2018 International Power Electronics Conference (IPEC-Niigata 2018 –ECCE Asia-)

Niigata, Japan
20-24 May 2018

Pages 1-758

IEEE Catalog Number: CFP1854I-POD
ISBN: 978-1-5386-4190-3

Copyright © 2018, IEEJ Industry Applications Society
All Rights Reserved

*** *This is a print representation of what appears in the IEEE Digital Library. Some format issues inherent in the e-media version may also appear in this print version.*

IEEE Catalog Number:	CFP1854I-POD
ISBN (Print-On-Demand):	978-1-5386-4190-3
ISBN (Online):	978-4-88686-405-5

Additional Copies of This Publication Are Available From:

Curran Associates, Inc
57 Morehouse Lane
Red Hook, NY 12571 USA
Phone: (845) 758-0400
Fax: (845) 758-2633
E-mail: curran@proceedings.com
Web: www.proceedings.com

2018 International Power Electronics Conference (IPEC-Niigata 2018 –ECCE Asia-)

Niigata, Japan
20-24 May 2018

Pages 1-758

IEEE Catalog Number: CFP1854I-POD
ISBN: 978-1-5386-4190-3

TABLE OF CONTENTS

THREE-PHASE INDUCTIVE POWER TRANSFER SYSTEM WITH 12 COILS FOR RADIATION NOISE REDUCTION 69
Keisuke Kusaka ; Jun-Ichi Itoh

SECONDARY-SIDE-ONLY CONTROL FOR SMOOTH VOLTAGE STABILIZATION IN WIRELESS POWER TRANSFER SYSTEMS WITH CONSTANT POWER LOAD 77
Giorgio Lovison ; Takehiro Imura ; Hiroshi Fujimoto ; Yoichi Hori

CONSTANT CURRENT CHARGING AND THE MAXIMUM SYSTEM EFFICIENCY TRACKING FOR WIRELESS CHARGING SYSTEMS EMPLOYING DUAL-SIDE CONTROL 84
Zhenjie Li ; Xiaoliang Huang ; Kai Song ; Jinhai Jiang ; Chunbo Zhu ; Zhijiang Du

ELECTRIC FIELD COUPLING TYPE HIGH POWER WIRELESS POWER TRANSFER WITH LEAKAGE ELECTRIC FIELD STRUCURE 88
Mitsuru Masuda

TRANSFER POWER ANALYSIS OF CAPACITIVELY ISOLATED OUTLET AND PLUG (CAPISOP) USING SERIES RESONANCE 94
Hirohito Funato ; Koki Amano ; Takuya Hatsumi ; Junnosuke Haruna

WIDE VOLTAGE GAIN RANGE LLC DC/DC TOPOLOGIES: STATE-OF-THE-ART 100
Qi Cao ; Zhiqing Li ; Haoyu Wang

DUAL HALF-BRIDGE LLC RESONANT CONVERTER WITH HYBRID-SECONDARY-RECTIFIER (HSR) FOR WIDE-OUPUT-VOLTAGE APPLICATIONS 108
Jae-Il Baek ; Chong-Eun Kim ; Keon-Woo Kim ; Min-Su Lee ; Gun-Woo Moon

A STUDY ON THE ANALYSIS AND CONTROL OF NO-LOAD CHARACTERISTICS OF LLC RESONANT CONVERTER FOR PLASMA PROCESS 114
Min-Jun Kwon ; Woo-Cheol Lee

MECHANISM OF CURRENT IMBALANCE IN LLC RESONANT CONVERTER WITH CENTER TAPPED TRANSFORMER 118
Mitsuru Sato ; Shingo Nagaoka ; Takeshi Uematsu ; Toshiyuki Zaitsu

PERFORMANCE STUDY OF HIGH-POWER HALF-BRIDGE INTERLEAVED LLC CONVERTER 123
Hung-I Hsieh ; Hui-Lung Chiu ; Guan-Chyun Hsieh

MULTI-CHIP SIC MOSFET POWER MODULES FOR STANDARD MANUFACTURING, MOUNTING AND COOLING 130
Alberto Castellazzi ; Asad Fayyaz ; Emre Gurpinar ; Abdallah Hussein ; Jianfeng Li ; Bassem Mouawad

AN ALTERNATIVE METHOD TO ACCURATELY DETERMINE THE THERMAL RESISTANCE OF SIC MOSFET STRUCTURES WITH DISCRETE DIODES 137
Andras Vass-Varnai ; Young Joon Cho ; Gabor Farkas ; Marta Rencz

HEAT-RESISTANT PACKAGING TECHNOLOGY FOR WIDE BANDGAP POWER DEVICES AND THERMAL RELIABILITY TESTING 142
K. Suganuma ; H. Zhang ; S. Nagao ; C. Chen ; T. Sugahara ; A. Shimoyama ; A. Suetake

VERIFICATION OF IDENTIFICATION ACCURACY OF LOSS CALCULATED BY INVERSE THERMAL ANALYSIS 148
Yuki Ikari ; Kazushige Nakao

PACKAGING ARCHITECTURES FOR SILICON CARBIDE POWER ELECTRONIC MODULES 153
H. Alan Mantooth ; Simon S. Ang

DEVELOPMENT OF A HOMO-POLAR BEARINGLESS MOTOR WITH CONCENTRATED WINDING FOR HIGH SPEED APPLICATIONS 157
Dai Suzuki ; Takaaki Oiwa

HIGH-SPEED SLOTLESS PERMANENT MAGNET MACHINES: MODELLING AND DESIGN FRAMEWORKS 161
S. Jumayev ; K.O. Boynov ; E.A. Lomonova ; J. Pyrhonen

DEVELOPMENT AND PERFORMANCE OF HIGH-SPEED SPM SYNCHRONOUS MACHINE 169
Kota Kawanishi ; Keisuke Matsuo ; Takayuki Mizuno ; Koji Yamada ; Takashi Okitsu ; Kouki Matsuse

1.2KW 100,000RPM HIGH SPEED MOTOR FOR AIRCRAFT 177
Takehiro Jikumaru ; Gen Kuwata

COMPARATIVE EVALUATION OF Y-INVERTER AGAINST THREE-PHASE TWO-STAGE BUCK-BOOST DC-AC CONVERTER SYSTEMS 181
Michael Antivachis ; Dominik Bortis ; David Menzi ; Johann W. Kolar

DC-POWERED OFFICE BUILDINGS AND DATA CENTRES : THE FIRST 380 VDC MICRO GRID IN A COMMERCIAL BUILDING IN GERMANY 190
Tilo Pueschel

RECENT TREND IN POWER ELECTRONICS FOR ICT SYSTEMS 196
Hiroshi Nakao ; Yu Yonezawa ; Yoshiyasu Nakashima

GREEN BASE STATION USING ROBUST SOLAR SYSTEM AND HIGH PERFORMANCE LITHIUM ION BATTERY FOR NEXT GENERATION WIRELESS NETWORK (5G) AND AGAINST MEGA DISASTER 201
M. Nakamura ; K. Takeno

OPTIMIZATION OF MAINTENANCE BY FAILURE PREDICTION CONSIDERING INSTANTANEOUS AND CUMULATIVE EFFECTS OF EXTERNAL ENVIRONMENTS 207
Kaisei Kanetani ; Masahiro Yamazaki ; Tadatoshi Babasaki ; Hideaki Kim ; Tatsushi Matsubayashi

HYBRID CONVERTERS WITH REDUCED INDUCTOR LOSS FOR INTEGRATABLE POWER CONVERSION 213
Gab-Su Seo ; Hanh-Phuc Le

ENERGY SAVING SYSTEM TREND FOR HARBOR CRANE WITH LITHIUM ION BATTERY 219
Hidemasa Yoshihara

INVERTER DRIVE OF DYNAMOMETERS FORAUTOMOTIVE EVALUATION SYSTEM 227
Shizunori Hamada ; Toshimichi Takahashi ; Nobutaka Kezuka ; Masaju Kouketsu ; Shingo Ishigaki

EXPERIMENTAL INVESTIGATION OF PROTOTYPE ALL-SIC CONVERTER FOR ULTRA-HIGH-SPEED ELEVATOR 233
Kazuhisa Mori ; Kaoru Katoh ; Yohei Matsumoto ; Tatsushi Yabuuchi ; Naoto Ohnuma

HIGH-VOLTAGE, LARGE-CAPACITY CONVERTER TECHNOLOGIES AND THEIR APPLICATIONS 238
Daisuke Yoshizawa ; Paul Bixel ; Masahiko Tsukakoshi

HIGHER RADIAL SUSPENSION FORCE OF MAGNETIC BEARING ON CENTRIFUGAL COMPRESSOR FOR HVAC 244
Yuji Nakazawa ; Yusuke Irino ; Atsushi Sakawaki ; Kazunobu Ohyama

NOVEL SWITCHING CONTROL METHOD FOR FULL-BRIDGE DC-DC CONVERTERS FOR IMPROVING LIGHT-LOAD EFFICIENCY USING REVERSE RECOVERY CURRENT 250
Fumihiro Sato ; Takae Shimada ; Takayuki Ouchi

A 800V/14V SOFT-SWITCHED CONVERTER WITH LOW-VOLTAGE RATING OF SWITCH FOR XEV APPLICATIONS 256
Byeongwoo Kim ; Kangsan Kim ; Sewan Choi

HIGH SPEED CONTROL METHOD FOR SUPERPOSING HIGH-FREQUENCY-HIGH-SINUSOIDAL-CURRENT WITH DC CURRENT TO ANALYZE BATTERY AC IMPEDANCE 261
Jin Xu ; Toshihiko Kishimoto ; Noboru Shimosato

EV BMS WITH TIME-SHARED ISOLATED CONVERTERS FOR ACTIVE BALANCING AND AUXILIARY BUS REGULATION 267
Z. Gong ; B.A.C. Van De Ven ; Y. Lu ; Y. Luo ; K. Gupta ; C. Da Silva ; H.J. Bergveld ; O. Trescases

A DRIVING CIRCUIT WITH PARTIAL POWER REGULATION FOR RGB LED LAMPS 275
You-Chun Huang ; Yu-Jen Chen ; Yong-Jyun Li ; Chin-Sien Moo

FPGA-BASED DYNAMIC DUTY CYCLE AND FREQUENCY CONTROLLER FOR A CLASS-E2DC-DC CONVERTER 282
Sanghyeon Park ; Juan Rivas-Davila

DESIGN METHODOLOGY OF 3 KW INDUCTION HEATING SYSTEM FOR BOTH LOW RESISTANCE AND HIGH RESISTANCE CONTAINERS IN A SINGLE BURNER 289
Si-Hoon Jeong ; Hwa-Pyeong Park ; Jee-Hoon Jung

MULTI-RESONANT INVERTER REALIZING DOWNSIZING AND LOSS REDUCTION FOR ALL-METALLIC IH COOKTOP 296
Takayuki Hirokawa ; Makoto Imai ; Atsushi Fujita

TEMPERATURE ESTIMATION OF ALUMINUM ELECTROLYTIC CAPACITOR UNDER ACTUAL CIRCUIT OPERATION 302
Kazuki Urata ; Toshihisa Shimizu

DESIGN AND EVALUATION OF CURRENT DISTRIBUTION IN POWER MODULE 309
Takaaki Ibuchi ; Eisuke Masuda ; Tsuyoshi Funaki

DEVELOPMENT OF IMPEDANCE-SOURCE INVERTER USING SIC-MOSFET 313
Ryuji IIjima ; Thilak Senanayake ; Takanori Isobe ; Hiroshi Tadano

CONTROL METHODOLOGY FOR REALIZATION OF 100KW HEECS CHOPPER WITH 99.5% EFFICIENCY 318
Yukinori Tsuruta ; Atsuo Kawamura

IRON LOSS REDUCTION IN THE CORES OF INDUCTION HEATING COILS FOR SMALL-FOREIGN-METAL PARTICLE DETECTOR WITH A 400-KHZ SIC-MOSFETS HIGH-FREQUENCY INVERTER 324

Takuya Shijo ; Yuki Uchino ; Yujiro Noda ; Hiroaki Yamada ; Toshihiko Tanaka

FREQUENCY TRACKING BURST-MODE PDM-CONTROLLED CLASS-D ZERO VOLTAGE SOFT-SWITCHING RESONANT CONVERTER FOR INDUCTIVE POWER TRANSFER APPLICATIONS 329

Yoichiro Tabata ; Tomokazu Mishima ; Tatsuya Kido

REDUCED-ORDER DYNAMICAL MODELS OF TUNED WIRELESS POWER TRANSFER SYSTEMS 337

Hongchang Li ; Jingyang Fang ; Yi Tang

DYNAMIC MODELLING AND CLOSED LOOP CONTROL OF TRANSMITTER PARALLEL AND RECEIVER SERIES COMPENSATED IPT TOPOLOGY FOR EV APPLICATIONS 342

Suvendu Samanta ; Akshay Kumar Rathore

DEVELOPMENT OF INDUCTIVE POWER TRANSFER SYSTEM FOR EXCAVATOR UNDER LARGE LOAD FLUCTUATION : CONSIDERATION OF RELATIONSHIP BETWEEN LOAD VOLTAGE AND RESONANCE PARAMETER 348

Jun-Ichi Itoh ; Kent Inoue ; Keisuke Kusaka

WIRELESS POWER TRANSFER SYSTEM USING THREE-PHASE TO SINGLE-PHASE MATRIX CONVERTER 356

Yuji Hayashi ; Hiromasa Motoyama ; Takaharu Takeshita

DESIGN OF A REDUCED-ORDER OBSERVER FOR SENSORLESS CONTROL OF DUAL-ACTIVE-BRIDGE CONVERTER 363

Nguyen Duy Dinh ; Goro Fujita

IMPROVED LOAD TRANSIENT RESPONSE OF A DUAL-ACTIVE-BRIDGE CONVERTER 370

Sheng-Zhi Zhou ; Chuan Sun ; Song Hu ; Guo Chen ; Xiaodong Li

MODULATION AND ACTIVE MIDPOINT CONTROL OF A THREE-LEVEL THREE-PHASE DUAL-ACTIVE BRIDGE DC-DC CONVERTER UNDER NON-SYMMETRICAL LOAD 375

Philipp Joebges ; Anton Gorodnichev ; Rik W. De Doncker

A NOVEL SWITCHING ALGORITHM TO IMPROVE EFFICIENCY AT LIGHT LOAD CONDITIONS FOR THREE-PHASE DAB CONVERTER IN LVDC APPLICATION 383

Hyun-Jun Choi ; Si-Hoon Jung ; Jee-Hoon Jung

DESIGN OF A HIGH-FREQUENCY DUAL-ACTIVE BRIDGE CONVERTER WITH GAN DEVICES FOR AN OUTPUT POWER OF 3.7 KW 388

Philipp Schülting ; Christian Winter ; Rik W. De Doncker

EXPLORATION OF THE DESIGN AND PERFORMANCE SPACE OF A HIGH FREQUENCY 166 KW/10 KV SIC SOLID-STATE AIR-CORE TRANSFORMER 396

Piotr Czyz ; Thomas Guillod ; Florian Krismer ; Johann W. Kolar

NOVEL CALCULATION METHOD OF IRON LOSS OF GAPPED INDUCTORS USING LOSS MAP 404

Yoshihiro Miwa ; Toshihisa Shimizu

VERIFICATION OF THE REDUCTION OF THE COPPER LOSS BY THE THIN COIL STRUCTURE FOR INDUCTION COOKERS 410

Morimasa Hataya ; Koki Kamaeguchi ; Eiji Hiraki ; Kazuhiro Umetani ; Takayuki Hirokawa ; Makoto Imai ; Hideki Sadakata

CONDITION MONITORING OF ELECTROLYTIC CAPACITOR BASED ON ESR ESTIMATION AND THERMAL IMPEDANCE MODEL USING IMPROVED POWER LOSS COMPUTATION 416

Sundararajan Prasanth ; Mohamed Halick ; Mohamed Sathik ; Firman Sasongko ; Tan Chuan Seng ; Peng Yaxin ; Rejeki Simanjorang

TEST SETUP FOR CHARACTERISATION OF BIASED MAGNETIC HYSTERESIS LOOPS IN POWER ELECTRONIC APPLICATIONS 422

Min Luo ; Drazen Dujic ; Jost Allmeling

A FAST OPEN-CIRCUIT FAULT DIAGNOSIS SCHEME FOR MODULAR MULTILEVEL CONVERTERS WITH MODEL PREDICTIVE CONTROL 428

Dehong Zhou ; Shunfeng Yang ; Yi Tang

AN ONLINE OPEN-CIRCUIT FAULT DIAGNOSIS AND FAULT TOLERANT SCHEME FOR THREE-PHASE AC-DC CONVERTERS WITH MODEL PREDICTIVE CONTROL 434

Dehong Zhou ; Yi Tang

THE LIFETIME ASSESSMENT OF A MICRO-INVERTER FOR PV APPLICATIONS 439

Tohihiro Shimao ; Koji Kato ; Youichi Ito ; Akio Iwabuchi ; Yongheng Yang ; Frede Blaabjerg

ONLINE HEALTH MONITORING OF MULTIPLE MOSFETS IN A GRID-TIED PV INVERTER USING SPREAD SPECTRUM TIME DOMAIN REFLECTOMETRY (SSTDR) 446
Sourov Roy ; Faisal Khan

AN IMPROVED EQUIVALENT MODEL FOR A LONG PV STRING UNDER PARTIAL SHADING CONDITIONS 453
Xiaoyang Wang ; Huiqing Wen ; Xingshuo Li

OPTIMIZED FLUX-WEAKENING CONTROL OF INDUCTION MOTOR FOR TORQUE ENHANCEMENT IN VOLTAGE EXTENSION REGION 459
Zhen Dong ; Yong Yu ; Bo Wang ; Qinghua Dong ; Dianguo Xu

IMPROVED PERFORMANCE OF CFTC-BASED DIRECT TORQUE CONTROL OF INDUCTION MACHINES BY INCREASING TORQUE LOOP BANDWIDTH 466
Ibrahim Mohd Alsofyani ; June-Hee Lee ; Byung-Moon Han ; Kyo-Beum Lee

μ-ANALYSIS EVALUATION OF A NOVEL COMBINED CURRENT-AND-SPEED CONTROL FOR INDUCTION MOTORS VIA ILQ DESIGN METHOD 471
Shuto Omori ; Hiroshi Takami ; Masashi Nakamura

LOSS MINIMIZATION CONTROL OF SENSORLESS SCALAR-CONTROLLED INDUCTION MOTOR DRIVES CONSIDERING IRON LOSS 478
Nguyen Anh Tan ; Dong-Choon Lee

TUNING OF INDUCTION MOTOR DRIVE WITH TORQUE SENSOR 483
Hajime Kubo ; Yugo Tadano

QUASI-TWO-LEVEL CONVERTER FOR OVERVOLTAGE MITIGATION IN MEDIUM VOLTAGE DRIVES 488
F. Bertoldi ; M. Pathmanathan ; R. S. Kanchan ; K. Spiliotis ; J. Driesen

A MEDIUM-VOLTAGE THREE-PHASE AC-DC CONVERTER CONSISTING OF CASCADED THREE-LEVEL BOOST-TYPE RECTIFIERS AND AN OPEN-END WINDING TRANSFORMER 495
Ryoji Tsuruta ; Hiromitsu Suzuki ; Ritaka Nakamura

A FAULT TOLERANT CONTROL STRATEGY FOR THE DELTA-CONNECTED CASCADED CONVERTER 503
Ping-Heng Wu ; Po-Tai Cheng

COOLING PERFORMANCE IMPROVEMENT OF HEAT SINK BY OSCILLATING HEAT PIPE ADDITION AND DESIGN FOR ENVIRONMENT OF OSCILLATING HEAT PIPE REFRIGERANT 511
Kuan-Chung Tey ; Kenichiro Suzuki

COMPACT LARGE CAPACITY GAS TURBINE STATIC STARTER 517
Hironori Kawaguchi ; Shigeyuki Nakabayashi ; Akinobu Ando ; Hiroshi Ogino ; Yasuaki Matsumoto ; Ikuto Udagawa ; Takahiro Ohta

VOLTAGE REFERENCE MODIFICATION SCHEME FOR RESONANCE SUPPRESSION IN LCL-FILTERED INVERTERS WITH DISCONTINUOUS PWM METHOD 521
Hyeon-Sik Kim ; Seung-Ki Sul

PARAMETRIC ROBUSTNESS ANALYSIS FOR PARALLEL FEEDFORWARD COMPENSATION BASED ACTIVE DAMPING OF LCL GRID CONNECTED INVERTER 528
Muhammad Talib Faiz ; Muhammad Mansoor Khan ; Xu Jianming ; Muhammad Ali ; Houjun Tang

OPEN-LOOP-BASED ISLAND-MODE VOLTAGE CONTROL METHOD FOR SINGLE-PHASE GRID-TIED INVERTER WITH MINIMIZED LC FILTER 534
Satoshi Nagai ; Jun-Ichi Itoh

EXPERIMENTAL VALIDATION OF ADAPTIVE CURRENT INJECTING METHOD FOR GRID-SYNCHRONIZATION IMPROVEMENT OF GRID-TIED REGS DURING SHORT-CIRCUIT FAULT 542
Shaokang Ma ; Hua Geng ; Geng Yang ; Bo Liu

ADAPTIVE CONTROL OF GRID-VOLTAGE FEEDFORWARD FOR GRID-CONNECTED INVERTERS BASED ON REAL-TIME IDENTIFICATION OF GRID IMPEDANCE 547
Roni Luhtala ; Tuomas Messo ; Tomi Roinila

MODEL BASED TUNING OF PROPORTIONAL RESONANT CONTROLLERS FOR VOLTAGE SOURCE INVERTERS 555
Stefan Almér ; Thomas Besselmann ; Mario Schweizer

AN SOC-BASED PLATFORM FOR INTEGRATED MULTI-AXIS MOTION CONTROL AND MOTOR DRIVE 560
Yongping Sun ; Ming Yang ; Yangyang Chen ; Wangpin He ; Dianguo Xu

VARIABLE SWITCHING FREQUENCY STRATEGY FOR ENHANCED SETTLING PERFORMANCE OF POSITION CONTROL WITHIN INVERTER LOSS LIMIT 565
Choongin Lee ; Jung-Ik Ha

TWO-WHEEL CANE FOR WALKING ASSISTANCE..571
Phi Van Lam ; Yasutaka Fujimoto

FALL PREVENTION AND VIBRATION SUPPRESSION OF WHEELCHAIR USING RIDER MOTION STATE..575
Isseki Takahashi ; Toshiyuki Murakami

STABILIZATION METHOD FOR RESIDENTIAL DC SYSTEM BASED ON PASSIVITY CRITERION..583
Hiroaki Kakigano

A NOVEL CONTROL APPROACH TO MULTI-TERMINAL POWER FLOW CONTROLLER FOR NEXT-GENERATION DC POWER NETWORK..588
Kenji Natori ; Yuta Nakao ; Yukihiko Sato

DC MICROGRID FOR TELECOMMUNICATIONS SERVICE AND RELATED APPLICATION..593
Keiichi Hirose

MVDC DISTRIBUTION GRIDS FOR ELECTRIC VEHICLE FAST-CHARGING INFRASTRUCTURE..598
Marco Stieneker ; Benedict J. Mortimer ; Arne Hinz ; Adolf Müller-Hellmann ; Rik W. De Doncker

REVIEW OF RESONANT GATE DRIVER IN POWER CONVERSION..607
Bainan Sun ; Zhe Zhang ; Michael A.E. Andersen

A LOW PROFILE HIGH FREQUENCY LED DRIVING SYSTEM BASED ON AIRCORE PLANAR INDUCTOR..614
Yueshi Guan ; Xihong Hu ; Shu Zhang ; Yijie Wang ; Dianguo Xu ; Wei Wang

ANALYSIS AND COMPENSATION OF DEAD-TIME EFFECT IN SIC-DEVICE-BASED HIGH-SWITCHING-FREQUENCY INVERTERS..619
Qingzeng Yan ; Xibo Yuan ; Xiaojie Wu ; Yiwen Geng

CONTROL AND PERFORMANCE OF NEW ASYMMETRICAL OPERATION FOR SWITCHED-CAPACITOR-BASED RESONANT CONVERTERS..626
Hadi Setiadi ; Hideaki Fujita

HIGH-FREQUENCY RESONANT CONVERTER WITH SYNCHRONOUS RECTIFICATION FOR HIGH CONVERSION RATIO AND VARIABLE LOAD OPERATION..632
Lei Gu ; Kawin Surakitbovorn ; Juan Rivas-Davila

SMART PV INVERTERS FOR SMART GRID APPLICATIONS..639
Cheng-Jhen Yang ; Terng-Wei Tsai ; Yi-Chan Li ; Cheng-Yu Tang ; Yaow-Ming Chen ; Yung-Ruei Chang

HIGH-VOLTAGE BI-DIRECTIONAL HALF-BRIDGE THREE-LEVEL SERIES RESONANT CONVERTER WITH FREQUENCY MODULATION CONTROL..645
Lee Sih-Yi ; Jhang Jynu-Jhe ; Lin Jing-Yuan ; Hsieh Yao-Ching ; Chiu Haung-Jen

A CONTROL STRATEGY FOR FLYING-START OF SHAFT SENSORLESS PERMANENT MAGNET SYNCHRONOUS MACHINE DRIVE..651
Zih-Cing You ; Sheng-Ming Yang

CONTACTLESS EV POWER TRACK SYSTEM WITH SEGMENT-EXCITED INDUCTIVELY COUPLED STRUCTURE..657
Jia-You Lee ; Yu-Chi Wang ; Chih-Yi Liao

DRIVING TEST EVALUATION OF SENSORLESS VEHICLE DETECTION METHOD FOR IN-MOTION WIRELESS POWER TRANSFER..663
Katsuhiro Hata ; Kensuke Hanajiri ; Takehiro Imura ; Hiroshi Fujimoto ; Yoichi Hori ; Motoki Sato ; Daisuke Gunji

A SYSTEM DESIGN METHOD OF HIGH-FREQUENCY CLASS-D INVERTER FOR WIDEBAND CURRENT CONTROL..669
Hiroki Kurumatani ; Seiichiro Katsura

ANALYSIS OF INTERIOR PERMANENT MAGNET TWO DEGREES OF FREEDOM MOTOR BASED ON CROSS-COUPLED STRUCTURE..675
Yoshiyuki Hatta ; Tomoyuki Shimono

STUDY COMPARISON BETWEEN FIREFLY ALGORITHM AND PARTICLE SWARM OPTIMIZATION FOR SLAM PROBLEMS..681
Mounia Janah ; Yasutaka Fujimoto

BANDWIDTH LIMITATIONS IN FORCE CONTROL OF A SERIES ELASTIC ACTUATOR WITH BACKLASH AND QUANTIZATION..688
Hanul Jung ; Chan Lee ; Sehoon Oh

ROTOR SHAPE OPTIMIZATION OF INTERIOR PERMANENT MAGNET SYNCHRONOUS MOTORS WITH CONCENTRATED WINDINGS BY CONSIDERING END-LEAKAGE FLUX..693
Katsumi Yamazaki ; Hiroki Narushima

LOSS ANALYSIS OF PERMANENT-MAGNET SYNCHRONOUS MACHINES CONSIDERING IN-PLANE EDDY CURRENT IN ELECTRICAL STEEL SHEETS...........699
Hideki Ohguchi ; Satoshi Imamori ; Katsumi Yamazaki ; Haiyan Yui ; Masao Shuto

STUDY ON INFLUENCE OF DIFFERENCE IN STRUCTURE OF CONCENTRATED WINDING IPMSMS OBTAINED BY AUTOMATIC DESIGN...........704
A. Ura ; M. Sanada ; S. Morimoto ; Y. Inoue

CARRIER HARMONIC LOSS REDUCTION TECHNIQUE ON DUAL THREE-PHASE PERMANENT-MAGNET SYNCHRONOUS MOTORS WITH PHASE-SHIFT PWM...........711
Yoshihiro Miyama ; Haruyuki Kometani ; Kan Akatsu

FLUX INTENSIFYING PM-MOTOR WITH VARIABLE LEAKAGE MAGNETIC FLUX TECHNIQUE...........718
Masahiro Aoyama ; Toshihiko Noguchi

CONTINUOUS OPERATION CONTROL OF PMSM IN THE CASE OF DC POWER SUPPLY LOSS...........726
Jongwon Heo ; Keiichiro Kondo

MODEL PREDICTIVE CONTROL FOR MULTIPHASE MOTOR DRIVES – A TECHNOLOGY STATUS REVIEW...........732
A. Tenconi ; S. Rubino ; R. Bojoi

INFLUENCE OF FAST SWITCHING SEMICONDUCTORS ON THE WINDING INSULATION SYSTEM OF ELECTRICAL MACHINES...........740
Kay Hameyer ; Andreas Ruf ; Florian Pauli

CENTRALIZED CONTROL OF MODULAR MULTI RECTIFIER FOR MOTOR DRIVE APPLICATIONS UNDER UNBALANCED GRID...........746
Yipeng Song ; Pooya Davari ; Frede Blaabjerg

VECTOR CONTROL OF MAGNETICALLY MODULATED MOTOR FOR POWER SPLITTING OF HEV APPLICATION...........753
Toshihiko Noguchi ; Sawanth Krishna Machavolu ; Masahiro Aoyama ; Yuto Motohashi

IMPEDANCE-BASED STABILITY EVALUATION OF VIRTUAL SYNCHRONOUS MACHINE IMPLEMENTATIONS IN CONVERTER CONTROLLERS...........759
Eneko Unamuno ; Atle Rygg ; Mohammad Amin ; Marta Molinas ; Jon Andoni Barrena

STABLE POWER SUPPLY METHOD FOR HOUSEHOLD APPLIANCES VIA VIRTUAL SYNCHRONOUS GENERATOR IN SINGLE-PHASE THREE-WIRE MICROGRID...........767
Yuko Hirase ; Hidehiko Nakagawa ; Eiji Yoshimura ; Shogo Katsura ; Kensho Abe ; Osamu Noro ; Kazushige Sugimoto ; Kenichi Sakimoto

A NOVEL OSCILLATION DAMPING METHOD OF VIRTUAL SYNCHRONOUS GENERATOR CONTROL WITHOUT PLL USING POLE PLACEMENT...........775
Jia Liu ; Yushi Miura ; Toshifumi Ise

OPERATION OF A MODULAR MULTILEVEL CONVERTER CONTROLLED AS A VIRTUAL SYNCHRONOUS MACHINE...........782
Salvatore D'arco ; Giuseppe Guidi ; Jon Are Suul

ASSESSMENT OF VIRTUAL SYNCHRONOUS MACHINE BASED CONTROL IN GRID-TIED POWER CONVERTERS...........790
Chi Li ; Igor Cvetkovic ; Rolando Burgos ; Dushan Boroyevich

RESEARCH ON THE BLOCKCHAIN-BASED INTEGRATED DEMAND RESPONSE RESOURCES TRANSACTION SCHEME...........795
Shengnan Zhao ; Yang Li ; Beibei Wang ; Huiling Su

INDIRECT CURRENT CONTROL FOR SEAMLESS TRANSFER OF UTILITY INTERACTIVE INVERTER...........803
Kyungbae Lim ; Injong Song ; Jaeho Choi

STUDY OF AC POWER INTERCHANGE AND DC POWER INTERCHANGE FOR MICRO GRID SYSTEMS...........809
Kazuto Yukita ; Daiki Owaki ; Shunsuke Horie ; Toshiro Matsumura ; Yasuyuki Goto

STABILITY ENHANCEMENT STRATEGY FOR ISLANDING MICROGRID WITH MULTI-TYPE INVERTERS BASED ON HYBRID IMPEDANCE MODELLING...........815
Meiqin Mao ; Yong Ding ; Yatao Shen ; Liuchen Chang

DC POWERED DATA CENTER WITH 200 KW PV PANELS...........822
Keiichi Hirose

INFLUENCES OF DETERIORATION IN CAPACITOR AND INDUCTOR ON CURRENT SENSORLESS STATIC MODEL DC-DC CONVERTER...........826
Fujio Kurokawa ; Masashi Taguchi ; Jizhe Wang ; Hidenori Maruta ; Nobumasa Matsui

CAPACITIVE DIVIDER BASED PASSIVE START-UP METHODS FOR FLYING CAPACITOR STEP-DOWN DC-DC CONVERTER TOPOLOGIES......831

Michael Halamicek ; Tom Moiannou ; Nenad Vukadinovic ; Aleksandar Prodic

HIGH VOLTAGE GAIN INTERLEAVED ACTIVE-CLAMP FORWARD (IACF) CONVERTER HAVING REDUCED PRIMARY CONDUCTION LOSS......838

Yeonho Jeong ; Mu-Hyun Park ; Gun-Woo Kim ; Byoung-Hee Lee ; Gun-Woo Moon

CONTROL OF SWITCHING-CAPACITOR BASED BUCK-BOOST CONVERTER......845

M. Veerachary ; Vasudha Khubchandani

IMPROVEMENT OF UPLOAD TRANSIENT RESPONSES FOR ULTRA HIGH STEP-DOWN CONVERTER......851

Y.T. Yan ; K.I. Hwu

POWER ELECTRONICS AND CONTROL TECHNOLOGIES FOR HOUSEHOLD WASHER......856

Toru Niki

DEVELOPMENT OF ROOM AIR CONDITIONER WITH TWIN-PROPELLER FANS......860

Takamasa Uemura ; Tomoya Fukui ; Kenichi Sakoda

ELECTROLYTIC CAPACITOR-LESS SINGLE-PHASE TO THREE-PHASE INVERTER WITH HARMONICS SUPPRESSION CONTROL FOR AIR CONDITIONER......866

Nobuo Hayashi ; Takuro Ogawa ; Tomoisa Taniguchi ; Morimitsu Sekimoto

LATEST DEVELOPMENT OF SIC POWER MODULE-BASED SINGLE-STAGE AC-AC RESONANT CONVERTER FOR HIGH-FREQUENCY INDUCTION HEATING APPLICATIONS......872

Tomokazu Mishima

AN OPTIMIZED CONTROL STRATEGY TO IMPROVE THE CURRENT ZERO-CROSSING DISTORTION IN BIDIRECTIONAL AC/DC CONVERTER BASED ON V2G CONCEPT......878

Lei Jing ; Xiaoqing Wang ; Bodong Li ; Maohang Qiu ; Bo Liu ; Min Chen

PER-PHASE CONTROL STRATEGY OF THE THREE-PHASE FOUR-WIRE INVERTER......883

Yi-Chan Li ; Terng-Wei Tsai ; Cheng-Jhen Yang ; Yaow-Ming Chen ; Yung-Ruei Chang

OPPORTUNITIES FOR PERFORMANCE IMPROVEMENT OF SINGLE-PHASE POWER CONVERTERS THROUGH ENHANCED AUTOMATIC-POWER-DECOUPLING CONTROL......889

Huawei Yuan ; Sinan Li ; Wenlong Qi ; Siew-Chong Tan ; S. Y. Ron Hui

ZERO VOLTAGE SWITCHING SCHEME FOR FLYBACK CONVERTER TO ENSURE COMPATIBILITY WITH ACTIVE POWER DECOUPLING CAPABILITY......896

Hiroki Watanabe ; Jun-Ichi Itoh

MODEL PREDICTIVE FAULT TOLERANT CONTROL OF BIDIRECTIONAL AC/DC CONVERTER WITH VOLTAGE BALANCE OF SPLIT CAPACITOR......904

Nan Jin ; Chongyan Zhao ; Leilei Guo

PWM STRATEGY FOR PARALLEL OPERATION OF THREE PHASE CONVERTERS TIED TO GRID......911

Hyun-Sam Jung ; Seung-Ki Sul

PRACTICAL ISSUES AND IMPLEMENTATION CIRCUITS OF THE DIGITAL-ANALOG HYBRID FULL FEED-FORWARD METHOD WITH UNIPOLAR AND BIPOLAR MODULATIONS......917

Xin Zhang ; Henry S. H. Chung ; Zhixun Ma

AN AC-DC POWER CONVERTER FOR ELECTROLYTIC CAPACITOR-LESS LED DRIVER WITH HIGH LUMINOUS EFFICACY......922

Kwon-Sik Park ; Byuong-Jun Seo ; Kyoung-Suk Kang ; Eui-Cheol Nho

AN IMPROVED CASCADED DUAL-BUCK INVERTER......927

Usman Ali Khan ; Honnyong Cha ; Ashraf Ali Khan ; Heung-Geun Kim ; Wilson Eberle ; Liwei Wang

A SINGLE-SWITCH INTEGRATED-STAGE LED DRIVER BASED ON CUK AND CLASS-E CONVERTER......934

Shu Zhang ; Yijie Wang ; Xiaosheng Liu ; Yan Zhou ; Dianguo Xu

A FAULT-TOLERANT PARALLEL INVERTER APPLIED TO MICRO-GRID......939

Xiangyue Shi ; Jinjie Peng ; Zhifeng Qiu ; Wei Xiong

STABILITY ANALYSIS OF GRID-CONNECTED CONVERTERS WITH ADD-ON VOLTAGE SUPPORT FUNCTIONALITY USING REPETITIVE CONTROL......946

Y. Zhang ; M. G. L. Roes ; M. A. M. Hendrix ; J. L. Duarte

ADAPTIVE SERIES STABILIZER MODULE FOR THE GRID CONNECTED INVERTER UNDER VARIABLE GRID CONDITIONS......953

Xin Zhang

AN IMPROVED DROOP CONTROL BASED SMOOTH TRANSFER CONTROL STRATEGY......957

Xin Meng ; Jinjun Liu ; Zeng Liu ; Ronghui An

FREQUENCY RESPONSE ANALYSIS OF LOAD EFFECT ON DYNAMICS OF GRID-FORMING INVERTER 963

Matias Berg ; Tuomas Messo ; Teuvo Suntio

A NEW CONTROL METHOD FOR TRIPLE-ACTIVE BRIDGE CONVERTER WITH FEED FORWARD CONTROL 971

Takanobu Ohno ; Nobukazu Hoshi

ANALYSIS OF PFM OPERATION MODEL FOR CAPACITOR CHARGER RESONANT TOPOLOGY WITH ENERGY DOSAGE 977

Pengyu Jia ; Yiqin Yuan ; Shengwen Fan ; Zhenyu Shan

AN ACTIVE-CLAMPED CURRENT-FED HALF-BRIDGE DC-DC CONVERTER WITH THREE SWITCHES 982

Truong-Duy Duong ; Minh-Khai Nguyen ; Young-Cheol Lim ; Joon-Ho Choi

A HIGH GAIN QUASI SINGLE STAGE LLC RESONANT DC/DC CONVERTER WITH COUPLED INDUCTOR AND PARTIAL ACTIVE CLAMP 987

Chongcan Huo ; Xiaogao Xie ; Shuai Jiang ; Hanjing Dong

SUPPRESSION OF RIPPLE CURRENT IN HIGH STEP-UP DC-DC CONVERTER UTILIZING COCKCROFT-WALTON CIRCUIT WITH INDUCTOR 992

Takumi Yasuda ; Masataka Minami ; Shin-Ichi Motegi ; Masakazu Michihira

AN OPTIMAL DESIGN METHOD CONSIDERING TRANSFORMER PARASITIC CAPACITANCE OF LLC RESONANT CONVERTERS 998

Naizeng Wang ; Xu Yang ; Mofan Tian ; Haiyang Jia ; Guangzhao Xu ; Zhenwei Li

COMPARISON OF HARMONIC LINEARIZATION AND HARMONIC STATE SPACE METHODS FOR IMPEDANCE MODELING OF MODULAR MULTILEVEL CONVERTER 1004

Jing Lyu ; Xin Zhang ; Jingjing Huang ; Jianwen Zhang ; Xu Cai

AN IMPROVED PHASE-SHIFTED PWM FOR A FIVE-LEVEL HYBRID-CLAMPED CONVERTER 1010

Kui Wang ; Nianzhou Liu ; Zedong Zheng ; Yongdong Li

INTEGRATED CONTROL METHODS FOR ASYMMETRICAL CASCADED H-BRIDGE RECTIFIER 1015

Wenjing Dai ; Jie Chen ; Xin Chen ; Chunying Gong

TRANSIENT VOLTAGE STRESS MODELING FOR SUBMODULES OF MODULAR MULTILEVEL CONVERTERS UNDER GRID VOLTAGE SAGS 1021

Zhijian Yin ; Yongheng Yang ; Huai Wang

SVPWM STRATEGY BASED ON MULTILEVEL 3LNPC-CR 1027

Xiaoqiong He ; Pengcheng Han ; Xiaolan Lin ; Yi Wang ; Xu Peng

THE MULTIPLE DEGREE OF FREEDOM BASED NEUTRAL POINT POTENTIAL CONTROL OF THREE LEVEL NEUTRAL POINT CLAMPED CONVERTERS 1032

Bo Guan ; Shinji Doki

A MODIFIED PHASE-SHIFTED PWM TECHNIQUE FOR THE GRID-CONNECTED HYBRID CASCADED CONVERTER 1038

Yu-Chen Su ; Po-Tai Cheng

NOVEL T-TYPE DUAL-BUCK INVERTER WITH MINIMUM NUMBER OF INDUCTORS 1046

Tien-The Nguyen ; Honnyong Cha ; Bang Le-Huy Nguyen ; Heung-Geun Kim

CONTROL OF DIRECT AC/AC MODULAR MULTILEVEL CONVERTER IN RAILWAY POWER SUPPLY SYSTEM 1051

Shuguang Song ; Jinjun Liu ; Shaodi Ouyang ; Xingxing Chen ; Baojin Liu

WIRELESS POWER TRANSFER: CRITICAL REVIEW OF RELATED STANDARDS 1062

Mohamad Abou Houran ; Xu Yang ; Wenjie Chen ; Mehdi Samizadeh

COMPARATIVE STUDY OF SINGLE-PHASE FUNDAMENTAL COMPONENT FREQUENCY ESTIMATION SCHEMES UNDER TIME-VARYING HARMONIC DISTORTION OPERATION 1067

E. B. Kapisch ; J. L. Duarte ; C. A. Duque

A COMPREHENSIVE DEAD-TIME COMPENSATION METHOD FOR A THREE-PHASE DUAL-ACTIVE BRIDGE CONVERTER WITH HYBRID MODULATION SCHEMES 1073

Jingxin Hu ; Zhiqing Yang ; Rik W. De Doncker

EVALUATION OF A HIGH-FREQUENCY REACTOR WITH A NEW WIRE GUIDE FOR A TOROIDAL CORE 1080

Hideki Ayano ; Akira Fujimura ; Yoshihiro Matsui

CORE LOSS EVALUATION IN POWDER CORES: A COMPARATIVE COMPARISON BETWEEN ELECTRICAL AND CALORIMETRIC METHODS 1087

Yuki Ishikura ; Jun Imaoka ; Mostafa Noah ; Masayoshi Yamamoto

MODELING, MAGNETIC DESIGN, AND SIMULATION METHODS CONSIDERING DC SUPERIMPOSITION CHARACTERISTIC OF POWDER CORES USED IN POWER CONVERTERS 1095

Jun Imaoka ; Kenkichiro Okamoto ; Masahito Shoyama ; Yuki Ishikura ; Mostafa Noah ; Masayoshi Yamamoto

MODELLING AND DESIGN OF A MEDIUM FREQUENCY TRANSFORMER FOR HIGH POWER DC-DC CONVERTERS 1103

Miloš Stojadinovic ; Jürgen Biela

EVALUATION OF INDUCTOR LOSSES ON Z-SOURCE INVERTER CONSIDERING AC AND DC COMPONENTS 1111

Ryuji IIjima ; Naoki Kamoshida ; Rene Alexander Barrera Cardenas ; Takanori Isobe ; Hiroshi Tadano

AN INTEGRATING STRUCTURE OF OUTPUT FILTER FOR GRID CONNECTED INVERTER BASED ON FMLF TECHNIQUE 1118

Jie Ma ; Yenan Chen ; Pingping Chen ; Wenxing Zhong ; Dehong Xu

NEW SCREENING METHOD FOR IMPROVING TRANSIENT CURRENT SHARING OF PARALLELED SIC MOSFETS 1125

Junji Ke ; Zhibin Zhao ; Peng Sun ; Huazhen Huang ; James Abuogo ; Xiang Cui

PSPICE MODELING AND APPLICATION FOR SIC POWER MOSFET TO EVALUATE THE POWER LOSS IN FULL-BRIDGE CONVERTER 1131

Juan Wei ; Fei Lin ; Zhongping Yang ; Xianjin Huang ; Chanjuan Xiao ; Hao Zhang ; Wencai Liang

ALL-SIC MODULE PACKAGING TECHNOLOGY 1137

Kento Shirata ; Norihiro Nashida ; Hideyo Nakamura ; Yoshitaka Nishimura

A NEW SMALLEST 1200V INTELLIGENT POWER MODULE FOR THREE PHASE MOTOR DRIVES 1141

Minsub Lee ; Miran Baek ; Junbae Lee ; Daewoong Chung

DESIGN AND ENHANCEMENT OF ESD RELIABILITY IN CIRCULAR UHV 300-V NLDMOS POWER COMPONENTS 1145

Shen-Li Chen ; Yi-Hao Chao ; Chih-Ying Yen ; Jen-Hao Lo ; Chun-Ting Kuo ; Yu-Lin Lin ; Yi-Hao Chiu ; Pei-Lin Wu ; Yu-Lin Jhou

A TECHNOLOGY ANALYSIS OF VOLTAGE SHARING IN SERIES CONNECTED POWER DEVICES 1149

Z Davletzhanova ; O Alatise ; R Bonyadi ; J Ortiz-Gonzalez ; T Dai ; M Jennings ; L Ran ; P Mawby

FAILURE MECHANISM ANALYSIS AND PHYSICS-OF-FAILURE LIFETIME PREDICTION METHOD FOR PRESS-PACK THYRISTOR OF CONVERTER VALVE 1157

Ning Liang ; Zhigang Zhang ; Yating Gou ; Cuicui Liu ; Zebin Yang ; Jiangnan Chen ; Fang Zhuo ; Feng Wang

SURGE VOLTAGE ABSORPTION BY A SILICON CARBIDE AVALANCHE-DIODE WITH P-N STRUCTURE 1162

K. Koseki ; Y. Tanaka

CALCULATION OF THYRISTOR RELIABILITY PARAMETER OF UHVDC CONVERTER VALVE IN HEMP ENVIRONMENT 1167

Zhigang Zhang ; Yating Gou ; Cuicui Liu ; Zebin Yang ; Xiaotong Du ; Jiangnan Chen ; Fang Zhuo ; Feng Wang ; Yuanliang Lan ; Caiwang Sheng

GENERALIZED STACKELBERG GAME-THEORETIC APPROACH FOR JOINTED ENERGY AND RESERVE COORDINATION OF ELECTRIC VEHICLES 1172

Tianyang Zhao ; Xuewei Pan ; Lei Li ; Fei Zhao ; Can Wang

IMPEDANCE INFLUENCE ANALYSIS OF PHASE-LOCKED LOOPS ON THREE-PHASE GRID-CONNECTED INVERTERS 1177

Yuncheng Wang ; Xin Chen ; Yang Zhang ; Jie Chen ; Chunying Gong

PULSE-INJECTION-BASED SENSORLESS CONTROL METHOD WITH IMPROVED DYNAMIC CURRENT RESPONSE FOR PMSM 1183

Hechao Wang ; Kaiyuan Lu ; Dong Wang ; Frede Blaabjerg

INFLUENCE OF PARAMETER VARIATIONS ON OPERATING CHARACTERISTICS OF MTPF CONTROL FOR DTC-BASED PMSM DRIVE SYSTEM 1189

Keisuke Fujii ; Yukinori Inoue ; Shigeo Morimoto ; Masayuki Sanada

A QUIET POSITION SENSORLESS CONTROL FOR AN IPMSM BASED ON EXTENDED EMF AND VOLTAGE INJECTION SYNCHRONIZED WITH PWM CARRIER 1196

Yuki Ishii ; Hiroki Yamashita ; Hisao Kubota

STUDY OF TORQUE RIPPLE REDUCTION AND TORQUE BOOST BY MODIFIED TRAPEZOIDAL MODULATION 1202

Satoshi Joryo ; Kazuto Tatsumi ; Toshimitsu Morizane ; Katsunori Taniguchi ; Noriyuki Kimura ; Hideki Omori

FAULT DIAGNOSIS METHOD OF CURRENT SENSOR FOR PERMANENT MAGNET SYNCHRONOUS MOTOR DRIVES 1206

Guoqiang Zhang ; Guoxin Wang ; Gaolin Wang ; Junya Huo ; Lianghong Zhu ; Dianguo Xu

SENSORLESS SPEED CONTROL OF DIESEL-GENERATOR SYSTEMS BASED ON MULTIPLE SOGI-FLLS 1212
Ngoc Dat Dao ; Dong-Choon Lee ; Dae-Sik Lim

ROBUSTNESS OF SIMPLIFIED SPEED-SENSORLESS VECTOR CONTROL FOR INDUCTION MOTOR 1217
Naoki Akao ; Mineo Tsuji ; Shin-Ichi Hamasaki

MAXIMUM TORQUE CONTROL REFERENCE FRAME BASED ON A TORQUE MAP FOR IPMSMS WITH LARGE INDUCTANCE VARIATION 1223
Kazuki Ohta ; Takumi Ohnuma ; Shinji Doki

PMSM MODEL DISCRETIZATION IN CONSIDERATION OF PARK TRANSFORMATION FOR CURRENT CONTROL SYSTEM 1228
Masamichi Inoue ; Shinji Doki

PSEUDO-RANDOM HIGH-FREQUENCY SINUSOIDAL VOLTAGE INJECTION BASED SENSORLESS CONTROL FOR IPMSM DRIVES 1234
Guoqiang Zhang ; Huiying Wang ; Gaolin Wang ; Junya Huo ; Lianghong Zhu ; Dianguo Xu

AT-NPC 3-LEVEL INVERTER-FED INDUCTION MOTOR VECTOR CONTROL WITH NEUTRAL POINT VOLTAGE CONTROL 1240
K. Sudo ; M. Tsuji ; S. Hamasaki ; T. Fukuoka ; H. Ichinose

INVESTIGATION OF VARIOUS POSITION ESTIMATION ACCURACY ISSUES IN PULSE-INJECTION-BASED SENSORLESS DRIVES 1246
Hechao Wang ; Kaiyuan Lu ; Dong Wang ; Frede Blaabjerg

POSITION SENSORLESS CONTROL OF SWITCHED RELUCTANCE MOTOR USING ESTIMATED PWM PHASE VOLTAGE 1253
Y. Nakazawa ; K. Ohyama ; H. Fujii ; H. Uehara ; Y. Hyakutake

EXPERIMENTAL CONFIRMATION OF THRUST AND ATTRACTIVE FORCE CONTROL OF LINEAR INDUCTION MOTOR BY TWO DIFFERENT FREQUENCY COMPONENTS 1259
Kenta Sannomiya ; Toshimitsu Morizane ; Noriyuki Kimura ; Hideki Omori

GA BASED OPTIMIZED TRAJECTORIES OF ROTATING SPEED AND D-Q AXIS CURRENTS FOR AN IPMSM 1264
Shuta Kumagai ; Kaoru Inoue ; Toshiji Kato

2-DEGREE-OF-FREEDOM DEADBEAT CONTROL WITH DISTURBANCE COMPENSATION FOR PMSM DRIVE SYSTEM USING FPGA 1270
Arata Takahashi ; Shotaro Takakura ; Tomoki Yokoyama

EXTENDED EMF-BASED SIMPLE IPMSM SENSORLESS VECTOR CONTROL USING COMPENSATED CURRENT CONTROLLER 1276
Takatoshi Inoue ; Yasumasa Hamabe ; Mineo Tsuji ; Shin-Ichi Hamasaki

FULL-BAND OUTPUT IMPEDANCE MODEL OF VIRTUAL SYNCHRONOUS GENERATOR IN DQ FRAMEWORK 1282
Li Wenbing ; Wang Jianhua ; Song Jingyu ; Luo Fangfang ; Gao Shang ; Wu Zaijun

AN MTPA CONTROL METHOD OF A PMSM AND A SYNRM BASED ON A DTC IN THE STATOR FLUX LINKAGE SYNCHRONOUS FRAME 1289
Gimpei Itoh ; Yukinori Inoue ; Shigeo Morimoto ; Masayuki Sanada

EEMFS EXCITED BY SIGNAL INJECTION FOR POSITION SENSORLESS CONTROL OF PMSMS AND THEIR PERFORMANCE COMPARISON BY USING IMAGINARY ELECTROMOTIVE FORCE 1295
Takumi Nimura ; Shota Kondo ; Shinji Doki ; Mutuwo Tomita

HARMONIC CURRENT CANCELLATION METHOD FOR PMSM DRIVE SYSTEM USING RESONANT CONTROLLERS 1301
Dongsheng Li ; Yoshitaka Iwaji ; Yasuo Notohara ; Ken Kishita

ESTIMATION ERROR ANALYSIS OF STATOR FLUX OBSERVER FOR DTC-BASED PMSM DRIVES 1308
Atsushi Shinohara ; Kichiro Yamamoto

APPLICATION OF FICTITIOUS REFERENCE ITERATIVE TUNING TO CONTROLLER DESIGN FOR VARIOUS MACHINES 1315
Hidehiro Ikeda ; Kazuya Goto ; Feili Zhang ; Kazuya Kayashima ; Tsuyoshi Hanamoto

HIGH EFFICIENCY CONTROL FOR PERMANENT MAGNET MOTOR DRIVE SYSTEM WITH FUEL CELLS CONNECTED IN SERIES WITH ELECTRIC DOUBLE-LAYER CAPACITORS 1322
Kichiro Yamamoto ; Fumiya Ohdera ; Atsushi Shinohara

COMPARATIVE STUDY OF SPEED RIPPLE REDUCTION BY VARIOUS CONTROL METHODS IN PMSM DRIVE SYSTEMS WITH PULSATING LOAD 1329
Yuma Komaru ; Yukinori Inoue ; Shigeo Morimoto ; Masayuki Sanada

ESTIMATION OF THE PARAMETERS OF THE SERVO DRIVE SYSTEM USING PARTICLE SWARM OPTIMIZATION ALGORITHM 1336

Helin Zhu ; Jae Hyuk Choi ; Sang Uk Park ; Jusuk Lee ; Hyong Gun Lee ; Hyung Soo Mok

A PROGRAMMABLE BATTERY TEST SYSTEM WITH ENERGY RECYCLING FEATURE BASED ON SINUSOIDAL LOADING TECHNIQUE 1341

Chang-Hua Lin ; Guan-Jung Chen ; Hwa-Dong Liu ; Kun-Feng Chen

DEVELOPMENT OF LARGE-CAPACITY CONVERTER FOR BATTERY ENERGY STORAGE SYSTEMS 1346

Hiroyoshi Komatsu ; Tatsuji Katayama ; Noriko Kawakami

ANALYSIS AND COMPARISON OF DC/DC TOPOLOGIES IN PARTIAL POWER PROCESSING CONFIGURATION FOR ENERGY STORAGE SYSTEMS 1351

Maria C. Mira ; Zhe Zhang ; A. E. Michael Andersen

TWO-STAGE PROTECTION FOR MULTI-CHANNEL POWER ELECTRONIC CONVERTERS FED LARGE ASYNCHRONOUS HYDRO-GENERATING UNIT 1358

R. R. Semwal ; Anto Joseph

CURRENT SHARING CONTROL FOR SERIES-PARALLEL CHANGEOVER USING BATTERY AND ELECTRIC DOUBLE-LAYER CAPACITOR BANK 1364

Taisei Nishino ; Keisaku Isozaki ; Naoki Kogai ; Kyungmin Sung

CONTROL METHOD OF ENERGY STORAGE SYSTEM TO IMPROVE OUTPUT POWER OF PCS 1370

Mikiya Ishibashi ; Hitoshi Haga ; Kenji Arimatsu ; Koji Kato

A CONTROL STRATEGY OF MMC BATTERY ENERGY STORAGE SYSTEM BASED ON ARM CURRENT CONTROL 1376

Liu Danqing ; Wang Guangzhu ; Ou Zhujian ; Liu Jiaxing

EQUIVALENT RESISTANCE CONTROL FOR MAXIMUM POWER TRANSFER METHOD OF PIEZOELECTRIC ELEMENT IN VIBRATION POWER GENERATION 1381

Kenya Takamura ; Hiroaki Yamada ; Toshihiko Tanaka ; Tomoharu Yada ; Hajime Fujiwara

DC BUS VOLTAGE STABILIZATION FOR CASCADED POWER CONVERTER BY INTEGRATING AN EXTRA PORT INTO LOAD SIDE PSFB 1386

Jiang You ; Weiyan Fan ; Mengyan Liao

COMMON MODE CURRENT REDUCTION OF THREE-PHASE CASCADED MULTILEVEL TRANSFORMERLESS INVERTER FOR PV SYSTEM 1391

Wenjie Wang ; Ke Chen ; Lijun Hang ; Anping Tong ; Yiliang Gan

CURRENT SHARING/VOLTAGE SHARING CONTROL STRATEGY FOR CASCADED DC/DC CONVERTER IN PHOTOVOLTAIC DC COLLECTION SYSTEM 1397

Bo Chen ; Yi Wang ; Yanjun Tian ; Shilei Wei

PCC VOLTAGE COMPENSATION OF PV INVERTER WITH ACTIVE POWER DECOUPLING CIRCUIT 1403

Duck-Hwan Hwang ; Jung-Yong Lee ; Younghoon Cho

A NOVEL PARTIAL SHADING DETECTION ALGORITHM UTILIZING POWER LEVEL MONITORING OF PHOTOVOLTAIC PANELS 1409

Thusitha Randima Wellawatta ; Sung-Jin Choi

BOOST INTEGRATED THREE-PHASE SOLAR INVERTER USING CURRENT UNFOLDING AND ACTIVE DAMPING METHODS 1414

N. Ha Pham ; Tomoyuki Mannen ; Keiji Wada

LINEAR ACTIVE DISTURBANCE REJECTION CONTROL FOR ISOLATED THREE-PORT CONVERTER 1421

Jiang You ; Mengyan Liao ; Weiyan Fan

STABILITY CONSTRAINED GAIN OPTIMIZATION OF DROOP CONTROLLED CONVERTERS IN DC NANOGRIDS 1426

Soumya Bandyopadhyay ; Laura Ramirez-Elizondo ; Pavol Bauer

SIC BASED SSPC FOR HIGH VOLTAGE SPACE APPLICATIONS 1435

D. Marroquí ; A. Garrigós ; José M. Blanes ; R. Gutiérrez

AN IMPROVED VOLTAGE-TYPE GRID-CONNECTED CONTROL STRATEGY FOR COMPENSATING UNBALANCED VOLTAGE 1442

Liu Hongpeng ; Zhou Jiajie ; Wang Wei

DUAL TWO-STAGE ISOLATED BIDIRECTIONAL DC-DC CONVERTER FOR DC GRID STORAGE 1447

Gabriel Tibola ; Jorge L. Duarte

MODULAR MULTILEVEL CONVERTER WITH CAPACITOR VOLTAGE SELF-BALANCING USING REDUCED NUMBER OF VOLTAGE SENSORS 1455

Taiyuan Yin ; Yue Wang ; Xiaolei Wang ; Shiyuan Yin ; Shumin Sun ; Guanglei Li

PLUG AND OUTLET IN HOUSEHOLD DC LOW VOLTAGE MICRO-GRID POWER DISTRIBUTION 1460
Worapong Pairindra ; Surin Khomfoi

PERFORMANCE PROGRAMMING TECHNIQUE FOR MULTI-STAGE DC POWER DISTRIBUTION SYSTEMS 1465
Syam Kumar Pidaparthy ; Hansang Kim ; Yeonjung Kim ; Byungcho Choi

COORDINATION CONTROL FOR PARALLELED INVERTERS BASED ON VSG FOR PV/BATTERY MICROGRID 1472
Meiqin Mao ; Cheng Qian ; Liuchen Chang ; Yan Du

ADAPTIVE VOLTAGE CONTROL SCHEME FOR DAB BASED MODULAR CASCADED SST IN PV APPLICATION 1478
Tao Liu ; Yang Xuan ; Xu Yang ; Peng Xu ; Yang Li ; Lang Huang ; Xiang Hao

SIX-STEP MMC-BASED HIGH POWER DC-DC CONVERTER 1484
Stefan Milovanovic ; Dražen Dujic

COMBINED DC POWER FLOW CONTROLLER FOR DC GRID 1491
Yongning Chi ; Xizhou Du ; Siqi Liu ; Xu Cai

AN APPROACH FOR THE EMULATION OF DC GRID ADMITTANCES: IMPLEMENTATION ON A BUCK CONVERTER 1498
Enrique Rodriguez-Diaz ; Fracisco D. Freijedo ; Drazen Dujic ; Juan C. Vasquez ; Josep M. Guerrero

A COMPOUND CONTROLLER FOR POWER FLOW AND SHORT-CIRCUIT FAULT IN DC GRID 1504
Han Ye ; Wu Chen ; Pengpeng Pan ; Xiaokun He

DESIGN PROCEDURE AND CONTROL OF A HYBRID CIRCUIT BREAKER WITH ADAPTABLE PULSE CURRENT INJECTION 1509
Andreas Jehle ; Jürgen Biela

A PRAGMATIC SOH AND SOC CO-ESTIMATOR FOR LITHIUM-ION BATTERIES IN SMART GRID APPLICATIONS 1517
Kaiyuan Li ; King Jet Tseng ; Feng Wei ; Boon-Hee Soong

MODELING AND STABILITY ANALYSIS OF PARALLEL DROOP-CONTROLLED AND CURRENT-CONTROLLED INVERTERS 1524
Shike Wang ; Zeng Liu ; Jinjun Liu ; Ronghui An

DIRECT WIRELESS BATTERY CHARGING SYSTEM 1530
Woo-Seok Lee ; Jin-Hak Kim ; Shin-Young Cho ; Il-Oun Lee

AN IMPROVED PWM SCHEME TO ACHIEVE ZERO-VOLTAGE SWITCHING FOR ALL DEVICES IN THREE-PHASE ISOLATED MATRIX RECTIFIER 1537
Xuerui Lin ; Yunwei Ryan Li ; Jahangir Afsharian ; Dewei David Xu

FIXED-FREQUENCY HF GATE DRIVER BY A PUSH-PULL SELF-EXCITATION LC OSCILLATOR HAVING A CAPACITANCE TRANSISTOR 1543
Naoyuki Ishibashi ; Takuya Mizushima ; Masahiko Hirokawa ; Akihiko Katsuki

A FLEXIBLE REDUCED CAPACITOR VOLTAGES STRATEGY FOR VARIABLE-SPEED DRIVES WITH MODULAR MULTILEVEL CONVERTER 1549
Fangzhou Zhao ; Guochun Xiao ; Daoshu Yang ; Zhiqian Wu ; Xin Meng

A LEAKAGE FLUX CANCELLATION TECHNIQUE FOR SERIES-PARALLEL COMBINED RESONANT CIRCUITS WITH ASYMMETRIC ROTARY TRANSFORMERS USED FOR ULTRASONIC SPINDLE DRIVE 1554
Jun Imaoka ; Masahito Shoyama

A NOVEL STRUCTURAL HEALTH MONITORING SYSTEM WITH WIRELESS POWER AND BI-DIRECTIONAL DATA TRANSFER 1562
Yujin Jangs ; Keon-Woo Kim ; Moo-Hyun Park ; Nayoung Lee ; Gun-Woo Moon

CONTROL STRATEGY FOR STARTER GENERATOR IN UAV WITH MICRO JET ENGINE 1567
Jun-Ichi Itoh ; Kazuki Kawamura ; Hiroyuki Koshikizawa ; Kazuyuki Abe

STUDY ON THE INFLUENCE OF VOLTAGE VARIATIONS FOR NON-INTRUSIVE LOAD IDENTIFICATIONS 1575
Yu-Hsiu Lin ; Shun-Kang Hung ; Men-Shen Tsai

BASIC EXPERIMENT OF A MAGLEV SYSTEM FOR A FLEXIBLE STEEL PLATE WITH CURVATURE: FUNDAMENTAL CONSIDERATION ON LEVITATION STABILITY UNDER DISTURBANCE 1580
Makoto Tada ; Kazuki Ogawa ; Takayoshi Narita ; Hideaki Kato ; Hiroyuki Moriyama

PERFORMANCE OF HYBRID MAGNETIC LEVITATION CONTROL SYSTEM FOR THIN STEEL PLATE BY EMS AND PMS: EXPERIMENTAL EVALUATION OF APPLYING OPTIMAL GAP AND ARRANGEMENT OF PMS 1586

Yasuaki Ito ; Yoshiho Oda ; Kengo Okuno ; Toshiki Suzuki ; Masahiro Kida ; Takayoshi Narita ; Hideaki Kato ; Hiroyuki Moriyama

A PRACTICAL LITHIUM-ION BATTERY MODEL BASED ON THE BUTLER-VOLMER EQUATION 1592

Kaiyuan Li ; King Jet Tseng ; Feng Wei ; Boon-Hee Soong

BONDING TECHNOLOGY USING COLD-ROLLED AG SHEET IN DIE-ATTACHMENT APPLICATIONS 1598

Seungjun Noh ; Chanyang Choe ; Chuantong Chen ; Hao Zhang ; Katsuaki Suganuma

HIGH-FREQUENCY SELF-DRIVEN SYNCHRONOUS RECTIFIER CONTROLLER FOR WPT SYSTEMS 1602

Akihiro Konishi ; Kazuhiro Umetani ; Eiji Hiraki

AUTOMATIC RESONANCE FREQUENCY TUNING METHOD FOR REPEATER IN RESONANT INDUCTIVE COUPLING WIRELESS POWER TRANSFER SYSTEMS 1610

Masataka Ishihara ; Kazuhiro Umetani ; Eiji Hiraki

INDUCTIVE POWER TRANSFER FOR T5 FLUORESCENT LAMP LIGHTING SYSTEM 1617

Chung-Chuan Hou ; Tang-Jung Chen ; Ching-Chen Chen ; Chen-Wei Chang ; Po-Wei Wang

AN IMPLEMENT 1.5 MHZ OF INDUCTION HEATING FOR ALUMINUM BASED ON VACUUM TUBE OSCILLATOR CIRCUIT 1622

A. Bilsalam ; P. Chanmontree ; S. Supanyapong ; V. Chunkag

SINGLE-INDUCTOR MULTIPLE-OUTPUTS DIMMABLE LED DRIVER WITH BUCK CONVERTER 1626

Ta-Wei Huang ; Wei-Jing Tseng ; Jun-Xian Huang

A SOFT-SWITCHED THREE-LEVEL T-TYPE INVERTER WITH AUXILIARY COMMUTATED POLES 1634

Apollo Charalambous ; Xibo Yuan

CARRIER-BASED REALIZATION OF ARBITRARY SPACE-VECTOR PWM METHODS FOR THREE-LEVEL INVERTERS 1642

Somboon Sangwongwanich ; Supakorn Paiboon

MULTI-LEVEL TOPOLOGY BASED LINEAR AMPLIFIER FAMILY FOR REALIZATION OF NOISE-LESS INVERTERS 1649

Hidemine Obara ; Tatsuki Ohno ; Atsuo Kawamura

A NEW ZERO-VOLTAGE SWITCHING THREE-LEVEL CONVERTER WITH REDUCED RECTIFIER VOLTAGE STRESS 1655

Keon-Woo Kim ; Cheon-Yong Lim ; Dong-Kwan Kim ; Yu-Jin Jang ; Gun-Woo Moon

MODEL PREDICTIVE CONTROL OF A THREE-LEVEL NPC RECTIFIER WITH A SLIDING MANIFOLD TERM 1661

Xiaonan Gao ; Wei Tian ; Xicai Liu ; Zhenbin Zhang ; Ralph Kennel

H∞ CONTROL-BASED VIBRATION SUPPRESSION IN ROBOT ARM WITH STRAIN WAVE GEARING 1666

Tran Vu Trung ; Makoto Iwasaki

FINE FORCE SENSORLESS FORCE CONTROL BASED ON FRICTION-FREE DISTURBANCE OBSERVER 1673

Ohishi Kiyoshi ; Naoki Kamiya ; Toshimasa Miyazaki ; Yuki Yokokura

KINEMATICS AND TRACKING CONTROL OF A FOUR AXIS ANTENNA FOR SATCOM ON THE MOVE 1680

Oguz Kaan Hancioglu ; Mustafa Celik ; Ugur Tumerdem

POSITION SENSORLESS POSITION CONTROL FOR DUAL SOLENOID ACTUATOR 1687

Sakahisa Nagai ; Atsuo Kawamura

CAE TECHNOLOGY APPLICATION TREND FOR LARGE-CAPACITY POWER ELECTRONICS DEVELOPMENT 1692

Teruo Yoshino ; Kuniaki Nagasaka ; Shigeaki Nakabayashi ; Ikuto Udagawa ; Isamu Tominaga ; Junya Konno

XILINX SYSTEM GENERATOR BASED MODELLING OF FINITE STATE MPC 1698

Vijay Kumar Singh ; Ravi Nath Tripathi ; Tsuyoshi Hanamoto

POWER HARDWARE-IN-THE-LOOP SETUP FOR STABILITY STUDIES OF GRID-CONNECTED POWER CONVERTERS 1704

Tommi Reinikka ; Henrik Alenius ; Tomi Roinila ; Tuomas Messo

PASSIVITY-BASED LCL FILTER DESIGN OF GRID-CONNECTED VSCS WITH CONVERTER SIDE CURRENT FEEDBACK 1711

Shih-Feng Chou ; Xiongfei Wang ; Frede Blaabjerg

ADAPTIVE CONTROL OF DC POWER DISTRIBUTION SYSTEMS: APPLYING PSEUDO-RANDOM SEQUENCES AND FOURIER TECHNIQUES..........1719

Tomi Roinila ; Hessamaldin Abdollahi ; Silvia Arrua ; Enrico Santi

AN IMPROVED FINITE-SET MODEL PREDICTIVE TORQUE CONTROL FOR INTERIOR PERMANENT MAGNET SYNCHRONOUS MOTOR DRIVES..........1724

Xinan Zhang ; Gilbert Foo ; Tung Ngo

PREDICTIVE TORQUE CONTROL FOR FIVE PHASE INDUCTION MOTOR DRIVE WITH COMMON MODE VOLTAGE REDUCTION..........1730

Apekshit Bhowate ; Mohan Aware ; Sohit Sharma ; Yogesh Tatte

INDIRECT MATRIX CONVERTER FOR PERMANENT-MAGNET-SYNCHRONOUS-MOTOR DRIVES BY IMPROVED TORQUE PREDICTIVE CONTROL..........1736

Yun Jang ; Yeongsu Bak ; Kyo-Beum Lee

PREDICTIVE DC-LINK CURRENT CONTROL BASED ON IPMSM DISCRETE STATE EQUATION FOR INVERTER WITHOUT INDUCTOR OR ELECTROLYTIC CAPACITOR..........1741

Yousuke Akama ; Kodai Abe ; Kiyoshi Ohishi ; Yuki Yokokura ; Koji Kobayashi ; Tatsuki Kashihara

NEW SEARCH ALGORITHM OF MODEL PREDICTIVE CONTROL TO REDUCING CALCULATION AMOUNT FOR IMPROVING STEADY CURRENT CONTROL PERFORMANCE..........1747

Masahiro Shimaoka ; Shinji Doki

DISTRIBUTED POWER SHARING STRATEGY FOR ISLANDED MICROGRIDS WITHOUT FREQUENCY AND VOLTAGE DEVIATIONS..........1752

Tuan V. Hoang ; Hong-Hee Lee

LIFETIME-ORIENTED DROOP CONTROL STRATEGY FOR AC ISLANDED MICROGRIDS..........1758

Yanbo Wang ; Dong Liu ; Fujin Deng ; Dao Zhou ; Zhe Chen

EXPERIMENT ON HIERARCHICAL CONTROL BASED POWER QUALITY ENHANCEMENT FOR STANDALONE MICROGRID..........1764

Darith Leng ; Sompob Polmai ; Kittichot Soontorntaweesub

A DISTRIBUTED PREDICTIVE CONTROL STRATEGY BASED ON STATE ESTIMATOR FOR ISLANDED MICROGRID..........1771

Mi Dong ; Li Li ; Xiaoyu Tian

MAXIMUM POWER POINT TRACKING METHOD FOR PV MODULE UNDER WIDE RANGE VARYING IRRADIANCE LEVELS..........1777

Hwa-Dong Liu ; Chang-Hua Lin

DUAL MPPT CONTROL AND FIELD TESTING FOR SWITCHED CAPACITOR-BASED CELL-LEVEL POWER BALANCING UTILIZING DIFFUSION CAPACITANCE OF PHOTOVOLTAIC CELLS..........1782

Masatoshi Uno ; Yota Saito ; Masaya Yamamoto ; Shinichi Urabe

SERIES RESONANT DC-DC CONVERTER WITH DUAL-MODE RECTIFIER FOR PV MICROINVERTERS..........1788

Yanfeng Shen ; Huai Wang ; Zhan Shen ; Yongheng Yang ; Frede Blaabjerg

VOLTAGE-REFERENCE ACTIVE POWER DECOUPLING BASED ON BOOST CONVERTER FOR SINGLE-PHASE BRIDGE INVERTER..........1793

Shuang Xu ; Meiqin Mao ; Riming Shao ; Liuchen Chang

A SINGLE-PHASE COMMON GROUND BOOST INVERTER FOR PHOTOVOLTAIC APPLICATIONS..........1799

Tan-Tai Tran ; Minh-Khai Nguyen ; Young-Cheol Lim ; Joon-Ho Choi

STUDY FOR FURTHER INTRODUCTION OF THE ELECTRONIC FREQUENCY CONVERTERS TO THE TOKAIDO SHINKANSEN..........1803

Toshimasa Shimizu ; Ken Kunomura ; Masahiko Kai ; Hiroki Miyajima ; Teruhisa Matsui

COUNTERMEASURE FOR PARTIAL TURN-OFF OF THYRISTOR CHANGEOVER SWITCH INTRODUCED TO TOHOKU SHINKANSEN SHIN-YONO SECTIONING POST..........1810

Yuki Mizumoto ; Nobuhito Kurosawa

HARDWARE–IN–THE–LOOP REAL–TIME SIMULATION EXPERIMENT PLATFORM FOR TRACTION POWER SUPPLY SYSTEM BASED ON DSPACE-XSIM..........1816

Runze Zhang ; Fei Lin ; Zhongping Yang ; Hu Cao ; Yuping Liu

EVALUATING THE NON-SINUSOIDAL AND NON-SYMMETRIC REGIMES FROM A RAILWAY SUPPLYING SUBSTATION..........1822

Ileana-Diana Nicolae ; Petre-Marian Nicolae ; Radu-Florin Marinescu

A FUNDAMENTAL TRAIN RUNNING EXPERIMENT FOR A BASIC PERFORMANCE VERIFICATION OF A TRAIN POWER DEMAND CONTROL SYSTEM BY DECENTRALIZED CONTROL ALGORITHM..........1828

Yusuke Oki ; Tomoyuki Ogawa ; Yoko Takeuchi ; Tatsuhito Saito ; Jun'ichiro Kawaguchi

VERIFICATION OF SIC BASED MODULAR MULTILEVEL CASCADE CONVERTER (MMCC) FOR HVDC TRANSMISSION SYSTEMS 1834
Y. Ishii ; T. Jimichi

CONTROL OF A 6.6-KV TRANSFORMERLESS STATCOM BASED ON THE MMCC-SDBC USING SIC MOSFETS 1840
Laxman Maharjan ; Toshihisa Tajyuta ; Hiroshi Shinohara ; Akio Suzuki ; Akio Toba

ISOLATED THREE–PHASE AC/DC CONVERTER USING A SOFT–SWITCHING TECHNIQUE FOR BATTERY CHARGER 1847
Yuto Matsui ; Kazuma Suzuki ; Takaharu Takeshita ; Wataru Kitagawa

IMPLEMENTATION OF A MINIATURIZED SIC INVERTER 1854
Hideaki Fujita ; Cristian Andres Garces Guajardo

DESIGN CONSIDERATION OF FLYING CAPACITOR MULTILEVEL INVERTERS USING SIC MOSFETS 1860
Yukihiko Sato ; Kenji Natori

A CONTROL METHOD OF OVERVOLTAGE SUPPRESSION ACROSS THE DC CAPACITOR IN A GRID-CONNECTION CONVERTER USING LEG SHORT-CIRCUIT OF POWER MOSFETS DURING THE INITIAL CHARGE 1866
Tomoyuki Mannen ; Keiji Wada

THE ESSENTIAL RELATIONSHIP BETWEEN DEADBEAT PREDICTIVE CONTROL AND CONTINUOUS-CONTROL-SET MODEL PREDICTIVE CONTROL FOR PWM CONVERTERS 1872
Bi Liu ; Tao Chen ; Wensheng Song

DEADBEAT CONTROL FOR MULTI-LEVEL INVERTER USING 1MHZ MULTISAMPLING METHOD FOR UTILITY INTERACTIVE SYSTEM 1877
Ryosuke Kikuchi ; Ryunosuke Araumi ; Tomoki Yokoyama

1MHZ MULTISAMPLING DEADBEAT CONTROL WITH DISTURBANCE COMPENSATION METHOD FOR THREE PHASE PWM INVERTER 1883
Hiroaki Ueta ; Tomoki Yokoyama

MODULAR MULTILEVEL CONVERTER REPLACED ONE MODULE WITH HIGH VOLTAGE IGBT 1890
Kazunobu Oi ; Kenta Takasho ; Yugo Tadano

INCREASED EFFICIENCY AND REDUCED REALIZATION EFFORT OF DSBC AND DSCC MODULAR MULTILEVEL CONVERTERS (MMCS) 1896
A. Hillers ; J. Biela

COMMON-MODE VOLTAGE INJECTION TECHNIQUES FOR QUASI TWO-LEVEL PWM-OPERATED MODULAR MULTILEVEL CONVERTERS 1904
Jakub Kucka ; Axel Mertens

CURRENT TRACKING AND CELL-VOLTAGE LIMITATIONS OF MODULAR MULTILEVEL CONVERTERS WITH DIRECT DIGITAL CONTROL 1912
T.-F. Wu ; T.-C. Chou ; K.-E. Lin ; T.-Y. Li

SWITCHING LOSS ANALYSIS OF SIC-MOSFET BASED ON STRAY INDUCTANCE SCALING 1919
Keiji Wada ; Masato Ando

MODELING AND OPTIMIZATION OF DISPLACEMENT WINDINGS FOR TRANSFORMERS IN DUAL ACTIVE BRIDGE CONVERTERS 1925
Zhan Shen ; Yanfeng Shen ; Zian Qin ; Huai Wang

OPTIMIZED SELECTION AND UTILIZATION OF DC-LINK CAPACITOR IN A SINGLE-PHASE PV GRID INVERTER SYSTEM 1931
Caspar Collins ; Li Ran

AN EVALUATION CIRCUIT FOR DC-LINK CAPACITORS USED IN A HIGH-POWER THREE-PHASE INVERTER WITH CONDITION MONITORING 1938
Kazunori Hasegawa ; Ichiro Omura ; Shin-Ichi Nishizawa

RECENT MARKET AND TECHNICAL TRENDS IN COPPER ROTORS FOR HIGH-EFFICIENCY INDUCTION MOTORS 1943
Daniel Liang ; Victor Zhou

OVERVIEW OF THE LATEST RESEARCH AND DEVELOPMENT FOR COPPER DIE-CAST SQUIRREL-CAGE ROTORS 1949
Shu Yamamoto

A NOVEL HEAT-RESISTANT INSULATION-PROCESSING AGENT APPLICABLE TO COPPER DIE-CAST SQUIRREL-CAGE ROTORS 1955
Junichi Uchida ; Yuki Sueuchi ; Naosumi Kamiyama

INSULATION-PROCESSING OF COPPER DIE-CAST SQUIRREL-CAGE ROTOR ON MOTOR EFFICIENCY IN HIGH-SPEED OPERATION OVER 10,000 R/MIN 1960
Hideaki Hirahara ; Akira Tanaka ; Shu Yamamoto

HIGH-PRECISION ROTOR POSITION ESTIMATION FOR HIGH-SPEED SPMSM DRIVE BASED ON STATE OBSERVER AND HARMONIC ELIMINATION 1966

Peng Yang ; Xi Xiao ; Meng Zhang ; Shkodyrev Vyacheslav

HARMONIC LOSS REDUCTION IN HIGH SPEED MOTOR DRIVE SYSTEMS BY FLYING CAPACITOR MULTILEVEL INVERTER 1972

Anudari Tumurbaatar ; Sae Mochidate ; Koji Yamaguchi ; Tomohiro Matsuda ; Yukihiko Sato

CURRENT SOURCE TYPE PMSG WIND TURBINE SYSTEM WITH THREE-PHASE THREE-SWITCH BUCK-TYPE RECTIFIER FOR MACHINE-SIDE CONVERTER 1977

Beomseok Chae ; Tahyun Kang ; Yongsug Suh

A STUDY OF 10MW LOAD COMMUTATED INVERTER FOR GAS-TURBINE START-UP 1985

An Hyunsung ; Cha Hanju

PROTOTYPING OF 500 KVA MEDIUM FREQUENCY TRANSFORMER FOR OFFSHORE DIRECT-CURRENT COLLECTION GRID 1991

Tomoyuki Hatakeyama ; Naoyuki Kurita ; Mamoru Kimura

PSCAD/EMTDC AND RTDS SIMULATION ANALYSIS OF MULTIVENDOR MULTI-TERMINAL HVDC SYSTEM CONNECTED TO OFFSHORE WINDFARMS 1997

Hiroshi Suwa ; Takuro Arai ; Takahiro Ishiguro ; Tohru Yoshihara ; Mamoru Kimura ; Tsuneshisa Wachi ; Takahiro Horikoshi ; Tatsuhito Nakajima

INTEROPERABILITY OF MODULAR MULTILEVEL CONVERTERS AND 2-LEVEL VOLTAGE SOURCE CONVERTERS IN A LABORATORY-SCALE MULTI-TERMINAL DC GRID 2003

Salvatore D'arco ; Atsede G. Endegnanew ; Giuseppe Guidi ; Jon Are Suul

PRINCIPLE EXPERIMENT OF CURRENT COMMUTATED HYBRID DCCB FOR HVDC TRANSMISSION SYSTEMS 2011

Ryuta Hasegawa ; Kazuhisa Kanaya ; Yushi Koyama ; Toshiaki Matsumoto ; Takahiro Ishiguro

A THREE-INPUT CENTRAL CAPACITOR DC/DC CONVERTER 2016

Jiaxin Liu ; Feng Gao

SERIES/PARALLEL SWITCHING CIRCUITS USING POWER MOSFETS FOR PHOTOVOLTAIC MODULES 2022

Masamichi Tanemo ; Koki Matsudate ; Shinichi Nomura

MODULARIZED EQUALIZATION ARCHITECTURE BASED ON SWITCHED CAPACITOR CONVERTER TO VIRTUALLY UNIFY MISMATCHED PHOTOVOLTAIC PANEL CHARACTERISTICS 2030

Masatoshi Uno ; Masaya Yamamoto

BUCK-BOOST TYPE MPPT CIRCUIT SUITABLE FOR PHOTOVOLTAIC GENERATION OF VEHICLE INSTALLATION 2036

Fumihisa Kano ; Yuji Kasai ; Hideki Kimura ; Kouhei Sagawa ; Junnosuke Haruna ; Hirohito Funato

VERIFICATION TEST OF ENERGY-EFFICIENT OPERATIONS AND SCHEDULING UTILIZING AUTOMATIC TRAIN OPERATION SYSTEM 2042

Shoichiro Watanabe ; Yasuhiro Sato ; Takafumi Koseki ; Eisuke Isobe ; Jun Kawashita

THE DIRECT BENEFIT OF SIC POWER SEMICONDUCTOR DEVICES FOR RAILWAY VEHICLE TRACTION INVERTERS 2047

Shingo Makishima ; Kazuki Fujimoto ; Keiichiro Kondo

THE LOSS CHARACTERISTICS OF PSFB ZVS DC-DC CONVERTER APPLIED TO THE AUXILIARY POWER SYSTEM 2051

Xianjin Huang ; Juan Zhao ; Fei Lin

SURVEY ON ELECTROMAGNETIC INTERFERENCE ANALYSIS FOR TRACTION CONVERTERS IN RAILWAY VEHICLES 2058

Zhichang Yang ; Hong Li ; Chao Feng ; Yanfeng Jiang ; Fei Lin ; Zhongping Yang

DEVELOPMENT OF TRACTION MOTOR FOR NEW ZERO - EMISSION VEHICLE 2066

Akinobu Iwai ; Satoshi Honjo ; Hirofumi Suzumori ; Toshio Okazawa

EMC DESIGN AND DEVELOPMENT METHODOLOGY FOR TRACTION POWER INVERTERS OF ELECTRIC VEHICLES 2073

Isao Hoda ; Jia Li ; Hiroki Funato

SIMULATION-DRIVEN DESIGN OPTIMIZATION OF A MULTILAYER EMC INPUT FILTER 2078

Fatou Diouf ; Nadim Sakr ; Anna Gheonjian

EV TRACTION INVERTER EMPLOYING DOUBLE-SIDED DIRECT-COOLING TECHNOLOGY WITH SIC POWER DEVICE 2082

Takashi Hirao ; Masami Onishi ; Yusuke Yasuda ; Akihiro Namba ; Kinya Nakatsu

AN OVERVIEW OF STABILITY IMPROVEMENT METHODS FOR WIDE-OPERATION-RANGE FLYBACK CONVERTER WITH VARIABLE FREQUENCY PEAK-CURRENT-MODE CONTROL 2086
Ching-Hsiang Cheng ; Ching-Jan Chen ; Shinn-Shyong Wang

DESIGN AND IMPLEMENTATION OF A HIGH POWER DENSITY ACTIVE-CLAMPED FLYBACK CONVERTER 2092
Yu-Chen Liu ; Bing-Siang Huang ; Cheng-Hung Lin ; Katherine A. Kim ; Huang-Jen Chiu

OPTIMIZED VARIABLE ON-TIME CONTROL FOR LED LIGHTING DRIVER 2097
Jizhe Wang ; Haruhi Eto ; Fujio Kurokawa

DESIGN OF MULTIMODE BATTERY CHARGER WITH DYNAMIC VOLTAGE TRACKING CONTROL 2102
Pang-Jung Liu ; Lin-Hao Chien ; Song-Kai Lee ; Ang-Tung Chen

DUAL-SLOT POWER-PICKUP STRUCTURE FOR CONTACTLESS STRIP INDUCTIVE POWER TRACK SYSTEM 2107
Jia-You Lee ; I-Lin Chen ; Chien-Tzu Ko

DISCONTINUOUS SVM TECHNIQUE FOR THREE-LEG VSI FED BALANCED/UNBALANCED TWO-PHASE LOADS 2113
Supanut Charoensuksirikul ; Yuttana Kumsuwan

REDUCTION OF POWER LOSSES BASED ON GENERALIZED TWO-LEVEL PWM ALGORITHM FOR A NINE-SWITCH VSI 2121
Neerakorn Jarutus ; Yuttana Kumsuwan

SIC-BASED THREE-PHASE QUASI-Z-SOURCE INVERTER VERSUS THE TWO-STAGE TOPOLOGY - A COMPARISON 2129
Kornel Wolski ; Mariusz Zdanowski ; Jacek Rabkowski

DC-SIDE CIRCUIT IMPLEMENTATION OF A THREE-PHASE INVERTER FOR BALANCING PHASE-LEG CAPACITOR CURRENTS 2137
Takashi Hirao ; Keiji Wada ; Toshihisa Shimizu

A THREE-PHASE HYBRID SWITCHED-BOOST INVERTER 2145
Minh-Khai Nguyen ; Tan-Tai Tran ; Hoan-Tien Luong ; Kyoung-Won Lee ; Youn-Ok Choi ; Geum-Bae Cho

THE EFFECT OF BUILT-IN CR SNUBBER CAPACITOR INTO THE POWER MODULE 2149
Ryotaro Hata ; Shigeki Nishiyama

EVALUATION OF NOVEL HYBRID PROTECTION BASED ON PYROSWITCH AND FUSE TECHNOLOGIES 2153
Tomokazu Sakuraba ; Rémy Ouaida ; Song Chen ; Thibaut Chailloux

OPTIMAL DESIGN OF A MAGNETICALLY COUPLED FILTER FOR HIGH EFFICIENCY, LOW COST AND LOW VOLUME DC-DC BATTERY STORAGE CONVERTER 2158
Timothé Delaforge ; Robert Pasterczyk ; Mickaël Robert ; Hervé Chazal ; Jean-Luc Schanen ; Sébastien Mariethoz

HIGH POWER/CURRENT INDUCTOR LOSS MEASUREMENT WITH SHUNT RESISTOR CURRENT-SENSING METHOD 2165
Pin Yu Huang ; Toshihisa Shimizu

SENSITIVITY ANALYSIS OF MEDIUM FREQUENCY TRANSFORMER DESIGN 2170
Marko Mogorovic ; Drazen Dujic

STANDARD MODELS FOR POWER ELECTRONIC SYSTEM SIMULATION 2176
Koichi Shigematsu ; Hiroki Ishikawa ; Taku Noda ; Kentarou Fukushima ; Yoichi Sekiba ; Yusuke Kouno ; Takashi Abe ; Takayuki Sekisue ; Shinji Katoh

MODELING AND MODEL PARAMETER EXTRACTION OF WIDE BANDGAP POWER SEMICONDUCTOR DEVICE, PACKAGE, AND CIRCUIT FOR SIMULATING FAST SWITCHING BEHAVIOR 2181
Tsuyoshi Funaki

STABILITY ANALYSIS METHODS OF A GRID-CONNECTED INVERTER IN TIME AND FREQUENCY DOMAINS 2186
Toshiji Kato ; Kaoru Inoue ; Taiki Sakiyama

FINITE ELEMENT METHODS FOR MULTI-OBJECTIVE OPTIMIZATION OF A HIGH STEP-UP INTERLEAVED BOOST CONVERTER 2193
Wilmar Martinez ; Camilo Cortes ; Ahmad Bilal ; Jorma Kyyra

HIGH FIDELITY REAL-TIME SIMULATION OF MULTI-LEVEL CONVERTERS 2199
Jost Allmeling ; Niklaus Felderer ; Min Luo

AN ENHANCED HIGH FREQUENCY PULSATING VOLTAGE INJECTION METHOD BASED ON IMMUNE ALGORITHM FOR SENSORLESS IPMSM DRIVES 2204
Yanping Zhang ; Zhonggang Yin ; Chao Du ; Youyun Wang ; Xiangdong Sun

POSITION ESTIMATION ACCURACY IMPROVEMENT FOR MAGNETIC SALIENCY BASED SENSORLESS CONTROL INCLUDING CROSS-COUPLING FACTOR 2210

Keita Shimamoto ; Shinya Morimoto

SENSORLESS DRIVE IN THE LOW SPEED REGION AND AUTO-TUNING METHOD FOR PERMANENT MAGNET SYNCHRONOUS MOTORS 2216

Naofumi Nomura ; Shinichi Higuchi

HIGH STABILITY V/F CONTROL OF PMSM USING STATE FEEDBACK CONTROL BASED ON N-T COORDINATE SYSTEM 2224

Yosuke Matsuki ; Shinji Doki

STABILIZATION METHOD USING EQUIVALENT RESISTANCE GAIN BASED ON V/F CONTROL FOR IPMSM WITH LONG ELECTRICAL TIME CONSTANT 2229

Jun-Ichi Itoh ; Takato Toi ; Koroku Nishizawa

SINGLE-PHASE SOLID-STATE TRANSFORMER USING MULTI-CELL WITH AUTOMATIC CAPACITOR VOLTAGE BALANCE CAPABILITY 2237

Jun-Ichi Itoh ; Kazuki Aoyagi ; Keisuke Kusaka ; Masakazu Adachi

A DEVELOPED DUAL MMC ISOLATED DC SOLID STATE TRANSFORMER AND ITS MODULATION STRATEGY 2245

Yan Li ; Chao Liu ; Xu Cai

DC FAULT RIDE-THROUGH OF A THREE-PHASE DUAL-ACTIVE BRIDGE CONVERTER FOR DC GRIDS 2250

Jingxin Hu ; Shenghui Cui ; Rik W. De Doncker

A COMPOUND 10KV DVR SYSTEM BASED ON SOLID STATE TRANSFORMER STRUCTURE 2262

Yaqian Zhang ; Jianzhong Zhang ; Xing Hu ; Zakiud Din

A DUAL-ENERGY-SOURCE UNINTERRUPTIBLE POWER SUPPLY (UPS) 2270

Hao Wang ; Dehong Xu ; Binci Xu ; Haijin Li ; Ye Zhu

INFLUENCE OF WIND POWER FORECASTS ON EQUITABLE DISTRIBUTION METHOD OF WIND POWER CURTAILMENT 2278

Daisuke IIoka ; Hiroumi Saitoh

COMPARISON OF OPTIMIZED DEMAND OF EGS FOR MINIMIZING FUEL CONSUMPTION AND EGS MODEL WITH POWER GRID FREQUENCY USING A HPSPITAL LOAD WITH PV 2283

Yuji Mizuno ; Teppei Baba ; Fujio Kurokawa ; Nobumasa Matsui

COORDINATED DFIG WIND TURBINES AND SOLAR PV GENERATORS FOR INTER-AREA OSCILLATION DAMPING 2287

Tossaporn Surinkaew ; Issarachai Ngamroo

ENERGY MANAGEMENT USING A QUICK CHARGER WITH STORAGE BATTERIES FOR ELECTRIC VEHICLES 2292

Taku Ishibashi ; Toyonari Shimakage ; Norikazu Takeuchi ; Takaaki Kikuchi ; Midori Nonogaki

A METHOD FOR JUNCTION TEMPERATURE ESTIMATION UTILIZING TURN-ON SATURATION CURRENT FOR SIC MOSFET 2296

Hui-Chen Yang ; Rejeki Simanjorang ; Kye Yak See

FIELD BUS FOR DATA EXCHANGE AND CONTROL OF MODULAR POWER ELECTRONIC SYSTEMS WITH HIGH SYNCHRONISATION ACCURACY 2301

Stefan Rietmann ; Simon Fuchs ; André Hillers ; Jürgen Biela

ANALYTICAL INVESTIGATION ON ASYMMETRIC LCC COMPENSATION CIRCUIT FOR TRADE-OFF BETWEEN HIGH EFFICIENCY AND POWER 2309

Kodai Takeda ; Takafumi Koseki

PROBABILISTIC PCA-SUPPORT VECTOR MACHINE BASED FAULT DIAGNOSIS OF SINGLE PHASE 5-LEVEL CASCADED H-BRIDGE MLI 2317

Nagendra Vara Prasad Kuraku ; Yigang He ; Murad Ali

A STUDY ON EDGE SUPPORTED ELECTROMAGNETIC LEVITATION SYSTEM: FUNDAMENTAL CONSIDERATION ON LEVITATION PERFORMANCE OF THIN STEEL PLATE 2324

Yoshiho Oda ; Yasuaki Ito ; Kengo Okuno ; Masahiro Kida ; Toshiki Suzuki ; Takayoshi Narita ; Hideaki Kato ; Hiroyuki Moriyama

APPLICATION OF FACTS DEVICES FOR A DYNAMIC POWER SYSTEM WITHIN THE USA 2329

Jan Paramalingam ; Fuminori Nakamura ; Akihiro Matsuda ; Daisuke Yamanaka ; Taichiro Tsuchiya

CAPACITOR VOLTAGE BALANCING IN SEMI-FULL-BRIDGE SUBMODULE WITH DIFFERENTIAL-MODE CHOKE : (INVITEDPAPER) 2335

Kalle Ilves ; Yuhei Okazaki ; Nan Chen ; Muhammad Nawaz ; Antonios Antonopoulos

RESEARCH ON KEY TECHNOLOGY AND EQUIPMENT FOR ZHANGBEI 500KV DC GRID 2343

Hui Pang ; Xiaoguang Wei

WHAT LED TO SUCCESS IN ACADEMIC RESEARCH ON THE FAMILY OF MODULAR MULTILEVEL CASCADE CONVERTERS?..2352

Hirofumi Akagi

OPERATING PRINCIPLE OF CURRENT RESONANT CONVERTER USING AIR CORE TRANSFORMER FOR ISOLATED POWER SUPPLY ON CHIP..2360

Seiya Abe ; Hikaru Kaishakuji ; Satoshi Matsumoto

ANALYSIS FOR HIGH-FREQUENCY LLC RESONANT CONVERTER WITH PLANAR TRANSFORMER AT LIGHT-LOAD CONDITION..2365

Keon-Woo Kim ; Jae-Il Baek ; Yeonho Jeong ; Ki-Mok Kim ; Gun-Woo Moon

A NOVEL FULL DIGITAL CONTROL H-BRIDGE DC-DC CONVERTER FOR POWER SUPPLY ON CHIP APPLICATIONS...2370

Shigeki Nakano ; Toshiomi Oka ; Seiya Abe ; Satoshi Matsumoto

A HIGH-EFFICIENCY POWER SUPPLY FROM MAGNETIC ENERGY HARVESTERS...............2376

Cheon-Yong Lim ; Yeonho Jeong ; Keon-Woo Kim ; Feel-Soon Kang ; Gun-Woo Moon

OPPORTUNITIES FOR LEVERAGING LOW-VOLTAGE GAN DEVICES IN MODULAR MULTI-LEVEL CONVERTERS FOR ELECTRIC-VEHICLE CHARGING APPLICATIONS...........................2380

Mojtaba Ashourloo ; Mohammad Shawkat Zaman ; Miad Nasr ; Olivier Trescases

A NEW CONTROL STRATEGY FOR MODULAR MULTILEVEL CONVERTER OPERATING IN QUASI TWO-LEVEL PWM MODE...2386

Chao Wang ; Kui Wang ; Zedong Zheng ; Yongdong Li

A CURRENT-SOURCE TYPE MMC WITH DELTA-CONNECTED ARMS FOR SMES...................2393

Yushi Miura ; Toshifumi Ise

NEW MODULE WITH ISOLATED HALF BRIDGE OR ISOLATED FULL BRIDGE FOR MODULAR MEDIUM VOLTAGE CONVERTER...2400

Yunpeng Si ; Yifu Liu ; Qin Lei

DEVELOPMENT OF A 700-V-CLASS REVERSE-BLOCKING IGBT FOR ADVANCED T-TYPE NEUTRAL POINT-CLAMPED POWER CONVERSION SYSTEM...2404

Hiroki Wakimoto ; Haruo Nakazawa ; David H. Lu ; Takashi Matsumoto ; Yoichi Nabetani

CERAMIC EMBEDDING AS PACKAGING SOLUTION FOR FUTURE POWER ELECTRONIC APPLICATIONS...2410

Hoang Linh Bach ; Tobias Maximilian Endres ; Daniel Dirksen ; Sigrid Zischler ; Christoph Friedrich Bayer ; Andreas Schletz ; Martin März

MICROELECTROMECHANICAL SYSTEM (MEMS) RESONATOR: A NEW ELEMENT IN POWER CONVERTER CIRCUITS FEATURING REDUCED EMI...2416

A N M Wasekul Azad ; Sourov Roy ; Abu Saleh Imtiaz ; Faisal Khan

A LUMPED THERMAL MODEL INCLUDING THERMAL COUPLING EFFECTS AND BOUNDARY CONDITIONS FOR CAPACITOR BANKS...2421

Qiusheng Wang

HYSTERESIS MODELING OF MAGNETIC DEVICES BASED ON RELUCTANCE NETWORK ANALYSIS...2426

Yoshiki Hane ; Kenji Nakamura

OPTIMAL SIZING AND PLACEMENT OF SOLAR POWERED CHARGING STATION UNDER EV LOADS PENETRATION USING ARTIFICIAL BEE COLONY TECHNIQUE........................2430

Yuttana Kongjeen ; Kulsomsup Yenchamchalit ; Krischonme Bhumkittipich

A COMPARISON OF AVERAGE MODEL, SAMPLED-DATA MODEL AND MULTI-FREQUENCY MODEL BASED ON DC/DC CONVERTERS...2435

Xiangpeng Cheng ; Jinjun Liu ; Zeng Liu ; Yiming Tu ; Danhong Xue

SMALL-SIGNAL DISCRETE-TIME MODELING AND DIGITAL CONTROL OF THE BI-DIRECTIONAL DC/DC CONVERTERS..2441

Jia Yaoqin ; Xu Yingchun ; Hou Yijie

ENERGY MANAGEMENT OF HYDROGEN-STORAGE PHOTOVOLTAIC GENERATION SYSTEM WITH A FUNCTION OF SUPPRESSING SHORT-PERIOD COMPONENTS...............2449

Yuuki Machida ; Akihisa Goto ; Akiko Takahashi ; Shigeyuki Funabiki

A DYNAMIC BATTERY CHARGING APPROACH FOR ENERGY TRADING IN THE SMART GRID...2456

Avinash Sharma ; Akshay Kumar Rathore ; Rajesh Kumar

A FORCED COMMUTATION METHOD OF THE SOLID-STATE TRANSFER SWITCH IN THE UNINTERRUPTED POWER SUPPLY APPLICATIONS..2462

Meng-Jiang Tsai ; Jiuyang Zhou ; Po-Tai Cheng

ONLINE INTERNAL IMPEDANCE MEASUREMENTS OF LI-ION BATTERY USING PRBS BROADBAND EXCITATION AND FOURIER TECHNIQUES: METHODS AND INJECTION DESIGN................2470

Jussi Sihvo ; Tuomas Messo ; Tomi Roinila ; Roni Luhtala

A DC CURRENT FLOW CONTROLLER FOR MESHED HVDC GRIDS................2476

Viktor Hofmann ; Mark-M. Bakran

AN ISOLATED SOFT-SWITCHING HYBRID-SOURCE DC-DC CONVERTER FOR DC OFFSHORE WIND FARMS................2484

Shenghui Cui ; Jingxin Hu ; Marco Stieneker ; Rik W. De Doncker

A TRANSFORMERLESS MULTI-CELL SOLID-STATE FAULT CURRENT LIMITER FOR MEDIUM VOLTAGE POWER SYSTEM................2490

Pantarote Techama ; Sompob Polmai ; Chanin Bunlaksananusorn

A NOVEL DC POWER FLOW CONTROLLER FOR HVDC GRIDS WITH DIFFERENT VOLTAGE LEVELS................2496

Ya'nan Wu ; Han Ye ; Wu Chen ; Xiaokun He

DESIGN AND CONTROL OF SINGLE-PHASE GRID-CONNECTED PHOTOVOLTAIC MICROINVERTER WITH REACTIVE POWER SUPPORT CAPABILITY................2500

Geon-Hong Min ; Kyung-Hwan Lee ; Jung-Ik Ha ; Myong Hwan Kim

OPTIMAL SIZE AND MULTI-OBJECTIVE CONTROL OF BATTERY ENERGY STORAGES IN DISTRIBUTION SYSTEM WITH HIGH PENETRATION OF DISTRIBUTED PV GENERATORS................2505

Meiqin Mao ; Lei Zhou ; Yangyang Wang ; Liuchen Chang

MISSION PROFILE-ORIENTED CONTROL FOR RELIABILITY AND LIFETIME OF PHOTOVOLTAIC INVERTERS................2512

Ariya Sangwongwanich ; Yongheng Yang ; Dezso Sera ; Frede Blaabjerg

DISCONTINUOUS CURRENT MODE CONTROL FOR MINIMIZATION OF THREE-PHASE GRID-TIED INVERTER IN PHOTOVOLTAIC SYSTEM................2519

Hoai Nam Le ; Jun-Ichi Itoh

A THEORETICAL ANALYSIS ON STATIC CHARACTERISTICS OF VOLTAGE BASED CONTROL METHOD AND CURRENT BASED CONTROL METHOD FOR THE WAYSIDE ENERGY STORAGE SYSTEM IN DC-ELECTRIFIED RAILWAY................2527

Hiroyasu Kobayashi ; Keiichiro Kondo ; Diego Iannuzzi

IMPROVEMENT OF A DC ELECTRICAL RAILWAY SIMULATOR USING ARTIFICIAL INTELLIGENCE................2534

Alvaro J. Lopez-Lopez ; Ramon R. Pecharroman ; Antonio Fernandez-Cardador ; Asuncion P. Cucala

FEEDING-LOSS REDUCTION BY HIGHER-VOLTAGE DC RAILWAY FEEDING SYSTEM WITH DC-TO-DC CONVERTER................2540

Hidenori Shigeeda ; Hiroaki Morimoto ; Kazuhiko Ito ; Toshiyuki Fujii ; Naoki Morishima

MODELING AND SIMULATION OF NOVEL RAILWAY POWER SUPPLY SYSTEM BASED ON POWER CONVERSION TECHNOLOGY................2547

Minwu Chen ; Ruofei Liu ; Shaofeng Xie ; Xiaofang Zhang ; Yimin Zhou

COMPARATIVE STUDY ON FRONT-END PARAMETER IDENTIFICATION METHODS FOR WIRELESS POWER TRANSFER WITHOUT WIRELESS COMMUNICATION SYSTEMS................2552

Sinan Li ; S. Y. Ron Hui

A NEW TYPE OF WIRELESS V2X SYSTEM WITH A DUAL-ACTIVE BIDIRECTIONAL SINGLE-ENDED CONVERTER AND OPTIMIZED SIC-MOSFET................2558

Hideki Omori ; Aoto Yamamoto ; Naoki Mukaiyama ; Masahito Tsuno ; Kenji Fukuda ; Hisato Michikoshi ; Noriyuki Kimura ; Toshimitsu Morizane

METAL OBJECT DETECTION SYSTEM WITH PARALLEL-MISTUNED RESONANT CIRCUITS AND NULLIFYING INDUCED VOLTAGE FOR WIRELESS EV CHARGERS................2564

Seog Y. Jeong ; Van X. Thai ; Jun H. Park ; Chun T. Rim

WIRELESS EV CHARGING SYSTEM WITHOUT AIR-GAP AND MISALIGNMENT................2569

Wenxing Zhong ; Dehong Xu

FIXED SLOPE CARRIER PWM FOR INDIRECT MATRIX CONVERTER................2576

Tzung-Lin Lee ; Chun-Yao Hung ; Yen-Wen Chen ; Wen-Mei Huang

CARRIER-BASED OVERMODULATION STRATEGY FOR MATRIX CONVERTERS................2581

Paiboon Kiatsookkanatorn ; Somboon Sangwongwanich

THREE-PHASE TO HIGH-FREQUENCY SINGLE-PHASE MATRIX CONVERTER : A FREQUENCY CONTROL SUITABLE FOR SOFT SWITCHING................2589

Wataru Kodaka ; Satoshi Ogasawara ; Koji Orikawa ; Masatsugu Takemoto ; Takashi Hyodo ; Hiroyuki Tokusaki

TWO-STEP COMMUTATION FOR ISOLATED DC-AC CONVERTER WITH MATRIX CONVERTER................2596

Shunsuke Takuma ; Jun-Ichi Itoh

A DC-LINK CAPACITOR VOLTAGE OSCILLATION REDUCTION METHOD FOR A MODULAR MULTILEVEL CASCADE CONVERTER WITH SINGLE DELTA BRIDGE CELLS (MMCC-SDBC) ... 2604

Takaaki Tanaka ; Huai Wang ; Frede Blaabjerg

OPTIMIZED DECOUPLING CONTROL OF FLYING CAPACITOR IN ANPC FIVE-LEVEL INVERTER ... 2611

Fusheng Wang ; Deyou Zheng ; Jianing Wang ; Fei Li ; Fang Liu ; Shuying Yang ; Zhen Xie

CASCADED DUAL-BUCK AC-AC CONVERTER USING COUPLED INDUCTORS ... 2619

Sanghun Kim ; Duekjin Jang ; Heung-Geun Kim ; Honnyong Cha

INSTANTANEOUS POWER LOSS CALCULATION FOR MMC BASED ON VIRTUAL ARM MATHEMATICAL MODEL ... 2625

Yin Shiyuan ; Wang Yue ; Yin Taiyuan ; Nie Cheng ; Duan Guozhao ; Wang Zhang

COMPARISON OF CURRENT CONTROL STRATEGIES IN MODULAR MULTILEVEL CONVERTER ... 2630

Jianzhao Wei ; Anirudh Budnar Acharya ; Lars Norum ; Pavol Bauer

MODEL PREDICTIVE CONTROL OF A MODULAR MULTILEVEL CONVERTER WITH AN IMPROVED CAPACITOR BALANCING METHOD ... 2638

Shichong Zhang ; Baodong Bai ; Dezhi Chen

HIGH STEP-UP DC-DC CONVERTER BASED ON MULTI-CELL COUPLED INDUCTOR DIODE-CAPACITOR NETWORK ... 2646

Xinying Li ; Yan Zhang ; Jinjun Liu ; Pengxiang Zeng

NOVEL ACTIVE CLAMPING STEP-DOWN DC-DC CONVERTER WITH LOWER VOLTAGE STRESS ... 2653

Chi-Hsuan Hsu ; Jun-Min Jian ; Jiann-Fuh Chen ; Hsuan Liao

DESIGN AND EVALUATION OF A MAGNETICALLY-LOOSELY-COUPLED INDUCTOR FOR A FOUR-PHASE INTERLEAVED BOOST CHOPPER ... 2660

Hiroki Kowatari ; Toshinori Kitamura ; Nobukazu Hoshi

A SYNCHRONOUS-REFERENCE-FRAME I-V DROOP CONTROL METHOD FOR PARALLEL-CONNECTED INVERTERS ... 2668

Mingshen Li ; Yonghao Gui ; Zheming Jin ; Yajuan Guan ; Josep M. Guerrero

TRANSIENT STABILITY IMPACT OF THE PHASE-LOCKED LOOP ON GRID-CONNECTED VOLTAGE SOURCE CONVERTERS ... 2673

Heng Wu ; Xiongfei Wang

COMPREHENSIVE ANALYSIS OF VIRTUAL IMPEDANCE-BASED ACTIVE DAMPING FOR LCL RESONANCE IN GRID-CONNECTED INVERTERS ... 2681

Teng Liu ; Zeng Liu ; Jinjun Liu ; Yiming Tu ; Zipeng Liu

A COMPARATIVE STUDY OF THE TRADITIONAL FS-MPC AND THE PROPOSED CSF-PCC FOR THE THREE-PHASE GRID-CONNECTED INVERTERS ... 2688

Zhixun Ma ; Xin Zhang ; Jingjing Huang

CONSTANT SWITCHING-FREQUENCY PREDICTIVE- CURRENT-CONTROL METHOD WITH A DICHOTOMY SOLUTION FOR THE GRID-TIED INVERTERS ... 2692

Zhixun Ma ; Xin Zhang ; Jingjing Huang ; Zhao Bin ; Lyu Jing

OBSERVER-BASED ACTIVE DAMPING FOR GRID-CONNECTED CONVERTERS WITH LCL FILTER ... 2697

Y. Zhang ; M. G. L. Roes ; M. A. M. Hendrix ; J. L. Duarte

CONDUCTION LOSS ANALYSIS AND OPTIMIZATION DESIGN OF FULL BRIDGE LLC RESONANT CONVERTER ... 2703

Yugang Yang ; Lifei Zhang ; Tianshu Ma

FULL-BRIDGE T-TYPE ISOLATED DC/DC CONVERTER WITH WIDE INPUT VOLTAGE RANGE ... 2708

Dong Liu ; Yanbo Wang ; Fujin Deng ; Zhe Chen

RESEARCH ON HIGH EFFICIENCY LLC DC-DC CONVERTER BASED ON SIC MOSFET ... 2714

Pengcheng Han ; Xiaoqiong He ; Haijun Ren ; Zhiqing Zhao ; Xu Peng

AN IMPROVED DUAL PHASE SHIFT CONTROL STRATEGY FOR DUAL ACTIVE BRIDGE DC-DC CONVERTER WITH SOFT SWITCHING ... 2718

Miao Hong ; Gao Xuanjie ; Zeng Chengbi ; Duan Shujiang

DEVELOPMENT OF AN SIC HIGH-FREQUENCY PWM INVERTER USING A THICK MULTILAYER PCB TO MINIMIZE STRAY INDUCTANCE ... 2725

Kohsuke Ishikawa ; Satoshi Ogasawara ; Masatsugu Takemoto ; Koji Orikawa

FAST SWITCHING PLANAR POWER MODULE WITH SIC MOSFETS AND ULTRA-LOW PARASITIC INDUCTANCE ... 2732

Arash Edvin Risseh ; Hans-Peter Nee ; Konstantin Kostov

EXPERIMENTAL EVALUATION OF INVERTER SYSTEM CONSISTING OF 4-PARALLEL GAN DEVICES UNIT..2738

Yoshiya Ohnuma ; Satoshi Miyawaki ; Fumiya Hattori ; Masayoshi Yamamoto

IMPACT OF THE THERMAL-INTERFACE-MATERIAL THICKNESS ON IGBT MODULE RELIABILITY IN THE MODULAR MULTILEVEL CONVERTER..2743

Yi Zhang ; Huai Wang ; Zhongxu Wang ; Yongheng Yang ; Frede Blaabjerg

NANOSCALE INVESTIGATION OF THE POWER MOSFET BY THE AFM/KFM/SCFM................................2750

Mizuki Nakajima ; Yuuki Uchida ; Nobuo Satoh ; Hidekazu Yamamoto

SIMULATION ANALYSIS OF OPTIMUM GATE DRIVING CONDITIONS OF IGBTS................................2756

Satoshi Sugahara ; Masaki Kawakami ; Kousuke Kamakura

IMPROVEMENT OF THE I2T CAPABILITY FOR XEV ACTIVE SHORT CIRCUIT PROTECTION BY COMBINATION OF RC-IGBT AND LEADFRAME TECHNOLOGIES................................2764

Keiichi Higuchi ; Hayato Nakano ; Akihiro Osawa ; Akio Kitamura ; Shunji Takenoiri ; Daisuke Inoue ; Souichi Yoshida ; Hiromichi Gohara

INVESTIGATION OF SWITCHING BEHAVIOR OF AN IGBT UNDER SOFT TURN-OFF IN APPLICATION FOR DUAL-ACTIVE BRIDGE CONVERTERS................................2768

Eri Ogawa ; Yuichi Onozawa ; Rik W. De Doncker

600 V HIGH VOLTAGE GATE DRIVER IC (HVIC) WITH 1.0 MHZ HIGH FREQUENCY OPERATION FOR LLC CURRENT RESONANT POWER SUPPLY................................2774

Masaharu Yamaji ; Masashi Akahane ; Takahide Tanaka ; Akihiro Jonishi ; Hidetomo Ohashi ; Masahiro Sasaki ; Hitoshi Sumida

AN INTEGRATED VOLTAGE AND CURRENT BALANCING STRATEGY OF SERIES-PARALLEL CONNECTED IGBTS................................2780

Xiaotong Du ; Fang Zhuo ; Haotian Sun ; Hao Yi ; Yanlin Zhu

THERMAL DESIGN AND ANALYSIS OF A CABLE CHARGER USED FOR PORTABLE ELECTRONICS................................2785

Mofan Tian ; Xu Yang ; Naizeng Wang ; Yang Chen ; Laili Wang

PARASITIC INDUCTANCE DESIGN CONSIDERATIONS TO SUPPRESS GATE VOLTAGE OSCILLATION OF FAST SWITCHING POWER SEMICONDUCTOR DEVICES................................2789

Yusuke Sugihara ; Kimihiro Nanamori ; Masayoshi Yamamoto ; Yasuki Kanazawa

THE EXAMINATION OF INCREASING OPERATION SPEED OF CONSEQUENT POLE TYPE AXIAL GAP MOTOR FOR HIGHER OUTPUT POWER DENSITY................................2796

Toru Ogawa ; Tomohira Takahashi ; Masatsugu Takemoto ; Satoshi Ogasawara ; Hideaki Arita ; Akihiro Daikoku

BASIC STUDY OF PMASYNRM WITH BONDED MAGNETS FOR TRACTION APPLICATIONS................................2802

Marika Kobayashi ; Shigeo Morimoto ; Masayuki Sanada ; Yukinori Inoue

STUDY ON ROTOR STRUCTURE SUITABLE FOR IMPROVING POWER DENSITY AND EFFICIENCY IN IPMSMS FOR AUTOMOTIVE APPLICATIONS................................2808

R. Imoto ; M. Sanada ; S. Morimoto ; Y. Inoue

EXAMINATION OF THE DEMAGNETIZATION SUPPRESSION EFFECT OF PLACING FLUX BARRIERS IN AN IPMSM USING RARE-EARTH BONDED MAGNETS................................2814

Takashi Umeda ; Masayuki Sanada ; Shigeo Morimoto ; Yukinori Inoue

A NOVEL POLE-CHANGING METHOD WITH A MULTIPLE THREE-PHASE INVERTER................................2820

Yuki Hidaka ; Taiga Komatsu ; Hideaki Arita

STARTING CHARACTERISTICS OF AN ULTRA-LIGHTWEIGHT MOTOR USING MAGNETIC RESONANCE COUPLING................................2826

Kenta Takishima ; Kazuto Sakai

DESIGN AND BASIC CHARACTERISTICS ANALYSIS OF TOROIDAL WINDING AXIAL GAP INDUCTION MOTOR................................2832

Ryosuke Sakai ; Yukihiro Yoshida ; Katsubumi Tajima

MAGNET ARRANGEMENT SUITABLE FOR LARGE AIR GAP LENGTH IN LINEAR PM VERNIER MOTOR................................2836

Tatsuya Ninomiya ; Abdulaziz Gasim ; Shoji Shimomura

MICRO ELECTROMAGNETIC VIBRATION ENERGY HARVESTER WITH MECHANICAL SPRING AND IRON FRAME FOR LOW FREQUENCY OPERATION................................2842

Yecheng Shen ; Kaiyuan Lu ; Yongming Xia

MEASUREMENT OF TWO-LEVEL INVERTER INDUCED CURRENT SLOPES AT HIGH SWITCHING FREQUENCIES FOR CONTROL AND IDENTIFICATION ALGORITHMS OF ELECTRICAL MACHINES................................2848

Simon Decker ; Andreas Liske ; Daniel Schweiker ; Johannes Kolb ; Michael Braun

A NEW TOPOLOGY OF SWITCHED-CAPACITOR MULTILEVEL INVERTER FOR SINGLE-PHASE GRID-CONNECTED WITH ELIMINATING LEAKAGE CURRENT 2854

Mehdi Samizadeh ; Xu Yang ; Bagher Karami ; Wenjie Chen ; Mohamad Abou Houran ; Adib Abrishamifar ; Abdolreza Rahmati

AN INTERLEAVED BUCK-CASCADED BUCK-BOOST INVERTER FOR PV GRID-CONNECTION APPLICATIONS 2860

Chien-Hsuan Chang ; Chun-An Cheng ; Hung-Liang Cheng

A NOVEL PV ARRAY CONNECTION STRATEGY WITH PV-BUCK MODULE TO IMPROVE SYSTEM EFFICIENCY 2866

Chi Shao ; Wenjie Wang ; Lijun Hang ; Anping Tong ; Shitao Wang

A COMMON-MODE VOLTAGE REDUCTION FOR TWO-STAGE THREE-PHASE TRANSFORMERLESS PV INVERTERS 2871

Adisak Promyoo ; Surapong Suwankawin

A GRID-CONNECTED PV-ENERGY STORAGE SYSTEM WITH SYNCHRONOUS GENERATOR CHARACTERISTICS 2877

Huadian Xu ; Jianhui Su ; Ning Liu ; Yong Shi ; Yan Du

A TRANSFORMERLESS BIDIRECTIONAL DC-DC CONVERTER BASED ON POWER UNITS WITH UNIPOLAR AND BIPOLAR STRUCTURE FOR MVDC INTERCONNECTION 2882

Lejia Sun ; Fang Zhuo ; Feng Wang ; Hao Yi ; Baohui Ma

NEW MODULATION CONTROL OF CONVERTER SYSTEM APPLIED FOR OFFSHORE WIND FARMS 2887

Naoki Kawabata ; Noriyuki Kimura ; Toshimitsu Morizane ; Hideki Omori

SPHERE DECODING BASED LONG-HORIZON PREDICTIVE CONTROL OF THREE-LEVEL NPC BACK-TO-BACK PMSG WIND TURBINE SYSTEMS 2895

Ferdinand Grimm ; Zhenbin Zhang ; Ralph Kennel

BASED ON PCHD AND HPSO SLIDING MODE CONTROL OF D-PMSG WIND POWER SYSTEM 2901

Lijun Hou ; Xuemei Zheng ; Chao Wang ; Yangman Li ; Haoyu Li

ESTABLISHMENT AND DYNAMIC CONTROL OF WIND INDUCTION GENERATOR 2907

M. Z. Lu ; V. K. Ganisetti ; C. M. Liaw

MIDDLE FREQUENCY SOLID STATE TRANSFORMER FOR HVDC TRANSMISSION FROM OFFSHORE WINDFARM 2914

Noriyuki Kimura ; Toshimitsu Morizane ; Isao Iyoda ; Kazushige Nakao ; Tomoki Yokoyama

SIMULATION OF WIND POWER GENERATION SYSTEM USING SWITCHED RELUCTANCE GENERATOR AND CAPACITOR-LESS AC-AC CONVERTER 2921

Guyuan Ji ; Kazuhiro Ohyama

VARIABLE FREQUENCY CONTROL AND FILTER DESIGN FOR OPTIMUM ENERGY EXTRACTION FROM A SIC WIND INVERTER 2932

Abdallah Hussein ; Alberto Castellazzi

EXPERIMENTAL VERIFICATIONS OF UPFC USING DEADBEAT CONTROL WITH 3-PHASE UNBALANCED COMPENSATION 2938

Shin-Ichi Hamasaki ; Hiroto Fukuda ; Syohei Tokumaru ; Mineo Tsuji

A CONTROL METHOD FOR TWO TYPES OF THREE-PHASE TRANSFORMERLESS UNIFIED POWER QUALITY CONDITIONER 2944

Fujian Li ; Guochun Xiao ; Fangzhou Zhao ; Shuai Zhang ; Baojin Liu

DESIGN OF CUSTOMER-END CONVERTER SYSTEMS FOR LOW VOLTAGE DC DISTRIBUTION FROM A LIFE CYCLE COST PERSPECTIVE 2948

A. Mattsson ; P. Nuutinen ; T. Kaipia ; P. Peltoniemi ; J. Karppanen ; V. Tikka ; A. Lana ; P. Pinomaa ; P. Silventoinen ; J. Partanen

A CONTROL METHOD OF DC CAPACITOR VOLTAGE IN MMC FOR HVDC SYSTEM USING NEGATIVE SEQUENCE CURRENT 2956

Hanis Afiqah Binti Jaffar ; Ahmad Arif Bin Abd Rahman ; Hiroaki Kakigano

A COORDINATE AND DISTRIBUTED CONTROL SCHEME FOR MULTILEVEL AND MULTI-STAGE MEDIUM VOLTAGE SOLID STATE TRANSFORMER 2963

Jintong Nie ; Liqiang Yuan ; Qing Gu ; Jianning Sun ; Zhengming Zhao

AN IMPROVED HARMONIC POWER SHARING SCHEME OF PARALLELED INVERTER SYSTEM 2969

Liu Hongpeng ; Liu Xiaoxi ; Zhang Wei ; Wang Wei

THE GRID IMPEDANCE ADAPTATION DUAL MODE CONTROL STRATEGY IN WEAK GRID 2973

Ming Li ; Xing Zhang ; Ying Yang ; Pengpeng Cao

TRANSMISSION POWER ANALYSIS AND CONTROL OF THE DC TRANSFORMER IN HYBRID AC/DC MICROGRID 2980
Jingjin Huang ; Xin Zhang ; Tengfei Zhang

A NOVEL FLEXIBLE INTERCONNECTION SCHEME FOR MICROGRID TO OPTIMIZE THE CAPACITY OF ENERGY STORAGE SYSTEM (ESS) 2986
Zhou Jianqiao ; Zhang Jianwen ; Cai Xu ; Li Zhuyong ; Wang Jiacheng ; Zang Jiajie

VSC CONTROL AND PARAMETERS DESIGN BASED ON VIRTUAL SYNCHRONOUS GENERATOR 2992
Fang Liu ; Meng Wang ; Zhen Xie ; Fusheng Wang ; Jinxin Deng ; Xing Zhang

MULTI-TARGET VIRTUAL RESISTANCE CONTROL STRATEGY IN A 400 HZ LOW VOLTAGE MICROGRID 2997
Yuze Li ; Xuejun Pei ; Zhi Chen ; Hanyu Wang ; Yong Kang

AN ADAPTIVE POWER COMPENSATION STRATEGY FOR THE VOLTAGE STABILIZATION OF LCL-VSC BASED MICROGRIDS 3002
Sheng Xu ; Wu Cao ; Dongchen Fan ; Jianfeng Zhao ; Shunyu Wang

RESONANCE DETECTION STRATEGY FOR MULTIPLE GRID-CONNECTED INVERTERS-BASED SYSTEM USING CASCADED SECOND-ORDER GENERALIZED INTEGRATOR 3010
Wu Cao ; Dongchen Fan ; Kangli Liu ; Jianfeng Zhao ; Liheng Ruan ; Xiaojun Wu

HARMONIC STABILITY ASSESSMENT BASED ON GLOBAL ADMITTANCE FOR MULTI-PARALLELED GRID-CONNECTED VSIS USING MODIFIED NYQUIST CRITERION 3015
Wu Cao ; Dongchen Fan ; Kangli Liu ; Jianfeng Zhao ; Liheng Ruan ; Xiaojun Wu

THE AC TRACTION POWER SUPPLY SYSTEM FOR URBAN RAIL TRANSIT BASED ON NEGATIVE SEQUENCE CURRENT COMPENSATOR 3020
Tianshu Zhao ; Xu Peng

GRID CONNECTED POWER GENERATION CONTROL METHOD FOR Z-SOURCE INTEGRATED BIDIRECTIONAL CHARGING SYSTEM 3025
Xu Jia ; Guoming Chuai ; Haonan Niu ; Qianfan Zhang

AN ISOLATED PFC CONVERTER WITH HARMONIC MODULATION TECHNIQUE FOR EV CHARGERS 3030
Byung-Kwon Lee ; Jun-Young Lee ; Dong-Hun Kang

HIGHLY DYNAMIC SWITCHING FREQUENCY-BASED CALCULATION OF POWER QUANTITIES, FUNDAMENTAL WAVEFORMS, AND RMS VALUES OF INVERTER-FED ELECTRICAL MACHINES 3034
Alexander Stock ; Johannes Teigelkötter ; Johannes Büdel

DESIGN AND ANALYSIS OF HIGH VOLTAGE POWER SUPPLY FOR INDUSTRIAL ELECTROSTATIC PRECIPITATORS 3040
Shengwen Fan ; Yiqin Yuan ; Pengyu Jia ; Zhigang Chen ; Haisi Li

LOAD SHARING OPERATION IN N+1 UPS SYSTEM BY USING HARMONIC SHARING CONTROL METHOD 3046
Prashant Patel ; Sagar Naina ; Utsav Patel ; Premal Patwa

RESEARCH ON CAPACITY OPTIMIZATION OF PV-WIND-DIESEL-BATTERY HYBRID GENERATION SYSTEM 3052
Cailing Zhu ; Furong Liu ; Sheng Hu ; Shu Liu

A NUMERICAL ANALYSIS AND IMPROVEMENT OF OUTPUT CHARACTERISTICS IN DIFFERENT PASSIVE RECTIFIERS BASED ON VIBRATION GENERATORS 3058
Tomoki Sakabe ; Masataka Minami ; Shin-Ichi Motegi ; Masakazu Michihira

CIRCUIT MODELING APPROACH FOR ANALYZING TRIBOELECTRIC NANOGENERATORS FOR ENERGY HARVESTING 3063
Bo-Kyung Yoon ; Jeong Min Baik ; Katherine A. Kim

GENERAL POWER ELECTRIC CONVERTER MODEL 3069
Jingwen Xie

A MODULAR CONVERTER- AND SIGNAL-PROCESSING-PLATFORM FOR ACADEMIC RESEARCH IN THE FIELD OF POWER ELECTRONICS 3074
Rüdiger Schwendemann ; Simon Decker ; Marc Hiller ; Michael Braun

CONTROL IC FOR BOOST-FLYBACK CONVERTER FOR ENERGY HARVESTING APPLICATIONS 3081
Jhih-Sian Li ; Kai-Hui Chen ; Jui-Hung Lai ; Jun-Xian Huang

NEW CONCEPT OF THE DC-DC CONVERTER CIRCUIT APPLIED FOR THE SMALL CAPACITY UNINTERRUPTIBLE POWER SUPPLY 3086
Dang Minh Huynh ; Yoichi Ito ; Shinji Aso ; Koji Kato ; Kenji Teraoka

COMPARATIVE STUDY ON THE PERFORMANCE OF DUAL-PHASE TAPPED-INDUCTOR BOOST CONVERTER AND INTERLEAVED BOOST PARALLEL-INPUT SERIES-OUTPUT CONVERTER IN 40 TO 400V APPLICATIONS 3092

Niño Christopher Ramos ; Tsuyoshi Funaki

A NEW STANDBY STRUCTURE INTEGRATED WITH BOOST PFC CONVERTER FOR SERVER POWER SUPPLY 3100

Jae-Il Baek ; Jae-Kuk Kim ; Jae-Bum Lee ; Moo-Hyun Park ; Gun-Woo Moon

NONISOLATED TWO-CHANNEL LED DRIVER WITH SIMPLE SNUBBER 3107

Jong-Woo Kim ; Jung-Kyu Han ; Jih-Sheng Lai

DESIGN AND IMPLEMENTATION OF SINGLE-PHASE ASYMMETRIC MULTILEVEL STATCOM 3112

Hao Chen ; Yang Han ; Ping Yang ; Congling Wang ; Josep M. Guerrero

SUBMODULE VOLTAGE BALANCING AND LOSS EQUALISATION IN ALTERNATE ARM CONVERTERS BASED ON VIRTUAL VOLTAGES 3117

Georgios Konstantinou ; Harith R. Wickramasinghe ; Salvador Ceballos ; Josep Pou

BALANCED CONDUCTION LOSS DISTRIBUTION AMONG SMS IN MODULAR MULTILEVEL CONVERTERS 3123

Zhongxu Wang ; Huai Wang ; Yi Zhang ; Frede Blaabjerg

SIMPLIFICATION OF MODEL PREDICTIVE CONTROL FOR MODULAR MULTILEVEL CONVERTER THROUGH DIRECT VOLTAGE LEVEL SELECTION 3129

Xingxing Chen ; Jinjun Liu ; Shaodi Ouyang ; Shuguang Song ; Rui Luo

FAMILY OF INTEGRATED MULTI-INPUT MULTI-OUTPUT DC-DC POWER CONVERTERS 3134

Bang Le-Huy Nguyen ; Honnyong Cha ; Tien-The Nguyen ; Heung-Geun Kim

LOW-COMPLEXITY STATE-SPACE BASED SYSTEM IDENTIFICATION AND CONTROLLER AUTO-TUNING METHOD FOR MULTI-PHASE DC-DC CONVERTERS 3140

Marc Kanzian ; Harald Gietler ; Christoph Unterrieder ; Matteo Agostinelli ; Michael Lunglmayr ; Mario Huemer

A PHASE-SHIFT DOUBLE FULL-BRIDGE (PSDB) CONVERTER WITH THREE SHARED LEADING-LEGS 3145

Junjie Zhu ; Qinsong Qian ; Shengli Lu ; Weifeng Sun ; Le Zhang

DUAL ACTIVE BRIDGE SYNCHRONOUS RECTIFIED STEP-DOWN CONVERTER 3151

Chien-Chun Huang ; Chang-Lin Tsai ; Tsung-Lin Tsai ; Yao-Ching Hsieh ; Huang-Jen Chiu ; Jing-Yuan Lin

ACCURATE IMPEDANCE MODEL OF GRID-CONNECTED INVERTER FOR SMALL-SIGNAL STABILITY ASSESSMENT IN HIGH-IMPEDANCE GRIDS 3156

Tuomas Messo ; Roni Luhtala ; Aapo Aapro ; Tomi Roinila

MODELING OF UNBALANCED THREE-PHASE GRID-CONNECTED CONVERTERS WITH DECOUPLED TRANSFER FUNCTIONS 3164

Wei Liu ; Xiongfei Wang ; Frede Blaabjerg

PREDICTING VOLTAGE CHARACTERISTIC OF CHARGING MODEL FOR LI-ION BATTERY WITH ANN FOR REAL TIME DIAGNOSIS 3170

Minella Bezha ; Naoto Nagaoka

IMPEDANCE MODELING AND STABILITY ANALYSIS OF THE CASCADED THREE-PHASE SYMMETRIC SYSTEMS USING COMPLEX TRANSFER FUNCTIONS 3176

Teng Liu ; Zeng Liu ; Jinjun Liu ; Yiming Tu ; Zipeng Liu

ACOUSTIC NOISE REDUCTION OF 12/8 POLES SRM WITHOUT EFFICIENCY DROP USING SIMPLE CURRENT WAVEFORMS 3182

Kyohei Kiyota ; Kenji Amei ; Takahisa Ohji ; Jun Jisaki ; Masanobu Nakai

STUDY OF SWITCHED RELUCTANCE MOTOR DIRECTLY DRIVEN BY COMMERCIAL THREE-PHASE POWER SUPPLY 3186

Masaki Takahashi ; Kohei Aiso ; Kan Akatsu

DOUBLE STATOR AXIAL-FLUX SWITCHED RELUCTANCE MOTOR FOR ELECTRIC CITY COMMUTERS 3192

Hiroki Goto

TORQUE RIPPLE REDUCTION USING ASYMMETRIC FLUX BARRIERS IN SYNCHRONOUS RELUCTANCE MOTOR 3197

Yuuto Yamamoto ; Shigeo Morimoto ; Masayuki Sanada ; Yukinori Inoue

ON-BOARD SINGLE-PHASE ELECTRIC VEHICLE CHARGER WITH ACTIVE FRONT END 3203

Theodore Soong ; Peter W. Lehn

A BIDIRECTIONAL BUFFERED CHARGING UNIT FOR EV'S (BBCU) 3209

Gabriel Fernandez

RECONFIGURABLE CONVERTER WITH MULTIPLE-VOLTAGE MULTIPLE-POWER FOR E-MOBILITY CHARGING 3215

Mohamed S A Dahidah ; He Liu ; Vassilios G. Agelidis

DEVELOPMENT OF A SERIES HYBRID ELECTRIC VEHICLE LABORATORY TEST BENCH WITH HARDWARE-IN-THE-LOOP CAPABILITIES..3223
Poria Fajri ; Nima Lotfi ; Mehdi Ferdowsi

NEW THREE-PHASE STATIC TRANSFER SWITCH USING AC SSCB..3229
Seung-Min Song ; Jin-Young Kim ; In-Dong Kim

HARMONICS COMPENSATION IN HIGH FREQUENCY RANGE OF ACTIVE POWER FILTER WITH SIC-MOSFET INVERTER IN DIGITAL CONTROL SYSTEM..3237
Shin-Ichi Hamasaki ; Kengo Nakahara ; Mineo Tuji

CONTROL OF BUCK-BOOST DIRECT MATRIX CONVERTER WITH LOW VOLTAGE RIDE-THROUGH CAPABILITY..3243
Nico Remus ; Martin Leubner ; Wilfried Hofmann

AN IMPROVED PLL BASED SEAMLESS TRANSFER CONTROL STRATEGY................................3251
Xin Meng ; Jinjun Liu ; Zeng Liu ; Ronghui An

EFFICIENT URBAN RAILWAY DESIGN INTEGRATING TRAIN SCHEDULING, ONBOARD ENERGY STORAGE, AND TRACTION POWER MANAGEMENT..3257
Warayut Kampeerawar ; Takafumi Koseki ; Fulin Zhou

OPTIMAL CONTROL METHOD OF AN ENERGY STORAGE SYSTEM FOR ENERGY SAVING................3265
Yoko Takeuchi ; Tomoyuki Ogawa ; Keisuke Sato ; Hiroaki Morimoto ; Tatsuhito Saito

START-UP AND TRANSIENT OPERATION OF A BIDIRECTIONAL CHOPPER WITH AN AUXILIARY CONVERTER..3273
Hamzeh J. Ahmad ; Haruna Ohnishi ; Makoto Hagiwara

EXPERIMENTAL RESULTS OF QUASI-OPTIMAL CHARGING CURRENT PATTERNS TO REDUCE THE INTERNAL HEAT GENERATION OF THE LITHIUM-ION BATTERY................................3280
Yoshiaki Taguchi ; Gaku Yoshikawa

DEVELOPMENT OF TEST METHODS AND EVALUATION RESULTS FOR 500KV HVDC CONVERTER..3286
Keisuke Hattori ; Asuka Ohtake ; Takayoshi Kamejima ; Haruhisa Wada

DISSIPATION LOOP FOR SHOOT-THROUGH FAULTS IN HVDC CONVERTER CELLS................3292
Keijo Jacobs ; Staffan Norrga ; Hans-Peter Nee

A SUPPRESSION METHOD OF HARMONIC INSTABILITY IN LINE-COMMUTATED CONVERTERS APPLYING ACTIVE HARMONIC FILTERS..3299
Kenichiro Sano ; Toshiaki Kikuma ; Tatsuhito Nakajima ; Junya Kanno

EXPERIMENT OF SEMICONDUCTOR BREAKER USING SERIES-CONNECTED IEGTS FOR HYBRID DCCB..3304
Kazuyasu Takimoto ; Hiroshi Takenaka ; Toshiaki Matsumoto ; Takahiro Ishiguro

STUDY OF EMI CAUSED BY BUCK CONVERTER ON CONTROLLER AREA NETWORK................3309
Ryo Shirai ; Toshihisa Shimizu

A STUDY ON REDUCTION TECHNIQUES OF A WIDEBAND COMMON-MODE VOLTAGE PRODUCED BY A PWM INVERTER..3315
Shotaro Takahashi ; Satoshi Ogasawara ; Masatsugu Takemoto ; Koji Orikawa ; Michio Tamate

A MODIFIED DISCONTINUOUS PWM FOR COMMON-MODE VOLTAGE ELIMINATION IN 3-LEVEL 4-LEG PWM CONVERTER SYSTEM..3323
Seon-Ik Hwang ; Jun-Hyung Jung ; In-Ho Cho ; Jang-Mok Kim ; Yung-Deug Son

EMI ANALYSIS OF FULL-SIC INTEGRATED POWER MODULE..3329
Xiliang Chen ; Wenjie Chen ; Yu Ren ; Liang Qiao ; Yilin Sha ; Xu Yang

EXPERIMENTAL VERIFICATION OF COUPLING EFFECT AND POWER TRANSFER CAPABILITY OF DYNAMIC WIRELESS POWER TRANSFER..3332
Chan Anyapo ; Nithiphat Teerakawanich ; Chowarit Mitsantisuk ; Kiyoshi Ohishi

NEIGHBORING EFFECTS ON THE DEACTIVATED INVERTER IN A SEGMENTED DYNAMIC WIRELESS EV CHARGING SYSTEM..3338
Qingwei Zhu ; Yanjie Guo ; Lifang Wang ; Shufan Li ; Chenglin Liao

MULTIPLE EXCITING VOLTAGE CONTROL FOR MAXIMIZATION OF MULTI-HOP WIRELESS POWER TRANSFER EFFICIENCY..3344
Masato Sasaki ; Masayoshi Yamamoto

GENERAL ANALYTICAL MODEL FOR INDUCTIVE POWER TRANSFER SYSTEM WITH EMF CANCELING COILS..3349
Keita Furukawa ; Keisuke Kusaka ; Jun-Ichi Itoh

STABILITY INFLUENCE OF FILTER COMPONENTS PARASITIC RESISTANCE ON LCL-FILTERED GRID CONVERTERS..3357
Hiroaki Matsumori ; Toshihisa Shimizu ; Frede Blaabjerg ; Xiongfei Wang ; Dongsheng Yang

REAL-TIME ESTIMATION CONTROL OF INDUCTANCE PARAMETERS USING DUST CORE MATERIALS FOR PWM INVERTER 3363
Kazu Imai ; Takuma Yoshino ; Ohasi Shunsuke ; Tomoki Yokoyama

CONTROL DESIGN OF OUTPUT-STAGE FILTERLESS SINUSOIDAL-WAVE INVERTER 3369
Shinichi Hiroshige ; Kenji Yamanaka ; Masahide Hojo

SERIES REACTIVE POWER COMPENSATOR WITH REDUCED CAPACITANCE FOR HYBRID TRANSFORMER 3375
Yuki Takahashi ; Takanori Isobe ; Hiroshi Tadano

AN INSIGHT INTO THE VOLTAGE RISING BEHAVIOR DURING TURN-OFF PROCESS OF SERIES CONNECTED SIC MOSFETS ON CIRCUIT LEVEL 3383
Panrui Wang ; Feng Gao ; Yang Jing ; Yufeng Chen ; Lei Zhang

PARALLELING SIX 320A 1200V ALL-SIC HALF-BRIDGE MODULES FOR A LARGE CAPACITY POWER STACK 3390
David Hongfei Lu ; Hiromu Takubo ; Sho Takano ; Yuhei Suzuki

3.3KV ALL-SIC MODULE FOR ELECTRIC DISTRIBUTION EQUIPMENT 3396
Ryohei Takayanagi ; Katsumi Taniguchi ; Satoshi Kaneko ; Naoyuki Kanai ; Keishirou Kumada ; Motohito Hori ; Yoshinari Ikeda ; Kouji Maruyama ; Itsuo Kawamura

PRESENT STATUS OF SIC BASED POWER CONVERTERS AND GATE DRIVERS – A REVIEW 3401
Abhijit Choudhury

METHOD OF APPLYING FORCE DISTRIBUTION FUNCTION FOR LINEAR SWITCHED RELUCTANCE MOTOR DRIVEN BY CURRENT SOURCE INVERTER 3406
Tadashi Hirayama ; Shuma Kawabata

A NOVEL DRIVE CIRCUIT FOR SWITCHED RELUCTANCE MOTORS WITH BIPOLAR CURRENT DRIVE 3412
Hiroki Ishikawa ; Yuma Uesugi ; Seiya Sakurai

TORQUE RIPPLE MINIMIZATION CONTROL OF SRM BASED ON NOVEL MOTOR MODEL CONSIDERING MUTUAL COUPLING EFFECT 3418
Sungyong Shin ; Naruse Hikaru ; Takashi Kosaka ; Nobuyuki Matsui

COMPARISON OF HIGH FREQUENCY VOLTAGE INJECTION METHODS FOR SHAFT SENSORLESS CONTROL OF WOUND-FIELD FLUX SWITCHING MACHINE 3426
Hong-Quan Nguyen ; Sheng-Ming Yang

DESIGN AND EXPERIMENTAL VERIFICATION OF A DAB MEDIUM FREQUENCY TRANSFORMER FOR A 6.6KV/200V SOLID STATE TRANSFORMER 3431
Rene Barrera-Cardenas ; Takanori Isobe ; Terazono Katsushi ; Tadano Hiroshi

RESEARCH ON THE UNBALANCED COMPENSATION RANGE OF DELTA-CONNECTED CASCADED H-BRIDGE MULTILEVEL SVG 3439
Rui Luo ; Yingjie He ; Yiming Tu ; Xingxing Chen ; Jinjun Liu

STATIC SYNCHRONOUS COMPENSATOR TO STABILIZE GRID VOLTAGE FOR WIND AND PHOTOVOLTAIC POWER PLANT 3450
Ryota Okuyama ; Naoki Morishima ; Yusuke Ashizaki ; Yohei Itaya

LARGE EQUALIZATION CURRENT CONTROL STRATEGY FOR SERIES CONNECTED BATTERY PACKS BASED ON BUCK-BOOST CONVERTER 3455
Xinbo Liu ; Zhuo Gao ; Xuehao Huang ; Yaohan Zou

A MULTI-PORT BIDIRECTIONAL POWER CONVERSION SYSTEM FOR REVERSIBLE SOLID OXIDE FUEL CELL APPLICATIONS 3460
Xiang Lin ; Kai Sun ; Jin Lin ; Zhe Zhang ; Wei Kong

SELF-PREHEATING METHOD FOR LI-ION BATTERY USING BATTERY IMPEDANCE ESTIMATOR 3466
Dong-Kwan Kim ; Young-Dal Lee ; Sang-Hyun Ha ; Yu-Jin Jang ; Gun-Woo Moon

ACTIVE ANTI-ISLANDING TECHNIQUE WITH REDUCED NON-DETECTION ZONE FOR CENTRALIZED INVERTERS 3471
Prashant Jain ; Vivek Agarwal ; Bishnu Prasad Muni ; Eswar Rao ; Deepak Gehlot ; S. Gautam Kumar

DEVELOPMENT OF SIC APPLIED TRACTION SYSTEM FOR SHINKANSEN HIGH-SPEED TRAIN 3478
Kenji Sato ; Hirokazu Kato ; Takafumi Fukushima

DEVELOPMENT OF A HIGH POWER DENSITY AUXILIARY CONVERTER BASED ON 1700V 225A SIC MOSFET FOR TRAMS 3484
Liu Hao ; Fei Lin ; Zhongping Yang ; Hu Cao ; Meng Xia

EXPERIMENTAL TESTS RESULTS OF DAMPING CONTROL WITH OVER VOLTAGE RESISTOR FOR REGENERATIVE BRAKE CONTROL OF RAILWAY VEHICLE 3490
Natsuki Kawagoe ; Febry Pandu Wijaya ; Hiroyasu Kobayashi ; Keiichiro Kondo ; Tetsuya Iwasaki ; Akihiko Tsumura ; Takumi Nagashima ; Yoshinori Yamashita ; Ryota Gondo

COILS LAYOUT OPTIMIZATION OF DYNAMIC WIRELESS POWER TRANSFER SYSTEM TO REALIZE OUTPUT VOLTAGE STABLE.................3495
Yi Wang ; Fei Lin ; Zhongping Yang ; Panpan Cai ; Zhiyuan Liu

QUICK CHARGER FOR A BATTERY USING MODULAR MATRIX CONVERTER (MMXC).................3501
Kazuma Suzuki ; Takaharu Takeshita

VARIABLE OUTPUT VOLTAGE CONTROL OF AN ISOLATED BI-DIRECTIONAL AC/DC CONVERTER WITH A SOFT-SWITCHING TECHNIQUE.................3507
Takumi Hamaguchi ; Kazuma Suzuki ; Wataru Kitagawa ; Takaharu Takeshita

A NEW MODULATION METHOD APPLYING OPTIMAL DUTY CYCLE AND PHASE SHIFT FOR BIDIRECTIONAL ISOLATED THREE-PHASE AC/DC CONVERTER BASED ON MATRIX CONVERTER.................3514
Koji Shigeuchi ; Jin Xu ; Noboru Shimosato ; Yukihiko Sato

DECOUPLING CONTROL METHOD FOR ELIMINATING DC BIAS FLUX OF HIGH FREQUENCY TRANSFORMER IN A BIDIRECTIONAL ISOLATED AC/DC CONVERTER.................3522
Kensuke Sakuma ; Koji Shigeuchi ; Jin Xu ; Noboru Shimosato ; Yukihiko Sato

INTERLEAVED VOLTAGE-DOUBLER BOOST CONVERTER FOR POWER FACTOR CORRECTION.................3528
Bo-Jia Huang

ZVS INTERLEAVED TOTEM-POLE BRIDGELESS PFC CONVERTER WITH PHASE-SHIFTING CONTROL.................3533
Moo-Hyun Park ; Jae-Il Baek ; Jung-Kyu Han ; Cheon-Yong Lim ; Gun-Woo Moon

A ZERO-VOLTAGE-SWITCHING TOTEM-POLE BRIDGELESS BOOST POWER FACTOR CORRECTION RECTIFIER HAVING MINIMIZED CONDUCTION LOSSES.................3538
Young-Dal Lee ; Chong-Eun Kim ; Jae-Il Baek ; Dong-Kwan Kim ; Gun-Woo Moon

POWER-FACTOR-CORRECTION WITH POWER DECOUPLING FOR AC-TO-DC CONVERTER.................3544
Wan-Jung Chen ; Tsung-Hsi Wu ; Yao-Ching Hsieh ; Chin-Sien Moo ; Po-Hsiang Wen

DESIGN AND ANALYSIS OF THE DISTRIBUTED CONTROLLER FOR THE MODULAR MULTILEVEL CASCADED CONVERTER.................3549
Ping-Heng Wu ; Yu-Chen Su ; Po-Tai Cheng

ASYMMETRIC MIXED MODULAR MULTILEVEL CONVERTER TOPOLOGY IN HYBRID BIPOLAR HVDC TRANSMISSION SYSTEMS.................3557
Joon-Hee Lee ; Jae-Jung Jung ; Seung-Ki Sul

HIGH POWER MEDIUM VOLTAGE 10 KV SIC MOSFET BASED BIDIRECTIONAL ISOLATED MODULAR DC–DC CONVERTER.................3564
Sayan Acharya ; Ritwik Chattopadhyay ; Anup Anurag ; Satish Rengarajan ; Yos Prabowo ; Subhashish Bhattacharya

MULTI-LEVEL POWER CONVERTER USING SERIES-CONNECTED SOLID-STATE TRANSFORMERS.................3572
Yuichi Mabuchi ; Yuki Kawaguchi ; Kimihisa Furukawa ; Mitsuhiro Kadota ; Mizuki Nakahara ; Akihiko Kanoda

CAPACITOR VOLTAGE CONTROL OF MMC-STATCOM DURING UNBALANCED AC SYSTEM FAULT.................3578
Kaho Nada ; Takeshi Kikuchi ; Tsuguhiro Takuno ; Toshiyuki Fujii ; Ryosuke Uda ; Takashi Sugiyama

SIC BASED POWER SEMICONDUCTOR IN APPLICATIONS - ASPECTS AND PROSPECTS.................3584
Peter Friedrichs

ELECTROMAGNETIC MODELING APPROACHES TOWARDS VIRTUAL PROTOTYPING OF WBG POWER ELECTRONICS.................3588
Ivana Kovacevic-Badstübner ; Daniele Romano ; Giulio Antonini ; Jonas Ekman ; Ulrike Grossner

SILICON BASED DEVICES FOR DEMANDING HIGH POWER APPLICATIONS.................3596
A. Kopta ; J. Vobecky ; M. Rahimo ; T. Wikström ; U. Vemulapati ; C. Papadopoulos ; C. Corvasce ; M. Andenna ; F. Dugal ; F. Fischer ; S. Hartmann

RECENT PROGRESS IN HIGH TO ULTRA-HIGH-VOLTAGE SIC POWER DEVICES: DEVELOPMENT AND APPLICATION.................3603
Y. Yonezawa

DYNAMIC DRIFT EFFECTS IN GAN POWER TRANSISTORS: CORRELATION TO DEVICE TECHNOLOGY AND MISSION PROFILE.................3607
Joachim Würfl ; Eldad Bahat-Treidel ; Oliver Hilt ; Maria Troppenz ; Mihaela Wolf ; Jan Böcker ; Carsten Kuring ; Sibylle Dieckerhoff

COMPENSATION METHOD OF RADIAL UNBALANCE FORCE AT FAILURE OF A MOTOR SECTION IN A D-Q AXIS CURRENT CONTROL BEARINGLESS MOTOR.................3613
Masahide Ooshima

A BEARINGLESS SYNCHRONOUS RELUCTANCE SLICE MOTOR WITH ROTOR FLUX BARRIERS 3619

Thomas Holenstein ; Thomas Nussbaumer ; Johann W. Kolar

PARAMETER IDENTIFICATIONS OF CURRENT-FORCE FACTOR AND TORQUE CONSTANT IN SINGLE-DRIVE BEARINGLESS MOTORS 3627

Hiroya Sugimoto ; Akira Chiba

DAMPENING OF AXIAL VIBRATIONS IN A BEARINGLESS FLUX-SWITCHING SLICE MOTOR BY FIELD CURRENT REGULATION 3632

Bianca Klammer ; Karlo Radman ; Wolfgang Gruber

ANALYSIS AND DESIGN OF A BEARINGLESS AXIAL-FORCE/TORQUE MOTOR WITH FLEX-PCB WINDINGS 3640

Nobuyuki Kurita ; Walter Bauer ; Gerald Jungmayr ; Wolfgang Gruber ; Wolfgang Amrhein

A PLOTTER-BASED AUTOMATIC MEASUREMENT AND STATISTICAL CHARACTERIZATION OF MULTIPLE DISCRETE POWER DEVICES 3644

Michihiro Shintani ; Benjamin Dauphin ; Kazuki Oishi ; Masayuki Hiromoto ; Takashi Sato

A NOVEL HIGH-SPEED SIC MOSFET DRIVER WITH A LOW SWITCH-VOLTAGE STRESS 3650

Xiuqin Wei ; Yuchong Sun ; Hiroo Sekiya

ENHANCEMENT OF DRIVING CAPABILITY OF GATE DRIVER USING GAN HEMTS FOR HIGH-SPEED HARD SWITCHING OF SIC POWER MOSFETS 3654

Takafumi Okuda ; Takashi Hikihara

DESIGN AND EXPERIMENTAL VERIFICATION OF ROBOT ARM OPERATION FOR POWER PACKET DISPATCHING SYSTEM 3658

Tomoki Yokoyama ; Ryunosuke Araumi ; Kazunori Asada ; Takashi Ando

A RESOURCE SHARING MODEL IN A POWER PACKET DISTRIBUTION NETWORK 3665

H. Ando ; R. Takahashi ; S. Azuma ; M. Hasegawa ; T. Yokoyama ; T. Hikihara

DECOUPLED DSOGI-PLL FOR IMPROVED THREE PHASE GRID SYNCHRONISATION 3670

A. A. Nazib ; D. G. Holmes ; B. P. Mcgrath

A DEVIATION ELIMINATION CONTROL BASED ON AUTONOMOUS CURRENT-SHARING CONTROLLER FOR THE PARALLEL-CONNECTED INVERTERS IN AC MICROGRIDS 3678

Yajuan Guan ; Wei Feng ; Baoze Wei ; Wenzhao Liu ; Mingshen Li ; C. Juan Vasquez ; M. Josep Guerrero

SISO TRANSFER FUNCTIONS FOR STABILITY ANALYSIS OF GRID-CONNECTED VOLTAGE-SOURCE CONVERTERS 3684

Hongyang Zhang ; Lennart Harnefors ; Xiongfei Wang ; Jean-Philippe Hasler ; Hans-Peter Nee

A COMMUNICATION-INDEPENDENT REACTIVE POWER SHARING SCHEME WITH ADAPTIVE VIRTUAL IMPEDANCE FOR PARALLEL CONNECTED INVERTERS 3692

Ronghui An ; Zeng Liu ; Jinjun Liu ; Shike Wang

DESIGN AND INTEGRATION OF THE BI-DIRECTIONAL ELECTRIC VEHICLE CHARGER INTO THE MICROGRID AS EMERGENCY POWER SUPPLY 3698

Yang Song ; Pengcheng Li ; Yuanliang Zhao ; Shuai Lu

STABILITY IMPACT OF PV INVERTER GENERATION ON MEDIUM VOLTAGE DISTRIBUTION SYSTEMS 3705

Ye Tang ; Rolando Burgos ; Chi Li ; Dushan Boroyevich

1MW POWER CONDITIONING SYSTEM WITH MULTIPLE DC INPUTS FOR PVS AND BATTERIES 3711

Yasuaki Furusho ; Yasuyuki Noto ; Kansuke Fujii

A ROBUST AND FLEXIBLE DC-LINKED 3-PHASE ENERGY MANAGEMENT SYSTEM WITH ADAPTIVE DROOP CONTROL STRATEGY 3717

Yue Ma ; Yuki Ishikura ; Hitoshi Tsuji ; Kazuaki Mino

MAXIMUM POWER POINT TRACKING CONTROL FOR SMALL HYDROELECTRIC GENERATION 3723

Kazuya Azegami ; Masashi Takiguchi ; Junya Yano ; Hirohiko Tsutsumi ; Toshitake Masuko

DESIGN AND EXPERIMENTAL VERIFICATION OF A THREE-PHASE DUAL-ACTIVE BRIDGE CONVERTER FOR OFFSHORE WIND TURBINES 3729

Takushi Jimichi ; Murat Kaymak ; Rik W. De Doncker

OPTIMIZED BIDIRECTIONAL PFC RECTIFIERS & INVERTERS - SI VS. SIC VS. GAN IN 2L AND 3L TOPOLOGIES - 3734

Jonas Wyss ; Jürgen Biela

A STANDARD BLOCK OF "SERIES CONNECTED SIC MOSFET" FOR MEDIUM/HIGH VOLTAGE CONVERTER 3742

Qin Lei ; Chunhui Liu ; Yunpeng Si ; Yifu Liu

DESIGN AND TESTING OF 1 KV H-BRIDGE POWER ELECTRONICS BUILDING BLOCK BASED ON 1.7 KV SIC MOSFET MODULE 3749
Jun Wang ; Rolando Burgos ; Dushan Boroyevich ; Zeng Liu

A FLYBACK CONVERTER WITH SIC POWER MOSFET OPERATING AT 10 MHZ: REDUCING LEAKAGE INDUCTANCE FOR IMPROVEMENT OF SWITCHING BEHAVIORS 3757
Kazuki Hashimoto ; Takafumi Okuda ; Takashi Hikihara

A STUDY ON LOAD FLUCTUATION OF ISOLATED DC-DC CONVERTER WITH CLASS PHI-2 INVERTER USING GAN-HFET 3762
Yuta Yanagisawa ; Yushi Miura ; Hiroyuki Handa ; Tetsuzo Ueda ; Toshifumi Ise

SINGLE-INDUCTOR MULTIPLE-OUTPUT CURRENT-SOURCE CONVERTER WITH IMPROVED CROSS REGULATION AND SIMPLE CONTROL STRATEGY 3768
Zheng Dong ; Xiaolu Lucia Li ; Chi K. Tse

LIMIT OPERATING FREQUENCY OF PEAK CURRENT-MODE CONTROL DC-DC CONVERTER CONSIDERING TURN-OFF DELAY TIME 3773
Ryo Ute ; Kazuya Fujiwara ; Jun Imaoka ; Masahito Shoyama

A NOVEL SINGLE SWITCH HIGH FREQUENCY DC/DC CONVERTER AND ITS MATHEMATIC MODEL 3780
Yueshi Guan ; Xihong Hu ; Shu Zhang ; Yijie Wang ; Dianguo Xu ; Wei Wang

ANALYSIS OF CLOSED LOOP OPERATION OF AN ISOLATED BIDIRECTIONAL DAB DC-DC CONVERTER WITH LC COUPLING 3785
Bruno Yukio Enomoto ; Kelly C. M. Carvalho ; Lourenço Matakas Junior ; Wilson Komatsu

ISOLATED AC/DC CONVERTER USING SIMPLE PWM STRATEGY 3791
Naoki Hirose ; Yuto Matsui ; Takaharu Takeshita

ANALYSIS OF ONE PHASE LOSS OPERATION OF THREE-PHASE ISOLATED BUCK MATRIX-TYPE RECTIFIER WITH EIGHT-SEGMENT PWM SCHEME 3797
Jahangir Afsharian ; Dewei David Xu ; Bin Wu ; Bing Gong ; Zhihua Yang ; Jun-Ichi Itoh

NOVEL ISOLATED BIDIRECTIONAL INTEGRATED DUAL THREE-PHASE ACTIVE BRIDGE (D3AB) PFC RECTIFIER 3805
F. Krismer ; E. Hatipoglu ; J. W. Kolar

LOAD VOLTAGE REGULATION METHOD FOR AN ISOLATED AC-DC CONVERTER WITH POWER DECOUPLING OPERATION 3813
Shohei Komeda ; Hideaki Fujita

OPTIMAL DESIGN OF A LOW COST 20KW 99.1% EFFICIENCY ACTIVE ZCS ISOLATED DC-DC CONVERTER 3820
Timothé Delaforge ; Sébastien Mariéthoz

SOFT-SWITCHING ANALYSIS AND PFM CONTROL METHOD OF BIDIRECTIONAL DC/DC CONVERTER TOPOLOGY 3825
Yijie Wang ; Haoyu Wang ; Hongyu Song ; Dianguo Xu

A FULLY SOFT-SWITCHED PWM DC-DC CONVERTER USING AN ACTIVE-SNUBBER-CELL 3833
Hai N. Tran ; Adhistira M. Naradhipa ; Sunju Kim ; Ali Tausif

FLYING CAPACITOR RESONANT POLE INVERTER WITH DIRECT INDUCTOR CURRENT FEEDBACK 3840
Sjef J. Settels ; Jorge L. Duarte ; Jeroen Van Duivenbode

DESIGN OF A GAN-BASED WIRELESS POWER TRANSFER SYSTEM AT 13.56 MHZ TO REPLACE CONVENTIONAL WIRED CONNECTION IN A VEHICLE 3848
Kawin Surakitbovorn ; Juan Rivas-Davila

EFFICIENCY MAXIMIZATION OF INDUCTIVE POWER TRANSFER SYSTEM BY IMPEDANCE AND SWITCHING FREQUENCY CONTROL IN SECONDARY-SIDE CONVERTER 3855
Ryosuke Ota ; Dannisworo S. Nugroho ; Nobukazu Hoshi

ANALYSIS OF OPTIMAL OPERATION FREQUENCY RANGE FOR BATTERY CHARGING IN WPT SYSTEM 3863
Yongbin Jiang ; Min Wu ; Junwen Liu ; Yue Wang ; Laili Wang ; Hailong Zhang

INITIAL CURRENT INJECTION METHOD OF A DIRECT THREE-PHASE TO SINGLE-PHASE AC/AC CONVERTER FOR INDUCTIVE CHARGER 3870
Ferdi Perdana Kusumah ; Jorma Kyyrä

MISSION PROFILE EMULATOR FOR PERMANENT MAGNET SYNCHRONOUS MACHINE BASED ON THREE-PHASE POWER ELECTRONIC CONVERTER 3877
Yubo Song ; Ran Cheng ; Ke Ma

A VARIABLE DC BUS VOLTAGE BASED POWER HARDWARE-IN-THE-LOOP EMULATION OF ELECTRIC MOTORS WITH WIDE VARIATION IN INTERFACE FILTER INDUCTANCE 3884
Tsai-Fu Wu ; Mitradatta Misra ; Ying-Yi Jhang ; Chang-Jun Yang ; Yin-Chi Xu

COPPER LOSS MINIMIZATION CONTROL AT ZERO OUTPUT VOLTAGE FOR ELECTROLYTIC CAPACITOR-LESS INVERTER 3890
Kodai Abe ; Haruya Kada ; Kiyoshi Ohishi ; Hitoshi Haga ; Yuki Yokokura

ARMATURE TEMPERATURE ESTIMATION INSENSITIVE TO ROTOR FLUX VARIATION FOR SPMSM 3896
Toshiki Sano ; Kiyoshi Ohishi ; Yuki Yokokura ; Hiroki Iwata ; Yuji Ide ; Daigo Kuraishi ; Akihiko Takahashi

VIRTUAL SYNCHRONOUS GENERATOR CONTROL WITH RELIABLE FAULT RIDE-THROUGH CAPABILITY BY ADOPTING MODEL PREDICTIVE CONTROL 3902
Jonggrist Jongudomkarn ; Jia Liu ; Toshifumi Ise

RESHAPING QUADRATURE-AXIS IMPEDANCE OF THREE-PHASE GRID-CONNECTED CONVERTERS FOR LOW-FREQUENCY STABILITY IMPROVEMENT 3910
Yi Tang ; Jingyang Fang ; Xiaoqiang Li ; Hongchang Li

COMPARISON BETWEEN TRADITIONAL DROOP AND A NEW AUTONOMOUS CONTROL SCHEME FOR PARALLEL INVERTERS 3916
Mohammad Bani Shamseh ; Teruo Yoshino ; Atsuo Kawamura

A NOVEL MICROGRID POWER SHARING SCHEME ENHANCED BY A NON-INTRUSIVE FEEDER IMPEDANCE ESTIMATION METHOD 3924
Baojin Liu ; Zeng Liu ; Jinjun Liu ; Ronghui An ; Shuguang Song

DEVELOPMENT OF A 3.2MW PHOTOVOLTAIC INVERTER FOR LARGE-SCALE PV POWER PLANTS 3929
Naoya Shibata ; Tsuguhiro Tanaka ; Masahiro Kinoshita

IMPEDANCE-BASED STABILITY ANALYSIS OF LARGE-SCALE PV STATION UNDER WEAK GRID CONDITION CONSIDERING SOLAR RADIATION FLUCTUATION 3934
Yiming Tu ; Jinjun Liu ; Teng Liu ; Xiangpeng Cheng

EXPERIMENTAL VERIFICATION OF GRID-CONNECTION OF A PV CONVERTER USING A SYMMETRICALLY CONNECTED BOOST CONVERTER FOR A HIGH-LEG DELTA TRANSFORMER 3940
Daiki Yamaguchi ; Hideaki Fujita

A NOVEL SINGLE- STAGE HIGH-FREQUENCY BOOST INVERTER CASCADED BY RECTIFIER-INVERTER SYSTEM FOR PV GRID-TIE APPLICATIONS 3945
Hamdy Radwan ; Mahmoud A. Sayed ; Takaharu Takeshita ; Adel A. Elbaset ; G. Shabib

NINE SWITCHES MATRIX CONVERTER USING BI-DIRECTIONAL GAN DEVICE 3952
Takashi Hirota ; Kentaro Inomata ; Daisuke Yoshimi ; Masato Higuchi

A MODEL PREDICTIVE DUAL CURRENT CONTROL METHOD FOR INDIRECT MATRIX CONVERTER FED INDUCTION MOTOR DRIVES 3958
Mei Yang ; Chen Lisha ; Liang Wang ; Yunwei Li

FAULT TOLERANT PREDICTIVE CONTROL OF THREE-LEVEL NEUTRAL-POINT-CLAMPED BACK-TO-BACK POWER CONVERTERS 3965
Zhenbin Zhang ; Xicai Liu ; Kejun Cai ; Feng Gao ; Ralph Kennel

TWO-STAGE OPTIMIZATION BASED PREDICTIVE TORQUE CONTROL WITH REDUCED COMPLEXITY FOR A THREE-LEVEL INVERTER DRIVEN INDUCTION MOTOR 3971
Ilham Osman ; Dan Xiao ; Faz Rahman

DESIGN CHALLENGES OF SIC DEVICES FOR LOW- AND MEDIUM-VOLTAGE DC-DC CONVERTERS 3979
Georges Engelmann ; Alexander Sewergin ; Markus Neubert ; Rik W. De Doncker

DESIGN AND TESTING OF 6 KV H-BRIDGE POWER ELECTRONICS BUILDING BLOCK BASED ON 10 KV SIC MOSFET MODULE 3985
Jun Wang ; Slavko Mocevic ; Jiewen Hu ; Yue Xu ; Christina Dimarino ; Igor Cvetkovic ; Rolando Burgos ; Dushan Boroyevich

HIGH POWER MEDIUM VOLTAGE CONVERTERS ENABLED BY HIGH VOLTAGE SIC POWER DEVICES 3993
Sanket Parashar ; Ashish Kumar ; Subhashish Bhattacharya

SOFT-SWITCHING – THE KEY TO HIGH POWER WBG CONVERTERS 4001
Deepak Divan ; Zheng An ; Prasad Kandula

SIC: TECHNOLOGY ENABLER FOR MV DC/DC GALVANICALLY INSULATED MODULAR CONVERTERS 4009
S. Alvarez ; M. Bellini ; U. Vemulapati ; F. Canales ; M. Rahimo

A BEARINGLESS SLICE MOTOR WITH A SOLID IRON ROTOR FOR DISPOSABLE CENTRIFUGAL BLOOD PUMP 4016
Tadahiko Shinshi ; Ryo Yamamoto ; Yoshiki Nagira ; Junichi Asama

REDUCED HARDWARE PARALLEL DRIVE FOR NO VOLTAGE BEARINGLESS MOTORS 4020
Eric L. Severson

DUAL FIELD-ORIENTED CONTROL OF BEARINGLESS MOTORS WITH COMBINED WINDING SYSTEM 4028
Wolfgang Gruber ; Siegfried Silber

OPEN-CIRCUIT FAULT TOLERANT STUDY OF BEARINGLESS MULTI-SECTOR PERMANENT MAGNET MACHINES 4034
G. Valente ; L. Papini ; A. Formentini ; C. Gerada ; P. Zanchetta

BALANCE CONTROL OF SPLIT CAPACITOR POTENTIAL FOR MAGNETICALLY LEVITATED MOTOR SYSTEM USING ZERO-PHASE CURRENT 4042
Takaaki Oiwa

ASYMMETRICAL HALF-BRIDGE CONVERTER WITH ZERO DC-OFFSET CURRENT IN TRANSFORMER USING NEW RECTIFIER STRUCTURE 4049
Jung-Kyu Han ; Jong-Woo Kim ; Seung-Hyun Choi ; Jih-Sheng Lai ; Gun-Woo Moon

CIRCULATING CURRENT-LESS PHASE-SHIFTED FULL-BRIDGE CONVERTER WITH NEW RECTIFIER STRUCTURE 4054
Jung-Kyu Han ; Gun-Woo Moon

A BI-DIRECTIONAL CURRENT DETECTION USING CURRENT TRANSFORMERS FOR BI-DIRECTIONAL DC-DC CONVERTER 4059
Seiji Iyasu ; Yuji Hahashi ; Yuuichi Handa ; Kimikazu Nakamura ; Keiji Wada

A 10 MHZ GANFET BASED ISOLATED HIGH STEP-DOWN DC-DC CONVERTER 4066
Prasanth Thummala ; Dorai Babu Yelaverthi ; Regan Zane ; Ziwei Ouyang ; Michael A. E. Andersen

ANALYSIS AND DESIGN OF A PARALLEL RESONANT CONVERTER FOR CONSTANT CURRENT INPUT TO CONSTANT VOLTAGE OUTPUT DC-DC CONVERTER OVER WIDE LOAD RANGE 4074
Tarak Saha ; Hongjie Wang ; Baljit Riar ; Regan Zane

NOVEL SINUSOIDAL INPUT CURRENT SINGLE-TO-THREE-PHASE Z-SOURCE BUCK+BOOST AC/AC CONVERTER 4080
M. Haider ; D. Bortis ; J. W. Kolar ; Y. Ono

SIMPLE PWM STRATEGY OF A MATRIX CONVERTER FOR MINIMIZING OUTPUT VOLTAGE HARMONICS 4088
Takuya Oshima ; Takaharu Takeshita

NOVEL THREE-LEVEL BACK-TO-BACK CONVERTERS: STRUCTURE, MODULATION METHOD, AND EXPERIMENT 4096
S. Sangwongwanich ; K. Niyomsatian ; S. Samermurn ; S. Nuchnoi ; S. Suwankawin

MODEL PREDICTIVE CONTROL USING SUBDIVIDED VOLTAGE VECTORS FOR CURRENT RIPPLE REDUCTION IN AN INDIRECT MATRIX CONVERTER 4104
Keon Young Kim ; Yeongsu Bak ; Jin-Hyuk Park ; Kyo-Beum Lee

DC-LINK RIPPLE CURRENT REDUCTION IN BACK-TO-BACK CONVERTERS WITH DPWM 4109
Anatolii Tcai ; Kyo-Beum Lee

AN ANALYSIS OF CLASS DE VOLTAGE-SOURCE PARALLEL RESONANT INVERTER 4114
Takeshi Kondo ; Tsuyoshi Inaba ; Yoshikazu Sakai ; Hirotaka Koizumi

AN IMPROVEMENT ON EXTENDED IMPEDANCE METHOD TOWARDS EFFICIENT STEADY-STATE ANALYSIS OF HIGH-FREQUENCY CLASS-E RESONANT INVERTERS 4122
Junrui Liang

OUTPUT POWER CAPABILITY COMPARISONS OF CLASS-E POWER AMPLIFIERS WITH HARMONIC RESONANCE 4127
Hiroo Sekiya ; Xiuqin Wei ; Yuchong Sun

A CLASS Φ2 RESONANT BUCK CONVERTER WITH RIPPLE INJECTION BURST CONTROL METHOD 4133
Min Lin ; Masahiko Hirokawa

PRACTICAL DESIGN TECHNIQUE FOR HIGH POWER DENSITY LLC RESONANT CONVERTER 4139
Shingo Nagaoka ; Hiroyuki Onishi ; Koji Takatori ; Toshiyuki Zaitsu ; Takeshi Uematsu

OPERATIONAL STUDY AND PROTECTION OF A SERIES RESONANT CONVERTER WITH DC CURRENT INPUT APPLIED IN DC CURRENT DISTRIBUTION SYSTEMS 4145
Hongjie Wang ; Tarak Saha ; Baljit Riar ; Regan Zane

A STUDY ON IMPROVEMENT OF POWER UTILIZATION RATE OF ENERGY SYSTEMS WITH PVS AND BATTERIES 4151
Hiroaki Endo ; Masakatsu Kurisaka ; Tsutomu Ueno ; Yusuke Yoshioka ; Kaoru Inoue ; Toshiji Kato

A NOVEL DC DISTRIBUTION NETWORK WITH MULTI-LEVEL BUS VOLTAGES AND ITS ENERGY MANAGEMENT SYSTEM DESIGN 4157
Jingjin Huang ; Xin Zhang ; Zhixun Ma ; Jianfang Xiao

A NOVEL DC-SIDE-PORT IMPEDANCE MODELING OF MODULAR MULTILEVEL CONVERTERS BASED ON HARMONIC STATE SPACE METHOD 4162
Jing Lyu ; Xin Zhang ; Zhixun Ma ; Xu Cai

AN IMPROVED MASTER-SLAVE CONTROL FOR THREE-PORT CONVERTER BASED DISTRIBUTED DC GRID-CONNECTED PV SYSTEM 4168
Siyue Jiang ; Kai Sun ; Hongfei Wu ; Haixu Shi ; Xiaofeng Dong ; Syed Muhammad Raza Kazmi

SENSORLESS POSITION ESTIMATION, PARAMETER IDENTIFICATION AND CONTROL INTEGRATION FOR PERMANENT MAGNET SYNCHRONOUS MACHINES USING CURRENT DERIVATIVE MEASUREMENTS 4174
M.X. Bui

DYNAMIC PERFORMANCE IMPROVEMENT OF BIDIRECTIONAL SWITCHED-CAPACITOR DC/DC CONVERTER BY RIGHT-HALF-PLANE ZERO ELIMINATION 4181
Ding Kaicheng ; Zhang Yan ; Liu Jinjun ; Zeng Pengxiang ; Zhang Jinshui

A MATRIX BASED ISOLATED BIDIRECTIONAL AC-DC CONVERTER WITH LCL TYPE INPUT FILTER FOR ENERGY STORAGE APPLICATION 4186
Prathamesh Pravin Deshpande ; Amit Kumar Singh ; Sanjib Kumar Panda

ON A STUDY OF VOLTAGE DIVIDING CLASS Φ AMPLIFIER 4193
Katsutoshi Hirayama ; Tadashi Suetsugu ; Yudai Furukawa ; Fujio Kurokawa

A DPWM BASED CONTROL STRATEGY TO INTEGRATE PHOTOVOLTAIC SYSTEM AND BATTERY STORAGE USING GRID CONNECTED THREE-LEVEL T-TYPE INVERTER 4198
Mohammad M. Hashempour ; Yue-Ting Tsai ; T. L. Lee

IMPEDANCE MEASUREMENT OF MEGAWATT-LEVEL RENEWABLE ENERGY INVERTERS USING GRID-FORMING AND GRID-PARALLEL CONVERTERS 4205
Matias Berg ; Tuomas Messo ; Tomi Roinila ; Henrik Alenius

IMPROVED VIRTUAL INDUCTANCE BASED CONTROL STRATEGY OF DFIG UNDER WEAK GRID CONDITION 4213
Ran Fang ; Wenjia Chen ; Xueguang Zhang ; Dianguo Xu

CONTROL OF VSC-HVDC FOR WIND FARM INTEGRATION WITH REAL-TIME FREQUENCY MIRRORING AND SELF-SYNCHRONIZING CAPABILITY 4220
Renxin Yang ; Chen Zhang ; Xu Cai ; Gang Shi ; Jing Lyu

A STUDY ON STEADY-STATE CHARACTERISTICS OF SERIES-CONNECTED WIND FARM USING AN EXPERIMENTAL SET OF LABORATORY SIZE 4227
Fujio Tatsuta ; Shoji Nishikata

A NOVEL ISLANDING DETECTION METHOD WITH TWO-PHASE MAGNIFICATION INSPECTION 4233
Jian-Tang Liao ; Shun-Hao Yeh ; Hong-Tzer Yang

Author Index

The 2018 International Power Electronics Conference -ECCE Asia-

IPEC 2018 ECCE ASIA

Niigata

May 20- 24, 2018
TOKI MESSE Niigata Convention Center
Sponsored by IEEJ Industry Applications Society

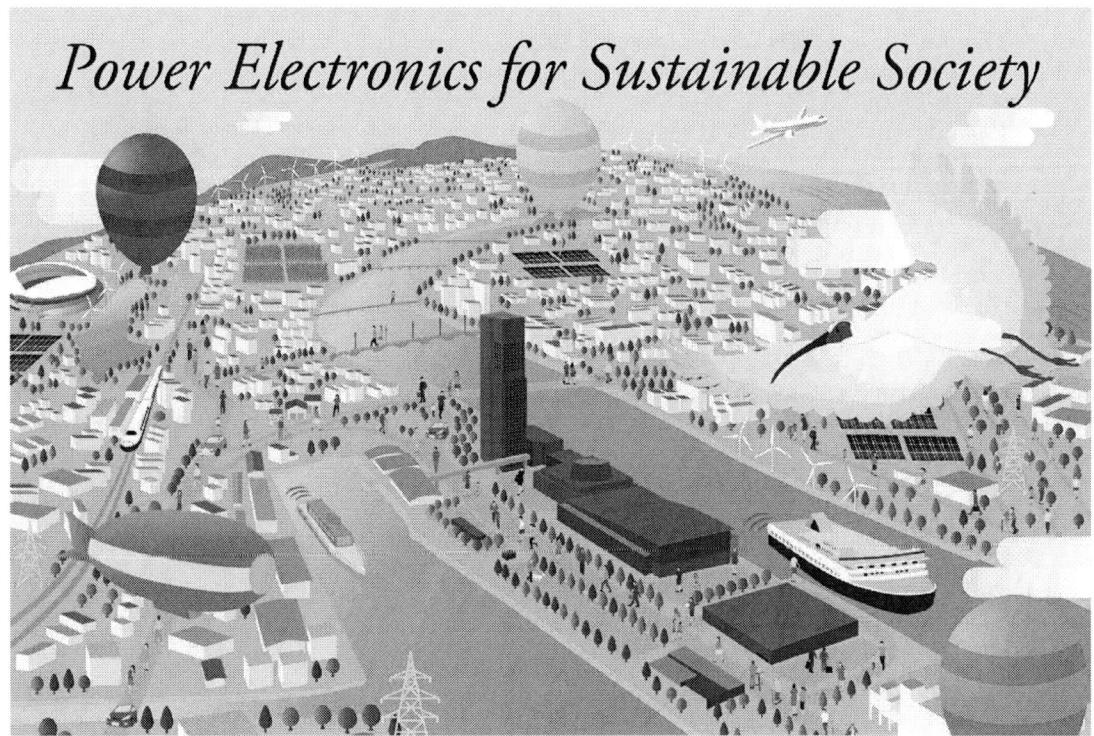

Power Electronics for Sustainable Society

ECCE Asia Cooperation

IEEJ Industry Applications Society

The Korean Institute of Power
Electronics (KIPE)

China Electrotechnical Society
(CES)

Co-Sponsorship

IEEE Power Electronics Society
(IEEE-PELS)

IEEE Industry Applications Society
(IEEE-IAS)

In Cooperation with

European Power Electronics and
Drives Association (EPE)

Welcome Message

On behalf of the Organizing, Steering and Technical Program Committees, we sincerely welcome you to the 2018 International Power Electronics Conference -ECCE Asia-, which we simply call IPEC-Niigata 2018. The first IPEC was held in 1983, and since then, in 1990, 1995, 2000, and 2005, IPECs have been held in different locations, sponsored by IEEJ (Institute of Electrical Engineers of Japan). Since 2010, IPEC has had the additional function of hosting the power electronics conference series, which is held in the three-country rotation of China, Japan, and Korea, as one of the ECCE Asia conferences through Co-sponsorship with the IEEE Power Electronics Society and the IEEE Industry Applications Society.

The eighth International Power Electronics Conference, IPEC-Niigata 2018 -ECCE Asia-, is held from May 20 to May 24, 2018, in Niigata, Japan. The conference venue is the Niigata Convention Center, Toki Messe, which faces to the estuary of the Shinano River, the longest river in Japan. Power electronics has been providing numerous new technologies in the fields of electric energy conversion and motor drive systems for more than 60 years. In recent years, global energy and environmental issues are becoming more serious and power electronics is expected to play a key role in solving such problems. The IPEC Niigata 2018 -ECCE Asia- will provide a unique opportunity for researchers, engineers, and academics from all over the world to present and exchange the latest information on power electronics, motor drives, and related subjects.

The Technical Program Committee received 930 extended summaries submitted from 40 countries and areas. Finally, 509 Regular Session papers and 168 Organized Session papers were accepted for presentation. We would like to thank the committee members and reviewers for their support of the IPEC.

IPEC-Niigata 2018 will start with an Industrial Seminar in the afternoon of May 20, followed by a welcome reception. Prior to the Industrial Seminar, a Students and Young Engineers Meeting will be held in conjunction with the IPEC Steering Committee, Ph.D. Candidates of Power Electronics in Japan (PPEJ), and the IEEE PELS and IAS in the morning of May 20. The Opening Ceremony will be held in the morning of May 21. At that time, winners of the Isao Takahashi Power Electronics Award and the Prize Paper Award will be announced, followed by the former part of The Plenary Session. The latter part of The Plenary Session will be held in the morning of May 22. The Regular Sessions consist of both Oral and Poster presentations, and they are open from afternoon of May 21 to 24. The Organized Session will present invited papers. A banquet is scheduled in the evening of May 23.

Niigata is the capital city of Niigata Prefecture and located at 300 kilometers (186 miles) northwest of Tokyo in the main land of Japan. Niigata is a major port city with more than 150 years of history. Echigo Plain which surrounds Niigata city is fertile. Due to these good circumstances, Niigata is very rich in fresh seafood, finest rice, a wide variety of fruits, vegetables, and Sake (Japanese traditional rice wine). Everyone will enjoy the beautiful landscape of the Shinano River, and can also savor Japanese traditional foods. If you need any help during the conference, please let us know or notify any member of the Steering Committee. We would be very happy to serve you and help you enjoy the conference.

Finally, we would like to express special thanks to the many companies that supported the conference as financial sponsors or exhibitors. We would also like to thank the Japan Society for the Promotion of Science (Grant-in Aid for Scientific Research) for financial support. In addition, we would like to thank the Niigata Visitors & Convention Bureau for helpful support in local arrangement.

Toshihisa Shimizu
Chairperson, Organizing Committee

Yukihiko Sato
Chairperson, Steering Committee

Hirohito Funato
Chairperson, Technical Program Committee

Toshihisa Shimizu
Chairperson
Organizing Committee

Yukihiko Sato
Chairperson
Steering Committee

Hirohito Funato
Chairperson
Technical Program Committee

Gap in pagination due to formatting issues.

Pages 1-68

The 2018 International Power Electronics Conference

Three-phase Inductive Power Transfer System with 12 coils for Radiation Noise Reduction

Keisuke Kusaka[1][*] and Jun-ichi Itoh[2]

1 Department of Electrical, Electronics and Information Engineering, Nagaoka University of Technology, Nagaoka, Japan
2 Department of Science of Technology Innovation, Nagaoka University of Technology, Nagaoka, Japan

*E-mail: kusaka@vos.nagaokaut.ac.jp

Abstract— A three-phase inductive power transfer system with six transmission coils and six receiving coils is proposed in this paper. The proposed system achieves a reduction in radiation noise by canceling the noise using multiple coils, which are placed opposite to each other. However, in IPT systems with multiple coils, a magnetic interference among the multiple coils decreases a transmission efficiency due to the occurrence of circulating current. The proposed three-phase IPT system achieves cancellation of the interference using six coils on each side. First, the IPT system with the 12 coils is proposed. Then, a canceling condition of the magnetic interference among the coils is mathematically introduced from the voltage equation on the coils. Finally, the proposed IPT system is experimentally demonstrated with the 3-kW prototype. The experimental result shows that the radiation noise at the fundamental frequency is suppressed to 1/6 (from 12.2dBµA to 2.1dBµA).

Keywords— inductive power transfer, wireless power transfer, three-phase transmission

I. INTRODUCTION

In recent years, inductive power transfer (IPT) systems for electrical vehicles (EVs) are actively studied [1–11] because IPT systems are capable of improving the usability of EV users. In order to put the IPT system for practical use, a standardization of the IPT system for EVs is ongoing [12–13]. According to the standards, which will be published by IEC/ISO, the IPT systems will be classified by the maximum input power into four class; the maximum input power is ≤ 3.7 kW for WPT1, the maximum input power is > 3.7 kW and ≤ 7.7 kW for WPT2, the maximum input power is > 7.7 kW and ≤ 11 kW for WPT3, the maximum input power is > 11 kW and ≤ 22 kW for WPT4, and the maximum input power is > 22 kW for WPT5 [12–13].

The standardization of WPT1 and WPT2 are currently in progress. However, the standardization of high-power IPT systems such as WPT5 has not been discussed well despite strong demand for higher power IPT system for heavy-duty vehicles because high-power IPT systems may cause two problems. One of the problems is a decrease in transmission efficiency. Copper loss and iron loss on transmission coils may increase because the current in the transmission coils increases [14]. The second problem is radiation noise. The radiation noise caused by the IPT systems must satisfy the regulations, which are published by CISPR or legislated in each country or region. However, the satisfaction of regulations becomes further difficult because the radiation noise caused by a loop coil is proportional to the conduction current.

In previous studies, a radiation noise reduction technique using a duplicated 44-kW IPT system for heavy-duty EVs has been proposed [14–16]. In [15], the radiation noise is canceled by duplicated transmission coils. The two transmission coils with differential current are placed to cancel out the leakage radiation noise. However, the duplicated IPT system limits freedom of coil positions because the duplicated coils cause a magnetic interference between the duplicated coils [15]. The magnetic interference decreases power factor regarding an inverter output.

Meanwhile, three-phase IPT systems have been proposed to increase the output power of the IPT system [17-20]. The three-phase IPT systems allow the reduction of the current on each transmission coil. Thus the three-phase IPT is an effective method to reduce the copper loss and the iron loss. However, in the conventional three-phase IPT system, a magnetic interference among the additional coils on each side occurs. The magnetic interference causes a degrade the efficiency because the magnetic interference leads to circulating currents.

This paper proposes a three-phase IPT system with 12 solenoid coils, which consists of six pairs of coils placed opposite to each other. The proposed IPT system contributes reducing the current on the winding without a byproduct such as a magnetic interference. A contribution of this paper is offering the new transmission coil structure, which reduces the radiation noise with a small effect on efficiency. In the proposed system, the radiation noise is canceled out by the countercurrent in the opposite coil. Moreover, the primary coils and the secondary coils are placed at an angle of 60 degrees to each other. Consequently, the induced voltage caused by magnetic interference among the 12 coils are canceled out. First, a system configuration of the proposed IPT system is unveiled. Next, a design procedure of the proposed three-phase IPT system with a star-star winding and a delta-delta winding is explained. Then, the canceling method of the magnetic interference among the multiple coils is mathematically introduced. Finally, the proposed IPT system with star-star winding is tested with a 3-kW prototype. Then, the radiation noise is assessed and compared between the conventional three-phase IPT system and the proposed three-phase IPT system with 12 coils.

II. THREE-PHASE IPT SYSTEM WITH 12 COILS

In this chapter, the three-phase IPT system is proposed. The proposed IPT system achieves radiation noise reduction using 12 coils without magnetic interference among multiple coils. In other words, the IPT system transmits power with unity power factor regarding inverter output.

A. System Configuration

Figure 1 shows the proposed three-phase IPT system with 12 coils. The three-phase coils can be connected with not only star-star winding but also star-delta, delta-star, and delta-delta windings. In this paper, only star-star and delta-delta windings are discussed due to the page limitation. Power is inductively transmitted through the three-phase coils. Each phase has two coils connected in series. These coils connected in series cancel out the radiation noise because of the countercurrent flowing on the coils [14].

Besides, the three-phase inverter in the primary side and the three-phase rectifier in the secondary side are used. The primary inverter output square-wave voltage in each phase. The phase difference of each phase are 120 degrees. Furthermore, resonant capacitors are connected to the output of the primary inverter and the input of the rectifier in series. The series-series compensation technique [21] is applied to the proposed system in order to cancel out the leakage inductance due to the weak magnetic coupling. Using three-phase IPT and the series-series compensation technique, copper loss of the windings can be reduced because the current is shunted into six coils. Therefore, the proposed IPT system is effective to improve the heat dissipation of the transmission coils.

Figure 2 shows the position and connection of the transmission coils. In the proposed IPT system, six solenoid coils are used in each of the primary side and the secondary side. Fig. 3 shows the mechanism of the reduction in radiation noise. Two transmission coils connected in series (e.g., L_{uv1A} and L_{uv1B}) are differentially connected and are placed in an opposite place. These opposite coils cancel out radiation noise at a measurement point, which is typically 10 m from the IPT system [22]. Other pairs of coils are placed in an angle of 120 degrees to the coils on phase-u each other. As a drawback of the reduction in radiation noise using opposite coils, magnetic interference occur among the six coils on each the primary side and the secondary side. In

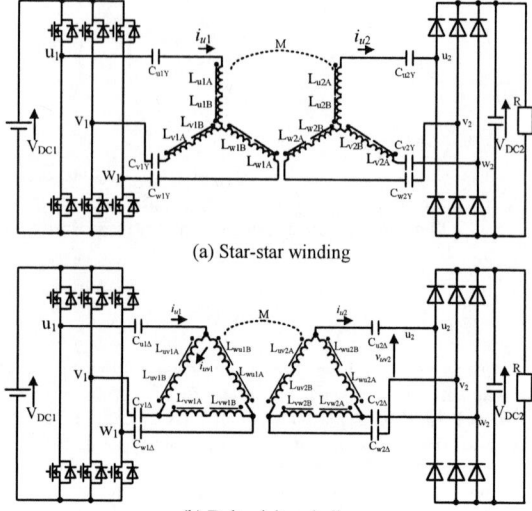

(a) Star-star winding

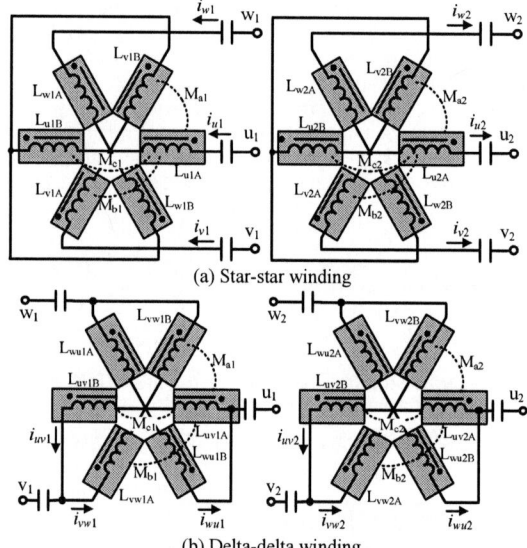

(b) Delta-delta winding

Fig. 1. Proposed three-phase IPT system with 12 coils.

(a) Star-star winding

(b) Delta-delta winding

Fig. 2. Placement and connection of coils.

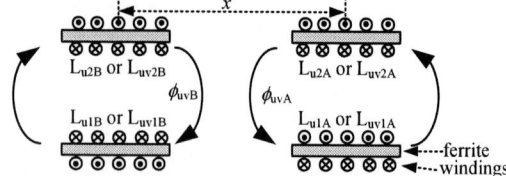

Fig. 3. Mechanism of reduction in radiation noise using solenoid coils.

order to cancel out the magnetic interference, the position relationship and figure of the coils have to be adjusted. The design method of the coils will be explained in section C.

B. Mathematical Expression of Three-phase Coils

Equations (1) and (2) represents the induced voltage of 12 coils on the star-star winding system and the delta-

$$
\begin{pmatrix} v_{u1A} \\ v_{u1B} \\ v_{v1A} \\ v_{v1B} \\ v_{w1A} \\ v_{w1B} \\ v_{u2A} \\ v_{u2B} \\ v_{v2A} \\ v_{v2B} \\ v_{w2A} \\ v_{w2B} \end{pmatrix} = \begin{pmatrix} L_{1Ys} & M_{c1} & M_{b1} & M_{a1} & M_{b1} & M_{a1} & M & 0 & 0 & 0 & 0 & 0 \\ M_{c1} & L_{1Ys} & M_{a1} & M_{b1} & M_{a1} & M_{b1} & 0 & M & 0 & 0 & 0 & 0 \\ M_{b1} & M_{a1} & L_{1Ys} & M_{c1} & M_{b1} & M_{a1} & 0 & 0 & M & 0 & 0 & 0 \\ M_{a1} & M_{b1} & M_{c1} & L_{1Ys} & M_{a1} & M_{b1} & 0 & 0 & 0 & M & 0 & 0 \\ M_{b1} & M_{a1} & M_{b1} & M_{a1} & L_{1Ys} & M_{c1} & 0 & 0 & 0 & 0 & M & 0 \\ M_{a1} & M_{b1} & M_{a1} & M_{b1} & M_{c1} & L_{1Ys} & 0 & 0 & 0 & 0 & 0 & M \\ M & 0 & 0 & 0 & 0 & 0 & L_{2Ys} & M_{c2} & M_{b2} & M_{a2} & M_{b2} & M_{a2} \\ 0 & M & 0 & 0 & 0 & 0 & M_{c2} & L_{2Ys} & M_{a2} & M_{b2} & M_{a2} & M_{b2} \\ 0 & 0 & M & 0 & 0 & 0 & M_{b2} & M_{a2} & L_{2Ys} & M_{c2} & M_{b2} & M_{a2} \\ 0 & 0 & 0 & M & 0 & 0 & M_{a2} & M_{b2} & M_{c2} & L_{2Ys} & M_{a2} & M_{b2} \\ 0 & 0 & 0 & 0 & M & 0 & M_{b2} & M_{a2} & M_{b2} & M_{a2} & L_{2Ys} & M_{c2} \\ 0 & 0 & 0 & 0 & 0 & M & M_{a2} & M_{b2} & M_{a2} & M_{b2} & M_{c2} & L_{2Ys} \end{pmatrix} \frac{d}{dt} \begin{pmatrix} i_{u1} \\ -i_{u1} \\ i_{v1} \\ -i_{v1} \\ i_{w1} \\ -i_{w1} \\ i_{u2} \\ -i_{u2} \\ i_{v2} \\ -i_{v2} \\ i_{w2} \\ -i_{w2} \end{pmatrix} \tag{1}
$$

$$
\begin{pmatrix} v_{uv1A} \\ v_{uv1B} \\ v_{vw1A} \\ v_{vw1B} \\ v_{wu1A} \\ v_{wu1B} \\ v_{uv2A} \\ v_{uv2B} \\ v_{vw2A} \\ v_{vw2B} \\ v_{wu2A} \\ v_{wu2B} \end{pmatrix} = \begin{pmatrix} L_{1\Delta s} & M_{c1} & M_{b1} & M_{a1} & M_{b1} & M_{a1} & M & 0 & 0 & 0 & 0 & 0 \\ M_{c1} & L_{1\Delta s} & M_{a1} & M_{b1} & M_{a1} & M_{b1} & 0 & M & 0 & 0 & 0 & 0 \\ M_{b1} & M_{a1} & L_{1\Delta s} & M_{c1} & M_{b1} & M_{a1} & 0 & 0 & M & 0 & 0 & 0 \\ M_{a1} & M_{b1} & M_{c1} & L_{1\Delta s} & M_{a1} & M_{b1} & 0 & 0 & 0 & M & 0 & 0 \\ M_{b1} & M_{a1} & M_{b1} & M_{a1} & L_{1\Delta s} & M_{c1} & 0 & 0 & 0 & 0 & M & 0 \\ M_{a1} & M_{b1} & M_{a1} & M_{b1} & M_{c1} & L_{1\Delta s} & 0 & 0 & 0 & 0 & 0 & M \\ M & 0 & 0 & 0 & 0 & 0 & L_{2\Delta s} & M_{c2} & M_{b2} & M_{a2} & M_{b2} & M_{a2} \\ 0 & M & 0 & 0 & 0 & 0 & M_{c2} & L_{2\Delta s} & M_{a2} & M_{b2} & M_{a2} & M_{b2} \\ 0 & 0 & M & 0 & 0 & 0 & M_{b2} & M_{a2} & L_{2\Delta s} & M_{c2} & M_{b2} & M_{a2} \\ 0 & 0 & 0 & M & 0 & 0 & M_{a2} & M_{b2} & M_{c2} & L_{2\Delta s} & M_{a2} & M_{b2} \\ 0 & 0 & 0 & 0 & M & 0 & M_{b2} & M_{a2} & M_{b2} & M_{a2} & L_{2\Delta s} & M_{c2} \\ 0 & 0 & 0 & 0 & 0 & M & M_{a2} & M_{b2} & M_{a2} & M_{b2} & M_{c2} & L_{2\Delta s} \end{pmatrix} \frac{d}{dt} \begin{pmatrix} i_{uv1} \\ -i_{uv1} \\ i_{vw1} \\ -i_{vw1} \\ i_{wu1} \\ -i_{wu1} \\ i_{uv2} \\ -i_{uv2} \\ i_{vw2} \\ -i_{vw2} \\ i_{wu2} \\ -i_{wu2} \end{pmatrix} \tag{2}
$$

delta winding system, respectively where L_{1Ys} is the self-inductance of each coil on the star-star winding, $L_{1\Delta s}$ is the self-inductance of each coil on the delta-delta winding.

The mutual inductance M represents the mutual inductances between the primary side and the secondary side, which contributes to the power transmission, e.g., the mutual inductance between L_{u1A} and L_{u2A}. M_a represents the mutual inductance between the adjacent coils, e.g., the mutual inductance between L_{u1A} and L_{v1B}. The mutual inductance M_b represents the mutual inductance between the coils, which are placed apart from 120 degrees, e.g., the mutual inductance between L_{u1A} and L_{w1A}. The mutual inductance M_c represents the mutual inductance between the opposite coils, e.g., the mutual inductance between L_{u1A} and L_{u1B}. The suffix "1" indicates inductances on the primary side, and "2" means inductances on the secondary side. Note that other unwanted mutual inductances, e.g., the mutual inductance between L_{u1A} and L_{u2B} are ignored in Eq. (1).

It is clear from Eq. (1) that the unwanted mutual inductances cause an unwanted induce voltage on each of the transmission coils. This induced voltage decreases the efficiency due to a circulating current. The canceling method of these unwanted induced voltages will be explained in the next section.

C. Cancellation of Magnetic Interference Coupling

In this section, the canceling method of unwanted induced voltage caused by the magnetic interference is explained. The cancellation method of the magnetic interference coupling is explaind with the star-star winding. However, the cancellation technique can be used for the delta-delta winding.

The induced voltage on L_{u1} is derived from the first column of Eq. (1),

$$
v_{u1A} = L_{1Ys}\frac{di_{u1}}{dt} + M\frac{di_{u2}}{dt} - M_{c1}\frac{di_{u1}}{dt}
$$
$$
- M_{a1}\left(\frac{di_{v1}}{dt} + \frac{di_{w1}}{dt}\right) + M_{b1}\left(\frac{di_{v1}}{dt} + \frac{di_{w1}}{dt}\right) \tag{3}
$$

The first term of the right side in Eq. (3) is the induced voltage by self-inductance, the second term is the mutual inductance caused by the coil on the secondary side. The second term contributes to the power transmission. The third to fifth terms are the induced voltage caused by the interference coupling. If the third to fifth terms of Eq. (3) is zero, the relationship between the primary coil L_{1Ys} and the secondary coil L_{2Ys} is seen as the same as a transmission with one-by-one.

Assuming the three-phase equilibrium, Eq. (3) is transformed as

$$
v_{u1A} = L_{1Ys}\frac{di_{u1}}{dt} + M\frac{di_{u2}}{dt}
$$
$$
- \omega I_m \left\{ M_{c1}\cos\omega t + (M_{a1} - M_{b1})\cos\left(\omega t - \frac{2}{3}\pi\right) \right.
$$
$$
\left. (M_{a1} - M_{b1})\cos\left(\omega t - \frac{4}{3}\pi\right) \right\} \tag{4},
$$

where I_m is the maximum value of the primary current i_{u1}, which flows in each of the primary coils.

From Eq. (4), it is found that the unwanted induced voltage does not occur when

$$M_{a1} = M_{b1} + M_{c1} \qquad (5)$$

is satisfied.

Because the self-inductance of the primary coils L_{1Ys} is equal to each other, the canceling condition of the unwanted induced voltage is

$$k_{a1} = k_{b1} + k_{c1} \qquad (6).$$

Therefore, the unwanted coupling can be canceled by designing the magnetic coupling according to Eq. (6) on the primary side.

Similarly, the unwanted coupling on the secondary side can be canceled by designing the magnetic coupling according to Eq. (7).

$$k_{a2} = k_{b2} + k_{c2} \qquad (7)$$

Figure 4 shows the simulated relationship between the inverter output power factor and the magnetic interference couplings k_a, k_b, and k_c. The dotted line on the graph represents Eq. (6). Fig. 4 (a), (b), and (c) show the relation when the magnetic interference coupling k_b is zero, 0.1 and 0.2, respectively. When the magnetic interference coupling is small enough to be negligible, the power factor closes to unity. The power factor decreases when one of the interference coupling increases. However, the power factor remains at high even when the interference couplings increase under as long as Eq. (6) is satisfied. On the other words, these results show that the magnetic interferences among the transmission coils do not affect the power factor when the relation of the interference coupling fits on Eq. (6).

III. DESIGN OF THREE-PHASE IPT SYSTEM

In this chapter, the design method of proposed three-phase IPT system with the star-tar winding and the delta-delta winding is described. In this design procedure, the induced voltage generated by the magnetic interference among the multiple coils can be ignored with assuming Eq. (6–7).

A. Star-star winding

This section represents the design method of the proposed three-phase IPT system with the star-star winding shown in Fig. 1 (a) with.

1) Equivalent AC resistance

The equivalent AC resistance is calculated from the rated output voltage V_{DC2}, and rated output power P. The equivalent AC resistance R_{eq} can be calculated by expanding a derivation of an equivalent AC resistance of a single-phase rectifier to the three-phase rectifier. The equivalent AC resistance for a certain phase is expressed by

$$R_{eq} = \frac{8}{\pi^2} \frac{\left(V_{DC2}/2\right)^2}{P/3} = \frac{6}{\pi^2} \frac{V_{DC2}^2}{P} \qquad (8).$$

2) Secondary inductance

Maximizing a transmission efficiency, an impedance of the secondary inductance should be equal to the impedance of the equivalent AC resistance. By ignoring a

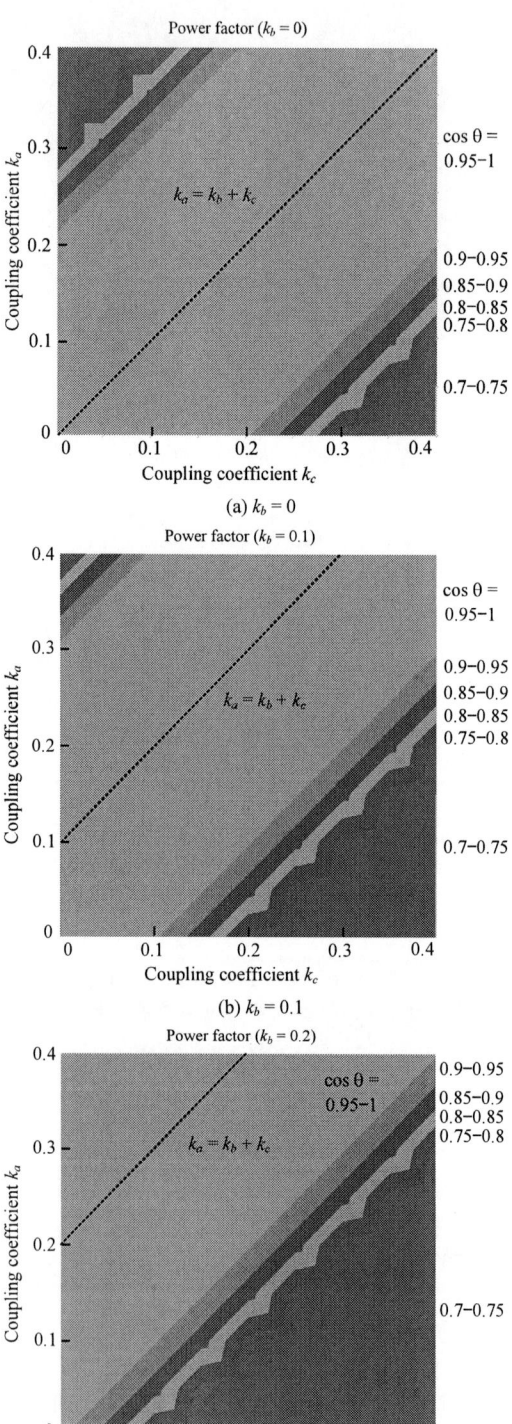

Fig. 4. Effect of magnetic interference coupling on power factor.

winding resistance and assuming the following resonance condition, the secondary self-inductance for a certain phase should be

$$L_{2Y} = L_{u2} = L_{v2} = L_{w2} = \frac{6}{\pi^2 k \omega} \frac{V_{DC2}^2}{P} \qquad (9),$$

where ω is the transmission angular frequency, k is the coupling coefficient.

The secondary inductances have to be half because two transmission coils are connected in each phase in the proposed IPT system. Thus, inductance for a certain coil is $L_{2Ys} = L_{2Y} / 2$.

3) Primary inductance

From the voltage ratio between the input DC voltage V_{DC1} and the output DC voltage V_{DC2} under the resonance condition, a primary self-inductance for obtaining the desired rated voltage is expressed by

$$L_{1Y} = L_{u1} = L_{v1} = L_{w1} = L_{2Y} \left(\frac{V_{DC1}}{V_{DC2}} \right)^2 = \frac{6}{\pi^2 k \omega} \frac{V_{DC1}^2}{P} \qquad (10),$$

where the primary inverter is operated with a square-wave operation. In the proposed IPT system, the transmission coils are divided and connected in series. The inductances represented by Eq. (10) have to be half. Thus, the inductance of primary coil is $L_{1Ys} = L_{1Y} / 2$.

4) Resonance capacitors

Besides, the resonance capacitors are connected to the primary inverter and the rectifier in series. The resonance capacitors are designed to resonate with the self-inductance at the transmission angular frequency ω. Thus, the resonant capacitors can be designed as

$$C_{u1Y} = C_{v1Y} = C_{w1Y} = \frac{1}{\omega^2 L_{1Y}} \qquad (11),$$

$$C_{u2Y} = C_{v2Y} = C_{w2Y} = \frac{1}{\omega^2 L_{2Y}} \qquad (12).$$

Note that, the operating frequency is slightly adjusted to achieve a zero-voltage switching of the MOSFETs in the inverter.

B. Delta-delta winding

This section describes the design method of the proposed three-phase IPT system with the delta-delta winding shown in Fig. 1 (b).

The equivalent resistance for the star connection is as same as the star-star winding.

1) Secondary inductance

The secondary inductance $L_{2\Delta}$ should be equal to the impedance of the equivalent AC resistance. Thus the secondary inductance with a star winding is expressed by

$$L_{2Y} = \frac{R_{eq}}{k \omega} = \frac{6}{\pi^2 k \omega} \frac{V_{DC2}^2}{P} \qquad (13)$$

with assuming small winding resistance under the resonance condition.

The secondary inductance expressed by Eq. (13) have be transformed into the inductance for the star-star winding. The secondary inductance is expressed by

$$L_{2\Delta} = L_{uv2} = L_{vw2} = L_{wu2} = 3L_{2Y} = \frac{18}{\pi^2 k \omega} \frac{V_{DC2}^2}{P} \qquad (14).$$

The inductance expressed in Eq. (14) have to be half in the proposed IPT system because the transmission coils are divided and connected in series.

2) Primary inductance

The primary inductance is calculated by Eq. (15) from the voltage ratio between the primary DC voltage and the secondary DC voltage when the primary inverter is operated in square-wave operation mode.

$$L_{1\Delta} = L_{uv1} = L_{vw1} = L_{wu1} = L_{2\Delta} \left(\frac{V_{DC1}}{V_{DC2}} \right)^2 = \frac{18}{\pi^2 k \omega} \frac{V_{DC1}^2}{P} \qquad (15)$$

The primary inductances also have to be half.

3) Resonance capacitors

The resonance capacitors are connected in series to the output of the inverter and the input of the rectifier. The resonance capacitors are designed to resonate with the primary and the secondary inductance at the transmission angular frequency ω.

$$C_{u1\Delta} = C_{v1\Delta} = C_{w1\Delta} = \frac{3}{\omega^2 L_{1\Delta}} \qquad (16)$$

$$C_{u2\Delta} = C_{v2\Delta} = C_{w2\Delta} = \frac{3}{\omega^2 L_{2\Delta}} \qquad (17)$$

IV. EXPERIMENTAL VERIFICATION

Figure 5 and Table I show the prototype and its specifications. In this paper, the proposed IPT system with the star-star winding is experimentally demonstrated. The magnetic interference couplings are $k_a = 0.048$, $k_b = 0.011$, and $k_c = 0.003$. The each transmission coils have ferrite plates of $245 \times 215 \times 10$ mm. The ferrite plates are put into boxes made from acrylic. As windings for the transmission coils, a litz-wire is used.

Figure 6 shows the operation waveforms with the 3-kW prototype. Fig. 6 (a) shows the inverter output voltage v_{uv1}, output current i_{u1}, rectifier input current i_{u2} and output voltage V_{DC2}. Note that the line voltage v_{uv1} as the inverter output voltage is observed. Thus, the inverter output current lags 30 degrees with respect to the inverter output line-volage. It means that the power factor correction using resonance is achieved despite the magnetic interference coupling. Moreover, it is confirmed that power is transmitted from the primary side to the secondary side.

Figure 7 shows the voltage gain between the primary DC voltage and the secondary DC voltage. The dotted line is the theoretical curve, which is calculated by ignoring the magnetic interference among the coils. The experimental results almost agree with the theoretical curve except for low load region. In the low-load area,

The 2018 International Power Electronics Conference

(a) Three-phase inverter

(b) Transmission coils
Fig. 5. 3-kW prototype.

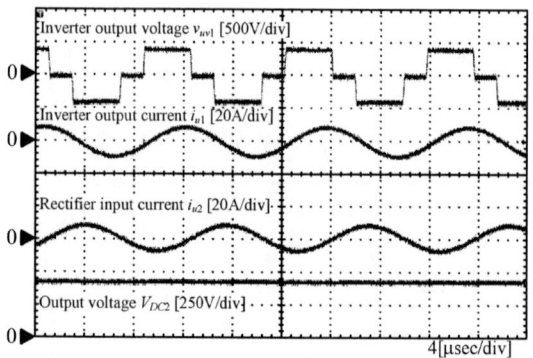

(a) Inverter output voltage, output current, rectifier input current and output voltage

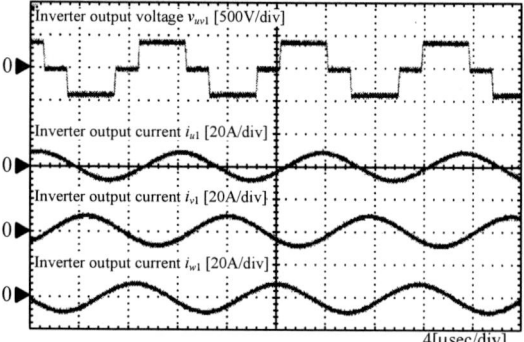

(a) Inverter output voltage and output currents on the primary side
Fig. 6. Operation waveforms.

Table 1. Specifications of the prototype.

Item	Symbol	Value	
Primary DC voltage	V_{DC1}	400	V
Secondary DC voltage	V_{DC2}	400	V
Rated power	P	3.0	kW
Transmission frequency	f	85.6	kHz
Coupling coefficient	k	0.26	
	k_a	0.048	
Interference coupling	k_b	0.011	
	k_c	0.003	
Primary inductance	L_{1Ys}	106	uH
Secondary inductance	L_{2Ys}	106	uH
Primary capacitance	$C_{u1Y}, C_{v1Y}, C_{w1Y}$	16.5	nF
Secondary capacitance	$C_{u2Y}, C_{v2Y}, C_{w2Y}$	16.5	nF
MOSFETs	SCT3030AL (ROHM)		
Diodes	VS-20ETF06-M3 (Vishay)		

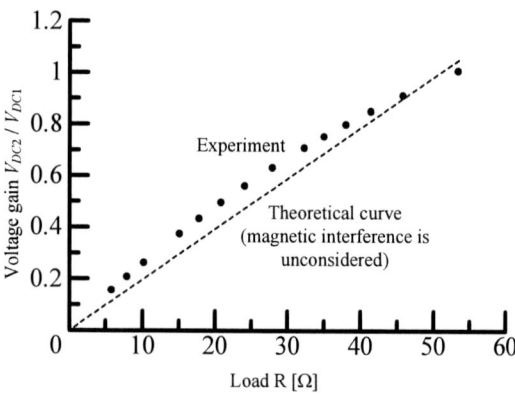

Fig. 7. Voltage gain between primary voltage and secondary voltage.

the hard switching occurs because the current during the dead time is not enough to achieve zero-voltage switching. Due to the hard switching, the voltage gain is slightly different to the theoretical curve.

Figure 8 shows the measured efficiency characteristic from the primary DC voltage to the secondary DC voltage. In fact, the efficiency takes into account the power loss in the inverter and the rectifier. The maximum efficiency reaches 91%. The reason why the efficiency is not so high because the transmission coils of the prototype are a scaled-down model. Thus, the efficiency is expected to be improved in an actual application.

Figure 9 shows the measuring point of the radiation

noise. The radiation noise is measured at 2 m from the coil edge. The radiation noise is measured in a shielded room. As the conventional system, three-phase IPT system with three coils is tested for the comparison on the radiation noise. When the conventional method is tested, the six coils (coil B in Fig. 2) are removed. The transmission coils of phase-v and phase-w are placed in an angle of 120 degrees to the coils on phase-u.

Moreover, the radiation noise is compared under the same current condition. In order to remove the effect of the radiation noise from the inverter, the inverter is placed out of the shield room.

Figure 10 shows the radiation noise. Fig. 10 (a) is the

The 2018 International Power Electronics Conference

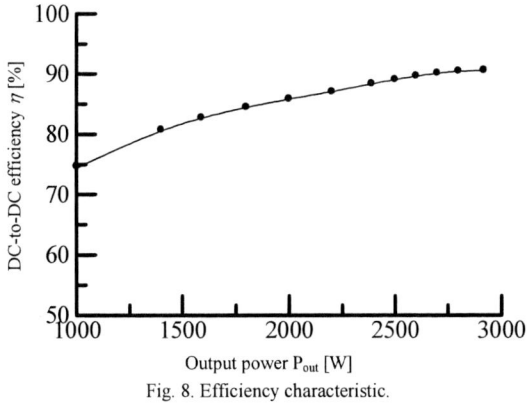

Fig. 8. Efficiency characteristic.

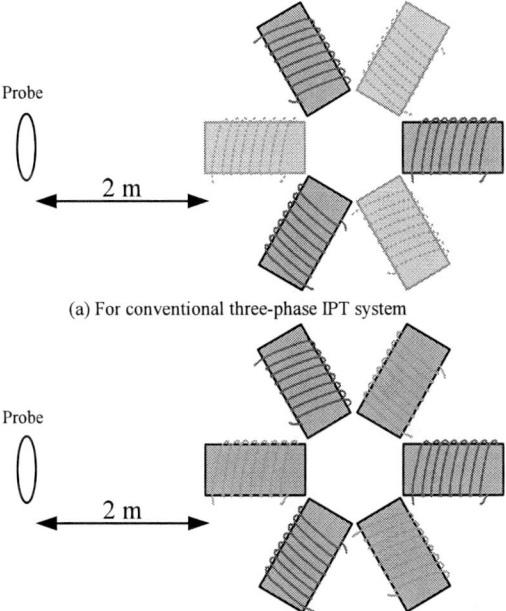

(a) For conventional three-phase IPT system

(b) For proposed three-phase IPT system

Fig. 9. Measurement of radiation noise.

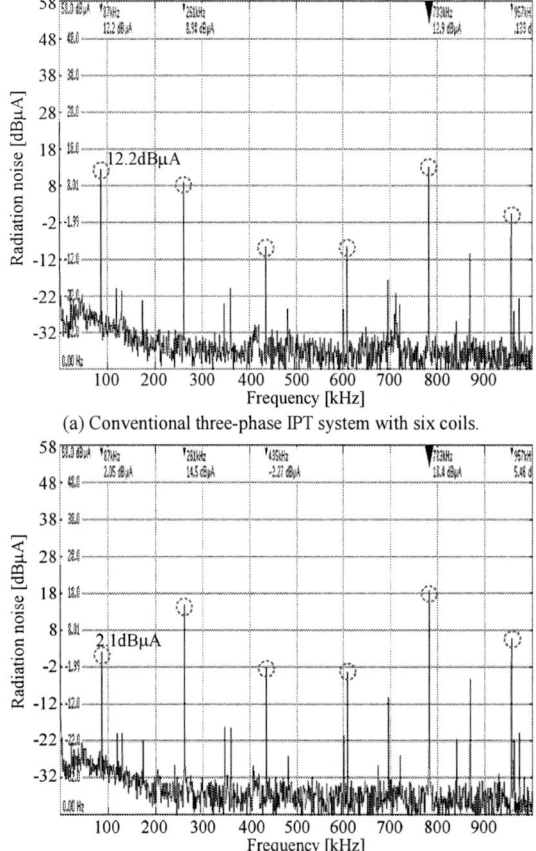

(a) Conventional three-phase IPT system with six coils.

(a) Proposed three-phase IPT system with 12 coils.

Fig. 10. Radiation noise at 2 m from the coil edge.

radiation noise generated by the conventional three-phase IPT system with six coils where the coils B in Fig. 2 are removed. Fig. 10 (b) is the radiation noise generated by the proposed IPT system with 12 coils. The measurement distance and conditions are same as those in Fig. 10 (a). In both the experiments, common transmission coils are used. In order to have a fair comparison, the conduction currents on each coil are adjusted to be same by adjusting the input DC voltage. Note that the inverter and the rectifier are put outside of the shielded room in order to measure only the radiation noise from the transmission coils.

The radiation noise at the fundamental frequency is suppressed from 12.2dBμA to 2.1dBμA by the proposed IPT system. It is confirmed that the proposed three-phase IPT system is effective to reduce the fundamental radiation noise. However, the radiation noise on the third-order harmonics of the proposed system slightly increases.

V. CONCLUSION

This paper proposed three-phase inductive power transfer system. The proposed system has six primary coils and six secondary coils. The opposite coils, which are in series and are differentially connected, cancel out the radiation noise. Moreover, the six coils on each side allow to cancel out the magnetic interference among the multiple coils. Due to the cancel of the magnetic interference, the IPT system can be operated with an unity power factor regarding the inverter output.

In this paper, the effect of the magnetic interference and its cancellation method is mathematically introduced. Then, the proposed scaled-mode was experimentally tested with the 3-kW prototype. The experimental results shows that the maximum DC-to-DC efficiency is 91%. Finally, the radiation noise is evaluated and compared to the conventional three-phase IPT system with six coils. The proposed method suppresses the radiation noise, which is measured at 2 m from the coil edge, from 12.2dBμA to 2.1dBμA on the fundamental frequency.

REFERENCES

[1] A. Kurs, R. Moffatt, M. Soljacic, "Simultaneous mid-range power transfer to multiple devices," *Applied Physics Letters*, No. 044102, pp. 96-98 (2009)

[2] A. Kurs, A. Karalis, R. Moffatt, J. D. Joannopoulos, P. Fisher, and M. Soljacic, "Wireless Power Transfer via Strongly Coupled Magnetic Resonances," *Science*, Vol. 317, pp. 83-86 (2007)

[3] S. Weearsinghe, D. J. Thrimawithana, and U. K. Madawala, "Modeling Bidirectional Contactless Grid Interfaces With a Soft DC-Link," *IEEE Trans. On Power Electronics*, Vol. 30, No. 7, pp. 3528-3541 (2015)

[4] F. Y. Lin, G. A. Covic, and J. T. Boys, "Evaluation of Magnetic Pad Sizes and Topologies for Electric Vehicle Charging," *IEEE Trans. On Power Electronics*, Vol. 30, No. 11, pp. 6391-6407 (2015)

[5] M. Budhia, J. T. Boys, G. A. Covic, and C. Huang, "Development of a Single-Sided Flux Magnetic Coupler for Electric Vehicle IPT Charging Systems," *IEEE Trans. On Industrial Electronics*, Vol. 60, No. 1, pp. 318-328 (2013)

[6] C. Huang, J. T. Boys, and G. A. Covic, "LCL pickup Circulating Current Controller for Inductive Power Transfer Systems," *IEEE Trans. On Power Electronics*, Vol. 28, No. 4, pp. 2081-2093 (2013)

[7] T. Koyama, K. Umetani, and E. Hiraki, "Design Optimization Method for the Load Impedance to Maximize the Output Power in Dual Transmitting Resonator Wireless Power Transfer System," *IEEJ Journal of Industry Applications*, Vol. 7, No. 1, pp. 49-55 (2018)

[8] T. Yamamoto, Y. Bu, T. Mizuno, Y. Yamaguchi, and T. Kano, "Loss Reduction of Transformer for LLC Resonant Converter Using a Magnetoplated Wire," *IEEJ Journal of Industry Applications*, Vol. 7, No. 1, pp. 43-48 (2018)

[9] G. Iovison, D. Kobayashi, M. Sato, T. Imura, and Y. Hori, "Secondary-side-only Control for High Efficiency and Desired Power with Two Converters in Wireless Power Transfer Systems," *IEEJ Journal of Industry Applications*, Vol. 6, No. 6, pp. 473-481 (2017)

[10] K. Kusaka. J. Itoh, "Development Trends of Inductive Power Transfer Systems Utilizing Electromagnetic Induction with Focus on Transmission Frequency and Transmission Power," *IEEJ Journal of Industry Applications*, Vol. 6, No. 5, pp. 328-339 (2017)

[11] R. Ota, N. Hoshi, and J. Haruna, "Design of Compensation Capacitor in S/P Topology of Inductive Power Transfer System with Buck or Boost Converter on Secondary Side," *IEEJ Journal of Industry Applications*, Vol. 4, No. 4, pp. 476-485 (2015)

[12] International Organization for Standardization, "Electrically propelled road vehicles — Magnetic field wireless power transfer — Safety and interoperability requirements," Publicly available specification: ISO/PAS19363:2017(E)

[13] International Electrotechnical Commission, "Electric vehicle wireless power transfer (WPT) systems – Part 1: General requirements," International standard: IEC 61980-1 (2015)

[14] T. Shijo, K. Ogawa, M. Suzuki, Y. Kanekiyo, M. Ishida, and S. Obayashi, "EMI Reduction Technology in 85 kHz Band 44 kW Wireless Power Transfer System for Rapid Contactless Charging of Electric Bus," *IEEE Energy Conversion Congress & Expo 2016*, No. EC-0641 (2016)

[15] M. Suzuki, K. Ogawa, F. Moritsuka, T. Shijo, H. Ishihara, Y. Kanekiyo, K. Ogura, S. Obayashi, and M. Ishida, "Design method for low radiated emission of 85 kHz band 44 kW rapid charger for electric bus," *IEEE Applied Power Electronics Conference and Exposition 2017*, pp. 3695-3701 (2017)

[16] T. Shijo, K. Ogawa, F. Moritsuka, M. Suzuki, H. Ishihara, Y. Kanekiyo, K. Ogura, M. Ishida, S. Obayashi, S. Shimmyo, K. Maki, F. Takeuchi, and N. Tada, "85 kHz band 44 kW wireless power transfer system for rapid contactless charging of electric bus," *International Symposium on Antennas and Propagation 2016*, pp. 38-39 (2016)

[17] D. J. Thrimawithana, U. K. Madawala, A. Francis, and M. Neath: "Magnetic Modeling of a High-Power Three Phase Bi-Directional IPT System," *37th Annual Conference of the IEEE Industrial Electronics Society*, pp. 1414-1419 (2011)

[18] G. A. Covic, J. T. Boys, M. L. G. Kissin, and H. G. Lu: "A Three-Phase Inductive Power Transfer System for Roadway-Powered Vehicles," *IEEE Trans. on Industrial Electronics*, Vol. 54, No. 6, pp. 3370-3378 (2007)

[19] A. Laka, J. A. Barrena, J. Chivite-Zabalza, M. A. Rodriguez, and P. Izurza-Moreno: "Isolated Double-Twin VSC Topology Using Three-Phase IPTs for High-Power Applications," *IEEE Trans. on Power Electronics*, Vol. 29, No. 11, pp. 5761- 5769 (2014)

[20] M. Kim, S. Ahn, and H. Kim: "Magnetic Design of a Three-Phase Wireless Power Transfer System for EMF Reduction," *IEEE Conference on Wireless Power Transfer Conference 2014*, pp. 17-20 (2014)

[21] Y. H. Sohn, B. H. Choi, E. S. Lee, G. C. Lim, G. Cho, and C. T. Rim: "General Unified Analyses of Two-Capacitor Inductive Power Transfer Systems: Equivalence of Current-Source SS and SP Compensations," *IEEE Trans. On Power Electronics*, Vol. 30, No. 11, pp. 6030-6045 (2015)

[22] International Special Committee on Radio Interference, "Industrial, scientific and medical equipment — Radio-frequency disturbance characteristics — Limits and methods of measurement," CISPR 11: 2015 (2015)

The 2018 International Power Electronics Conference

Secondary-side-only Control for Smooth Voltage Stabilization in Wireless Power Transfer Systems with Constant Power Load

Giorgio Lovison[1*], Takehiro Imura[2], Hiroshi Fujimoto[1], Yoichi Hori[1]

1 Department of Advanced Energy, The University of Tokyo, Tokyo, Japan

2 Department of Electrical Engineering and Information Systems, The University of Tokyo, Tokyo, Japan

*E-mail: giorgio.lovison@gmail.com

Abstract—**In recent years, wireless power transfer technology has received considerable attention because of its wide range of applications. Almost all literature focuses on resistance load or constant voltage load, but constant power load is never considered. The constant power load voltage is unstable, thus stabilization is required. In order to perform it without resorting to discontinuous operation and avoid big voltage transients, this paper proposes a control strategy for a secondary side with a single converter. It is based on the combination of synchronous rectification and symmetric phase shift, without communication with the primary side. While the primary side is not manipulated, the AC/DC converter regulates the amplitude of the secondary coil voltage and stabilizes the constant power load voltage on the DC side. In this paper, the control concept, the design and the stability analysis are provided. The proposed control is verified through experimental results both in a static scenario.**

Keywords—*Wireless power transfer, Constant Power Load (CPL), secondary-side-only, symmetric phase shift.*

I. INTRODUCTION

Wireless power transfer (WPT) technology recently has become appealing for both industrial and automotive applications because it simplifies the powering process and eliminates potential dangers for the user. WPT by magnetic resonant coupling achieves high efficiency and is capable of transmitting high power [1]. There are different types of WPT tuning by magnetic resonant coupling, depending on the circuit topology [2]: in particular, the series-series (SS) compensation allows high power to flow on the secondary side with high efficiency. These characteristics are favorable for applications such as Electric Vehicles (EVs) and related literature is abundant [2]–[5]. Currently, research is pointing towards in-motion WPT because it is considered a way to reduce both battery weight and range anxiety; recently, some trial units are been experimented upon and achieved promising results [6]–[8].

In the case of an EV, the main objective is charging the onboard battery; however, it can be necessary to power wirelessly the motor, either from the battery or directly from the ground facilities. While research about constant power load (CPL) has been performed for other application such as distributed grids regulation or automotive [9], in WPT application it is very seldom considered. In fact, past literature considers mostly resistance loads [10][11] or voltage loads [13][12]. One notable case is a battery-less vehicle powered directly by capacitive power transfer [14][15], which achieved 80% efficiency and regulated by a double current HF rectifier and followed by a DC/DC converter for regulating the current. Past research as for inductive power transfer has two cases: in the first one [16], a discontinuous operation with a semi-bridgeless rectifier was proposed; in the second case a coordinated control between the two sides [17] reported a successful operation. However, it is necessary to consider the weak points of the aforementioned research. In the first case, discontinuous operation, while having high efficiency, generates big current transients in the switching between OFF state and ON state, resulting in increased EMI and stress to the converter components. On the other hand, in the second case, the secondary side AC/DC converter output current is estimated by the primary side, whose voltage is thus regulated in order to always have sufficient margin to maintain stability. This results in a smooth operation, but if primary side regulation is not possible then stability is not achievable. Now, primary side voltage regulation is a popular choice for power flow control [18] and may be a necessity in case of particular applications (such as medical implants) to limit the secondary side heating. However, in case of multiple different loads being supplied at the same time, actively controlling the secondary side is the only option. Furthermore, the ON/OFF operation forces the primary coil current to a very low value, thus making it impractical for real applications with multiple loads.

Generally, in the secondary side of a WPT system, an AC/DC converter and a DC/DC converter are present. The AC/DC converter should not be a passive diode bridge rectifier but must include active devices. The DC/DC converter is generally used to regulate either the load current or the DC link voltage. However, considering a real application in which the CPL is interfaced with the DC bus by a three-phase inverter, the addition of a DC/DC converter is not recommended because of space constraints. Thus, the voltage stabilization of constant power load should be performed only with the AC/DC converter and without communication. This paper proposes a control for AC/DC converter based on symmetric phase shift. The control concept and the stability analysis are provided, and its effectiveness is verified by experiments.

This paper is organized into seven sections. Section II intro-

The 2018 International Power Electronics Conference

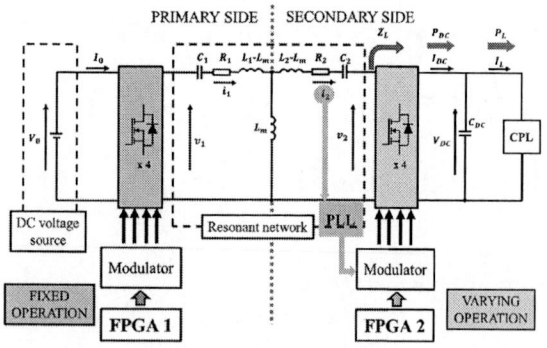

Fig. 1.   Reference circuit.

duces the case of study and describes the power characteristics of WPT. Section III explains the proposed control concept and design and Section IV presents the experimental results. Section V explains the stability conditions depending on the parameters and Section VI reports the results in case of change of mutual inductance. Section VII finally draws the conclusions and indicates the future works.

II.   CASE OF STUDY

This study focuses on the secondary side of a WPT system with SS compensation, as shown in the equivalent circuit of Fig. 1. It includes one converter per side, which is a configuration allowing bidirectional power flow. In the primary side there is an inverter used to produce the high frequency waveforms, while in the secondary side there are a full bridge active rectifier and the CPL such as a motor driven by a three-phase inverter as in [16]. The power transmitted to the secondary side is equal to:

$$P_{DC} = \frac{(\omega L_m)^2 Z_L}{[R_1(R_2 + Z_L) + (\omega L_m)^2]^2} V_1^2 \quad (1)$$

with V_1 as RMS value of v_1 and Z_L as load input impedance. Considering a system in resonance (i.e. with unity power factor), the input impedance is regarded as a pure resistance. The power in (1) is thus real power. Furthermore, in condition of resonance it is allowed to approximate the AC voltages (v_1 and v_2) and currents (i_1 and i_2) by considering only the fundamental wave components.

Past research quite never considered the CPL because, currently, the main application for EV is battery charger. However, supplying the battery but not the motor means that the power must reach the motor from the battery again with losses on the way. From a pure efficiency point of view, this is not an optimal solution. Therefore, it makes sense to think about powering directly the motor and research on this theme. In [16], the voltage stabilization by the secondary side is performed by using the ON/OFF control as an hysteresis to keep the desired value of DC link voltage V_{DC}. With respect to change of parameters, such as load impedance or mutual inductance, the secondary side transmits by wireless communication the power flow regulation command and a

chopper located before the inverter converts boosts the battery voltage in order to create some margin of deliverable power to maintain stability. In so doing, the primary side cooperates with the secondary side and the control has a very good performance, with the only disadvantage represented by the big current transients occurring because of the switching from rectification mode to short mode operation. However, in case of wireless communication malfunction or in case of lack of primary side chopper, the above mentioned control scheme is not available anymore and the hysteresis control alone would keep the disadvantage of big transients, making it a simple but not very good choice for voltage stabilization.

Therefore, in this paper, another controller is proposed. The control is applied to the secondary side full active rectifier and uses the combination of synchronous rectification (SR) and symmetric phase shift explained in the previous chapter. The controller is a simple PI feedback. The advantage of the proposed control is that the severe transients of hysteresis control are mitigated and a smoother voltage stabilization is achieved. In order to design the controller, however, it is first necessary to analyze the load plant in case of CPL; then, a sufficiently fast controller to counteract the slow pole typical of CPL can be designed.

A.   Symmetric phase shift by use of digital PLL with FPGA

Since the primary side is considered fixed, the control must be performed on the secondary side rectifier. In other words, the rectifier has to be equipped with MOSFETs, which are controllable and are more suited to high frequency operation rather than IGBTs. The load is a CPL, which requires voltage stabilization because its open loop plant is unstable. In particular, having only one rectifier on the secondary side, the control must perform at the same time synchronization to the primary side and load voltage stabilization without recurring to discontinuous operation or primary side regulation. Hence, in this paper the chosen control method is SR with symmetric phase shift as in Fig. 2(b). It is then necessary to adopt an analog polarity circuit or implement a digital PLL. Given the high frequency environment, an analog circuit is preferrable; however, it is an additional circuitry that occupies space. Therefore, coping with the constraint condition, the only solution is to use a digital phase locked loop (PLL). In this sense, a high performance control board with fast sampling time and calculation time is necessary: in other words, a field programmable gate array (FPGA) is mandatory.

With normal SR of Fig. 2(a), the output voltage of the converter is a square wave and the converter duty cycle is fixed at 0.5 because of the positive and negative halfwave. However, by modifying the parameter t_α, v_2 becomes the three-level waveform represented in Fig. 3 and symmetric phase shift operation is achieved. In so doing, the amplitude of the fundamental waveform component of the output voltage V_2, which depends on V_{DC}, changes accordingly to t_α. For square wave operation, t_α is equal to zero. The duty cycle in case of symmetric phase shift is still equal to 0.5, but the phase

78

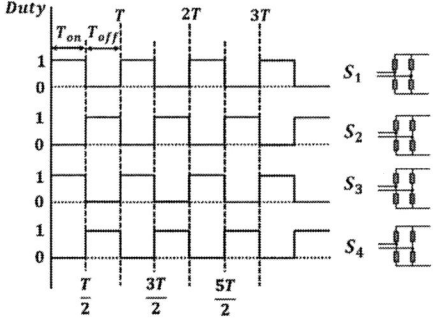

(a) Synchronous rectification (SR).

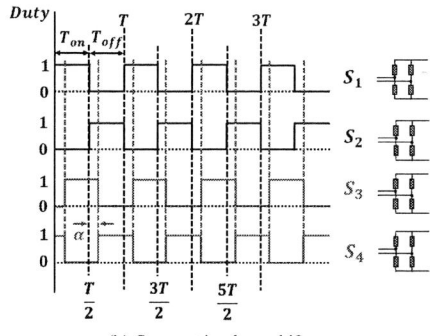

(b) Symmetric phase shift.

Fig. 2.   Duty pattern of SR and symmetric phase shift.

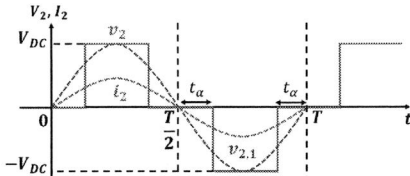

Fig. 3.   Three-level voltage waveform of v_2.

shift ratio is different. It can be expressed by:

$$d_{cr} = \frac{t_\alpha}{T} \tag{2}$$

By adopting the Fourier expansion series to the fundamental value of V_2, the following formula is obtained:

$$V_2 = \frac{2\sqrt{2}}{\pi} V_{DC} \cos(2\pi d_{cr}) \tag{3}$$

with T as the period of the waveform and V_{DC} as the secondary DC link voltage. It is then possible to identify the conversion ratio α, given by:

$$\alpha = \cos(2\pi d_{cr}) \tag{4}$$

In case of the secondary side active rectifier and its output current, the normal square wave operation current I_2 (neglecting the internal diodes' forward voltage drop) is given by :

$$I_2 \simeq \frac{\omega L_m V_1 - R_1 V_2}{R_1 R_2 + (\omega L_m)^2} \tag{5}$$

Hence, the related DC current $I_{DC,SW}$ is given by:

$$I_{DC,SW} \simeq \frac{2\sqrt{2}}{\pi} I_2 \tag{6}$$

On the other hand, in case of a phase shifting operation, (6) becomes the following expression:

$$I_{DC,PS} = I_{DC}(\alpha) = \frac{2\sqrt{2}}{\pi} \frac{\omega L_m V_1 - R_1 V_2 \alpha}{R_1 R_2 + (\omega L_m)^2} \alpha \tag{7}$$

The second term in the denominator $(R_1 V_2)$ is much smaller than the other $(\omega L_m V_1)$ and therefore can be ignored. By phase shifting, $I_{DC}(\alpha)$ is modified in such a way that stabilization is ensured.

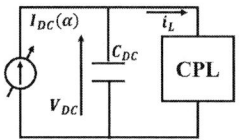

Fig. 4.   Equivalent circuit of CPL.

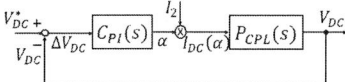

Fig. 5.   Proposed controller.

III.   CONTROLLER DESIGN FOR CPL

The equivalent circuit for the constant power load considered in this circuit is shown in Fig. 4. Its characteristic equations is as following:

$$i_L = I_{DC}(\alpha) - C_{DC} \frac{dV_{DC}}{dt} \tag{8}$$

$$|Z_L| = \frac{V_{DC}^2}{p_L}, \quad p_L = V_{DC} i_L$$

where C_{DC} is the DC smoothing capacitor. These characteristic equations are non-linear and therefore need to be linearized. The linearization is performed by setting an equilibrium point at voltage V_{eq} and current I_{eq} and considering the variation around it as following:

$$V_{DC} = V_{eq} + \Delta V_{DC} \tag{9}$$

$$I_{DC}(\alpha) = I_{eq} + \Delta I_{DC}(\alpha) \tag{10}$$

Here, the voltage V_{eq} and current I_{eq} are the load voltage and current at the operation point. The voltage reference can be selected based on the requirements of the CPL. However, it is noted that the V_{eq} can be the voltage value related to optimal input load impedance Z_{Lopt}. Hence, the following

The 2018 International Power Electronics Conference

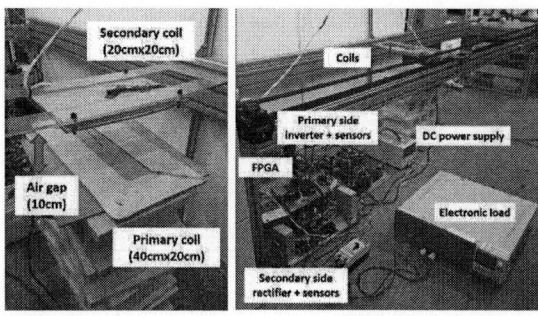

Fig. 6. Experimental setup.

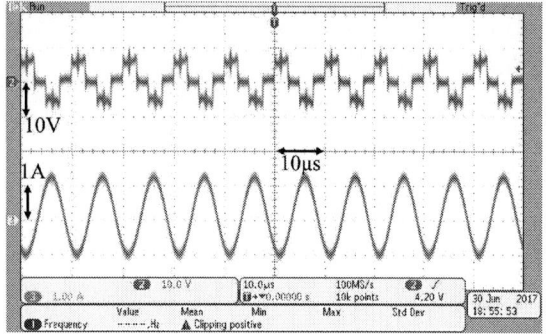

Fig. 7. AC variables (red:voltage v_2; green:current i_2) with $V_{DC} = 4V$, $p_L = 1W$.

approximated open loop transfer function is obtained:

$$P_{CPL}(s) = \frac{\Delta V_{DC}(s)}{\Delta I_{DC}(\alpha)(s)} \simeq \frac{1}{C_{DC}(s - \frac{p_L}{C_{DC}V_{eq}^2})} \quad (11)$$

In (11), it is clear that the pole resides in the right half plane. More in detail, it depends on load power p_L, DC link voltage V_{DC} and smoothing DC capacitor C_{DC}. For bigger p_L, smaller V_{DC} and smaller C_{DC}, the poles are faster. In this paper, the stabilization of V_{DC} is performed by a high bandwith PI controller as shown in Fig. 5. Thus, the controller regulates the value of V_2 by adapting t_α. The controller gains are chosen by pole placement method. From the Routh stability criterion the following formulations must be satisfied in order to maintain stability:

$$k_P \geq \frac{p_L}{V_{eq}^2} \quad (12)$$

$$k_I \geq 0 \quad (13)$$

Then, choosing the damping factor ψ equal to one, the gains can be chosen according to:

$$k_P = \frac{2mZ_L^* C_{DC} - 1}{Z_L^*} \quad (14)$$

$$k_I = m^2 C_{DC} \quad (15)$$

TABLE I. SYSTEM PARAMETERS

Parameter	Value
Primary side DC voltage source V_0	10 V
Primary side coil resistance R_1	1.83 Ω
Secondary side coil resistance R_2	1.683 Ω
Primary side coil inductance L_1	417.8 μH
Secondary side coil inductance L_2	208.3 μH
Primary side coil capacitance C_1	6.03 nF
Secondary side coil capacitance C_2	12.15 nF
Operation frequency f	100 kHz
Mutual inductance L_m (best alignment)	37.9 μH
Coil gap (best alignment)	100 mm
Smoothing capacitor C_{DC}	1000 μF

where m is the pole in closed loop. In (14), Z_L^* is determined by the desired operating voltage V_L^* and the load power p_L as follows:

$$Z_L^* = \frac{V_{DC}^{*2}}{p_L} \quad (16)$$

It is clear that the proportional gain is variable. In any case, m must be chosen high enough to avoid frequency bifurcation and other phenomena occurring at slow pole position. Clearly, that by using a two degree of freedom PI control, which include the feedforward part, it is easier to design the controller. However, in this research, the basic case is introduced, not to mention that the feedback controller design is the very same.

IV. EXPERIMENTAL RESULTS

The experimental setup is shown in Fig. 6 and the circuit parameters are reported Table I. As in Fig. 1, both the converters in the systems are controlled independently by a FPGA. All the current and voltage sensors are in the DC side, except for a single AC current sensor in the secondary coil, necessary to the PLL. The primary side voltage V_1 is fixed. In the experiments, the stability of the proposed control is verified by changing step-wise the voltage reference V_{DC}^*. The results are presented in Fig. 7, Fig. 8 and Fig. 9. In Fig. 7, the three-level operation explained in Fig. 3 is confirmed. The little surges on v_2 depend on the fact that a single FPGA is used to control at the same time both converters in primary and secondary side. It is noticed that the noise is generated at the switching of the primary side, therefore it must have found a path through the FPGA since the resonance network in case of SS compensation has bandpass filter properties. Nevertheless, the control effectiveness is not affected by this issue, as the voltage is stable. In Fig. 8, the change in V_{DC} and i_L is shown; on the other hand, in Fig. 9 the change in α is shown, starting from t=0. Moreover, in Fig. 8(a) and Fig. 9(a) the power p_L is 1 W, while in Fig. 8(b) and Fig. 9(b) p_L is 1.5W. Furthermore, in Fig. 8(a) and Fig. 9(a) the V_{DC}^* changes from 3V to 4V, while in Fig. 8(b) and Fig. 9(b) the V_{DC}^* varies from 3.5V to 4V. As it can be seen, in Fig. 8(a) and Fig. 8(b) the voltage stability is maintained. In Fig. 9 the variation of α is shown. In both cases, the secondary side conversion ratio α varies in two phases, starting from t=0. First, from its initial

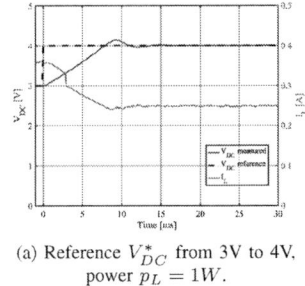
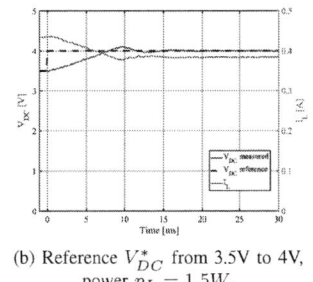

(a) Reference V_{DC}^* from 3V to 4V, power $p_L = 1W$.

(b) Reference V_{DC}^* from 3.5V to 4V, power $p_L = 1.5W$.

Fig. 8. Experimental results of proposed control: profile of V_{DC} and i_L when the voltage reference is changed.

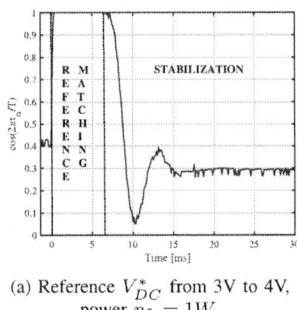
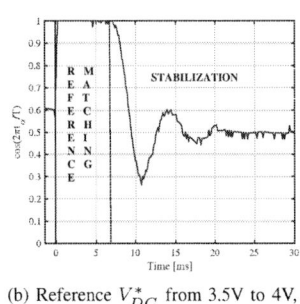

(a) Reference V_{DC}^* from 3V to 4V, power $p_L = 1W$.

(b) Reference V_{DC}^* from 3.5V to 4V, power $p_L = 1.5W$.

Fig. 9. Experimental results of proposed control: profile of secondary side conversion ratio α when the voltage reference is changed.

value α increases up to one in order to let the DC capacitor charge up to the new voltage reference. Then, after about 7ms, when the reference voltage has been matched, α is controlled to drop to its final value. In this phase, the voltage V_{DC} is stabilized and the dynamic response occurs according to the pole placement frequency. Thus, from these experiments it is verified that stability is achieved.

V. STABILITY CONDITIONS DEPENDING ON OPERATION POINT PARAMETERS

The controller design presented grants stability in the wide operation area because of the adaptive gains calculated from pole placement. However, there are some physical limitation deriving from the system parameters that are unavoidable (e.g. conversion ratio of a converter can not be higher than one, $\alpha \leq 1$). This applies in the case of WPT systems for CPLs.
In fact for a determined load power, by considering (8), the maximum current $I_{DC}(\alpha)$ is given by:

$$I_{DC}(\alpha)|_{\alpha=1} = I_{DC,SW} = \frac{\omega L_m V_1 - R_1 V_2}{R_1 R_2 + (\omega L_m)^2} \quad (17)$$

and therefore, because of (8), also the voltage V_{DC} is limited. Its value is expressed by:

$$V_{DC}(\alpha)|_{\alpha=1} = \frac{P_L}{I_{DC,SW}} \quad (18)$$

which is the minimum value possible for a given load power. In this case, a secondary converter for physical decoupling and further control is not available, therefore I_{DC} and V_{DC} will

not undergo further conversions. Thus, (17) and (18) hold for i_L as well.

This can be verified with an experiment by showing how the current limit changes by varying the mutual inductance L_m and the DC source voltage V_0. The experiments are shown in Fig. 10. The experiment is performed by changing the voltage reference V_{DC}^* to a lower value in order to increase the load current as the load is CPL. At $t = 13ms$, the reference change command is given and the behaviour of the current $I_{DC}(\alpha)$ is observed. As a starting condition, the parameters adopted are the same as Table I and the reference power is 1.5W. In Fig. 10(a) it can be seen that the current limit calculated for the abovementioned conditions is 0.55A. After the change in voltage reference, the current runs away at the moment it surpasses the limit and stops at 1.5A because the voltage V_{DC} is nearly zero. Needless to say, the system stability is lost. Interestingly, since the PI controller is still working and trying to have a lower voltage to match the reference V_{DC}^*, the current meets the limit and continues to hold that value. This means that the secondary side conversion ratio α is trying to be higher than one but it is not possible due to physical limitations. Hence, it is confirmed that the stability is maintained as long as α is less than one.
In Fig. 10(b), the mutual inductance L_m is decreased from 37.9 μH to 29μH, meaning that the coupling is more loose, the efficiency is lower and the available transmittable power is higher. All the other parameters are the same as the first case. In this case, it is expected that the current limit is higher: in

The 2018 International Power Electronics Conference

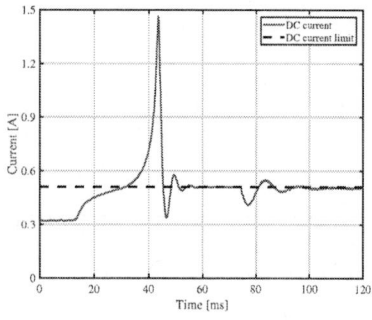
(a) Verification of current limit for $p_L = 1.5W$ with $L_m = 37.9\mu H$ and $V_0 = 14V$.

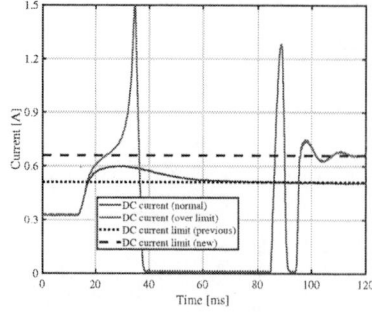
(b) Verification of current limit for $p_L = 1.5W$ with $L_m = 29\mu H$ and $V_0 = 14V$.

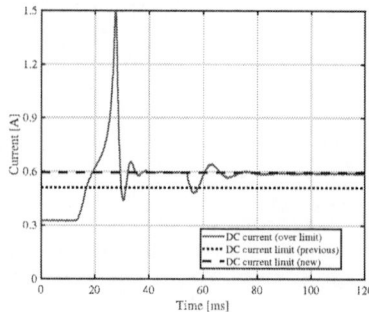
(c) Verification of current limit for $p_L = 1.5W$ with $L_m = 37.9\mu H$ and $V_0 = 16V$.

Fig. 10. Verification of current $I_{DC}(\alpha)$ limit.

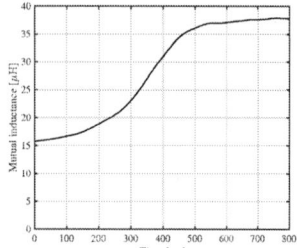
(a) Mutual inductantance obtained from fitting of measurements.

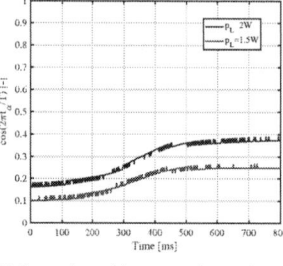
(b) Secondary side conversion ratio α.

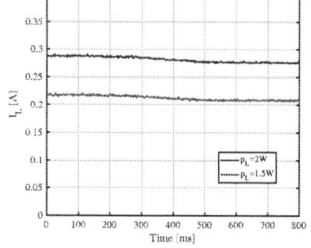
(c) Secondary side current $I_{DC}(\alpha)$.

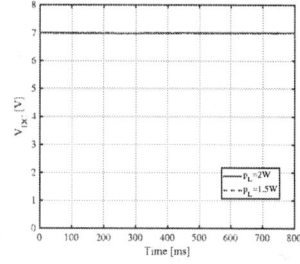
(d) Secondary side voltage V_{DC}.

Fig. 11. Experimental results with mutual indutance variation.

fact, from the calculation it results that the limit is increased from 0.55A to 0.66A. By performing the very same change of V_{DC}^* as before, it is shown that the current does not approach the limit becuase the maximum transient value is 0.6A; then, the stability is maintained. On the other hand, for a bigger change of V_{DC}^*, the current exceeds the limit and the stability is lost. After losing the stability, a different behaviour from Fig. 10(a) is observed at first, but in the end the current converges to the limit as the controller keeps α to be equal to one.

Finally, in Fig. 10(c), the DC voltage source V_0 is increased from 14V to 16V, meaning that the efficiency is slightly lower and the available transmittable power is higher. All the other parameters are the same as the first case. In this last case, too, the current limit is higher with respect to the first case: the new limit value is 0.59A, while in the previous case it was 0.55A. It is demonstrated once more that, when the voltage reference V_{DC}^* is big, the current $I_{DC}(\alpha)$ will run away at the moment the current limit is exceeded and the stability is lost because the α is trying to be bigger than one but it is impossible. By these results, the unavoidable limits of secondary-side-only control for voltage stabilization of CPL have been clarified.

VI. EFFECT OF MUTUAL INDUCTANCE VARIATION

Since one of WPT's main contribution is the enhanced mobility, it is expected that in target applications there are cases in which the coils are moving. In these condition of dynamic WPT, the mutual inductance L_m changes, and since the mutual inductance is relevant to many other parameters, it is necessary to ascertain that its change does not affects them negatively. Therefore, in the present investigation, it is necessary to verify that even with change in the mutual inductance the voltage V_{DC} does not change. The experiment results are reported in Fig. 11. In these experiments, the voltage reference V_{DC}^* is equal to 7V, therefore it is expected that V_{DC} will be always 7V notwithstanding any change in L_m.

In Fig. 11(a), the change of L_m is shown. In total, the mutual inductance increases from $16\mu H$ to $37\mu H$ in 700 milliseconds. As stated in a previous chapter, the mutual inductance can not be measured during the coil movement, therefore this graphic is obtained fitting the measured values at fixed operation points with some misalignment from the center, which is the best condition. In Fig. 11(b), the variation of the secondary side converter ratio α is reported for two different power levels: one is 2W, the other one is 1.5W. It is possible to see that α follows the same pattern as the mutual inductance. This is because in WPT systems by magnetic resonant coupling with SS compensation, as the mutual inductance increases, the power delivered to the secondary side is reduced. Consequently, α correctly becomes large in order to let more current flow in the effort of keeping the voltage V_{DC} fixed to the reference. Furthermore, for higher power α is higher because, at parity of voltage, the current must be higher.

In Fig. 11(c), the current $I_{DC}(\alpha)$ is shown. Its value changes only slightly, becoming about 5% lower than the initial value. Finally, in Fig. 11(d), the voltage V_{DC} is shown. It is immediate to notice that the DC voltage V_{DC} is equal to 7V, matching the reference and being unaffected from the change of L_m. The control is smooth and no transients are observed. From these results it is apparent that the proposed controller copes well with changes of L_m and can be suitable for dynamic WPT applications.

VII. CONCLUSION

This paper proposes a control for voltage stabilization for WPT systems with CPL. The proposed control is performed only with one rectifier on the secondary side, without communication to and regulation thereof of the primary side. The control consists in symmetric phase shift to change the amplitude of v_2: this causes smoother transient with less noise. Furthermore, the controller is a simple PI feedback with high bandwith applied to t_α. The controller gains derivation method is described.

Experimental results confirm the validity of the proposed method, covering different parameters variations in order to prove the stability of the load voltage. Given the lack of other converters for decoupling, there are intrinsic limits of stability depending on the circuit parameters, the DC source voltage and the mutual inductance. If the secondary current or the secondary voltage surpasses those limits, the stability cannot be maintained because of the unavoidable limitation of the secondary side conversion ratio α, which can not be higher than one. Higher power experiments and better transient analysis are the main direction of further investigation.

VIII. ACKNOWLEDGMENTS

This work is partly supported by JSPS KAKENHI grant number 15H02232 and number 17H04915 and the JST-CREST grant number JPMJCR15K3. Moreover, the authors would like to express their deepest gratitude to Dr. Giuseppe Guidi for the precious discussions about the FPGA operation and settings.

REFERENCES

[1] A. Kurs, A. Karalis, R. Moffatt, J.D. Joannopoulos, P. Fisher and M. Soljacic, "Wireless power transfer via strongly coupled magnetic resonance", *Science Express*, vol. 317, no. 5834, pp. 83-86 (2007).

[2] S. Li and C.C. Mi, "Wireless power transfer for electric vehicle applications", IEEE Journ. of Em. and Select. Top. in Pow. El., vol. 3, issue 1, pp. 4-17 (2015).

[3] C.C. Mi, G. Buja, S.Y. Choi, and C.T. Rim, "Modern advances in wireless power transfer systems for roadway powered electric vehicles ", IEEE Trans. on Ind. El., vol. 63, no. 10, pp. 6533-6545 (2016).

[4] Y. Nagatsuka, N. Ehara, Y. Kaneko, S. Abe and T. Yasuda, "Compact contactless power transfer system for electric vehicles", in Int. Pow. El. Conf. (IPEC), pp. 807-813 (2010).

[5] M. Budhia, J.T. Boys, G.A. Covic and C. Huang, "Development of a Single-Sided Flux Magnetic Coupler for Electric Vehicle IPT Charging Systems", IEEE Trans. on Ind. El., vol. 60, no. 1 pp. 318-328 (2013).

[6] K. Throngnumchai, A. Hanamura, Y. Naruse and K. Takeda, "Design and evaluation of a wireless power transfer system with road embedded transmitter coils for dynamic charging of electric vehicles", in El. Veh. Symp. and Exhib. (EVS27), pp. 1-10 (2013).

[7] J. Huh, S.W. Lee, W.Y. Lee, G.H. Cho and C.T. Rim, "Narrow-width inductive power transfer system for on-line electrical vehicles", IEEE Trans. on Pow. El., vol. 26, no. 12, pp. 3666-3679 (2011).

[8] J.M. Miller, O.C. Onar, C. White, S. Campbell, C. Coomer, L. Seiber, R. Sepe and A. Steyerl, "Demonstrating Dynamic Wireless Charging of an Electric Vehicle: The Benefit of Electrochemical Capacitor Smoothing", IEEE Pow. El. Mag., vol. 1, no. 1, pp. 12-24 (2014).

[9] A. Emadi, A. Khaligh, C.H. Rivetta, G.A. Williamson, "Constant Power Loads and Negative Impedance Instability in Automotive Systems: Definition, Modeling, Stability, and Control of Power Electronic Converters and Motor Drives", IEEE Trans. on Veh. Tech., vol. 55, no. 4, pp. 1112-1125 (2006).

[10] K. Colak, M. Bojarski, E. Asa and D. Czarkowski, "A constant resistance analysis and control of cascaded buck and boost converter for wireless EV chargers", in IEEE Appl. Pow. El. Conf. and Expo. (APEC), pp. 3157-3161 (2015).

[11] D. Ahn, S. Kim, J. Moon, and I.K. Cho, "Wireless Power Transfer With Automatic Feedback Control of Load Resistance Transformation", IEEE Trans. on Pow. El., vol. 31, no. 11, pp. 7876-7886 (2016).

[12] G. Guidi and J.A. Suul, "Minimizing Converter Requirements of Inductive Power Transfer Systems With Constant Voltage Load and Variable Coupling Conditions", IEEE Trans. on Ind. El., vol. 63, no. 11, pp. 6835-6844 (2016).

[13] Y. Zhang, K. Chen, F. He, Z. Zhao, T. Lu and L. Yuan, "Closed-Form Oriented Modeling and Analysis of Wireless Power Transfer System With Constant-Voltage Source and Load ", IEEE Trans. on Pow. El., vol. 31, no. 5, pp. 34723480 (2016).

[14] N. Sakai, D. Itokazu, Y. Suzuki, S. Sakihara, T. Ohira, "One-kilowatt capacitive Power Transfer via wheels of a compact Electric Vehicle", 2016 IEEE Wireless Power Transfer Conference (WPTC), pp. 1-3 (2016).

[15] T. Ohira, "A battery-less electric roadway vehicle runs for the first time in the world", 2017 IEEE MTT-S International Conference on Microwaves for Intelligent Mobility (ICMIM), pp. 75-78 (2017).

[16] M. Sato, G. Yamamoto, D. Gunji, T. Imura and H. Fujimoto, "Development of Wireless In-Wheel Motor Using Magnetic Resonance Coupling", IEEE Trans. on Pow. El., vol. 31, no. 7, pp. 5270-5278 (2016).

[17] D. Gunji, T. Imura and H. Fujimoto, "Stability analysis of constant power load and load voltage control method for Wireless In-Wheel Motor", in Proc. 9th Int. Conf. on Pow. El. and ECCE Asia (ICPE-ECCE Asia), pp. 1944-1949 (2015).

[18] H.L. Li, A.P. Hu, G.A. Covic and C. Tang, "A new primary power regulation method for contactless power transfer", in IEEE Int. Conf. on Ind. Tech. (ICIT), pp. 1-5 (2009).

Constant Current Charging and the Maximum System Efficiency Tracking for Wireless Charging Systems Employing Dual-side Control

Zhenjie Li[1], Xiaoliang Huang[2], Kai Song[1], Jinhai Jiang[1], Chunbo Zhu[1] and Zhijiang Du[1]

[1]Harbin Institute of Technology, Harbin, China

[1]E-mail: kaisong@hit.edu.cn and [2]E-mail: xiaoliang.huang@ieee.org

Abstract-The system efficiency and charging current highly dependents on load conditions, which varies in a wide range for practical applications, such as the supercapacitor charge. To achieve the maximum system efficiency (MSE) tracking and constant current (CC) charging, this paper designs the dual-side controlled wireless power transfer (WPT) system. The fundamental harmonic analysis based on the mutual inductance model is built to illustrate the system characteristics. Based on the algorithm of perturbation observation, the primary-side Buck converter realizes the MSE tracking through measuring the minimum system dc input current. The secondary-side PI controlled semi-active rectifier achieves CC charging for the supercapacitor. The simulation results verify the proposed control method. During the CC charging of 2 A, the system efficiency is significantly improved.

Keywords- Constant current charging, system efficiency, semi-active rectifier, wireless power transfer.

I. INTRODUCTION

Wireless power transfer system has the many advantages, such as the convenience of being cordless, insusceptible to weather conditions and fully automatic charging [1]. It can transfer electric power to the battery or supercapacitor powered portable devices and electric vehicles (EVs). Generally, the constant current (CC) charging is required to realize quick charge and the maximum system efficiency (MSE) tracking is essential to fully utilize the system input power. However, the equivalent load resistance of the supercapacitor varies during the charging process, which influences the charging current and system efficiency inevitably. Therefore, the goal of this paper is to fulfill the analysis and design of a system that realizes CC charging and the MSE tracking for the supercapacitor.

Generally, the closed-loop control methods for CC charging can be classified as three types: primary-side control, secondary-side control and dual-side control [2]-[4]. For the primary-side control methods, the dc-dc converter and frequency adjustment/phase shift H-bridge inverter are always used. For the secondary-side control methods, the semi-active rectifier, dc-dc converter and on-off control can be used. The dual-side control methods are the combination of the above two control methods. In this paper, the secondary-side semi-active rectifier is used to achieve the CC charging through the designed PI controller. It reduces one inductor and capacitor when compared with the commonly used dc-dc converter, such as Buck converter.

For the methods of system efficiency improvement, the basic idea is based on the impedance matching and load transformation. Firstly, with optimizing the parameters of the WPT system, the system losses should be as small as possible. Once the WPT system is fabricated, the control methods are also essential to further improve the system efficiency. In practice, the impedance matching networks that consist of capacitor, inductor or their combination are used to improve the system efficiency for the high frequency and low power WPT systems [5]. In addition, the optimization of magnetic coupler and compensation topology are also able to improve the system efficiency. However, its adjustment range is always limited when the load variation is large [6]. Therefore, the primary-side dc-dc converter is proposed to realize the MSE tracking for the supercapacitor.

As the WPT system is a multi-parameter correlation system, the interaction between primary-side and secondary-side control are analyzed in-depth in this paper. It proves that the required combination adjustment of dc-dc converter and semi-active rectifier achieves CC charging and the MSE tracking for supercapacitor. The rest sections are organized as follows. Section II introduces the system structure and theoretical analysis. Section II analyzes and simulates the proposed dual-side control method. Finally, the conclusions and future works are drawn in Section IV.

II. SYSTEM STRUCTURE AND ANALYSIS

A. System Structure

The WPT systems consist of two electrically isolated parts that are named as primary side and secondary side, as shown in Fig. 1. On the primary side, the system dc input voltage is regulated by Buck converter, which realizes the maximum system efficiency (MSE) tracking through the algorithm of perturbation observation. Its output dc voltage is inverted by H-bridge inverter. Then, the ac voltage is transferred from primary-side to secondary-side. The series-series compensation is used to reduce the reactive power and power volume of the H-bridge inverter. On the secondary side, the obtained ac voltage is rectified and filtered by semi-active rectifier and capacitive filter. Then, the CC charging is achieved by the designed PI controller. Finally, the power is transferred to supercapacitor. The system dc input current and charging current are measured by the current sensors.

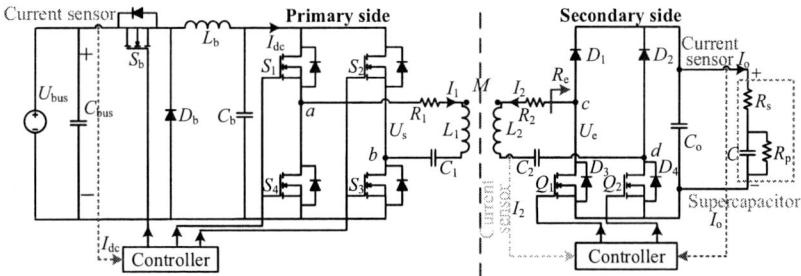

Fig. 1. System structure.

From Fig. 1, the metallic oxide semiconductor field effect transistor (MOSFET) S_b, diode D_b, inductor L_b and filter C_b constitutes the Buck converter. Four MOSFETs S_1~S_4 constitutes the H-bridge inverter. The L_i, R_i, C_i and I_i (i=1, 2) are self-inductance, internal resistance, resonant capacitor and tank current of the primary and secondary side coil, respectively. The secondary-side semi-active rectifier consists of two MOSFETs Q_1~Q_2 and two diodes D_1~D_2. The capacitor C_o is used as the capacitive filter. The supercapacitor model constitutes series resistance R_s, ideal capacitor C and parallel resistance R_p. The R_e and U_e are equivalent input resistance and voltage seen into the semi-active rectifier, respectively. The system dc input voltage is U_{bus} and M is the mutual inductance. The current signals used for dual-side control are obtained by the hall current sensor LEM LV25-P.

B. System Efficiency Analysis

The mutual inductance model is used to perform the circuit analysis, as shown in Fig. 2.

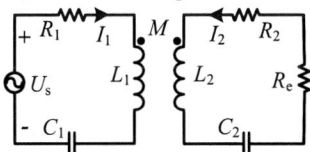

Fig. 2. Mutual inductance model.

According to the Kirchhoff's Voltage Law (KVL), the system is expressed as:

$$\begin{cases} \left(R_1 + j\omega L_1 + \dfrac{1}{j\omega C_1}\right)\cdot \dot{I}_1 + j\omega M \cdot \dot{I}_2 = \dot{U}_s \\ j\omega M \cdot \dot{I}_1 + \left(R_2 + j\omega L_2 + \dfrac{1}{j\omega C_2} + R_e\right)\cdot \dot{I}_2 = 0 \end{cases} \quad (1)$$

where, $\dot{U}_s$, $\dot{I}_1$ and $\dot{I}_2$ are phasors of equivalent output voltage of H-bridge inverter, primary and secondary side resonant current, respectively. The ω is system operating angular frequency and j is imaginary unit. Assuming the WPT system operates at resonant state, the system efficiency η is given by

$$\eta = \frac{(\omega M)^2 R_e}{(\omega M)^2 (R_2 + R_e) + R_1 (R_2 + R_e)^2} \quad (2)$$

It is obvious that η is influenced by M and R_e. When

the magnetic coupler is fixed, an optimal value R_{e_opt} that achieve the MSE point η_{max} exists. Through the derivation of η, the R_{e_opt} is deduced as

$$R_{e\_opt} = R_2 \sqrt{1 + \frac{(\omega M)^2}{R_1 R_2}}$$
$$\approx (\omega M)\sqrt{\frac{R_2}{R_1}} \quad for \quad \frac{(\omega M)^2}{R_1 R_2} \gg 1 \quad (3)$$

The simulation results of (3) are plotted in Fig. 3. It shows that for either small or large R_e that deviates from R_{e_opt}, the η decreases rapidly and η_{max} exists for a fixed M. The larger the M is, the larger the η_{max} is.

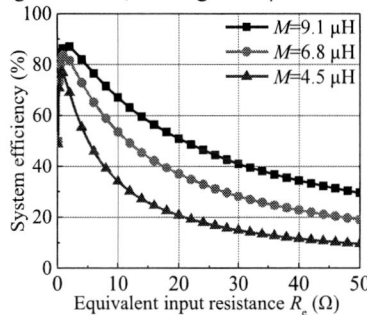

Fig. 3. System efficiency simulations with different R_e and M.

C. Charging Current Analysis

Generally, the Fundamental Harmonic Analysis (FHA) has sufficient accuracy for steady-state analysis [8]. The Root Mean Square (RMS) value of H-bridge inverter output voltage U_s is deduced as

$$U_s = \left|\dot{U}_s\right| = \frac{2\sqrt{2}}{\pi} U_{in} \quad (4)$$

where, U_{in} and U_s are the dc input voltage and ac output voltage of H-bridge inverter. Further, the U_{in} is adjusted by the primary-side Buck converter and expressed as

$$U_{in} = D U_{bus} \quad (5)$$

where, D is duty cycle of Buck converter. Based on (1), (4) and (5), the RMS value of secondary-side resonant current is given by

$$I_{2\_RMS} = \frac{(\omega M)}{R_1(R_2 + R_e) + (\omega M)^2}\frac{2\sqrt{2}}{\pi} D U_{bus} \quad (6)$$

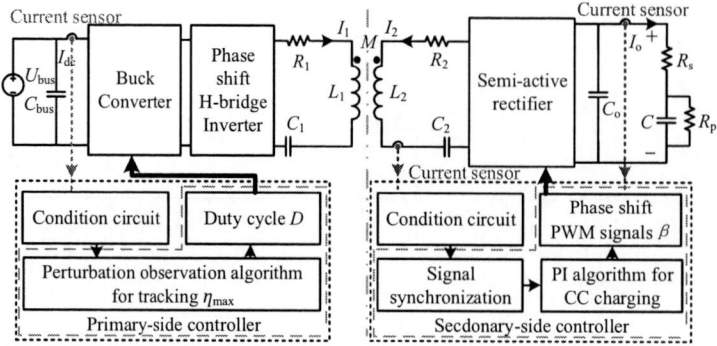

Fig. 4. Closed-loop control diagram.

As shown in Fig. 1, for secondary-side semi-active rectifier that combines with capacitive filter, its input equivalent input resistance R_e is expressed as

$$R_e = \frac{8}{\pi^2} R_o \cos^2 \frac{\beta}{2} \qquad (7)$$

where, R_o is equivalent load resistance of supercapacitor. According to the power balance equation, the charging current I_o is given by

$$I_o = \frac{8DU_{\text{bus}}(\omega M)}{R_1\left(R_2\pi^2 + 8R_o\cos^2\dfrac{\beta}{2}\right) + (\omega M)^2 \pi^2} \cos\frac{\beta}{2} \qquad (8)$$

Based on (7), the (2) is further deduced as

$$\eta = \frac{(\omega M)^2 \, 8\pi^2 R_o \cos^2 \dfrac{\beta}{2}}{\left(R_2\pi^2 + 8R_o\cos^2\dfrac{\beta}{2}\right)^2 R_1 + B} \qquad (9)$$

$$B = (\omega M)^2 \left(R_2\pi^4 + 8\pi^2 R_o \cos^2\frac{\beta}{2}\right)$$

From (8) and (9), it shows that I_o is a function of D, β and R_o, and η is a function of β and R_o, which means I_o is also a function of η. Then, the adjustment of both I_o and η should be coordinated. Further, when the $\eta_{\max}$ is tracked for pre-set I_{o_set}, the D is given by

$$D = \frac{I_{o\_set}\left(A + (\omega M)^2 \pi^2\right)}{8U_{\text{bus}}(\omega M)\sqrt{\dfrac{\pi^2 R_2}{8R_o}}\sqrt{1 + \dfrac{(\omega M)^2}{R_1 R_2}}} \qquad (10)$$

$$A = \pi^2\left(R_1 R_2 + \sqrt{(R_1 R_2)^2 + (R_1 R_2)(\omega M)^2}\right)$$

In practice, the system parameters should be welly designed and optimized to ensure D is available during the dual-side control process for supercapacitor charge.

III. Analysis and Simulations of the Proposed Controller

The proposed closed-loop control diagram is shown in Fig. 4. On the primary side, the system dc input current for Buck converter is measured to track the MSE point through the algorithm of perturbation observation. On the secondary side, the charging current is measured and sent

to the controller of semi-active rectifier, then CC charging is achieved by the PI control algorithm. The secondary-side resonant current is used to synchronize the PWM signals for semi-active rectifier. The feasibility of the proposed dual-side control method is verified by Simulink software and parameters are listed in Table I.

TABLE I
SYSTEM PARAMETERS

Symbol	Parameter	Value
L_1	Primary-side coil inductance	56.9 μH
R_1	Primary-side coil resistance	0.16 Ω
ϕ_1	Primary-side transmitter diameter	10 cm
L_2	Secondary-side coil inductance	36.2 μH
R_2	Secondary -side coil resistance	0.1 Ω
ϕ_2	Secondary-side transmitter diameter	7 cm
d	Rated transfer distance	3 cm
k	Rated coupling coefficient	0.15
f_s	System operating frequency	85 kHz
U_{bus}	System dc input voltage	48 V
I_o	Charging current	2 A

A. Secondary-side Control for CC Charging

The secondary-side control circuit is shown in Fig. 5. It composes of PWM synchronize unit that generates two PWM signals for semi-active rectifier and PI controller unit that generates phase shifted angle β for PWM signals.

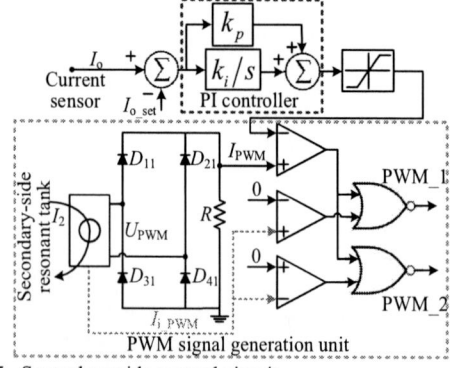

Fig. 5. Secondary-side control circuit.

As shown in Fig. 6 (a), when the supercapacitor voltage increases from 12.5 V to 14.5 V, the charging current is maintained as 2 A through adjusting β. Fig. 6 (b) shows that the secondary-side resonant current is in-phase with the input voltage of semi-active rectifier.

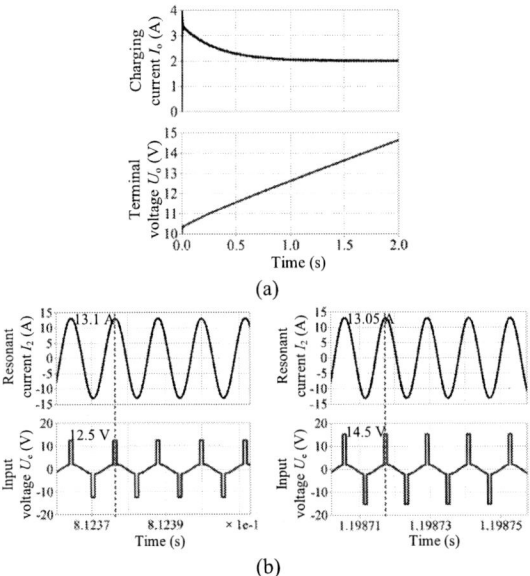

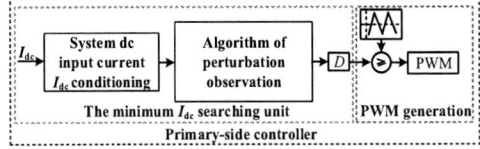

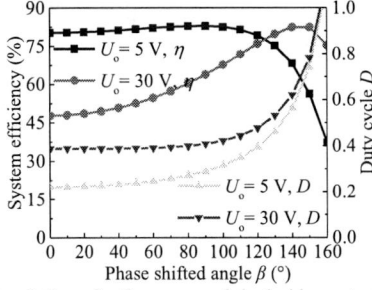

(a)

(b)

Fig. 6. Simulations of secondary-side control. (a) Charging process and (b) Input voltage/current of the semi-active rectifier.

B. Primary-side Control for the MSE tracking

The primary-side control circuit consists of algorithm of perturbation observation unit and PWM generate unit, as shown in Fig. 7.

Fig. 7. Primary-side control circuit.

It is obvious that CC charging is realized by the PI controlled semi-active rectifier. Assume that output power is fixed when executes the algorithm of perturbation observation, the minimum system dc input current corresponds to the MSE point. Combined with the secondary-side PI controller, the duty cycle D of Buck converter is adjusted to track the MSE. The simulation results are shown in Fig. 8. It is obvious that system parameters should be optimized to ensure D is available during the search of the MSE.

Fig. 8. Simulations for the proposed dual-side control method.

C. Verification of the proposed dual-side control method

To verify the feasibility of the proposed control method, the system efficiency with/without using the MSE

tracking is analyzed. Fig. 9 shows that the proposed control method can improve the system efficiency.

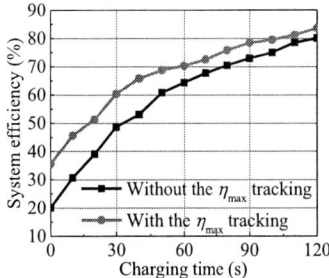

Fig. 9. Simulation analysis of the system efficiency.

IV. CONCLUSIONS

A dual-side control method that consists of primary-side dc-dc converter and secondary-side semi-active rectifier is proposed to realize CC charging and the MSE tracking for supercapacitor charge. Based on the mutual inductance model, the adjustment of charging current and system efficiency is analyzed. Through measuring the system dc input current, the CC charging is achieved by the PI controlled semi-active rectifier. Without the dual-side wireless communication link, through executing the algorithm of perturbation observation, the minimum system dc input current is measured and then the MSE tracking is realized by adjusting the duty cycle of Buck converter. The feasibility of proposed control method is verified by simulation results. During CC charging (2 A) process, the system is significantly improved.

V. ACKNOWLEDGMENT

This work was supported by Natural Science Foundation of China under Project 51677032 and 51577034. Natural Science Foundation of Heilongjiang Province No. E2017045. Harbin Science and Technology Innovation Talents Special Fund Project under Grant No.2016RAQXJ002. China Postdoctoral Science Foundation under Grant No.2014M560254 and 2015T80338.

REFERENCES

[1] S. Li and C. Mi, "Wireless power transfer for electric vehicle applications," *IEEE J. Emerging Sel. Topics Power Electron.*, vol. 3, no. 1, pp. 4–17, Mar. 2015.

[2] G. Buja, M. Bertoluzzo, and K.N. Mude, "Design and Experimentation of WPT Charger for Electric City-Car", *IEEE Trans. Ind. Electron.*, vol. 62, no. 62, pp. 7436-7447, Dec. 2015.

[3] S. Kim, G. Covic, and J. Boys, "Tripolar pad for inductive power transfer systems for EV Charging," *IEEE Trans. Power Electron.*, vol. 32, no. 7, pp. 5045–5057, Jul. 2017.

[4] Z. Li, C. Zhu, J. Jiang, K. Song, and G. Wei, "A 3 kW wireless power transfer system for sightseeing car supercapacitor charge," *IEEE Trans. Power Electron.*, vol. 32, no. 5, pp. 3301–3316, Jun. 2017.

[5] Y. Lim, H. Tang, S. Lim and J. Park, "An adaptive impedance-matching network based on a novel capacitor matrix for wireless power transfer," *IEEE Trans. Power Electron.*, vol. 29, no. 8, pp. 4403-4413, Aug. 2014.

[6] H. Feng, T. Cai, S. Duan, J. Zhao, X. Zhang, and C. Chen, "An LCC compensated resonant converter optimized for robust reaction to large coupling variation in dynamic wireless power transfer," *IEEE Trans. Ind. Electron.*, vol. 63, no. 10, pp. 6591–6601, Oct. 2016.

Electric field coupling type high power wireless power transfer with leakage electric field strucure.

Mitsuru MASUDA[1]

1 FURUKAWA ELECTRIC CO.,LTD. Automotive System & Electronics Laboratories, Kanagawa, JAPAN
*E-mail: Mitsuru.yhpe.masuda@furukawaelectric.com

Abstract— **In consideration of the electric field in the near region, we are conducting a study with the purpose of achievement of a wireless power transmission system using the electric field resonance. In this report, a comparison of the electric field coupling type and the magnetic field coupling type has been conducted and it was demonstrated by the equivalent circuit analysis that the series resonance structure in the electric field coupling type makes it possible to transmit the electric power with high efficiency even at a remote distance. Furthermore, we have mounted the system in a mobility car and have demonstrated that it is possible to transmit the power in the kW order.**

Keywords— The authors shall provide up to 4 keywords or phrases (in alphabetical order and separated by commas) to help identify the major topics of the paper.

I. INTRODUCTION

Currently, the researches on the wireless power transmission using the magnetic field coupling are active. This technology has been put to practical use in the electric home appliances such as in an electric toothbrush, in an electric shaver, etc. that are used around the water and a smart phone, etc. The magnetic field coupling type wireless power supply uses the frequencies around 100 kHz and the devices with improved power factor by adding a capacitance component to the inductance component of the power transmission and reception part have been produced and the commercialization is proceeding. Furthermore, in 2006, the technology named "Magnetic Field Resonance" was presented by the Massachusetts Institute of Technology in the United States of America[1], [2], and the research on the wireless power transmission has been active. In the EV field, the standardization of the wireless power supply using the 85 kHz band is under way. In the electric circuit, the magnetic field and the electric field are closely related and what is made possible in the magnetic field is also made possible in the electric field. In the wireless electric power transmission using the electric field coupling, the researches of the high electric power supply from the electrode buried in the asphalt to the steel belt in the tire[3] and the power supply to the multimedia transmission line[4] are under way, but due to the constraint of the breakdown voltage in the air, etc., the number of the research example is fewer compared to the one of the magnetic coupling. In this report, while contrasting with the magnetic field coupling type, we analyze the fact that

the electric power transmission is possible by using the electric field coupling even if the transmission and reception distance is long.

II. THE WIRELESS POWER TRANSMISSION USING ELECTRIC FIELD COUPLING

As a coupler structure using the wireless electric power transmission by means of the electric field coupling, two types of systems, a Meander line shape with the tip of the power transmission coil opened5) and a flat plate electrode plate shape6) are being studied: In the Meander line shape, there is an advantage that a complete planar structure can be easily obtained. However, since the transmission line length is long, the loss is increased due to the influence of the skin effect at high frequencies similar to the magnetic field coupling type. On the other hand, in the system using the flat electrode, even if a skin effect occurs, the loss due to the high frequency is small because its surface area is large. In addition, it is possible to constitute an electrode surface material from an inexpensive metal material such as aluminum, etc. other than copper. As a weak point of the electric field coupling type, there is a limitation of the power supply due to the breakdown voltage value. In addition, since the electric field is a divergent field, the leakage electric field around the coupler tends to be generated as a noise. On the other hand, in the magnetic field coupling type, this limitation exists in the withstand voltage of the resonance capacitor for improving the power factor and between the coil windings, but this restriction does not occur between the primary side transmission line and the secondary side transmission line.

III. THE STRUCTURE AND THE OPERATION OF THE ELECTRIC FIELD COUPLING TYPE WIRELESS POWER TRANSMISSION COULER

At first, we will explain the operation of the electric field coupling type wireless power transmission with a flat plate electrode structure. The circuit has a series resonance structure in which the resonance coil and the flat plate electrodes for transmitting and receiving the electric power are connected in series. When the high frequency power is applied to this circuit, the electric flux lines are generated. A high electric field is generated at the end where the electrode plates are close to each other and a lower electric field is generated at the separated portion (Figure 1 (a)). At this time, if the coupler having the same resonance frequency is brought close to it, this electric flux line faces the direction of the power transmission and reception (Fig. 1 (b)). This enables the power transmission between the transmission and reception electrodes. In addition, at this time, a curved electric flux line is generated by the fringe effect at the end of the coupler. This bent state of the electric flux lines generates a high electric field intensity region in the vicinity of the coupler. This characteristic gives a high degree of freedom to the positional characteristics between the transmission and reception electrodes, but also creates the coupling with the low electric potential objects. This measure will be described in Section 7.

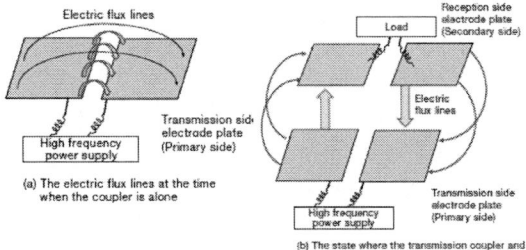

Figure 1 .The image of the power transmission using the electric field resonance.

IV. THE COMPARISON OF THE EQUIVALENT CIRCUIT MODEL ELECTRIC FIELD COUPLING

An equivalent circuit model of the electric field coupling type and of the magnetic field coupling type contactless power supply system are shown in Figure 2. In the electric field coupling type, a π type equivalent circuit model is used from the configuration system and in the magnetic field coupling type, the T type equivalent circuit model is used. In the model shown, a matched load Z0 is connected. R is the parasitic resistance such as in the capacitance, the internal resistance of the coil, the wiring, etc. Therefore, the capacitance and the inductance on the model can be treated as an ideal element without the loss.

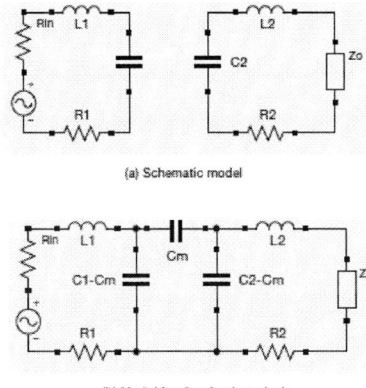

Figure 2. The model for the electric field coupling type equivalent circuit.

V. THE ANALYSIS ON THE ELECTRIC FIELD COUPLING TYPE AND THE MAGNETIC FIELD COUPLING TYPE NON-CONTACT POWER SUPPLY

A. The Analysis Equation for the Electric Field Coupling Type

The analysis is started by transforming Figure 2 of the electric field coupling type equivalent circuit model into that of Figure 4. We analyze under the assumption that the loaded end is a short circuit.

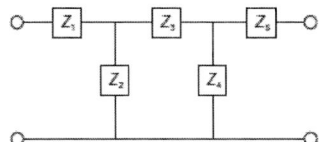

Figure 4.The model for the electric field type analysis.

The impedance seen from the power supply is expressed by the equation (1) and if transformed, it becomes the equation (2).

$$Z = Z_1 + \cfrac{1}{Z_2^{-1} + \cfrac{1}{Z_3 + \left(Z_4^{-1} + Z_5^{-1}\right)}} \tag{1}$$

$$Z = Z_1 + \frac{Z_2\left(Z_3 Z_4 + Z_4 Z_5 + Z_5 Z_3\right)}{Z_3 Z_4 + Z_4 Z_5 + Z_5 Z_3 + Z_2 Z_4 + Z_2 Z_5} \tag{2}$$

Since the equivalent circuit model is a symmetric system, the equation (2) becomes the equation (3) if $Z1 = Z5$ and $Z2 = Z4$.

$$Z = Z_1 + \frac{Z_2\left(Z_1 Z_2 + Z_2 Z_3 + Z_3 Z_1\right)}{2 Z_1 Z_2 + Z_2 Z_3 + Z_3 Z_1 + Z_2^{2}} \tag{3}$$

B. The Analysis in the ideal State

We analyze in an ideal state where the circuit configuration element, the internal resistance and the load of the power supply are 0 Ω. Using the equation (3), input

$Z1 = j\omega L$, $Z2 = 1 / j\omega(C - Cm)$, $Z3 = 1 / j\omega Cm$

The equation (4) is obtained.

$$Z = j\frac{\omega^{4} L^{2}\left(C^{2} - C_m^{2}\right) - \omega^{2} 2LC + 1}{\omega^{2} L\left(C^{2} - C_m^{2}\right) - \omega C} \tag{4}$$

Since the resonance mode appears at each frequency where the imaginary part becomes zero, only the numerator of the equation (5) is taken out,

$$\omega^{4} L^{2}\left(C^{2} - C_m^{2}\right) - \omega^{2} 2LC + 1 = 0 \tag{5}$$

and solve this equation (5). Therefore, the solution is

$$\omega_{1,2} = 1 / \sqrt{LC\left(1 \pm k_\varepsilon\right)} \tag{6}$$

Here, $ke = C / Cm$ (coupling coefficient).

C. The Analysis Including the Loss Term

In the actual experiment system, it is necessary to consider the impedance, the internal resistance and the load of the power supply. Here, we perform the analysis including such a loss term
Using the equation (3),
input $Z1 = R + j\omega L$, $Z2 = 1 / j\omega(CCm)$,
$Z3 = 1 / j\omega Cm$ and proceed with the analysis. The real part is shown in the equation (7) and the imaginary part is shown in the equation (8).

Here, R is assumed to include all loss terms included in the actual equipment. Here, R is assumed to include all loss terms included in the actual equipment.

$$\mathrm{Re}[Z] = R + \frac{\frac{K_3}{\omega^2(C-C_m)^2}R}{(RK_1)^2 + \left(\omega L K_1 - \frac{K_2}{\omega(C-C_m)}\right)^2} \quad (7)$$

$$\mathrm{Im}[Z] = \omega L + \frac{\frac{R^2 K_1 K_2}{\omega(C-C_m)}}{(RK_1)^2 + \left(\omega L K_1 - \frac{K_2}{\omega(C-C_m)}\right)^2} +$$

$$\frac{\left\{\frac{LK_2}{C-C_m} - \frac{1}{\omega^2(C-C_m)^2}\right\}\left\{\omega L K_1 - \frac{K_2}{\omega(C-C_m)}\right\}}{(RK_1)^2 + \left(\omega L K_1 - \frac{K_2}{\omega(C-C_m)}\right)^2} \quad (8)$$

Here, $K1 = (C+Cm) / (C - Cm)$, $K2 = C / (C - Cm)$,

$K3 = Cm /(C - Cm)$

Since the resonance mode appears at each frequency where the imaginary part becomes zero, it becomes the equation (9) if only the numerator of the equation (8) is taken out

$$\omega^6 + \alpha \omega^4 + \beta \omega^2 + \gamma = 0 \quad (9)$$

Here, α, β and γ are as follows.

$$\alpha = \frac{(C-C_m)R^2 K_1 - 3LK_2}{L^2(C-C_m)LK_1}$$

$$\beta = \frac{2LK_2^2 + LK_1 - (C-C_m)R^2 K_1 K_2}{L^3(C-C_m)^2 K_1^2}$$

$$\gamma = -\frac{K_2}{L^3(C-C_m)^3 K_1^2}$$

By factorizing the equation (9), the equation (10) is obtained.

$$\left\{1 - \omega^2 LC(1+k_e)(1-k_e)\right\}\left[\omega^4 - \frac{1}{(1+k_e)(1-k_e)}\right.$$

$$\left.\left\{\frac{2}{LC} - \frac{R^2}{L^2}(1+k_e)(1-k_e)\right\}\omega^2 + \frac{1}{L^2 C^2(1+k_e)(1-k_e)}\right] = 0 \quad (10)$$

From the equation (10), the solution is as follows.

$$\omega_0 = 1/\sqrt{LC(1+k_e)(1-k_e)}$$

$$\omega_{12} = \sqrt{\frac{1}{(1+k_e)(1-k_e)}\left\{\frac{1}{LC} - \frac{R^2}{2L^2}(1+k_e)(1-k_e)\right\}} \quad (11)$$

$$\overline{(1\pm\sqrt{1-\kappa_e})}$$

Here, Ke becomes as follows.

$$\kappa_e = \frac{(1+k_e)(1-k_e)}{1 + \frac{CR^2}{L}\left(\frac{CR^2}{4L} - 1\right)(1+k_e)(1-k_e)} \quad (12)$$

D. The Considerration of the Analsis Equation

The analysis of the electric field coupling type and the magnetic field coupling type including the ideal state and the loss term is carried out in the equivalent circuit model used for analyzing the wireless power transmission system so that in the ideal state, the frequency split that depends on the coupling coefficient as the center of the resonance frequency $\omega 0 = 1 / \sqrt{LC}$ could be derived by the mathematical equation for both types. In addition, If the loss term is included, there is a difference between the electric field coupling type and the magnetic field coupling type and as far as the resonance frequency $\omega 0$, it does not change from the ideal state in the magnetic field coupling type, but it is understood that the coupling coefficient contributes to the electric field coupling type. It is considered that this is due to the fact that the mutual capacitors contribute serially from the equivalent circuit model. For this reason, in the electric field coupling type, the Q value can be increased by the series resonance structure and the distance between the power transmission and reception electrodes can be increased with high efficiency, but a matching circuit on the power transmission side and the power reception side is needed because of the influences of the output impedance of the power supply and load impedance.

VI. THE CIRCUIT CONFIGURATION METHOD IN THE ELECTRIC FIELD RESONANCE

The Interpretation of the Coupling with the Electric Field

We introduce the electric flux in consideration of the electric field coupling. The electric flux is proportional to the amount of the charge accumulated on the electrode plate. In the case of the electric field, an attention is required since it is the divergent field rather than the rotating field like the magnetic field. If there are two pairs of the electrodes that are positively charged and negatively charged, the electric flux terminates between the two electrodes and a potential difference based on either one can be defined and a proportional constant capacitance can be defined between the charge amounts accumulated between the electrodes and the potential difference. If there are two pairs of electrodes as well as the case of the magnetic field coupling, it can be divided into the electric flux that terminates between a pair of electrodes and the electric flux that terminates via another set of electrodes (Figure 5 (a)). If assigning the term with the electric field, the former is the leakage capacitance and the latter is the mutual capacitance. The electric flux generated from the accumulated charge is constant.

The 2018 International Power Electronics Conference

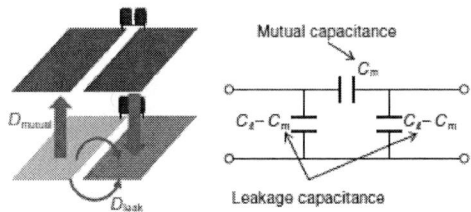

(a) The direction of the electric field.　(b) The equivalent circuit.

Figure 5. The coupling with the electric field and the equivalent circuit.

The equivalent circuit of the electric field resonance can be represented by a π type equivalent circuit (Figure 5 (b)). If the degree of the coupling is small, the leakage capacitance is large and the mutual capacitance is small. So the voltage drop increases and the voltage on the secondary side cannot be picked up (Figure 6 (a)). If the coupling is large, the leakage capacitance is small and the mutual capacitance is large. So a voltage can be taken on the secondary side (Figure 6 (b)).

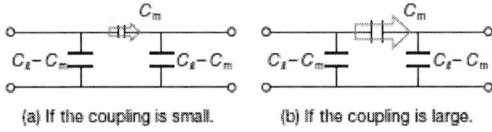

(a) If the coupling is small.　(b) If the coupling is large.

Figure 6. The image diagram for the equivalent circuit based on the magnitude of the coupling (the electric field resonance).

The series resonance coupling type of the electric field resonance system.

In the electric field resonance type of the series resonance system, the resonance inductance is connected to the electrode in series (Figure 7 (a)). This resonance inductance generates the Q value multiplied voltage of the power supply voltage at both ends of the leakage capacitance by resonating with the leakage capacitance. For this reason, even if the voltage drops because the mutual capacitance is large, the electric power can be transmitted to the secondary side (Figure 7 (b)). Therefore, it is considered that electric power can be transmitted even if the power transmission distance is large. We are researching and developing this series resonance type of the electric field resonance system.

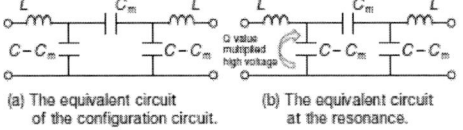

(a) The equivalent circuit of the configuration circuit.　(b) The equivalent circuit at the resonance.

Figure 7. The equivalent circuit for the parallel resonance type of the electric field resonance system.

The parallel resonance coupling type of the electric field resonance system.

In the parallel resonance type of the electric field

resonance system, the resonance inductance is connected to the electrode in parallel (Figure 8 (a)). This resonance inductance can cancel the leakage capacitance component by the resonance. Since the impedance at the parallel resonance can be considered as open, the equivalent circuit at the resonance is only the mutual capacitance (Figure 8 (b)). Even at the resonance, the voltage drop corresponding to the mutual capacitance becomes a loss, so that the transmission distance cannot be increased and the transmission distance becomes short. In the resonance coupling system of the magnetic field and the electric field, the parallel resonance type is the same as the electromagnetic induction in the case of the magnetic field resonance but since the Q value multiplied current is flowing, the voltage which can be transmitted to the secondary side can be increased by increasing the number of windings. In the case of the electric field resonance type, it is possible to increase the transmission voltage on the secondary side by increasing the mutual capacitance and decreasing the voltage drop by interposing a dielectric material.

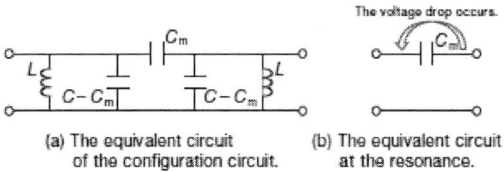

(a) The equivalent circuit of the configuration circuit.　(b) The equivalent circuit at the resonance.

Figure 8. The equivalent circuit for the parallel resonance type of the electric field resonance system.

From the viewpoint of the circuit theory, the electric field resonance system corresponding to the series magnetic field resonance system is parallel type. Unlike the serial magnetic field resonance system, the mutual capacitance remains between the power supply and the load in the parallel type electric field resonance system. Therefore, in the view of the conventional electric field resonance system, it has been considered that it can be used only in a short distance. However, by applying the series type, the electric power can be practically transmitted over a long distance, even with the electric field resonance system because the electric field resonance coupler can be driven with the Q value multiplied power supply voltage. This can similarly be said for the parallel type magnetic field resonance, but in the case of the magnetic field resonance, since the current is multiplied by the Q value, it is considered that the advantage is stronger when the power is transmitted mainly by the current, not by increasing the number of windings of the coil or by increasing the voltage

VII. THE MEASURES TO REDUCE THE PERIPHERAL ELECTRIC FIELD OF THE ELECTRIC FIELD RESONANCE COUPLER

Electric flux lines in a bent state are generated due to the fringe effect generated at the end of the electric field resonance coupler and a high electric field intensity region is generated in the vicinity of the coupler. As a countermeasure against this, we studied a structure that suppresses the unnecessary coupling by installing a coupler in a metal case where the power transmission surface is open. Figure 9 and 10 show the characteristics of the electric field near the coupler analyzed by the electromagnetic field simulation (moment method) when the power is applied the level of 1 kW and the couplers with the same configuration are opposed for the power transmission and reception. Figure 9 is a diagram showing the X axis in the power transmission direction, the Y axis in the direction parallel to the electric field vector and the Z axis in the power transmission and reception direction if the circumference of the coupler is free space. Figure 10 shows the characteristics if a metal case is provided for this coupler. An attenuation in the Y axis and the Z axis can be confirmed by installing the metal case. In this case, since the distance between the power transmission surface and the shield plate is required to be about the same as the power supply distance, thinning this equipment is a future subject.

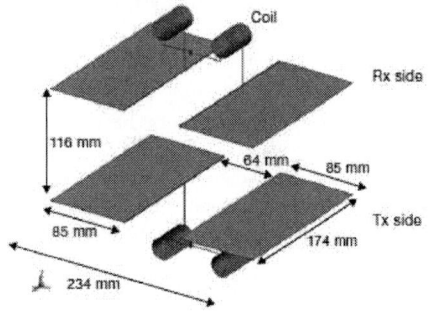

(a) The model with the coupler alone

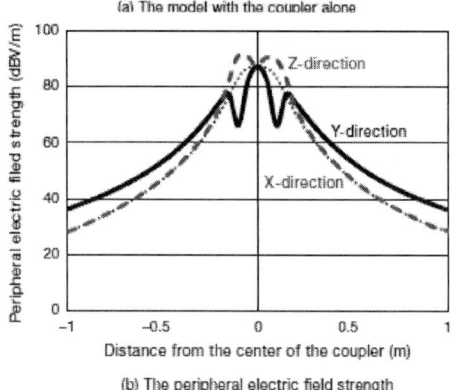

(b) The peripheral electric field strength

Figure 9. The peripheral electric field strength of the electric field resonance coupler.

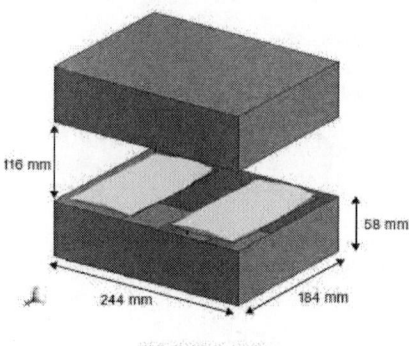

(a) Schematic model

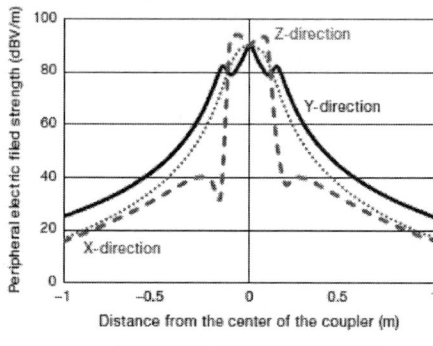

(b) The peripheral electric field strength

Figure 10. The peripheral electric field strength of the electric field resonance coupler with a shield box

VIII. THE DEPLOYMENT IN THE MOBILITYCAR

We have developed a system that provides the wireless power transmission by mounting the electric power resonance type coupler with the series resonance structure in a mobility car. Figure 11 shows a demonstration equipment of an electric field resonance coupler with a shield box mounted in a mobility car. In the coupler configuration, two electrodes are provided on the insulating plate and the resonance coil is connected to them. The electrode size is 458 mm × 220 mm and 2 electrode plates are set 18 mm apart. The material of the electrode plate is an Aluminum plate of A1100 and is placed in a shield case of 480 mm × 480 mm × 86 mm. The impedance of the transmission and reception coupler is 50 Ω and the power supply distance showing the maximum efficiency is 70 mm. The coupler is capable of the wireless power transmission of up to 1 kW in a natural air cooling condition.

Figure 11. The electric field resonance coupler with shield box mounted in a mobility car.

IX. CONCLUSIONS

We compared and considered the electric field coupling type and the magnetic field coupling type for the wireless power transmission and demonstrated that even the electric field coupling type can transmit the power at a remote distance by using the series resonance structure. This technology was mounted in a mobility car and verified that even the electric field resonance type can transmit the power in kW order. Conclusions are one of the most important parts of a paper. Please give careful consideration to this section.

REFERENCES

List only one reference per reference number according to the following samples:

[1] M. Young, "The PWM strategy on DC-DC converter," *IEEJ Journal of Industry Applications*, vol. 28, no. 15, pp. 123-129, 1989.

[2] G. Eason, B. Noble, and I. N. Sneddon, "On certain integrals of Lipschitz-Hankel type involving products of Bessel functions," *IEEE Trans. on Power Electronics*, vol. 247, no. 8, pp. 529-551, 1995.

[3] J. Clerk Maxwell, "A treatise on electricity and magnetism," *IEEE Trans. on Industry Applications*, vol. 589, no. 2, pp. 68-73, 2010.

[4] G. Eason, B. Noble, and I.N. Sneddon, "On certain integrals of Lipschitz-Hankel type involving products of Bessel functions," *Phil. Trans. Roy. Soc. London*, vol. A247, pp. 529-551, April 1955.

[5] J. Clerk Maxwell, *A Treatise on Electricity and Magnetism*, 3rd ed., vol. 2. Oxford: Clarendon, 1892, pp. 68-73.

[6] I.S. Jacobs and C.P. Bean, "Fine particles, thin films and exchange anisotropy," *in Magnetism*, vol. III, G.T. Rado and H. Suhl, Eds. New York: Academic, 1963, pp. 271-350.

[7] K. Elissa, "Title of paper if known," unpublished.

[8] R. Nicole, "Title of paper with only first word capitalized," *J. Name Stan. Abbrev.*, in press.

Transfer Power Analysis of Capacitively Isolated Outlet and Plug (CapIsOP) using Series Resonance

Hirohito Funato[1*], Koki Amano[1], Takuya Hatsumi[1] and Junnosuke Haruna[1]

1 Department of Electrical and Electronic Engineering, Utsunomiya University, Utsunomiya, Tochigi, Japan
*E-mail: funato@utsunomiya-u.ac.jp

Abstract— Authors have proposed capacitively isolated outlet and plug (CapIsOP) suitable for DC power distribution. In the previous studies, comb type capacitor is used to realize isolated outlet and plug. Comb type capacitor can realize large capacitance which results in large transfer power. In the previous studies, series resonance is employed for simple structure and both choke-input and capacitor-input rectifiers are used as load side rectifier. In this paper, analysis of transfer power using both choke-input rectifier and capacitor-input rectifier will be done for selection of circuit topology to realize more simple structure of capacitive couplings and enhancement of transfer power. In the experiment based on the analysis, power of 41.5W was successfully transferred using capacitor-input rectifier. In addition, it is experimentally confirmed that the plug from outlet was safely removed from the outlet.

Keywords— Capacitive Power Transfer, Isolated Outlet and Plug, Comb Capacitor

I. INTRODUCTION

In recent years, DC power distribution has been studied[1]. In DC power distribution, there is a danger of arcing when the plug is removed from the outlet. Therefore, from the view point of safety, isolated outlet and plugs have been proposed[2][3]. Since isolated outlet and plugs are electrically insulated, there is no possibility of dangerous arcing. In [2], power of 1kW is successfully transferred using inductive coupling type isolated outlet and plug, based on the inductive power transfer (IPT).

On the other hand, there is another method of wireless power transfer using capacitive coupling, called capacitive power transfer (CPT)[4][5]. In [4], 838W power transfer with a gap of 20cm was successfully realized. On the other hand, smaller gap results in larger capacitance in capacitive couplings. There are several studies of capacitive power transfer with small gap[6]-[12]. In [6], battery charger for a soccer playing robot is proposed using CPT. Lead zirconate titanate with thickness of 1mm is used as a dielectric between capacitive couplings. Capacitively coupled contactless charging platform using matrix charging pad was proposed in [7]. The proposed method employs capacitive couplings with 0.5mm gap filled with plastics. . In [8], Characteristic of coupling capacitance for CPT coated with titanium dioxide as a dielectric is analyzed. In [9], power transfer to a rotating device using a hydrodynamic coupling capacitor is proposed. In [10], CPT is used to provide power to slip rings of synchronous machines. Capacitor with air gap of 0.115 mm is formed between the rotor and the stator. Capacitive power transfer system using non-resonant power circuit was proposed in [11]. In this system, capacitive couplings with gap of 1mm filled with barium titanium were used. In [12], power transfer to electric vehicles using window glass as a dielectric is proposed. It achieves transfer of 200 W via glass with a thickness of 2 mm.

In outlet and plugs, it is desirable to place plug and outlet very close together. This feature is suitable for capacitive power transfer because larger coupling capacitance can be obtained. Therefore, authors have proposed capacitively isolated outlet and plug using CPT[13]. Fig. 1 shows concept of comb type capacitor proposed as a power supply connector. Fig. 2 shows Prototype of outlet and plug. The design policies are as follows.

(1) Size

It is desirable that it is not too large in order to ensure the same convenience as the conventional outlet socket and plug. Therefore, target value of size is decided as 120mm x 75mm x 45mm.

(2) Comb structure

To obtain wide area, comb type structure is adopted. As shown in Fig. 3, since the coupling part is filled with the dielectric, decrease of electromagnetic interference can be expected. This is superior to IPT.

(3) Electrode plate and

The material of the electrode plate is a phosphor bronze plate with a thickness of 0.1 mm. Since the phosphor bronze plate is excellent in spring property, it is expected to realize close contact.

(4) Dielectric

The dielectric is a Teflon sheet with a thickness of 0.08 mm. Its dielectric constant is 2.1. The breakdown voltage is 8 kV, which is sufficient value for power transfer. Since the surface of the Teflon sheet is slippery, it is expected to realize easily detachability.

In the previous studies, series resonance is employed for simple structure and both choke-input and capacitor-input rectifiers are used as load side rectifier. In this paper, transfer power using both choke-input rectifier and capacitor-input rectifier will be calculated theoretically and verified simulations in order to select suitable circuit topology to realize more simple structure of capacitive couplings and enhancement of transfer power. In the experiment based on the analysis, power of 41.5W was successfully transferred using capacitor-input rectifier.

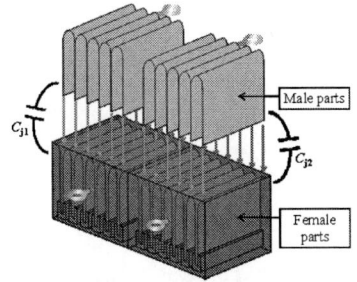

Fig. 1: Concept of outlet and plug using comb type structure

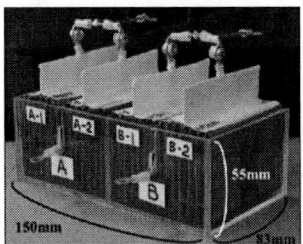

Fig. 2: Prototype of outlet and plug

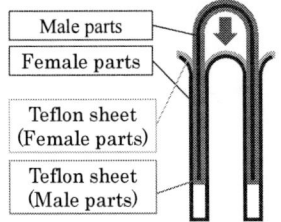

Fig. 3: One pair of comb type outlet and plug

II. PRINCIPLE AND ANALYSIS OF CAPACITIVELY ISOLATED OUTLET AND PLUG

In this section, concept of the proposed capacitively isolated outlet and plug (CapIsOP) using series resonance is introduced then the transfer power is theoretically analyzed. Fig. 4 shows conceptual block diagram of the proposed outlet and plug. In this figure, capacitance C represents capacitive couplings for outlet and plug.

Fig. 5 shows general equivalent circuit model of capacitive coupling used in the capacitive power transfer system. In the proposed system, the source side plate (plate1 or 2) and load side plate (plate 3 or 4) are placed very closely and two plates of each side are usually placed physically separated so that C_{12}, C_{14}, C_{23}, C_{34} can be negligible. Therefore, the capacitive coupling in the proposed system can be expressed as C_{j1}, C_{j2} shown in

Fig. 4 with simple series capacitance. The capacitance of couplings is nF order so that it is required to use resonance to enhance transfer power. In the proposed system simple series resonance is employed because of simplicity.

Choke-input and capacitor-input rectifiers in load side are considered as shown in Fig. 4 (a) and (b). This circuit contains series resonance so that flowing current becomes sinusoidal under resonant condition. For this reason, capacitor-input rectifier is suitable, however, transfer power of both circuits are calculated in the following part in order to consider various load condition.

(a) choke-input rectifier

Assuming that the current in series resonance circuit i_{cj} is sinusoidal and load side current is constant due to large inductance in rectifier, amplitude of i_{cj} becomes equal to load current I_{load}. Therefore, I_{cj}, rms value of i_{cj} can be calculated as following equation,

$$I_{cj} = \frac{I_{load}}{\sqrt{2}} . \tag{1}$$

The fundamental component of inverter output voltage can be expressed as the following equation when duty ratio is 0.5.

$$v_{S\_e} = \frac{2E}{\pi} \sin 2\pi f t , \tag{2}$$

where f is switching frequency. Therefore, provide power from power source can be calculated as the following equation.

$$P_{in} = \left(\frac{I_{load}}{\sqrt{2}} \right) \cdot \left(\frac{1}{\sqrt{2}} \cdot \frac{2E}{\pi} \right) \tag{3}$$

When the loss component of inductor R_L can be negligible, P_{in} is equal to consumption power of load P_{load}, then the following equation is obtained

$$P_{load} = R_{load} I_{load}^2 = \frac{E^2}{\pi^2 R_{load}} \tag{4}$$

Next, equivalent circuit shown in Fig. 6 (a) is considered. In this equivalent circuit, inverter and load is represented by sinusoidal voltage and ac resistance respectively. C_j is synthesized capacitance of capacitive couplings C_{j1}, C_{j2}. Using this equivalent circuit, consumption power P_{load_e} at load resistance R_{load_e} can be expressed as follows;

$$P_{load\_e} = \frac{\left(\frac{1}{\sqrt{2}} \cdot \frac{2E}{\pi} \right)^2}{R_{load\_e}} \tag{5}$$

Comparing Eq. (4) and Eq. (5), R_{load_e} is obtained as follows because P_{load_e} is equal to P_{load}.

$$R_{load\_e} = 2R_{load} \tag{6}$$

Eq. (6) can be applied when load resistance is relatively small.

(b) Capacitor-input rectifier

Using capacitor-input rectifier, the circuit configuration is equal to series resonant converter, so that transfer power is calculated using the same way of series resonant converter. Fig. 6 (b) is equivalent circuit to calculate transfer power. Assuming that V_{load} in Fig. 4 (b) is constant, v_2 becomes rectangle voltage. On the other hands, the voltage of equivalent resistance R_{load_e} becomes sinusoidal because of series resonance. The fundamental component of rectangle voltage is obtained as shown in Fig. 7. The rectangle voltage shown in Fig. 7 (a) is applied to R_{load} in Fig. 4 (b), while the sinusoidal voltage shown in Fig. 7 (b) is applied to R_{load_e} in Fig. 6 (b). Because consumption power in of R_{load_e} should be equal to that in R_{load}, the following equation is obtained,

$$\frac{\left(\frac{1}{\sqrt{2}} \cdot \frac{4}{\pi} V_{load}\right)^2}{R_{load\_e}} = \frac{V_{load}^2}{R_{load}} . \tag{7}$$

Solving Eq. (7), R_{load_e} is derived as follows,

$$R_{load\_e} = \frac{8}{\pi^2} R_{load} \tag{8}$$

Transfer power P_{load} can be calculated as the following equation from equivalent circuit in Fig. 7 (b) substituting Eq. (8) under resonant condition.

$$P_{load} = \left(\frac{8}{\pi^2} R_{load}\right) \cdot \left(\frac{\frac{\sqrt{2}}{\pi} E}{R_L + \frac{8}{\pi^2} R_{load}}\right)^2 \tag{9}$$

If R_L can negligible, Eq. (9) becomes;

$$P_{load} = \frac{E^2}{4 R_{load}} \tag{10}$$

From Eq. (4) and Eq. (10), transfer power can be estimated.

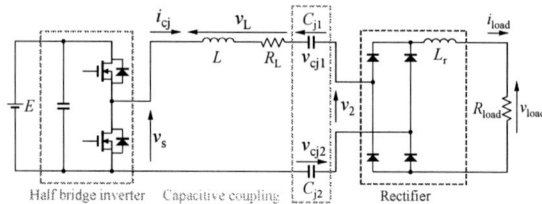

(a) With choke-input rectifier

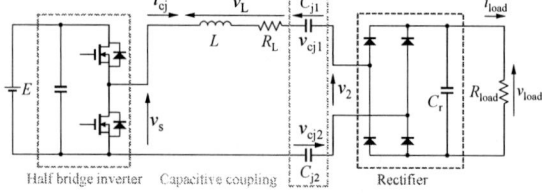

(b) With capacitor-input rectifier

Fig. 4: Circuit diagram of proposed capacitively isolated outlet and plug

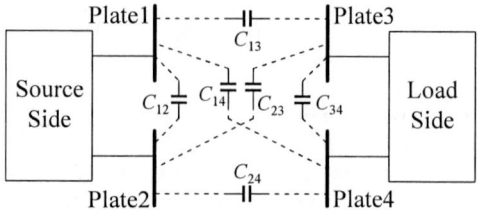

Fig. 5: Model of coupling part

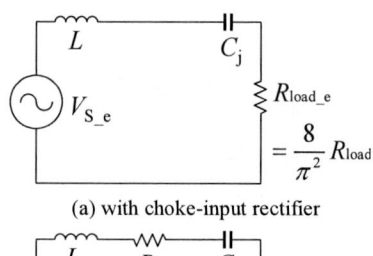

(a) with choke-input rectifier

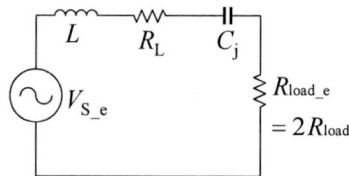

(b) with capacitor-input rectifier

Fig. 6: Equation circuit of series resonance type

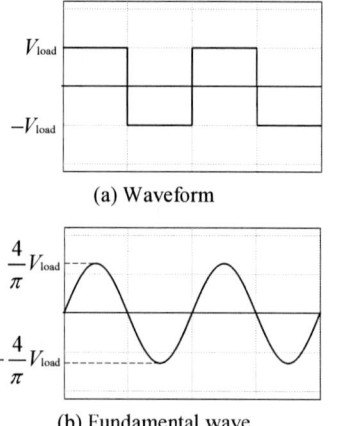

(a) Waveform

(b) Fundamental wave

Fig. 7 Waveform of v_2 with capacitor-input rectifier

III. SIMULATION ANALYSIS

In this section, transfer power of actual system shown in Fig. 4 is analyzed using simulations. Table I shows parameters used in the simulations. DC source voltage E is not so large because of the limitation of experimental equipment. Capacitances of couplings C_{j1}, C_{j2} are measured value of comb type capacitor shown in Fig.1. Operating frequency is selected as 400 kHz.

Fig. 8 shows simulation results using choke-input rectifier. Transfer power becomes 23.8 W, while theoretical power becomes 25.3 W using Eq. (4). Total efficiency from DC source to load becomes 93.0%. There

The 2018 International Power Electronics Conference

are two reasons of difference between theoretical value and simulated value. One is loss component R_L is neglected in theoretical analysis. The other is current flowing in the series resonant circuit does not become sinusoidal. Fig. 9 shows current waveform and current path in the CapIsOP with choke-input rectifier. After changing v_s, i_{cj} increases due to resonance (mode 1). In this mode 1, load current circulating in the rectifier as shown in the upper diagram of Fig. 9 (b). When i_{cj} arrive at i_{load}, D_2 and D_3 turned off. Therefore, resonant current is clumped by load current. If load resistance is small, mode 2 becomes short, so that i_{cj} becomes closer to sinusoidal.

Fig. 10 shows simulation results using capacitor-input rectifier. Transfer power becomes 47.3W and total efficiency becomes 86.9%. Theoretical power also becomes 47.3 W using Eq. (9). The simulated power well agrees with theoretical power.

The above analysis clarifies that the transfer power is higher in capacitor-input rectifier. The next section shows the experiment based on this analysis.

Table I: Parameters used in simulations

Symbol	Meaning	Value
f [kHz]	Resonance frequency	400
E [V]	Source voltage	50
C_{i1} [nF]	Coupling capacitance 1	13.3
C_{i2} [nF]	Coupling capacitance 2	10.9
L [μH]	Resonance inductance	26.4
R_L [Ω]	Winding resistance	1.21
R_{load} [Ω]	Load resistance	10
L_r [μH]	Choke inductance	250
C_r [μF]	Smoothing capacitance	1

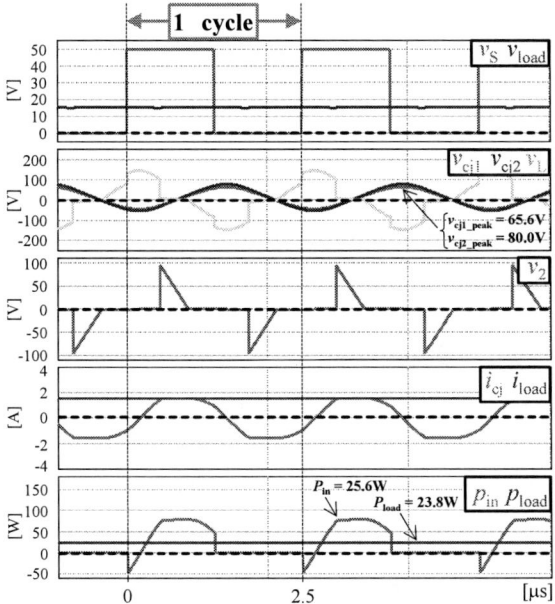

Fig. 8: With choke-input rectifier

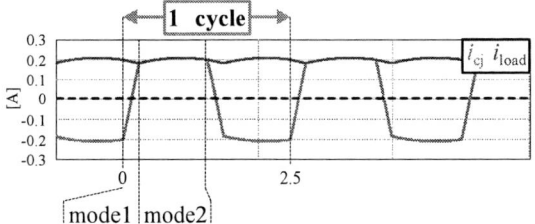

(a) Waveforms of i_{cj} and i_{load} when R_{load} is 100Ω

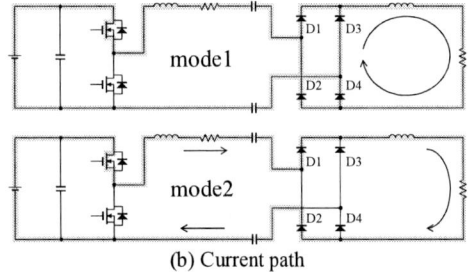

(b) Current path

Fig. 9: i_{cj} and i_{load} when the load resistance is large (with choke-input rectifier)

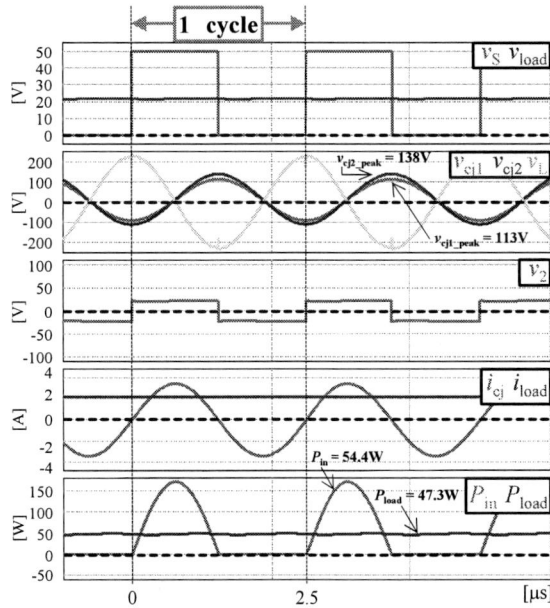

Fig. 10: With capacitor-input rectifier

IV. EXPERIMENTAL RESULTS

Fig. 11 shows experimental system. In this figure, the blue part with indication of "Capacitive coupling" is isolated outlet and plug using comb capacitor. The circuit diagram of the experimental system is the same as in Fig. 4(b). In this experiment, capacitor-input rectifier is used to transfer more power. Parameters used in experiments are shown in Table II. The resonant inductance is adjusted to resonate at 400 kHz. Winding resistance is experimental measured value at 400 kHz. The load is non-inductive resistance. The switching device of the inverter is GaN MOSFET (GS66508B), and the dead time is set to 50 ns.

Fig. 12 shows experimental results. Fig. 12(a) shows experimental waveforms with 53 V dc power supply. In this case, power transfer of 41.5 W is successfully obtained. In the previous experiment using the choke-input rectifier, 15 W power transfer was obtained with 49 V dc power source[13]. Therefore, enhancement of 36.5 W is confirmed. The DC difference of power supply voltage is due to the limitation of experimental system. The theoretical value of the transfer power described in section 3 is 54.4 W. The reason why the experimental value is smaller than the theoretical value is considered to be the forward drop voltage of the diode and the dead time of the inverter. Experimental transfer power approaches the theoretical value by using diodes with small forward drop voltage.

Fig. 12 (b) shows the voltage of each part on the source side and resonance current i_{cj}. The DC power supply voltage is set to 35 V due to the limitation of measurement. v_{cj1}, v_{cj2}, i_{cj} are sinusoidal waveforms and well agree with the simulation waveforms in Fig. 10. However, the shape of peak point of v_L is slight different from the simulation waveform. This is because the switching frequency of the inverter is slightly higher than the resonance frequency of the circuit.

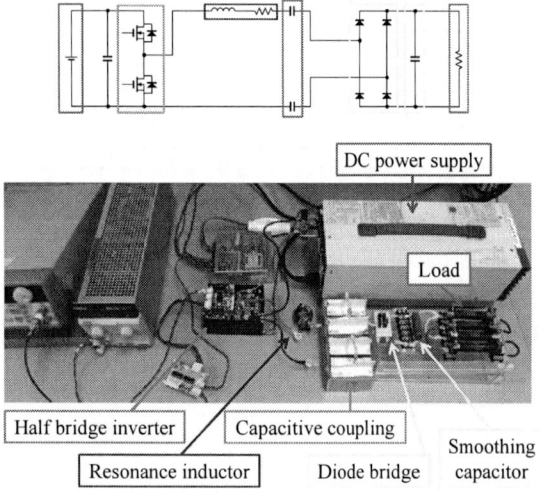

Fig. 11: Experimental setup

Table II: Parameters used in experiment

Symbol	Meaning	Value
f [kHz]	Resonance frequency	400
E [V]	Source voltage	53.0, 35.0
C_{i1} [nF]	Coupling capacitance 1	13.3
C_{i2} [nF]	Coupling capacitance 2	10.9
L [μH]	Resonance inductance	26.4
R_L [Ω]	Winding resistance	1.21
R_{load} [Ω]	Load resistance	9.7
C_s [μF]	Smoothing Capacitance	1.00
v_f [V]	Forward drop voltage of diode	1.4
t_d [ns]	Dead time	50

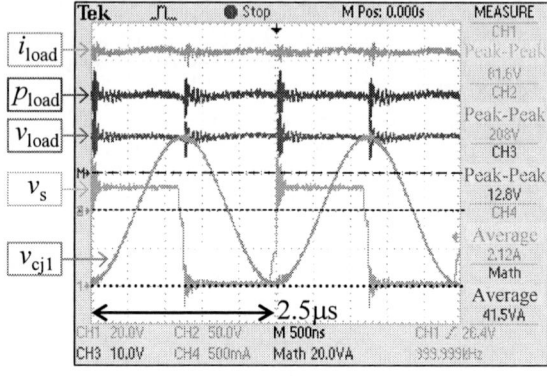

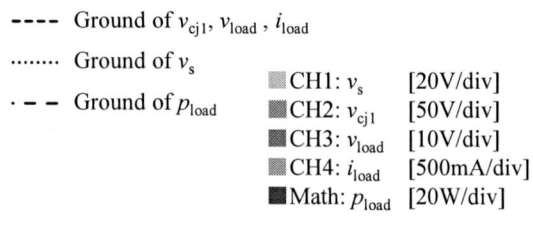

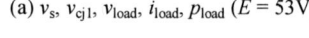

---- Ground of v_{cj1}, v_{load}, i_{load}

.... Ground of v_s

·－－ Ground of p_{load}

CH1: v_s [20V/div]
CH2: v_{cj1} [50V/div]
CH3: v_{load} [10V/div]
CH4: i_{load} [500mA/div]
Math: p_{load} [20W/div]

(a) v_s, v_{cj1}, v_{load}, i_{load}, p_{load} (E = 53V)

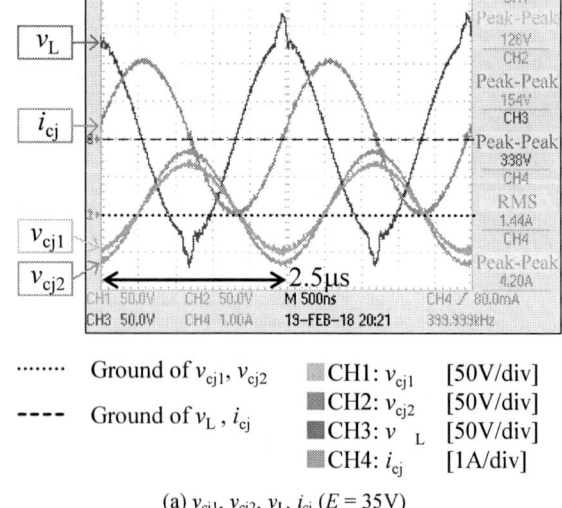

.... Ground of v_{cj1}, v_{cj2}

---- Ground of v_L, i_{cj}

CH1: v_{cj1} [50V/div]
CH2: v_{cj2} [50V/div]
CH3: v_L [50V/div]
CH4: i_{cj} [1A/div]

(a) v_{cj1}, v_{cj2}, v_L, i_{cj} (E = 35V)

Fig. 12: Waveforms of power transfer experiment

Insertion and removal characteristic when energized is an important analysis from the viewpoint of safety of isolated outlet and plug. Experimental analysis of circuit operation when male part of the comb capacitor is removed is completed. Fig. 13 shows how to remove the mail part of the comb capacitor. In this experiment, the mail part of C_{j1} in Fig. 4(b) is removed. DC power supply voltage is set to 35V. The experimental waveform during removal is shown in Fig. 14. In this figure, resonance current i_{cj} and load current i_{load} are safely shut off when C_{j1} is removed. In the case of insertion, it is considered to be the reverse waveform. Therefore, the safety of the capacitively isolated outlet and plug is verified.

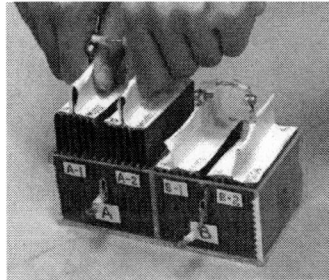

Fig. 13: Removal of male part of comb capacitor

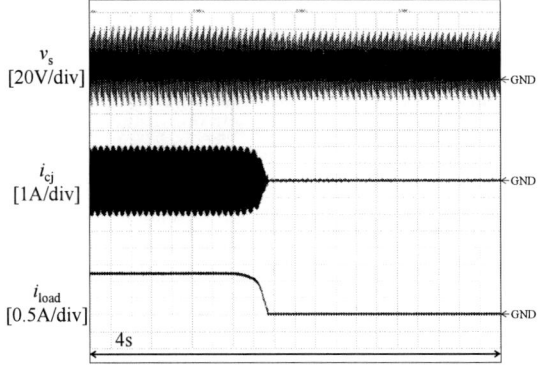

Fig. 14: Waveforms of removal experiment when energized

V. CONCLUSIONS

In this paper, transfer power of proposed capacitively isolated outlet and plug (CapIsOP) is analyzed using theoretical analysis and simulations. For the proposed isolated outlet socket and plug system, two types of rectifier in load side were considered. Analysis using the equivalent circuit theoretically clarified that the transfer power with capacitor-input rectifier is larger than that with choke-input rectifier. Experiments are also completed. In the experiment using the capacitor-input rectifier, power of 41.5W was successfully transferred. Further enhancement of transfer power will be considered to realize final target transfer power of 1kW. The next mission is to develop new smaller connector with higher dc voltage and higher frequency.

REFERENCES

[1] Po-Hsu Huang, Weidong Xiao, Mohamed Shawky El Moursi, "A Practical Load Sharing Control Strategy for DC Microgrids and DC Supplied Houses", *39th Annual Conference of the IEEE Industrial Electronics Society (IECON 2013)*, pp. 7122-7126, 2013

[2] Satoshi Ojika, Yushi Miura, Toshifumi Ise, "Inductive Contactless Power Transfer System with Coaxial Coreless Transformer for DC Power Distribution", *2013 IEEE ECCE Asia Downunder*, pp. 1046-1051, 2013

[3] Kenzo Kakinuma, Yasuyoshi Kaneko, "Development and Evaluation of Insertion Type Wireless Power Transformer", *the 2016 Annual Meeting of the Institute of Electrical Engineers of Japan* ,4-174, pp.292-293, 2016 (in Japanese)

[4] Mitsuru Masuda, Masahiro Kusunoki, Daiki Obara, Yujiro Nakayama, Hiroki Hamada, Shoichi Negami,"Wireless Power Transfer via Electric Coupling", *The Institute of Electronics Information and Communication Engineers Technical Report*, WPT2013-20, pp.15-19, 2013 (in Japanese)

[5] Fei Lu, Hua Zhang, and Chris Mi, "A Two-Plate Capacitive Wireless Power Transfer System for Electric Vehicle Charging Applications", *IEEE Transactions on Power Electronics*, vol. 33, No. 2, pp.964-969, 2018

[6] Aiguo Patrick Hu, Chao Liu, Hao Leo Li: "A Novel Contactless Battery Charging System for Soccer Playing Robot", *15th IEEE International Conference, Mechatronics and Machine Vision in Practice*, pp.623-626, 2008

[7] Chao Liu, Aiguo Patrick Hu, Bob Wang, and N. Nair, "A Capacitively Coupled Contactless Matrix Charging Platform With Soft Switched Transformer Control," *IEEE Transactions on Industrial Electronics*, vol. 60, no. 1, pp. 249-260, 2013

[8] Baoyun Ge, Daniel C. Ludois, Rodolfo Perez, "The Use of Dielectric Coatings in Capacitive Power Transfer Systems", *Energy Conversion Congress and Exposition (ECCE)*, pp, 2193-2199, 2014

[9] Daniel C. Ludois, Micah J. Erickson, Justin K. Reed, "Aerodynamic Fluid Bearings for Translational and Rotating Capacitors in Noncontact Capacitive Power Transfer Systems", *IEEE Transactions on Industry Applications*, vol. 50, No. 2, pp. 1025-1033, 2014

[10] Antonio Di Gioia, Ian P. Brown, Yue Nie, Ryan Knippel, Daniel C. Ludois, Jiejian Dai, Skyler Hagen, Christian Alteheld, "Design and Demonstration of a Wound Field Synchronous Machine for Electric Vehicle Traction with Brushless Capacitive Field Excitation, *IEEE Transactions on Industry Applications*, vol. pp, No. 99, 13pages, 2017

[11] Hirohito Funato and Yuki Chiku, "Capacitve Power Transfer System using a Cascaded Improved One-Pulse Switching Active Capacitor", *IEEJ Journal of Industry Applications*, vol.4, No.6, pp.714-721, 2015

[12] Kang Hyun Yi, JaeYup Jung, Byoung-Hee Lee, YoungSoo You, "Study on a capacitive coupling wireless power transfer with electric vehicle's dielectric substrates for charging an electric vehicle", *EPE'17 ECCE Europe*, 7pages, 2017

[13] Koki Amano, Hirohito Funato, and Junnosuke Haruna, "Proposal of Isolated Outlet Socket and Plug using Capacitive Power Transfer", *IEEE IFEEC 2017 -ECCE Asia*, No.1487, 6pages, 2017

Wide Voltage Gain Range LLC DC/DC Topologies: State-of-the-Art

Qi Cao, Zhiqing Li and Haoyu Wang
School of Information Science and Technology
ShanghaiTech University, Shanghai, China, 201210
wanghy@shanghaitech.edu.cn

Abstract— LLC resonant converter is a prevalent isolated dc/dc topology due to its high efficiency, high power density, and simple structure. However, it is challenging to optimize the design of conventional LLC converter in wide gain range applications. Therefore, modifications either on the circuit structure or control strategy are necessary to improve its voltage regulation performance. This paper presents a comprehensive overview of recent evolvements of LLC topology in wide voltage gain applications. The broad applications of wide gain LLC converters, which include hold-up operation, wide input voltage range, and wide output voltage range, are investigated. State-of-the-art wide gain LLC solutions are reviewed and classified into four categories. Benefits and constraints of different solutions are addressed in detail. Finally, the circuit selection principles for different application backgrounds are summarized with a comparative study of different configurations.

Index Terms—LLC topology, review, wide voltage range.

I. INTRODUCTION

Dc/dc converters with wide voltage gain capability are widely used in different power conversion applications. Typically, those applications are featured with one of those characteristics: 1) hold-up operation, 2) wide input voltage range, and 3) wide output voltage range.

Hold-up operation is required in applications with demanding requirements on power supply continuity and reliability, such as sever power supply and telecommunication systems [1]–[3]. When a short duration of grid blackout occurs, the dc bus voltage between the ac/dc power-factor-correction stage and the front-end dc/dc stage drops substantially [4]. However, the dc/dc converter is expected to maintain a stable output voltage to ensure the proper function of the equipment [5]. Therefore, the dc/dc converter needs to fit the wide input voltage range. The hold-up operation lasts about 20 milliseconds [6]–[8]; hence, the converter mainly operates in normal mode, while the high gain mode rarely occurs [9], [10]. In those scenarios, research is mainly focused on maintaining high efficiency in normal mode. Thus, relatively low priority is assigned to the efficiency performance at high voltage gain mode for hold-up operation.

Wide input range applications regularly require a wide gain range to maintain a constant output under different input voltages. The most common scenario is the photovoltaic systems, where a dc/dc converter is required to couple the renewable sources with the grid [11]. Renewable energy sources such as solar panels usually have a wide terminal voltage range and low output power (below 500 W) [12], [13]. In those scenarios, researches are mainly focused on optimizing the efficiency over the entire input voltage range.

For wide output voltage applications, the input voltage is usually constant, while the output voltage varies in a wide range. The efficiency optimization over entire output range must be considered. For example, in battery-charging systems, constant current (CC) stage and constant voltage (CV) stage are enforced in the charging profile [14], [15]. In CC stage, battery pack's terminal voltage demonstrates a wide variation range [16]. The dc/dc stage of the battery charger needs to match its output voltage with the battery pack terminal voltage [17].

Frequency modulated LLC resonant converter is a prevalent isolated dc/dc topology and has attracted wide research focus in recent years [18]–[22]. This is mainly due to its appealing merits: 1) high efficiency,2) wide zero-voltage-switching (ZVS) and zero-current-switching(ZCS) range, 3) simple circuit structure, and 4) high power density [23]. How to design conventional LLC topology to fit the wide gain range applications has been explored in the literatures [15], [24]–[26]. However, when the conventional LLC topology is deployed in wide gain range applications, switching frequency range will be extra wide. This leads to several drawbacks: 1) constrained ZVS and ZCS range, 2) increased core size determined by lowest f_s, and 3) limited light-load regulation ability, especially considering secondary-side equivalent parasitic capacitance [5], [9], [19]. To overcome these drawbacks, researchers have proposed various modifications and corresponding design considerations to the conventional LLC structure. The main purpose is to make the LLC-based converter achieve a wide voltage range and meanwhile to maintain a high overall efficiency.

This paper presents a comprehensive overview of the recent evolvements of the LLC type topologies adapted to wide gain range applications. Section II analyzes the conventional LLC structure. Section III reviews the modified LLC topologies and proposed control strategies. A comparison of these solutions and summary of topology selection principles for different applications are given in Section IV. Conclusion is drawn in Section V.

II. CONVENTIONAL LLC RESONANT CONVERTER

Fig. 1 shows the block diagram of a typical LLC resonant converter. It consists of four parts: 1) a primary-side switch network, 2) a resonant tank, 3) a high-frequency transformer (Tx), and 4) a secondary-side rectifier. The primary-side switch network is either a full-bridge (FB) or half-bridge (HB) inverter. A square wave is generated and fed into the resonant tank. The resonant tank consists a resonant capacitor C_r, resonant inductor L_r and magnetizing inductor L_m. The high-frequency

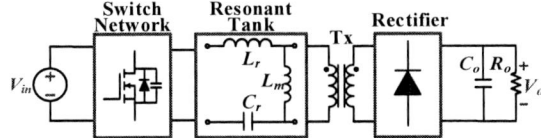

Fig. 1. Block diagram of typical LLC resonant converter.

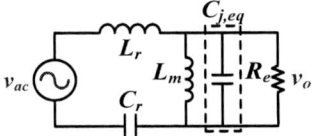

Fig. 2. Ac equivalent circuit model of conventional LLC resonant converter.

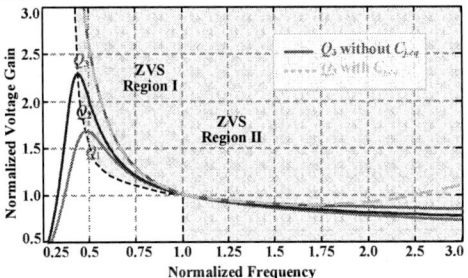

Fig. 3. DC gain curves of conventional LLC resonant converter.

transformer promises electrical isolation and delivers energy to the output. The secondary-side rectifier could be a FB rectifier or a center-tapped synchronous rectifier (SR).

Among mainstream isolated dc/dc topologies, LLC topology distinguishes itself with its good circuit performance. In LLC topology, ZVS for primary-side switches and ZCS for diodes can be maintained in a wide load range. The soft switching feature enables a high switching frequency, which facilitates a compact transformer [23]. Thus, high power density can be achieved. Besides, on the primary side, L_r and L_m can be integrated into one single magnetic core; while on the secondary side, only one capacitor filter is required. Hence, the circuit complexity is reduced [27].

To design and analyze an LLC resonant converter, a voltage transfer function should be determined. The widely used first harmonic approximation (FHA) method approximates the square wave with its fundamental harmonic while ignoring the higher order components. Fig. 2 shows the ac equivalent circuit model, where v_{ac} is the output of primary side switch network and R_e is the equivalent load resistance. $C_{j,eq}$ is the equivalent junction capacitor of secondary side rectifier and transformer parasitic capacitor. The FHA method provides acceptable accuracy if f_s is at the vicinity of f_r, since the first harmonic part of the resonant current dominates. Ignoring $C_{j,eq}$, the normalized voltage gain function can be determined as

$$M_g = \frac{nV_o}{V_{in}} = \left| \frac{L_n \cdot f_n^2}{\left[(L_n+1)f_n^2 - 1 \right] + j\left[(f_n^2-1) \cdot f_n \cdot Q_e \cdot L_n \right]} \right| \tag{1}$$

where $f_n = f_s / f_r$ is normalized frequency, $L_n = L_m / L_r$ is inductance ratio, Q_e is the quality factor, which is defined as

$$Q_e = \sqrt{L_r/C_r} / R_e \tag{2}$$

and n is the turns ratio of Tx. The output voltage can be regulated by frequency modulation (FM) since M_g varies with f_s.

Meanwhile, either decreasing L_n or Q_e results in a larger peak gain. The gain curves are plotted in Fig. 3. As shown, there always exists a combination of L_n and Q_e to fulfill the design requirements. However, there are some critical issues and restrictions, especially in wide gain range applications. As shown in Fig. 3, the regulation ability of FM degrades when f_s is beyond f_r, especially in light load condition [28]–[30]. In light load condition, the influence of $C_{j,eq}$ increases the voltage gain when f_s is above 2-2.5 times of f_r [31]. This phenomenon is also illustrated in Fig. 3. The efficiency with a wide frequency range degrades as well since the optimal operation point only occurs in the vicinity of f_r [32], [33]. When f_s deviates beyond f_r, the secondary-side diodes lose ZCS, and the MOSFET turning-off current increases [34]. This leads to increased switching loss and the diodes' recovery issues [35]. When f_s deviates below f_r, the circulating power increases. This leads to increased conduction loss and current stresses.

Moreover, a wide switching frequency range leads to a large transformer core, high core loss, and low power density [9], [36]. Also, a high peak gain requires a small L_n, which results in increased circulating power and degraded efficiency [10]. Besides, the accuracy of FHA method will also be undermined [11]. Therefore, the conventional LLC structure is not an optimal solution for wide voltage range applications.

III. MODIFIED LLC TOPOLOGIES FOR WIDE VOLTAGE RANGE

This section provides a comprehensive review of recent research efforts to narrow down the f_s range of LLC topology. According to the specific stage where the circuit modification is enforced, the solutions for LLC topology extension are classified into four categories.

A. Reconfiguration of Resonant Tank

This kind of modification mainly reconfigures the structure of resonant tank (C_r, L_r, and L_m) and transformer's turns ratio n. The equivalent value of those parameters can be tuned by either mode transition or switching frequency variation. Therefore, the voltage gain can be extended to a wide range with narrow frequency window. Typically, the control strategy is simple, while the parameter design is more complicated.

Several modifications are proposed to change L_m. In conventional LLC structure, the trade-off between wide voltage range and small circulating current needs to be addressed when designing L_m. Thus, a variable L_m is beneficial to extend the gain range with improved overall efficiency. The modification includes linking an extra transformer, adding an auxiliary transformer as a variable L_m and auxiliary LC structures [2], [5], [10], [11]. Figures 4-7 shows the topologies. The first method changes L_m by controlling a bi-directional switch, which endures overshoot problem during mode transition. Besides, an extra FB rectifier increases its volume. The second method changes the equivalent L_m by controlling dc current bias of auxiliary transformer. In the third method, equivalent L_m can be changed adaptively with frequency.

Taking the LCLC structure proposed in [2] as the example, the equivalent L_m can be expressed as,

$$L_{m\_eq}(f_s) = L_p - \frac{1}{(2\pi f_s)^2 C_p} \tag{3}$$

The 2018 International Power Electronics Conference

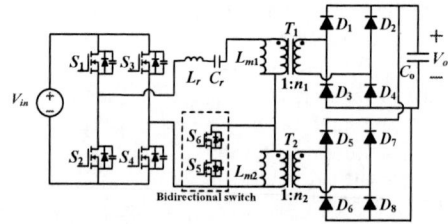

Fig. 4. Two-transformer LLC converter [11].

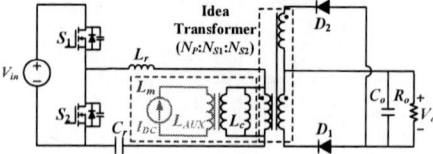

Fig. 5. Modified LLC converter with variable inductance [10].

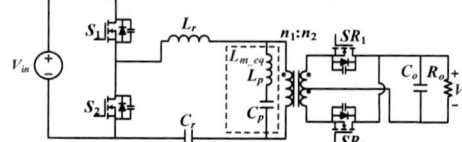

Fig. 6. Topology of LCLC converter [2].

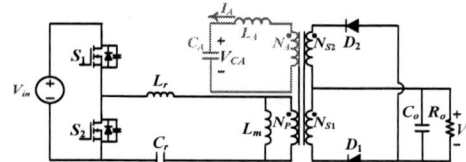

Fig. 7. Modified LLC converter with auxiliary LC circuit [5].

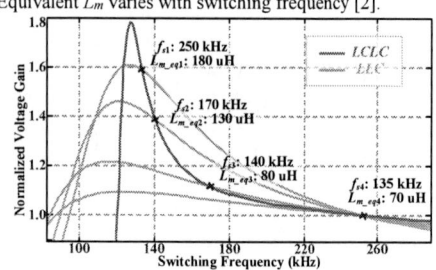

Fig. 8. Equivalent L_m varies with switching frequency [2].

Fig. 9. Gain curves of LCLC and equivalent LLC [2].

Figures 8-9 illustrates operation principles in [2], where L_p = 230 uH and C_p = 9.4 nF. The circuit reliability can be enhanced since there is no need for extra control circuits. Moreover, the abrupt mode transition is avoided; hence, the overshoot problem no longer exists.

Alternatively, in [1], equivalent C_r can be changed by paralleling an extra capacitor. In [17], [37], transformer's turns ratio can be tuned by connecting extra coils or switching between series and parallel connected structure of two

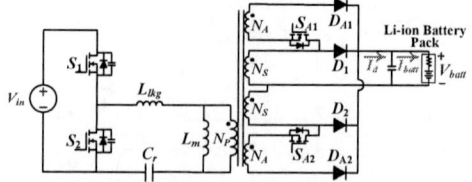

Fig. 10. Reconfigured LLC converter with adaptive turns ratio [17].

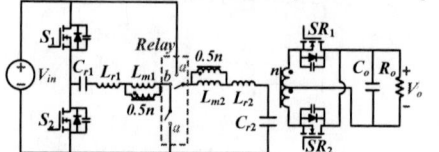

Fig. 11. Reconfigured LLC converter with adaptive turns ratio [37].

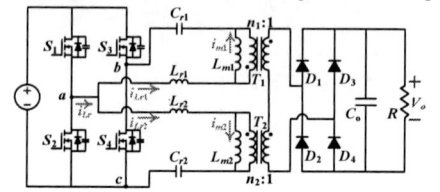

Fig. 12. Modified LLC converter with two split resonant tank [38].

primary-side coils. Figures 10-11 shows the circuit structures.

As shown in Fig. 12, , an extra resonant tank is added in [38]., Smooth mode transition between HB/FB can be realized by different PWM pattern of S_{1-4}. This method restrains the overshoot problem.

B. Modification on Primary-side Switch Network

The modifications on the primary-side switch network mainly follow those principles: 1) regulation of equivalent input voltage of the resonant tank, and 2) modification for specific control strategies.

The primary-side switch network generates an ac voltage (v_{ac}) to feed the resonant tank. Regulating the amplitude and pulse width of v_{ac} changes its root-mean-square (RMS) voltage. This improves regulation ability of the voltage gain. Moreover, the design procedure of resonant parameters can be simplified. However, this method usually requires some extra switches, which increases switching loss, complexity of control strategy, and the volume of converter.

The modifications include fixed frequency dual bridge structure, variable frequency multiplier (VFX) with three-level structure, three-level full-bridge structure, buck-boost cascaded structure, and Interleaved Boost-Integrated LLC (IBI-LLC) [23], [35], [39]–[42]. Figures 13-16 shows their topologies.

The first structure realizes v_{ac} regulation by transition between HB and FB mode in primary side. The second and third structures generate various waveforms of v_{ac} by controlling duty cycles and phase-shift (PS) angles of primary-side switches. In the third structure, the v_{ac} waveforms in three modes are shown in Fig. 17, where α, β, and γ are control variables. In this structure, lower voltage-rating devices can be used since the voltage stress on each MOSFET is reduced. The last two structures combine a buck-boost or interleaved boost stage with LLC stage. The equivalent input voltage V_{CLINK} can be regulated by duty cycle control. However, its non-symmetric operation results to dc offset current and a larger size of transformer.

102

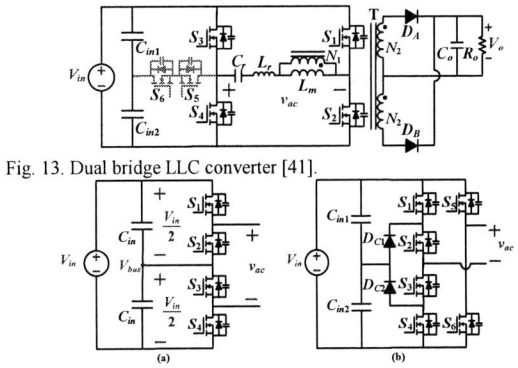

Fig. 13. Dual bridge LLC converter [41].

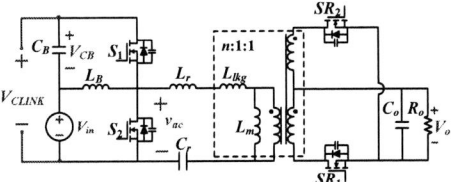

Fig. 14. Topology of modified primary-side inverter. (a) Three-level bridge structure [23]. (b) Three-level full-bridge structure [35].

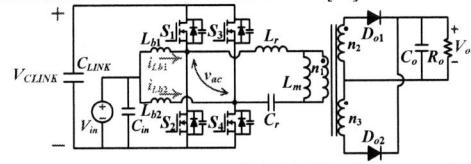

Fig. 15. Buck-boost + LLC cascaded converter [39].

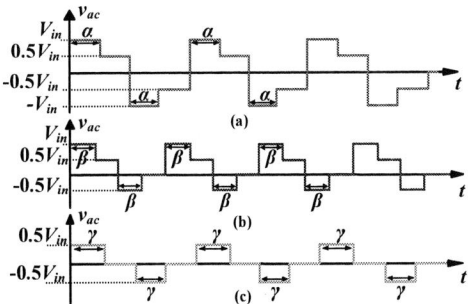

Fig. 17. Waveforms of v_{ac} proposed in [40]. (a) HG mode. (b) MG mode. (c) LG mode.

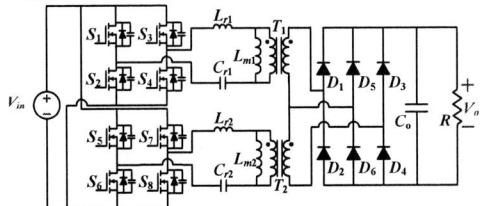

Fig. 18. Topology of interleaved LLC converter [44].

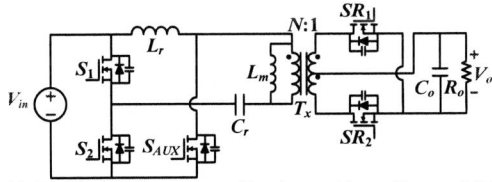

Fig. 19. Modified LLC converter with primary-side auxiliary switch [45].

Differently, the main purpose of modification proposed in [43] and [44] is to realize their unique control strategies that work complementarily with FM to squeezes the f_s range. The structure proposed in [43] combines an identical structure shown in Fig. 14(a) and a secondary-side SR. The input voltage range is extended by PS control between primary-side and secondary-side switches. The structure proposed in [44] is shown in Fig. 18. Once $f_s > f_r$, voltage gain can be further extended by controlling PS angle between two resonant tanks.

In [45], a different method to extend the voltage range is proposed by injecting more energy into resonant tank by controlling ON time of S_{AUX} in each switching cycle. Fig. 19 illustrates this topology.

C. Modification on control strategies

Instead of topological modifications, customizing the control strategies is also considered as a feasible solution to extend the gain range of LLC converter. Reported works are mainly based on those two principles: 1) modifying the control strategies on the primary-side switch network, and 2) regulating the dc link voltage.

The first type solutions include fixed frequency phase shift (FFPS) approach, HB/FB transition, and asymmetric (APWM) control of HB inverter [46]–[48]. In [46], the HB/FB transition is proposed based on a FB inverter. Nevertheless, this abrupt mode transition leads to a severe overshoot issues. In [47], APWM control works as a supplement to FM, which can further increase the voltage gain by adjusting the duty ratio of the HB. The asymmetric operation introduces offset current, which

leads to saturation problem of magnetic component. Typically, the size of transformer will be increased to avoid saturation. However, its volume decreases due to a squeezed frequency range. Conclusively, the low conduction loss and high power density are also retained.

For the second principle, since an isolated battery charger typically consists a front-end ac/dc stage for power factor correction (PFC) and a second dc/dc stage for voltage regulation, the variable dc link strategy can be realized by the modulation of the PFC stage [49]–[51]. The control methods in these three papers are similar, which can be summarized as, 1) to ensure the power factor, and 2) to regulate the dc-link voltage following battery voltage. This strategy can keep the LLC converter always operating at f_r with symmetric operation of primary-side switches, thus the switching and conduction losses are decreased. However, the PFC stage has a limited range of voltage variation.

D. Modification on Secondary-side Rectifier

Regarding the modifications on the secondary-side rectifier, the variable-structure voltage multipliers are typically adopted in [31], [33], [52], [53].

In [31], [53], the secondary side can be switched between a FB rectifier and a voltage-doubler rectifier (VDR). In [52], it can operate as a VDR or a voltage-quadrupler rectifier, which is called semi-active variable-structure rectifier (SA-VSR). Fig. 20 shows their structures. However, these three methods require mode transition between two structures, which ensures that the maximum voltage gain should be at least twice of minimum

103

voltage gain. Hence, the complexity of parameter design increases.

Applying the same concept shown in Fig. 20(b), PWM control is applied on the extra switch [33]. Thus, the circuit could always operate at f_r and output voltage could be regulated by duty cycle. Meanwhile, it can further widen voltage gain range by PWM+FM hybrid control.

In [9], one leg of diode bridge is replaced by synchronous switches. Fig. 21 shows the secondary-side circuit and critical operation mode. During this mode, large amount of energy is stored in L_r when the secondary side is shorted. The principle is similar to that in [45].

IV. COMPARISON AND SUMMARY OF MODIFICATION PRINCIPLES BASED ON DIFFERENT APPLICATIONS

In this section, Table I shows the comparisons of the various solutions for different applications. Modification, control strategies, and performance of various topologies are listed. Then, some circuit selection principles are summarized for different applications.

A. Hold-up operation

The hold-up operation rarely occurs and lasts for a short time duration. Thus, improving the efficiency during normal mode is more important. Typically, normal mode and extra hold-up compensation mode are designed. To design the hold-up compensation mode, less extra components and simpler control strategies are preferred. As shown in Table I, modification of the resonant tank is a popular method. The extra operation mode can be simply implemented by an extra resonant component and a switch for mode transition [1], [9], [10], [45]. Some topologies even realize automatic mode transition without switch control [2], [5].

Moreover, relatively low efficiency in hold-up compensation mode is acceptable. For example, the asymmetric PWM control in [47] results in a large current stress and transformer dc offset current, which increases core loss and is not preferable for high power cases [2], [5]. The solutions proposed in [9], [45] increase the current stress on primary-side components. However, the server power supply and telecommunication system usually have a relatively low rated power under 500 W [2], [10], [45]. Thus, the increased current stress is acceptable during the short hold-up operation.

B. Wide Input Voltage Applications

Solar energy system is a typical example of wide input voltage applications, which features a wide output voltage range and large current ripple [42], [54]. Therefore, high efficiency over the entire input voltage range is required, which is different from the hold-up operation.

Mode transition with extra resonant components is no longer an optimal choice, since frequent mode transitions damage the overall efficiency and slow down the transient response. Moreover, in some structures such as [10], [11], [45], the abrupt mode transition leads to current overshoot problem. However, applying the smooth mode transition technique can be a better solution, the solution proposed in [38] is an example.

Among the modifications reviewed in section III, most solutions are suitable for this application by regulating the RMS

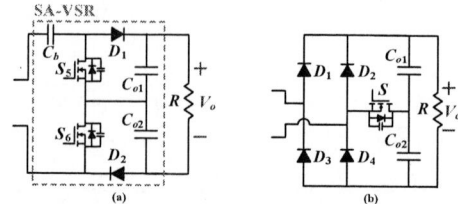

Fig. 20. Topology of modified secondary-side rectifier. (a) SA-VSR structure [52]. (b) Modified voltage doubler structure [31].

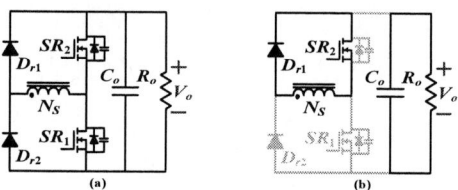

Fig. 21. (a) Modified secondary-side rectifier proposed in [9] (b) Its critical operation mode.

voltage of v_{ac} [23], [38]–[42]. These methods have three advantages: 1) the voltage variation in resonant tank is repressed, so the switching frequency range is narrow and may even be fixed at f_r, 2) with the squeezed frequency range, parameter design of resonant components is easier and 3) L_m can be larger so the primary-side circulating current is reduced.

Besides, for this application, that realized by cascading a buck-boost or an interleaved boost stage with LLC are also preferable methods [39], [42]. First, wide input voltage range is realized, meanwhile sudden mode transition is avoided. Secondly, the cascaded buck-boost or interleaved boost stage can effectively reduce input current ripple. However, these two methods apply PWM control on primary-side switches, which requires non-symmetric operation. This results in a dc offset current and leads to a larger size of transformer.

Generally, the structure modifications and control strategies for this application are usually more complex or contain more extra switches to realize the v_{ac} regulation.

C. Wide Output Voltage Applications

For wide output voltage applications, variable structure on secondary side are suitable than those on primary side. When the voltage gain is regulated on the secondary side, it's easier to optimize resonant parameters and to reduce circulating current. Besides, among the modifications on primary-side, regulating the RMS value of v_{ac} or the dc-link voltage are feasible [35], [48], [50], [51]. Resonant tank has a small variation, which is helpful to reduce circulating currents. The dc-link voltage regulation can also reduce the switching and conduction loss of LLC stage. However, it can't be applied to renewable energy applications such as PV or fuel cell systems.

Battery charger is a common example of this kind of application. The ripple-free output current is necessary for this case since it provides a high-quality charging current and reliable operation of battery management system. For example, in [48], the proposed FFPS technique provides a ripple-free charging current to eliminate burst mode oscillation. The control algorithm in dc-link voltage regulation methods can also suppress the output current ripple [50], [51].

TABLE I
COMPARISON OF MODIFICATIONS FOR DIFFERENT APPLICATIONS

MODIFICATION		CONTROL			PERFORMANCE		
STRUCTURE	TYPE & COMPLEXITY	MODULATION	OPERATION MODE	RATED POWER	VOLTAGE RANGE	RANGE OF SWITCHING FREQ.	RESONANT FREQ. f_r
Modifications for Hold-up Operation							
Auxiliary LC Circuit [5]	Resonant Tank, Simple	FM	-	56 V/350 W	330-390 V	60- ~110 kHz	100 kHz
LCLC Structure [2]	Resonant Tank, Simple	FM	-	12 V/500 W	250-400 V	135-250 kHz	250 kHz
Auxiliary Capacitor [1]	Resonant Tank, Simple	FM	Mode Transition	56 V/350 W	325-385 V	34-106 kHz	90 kHz
Auxiliary Switch [45]	Resonant Tank, Simple	FM + PWM	Hybrid Modulation + Mode Transition	12 V/300 W	250-400 V	150-260 kHz	260 kHz
Variable Inductance [10]	Resonant Tank, Medium	FM	Mode Transition	56 V/350 W	290-405 V	~75-110 kHz	110 kHz
Two secondary-side SR [9]	Secondary-side, Simple	FM + PS	Hybrid Modulation	200 V/300 W	250-400V	35-75 kHz	75 kHz
APWM Approach [47]	No modification on topology	FM + APWM	Mode Transition	18.5 V/85 W	300-400 V	61-100 kHz	100 kHz
Modifications for Wide Input Voltage Range							
FB/HB Switch [46]	No modification on topology	FM	Mode Transition	500 V/2 kW	125-550 V	40-120 kHz	60 kHz
Dual-Bridge LLC [41]	Primary-side, Medium	PWM	-	24 V/480 W	120-240 V	Fixed @ f_r	100 kHz
Buck-boost + LLC [39]	Primary-side, Medium	FM + APWM	Hybrid Modulation	15 V/300 W	36-72 V	Range: ~20 kHz	100 kHz
Interleaved Boost+LLC [42]	Primary-side, Medium	PWM	-	24 V/600 W	120-240 V	Fixed @ f_r	100 kHz
Three-level Bridge with VFX [23]	Primary-side, Complex	FM	Mode Transition	20 V/50 W	85-340 V	~240-~500 kHz	500 kHz
Three-level Full Bridge [40]	Primary-side, Complex	PWM + PS	Hybrid Modulation + Mode Transition	30-60 V/1 kW	240-480 V	Fixed @ f_r	100 kHz
Three-level Bridge [43]	Primary-side, Complex	FM + PS	Hybrid Modulation	48 V/1 kW	200-400 V	Around f_r	~43 kHz
Two Split Resonant Tank [38]	Resonant Tank, Complex	FM	Mode Transition by PWM	400 V/1 kW	80-200 V	80-160 kHz	140 kHz
Auxiliary Tx, Serial Connected [11]	Resonant Tank, Medium	FM	Mode Transition	210 V/250 W	22-65 V	80-140 kHz	140 kHz
Modifications for Wide Output Voltage Range							
FFPS Approach [48]	No modification on topology	FM + PS	Mode Transition	400 V/3 kW	120-180 V	100-200 kHz	122 kHz
Variable DC-link [50]	No modification on topology	FM	-	269-352 V/ 1 kW	320-420 V	Around f_r	200 kHz
Variable DC-link [51]	No modification on topology	FM	-	500-800 V/ 6.6 kW	250-420 V	Around f_r	500 kHz
Three-level Full-bridge [35]	Primary-side, Complex	PS + PWM	Hybrid Modulation + Mode Transition	385 V/6.6 kW	225-378 V	Fixed @90 kHz	78 kHz
Adaptive Turns Ratio [17]	Resonant Tank, Medium	FM	Mode Transition	311 V/300 W	25-42 V	~53-130 kHz	130 kHz
Interleaved LLC [44]	Two Resonant Tank, Complex	FM + PS	Mode Transition	400 V/3.5 kW	150-500 V	~45-100 kHz	100 kHz
SA-VSR [52]	Secondary-side, Medium	FM + PWM	Hybrid Modulation + Mode Transition	400 V/1.5 kW	100-500 V	70-150 kHz	100 kHz
Modified Voltage Doubler [31]	Secondary-side, Medium	FM	Mode Transition	400 V/1.5 kW	100-420 V	94.37-236.6 kHz	170 kHz
Modified Voltage Doubler [33]	Secondary-side, Medium	PWM	PWM+FM hybrid control is available	390 V/1 kW	250-420 V	Fixed @ f_r	100 kHz

V. CONCLUSIONS

In this paper, a comprehensive overview of LLC type topologies suitable for wide gain range applications are presented. The state-of-the-art wide gain LLC solutions based on conventional structure mainly aim to squeeze f_s range. The modifications are analyzed and classified into four categories based on circuit structure: modification on the resonant tank, primary-side switch network, control strategies and secondary-side rectifier. The first type usually adds extra resonant components. It features a simple control strategy. However, the parameter design of resonant components is challenging. The second type utilizes more extra switches on primary-side to realize v_{ac} regulation. Nonetheless, the control of switch patterns is complicated and some might cause extra switching losses. The third type modifications apply various control strategies without extra components, while they may introduce some problems like circuit imbalance and dc bias current through L_m. The last type distinguishes itself with variable rectifier structure; hence, the control and design is simple. However, mode transition problem, extra switches and conduction losses are introduced.

Besides, the comparison of reviewed topologies for different applications from modification, control, and performance is conducted. For hold-up operation, since it rarely occurs and time duration is short, introducing an extra compensation mode is preferred, which features high efficiency in normal mode and simple control [1], [2], [5], [10], [45], [47]. For wide input voltage applications, overall high efficiency is required, input current ripple should be suppressed, and sudden mode transition should be avoided. v_{ac} regulation is one of the suitable solution [23], [38]–[42]. For wide output voltage applications, overall high efficiency and restrained output current ripple are necessary. Variable-structure on secondary side is widely adopted for this application [31], [33], [52].

As described, the requirements of various applications are different. The review and comparison in this paper give a design reference for different LLC-based wide gain applications. Besides, it can bring some insights to the further improvements of LLC resonant converter.

ACKNOWLEDGEMENT

This work was supported in part by the National Natural Science Foundation of China under Grant 51607113, and in part by the Shanghai Sailing Program under Grant 16YF1407600.

REFERENCES

[1] J.-B. Lee, J.-K. Kim, J.-I. Baek, J.-H. Kim, and G.-W. Moon, "Resonant Capacitor On/Off Control of Half-Bridge LLC Converter for High-Efficiency Server Power Supply," *IEEE Trans. Ind. Electron.*, vol. 63, no. 9, pp. 5410–5415, Sep. 2016.

[2] Y. Chen, H. Wang, Z. Hu, Y.-F. Liu, J. Afsharian, and Z. A. Yang, "LCLC resonant converter for hold up mode operation," in *Proc. IEEE Energy Conversion Congress and Exposition (ECCE)*, Monteral, Canada, Spet. 2015, pp. 556–562.

[3] Y.-S. Lai, Z.-J. Su, and W.-S. Chen, "New Hybrid Control Technique to Improve Light Load Efficiency While Meeting the Hold-up Time Requirement for Two-Stage Server Power," *IEEE Trans. Power Electron.*, vol. 29, no. 9, pp. 4763–4775, Sep. 2014.

[4] Jung-Kyu Han, Jong-Woo Kim, Yujin Jang, Byunggu Kang, Jaewon Choi, and Gun-woo Moon, "Efficiency optimized asymmetric half-bridge converter with hold-up time compensation," in *Proc. IEEE 8th International Power Electronics and Motion Control Conference (IPEMC-ECCE Asia)*, Hefei, China, May 2016, pp. 2254–2261.

[5] D. Kim, S. Moon, C. Yeon, and G. Moon, "High Efficiency LLC Resonant Converter with High Voltage Gain Using Auxiliary LC Resonant Circuit," *IEEE Trans. Power Electron.*, vol. 31, no. October 2016, pp. 6901–6909, 2016.

[6] Y.-D. Kim, K. Cho, D.-Y. Kim, and G.-W. Moon, "Wide-Range ZVS Phase-Shift Full-Bridge Converter With Reduced Conduction Loss Caused by Circulating Current," *IEEE Trans. Power Electron.*, vol. 28, no. 7, pp. 3308–3316, Jul. 2013.

[7] In-Ho Cho, Kyu-Min Cho, Jong-Woo Kim, and Gun-Woo Moon, "A New Phase-Shifted Full-Bridge Converter With Maximum Duty Operation for Server Power System," *IEEE Trans. Power Electron.*, vol. 26, no. 12, pp. 3491–3500, Dec. 2011.

[8] In-Ho Cho, Young-Do Kim, and Gun-Woo Moon, "A Half-Bridge LLC Resonant Converter Adopting Boost PWM Control Scheme for Hold-Up State Operation," *IEEE Trans. Power Electron.*, vol. 29, no. 2, pp. 841–850, Feb. 2014.

[9] J.-W. Kim and G.-W. Moon, "A New LLC Series Resonant Converter with a Narrow Switching Frequency Variation and Reduced Conduction Losses," *IEEE Trans. Power Electron.*, vol. 29, no. 8, pp. 4278–4287, Aug. 2014.

[10] Y. Jeong, G. Moon, and J.-K. Kim, "Analysis on half-bridge LLC resonant converter by using variable inductance for high efficiency and power density server power supply," in *Proc. IEEE Applied Power Electronics Conference and Exposition (APEC)*, Tampa, FL, Mar. 2017, pp. 170–177.

[11] H. Hu, X. Fang, F. Chen, Z. J. Shen, and I. Batarseh, "A Modified High-Efficiency LLC Converter With Two Transformers for Wide Input-Voltage Range Applications," *IEEE Trans. Power Electron.*, vol. 28, no. 4, pp. 1946–1960, Apr. 2013.

[12] F. Edwin, W. Xiao, and V. Khadkikar, "Topology review of single phase grid-connected module integrated converters for PV applications," in *Proc. 38th Annual Conference on IEEE Industrial Electronics Society (IECON)*, Montreal, Canada, Oct. 2012, pp. 821–827.

[13] S.-M. Chen, T.-J. Liang, L.-S. Yang, and J.-F. Chen, "A Safety Enhanced, High Step-Up DC-DC Converter for AC Photovoltaic Module Application," *IEEE Trans. Power Electron.*, vol. 27, no. 4, pp. 1809–1817, Apr. 2012.

[14] J.-H. Kim, I.-O. Lee, and G.-W. Moon, "Integrated Dual Full-Bridge Converter With Current-Doubler Rectifier for EV Charger," *IEEE Trans. Power Electron.*, vol. 31, no. 2, pp. 942–951, Feb. 2016.

[15] F. Musavi, M. Craciun, D. S. Gautam, W. Eberle, and W. G. Dunford, "An LLC Resonant DC-DC Converter for Wide Output Voltage Range Battery Charging Applications," *IEEE Trans. Power Electron.*, vol. 28, no. 12, pp. 5437–5445, Dec. 2013.

[16] I.-O. Lee, "Hybrid PWM-Resonant Converter for Electric Vehicle On-Board Battery Chargers," *IEEE Trans. Power Electron.*, vol. 31, no. 5, pp. 3639–3649, May 2016.

[17] H.-G. Han, Y.-J. Choi, S.-Y. Choi, and R.-Y. Kim, "A High Efficiency LLC Resonant Converter with Wide Ranged Output Voltage Using Adaptive Turn Ratio Scheme for a Li-Ion Battery Charger," in *Proc. IEEE Vehicle Power and Propulsion Conference (VPPC)*, Zhejiang, China, Oct. 2016, pp. 1–6.

[18] C. Cecati, H. A. Khalid, M. Tinari, G. Adinolfi, and G. Graditi, "DC nanogrid for renewable sources with modular DC/DC LLC converter building block," *IET Power Electron.*, vol. 10, no. 5, pp. 536–544, Apr. 2017.

[19] C.-O. Yeon, J.-W. Kim, M.-H. Park, I.-O. Lee, and G.-W. Moon, "Improving the Light-Load Regulation Capability of LLC Series Resonant Converter Using Impedance Analysis," *IEEE Trans. Power Electron.*, vol. 32, no. 9, pp. 7056–7067, Sep. 2017.

[20] U. Kundu and P. Sensarma, "A Unified Approach for Automatic Resonant Frequency Tracking in LLC DC–DC Converter," *IEEE Trans. Ind. Electron.*, vol. 64, no. 12, pp. 9311–9321, Dec. 2017.

[21] Z. Fang, J. Wang, S. Duan, K. Liu, and T. Cai, "Control of an LLC Resonant Converter Using Load Feedback Linearization," *IEEE Trans. Power Electron.*, vol. 33, no. 1, pp. 887–898, Jan. 2018.

[22] C. Fei, F. C. Lee, and Q. Li, "High-Efficiency High-Power-Density LLC Converter With an Integrated Planar Matrix Transformer for High-Output Current Applications," *IEEE Trans. Ind. Electron.*, vol. 64, no. 11, pp. 9072–9082, Nov. 2017.

[23] W. Inam, K. K. Afridi, and D. J. Perreault, "Variable Frequency Multiplier Technique for High-Efficiency Conversion Over a Wide Operating Range," *IEEE J. Emerg. Sel. Top. Power Electron.*, vol. 4, no. 2, pp. 335–343, Jun. 2016.

[24] Z. Fang, T. Cai, S. Duan, and C. Chen, "Optimal Design Methodology for LLC Resonant Converter in Battery Charging Applications Based on Time-Weighted Average Efficiency," *IEEE Trans. Power Electron.*, vol. 30, no. 10, pp. 5469–5483, Oct. 2015.

[25] Junjun Deng, Siqi Li, Sideng Hu, C. C. Mi, and Ruiqing Ma, "Design

Methodology of LLC Resonant Converters for Electric Vehicle Battery Chargers," *IEEE Trans. Veh. Technol.*, vol. 63, no. 4, pp. 1581–1592, May 2014.

[26] R. Beiranvand, B. Rashidian, M. R. Zolghadri, and S. M. Hossein Alavi, "A Design Procedure for Optimizing the LLC Resonant Converter as a Wide Output Range Voltage Source," *IEEE Trans. Power Electron.*, vol. 27, no. 8, pp. 3749–3763, Aug. 2012.

[27] G. Yang, P. Dubus, and D. Sadarnac, "Double-Phase High-Efficiency, Wide Load Range High- Voltage/Low-Voltage LLC DC/DC Converter for Electric/Hybrid Vehicles," *IEEE Trans. Power Electron.*, vol. 30, no. 4, pp. 1876–1886, Apr. 2015.

[28] Yiqing Ye, Chao Yan, Jianhong Zeng, and Jianping Ying, "A novel light load solution for LLC series resonant converter," in *Proc. 29th International Telecommunications Energy Conference (INTELEC)*, Rome, Italy, Oct. 2007, no. 1, pp. 61–65.

[29] B.-H. Lee, M.-Y. Kim, C.-E. Kim, K.-B. Park, and G.-W. Moon, "Analysis of LLC Resonant Converter considering effects of parasitic components," in *Proc. 31st International Telecommunications Energy Conference (INTELEC)*, Incheon, South Korea, Oct. 2009, vol. 32, no. 9, pp. 1–6.

[30] J.-H. Kim, C.-E. Kim, J.-K. Kim, and G.-W. Moon, "Analysis for LLC resonant converter considering parasitic components at very light load condition," in *Proc. 8th International Conference on Power Electronics (ECCE Asia)*, Jeju, South Korea, May 2011, pp. 1863–1868.

[31] M. I. Shahzad, S. Iqbal, and S. Taib, "A Wide Output Range HB-2LLC Resonant Converter With Hybrid Rectifier for PEV Battery Charging," *IEEE Trans. Transp. Electrif.*, vol. 3, no. 2, pp. 520–531, Jun. 2017.

[32] M. Shang and H. Wang, "A LLC type resonant converter based on PWM voltage quadrupler rectifier with wide output voltage," in *Proc. IEEE Applied Power Electronics Conference and Exposition (APEC)*, Tampa, FL, Mar. 2017, pp. 1720–1726.

[33] H. Wang and Z. Li, "A PWM LLC Type Resonant Converter Adapted to Wide Output Range in PEV Charging Applications," *IEEE Trans. Power Electron.*, vol. 33, no. 5, pp. 3791–3801, May 2018.

[34] C.-C. Hua, Y.-H. Fang, and C.-W. Lin, "LLC resonant converter for electric vehicle battery chargers," *IET Power Electron.*, vol. 9, no. 12, pp. 2369–2376, Oct. 2016.

[35] H. Haga and F. Kurokawa, "Modulation Method of a Full-Bridge Three-Level LLC Resonant Converter for Battery Charger of Electrical Vehicles," *IEEE Trans. Power Electron.*, vol. 32, no. 4, pp. 2498–2507, Apr. 2017.

[36] Y. Shen, H. Wang, Z. Qin, F. Blaabjerg, and A. Al Durra, "A reconfigurable series resonant DC-DC converter for wide-input and wide-output voltages," in *Proc. IEEE Applied Power Electronics Conference and Exposition (APEC)*, Tampa, FL, Mar. 2017, pp. 343–349.

[37] C.-E. Kim, J.-I. Baek, and J. Lee, "High-Efficiency Single-Stage LLC Resonant Converter for Wide-Input-Voltage Range," *IEEE Trans. Power Electron.*, vol. 8993, no. c, pp. 1–1, 2017.

[38] W. Sun, Y. Xing, H. Wu, and J. Ding, "Modified High-efficiency LLC Converters with Two Split Resonant Branches for Wide Input-Voltage Range Applications," *IEEE Trans. Power Electron.*, vol. 8993, no. c, pp. 1–1, 2017.

[39] Y. Jeong, J.-K. Kim, J.-B. Lee, and G.-W. Moon, "An Asymmetric Half-Bridge Resonant Converter Having a Reduced Conduction Loss for DC/DC Power Applications With a Wide Range of Low Input Voltage," *IEEE Trans. Power Electron.*, vol. 32, no. 10, pp. 7795–7804, Oct. 2017.

[40] T. Jiang, J. Zhang, X. Wu, K. Sheng, and Y. Wang, "A bidirectional three-level LLC resonant converter with PWAM control," *IEEE Trans. Power Electron.*, vol. 31, no. 3, pp. 2213–2225, 2016.

[41] X. Sun, X. Li, Y. Shen, B. Wang, and X. Guo, "Dual-Bridge LLC Resonant Converter With Fixed-Frequency PWM Control for Wide Input Applications," *IEEE Trans. Power Electron.*, vol. 32, no. 1, pp. 69–80, Jan. 2017.

[42] X. Sun, Y. Shen, Y. Zhu, and X. Guo, "Interleaved Boost-Integrated LLC Resonant Converter With Fixed-Frequency PWM Control for Renewable Energy Generation Applications," *IEEE Trans. Power Electron.*, vol. 30, no. 8, pp. 4312–4326, Aug. 2015.

[43] S. M. S. I. Shakib and S. Mekhilef, "A Frequency Adaptive Phase Shift Modulation Control Based LLC Series Resonant Converter for Wide Input Voltage Applications," *IEEE Trans. Power Electron.*, vol. 32, no. 11, pp. 8360–8370, 2017.

[44] H. Wu, X. Zhan, and Y. Xing, "Interleaved LLC Resonant Converter With Hybrid Rectifier and Variable-Frequency Plus Phase-Shift Control for Wide Output Voltage Range Applications," *IEEE Trans. Power Electron.*, vol. 32, no. 6, pp. 4246–4257, Jun. 2017.

[45] H. Wang, Y. Chen, P. Fang, Y.-F. Liu, J. Afsharian, and Z. Yang, "An LLC Converter Family With Auxiliary Switch for Hold-Up Mode Operation," *IEEE Trans. Power Electron.*, vol. 32, no. 6, pp. 4291–4306, Jun. 2017.

[46] Z. Liang, R. Guo, G. Wang, and A. Huang, "A new wide input range high efficiency photovoltaic inverter," in *Proc. IEEE Energy Conversion Congress and Exposition (ECCE)*, Atlanta, GA, Spet. 2010, pp. 2937–2943.

[47] B.-C. Kim, K.-B. Park, and G.-W. Moon, "Asymmetric PWM Control Scheme During Hold-Up Time for LLC Resonant Converter," *IEEE Trans. Ind. Electron.*, vol. 59, no. 7, pp. 2992–2997, Jul. 2012.

[48] N. Shafiei, M. Ordonez, M. Craciun, C. Botting, and M. Edington, "Burst Mode Elimination in High-Power LLC Resonant Battery Charger for Electric Vehicles," *IEEE Trans. Power Electron.*, vol. 31, no. 2, pp. 1173–1188, Feb. 2016.

[49] B.-C. Kim, K.-B. Park, C.-E. Kim, B.-H. Lee, and G.-W. Moon, "LLC Resonant Converter With Adaptive Link-Voltage Variation for a High-Power-Density Adapter," *IEEE Trans. Power Electron.*, vol. 25, no. 9, pp. 2248–2252, Sep. 2010.

[50] H. Wang, S. Dusmez, and A. Khaligh, "Maximum Efficiency Point Tracking Technique for LLC-Based PEV Chargers Through Variable DC Link Control," *IEEE Trans. Ind. Electron.*, vol. 61, no. 11, pp. 6041–6049, Nov. 2014.

[51] B. Li, F. C. Lee, Q. Li, and Z. Liu, "Bi-directional on-board charger architecture and control for achieving ultra-high efficiency with wide battery voltage range," in *Proc. IEEE Applied Power Electronics Conference and Exposition (APEC)*, Tampa, FL, Mar. 2017, pp. 3688–3694.

[52] H. Wu, Y. Li, and Y. Xing, "LLC Resonant Converter With Semiactive Variable-Structure Rectifier (SA-VSR) for Wide Output Voltage Range Application," *IEEE Trans. Power Electron.*, vol. 31, no. 5, pp. 3389–3394, May 2016.

[53] Yilei Gu, Lijun Hang, and Zhengyu Lu, "A Flexible Converter With Two Selectable Topologies," *IEEE Trans. Ind. Electron.*, vol. 56, no. 12, pp. 4854–4861, Dec. 2009.

[54] U. R. Prasanna and A. K. Rathore, "Analysis, Design, and Experimental Results of a Novel Soft-Switching Snubberless Current-Fed Half-Bridge Front-End Converter-Based PV Inverter," *IEEE Trans. Power Electron.*, vol. 28, no. 7, pp. 3219–3230, Jul. 2013.

The 2018 International Power Electronics Conference

Dual Half-Bridge *LLC* Resonant Converter with Hybrid-Secondary-Rectifier (HSR) for Wide-Ouput-Voltage Applications

Jae-Il Baek[1*], Chong-Eun Kim[2], Keon-Woo Kim[1], Min-Su Lee[1], and Gun-Woo Moon[1]
1 School of Electrical Engineering, KAIST, Daejeon, Republic of Korea
2 Power R&D Team, SoluM, Youngin-si, Republic of Korea
*E-mail: dpi1067@kaist.ac.kr

Abstract— In this paper, a dual half-bridge (HB) *LLC* resonant converter with a new hybrid secondary rectifier (HSR) is proposed for wide-output-voltage applications. In the proposed converter, the HSR can operate like a voltage doubler rectifier or voltage quadrupler rectifier according to the switching strategy of the dual HB *LLC* resonant converter. For low output voltage range, the dual HB *LLC* resonant converter are controlled by 180 degree phase-shifted gate signals. Thus, it can achieve low voltage gain with a voltage doubler rectifier. Whereas, in high output voltage range, in-phase gate signals are applied to the dual HB *LLC* resonant converter. Thus, the proposed converter can obtain high voltage gain with a voltage quadrupler rectifier. As a result, the proposed converter is able to be designed with narrow switching frequency variation, which results in high efficiency over the entire output voltage range. The validity of the proposed converter is confirmed by a 750W prototype with 400V input, 100-300V output.

Keywords— *LLC resonant converter, secondary rectifier, voltage doubler, voltage quadrupler, wide-output-voltage range.*

I. INTRODUCTION

Recently, *LLC* resonant converters have been widely researched in many applications due to their several advantages; 1) simple control and structure, 2) zero voltage switching (ZVS) for primary MOSFETs and zero current switching (ZCS) for secondary diodes, 3) reduced electromagnetic interferences [1]-[6]. However, in wide-output-voltage applications such as battery charger and LED driver, etc., they have critical challenges [7]-[16].

Typically, when the *LLC* resonant converter is designed under narrow-output-voltage range, it can operate around the resonant frequency. Thus, it can easily achieve high efficiency with small conduction, switching, and core losses. However, the *LLC* resonant converter has degraded efficiency in wide-output-voltage applications. This is because its operating frequency has to swing in a wide range and deviates from the resonant frequency. For example, provided that the operating frequency of the *LLC* resonant converter is higher than the resonant frequency, i.e., above region, the *LLC* resonant converter suffers from large primary turn-off switching loss and reverse recovery problems caused by the secondary diode. In the opposite case, i.e., below region, the conversion efficiency is degraded due to large magnetic core and conduction losses. Therefore, the *LLC* resonant converter has challenges to be adopted in the wide-output-voltage applications [7]-[16].

As aforementioned reasons, many approaches have been researched to achieve high efficiency *LLC* resonant converter under wide-output-voltage range [12]-[16]. First, in [12]-[13], improved design methodologies are presented to optimize the *LLC* resonant converter. Thus, they can improve the efficiency and power density by drawing full capabilities of the *LLC* resonant converter. However, they still have low efficiency once the switching frequency is far away from the resonant frequency. Next, from the viewpoint of the topology improvement, several improved *LLC* resonant converters are proposed in [14]-[16]. In [14], semiactive variable-structure rectifier (SA-VSR) is proposed for the *LLC* resonant converter. Thus, the converter proposed in [14] can reduce the switching frequency variation because the SA-VSR can operate as the voltage doubler rectifier or voltage quadrupler rectifier according to the output voltage range. However, it requires two extra switches as well as pulse-frequency-modulation (PFM) and pulse-width-modulation (PWM) controls for the SA-VSR. Moreover, high turn-off switching loss and reverse recovery problems are caused by two extra switches, which results in low efficiency. To reduce the number of additional switch, in [15], a hybrid rectifier is proposed for the HB-2LLC resonant converter. In this converter, the hybrid rectifier can operate like full bridge rectifier or voltage doubler rectifier according to ON/OFF operation of one relay. Thus, it can be also designed with narrow switching frequency variation. However, it still requires one relay resulting in large volume and complex control. Moreover, it has low utilization of two rectifier diodes in voltage doubler rectifier mode. Finally, a modified PWM *LLC* type resonant converter is proposed in [16]. In this converter, since the output voltage is regulated by modulating the duty cycle of a secondary side additional switch, the switching frequency of the primary switches can be constant. Thus, it can achieve high efficiency. However, the secondary switch should be driven with the floating gate driver. In addition, it suffers from high turn-off switching loss. Above all, it still requires wide switching frequency variation in the applications requiring more than twice output voltage range.

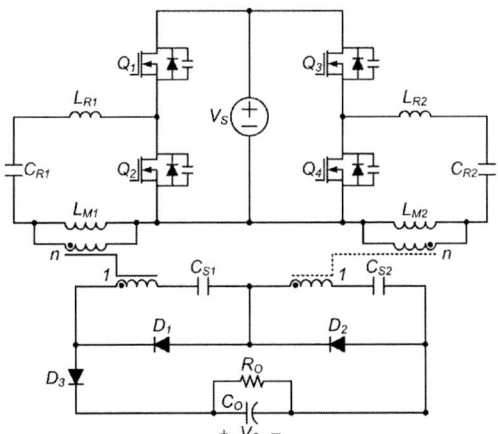

Fig. 1. Circuit diagram of proposed converter.

To relieve the drawback of previous research, a new hybrid-secondary-rectifier (HSR) and primary switching strategy are proposed for the dual half-bridge (HB) *LLC* resonant converter to achieve high efficiency in wide-output-voltage range. In low output voltage range, the HSR can operate like the voltage doubler rectifier by adopting 180 degree phase-shifted gate signals to the dual HB *LLC* resonant converter. In high output voltage range, the HSR is able to be the voltage quadrupler rectifier with in-phase gate signals. Therefore, the proposed converter can extend the output voltage range without additional switch by adaptively changing the operation mode of the HSR. As a result, the proposed converter can be designed with narrow switching frequency variation.

II. DESCRIPTION OF PROPOSED CONVERTER

A. Topology of Proposed Converter

Fig. 1 shows the circuit diagram of the proposed converter which is consist of the dual HB *LLC* resonant converter and hybrid-secondary-rectifier (HSR). From this figure, the dual HB *LLC* resonant converter is composed of two parallel-connected HB *LLC* resonant converters. In case of the secondary side, the HSR has no active devices and it is composed of two capacitors (C_{S1} and C_{S2}), three diodes (D_1, D_2, and D_3). According to the switching strategy of the dual HB *LLC* resonant converter, two capacitors and output voltage (V_O) can be differently connected. Provided that one of two capacitors is only connected with V_O, the voltage across the capacitor (V_{CS1} and V_{CS2}) is half of V_O, which is the operation of the voltage doubler rectifier. On the other hand, if both two capacitors are connected with V_O in series, V_{CS1} and V_{CS2} are one fourth of V_O. As a result, the HSR operates like the voltage quadrupler rectifier

B. Operational Principles of Proposed Converter

In the proposed converter, the frequency control is basically used to regulate the output voltage like the conventional *LLC* resonant converter. It means that all primary switches (Q_1-Q_4) have the constant duty cycle of 0.5. Moreover, since the operational principles of the proposed converter are similar to the conventional *LLC*

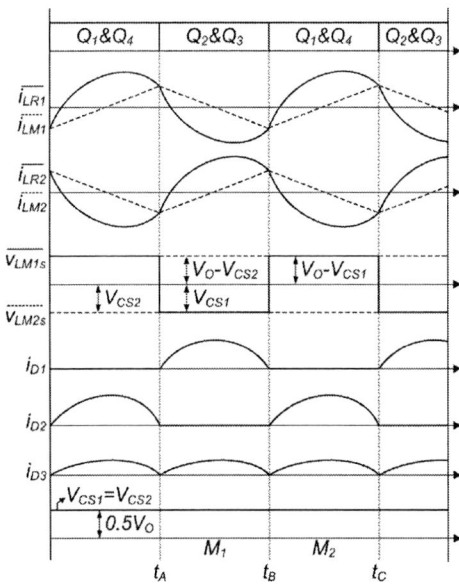

Fig. 2. Simplified waveforms of voltage doubler rectifier mode.

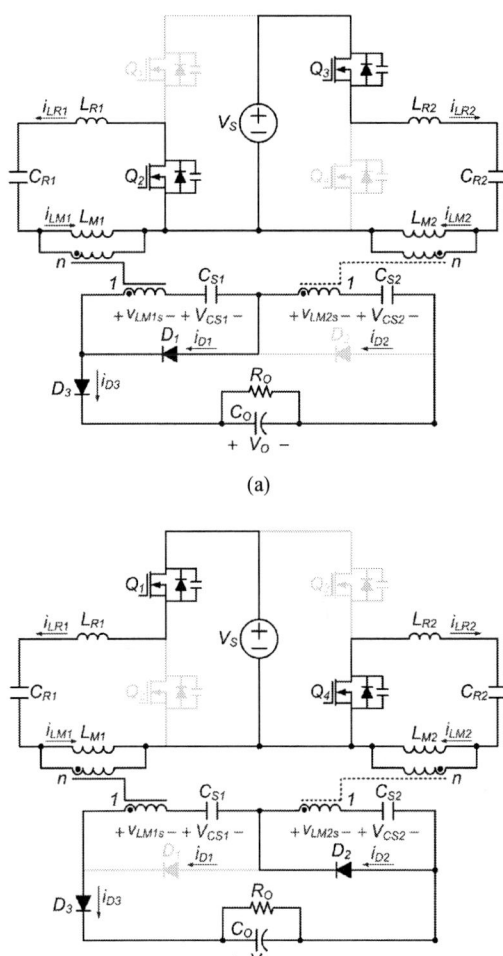

(a)

(b)

Fig. 3. Operating circuit of voltage doubler rectifier mode. (a) Mode 1 (t_A-t_B). (b) Mode 2 (t_B-t_C).

109

resonant converter [17]-[18] except for the operation of the HSR. Thus, this part focuses on the operation of the HSR modes; 1) voltage doubler rectifier mode, 2) voltage quadrupler rectifier mode.

To perform the mode analysis, several assumptions are made as follows; 1) V_{CS1}, V_{CS2}, and V_O are constant, 2) dead time between primary gate-signals is negligibly short, 3) all parasitic components are ignored except for those specified in Fig. 1, 4) all components of dual half-bridge converters are matched ($L_{R1}=L_{R2}$, $C_{R1}=C_{R2}$, $L_{M1}=L_{M2}$, $C_{S1}=C_{S2}$).

1)Voltage Doubler Rectifier Mode:

In low output voltage range, primary switches of the dual HB *LLC* resonant converter are driven in the interleaved fashion (with 180 degree phase shift), as shown in Fig. 2. There are two simplified stages (t_A-t_B, t_B-t_C) in one switching cycle. Fig. 3 shows topological states of the proposed converter.

Mode 1 [t_A-t_B, Fig. 3(a)]: At time t_A, Q_1 and Q_4 are turned off, and Q_2 and Q_3 are turned-on. In this mode, the power is transferred from the resonance between L_{R2} and C_{R2} to V_O through C_{S2}, D_1, and D_3. On the other hand, the other resonant current of L_{R1} and C_{R1} charges C_{S1}, as shown in Fig. 3(a). Moreover, the voltages applied on the secondary windings of the transformers (v_{LM1s} and v_{LM2}) are can be expressed as follows:

$$v_{LM1s} = -V_{CS1}, \tag{1}$$
$$v_{LM2s} = V_O - V_{CS2}. \tag{2}$$

Mode 2 [t_B-t_C, Fig. 3(b)]: At time t_B, Q_2 and Q_3 are turned off, and Q_1 and Q_4 are turned-on. This mode is similar to Mode 1 except direction of secondary resonant currents. Thus, C_{S2} is charged, and the stored energy of C_{S1} is delivered to V_O with the transferred energy from the resonance between L_{R1} and C_{R1}. As a result, the v_{LM1s} and v_{LM2s} are given by

$$v_{LM1s} = V_O - V_{CS1}, \tag{3}$$
$$v_{LM2s} = -V_{CS2}. \tag{4}$$

2) Voltage Quadrupler Rectifier Mode:

In high output voltage range, in-phase gate signals are applied to the primary switches in the dual HB *LLC* resonant converter. Fig. 4 shows the key waveforms of voltage quadrupler mode in the proposed converter. From this figure, there are also two simplified stages (t_A-t_B, t_B-t_C) in one switching cycle. Fig. 5 shows topological states of the proposed converter.

Mode 1 [t_A-t_B, Fig. 5(a)]: At time t_A, Q_1 and Q_3 are turned off, and Q_2 and Q_4 are turned on. Unlike the voltage doubler rectifier mode, since the directions of secondary resonant currents are negative, the transferred energy from the primary side is used to charge C_{S1} and C_{S2}, as shown in Fig. 5(a). Thus, v_{LM1s} and v_{LM2s} can be expressed as like(4) and (5), respectively. Moreover, the voltage stress of D_3 is V_O.

$$v_{LM1s} = -V_{CS1}, \tag{5}$$
$$v_{LM2s} = -V_{CS2}. \tag{6}$$

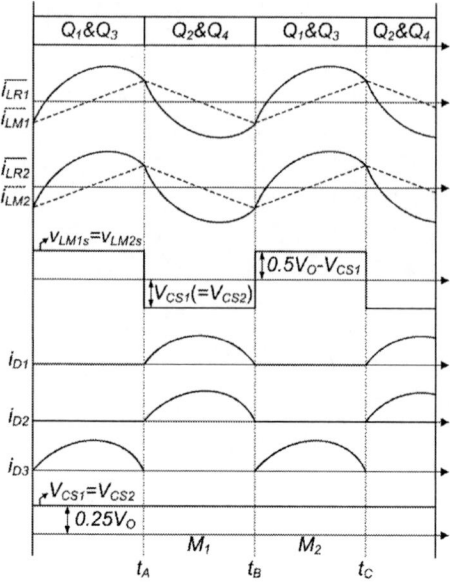

Fig. 4. Simplified waveforms of voltage quadrupler rectifier mode.

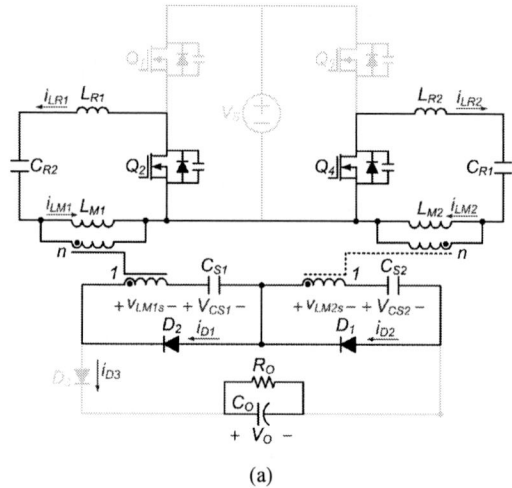

(a)

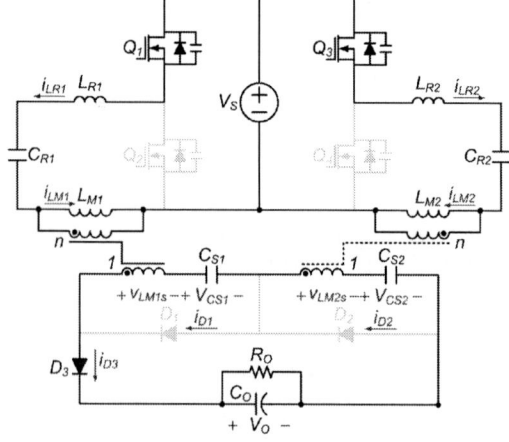

(b)

Fig. 5. Operating circuit of voltage quadrupler rectifier mode. (a) Mode 1(t_A-t_B). (b) Mode 2(t_B- t_C).

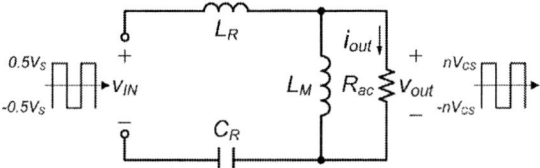

Fig. 6. AC equivalent circuit of the proposed converter.

Mode 2 [t_B-t_C, Fig. 5(b)]: At time t_B, Q_2 and Q_4 are turned off, and Q_1 and Q_3 are turned on. During this mode, the energy stored in C_{S1} and C_{S2} are delivered to V_O with the transferred energy from the primary side. Thus, assuming the impedance of each secondary windings are same, v_{LM1s} and v_{LM2s} can be expressed as follows:

$$v_{LM1s} = 0.5V_O - V_{CS1}, \tag{7}$$

$$v_{LM2s} = 0.5V_O - V_{CS2}. \tag{8}$$

From Fig. 5(b), since D_1 and D_2 are connected in series, the voltage stresses of D_1 and D_2 are $0.5V_O$, which is half of the voltage stress of the conventional LLC resonant converter with voltage doubler.

III. VOLTAGE GAIN OF PROPOSED CONVERER

In order to obtain the voltage gain of the proposed converter, the fundamental harmonic approximation (FHA) analysis is adopted. Thus, the transfer power from the input source to output load is based on the fundamental switching frequency, and the harmonics of the switching frequency are neglected in this chapter.

In the proposed converter, since all components of dual half-bridge converter are matched, the voltage gain of the proposed converter can be derived by considering only one half-bridge converter. Therefore, Fig. 6 shows the AC equivalent circuit of one half-bridge converter. From this figure and FHA analysis, V_S to nV_{CS} conversion ratio (M_{LLC}) can be expressed as follows:

$$
\frac{V_{CS}}{V_S} = M_{LLC} = \frac{(L_M \parallel R_{ac})}{2n\left[sL_R + \dfrac{1}{sC_R} + (L_M \parallel R_{ac})\right]}
$$
$$
= \frac{k}{2n\sqrt{\left(1 + k - f_N^{-2}\right)^2 + k^2 Q^2 (f_N - f_N^{-1})^2}}, \tag{9}
$$

where $k = L_M/L_R$, $Q = (L_R/C_R)^{0.5}/R_{ac}$, $f_N = f_S/f_R$.

From (7), to derive the voltage gain of the proposed converter (V_O/V_S), V_{CS} should be expressed as V_O, which can be obtained by applying the volt-second balance of the transformer. In case of the voltage doubler mode, according to (1), (2), (3), and (4), the volt-second balance of two transformers can be expressed as follows:

$$V_O - V_{CS1} - V_{CS1} = 0, \tag{10}$$

$$V_O - V_{CS2} - V_{CS2} = 0. \tag{11}$$

Thus, as mentioned previously, V_{CS1} and V_{CS2} are $0.5V_O$ like the voltage doubler rectifier. Similarly, when the HSR operates like the voltage quadrupler mode, according to (5)-(8), V_{C1} and V_{CS2} can be calculated as $0.25V_O$. Therefore, the voltage gain of the proposed converter can be derived as follows:

TABLE I
DESIGNED PARAMETERS OF PROTOTYPES

Items	Conventional Converter (FB LLC with VDR)	Proposed LLC Converter
Primary switches (Q_1-Q_4)	IPA60R280C6 (650V, 280mΩ)	
Secondary diodes	MUR1540G (V_{RM}=400V, V_F=0.9V)	D_1 & D_2 : MBR40250TG (V_{RM}=250V, V_F=0.6V) D_3 : MUR1540G (V_{RM}=400V, V_F=0.9V)
Transformer	Core : PQ3535 (V_e : 43327mm³) L_M=1.2mH, L_{lkg}=27µH N_P : N_S = 104:13 (n=8)	Core : PQ3220*2EA (V_e : 20450mm³*2EA) L_{M1}=404.0µH, L_{lkg1}=6.7µH, L_{M2}=404.6µH, L_{lkg2}=7.1µH, N_P : N_S = 48:12 (n=4)
Resonant inductor	Core : CM229060 (V_e : 1877mm³) L_R : 122µH, 53T	Core : CM229060*2EA (V_e : 960mm³*2EA) L_{RI} : 107.4µH, 50T, L_{RI} : 109.3µH, 50T
Resonant capacitor	C_R : 16.5nF (V_e : 2970mm³)	C_{R1} : 20nF, C_{R2} : 20nF (V_e : 687.5mm³*2EA)

$$
M_{pro} = \frac{V_O}{V_S} = \left(\begin{array}{l} 2M_{LLC}, Voltage\,Doubler \\ 4M_{LLC}, Voltage\,Quadrupler \end{array} \right), \tag{10}
$$

where R_{ac} is $4n^2R_O/\pi^2$ and n^2R_O/π^2 at voltage doubler mode and voltage quadrupler mode respectively, and R_O is the output resistor.

IV. EXPERIMENTAL RESULTS

To confirm the validity of the proposed converter, a 750W prototype with the specifications of 400V input, 100-300V/2.5A output, and 100kHz resonant frequency has been built and tested. Moreover, for the comparison, the prototype of the conventional full-bridge (FB) LLC converter with voltage doubler rectifier is also implemented. The designed parameters are presented in Table I. From this table, although the proposed converter should use two transformers and resonant inductors, since the voltage and current stresses are half of the conventional converter, the size of transformer and resonant inductor is nearly half of the conventional converter. Moreover, the size of resonant capacitor is reduced. Furthermore, the proposed converter can employ the schottky diode which generally has better performance compared to the fast recovery diode for D_1 and D_2 because the voltage stress of D_1 and D_2 are half of that of the conventional converter.

Fig 7 shows the experimental key waveforms of the proposed converter at 100V, 200V, and 300V output voltage. From Fig. 7(a), when V_O is 100V, the HSR of the proposed converter operates like the voltage doubler rectifier. Thus, primary currents (i_{LR1} and i_{LR2}) of the proposed converter has 180 degree phase shift. Next, from Figs. 7(b) and 7(c), since in-phase gate signals are adopted to the proposed converter, the HSR operates like the voltage quadrupler rectifier. Therefore, the proposed converter can operate at resonant frequency despite of V_O=200V. Moreover, to regulate V_O as 300V, the

The 2018 International Power Electronics Conference

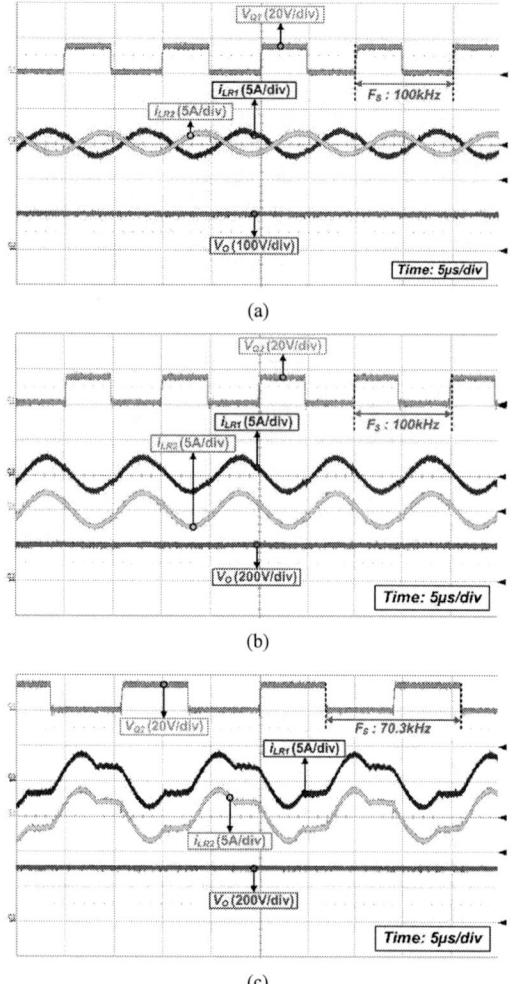

(a)

(b)

(c)

Fig. 7. Experimental key waveforms of proposed converter. (a) Voltage doubler mode at 100V output voltage. (b) Voltage quadrupler mode at 200V output voltage. (c) Voltage quadrupler mode at 300V output voltage.

switching frequency of the proposed converter is 70.3kHz, which is much higher than that of the conventional converter (40kHz). Therefore, the proposed converter can achieve narrow switching frequency range compared to the conventional FB LLC converter.

Fig. 8 shows the measured efficiency of the conventional and proposed converters. The efficiency was measured with a power analyzer, i.e., Yokogawa WT1600. From this figure, as the output voltage increase, the conventional converter has lower efficiency because its switching frequency is further away from the resonant frequency. On the other hand, the proposed converter has higher efficiency over the entire output voltage range compared to the conventional converter. Under 175V output voltage, the voltage doubler mode of the proposed converter can achieve high efficiency because it has low conduction loss and core loss. Over 175V output voltage, the quadrupler mode of the proposed converter can achieve high efficiency by operation near resonant switching frequency. Thus, it is prefer to select the

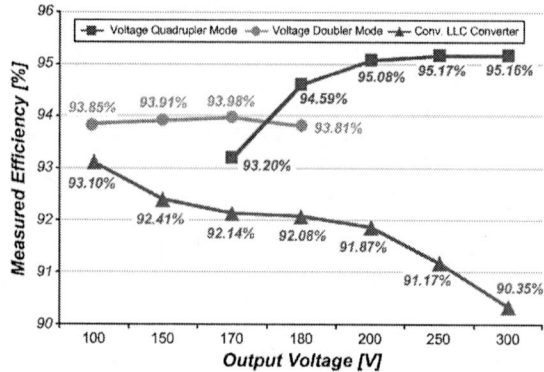

Fig. 8. Measured efficiency.

threshold voltage as 175V.

V. CONCLUSION

In this paper, a dual half-bridge (HB) *LLC* resonant converter with a new hybrid secondary rectifier (HSR) is proposed. In the proposed converter, without additional switch, it can achieve high efficiency under wide-output-voltage range only by changing the primary switching strategy. For example, in low output voltage range, the HSR can operate as the voltage doubler rectifier due to interleaved primary gate signals (180 degree phase shift). On the other hand, in high output voltage range, by adopting in-phase gate signals, the HSR can be used as the voltage quadrupler rectifier. Moreover, all semiconductor devices in the proposed converter can achieve the soft switching operation. As a result, the proposed converter can achieve high efficiency over the entire output voltage range due to narrow switching frequency variation, which results in small magnetic core, switching, and conductions losses. Experimental results on a 750W prototype with 400V input and 100-300V output verify the effectiveness and feasibility of the proposed converter.

ACKNOWLEDGMENT

This work was supported by the National Research Foundation of Kore (NRF) grant funded by the Korea government (MSIP) (No. 2016R1A2B2010328).

REFERENCES

[1] W. Li, Q. Luo, Y. Mei, S. Zong, X. He, and C. Xia, "Flying-Capacitor-Based Hybrid LLC Converters With Input Voltage Autobalance Ability for High Voltage Applications," *IEEE Trans. Power Electron.*, vol. 31, no. 3, pp. 1908–1920, Mar. 2016.

[2] H. Nguyen, R. Zane, and D. Maksimovic, "ON/OFF Control of a Modular DC-DC Converter Based on Active-Clamp LLC Modules," *IEEE Trans. Power Electron.*, vol. 30, no. 7, pp. 3748-3760, Jul. 2015.

[3] X. Sun, Y. Shen, Y. Zhu, and X. Guo, "Interleaved Boost-Integrated *LLC* Resonant Converter With Fixed-Frequency PWM Controls for Renewable Energy Generation Applications," *IEEE Trans. Power Electron.*, vol. 30, no. 8, pp. 4312-4326, Aug. 2015.

[4] D. K. Kim, S. C. Moon, C. O. Yeon, and G. W. Moon, "High-Efficiency *LLC* Resonant Converter With High Voltage Gain Using an Auxiliary *LC* Resonant Circuit," *IEEE Trans. Power Electron.*, vol. 31, no. 10, pp. 6901–6909, Oct. 2016.

[5] J. B. Lee, J. K. Kim, J. I. Baek, J. H. Kim, and G. W. Moon, "Resonant Capacitor On/Off Control of Half-Bridge *LLC*

112

Converter for High-Efficiency Server Power Supply," *IEEE Trans. Ind. Electron.*, vol. 63, no. 9, pp. 5410–5415, Sep. 2016.

[6] J. I. Baek, J. K. Kim, J. B. Lee, H. S. Youn, and G. W. Moon, "A Boost PFC Stage Utilized as Half-Bridge Converter for High-Efficiency DC-DC Stage in Power Supply Unit ," *IEEE Trans. Power Electron.*, vol. 32, no. 10, pp. 7449–7457, Oct. 2017.

[7] X. Chen, D. Huang, Q. Li, and F. C. Lee, "Multichannel LED Driver With *CLL* Resonant Converter," *IEEE J. Emerg. Sel. Topics Power Electron.*, vol. 3, no. 3, pp.589-598, Sep. 2015.

[8] J. I. Baek, J. K. Kim, J. B. Lee, H. S. Youn, and G. W. Moon, "Integrated Asymmetrical Half-Bridge Zeta(AHBZ) Converter for DC/DC Stage of LED Driver With Wide Output Voltage Range and Low Output Current," *IEEE Trans. Ind. Electron.*, vol. 62, no. 10, pp. 7489–7498, Dec. 2015.

[9] F. Musavi, M. Craciun, D. Gautam, and W. Eberle, "Control Strategies for Wide Output Voltage Range LLC Resonant DC-DC Converters in Battery Chargers," *IEEE Trans. Veh. Technol.*, vol. 63, no. 3, pp. 1117–1125, Mar. 2014.

[10] M. Shang, H. Wang, "A LLC Type Resonant Converter Based on PWM Voltage Quadrupler Rectifier with Wide Output Voltage," in *Proc. Appl. Power Electron. Conf. Expo.*, Mar. 2017, pp. 1720-1726.

[11] H. Haga and F. Kurokawa, "Modulation Method of a Full-Bridge Three-Level *LLC* Resonant Converter for Battery Charger of Electrical Vehicles," *IEEE Trans. Power Electron.*, vol. 32, no. 4, pp. 2498–2507, Apr. 2017.

[12] Z. Fang, T. Cai, S. Duan, and C. Chen, "Optimal Design Methodology for *LLC* Resonant Converter in Battery Charging Applications Based on Time-Weighted Average Efficiency," *IEEE Trans. Power Electron.*, vol. 30, no. 10, pp. 5469–5483, Oct. 2015.

[13] C. Buccella, C. Cecati, H. Latafat, P. Pepe, and K. Razi, "Observer-Based Control of *LLC* DC/DC Resonant Converter Using Extended Describing Functions," *IEEE Trans. Power Electron.*, vol. 30, no. 10, pp. 5881–5891, Oct. 2015.

[14] W. Hongfei, L. Yuewei, and X. Yan, "LLC Resonant Converter With Semiactive Variable-Structure Rectifier (SA-VSR) for Wide Output Voltage Range Application," *IEEE Transactions on Power Electronics*, vol. 31, pp. 3389-3394, 2016.

[15] M. I. Shahzad, S. Iqbal, and S. Taib, "A Wide Output Range HB-2LLC Resonant Converter With Hybrid Rectifier for PEV Battery Charger," *IEEE Trans. Transp. Electrific.*, vol. 3, no. 2, pp. 520-531, Jun. 2017.

[16] H. Wang and Z. Li, "A PWM LLC Type Resonant Converter Adapted to Wide Output Voltage Range in PEV Charging Applications," *IEEE Trans. Power Electron.*, vol. 33, No. 5, pp. 3791-3801, May 2018.

[17] Y. Gu, L. Hang, Z. Lu, Z. Qian, and D. Xu, "Voltage Doubler Application in Isolated Resonant Converter," in *Proc.* IECON 2005, Nov. 2005, pp. 1184-1188.

[18] B. R. Lin, W. R. Yang, J. J. Chen, C. L. Huang, and M. H. Yu, "Interleaved LLC Series Converter with Output Voltage Doubler," in *Proc.* IPEC 2010, Jan. 2010, pp. 92-98.

A Study on the Analysis and Control of No-load Characteristics of LLC Resonant Converter for Plasma Process

Min-Jun Kwon[1] and Woo-Cheol Lee[1*]
Department of Electrical, Electronic and Control Engineering
Hankyong National University
Anseong, Republic of Korea
*E-mail: woocheol@hknu.ac.kr

Abstract-This study applied an LLC resonant converter with various advantages as a power supply for a plasma process. Owing to the nature of the process, it was necessary to analyze the no-load characteristics of the LLC resonant converter because the plasma power supply has an initial no-load operation. The conventional method for analyzing the characteristics of an LLC resonant converter shows the characteristics of the load condition well, but it is difficult to analyze the characteristics of the no-load condition. Therefore, this study analyzed the no-load characteristics of an LLC resonant converter as a method to consider parasitic components and analyzed them in the time domain. It also describes the control method with a duty ratio by the phase shift for operating the LLC resonant converter as a power supply for the plasma process through the analysis results.

Keywords— LLC converter, Plasma process, No-laod.

I. INTRODUCTION

Recently, the plasma process has attracted attention in the IT, medical, and textile fields. As interest in the plasma process increases, research on a power supply for plasma is actively under way. An LLC resonant converter that can achieve high efficiency and power density can be applied as a power supply for the plasma process. The power supply for a plasma process must be able to operate under no-load conditions to generate the plasma because of the characteristics of the process. Therefore, it is necessary to analyze the no-load characteristics of the power supply [1-4]. The most representative method used to analyze the characteristics of LLC resonant converter is first harmonic approximation (FHA). FHA analysis is a method of analyzing an equivalent circuit considering only the fundamental wave in the frequency domain and has the advantage of showing the characteristics of an LLC resonant converter because it can the operation close to the series resonance. FHA analysis has effectively shown the characteristics of an LLC resonant converter in a load condition but has confirmed that the characteristics in a no-load condition are different from the actual operation [4-6]. In this study, to apply an LLC resonant converter as a power supply for a plasma process, it analyzed the no-load characteristics of the converter and conducted a study for controlling the operation in a no-load condition. It analyzed the no-load characteristics of the LLC resonant

converter by identifying and complementing the limitations of FHA analysis. The validity of the analysis and control method was investigated by simulation.

II. ANALYSIS OF NO-LOAD CHARACTERISTICS

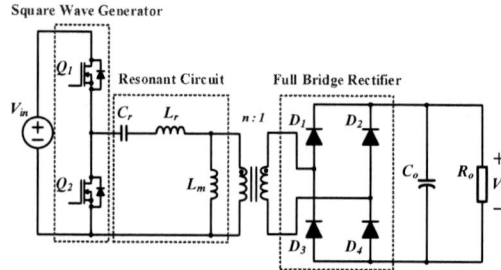

Fig. 1. LLC resonant converter circuit.

Fig. 1 shows the basic circuit of the LLC resonant converter. The converter consists of a square wave generator, a resonant circuit, and a rectifier. A full bridge rectifier may be applied for depending on the application. The LLC resonant converter has the advantage of achieving a high efficiency through zero voltage switching (ZVS). The resonant frequency f_r of the LLC resonant converter is as follows [3][7].

$$f_r = \frac{1}{2\pi\sqrt{L_r C_r}} \qquad (1)$$

When analyzing the no-load characteristics of an LLC resonant converter through FHA analysis, it can be confirmed that the gain in a range where the switching frequency f_s is larger than the resonance frequency f_r has a constant value and is linear regardless of the switching frequency. However, the actual operation of the LLC resonant converter in a no-load condition is nonlinear and has no constant value. This is a limitation of the FHA analysis method that neither considers harmonic components nor parasitic components. The converter circuit is not an ideal no-load circuit because of the resistance values in the analog to digital converter (ADC) circuit or the snubber circuit when constructing the circuit. Therefore, to analyze actual no-load behavior, it is necessary to analyze the very low light load condition rather than an ideal no-load condition. To compensate for the limitation of FHA analysis, this study considered the parasitic components of the circuit and analyzed them in the time domain and also confirmed that the diode junction capacitance C_j among various parasitic components affects the no-load characteristics [8].

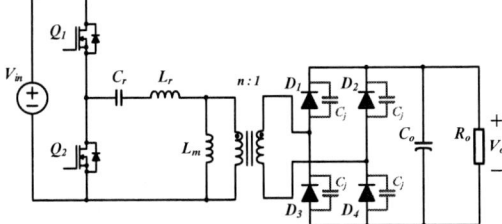

Fig. 2. LLC resonant converter circuit considering C_j.

Fig. 2 shows the LLC resonant converter considering the diode junction capacitance C_j.

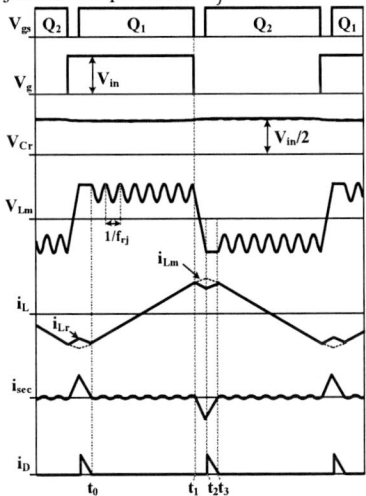

Fig. 3. Waveform considering C_j in no-load condition.

Fig. 3 shows the waveform of the LLC resonant converter considering C_j in a no-load condition. The waveform consists of V_{gs}, the square wave input voltage V_g, the resonant capacitance voltage V_{Cr}, the magnetizing inductance voltage V_{Lm}, the magnetizing inductance current i_{Lm}, resonant inductance current i_{Lr}, the secondary current i_{sec}, and the current delivered to the output through the diode, i_D.

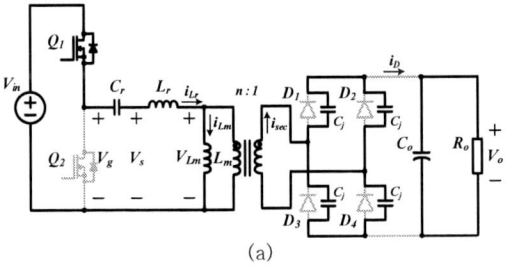

(a)

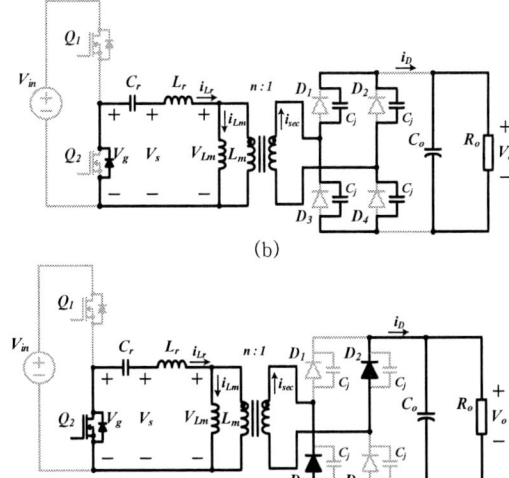

(b)

(c)

Fig. 4. Circuit operation considering C_j in no-load condition: (a) Mode 1 ($t_0 - t_1$), (b) Mode 2 ($t_1 - t_2$), and (c) Mode 3 ($t_2 - t_3$)

The no-load operation of the LLC resonant converter considering the C_j analyzed in the time domain can be largely classified into three modes.

Mode 1 ($t_0 - t_1$): In Mode 1, switch Q1 turns on, and resonance occurs between the parameters. The magnitude of the square wave input voltage V_g is V_{in}, and the V_{Cr} is almost constant at $V_{in}/2$. Therefore voltage V_s becomes $+V_{in}/2$. The VLm causes a resonance ripple between the parameters, and the i_D current does not occur. The resonant ripple frequency f_{rj} is

$$f_{rj} = \frac{1}{2\pi\sqrt{\frac{L_r L_m}{L_r + L_m} \times \frac{C_j}{n^2}}} \qquad (2)$$

The voltage $V_{Lm}(t_1)$ at t_1 when Mode 1 ends can be calculated as

$$V_{Lm}(t_1) = \frac{L_m}{L_r + L_m} V_s + \left(V_{Lm}(t_0) - \frac{L_m}{L_r + L_m} V_s\right)\cos(\omega_r t) \qquad (3)$$

where $\omega_r = 1/\sqrt{L_r L_m C_j/((L_r + L_m)n^2)}$.

Mode 2 ($t_1 - t_2$): Mode 2 is the charge section, and switch Q1 turns off. When Q1 turns off at t1, the body diode of Q2 turns on and voltage V_g decreases from V_{in} to 0 V. The voltage V_s decreases rapidly from $+V_{in}/2$ to $-V_{in}/2$, while the voltage V_g decreases to 0 V. V_{Lm} also starts reversing rapidly as voltage V_s decreases. $V_{Lm}(t_2)$ can be calculated as shown in Equation (3) because the circuits, except for the V_g and V_s, are the same as those of Mode 1.

$$V_{Lm}(t_2) = \frac{L_m}{L_r + L_m} V_s + \left(V_{Lm}(t_1) - \frac{L_m}{L_r + L_m} V_s\right)\cos(\omega_r t) \qquad (4)$$

The current i_{sec} flowing across the diode junction capacitance C_j is calculated as

$$i_{sec}(t_2) = \frac{1}{n}C_j\omega_r\left(\frac{L_m}{L_r+L_m}V_s - V_{Lm}(t_1)\right)\sin(\omega_r t) \quad (5)$$

Mode 3 ($t_2 - t_3$): Mode 3 is the section in which charging is completed and power is supplied, and rectifier diodes D2 and D3 turn on. As the rectifier diodes turn on, the i_{sec} current is delivered to the output. In Mode 3, the i_D current is the same as the i_{sec} current. The time t_a when i_D current flows can be calculated as the time the current flowing across the output filter capacitance C_o

$$t_a = \sin^{-1}\left(\frac{i_{sec}(t_2)}{\frac{1}{n}C_o\omega_o\left(\frac{L_m}{L_r+L_m}V_s - V_{Lm}(t_1)\right)}\right)\frac{1}{\omega_o} \quad (6)$$

where $\omega_o = 1/\sqrt{L_rL_mC_o/((L_r+L_m)n^2)}$. Through the analysis, the output voltage of the LLC resonant converter can be calculated as

$$V_o = i_{sec}(t_2)t_aR_oF_s \quad (7)$$

Fig. 5 shows the gain curve of the LLC resonant converter in a no-load condition. It can be confirmed that the gain curve of the LLC resonant converter in a no-load condition was not constant and was nonlinear. This gain curve was the result of the resonance ripple caused by C_j.

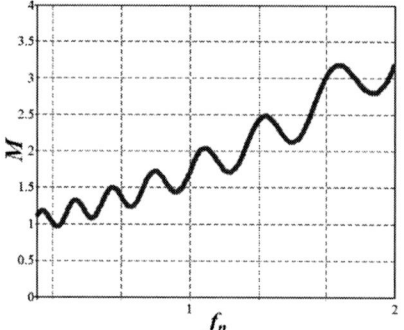

Fig. 5. Gain curve of LLC resonant converter in no-load condition.

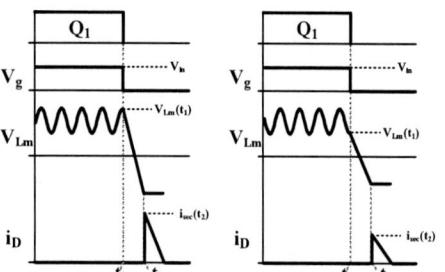

Fig. 6. Waveform according to t_1 by switching frequency.

The magnitude of i_{sec} (t_2), was affected by the magnitude of voltage V_{Lm} (t_1), as shown in Equation (5). And the magnitude of the output voltage, was affected by the magnitude of i_{sec} (t_2). As shown Fig. 6, the C_j caused the resonance ripple by V_{Lm}, so that the magnitude of voltage V_{Lm} (t_1) was determined according to t_1 by the

switching frequency. As a result, the resonance ripple of V_{Lm} (t_1) caused the ripple in the gain curve.

III. CONTROL METHOD IN NO-LOAD CONDITION

The power supply for a plasma process requires no-load operation for plasma generation. The LLC resonant converter does not operate linearly under a no-load condition and has a non-linear characteristic in that the increase and decrease of the output repeat according to the switching frequency. In order to apply the LLC resonant converter as the plasma power supply, it is necessary to output a high gain during the no-load operation and control the no-load operation.

When the square wave generator of an LLC resonant converter contains a full bridge, a voltage gain higher than that of a half bridge can be obtained, and the gain can be controlled by adjusting the duty ratio through the phase shift with a fixed switching frequency. Therefore, a full bridge structure is more suitable than that of a half bridge when an LLC resonant converter is applied as a power supply for a plasma process.

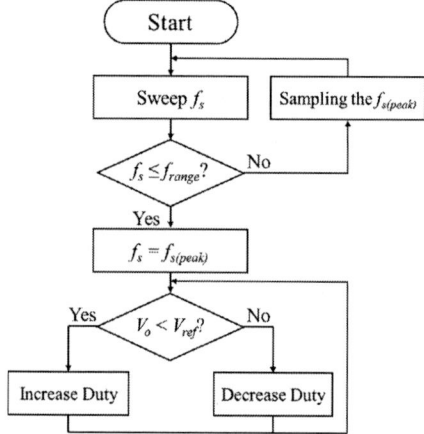

Fig. 7. Control algorithm in no-load condition.

Fig. 7 shows the control algorithm of an LLC resonant converter in no-load condition for a plasma process. First, the switching frequency f_s start at maximum frequency $f_{s(max)}$. The $f_{s(peak)}$ at the maximum gain is determined by sweeping the switching frequency in the direction of decreasing frequency. The switching frequency f_s should sweep by at least f_{range} because gain increases and decreases are related to the resonant ripple frequency f_r, where $f_{range} = f_{s(max)}f_r/(2f_{s(max)} + f_r)$. Once $f_{s(peak)}$ is determined, the converter controls the gain by changing the duty at the $f_{s(peak)}$. The converter can be controlled and achieve maximum gain through this algorithm in a no-load condition.

IV. SIMULATION RESULTS

Table 1 shows simulation system parameters. The operating range of the LLC resonant converter was 200 kHz to 100 kHz, and the circuit contained a full bridge. This study used a 120 kΩ snubber circuit. The simulation was carried out through the PSIM program. The output waveform was confirmed by sweeping from 200 kHz to 100 kHz using the DDL function of the PSIM program.

TABLE I SYSTEM PARAMETERS

Parameter	Value
Input Voltage, V_{in}	100 V
Resonant Inductance, L_r	71.2 uH
Magnetizing Inductance, L_m	274 uH
Resonant Capacitance, C_r	22 nF
Diode Junction Capacitance, C_j	61 pF
Load Resistance, R_o (snubber)	120 kΩ

Fig. 8 shows the simulation results in the no-load condition. Fig. 8(a) shows the output voltage V_o according to the switching frequency F_s. It was confirmed that the output voltage of the converter was nonlinear in the no-load condition. Fig. 8(b) and (c) show the voltage V_{Lm} at maximum and minimum gains, respectively. Resonant ripple as shown in Equation (2) occurs at the V_{Lm}. As shown in Fig. 8(d), the search for the $F_{s(peak)}$ occurred while sweeping the switching frequency, and the V_o followed the V_{ref} by controlling the duty ratio.

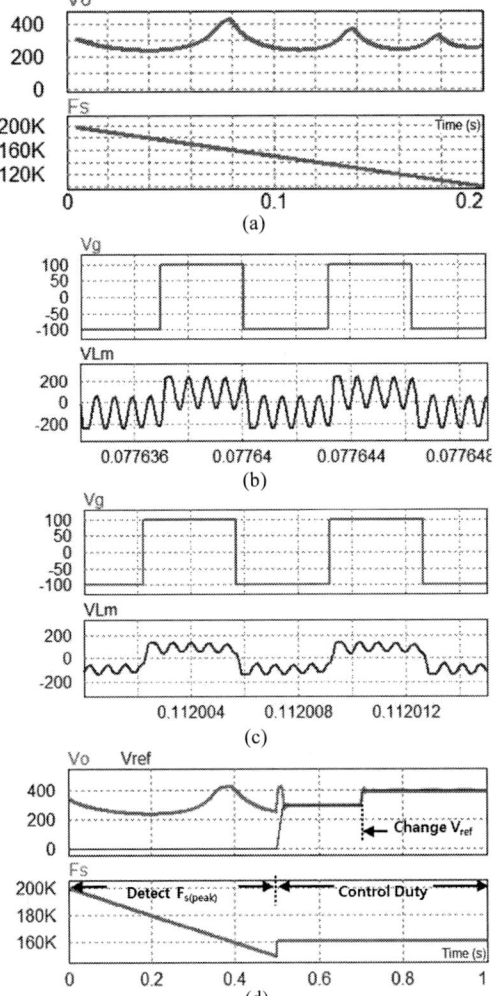

Fig. 8. Simulation results: (a) V_o according to F_s, (b) V_g and V_{Lm} at maximum gain, (c) V_g and V_{Lm} at minimum gain, and (d) control in no-load condition.

It was confirmed that the output voltage of the converter was nonlinear in a no-load condition, and the proposed control was confirmed.

V. CONCLUSION

When analyzing an LLC resonant converter with the FHA method, the characteristics were shown well in the load condition, but the characteristics in the no-load condition were different from those in the actual operation. To compensate for the limitation of the FHA method, it confirmed the no-load characteristics by considering the diode junction capacitance, which affects the no-load characteristics, and by analyzing the LLC resonant converter in the time domain. It was confirmed that the LLC resonant converter in a no-load condition has nonlinear gain because of the resonance ripple frequency caused by the diode junction capacitance. To control the LLC resonant converter in a no-load condition, this study used a duty ratio by the phase shift of a full bridge structure with a fixed switching frequency. The analysis and control methods were verified by simulation.

ACKNOWLEDGMENT

This research was supported by Basic Science Research Program through the National Research Foundation of Korea(NRF) funded by the Ministry of Education(No. NRF-2017R1D1A1B03031532.

REFERENCES

[1] W. Somkhunthot, "Bipolar Pulsed-DC Power Supply for Magnetron Sputtering and Thin Films Synthesis" Elektrika Journal of Electrical Engineering Vol. 9, no. 2, pp20-26

[2] F. Canales, P. Barbosa, F. C. Lee "A wide input voltage and load output variations fixed-frequency ZVS DC/DC LLC resonant converter for high-power applications" in Conf Rec Industry Applications 2002, Vol. 4, pp. 2306-2313, 13-18 Oct 2002.

[3] B. Yang, F. C. Lee, A. J. Zhabg, and G. Huang, "LLC Resonant Converter for Front End DC/DC Conversion," in Proc. IEEE APEC'02, pp. 1108-1112, 2002.

[4] B. Lu, W. Liu, Y. Liang, F. C. Lee, and J. D. Van Wyk, "Optimal Design Methodology for LLC Resonant Converter," in Proc. IEEE APEC' 06, pp. 533-538, 2006 .

[5] T. Duerbaum, "First harmonic approximation including design constraints," in Proc. INTELEC' 98, pp. 321-328, Oct. 1998.

[6] H. K. Lee, E. S. Kim, D. Y. Huh, G. S. Lee, B. G. Chung, and S. I. Kang, "Operating Characteristics of LLC Series Resonant Converter Using a LLT Transformer." The Transactions of the Korea Institute of Power Electronics. vol. 11, no. 5, pp. 409-416, 2006.

[7] G. Pledl, M. Tauer, and D. Buecherl, "Theory of operation, design procedure and simulation of a bidirectional LLC resonant converter for vehicular applications," in Proc. IEEE Vehicle Power Propulsion Conf., Sep. 2010,pp. 1–5.

[8] B. H. Lee, M. Y. Kim, C. E. Kim, K. B. Park and G. W. Moon, "Analysis of LLC Resonant Converter Considering Effect of Parasitic Components" in Proc. INTELEC'09, pp.1-6, Oct. 2009.

Mechanism of Current Imbalance in LLC resonant converter with Center Tapped Transformer

Mitsuru Sato, Shingo Nagaoka, Takeshi Uematsu, and Toshiyuki Zaitsu
OMRON Corporation, Kyoto, Japan
E-mail: mitsuru_sato@omron.co.jp

Abstract— In recent years, the LLC resonant converter with a center-tapped transformer has been widely used due to its high power density and high efficiency features in the Factory Automation which requires the low output voltage and high output current. However, the imbalanced secondary current for each half cycle occurs, and it cause a problem of reliability degradation associated with the high current. In this paper, the mechanism of imbalanced current is investigated focusing on the effects of coupling coefficient of the center-tapped transformer. This mechanism has been confirmed by experiments.

Keywords— LLC resonant converter, center-tapped transformer, imbalanced current,

I. INTRODUCTION

High power density of power supplies for Factory Automation (FA) has been required for space optimization of control panels, because the introduction of IoT and AI of control equipment. As a result, higher output current has been desired for the power supplies in order to achieve high power density. Hence, LLC resonant converters, which have the features of high power density and high efficiency, are used extensively as isolated DC-DC converter [1] [2]. LLC resonant converter is able to be miniaturized since it can utilize the leakage inductance of the transformer as the resonance inductance and does not need to have the smoothing inductance for output.

In the case of step-down from high to low voltage, the LLC resonant converter is generally designed with the center tapped transformer as shown in Fig. 1 to reduce the conduction loss of the rectifying element [3] - [5]. However, in the center tapped transformer, imbalanced current likely to occur which leads to the degradation reliability of the device. Especially, in the case of low voltage and high current output, the secondary windings of the transformer are designed with a small number of windings such as 1-3 turns to reduce copper losses. Therefore, the current imbalance from the variations of the transformer characteristics is likely to occur. Previously, a winding structure for suppressing the characteristic variation of the center tapped transformer and technique to reduce the secondary current imbalance

in the LLC resonant converter have been studied [3] [4]. However, the mechanism of the current imbalance of the LLC resonant converter with the center tapped transformer has not been examined adequately.

In this paper, we investigated the imbalanced current mechanism with focusing on the difference of the coupling coefficient in the center tapped transformer of the LLC resonant converter. This paper is organized as follows. Chapter □ introduces operation of the LLC resonant converter with imbalanced current. Chapter □ describes the proposed mechanism. Chapter IV mentions simulation and experimental results. Chapter V concludes this paper.

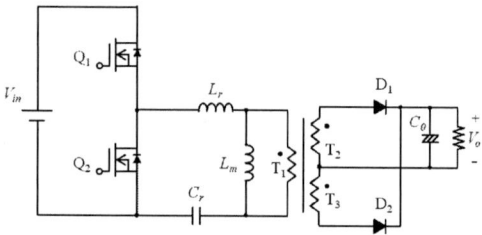

Fig. 1. Block diagram of LLC resonant converter with center-tapped transformer

II. IMBALANCED CURRENT OF LLC RESONANT CONVERTER WITH CENTER-TAPPED TRANSFORMER

Fig. 2 shows the four operation modes of the LLC resonant converter with center tapped transformer. A square wave of 50% duty generated by the primary switches, Q_1 and Q_2, is applied to resonate the resonant capacitor C_r and the leakage inductance L_r of the transformer. Resonant current i_r flows to the load as rectified currents i_{d1} and i_{d2}. LLC resonant converter has two patterns of resonant circuits that depends on when the switches Q_1 and D_1 are on, and when the switches Q_2 and D_2 are on, as shown in Fig. 2(a) and (c). In Fig. 2(a), the primary winding of the transformer T_1 is coupled with the secondary winding T_2. Fig. 2(c) shows the resonant circuit where the primary winding of the transformer T_1 is coupled with the secondary winding T_3. M_{12} is mutual inductance between T_1 and T_2. M_{13} is mutual inductance between T_1 and T_3.

The series resonant frequency ω_{sr1} of the Fig. 2(a) operation and the series resonance frequency ω_{sr2} of the Fig. 2(c) operation can be obtained by the following equations, respectively.

$$\omega_{sr1} = \frac{1}{\sqrt{(1 - k_{12}{}^2)L_{T1}C_r}} \qquad (1)$$

$$\omega_{sr2} = \frac{1}{\sqrt{(1 - k_{13}{}^2)L_{T1}C_r}} \qquad (2)$$

Where L_{T1} is the inductance of the primary winding, k_{12} and k_{13} are coupling coefficient between T_1-T_2 and T_1-T_3, respectively. When the switching frequency f_{SW} is below series resonant frequency, voltage gain is always obtained over nV_o/V_{in} as boost mode where n is turn-ratio between primary and secondary windings.

In Fig. 2(b) and (d), i_r flows through primary circuit as the resonant current which is parallel resonant frequency ω_{pr} obtained in Eq. (3) while D_1 and D_2 are off.

$$\omega_{pr} = \frac{1}{\sqrt{L_{T1}C_r}} \qquad (3)$$

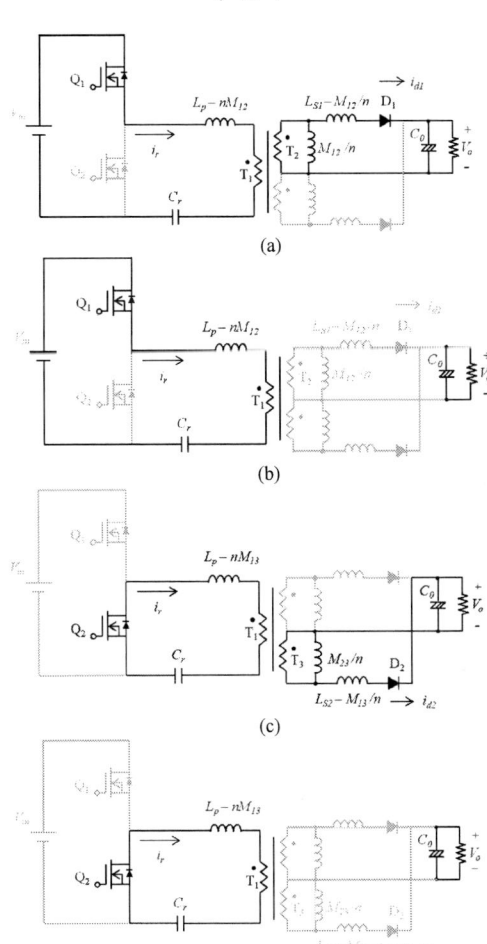

Fig. 2 Operation mode

When k_{12} and k_{13} are different, imbalance of the resonant current is presumed to occur because it operates in different resonant circuits for each half cycle. Imbalanced current i_r becomes the imbalanced secondary current for each cycle.

Fig. 3 shows a simulated waveform of a steady state of the LLC resonant converter with center tapped transformer at boost mode. In the case of $k_{12} = k_{13}$, the diode D_1 current i_{d1} and the diode D_2 current i_{d2} are balanced as shown in Fig. 3(a). This results in a symmetrical waveform of the resonant current i_r. On the contrary, in the case of $k_{12} \neq k_{13}$, the peak values of i_{d1} and i_{d2} are imbalanced as shown in Fig. 3(b), and i_r is the positive-negative asymmetric.

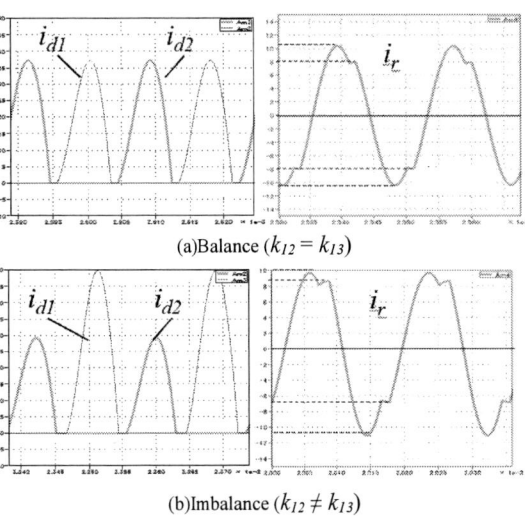

(a)Balance ($k_{12} = k_{13}$)

(b)Imbalance ($k_{12} \neq k_{13}$)

Fig. 3. Key waveforms when imbalanced current occurs

III. MECHANISM OF IMBALANCED CURRENT

In this chapter, we examine four states in order to study the imbalanced current in boost mode operation. Fig. 4 shows the waveform at the heavy load of i_r, i_{d1}, i_{d2}, and the magnetizing current i_m, when the secondary current is imbalanced. For simplicity, dead time for achieving ZVS of Q_1 and Q_2 is not taken into consideration here. In addition, it is assumed that the output voltage, V_o is the constant as voltage source, and the inductance L_{t2} and L_{t3} for the secondary windings is same.

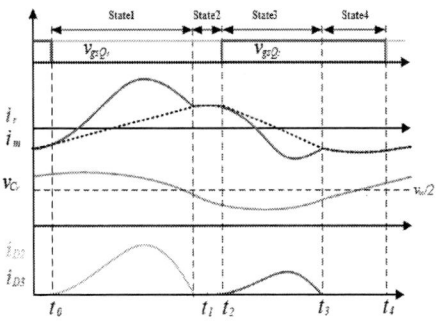

Fig. 4. Key waveforms at the heavy load

State 1 ($t_0 < t < t_1$) [Fig.2 (a)]

Q_1 and D_1 are on, and Q_2 and D_2 are off. At t_0, i_r becomes larger than i_m, consequently D_1 turns on. During this period, i_m increases linearly, and the resonance characteristics are determined by the coupling coefficient between the transformer T_1 and T_2 and C_r. i_r, i_{d1}, i_{d2}, i_m are expressed in Eq. (4)

$$
\begin{cases}
i_r(t) = \alpha_1 \sin \omega_{sr1}(t - t_0) + \beta_1 \cos \omega_{sr1}(t - t_0) \\
i_{d1}(t) = k_{12}\sqrt{\dfrac{L_{T1}}{L_{T2}}}\left\{\alpha_1 \sin \omega_{sr1}(t - t_0) + \beta_1\left(\cos \omega_{sr1}t - 1\right)\right\} \\
\qquad\qquad\qquad\qquad\qquad - \dfrac{V_O}{L_{T2}}(t - t_0) \\
i_{d2}(t) = 0 \\
i_m(t) = (1 - k_{12})\alpha_1 \sin \omega_{sr1}(t - t_0) + (1 - k_{12})\beta_1 \cos \omega_{sr1}(t - t_0) \\
\qquad\qquad\qquad\qquad\qquad + \dfrac{k_{12}V_O}{M_{12}}(t - t_0) \\
\alpha_1 = \dfrac{C_r \omega_{sr1}\left(L_{T2}V_{in} - M_{12}V_O - L_{T2}v_{cr}(t_0)\right)}{L_{T2}} \\
\beta_1 = i_r(t_0)
\end{cases}
\tag{4}
$$

State 2 ($t_1 < t < t_2$) [Fig.2 (b)]

Q_1 is on, and Q_2, D_1 and D_2 are off. At t_1, i_r corresponds with i_m, consequently D_1 turns off. During this period, i_r and i_m are equal, and a resonant current i_r of L_{T1} and C_r flows through primary side only. i_r, i_{d1}, i_{d2}, i_m are expressed in Eq. (5)

$$
\begin{cases}
i_r(t) = \dfrac{v_{Cr}(t_1)}{L_{T1}\omega_{pr}} \sin \omega_{pr}(t - t_1) + I_r(t_1)\cos \omega_{pr}(t - t_1) \\
i_{d1}(t) = 0 \\
i_{d2}(t) = 0 \\
i_m(t) = i_r(t)
\end{cases}
\tag{5}
$$

State 3 ($t_2 < t < t_3$) [Fig.2 (c)]

Q_2 and D_2 are on, and Q_1 and D_1 are off. This operation is symmetrical to State 1. The resonant characteristics are expressed in Eq. (6)

$$
\begin{cases}
i_r(t) = \alpha_2 \sin \omega_{sr2}(t - t_2) + \beta_2 \cos \omega_{sr2}(t - t_2) \\
i_{d1}(t) = 0 \\
i_{d2}(t) = k_{13}\sqrt{\dfrac{L_{T1}}{L_{T2}}}\left\{\alpha_2 \sin \omega_{sr2}(t - t_0) + \beta_2\left(\cos \omega_{sr2}t - 1\right)\right\} \\
\qquad\qquad\qquad\qquad\qquad - \dfrac{V_O}{L_{T3}}(t - t_2) \\
i_m(t) = (1 - k_{13})\alpha_2 \sin \omega_{sr1}(t - t_2) + (1 - k_{13})\beta_2 \cos \omega_{sr1}(t - t_2) \\
\qquad\qquad\qquad\qquad\qquad + \dfrac{k_{13}V_O}{M_{13}}(t - t_2) \\
\alpha_2 = \dfrac{C_r \omega_{sr2}\left(-L_{T3}V_{in} + M_{13}V_O + L_{T3}v_{cr}(t_2)\right)}{L_{T3}} \\
\beta_2 = i_r(t_0)
\end{cases}
\tag{6}
$$

State4 ($t_3 < t < t_4$) [Fig.2 (d)]

Q_2 is on, and Q_1, D_1 and D_2 are off. This operation is symmetrical to State 2. The resonant current of L_{T1} and C_r flows through primary side only. i_r, i_{d1}, i_{d2}, i_m are expressed in Eq. (7)

$$
\begin{cases}
i_r(t) = -\dfrac{v_{Cr}(t_3)}{L_{T1}\omega_{pr}} \sin \omega_{pr}(t - t_3) - I_r(t_3)\cos \omega_{pr}(t - t_3) \\
i_{d1}(t) = 0 \\
i_{d1}(t) = 0 \\
i_m(t) = i_r(t)
\end{cases}
\tag{7}
$$

From equations (4)-(7), it can be seen that the resonant capacitor voltage and the resonant current at the end of the previous state is set to the initial value in each state.

In the steady state, the magnetizing current does not diverge and thus satisfies the following equation.

$$
i_m(t_2) - i_m(t_0) = i_m(t_4) - i_m(t_2)
\tag{8}
$$

Moreover, the resonant capacitor behaves as a DC blocking capacitor to compensate DC offset of resonant capacitor voltage in each half cycle; Therefor, Eq. (9) must be satisfied.

$$
\int_{t_0}^{t_4} i_r(t)\,dt = 0
\tag{9}
$$

From Eq. (4) - (9), the magnetizing current is balanced when the duration of State 1 and State 3 as $i_m(t_1)$-$i_m(t_0)$=$i_m(t_3)$-$i_m(t_2)$. In case of $k_{12}{\neq}k_{13}$, the absolute value of $i_m(t_1)$-$i_m(t_0)$ and $i_m(t_3)$-$i_m(t_2)$ are not same due to the different coupling coefficient.

Hence, i_m becomes the positive-negative asymmetric waveform, because initial conditions of State2 and State4 are not symmetric. As a result, i_{d1} and i_{d2} are imbalanced to satisfy Eq. (9), and an offset occurs in i_m since $i_m(t_2)$+ $i_m(t_4) \neq 0$.

Fig. 5 shows the key waveforms in the light load condition which is divided into 6 states. When the load current becomes smaller, the duration of state2' and state 5' becomes shorter. As a result, D1 and D2 turn on after Q1 and Q2 turn on, respectively.

At the no load, $i_r(t'_0)$ is equal to $i_r(t'_3)$, consequently state 2 'and state 5' do not exist. Hence, resonant current i_r of L_{T1} and C_r flows continuously in the switching period. L_{T1} of Center-tapped transformer and C_r are same in the all-state, respectively. So the offset of i_m does not occur despite of $k_{12}{\neq}k_{13}$ in the light load condition. Therefore, it found that the offset of magnetizing current depends on the load condition. When the duration of state 2 'and state 5' increases in the heavy load, current imbalance becomes larger by the difference of coupling coefficient.

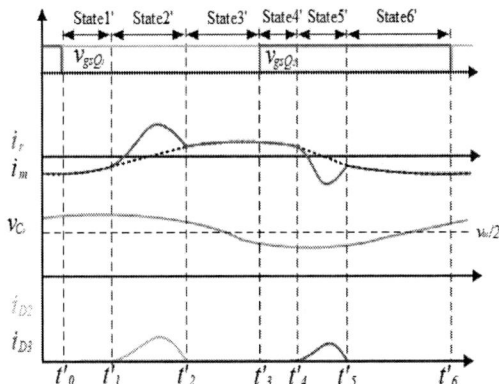

Fig.5. Key waveforms at the light load

IV. SIMULATION AND EXPERIMENTS

A. Transformer design and simulation

In order to assess the validity of the imbalanced current mechanism, prototype of the transformers have been fabricated for testing and compared with the simulations.

Fig. 6 shows three transformer structures with different coupling coefficients. Table I shows magnetic simulations and the measurement results of the secondary winding structures. The simulation conditions are 55 kHz for the operating frequency. The core type is PC95PQ3535. Primary and secondary winding are used $\Phi 0.1x180$ (litz wire).The primary and secondary number of turns is 18T and 3T, respectively. The gap of 4mm is inserted between primary and secondary windings because the coupling coefficient is determined by position of the windings.

In Fig. 6(a), each secondary winding are winded vertically on the bobbin so that the distance between each secondary and primary winding can be close as possible. In Fig. 6(b), each secondary winding are winded horizontally on the bobbin so that the distance between each secondary and primary winding can vary. In Fig. 6(c), in order to cause larger variations of coupling coefficient than Fig. 6(b), the gap of 2.2mm is inserted between each secondary winding.

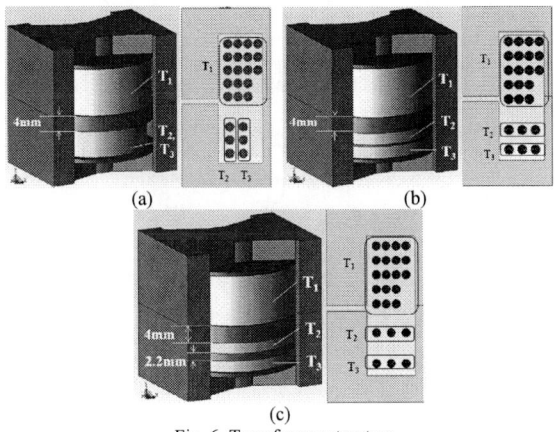

Fig. 6. Transformer structure

Table I   Magnetizing simulation and measurement result

	(a)		(b)		(c)	
	Sim	Exp	Sim	Exp	Sim	Exp
Inductance of T_1 L_{T1}[μH]	88.01	88.17	87.62	87.97	87.38	89.10
Coupling Coefficient k_{12}	0.814	0.822	0.851	0.833	0.861	0.859
Coupling Coefficient k_{13}	0.811	0.819	0.815	0.792	0.792	0.796
Mutual inductance M_{12}[μH]	12.42	12.43	12.65	12.83	12.49	12.93
Mutual inductance M_{13}[μH]	12.55	12.59	12.63	12.87	12.45	12.95
Error between k_{12} and k_{13} Δk [%]	0.2	0.1	2.2	2.5	4.2	3.8

B. Experimental result

In order to validate the current imbalance mechanism, a prototype of 480W LLC resonant converter was implemented using the transformer shown in Fig. 6 (a), (b) and (c). Table II shows experimental conditions.

Fig. 7 shows the waveforms with different coupling coefficient at output current 20A as heavy load. It is observed that there increases offset in resonant current i_r which include magnetizing current by increasing the difference of the coupling coefficient. And secondary current imbalance is observed as well.

Table II.   Experimental condition

Input Voltage V_{in}[V]	200
Output Voltage V_o[V]	24
Rated output Power P_o[W]	480
Resonant capacitor C_r [nF]	220

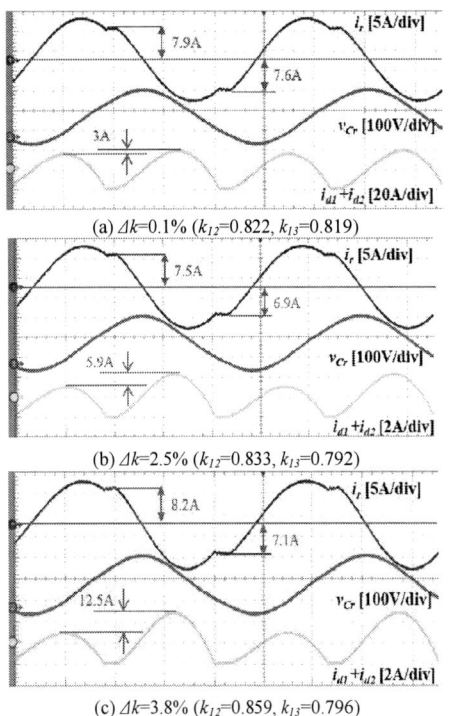

Fig. 7. Experimental waveform at output current = 20A with different coupling coefficient

Fig. 8 shows the waveforms at output current 1A as light load. Resonant current almost becomes the positive-negative symmetric waveform. Therefore, offset of magnetizing current become smaller than the case of output current 20A. Similarly, secondary current imbalance is small since rectification period is short.

Fig. 9 shows a plot of the relationship between the output current and the offset of magnetizing current. We saw the offset depends on the load condition.

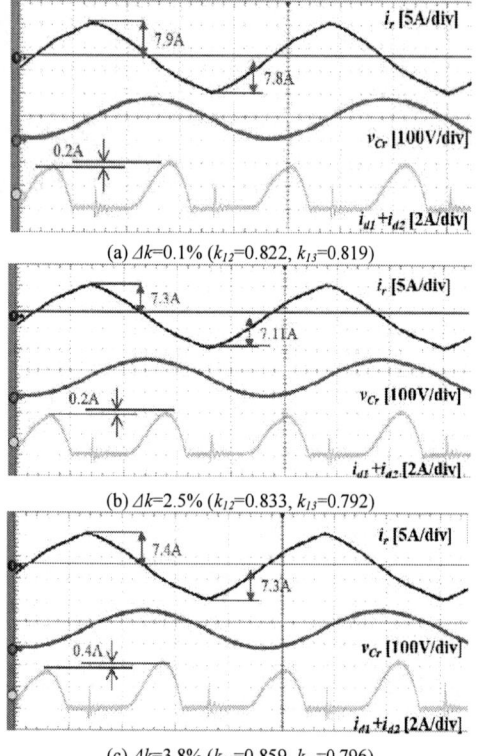

(a) Δk=0.1% (k_{12}=0.822, k_{13}=0.819)

(b) Δk=2.5% (k_{12}=0.833, k_{13}=0.792)

(c) Δk=3.8% (k_{12}=0.859, k_{13}=0.796)

Fig. 8. Experimental waveform at output current =1A with different coupling coefficient

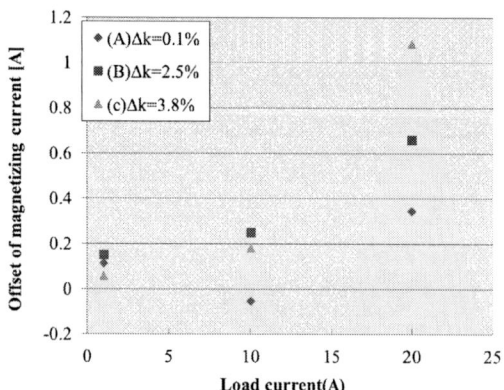

Fig. 9. Offset of resonant current at different loads

V. CONCLUSION

In this paper, the mechanism of the current imbalance of the LLC resonant converter with a center tapped transformer has been investigated.

We built the hypothesis that the imbalanced current mechanism is come from the difference of coupling coefficient in the transformer windings. This causes the magnetizing current imbalance in each half cycle which generates the DC offset, finally leads to imbalanced resonant current.

In order to verify this mechanism, three types of transformers which have different coupling coefficient by different winding structure were fabricated. And the board of LLC resonant converter was implemented and tested using the above different types of transformers. As a result, the offset of primary resonant current and imbalanced secondary current were observed as expected. It is confirmed that the difference of coupling coefficient causes the current imbalance in the LLC converter with center tapped transformer.

REFERENCES

[1] B. Yang, F.C. Lee, A.J. Zhang, and G. Huang, "LLC Resonant Converter for Front End DC/DC Conversion," in Proc. IEEE APEC02. pp. 1108-1112, Boston, 2002.

[2] S. De Simone, C. Adragna and C. Spini, "Design guideline for magnetic integration in LLC resonant converters," in Proc SPEEDAM 2008. pp. 950-957, Ischia, 2008.

[3] M. Li, Q. Chen, X. Ren, Y. Zhang, K. Jin and B. Chen, "The integrated LLC resonant converter using center-tapped transformer for on-board EV charger," in Proc. IEEE ECCE15, Montreal, QC, 2015, pp. 6293-6298.

[4] J. H. Jung, "Bifilar Winding of a Center-Tapped Transformer Including Integrated Resonant Inductance for LLC Resonant Converters," IEEE Trans. Power Electronics, vol. 28, no. 2, pp. 615-620, Feb. 2013.

[5] K. B. Park, B. H. Lee, G. W. Moon and M. J. Youn, "Analysis on Center-Tap Rectifier Voltage Oscillation of LLC Resonant Converter," in IEEE Trans. Power Electronics, vol. 27, no. 6, pp. 2684-2689, June 2012.

[6] B. A. McDonald and D. Freeman, "Practical transformer modeling and characterization for a high performance LLC converter," in Proc. APEC13, pp. 2582-2589, Long Beach, CA, USA, 2013.

[7] J. F. Lazar and R. Martinelli, "Steady-state analysis of the LLC series resonant converter," in Proc. IEEE APEC01, pp. 728-735 vol.2. Anaheim. 2001.

[8] Bing Lu, Wenduo Liu, Yan Liang, F. C. Lee and J. D. van Wyk, "Optimal design methodology for LLC resonant converter," in Proc. IEEE APEC 06. pp. 533-538 2. Dallas. 2006.

[9] M. P. Foster, C. R. Gould, A. J. Gilbert, D. A. Stone and C. M. Bingham, "Analysis of CLL Voltage-Output Resonant Converters Using Describing Functions," IEEE Trans. Power Electronics, vol. 23, no. 4, pp. 1772-1781, July 2008.

[10] X. Fang, H. Hu, Z. J. Shen and I. Batarseh, "Operation Mode Analysis and Peak Gain Approximation of the LLC Resonant Converter," in IEEE Trans. Power Electronics, vol. 27, no. 4, pp. 1985-1995, April 2012.

[11] B. H. Lee, M. Y. Kim, C. E. Kim, K. B. Park and G. W. Moon, "Analysis of LLC Resonant Converter considering effects of parasitic components," in Proc. INTELEC 2009, pp. 1-6, Incheon, 2009.

Performance Study of High-Power Half-Bridge Interleaved LLC Converter

Hung-I Hsieh
Department of Electrical Engineering
National Chiayi University
Chiayi, Taiwan 60004 (R.O.C.)
hihsieh@mail.ncyu.edu.tw

Hui-Lung Chiu and Guan-Chyun Hsieh
Department of Electrical Engineering
Chung Yuan Christian University
Chung-Li, Taiwan 320 (ROC)
gchsieh@dec.ee.cycu.edu.tw

Abstract-Equalizing and minimizing output current ripple in an interleaved half-bridge LLC converter is explored to raise output power with high efficiency. A strategy to let the half-bridge LLC work in regions 1, $\omega > \omega_{r1}$, is implemented to ensure the two interleaved duty cycle close to 50%, which makes the output current width continuous link without lag and also reduce the component size. In addition, the magnetizing currents can also be equal to mitigate the circulation current in the MOSFETs. In transformer design, the secondary winding is wound with equal-length wire to lead the amplitudes of the full-wave output current ripple almost the same, which equalizes the wire thermal to improve the power efficiency. Further, with sharing the output current, an interleaved half-bridge LLC converter can perform a power conversion up to 2.7 kW with the highest efficiency close to 98%. Besides, in-phase and 90°-phase-shift output currents merging to reduce the output ripple currents are respectively explored in experiment. In this paper, a simple modeling of the interleaved LLC is also explored and a design consideration for implementing a 2.7 kW high-power interleaved half-bridge LLC converter is described. All experimentations including the switching statuses and temperature distribution are displayed and discussed in details.

I. INTRODUCTION

LLC converter has been come forward in extending the applications involving power server, charger for vehicle, energy conversion for energy storage and so on due to its inherent advantage including low noise, high efficiency, and wide control range adapting to load change [1-4]. In order to extend the power output ability, the half-bridge LLC converter with interleaved topology is explored, which can output power up to 2 kW with high efficiency close to 95% [5-9]. However, the issue how to equalize and minimize the magnitude of the output ripple current to raise the output power is always the design goal. In addition, how to place the operation point for optimal design is also the vital design consideration. In this paper, a design consideration to explore how to place the operation point and equalize the magnitude of the output ripple current so as to increasing the power efficiency is conducted. In addition, a simple modeling for describing the interleaved half-bridge LLC converter is presented. Finally, an interleaved half-bridge LLC converter is designed and implemented to verify the proposed theme, which can produce power of 2.7 kW with highest power efficiency up to 97.51% at 35 A and

96.14% at 50 A. Besides, all output range from 10 A to 50 A can gain power efficiency from 95.2% to 97.51%. Even at light load situation of 5 A, the efficiency can also kept at 92.78%. In addition, in-phase and 90°-phase-shift output currents merging to reduce the output ripple currents are respectively explored in experiment. The measured ripple currents flowing through the filter capacitor C_L at heavy load condition of 50 A for in-phase and 90°-phase-shift current merging cases are 20.2 A_{rms} and 6.8 A_{rms}, respectively. The ripple current in the 90o-phase-shift case is just only 37% of that in-phase merge case, which dramatically reduces the current stress in the output capacitor. Meanwhile, the temperatures without/with fan for the transformer and the MOSFETs are respectively measured for comparison under room temperature 25°C.

II. DESCRIPTION OF AN INTERLEAVED HALF-BRIDGE LLC CONVERTER

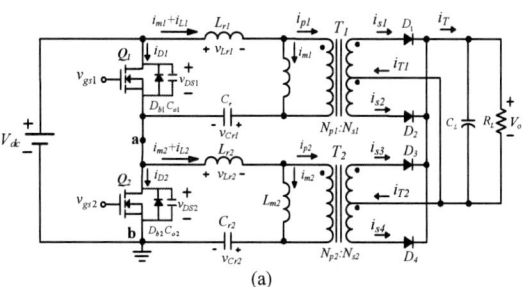

(a)

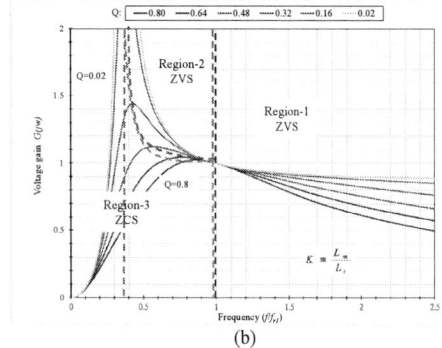

(b)

Fig. 1. Circuit scheme of an interleaved half-bridge LLC converter.

Fig. 1(a) shows a typical circuit topology of an interleaved half-bride LLC converter, in which two transformers with two resonant L_r-C_r loops are built across two MOSFETs, respectively. The two outputs at secondary side are connected in parallel. MOSFETs Q_1 and Q_2 serve as the interleaved half-bridge switches that alternately supplies energy to the two series-resonant tanks, which are respectively formed by two resonant inductors L_{r1} and L_{r2}, magnetizing inductances L_{m1} and L_{m2}, and resonant capacitors C_{r1} and C_{r2}. The frequency response of the LLC with different Q's is shown in Fig. 1(b). In this interleaved case, the inherent resonant frequency ω_{r1} at $f/f_{r1} = 1$ is function of L_r and C_r and the second ω_{r2} is set such at $f/f_{r1} = 0.63$, which is primarily dependent on the magnetizing inductance L_m. The switching statuses of the two MOSFETs in Region 1 and Region 2 are predicted in Fig. 2, respectively. When working in $\omega_{r2} < \omega < \omega_{r1}$ for Region 2 as in Fig. 2(a), the output ripple currents exist a little break each other, which may ensure the output rectifiers switching at zero-current-switching (ZCS) to reduce switching loss.

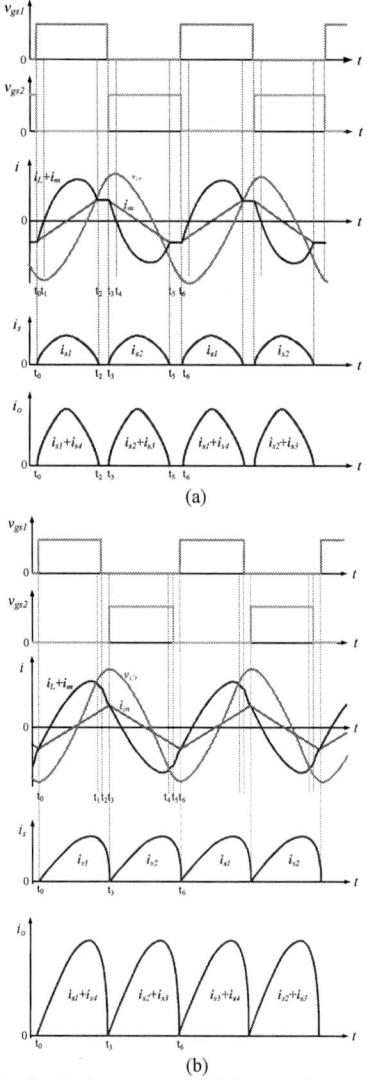

Fig. 2 Predicted switching statuses of LLC conversion operation in (a) Region 2, $\omega_{r2} < \omega < \omega_{r1}$ and (b) Region 1, $\omega > \omega_{r1}$.

As shown in Fig. 2(b), the output ripple currents in Region 1 are link continuously without break, which can sustain the output power supplying continuously but still keep the output rectifiers switching at near ZCS almost in boundary-conduction mode (BCM) to reduce switching loss. In shorts, the LLC converter can always sustain low power loss for the output rectifier as one of the merits. In order to acquire high power output with high efficiency, the operation point in this approach is operated above the inherent frequency ω_{r1} in the Region 1.

III. MODELLING OF THE INTERLEAVED HALF-BRIDGE LLC CONVERTER

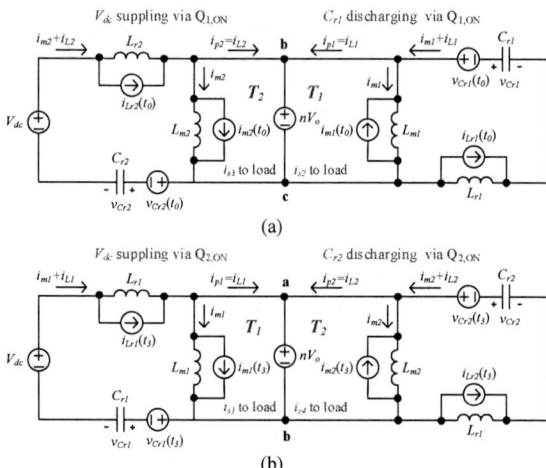

Fig. 3 Modeling of the interleaved statuses, (a) phase 1 for Q1 on and Q2 off, and phase 2 for Q1 off and Q2 on.

The modeling of the interleaved LLC in Fig. 1(a) for switching phases 1 and 2 can be respectively described in Fig. 3. Since under switching period the voltage on the primary windings of the transformers T1 and T2 are almost clamped with a voltage nV_o, which ensures the magnetizing current i_{m1} and i_{m2} are always periodically circulating in the loop of resonant tank but only through the MOSFETs. Thus, only the resonant current i_{L1} and i_{L2} can pass through the transformers to the secondary side. For convenience in analysis, both two parameters of T_1 and T_2 and the tank components are presumed to be the same; i.e. $L_{r1} = L_{r2} = L_r$, $C_{r1} = C_{r2} = C_r$, $L_{m1} = L_{m2} = L_m$, $N_{p1} = N_{p2} = N_p$, $N_{s1} = N_{s2} = N_s$ and turn ratio $n = N_p/N_s$ etc.

(A) *Phase 1*: ($t_0 \le t \le t_3$)

In this phase, Q_1 turns on and Q_2 is in off state. The energy initially supplies to resonant tank 2, in which the energy previously stored in C_{r1} will release to tank 1 via the Q_1. We assume the initial states in tank 1 and tank 2 in steady-state will be $i_{L2}(t_0) = i_{L1}(t_0) = I_{r0} \approx 0$, $v_{Cr2}(t_0) = v_{Cr1}(t_0) = V_{t0}$, $i_{m1}(t_0) = I_{m2t0}$, $i_{m2}(t_0) = I_{m2t0}$. Since all magnetizing currents i_{m2} and i_{m2} previously mentioned are circulation currents with zero mean, the magnetizing current can then be obtained by

124

$$i_m(t) = -\frac{nV_o}{L_m} \cdot \frac{T}{4} cos(\omega t) \tag{1}$$

In the left-hand side of Fig. 3(a), we will have

$$L_{r2}\frac{di_{L2}}{dt} = V_{dc} - nV_o - v_{cr2} \tag{2}$$

and

$$i_{L2} = C_{r2}\frac{dv_{cr2}}{dt} \tag{3}$$

The resonant inductor current $i_{L2}(t)$ and resonant capacitor voltage $v_{Cr2}(t)$ can then be given by

$$\begin{aligned}i_{L2}(t) &= I_{t0}\cos\omega(t-t_0) - \omega C_{r2}[V_{t0}-(V_{dc}-nV_o)]\sin\omega(t-t_0) \\ &\approx -\omega C_{r2}(V_{t0}-nV_o)\sin\omega(t-t_0)\end{aligned} \tag{4}$$

and

$$\begin{aligned}v_{cr2}(t) &= [V_{t0}-(V_{dc}-nV_o)]\cos\omega(t-t_0) \\ &+ \frac{I_{t0}}{\omega_1 C_{r2}}\sin\omega(t-t_0) + (V_{dc}-nV_o) \\ &\approx [V_{t0}-(V_{dc}-nV_o)]\cos\omega(t-t_0)+(V_{dc}-nV_o)\end{aligned} \tag{5}$$

In the right-hand side of Fig. 3(a), we will have

$$V_{cr1} = nV_o + L_{r1}\frac{di_{L1}}{dt} \tag{6}$$

and

$$i_{L1} = C_{r1}\frac{dv_{cr1}}{dt} \tag{7}$$

Then, we have

$$\begin{aligned}v_{cr1}(t) &= (V_{t0}-nV_o)\cos\omega(t-t_0) \\ &+ \frac{I_{t0}}{\omega C_{r1}}\sin\omega(t-t_0) + nV_o \\ &\approx (V_{t0}-nV_o)\cos\omega(t-t_0)+nV_o\end{aligned} \tag{8}$$

and

$$\begin{aligned}i_{L1}(t) &= I_{t0}\cos\omega(t-t_0) - \omega C_{r1}(V_{t0}-nV_o)\sin\omega(t-t_0) \\ &\approx -\omega C_{r1}(V_{t0}-nV_o)\sin\omega(t-t_0)\end{aligned} \tag{9}$$

The sub-current sum $i_{p,1}(t)$ at the primary sides of T_1 and T_2 for phase 1 will be

$$\begin{aligned}i_{p,1}(t) &= i_{p1}(t) + i_{p2}(t) \\ &= \left[-i_{s2}(t) + i_{s3}(t)\right]/n\end{aligned} \tag{10}$$

where $C_{r1} = C_{r2} = C_r$ has been assumed previously. In (10), the $i_{p,1}(t)$ for phase 1 corresponds to the sum of currents the $i_{s2}(t)$ of T_1 and $i_{s3}(t)$ of T_2. From (4)-(5) and (7)-(8), we then have

$$i_{p,1}(t) = -2\omega C_r(V_{t0}-nV_o)\sin\omega(t-t_0) \tag{11}$$

In (11), $i_{p,1}(t)$ corresponds sum of double half-sinusoidal wave at secondary side of the two transformers T_1 and T_2 during half switching period (t_0, t_3) as shown in Fig. 2.

(B) *Phase 2*: $(t_3 \le t \le t_6)$

As shown in Fig. 3(b), in this phase, Q1 turns off and Q2 is in on state. The analysis results are the same as those in phase 1. Accordingly, the $i_{p,2}(t)$ at the primary sides of T_1 and T_2 for phase 2 will be

$$\begin{aligned}i_{p,2}(t) &= i_{p1}(t) + i_{p2}(t) \\ &= \left[i_{s1}(t) - i_{s4}(t)\right]/n\end{aligned} \tag{12}$$

The $i_{p,2}(t)$ for phase 2 corresponds to the sum of currents the $i_{s1}(t)$ of T_1 and $i_{s4}(t)$ of T_2. The resonant inductor current $i_{L1}(t)$ and resonant capacitor voltage $v_{Cr1}(t)$ via Q2 can then be given by

$$\begin{aligned}i_{L1}(t) &= I_{t3}\cos\omega(t-t_3) \\ &- \omega C_{r1}[V_{t3}-(V_{dc}-nV_o)]\sin\omega(t-t_3) \\ &\approx -\omega C_{r1}(V_{t3}-nV_o)\sin\omega(t-t_3)\end{aligned} \tag{13}$$

and

$$\begin{aligned}v_{cr1}(t) &= [V_{t3}-(V_{dc}-nV_o)]\cos\omega(t-t_3) \\ &+ \frac{I_{t3}}{\omega_1 C_{r3}}\sin\omega(t-t_3) + (V_{dc}-nV_o) \\ &\approx [V_{t3}-(V_{dc}-nV_o)]\cos\omega(t-t_3)+(V_{dc}-nV_o)\end{aligned} \tag{14}$$

where $v_{Cr1}(t_3) = V_{t3}$. and

$$\begin{aligned}i_{p,2}(t) &= i_{p1}(t) + i_{p2}(t) \\ &= -2\omega C_r(V_{t3}\ nV_o)\sin\omega(t-t_3)\end{aligned} \tag{15}$$

In (15), $i_{p,2}(t)$ corresponds sum of double half-sinusoidal wave at secondary side of the two transformers T_1 and T_2 during half switching period (t_3, t_6) as shown in Fig. 2.

(C) *Output Current State*

During phase 1 and phase 2, total current $i_p(t)$ equivalently summing at the two primary terminals of the transformer T_1 and T_2 can be described as, from (11) and (15),

$$i_p(t) = i_{p,1}(t) + i_{p,2}(t) \tag{16}$$

Total average output current at the output terminal for interleaved phases can then be easily estimated from either (11) or (15), by

$$\begin{aligned}I_T &= \frac{2}{\pi} \cdot n \cdot 2\omega C_r (V_{t3}-nV_o) \\ &= \frac{4n\omega C_r (V_{t3}-nV_o)}{\pi}\end{aligned} \tag{17}$$

and the output power P_o can be easily given by

$$P_o = V_o I_T \qquad (18)$$

At input side, the input power drawn from the V_{dc} will be

$$
\begin{aligned}
P_{in} &= 2 \cdot V_{dc} \left(i_{Lr} + i_m \right)_{av} \\
&= 2 \cdot V_{dc} \cdot \frac{1}{T} \int_0^{T/2} \left[\frac{\pi}{2} \cdot \frac{1}{n} \cdot \frac{I_T}{2} sin\left(\omega t \right) - \frac{nV_o}{L_m} \cdot \frac{T}{4} cos\left(\omega t \right) \right] dt \\
&= V_{dc} \cdot 2 \cdot \frac{\pi I_T}{4n} \cdot \frac{1}{\pi} \\
&= V_{dc} \cdot \frac{I_T}{2n}
\end{aligned}
$$

$$(19)$$

In (19), the average of the magnetizing current during half-period of the switching period T is zero since it is a steady-state circulation current through the resonant tank without consuming any power from the V_{dc}. In addition, the input power in (19) is for the interleaved LLC converter and is twice that of the general LLC. The power efficiency η can then be given by

$$
\begin{aligned}
\eta &= \frac{P_o}{P_{in}} \\
&= \frac{2nV_o}{V_{dc}}
\end{aligned}
$$

$$(20)$$

Equation (20) is interestingly close to ideal case. However, this estimation is quite close to the experimental result in our study.

IV. DESIGN, IMPLEMENT, AND RESULTS

In this paper, a prototype of a 2.7 kW high-power interleaved half-bridge LLC converter is designed and implemented, in which the full output is specified as 54 V/ 50 A for server use and the $V_{dc} = 390$ V is from a PFC. In-phase and 90°-phase-shift output currents merging at the output terminal to reduce the output ripple currents are respectively explored in experiment.

Table 1 Parameters Estimation for each resonant tank

L_m (µH)	280
L_r (µH)	42
C_r (nF)	94
n	3.5 (21:6)
f_r (kHz)	80.1
f_m (kHz)	28.9
$K = L_m / L_r$	6.667
Gain	0.969
V_{in} (V)	390
V_{out} (V)	54
I_{out} (A)	25

To equalize and minimize the output current ripple, an equal-length wire for the secondary winding is wound and the operation point is placed in the Region 1 such that the neighbor output current ripple without break.

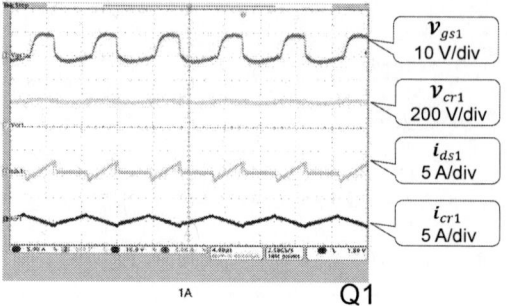

(a-1)

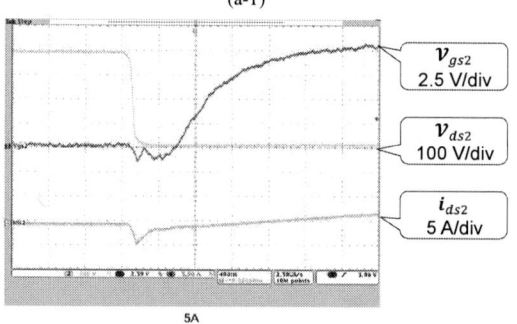

(a-2)

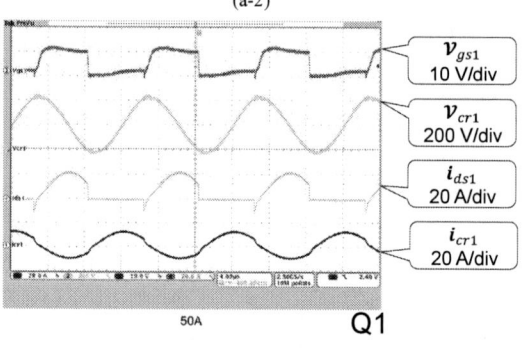

(b-1)

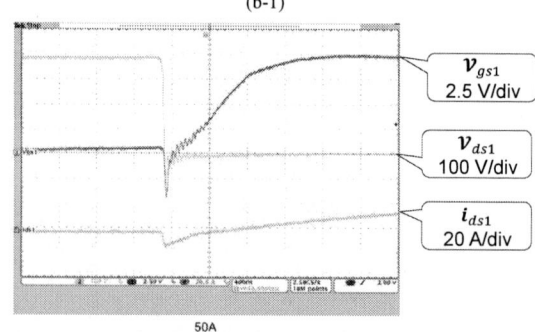

(b-2)

Fig. 4. The switching states of MOSFET Q1 working in phase 1, (a) light load of 1 A, ZVS exits as shown in (a-2); (b) heavy load of 50 A, ZVS behavior as shown in (b-2).

The 2018 International Power Electronics Conference

In this design, the inherent resonant frequency f_{r1} is 80.1 kHz and the second resonant frequency f_{r2} is 28.9 kHz with the resonant inductance $L_r = 42$ μH and magnetizing inductance $L_m = 280$ μH, all parameters are listed in Table 1. In implementation, the cool MOSFET IPW60R041P6 is used. The operation frequency spans from 142 kHz to 82 kHz in the Region 1 for the load change from 1A to 50 A. Figure 4 shows the states of the MOSFET Q1 for load currents at 1 A (light load) and 50 A (heavy load). In Fig. 4(a-1), at light load situation, the i_{ds1} is almost the magnetizing current; however, the zero-voltage switching (ZVS) still exists as shown in the Fig. 4(a-1).

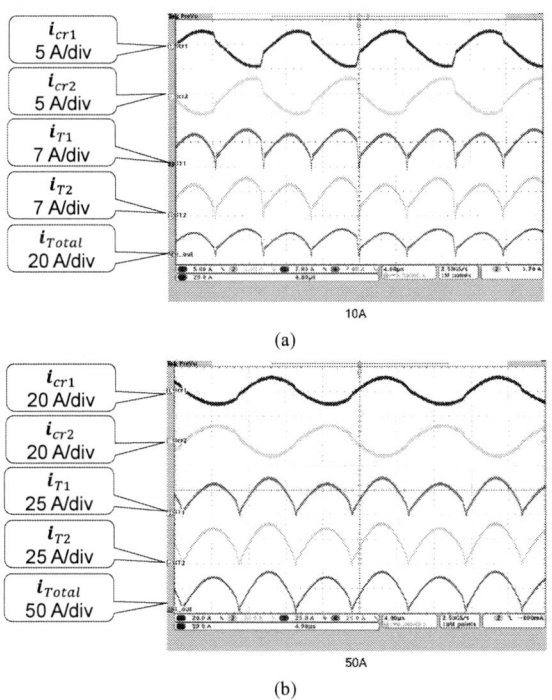

Fig. 5 The output current ripple behaviors, including the currents i_{T1} and i_{T2} at the secondary's of T1 and T2; total sum of current i_T before filtering, under the output currents at (a) 10 A and 50 A.

As for the heavy load condition, the i_{ds1} shown in Fig. 4(b-1) is quite as the prediction in Fig. 2(b); the ZVS is certainly well done as shown in Fig. 4(b-2).The responses of the output current ripples i_{T1} and i_{T2} at the secondary's of the transformers T1 and T2, and the total output current i_T before filtering, under the load of 10A and 50 A are respectively shown in Fig. 5. Both two current ripple trains of i_{T1} and i_{T2} are almost equal peak magnitude. Only a little peak droop happens during the discharging period of the resonant capacitor. This is because the capacitor energy release to the resonant tank may be a little bit tilt voltage but not a constant V_{dc}.

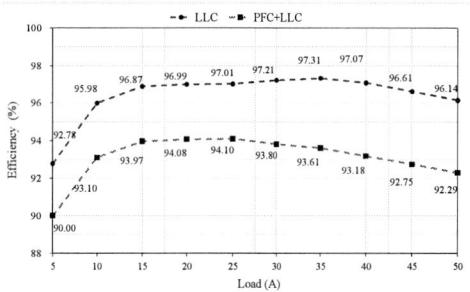

Fig. 6 Measured efficiency of the 2.7 kW interleaved half-bridge LLC converter

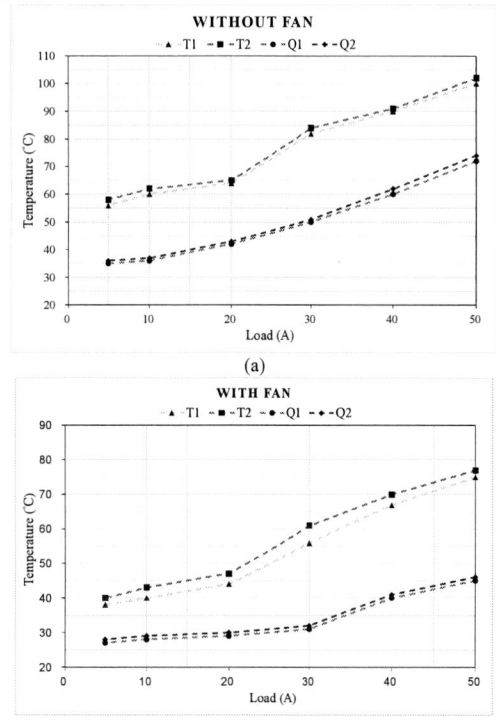

Fig. 7 Measured temperature response of the transformer and the power MOSFETs (a) without fan and (b) with fan cooling, all measured at 25°C.

The power efficiency is measured in Fig. 6, in which the highest efficiency 97.31% occurs at 35 A and 96.14% at heavy load 50 A. From 10 A, the efficiencies are above 96% for all load currents above 5 A; even at light load of 5A, the efficiency still reaches 92.79%. Including the PFC, all efficiencies can also sustain between 90% and 94.1% for all load range. The temperature distributions in the transformers and MOSFETs are shown in Fig. 7, in which the highest temperatures on transformers T_1 and T_2 at heavy load of 50 A without fan cooling are around 101°C and are 75°C with fan cooling. As for MOSFETs at heavy load of 50 A, the highest temperature without fan cooling are 73°C and are 46°C with fan cooling. Comparison of the current ripples in the output capacitor is measured in Fig. 8 under heavy load condition of 50 A.

127

The 2018 International Power Electronics Conference

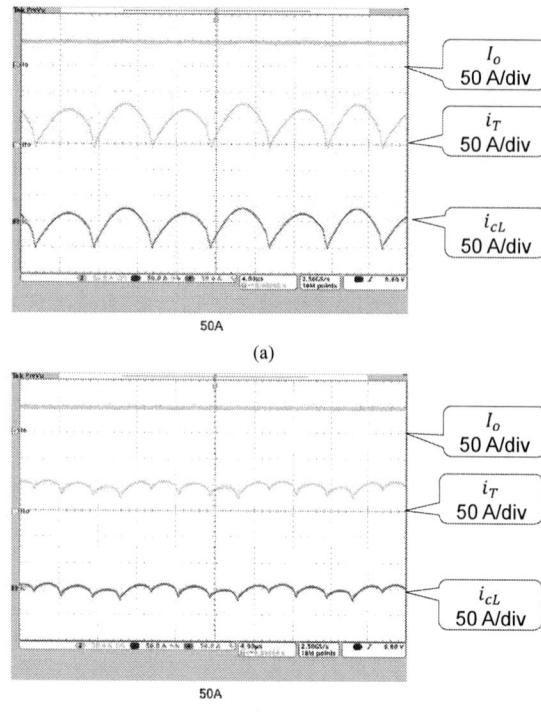

(a)

(b)

Fig. 8 The behaviors of ripple current flowing through the output filter capacitor C_L, measured under heavy load condition of 50 A: (a) in in-phase current merge case, the ripple current is as high as 79.6 A_{p-p} or 20.2 A_{rms}; (b) in 90°-phase-shift current merge case, 30.8 A_{p-p} 6.8 A_{rms}.

The ripple current flowing through the output filter capacitor C_L is as high as 79.6 A_{p-p} or 20.2 A_{rms}, in in-phase current merge case; is 30.8 A_{p-p} or 6.8 A_{rms}, in 90°-phase-shift current merge case, respectively. Obviously, the expected performance of the proposed approach is close to the experimental results. Fig. 9 shows the prototype of 2.7 kW interleaved half-bridge LLC converter including PFC, in which the power density is 17.9 W/in³ suitable for industry application.

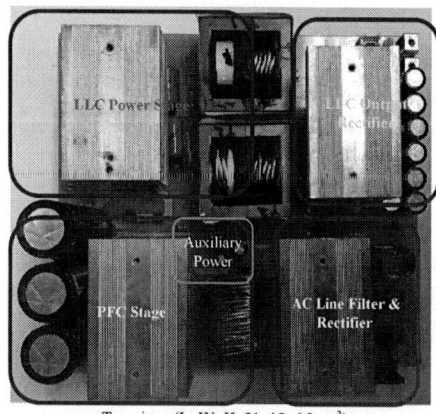

Top view: (L×W×H: 21×19×6.2 cm³)
(a)

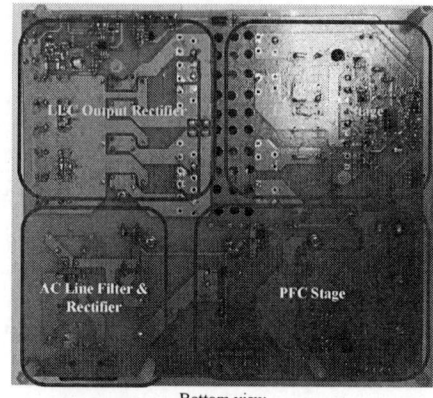

Bottom view
(b)

Fig. 9 The prototype of the 2.7 kW interleaved half-bridge LLC converter including PFC

V. CONCLUSIONS

This paper presents a simple modeling for describing the interleaved half-bridge LLC converter. The approach to equalize and minimize the output current ripple is performed by using equal-length wire for secondary winding and placing the operation point in the Region 1, which makes all output current ripple magnitudes almost equal without any break between neighbor current ripples. In addition, a 90°-phase-shift ripple current merge at output terminal is also studied and experimented to compare the ripple current flowing via the filter capacitor in the in-phase merge case. Under heavy load condition of 50 A, the ripple current in the output filter capacitor C_L for the 90°-phase-shift case is just only 37% of that in the in-phase merge case, which dramatically reduces the current stress in the output capacitor.

REFERENCES

[1] R. Liu, and C.Q. Lee, "Analysis and design of LLC-type series resonant converter," *IEE Electron. Letter*, vol. 24, no. 24, pp. 1517-1519, 1988.

[2] R. Liu and C.Q. Lee, "The LLC-type series resonant converter-variable switching frequency control," in *Proc. 32th Midwest Symp. Circuits and System*, 1990, pp. 509-512.

[3] B. Yang, F. C. Lee, A. J. Zhang and G. Huang, "LLC resonant converter for front end dc/dc conversion," in *Proc. IEEE Applied Power Electronics Conference (APEC)*, pp. 1108-1112, Mar. 2002.

[4] B. Lu, W. Liu, Y. Liang, F. C. Lee and J. D. van Wyk, "Optimal design methodology for LLC resonant converter," in Proc. IEEE Applied Power Electronics Conference (APEC), pp. 533-538, Mar. 2006.

[5] B.-Chul Kim, K.-Bum Park, C.-Eun Kim, and G.-Woo Moon, "Load sharing characteristic of two-phase interleaved LLC resonant converter with parallel and series input structure," in *Proc. IEEE Energy Conversion Congress and Exposition (ECCE)*, pp. 750-753, Sep. 2009.

[6] E. Orietti, P. Mattavelli, G. Spiazzi, C. Adragna and G. Gattavari,"Two-phase interleaved LLC resonant converter with current controlled inductor," in *Proc. IEEE Brazilian Power Electronics Conference (COBEP)*, pp. 289-304, Sep. 2009.

[7] Z. Hu, Y. Qiu, Y. Fei Liu, and P. C. Sen, "An interleaving and load sharing method for multiphase LLC converters," in *Proc. IEEE Applied Power Electronics Conference (APEC)*, pp. 1421-1428, Mar. 2013.

[8] K. Murata and F. Kurokawa, "A Novel Interleaved LLC Resonant Converter with Phase Shift Modulation," in *Proc. IEEE Energy*

128

Conversion Congress and Exposition (ECCE), pp. 750-753, Sep. 2014.

[9] Z. Hu, Y. Qiu, L. Wang, and Y.F. Liu, "An Interleaved LLC Resonant Converter Operating at Constant Switching Frequency," *IEEE Trans. Power Electron.* vol. 29, no. 6, 2014.

The 2018 International Power Electronics Conference

Multi-Chip SiC MOSFET Power Modules for Standard Manufacturing, Mounting and Cooling

Alberto Castellazzi[1*], Asad Fayyaz[1], Emre Gurpinar[2], Abdallah Hussein[1], Jianfeng Li[1], Bassem Mouawad[1]

[1] Power Electronics, Machines and Control Group, University of Nottingham, Nottingham, UK
[2] Electrical and Electronics Systems Research Division, Oak Ridge National Laboratory, Oak Ridge, Tennessee, USA
*E-mail: alberto.castellazzi@nottingham.ac.uk

Abstract— **Taking full advantage of the superior characteristics of SiC Power MOSFETs in the application requires the development of bespoke packaging solutions. Their design needs to thoroughly encompass electro-magnetic and electro-thermal aspects to yield major system-level benefits. New design approaches are needed in particular for parallel multi-chip structures at higher voltage ratings. With the aim of enabling full exploitation of the disruptive potential of SiC technology, this paper proposes a review of learnings made in the development of SiC bespoke power modules, focusing in particular on module designs compatible with the most widely established manufacturing, converter assembly and thermal management solutions for large volume applications.**

Keywords— *The authors shall provide up to 4 keywords or phrases (in alphabetical order and separated by commas) to help identify the major topics of the paper.*

I. INTRODUCTION

The analysis of a very broad range of power electronic converters can be reduced, in terms of the operation of the semiconductor devices within them, to the study of a half-bridge switch (HBS) architecture with free-wheeling diodes, a characteristic building block of a large category of inverters and dc-dc converters. In the case of power MOSFETs, a typical arrangement is as depicted in Fig. 1 a), which is most often operated with a switching sequence as in Fig. 1 b): between turn-off of the high-side transistor (HST) and turn-on of the low-side one (LST), and vice-versa, a dead-time is interposed to avoid shoot-through phenomena; at least during the dead-times, in the most typical situation of inductive load, continuous current conduction needs to be ensured by the high-side and low-side free-wheeling diodes (HFD and LFD, respectively). After that, synchronous rectification over the transistors channel is a usually implemented option, given the good third quadrant conduction characteristics of MOSFETs. So, the dead-time can be made a more or less significant portion of the overall transistor conduction time in any given modulation scheme, depending primarily on the switching speed capability of the devices. Interest exists to minimize its duration, as this is beneficial from both an efficiency point of view and, in AC systems, for the harmonic signature of the output waveforms [1]. Depending on the specific application and design targets, the freewheeling diodes in

the HBS can be intrinsic to the MOSFET (body-diode) or external components can be used (with SiC, the most typical solution is to use Schottky diodes): whenever possible, use of the MOSFET intrinsic diode is looked at as an opportunity to contain costs and increase power density. SiC MOSFET technology has progressed significantly over the last few years and most available devices nowadays, enable a reliable use of the MOSFET body-diode, having solved long-standing issues such as stacking faults and bipolar degradation in the SiC crystal [2].

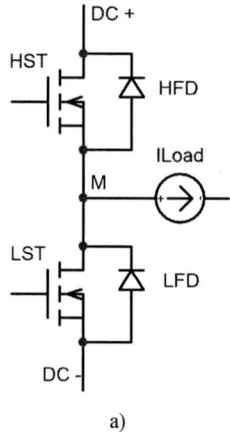

a)

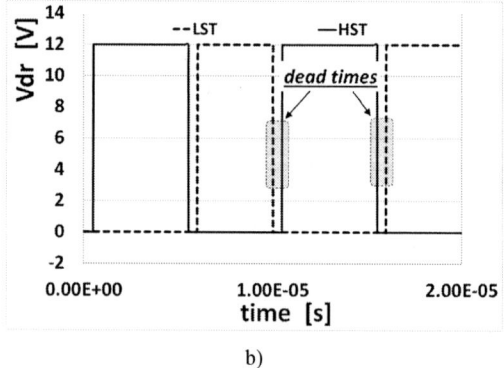

b)

Fig. 1. a), schematic diagram of a half-bridge switch powering an inductive load; b), corresponding switching sequence implementing synchronous rectification with variable dead-times.

In physically realizing the switch topology of Fig. 1, it is very important to optimize the design of the current commutation loop between HST and LFD and LST and HFD to minimize parasitic inductance components which can result in undesirable voltage overshoots and ringing during the switching transitions. This is best achieved if the devices are co-packaged within the same module, a realization which has recently led to the development of new half-bridge type industrial package standards [3]. SiC MOSFETs can easily be switched ten times faster than Si IGBTs in the same voltage class. So, containment of parasitic inductance associated with layout and assembly is paramount for the achievement of good switching performance.

II. 1.2 kV POWER MODULE

Fig. 2 shows a 3-phase all SiC MOSFET half-bridge power module, developed for use in a 2-level chopper/inverter topology. Fig.2 a) shows a detailed view of the individual switch: only MOSFETs were used, rated at 1.2 kV-30 A, without nay external free-wheeling diode.

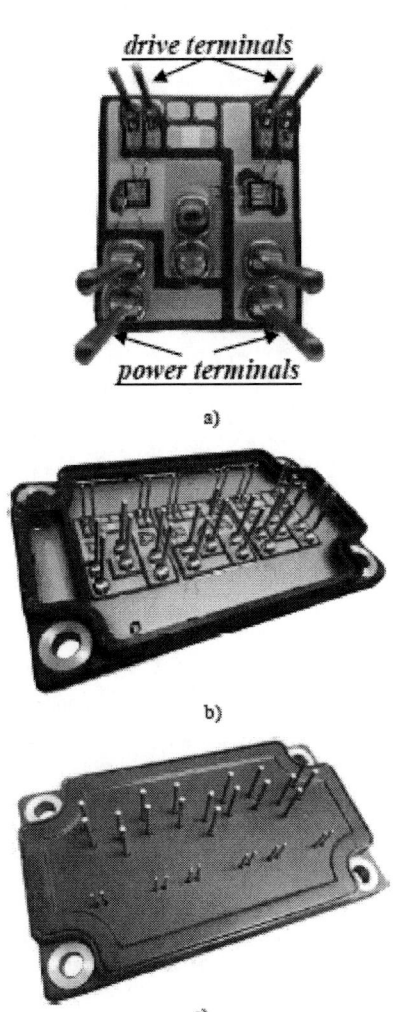

Fig. 2. Assembled switches a), full power module b) and encapsulated device c).

The main objectives in the development and testing of this module were: to assess the reliability of a small footprint/high-power density lightweight solution with thinned baseplate; investigate the electro-thermal and electro-magnetic operational limits. Extensive design and experimental characterization results are reported in [4, 5].

The design feature which mainly distinguish it from established solutions developed for Si are:

-        built-in reliability design approach, with optimum matching of substrate, base-plate and solder materials and thicknesses to yield minimum thermo-mechanical stresses and the required target reliability, using a thin baseplate (2mm vs. the more traditional 3mm);

-        low-parasitic inductance by layout design and increased power connector sizes;

-        full separation of the gate-drive and power loops, so as to avoid delayed switching and higher power dissipation due voltage drops on the gate-loop internal source terminal parasitic inductance.

The module was instrumental in demonstrating extremely high switching speeds, well below 100 ns over a very broad range of bias conditions. Fig. 3 shows results for the turn-on and turn-off waveforms at 540 V input voltage and 20 A peak output current.

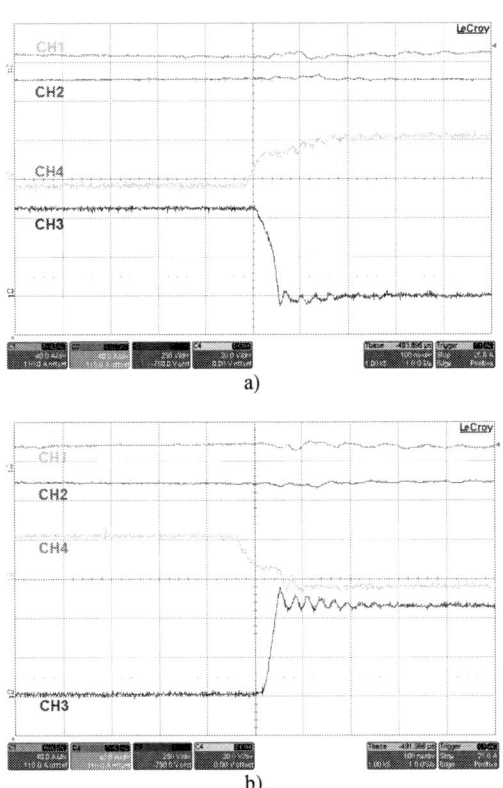

Fig. 3. Turn-on, a), and turn-off waveforms, b), at fsw=100 kHz, full-load (CH1 and CH2: Current in phase A and B, 40A/div,   CH3:Vds=250V/div,   CH4:Vgs=20V/div, timebase=100ns/div).

It was also used to illustrate the drastic reduction in output filter inductance that can be achieved by the increased switching frequencies, in particular going up to 100 kHz at an output load of 6 kW. Fig. 4 a) shows the phase output currents with a filter inductance of 3 mH; Fig. 4 b) shows the phase output current without any output filter inductance other than that associated with connection cables and resistors.

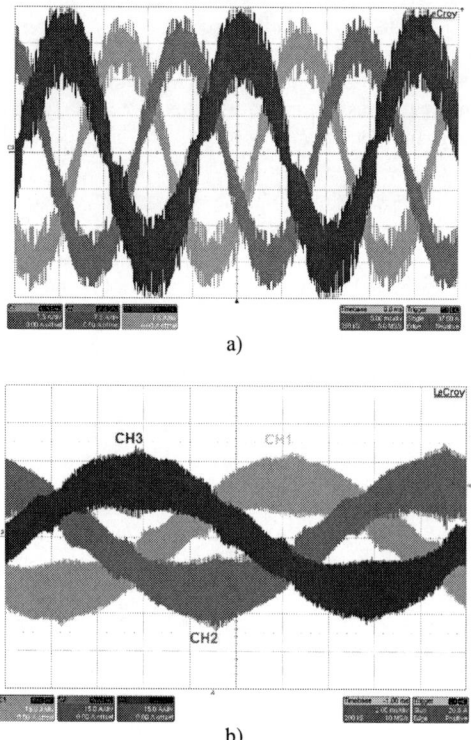

a)

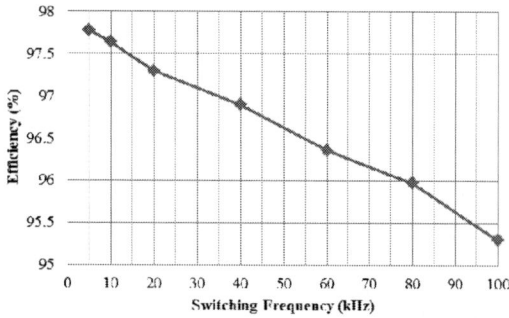

b)

Fig. 4. Output current waveforms at f_s=100 kHz, full-load (CH1= Phase A current 15 A/div, CH2= Phase B current 15 A/div, CH3= Phase C current 15A/div, timebase= 2ms/div).

Fig. 5 shows the measured efficiency over switching frequency at 6 kW output power: as can be seen, the performance at higher switching frequencies remains credible and viable, especially for motor-drive type applications, where the overall system efficiency is in any case not primarily determined by the power electronics.

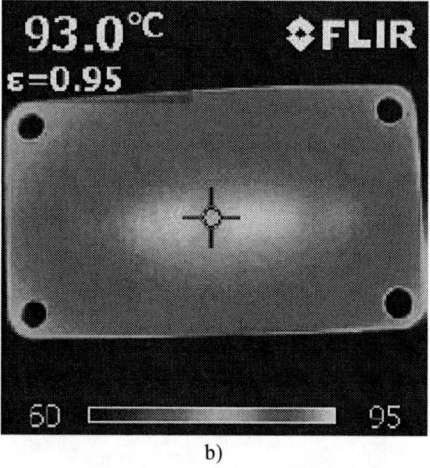

Fig. 5. Efficiency versus switching frequency for an output power of 6 kW.

In terms of handling the higher losses at module level, it is interesting to also point out that the module was operated safely without any heatsink, reaching a baseplate temperature of 90 °C at 2.5 kW output load for a 5 kHz switching frequency and 1.5 kW output load for an 80 kHz switching frequency. Fig. 6 shows the corresponding thermal maps of the module baseplate.

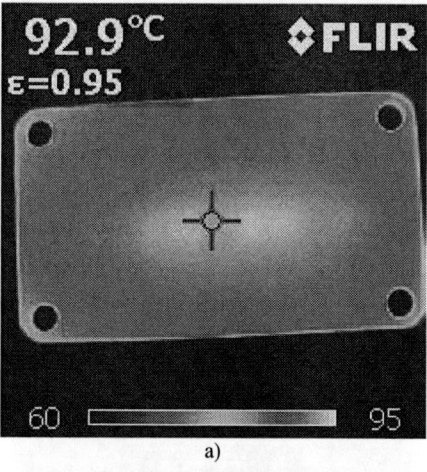

a)

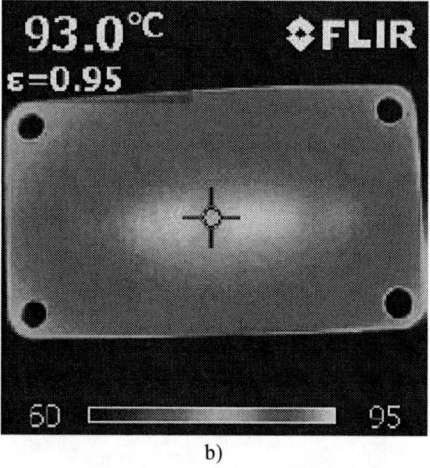

b)

Fig. 6. Thermal maps of the module baseplate: a), P$_{OUT}$=2.5 kW and fs=5kHz; b), P$_{OUT}$=1.5kW and fs=80 kHz.

The main limitation in deriving a markedly disruptive performance in the application of the module was due relatively large common mode currents flowing at the higher end of the switching frequency range. In a power module, common mode currents are mainly contributed by the parasitic capacitance associated with terminal M in the schematic illustration of Fig. 1 a). In a SiC MOSFET HBS, this node swings between DC+ and DC- voltages at high frequencies and relatively high voltages, with very high dV/dt's. So, it is crucial to contain the value of parasitic capacitance between this node and ground if common mode currents are to be limited. This aspect was duly taken into account in the design of a second module generation, populated this time with state-of-the-art 3.3 kV SiC MOSFETs [6].

132

III. 3.3 kV POWER MODULE

Here, a single-phase HBS was built, but using two parallel devices for both the HSM and LSM to achieve a nominal 100 A current rating and still relying only on the transistors body-didoe for the current-freewheeling action. The module design is illsutrated in Fig. 7 a). Many of the design features of the previous version are kept for minimsing parasitic inductance and ensuring a fast switching symmetrically balanced electro-magnetic circuit. However, the extension of the mounting track for the LSM collector is intentionally reduced to minise commone mode capacitance towards the baseplate underneath (grounded in the application). The amount of metal track size reduction that can be realistically achieved is determined by the need for trading-off on thermal performance and esnuring uniform temeprature distribution and symmetrical electro-thermal performance for all devices. Design optimisation was based on extensive structural simulation, of the kind shown in Fig. 7 b) [7].

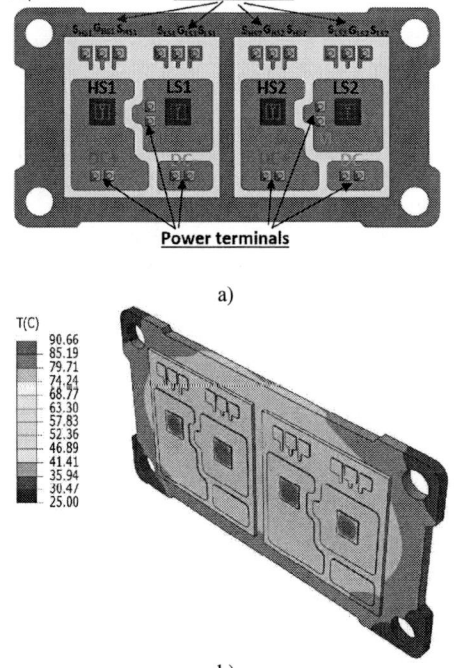

a)

b)

Fig. 7. Module design of the 3.3 kV-100A half-bridge module: a), module layout; b), thermal simulation showing uniform temperature distribution even with asymmetrical sizing of the copper tracks used for mounting and cooling the chips.

Fig. 8 shows the module hardware prototype, fully covered in insulating gel, but still without the lid. The module was tested extensively for switching performance of the single and parallel dies, including a parametric analysis of switching energy over temperature, load current and input voltage. Reported in Fig. 9 are representative voltage and current switching waveforms at 1.8 kV and 30 A (per transistor), which show the ability to operate with significant dV/dt's values.

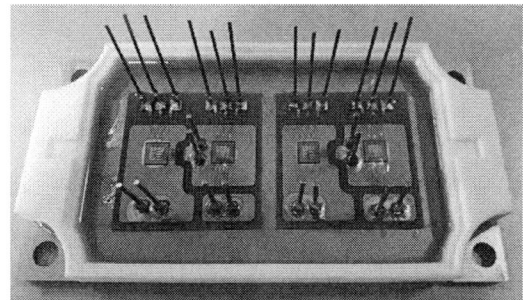

Fig. 8. 3.3 kV -100 A SiC MOSFET power module prototype.

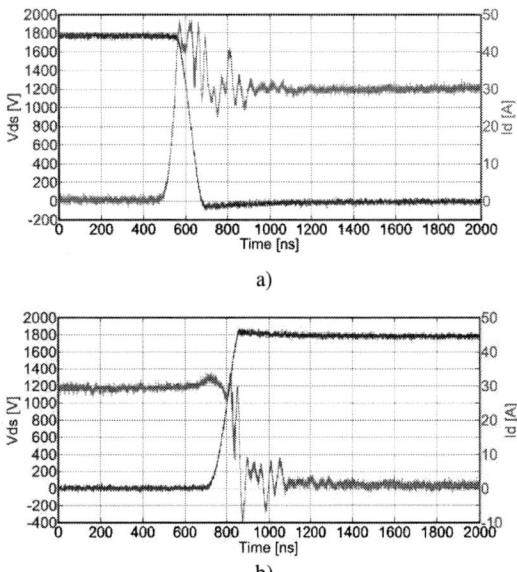

a)

b)

Fig. 9. Representative voltage and current waveforms at turn-on, a), and turn-off, b).

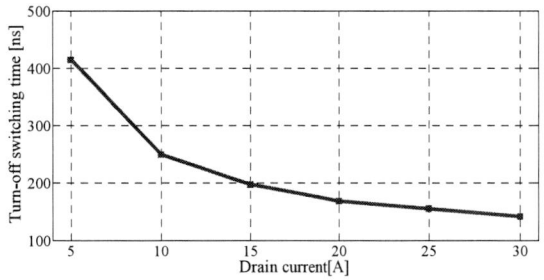

Fig. 10. Turn-off switching time as a function of load current.

Interesting in relation to this are the results of Fig. 10, which report a marked dependence of the turn-off switching time as a function of the load current, an effect which was not highlighted in the testing of devices with lower voltage rating and which needs to be duly taken into account in the optimization of the design parameters of a power converter. The longer switching times do not adversely affect switching energy, since they are associated with lower load current values. Moreover, since realistically the target switching frequencies at higher voltages are reduced compared to lower voltage applications, the maximum measured values are deemed fully suitable for the achievement of excellent performance.

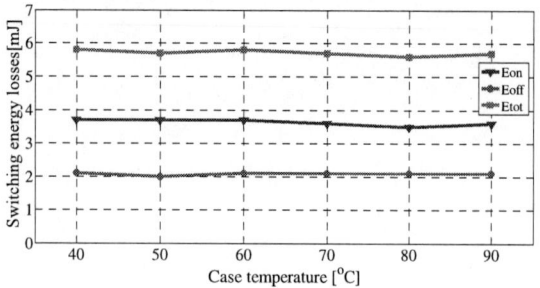

Fig. 10. Turn-off switching time as a function of load current.

IV. MULTI-CHIP POWER MODULE ROBUSTNESS

In power electronics applications, semiconductor power devices are intended mainly to operate as high-frequency switches, which are either fully on, that is, conducting current with virtually zero voltage drop between their terminals, or fully off, that is, blocking voltage with virtually zero current through them. The on and off bias points and the switching trajectories between the two points need to be fully contained within the so-called *Safe Operating Area* (SOA) of the transistors or diodes. The situation is illustrated in Fig. 11.

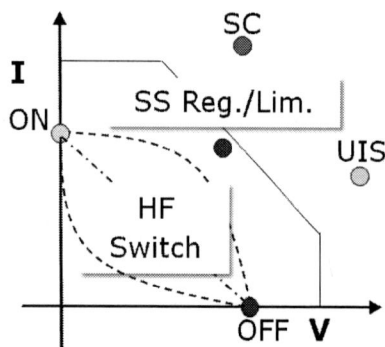

Fig. 11. Illustration of power device operation against its nominal SOA.

As depicted, next to their nominal operation, the devices are also expected to safely withstand a number of *Edge-of-* or *Out-of-*SOA transient events: short circuit (SC) is probably the most important one, being required in virtually all motor drives applications (a very large percentage of all power electronics applications); for the specific case of SiC MOSFETs, unclamped inductive switching (UIS) is also of interest, since the material properties and device features enable dissipation of significant energy values in the avalanche regime [8]; solid-state (SS) current regulation and limiting are gaining more and more relevance. Although some of these operational situations may be non-intentional, that

does not necessarily mean that they are infrequent during system operation. So, relevant are both the devices *single event* withstand capability and its aging as a result of *repetitive stress* application. The performance, robustness and failure mechanisms of single SiC devices have been exhaustively explored in relatively recent studies [1, 8-9], which have generally highlighted steady progress in technology development, with the recent overcoming of long-standing issues, such as threshold voltage, *Vth*, stability of SiC MOS interfaces or body-diode degradation as a result of stacking-faults and bipolar degradation. Also, short-circuit and avalanche robustness have been investigated in depth, pointing out limitations and ways to improve robustness where strictly necessary [10-11]. Thorough investigations of parallel chips robustness are, on the other hand, still wanting, with very few initial studies published on the subject [12].

For example, Fig. 12 a) reports a measure of the threshold-voltage value, *Vth*, on a number of commercial 1.2 kV devices (not of latest generation): the values are contained within data-sheet specification, but the spread is much more significant than in silicon and in view of the switching speed of SiC MOSFETs. Moreover, the quantitative spread is temperature dependent, as Fig. 12 b) shows, so that transistors with a Vth difference of about 250 mV at ambient temperature may end up having nearly twice as much as the temperature is increases towards the nominal maximum of 150 °C (it is worth considering that during transient operation, the actual device junction temperature can rise well beyond the maximum nominal steady-state value, which is mainly package related).

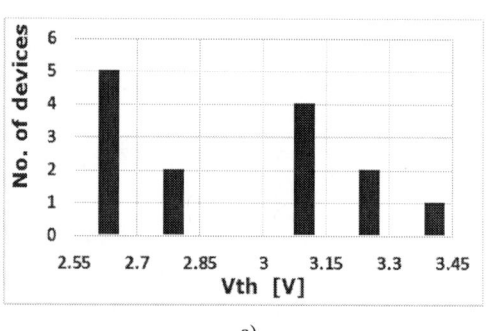

a)

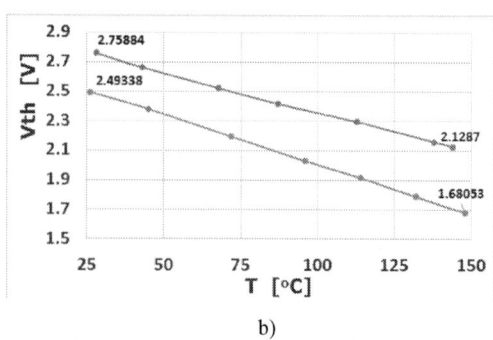

b)

Fig. 12. Measured *Vth* distribution on a sample population of 14 SiC MOSFETs (1200V – 80 mΩ; data-sheet specification is 2-4 V), a), and *Vth* temperature variation for two devices, b).

As further shown in Fig. 13, which reports the measured short circuit current waveforms for some of the devices from the lot of Fig. 12, the implication of the *Vth* difference alone can be, for instance, that devices used in parallel within a module are actually exposed to very different stress levels during short-circuit events, with a significant difference in the maximum chip temperature in terms of module reliability and lifetime and inherent presence of a *weak-spot* for parallel use, requesting the introduction of bespoke derating measures.

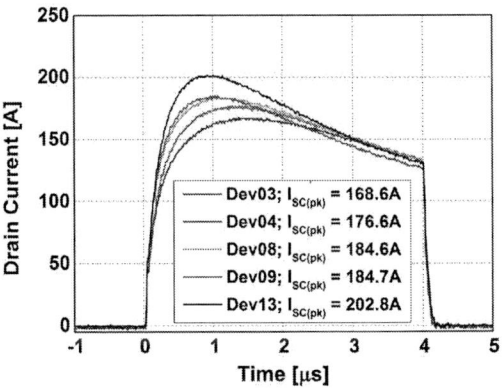

Fig. 13. Experimental short-circuit waveforms for *Vth* ranging between 2.55 and 3.45 V (in these tests, V_{DS} = 600 V, T_{CASE} = 300 K and V_{GS} = 18 V).

In relation to unclamped inductive switching and avalanche energy dissipation capability, Fig. 14 reports the distribution of measured breakdown voltage, V_{BD}, against *Vth*. Here, the spread is quite contained, compatible with statistical device design and processing parameters spread. It is worth stressing that although the results of Fig. 14 seem to indicate a correlation between V_{BD} and *Vth*, specifically, that devices with lower *Vth* exhibit higher V_{BD}, the present interpretation is actually only that probably devices from two distinct manufacturing batches were tested here, one group having somewhat higher V_{BD} and one lower and each group featuring some spread in their *Vth* value.

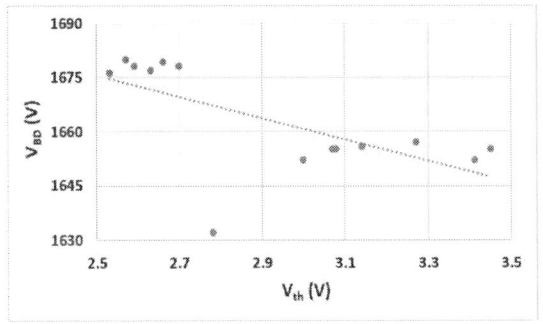

Fig. 14. Measured breakdown voltage, V_{BD}, over threshold voltage, *Vth*, for the same devices used for the results of Fig. 12.

Results of dynamic UIS tests run with different device pairs connected in parallel, have consistently indicated a behavior as illustrated by the results of Fig. 15, which

shows representative drain-source voltage, V_{DS}, and drain current, I_D, for two transistors (indicated as Dev06 and Dev14, respectively) during avalanche breakdown. As can be seen, when the SiC MOSFETs are first turned-off, at time t=0 µs, one of the two (Dev06) takes on the full current, while the other (Dev14), turns-off: this is attributed to Dev14 having a higher Vth value and thus turning-off sooner than Dev06. Dev06 is subsequently also the first to go into avalanche breakdown, having also slightly lower value of V_{BD}; however, shortly after avalanche has commenced, V_{BD}, of Dev06 increases as a result of temperature and soon causes Dev14 to also avalanche. Thus, the load current tends to be shared uniformly and tests run at higher current values have confirmed such behavior and the possibility to robustly use devices in parallel for energy dissipation of UIS.

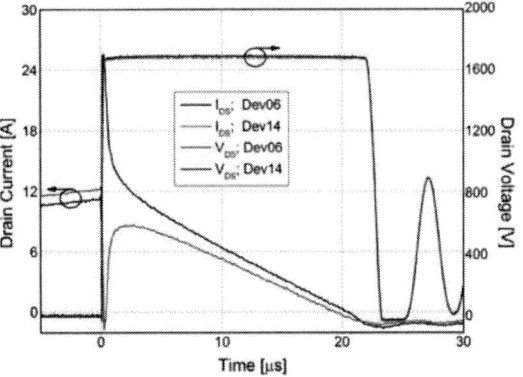

Fig. 15. Measured drain current, I_D, and drain-source voltage, V_{DS}, of two parallel SiC MOSFETs during a UIS event.

Functional tests on parallel devices in the higher voltage class, 3.3 kV, have also indicated very good parallel performance, with very contained spread of the main device parameters [13]. For illustration, Fig. 16 proposes the body-diode currents of two parallel MOSFETs during free-wheeling and subsequent turn-off: the only apparent difference is in the value of the reverse recovery current peak.

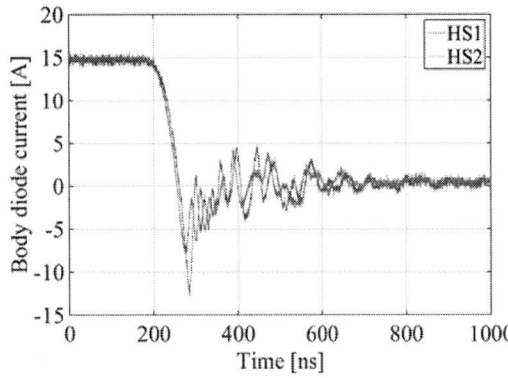

Fig. 16. Measured breakdown voltage, V_{BD}, over threshold voltage, *Vth*, for the same devices used for the results of Fig. 12.

V. CONCLUSION

This paper has presented and discussed the development of multi-chip power modules using well established design solutions and assembly methodologies for the specific case of fast-switching SiC power MOSFETs of different voltage classes. Next to the broadly acknowledged need to minimize parasitic inductance, largely shared with Si power modules, key aspects of performant design in the case of SiC are:

- separation of gate drive and load current switching loops;
- minimization of parasitic capacitance and common mode currents;
- symmetry of layout and device screening for device parameter matching in parallel operation.

Hints at the need for bespoke investigations of SiC MOSFET paralleling derating guidelines have also been proposed and, in the near future, it is expected that comprehensive evaluation of higher voltage (3.3 kV, 6.5 kV) device robustness towards short-circuit and other stressful events will also commence. Particular attention in this regard is dedicated to avalanche robustness and unclamped inductive switching capability: this feature is only exploited in Si at low voltages (indicatively, below 100-200 V) and has come as the result of significant effort invested in the improvement of device design and manufacturing to avoid the activation of the parasitic bipolar BJT in MOSFETs. In SiC, the ability to withstand significant avalanche energy dissipation comes as a result of the material properties (the wider band-gap) and can in principle be taken advantage of even in high voltage devices. Indeed, this might very well be one feature enabling a disruptive revision of traditional power system design in the high voltage high power class (e.g., railway, naval, surface traction; wind), revisiting established voltage de-rating rules and snubbering/protection circuit deployment towards more competitive system development. As such, ability to exploit this particular feature could play a key role in the discussion on compensating the higher purchase price of the semiconductors with cost reductions in other parts of the system. For this to happen, however, the margin between nominal and actual breakdown voltage will need to decrease or shift towards multi-level topologies employing devices with lower voltage rating should be considered. Concurrent improvements in passive device technology will be crucial for enabling more marked improvements and achieving best performance in future SiC based power converters.

Finally, alternative mounting schemes characterized in particular by the removal of the base-plate or its replacement with either ceramic or Insulated Metal Substrate (IMS) components have also been presented, which enable more performant or more cost-effective design, contributing the possibility of counterbalancing higher semiconductor material price, with savings elsewhere in the system [14, 15].

ACKNOWLEDGMENT

The authors wish to acknowledge Mr. Takui Sakaguchi, Dr. Takashi Nakamura and Mr. Masaharu Nakanishi of ROHM Semiconductors, Japan, for supplying the 3.3 kV MOSFETs and information about them and Dr. Prasad Bhalerao of ROHM Semiconductors GmbH, Germany, for effectively managing the communication and information exchange.

REFERENCES

[1] A. Castellazzi, A. Fayyaz, G. Romano, L. Yang, M. Riccio, A. Irace, *SiC power MOSFETs performance, robustness and technology maturity*, Microelectronics Reliability Volume 58, March 2016, Pages 164-176 (Invited paper).

[2] T. Kimoto, *Material science and device physics in SiC technology for high-voltage power devices*, Japanese Journal of Applied Physics, Volume 54, Number 4.

[3] D. Kawase, M. Inaba, K. Horiuchi, K. Saito, *High voltage module with low internal inductance for next chip generation - next High Power Density Dual (nHPD2)*, in Proc. PCIM Europe 2015, Nuremberg, Germany, 19-20 May 2015.

[4] J. Li, E. Gurpinar, S. Lopez-Arevalo, A. Castellazzi, L. Mills, *Built-in reliability design of a high-frequency SiC MOSFET power module*, in Proc. IPEC2014, Hiroshima, Japan, 18-21 May 2014.

[5] E. Gurpinar, S. Lopez-Arevalo, J. Li, D. De, A. Castellazzi, L. Mills, *Testing of a Lightweight SiC Power Module for Avionic Applications*, in Proc. PEMD2014, Manchester, UK, 8-10 April 2014.

[6] T. Sakaguchi, M. Aketa, T. Nakamura, M. Nakanishi, M. Rahimo, *Characterization of 3.3kV and 6.5kV SiC MOSFETs*, in Proc. PCIM2017, Nuremberg, Germany, 2017.

[7] B. Mouawad, A. Hussein, A. Castellazzi, *A 3.3 kV SiC MOSFET Half-Bridge Power Module*, in Proc. CISP2018, Stuttgart, Germany, 20-22 March, 2018.

[8] A.Fayyaz, L.Yang, M.Riccio, A.Castellazzi, A.Irace, *Single pulse avalanche robustness and repetitive stress ageing of SiC power MOSFETs*, Microelectronics Reliability Volume 54, Issues 9–10, September–October 2014, Pages 2185-2190.

[9] A. Castellazzi, A. Fayyaz, G. Romano, M. Riccio, A. Irace, J. Urresti-Ibanez, N. Wright, *Transient Out-of-SOA Robustness of SiC Power MOSFETs*, in Proc. IRPS2017, Monterey, California, USA, 2-5 April, 2017.

[10] G. Romano, A. Fayyaz, M. Riccio, L. Maresca, G. Breglio, A. Castellazzi, A. Irace, *A Comprehensive Study of Short-Circuit Ruggedness of Silicon Carbide Power MOSFETs*, IEEE Journal of Emerging and Selected Topics in Power Electronics (Volume: 4, Issue: 3, Sept. 2016).

[11] A. Fayyaz, G. Romano, J. Urresti, M. Riccio, A. Castellazzi, A. Irace, N. Wright, *A Comprehensive Study on the Avalanche Breakdown Robustness of Silicon Carbide Power MOSFETs*, Energies 2017, 10(4), 452.

[12] A. Castellazzi, A. Fayyaz, R. Kraus, *SiC MOSFET Device Parameter Spread and Ruggedness of Parallel Multichip Structures*, in Proc. ICSCRM2017, Washington D.C., USA, 17-22 Sep., 2017.

[13] A. Hussein, B. Mouawad, A. Castellazzi, *Dynamic performance analysis of a 3.3 kV SiC MOSFET half-bridge module with parallel chips and body-diode freewheeling*, in Proc. ISPSD2018, Chicago, Illinois, USA, May 2018.

[14] B. Mouawad, J. Li, A. Castellazzi, P. Friedrichs, M. Jonnsohn, *Low parasitic inductance multi-chip SiC devices packaging technology*, in Proc. EPE2016, Karlsruhe, Germany, 5-9 Sept. 2016.

[15] B. Mouawad, Z. Wang, J. Buettner, A. Castellazzi, *3.3 kV SiC JBS diode configurable rectifier module*, in Proc. EPE2017, Warsaw, Poland, 11-14 Sept. 2017.

An Alternative Method to Accurately Determine the Thermal Resistance of SiC MOSFET Structures with Discrete Diodes

Andras Vass-Varnai[1*], Young Joon Cho[1], Gabor Farkas[2] and Marta Rencz[2,3]

[1] Mechanical Analysis Division, Mentor Graphics, Seoul, Korea
[2] Mechanical Analysis Division, Mentor Graphics, Budapest, Hungary
[3] Department of Electron Device, Budapest University of Technology and Economics, Budapest, Hungary
*E-mail: andras_vass-varnai@mentor.com

To determine the thermal properties of power semiconductor devices and structures, the JEDEC JESD 51-1 static, electrical test method is a well-known and industry-wide accepted technique. The approach provides accurate and repeatable results in case of silicon based transistors in all cases. For certain compound semiconductor components, such as SiC MOSFET-s and GaN HEMT structures, the application of the electrical test method becomes in some cases challenging. If traditional test setups are used, in the unit step response function, due to parasitic effects, an electric signal may superpose on the thermal signal of interest, making it hard or even impossible to analyze the test results. If the structure has a physical diode as well, it can be used to understand the thermal properties of the package and its layers. This information can be applied in another step to gather the thermal properties from die transistors' point of view as well, without measuring it. In this article we show a combined measurement and simulation based method, which allows the accurate thermal characterization of such components, even in cases when other approaches may fail.

Keywords— Thermal transient testing, structure functions, power semiconductor

I. INTRODUCTION

In the recent years a significant increase can be witnessed in the commercial application of compound power semiconductor devices, especially in the automotive sector. SiC structures have a more than two times wider bandgap than Si (3.26eV), high thermal conductivity (~120 W/mK) and they are chemically inert. They are usually applied in power conversion due to unquestionable benefits over the traditional silicon counterparts. Because of their wide bandgap SiC MOSFETs have high breakdown voltage, comparable to silicon IGBTs, but provide lower power losses, higher switching frequencies and operating temperatures [1].

On the other hand, the gate region of the SiC devices, the SiO_2 – SiC transition, may contain trapped charge carriers due to the large concentration of crystalline errors at the interface [2]. Even though some techniques, such as post-oxidation annealing of the gate oxide in nitric or nitrous oxide (NO or N_2O) may improve the device performance [3], in some structures the movement

of these trapped charges cause electrical disturbances up to the several seconds range after the switching [4]. For this reason thermal transient tests should be carried out in connection modes, where the gate potential remains unchanged during the process.

This makes common test procedures, such as the "MOS diode" setup and the common gate arrangement unsuitable for testing SiC devices. Measuring the thermal response of the device on the channel is not possible due to the low R_{DSon} resistance, characteristic to these devices, unless large sensor current is used to gain a measurable signal.

In the following figure a SiC MOSFET's thermal response can be observed at elevated sensor current level.

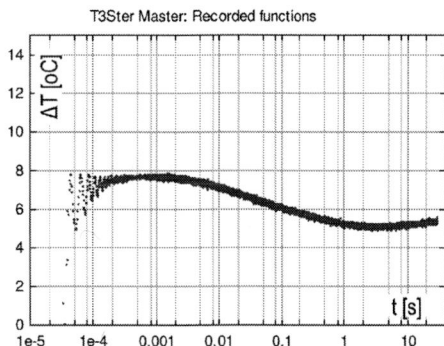

Fig. 1. SiC MOSFET measured with 20A sensor current, 240 A heating current and 15V VGS

The test shown in Fig. 1. was carried out using $I_{sense} = 20A$ sensor current and an open gate at $V_{GS} = 15V$ gate voltage. One can observe that the response curve is not monotonous, which indicates that an electric response is superposed on the thermal one.

In another experiment the gate is regulated to $V_{GS} = 10V$ to elevate the channel resistance and increase the signal. This time the test results look monotonous and the signal strength is indeed increased. In order to prove that the signal is purely thermal, dual interface tests were carried out, following the JEDEC JESD 51-14 standard.

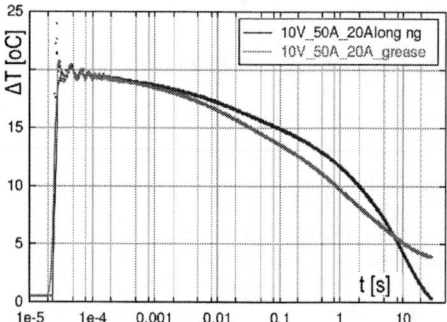

Fig. 2. SiC MOSFET measured with 20A sensor current, 5A heating current and 10V VGS

The resulting curves fit well below 100 ms, but they start to separate long before the expected location of the case, then they cross in the seconds range. This behavior is again showing a not thermal, but electrical nature.

As the issues demonstrated above most likely correspond to a gate charge related phenomena, the SiC diodes are not affected, thus their thermal measurement is quite simple and straightforward, see Fig. 3.

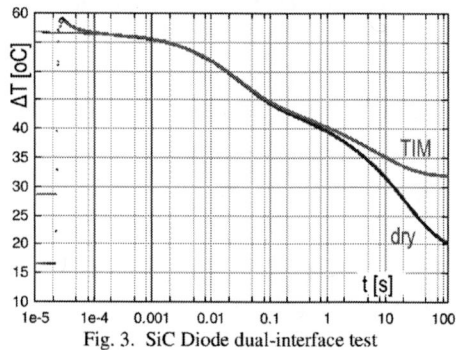

Fig. 3. SiC Diode dual-interface test

As most bridge arrangements have a separate freewheeling diode in close proximity to the transistor, the fact that SiC diodes can be tested fairly easily provides an opportunity to determine the thermal characteristics from the transistors' point of view as well:

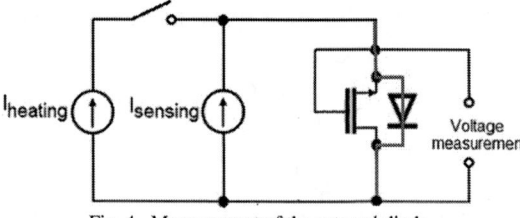

Fig. 4. Measurement of the external diode

The first way to get a valid thermal signal is to measure the external diode only, following the schematic in Fig. 4.

The thermal measurement of the diode alone would not provide sufficient information, as the diode's heated area and the die volume are both different from the transistor's characteristic values. In Fig. 5 we demonstrate the difference using structure functions. The

diode has a lower initial capacitance, and a higher die attach resistance than the transistor, in direct relationship with the difference in their geometry. On the other hand, after the die attach area the package parameters look similar in both cases. This explains well that calibrating the thermal model of the package structure through the signal measured at the diode and simulating the transistor is a feasible option.

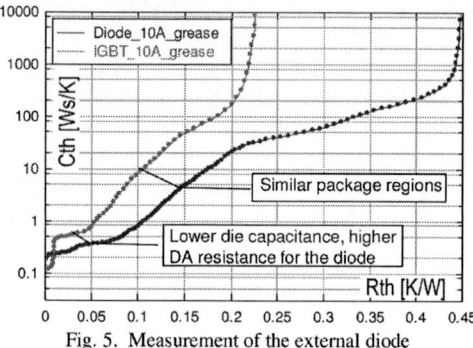

Fig. 5. Measurement of the external diode

Another measurement option is shown in Fig. 6. In this case we heat the channel of the transistor as in the previous experiments, but upon switching we turn the transistor off and initiate a negative sensor current to sense the thermal signal of the diode. This will be a transfer effect between the heated MOSFET and the measured diode.

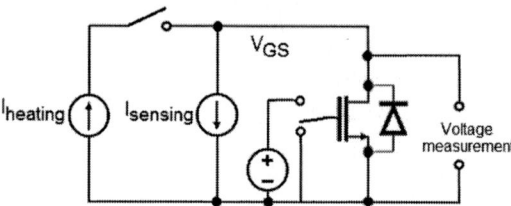

Fig. 6. Heating the channel and sensing the diode

A cross-heating example is demonstrated in Fig. 7, on Zth curves (temperature change divided by power step). The blue color curve corresponds to the cooling of the transistor (simulation result) and the red curve is the result of the cross-heating effect captured on the diode.

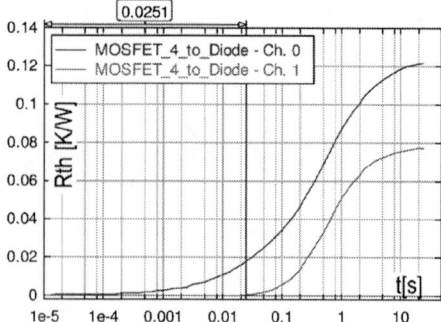

Fig. 7. Self and cross heating effects in the power package

The propagation delay can clearly be observed, the

temperature of the diode starts changing approximately 25 ms after the power step is initiated on the MOSFET.

Even if the self-impedance of the transistor cannot be measured in some cases, the transfer curve is useful to verify the simulated results.

II. EXPERIMENTAL RESULTS

In case none of the above introduced techniques work, a detailed thermal analysis of the package is still possible using a combined measurement and simulation technique. The ability to find a measurable structure allows us to create a simulation model of the package and calibrate it against the measured thermal signal [5]. As even in case of SiC devices the measurement of the reverse diode is normally possible without any issue, its thermal transient signal can be applied to calibrate the thermal model describing the package. Once the package model has been calibrated based on the diode's signal, the thermal characterization of the SiC MOSFET becomes possible in a simulation environment, if the package model will behave as the real package would for a wide range of time constants after calibration.

To demonstrate this, we selected a Si IGBT package with embedded reverse diodes. This allows us not only to do the calibration using the diode and characterize the IGBT's thermal performance in a CFD simulator, but as the IGBT chip is easy to test using traditional ways, we can also verify the simulation results by an additional measurement step. This may not be possible in case of SiC MOSFET components.

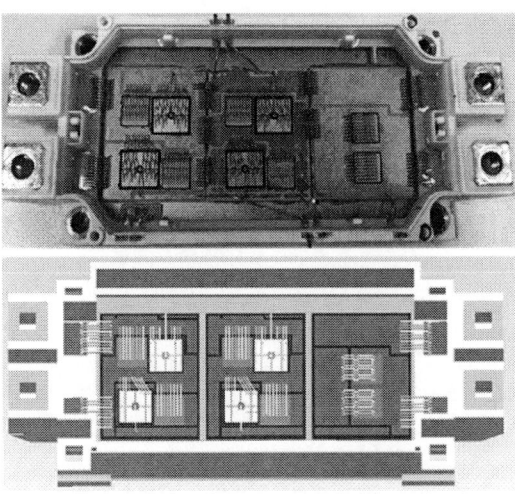

Fig. 8. Image and simulation model of the analyzed arrangement.

We have selected a commercially available IGBT inverter as shown in Fig. 8. with a high and a low side IGBT, physically two times two chips connected in parallel. The IGBTs are all equipped with the equal number of reverse diodes. Except for the standard datasheet information, the detailed structure of the package was not known to us; therefore we had to measure the physical parameters of the DBC layer

structure. The analyzed arrangement is shown in Fig. 9. For building up the initial thermal model, beside the geometry, the material properties of the layers were unknown parameters as well.

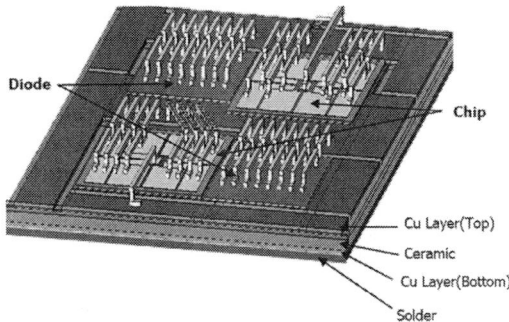

Fig. 9. Layer structure of the tested device

The calibration process started with a thermal transient test to obtain baseline information on the behavior of the selected semiconductor component, in our case the diode (two chips connected in parallel, thus heated and measured at the same time). After creating the first version of the thermal model of the component, applying the same power step on the same dies in the simulation environment, the resulting transient curves, or the corresponding structure functions can be compared.

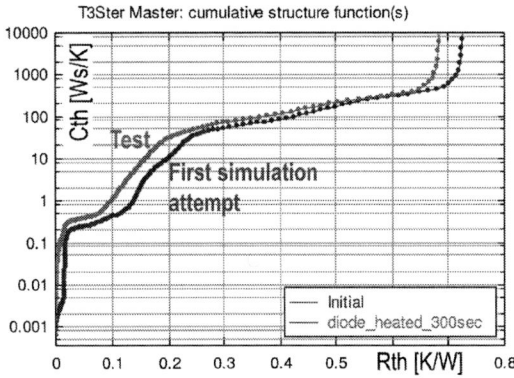

Fig. 10. Structure function comparison of the test data and the first simulation model

In Fig. 10 the shape of the test and simulation-based structure functions shows similarities, however the magnitude of certain partial thermal resistance sections does not match perfectly. As some of the material parameters were best guess values in the beginning, this is a normal situation. To overcome this problem a set of experiments were designed, mainly focusing on the variation of the thermal conductivity coefficients of materials in the model. Normally they have the highest influence on the results.

Another important parameter we investigated was the active area size, the coverage of the heated area on the die. Table 1. summarizes the set of variables applied in this example:

Table 1: Set of calibration parameters

	Unit	Initial value	Parameter range	Calibrated value
Active area	mm^2	81	64 ~ 81	79
Die adhesive	W/mK	33	30 ~ 35	33
Ceramic	W/mK	25	25 ~ 35	34
Solder	W/mK	40	35 ~ 45	45

The simulation software was set to create scenarios within the pre-defined parameter ranges and run a set of simulation experiments to check the effect of these parameters on the final results. Based on the analysis of these simulation experiments we could select the best set of parameters which make the tested and simulated curves overlap.

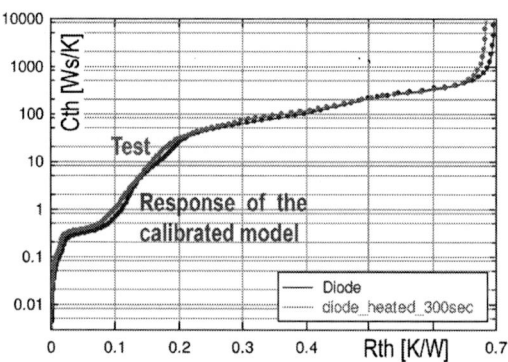

Fig. 11. Structure function comparison of the test data and the calibrated model

Fig. 11. demonstrates the final, good fit between the measurement and the simulation response in structure function space. By increasing the thermal conductivity coefficients of the ceramic and the solder layers, the points of the simulated structure function shifted towards the origin over the R_{th} scale and finally the two structural models showed alignment.

At this stage the package structure is calibrated, so a thermal transient simulation can be carried out on the transistor as well. The simulation should not need any further model alignment, however it is important to set the heated area of the transistor correctly as well. After performing the simulation we could compare the data to the test data of the IGBT to obtain immediate verification of this method, see Fig. 12.

Fig. 12. confirms that if it is not possible to physically test a component for any reason, it still makes sense to look for another one which allows a proper thermal transient test and use a combined test and simulation method to characterize all heat sources of interest.

A calibrated simulation model is useful anyway, as it can help to obtain characteristic package metrics, or it can help to verify them if they were previously defined using tests, such as by the JEDEC JESD 51-14 method (although it is defined for discrete, single heat-source packages).

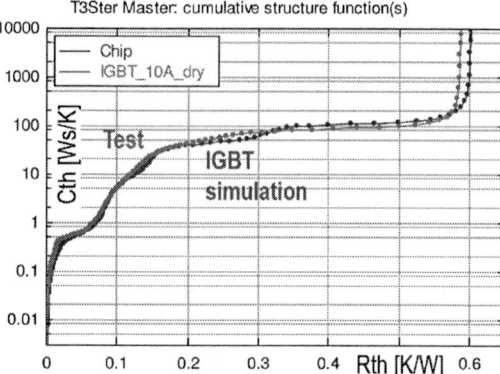

Fig. 12. Structure function comparison of the test data and the first simulation model

In case of this particular sample we performed the R_{thJC} measurement of the IGBT using a combination of tests and simulation. First so-called thermal regions were defined over the IGBT die surface and over the package surface, and then we used simulation to determine the average temperatures in both regions.

Fig. 13. Baseplate temperature plot of the package, with component outlines turned on.
Thick white lines indicate the tested regions

Fig. 13 illustrates the distribution of the temperature over the package surface. The final R_{thJC} will definitely depend on the selection of the regions, or perhaps individual isotherms (which is even harder to control). In case of this package the following data was obtained:

Table 2: Maximum and average region temperatures

Temperatures	Maximum	Mean
Base plate [°C]	70.34	62.90
IGBT active area [°C]	75.83	74.29
R_{thJC} [K/W]	**0.079**	**0.163**

The power step used for the calculations was approximately 69.5W. Not surprisingly, the difference between using the maximum temperature values of the chip and the base plate and using the mean values is large, about 1:2 ratio in this case. Using the maximum temperature values will result in a clear underestimation of the R_{thJC} value, while the mean values depend on the actual size of the base plate region selected for the averaging.

140

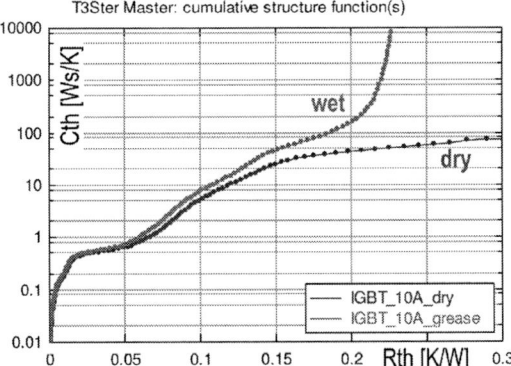

Fig. 14. Dual interface results taken on the package
from the IGBT-s point of view
(wet with grease, dry case without thermal grease)

Fig. 14. shows the divergence of the structure functions if the dual interface method is applied.

The separation of the two curves is monotonous, starting from app. 0.05 K/W to a very explicit separation around 0.17 K/W. This is a commonly seen phenomenon in case of large area, multi heat-source packages [6].

Looking at Fig. 13. it is obvious that there is a wide range of isotherms crossing the case, therefore the selection of a proper value (min, max or average) has to be defined by individual companies based on their own standards.

A calibrated simulation model helps to set this standard, even if the measurements are possible on their own. Once the standard is set, it helps to read the correct value of the R_{thJC} from the structure functions in a reproducible way.

III. CONCLUSIONS

In this article we demonstrated that if in case of some novel compound semiconductor structures the standard thermal test methods do not work, it is a good way to look for another, measurable component, such as a reverse diode.

Based on the thermal transient response of this component the package structure can be identified.

If a thermal simulation environment is available, the simulation model can be calibrated to the test-based structure functions, allowing the model to respond correctly in a wide range of time constants.

Using this calibrated model the thermal properties of all heat sources in the package can be simulated. In a worked example we demonstrated that the simulated thermal response of and IGBT fits the measured response very well, even though the package was calibrated to the diode's response only.

Having a calibrated package model has further benefits. It helps to identify a suitable way to interpret the separation point of structure functions if the separation is not a clear single point, but shows up rather like a continuous region.

REFERENCES

[1] J. A. Cooper, M. R. Melloch, R. Singh, A. Agarwal and J. W. Palmour, "Status and prospects for SiC power MOSFETs," in IEEE Transactions on Electron Devices, vol. 49, no. 4, pp. 658-664, Apr 2002. doi: 10.1109/16.992876G

[2] J. Biela, M. Schweizer, S. Waffler and J. W. Kolar, "SiC versus Si—Evaluation of Potentials for Performance Improvement of Inverter and DC–DC Converter Systems by SiC Power Semiconductors," in IEEE Transactions on Industrial Electronics, vol. 58, no. 7, pp. 2872-2882, July 2011. doi: 10.1109/TIE.2010.2072896

[3] G. Y. Chung et al., "Improved inversion channel mobility for 4H-SiC MOSFETs following high temperature anneals in nitric oxide," in IEEE Electron Device Letters, vol. 22, no. 4, pp. 176-178, April 2001. doi: 10.1109/55.915604

[4] G. Farkas, G. Simon, "Thermal transient measurement of insulated gate devices using the thermal properties of the channel resistance and parasitic elements" in Microelectronics Journal, Volume 46 Issue 12, December 2015 Pages 1185-1194

[5] A. Vass-Varnai, R. Bornoff, S. Ress, Z. Sarkany, S. Hodossy and M. Rencz, "Accurate thermal characterization of power semiconductor packages by thermal simulation and measurements," 2011 Symposium on Design, Test, Integration & Packaging of MEMS/MOEMS (DTIP), Aix-en-Provence, 2011, pp. 324-329.

[6] András Vass-Várnai, Shan Gao, Zoltán Sárkány, Jongman Kim, Seogmoon Choi, Gábor Farkas, András Poppe, Márta Rencz: 'Issues in junction-to-case thermal characterization of power packages with large surface area.'In: Proceedings of the 26th IEEE Semiconductor Thermal Measurement and Management Symposium (SEMI-THERM'10).Santa Clara, USA, 2010.02.21-2010.02.25. pp. 158-164.

Heat-resistant packaging technology for wide bandgap power devices and thermal reliability testing

K. Suganuma, H. Zhang, S. Nagao, C. Chen, T. Sugahara, A. Shimoyama, and A. Suetake

The Institute of Scientific and Industrial Research, Osaka University
8-1 Mihogaoka, Ibaraki, Osaka, 567-0047, Japan
e-mail: suganuma@sanken.osaka-u.ac.jp

Abstract- **SiC and GaN are attractive wide bandgap semiconductors for the new generation power devices. In order to realize the application in the market, the packaging materials, especially die-attach, has one of the key role as well as the packaging design structure. Ag sinter joining is one of the promising die-attach technologies. The advantages of Ag sintering are excellent heat-resistance, high bonding strength, high heat conductivity, easy handling capability (ambient atmosphere, low temperature and low pressure), and affordability without any nano materials.**

SiC die-attached made by Ag sinter joining exhibits excellent reliability in power cycling beyond 250 °C (T_{jmax}). Low temperature and low pressure Ag sintering utilizes its unique reaction with oxygen around 200 °C. The metallization structure both for a SiC die and for a substrate is also critical due to thermal stability and to oxidation. Some features of reliability testing methods will be mentioned in the presentation.

Keywords - die-attach, sinter joining, Ag, reliability

I. INTRODUCTION

Die-attach is one of the essential technologies to provide high reliability and excellent performance of WBG power devices. There are several candidate materials developed in the past decade for WBG die-attach [1, 2, 3]. Those candidates are Bi based alloys [4], Zn-Al [5], Zn-Sn [6] and pure Zn [7] as well as the conventional Au alloys [6]. Transient liquid phase bonding has been also examined for off-eutectic Au or Ag alloys [8, 9] and for Cu-Sn and Ni-Sn based alloys [10].

In contrast, Ag sinter joining has attracted much attention due to its excellent high performance and high temperature reliability [2, 3]. This paper reviews the current status of sinter joining including both Ag and Cu and a new method utilizing film stress migration bonding developed by the present authors [11, 12].

II. SINTER JOINING

Sinter joining has a long history back to 1980s. One of the present author reported the excellent potentials of sinter joining for joining ceramics and metals [13]. Al_2O_3 and steels was successfully bonded by inserting the mixture of Al_2O_3 and Fe powers, which was named as

"functionally gradient method" afterwards. Nevertheless, the joining requires high pressure beyond 100 MPa to make a dense bonding layer, which restricted the practical adoption of the sinter joining.

Ag powder paste wiring method is a well-known method for Si solar cells. This Ag paste needs glass addition to make proper electric contact with Si wafer and firing temperature is very high reaching 900 °C. The author demonstrated the potential of Ag nanoparticles sinter joining even at room temperature [14]. Because of the active nature of nanoparticles, sintering of nanoparticles can be achieved even at room temperature when the surface protective polymer/monomer coating on nanoparticles can be effectively removed. Sinter joining with nanoparticles, however, has serious drawbacks of the inhomogeneity of the bonding layer even with applied high pressure beyond a few MPa and remained organic or carbon residue when they are heated at low temperature below 200 °C because of difficulty in removing the surface protective coating. In addition, the high cost of nanoparticles also limits the wide application of nanoparticle sinter joining

In contrast, sinter joining with micron-size Ag hybrid particle pastes provides a stable bonding structure even at 200 °C without any high applied pressure [3]. It can fulfil the requirements such as printability of pastes to make a uniform layer, and low applied pressure below 1 MPa as well as excellent affordability.

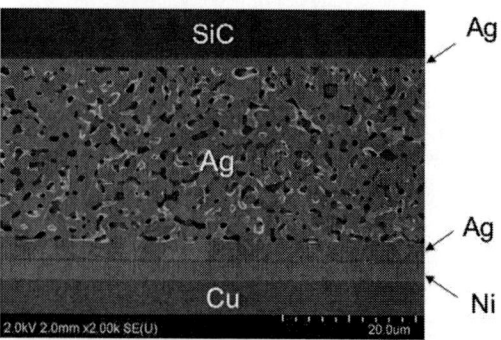

Fig. 1. (a) GaN die-attach structure with Ag hybrid paste and (b) the bonding layer microstrcutre (SEM) [1].

Fig. 1 shows an example of a LED die-attach with the hybrid paste. The presence of oxygen plays a key role to form stable bonding around 200 °C in air. The Ag sinter joining provides the microporous interlayer, which is strong enough just after the fabrication and can provide stress relaxation between dies and substrates due to the low Young's modulus less than 50 GPa, which can be adjusted to optimal value by sintering conditions. It was found that bimodal size distribution of sintering particles can increase sintering ability due to its initial dense packing of particles.

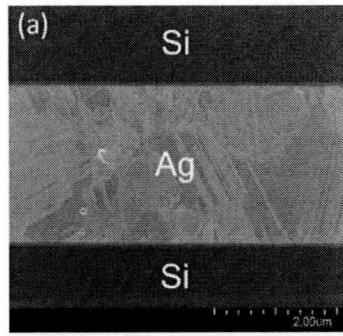

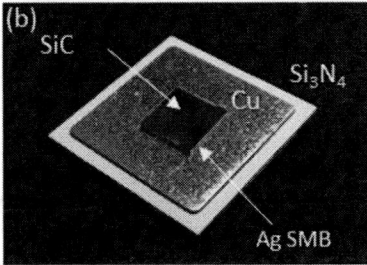

Fig. 2. (a) Si-to-Si joint interface by SMB mthod bonded at 250 °C in air without any pressure (SEM) and (b) SiC die attached on DBC [12].

Recently, the Ag thin film stress migration bonding (SMB) method has been developed, providing a perfect bonding without any large voids in ambient atmosphere at 250 °C as shown in Fig. 2. SMB can be also applicable to fine pitch interconnection such as flip-chip bonding or TSVs.

III. LOW TEMPERATURE SINTERING MECHANISM OF AG

It is very interesting to note that Ag sinter joining temperature and the SMB can be accomplished even at far lower temperature than the effective sintering temperature of Ag, which is usually considered to be the half of melting temperature of a metal. When metals are heated beyond the half of its melting temperature, self-diffusion is efficiently activated resulting in necking formation among particles and sintering. The half of melting temperature of Ag is about 350 °C. The sintering temperature 200 – 250 °C is far lower than this diffusion activation temperature for Ag. No activation driven by the surface energy of nanoparticles cannot be expected for large particles beyond a few hundred nm

and even for Ag films. Thus, there must be certain diffusion activation mechanism for bonding with Ag.

In the SMB process, numbers of Ag hillocks appear on the surface of Ag film as shown in Fig. 3 [12]. The neck region of a hillock shows a characteristic appearance of Ag grain aggregation in sintering process as indicated by the thick arrow in the figure.

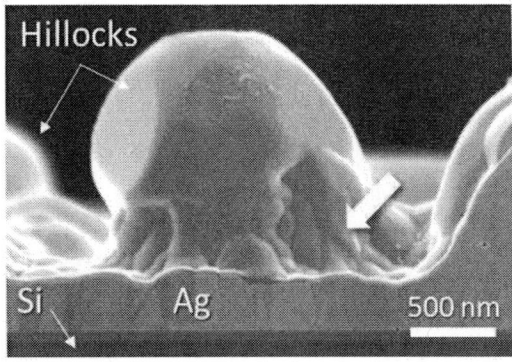

Fig. 3 Ag hillocks formed on the free surface of a Ag film on a Si die heated at 250 °C in air (SEM) [12].

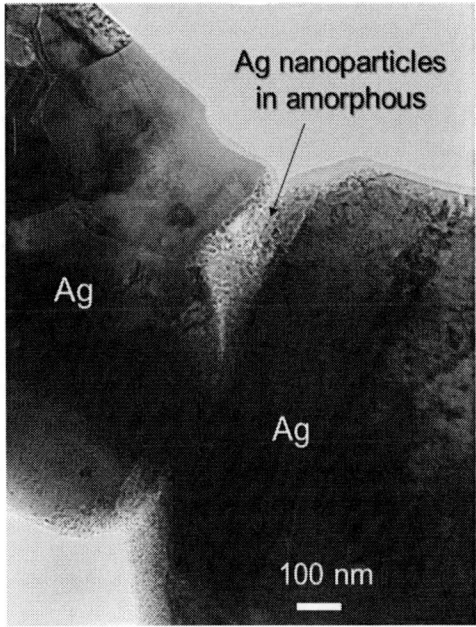

Fig. 4 TEM of two Ag hillocks contact showing the formation of Ag nanoparticles in Ag amorphous layer on the surface of a Ag film [15].

Fig. 4 shows the TEM photograph of two hillocks contacting each other in the SMB process. Interestingly, there are numbers of Ag nanoparticles filling the gap between two hillocks. These particles are dispersed in a Ag amorphous layer formed on the Ag film. Metallic amorphous cannot be normally formed because of the thermodynamic instability. However, in joining Ag particles and Ag films at around 200 °C in air, the formation of Ag amorphous was identified by a series of high resolution TEM observation without oxygen. Oxygen was locally identified in only some region. Inside the amorphous layer, Ag clusters and nanoparticles were also found. Thus, the microstructural observations revealed that the dense hillock formation on two mating Ag film surfaces with the self-generation of Ag amorphous promotes sintering and bonding, followed by the formation of nanoparticles in the contact area among Ag micron particles or between hillocks on Ag films, these nanoparticles fills the gaps effectively. Beyond 200 ºC, sintering nanoparticles can easily proceed.

A thermodynamic simulation with the aid of first principal calculation revealed that, at such low temperature, Ag can absorb oxygen along grain boundaries and that Ag-O in grain boundary becomes liquid under a high partial pressure [15]. Since the Ag film gains compressive stress by thermal expansion mismatch with a Si substrate, the Ag-O liquid squeeze out like eruption, which the present author named as "Nano-volcanic Eruption", as schematically shown in Fig. 5.

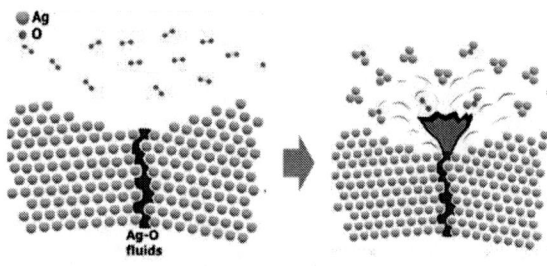

Fig.5 Nano-volcanic eruption of Ag in air [15].

In sintering Ag submicron and micron particles, the same "Nano-volcanic Eruption" reaction was observed by a series of high resolution TEM. Thus, Ag is the special materials that can be sintered at very low temperature in the presence of oxygen. The lowest oxygen concentration is about 1 % [11]. The lowest sintering temperature for the mechanism is 145 °C [15], which temperature is low enough for bonding and wiring for most cases. Since 250 °C is the current target operation temperature for WBGs, bonding and wiring should be carried out higher than 250 °C. in order to obtain good bonding, the metallization both of a die and of a substrate should be Ag to enhance bonding from both sides [3].

IV. POLYOL SYNTHESIS OF BIMODAL Ag PARTICLES

As mentioned above, mixing two size distribution of Ag particles is effective for sinter joining due to its particles packing nature. The polyol synthesis of Ag particles to form bimodal distribution particles was found effective to increase sintering ability [16]. A one-pot synthesis method was developed to obtain the ideal bimodal submicron/micron particles for Ag and also for Cu. The size and size distribution of metal particles have

been freely tailored by adjusting the reaction temperature, time and using different additive agents. Fig. 6 shows the SEM images of Cu and Ag particles and the shear strength of the two metal pastes applied in bonding process. High bonding strength has been achieved with the two metal pastes under a very low pressure of 0.4 MPa.

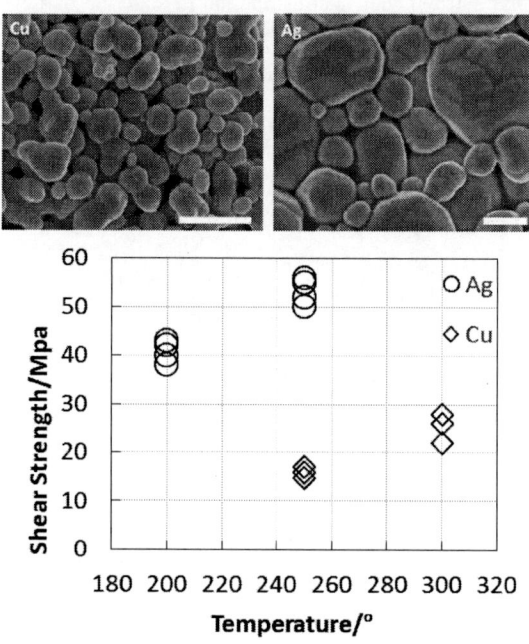

Fig. 6 SEM images of Ag and Cu particles prepared with polyol process and the shear strength [16].

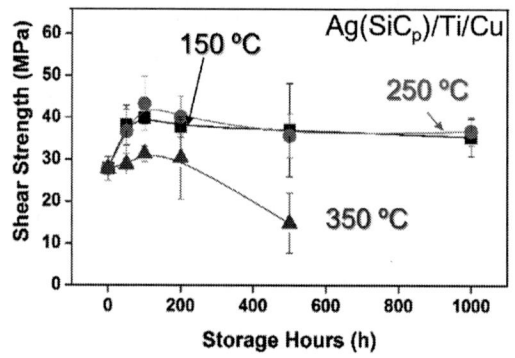

Fig. 7 Bonding shear strength change exposed to high temperature [17].

V. IMPROVEMENT OF HIGH TEMPERATURE STABILITY

When a die-attached assembly is exposed to high temperature of 250 ºC, the microporous sintered layer exhibits grain coarsening. The coarsening itself does not influence on the bonding strength even up to 1000 hours. The addition of ceramic submicron particles effectively can stabilize the microporous joint structure [17]. Excellent heat-resistance up to 250 ºC was proved as shown in Fig. 7. At 350 ºC, the shear strength decreased down to the half of the initial value. This degradation can be attributed to the severe oxidation of the Cu substrate not by the coarsening of the Ag sintered layer as shown in Fig. 8.

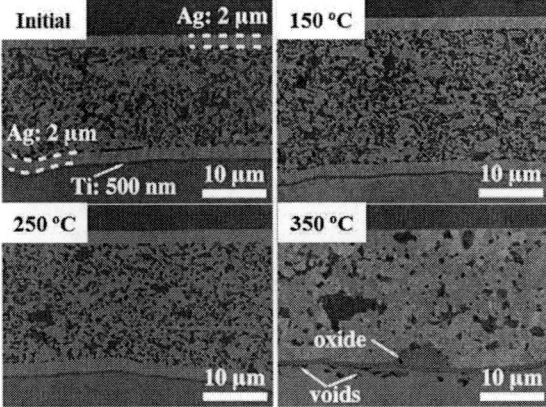

Fig. 8 SEM microstructure of Cu-to-Cu joints bonded with Ag hybrid paste with 2% SiC submicron particles as a function of exposure temperature for 1000 h.

The interface metallization stability is also the key to fabricate a sound structure. Because of the intrinsic nature of Ag as mentioned above, oxygen can diffuse very fast along grain boundaries of Ag. This sometimes cause oxidation of the metallization layer such as Ni plating as well as a Cu substrate, which sometimes degrades interface bonding. To prevent this interface degradation due to oxidation, a protective coating is required. Fig. 9 shows the ideal Ag sinter joint structure for high temperature uses.

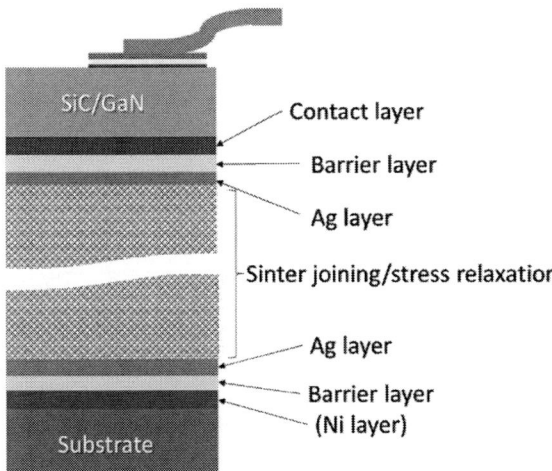

Fig. 9 Ideal joint structure for Ag sinter joining.

Since Ag sinter joining can be achieved in the specific reaction with oxygen as mentioned in section III, strong interface can be formed onto Ag metallization on substrates and dies. Bonding onto Au and the other metallization surfaces requires much higher temperature, longer time or high applied pressure. Recently, the author overcome this limitation by using certain kinds of solvents, which accelerates metal interface reaction [18].

VI. 3D POWER WIRING

Wiring method and materials have another key role for power device reliability as well as functions. Al wiring has been widely adopted to power device while the other new methods have been applied for high power application such as Al ribbon, Al/Cu clad, Cu wiring/ribbon and clip bonding. Especially, clip bonding not only can carry high current but also can dissipate substantial heat from a power semiconductor.

Recently the present authors group proposed 3D wirings for SiC/GaN power devices instead of clip bonding. Two methods have been proposed. A direct printing Ag paste can form thick, wide and flexible wiring. The resistivity of Ag printed wiring tracks can reach about 3.4×10^{-6} $\Omega \cdot$ cm sintered at 250 °C, which is about twice of bulk Ag. Heat-resistance and electromigration resistance of Ag printed wiring was proved to be excellent. The other method is bonding Ag or Ag coated plate direct bonding with no/low pressure as shown in Fig. 10. Ag surface can be activated by grinding or polishing by introducing defect structure that enhances Ag-O reaction at 200 °C.

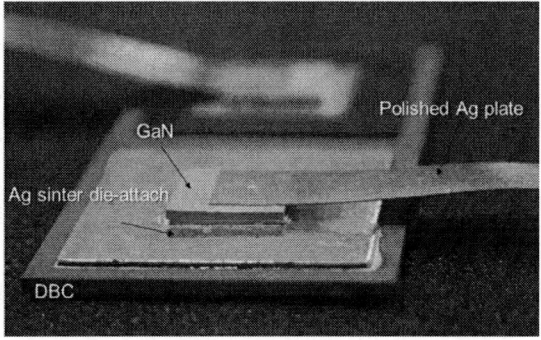

Fig.10 Ag plate bonding on GaN die by surface activation of Ag wiring plate.

Electromigration is another reliability concern for advanced power electronics. It is well known that very large current sometimes causes wiring failure for semiconductors. For many cases of WBG devices, high current density beyond 1 kA/cm² is expected. Such high current density may cause atomic diffusion in wiring and across interfaces. Fig. 11 shows our experimental set-up for electromigration for high power devices. For wiring dimension of width 600 μm, height 130 μm, the current density of 2.7×10^4 A/cm² can be applied.

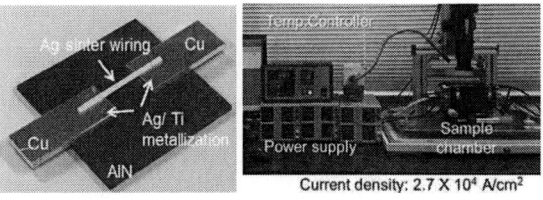

Fig.11 Electromigration test sample and set-up for Ag sinter wiring.

It was proved that Ag wiring exhibited excellent stability at 250 °C as shown in Fig. 12. Initially, the resistivity decreased a little while no severe degradation after 20 hours. This little decrease in the initial stage can be attributed to the additional sintering of Ag wiring since the printed Ag tracks were sintered at 250 °C.

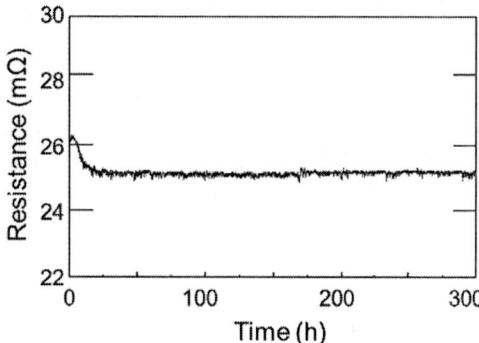

Fig. 12 Resistivity change of Ag sinter wiring as a function of exposure time. The current density was 2.7 x 10^4 A/cm^2 at 250 °C.

Some grain growth was observed in Ag wiring tracks as shown in Fig. 13. No serious cracking or void formation was observed in the wiring tracks. In contrast, the interfaces showed a certain microstructural change, especially at the anode side, while it was no influence on the wiring performance as shown in Fig.12. At the anode side, Ag diffused across the interface into Cu substrate.

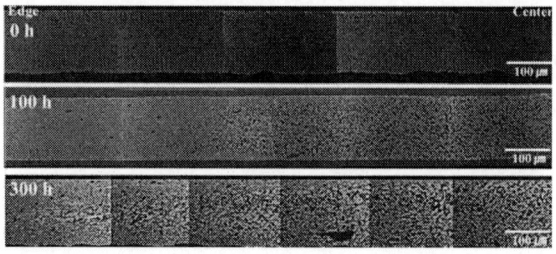

Fig. 13 Cross sections of Ag wiring tracks during EM test 2.7 x 10^4 A/cm^2 at 250 °C.

VII. FINE PITCH INTERCONNECTION

Ag sinter joining can be also applied to fine pitch interconnection such as flip-chip and TSV assemblies [19]. Fig. 14 shows an example of Si die stacking fabricated at 250 °C without any applied pressure. First 200 μm square Cu pads are coated with Ag, two Si die were bonded together. Two Si dies were successfully bonded and its bond line did not show any significant void. The bonding strength is extremely high because of this dense bond structure. Thus, SMB provides a good replacement to brittle IMC interconnection with soft Ag interlayer.

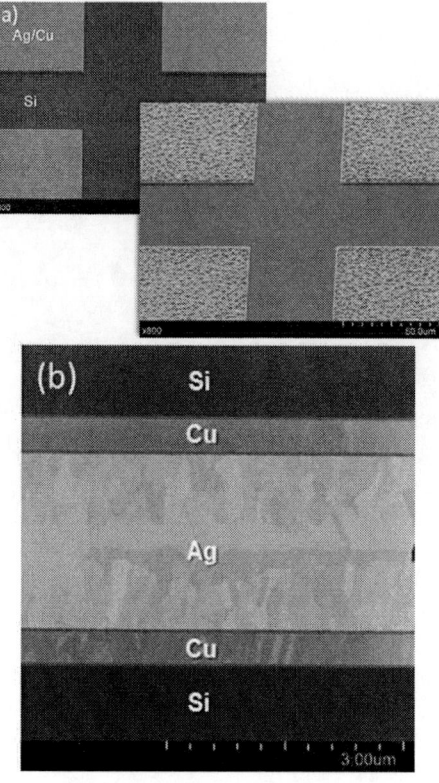

Fig. 14 Application of SMB to flip chip bonding [19]. (a) Ag film bumps before and after heating, and (b) bonded interface microstructure.

VIII. SUMMARY

This paper briefly describes the current status of Ag sinter joining and Ag film stress migration bonding. Ag has great advantages in both the surface reaction in air benefitting joining quality and the excellent electric/thermal properties. Ag sinter joining has already exhibited a great potential in the market as a high temperature interconnection technology, while our Ag film stress-migration bonding is expected to provide an alternative route. Similar bonding methods using Cu or other metallic materials instead of Ag would be explored as cost-effective interconnections in future.

ACKNOWLEDGMENT

The authors are grateful partial for the support from the JSPS (Grant-in-Aid for Scientific Research, Grant No. 24226017) and for by the JST Advanced Low Carbon Technology Research and Development Program (ALCA) project "Development of a high frequency GaN power module package technology".

REFERENCES

[1] K.Suganuma et al, JOM, 61, 64-71(2009).

[2] K. S. Siow, J. Alloys Compd., 514, 6-19 (2012).

[3] K. Suganuma et al, Microelectron. Reliab., 52, 375-380 (2012).

[4] M. Rettenmary et al, Adv. Engineer. Mater., 7, 965-969 (2005).

[5] S.-J. Kim at al, Mater. Trans., 49, 1531-1536 (2008).

[6] S. Kim et al, J. Electron. Mater., 38, 2668-2675 (2009).

[7] K. Suganuma et al, IEEE EDL, 31, 1467-1469 (2010).

[8] H. A. Mustain et al, IEEE Trans. CPT, 33, 563-570 (2010).

[9] A. Sharif et al, J. Alloys Compd., 587, 365–368 (2014).

[10] S. W. Yoon, J. Micromech. Microeng. 23, 015017 (2013).

[11] M. Kuramoto et al, IEEE Trans.CPMT, 33, 801-808 (2010).

[12] C. Oh et al, Appl. Phys. Letters, 104, 161603 (2014).

[13] K.Suganuma et al, J. Amer. Ceram. Soc., 66, c117-118 (1983).

[14] D. Wakuda et al, IEEE Trans.CPMT, 33, 437-442(2010).

[15] S.-K. Lin et al, Scientific Report, 6, 34769(2016).

[16] J. Jiu et al, J. Mater. Sci.: Mater. Electron., 26, 7183-7191 (2015).

[17] H. Zhang et al, J. Electron. Mater., 44, 3896-3903 (2015).

[18] T. Fan et al, J. Alloys Compds, 731, 1280-1287 (2018).

[19] C. Oh et al, Appl.Phys.Letters, 104, 161603 (2014).

Verification of Identification Accuracy of Loss calculated by Inverse Thermal Analysis

Yuki Ikari[1], Kazushige Nakao [1]*
1 Fukui University of Technology, Fukui, Japan
*E-mail: nakao@fukui-ut.ac.jp

Abstract — **In power electronics system, it is very important for an engineer to design a cooling system and mount power module for aiming high power density. To realize an optimum system, we need to identify loss and heat source of power device.**

So, we conduct an approach using inverse thermal analysis instead of a convectional calorie method and electromagnetic analysis. This proposed approach is method that heat source and heat loss are calculated inversely from the measured temperature values.

At first, we conduct heater experiment in wind tunnel with fully controlled velocity to provide a high-accuracy order analysis model.

Next, the measured temperature values are set as an initial condition in inverse thermal fluid analysis. As a result, the heater loss, temperature profile and heat source are calculated, and these analytical values are compared with experimental values.

We found the fact that the heater loss calculated by inverse analysis was good agreement with experimental values within errors of ±7%. Also, the position of heat source and temperature profile are specified with this analysis. Then the effectiveness of proposed analysis has been confirmed.

Keywords — *Inversed thermal analysis, loss, temperature measurement, induction equipment*

I. INTRODUCTION

In power electronics system, it is very important for an engineer to design a cooling system and mount power module for aiming high power density [1][2]. To realize an optimum system, we need to identify loss and heat source of power device such as induction equipment[3][4] .

So, we conduct an approach using an inverse thermal analysis instead of a convectional calorie method and electromagnetic analysis. This proposed approach is method to find inversely heat source and heat loss from measured temperature values.

At first, we conduct heater experiment in wind tunnel with fully controlled velocity to provide a high-accuracy order analysis model.

Next, the measured temperature values are set as an initial condition in inverse thermal fluid analysis. As a result, the heater loss, temperature profile and heat source are calculated and, these analytical values are compared with experimental values.

II. HEATING EXPERIMENT WITH THE HEATER BLOCK AND EXPERIMENTAL METHOD

Fig.1 shows an experimental apparatus. This apparatus consists of wind tunnel (□290 mm×390 mm) embedded flow velocity control grid and cooling fan. Air velocity is controlled from 0.4 m/s to 3.1 m/s by varying inverter frequency from to 22 Hz to 60Hz.

Fig.2 shows the detailed picture of heater block (110 mm×80 mm×25 mm) and heater cooling system. The heater block is equipped three heaters in parallel along air flow direction.

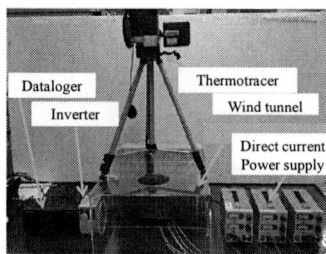

Fig.1. Experimental apparatus

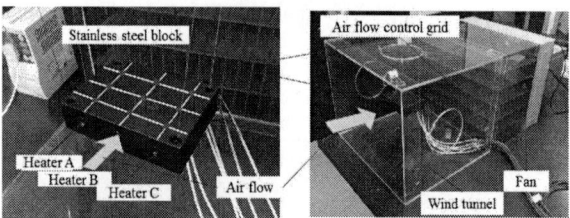

Fig.2. Heater blocks and embedded wind tunnel

Fig.3. Velocity measurement point at inlet of wind tunnel

Fig.3 shows an air velocity measurement points. A uniform flow is formed by air flow control grid. This uniformity is an essential to increase calculation accuracy. The air velocity is measured at points showed in the Fig.3

Fig.4 shows the velocity profile of wind tunnel open space at frequency of 40 Hz. The uniformity of velocity is satisfied within errors of ±10% for the average velocity. The under mentioned experiment is conducted on the condition of this velocity.

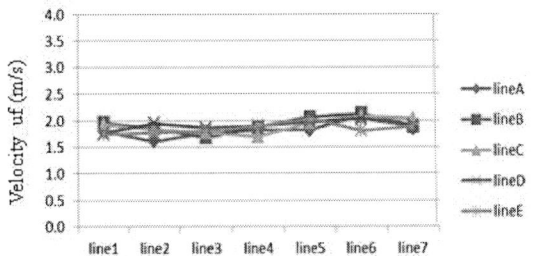

Fig.4.　Inlet velocity of wind tunnel at inverter frequency 40 Hz

Table 1 shows six heating patterns in heating experiment. Then, the block surface temperature at measurement point(red point ●) showed in Fig.2 are measured.

Table 1　Heating pattern

	Heater A (W)	Heater B (W)	Heater C (W)
Pattern ①	8	8	0
Pattern ②	8	0	8
Pattern ③	8	0	5
Pattern ④	8	0	2
Pattern ⑤	8	0	0
Pattern ⑥	0	8	0

III. Order analysis and inverse analysis

When an order analysis model is formed by using FlowDesigner which is heat and fluid commercial software, the calculation is conducted under the heating conditions showed in Table 1, and velocity profile at wind tunnel inlet and temperature profile on the heater surface is calculated. Construction of order analysis model is done repeatedly until analytical results are good agreement with experimental results of temperature profile and air flow velocity profile.

At first, we explain about the process of inverse analysis. The nine measured values showed in Fig.2, is set as an input condition of completed model.

Secondary, an inverse analysis is conducted and the heater loss, temperature profile and heat source are output.

Finally, these analytical values are compared with experimental values.

A. Order analysis

Fig.5 shows the analytical model of total system. Fig.6 shows the detailed model of heater. In addition, Fig.7(a) shows a cross section of real heater. Fig.7(b) shows the analytical model of that.

The heater model consists of lead line for conducting current, annular nichrome wire for heating, MgO for insulating between nichrome wire and stainless-steel (SUS 304) tube.

B. Inverse thermal analysis

Table 2 shows heater surface temperature measured by infrared camera in the case of heating pattern ①. These

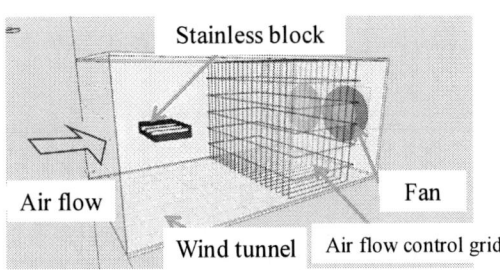

Fig.5.　Heater block and cooling system in a wind tunnel

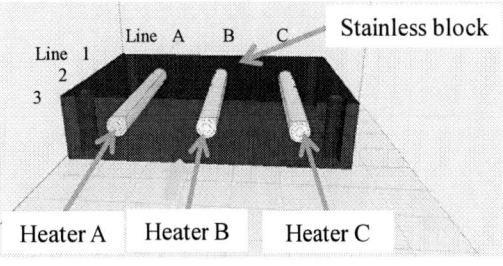

Fig.6.　Heater block model

values are input in inverse model. The temperature profile of total heater block is calculated. As a result, heat source and heating rate is specified.

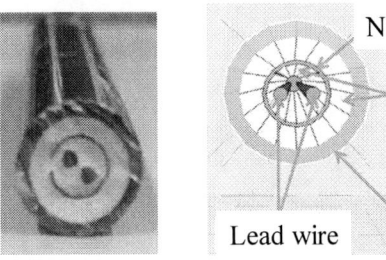

(a) Stainless heater　　　(b) Heater model

Fig.7.　Cross section of stainless heater and analytical model

Table 2　Target temperature [℃]

	Line A	Line B	Line C
Line 1	33.8	31.8	28.1
Line 2	34.4	33.4	28.8
Line 3	35.1	33.9	29.3

IV. CONSIDERATIONS BY COMPARISON BETWEEN EXPERIMENTAL RESULTS AND ANALYTICAL RESULTS

Fig.8, 9 shows temperature profile on heater block surface measured by experiment and calculated by order analysis. From the comparison of both figures, we found that an analytical temperature profile is good agreement with that of experimental profile. The surface temperature has the tendency of $t_A > t_B > t_C$. Because heater A receives heat interference from heater B, the temperature of heater A is highest, and that of heater B is nearly equal or slightly lower because heater B also receives heat interference from heater B, but there is no heat interference from heater C. From good agreement of results, we confirmed that an order analytical model was formed efficiently with good accuracy.

Fig.10 shows surface temperature on the heater block. Fig.10(a) shows the temperature profile at starting state of time T=0 sec measured by infrared camera. Fig.10(b) shows that of thermal steady state of T=3600 sec. From these data, 9 temperature values showed in Fig.2 and Table 2, are got as a target temperature.

Next, the above-mentioned temperature values are set as an input boundary condition to analytical model.

So, the inverse analysis is conducted, and the heater loss profile and temperature profile are output. Table 3 shows the heat rate output by analysis and experiment. Both values coincide with errors of 6.5 % for heater A and that of−7.0% for heater B.

Fig.11 shows surface temperature profile of heater A, B, C. Fig.11(a) shows temperature profile at A= 0 W, B=8 W, C = 0 W. Fig.11(b) shows that at A= 8W, B=8 W, C = 0 W. Fig.12 shows another perspective view of temperature profile of Fig.11(b).

In Fig.11(a), as only heater B is heated, its surface temperature profile is symmetrical. Also, the temperature of heater middle section in flow direction is higher than end section of heater. In Fig.11(b), heater A is heated at 8 W and heater C isn't heated. As a result, heater B surface temperature is asymmetrical profile which left side temperature of heater B is higher than right side of that. These phenomena are influenced to thermal interface from heater A.

Fig.13 shows temperature profile of heater cross section calculated by inverse thermal analysis. Fig.13 (a) shows the perpendicular cross section profile. Fig.13(b) shows flow directional cross section profile.

In Fig.13(a), we found that heat source is formed an annular section where nichrome wire generates Joule's heat. Also, in Fig.13(b), the heater temperature become higher and higher along flow direction. As a result, the position of heat source and temperature profile is specified with inverse analysis.

Then, the effectiveness of proposed inverse thermal analysis has verified.

Table 3 Comparison between experimental values and analytical values

	Heater A	Heater B	Heater C
Experimental values (W)	8.00	8.00	0
Inverse analysis values (W)	8.52	7.44	0
Error (%)	6.5	-7.0	0

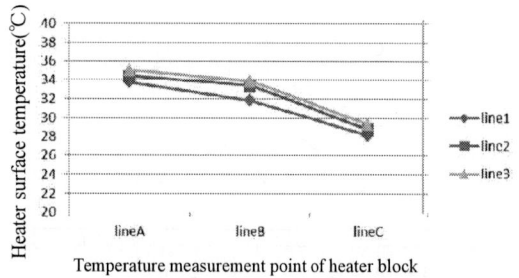

Fig.8. Temperature profile measured by experiment

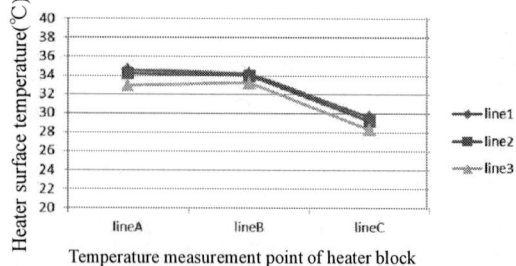

Fig.9. Temperature profile calculated by analysis

The 2018 International Power Electronics Conference

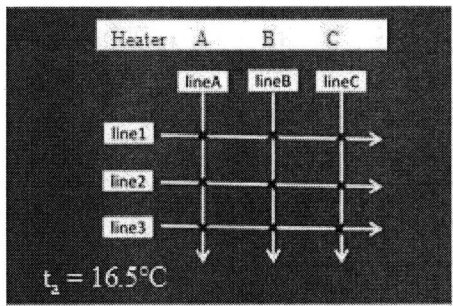 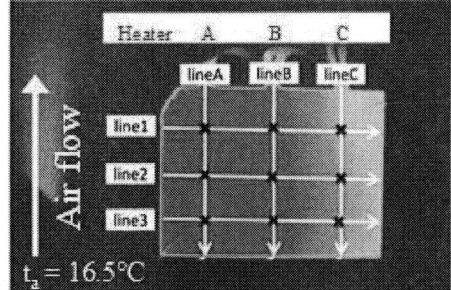

(a) Initial temperature profile
at t = 0 sec

(b) Temperature profile
at t = 3600 sec

Fig.10. Temperature profile measured by infrared camera at heater A= 8 W, B = 8 W, C = 0 W

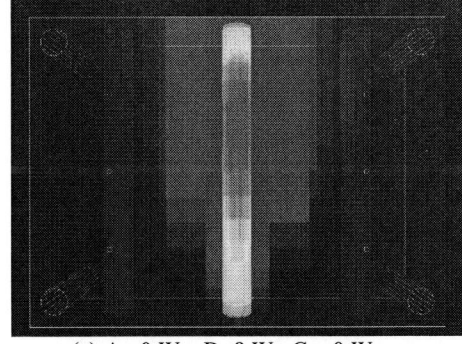 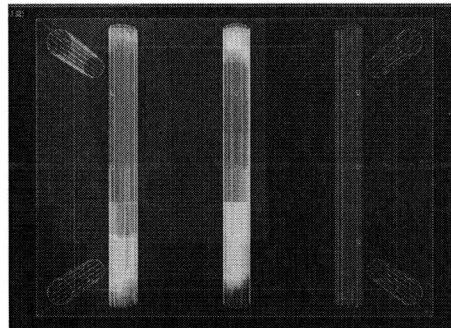

(a) A= 0 W B=8 W C = 0 W

(b) A= 8W B=8 W C = 0 W

Fig.11. Three-dimensional temperature profile

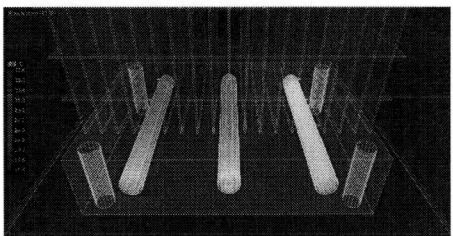

Fig.12. Perspective view of temperature profile of Fig.12(b)

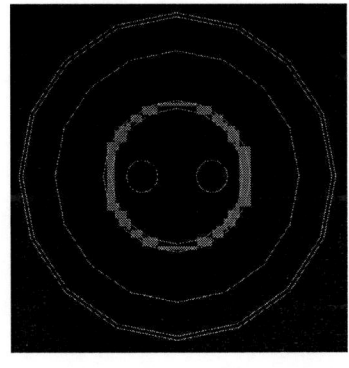 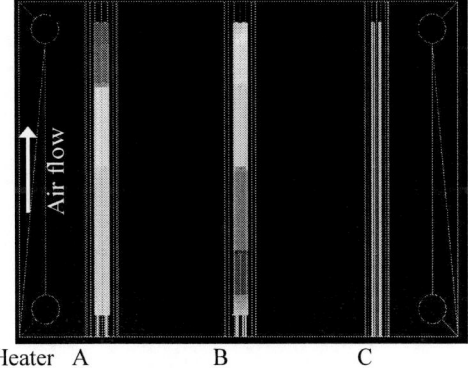

(a) Perpendicular cross section profile

(b) Flow directional profile

Fig.13. Temperature profile calculated by inverse thermal analysis

V. CONCLUSIONS

We found the fact that the heater loss calculated by an inverse analysis was agreement with that found by heater experiment within errors of ±7%.

Also, the position of heat source and temperature profile is specified with inverse analysis.

As a perspective about research, we will conduct inverse analysis to specify a loss and heat source of power module and electromagnetic device such as transformer and induction heating device.

REFERENCES

[1] Kazushige Nakao, "Review: Power Electronics System Integration for increasing Power Density", The Annual Meeting record I.E.E. Japan (Tohoku university), 2016.

[2] Bernd Eckardt, "Cool Systems with SiC and GaN" ECPE Workshop Advances in thermal Materials and System for Electronics, 2015.

[3] G.Ortiz,M.Leibl,J.W.Kolar and O.Apeldoorn,"Medium Frequency Transformers for Solid-State-Transformer Applications-Design and Experimental Verification" Proceedings of the 10th IEEE International Conference on Power Electronics and Drive Systems (PEDS 2013), Kitakyushu, Japan, April 22-25, 2013.

[4] G. Ortiz, J. Biela and J. W. Kolar, "Optimized Design of Medium Frequency Transformers with High Isolation Requirements" , IECON 2010 - 36th Annual Conference on IEEE Industrial Electronics Society , 7-10 Nov. 2010.

[5] Thiemo Kleeb, Benjamin Dombert, Samuel Araújo, Peter Zacharias, "Loss Measurement of Magnetic Components under real Application Conditions" EPE'13 ECCE Europe, 2013

Packaging Architectures for Silicon Carbide Power Electronic Modules

H. Alan. Mantooth* and Simon S. Ang
Department of Electrical Engineering, University of Arkansas, Fayetteville, USA
*E-mail: mantooth@uark.edu

Abstract-High-density power electronic systems require the integration of multiple power semiconductor devices into single power electronic modules. One of the major advances to power electronic system integration has been the commercial availability of silicon carbide (SiC) power semiconductor devices. Increased switching speeds, lower on-state resistance, higher breakdown voltage, and higher operating temperatures of these SiC power devices lead to higher performance and greater miniaturization when compared to those of silicon power devices. At present, these SiC devices are limited by their current handling capability. As such, paralleling of multiple SiC devices is commonly used to increase current handling capability in SiC power electronic modules. This paper describes techniques to address performance limitations while achieving the potential merits of wide bandgap device technology. A great deal of this activity involves wire bondless packaging techniques to achieve lower inductances and higher packaging density.

Keywords— 3D power modules, parasitic inductance, silicon carbide power modules, wire-bondless power modules.

I. INTRODUCTION

Considerable advances in bulk and epitaxial semiconductor quality, device processing, and packaging technology have made silicon carbide (SiC) power semiconductors readily commercially available in recent years. Their high power density at high junction temperatures, coupled with faster switching speeds as compared with their silicon (Si) counterparts [1] have made these SiC devices very attractive in many power electronic applications. In addition, the switching losses associated with SiC power devices are much lower than those reported for silicon power devices. The faster switching speeds of SiC devices reduce sizes of inductors and capacitors with a resultant size and volume reduction for their power electronic systems.

Power electronic modules require the integration of multiple materials, each with different thermo-mechanical properties. When large electrical current flows through these power semiconductor devices, Joule heating is generated in the active junctions of the power semiconductor devices and creates a temperature gradient within these different layers. Due to coefficient of thermal expansion (CTE) mismatch among different layers, reliability issues caused by warpage or delamination are a concern [2]. Therefore, in order to improve the reliability of these power electronic modules, it is necessary to avoid excessive heat concentration and thermo-mechanical stresses. For high power density

modules, it is necessary to remove heat, not only from top and bottom as in double-sided cooling, but also from all sides to avoid hot spots within the power module.

For high current ratings, several SiC power devices are connected in a single switching position. Paralleling of multiple high-speed SiC power devices present many challenges in power module design and packaging technology. Due to their high switching speeds, the performance of these paralleled SiC devices is becoming very sensitive to parasitic circuit elements. It is conceivable that each of the paralleled SiC devices may have different parasitic inductances due to their interconnections and geometrical locations. Hence, these paralleled SiC devices may have different switching characteristics.

Wire bonds are commonly used to interconnect power devices to the interconnection traces on the packaging substrate and from the interconnection traces to the package pins. A single wire bond can contribute a few nH of parasitic inductance. Usually multiple wire bonds are used in practice to reduce this parasitic inductance. For high current power modules, it is not unusual to have hundreds of bond wires for the purposes of decreasing parasitic inductance and increasing current handling capability. Bond wires are also the weak link in reliability when subjected to thermal cycling or thermal shock [2]. Hence, eliminating wire bonds will reduce parasitic inductances and increase dimensional integration, from 2D to 2.5D and 3D, of SiC power devices possible in power electronic modules. The immediate benefits are not only reduction in parasitics but also enablement of efficient distributed cooling schemes, a reduction in footprint, and consequent increase in power density. This paper presents several packaging architectures to achieve lower parasitic inductances and higher packaging density for SiC power electronic modules.

II. HIGH DENSITY POWER MODULE PACKAGING

A conventional half-bridge power module is shown in Fig. 1. As can be seen, the module consists of 15 paralleled SiC MOSFETs in the upper and lower switching positions, respectively [3]. The multiple bond wires are to reduce parasitic inductances and to increase current handling capability. These paralleled SiC MOSFETs and diodes are packaged in a conventional rectangular power package. Three separate direct bond copper (DBC) substrates are used to house these SiC MOSFETs to reduce the potential bowing problem due to

CTE mismatch of the DBC to the copper base plate. The three separate DBC substrates are arranged in a rectangular row because of the geometrical constraints. They are then connected by wire bonds to each other as shown. This highly asymmetrical layout scheme creates several challenging issues for paralleled SiC devices. The gate length for each device has a different length to its source terminal and this creates an asymmetric gate turn on for each SiC MOSFET. Besides, the source and drain parasitic circuit elements are different for each power device.

Fig.1. A rectangular SiC power module.

Fig. 2 shows two different layouts to illustrate the effects of different layout topologies on the parasitic inductances of wire-bonded power electronic modules.

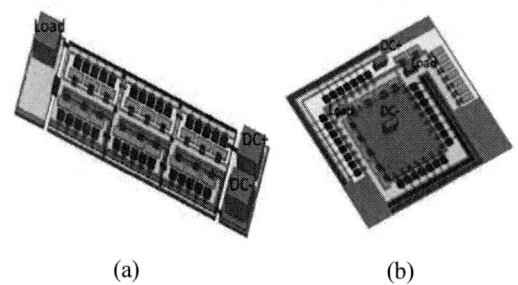

(a)   (b)

Fig. 2. (a) Rectangular layout versus (b) square layout of a SiC power module.

Table 1 shows the comparison of the parasitic inductances for these two modules. The conventional rectangular module package has an interconnection trace with a maximum parasitic inductance of 69 nH in its DC+ to load circuit path compared to a maximum parasitic inductance of 20 nH in its DC+ to DC- circuit path for the square module package. The square module adopts the N/cell/P-cell power device layout scheme to further reduce the parasitic inductance. This illustrates the module geometrical effect on the parasitic inductances, and as such, conventional rectangular geometry may not be the optimum package geometry for modules with many paralleled SiC power devices.

Table 1. Comparison of Parasitic Inductances for Layouts of Fig. 2.

Parasitic Path Inductance	Rectangular Module	Square Module
DC+ to DC-	31 nH	20 nH
DC+ to Load	69 nH	14 nH
DC- to Load	62 nH	10 nH

Three different wire bondless packaging schemes, each with different packaging configurations, have been investigated. Fig. 3 shows a wire bondless package of one or several SiC power devices in a single package [4]. A commercially available bare die SiC MOSFET was processed into a flip-chip capable metallized die and, along with an L-shaped copper connector to the drain electrode on the backside of the die, attached to the substrate through solder ball attachment as shown in Fig. 5 [4]. The drain connector also provides a low thermal resistance path to the heat sink. Notice that the flip-chip SiC device is directly attached to the base plate without a substrate. This feature decreases both the thermal path and the number of different material interfaces in the package. The solder balls serve as low-inductance interconnects to the power electronic circuitry. It was demonstrated in this study that the flip-chip interconnection reduced the parasitic loop inductance by 3x as compared with a conventional wire bonded module. A 24% reduction in the on-state resistance was also reported [5].

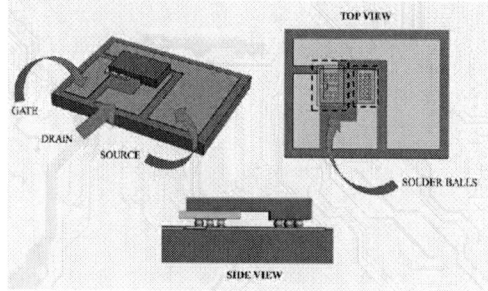

Fig. 3. A flip-chip SiC power package [4].

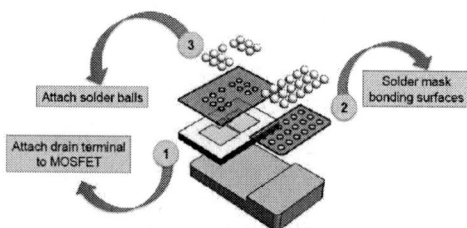

Fig. 4. A schematic depicting the process flow [4].

Fig. 5 shows a wire-bonded half-bridge power module with two parallel SiC devices in parallel in each switching position and a similar half-bridge wire-bondless power module. The most common layout technique for the wire bonded module is to have both paralleled power devices next to each other to facilitate shorter interconnections between their respective source and gate electrodes for power MOSFETs. Of course, their proximities depend on thermal considerations to avoid intense heat generated in a confined area. As such, larger parasitic inductances are expected. The wire-bondless SiC power module has their parallel-connected SiC devices stacked on top of each other to accomplish a 3D packaging architecture. Compared to conventional wire-bonded integration, the wire-bondless assembly with

154

shorter interconnect lengths will reduce parasitic inductance, decrease on-state resistance, and increase the power density, which translates into performance improvements and a reduction in power consumption. 3-D packaging has been suggested by many as an attractive method to meet the demands of higher functionality in ever-smaller and lightweight packages. On the other hand, the closer distances between power devices in the 3-D packages make the heat density inside to be much higher than 2-D packages. Efficient heat dissipation becomes a key issue to guarantee the electrical and reliability performances of the 3-D modules. The low temperature co-fired ceramic (LTCC), whose CTE closely matches that of SiC, can combine with high thermally conductive materials such as nano diamond particles to improve the thermal dissipation near power devices in power modules, and works as interposer to integrate both electrical and thermal power routes [6].

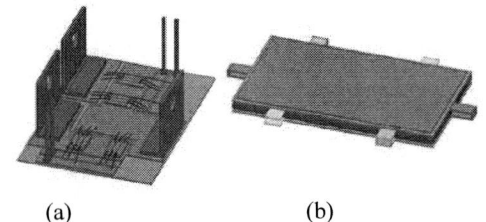

(a)   (b)

Fig. 5. A wire-bonded power module (a) and a wire-bondless power module (b) that house a similar half bridge switching leg.

The proposed 3-D stacked wire bondless module and its exploded view are shown below in Fig. 6. As can be seen from the exploded view, SiC devices are flip-chip bonded onto the LTCC interposer surfaces face-to-face to form parallel connections, electrical interconnections are achieved through conductive vias in the LTCC interposer. Power and signal terminals are soldered onto the metallized interposer or DBCs, respectively, for external connections. DBC substrates at top and bottom provide drain connections for the die as well as a vertical thermal path. Besides a reduction in the parasitic inductance of about 80%, the wire-bondless module also enables a volumetric reduction with an increase in power density. Moreover, a double-sided cooling scheme can be implemented for this wire-bondless power module.

A wire-bondless package architecture can also be implemented by stacking together two individual power modules as shown in Fig. 7 [7]. As can be seen, power modules 1 and 2 are interconnected by spring-loaded copper pins using a ceramic fixture for mechanical compliance. Fig. 8 illustrates the interconnection scheme for the top and bottom power modules [8]. For this packaging architecture, many similar power modules can be stacked on top of each other to series connect the individual power modules to increase their breakdown voltage to achieve a high-voltage module such as the ABB high voltage IGBT valve [9]. The fundamental difference between this series-connected stacked power module and the ABB IGBT high-voltage IGBT valve is that the spring-loaded interconnects serve as interconnects between the two power modules which are individually packaged in a wire-bondless module. In press-pack IGBT modules, the gate and emitter electrodes of the IGBTs are connected by individual spring-loaded pins placed directly on them to form electrical contacts [10]. However, when subject to wide range of thermal cycles and high junction temperatures during normal operation of these modules, stress and strain in the contact materials occur due to different coefficients of thermal expansion (CTE), thermal conductivity, and mechanical contact properties between the materials. As such, the contact metallization on the power devices gradually wears out over a period of time and leads to a condition known as fretting failure [11]. Double-pulse tests of this stacked half-bridge switching position reveal both a peak current overshoot reduction during turn-on and a peak voltage overshoot reduction during turn-off, with minimum ringing indicating low parasitic inductances within the power module. Moreover, a parasitic inductance mitigation scheme of utilizing current return path cancellation can be implemented within the individual wire-bondless power modules. All these lead to the mitigation of electromagnetic interference (EMI) in this stacked spring-loaded interconnected power module [12-14].

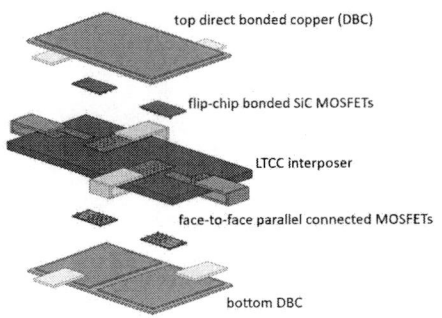

Fig. 6. Exploded view of the 3D wire-bondless package.

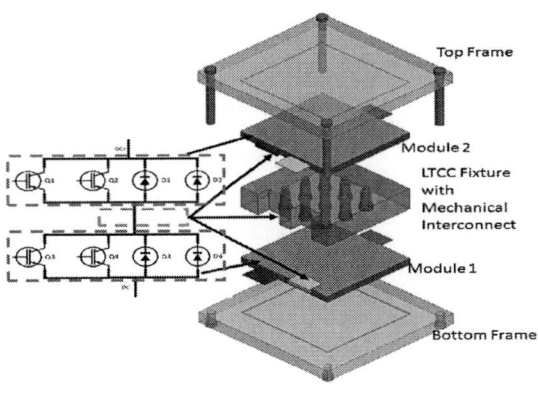

Fig. 7. A spring-loaded interconnected power module [7].

155

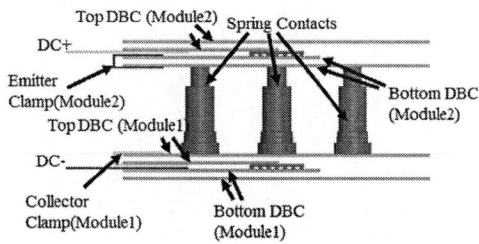

Fig. 8. Interconnection scheme of two power modules [8].

III. CONCLUSIONS

Several packaging architectures to realize high-density power modules for silicon carbide power semiconductor devices are discussed. Wire-bondless and 3D package configurations to achieve lower parasitic inductances and higher packaging density are keys to address performance limitations while achieving the merits of silicon carbide device technology. Of course, thermal and mechanical aspects of packaging architectures must also be considered for ultimate reliable power modules for wide bandgap power devices.

ACKNOWLEDGMENT

The authors would like to recognize the research contributions of many of their graduate students, in particular, Atanu Dutta, Si Huang, and Sayan Seal, in power electronic modules research. They also would like to express their sincere thanks to the professional staff at the High Density Electronics Center (https://high-density-electronics.uark.edu/) at the University of Arkansas for their excellent technical support in fabricating power electronic modules and offering an excellent power packaging prototyping and research facility.

REFERENCES

[1] J. Rabkowski, D. Peftitsis, H.P. Nee, "Silicon carbide power transistors: A new era in power electronics is initiated," *IEEE Industrial Electronics Magazine,* 6, pp. 17–26, 2012.

[2] S. S. Ang, H. A. Mantooth, "Reliability of power electronic packaging," in *Reliability of Power Electronic Converter Systems,* edited by Henry Shu-hung Chung, Huai Wang, Frede Blaabjerg and Michael Pecht, IET Research Book, September 2015.

[3] Atanu Dutta, Shijie Wang, Jinchang Zhou, Simon S. Ang, June-Chien Chang, and Chang-Sheng Chen, "The Design and Fabrication of a 50KVA 450A Silicon Carbide Power Electronic Module," *The 4th International Symposium on Power Electronics for Distributed Generation Systems,* July 8-11, 2013.

[4] S. Seal, "The Development of Novel Interconnection Technologies for 3D Packaging of Wire Bondless Silicon Carbide Power Modules," *Doctoral Dissertation, University of Arkansas,* June 2017.

[5] Sayan Seal, Michael D. Glover, and H. Alan Mantooth, "3D Wire Bondless Switching Cell Using Flip-chip Bonded Silicon Carbide Power Devices," *DOI 10.1109/TPEL.2017.2782226, IEEE Transactions on Power Electronics.*

[6] Si Huang, Z. Xu, Fang Yu, Simon S. Ang, "Impact of Nano-diamond Composites on Low-Temperature Co-fired Ceramic Interposer for Wide Bandgap Power Electronic Module

Packages", *2016 IEEE 4th Workshop on Wide Bandgap Power Devices and Applications (WiPDA),* Fayetteville, AR, pp. 314-318, 2016.

[7] A. Dutta and S. S. Ang, "A 3-D stacked wire bondless silicon carbide power module," *2016 IEEE 4th Workshop on Wide Bandgap Power Devices and Applications (WiPDA),* Fayetteville, AR, pp. 11-16, 2016.

[8] A. Dutta, "A Low Temperature Co-Fired Ceramic (LTCC) Based Three-Dimensional Stacked Wire Bondless Power Module," *Doctoral Dissertation, University of Arkansas,* June 2017.

[9] Jan R. Svensson, "Efficient power transfer with HVDC Light," https://library.e.abb.com/public/ 2fdb92b24fec4e0f946b30434c1c514b/24- 26%203m6043_EN_72dpi.pdf

[10] S. Gunturi and D. Schneider, "On the operation of a press pack IGBT module under short circuit conditions," *IEEE Transactions on Advanced Packaging,* vol. 29, no. 3, pp. 433–440, 2006. doi: 10.1109/TADVP.2006.875090.

[11] R. Wu, F. Blaabjerg, H. Wang, M. Liserre and F. Iannuzzo, "Catastrophic failure and fault-tolerant design of IGBT power electronic converters - an overview," *IECON 2013 - 39th Annual Conference of the IEEE Industrial Electronics Society,* Vienna, pp. 507-513, 2013.

[12] A. Dutta and S. S. Ang, "Electromagnetic Interference Simulations for Wide-Bandgap Power Electronic Modules," *IEEE Journal of Emerging and Selected Topics in Power Electronics,* vol. 4, no. 3, pp. 757-766, 2016.

[13] Atanu Dutta and Simon Ang, "Effects of parasitic parameters on electromagnetic interference of power electronic modules," *2017 IEEE Applied Power Electronics Conference and Exposition,* 2017

[14] A. Dutta and S. S. Ang, "Electromagnetic interference simulations of power electronic modules," *Integrated Power Packaging (IWIPP), 2015 IEEE International Workshop,* pp. 83-86, 2015.

Development of a Homo-Polar Bearingless Motor with Concentrated Winding for High Speed Applications

Junichi Asama*[1], *Member, IEEE*, Dai Suzuki[1], Takaaki Oiwa[1], and Akira Chiba[2], *Fellow, IEEE*

1 Department of Mechanical Engineering, Shizuoka University, Hamamatsu, Japan
2 Department of Electrical and Electronic Engineering, Tokyo Institute of Technology, Tokyo, Japan
*E-mail: asama@shizuoka.ac.jp

Abstract— This paper introduces a six-pole homo-polar bearingless motor intended for high-speed applications. This bearingless motor has four degrees of freedom active positioning control. The rotor includes two salient iron cores without permanent magnets, and hence it has an advantage of rigid rotor structure. The stator consists of two units in tandem. Between these two cores, a permanent magnet with axial magnetization is located, and hence it provides the magnetic flux bias to polarized the rotor cores. We propose a suspension winding configuration for the six-pole homo-polar bearingless motor with six salient poles, although conventional homo-polar bearingless motors have eight-pole structure. The suspension winding arrangement was designed to reduce the suspension force variation without mechanical angle detection. This paper shows the motor structure, principle of suspension force generation, and results of the finite-element-method calculations and experiments.

Keywords— Bearingless motor, homo polar, four degrees of freedom, high speed

I. INTRODUCTION

A motor output is the product of rotational speed and torque, and the torque is proportional to the product of the rotor diameter squared and the axial length. Hence, the power density, which is the mechanical output per the motor volume/weight, can increase with an increase in rotational speed, or the motor size can be reduced when the power density increases. Many previous works have investigated high-speed permanent magnet (PM) motor drive [1-5], but the rotor is suspended with bearings that may cause significant mechanical friction and reduce the long term operation due to the bearing failure.

Active/passive magnetic bearings (AMB/PMB) can support the rotor without any mechanical contact [6]. The AMB generates suspension force for magnetic levitation, and the PMB has self-positioning function due to the magnetic attractive or repulsive force. The non-contact operation reduces mechanical loss significantly and enables continuous high-speed operation with less maintenance. However, the magnetic bearing system needs additional motor part, for instance, the motor unit is installed between the AMB components in tandem. This causes a long shaft length, and hence, the maximum

rotational speed is limited due to the rotor flexible vibration at high-speed region.

Bearingless motor technique is one possible solution. It combines functions of motoring toque generation and suspension force generation in one unit, or the non-contact levitation function is magnetically integrated in the motor component [7]. Hence, the rotor axial length can be reduced and the maximum rotational speed increases in comparison with the magnetic bearing system. For high-speed application of the bearingless motor, angle detection for suspension force regulation should be avoided, because the active positioning control becomes unstable if the angle detection has significant error. A consequent-pole bearingless motor needs no angle detection [8, 9], however, magnet pieces are inset at the rotor surface. The rigid rotor structure withstands centrifugal force due to the high-speed rotation. Hence, a homo-polar bearingless motor (HPBM) [10] is a possible candidate for high-speed application. The HPBM rotor consists of only iron cores. In addition, no angle detection is needed for magnetic suspension, if the suspension winding is well designed.

In previous work [8, 10], eight-pole bearingless motors with distributed winding have been introduced, and it is concluded that a consequent-pole or homo-polar rotor with four or more pole-pair numbers enables little suspension force variations. On the other hand, we have proposed the six-pole HPBM with concentrated winding to reduce the driving frequency. This six-pole HPBM structure is new and challenging. In our previous work [11], the winding arrangement has been numerically investigated. Section II provides brief introduction of the structure and working principle for this HPBM. A prototype was built and tested. The experimental results are shown in section III.

II. PROPOSED HPBM STRUCTURE AND PRINCIPLE

Fig. 1 shows a structure of the eight-pole HPBM with PM flux bias. Two stators are located in tandem, where an axially magnetized PM is sandwiched. The rotor consists of salient pole iron cores. The PM bias flux contributes to the polarization of the rotor iron cores. In the figure, the magnetic flux flows into the left rotor core through a ferromagnetic shaft and goes out from the right core. Hence, the rotor iron cores are polarized and only one pole

---

This work was supported by JSPS (Japan Society for the Promotion of Science) KAKENHI Grant Numbers 17H03203.

appears in the same radial plane. The rotor salient cores are twisted by half pitch, and thus, the HPBM behaves like a conventional PM motor. Due to the tandem structure, this HPBM has four degrees of freedom (DOF) active positioning.

A conventional HPBM has eight-pole structure with distributed winding to achieve sinusoidal distribution of magneto-motive force and thus low suspension force variation [10]. To reduce the driving frequency, we have proposed the 6-pole/9-slot structure [11], but reduction in pole pair number may cause crucial suspension force variation and interference [8]. For reduction in the force variation and interference without mechanical angle detection, the suspension winding arrangement for the 6-pole/9-slot HPBM with concentrated windings has been investigated and determined as shown in Fig. 2. The three-phase suspension winding generates 2-pole field for suspension force production. The three-phase motor winding is separately wound in the same stator. The ratio of the number of turns of each tooth in one suspension winding phase is 1: 0.6: 0.4, as shown in Fig. 2 (a). The V-phase and W-phase are shifted by ±120° from the U-phase. This unique arrangement of the suspension winding contributes to little force variation and interference.

When the suspension current is provided, two pole field is generated in the air gap and is overlapped with the bias eight-pole field. The resulting flux unbalance causes the radial suspension force generation. As explained before, the suspension winding arrangement is well designed so as not to detect the rotor angle for suspension force regulation. The axial movement of the rotor is passively stabilized for this four-DOF controlled HPBM. When the rotor moves to the axial direction, the restoring force is generated.

III. EXPERIMENTS

A. Test Machine

Fig. 3 shows a fabricated test machine of the 9-slot/6-pole HPBM. It has a three-phase 2-pole suspension winding and a three-phase 6-pole torque production winding. The motor frame/bearing bracket has an outer diameter of 74 mm and an axial length of 94 mm. The outer diameter, inner diameter, and axial length of the stator are 54 mm, 22 mm, and 15 mm, respectively. The outer diameter, inner diameter, and axial length of the rotor are 20 mm, 10 mm, and 15 mm, respectively. Hence, the air-gap between the rotor and stator iron cores is 1 mm. The core material for the stator and rotor is 6.5 % Si Steel (10JNEX900, JFE Steel., Japan). The outer diameter, inner diameter, and axial length of the PM (N40 garade) located between the stators are 56.5 mm, 50mm, and 6 mm, respectively.

The stator has two separated windings for torque and suspension force generations with concentrated winding, as shown in Fig. 3 (b). The number of turns of the torque producing winding in each tooth is 10 with a copper diameter of 0.7 mm. The numbers of turns for suspension winding in each phase, as shown in Fig. 2 (a), are 20, 12, and 8 turns. The motor windings in the left and right units are connected in series, and hence one inverter is needed for rotation. Eddy current type displacement sensors are used for rotor position detection. Hall elements are embedded between the stator teeth for detection of the rotor polarization.

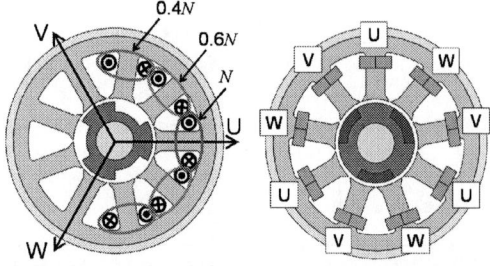

Fig. 1. Structure of the eight-pole HPBM with PM flux bias.

(a) 2-pole suspension winding (U-phase is only shown).   (b) 6-pole torque production winding.

Fig. 2. Winding arrangement of the proposed 6-pole/9-slot HPBM.

(a) Assembled HPBM prototype (motor frame diameter: 74 mm).   (b) Stator (core outer and inner diameters: 54 mm and 22 mm) with Hall sensors.

(c) Rotor (core diameter: 20 mm) with sensor targets (diameter: 20 mm) for eddy current displacement sensor.

(d) Short (left) and long (right) bearing brackets.

Fig. 3. Fabricated prototype.

As shown in Fig. 3 (c), two rotors are twisted by half pitch, and aluminum targets for shaft displacement detection are attached. The shaft has two supporting parts with different diameters of 5.4 mm and 6 mm. Two bearing brackets were prepared with different axial length (long and short), as shown in Fig. 3 (d). The same ball bearings with an inner race diameter of 6 mm are used for both brackets. For only rotation without magnetic levitation, the short bearing brackets are used to support the shaft mechanically with bearings. The ball bearing, when the long bracket is used, works as auxiliary bearing with a radial touch-down gap of 0.3 mm.

B. Static Rotor Levitation

A proportional derivative integral (PID) controller is adopted for the HPBM active magnetic suspension. A three-phase pulse width modulation (PWM) voltage source inverter is employed to regulate the suspension currents. For four-DOF active positioning and rotation, totally three PWM inverters are used. Non-contact magnetic suspension of the shaft was realized by tuning PID control parameters. Then, suspension force and its error angle were statically measured at 0 rpm. The test results are also shown in Fig. 4, in comparison with finite element method (FEM) simulation (JMAG, JSOL Corp., Japan). The measured suspension force corresponded well with calculated results. The maximum error angle was successfully suppressed within 4.5°. A tendency of the experimental result agreed well with calculated results. It is found that the proposed winding configuration, shown in Fig. 2 [11], is useful for reduction of the suspension force interference.

C. High Speed Rotation with Ball Bearing

The brackets were changed, and the rotor was driven with ball bearings without magnetic levitation to verify only high speed operation. Fig. 5 shows how to estimate the rotational angle for this motor. Hall elements have analog output signals, and these are converted to digital signal by using a hysteresis comparator. Then, the pulse per electrical cycle can be measured, and the clock pulses are counted at high speed operation in one electrical cycle, denoted by M counts. It is assumed that the number of pulses in the current cycle is same with the previous one, and hence the rotational angle, θ_r, can be estimated by using the current counted pulse, k, as $\theta_r = 120\ k/M$ degrees. The DC voltage of the inverter was 36 V, and the carrier frequency was 20 kHz. The dq vector control was applied with $i_d = 0$ A.

The rotor was driven at up to 60000 rpm with a driving frequency of 1 kHz (due to the 3-pole pair rotor) with ball bearings. Fig. 6 shows the measured three-phase motor currents at 60000 rpm. As can be seen, the current ripple can be observed. This may be caused due to the low inductance value of the motor winding of 0.127 mH. When the rotor speed increases at more than that speed, undesirable current pulse spike more than 6 A occurred and then the inverter automatically stopped due to the protection mode.

D. Magnetic Suspension and Rotation

When the rotor was magnetically suspended, the rotor was driven at up to 10000 rpm. Fig. 7 shows the measured (a) radial displacement and (b) tilting angles of the rotor.

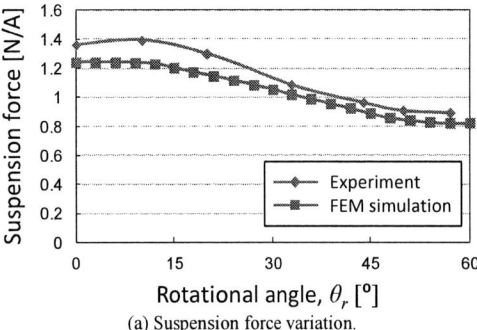

(a) Suspension force variation.

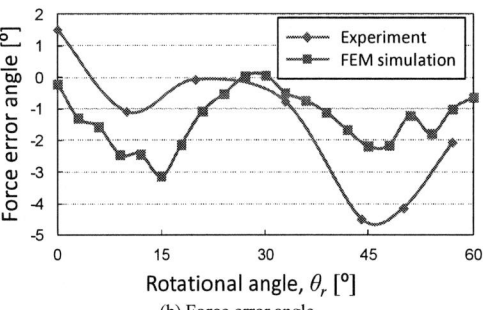

(b) Force error angle.

Fig. 4. Measured and calculated suspension force variation and force error angle under static levitation condition at 0 rpm.

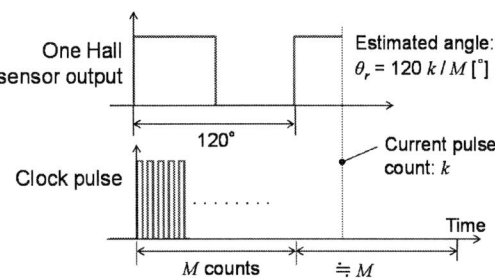

Fig. 5. Rotational angle detection using Hall elements.

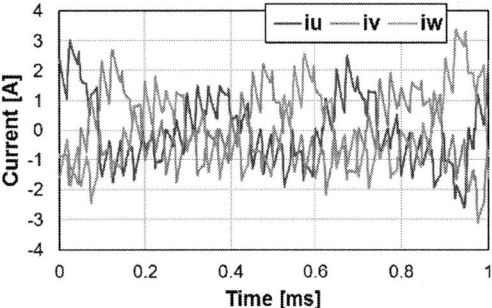

Fig. 6. Three-phase motor currents at 60000 rpm (driving frequency: 1 kHz due to 6-pole rotor) with ball bearings.

The amplitudes of the radial and tilting vibrations were 15 μm and 0.12 mrad, respectively. The primary component of the vibration is fundamental frequency, and hence this is caused by the mechanical unbalance. This unbalance force can be removed by the unbalance compensation control [12] or rotation at the estimated center of inertia [13], and we are now addressing this issue.

IV. CONCLUSIONS

This paper introduces a 6-pole/9-slot HPBM, where the rotor consists of only salient iron cores, and hence this motor is suitable for high speed applications. The suspension winding arrangement was well designed to reduce the suspension force variation and interference without mechanical angle detection. The rotor was magnetically levitated at 0 rpm and the measured force variation and angle error were small and enough for stable magnetic suspension. The rotor was driven at 60000 rpm with ball bearings. However, when the rotor was magnetically suspended, the maximum speed was 10000 rpm. The mass unbalance is primary reason for the speed limitation and hence we are now addressing this issue.

ACKNOWLEDGMENT

The authors gratefully acknowledge the significant contributions of Mr. Hisao Fukuhara, Mr. Ryuichi Natsume, and Mr. Kazuki Tsushima, former students, in the Department of Mechanical Engineering, Shizuoka University, Japan, for his work.

REFERENCES

[1] I. Takahashi, T. Koganezawa, G. Su, and K. Ohyama, "A Super High Speed PM Motor Drive System by a Quasi-Current Source Inverter," *IEEE Trans Industry Applications*, vol. 30, no. 3, 1994, pp.683-690.

[2] T. Noguchi, Y. Takata, Y. Yamashita, and S. Ibaraki, "160,000-r/min, 2.7-kW Electric Drive of Supercharger for Automobiles," *Proc IEEE PEDS*, 2005, pp. 1380-1385.

[3] N. Bianchi, S. Bolognani, and F. Luise, "High Speed Drive Using a Slotless PM Motor," *IEEE Trans Power Electronics*, vol. 21, no. 4, July 2006, pp. 1083-1090.

[4] L. Zhao, C. Ham, L. Zheng, T. Wu, K. Sunaram, J. Kapat and L. Chow, "A Highly Efficient 200000 RPM Permanent Magnet Motor System, " IEEE Trans on Magnetics, vol. 43, no. 6, 2007.

[5] C. Zwyssig, S. D. Round and J. W. Kolar, "An Ultrahigh-speed, Low Power Electrical Drive System," *IEEE Trans. Industrial Electronics*, Vol. 55, No. 2, pp. 577-585, 2008.

[6] G. Schweitzer and E. H. Maslen, "Magnetic Bearings", Springer.

[7] A. Chiba, T. Fukao, O. Ichikawa, M. Oshima, M. Takemoto, and D. G. Dorrell, "Magnetic Bearings and Bearingless Drives", Elsevier.

[8] J. Amemiya, A. Chiba, D. G. Dorrell, and T. Fukao, "Basic Characteristics of a Consequent-Pole-Type Bearingless Motor," IEEE Trans. Magnetics, vol. 41, pp. 82-89, Jan. 2005.

[9] J. Asama, M. Amada, N. Tanabe, N. Miyamoto, A. Chiba, S. Iwasaki, M. Takemoto, T. Fukao, and M. A. Rahman, "Evaluation of a Bearingless PM Motor with Wide Magnetic Gaps", IEEE Transactions on Energy Conversion, Vol. 25, No. 4, pp. 957-964, December 2010.

[10] O. Ichikawa, O, A. Chiba, and T. Fukao, "Inherently decoupled magnetic suspension in homopolar-type bearingless motors", IEEE Transactions on Industry Applications, Vol. 37, Issue 6, pp. 1668-16674, 2002.

[11] J. Asama, R. Natsume, H. Fukuhara, T. Oiwa, and A. Chiba, "Optimal Suspension Winding Configuration in a Homo-polar Bearingless Motor", IEEE Transactions on Magnetics, Vol. 48, Issue 11, pp. 2973-2976 , 2012.

[12] T. Mizuno and T. Higuchi, "Compensation for Unbalance in Magnetic Bearing Systems", Transactions of the Society of Instrument and Control Engineering, Vol. 20, No. 12, pp. 1095-1101, 1984.

[13] J. Asama, T. Shibata, T. Oiwa, T. Shinshi, and A. Chiba, "Performance Improvement of a Bearingless Motor by Rotation about an Estimated Center of Inertia", Proceedings of the 11th International Symposium on Linear Drives for Industry Applications (LDIA), 2017.

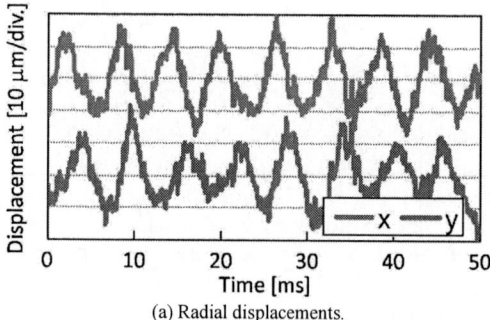

(a) Radial displacements.

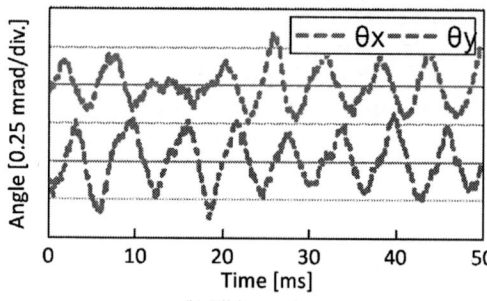

(b) Tilting angles.

Fig. 7. Measured rotor vibrations in the radial and tilting directions at 10000 rpm when the rotor is magnetically suspended.

High-speed Slotless Permanent Magnet Machines: modelling and design frameworks

S. Jumayev[*], K.O. Boynov[*], E.A. Lomonova[*], and J. Pyrhönen[†]

[*]Electrical Engineering Department, Eindhoven University of Technology, Eindhoven 5612 AZ, The Netherlands
[†]Electrical Engineering Department, Lappeenranta University of Technology, Lappeenranta 53850, Finland
e-mail: s.jumayev@tue.nl

Abstract—**This paper presents a design framework for high-speed slotless permanent magnet machines based on extended harmonic modeling (HM) technique to predict various electromagnetic properties and torque distribution. The developed models for generic design framework are able to evaluate slotless PM machines' topologies with a wide range of 3D slotless windings, (including those with skewing), and can be also used for future design optimization routines.**

Keywords—*3D harmonic modeling, 3D slotless winding, BLDC PM machine, high-speed.*

I. INTRODUCTION

Small-sized high-speed machines are also widely used in medical applications, such as surgery or dental drilling tools. Another example of medical small-sized high-speed machine applications is a portable medical ventilator that provides breathing assistance for people suffering from a respiratory system malfunction. Nowadays, there is a high interest in portable medical ventilators for stationary and autonomous operation. Current medical ventilators, which are usually based on closed systems with an external pressurized air or oxygen source, are not suitable for these purposes. Moreover, they are usually energy inefficient and heavy, and require extensive maintenance. The market demands new products intended for specific applications and homecare. The general aim is to develop a compact, energy-efficient, low life-cycle cost, and reliable breathing support system for innovative medical ventilators, which can be universally employed for stationary use (both neonates and adults, Fig. 1), during emergency transportation and independent of medical gases. The dynamic operating profile requires extra power from the machine for acceleration and deceleration of the overall system inertia. From the electrical drive system perspective, on the one hand, the drive has to satisfy the required speed profiles, and on the other hand, consume a minimum amount of energy from the battery supply to extend the autonomous operating time of the portable ventilator. Various machine topologies (Fig. 2) can be used in small-sized high-speed machines; however, the only feasible solution is a permanent magnet (PM) synchronous machine or a brushless DC machine, when high efficiency is needed [1–3]. The permanent magnet excitation enables, in principle, lossless excitation of the machine, which is the key to a high efficiency. Moreover, PM machines demonstrate a higher power density and a higher power factor [4]. In this paper, the generic

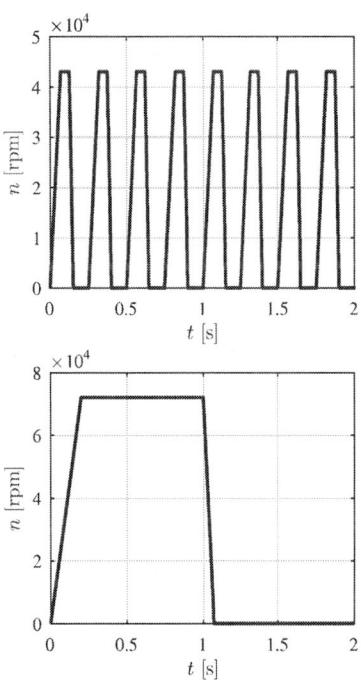

Fig. 1: Periodic motion profiles of two extreme ventilation cases: extreme neonatal (top) and extreme adult (bottom).

approach for design, modeling and verification of the small-size high speed PM motors with various windings configurations is extensively treated.

II. ELECTROMAGNETIC FIELD MODELING FRAMEWORK

The slotless winding embodiments enable a wide variety of complex winding geometries with skewing, which mainly result in lower-cost manufacturing of small sized machines. The electromagnetic field modeling of high-speed PM machines with these slotless windings lacks a unified, fast, and accurate approach. This paper presents an extended harmonic modeling (HM) technique to model the electromagnetic field, covering the induced eddy-current phenomenon, in 3D problems in a cylindrical coordinate system. The developed models, summarized in Tables I-III, are able to evaluate slotless PM machines

The 2018 International Power Electronics Conference

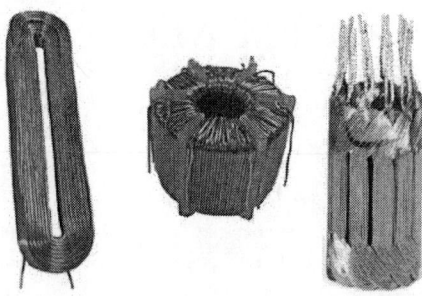

Fig. 2: Examples of slotless windings for high speed motors: a coil of an AC concentrated winding [5], AC toroidal winding, and knitted winding for a DC motor [6], respectively from left to the right.

with a wide range of slotless windings, including those with skewing. The extension of the harmonic modeling technique incorporates the 2D HM (Table II), formulated by the vector potential, and the 3D HM, expressed by the second-order vector potential (SOVP) (Table III). The concept of the SOVP to model 3D electromagnetic problems including the induced eddy currents has previously been implemented by a few researchers, in particular in [7–11]. However, these models do not provide an option to implement the 3D HM for a wide range of winding geometries. Another approach has been chosen because of the space distribution of the magnetic field source given by the linear current density, which is described by a 2D Fourier series and contains linear current density components with single and double space coordinate dependence [12]. The obtained electromagnetic field modeling framework (Table I) has been verified by using 3D FEM models of the benchmark slotless PM machines with Faulhaber, rhombic, and diamond windings (Fig. 3) for a comparison of the magnetic (Fig. 4), and electric (Fig. 5) fields. The maximum difference obtained for the magnetic field comparison does not exceed 10% of the peak value of the magnetic field distribution. The comparison of the electric field strength on the rotor surface gives a higher maximum difference, which is about 20%. This high value of discrepancy in the case of electric field strength can be caused by the limited amount of space harmonics and the Gibbs phenomenon, and does not have a strong influence on the prediction of the rotor eddy-current losses.

The developed HM simulation framework has limitations concerning the actual physical properties of certain regions. The main limitation is the assumption of the infinitely permeable soft-magnetic material and omission of its nonlinear behavior. This assumption, however, has only a slight impact in the modeling of PM machines, especially with slotless windings. This is explained by the large effective air gap. As it has been demonstrated in the benchmark machines with the M270-35A iron material, the decrease in the relative permeability value from $\mu_r = 10^5$ to $\mu_r = 10^2$ causes only a 1% error in the magnetic field calculation.

TABLE I: General modeling framework arrangement of armature and PM field modeling in 2D and 3D problems

	2D problem	3D problem
Armature field		
Modeling approach	2D HM	2D and 3D HM
Coordinate system	Polar (r, θ)	Cylindrical (r, θ, z)
Space current components	$\vec{I}_z$	$\vec{I}_z$ and $\vec{I}_\theta$
Source description	1D Fourier series	2D Fourier series
Potential formulation	$\vec{A}$	$\vec{A}$ and $\vec{W}$
Periodicity	θ-direction	θ- and z-directions
PM field		
Modeling approach	2D HM	
Coordinate system	Polar (r, θ)	
Magnetization direction	$\vec{M}_r$ and $\vec{M}_\theta$	
Source description	1D Fourier series	
Potential formulation	$\vec{A}$	
Periodicity	θ-direction	

TABLE II: Overview of 2D harmonic modeling

Governing equations	
Air (vacuum) region	$\nabla^2 \vec{A}_z = 0$
Winding region	$\nabla^2 \vec{A}_z = -\mu \vec{J}$
PM region	$\nabla^2 \vec{A}_z = -\mu_0 \nabla \times \vec{M}_0$
Conducting region	$\nabla^2 \vec{A}_z = \mu\sigma \dfrac{\partial \vec{A}_z}{\partial t}$
Boundary conditions	
Continuous	$B_{r,1} - B_{r,2} = 0,$ $\vec{H}_{\theta,1} - \vec{H}_{\theta,2} = 0$
Dirichlet	$A_z\|_{r=0}\|_{r\to\infty} = 0$
Magnetic field source	
Current density	$\vec{J} = \vec{J}_z(\theta, t) \cdot \vec{e}_z$
Magnetization	$\vec{M}_0 = \vec{M}_r(\theta) \cdot \vec{e}_r + \vec{M}_\theta(\theta) \cdot \vec{e}_\theta$

TABLE III: Overview of 3D harmonic modeling

Governing equations	
Air region	$\nabla^2 \vec{W} = 0$
Conducting region	$\nabla^2 \vec{W}_1 = \mu\sigma \dfrac{\partial W_1}{\partial t},$ $\nabla^2 \vec{W}_2 = \mu\sigma \dfrac{\partial W_2}{\partial t}$
Boundary conditions	
Continuous	$B_{r,1} - B_{r,2} = 0,$ $\vec{H}_{\theta,1} - \vec{H}_{\theta,2} = \vec{K}_\theta,$ $\vec{H}_{z,1} - \vec{H}_{z,2} = \vec{K}_z,$ $\vec{E}_{z,1} - \vec{E}_{z,2} = 0 \text{ or } \vec{E}_{\theta,1} - \vec{E}_{\theta,2} = 0$
Dirichlet	$\vec{W}_1\|_{r=0}\|_{r\to\infty} = 0,$ $\vec{W}_2\|_{r=0}\|_{r\to\infty} = 0$
Source description	
Linear current density	$\vec{K} = \vec{K}_z \vec{e}_z + \vec{K}_\theta \vec{e}_\theta$

Besides, if a laminated structure is used, the iron magnetic properties become anisotropic, meaning that the relative permeability is significantly lower in the axial direction because of the gaps between the laminations. This influences the modeling of the axial field component. However, the impact is minimal because of the large effective air gap in slotless PM machines. Furthermore, in the modeling of the rotor eddy currents, the conducting parts of the rotor are assumed solid. The assumption has a strong effect on the modeling of the eddy-current phenomenon in rotors with segmented PMs, which is usually not the case for small-sized high speed slotless PM machines. The extended HM for 3D [12] electromagnetic problems can also take into account the end effects of the induced rotor eddy currents. These effects are usually

162

neglected in problems modeled by using the 2D HM. This is especially beneficial for machines with a short active length, where the end path of the eddy currents is not negligible.

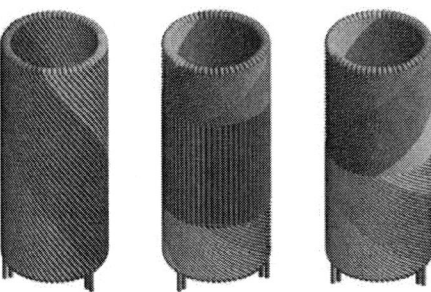

Fig. 3: Windings of the slotless high speed machines under investigation: Faulhaber, diamond, rhombic, respectively from left to right.

III. DERIVATION OF ELECTROMAGNETIC QUANTITIES

The electromagnetic field distribution is further used to derive the electromagnetic quantities of slotless PM machines, namely the developed electromagnetic torque, the emf induced by the PM field, and the synchronous inductance [13]. The generalized expression to calculate the electromagnetic torque has been derived using the armature and PM field solutions in the air gap for the 2D and 3D electromagnetic problems applying Maxwell's stress tensor.

A. Force calculation based on the 2D harmonic modeling

To obtain the general expression of force calculation for the 2D case, the following vector potential of the PM field given in the stator reference frame is assumed

$$A_{z,PM} = \sum_{\nu=1}^{\infty} (C_{PM} r^{-\nu} + D_{PM} r^{\nu}) \sin(\nu(\theta - \omega t) + \theta_0),$$
(1)

where θ_0 is the initial rotor position [rad]. The expression of the vector potential is given in the real number domain in order to simplify the derivation of the force calculation expression. Additionally, it is assumed that the vector potential contains only sine terms, which can be obtained by choosing a proper origin of the coordinate system while deriving the PM magnetization distribution. The radial and circumferential components of the PM flux density are written as

$$B_{r,PM} = \frac{1}{r} \sum_{\nu=1}^{\infty} \Lambda_{PM,r} \cos(\nu(\theta - \omega t) + \theta_0),$$
(2)

$$B_{\theta,PM} = \sum_{\nu=1}^{\infty} \Lambda_{PM,\theta} \sin(\nu(\theta - \omega t) + \theta_0),$$
(3)

where $\Lambda_{PM,r}$ and $\Lambda_{PM,\theta}$ are defined as PM-field constants.

Similarly to the PM field, the general expression for the armature reaction vector potential in the stator reference frame can be given as

$$\bar{A}_{z,AR} = \sum_{\substack{k=1 \\ L=-\infty}}^{\infty} \left[\left(\bar{C}_{s,AR} r^{-(3L-k)} + \right. \right.$$
$$\bar{D}_{s,AR} r^{(3L-k)} \Big) e^{j((3L-k)\theta + k\omega t)} +$$
$$\left. \left(\bar{C}_{c,AR} r^{-(3L-k)} + \bar{D}_{c,AR} r^{(3L-k)} \right) e^{j((3L-k)\theta + k\omega t)} \right],$$
(4)

where $\bar{C}_{s,AR}$ and $\bar{D}_{s,AR}$ are complex unknown constants related to the sine term of the vector potential, and $\bar{C}_{c,AR}$ and $\bar{D}_{c,AR}$ are complex unknown constants related to the cosine term of the vector potential. In order to obtain these unknown coefficients, initial and boundary conditions between domains have to be defined. For the regions with magnetic field sources, to comply with the periodic part of the solution, these sources are represented by a Fourier series. With certain assumptions, as discussed in [12], this method is very suitable to derive an application-specific solution of Maxwell's equations. The radial and circumferential components of the armature reaction flux density are found as

$$B_{r,AR} = \frac{1}{r} \sum_{\substack{k=1 \\ L=-\infty}}^{\infty}$$
$$\left[\text{Im}(\bar{A}_{s,r} e^{j((3L-k)\theta + k\omega t)}) + \text{Re}(\bar{A}_{c,r} e^{j((3L-k)\theta + k\omega t)}) \right],$$
$$B_{\theta,AR} = \sum_{\substack{k=1 \\ L=-\infty}}^{\infty}$$
$$\left[\text{Im}(\bar{A}_{s,\theta} e^{j((3L-k)\theta + k\omega t)}) + \text{Re}(\bar{A}_{c,\theta} e^{j((3L-k)\theta + k\omega t)}) \right],$$
(5)

where $\bar{A}_{s,r}$, $\bar{A}_{c,r}$, $\bar{A}_{s,\theta}$, and $\bar{A}_{c,\theta}$ are defined by armature reaction field constants.

The circumferential force is expressed as

$$F_\theta = \frac{1}{\mu_0} \oint_{S_1} B_\theta B_r dS_1,$$
(6)

where

$$B_r = B_{r,PM} + B_{r,AR},$$
$$B_\theta = B_{\theta,PM} + B_{\theta,AR}$$
(7)

are radial and circumferential flux densities of the total field, including PM and armature fields.

In this case, the armature field also contains the reaction of the rotor eddy currents. By substituting (7)

The 2018 International Power Electronics Conference

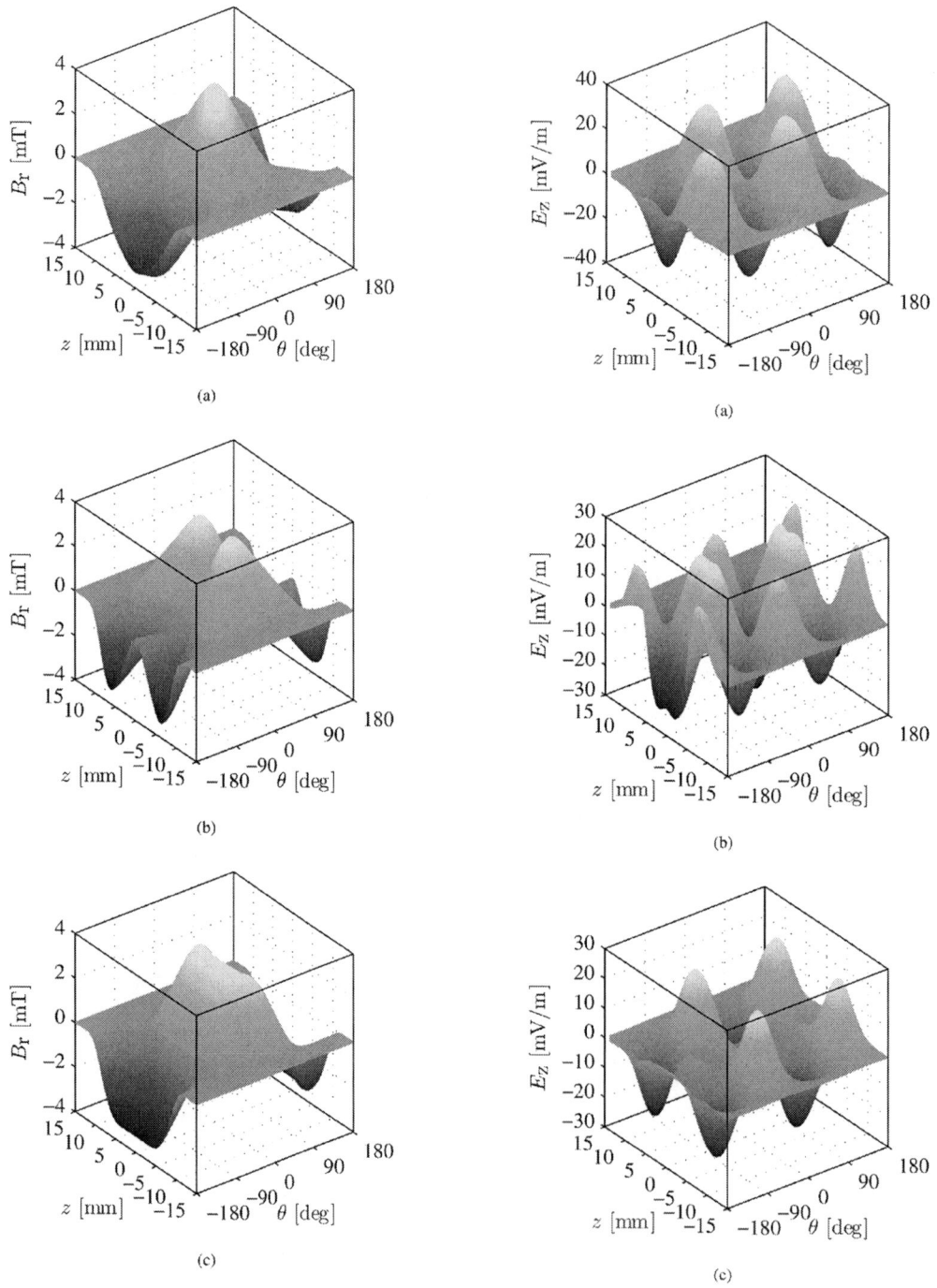

Fig. 4: Radial component of the armature reaction magnetic flux density of the benchmark PM machines with a) Faulhaber, b) rhombic, and c) diamond slotless winding types at $r = r_{\mathrm{m}}$ (results of the harmonic modeling approach).

Fig. 5: Axial component of the electric field strengths of the benchmark PM machines with a) Faulhaber, b) rhombic, and c) diamond slotless winding types at $r = r_{\mathrm{m}}$ (results of the harmonic modeling approach).

164

into (6), the tangential force expression can be rewritten as

$$F_\theta = \frac{lr}{\mu_0} \int_{-\pi}^{\pi} (B_{\mathrm{r,PM}} B_{\theta,\mathrm{AR}} + B_{\theta,\mathrm{PM}} B_{\mathrm{r,AR}} + B_{\mathrm{r,AR}} B_{\theta,\mathrm{AR}} + B_{\mathrm{r,PM}} B_{\theta,\mathrm{PM}}) \mathrm{d}\theta. \quad (8)$$

It should be noted that the radial and tangential PM field components are always shifted by $\pi/2$, because no circumferential permeance variation in the stator is present. This means slotless PM machines do not produce any cogging torque.

B. Force calculation based on the 3D harmonic modeling

The calculation of the force and torque in a 3D electromagnetic problem is similar to the 2D one. To derive the tangential force for the 3D case, the armature field in the air gap in terms of second-order vector potential can be generally assumed as

$$\bar{W}_{\mathrm{AR}} = \sum_{\substack{m=0 \\ k=1 \\ L=-\infty}}^{\infty}$$

$$[(\bar{C}_{\mathrm{ss}} I_{(3L-k)}(\omega_z r) + \bar{D}_{\mathrm{ss}} K_{(3L-k)}(\omega_z r)) \mathrm{e}^{\mathrm{j}((3L-k)\theta + k\omega t)} \sin(\omega_z z + \theta_z) + (\bar{C}_{\mathrm{sc}} I_{(3L-k)}(\omega_z r) + \bar{D}_{\mathrm{sc}} K_{(3L-k)}(\omega_z r)) \mathrm{e}^{\mathrm{j}((3L-k)\theta + k\omega t)} \cos(\omega_z z + \theta_z) + (\bar{C}_{\mathrm{cs}} I_{(3L-k)}(\omega_z r) + \bar{D}_{\mathrm{cs}} K_{(3L-k)}(\omega_z r)) \mathrm{e}^{\mathrm{j}((3L-k)\theta + k\omega t)} \sin(\omega_z z + \theta_z) + (\bar{C}_{\mathrm{cc}} I_{(3L-k)}(\omega_z r) + \bar{D}_{\mathrm{cc}} K_{(3L-k)}(\omega_z r)) \mathrm{e}^{\mathrm{j}((3L-k)\theta + k\omega t)} \cos(\omega_z z + \theta_z)], \quad (9)$$

where the first letter in the subscript of the unknown coefficient relates to the sine or cosine terms of circumferential spatial distribution, and the second letter in the subscript of the unknown coefficient relates to the sine or cosine terms of axial spatial distribution. The real-number representation of the radial and circumferential components of the flux density can be derived as

$$B_{\mathrm{r,AR}} = \sum_{\substack{m=0 \\ k=1 \\ L=-\infty}}^{\infty}$$

$$[\mathrm{Re}(\bar{A}_{\mathrm{ss,r}} \mathrm{e}^{\mathrm{j}((3L-k)\theta + k\omega t)} \cos(\omega_z z + \theta_z)) + \mathrm{Re}(\bar{A}_{\mathrm{sc,r}} \mathrm{e}^{\mathrm{j}((3L-k)\theta + k\omega t)} \sin(\omega_z z + \theta_z)) + \mathrm{Im}(\bar{A}_{\mathrm{cs,r}} \mathrm{e}^{\mathrm{j}((3L-k)\theta + k\omega t)} \cos(\omega_z z + \theta_z)) + \mathrm{Im}(\bar{A}_{\mathrm{cc,r}} \mathrm{e}^{\mathrm{j}((3L-k)\theta + k\omega t)} \sin(\omega_z z + \theta_z))], \quad (10)$$

$$B_{\theta,\mathrm{AR}} = \sum_{\substack{m=0 \\ k=1 \\ L=-\infty}}^{\infty} \frac{1}{r}$$

$$[\mathrm{Re}(\bar{A}_{\mathrm{ss,\theta}} \mathrm{e}^{\mathrm{j}((3L-k)\theta + k\omega t)} \cos(\omega_z z + \theta_z)) + \mathrm{Re}(\bar{A}_{\mathrm{sc,\theta}} \mathrm{e}^{\mathrm{j}((3L-k)\theta + k\omega t)} \sin(\omega_z z + \theta_z)) + \mathrm{Im}(\bar{A}_{\mathrm{cs,\theta}} \mathrm{e}^{\mathrm{j}((3L-k)\theta + k\omega t)} \cos(\omega_z z + \theta_z)) + \mathrm{Im}(\bar{A}_{\mathrm{cc,\theta}} \mathrm{e}^{\mathrm{j}((3L-k)\theta + k\omega t)} \sin(\omega_z z + \theta_z))], \quad (11)$$

where $\bar{A}_{\mathrm{ss,r}}$, $\bar{A}_{\mathrm{sc,r}}$, $\bar{A}_{\mathrm{cs,r}}$, $\bar{A}_{\mathrm{cc,r}}$, $\bar{A}_{\mathrm{ss,\theta}}$, $\bar{A}_{\mathrm{sc,\theta}}$, $\bar{A}_{\mathrm{cs,\theta}}$, and $\bar{A}_{\mathrm{cc,\theta}}$ are the armature reaction field constants for the 3D case.

The equation for the circumferential force calculation for a 3D case can be written as

$$F_\theta = \frac{r}{\mu_0} \int_{-\pi}^{\pi} \int_{-l/2}^{l/2} (B_{\mathrm{r,PM}} B_{\theta,\mathrm{AR}} + B_{\theta,\mathrm{PM}} B_{\mathrm{r,AR}} + B_{\mathrm{r,AR}} B_{\theta,\mathrm{AR}} + B_{\mathrm{r,PM}} B_{\theta,\mathrm{PM}}) \mathrm{d}\theta \mathrm{d}l. \quad (12)$$

The torque expression can be used to calculate the time and position-dependent torque profiles. It has been validated for the benchmark PM machine with the Faulhaber winding by means of the 3D FEM model, where the relative error is less than 4%. The emf induced by the PM field is calculated by integrating the magnetic field over the winding spatial distribution. The relative error of 4% is obtained in the emf calculation in comparison with the 3D FEM. The calculation of the synchronous inductance using the field predicted by the HM is able to take into account the inductance reduction produced by the rotor eddy currents. This frequency-dependent synchronous inductance value is further used to estimate the current ripple caused by the PWM voltage. The analytically evaluated synchronous inductance value demonstrates a good matching, where the maximum relative difference from the 3D FEM model does not exceed 5%.

IV. ELECTROMAGNETIC LOSSES IN HIGH-SPEED SLOTLESS PM MACHINES

Accurate modeling of losses in electrical machines is of primary importance, as it facilitates accurate prediction of the machine performance and thermal stresses on vulnerable machine parts. In order to complete the electromagnetic modeling framework, the lacking parts of the electromagnetic loss calculation in high-speed slotless PM machines have been incorporated. By using the electromagnetic field distribution in the machine, the rotor eddy-current losses have been simulated employing the Poynting vector. The rotor eddy-current losses assessed using the HM and 3D FEM demonstrate a good matching, where the maximum relative difference does not exceed 5%.

Furthermore, the strand-level proximity-effect loss calculation model for the round conductor cross-section (2D case) has been extended to the oval conductor cross sections, which is the case for the slotless windings with skewing. The strand-level proximity-effect loss calculation model for the oval conductor cross-section has been validated by using the FEM with a relative error that does not exceed 10%. Additionally, the approach to assess the impact of the PWM voltage on the rotor eddy-current losses in slotless PM machines has been demonstrated. This method is based on the solution of the electrical circuit of the slotless PM machines with frequency-dependent circuit parameters, where the PWM voltage is defined as a Fourier series over the frequency. Performing the rotor eddy-current loss calculation over each time harmonic of the current ripple results in the overall impact of the PWM on the rotor eddy-current losses. This approach to obtain the current ripple content has been verified by measurements on the test machine having a toroidal winding. Two voltage-source converters, one with sinusoidal and another with space vector PWM, have been used. The comparison shows a reasonable accuracy of the proposed approach, where the maximum obtained relative error is 39%. The error in the measurement can be explained by the measurement error, resulting from the high frequencies and parasitic impedance present in the circuit, and error in the estimation of the electric circuit parameters of the machine.

V. DESIGN OF HIGHLY DYNAMIC HIGH-SPEED SLOTLESS PM MACHINES

One of the objectives of this paper is to present the results of a high-speed slotless PM machine design research for the ventilator systems with a highly dynamic speed profile (Fig. 1) and minimum power consumption. By using the extreme cases of the motion profiles, the required rated electromagnetic torque of the machine has been defined as the maximum RMS torque value over one period. This torque value has to be updated with the change in the rotor dimensions as a function of rotor inertia. Different slotless PM machine topologies (Table V) have been analyzed with the aim to determine suitable rotor and stator combinations for the application in question. It has been found that a shaftless rotor results in lower rotor inertia with the same magnetic loading. Furthermore, diametrical rotor magnetization is advantageous as it produces a purely sinusoidal magnetic field. The stator configuration is mainly defined by the slotless winding topology, and complicates the selection of the possible slotless winding options. Therefore, it has been decided to design slotless PM machines with three different winding configurations, namely toroidal, concentrated, and Faulhaber ones. The design optimization framework for high-speed slotless PM machines intended for highly dynamic operation has been developed. This framework covers the electromagnetic, thermal, and mechanical models of slotless PM machines.

The parametric search optimization of the slotless PM machines with the chosen configurations is performed over the variable (varied parameter) ranges specified in Table IV. The required rated electromagnetic power of the machine for different rotor dimensions is depicted in Fig. 6. This plot is not made over the full PM diameter range, as specified in Table IV, because the

TABLE IV: Ranges and steps of the variables applied in the optimization procedure

Parameter, [unit]	Range	Step
PM diameter, [mm]	$4 \leq d_{\mathrm{m}} \leq 6.5$	0.1
Active length, [mm]	$15 \leq l \leq 35$	1
Conductor current density, [A/mm^2]	$6 \leq J \leq 24$	2
Number of parallel strands, [-]	$1 \leq a \leq 4$	1

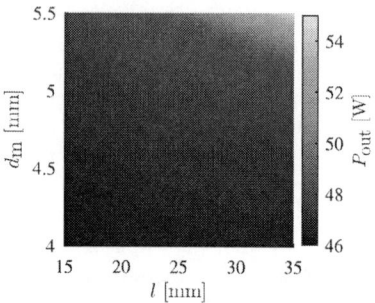

Fig. 6: Required rated electromagnetic power of the motor over the rotor length and the PM diameter.

machine designs obtained in the concealed region are not feasible. Figure 6 shows that with an increase in the rotor dimensions, also the required electromagnetic power increases. An increase in the rotor dimensions affects the inertia, and thereby, results in a higher power required during dynamic operation. The designs of slotless PM machines with toroidal, concentrated, and Faulhaber winding configurations are determined in accordance with this required power profile. As it was already mentioned, the optimal machine design is a design with minimum required input power as the drive assumes autonomous operation.

As the resulting multi-physical model is fast, determination of the optimal machine design by parametric search has been employed as the optimization method. In the developed design framework, iteration between the electromagnetic and thermal models is not implemented, which requires a close guess of the material temperature of the machine parts to take into account the thermal influence. The iteration process, however, can be implemented and evaluated by the proposed approach.

VI. PROTOTYPE IMPLEMENTATION AND MEASUREMENTS

A prototype of the high-speed slotless PM machine (see Figs. 7, 8 and Table VI) with a Faulhaber winding has been manufactured and tested. The measurements have shown that the model gives a satisfactory agreement in the prediction of the torque constant, DC resistance, and synchronous inductance of the prototype machine with relative errors smaller than 1%, 4%, and 14%, respectively. The rotor eddy-current loss calculation model has

The 2018 International Power Electronics Conference

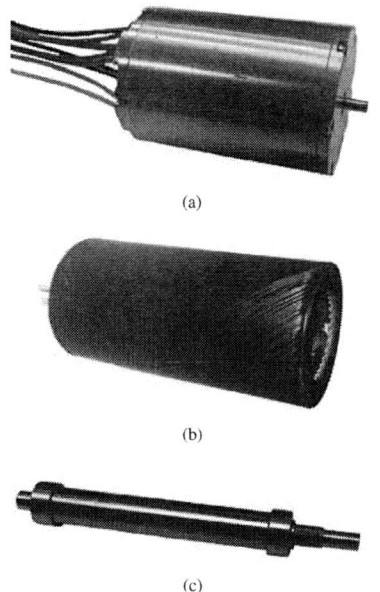

(a)

(b)

(c)

Fig. 7: Photos of a) assembled prototype, b) Faulhaber winding, and c) shaftless rotor.

TABLE V: Optimal designs specifications of slotless PM machines with a) toroidal, b) concentrated, and c) Faulhaber windings for centrifugal compressor load.

Parameter, symbol [unit]	Toroidal	Concentrated	Faulhaber
Input power, P_{in} [W]	51.1	52.2	53.2
Output power, P_{out} [W]	45.5	46.0	46.1
Rated output torque, T [mNm]	5.4	5.5	5.5
Phase current, RMS, I [A_{RMS}]	2.0	2.0	1.5
Phase resistance, R [Ω]	0.22	0.33	0.37
Efficiency, η [%]	89	88	87
Current density, J [A_{RMS}/mm^2]	6	10	10
Phase turn number, N [-]	18	32	41
Number of parallel strands, a [-]	3	3	4
Conductor diameter, d_c [mm]	0.355	0.28	0.25
Active length, l [mm]	32	34	35
PM diameter, d_m [mm]	4.5	4.6	4.6
Iron inner radius, r_{si} [mm]	5.4	6.2	7.2
Iron outer radius, r_{so} [mm]	10.4	11.2	12.2
Maximum PM temp., T_{PM} [^oC]	89	88	93
Maximum winding temperature, T_w [^oC]	89	89	99
Sleeve material	No	No	No
Maximum PM reference stress, $\sigma_{refm,max}$ [MPa]	1.08	1.13	1.13

TABLE VI: Initial and final (resultant) design specifications of the slotless PM machine with the Faulhaber winding.

Parameter, symbol [unit]	Initial	Final
Output power, P_{out} [W]	46.1 (120^oC)	45.7 (20^oC)
Rated output torque, T [mNm]	5.5 (120^oC)	5.5 (20^oC)
Phase current, RMS, I [A]	1.5	2.2
Phase resistance, R [Ω]	0.37	0.28
Phase turn number, N [-]	41	35
Conductor diameter , d_c [mm]	0.25	0.2
Active length, l [mm]	35	28
PM diameter, d_m [mm]	4.6	5.5
Winding thickness, h_w [mm]	3.7	3
Yoke inner radius, r_{si} [mm]	7.2	7.3
Yoke outer radius, r_{so} [mm]	12.2	12
PM remanent flux density, B_{rem} [T]	1.2 (120^oC)	1.23 (20^oC)
PM relative permeability, μ_{rPM} [-]	1.035	1.088

been validated by conducting indirect measurements by no-rotor and locked-rotor tests. In order to reduce the influence of the other possible losses, a measurement was conducted with the Faulhaber winding without the back iron and a dummy cylindrical PM. The results of the measurements have proven the validity of the developed models by demonstrating a maximum relative error of about 10%. As losses caused by the rotor motion cannot be measured separately, a no-load test of the prototype was conducted to measure them together. To separate the losses caused by the rotor motion, the losses caused by the applied current (dc Joule losses), and the rotor eddy-current losses, they are subtracted from the no-load losses. The obtained relative error between the measured and calculated losses caused by the rotation is about 30%.

The significant error in the results can be explained by the poor accuracy of the predicted bearing loss modeling and stray losses, produced by the eddy currents induced in the conducting parts of the machine. Based on the implementation of the prototype, it can be concluded that an initial Faulhaber winding type is not the best candidate for a large winding thickness because of the manufacturing limitations. It is identified that a three-layer Faulhaber winding had to be manufactured to meet the thickness requirements.

VII. DRIVE-LEVEL MODELING AND DESIGN FRAMEWORK

High-speed electrical machines are inherently employed with power electronic converters and are driven by certain control strategies. The common practice during the

drive design is to develop each of the drive components, namely the electrical machine and converter, separately. However, such an approach easily results in an unpractical solution as the mutual influence of the drive components is not taken into account. Therefore, to make a further step in high-speed drives design, a system-level approach has to be adopted.

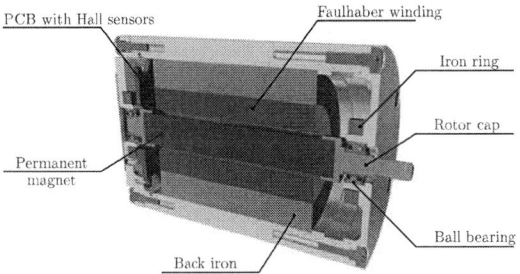

Fig. 8: 3D cross-sectional view of the prototyped machine.

VIII. Conclusions

The obtained electromagnetic field modeling framework has been verified by using 3D FEM models of the benchmark slotless PM machines with Faulhaber, rhombic, and diamond windings for a comparison of the magnetic and electric fields. The maximum difference obtained for the magnetic field comparison does not exceed 10% of the peak value of the magnetic field distribution. The comparison of the electric field strength on the rotor surface gives a higher maximum difference, which is about 20%.

Three optimal designs of the small-sized high-speed slotless PM machine with either toroidal, concentrated, or Faulhaber windings have been obtained by using the parametric search optimization framework. These identified optimal machine designs with three different windings have similar machine performance indices (input power and efficiency) and dimensions of the active part, excluding the end windings. Therefore, it is not obvious how to select the most suitable winding topology, which is, in practice, mainly defined by the manufacturing capabilities. Meanwhile, it has been observed that the common tendency is to select a decreased rotor diameter and an increased rotor length, which lead to decreased rotor inertia and, consequently, lower power consumption in this dynamic application. The decreased rotor diameter gives a lower magnetic loading. Therefore, to maintain the required torque, the electrical loading should be increased, which can be identified in the relatively large thickness of the winding. It has also been observed that with the increased winding thickness, the allowed conductor current density is decreased because of the poorer cooling capabilities. Furthermore, with the given PWM type (asymmetrical space vector), the optimal machine designs for all three winding types result in a rotor without any sleeve material, as it gives the lowest rotor eddy-current losses.

References

[1] D. Gerada, A. Mebarki, N. Brown, C. Gerada, A. Cavagnino, and A. Boglietti, "High-speed electrical machines: Technologies, trends, and developments," *IEEE Transactions on Industrial Electronics*, vol. 61, pp. 2946–2959, June 2014.

[2] A. Binder and T. Schneider, "High-speed inverter-fed AC drives," in *International Aegean Conference on Electrical Machines and Power Electronics, 2007*, pp. 9–16, Sept 2007.

[3] A. Borisavljevic, S. Jumayev, and E. Lomonova, "Toroidally-wound permanent magnet machines in high-speed applications," in *2014 International Conference on Electrical Machines (ICEM)*, pp. 2588–2593, Sept 2014.

[4] C. Zwyssig, S. Round, and J. Kolar, "An ultrahigh-speed, low power electrical drive system," *Industrial Electronics, IEEE Transactions on*, vol. 55, pp. 577–585, Feb 2008.

[5] P.-D. Pfister and Y. Perriard, "Slotless permanent-magnet machines: General analytical magnetic field

calculation," *IEEE Transactions on Magnetics*, vol. 47, pp. 1739–1752, June 2011.

[6] Maxon Academy, www.maxonmotor.com, *Maxon DC motor: Permanent magnet DC motor with coreless winding*.

[7] T. Theodoulidis, C. S. Antonopoulos, and E. E. Kriezis, "Analytical solution for the eddy current problem inside a conducting cylinder using the second order magnetic vector potential," *COMPEL - The International Journal for Computation and Mathematics in Electrical and Electronic Engineering*, vol. 14, no. 4, pp. 45–48, 1995.

[8] H. von Markus, *Uber die dreidimensionale analytische Berechnung der elektrischen Wirbelstrome in Kreiszylinderschalen*. PhD thesis, Technischen Universitat Munchen, 1977.

[9] K. Weigelt, *Uber die dreidimensionale analytische Berechnung von Wirbelstromen in gekoppelten Kreiszylinderschalen*. PhD thesis, Technischen Universitat Munchen, 1986.

[10] K. Pipis, *Eddy-current testing modeling of axisymmetric pieces with discontinuities along the axis by means of an integral equation approach*. PhD thesis, Universite Paris-Saclay, 2015.

[11] S. Jumayev, J. J. H. Paulides, K. O. Boynov, J. Pyrhönen, and E. A. Lomonova, "Three-dimensional analytical model of helical winding PM machines including rotor eddy-currents," *IEEE Transactions on Magnetics*, vol. 52, pp. 1–8, 2016.

[12] S. Jumayev, K. O. Boynov, J. J. H. Paulides, E. A. Lomonova, and J. Pyrhönen, "Slotless pm machines with skewed winding shapes: 3-D electromagnetic semianalytical model," *IEEE Transactions on Magnetics*, vol. 52, pp. 1–12, Nov 2016.

[13] S. Jumayev, A. Borisavljevic, K. Boynov, J. Pyrhönen, and E. A. Lomonova, "Inductance calculation of high-speed slotless permanent magnet machines," *COMPEL - The international journal for computation and mathematics in electrical and electronic engineering*, vol. 34, no. 2, pp. 413–427, 2015.

Development and Performance of High-Speed SPM Synchronous Machine

Kota Kawanishi, Keisuke Matsuo, Takayuki Mizuno, Koji Yamada & Takashi Okitsu,
High-speed Motor Development Project Meidensha Corporation, Tokyo, Japan
Kouki Matsuse
Meiji University Tokyo, Japan

Abstract—This paper presents a performance evaluation for a 250 kW – 20,000 min^{-1} high-speed permanent magnet (PM) motor for high-speed applications such as industrial blowers and compressors. The utilization of a high-speed motor for a direct drive system can contribute to making the system size more compact when compared to a geared system. However, this could cause the motor loss density to be high due to the high output power density in the high-speed motor. At this time, we evaluated the electrical characteristics and efficiency of the prototype 250 kW – 20,000 min^{-1} high-speed PM motor and measured the temperature increase by carrying out a heat run test. Furthermore, loss separation was performed to improve the efficiency and design accuracy.

I. Introduction

Permanent magnet (PM) synchronous motors have been widely used in various industrial applications due to their compact size and high efficiency [1, 2]. For high-speed applications such as industrial blowers and compressors, we have developed a 250 kW – 20,000 min^{-1} high-speed PM motor using an active magnetic bearing (AMB) [18–21].

When the high-speed motor is utilized in such a way, the system can be replaced by using a speed increasing gear with the direct drive system to eliminate the gear. Thus, in this way, the system size can be made compact, costs can be lowered, system efficiency can be improved, and maintenance time can be reduced.

On the other hand, the high output power density that is produced due to compacting the size while maintaining output power via high-speed rotation can bring about high loss density or high heat generating density. Therefore, it is necessary to reduce the motor loss and choose an appropriate cooling method.

Therefore, many studies have been implemented, including development of various types of high-speed motors [3–10], examination of loss reduction [11–15], and research on magnetic bearings [16–17]. In this paper, we present the characteristics of the prototype (250 kW – 20,000 min^{-1}) and the results of the loss separation.

II. DEVELOPMENT AND SPECIFICATION OF THE PROTOTYPE MOTOR

Table 1 lists the specifications and Fig. 1 shows the general view of the prototype motor [18].

Table 1 Specifications of the prototype motor

Motor type	PM motor
Rated output [kW]	250
Rated rotational Speed [min^{-1}]	20,000
Number of poles	2
Frequency [Hz]	333.3
Rated torque [N·m]	119.4
Cooling system	Forced air cooling system
Bearing	Active magnetic bearing
Magnet type	Nd-Fe-B

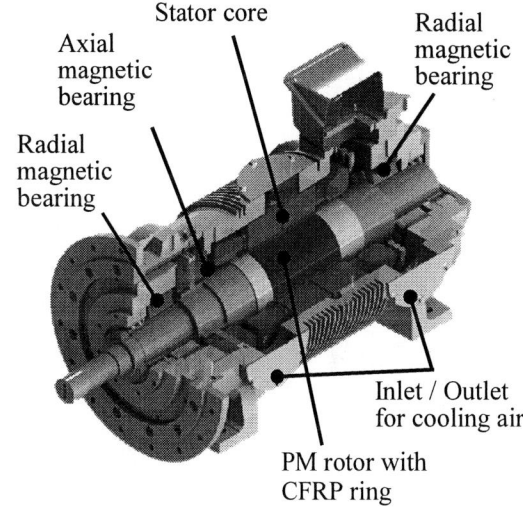

Fig. 1 Cross-sectional configuration of the High speed moto

The rated output power is 250 kW at a maximum rotational speed of 20,000 min^{-1} and there are two poles. Forced air cooling (via a separately-installed air blower) was utilized to cool the motor. As the rotor shaft is levitated by the AMB, contactless rotation can be realized. In addition, a surface permanent magnet (SPM) has been adopted for the rotor structure and the PMs mounted on the surface of the shaft have been retained by the sleeve. In this prototype motor, Carbon Fiber Reinforced Plastic (CFRP), which has features of high stress with light weight and low conductivity, has been applied as the material for the sleeve for enduring centrifugal force at 160 m/s over the circumferential speed to reduce the rotor eddy current losses.

Fig.2 shows the AMB structure. Radial magnetic bearings with floating RT that follow electromagnetic principle make it possible to rotate so as to avoid contact. Also, the axial magnetic bearing controls the axial position of RT based on the same principle. These RT positions of Radial and Axial magnetic bearings are detected by a position sensor. The sensor is controlled by an active magnetic bearing. (Fig. 3) In case of an emergency, touch-down bearings are mounted on the outside of the radial magnetic bearings.

Fig. 4 is high frequency inverter manufactured by MEIDEN. This inverter is a sister model of the high functional universal inverter VT240S, with its highest frequency enhanced to 800 Hz (200 V/400 V system). It enables sensorless control of high-speed PM motors. In case of two pole motors, the maximum rotation speed is 48000 rpm and the maximum rated output is 475 kW.

An LC filter is inserted between the inverter and motor to reduce the motor loss by smoothening the current waveform.

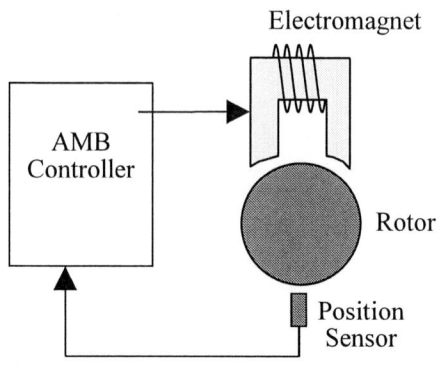

Fig. 3 AMB control principle

Inverter Specification

Voltage	Output	Frequency
200V	up to 90 kW	up to 800 Hz
400V	up to 475 kW	up to 800 Hz

Fig. 4 High frequency Inverter
(THYFREC-VT240S)

Fig. 5 is system configuration diagram from a commercial power supply to an inverter and motor.

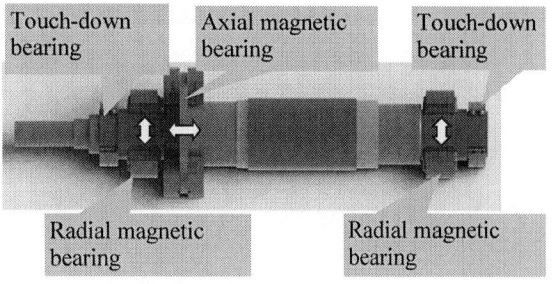

Fig. 2 AMB structure

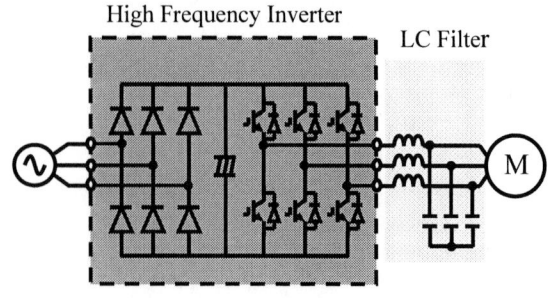

Fig. 5 Circuit Configuration

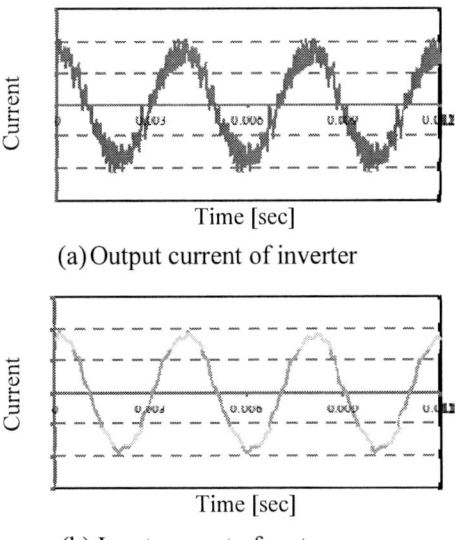

(a) Output current of inverter

(b) Input current of motor

Fig. 6 Current waveform
(15,000 min^{-1}, 100%_load)

Fig. 6 is an example of the current waveform in the prototype motor. From this figure, it can be seen that clean sinusoidal current flows through the motor.

III. PERFORMANCE EVALUATION

The motor loss consists of copper loss, iron loss, mechanical loss, and stray load loss. With high-speed motors, an increase in stator iron loss and mechanical loss is expected as the frequency increases. In examining various characteristics such as motor efficiency and temperature, these loss evaluations are considered to be extremely important themes.

Here, to create basic data for pursuing an effective loss reduction measure, each loss ratio will be experimentally clarified by the method shown below.

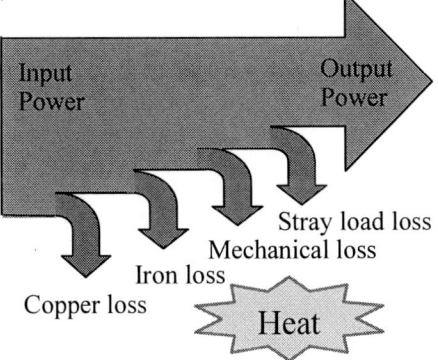

Fig. 7 Motor Loss

◆ copper loss: 3 x (current effective value)^2 x
   winding resistance
◆ Iron loss: No load iron loss
   (separate from single motor no-load input)
◆ Mechanical loss: Measure with a non-
   magnetized rotor
   (windage loss, AMB iron loss)
◆ Stray load loss: (input current at motor load)-
   (copper loss, iron loss, mechanical loss,)

The load test is conducted with the same motor connected directly.

A. The loss separation in no-load test
We performed the loss separation in a no-load test. When driving a PM motor in a no-load state, the iron loss due to only the magnetic flux generated by the PM will occur in the stator core. Therefore, the mechanical loss can be determined by the difference of two input powers; the one for the single motor is measured in a no-load state and the other from the total amount of two identical coupled motors. The one rotor is magnetized and the other one is non-magnetized, measured in a back-to-back state. (See Fig. 8)

Here, as the rotor shaft is levitated by the magnetic force generated by the AMB, there will not be any friction loss in this motor due to contact between the rotor shaft and bearing. Hence, the mechanical loss will include the windage loss in the air gap and the iron loss that occurred in the AMB. Furthermore, the iron loss in the stator core can be determined by subtracting the mechanical loss and the measured copper loss (I^{2}R) in the winding coil from the no-load loss of the single motor.

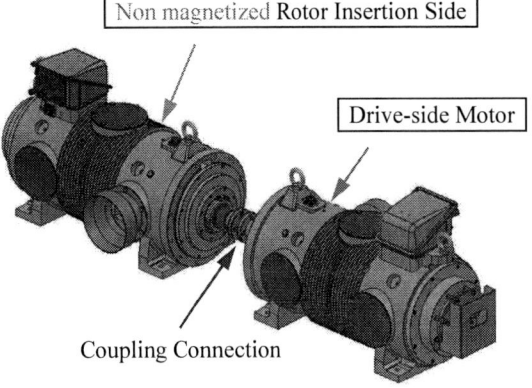

Fig. 8 Mechanical Loss Measurement Test

Fig. 9 shows the result of the loss separation at the speed of 5,000 min⁻¹ intervals in the no-load test. In Fig. 9, each loss was standardized with the loss at a maximum rotational speed of 20,000 min⁻¹. As a result of the loss separation, the ratio of the mechanical loss increases as the rotational speed becomes higher, and is about 64% at a speed of 20,000 min⁻¹. Note that the ratio of the copper loss is very small because the current is small in the no-load state. Accordingly, it will be an issue of reducing the mechanical loss for improving the efficiency of the high-speed motor.

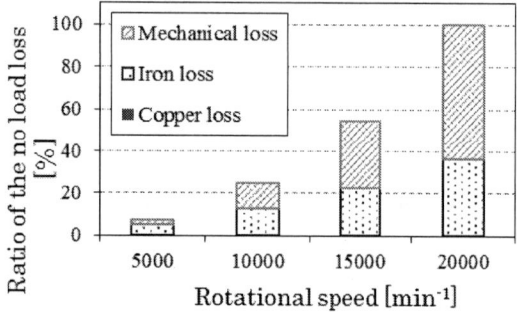

Fig. 9 Result of the loss separation in no-load test

B. Load test by back-to-back test

By coupling the two same prototype motors, we carried out the back-to-back test to measure the electrical characteristics and calculate the motor efficiency. Fig. 10 shows a view of the back-to-back test. In the test, we measured the input power W_{in} of the drive motor and the regenerated power W_{out} of the regeneration motor by using the power meter and the relation of W_{in} and W_{out} with the output power W_{trq} would be expressed as follows;

$$W_{in} = W_{trq} + (\text{loss of the drive motor}) \quad \cdots (1)$$
$$W_{out} = W_{trq} - (\text{loss of the regeneration motor}) \quad \cdots (2)$$

Here, assuming each loss generated in both drive and regeneration motor would be almost the same, W_{trq} is derived as the following equation;

$$W_{trq} = (W_{in} + W_{out})/2 \quad \cdots (3)$$

By the above method, we measured the electric characteristics while changing the rotational speed and load factor and calculated the motor efficiency.

Fig. 11 shows the efficiency map of the drive motor. The efficiency at the rated point is 96.6%

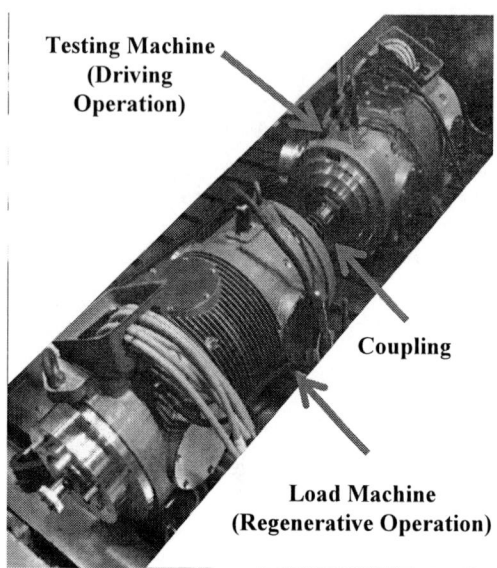

Fig. 10 Setup of the back-to-back test

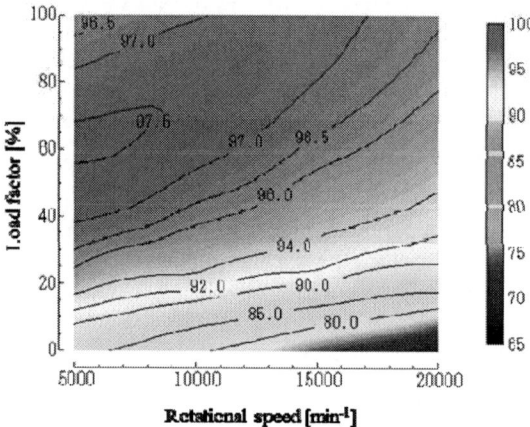

Fig. 11 Efficiency map of the prototype motor

and it is confirmed that the efficiency tends to decrease in the region of higher rotational speed.

Based on the test results from the no-load test and load test, the results of loss separation at the rated load are shown in Fig. 12. The mechanical loss accounts for about 55% of the total, and in order to achieve high efficiency, it is necessary to clarify the factors causing mechanical loss and to take measures to reduce the same.

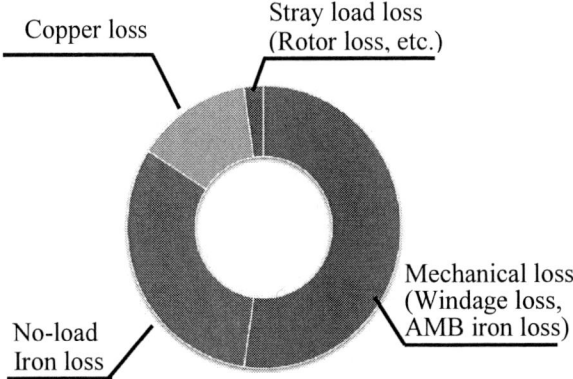

Fig. 12 Loss Separation Results at Rated Load
(250kW – 20,000 min⁻¹)

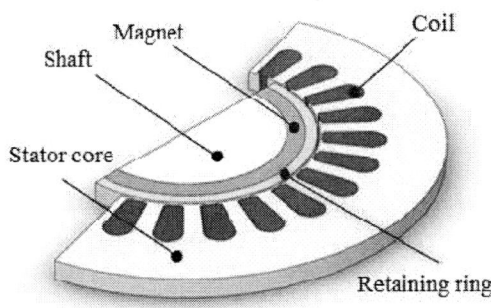

Fig. 13 Magnetic Field Analysis Model

Table 2. Loss Analysis pattern

	Ring material	No. of Ring division	Conductivity [×10⁵ S/m]
A	Titanium alloy (Ti-6Al-4V)	20	5.8
B	CFRP	5	// : 0.23 ⊥ : 0.000046

C. Consideration of rotor retaining ring [20]

The retaining ring is located at the outermost part of the rotor. In case of the conductivity ring, It has a possibility of decreasing efficiency or demagnetization by heating because an eddy current is caused by space harmonics due to stator slots or time harmonics due to the distorted current waveform.

Therefore, it is necessary to quantitatively grasp the loss of the retaining ring and the influence of the temperature. [20].

The authors conducted a magnetic field analysis on a holding ring (hereinafter referred to as a titanium ring) made of a titanium alloy (Ti - 6Al - 4V) having a high specific strength and a low conductivity and a holding ring made of CFRP by comparing actual measurements. Losses were compared and evaluated when applied to high speed motors. Therefore, the authors compared and evaluated the losses from the retaining ring made of titanium alloy (Ti-6Al-4V) and the holding ring made of CFRP by utilizing magnetic field analysis and actual measurements.

C-1 Loss analysis of retaining ring

Fig. 13 shows a magnetic field analysis model for the retaining ring loss analysis. Loss generated in each part of the rotor was analyzed and compared by changing the material of the retaining ring to titanium, CFRP. When using the titanium ring, the axial division number of the retaining ring is increased so that the eddy current loss occurring in the retaining ring can be reduced. (See Table 2.)

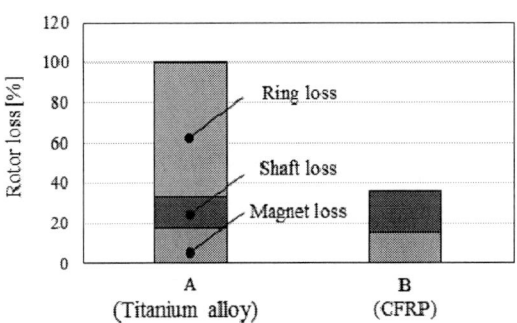

Fig. 14 Analysis result of rotor loss

Fig. 14 shows the analysis results of the loss generated in the rotor at the rated load (250 kW– 20,000 min⁻¹).

The ring conductivity differs by a parallel or vertical direction from the carbon fiber. As resin is filled around the fiber, the value gets much lower than that of a titanium ring.

The loss represents the loss of the rotor of condition A as 100%. In Condition A, significant eddy current loss occurs in the titanium ring, and it can be confirmed that the ring loss occupies the value of 67% of the total rotor loss. In condition B, the conductivity of the CFRP ring is lower than that of the titanium ring as described above. Therefore, it can be confirmed that the eddy current loss generated in the ring is very small. From the comparison of both conditions, the loss of the rotor for condition B is about 35% of the loss of the rotor for condition A.

From the above results, the CFRP ring seems to be superior in terms of loss reduction.

C-2 Ring loss measurement test

According to the method of loss separation described above, loss at the rated load was measured with a motor using each ring. From the result, the stray load loss W_{str}, including the loss of the retaining ring, was calculated by equation (4).

$$W_{str} = W_{total} - W_m - W_c - W_{ni} \quad \cdots (4)$$

W_{total} is total loss at load, W_m is mechanical loss, W_c is copper loss, and W_{ni} is no-load core loss.

According to the result shown in Fig. 14, loss occurring at the magnet and shaft is respectively equal. So, if you suppose that the CFRP ring's eddy current loss is 0, the eddy current loss of the titanium ring can be calculated as a difference of each stray load loss by the following formula (5).

$$W_{Ti} = W_{str\_Ti} - W_{str\_Cf} \quad \cdots (5)$$

W_{Ti} shows the eddy current loss of the titanium ring. W_{str_Ti} and W_{str_Cf} shows the stray load loss of the motor when the titanium or CFRP ring is respectively applied.

Fig. 15 shows the eddy current loss calculated by the formula (5) using the measured value and eddy current loss acquired by the magnetic field analysis on the titanium ring. Fig. 15 indicates that the eddy current loss of the titanium ring is around 15% larger in the analysis compared to the measurement. A possible reason for this would be an error caused by an assumption when formula (5) was derived or due to alteration of the conductivity caused by the temperature increase when measured.

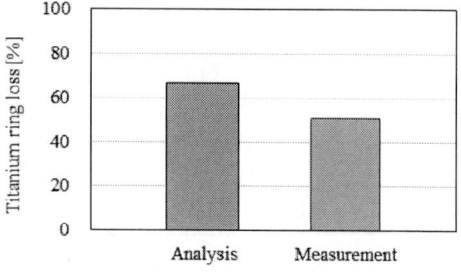

Fig. 15 Comparison of titanium ring loss

C-3 Temperature Evaluation of the Retaining Ring

To evaluate the influence of the retaining ring on the temperature, we performed a continuous operation under the rated load condition and checked the temperature increase of the magnet during steady operation. The magnet temperature was estimated using the change rate of the induced voltage both before and after the operation as well as the temperature coefficient of the residual flux density of the magnet. As a result, the CFRP ring motor took a value whose temperature increase from the magnet was around 30 K lower than that of the titanium ring motor. This result shows that the heating caused by the eddy current loss of the retaining ring has a significant influence on the magnet temperature. Therefore, the CFRP ring can be considered to be suitable for rotor retention of the high-speed motor because its heating and eddy current are smaller compared to the titanium ring.

D. Cooling system and heat run test

D-1 Cooling System

We carried out the heat run test at the rated point for measuring steady temperature inside the motor. Fig. 16 shows the configurations of the air cooling system in the test. At this time, two cooling systems were evaluated by installing one, 3.7 kW–50 Hz, cooling blower for exhausting air. In system A, inlet air would be flowing through to a ventilation hole of the direct-coupled side and passing through the other ventilation hole of the non-direct-coupled side.

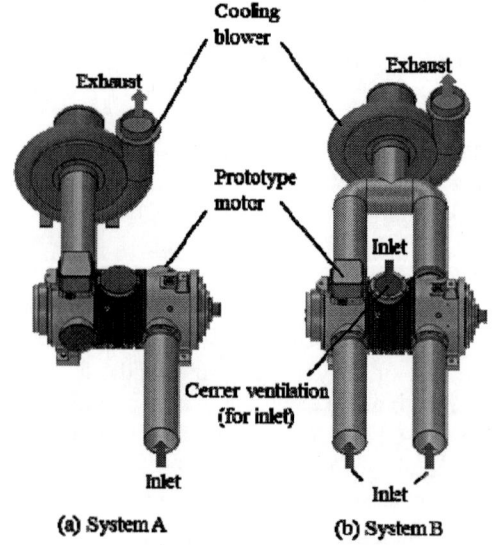

Fig. 16 Air cooling system in the heat run test

In system B, in contrast, inlet air would be flowing through to ventilation holes on both the direct-coupled side and the non-direct-coupled side and passing through ventilation holes opposite both sides. Additionally, the center ventilation hole was opened for inlet air in system B.

D-2 Temperature Test Result

To measure the coil temperature during the operation, a plurality of thermocouples was mounted inside the coil as shown in Fig. 17 (a). Furthermore, the temperature of the rotor magnet was estimated as an average temperature by calculation with the temperature coefficient of residual flux density %/K of the magnet and the decrease rate in the induced voltage was measured before and after the operation. Fig. 18 (b) shows the temperature increases in the coils and the magnet. As a result of measurements, the temperature increases in both the coils and the magnet in system B is lower than that in system A.

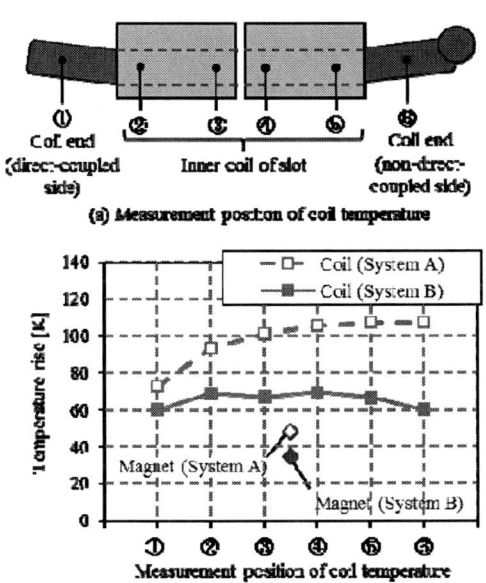

(a) Measurement position of coil temperature

Fig. 17 Temperature increase in the heat run test

E. Flow analysis

Flow in the ventilation system has a significant influence on the temperature increase of each part of the motor. We conducted computer fluid analysis (CFD) to evaluate the cooling system. Fig.18 shows the analysis result for the air flow inside motor. From this analysis result, it is possible to clarify the state of air flow inside motor.

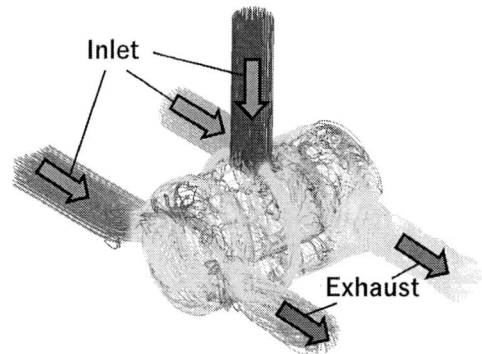

Fig. 18 Air flow inside motor

IV. CONCLUSION

In this paper, we evaluated the performance of the prototype for a 250 kW–20,000 min^{-1} high-speed PM motor. As the result of the evaluation, it was found that the motor efficiency is 96.6% at the rated point and the mechanical loss could be considered to be dominant loss, especially in the high-speed region. Furthermore, we compared two air cooling systems by carrying out a heat run test at the rated point and confirmed that system B could have sufficient cooling performance for the motor.

In the future, we will verify the design accuracy by performing comparisons with the results from loss separation found at this time to improve efficiency and work on development of further high-speed and large output power motors.

REFERENCES

[1] Investigating R&D Committee on New Technologies for Industry Applications of PM motors, "Trend of New Technologies toward their Spread" IEEJ Technical Report, No.1207, 2010 (in Japanese)

[2] IEEJ, "Current Situations and Future Issues of Adjustable-Speed AC Drives," Technical Report No.1326, p.64, 12, 2014 (in Japanese)

[3] T. Fukao and A. Chiba, "Super High-Speed Drives-Introduction", Proceeding of National Convention of IEEJ, S.21.1-4,3.1998
(in Japanese)

[4] R. R. Moghaddam, "High Speed Operation of Electrical Machines, a Review on Technology, Benefits and Challenges", IEEE ECCE 2014, pp.5539-5546

[5] M. Rahman, A. Chiba and T. Fukao, "Super High Speed Electrical Machines-Summary", IEEE, Power Engineering Society General Meeting 2004, vol.2, pp.1272-1275

[6] B. Sarlioglu, T. Jahns and D. Ionel, "Design and Manufacturing of PM Electric Machines", IEEE ECCE 2015, Tutorial, p.196

[7] E. Schubert and B. Sarlioglu, "Mechanical Design Method for a High-Speed Surface Permanent Magnet Rotor", IEEE ECCE 2016, p.6

[8] D. Matsuhashi and K. Matsuse, "High Power High Speed Motor Drives for Industry Applications", IEEE PEDES 2016, Plenary speech, India, Dec.16, 2016

[9] K. Matsuse and D. Matsuhashi, "New Technical Trends on Adjustable Speed AC Motor Drives", Chinese Journal of Electrical Engineering, Vol.3, No.1, pp.1-9, June 2017

[10] C. Babetto, G. Bacco, G. Berardi and N. Bianchi, "High Speed Motors: A Comparison between Synchronous PM and Reluctance Machines", IEEE ECCE 2017, pp3927-3934

[11] B. Sarlioglu and T. Wu, "Design and Analysis of Electrical Machines including High-Speed Types", IEEE IECON Tutorial 2014, p.176

[12] J. Luomi, C. Zwyssig, A. Looser, and J. W. Kolar, "Efficiency Optimization of a 100-W 500000 - r/min Permanent-Magnet Machine Including Air-Friction Losses", IEEE Trans. On I.A, vol.45, No.4, pp.1368-1377, 2009

[13] H. Fang, R. Qu and J. Li, "Rotor design for a high-speed high-power permanent-magnet synchronous machine", IEEE ECCE 2015, pp4405-4412

[14] P. Mellor, R. Wrobel, D. Salt and A. Griffo, "Experimental and Analytical Determination of Proximity Losses in a High-Speed PM Machine" IEEE ECCE 2013, pp.3504-3511

[15] P. B. Reddy, T. M. Jahns and T. P. Bohn, "Modeling and analysis of Proximity Losses in High–Speed Surface Permanent Magnet Machines with Concentrated Winding" IEEE, ECCE 2010, pp.996-1003

[16] A. Smirnov, N. Uzhegov, T. Sillanpää, J. Pyrhönen and O.Pyrhönen, "High-Speed Electrical Machine with Active Magnetic Bearing System Optimization", IEEE Tran. On I.E., Vol.64, No.12, pp.9876-9885, December 2017

[17] Z. Fu, D. Jiang and R. Qu, "Design of four-axis magnetic bearing for high speed motor", IPEMC-ECCE Asia, 2016

[18] T. Okitsu, S. Uchiyama, K. Matsuo and D. Matsuhashi, "Performance Evaluation of High Speed PM Motor 250kW-20,000rpm", Meiden-jiho, Vol.352, No.3, pp.30-33, 2016 (in Japanese)

[19] S. Uchiyama, K. Matsuo, T. Onishi, T. Okitsu, and D. Matsuhashi, "Characteristic evaluation of 250kW -20,000min^{-1} High-speed Motor" in Proc. IEEJ 2016, Sendai, 2016, vol. 5, pp. 35-36 (in Japanese)

[20] M. Omura, K. Matsuo, S. Uchiyama, T. Okitsu and D. Matsuhashi, "Loss comparison of rotor retaining ring in a High-speed motor" in Proc. IEEJ 2017, Toyama, 2017, vol. 5, pp.55-56 (in Japanese)

[21] Meidensha Catalogue, "Ultra-High Speed PM & Medium Voltage Drive", Meidensha Corporation, Apr. 2016

1.2kW 100,000rpm high speed motor for aircraft

Takehiro JIKUMARU[1], Gen KUWATA[1]
1 IHI corporation, Japan
*E-mail:takehiro_jikumaru@ihi.co.jp

Abstract— **This paper presents high speed motor for electric turbo blowers used in aircraft. In this application, equipment specifications such as light weight and high power density are strongly required. Therefore, it is necessary to increase the rotational speed of motors. We have developed 1.2kW/100krpm high speed motor. This paper also shows experimental results in the prototype.**

Keywords— Power density, high speed motors, Surface Permanent Magnet Synchronous Motor , Electric turbo blower

I. INTRODUCTION

In recent years, developments of technologies for More Electric Aircraft (MEA) have been actively studied. [1][2] Furthermore, the feasibility of All Electric Aircraft (AEA) is also being studied, and it is expected that the trend of MEA and AEA will accelerate in the future. [3] In order to realize these, it is necessary to suppress the increasing heat generation in electric equipment. In the most advanced electric aircraft Boeing 787, a liquid cooling heat exhaust system is adopted for cooling large electric power converter, however the poor maintainability is pointed out. We have studied an air cooling system using an electric turbo blower as an exhaust heat system with better maintainability. [4] Therefore, high speed motors such as light-weight and high power density are strongly required for electric turbo blowers used in aircraft.

There are many researches on high-speed motors applied to compressors and pump and so on[5]-[7], but there are few reports on application to aircraft. In this paper, 1.2kW/100krpm high-speed motor for aircraft achieves light-weight and high power density specifications. The motor was designed using FEA and a prototype was built to verify FEA results.

II. MECHNICAL DESIGN

As shown in FIG. 1, the impeller of the electric turbo blower is direct drive by the high speed motor. The high speed motor is 2 poles Surface Permanent Magnet Synchronous Motor (SPMSM). The permanent magnets (PMs) are made of Nd-Fe-B. The rotor is made up of hollow ring magnets, which is fitted on a permeable shaft and retained by sleeve. PMs are parallel magnetized to

reduce the space harmonics in air gap. there is an advantage that the eddy current loss on sleeve and magnet is reduced by large air gap. On the other hand, there is a disadvantage that the armature current increases. Thus, the air gap has to select the optimum value in consideration of the loss and the cool in the rotor. In high speed motor, the heat generated on the rotor is cooled with the air flow through the air gap. The stator with 24 slots is equipped with distributed winding. By connecting the windings in parallel, it is possible to suppress the volume of the neutral point in the coil-end and to prevent an increase in the total stator volume. An electrical steel is using thin sheets to reduce iron loss.

The output and losses of the high speed motor are evaluated using the commercial 2D FEA software Maxwell. The final geometry dimension of high speed motor is shown in FIG. 2. The power density is about 20kW/L without coil-end. The magnetic flux density in stator yoke and teeth in FIG. 3 is maximum 1.6T and 1.2T, respectively. It is seen that the magnetic flux density concentration occurs in stator yoke.

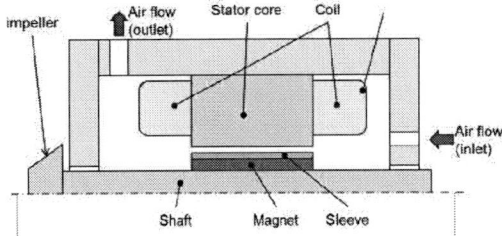

FIG.1. Cross section on electric turbo blower

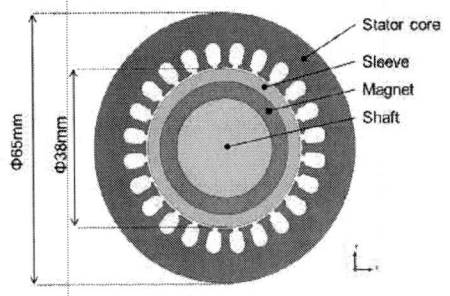

FIG2. Geometry dimension in high speed motor

The 2018 International Power Electronics Conference

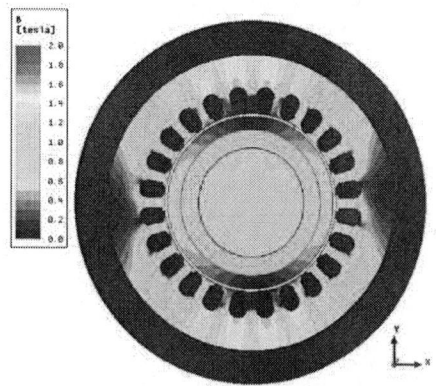

FIG.3 Field distribution of the high speed motor at rated power

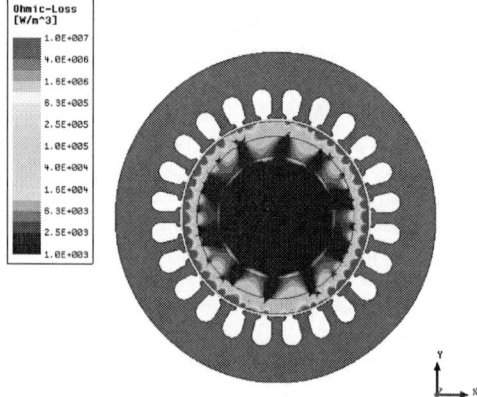

FIG.4 Contour of eddy current loss at rated power

FIG. 4 shows the contour of the eddy current loss in the rotor. The eddy current loss in the magnet is smaller than the eddy current loss in the sleeve and it is found that the loss concentrates on the surface of sleeve.

III. MEASUREMENTS

High-speed motor was evaluated using an aerodynamic load, since there is no method to directly measure the torque. FIG. 5(a) shows system of piping diagram. The impeller power is calculated from the following formula.

$$P_o = C_p \times G \times \Delta T$$

Where C_p is specific heat at constant pressure, G is mass flow rate, ΔT is temperature difference of inlet and outlet in impeller or cooling fan. FIG 5(b) shows test rig exterior. In the test rig, in addition to the temperature sensor for calculating the impeller power, the temperature in each part such as the armature coil, stator, bearings and so on was measured.

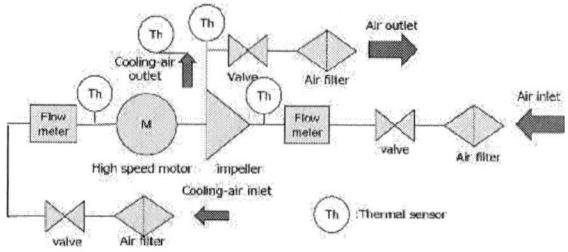

(a) System of piping diagram

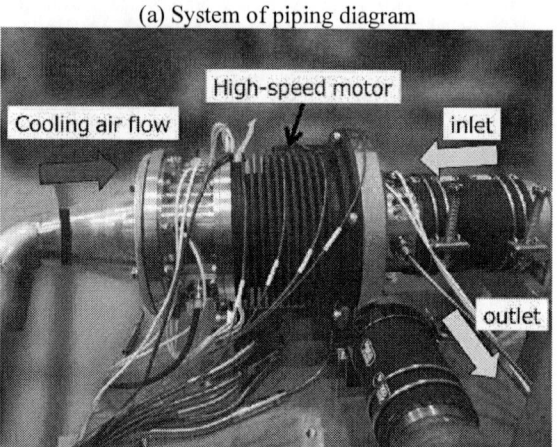

(b) Test rig exterior

FIG.5 Test rig system

High-speed motor with position sensorless control was driven by PWM converter using IGBT. The carrier frequency is set to 25kHz.

IV. EXPERIMENT RESULTS

This section introduces developed high speed motor in electric turbo blower with experimental results. FIG. 6 shows the current vs rotational speed characteristic of prototype. Although the experimental values as expected are obtained at 60krpm or more, there is a difference from the calculated value at 60krpm or less. Since low inductance in prototype, the current ripple becomes relatively large with respect to the fundamental current at 60krpm or less. Thus, the current effective value increases.

FIG. 7 shows frequency property of current on 80krpm, and the current value around 50 kHz is included 30% of the current value of the fundamental frequency. This frequency component increases the eddy current loss on the rotor surface.

FIG. 8 shows the voltage vs rotational speed characteristic of prototype. The experimental value is roughly on the calculation line. If it is operated with the rotational speed kept at 80krpm, the voltage value further decreased by 5% due to the temperature dependence of the magnet. (FIG.9) Since the temperature coefficient of the magnet is -0.1%/K, it is estimated that the magnet temperature rises until about 150 degree C.

178

The efficiency up to 100krpm is shown in FIG.10. This efficiency includes bearing loss and windage loss. When these losses are obtained from theoretical calculation, it becomes about 4.5% of the output at 100 krpm. So that means the efficiency of the high-speed motor at 100krpm is 89.5%.

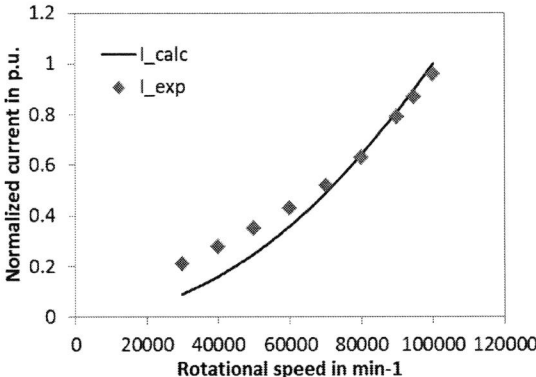

FIG.6 Characteristic of Current vs Rotational speed

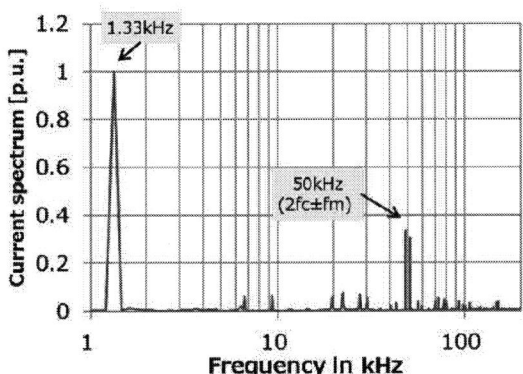

fm : Fundamental frequency   fc : Carrier frequency

FIG 7 Frequency property of current on 80krpm

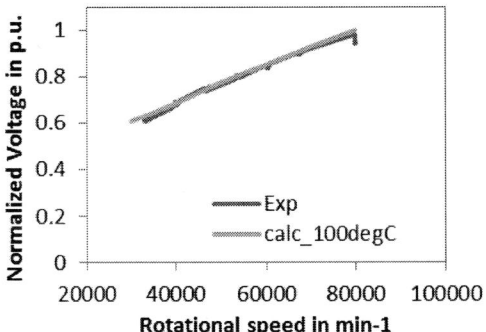

FIG.8 Characteristic of Voltage vs Rotational speed

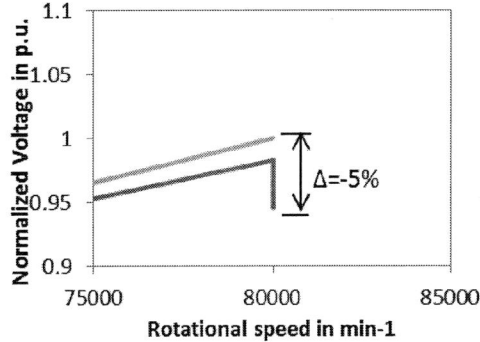

FIG.9 Characteristic of Voltage vs Rotational speed (enlarge in FIG.8)

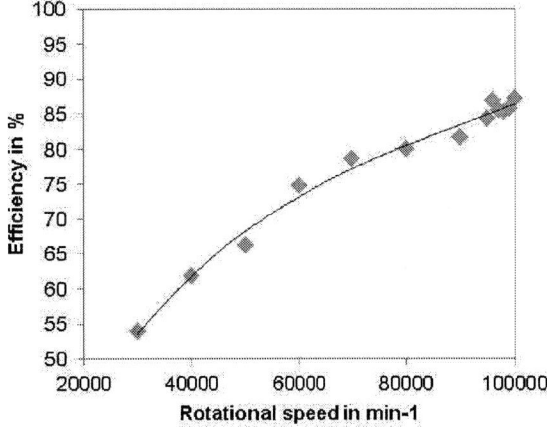

FIG. 10 Efficiency of high speed motor with mechanical loss

I. CONCLUSIONS

We confirmed that the design high speed motor has high power density with experimental results. The power density of 20kW/L is realized with air-cooling system, and this performance is comparable to a liquid-cooling system. However, the magnet temperature estimated from the voltage characteristics is about 150 degree C. The temperature coefficient of the NdFeB magnet is inferior to that of the SmCo magnet. Therefore, depending on the temperature of the magnet, it may be better to change the type of the magnet.

In order to properly select the type of the magnets, we will improve the estimation accuracy of magnet temperature and evaluate the extent of irreversible demagnetization. As a result, it will be possible to further improve the power density.

REFERENCES

[1] Davide Lusignani, Davide. Barater, Giovanni Franceschini, "A high-speed electric drive for the more electric engine" *Energy Conversion Congress and Exposition (ECCE)*, 2015
[2] P. THALIN, "Delivering innovative and optimized solutions for the more electric aircraft" *Electric & hybrid aerospace technology symposium*, 2015.
[3] Janet L. Kavandi "Powering the future with green propulsion technologies " *Electric & hybrid aerospace technology symposium*, 2015

[4] N. Morioka H. Saito, N. Takahashi, M. Seta, H. Oyori "Thermal Management System Concept with an Autonomous Air-Cooled System" *SAE 2014 Aerospace Systems and Technology Conference*, 2014

[5] K. Wang, M. J. Jin, J. X. Shen, and H. Hao,"Study on rotor structure with different magnet assembly in high speed sensorless brushless DC motors" *IET Proc. Elect. Power Appl.*, vol.4, no.4, pp.241-248,2014

[6] A. Gils; S. Tavernier, M. Gerber, C. Espanet, F. Dubas, D. Depernet, "Design of a cost-efficient high-speed high-efficiency PM machine for compressor applications" *Energy Conversion Congress and Exposition (ECCE)*, 2015

[7] Yu Fu, Masatsugu Takemoto , Satoshi Ogasawara , Koji Orikawa, "Investigation of a high speed and high power density bearingless motor with neodymium bonded magnet" *Electric Machines and Drives Conference (IEMDC)*, 2017

The 2018 International Power Electronics Conference

Comparative Evaluation of Y-Inverter against Three-Phase Two-Stage Buck-Boost DC-AC Converter Systems

Michael Antivachis, Dominik Bortis, David Menzi and Johann W. Kolar
Power Electronic Systems Laboratory
ETH Zurich, Switzerland
antivachis@lem.ee.ethz.ch

Abstract—Modern motor drives feature output filtering capability in order to protect the motor from high converter output voltage du/dt rates and provide a sinusoidal current to the machine in order to minimize the rotor losses. The incorporation of such motor drive into a fuel-cell (FC) application is challenging since the power electronics converter has to cope with the power dependent variation of the FC voltage. In this paper three candidate converter concepts are comparatively evaluated i.e. a voltage source inverter with front-end DC-DC boost converter (boost VSI), a current source inverter with front-end DC-DC buck converter (buck CSI), and the recently proposed buck-boost Y-inverter topology. In a first step, the three implementations are assessed based on fundamental scaling laws, i.e. the semiconductor losses and/or the required chip area and the inductive components volume, which constitute a major part of the total system losses and volume, are analytically derived. In a second step, the exact performance of the different motor drive solutions in terms of efficiency η and power density ρ is quantified by means of a comprehensive multi-objective optimization. The optimization results reasonably mach the scaling-laws approach and hence further verify the practical value of the analytic calculations. Both the scaling-laws based analysis and the accurate optimization indicate a clear power density advantage in favor of the Y-inverter.

Index Terms—High-speed drives, Scaling laws, Optimization, Modular Y-inverter

I. INTRODUCTION

Drive systems typically comprise a voltage source inverter (VSI) powered from a stabilized DC energy source followed by a motor as visualized in **Fig. 1(a)**. In case of a fuel-cell (FC) powered drive system, however, the input voltage strongly drops with increasing output power (cf. **Fig. 1(b.i)**) which greatly impacts the achievable performance and/or design of the following power electronics stage and the motor. For example, under low power conditions, the high input voltage $U_{FC,max}$ imposes a high voltage stress on the converter components, whereas during rated power operation, the low input voltage $U_{FC,min}$ leads to a high current stress (cf. **Fig. 1(b.ii)**), thus the drive system has to be inevitably overdimensioned. Furthermore, since the machine voltage linearly increases with the rotational speed while the FC voltage drops, in certain cases the machine can no longer be directly driven by a VSI. An application of this type is considered in the followng (cf. **Tab. I**). At a nominal rotational speed of 300 krpm [1], the motor of the underlying 1 kW compressor drive system exhibits a back-EMF phase voltage of $\tilde{U}_M = 30\,V_{RMS}$, whereas at the same time the FC voltage drops from $U_{FC,max} = 120\,V$ down to $U_{FC,min} = 60\,V$ at its nominal output power. Hence, the direct VSI can no longer drive the permanent magnet synchronous motor (PMSM) at its nominal output power and rotational speed, since in this case a minimum DC-link voltage of $U_{DC,min} = 74\,V$ would be needed.

One possibility to raise the low FC voltage at rated power operation is to place a boost-type DC-DC converter in front of the VSI (boost VSI - cf. **Fig. 2(a)**), which generates an intermediate DC supply voltage $U_{FC,min} \leq U_{DC}^* \leq U_{FC,max}$, that in return offers flexibility with respect to the rated motor voltage and reduces the current stress of the inverter under full power operation. Beneficially, the additional component count is low and thus the drive system still stays simple. However, due to the two-stage voltage conversion and the bulky boost inductors, both the efficiency and the power density drop.

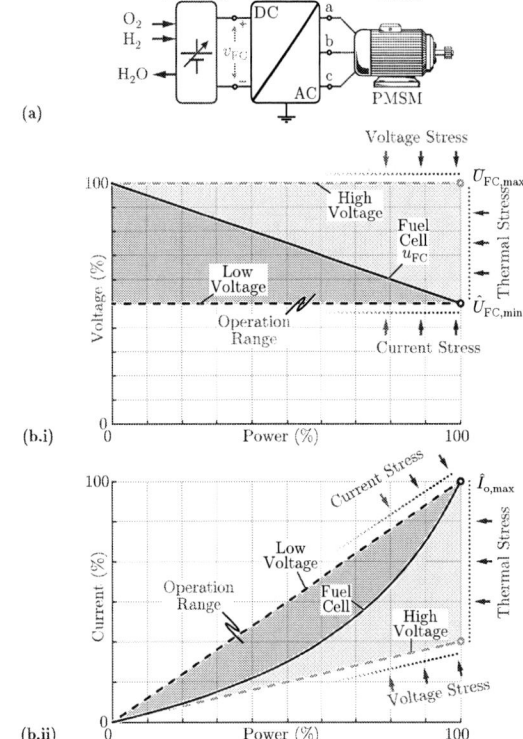

Fig. 1: (a) Fuel-cell (FC) powered motor drive employing a conventional voltage source inverter (VSI). The power dependent voltage variation of the FC is highlighted in **(b.i)** while the corresponding output current is plotted in **(b.ii)**.

Furthermore, as shown in **Fig. 2(a)**, an output filter is typically added to modern drive systems in order reduce down the high du/dt on the inverter output voltage resulting from today's wide bandgap semiconductor devices, which otherwise are directly applied to the motor windings [2]–[4], and to get smooth output voltages/currents which drastically reduce the radiated and conducted EMI noise, and thus also enables the use of unshielded motor cables (cost reduction) [5]–[7] and the reduction of rotor losses, that predominantly exist in high-speed motors [8], [9].

As an alternative to the boost VSI, a current source inverter (CSI) with integrated boost functionality can be used, which in addition benefits from an integrated output filter and therefore provides a continuous AC output voltage eliminating the need of a dedicated output filter. Moreover, as shown in **Fig. 2(b)**, the buck CSI incorporates only one inductive component which in addition is operated with a DC current, thus promises a compact overall design. As a drawback, however, the switches of the inverter stage have to be realized with

181

The 2018 International Power Electronics Conference

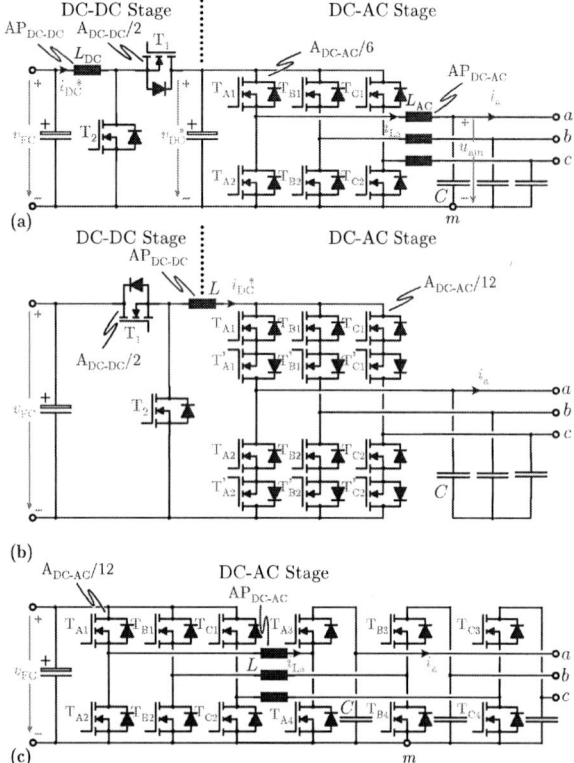

(a)

(b)

(c)

Fig. 2: Different fuel-cell (FC) powered motor drive implementations: **(a)** cascaded boost DC-DC converter with voltage source inverter (boost VSI), **(b)** current source inverter (CSI) based solution where a buck-type DC-DC converter precedes a CSI (buck CSI) and **(c)** Y-inverter, which enables wide input-output voltage range with single-stage energy conversion (i.e. no interface DC-DC converter required). In all converters the semiconductor chip areas A of each switch and the inductor area products $AP_{\mathrm{DC-DC}}$ and $AP_{\mathrm{DC-AC}}$ are highlighted.

e.g. two anti-series connected MOSFETs, which in consequence leads to higher conduction losses. Furthermore, due to the inherent boost functionality, the FC input voltage always has to be below the output voltage. Hence, at low output power, where the FC input voltage exceeds the needed machine voltage, the input voltage of the CSI has to be reduced by a precedent buck-stage (buck CSI), which can be bypassed at higher output power (T_1 is permanently on), hence only generates additional conduction losses at rated power.

The same buck-boost functionality with integrated output filter can also be achieved with the so-called Y-inverter (cf. **Fig. 2(c)**) as proposed by the authors in [10]. The Y-inverter represents an interesting alternative, since it manages to drive high EMF motors without any additional DC-DC converter. Furthermore, compared to the buck CSI, the Y-inverter features a phase-modular structure, i.e. three separate buck-boost DC-DC converters connected to a common ground, and doesn't need any anti-series connected switching devices, which results in a simpler converter and control structure as well as in lower conduction losses due to the reduced component count in the conduction path. Similarly to the buck CSI, depending on the input to output voltage ratio, either only the buck or the boost stage is switching while the high-side switch of the other stage is permanently on. Hence, compared to the boost VSI, the switching losses which

TABLE I: Fuel-cell powered motor drive specifications.

Parameter	Value
Motor Speed n	300 krpm
Motor EMF $\bar{U}_{\mathrm{M}}$	30V (Phase, RMS)
Fundamental freq. f_{o}	5 kHz
Fuel-cell voltage U_{FC}	60V...120V
Power P	1 kW

especially occur at high switching frequencies can drastically be reduced.

Nevertheless, each converter topology shown in **Fig. 2** has its advantages and disadvantages which at first glance are difficult to assess and in consequence no topology can be immediately discarded. Therefore, in this paper a comprehensive comparison of the above mentioned candidate topologies is performed. Firstly, analytic models based on fundamental scaling laws are derived for the semiconductor devices and the inductive components which are the major drivers of losses and volume respectively in **Sec. II**. This analytic approach enables an intuitive and fair preliminary comparison of the three converter topologies in **Sec. III**. In **Sec. IV**, the performance of the various solutions, including all the converter components (in addition to the inductors and semiconductor devices), is evaluated in terms of efficiency η and power density ρ by means of a comprehensive multi-objective optimization. There, a performance benefit in favor of the Y-inverter is deduced for the given fuel-cell application case at hand. Finally, the conclusions are drawn in **Sec. V** and an outlook on the continuation of this research is given.

II. ANALYTIC COMPARISON APPROACH

In order to identify the basic performance characteristics of the various converter concepts shown in **Fig. 2**, a comparison is performed based on fundamental scaling laws. Namely, the semiconductor devices and the inductive components, which are major drivers of losses and volume respectively, are modeled by analytic formulas. Subsequently, the derived models are applied to the considered converter concepts.

A. Chip Area and Losses of Semiconductor Devices

Typically different semiconductor devices must be selected in terms of voltage rating U_{r} and chip area A for the different converter implementations [11], [12]. In order to meaningfully compare the different converter options, a comprehensive and fair figure of merit (FOM) addressing the needs of industry (i.e. low cost) is required. To this end, the chip area based approach of [13], [14] is selected which determines the minimum required semiconductor chip area A_{min} and/or minimizes the semiconductor cost.

The chip area calculation algorithm is visualized in **Fig. 3(b)** and is described in the following. The chip is assumed to be mounted on a heatsink with a given surface temperature of $T_{\mathrm{h}} = 85^{\circ}$C via a thermal pad (cf. **Fig. 3(a)**). In addition the semiconductor device is considered to be operated at its junction temperature limit $T_{\mathrm{j,max}} = 120^{\circ}$C, a value widely used in industry. Hence, depending on the thermal resistances $R_{\theta\mathrm{jc}}$ (junction-to-case) and $R_{\theta\mathrm{ch}}$ (case-to-heatsink), which scale with the chip area A, the maximum allowed semiconductor losses are

$$P_{\mathrm{max}}(A) = \frac{T_{\mathrm{j,max}} - T_{\mathrm{h}}}{R_{\theta\mathrm{jc}}(A) + R_{\theta\mathrm{ch}}(A)}. \qquad (1)$$

If the chip area A is increased, the thermal resistance $R_{\theta\mathrm{jc}} + R_{\theta\mathrm{ch}}$ decreases as are the conduction losses because of the lower on-state resistance R_{ds}. In contrast, however, the switching losses which scale with the parasitic output capacitance C_{oss} and therefore with the

182

The 2018 International Power Electronics Conference

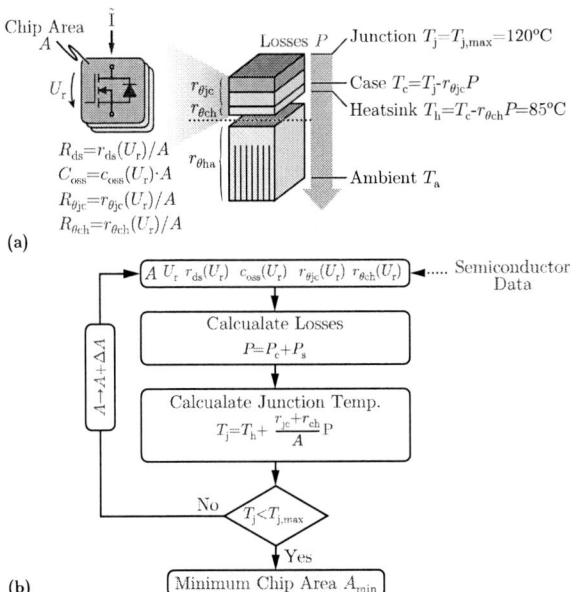

(a)

(b)

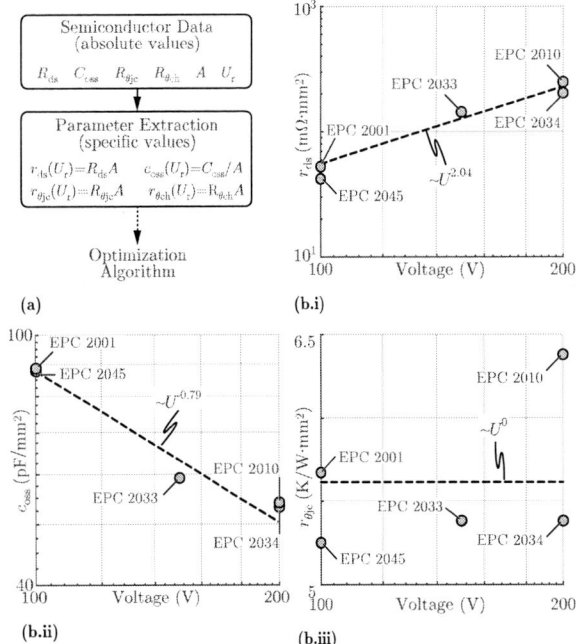

(a)

(b.i)

(b.ii)

(b.iii)

Fig. 3: (a) The employed MOSFET model is illustrated: The model is based on specific (i.e. normalized to the chip area A) electrical and thermal parameters $r_{dc}, c_{oss}, r_{\theta jc}, r_{\theta ch}$ and assumes a constant heatsink temperature of $T_h = 85^\circ$C. The iterative chip area based semiconductor optimization algorithm is highlighted in **(b)**. There starting from a small chip area value A the chip area dependent converter losses are calculated. Afterwards, the chip area is gradually increased $A \rightarrow A + \Delta A$ until the losses can be thermally dissipated opposite the heatsink temperature $T_h = 85^\circ$C without exceeding the maximum junction temperature $T_j = 120^\circ$C. Accordingly, the minimum chip area A_{min} is selected by the algorithm.

chip area A would increase. Hence, with too small chip areas the conduction losses and the thermal resistances are too high, which means the chip area has to be increased until the chip temperature decreases to the $T_{j,max} = 120^\circ$C. Thus optimum and/or minimum chip area A_{min} is defined as the chip area which satisfies the junction temperature sidecondition $T_j = T_{j,max}$. Further enlargement of the chip area above A_{min} would probably lead to lower losses (depending how the switching losses increase), however, the semiconductor costs would definitely increase. Therefore, the minimum chip area A_{min} is an intuitive FOM that enables comparison of fundamentally different converter topologies mainly concerning costs.

If the algorithm is applied to the example case of a buck DC-DC converter (where the total chip area A_{DC-DC} is assumed to be equally shared among the high and the lows side switches, resulting in a single semiconductor device chip area $A = \frac{A_{DC-DC}}{2}$), a lower bound for the semiconductor losses described in [12] is

$$P_{DC-DC} = \underbrace{\frac{r_{ds}(U_r)}{A} \tilde{I}_o^2}_{\text{Conduction losses}} + \underbrace{A c_{oss}(U_r) U_{in}^2 f_s}_{\text{Switching losses}}, \quad (2)$$

where r_{ds} (mΩ mm^2) is the specific on-state resistance, c_{oss} (pF/mm^2) is the specific parasitic output charge equivalent capacitance, U_{in} and $\tilde{I}_o$ are the input voltage and the RMS output current of the buck converter respectively, while $U_r = \frac{U_{in}}{0.7}$ is the rated voltage of the semiconductor devices. The thermal resistances are subsequently calculated based on the total semiconductor chip area A_{DC-DC} (i.e.

Fig. 4: (a) Extraction procedure of the specific (i.e. normalized to the chip area A) electrical and thermal parameters $r_{ds}, c_{oss}, r_{\theta jc}$ of GaN MOSFETs. The scaling of the specific on-state resistance r_{ds} for $T_j = 120^\circ$C junction temperature, the charge equivalent parasitic output capacitance c_{oss} and the thermal junction-to-case resistance $r_{\theta jc}$ with respect to the rated voltage U_r are plotted in **(b.i)-(b.iii)**, respectively.

not only considering a single device) and the junction temperature is derived as

$$T_j = T_h + \frac{r_{\theta jc} + r_{\theta ch}}{A_{DC-DC}} P_{DC-DC}, \quad (3)$$

where $r_{\theta jc}$ and $r_{\theta ch}$ (K mm^2 W^{-1}) are the specific junction-to-case and case-to-heatsink thermal resistances. As sown in **Fig. 3** the chip area A_{DC-DC} is gradually increased until the junction temperature constraint $T_j < 120^\circ$C is satisfied.

The presented algorithm relies heavily on the accuracy of the specific electrical and thermal MOSFET's parameters $r_{ds}, c_{oss}, r_{\theta jc}$ which in turn scale with the semiconductor device rated voltage U_r, as well as the case-to-heatsink isolation material thermal properties $r_{\theta ch}$. Those parameters were carefully extracted based on the latest GaN devices with voltage ratings $U_r \in [100\,\text{V}, 200\,\text{V}]$. The parameter extraction algorithm is visualized in **Fig. 4(a)**, while the analytic models are plotted against the commercial semiconductor devices data in **Fig. 4(b)**. The MOSFET's parameters derived as a function of the rated voltage are

$$r_{ds} = 4.54 \cdot 10^{-6} U_r^{2.04} \text{m}\Omega\,\text{mm}^2$$
$$c_{oss} = 3.3 \cdot 10^3 U_r^{-0.79} \text{pF}\,\text{mm}^{-2} \quad (4)$$
$$r_{\theta jc} = 5.61 \text{K}\,\text{mm}^2\,\text{W}^{-1} \quad r_{\theta ch} = 200 \text{K}\,\text{mm}^2\,\text{W}^{-1} \text{ [15]}.$$

B. Area Product and Volume of Inductive Components

The minimum achievable volume of inductors is derived analytically considering the area product (AP). The area product $A_c A_w$ which depends on the electrical parameters (i.e. inductance value L, peak $\hat{I}_L$ and RMS inductor current $\tilde{I}_L$), the core and winding material properties (i.e. core saturation flux density $\hat{B}$, maximum allowable

183

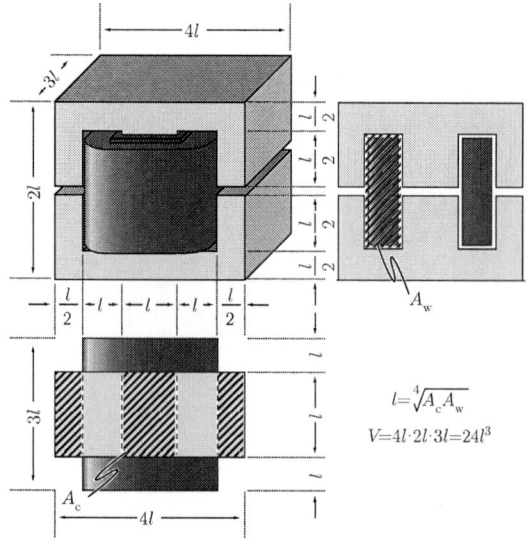

Fig. 5: Volume estimation model corresponding to an E core with area product $A_c A_w$. Assuming an E core according to the drawing with a characteristic core length $l = \sqrt[4]{A_c A_w}$ the total volume (i.e. including core and windings) can be approximated as $V = 24l^3$.

RMS current density $\tilde{J}$ and winding fill factor k) is

$$AP = A_c A_w = \frac{L \hat{I}_L \tilde{I}_L}{\hat{B} \tilde{J} k}. \tag{5}$$

Based on the AP (cm^4), which is independent of the core geometry, the characteristic core length $l = \sqrt[4]{A_c A_w}$ can be derived and subsequently utilized for the volume estimation of an E-core as $V \simeq 24l^3$ (cf. **Fig. 5**).

III. COMPARISON OF ANALYTIC RESULTS

The presented semiconductor and inductor models are applied to the considered converter implementations. With reference to a permanent magnet synchronous machine (PMSM) a sinusoidal motor current $i_a = \hat{I}_M \sin(\omega_o t)$ in phase with the motor AC voltage $u_{an} = \hat{U}_M \sin(\omega_o t)$ (i.e. low motor reactive power consumption) is assumed, while the reactive power consumption of the different passive components (i.e. capacitors and inductors) is considered negligible for the sake of simplicity ($\hat{U}_o = \hat{U}_M$, $\hat{I}_o = \hat{I}_M$). A common switching frequency f_s is assumed for all the power semiconductor stages and a safety margin of $a_u = 0.7$ is considered for the voltage rating U_r calculation of the semiconductor devices. A maximum allowed peak current ripple amplitude (i.e. half of the peak-to-peak value) to fundamental current ratio $a_i = 0.25$ is considered for the inductive components analysis.

A. Boost VSI

The boost VSI solution utilizes a chip area $A_{DC\text{-}DC}$ for its interface boost DC-DC converter and a chip area $A_{DC\text{-}AC}$ for the inverter stage, while the total chip area is $A = A_{DC\text{-}DC} + A_{DC\text{-}AC}$. The DC-DC and DC-AC parts can be optimized independently according to the minimum chip area calculation algorithm of **Sec. II**. To this end, the loss dissipation equations specific to each converter stage are determined. In a first step, the losses of the DC-DC stage are derived

as a function of its basic system operating parameters (cf. **Fig. 2**, DC-DC stage),

$$U_r = \frac{U_{FC,max}}{a_u}, \; U = U_{DC}^*, \; \tilde{I} = \frac{P}{U_{FC,min}}, \; A = \frac{A_{DC\text{-}DC}}{2}, \tag{6}$$

where the inverter supply voltage is selected as $U_{DC}^* = 2\hat{U}_o = 2\hat{U}_M$. The losses are then calculated as

$$P_{DC\text{-}DC} = \frac{r_{ds}(U_r)}{A} \tilde{I}^2 + A c_{oss}(U_r) U^2 f_s = \\ \frac{2 r_{ds}(U_r) P^2}{A_{DC\text{-}DC} U_{FC,min}^2} + \frac{1}{2} c_{oss}(U_r) A_{DC\text{-}DC} U_{DC}^{*2} f_s. \tag{7}$$

Subsequently, the losses of the DC-AC stage are derived (cf. **Fig. 2**, DC-AC stage) according to,

$$U_r = \frac{U_{FC,max}}{a_u}, \; U = U_{DC}^*, \; \tilde{I} = \frac{4P}{3\sqrt{2}U_{DC}^*}, \; A = \frac{A_{DC\text{-}AC}}{6}, \tag{8}$$

as

$$P_{DC\text{-}AC} = 3 \left[\frac{r_{ds}(U_r)}{A} \tilde{I}^2 + A c_{oss}(U_r) U^2 f_s \right] = \\ \frac{16 r_{ds}(U_r) P^2}{A_{DC\text{-}AC} U_{DC}^{*2}} + \frac{1}{2} c_{oss}(U_r) A_{DC\text{-}AC} U_{DC}^{*2} f_s. \tag{9}$$

The inductor volume of the DC-DC stage and the DC-AC stage are examined separately. First, the area product of the single DC-DC stage inductor is calculated based on

$$\hat{I}_L = \tilde{I}_L = \frac{P}{U_{FC,min}}, \; L = \frac{U_{FC,min}(U_{DC}^* - U_{FC,min})}{2U_{DC}^* f_s a_i \hat{I}_L}, \tag{10}$$

resulting in

$$AP_{DC\text{-}DC} = \frac{L \hat{I}_L \tilde{I}_L}{\hat{B}_{max} \tilde{J}_{max} k} = \frac{(U_{DC}^* - U_{FC,min})P}{2U_{DC}^* f_s a_i \hat{B} \tilde{J} k}, \tag{11}$$

and its volume $V_{DC\text{-}DC} = 24 AP_{DC\text{-}DC}^{3/4}$ is estimated based on the model of **Fig. 5**. Afterwards, the area product of one out of three in total DC-AC stage inductors is derived considering

$$\hat{I}_L = \frac{4P}{3U_{DC}^*}, \; \tilde{I}_L = \frac{4P}{3\sqrt{2}U_{DC}^*}, \; L = \frac{U_{DC}^*}{8 f_s a_i \hat{I}_L}, \tag{12}$$

which gives

$$AP_{DC\text{-}AC} = \frac{L \hat{I}_L \tilde{I}_L}{\hat{B}_{max} \tilde{J}_{max} k} = \frac{\sqrt{2}P}{12 f_s a_i \hat{B} \tilde{J} k}, \tag{13}$$

and the volume $V_{DC\text{-}AC} = 72 AP_{DC\text{-}AC}^{3/4}$ is evaluated. Finally, the total inductors volume is calculated as $V = V_{DC\text{-}DC} + V_{DC\text{-}AC}$.

B. Buck CSI

The buck CSI solution utilizes a chip area $A_{DC\text{-}DC}$ for its front-end buck DC-DC converter and a chip area $A_{DC\text{-}AC}$ for the inverter stage, resulting in a total chip area $A = A_{DC\text{-}DC} + A_{DC\text{-}AC}$. The loss dissipation equation of the DC-DC stage (cf. **Fig. 2(b)**, DC-DC stage) follows considering

$$U_r = \frac{U_{FC,max}}{a_u}, \; U = U_{FC,min}, \; \tilde{I} = \frac{P}{U_{FC,min}}, \; A = \frac{A_{DC\text{-}DC}}{2}, \tag{14}$$

as

$$P_{DC\text{-}DC} = \frac{r_{ds}(U_r)}{A} \tilde{I}^2 + A c_{oss}(U_r) U^2 f_s = \\ \frac{2 r_{ds}(U_r) P^2}{A_{DC\text{-}DC} U_{FC,min}^2} + \frac{1}{2} c_{oss}(U_r) A_{DC\text{-}DC} U_{FC,min}^2 f_s. \tag{15}$$

The CSI generates an AC current space vector (SV) by appropriately modulating the constant DC-link current I_{DC}^*, where the modulation is derived considering

$$\hat{I}_{\mathrm{L}} = \tilde{I}_{\mathrm{L}} = \frac{P}{U_{\mathrm{FC,min}}}, \quad M = \frac{\hat{I}_{\mathrm{o}}}{I_{\mathrm{DC}^*}} = \frac{2U_{\mathrm{FC,min}}}{3\hat{U}_{\mathrm{o}}}$$

$$L_{\mathrm{DC}} = \frac{U_{\mathrm{FC,min}}}{8f_s a_{\mathrm{i,CSI}}\hat{I}_{\mathrm{L}}}, \quad L_{\mathrm{AC}} = \frac{U_{\mathrm{FC,min}}\left(1 - \frac{\sqrt{3}}{2}M\right)}{2f_s a_{\mathrm{i,CSI}}\hat{I}_{\mathrm{L}}}, \tag{18}$$

and results as

$$AP_{\mathrm{DC\text{-}DC}} = \frac{(L_{\mathrm{DC}} + L_{\mathrm{AC}})\,\hat{I}_{\mathrm{L}}\tilde{I}_{\mathrm{L}}}{\hat{B}_{\max}\tilde{J}_{\max}k} = \frac{\left(\frac{5}{4} - \frac{\sqrt{3}}{2}M\right)P}{2f_s a_{\mathrm{i,CSI}}\hat{B}\tilde{J}k}, \tag{19}$$

resulting in a volume $V = V_{\mathrm{DC\text{-}DC}} = 24AP_{\mathrm{DC\text{-}DC}}^{3/4}$.

C. Y-Inverter

The Y-inverter shows a modular structure and employs three identical buck-boost DC-DC converter modules connected to a common star point [17]–[20] i.e. attached to the negative DC-rail m (cf. **Fig. 2(c)**). This arrangement is ideal for fuel-cell powered high-speed motor drives thanks to two key features. Firstly, it provides a continuous AC output voltage which eliminates the need of a dedicated output filter. Secondly, due to its buck-boost characteristic, the DC input voltage can be higher or lower than the AC output voltage with a single energy conversion stage, i.e. without requiring a interface DC-DC converter.

Each phase is comprised of two half-bridges connected to the opposite terminals of an inductor L, and an output capacitance C placed between the corresponding AC output terminal a, b, c and the negative DC-rail m, which, as already mentioned forms a common star (Y) point among the three phases. In order to generate the sinusoidal phase a motor voltage $u_{\mathrm{an}} = \hat{U}_{\mathrm{o}}\sin(\omega t) = M\frac{U_{\mathrm{FC,min}}}{2}\sin(\omega t)$, the converter phase module generates a strictly positive terminal voltage $u_{\mathrm{am}} = \hat{U}_{\mathrm{o}}\sin(\omega t) + \hat{U}_{\mathrm{o}} = M\frac{U_{\mathrm{FC,min}}}{2}(1 + \sin(\omega t))$, i.e. a sinusoidal voltage with a constant offset, where the modulation depth M can exceed value $M = 1$ (cf. **Fig. 6(b)**). The left half-bridge (T_{A1}, T_{A2}) of the phase module is dedicated to buck converter operation ($u_{\mathrm{am}} \leq U_{\mathrm{FC}}$, cf. **Fig. 6(a.ii)**), while the right hand side bridge (T_{A3}, T_{A4}) is exclusively used for boost operation ($u_{\mathrm{am}} > U_{\mathrm{FC}}$, cf. **Fig. 6(a.i)**). The buck and boost bridge-legs are operated in a mutually exclusive fashion, meaning that only one of the two bridge-legs is pulse width modulated (PWM) at a time, while the top side switch of the second bridge is clamped to an active on-state. Namely, the buck bridge-leg ($T_{\mathrm{A1}}, T_{\mathrm{A2}}$) is switched for the fraction $t_{\mathrm{A}} = \frac{2\pi - 2\phi_0}{2\pi}T_{\mathrm{o}}$ of the fundamental period T_{o}, while the boost bridge-leg ($T_{\mathrm{A3}}, T_{\mathrm{A4}}$) is switched for $t_{\mathrm{B}} = \frac{2\phi_0}{2\pi}T_{\mathrm{o}}$, where $\phi_0 = \cos^{-1}\left(\frac{2}{M} - 1\right)$. A detailed analysis and verification of the Y-inverter concept is given in [10].

The Y-inverter solution utilizes a total semiconductor chip area $A = A_{\mathrm{DC\text{-}AC}}$ which is assumed to be equally distributed to the buck and boost half-bridges. The semiconductor losses of the buck bridge-legs are calculated considering

$$U_{\mathrm{r}} = \frac{U_{\mathrm{FC,max}}}{a_{\mathrm{u}}}, \quad U = U_{\mathrm{FC,min}}, \quad \tilde{I} = \tilde{I}_{\mathrm{L}}, \quad A = \frac{A_{\mathrm{DC\text{-}AC}}}{12}, \tag{20}$$

and result in

$$P_{\mathrm{DC\text{-}AC,1}} = 3\left[\frac{r_{\mathrm{ds}}(U_{\mathrm{r}})}{A}\tilde{I}^2 + Ac_{\mathrm{oss}}(U_{\mathrm{r}})U^2 f_s \frac{2\pi - 2\phi_0}{2\pi}\right] =$$

$$\frac{36 r_{\mathrm{ds}}(U_{\mathrm{r}})\tilde{I}_{\mathrm{L}}^2}{A_{\mathrm{DC\text{-}AC}}} + \frac{1}{4}c_{\mathrm{oss}}(U_{\mathrm{r}})A_{\mathrm{DC\text{-}AC}}U_{\mathrm{FC,min}}^2 f_s\frac{2\pi - 2\phi_0}{2\pi}. \tag{21}$$

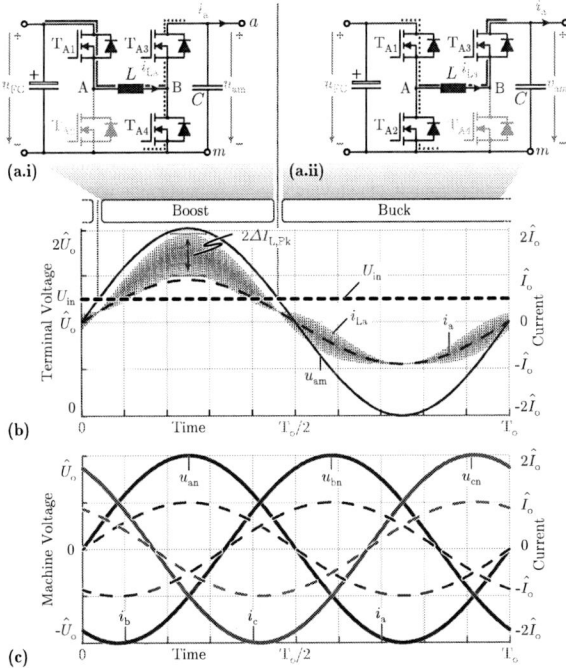

Fig. 6: Operating principle of a Y-inverter phase module and/or bridge-leg. In **(a.i)-(a.ii)**, the boost and buck operation are highlighted respectively. In **(b)** the generated voltage and current waveforms of phase-leg a are depicted, while the resulting sinusoidal motor currents and voltages are depicted in **(c)**.

depth of a CSI is defined as $M = \frac{\hat{I}_{\mathrm{o}}}{I_{\mathrm{DC}}^*} \in [0, 1]$. To do so, the CSI semiconductor arrangement must provide bidirectional voltage blocking, i.e. has to be implemented using anti-series connected semiconductor devices which result in twice the conduction losses compared to a single semiconductor device (cf. **Fig. 2(b)**). The losses of the DC-AC stage are calculated considering

$$U_{\mathrm{r}} = \frac{\sqrt{3}\hat{U}_{\mathrm{o}}}{a_{\mathrm{u}}}, \quad \tilde{I} = \frac{I_{\mathrm{DC}}^*}{\sqrt{3}} = \frac{P}{\sqrt{3}U_{\mathrm{FC,min}}}, \quad A = \frac{A_{\mathrm{DC\text{-}AC}}}{12}$$

$$U_1 = \sqrt{\frac{3}{\pi}\int_{-\pi/6}^{+\pi/6}\left[\sqrt{3}\hat{U}_{\mathrm{o}}\cos(\phi)\right]^2 d\phi} = \hat{U}_{\mathrm{o}}\frac{\sqrt{3(2\pi + 3\sqrt{3})}}{2\sqrt{\pi}}$$

$$U_2 = \sqrt{\frac{3}{\pi}\int_{-\pi/6}^{+\pi/6}\left[\sqrt{3}\hat{U}_{\mathrm{o}}\sin(\phi)\right]^2 d\phi} = \hat{U}_{\mathrm{o}}\frac{\sqrt{3(2\pi - 3\sqrt{3})}}{2\sqrt{\pi}}, \tag{16}$$

which results in

$$P_{\mathrm{DC\text{-}AC}} = 12\frac{r_{\mathrm{ds}}(U_{\mathrm{r}})}{A}\tilde{I}^2 + Ac_{\mathrm{oss}}(U_{\mathrm{r}})(U_1^2 + U_2^2)f_s =$$

$$\frac{48 r_{\mathrm{ds}}(U_{\mathrm{r}})P^2}{A_{\mathrm{DC\text{-}AC}}U_{\mathrm{FC,min}}^2} + \frac{1}{4}c_{\mathrm{oss}}(U_{\mathrm{r}})A_{\mathrm{DC\text{-}AC}}\hat{U}_{\mathrm{o}}^2 f_s. \tag{17}$$

The CSI approach benefits from a reduced number of passive components since only one DC side inductor is required. However, this inductor must maintain a low current ripple in order to ensure stable operation and reasonably low output voltage distortion. Namely, $a_{\mathrm{i,CSI}} = 7.5\%$ is considered as the upper bound of the peak current ripple to related DC inductor current ratio [16], which leads to a comparably large inductor volume. The area product inductor scaling

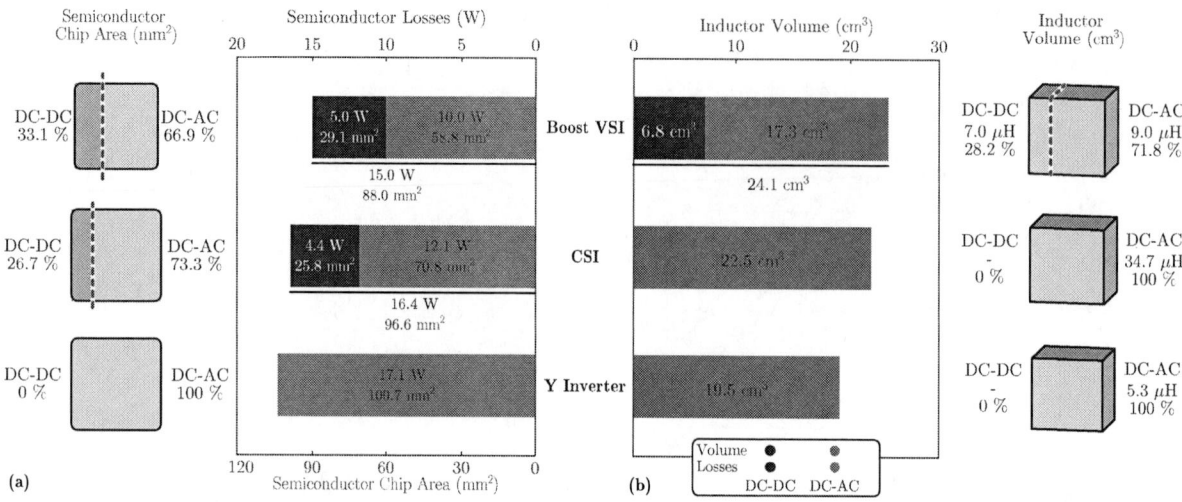

Fig. 7: Comparison of the conventional boost VSI and buck CSI motor drive implementations (cf. **Fig. 2(a)-(b)**) against the Y-inverter (cf. **Fig. 2(c)**) based on fundamental scaling laws of semiconductors and inductors. In **(a)** the minimum achievable semiconductor chip area of the different solution as well as the corresponding semiconductor losses are illustrated. In **(b)** the inductive components volume are shown. The parameter values $a_u = 0.7$, $a_i = 25\%$, $a_{i,CSI} = 7.5\%$, $\hat{B} = 350$ mT, $\tilde{J} = 5$ A mm^{-2}, $k = 60\%$ are considered for the calculations.

The loss dissipation of the boost bridge-legs follows with

$$U_r = \frac{2\hat{U}_o}{a_u}, \tilde{I} = \tilde{I}_L, A = \frac{A_{DC\text{-}AC}}{12}$$

$$U = \sqrt{\frac{1}{2\phi_0} \int_{-\phi_0}^{+\phi_0} \left[\hat{U}_o(1 + \cos(\phi))\right]^2 d\phi} = \hat{U}_o \sqrt{\frac{3\phi_0 + \sin(\phi_0)(\cos(\phi_0) + 4)}{2\phi_0}} \quad (22)$$

as

$$P_{DC\text{-}AC,2} = 3\left[\frac{r_{ds}(U_r)}{A}\tilde{I}^2 + Ac_{oss}(U_r)U^2 f_s \frac{2\phi_0}{2\pi}\right] = \frac{36 r_{ds}(U_r)\tilde{I}_L^2}{A_{DC\text{-}AC}} + \frac{1}{4}c_{oss}(U_r)A_{DC\text{-}AC}U^2 f_s \frac{2\phi_0}{2\pi}. \quad (23)$$

Finally, the total semiconductor losses are calculated as $P_{DC\text{-}AC} = P_{DC\text{-}AC,1} + P_{DC\text{-}AC,2}$. The filter inductors scaling is afterwards derived considering

$$\hat{I}_L = M\hat{I}_o = \frac{4P}{3U_{FC,min}}, \tilde{I}_L \simeq M\tilde{I}_o = \frac{4P}{3\sqrt{2}U_{FC,min}},$$

$$L = \begin{cases} \frac{U_{FC,min}}{8 f_s a_i \hat{I}_L}, M < \frac{4}{3} \\ \frac{(M-1)U_{FC,min}}{2M f_s a_i \hat{I}_L}, M > \frac{4}{3} \end{cases}, \quad (24)$$

as

$$AP_{DC\text{-}AC} = \begin{cases} \frac{\sqrt{2}P}{12 f_s a_i \hat{B}\tilde{J}k}, M < \frac{4}{3} \\ \frac{\sqrt{2}(M-1)P}{3M f_s a_i \hat{B}\tilde{J}}k, M > \frac{4}{3} \end{cases}, \quad (25)$$

where the modulation depth is defined as $M = \frac{\hat{U}_o}{\frac{U_{FC,min}}{2}}$. The inductor volume is hence $V = V_{DC\text{-}AC} = 72 AP_{DC\text{-}AC}^{3/4}$.

D. Comparison

The drive system implementations are compared for a switching frequency of $f_s = 300$ kHz and a relative peak current ripple of

$a_i = 25\%$ for all DC-DC and DC-AC converters except for the CSI where the peak current ripple has to be reduced to $a_{i,CSI} = 7.5\%$ for the reasons stated above. Utilizing the expressions **(6)-(25)**, the semiconductor losses and the volume of the inductive components are resulting as visualized in **Fig. 7**. As can be noted, the boost stage of the boost VSI roughly increases both the semiconductor losses and the inductor volume by 30% compared to the VSI stage. In contrast, for the buck CSI, the already existing boost inductor can be used for the precedent buck stage and only the buck half-bridge has to be added to the CSI. Hence, only the semiconductor losses are slightly increased by around 25%, which means that the performance - mainly in terms of efficiency - is only slightly decreased compared to the conventional CSI.

Comparing the different topologies, it can be noted that based on the given assumptions the boost VSI results in the smallest chip area (-14% compared to the Y-inverter) and therefore the lowest semiconductor losses and costs are achieved. Surprisingly, the chip area of the buck CSI and the Y-inverter are quasi identical, even if the number of semiconductors in the conduction path is higher for the CSI than for the Y-inverter. This can be explained by the larger switching losses of the Y-inverter: Namely, three hard switching transition per switching period f_s occur for the Y-inverter instead of only two for the CSI, while on average the Y-inverter is switching a higher output voltage (sinusoidal with DC offset) compared to the buck CSI (line-to-line sinusoidal motor voltage). On the other hand, with the Y-inverter the smallest overall inductor volume is achieved, whereas the inductor of the buck CSI is slightly bigger ($+15\%$). The largest inductor is resulting for the boost VSI, which compared to the Y-inverter increases the volume by roughly $+24\%$.

The Y-inverter and the VSI stage of the boost VSI could be operated with phase clamping modulation where always only two out of three phases are switching and one phase remains in a continuous on-state. This mode of operation would reduce the switching losses but is not considered here for the sake of simplicity.

In general, it can be noticed that the Y-inverter and the buck CSI achieve similar performance concerning power density and efficiency,

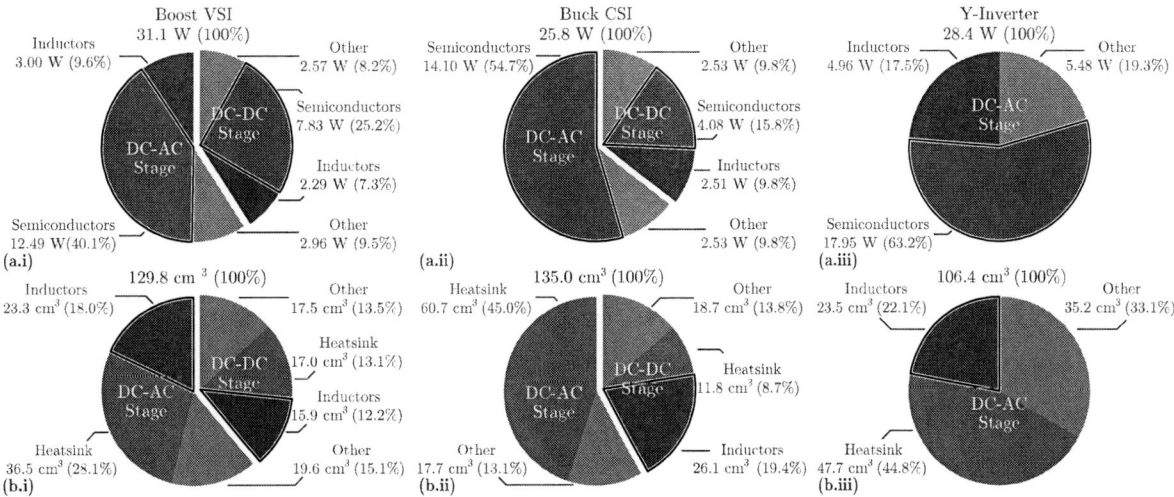

Fig. 8: Comparative break-down of (a) the volume and (b) the losses corresponding to the three converter approaches of Fig. 2.

which is not particularly surprising, since both topologies feature the same buck-boost structure with integrated output filter, whereas the Y-inverter basically only constitutes the phase modular approach of the buck CSI. Furthermore, also the boost VSI achieves a considerable high overall performance with highest efficiency, but lowest power density. Due to the fact that for the given criterion all topologies achieve similar performance, only general statements can be given, but no clear trend with the exclusion of one or several topologies is visible. Therefore, in a further step, the loss and volume calculations as well as the models of the semiconductors and inductors are refined, which means that e.g. also the high-frequency losses and the thermal aspects in the inductor winding and core are considered. Furthermore, also the additional circuit components like capacitors, control and measurement circuits, auxiliary supplies and heatsink are taken into account.

IV. MULTI OBJECTIVE OPTIMIZATION RESULTS

A. Minimum Chip Area, Area Product Optimization

As already discussed, the three converter topologies are optimized again for the same design criterion (i.e. minimum chip area, area product) and the same specifications (e.g. $f_s = 300 \, \text{kHz}$), whereas for the dimensioning of the individual components on the one hand all electric (e.g. high-frequency effects), magnetic (e.g. flux densities), mechanical (e.g. heatsink dimensions, interconnections, spacings between components or design rules) and thermal constraints (e.g. junction, core and winding temperatures) are considered [21], [22] and on the other hand only real and commercially available components (semiconductor devices with discrete voltage and current ratings, discrete core sizes and core types, solid and litz wire diameters with discrete numbers of strands, capacitors with discrete voltage and capacitance values) are used.

The comparative loss and volume breakdowns of the three designs are presented in Fig. 8(a),(b).

At first glance it can be noted that the major part (63 − 70%) of the overall losses is really generated in the semiconductors. The reason for this lies also in fact that the converters are optimized concerning a minimum chip area and therefore the optimization leads to a design with maximum allowable semiconductor losses per chip. In analogy to the analytical scaling laws, the total semiconductor

losses of the Y-inverter and the buck CSI are practically identical (18 W) and the losses generated in the buck stage of the buck CSI constitute approximately one quarter of the total semiconductor losses. The slight mismatch in the absolute values can be explained by the more detailed calculation of the switching losses, where the switching losses also scale with the switched current. Surprisingly, the highest semiconductor losses of around 20 W are found in the boost VSI, since the scaling laws predicted the lowest semiconductor losses for this topology. The reason for this is given by the fact that now in the accurate design only semiconductor components with discrete chip size can be used and therefore the investigated converters cannot reach their theoretically calculated chip area (cf. Fig. 7). The CSI for example employs 12 semiconductor devices with 150 V rating: The smallest commercially available 150 V MOSFET is the EPC2033 with chip area of $12 \, \text{mm}^2$. Therefore the smallest possible total CSI chip area is $A_{\text{DC-AC}} = 12 \cdot 12 = 144 \, \text{mm}^2$ which is twice the theoretical value of $70.8 \, \text{mm}^2$ (cf. Fig. 7). The larger chip area allows for lower conduction and therefore lower overall semiconductor losses for the buck CSI concept. In contrast the VSI employs 6 EPC2034 200 V rated switches with total chip area $A_{\text{DC-AC}} = 6 \cdot 12 = 72 \, \text{mm}^2$ which is much closer to the theoretically derived value of $58.8 \, \text{mm}^2$ (cf. Fig. 7). Therefore, the commercial devices chip area discretization artificially favors the buck CSI (as well as the Y-inverter) in terms of losses.

The relatively high semiconductor losses in all topologies are also reflected in the volume distribution, since for the cooling of the switching devices a relatively large heatsink is needed, which in all designs roughly consumes $1/3$ of the overall volume. The equal share of the heatsink volume confirms that it is permissible to neglect this component in the first calculation with the scaling laws. However, the heatsink volume could also be easily considered with a defined cooling system performance index (CSPI) [23].

The comparison of the precise calculation with the simple scaling laws also reveal that the inductor volume of the Y-inverter is the smallest, whereas the one of the buck CSI is around 11% larger. The largest inductor volume (+66% compared to the Y-inverter) is still consumed by the boost VSI, where 40% instead of 30% is accounted to the boost inductor. Nevertheless, it can be shown that these relative comparisons nicely match with the scaling laws. For the

The 2018 International Power Electronics Conference

given case, this is also true concerning the absolute inductor volumes, which means that the different inductor volumes calculated with the scaling laws only differ by $-15... - 25\%$ compared to the accurate calculation (expect the boost inductor deviates by -50%). There the major challenge of the inductor area product is to select a reasonable current density $\bar{J}$, since there are no thermal constraints and thus the current density can be arbitrarily chosen. Hence, the area product only gives a reliable statement concerning the relative inductor volumes, however, to get a prediction of the correct absolute inductor volumes is difficult and therefore these results should be treated with caution.

In general, the accurate calculation also reveals that for the considered output power of $1\,\mathrm{kW}$ in all converter topologies 20% of losses are generated by additional components like measurement and control circuits, auxiliary supplies, PCB and capacitors. Hence, in the analytic calculation based on the scaling laws, these losses can be neglected, since they only result in an offset in the loss balance. The same can be applied to the volume consumed by these components, which in all designs is roughly 35% of the overall volume.

In summary, with all topologies similar overall efficiencies and power densities are achieved. Interestingly, the buck CSI exhibits the highest efficiency (97.4%), while the Y-inverter is only slightly worse (97.2%). The difference is found in the inductor losses, because the inductor of the buck CSI only conducts a DC current and the inductors of the Y-inverter are excited with a sinusoidal current. The lowest efficiency (96.9%) is resulting for the boost VSI, but it has to be mentioned again that GaN devices with larger chip size had to be used. On the other hand, the Y-inverter features the highest power density ($9.3\,\mathrm{kW/dm^3}$). Compared to this, the volume of the buck CSI is around 30% larger ($7.4\,\mathrm{kW/dm^3}$) and even exceeds the volume of the boost VSI by 4% ($7.7\,\mathrm{kW/dm^3}$).

B. $\eta\rho$-Pareto Front Optimization

The previously discussed optimization concerning minimum chip area and area product fits the low cost needs of industry applications and hence the converter designs derived from such optimization are from now on denoted as industrial designs. The industrial designs of **Fig. 8** exhibit similar performances in terms of converter efficiency η and power density ρ. Now, the question arises how much these performance indexes can be increased compared to the industrial designs if further degrees of freedom (i.e. arbitrary chip size, switching frequency $f_s \in [200\,\mathrm{kHz}, 600\,\mathrm{kHz}]$, passive component values L, C) are considered. Therefore, a multi-objective optimization routine with respect to converter efficiency η and power density ρ is examined. The η-ρ Pareto limits of the different converter candidates are depicted in **Fig. 9**, whereas the following performance trends can be identified: The boost VSI quickly reaches a power density threshold because of the volume contribution related to the DC-DC stage. At the nominal operating point, where the boost stage must step-up the FC voltage to the greatest degree, the semiconductors are switching the DC voltage U_{DC}^* and the boost inductor is exposed to large voltage-time-areas resulting in high losses. Therefore, the boost VSI is outperformed by the two other solutions. Compared to the industrial design with minimum chip area, the efficiency can be increased by around 0.5% for the same power density. On the other hand, for the same efficiency the power density could be increased by 23%.

The buck CSI achieves a high efficiency by paralleling several MOSFETs per switch, which mitigate the conduction losses. Hence, the efficiency can be raised to roughly 98.3%, i.e. almost +1%, compared to an industrial design. However, this low loss profile is achieved at the expense of substantially increased semiconductor cost, which might be impractical for industry. Furthermore, the low DC-

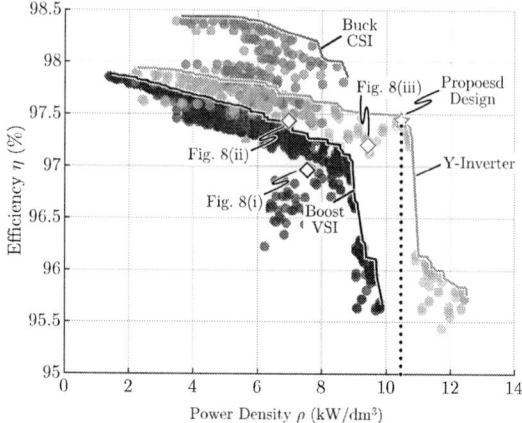

Fig. 9: The efficiency (η) power density (ρ) Pareto optimization results for the different candidate converter concepts of **Fig. 2**.

link current ripple condition ($a_{i,CSI} = 7.6\%$) leads to a bulky inductor design which compromises the overall power density.

Finally, the Y-inverter breaks through the power density barriers of the other systems, while a high efficiency can be maintained. At the same power density as the industrial design, the efficiency can be increased to around 97.5%. A reasonable Y-inverter benchmark design would achieve an efficiency of $\eta = 97.2\%$ and a power density of $\rho = 10.5\,\mathrm{kW/dm^3}$ as highlighted in **Fig. 9**. It should be mentioned that this performance is achieved with conventional PWM, which results in hard-switching. In a future step, a trapezoidal current modulation (TCM) as proposed in [24] could be implemented, which enables soft-switching operation at constant switching frequency and further reduces the needed volume of the inductors. Hence, besides the advantages of modularity, scalability and high power density also a high efficiency, comparable to the buck CSI, can be achieved.

V. CONCLUSIONS

A comparative evaluation of different converter concepts is performed within the context of a fuel-cell powered motor drive application. Namely, the recently proposed Y-inverter is compared against traditional voltage source (boost VSI) and current source inverter based (buck CSI) approaches. First, a preliminary analysis is performed based on fundamental scaling laws of semiconductor devices and inductive components, which are the main contributors of losses and volume respectively. The analytic nature of the underlying semiconductor and inductor models enables a fair and intuitive relative comparison of the investigated converters, however, the absolute volume and losses calculation have limited accuracy. For this reason, in order to identify the exact efficiency (η) vs. power density (ρ) Pareto limits of the investigated converter implementations (including the remaining system components in addition to semiconductor devices and inductors) a comprehensive converter optimization routine is employed. Both the analytic calculations and the precise Pareto optimization indicate a considerable power density gain in favor of the Y-inverter with a small decrease in efficiency compared to the traditional solutions. The reason behind the low Y-inverter volume is its integrated filter structure (with minimal inductive components number) and the buck-boost modular approach it follows. A trapezoidal current modulation (TCM) would be an effective measure for efficiency improvement of the Y-inverter that should be carefully analyzed in future research.

188

ACKNOWLEDGMENT

The authors gratefully acknowledge the financial support by the Swiss Federal Commission for Technology and Innovation (CTI) and the technical contributions of Celeroton AG.

REFERENCES

[1] Celeroton. (2018) Permanent-magnet motor cm-25-280. [Online]. Available: http://www.celeroton.com/fileadmin/user_upload/produkte/motoren/datasheets/Datasheet-CM-25-280.pdf

[2] D. Rendusara and P. Enjeti, "New inverter output filter configuration reduces common mode and differential mode dv/dt at the motor terminals in pwm drive systems," in *Proceedings of 28th Annual IEEE Power Electronics Specialists Conference (PESC)*, Jun. 1997, pp. 1269–1275.

[3] T. G. Habetler, R. Naik, and T. A. Nondahl. "Design and implementation of an inverter output lc filter used for dv/dt reduction," *IEEE Transactions on Power Electronics*, vol. 17, no. 3, pp. 327–331, May. 2002.

[4] K. Hatua, A. K. Jain, D. Banerjee, and V. T. Ranganathan, "Active damping of output lc filter resonance for vector-controlled vsi-fed ac motor drives," *IEEE Transactions on Industrial Electronics*, vol. 59, no. 1, pp. 334–342, Jan. 2012.

[5] A. von Jouanne and P. N. Enjeti, "Design considerations for an inverter output filter to mitigate the effects of long motor leads in asd applications," *IEEE Transactions on Industry Applications*, vol. 33, no. 5, pp. 1138–1145, Sep. 1997.

[6] A. F. Moreira, P. M. Santos, T. A. Lipo, and G. Venkataramanan, "Filter networks for long cable drives and their influence on motor voltage distribution and common-mode currents," *IEEE Transactions on Industrial Electronics*, vol. 52, no. 2, pp. 515–522, Apr. 2005.

[7] X. Chen, D. Xu, F. Liu, and J. Zhang, "A novel inverter-output passive filter for reducing both differential- and common-mode dv/dt at the motor terminals in pwm drive systems," *IEEE Transactions on Industrial Electronics*, vol. 54, no. 1, pp. 419–426, Feb. 2007.

[8] M. A. Rahman, A. Chiba, and T. Fukao, "Super high speed electrical machines - summary." in *Proceedings of IEEE Power Engineering Society General Meeting*, Jun. 2004, pp. 1272–1275.

[9] C. Zwyssig, S. D. Round, and J. W. Kolar, "An ultra-high-speed, low power electrical drive system," *IEEE Transactions on Industrial Electronics*, vol. 55, no. 2, pp. 577–585, Feb. 2008.

[10] M. Antivachis, D. Bortis, L. Schrittwieser, and J. W. Kolar, "Novel buck-boost inverter topology for fuel-cell powered drive systems," in *Proceedings of IEEE Applied Power Electronics Conference and Exposition (APEC)*, Mar. 2018, pp. 1492–1499.

[11] G. Deboy, O. Haeberlen, and M. Treu, "Perspective of loss mechanisms for silicon and wide band-gap power devices," *CPSS Transactions on Power Electronics and Applications*, vol. 2, no. 2, pp. 89–100, 2017.

[12] J. A. Anderson, L. Schrittwieser, C. Gammeter, G. Deboy, and J. W. Kolar, "Relating the figure of merit of power mosfets to the maximally achievable efficiency of converters," *under review for the CPSS Transactions on Power Electronics and Applications*.

[13] T. Friedli and J. W. Kolar, "A semiconductor area based assessment of ac motor drive converter topologies," in *Proceedings of 24th Annual IEEE Applied Power Electronics Conference and Exposition (APEC)*, Feb. 2009, pp. 336–342.

[14] M. Schweizer, T. Friedli, and J. W. Kolar, "Comparative evaluation of advanced three-phase three-level inverter/converter topologies against two-level systems," *IEEE Transactions on Industrial Electronics*, vol. 60, no. 12, pp. 5515–5527, Dec. 2013.

[15] The Bergquist Company. (2018) Gap pad 5000s35. [Online]. Available: http://www.bergquistcompany.com/pdfs/dataSheets/PDS_GP_5000S35_0711%20v2.pdf

[16] M. Baumann and J. W. Kolar, "A 5kw three-phase buck boost telecommunications power supply module input stage maintaining unity power factor under failure of a mains phase," in *Proceedings of the 9th European Power Quality Conference (PCIM Europe)*. May. 2003, pp. 291–598.

[17] R. Erickson and L. Colony, "Dc to three phase switched mode converters," Patent US 4,677,539, 1987.

[18] K. D. Ngo, S. Cuk, and R. D. Middlebrook, "A new flyback dc-to-three-phase converter with sinusoidal outputs," in *Proceedings of IEEE Power Electronics Specialists Conference (PESC)*, Jun. 1983, pp. 377–388.

[19] S. Mehrnami, S. K. Mazumder, and H. Soni, "Modulation scheme for three-phase differential-mode inverter," *IEEE Transactions on Power Electronics*, vol. 31, no. 3, pp. 2654–2668, Mar. 2016.

[20] A. Darwish, D. Holliday, S. Ahmed, A. M. Massoud, and B. W. Williams, "A single-stage three-phase inverter based on cuk converters for pv applications," *IEEE Journal of Emerging and Selected Topics in Power Electronics*, vol. 2, no. 4, pp. 797–807, Dec. 2014.

[21] R. M. Burkart, H. Uemura, and J. W. Kolar, "Optimal inductor design for 3-phase voltage-source pwm converters considering different magnetic materials and a wide switching frequency range," in *Proceedings of International Power Electronics Conference (IPEC - ECCE ASIA)*, May. 2014, pp. 891–898.

[22] P. Papamanolis, F. Krismer, and J. W. Kolar, "Minimum loss operation of high-frequency inductors," in *Proceedings of IEEE Applied Power Electronics Conference (APEC)*, Mar. 2018, pp. 1756–1763.

[23] D. Bortis, D. Neumayr, and J. W. Kolar, "$\eta\rho$-pareto optimization and comparative evaluation of inverter concepts considered for the google little box challenge." in *In Proceedings of IEEE 17th Workshop on Control and Modeling for Power Electronics (COMPEL)*, Jun. 2016, pp. 1–5.

[24] S. Waffler and J. W. Kolar, "A novel low-loss modulation strategy for high-power bidirectional buck + boost converters," *IEEE Transactions on Power Electronics*, vol. 24, no. 6, pp. 1589–1599, Jun. 2009.

DC-Powered Office Buildings and Data Centres
The First 380 VDC Micro Grid in a Commercial Building in Germany

Mr. Tilo Pueschel

Business Development Department., Bachmann GmbH, Stuttgart, Germany

E-Mail: tilo.pueschel@bachmann.com

Abstract - Nowadays, the use and storage of renewable energies on the one hand, coupled with demands to save energy on the other, mean that modern data centres and office buildings have to be designed and operated in a way that enables them to meet a number of totally new requirements. On top of this, the latest information and communication technologies are developing at breakneck speed, calling for a building's physical infrastructure to provide as much flexibility as possible.

This paper illustrates how a 380 VDC micro grid was installed in an office building and the data centre of a German company. However, the analysis is not just restricted to factors relating to energy efficiency, it also looks at safety-related aspects of personal and plant protection.

Keywords: Direct Current (DC), 380 VDC, Micro Grid, Battery, Efficiency, Office Building, Data Centre, Personal Protection

I. INTRODUCTION

Today's data centres and office buildings are supplied as standard with 230 VAC electricity (AC = alternating current). The reasons for this are historical; AC voltage was easier to convert to different voltage levels through transformers. Furthermore, it was also possible to transfer energy over long distances. Modern power electronics now offer numerous ways of also converting DC voltage (DC = direct current) with virtually no losses, making the process of transmitting and transforming energy much more efficient. At the same time, renewable energies can be called upon for the environmentally sound generation of direct current. This paper lays out the challenges to be faced and benefits to be gained when using renewable energies in combination with direct current within buildings.

II. HISTORY

In Germany, trends in electricity prices over the last few years have led to more and more private households investing in renewable energies, with the German government providing support in the form of public subsidy programmes such as the one created by the Renewable Energy Sources Act (feed-in tariff).

Into this mix comes another political decision, the "energy revolution", which essentially has the following aims:

- Primary energy consumption -20% by 2020 and -50% by 2050 (compared to 2008 levels)
- Heat requirements in buildings -20% by 2020 and -80% by 2050 (compared to 2008 levels)
- Electricity consumption -10% by 2020 and -25% by 2050 (compared to 2008 levels)
- Renewable energies (gross) +35% by 2020 and +80% by 2050

The energy revolution means that the owners or operators of office buildings also have to comply with ever more requirements when it comes to using renewable energies. Thus, owners and operators of office buildings have some decisions to make: should they install and use renewable energies? If the answer is yes, what kind of renewable energies should they choose? Architects and designers are not only installing photovoltaic systems on roofs, but are now increasingly integrating them into building façades as well.

© Andi Schmid, Munich (Germany)

III. INDIRECT USE

With this type of use, the photovoltaic energy generated is converted into AC voltage via an inverter, then either used within the building itself or fed into the public power grid as surplus energy. Furthermore, conventional solar-energy storage units can be used, which rectify the AC voltage produced by the inverter again and feed it to a storage cell (battery). If the photovoltaic installation stops producing energy due to a lack of solar irradiation, the user can then fall back on the energy in the storage cell. The stored energy is also adjusted to the standard mains frequency of 50 Hz by an inverter. With storage systems, consumers are able to increase the percentage of the power generated from renewable energies which they actually use themselves. However, as has been explained in this paper already, a number of AC/DC or DC/AC conversion processes are needed to ensure the functionality of the entire system. In this context, the AC voltage is only really used to transport energy within the overall system, since most devices (IT equipment, LEDs, smartphones, etc.) have their own power supply units, which convert the AC voltage into DC voltage and only then "process" it electronically.

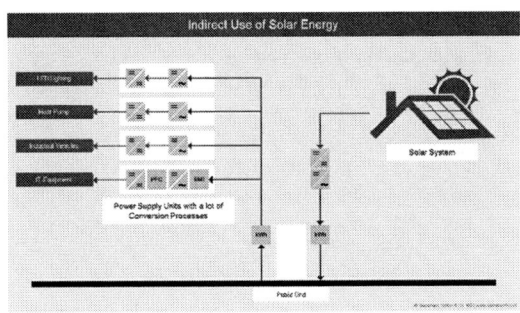

IV. DIRECT USE

Photovoltaic installations generate DC voltage. Electrical storage systems also process DC voltage and, as described above, devices also run on DC voltage internally. This leads inevitably to the idea of not converting the DC voltage produced into AC voltage, but of distributing and using it within the building directly instead. This shortens the process chain considerably, since many AC/DC and DC/AC conversion processes can be done away with.

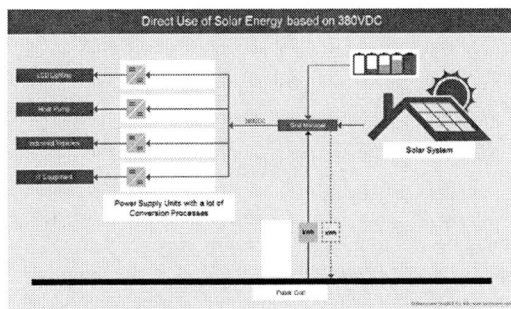

To achieve an exact comparison of a DC and an AC voltage installation, both systems would have to be operated in parallel and undergo technical measurements under the same conditions. However, the amount of expenditure involved render this course of action unprofitable and therefore infeasible. As a result, the comparison is restricted to the efficiency data provided by the various manufacturers of power supply units, UPS systems, inverters and so on. Based on the process chains already illustrated, this enables the efficiency of the entire system in question to be estimated. First of all, the process chain for an AC system is illustrated:

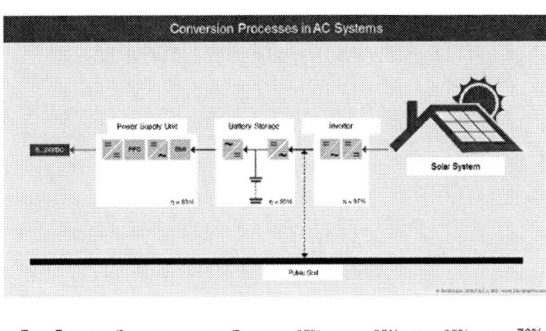

$$\eta = \eta_1 \times \eta_2 \times \ldots \times \eta_n = 97\% \times 95\% \times 85\% = 78\%$$

And now an illustration of the process chain for a DC system:

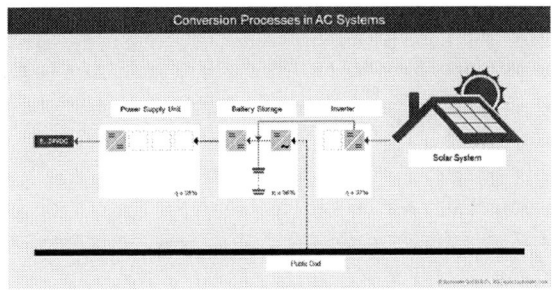

$$\eta = \eta_1 \times \eta_2 \times \ldots \times \eta_n = 97\% \times 95\% \times 95\% = 88\%$$

Comparing the two process chains, it is clear that a DC system is approximately 10% more efficient than an AC system, due to the omission of certain conversion elements.

V. TECHNOLOGICAL CHANGE

Modern companies are nowadays faced with new technologies and trends that are evolving ever faster. This applies both to personal equipment such as computers, laptops and phones, and to technological trends, which can have an impact on buildings and how they are used. Viewed statistically, the intervals between fundamental technical changes are becoming ever shorter, because the integration level of integrated circuits is doubling at an average of approximately every 18 months (Moore's law). This makes it increasingly difficult to procure IT equipment for office workers that is suitable for the long term and to integrate it into the existing infrastructure of office buildings. Good examples of this are constantly rising data rates and PoE (Power over Ethernet): in both cases, management has to respond quickly and flexibly to remain competitive. Added to this is the political pressure or conditions at play when it comes to meeting the demands of the energy revolution in Germany.

VI. THE TREND IS MOVING TOWARDS DC

Renewable energies are without doubt in the ascendancy and will be crucial in shaping how power is supplied in the future. How efficiently this shift to renewable energies happens will have a major impact on whether economic and environmental targets are achieved. Many data centres are the primary focus here, since they have very high energy requirements on the one hand, and, on the other hand, by their very essence need to run on DC voltage. For example, it is not unusual for a data centre of 1000 m² to have up to 8000 IT devices installed in it, most of which will run on 5 to 24 VDC internally. The common supply voltage for data centres is 400 V or 3-phase alternating current, which is the conventional standard. Once the power is fed into the building, the AC system process chain already illustrated performs a seemingly endless stream of conversions from AC to DC and back again. The large number of devices found in data centres, and the fact that they are usually housed in buildings with flat roofs, means photovoltaic technology is suitable as an additional source of renewable energy. Several large telecommunications corporations are already pursuing this trend. These companies have been using -48 VDC for their systems for years now, and therefore have no reservations about the new 380 VDC technology. Quite the opposite, in fact: 380 VDC offers even further potential savings compared to the -48 VDC used until now, and these savings can be made in the area of cabling. In the past, if 100 kW of power was to be transferred from a power source to a device, for example, this would require 20 copper cables, each with a

cross-section of 325 mm² – an enormous investment in terms of both expenditure and labour. If the same power was to be transferred via a 380 VDC system, only two copper cables would be needed, each with a cross-section of 200 mm². Fewer copper cables in the data centre means fewer cable runs above the server racks on the ceiling or inside the raised floor, with either option resulting in optimised ventilation and thus more efficient cooling of the data centre. This would be another synergistic effect, even though it is only related to DC voltage technology indirectly.

VII. SAFETY ASPECTS

The topic of DC voltage is not a fundamentally new one; it is already used in many applications today, such as telecommunications (-48 V), railway signalling technology (60 V), electric cars (200 to 800 V) and trams (700 V). Thus, experience has been gained in working with DC voltage; it is just the office building and data centre applications that are new. There are additional personal, installation and fire protection requirements to consider in this field, because end users are in very close proximity to the DC voltage when switching lights on and off or plugging electrical appliances into sockets, for instance. To the layman, this would not seem dangerous at all, since we are used to these scenarios in the familiar AC voltage environment. With AC voltage, the sine curve crosses the zero point 100 times per second, which makes switching lights, disconnecting plug connectors or tripping miniature circuit breakers easier. In contrast, DC voltage – as the name suggests – provides a constant voltage level, which presents a challenge when switching, disconnecting and tripping, due to the energy content. It is exactly this issue that standardisation committees around the world are grappling with at the moment, with corresponding device types undergoing tests in the first publicly funded projects. This sounds more like the future than reality. This is not, exactly, true. As has been covered above, a wealth of experience in working with DC voltage technology has already been collected, so there is already suitable equipment available for creating installations in data centres, for example. Depending on which power concept is selected (TN-S in office buildings and IT in data centres), comprehensive residual current monitoring or insulation monitoring is recommended; this is a familiar method, approved by experts, of continuously monitoring installations using measuring instruments. This enables even the tiniest status changes to be detected quickly, thus guaranteeing maximum safety for both the operator and the application. The TN-S concept was installed in the Bachmann GmbH Data Centre. The main reason for this earthing concept was the parallel installation of direct and alternating current in one building. All final current circuits were monitored with one residual current sensor each. The following threshold values were defined for monitoring purposes:

0-30 mA	Status OK
30-100 mA	Status NOK
> 100 mA	Status critical – switch-off

Switch-off was realised via shunt releases. As soon as the residual current sensor detects a threshold value of 100 mA, the superordinate monitoring system switches a floating contact. This floating contact ensures that the shunt release trips the miniature circuit breaker.

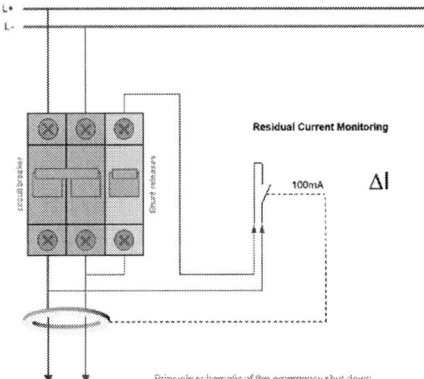

Principle schematic of the emergency shut down

In the event of escalation, just one of the relevant current paths is switched off in each case. (e.g. air conditioning system 1) In data centres designed with redundancy, functionality would not be endangered in this scenario.

VIII.   REALISATION PHASE

Work began on installing the first 380 VDC micro grid in an office building used for commercial purposes in 2014. All the factors mentioned above were taken into consideration and implemented. As well as connecting the photovoltaic system to the storage system directly, the IT equipment, the lighting and the air conditioning were also designed to run on 380 VDC. In order to ensure that the data centre is fail-safe, thus having maximum availability, all key components such as air conditioning systems and UPS systems were installed with 1:1 redundancy.

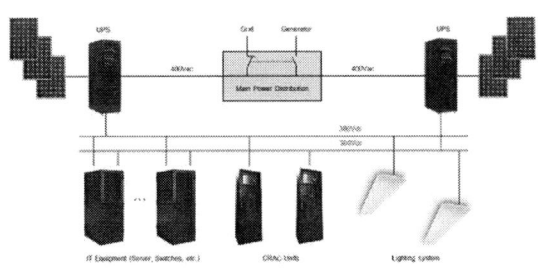

Prinziple schematic of the data center

The safety concept described, with residual current monitoring and switch-off at values >100 mA, was also realized. All key areas of the installation are monitored with separate RCM sensors. Even the power strips (PDUs – Power Distribution Units) in the server cabinets are each equipped with an RCM sensor.

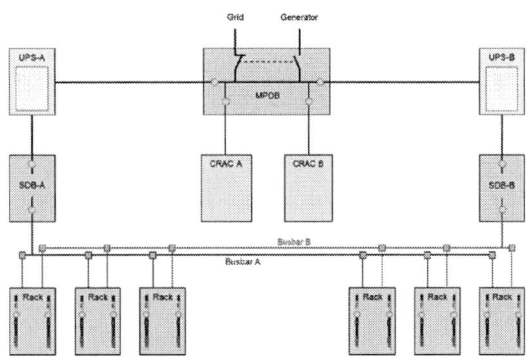

RCM sensors displayed in green

The RCM sensors used are all-current-sensitive (AC and DC) in compliance with the standard IEC 62020.

The direct current generated is not just used for the data centre, but also for the lighting in the entire administration building. Here, the 380 VDC electricity is used for power distribution from the ground floor up into the four floors. For safety reasons, the voltage is stepped down from 380 VDC to 48 VDC or 24 VDC on each floor. This 48 VDC/24 VDC electricity is used both for the lighting and as a power supply for the desks and workstations. The advantage of this is that the power adapters for PCs, laptops, etc. are no longer necessary, which saves energy, but also helps in realising the "clean desk" concept. The following schematic illustration shows the power distribution in the building and on the different floors.

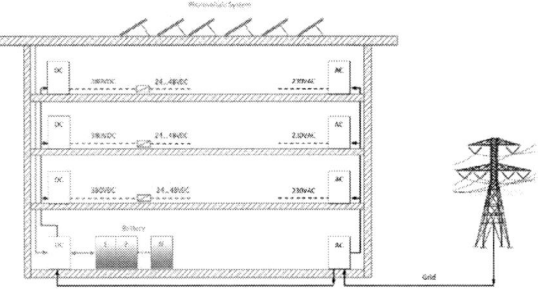

On each floor, there are two possible methods of distributing the energy. Either large converters are used in the sub-distribution stations, or local converters are used in the immediate vicinity of the desks and workstations.

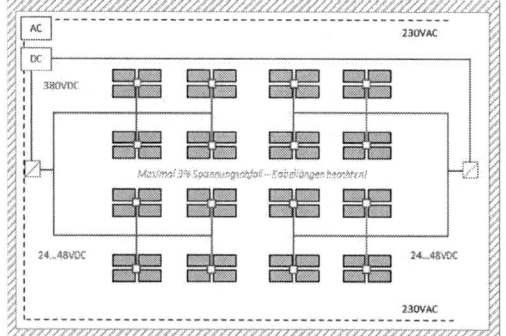

Energieverteilung mit zentralen Konvertern

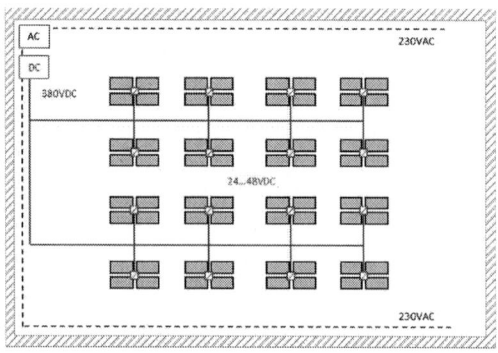

Energieverteilung mit dezentralen Convertern

IX. ENERGY EFFICIENCY

The 380 VDC micro grid has been running for two and a half years with no faults; therefore, resilient statements regarding the energy footprint can also be made. This primarily concerns the energy generated by the photovoltaic system. On sunny summer days, the photovoltaic installation provides approximately two thirds of the building's entire energy requirements around midday.

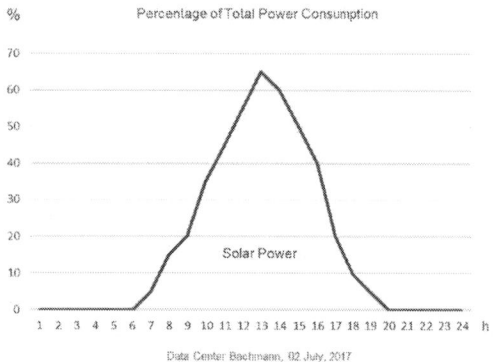

The energy footprint of the building in comparison to the years 2013, 2014 and 2015 shows that a total energy saving of 10-12% is achieved with the current system.

X. NEXT STEPS

Other solutions for the office environment have been developed in response to the experience gained with direct current. A desk with internal battery storage can be operated autonomously. This means that no cabling is required, which leads to infrastructure savings, easier relocations and greater flexibility in general.

Autarker Schreibtisch von Bachmann

The same applies to the autonomous monitor, which is ideal for conferences or trade fairs.

Autarker Monitor von Bachmann

XI. CONCLUSION

This paper presents the benefits of 380 VDC as the voltage supply of the future for office buildings and data centres. The author has illustrated the direct use of renewable energies, the potential savings associated with direct current, and possible applications. At the same time, safety aspects have been considered, potential solutions identified and the two and a half years of successful operation mentioned.

REFERENCES

[1] Bachmann GmbH - Application Report, Stuttgart (Germany), June 2017

[2] Bachmann Systems GmbH – Photos Work-Life Technology, Stuttgart (Germany), March 2017

[3] Bachmann GmbH – MarCom Department Stuttgart (Germany), June 2017

Recent Trend in Power Electronics for ICT Systems

Hiroshi Nakao[1,2], Yu Yonezawa[1]*and Yoshiyasu Nakashima[1]

1 Computer Systems laboratory, Fujitsu Laboratories LTD., Kawasaki-Shi, Japan
2 Graduate School of Engineering, Nagasaki University, Nagasaki-Shi, Japan
*E-mail: nakao.h@jp.fujitsu.com

Abstract— In recent years, the scale of datacenter and its computing power has become larger and larger, however, the total consumed power of datacenter is not increased so much, on the contrary, slightly decreased. This is a frits of the continuous effort for the energy conservation in datacenter or ICT system. In this paper, resent trend of power Electronics in ICT systems are reported.

Keywords—Recent trend of power electronics, ICT system, power delivery.

I. INTRODUCTION

In recent years, the scale of datacenters and their computing power have become larger and larger, however, the total consumed power of datacenter is not increased so much. Figure 1 shows the total energy consumption trend of datacenters in USA [1]. The gray dotted line shows trend line estimated from energy efficiency trend before 2010, energy consumption of datacenter increases continuously and future energy crisis was concerned. However, current trend (Black solid line) becomes slower or will be slightly decreased after 2015. This is obviously the flirts of continuous effort for energy conservation in datacenter.

The energy saving of datacenter is advancing in many fields, a cooling optimization, computer usage optimization with virtual machine installed in cloud system, improvement of electricity efficiency and so on [2]. In this paper, after mentioned briefly cooling and usage optimization, improvement of electricity efficiency is described.

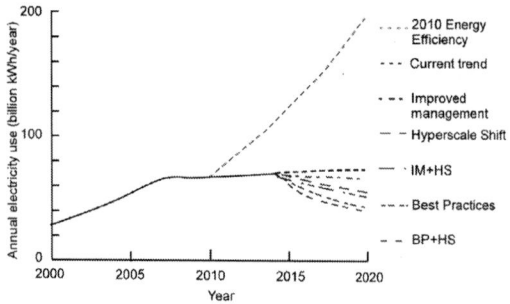

Fig. 1. The total energy consumption trend of datacenters in USA [1].

II. COOLING OPTIMIZATION AND PUE [2]

Power usage effectiveness or PUE[1] is an index of efficiency of energy usage of datacenter, which is defined as a ratio of the energy used by facility and the energy used by ICT equipment. Before the trend curve of energy consumption start slowing, PUE was very poor and most of datacenter PUE is over 2.0, less than half energy is used for actual computing. After around 2010, datacenter PUE is drastically improved and now some datacenter PUE become almost 1.0. This improvement is mainly occurred by cooling optimization, such as separation of cool and hot area using aisle structure, implement of the outside air cooling system, higher temperature operation of server room, and so on. However, I think detail explanation of these methods are out of IPEC scope. Please check reference for cooling optimization [2].

III. IMPROVEMENT OF ELECTRICITY EFFICIENCY

Of cause, efficiency of electricity, such as power supply unit or PSU for servers, power delivery system, uninterrupted power supply or UPS and so on (Fig. 2), are also improving together with cooling improvement. In this section topics of these are described.

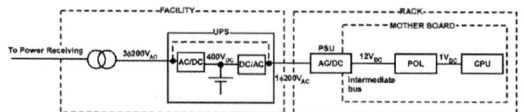

Fig.2 Example of power delivery system for conventional data center. Efficiency is improved on every stages of this system.

A. High efficiency power supply and 80 plus

From the mid of 2000s, high efficiency PSU certification of 80 plus was established [3]. 80 plus started certifying PSU with maximum efficiency over 80%, and now there are five category to over 96% efficiency (Table 1). Valuing higher efficiency for servers, 80 plus standard certification is not specified for

---

[1] From the definition of PUE, the efficiency improvement of PSU makes PUE worse. You need take care to use PUE as index of energy saving.

TABLE 1 80+ citification for AC 230 V input redundant PSUs

Loading	10%	20%	50%	100%
Bronze	-	81%	85%	81%
Silver	-	85%	89%	85%
Gold	-	88%	92%	88%
Platinum	-	90%	94%	91%
Titanium	90%	94%	96%	91%

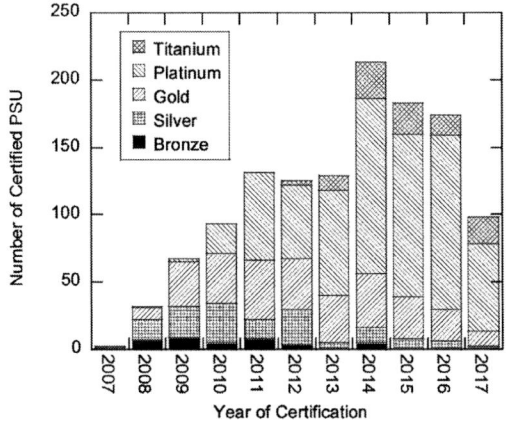

Fig. 3 Trend of 80 plus certificated PSUs.

230 V input redundant PSU. Figure 3 shows trend of 80 plus certified PSUs for servers. Before 2010, gold certificated PSUs was already dominant and now platinum certificated are become common. However, Titanium certificated PSU is not become major, mainly caused by its relatively poor cost-performance.

B. Optimization of datacenter power delivery

There are also many proposals to suppress power loss before PSU. High voltage DC power delivery or HVDC is one of the most popular proposal. Figure 4 shows schematic diagrams of HVDC 400 V system (Fig.4-a) [4] and its variations.

The HVDC system has two power loss suppression concepts. First concept is reducing the number of AC/DC and DC/AC power conversion stages. In conventional power delivery system (Fig. 2), there are three AC/DC, DC/AC power conversion stages, including two conversion stages in UPS. In the late 2000s, efficiency of

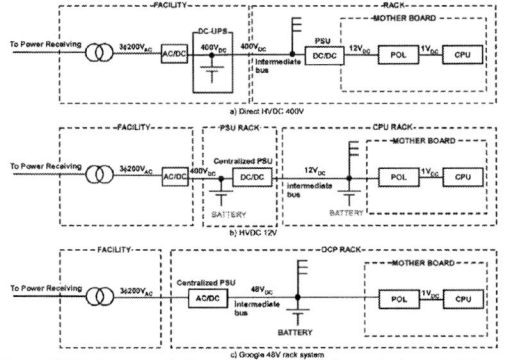

Fig. 4 HVDC power delivery system and its variation. a) Direct HVDC 400V, b) HVDC 12V, c) Google 48V Rack system.

UPS, which was already installed in datacenter[2], was relatively low as same as PSU and over 10 point improvement was expected by reducing number of power conversion stages. However, today this advantage is almost deceased by UPS efficiency improvement. For example, Fuji electric "UPS8000 series" achieved 98% efficiency by dual processing system; in which almost all power is delivered through by-pass AC line and inverter is used only for waveform shaping in normal state [5].

The second concept is reducing resistive loss by high voltage power delivery. By shorten low voltage and high current pass near the point of use, saving resistive loss and weight of Cu pattern or wiring. For this concept, in the HVDC 400 V, isolated 400 V bus bar was installed in server rack. However, safety and usability issues were pointed out and HVDC-12 V system is proposed [7]. In HVDC-12V system, 12 V bus bar is installed in server rack (Fig. 4-b). Two variations of backup battery connection are there, to 400V DC bus or to 12V DC bus. The later may enable distributed install of backup batteries and disuse of facility level UPS. Power leveling with local battery is also investigated. Especially in the cluster system, there is a possibility that all virtual machines installed in the same rack run synchronized to higher power and cause over power protection [7]. Peak cut using local battery is also a desirable function.

HVDC-12V is now being developed in Open Computing Project (OCP) [8]. First, OCP is proposed by Facebook and try making open specification of hardware for large scale datacenter. They are remarkable project in the point of ICT power delivery.

In OCP summit 2016, Google proposed 48V rack system (Fig. 4-c) [9]. Instead of 12V intermediate bus, 48 V bus was proposed. 48 V intermediate bus is common in telecommunication system, however, it is two step bus system; from 48V via 12 V to 1 V. So the efficiency advantage for conventional system is limited. 48 V rack system is characterized by the 48V to 1V direct power conversion point of load or POL converter, which is enabled in resent improvement of power electronics.

Resent topics for high performance power converter will be described in next section.

Appling renewal energy for ICT power delivery is also resent topics. Some ICT companies such as apple, google, Facebook, and so on joined RE100 and aiming 100% of their global electricity consumption from renewable sources [10]. However, they buy renewable source energy from electric power companies so far. Their own renewable energy sources are almost sell to the electric power company or used in non-critical facility such as lighting of datacenter building, and don't connect directly to servers. To meet the specifications required by ICT equipment, more stability of renewal power source is required.

---

[2] Unlike PSUs are used as expendable parts, UPS is treated as facility and its lifetime is over 10years.

C. Topics for high performance Power Supply

1) Soft switching

Now 48 V bus solutions are provided from many power supply chip makers. The one of the key technologies of direct conversion from 48 to 1 V is a soft switching.

For example, Vicor provide a sine amplitude converter with a fixed voltage conversion ratio. They named it as "Voltage Transformation Module" or VTM [11]. Using fully resonant high frequency (> 1 MHz) sine wave switching, zero volt switching (ZVS) and zero current switching (ZCS) with very high efficiency (~ 97%) is obtained. Combining this VTM with ZVS pre-regulator module (PRM), over 90% efficiency is obtained from 48 to 1V [12].

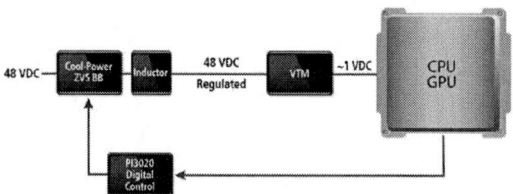

Fig.5 Vicor's 48 to 1 V direct conversion [12]

2) New materials for power electronics

Although almost all 48V solutions use resonant transformer, EPC proposed GaN based POL with both simple buck and transformer isolated topologies [13]-[14]. GaN have one order high break down electric field compared to Si and suitable for high speed and low on resistance transistor [15]. However, even though GaN based POL, their switching frequency and efficiency are not so high, 300 kHz-83% and 600 kHz-90%, respectively.

There seems to be two reasons. First issue is related to controller chip. For example, in the buck converter topology, only 2% of duty ratio and 67 ns of minimum on time are required for 48 to 1 V regulation at 300 kHz switching. Typical minimum on time of the conventional power supply controller is over several 10s-ns and this specification limits switching frequency of buck converter.

The second issue is related to the process rule of GaN

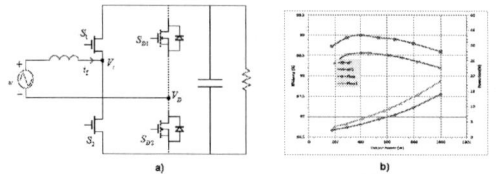

Fig. 6 Transphorm's totem pole PFC (a) and its efficiency (b) [17]

devices. EPC is only one maker developing less than 100 V breakdown voltage GaN devices. However, today's major wafer size of GaN on Si is 6 inϕ and minimum process rule for 6 inϕ wafer line is limited to 0.35 μm. 30 V power device is available using 1 um rule of Si [16], so 0.35μm rule is too large for sub 100 V GaN device.

On the other hand, using high voltage (> 600 V breakdown voltage) GaN device, high efficiency totem pole type PFC is considered (Figure 6) [17]. High speed and recovery less characteristics of GaN device is suitable for totem pole type PFC. However, application for DC-DC is not remarkable. It seems to be related with lack of low voltage GaN transistor suited for synchronous rectifier. 12 V, 5V, 3.5 V or less than 1V regulation are required for ICT power system. Low voltage, suitable for 10 V or less application, GaN device are desirable for ICT application.

3) CPU integrated voltage regulator

CPU embedded voltage regulator is also topic in ICT system. Intel Haswell [18] and IBM power 8 [19] CPU have integrated voltage regulator in its package. Voltage regulator topologies embedded in these two CPUs are very contrastive. Intel embedded multiphase switching regulator with coreless planar coil using PCB pattern. On the other hand, IBM power8 integrated LDO as embedded voltage regulator. IBM has already reported PowerSoC technology at 2012 [20], this is look like very conservative. In my personal view point, there seems to be difference of policy expected for embedded regulator.

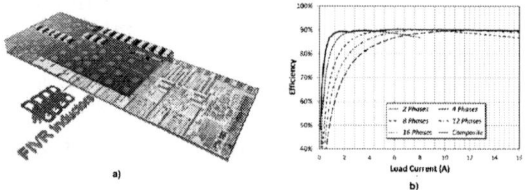

Fig. 6 Intel's a) embedded switching regulator and b) its efficiency [18].

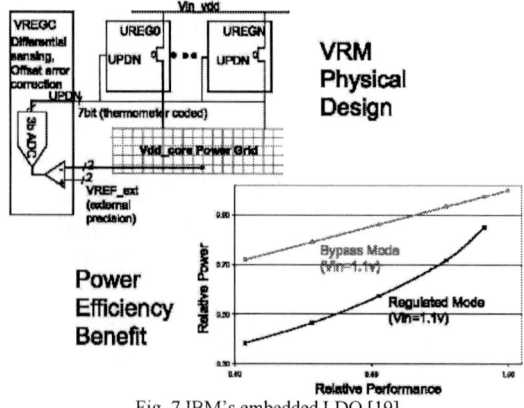

Fig. 7 IBM's embedded LDO [19]

Haswell aims very wide market from mobile to high-end computer, and expected relaxing restriction for mother board voltage supply. Power 8 has twelve cores in one chip and its market is almost high-end only, they expect certain power down of deactivated core using embedded LDO.

4) Model based design and falure prediction of power supply

Finally, Fujitsu laboratories' resent activity on power electronics is introduced. In recent years, we developed new scheme for developing digitally controlled power supply, model based design or MBD [21], [22]. Digitally controlled power supply has many advantage such as high efficiency, easy to configure using communication port online, easy to be monitored from cloud system, and so on. However, its development is much difficult because isolating faults between circuit and firmware is not so easy. Our MBD scheme enables debugging process of circuit and firmware separately.

Using this scheme, degradation detection/failure prediction method is being developing [23]-[25]. In large scale datacenter or base stations of cellular phone scattered in the deep mountain, these maintenance cost becomes an issue. We believe failure prediction can solve this issue. We demonstrated degradation of electrolytic capacitor installed DCDC converter can be detected using digital control oriented Low Stability Phenomenon.

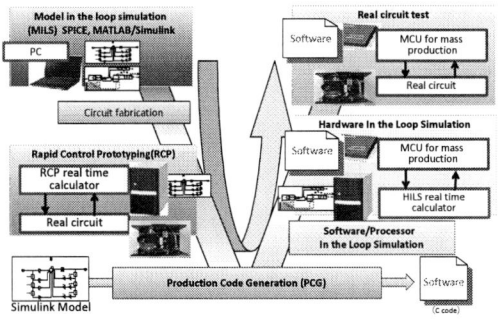

Fig.8 V process of our power supply MBD scheme.

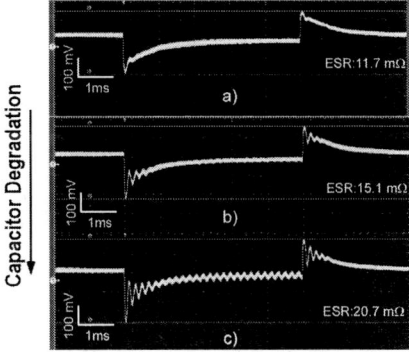

Fig. 9 Digital control oriented Low Stability Phenomenon.

IV. SUMMARY

Recent trend of power electronics in ICT system is briefly reported. High efficiency race in PSU is seems be saturated. However, there still exist some issues, such as power loss suppression in power delivery system, application of renewable energy in ICT system, maintenance cost suppression and so on.

References

[1] A. Shehabi, S. Smith, D. Sartor, R. Brown, M. Herrlin, J. Koomey, E. Masanet, N. Horner, I. Azevedo, and W. Lintner, "United States Data Center Energy Usage Report", LBNL-1005775, Ernest Orlando Lawrence Berkeley National Lab, USA, June, 2016.

[2] L. A. Barroso, J. Clidaras, and U. Hölzle, "The datacenter as a computer: an introduction to the design of warehouse-scale machines, second edition", Morgan&Claypool, July 2013.

[3] 80 PLUS [Online], http://www.plugloadsolutions.com/80PlusPowerSupplies.a spx, last accessed at October 18th 2017.

[4] T. Babasaki, T. Tanaka, Y. Nozaki, T. Tanaka, T. Aoki, and F. Kurokawa, "Developing of higher voltage direct-current power-feeding prototype system", in Proc. INTELEC 2009: 31st International telecommunications energy conference, Incheon, Korea, pp. 1-5, Oct. (2009).

[5] J. Hirose, M. Yatsu, and S. Fukuda, "*DAIYOURYOU UPS "UPS8000 series*", Fuji Gihou, PP458-462, Vol. 77, No. 6, 2004.

[6] F. Mura, "*KOREKARANO JISEDAI, CYOKURYU DATACENTER! "KOUKOURITSU: HVDC+DC12V"*", INTEROP TOKYO 2014, Tokyo, Japan, 2014 [online], http://www.ip-core.jp/doc/14/interop/03.pdf

[7] T. Sugimoto, "*SAVER NO SYOUHIDENNRYOKUKAIKOU TO PEEK CUT NO TORIKUMI*" [Online], http://techblog.yahoo.co.jp/infrastructure/power_peak_cutting/

[8] Open Computing Project [online], http://opencompute.org/

[9] N. Neilus O'Sullivan, "Optimized Power Delivery Architecture for Data Center Scale Server Applications", OCP US Summit 2016, Santa Clara, USA, Nov. 2016 [online], http://files.opencompute.org/oc/public.php?service=files&t=66a58 4dabe790c3bf4eb3d31f9789642

[10] RE100 [online], http://there100.org/

[11] "FPA Overview: An Introduction to FPA" whitepaper, [online], http://www2.vicorpower.com/l/5132/2015-05-28/2b31tp

[12] "48 V Direct to CPU" [online], http://www.vicorpower.com/industries-computing/48v-direct-to-cpu#Higher-Efficiency

[13] D. Reusch, J. Glaser, A. Lidow, "Re-evaluating 48 VIN Server Architectures with High Performance GaN Transistors", in Proc. APEC2017: 2017 the 32st annual IEEE applied power electronics conference and exposition, industrial Session IS10: Server Power Topics, Tampa, USA , Mar. 2017.

[14] D. Reusch, "Rethinking Server Power Architecture in a Post-Silicon World: Getting from 48 Vin – 1 Vout Directly" [online], http://epc-co.com/epc/GaNTalk/Post/13523/Rethinking-Server-Power-Architecture-in-a-Post-Silicon-World-Getting-from-48-Vin-1-Vout-Directly

[15] E.O. Johnson, "Physical Limitations on Frequency and Power Parameters of Transistors", RCA Review, pp.163-177, June 1965.

[16] K. Kumada and S. Yokoyama, "*KOUTAIATSU CMOS PROCESS GIJUTSU*", Fuji Gihou, PP456-458, Vol. 71, No. 8, 1998.

[17] L. Zhou,and Y.F. Wu, "99% Efficiency True-Bridgeless Totem-Pole PFC Based on GaN HEMTs", [online], http://www.transphormusa.com/sites/default/files/transphorm/new s/Totem-pole%20paper_0.pdf

[18] E. A. Burton G. Schrom, F. Paillet, J. Douglas, W. J Lambert, K. Radhakrishnan, and M. J. Hill., "FIVR – Fully Integrated Voltage

Regulators on 4th Generation Intel® Core™ SoCs", Proc. of APEC2014, pp.432-439.

[19] E. J. Fluhr, J. Friedrich, D. Dreps, V. Zyuban, G. Still, C. Gonzalez, A. Hall, D. Hogenmiller, F. Malgioglio, R. Nett, J. Paredes, J. Pille, D. Plass, R. Puri, P. Restle, D. Shan, K. Stawiasz, Z. T. Deniz, D. Wendel, and M. Ziegler, "POWER8TM: A 12-Core Server-Class Processor in 22nm SOI with 7.6Tb/s Off-Chip Bandwidth", in proc. ISSCC2014, San Francisco, USA.

[20] N. Wang, "Magnetics integration - from thin film heads to on chip inductors", PowerSoC2012 [online], http://pwrsocevents.com/wp-content/uploads/2012-presentations/session-3/3.3_Naigang%2[20]0Wang.pdf

[21] Y. Yonezawa, T. Sasaki, H. Hosoyama, H. Nakao, A. Manabe, J. Kaneko, Y. Nakashima and T. Maruyama, "Rapid control prototyping for server power supply with high-resolution PWM", in Proc. APEC2015: 2015 the 30th annual IEEE applied power electronics conference and exposition, pp. 2635-2641, Charlotte, USA , Mar. 2015.

[22] T. Sasaki, H. Hosoyama, Y. Yonezawa, A. Manabe, K. Huang, X. Liu, J. Chen, J. Kaneko, and Y. Nakashima, "Production code generation for server power supply controller", in Proc. APEC2015: 2015 the 30th annual IEEE applied power electronics conference and exposition, pp. 2656-2663, Charlotte, USA, Mar. 2015.

[23] H. Nakao, Y. Yonezawa, Y, Nakashima, and F. Kurokawa, "RCP evaluation of electrolytic capacitor degradation for SMPS failure prediction", in Proc. APEC2016: 2016 the 31st annual IEEE applied power electronics conference and exposition, pp. 754-758, Long Beach, USA, Mar. 2016

[24] H. Nakao, Y. Yonezawa, Y, Nakashima, and F. Kurokawa, "Failure Prediction Using Low Stability Phenomenon of Digitally Controlled SMPS by Electrolytic Capacitor ESR Degradation", in Proc. APEC2017: 2017 the 32st annual IEEE applied power electronics conference and exposition, pp. 2323-2328, Tampa, USA, Mar. 2017.

[25] H. Nakao, Y. Yonezawa, T. Sugawara, Y, Nakashima, and F. Kurokawa, "Online Evaluation Method of Electrolytic Capacitor Degradation for Digitally Controlled SMPS Failure Prediction", IEEE Trans. Power Electron., vol. 33, no. 3, pp. 2552-2558, Mar. 2018.

Green Base Station Using Robust Solar System and High Performance Lithium ion battery for Next Generation Wireless Network (5G) and against Mega Disaster

M. Nakamura[1] and K. Takeno[1]

1. Research Laboratories, NTT DOCOMO, INC.

DOCOMO R&D Center, 3-6 Hikarino-oka, Yokosuka-shi, Kanagawa, 236-8536 JAPAN

*E-mail: masaki.nakamura.ua@nttdocomo.com

Abstract— To secure wireless communication services, we are researching and developing disaster-resistant and environmentally friendly green base stations. One effective disaster countermeasure in carriers is to make backup time long for base stations during a power outage. Therefore, we have developed a photovoltaics (PV) system for green base stations to prolong the backup. In this paper, we propose a power control method that realizes long-term autonomous operation by PV and lithium-ion batteries (LiB) and regeneration operation by only PV for when commercial power is lost during a power outage and describe the results obtained at field test station.

Keywords— Green Base Station; Lithium- ion battery

I. INTRODUCTION

In the wake of recent power outages brought on by natural disasters and other events, enhancing power-backup measures for mobile communications is becoming increasingly important. Although radio base stations are already equipped with storage batteries for backup purposes during power outages in the utility grid, it is thought that this scheme can be further enhanced by changing from lead-acid batteries to LiB to increase full-charge battery capacity [1][2].

At the same time, studies are being performed on migrating to natural energy as an environmental contribution. In this regard, particular attention is being given to PV generation, which can generate power during daylight hours when demand is high and therefore reduce demand for commercial power during peak hours. Looking forward, the introduction of natural energy such as PV power in radio base stations can also be envisioned, and it is said that "green base stations" that can effectively combine storage batteries with natural energy and reduce environmental load can be an effective means of doing so [3].

To demonstrate the effectiveness and reliability of green base stations, NTT DOCOMO has so far installed ten stations for field-testing in the Kanto-Koshinetsu region (Tokyo and Kanagawa, Gunma, Ibaraki, Yamanashi, Nagano, and Niigata prefectures). Furthermore, based on the results obtained from these stations, 44 commercial green base stations have been installed throughout the country, which means that green base stations including the test stations are now operating in every prefecture in Japan.

A green base station improves the self-generation rate by providing power shift control that efficiently stores surplus power generated in the daytime [4]. However, given that the power generated by solar panels may be insufficient or excessive depending on weather conditions, either LiB capacity must be increased or the battery's State of Charge (SOC) must be brought sufficiently low beforehand to ensure charging without losing use of any surplus power generated. Yet, from the viewpoint of a disaster countermeasure, it is desirable to keep SOC as high as possible to ensure the provision of a power supply in the event of an emergency such as a power outage. However, increasing LiB capacity is constrained by costs and installation space, so keeping SOC low is necessary in practice, but this prevents an environmental contribution and a disaster countermeasure from being achieved together. To solve this issue, we have developed a technology called "weathercast-linked control." This paper describes the configuration of a green base station, gives an overview of this technology, and reports on the results of field trials conducted at a test station.

II. POWER CONTROL OF GREEN BASE STATIONS

A. Typical Configuration

A typical configuration of a green base station is shown in Figure 1 and Fig. 2. In addition to a rectifier, which serves as a power-supply facility in conventional radio base stations, a green base station introduces LiB in place of lead-acid batteries and connects the solar panels to a DC 48V bus (hereinafter referred to as "DC bus") in parallel so that generated power can be used at times of a power outage. By incorporating a power-supply control section for high-efficiency control of these three types of power supplies, a green base station achieves effective use of natural energy and deals with peak power demand.

Fig. 1. Green Base Station test site.

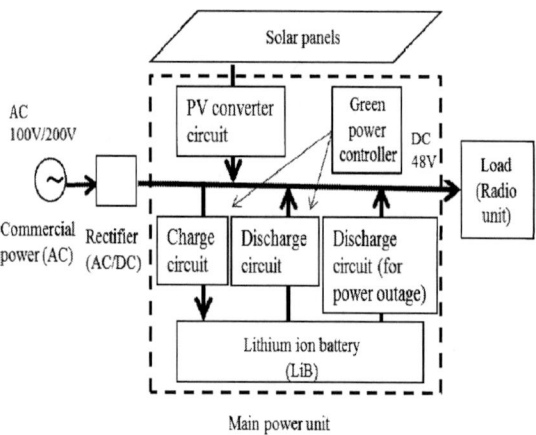

Fig. 2. Block diagram of Green Base Station.

The power-supply control section consists of a (1) photovoltaic converter, (2) charge/discharge section, (3) distribution board, and (4) general equipment controller as described below.

1) PV converter: The photovoltaic converter incorporates a Maximum Power Point Tracking (MPPT) function to generate power at maximum efficiency [5][6]. It also sets the output voltage after Direct Current-to-Direct Current (DC/DC) conversion higher than the rectifier and LiB voltage to give priority to photovoltaic power in the supply of power to the radio equipment.

2) Charge/discharge section: The charge/discharge section incorporates charge and discharge circuits and performs a specific amount of charging and discharging

based on instructions received from the control/monitoring section. The discharge circuit (for power-outage) includes a circuit for detecting voltage drops in the DC bus so that power can be supplied from the batteries without interruption at the time of a power outage.

3) Distribution board: The distribution board provides a connection to the DC bus and detaches individual equipment from the DC bus at the time of an emergency.

4) General equipment controller: The general equipment controller performs unified management of generated power measured at IV current/voltage measurement locations and of various types of measurement data such as battery-unit information. It includes a monitoring function for passing information to a remote base and a function for responding to remotely received control instructions.

B. Charge and Discharge Control of the Battery

Charge and discharge of the battery is controlled by the voltage regulator. For charging, the voltage regulator boosts the bus voltage above the battery voltage and supplies power to the battery. For discharging, the voltage regulator boosts the battery voltage above bus voltage and supplies power to the bus line. Conversely, applying a proper bus voltage caused by the rectifier output voltage to the battery voltage enables charge and discharge control of the battery. This method means that the rectifier output voltage is boosted above the battery voltage during charging and is lowered below the battery voltage during discharging. This charge and discharge method is shown in Fig. 3.

As we adjust the bus voltage, we focus on the features of the existing rectifier. The rectifier has a DC / DC converter to output power suitable for the allowable voltage of the base stations after converting AC into DC. Although the existing rectifier output voltage is fixed, the rectifier is used instead of the charge-discharge control device by changing the output voltage. In addition, without changing the hardware of the existing rectifier to adjust the output voltage, the firm-ware should be rewritten to output only a predetermined voltage in accordance with a command of the Green Power Controller. In this way, not only is the control unit cost reduced to virtually zero, but also high reliability and versatility can be maintained.

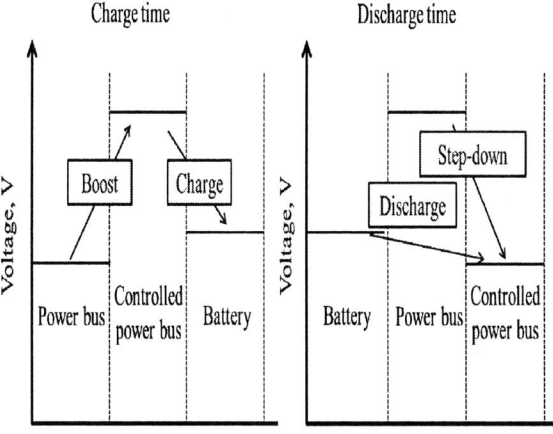

Fig. 3. Charge and discharge control by rectifier voltage control

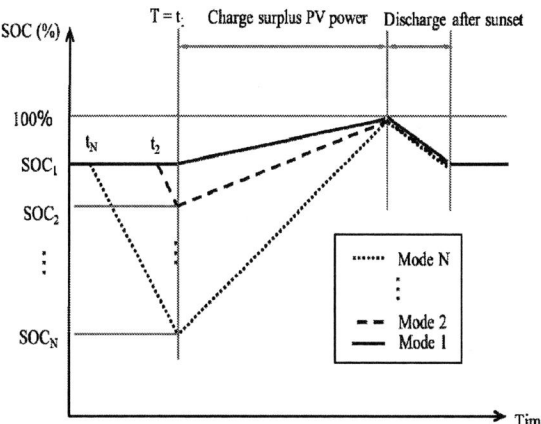

Fig. 4. Overview of sunlight control mode.

III. WEATHERCAST-LINKED CONTROL

The weathercast-linked control technology features two operation modes: sunlight control mode during normal times and autonomous/regeneration mode during power outages.

A. Sunlight Control Mode

Weather-dependent operation in sunlight control mode is shown in Fig. 4. In this mode, the amount of surplus power generated by the solar panels is estimated based on weather forecasts. Then, by discharging that amount by time T when surplus power generation begins, solar-panel generated power that was conventionally discarded can be effectively used even for batteries that cannot hold sufficient capacity other than that needed for backup during a power outage. Moreover, at the time of an emergency such as a power outage, a solar-panel backup effect achieved by switching to autonomous/regeneration mode enables communication services to be continued for a specific amount of time.

When the number of modes is n, SOCk and tk can be expressed the following equation (1) and (2) using the maximum surplus power P, power generation prediction error δ, full charge capacity (FCC) and load power Q.

$$SOC_k = SOC_1 - (SOC_1 - SOC_N) \times k/N \quad (1)$$

$$t_k = T - FCC \times (SOC_1 - SOC_k)/Q \quad (2)$$

where.

$$N = \lfloor P/\delta \rfloor \quad (3)$$

$$SOC_1 = 100 - \delta/FCC \quad (4)$$

$$SOC_N = 100 - P/FCC \quad (5)$$

The following describes the control technique in sunlight control mode. This technique begins by determining the weather-dependent SOC value (SOC-sunny, SOC-cloudy, SOC-rain) and the discharge start time (t sunny, t cloudy) based on information in the previous day's weather report. If rain is forecast, surplus power from the solar panels cannot be expected, so there is no need to discharge battery power beforehand. In this case, SOC-rain is determined to be the amount of backup needed for ensuring communications during a power outage. SOC-sunny, meanwhile, is determined to be a value such that the maximum amount of surplus power generated in the last month is available for battery charging. Finally, SOC-cloudy is determined from the average value of SOC-sunny and SOC-rain. Surplus-power-generation start time T is determined as the earliest time from among the actual start times for the last month, and discharge start time t is determined using full-charge battery capacity and the power load such that weather-dependent SOC is reached at surplus-power-generation start time T. Weather information for the base station is obtained before time t, and surplus power generated by the solar panels is used for battery charging after surplus-power-generation start time T. The evening-discharge start time is set from some time in the evening such as sunset when no surplus power is being generated, and on reaching this time, battery discharging begins so that SOC drops to SOC-rain (backup portion), at which time LiB enters a standby state. However, if the weather forecast includes the possibility of a natural disaster, LiB will be put into a fully charged state and operation will continue as usual until the disaster forecast is lifted. Additionally, if a disaster actually occurs and a state of power outage is entered, the radio equipment will be supplied with power from LiB, and when the power-outage state ends, LiB will again be charged up to the SOC-rain level.

B. Autonomous/Regeneration Mode

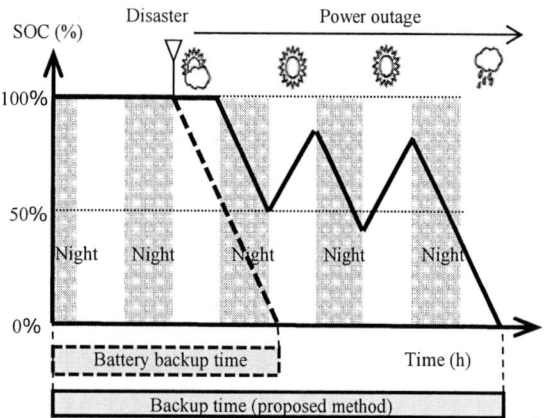

Fig. 5. Operation in autonomous/regeneration control mode.

Operation in autonomous/regeneration mode is shown in Fig. 5. In the past, the LiB SOC would decrease monotonously at the time of a power outage due to a natural disaster or other event, given that radio-equipment load was about constant. As a result, battery power would gradually diminish leading to a cut in service. In autonomous/regeneration mode, however, the power generated by the solar panels can be supplied to the radio equipment during a clear day and surplus power can be used to charge LiB thereby extending service time. Regeneration operation can be performed in the same way even after LiB depletes.

The following describes the control technique in autonomous/regeneration mode.

A charge adjusting method using hill climbing (HC) is shown in Fig. 6. Specifically, when LiB output is zero and surplus power is being generated, this method detects that the DC bus rises to the output voltage of the PV converter, so the method increases the amount of charging in very small increments. Then, if surplus power is not being generated, the method decreases the amount of charging in very small decrements. Repeating this process in such small steps enables surplus power to be estimated and only surplus power to be used for charging. Furthermore, while power generation can suddenly drop due to momentary shading by passing clouds on an otherwise clear day, the charge/discharge section can deal with this by switching to a discharge state from a charge state after detecting voltage drops in the DC bus so that power is uninterruptedly fed to the radio equipment.

C. Control Algorithm

A flowchart of the control is shown in Fig. 7. In this control, parameters are determined based on past data, and the current operation mode of the day is determined based on the weather forecast information of the previous day.

Table 1 shows a part of the control parameters of weather forecast linkage control used for the verification test.

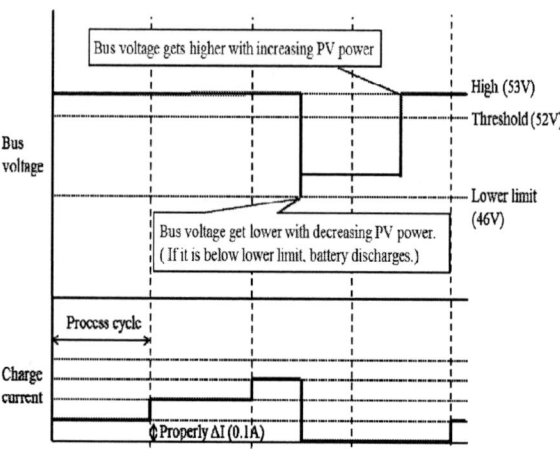

Fig. 6. Charge adjusting method using HC.

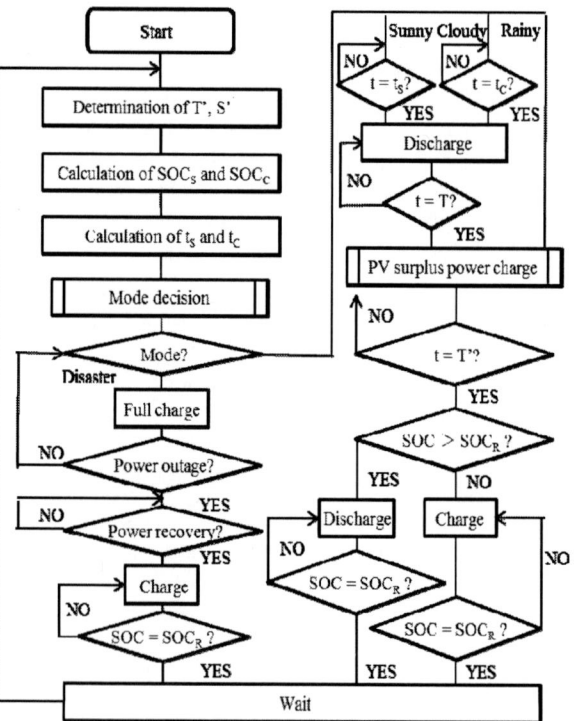

Fig. 7. Program flowchart.

TABLE I
PARAMETER OF PROPOSED CONTROL.

No.	Parameter	Description	Value
1	PV charge starting time	The Assumed PV charge starting time	8:00
2	Surplus discharge starting time	The discharge starting time which is the part beyond Base SOC	15:30
3	Maximum power of surplus charge	The Assumed maximum power of the surplus charge	2.16 kW
4	Base SOC(Start SOC (Rain))	Backup power standard SOC. The starting SOC which is at the rain forecast	80%
5	Start SOC(Fine)	The starting SOC which is at the fine forecast	42.4%
6	Start SOC(Cloudy)	The starting SOC which is at the cloudy forecast	61.2%
7	Discharge starting time(Fine)	The discharge starting time which is at the fine forecast	5:10
8	Discharge starting time(Cloudy)	The discharge starting time which is at the cloudy forecast	6:45
9	Blackout voltage threshold	Discharge starting by this voltage	46.5 V
10	Bus voltage threshold	The threshold by which this judges fluctuation of the charging rate.	52 V
11	Default charging current	The charge starting value.	1 A
12	Maximum charging current	Maximum charging rate.	30 A
13	Increase current (ΔI)	(Bus voltage > bus voltage threshold) The increased current value.	50 mA
14	Decrease current	(Bus voltage ≤ bus voltage threshold) The decreased current value.	50 mA
15	Process cycle	The setting, control, and watching cycle.	5 s

IV. EXPERIMENTAL RESULT

We installed a green base station providing 1.4 kW-rated photovoltaic generation at a radio base station in Gunma prefecture as a field test station. The power consumption of the radio equipment at this test station was approximately 0.5 kW. During the test, various types of measurement data regarding the power-supply system and any system alarms such as abnormal battery temperature were sent to a remote monitoring base over a communications circuit. The green base station was found to operate well with no system alarms generated during the testing period. Table II shows the specifications.

TABLE II
PARAMETER OF PROPOSED CONTROL.

Elements	Data
Latitude	North latitude 36°21′ 25″
Longitude	East longitude 139°14′ 9″
Altitude	93 m
PV installation direction	Due south
PV installation angle	30°
Load capacity	500 W
PV rated power	1.4 kW

A. Sunlight Control Mode

Test results for sunlight control mode are shown in Fig. 8 and Fig. 9. For this demonstration, we conducted a long-term test from December 4, 2015 to January 3, 2016 using LiB with a capacity of 3.75 kWh. The graphs in the figure overlay some results for this period. Examining the change in SOC in Fig. 6, it can be seen from the manner of falling SOC by early-morning discharging that control could be performed according to the three prescribed control patterns (clear, cloudy, rain). Additionally,

examining the change in generated power in Fig. 7, it can be seen that the amount of generated power was generally greater than the power consumption of the radio equipment during the day so that surplus power could be effectively used. These test results show that an improvement in the self-generation rate of approximately 11% can be expected compared with conventional power shift control under the same conditions.

B. Autonomous/Regeneration Mode

Test results for autonomous/regeneration mode are shown in Fig. 10. In this test, we used LiB with a capacity of 13.5 kWh and reproduced power-outage conditions by turning off the power supply to the rectifier in the field test station. The graphs in the figure show photovoltaic power generation and LiB SOC after the beginning of a power outage.

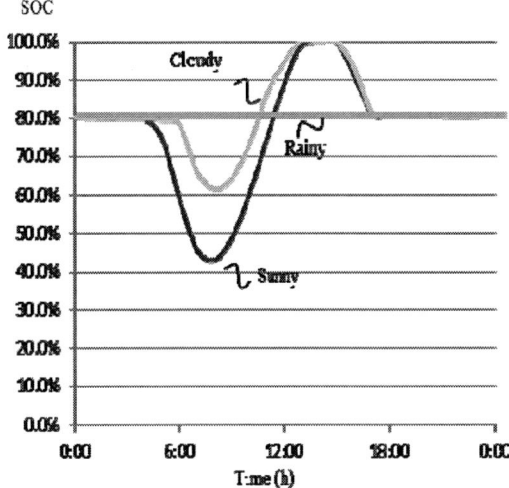

Fig. 8. Transition of SOC.

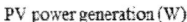
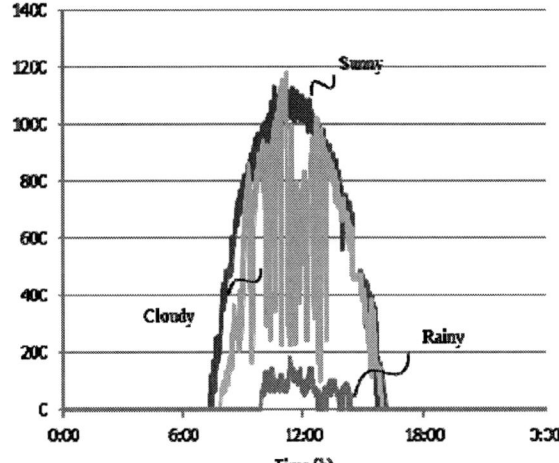

Fig. 9. Solar power characteristics.

The 2018 International Power Electronics Conference

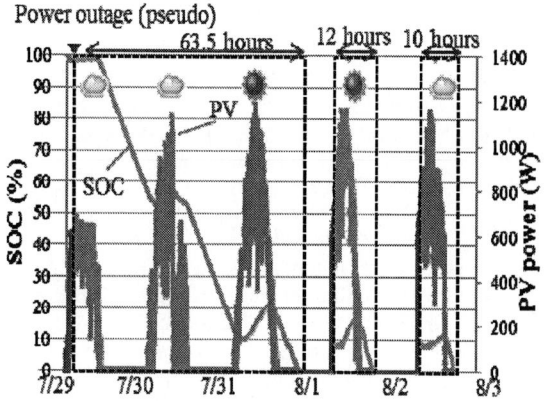

Fig. 10. Power outage test results.

It can be seen that LiB charging took place even during a power outage, which means that power generation that included surplus power greater than the amount of power consumed by the radio equipment was performed. These results show that the autonomous operation period was approximately 2.4 times that when using LiB of the same capacity as an emergency power supply. It can also be seen from these results that the radio base station could operate for a certain period of time through power supplied from LiB even after sunset depending on the amount of power stored during the day.

V. SUMMARY

This paper described weathercast-linked control that achieves both an environmental contribution and a disaster countermeasure in green base stations. The use of this control technology can be expected to improve the self-generation rate, reduce the amount of commercial power used, and extend service time during a power outage. Going forward, we plan to enhance the service-time extension effect during power outages without diminishing the self-generation rate by further dividing control/operation modes to raise the level of LiB backup time. We also plan to promote the introduction of this technology in parallel with the expansion of green base stations.

ACKNOWLEDGMENT

My heartfelt appreciation goes to Mr. Kimura who provided carefully considered feedback and valuable comments. Special thanks also goes to environmental technology research group whose opinions and information have helped me very much throughout the production of this study.

REFERENCES

[1] I. Suzuki, T. Shizuki and K. Nishiyama:"High power and long life lithium-ion battery for backup power sources, "Proc. of INTELEC '03, pp. 317-322, Yokohama, Oct. 2003.

[2] T. Matsushima: "Characteristics of Lithium-Ion Cells with Large-Capacity and DC-Power-Supply Systems for Telecommunications Applications," Transactions of the Institute of Electronics, Information and Communication Engineers (B), Vol. J91-B, No. 12, pp. 1725-1734, Dec. 2008.

[3] K. Takeno: "Deployment of Green Base Station," Journal of the Institute of Electrical Installation Engineers of Japan, Vol. 34, No. 2, pp. 110-113, Sep. 2014.

[4] M. Nakamura and K. Takeno: "Power Control Method of Mobile Base Stations Using Lithium-ion Battery and Solar Power," Transactions of the Institute of Electronics, Information and Communication Engineers (B), Vol. J100-B, No. 4, pp. 307-314, Apr. 2017.

[5] G.J. Yu, Y.S. Jung, J.Y. Choi and G.S. Kim: "A novel two-mode MPPT control algorithm based on comparative study of existing algorithms," Proc. of Solar Energy, Vol. 76, No. 4, pp. 455-463, 2004.

[6] T. Tafticht, K. Agbossou, M.L. Doumbia and A. Cheriti: "An improved maximum power point tracking method for photovoltaic systems," Proc. of Renewable Energy, Vol. 33, No. 7, pp. 1508-1516, Apr. 2008.

Optimization of maintenance by failure prediction considering instantaneous and cumulative effects of external environments

Kaisei Kanetani[1]*, Masahiro Yamazaki[1], Tadatoshi Babasaki[1], Hideaki Kim[2], Tatsushi Matsubayashi[2]

1 Research and Development Division, NTT Facilities, Inc., Tokyo, Japan
2 Service Evolution Laboratories, Nippon Telegraph and Telephone Corporation, Tokyo, Japan
*E-mail: kaneta24@ntt-f.co.jp

Abstract—We propose a method of determining power equipment renewal by using a method of failure prediction. Renewal is conventionally determined from equipment age by using a bathtub curve. However, it is impossible to make a renewal assessment in accordance with the deterioration level of individual units that are in different environment with the conventional method. It may be possible to renew equipment whose deterioration has not progressed more than its actual aging. Therefore, we propose a method of individually predicting the deterioration level of each unit by considering the instantaneous and cumulative effects of external environment simultaneously. We tested and verified the accuracy of this proposed method using real maintenance data of power equipment and reported the results.

Keywords—Failure prediction, Machine learning, Maintenance, Power equipment

I. INTRODUCTION

In recent years, many infrastructure facilities in Japan have become decrepit, resulting in the occurrence of serious accidents. To prevent such accidents, the facilities equipment is required to be properly maintained and renewed as necessary. Time based maintenance (TBM) that performs maintenance on the basis of a schedule has long been the most common management method in facility maintenance management. However, when managing a vast number of facilities, both maintenance and management increases in amount, complexity, and cost. One approach to solving that problem is condition-based maintenance (CBM), in which maintenance is performed in accordance with the state of the equipment as determined by remote monitoring or other such means rather than TBM. In addition to the simple monitoring of equipment, a failure prediction method [1] was proposed to achieve more efficient maintenance and management.

Prognostics and health management (PHM) is an academic field related to failure prediction. In PHM, the remaining useful life (RUL) of equipment is estimated for the purpose of predicting its maintenance time and potential extension of its life. For this purpose, two main approaches are used: physics of failure (PoF) and data-driven type. In the PoF approach, after specifying the failure mode of the equipment, its RUL is calculated by sensing the operating data and environment data of the equipment and inputting the obtained data into the PoF model. This approach is used when the failure mechanism of the equipment is known in detail. On the other hand, the data driven approach calculates the RUL statistically from the already available data mainly using machine learning. This approach is used when data related to the equipment's health condition is already stored.

In this paper, we propose a method of estimating power equipment renewal timing by using the data driven approach to determine its optimum timing. First, we describe our facility maintenance system and the data we have stored. Then, after describing the conventional method of facility renewal judgment, we propose a failure prediction model that considers the effects using the external environment of the equipment. Finally, we verify the accuracy of this proposal and report the result.

II. BACKGROUND

NTT FACILITIES is responsible for the construction and maintenance of buildings, power supply facilities, and facilities that enable communication services throughout Japan. Since stopping these facilities means stopping communication services, high reliability is guaranteed by the redundant and backup configurations. However, as the number of required facilities increases, the costs for construction and maintenance increases. To reduce these costs and continue to maintain and manage power supply facilities with high reliability and low costs, we have developed monitoring technology and data utilization technology.

We maintain approximately 200,000 power equipment units installed in approximately 10,000 buildings. We operate CBM using remote monitoring to efficiently maintain these units. Fig. 1 shows the equipment maintenance system based on the monitoring system. An operation center uses the system to monitor and manage the state of power-supply equipment in buildings and data centers located in various places. Local workers regularly maintain the equipment. The occurrence of abnormalities and failures is mainly found in the operations center, and this information is transmitted to the local workers so that they can respond to it. These correspondences are recorded in the system. Using this recorded information, ICT engineers and analysts compile and analyze the information on the related equipment.

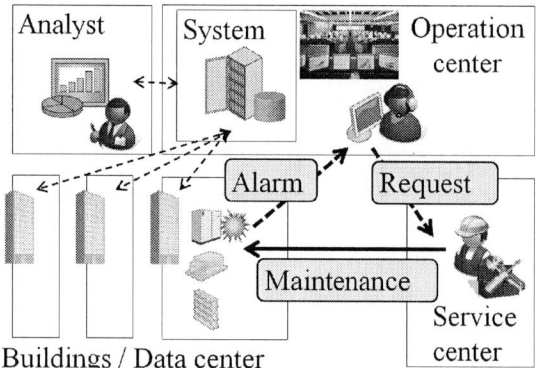

Fig. 1. Equipment maintenance by operation system

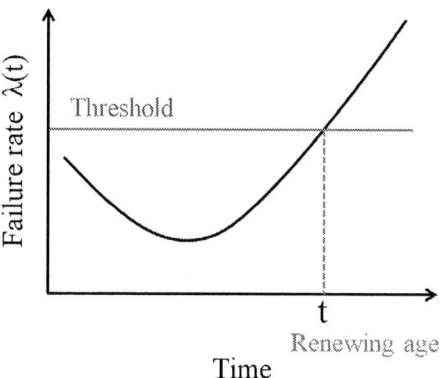

Fig. 2. Bathtub curve to determine equipment renewal

Through the operation of the monitoring system, we have stored different types maintenance data. Typical data features are as follows.

- Equipment data: The group name of the facility, the unique name of the facility, the installation place, etc. are recorded as management data for the facility. These data are entered and registered by the user.

- Alarm data: The system monitors the state of the equipment, and an alarm is issued when an abnormality occurs. The alarm is automatically transmitted from the facility and the operator is notified.

- Measurement data: The operation data of equipment measured are recorded at constant intervals. For example, in the case of power supply equipment, voltages, currents, etc., are recorded. It also measures environmental data such as temperature and humidity.

- Failure detail data: Data corresponding to failure is recorded. These data include abnormal equipment status, operator correspondence, worker correspondence, abnormal parts, cause of failure, etc. Most of the data is input by the user.

Hundreds of thousands of records related to equipment and failure data and hundreds of millions of records related to measurement data and alarm records have already been stored. We studied failure prediction by using these data. As a result, we developed lot abnormality detection technology [3] to detect defective lots, in which failure frequently occurs, early. In the method for detecting lot defects, a lot is defined by the combination of equipment type, maker, and manufacturing year and the method automatically detects and reports the failure rate quickly from the failure rate trend. This technology is intended to prevent the lots spread as early as possible by detecting lots that have defects.

III. CONVENTIONAL METHOD

To reduce facility costs, it is desirable to keep equipment running for as long as possible until just before failure. However, it is generally known that the equipment failure rate $\lambda(t)$ increases as the aging time t increases from the start of operation. Therefore, when formulating a facility renewal plan a few years ahead, we conventionally calculate the bathtub curve shown in Fig. 2 and plan to uniformly renew aged equipment units that were over the threshold. This threshold indicates the failure rate that is necessary to ensure reliability. This planning can determine the criteria for renewing equipment.

Many models have been proposed for statistical analysis using bathtub curves [4]. The virtual age model (VAM) is a inter-failure corrective maintenace using age-based model. For each corrective maintenance, the system's virtual age τ is determined by a variety of additive or multiplicative age-reduction factors. As shown in Fig. 3 (a), VAM [5] uses the virtual age τ which is refreshed and restored by the facility renewal. The failure rate is described as follows,

$$\lambda_{\mathrm{VA}}(t) = \lambda_{\mathrm{age}}(\tau(t)) , \qquad (1)$$

and

$$\tau(t) = \alpha T_K , \qquad (2)$$

and

$$T_k = \alpha T_{k-1} + \Delta t_k , \qquad (3)$$

where K is the number of facility renewals, Δt_k is the time duration from $k-1$ th to k th for renewal timing, $T_0 = 0$, and α is the restored rate ($0 < \alpha < 1$). The Proportional harzard model (PHM) [6] uses the gained rate β ($0 < \beta < 1$), and is described as follows,

$$\lambda_{\mathrm{PHM}}(t) = \beta^K \times \lambda_{\mathrm{age}}(\tau(t)) . \qquad (4)$$

For the PHM, it is difficult to refresh the equipment's aging for occurences such as overhaul maintenance. However, VAM can describe the aging progress using various maintenances or external environments. Therefore, we propose a new aging model for failure rate based on the VAM.

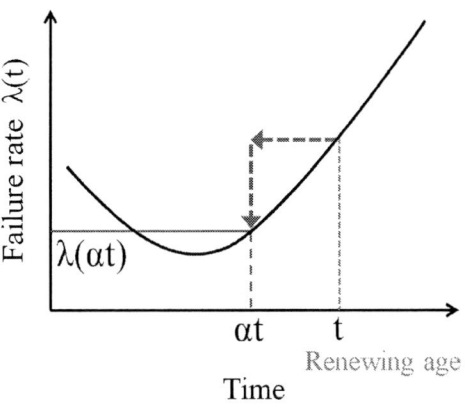

(a) Virtual age model

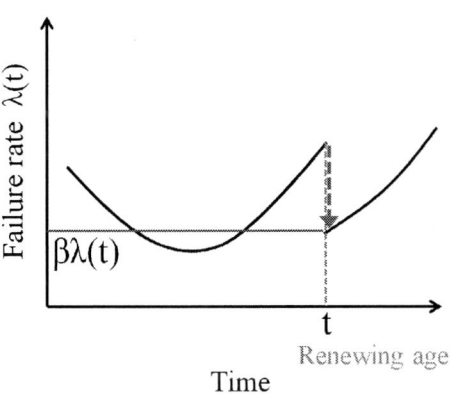

(b) Hazard rate model

Fig. 3. Analysis model using the bathtab curves

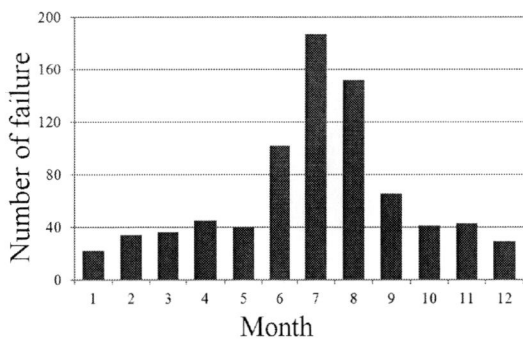

Fig. 4. Number of failures of inverter units by month

Although the renewal age for any type of equipment can be determined by using bathtub curves, they cannot be used to detect individual equipment deterioration. Even if the aging of equipment is the same, the progress of deterioration differs depending on the environment it is in. Therefore, it is not possible to renew equipment whose deterioration has progressed more than the actual aging due to this problem. It may be possible, however, to renew equipment whose deterioration has not progressed more than the actual aging. We propose a method of equipment renewal planning that takes into account the risk of failure due to external environments for equipment units to solve such problems.

IV. PROPOSED METHOD

We considered on the basis of the analysis of failure data and experience in equipment maintenance that the environment around equipment can affect failure. Furthermore, we consider that there are two types of effects of external environment, instantaneous and cumulative. Fig. 4 shows the number of failures of inverter units by month from 2006 to 2015. As can be seen in Fig. 4, the number of failures increases substantially from June to August. In Japan, this period is hot and humid, and we infer that failures due to temperature and humidity occur.

There are various external environments that affect failure. For example, coastal areas tend to have more equipment failures compared to inland areas due to the possibility of corrosion from salt damage. Failures near arterial roads can similarly be attributed to dust caused by exhaust gases from cars. When the number of failures is counted on a monthly basis in one year, it increases in summer from June to August. This is considered to be due to the influence of temperature and humidity. It is known that electrolytic capacitors and batteries deteriorate due to heat. Furthermore, the printed circuit boards of power equipment fail due to corrosion and the accumulation of dust.

The conceptual diagrams for the two types of effects are plotted in Fig. 5. We defined instantaneous effects as failures that resulted from a temporary increase in environmental factors. For example, shorts in printed circuit boards due to dewing at high humidity. The relation of air temperature to failure rate for inverter equipment as instantaneous effects is plotted in Fig. 6, where the horizontal axis indicates the temperature when the failure occurred and the vertical axis is the normalized failure rate, which indicates the ratio of the number of failures to the cumulative time for each temperature. Fig. 6 shows that high air temperatures were associated with relatively high failure rates. However, we defined cumulative effects as failures that were caused by damage that built up over time. For example, deterioration due to corrosion caused by salt damage. However, because the instantaneous and cumulative effects were mutually influenced, the two types of effects must be considered together to assume a situation that is closer to reality. Therefore, we propose a failure prediction model considering both the instantaneous and cumulative effects of external environments and a facility renewal judgment by the model. By using the aging speed function $G(H(t))$ including both effects, virtual age $\tau(t)$ can be described by integrating $G(H(t))$

The 2018 International Power Electronics Conference

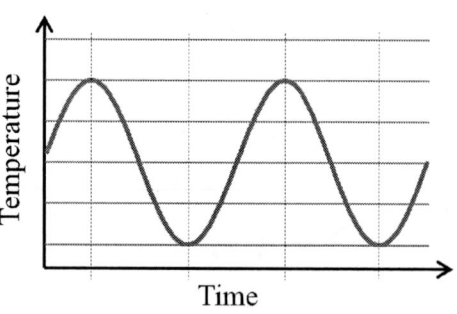

(a) Temperature changes

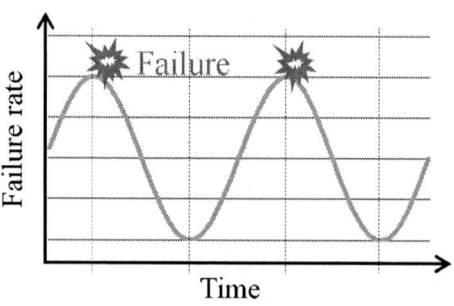

(b) Instantaneous effect

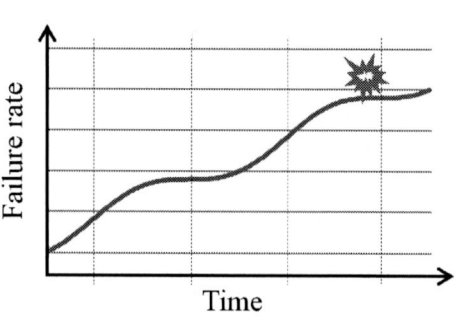

(c) Cumulative effect

Fig. 5.   Instantaneous and cumulative effects

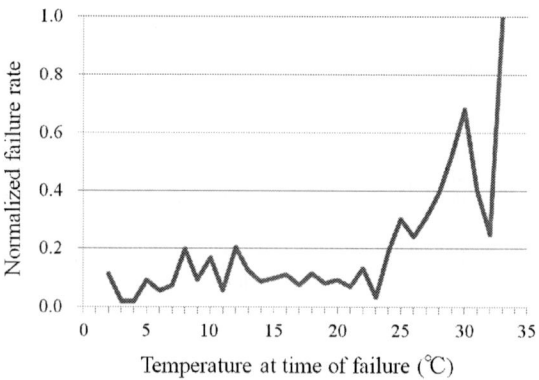

Fig. 6.   Relation of instantaneous effect of air temperature with failure rate

as follows,

$$\tau(t) = \int G(H(t))dt, \qquad (5)$$

where $H(t)$ is environmental cost (effected by temperature, moisture, dust, etc.) at real time t. Therefore, the failure rate is given as

$$\lambda_{\mathrm{GVAM}}(t) = \lambda_{\mathrm{age}}(\tau(t)) \times G(H(t)). \qquad (6)$$

We call the proposed model as a generalized virtual age model (GVAM). We used the concept of the virtual age model [7] in a failure prediction model that took into account both instantaneous and cumulative effects to

accomplish the proposed method. This model calculated the virtual age of all equipment units using data such as the measurements of the external environment, failures, and maintenance timing. Virtual age is an index-like age that takes into account progress in deterioration by external environments and recovery due to maintenance. Additionally, this model produces future failure probabilities for all units by using measurement data that have been estimated from future external environments. It is possible to determine the renewal of all units regardless of age by using these prediction results.

We explain the equipment renewal plan using a failure prediction model. Fig. 7 shows an overview of the failure prediction model. First, measurement data of the external environment of the equipment and the failure data are given as input data. For example, when temperature data around the equipment and failure data as shown in Fig. 7 are given, this model learns the temperature conditions the equipment will fail using the data recorded up to the point of failure (cumulative effect) and the data at the time the failure occurs (instantaneous effect). The learning result is output as a deterioration speed characteristic. The deterioration speed is a parameter that determines the progression of the virtual age by the external environment. For example, the model learns that the deterioration speed is low at 10°C, and increases at 30°C. In addition to deterioration due to aging, the virtual age progresses in accordance with the deterioration speed of the actual temperature experienced. For example, when the temperature is 30°C at which the deterioration speed increases as shown in the lower part of Fig. 7, the virtual age further increases. By giving information on the learning parameters (degradation speed) and predicted measured values of the external environment, the future virtual age and failure probability can be predicted for each equipment unit. Using this prediction result, an equipment list as shown in the lower part of Fig. 7 is output. Presenting the virtual age for each equipment unit makes it possible to judge the degree of deterioration in accordance with the environment. We hope to optimize the lifecycle cost and secure a high reliability maintenance system by preferentially renewing equipment with

210

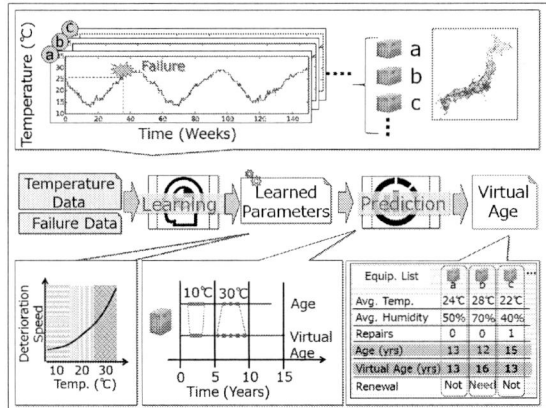

Fig. 7. Overview for failure prediction model

TABLE I.   CONDITON OF DATA SET

External environment	Room temperature
Measurement interval	1 day
Training data	From 2006 to 2014
Test data	2015
No. of predicted units (in 2015)	21076 units
No. of failures in 2015	101 failures
Failure rate in 2015	0.48%

a high virtual age.

The prediction accuracy of the model is important to successfully use the proposed method. Therefore, we tested and verified the accuracy of the model. The verification accuracy was tested by using the data set in Table I. The model learned each data item from 2006 to 2014. Furthermore, the model predicted units of equipment that failed in 2015. We compared these prediction results with the failure records for 2015 and evaluated the results.

V.   VERIFICATION RESULT

The prediction results for the failure prediction model are summarized in TableII. True positive (TP) is the number of failed units of equipment that were predicted to fail by the model. True negative (TN) is the number of non-failed units that were not predicted to fail by the model. The accuracy rate is the ratio of these sums to all the units. As a result, the accuracy rate was 98.9%. However, this index could not be used to determine how many failures could be predicted because the number of TNs was dominant. Therefore, we evaluated the model for the precision and recall rates that are used for the evaluation of machine learning models. The precision rate is the ratio of TPs to the number of units that were predicted to fail by the model. As a result, the precision rate was 6.8%. The recall rate is the ratio of TPs to the number of failed units. As a result, the recall rate was 10.9%. In comparison with the actual failure rate in 2015, which was 0.48%, the precision rate was large. Therefore, we considered that it was a good result.

Table III shows the results in the case where all 9-year-old equipment was renewed on the basis of the judg-ing renewal method with conventional equipment aging. Table IV shows the comparison result of each accuracy index of the conventional and proposed methods. In the case of the conventional method, the precision rate is 0.6%, which is roughly equivalent to the actual failure rate in 2015. In the comparison of accuracy indexes, the proposed method is superior for all indexes. In particular, the precision rate improved by 10 times or more, indicating that wasteful equipment renewal can be reduced.

Although the proposed method can be expected to further reduce facility costs compared to the conventional method, we considered that it was necessary to obtain higher precision and recall rates to validate the proposed method. In particular, since the precision rate is an important index for cost reduction, we intend to make a further improvement. We think that it is important to use external environmental data that are more highly correlated with failure of equipment than room temperature data to obtain higher rates. For example, humidity is considered to be the environmental factor that most affects failures in equipment with circuit board inside. Currently, we try to acquire it in order to carry out verification by humidity data.

TABLE II.   RESULTS FROM TEST AND VERIFICATION

		Prediction with model	
		2006	2007–2015
Result in 2015	Failures	10 (TP)	91 (FN)
	No Failures	1685 (FP)	19,299 (TN)

TABLE III.   RESULTS FROM CONVENTIONAL METHOD

		Prediction with model	
		Failures	No Failures
Result in 2015	Failures	11 (TP)	90 (FN)
	No Failures	151 (FP)	20,833 (TN)

TABLE IV.   COMPARISON RESULTS OF CONVENTIONAL AND PROPOSED

	Conventional	Proposed
Accuracy rate	91.6%	98.9%
Precision rate	0.6%	6.8%
Recall rate	9.9%	10.9%

VI.   CONCLUSIONS

We described failure prediction, which is a technology for low-cost maintenance, and explained studies and developments being advanced with this technology. It is important to focus attention on external environments, such as temperature and humidity, and instantaneous and cumulative effects to predict power equipment failures. We therefore proposed a failure prediction model and maintenance plan that introduced these effects, and tested and verified the accuracy of the model. On the basis of the verification results, we suggested that the proposed method can be expected to further reduce facility cost compared with the conventional method. The results indicated that it is necessary to obtain higher precision and recall rates to implement the proposed method. We will attempt to improve these rates in future work by using external environmental data that are highly correlated with equipment failures.

REFERENCES

[1] Japan Electronics and Information Technology Industries Association,"Investigative report about predictive maintenance technology," http://home.jeita.or.jp/upload_file/20120628102027_elgD9Ed5tM.pdf, Jan. 2012.

[2] K. Wang, L. Wu, J. Tian, A. Xu, and M. Pecht, "A review of Prognostics and health management for electronics," Transactions of the Japan Society of Mechanical Engineers,Vol.118,No.1160, July 2015.

[3] H. Hayasaka, K. Kanetani, S. Nakashima, M. Yamazaki, and T. Babasaki, "Method for detection of lot defects for maintenance of ICT power supplies and air conditioning equipment and verification results," IEEE The 38th International Telecommunications Energy Conference, Oct. 2017.

[4] K. A. Helmy Kobbacy, and D. N. Prabhakar Murthy, "Complex System Maintenance Handbook," Springer Series in Reliability Engineering, 2008.

[5] Jack, N. "Age-reduction models for imperfect maintenance," IMA Journal of Management Mathematics, pp.347-354, 1998.

[6] NY.Cox, David R. "Regression models and life-tables," Breakthroughs in statistics. Springer, pp.527-541, 1992.

[7] Kijima, "Some results for repairable systems with general repair," Journal of Applied Probability, pp.89-102, 1989.

The 2018 International Power Electronics Conference

Hybrid Converters with Reduced Inductor Loss for Integratable Power Conversion

Gab-Su Seo[1,2] and Hanh-Phuc Le[1]

[1] Colorado Power Electronics Center (CoPEC), University of Colorado, Boulder, Colorado, United States

[2] Power Systems Engineering Center, National Renewable Energy Laboratory, Golden, Colorado, United States

E-mail: {Gabsu.Seo, hanhphuc}@colorado.edu

Abstract—This paper discusses hybrid converters that achieve high efficiency with an inductor positioned at the lower current path to reduce the DC component of inductor current and thus significantly decrease inductor loss. The concept also features reduced inductance requirement by decreasing the voltage swing blocked by the inductor. These improvements bring benefits in both higher efficiency and better integration. The hybrid architecture exploits less voltage stress on switches for switching loss reduction and switching frequency increase. As a result, low-voltage-rated switches with better figure-of-merit can be used, and passive component size reduced. The hybrid converters can be realized for both step-down and step-up power conversion as well as bidirectional power flow. For simplicity and cost, some of switches can be replaced with passive switches such as diodes that simplifies the converter circuit implementation. To explore different applications, a possible extension of an original hybrid topology is discussed as well as utilization of power delivery cable as power conversion component and techniques for soft-switching and soft-charging switched capacitor operations.

Keywords—*inductor loss reduction, hybrid converter, S-Hybrid converter, switched capacitor.*

I. INTRODUCTION

Buck converter topology has been widely adapted in various DC-DC conversion applications due to its simplicity and higher efficiency than linear regulators that step down input voltage to output voltage by burning out power on a controllable resistor. In fact, Buck converter is extensively used in personal computer and server applications [1], [2]. From a bus voltage, e.g. 12 V, Buck converters supply power to loads, such as microprocessors. Buck converter is used in applications for mobile devices to charge Li-ion batteries at ~3.0 V to 4.5 V from a 5 V input.

As shown in Fig. 1(a), a Buck converter has a switch, a current rectifying diode, an inductor and output capacitor. In a more popular variation, the converter can have a switch, in place of diode, that is operated synchronously with the switch Q and form a half bridge implementation for synchronous rectification in order to achieve better efficiency. However, in some applications that require better integration and efficiency, conventional Buck converters are not suitable due to two drawbacks. The first drawback is the relatively large voltage swing across the inductor as the switch node v_x swings between zero to

input voltage V_{IN} as shown in Fig. 1(b). This large swing of v_x requires an inductor having a larger inductance value to achieve low inductor current ripple. In the same form factor, an increase in inductance leads to increased equivalent series resistance (ESR) of the inductor and thus increased copper loss. In other words, higher inductance requirement results in larger inductor size for high efficiency leading to lower power density. As the inductor carries the average output current I_o in the Buck converter, the inductor copper loss is significant since it is proportional to the square of the current expressed as

$$P_{cond,Buck} = R_L I_{L,rms}^2 = R_L I_{o,rms}^2 \quad (1)$$

where R_L and $I_{L,rms}$ are ESR of the inductor and the root mean square (RMS) value of the inductor current. The second drawback of Buck converter is that switches need to block the input voltage V_{IN}. This imposes the need for high breakdown voltage devices, which leads to either higher switch on-resistance or parasitic capacitance given the same form factor. This contributes to efficiency decrease of the switching regulator.

These two drawbacks of the conventional Buck converter are effectively addressed by a three-level Buck converter, such as illustrated in Fig. 2(a). By reducing the voltage swing at v_x to one half of the input voltage and doubling the effective inductor switching frequency because of the interleaving effect, the reliance on inductance is significantly reduced. As a result, the three-level Buck converter can enjoy improved efficiency and better integration as pointed out in [3] and [4]. In addition, by dividing the input voltage equally into two capacitors C_1 and C_2 as shown in Fig. 2(b), the voltage sustained by switches Q_1 and Q_2 and diodes D_1 and D_2 is one half of that of conventional Buck converter. This voltage stress reduction is beneficial to alleviating switching loss of semiconductor devices because the loss is expressed as

$$P_{switch} = \frac{1}{2} C_{DS} V_{DS}^2 f_s \quad (2)$$

where C_{DS}, V_{DS}, and f_s are the equivalent drain-to-source capacitance, voltage applied to a switch during the off times, and switching frequency. Therefore, three-level Buck converter could achieve better efficiency and shows better potential for integration.

In spite of the advantages in lower inductor voltage swing and semiconductor voltage stress, three-level Buck

The 2018 International Power Electronics Conference

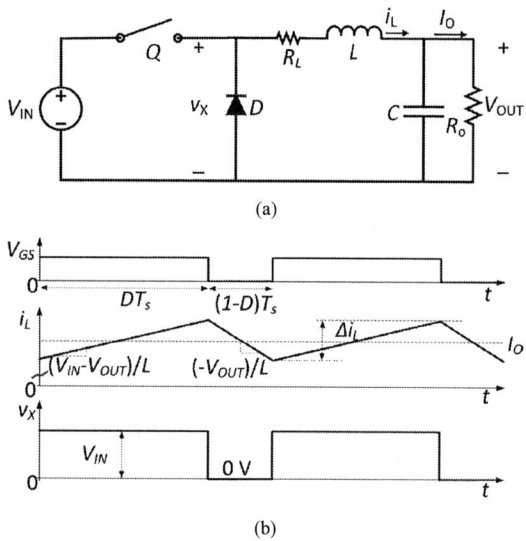

(a)

(b)

Fig. 1.   Buck converter: (a) converter schematic, and (b) operation.

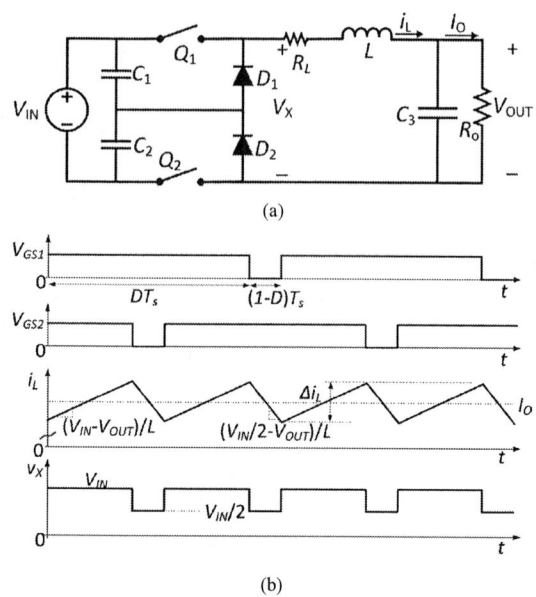

(a)

(b)

Fig. 2.   Three level Buck converter: (a) converter schematic, and (b) operation (in case of D>0.5).

converter has several demerits. First, even though three-level Buck converter can alleviate the inductor loss by reducing the required inductance, i.e. less copper loss with decreased ESR, it still undesirably has to carry the full output current and the related loss. Second, the minimum number of active switches required to operate the circuit is two, twice that of a conventional Buck converter. Three-level Buck converter needs one additional active switch to ensure the charge balance between C_1 and C_2 [5], increasing design complexity [6].

To overcome the limitations of switched inductor-based converters, switched-capacitor (SC) converters have been introduced [7]. SC converter features easier integration with only capacitors and no magnetic components. SC converter achieves high efficiency when the output voltage is a predetermined fraction of input voltage e.g. 1/2, 1/3, and 2/3. However, as the voltage conversion ratio changes, its efficiency decreases since it operates like a fixed conversion ratio linear regulator [8], [9].

To overcome the drawback of SC converters and Buck converters, a hybrid converter that cascades the two kinds of converters have been proposed [10]. Using the hybrid structure, the output voltage can be tightly regulated with reasonable efficiency. However, the increased number of components, particularly the number of switches, makes the converter circuitry more complex. In addition, efficiency degradation due to the cascaded power stages is not desirable.

In this paper, a new class of hybrid converters, named *S-Hybrid* converter, with an inductor positioned at the lower current path that significantly decreases inductor loss is presented. By reducing the voltage swing of the inductor switching node, the converter further reduce the

reliance of inductor or increase the converter efficiency. As the converter can be used not only for step-down power conversion but also step-up, it can be used for various applications including bidirectional circuits.

Section II discusses converter topologies for inductor loss and size reduction, recognizing their benefits and limitations, and presents S-Hybrid architecture with its derivation and technical aspects. Topology extension of the S-Hybrid circuits and techniques for further performance improvement is delivered in Section III.

II.   CONVERTERS WITH REDUCED INDUCTOR LOSS

As discussed in Section I, it is challenging to effectively relieve the reliance of a power conversion circuit on inductor which is one source of difficulties in converter thermal management and/or integration for high power density. Three-level Buck converter effectively improves this by reducing the voltage swing and doubling the effective switching frequency. However, the inductor is still positioned at the output similar to Buck converter, which forces it to carry full output current limiting the advantage. While this paper is focused on step-down conversion using Buck type circuits, the same argument can be made for step-up using a Boost type converter where the inductor (shorted to the input) experiences high voltage swing (zero to output voltage) and conducts high current (input current in this case), resulting in a poor switch utilization [11]. Therefore, a new converter architecture effectively relieving inductor reliance is desirable for better integration, thermal management, and higher power density.

214

A. Step-down Converter with Inductor Relocated at Front

In this new step-down converter architecture, the inductor is relocated to the input. Compared with the output position in Buck-type topologies, this position of the inductor is more beneficial since it allows the inductor to handle only a fraction of output current. Therefore, the inductor DC loss is reduced to

$$P_{cond,S-Hybrid} = R_L I_{L,rms}^2 = M^2 R_L I_{o,rms}^2 \qquad (3)$$

where $M = \frac{V_{OUT}}{V_{IN}}$, input-to-output voltage conversion ratio, and $M<1$ for step-down conversion, yielding M^2 less DC inductor loss.

Figure 3(a) depicts a simplified circuit to realize the new step-down converter. While the circuit provides a critical idea to develop a fully working hybrid topology, it should be noted that it needs additional circuitry to properly operate in steady state. To describe the operation, the two capacitor voltages v_{C1} and v_{C2} are assumed to be maintained equal, $v_{C1} = v_{C2} = v_{OUT}$. When Q_1 turns on, the inductor is charged by (V_{IN}-v_{OUT}) while it is discharged by ($2 \cdot v_{OUT}$ -V_{IN}) via Q_2 during the other phase. By the inductor voltage-second balance, the converter input-to-output voltage conversion ratio can be derived as

$$M_{S-Hybrid} = \frac{1}{2-D} \qquad (4)$$

where D is defined as the duty cycle of Q_1. The converter provides a step-down conversion with an inductor at the input side with less inductor DC loss. The inductor switching voltage is reduced to V_{OUT} from V_{IN} as illustrated in Fig. 5, reducing the reliance on inductor. In thi simplified circuit, however, the two capacitor voltages cannot be maintained equal since their charge-balance are not satisfied; specifically, capacitor C_1 only receives charge from the inductor but has no discharge phase. Therefore, the circuit needs an additional circuitry that will re-distribute the charge from C_1 and make sure the two capacitor voltages are maintained equal. For this purpose, an effective voltage equalizer is necessary.

A flying capacitor balancer, similar to one used in an active battery cell balancing circuit [12] is added in the complete circuit in Fig. 3(b). The circuit can be divided into two functional sub-circuits: 1) a flying-inductor step-down power conversion sub-circuit that processes power from source V_{IN} to load and 2) a voltage balancing sub-circuit that redistributes and balances the charge of C_1 to C_2. With the flying capacitor C_{fly} periodically forwarding the charge on C_1 to C_2, the two capacitor voltages can be maintained equal. Once the balancer works properly, the entire circuit can regulate output voltage by controlling the duty cycle of Q_1.

As illustrated in Fig. 3(b), compared to the conventional Buck converter, this configuration achieves power conversion at the expense and complexity of three active switches and two additional capacitors. In case of using synchronization switches for higher efficiency, six switches are required. Despite the advantages on inductor, it would be challenging to justify the increased complexity

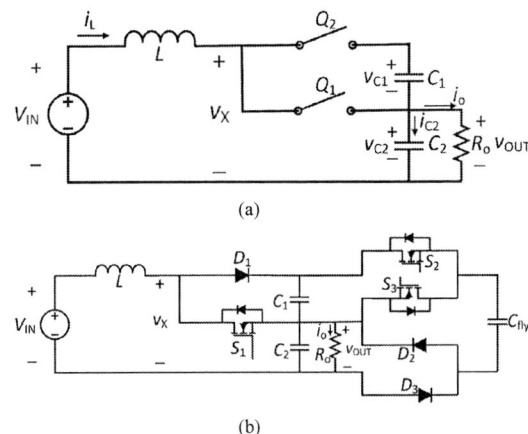

(a)

(b)

Fig. 3. Hybrid converter to allieviate the inductor reliance: (a) concept circuit with two identical voltage capacitors, and (b) complete circuit example with a ladder type voltage balancer.

and to manage the trade-off between switching frequency of balancer and passive component size.

B. S-Hybrid Converter

A recent approach, S-Hybrid converter architecture [13], shown in Fig. 4, effectively addresses the drawbacks in Buck-type converters and eliminates the need of voltage balancer of the inductor-front circuit in Fig. 3(a). In series with the inductor, the network of C_1, Q, D_1, and D_2, is operated to switch the switching node voltage between V_{OUT} and $2 \cdot V_{OUT}$ similar to the previous six switch circuit in Fig. 3(b) while requiring only one active switch Q to control the output voltage by modulating the duty cycle of Q. This basic S-Hybrid converter has exactly the same voltage swing at the switching node v_X as illustrated in Fig. 5. The S-Hybrid converter allows two capacitor voltages to be periodically equalized by nature with only one additional diode compared to conventional Buck converter. Similar to other topologies, diodes can be replaced with synchronous switches for high efficiency and/or reverse or bidirectional power flow.

By the inherent balancing operation, the converter achieves the same step-down conversion ratio in (4), featuring reduced switching voltages and lower inductor DC loss. These characteristics are beneficial for future integrations targeting more compact implementation. In addition, the voltage stress on switches is also reduced to v_{OUT} that is advantageous to utilizing low voltage device with better figure of merit, e.g. switches with the same on-resistance but smaller parasitic capacitance that enables higher frequency operation with high efficiency.

III. EXTENSION OF S-HYBRID CONVERTER

A. Further Step-down (Step-up) Conversion

Theoretical conversion ratio of the basic S-Hybrid converter as indicated in (4) is limited to 0.5 to unity, which is half of Buck converter. While its potential

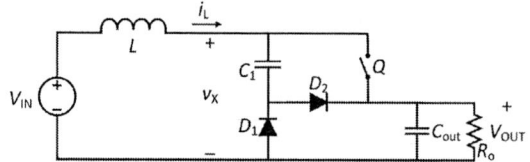

Fig. 4.   Schematic of S-Hybrid converter (N=1).

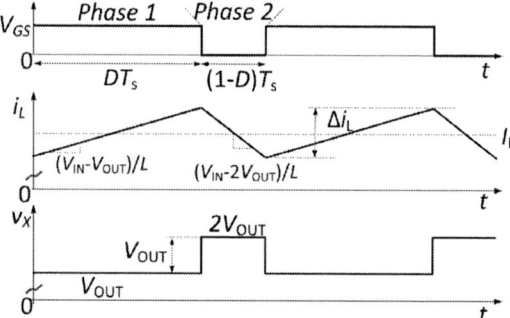

Fig. 5.   Operation of S-Hybrid converter (N=1).

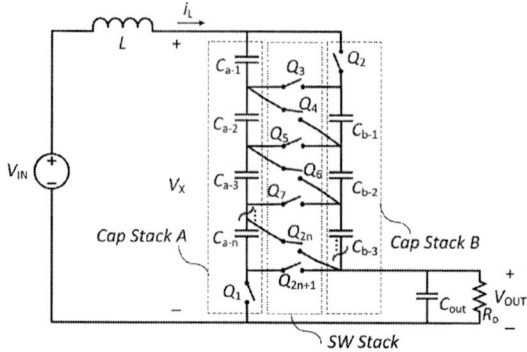

Fig. 6.   Generalized S-Hybrid converter to accommodate various conversion ratio with ladder type swtiched-capacitor network.

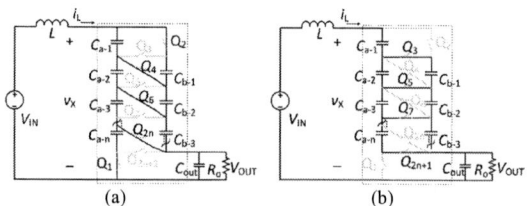

(a)   (b)

Fig. 7.   Circuit operation of generalized S-Hybrid converter: (a) inductor charges in phase 1, (b) inductor discharges in phase 2.

applications can be found matching the operable range, e.g. single cell battery charger from 5 V, it is desirable to identify possible extension of the topology for a wider range of applications.

To identify its extendibility, the concept of S-Hybrid converter can be generalized as illustrated in Fig. 6. It should be noted that the generalization is not the only way but many different generalized forms can be found using different switched-capacitor networks, which will be discussed later. Retaining the key concept of positioning the inductor at the lower current side and using synchronous switched-capacitor network to lower the inductor switching voltage, the basic S-Hybrid converter in Fig. 4 can be extended to accommodate different voltage conversion ratio. As dotted boxes highlight, the switched-capacitor network following the inductor can be divided into three parts: capacitor stack A consisting of N capacitors and Q_1 in series, capacitor stack B containing another series of (N—1) capacitors and Q_2, and ($2N$—1) switch stack interconnecting the two capacitor stacks. Therefore, the SC network contains ($2N$—1) capacitors and ($2N$+1) switches in total. Through this generalization, the three switch S-Hybrid circuit in Fig. 4 has N=1. The fundamental operation principle includes two phases, as shown in Fig. 7(a) and (b) and Fig. 8. During each of these phases, the inductor is either charged by $N \cdot v_{OUT}$ or discharged by (N+1)$\cdot v_{OUT}$ from V_{IN}, provided that V_{IN} is in between $N \cdot v_{OUT}$ and (N+1)$\cdot v_{OUT}$. Similarly, by voltage-second balance, the input-to-output voltage conversion ratio can be derived as

$$M_{S-Hybrid,ladder} = \frac{1}{(N+1)-D}.$$ (5)

By limiting the inductor voltage swing to v_{OUT}, the inductor current ripple can be effectively minimized. It can be derived as a function of duty cycle and other parameters

as

$$\Delta i_{L,S-Hybrid} = \frac{V_{IN}T_s}{L} \frac{(1-D)D}{(N+1-D)}$$ (6)

where T_s is switching period, while the inductor current ripple of Buck converter is expressed as

$$\Delta i_{L,Buck} = \frac{V_{IN}T_s(1-D)D}{L}.$$ (7)

Fig. 9 illustrates these current ripples normalized to the peak inductor current ripple of Buck converter at the conversion ratio of 0.5. In the figure, the ripple values are re-constructed as a function of conversion ratio M for comparison between topologies. Significant current ripple reduction as well as the conversion ratio range limit with different N values is observed.

It would be noteworthy that the generalized S-Hybrid circuit structure shown in Fig. 6 is constructed by a modified ladder topology [14]. The ladder switched-capacitor converters, similar with other switched-capacitor topologies, have to sacrifice significant efficiency drop to regulate output voltage off the predetermined value [8]. By having an inductor at the input rather than a switch, the switched capacitor converter becomes duty-cycle-controllable with high power conversion efficiency.

While the ladder-based S-Hybrid converter shows significantly reduced switching node voltage swing, it has relatively large number of capacitors and switches leading to high complexity [15]. To better trade-off the complexity and other aspects of circuit such as inductor voltage swing, other switched-capacitor topologies can be used to construct the switched-capacitor network, e.g. series-parallel, Dickson charge pump, or Fibonacci [16]. With different switched-capacitor network, the circuit can

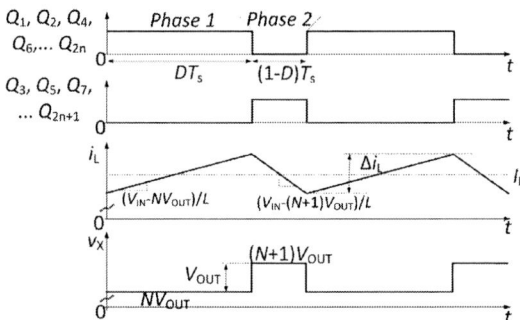

Fig. 8.   Timing diagram of ladder type S-Hybrid converter operation.

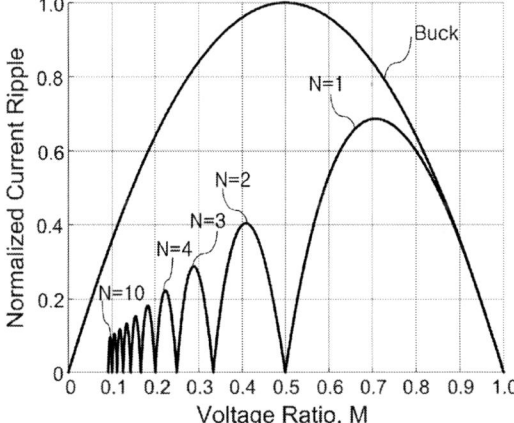

Fig. 9.   Inductor current ripple comparison of S-Hybrid converter to Buck converter.

achieve less complexity at the cost of increased inductor voltage swing and/or increased loss.

S-Hybrid topology can also be used for step-up and bidirectional power conversion by employing synchronous switches and interchanging input and output. As the benefit of the circuit still remains, the converter would be beneficial in applications where relieving the reliance on inductor is critical for high performance.

B. Utilization of Input or Output Cable Impedance

To complete an electrical power system, most application involves cables for power delivery. That led to discussion on utilizing cable as a power conversion component [13], [17]. Together with Buck converter that has the inductor at the output, S-Hybrid converter with its inductor at the input completes the possibility of replacing discrete inductors by parasitic inductance of input or output cable in step-down applications. The complementary topologies of these two, i.e. Buck converter and S-Hybrid step-up, can cover step-up conversions.

An inductor-less S-Hybrid converter was demonstrated to provide an efficient battery charger for mobile applications [13]. Removing inductor off the circuit board not only removes bulky magnetic but also leads to

significantly reduced thermal concern of inductor. It is reported that the on-board loss can be reduced by 45% compared to Buck converter solution for a 3.3V~3.8V single cell Li-ion battery charging application from 5V input.

In input cable utilization, variance in cable inductance should be considered if the system is not hard-wired but consists of mobile and exchangible cable such as USB powered system where cable length and configuration can vary. It is critical to ensure the system stability and dynamic performance with various cables that nececitates dynamic behavior modeling of the converter such as small signal modeling [18]. In some cases, it would be required to limit the type of cable used or a small inductor can be embeded in the power conversion circuitry to guarantee circuit operation in a worst case scenario.

In addition, electromagnetic interference (EMI) issues that could be caused by high-frequency current ripple in the cable should be considered. Shielding the cable would be beneficial at the cost of reduced equivalent inductance and increased parasitic capacitance.

C. Additional Techniques for Improvement

To further improve the system performance, several additional techniques can be utilized. Zero-voltage-switching technique can be used to reduce the switching loss [19]. With additional resonant circuitry, S-Hybrid circuit would achieve soft-switching. Boundary-conduction mode (BCM) can also be employed to achieve zero-voltage switching utilizing the resonance between inductor and drain-to-source capacitance of power switches without adding additional component [20].

Switched-capacitor topologies suffer from hard-charging capacitor loss which occurs when two or multiple capacitors with different voltages are connected in parallel [21]. S-Hybrid converter mitigates hard-charging loss since the inductor soft-charges capacitors in one phase (during $(1-D)T_s$ in Fig. 5) incuring no hard-charging loss. However, hard-charing in the capacitors still occurs in the beginning of phase 1 (during DT_s in Fig. 5). Achieving complete soft-charging capacitor operation would enable higher power density by better utilizing passive components and maintaining high efficiency as shown in [22]. Therefore, it would be desirable to add passive and/or active components to the S-Hybrid topology or to develop a new hybrid topology to completely eliminate the hard-charging loss.

IV. CONCLUSIONS

This paper has presented S-Hybrid architecture that relocates the inductor to the lower current side. With the inductor at low current side and reduced switching node voltage swing, the converter significantly reduces its reliance on inductor with lower inductor DC loss and smaller required inductance. The topology can be extended to cover different applications while retaining its key benefit. By properly addressing the possible issues in using cable, the circuit can contribute to high power density and efficiency with superior thermal management.

With additional techniques to further reduce the converter loss such as hard-switching loss and hard-charging loss, the converter performance can be improved and utilized for various applications.

ACKNOWLEDGMENT

The authors wish to acknowledge the students, faculty and sponsors of the Colorado Power Electronics Center (CoPEC), and the support of Texas Instruments and the University of Colorado Boulder. The authors would like to especially thank Dr. Dragan Maksimovic for helpful discussions and support.

REFERENCES

[1] P. Xu, J. Wei, and F. C. Lee, "Multiphase coupled-buck converter-a novel high efficient 12 V voltage regulator module," *IEEE Trans. Power Electron.,* vol. 18, no. 1, pp. 74-82, Jan. 2003.

[2] Y. Panov and M. M. Jovanovic, "Design considerations for 12-V/1.5-V, 50-A voltage regulator modules," *IEEE Trans. Power Electron.,* vol. 16, no. 6, pp. 776-783, Nov. 2001.

[3] W. Kim, D. Brooks, and G. Y. Wei, "A Fully-Integrated 3-Level DC-DC Converter for Nanosecond-Scale DVFS," *IEEE J. Solid-State Circuits,* vol. 47, no. 1, pp. 206-219, Jan. 2012.

[4] V. Yousefzadeh, E. Alarcon, and D. Maksimovic, "Three-level buck converter for envelope tracking applications," *IEEE Trans. Power Electron.,* vol. 21, no. 2, pp. 549-552, Mar. 2006.

[5] X. Ruan, B. Li, Q. Chen, S.-C. Tan, and C. K. Tse, "Fundamental considerations of three-level dc–dc converters: Topologies, analyses, and control," *IEEE Trans. Circuits Syst. I,* vol. 55, no. 11, pp. 3733-3743, Dec. 2008.

[6] Z. Ye, Y. Lei, Z. Liao, and R. C. N. Pilawa-Podgurski, "Investigation of capacitor voltage balancing in practical implementations of flying capacitor multilevel converters," in *Proc. IEEE Workshop on Control and Modeling for Power Electronics (COMPEL),* 2017, pp. 1-7.

[7] S. V. Cheong, H. Chung, and A. Ioinovici, "Inductorless DC-to-DC converter with high power density," *IEEE Trans. Ind. Electron.,* vol. 41, no. 2, pp. 208-215, Apr. 1994.

[8] V. W.-S. Ng, "Switched capacitor DC-DC converter: Superior where the buck converter has dominated," Ph. D. Dissertation, University of California, Berkeley, 2011.

[9] H. P. Le, M. Seeman, S. R. Sanders, V. Sathe, S. Naffziger, and E. Alon, "A 32nm fully integrated reconfigurable switched-capacitor DC-DC converter delivering 0.55W/mm2 at 81% efficiency," in *Proc. IEEE International Solid-State Circuits Conference (ISSCC),* 2010, pp. 210-211.

[10] R. C. Pilawa-Podgurski and D. J. Perreault, "Merged two-stage power converter with soft charging switched-capacitor stage in 180 nm CMOS," *IEEE J. Solid-State Circuits,* vol. 47, no. 7, pp. 1557-1567, Jul. 2012.

[11] S. R. Sanders, E. Alon, H.-P. Le, M. D. Seeman, M. John, and V. W. Ng, "The road to fully integrated DC–DC conversion via the switched-capacitor approach," *IEEE Trans. Power Electron.,* vol. 28, no. 9, pp. 4146-4155, Sep. 2013.

[12] J. Cao, N. Schofield, and A. Emadi, "Battery balancing methods: A comprehensive review," in *Proc. IEEE Vehicle Power and Propulsion Conf.,* 2008, pp. 1-6.

[13] G. S. Seo and H.-P. Le, "An inductor-less hybrid step-down DC-DC converter architecture for future smart power cable," in *Proc. IEEE Applied Power Electronics Conference and Exposition,* 2017, pp. 247-253.

[14] T. Van Breussegem and M. Steyaert, *CMOS integrated capacitive dc-dc converters*: Springer Science & Business Media, 2012.

[15] M. D. Seeman and S. R. Sanders, "Analysis and optimization of switched-capacitor DC–DC converters," *IEEE Trans. Power Electron.,* vol. 23, no. 2, pp. 841-851, Mar. 2008.

[16] G. S. Seo and H.-P. Le, "Hybrid Converter with Reduced Inductor Loss," PCT/US2017/035282, Jul. 12, 2017.

[17] R. Goel, G. S. Seo, and H. P. Le, "A smart-USB-cable buck converter with indirect control," in *Proc. IEEE Workshop on Control and Modeling for Power Electronics (COMPEL),* 2017, pp. 1-6.

[18] G. S. Seo and H. P. Le, "Small-signal analysis of S-hybrid step-down DC-DC converter," in *Proc. IEEE Workshop on Control and Modeling for Power Electronics (COMPEL),* 2017, pp. 1-6.

[19] K. H. Liu and F. C. Y. Lee, "Zero-voltage switching technique in DC/DC converters," *IEEE Trans. Power Electron.,* vol. 5, no. 3, pp. 293-304, Jul. 1990.

[20] G. S. Seo, J. W. Shin, B. H. Cho, and K. C. Lee, "Digitally controlled current sensorless photovoltaic micro-converter for dc distribution," *IEEE Trans. Ind. Inform.,* vol. 10, no. 1, pp. 117-126, Feb. 2014.

[21] H.-P. Le, S. R. Sanders, and E. Alon, "Design techniques for fully integrated switched-capacitor DC-DC converters," *IEEE J. Solid-State Circuits,* vol. 46, no. 9, pp. 2120-2131, Sep. 2011.

[22] Y. Lei, R. May, and R. Pilawa-Podgurski, "Split-Phase Control: Achieving Complete Soft-Charging Operation of a Dickson Switched-Capacitor Converter," *IEEE Transactions on Power Electronics,* vol. 31, no. 1, pp. 770-782, 2016.

The 2018 International Power Electronics Conference

Energy saving system trend
for harbor crane with lithium ion battery

Hidemasa Yoshihara
System Technology Dept. Engineering and Technology Div.,
Yaskawa Siemens Automation and Drives Corp., Yukuhashi, Japan
*E-mail: hidemasa.yoshihara@ysad.co.jp

Abstract— **This paper proposes several kinds of energy saving system for harbor crane such as Rubber Tyred Gantry Crane (RTGC) and Straddle Carrier (SC). Harbor crane carry the cargo not only horizontal direction but also vertical direction. Especially for vertical direction operation, power demand becomes greater in hoisting up a cargo. On the other hand, big regenerative power can be back in lowering a cargo. Even if big regenerative power is generated by motor so that RTGC and SC are powered by diesel engine generator, it is regenerated to not power source but braking resistor. Hence it can be reduced fuel consumption of diesel engine by utilizing its regenerative energy. To utilize regenerative energy it requires energy storage system. The proposed system adopts lithium ion battery (LiB) and its combined system is separated into three kinds of concept. Their proposed system accomplished 25 - 60% fuel consumption cut.**

Keywords— *Fuel Saving, Harbor Crane, Hybrid sytem, Lithium Ion Battery*

I. INTRODUCTION

The machine energized by fossil fuel is improving as reducing fuel consumption in many fields. In case that machine is constructed by drive system such as diesel engine generator, converter, inverter and motor, there are two kinds of method to improve the efficiency broadly. One is to consider each drive component efficiency [1-2], another is to consider total system efficiency [3-4]. In the present situation, harbor crane also has an identity of the target and requires common technology.

RTGC and SC are a kind of yard crane. In RTGC yard, quayside crane (herein after QC) perform loading/ unloading the container between ship and chassis. Chassis moves the container between quayside and container stacking area. Then RTGC perform mounting/offloading/shuffling the container between chassis and container stacking area. On the other hand, in SC yard, QC performs loading/unloading the container between ship and ground. SC moves the container between quayside and container stacking area and chassis, then mounting/offloading/shuffling the container to container stacking area or chassis by itself [5].

Typical system construction of RTGC is shown in Fig.1. RTGC figure is shown in Fig. 2. Main motor drives of RTGC are Hoist, Trolley and Gantry. These drives are powered by diesel engine generator. Generator output is 3 phase AC power. AC power is converted into DC power by diode rectifier. DC power energized each inverters, input of each inverter is connected each other, so called DC link. Each inverter supplies the power to each motor; Hoist, Trolley and Gantry. Hoist is vertical operation of the container or only spreader. Spreader is tool of lifting up the container. Trolley is a kind of a cart to move hanged spreader in horizontal direction. Gantry is horizontal operation of RTGC itself. Each inverter is given speed reference by PLC (Programmable Logic Controller), and controls each motor. In case that motors regenerate the power during lowering and decelerate operation of horizontal movement, the regenerative power returns to DC link, braking resistor works over set DC voltage, then braking resistor consumes the regenerative power.

TABLE I shows capacity of equipment about RTGC. Each motor peak power is roughly 1.5 times of rated capacity. And hoist can be operated with trolley simultaneously, then maximum load power of RTGC reaches over 350kW. Therefore typical diesel engine generator for RTGC can supply 400kW. However peak power last only 1 or 2 second. This period is less than 5% of 1 cycle. Thus typical RTGC requires diesel engine with big capacity for short duty.

Typical SC system is similar configuration except for trolley. SC figure is shown in Fig. 3. SC has no trolley drives. Each motor requires 1.5 times power of rated capacity. TABLE II shows capacity of equipment about SC.

In this paper, three kinds of energy saving systems with LiB are proposed. These systems are as followings; Hybrid RTGC, Battery RTGC, and Hybrid SC.

All systems have diesel engine generator and LiB for power supply.

A. Hybrid RTGC

Hybrid RTGC uses diesel engine generator for prime power, secondary power is LiB. Bidirectional DC/DC converter is connected between DC link and LiB, and it can charge and discharge to LiB. Recently, this system becomes quite popular, also different energy storage can be used [6-8].

B. Battery RTGC

Battery RTGC can be operated by only LiB. When remained LiB energy reduced, minimized diesel engine generator starts to supply power to load and LiB.

C. Hybrid SC

Variable speed diesel engine generator energizes to Hybrid SC. Diesel engine for Hybrid SC changes its speed considering with fuel consumption. LiB assists lack of generator power in case that engine increases its speed or load power changes rapidly.

II. FEATURE OF CONVENTIONAL EQUIPMENT

A. Duty of Yard Crane

Fig.4 shows RTGC container yard layout. QC moves container between ship and yard chassis. After QC moves container from ship to yard chassis, yard chassis brings container to RTGC place, then RTGC picks it up and put in container stacking area. This is offloading operation of RTGC. In the view point of RTGC operation, there are 3 kinds of operation mode; mounting, offloading, and shuffling. Offloading and mounting are operation between container stacking area and chassis. Shuffling is operation inside container stacking area, it is mainly preparation for next mounting or offloading operation.

Fig.5 shows SC container yard layout. QC moves container between ship and ground. After QC moves container from ship to ground, SC brings container to container stacking area and put it on there. This is offloading operation of SC. SC operation also has three kinds of operation mode; mounting, offloading, and shuffling.

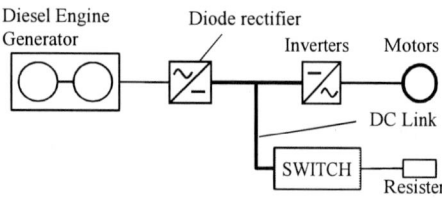

Fig. 1 Typical system construction of RTGC

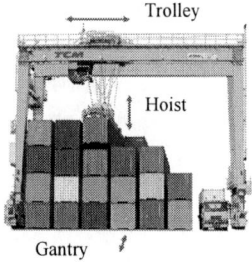

Fig. 2 RTGC figure

TABLE I
TYPICAL CAPACITY OF EQUIPMENT FOR RTGC

Equipment	Capacity[kW]
Hoist motor	200
Trolley motor	40
Gantry motor	120
Auxiliary load	40

Fig. 3 SC figure

TABLE II
TYPICAL CAPACITY OF EQUIPMENT FOR SC

Equipment	Capacity[kW]
Hoist motor	180
Gantry motor	150
Auxiliary load	20

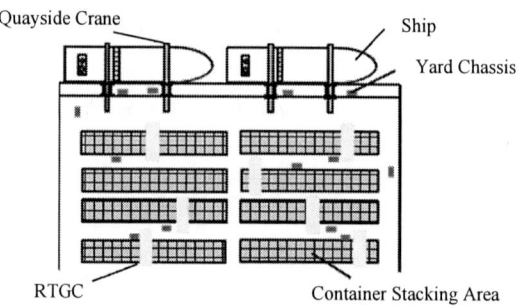

Fig. 4 RTGC container yard

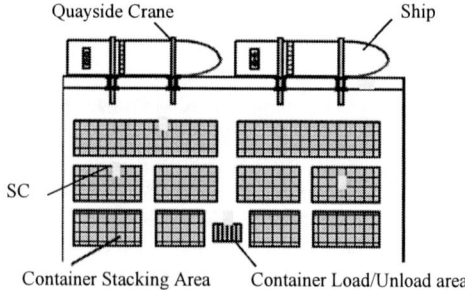

Fig. 5 SC container yard

B. Operation Pattern and Power Demand

Fig. 6 shows typical RTGC operation pattern by each step.
1. Hoist down with spreader
2. Hoist up with container
3. Trolley forward with container
4. Hoist down with container
5. Hoist up with spreader
6. Trolley backward with spreader

Spreader is tool to pick up container. Fig. 7 shows power demand time chart when RTGC works as like Hoist and trolley speed reference are given by trapezoid pattern with constant acceleration and deceleration. Hoist positive speed is hoist up, negative speed is hoist down. Trolley positive speed is forward, negative speed is backward. Constant speed is decided by container weight. Heavy container is slower than light container. Power demand of hoist is larger than power demand of trolley.

Fig. 8 shows typical SC operation pattern by each step.
1. Gantry with spreader
2. Hoist down with spreader
3. Hoist up with container
4. Gantry with container
5. Hoist down with container
6. Hoist up with spreader
7. Gantry with spreader

Fig. 9 shows power demand time chart when SC works as like Fig. 8. Hoist speed reference is as same as RTGC's one. Gantry speed reference is given by trapezoid pattern with variable torque limit in acceleration. So acceleration is not constant. Gantry positive speed is forward, negative speed is backward. Power demand of hoist is similar with power demand of gantry.

Hoist is important factor to consider power demand of RTGC. On the other hand, SC requires power demand consideration about gantry rather than hoist.

III. SYSTEM DESIGN WITH BATTERY

A. System Construction

Hybrid RTGC system construction is shown in Fig. 10. Hybrid RTGC consists of typical RTGC system with bidirectional DC/DC converter and LiB, and diesel engine generator can be replaced one third of the one. Current of bidirectional DC/DC converter is controlled by feedback control about DC link voltage and generator output power. LiB is selected to consider maximum current and continuous current, hi-rate type LiB is suitable for Hybrid RTGC.

Battery RTGC and Hybrid SC system construction is shown in Fig. 11. Battery RTGC and Hybrid SC consist of typical system with Step-up converter and LiB. Battery RTGC's diesel engine generator can be replaced less than one fourth of the one. According to LiB SOC (State of Charge), PLC controls diesel engine start and stop. When SOC is not low, RTGC is energized by only LiB and diesel engine is stop. When SOC becomes low, diesel engine starts up and generator starts to supply the power. Step-up converter controls DC link voltage with Step-up, voltage reference requires over LiB voltage to charge.

Hybrid SC's diesel engine functions in variable speed. Diesel engine changes own speed commanded by PLC. PLC sends engine speed reference to engine control unit (ECU), engine speed reference is considered load power. Generator output voltage and frequency is changed by engine speed. Step-up converter controls DC link voltage

with Step-up even if input ac voltage and frequency changes widely. According to engine speed feedback, PLC sends current limitation to Step-up converter.

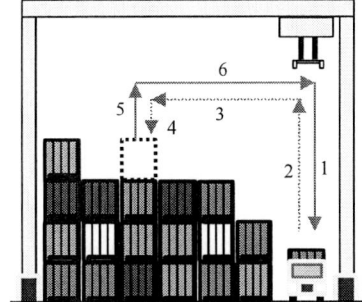

Fig. 6 typical RTGC operation pattern

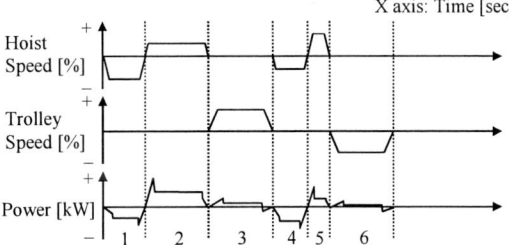

Fig. 7 RTGC power demand time chart

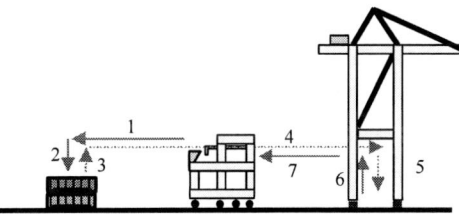

Fig. 8 typical SC operation pattern

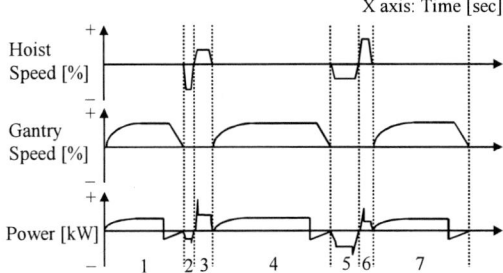

Fig. 9 SC power demand time chart

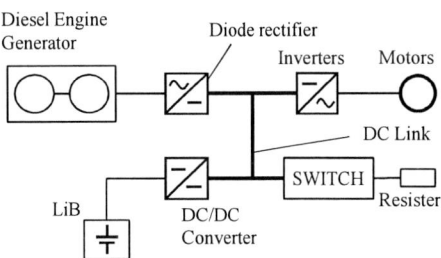

Fig. 10 System construction of Hybrid RTGC

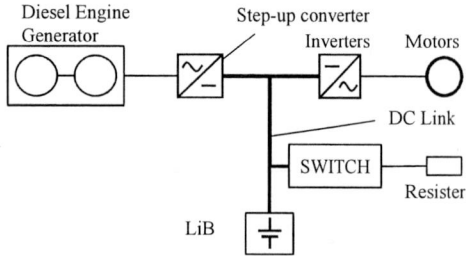

Fig. 11 System construction of Battery RTGC and Hybrid SC

B. Battery System Design

Battery system is designed by several steps as Fig. 12. In Fig. 12, "Select Crane Type" is to select Hybrid RTGC or Battery RTGC or Hybrid SC. Hybrid requires high current and doesn't require high energy. On the other hand, Battery RTGC requires high energy and doesn't require high current. So "Battery Type" becomes that hybrid requires high rate type battery, battery RTGC requires high energy type battery. From power source and load specification, battery system can be configured. In the next step, life time simulation is investigated with current cycle pattern in defined operation pattern. As a result, if life time estimation doesn't satisfy required life time, life time simulation will be required after increasing battery parallel number again. Fig. 13 shows how to decide battery series number. To decide battery series number, DC link voltage should be considered at initial state. Maximum DC link voltage is decided by AC input voltage or Step-up converter control voltage. When system doesn't use Step-up converter, AC input voltage is converted into DC voltage by diode rectifier. When system uses Step-up converter, DC voltage is decided by considering with and motor surge voltage. Thus, maximum DC link voltage is decided. On the other hand, minimum DC link voltage is decided by motor rated voltage. OCV (open circuit voltage) means voltage without any current flow as like open circuit. Fig. 14 shows battery equivalent circuit. Actually, maximum and minimum battery voltage requires voltage drop consideration with current flow. Relationship between battery OCV and battery terminal voltage can be written in

$$V_{lib\_terminal} = V_{lib\_OCV} + \frac{n_s}{n_p} \cdot r_{internal} \cdot I_{lib} \qquad (1)$$

where $V_{lib\_terminal}$ is the terminal voltage of LiB, V_{lib_OCV} is the OCV of LiB, $r_{internal}$ is the internal resistance of battery cell, I_{lib} is the input current of LiB, n_s is series number of battery cell, and n_p is parallel number of battery cell. $R_{internal}$ is total LiB internal resistance. Thus, maximum and minimum SOC toward maximum and minimum voltage can be decided in each series number of battery. Then, DOD (depth of discharge) can be gotten in each series number of battery. Finally, comparing with these DOD, series number of battery can be decided as DOD becomes widest.

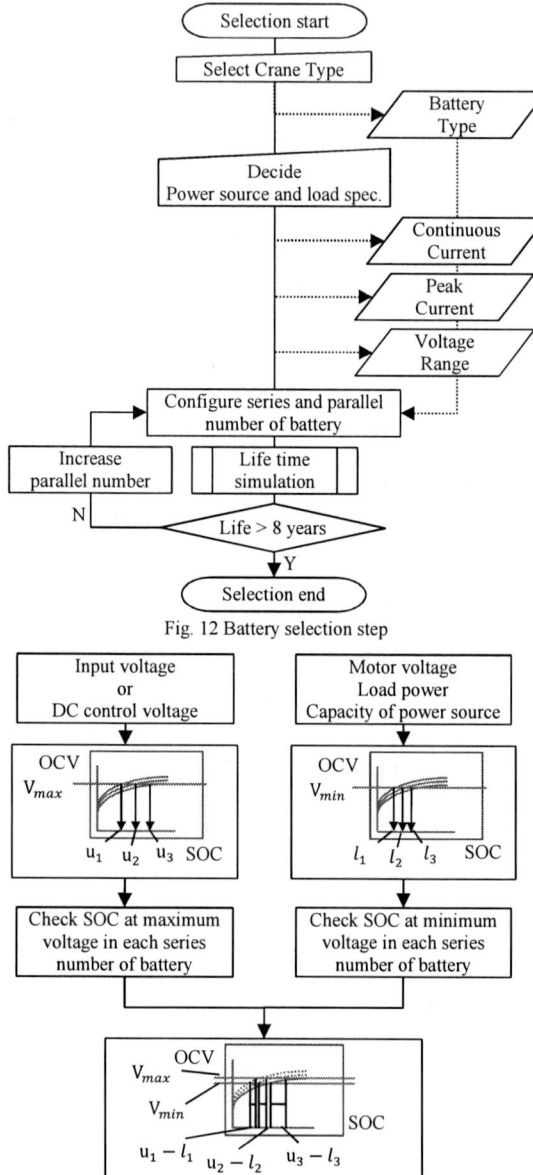

Fig. 12 Battery selection step

Fig. 13 Decision method of battery series number

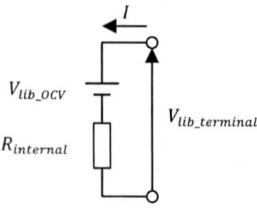

Fig. 14 Battery equivalent circuit

TABLE III
CONTROL DEVICE AND OBJECT

Equipment	Control Device	Control Object
Hybrid RTGC	DC/DC converter	AC power DC link voltage
Battery RTGC	Engine	AC power
Hybrid SC	Engine	Engine speed
	Step-up converter	Current

C. Control Device and Object

Table III shows control device and object. In Hybrid RTGC, it requires charge and discharge control of LiB to operate RTGC with small-sized engine generator. DC/DC converter performs charge and discharge to LiB. To avoid excessive supply request to engine generator, DC/DC converter has to be controlled considering with AC power supply. To charge regenerative power from motor, DC link voltage has to be considered. In Battery RTGC, it requires to control only battery mode and battery with engine generator mode. To control two modes, engine has to be controlled in ON or OFF considering with LiB SOC. In Hybrid SC, it requires charge and discharge control of LiB to operate SC with variable speed engine generator. To avoid excessive supply request to each speed engine generator, Step-up converter has to be controlled considering with AC power supply. To achieve AC power supply control, Step-up converter is controlled by current for each engine speed.

D. Control Flow

Fig. 15 shows Hybrid RTGC Control flowchart of changeover control mode for DC/DC converter. There are three control modes; power control, voltage control, and hybrid off. In power control mode, engine generator output power is controlled in power reference. Power reference is defined by constant value with accelerator. In voltage control mode, dc link voltage is controlled in voltage reference. Voltage reference is defined by constant value and it has to be set less than braking resister working voltage. In hybrid off mode, DC/DC converter never works and load works in limited speed. In this mode, engine generator supplies whole load power. When LiB SOC level is less than low level (LVL), DC/DC converter never works. When DC link voltage is over normal voltage of diode rectifier, it means that regenerative power goes back to DC link from motor. V_{dcH} is constant value to detect regenerative operation or not. Needless to say, when LiB SOC level is over high level (LVH), DC/DC converter doesn't work in charge direction.

Fig. 16 shows Battery RTGC Control flowchart of changeover control mode for engine. There are two control modes; engine use or no use. It's decided by LiB SOC level. When LiB has enough energy storage, only LiB energizes to load. Inversely, when LiB has not enough energy storage, LiB needs to charge from engine generator.

Fig. 17 shows Hybrid SC Control flowchart of changeover engine speed. There are three kinds of

commanding speed reference to engine. When LiB has not enough energy storage, engine speed reference becomes maximum speed so that LiB needs to charge.

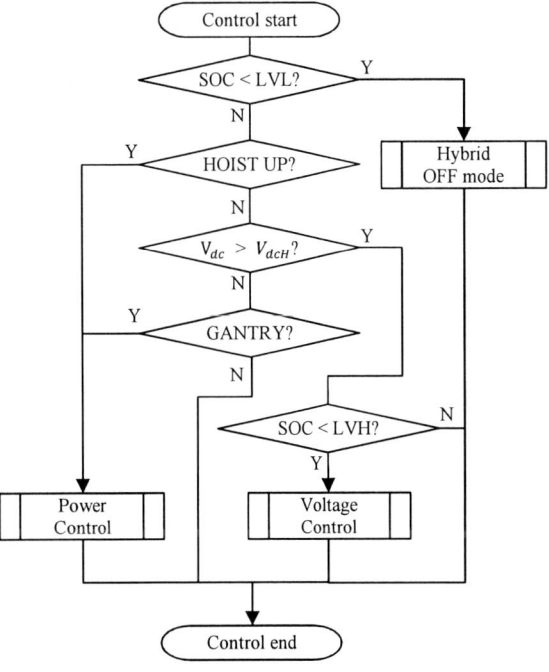

Fig. 15 Hybrid RTGC Control flowchart

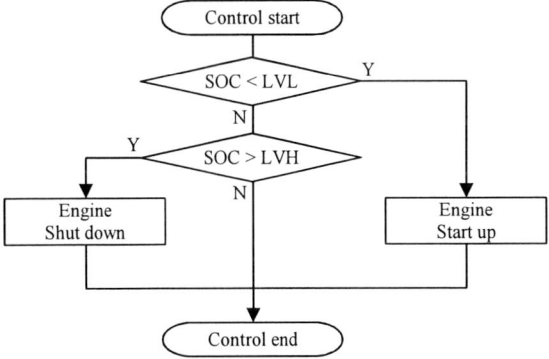

Fig. 16 Battery RTGC Control flowchart

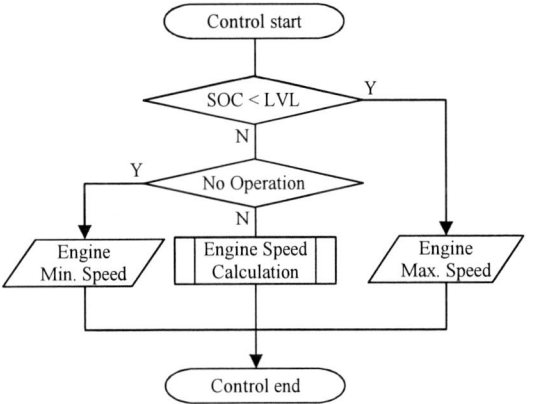

Fig. 17 Hybrid SC Control flowchart

During no gantry and hoist operation, engine can rotate in minimum speed to save fuel consumption. During gantry or hoist operation, engine is commanded calculated engine speed reference. Engine speed reference is decided by load power as Fig.18. Engine has limitation for power in each speed. When engine speed is low, engine can't supply big power. Therefore Hybrid SC requires not to request big power to engine generator while engine speed is not suitable for output power. In this system Step-up converter performs as like power limiter for engine generator with current limitation. Fig. 19 shows Step-up converter current limitation toward engine speed feedback.

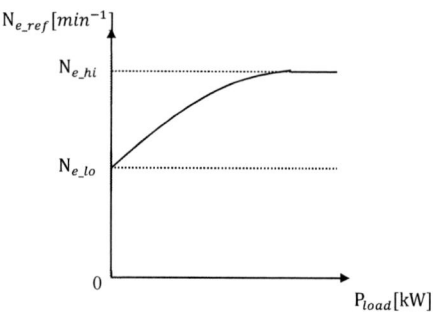

Fig. 18 Characteristics of load power and engine speed

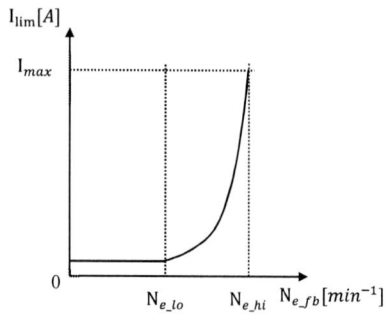

Fig. 19 Characteristics of engine speed feedback and current limit

IV. OPERATION DATA AND FUEL CONSUMPTION

A. Hybrid RTGC

Fig. 20 shows control data of Hybrid RTGC. V_{h_fb} is feedback of hoist motor speed, V_{dc} is feedback of DC link voltage, V_{lib} is LiB voltage, I_{lib} is current of DC/DC converter LiB side, P_{ac_ref} and P_{ac_fb} are reference and feedback about generator output power, SOC is percentage of charge about LiB, X axis is time by second and 1 division is 10 second. Positive of I_{lib} means charging, negative means discharging. Positive of V_{h_fb} means hoisting up, negative means lowering. (a) shows waveform of hoisting up with 40 ton container. (b)

shows waveform of lowering with only spreader. Spreader is around 10 ton. Hoist speed reference is generated in trapezoid pattern by PLC. During hoisting up, DC/DC converter discharges LiB energy to DC link with controlling generator output power in power reference sent by PLC. During lowering, DC link voltage increase voltage so that hoist motor is rotated by load weight and motor is operated as generator. DC/DC converter charges regenerative power to LiB with controlling DC link voltage so as not to increase over voltage reference sent by PLC.

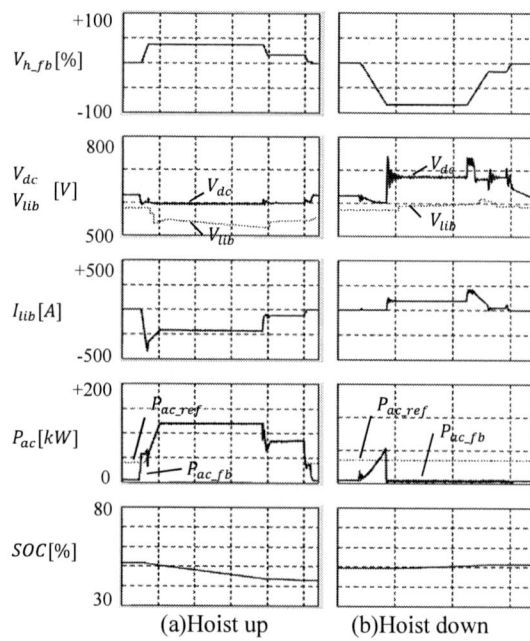

Fig. 20 Control data of Hybrid RTGC

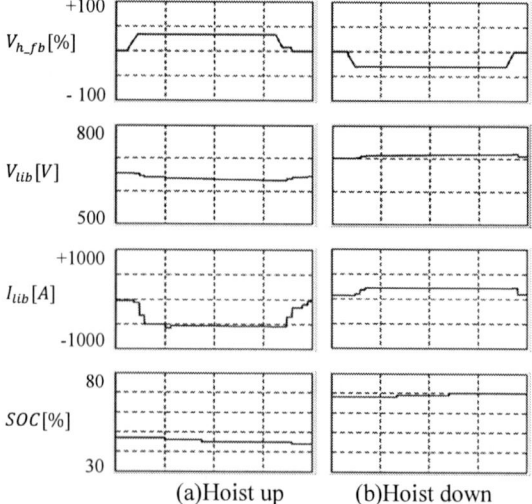

Fig. 21 Control data of Battery RTGC

B. Battery RTGC

Fig. 21 shows operation data of Battery RTGC. V_{h_fb} is feedback of hoist motor speed, V_{lib} is LiB voltage, I_{lib} is charge/discharge current of LiB, SOC is percentage of charge about LiB, X axis is time by second. (a) shows waveform of hoisting up with 30 ton container. (b) shows waveform of lowering with 30 ton container. V_{lib}, SOC, and I_{lib} are monitored value in battery management unit (BMU), and refreshing time is around 1 second. So waveform is not smooth.

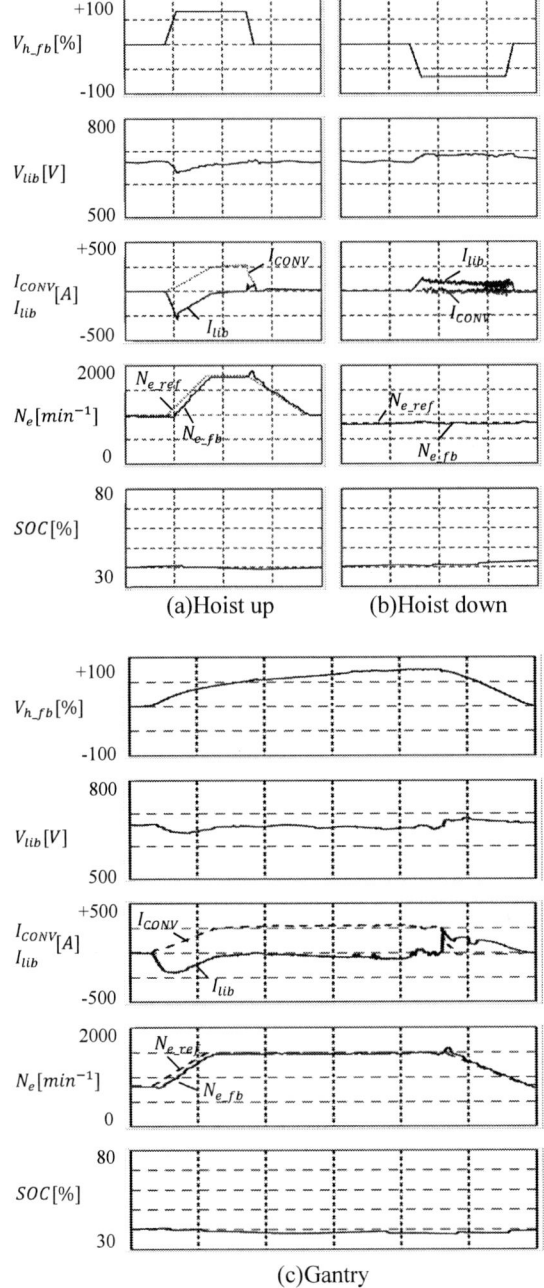

(a)Hoist up   (b)Hoist down

(c)Gantry

Fig. 22 Control data of Hybrid SC

C. Hybrid SC

Fig. 22 shows control data of Hybrid SC. V_{h_fb} and V_{g_fb} are feedback of hoist and gantry motor speed, V_{lib} is LiB voltage, I_{CONV} is feedback of converter current in DC link side, I_{lib} is charge/discharge current of LiB, N_{e_ref} and N_{e_fb} are reference and feedback of diesel engine speed, SOC is percentage of charge about LiB, X axis is time by second. (a) shows waveform of hoisting up with 40 ton container. (b) shows waveform of lowering with 40 ton container. (c) shows waveform of gantry with 40 ton container.

D. Fuel Consumption

Table IV shows fuel consumption data and its reduced rate toward conventional machine in real working RTGC. In case of RTGC, fuel consumption is related with engine capacity. The bigger engine capacity becomes, the more RTG consumes fuel. In case of SC, fuel consumption is related with engine speed. Generally speaking, the faster engine speed becomes, the bigger engine output becomes. But fuel consumption becomes worse in high engine speed.

TABLE IV
FUEL CONSUMPTION DATA

Equipment	Fuel Consumption [litter/h]	Reduced Rate [%]
Conventional RTGC	23.5	-
Hybrid RTGC	11.5	-51.1
Battery RTGC	9.5	-59.6
Conventional SC	27	-
Hybrid SC	20.3	-24.8

V. CONCLUSIONS

In this paper, three kinds of harbor crane system with Lithium ion Battery have been proposed. Each system has different solution. Hybrid RTGC requires intelligent DC/DC converter control to charge and discharge energy stored in LiB. Considering with RTGC operation, due to various load weight generative and regenerative power are unbalanced. In this situation, in case that quantity of battery assist is small, charge amount of regenerative power from motor will be more than amount of assist power from LiB, therefore LiB may not charge whole regenerative power. On the other hand, in case that quantity of battery assist is big, discharge amount of assist power from LiB will be more than amount of regenerative power from motor, therefore LiB may need to charge energy from engine generator. From a viewpoint of efficiency, one of key point is not to charge energy to LiB from engine generator as long as possible. Charging LiB and discharging LiB are including with heat loss of LiB and switching loss of DC/DC converter. And they are not small. So it's important to decide optimum assist power considering with LiB status. Most important thing for Battery RTG is to select most suitable LiB. Key points are wide range of DOD, low LiB internal resistance, and small margin for LiB current consideration. In case of Hybrid SC, it's important to how frequent engine can set low speed. In case that engine speed reduces frequently, LiB may have low voltage frequently. When LiB is low voltage, LiB requires to charge energy from engine generator and this situation makes fuel consumption worse. In consideration of these topics, Hybrid RTGC can cut fuel consumption to around 50% of conventional one, Battery RTG can cut fuel consumption to around 60% of conventional one and Hybrid SC can cut fuel consumption to around 25% of conventional one.

REFERENCES

[1] T. Kitamura, M. Yamada, S. Harada and M. Koyama, "Development of High-power-density Interleaved DC/DC Converter with SiC Devices," *IEEJ Trans. IA,* vol.134, No.11, pp.956-961, 2014.

[2] H. Tokoi, S. Kawamata and Y. Enomoto, "Study of High-Efficiency Motors Using Soft Magnetic Cores," *IEEJ Trans. IA,* vol.132, No.5, pp.574-580, 2012.

[3] M. Morimoto, "The First Hybrid Electric Vehicle in the World," *IEEJ Trans. IA,* vol.137, No.1, pp.69-74, 2017.

[4] N. Shiraki and K. Kondo, "Evaluation of Design Method for Engine Output and Battery Capacity for Lithium Ion-Battery Hybrid Diesel Railway Vehicles," *IEEJ Trans. IA,* vol.132, No.2, pp.178-184, 2012.

[5] R. Naicker and D. Allopi, "Evaluating Straddle Carriers And Rubber TyredGantrys To Determine Which Would Be The Most Suitable Container Handling Infrastructure Between The Quay And Stack Area At The Durban Container Terminal; Pier2," *IJISET.* vol.2, Issue 5, pp.134-142, 2015.

[6] S.M. Kim and S.K. Sul, "Control of Rubber Tyred Gantry Crane With Energy Storage Based on Supercapacitor Bank" *IEEE Trans. on Power Electronics*, vol.21, No.5, pp.1420-1427, 2006.

[7] H. Yoshihara, "Development of Energy-Saving type crane system," *IEEJ Trans. IA,* vol.MID-09, No.14-21, pp.15-20, 2009.

[8] H. Yoshihara, "Energy Solution for Loading and Unloading Machines," *YASKAWA TECHNICAL REVIEW*, vol. 76, No.296, pp.163-168, 2012.

Inverter Drive of Dynamometers
for Automotive Evaluation System

Shizunori Hamada[1*], Toshimichi Takahashi[2], Nobutaka Kezuka[3], Masaju Kouketsu[3] and Shingo Ishigaki[3]

1 Power Electronics Research Department, MEIDENSHA CORPORATION, Numazu, Japan
2 Dynamometer Systems Engineering Division, MEIDENSHA CORPORATION, Tokyo, Japan
3 Dynamometer System Factory, MEIDENSHA CORPORATION, Ota, Japan
*E-mail: hamada-sh@mb.meidensha.co.jp

Abstract- We delivered the dynamometer in 1920 first in Japan and since then has continued development of the dynamometer for 97 years. The purpose of the dynamometer system as testing equipment for automobiles is to reproduce the operating conditions of the components used in the automobile in the indoor testing condition. More realistic environment representing the actual operating conditions becomes necessary recently. We responded to such need with the system of improved performance and functionality. In this paper, increased capacity, improved current response, and higher accuracy of current control are explained.

Keywords— Current Response, Dynamometer, Inverter Drive and Speed Responce.

I. INTRODUCTION

The dynamometer system is equipment used for the testing and evaluation of motor vehicles and powertrain components. Recently, it has been used as a basic tool in diversified automotive-related industries, in addition to the motor vehicle manufacturers. The principle of the dynamometer is to absorb the power of the automobile and to measure torque and speed and then to determine the engine output power and/or the powertrain power loss multiplying the torque and speed.

Recently, requirements for testing of compliance with the environmental regulations, such as emission, fuel economy, and EMC (Electro Magnetic Compatibility) become more demanding of the dynamometer systems. To satisfy such requirements, the workload for the testing and verification of the automobiles and the components in development phase using our equipment is substantially increased.

For the engine and drivetrain testing system, engine simulation reproducing the changing torque due to engine combustion and testing the reproducing behavior of the vehicle's drivetrain using the dynamometer are required. To improve accuracy in reproducing such behavior, higher capacity, improved current response, and improved accuracy in control of motor torque are required [1].

II. INVERTER PERFORMANCE REQUIRED FOR DYNAMOMETER

Fig. 1 shows how the engine bench test and drivetrain test are conducted. In Fig. 1 (a) a dynamometer for engine test is shown. The engine and the dynamometer

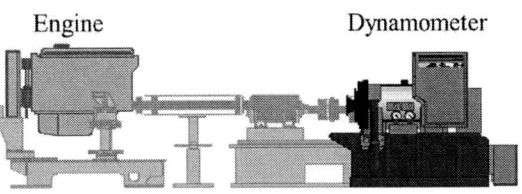

(a) Engine Evaluation system

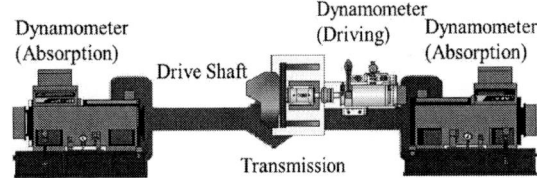

(b) Drive Train Evaluation system

Fig. 1. Setup for Automotive Component Testing

are connected via a shaft. The shaft comprises the clutch, transmission, and propeller shaft. The tested equipment is the engine, and the dynamometer is used to simulate the engine load. For this purpose, the torque response in a higher torque and higher frequency range is required for the dynamometer. The performance required for the dynamometer is high response to torque change and torque characteristics independent of the operating condition of the motor. Fig. 1(b) shows configuration of the drivetrain bench for the FWD (Front Wheel Drive) system. The drivetrain bench is a system used for various measurements and testing, such as the measurement of transmission efficiency of powertrain components like the various types of automotive transmissions and torque converters and endurance tests. Various configurations are used for the drivetrain bench depending on the drive system of the automobile. The two types of dynamometer used are one for simulating the engine connected to the transmission input shaft and the other is for simulating the road load connected to the driveshaft. Higher output, higher frequency characteristics, and wide constant output range are required for the dynamometer simulating road load. Simulation of the characteristics in engine idling and in starting on a slope is also required. A dynamometer that is robust enough against disturbances caused by external torque is required.

The IPMSM is commonly used for the dynamometer to produce higher output torque.

III. INCREASING INVERTER OF VOLTAGE

In this section, an explanation of the inverter THYFREC VT340DY-21K (hereinafter referred to as VT340DY-21K) used for the dynamometer is given [2],[3].

In the case of the common inverter, the carrier frequency is reduced because heat density increases when capacity is increased. On the other hand, a higher carrier frequency is required to improve the response for electric current control in the case of the inverter used in the dynamometer. Additionally, when the capacity of the motor is increased for the same voltage, the electric current increases. So, higher voltage of the inverter output is also required.

Accordingly, in the case of VT340DY-21K, series connection of the single-phase inverters is adopted. Fig. 2 shows the main circuit diagram. The harmonics of the input current are reduced by the multiple winding transformer installed at the input.

At the output, two single-phase inverters with regenerating functions are connected in series for each phase, and they are star connected. PS (Phase Shift) method is used for PWM. Fig. 3 shows the simulation waveform used in the PS method. PWM is realized using four carrier waves with the phases delayed by 90 degrees respectively. While the carrier frequency of the-single-phase inverter is 5 kHz, the carrier frequency is equal to 20 kHz when the carrier frequency of the entire system is converted to a 2-level inverter scheme. Fig. 4 shows the output line voltage of the VT340DY-21K. Compared with the two-level inverter, there are more levels, and the magnitude of the harmonics is small.

Table 1 shows the rating of the inverter. Because the maximum output voltage is 700 V, which provides higher allowance in output voltage, the response will not be impaired because of the voltage saturation caused by the steep current change. Current response is significantly improved from the conventional inverter by the combination of such inverters with high-speed current control using the FPGA.

However, because of the high carrier frequency, noise and heat generated in the electrolytic capacitors and IGBTs are problematic.

Therefore, modification of construction was made to improve heat rejection through the reduction of inductance at the DC links and cooling of large switching loss.

Fig. 5 shows Bode diagram of the current response of VT340DY-21K. Our low inertia dynamometer PCDY600 was used in measurement. Table 2 shows the specifications of PCDY600.In the case of VT340DY-21K, the gain characteristic of about -0.7 dB in frequency response of 2.0 kHz is achieved.

In addition, as shown in Fig. 5, a certain level of

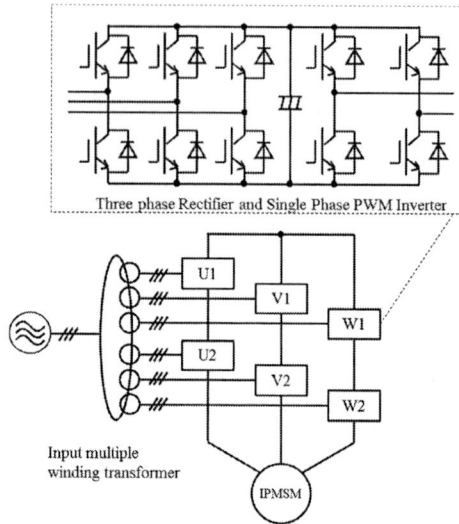

Fig. 2. Main Circuit of VT340DY-21K Configuration

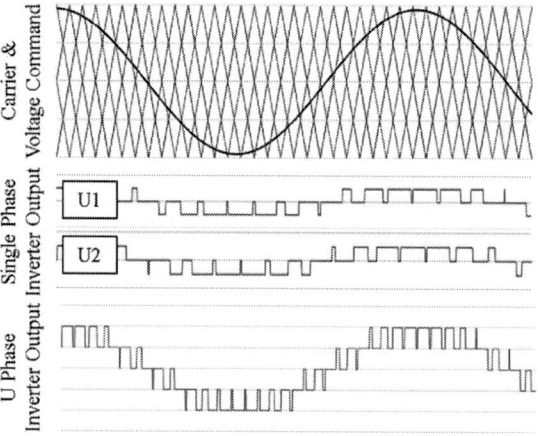

Fig. 3. Simulated waveform of PS method

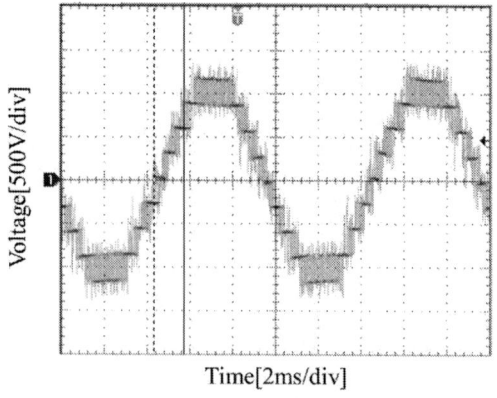

Fig. 4. Output Line Voltage of VT340DY-21K

response can be maintained even when the measurement conditions are different. The conditions of measurement in Fig. 5 were motor speed of 5000 min^{-1}, base load of 20%, and oscillation amplitude of 10%. Tests are

TABLE I
MAJOR SPECIFICATION OF VT340DY-21K

Item	Specification
Max. capacity	600kW
Max. output current	1800A
Max. output voltage	700V
Output rating	100% continuous No overload rating
Output frequency	0~440Hz

TABLE II
MAJOR SPECIFICATION OF PCDY600

Item	Specification
Rated continuous capacity	Absorption 600kW Driving 600kW
Rated continuous torque	1146Nm
Base/Max. revolution speed	5000/8000 min^{-1}
Moment of inertia	0.23kgm^2(single unit)

conducted by changing speed, base load, oscillation amplitude, and conditions of driving and power absorption. As explained above, the VT340DY-21K demonstrated a high current response and a response independent of the load owing to its high carrier frequency and high output voltage.

IV. INCREASING INVERTER OF CURRENT

In this section, an explanation of the inverter for the dynamometer THYFREC VT340DY-H15000 (hereinafter referred to as VT340DY-H15000) is given [4]. Fig. 6 shows the main circuit diagram. Since the VT340DY-H15000 is designed for 200% overloading, high capacity is realized through the parallel connection of the main circuits. The circuit consists of two inverters and interphase reactors with the same DC voltage from the rectifier. Coupled inductance of the interphase reactors produces impedance only for cross current and does not affect the output current. When cross current is suppressed only by the interphase reactors, heat is generated by the interphase reactors, which requires large interior volume in the equipment. Accordingly, the cross current needs to be controlled. Fig. 7 shows the block diagram of inverter current control.

I_{dq}^*: Current command of inverter [A]
I_{dq}: Output current of inverter [A]
I_1, I_2: Current of inverter[A]
I_c: Cross current [A]
V_{ce1}, V_{ce2}: Collector emitter voltage[V]

In Fig. 7, the cross current control and Vce compensation are additionally included in the current control. In cross current control, output current from the inverter is detected, and the voltage command is compensated to reduce cross current. In Vce compensation, the gate command and measured value of Vce are compared, and the difference is used to compensate the voltage command.

Respective control command are coordinated each other and no interference will occur, because of both commands are converted to voltage[5]-[7].

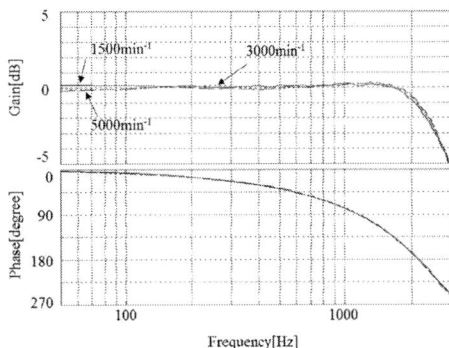

(a) Revolution speed dependency

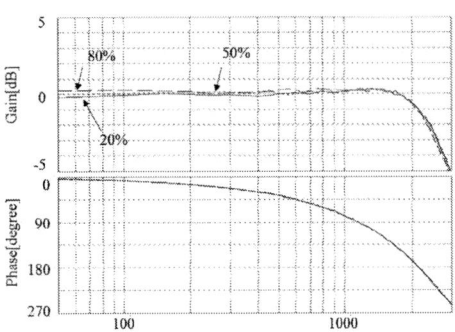

(b)Base load dependency

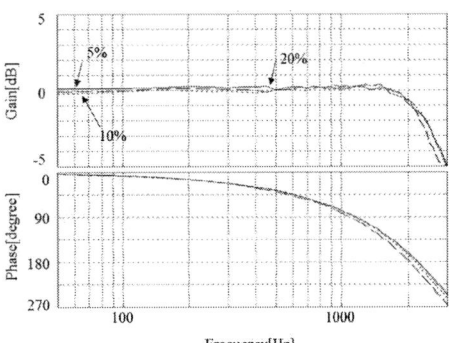

(c)Oscillation amplitude dependency

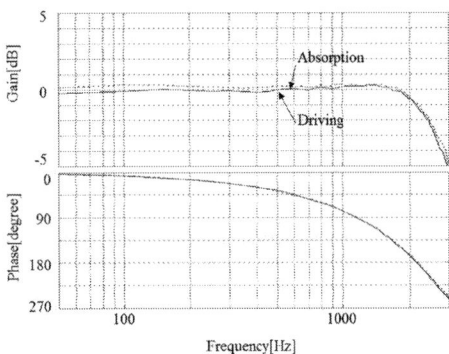

(d) Driving and absorption load dependency

Fig. 5. Current Response Bode Diagram

The 2018 International Power Electronics Conference

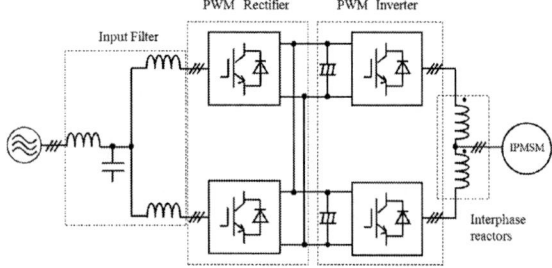

Fig. 6. Main Circuit of VT340DY-H15000

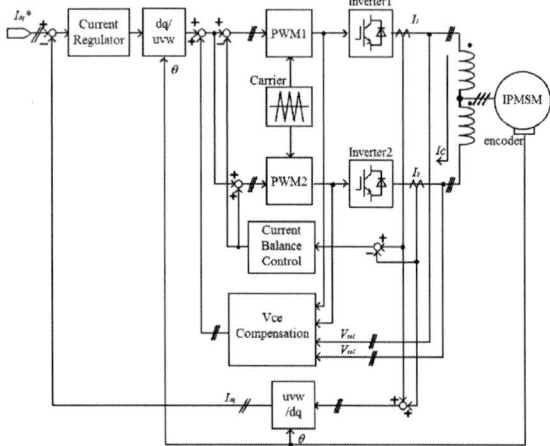

Fig. 7. Inverter Current Control Circuit Configuration

TABLE III
MAJOR SPECIFICATION OF VT340DY-H15000

Item	Specification
Capacity	652kVA (continuous) 1306kVA(60 seconds)
Output rating	100% continuous 200% for 60 seconds
Output frequency	0~533Hz

TABLE IV
MAJOR SPECIFICATION OF PCDY330

Item	Specification
Rated continuous capacity	Absorption 330kW Driving 330kW
Rated continuous torque	525Nm
Base/Max. revolution speed	6000/10000 min^{-1}
Moment of inertia	0.12kgm^2(single unit)

In addition, with respect to control of the PWM rectifier, robustness of the DC voltage is increased to improve accuracy of output voltage of the inverter [8].

Fig. 8 shows the waveform of the output line voltage and output current. Distortion of the output current waveform is reduced by combination of the high-speed current control and parallel control. Fig. 9 shows a Bode diagram of the current response of the VT340DY-H15000. Our low inertia dynamometer PCDY330 was used in measurement. Table 4 shows the specification of

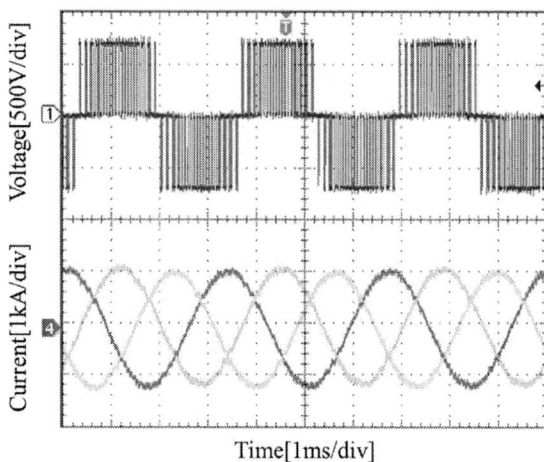

Fig. 8. Output Line Voltage & Current of VT340DY-H15000

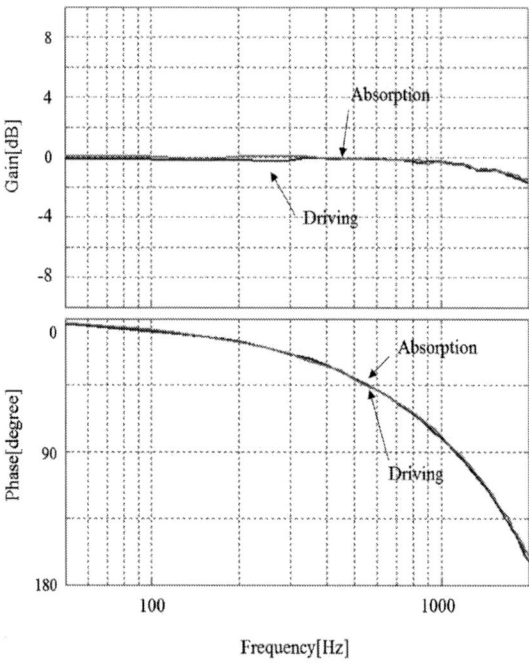

Fig. 9. Current Response Bode Diagram

PCDY330. As shown in Fig. 9, current control response more than 1.5 kHz is realized both for driving and power absorption cases.

Excellent frequency response of current more than 1.5 kHz is achieved by the VT340DY-H15000. Table 3 shows the specification of the VT340DY-H15000. In the case of the VT340DY-H15000, output up to the output frequency of 533 Hz is possible corresponding to the short time rating (200% overload 60 seconds). The system can also be used for the stall performance test (operation at 0 min^{-1}) that is frequently required in the market.

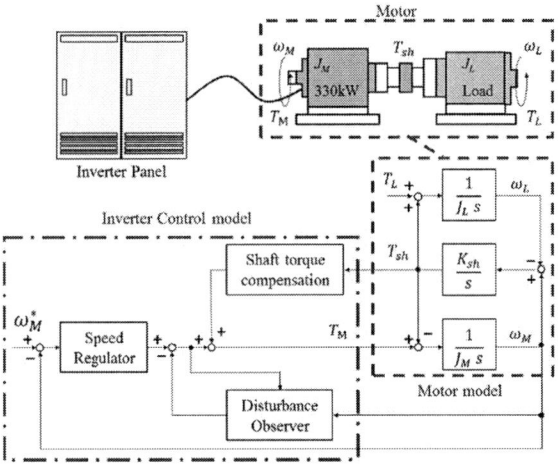

Fig. 10. Speed Control Circuit Configuration

TABLE V
MAJOR SPECIFICATION OF VT340DY-H8500

Item	Specification
Capacity	850kVA (continuous)
Output rating	100% continuous No overload rating
Output frequency	0~533Hz

V. ROBUSTNESS OF INVERTER FOR DYNAMOMETER

In this section, our approach for speed control is explained. Fig. 10 shows the speed control block diagram.

Assume two-inertia system model for the mechanical configuration of the dynamometer system.

J_L: Moment of inertia of the load motor [kgm^2]
J_M: Moment of inertia of the dynamometer [kgm^2]
T_L: Load torque [Nm]
ω_L: Angular velocity of the load motor [rad/s]
ω_M: Angular velocity of the dynamometer [rad/s]
K_{sh}: Torsional stiffness of the shaft [Nm/rad]

In the case of the two-inertia system model in Fig. 10 with improved speed control response, when the moment of inertia of the load J_L is greater than the moment of inertia of the dynamometer J_M, instabilities of hunting and divergence will become likely to occur because of the resonance of the mechanical system. Accordingly, two speed control mechanisms are additionally incorporated in order to improve robustness in speed control [9].

One is the shaft torque compensation mechanism using the measured torque to reduce mechanical resonance, and the other is the load torque disturbance observer. When these two mechanisms function at the same time, interference of operations by both mechanisms will occur. The filters are used to segregate the frequency response band of each mechanism to prevent any interference. Cutoff frequency setting of the filter for the shaft torque compensation mechanism is near the resonance frequency of the mechanical system and frequency setting of the load torque disturbance observer is in the frequency band lower than the resonance frequency. Such arrangement eliminates interference between two mechanisms.

The inverter THYFREC VT340DY-H8500 (hereinafter referred to as VT340DY-H8500) and the PCDY330 are used in measurements. VT340DY-H8500 is the system in which the main circuits of the VT340DY-H15000 are reduced to one system. The speed control system is the same in both systems.

Table 5 shows the principal specifications of the VT340DY-H8500. Fig. 11 shows the waveform of the speed and shaft torque when ±200 Nm torque step is applied when the system is running at constant speed of 1000 min^{-1}. Fig. 11 (a) represents the waveform without torque compensation and Fig. 11 (b) represents the waveform with torque compensation. As shown in these graphs, speed changes and the effect of the mechanical resonance because of the torque change caused by the load disturbance are small in Fig. 11 (b). Fig. 12 shows the revolution speed and shaft torque waveforms during acceleration and deceleration of the motor. In Fig. 12(b), overshooting of the speed and the effect of mechanical resonance on torque are small compared with those in Fig. 12(a). Fig. 13 shows the speed and shaft torque waveform at the revolution speed of 0 min^{-1}. In Fig. 13(b), speed ripple and torque ripple are smaller than in Fig. 13 (a).

From these results, the VT340DY-H8500 is capable of maintaining highly accurate speed control even with the load disturbance.

VI. CONCLUSION

Outline of the large capacity dynamometer control systems VT340DY-21K, VT340DY-H15000, and VT340DY-H8500 is explained.

We have taken the following three major steps to improve the performance of the dynamometer.

(1) Increase in the maximum output voltage of the inverter

(2) Increase in the inverter rated current

(3) Enhancement of the shaft torque robust performance

These inverters, having large capacity and high responsive current control features are the products that will be effectively used in the transient response test systems. Use of these inverters will also improve control performance such as in speed control, torque control, position control, etc.

Development of the THYFREC VT350DY series inverters is ongoing as the successor to the above models. Compared with the VT340DY, the new series will realize a smaller volume of equipment while maintaining performance.

We will continue to improve the performance and quality of the inverter system to deliver products that will meet customer needs.

*All product and company names mention in this paper are the trademarks and/or service of their respective owners.

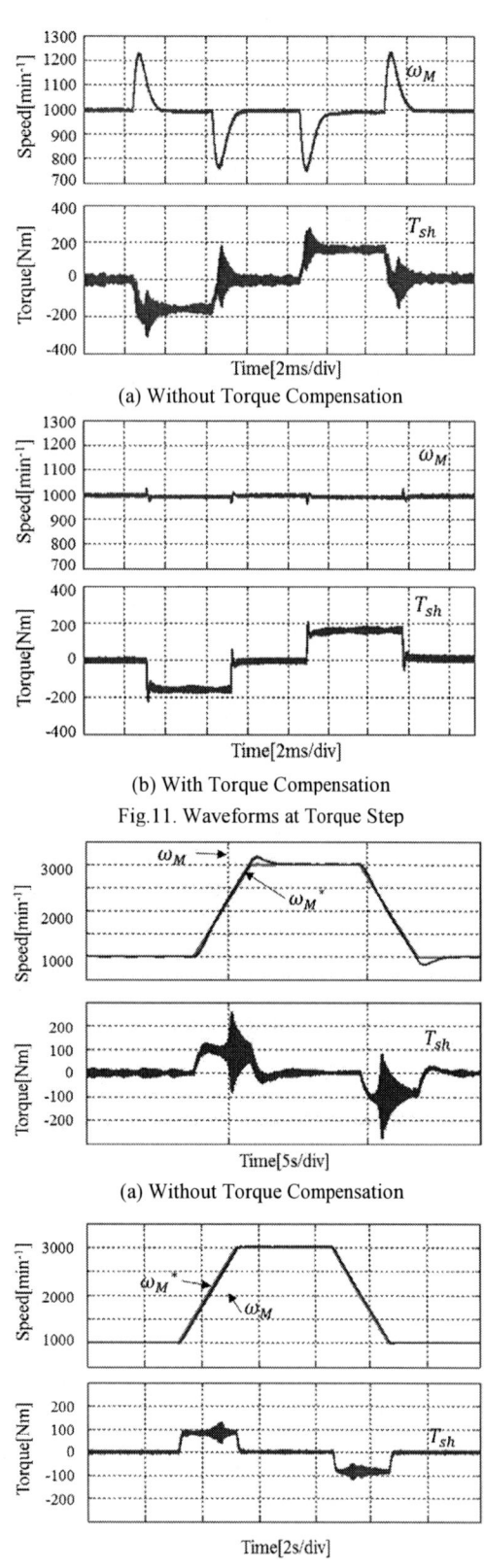

(a) Without Torque Compensation

(b) With Torque Compensation

Fig.11. Waveforms at Torque Step

(a) Without Torque Compensation

(b) With Torque Compensation

Fig.12. Waveforms acceleration and deceleration of the motor

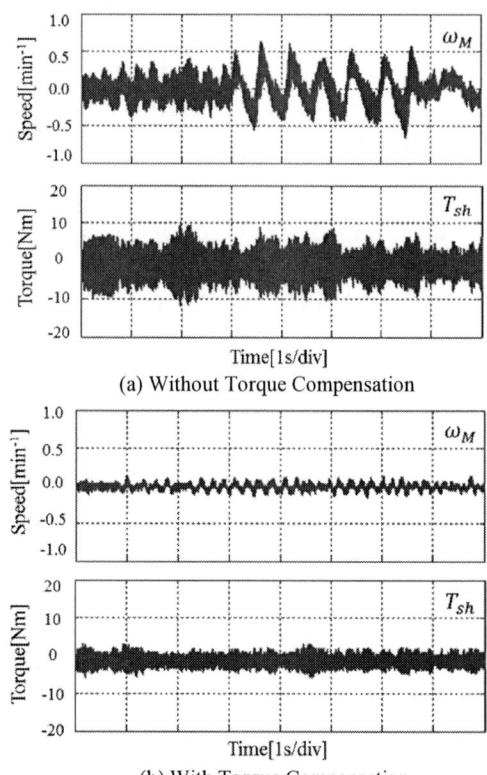

(a) Without Torque Compensation

(b) With Torque Compensation

Fig.13. Waveforms at 0 min^{-1}

REFERENCES

[1] Jiaqiang Yang, Jin Huang, Jian Ma, "Research on a Novel AC Variable-Frequency Dynamic Power Dynamometer", Electrical Machines and Systems 2005. ICEMS 2005. Proceedings of the Eighth International Conference on, vol. 2, pp. 1687-1691, 2005.

[2] K.Hirao,S.Ishigaki,K.Ogura, "Inverter Drive for Large-Capacity Low Inertia Dynamometers, THYFREC VT340DY-21K ",MEIDEN REVIEW ,No.157,pp.14-16(2011)

[3] T.Takahashi,S.Ishigaki,S.Hamada,M.Ejiri,K.Hirao :"High Response Current Control Using Direct Multiple Inverters",2011 National Meeting of The Institute of Electrical Engineers of Japan,No.4-108,2011

[4] T.Takahashi,S.Hamada,M.Ejiri,"Inverter Drive for Large Capacity Dynamometers, THYFREC VT340DY-H15000MDK",MEIDEN REVIEW ,No.157,pp.11-13(2011)

[5] S.Hamada,T.Takahashi, "Dead Time Compensation System of PWM Power Conversion Apparatus", Japanese Patent,2012-016232

[6] S.Hmada,T.Takahashi, "Parallel Operation Device and Parallel Operation Method of PWM Power Converter", Japanese Patent,2012-244674

[7] S.Hamada,T.Takahashi, "Parallel Operation Device of PWM Power Converter" ,Japanese Patent,2013-143880

[8] S.Hamada,T.Takahashi, "Voltage Control Device and Voltage Control Method for Electric Power Conversion Device" ,Japanese Patent,2013-172578

[9] T.Takahashi,"Dynamo Meter System",Japanese Patent,2013-179810

Experimental Investigation of Prototype All-SiC Converter for Ultra-High-Speed Elevator

Kazuhisa Mori[1*], Kaoru Katoh[1], Yohei Matsumoto[2], Tatsushi Yabuuchi[2] and Naoto Ohnuma[2]

1 Research & Development Group, Hitachi, Ltd., Hitachi, Japan

2 Building Systems Business Unit, Hitachi, Ltd., Hitachinaka, Japan

*E-mail: kazuhisa.mori.xz@hitachi.com

Abstract— As the population concentration in large cities steadily increases, buildings will require even more energy savings. Articles about the effect of converters using SiC, which is a low-loss power device, have been reported in various applications. We developed a prototype of a compact converter using SiC-MOSFETs for elevators and evaluated the energy savings in actual elevator operation.

In a former paper, we reported the following two results. The first was the prototype control panel in the all-SiC converter had a 57% smaller setting area than that of the conventional one using Si-IGBT. The other was an energy savings of approximately 17% was achieved in the total elevator system under a high temperature condition.

In this paper, we present our analysis of the power loss of the converter for the possibility of further energy savings. Approximately 70% of the total energy loss of the all-SiC module was found to be switching loss, that is, turn-on loss and turn-off loss. This result suggests that we can expect further energy savings using faster switching with a higher dv/dt to reduce switching loss. Overall, we estimated energy savings of about 20% in an elevator system under higher dv/dt switching.

Keywords— Converter, SiC-MOSFET, elevator

I. Introduction

The power semiconductors using the wide-band-gap materials, such as silicon carbide (SiC) and gallium nitride (GaN) ones, have gained considerable attention for use in various applications in power electronics systems because of their useful characteristics. For example, all-SiC converters using SiC metal-oxide-semiconductor field-effect transistors (SiC-MOSFETs) and Schottky barrier diodes (SiC-SBDs) for railways and for photovoltaic power generation have been reported [1–4].

The size of high-rise buildings has increased as urbanisation has progressed. Thus, demand is increasing for faster elevators and large-capacity elevators. Therefore, the power to drive these elevators is increasing. For example, converters with parallel connections of insulated gate bipolar transistors (IGBTs) have enabled higher speed in elevators [5]. In addition, demand exists for more compact machine rooms and for smaller converters to improve the amount of useful space in buildings. Furthermore, as the population concentration in large cities becomes more of a serious problem, buildings will require even more energy savings.

Therefore, we developed a prototype of a converter unit and control panel by applying an all-SiC module for an ultra-high-speed elevator. As a result, the control panel

(main part) had a 57% smaller setting area than that of the conventional one. We also evaluated the loss reduction effect in an actual elevator. An energy savings of approximately 17% was achieved in the total elevator system under a high temperature condition [6]. However, we suppressed dv/dt to the same level of the conventional one in that evaluation. Further energy savings using faster switching with a higher dv/dt are expected. In this paper, we present our analysis of the power loss of the converter and our investigation on the possibility of further energy savings.

II. Switching Characteristics of SiC Module

A. Specifications of all-SiC module

We used the 1200V/800A all-SiC module (FMF800DX-24A) for the converter unit. The specifications of this module and conventional Si-IGBT are shown in Table I. Even though the rated gate voltage is equal to that of Si-IGBT module, gate threshold-voltage of the all-SiC module is very lower than that of the Si-IGBT module. Therefore, to avoid the false turn-on, designs of the converter in consideration of the dv/dt at the switching are required.

TABLE I
COMPARISON OF POWER MODULE SPECIFICATIONS

	Si-IGBT [7]	SiC-MOSFET [8]
Module type	CM900DU-24NF	FMF800DX-24A
Rated voltage	1200V	1200V
Rated current	900A	800A
Rated gate voltage	±20V	±20V
Threshold voltage	7V	1V

B. Measured results of SiC module

Fig. 1 shows the turn-on and turn-off waveforms for the all-SiC module at normal temperature (Tj=25°C).

Fig. 2 shows the dv/dt in switching versus the switching energy loss (Esw) at normal temperature (Tj=25°C). Esw contains turn-on and turn-off energy losses (Esw = Eon + Eoff). In Fig. 2, Esw values are normalized on the basis of Si-IGBT's ones. Esw can be reduced by 75% in comparison with IGBT's when the gate resistance is the minimum of the recommended range (R$_G$=2.2 Ω). However, dv/dt extends to nearly 18 kV/µs. Even if we suppress dv/dt to the same level (below 10 kV/µs) of the conventional IGBT module's, Esw can be reduced by 54% in comparison with Si-IGBT's.

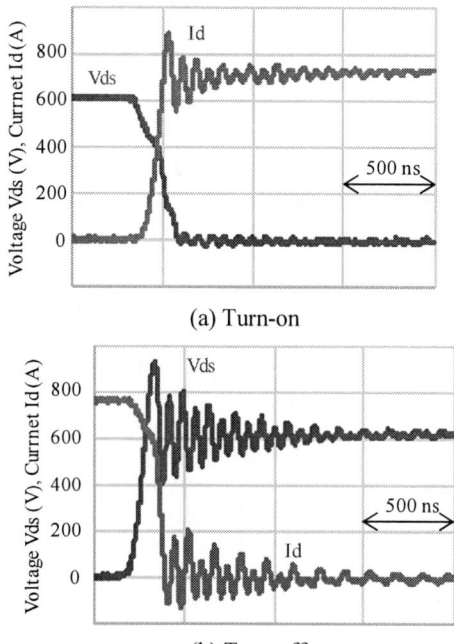

(a) Turn-on

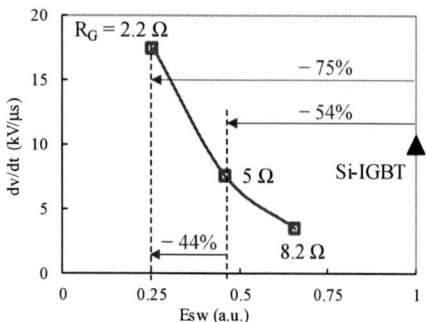

(b) Turn-off

Fig. 1. Switching waveforms of all-SiC module.
(Vdc=600 V, Id=720 A, Tj=25 °C, R_G=2.2 Ω)

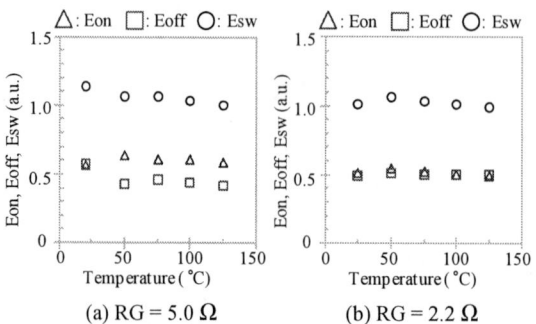

(a) RG = 5.0 Ω   (b) RG = 2.2 Ω

Fig. 3. Measured characteristics of Esw vs. temperature.
(normalized on Esw at Tj=125 °C)

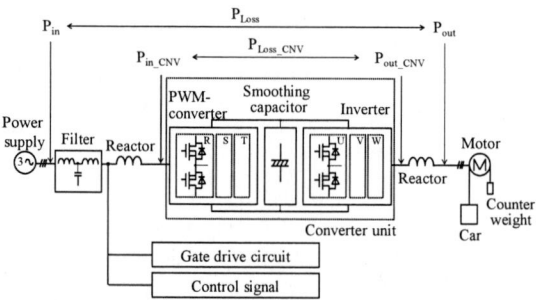

Fig. 4. Circuit diagram of converter for elevators.

Fig. 4 also shows the power measurement points, such as Pin, Pout, Pin_CNV, and Pout_CNV, mentioned later (in section IV). In addition, because elevator users and habitants demand reductions in noise, the switching frequency of the converter is about 8 or 10 kHz. Thus, the ratio of Esw in the losses of the module is high, which in turn means the reduction in Esw by applying all-SiC enables further loss reductions in the converter.

Fig. 2. Measured characteristics of Esw vs. dv/dt.
(normalized on Esw of Si-IGBT)

Fig. 3 shows the dependence on temperature of Eon and Eoff. Here, Eon and Eoff are numerically normalized on the basis of Esw at high temperature (125°C) in each gate resistance (R_G). In Fig. 3, Eon and Eoff hardly have dependence on temperature, and the differences between Eon and Eoff are small.

III. PROTOTYPE ALL-SiC CONVERTER

A. Features of converter for elevators

Fig. 4 shows the circuit diagram of the converter for elevators. Because the elevator system has a car and counter weight, the power to lift is reduced. The elevator system is driven and controlled by the control panel consisting of a filter, reactor, converter unit, and signal circuits. The filter and reactor are connected between the power supply and converter. The converter unit has a PWM converter, smoothing capacitor, and inverter.

B. Prototype Converter using all-SiC modules

Fig. 5 shows a prototype control panel which has an all-SiC converter unit. The loss reduction using all-SiC modules can make not only heat sink smaller but also less dead space. The volume of the all-SiC converter unit is reduced by 85% compared with the conventional one consisting of Si-IGBTs.

Fig. 6 shows the volume ratio of the main parts constituting a converter. The main parts are power modules, cooling parts (heat sinks and fans) and capacitors. Others include the path of cooling air. The cooling part has been greatly downsized. This is an effect of the large reduction in the power module loss. In addition, others occupied approximately 50% in the conventional converter, but it became about 1/8 using the high density assembly due to the downsizing of cooling parts. In this way, the loss reduction of the power modules not only enabled the downsizing of the power modules and the cooling parts but also reduced the space of others. This enabled the large downsizing of the converter, which in turn enabled putting the filter and the reactors (which conventionally were put in other panels) in one panel. As a result, the setting area of the control panel could be reduced by 57%.

IV. EVALUATION OF PROTOTYPE CONVERTER

A. Evaluation method

We evaluated the prototype in an actual elevator system. The elevator has the following specifications, the rated load weight is 1800 kg, the rated speed is 6 m/s, and the rated output power of the traction machine is 68 kW. We measured the power at the points shown in Fig. 4 when the elevator car moves during one round trip. The measured power values are the input power (Pin), the output of the control panel (Pout), the input of the converter (Pin_cnv), and the output of the converter (Pout_cnv). We derived loss in the control panel (P_{Loss}) with equation (1). The loss of the converter (P_{Loss_cnv}) was derived with equation (2).

$$P_{Loss} = Pin - Pout \quad (1)$$

$$P_{Loss\_CNV} = P_{in\_CNV} - P_{out\_CNV} \quad (2)$$

We calculated the following energy losses in the elevator operation. The energy loss (E_{Loss_1}) of the converter is derived with equation (3). The energy loss (E_{Loss_2}) in the control panel except for the converter is derived with equation (4). Here, this loss (E_{Loss_2}) involves the loss of the filter, reactors, and signal circuits.

$$E_{Loss\_1} = \int P_{Loss\_CNV} \cdot dt \quad (3)$$

$$E_{Loss\_2} = \int (P_{Loss} - P_{Loss\_CNV}) \cdot dt \quad (4)$$

$$E_{Loss\_3} = \int P_{out} \cdot dt \quad (5)$$

$$E_{tot} = \int P_{in} \cdot dt \quad (6)$$

Fig. 7 shows the measured input power of the converter (P_{in_CNV}) during one round trip. Here, input power is normalized at constant speed when moving up. In this case, because the elevator has a rated weight in the car, P_{in_CNV} is a positive value in the up mode, P_{in_CNV} is a negative one in the down mode.

In this operation, the potential energy after the trip is equal to the one before the trip because the elevator car has a constant weight during one round trip. Thus, the energy loss (E_{Loss_3}) derived from the output power of the control panel (Pout) with equation (5) is the energy loss in the motor and mechanical parts. Additionally, the energy (E_{tot}) derived by equation (6) is the total power consumption in the elevator system, and it is a key performance indicator of the energy savings.

When an elevator is frequently used, the temperature of the power module in the converter rises. The switching loss of the SiC module does not increase (in Fig. 3), though the Si-IGBT increases at a high temperature. Therefore, we operated the elevator continually and measured the losses at a high temperature. During this operation, we measured the temperature of the module case using a thermocouple attached to the module base plate.

While these power modules are permitted when the junction temperature (Tj) is up to 150°C, we cannot directly measure the junction temperature. Thus, we

measured it when the temperature (Tc) of the module was under 100°C and extrapolated it from the results of a measurement on temperature increase.

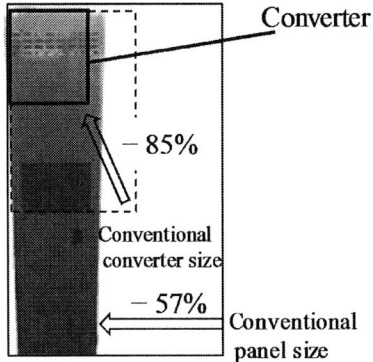

Fig. 5. Prototype control panel (main part).

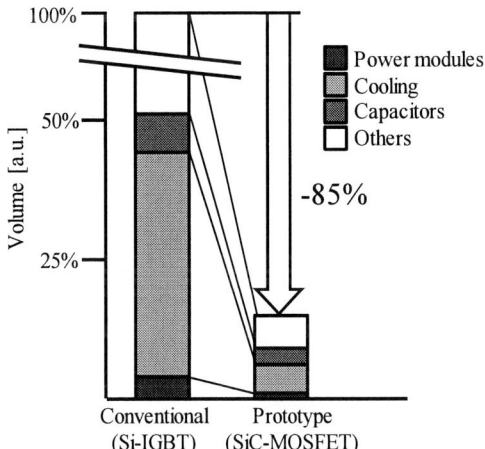

Fig. 6 Comparison of each part volume.

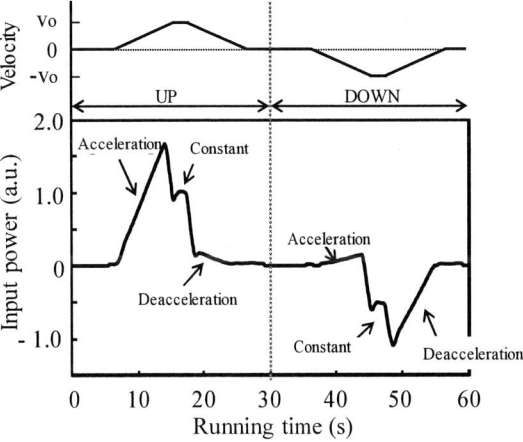

Fig. 7. Input power of the converter during one round trip.

The 2018 International Power Electronics Conference

B. Results of evaluation

Fig. 8 shows the dependence on temperature of the energy loss (E_{loss_1}) of the converter during one round trip. For Si-IGBT, Eoff increased at high temperature because the tail current increased at high temperature. In addition, because the conventional IGBT module had a PiN (p-intrinsic-n) diode as a free-wheeling diode, the reverse recovery current increased at high temperature, thereby increasing Eoff. Therefore, Esw showed a tendency toward a simple increase in temperature. In contrast, for the all-SiC module, Eon and Eoff did not increase at high temperature.

C. Analysis of loss and estimation

Fig. 9 shows each of the rates of the following losses in the all-SiC module. Here, the losses in the module are defined by equations (7)–(10). So, Psw is the switching loss, Pcd is the conducting loss, Pvf is the conducting loss of the free-wheeling diode, Prr is the recovery loss, and fsw is the switching frequency.

$$Psw = (Eon + Eoff) \cdot fsw \quad \text{(switching state)} \quad (7)$$
$$Pcd = Vds \cdot Id \quad \text{(on state)} \quad (8)$$
$$Pvf = Vsd \cdot Id \quad \text{(free-wheeling state)} \quad (9)$$
$$Prr = Err \cdot fsw \quad \text{(diode recovery state)} \quad (10)$$

Psw includes turn-on loss and turn-off loss, accounting for over 70%. Fig. 2 (in section II.) shows that we can reduce it by 44% at R_G=2.2 Ω in comparison with one at R_G=5 Ω. Though Prr is increased slightly by faster switching, it does not matter because the rate of Prr in the total loss of the module is very small.

Therefore, we estimated that the total loss in the all-SiC module is reduced by 30%. Fig. 10 shows the analysis results of losses in the following part of the elevator system. E_{loss_1} is the loss in the converter, E_{loss_2} is the loss in the control panel except for the converter, and E_{loss_3} is the loss in the motor and mechanical parts. In the case of a conventional Si-IGBT converter, the loss of the converter (E_{loss_1}) accounts for 30%. Thus, the losses except for those in the converter (E_{loss_2} and E_{loss_3}) in the total system account for 70%. The measured results of the prototype all-SiC converter mean that we can reduce the energy for the operating elevator by 17% at maximum.

Additionally, we estimated that faster switching can save energy by 20% in comparison with the conventional Si-IGBT converter. However, higher tolerance is needed for dv/dt at faster switching. As aforementioned, Fig. 2 shows the dv/dt value at R_G=2.2 Ω is approximately 18 kV/μs.

The electromagnetic interference (EMI) needs to be considered [9]. We derived the voltage amplitude spectra during turn-off. Fig. 11 shows a comparison of the voltage amplitude spectra in two cases of the gate resistance (R_G=2.2, 5 Ω).

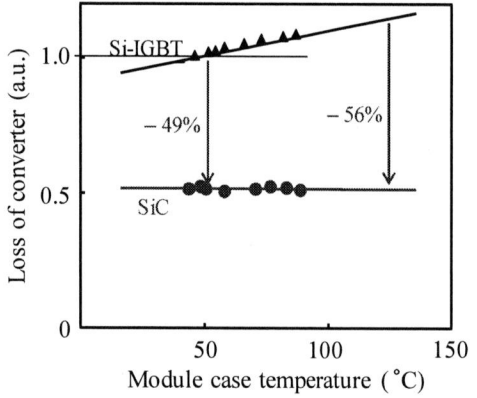

Fig. 8. Temperature characteristics of power loss of converter.

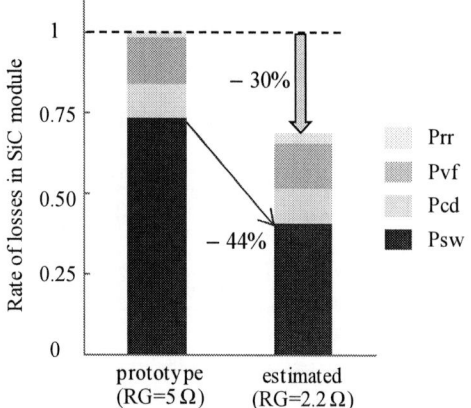

Fig. 9. Analysis of losses in all-SiC module.

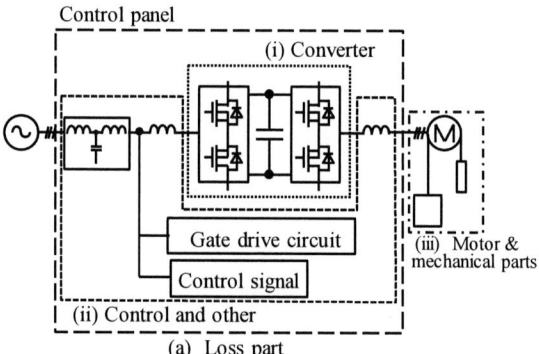

(a) Loss part

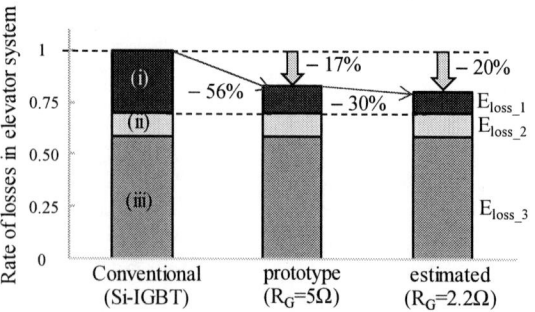

(b) Rate of power loss in each part

Fig. 10. Analysis of losses in elevator system.

236

In Fig. 11, the voltage amplitude is normalized on the basis of the value in the case of R$_G$=5Ω. For R$_G$=2.2Ω, the voltage amplitude spectrum is derived from Fig. 1 (in section II). In Fig. 1(b), we can see the ringing frequency is about 14 MHz. In Fig. 11, the ratio of voltage amplitude at this frequency is high. Also, the voltage amplitude in RG=2.2Ω is double for the voltage amplitude in RG=5Ω. This is only a comparison using the frequency analyses (Fourier transform) of the voltage waveform during turn-off, but care needs to be taken to avoid the problem of the EMI increasing as dv/dt increases.

Fig. 12 shows the dependence on temperature of the dv/dt value. Here, the dv/dt values are normalized at 25°C. At 125°C, dv/dt increased by 12%. Therefore, we used a device which has tolerance to high dv/dt (about 20 kV/μs) in the gate drive circuits and signal controller.

V. CONCLUSIONS

We developed a prototype all-SiC converter for ultra-high-speed elevators with a reduced volume of 15% compared with the conventional Si-IGBT converter. The setting area of the control panel with the compact converter was reduced by 57% in comparison with the conventional one. An energy-saving effect of approximately 17% was achieved in the same level of dv/dt in the switching period. Additionally, because further energy savings are expected using faster switching with the higher dv/dt, we investigated the possibility of further loss reductions. We

analyzed the losses, finding that switching loss accounts for about 70% in SiC module. We estimated the energy savings at higher dv/dt (about 20 kV/μs) were nearly 20% in comparison with a conventional Si-IGBT converter.

REFERENCES

[1] H. Akagi, T. Yamagishi, N. M. L. Tan, S. Kinouchi, Y. Miyazaki, and M. Koyama, "Power-loss breakdown of a 750-V, 100-kW, 20-kHz bidirectional isolated DC-DC converter using SiC-MOSFET/SBD dual modules," *IPEC-Hiroshima 2014 -ECCE-ASIA*, pp. 420–428, 2014.

[2] N. Nashida, Y. Hinata, M. Horio, R. Yamada, and Y. Ikeda, "All-SiC power module for photovoltaic power conditioner system," *Proc. ISPSD2014*, pp. 342–345, 2014.

[3] M. Shinbo, H. Abiko, H. Sonoda, K. Shibanuma, T. Ishida, and Y. Chiba, "Research of efficient main power equipment using SiC power device," *IPEC-Hiroshima 2014 -ECCE-ASIA*, pp. 634–639, 2014.

[4] A. Hatanaka, H. Kageyama, and T. Masuda, "A 160kW high-efficiency photovoltaic inverter with paralleled SiC-MOSFET modules for large-scale solar power," *Proc. INTELEC2015*, pp. 1263–1267, 2015.

[5] K. Mori, N. Ohnuma, T. Sakoda, Y. Matsumoto, and T. Yabuuchi, "Power converter with parallel connected IGBTs for ultra-high-speed elevators," *IEE Japan Trans. IA*, Vol. 137, No. 1, pp. 1–9, 2017 (in Japanese).

[6] K. Katoh, K. Mori, Y. Matsumoto, T. Yabuuchi, and N. Ohnuma, "Converter using SiC-MOSFET for ultra high-speed elevators," *IEE Japan Trans. IA*, Vol. 137, No. 4, pp. 334–341, 2017 (in Japanese).

[7] J. Yamada, Y. Yu, J.F. Donlon, and E.R. Motto, "New MEGA POWER DUAL™ IGBT module with advanced 1200V CSTBT chip," *2002 IEEE Industry Applications Conference*, pp. 2159–2164, 2002.

[8] E. Wiesner, K. Masuda, and M. Joko, "New 1200V full SiC module with 800 A rated current," *Proc. EPE'15 ECCE-Europe*, pp. 1–9, 2015.

[9] N. Oswald, P. Anthony, N. McNeill, and B. H. Stark, "An experimental investigation of the tradeoff between switching losses and EMI generation with hard-switched All-Si, Si-SiC, and All-SiC device combinations," *IEEE Trans. on Power Electronics*, Vol. 29, No.5, pp. 2393–2407, 2014.

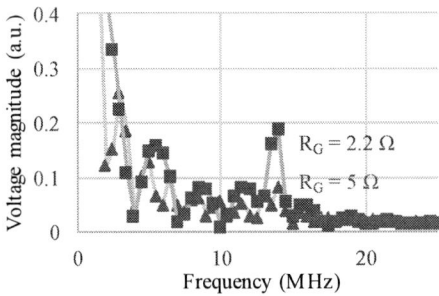

Fig. 11. Comparison using voltage amplitude spectra during turn-off. (normalized on the basis of R$_G$=5Ω)

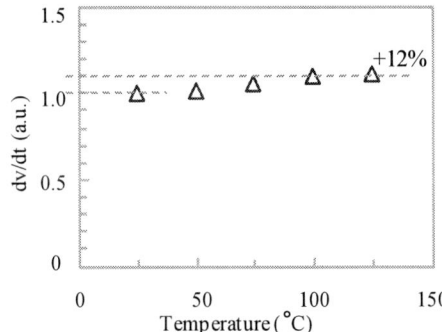

Fig. 12. Measured characteristics of dv/dt vs. temperature. (normalized at 25°C)

High-Voltage, Large-Capacity Converter Technologies and Their Applications

Daisuke Yoshizawa[1*], Paul Bixel[2] and Masahiko Tsukakoshi[1]

1 Toshiba Mitsubishi-Electric Industrial Systems Corporation, 1, Toshiba-cho, Fuchu-shi, Tokyo, 183-8511, JAPAN

2 TMEIC Corporation, 1325 Electric Road, Roanoke VA 24018 U.S.A

*E-mail: YOSHIZAWA.daisuke@tmeic.co.jp

Abstract-In this paper, three main technologies we apply to large-capacity power converters are described. The first one is electromagnetic field analysis by simulation with 3D-CAD models. This allows us to reduce inductance of busbars to the utmost limit by changing their layout, which ultimately leads to surge voltage reduction. The second one is adoption of aluminum instead of copper for lighter heat sinks. To supplement the lower heat conductivity of aluminum, the water-flow passages of the heat sinks are modified to keep the thermal resistance low. The final one is the fixed pulse pattern method used for pulse width modulation (PWM). In large-capacity power converters, torque ripples may increase depending on the PWM method being applied. To mitigate this problem, we use the fixed pulse pattern to reduce harmonics that generate torque ripples, as in the both simulated and experimental data, which are shown later in this paper.

I. INTRODUCTION

The common motor drives in the 1960s to 80s were thyristor Leonard direct current (DC) drive and cycloconverter as alternating current (AC) drive. Due to a lack of gate turn-off capability, however, those thyristor-controlled drives had problems of low power factor as well as integer and non-integer harmonics generated in the main power supply. In the 1990s, the insulated gate bipolar transistor (IGBT) with gate turn off capability, which does not require complex auxiliary circuits, became available and was adopted for various kinds of motor drive systems.

The 2000s saw an increase in world-wide energy consumption, boosting demand for lower-power-loss semiconductor devices for higher voltage and larger current applications. To satisfy this demand, the injection enhanced gate transistor (IEGT), a semiconductor device featuring small power loss, was developed. Since then, converters adopting IEGTs, which are suitable for efficient driving of several MW-class motors for compressors used in the iron and steel rolling mills as well as in the oil and gas facilities, have been developed over the last less than two decades[1][2]. Along with the advances in the semiconductor technology, its related technologies have also been developed. Among them are those for reducing harmonics and cooling down the semiconductor devices.

This paper reports the development of semiconductor devices applied to motor drives and their related technologies.

II. AC MOTOR DRIVE SYSTEMS

Operation of high-power motors of about 2MW to 100MW for drives used in oil and gas facilities as well as for main motor drives used in iron and steel rolling mills requires a drive system with large capacity, which has been achieved by multi-level inverters. The followings are examples of the circuit configurations of the commercially-available, large-capacity dives:

<2-1> 3-level inverter

The 3-level inverter shown in Fig.1 provides phase voltage of three levels, +E, 0, and -E, which means up to five levels of line-to-line output voltage. This circuit configuration is applicable to output of maximum 36MVA and has been adopted for many main motor drives for iron and steel rolling mills[3][4][5].

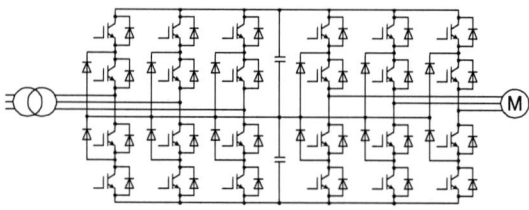

Fig. 1. 3-level inverter

Fig. 2 illustrates the topology for the common DC bus method in which multiple converters are shared by multiple inverter circuits, realizing a drive system with minimum configuration. This method is widely used for the drives for iron and steel mills [6].

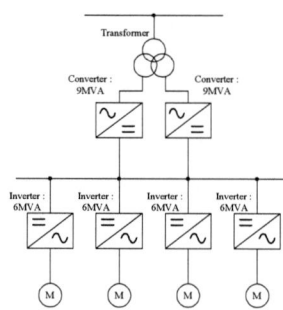

Fig.2 Common converter method

<2-2> 5-level inverter

Fig. 3 is a circuit configuration of the 5-level inverter. Greater output voltage has been realized with a cascade connection of three single-phase 3-level inverters, providing up to nine levels of line-to-line output voltage. Also, 36-pulse diode rectifier circuit reduces harmonics in the input side to a level where they comply with the IEEE-519 standard. This circuit configuration is applicable to output of maximum 120MVA and has been adopted for many main motor drives for oil and gas facilities [7][8].

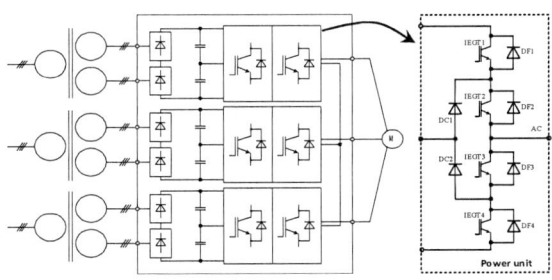

Fig.3 5-level inverter

III. FEATURES AND CHARACTERISTICS OF IEGTs

Since the emergence of large-capacity multi-level inverters, IEGTs with lower on-state voltage have been developed to meet the demand for large-capacity semiconductor devices. Fig. 4 is an outline of the 4500V-3000A type IEGT. The IEGT features improved thermal cycle resistance thanks to the wireless bonding of the multiple chips, as well as of the gate wiring. The chips, which are placed in the IEGT, are connected to one another in parallel for greater capacity of the semiconductor device, making the IEGT inverters applicable to main motor drives used in the iron and steel rolling mills, as well as to drives used in the oil and gas industry.

Fig. 4 4500V-3000A type IEGT

Table 1 lists main specifications of the 4500V-3000A type IEGT. Its improved gate structure contributes to lowering the on-state voltage, which is proven by the collector-emitter saturation voltage of approximately 3.4V (typ.) [5][6][7].

Table 1. IEGT specifications (4500V-3000A type)

Item	Rating	Condition
Collector-emitter voltage	4500V	VGE=-15V Tj=125℃
Maximum turn-off current of collector	3000A	Vcp = 4500V, Vcc = 3000V, VGE = ±15V, Tj≦125℃, Non-repetitive
Collector-emitter saturation voltage	3.4V (typ.)	Ic = 1500A, VGE = 15V, Tj = 125℃
Turn-on loss	15J	Vcc = 3000V, Ic = 1500 A, Tj = 125℃,VGE = ±15 V
Turn-off loss	10.5J (Vce: 3.4V)	

Vce : Maximum rated voltage
Vcc : Rated DC voltage

IV. MAIN CIRCUIT AND COOLING MECHANISM CONFIGURATIONS

Usually, a power converter adopting large-capacity semiconductor devices requires snubber circuits to suppress the higher surge voltage generated by switching. By reducing the inductance to the limit, however, the snubber circuits can be omitted.

In these days, a various simulation tools for electromagnetic field analysis have been applied to evaluate appropriate inductance values. In this study, it was also used one of the simulation tools which confirms the nearly equivalent values measured on actual power module.

Fig.5 shows the 3D-CAD model used for inductance simulation with a 3-level configuration. It was evaluated to obtain the target value or lower inductance for the bus bars connected with semiconductor devices and heat sinks Case A and Case B show the current flow path requiring inductance reduction on turn-off of IEGT1 (Q1) and IEGT2 (Q2), respectively for the same power module with different switching stages.

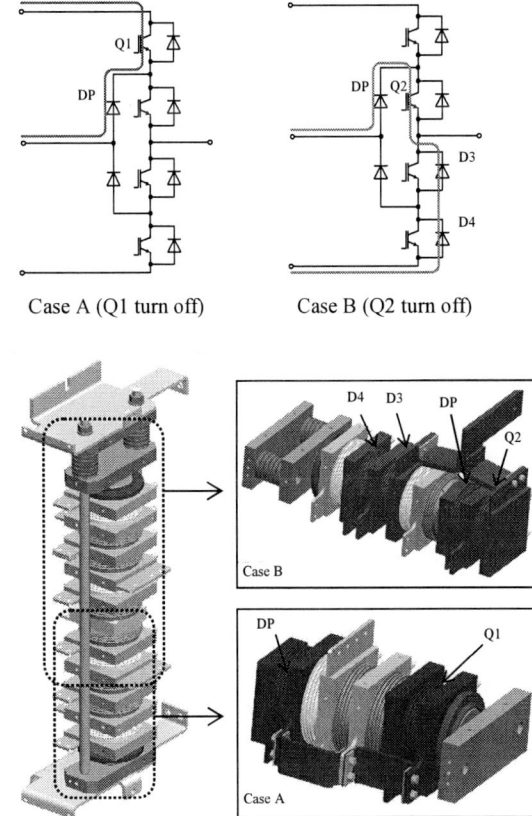

Case A (Q1 turn off)   Case B (Q2 turn off)

Fig. 5 3D-CAD models used for inductance simulation

The main circuit configuration of the 3-level inverter provided for iron & steel rolling mills as well as oil & gas facilities is shown in Fig. 6. The power unit (b) consists of IEGTs free-wheeling diodes and clamp diodes. The less complex configuration achieved by omission of the snubber circuits leads to reliability of the equipment.

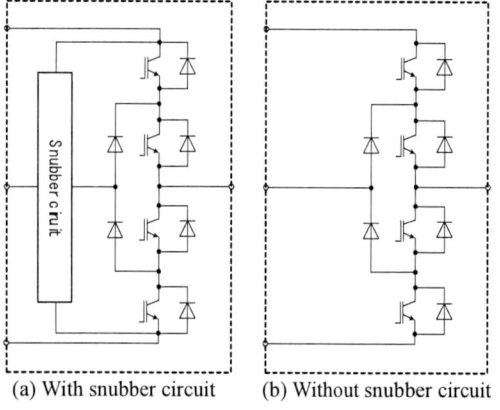

(a) With snubber circuit   (b) Without snubber circuit

Fig. 6 Power unit (Single phase)

Fig. 7 shows turn-off waveforms of an IEGT tested with the DC voltage of 3000V, and both the turn-off and turn-on currents of 2000A. As shown here, the low inductance configuration limits the peak surge voltage to approximately 3800V for the IEGT without a snubber circuit [6][7].

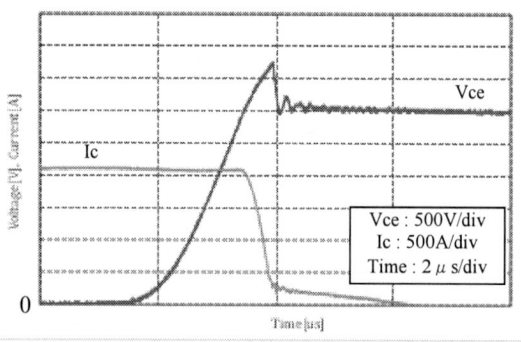

Fig. 7 Turn-off waveforms of IEGT

Fig. 8 is an outline of a power unit. A single power unit comprises IEGTs, diodes and gate drive boards. The power unit approximately 67% lighter than the conventional unit is achieved by adopting aluminum heat sinks and snubber less circuit. Also, mean-time-to repair (MTTR) for the equipment can be shortened, because the power unit can be replaced as a whole.

The problem is that an aluminum heat sink is less heat-conductive than a copper heat sink. The disadvantageous higher thermal resistance of the aluminum heat sinks is, however, lowered by approximately 40% by modifying the water-flow passages and expanding the water contact area [5][6][7]. Fig. 9 shows the thermal resistance ratio of a copper heat sink to aluminum heat sink. Fig.10 shows the heat sink structure.

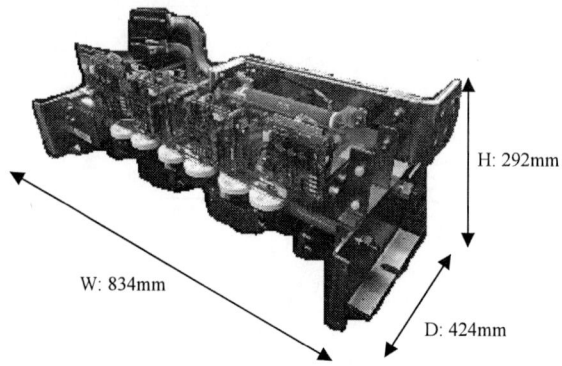

Fig. 8 Power unit

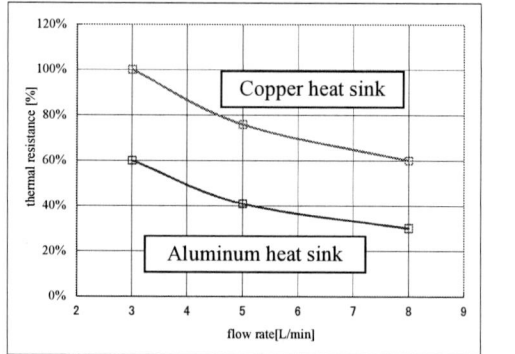

Fig. 9 Thermal resistances of the heat sink

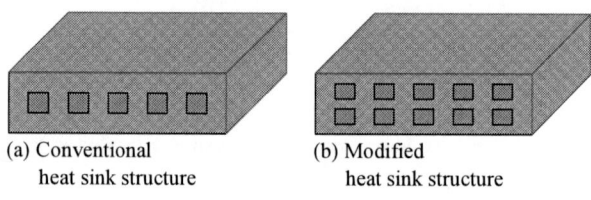

(a) Conventional   (b) Modified
    heat sink structure   heat sink structure

Fig. 10 Cross-sectional view of the fin structures

V. HARMONIC REDUCTION WITH THE FIXED PULSE PATTERN METHOD

The waveform of voltage output from PWM inverter contains harmonics along with the fundamental frequency. The torque ripples resulting from the presence of the harmonics resonate with natural frequencies of the motor couplings or other mechanics, and may eventually damage them. With the asynchronous carrier comparison PWM method, voltage harmonics, which can be determined using the following formula, are generated:

$$| \, k \cdot \mathrm{fsw} \, +/- \, l \cdot \mathrm{fm} \, | \qquad (1)$$

where:
- fsw is the carrier frequency;
- fm is the fundamental frequency;
- k and l are any integer.

With the asynchronous carrier comparison PWM method, as can be seen in the formula (1) above, the inverter outputs voltage whose waveform contains harmonic frequencies which are non-integer multiples of the fundamental frequency, possibly including those same as the natural frequencies of the mechanics. The voltage waveform output by the 5-level inverter adopting the fixed pulse pattern method, on the other hand, only includes frequencies which are integer multiples of the fundamental frequency. In this method, the voltage waveform which is symmetric in 1/4 cycle is generated, and therefore, non-integer harmonics cannot be created [2][9]. Fig. 11 is an example of the voltage waveforms generated with the fixed pulse pattern method. As shown in the figure, voltages are output in accordance with the specified phases (a1 to a3 and b1 to b3).

To avoid resonance with natural frequencies of the mechanics that are generally as low as several Hz to 100Hz, lower-order harmonics are required to be eliminated from the output frequency of the inverter. Here, another feature of the fixed pulse pattern method–creation of voltage waveforms that eliminate the low-order harmonics to theoretically zero –helps. As derived from the example formulas (2) and (3) for calculating 3-pulse output with the fixed pulse pattern method, voltage is output in accordance with the phase values a1 to b3 which are found by calculation to eliminate arbitrary low-order harmonics. Fig. 12 shows pulse pattern corresponding to the modulation rate [10].

$$V_n = 4 \times \frac{2}{n\pi}\{(cosna1 - cosna2 + cosna3) + (cosnb1 - cosnb2 + cosnb3)\} (2)$$

$$V_1 = m, V_5 = 0, V_7 = 0, V_{11} = 0, V_{13} = 0 \qquad (3)$$

m: modulation rate

Vn: N order voltage harmonics

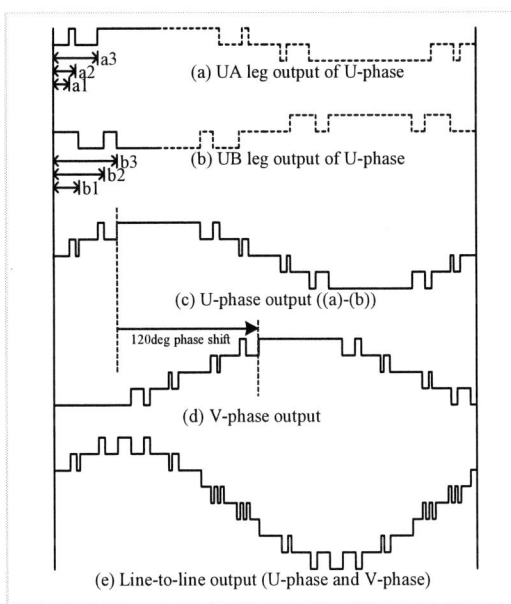

Fig. 11 Example of waveforms generated with the fixed pulse pattern method

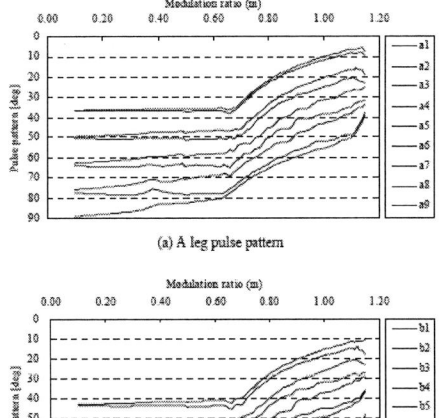

(a) A leg pulse pattern

(b) B leg pulse pattern

Fig. 12 Pulse pattern corresponding to the modulation rate

Fig. 13 and Fig. 14, respectively, are the waveform of the line-to-line output voltage and the fast Fourier transform (FFT) analysis result, observed in simulation on the 5-level inverter adopting the fixed pulse pattern method. Fig. 14 shows reduction in low-order harmonics (5th, 7th, 11th, and 13th harmonics).

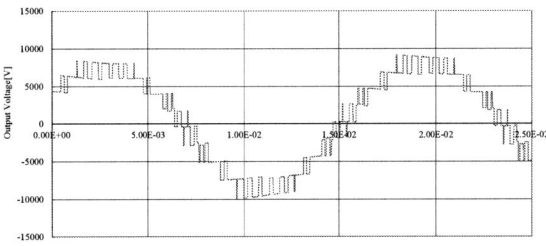

Fig. 13 Voltage waveform to which fixed pulse pattern applied (Simulation)

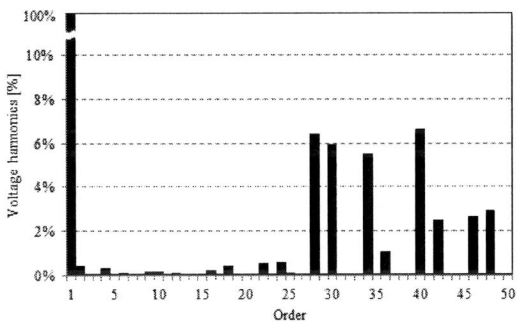

Fig. 14 Output voltage harmonics with 20MW load

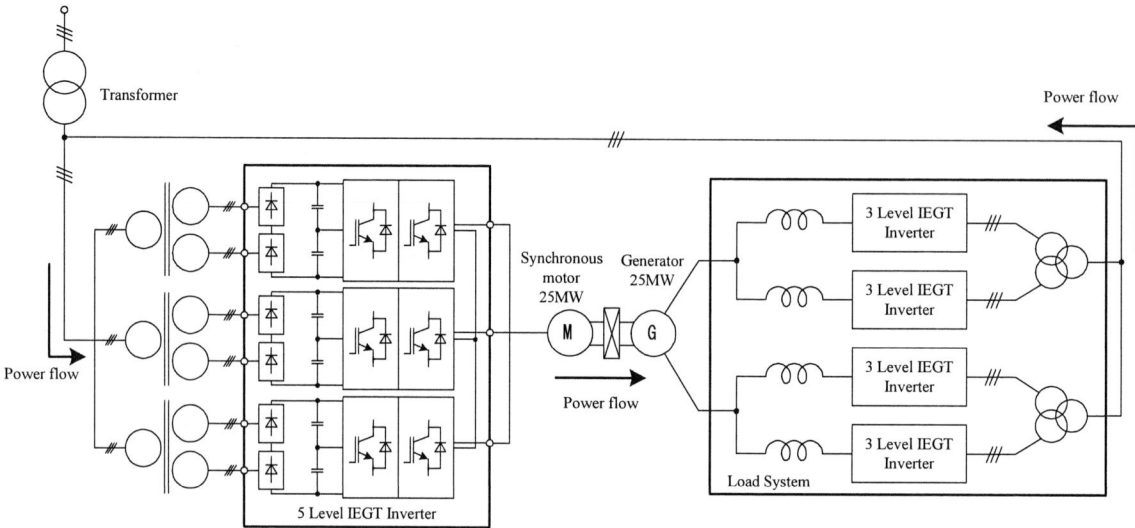

Fig 15 System configuration with 5-level inverter

VI. BACK-TO-BACK TEST

Fig. 15 shows a system configuration with a 5-level inverter under the back-to-back (BTB) test for evaluation of a large-capacity motor drive.

Power output from the 5-level IEGT inverter is supplied to a motor connected to a generator, and converted by four 3-level inverters connected in parallel, circulating in the system. Thus, this system as a whole, including the motor, can be evaluated with low power consumption supplementing power losses generated only by the motor and inverter. This test method allows evaluation of input/output waveforms and of motor operation equal to that of actual equipment [2][8].

Speed and torque controls are available for a motor connected to a 5-level inverter. Fig.16 shows system test on 5-level inverter.

![Fig. 16 System test on 5-level inverter]

Fig. 16 System test on 5-level inverter

In the test, a two-pole 25MW synchronous motor was connected to a generator via a gear. With the 5-level inverter, it was confirmed that torque ripples of low-order harmonics can be reduced by using the fixed pulse pattern method. The test was conducted using 5-level and 3-level IEGT inverters with output capacity of 20 MVA, output voltage of 6000 V, and output current of 1925 A.

Figs. 17, 18, and 19 are, respectively, the line-to-line output voltage and waveform test result and FFT analysis result of the line-to-line output voltage waveform. As in the evaluation by simulation, they show reduction in low-order harmonics achieved by the fixed pulse pattern method.

Fig. 17 shows the line-to-line output voltage. Due to the low motor torque harmonics, it was obtained the reduction in low-order output voltage harmonics shown in Fig. 19.

Fig. 18 shows the output current waveform which shows a very clean output of the inverter near to sinusoidal waveform. Without using any additional output filter of the inverter, it is possible to obtain very clean output with less noise.

Evaluation of torque ripples using an actual motor is important for large-capacity drive systems.

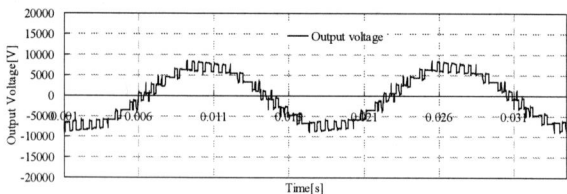

Fig. 17 Output voltage waveform (Line-to-line voltage)

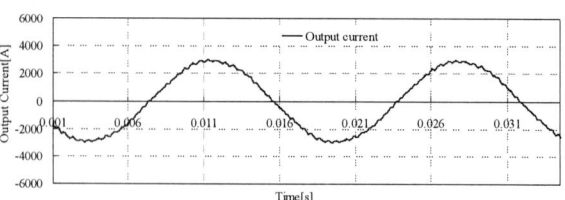

Fig. 18 Output current waveform

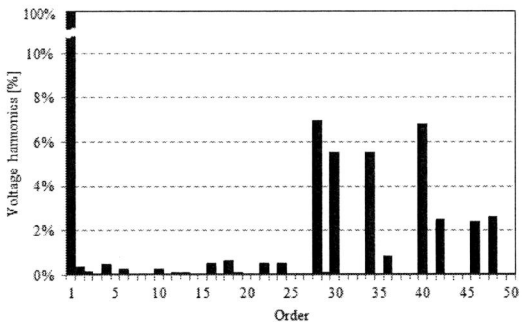

Fig. 19 Output voltage harmonics with motor connection (20 MW load)

Because 5-level inverters are mostly used for compressors or other applications that do not require regenerative operation, diode converters are used in the converter circuit of the IEGT. Regarding the harmonics generated in the converter circuit, the 35th and lower order harmonics can be reduced to meet the IEEE-519 standard by shifting each phase of the two serially-connected six-pulse rectifier diode converters by 30 degrees to obtain input current whose waveform is closer to a sine wave. Fig. 20 shows the input current harmonics on 20 MW-load operations [2].

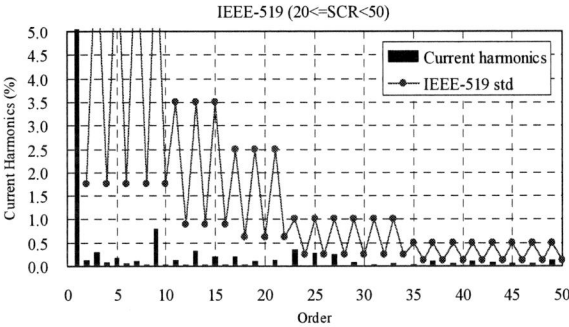

Fig. 20 Input current harmonics on 20 MW-load operations

Thus, by using a system configuration such as the BTB test, the whole system including a motor can be evaluated with the lowest power consumption. In combination with newly-developed, higher-efficient converters, a further reduction of power consumption has been possible in recent years.

VII. CONCLUSION

Thanks to their improved main circuit configuration achieving larger capacity and control method reducing low-order harmonics, the 3-level and 5-level inverters have been widely used for the motor drives for industry applications. The optimally designed motor drive realized the power efficiency of 99%.

In the recently, not only power efficiency but also maintenance improvements, safety of the converter, etc. are required.

Moreover it is important to develop drive units in accordance with various countries' industrial standards and of motor drives flexible in specification and optimal to meet market demands.

REFERENCES

[1] K.Fukuma, H.Hosoda, K.Oda, "Large Voltage Source Inverter for Hot Strip Mill", 2012 IEEE 7th International Power Electronics and Motion Control Conference – ECCE Asia, June 2-5,2012, p540-545

[2] M. A. Mamun, M. Tsukakoshi, K. Hashimura, "Introduction of a Large Scale High Efficiency 5-level IEGT Inverter for Oil and Gas Industry", IEEE Energy Conversion Congress and Exposition (ECCE2010), Session 126: Electric Drives, p4313-4320

[3] K.Oda, M.A Mamun, K.Fukuma, H.Hosoda, M.Lara, "A Large Capacity 3-Level IEGT Inverter for Rolling Mill Application", AISTech 2014 Proceedings, ELECTRICAL APPLICATIONS, Drive system, p3237-3245

[4] M. A. Mamun, D.Yoshizawa, M. Mukunoki, "Performance Evaluation of a Large Capacity 3-Level IEGT Inverter", ECCE-Asia 2013, Session #3-2, p201-207

[5] D.Yoshizawa, M.Mukunoki, K.Omote, M.Hayashi, T.Ishida, "Main circuit technology of a large capacity 3-level IEGT inverter", 2014-IEE Japan, 3-S1-5, Ⅲ-21～26.

[6] D.Yoshizawa, M.Mukunoki, K.Omote, M.Hayashi, T. Ishida, "A Large Capacity 3-Level IEGT Inverter", The 2014 International Power Electronics Conference – ECCE Asia, 20A4-5, p1950-1955

[7] D.Yoshizawa, H.Higure, K.Omote, M.Mukunoki, "The large capacity 5-level IEGT inverter for OIL & GAS application", 2016-IEE Japan, 1-66, I-223～226.

[8] D.Yoshizawa, M.Mukunoki, O.Tanaka, " The large Capacity five-level IEGT Inverter", 2010-IEE Japan, Vol.4, p.91

[9] Y.Ishimaru, M. Adachi, M. Tsukakoshi, R. Nakamura, H. Masuda, Y. Ogashi, Y.Tsuboi, "Testing Facility Using Large Capacity Inverter", The 2014 International Power Electronics Conference – ECCE Asia, 19P1-5, p92-96

[10] M.Tsukakoshi, M.A Mamun, K.Hashimura, H.Hosoda, T.Kojima, "Performance Evaluation of a Large Capacity VSD System for Oil and Gas Industry", 2009 – IEEE ECCE, p3485-3492

Higher Radial Suspension Force of Magnetic Bearing on Centrifugal Compressor for HVAC

Yuji Nakazawa[1*], Yusuke Irino[1], Atsushi Sakawaki[1]and Kazunobu Ohyama[1]

Technology and Innovation Center, DAIKIN INDUSTRIES, LTD, Osaka, Japan

*E-mail: yuuji.nakazawa@daikin.co.jp

Abstract— The implementation of a magnetic bearing within a high-speed, high-capacity SPMSM (Surface Permanent Magnet Synchronous Motor) (450kW, 17,000min[-1]) and the subsequent elimination of the step up gear from the centrifugal compressor have achieved increased efficiency, size reduction, and oil-free design of large HVAC chillers. This report examines the characteristics of magnetic bearings and describes the higher radial suspension force technology that allows for a shorter bearing length, key to achieving higher speeds and compressor size reduction.

Keywords— Magnetic Bearing, Centrifugal Compressor, High speed, Direct drive, Oil-free

I. INTRODUCTION

In the world-largest North American air conditioner market accounting for about 30% of the world's total sales of air conditioning equipment, large buildings normally use central air conditioning systems with a concentrated configuration of heat source equipment (water cooled chillers) of more than several hundred kW in total comprising large-capacity centrifugal compressors.

Recent increasing energy-saving consciousness in the market has led to the quick spread of inverter controls in the 2010s and later, and magnetic bearings as well. A configuration of a centrifugal compressor, magnetic bearings and a high-speed SPMSM with a gearless, direct drive of the impeller is a very highly efficient, downsized, oil-free system [1].

Chapter 2 of this paper describes features and advantages of magnetic bearings that support a high-speed, large-capacity SPMSM (450kW, 17,000min[-1]) of a centrifugal compressor. Chapter 3 describes the higher radial suspension force for our newly developed magnetic bearing. To directly drive a centrifugal compressor's impeller without using a step up gear, it is necessary to increase the radial force per the unit length of the magnetic bearing so that the rotor shaft can be shortened. Using a specially designed structure that permits the use of shorter coil ends and can suppress magnetic saturation has led to a larger radial force by 27% compared with the ordinary design of magnetic bearings. Chapter 4

describes the structure and technical specifications of our newly developed magnetic bearing system. Also, by applying magnetic bearing with higher suspension force, it is shown the orbit of the shaft during rotating stall area could be kept within the allowable range with measurement waveform.

II. FEATURES AND ADVANTAGES OF A CENTRIFUGAL COMPRESSOR WITH MAGNETIC BEARINGS

An ordinary centrifugal compressor comprises a drive system of a large-capacity IM (Induction Motor) connected to the compressor via a step up gear to obtain high-speed revolution of the refrigerant compressing impeller, as shown in Fig.1(a). However, such high-speed systems suffered large friction losses due to the gear and bearings, and needed a lubrication circulation system to lubricate these devices.

To eliminate these disadvantages of the existing centrifugal compressor systems, a configuration using magnetic bearings has been proposed (see Fig.1(b)). The magnetic bearings provide a non-contact, frictionless support of the rotating axis, which means that the motor can be rotated at higher speed with direct drive of the impeller. Therefore, the required high-speed revolution of the compressor impeller can be obtained without using a step up gear.

Therefore, advantages of using magnetic bearings include:

(1) Mechanical losses decrease whereby the compressor's operating efficiency increases
(2) Disuse of gears and bearings subject to wear eliminates overhaul inspection of the compressor
(3) Downsizing due to higher speed
(4) Disuse of the lubrication system eliminates maintenance work for the system

How much mechanical losses attributable to gears and bearings can be reduced by using magnetic bearings is described in the literature of a Danfoss make compressor Turbocor®. This compressor is directly coupled to a 120HP PMSM (Permanent Magnet Synchronous Motor) with magnetic bearings to provide completely contactless

The 2018 International Power Electronics Conference

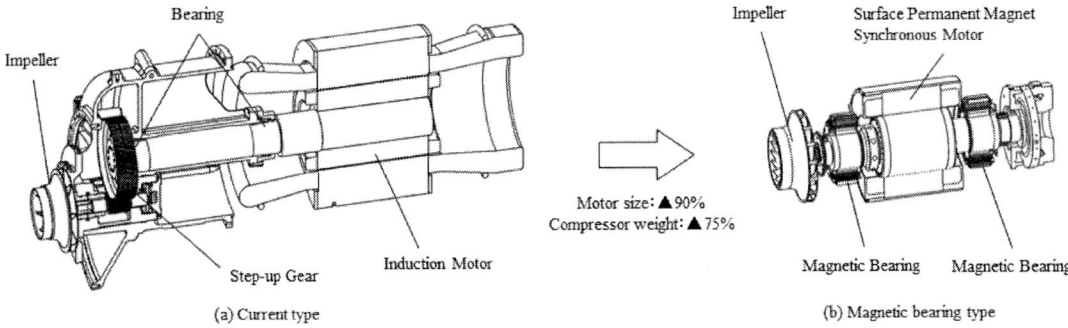

Motor size: ▲90%
Compressor weight: ▲75%

(a) Current type

(b) Magnetic bearing type

Fig.1. Compressor structure.

support of the axis during operation at 48000 min⁻¹, reducing mechanical losses to 180W compared to about 10kW for conventional configurations with ordinary bearings [1].

Our company has released, in 2009, Magnitude® series central air conditioning water-cooled chiller systems (cooling capacity = 500 to 700USRT) comprising a centrifugal compressor with magnetic bearings intended for North American markets. The large-capacity 450kW SPMSM can be operated at as high as 17000 min⁻¹ thanks to the magnetic bearing's completely noncontact support of the motor axis, which enables a direct drive of the high-speed compressor.

The noncontact, direct drive system achieved by using magnetic bearings reduced mechanical losses from 13.5kW to 0.4kW. Changing motor types from IM to PMSM led to improvement in motor efficiency in rated output from 95% to 98%. As a result, the product efficiency (representative IPLV value) rose by 1.5 points to an almost industry-highest value of 11. Comparison in performance between centrifugal compressors with magnetic bearings and those with ordinary bearings is shown in TABLE I.

TABLE I
PERFORMANCE OF THE CENTRIFUGAL AND COMPRESSOR

	Conventional	Magnetic Bearing
Cooling Capacity	500 – 600USRT	500 – 600USRT
Motor power	450kW	450kW
Motor Speed	3600min⁻¹	17000min⁻¹
Motor efficiency	95%	98%
Motor size	D508mm L788mm	D290mm L260mm
Compressor weight	1440kg	380kg
Losses (Gear + Bearing)	13.5kW	0.4kW
Production efficiency (IPLV typical)	9.7	11.2

III. BEARING FORCE CAPABILITY OF VARIOUS TYPES OF RADIAL MAGNETIC BEARINGS

It is needed to adapt several usage of air conditioner for each region in order to sell Magnetic Bearing Centrifugal Compressors all over the world. In the US, Cooling capacity increases or decreases in accordance with outdoor temperature since the air conditioner is mainly operated for whole room. On the other, in China or other Asia countries, the air conditioners are sometimes operated with low cooling capacity since users in those countries adjust each room temperature individually, and it is required to conserve power consumption by stopping air conditioners in unused room.

Fig.2 shows relations between operation area of water-cooled centrifugal compressor, cooling capacity and condenser temperature (≈Outdoor temperature).

The shaded area in upper left shows operation area required additionally for China and other Asian countries.

However, in this operation area, stall phenomenon that causes exfoliation that refrigerant gas does not flow along the vane and decrease of aerodynamic lift happens in some of vanes since the refrigerant flow is lowered with high ratio of suction and discharge pressure of compressor and angle of incidence of the vane inside an

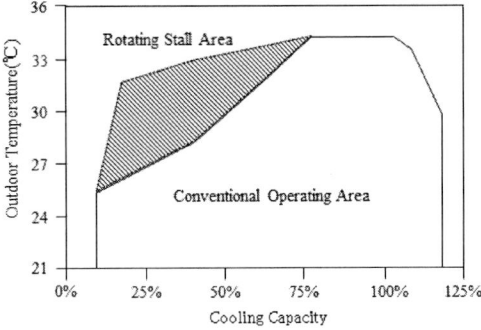

Fig.2. Operation area of water-cooled centrifugal compressor.

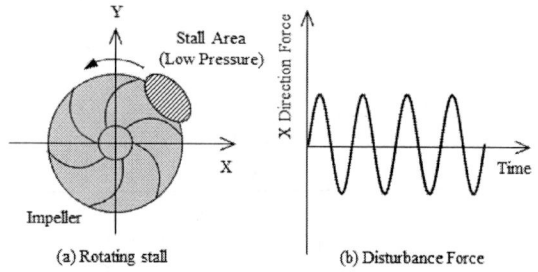

(a) Rotating stall   (b) Disturbance Force

Fig.3. Rotating stall and disturbance force.

impeller becomes bigger than the refrigerant flow relatively.

That is called rotating stall what the stall area moves vane to vane in rotational direction since each vane is affected to next vane of a Impeller [2]-[3]. (Fig.3(a))

As a result, disturbance force oscillating a rotor is generated as shown in Fig.3(b) since fluid is not composed in the staled vanes and that causes imbalance of pressure distribution. In the conventional magnetic bearing, it was impossible to suppress this oscillating disturbance force. Therefore, increasing of radial magnetic force is required.

The axis-supporting capacity of a magnetic bearing can be increased easily by increasing the lamination thickness. However, such modifications leading to a longer axis may cause bending resonance related problems to occur. Therefore, we developed a magnetic bearing having a larger axis-supporting capacity per the unit length of the axis plus coil ends.

Existing types of radial magnetic bearings include a hetero-polar type bearing that generates attracting power using a stator core and coil only, and homo-polar type

bearing consisting of a stator core, coil and magnet [4]-[5]. The latter uses a magnet to reduce the electric power needed for controlling the magnet bearing. However, it may have to be complicated in structure and larger in size. Therefore, the hetero-polar type magnetic bearing better suits the need for larger load-bearing capacity. Possibilities of increasing the load-bearing capacity of hetero-polar type magnetic bearings, and the load-bearing capacity of various derivatives of the hetero-polar type magnetic bearing as well, is discussed below.

Implementations of hetero-polar type magnetic bearings reported to date include those having an equidistant C-type core and those having an E-type core. Fig.4(a) shows the equidistant C-type core. Teeth are arranged radially at equal distance and connected in series so that magnetic flux flows oppositely between adjacent two teeth. The configuration is named a C-type core after the C-shaped single magnetic path formed as described above. Fig.4(a) shows an implementation using two C cores per axis.

Fig.4(b) shows the E-type core. The core has a coil on the E-shaped core's mid tooth only. E-type cores are independent each other and arranged so that they have a constant distance with others for the purpose of shielding against magnetic flux leak protection and axis-to-axis interference. Therefore, it is possible to reduce the distance between adjacent teeth and resultantly increase the width of the teeth. The E-type core is better than the equidistant C-type core in that the bearing capacity per the unit length of the axis can be increased more. However, a disadvantage of the E-type core is that the coil end becomes longer since the wire is wound on the

	(a) Equidistant C-type core	(b) E-type core	(c) Non-Equidistant C-type core
Stator core length [mm]	55.3	43.3	55.3
Coil end length [mm]	20	32	20
Total axial length [mm]	75.3	75.3	75.3
Radial force @16A [N]	1502	1499	1902

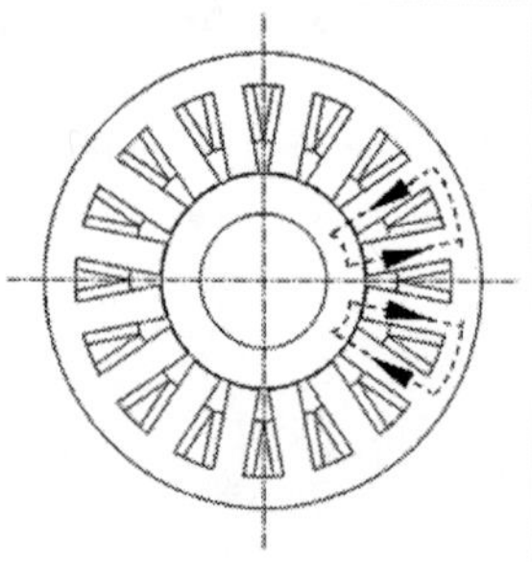

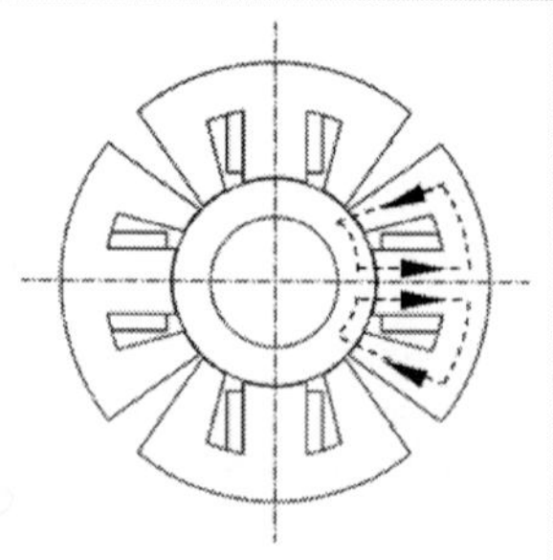

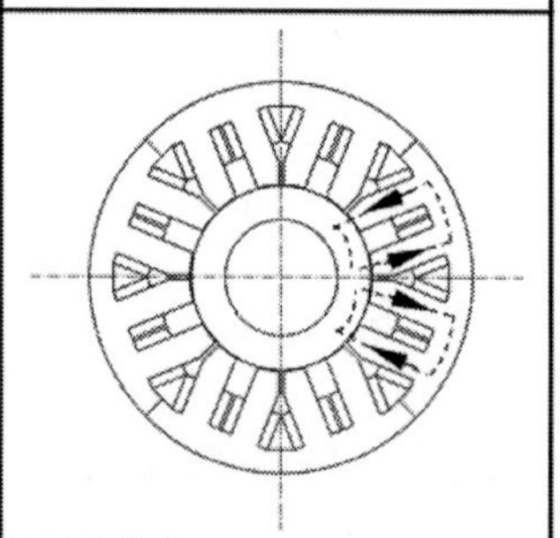

Fig.4. Various type of Hetero polar Radial Magnetic Bearing.

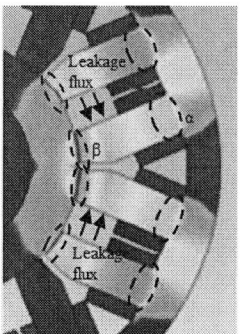

Fig.5. Analysis result of Electromagnetic field.

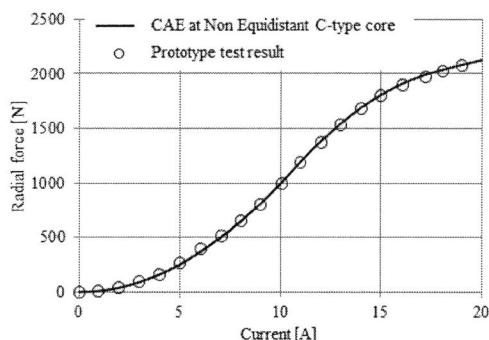

Fig.7. Analysis and measurement result of Magnetic Force.

mid tooth only.

Therefore, we developed a non-equidistant C-type core that, providing the same per-unit-length bearing capacity as the E-type core, has a shorter coil end length achieved as a result of winding dispersedly on all teeth. Fig.4(c) shows the structure of the non-equidistant C-type core.

Fig.5 shows the result of electromagnetic field analysis to show the distribution (contour) of magnetic flux density. This new core design causes the flux density to become larger at the outer end (root) of the teeth (regions denoted by α in Fig.5) than the inner end (tooth tip) (regions denoted by β in Fig.5) due to the leakage flux between teeth. Therefore, we found it was possible to increase the bearing capacity by increasing the width of the teeth at the outer end (root) so that magnetic saturation could be reduced and decreasing their width at the inner end (tooth tip) so that the flux density in the gap could be increased.

Fig.6 shows the bearing capacity of various types of radial magnetic bearing. These magnetic bearings have the same maximum height measured inclusive of the inside and outside diameter of stator, inside and outside diameter of rotor and coil end length to facilitate

comparison between different structures of magnetic bearing.

Within a maximum electric current of 16A that corresponds to the radial force requirements prevailing in the US market, the equidistant C-type core and E-type core are comparable with respect to the radial bearing capacity while the non-equidistant C-type core shows a larger bearing capacity by 27%. The non-equidistant C-type core can satisfy the requirements for radial bearing capacity for magnetic bearings intended for Chinese and other Asian markets. Measurements of radial bearing capacities using a prototype of the non-equidistant C-type core is shown in Fig.7. These measurements agree well with analytical results.

IV. OUR PROPRIETARY MAGNETIC BEARING SYSTEM

The diagram and specification of magnetic bearing is shown in Fig.8 and TABLEII.

To rotate the impeller at high speed, the compressor is rotated by SPMSM connected directly thereto. The SPMSM is driven by a voltage-source PWM inverter and the compressor system can be operated at variable speed.

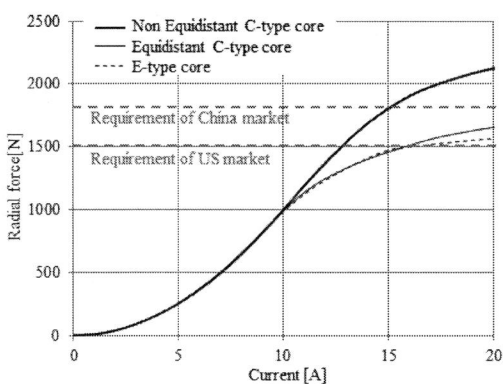

Fig.6. Analysis result of Radial Force.

TABLE II
SPECIFICATIONS FOR MAGNETIC BEARINGS.

Cooling Capacity	500 - 700USRT
Motor Power	450kW
Motor Speed	17000min^{-1}
Rotor weight	48kg
Maximum Radial Force	1800N
Clearance at Touchdown Bearing	0.25mm
Number of Control Axes	5
Cycle for Position Control	25kHz
Power Loss at Magnetic Bearing	Maximum 0.4kW
Power Supply Voltage	DC350V

The 2018 International Power Electronics Conference

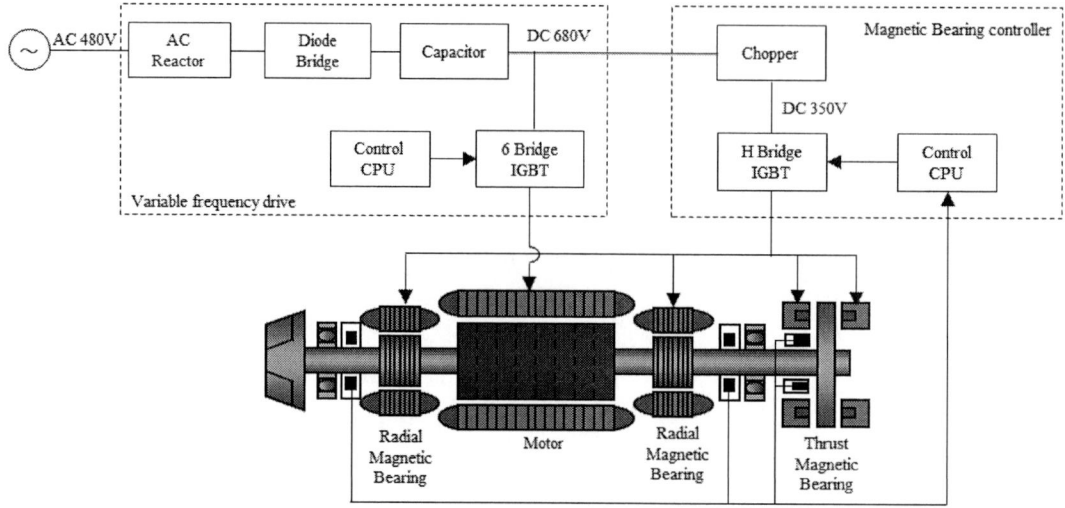

Fig.8. Magnetic Bearing System Diagram.

In addition, refrigerant gas flow is controlled by guide vane to expanding operation area [6].The magnetic bearings applied for high-speed and direct drive systems are composed of radial magnetic bearings supporting radial direction and thrust magnetic bearings supporting axial direction. The magnetic bearing controller implements the feedback control by using the signal of the position sensor for active control of 5 axes. FIG.9 is a photograph of the Compressor rotor and Radial magnetic bearing stators of prototype model. Magnetic force with radial magnetic bearings and thrust magnetic bearings levitate the Rotor which weighs is 48kg to realize non-contact completely.

Our magnetic bearing system can use the compressor motor as a generator in the event of power failure whereby the magnetic bearing controller can be kept active so that the motor axis can be kept levitated during rotation. Moreover, it has a touchdown bearing that works when the rotor is not levitated or when the magnetic bearing or control circuit fails.

Fig.10 shows the Rotor orbit under maximum load condition in conventional operating area. The dashed circle in Fig.10 indicated the inner diameter of the touchdown bearing. Since the Rotor orbit is lower than approximately 40 um and smaller than the inner diameter of the touchdown bearing. The rotor can be stably levitated.

Fig.11 shows the current waveform under same condition at the one of Fig10. The dashed line in the Fig.11 shows the limit of the current effective value depend on magnetic bearings controller. Since a disturbance force to the rotor is small, current is smaller than the limit value as shown in Fig11.

Fig.12 shows the orbit under the rotating stall

Fig.9. Rotor shaft and Radial Magnetic Bearing.

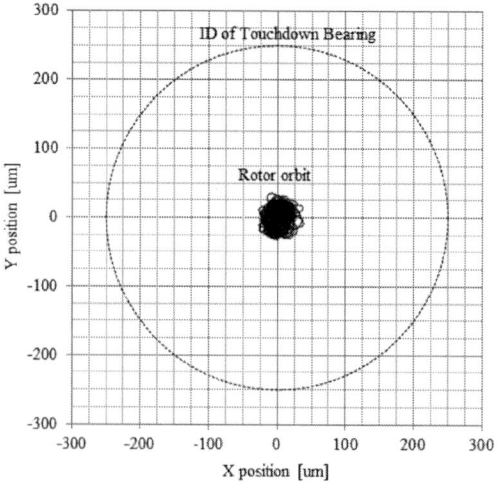

Fig.10. Orbit of the rotor under normal operation.

248

The 2018 International Power Electronics Conference

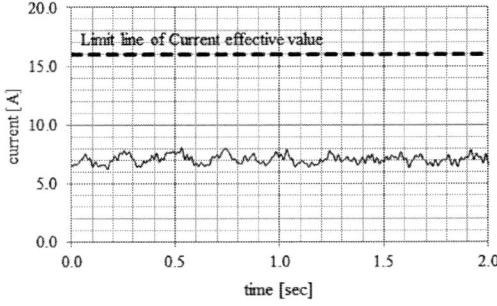

Fig.11 Current waveform of the Radial Magnetic Bearing under normal operation

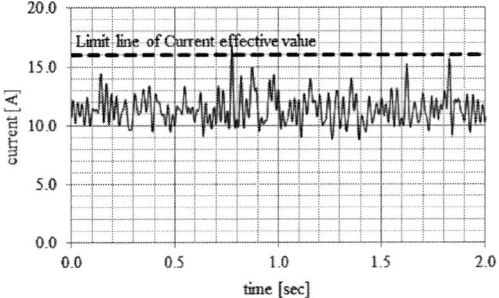

Fig.13 Current waveform of the Radial Magnetic Bearing under rotating stall.

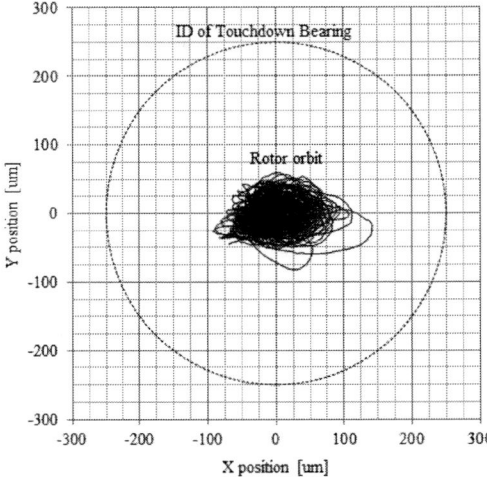

Fig.12 Orbit of the rotor under rotating stall

V. SUMMARY

This paper described our proprietary technologies for higher radial suspension force of magnetic bearings applicable for high-efficiency centrifugal compressors, and features of the water-cooled chiller unit comprising our newly developed magnetic bearings.

Large-capacity water-cooled chiller systems that have now been rather outdated in that they have yet to make full use of the recent power electronics technologies would recover their commercial values in the world market using these magnetic bearing technologies that can contribute to higher operating efficiency, downsizing, and oil-free.

condition. The Rotor orbit is larger than that shown in FIG.10, nevertheless it shows that the rotor is kept levitated without contact touchdown bearing.

Fig.13 shows the current waveform under same condition at the one of Fig12. Since the current is oscillating, disturbance force is applied to the rotor under this condition. The magnetic bearing force is sufficient for the disturbance force, because the current effective value does not exceed the limit (16A).

Hence, it could be proved experimentally that the developed Radial magnetic bearing can levitate the Rotor during disturbance force of the rotating stall.

REFERENCES

[1] Ron Conry, Len Whelan, John Ostman, "Magnetic Bearings, Variable Speed Centrifugal Compression And Digital Controls Applied In A Small Tonnage Refrigerant Compressor Design", (2002) International Compressor Engineering Conference. Paper 1500.

[2] Turbomachinery Society of Japan, "Turbomachinery", Japan Industrial Publishing Co.LTD, ISBN978-4-8190-1711-4 (IN Japanese)

[3] Jan Tommy Gravdahl, Olav Egeland, " Compressor Surge and Rotating Stall: Modeling and Control 1st", Springer Publishing Company, ISBN:1447112113 9781447112112

[4] T.Azuma, Y.Kanemitsu, Y.Fukushima, O.Matsushita, "Magnetic Bearing Guidebook for Rotating Machine Designers", Japan Industrial Publishing Co.LTD, ISBN4-8190-1606-7 (IN Japanese)

[5] T.SHINSHI, "Magnetic Bearing – Fundamentals and Applications-", Journal of the Japan Society of Precision Engineering vol.78, No12, 2012 (IN Japanese)

[6] Joost Brasz, Lee Tetu, "Variable-Speed Centrifugal Chiller Control for Variable Primary Flow (VPF) Applications", (2008) International Compressor Engineering Conference. Paper 1922.A

The 2018 International Power Electronics Conference

Novel Switching Control Method for Full-Bridge DC-DC Converters for Improving Light-Load Efficiency Using Reverse Recovery Current

Fumihiro Sato[1*], Takae Shimada[1], and Takayuki Ouchi[1]

1 Power Electronic Systems Research Department, Hitachi, Ltd., Hitachi, Ibaraki, Japan

*Email: fumihiro.sato.yp@hitachi.com

Abstract— The phase-shift full-bridge DC-DC converter has been widely used for electric vehicles. However, the conversion efficiency is degraded under light-load conditions due to switching losses of the semiconductor switches on the high-voltage side of the converter. In this paper, a novel switching control method for full-bridge converters is proposed for improving the light-load efficiency. To reduce switching losses with the method, lagging-leg switches stop switching in the off-state under light-load. The proposed converter exhibits zero voltage switching for all switches with the use of a reverse recovery current in the body diode of lagging-leg switches. The effectiveness of the method was experimentally verified by using a 2.8-kW prototype converter. Experimental results show that the proposed method increases light-load efficiency by 33% at a load ratio of 3% compared with the conventional phase-shift control.

Keywords— *Full-bridge DC-DC converter, reverse recovery of body diode, light-load efficiency, zero voltage switching*

I. INTRODUCTION

In recent years, due to the growing awareness of the need for global environmental conservation, the spread of electric vehicles (EVs) and plug-in hybrid vehicles (PHEVs) has been desired. A DC-DC converter is installed in these vehicles to transfer energy from the main high-voltage battery to an auxiliary low-voltage battery [1], [2]. Such converters need to be highly efficient. In particular, the light-load efficiency of a DC-DC converter should be considered since the converter operates under a light-load while charging the main high-voltage battery from a power supply. In this case, the converter supplies a small amount of energy to control circuit and auxiliary components with a load ratio of less than 5%.

A full-bridge converter that uses phase-shift control [3] is most commonly used for EVs and PHEVs. Zero voltage switching (ZVS) can be implemented for all switches by using a phase-shift control method under heavy-load conditions. However, under light-load conditions, ZVS is not maintained due to the small current flowing in the circuit of phase-shift converter, which is needed to discharge the parasitic capacitor.

To improve the light-load efficiency, several circuit topologies have been proposed [4]-[6]. A full-bridge converter was proposed for achieving ZVS for leading-leg switches and zero current switching (ZCS) for lagging-leg switches [4]. However, the converter requires an additional auxiliary circuit that consists of an inductor, capacitor, and two diodes. A full-bridge converter that uses a voltage complementary auxiliary network on the primary side for ZVS over the entire load range was proposed [5]. However, a full-bridge ZVS converter is added to the auxiliary energy storage circuit that consists of one coupled inductor and two auxiliary capacitors. A full-bridge converter that offers ZVS over a wide range of output loads was proposed [6]. However, this topology requires an additional auxiliary transformer.

Several control methods without additional auxiliary components have been proposed [7]-[9]. A full-bridge converter that switches between a phase-shift topology and a forward topology in response to load was proposed [7]. It improves efficiency in 11% under light-load conditions as compared with phase-shift control, but it does not achieve ZVS in switching devices, so the reduction in loss is insufficient. A switching control method for full-bridge converters was proposed that consists of three control modes: phase-shifted switching, pulse-width-modulated (PWM) switching, and PWM switching with burst mode [8]. It improves efficiency in 26% under light-load conditions as compared with phase-shift control. However, it requires complex control-mode switching. A dead-time adjustment strategy for improving the efficiency of the light-load conversion of phase-shift converters was proposed [9]. It can adjust the dead time for each switch simultaneously in accordance with the load current, improving efficiency in 5.75% as compared with phase-shift converters without dead-time adjustment. However, under light-load conditions in which enough energy cannot be stored to enable ZVS for leading-leg switches in resonant inductors, the switches cannot achieve ZVS even if the dead time of the phase-shift converter is extended. Therefore, adjusting the dead time has a limited effect on increasing the efficiency under light-load conditions.

We propose a novel switching control method for achieving ZVS for all switches under light-load conditions without additional auxiliary components. The proposed method stops the switching of lagging-leg switches in the off-state under light-load conditions. It effectively uses the reverse recovery current of the body diodes in lagging-leg switches for power conversion.

II. Circuit Topology

The circuit topology of a full-bridge phase-shift converter is shown in Fig. 1, where V_1 indicates a high-voltage battery, and V_2 indicates a low-voltage one. The high voltage side has a voltage-fed full-bridge circuit that consists of leading source switches (H_1 and H_2) and lagging-leg switches (H_3 and H_4). Low on-resistance MOSFETs, such as super junctions, are used for all H_1 to H_4 at a high switching frequency. A series resonant inductor L_r is connected to the primary winding N_1 of the transformer Tr. The low-voltage side has a current inverter that uses switching devices S_1 to S_4 and smoothing inductor L_{out} with an active clamp circuit that consists switching devices S_3 and S_4 [10], [11].

III. Control Method

A. Conventional Phase-Shift Control

Fig. 2 shows the operational waveforms of the conventional phase-shift converter under a light-load. V_{gsH1}–V_{gsH4} are the gate-source voltages of H_1–H_4, V_{dsH1}–V_{dsH4} are the drain-source voltages of H_1–H_4, V_{N1} is the voltage applied to N_1, and I_{Lr} and I_{Lout} are the currents in L_r and L_{out}. Each of the legs (H_1 and H_2, H_3 and H_4) are switched at a duty of 50% and 180 degrees out of phase with each other at a constant frequency. The phase shift value decides the amount of energy transferred from V_1 to V_2.

The phase-shift converter enables ZVS under a heavy load since a wide voltage pulse is applied to N_1, and, thus, L_r stores enough energy to achieve ZVS in switches on the high-voltage side. However, since a narrow voltage pulse is applied to N_1, L_r cannot store enough energy to enable ZVS under a light-load. In this case, the charge-discharge of the parasitic capacitance of H_1 to H_4 is insufficient due to the small current flowing on the primary side of the Tr. Turning on high voltage switches while this charge-discharge is insufficient would make hard switching, causing the increasing of switching loss and decreasing of efficiency.

B. Proposed Control

Fig. 3 shows switching stages on the primary side of the Tr with the proposed control, and Fig. 4 shows operational waveforms of the proposed control. The proposed control can be applied to the conventional full-bridge circuit topology without any change in power stages. Under a heavy load, the proposed control requires a converter to operate in general phase-shift control, and under a light-load, H_3 and H_4 stop switching in the off-state, and H_1 and H_2 are driven by complementary signals with a dead time. The most significant point of the proposed control is that it utilizes the reverse recovery current of the body diodes in H_3 and H_4.

[Stage (a)] On this stage, H_1 is in the on-state, and the H_4 body diode conducts a current. The current flowing in L_r decreases gradually since V_1 is applied to L_r.

[Stage (b)] When the current flowing in L_r decreases to zero, the circuit state switches to stage (b). The current flowing in L_r increases in the positive direction due to the reverse recovery current flowing in the H_4 body diode. At this stage, despite H_4 being in the off-state, the circuit behaves as if H_4 is in the on-state.

[Stage (c)] After the reverse recovery time of the H_4 body diode has elapsed, the current flowing in L_r becomes a charge current to the H_4 parasitic capacitor and a discharge current to the H_3 parasitic capacitor.

[Stage (d)] After the H_3 parasitic capacitor discharges, the circulating current flows through the H_1, L_r, N_1, and H_3 body diodes.

[Stage (e)] At this stage, H_1 is turned off. The current flowing in L_r becomes a charge current to the H_1 parasitic capacitor and discharge current to the H_2 parasitic capacitor.

[Stage (f)] After the H_2 parasitic capacitor discharges completely, the H_2 body diode conducts a current.

[Stage (g)] At this stage, H_2 is turned on. The voltage of the H_2 parasitic capacitor is zero due to the conduction of the H_2 body diode; thus, H_2 achieves ZVS.

The circuit behavior in stage (g) is symmetrical with that of stage (a). Subsequently, after stages that are symmetrical with stage (b) to (f) have been completed, the operation cycle returns to stage (a).

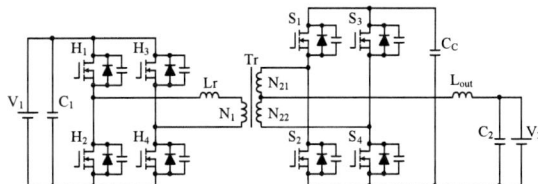

Fig. 1. Circuit topology for full-bridge phase-shift DC-DC converter.

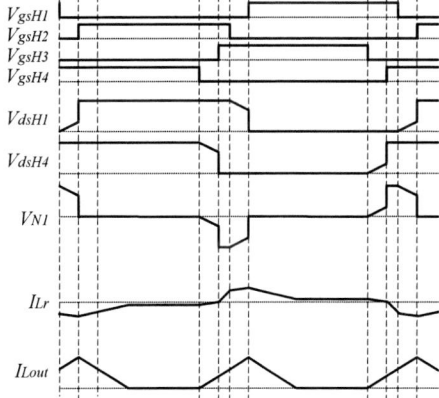

Fig. 2. Operational waveforms of conventional phase-shift control under light-load.

The 2018 International Power Electronics Conference

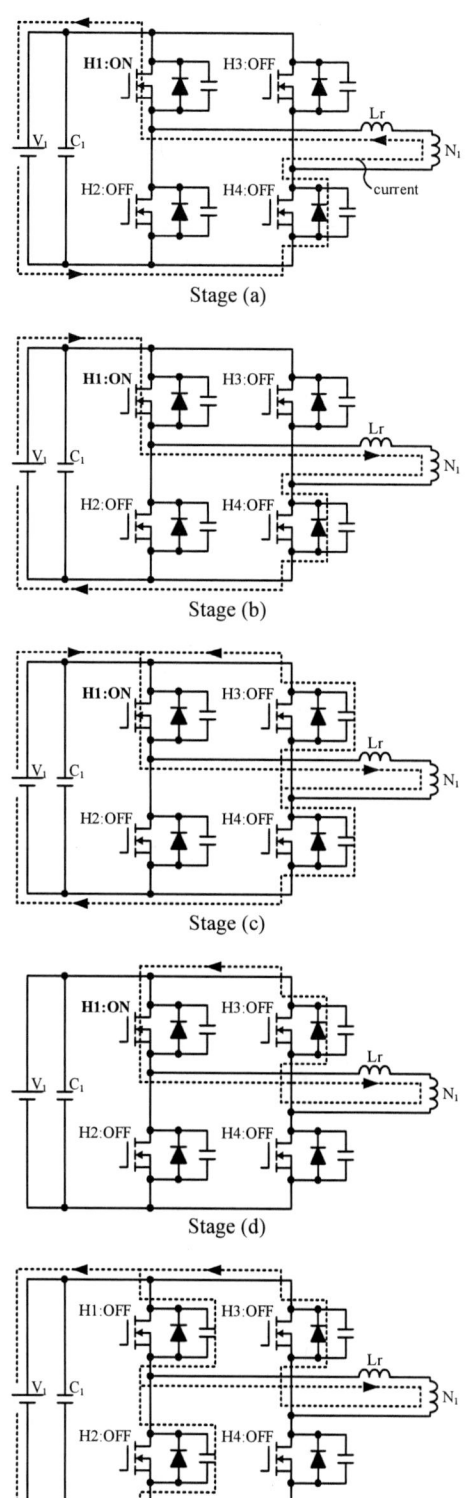

Stage (a)

Stage (b)

Stage (c)

Stage (d)

Stage (e)

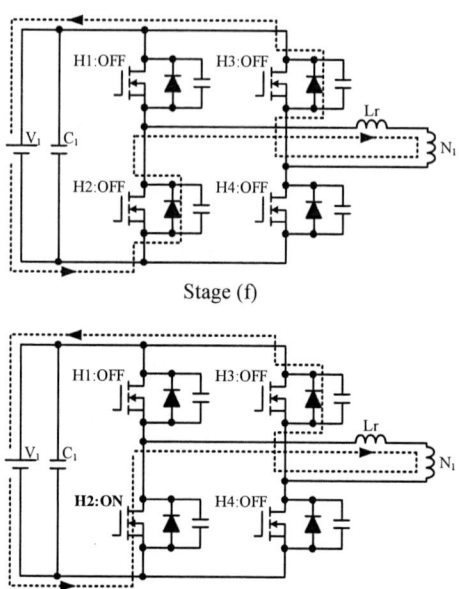

Stage (f)

Stage (g)

Fig. 3. Switching stages of proposed control method.

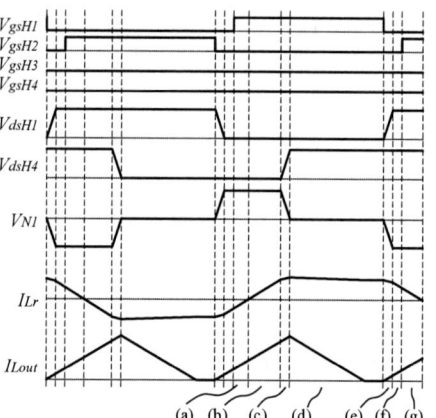

Fig. 4. Operational waveforms of proposed control method.

Fig. 5 and Fig. 6 show the operation waveforms of the proposed control method in detail. Fig. 5 shows the switching stages with/without a reverse recovery current in lagging-leg switches. In the proposed method, the converter is able to transfer a small amount of energy from V_1 to V_2 for utilizing the charge-discharge current of the parasitic capacitor in each switch since the voltage of N_1 enlarges at stage (e) as shown in Fig. 5(a). To increase the output power, it is preferable to adopt switching devices that have a body diode with a large reverse recovery charge for H_3 and H_4 because doing so would make it possible to extend the period when V_1 is applied to N_1, as shown in Fig. 5(b). The voltage pulse width applied to N_1 is determined by the output capacitances of each switch and the reverse recovery charge of the lagging-leg switches with proposed method.

Fig. 6 shows the operational waveforms in a high frequency state and low frequency state with the proposed method. The L_{out} current increases during the

252

powering period, which means that V_1 is applied to N_1 [stages (f) to (b)] and the charging time of the parasitic capacitances of H_1 to H_4 [stages (e), (c)]. Since the transferred energy per switching cycle is at a constant value determined by the length of the powering period, output power control for the proposed converter can be achieved by varying the switching frequency of H_1 and H_2. That is, by increasing the frequency, the number of powering period increases, and thus, the output power can be increased.

In the phase-shift control, the duration of the period in which V_1 is applied to N_1 is determined by a duty ratio, so under a light-load, the current in L_r for discharging the parasitic capacitances of each switch is not sufficient enough to achieve ZVS because of the narrow pulse. In comparison, in the proposed control, the duration of the period in which V_1 is applied to N_1 is at a constant value as described above. That is, the current for the charge-discharge to the H_1 and H_2 parasitic capacitances increases more compared with a phase-shift converter under a light-load. Consequently, ZVS becomes easier with the proposed control than phase-shift control under a light-load.

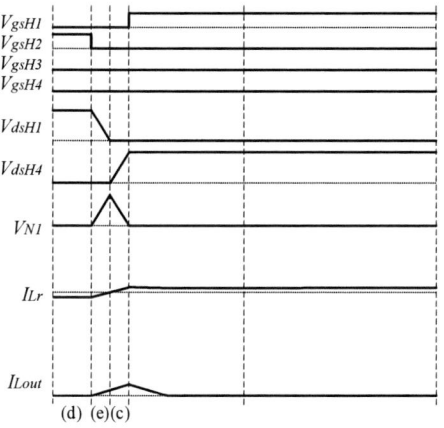

(a) without reverse recovery current

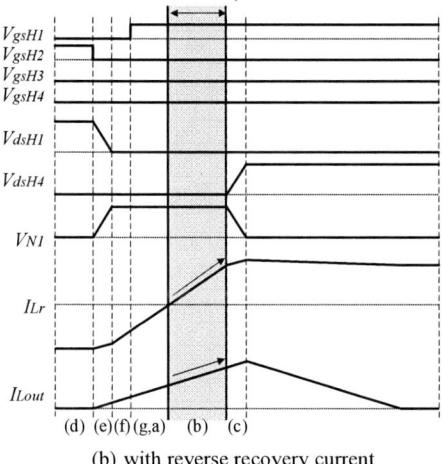

(b) with reverse recovery current

Fig. 5. Operational waveforms with/without reverse recovery current in body diode of lagging-leg switches

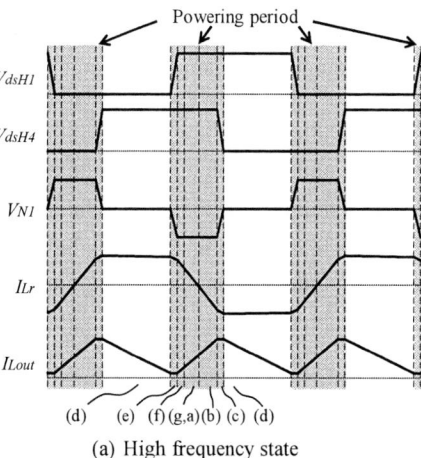

(a) High frequency state

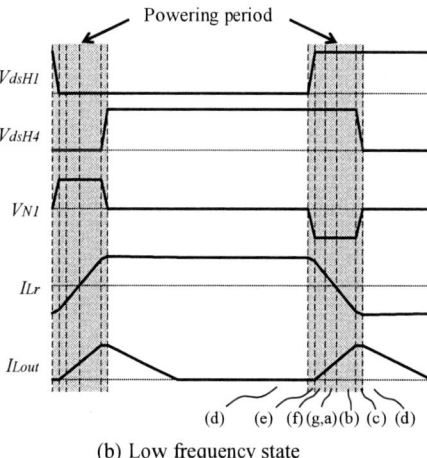

(b) Low frequency state

Fig. 6. Operational waveforms for high-switching-frequency operation and low-switching-frequency operation.

IV. EVALUATION

A 2.8-kW prototype full-bridge DC-DC converter with phase-shift control and the proposed control method were constructed for evaluation. Table I shows the parameters for the circuit of the prototype. H_1–H_4 are super junction MOSFETs.

Fig. 7 shows experimental waveforms of the proposed control with switching frequencies in the leading-leg switches, of 100 kHz [Fig. 7(a)] and 50 kHz [Fig. 7(b)]. The conditions were an input voltage of 320 V, an output voltage of 14 V. As shown in Fig. 7(a), it was revealed that the converter with the proposed control achieved ZVS in H_1 and H_2 without switching H_3 and H_4 by utilizing the reverse recovery current and charge-discharge current of the parasitic capacitances.

It was also confirmed that the length of the powering period was at a constant value in the proposed control. Fig. 8 shows the relationship between the output power and switching frequency of the leading-leg switches at the input voltage of 410 V and the output voltage of 14 V. It was confirmed that the output power linearly increased in relation to an increase in the switching frequency since the number of powering periods increased linearly.

Fig. 9 shows the measured efficiency, and Fig. 10 shows the measured converter losses at 410 V input and 14 V output. The switching frequency of the proposed converter varied from 40 to 100 kHz. The conventional phase-shift converter operated at a switching frequency of 100 kHz, and the output power was controlled by the amount of phase shift. It was clear that the light-load efficiency between the phase-shift control and proposed control improved in 33%, and the loss was reduced in 77% at a load ratio of 3% compared with the conventional phase-shift control.

TABLE I
PARAMETERS FOR PROTOTYPE CONVERTER

Symbol	Parameter	Value
H_1–H_4	High-voltage-side MOSFETs	650 V
S_1–S_4	Low-voltage-side MOSFETs	150 V
N_1:N_{21},N_{22}	Transformer turns	11:1:1
L_r	Resonant inductor	6.5 µH
L_{out}	Smoothing inductor	1.61 µH

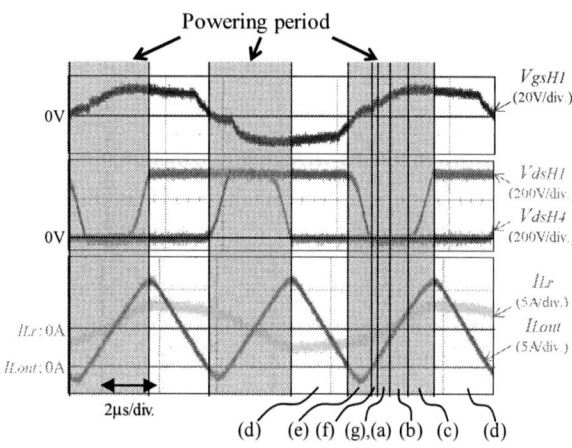

(a) Switching frequency: 100 kHz

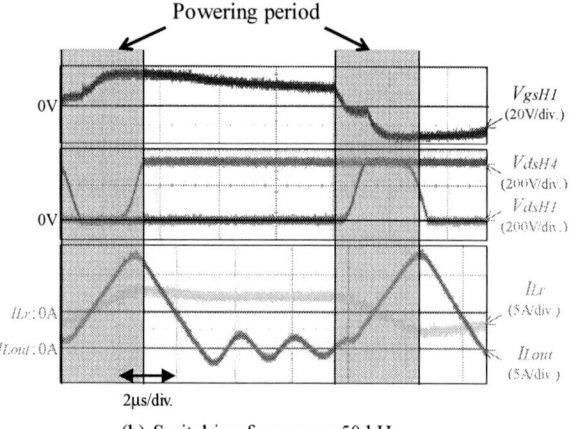

(b) Switching frequency: 50 kHz

Fig. 7. Experimental waveforms of V_{gsH1}, V_{dsH1}, V_{dsH4}, I_{Lout}, and I_{Lr} ($V_1 = 320$ V, $V_2 = 14$ V)

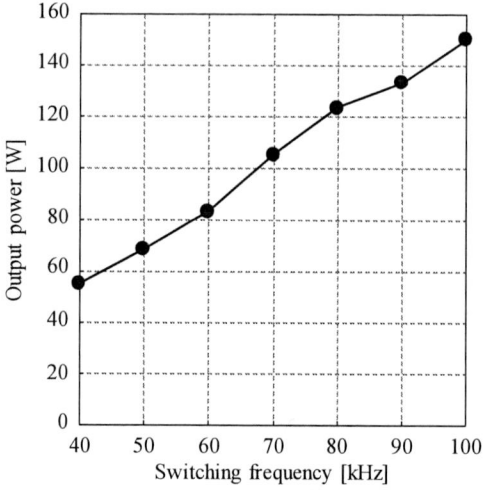

Fig. 8. Output power vs. switching frequency ($V_1 = 410$ V, $V_2 = 14$ V)

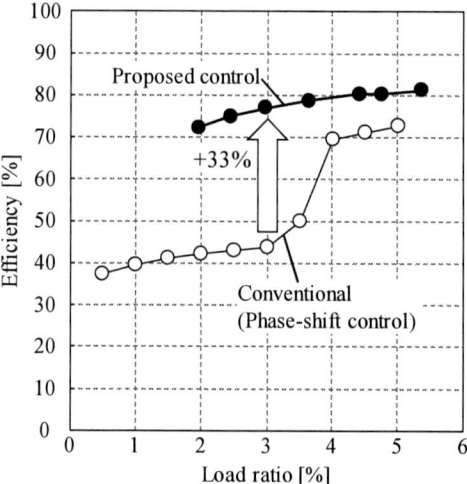

Fig. 9. Measured efficiency vs. load ratio.

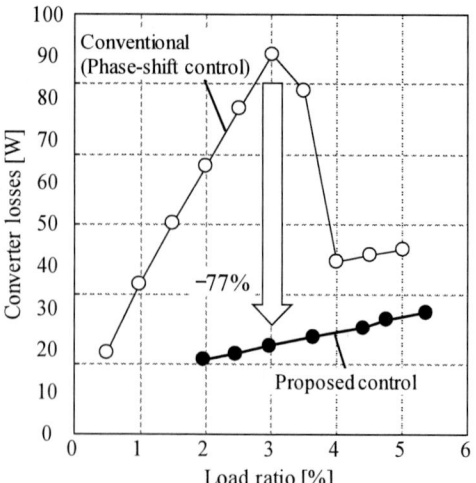

Fig. 10. Measured converter losses vs. load ratio.

V. CONCLUSION

A novel control method for a DC-DC converter with which switching loss decreases more compared with the conventional phase-shift control was developed. With the proposed method, lagging-leg switches stop switching in the off-state, and the reverse recovery current in the body diodes of the switches are used for power conversion. The proposed method exhibited high efficiency under light-load conditions, an increase in 33%, and the loss decreased 77% at a load ratio of 3% compared with the conventional phase-shift control.

REFERENCES

[1] A. Emadi, K. Rajashekara, S. S. Williamson, and S. M. Lukic, "Topological Overview of Hybrid Electric and Fuel Cell Vehicular Power System Architectures and Configurations," *IEEE Trans. On Vehicular Technology*, vol. 54, no. 3, pp. 763–770, 2005.

[2] A. Emadi, Y. J. Lee, and K. Rajashekara, "Power Electronics and Motor Drives in Electric, Hybrid Electric, and Plug-In Hybrid Electric Vehicles," *IEEE Trans. On Industrial Electronics*, vol. 55, no. 6, pp. 2237–2245, 2008.

[3] U. Badstuebner, J. Biela, D. Christen, and J. W. Kolar, "Optimization of a 5-kW Telecom Phase-Shift DC-DC Converter with Magnetically Integrated Current Doubler," *IEEE Trans. On Industrial Electronics*, vol. 58, no. 10, pp. 4736–4745, 2011.

[4] X. Wu, X. Xie, C. Zhao, and Z. Qian, "Low Voltage and Current Stress ZVZCS Full Bridge DC-DC Converter Using Center Tapped Rectifier Reset," *IEEE Trans. On Industrial Electronics*, vol. 55, no. 3, pp. 1470–1477, 2008.

[5] Y. Jang and M. M. Jovanovic, "A New PWM AVS Full-Bridge Converter," *IEEE Trans. On Power Electronics*, vol. 22, no. 3, pp. 987–994, 2007.

[6] Z. Chen, F. Ji, B. Ji, and X. Zhang, "A Novel Soft-Switching Full-Bridge Converter for ZVS in Light and Full Load Conditions with Current-Doubler Rectifier," *Proc. of IEEE Power Electronics Conference*, pp. 2339–2346, Sapporo, 2010.

[7] K. I. Hwu, Y. T. Yau, and T. H. Chen, "Improvement in Efficiency of the Phase-shift Current-doubler-rectification ZVS Full-bridge DC-DC Converter," *Proc. of Applied Power Electronics Conference*, pp. 991–997, Anaheim, CA, 2007.

[8] B. Y. Chen and Y. S. Lai, "Switching Control Technique of Phase-Shift Controlled Full-Bridge Converter to Improve Efficiency Under Light-Load and Standby Conditions Without Additional Auxiliary Components," *IEEE Trans. On Power Electronics*, vol. 25, no. 4, pp. 1001–1012, 2010.

[9] B. H. Liu, J. H. Teng, M. Y. Lin, and C. C. Huang, "Light-Load Conversion Efficiency Improvement Strategy for Phase-Shift Full-Bridge Converters," *Proc. of International Future Energy Electronics Conference and ECCE Asia*, pp. 488–493, Kaohsiung, 2017.

[10] T. Shimada, H Shoji, and K. Taniguchi, "Two Novel Control Methods Expanding Input-output Operating Range for a Bi-directional Isolated DC-DC Converter with Active Clamp Circuit," *Proc. of Energy Conversion Congress and Exposition*, pp. 2537–2543, Raleigh, NC, 2012.

[11] T. Shimada, H Shoji, and K. Taniguchi, "A Novel Scheme for a Bi-directional Isolated DC-DC Converter with a DC-link Diode Using Reverse Recovery Current," *Proc. of European Conference on Power Electronics and Applications*, pp. 1–7, Lappeenranta, 2014.

The 2018 International Power Electronics Conference

A 800V/14V Soft-switched Converter with Low-Voltage Rating of Switch for xEV applications

Byeongwoo Kim, Kangsan Kim and Sewan Choi
Department of Electrical and Information Engineering
Seoul National University of Science and **Tech**nology
Seoul, Korea
E-mail: schoi@seoultech.ac.kr

Abstract- **This paper proposes 800V/14V soft-switched dc/dc converter with low-voltage rating of switch for xEV application. The proposed soft-switched converter has theoretically zero dc magnetizing current offset. Also, low-voltage rating of switches makes it possible to achieve low conduction loss and switching loss. It also achieves ZVS turn-on of all switches over the entire load and input-voltage variation. The proposed converter is compared to some low-voltage dc/dc converters. Experimental results form a 2-kW preliminary lab prototype of the proposed converter will be presented to verify the validity of the proposed concept and operation for xEV applications.**

I. INTRODUCTION

Recently, eco-friendly vehicles such as electric vehicles (EVs), hybrid electric vehicles (HEVs) and plug in hybrid electric vehicles (PHEVs) are attracting increasing attention as a viable solution of greenhouse gas emissions, fossil-fuel consumption. Configurations of several types of HEV and PHEV power train systems are delineated in [1], and block diagram of a typical xEV power train is shown in Fig. 1. The low- voltage dc/dc converter (LDC) with power range of a few kilowatts(1-5kW) is used to charge power to 14-V loads such as head lamps, navigation, audio, interior light etc., and charges a 14-V auxiliary battery from a high-voltage battery [2]. The voltage range of the high-voltage xEV battery is usually around 200V to 400V [2]. It takes 5 to 8 hours to fully charge the battery if the battery is charged by 3.3kW onboard battery charger. In order to increase the charging speed and extend the driving range, xEVs with 800V battery pack have been investigated [3]-[4]. In this case a 800V/14V LDC needs to be developed.

In the meantime, the phase-shift full-bridge (PSFB) converter is considered as the most popular topology for the 400V/14V LDC due to its simple structure, wide input range and high reliability [6]-[9], where 600V Si MOSFETs is selected as switching devices of the PSFB converter.

However, the PSFB converter could not be a suitable topology for the 800V/14V LDC since Si based MOSFETs with voltage rating higher than 1000V are seldom found in market, and MOSFETs with high voltage ratings typically have not only high on-resistances $R_{ds(on)}$ resulting in high conduction losses, but also high switching losses due to high blocking voltage.

In this paper, a dc/dc converter with low-voltage rating of switch is proposed for 800V/14V LDC. The proposed

converter has the following features: 1) the switch blocking voltage of half of input voltage; 2) ZVS turn-on of all switches over the entire load and input-voltage range; 3) ZCS turn-off of all diodes without voltage surge related to reverse recovery; and 4) zero magnetizing current offset due to the use of dc blocking capacitors in series with winding of the transformer. To verify the proposed concept and operation for xEV application, experimental results on a 100 kHz, 2-kW preliminary lab prototype are provided.

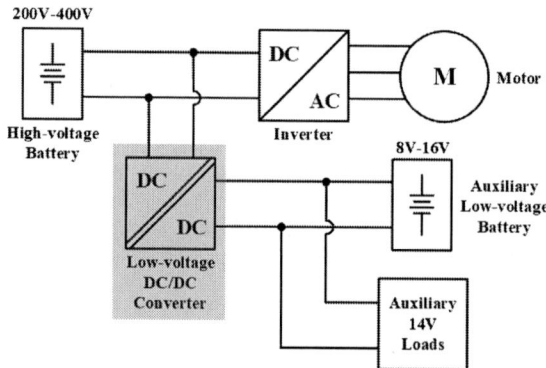

Fig. 1. Block diagram of a typical xEV power train.

II. PROPOSED SOFT-SWITCHED CONVERTER

Fig. 2 shows the circuit configuration of the proposed 800V/14V soft-switched converter with low-voltage rating of switch. Two half-bridge cells in the primary side are connected with the inputs in series, and the secondary side consists of two single diode for rectification and an LC filter of the output. The battery current and voltage of the proposed dc/dc converter is controlled by variable duty ratio.

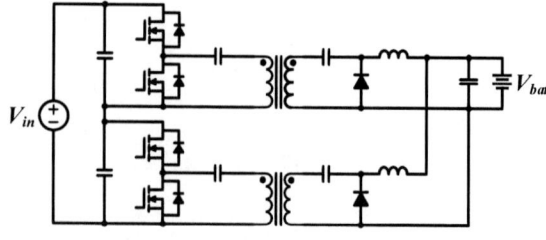

Fig. 2. Circuit configuration of the proposed 800V/14V soft-switched converter.

256

A. Operating Principle

Fig. 3 and Fig. 4 show operating modes and key waveforms of the proposed converter, respectively. The operation of the proposed converter can be divided into five modes, as shown in Fig. 2. In order to simplify the operating mode, it is assumed that magnetizing inductor current is very small so that it can deal with zero current during a switching period.

The half-bridge cell in the primary side is operated with asymmetrical complementary switching and the two switch pair of the half-bridge cells are in interleaved with 180° phase shift, which leads to an increased effective switching frequency of the output side. Therefore, battery current ripple is reduced. It can be seen from Fig. 4 that the proposed converter achieves ZVS turn on of all switches. Fig. 5 shows switch turn-on current waveforms of the switches according to variation of duty ratio (input voltage) and load magnitude.

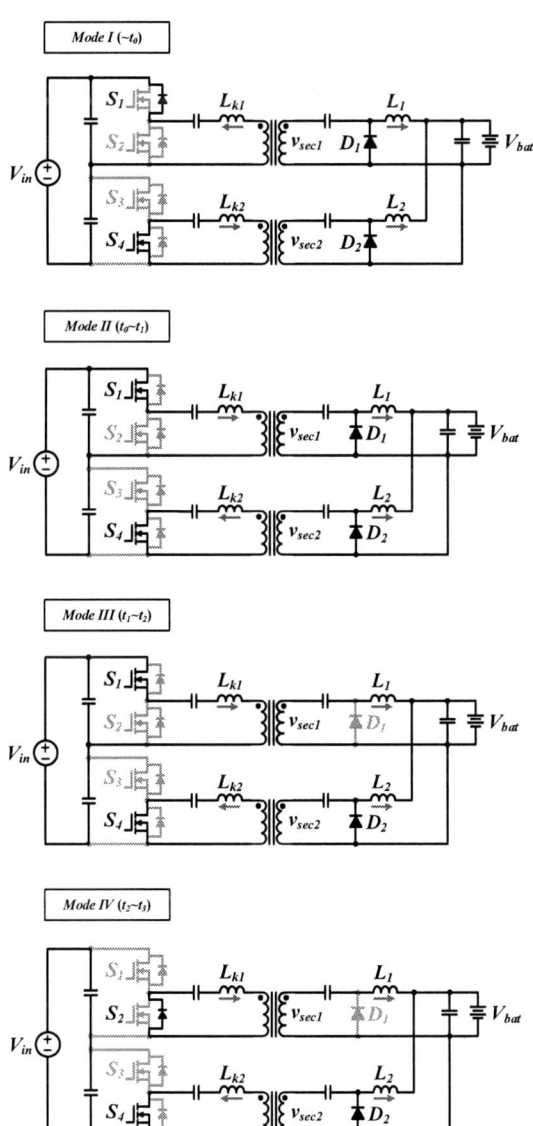

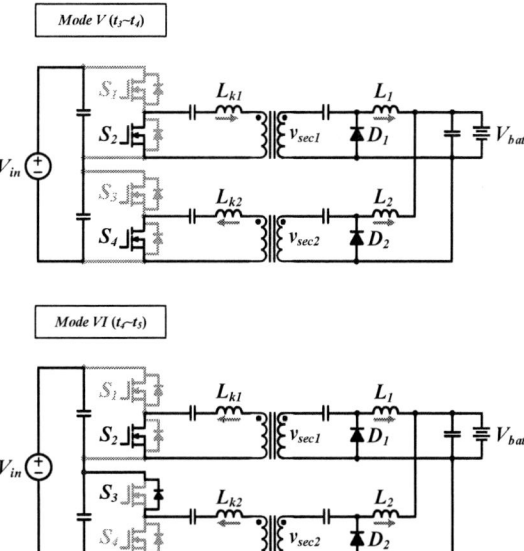

Fig. 3. Operating modes of the proposed converter.

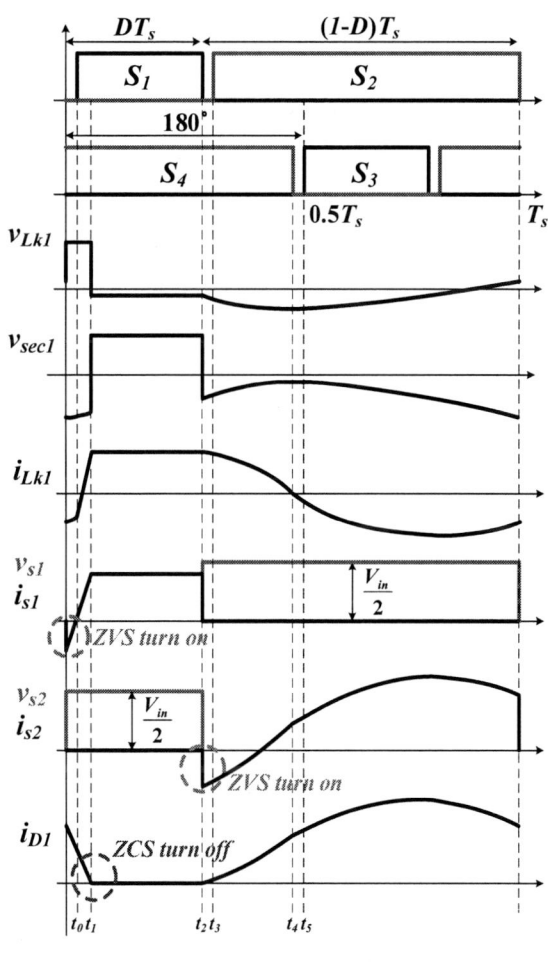

Fig. 4. Key waveforms of the proposed converter.

TABLE I
PERFORMANCE COMPARISON OF DIFFERENT TOPOLOGIES

	PSFB converter[5]		Active-clamp forward (ACF) converter[11]		Proposed converter
Circuit diagram	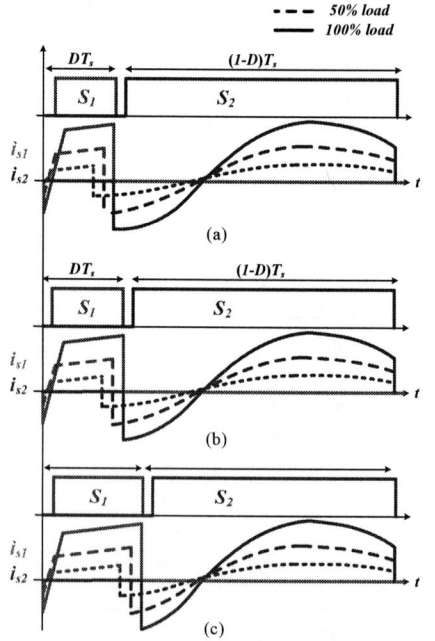				
Turn ratio	Design A	Design B	Design A	Design B	6:1
	11:1:1	24:1:1	2:1	9:1	
DC magnetizing current offset	X	X	O (>9A)	O (>0.5A)	X
	$i_{Lm}=0$		$i_{Lm}\neq 0$		$i_{Lm}=0$
switch — Voltage	800V		450V	620V	400V
switch — Current	S₁=14A (rms) S₂=14A (rms)	S₁=8A (rms) S₂=8A (rms)	S₁=9A (rms) S₂=12A (rms)	S₁=5A (rms) S₂=2A (rms)	S₁=7.5A (rms) S₂=10A (rms)
switch — Turn-off current	12A	9A	S₁=30A S₂=9A	S₁=8.8A S₂=3A	S₁=8A (rms) S₂=13A (rms)
Switching characteristics	ZVS turn on is failed at light load	Hard switching	ZVS turn on is failed at light load	Hard switching	Always ZVS turn on
Diode — Voltage	145V	60V	40V	23V	75V
Diode — Current	71A		71A		71A
Additional capacitors	X		O		O

(switch — Current: $S_1=14A$ (rms), $S_2=14A$ (rms) | $S_1=8A$ (rms), $S_2=8A$ (rms) | $S_1=9A$ (rms), $S_2=12A$ (rms) | $S_1=5A$ (rms), $S_2=2A$ (rms) | $S_1=7.5A$ (rms), $S_2=10A$ (rms))

III. COMPARISON

In this section, the PSFB, active-clamp forward and proposed converter are designed under a specification and their performances are compared. The design specification is given as follow: P_O =2kW, V_{in} = 400-800V, V_o = 14V and f_s = 100 kHz. Table I shows a comparison result of the three converters. Zero magnetizing offset of the PSFB converter is not guaranteed, therefore this converter often requires use of dc blocking capacitor in series with transformer winding. The proposed converter has theoretically zero dc magnetizing current offset. The switch voltage rating of the proposed converter is 400V which is half of the input voltage, so that 600V MOSFET can be used while the switch voltage rating of PSFB converter is the same as input voltage which is 800V. The switch voltage rating of the ACF can be designed as 440V, but the dc offset magnitude is 9A. Also, ZVS turn on of the PSFB and ACF converters is failed at light load while the proposed converter is achieved over the entire voltage and load range. Additionally, the turn-off current of the switches can be reduced depending on the value of dc blocking capacitor. It can be seen that the switching loss of the proposed converter switches is reduced. Thereby, the proposed converter can achieve high efficiency.

Fig. 5. Switch turn-on current waveforms according to variation of input-voltage and load. (a) V_{in}=800V, (b) V_{in}=700V, (c) V_{in}=600V.

The 2018 International Power Electronics Conference

IV. EXPERIMENTAL RESULT

A 2-kW prototype of the proposed converter has been built and tested to verify the operating principle, and the experimental results are provided. The proposed soft-switched converter with low-voltage rating of switch specification used in the experiment as shown in Table II. Fig. 7 and 8 shows experimental waveforms of the proposed converter. Figs. 7(a), 7(b), 7(c) and 7(d) show waveforms of switch voltage and current. It can be seen that all switches are turned on under ZVS condition over the entire input voltage and load range. Figs. 8(a) and 8(b) show that all diodes are turned OFF under ZCS condition.

TABLE II
EXPERIMENTAL SPECIFICATIONS

Symbol	Meaning	Value
P_o	Output power	2 kW
V_{in}	Input voltage	400-800 V
V_o	Output voltage	14 V
$N_p{:}N_s$	Turn ratio	6:1
f_s	Switching frequency	100 kHz
L_{k1}, L_{k2}	Leakage inductance	14 μH
C_{c1}, C_{c2}	Dc blocking capacitance of the primary side	10 μF
C_{b1}, C_{b2}	Dc blocking capacitance of the secondary side	6.6 μF
L_1, L_2	Output filter inductance	10 μH
C_o	Output filter capacitance	47 μF

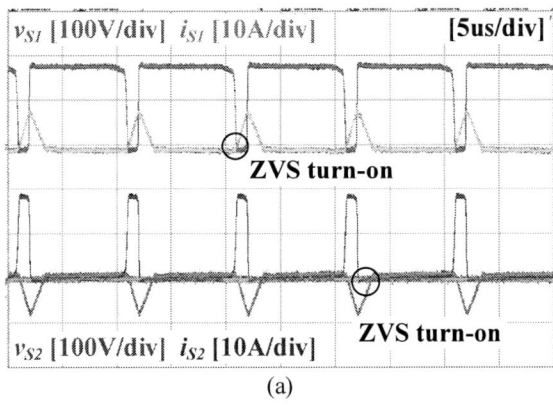

(a)

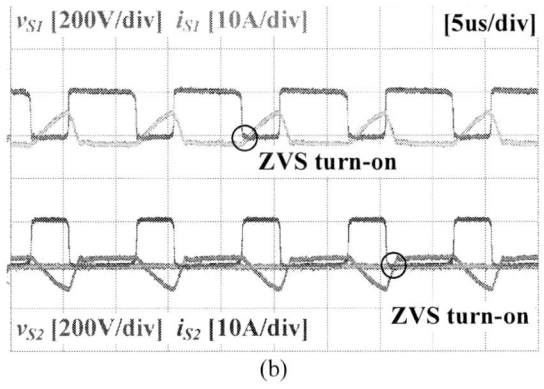

(b)

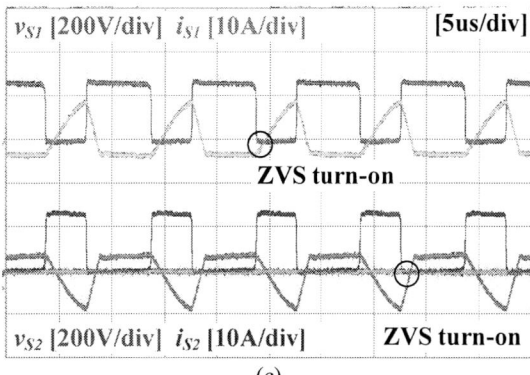

(c)

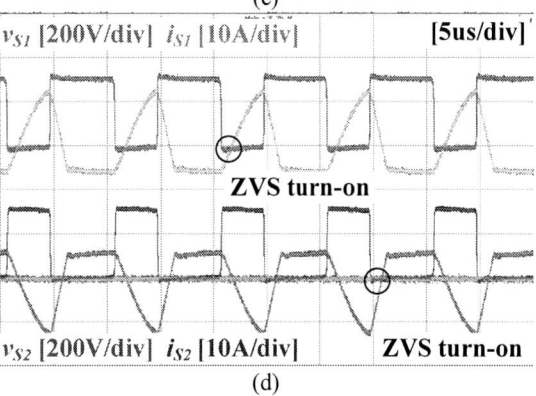

(d)

Fig. 7. Experimental waveforms of the switch. (a) Switch voltage and current at V_{in}=400V, P_o=150W. (b) Switch voltage and current at V_{in}=400V, P_o=350W. (c) Switch voltage and current at V_{in}=500V, P_o=700W. (d) Switch voltage and current at V_{in}=600V, P_o=1kW.

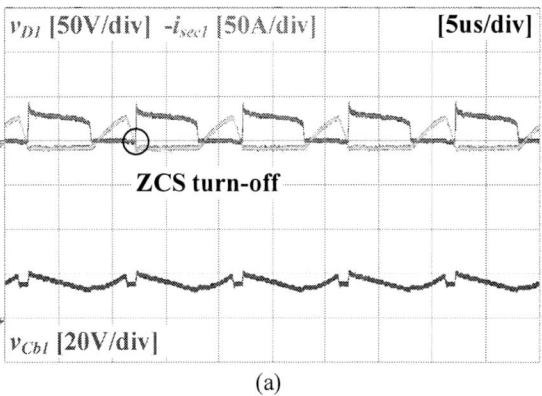

(a)

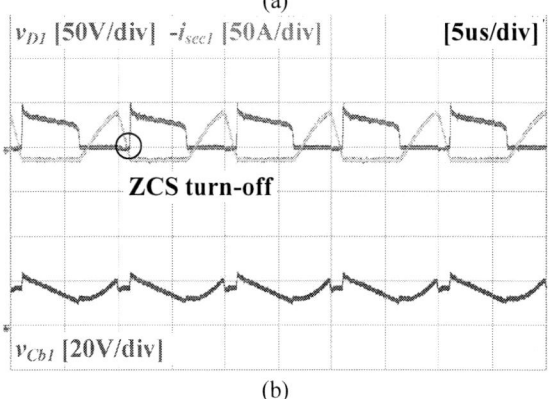

(b)

259

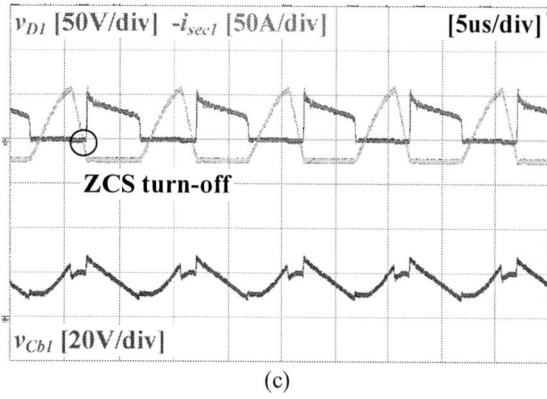

(c)

Fig. 8. Experimental waveforms of the diode and dc blocking capacitor of the secondary side. (a) Diode voltage and current at V_{in}=600V, P_o=200W. (b) Diode voltage and current at V_{in}=600V, P_o=550W. (c) Diode voltage and current at V_{in}=600V, P_o=800W.

V. CONCLUSION

In this paper, a soft-switched converter with low-voltage rating of switch is proposed for 800V/14V LDC. The soft-switched converter achieves ZVS turn-on over the entire input voltage and load range resulting in reduced switching loss. It also has low voltage rating of the switch which is half of the input voltage. Consequently, 600V MOSFET can be used which has less conduction loss. In addition, the turn-off current of the switches can be reduced depending on the value of dc blocking capacitor. The proposed converter has theoretically zero dc magnetizing current offset. Experimental results on a 100 kHz, 2-kW laboratory prototype are provided to validate the proposed concept.

ACKNOWLEDGEMENT

This work was supported by the National Research Foundation of Korea (NRF) grant funded by the Korea Government (MSIT) (2017R1A2A2A05001054)

REFERENCES

[1] A. Emadi, L. Young Joo, and K. Rajashekara, "Power electronics and motor drives in electric, hybrid electric, and plug-in hybrid electric vehicles," *IEEE Trans. Ind. Electron.*, vol. 55, no. 6, pp. 2237–2245, Jun. 2008.

[2] A. Emadi, S. S.Williamson, and A. Khaligh, "Power electronics intensive solutions for advanced electric, hybrid electric, and fuel cell vehicular power systems," *IEEE Trans. Power Electron.*, vol. 21, no. 3, pp. 567–577, May 2006.

[3] Porsche, e-mobility. New possibilities with 800-volt charging, Porsche 2016. [Online]. Available: https://newsroom.porsche.com/en/technology/porsche-engineering-epower-electromobility-800-volt-charging-12720.html [Accessed: 10-Oct-2017].

[4] "BRUSA: Auxiliary DC/DC." [Online]. Available: http://www.brusa.biz/en/products/dcdc-converter/hvlv-800-v/bsc628.html [Accessed: 10-Jul-2017].

[5] D. Moon, J. Park and S. Choi, "New Interleaved Current-Fed Resonant Converter With Significantly Reduced High Current Side Output Filter for EV and HEV Applications," in *IEEE Transactions on Power Electronics*, vol. 30, no. 8, pp. 4264-4271, Aug. 2015.

[6] A. Kawahashi, "A new-generation hybrid electric vehicle and its supporting power semiconductor devices," in Proc. International Symposium Power Semiconductor Devices and ICs, Kitakyushu, Japan, May. 2004, pp. 23-29. D

[7] F. Krismer and J.W. Kolar, "Efficiency-optimized high-current dual active bridge converter for automotive applications," IEEE Trans. Ind. Electron., vol. 59, no. 7, pp. 2745–2760, Jul. 2012.

[8] M. Pahlevaninezhad, J. Drobnik, P. K. Jain, and A. Bakhashai, "A load adaptive control approach for a zero-voltage-switching dc/dc converter used for electric vehicles," IEEE Trans. Ind. Electron., vol. 59, no. 2, pp. 920–933, Feb. 2012.

[9] U. Badstuebner, J. Biela, D. Christen, and J. W. Kolar, "Optimization of a 5-kW telecom phase-shift dc–dc converter with magnetically integrated current doubler," IEEE Trans. Ind. Electron., vol. 58, no. 10, pp. 4736–4745, Oct. 2011.

[10] C. Park and S. Choi, "Quasi-Resonant Boost-Half-Bridge Converter With Reduced Turn-Off Switching Losses for 16 V Fuel Cell Application," in *IEEE Transactions on Power Electronics*, vol. 28, no. 11, pp. 4892-4896, Nov. 2013.

[11] S. J. Chen, S. P. Yang and M. F. Cho, "Analysis and implementation of an interleaved series input parallel output active clamp forward converter," in IET Power Electronics, vol. 6, no. 4, pp. 774-782, April 2013.

[12] J. Lee, M. Kim, H. Jeong and S. Choi, "Single switch ZCS resonant converter with high step-up ratio," *2016 IEEE 8th International Power Electronics and Motion Control Conference (IPEMC-ECCE Asia)*, Hefei, 2016, pp. 3495-3500.

[13] M. Kim and S. Choi, "A Fully Soft-Switched Single Switch Isolated DC - DC Converter," in *IEEE Transactions on Power Electronics*, vol. 30, no. 9, pp. 4883-4890, Sept. 2015.

High Speed Control Method for Superposing High-Frequency-High-Sinusoidal-Current with DC Current to Analyze Battery AC Impedance

Jin Xu*, Toshihiko Kishimoto and Noboru Shimosato

Power Supply Systems Department, Myway Plus Corporation, Yokohama, Japan

*E-mail: jin_xu@myway.co.jp

Abstract-In this paper, for improving battery performance and exactly analyzing the characteristics of battery pack, a high speed control method for superposing high sinusoidal current at high frequency with DC current is proposed. The design and implemental method is presented in detail. The validity is verified by experiment. Using this method, a 100App/20kHz (pp: peak to peak) sinusoidal current superposed with several hundred DC current can be regulated instantaneously at 600kHz control frequency to charge and discharge the lithium-ion (Li-ion) battery pack.

Keywords—high speed control, high frequency sinusoidal current, superpose, battery AC impedance.

I. INTRODUCTION

With the rapid growth of electric/hybrid vehicles (EV/HEV) and energy storage systems, the Li-ion battery packs in series/parallel-connected with power converters are widely used. For improving the battery performance and analyzing the battery characteristic such as the state-of-charge (SOC), state-of-health (SOH) under the influence of the switching frequency current [1,2], the AC impedance as shown in Fig.1 needs to be understood and measured exactly.

The battery AC impedance as shown in Fig.1 can be represented as: parasitic inductance L_s, ohmic resistance R_o, charge transfer resistance R_{ct} as a result of activation polarization, double-layer capacitor C_{dl} as a result of activation polarization, warburg impedance Z_w as a result of concentration polarization [3,4]. At the low frequency range (~several Hz), the small L_s can be neglected and the large C_{dl} can be seen as an open circuit. Thus, the impedance is decided by R_o, R_{ct} and Z_w. At the middle frequency range (several Hz ~ several kHz), the impedance is decided by R_o, R_{ct} and C_{dl}. At the high frequency range (several kHz ~), the C_{dl} is seen as a short circuit and L_s is too large, so the impedance is decided by L_s and R_o.

In order to measure the battery AC impedance, the sinusoidal current superposed with DC current is necessary [2,5]. Most commonly used battery impedance testing methods available require small signal testers, such as electrochemical impedance spectroscopy, potentiostat [3,4,6,7]. These testers only can be used to test the battery cell impedance, but cannot be suitable for testing the battery pack at high-frequency-high-current conditions. One reason is that for charging and discharging the battery pack, several hundred voltage (EV: about 200V~400V, EV Bus: about 500V~700V) and several hundred current is necessary. Another reason is that these testers are using linear amplifier technology but not using the switching technology, so they can only generate small AC currents and frequencies less than several kHz.

Thus, in order to test and analyze the battery pack impedance, it is necessary to develop a system that can generate high sinusoidal current at high frequency superposed with DC current. According to our research results, there are some researches that using switching circuit to generate high-frequency-high-current. But, in these researches, the battery current cannot be regulated instantaneously and dynamically [2,5,8,9].

In this paper, a high speed control method implemented in FPGA is proposed. Using this method, the following functions can be realized.

• Low distortion rate sinusoidal current at high frequency from DC to 20kHz.

• High speed current controller at 600kHz control frequency to regulate the battery current.

Firstly, the system configuration and the coordination control principle is described. Next, the design and implemental method of the high speed control is proposed and described in detail. The communication data between CPU and FPGA, floating-point calculation

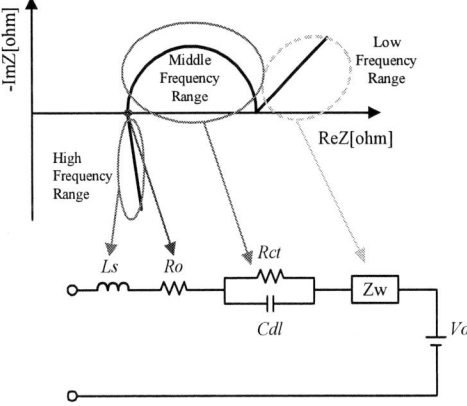

Fig.1. Equivalent battery AC impedance model

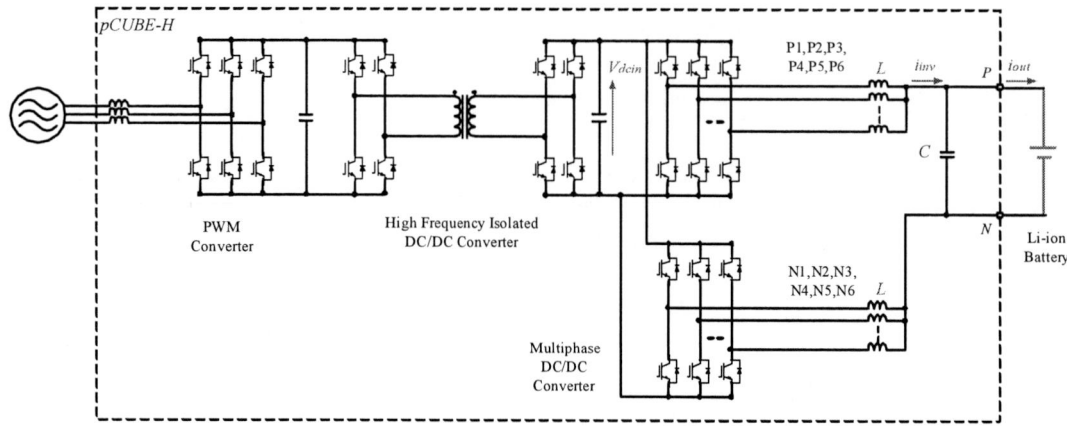

Fig.2. System configuration

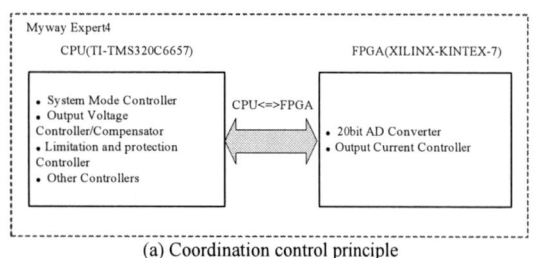

(a) Coordination control principle

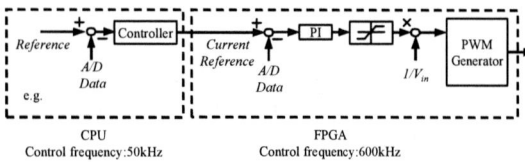

(b) Brief control blocks in CPU and FPGA

Fig.3. Control configuration

method, data and timing flowchart from data input to PWM signal output in FPGA is presented clearly. The results of high speed control and calculation time are confirmed by a hardware in loop simulator. Then, the parameter affecting the superposed sinusoidal current is extracted and analyzed in detail, and a solution is proposed. Finally, maximum 100App at 20kHz superposed with several hundred DC current is generated and the proposed high speed control method is verified in a real bidirectional DC power supply and battery system.

II. System Configuration and Coordination Control Principle between CPU and FPGA

In order to test battery packs, our company developed a 100kW bidirectional DC power supply. This system can generate a 100App and maximum 20kHz sinusoidal current superposed with several hundred DC current.

Fig.2 shows the system configuration of the bidirectional DC power supply. This system is composed of three-phase PWM converter to connect AC grid, high frequency isolated DC/DC converter using the Dual Active Bridge (DAB) circuit, multiphase DC/DC converter and output LC filter. In this system, SiC

devices are used for all switching devices. The switching frequency of the output DC/DC converter is 50kHz, and the total switching frequency is 600kHz by means of the multiphase circuit and phase shift PWM principle.

For superposing high-frequency-high-sinusoidal-current with DC, a high speed control method to instantaneously regulating current is necessary. In conventional controller, all voltage and current controllers are implemented in CPU and only PWM generator is implemented in FPGA. In a complex and multifunctional product as shown in Fig.2, the control frequency can only be raised to several ten kHz. Thus, it is unable to regulate a maximum 20kHz sinusoidal current.

Thus, we proposed a high speed control method implemented in FPGA. Fig.3(a) shows the coordination control principle between CPU and FPGA. The system mode controller, for controlling the system mode such as initialization, start/stop sequence, etc., output voltage controller/compensator, output current limitation and protection functions are implemented into CPU (TI TMS320C6657). The high speed controller for regulating the output current and 20bit/0.1%accuracy A/D (analog/digital) converter are implemented in FPGA (XILINX KINTEX-7). Fig.3(b) shows the brief control blocks in CPU and FPGA. Here, the control frequencies are 50kHz in CPU, and 600kHz in FPGA.

III. High Speed Control Method

In this chapter, the high speed controller method is proposed and described in detail, and the debug results using Hardware in Loop simulation (Typhoon HIL) are confirmed.

1) High speed controller configuration

Fig.4 shows the high speed controller configuration in FPGA. A PI controller as shown in Fig.3(b) is implemented in FPGA. Here, we will introduce the data flowchart in control loop, the timing flowchart and clock setting method of each block in detail.

a) Input parameters from CPU to FPGA:

For communicating well with FPGA, as shown in Fig.4, the DC current reference, AC current frequency and

The 2018 International Power Electronics Conference

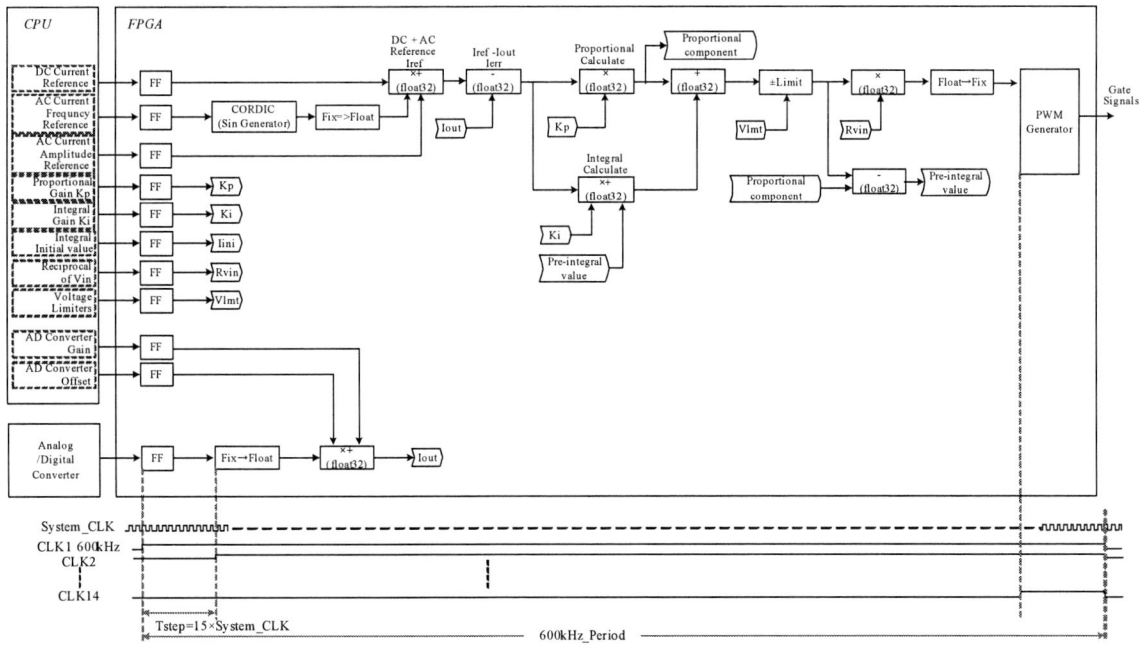

Fig.4. High speed control configuration in FPGA
Data and timing flowchart

TABLE I
FPGA BLOCKS AND FUNCTIONS

Block Name	Function	Xilinx IP/VHDL
FF	Flip-Flop	RAM-based shift register
CORDIC	Sin generator	Coordinate Rotational Digital Computer
Fix→Float	Fixed-point to Float 32bit conversion	Floating-point
Float→Fix	Float32bit to Fixed-point conversion	Floating-point
×+ Float	Float32bit multiply and add	Floating-point
+ Float	Float32bit add	Floating-point
- Float	Float32bit subtract	Floating-point
±Limit	Limit operate	VHDL coder

amplitude reference, proportional gain Kp, integral gain Ki, integral initial value, output voltage limiters, AD converter gain and offset are transferred from CPU to FPGA. Then, for holding the input data in 1 control period, the flip-flops (FF) are used. Although not shown in Fig.4, the internal data in FPGA such as AD converter data and output data (proportional component, etc.) of each blocks are transferred from FPGA to CPU by a slow speed data bus that has no effect on the control loop speed.

b) Function blocks:

As shown in Fig.4, the high speed controller in FPGA uses floating-point operator, and the functions of blocks are shown in TABLE I.

Here, in order to generate a sinusoidal wave as reference data, a CORDIC function block is used. The CORDIC is generated according to the input frequency and the counter value in FPGA. Using this function, high resolution at arbitrary frequency can be realized compared with using lookup table method.

c) Clock setting and Calculation time of function blocks:

For regulating a 20kHz sinusoidal current, the control frequency must be at least 200kHz. In the proposed high speed control method, the control frequency synchronizing the top of phase-shift-PWM signals is 600kHz. Thus, the current will be calculated and adjusted 30 times that is enough points to generate a 20kHz sinusoidal wave.

2) Debug results

In order to complete the current regulation in 600kHz, the calculation time of all function blocks should be clarified and controlled.

Fig.5 shows the debug results of Integrated Logic Analyzer (ILA) of Vivado Design Suite using Hardware in Loop Simulation. The main circuit as shown in Fig.2 is implemented in HIL, and the control boards of product are used. From these results, we confirmed that the maximum calculation time is about 13 times of system clock period. Thus, from input FFs to the output PWM generator, the interval time between two blocks (e.g.: clock1 to clock2) is set as 15 times of system clock period. Here, the system clock is 200MHz. The total calculation time is approximately 975nsec (= 13*15*5nsec) and confirmed by debug results.

263

The 2018 International Power Electronics Conference

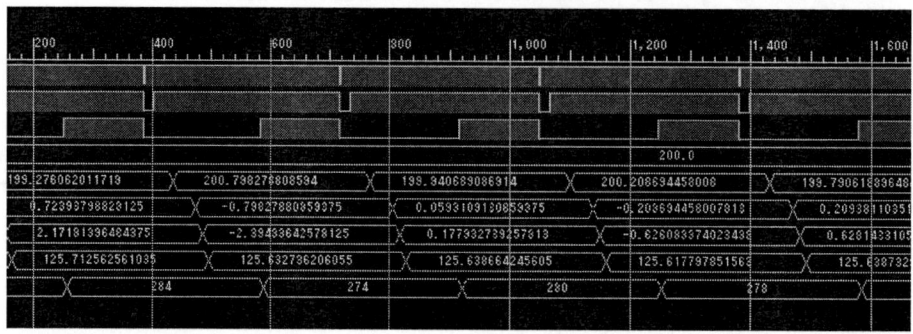

Fig.5. Debug results of Integrated Logic Analyzer (ILA) of Vivado Design Suite

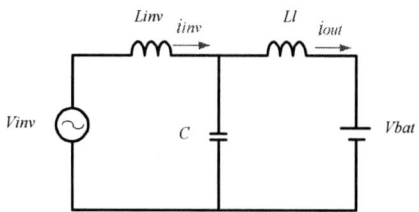

Fig.6. Equivalent circuit

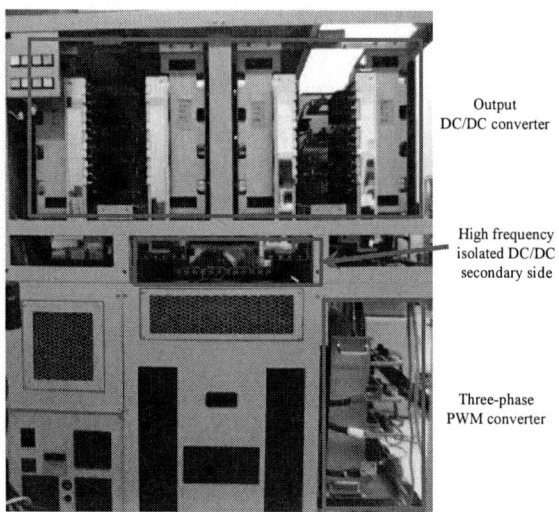

Output
DC/DC converter

High frequency
isolated DC/DC
secondary side

Three-phase
PWM converter

Fig.7. Experimental system

IV. PARAMETERS AFFECT SUPERPOSED SINUSOIDAL CURRENT

When superposing the sinusoidal current with DC, the inductance and the parasitic impedance of battery will affect the sinusoidal current. The equivalent circuit as shown in Fig.2 can be described as Fig.6.

Here, V_{inv} is the output voltage of the DC/DC converter. L_{inv} is the equivalent inductance of one phase.

L_l is the inductance of the cable and the parasitic inductance of the battery.

The impedances of L_{inv}, C, Ll, and the output current can be given as follows.

$$Z_c(s) = \frac{1}{Cs}, Z_{Linv}(s) = \frac{1}{L_{inv}s}, Z_{Ll}(s) = \frac{1}{L_l s}, \tag{1}$$

$$\frac{I_{out}}{V_{inv}} = \frac{Z_c(s)}{Z_{Linv}(s)(Z_c(s) + Z_{Ll}(s) + Z_{bat}(s)) + Z_c(s)(Z_{Ll}(s) + Z_{bat}(s))}$$

As shown in Fig.1, according to the battery AC impedance at each frequency range, the battery impedance $Z_{bat}(s)$ can be seen as 0, C or L. Thus, the equivalent circuit can be expressed as *LCL* or *LCLC* circuit. In addition, according to the transfer function of I_{out}/V_{inv}, there has a resonant and decay phenomenon. In order to solve this problem, not only the current i_{inv} but also the current i_{out} are regulated in a close control loop. For controlling the amplitude of sinusoidal current following the reference, the peak to peak value of the output current i_{out} is detected, calculated and compensated by CPU and FPGA.

V. EXPERIMENTAL RESULTS

In this chapter, the proposed high speed control method implemented in FPGA is verified by experiment.

Fig.7 shows the experimental system (pCUBE-H: the bidirectional DC Power Supply). The system is composed of PWM converter, high frequency isolated DC/DC converter, multiphase DC/DC converter as shown in Fig.2. The output power of this power supply is 100kW, the output maximum voltage and current are 600V and 400A, respectively. Because we have not a 600V battery pack, the functions and performances of this power supply will be tested by another pCUBE-H used as an electronic load and a 33V battery.

A. Electronic Load

Using the electronic load, we confirmed the validity of the proposed high speed control method and the functions and performances of the power supply at rated conditions.

Fig.8, 9 show the experimental results when the output voltage is 250V. In Fig.8, the output current is DC 350A with superposing a 100App/20kHz sinusoidal current. In Fig.9, the output is DC -350A with superposing a 100App/20kHz sinusoidal current.

Fig.10, 11 show the experimental results when the

264

The 2018 International Power Electronics Conference

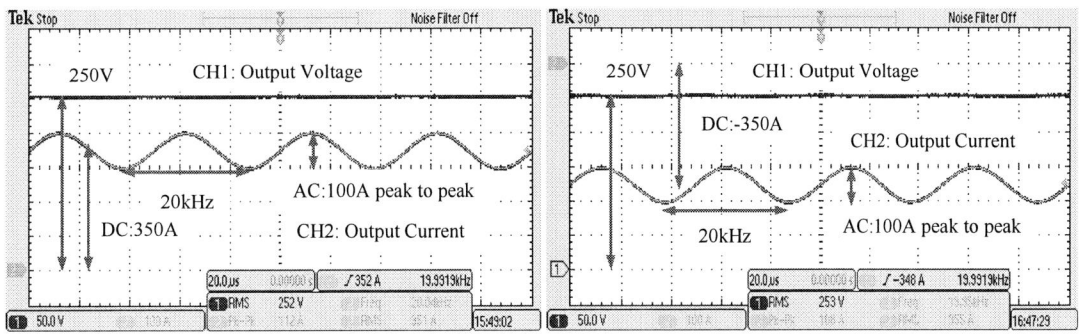

Fig.8. DC350A with 100App/20kHz when output voltage is 250V   Fig.9. DC-350A with 100App/20kHz when output voltage is 250V

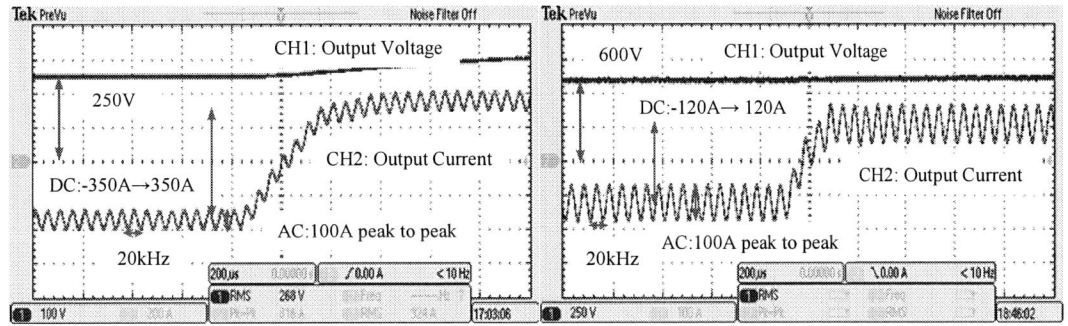

Fig.10. DC-350A to DC350A with 100App/20kHz   Fig.11. DC-120A to DC120A with 100App/20kHz
at stepwise change when output voltage is 250V   at stepwise change when output voltage is 600V

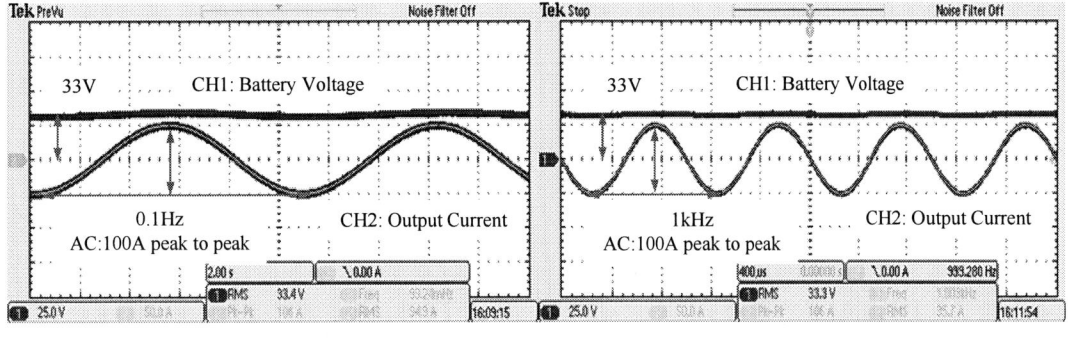

Fig.12. 100App/0.1Hz using battery load   Fig.13. 100App/1kHz using battery load

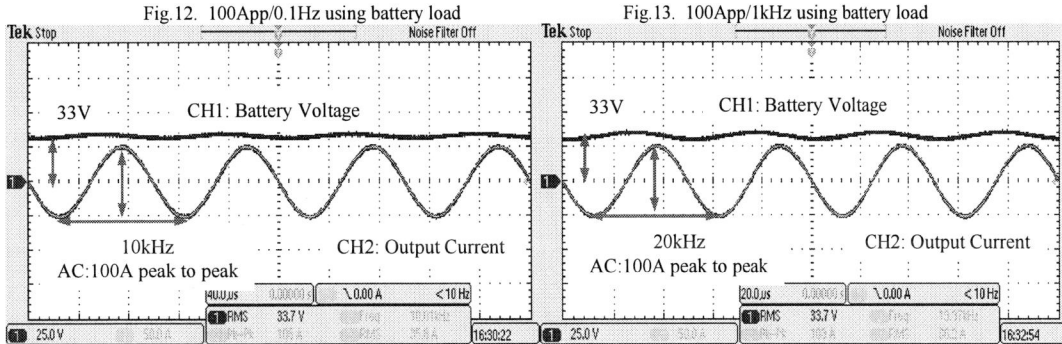

Fig.14. 100App/10kHz using battery load   Fig.15. 100App/20kHz using battery load

output current is changed at stepwise. The output current is changed from DC-350A to DC350A with a

100App/20kHz current when the output voltage is 250V. Here, the output voltage is unstable, because there is only

265

a small capacitor between the power supply side and the electronic load side. In Fig.11, the current is changed from DC-120A to DC120A when the output voltage is 600V.

From these results, it can be seen that the high frequency high sinusoidal current superposed with high DC current is regulated instantaneously and the validity of the proposed high speed control method is verified as designed.

B. Battery Load

Next, we confirmed the proposed method and the functions using a battery load. Because the limitation of the battery, only the sinusoidal current from 0.1Hz to 20kHz is generated. Here, we measured the AC impedance of the battery and analyzed the THD of the sinusoidal current.

Fig.12~15 show the experimental results when the frequency of the sinusoidal current are 0.1 Hz, 1kHz, 10kHz and 20kHz. The amplitude is 100A peak to peak. From these results, the high frequency sinusoidal current can also be realized using battery load.

Fig.16 shows the THDs of the sinusoidal current at 0.1Hz, 1kHz, 5kHz, 10kHz, 15kHz and 20kHz. From these results, the THDs can be suppressed under 5% and is not related to the frequencies.

Fig.17 shows the measured AC impedance of the battery load. Compared with Fig.1, we can say that the sinusoidal current, generated by the proposed high speed control method, can satisfy the demands of testing and analyzing the battery AC impedance.

VI. Conclusion

In this paper, a high speed control method to superpose 100App/20kHz sinusoidal current with DC current is proposed and verified by experiment. Using this method, the achievements can be summarized as follow.

- The design method of high speed control is proposed and verified by HIL simulation and experiment.
- The parameter affecting the sinusoidal current is exacted and investigated in detail.
- The sinusoidal current at high frequency from DC to 20kHz can be regulated instantaneously at a 600kHz control frequency.

References

[1] Y.C Zhang, R.X Zhao, J. Dubie, T. Jahns, L. Juang, "Investigation of Current Sharing and Heat Dissipation in Parallel-Connected Lithium-Ion Battery Packs," *2016 IEEE Energy Conversion Congress and Exposition (ECCE)*, pp. 1-8, 2016.

[2] K. Zou, S. Nawrocki, R.X. Wang, J. Wang, "High Current Battery Impedance Testing for Power Electronics Circuit Design," *2009 IEEE Vehicle Power and Propulsion Conference*, pp. 531-535, 2009.

[3] S.Y. Cho, Il.O. Lee, J.Il. B, G.W. Moon, "Battery Impedance Analysis Considering DC Component in Sinusoidal Ripple-Current Charging," *IEEE Trans. on Industrial Electronics*, vol. 63, no. 3, pp. 1561-1573, 2016.

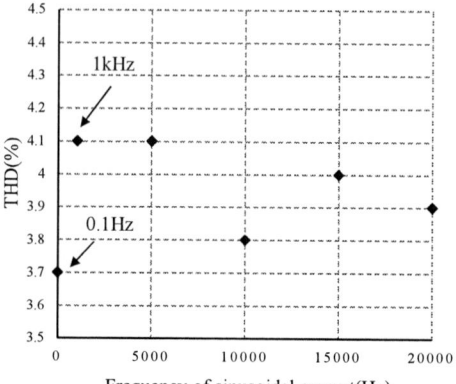

Fig.16. THD of sinusoidal current from 0.1Hz to 20kHz

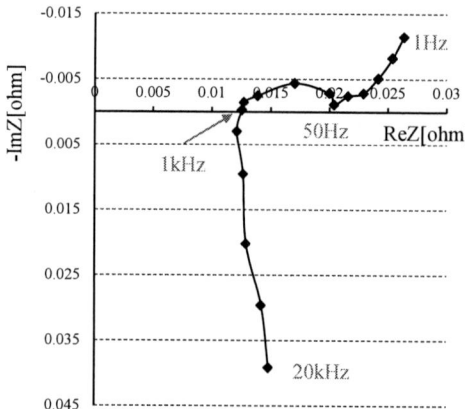

Fig.17. Battery AC impedance when 100App from 1Hz to 20kHz

[4] L.W. Juang, P.J. Kollmeyer, R.X. Zhao, T. M. Jahns, R. D. Lorenz, "The Impact of DC Bias Current on the Modeling of Lithium Iron Phosphate and Lead-Acid Batteries Observed using Electrochemical Impedance Spectroscopy," *2014 IEEE Energy Conversion Congress and Exposition (ECCE)*, pp. 2575-2581, 2014.

[5] J. Wang, K. Zou, C.C. Chen, L.H. Chen, "A High Frequency Battery Model for Current Ripple Analysis," *2010 Twenty-Fifth Annual IEEE Applied Power Electronics Conference and Exposition (APEC)*, pp. 676-680, 2010.

[6] G. Pérez, I. Gandiaga, M. Garmendia, J.F. Reynaud, U. Viscarret, "Modelling of Li-ion Batteries Dynamics using Impedance Spectroscopy and Pulse Fitting: EVs Application," *Electric Vehicle Symposium and Exhibition (EVS27)*, pp. 1-9, 2013.

[7] J.Y. Choi, J.B. Jeong, H.J. Lee, D.H. Shin, "Development and Verification of Impedance Measurement Equipment of High-Voltage Battery Pack," *2016 IEEE Transportation Electrification Conference and Expo, Asia-Pacific*, pp. 828-831, 2016.

[8] Y.D. Lee, S.Y Park, S.B Han, "On-line optimal ion conductivity control of Li-ion battery," *Energy Conversion Congress and Exposition (ECCE)*, pp. 4493-4500, 2012.

[9] Y.D. Lee, S.Y Park, "Electrochemical State-Based Sinusoidal Ripple Current Charging Control," *IEEE Trans. on Power Electronics*, vol. 30, no. 8, pp. 4232-4243, 2015.

The 2018 International Power Electronics Conference

EV BMS with Time-Shared Isolated Converters for Active Balancing and Auxiliary Bus Regulation

Z. Gong[1*], B.A.C. van de Ven[1,2], Y. Lu[1], Y. Luo[1], K. Gupta[1], C. da Silva[1], H.J. Bergveld[2], O. Trescases[1]

1 Electrical and Computer Engineering, University of Toronto, Toronto, Canada
2 Electrical Engineering, Eindhoven University of Technology, Eindhoven, The Netherlands
*E-mail: zhe.gong@mail.utoronto.ca

Abstract—Improved utilisation of the total energy storage in Electric Vehicle (EV) battery systems can be achieved through balancing of the series-connected battery units based on parameters such as the terminal voltage and State-of-Charge (SOC). This paper proposes a BMS power architecture where at any given time, an isolated converter connects either a module or one of its constituent sub-modules to the vehicle auxiliary bus, where a 12V lead-acid battery is present. The converters operate in burst-mode with a period of 10 s to simultaneously balance the sub-modules and regulate the auxiliary bus voltage. The use of module and sub-module input modes to the converters enables the supply of high-power auxiliary loads without an increase in converter input current rating. Simulations of one rule-based and one variable-priority control algorithm, both using SOC as the balancing parameter, are shown over a 6 hour load profile and 5% maximum initial SOC imbalance, for a 4 kWh liquid-cooled battery module prototye. Measurements using the same prototype are shown to match the simulation results. The simulation and experimental results highlight the necessary trade-off, in the system control, between auxiliary bus voltage regulation and balancing rate.

Keywords—*battery management system, electric vehicle, cell balancing, active balancing*

I. INTRODUCTION

In 2015 the global stock of Electric Vehicles (EVs) exceeded 1 million for the first time [1]. As EV production volumes increase, it is becoming increasingly important to develop cost-saving battery system technologies that maintain reliability and performance. One way to achieve this is through improved utilisation of the total energy storage in the pack through balancing of the series-connected battery units, which are referred to as sub-modules in this work. It is well known that mismatches in sub-module characteristics, such as capacity and impedance, can result from issues in manufacturing, temperature gradients across the pack, and parasitic currents in battery management electronics [2]. These mismatches can negatively affect the pack performance. For example, the lowest State-of-Charge (SOC) sub-module limits the capacity during discharge, and the highest impedance sub-module limits the peak current. It is therefore necessary

The authors thank Tony Han and Havelaar Canada for their support of the UofT Electric Vehicle Research Centre. This work was also supported by The Natural Sciences and Engineering Research Council of Canada (NSERC).

for a Battery Management System (BMS) to balance sub-modules based on parameters such as the terminal voltage and SOC. Extensive reviews of sub-module balancing methods and architectures are provided in [3], [4].

Active balancing involves using power converters to transfer energy between sub-modules. In some active balancing architectures, each sub-module is connected to a common bus through an isolated converter, which controls the energy transfer [5]–[7]. In [8], [9], this is extended to EV applications by using the vehicle's 12V auxiliary bus as the common bus. The control objective is to simultaneously balance the sub-modules and regulate the bus voltage, and the system does not consider the lead-acid battery, which is typically present on the auxiliary bus for operation of auxiliary systems in case of main pack failure. The major strength of this architecture is cost reduction through the incorporation of the high-voltage (>400V) to low-voltage (12V) dc-dc converter, normally present in the system as a separate liquid-cooled unit, into the battery system. Two weaknesses are that: 1) the cost of having one converter per sub-module is high, and 2) disregarding the lead-acid battery in the system control means its energy storage cannot be used to buffer the auxiliary bus voltage while the controller prioritises the balancing speed or low-loss operation.

This paper proposes a BMS power architecture where at any given time, one isolated converter connects either a module or one of its constituent sub-modules to the vehicle's auxiliary bus, where a conventional lead-acid battery is connected. The converters simultaneously balance the sub-modules and regulate the auxiliary bus voltage. Compared to systems with one converter per sub-module, cost reduction is achieved by using fewer converters, and drawing the auxiliary power at higher voltage using the module connection. A new control scheme is proposed, which allows trade-offs to be made between the balancing speed and bus regulation integrity. The system is designed for custom battery modules with liquid thermal management to be deployed in an electric pickup truck, which are both shown in Fig. 1.

II. BMS POWER ARCHITECTURE

The electric pickup truck battery system is composed of 18 series-connected modules, each consisting of 44 Ah lithium Nickel-Manganese-Cobalt (NMC) pouch cells in a 6S4P configuration. The simplified BMS architecture is shown

The 2018 International Power Electronics Conference

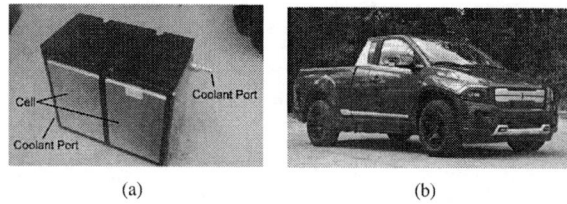

(a)   (b)

Fig. 1. (a) Custom liquid-cooled battery module used in this work. (b) Electric pick-up truck prototype.

in Fig. 2(a). The electrical nodes of the module and the converter input terminal are connected to a switch matrix, which engages connection between the converter and module or sub-modules. Each battery module is connected to a BMS module which performs: 1) temperature and voltage sensing, 2) switch matrix actuation, 3) dc-dc converter control, and 4) online Electrochemical Impedance Spectroscopy (EIS) measurements [10]–[13]. A single supervisor aggregates data sensed at the sub-module level and computes the sub-module states using a state observer. The battery model used for the observer is a second order equivalent circuit model, and the parameter identification is beyond the scope of this paper. The supervisor computes and issues the balancing commands.

A. Principle of Operation

In module mode, the converter input V_{in} is connected across the entire module. In sub-module mode, V_{in} is instead connected across any of the sub-modules within the module. The switch matrix configuration is controlled by the sw_sel signal. To actuate the switch matrix, the BMS module periodically updates sw_sel, based on the supervisor's periodic balancing command. Enabling module or sub-module connections to the dc-dc converter presents a design challenge: the converter must operate at two input voltage ranges and its efficiency must be optimised for both. In this work, V_{in}, the dc-dc input voltage, varies between 3-4.2 V in sub-module mode, and 18-25.2 V in module mode. The converter is implemented with an isolated Ćuk topology, as described in [14], switching at 250 kHz, and is controlled using average current mode control, at the input side, with a PI compensator. The converter and compensator are shown in Fig. 2(b).

In order to reduce the system cost, the converter operates in uni-directional power transfer mode. To maximise efficiency, the converter is operated at only the peak efficiency current at each input voltage. Burst-mode operation is used to control the average current, and one burst-mode operation cycle is shown in Fig. 3. The burst-mode period is T_{cycle}. Two control parameters are used to vary the total average output current of all converters, $I_{out,avg}$: 1) each converter's on-time, t_{on}, and 2) the switch matrix configuration sw_sel. Within each burst-mode cycle, the configuration phase T_{config} is when sw_sel is first updated and the converter is then activated. Since T_{config} is short compared to T_{cycle}, it can be neglected for calculation of $I_{out,avg}$. The next cycle's t_{on} and sw_sel are determined during T_{calc}, based on the control algorithm, and this period can overlap with t_{on}.

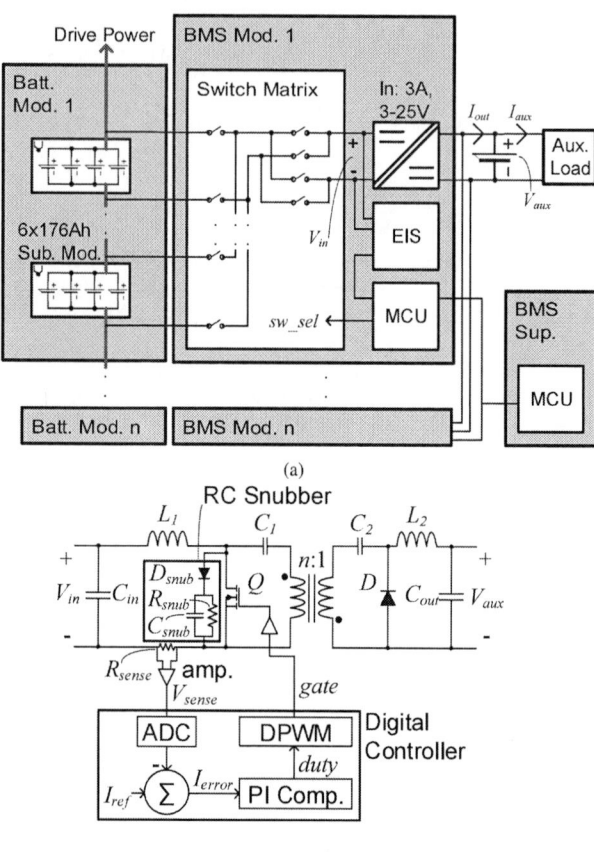

(a)

(b)

Fig. 2. BMS architecture: (a) System schematic. (b) Isolated Ćuk converter and compensator.

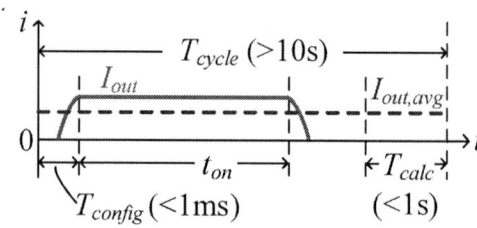

Fig. 3. Timing diagram of the burst-mode operating cycle of each converter in the system.

III. RULE-BASED CONTROL

One simple method for simultaneously performing active balancing and auxiliary bus regulation is to apply a closed-loop controller, which calculates a current command for auxiliary bus regulation, and uses a set of rules to determine sw_sel. A model for the rule-based control system is shown in Fig. 4.

The current source, $I_{out,avg}$, represents the same current

268

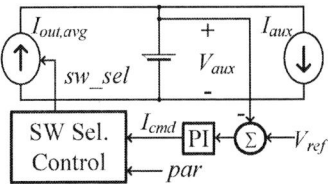

Fig. 4. Auxiliary bus model and control system for the rule-based algorithm.

as in Fig. 3, and is defined as follows:

$$
\begin{aligned}
I_{out,avg} = \frac{t_{on}}{T_{cycle}}(&\eta_{mod}n_{mod}I_{mod}\frac{V_{mod}}{V_{bus}} \\
&+\eta_{submod}n_{submod}I_{submod}\frac{V_{submod}}{V_{bus}}),
\end{aligned} \tag{1}
$$

where n_{mod} and n_{submod} are the number of converters operating in module and sub-module input mode, I_{mod} and I_{submod} are the converter module and sub-module input mode currents, and η_{mod} and η_{submod} are the module and sub-module input mode efficiencies of the dc-dc converters at the module and sub-module input mode currents. The 12V lead-acid battery is modelled according to [15]. The auxiliary bus load is modelled as a current source operating with time-varying current I_{aux}. The auxiliary bus voltage, V_{bus}, is regulated to the reference voltage, V_{ref}, by varying $I_{out,avg}$ through a current command, I_{cmd}, calculated using a PI compensator. The current command is converted into *sw_sel* using the *SW Sel. Control* block. The signal *par* is an arbitrary balancing parameter, for example SOC or the terminal voltage, which can be issued by a higher-level vehicle controller. The *SW Sel. Control* block operates according to the following rules:

1) Set t_{on} to T_{cycle}.
2) Use eq. (1) to find n_{submod}, n_{mod} combination that meets or just exceeds current command and maximises n_{submod}, to maximise sub-module balancing.
3) Use eq. (1) to find t_{on} that meets I_{cmd} with n_{submod} and n_{mod} from step 2).
4) Choose the n_{mod} highest-minimum-*par*-modules in the pack to operate in module input mode.
5) Choose the highest-*par*-sub-module in each remaining module to operate in sub-module input mode.
6) Issue *sw_sel* and begin converter operation.
7) Wait for new current command, return to step 1.

IV. VARIABLE-PRIORITY CONTROL

The primary weakness of the rule-based algorithm is its inability to optimise the trade-off between the balancing rate and auxiliary bus voltage regulation. The presence of energy storage on the auxiliary bus allows the supply current $I_{out,avg}$ to deviate from the load current I_{aux} periodically, while maintaining V_{aux} within acceptable limits. It is thus desirable to model the sub-module balancing dynamics in addition to the auxiliary bus voltage, such that trade-offs can be made between the two parameters. The updated Variable-Priority

Control (VPC) scheme is shown in Fig. 5. The objective is to control the system according to the vehicle-level parameter *Opt. mode*. This parameter can be selected according to the operating scenario, for example to favour SOC balancing in order to maximise the total charging phase input energy in the presence of SOC imbalance. The proportional controller outputs a current command, I_{cmd}, that is used to generate a table of e_Q-$\dot{e}_{par}$ pairs, each associated with a switch matrix configuration and t_{on}, where e_Q is the total auxiliary charge error, obtained from

$$
e_Q = \int (I_{cmd}(t) - I_{out,avg}(t))dt, \tag{2}
$$

and $\dot{e}_{par}$ is the rate of imbalance decrease, obtained from

$$
\dot{e}_{par} = e_{par}(t)\frac{d}{dt}(par(t) - par_{ref}(t)), \tag{3}
$$

where e_{par} is used as a ranking factor to prioritise sub-modules with higher imbalance, and the variable *par* is the same as in Section III. For example, if *par* is the sub-module SOC, $\dot{e}_{par}$ would be proportional to the balancing current. If *par* were instead the sub-module terminal voltage, $\dot{e}_{par}$ would be proportional to the local slope of the electro-motive force curve and the balancing current.

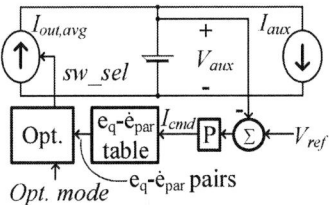

Fig. 5. Auxiliary bus model and control system for the VPC algorithm.

A cost function is used in the *Opt.* block, such that the system can choose a *sw_sel* and set of converter t_{on} values that decreases e_Q and increases $\dot{e}_{par}$ according to *Opt. mode*. The *Opt. mode* parameter is thus specified as tuning of the cost in order to bias the controller toward either decreasing e_Q or increasing $\dot{e}_{par}$.

A. SOC Balancing Implementation

To highlight one example implementation of the VPC, the SOC is selected as the balancing parameter, and the cost function and model for $\dot{e}_{par}$ are presented. According to the ideal description of the VPC, a set of e_Q-$\dot{e}_{par}$ pairs is required to be generated for every possible switch matrix configuration. In the Bison battery system, where 18 modules of 6 series sub-modules are present, 8 switch matrix configurations are available: the no-connection case, any of the sub-modules, or the entire module. This leads to a total of 8^{18} possible switch matrix configurations to evaluate, which is not practical for implementation in the BMS. One method for reduction of the computation requirement is to limit sub-module balancing to a module-level reference, and only balance modules according to a pack-level reference. This reduces the global balancing problem to multiple instances of intra-module balancing problems, and one inter-module balancing problem. The total number of switch matrix configurations to evaluate in this scheme is 8×18. A duty cycle, D, is used to represent the burst-mode on-time, t_{on}, and in this work, duty cycle steps of 0.01 are used, thus the total number of $e_Q\text{-}\dot{e}_{par}$ pairs to compute per burst-mode cycle is 14,400, which can be accomplished in short time (less than 1s) with a modern microprocessor or FPGA.

The SOC balancing implementation uses a quadratic cost function, expressed as

$$J(i,j) = x(i,j)^\top Q x(i,j) + u(i,j)^\top R u(i,j), \qquad (4)$$

where i is the module number and j represents a combination of one switch matrix configuration, sw_sel, and one duty cycle value, D. The balancing system states $x(i,j)$ are expressed in the cost function as

$$x(i,j) = \begin{bmatrix} \frac{1}{\dot{e}_{SOC}(i,j)} \\ |e_{Q,predicted}(i,j)| \end{bmatrix}, \qquad (5)$$

and the cost function input $u(i,j)$ is expressed as

$$u(i,j) = D(i,j). \qquad (6)$$

The first state of the system is the balance rate, expressed as

$$\dot{e}_{SOC}(i,j) = \frac{e_{SOC}(i,j)D(i,j)I_{in}}{3600 Q_{nom}(i,j)}, \qquad (7)$$

where Q_{nom} is the nominal cell capacity, and I_{in} is the fixed input current of the converter. Due to the selective balancing scope for computation reduction, e_{SOC} is expressed differently for sub-module and module balancing. In the sub-module case, the expression is

$$e_{SOC} = SOC_{submod} - SOC_{min,module}, \qquad (8)$$

where SOC_{submod} is the SOC of the sub-module being considered for balancing, and $SOC_{min,module}$ is the minimum sub-module SOC within the immediate module. In the module balancing case, the expression is

$$e_{SOC} = SOC_{avg,module} - SOC_{avg,min}, \qquad (9)$$

where $SOC_{avg,module}$ is the average sub-module SOC in the module being considered for balancing, and $SOC_{avg,min}$ is the minimum of the average sub-module SOCs of all modules.

The second state of the system is the cumulative predicted auxiliary battery charge error at the next computation step. The prediction is performed using the average auxiliary load current over the vehicle life, in order to prevent erratic controller behaviour under high transient auxiliary load currents. This is similar to the integrator portion of a PI compensator. Due to the reduced balancing scope, it is necessary to evaluate the cumulative charge error at the module level. This is performed by dividing the auxiliary load among the modules according to the values of $SOC_{avg,module}$. As opposed to the evenly-divided case, using the $SOC_{avg,module}$ biases the controller to favour balancing. This is done to avoid cases where modules

with low $SOC_{avg,module}$ are forced to operate in module mode under high auxiliary load, which dramatically increases the level of global imbalance. The cumulative charge error is expressed as

$$e_{Q,predicted}(i,j) = n(i)e_{Q,real}(t) + e_{dQ}(i,j), \qquad (10)$$

where $e_{Q,real}$ is the real instantaneous charge error, expressed as

$$e_{Q,real}(t) = \int_0^t (I_{cmd}(t) - I_{out,avg}(t))dt, \qquad (11)$$

where I_{cmd} and I_{out} are as defined in Fig. 5. The current-step predicted charge error, e_{dQ}, is expressed as

$$e_{dQ}(i,j) = \int_t^{t+T_{cycle}} (n(i)I_{average}(t) - D(i,j)I_{out,avg}(i,j))dt, \qquad (12)$$

where $I_{average}$ is the average load current. The auxiliary load division factor $n(i)$ is calculated for each module, and is expressed as

$$n(i) = \frac{\text{mean}(SoC(i,:))}{\sum_{x=1}^{N} \text{mean}(SoC(x,:))}. \qquad (13)$$

By manipulating the cost matrix Q in cost function (4), the bias between cell balancing or bus regulation can be controlled. This is done by changing the balance between the two values on the diagonal of the cost matrix Q, where a higher value will lead to more penalisation being brought upon high values of the associated state, e.g. a low charge error $e_{Q,predicted}$ or a high $\dot{e}_{SoC}$ will have a low associated cost, and will hence be preferred. On the other hand R is tuned to add a weight to a higher t_{on}, in order to allow for efficiency optimisation in case the converter module and sub-module input efficiencies are different.

The final values for t_{on} and sw_sel are chosen according to the minimum cost among all computed options, and is expressed as

$$j_{opt}(i) = \underset{j}{\text{argmin}}\, J(i,j). \qquad (14)$$

V. Simulation Results

The simulations of the two control algorithms, as described in Sections III and IV, are performed using a scaled-down battery system with the following hardware configuration: two lithium NMC battery modules of 6S2P cell configuration, two BMS modules operating at 3 A in both sub-module and module modes, and a 60 Ah lead-acid auxiliary battery. The auxiliary load profile contains a 1.7 hour drive phase, where the current is higher and variable, and a 4.3 hour idle phase, where the current is lower and relatively static. The same profile is used for all simulations, and current is scaled down by a factor of 18 from 800 W average while driving, and 200 W average while idle, to account for the reduced module count and sub-module capacity. Operation of essential auxiliary bus loads including electronics involved in management of thermal, chassis, body, and interior systems were analysed to develop the load profile. The balancing parameter par is the sub-module SOC and the initial chosen sub-module imbalance varies over a range of 5%. The pack power would impact SOC imbalance and state observer estimation of the SOC, but it is not necessary for initial demonstration of the control algorithms, therefore it is not considered in this work. The auxiliary bus reference V_{ref} =12.5 V. For both control algorithms, T_{cycle} =10s.

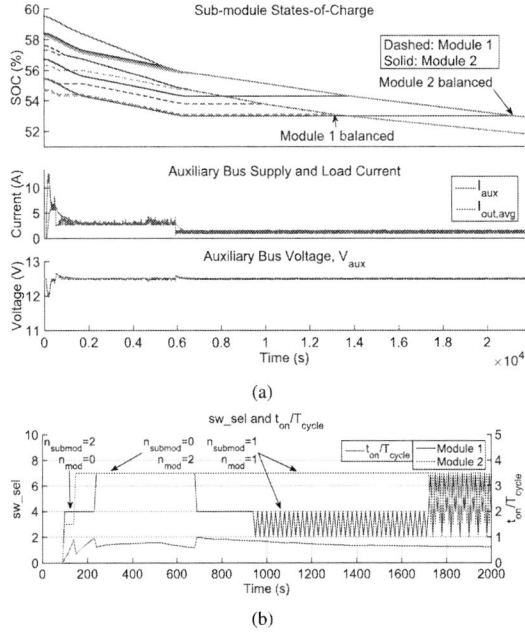

(a)

(b)

Fig. 6. Simulation results of the rule-based control from Section III with 5% maximum initial sub-module imbalance: (a) simulation data, and (b) the switch matrix configuration and t_{on}/T_{cycle} for the first 2000 s of operation: positions 1-6 refer to sub-module input, and 7 refers to module input, n_{submod} and n_{mod} are labelled for relevant periods.

The operation using the rule-based algorithm is shown in Fig. 6(a). The sub-module SOCs from each module converge to the same point, however there is still a difference in the average SOC of the modules. Since I_{aux} is relatively low beyond 6000s, the rule-based algorithm operates with $n_{submod} = 2$ and $n_{mod} = 0$. To maintain SOC balance, the sub-modules in each module are discharged in sequence, and the algorithm must wait for I_{aux} to increase again before further reducing the imbalance.

The operation using the VPC, tuned for fastest balancing speed and best bus regulation, are shown in Figs. 7 and 8, respectively. Tuning is achieved by selecting the sw_sel resulting in the highest $\dot{e}_{par}$ for balancing speed, or lowest e_Q for bus regulation, at every step of the control loop. In the balancing speed case, it can be seen that the system is capable of achieving balancing before the test finishes, but the bus regulation is visibly worse than in the rule-based simulation from Fig. 6(a). In the best regulation case, it can be seen that the bus voltage is regulated more closely to V_{ref} compared to both the rule-based and fastest balancing simulations in Figs. 6(a) and 7, but the sub-modules are less balanced after 6 hours of operation.

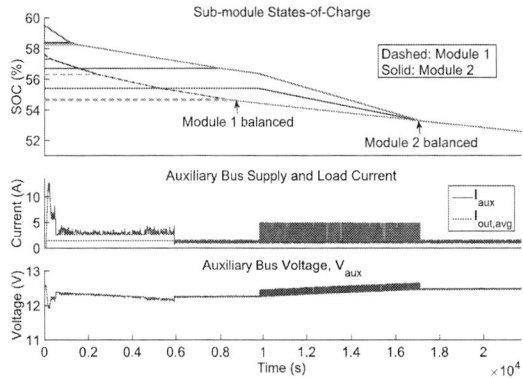

Fig. 7. Simulation results of the VPC from Section IV, tuned to achieve maximum balancing speed with 5% maximum initial sub-module imbalance.

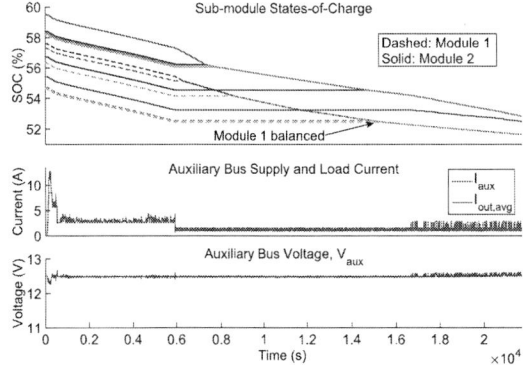

Fig. 8. Simulation results of the VPC from Section IV, tuned to achieve best auxiliary bus regulation with 5% maximum initial sub-module imbalance.

VI. EXPERIMENTAL RESULTS

A. Isolated Ćuk Converter

The isolated Ćuk converter operates at 250 kHz and 15μH inductors as L_1 and L_2. The objective in implementation is to use automotive-qualified, commercially-available components where possible. Due to the high dependence of the converter steady-state and dynamic operation on the transformer turns ratio, the first step in converter implementation is to source an off-the-shelf transformer with appropriate magnetic specifications, minimal conduction loss, and turns ratio between 1-2 to maintain practical duty cycles and currents in the converter. All capacitors in the design are ceramic. The converter primary and secondary-side switch node waveforms are shown in Fig. 9.

Start-up and shut-down transient waveforms for sub-module mode are shown in Fig. 10. Both the positive and negative input terminals of the converter are shown. The switch matrix is configured to connect the sixth sub-module, therefore the common-mode voltage of the two terminals is high, but the differential voltage is that of the sub-module. During the start-up sequence, the PI compensator slowly ramps the input current to the reference value of 3.7 A over 100 ms. During the

The 2018 International Power Electronics Conference

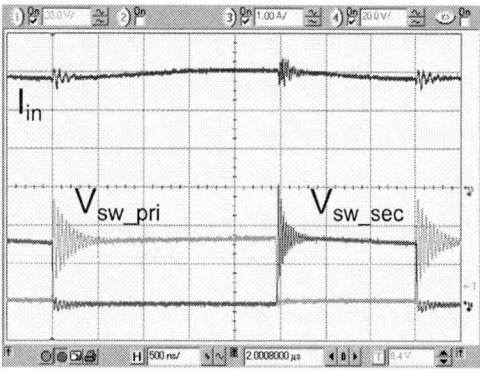

Fig. 9. Isolated Ćuk converter primary and secondary-side switch nodes when operating with $V_{in} = 20$V and $I_{in} = 3$A.

shut-down sequence, the duty cycle is ramped down to zero from its operating value over 1 ms.

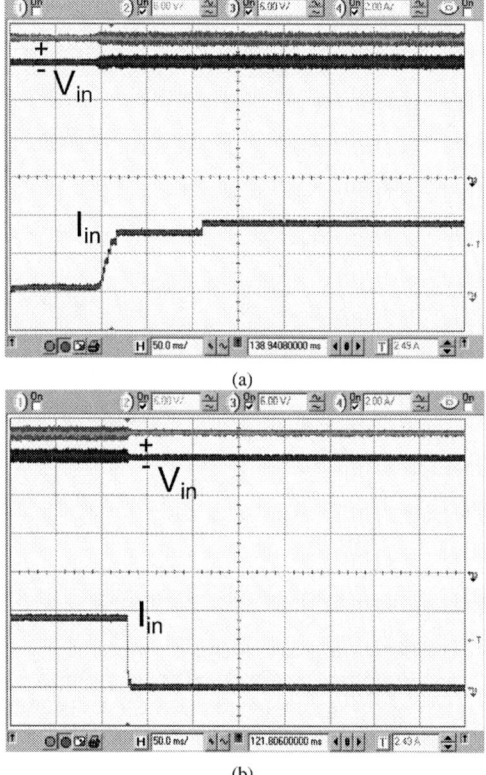

(a)

(b)

Fig. 10. Isolated Ćuk converter experimental waveforms: a) V_{in} and I_{in} at start-up, and b) V_{in} and I_{in} at shut-down.

The converter input voltage and current from one burst cycle is shown in Fig. 11, with $T_{cycle} = 10$s and $T_{on} = 5$s. It can be seen that T_{config} occupies more than 1 s, which is 10 % of the T_{cycle} shown. The configuration time can be significantly reduced with further tuning of the PI compensator and optimisation of the MCU task management. A transition

of the switch matrix configuration from module mode to the fourth sub-module is shown by the change in V_{in} during T_{config}. In-rush current to the converter input capacitors can be seen at the switch matrix transition instant.

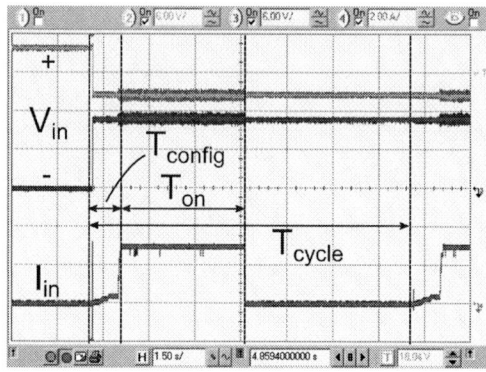

Fig. 11. Isolated Ćuk converter V_{in} and I_{in} over a single burst cycle.

The converter efficiency at the two nominal operating voltages is shown in Fig. 12. The efficiency is obtained with a laboratory power supply at the input and electronic load at the output, to simulate the sub-module, module, and auxiliary batteries.

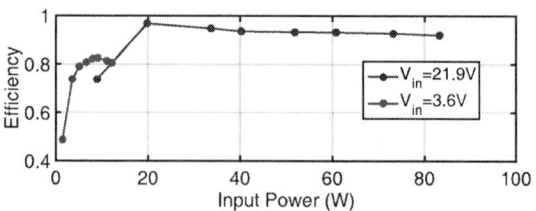

Fig. 12. Isolated Ćuk converter efficiency at the two nominal operating voltages. The converter operates at 10W in sub-module mode, and 65W in module mode.

B. Rule-Based Control

In the experimental setup, the auxiliary bus dynamic load is simulated using a *Chroma 17020* battery cycler, and the BMS operates with the rule-based control algorithm. The battery modules are the same as that shown in Fig. 1(a), and the lead-acid battery is a conventional 60Ah automotive battery. The BMS supervisor is a real-time embedded hardware platform from *National Instruments*. An image of one BMS module is shown in Fig. 13. The circuit board occupies below 1% of the battery module volume with which it is associated. The measurements from the BMS are shown in Fig. 14. Both the SOC evolution over time and bus regulation match well with the rule-based algorithm simulation, demonstrating successful operation of the hardware.

C. Variable-Priority Control

The VPC, tuned for fastest balancing, is tested using the same hardware setup as for the rule-based control, and the test

The 2018 International Power Electronics Conference

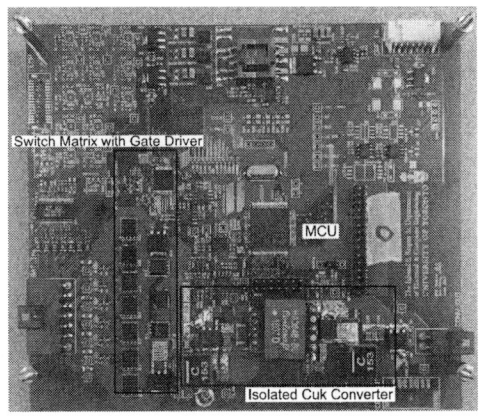

Fig. 13. One of the custom BMS modules used in the experimental setup.

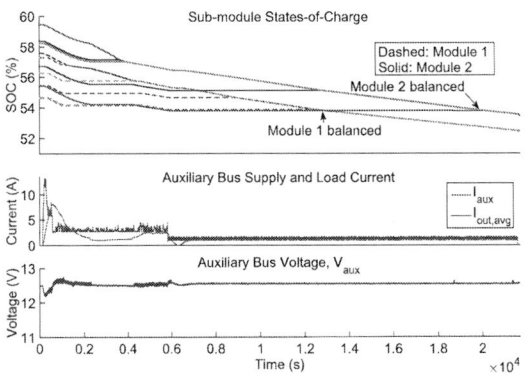

Fig. 14. Measured results from a scaled-down main battery test with the rule-based control, dynamic load on the auxiliary bus, and 5% maximum initial sub-module imbalance.

results are shown in Fig. 15. It can be seen that compared to the simulation from Fig. 7, the system uses less time to balance module 1, and more time for module 2. Compared to the experimental rule-based algorithm results in Fig. 14, it can be seen that the VPC achieves faster balancing, but provides a less regulated auxiliary bus voltage. Overall, the experimental VPC is demonstrated to function as designed, when tuned for fastest balancing speed.

VII. CONCLUSIONS

A new EV BMS, with time-shared isolated converters that achieves hardware cost-reduction, without sacrificing reliability or performance, is demonstrated. The use of module and sub-module input modes to the converters enables the supply of high-power auxiliary loads without increasing the converter input current rating. A good match is observed between simulation and experimental results under the rule-based control. Simulation of the variable-priority control, tuned for balancing speed and bus regulation, demonstrates the capability to prioritise performance parameters dynamically. The VPC, tuned for balancing speed, is demonstrated experimentally to match the

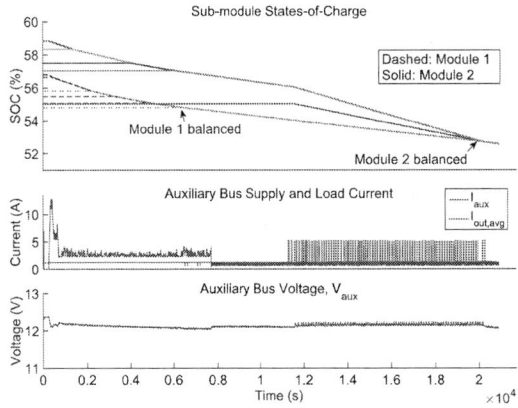

Fig. 15. Measured results from a scaled-down main battery test with the VPC, dynamic load on the auxiliary bus, and 5% maximum initial sub-module imbalance.

simulation results and achieve faster sub-module balance than the rule-based control. The VPC enables the use of the lead-acid battery's energy storage in meeting the multidimensional control objective.

ACKNOWLEDGEMENTS

The authors would like to thank Mazhar Moshirvaziri and Theo Soong, who provided valuable technical guidance in this project.

REFERENCES

[1] "Global ev outlook 2016," 2016, p.52. [Online]. Available: https://www.iea.org/publications/freepublications/publication/global-ev-outlook-2016.html

[2] I. Aizpuru, U. Iraola, J. M. Canales, E. Unamuno, and I. Gil, "Battery pack tests to detect unbalancing effects in series connected Li-ion cells," *4th International Conference on Clean Electrical Power: Renewable Energy Resources Impact, ICCEP 2013*, pp. 99–106, 2013.

[3] J. Cao, N. Schofield, and A. Emadi, "Battery balancing methods: A comprehensive review," *2008 IEEE Vehicle Power and Propulsion Conference, VPPC 2008*, pp. 3–8, 2008.

[4] J. Qi and D. D.-C. Lu, "Review of battery cell balancing techniques," in *2014 Australasian Universities Power Engineering Conference (AUPEC)*, Sept 2014, pp. 1–6.

[5] M. Einhorn, W. Guertlschmid, T. Blochberger, R. Kumpusch, R. Permann, F. V. Conte, C. Kral, and J. Fleig, "A current equalization method for serially connected battery cells using a single power converter for each cell," *IEEE Transactions on Vehicular Technology*, vol. 60, no. 9, pp. 4227–4237, 2011.

[6] G. Altemose, P. Hellermann, and T. Mazz, "Active cell balancing system using an isolated share bus for Li-Ion battery management: Focusing on satellite applications," *2011 IEEE Long Island Systems, Applications and Technology Conference, LISAT 2011*, 2011.

[7] H. D. Gui, Z. Zhang, D. J. Gu, Y. Yang, Z. Lu, and Y. F. Liu, "A hierarchical active balancing architecture for Li-ion batteries," *Conference Proceedings - IEEE Applied Power Electronics Conference and Exposition - APEC*, vol. 2016-May, no. 4, pp. 1243–1248, 2016.

273

[8] M. M. U. Rehman, M. Evzelman, K. Hathaway, R. Zane, G. L. Plett, K. Smith, E. Wood, and D. Maksimovic, "Modular approach for continuous cell-level balancing to improve performance of large battery packs," *2014 IEEE Energy Conversion Congress and Exposition (ECCE)*, pp. 4327–4334, 2014. [Online]. Available: http://ieeexplore.ieee.org/lpdocs/epic03/wrapper.htm?arnumber=6953991

[9] M. Evzelman, M. M. U. Rehman, K. Hathaway, R. Zane, D. Costinett, and D. Maksimovic, "Active balancing system for electric vehicles with incorporated low-voltage bus," *IEEE Transactions on Power Electronics*, vol. 31, no. 11, pp. 7887–7895, Nov 2016.

[10] S. Buller, M. Thele, R. W. A. A. D. Doncker, and E. Karden, "Impedance-based simulation models of supercapacitors and li-ion batteries for power electronic applications," *IEEE Transactions on Industry Applications*, vol. 41, no. 3, pp. 742–747, May 2005.

[11] A. Moshirvaziri, J. Liu, Y. Arumugam, and O. Trescases, "Modelling of temperature dependent impedance in lithium ion polymer batteries and impact analysis on electric vehicles," in *IECON 2014 - 40th Annual Conference of the IEEE Industrial Electronics Society*, Oct 2014, pp. 3149–3155.

[12] E. Din, C. Schaef, K. Moffat, and J. T. Stauth, "A scalable active battery management system with embedded real-time electrochemical impedance spectroscopy," *IEEE Transactions on Power Electronics*, vol. 32, no. 7, pp. 5688–5698, July 2017.

[13] J. P. M. v. Lammeren, "Battery impedance detection system, apparatus and method," Apr. 26 2017, eP Patent App. EP20,120,169,313. [Online]. Available: http://www.google.com/patents/EP2530480A3?cl=en

[14] R. W. Erickson and D. Maksimovic, *Fundamentals of Power Electronics, 2nd edition.* Springer US, 2001.

[15] O. Tremblay, L. A. Dessaint, and A. I. Dekkiche, "A generic battery model for the dynamic simulation of hybrid electric vehicles," in *2007 IEEE Vehicle Power and Propulsion Conference*, Sept 2007, pp. 284–289.

The 2018 International Power Electronics Conference

A Driving Circuit With Partial Power Regulation for RGB LED Lamps

You-Chun Huang[1]*, Yu-Jen Chen[2]; Yong-Jyun Li[3] and Chin-Sien Moo[1]

1 Department of Electrical Engineering, National Sun Yat-sen University, Kaohsiung, Taiwan
2 Green Energy & Environment Research Laboratories, Industrial Technology Research Institute, Hsinchu, Taiwan
3 Research & development department II, Gemtek Technology Co., Ltd, Hsinchu, Taiwan
*E-mail: chunyen732@gmail.com

Abstract—An efficient dimmable light-emitting diode (LED) driving circuit with partial power regulation is proposed for the lamp with three primary-color (red-green-blue, RGB) LED strings. The currents on RGB LED strings are individually regulated by a flyback converter with three output ports. A multi-phase pulse-width modulation (PWM) control scheme is introduced to operate the LED lamp at the desired irradiance and color. A driving circuit is designed for a 45-W RGB LED lamp supplied from a 24-V dc voltage source. The detailed circuit analyses for parameter design are provided. Experimental results have shown that a better circuit efficiency of the LED driver with partial power regulation over the dimmable range.

Keywords—light emitting diode (LED); LED driver; multi-phase pulse-width modulation; partial power regulation

I. INTRODUCTION

The high-brightness light-emitting-diode (LED) with high luminous efficacy, small size and long lifetime has shown its great potential for energy saving in many lighting applications [1]. There is a tendency to substitute it for the existing lighting sources, such as automotive lighting, liquid crystal display backlight, and general-purpose lighting [2]-[4]. A considerable lighting energy has been decreased since the technique of LED drivers have been improved, including a demanding power conversion efficiency, the flexible control ability of the color temperature and the luminous intensity, and the excellent mechanism for heat dissipation. Many reports on the human visual performance with the LED characteristics, including light color, radiation wavelength, luminous intensity, and others have provided a reference for the designers, standard certifications, and users [5]-[7].

Among the lighting applications, a huge amount of energy is consumed by the residential and commercial buildings, which need a variety of white lightings. To replace the conventionally used artificial white light sources, the phosphor-converted blue-light LEDs are generally used [5]. This method is simple and of low cost, but with poor lighting performance. Consequently, an alternative white light generation is obtained by mixing the lights of three-primary-color (red, green, blue, RGB) LEDs. It offers a better color control flexibility than the phosphor-converted LEDs [8]-[10]. Moreover, a

dimmable RGB LED lamp with adjustable color temperature also provides variable lighting environment with the care of mental and health. Conventionally, an RGB LED driver is composed of three conversion circuits, for each color LED string [11], [12]. This approach, however, is obviously at expense of more complex circuit and thus of poor efficiency [13]-[16].

In this paper, an efficient driving circuit with partial power regulation is proposed for the RGB LED lamp. A flyback converter with three output ports is used. With multi-phase control, the LED strings can be individually dimmed with a relatively higher efficiency.

II. CIRCUIT CONFIGURATION

The circuit topology of an LED driver with partial power regulation for a string of LEDs is illustrated in Fig. 1. The LED string is powered by a dc voltage source, v_i, in series with a power converter for output power regulation. The power converter is with an adjustable voltage, v_x, depending on the required current, i_o, which flows through the LED string. With such an arrangement, the LED string is driven by the two power sources with voltages, v_i and v_x, respectively. In other words, the output power, p_o, delivered to LEDs is the sum of the powers from the dc voltage source and the power regulator.

$$p_o = (v_i + v_x)i_o \qquad (1)$$

By specifying v_x within a range much smaller than v_i, the LED power is mostly supplied directly from the dc voltage source. By processing a small amount of the input power, the conversion loss in the power converter can be minor and the overall efficiency can thus be improved effectively.

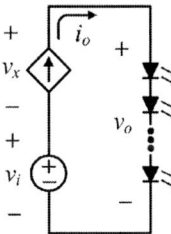

Fig. 1. Partial power regulation for an LED string.

A step-up LED driver with partial power regulation is illustrated in Fig. 2 [16]. The power regulator with an adjustable voltage for partial power regulation is realized by a flyback converter. The primary side of the flyback transformer is parallel-connected to the dc voltage source, and the secondary side is associated with a filter capacitor which is interposed between the dc source and the LED string. The primary and secondary inductances of the flyback transformer are denoted by L_p and L_s, respectively. With such an arrangement, the flyback converter draws a part of power from the dc voltage source and then converts it to the filter capacitor, C_f. By varying the duty-ratio of the power switch, Q_p, the average voltage on C_f and thus the LED current can be regulated. As a result, the LED string is powered from the dc voltage source directly with a partially processed power by the flyback converter.

The proposed driving circuit is shown in Fig. 3. The LED lamp is formed by three RGB LED strings. A flyback converter is with three secondary windings, L_{sR}, L_{sG}, and L_{sB}. Each winding is associated with an active power switch to conduct the stored energy in the coupled inductors. The average currents on LED strings are powered from the secondary windings, which can be regulated with the controllable duty-ratio of the corresponding active power switch. Accordingly, a dimming control scheme can be achieved with desired luminance, color temperature, and color rendering.

III. MULTI-PHASE PWM DIMMING CONTROL

Fig. 4 illustrates the timing sequence of the multi-phase pulse-width modulator for gating the active power switches of the multi-output flyback converter. The active power switches on the secondary sides associated with the corresponding coupled inductors conduct the stored energy from the flyback transformer. Each LED string current is regulated by switching the active power switch on the primary side as well as the corresponding active power switch on the secondary side with complementary signals in a switching period of T_s. An overall three-phase modulation is completed in a period of $3T_s$.

The flyback converter is designed to be operated at the discontinuous-conduction mode (DCM). The energy in the flyback transformer is completely released before turning Q_p on subsequently for the following LED current regulation. Then, the RGB LED currents can be sequentially modulated with non-overlapping PWM signals for individual regulation.

IV. POWER CONVERSION EFFICIENCY

In the RGB LED driver, the major power directly supplied by the dc voltage source, V_i, is expressed as

$$P_d = V_i(I_{oR} + I_{oG} + I_{oB}) \tag{2}$$

where I_{oR}, I_{oG}, and I_{oB} are the average currents of the red, green, and blue LED strings, respectively.

The partially processed power P_t is

$$P_t = V_i I_{pR} + V_i I_{pG} + V_i I_{pB} \tag{3}$$

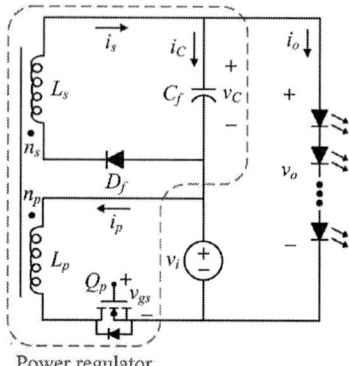

Fig. 2 LED driving circuit with partial power regulation [16].

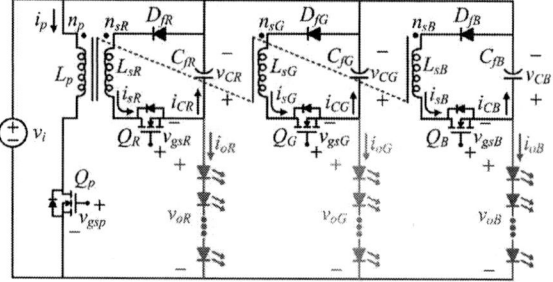

Fig. 3 Driving circuit for RGB LED strings.

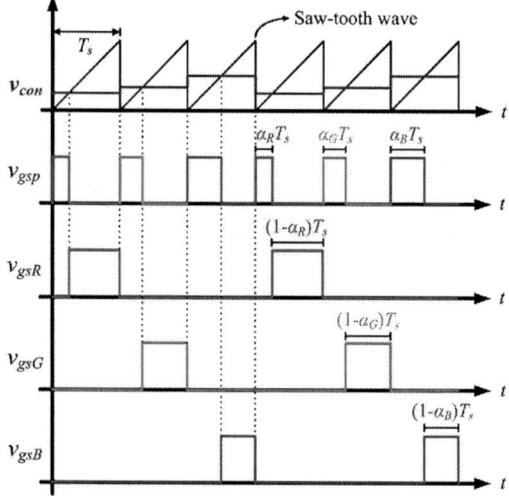

Fig. 4. Timing sequence of non-overlapping PWM.

where I_{pR}, I_{pG}, and I_{pB} are the average primary currents of the flyback transformer.

The regulated power by the flyback converter is given as

$$P_r = V_{CR}I_{oR} + V_{CG}I_{oG} + V_{CB}I_{oB} \tag{4}$$

where V_{CR}, V_{CG}, and V_{CB} are the average voltages on the filter capacitors, C_{fR}, C_{fG}, and C_{fB}, respectively.

The total output power is the sum of the major power directly supplied by V_i and the regulated power of the flyback converter with a conversion efficiency of η_t.

$$P_o = P_d + P_r = P_d + \eta_t P_t \tag{5}$$

As a result, the overall circuit efficiency, η_o, of the driver is

$$\eta_o = \frac{p_o}{p_d + p_t} = \frac{p_d + \eta_t p_t}{p_d + p_t} \tag{6}$$

This equation indicates that the overall conversion efficiency can be always much higher than η_t.

V. CIRCUIT OPERATION

The steady-state operation of the RGB LED driver can be described by three operation stages as illustrated in Fig. 5 with the theoretical waveforms shown in Fig. 6.

A. Stage I

In Stage I, all LED currents are supplied by the dc voltage source v_i and the filter capacitors, which are charged by the secondary inductors at the previous stage. Stage I ends when the active power switch Q_p is switched on. In the case that Q_p has been turned off for a relatively long time, the capacitors may be discharged to zero voltage and then is reversely charged to cut off the LED currents.

B. Stage II

At the beginning of Stage II, Q_p is turned on, and L_p is charged by v_i. The primary current, i_p, rises linearly. At the same time, v_i and the filter capacitors, C_{fR}, C_{fG}, and C_{fB}, supplies the currents, i_{oR}, i_{oG}, and i_{oB}, of the RGB LED strings. Stage II ends when Q_p is switched off and one of the secondary-side switches is turned on.

C. Stage III

With multi-phase PWM control, Stage III is classified into three sub-stages, for three LED strings, as shown in Figs. 5(c) to 5(e). In this stage, the corresponding switch on the secondary-side is activated cyclically and sequentially to conduct the energy stored in the flyback transformer at the Stage II. Then, the associated filter capacitors supply the regulated power to the LED strings. This stage ends when the currents on the secondary sides decline to zero, meaning that the stored energy has been completely released.

VI. CIRCUIT ANALYSIS

The flyback converter of the RGB LED driver with the multi-phase PWM is designed to be operated at the DCM. To simplify the circuit analysis, each LED string is represented by its cut-in voltage in series with an equivalent resistance. The voltage on the LED string can be expressed as

$$v_{ok} = V_{Fk} + i_{ok} R_{Ek} \tag{7}$$

The subscript k in the cut-in voltage, V_{Fk}, and the equivalent resistance, R_{Ek}, represents the color of red-green-blue LEDs.

During Stage II, i_p rises linearly from zero when Q_p is switched on and then reaches the peak as Q_p is turned off. The peak of i_p can be expressed as

$$I_{pk} = \frac{v_i \alpha_k}{L_p f_s} \tag{8}$$

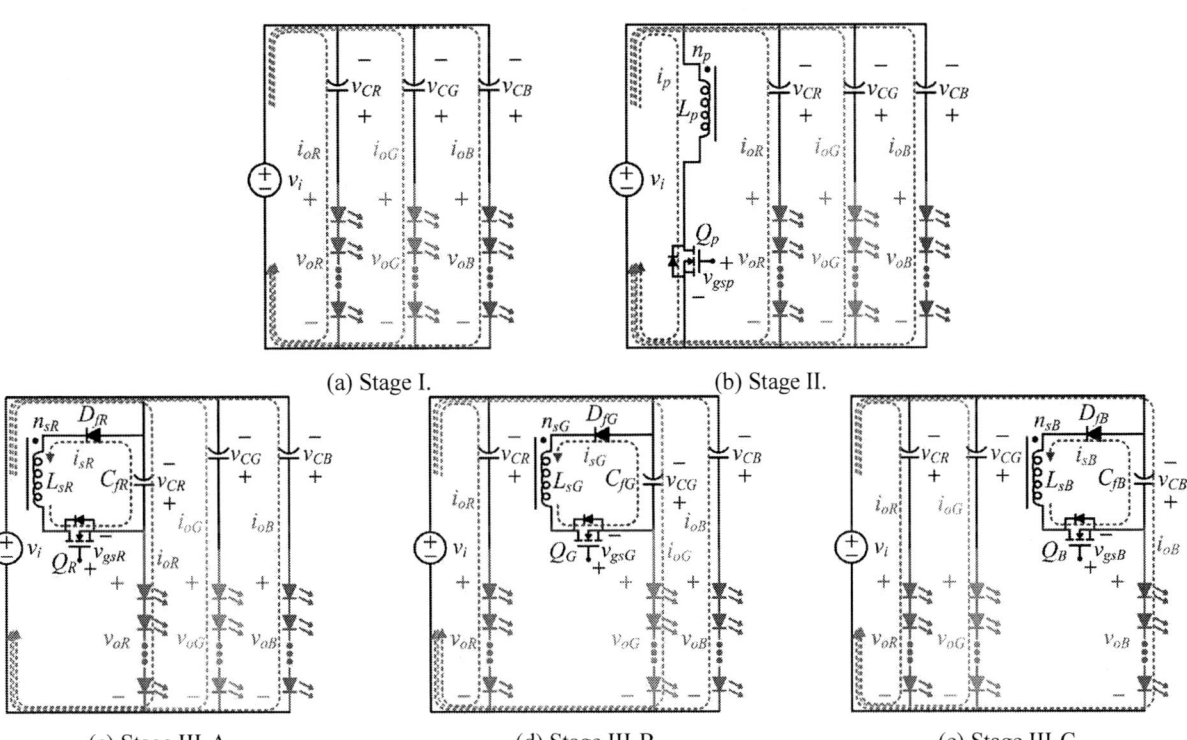

(a) Stage I. (b) Stage II.

(c) Stage III-A. (d) Stage III-B. (e) Stage III-C.

Fig. 5. Operation stages of the RGB LED driver

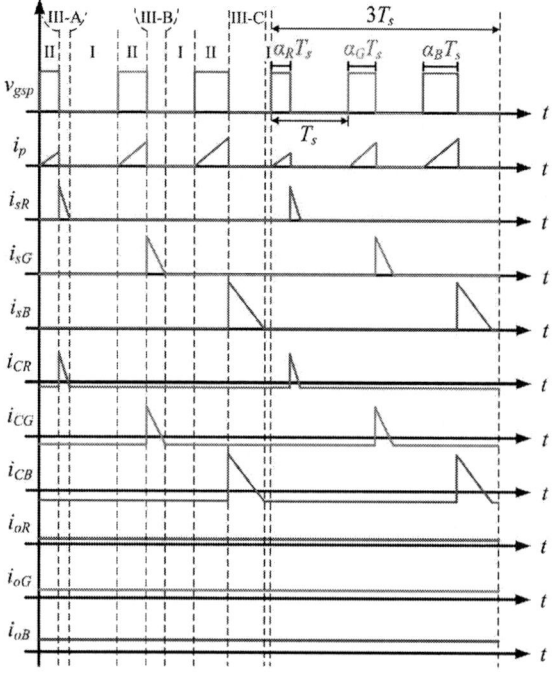

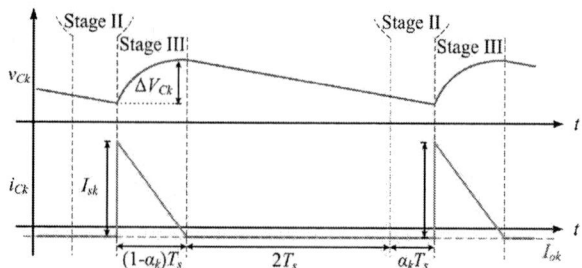

Fig. 7. Theoretical waveforms on the filter capacitor.

$$\Delta V_{Ck} \approx \frac{I_{ok}(\alpha_k + 2)T_s}{C_{fk}} \tag{13}$$

The maximum of the inductance of L_p can be derived from (10), (11), and (12).

$$L_{p(k)} \leq \frac{(v_i \alpha_k)^2}{6 f_s (v_{ok} - v_i) I_{ok}} \tag{14}$$

On the other hand, to ensure that the power regulator is always operated at the DCM for all LED strings, the minimum L_p is chosen by

$$L_p \leq \min\{L_{p(R)}, L_{p(G)}, L_{p(B)}\} \tag{15}$$

By substituting (7) into (14), I_{ok} is derived as

$$I_{ok} = \frac{\sqrt{(V_{Fk} - v_i)^2 + \frac{2 R_{Ek}(v_i \alpha_k)^2}{3 L_p f_s}} + (v_i - V_{Fk})}{2 R_{Ek}} \tag{16}$$

Therefore, the LED current can be dimmed by adjusting α_k with multi-phase PWM.

VII. EXPERIMENTAL VERIFICATIONS

An LED driver supplied from a 24-V dc voltage source is designed for three strings of RGB LEDs. Each string is composed of 15 1-W high-brightness LEDs connected in series. As illustrated in Fig. 8, all LEDs are rated at 350 mA with maximum luminance at different rated output voltages of 33.3 V, 45.5 V, and 45.2 V, respectively. Among which, the cut-in voltage of the red LED, V_{FR}, is the smallest one. With the same number in RGB LED strings, the cut-in voltage of the red LED string is designated to be slightly higher than the dc source voltage to ensure that all LEDs are cut off when the power regulator is inactive. With cut-in voltages of 24.0 V, 37.0 V, and 36.0 V, the equivalent resistances, R_{ER}, R_{EG}, and R_{EB}, calculated by (7) are 26.57 Ω, 24.29 Ω, and 23.43 Ω, respectively.

The switching frequency, f_s, is selected at 60 kHz. The maximum primary-side inductances for the RGB LED strings calculated by (14) are 176.96 µH, 76.55 µH, and 77.63 µH, respectively. Then, the maximum inductance of L_p is selected at 76.55 µH with (15), which ensures that the multi-output flyback converter is always operated at

Fig. 6. Theoretical waveforms.

where α_k and f_s are the duty-ratio and the switching frequency of Q_p, respectively.

The turn-ratio, N_k, can be expressed with the inductances of the primary and secondary of the flyback transformer.

$$N_k = \frac{n_p}{n_{sk}} = \sqrt{\frac{L_p}{L_{sk}}} \tag{9}$$

The current, i_{sk}, on the secondary side decreases from the peak to zero for the DCM operation. Then, the average current of i_{sk} is calculated as

$$I_{sk} = \frac{N_k^2 (v_{ok} - v_i) \alpha_k'}{L_p f_s} \tag{10}$$

where α_k' is the effective duty-ratio of Q_k.

The turn-ratio of the flyback transformer N_k can be obtained by

$$N_k \geq \frac{v_i \alpha_k}{(1 - \alpha_k)(v_{ok} - v_i)} \tag{11}$$

Fig. 7 depicts the voltage and current waveforms on C_{fk}, showing that the capacitor is charged by i_{sk} and discharged by i_{ok}. The LED string current, I_{ok}, is assumed to be a constant in a switching period of $3T_s$. An equation for charge balance on the capacitor can be obtained as

$$\frac{I_{sk}(1 - \alpha_k)T_s}{2} = 3 T_s I_{ok} \tag{12}$$

Then, currents flowing into and out from the filter capacitor cause a voltage ripple, ΔV_{Ck}, on the LED string can be calculated.

the DCM. The voltage ripple is set to be less than 0.5 % at the rated output voltage. Then, the capacitances, C_{fR}, C_{fG}, and C_{fB}, calculated by (13) are 91.09 μF, 66.67 μF, and 67.11 μF. Since L_p is obtained according to the specified green LED string by (15), the inductances associated with the red and blue LED stings, L_{sR} and L_{sB}, should be modified by (9) and (11). The driver specifications and the circuit parameters of the LED driver are listed in Table I.

The voltage and current waveforms of the RGB LED driver are shown in Figs. 9, 10 and 11. The LED currents, I_{oR}, I_{oG}, and I_{oB}, are regulated at 100 %, 60 %, and 20 % of the rated current, respectively. Fig. 9 shows the gating signals for multi-phase PWM. The non-overlapping secondary-side of the flyback converter ensures that Q_R, Q_G, and Q_B would not be conducted simultaneously.

Fig. 10 illustrates the current waveforms of the flyback transformer. When the primary-side switch, Q_p, is switched on, the primary-side current, i_p, rises linearly from zero and reaches to I_{pk}. The peak of the inductor current for the red LED string, I_{pR}, is the highest, meaning that the power regulator processes the most power from the dc source for the red string. Each secondary-side inductor current decreases linearly from the peak current to zero before Q_p is turned on again.

Fig. 11 shows the output current waveforms of the RGB LED strings. The LED currents, I_{oR}, I_{oG}, and I_{oB}, are regulated at 350 mA, 210 mA, and 70 mA, respectively. Fig. 12 illustrates the output current variations when dimming the green LED string. The green LED current, I_{oG}, is dimmed from 350 mA to 35 mA by adjusting the corresponding duty-ratio, α_G, from 0.611 to 0.166. Meanwhile, the red and blue LED currents, I_{oR} and I_{oB}, are maintained at 350 mA and 210 mA, respectively. Fig. 13 shows the output current variations as the red and blue LED currents, I_{oR} and I_{oB}, are dimmed from 350 mA to 35 mA simultaneously by adjusting α_R and α_B. The green LED current, I_{oG}, is kept at the 350 mA during the dimming of other two LED strings. These tested results indicate that the currents on the RGB LED strings can be adjusted individually.

TABLE I

SPECIFICATIONS OF THE DRIVING CIRCUIT

DC input voltage, v_i	24.0 V
Cut-in voltages of the LED strings, V_{FR}, V_{FG}, V_{FB}	24.0 V, 37.0 V, 36.0 V
Rated voltages of the LED strings, V_{rR}, V_{rG}, V_{rB}	33.3 V, 45.5 V, 45.2 V
LED rated currents, I_{rR}, I_{rG}, I_{rB}	350 mA
Number of LEDs in each string, n	15
Switching frequency, f_s	60 kHz
Primary-side inductance of the coupled inductors, L_p	61.5 μH
Secondary-side inductances of the coupled inductors, L_{sR}, L_{sG}, L_{sB}	4.7 μH, 24.4 μH, 23.9 μH
Filter capacitances, C_{fR}, C_{fG}, C_{fB}	100 μF
Active switches, Q_p, Q_R, Q_G, Q_B	IRFB4019PbF
Power diodes, D_{fR}, D_{fG}, D_{fB}	STTH602C

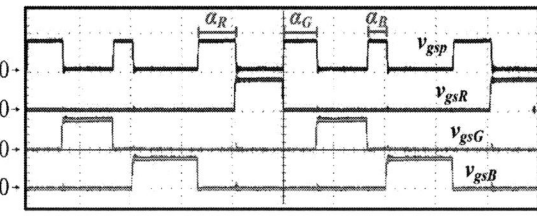

(v_{gsp}, v_{gsR}, v_{gsG}, v_{gsB}: 20 V/div; time: 10 μs/div)

Fig. 9. Gating signals with multi-phase PWM.

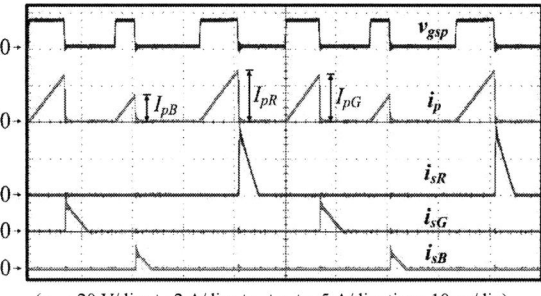

(v_{gsp}: 20 V/div; i_p: 2 A/div; i_{sR}, i_{sG}, i_{sB}: 5 A/div; time: 10 μs/div)

Fig. 10. Current waveforms of the flyback transformer.

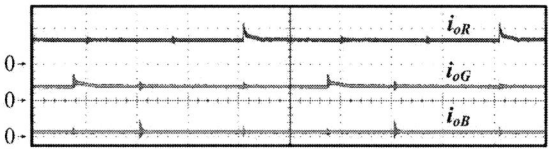

(i_{oR}, i_{oG}, i_{oB}: 500 mA/div; time: 10 μs/div)

Fig. 11. Regulated LED currents.

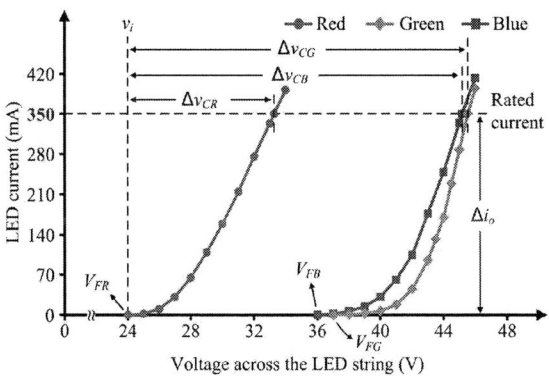

Fig. 8. Characteristic curves of the RGB LED strings.

Fig. 14 shows the overall efficiency comparison between the proposed LED driver with the multi-phase PWM control and three flyback converters conventionally used for the RGB LED lamp. The power conversion efficiency of the proposed driver is much higher than that of the driver with three flyback converters. The overall efficiency of the proposed LED driver is 91.3 % at the rated output of 45 W. As the string currents are all dimmed at 35 mA, the efficiency of the driver can be higher as 96.5 %. The efficiency of the conventional flyback converters is 87.4 % at rated output current of 350 mA, and the highest efficiency is 91.5 % at 70 mA.

The experimental results show that the proposed RGB LED lamp driver with the partial power regulation has a significant improvement in the circuit efficiency, especially when the LED lamp is dimmed at a relatively lower power.

VIII. CONCLUSION

An efficient RGB LED driver with partial power regulation has been proposed. The partial power regulation is realized by a multi-output flyback converter with multi-phase PWM control. A laboratory circuit has been designed for a 45-W RGB LED lamp supplied from a 24-V dc source to verify the theoretical analyses. An excellent power conversion efficiency of the proposed LED driver has been demonstrated by the experimental results over a wide dimmable range. In addition, the control flexibility is confirmed by the dimming feature on individual LED string. As compared with the conventionally used flyback converters for an RGB LED lamp, the proposed driver circuit is advantageous of a much higher efficiency with a simpler circuit.

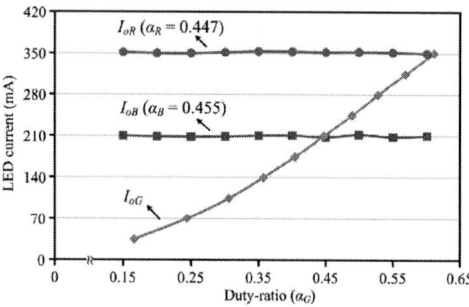

Fig. 12. LED currents with dimmed green LED string.

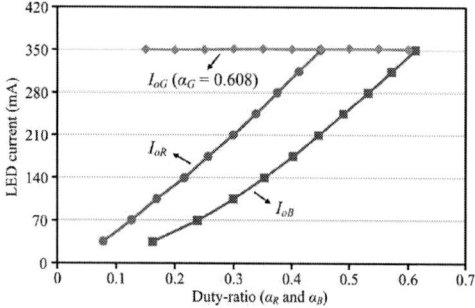

Fig. 13. LED currents with dimmed red and blue LED strings.

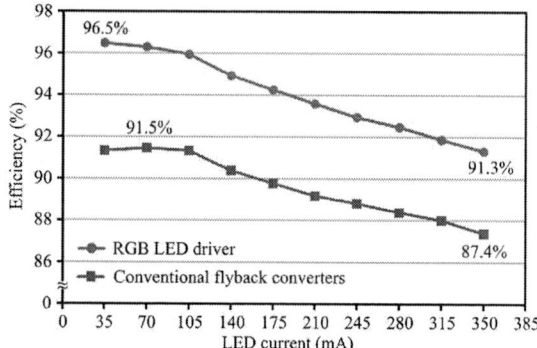

Fig. 14. Power conversion efficiencies of LED drivers.

ACKNOWLEDGMENT

This work was supported by the Ministry of Science and Technology of Taiwan under the grant number of 106-2218-E-006-024.

REFERENCES

[1] J. Peck, G. Ashburner, and M. Schratz, "Solid state LED lighting technology for hazardous environments: Lowering total cost of ownership while improving safety, quality of light and reliability," in *Proc. IEEE PCIC EUROPE*, Jun. 2011, pp. 1–8.

[2] L. Corradini and G. Spiazzi, "A high-frequency digitally controlled LED driver for automotive applications with fast dimming capabilities," *IEEE Trans. Power Electron.*, vol. 29, no. 12, pp. 6648–6659, Dec. 2014.

[3] Y. L. Lin, H. J. Chiu, Y. K. Lo, and C. M. Leng, "LED backlight driver circuit with dual-mode dimming control and current-balancing design," *IEEE Trans. Ind. Electron.*, vol. 61, no. 9, pp. 4632–4639, Sep. 2014.

[4] M. F. Melo, W. D. Vizzotto, P. J. Quintana, A. L. Kirsten, M. A. Dalla Costa, and J. Garcia, "Bidirectional grid-tie flyback converter applied to distributed power generation and street lighting integrated system," *IEEE Trans. Ind. Appl.*, vol. 51, no. 6, pp. 4709–4717, Nov. 2015.

[5] K. H. Loo, Y. M. Lai, S. C. Tan, and C. K. Tse, "On the color stability of phosphor-converted white LEDs under DC, PWM, and bilevel drive," *IEEE Trans. Power Electron.*, vol. 27, no. 2, pp. 974–984, Feb. 2012.

[6] F. Rahman, A. F. George, and R. Drinkard, "Short- and long-term reliability studies of broadband phosphor-converted red, green, and white light-emitting diodes," *IEEE Trans. Device Mater. Rel.*, vol. 16, no. 1, pp. 1–8, Mar. 2016.

[7] D. Gacio, J. M. Alonso, J. Garcia, D. Garcia-Llera, and J. Cardesin, "Study on passive self-equalization of parallel-connected LED strings," *IEEE Trans. Ind. Appl.*, vol. 51, no. 3, pp. 2536–2543, May/Jun. 2015.

[8] X. Qu, S. C. Wong, and C. K. Tse, "Temperature measurement technique for stabilizing the light output of RGB LED lamps," *IEEE Trans. Instrum. Meas.*, vol. 59, no. 3, pp. 661–670, Mar. 2010.

[9] S. K. Ng, K. H. Loo, Y. M. Lai, and C. K. Tse, "Color control system for RGB LED with application to light sources suffering from prolonged aging," *IEEE Trans. Ind. Electron.*, vol. 61, no. 4, pp. 1788–1798, Apr. 2014.

[10] Y. F. Cheung and H. W. Choi, "Color-tunable and phosphor-free white-light multilayered light-emitting diodes," *IEEE Trans. Electron Devices*, vol. 60, no. 1, pp. 333–338, Jan. 2013.

[11] H. J. Chiu, Y. K. Lo, T. P. Lee, S. C. Mou, and H. M. Huang, "Design of an RGB LED backlight circuit for liquid crystal display panels," *IEEE Trans. Ind. Electron.*, vol. 56, no. 7, pp. 2793–2795, Jul. 2009.

[12] Y. H. Liu, Z. Z. Yang, and S. C. Wang, "A novel sequential-color RGB-LED backlight driving system with local dimming control and dynamic bus voltage regulation," *IEEE Trans. Consum. Electron.*, vol. 56, no. 4, pp. 2445–2452, Nov. 2010.

[13] C. S. Moo, Y. J. Chen, and W. C. Yang, "An efficient driver for dimmable LED lighting," *IEEE Trans. Power Electron.*, vol. 27, no. 11, pp. 4613–4618, Nov. 2012.

[14] J. M. Alonso, J. Vina, D. G. Vaquero, G. Martinez, and R. Osorio, "Analysis and design of the integrated double buck–boost converter as a high-power-factor driver for power-LED lamps," *IEEE Trans. Ind. Electron.*, vol. 59, no. 4, pp. 1689–1697, Apr. 2012.

[15] J. Zhang, J. Wang, and X. Wu, "A capacitor-isolated LED driver with inherent current balance capability," *IEEE Trans. Ind. Electron.*, vol. 59, no. 4, pp. 1708–1716, Apr. 2012.

[16] C. S. Moo, Y. J. Chen, Y. J. Li, and H. C. Yen, "A dimmable LED driver with partial power regulation," in *Proc. IEEE IECON*, Nov. 2015, pp. 672–677.

FPGA-based Dynamic Duty Cycle and Frequency Controller for a Class-E^2 DC-DC Converter

Sanghyeon Park and Juan Rivas-Davila
Department of Electrical Engineering
Stanford University
Stanford, CA 94305, USA
Email: spark15@stanford.edu

Abstract—**This paper presents a controller that achieves zero-voltage and zero-dv/dt switching (ZVS and ZVDS) for a class-E^2 dc-dc converter under a wide variation in the input and output voltages. One of the major problems in using the class-E topology for dc-dc conversion is that the increased switching loss significantly degrades the efficiency when the voltage condition deviates from the point which the circuit is designed for. The proposed controller modulates the duty cycle and the frequency of the gate driving signal so that ZVS and ZVDS are always achieved under the voltage variation, while regulating the output voltage by on-off control. We demonstrate an FPGA-based prototype controller in conjunction with a 2 MHz isolated class-E^2 dc-dc converter that minimizes the efficiency drop due to the switching loss for the input voltage from 80 V to 200 V and the output voltage from 5 V to 20 V.**

I. INTRODUCTION

The ever-increasing market of portable devices calls for chargers with less weight and smaller size. We can meet the demand by developing small, efficient power converters with multi-level outputs that fits various mobile devices such as smartphones, tablet computers and laptops. This paper discusses design of a controller that efficiently drives a high-frequency class-E^2 resonant converter over wide voltage ranges.

A class-E^2 dc-dc converter, consisting of a class-E inverter and a class-E rectifier [1]–[3], features low switching loss due to the zero-voltage switching (ZVS) and zero-dv/dt switching (ZVDS). This low switching loss makes the class-E^2 topology a good candidate for high-frequency high-power-density dc-dc conversion. The relatively high voltage stress across switching devices, one of the biggest disadvantages of the class-E topology, is becoming an easier problem to deal with thanks to recent advances in wide-bandgap (WBG) semiconductor technologies and many WBG devices recently launched in the market.

The major hurdle in developing a class-E^2 converter for a mobile charger application is the circuit's sensitivity to the input and output voltages. When the circuit design, switching frequency, and duty cycle of the switching signal are fixed, ZVS and ZVDS are no longer possible if the input or output voltages change from the single value the converter is designed for. The disappearance of soft switching not only degrades

the efficiency of the converter but also possibly damages the switching device as a consequence of high power loss induced on the device.

One can achieve ZVS and ZVDS under varying input and output voltage conditions by changing the switching frequency and the duty cycle of the switching signal [4]. Previous publications that analyze the class-E topology [5]–[7] reveal that the behavior of the circuit can be fully described as a function of two variables: the duty cycle and the normalized switching frequency (switching frequency divided by the inductance and capacitance of the class-E inverter or rectifier). This analysis implies that, when the duty cycle and the frequency can be freely adjusted, ZVS and ZVDS can be achieved under any variations in circuit parameters including the input and the output voltages.

This paper presents a design and implementation of a controller that changes the frequency and the duty cycle to meet soft switching conditions under wide variations of the input and output voltages. The FPGA-based controller receives the voltage information through analog-to-digital converters (ADCs). The controller uses an on-off control method to regulate the output voltage under load variations. Since the target application of this development is a universal power supply unit for mobile applications, the circuit is designed to be compatible with the wall voltage inputs, namely from 80 V_{dc} to 200 V_{dc}, as well as the output voltage range that is required by various types of mobile devices, namely from 5 V_{dc} to 20 V_{dc}.

II. CLASS-E^2 DC-DC CONVERTER

Fig. 1a is the schematic of the dc-dc converter of our interest which consists of a class-E inverter and a class-E rectifier [3]. Both the inverter and the rectifier have finite dc-feed inductance [2] in order to reduce the size and weight of the circuit. Assuming that only a sinusoidal ac current flows between the inverter and the rectifier, the circuit of Fig. 1a can be divided into two parts as illustrated in Fig. 1b and Fig. 1c. The inverter [Fig. 1b] drives a load impedance $R_{load} + jX_{load}$ with a sinusoidal load current of I_s, and the rectifier [Fig. 1c] presents an input impedance $R_{rect} + jX_{rect}$ and is driven by a sinusoidal current source with I_s. Fig. 1d depicts waveforms

The 2018 International Power Electronics Conference

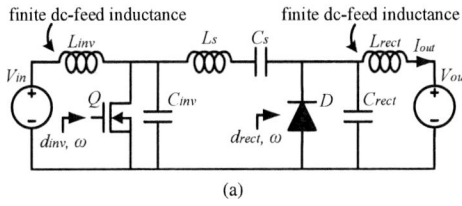

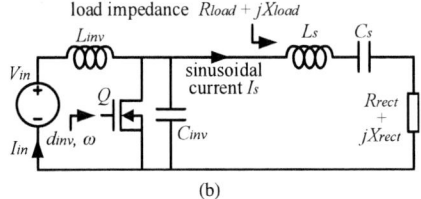

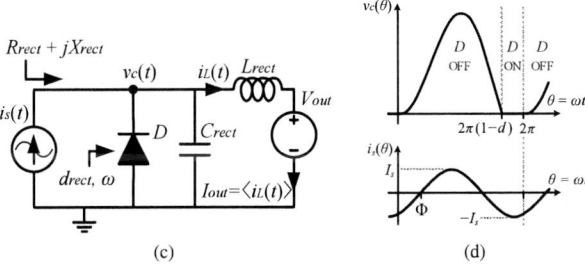

Fig. 1. The class-E^2 dc-dc converter. Instead of an (ideally) infinitely large choke inductance [1], [3], a finite dc-feed inductance [2] is used for the reduction in volume and weight of the converter. (a) A class-E^2 dc-dc converter. The converter consists of a class-E inverter and a class-E rectifier [3] with a series LC filter in between them. (b) The class-E inverter. We assume that only sinusoidal ac current flows into the LC filter branch. (c) The class-E rectifier [2], [3]. We assume that the rectifier is driven by a sinusoidal ac current. (d) Waveforms of the capacitor voltage $v_c(t)$ and the sinusoidal input current $i_s(t)$ of the class-E rectifier [Fig. 1c] over one cycle of the periodic operation.

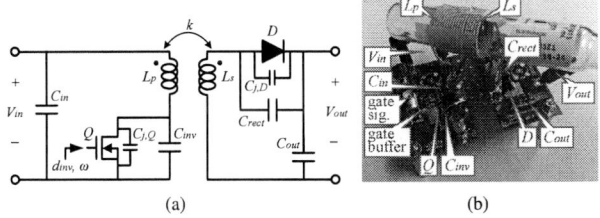

Fig. 2. The class-E^2 converter implementation. (a) The schematic. (b) The implementation.

of the capacitor voltage $v_c(t)$ and the sinusoidal input current $i_s(t)$ of the class-E rectifier [Fig. 1c] over one switching cycle.

Fig. 2 shows the schematic and picture of the class-E^2 converter that was implemented. Table I lists the relevant parameters. We designed the circuit by following the method described in [8].

TABLE I. IMPLEMENTATION DETAILS OF THE CONVERTER IN FIG. 2B.

Parameter	Value	Description
C_{inv}	3 nF	Mica capacitor
C_{rect}	6 nF	Mica capacitor
L_p	2000 nH	675×48 AWG Litz; air-core
L_s	499 nH	675×48 AWG Litz; air-core
k	0.257	air-core coupling
Q	GS66502B	GaN transistor
D	MBR5H100MFS	Si Schottky diode
C_{in}	0.2 μF	X7R capacitor
C_{out}	0.3 μF	X7R capacitor
$C_{j,Q}$	36 pF	approx. of nonlinear cap.
$C_{j,D}$	120 pF	approx. of nonlinear cap.

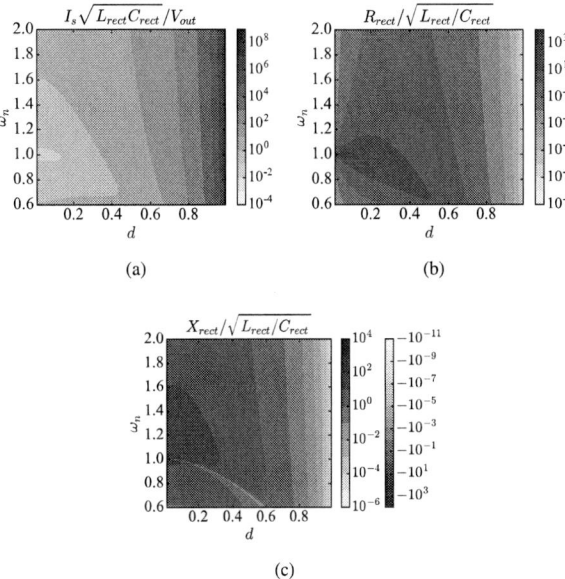

Fig. 3. General solution maps of the class-E rectifier in Fig. 1c. The derivation follows a similar manner as that in previous works [5]–[7]. (a) The normalized sinusoidal ac current. Variables here are circuit parameters normalized by the dc voltage, resonant inductance and capacitance. (b) The normalized dc current. (c) The equivalent resistance. (d) The equivalent reactance.

III. DUTY CYCLE AND FREQUENCY FOR ZVS AND ZVDS

Regarding the rectifier in Fig. 1c, we define the normalized switching frequency ω_n as $\omega\sqrt{L_{rect}C_{rect}}$ and the duty cycle d as d_{rect}. Previous publications [5]–[7] reveal that when ω_n and d values are given, the normalized ac current $I_s\sqrt{L_{rect}C_{rect}}/V_{out}$, normalized input resistance $R_{rect}/\sqrt{L_{rect}/C_{rect}}$ and reactance $X_{rect}/\sqrt{L_{rect}/C_{rect}}$ values are determined. The solution maps are drawn in Fig. 3.

Fig. 3a shows that each set of the normalized ac current and ω_n values uniquely determines every different point on the (d, ω_n) plane. We use this property to remap Fig. 3b and Fig. 3c onto a $(I_s\sqrt{L_{rect}C_{rect}}/V_{out}, \omega_n)$ plane. The reconstructed solution maps are shown in Fig. 4.

The class-E inverter operation is time-reversal of the class-E rectifier operation [9], [10]. Therefore, the solution maps for the rectifier [Fig. 4] apply equally well for the class-E

283

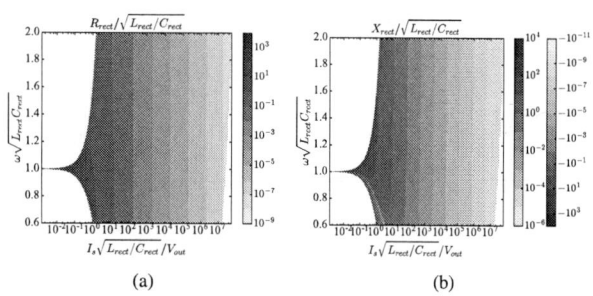

(a)   (b)

Fig. 4. Solution maps for the normalized input impedance of the rectifier, reconstructed from Fig. 3b and Fig. 3c using the information in Fig. 3a. (a) The normalized input resistance. (b) The normalized input reactance.

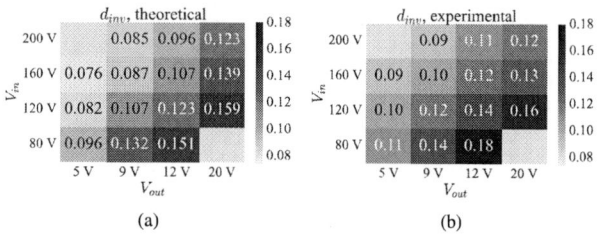

(a)   (b)

Fig. 5. Duty cycles of the inverter switch Q for ZVS and ZVDS at different input and output voltages. (a) Theoretical. (b) Experimental.

inverter [Fig. 1b] when we re-define the normalized switching frequency ω_n as $\omega\sqrt{L_{inv}C_{inv}}$ and replace all the other parameters by their counterparts of the inverter, except that the sign of X_{inv} is flipped to negative due to the time-reversal relation.

The switching frequency ω and the ac current I_s are shared between the inverter [Fig. 1b] and the rectifier [Fig. 1c]. Therefore, as long as there exists a point on the (I_s, ω) plane where $R_{load} = R_{rect}$ and $X_{load} = X_{rect}$, there exists a condition under which ZVS and ZVDS are achieved for both the inverter and the rectifier. Such point can be easily found by using analysis tools such as MATLAB or Python [4].

We experimentally obtain the duty cycle and frequency for ZVS and ZVDS and compare them with the calculation. As shown in Fig. 5 and Fig. 6, theoretical predictions closely match experimental results, verifying the analysis presented in this section. Fig. 7 depicts some of the measured waveforms of

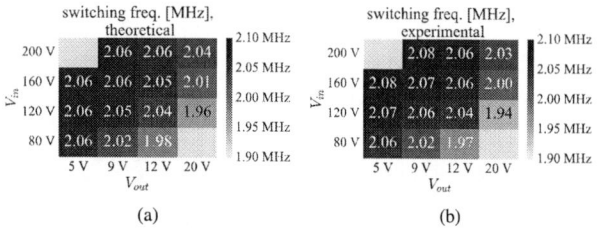

(a)   (b)

Fig. 6. Switching frequencies for ZVS and ZVDS at different input and output voltages. (a) Theoretical. (b) Experimental.

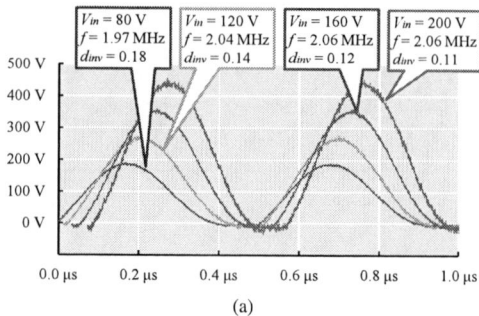

(a)

Fig. 7. Experimental waveforms of the voltage across the inverter switch Q when we fix V_{out} to 12 V and change V_{in} from 80 V to 200 V while modulating the duty cycle and frequency as indicated in Fig. 5 and 6.

the voltage across the inverter switch Q where soft switching always occurs even when the input voltage varies from 80 V to 200 V.

IV. CONTROLLER DESIGN

We design a controller that achieves soft switching under various input and output voltages for the class-E^2 topology. As discussed in the previous section, ZVS and ZVDS occur when the duty cycle and switching frequency are adjusted accordingly to the input and output voltage levels. This requires the controller to be capable of reading the input and output voltage and be able to finely adjust the pulse width and period of its output signal that drives the transistor gate. The controller should also be able to regulate the output voltage under load variations.

We choose Mojo V3 development board from Embedded Micro for the controller implementation. Mojo board is equipped with Xilinx Spartan-6 XC6SLX9 FPGA and an eight-channel ADC that is interfaced to the FPGA. The FPGA can operate at 300 MHz clock speed which allows better-than-1% resolution of the duty cycle and frequency for the power converter when the nominal switching frequency is less than 3 MHz.

The controller programmed on the FPGA largely consists of the following three parts: the lookup table, the pulse-width modulation (PWM) module, and the on-off control gate. The connections between modules are illustrated in Fig. 8a. The lookup table [Fig. 8b] provides *switching period* and *on time* that correspond to the frequency and the duty cycle displayed in Fig. 5b and Fig. 6b. The PWM module [Fig. 8c] contains a counter that counts FPGA clock cycles and resets when the count value reaches the *switching period*. The *PWM signal* becomes HIGH only when the count value is less than *on time*; otherwise *PWM signal* is zero. The on-off control module [Fig. 8d] compares the output voltage V_{out} with the desired output voltage level $V_{out,target}$, and passes the *PWM signal* only when V_{out} is less than $V_{out,target}$. The resulting output, *PWM on-off signal*, is fed to the gate driver of the inverter-side switch in the dc-dc converter.

Fig. 9 is the schematic of the controller and auxiliary

The 2018 International Power Electronics Conference

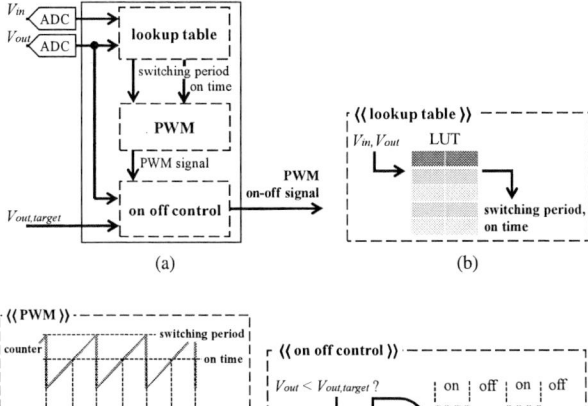

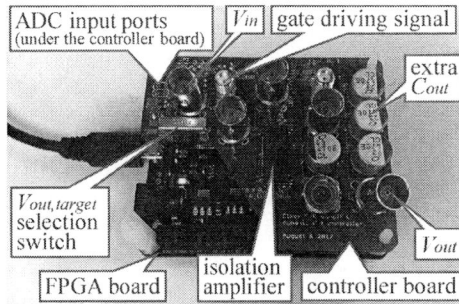

Fig. 10. The controller implementation following the schematic in Fig. 9.

Fig. 8. The controller design programmed on the FPGA. (a) The top module. (b) The lookup table module. (c) The PWM module. (d) The on-off control module.

Fig. 11. Experimental setup to test the controller. (a) The dc-dc converter and the controller connected to power supplies, the electronic load, and measuring equipment. (b) Close-up picture of the converter and the controller.

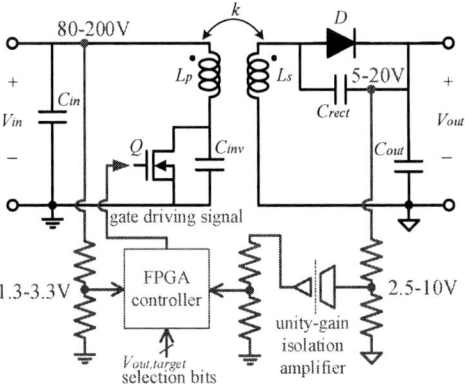

V. EXPERIMENTAL RESULTS

Fig. 11a shows the entire experimental setup to test the controller together with the class-E^2 dc-dc converter. Fig. 11b is the close-up picture of the converter and the controller. Out of all cables in the picture, two BNC cables deliver the input and output voltages from the converter to the controller, one SMA cable delivers the gate driving signal to the converter, and the rest provide dc power to the isolation amplifier and buffers.

A. Step Change in Output Voltage

We change the target output voltage from 5 V to 9 V, 12 V, and 20 V by sequentially moving the slider of the $V_{out,target}$ selection switch [Fig. 10] while fixing the input voltage to 120 V. Fig. 12 shows voltage waveforms. 10 Ω load is used for Fig. 12a and Fig. 12b, and 20 Ω for Fig. 12c. As indicated in Table II, the output voltage settles to the set value within 20 ms with an error in the average value less than 5 %. The transient response can be made faster by reduction in the output filter capacitance at the cost of possible increase in the output voltage ripple.

Fig. 13 shows close-up waveforms associated with multiple output voltages of Fig. 12. Depending on the detected output voltage level, the controller changes the duty cycle and frequency of its gate driving signal by following the lookup table of Fig. 5b and Fig. 6b. The waveforms demonstrate that

Fig. 9. Schematic of the class-E^2 dc-dc converter (black) [Fig. 2] and the controller with auxiliary circuitry (red).

circuitry connected to the class-E^2 dc-dc converter previously shown in Fig. 2. Resistive voltage dividers scale down the input and the output voltages to the level that is within the input range of the ADC, namely from 0 V to 3.3 V. The isolation amplifier (ISO124 from Texas Instruments) delivers the output voltage information without breaking the galvanic isolation of the converter. The user selects the desired output voltage $V_{out,target}$ by means of a multi-bit digital input.

Fig. 10 shows the FPGA-based controller implementation following the design in Fig. 8 and Fig. 9. To minimize the effect of on-off control on the power conversion efficiency, we slow down the change rate of the output voltage by adding extra filter capacitance C_{out} of 1.88 mF (4 units of 470 μF electrolytic capacitors).

285

The 2018 International Power Electronics Conference

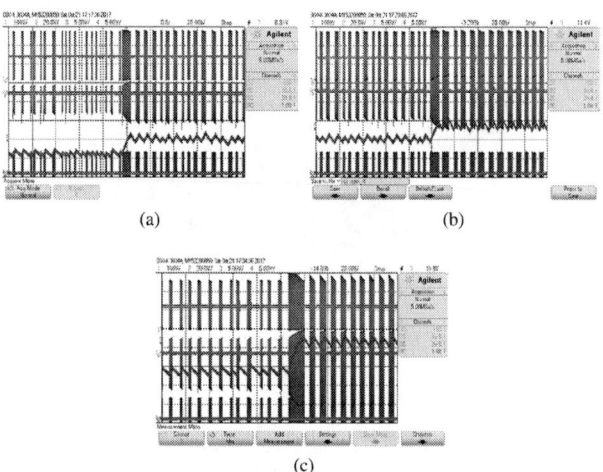

(a)   (b)

(c)

Fig. 12. Voltage waveforms during step changes in the target output voltage $V_{out,target}$. The horizontal scale is 20 ms/div. From top to bottom: the voltage across the inverter transistor (yellow, 100 V/div); the voltage across the rectifier diode (green, 20 V/div); the output voltage (blue, 5 V/div); and the gate driving signal (red curve at the bottom, 5 V/div). (a) $V_{out,target}$ is changed from 5 V to 9 V. (b) $V_{out,target}$ is changed from 9 V to 12 V. (c) $V_{out,target}$ is changed from 12 V to 20 V.

TABLE II.   MEASUREMENTS FROM WAVEFORMS IN FIG. 12.

$V_{out,target}$ Set Value	Measured Average Output Voltage	Peak-to-Peak Ripple
5 V	4.9 V	2.6 V
9 V	8.9 V	4.0 V
12 V	12.4 V	4.5 V
20 V	20.3 V	3.0 V

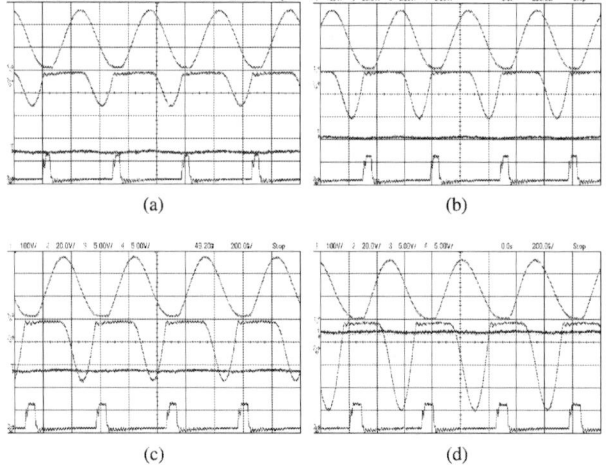

(a)   (b)

(c)   (d)

Fig. 13. Close-up waveforms associated with multiple output voltages of Fig. 12. The horizontal scale is 200 ns/div. From top to bottom: the voltage across the inverter transistor (yellow, 100 V/div); the voltage across the rectifier diode (green, 20 V/div); the output voltage (blue, 5 V/div); and the gate driving signal (red curve at the bottom, 5 V/div). (a) When $V_{out,target}$ is 5 V. The gate driving signal is 0.10 duty cycle and 2.07 MHz frequency. (b) When $V_{out,target}$ is 9 V. The gate driving signal is 0.12 duty cycle and 2.06 MHz frequency. (c) When $V_{out,target}$ is 12 V. The gate driving signal is 0.14 duty cycle and 2.04 MHz frequency. (d) When $V_{out,target}$ is 20 V. The gate driving signal is 0.16 duty cycle and 1.94 MHz frequency.

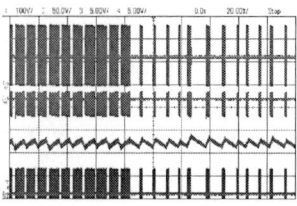

Fig. 14. Voltage waveforms during step changes in the load resistance from 7 Ω to 30 Ω. The horizontal scale is 20 ms/div. From top to bottom: the voltage across the inverter transistor (yellow, 100 V/div); the voltage across the rectifier diode (green, 50 V/div); the output voltage (blue, 5 V/div); and the gate driving signal (red curve at the bottom, 5 V/div).

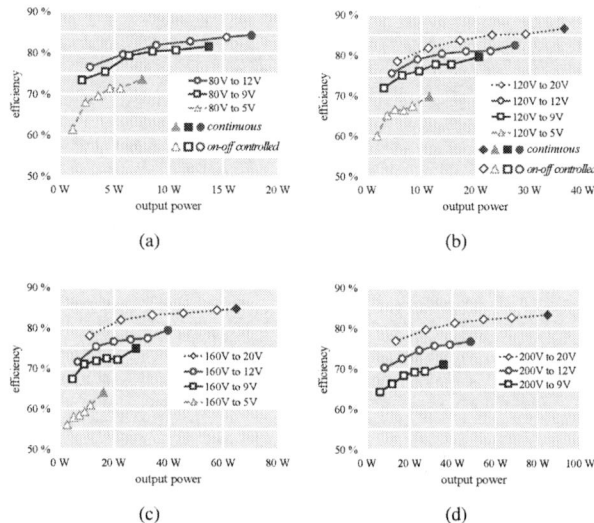

(a)   (b)

(c)   (d)

Fig. 15. Power conversion efficiency at different output power levels. (a) When the input voltage is 80 V. (b) When the input voltage is 120 V. (c) When the input voltage is 160 V. (d) When the input voltage is 200 V.

the inverter transistor Q and the rectifier diode D achieve ZVS and ZVDS for all output voltage conditions.

B. Step Change in Load Resistance

We step-change the load resistance from 7 Ω to 30 Ω, which is equivalent to changing the output power from 80 % full power to 20 % full power. The input voltage and the target output voltage are fixed at 120 V and 12 V, respectively. The step change in the load is equivalent to 80 % to Fig. 14 shows measured waveforms in which the average output voltage remains unchanged at 12 V under the load variation.

C. Efficiency at Different Output Powers

Fig. 15 depicts the efficiency of the converter at different output power levels. The power consumption of the gate driver chip and the controller was not included in the efficiency calculation. The plots show that the efficiency is generally higher at the higher power and the lower voltage gain.

286

The 2018 International Power Electronics Conference

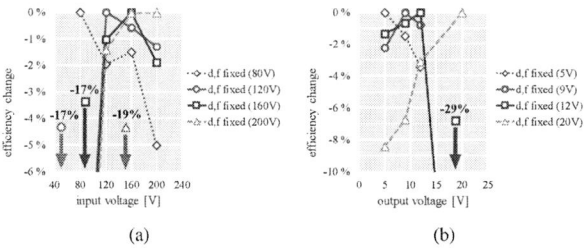

(a)   (b)

Fig. 16. Efficiency drops when the duty cycle and frequency are fixed to a certain value instead of being dynamically changed. The load resistance is 20 Ω for all data points. (a) When the duty cycle and switching frequency are fixed to the value for the output voltage of 12 V and the input voltage of 80 V, 120 V, 160 V, and 200 V. (b) When the duty cycle and switching frequency are fixed to the value for the input voltage of 120 V and the output voltage of 5 V, 9 V, 12 V, and 20 V.

D. Comparison With and Without Dynamic Duty Cycle and Frequency

We operate the converter with and without the duty cycle and frequency modulation, and compare these two cases in regard to the efficiency and the switching loss. Fig. 16 shows by how much the efficiency drops when the duty cycle and frequency are fixed to a certain value instead of being dynamically changed. In all cases, the efficiency peaks when the voltage condition matches what the set values are tuned for, and drops by as much as 29 % when the voltage condition deviates from what is intended for the duty cycle and frequency in use.

Thermal images and waveforms in Fig. 17 further confirms that the switching loss is indeed reduced by the proposed method. We use the input voltage of 120 V, the output voltage of 12 V, and the load resistance of 20 Ω. Fig. 17a and Fig. 17b are thermal images of the dc-dc converter when the duty cycle and frequency are set for 12 V and 5 V output voltages, respectively. Fig. 17c is shown as a reference picture for thermal images. The higher temperature in the inverter transistor indicates the increased switching loss due to the lack of ZVS. Fig. 17d shows the voltage waveform of the inverter transistor, the yellow curve at the top, recorded during the thermal measurement of Fig. 17b. The waveform captures 30 V of voltage across the switch at the turn-on moment which is associated with the increased power loss at the transistor and reduced efficiency in Fig. 16b.

VI. Conclusion

This paper presented a controller for efficient operation of a class-E^2 dc-dc converter. Using a development board equipped with an FPGA and ADC, we implemented a controller that changes the duty cycle and frequency of its gate driving signal to always achieve ZVS and ZVDS under wide variations of input and output voltages. We briefly analyzed the operation of class-E^2 topology and explained a method to figure out the condition for soft switching. We experimentally demonstrated the effectiveness of the proposed controller design in keeping the switching loss low and maintaining a high power conversion efficiency over a wide range of input and output voltages.

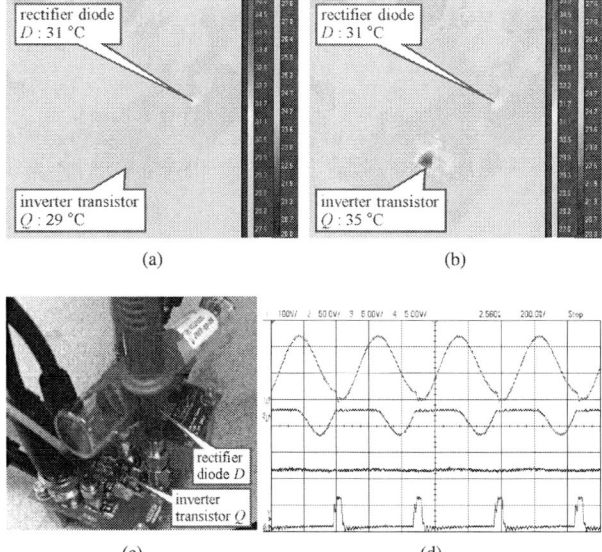

(a)   (b)

(c)   (d)

Fig. 17. Measurements that verify the reduction in the switching loss by the proposed method. The input voltage of 120 V, the output voltage of 12 V, and the load resistance of 20 Ω are used. (a) Thermal image of the dc-dc converter when the duty cycle and frequency are set for 12 V output voltage. (b) Thermal image of the dc-dc converter when the duty cycle and frequency are intentionally mistuned to 5 V output voltage. (c) Reference picture for thermal images. (d) Waveforms recorded during the thermal measurement of Fig. 17b. Note the voltage waveform of the inverter transistor, the yellow curve at the top, which shows 30 V of voltage at the switch turn-on moment which is associated with the increased power loss. The horizontal scale is 200 ns/div. From top to bottom: the voltage across the inverter transistor (yellow, 100 V/div); the voltage across the rectifier diode (green, 50 V/div); the output voltage (blue, 5 V/div); and the gate driving signal (red curve at the bottom, 5 V/div).

Acknowledgment

This work was supported in part by Huawei through the FMA program of Stanford SystemX Alliance.

References

[1] N. O. Sokal and A. D. Sokal, "Class E-a new class of high-efficiency tuned single-ended switching power amplifiers," *IEEE Journal of Solid-State Circuits*, vol. 10, no. 3, pp. 168–176, Jun 1975.

[2] R. Zulinski and J. Steadman, "Class E power amplifiers and frequency multipliers with finite dc-feed inductance," *IEEE Transactions on Circuits and Systems*, vol. 34, no. 9, pp. 1074–1087, Sep 1987.

[3] M. K. Kazimierczuk and J. Jozwik, "Resonant dc/dc converter with class-E inverter and class-E rectifier," *IEEE Transactions on Industrial Electronics*, vol. 36, no. 4, pp. 468–478, Nov 1989.

[4] S. Park and J. R. Davila, "Duty cycle and frequency modulations in class-e dc-dc converters for wide input and output voltage ranges," *IEEE Transactions on Power Electronics*, in press.

[5] M. Acar, A. J. Annema, and B. Nauta, "Analytical design equations for class-E power amplifiers," *IEEE Transactions on Circuits and Systems I: Regular Papers*, vol. 54, no. 12, pp. 2706–2717, Dec 2007.

[6] K. H. Lee, E. Chung, G. S. Seo, and J. I. Ha, "Design of GaN transistor-based class E dc-dc converter with resonant rectifier circuit," in *2015 IEEE 3rd Workshop on Wide Bandgap Power Devices and Applications (WiPDA)*, Nov 2015, pp. 275–280.

[7] Y. Guan, Y. Wang, W. Wang, and D. Xu, "Analysis and design of high frequency dc/dc converter based on resonant rectifier," *IEEE Transactions on Industrial Electronics*, vol. PP, no. 99, pp. 1–1, 2017.

[8] S. Park and J. Rivas-Davila, "Isolated resonant dc-dc converters with a loosely coupled transformer," in *2017 IEEE 18th Workshop on Control and Modeling for Power Electronics (COMPEL)*, July 2017, pp. 1–7.

[9] D. C. Hamill, "Time reversal duality and the synthesis of a double class E dc-dc converter," in *21st Annual IEEE Conference on Power Electronics Specialists*, 1990, pp. 512–521.

[10] J. J. Jozwik and M. K. Kazimierczuk, "Analysis and design of class-E2 dc/dc converter," *IEEE Transactions on Industrial Electronics*, vol. 37, no. 2, pp. 173–183, Apr 1990.

Design Methodology of 3 kW Induction Heating System for both Low Resistance and High Resistance Containers in a Single Burner.

Si-hoon Jeong[1], Hwa-pyeong Park[1], Jee-hoon Jung[1*]
1 Electrical Engineering, Ulsan National Institute of Science and Technology (UNIST)
Ulsan, Republic of Korea
*jhjung@unist.ac.kr

Abstract- Conventional resonant converters for IH applications dedicated to IH-Only containers have been designed for heating the high-resistance containers made by specific manufacturers such as All-Clad and …. Because the IH-Only container made by the other manufacturers has relatively low-resistance which induces large resonant current to power switches in series resonant IH inverters. Hence, the rated power cannot be transferred to the container due to overcurrent passing through the low impedance container , which is higher than the rated switch current. However, the heating capability for various materials is significant to the usability of IH products. In this paper, the boost-type half-bridge series resonant converter is adopted to enlarge the heating capability for the various IH-Only containers which have different impedance to each other. The boost PFC can change the magnitude of a DC-link voltage according to the impedance of the IH-Only container. In this paper, the design methodology of the resonant converter and the IH coil are proposed to properly heat the IH-Only containers which have various impedance. The validity of the design method and the control algorithm is experimentally verified using a 3 kW prototype series resonant half-bridge inverter with the IH coil.

Keywords— Induction Heating, Series Resonant Inverter, IH coil design, Power transmission

I. INTRODUCTION

The IH power conversion systems have been developed rapidly for home appliances and industry due to its fast heating speed, cleanliness, and safety. Conventional IH applications have an input rectifier, a half-bridge structure, a series resonant tank, and a planar type working coil. The half-bridge series resonant inverter is the most used due to its simplicity and its cost-effectiveness. The resonant load consists of the container and the planar type working coil. This resonant tank with resonant capacitor makes an alternating magnetic field, which causes eddy currents and magnetic hysteresis, heating up the container.

The IH resonant inverter dedicated to IH-Only container has been developed to heat the ferromagnetic container. IH-Only container is made from the ferromagnetic material which has high resistance with low skin-depth and high permeability. However, IH-Only container has different resistance according to manufacturer. Fig. 1 shows the comparative data of the

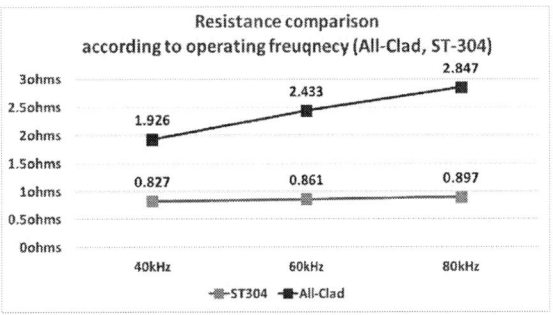

Fig. 1. Comparative analysis of the containers' resistance according to the manufacturers (All-Clad, STS-304)

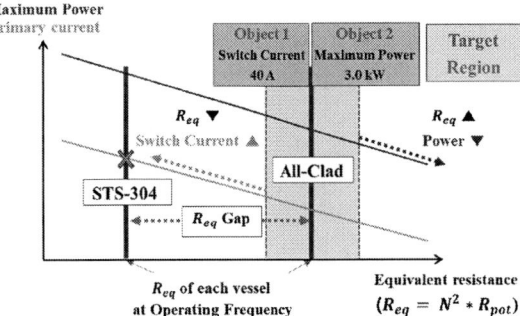

Fig. 2. Problems of the different load conditions in the conventional IH power conversion systems

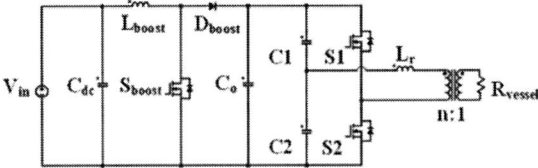

Fig. 3. Boost type series resonant converter topology

resistance according to manufacturers. Some IH-Only container has the relatively low resistance which induces the high resonant current for the same transmission power compared with the high resistance IH-Only container such as All-Clad. The low-resistance IH-Only container is STS-304. It shows that the low resistance IH-Only container has 2 times smaller resistance than the high resistance IH-Only container at the same operating frequency. Fig. 2 shows the problems of the different load conditions in the conventional IH power conversion systems. Conventional design method has been developed to heat the high-resistance IH-Only container. The low-resistance IH-Only container which induces a relatively high resonant current cannot transfer the desired power. However, in order to improve the functionality and convenience, the IH resonant inverter has been required to heat the various IH-Only container according to the manufacturers.

Several research have been proposed to heat the containers made from various materials. Fig. 3 shows the boost type series resonant converter topology. The boost type active PFC was proposed to cope with various load conditions, but the working coil design methodology and control algorithm are not clearly demonstrated. Analysis of the design method and control algorithm can be applied to the various IH systems.

This paper proposes a design methodology of a working coil that can heat both high- and low resistance IH-Only containers to 3 kW rated power in a single burner and the operational algorithm that controls the DC-link voltage. The validity of the proposed design method will be verified with simulation and experimental results using the 3 kW prototype series resonant half bridge inverter.

II. COIL DESIGN CONSIDERATIONS

A. Voltage Induced to Working Coil

The transmission power of the induction heating power conversion system is determined by the load resistance and the voltage induced to a working coil. The voltage induced to a working coil is described as follows:

$$V_{coil} = \frac{\sqrt{2}}{\pi} G_v V_{DC-link} \qquad (1)$$

Where G_v is the resonant tank voltage gain, $V_{DC-link}$ is the DC- link voltage of the boost PFC.

B. Conventional Load Calculation

The resistance of the load can be approximated to a transformer model. The primary side of the transformer is a working coil and the secondary side of the transformer is the container. The resistance of the container is described as follows:

$$R_{eq} = n^2 R_{container} \qquad (2)$$

where R_{eq} is the equivalent resistance of the working coil and container, n is the number of turns in primary, $R_{container}$ is the resistance of the container.

C. Primary Side Switch Current

In an induction heating converter, the primary side switch current is a factor in determining the rated current

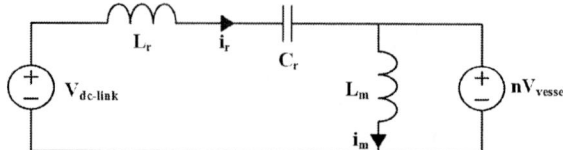

Fig. 4. Equivalent circuit of resonant tank at resonant frequency.

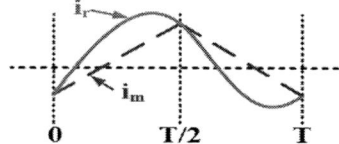

Fig. 5. Resonant tank current waveform at resonant frequency.

of the switch and the range of the load resistance. The current flowing through the resonant tank can be approximated by the primary side switch current. Fig. 4 shows the equivalent circuit of the resonant tank at the resonant frequency and Fig. 5 shows the waveform of the resonant tank current. The peak current of the magnetizing inductor is derived as (3). And the resonant tank current at the resonant frequency is described as (4). Where n is the number of turns in primary, $V_{container}$ is the voltage induced to the container, L_m is magnetizing inductance, L_r is the resonant inductance of the load, T_s is the switching period, I_{rms} is the RMS current of the resonant tank.

$$I_m = \frac{4nV_{container}}{L_m T_s} \qquad (3)$$

$$I_r = \sqrt{2} I_{rms} \sin(2\pi f_r + \phi) \qquad (4)$$

There is a point where the resonant tank current and the magnetizing inductor current are the same at each half-cycle according to the current waveform in Fig 5. Thus, (5) is derived as follows:

$$\sqrt{2} I_{rms} \sin(\phi) = \frac{4nV_{container}}{L_m T_s} \qquad (5)$$

The difference between the resonant tank current and the magnetizing inductor current is equal to the current delivered to the load.

$$\int_0^{\frac{T_s}{2}} (i_r - i_m) dt = \frac{T_s V_{container}}{2nR_{container}} \qquad (6)$$

The RMS current of the resonant tank is derived by summarizing the above equations, which can be described as follows:

$$I_r = \frac{1}{8} \frac{V_{container}}{nR_{container}} \sqrt{\frac{2n^4 R_{container}^2 T_s^2}{L_m^2} + 8\pi^2} \qquad (7)$$

where L_m is magnetizing inductance, T_s is the switching period. Thus, the minimum value of the load resistance range can be calculated by using (7).

D. Target Resistance Range

Equation (8) is the power transferred to the container. Equations (7) and (8) are used to estimate the target resistance range that meets rated power and switch rated current. The target resistance range for each container is calculated since the DC link voltage to be induced is different for each container.

$$P_{transfer} = \frac{V_{coil}^2}{R_{eq}} = \frac{\left(\frac{\sqrt{2}}{\pi} G_v V_{dc-link} \right)^2}{n^2 R_{container}} \tag{8}$$

E. Permeability and Operating Frequency

The IH-Only container has inherent resistivity and permeability depending on the material, the skin depth of each container changes accordingly.

$$\delta_s = \sqrt{\frac{1}{\pi f \mu \sigma}} \tag{9}$$

$$R_{eff} = \sqrt{\frac{L\rho}{\pi D \delta_s}} \tag{10}$$

Equations (9) and (10) show the skin depth and the container resistance. where f is the operating frequency, μ is the permeability, σ is the conductivity, R_{eff} is the effective container resistance, δ_s is the skin-depth, D is the litz wire's diameter, L is the litz wire's length, ρ is the resistivity. Using (9) and (10), the load resistance can be calculated according to the operating frequency. The equivalent resistance of each container can be varied by adjusting the operating frequency of each container.

III. PROPOSED DESIGN METHODOLOGY

A. Coil Design Procedures

In this section the proposed coil design concepts are presented.

First, calculate the equivalent resistance of high resistance container. The equivalent resistance of high resistance container is a factor that is the basis for determining the number of the working coil turns in primary. If the high resistance container be heated to the desired power at the highest DC-link voltage that can be applied, the low resistance container can also be heated by adjusting the DC-link voltage. The equivalent resistance of the high resistance container is calculated by using (1) and (8).

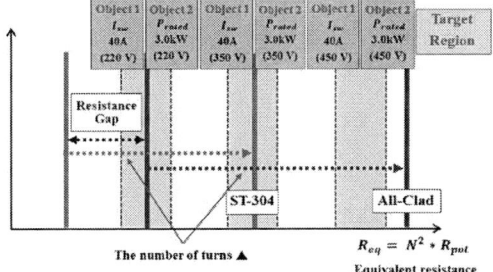

Fig. 6. Variation of the containers' resistance by increasing the number of coil turns.

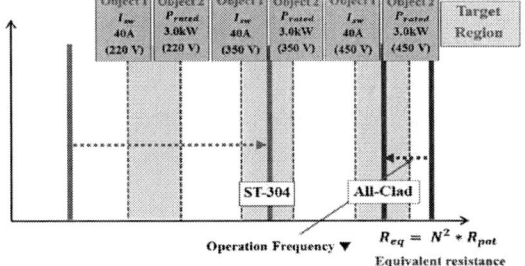

Fig. 7. Variation of the container's resistance by decreasing the operating frequency.

Second, select a working coil that satisfy the equivalent resistance of the high resistance container. And the equivalent resistance of the working coil and the high resistance container should be higher than the calculated equivalent resistance at 20 kHz.

Third, calculate resonant capacitance. After measuring the resonant frequency satisfying the calculated equivalent resistance of the high resistance container, measure the equivalent inductance of a working coil and the container at the resonant frequency. Next, the resonant capacitance is calculated using (11). Equation (11) is the resonant frequency, and (12) is the resonant capacitance, which is described as follows:

$$f_r = \frac{1}{2\pi \sqrt{L_r C_r}} \tag{11}$$

$$C_r = \frac{1}{4\pi^2 f_r^2 L_r} \tag{12}$$

where f_r is resonant frequency, L_r is leakage inductance, C_r is series resonant capacitance.

Fourth, measure the equivalent resistance of the low resistance container. Procedure 1-3 is repeated using the lowest DC-link voltage value of the PFC to calculate the equivalent resistance of the low resistance container, and measure the equivalent inductance and resistance of the low resistance container.

Finally, calculate DC-link voltage. The DC-link voltage to be applied when heating the low-resistance container is derived by using the voltage equation applied to the working coil. Equations (13), (14), and (15) are derived by using (8). Thus, the DC-link voltage to be applied when

The 2018 International Power Electronics Conference

Calculate equivalent resistance of high resistance vessel ($R_{high-calc}$)

- $P_{transfer} = \dfrac{V^2_{coil}}{R_{high-calc}} = \dfrac{\left(\left(V_{dc-link}*\frac{\sqrt{2}}{\pi}\right)*G_v\right)^2}{N^2*R_{vessel}}$

- $V_{coil} = \left(V_{dc-link} * \dfrac{\sqrt{2}}{\pi}\right) * G_v$

- $R_{high-calc} = \dfrac{V_{coil}^2}{P_{trasfer}}$

↓

Select working coil

- $R_{eq} = n^2 \times R_{vessel}$
- R_{eq} is lower than $R_{high-calc}$ at 20 kHz
- Select a number of turns in primary

↓

Calculate resonant capacitance

- Measure resonance frequency which satisfy $R_{high-calc}$
- Measure resonance inductance L_{r-high}
- $C_r = \dfrac{1}{4\pi^2 f_r^2 L_{r-high}}$
- Calculate resonance capacitance

↓

Measure equivalent resistance of low resistance vessel ($R_{low-meas}$)

- Measure equivalent resistance of low resistance vessel $R_{low-calc}$
- Measure resonance inductance L_{r-low}

↓

Calculate DC link voltage

- $P_{transfer} = \dfrac{V^2_{coil}}{R_{low-meas}}$
- $V_{coil} = \sqrt{P_{transfer} \times R_{eq}}$
- $V_{coil} = \left(V_{dc-link} * \dfrac{\sqrt{2}}{\pi}\right) * G_v$
- $V_{dc-link} = \dfrac{V_{coil}\times\pi}{G_v\times\sqrt{2}}$

Fig. 8. Flow chart of IH coil design procedures

$$V_{coil} = \sqrt{P_{transfer} R_{eq}} \tag{13}$$

$$V_{coil} = \frac{\sqrt{2}}{\pi} V_{DC-link} G_v \tag{14}$$

$$V_{DC-link} = \frac{\pi V_{coil}}{\sqrt{2} G_v} = \frac{\pi \sqrt{P_{transfer} R_{eq}}}{\sqrt{2} G_v} \tag{15}$$

heating the low-resistance container is calculated by substituting the equivalent resistance calculated in Procedure 4 into (15).

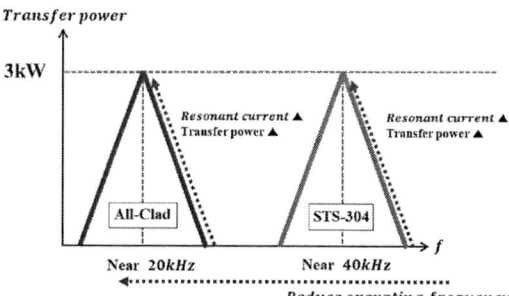

Fig. 9. Power variation according to the containers by decreasing the operating frequency

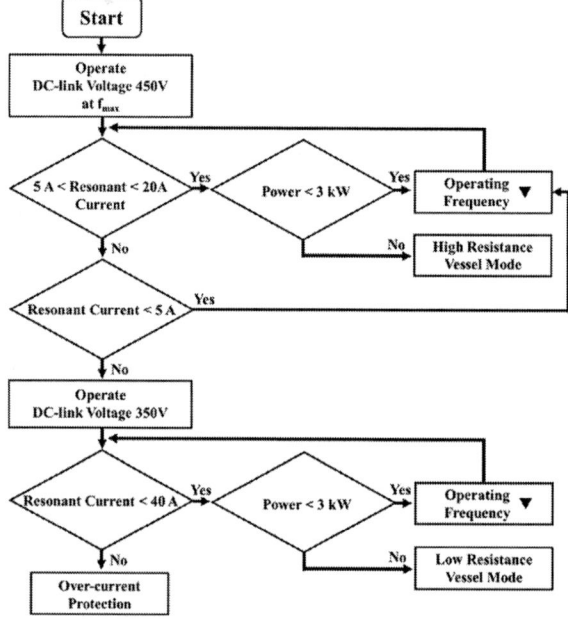

Fig. 10. Block diagram of operational mode selection sequence

B. Operational Method Design

The operational mode is determined by the resonant current and the transmission power. The equivalent resistance of the working coil and the container is inversely proportional to the resonant current. Thus, the resonant current which be induced by heating the high resistance container is twice lower than the resonant current which be induced by heating the low resistance container. It is possible to distinguish the operation mode by using the relationship of resonant current to a transmission power. Fig. 9 shows the power variation according to the containers by decreasing the operating frequency in the proposed prototype. When the operating frequency is decreased from the high frequency outside the power generation range, the transmission power and the resonant current increase as the resonant frequency of each container becomes closer. When heating the high resistance container, the transmission power first reaches 3 kW rather than when heating the low resistance container. Fig. 10 shows the block diagram of operational mode selection sequence. Therefore, the DC-link voltage starts

292

TABLE I
TARGET SPECIFICATIONS

Symbol	Meaning	Value
V_{in}	Input voltage(AC)	220 V
I_{sw}	Switch maximum current	40 A (RMS)
P_{rated}	Rated power	3,000 W

TABLE II
CONVENTIONAL DESIGN SPECIFICATIONS

Symbol	Meaning	Value
n	Coil Turn Numbers	16 turns
R_{eq-All}	Equivalent resistance of All-Clad	3.30 Ω at 45 kHz
R_{eq-STS}	Equivalent resistance of STS-304	1.64 Ω at 66 kHz
f_{sw-All}	Operating Frequency (All-Clad)	45 kHz
f_{r-All}	Resonant Frequency (All-Clad)	39 kHz
f_{sw-ST}	Operating Frequency (ST-304)	66 kHz
f_{r-ST}	Resonant Frequency (ST-304)	50 kHz

TABLE III
CONVENTIONAL DESIGN EXPERIMENTAL RESULTS

Symbol	Meaning	Value
P_{All}	All-Clad input power	2,840 W
I_{sw-All}	All-Clad switch current	34.51 A
P_{ST}	ST-304 input power	1,640 W
I_{sw-ST}	ST-304 switch current	38.35 A

TABLE IV
PROPOSED DESIGN SPECIFICATIONS

Symbol	Meaning	Value
n	Coil Turn Numbers	28 turns
V_{DC-All}	DC-link voltage (All-Clad)	450 V
V_{DC-STS}	DC-link voltage (STS-304)	350 V
R_{eq-All}	Equivalent resistance of All-Clad	12.15Ω at 40kHz
R_{eq-STS}	Equivalent resistance of STS-304	6.20Ω at 48 kHz
f_{sw-All}	Operating Frequency (All-Clad)	40 kHz
f_{r-All}	Resonant Frequency (All-Clad)	36 kHz
f_{sw-ST}	Operating Frequency (ST-304)	48 kHz
f_{r-ST}	Resonant Frequency (ST-304)	42 kHz

TABLE V
PROPOSED DESIGN SIMULATION RESULTS

Symbol	Meaning	Value
P_{All}	All-Clad input power	3,150 W
I_{sw-All}	All-Clad switch current	20.15 A
P_{ST}	ST-304 input power	3,010 W
I_{sw-ST}	ST-304 switch current	38.10 A

TABLE V
PROPOSED DESIGN EXPERIMENTAL RESULTS

Symbol	Meaning	Value
P_{All}	All-Clad input power	3,050 W
I_{sw-All}	All-Clad switch current	18.21 A
P_{ST}	ST-304 input power	2,940 W
I_{sw-ST}	ST-304 switch current	37.20 A

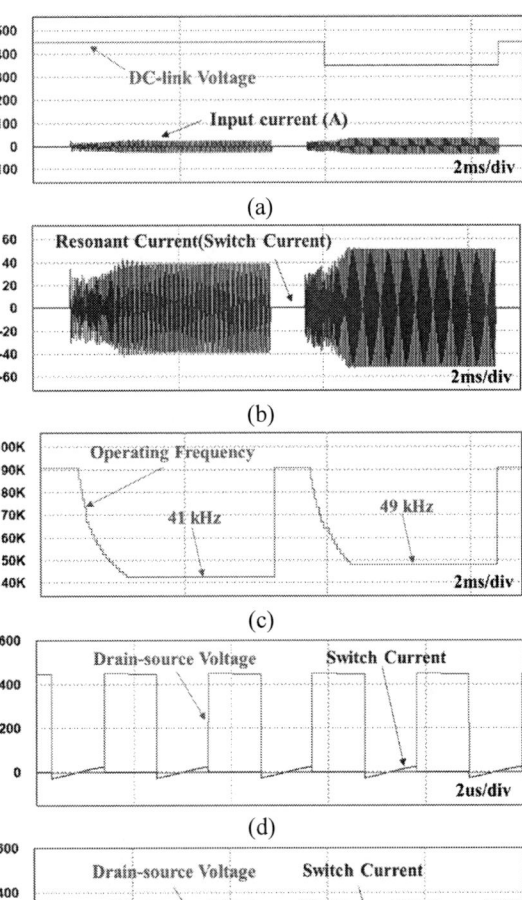

Fig. 11. Simulation results of proposed design: (a) Operational waveform of the mode selection algorithm (b) Operational waveform of the switch current (c) Operational waveform of the operating frequency control (All-Clad) (d) ZVS operation of the proposed design (All-Clad) (e) ZVS operation of the proposed design (STS-304)

driving based on the high resistance container, and the DC-link voltage is lowered when it does not reach 3 kW under the low resonant current conditions. At lower DC-link voltage conditions, the operating frequency continually decrease toward the resonant frequency of the low resistance container, and drive the steady state operation when the transmission power reaches 3 kW. If the resonant current exceeds 40 A, over-current protection is implemented.

The 2018 International Power Electronics Conference

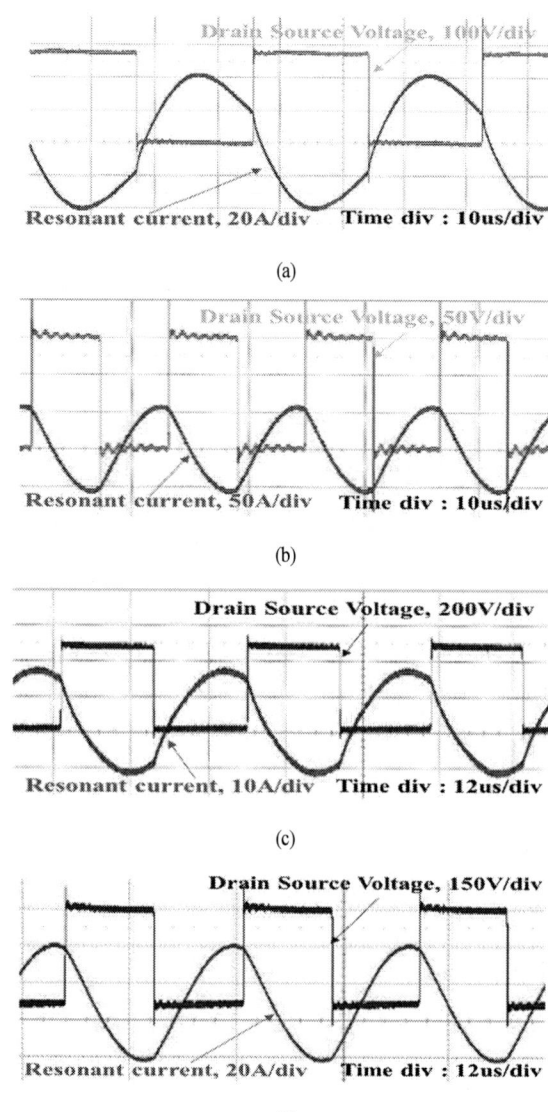

(a)

(b)

(c)

(d)

Fig. 12. Experimental results of drain-source voltage & resonant current: (a) Conventional prototype (All-Clad) (b) Conventional prototype (STS-304) (c) Proposed prototype (All-Clad) (d) Proposed prototype (STS-304)

IV. SIMULATION AND EXPERIMENTAL RESULTS

The design specification of the proposed resonant inverter is described in Table IV. The two containers (All-Clad, STS-304) are used to implement the experiment. The simulation results which uses the same specifications as the experiment is obtained to use a PSIM software. Fig. 11 and Fig. 12 show the simulation and experimental results of 3 kW power conversion according to the containers. Fig. 11 shows the simulation results of 3 kW power conversion according to the operational modes. Fig. 11 (a) shows that the different DC-link voltages are applied according to the container. The first was the All-Clad container, and the

second was the STS-304 container. Fig. 11 (b) shows the sequence of operation modes. When the detection mode starts, it operates from a high frequency to a low frequency, and the switch current increases. If the resonant current of the second container exceeds 20A before reaching 3kW output power, it is determined as STS-304 and the DC-link voltage is lowered to 350V. Fig. 11 (c) shows the variation of the operating frequency. Both the operational modes obtain the soft switching capability and the desired power 3 kW as shown Fig 11 (d) and (e). Fig. 12 shows the experimental waveforms of the conventional and proposed methods. The transmission power and resonant current measurements of the proposed method is presented in Table V.

V. CONCLUSIONS

In this paper, the design procedures of a working coil for induction heating and the operational algorithm is proposed to heat the wide induction heating capability for the containers according to the manufacturers. The series resonant IH inverter with boost type PFC is adopted as the topology of the proposed design method. The boost type PFC can provide large DC-link voltage which can heat the 3 kW rated power under the large turn numbers of IH coil. It can heat the low resistance container by reducing the resonant current, and the high resistance container by boosting DC-link voltage. The proposed operational algorithm includes the container detection, the DC-link voltage and operating frequency variation. The containers were detected according to the ratio of the transmission power to the resonant current.

The validity of the proposed coil design methodology and the operational mode design methodology were verified with experimental results using a 3 kW prototype series resonant converter.

ACKNOWLEDGMENT

This work was supported by the 'Development of pilot-level facility and testing board for Seawater Battery' Research Fund of EWP (KOREA East-West Power co., LTD.).

REFERENCES

[1] Jee-hoon Jung and Joong-gi Kwon, "Theoretical Analysis and Optimal Design of LLC Resonant Converter", IEEE International Conference on Power and Energy (PECon)

[2] ME ME KHANG, SOE SANDAR AUNG, "Design and Comparison of Conductor Size for Inductino Cooker Coil", International Journal of Scientific Engineering and Technology Research, Vol. 03, pp. 1240-1244, May-2014

[3] Soe Sandar Aung, Han Phyo Wai, and Nyein Nyein Soe, "Design Calculation and Performance Testing of Heating Coil in Induction Surface Hardening Machine", World Academy of Science, Engineering and Technology, vol. 2, pp. 1134-1138, Nov-2008.

[4] Lahore. Pakistaan, Umar Shami., University of Engineering and Technology, "Design and Development of a efficient coil for a resonant high frequency inverter for induction heating"

[5] Hwa-Pyeong Park and Jee-Hoon Jung, "Improved Control Starategy of 1MHz LLC Converter for High Frequency

Resolution", 2016 IEEE Applied Power Electronics Conference and Exposition(APEC), pp. 3213-3218, March-2016.

[6] Hwa-Pyeong Park and Jee-Hoon Jung, "PWM and PFM Hybrid Control Method for LLC Resonant Converters in High Switching Frequency Operation", IEEE Transactions on Industrial Electronics, vol. 64, pp. 253-263, Jan-2017.

[7] Hector Sarnago, Oscar Lucia, Arturo Mediano and Jose. M. Burdio, "Class-D/DE Dual-Mode-Operation Resonant Converter for Improved-Efficiency Domestic Induction Heating System", IEEE TRANSACTIONS OF POWER ELECTRONICS, vol. 28, pp. 1274-1285, March-2013.

[8] J. Acero, J. Burd"L₁o, L. Barrag"Lan, D. Navarro, R. Alonso, J. Garcia, F. Monterde, P. Hernandez, S. Llorente, and I. Garde, "The domestic induction heating appliance: An overview of recent research," in Applied Power Electronics Conference and Exposition, 2008. APEC 2008. Twenty-Third Annual IEEE, pp. 651–657, Feb 2008.

[9] J. Acero, J. M. Burdio, L. A. Barragan, and R. Alonso, "A model of the equivalent impedance of the coupled windingload system for a domestic induction heating application," in 2007 IEEE International Symposium on Industrial Electronics, pp. 491–496, June 2007.

[10] H. Koertzen, J. Van Wyk, and J. Ferreira, "Design of the half-bridge, series resonant converter for induction cooking," in Power Electronics Specialists Conference, 1995. PESC '95 Record., 26th Annual IEEE, vol. 2, pp. 729–735 vol.2, Jun 1995.

[11] Y. Yu, Y. Zou, M. Jiang, and D. Zhang, "Investigation on conductivity invariance in eddy current ndt and its application on magnetic permeability measurement," in 2015 IEEE Far East NDT New Technology Application Forum (FENDT), pp. 257– 262, May 2015.

[12] Yong-Ju Kim, Dae-Cheul Shin, Kee-Hwan Kim, Y.Uchihori, Y. Kawamura, "Fluid Heating System using High-Frequency Inverter Based on Electromagnetic Indirect Induction Heating", ICPE'01, pp.69-74, 2001,10.

[13] J. Acero, R. Alonso, L. A. Barragan, C. Carretero, O. Lucia, I. Millan, J.M. Burdio, "Domestic induction heating impedance modeling including windings, laod, and ferrite substrate", 2009 13th European Conference on Power Electronics and Applications, pp.1-10.

The 2018 International Power Electronics Conference

Multi-resonant Inverter Realizing Downsizing and Loss Reduction for All-metallic IH Cooktop

Takayuki Hirokawa[1*], Makoto Imai[1] and Atsushi Fujita[1]

Appliances Company, Panasonic Corporation, 2-3-1-2, Noji-higashi, Kusatsu, Shiga 525-8555, Japan

*E-mail: hirokawa.takayuki@jp.panasonic.com

Abstract— Induction heating (IH) cooktops require no flame, and are popular with consumers. They have a drawback, though, in that the volume of the circuit becomes very bulky if modified to carry the high currents required to heat cooking utensils made of aluminum, copper, and other non-magnetic metals. This paper discusses the application of a multi-resonant inverter that achieves a small, low-loss IH cooktop suitable for heating non-magnetic cookware. Applying this multi-resonant inverter to the IH cooktop allows losses in power semiconductors to be reduced by 69% when heating an aluminum pot and by 13% for an iron pot, and also allows a 27% reduction in the size of the conventional heating coil and a 36% cut in the size of the high-frequency inverter. This paper also describes the successful development of an input power control for application to an IH cooktop of this type.

Keywords— *Soft-switching, Induction heating, High-frequency inverter, Multi-resonant, Downsizing, Loss reduction.*

I. INTRODUCTION

IH cooktops can be roughly classified into two types. One is capable of heating only cooking pots made of magnetic materials; the other is able to heat any kind of metallic cookware. Non-magnetic materials, such as aluminum and copper, have a resistivity of about 1/10 that of magnetic materials such as iron and magnetic stainless steel. To satisfactorily heat non-magnetic materials, therefore, the resonance frequency, or the number of turns of the wires making up a heating coil, or the coil current must be increased [1-4]. For this reason, any model designed to heat all types of metallic cookware faces a problem in that the volume of its circuit, including the heating coil, needs to be far greater than for models designed for heating only cooking pots made of a magnetic material. Some next-generation power semiconductors, such as SiC and GaN types, can be used to reduce the size and power loss of the circuit [5], but their cost tends to be high.

This paper discusses an IH cooktop capable of heating all kinds of metallic cookware, but which uses ordinary power semiconductors and a smaller and more efficient heating coil and high-frequency inverter. A heating coil can be made smaller than the conventional type by reducing the number of turns of its wires. As a result, the current in the heating coil becomes higher to maintain the ampere-turns. A high-frequency inverter using a heating coil with fewer turns causes power semiconductors to lose a larger amount of power, thus making it an unsuitable strategy for size reduction. To solve the problems described above, we adopt a multi-resonant inverter as a high-frequency inverter for an IH cooktop that is capable of heating all kinds of metallic cookware in order to confirm the effectiveness of this device when incorporated in an actual IH cooktop. We also verify a power control method for the circuit making up the inverter.

II. TECHNIQUES FOR REDUCTION OF SIZE AND LOSS

A. Reduction in size of heating coil

Let us consider how the number of turns of wires N of the heating coil is determined when heating a pot made of a non-magnetic material that requires a large coil current. The heating coil L_1 and a pot (L_2, R) shown in Figure 1(a) can also be described as an equivalent series inductance L_S and an equivalent series resistance R_S, as shown in Figure 1 (b). M in Figure 1(a) denotes mutual inductance between the heating coil and the pot. A and B denote the terminals on both ends of the heating coil. L_S and R_S are described by equations (1) and (2), respectively.

$$L_S = \omega(L_1 - \frac{\omega^2 \cdot M \cdot L_2}{R^2 + \omega^2 \cdot L_2^2}) \tag{1}$$

$$R_S = \frac{\omega^2 \cdot k^2 \cdot L_1 \cdot L_2 \cdot R}{R^2 + \omega^2 \cdot L_2^2} \tag{2}$$

These equations indicate that L_S and R_S are proportional to L_1. Table 1 shows measurements of L_S and R_S for different numbers of turns of wires. In the table, the power loss and the number of wires are normalized to be 1 at 40 T, which is the currently adopted

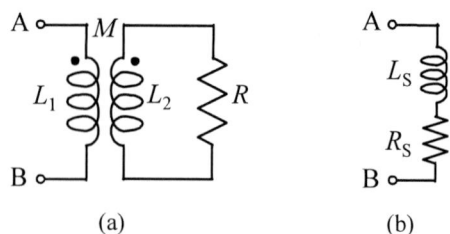

(a)   (b)

Fig.1. Load model of IH cooktop.

(a) Transformer Model. (b) Equivalent Model.

number of turns of wires. The aluminum pot is placed on the heating coil, and the measurement is carried out at both ends of the heating coil. In the table 1, frequency is fixed 90 kHz. The heating coil can be made thinner by reducing the number of turns of its wires. This, however, gives a drastically smaller than R_S, and as a result, to maintain the same power output, the heating coil shows a higher power loss than seen in conventional heating coils. By increasing the number of wires by 25% over the conventional type, the resistance of the heating coil is reduced, bringing its power loss down to that of conventional heating coils. The number of turns of wires N of the heating coil is thus determined to be 27 (9 turns and 3 layers). This reduces the thickness of the heating coil by 20% compared with conventional heating coils.

B. Configuration and operation of proposed circuit

If it has fewer turns, a heating coil will need to carry more current than a conventional heating coil. According to the circuit configuration of the conventional inverter shown in Figure 2, therefore, the power semiconductor will have a higher power loss. A multi-resonant inverter shown in Figure 3 was used as the high-frequency inverter in the IH cooktop to achieve a heating coil with fewer turns. We then examined whether this IH cooktop solves the above problems. The circuit is formed by adding a resonant inductor L_{1a} and a resonant capacitor C_{1a} to the circuit of a conventional inverter. Figure 4 depicts the waveforms of currents and voltages flowing through respective parts of the circuit when a non-magnetic material is heated. Figure 5 depicts current paths that result in different operation modes indicated in Fig. 4, respectively. The circuit operation will be described referring to Figs. 4 and 5.

(Mode I) Period in which Q_1 to Q_4 are turned off

After transistors Q_2 and Q_3 are turned off, current flowing through a coil L_{1a} proceeds to a capacitor C_{S1}, which accumulates electric energy. Energy accumulation at the C_{S1} slows down the time-dependent change of a voltage applied to the Q_2, thus reducing turn-off loss. Meanwhile, a capacitor C_{S2} discharges its accumulated energy. At this time, current flows through the circuit by taking these two paths: C_{1a} - C_1 - L_1; C_{S2} - C_1 - L_1 - L_{1a} - C_{S1}. The resonance frequency of the circuit is given by equations (3) and (4).

$$f_{r1} = \frac{1}{2\pi\sqrt{\frac{C_{1a}\cdot C_1}{C_{1a}+C_1}\cdot L_1}} \text{ [Hz]} \qquad (3)$$

$$f_{r2} = \frac{1}{2\pi\sqrt{\frac{C_1\cdot C_{S1}\cdot C_{S2}}{C_1\cdot C_{S1}+C_1\cdot C_{S2}+C_{S1}\cdot C_{S2}}\cdot(L_1+L_{1a})}} \text{ [Hz]} \qquad (4)$$

Table.1 Measurements of L_S and R_S (at 90 kHz)

Turns	L_S	R_S	Loss ratio	Wire number ratio
20 T	48 μH	0.80 Ω	1.69	1.00
27 T	85 μH	1.4 Ω	1.01	1.25
40 T	188 μH	2.7 Ω	1.00	1.00

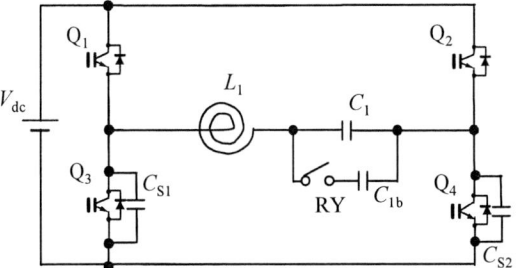

Fig. 2. Conventional inverter.

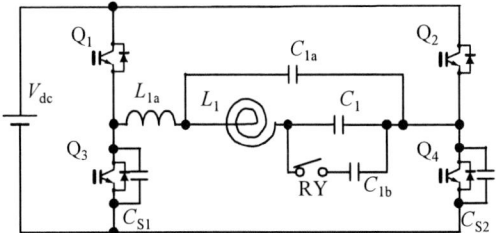

Fig. 3. Multi-resonant inverter.

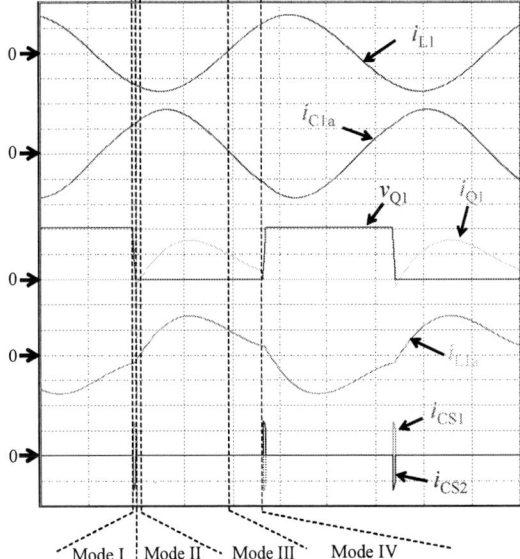

Fig. 4. Operating waveforms (non-magnetic material).

(Mode II) Period in which the anti-parallel diodes of Q_1 and Q_4 are conductive

After C_{S1} and C_{S2} discharge their accumulated energy, the anti-parallel diodes of the transistors Q_1 and Q_4 are conducted until L_{1a} finishes discharging its electromagnetic energy. As a result, current flows through the circuit by taking these two paths: C_{1a} - C_1 - L_1; C_1 - L_1 - L_{1a} - V_{dc}. The resonance frequency of the circuit in this case is given by equations (3) and (5).

$$f_{r3} = \frac{1}{2\pi\sqrt{(L_1+L_{1a})\cdot C_1}} \text{ [Hz]} \qquad (5)$$

297

(Mode III) Period in which Q_1 and Q_4 are turned on

After the L_{1a} discharges energy, the period of energy accumulation follows. In this period, a gate signal is applied to the Q_1 and Q_4 to switch them on, which turns current flowing through the anti-parallel diodes of the Q_1 and Q_4 into their collector current. This achieves Zero Voltage Switching (ZVS), thus reducing switching loss resulting from turning on of the transistors to zero. In this case, current flows through a signal path: C_{1a} - C_1 - L_1 in the circuit, of which the resonance frequency is given by equation (3).

(Mode IV) Period in which the capacitor C_{1a} discharging

Under the influence of the resonance frequency created by C_{1a}, L_1, and C_1, the capacitor C_{1a} discharges its accumulated energy, which changes the direction of current flow in the heating coil to the direction opposite that of Mode III. Meanwhile, the coil L_{1a} in its charge period allows current to flow in the same direction as in Mode III. As a result, current flows through the circuit by taking these two paths: C_{1a} - C_1 - L_1; V_{dc} - L_{1a} - L_1 - C_1. The resonance frequency of the circuit in this case is given by equations (3) and (5).

Afterward, Mode V, Mode VI, ... Mode VIII develop, in which the circuit operation is opposite that in Modes I-IV and therefore will not be described further.

Current supplied from the power supply V_{dc} meets a great deal of impedance at the coil L_{1a}, at which a current flow in the circuit becomes significantly smaller than the same in the path: C_{1a} - C_1 - L_1. Hence, resonance frequency of the circuit in this case is determined by equation (3).

C. Design of circuit parameter

The operating frequency of the circuit is then determined. Figure 6 depicts a resonance section formed of a set of midpoints Q_1 and Q_3 and a set of midpoints Q_2 and Q_4. In Figure 6, R_S is not depicted for reasons of simplicity. Impedances at both ends of this resonance section are calculated to determine the series resonance frequency and the parallel resonance frequency, which are given by equations (6) and (7), where $\alpha = L_S \times C_1 + L_{1a} \times C_1 + L_{1a} \times C_{1a}$ and $\beta = L_S \times L_{1a} \times C_1 \times C_{1a}$.

$$f_{r4} = f_{r5} = \frac{1}{2\pi} \cdot \sqrt{\frac{\alpha \pm \sqrt{\alpha^2 - 4\beta}}{2\beta}} \text{ [Hz]} \quad (6)$$

$$f_{r6} = f_{r1} = \frac{1}{2\pi} \cdot \sqrt{\frac{C_1 + C_{1a}}{L_S \cdot C_1 \cdot C_{1a}}} \text{ [Hz]} \quad (7)$$

The calculation result indicates that this circuit resonates at series resonance frequencies f_{r4} and f_{r5} and a parallel resonance frequency f_{r6}, thus showing the operating characteristics illustrated in Figure 6, which are different from those of a conventional inverter. In Figure 6 (b), of the two series resonance frequencies f_{r4} and f_{r5}, f_{r5} has a higher frequency ($f_{r4} < f_{r5}$) close to the parallel resonance frequency f_{r6}. When the circuit operates at frequency near f_{r5}, therefore, the impedance as seen from V_{dc} reaches a peak. In this case, the heating coil is thus allowed to carry a high current flow through the C_{1a} - C_1 -

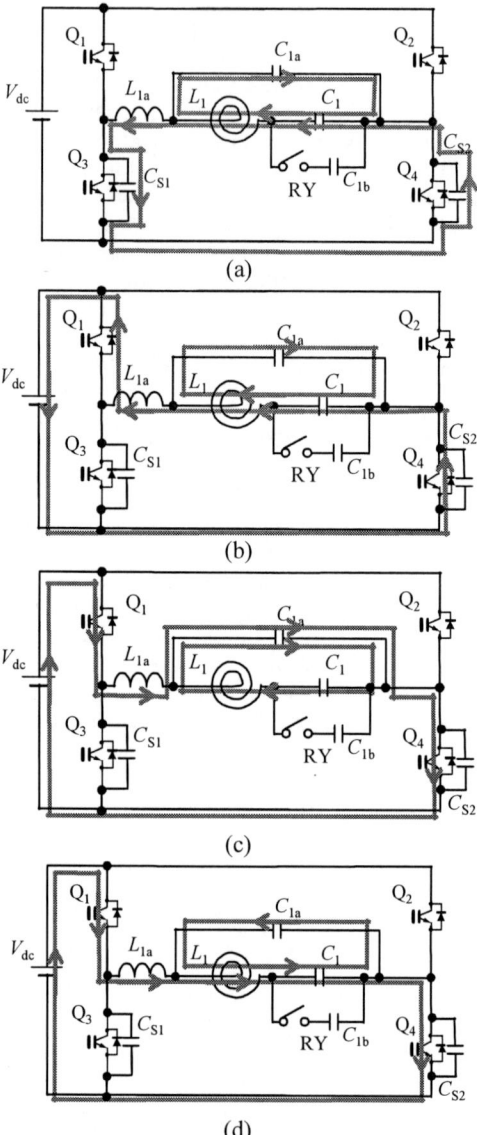

Fig. 5 Current path in each mode.
(a) Mode I. (b) Mode II. (c) Mode III. (d) Mode IV

L_1 closed loop circuit because the current flow through the power semiconductor is reduced. Circuit constant design is carried out according to the following steps.

C-1. Heating an aluminum pot

(i) To reduce the heating coil current by utilizing the skin effect, both the switching frequency and the resonance frequency in equation (3) were set at about 90 kHz.

(ii) Because the main current path provided by the C_{1a} - C_1 - L_1 loop circuit for the heating coil current, the capacitance relationship between C_{1a} and C_1 was set such that $C_{1a} > C_1$, so that the capacitor C_{1a} served as a high-frequency voltage source.

(iii) Based on the impedance of $L_1(L_S, R_S)$ - C_1 determined by the above step (ii), we set the voltage V_{C1a} to the capacitor C_{1a} that is necessary to cause a current I_{L1} required for the rated power output to flow through the heating coil.

C-2. Heating an iron pot

(i) The heating coil L_1 (L_S, R_S) needed for heating an iron pot is larger than that for heating an aluminum pot. For this reason, when the cooktop is operated at the same resonance frequency used for heating the aluminum, V_{dc} must be higher. To allow the cooktop to operate at a constant V_{dc}, therefore, the switching frequency and the resonance frequency are set lower.

(ii) Closing relay RY connects C_1 and C_{1b} to form a combined capacitance. C_{1b} is a capacitance that allows circuit operation at constant V_{dc} and input of the rated power of 3.0 kW.

Circuit parameters determined according to the above steps are shown in Table 2. The switching frequency is set to be close to the parallel resonance frequency. To allow zero-voltage switching (ZVS), the resonance frequencies and the switching frequency are shifted so that they do not match.

III. TEST RESULTS

A. Operation Waveforms

Figures 7 and 8 depict the operation waveforms of the multi-resonant and the conventional circuit. Figures 7 (a) and (b) depict the operation waveforms of the circuits that heat the aluminum and iron pot, respectively. Figure 7 (a) indicates that because the heating coil of the multi-resonant circuit has fewer wire turns than a conventional heating coil, the I_{L1} of the heating coil current is 1.4 times than that of the conventional circuit in Figure 8 (a). However, because the proposed circuit is designed such that its impedance seen from V_{dc} is higher, and a high current is allowed to flow through the C_{1a} - L_1 - C_1 closed circuit, the effective value of its collector current I_C is reduced to about 1/3 of a conventional circuit. Meanwhile, when heating the iron pot shown in Figure 7 (b), I_{L1} turns out to be larger than the I_{L1} of a conventional circuit, as I_C is reduced in the same

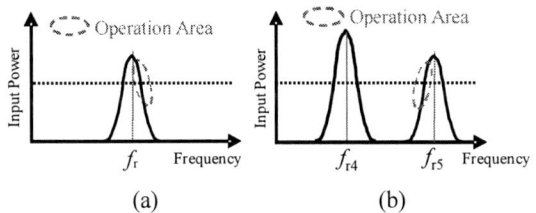

(a)   (b)

Fig. 6. Power characteristics.
(a) Conventional circuit. (b) Multi-resonant circuit.

Table.2 Circuit constants

L_1	@90kHz (Aluminum)	85 μH / 1.4 Ω
(L_S/R_S)	@50kHz (Iron)	120 μF / 11 Ω
C_1		46 nF
C_{1a}		0.2 μF
L_{1a}		60 μH
C_{1b}		0.18 μF
C_{S1}, C_{S2}		1.0 nF
f_{SW} (Aluminum / Iron)		90.3 kHz / 44 kHz
V_{dc}		330 Vdc

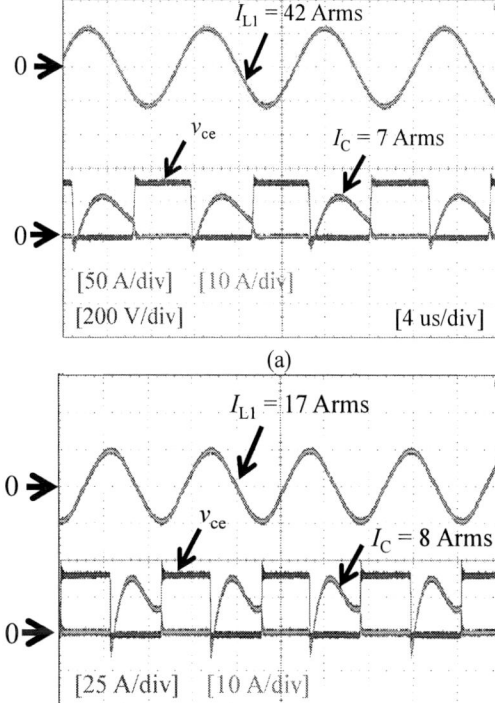

(a)

(b)

Fig. 7. Multi-resonant waveforms.
(a) Aluminum pot heating (Pin = 2.5 kW). (b) Iron pot heating (Pin = 3.0 kW)

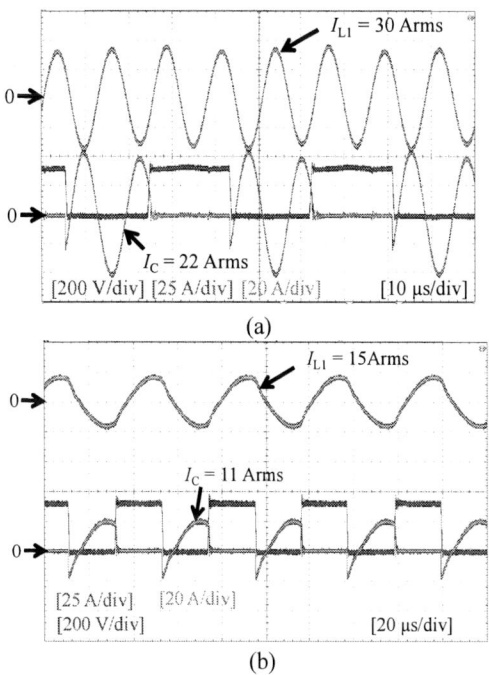

(a)

(b)

Fig. 8. Conventional resonant waveforms.
(a) Aluminum pot heating (Pin = 2.5 kW). (b) Iron pot heating (Pin = 3.0 kW)

parameter design when heating the aluminum pot. These collector current waveforms when heating the aluminum pot and iron pot demonstrate that ZVS is taking place, suggesting the potential for minimizing switching loss.

The above results lead us to the conclusion that our multi-resonant circuit, when operated at the rated power input, can significantly reduce the collector current compared to a conventional circuit. The multi-resonant circuit is thus expected to reduce the loss of the power semiconductor, both when heating an aluminum pot and an iron pot.

B. Evaluating power semiconductor loss

Figures 9(a), 9(b), 10(a), and 10(b) depict switching waveforms resulting from turning off power semiconductors for the cases of aluminum pot heating and iron pot heating. The same power semiconductors are used in the multi-resonant circuit and the conventional circuit. Figure 8 (a) indicates that the conventional circuit, whose switching frequency is 1/3 that of the resonance frequency, suffers switching losses 1/3 that of the multi-resonant circuit. Turn-off losses are calculated from voltage and current waveforms resulting from turning off power semiconductors and turn-on losses are estimated at zero because of ZVS actions. Meanwhile, conduction losses are calculated from measured current values and data indicated on a power semiconductor data sheet. Figure 11 depicts the results of measurement of losses in all of the normalized power semiconductors. As shown in Fig. 11 (a), comparing the conventional circuit with the multi-resonant circuit in terms of losses in power semiconductors for the case of heating the aluminum pot reveals that the multi-resonant circuit reduces the losses by 69%. Particularly, this figure gives us a confirmation that conduction losses are reduced significantly. However, we have observed a small reduction in turn-off losses because of the low switching frequency of the conventional circuit that is 1/3 of the resonance frequency. In the case of heating the iron pot, as shown in Fig. 11(b), losses in power semiconductors are reduced by 13%. Since the resistance of the iron pot is larger than that of the aluminum pot, however, the conduction loss reduction effect turns out to be small in this case. Turn-off losses have rather increased because the switching frequency of the multi-resonant circuit is higher than that of the conventional circuit.

The above results indicate that losses in power semiconductors have been reduced widely for the case of aluminum pot heating, which allows the use of the same heat sink as that used for iron pot heating. This allows for a 36% reduction in the height of the heat sink under the same cooling conditions. Less loss in power semiconductors gives us hope for higher heating efficiency as well.

C. Power control characteristics

Figures 12 depict the power characteristics of different materials in a case where the operation of the multi-resonant and conventional inverter is controlled by duty. This figure demonstrates that power control over loads widely different in characteristics, i.e. the aluminum pot and iron pot, is successfully carried out. "Duty" noted

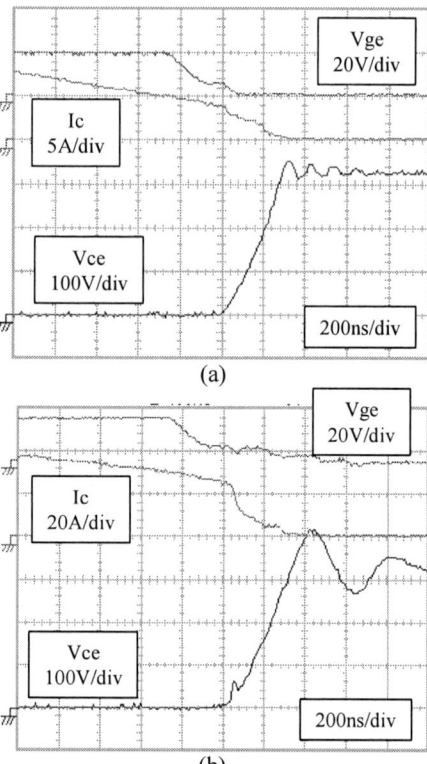

Fig. 9. Switching waveforms (Aluminum pot)
(a) Multi-resonant. (b) Conventional.

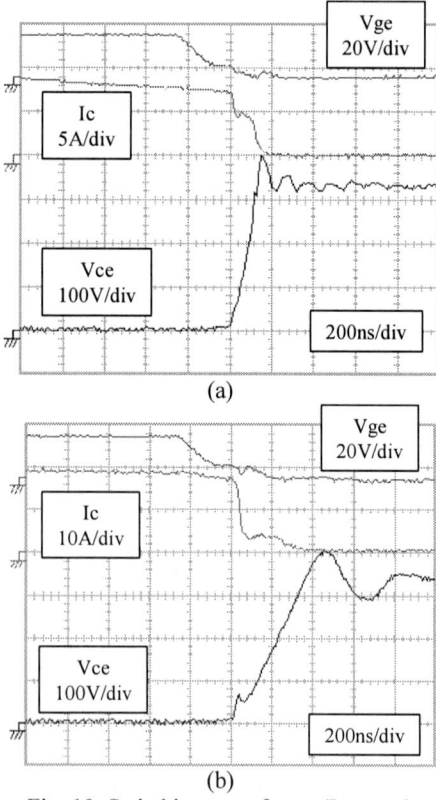

Fig. 10. Switching waveforms (Iron pot)
(a) Multi-resonant. (b) Conventional.

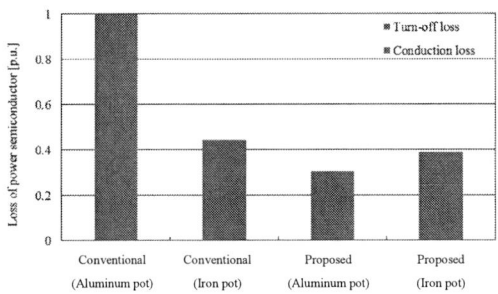

Fig.11. Comparisons of power semiconductor loss.

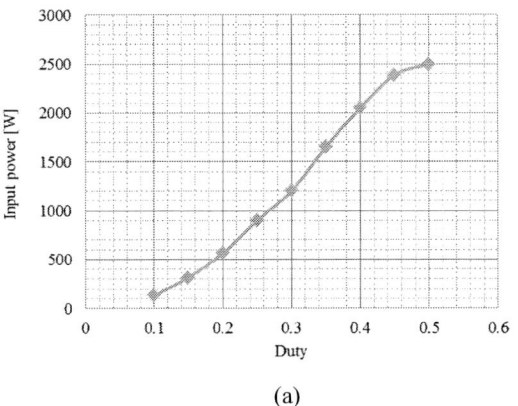

(a)

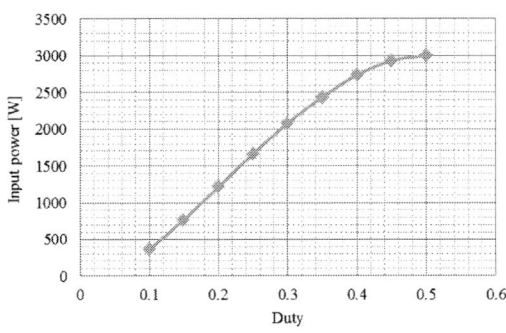

(b)

Fig. 12. Duty power control.
(a) Aluminum heating. (b) Iron heating.

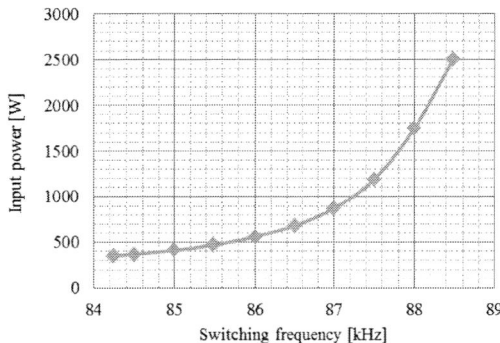

Fig. 13. Frequency power control (Aluminum heating).

IV. CONCLUSIONS

In this paper we have investigated the potential for reduction in the size and loss of a high-frequency inverter incorporated in an IH cooktop that is capable of heating various types of cooking pots. In the conventional circuit, the reduction in the number of turns of the heating coil wires raises the currents passing through the heating coil and power semiconductor. Using a multi-resonant circuit provides a current path that bypasses the power semiconductor, thus reducing the loss of the power semiconductor and allowing a high current to flow through the heating coil. This effect was confirmed in tests on an actual device. An evaluation of the loss of the power semiconductor confirmed that such loss is reduced by 69% when heating an aluminum pot and by 13% for an iron pot. These results have led to a 27% reduction in the size of the heating coil and of 36% of the size of the heat sink compared to that of a conventional heat sink. We also confirmed that the multi-resonant circuit allows input power control.

REFERENCES

[1] H. Sadakata, A. Fujita, S. Sumiyoshi, H. Omori, B. Saha, T. Ahmed, and M. Nakaoka: "Latest Practical Developments of Triplex Series Load Resonant Frequency-Operated High Frequency Inverter for Induction-Heated Low Resistivity Metallic Appliances in Consumer Built-In Cooktops", APEC 2010, pp. 1825-1832, 2010.

[2] T. Hirokawa, E. Hiraki, T. Tanaka, M. Okamoto, and M. Nakaoka: "The Practical Evaluations of Time-Sharing High-Frequency Resonant Soft-Switching Inverter for All Metal IH Cooking Appliances", IECON2012, pp.3284-3289, 2012.

[3] S. Hiroyuki, U. Junpei, I. Masayuki, and Y. Takeshi: "Buck-Boost-Full-Bridge Inverter for All-Metals Induction Heating Cookers", IEEJ Journal of Industry Applications, Vol.5, No.3, pp.289-295, 2016.

[4] I. Millan, J.M. Burdio, J. Acero, O. Lucia, and S. Llorente: "Series resonant inverter with selective harmonic operation applied to all-metal domestic induction heating", IET Power Electron, Vol.4, Iss.5, pp.587-592, 2011.

[5] L. Sun, Z. Ding, D. Guo, and L. Yan: "High Efficiency High Density Telecom Rectifier with GaN Device", Proc. INTELEC2015, 2015.

under the horizontal axis of Fig. 12 represents the duty power of the transistors Q_1 and Q_4. The transistors Q_2 and Q_3 have duty power symmetrical with that of the transistors Q_1 and Q_4. Figure 13 depicts the power characteristics of the aluminum pot that result when the multi-resonant inverter is subjected to pulse frequency modulation. This figure demonstrates that power control over the aluminum pot can be carried out even if the multi-resonant inverter is subjected to pulse frequency modulation. Power control over the iron pot through the multi-resonant inverter subjected to frequency modulation fails to reduce power sufficiently. This because the Q value of the aluminum pot is sufficiently larger than that of the iron pot, which eliminates a frequency that reduces power to zero in the area between f_{r4} and f_{r5} shown in Fig. 6.

The 2018 International Power Electronics Conference

Temperature Estimation of Aluminum Electrolytic Capacitor under Actual Circuit Operation

Kazuki Urata and Toshihisa Shimizu*
Tokyo Metropolitan University
1-1, Minamiosawa, Hachioji, Tokyo, 192-0397, Japan
*E-mail: shimizut@tmu.ac.jp

Abstract— To increase the power density and lifetime of power converters, the loss evaluation of passive components is crucial. The lifetime and reliability of electrolytic capacitors are strongly influenced by temperature. To calculate power loss and temperature rise, an accurate loss calculation method is proposed herein. First, a novel measuring method of the equivalent series resistance (ESR) of a capacitor by utilizing a B-H analyzer is introduced. The measured results show that the ESR of the electrolytic capacitor has strong dependency on the temperature, current frequency, etc. By considering the characteristics of the ESR and the thermal resistances in several portions of the capacitor, the power loss of the electrolytic capacitor in a practical operating condition, for instance in a non-sinusoidal current waveform and high temperature, is calculated. The experimental results agree with the calculated ones.

Keywords— Aluminum electrolytic capacitor, Lifetime, Passive component, Temperature rise

I. INTRODUCTION

The optimum design of passive components is important to increase the power density of power converters [1]–[3]. Although aluminum electrolytic capacitors have been used as DC-link capacitors in power converters, their failure rate is still high [4], [5]. The lifetime of the electrolytic capacitors is highly susceptible to the temperature; the Arrhenius law states that the lifetime will be reduced by half when the temperature rises by 10 °C. Thus, the accurate calculation of power loss and temperature rise of the electrolytic capacitor is extremely important to evaluate the reliability of the electrolytic capacitors [6]. The equivalent series resistance (ESR) is a key factor for the evaluation of power loss. Generally, the ESR values of electrolytic capacitors have been measured using an impedance meter, and the power loss and resultant temperature rise have been calculated using the measured ESR values. However, the temperature rise calculated using the ESR often deviates from the measured value. Therefore, the actual ESR value of the electrolytic capacitor operating in the power converter may not coincide with the ESR value measured using an impedance analyzer. Although, an impedance analyzer measures the ESR on a very low voltage and current with the sinusoidal

waveform condition, high voltage stresses and high currents with non-sinusoidal current waveforms are induced in the capacitor of the power converter system. Hence, the actual ESR value on a high-voltage and high-current condition must be identified. Recently, the power loss estimation on a practical current condition has been reported [7]. However, to the best of our knowledge, power loss calculation based on ESR values measured by the high-current condition has not been reported.

Recently, the authors have proposed a novel ESR measuring method on a high-current and high-voltage condition using a B-H analyzer, SY-8218 (IWATSU). In this method, accurate ESR values can be measured on a wide frequency band of up to 10MHz, a high current condition of up to 6 A, and a high DC-biased condition as well [8], [9]. From the measured results, we found that not only the ESR value but also the capacitance value vary depending on several parameters such as current, temperature, and frequency. Hence, power loss and temperature rise cannot be calculated by simple calculations [10], [11]. Many studies on the online measurement of ESR have been conducted to evaluate the life span of electrolytic capacitors [12], [13]. Those may be effective for the evaluation process of the power converters that are already manufactured, but is not applicable to the general design process of power converters.

This paper focuses on the accurate calculation method of the power loss and temperature rise of the aluminum electrolytic capacitor under the practical operating conditions of power converters. First, the measuring method of ESR and the capacitance of the general capacitors on the high-current condition using a B-H analyzer are reviewed, and the frequency and temperature dependencies of the ESR of the capacitors are shown. Additionally, the temperature distribution of the capacitor is measured.

Next, a method to estimate the capacitor temperature and the loss from the current waveform in the offline state using simple thermal resistance modeling, as well as the temperature and frequency characteristics data, is presented.

To validate the proposed temperature estimation

method, the estimated temperature and the measured temperature in the actual circuit is compared.

II. CAPACITOR MEASUREMENT SYSTEM

Herein, a B-H analyzer is used to measure the capacitor's characteristics. The expressions for magnetic flux density, B and magnetic field strength, H of the magnetic material, and the expressions for electric flux density, D and the electric field intensity, E of the dielectric material are defined as

$$B(t) = \frac{1}{NS} \int v_L(t) dt \quad \cdots\cdots\cdots\cdots (1)$$

$$H(t) = \frac{N}{l} i_L(t) \quad \cdots\cdots\cdots\cdots (2)$$

$$D(t) = \frac{1}{s} \int i_C(t) dt \quad \cdots\cdots\cdots (3)$$

$$E(t) = \frac{1}{d} v_C(t), \quad \cdots\cdots\cdots\cdots (4)$$

where N is the number of windings, A is the effective sectional area, l is the effective magnetic path length, S is the electrode plate area, and d is the distance between the plates. A B-H analyzer is a high-precision loss-measuring instrument for an inductor. It magnetizes the inductor and measures the B-H curve and iron loss characteristics using (1) and (2) from the voltage and current waveform. The current and the voltage waveforms are acquired by the B-H analyzer, capacitance and ESR of the capacitor, of which all obtained by the external calculation using the duality of the relationship between current and voltage in equations (1), (2) and (3), (4).

Fig. 1 shows the measurement circuit using a B-H analyzer: (a) is the B-H analyzer, and (b) is the capacitor that is the device under test (DUT). The equivalent circuit of the capacitor is expressed by the pure capacitance and ESR. An equivalent series inductance (ESL) can be neglected because the ESR value is measured on a frequency that is lower than the resonant frequency of the capacitor. The sinusoidal wave signal from the B-H analyzer and the DC bias voltage are amplified by the two-input high-fidelity power amplifier (NF HSA 4014). The P1, P2, S1, and S2 are the measuring terminals (POD) of the B-H analyzer that detect the voltage and current of the DUT. C_1, C_2 are the DC block capacitors, and 6.8-μF metallized polypropylene film capacitors are used. The B-H analyzer is advantageous in that it provides precise loss measurements under accurate current and voltage conditions. The maximum voltage and current are ± 200 V and ± 6 A, respectively. In addition, the ambient temperature of the DUT can be changed by synchronizing the B-H analyzer with a thermostat scanner system, SY-320A (IWATSU Corporation).

III. CHARACTERISTICS OF ELECTROLYTIC CAPACITOR

Herein, an aluminum electrolytic capacitor (capacitance 47 μF, withstand voltage 50 V, rated current ripple 155 mA$_{rms}$, diameter 6.3 mm) is used as the DUT. Each characteristic of the capacitor is measured using the B-H analyzer. In this section, the characteristics that should be considered in this study is discussed.

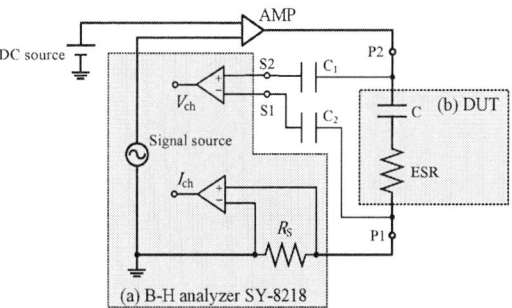
Fig. 1. Measurement circuit using B-H analyzer.

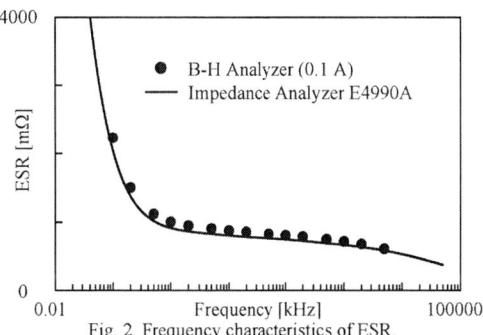
Fig. 2. Frequency characteristics of ESR.

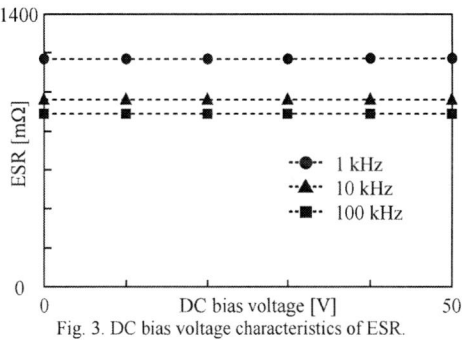
Fig. 3. DC bias voltage characteristics of ESR.

A. Frequency characteristics of ESR

Fig. 2 shows two measurement results: frequency characteristics of ESR measured using an impedance analyzer and B-H analyzer. The ESR decreases as the frequency increases. This can be attributed to the electrolytic capacitor expressed by an RC ladder circuit that is composed of an oxide film dielectric and the electrolytic solution resistance due to of the unevenness arising from the surface etching [14].

B. DC bias voltage characteristics of ESR

Fig. 3 shows the DC bias voltage characteristics of the ESR on the current condition of 0.1 A with three frequencies: 1 kHz, 10 kHz, and 100 kHz. Since the ESR shows a constant value in each frequency, the effect of DC bias voltage on the ESR is considered little.

C. Temperature characteristics of ESR

Fig. 4 shows the temperature characteristics of the ESR

measured using a B-H analyzer and thermostat scanner system. The measured current and frequency are 0.1A and 10 kHz, respectively. The ESR value decreases as the temperature increases. This is because the ESR value is governed by the electrical conductivity of the electrolyte and its electrical conductivity increases with the temperature rise.

D. Current characteristics of ESR

Fig. 5 shows the current characteristics of ESR measured from 0.1 A to 3 A. The current characteristics measurement is one of the advantages of using a B-H analyzer. The measurement frequency is 10 kHz, and the current is applied to the DUT by the B-H analyzer. The measurements are performed after a sufficient time elapses. The values inside the graph in Fig. 5 are capacitor surface temperature when measuring. Although the ESR value decreases as current increases, the temperature increases on the high-current condition simultaneously. Fig. 6 shows a comparison of the ESR values on both the self-heated condition by high-current, and on the high ambient temperature condition of low-current condition. Both ESR characteristics coincide well, thus the current characteristics of the ESR are considered to be caused by the temperature rise. The small difference between these two ESR values is caused by the difference in temperature of the capacitor element, which is covered by the case, and the temperature of the case surface. This will be described in the next section.

IV. TEMPERATURE DISTRIBUTION AND THERMAL RESISTANCE OF ELECTROLYTIC CAPACITOR

Since the ESR of the capacitor is strongly dependent on the temperature, the temperature distribution of the capacitor must be identified to calculate the temperature rise. A simplified structure of the electrolytic capacitor is shown in Fig. 7. The capacitor element body is enclosed in an aluminum case, and a small hole at the center of the capacitor element is present because of the winding process step. An equivalent thermal circuit is expressed in Fig. 8, where the element center temperature is T_{e-in}, the element surface temperature is T_{e-surf}, the case temperature is T_c, and the ambient temperature is T_a. To obtain the temperature distribution, the temperature of each part of the electrolytic capacitor with a case diameter of 32 mm was measured using a thermocouple, and the result is shown in Fig. 9. The result shows that the capacitor element has a near-uniform temperature distribution, and it means that $R_{th(ei-es)}$ is small compared to $R_{th(es-c)}$ and $R_{th(c-a)}$. This is because the capacitor element is constructed by winding an aluminum electrode plate that has extremely high thermal conductivity compared to air. Hence, herein, the capacitor element is treated as having a uniform temperature distribution.

The relationship between temperature rise and loss is expressed by the following equation:

$$\Delta T_{cap} = \left(R_{th(es-c)} + R_{th(c-a)} \right) \cdot P_{loss} \quad \cdots\cdots\cdots\cdots (5)$$

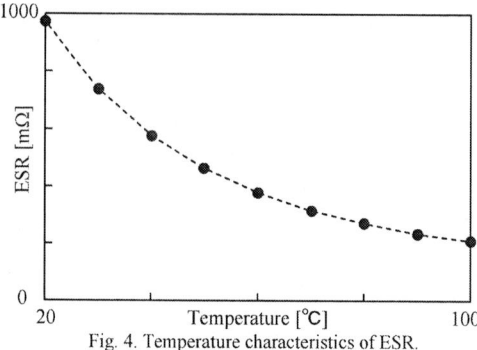

Fig. 4. Temperature characteristics of ESR.

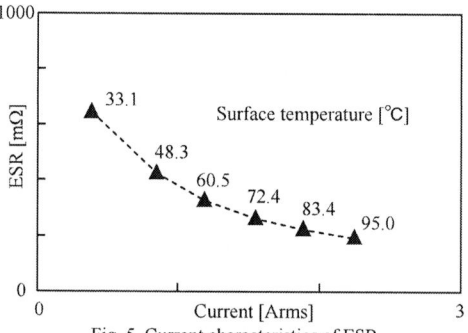

Fig. 5. Current characteristics of ESR.

Fig. 6. Comparison of ESRs under different current conditions.

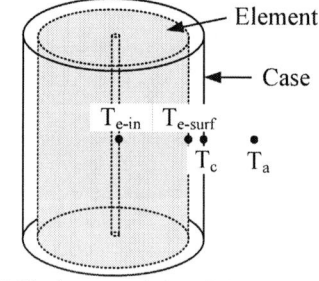

Fig. 7. Simple structural view of electrolytic capacitor.

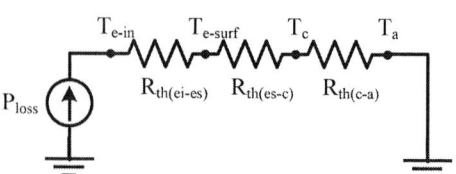

Fig. 8. Thermal resistance for each part of the capacitor.

$R_{th(c-a)}$ can be obtained by measuring the case surface temperature and ambient temperature using infrared thermography. To obtain $R_{th(es-c)}$, the element body temperature is required, and can be measured by inserting a thermocouple; however, a small hole has to be drilled to insert a thermocouple wire. The hole made in the aluminum case causes a problem in that the characteristics are changed due to the volatilization of the electrolytic solution. Moreover, if the surface area of the element is small, the temperature changes owing to the heat conduction from the thermocouple, making accurate measurements difficult. Therefore, this paper proposes a method to obtain $R_{th(es-c)}$ nondestructively by measuring the ESR. Because the element body has a uniform temperature distribution, the ESR is determined depending on the element body temperature. Therefore, the element body temperature can be estimated by the reverse calculation of the measured ESR value using the relationship between the ESR and the temperature measured using the thermostat scanner system. $R_{th(es-c)}$ can be obtained from the estimated element body temperature, the actual measured case surface temperature and the loss at the time. By simultaneously measuring the loss, ESR, case temperature, and element temperature using the B-H analyzer and infrared thermography, all thermal resistances are obtained.

V. Temperature Rise Calculation Considering Temperature and Frequency Characteristics

In this section, the temperature rise calculation method of an electrolytic capacitor using the ESR characteristics obtained in the previous section is explained.

A. Functionalization of ESR

The loss of the capacitor P_{loss} is defined as

$$P_{loss} = ESR \cdot I_{RMS}^2, \quad\cdots\cdots\cdots\cdots\cdots\cdots\cdots (6)$$

where I is the current of the capacitor. From (5), the element temperature is changed with the loss. Further, since the ESR is temperature dependent, the loss changes according to temperature. Thus, Predicting loss by only (6) is difficult. Therefore, herein, the ESR is treated as a function of frequency and temperature: $ESR(f_c, T_c)$. In [15], an approximate expression for ESR considering temperature is proposed. However, as the frequency of the current changes, each parameter of the approximate expression also changes; thus, the data for each frequency to be used must be acquired. Herein, the ESR when frequency and temperature are changed is measured, and an ESR function is obtained by approximating the map data with a fifth-order polynomial using the MATLAB curve-fitting tool, as shown in Fig. 10. The error is within 4.2% at the maximum within the measurement range of the data, hence the ESR can be calculated from the current frequency and element temperature using the obtained functional formula.

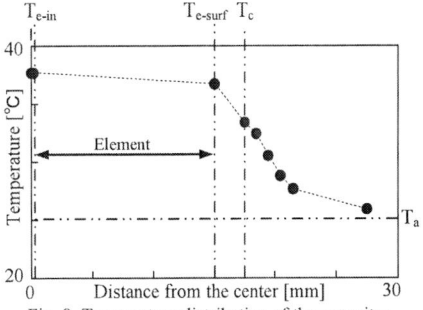

Fig. 9. Temperature distribution of the capacitor.

Fig. 10. Map data for frequency and temperature of ESR and the approximate surface.

B. The loss calculation from sinusoidal current

Using the functionalized ESR, P_{loss} when a sinusoidal current flows can be expressed as

$$P_{loss} = ESR(f_1, \Delta T_{cap} + T_a) \cdot I_{RMS}^2, \quad\cdots\cdots\cdots (7)$$

where T_a is the ambient temperature. With reference to (5) and (7), the temperature rise of a capacitor is calculated as

$$\Delta T_{cap} = \left(R_{th(es-c)} + R_{th(c-a)}\right) \\ \cdot ESR(f_1, \Delta T_{cap} + T_a) \cdot I_{RMS}^2 \quad\cdots\cdots (8)$$

Therefore, the temperature rise of the capacitor can be calculated when the functionalized ESR, ambient temperature, and thermal resistance is known. Furthermore, the loss of the capacitor can be calculated under practical temperature conditions from (5). However, since the solution of $\Delta T_{cap} + T_a$ in the functionalized ESR (f, T) is a fifth-order polynomial, the solution processes are complicated. Thus, in the actual calculation, by focusing on the range of temperature rise, ΔT_{cap} is an unknown variable only within the range of 0 °C to 80 °C, and values are assigned to the variable in increments of 0.1. The solution is decided as the value that minimizes the difference between the right side and left side of the equation.

C. The loss calculation from non-sinusoidal current

In power converters, a current with non-sinusoidal waveforms flows into the power capacitor. The loss when a current contains multiple-frequency flow components has been reported to be different from the loss when a single sinusoidal current flow [9]. This can be attributed to the ESR changing according to the sinusoidal frequency, as shown in Section 3. In a ferroelectric capacitor, a current waveform distortion occurs with respect to the sinusoidal voltage. However, in an aluminum electrolytic capacitor, distortion hardly occurs. Therefore, for the non-sinusoidal current, the loss can be calculated separately for each frequency component by the Fourier transform as

$$I(t) = \sum_{n=1}^{N} I_n \cdot \sin\omega_n t , \quad\cdots\cdots\cdots\cdots\cdots\cdots\cdots (9)$$

in which N is the number of sinusoidal frequency components included in the current. By the orthogonality of the trigonometric function, the effective value of the current is calculated as

$$
\begin{aligned}
I_{RMS}^2 &= \frac{1}{T}\int_0^T I(t)^2 dt \\
&= \frac{1}{T}\int_0^T (\sum_{n=1}^{N} I_n \cdot \sin\omega_n t)^2 dt \\
&= \sum_{n=1}^{N} I_{nRMS}^2 \quad\cdots\cdots\cdots\cdots\cdots\cdots\cdots (10)
\end{aligned}
$$

Therefore, in consideration of the frequency dependency of the ESR, the loss in (6) is expressed as

$$
\begin{aligned}
P &= ESR \cdot (I_{1RMS}^2 + I_{2RMS}^2 + I_{3RMS}^2 + \cdots) \\
&= \sum_{n=1}^{N} ESR_{f_n} \cdot I_{nRMS}^2 \quad\cdots\cdots\cdots\cdots\cdots\cdots (11)
\end{aligned}
$$

Subsequently, by substituting equation (11) into equation (8), the temperature rise of the capacitor is expressed as

$$
\begin{aligned}
\Delta T_{cap} &= \left(R_{th(es-c)} + R_{th(c-a)}\right) \\
&\quad \cdot \sum_{n=1}^{N} ESR(f_n, \Delta T_{cap} + T_a) \cdot I_{nRMS}^2 \quad\cdots (12)
\end{aligned}
$$

By solving equation (12), the temperature rise and the loss of the capacitor can be calculated, in consideration of the frequency and temperature characteristics. Fig. 12 shows the calculation process for obtaining the temperature rise from the current waveform.

VI. EXPERIMENTAL VERIFICATION

In this section, the results and discussion are described, on the actual verification experiments of the temperature rise calculation method described in Chapter 5.

A. Experimental verification using the simplified circuit

As shown in Fig. 12, the current is passed through the DUT using an amplifier, the current waveform and the ambient temperature are measured to calculate the temperature rise, and simultaneously the case surface temperature is measured using infrared thermography. Comparison is made between the actual measurement and calculated value. The measurements were performed with a single sinusoidal wave at 1 kHz and 10 kHz, with a non-sinusoidal wave superimposed with two sinusoidal waves for each frequency. Fig. 13 shows the measurement results of the single sinusoidal wave current. The measured value and the estimated value calculated from the current waveform are within 10% of the error, thus the validity of the temperature calculation using the thermal resistance was confirmed. Fig. 14 shows the measurement results of the superposed current of the 1-kHz and 10-kHz sinusoidal waves.

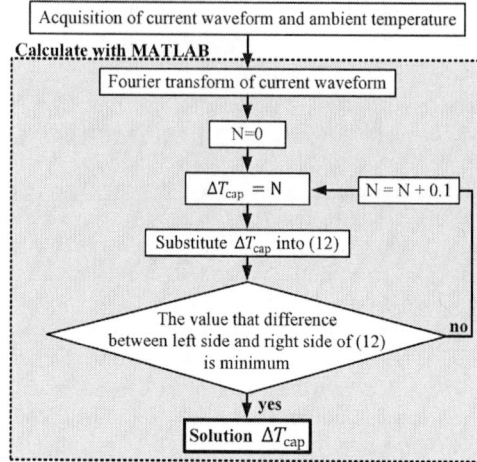

Fig. 11. Calculation flowchart of temperature estimation.

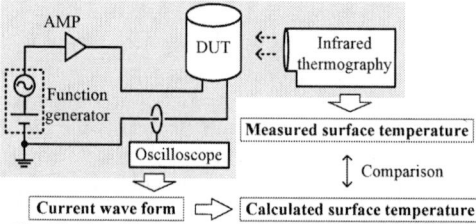

Fig. 12. Simplified measurement schematic for verification.

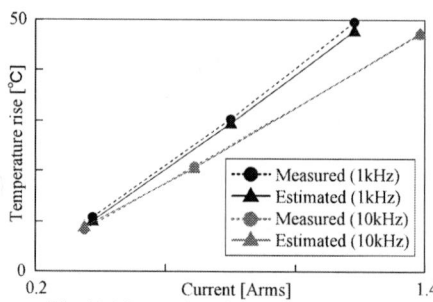

Fig. 13. Measured results on sinusoidal wave.

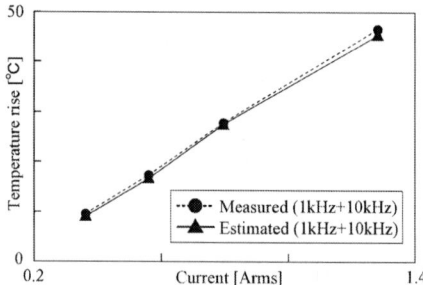

Fig. 14. Measured results on sinusoidal superimposed wave.

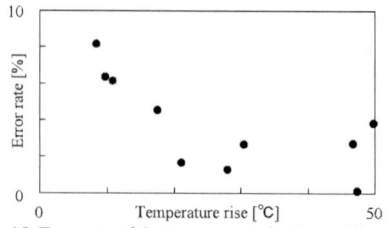

Fig. 15. Error rate of the temperature rise for verification.

The measured value and the estimated value are within 10% of the error, thus the calculation by (12) using the Fourier transform is considered to be appropriate. Fig. 15 shows the error rate of the estimated value against the temperature rise in the verification experiment by the simplified measurement circuit, and the error rate tends to increase as the temperature rise becomes smaller. This is because when the temperature rise is small, the stationary measurement error during temperature measurement becomes dominant.

B. Experimental verification using buck-chopper circuit

For the input capacitor and the output capacitor on the buck-chopper circuit shown in Fig. 16, the estimation of temperature rise was made from the current and ambient temperature. TABLE I shows the specifications of the buck-chopper circuit. Inductor L_1 is inserted for the evaluation of the input capacitor. Fig. 17 shows the actual current that flows to the capacitors, in which both are non-sinusoidal currents. Fig. 18 shows the results of the temperature rise calculated using the proposed method and the measured temperature rise. The data plotted with squares is the result of the temperature rise calculated using only the ESR at room temperature and the sinusoidal wave of 10 kHz without considering the frequency and temperature characteristics. In both (a) and (b), the value calculated using the proposed method agrees with the measured value; therefore, the calculation of the proposed method that considered the temperature and frequency characteristics is considered to be effective during actual circuit operation.

VII. CONCLUSION

A novel measuring method of an equivalent series resistance (ESR) and the power loss of a capacitor by utilizing a B-H analyzer is introduced. The measured results show that the ESR of the electrolytic capacitor has strong dependency on the temperature, current frequency, and current waveform. The measured temperature distribution of the capacitor body showed that the temperature is almost uniform. From the results above, the capacitor temperature is calculated by considering the thermal resistances in several portions of the capacitor. To express the ESR characteristics, the ESR is functionalized by a curved surface approximation of the map data. Using the function of the ESR, the calculation of temperature rise and power loss of the electrolytic capacitor on a sinusoidal current waveform and also an arbitrary current waveform are proposed. The experimental results agree with the calculated ones. It is verified that the proposed method is valuable in evaluating the power loss and temperature rise of the electrolytic capacitor on a practical operating condition.

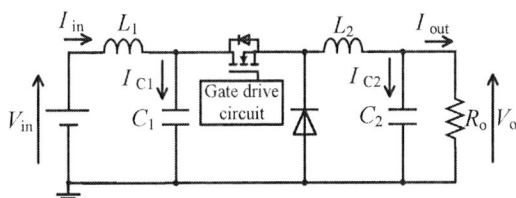

Fig. 16 Buck-chopper circuit.

TABLE I
SPECIFICATION OF THE BUCK-CHOPPER CIRCUIT.

Items	Symbol	Values
Input voltage	V_{in}	0–50 V
Output voltage	V_o	0–25 V
Switching frequency	f_{sw}	10 kHz
Toroidal inductor	L_1	322 µH
Toroidal inductor	L_2	329 µH
Road resistor	R_o	10 Ω
Electrolytic capacitor	C_1	47 µF
Electrolytic capacitor	C_2	47 µF

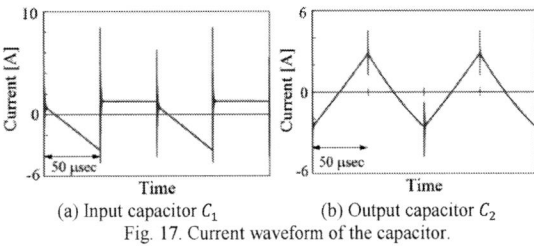

(a) Input capacitor C_1 (b) Output capacitor C_2
Fig. 17. Current waveform of the capacitor.

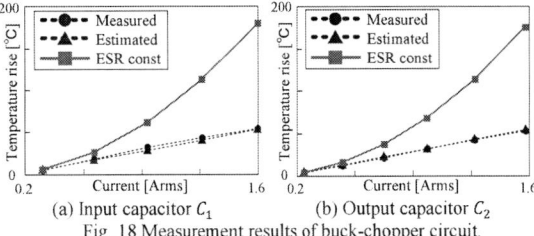

(a) Input capacitor C_1 (b) Output capacitor C_2
Fig. 18 Measurement results of buck-chopper circuit.

ACKNOWLEDGMENT

This work was supported by Council for Science, Technology and Innovation (CSTI), Cross-ministerial Strategic Innovation Promotion Program (SIP), "Next-generation power electronics" (funding agency: NEDO)

REFERENCES

[1] A. B. Ponniran, K. Orikawa, J.-I. Itoh, "Minimization of passive components in multi-level flying capacitor DC-DC converter," IEEJ J. Ind. Appl., vol. 5 no. 1 pp. 10-11, 2016.

[2] H. Obara, Y. Sato, "Selection criteria of capacitors for flying capacitor converters," IEEJ J. Ind. Appl., vol. 4 no. 2 pp. 105-106, 2015.

[3] J. W. Kolar U. Drofenik J. Biela M. Heldwein H. Ertl T. Friedli S. Round "PWM converter power density barriers," IEE Japan Trance. Ind. Appl. vol. 128, no. 4 pp. 468-480 2008.

[4] Huai Wang, Frede Blaabjerg, "Reliability of capacitors for DC-Link applications in power electronic converters—An overview," IEEE Transactions on Industry Applications, vol. 50, no. 5, September/October 2014.

[5] Power supply failure survey part II. [Online]. Available: http://www.rsonline.com/designspark/electronics/knowledge-item/power-supply-failure-survey-part-ii

[6] A. Albertsen, "Electrolytic capacitor lifetime estimation," Jianghai Capacitor Technical Note. [Online]. Available: http://jianghai-america.com/uploads/technology/JIANGHAI_Elcap_Lifetime_-_Estimation_AAL.pdf

[7] Kazunori Hasegawa, Ichiro Omura, and Shin-ichi Nishizawa, "A new evaluation crcuit for DC-link capacitors used in a three-phase inverter'', Japanese Industry Applications Society Conference (JIASC), 1-73, 2015.

[8] Pin-Yu Huang, Hironori Nagasaki, and Toshihisa Shimizu, "Characteristics of capacitor measurement set up by using B-H analyzer in power converters," ECCE Asia, 2016.

[9] H. Nagasaki, P. Huang and T. Shimizu, "Characterization of power capacitors under practical current condition using capacitor loss analyzer," ECCE, 2016.

[10] P. Venet, F. Perisse, M. El-Husseini, and G. Rojat, "Realization of a smart electrolytic capacitor circuit," IEEE Ind. Appl. Mag., vol. 8, no. 1, pp. 16-20, Jan. 2002.

[11] "Judicious use of aluminum electrolytic capacitors" Nippon Chemi-Con Corporation of the internet. [online] Available: https://www.chemi-con.co.jp/e/catalog/pdf/al-e/al-sepa-e/001-guide/al-technote-e-171001.pdf

[12] E. Aeloiza, J.-H. Kim, P. Enjeti, and P. Ruminot, "A real time method to estimate electrolytic capacitor condition in PWM adjustable speed drives and uninterruptible power supplies," in Proc. IEEE 36th Power Electron. Spec. Conf., Jun. 2005, pp. 2867-2872.

[13] Y. Yang K. Ma H. Wang F. Blaabjerg "Instantaneous thermal modeling of the dc-link capacitor in photovoltaic systems," 2015 IEEE Applied Power Electronics Conference and Exposition (APEC), pp. 2733-2739, March 2015.

[14] R. H. Broadbent, "Alternating-current properties of aluminum foil electrolytic capacitors," Electrochem. Technology, May-June 1968, pp. 163-166.

[15] M.L. Gasperi, "Life prediction model for aluminum electrolytic capacitors," in Conf. Rec. IEEE 1996 IAS Annu. Meeting, vol. 3, pp. 1347-1351.

Design and Evaluation of Current Distribution in Power Module

Takaaki Ibuchi*, Eisuke Masuda and Tsuyoshi Funaki

Osaka University

Div. of Electrical, Electronic and Information Eng., Graduate school of Engineering,
Suita, Osaka 565-0871, Japan

*E-mail: ibuchi@eei.eng.osaka-u.ac.jp

Abstract— This report presents the design and evaluation of current distribution in a power module based on near magnetic field scanning. This study focuses on the noise current distribution in a power module for its switching operation. This current distribution can be used for optimizing the wiring pattern design of a power module. This measurement methodology can visualize the practical current distribution on a wiring pattern and it applicable to realize compact and low-electromagnetic interference (EMI) power module design.

Keywords— *Current density, Electromagnetic interference (EMI), Near magnetic field intensity, Power module design*

I. INTRODUCTION

Power conversion circuits in electric vehicles and renewable energy conversion systems are required to be high power density and high reliability in the harsh environment. The characteristics of wide-bandgap SiC power semiconductor devices are especially attractive as high power density with fast switching operation of high voltage and large current [1]. However, these fast transitions also lead to electromagnetic interference (EMI) noise problems by interacting with circuit parasitic components [2]. It requires a more comprehensive understanding of the effects of parasitic elements, to achieve both low switching losses and low EMI characteristics. Previous works generally analyze the parasitic components and its current distribution using 3-D electromagnetic analysis [3], and evaluate their influence on the switching and the EMI characteristics [4]. This study focuses the current distribution in a power module, identified with the measured near magnetic field intensity [5, 6], for optimizing layout and packaging design of a power module.

II. EVALUATION OF CURRENT DISTRIBUTION USING NEAR MAGNETIC FIELD MEASUREMENT

The device under test (DUT) is a commercially available SiC half-bridge power module (ROHM, BSM080D12P2C008, 1200 V, 80 A), as shown in Fig. 1. There are three MOSFETs with antiparallel SiC schottky barrier diodes in each of the upper and lower side. Fig. 1

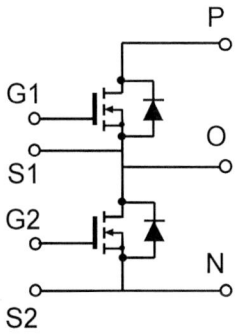

(a) Half-bridge configuration

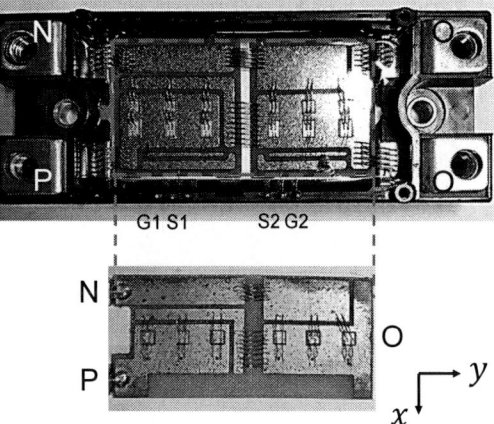

(b) Studied SiC power module

Fig. 1. Wiring-pattern of the studied power modules.

also shows the printed circuit board (PCB) that modeled wiring-pattern of the studied power module.

A. Static near magnetic field scanning

Figure 2 shows the near magnetic field scanning system. A magnetic probe (NEC Engineering, MP-10L) was connected to a spectrum analyzer (Tektronix, RSA3308B) and used to measure the near magnetic field over the wiring pattern of the DUT. The magnetic field were measured in two orthogonal directions (H_x, H_y) at each measurement point and magnetic field amplitude H_{xy} was calculated.

The 2018 International Power Electronics Conference

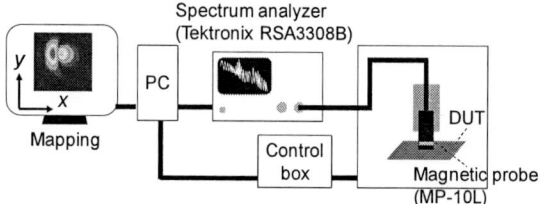

Fig. 2. System configuration of near magnetic field evaluation.

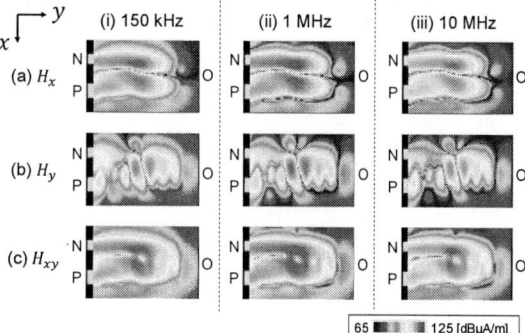

65 ▮▮▮ 125 [dBµA/m]

Fig. 3. Measured near magnetic field distribution
for a studied power module.

The measurement result of the near magnetic field on the wiring pattern depicts in Fig. 3. The single frequency sinusoidal voltage (V_{pp}= 10 V, Freq. = 150 kHz, 1 MHz, and 10 MHz) was applied on P-terminal in Fig. 1. N-terminal was terminated with 50 Ω. A magnetic probe scanned on the copper foil above 6mm, with 1 mm pitch. The result suggests that the current concentrates in the inner side wiring-pattern.

B. Dynamic near magnetic field scanning

This study focuses on the noise current distribution in a power module for its switching operation. Fig. 4 shows the system configuration of dynamic near magnetic field evaluation. A mixed domain oscilloscope (Tektronix, MDO4104C-6) was used. A magnetic field probe was connected to the RF input of the built-in spectrum analyzer. A voltage probe measured the switching gate voltage of the MOSFET and it realizes the evaluation of the time-synchronized frequency spectrum of the near magnetic field in the power module.

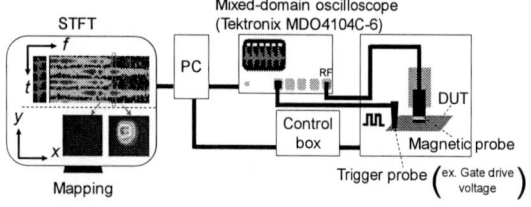

Fig. 4. System configuration of dynamic near magnetic field evaluation.

The test circuit configuration for visualizing the noise current in power module is shown in Fig. 5 (a). The switching frequency was set to 148.5 kHz and duty ratio D= 0.465. The input DC voltage V_{in}= 600 V, and 100 µH inductive load current is ±10 A. Fig. 5 (b) depicts the

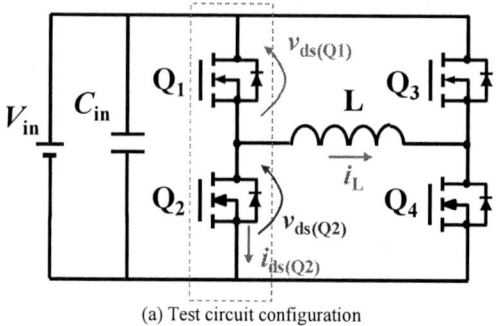

(a) Test circuit configuration

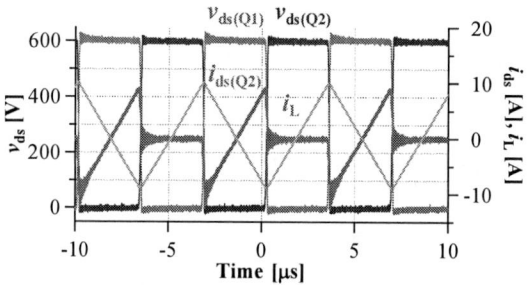

(b) Time response of the voltage and current in the test circuit
Fig. 5. Single-phase full-bridge circuit.

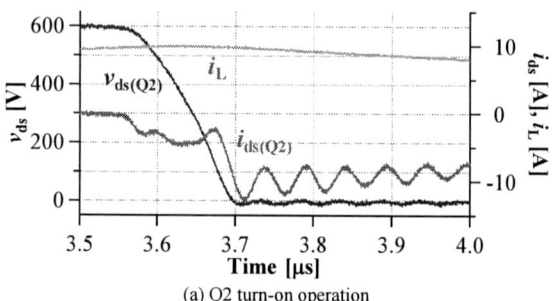

(a) Q2 turn-on operation

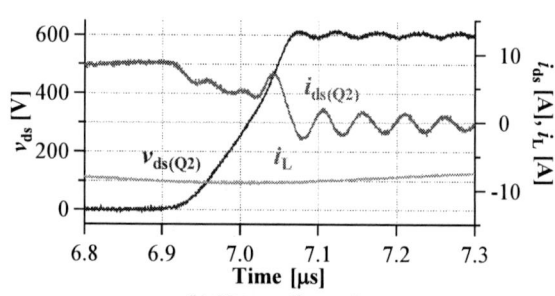

(b) Q2 turn-off operation
Fig. 6. Switching characteristics of voltage and current.

measured time response of the voltage and current in the tested single-phase full-bridge circuit.

Figure 6 shows the switching characteristics of inductor current, Q2 drain voltage and current. 19 MHz ringing oscillation were appeared in i_{dsQ2} both turn-on and turn-off operations. The load current, however, does not show the ringing oscillation.

A mixed-domain oscilloscope was triggered on the v_{gsQ2} and simultaneously captured a frequency spectrum

310

The 2018 International Power Electronics Conference

(a) Measuring point of near magnetic field

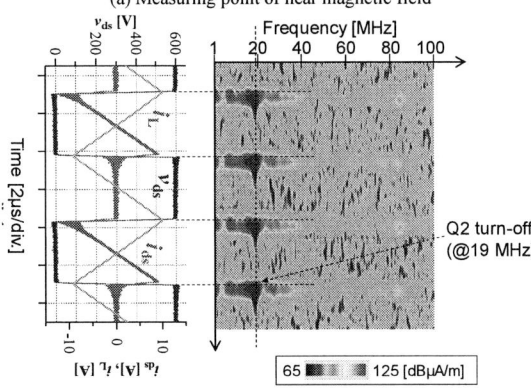

(b) Spectrogram of near magnetic field at a measuring point
Fig. 7. Near magnetic field evaluation in a power module for switching operation in the tested circuit.

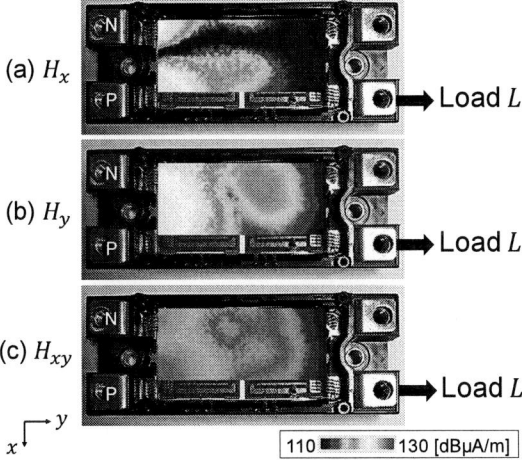

Fig. 8. Evaluation of noise current distribution in a power module (19 MHz, at Q2 turn-off operation).

of near magnetic field H_y at the switching operation. Fig.7 shows a measuring point of near magnetic field and the spectrogram of up to 100 MHz. The horizontal axis is frequency, the vertical axis is time, and the color scale represents the magnetic field intensity. Fig. 7 (b) also shows the corresponding Q2 voltage and current waveform. The near magnetic field on the MOSFET chip exhibits the highest spectrum peak at 19 MHz in the switching operations. The measured 19 MHz noise current distribution in a power module at Q2 turn-off operation are shown in Fig. 8. The results suggest that the noise current, which was caused by the switching operations flows the Q1, Q2 and input smoothing capacitor C_{in}, and it does not flow out the load inductor.

III. DESIGN OF WIRING PATTERN OF POWER MODULE

The developed power module wiring, which is based on the evaluation results of current distribution shown in the previous section, is depicted in Fig. 9. The three MOSFETs without antiparallel SBDs are parallel connected in each of the upper and lower side. The length of module is about 2/3 of the tested conventional module (Fig. 1). The frequency characteristics of interconnect impedance across P and N terminals in a module was evaluated with impedance analyzer (Agilent, 4294A). The measured results shown in Fig. 10 indicate that the parasitic inductance in a developed power module is about 1/2 of the conventional module.

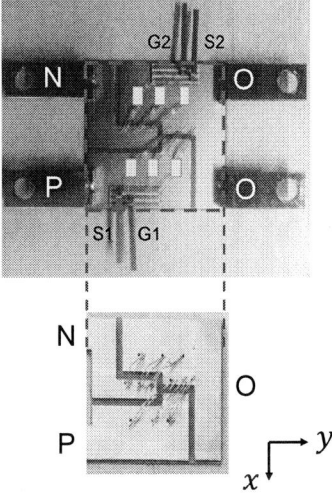

Fig. 9. Wiring pattern of the developed power module.

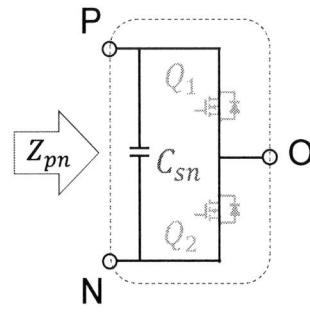

(a) Impedance measurement across P and N terminals

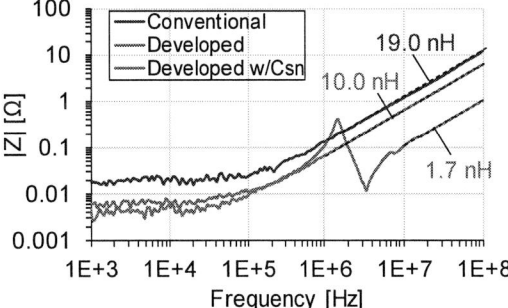

(b) Frequency characteristics of impedance $|Z_{pn}|$
Fig. 10. Evaluation of parasitic inductance across P and N terminals.

311

The 2018 International Power Electronics Conference

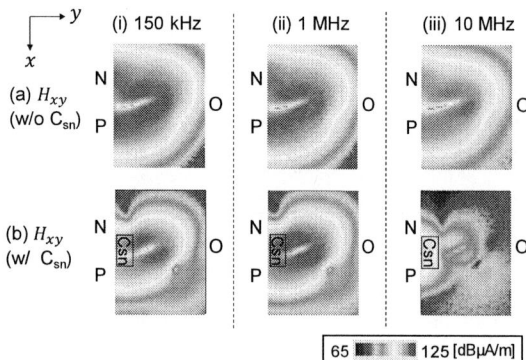

Fig. 11. Measured near magnetic field distribution
for developed power modules.

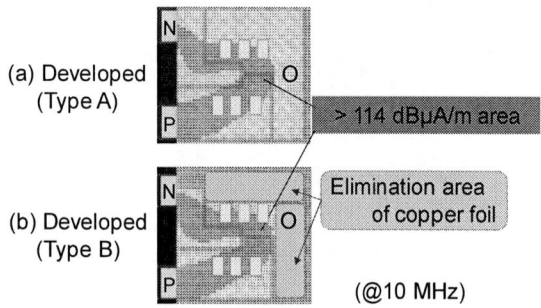

Fig. 12. Near magnetic field distribution over 114 dBµA/m
for developed power module wiring pattern.

Figure 11 are the measured results of current distribution for developed power module (type A). The current distribution in all parallel dies were well balanced on the shortest path across P and N terminals. This study also evaluates another wiring pattern design (type B) that eliminate relatively small magnetic field intensity. The red-painted area for type B in Fig. 12, where the magnetic field intensity is higher than 114 dBµA/m, is almost identical to that of type A. It is also confirmed that the parasitic inductance of type B is almost identical to that of type A. Therefore, the relatively small magnetic field intensity area has little influence on the current distribution and loop inductance. It leads to smaller package and enhance the function of power module with integrating other components in the spare space.

Figure 11 also shows the current distribution in a power module with and without embedded DC-link snubber capacitor. The results show that 10 MHz current concentrates around snubber capacitor. Fig. 13 depicts an example of measured switching characteristic in turn-off operation. The voltage overshoot for without snubber capacitor is caused by interaction with large di/dt and commutation loop inductance. While the built-in snubber capacitor can provide low-inductance commutation path and can suppress the turn-off surge voltage with comparable switching speed.

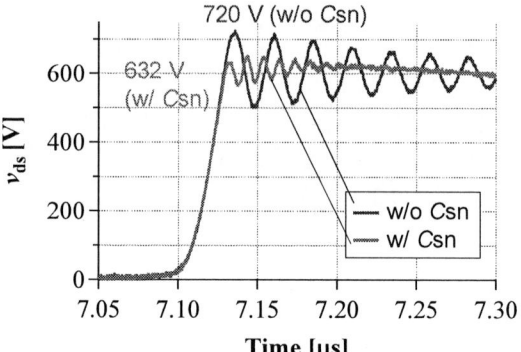

Fig. 13. Switching characteristics of developed power modules
in turn-off operation.

IV. CONCLUSIONS

This report studies design and evaluation of power module using near magnetic field measurement. The results shown in this report indicates that the near magnetic field scanning is one of the effective tool to realize the wiring layout and packaging design of novel power modules in the early design stage.

ACKNOWLEDGMENT

All the SiC power modules evaluated in this paper were manufactured and provided by ROHM Co., LTD (Japan). This work was partially supported by Council for Science, Technology and Innovation (CSTI), Cross-ministerial Strategic Innovation Promotion Program (SIP), "Next-generation power electronics" (funding agency: NEDO).

REFERENCES

[1] Z. Liang, P. Ning, F. Wang, "Development of Advanced All-SiC Power Modules," *IEEE Trans. Power Electron.*, vol. 29, no. 5, pp. 2289-2295, 2014.

[2] N. Oswald, P. Anthony, N. McNeill, B.H. Stark, "An Experimental Investigation of the Tradeoff between Switching Losses and EMI Generation With Hard-Switched All-Si, Si-SiC, and All-SiC Device Combinations," *IEEE Trans. Power Electron.*, vol. 29, no. 5, pp. 2393-2407, 2014.

[3] C. Martin, J-L Schanen, J-M Guichon and R. Pasterczyk, "Analysis of Electromagnetic Coupling and Current Distribution Inside a Power Module," *IEEE Trans. Ind. Appl.*, vol. 43, no. 4, pp. 893-901, 2007.

[4] A. Dutta and S. S. Ang, "Electromagnetic Interference Simulations for Wide-Bandgap Power Electronic Modules," *IEEE Journal of Emerging and Selected Topics in Power Electronics*, vol. 4, no. 3, pp. 757-766, 2016.

[5] Y. Vives-Gilabert, C. Arcambal, A. Louis, F. de Daran, P. Eudeline and B. Mazari, "Modeling Magnetic Radiations of Electronic Circuits Using Near-Field Scanning Method," *IEEE Trans. EMC.*, vol. 49, no. 2, pp. 391-400, 2007.

[6] H. Weng, D. G. Beetner, R. E. DuBroff, J. Shi, "Estimation of High-Frequency Currents From Near-Field Scan Measurements," *IEEE Trans. EMC.*, vol. 49, no. 4, pp. 805-815, 2007.

Development of Impedance-Source Inverter Using SiC-MOSFET

Ryuji Iijima, Thilak Senanayake, Takanori Isobe and Hiroshi Tadano*
University of Tsukuba, Tsukuba, Japan
*E-mail: tadano.hiroshi.fn@u.tsukuba.ac.jp

Abstract— Impedance-source inverter using SiC-MOSFET was studied. A 3-kW laboratory level inverter had the efficiency of about 96 % at the condition of output boost ratio G of 1.77 (boost ratio B of 2.5), output power of 3.16 kW and input voltage of 240 V. At lower G case of 1.32 (B of 1.61), the efficiency of impedance-source inverter was increased by 97.4 % due to low conduction loss of SiC-MOSFET.

Keywords— Impedance-Source Inverter, SiC-MOSFET, Boost, SVM, Short-Through .

I. INTRODUCTION

The 3-phase impedance-source inverter (Z-source inverter; ZSI) [1] has a boost function and inverter function using short-through mode, and is expected to apply to a traction motor driving system [2]-[6] and photovoltaic system [7]-[10]. On the other hand, SiC-MOSFET is expected as a high-performance next-generation power device due to low on-state voltage, high switching speed and high temperature operation. Therefore, as for the inverter using SiC-MOSFET, the application to a motor drive system is expected as an inverter with the voltage boost up function which read to higher rotation and lower volume of traction motor.

The distinctive characteristic of Z-source inverter is that has short-through mode and no dead time. Therefore, body diode of SiC-MOSFET does not conduct at any time. The ZSI with SiC-MOSFET needs not anti-parallel diodes [4] which are used in conventional ZSI with Si-IGBT, and that results in size reduction of inverter module. The previous paper [6] has reported to have the efficiency of 95.6 % at the condition of boost ratio of 1.9, where SiC-MOSFET and PWM modulation method were used. In this paper, it was reported the output boost ratio of 1.4 under the condition of the boost ratio of 1.9. In order to apply the ZSI to traction motor control for EV or FCEV, it is expected to boost the input DC voltage up to about 2 in order to rotate the motor at high speed.

In this paper, experimental results with a laboratory level 3-kW ZSI system using SiC-MOSFETs are reported, where Space Vector Modulation (SVM) method is applied to get sinusoidal output waveforms. And efficiency of ZSI at the condition of high output boost ratio is discussed.

II. ZSI USING SiC-MOSFET

Fig. 1 shows the ZSI circuit. In the ZSI, upper MOSFET and lower MOSFET can be turn on at the same time, that is a short-through mode, and the current of inductor during in this mode is increased to boost up the DC input voltage.

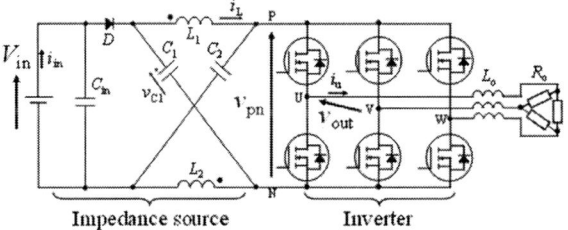

Fig. 1. Circuit diagram of Z source inverter using SiC-MOSFET.

SiC-MOSFETs with low on-resistance and high switching speed characteristics can be used in this ZSI. In this ZSI, the diode conduction does not occur. Fig. 2 shows the switching mode of ZSI. Fig. 2-(a), (b) and (c) are active mode, zero-vector mode and short-through mode, respectively. During an active mode and zero-vector mode, inverter applies the output voltage and zero voltage to the load, respectively, as same as conventional voltage source inverter. In a short-through mode, all SiC-MOSFETs are turned on and inductor currents of ZSI are increased to make boosted voltage with this short-through time period T_{sh}. The boosted average PN voltage v_{pn_ave} and boosted output line to line average voltage v_{out_ave} are controlled with this short-thorough time period, as shown in equation (1) and (2).

$$v_{pn\_ave} = \frac{1-\frac{T_{sh}}{T_{sw}}}{1-2\cdot\frac{T_{sh}}{T_{sw}}} \cdot V_{in} , \qquad (1)$$

$$v_{out\_ave} = \frac{T_{active}}{T_{sw}} \cdot \frac{1}{1-2\cdot\frac{T_{sh}}{T_{sw}}} \cdot V_{in} , \quad (2)$$

where, T_{active} and T_{sw} are an active mode time period and a switching cycle time, respectively, and V_{in} is input voltage.

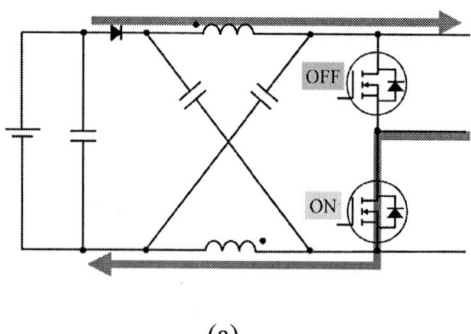

(a)

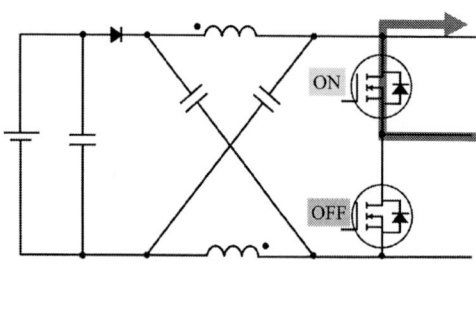

(b)

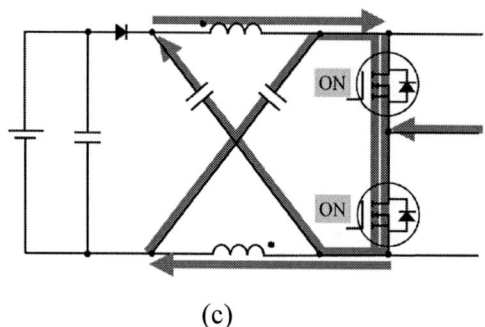

(c)

Fig. 2. Example of current flow of Z source inverter,
(a) active mode,
(b) zero vector mode,
(c) short-through mode.

Fig. 3 is a fabricated 3-kW ZSI using a full SiC-MOSFET 6 in 1 module. ZSI used in this experiments has no free-wheeling diode because ZSI has no dead-time. And also there is no conduction mode of body diode in SiC-MOSFET which read to degradation of SiC-MOSFET characteristics. The circuit parameters used in this experiment are shown in Table I.

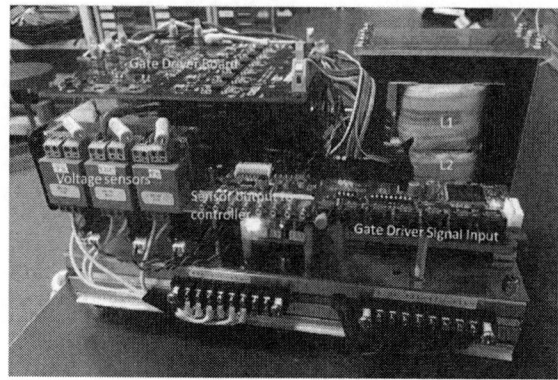

Fig. 3. Photograph of fabricated lab level Z source inverter.

Table I.   Circuit parameters of 3-kW ZSI.

SiC 6 in 1 module		CCS050M12CM2 1200 V, 50 A
Diode	D	SCS240KE2 1200 V, 40 A
Inductor	L_1, L_2	203 µH
Capacitor	C_1, C_2	60 µF
Input Capacitor	C_{in}	50 µF
Control frequency	$f_{control}$	10 - 24 kHz

III.   EXPERIMENTAL RESULTS

Fig. 4 shows a SVM diagram used in this experiments. The six short through periods were inserted in one switching control cycle. At each short-through modes, all SiC-MOSFETs in the inverter ware turned on to make the boosted voltage and to reduce the SiC-MOSFET conduction losses by dividing the large short-through current into 3 legs. The boost ratio B and output boost ratio G of ZSI are defined as equation (3) and (4), respectively.

$$B = {v_{pn\_max}}/{V_{in}} \quad , \quad (3)$$
$$G = {v_{out\_max}}/{V_{in}} \quad , \quad (4)$$

where, v_{pn_max} is the maximum boosted voltage of dc-link voltage v_{pn}, v_{out_max} is the maximum output line to line voltage and V_{in} is the input DC voltage.

Fig. 5 shows a dc-link voltage (v_{pn}) and output line currents of fabricated ZSI. The input DC voltage V_{in} is 150 V and peak PN voltage v_{pn_max} boosted with short-through mode of ZSI is 420V. The boost ratio B in this case is calculated as 2.8. From this figure, it is confirmed that the ZSI can boost the input DC voltage and output the AC voltage successfully using boosted DC voltage.

The 2018 International Power Electronics Conference

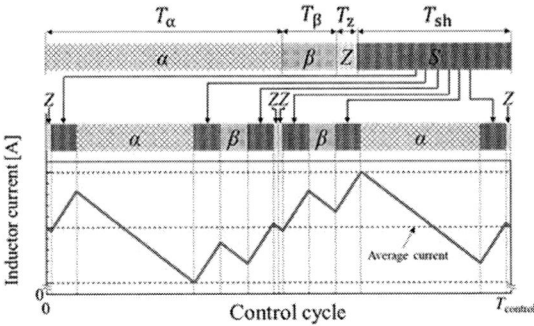

Fig. 4. SVM switching pattern used in this experiments. Six short-through were inserted in one control cycle.

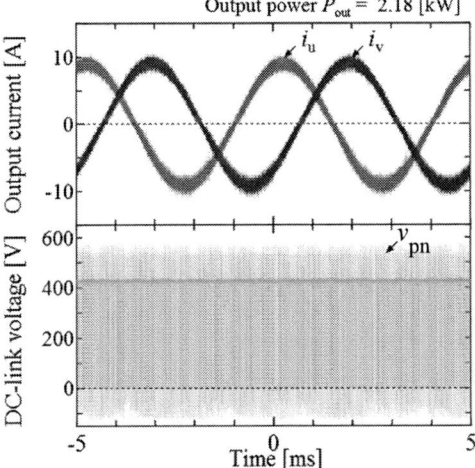

Fig. 5. Output line currents and DC-link voltage of Z-source inverter. Output power is 2.18 kW.

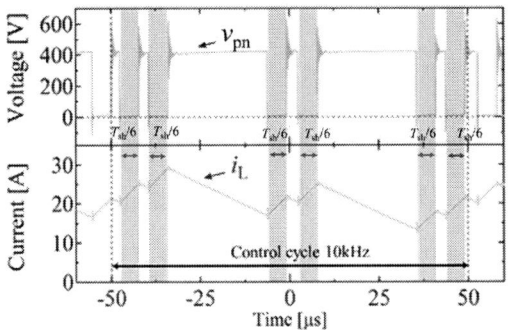

Fig. 6. Switching waveform of DC-link voltage and inductor current. $T_{sh}/6$ is the short-through period.

Fig. 6 is the expanded waveforms of the PN voltage and the inductor current i_L of impedance network. This shows one control cycle of ZSI and 6 short-through modes are

inserted in this one control cycle that are indicate with $T_{sh}/6$ in this figure. In these short-through periods, all devices of 3 legs are turned on to increase the inductor current to boost the voltage. Therefore, each SiC-MOSFET was switched by 3 times in one control cycle. In this experiments, 10 kHz and 20 kHz control frequency were used and effective switching frequencies of SiC-MOSFET were 30 kHz and 60 kHz, respectively.

IV. EFFICIENCY

The power loss of ZSI was measured using power meter WT1800. Fig. 7 shows the total loss of ZSI in case of input DC voltage of 240 V and output average AC voltage of 300 V at the control frequencies of 10, 14, 20 and 24 kHz. In these cases, the output boost ratio G is about 1.77. The power conversion efficiency of about 96 % from DC to AC was obtained at the condition of the input DC voltage of 240 V and the output average AC voltage of 300 V with the control frequency of 10 kHz (Fig. 8).

Fig.9 shows the ZSI losses depended on control frequency. As shown in this figure, the total power loss depends on the control frequency, and also has the loss not to depend on the frequency. The frequency independent loss was about 80 W which was 2.67% of the output power of 3000 W.

The some amount of power loss was caused by inductors and diode. Then, the efficiency of ZSI will be increased using low loss inductor and synchronous rectification of diode used in the impedance source network.

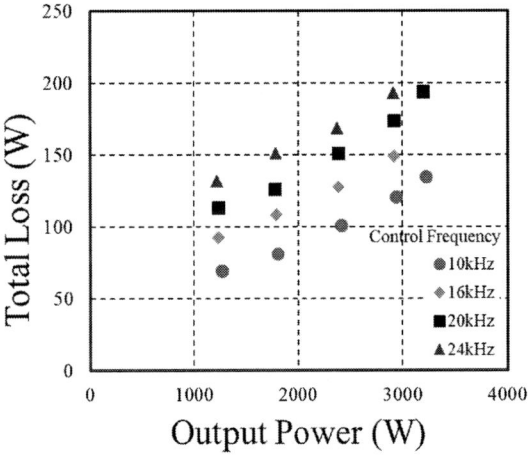

Fig. 7. Total losses depended on the output power. Control frequencies are 10, 16, 20 and 24 kHz.

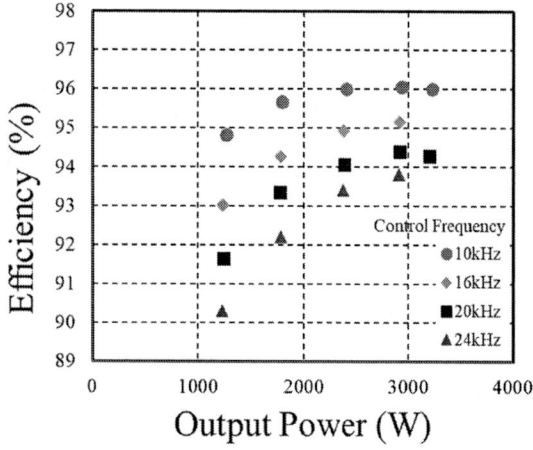

Fig. 8. Efficiencies of fabricated Z source inverter. The input voltage is 240 V, and the output average voltage is 300 V (*G*=1.77).

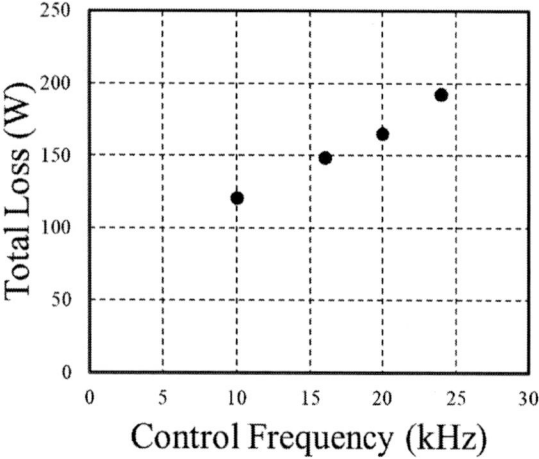

Fig. 9. Total loss dependency on control frequency.

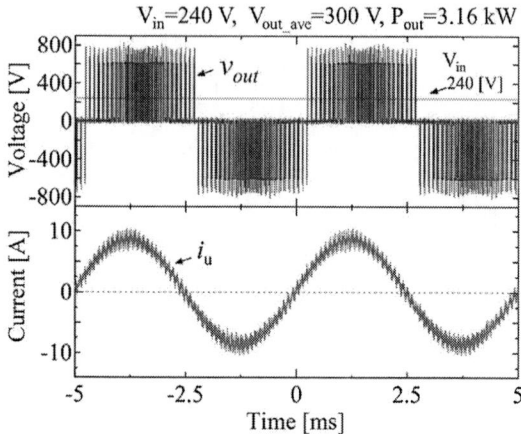

Fig. 10. Switching waveforms at the output power of 3.16 kW. (*B* = 2.58, *G* = 1.77)

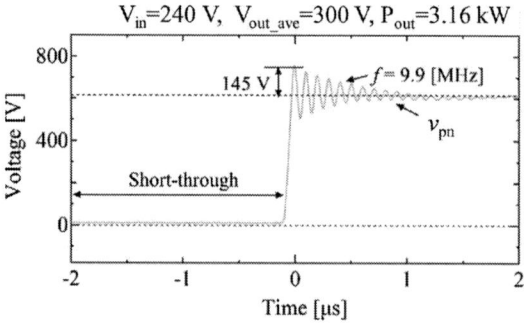

Fig. 11. Expanded switching waveform at the output power of 3.16 kW.

Fig. 10 is the switching waveforms in the case of output power of 3.16 kW, input voltage of 240 V and output average AC voltage of 300 V. In this figure, there are some surge waveforms in PN voltage due to short-through operation and parasitic inductance in the short-through circuit. Fig. 11 shows the expanded PN voltage of figure 10. After the short-through, the surge voltage with the oscillation frequency of 9.9 MHz was observed. In this period, ZSI acts as inverter mode, then the total capacitance of inverter module assume as about 3 times of output capacitance of SiC-MOSFET (total capacitance of 1500 pF). Therefore, the parasitic inductance in this circuit is assumed as 170 nH. In order to decrease this voltage surge, it is effective to connect the capacitance with high frequency responsibility at the suitable points in the circuit.

Fig. 12 shows *G* depended efficiencies. In this figure, output voltage was set at 300 V, and input voltages were changed. At lower *G* case of 1.32 (*B* of 1.61), the efficiency is increased to 97.4% at the output power of 3000 W. Fig. 13 shows *B* dependencies. In these cases, the input voltage was fixed at 240 V and the output voltage was changed. As shown in this figure, the efficiency is increased by decreasing *B*. The efficiency in this experiment was higher than that reported in Ref. [6].

The 2018 International Power Electronics Conference

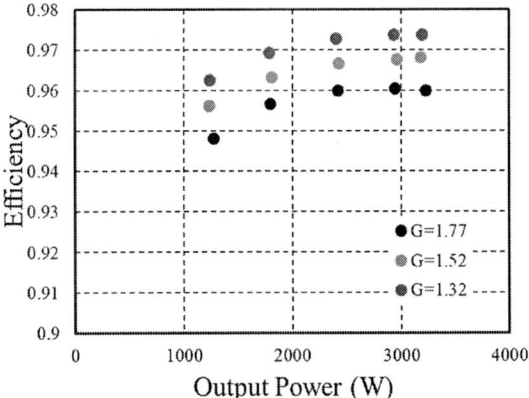

Fig. 12. Efficiencies of Z-source inverter at the G of 1.77, 1.52 and 1.32.

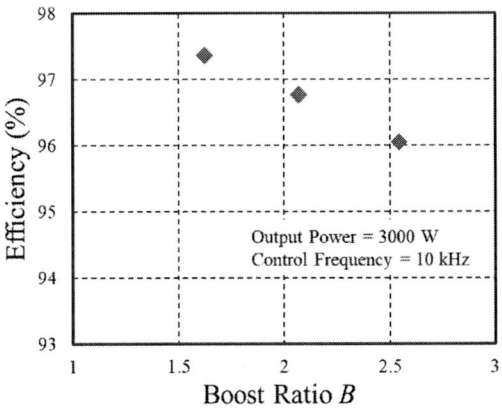

Fig. 13. Efficiency dependency on boost ratio B.

V. CONCLUSIONS

In this paper, high efficiency Z-source inverter using SiC-MOSFET without body diode conduction was studied. Fabricated 3-kW ZSI had the power conversion efficiency of 96 % at the condition of 3.1 kW of output power and output boost ratio G of 1.77. In case of lower boost ratio B of 1.61, the efficiency of ZSI was increased by 97.4 % due to low conduction and switching losses of SiC-MOSFET. ZSI using high performance SiC-MOSFET was the high power conversion efficiency without body-diode conduction mode. Then it would have high reliability due to no bipolar degradation of SiC-MOSFET and would be able to be applied to traction motor driving of EV at high rotating speed.

This work was supported by Council for Science, Technology and Innovation (CSTI), Cross-ministerial Strategic Innovation Promotion Program (SIP), Next-generation power electronics (funding agency: NEDO)

REFERENCES

[1] E. Z. Peng, "Z-Source Inverter," IEEE Trans. Ind., vol.39, no.2, pp.504-519, 2003.

[2] M. Ghasem Hosseini Aghdam, "Z-Source Inverter with SiC Power Semiconductor Devices for Fuel Cell Vehicle Applications", Journal of Power Electronics, Vol. 11, No. 4, pp. 606-611, 2011.

[3] F. Guo, L. X. Fu, C. H. Lin, C. Li,W. Choi, and J.Wang, "Development of an 85-kW bidirectional quasi-Z-source inverter with DC-link feed-forward compensation for electric vehicle applications," *IEEE Trans. Power Electron.*, vol.28, no.12, pp. 5477–5488, 2013.

[4] R. Iijima, T. Isobe and H. Tadano, "Investigation of Eliminating Free-wheeling Diode Conduction of Z-source Inverter using SiC-MOSFET," IECON2015-Yokohama, 2015.

[5] R. Iijima, T. Isobe and H. Tadano, "Loss analysis of Z-source inverter using SiC-MOSFET from the perspective of current path in the shortthrough mode," 18th European Conference on Power Electronics and Applications, pp. 1-10, Sep. 2016.

[6] M. Zdanwski, D. Peftisis, S. Piasecki and J. Rabkowski, "On the Design Process of 6-kVA Quasi-Z-inverter Employing SiC Power Devices," IEEE Trans. on Power Elec., vol.31, no.11, pp.7499-7508, 2016.

[7] Y. Liu, H. Abu Rub and G. Baoming, "Z-source/quasi-Z-source inverter: Derived networks, modulations, controls and emerging applications to photovoltaic conversion," IEEE Ind. Electron. Mag., vol.8, no.4, pp.32-44, 2014.

[8] Y. Li, J. Anderson, F. Z. Peng and D. Liu, "Quasi-Z-source inverter for photovoltaic power generation systems," in *Proc. 24th Annu. IEEE Appl. Power Electron. Conf. Expo.*, Feb. 15–19, 2009.

[9] B. Ge, H. Abu-Rub, F. Peng, Q. Lei, A. de Almeida, F. Ferreira, D. Sun and Y. Liu, "An energy stored quasi-Z-source inverter for application to photovoltaic power system," *IEEE Trans. Ind. Electron.*, vol.60, no.10, pp. 4468–4481, 2013.

[10] Y. Liu, B. Ge, H. Abu-Rub, and F. Z. Peng, "An effective control method for quasi-Z-source cascade multilevel inverter based grid-tie single-phase photovoltaic power system," *IEEE Trans. Ind. Informat.*, vol.10, no.1, pp. 399–407, Feb. 2014.

Control Methodology for Realization of 100kW HEECS Chopper with 99.5% Efficiency

Yukinori Tsuruta [1*] and Atsuo Kawamura [1]

1 Electrical and Computer Engineering, Yokohama National University, Yokohama, Japan

*E-mail: tsuruta-yukinori-px@ynu.ac.jp

Abstract— This paper describes a study for extremely high precision efficiency measurement by build of back to back (BTB) system of high power DC-DC converter aiming at realization of 100 kW high efficiency energy conversion system (HEECS) chopper with 99.5% efficiency. The proposed new water-cooled 2-phase HEECS BTB chopper system offers solutions in saving reduction of input power and high precision efficiency measurement of high power DC-DC converter with extremely high efficiency of 99.5%.

Keywords— **Partial Boost Method, Dual Active Bridge Converter, SiC-MOSFET, 2-Phase Back to Back System**

I. Introduction

A very high efficiency chopper is expected to enhance the use of power conversion technology in various fields in the electric power based modern society. Authors have already proposed "HEECS" chopper, in which a partial power conversion is the key concept as shown in [1], [12], and [13]. The aim of this HEECS chopper is realization of a very high efficiency, assuming that the output load voltage variation is not large, or the input voltage fluctuation is not so large.

The efficiency of HEECS is given in (1).

$$\eta_{system}=\frac{W_{out}}{W_{in}}=\frac{W_1+\eta_{DAB}W_2}{W_1+W_2}=1-\frac{1}{(1+n_S)}(1-\eta_{DAB})\ \cdots\cdots\cdots (1)$$

All symbols in this equation are shown in Fig.1. For example, if efficiency of Dual Active Bridge (DAB) converter η_{DAB}=94.8%, W_1: W_2=n_S : 1, and n_S=6.5 as for system efficiency η_{system}, extremely high efficiency η_{system}=99.3% can be obtained.

In this paper new 600 V, 20 kHz, 100 kW 2 phase HEECS BTB system using DAB converter has been designed, built and tested. Efficiency of 99.57 % at maximum output power of 100.05 kW with boost mode (input: V_{in}=537.27 V /output: V_S=607.62 V) was obtained.

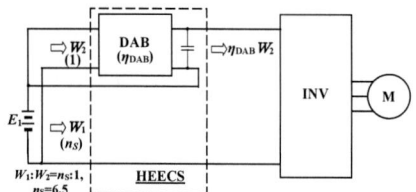

Fig. 1. Principle of extremely high efficiency by HEECS.

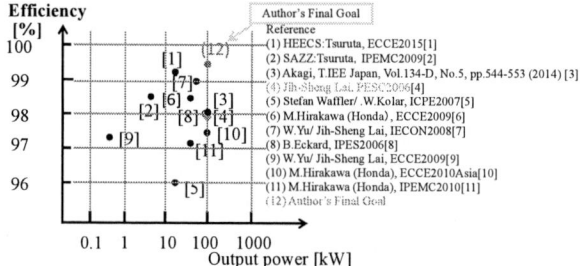

Fig. 2. Power vs. efficiency of DC-DC converter by literature search.

The survey of the high efficiency literatures [1]- [11] is shown in Fig.2. There is no example of BTB system with extremely high efficiency of 99.5%. Viewed from the point of high precision efficiency measurement for high power DC-DC converter, test result is discussed in following paragraphs as follows: II. BTB Methodology for Efficiency Measurement of High Power DC-DC Converter, III. Design and Build of BTB system, IV. Experimental Results, V. Discussion on Highly Precise Measurement of Efficiency by HEECS-BTB system.

II. BTB Methodology for Efficiency Measurement of High Power DC-DC Converter

A. Conventional BTB System

Fig.3 shows the principal circuit configuration of conventional high power BTB system. The BTB system has two groups of line-frequency transformers T_1, T_2 and two groups of converter units AC/DC-1, DC/AC-1 with a common dc-link reactor L_0. The BTB converter unit can control active power P as well as reactive power Q.

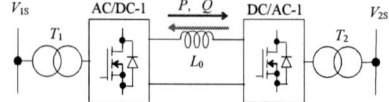

Fig. 3. Conventional circuit configuration of high power BTB converter.

B. BTB System for High Precision Measurement of High Power DC-DC Converter

Fig.4 shows a block diagram of conventional test setup for DC-DC converter. Input power source V_{IN} needs to supply rating power P_R corresponding to the rating power of DC-DC converter. Therefore, viewed from the point of high precision efficiency measurement for high power

DC-DC converter, the way of extremely high system efficiency and high precision measurement of efficiency was studied by the function of regenerating power of BTB system.

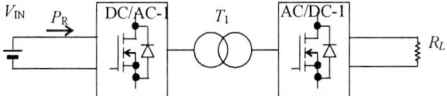

Fig. 4. Conventional test setup for DC-DC converter.

Fig.5(a) shows the conventional BTB system with transformer turn ratio 1:1 [3]. The output power of DC-DC converter is regenerated back to the input side. Input power of V_{IN} is reduced surely. However, this system configuration is the same as conventional test setup, so total system efficiency is restricted to efficiency of DC-DC converter itself. Fig.5(b) shows another way of conventional BTB system by 2-phase high power DC-DC converter. Reduction of input power is obtained by traction mode operation of the first phase DC-DC converter and regenerating operation of the second phase DC-DC converter. However, system efficiency becomes DC-DC converter itself.

Corresponding to these restriction, a new HEECS BTB system with any transformer ratio n:1, which can operate extremely high efficiency and can obtain high precision measurement of system efficiency, shown in Fig.5 (c) was proposed. Total rating power is 2 DC-DC converters and total system efficiency can be measured in high precision by high precision measurement of input power, which is total system loss.

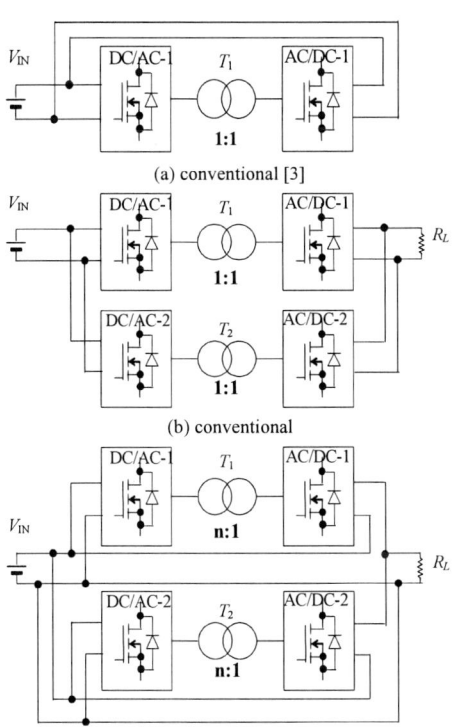

(a) conventional [3]

(b) conventional

(c) proposed

Fig. 5. BTB systems for high power DC-DC converter.
(a)(b)conventional, (c)proposed

III. DESIGN AND BUILD OF HEECS-BTB SYSTEM

A. Fabrication of 2 Phase HEECS BTB Chopper

We decided to make water cooled 600VDC-100kW 2 phase HEECS BTB test set-up. Table 1 shows the specification of HEECS BTB system. Fig.6 shows main circuit configuration of 100 kW 2 phase HEECS-BTB chopper. Fig.7 shows control system chart of 100 kW 2 phase HEECS BTB test set-up.

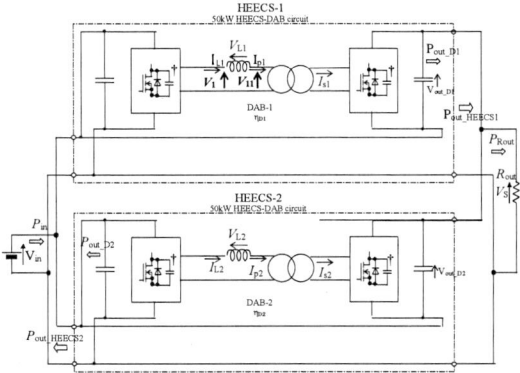

Fig. 6. Main circuit configuration of 100 kW
2 phase HEECS BTB chopper.

† Products of laminated ceramic capacitor by Murata Manufacturing Co., Ltd. are applied to snubber capacitors.

TABLE I
SPECIFICATION OF HEECS BTB SYSTEM

Input	V_{in}	520 V
Output	V_S	600 V
Load	R_L	160 Ω
Frequency	F_S	20 kHz
Rating power	P_S	100 kW (=$P_{out\ HEECS1}$+$P_{out\ HEECS2}$)
Efficiency	η_{System}	Over 99.3 [%]
Turn ration	n	6.5
Primary snubber capacitor		† 330pF, 1000Vdc, RCE7U3A331J2K1H03B, laminated ceramic capacitor by Murata Manufacturing Co., Ltd.
Secondary snubber capacitor		† 15000pF, 630Vdc, RCE7U2J153J4K1H03B, laminated ceramic capacitor by Murata Manufacturing Co., Ltd.
	C_{S2a}	0.33μF, 630Vdc, SNFPJ033306FD2JSSD, WIMA

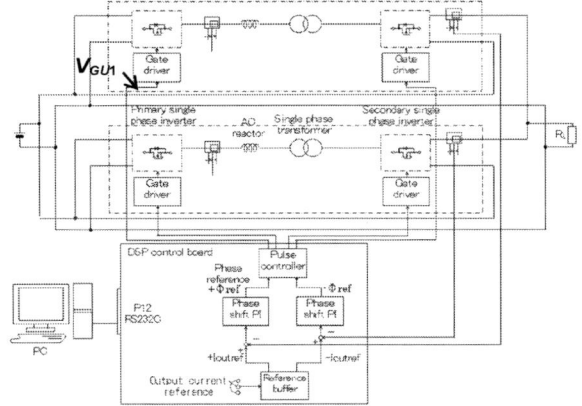

Fig. 7. Control system chart for 2 phase HEECS BTB test set-up.

319

B. Build of HEECS- BTB System for Testing

Fig.8 shows power flow block diagram of HEECS BTB system. Whole power flow in HEECS BTB system can be expressed by balance rule on node2 as

$$P_2 = P_3 + P_4 \quad \dots\dots\dots\dots\dots\dots\dots\dots (2)$$

where P_2: output power of HEECS-1, P_3: power dissipation of resistance R_L, P_4: input power of HEECS-2.

HEECS-1 power conversion ratio η_1 is shown as

$$P_2 = \eta_1 P_1 \quad \dots\dots\dots\dots\dots\dots\dots\dots (3)$$

where P_1: input power of HEECS-1.

Similarly, HEECS-2 power conversion ratio η_2 is

$$P_5 = \eta_2 P_4 \quad \dots\dots\dots\dots\dots\dots\dots\dots (4)$$

where P_5: output power of HEECS-2.

From law of conservation of energy for BTB system

$$P_{in} = (1-\eta_1)P_1 + (1-\eta_2)P_4 + P_3 \quad \dots\dots\dots (5)$$

where P_{in}: total HEECS input power.

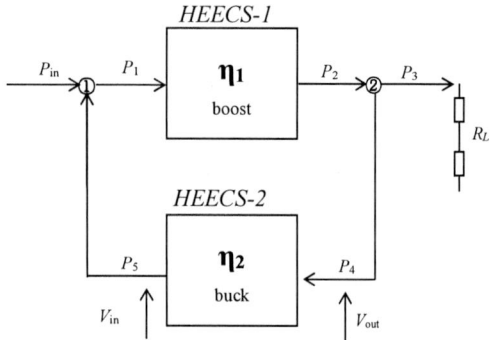

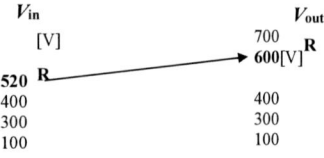

Fig. 8. Power flow block diagram of HEECS-BTB system.

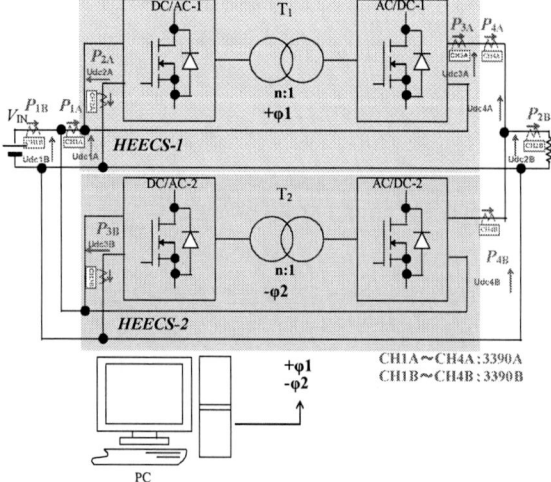

Fig. 9. Block diagram of 100 kW HEECS BTB system for measurement of total system efficiency η_{HEECS}.

Total HEECS output power P_{out_HEECS} is sum of output power P_2 of HEECS-1 and output power P_5 of HEECS-2. Thus,

$$P_{out\_HEECS} = P_2 + P_5 \quad \dots\dots\dots\dots\dots\dots (6)$$

Total HEECS system efficiency η_{HEECS} is

$$\eta_{HEECS} = \frac{P_2 + P_5}{P_2 + P_5 + P_{HEECS-loss}} \quad \dots\dots\dots\dots (7)$$

where $P_{HEECS-loss}$: total HEECS system loss.

Fig. 9 shows block diagram of 100 kW HEECS-BTB system for measurement of total system efficiency η_{HEECS}. To circulate 50 kW power in test system, HEECS-1 operates in traction mode with phase shift angle of $+\Phi_1$ and HEECS-2 operates in regeneration mode with phase shift angle of $-\Phi_1$ corresponding to digital control of phase shift reference through desktop personal computer PC keyboard. The measuring instruments are 2 digital power meter HIOKI 3390, which are distinguished by suffix A and B.

IV. EXPERIMENTAL RESULT

A. Noise Problem during Setup from 77 kW to 100 kW

Fig.10(a) and Fig.10(b) show difference of waveforms with increased surge voltage and gate noise during setup from 77 kW to 87 kW. V_1 is output voltage of primary inverter and V_{11} is primary winding voltage of transformer T_1 as shown in Fig. 6. New surge voltage at back edge of V_{11} appeared over 87 kW and adds to surge voltage at front edge of V_{11} during setup until 77 kW. Surge voltage at front edge is result from resonance between snubber C_S, stray C_{ST} and leakage inductance L_w of transformer T_1. Back edge is result from turn-off switching of increasing current.

Fig.10(c) shows phase timing between surge voltage and gate noise of 1-phase primary U-phase gate signal V_{GU1} from DSP controller which is shown in Fig.7. Gate noise occurred at front edge and back edge of 1-phase transformer primary voltage. One of noise measures by aluminum case for DSP control board against increased noise mentioned above is shown in Fig.11(b).

B. HEECS-BTB Recirculation Power Test at 100 kW

Table II shows test result of HEECS BTB recirculation power test at rating condition of 100 kW. Total HEECS output power $P_{out\text{-}HEECS}$ of 100.05 kW was achieved with only 2.76 kW of input power P_{in}. P_{in} is sum of total HEECS-BTB system loss $P_{HEECS\text{-}loss}$ and road resistance R_L loss. Fig.12 shows its readings of 2 digital power meter HIOKI 3390A and 3390B. Total HEECS efficiency η_{HEECS}=99.57 % was measured at high precision with measurement error of ±0.0055 % as discussed in chapter V. Fig.13 shows waveforms of rating 100 kW (measured vs. simulation). Circuit operation at rating condition of 100 kW, 600VDC was verified by comparison between experiment and simulation. Fig.14 shows efficiency measurement results. It was measured in ΔV_{in}=50V steps from P_{out_HEECS}=130W to rating power P_{out_HEECS}=100.05 kW. Extremely high efficiency of 99.57 % at 100.05 kW was obtained.

320

The 2018 International Power Electronics Conference

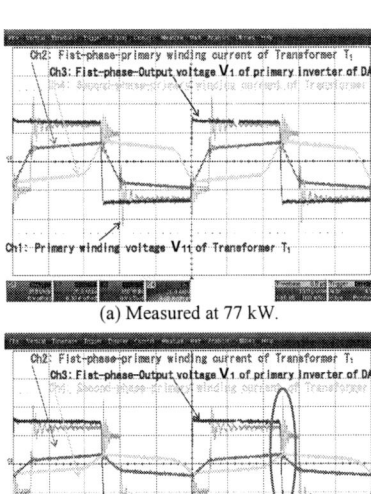

(a) Measured at 77 kW.

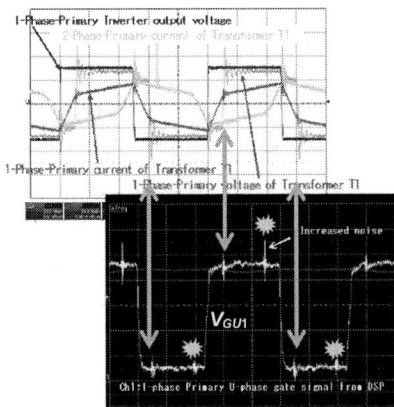

(b) Measured at 87 kW.

(c) Phase timing between surge voltage and gate noise at 87 kW

Fig. 10. Waveforms with increased surge voltage and gate noise from 77 kW to 87 kW.

TABLE II

TEST RESULT ON HEECS- BTB SYSTEM

Input power	P_{in}	2.76 kW
		$(P_{in} = P_{1B})$
Output dissipation power	P_3	2.32 kW
		$(P_3 = P_{2B})$
Output power of HEECS-1	P_2	51.45 kW
		$(P_2 = P_{4A})$
Output power of HEECS-2	P_5	48.60 kW
		$(P_5 = P_{1A} - P_{1B})$
Total HEECS output power	P_{out_HEECS}	100.05 kW
		$(P_{out\_HEECS} = P_{out\_HEECS1} + P_{out\_HEECS2} = P_2 + P_5)$
Total HEECS system loss	P_{HEECS_loss}	436.2 W
		$(P_{HEECS\ loss} = P_{1B} - P_{2B})$
Total HEECS efficiency	η_{HEECS}	99.57% > Over 99.3%
		$(\eta_{HEECS} = (P_2 + P_5)/(P_2 + P_5 + P_{HEECS\ loss})$

Fig. 12. Test result at rating of 100 kW.

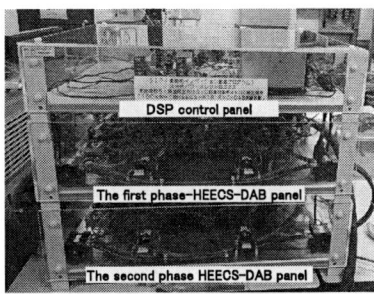

(a) before noise measure

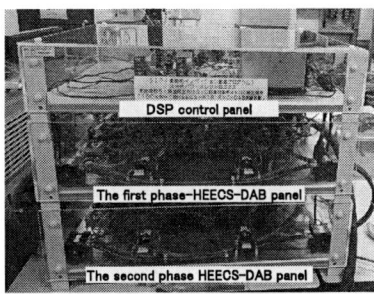

(b) after noise measure

Fig.11. Noise measure by aluminum case for DSP control board.

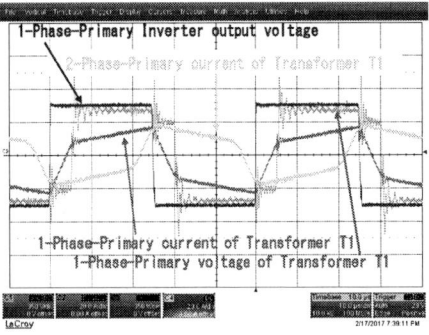

(a) Measured at rating of 100 kW.

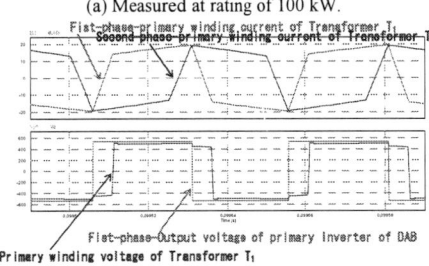

(b)Simulation at rating of 100 kW

Fig. 13. Waveforms at rating of 100 kW (Measured vs. Simulation).

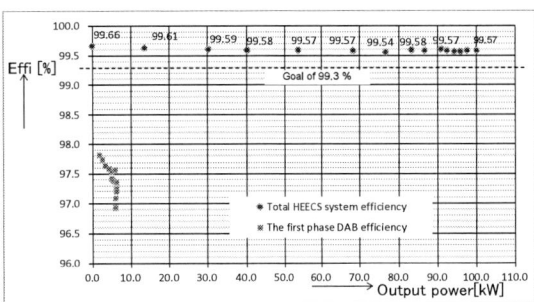

Fig. 14. Efficiency measurement results.

321

V. DISCUSSION ON HIGHLY PRECISE MEASUREMENT OF EFFICIENCY BY HEECS-BTB SYSTEM

A. Principle of Efficiency Measurement by Conventional System

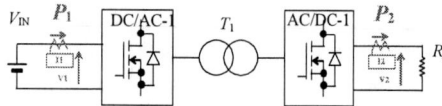

Fig. 15. Conventional efficiency measurement for DC-DC converter.

Fig.15 shows conventional efficiency measurement for DC-DC converter. Efficiency: η can be measured by input power: P_1 and output power: P_2 depending on precision of digital power meter. Power dissipation P_{loss} of DC-DC converter with efficiency $\eta=P_2/P_1=99.5$ %, output power P_2=100 kW is shown as

$$P_{loss} = P_1 - P_2 = \frac{P_2}{\eta} - P_2 = P_2 \bullet \frac{1-\eta}{\eta} \quad \text{..............(8)}$$

where η=0.995, P_2=100 kW. Thus from equation (8), power dissipation P_{loss} of DC-DC converter becomes P_{loss}=502.51 W. .... (9)

Measurement error εP_{loss} of power dissipation is

$$\varepsilon P_{loss} = P_1 \bullet (1+\varepsilon) - P_2 \bullet (1-\varepsilon) - P_{loss} \quad \text{.................(10)}$$

where power meter precision ε=0.16 %, output power P_2=100 kW, efficiency $\eta=P_2/P_1$=99.5 %, from equation (9), (10), measurement error εP_{loss} of power dissipation becomes

$$\varepsilon P_{loss} = 320.79\,W \quad \text{.................................. (11)}$$

Thus measurement error of efficiency: $\varepsilon\eta$ is shown as

$$\varepsilon\eta = \frac{P_2}{P_2 + P_{loss} - \varepsilon P_{loss}} \times 100 \sim \frac{P_2}{P_2 + P_{loss} + \varepsilon P_{loss}} \times 100 \quad \text{.......(12)}$$

Thus from equation (9), (11) and (12), measurement error of efficiency: $\varepsilon\eta$ becomes

$$\varepsilon\eta = 99.5\% \pm 0.319\% \quad \text{...(13)}$$

B. Principle of Highly Precise Measurement by HEECS-BTB system

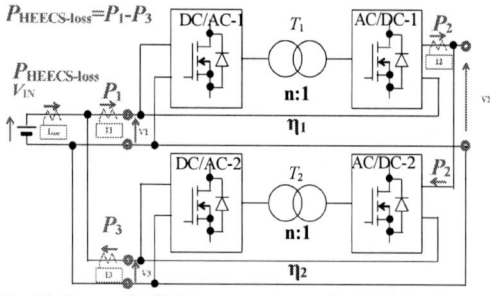

Fig. 16. Proposed efficiency measurement for DC-DC converter by HEECS-BTB system.

Fig.16 shows proposed efficiency measurement for DC-DC converter by HEECS-BTB system. Total HEECS system efficiency η_{HEECS} by 2-sets HEECS is shown as

$$\eta_{HEECS} = \eta_1 \bullet \eta_2 = 0.995 \bullet 0.995 = 0.990025 \quad \text{..........(14)}$$

where $\eta_1=\eta_2$=99.5%.

$P_{HEECS\text{-}loss}$: total HEECS system loss by 2-sets HEECS with total output power P_3+P_2=100 kW is shown as

$$P_{HEECS-loss} = P_1 - P_2 + P_2 - P_3 = P_1 - P_3 = P_3 \bullet \frac{1-\eta_{HEECS}}{\eta_{HEECS}} \quad \text{...........(15)}$$

where η_{HEECS}=0.990025

Power distribution between P_3 and P_2 of 2-sets HEECS with total output power P_3+P_2=100 kW can be obtained by solving following 3 equations (16), (17), (18).

$$P_2 + P_3 = 100\,kW \quad \text{..(16)}$$

$$P_2 \bullet \eta_2 = P_3 \quad \text{..(17)}$$

$$P_1 \bullet \eta_1 = P_2 \quad \text{..(18)}$$

where $\eta_1=\eta_2$=99.5%.

From equations (16), (17), (18), power distribution between P_1, P_2 and P_3 becomes

$$P_1 = 50.3771\,kW \quad \text{...(19)}$$

$$P_2 = 50.1253\,kW \quad \text{...(20)}$$

$$P_3 = 49.8747\,kW \quad \text{...(21)}$$

From equations (19), (20), (21), $P_{HEECS\text{-}loss}$: total HEECS system loss becomes

$$P_{HEECS-loss} = 502.4\,W \quad \text{...(22)}$$

Total HEECS system loss $P_{HEECS\text{-}loss}$ can be directly measured by voltage V_{IN} and current I_{loss} of input power source as shown in Fig.16. So measurement error εP_{loss} of total HEECS system loss becomes negligible small

$$\varepsilon P_{loss} = (P_1 - P_3) \bullet \varepsilon \quad \text{..........................(23)}$$
$$= 502.4\,W \bullet 0.16\% = 0.80384\,W$$

corresponding to 320.79 W of conventional measurement. Thus measurement error of efficiency: $\varepsilon\eta$ becomes negligible small as shown as following equation

$$\varepsilon\eta = \frac{P_2 + P_3}{P_2 + P_3 + P_{HEECS-loss} - \varepsilon P_{loss}} \times 100 \quad \text{..........(24)}$$
$$\sim \frac{P_2 + P_3}{P_2 + P_3 + P_{HEECS-loss} + \varepsilon P_{loss}} \times 100$$

where P_2+P_3=100 kW, $P_{HEECS\text{-}loss}$=502.4 W, εP_{loss}=0.80384 W

Thus from equation (16), (22) and (23), measurement error of efficiency: $\varepsilon\eta$ becomes

$$\varepsilon\eta = 99.5\% \pm 0.00091\% \quad \text{...(25)}$$

So corresponding to conventional measurement error of efficiency ±0.319%, very high precision measurement of efficiency with measurement error: ±0.00091% can be obtained.

C. Verification of experimental result

Measurement error εP_{loss} of total HEECS system loss is as follows

$$\varepsilon P_{loss} = 2.76\,kW \text{ of reading } P_{1B} \text{ on CH1B}(= P_{Rout} + P_{HEECS-loss}) \times \varepsilon \quad (26)$$

where ε=0.16%

From experimental result of direct measurement by channel CH1B of digital power meter 3390 as shown in Fig.12 and equation (26), Measurement error εP_{loss} of total HEECS system loss becomes

$$\varepsilon P_{loss}=4.416\,W \quad \text{.......................................(27)}$$

Thus from reading values of 2-set of digital power meter 3390A and 390B, measurement error of efficiency: $\varepsilon\eta$ becomes

$$\varepsilon\eta = 99.5659\% \; of \; \textit{effi} \pm 0.0055\% \quad\text{............................ (28)}$$

where *effi* is

$$\textit{effi} = \frac{(P1A - P1B + P4A)}{(P1A - P1B + P4A) + (P1B - P2B)} \quad\text{..................... (29)}$$

Corresponding to conventional measurement error of efficiency ±0.319% and principal high precision measurement of efficiency with measurement error: ±0.00091% by HEECS-BTB system, experimental result of ±0.0055% was verified.

VI. CONCLUSIONS

The proposed new water-cooled 2-phase HEECS BTB chopper system offers solutions in saving reduction of input power and high precision efficiency measurement. 100 kW HEECS-BTB converter with extremely high efficiency of 99.5% (input: V_{in}=537.27 V /output: V_S=607.62 V) was verified.

ACKNOWLEDGMENT

This work was supported by Council for Science, Technology and Innovation (CSTI), Cross-ministerial Strategic Innovation Promotion Program (SIP), "Next-generation power electronics" (funding agency: NEDO)

REFERENCES

[1] Y. Tsuruta and A. Kawamura, "Principle Verification Prototype Chopper Using SiC MOSFET Module Developed for Partial Boost Circuit System", *ENERGY CONVERSION & EXPO, ECCE-2015-Canada*, Poster Session I P904, pp.1421-1426, 2015.

[2] Y. Tsuruta, M. Pavlovsky and A. Kawamura, "Very High Efficiency SAZZ Chopper Using High Speed IGBT", *Proc. of 2009 IEEE 6th International Power Electronics and Motion Control Conference (IPEMC 2009)*, pp.573-579, 2009.

[3] T. Yamagishi, H. Akagi, S. Kinouchi, Y. Miyazaki and M. Koyama, "A 750-V, 100-kW, 20-kHz Bidirectional Isolated DC/DC Converter Using SiC-MOSFET/SBD Modules", *IEEJ Transactions on Industry Applications*, Vol.134-D, No.5, pp.544-553 (2014) (in Japanese).

[4] J. Zhang, R.-young Kim and J.-Sheng Lai, "High-Power Density Design of a Soft Switching High-Power Bidirectional DC-DC Converter" in *Proc. Of IEEE 37th Annual Power Electronics Specialists Conference (PESC)*, WeA2-1, pp.2119-2125, 2006.

[5] Stefan Waffler and Prof. Johann W. Kolar, "A Novel Low-Loss Modulation Strategy for High-Power Bi-directional Buck+Boost Converters", *2007 7th International Conference on Power Electronics (ICPE)*, pp.889-894, 2007.

[6] M. Hirakawa, M. Nagano, Y. Watanabe, K. Andoh, S. Nakatomi and S. Hashino, "High Power Density DC/DC Converter using the Close-Coupled Inductors", *European Conference on Cognitive Ergonomics 2009 (ECCE-2009)*, pp.1760-1767, 2009.

[7] Wensong Yu, Hao Qian, and Jih-Sheng Lai, "Design of High-Efficiency Bidirectional DC-DC Converter and High-Precision Efficiency Measurement", *Proc. of the 34th Annual Conference of the IEEE Industrial Electronics Society (IECON 2008)*, pp. 685-690, 2008.

[8] B. Eckard, A. Hofman, S. Zeltner, and M. Maerz, "Automotive Powertrain DC/DC Converter with 25kW/dm3 by using SiC Diodes", *4th International Conference on Integration of Power Electronics System*, pp.1-6, 2006.

[9] Wensong Yu,Chris Hutchens,Jih-Sheng Lai,Jianhui Zhang , Gianpaolo Lisi, Ali Djabbari,Greg Smith, and Tim HEGARTY, "High Efficiency Converter with Charge Pump and Coupled Inductor for Wide Input Photovoltanic AC Module Applications", *Proc. of IEEE ECCE2009*, pp. 3895-3900, 2009.

[10] M. Hirakawa, M. Nagano, Y. Watanabe, K. Andoh, S. Nakatomi, S. Hashino and T. Shimizu, "High Power DC/DC Converter using

Extreme Close-Coupled Inductors aimed for Electric Vehicles", *The 2010 International Power Electronics Conference ECCE ASIA-*, Japan, pp. 2941-2948, 2010.

[11] M. Hirakawa, M. Nagano, Y. Watanabe, K. Andoh, S. Nakatomi, S. Hashino and T. Shimizu, "High Power Density 3-level Converter with Switched Capacitors aimed for HEV", *14th International Power Electronics and Motion Control Conference (IPEMC)*, p. T9.27-33, 2010.

[12] K. Aoyama, N. Motoi, Giuseppe Guidi, Y. Tsuruta and A. Kawamura: "High Efficiency Battery Voltage Compensation for Electric Vehicles using Partial Boost Circuit", *IEEJ/JIASC 2013*, No.1-20, pp.125-130 (2013) (in Japanese).

[13] K. Aoyama, N. Motoi, Giuseppe Guidi, Y. Tsuruta and A. Kawamura: "Ultra High Efficient Battery Voltage Compensation against Decrease in the Terminal Voltage of Electric Vehicles", Industrial Electronics Society, *IECON2013-39th Annual Conference of IEEE*, pp.7280-7285, 2013.

[14] Martin Pavlovsky, Y. Tsuruta and A. Kawamura, "Evolution of Automotive Chopper Circuits Towards Ultra High Efficiency and Power Density", *IEEJ Transactions on Industry Applications*, Vol.131, No.10, pp.1-8, 2011.

[15] K. Aoyama, N. Motoi, Y. Tsuruta and A. Kawamura, "High Efficiency Energy Conversion System for Decreases in Electric Vehicle Battery Terminal Voltage", *IEEJ Journal of Industry Applications* Vol.5 No.1 pp.12-19, 2016.

[16] Y. Tsuruta and A. Kawamura, "Loss Minimization by Partial SAZZ for High-Power Chopper", *IEEJ Journal of Industry Applications*, Vol.3 No.5 pp.388-394, 2014.

[17] Y. Tsuruta and A. Kawamura, "Loss Analysis of High Power Chopper by Under-Zero Current Switching", *IEEJ Journal of Industry Applications*, Vol.4 No.1 pp.31-39, 2015.

Iron Loss Reduction in the Cores of Induction Heating Coils for Small-Foreign-Metal Particle Detector With a 400-kHz SiC-MOSFETs High-Frequency Inverter

Takuya Shijo, Yuki Uchino, Yujiro Noda, Hiroaki Yamada, and Toshihiko Tanaka

Department of Electrical and Electronic Engineering, Yamaguchi University, Ube, Japan

*E-mail: totanaka@yamaguchi-u.ac.jp

Abstract—This paper presents iron loss reduction in the cores of induction heating (IH) coils for a small-foreign-metal particle (SFMP) detector with a 400-kHz SiC-MOSFETs high-frequency inverter. A new IH coil shape, which can reduce iron loss, is proposed for the stable and continuous operation of the SFMP detector. Magnetic field analysis results using JSOL JMAG software, which is a 3D full-wave electromagnetic field simulation software, demonstrate that the iron loss caused by the newly proposed IH coils is decreased by 71.8 % compared to the previously proposed IH coils. The stable and continuous operation of the SFMP detector with the newly proposed IH coils can be achieved. It is also shown that SFMP represented by 0.3-mm-diameter stainless steel balls (SUS304) on the films can be detected with the newly proposed IH coils using JSOL JMAG software. A prototype experimental setup of the SFMP detector with the newly proposed IH coils is constructed and tested. Experimental results demonstrate that a 400-kHz SiC-MOSFETs high-frequency-inverter-based SFMP detector with the newly proposed IH coil can heat SFMPs, which can be observed by a thermographic camera. Therefore, the authors concluded that the 400-kHz SiC-MOSFETs high-frequency-inverter-based SFMPs detector with the newly proposed IH coil is applicable for high-performance chemical film production lines.

Keywords—SiC-MOSFET, small-foreign-metal particle detector, induction heating, iron loss

I. INTRODUCTION

High-performance chemical films (HPCFs) are widely used in lithium-ion batteries (LiBs) and flat panel displays of notebook computers, notepads, and mobile phones. For example, they are used as the separators of LiBs. The film material is plastic, and the required thickness is less than 0.01 mm. However, during the manufacturing process of the HPCFs, small-foreign-metal particles (SFMPs) produced by the manufacturing machines occasionally adhered to them. The SFMPs adhering to the HPCFs are typically made up of stainless steel or iron, and their diameter is approximately 0.1 mm. It is well known that SFMPs adhering to HPCFs occasionally rarely construct a short circuit between the positive electrode and negative electrode through the separator, which comprising the HPCFs. Overheating of the LiBs occurs due to this short circuit [1], [2]. Therefore, there is a strong requirement

in industry for practical methods of detecting SFMPs. S. Tanaka *et al.* proposed an SFMP detection method with a two-channel high-temperature scanning superconductive quantum interference device (HTS SQUID) gradiometer system [3]. They demonstrated that small iron particles with a diameter of less than 100 μm can be detected with the HTS SQUID. However, the SQUID needs a highly reliable cooling system to achieve superconductivity. Thus, a more practicable SFMP detector is needed for industrial applications. The present authors proposed an IH-based SFMPs detection method with a high-frequency inverter [4]- [6]. In [4], the winding locations on the core of the induction heating (IH) coils were discussed in detail. The magnetic analysis results with JSOL JMAG software, which is a 3D full-wave electromagnetic field simulation software, demonstrated that placing the windings for the core close to the gap is effective in reducing iron loss in the core of the IH coils. The validity and high practicability of the IH-based SFMP detector with the new winding locations on the IH coils were experimentally confirmed. Experimental results demonstrated that a 0.15-mm-diameter SFMP can be heated using a 400-kHz high-frequency inverter with SiC-MOSFETs and observed with a thermographic camera. However, this proposed IH-based SFMP detection method with a high-frequency inverter cannot be applied to actual continuous HPCF production lines because the SFMP was located in the gap of the IH coils. The present authors previously proposed a detection method using a leakage flux on the core gap [5], [6]. Fig. 1 shows a block diagram of the previously proposed SFMP detector using the SiC-MOSFETs high-frequency inverter. Fig. 2 shows the dimensions of the core used in the arrangement illustrated in Fig. 1. A leakage magnetic flux was used to heat and detect a 0.3-mm-diameter SFMP. The IH coils generate a high-frequency magnetic flux, which induces eddy currents in the SFMPs. The induced eddy current raises the temperature of the SFMPs. Subsequently, the heated SFMPs expand their own heat into the ambient resin material by heat diffusion and are thus easily observed with a thermographic camera. However, core cracking of the previously proposed IH coils in Fig. 2 often occurred owing to thermal stress caused by the iron loss. Therefore, the SFMP detector with the previously proposed IH

The 2018 International Power Electronics Conference

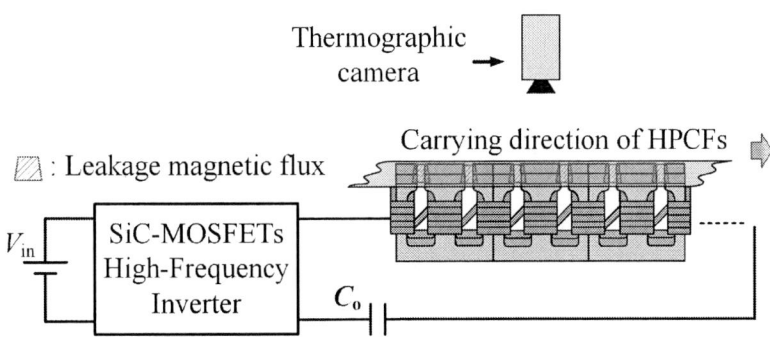

Fig. 1. Block diagram of the previously proposed SFMP detector using SiC-MOSFETs high-frequency inverter.

coils in Fig. 1 cannot achieve the stable and continuous operation that is desired in an HPCF production line.

In this paper, a new IH coil shape with a 400-kHz SiC-MOSFETs high-frequency-inverter-based SFMP detector is proposed, designed to reduce iron iron loss, for stable and continuous operation. A magnetic-field analysis is performed to confirm the validity and practicability of the newly proposed IH coil using JSOL JMAG software. Magnetic-field analysis results demonstrate that the iron loss in the newly proposed core is reduced by 71.8 % compared to that in a previously proposed core increasing the magnetic flux density to heat the SFMPs by 20 %. Reducing the iron loss can avoid the temperature rise in the newly proposed core. This means that core cracking, which often occurred in [5], [6], is also avoided. A prototype experimental model of the high-frequency SiC-MOSFETs based SFMP detector with the newly proposed IH coils is constructed and tested. Experimental results demonstrate that a 400-kHz SiC-MOSFETs high-frequency-inverter-based SFMP detector with the newly proposed IH coil can heat 0.3-mm-diameter SFMPs, and the heated SFMPs can be observed by a thermographic camera. From the experimental results, the authors conclude that the 400-kHz SiC-MOSFETs high-frequency inverter-based SFMPs detector with the newly proposed IH coil is applicable to the production of HPCFs.

II. NEWLY PROPOSED IH COIL DESIGN FOR IRON LOSS REDUCTION

Fig. 3 shows a block diagram of the SFMP detector with the newly proposed IH coil. A 400-kHz SiC-MOSFETs high-frequency inverter is also used with the new IH coil to generate high-frequency magnetic flux. Fig. 4 shows the shape of the new core with reduced iron loss. The newly proposed IH coil comprising of an E-shaped core, an I-shaped core, and a copper plate, with a high-frequency magnetic flux generated in the core gap between the E-shaped core and I-shaped core. The HPCFs pass through the gap. The gap length between the E-shaped core and I-shaped core is 5 mm. A Litz wire can be wound around in the E-shaped core. Therefore, iron loss can be reduced. The E-shaped and I-shaped cores shown in Fig. 4 are connected in multiple series and parallel, respectively, as shown in Fig. 3. In [5], [6], the frequent occurrence of core cracking was reported,

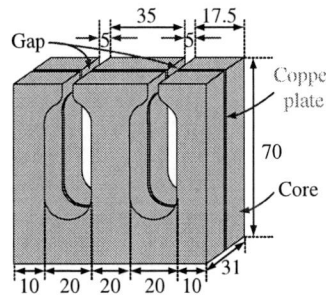

Fig. 2. IH core dimensions of the previously proposed SFMP detector.

caused by temperature differences in the cores. To sink the temperature difference in the cores, a copper plate of 1-mm thickness is inserted in the cores, effectively avoiding core cracking. The turn-number of windings for each core leg is five turns. Newly proposed HPCFs are suspended at a height of 3 mm from the upper surface of the E-shaped cores. As the IH coil produces a magnetic field in the core gap, the SFMPs on the film are heated and then observed using a thermographic camera. The number of multiple-series-parallel connections depends on the width of the manufactured HPCFs. Magnetic-field analysis is performed to confirm the reduced effect of the iron loss in the newly constructed IH coil using JSOL JMAG software. The frequency of the coil current is 400 kHz, and its value of RMS is 30 Arms. Fig. 5 shows the magnetic-field analysis results of the newly proposed IH coil in Fig. 4 using JSOL JMAG software. Fig. 5(a) shows a finite element method (FEM) model, and Fig. 5(b) shows the magnetic-field analysis results of the magnetic flux density distribution at 3 mm from the upper surface of the newly constructed IH coil, where the core material is PC40 (TDK Co., Ltd.). As shown in Fig. 5(b), the magnetic flux density near the core gap is approximately 36 mT with an its RMS value of coil current of 30 Arms, where the gap length between the E-shaped and I-shaped cores is 5 mm. In [5], the leakage magnetic flux density was 30 mT with an RMS value of coil current of 30 Arms. Therefore, the magnetic flux density for heating the SFMPs is increased by 20 % using the newly proposed IH coil.

325

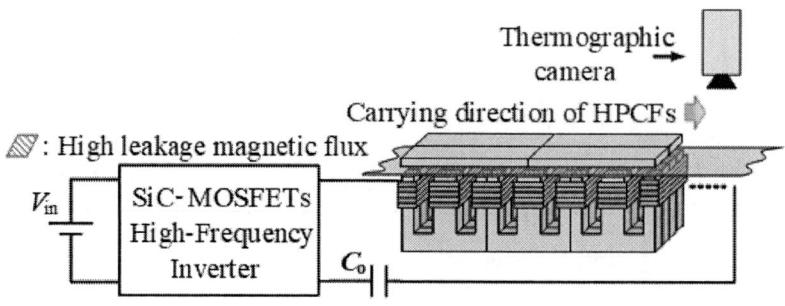

Fig. 3. Block diagram of the SFMP detection using newly proposed IH coils.

Table I presents the iron loss comparison between the previously proposed and newly proposed cores, where three cores are connected in series. It is well known that the iron losses in the cores of IH coils are caused by Joule and hysteresis losses. In the previously proposed cores, the Joule loss is 2686 W, and the hysteresis loss is 1402 W. Thus, the total iron loss is 4088 W while the total iron loss in the newly proposed core is 1153 W. The iron loss in the newly proposed core is reduced by 71.8 % compared to that in the previously proposed core increasing the magnetic flux density to heat SFMPs by 20 %. Reducing the iron loss in the newly proposed core can avoid the temperature-rise in the core. This means that core cracking, which often occurred in [5], [6], is also avoided in the new core design. It is thus concluded that stable and continuous operation of the high-frequency SiC-MOSFETs based SFMP detector with the newly proposed IH coils is possible, and the detector can therefore be applied to HPCF production lines.

III. EXPERIMENTAL RESULTS OF SFMP HEATING

A prototype experimental model of the high-frequency SiC-MOSFETs-based SFMP detector with the newly proposed IH coil is constructed and tested. Fig. 6 shows a picture of the newly proposed IH coil, which is used in the prototype experimental model. Fig. 7 shows a power circuit diagram of the SiC-MOSFETs high-frequency inverter, which supplies the high-frequency current to the IH coils, where the C2M0025120D (CREE Co., Ltd) is used. Table II shows circuit constants for Fig. 7. R_o and L_o are the equivalent resistor and inductor of the newly proposed IH coil, respectively, shown in Fig. 6. A resonant capacitor C_o is connected in series to the IH coil. To avoid the cracking of the IH cores caused by a spark between the wound coils and cores, a self-fusion silicon rubber tape (Nitto Denko Co., Ltd.) is wrapped around the Ritz wires. The self-fusion silicon rubber tape has a thickness of 5 mm, a heat-resistance temperature of 200 °C and a breakdown voltage of 56.6 kV/mm. For the SFMPs, 0.3-mm-diameter stainless steel balls (SUS304) are used. Fig. 8 shows the locations of the SFMPs and the experimental results. Fig. 8(a) shows the locations of the 0.3-mm-diameter stainless steel balls that used as model SFMPs. The gap between the E-shaped core and I-shaped core is 5 mm. The film is placed in the gap between the E-shaped core and I-shaped core, as shown in

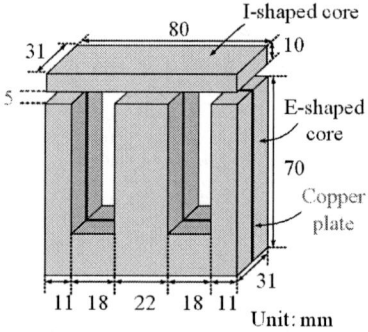

Fig. 4. IH core dimensions of the newly proposed SFMP detector.

TABLE I.   COMPARISON OF IRON LOSS.

	Previously proposed IH cores	Newly proposed IH cores
Iron loss W	4088	1153
Joule loss W	2686	702
Hysteresis loss W	1402	451

Fig. 6 to heat the model SFMPs. After an exposure of 5 s, an increase in the temperature rise of the stain less steel balls is observed using a thermographic camera (R300SR, Nippon Avionics Co. LTD.). Fig. 8(b) shows the observed SFMPs using the thermographic camera. The temperature difference $\Delta T_{\max}$ between the heated stainless steel balls and room temperature is 6.6 °C. From the experimental results shown in Fig. 8, the authors conclude that the high-frequency SiC-MOSFETs-based SFMP detector with the newly proposed IH coil is useful for industrial continuous production lines of HPCFs.

IV. CONCLUSION

In this paper, a new IH coil shape with a 400-kHz SiC-MOSFETs high-frequency-inverter-based SFMP detector, which can reduce iron loss, has been proposed for a stable and continuous operations to detect SFMPs. A magnetic field analysis has been performed to confirm the validity and practicability of the newly proposed IH coil using JSOL JMAG. The magnetic-field analysis results demonstrated that iron loss in the newly proposed core is reduced by 71.8 % compared to that in the previously proposed core increasing the magnetic flux density to heat

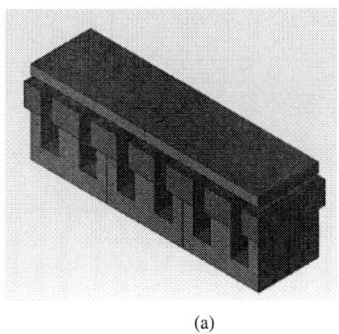
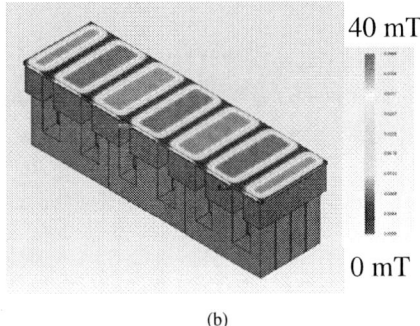

40 mT

0 mT

(a)　　　　　　　　　　　　(b)

Fig. 5.　Magnetic-field analysis results of the magnetic flux density near the core gap modelled using with JMAG software. (a) FEM model. (b) Magnetic-field analysis result.

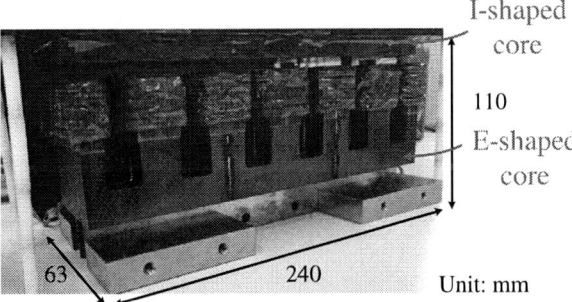

I-shaped core

110

E-shaped core

63　　　　240　　　Unit: mm

Fig. 6.　Prototype of newly proposed IH coils.

the SFMPs by 20 %. Reducing iron loss in the newly proposed core results in a stable and continuous operation owing to lowered thermal stress damage. A prototype experimental model of the high-frequency SiC-MOSFETs inverter-based SFMP detector with the newly proposed IH coil has been constructed and tested. Experimental results have demonstrate that a 400-kHz SiC-MOSFETs high-frequency inverter-based SFMP detector with the newly proposed IH coils can heat 0.3 mm-diameter SFMPs, and the heated SFMPs can be observed using a thermographic camera. Based on the experimental results, the authors have concluded that the 400-kHz SiC-MOSFETs high-frequency inverter-based SFMP detector with the newly proposed IH coil can be applied to HPCF production lines.

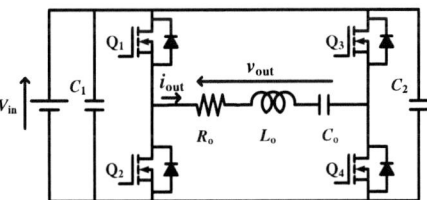

Fig. 7.　Circuit diagram of SiC-MOSFETs full-bridge inverter.

TABLE II.　CIRCUIT CONSTANTS FOR FIG. 7.

Item	Symbol	Value
Input voltage	V_{in}	180 Vdc
Coil current	I_{out}	30 Arms
Switching frequency	f_{SW}	400 kHz
Bypass capacitance	C_1, C_2	1.12 μF
Equivalent resistance of IH coils	R_{o}	4 Ω
Inductance of IH coils	L_{o}	120.2 μH
Resonant capacitor	C_{o}	1.3 nF
Exposure time	t	5 s

REFERENCES

[1]　D. Capozzo, S. Fleming, B. Foley, and M. Macri, "Lithium ion battery safety," Worcester Polytechnic Institute, pp. 1-169, Dec. 2006.

[2]　M. Arakawa, M. Ichimura, "Latest safety test technologies of lithium-ion batteries and construction of battery safety evaluation site",*NTT Facilities Research Institute, ,*Annual Report No. 20, pp. 1-7, June 2009.

[3]　S. Tanaka, T. Akai, Y. Kitamura, "Two-Chanel HTS SQUID Gradiometer System for Detection Metallic Contaminants in Lithium-Ion Battery", *IEEE Transactions on Applied Superconductivity,* Vol.21, Issue. 3, pp. 424-427, June 2011.

[4]　S. Kurachi, N. Yamamoto, H. Yamada, T. Tanaka, E. Hiraki, Y. Yamada, T. Nagao, Y. Miyake, and Y. Noda, "High-frequency induction heating for tiny foreign metals," in *Proc. 9th Annu. IEEE Int. Energy Convers. Congr. Exhi. (ICPE)*, Jul. 2015, pp. 2203-2208.

[5]　T. Shijo, S. Kurachi, Yuki Uchino, Y. Noda, H. Yamada, and T. Tanaka, "High-frequency induction heating for small-foreign-metal particles using SiC-MOSFETs inverter," *in Proc. of IEEE International Future Energy Electronics Conference (IFEEC 2017-*

The 2018 International Power Electronics Conference

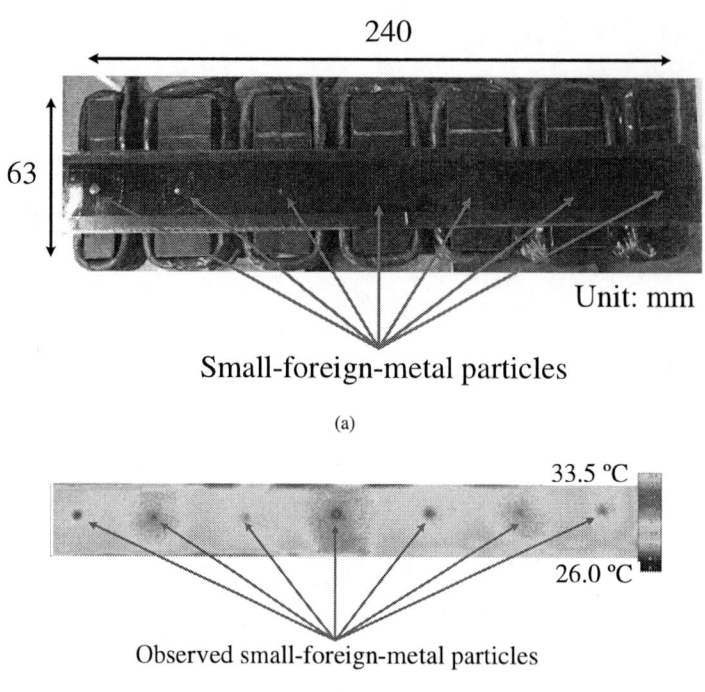

Small-foreign-metal particles

(a)

Observed small-foreign-metal particles

(b)

Fig. 8.   Prototype of newly proposed IH coil (a) Top view of IH cores. (b) Thermal image using thermographic camera

ECCE-Asia), O15, 1532, Jun 2017.

[6]   T. Shijo, S. Kurachi, Yuki Uchino, Y. Noda, H. Yamada, and T. Tanaka, "High-frequency induction heating for small-foreign-metal particles detection using 400 kHz SiC-MOSFETs inverter," *in*

Proc. of IEEE Energy Conversion Cogress and Expo (ECCE), Oct. 2017.

Frequency Tracking Burst-Mode PDM-controlled Class-D Zero Voltage Soft-Switching Resonant Converter for Inductive Power Transfer Applications

Yoichiro Tabata, Tomokazu Mishima*, and Tatsuya Kido

Dept. of Marine Engineering, Graduate School of Maritime Sciences, Kobe University

5-1-1, Higashinada, Kobe, Hyogo 658–0022, Japan

E-mail: mishima@maritime.kobe-u.ac.jp

Abstract—A newly developed prototype of zero-voltage soft-switching (ZVS) high-frequency resonant for inductive power transfer (IPT) applications is presented in this paper. By adopting the burst mode pulse-density-modulation (PDM) scheme with resonant frequency tracking, the load power can be continuously regulated under the conditions of full-range soft-switching due to the undamped resonant currents through the coils, thereby the current surges which appears with the conventional PDM scheme can be eliminated effectively. The essential performances on the output power regulation and soft-switching operations are demonstrated in experiment using a class of 400 W-500 kHz prototype, and the validity is evaluated from the practical point of view.

Keywords—*Burst mode pulse-density-modulation (PDM), Class-D inverter, Inductive power transfer (IPT), resonant converter, soft-switching.*

I. INTRODUCTION

IPT technology utilizes magnetic coupling to transfer power across a large air gap, and nowadays its applications are widely expanded from the battery-powered mobility such as electric vehicles (EV) as in Fig. 1 and induction heating for metal surface treatments. Highly efficient power conversion and high-frequency switching operation are essential for achieving high-power density. In addition to the power converter topology and the sending / receiving coils design, the power control of the high frequency (HF) inverter is a critical issue for the IPT[1]-[9]. The circuit topologies and modulation strategies applicable for IPT are summarized in TABLE I. The pulse density modulation (PDM) provides a simple but practical solution for load power control with full-range of soft switching.

Strategies of PDM are sorted by several type: as predefined switching pattern, delta-sigma modulation, and burst mode[10][11]. The burst mode PDM is most simple and easy-to-implement for contactless power and signal communications in IPT systems. The technical issues of conventional burst-mode PDM is illustrated in Fig. 2: while holding the charge of the lossless snubber capacitor in parallel with the switch, it is forced to discharge when

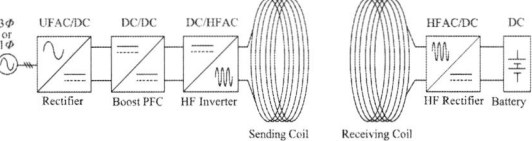

Fig. 1. Energy and power process of IPT battery charger applications.

the inverter gets back to switching action, and then the current surge occurs.

The proposed burst-mode PDM naturally avoids the surge and related ringing by making use of the second-order undamped oscillation through sending and receiving coils. The original idea of the proposed PDM exist in the incorporation of resonant frequency tracking into the burst mode PDM for keeping the high power factor and wide range of ZVS in accordance with variations of coupling coefficient between the coils. The frequency tracking allows for maximization of the transferred power in the situations of misalignment of coil positions and variations of air gap length[12].

This paper is organized as follows: the circuit operation with the burst mode PDM scheme is described in Section II. The steady-state analysis based on the frequency domain equivalent circuit is demonstrated in Section III, thereby the effectiveness of the frequency tracking PDM scheme is clarified in principle. The experimental verifications are presented in Section IV with comparison of operating waveforms and efficiency, after which the performances and features of the proposed converter are summarized in Section V.

II. CIRCUIT TOPOLOGY AND PULSE MODULATION

The circuit topology of the proposed converter is presented in Fig.3. The primary-side, i.e. sending coils-side high-frequency resonant (HF-R) inverter generates the resonant current through the series compensation network L_1-C_1. The class-D rectifier receives the HF-R current through the series compensation network L_2-C_2 in the secondary-side, i.e. receiving side, and then transfer the power to the dc load which is represented by

The 2018 International Power Electronics Conference

TABLE I. CIRCUIT TOPOLOGIES AND POWER CONTROL SCHEMES FOR IPT SYSTEMS

Circuit topology	Power range	Control	Pros and cons
Class-D inverter	Low-Medium power	Asymmetry PWM, PFM, PDM	Low voltage-stress
Class-E inverter	Low power	PAM, PDM	Excessive voltage-stress
Full-bridge inverter	Medium-High power	PS-PWM	Increase of devices and components

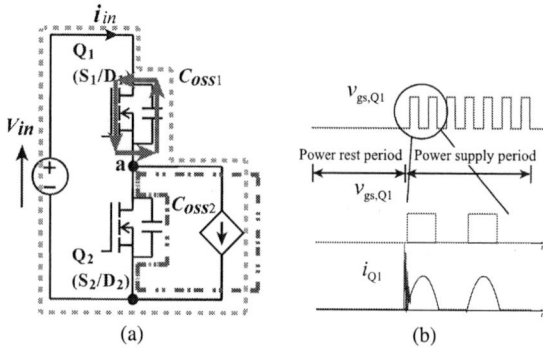

(a)   (b)

Fig. 2. Surge current at the turn-on transition in the conventional burst-mode PDM: (a) short circuit loop, and (b) resultant surge and ringings.

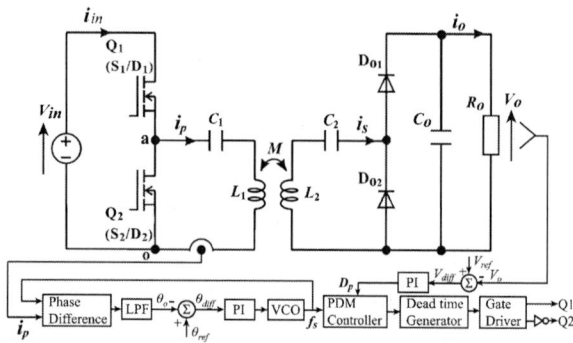

Fig. 3. Class-D ZVS resonant DC-DC converter for IPT systems.

R_o. The active switches Q_1-Q_2 operate under the edge-resonant ZVS utilizing a parasitic output capacitance C_{oss} of switching power device.

The logic circuit of pulse generator and key waveforms are compared in Fig.4 and Fig.5, where the PDM duty factor D_p is defined by

$$D_p = \frac{T_{p,on}}{T_p} \quad (1)$$

Note here the PDM cycle $f_{dp}(=1/T_p)$ should be set as the specified number of fraction with respect to the switching frequency f_s.

The key idea of the proposed burst mode PDM is to keep the ON-state of the low-side switch for suspending the power transfer interval from the input dc source to the sending coils by sustaining the undamped resonant current due to the high load quality factor of WPT system. Accordingly, the oscillation between the parasitic

output capacitances and stray inductances of the HF-R inverter are dramatically suppressed. The switching operation resumes by turning the low-side switch Q_2 off. Then, the capacitive energy of the high-side switch Q_1 is completely discharged, consequently no surge current appears at the turn-on transition of the high-side switch, and the power dissipation is reduced under the high frequency condition. Another feature is the switching frequency of the HF-R inverter corresponds to the natural frequency of the converter. Therefore, no high sensitivity of high-frequency current sensor is required just for implementing the proposed PDM, which is advantageous over the alternative solution reported in [10].

The corresponding mode-transitions and equivalent circuits are shown in Fig.7. They are divided into the switching and free-oscillation intervals in the burst mode PDM.

Taking the winding resistance R_1 of the sending coil and the winding resistance R_2 of the receiving coil into consideration, the impedance parameters are expressed by

$$\begin{bmatrix} \dot{V}_1 \\ 0 \end{bmatrix} = \begin{bmatrix} \dot{Z}_1 & -j\omega L_m \\ -j\omega L_m & \dot{Z}_2 \end{bmatrix} \begin{bmatrix} \dot{I}_1 \\ \dot{I}_2 \end{bmatrix}$$

$$(2)$$

where the primary and secondary-side impedances are represented respectively by $\dot{Z}_1 = R_1 + j(\omega L_1 - \frac{1}{\omega C_1})$ and $\dot{Z}_2 = R_2 + R_L + j(\omega L_2 - \frac{1}{\omega C_2})$. Transforming (2) gives birth to the admittance parameters as

$$\begin{bmatrix} \dot{I}_1 \\ \dot{I}_2 \end{bmatrix} = \frac{1}{\dot{Z}_1 \dot{Z}_2 + (\omega L_m)^2} \begin{bmatrix} \dot{Z}_2 & j\omega L_m \\ j\omega L_m & \dot{Z}_1 \end{bmatrix} \begin{bmatrix} \dot{V}_1 \\ 0 \end{bmatrix}.$$

$$(3)$$

The input power is decided by $P_1 = \{|\dot{V}_1||\dot{I}_1| = V_1^2 |\dot{Z}_2|/\dot{Z}_1\dot{Z}_2 + (\omega L_m)^2\}$, and the output power is expressed by $P_2 = R_L|\dot{I}_2|^2 = V_1^2(\omega L_m)^2 R_L/\{\dot{Z}_1\dot{Z}_2 + (\omega L_m)^2\}$. Therefore, the power conversion efficiency is defined as

$$\eta = \frac{(\omega L_m)^2 R_L}{(R_L + R_2)\{R_1(R_L + R_2) + (\omega L_m)^2\}}. \quad (4)$$

It is understood from (4) that the magnetizing inductance L_m varies with the position shift and gap-length between the sending and the receiving coils, so that the power transmission efficiency can keep as close to maximum by adjusting the operation frequency around the resonance frequency.

330

The 2018 International Power Electronics Conference

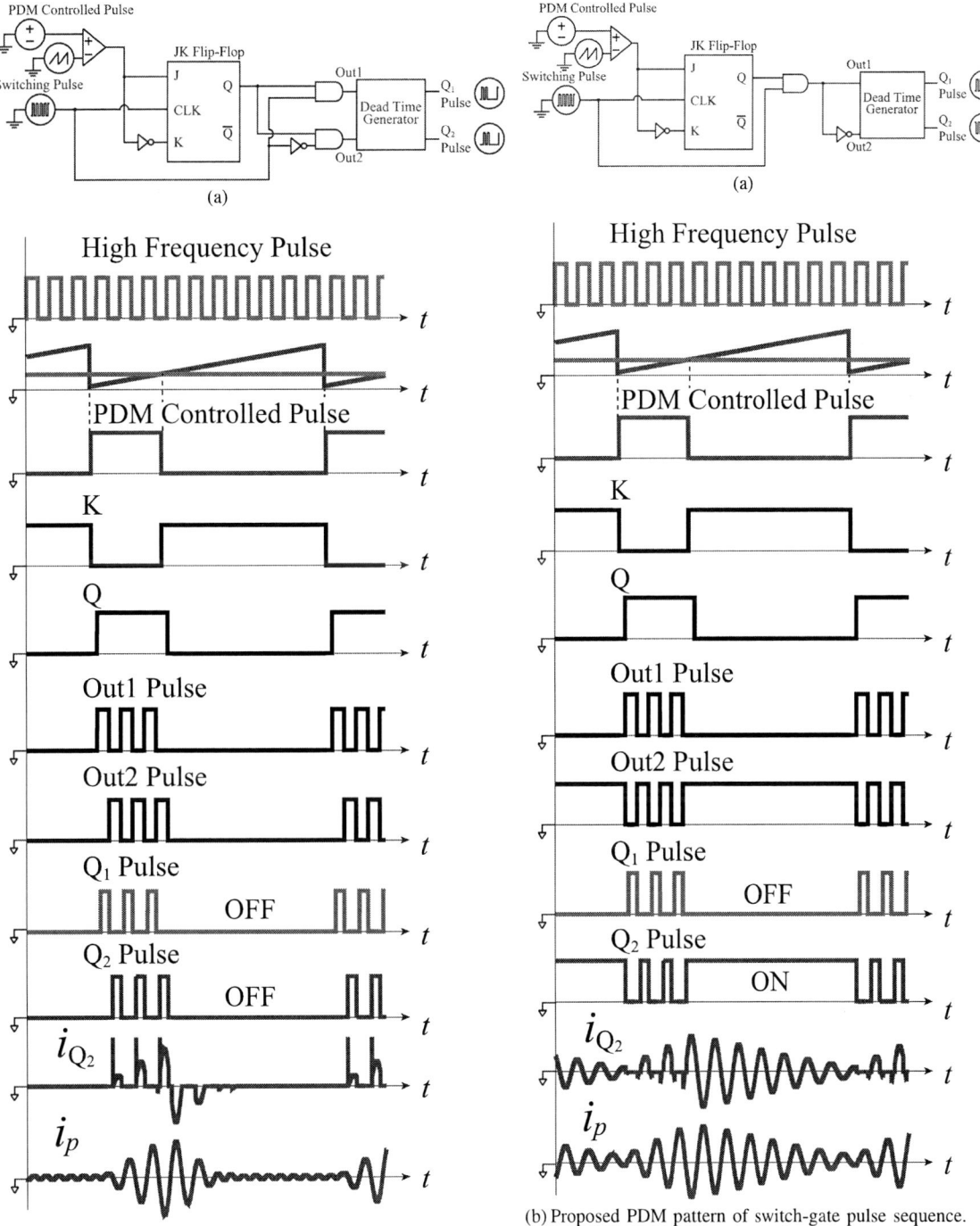

Fig. 4. Conventional PDM pattern of switch-gate pulse:(a) logic circuit diagram, and (b) pulse sequences.

Fig. 5. Proposed PDM pattern of switch-gate pulse: (a) pulse sequences, and (b) pulse sequence.

III. STEADY-STATE ANALYSIS

The frequency domain analysis is described by the simplified equivalent circuit of the proposed HF-R inverter as illustrated in Fig.8. The leakage and magnetizing inductances which depend on the gap length between the two coils are uniquely expressed by $L_r = (1 - k^2)L_1$

and $L_m = k^2 L_1$. The fundamental-harmonics approximation (FHA) for the proposed converter justifies the adoption of equivalent ac resistance $R_{ac} = 2aR_o/\pi^2$ for the sake of simplicity.

The voltage conversion ratio depicted in Fig. 8 can be expressed with the HF transformer winding turns ratio

331

The 2018 International Power Electronics Conference

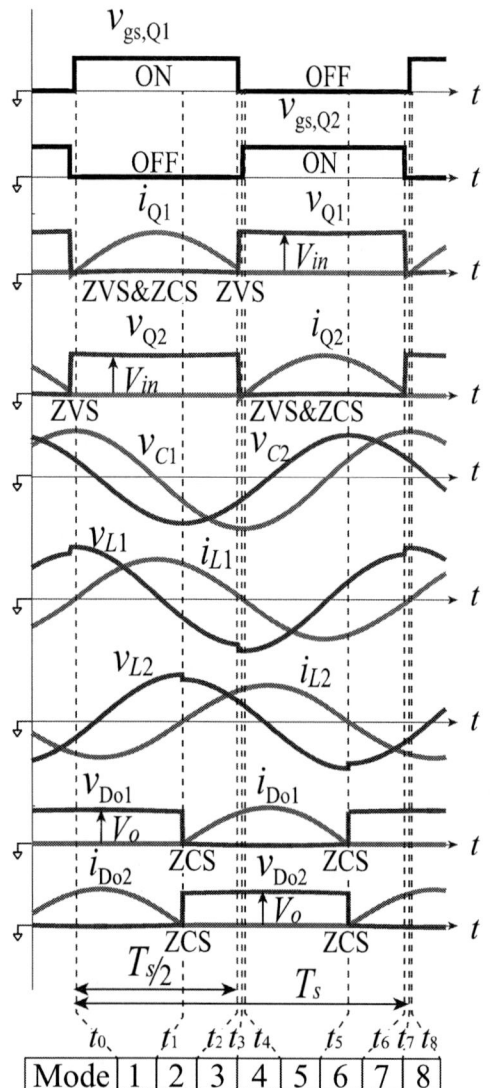

Fig. 6. Voltage and current waveforms during switching one cycle.

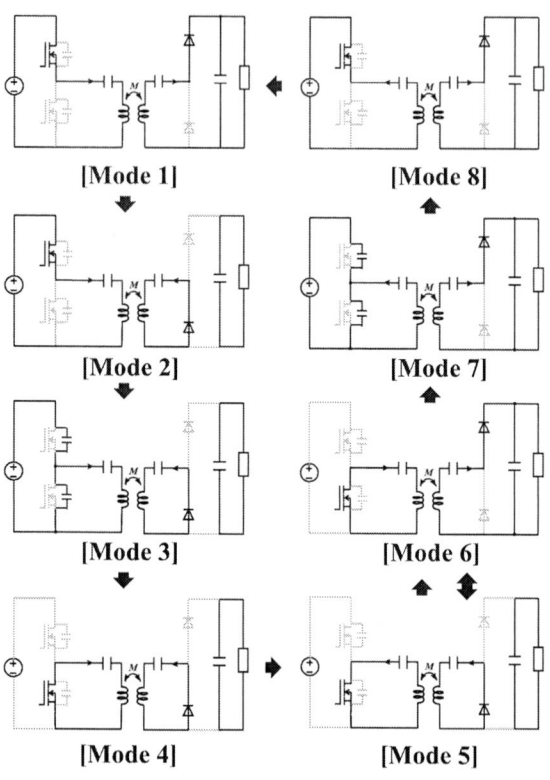

Fig. 7. Voltage and current waveforms during switching one cycle.

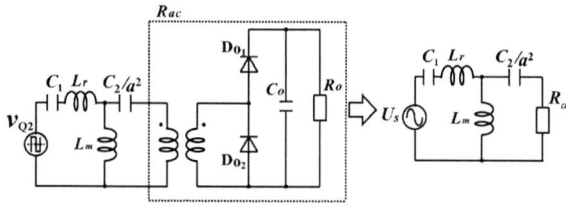

Fig. 8. Simplified equivalent circuits based on FHA method.

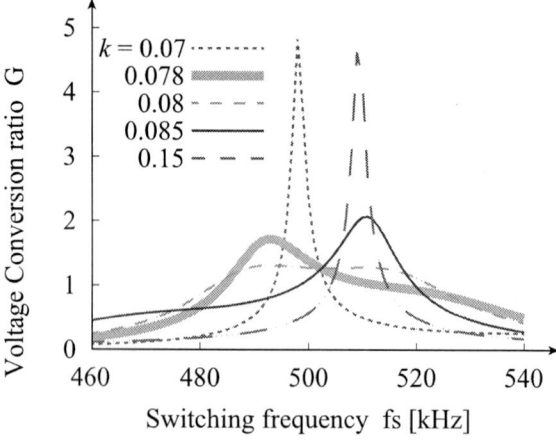

Fig. 9. Voltage conversion ratios versus switching frequency.

$a = k\sqrt{L_1/L_2}$ as

$$G = \frac{1}{a\sqrt{\left\{1 + \frac{1}{S} - \frac{1}{S}\left(\frac{f_{r1}}{f_s}\right)^2\right\}^2 + \left(\frac{f_{ms}}{f_s}\right)^2 (Q_r\,\xi - Q_{ms})^2}} \tag{5}$$

$$\xi = \left(\frac{f_s}{f_{r1}} - \frac{f_{r1}}{f_s}\right)\left(\frac{f_s}{f_{ms}} - \frac{f_{ms}}{f_s}\right) \tag{6}$$

where $S = L_m/L_r, Z_{ms} = \sqrt{a^2 L_m/C_2}, Z_{ms} = \sqrt{a^2 L_m/C_2}, Z_r = \sqrt{L_r/C_1}, Q_{ms} = Z_{ms}/R_{ac}$ and $f_{ms} = 1/(2\pi\sqrt{L_m C_2/a^2})$. Referring to the characteristic impedance ratio $\lambda_{ms} = Z_{ms}/Z_r$ obtained from Fig. 9, the three resonance frequencies are defined respectively under the condition of loose coupling coefficient by

$$f_r = \frac{1}{2\pi\sqrt{L_1 C_1}} \tag{7}$$

332

$$f_{r1} = \sqrt{\frac{\Gamma - \sqrt{\Gamma^2 - 4f_r{}^2 f_{ms}{}^2}}{2}} \qquad (8)$$

$$f_{r2} = \sqrt{\frac{\Gamma + \sqrt{\Gamma^2 - 4f_r{}^2 f_{ms}{}^2}}{2}} \qquad (9)$$

where $\Gamma = f_r{}^2 + \lambda_{ms} f_r f_{ms} + f_{ms}{}^2$. Furthermore, considering the energy balance between the leakage inductance of the sending and receiving coils and the switch parasitic capacitances, the turn-off current $i_{\mathrm{Q}x, off}$ (the suffix x denotes $1/2$ for Q_1/Q_2) which is relevant to the ZVS condition is given by

$$\frac{1}{2} L_1 i_{\mathrm{Q}x, off}^2 > \frac{1}{2}(C_{oss1} + C_{oss2})V_{in}^2. \qquad (10)$$

As a result, the lower limit value by which the turn-off dv/dt of Q_1 and Q_2 can be determined as

$$i_{\mathrm{Q}, off} > \frac{V_{in}}{\sqrt{\frac{L_1}{(C_{oss1}+C_{oss2})}}} \qquad (11)$$

$$= \frac{100 \,[\mathrm{V}]}{\sqrt{\frac{19\,[\mu\mathrm{H}]}{(100\,[\mathrm{pF}]+100\,[\mathrm{pF}])}}} \qquad (12)$$

$$= 0.32\,[\mathrm{A}] \qquad (13)$$

It is possible to satisfy (13) and perform the ZVS operation using the resonant current of primary coils at the zero crossing point with the frequency tracking controller.

IV. EXPERIMENTAL VERIFICATION

The practical effectiveness of the proposed converter is investigated by experiment. The exterior appearance of a 400 W-510 kHz prototype is shown in Fig.10. The circuit parameters and specifications are summarized in TABLE II.

The observed voltage and current waveforms are displayed in Fig.11, where zero-voltage and zero-current soft-switching (ZVZCS) turn-on and ZVS turn-off operations can be confirmed for the primary-side active switches. The operating waveforms with the conventional and proposed PDM are depicted in Figs.12–14, respectively. It can be confirmed that the inverter current i_{L_1} through the primary-side coil is continuously regulated in accordance with D_p. The surge currents occur at the turn-on transition of each active switch due to the residual capacitive energy in C_{oss1} and C_{oss2}, which is the phenomenon inherent to the conventional PDM-controlled voltage source class-D inverter. In contrast to that, no surge current emerges at the turn-on transitions with the proposed burst mode PDM.

The steady-state characteristics of the output power versus the PDM duty ratio are presented in Fig.15, where the wide range of power regulation is verified for the proposed converter. The actual efficiency curve is presented in Fig. 16 and the maximum efficiency is obtained as 72.4 % at $P_o = 220$ W with gap-length $g = 15$ cm.

The characteristics for the load-resistance variations are depicted in Fig.17 when the load resistance of the prototype is changed from $25\,\Omega$ to $55\,\Omega$. It can be seen

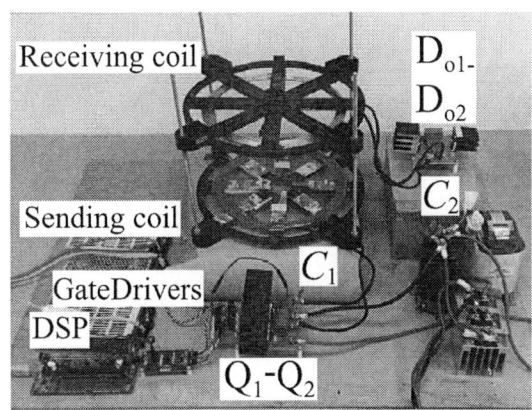

Fig. 10. Exterior appearance of the prototype converter.

TABLE II. EXPERIMENTAL CIRCUIT PARAMETERS

Item	Symbol	Value [unit]
DC input voltage	V_{in}	100[V]
Switching frequency	f_s	510[kHz]
Output power rating	P_o	410[W]
Input smoothing capacitor	C_{in}	10[μF]
Resonant capacitors	C_1, C_2	5.3[nF]
Output parasitic capacitance	C_{oss}	100 [pF]
Output smoothing capacitor	C_o	10[μF]
Dead time interval	T_d	150 [ns]
Load resistor	R_o	40[Ω]
Winding turns	w_1/w_2	5/5[turn]
Self inductance of sending coil	L_1	19 [μH]
Self inductance of receiving coil	L_2	19 [μH]
Mutual inductance of L_1 and L_2	M	1.52 [μH]
Coupling coefficient of L_1 and L_2	k	0.08
Air gap length between L_1 and L_2	g	15 [cm]

$*$ Super Junction-MOSFET : IPW50R190CE, 550[V], 18.5[A]

$*$ D_{o1}-D_{o2} : C3D20060D, 600[V], 28[A]

from the result that both the output voltage and power can be continuously adjusted for a wide range of load variations.

The results of power loss analysis at the rated output are shown in Fig.18. It can be seen from the result that the switching loss is suppressed by the effect of edge-resonant ZVS. Thus, it becomes clear that it is necessary to reduce the copper loss of the high frequency LITZ wire in the power sending and receiving coils besides suppressing the conduction loss of the switch. Power loss analysis of the prototype converter is revealed in Fig. 18. It can be confirmed from the breakdown that the power consumption of switches are reduced than the sending and receiving coils, so the elimination of surge and ringing is proven by experimental data.

The effectiveness of resonance frequency tracking is confirmed by the open-loop controller against the gap length variation. The efficiency versus gap length curves are presented in Fig.19 with the frequency variation between 505 kHz–530 kHz for the unique load resistance. The resonant frequency slightly declines as the air-gap expands and the coupling coefficient aggravates as

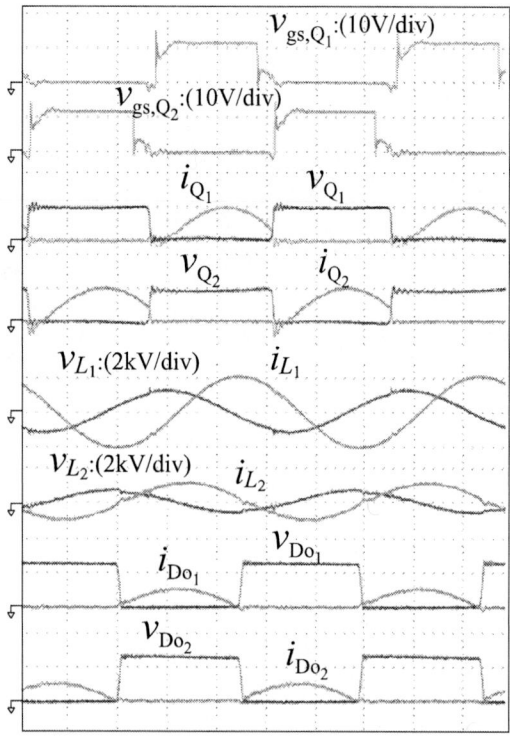

Fig. 11.   Observed waveforms during one switching-cycle at P_o = 410 [W] (100 V/div, 20 A/div, 400 ns/div).

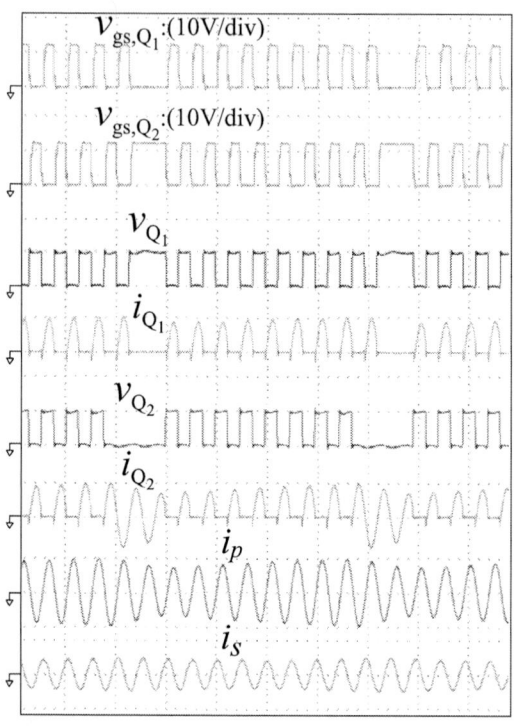

Fig. 13.   Observed voltage and current waveforms at D_p = 0.8 with proposed PDM control (100 V/div, 20 A/div, 4 μs/div).

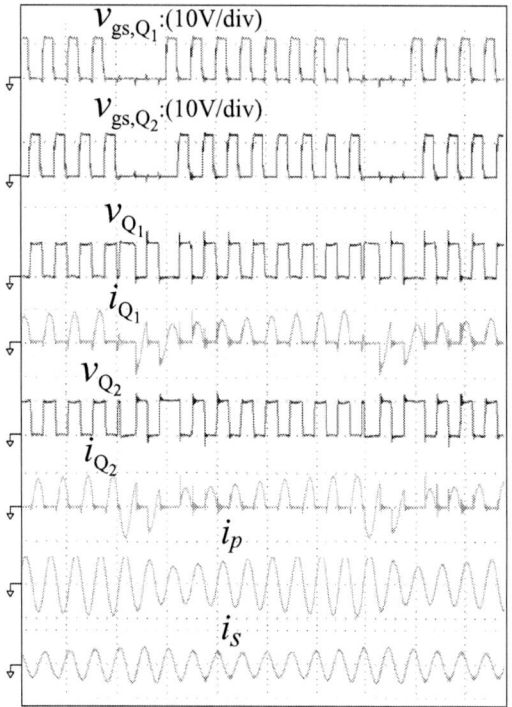

Fig. 12.   Observed voltage and current waveforms at D_p = 0.8 with conventional PDM control (50 V/div, 10 A/div, 4 μs/div).

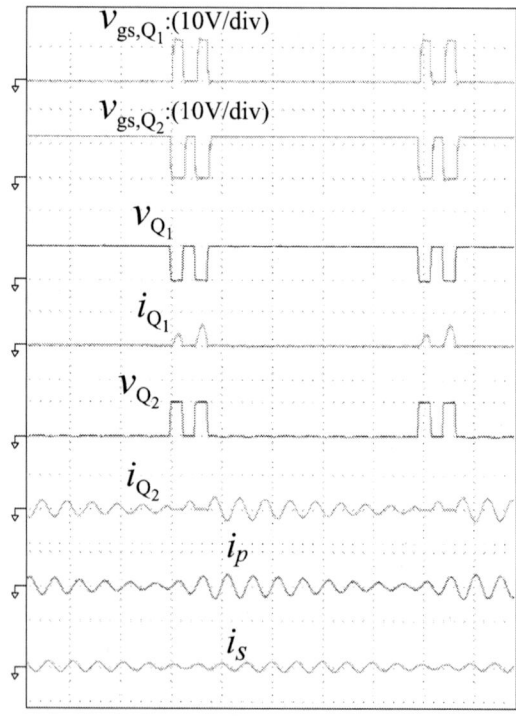

Fig. 14.   Observed voltage and current waveforms at D_p = 0.2 with proposed PDM control (100 V/div, 20 A/div, 4 μs/div).

The 2018 International Power Electronics Conference

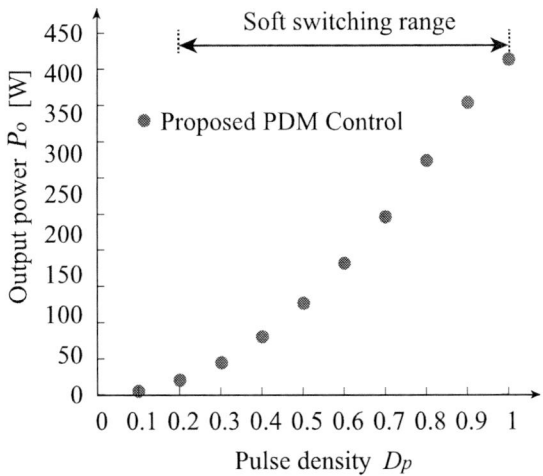

Fig. 15.   Characteristics of output power vs. pulse density.

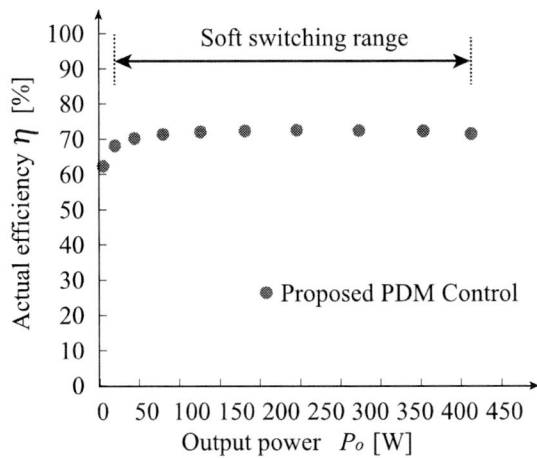

Fig. 16.   Actual efficiency curves of DC-DC power conversion.

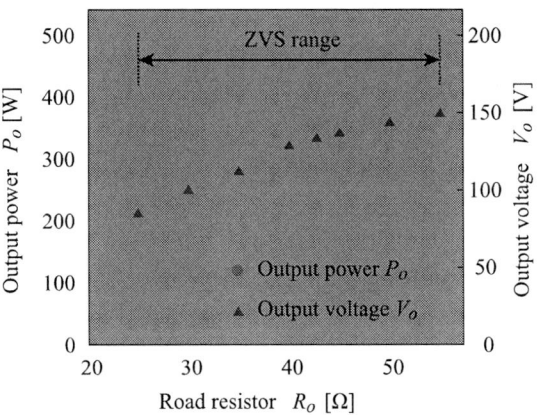

Fig. 17.   Characteristic the load variation under same gap.

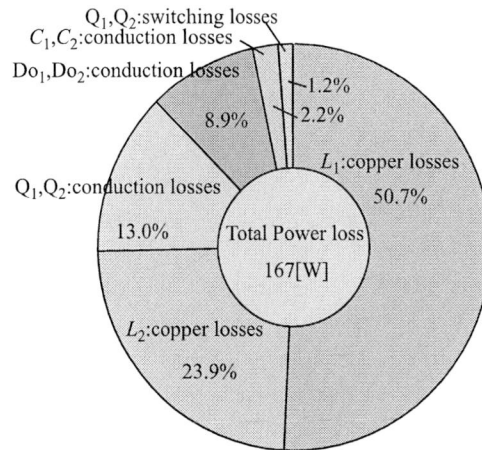

Fig. 18.   Experimental power loss analysis of IPT ZVS-prototype.

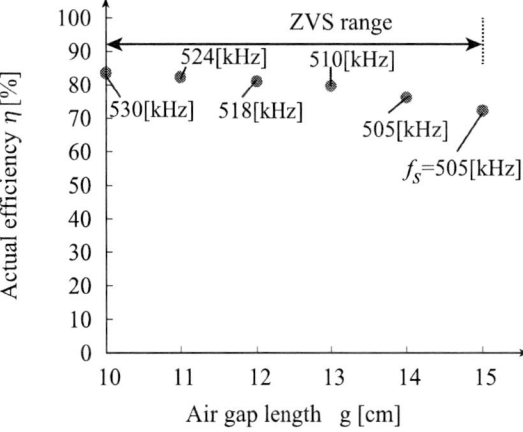

Fig. 19.   Actual efficiency versus the air gap length.

reverse conduction intervals while maintaining the edge-resonant ZVS operation as mentioned above.

The actual efficiency characteristics are compared in Figs.20 and 21 for the conventional and proposed burst mode PDM schemes. The higher power and higher efficiency can be observed over the whole range under the same duty cycle of D_p by the proposed PDM. The efficiency is improved with the proposed PDM, especially 29.4 % of efficiency escalation can be observed as maximum at the pulse $D_p = 0.5$. Thus, the practical effectiveness of the proposed PDM control is actually verified herein.

V.   CONCLUSION

A burst mode PDM controlled ZVS class-D resonant converter for IPT systems has been proposed, which is featured by resonant frequency tracking for high power factor and wide range of soft switching. The validity of the proposed converter has been demonstrated by experiment, and the advantageous technologies over the conventional PDM scheme has been clarified. The power

demonstrated in Fig. 9. This graph shows the effectiveness of the frequency tracking that can suppress the MOSFET

The 2018 International Power Electronics Conference

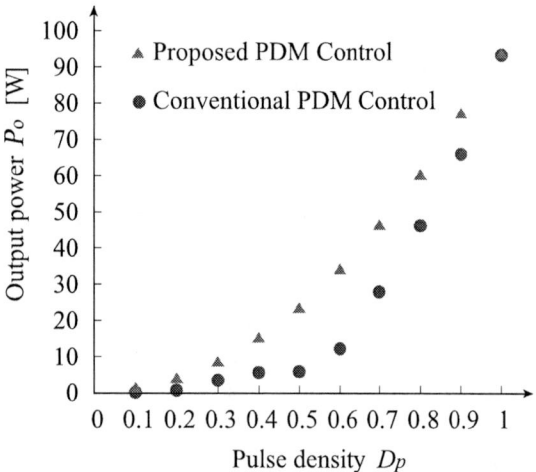

Fig. 20. Characteristics of output power vs. pulse density.

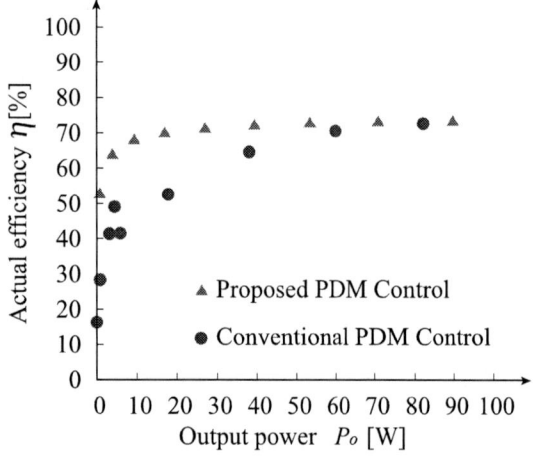

Fig. 21. Actual efficiency curves of DC-DC power conversion.

conversion efficiency of dc-dc power stage is recorded as 72.4 % at $P_o = 220\,\mathrm{W}$, and almost 30 % of efficiency improvement attains by means of the proposed modulation scheme. The validity of resonant frequency tracking has been confirmed by open loop controller, and the adaptability for misalignment of sending and receiving coils are clarified.

REFERENCES

[1] G.A. Covic, and J.T. Boys, "Inductive Power Transfer," *Proc. The IEEE*, vol.101. no.6, pp.1276-1289, Jun. 2013.

[2] S.Y.R. Hui, W. Zhong, and C.K. Lee, "A critical review of recent progress in mid-range wireless power transfer," *IEEE Trans. Power Electron.*, vol.29, No.9, pp.4500–4511, Sep. 2014.

[3] U.K. Madawala, and D.J. Thrimawithana, "A bidirectional inductive power interface for electric vehicles in V2G systems," *IEEE Trans. Ind Electron.*, vol.58, no,10, pp.4789-4796, Oct. 2011.

[4] G. Buja, M. Bertoluzzo, and K.N. Mude, "Design and experimentation of WPT charger for electric city car," *IEEE Trans. Ind Electron.*, vol.62, no.12, pp.7436-7447, Dec. 2015.

[5] W. Li, H. Zhao, S. Li, J. Deng, T. Kan, and C.C. Mi, "Integrated *LCC* compensation topology for wireless charger in electric and plug-in electric vehicles," *IEEE Trans. Ind Electron.*, vol.62, no,7, pp.4215-4225, Jul. 2015.

[6] K. Yan, Q. Chen, J. Hou, X. Ren, and X. Ruan, "Self-oscillating contactless resonant converter with phase detection contactless current transformer," *IEEE Trans. Power Electron.*, vol.29, No.8, pp.4438–4449, Aug. 2014.

[7] R. Haldi, and K. Schenk, "A 3.5kW wireless charger for electric vehicles with ultra high efficiency." *Proc. 2014-ECCE*, pp.668-674, Sep. 2014.

[8] B. Esteban, M. Sid-Ahmed, and N.C. Kar, "A comparative study of power supply architectures in wireless EV charging systems," *IEEE Trans. Power Electron.*, vol.30, No.11, pp.6408–6422, Sep. 2015.

[9] T. Mishima and E. Morita, "High-frequency bridgeless rectifier-based ZVS multi-resonant converter for inductive power transfer featuring high-voltage Gan-HFET", *IEEE Trans. Ind. Electron.*, pp.9155-9164, Nov. 2017.

[10] H. Y. Leung, D. McCormick, D. M. Budgett, and A. P. Hu, "Pulse density modulated control patterns for inductively powered implantable devices based on energy injection control", *IET Power Electronics*, vol. 6, issue. 6, pp. 1051-1057, 2013.

[11] H. Li, J. Fang, S. Chen, K. Wang, and Y. Tang, "Pulse density modulation for maximum efficiency point tracking of wireless power transfer systems," *IEEE Trans. Power Electron.*, vol.33, no.4, pp.3595-3603, Arp. 2018.

[12] J. Tian and A.P. Hu, "A dc-voltage-controlled variable capacitor for stabilizing the ZVS frequency of a resonant converter for wireless power transfer," *IEEE Trans. Power Electron.*, vol.32, no.3, pp.2312-2318, Mar. 2017.

The 2018 International Power Electronics Conference

Reduced-Order Dynamical Models of Tuned Wireless Power Transfer Systems

Hongchang Li[1*], Jingyang Fang[2], and Yi Tang[2]

1 Energy Research Institute, Nanyang Technological University, Singapore

2 School of Electrical and Electronic Engineering, Nanyang Technological University, Singapore

*E-mail: hongchangli@ntu.edu.sg

Abstract—Dynamical models are of primary interest when studying the dynamical control of wireless power transfer (WPT) systems. Most existing dynamical models of WPT systems are derived using the generalized state space averaging method, extended describing functions, or the concept of coupled modes. These models are applicable to both tuned and detuned WPT systems but suffer from high orders and complex formulae. This paper focuses on tuned WPT systems and proposes the reduced-order dynamical models, which are derived from the energy point of view. The proposed models have much lower orders and simpler formulae as compared to the existing models. Experimental results are presented for verification.

Keywords— Dynamical models, reduced-order, tuned, wireless power transfer (WPT).

I. INTRODUCTION

Wireless power transfer (WPT) technologies have been considerably improved over the past decades. Remarkable research work can be found in the hardware designs [1], the steady-state control strategies [2], and the extensions of application areas [3]. However, the study on the dynamical modeling is still at the early stage.

Most existing dynamical models of WPT systems are derived using the generalized state space averaging method and extended describing functions [4-9], which are originally used to model resonant converters [10, 11]. The major disadvantage of these modeling methods is that the order of the derived model is increased as compared with the number of energy storage elements. For example, the WPT system in [8] has 5 energy storage elements (4 resonant elements and 1 filter capacitor), while, the order of the derived model is 9.

To overcome the order increase problem, the dynamical modeling method based on the concept of coupled modes was proposed in [12]. By using this method, the model's order equals the number of energy storage elements. However, the formulae of the model are still too complicated because of the intrinsic complexity of WPT systems.

To further simply the dynamical models, this paper focuses on tuned WPT systems because their dynamical behaviors may be much simpler than those of detuned WPT systems. Moreover, tuned WPT systems as the simplest ideal systems have been well investigated from

different aspects, and their dynamical models will also have special theoretical values.

II. TUNED WPT SYSTEM

Fig. 1 shows the schematic of a dual-side pulse density modulated WPT system [13]. The system consists of a dc voltage source, a half-bridge inverter with switches S_1 and S_2, a transmitter resonator with inductance L_1, capacitance C_1, equivalent series resistance (ESR) R_1, and resonant frequency ω_1, a receiver resonator with inductance L_2, capacitance C_2, ESR R_2, and resonant frequency ω_2, a synchronous half-bridge rectifier with switches S_3 and S_4, a filter capacitor C_f, and a load resistor R_L.

The inverter converts the dc input voltage v_{in} into pulses u_1 to inject energy into the transmitter resonator. The transmitter resonant current i_{L1} induces the receiver resonant current i_{L2} through the mutual inductance M. The rectifier converts the dc output voltage v_o into pulses u_2 to absorb energy from the receiver resonator. The absorbed energy is filtered by C_f and consumed by R_L. The pulse density of u_1 is modulated to d_1 based on the reference pulses of frequency ω_s generated by the transmitter controller. The pulse density of u_2 is modulated to d_2 based on the pulses synchronized with i_{L2}.

Fig. 2 shows the ac equivalent circuit of the system, where U_1, U_2, I_{L1}, and I_{L2} are the slow-varying dynamic phasors of u_1, u_2, i_{L1} and i_{L2}, respectively, P_1, P, and P_2 are the injected, transferred, and absorbed average power during a switching cycle $2\pi/\omega_s$, respectively.

The system is said to be tuned when u_2 is synchronized with i_{L2}, and the frequency of the reference pulses equals the resonant frequencies, i.e.

$$\omega_s = \omega_1 = \omega_2. \tag{1}$$

The system is controlled by the pulse densities d_1 and d_2 as described in [13]. The power flows are given by

$$\begin{cases} P_1 = \dfrac{\sqrt{2}}{\pi} d_1 v_{in} I_{L1} \\[2mm] P = \omega_s M I_{L1} I_{L2} \\[2mm] P_2 = \dfrac{\sqrt{2}}{\pi} d_2 v_o I_{L2} \operatorname{sgn} I_{L2} \end{cases} \tag{2}$$

337

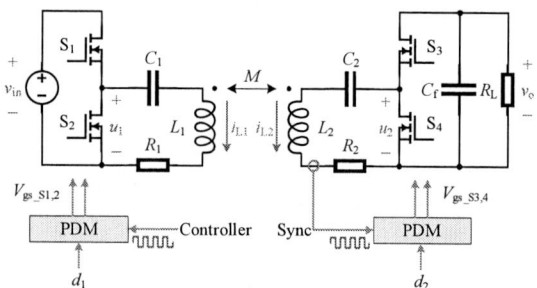

Fig. 1. Schematic of a tuned PDM WPT system.

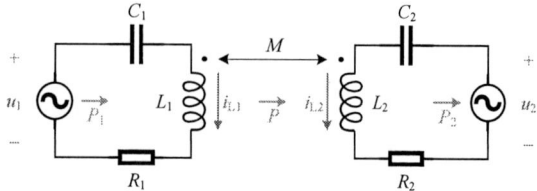

Fig. 2. The ac equivalent circuit.

where I_{L1} and I_{L2} are positive or negative real numbers because the phases of i_{L1} and i_{L2} are both binary when the system is tuned. More specifically, the phase of i_{L1} relative to u_1 is either 0° or 180°, the phase of i_{L2} relative to i_{L1} is either 90° or −90°, other phases will not occur.

III. Large Signal Model

The large signal model of the system can be intuitively derived from the energy point of view.

The average energies stored in the two resonators in a switching cycle $2\pi/\omega_s$ can be denoted by $W_1(t)$ and $W_2(t)$ and approximately expressed by the amplitudes of the two resonant currents:

$$\begin{cases} W_1(t) = L_1 I_{L1}{}^2(t) \\ W_2(t) = L_2 I_{L2}{}^2(t) \end{cases} \tag{3}$$

if

$$\begin{cases} \dfrac{2\pi}{\omega_s}\left| P_1(t) - P(t) \right| << W_1(t) \\ \dfrac{2\pi}{\omega_s}\left| P(t) - P_2(t) \right| << W_2(t) \end{cases} \tag{4}$$

The derivatives of $W_1(t)$ and $W_2(t)$ equal the differences between the average input and output powers minus the resonator ESR consumptions, namely

$$\begin{cases} \dfrac{dW_1(t)}{dt} = P_1(t) - P(t) - R_1 I_{L1}{}^2(t) \\ \dfrac{dW_2(t)}{dt} = P(t) - P_2(t) - R_2 I_{L2}{}^2(t) \end{cases} \tag{5}$$

In addition, the dynamics of C_f can be described by

$$C_f \frac{dv_o(t)}{dt} = \frac{\sqrt{2}}{\pi} d_2(t) I_{L2}(t)\, \mathrm{sgn}\, I_{L2}(t) - \frac{v_o(t)}{R_L}. \tag{6}$$

Substituting (2) and (3) into (5) and combining with (6), the large signal model of the system yields:

$$\begin{cases} \dfrac{dI_{L1}(t)}{dt} = -\dfrac{R_1}{2L_1} I_{L1}(t) - \dfrac{\omega_s M}{2L_1} I_{L2}(t) \\ \qquad\quad + \dfrac{\sqrt{2}}{2\pi L_1} d_1(t) v_{in}(t) \\[4pt] \dfrac{dI_{L2}(t)}{dt} = \dfrac{\omega_s M}{2L_2} I_{L1}(t) - \dfrac{R_2}{2L_2} I_{L2}(t) \\ \qquad\quad - \dfrac{\sqrt{2}}{2\pi L_2} d_2(t) v_o(t)\, \mathrm{sgn}\, I_{L2}(t) \\[4pt] \dfrac{dv_o(t)}{dt} = \dfrac{\sqrt{2}}{\pi C_f} d_2(t) I_{L2}(t)\, \mathrm{sgn}\, I_{L2}(t) - \dfrac{v_o(t)}{R_L C_f} \end{cases} \tag{7}$$

This is a 3rd order nonlinear model. The nonlinearity is caused by the sign functions.

IV. Equilibrium Point

The unique equilibrium point of the large signal model is derived by setting the derivatives in (7) to zeros and expressed as

$$\begin{cases} I_{L1} = \dfrac{\sqrt{2} d_1 v_{in}}{\pi(R_1 + R_r)} \\[6pt] I_{L2} = \dfrac{\omega_s M}{R_2 + R_e} I_{L1} . \\[6pt] v_o = \dfrac{\sqrt{2}}{\pi} d_2 R_L I_{L2} \end{cases} \tag{8}$$

where I_{L1} and I_{L2} are both positive, R_e and R_r are the equivalent load resistance and the reflected resistance, respectively:

$$\begin{cases} R_e = \dfrac{2}{\pi^2} d_2{}^2 R_L \\[6pt] R_r = \dfrac{(\omega_s M)^2}{R_2 + R_e} \end{cases} \tag{9}$$

The equilibrium point given by (8) equals the steady-state operating point derived from the fundamental harmonic analysis.

V. Small Signal Model

The large signal model (7) can be linearized at the equilibrium point and gives the small signal model:

$$\frac{d}{dt}\begin{bmatrix} \hat{I}_{L1}(t) \\ \hat{I}_{L2}(t) \\ \hat{v}_o(t) \end{bmatrix} = A \begin{bmatrix} \hat{I}_{L1}(t) \\ \hat{I}_{L2}(t) \\ \hat{v}_o(t) \end{bmatrix} + B \begin{bmatrix} \hat{v}_{in}(t) \\ \hat{d}_1(t) \\ \hat{d}_2(t) \end{bmatrix} \tag{10}$$

where $\hat{I}_{L1}(t)$, $\hat{I}_{L2}(t)$, and $\hat{v}_o(t)$ are the small signals of $I_{L1}(t)$, $I_{L2}(t)$, and $v_o(t)$, respectively

$$
A = \begin{bmatrix} -\dfrac{R_1}{2L_1} & -\dfrac{\omega_s M}{2L_1} & 0 \\[2mm] \dfrac{\omega_s M}{2L_2} & -\dfrac{R_2}{2L_2} & -\dfrac{\sqrt{2}d_2}{2\pi L_2} \\[2mm] 0 & \dfrac{\sqrt{2}d_2}{\pi C_f} & -\dfrac{1}{R_L C_f} \end{bmatrix} \tag{11}
$$

and

$$
B = \begin{bmatrix} \dfrac{\sqrt{2}d_1}{2\pi L_1} & \dfrac{\sqrt{2}v_{in}}{2\pi L_1} & 0 \\[2mm] 0 & 0 & -\dfrac{\sqrt{2}v_o}{2\pi L_2} \\[2mm] 0 & 0 & \dfrac{\sqrt{2}I_{L2}}{\pi C_f} \end{bmatrix}. \tag{12}
$$

The transfer functions from $\hat{v}_{in}(t)$, $\hat{d}_1(t)$, and $\hat{d}_2(t)$ to $\hat{I}_{L1}(t)$, $\hat{I}_{L2}(t)$, and $\hat{v}_o(t)$ are derived from (10):

$$
\begin{bmatrix} \dfrac{\hat{I}_{L1}(s)}{\hat{v}_{in}(s)} & \dfrac{\hat{I}_{L1}(s)}{\hat{d}_1(s)} & \dfrac{\hat{I}_{L1}(s)}{\hat{d}_2(s)} \\[2mm] \dfrac{\hat{I}_{L2}(s)}{\hat{v}_{in}(s)} & \dfrac{\hat{I}_{L2}(s)}{\hat{d}_1(s)} & \dfrac{\hat{I}_{L2}(s)}{\hat{d}_2(s)} \\[2mm] \dfrac{\hat{v}_o(s)}{\hat{v}_{in}(s)} & \dfrac{\hat{v}_o(s)}{\hat{d}_1(s)} & \dfrac{\hat{v}_o(s)}{\hat{d}_2(s)} \end{bmatrix} = (sI - A)^{-1} B \tag{13}
$$

where I is a 3×3 identity matrix. In (13), the transfer functions from $\hat{d}_1(t)$ and $\hat{d}_2(t)$ to $\hat{v}_o(t)$ are the control-to-output transfer functions.

VI. EXPERIMENT

The experimental system was as described in [13] but had two differences: 1) the power transfer distance was changed to 0.45 m, as shown in Fig. 3, and 2) the system was tuned by finely adjusting the resonant frequencies to ω_s and setting the phase difference between u_2 and i_{L2} to 180°. These changes sacrificed the soft switching feature, and therefore, the operating power of the system in this paper must be reduced for safety. The parameters and the equilibrium point of the system are listed in Table I and Table II, respectively.

Fig. 4 shows the measured operating waveforms at the equilibrium point. The phase differences between u_1 and i_{L1}, u_2 and i_{L2}, and i_{L1} and i_{L2} were about 0°, 180°, and 90°, respectively, indicating that the system was well tuned. Spikes can be seen on the voltage waveforms because of the hard switching.

Fig. 5 shows the resonant current and output voltage responses when stepping the pulse densities d_1 and d_2 between 0.5 and 1. The experimental results coincided with the predictions given by (7), and therefore, the large signal model was verified.

TABLE I
PARAMETERS OF THE EXPERIMENTAL SYSTEM

Symbol	Quantity	Value
V_{in}	Input voltage	9.9 V
ω_s	Switching frequency	5.71 Mrad/s
L_1	Transmitter resonant inductance	76.6 μH
C_1	Transmitter resonant capacitance	400 pF
R_1	Transmitter ESR	1.1 Ω
M	Mutual inductance	0.96 μH
L_2	Receiver resonant inductance	76.6 μH
C_2	Receiver resonant capacitance	400 pF
R_2	Receiver ESR	0.9 Ω
C_f	Filter capacitance	1 μF
R_L	Load resistance	92 Ω

TABLE II
EQUILIBRIUM POINT

Symbol	Quantity	Value
d_1	Pulse density of inverter	0.5
d_2	Pulse density of rectifier	0.5
I_{L1}	RMS of transmitter resonant current	0.34 A
I_{L2}	RMS of receiver resonant current	0.34 A
v_o	Output voltage	7 V

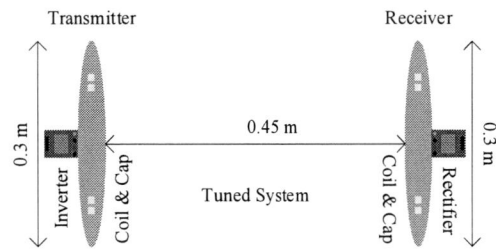

Fig. 3. Configuration of the experimental system.

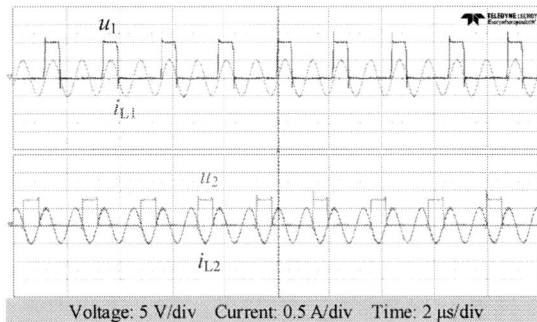

Voltage: 5 V/div   Current: 0.5 A/div   Time: 2 μs/div

Fig. 4. Operating waveforms at the equilibrium point.

Fig. 6 shows the resonant current and output voltage responses when injecting a 10 kHz low-magnitude (0.1) sine wave into d_1 and d_2 at the equilibrium point. The magnitudes and phases of the control-to-output transfer functions at 10 kHz can be calculated by taking the Fourier transformations of the recorded waveform data.

By injecting sine waves with various frequencies, the system control-to-output frequency characteristics were measured and the results are shown in Fig. 7, which coincided with the predictions given by the small signal transfer functions (13), and therefore, the small signal model was verified.

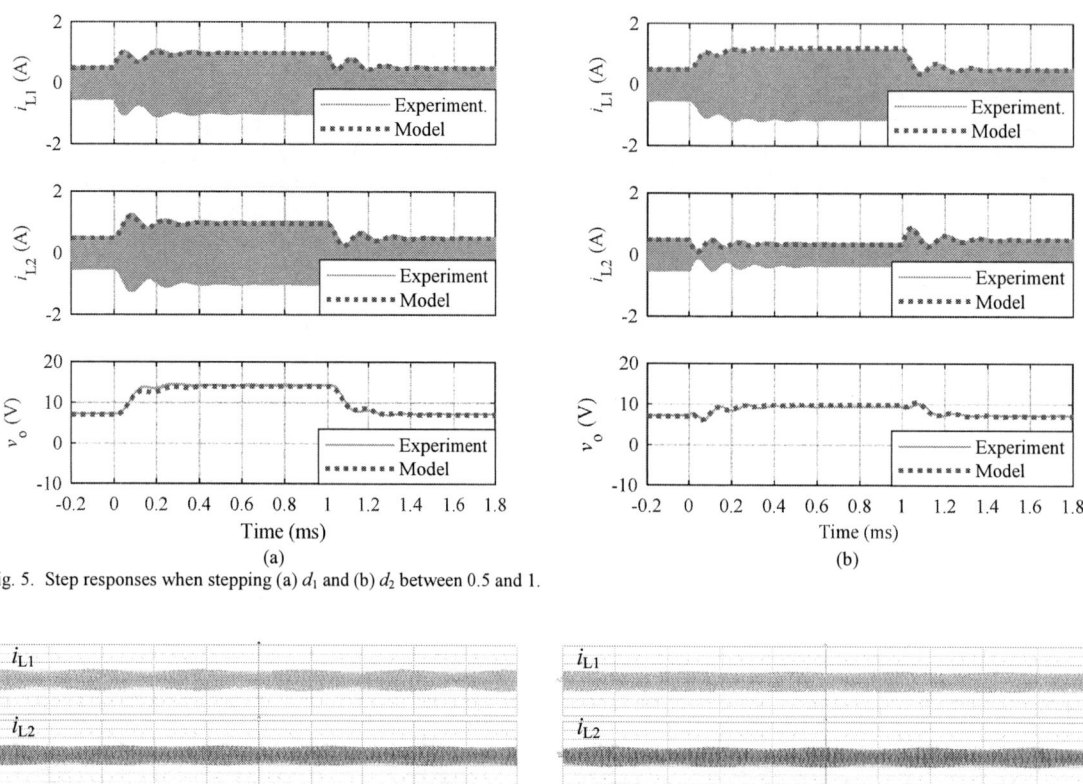

Fig. 5. Step responses when stepping (a) d_1 and (b) d_2 between 0.5 and 1.

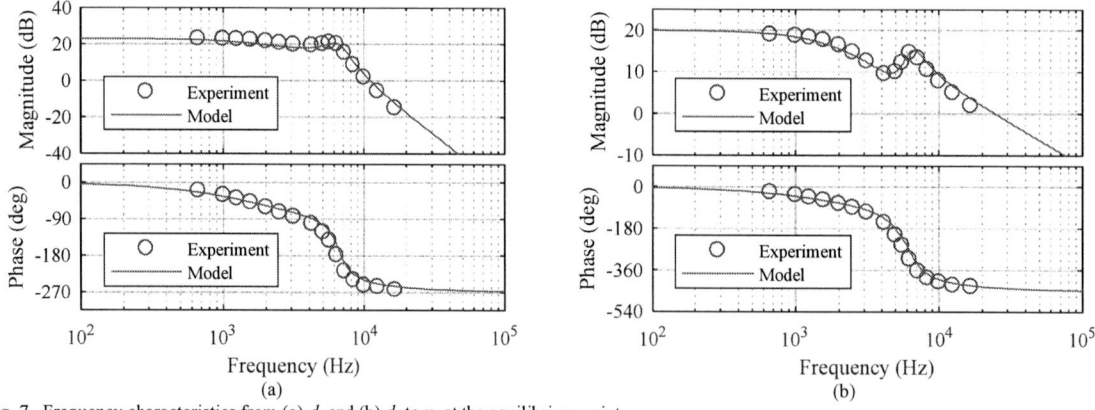

Fig. 6. Frequency responses when injecting a 10 kHz sine wave into (a) d_1 and (b) d_2 at the equilibrium point.

Fig. 7. Frequency characteristics from (a) d_1 and (b) d_2 to v_o at the equilibrium point.

VII. CONCLUSION

The dynamical behaviors of tuned WPT systems are much simpler than those of detuned WPT systems, and can be described by the reduced-order dynamical models. This paper modeled a tuned PDM WPT system as an example. The orders of the derived large and small signal models are both 3, which is smaller than the number of energy storage elements.

The limitation of the proposed models is that they are not applicable to detuned WPT systems. However, the authors believe that practical tuned WPT systems with soft switching features will come soon.

REFERENCES

[1] S. Kim, G. A. Covic, and J. T. Boys, "Tripolar Pad for Inductive Power Transfer Systems for EV Charging," *IEEE Transactions on Power Electronics*, vol. 32, no. 7, pp. 5045-5057, 2017.

[2] H. Li, J. Li, K. Wang, W. Chen, and X. Yang, "A Maximum Efficiency Point Tracking Control Scheme for Wireless Power Transfer Systems Using Magnetic Resonant Coupling," *IEEE Transactions on Power Electronics*, vol. 30, no. 7, pp. 3998-4008, Jul 2015.

[3] C. C. Mi, G. Buja, S. Y. Choi, and C. T. Rim, "Modern Advances in Wireless Power Transfer Systems for Roadway Powered Electric Vehicles," *IEEE Transactions on Industrial Electronics*, vol. 63, no. 10, pp. 6533-6545, 2016.

[4] A. P. Hu, "Modeling a contactless power supply using GSSA method," in *IEEE International Conference on Industrial Technology*, 2009, pp. 1-6.

[5] H. Hao, G. A. Covic, and J. T. Boys, "An Approximate Dynamic Model of LCL-T-Based Inductive Power Transfer Power Supplies," *IEEE Transactions on Power Electronics*, vol. 29, no. 10, pp. 5554-5567, Oct 2014.

[6] Z. Huang, S. C. Wong, and C. K. Tse, "Control Design for Optimizing Efficiency in Inductive Power Transfer Systems," *IEEE Transactions on Power Electronics*, vol. PP, no. 99, pp. 1-1, 2017.

[7] Z. U. Zahid, Z. Dalala, and J. S. J. Lai, "Small-signal modeling of series-series compensated induction power transfer system," in

[8] Z. U. Zahid et al., "Modeling and Control of Series-Series Compensated Inductive Power Transfer System," *IEEE Journal of Emerging and Selected Topics in Power Electronics*, vol. 3, no. 1, pp. 111-123, Mar 2015.

[9] H. Feng, T. Cai, S. Duan, X. Zhang, and H. Hu, "Modeling and analysis of phase-shift controlled LCL resonant converter in wireless charging systems," in *IEEE Applied Power Electronics Conference and Exposition*, 2017, pp. 3714-3719.

[10] S. R. Sanders, J. M. Noworolski, X. Z. Liu, and G. C. Verghese, "Generalized averaging method for power conversion circuits," *IEEE Transactions on Power Electronics*, vol. 6, no. 2, pp. 251-259, 1991.

[11] E. X. Yang, F. C. Lee, and M. M. Jovanovic, "Small-signal modeling of series and parallel resonant converters," in *IEEE Applied Power Electronics Conference and Exposition*, 1992, pp. 785-792.

[12] H. Li, K. Wang, L. Huang, W. Chen, and X. Yang, "Dynamic Modeling Based on Coupled Modes for Wireless Power Transfer Systems," *IEEE Transactions on Power Electronics*, vol. 30, no. 11, pp. 6245-6253, Nov 2015.

[13] H. Li, J. Fang, S. Chen, K. Wang, and Y. Tang, "Pulse Density Modulation for Maximum Efficiency Point Tracking of Wireless Power Transfer Systems," *IEEE Transactions on Power Electronics*, to be published, 2017.

Dynamic Modelling and Closed Loop Control of Transmitter Parallel and Receiver Series Compensated IPT Topology for EV Applications

Suvendu Samanta, *Student Member, IEEE*
Concordia University
Montreal, QC, Canada H3G 1M8
suvendu.cil@gmail.com

Akshay Kumar Rathore, *Senior Member, IEEE*
Concordia University
Montreal, QC, Canada H3G 1M8
arathore@encs.concordia.ca

Abstract – **Usually, parallel compensated IPT topologies are controlled either by frequency modulation or by dynamically varying tank capacitance. The dynamic load demand is usually met by an additional dc-dc chopper at load side. This paper presents dynamic modelling and closed loop control of a IPT topology with parallel compensation in transmitter side and series compensation in receiver side. Complete control is carried out though two-loop method, where the inner input current loop controls source current, and outer output current loop controls load current. Compare with conventional parallel compensated topology controlled by an additional dc-dc chopper for meeting load requirements, the proposed system achieves it directly by IPT inverter. The proposed dynamic analysis and closed loop control are validated by simulation results obtained from PowerSIM and experimental results performed in an 800W scale-down lab porotype.**

Keywords: **Wireless power transfer, Inductive power transfer, Dynamic modelling, and Closed loop etc.**

I. INTRODUCTION

Wireless inductive power transfer (IPT) is feasible due to wide availability of higher voltage and current rated semiconductor devices at high switching frequency. IPT technology is getting wide acceptance because it is very convenient, safe, and provides galvanic isolations. Power transfer is unaffected in snow, water, dirt, and hazardous chemical environments [1] [2] [3]. IPT finds applications in EVs, electronic gadgets, lighting, and biomedical implants etc. [1] [4] [5] [6] [7] [8].

Owing to very weak coupling between the IPT coils, compensation in both primary and secondary sides of IPT coil are required. A series compensated primary has load independent resonance, where parallel compensated is reported to be load dependent [9]. The parallel compensated topology requires a current-source inverter (CSI) to be compatible, whereas series, or *LCL*, or *LCCL* topologies require voltage source inverter (VSI). Parallel compensated topology has several merits including high quality coil current, lower device current rating and zero voltage switching [10], [11]. Unlike series compensated IPT topology, parallel compensated IPT topology at primary side does not face instability issues in the

absence of load or secondary circuitry [19]. Although, CSI requires an additional inductor at dc input, but it is quite small because of high switching frequency. *LCL* or *LCCL* tanks networks with VSI topology also require additional inductor at the ac side. Effective wireless power transfer is highly sensitive to the variation of tank inductor of *LCL* or *LCCL* tanks. However, the parameter variation of dc inductor in CSI topology does not affect effective wireless power transfer, because it is not a part of the tank network. Although, the dc inductor at the input of CSI has higher inductance than the ac inductor of *LCL* or *LCCL* tanks, but the peak current rating of CSI dc inductor is comparatively much lesser.

However, due to load dependent resonance of parallel tuned IPT topology, converter control is usually carried out either by dynamically varying inverter switching frequency or by dynamically tuning circuit parameter values to reach exact tuning [9], [12]. Owing to this control constraint, dynamic load demand is fulfilled by additional dc-dc chopper at load side. Recent research shows that these topologies are load independent if designed properly [13]. This paper follows this load independent design technique for transmitter-parallel and receiver-series compensated IPT topology. This eliminates the requirement for dynamic tuning of tank circuit due to load change. This paper proposes dynamic model and closed loop control of this topology. The complete control has two loops, where inner input current loop controls source current, and outer output current loop controls load current. Owing to load independent tuning of resonant tank, and direct control of load power requirements by CSI, additional chopper stage at load side is eliminated.

II. PARALLEL/ SERIES IPT TOPOLOGY

Fig. 1a shows a parallel compensated primary and series compensated secondary IPT topology, where the inverter is full bridge current-fed type and rectifier is voltage doubler. The series tank at secondary ensures least number of components onboard of EV. The inverter provides a quasi-square shaped current to the primary tank, whereas tank input voltage profile is sinusoidal. The output current of parallel/series tank, i_2 is rectified with voltage doubler circuit and fed to load. To achieve direct control of load through duty cycle modulation of inverter, unipolar pulse width modulation scheme is selected, i.e., S_1' and S_2' are complementary of S_1 and S_2 respectively. S_1 and S_2 receives same duty cycle with 180^0 phase displacement. Detail of steady-state operation and operating waveforms are not included and it can be found in [10], [14]. However, one difference in steady-state operation of this paper is that the CSI modulation is unipolar type, whereas the existing works are reported with bipolar modulation technique.

To achieve load independent tuning with parallel/ series IPT topology, a transformer equivalent circuit of coils is considered as shown in Fig. 1b, where the leakage and magnetizing inductances are related with coil turns ratio (n:1), self and mutual inductances as

$$L_{1k} = L_1 - nM, \quad L_m = nM, \quad L_{2k} = L_2 - M/n. \quad (1)$$

The load independent compensation circuit parameters are derived by impedance analysis of Fig. 1b circuit and the design expressions are given as [13]

$$C_p = 1/\omega_o^2 L_1 \quad (2)$$

$$C_2' = \left[\omega_o \left(X_{2k}' - \frac{X_m X_{1k}}{X_m + X_{1k}}\right)\right]^{-1} \quad (3)$$

where, ''' indicates secondary circuit parameters referred to primary, ω_o is inverter switching (resonance) frequency and 'X' shows impedance of corresponding circuit element e.g. $X_{2k}' = \omega_o L_{2k}'$.

III. Dynamic Modelling of the Converter

The converter is controlled through two-loop method, where the inner loop controls source current, and outer loop meets dynamic load demand. Similar two-loop control of non-resonant converters with CSI topology can be found in [15], [16]. Considering duty cycle of inverter device S_1 as d, the dynamic expression of input inductor current based on switching cycle average is given as

$$L_d \frac{di_d}{dt} = v_d - v_x. \quad (4)$$

Considering the range of duty cycle of inverter device S_1 as $0 \leq d \leq 0.5$, the instantaneous input voltage of the inverter in terms of output RMS voltage is derived considering i_i is quasi-square and v_i is sinusoidal. It is given as

$$v_x = \frac{2\sqrt{2}}{\pi} V_i \times \cos \alpha \times \sin \pi d. \quad (5)$$

Using this expression, the dynamic expression of the inner loop system is directly given as

$$L_d \frac{di_d}{dt} = v_d - \frac{4 A_t^2}{\pi^2 R_{eq}} \times i_d (1 - \cos 2\pi d). \quad (6)$$

where, $A_t = V_r/I_i$ = gain of the tank network at operating frequency. Introducing small perturbations ($\tilde{i}_d, \tilde{v}_d, \tilde{d}$) around an equilibrium point (I_d, V_d, D), and applying Laplace transformation, inner loop plant transfer function is derived as

$$G_i(s) = \frac{\tilde{i}_d(s)}{\tilde{d}(s)} = -\frac{\frac{8 A_t^2}{\pi R_{eq}} I_d \sin 2\pi D}{s L_d + \frac{4 A_t^2}{\pi^2 R_{eq}} (1 - \cos 2\pi D)} \quad (7)$$

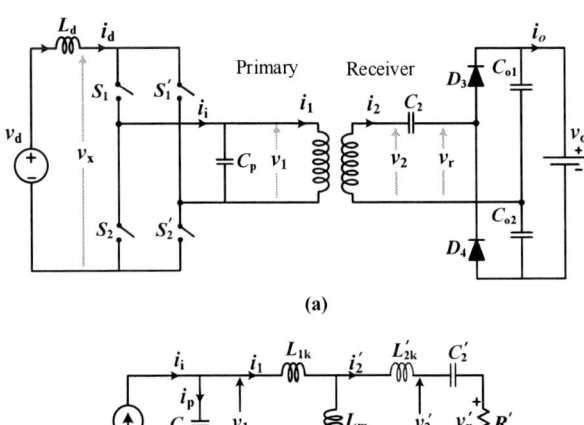

(a)

(b)

Fig. 1 (a) Parallel/ series tuned IPT topology with full-bride current-source inverter and voltage doubler rectifier (b) equivalent circuit

Table 1 Selected Circuit Parameters

Parameters	Values
Self-inductances of TC and RC	131.5 μH, 137.5 μH
Mutual inductance, M	30 μH
Compensation capacitors, C_p, C_2	77 nF, 81.2 nF
Switching frequency, f_s	50 kHz
Rated load power, P_o	800 W
Rated input voltage, v_d	200 V
Rated output voltage, v_o	160 V
Input inductor, L_d	1.4 mH
Output capacitors, C_{o1}, C_{o2}	15 μF, 15 μF

343

An integral controller with integral gain 25 is used to compensate the system. The bode plots of both compensated and uncompensated plants, and stability margins are shown in Fig. 2. The circuit parameters for these plots are listed in Table 1. The phase margin of compensated system is 88.6^0 and gain crossover frequency 1142rad/s. This indicates a closed loop settling time around 3.5ms.

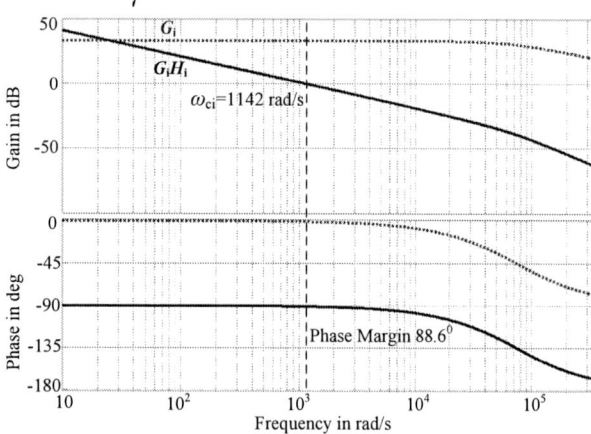

Fig. 2 Bode plot of inner input current loop

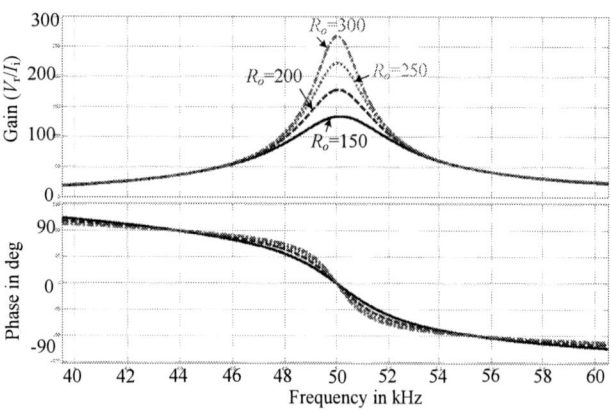

Fig. 3 Resonant tank gain for different load

The input to the outer loop is input inductor current (i_d), and the output is load current. The plant transfer function of outer loop system can be split as

$$G_o(s) = \frac{\tilde{i}_o(s)}{\tilde{i_d}(s)} = \frac{\tilde{\tilde{I}}_i(s)}{\tilde{i_d}(s)} \cdot \frac{\tilde{\tilde{V}}_r(s)}{\tilde{\tilde{I}}_i(s)} \cdot \frac{\tilde{i}_o(s)}{\tilde{V}_r(s)} = G_{inv} A_t G_{rec}. \quad (8)$$

where, all the ac side variables, i.e., $\hat{I}_i$, $\hat{V}_r$ indicate their peak values. The CSI transfer function is easily derived as

$$G_{inv} = \frac{\tilde{\tilde{I}}_i(s)}{\tilde{i_d}(s)} = \frac{4}{\pi} \sin \pi D \quad (9)$$

The resonant tank shown in Fig. 1b is a liner circuit. Therefore, applying KCL and KVL, the tank gain is easily derived as

$$A_t = \frac{\tilde{\tilde{V}}_r(s)}{\tilde{\tilde{I}}_i(s)} = \left[(1 + sC_p sL_{1k}) \left\{ \frac{1}{R'_{eq}} \right. \right.$$
$$+ \frac{1}{sM} \left(1 + \frac{sL'_{2k} + 1/sC'_2}{R'_{eq}} \right) \right\} \quad (10)$$
$$\left. + sC_p \left(1 + \frac{sL'_{2k} + 1/sC'_2}{R'_{eq}} \right) \right]^{-1},$$

The voltage doubler circuit transfer function is derived by applying power balance between input and output of rectifier. It is given as

$$G_{rec} = \frac{\tilde{i}_o(s)}{\tilde{V}_r(s)} = \frac{2}{\pi} \times \frac{1/R_{eq}}{sC_o r_b + 2}. \quad (11)$$

Therefore, overall outer loop transfer function is derived as

$$G_o(s) = \frac{\tilde{i}_o(s)}{\tilde{V}_r(s)} = \frac{8}{\pi^2} \frac{B}{R_{eq}} \sin \pi d \times \frac{1}{sC_o r_b + 2}. \quad (12)$$

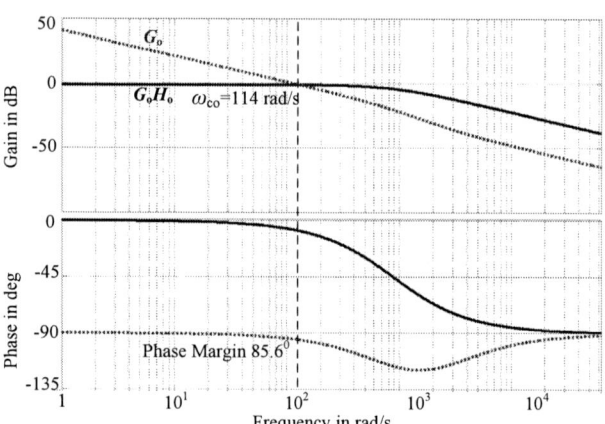

Fig. 4 Bode plot of outer loop

Fig. 3 shows resonant converter gain A_t for different load impedances. Fig. 4 shows bode plot of outer loop plant transfer function. A PI Controller, [0.05+130/s] is used to get desired performance of the closed loop system. The phase margin of compensated system is 85.20 and gain crossover frequency 114.3rad/s indicates a closed loop settling time 35ms. Compensated system bode plot is shown in Fig. 4.

The 2018 International Power Electronics Conference

IV. RESULTS

To verify the proposed dynamic modelling and closed loop control, simulation is performed in PowerSIM. Also, experimental results obtained from an 800W scale down lab-prototype is reported for farther verification. The experimental set-up diagram is shown in Fig. 5, where IPT coils are circular type and corresponding circuit parameters are listed in Table 1.

Fig. 6, Fig. 7, and Fig. 8 show steady state performance results of $(L)(C)$ transmitter and (LC) receiver compensated IPT topology. Fig. 6, Fig. 7, and Fig. 8 show results for 800W (100%), 400W (50%) and 160W (20%) power outputs, respectively. Clearly, the inverter output current is quasi-square. Zero state of this quasi-square wave appears due to boosting current through the input dc inductor, and it increases with higher power. This is exactly like a conventional boost-chopper. The inverter output voltage, v_i is very close to sinusoidal, even

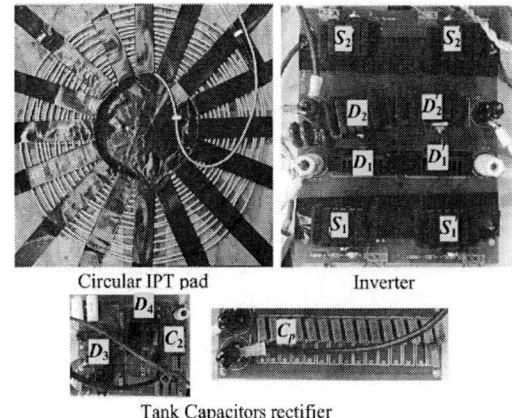

Circular IPT pad   Inverter

Tank Capacitors rectifier

Fig. 5 Experimental set-up

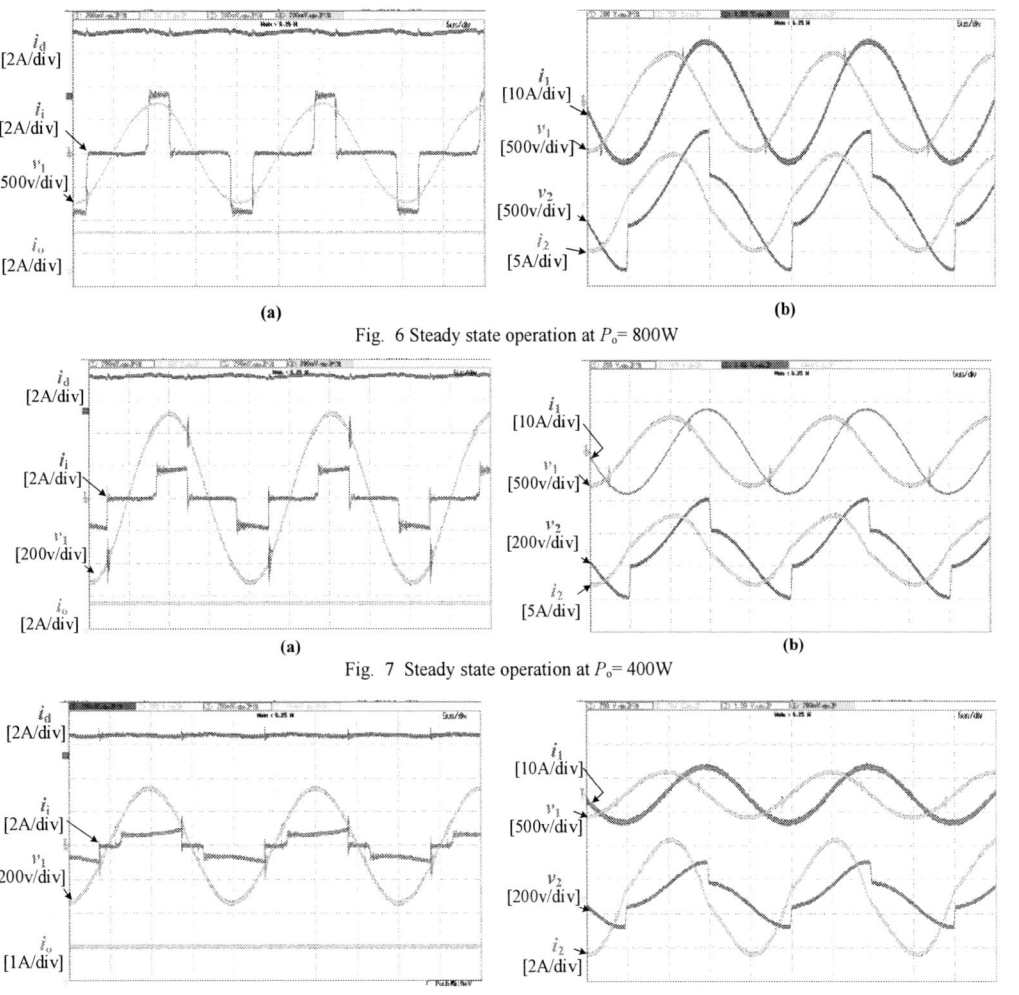

(a)   (b)

Fig. 6 Steady state operation at P_o= 800W

(a)   (b)

Fig. 7 Steady state operation at P_o= 400W

Fig. 8 Steady state operation at P_o= 160W

The 2018 International Power Electronics Conference

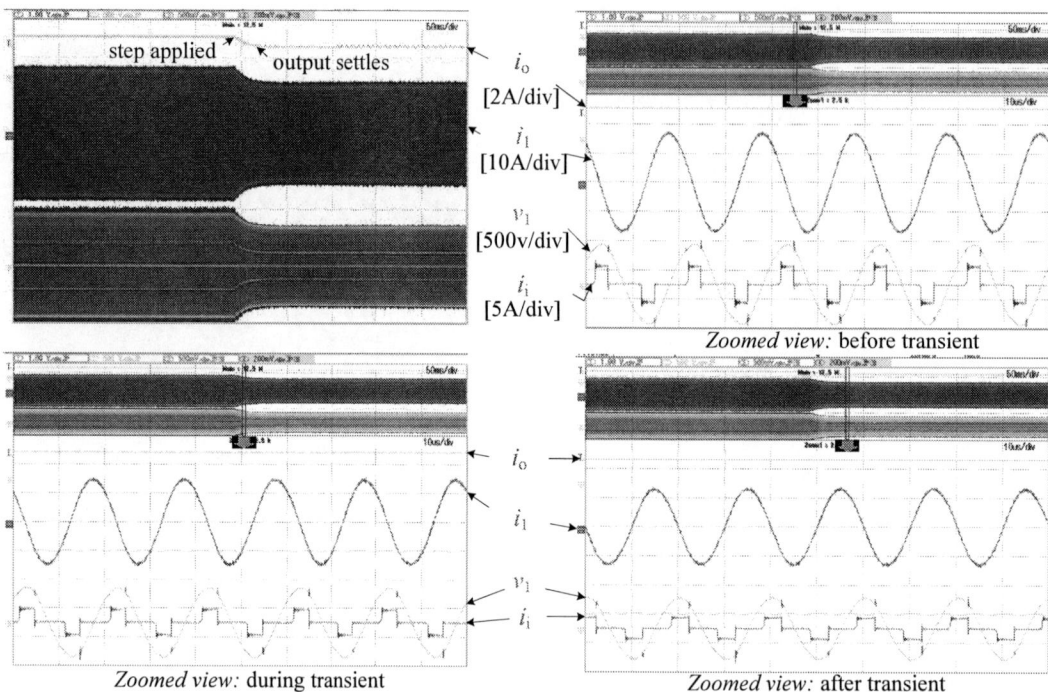

Fig. 9 Step response in presence of a step-down command applied to output current reference

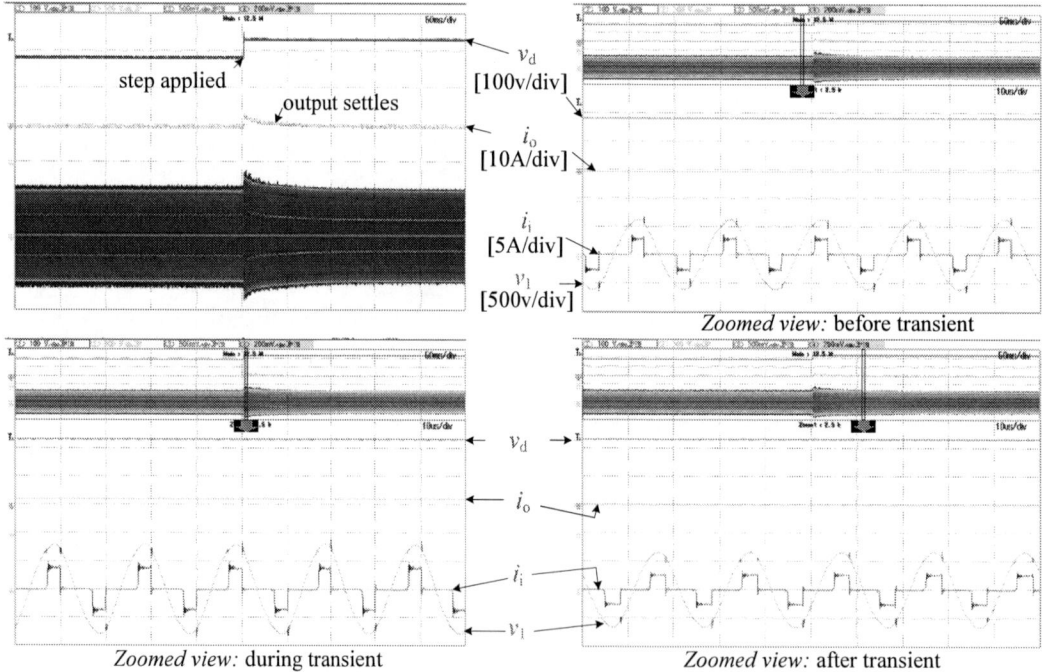

Fig. 10 Closed loop performance in presence of 25% input voltage disturbances (175 to 225V)

with wide load variations. Therefore, the assumption in derivation of plant transfer-function that only fundamental component of i_i in involved in active power transfer is validated. Also, the relative phase difference between i_i (fundamental) and v_i are zero, which verifies the effectiveness of load independent tank design. Therefore, the inverter need to supply the least amount of VA to tank network for given load power. This ensures least voltage and current ratings of inverter devices, thereby enabling lowest conduction loss of devices and compact size of converter.

Fig. 6b, Fig. 7b, and Fig. 8b show steady state results of TC and RC voltages and currents. Clearly, like earlier, the TC coil voltage is very close to sinusoidal due to presence of parallel capacitor, C_t. The sinusoidal profile of v_1 results in sinusoidal profiles of i_1, v_2 and i_2. This is very important in IPT systems to reduce unnecessary losses and electromagnetic interferences causing due to harmonics in coil currents.

Fig. 9 shows dynamic performance of the closed loop system, when a step-down command is applied to output current reference such that power reduces from 800W to 400W. The output reaches to desired new operating point with zero steady state error and settles in around 35ms after the step command. The zoomed views of this result before, during, and after the transient show fundamental component. Clearly, inverter output voltage is sinusoidal, and the current is quasi-square. This verifies the assumption that active power flow occurs predominantly due to fundamental component. Also, unity displacement power factor at inverter output with this wide load variations clearly shows the effectiveness of load independent resonant tank design. Primary coil current remains very close to sinusoidal throughout this operation as shown in Fig. 9. Fig. 10 shows performance of the closed loop system in presence of an input disturbance. This is done by abruptly increasing input voltage of the converter by 25%, i.e., from 175V to 225V. Immediately, a slight increase in output current observed due to more power injection from input. However, the output current is restored to its original value within the designed settling time of outer loop. This verifies the performance of the closed loop system. The zoomed views of this result validate the effectiveness of load independent design of converter.

V. CONCLUSIONS

A detailed dynamic modelling and closed loop control of a parallel primary and series secondary compensated IPT topology is reported. Overall control is carried out with two loop method, where inner loop controls source current and outer loop meets dynamic load demand. Compare with conventional parallel compensated IPT topology, where load demand is fulfilled with additional chopper at output side, this system directly fulfills load requirements by IPT inverter. Therefore, number of power conversion stages are reduced. Selected load independent tuning is effective and eliminates the dynamic tuning circuitry of parallel compensated topology. Experimental

results obtained from an 800W lab-porotype verifies the modelling, and closed loop control of converter.

REFERENCES

[1] C. C. Mi, G. Buja, S. Y. Choi and C. T. Rim, "Modern advances in wireless power transfer systems for roadway powered electric vehicles," *IEEE Trans. Ind. Electron.*, vol. 63, no. 10, p. 6533–6545, Oct. 2016.

[2] A. Khaligh and S. Dusmez, "Comprehensive Topological Analysis of Conductive and Inductive Charging Solutions for Plug-In Electric Vehicles," *IEEE Trans. Veh. Tech.*, vol. 61, no. 8, pp. 3475-3489, Oct. 2012.

[3] G. Buja, M. Bertoluzzo and K. N. Mude, "Design and Experimentation of WPT Charger for Electric City Car," *IEEE Trans. Ind. Electron.*, vol. 62, no. 12, pp. 7436-7447, December 2015.

[4] K. Colak, E. Asa, M. Bojarski, D. Czarkowski and O. C. Onar, "A Novel Phase-Shift Control of Semibridgeless Active Rectifier for Wireless Power Transfer," *IEEE Trans. Power Electron.*, vol. 30, no. 11, pp. 6288-6297, Nov 2015.

[5] M. Budhia, G. A. Covic and J. T. Boys, "Design and optimization of circular magnetic structures for lumped inductive power transfer systems," *IEEE Trans. Power Electron.*, vol. 26, no. 11, pp. 3096-3108, November 2011.

[6] D. Patil, M. Sirico, L. Gu and B. Fahimi, "Maximum efficiency tracking in wireless power transfer for battery charger: Phase shift and frequency control," *2016 IEEE Energy Conversion Congress and Exposition (ECCE), Milwaukee, WI*, pp. 1-8, 2016.

[7] K. Aditya and S. S. Williamson, "A Review of Optimal Conditions for Achieving Maximum Power Output and Maximum Efficiency for a Series–Series Resonant Inductive Link," *IEEE Trans. Transport. Electrific.*, vol. 3, no. 2, pp. 303-311, Jun. 2017.

[8] K. Aditya and S. S. Williamson, "Design considerations for loosely coupled inductive power transfer (IPT) system for electric vehicle battery charging - A comprehensive review," *Transportation Electrification Conference and Expo (ITEC), 2014 IEEE*, pp. 1-6, June 2014.

[9] G. A. Covic and J. T. Boys, "Inductive power transfer," *Proceedings of the IEEE*, vol. 101, no. 6, pp. 1276-1289, June 2013.

[10] S. Samanta and A. K. Rathore, "Wireless power transfer technology using full-bridge current-fed topology for medium power applications," *IET Power Electron.*, vol. 9, no. 9, pp. 1903-1913, Jul. 2016.

[11] S. Samanta and A. K. Rathore, "A New Current-Fed CLC Transmitter and LC Receiver Topology for Inductive Wireless Power Transfer Application: Analysis, Design, and Experimental Results," *IEEE Trans. Transport. Electrific.*, vol. 1, no. 4, pp. 357-368, Dec 2015.

[12] A. Kamineni, G. A. Covic and J. T. Boys, "Self-Tuning Power Supply for Inductive Charging," *IEEE Trans. Power Electron.*, vol. 32, no. 5, pp. 3467-3479, May 2017.

[13] S. Samanta and A. K. Rathore, "Analysis and Design of Load Independent ZPA Operation for P/S, PS/S, P/SP and PS/SP Tank Networks in IPT Applications," *IEEE Trans. Power Electron.*, Early access article DOI: 10.1109/TPEL.2018.2794623.

[14] S. Samanta and A. K. Rathore, "Analysis and design of current-fed (L)(C) (LC) converter for inductive wireless power transfer," *IEEE ECCE 2015*, pp. 5724-5731, Sep. 2015.

[15] P. Xuewei and A. K. Rathore, "Small-Signal Analysis of Naturally Commutated Current-Fed Dual Active Bridge Converter and Control Implementation Using Cypress PSoC," *IEEE Trans. Vehicular Tech.*, vol. 64, no. 11, pp. 4996-5005, November 2015.

[16] U. R. Prasanna and A. K. Rathore, "Small-Signal Modeling of Active-Clamped ZVS Current-Fed Full-Bridge Isolated DC/DC Converter and Control System Implementation Using PSoC," *IEEE Trans. Ind. Electron.*, vol. 61, no. 3, pp. 1253-1261, Mar. 2014.

The 2018 International Power Electronics Conference

Development of Inductive Power Transfer System for Excavator under Large Load Fluctuation

-Consideration of relationship between load voltage and resonance parameter-

Jun-ichi Itoh[1], Kent Inoue[2]*and Keisuke Kusaka[2]

1 Department of Science of Technology Innovation Engineering
2 Department of Electrical, Electronics and Information Engineering
Nagaoka University of Technology
Nagaoka, Niigata, Japan
*E-mail: k_inoue@stn.nagaokaut.ac.jp

Abstract— An inductive power transfer system for an excavator, which is operated under an air pressure environment, is designed and developed by considering the load voltage fluctuation. In the conventional excavator systems, the power is supplied via the contact wires, which may cause fire when a spark occurs because a working chamber environment is under high air pressure. In the proposed system, the series-parallel compensation is applied to cancel out the leakage inductance. By using the series-parallel compensation, the load voltage is ideally constant regardless of load fluctuation. However, the constant-voltage characteristic degrades due to winding resistance and an error of the resonance parameter. Thus, the resonance parameters have to be designed considering the error. In this paper, the design method of the resonance parameter is proposed with the voltage ratio maps considering the error of the parameter including the winding resistance. In the experiments with a developed 15-kW IPT system, the voltage fluctuation is smaller than 4.3%. Furthermore, the constant-voltage characteristic is maintained even when the output power of an induction motor changes from 5 kW to 15 kW and vice versa.

Keywords— Inductive power transfer, Excavator, Load fluctuation, Constant voltage characteristic

I. INTRODUCTION

A pneumatic caisson method is used in many structures: foundations of bridges and buildings, shafts for insertion of shield tunneling machines, tunnel and railways, e.g., the Chuo Shinkansen [1]. Figure 1 shows the schematic of the pneumatic caisson method. First, a reinforced concrete caisson is constructed on the ground. Second, an airtight working chamber is formed at the bottom of the caisson. Finally, the caisson is immersed at a predetermined depth and the pressurized air is supplied into the working chamber in order to prevent the underground water from coming into the chamber. Therefore, the working chamber environment is under high air pressure.

Figure 2 shows the schematic of the charging system. Fig. 2 (a) and (b) show the conventional charging system

and the proposed charging system, respectively. An excavator hangs onto the ceiling of the working chamber and moves along a traveling rail. Meanwhile, an electric hydraulic pump is used in order to move and operate the excavator. The power for an electric motor which drives hydraulic pump is supplied via an insulated trolley wire which is placed into the traveling rail. However, due to the movement of the excavator along the traveling rail, there is a threat of a spark which occurs at connection points. Even a small spark may cause a large-scale fire because the working chamber is under high air pressure environment. Therefore, in order to reduce the risk of the fire which is caused by the spark, the application of an inductive power transfer for the excavator has been proposed.

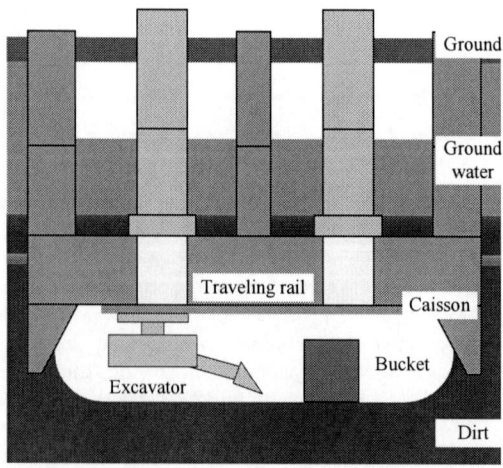

Fig. 1. Schematic of pneumatic caisson method which is immersed at predetermined depth and pressurized air is supplied into working chamber in order to prevent underground water from coming into chamber.

In the conventional method which is connected the capacitor in series to the primary side and the secondary side (SS compensation method), the secondary side features a constant-current characteristic when the primary side is driven at a constant voltage [2]–[4]. Therefore, the additional circuits, e.g., the buck converter and the boost converter, on the secondary side are required for the load voltage regulation. In addition, in the SS compensation method, the high-speed communication is needed between the primary side and the secondary side. On the other hand, the method which is connected the capacitor in series to the primary side and in parallel to the secondary side (SP compensation method) is regulated the load voltage without the additional circuits [5] [6]. Therefore, in the SP compensation method, the high-speed communication is unnecessary.

In this paper, the inductive power transfer system [7]–[14] for the excavator, which ensure the large load fluctuation, is developed. In the proposed system, the SP compensation is employed in order to cancel out the leakage inductance and a constant output voltage is ideally supplied even under the large load fluctuation without the

additional circuit. However, the constant-voltage characteristic degrades due to the winding resistance and the error of the resonance parameter.

The new contribution in this paper is providing an analysis how to develop the robust system against the voltage fluctuation. Analysis which is indicated this paper takes the relationship among the load fluctuation, the resonance parameter and the winding resistance takes into account in order to optimize parameters which is influenced the load voltage.

Based on the analysis on the voltage characteristic, a 15-kW inductive power transfer system is developed and tested. The constant load voltage characteristic is evaluated using a resistance and an induction motor load.

II. INDUCTIVE POWER TRANSFER SYSTEM FOR EXCAVATOR

A. System Configuration

Figure 3 shows the system configuration of the inductive power transfer system for the excavator. The proposed system consists of a converter with pulse-width

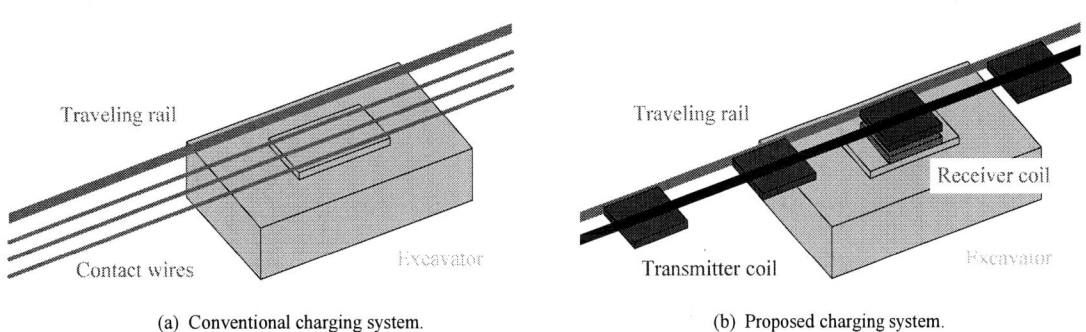

(a) Conventional charging system. (b) Proposed charging system.

Fig. 2. Schematic of charging system which hangs onto ceiling of working chamber.

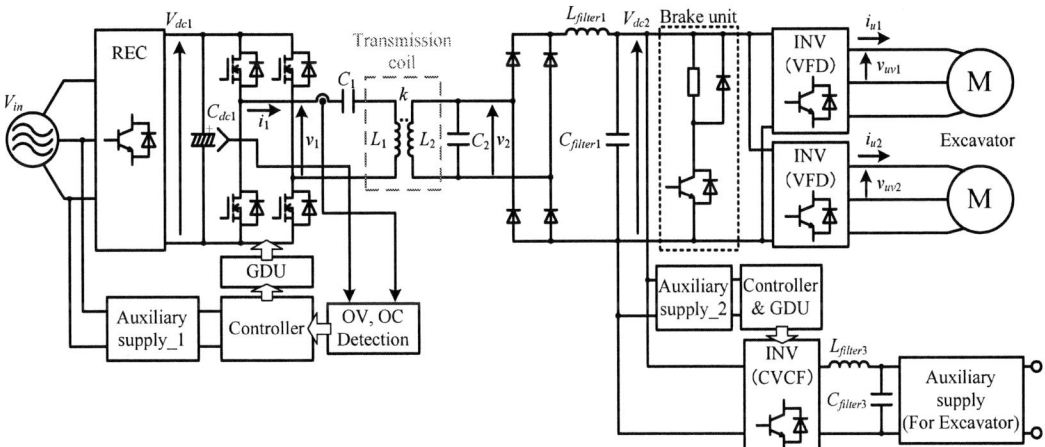

Fig. 3. Inductive power transfer system for excavator.

349

modulation (PWM) and a single-phase inverter in the primary side, a rectifier, a brake unit, a three-phase variable frequency drive (VFD) inverter and a single-phase constant voltage constant frequency (CVCF) inverter in the secondary side. The three-phase VFD inverters are used for adjustable speed driving an induction motors for an electric hydraulic pumps. The single-phase CVCF inverter is used as an auxiliary supply which is needed for an excavator control. The power is transferred from the primary side to the secondary side through the transmission coils. The SP compensation is employed in the system. Therefore, the secondary DC voltage is constant in the ideal conditions because the secondary side output features a constant-voltage characteristic regardless of a load fluctuation even if the primary side is driven at constant voltage. However, the winding resistance and an error of the resonance parameter practically degrade the constant-voltage characteristic.

B. Compensation Method of Reactive Power by Leakage Inductance of Transmission Coil

Figure 4 shows the typical compensation methods which is connected a resonance capacitor in series or parallel to the primary side coil and the secondary side coil in order to cancel out the reactive power [15]. Reactance components by leakage inductance are canceled out because of the resonance with the transmission coil. Consequently, the power factor seen from the primary side is.

Fig. 4 (a) shows the SS compensation method, which features a constant current characteristic at the secondary side when the primary side is driven at a constant voltage. Therefore, the SS compensation is unsuitable for the existing system because the additional circuit is needed in order to convert the constant current characteristic into the constant-voltage characteristic. In addition, the SS compensation is undesirable for the existing system because the high-speed communication is needed between the primary side and the secondary side. On the other hand, Fig. 4 (b) shows the SP compensation method, which features a constant voltage characteristic at the secondary side when the primary side is driven at a constant voltage. Therefore, the SP compensation is suitable to be applied to the proposed system because the inductive power transfer is possible to be applied into the existing excavator system without any modifications. However, the error of the resonance parameter needs to be considered.

III. DESIGN OF TRANSMISSION COIL

A. Specifications

Figure 5 shows the transmission coil. An excavator system hangs onto the ceiling of the working chamber and moves along a traveling rail. Therefore, the upper side is the transmitter coil, the lower side is the receiver coil. In the proposed system, the solenoid coil is employed to obtain higher magnetic coupling in comparison with a circular coil. The cores are employed with PC40 manufactured by TDK. The size of the core is W237 × H210 × D20 mm. Meanwhile, the transmission distance is

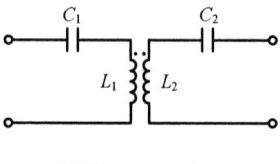

(a) SS compensation.

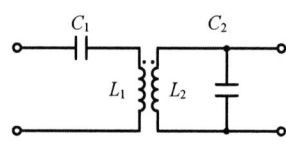

(b) SP compensation.

Fig. 4. Typical compensation method which is connected capacitor in series or parallel in order to cancel out leakage inductance

Fig. 5. Transmission coils which is used core with PC40 manufactured by TDK. Upper side is transmitter coil, lower side is receiver coil. Coil size is W237 × H210 × D20 mm. Transmission distance is 50 mm.

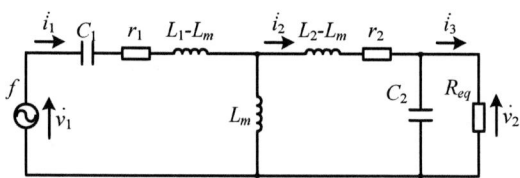

Fig. 6. Equivalent circuit of IPT system with SP compensation.

50 mm.

B. Parameters Design

Figure 6 shows the equivalent circuit for designing the IPT system. The circuit equations of the equivalent circuit which is shown Figure 6 are calculated as

350

$$V_1 = \left\{ r_1 + j\omega \left(L_1 - \frac{1}{\omega^2 C_1} \right) \right\} I_1 - j\omega L_m I_2 \quad \text{...............} \quad (1),$$

$$0 = -j\omega L_m I_1 + \left\{ r_2 + j\omega \left(L_2 - \frac{1}{\omega^2 C_2} \right) \right\} I_2 + j\frac{1}{\omega C_2} I_3$$

$$\text{..............................} \quad (2),$$

$$0 = j\frac{1}{\omega C_2} I_2 + \left(R_{eq} - j\frac{1}{\omega C_2} \right) I_3 \quad \text{...................} \quad (3),$$

where V_1 is the primary voltage, R_{eq} is the equivalent load resistance, r_1 is the equivalent series resistance of the primary winding, r_2 is the equivalent series resistance of the secondary winding, L_1 is the primary inductance, L_2 is the secondary inductance, C_1 is the primary compensation capacitor, C_2 is the secondary compensation capacitor, L_m is the mutual inductance, and ω is the angular frequency of the power supply.

The currents I_1, I_2 and I_3 are calculated by (4), (5) and (6) when an input voltage V_1 is applied into the primary side.

$$I_1 = \frac{V_1}{\Delta} \left[\left\{ r_2 + j \left(x_2 + x_0 - x_p \right) \right\} \left(R_{eq} - jx_p \right) + x_p^2 \right] \text{..} \quad (4),$$

$$I_2 = \frac{V_1}{\Delta} jx_0 \left(R_{eq} - jx_p \right) \quad \text{........................} \quad (5),$$

$$I_3 = \frac{V_1}{\Delta} x_0 x_p \quad \text{.....................................} \quad (6),$$

$$\Delta = \begin{vmatrix} r_1 + j\omega \left(L_1 - \dfrac{1}{\omega^2 C_1} \right) & -j\omega L_m & 0 \\ -j\omega L_m & r_2 + j\omega \left(L_2 - \dfrac{1}{\omega^2 C_2} \right) & j\dfrac{1}{\omega C_2} \\ 0 & j\dfrac{1}{\omega C_2} & R_{eq} - j\dfrac{1}{\omega C_2} \end{vmatrix}$$

$$\text{...} \quad (7).$$

Note that the voltage V_1 is the fundamental component of the output voltage of the inverter.

The parameters of the transmission coil are designed with the equivalent circuit. The resistance R_{eq} indicates that equivalent load resistance considering the full-bridge rectifier. Then the equivalent load resistance is given by [16]

$$R_{eq} = \frac{\pi^2}{8} \frac{V_{dc,2}^2}{P_2} \quad \text{...} \quad (8),$$

where $V_{dc,2}$ is the DC voltage on the secondary side and P_2 is the output power.

The inductances of the primary and the secondary coils are designed according to the following equations

$$L_2 = \frac{R_{eq}}{\omega} \frac{k}{\sqrt{1+k^2}} \quad \text{...............................} \quad (9),$$

$$L_1 = L_2 \left(\frac{8}{\pi^2 k} \frac{V_{dc,1}}{V_{dc,2}} \right)^2 \quad \text{..............................} \quad (10),$$

where $V_{dc,1}$ is the DC voltage on the primary side and k is the coupling coefficient.

The compensation capacitors are selected in order to cancel out the reactive power at the input frequency. Thus, the value of the compensation capacitors is calculated as

$$C_1 = \frac{1}{\omega^2 L_1 \left(1 - k^2 \right)} \quad \text{..............................} \quad (11),$$

$$C_2 = \frac{1}{\omega^2 L_2} \quad \text{.......................................} \quad (12).$$

C. Influence of Parameter Error and Winding Resistance

Figure 7 shows the voltage ratio v_2/v_1 against the error of the resonance parameter including the winding resistance. Fig. 7 (a) and (b) show the v_2/v_1 ratio against the error of L_1 and C_1 and the error of L_2 and C_2 from 80% to 120%, respectively. Figure 7 represents the v_2/v_1 ratio which is calculated by (13) and (7).

$$\frac{v_2}{v_1} = \frac{R_{eq} \omega^2 L_m \left(L_1 - L_m \right)}{\Delta} \quad \text{.....................} \quad (13)$$

As a result of Fig. 7 (a), it is confirmed that v_2/v_1 ratio is high in a large area of the primary inductance L_1 and a small area of the primary capacitor C_1. As a result of Fig. 7 (b), it is confirmed that v_2/v_1 ratio is high in a large area of the secondary inductance L_2. Thus, in the resonance parameter of the secondary side, v_2/v_1 ratio depends on the secondary inductance L_2. From the results, it is confirmed that the error of the resonance parameter affects v_2/v_1 ratio, i.e., the secondary DC voltage.

D. Coil size design

The size of the coil is decided based on the desired coupling coefficient and the transmission distance, which is obtained from the coupling coefficient maps [17]. The

The 2018 International Power Electronics Conference

core length is decided larger than the core depth because the coupling coefficient may be smaller than the design value.

IV. EXPERIMENTAL RESULTS

A. Experimental Conditions

Table I shows the experimental conditions. The rated power is 15 kW. The self-inductance of the primary side and the secondary side in Table I is measured value. In this experiment, a resistance or an induction motor is used as the load.

B. Resistance Load

Figure 8 shows the operation waveforms with the resistance load. Fig. 8 (a) and (b) show the waveforms obtained at an output of 5 kW and 10 kW, respectively. In the figure, the secondary voltage is constant against the load power. Though the resonance condition is correct, the low-order harmonics affect the primary current in the light load. Therefore, the primary voltage and current

waveforms are misaligned from the resonance point.

Figure 9 shows the frequency characteristics of the voltage gain. Fig. 9 (a) and (b) show the input/output

TABLE 1   EXPERIMENTAL CONDITIONS.

	Symbol	Value
Switching frequency	f	20 kHz
Rated power	P	15 kW
Coupling coefficient	k	0.41
Primary inductance	L_1	393 μH
Secondary inductance	L_2	113 μH
Primary capacitance	C_1	198 nF
Secondary capacitance	C_2	582 nF
Primary winding resistance	r_1	107 mΩ
Secondary winding resistance	r_2	70 mΩ
MOSFETs		BSM120D12P2C005
Diodes		DH 2X61-18A

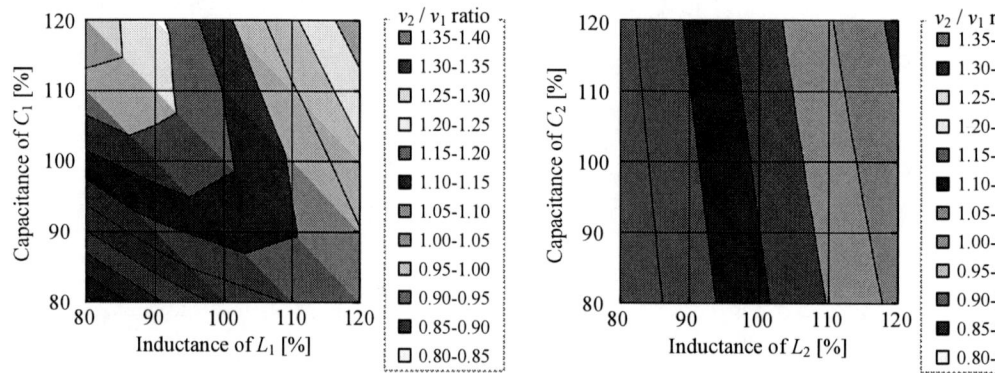

(a) Error of primary inductance L_1 and primary capacitor C_1 from 80% to 120%. (b) Error of secondary inductance L_2 and secondary capacitor C_2 from 80% to 120%.

Fig. 7. v_2 / v_1 ratio against error of resonance parameter including winding resistance. Graph legends shows v_2 / v_1 ratio.

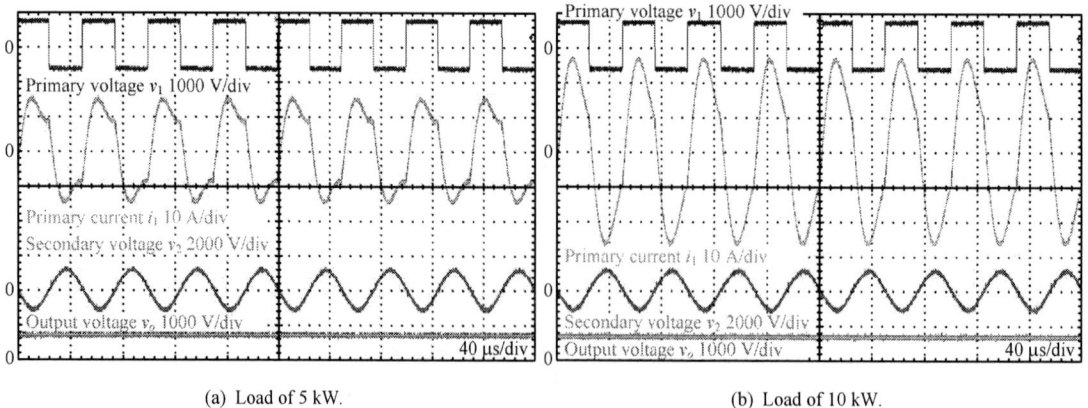

(a) Load of 5 kW.   (b) Load of 10 kW.

Fig. 8. Operation waveforms with resistance load.

352

The 2018 International Power Electronics Conference

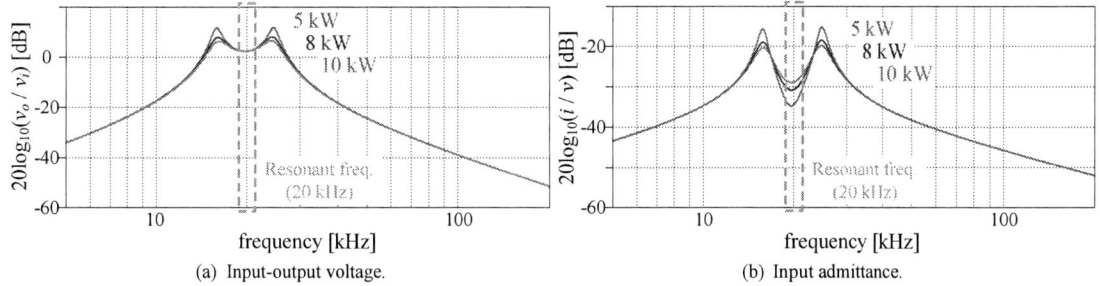

(a) Input-output voltage.

(b) Input admittance.

Fig. 9. Frequency characteristics of input-output voltage and input admittance gain

voltage ratio characteristic and the input admittance characteristic, respectively. The frequency characteristics in Fig. 9 are obtained in simulation in order to evaluate the effect of harmonics on the load. Around the fundamental frequency of the voltage gain frequency responses, the gain is same value regardless of the output power. On the other hand, it is concluded from the frequency characteristic of the input admittance; the gain around the fundamental frequency decreases at light load because low-order harmonics component of the input current is relatively large.

Figure 10 shows the frequency characteristics of the primary admittance. The gain of the fundamental component becomes smaller at light load, which has been expected from the frequency characteristics which is shown in Fig. 9 (b). Therefore, the distortion of the primary current at 5 kW is larger than the waveforms at 10 kW because low-order harmonics component of the primary current are relatively large at the light load.

Figure 11 shows the secondary/primary DC voltage ratio characteristic against the output power. The blue line in Fig. 11 represents the calculation value which is calculated without consideration of the resistance components in the rectifier. In the proposed system, the secondary/primary voltage ratio characteristic is expected to be constant because the SP compensation is employed in order to cancel out the leakage inductance. However, the secondary/primary DC voltage ratio decreases by 1.9% when the output power increases because the effect of voltage drop which is caused by the winding resistance is large at the high output power.

Figure 12 shows the secondary/primary voltage ratio characteristic against the output power. The blue line in Fig. 12 represents the calculation value by (13) and (7) using actual parameters considering the error from a nominal value. The green line in Fig. 12 represents calculation value which is calculated using (13) and (7) with design parameters (theoretical value). The experimental results agree with the calculation value of the prototype model with a small error (less than 0.5%). The error is caused by the difference between the nominal values and the actual values in the process of the resonance parameter design. Nevertheless, the secondary/primary voltage ratio characteristic is constant regardless of the output power. Therefore, it is confirmed that the decrease

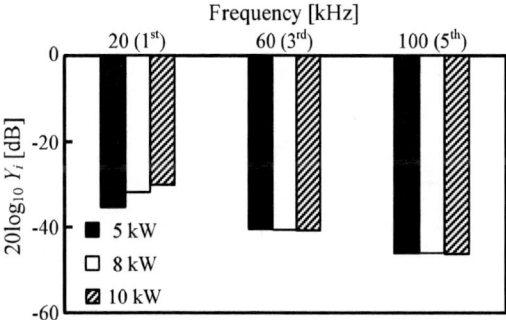

Fig. 10. Frequency characteristics of primary admittance. 1st: fundamental harmonic component (20 kHz), 3rd: triple harmonics component (60 kHz), 5th: fifth-order harmonics component (100 kHz)

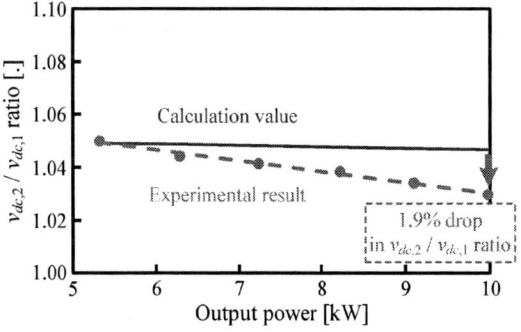

Fig. 11. Secondary/Primary DC voltage ratio. Red point is experimental results. Blue line is calculation value without consideration of resistance components in rectifier.

in the output/input voltage ratio at the high output power is mainly caused by the post-stage conversion, i.e., the rectifier.

Figure 13 shows the operation waveforms of the transient characteristic with the resistance load. Fig. 13 (a) and (b) show the step load response from 10 kW to 5 kW, and vice versa, respectively. The secondary DC voltage is maintained at constant even when the load step occurs. Consequently, it is confirmed that the secondary DC

353

The 2018 International Power Electronics Conference

voltage is constant regardless of the output power by using the SP compensation as a leakage inductance canceling method. In particular, the secondary DC voltage is constant even when a large load fluctuation occurs.

C. Induction Motor Load

Figure 14 shows the operation waveforms with the induction motor load which is connected to the pump. The output power is 15 kW. Fig. 14 (a) shows the primary voltage and current waveforms whereas Fig. 14 (b) shows the secondary DC voltage, output voltage and current waveforms. The primary current is confirmed that the waveform distortion is small because the fundamental components is relatively large, i.e. the influence of low-harmonics components is relatively small.

Figure 15 shows the operation waveforms of transient characteristic in the induction motor load. Fig. 15 (a) and (b) show the step response from 5 kW to 15 kW and vice versa, respectively. The secondary DC voltage is maintained at the constant value even when the load step occurs. Consequently, it is confirmed that the secondary

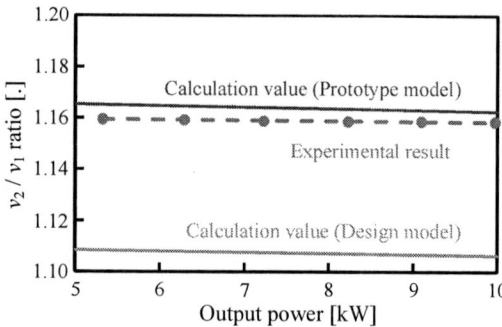

Fig. 12. Secondary/primary voltage ratio of transmission coil. Red point is experimental results. Blue line is calculation value by (13) and (7) using actual parameters considering error from nominal value. Green line is calculation value by (13) and (7) with design parameters.

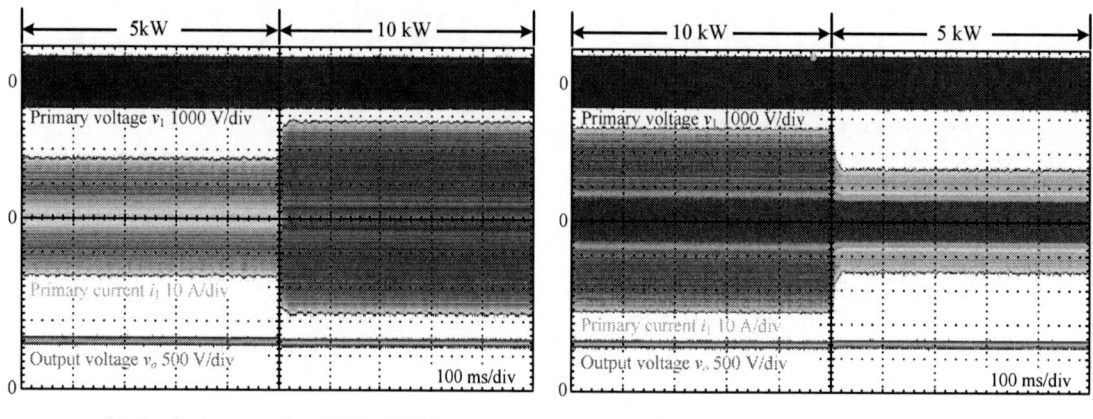

(a) Step load response from 5 kW to 10kW.

(b) Step load response from 10 kW to 5 kW.

Fig. 13. Operation waveforms of transient characteristic with resistance load.

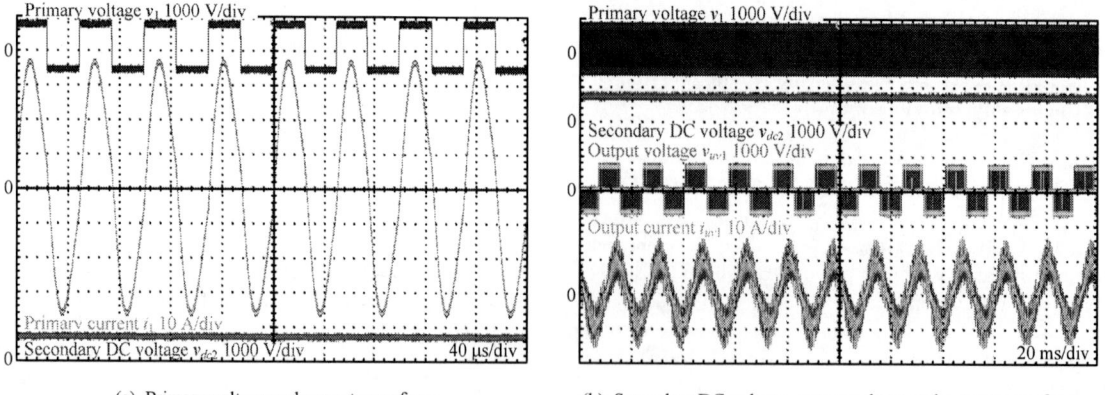

(a) Primary voltage and current waveform.

(b) Secondary DC voltage, output voltage and current waveform.

Fig. 14. Operation waveforms with induction motor load.

354

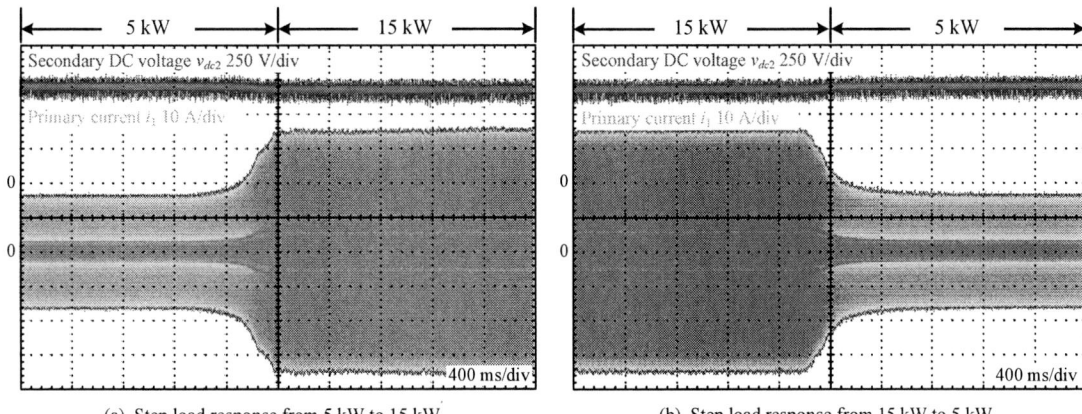

(a) Step load response from 5 kW to 15 kW.   (b) Step load response from 15 kW to 5 kW.

Fig. 15. Operation waveforms of transient characteristic with induction motor load.

DC voltage is constant regardless of the output power by using the SP compensation. In particular, the secondary DC voltage is constant even when a large load fluctuation occurs.

V. CONCLUSIONS

In this paper, the inductive power transfer system was applied in the existing excavator in order to reduce the fire risk which was caused by the spark of contact charging. In the proposed system, the load voltage against the load fluctuation should be stabilized without the additional circuit. In order to stabilize the load voltage against the load fluctuation, the SP compensation method was applied as the method in order to cancel out the leakage inductance. Moreover, the constant-voltage characteristic depending on the error of the resonance parameter and the winding resistance is theoretically analyzed in order to obtain the constant-voltage characteristic. In the experiments with an output power of 15 kW, the voltage fluctuation was smaller than 4.3%. The constant voltage is maintained even when the load step of the induction motor occurs, i.e. from 5 kW to 15 kW operation and vice versa. From the experimental results, the constant secondary DC voltage characteristic was confirmed, i.e. the constant output voltage.

REFERENCES

[1] K. Kodaki, M. Nakano, S. Maeda: "Development of the automatic system for pneumatic caisson", ELSEVIER Automation in Construction, Vol. 6, No. 3, pp. 241-255, (1997).

[2] K. Hata, T. Imura, Y. Hori: "Maximum Efficiency Control of Wireless Power Transfer via Magnetic Resonant Coupling Considering Dynamics of DC–DC Converter for Moving Electric vehicles", IEEE Applied Power Electronics Conference and Exposition, pp. 3301-3306, (2015)

[3] M. Kato, T. Imura, Y. Hori: "Study on Maximize Efficiency by Secondary Side Control Using DC-DC Converter in Wireless Power Transfer via Magnetic Resonant Coupling", The International Electric Vehicle Symposium & Exhibition, (2013)

[4] M.. Sato, G.Guidi, T. Imura, H. Fujimoto: "Model for Loss Calculation of Wireless In-Wheel Motor Concept Based on

Magnetic Resonant Coupling", IEEE Workshop on Control and Modeling for Power Electronics, No. 16267987, (2016)

[5] T. Fujita, Y. Kaneko, S. Abe: "Contactless Power Transfer Systems using Series and Parallel Resonant Capacitors", IEEJ Trans. of Industry Applications, Vol. 127, No. 2, pp. 174-180, (2007).

[6] R. Ota, N. Hoshi, J. Haruna: "Design of Compensation Capacitor in S/P Topology of Inductive Power Transfer System with Buck or Boost Converter on Secondary Side", IEEJ Journal of Industry Applications, Vol .4, No. 4, pp. 476-485, (2015)

[7] K. Kusaka, J. Itoh: "Development Trends of Inductive Power Transfer Systems Utilizing Electromagnetic Induction with Focus on Transmission Frequency and Transmission Power", IEEJ Transactions on Industry Applications, Vol. 137, No. 5, pp. 445-457, (2017).

[8] S. Li, C. C. Mi: "Wireless Power Transfer for Electric Vehicle Applications", IEEE Journal, Vol. 3, No. 1, pp. 4-17, (2015)

[9] D. Shimode, T. Murai, S. Fujiwara: "A Study of Structure of Inductive Power Transfer Coil for Railway Vehicles", IEEJ Journal of Industry Applications, Vol. 4, No. 5, pp. 550-558, (2015)

[10] T. Mizuno, T. Ueda, S. Yachi, R. Ohtomo, Y. Goto: "Dependence of Efficiency on Wire Type and Number of Strands of Litz Wire for Wireless Power Transfer of Magnetic Resonant Coupling", IEEJ Journal of Industry Applications, Vol. 3, No. 1, pp. 35-40, (2014)

[11] T. Koyama, K. Umetani, E. Hiraki: "Design Optimization Method for the Load Impedance to Maximize the Output Power in Dual Transmitting Resonator Wireless Power Transfer System", IEEJ Journal of Industry Applications, Vol. 7, No. 1, pp. 49-55, (2018)

[12] S. Li, C. C. Mi: "Wireless Power Transfer for Electric Vehicle Applications", IEEE Journal, Vol. 3, No. 1, pp. 4-17, (2015)

[13] A. Kurs, A. Karalis, R. Moffatt, J. D. Joannopoulos, P. Fisher, M. Soljacic: "Wireless Power Transfer via Strongly Coupled Magnetic Resonances", SCIENCE, Vol. 317, pp. 83-86, (2007)

[14] K. Inoue, K. Kusaka, J. Itoh: "Reduction in Radiation Noise Level for Inductive Power Transfer Systems using Spread Spectrum Techniques", IEEE Transaction on Power Electronics, Vol. 33, No. 4, pp. 3076-3085, (2018)

[15] T. Imura, Y. Hori: "Unified Theory of Electromagnetic Induction and Magnetic Resonant Coupling", IEEJ Trans. of Industry Applications, Vol. 135, No. 6, pp. 697-710, (2015).

[16] R. Bosshard, J. W. Kolar, J. Muhlethaler, I. Stevanovic, B. Wunsch, F. Canales: "Modeling and η-α-Perato Optimization of Inductive Power Transfer Coils for Electric Vehicles", IEEE Transactions, Vol. 3, No. 1, pp. 50-64, (2015).

[17] K. Inoue, K. Kusaka, D. Sato, J. Itoh: "Coupling Coefficient Maps for Wireless Power Transfer Using Solenoid Type Coil", IEICE Workshop on EE and WPT, No. WPT2016-34, pp. 85-90, (2016)

The 2018 International Power Electronics Conference

Wireless Power Transfer System Using Three-phase to Single-phase Matrix Converter

Yuji Hayashi[1], Hiromasa Motoyama[1]*, Takaharu Takeshita[1]

1 Nagoya Institute of Technology, Nagoya, Japan

*E-mail: 29413203@stn.nitech.ac.jp

Abstract—This paper presents the wireless power transfer system using three-phase to single-phase matrix converter. The operation principle and control method of the proposed system are explained. The effectiveness of the proposed circuit is verified by experiments.

Keywords—AC-AC converter, electric vehicle, inductive power transfer, matrix converter, wireless charging systems.

I. INTRODUCTION

In recent years, the demand of electric vehicles and plug-in hybrid vehicles has increased due to environmental problems such as global warming and resource depletion. These charging methods are contact types that connects a power source to a car body using a charging cable and a connector. However, it is a danger of electric shock or leakage in this system and there is a necessary to contact the cable to the car body every time for charging. Therefore, this system has safety and convenience problems. As a method for solving these problems, there is a non-contact method using an electromagnetic induction system. Application to automotive charging systems is being studied in recent years [1] [2]. This method has no mechanical contact. So safety and convenience problems are improved. Also, it may be possible to charge in driving by using the wireless transfer system. Therefore, it is possible to reduce the capacity of the storage battery [3].

A typical wireless power transfer system is shown in Fig.1 [4][5]. The primary side is composed of an AC source, a rectifier, an inverter and a resonant network to compensate the primary coil inductance. The secondary side is composed of a resonant compensation network and a power regulator, which can be a rectifier in the simplest form [6]. Therefore, in order to generate high

frequency current in the primary coil, two power conversion stages are required on the primary side. In addition, the conventional circuit is required a large electrolytic capacitor on the primary side in order to smooth the DC voltage. Because the capacitor is very large volume and its life time is short, it is difficult to achieve compact size and high reliability of the system [7][8]. Consequently, the system suffers from disadvantages such as lower reliability, increase in power losses, cost and overall physical size.

In order to reduce the number of conversion stages and thereby alleviate some of the above disadvantages, wireless power transfer systems using matrix converters have been proposed [10][11]. However, complicated commutation control is required for the matrix converter. In particular, the control complexity of the wireless power transfer system having a high frequency resonance circuit is remarkably increased. In the conventional proposal, it is only the control focusing on the high-frequency voltage application to the power transmission coil on the primary side, The input current control method for improving the current distortion and the power factor of the three-phase AC power source has not yet been clarified [12][13].

In order to solve these problems, the authors propose a wireless power transfer system using a matrix converter that directly converts from three-phase AC voltage to single-phase AC voltage. In addition, we propose a control method that achieves stable power transmission to the secondary side and reduction of current distortion of three-phase AC input current. In this paper, the switching pattern and duty of the six bidirectional switches of the primary side matrix converter are clarified, and the effectiveness of the proposed control method is verified by experiment using 2 kW prototype system.

II. PROPOSED SYSTEM CONFIGURATION

A. Main Circuit Configuration

Fig.2 shows the main circuit configuration of the proposed system. The three-phase AC voltages e_{su}, e_{sv}, e_{sw} are connected to the matrix converter through a filter circuit composed of reactors L_f and a capacitors C_f in order to prevent harmonic current flowing into the power supply. The matrix converter is composed of six bidirectional switches S_{ug}-S_{wh} that connect two MOSFETs in reverse series, and converts three-phase AC voltage to single-phase AC voltage v_1 directly. In the wireless power transfer, the coupling coefficient is small and the leakage reactance is large because the long distance between the

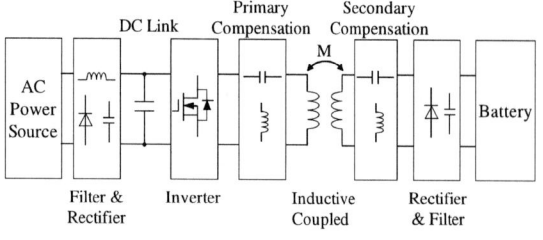

Fig. 1. A tyipical wireless power transfer system

The 2018 International Power Electronics Conference

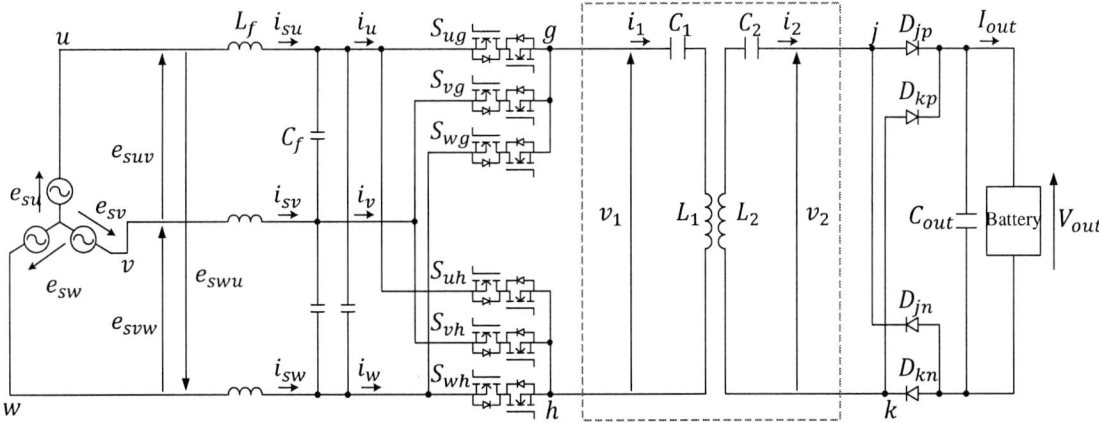

Fig. 2. Proposed system configuration.

coils. Therefore, the resonant capacitors are connected to the primary and secondary sides for the leakage reactance compensation. As the battery for automobiles is often charged at constant current, this circuit uses an SS topology that connects the resonant capacitors C_1 and C_2 to the primary coil and the secondary coil in series, respectively [9]. On the secondary side, the high frequency AC power transmitted from the primary side is rectified by four diodes D_{jp}-D_{kn} and the DC voltage is generated.

B. Equivalent Circuit of SS Topology

Fig.3 shows the equivalent circuit focusing on the switching frequency in the dashed line in Fig.2. Because LC resonance is used, circuit operation with switching frequency components becomes dominant, and the influence of harmonic components can be neglected [14]. In Fig.3, C_1, C_2 are the series resonant capacitors of the primary and the secondary, L_1, L_2 are the self-inductance of the primary side coil and the secondary side coil and r_1, r_2 is the winding resistance of the primary side coil and the secondary side coil. $M(=k\sqrt{L_1 L_2})$ represents mutual inductance between the primary side coil and the secondary side coil. The series resonance capacitor C_1, C_2 resonate with the self-inductance of coils L_1, L_2 at resonance frequency $\omega_0 (=2\pi f_0)$, respectively. Therefore,

the following equation is obtained.

$$
\begin{cases}
\omega_0 L_1 = \dfrac{1}{\omega_0 C_1} \\[2mm]
\omega_0 L_2 = \dfrac{1}{\omega_0 C_2}
\end{cases}
\tag{1}
$$

In Fig.3, the primary voltage V_1 and the secondary voltage V_2 are expressed by the following equation. However, winding resistances r_1 and r_2 are negligible because they are sufficiently smaller than the impedances of the coils [15][16].

$$
\begin{cases}
V_1 = \dfrac{I_1}{j\omega_S C_1} + j\omega_S L_1 I_1 - j\omega_S M I_2 \\[2mm]
V_2 = -j\omega_S L_2 I_2 + j\omega_S M I_1 - \dfrac{I_2}{j\omega_S C_2}
\end{cases}
\tag{2}
$$

Normally, switching frequency $\omega_S (= 2\pi f_S)$ is set to resonance frequency ω_0. Therefore, if ω_S and ω_0 are equal, the primary voltage V_1 and the secondary voltage V_2 are obtained from (1) and (2) as follows.

$$
\begin{cases}
V_1 = -j\omega_0 M I_2 \\[2mm]
V_2 = j\omega_0 M I_1
\end{cases}
\tag{3}
$$

From equation (3), the proposed system can control the secondary current I_2 in proportion to the primary voltage V_1.

The input impedance Z_{in} is expressed as follows.

$$
\begin{aligned}
Z_{in} &= \frac{V_1}{I_1} \\
&= j\left(\omega_S L_1 - \frac{1}{\omega_S C_1}\right) \\
&\quad + \frac{(\omega_S M)^2}{R_L + j\left(\omega_S L_2 - \dfrac{1}{\omega_S C_2}\right)}
\end{aligned}
\tag{4}
$$

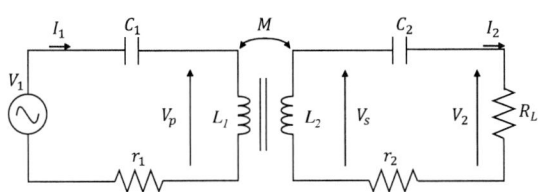

Fig. 3. Equivalent circuit of SS topology

The equivalent load resistance R_L in the above equation is obtained by using the output voltage V_{out} and the output

357

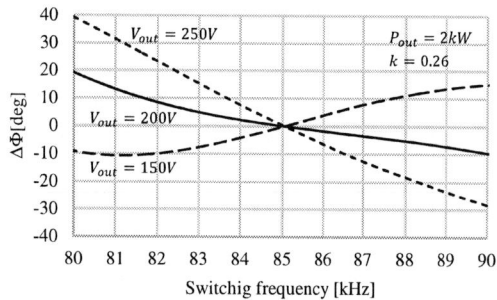

Fig. 4. Frequency characteristics of phase difference.

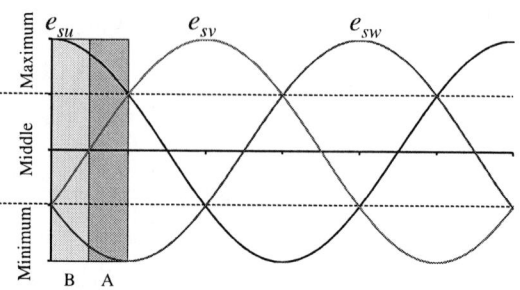

Fig. 5. Level of the grid phase voltage.

power P_{out} as follows;

$$R_L = \frac{\left(\frac{2\sqrt{2}}{\pi}V_{out}\right)^2}{P_{out}} \quad (5)$$

Equation (4) is as follows when ω_S is ω_0.

$$Z_{in} = \frac{V_1}{I_1} = \frac{(\omega_0 M)^2}{R_L} \quad (6)$$

From equation (5), the load resistance R_L is only the resistance component. Therefore, the input impedance Z_{in} is only the real part, and the primary voltage V_1 and the primary current I_1 are in phase [17]. On the other hand, if the switching frequency ω_S is different from the resonance frequency ω_0, Since the input impedance Z_{in} has real part and imaginary part, the phase difference $\Delta\phi$ is generated between the primary voltage V_1 and the primary current I_1. As an example, the relationship between the switching frequency and the phase difference $\Delta\phi$ in the case of $L_1=L_2=97\mu$H, $C_1=C_2=36$nF is shown in Fig.4. Switching frequency ω_S and resonance frequency ω_0 do not always match in real system. When switching frequency ω_S and resonance frequency ω_0 do not match, a phase difference $\Delta\phi$ is generated between the primary voltage V_1 and the primary current I_1 according to the switching frequency.

III. COMMUTATION STRATEGY

A. Proposed Switching Pattern

The grid three-phase voltages are continuously varying among three levels, as shown in Fig.5. Therefore, each phase voltage may be in the maximum, middle or minimum level that varies six times during each cycle. The proposed switching pattern of the matrix converter depends mainly on the grid phase voltage levels. The phase voltage with middle level may be positive or negative, as shown in the shaded areas A and B in Fig.5. For positive and negative intermediate phase voltages, two different PWM switching patterns for matrix converters have been proposed.

Fig.6 (a) shows the switching pattern of the matrix converter in the case of $e_{su} > e_{sv} > 0 > e_{sw}$, region A in Fig.5. During the half-period T_s of the positive primary

voltage in this case, the conducting switches are shifted as $S_{wg} \rightarrow S_{ug} \rightarrow S_{vg} \rightarrow S_{wg}$ in the g-phase. In the h-phase, the switch S_{wh} is always turned on. At this time, the primary voltage v_1 becomes three levels of square wave voltages of 0, e_{uw}, e_{vw}, as shown in Fig.6 (a). Fig.6 (b) shows the switching pattern of the matrix converter in the case of $e_{su} > 0 > e_{sv} > e_{sw}$, region B in Fig.5. During the half-period T_s of the positive primary voltage in this case, the conducting switches are shifted as $S_{uh} \rightarrow S_{wh} \rightarrow S_{vh} \rightarrow S_{uh}$ in the h-phase. In the g-phase, the switch S_{ug} is always turned on. At this time, the primary voltage v_1 becomes three levels of square wave voltages of 0, e_{uv}, e_{uw}, as shown in Fig.6 (b). On the other hand, the switching pattern during the half-period Ts of the negative primary voltage is obtained by replacing g-phase and h-phase in the switching pattern in the case of the positive primary voltage. The secondary voltage v_2 becomes a square waveform shifted by 90 degrees against the primary voltage. The primary and secondary currents i_1 and i_2 are sinusoidal waveforms in phase with the primary and secondary voltages, respectively.

B. Input Current References

Because the voltage drop in the reactor L_f is sufficiently small compared with the source voltages e_{su}, e_{sv} and e_{sw}, the input voltages e_u, e_v and e_w can be assumed to be equal to the source voltages. The source voltages are given by using the source line voltage effective value E and the phase angle θ as follows;

$$\begin{bmatrix} e_{su} \\ e_{sv} \\ e_{sw} \end{bmatrix} = \sqrt{\frac{2}{3}}E \begin{bmatrix} \cos\theta \\ \cos(\theta - 2\pi/3) \\ \cos(\theta + 2\pi/3) \end{bmatrix} \quad (7)$$

The input current references i_u^*, i_v^* and i_w^* are given by using the source current effective value I and the power factor angular reference φ^* as follows;

$$\begin{bmatrix} i_u^* \\ i_v^* \\ i_w^* \end{bmatrix} = \sqrt{2}I \begin{bmatrix} \cos(\theta + \varphi^*) \\ \cos(\theta + \varphi^* - 2\pi/3) \\ \cos(\theta + \varphi^* + 2\pi/3) \end{bmatrix} \quad (8)$$

The input instantaneous power are given by using (7) and (8) as follows;

$$p_{in} = e_u i_u^* + e_v i_v^* + e_w i_w^* = \sqrt{3}EI\cos\varphi^* \quad (9)$$

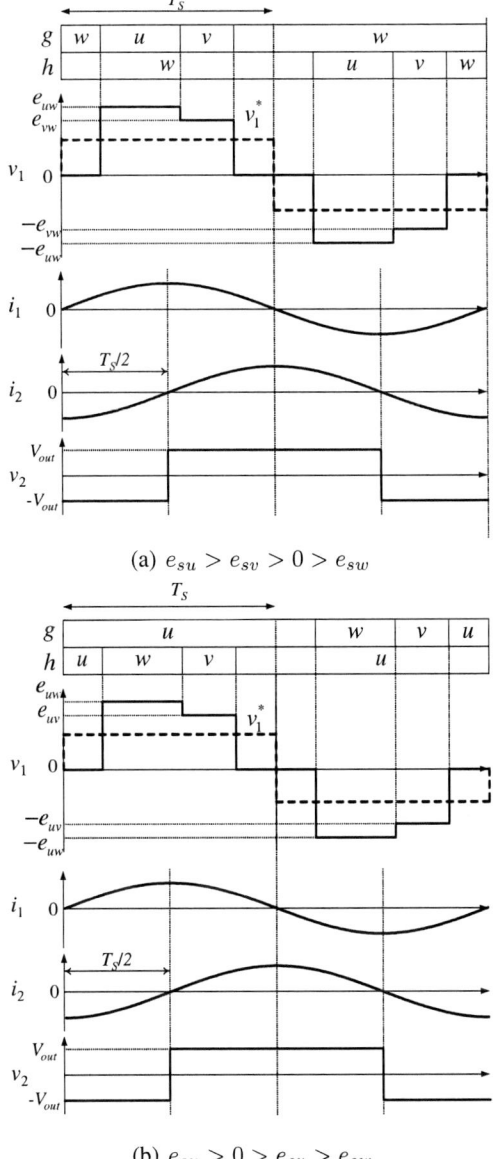

(a) $e_{su} > e_{sv} > 0 > e_{sw}$

(b) $e_{su} > 0 > e_{sv} > e_{sw}$

Fig. 6. Switching pattern of matrix converter

Assuming that the input power is equal to the primary power, the input current effective value is expressed by using (9) as follows;

$$I = \frac{v_1^* i_1}{\sqrt{3}E\cos\varphi^*} \qquad (10)$$

The input current references i_u^*, i_v^* and i_w^* are obtained by using (8) and (10).

C. Duty Cycles

The procedure for calculating the duty cycles d_{ug}-d_{wh} of the switches S_{ug}-S_{wh} of the matrix converter is as follows. First, only one of three switches on each phase is always turned on in order to prevent the short circuit and

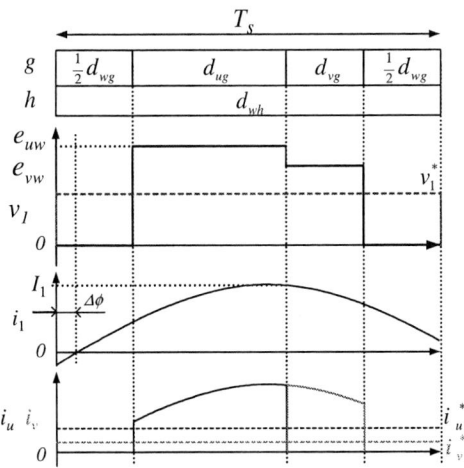

Fig. 7. Primary side voltage and current waveforms of contactless power transformer

current discontinuity. Therefore, the following equations are obtained from this condition.

$$d_{ug} + d_{vg} + d_{wg} = 1 \qquad (11)$$
$$d_{uh} + d_{vh} + d_{wh} = 1 \qquad (12)$$

Next, the following equation is established because the primary voltage v_1 is equal to the primary voltage reference v_1^* during the half-period T_s.

$$v_1^* = (d_{ug}-d_{uh})e_{su}+(d_{vg}-d_{vh})e_{sv}+(d_{wg}-d_{wh})e_{sw} \qquad (13)$$

In the case of $e_{su} > e_{sv} > 0 > e_{sw}$ and $v_1^* > 0$, the switch S_{wh} is always turned on. Therefore, the duty cycles of the h-phase d_{uh}-d_{wh} are obtained as follows;

$$d_{uh} = 0, \ d_{vh} = 0, \ d_{wh} = 1 \qquad (14)$$

the primary voltage reference v_1^* is obtained by using (11), (13) and (14) as follows;

$$v_1^* = e_{uw}d_{ug} + e_{vw}d_{vg} \qquad (15)$$

Next, in order to make the source current a sinusoidal waveform, the input current references i_u^*, i_v^* and the input current averages $\overline{i_u}, \overline{i_v}$ satisfy the following equation.

$$i_u^* = \overline{i_u}, \ i_v^* = \overline{i_v} \qquad (16)$$

As described in Section III-A, the primary current is ideally sinusoidal waveform in phase with the fundamental frequency component of the primary voltage reference. However, as shown in Fig.7, the phase difference $\Delta\phi$ is generated due to the mismatch between the switching frequency and the resonance frequency. The primary current model i_{1M} is given by using $\Delta\phi$ as follows;

$$i_{1M} = \sqrt{2}I_1 \sin(2\pi\frac{1}{2T_s}t + \Delta\phi) \qquad (17)$$

The input current averages $\overline{i_u}$ and $\overline{i_v}$ are obtained by the following equations.

$$
\begin{aligned}
\overline{i_u} &= \frac{1}{T_s} \int_{T_s \frac{1}{2} d_{wg}}^{T_s\left(\frac{1}{2} d_{wg} + d_{ug}\right)} i_{1M}\, dt \\
&= \frac{\sqrt{2} I_1}{\pi} \left\{ \cos\left(\frac{\pi}{2} d_{wg} + \Delta\phi\right) \right. \\
&\quad \left. - \cos\left(\frac{\pi}{2} d_{wg} + \pi d_{ug} + \Delta\phi\right) \right\}
\end{aligned}
\tag{18}
$$

$$
\begin{aligned}
\overline{i_v} &= \frac{1}{T_s} \int_{T_s\left(\frac{1}{2} d_{wg} + d_{ug}\right)}^{T_s\left(1 - \frac{1}{2} d_{wg}\right)} i_{1M}\, dt \\
&= \frac{\sqrt{2} I_1}{\pi} \left\{ \cos\left(\frac{\pi}{2} d_{wg} - \Delta\phi\right) \right. \\
&\quad \left. + \cos\left(\frac{\pi}{2} d_{wg} + \pi d_{ug} + \Delta\phi\right) \right\}
\end{aligned}
\tag{19}
$$

The following equation is established from (16), (18) and (19).

$$
\begin{aligned}
d_{ug} &= \frac{1}{\pi} \cos^{-1} \left\{ \frac{i_u^*}{i_w^*} \cos\left(\frac{\pi}{2} d_{wg} - \Delta\phi\right) \right. \\
&\quad \left. - \frac{i_v^*}{i_w^*} \cos\left(\frac{\pi}{2} d_{wg} + \Delta\phi\right) \right\} \\
&\quad - \frac{1}{2} d_{wg} - \frac{\Delta\phi}{\pi}
\end{aligned}
\tag{20}
$$

This equation can't be solved anymore because it is nonlinear. Equation (20) is approximated to a linear function using Taylor expansion as follows;

$$
\begin{aligned}
d_{ug} &= \frac{1}{2} \left(\frac{B}{C} - 1\right) d_{wg} \\
&\quad + \frac{1}{\pi} \cos^{-1}\left(-\frac{A}{\sqrt{2}}\right) - \frac{B}{4C} - \frac{\Delta\phi}{\pi}
\end{aligned}
\tag{21}
$$

A, B and C are expressed as follows;

$$
\begin{cases}
A = \dfrac{(i_u^* + i_v^*)\sin\Delta\phi + (i_u^* - i_v^*)\cos\Delta\phi}{i_u^* + i_v^*} \\[3mm]
B = \dfrac{(i_u^* + i_v^*)\sin\Delta\phi - (i_u^* - i_v^*)\cos\Delta\phi}{i_u^* + i_v^*} \\[3mm]
C = \sqrt{2 - A^2}
\end{cases}
\tag{22}
$$

d_{wg} is obtained by using (15) and (21) as follows;

$$
\begin{aligned}
d_{wg} &= \frac{2C\left(v_1^* - e_{vw}\right)}{\{(B - C)\, e_{uv} - 2C e_{vw}\}} \\
&\quad - \frac{\left\{4C\cos^{-1}\left(-\dfrac{A}{\sqrt{2}}\right) - \pi B - 4C\Delta\phi\right\} e_{uv}}{2\pi\{(B - C)\, e_{uv} - 2C e_{vw}\}}
\end{aligned}
\tag{23}
$$

Duty cycle d_{ug} is obtained by using (20) and (23). Next, d_{vg} is obtained by substituting d_{wg} and d_{ug} into equation (11). The duty cycles during the half-period T_S of the

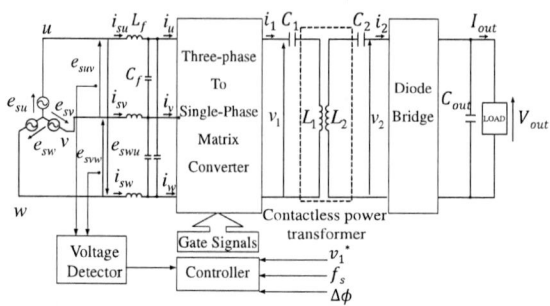

Fig. 8. Experimental system configuration

TABLE I. SPECIFICATIONS OF EXPERIMENTAL SYSTEM

Source voltage E, ω	200V , $2\,\pi \times 60$ rad/s
Output voltage V_{out}	$150 - 250$V
Output power P_{out}	2kW
Input filter L_f, C_f	300μH , 8.8μF
Inductance L_1, L_2	97μH , 97μH
Winding resistance r_1, r_2	141mΩ , 140mΩ
Number of turns $N_1 = N_2$	14
Compensated capacitance C_1, C_1	36nF , 36nF
Resonance frequency f_0	85.2kHz
Switching frequency f_S	85 ± 5kHz
Coupling coeffiecient k	0.26

negative primary voltage is obtained by replacing duty cycles of g-phase and h-phase in the case of the positive primary voltage.

IV. EXPERIMENTAL RESULTS

A. Experimental Conditions

Fig.8 shows the experimental system configuration and Table I shows the experimental conditions. The effective value of the source line voltage is 200 V and the frequency is 60 Hz. The output power is 2 kW. The output voltage V_{out} is in the range of 150-250V. The coupling coefficient k is 0.26, the inductances of L_1 and L_2 are 97 μH, the capacitance of the resonant capacitor C_1 and C_2 are 36 nF, the quality factors Q_1 and Q_2 are 360, the resonance frequency is 85 kHz and the trun ratio of wireless transformer is 1:1. The bidirectional switches of the matrix converter on the primary side connect two SiC-MOSFETs (Rohm, SCH2080KE) in reverse series. Also, the duty cycles are determined by detecting the source line voltages e_{suv}, e_{svw} and giving the primary voltage references v_1^*. In this experiment, primary voltage reference v_1^*, swithcing frequency f_S and phase difference $\Delta\phi$ between primary voltage v_1 and primary current i_1 are controlled in an open loop.

B. Experimental Waveforms

Fig.9 shows the experimental waveforms when the output voltage V_{dc} is 200 V. In this case, the switching frequency f_S is 85kHz, and phase difference $\Delta\phi$ is 0 deg. The waveforms are the source line voltage e_{suv}, the source current i_{su}, the primary voltage v_1, the primary current i_1, the secondary voltage v_2, the secondary current i_2, the output voltage V_{out} and the output current I_{out}.

The 2018 International Power Electronics Conference

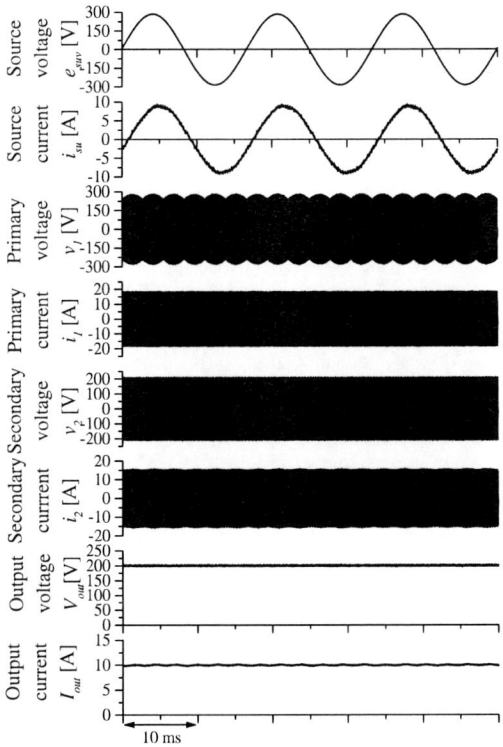

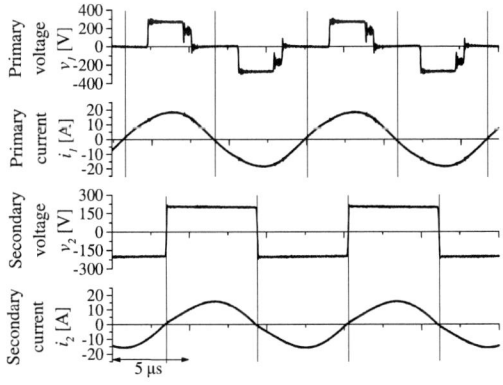

Fig. 9. Experimental waveforms at V_{out}=200V,f_S=85kHz

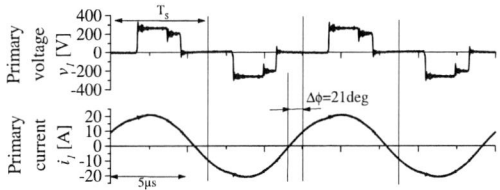

Fig. 11. Zoomed voltage and current waveforms of contactless power transformer at V_{out}=200V, f_S=80kHz

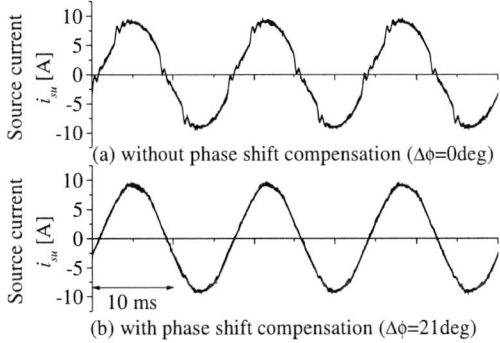

Fig. 12. Comparison of supply current waveforms

Fig. 10. Magnification waveforms

The output voltage V_{out} is controlled to the reference voltage of 200 V and the sinusoidal source current i_{su} is obtained. Fig.10 shows the magnification waveforms of the high frequency part. The waveforms are the primary voltage v_1, the primary current i_1, the secondary voltage v_2 and the secondary current i_2. Although the primary and secondary voltages v_1 and v_2 are pulsed waveforms, the primary and secondary currents i_1 and i_2 are the sinusoidal waveforms. In this case, the switching frequency f_S is set to 85kHz which is almost equal to the resonance frequency f_0. Therefore, the fundamental wave components of the primary voltage v_1 and the primary

current i_1 are almost in phase. Fig.11 shows the zoomed waveforms of the primary voltage v_1 and the primary current i_1 when the output voltage V_{out} is 200 V and the switching frequency f_S is 80 kHz. The phase difference $\Delta\phi$ between the primary voltage v_1 and the primary current i_1 is 21 deg. This result matches the value of Fig.4. In this case, supply current i_{su} is shown in Fig.12. By considering the phase difference $\Delta\phi$, The distortion of the power supply current i_{su} is significantly reduced.

C. THD of Source Current

Table II shows the measurement result of the source current THD with or without compensation for the phase difference $\Delta\phi$ of the primary voltage v_1 and the primary current i_1 when the switching frequency f_S is varied. In the case of no compensation for the phase difference $\Delta\phi$, THD is getting worse as the phase difference $\Delta\phi$ increases. On the other hand, the compensation for the phase difference $\Delta\phi$ can suppress the error between the actual primary current i_1 and the controller internal model current i_{1M}. Therefore, the THD is greatly reduced. From this result, it can be verified that even when the resonance frequency and the switching frequency are different, source current distortion can be greatly suppressed.

D. System Efficiency

Fig.13 shows the measurement results of the system efficiency from the three-phase voltage source (AC) to DC load (DC) at 2 kW output. For comparison, the system efficiency of the conventional system in Fig.1 is shown in

361

TABLE II. Source current THD reduction results by phase shift compensation

f_S[kHz]	$\Delta\phi$[deg]	THD[%]	
		w/o phase shift compensation	w/ phase shift compensation
80	21	7.44	2.0
85	0	1.45	1.45
90	9	3.52	1.75

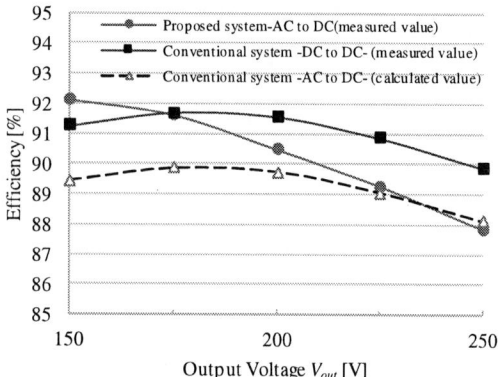

Fig. 13.   System efficiency measurement result

Fig.13. The system efficiency of the conventional system (dotted line) is obtained by multiplying the measured efficiency (solid line) between the inverter input and the DC output by the efficiency estimate value (98%) between the power supply and the inverter input shown in Fig.1. The inverter of the conventional system is a general H bridge configuration, and the same hardware configuration including used element and wireless power transfer coil is used as the proposed system. In Fig.13, the proposed system is more efficient than the conventional system. In the proposed system, when the output voltage V_{out} is 150 V, the highest efficiency of 92.1 % is obtained. The efficiency decreases as the output voltage V_{out} increases. Because the primary current i_1 increases as the output voltage V_{out} increases, and the conduction loss of the bidirectional switch of the primary side matrix converter increases. Therefore, as the output voltage V_{out} increases, the superiority in terms of efficiency decreases.

V.   Conclusion

This paper presents a circuit configuration of a wireless power transfer system using a three-phase to single-phase matrix converter and a control method for charging the load and reducing supply current distortion. Even when the switching frequency and the resonance frequency are different, the supply current can be controlled to sinusoidal current of THD 3 % or less by experiment. In addition, it is verified that the constant power of 2 kW can be supplied to the load in the range of the output voltage from 150 to 250 V by controlling the primary voltage of the wireless power transfer coil. Furthermore, a high efficiency of 92 % at the rated output 2 kW is obtained in the experiment, confirming the advantage in terms of efficiency over the conventional system.

References

[1]   G. A. Covic, J. T. Boys, "Modern trends in inductive power transfer for transportation applications", *IEEE J. Emerging Sel. Topics Power Electron.*, vol. 1, no. 1, pp. 28-41(2013)

[2]   B. M. Frauk, U. S. Jawarkar and T. G. Pal, A. S. Gugliya, "Wireless Power Transfer Electric Vehicle", International Advanced Reserch Journal in Science, Engineering and Technology, Vol. 4, Special Issue 3, pp. 111-119 (2017)

[3]   S. Abe and Y. Kaneko, "Technology Trends of Contactless Power transfer Systems for Electric Vehicle", IEEJ Industry Applications, 2-S11-6, ppⅡ-205-210 (2011) (in Japanese)

[4]   K. Kusaka and J. Ito, "Development Trend of Inductive Power Transfer Systems", IEEJ SPC-16-121, MD-16-085 (2016) (in Japanese)

[5]   T. Imura, H. Okabe and Y. Hori, "Basic Experimental Study on Helical Antennas of Wireless Power Transfer for Electric Vehicles by using Magnetic Resonant Couplings", IEEE Vehicle Power and Propulsion Conference, pp. 936-940 (2009)

[6]   T. Fujita, Y. Kaneko and S. Abe, "Contactless Power Transfer Systems using Sries and Paralell Resonant Capacitors", IEEJ Trans. IA, Vol. 127, No. 2, pp. 174-180 (2007) (in Japanese)

[7]   S. Chopra and P. Bauer, "Analysis and Design Considerations for a Contactless Power Transfer System", IEEE 33rd International Telecommunivations Energy Conference (INTELEC) (2011)

[8]   M. Vogelsberger, T. Wiesinger, H. Ertl, "Life-cycle monitoring and voltage-managing unit for DC-link electrolytic capacitors in PWM converters", *IEEE Trans. Power Electron.*, vol. 26, no. 2, pp. 493-503(2011)

[9]   K. Toshihiro and K. Throngnumchai, "A Study on Receiver Circuit Topology of Non-contact Charger for Electric Vehicle", *IEEJ Trans. Ind. Appl.*, Vol. 132, No. 11, pp. 1048-1054 (2012) (in Japanese)

[10]   L. L. Hao, A. P. Hu, G. A. Covic, "A direct AC?AC converter for inductive power-transfer systems", *IEEE Trans. Power Electron.*, vol. 27, no. 2, pp. 661-668(2012)

[11]   D. J. Thrimawithana, U. K. Madawala, "A novel matrix converter based bi-directional IPT power interface for V2G applications", *Proc. IEEE Int. Energy Conf. Exhib.*, pp. 495-500,(2010)

[12]   X. B. Nguyen, D. M. Vilathgamuwa, U. K. Madawala, "A SiC based matrix converter topology for inductive power transfer system", *IEEE Trans. Power Electron*, vol. 29, no. 8, pp. 4029-4038(2014)

[13]   D. S. B Weerasinghe, U. K. Madawala, D. J. Thrimawithana, D. M. Vilathgamuwa, "A three-phase to single-phase matrix converter based bi-directional IPT system for charging electric vehicles", *ECCE Asia Downunder (ECCE Asia)*, pp. 1240-1245(2013)

[14]   H. Abe, T. Akiyama, M. Ozaki, and H. Kohara, "Simple Equivalent Circuit for a Wireless Power Transfer System Using a Repeating Coil and Effects Confirming the Simplification in the Output Voltage Estimation", IEEJ Trans. IA, Vol.135, No.6, pp.679-688(2014)(in Japanese)

[15]   C. S. Wang, G. A. Covic, and O. H. Stielau, "Power Transfer Capability and Bifurcation Phenomena of Loosely Coupled Inductive Power Transfer System," IEEE Transactions on Industrial Electronics, Vol. 51, No. 1, pp. 148-157 (2004)

[16]   T. Imura, H. Okabe, Y. Hori, "Basic experimental study on helical antennas of wireless power transfer for electric vehicles by using magnetic resonant couplings", Proc. IEEE Vehicle Power and Propulsion Conference(VPPC'09), pp. 936-940(2009)

[17]   T. Imura, Y. Hori:, "Unified Theory of Electromagnetic Induction and Magnetic Resonant Coupling", IEEJ Trans. Ind. Appl, Vol. 135, No. 6, pp. 697-710(2015)(in Japanese)

Design of a reduced-order observer for Sensorless control of Dual-Active-Bridge converter

NGUYEN Duy Dinh[1,2] and Goro Fujita[1]

[1]Shibaura Institute of Technology, 3-7-5 Toyosu, Koto, Tokyo 135-8548, Japan
[2]Hanoi University of Science and Technology, Hanoi, Vietnam
Email: nguyen.duy.dinh.g7@shibaura-it.ac.jp

Abstract—A reduced-order observer is developed in this paper. The target is for sensorless current control of Dual Active Bridge (DAB) converters. Inherent from some relevance researches, this research also uses the Fundamental Harmonic Approximation analysis to establish the observer model. The observer uses the Terminal 2 voltage as the only feedback signal. No current sensors are required. Experimental results show that the load current can be estimated with high accuracy. The load angle which represent the reactive power can also be estimated with the error less than ±5 degrees. The observer can be used for current mode control or voltage mode plus current feedforward control of DAB converters.

Index Terms—Dual-active-bridge converter, reduced-order observer, frequency domain analysis, sensorless current control, single phase shift

I. INTRODUCTION

In micro-grid applications, a bidirectional converter is usually necessary to interconnect between the micro-grid and the large-scale power system. Conventionally, line frequency transformer (LFT) is employed. However, as the penetration of renewable energy becomes higher, negative effects such as, voltage fluctuation, frequency variation, etc. might be caused due to the intermittence of renewable resources. In such the circumstances, Solid-State-Transformer (SST) [1] depicted in Fig. 1 can solve the problem well.

Similar to the LFTs, SSTs also allow galvanic isolation and bidirectional power transmission. However, unlikes LFTs, SSTs can be used not only to convert from AC to AC, but all kinds of conversion. Furthermore, they can perform asynchronous interconnection, voltage/load regulation, protection, etc. The key factor that enables all the capabilities above is the Dual-Active-Bridge (DAB) converter in the construction of a SST.

DAB converter is illustrated in Fig. 1. A high-frequency isolated transformer is used to separate two DC sides. Each winding of the transformer connects to an inverter. Since the switching frequency is usually from several kHz to hundreds of kHz, the overall size could be reduced and power density could be increased.

Assuming a power transmission from medium voltage (MVDC) side to the low voltage (LVDC) side. When the LVAC voltage is not available, the DAB converter operates as a voltage source to regulate the LVDC bus. On the other hand, in the interconnected mode, the LVDC bus is regulated

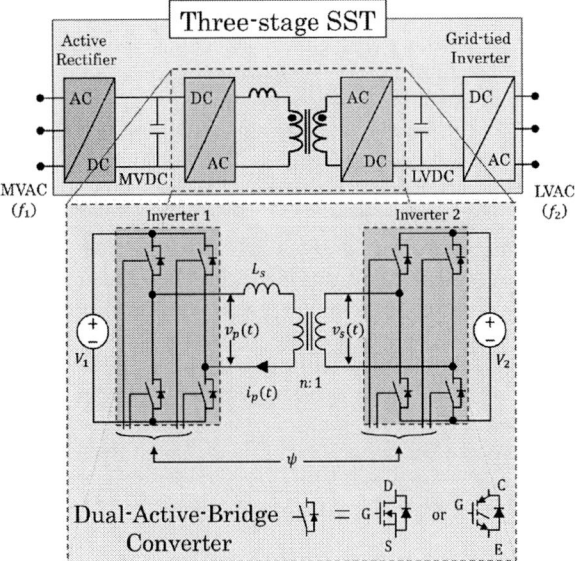

Fig. 1: Three-stage Solid State Transformer.

by the grid-tied inverter. Thus, the DAB converter plays as a current source. In this context, the current at LVDC bus is usually measured for current control purpose [2], [3]. Even in the voltage mode, that current is still useful for enhancing system dynamics [4], [5].

All the aforementioned studies required the information of the terminal current or the transmission current for the control system. However, in order to measure such the currents, expensive, high bandwidth current transducer is necessary. Besides, high sampling rate Analog-to-Digital conversion is also required to process the measured data. In order to ensure the accuracy, the sampling rate is usually at least four times faster than the switching frequency, that is sometime impossible when the switching frequency is in the range of hundreds kHz. The use of the sensor also introduces noise into the system that requires effort to calibrate and filter out. Therefore, reducing the number of sensors can simplify the signal processing procedures to save system resources for modulation and control algorithms.

It is possible to eliminate the need of current sensors. The technique current sensorless control has been popularly applied for DC/DC converters [6]–[8]. Recently, it has also been applied for controlling DAB converter [9]–[11]. In [9], the load current is estimated by a nonlinear disturbance observer then fed-forward to the voltage control loop to improve the dynamic performance. The study in [10] tried to get the data by communicating with other conversion stage. While Ge *et.al* in [11] established a set of mathematic equations to calculate that current.

The techniques reported in [12], [13] used observer to estimate active and reactive powers, then a decoupled current controller was employed to regulate the two power components individually. Dead-time was also regarded when deriving the reference for the reactive power control loop. Nevertheless, the observer required terminal current [12] or the transmission AC current [13] as input information. Furthermore, the observer model was established based on some approximations of the peak of the fundamental current component from the feedback signal. When the voltage ratio is different from unity, the accuracy of the approximation decreased due to the distortion of the current waveform.

This paper further improves the observer-based control strategies presented in [12] and [13]. The current sensors could be removed then an observer is developed to estimate the Terminal 2 current. In order to reduced the computation amount, reduced-order observer is preferred since the voltage can be measured directly with a transducer. After that, a current controller is designed to regulate the terminal current to demonstrate one application of the proposed observer.

II. LARGE SIGNAL MODELING

Let $v_1(t)$ and $v_2(t)$ be the voltages at each terminal. The conventional SPS modulation (both inverter switch with the same, constant duty cycle of approximately 50%) with the bridge shift angle of ψ is used in analysis. In order for simplification, primary referred equivalent circuit of fundamental components illustrated in Fig. 2 is employed. In the diagram, L_s and R_s are the total leakage inductance and resistance of the transformer referred to the primary side; $v_p(t)$ and $v_s'(t)$ are the fundamental components of the instantaneous voltage across the primary and secondary windings of the transformer:

$$\begin{cases} v_p(t) \approx \dfrac{4}{\pi} v_1(t) \cos(\omega_s t) \\ v_s'(t) \approx \dfrac{4n}{\pi} v_2(t) \cos(\omega_s t - \psi) \end{cases} \tag{1}$$

where ω_s is the angular switching frequency of the converter; n is the transformer winding ratio; $v_1(t)$ and $v_2(t)$ are the voltages at two DC terminals.

Let $i_d(t)$ and $i_q(t)$ be the cosine and sine components of the primary current, we have:

$$i_p(t) = i_d(t) \cos(\omega_s t) + i_q(t) \sin(\omega_s t) \tag{2}$$

Ignoring the power loss on Inverter 2, terminal 2 power can

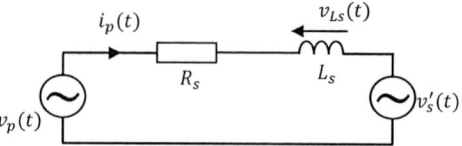

Fig. 2: Primary referred equivalent circuit.

be calculated by:

$$\begin{aligned} P_2 &= v_2(t) i_2(t) \\ &\approx \frac{2n}{\pi} v_2(t) (i_d(t) \cos\psi + i_q(t) \sin\psi) \end{aligned} \tag{3}$$

From (3), we have:

$$i_2(t) \approx \frac{2n}{\pi} \big(i_d(t) \cos\psi + i_q(t) \sin\psi \big) \tag{4}$$

According to [12] and from (4), the large signal model can be expressed as:

$$\begin{cases} \dfrac{di_d(t)}{dt} = -\omega_p i_d(t) - \omega_s i_q(t) + \dfrac{4v_1(t)}{\pi L_s} - \dfrac{4n v_2(t)}{\pi L_s} \cos\psi \\ \dfrac{di_q(t)}{dt} = \omega_s i_d(t) - \omega_p i_q(t) - \dfrac{4n}{\pi L_s} v_2(t) \sin\psi \\ \dfrac{dv_2(t)}{dt} = \dfrac{2n}{\pi C_2} \big(i_d(t) \cos\psi + i_q(t) \sin\psi \big) - \dfrac{v_2(t)}{R C_2} \end{cases} \tag{5}$$

where C_2 is the capacitance of the DC capacitor at Terminal 2, R is the load resistance and $\omega_p = \frac{R_s}{L_s}$.

It has been discussed in [12]–[14] that:

- $i_d(t)$ is proportional to the active power, and $i_q(t)$ is proportional to the circulating (reactive) power,
- the so-called *load angle* ϕ, defined by:

$$\phi(t) = \arctan \frac{i_q(t)}{i_d(t)}$$

is a good estimation of the phase different between the primary current and voltage,

- by applying an appropriate control algorithm, the load angle ϕ can be kept around a suitable value to reduced the conduction loss while regulating the output voltage or current.

However, the estimation of $i_d(t)$ and $i_q(t)$ required the information of either the terminal current [12] or the transmission current [13]. The performance of the observer designed in [13], [14] was relied on the approximation of the peak fundamental current. Since the current is not sinusoidal, the accuracy of the approximation decreased when the current becomes more distorted as the voltage ratio is further different from unity. From the large signal model (5), since Terminal 2 voltage is measurable, all the system states can be determined by an observer. The design of the observer is presented in the next section.

III. OBSERVER DESIGN

Similar to the studies reported in [12]–[14], switching frequency is also chosen as the additional control variable. Notes that, the observer technique based on the FHA method can also be applied for other modulation methods such as Dual-Phase-Shift, Enhanced-Phase-Shift or Triple-Phase-Shift.

By linearizing the large signal model (5) around one nominal operation point, the small signal model (6) can be obtained:

$$
\frac{d}{dt}\begin{bmatrix} i_d \\ i_q \\ v_2 \end{bmatrix} = \begin{bmatrix} -\Omega_p & -\Omega_s & -\rho_1\cos\psi \\ \Omega_s & -\Omega_p & -\rho_1\sin\psi \\ \rho_2\cos\psi & \rho_2\sin\psi & -\frac{1}{RC_2} \end{bmatrix} \begin{bmatrix} i_d \\ i_q \\ v_2 \end{bmatrix}
$$
$$
+ \begin{bmatrix} -\Omega_s I_q & \rho_1 V_2\sin\psi \\ \Omega_s I_d & -\rho_1 V_2\cos\psi \\ 0 & \rho_2 I_\psi \end{bmatrix} \begin{bmatrix} f_x \\ \varphi \end{bmatrix} + \frac{4}{\pi L_s}\begin{bmatrix} 1 \\ 0 \\ 0 \end{bmatrix} v_1 \tag{6}
$$

where $\rho_1 = \frac{4n}{\pi L_s}$; $\rho_2 = \frac{2n}{\pi C_2}$; $I_\psi = -I_d\sin\psi + I_q\cos\psi$; R is the equivalent load resistance; the low case characters imply small signal quantities and the upper case ones express values at equilibrium.

In (6), v_1 is a system disturbance. Usually, $v_1(t)$ is measured in order for the front-end active rectifier to regulate the MVDC bus. A large capacitor is usually present at the input side of Inverter 1 to eliminate the double line frequency oscillation caused by the rectifier. Hence, in one sampling period, the small variation v_1 can be assumed to be zero without losing the generality.

The state vector $\begin{bmatrix} i_d & i_q & v_2 \end{bmatrix}^T$ in (6) can be determined by employing an observer. As the characteristic matrix in (6) has the size of 3×3, and because $v_2(t)$ is measurable, a reduced-order observer is preferred in order to reduce the computation amount and to save some system resources. Let $\hat{\boldsymbol{x}} = \begin{bmatrix} \hat{i}_d & \hat{i}_q \end{bmatrix}^T$ be the estimated states, we aim to design the reduced-order observer as (7):

$$
\begin{cases} \dot{\hat{\boldsymbol{x}}} = \boldsymbol{A}_{11}\hat{\boldsymbol{x}} + \boldsymbol{A}_{12}v_2 + \boldsymbol{B}_1\boldsymbol{u} + \mathcal{L}(y - \hat{y}) \\ \hat{y} = \boldsymbol{A}_{21}\hat{\boldsymbol{x}} = \dot{v}_2 - \boldsymbol{A}_{22}v_2 - \boldsymbol{B}_2\boldsymbol{u} \end{cases} \tag{7}
$$

where $\boldsymbol{u} = [f_x \ \varphi]^T$; $\boldsymbol{A}_{11} = \begin{bmatrix} -\Omega_p & -\Omega_s \\ \Omega_s & -\Omega_p \end{bmatrix}$;

$\boldsymbol{A}_{12} = \begin{bmatrix} -\rho_1\cos\psi \\ -\rho_1\sin\psi \end{bmatrix}$; $\boldsymbol{A}_{21} = \begin{bmatrix} \rho_2\cos\psi & \rho_2\sin\psi \end{bmatrix}$;

$\boldsymbol{A}_{22} = -\frac{1}{RC_2}$; $\boldsymbol{B}_1 = \begin{bmatrix} -\Omega_s I_q & \rho_1 V_2\sin\psi \\ \Omega_s I_d & -\rho_1 V_2\cos\psi \end{bmatrix}$;

$\boldsymbol{B}_2 = \begin{bmatrix} 0 & \rho_2(-I_d\sin\psi + I_q\cos\psi) \end{bmatrix}$; and $\mathcal{L} = \begin{bmatrix} \ell_1 \\ \ell_2 \end{bmatrix}$; $\mathcal{L}$ is the observer gain.

Let us check the observability of system (7) by considering the observability matrix $\mathcal{O}(\boldsymbol{A}_{21}, \boldsymbol{A}_{11})$, where $\mathcal{O}(\boldsymbol{A}_{21}, \boldsymbol{A}_{11}) = [\boldsymbol{A}_{21} \ \boldsymbol{A}_{21}\boldsymbol{A}_{11}]^T$. With system parameters listed in Table I, we have:

$$
\text{rank}\big(\mathcal{O}(\boldsymbol{A}_{21}, \boldsymbol{A}_{11})\big) = 2
$$

Since $\boldsymbol{A}_{11} \in \mathbb{R}^{2\times 2}$, system (7) is observable.

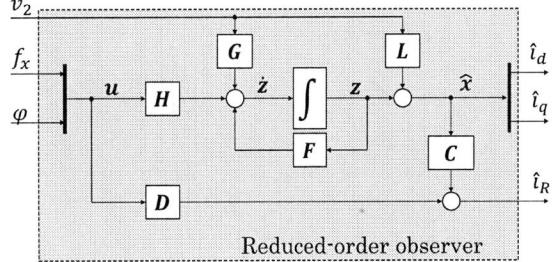

Fig. 3: Block diagram of the proposed observer.

Denoting the observer error as $\boldsymbol{e} = \boldsymbol{x} - \hat{\boldsymbol{x}}$, from (6) and (7), derivative error is calculated as:

$$
\dot{\boldsymbol{e}} = (\boldsymbol{A}_{11} - \mathcal{L}\boldsymbol{A}_{21})\boldsymbol{e} \tag{8}
$$

Next, we have to find value for $\mathcal{L}$ to make $(\boldsymbol{A}_{11} - \mathcal{L}\boldsymbol{A}_{21})$ stable. This can be done using Ackerman method by by placing all the eigenvalues of system (8) at twice to six times of system poles. And thus, observer error will vanish in limited time.

After determining the gain $\mathcal{L}$, the observer can be deployed by applying coordinator transformation (9) with the new state variable $\boldsymbol{z}$ to avoid the derivative $\dot{v}_2$ in (7):

$$
\hat{\boldsymbol{x}} = \boldsymbol{z} + \mathcal{L}v_2 \tag{9}
$$

Substituting (9) into (7), and denoting $\boldsymbol{F} = \boldsymbol{A}_{11} - \mathcal{L}\boldsymbol{A}_{21}$; $\boldsymbol{G} = \boldsymbol{A}_{12} - \mathcal{L}\boldsymbol{A}_{22} + \boldsymbol{F}\mathcal{L}$; and $\boldsymbol{H} = \boldsymbol{B}_1 - \mathcal{L}\boldsymbol{B}_2$, we have:

$$
\dot{\boldsymbol{z}} = \boldsymbol{F}\boldsymbol{z} + \boldsymbol{G}v_2 + \boldsymbol{H}\boldsymbol{u} \tag{10}
$$

Finally, Terminal 2 current can be derived by linearizing (4):

$$
i_2 \approx \hat{i}_2 = \boldsymbol{C}\hat{\boldsymbol{x}} + \boldsymbol{D}\boldsymbol{u} \tag{11}
$$

where $\boldsymbol{C} = \frac{2n}{\pi}\begin{bmatrix}\cos\psi & \sin\psi\end{bmatrix}$; and $\boldsymbol{D} = \frac{2n}{\pi}\begin{bmatrix} 0 & I_\psi \end{bmatrix}$.

The block diagram of the observer is depicted in Fig. 3. Obviously, the proposed observer can estimate the two current components, and also determine the load current without any current sensors. Those who prefers feedforward control like studies reported in [4], [9] or decoupled current control for simultaneously managing active and reactive powers as presented in [12] can take this advantage of the proposed observer. In this paper, a feedback control system for regulating Terminal 2 current is designed to demonstrate one application of the proposed observer.

IV. CONTROL SYSTEM DESIGN

Fig. 4 describes the example control system. Only current loop is examined in order to demonstrate one example application of the proposed observer. The current controllers $G_c(s)$ is designed with the following specifications:

- Type: Proportional - Integral
- Bandwidth: 2.0 kHz
- Phase margin: 75 degrees

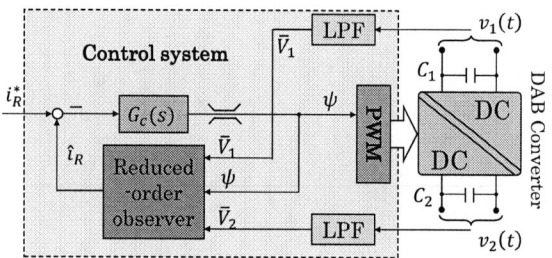

Fig. 4: Current control system.

V. EXPERIMENTAL RESULTS

The laboratory scaled experiment system is depicted in Fig. 5. Summary of system parameters is shown in Table I. A programmable power supply configured at constant voltage mode is connected to Terminal 1, whereas a DC electronic load is connected to Terminal 2. The winding ratio of the transformer is $1:1$. The primary referred leakage inductances and resistance measured at 20 kHz is 67.5 µH and 50 mΩ, respectively.

Aiming to compare among methods, the terminal currents are also measured by using two shunt resistors. The voltages across the resistors are then amplified by isolation amplifiers and filtered out low pass filter (LPF). Terminal voltages are also determined in the same manner using voltage divider, isolation amplifier and LPF. The amplifier used in the experiment is ACPL-C79A with 200 kHz bandwidth, hence, the amplitude and phase of the measured voltage and current could be ensured.

A TMS320F28335 experiment kit is used to deploy the proposed observer and the control system. Eigenvalues of the observer are placed at twice of system poles. Modulation frequency range is from 10 kHz to 30 kHz. Sampling rate is fixed at 20 kHz. In order to ensure the accuracy of the measured terminal current (for comparison purpose), the Analog-to-Digital rate is set to 100 kHz. The data is then taken average every 5 sampling.

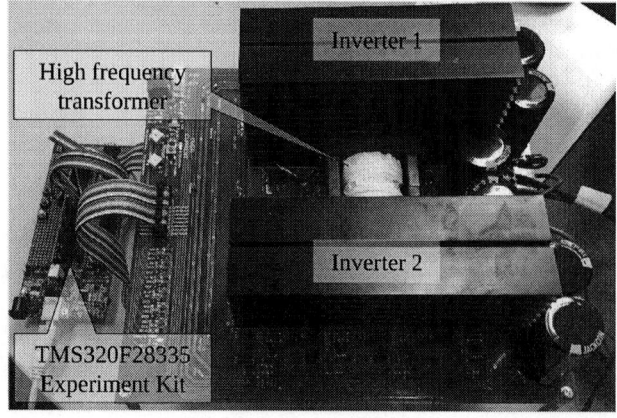

Fig. 5: Laboratory-scaled experiment system.

A. Open-loop evaluation

Fig. 6 and Fig. 7 illustrate the performance of the proposed observer (RBOS2) in term of Terminal 2 current and load angle estimations in comparison with the actual values and with the observer designed in [12] (ROBS1) based on the Terminal 1 current in some different operation conditions. The actual load angle is determined by measuring the time difference between the zero crossing points of primary voltage and current, then converting into degree scale. The estimated load angle $\hat{\phi}(k)$ is calculated by taking arctangent of $\hat{i}_q(k)$ over $\hat{i}_d(k)$, $\hat{\phi}(k) = \arctan\left(\hat{i}_q(k)/\hat{i}_d(k)\right)$. There are three types of experiments to evaluate the effect of the variation of ψ, F_s and V_2 on the observation results consequently.

When ψ varies $\pm 33\%$ around its nominal value (30 degrees) while fixing both Terminal voltages and switching frequency, the error between the estimated value $\hat{I}_{R,ROBS_2}$ and the measured one is up-to-about ± 0.1 A (Fig. 6(a)). Although that error is pretty small, it is equivalent to $\pm 8\%$ of the actual current. For the same experiment, ROBS1 can help achieve a slightly better performance with the error up-to-about ± 0.05 A. Regarding load angle estimation, both can provide good approximation of the actual one with the error around 2 degrees (Fig. 7(a)). This results show that, the both observers have good robustness against phase shift variation.

In the next experiment, we fix the voltage and the phase shift at their nominals and change the switching frequency in the range of $\pm 50\%$ its designated value of 20 kHz. The results are illustrated in Fig. 6(b) and Fig. 7(b). Both observers perform not as good as expectation. The estimation error by ROBS1 in the load current is from -0.235 A to 0.369 A , while that by ROBS2 is even higher, from -0.309 A to 0.467 A (i.e. -23.3% to 42.1% of the measured data). ROBS2 can do better when deriving the load angle as the error is in the range of ± 5 degrees. The results imply that the proposed method is quite sensitive to the change of switching frequency.

Finally, we try to investigate the effect of Terminal 2 voltage on the observation performance. The voltage is increased gradually from 20 V to 30 V ($\pm 20\%$ of the nominal value) while F_s and ψ are kept at their nominals. In term of Terminal current, ROBS2 can provide a little bit better estimation results than ROBS1 as the error is up-to-about ± 0.12 A (see Fig. 66c). Nevertheless, it is not so successful in detecting the load angle as seen in Fig. 7(c). While ROBS1 can obtain the load angle error of no bigger than ± 7.5 degrees, ROBS1 does the estimation with the error of more than ± 12 degrees. This can be explained by the phase drift due to the distortion of the current waveform when the Terminal voltages are not match. In the next section, we propose a procedure to compensate for that phase drift phenomenon.

B. Phase drift compensation

According to [15], the load angle ϕ could be estimated by a simple equation:

$$\phi_{est} = \frac{nV_2}{V_1 + nV_2}\left(\psi + \frac{V_1 - nV_2}{nV_2} \times \frac{\pi}{2}\right) \qquad (12)$$

The 2018 International Power Electronics Conference

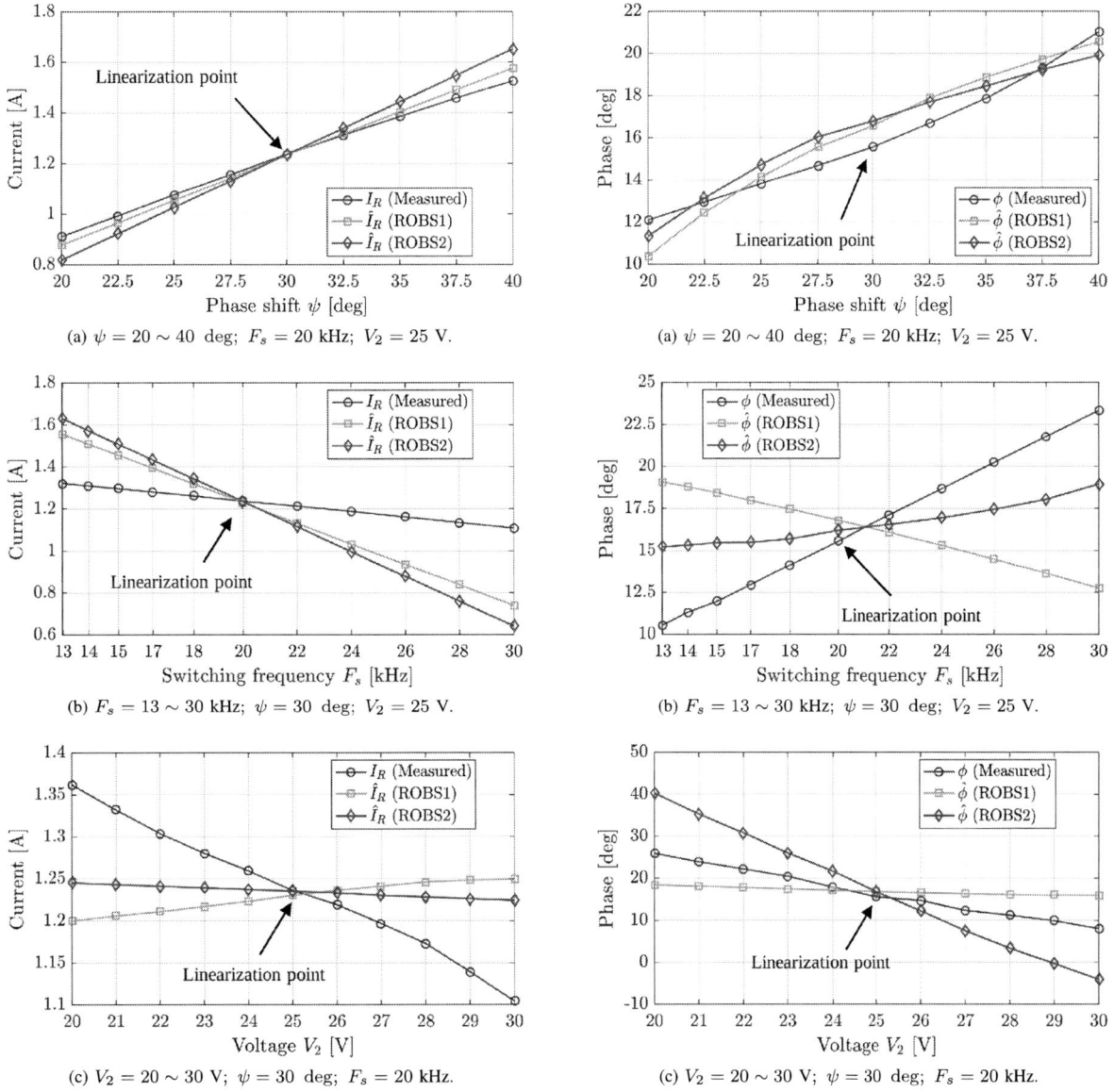

(a) $\psi = 20 \sim 40$ deg; $F_s = 20$ kHz; $V_2 = 25$ V.

(b) $F_s = 13 \sim 30$ kHz; $\psi = 30$ deg; $V_2 = 25$ V.

(c) $V_2 = 20 \sim 30$ V; $\psi = 30$ deg; $F_s = 20$ kHz.

Fig. 6: Load current I_R estimation performance.

(a) $\psi = 20 \sim 40$ deg; $F_s = 20$ kHz; $V_2 = 25$ V.

(b) $F_s = 13 \sim 30$ kHz; $\psi = 30$ deg; $V_2 = 25$ V.

(c) $V_2 = 20 \sim 30$ V; $\psi = 30$ deg; $F_s = 20$ kHz.

Fig. 7: Load angle ϕ estimation performance.

As experimented in [15], (12) can be used to calculate ϕ very accurately with the error less than 2 degrees.

Now, denoting $\hat{\phi}$ as the load angle obtained by the designed observer:

$$\hat{\phi} = \arctan \frac{\hat{i}_q}{\hat{i}_d} \tag{13}$$

Since both ϕ_{est} and $\hat{\phi}$ can be derived easily, there is no difficulty in calculating:

$$\delta = \hat{\phi} - \phi_{est} \tag{14}$$

After that, a vector rotation is applied to correct the error in the observed data. Fig. 8 demonstrates the compensation

principle. The rotation is accomplished using (15).

$$\begin{bmatrix} \hat{i}_{d,com} \\ \hat{i}_{q,com} \end{bmatrix} = \begin{bmatrix} \cos\delta & \sin\delta \\ -\sin\delta & \cos\delta \end{bmatrix} \begin{bmatrix} \hat{i}_d \\ \hat{i}_q \end{bmatrix} \tag{15}$$

Fig. 9 shows the observation performance before and after applying the compensator (15). In both sub-figures, the compensated data is represented as orange, triangle-marked curves. As seen, the observation performance is much improved. The error in terminal current decreases to only up-to -0.06 A, while the load angle is mostly coincident with the actual one.

Although containing some division and trigonometric functions, the compensator (15) executes quite fast on DSP F28335. At 150 MHz system clock, the observer (including

367

The 2018 International Power Electronics Conference

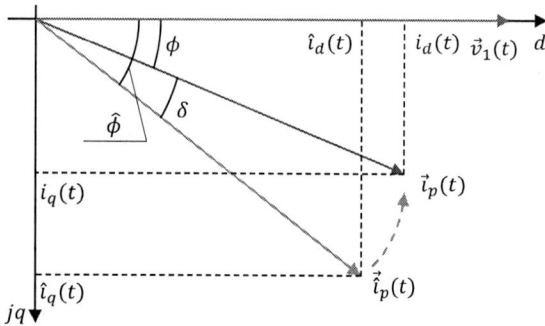

Fig. 8: Vector rotation for compensating the phase drift.

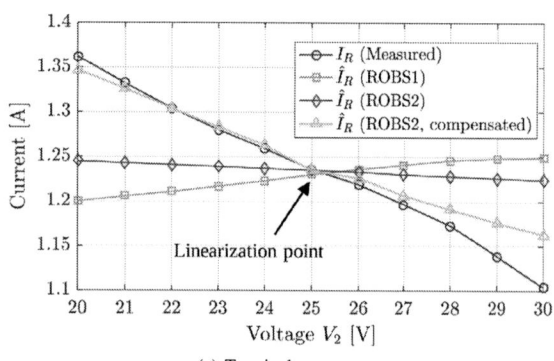

(a) Terminal current.

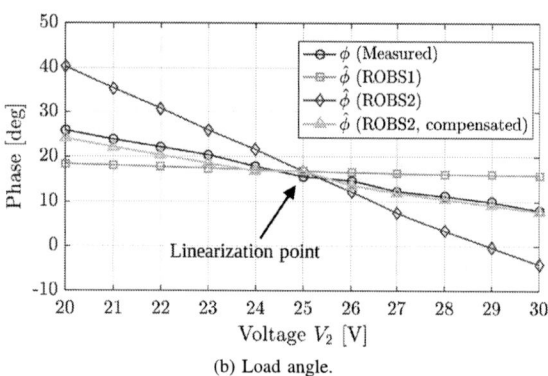

(b) Load angle.

Fig. 9: Load angle and Terminal 2 current before and after compensation, $V_2 = 20 \sim 30$ V; $\psi = 30$ deg; $F_s = 20$ kHz.

compensation procedures) take only 2.4 µs to accomplished. Since the sampling time is 50 µs, there is plenty of time or other functions and algorithms. This also allows to increase the sampling speed to more than 100 kHz if required.

C. Closed-loop evaluation

Fig. 10 expresses the Terminal 2 current response when deploying the observer-based current controller designed above. The voltage at Terminal 1 is 25 V, whereas that of Terminal 2 is 22 V. At first, the referent current is preset at 1 A. After some time, it is suddenly stepped up to 1.5 A.

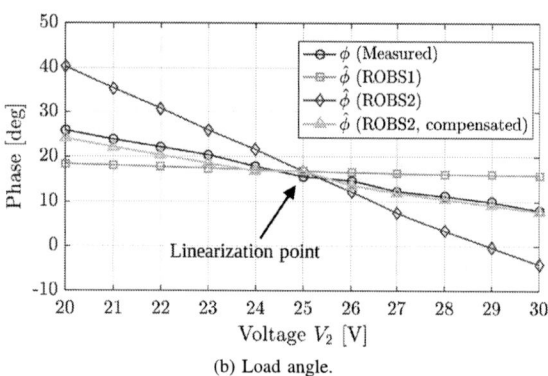

Fig. 10: Terminal current response.

As seen, the current is stabilized well at the first stage. The expected current is 1 A, however, there is a steady state error of 0.1 A. At the second stage, the current is stabilized at 1.45 A, mostly close to the desired one with a minor error of 0.05 A. There is an overshoot of 6%. The rising time is less than 0.8 ms. The response time could be lessened by carefully tuning the PI controller. And the steady state error can be compensated easily by the outer voltage controller.

The results above shows that, the observer serves well.

VI. CONCLUSION

This paper developed a reduced-order observer for sensorless current control of DAB converter. Although using only information of Terminal voltages, the proposed ROBS2 can predict Terminal 2 current quite well. Besides, it can also provide meaningful information about the load angle, which is a tool to express the reactive power in the system. The ROBS2 seems robust against the change of phase shift, but sensitive to the frequency modulation. When the voltage ratio is other than unity, the phase drift phenomenon causes estimation error on the observer. It has been resolved with the phase drift compensation tool proposed in this paper.

It is shown that, the ROBS2 can be used for feed-back current control with reasonable accuracy. The available information of the direct and quadrature current can also be used for other purposes such as feedforward current mode control, decoupled-current control, etc. Besides, the design of the observer is very simple and uses few system resources. And because the need of current sensors is eliminated, it is possible to reduce the system cost by using the proposed technique.

APPENDIX

REFERENCES

[1] X. She, A. Q. Huang, and R. Burgos, "Review of solid-state transformer technologies and their application in power distribution systems," *IEEE Journal of Emerging and Selected Topics in Power Electronics*, vol. 1, no. 3, pp. 186–198, 2013.

[2] F. Krismer and J. W. Kolar, "Accurate small-signal model for the digital control of an automotive bidirectional dual active bridge," *IEEE Trans. Power. Electron.*, vol. 24, no. 12, pp. 2756–2768, 2009.

The 2018 International Power Electronics Conference

TABLE I: System Parameters.

Parameter	Symbol	Value	Unit
Nom. Terminal 1 voltage	V_1	25	V
Nom. Terminal 2 voltage	V_2	25	V
Nom. Load resistance	R	20	Ω
Transformer ratio	n	1:1	
Total inductance	L_s	67.5	μH
Total resistance	R_s	50	mΩ
Nom. switching frequency	F_s	20	kHz
Nom. phase shift	Ψ	30	deg
Sampling time	T_z	50	μs
Dead-time	T_d	1000	ns
DC capacitors	C_1, C_2	1000	μF

[3] S. Dutta, S. Hazra, and S. Bhattacharya, "A digital predictive current-mode controller for a single-phase high-frequency transformer-isolated dual-active bridge dc-to-dc converter," *IEEE Trans. Ind. Electron.*, vol. 63, no. 9, pp. 5943–5952, 2016.

[4] D. Segaran, D. G. Holmes, and B. P. McGrath, "Enhanced load step response for a bidirectional dc–dc converter," *IEEE Trans. Power. Electron.*, vol. 28, no. 1, pp. 371–379, 2013.

[5] W. Song, N. Hou, and M. Wu, "Virtual direct power control scheme of dual active bridge dc-dc converters for fast dynamic response," *IEEE Transactions on Power Electronics*, 2017.

[6] M. Pahlevaninezhad, S. Eren, H. Pahlevani, I. Askarian, and S. Bagawade, "Digital current sensorless control of current-driven full-bridge dc/dc converters," *IEEE Transactions on Power Electronics*, vol. PP, no. 99, pp. 1–1, 2017.

[7] Q. Zhang, R. Min, Q. Tong, X. Zou, Z. Liu, and A. Shen, "Sensorless predictive current controlled dc–dc converter with a self-correction differential current observer," *IEEE Transactions on Industrial Electronics*, vol. 61, no. 12, pp. 6747–6757, 2014.

[8] G. Cimini, G. Ippoliti, G. Orlando, and M. Pirro, "Current sensorless solution for pfc boost converter operating both in dcm and ccm," in *Control & Automation (MED), 2013 21st Mediterranean Conference on*. IEEE, 2013, pp. 137–142.

[9] F. Xiong, J. Wu, Z. Liu, and L. Hao, "Current sensorless control for dual active bridge dc-dc converter with estimated load-current feedforward," *IEEE Transactions on Power Electronics*, 2017.

[10] X. She, A. Q. Huang, and X. Ni, "Current sensorless power balance strategy for dc/dc converters in a cascaded multilevel converter based solid state transformer," *IEEE Transactions on Power Electronics*, vol. 29, no. 1, pp. 17–22, 2014.

[11] J. Ge, Z. Zhao, L. Yuan, and T. Lu, "Energy feed-forward and direct feed-forward control for solid-state transformer," *IEEE Transactions on Power Electronics*, vol. 30, no. 8, pp. 4042–4047, 2015.

[12] D. D. Nguyen, G. Fujita, Q. Bui-Dang, and M. C. Ta, "Reduced-order observer-based control system for dual-active-bridge dc/dc converter," *IEEE Transactions on Industry Applications*, vol. PP, no. 99, pp. 1–1, 2018.

[13] D. D. Nguyen and G. Fujita, "Observer-based decoupling power control for frequency modulated dual-active-bridge converter," in *Power Electronics and Motion Control Conference (IPEMC-ECCE Asia), 2016 IEEE 8th International*. IEEE, 2016, pp. 754–760.

[14] D. D. Nguyen, M. L. Nguyen, T. Nguyen-Duc, and G. Fujita, "Observer-based nonlinear control for frequency modulated dual-active-bridge converter," in *Energy Conversion Congress and Exposition (ECCE), 2016 IEEE*. IEEE, 2016, pp. 1–8.

[15] D. D. Nguyen, D. T. Nguyen, and G. Fujita, "New modulation strategy combining phase shift and frequency variation for dual-active-bridge converter," *IEEJ Journal of Industry Applications*, vol. 6, no. 2, pp. 140–150, 2017.

The 2018 International Power Electronics Conference

Improved Load Transient Response of a Dual-Active-Bridge Converter

Sheng-Zhi Zhou, Chuan Sun, Song Hu, Guo Chen, Xiaodong Li*

Faculty of Information Technology, Macau University of Science and Technology, Macau, China

*E-mail: xdli@must.edu.mo

Abstract—In this work, the transient response of a dual-active-bridge (DAB) converter to load-change command is investigated under dual-phase-shift (DPS) control. During the transient process, there might be dc bias current and overshoot current arising in the inductor and transformer due to improper transient modulation. A dedicated transient modulation method is then proposed, which adjusts all phase-shift ratios by referring to a floating reference gating signal. Consequently a fast transient response can be obtained without dc bias current and overshoot current. The dynamic movement of the floating reference signal can be determined easily by a uniformed expression irrespective of the different steady state operation modes. The proposed transient control has been proved through simulation and experimental results.

I. INTRODUCTION

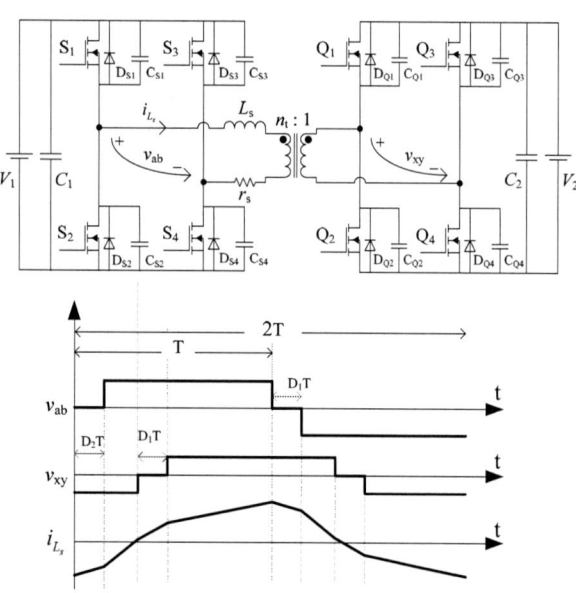

Fig. 1. A DAB converter and its steady-state waveforms with DPS control scheme.

As a typical topology of isolated bidirectional dc-dc converter, a dual-active-bridge (DAB) converter has been widely used in different applications due to high efficiency, high power density and simple phase-shift control [1]–[5]. The typical schematics of a DAB converter and its steady state waveforms using dual-phase-shift (DPS) scheme are presented in Fig. 1. Two switches in each bridge arm are always operated complementarily with a 50% duty cycle. It can be seen that

DPS scheme is a three-dimensional gating scheme with an identical inner phase-shift ratio D_1 for two bridges and one outer phase-shift ratio D_2 between two bridges. The converter may work in four different steady-state modes with distinctive power characteristics for positive power flow from V_1 to V_2. There are also four symmetric steady-states modes for reverse power flow.

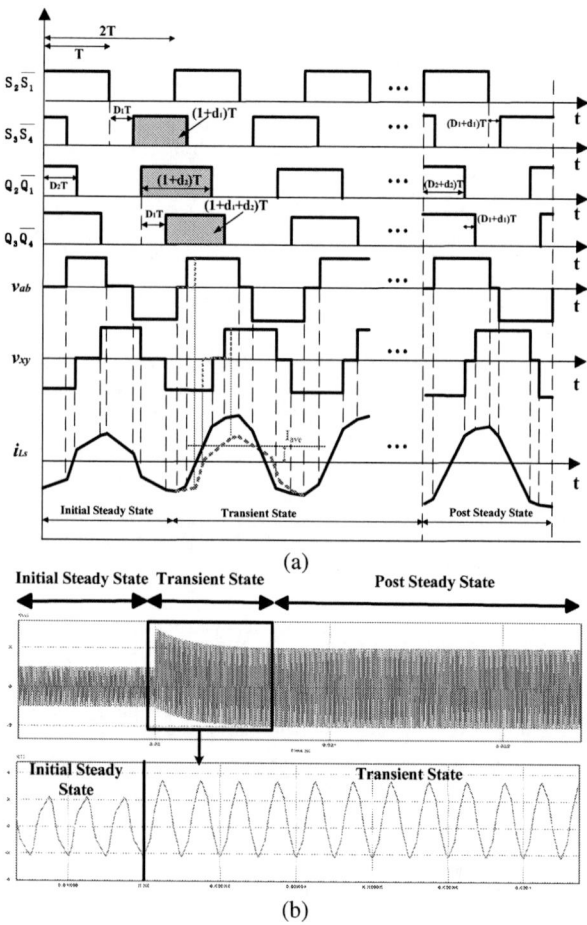

Fig. 2. Transient state in a DAB converter with a fixed reference: (a) theoretical waveforms; (b) simulated waveforms.

To adjust the bidirectional power level, some or all phase-shift ratios are to be changed according to some preset control strategies. However, the way to implement those changes of the phase-shift ratios can have quite different influence on the transient process. In an example shown in Fig. 2(a), those phase-shift ratios are to be changed from D_1, D_2 to $D_1 + d_1, D_2 + d_2$. By referring to the fixed gating signal $S_2, \bar{S}_1$, the gating signals $S_3, \bar{S}_4$ are shifted by d_1, the gating signals $Q_2, \bar{Q}_1$ are shifted by d_2, the gating signals $Q_3, \bar{Q}_4$ are shifted by $d_1 + d_2$. The direct effect is that a non-zero dc bias current I_{ave} emerging instantly in the power inductor L_s and the high-frequency (HF) transformer once the modifications are done. Considering the parasitic parameters in practise, this temporal dc bias current will decay gradually through parasitic resistance r_s in the converter as shown from the simulation results in Fig. 2(b). The long transient process with extra loss or potential saturation problem should be paid attention to and be prevented in a proper way. Similar phenomena also exist when other phase-shift gating schemes are used. In reported works, there are several solutions reported to depress such a dc bias current and speed up the whole transient process [2]–[5]. However, those solutions were presented in the scope of one dimensional scheme (single-phase-shift –SPS) or two dimensional scheme (extended-phase-shift –EPS). As a simple version of three dimensional scheme, DPS provides a balance of flexibility in control and complexity of implementation. Therefore, the focus of this paper is put on how to achieve a dc bias-free transient response of a DAB converter under DPS control by means of an optimized transient modulation (OTM). Besides, the obtained results can be extended to the most complicated scheme – triple-phase-shift control (TPS) with necessary modification in the future.

II. Optimized Transient Modulation for Load Transient Response

The basic principles of the proposed OTM are concluded as: (1) a gating signal is selected as a floating reference whose position is assumed to be shifted by a dynamic phase-shift ratio x during the transient process; (2) other gating signals will be shifted to achieve the required modification of phase-shift ratio by referring to the floating reference; (3) assuming that the value of x is an optimized one, the post-transient average current should be zero or the post-transient instantaneous current should match the corresponding current of the destined steady state. Therefore, the value of x can be solved according to the assumption.

Due to the existence of different steady state modes under DPS scheme, the number of different types of load transition will be high. Due to the limited space, the application of the proposed OTM will be exemplified by two different case in the following part. One example is for an inner-mode transition (the initial and final steady state are same), the other example is for an inter-mode transition (the initial and final steady state are different).

A. Inner-Mode Transition

An inner-mode transition using OTM is exemplified by a transition case within a particular forward power mode, which is featured with $D_2 > D_1 > 0$; $D_1 + D_2 < 1$. Fig. 3(a) shows the theoretical diagram of the transient response inside

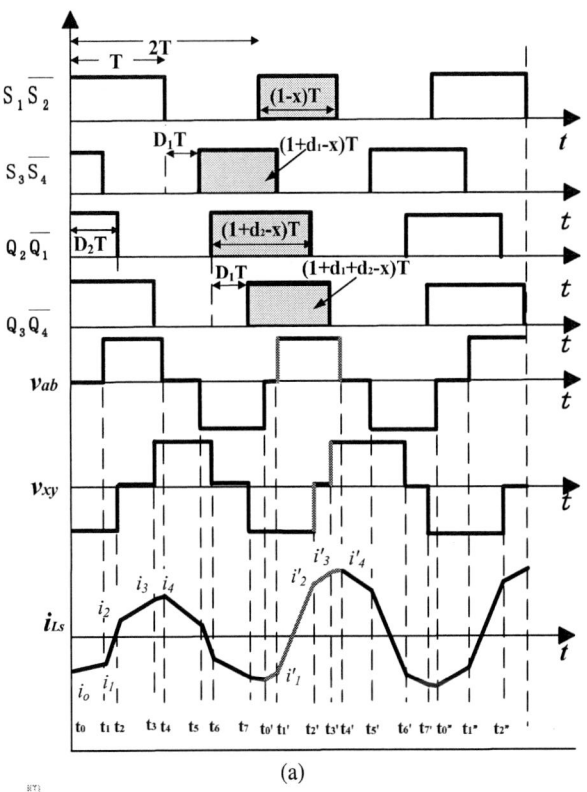

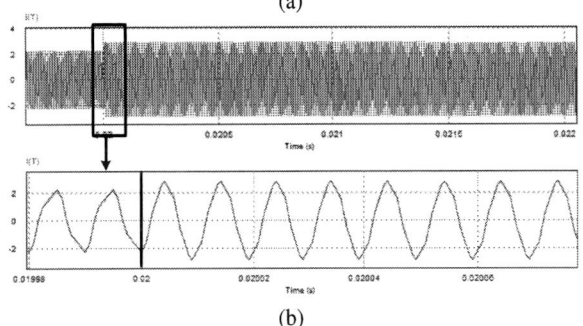

Fig. 3. Transient response in a DAB converter with OTM for an inner-mode transition: (a) theoretical waveforms; (b) simulated waveforms.

the mode. Before t_0' the DAB converter is in the initial steady state. The transient state lasts from t_0' to t_4'.

Neglecting the parasitic resistance r_s, the instantaneous currents in the initial steady state can be calculated as following:

$$\begin{cases} i_1 = i_o + \frac{n_t V_2}{L_s} D_1 T \\ i_2 = i_1 + \frac{V_1 + n_t V_2}{L_s}(D_2 - D_1)T \\ i_3 = i_2 + \frac{V_1}{L_s} D_1 T \\ i_4 = i_3 + \frac{V_1 - n_t V_2}{L_s}(1 - D_2 - D_1)T = -i_o. \end{cases} \quad (1)$$

Therefore, i_4 can be found as:

$$i_4 = -i_o = \frac{1}{2 f_s L_s}(V_1(1 - D_1) + n_t V_2(2D_2 + D_1 - 1)). \quad (2)$$

During the transient state, the following operations are applied on the gating signals according to the proposed transient modulation:

1) The turn-off moment of S_1 is selected as the reference, which will be shifted earlier by $x \cdot T$, i.e. the on-state duration of S_1 is shortened to $(1 - x)T$.

2) The turn-off moment of S_3 is shifted forward by $(d_1 - x)T$, i.e. the on-state duration of S_3 is changed to $(1 + d_1 - x)T$.

3) The turn-off moment of Q_1 is shifted forward by $(d_2 - x)T$, i.e. the on-state duration of Q_1 is changed to $(1 + d_2 - x)T$.

4) The turn-off moment of Q_3 is shifted forward by $(d_1 + d_2 - x)T$, i.e. the on-state duration of Q_3 is changed to $(1 + d_1 + d_2 - x)T$.

5) The on-state duration of all other switches in each HF period are still kept at T.

Therefore, the instantaneous currents in transient state are calculated as following:

$$
\begin{cases}
i_1' &= i_o + \frac{n_t V_2}{L_s}(D_1 + d_1 - x)T \\
i_2' &= i_1' + \frac{V_1 + n_t V_2}{L_s}(D_2 + d_2 - D_1 - d_1)T \\
i_3' &= i_2' + \frac{V_1}{L_s}(D_1 + d_1)T \\
i_4' &= i_3' + \frac{V_1 - n_t V_2}{L_s}(1 - D_2 - d_2 - D_1 - d_1)T
\end{cases}
\tag{3}
$$

From (3), i_4' can be found as:

$$
i_4' = \frac{1}{2f_s L_s}(V_1(1 - D_1 - 2d_1) \\
+ n_t V_2(2D_2 + 4d_2 + D_1 + 2d_1 - 1 - 2x)).
\tag{4}
$$

According to (2), the instant current located at t_4' in the post steady state is expected to be:

$$
i^* = \frac{V_1(1 - D_1 - d_1) + n_t V_2(2(D_2 + d_2) + D_1 + d_1 - 1)}{2f_s L_s}
\tag{5}
$$

In order to make the average value of inductor current become zero after the transient state, $i_4' = i^*$ is expected to be true. Therefore, the dynamic phase-shift angle x can be found:

$$
x = d_2 + \frac{M - 1}{2M}d_1
\tag{6}
$$

where the converter gain is defined as $(M = \frac{n_t V_2}{V_1})$.

Fig. 3(b) gives the simulation plots of such an inner-mode transient response using OTM. It can be seen that the duration of total transient state is less than one HF period. The converter seems be able to jump from one load level to another without introducing any disturbance in the transformer current.

B. Inter-Mode Transition

An inter-mode transient response using OTM is exemplified by a case between two forward power modes. The initial mode same as the one in last example. And the destinated mode is featured with $D_2 > D_1 > 0$; $D_1 + D_2 > 1$.

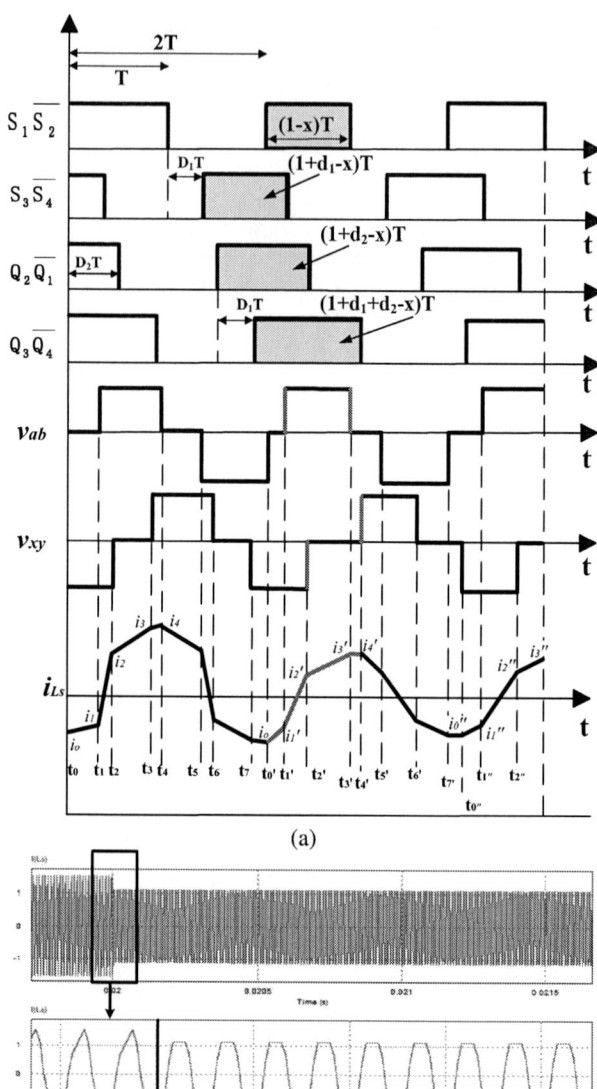

(a)

(b)

Fig. 4. Transient response in a DAB converter with OTM for an inter-mode transition: (a) theoretical waveforms; (b) simulated waveforms.

Following the principle of OTM, the same operation on each gating signals are implemented with a floating reference – the turn-off moment of S_1. During the transient state, the variation of i_{Ls} can be predicted as:

$$
\begin{cases}
i_1' &= i_o + \frac{n_t V_2}{L_s}(D_1 + d_1 - x)T_{hc} \\
i_2' &= i_1' + \frac{V_1 + n_t V_2}{L_s}(D_2 + d_2 - D_1 - d_1)T_{hc} \\
i_3' &= i_2' + \frac{V_1}{L_s}(1 - D_2 - d_2)T_{hc} \\
i_4' &= i_3'
\end{cases}
\tag{7}
$$

where i_o is same as the one from (2). At the end of the transient

372

state, i_4' is found to be:

$$i_4' = \frac{(V_1(1 - D_1 - 2d_1) + n_t V_2(1 - D_1 + 2d_2 - 2x))}{2f_s L_s} \quad (8)$$

In the final steady state, the instantaneous current at the positive rising edge of v_{xy} is expected to be:

$$i^* = \frac{(V_1(1 - (D_1 + d_1)) + n_t V_2(1 - (D_1 + d_1)))}{2f_s L_s}. \quad (9)$$

If the value of x is an optimized one, the current i_4' should be same as i^*, which indicates a seamless transition. Consequently, the dynamic movent x of the floating reference can be solved from $i_4 = i^*$, which yields to a same expression as (6).

There are still other different types of load transient responses involving different initial and final steady state modes. Although the calculation process are not same, the expression of x is always same as (6), wihc make it easy for the OTM to be implemented in a micro-controller-based platform.

III. VERIFICATION THROUGH EXPERIMENTAL TEST

To verify the feasibility and correctness of the theoretical analysis in the previous section, the load transient response of a DAB converter with DPS control is tested on a lab prototype converter. The parameters of the prototype converter is $V_1 = 100V$, $V_2 = 60V$, $n_t = 1:1$, $M = 0.6$, $L_s = 121.8\mu H$. The switching frequency is set at 100 kHz, i.e. $T = 5\mu s$.

The measured waveforms of an inner-mode transition is presented in Fig. 5. The initial steady state has: $D_1 = 0.06$; $D_2 = 0.13$, the final steady state has: $D_1 + d_1 = 0.13$; $D_2 + d_2 = 0.208$. With $d_1 = 0.07$, $d_2 = 0.078$, the dynamic x is calculated as 0.063. Fig. 5(a) shows the case with a fixed reference, in which the current amplitude jumps from 1.05 A to 1.77Aand takes almost 10 HF periods to decay to 1.4 A, which indicates a non-zero dc bias current. When the OTM is applied shown in Fig. 5(b), the transition is almost finished instantly without noticeable unequal positive/negative current amplitude in transient response.

The measured waveforms of an inter-mode transition is presented in Fig. 6(a). The initial steady state has: $D_1 = 0.13$; $D_2 = 0.13$, the final steady state has: $D_1 + d_1 = 0.36$; $D_2 + d_2 = 0.38$. With $d_1 = 0.23$, $d_2 = 0.244$, the dynamic x is calculated as 0.167. When the OTM is applied, the current amplitude decreased from 1.25A to 0.95A immediately. Thus, the transient response is finished in half period without dc bias current. In Fig. 6(b) another example of inter-mode transition is given. The initial steady state has: $D_1 = 0.10$; $D_2 = 0.13$, the final steady state has: $D_1 + d_1 = 0.184$; $D_2 + d_2 = 0.26$. With $d_1 = 0.084$, $d_2 = 0.13$, the dynamic x is calculated as 0.092. The current amplitude jumped from 1.05A to 1.20A immediately in the transient process.

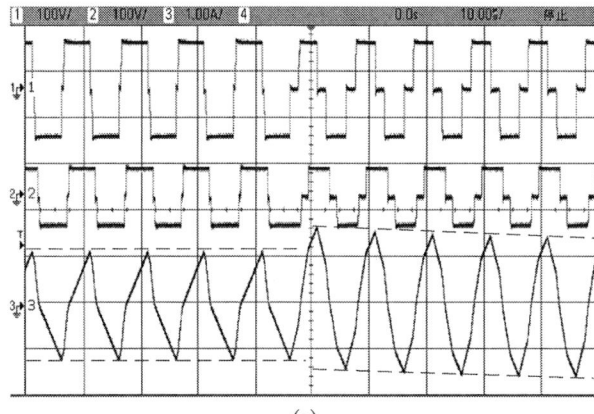

(a)

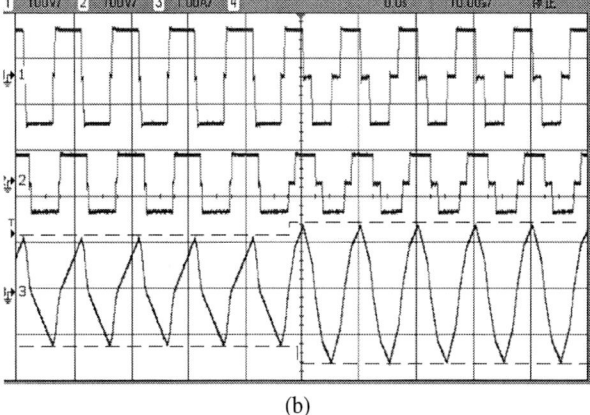

(b)

Fig. 5. Experimental plots of inductor current during load transient response for an inner mode transition (a) using a fixed reference (b)using OTM.

IV. CONCLUSIONS

In this paper, a transient modulation on the gating signals is proposed for the load transition in a DAB converter. In the scenario of DPS scheme, the possible dc bias current during the transient process can be depressed and the whole transient duration is independent on the parasitic resistance. The advantage is that the calculation to search for the dynamic parameter for the floating reference is quite simple, which only requires the converter gain and the increments of the two phase-shift ratios.

V. ACKNOWLEDGEMENT

This work was supported by Science and Technology Development Fund (FDCT) of Macau SAR under Grant Agreement no. 060/2017/A.

REFERENCES

[1] R. W. De Doncker, D. M. Divan and M. H. Kheraluwala, "A three-phase soft-switched high power density DC/DC converter for high power applications", *IEEE Trans. Ind. Applicat.*, vol. 27, no. 1, pp. 63-73, Jan./Feb. 1991.

[2] X. Li and Y.-F. Li, "An optimized phase-shift modulation for fast transient response in a dual-active-bridge converter". *IEEE Trans. on Power Electronics*, Vol. 29, No. 6, pp.2661-2665, June 2014.

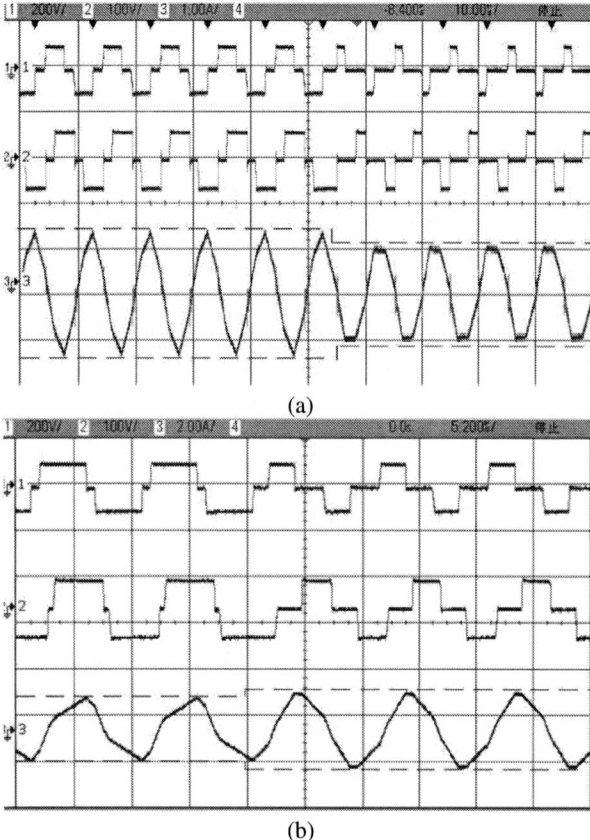

(a)

(b)

Fig. 6. Experimental plots of inductor current during load transient response for inter-mode transitions using OTM.

[3] B. Zhao, Q. Song, W. Liu and Y. Zhao, "Transient DC Bias and Current Impact Effects of High-Frequency-Isolated Bidirectional DC-DC Converter in Practice', *IEEE Trans. Power Electr.*, vol.31, no.4, pp. 3203-3216, 2016

[4] K. Takagi and H. Fujita, "Dynamic control and performance of an isolated dual-active-bridge DC-DC converter", *Power Electronics and ECCE Asia (ICPE-ECCE Asia), 2015 9th International Conference on*, pp. 1521-1527.

[5] S. T. Lin , X. Li, C. Sun and Y. Tang. "Fast transient control for power adjustment in a dual-active-bridge converter". *Electronics Letters*, 2017, 53(16): 1130-1132.

The 2018 International Power Electronics Conference

Modulation and Active Midpoint Control of a Three-Level Three-Phase Dual-Active Bridge DC-DC Converter under Non-Symmetrical Load

Philipp Joebges, Anton Gorodnichev, Rik W. De Doncker
Institute for Power Generation and Storage Systems,
E.ON Energy Research Center, FEN Research Campus, RWTH Aachen University, Aachen, Germany
E-mail: post_pgs@eonerc.rwth-aachen.de

Abstract—This paper discusses the operation of a three-phase three-level dual-active bridge (3L-DAB3) dc-dc converter in medium-voltage dc (MVDC) grids with a symmetrical monopole configuration. Due to the demand of bipolar mode operation, active control of the midpoint potential is essential to ensure balanced grid voltages even under non-symmetrical loads. The proposed solution is achieved with a novel modulation scheme, which allows a power and midpoint control to enable voltage balancing. Due to the large number of degrees of freedom, look-up tables (LUTs) are generated and used to operate with a maximum converter efficiency. The proposed method is implemented and validated in simulations and on a low-power prototype setup.

I. INTRODUCTION

Power generation from renewable energy sources has seen a rapid increase over the last years [1]–[3]. Flexible medium-voltage dc (MVDC) grids show a promising approach for the efficient distribution of electrical energy between the various sources and consumers [4], [5]. In order to increase the power transfer capability of a MVDC cable, a bipolar approach with positive and negative voltage levels is promising as it doubles the total pole-to-pole voltage while keeping the pole-to-ground voltage, which defines the insulation effort. To connect such bipolar MVDC systems, two bipolar dc-dc converters (e.g. two-level dual-active bridge (2L-DAB3)) can be combined as depicted in Fig. 1a. Each converter is connected to one pole and the midpoint of the bipolar configuration. The converters can be controlled independently and therefore balance the positive and negative voltage at the output under non-symmetrical conditions. An alternative solution is the use of a three-level dual-active bridge (3L-DAB3) with a controllable neutral point potential. This approach enables a midpoint control with one converter, shown in Fig. 1b, and is investigated within this work.

II. THREE-PHASE THREE-LEVEL DUAL-ACTIVE BRIDGE CONVERTER

In order to decrease the required blocking voltage of the semiconductor devices and to provide a controllable midpoint, a three-phase dual-active bridge can be implemented with neutral-point-clamped (NPC) converter phase legs. The topology, shown in Fig. 2, consists of a primary and a secondary NPC-bridge which are linked by a three-phase medium-frequency transformer in Y-Y configuration. In this work, the

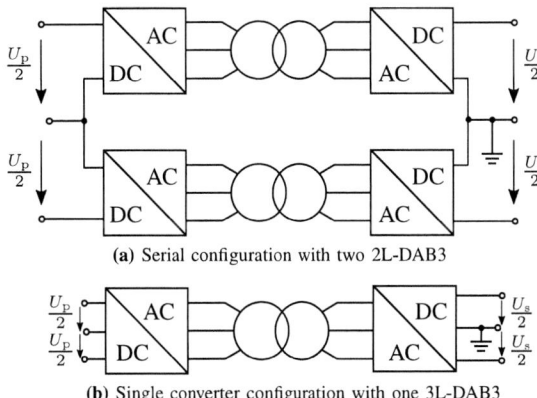

(a) Serial configuration with two 2L-DAB3

(b) Single converter configuration with one 3L-DAB3

Fig. 1: DC substation topologies to connect bipolar dc grids

primary side of the converter is connected to a dc grid with a constant input voltage $U_\mathrm{p}/2 = 2.5\,\mathrm{kV}$. The 3L-DAB3 with a nominal power of $5\,\mathrm{MW}$ is used to connect two bipolar medium-voltage dc grids with midpoint conductor. A detailed analysis of this topology under symmetrical dc-link conditions is carried out in [6] and [7] focusing on modulation strategies, power-transfer, soft-switching characteristics and rms-current reduction. Further, the primary and secondary side duty cycles are defined in [7] as modulation parameters. The parameter k_1 indicates the duty cycle of the primary side outer switches $S_{1,5,9}$ and $S_{4,8,12}$, while the inner switches $S_{3,7,11}$ and $S_{2,6,10}$ are operated with inverse gate signals (cf. Fig. 2). In analogy to k_1, the control parameter k_2 is defined as the duty cycle of outer power switches of the secondary side. For $k_1 = k_2 = \frac{1}{2}$ the voltage waveforms approach those of a 2L-DAB3 operated under single-phase-shift modulation as introduced in [8].

III. MODULATION SCHEME FOR ACTIVE NEUTRAL-POINT BALANCING

The control of the neutral-point voltage for NPC inverters has been the subject of many studies [9]–[12]. The space vector pulse-width modulation (SVPWM) is a state-of the art method to control the inverter current and thus the output voltage. For NPC inverters, the application of small space vectors in redundant switching states can be used to control

375

The 2018 International Power Electronics Conference

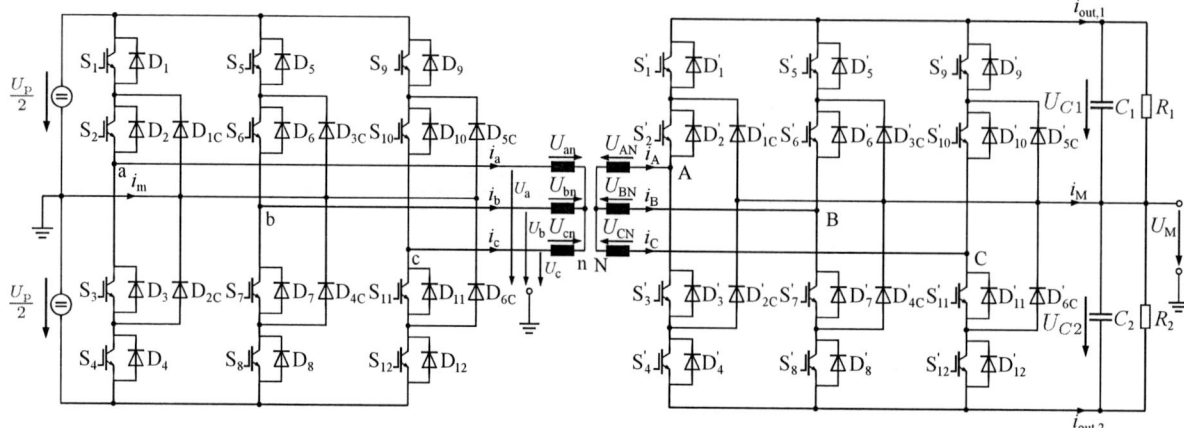

Fig. 2: Three-phase three-level dual-active bridge converter under NPC configuration

the neutral-point current and thus balance the dc-link voltages [11]. Unfortunately, this control method cannot be applied to a 3L-DAB3, since it demands a high ratio between the switching frequency and the fundamental frequency.

For a three-level single-phase dual-active bridge, a capacitor voltage balancing control was proposed in [13]. This approach suggests modifying the phase voltage waveforms by adjusting the switching instances. A three-phase topology increases the modulation complexity, since all three staircase phase voltages have to be modified while being phase-shifted by $\frac{2\pi}{3}$. However, an essential requirement is, that the average of the applied transformer voltage is zero, hence with no dc offset which would lead to transformer saturation and finally damage of the converter.

A. Active Control of the Midpoint Current

In order to suppress the neutral-point voltage U_M on the secondary side for non-symmetrical loads ($R_1 \neq R_2$), the modulation scheme has to provide a degree of freedom to control the average midpoint current i_M. This leads to an uneven average current distribution ($\bar{i}_{out1} \neq -\bar{i}_{out2}$) of the converter output in Fig. 2. For balanced output voltages, $U_{C1} = U_{C2} = \frac{U_s}{2}$ must be fulfilled. In this case, the required average neutral-point current is

$$\bar{i}_M = -\left(\bar{i}_{out1} + \bar{i}_{out2}\right) = \frac{U_s}{2} \cdot \left(\frac{1}{R_2} - \frac{1}{R_1}\right) \quad (1)$$

In the symmetrical load condition, positive and negative pulses of the midpoint current are canceled out in average. Therefore, $\bar{i}_M$ is zero (cf. Fig 4a).

However, for a non-symmetrical load, the voltage splits unsymmetrically over the dc-link capacitors, if no further measures are taken. To adjust the average midpoint current in order to control $U_M = U_{C1} - U_{C2} = 0$, the current injected into the neutral point during a positive pulse must be different from the current withdrawn during the negative pulse. This leads to an average current flow and therefore results in charging or discharging of C_1 and C_2, respectively.

Therefore, a new parameter λ is defined, which delays the turn-on transition of the lower outer switches $[S_4, S_8, S_{12}]_{\text{primary}}$, $[S'_4, S'_8, S'_{12}]_{\text{secondary}}$ and the inversely operated upper inner switches $[S_2, S_6, S_{10}]_{\text{primary}}$, $[S'_2, S'_6, S'_{10}]_{\text{secondary}}$ of the converter. The impact of λ on the transformer voltage is shown in Fig. 3b for one phase of the secondary side. It can be seen, that the turn-on times of the lower outer switches $S'_{4,8,12}$ and inner top switches $S'_{2,6,10}$, which are controlled inversely, are delayed.

In Fig. 4, the transformer and midpoint currents are compared for $\lambda = 0$ and $\lambda > 0$. Time intervals that allow an influence on the midpoint at the converter output are denoted $T'_{i \to j}$. These intervals indicate the time between the turn-off of S'_i and the turn-on of S'_j. S'_i and S'_j are the outer switches of the related phase leg. In Fig. 4, these time intervals are marked by gray areas for a particular phase. For the given waveform, a positive average non-zero midpoint current can be achieved by increasing the transition time intervals $[T'_{4 \to 1}, T'_{8 \to 5}, T'_{12 \to 9}]$ which are the times of change in switching state from the negative voltage to positive and reducing the intervals $[T'_{1 \to 4}, T'_{5 \to 8}, T'_{9 \to 12}]$ from positive to negative vise versa. During the time intervals $T'_{1 \to 4}$ and $T'_{4 \to 1}$, the phase current i_A is equal to the midpoint current i_M. Additional pulses are produced by the other phase currents in the same manner.

In oder to further influence the midpoint current, the parameters k_1 and $k_2 \in \left[\frac{1}{3}, \frac{1}{2}\right]$ as introduced in the previous section are used. In general, for $k_1 = k_2$ under a voltage ratio $d = \frac{U'_s}{U_p} = 1$, the voltage waveforms U_{an} and U_{AN} are identical and only shifted by the control variable φ which is the relative shift between primary and secondary switching signals. This results in relatively small charge transmitted during one midpoint current pulse as during the charging intervals $T'_{i \to j}$ the according phase current is close to zero (cf. Fig. 4a). Therefore, the impact of λ on the average midpoint current $\bar{i}_m$ is relatively small for $k_1 = k_2$ and $d = 1$ (cf. Fig. 4b).

The 2018 International Power Electronics Conference

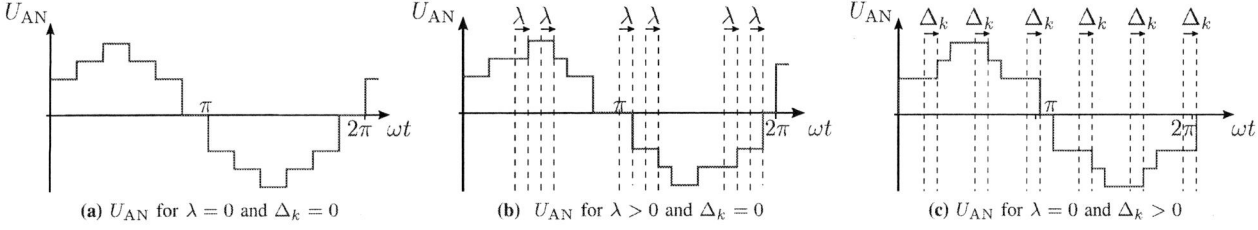

(a) U_{AN} for $\lambda = 0$ and $\Delta_k = 0$ **(b)** U_{AN} for $\lambda > 0$ and $\Delta_k = 0$ **(c)** U_{AN} for $\lambda = 0$ and $\Delta_k > 0$

Fig. 3: Impact of λ and Δ_k on the secondary-side transformer winding voltage U_{AN}

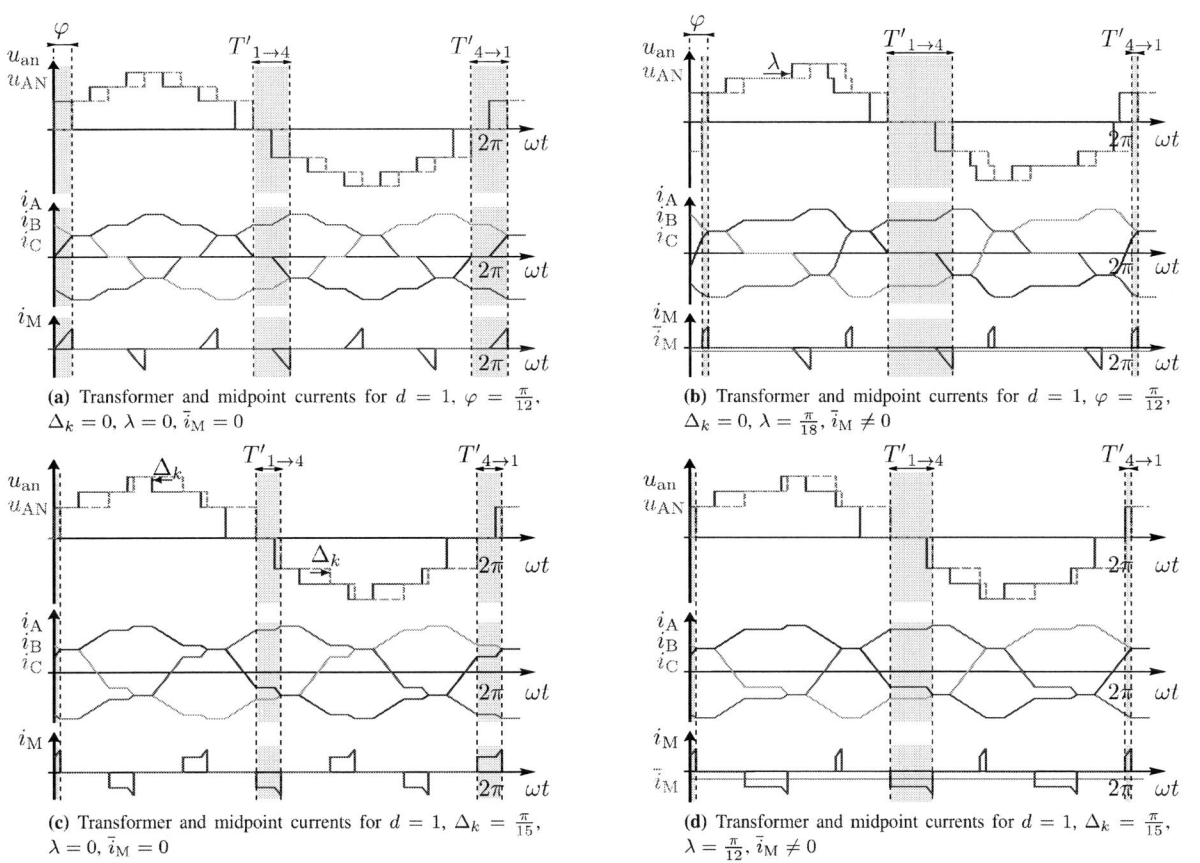

(a) Transformer and midpoint currents for $d = 1$, $\varphi = \frac{\pi}{12}$, $\Delta_k = 0$, $\lambda = 0$, $\bar{i}_M = 0$

(b) Transformer and midpoint currents for $d = 1$, $\varphi = \frac{\pi}{12}$, $\Delta_k = 0$, $\lambda = \frac{\pi}{18}$, $\bar{i}_M \neq 0$

(c) Transformer and midpoint currents for $d = 1$, $\Delta_k = \frac{\pi}{15}$, $\lambda = 0$, $\bar{i}_M = 0$

(d) Transformer and midpoint currents for $d = 1$, $\Delta_k = \frac{\pi}{15}$, $\lambda = \frac{\pi}{12}$, $\bar{i}_M \neq 0$

Fig. 4: Impact of different values of λ and Δ_k on transformer and midpoint currents for $d = 1$.

To increase the voltage-time area per pulse, different values for the parameters k_1 and k_2 can be chosen, which reduces the symmetry of the phase currents, causing i_M to further deviate from zero during the described transition intervals (cf. Fig. 4d).

In order to reduce the complexity of the modulation scheme, a new parameter Δ_k is defined which describes the relationship between k_1 and k_2:

$$\Delta_k = (k_2 - k_1) \cdot \pi \qquad (2)$$

$$k_1 = \frac{5}{12} - \frac{\Delta_k}{2\pi} \qquad (3)$$

$$k_2 = \frac{5}{12} + \frac{\Delta_k}{2\pi} \qquad (4)$$

From (3) the symmetric influence of Δ_k on k_1 and k_2 is obtained, which increased the influence of the shifting parameter λ.

For the first phase, a positive value of Δ_k reduces the duty cycles of S_1 and S_4 at the primary side and increases the duty cycles of S'_1 and S'_4 at the secondary side of the converter. The impact on the secondary-side transformer voltage can be seen in Fig. 3c where the waveform is manipulated by delaying the transitions. The effect of a positive Δ_k on the secondary side first phase transition times $T'_{1\rightarrow4}$ and $T'_{4\rightarrow1}$ can be seen by in Fig. 4c. The applied voltage between primary and secondary side during the transition is increased non-symmetrically and thus a higher current is flowing. The

377

The 2018 International Power Electronics Conference

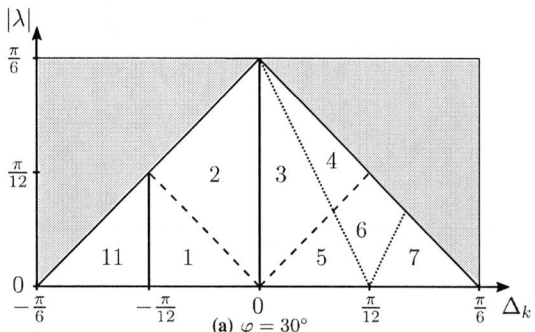
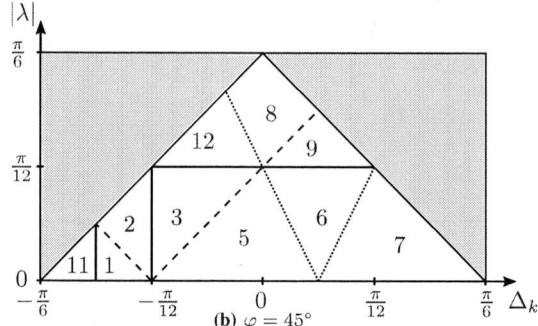

Fig. 5: Operational cases and boundaries depending on Δ_k and $|\lambda|$ for two exemplary values of φ

same principal is applied at the other two phases. It has to be avoided, that both outer switches of either primary or secondary side are switched on simultaneously, which would results that one of the transition time intervals $T_{i \to j}$ will become negative. Thus, the sum of $|\lambda|$ and $|\Delta_k|$ is not allowed to be larger than $\frac{\pi}{6}$, which is the length of one transition sequence for $k_1 = k_2 = \frac{5}{12}$ and $\lambda = 0$. Hence, the allowed operating range of λ is defined depending on Δ_k as

$$|\lambda| \le \frac{\pi}{6} - |\Delta_k| \qquad (5)$$

Comparing $\bar{i}_M$ in Fig. 4b with the one in Fig. 4d, it can be seen that applying a $\Delta_k \ne 0$ can increase the average midpoint current. In total, three degrees of freedom exists as already introduced above in order to control output power and neutral-point voltage, which are summarized in the following:

φ represents the phase shift between the primary and the secondary side of the converter and is used to control the total output power.

λ implements an additional delay between the upper and the lower power switches within one phase leg of the converter and thus introduces a non-zero average midpoint current at the converter output which is therefore indispensable for voltage balancing under asymmetrical loads. λ is applied evenly to the primary and the secondary side of the converter in order to reduce the effect on the total output power.

Δ_k has an impact on the total power transmitted through the converter. It represents the difference between the secondary side duty cycle k_2 and the primary side duty cycle k_1 in equivalent radians. It can increase the influence of the shift parameter λ.

In contrast to λ, the parameter Δ_k has a significant impact on the output power, which is discussed in the following paragraph.

B. Operation Cases of the 3L-DAB3 with λ-Δ-Modulation

The modulation of the converter is extended by the use of the two control parameters λ and Δ_k, named as λ-Δ-modulation in the following.

In order to identify the operation case of the inverter, all possible switching states and transition timings are analyzed.

During one switching period T_s, 24 different switching transitions occur. The sequence of the transitions is affected by φ, λ and Δ_k, which results in up to 13 separate operational cases where a different analytical equation of output power P_{out}, power ratio r_{power} and average midpoint current $\bar{i}_M$ is valid. Out of that, 12 different operational cases occur for $\varphi \le 45°$. Each case is precisely defined by a specific sequence of switching transitions with respect to the operating range of φ, λ and Δ_k related to (3) - (5):

$$0 \le \varphi \le \frac{\pi}{3} \qquad (6a)$$

$$-\frac{\pi}{6} \le \Delta_k \le \frac{\pi}{6} \qquad (6b)$$

$$\frac{\pi}{6} \ge |\lambda| + |\Delta_k| \qquad (6c)$$

The obtained borders and the related cases of operation are depicted in Fig. 5 for two exemplary phase-shifts. For each combination of the three control parameters φ, λ and Δ_k a related operational case can be found. For every case of operation, a different analytical solution is derived to calculate the converter's output power P_{out}, the power ratio $r_{power} = \frac{P_{R2}}{P_{R1}}$, which describes the ratio between the output power of the lower and upper part of the output dc-link, and the average midpoint current $\bar{i}_M$. The validity of every equation is proven by simulative results. In the following, operational case number one is given as an example:

$$P_{out} = \frac{d U_p^2 \left(7\pi (\varphi + \Delta_k) - 3 (\varphi + \Delta_k)^2 - 3\Delta_k^2 \right)}{12\pi\omega L_s} \qquad (7a)$$

$$r_{power} = \frac{\left(\begin{array}{c} 7\pi\Delta_k + 7\pi\varphi - 6\Delta_k\varphi - 3\varphi^2 \\ -6\Delta_k^2 + 3\lambda(\pi - \pi d - 2d\Delta_k) \end{array} \right)}{\left(\begin{array}{c} 7\pi\Delta_k + 7\pi\varphi - 6\Delta_k\varphi - 3\varphi^2 \\ -6\Delta_k^2 - 3\lambda(\pi - \pi d - 2d\Delta_k) \end{array} \right)} \qquad (7b)$$

$$\bar{i}_M = \frac{U_p}{\omega L_s} \cdot \frac{\lambda(\pi - \pi d - 2d\Delta_k)}{2\pi} \qquad (7c)$$

From (7a) it is apparent that P_{out} is not dependent on the parameter λ for the first operational case. Furthermore, the influence of φ and Δ_k on the total output power is almost equal. Even though the output power P_{out}, the output power ratio r_{power} and the average midpoint current $\bar{i}_m$ are calculated

378

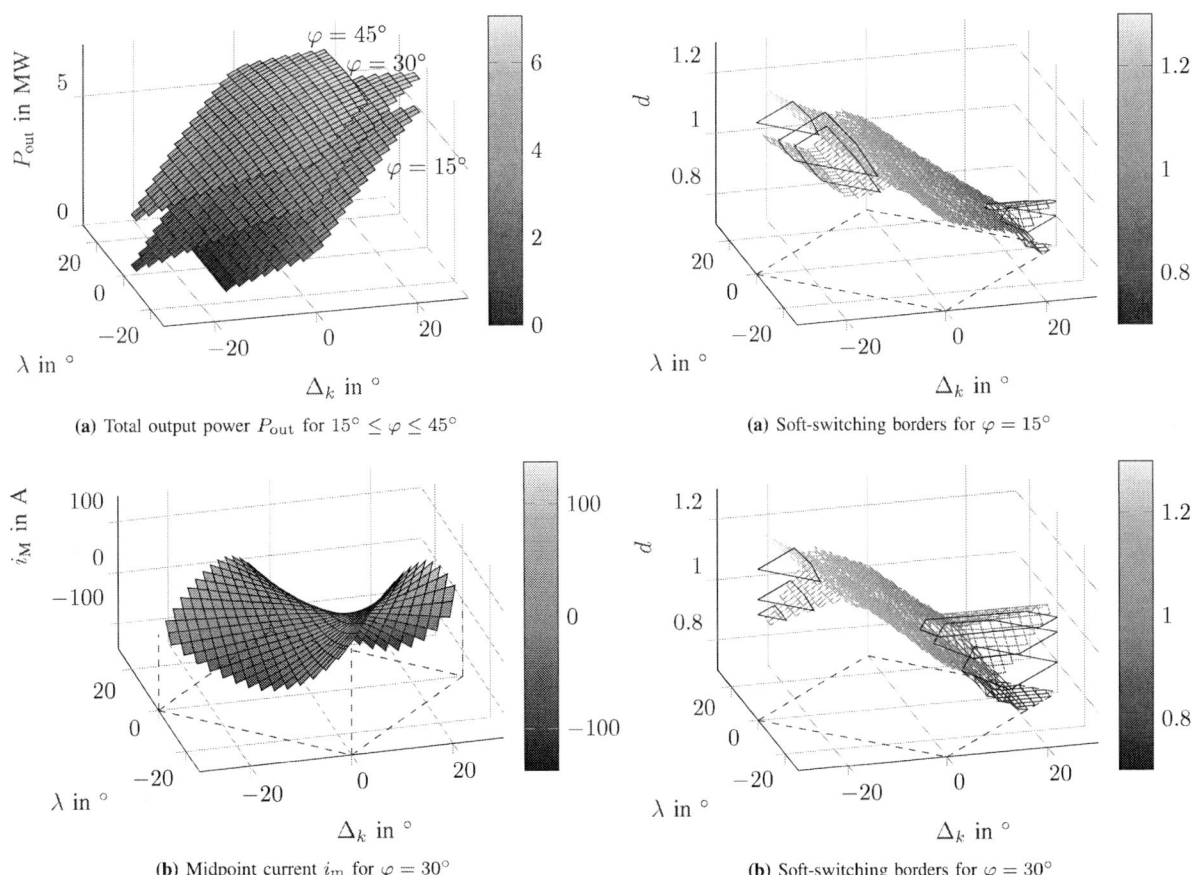

(a) Total output power P_{out} for $15° \leq \varphi \leq 45°$

(b) Midpoint current i_m for $\varphi = 30°$

Fig. 6: Output power and midpoint current as function of λ and Δ_k for $d = 1.0$ and exemplary values of φ

(a) Soft-switching borders for $\varphi = 15°$

(b) Soft-switching borders for $\varphi = 30°$

Fig. 7: Soft-switching borders for the primary (top mesh) and secondary bridge (bottom mesh). Area between the meshes is defined as the ZVS operating area and framed by black edges.

with 13 different equations, these quantities are continuous and continuously differentiable over the complete operating range of φ, λ and Δ_k, as can be seen in Fig. 6. P_{out} is monotonically increasing with φ and Δ_k over the defined operating range.

The influence of the voltage ratio d on the output power ratio is complex and differs for every operational case. For case 1, the equation for r_{power} in dependence of d, λ, φ and Δ_k is stated in (7b). The impact of λ on P_{out} is very small since this parameter effects the staircase voltages of the primary and the secondary side equally. The effect of λ on the midpoint current i_m, depicted in Fig. 6b, is clearly visible which is described analytically in (7c) for case one.

The maximal output power ratio is influenced by the midpoint current that for $r_{power} = 1$ can be seen as a pulsating current with a mean value $\bar{i}_M = 0$. A higher transmitted charge during one current pulse in relation to the output power allows a higher maximal power ratio balancing. Additionally, reactive power is transferred between the primary side dc-link and the output capacitors at the secondary side through the medium-frequency transformer of the converter. This degree of freedom allows an increase of the maximal adjustable level

of r_{power} for low output powers values. For a low active output power, a relatively high internal reactive power, as well as a high reactive current can be adjusted inside the converter. By choosing $\lambda > 0$ and $\Delta_k < 0$ with $|\lambda| > |\Delta_k|$, a high internal reactive power can be applied without increasing the active output power at the converter output.

C. Soft-Switching Analysis

To achieve a highly efficient operation of the converter, a wide soft-switching operation range is desirable. Soft-switching is ensured, if the power switch is turned-on while the current is carried by the anti-parallel diode. To analyze the zero-voltage switching (ZVS) region of the whole converter, the borders of the primary and the secondary side ZVS range, depending on the modulation parameters and voltage ratio are calculated and plotted as shown in Fig. 7. The two depicted meshes show the soft-switching borders. For the primary side, all operating points below the top mesh are under ZVS. Vise versa, all operating points above the lower mesh are secondary-side ZVS points. The region between the surfaces depicts the overall soft-switching range of the 3L-DAB3 and is marked

The 2018 International Power Electronics Conference

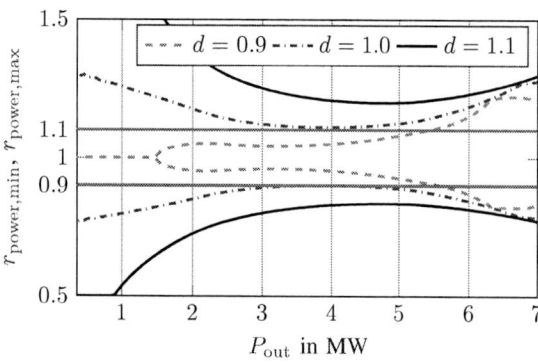

(a) ZVS range for the secondary side

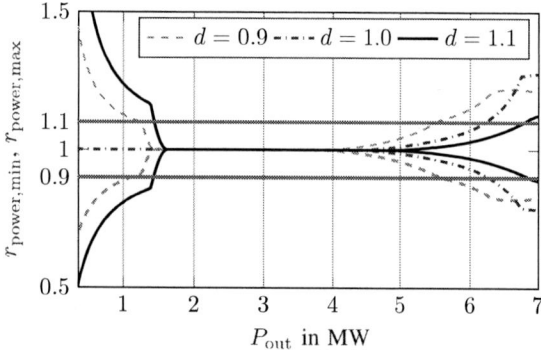

(b) ZVS range for primary and the secondary side

Fig. 8: Maximal and minimal power ratio power ratio r_{power} over P_{out} that can be balanced under ZVS operation

with black spanned edges. Apparently, the ZVS operation is limited to a few regions in the operating area of the converter. Considering the soft-switching regions of the primary and the secondary side, it becomes clear that for most operating points only one side of the converter can be operated at ZVS. A soft-switching operation of the converter is possible for higher values of φ and Δ_k in combination with a relatively low voltage ratio d (cf. Fig. 7b). Further it can be seen in Fig. 7a and 7b, that for $d = 1.0$ and $\varphi < 30°$, the soft-switching operating region for the primary side is given by $\Delta_k < 0°$ and is not further dependent on λ.

To summarize the ZVS range over the output power with regard to the voltage ratio d, the corresponding soft-switching boundaries of the power ratios $r_{\text{power,min}} < 1$ and $r_{\text{power,max}} > 1$ depending on the output power P_{out} are depicted in Fig. 8. All operating points (coordinates of P_{out} and r_{power}) between the drawn boundary lines are under ZVS. An operating area with $10\,\%$ voltage variation is depicted with red lines. A maximal operating range for P_{out} and r_{power} under ZVS operation of the secondary side of the 3L-DAB3 is depicted in Fig. 8a. A soft-switching operation of all 24 power power switches is possible only for a limited output power range. The maximal and minimal power ratio r_{power}

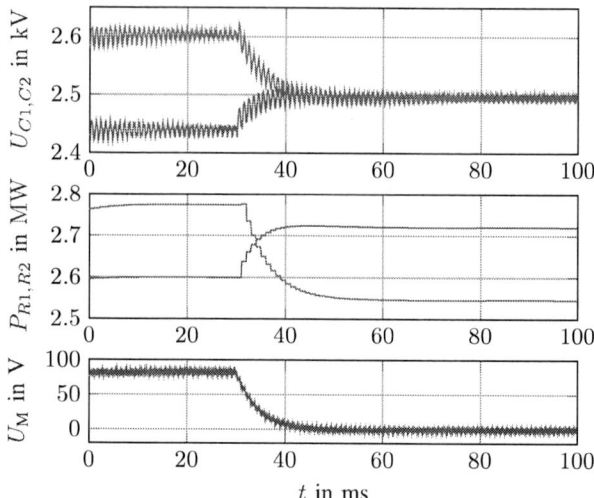

Fig. 9: Simulation results for the capacitor voltage balancing, activation at $30\,\text{ms}$

over P_{out} that can be achieved under ZVS operation of the entire converter for various d is shown in Fig. 8b.

IV. CONTROL WITH LOOK-UP TABLES

In order to generate a constant symmetrical output voltage of the converter under various loads, a closed loop control is necessary. To ensure the setting of an optimal operating point, utilization of look-up tables (LUTs) are a promising approach for such complex and highly non-linear systems [14], [15]. Look-up tables are created, to allow optimal control and output voltage balancing with different values of the parameters φ, λ and Δ_k. The λ-Δ-modulation offers additional degrees of freedom, which lead to an infinite number of parameter combinations to achieve the required reference power. These additional degrees of freedom can be used to optimize the LUTs for specific constraints, such as maximal converter efficiency. The calculation of converter efficiencies can be carried out for every operating point. To obtain the LUTs for the maximal converter efficiency, the total losses were calculated for every combination of the control parameters φ, λ and Δ_k. Parameter combinations that generate the same power at the converter output were compared and for every operating point the combination leading to the minimal losses was chosen. The optimal parameter combinations to achieve maximal converter efficiency can be summarized in LUTs. Three separate LUTs, one for each control parameter, can be obtained:

$$\varphi\left(P_{\text{out}}, r_{\text{power}}\right) \tag{8a}$$

$$\lambda\left(P_{\text{out}}, r_{\text{power}}\right) \tag{8b}$$

$$\Delta_k\left(P_{\text{out}}, r_{\text{power}}\right) \tag{8c}$$

The LUTs operate with normalized power values. This offers the advantage, that the LUTs are not depending on the concrete converter specifications (dc voltages, switching frequency,

380

The 2018 International Power Electronics Conference

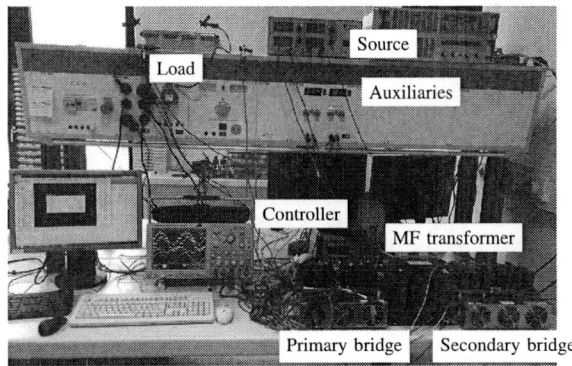

Fig. 10: Laboratory hardware setup for test and verification

transformer parameters). The LUTs are optimized for maximal efficiency only under a defined output power configuration. For a deviating power range, the LUTs can still be used for voltage balancing, but as the loss optimization is non-linearly with output power, a non-optimal operation point might be selected in that case.

In oder to validate the functionality of the implemented LUTs, several converter simulations under different load configurations are carried out. Figure 9 shows the effect of the proposed modulation, which is activated at $t = 30\,$ms, under non-symmetrical load with non-equal resistances R_1 and R_2. As can be seen, the dc-link voltages U_{C1} and U_{C2} are equalized and hence the neutral-point voltage U_M becomes zero. Since the loads are completely passive, the dissipated power P_{R1} and P_{R2} in the corresponding resistance changes according to their voltages. However, the presented approach is not depending on the type of load and also valid for general passive and active load characteristics.

V. EXPERIMENTAL SETUP AND VALIDATION

The proposed modulation has been successfully implemented and tested on a low power prototype of a three-phase three-level DAB based on 3L NPC MiniSKiiP modules [16]. The test setup, depicted in Fig. 10, is operated under an external input dc voltage of 60 V. The waveforms of primary and secondary transformer voltages and the resulting ac current are depicted in Fig 11, visualizing the influence of the shifting parameter λ. It can clearly be seen, that the influence of λ on the ac voltage waveforms appears in

TABLE I: Characteristic values of the experimental setup

(a) Hardware parameters

U_p	$U_{C1} + U_{C2}$	C_1	C_2	d	f_SW	L_σ
60 V	60 V	2 mF	2 mF	1	1 kHz	425 µH

(b) Operating point for voltage balancing (cf. Fig. 12)

R_1	R_2	P_total	r_power	d	φ	λ	Δ_k
10 Ω	9 Ω	190 W	0.9	1	19°	−10°	6°

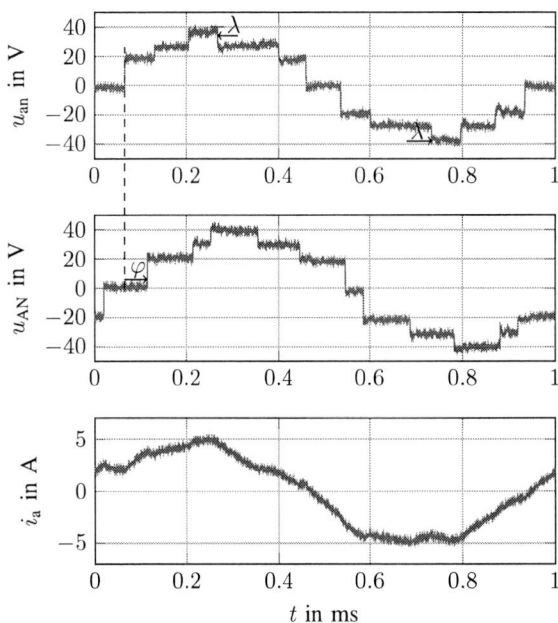

Fig. 11: Measured ac waveforms with exemplary markings of phase shift $\varphi = 19°$ and shifting parameter $\lambda = -10°$.

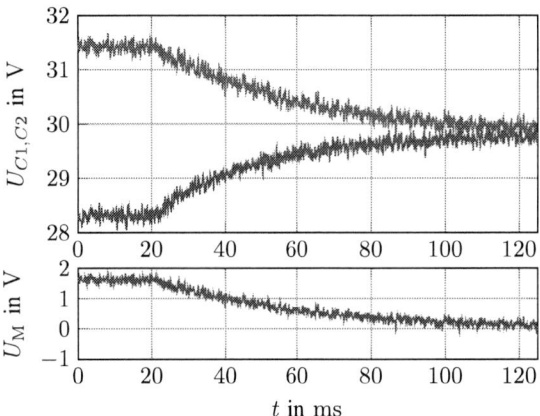

Fig. 12: Capacitor and neutral-point voltage with unequal loads after activation of the proposed modulation at 20 ms

hardware measurement as theoretically explained and depicted in section III-A.

To evaluate the effect of the converter's dc-link output, the prototype setup was equipped with two different load resistances R_1 and R_2 according to Fig. 2. The load parameters and the respective operating point is given in Table I.

Figure 12 shows the output dc-link voltages influenced by the modulation parameters given in the right part of Table I(b). The new modulation parameters are set at 20 ms and subsequently the dc-link voltages U_{C1} and U_{C2} are equalized and thus also the midpoint voltage U_M reduces to zero. This experimental voltage characteristic is consistent with the analytical and simulative results depicted in Fig. 9.

VI. CONCLUSION

In order to achieve dc-link voltage symmetry under non-symmetrical dc loads, a new modulation scheme has been introduced and investigated. It enables an active control of the positive and negative dc-link output voltage individually by influencing the dc midpoint current. All operating cases and their respective borders are shown, which can be used to generate analytical solutions for output power, power ratio and average midpoint current in all operating points. The effect of the modulation parameters on the output power and midpoint current has been discussed. Further the power ratio limitations regarding soft-switching of the converter are discussed and visualized. To allow an optimized control, look-up tables are used to set the matching modulation parameters.

The proposed method is implemented on a low power 3L-DAB3 prototype and the influence on the ac waveforms is shown. The effectiveness of balancing the dc-link voltages for unbalanced loads on the prototype is demonstrated by applying non-symmetrical resistive loads to the converter output.

ACKNOWLEDGMENT

Funded by the Federal Ministry of Education and Science (BMBF, FKZ03SF0491A), Flexible Electrical Networks (FEN) Research Campus.

REFERENCES

[1] A. Zervos and C. Lins, "Renewables global status report 2016," *REN21*, p. 17, 2016.

[2] F. S. Rabia Ferroukhi, Janet Sawin, "Rethinking energy 2017: Accelerating the global energy transformation," *IRENA International Renewable Ebergy Agency*, 2017.

[3] J. Conti, P. Holtberg, J. Diefenderfer, A. LaRose, and J. T. Turnure, "International energy outlook 2016," *US Energy Information Administration*, 2016.

[4] R. W. D. Doncker, "Power electronic technologies for flexible DC distribution grids," in *Proc. Int. Power Electronics Conf. (IPEC-Hiroshima 2014 - ECCE ASIA)*, May 2014, pp. 736–743.

[5] M. Stieneker and R. W. D. Doncker, "Medium-voltage DC distribution grids in urban areas," in *Proc. IEEE 7th Int. Symp. Power Electronics for Distributed Generation Systems (PEDG)*, Jun. 2016, pp. 1–7.

[6] N. H. Baars, J. Everts, C. G. E. Wijnands, and E. A. Lomonova, "Evaluation of a high-power three-phase dual active bridge dc-dc converter with three-level phase-legs," in *2016 18th European Conference on Power Electronics and Applications (EPE'16 ECCE Europe)*, Sept 2016, pp. 1–10.

[7] ——, "Modulation strategy for wide-range zvs operation of a three-level three-phase dual active bridge dc-dc converter," in *2017 IEEE Applied Power Electronics Conference and Exposition (APEC)*, March 2017, pp. 3357–3364.

[8] R. W. A. A. D. Doncker, D. M. Divan, and M. H. Kheraluwala, "A three-phase soft-switched high-power-density dc/dc converter for high-power applications," *IEEE Transactions on Industry Applications*, vol. 27, no. 1, pp. 63–73, Jan 1991.

[9] F. Wang, "Sine-triangle vs. space vector modulation for three-level pwm voltage source inverters," in *Conference Record of the 2000 IEEE Industry Applications Conference. Thirty-Fifth IAS Annual Meeting and World Conference on Industrial Applications of Electrical Energy (Cat. No.00CH37129)*, vol. 4, Oct 2000, pp. 2482–2488 vol.4.

[10] C. Newton and M. Sumner, "Neutral point control for multi-level inverters: theory, design and operational limitations," in *Industry Applications Conference, 1997. Thirty-Second IAS Annual Meeting, IAS '97., Conference Record of the 1997 IEEE*, vol. 2, Oct 1997, pp. 1336–1343 vol.2.

[11] D. H. Lee, S. R. Lee, and F. C. Lee, "An analysis of midpoint balance for the neutral-point-clamped three-level vsi," in *PESC 98 Record. 29th Annual IEEE Power Electronics Specialists Conference (Cat. No.98CH36196)*, vol. 1, May 1998, pp. 193–199 vol.1.

[12] N. Celanovic and D. Boroyevich, "A comprehensive study of neutral-point voltage balancing problem in three-level neutral-point-clamped voltage source pwm inverters," *IEEE Transactions on Power Electronics*, vol. 15, no. 2, pp. 242–249, Mar 2000.

[13] A. Filb-Martnez, S. Busquets-Monge, and J. Bordonau, "Modulation and capacitor voltage balancing control of a three-level npc dual-active-bridge dc-dc converter," in *IECON 2013 - 39th Annual Conference of the IEEE Industrial Electronics Society*, Nov 2013, pp. 6251–6256.

[14] P. Falkowski, K. Kulikowski, and R. Grodzki, "Predictive and look-up table control methods of a three-level ac-dc converter under distorted grid voltage," *Bulletin of the Polish Academy of Sciences Technical Sciences*, vol. 65, no. 5, pp. 609–618, 2017.

[15] A. Godlewska, R. Grodzki, P. Falkowski, M. Korzeniewski, K. Kulikowski, and A. Sikorski, "Advanced control methods of dc/ac and ac/dc power converterslook-up table and predictive algorithms," in *Advanced Control of Electrical Drives and Power Electronic Converters*. Springer, 2017, pp. 221–302.

[16] Semikron, *SEMIKRON three level (3L) evaluation inverter*, Semikron, miniSKiiP MLI EVA Inverter: 91 28 70 01.

A Novel Switching Algorithm to improve Efficiency at light load conditions for Three-Phase DAB Converter in LVDC Application

Hyun-jun Choi[1], Si-hoon Jung[1], Jee-hoon Jung[1*]
1 Electrical Engineering, Ulsan National Institute of Science and Technology (UNIST),
Ulsan, Republic of Korea
*jhjung@unist.ac.kr

Abstract- **In this paper, a novel switching algorithm for a three-phase dual active bridge (3P-DAB) converter in LVDC applications is proposed to improve the efficiency at light load conditions. The 3P-DAB converter is one of popular topologies in high-power applications due to small filtering size, the lower conduction losses with interleaved construction, the lower switching losses with inherent soft switching capability and seamless control in bi-directional power flow. However, the conventional power control method in 3P-DAB, single phase shift (SPS), does not effective due to possibility of ZVS failure in the light load conditions. In this paper, a novel control strategy for 3DAB converter to realize ZVS in the light load by using asymmetrical pulse width modulation (APWM) cascaded with SPS control algorithm. The transferred powers are calculated in both APWM and SPS mode, respectively. Also, the soft switching range also is also defined in proposed control strategy. A 3-kW prototype is implemented and tested to demonstrate the effectiveness of the proposed methods.**

Keywords— Three phase dual acitve bridge converter, Low voltage direct current (LVDC) system, modulation strategy, efficinecy improvement.

I. INTRODUCTION

DC microgrids (MG) have become an alternative to conventional AC power distribution systems. The DC MG can overcome the limitations of the AC systems, such that lower system efficiency is caused by many power conversion steps, and additional controls for frequency, phase, and reactive power are required [1].

The low voltage direct current (LVDC) distribution system is required to realize the DC MG, which achieve the requirements of the new electrical network. The target of the LVDC network is to generate higher quality power to the customers than in the traditional distribution. Fig. 1. shows the common DC distribution system. As shown in Fig. 1, the LVDC converter is required to convert the bi-polar very high voltage to application voltage level, 380V. Therefore, converter for LVDC system should handle the high voltage and high power with bi-directional capability.

A suitable dc–dc converter for this application is the three-phase dual active bridge (3P-DAB) converter in Fig. 2. Due to the inherent zero voltage switching (ZVS)

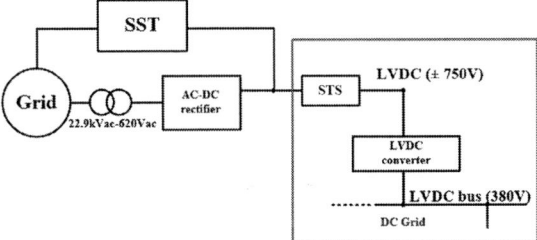

Fig. 1. Example of DC distribution system.

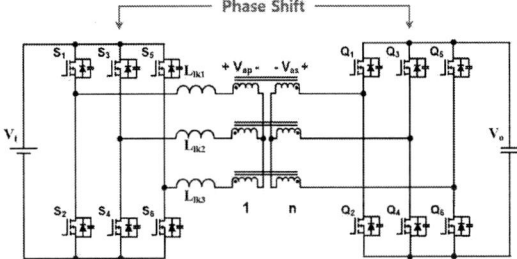

Fig. 2. Three Phase Dual Active Bridge Converter scheme.

capability and smooth bi-directional operation under the seamless manners. Specially, 3P-DAB converter is applied in high power application due to reduction of conduction losses for interleaved structure and reduction of passive components size. However, 3P-DAB converter also has some demerits such as ZVS limitation in light load conditions. In 3p-DAB converter, the reactive current is essential to make soft switching. However, at the light load conditions, the amount of reactive current is not enough, resulting the hard commutation in power switches. Several methods such as simultaneous PWM control and phase shift modulation with fixed asymmetrical duty are proposed [2,3]. However, these are not only very complex to control the output power, but also analyzed only in ideal case. Also, the triangular current mode and trapezoidal current mode are applied to 3P-DAB converter as the single DAB converter's operation strategy, but the volume and cost of filter and transformer of 3P-DAB converter are limited due to the non-interleaved operation

In this paper, asymmetrical pulse width modulation (APWM) control cascaded with single phase shift (SPS) control algorithm for the 3P-DAB converter in LVDC applications is proposed to improve the efficiency at light

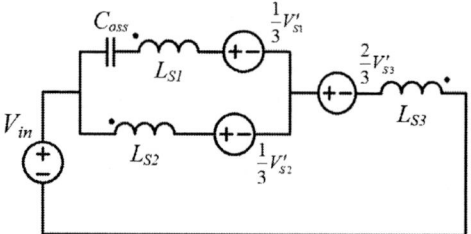

Fig. 3. Equavlent circuit of switches S_1 and S_2 turn-on and-off mechanism.

load conditions. In Section II, the SPS method as the conventional control algorithm for 3DAB converter is briefly introduced. In Section III, the proposed control algorithm is presented based on the mathematical analysis, and this method will be compared with SPS modulation. Finally, through the simulation and experimental results of 3-kW DAB prototype, the effectiveness of proposed method is verified.

II. SINGLE PHASE SHIFT CONTROL IN THREE-PHASE DUAL ACTIVE BRIDGE CONVERTER

As shown in Fig. 2, the 3P-DAB converter consists of two three-phase bridges in both sides, and these are connected through transformer with the coupling inductance, which combine external inductance with leakage inductance in transformer. The SPS control is a conventional manner to operate three phase dual active bridge converter, which is applied with a fixed duty cycle as 50 %. The SPS control is used to make easy operate and simple. The transmission power is controlled by phase different angle, Φ, between primary and secondary sides. The amounts of phase difference induce the voltage difference across the coupling inductance and control the transferred power, which causes six-step waveform in each phase voltage in both sides.

To derive accurate soft switching region, the current should flow through the anti-parallel diode before the switches are turn on/off. Soft switching condition can be expressed as (1), which is basic requirement for ZVS in 3P-DAB converter [4].

$$i_{L\_ZVS} = \begin{cases} i_L(0) \leq 0 \\ i_L(\phi) \geq 0 \end{cases} \qquad (1)$$

In [4], the ZVS condition is already presented as (1), however; it is not accurate because this condition ignores the parasitics in the switches. In practice, due to the output capacitors in power MOSFET (C_{OSS}), the commutation makes resonance between C_{OSS} and coupling inductance. Fig. 3 shows an equivalent circuit of one of the steady states. Because the 3P-DAB converter includes the three-phase transformer, equivalent components are required in each phase to solve the resonant circuit. Assume that equivalent output capacitance, C_{OSS}, is the same and is connected with power switches S1 and S2 parallelly. By applying the superposition rule to circuit in Fig. 3, the resonance can

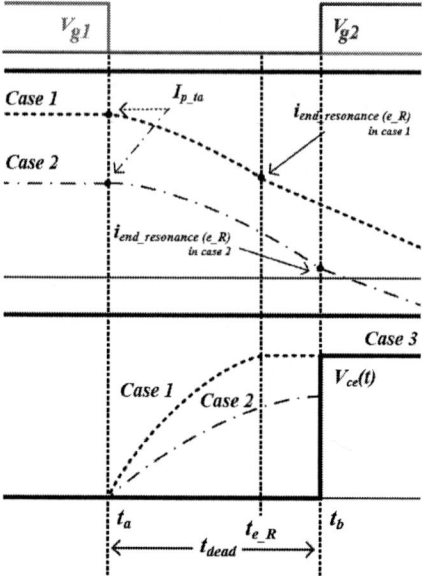

Fig. 4. Detail turn-on mechanism of three phase DAB converter in Fig. 4: Switches S_1 and S_2 turn on and off transient waveform

be analyzed. The resonant phase current, $i_{P\_R}(t)$ charged by $i_{P\_R}(t)$, and the output capacitor voltage are expressed as (2) and (3), respectively.

$$i_{p\_R}(t) = 3I_{p\_ta} + \frac{V_{in}}{L}\left(\frac{d}{3}-1\right)(t_a-t) +$$
$$I_{p\_ta}\cos\left[\sqrt{2}\omega_0(t_a-t)\right] - \frac{\sqrt{2}V_o}{3nZ_0}\sin\left[\sqrt{2}\omega_0(t_a-t)\right] \qquad (2)$$

$$V_{DS\_1}(t) = I_{p\_ta}(t_a-t) + \frac{V_{in}}{2L}\left(\frac{d}{3}-1\right)(t_a-t)^2 +$$
$$Z_0 I_{p\_ta}\frac{\sin\left[\sqrt{2}\omega_0(t_a-t)\right]}{2} + \frac{\sqrt{2}V_o}{3n}\frac{\cos\left[\sqrt{2}\omega_0(t_a-t)\right]-1}{2} \qquad (3)$$

Where $\omega_0 = \frac{1}{\sqrt{L_sC_{oss}}}, Z_0 = \frac{\sqrt{L_s}}{\sqrt{C_{oss}}}$

Fig. 4 shows the detail turn-on/off transient waveforms of 3P-DAB converter of S_1 and S_2 in Fig. 2, where the dead time and resonant transition are included. The conducting switch, S_1 in primary side, is turned off at t_a, and the value of the phase current is assumed being constant of 3P-DAB converter of S_1 and S_2 in Fig. 2, where the dead time and resonant transition are included. The conducting switch, S_1 in primary side, is turned off at t_a, and the value of the phase current is assumed being constant value as I_{p_ta} before t_a. Also, it is assumed that all the value of switches and inductances are the same.

In case the initial phase current, $I_{p\text{-}ta}$, is positive, the ZVS achieve when S_2 is turned off. The coupling inductor resonant with output capacitance of S_1 and S_2. The initial voltage condition of output capacitance of S_1 and S_2 is zero and V_{in}, respectively. As shown in Fig. 4, in the case I, the drain-source voltage of S_1, $V_{DS\_1}(t)$, is zero

384

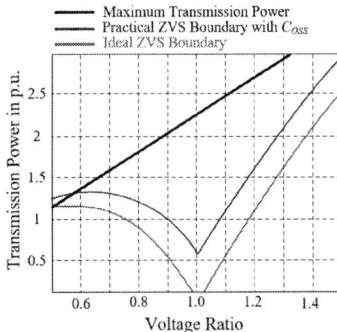

Fig. 5. ZVS boundaries comparison between ideal and practical case.

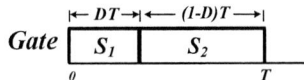

Fig. 6. Concept of Asymmetrical pulse width modulation algorithm

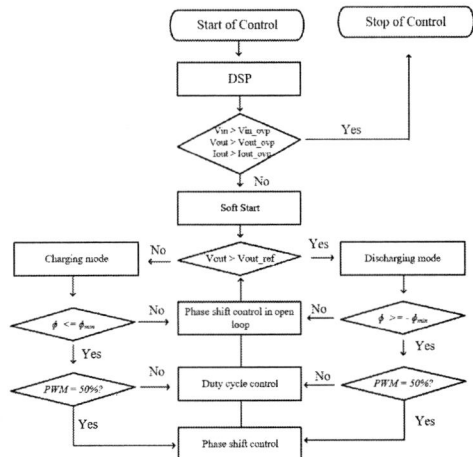

Fig. 7. Control Algorithm diagram of proposed cascade asymmetrical PMW plus phase shift mode control.

in the initial state, and $V_{DS\_1}(t)$ reaches the input voltage, Vin, by the end of dead time. The switch conducts current through a parallel diode in the reverse direction to achieve Full ZVS.

If the magnitude of I_{p_ta} is greater than zero, but I_{p_ta} is not enough to achieve the ZVS, switching losses can be caused at the light load conditions. As shown case II in Fig. 4, the initial state of I_{p_ta} seems to satisfy the ZVS condition. However, the complete ZVS is not realized, because the phase-current is insufficient to charge/discharge the $V_{DS\_1}(t)/V_{DS\_2}(t)$ to V_{in}/zero fully by the end of the dead time. Since the drain-source voltage of S_2 is not fully discharged, the remaining voltage in S_2 is instantaneously discharged leading to the high current spike. Therefore, the condition of ZVS boundaries with COSS should be expressed as (4)

$$i_{L\_ZVS} = \begin{cases} i_L(0) \le I_P \\ i_L(\phi) \ge -I_S/\text{n} \end{cases} \quad (4)$$

Where $I_P = I_S = \dfrac{\sqrt{3V_{in}V_{out}}}{2Z_O}, Z_O = \dfrac{\sqrt{L_S}}{\sqrt{C_{OSS}}}$.

Fig. 5. shows the comparison of ZVS boundaries of 3DAB converter between with and without C_{OSS}. Fig. 5 indicates even though the voltage ratio is close to the '1', the failure of ZVS can happen.

III. ASYMMETRICAL CASCADE PWM PLUS PHASE SHIFT MODE CONTROL

In terms of single-phase DAB converter (1DAB), a lot of control algorithms such as PWM plus Phase shift, trapezoidal strategy and triangular method are proposed to overcome low efficiency at the light lad conditions [5]. However, in the 3DAB converter, it is relatively difficult to modulate the switching algorithm for reducing the switching losses at the light load condition because of the 120° phase shift angle among three phase bridges. To reduce the switching losses, the ZVS region should be increased in the light load condition. As mentioned, the certain level of reactive current is required, and reactive current can be increased based on the APWM method as shown in Fig. 6. The target of APWM cascaded SPS method is smoothly that enlarged ZVS region based on

TABLE I
DESIGN SPECIFICATION

Meaning	Value
Input Voltage	500 V
Output Voltage	252 V
Target Power	3 kW
Switching Frequency	100 kHz
Power switches	C3M0065090D
Coupling inductance	65 uH
Turn ratio	0.5
Dead time	0.1 $uSec.$
Φ_{min}	8°

the APWM control manner is connected to the ZVS region of conventional SPS control mode. Therefore, the minimum phase shift angle ($\mp\Phi_{min}$) is predetermined. Using the analysis of accurate ZVS region in the previous section, Φ_{min} is determined when ZVS condition is achieved with SPS modulation. Fig. 7 shows the power control diagram. The modes are divided into three; open loop control, asymmetrical duty cycle, and phase shift control modes, and the control variable is only one in each mode. When the 3P-DAB converter operate, Φ_{min} is applied initially in the open loop mode where not power transfer region. After reaching for the Φ_{min}, the power is controlled by asymmetrical duty control with PI controller. In this region, the power is transferred with higher reactive current compared with SPS, which make soft switching and enlarge the ZVS region. As transferred power increase, the duty (DT) in Fig. 6 increases. If the DT is reach for 0.5T, then conventional SPS control is applied.

Because inductor is power transfer component, the transmission power can be derived based on the analysis of inductor voltage and current.

385

The 2018 International Power Electronics Conference

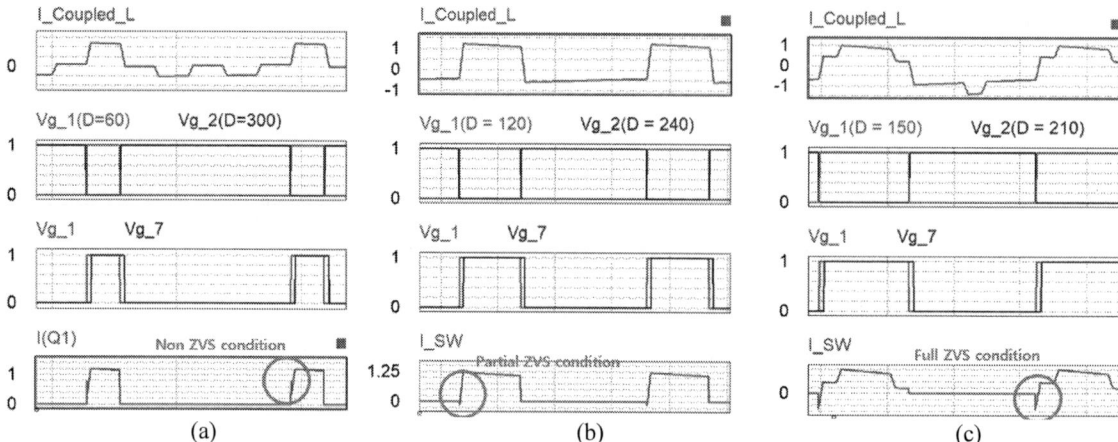

Fig. 8. Waveform by simulation results under the proposed APWM cascaded SPS control mothed according to output power: (a) 180W, (b) 320W, (c) 450W

$$L_s \frac{di_{LS}(t)}{dt} = V_P(t) - V_S'(t) = V_P(t) - \frac{V_S}{N}(t) \quad (5)$$

Based on the phase voltage and phase current, the transmission P_{out} can be calculated according to duty, D, as (6).

$$P_{out} = \begin{cases} \dfrac{V_{in}^2}{\omega L} \dfrac{wdD^2}{f_s}, & 0 < D < \dfrac{\phi_{min}}{2\pi} \\[2ex] \dfrac{V_{in}^2}{\omega L}(2dD\phi_{min} - \dfrac{d\phi_{min}^2}{2\pi}), & \dfrac{\phi_{min}}{2\pi} < D < \dfrac{1}{3} - \dfrac{\phi_{min}}{\pi} \\[2ex] \dfrac{V_{in}^2}{\omega L}(\dfrac{-d\pi(3D-1)^2}{9} + \phi_{min}(\dfrac{3\phi_{min}}{4\pi} - D - \dfrac{1}{3})), & \dfrac{1}{3} - \dfrac{\phi_{min}}{\pi} < D < \dfrac{1}{3} + \dfrac{\phi_{min}}{2\pi} \\[2ex] \dfrac{V_{in}^2}{\omega L}(\dfrac{2}{3}d\phi_{min} - \dfrac{1}{2}d\phi_{min}^2), & \dfrac{1}{3} + \dfrac{\phi_{min}}{2\pi} < D < \dfrac{1}{2} \\[2ex] \dfrac{V_{in}^2}{\omega L}d(\dfrac{2}{3} - \dfrac{\phi}{2\pi})\phi, & D = \dfrac{1}{2} \end{cases} \quad (6)$$

Where D is duty, $w = 2\pi f_S$, f_S is switching frequency, d is conversion ratio between primary and secondary bridge.

IV. SIMULATION RESULTS

The table 1 shows the design specifications of 3P-DAB prototype. Based on the consideration of switching and conduction losses, the coupling inductance is designed as 65 *uH*, designed by (5).

Simulation results are shown in Fig. 8. In Fig. 8 (a), (b) and (c) indicate the phase current and phase voltage waveform under proposed mode when output power is 180W, 320W and 450W, respectively. As shown in Fig. 8, the duty is controlled when output power is changed with fixed minimum phase shift angle, Φ_{min}. As mentioned in previous, the Φ_{min} is determined by criteria of ZVS region under the SPS control in 3P-DAB converter. Although (a) in Fig. 8 cannot achieve soft-switching, after 300W, the ZVS can be attained under the proposed control manner. These results show when the duty is controlled asymmetrically and changed according to the output power, the proposed control algorithm enlarge ZVS region compared with conventional SPS control manner with easy operation. The green circle in Fig. 8 shows the

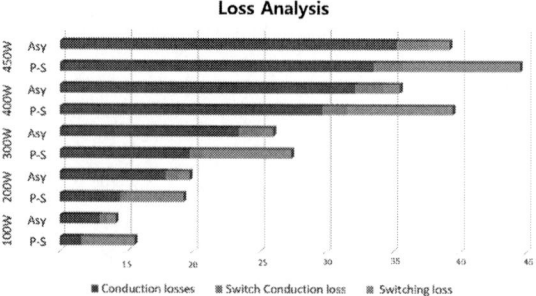

Fig. 9. Loss Comparison between SPS control manner and proposed APWM cascade SPS control method at the light load condition

negative current, which reduces switching losses. In the SPS with design specification in Table I, the full ZVS can occur at the 520W. The Fig. 9 shows the loss evaluation based on switching losses in power switches and conduction losses, which include core losses. The core losses are derived based on the [6-8]. As shown in Fig. 9, in proposed method, the ZVS can be obtained under the 100W. Although it increases the conduction losses compared with conventional method, most conversion efficiency at the light load is improved.

V. CONCLUSION

In this paper, the control strategy of the 3P-DAB converter design is proposed to improve the light load efficiency. The 3P-DAB converter is one of useful bi-directional converters with galvanic isolation, high efficiency, high performance, and ZVS capability; however, the 3P-DAB also has limitation due to the low efficiency at the light load conditions when conventional method is applied. In the paper, analysis and problem statement of the conventional SPS control manner are presented. As a solution, the APWM cascaded SPS control manner is proposed. The proposed algorithm is divided into three parts; open loop mode, APWM modulation mode and SPS mode. Basically, the SPS

386

mode is the same as conventional method, however; APWM mode can increase the ZVS region and reduce the switching losses. The accurate analysis in APWM mode according to duty at the light load condition is presented based on the mathematical method. the proposed design methodology and its effectiveness are verified by simulation results through the 3-kW 3P-DAB converter.

VI. DISCUSSION

Although the proposed algorithm is effective at the light load condition, as shown in Fig. 9, the conduction losses is increased, which means that this method should be applied for the higher frequency application. Therefore, the hybrid control algorithm should be added to reduce peak current when the proposed switching strategy is applied at the light load condition.

ACKNOWLEDGMENT

This research was supported by the KEPCO under the project entitled by "Demonstration Study for Low Voltage Direct Current Distribution Network in an Island." (D3080)

REFERENCES

[1] P. Chiradeja and R. Ramakumar, "An approach to quantify the technical benefits of distributed generation," *IEEE Transactions. Energy Convers.*, vol.19, no.4, pp. 764-773, Dec. 2004.

[2] F. Krismer and J. W. Kolar, "Efficiency-optimized high-current dual active bridge converter for automotive applications," IEEE Transactions on Industrial Electronics, vol. 59, no. 7, pp. 2745–2760, July 2012

[3] H. van Hoek, M. Neubert, and R. De Doncker, "Enhanced modulation strategy for a three-phase dual active bridge-boosting efficiency of an electric vehicle converter," Power Electronics, IEEE Transactions on, vol. 28, no. 12, pp. 5499 – 5507, March 2013

[4] J. Huang, Y. Wang, Z. Li, and Y. Jiang, "Simultaneous pwm control to operate the three-phase dual active bridge converter under soft switching in the whole load range," in Applied Power Electronics Conference and Exposition (APEC), 2015 IEEE, March 2015, pp. 2885 – 2891.

[5] H. van Hoek, M. Neubert, A. Krober, and R. W. De Doncker" Enhanced modulation strategy for a three-phase dual active bridge boosting efficiency of an electric vehicle converter," IEEE *Transactions Power Electronics*, vol.28, no 12, pp. 5499-5507, Dec.2013

[6] W. A. Roshen, "A Practical, Accurate and Very General Core Loss Model for Nonsinusoidal Waveforms," in *IEEE Transactions on Power Electronics*, vol. 22, no. 1, pp. 30-40, Jan. 2007

[7] P. L. Dowell, "Effects of eddy currents in transformer windings, Electrical Engineers, Proceedings of the Institution of, vol. 113, no. 8, pp. 1387–1394, August 1966.

[8] G. G. Oggier, G. O. García and A. R. Oliva, "Switching Control Strategy to Minimize Dual Active Bridge Converter Losses," in *IEEE Transactions on Power Electronics*, vol. 24, no. 7, pp. 1826-1838, July 2009.

The 2018 International Power Electronics Conference

Design of a High-Frequency Dual-Active Bridge Converter with GaN Devices for an Output Power of $3.7\,\mathrm{kW}$

Philipp Schülting, Christian Winter, Rik W. De Doncker
Institute for Power Electronics and Electrical Drives
RWTH Aachen University, Aachen, Germany
Email: post@isea.rwth-aachen.de

Abstract—In the automotive industry weight and volume are important issues to design power electronics besides costs. With the upcoming of wide band-gap devices like Gallium Nitride (GaN) devices high switching frequencies become a potential for converters. Operating at high switching frequencies reduces the volume of passive components e.g., transformer significantly. Auxiliary supplies or on-board charging systems for electric vehicles are typical areas of application for a reduction of volume and weight. For this kind of application dc-dc converters like a Dual Active Bridge converter can be used. This paper describes a detailed electrical analysis of a compact single phase Dual Active Bridge converter operated at a high switching frequency of $500\,\mathrm{kHz}$ at a dc voltage of $400\,\mathrm{V}$ on both full bridges to achieve a power transfer of $3.7\,\mathrm{kW}$. In this paper it is shown how a fast switching and compact dual active bridge converter is designed that operates with a high efficiency of about $96\,\%$ at nominal power.

I. INTRODUCTION

This paper focuses on design considerations of a high-frequency single-phase Dual-Active-Bridge (DAB) converter as part of a bidirectional on-board battery charger for electric vehicles with a rated power of $3.7\,\mathrm{kW}$. The topology of the charging system is depicted in Fig. 1. The system is connected to a single-phase grid V_{grid} via an ac-dc converter which includes the grid filter to comply with EMI regulations. The dc-link voltage V_{dc} is connected to the battery voltage V_{batt} via a DAB converter. This allows the operation at a constant dc-link voltage which is independent of the state-of-charge of the battery.

Nowadays, bidirectional battery chargers are not common but can provide important features in the future [1]. It has been shown in [2] that the lifetime of the battery can be increased by partially discharging the battery during a rest period since a fully charged battery ages the fastest. Thus, the batteries of electric vehicles could be used as distributed storage systems.

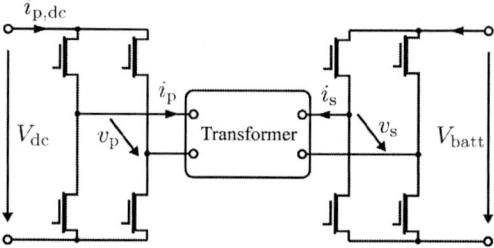

Fig. 2: Topology of a Dual-Active Bridge converter with a universal transformer as a coupling network

Then it is possible to supply and stabilize the grid with active and reactive power. A third argument for bidirectional chargers is the possibility to earn money by charging the battery at low electricity cost and discharging at higher costs.

If a charging system is integrated into a vehicle the main aspects are volume and weight. An increased switching frequency leads to a reduction in volume of the passive components [3]. Due to low switching energy of Gallium Nitride devices (GaN) devices higher switching frequencies can be used without resulting in higher switching losses compared to silicon devices operated at lower frequency. In this paper GaN devices [4] are considered to design a DAB converter for a nominal power of $3.7\,\mathrm{kW}$ at a switching frequency of $500\,\mathrm{kHz}$. The topology of the DAB converter is depicted in Fig. 2 with a transformer as a coupling network between both full bridges. The primary voltage at the clamps of the transformer is given with v_{p}, the voltage on the secondary side is v_{s}. The currents flowing into the transformer are i_{p} and i_{s}.

The method introduced in this paper, which was originally published in [5], allows arbitrary linear models for the transformer to analyze the system behavior [6]. Thus, it is possible to consider parasitic elements like winding capacitances of the transformer. Additionally at a switching frequency of $500\,\mathrm{kHz}$ low order harmonics, e.g., third and fifths harmonics, can induce significant losses in magnetics which has to be investigated, too [7].

In the next section it will be shown how a linear equivalent circuit of the transformer can be used to analyze the DAB converter in the frequency domain to consider harmonic influence and parasitic impacts.

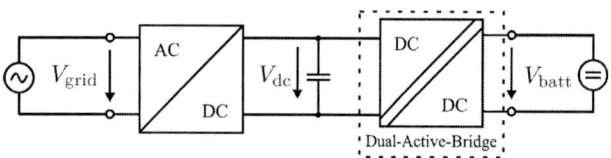

Fig. 1: Topology of a two-stage battery-charging system

388

II. Dual-Active-Bridge Converter in the Frequency Domain

There are two methods to investigate the behavior of a DAB converter. The analysis can be done in the time domain like in [8] or in the frequency domain [5]. In the time domain the fundamental wave is considered which gives insight in the behavior of the DAB converter comprehensively. Voltage square-waves are applied to the transformer on the primary and secondary side. According to Fourier analysis the applied voltage waveform with the amplitude V and the switching frequency ω_{sw} is given in (1) with n the nth harmonic and has strong third and fifth harmonic content. At high switching frequencies these harmonics result in a significant increase of losses in magnetics [7, 9, 10].

$$v_{\text{rect}}(t) = \frac{4V}{\pi} \sum_{n=1}^{\infty} \frac{\sin((2n-1)\omega_{\text{sw}}t)}{2n-1} \tag{1}$$

At a switching frequency of $500\,\text{kHz}$ those harmonics have to be considered to design the core and the cooling system. To get insight of the impact of harmonics the analysis and design procedure for the DAB converter are done in the frequency domain according to [5].

In single-phase-shift operation the duty cycle of both full bridges is equal to $50\,\%$ and both bridges are shifted by the angle δ to each other to apply a voltage across the stray inductance of the transformer which results in power transfer [8]. This operation principle is shown in Fig. 3. The values v_{p} and i_{p} are the voltage and current on the primary side and v_{s} and i_{s} correspond to the secondary side. In contrast to a lot of publications $\delta = 0\,\text{rad}$ is center aligned with the primary side and not aligned to the rising edge. This simplifies the analysis in the frequency domain because of symmetrical waveforms according to the phase shift δ.

It is also possible to deviate from the fixed duty cycle of $50\,\%$ for both bridges. For arbitrary duty cycles on both sides the variables α and β are introduced. Thus, α is the duty cycle of the primary side and β of the secondary side. In this paper the primary side corresponds to the dc link and the secondary side to the battery. Considering 1, both applied voltages to the transformer are described as (2) and (3) where n is the nth harmonic.

$$v_{\text{p}}(t) = V_{\text{dc}} \frac{2}{\pi} \sum_{n=1}^{\infty} \left(\frac{\sin\left[(2n-1)(\omega_{\text{sw}}t + \alpha/2)\right]}{2n-1} \right.$$
$$\left. - \frac{\sin\left[(2n-1)(\omega_{\text{sw}}t - \alpha/2)\right]}{2n-1} \right) \tag{2}$$

$$v_{\text{s}}(t) = V_{\text{batt}} \frac{2}{\pi} \sum_{n=1}^{\infty} \left(\frac{\sin\left[(2n-1)(\omega_{\text{sw}}t + \beta/2 - \delta)\right]}{2n-1} \right.$$
$$\left. - \frac{\sin\left[(2n-1)(\omega_{\text{sw}}t - \beta/2 - \delta)\right]}{2n-1} \right) \tag{3}$$

In order to calculate the transformer currents i_{p} and i_{s} which are shown in Fig. 2 and Fig. 3 the analysis is done

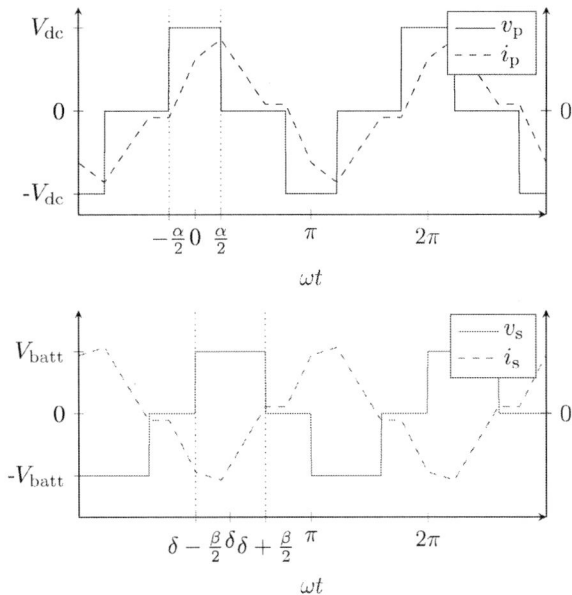

Fig. 3: General switching pattern of a single phase Dual-Active Bridge converter with single phase shift operation

completely in the frequency domain. In the end the results can be transformed back to the time domain.

To calculate the transformer currents the admittance matrix of the transformer (4) is used. For simplicity $m = 2n - 1$ is defined. With this equation the transformer currents for each harmonic are calculated.

$$\begin{pmatrix} I_{\text{p}}^{m} \angle \theta_{\text{p}}^{m} \\ I_{\text{s}}^{m} \angle \theta_{\text{s}}^{m} \end{pmatrix} = \begin{pmatrix} Y_{\text{p,p}}^{m} \angle \gamma_{\text{p,p}}^{m} & Y_{\text{p,s}}^{m} \angle \gamma_{\text{p,s}}^{m} \\ Y_{\text{s,p}}^{m} \angle \gamma_{\text{s,p}}^{m} & Y_{\text{s,s}}^{m} \angle \gamma_{\text{s,s}}^{m} \end{pmatrix} \cdot \begin{pmatrix} V_{\text{p}}^{m} \angle \phi_{\text{p}}^{m} \\ V_{\text{s}}^{m} \angle \phi_{\text{s}}^{m} \end{pmatrix} \tag{4}$$

If voltages in the time domain are described as $\Im\left(\hat{V}\angle \cdot e^{j\omega t}\right)$ the amplitude and phase of (2) and (3) can be seen directly as shown in (5).

$$V_{\text{p}}^{m} \angle \phi_{\text{p}}^{m} = V_{\text{dc}} \frac{\sqrt{2}}{m\pi} \left\{ 1\angle m(\alpha/2) - 1\angle m(-\alpha/2) \right\}$$
$$V_{\text{s}}^{m} \angle \phi_{\text{s}}^{m} = \tag{5}$$
$$V_{\text{batt}} \frac{\sqrt{2}}{m\pi} \left\{ 1\angle m(\beta/2 - \delta) - 1\angle m(-\beta/2 - \delta) \right\}$$

With the combination of (4) and (5) the currents in the time domain can be derived by an inverse Fourier transformation. The result is given in the equations (6) and (7).

$$i_{\mathrm{p}}(\omega_{\mathrm{sw}}t) =$$
$$V_{\mathrm{dc}}\frac{2}{\pi}\sum_{n=1}^{\infty}\left(\frac{1}{m}\left\{Y_{\mathrm{p,p}}^{m}\left\{\sin\left(m\left(\omega_{\mathrm{sw}}t+\alpha/2\right)+\gamma_{\mathrm{p,p}}^{m}\right)\right.\right.\right.$$
$$\left.-\sin\left(m\left(\omega_{\mathrm{sw}}t-\alpha/2\right)+\gamma_{\mathrm{p,p}}^{m}\right)\right\}$$
$$+\,dY_{\mathrm{p,s}}^{m}\left\{\sin\left(m\left(\omega_{\mathrm{sw}}t+\beta/2-\delta\right)+\gamma_{\mathrm{p,s}}^{m}\right)\right.$$
$$\left.\left.\left.-\sin\left(m\left(\omega_{\mathrm{sw}}t-\beta/2-\delta\right)+\gamma_{\mathrm{p,s}}^{m}\right)\right\}\right\}\right) \tag{6}$$

$$i_{\mathrm{s}}(\omega_{\mathrm{sw}}t) =$$
$$V_{\mathrm{batt}}\frac{2}{\pi}\sum_{n=1}^{\infty}\left(\frac{1}{m}\left\{Y_{\mathrm{s,p}}^{m}\left\{\sin\left(m\left(\omega_{\mathrm{sw}}t+\alpha/2\right)+\gamma_{\mathrm{s,p}}^{m}\right)\right.\right.\right.$$
$$\left.-\sin\left(m\left(\omega_{\mathrm{sw}}t-\alpha/2\right)+\gamma_{\mathrm{s,p}}^{m}\right)\right\}$$
$$+\,dY_{\mathrm{s,s}}^{m}\left\{\sin\left(m\left(\omega_{\mathrm{sw}}t+\beta/2-\delta\right)+\gamma_{\mathrm{s,s}}^{m}\right)\right.$$
$$\left.\left.\left.-\sin\left(m\left(\omega_{\mathrm{sw}}t-\beta/2-\delta\right)+\gamma_{\mathrm{s,s}}^{m}\right)\right\}\right\}\right) \tag{7}$$

With this method it is possible to calculate the currents for arbitrary transformer models. For instance, ohmic resistances or parasitic capacitances can be considered which allows a deep insight of the system behavior. As an example, measurement and simulation results are compared in Fig. 4. The equivalent circuit of a transformer and two external series inductances are given in Fig. 4a. One series inductor is connected to the primary side and the second one to the secondary side.

As an example for the effectiveness of this method a measurement result of a DAB converter is compared to a simulation which is shown in Fig. 4. The equivalent circuit including the transformer and two series inductances is shown in Fig. 4a with R_{coil} and C_{coil} the parasitic elements of the series inductor L_{coil} as well as the series resistance R_{s}, the stray inductance L_{σ} and the main inductance L_{m} of the transformer. The measurement result including the transformer current i_{p} as well as the primary and secondary voltage is shown in Fig. 4b whereas the simulation result is given in Fig. 4c. It can be seen that the simulations result fits the measurement good. For this simulation the used transformer and the series inductor have been measured with an impedance analyzer in advance. The resulting parasitic elements of the devices have been considered in the linear model of the transformer.

In the next section a closer look to the electrical design including the transformer and the series inductor are given.

III. ELECTRICAL DESIGN

Considering the battery-charging system, the high frequency components of voltage and current of the DAB should not be seen either by the grid-connected inverter or the battery. This could lead to EMI [11] or voltage spikes either at the battery clamps or the dc link due to its inductive behavior for high frequency components [3]. Therefore an LC-lowpass filter is connected to both sides of the DAB. The lowpass filter is depicted in Fig. 5.

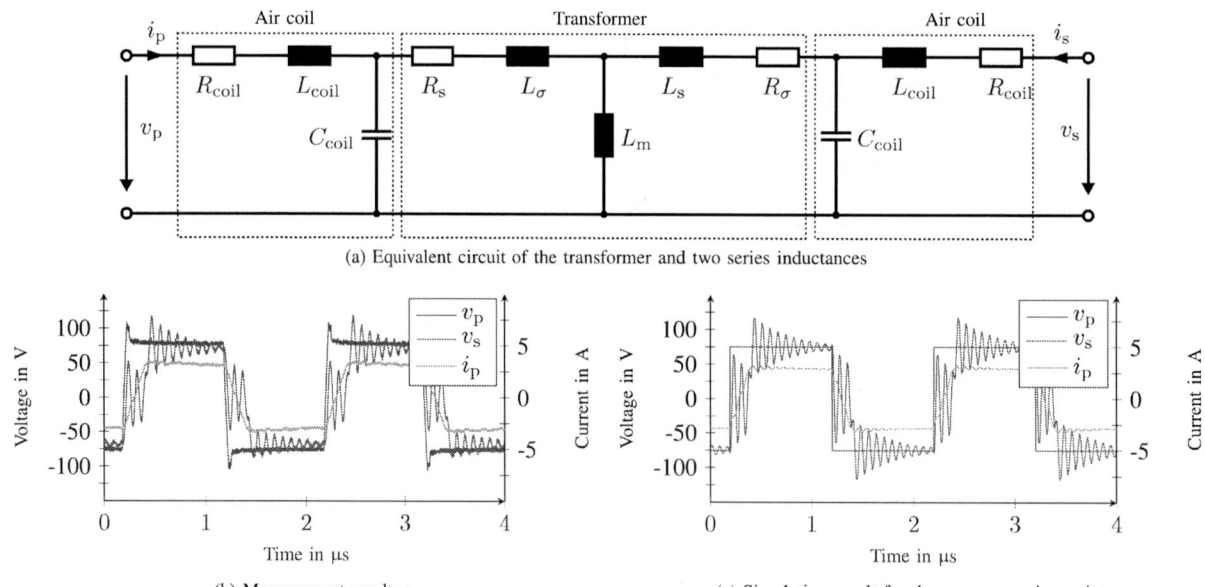

(a) Equivalent circuit of the transformer and two series inductances

(b) Measurement result

(c) Simulation result for the same operating point

Fig. 4: Comparison of measurement result with a simulation of the proposed design concept in the frequency domain. A low damped coupling network has been chosen to demonstrate the concept.

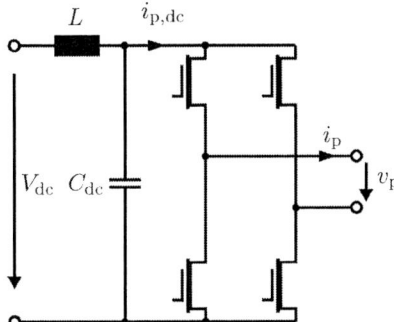

Fig. 5: *LC*-lowpass filter for the dc link

The design of this lowpass filter is done in the next subsection. After that, electrical design considerations for the high frequency transformer and the additional series inductance are done to give insight into the system design and dynamics.

A. Input Filter

Designing a suitable lowpass filter for the current $i_{p,dc}$ needs knowledge about the harmonic content of the current i_p and i_s [6]. As an example the design considerations are done for an *LC* filter for the dc link to demonstrate the concept. The current $i_{p,dc}$ is depicted in Fig. 5 and will be analyzed for the filter design. The filter for the battery side is designed the same way.

During the time interval $-\alpha/2 \leq \omega_{sw}t \leq \alpha/2$ the current $i_{p,dc}$ is identical to i_p and during $-\alpha/2 + \pi \leq \omega_{sw}t \leq \alpha/2 + \pi$ the current $i_{p,dc}$ is identical to $-i_p$. With the complex Fourier transform in (8) the coefficients of $i_{p,dc}$ can be calculated to get insight of the harmonic content.

$$
\underline{c}_{p,dc}^k = \frac{1}{2\pi} \left(\int_{-\frac{\alpha}{2}}^{\frac{\alpha}{2}} i_p(\omega_{sw}t) e^{-jk\omega_{sw}t} d(\omega_{sw}t) \right.
$$
$$
\left. - \int_{-\frac{\alpha}{2}+\pi}^{\frac{\alpha}{2}+\pi} i_p(\omega_{sw}t) e^{-jk\omega_{sw}t} d(\omega_{sw}t) \right) \tag{8}
$$

The current i_p itself can also be represented as its Fourier series (9). With the use of the Euler equation this equation can be rewritten to (10) which can be calculated analytically.

$$
\underline{c}_{p,dc}^k = \frac{1}{2\pi} \sum_{n=1}^{\infty}
$$
$$
\hat{I}_p^m \left(\int_{-\frac{\alpha}{2}}^{\frac{\alpha}{2}} \sin(m\omega_{sw}t + \theta_p^m) e^{-jk\omega_{sw}t} d(\omega_{sw}t) \right. \tag{9}
$$
$$
\left. - \int_{-\frac{\alpha}{2}+\pi}^{\frac{\alpha}{2}+\pi} \sin(m\omega_{sw}t + \theta_p^m) e^{-jk\omega_{sw}t} d(\omega_{sw}t) \right)
$$

TABLE I: Voltage fluctuation in the dc link depending on the dc-link capacitor

ΔV_{max}	0.5 V	1 V	2 V	4 V
C_{dc}	4.84 μF	2.42 μF	1.21 μF	0.61 μF

$$
\underline{c}_{p,dc}^k = \begin{cases} \dfrac{1}{2\pi} \displaystyle\sum_{n=1}^{\infty} \hat{I}_p^m \left(\dfrac{\left(e^{j(m-k)\frac{\alpha}{2}} - e^{-j(m-k)\frac{\alpha}{2}}\right) e^{j\theta_p^m}}{k-m} \right. \\ \qquad \left. - \dfrac{\left(e^{-j(m+k)\frac{\alpha}{2}} - e^{-j(m+k)\frac{\alpha}{2}}\right) e^{j\theta_p^m}}{k+m} \right), \text{ even } k \\ 0, \text{ odd } k \end{cases}
$$
$$
\tag{10}
$$

By knowing the complex Fourier coefficients the current $i_{p,dc}$ can be described with (11). Though, $\underline{c}_{p,dc}^0$ is the dc component of $i_{p,dc}$ which allows the calculation of the transferred power by $P = \underline{c}_{p,dc}^0 V_{dc}$.

$$
i_{p,dc} = \sum_{-\infty}^{\infty} \underline{c}_{p,dc}^k e^{jk\omega_{sw}t} \tag{11}
$$

With the known current waveform of the harmonics the voltage ripple ΔV in the dc link can be calculated with (12) at a known dc-link capacitor C_{dc}.

$$
\Delta V = \sqrt{\sum_{k=1}^{\infty} \frac{2\underline{c}_{p,dc}^k \underline{c}_{p,dc}^{-k}}{(k\omega_{sw}C_{dc})^2}} \tag{12}
$$

The dc current is neglected in this equation because the dc current is fed by the inverter of the battery-charing system. Thus, the high frequency current has to be supplied by dc-link capacitors. If there is a limitation in the voltage drop ΔV caused by the current harmonics, equation (12) can be rearranged to (13) to calculated the minimum needed capacitance C_{dc}.

$$
C_{dc} \geq \frac{1}{\Delta V} \sqrt{\sum_{k=1}^{\infty} \frac{2\underline{c}_{p,dc}^k \underline{c}_{p,dc}^{-k}}{(k\omega_{sw})^2}} \tag{13}
$$

A summary of a needed minimum capacitance is given in Tab. I. It can be seen that several μF can keep the voltage dip to less than 1 V caused by current harmonics. At a switching frequency of 500 kHz ceramic capacitors are suitable due to their low parasitic stray inductance which allows them to supply current at high frequencies. For this design, a ceramic based capacitor with 5 μF has been chosen to suppress the voltage dip nearly completely.

The resulting *LC*-lowpass filter can be designed by setting the resonance frequency f_{res} of this filter below the switching frequency f_{sw}. For the design a resonance frequency of $f_{res} = 100$ kHz has been chosen which results in an inductance of approximately $L = 500$ nH as it is calculated in (14).

$$
L = \frac{1}{(2\pi f_{res})^2 C} \approx 0.51 \, \mu H \tag{14}
$$

The Bode diagram of the realized filter is shown in Fig. 6. It has to be mentioned that the attenuation will vary at higher frequencies due to parasitic elements of the filter elements since ideal elements have been assumed for filter design. The influence of parasitic elements is analyzed in [12]. For frequency components less than the resonance frequency the voltage will not have any attenuation. For frequencies above the resonance frequency the attenuation increseases with $40\,\mathrm{dB}/\mathrm{decade}$ which results in an attenaution of $-33\,\mathrm{dB}$ at the switching frequency of $500\,\mathrm{kHz}$ and $-33\,\mathrm{dB}$ at the third harmonic of $1.5\,\mathrm{MHz}$.

B. Transformer

In general the volume of a transformer decreases if the switching frequency is increased at the same power [9]. A comparison of a $50\,\mathrm{kHz}$ transformer with the realized $500\,\mathrm{kHz}$ transformer with nearly the same power rating is depicted in Fig. 8. The corresponding dimensions can be found in Tab. II. It can be seen that the volume has been decreased by a factor of 6 by increasing the switching frequency by a factor of 10. Both transformer are realized in planar technology. The realized transformer consists of several PCBs with 7 turns and a core allowing high frequency operation. Due to geometrical issues, skin and proximity effect can be reduced compared to other technologies [13, 14] which allows high frequency operation. On the other hand primary and secondary side are close to each other which increases the coupling capacitance C_{coupling} between both sides. Measurement results of the transformer are depicted in Fig. 7.

The stray inductance L_σ is closely constant over a wide frequency range of about $200\,\mathrm{nH}$. This reduces the losses in the transformer and enables the operation at the given power and frequency. To operate the DAB securely, a certain stray inductance is needed to limit the current slope. In operation the stray inductance should be [8]

$$L_\sigma \leq \frac{V_{\mathrm{dc}} V_{\mathrm{batt}} \pi}{4 \omega_{\mathrm{sw}} P_{\mathrm{peak}}} \tag{15}$$

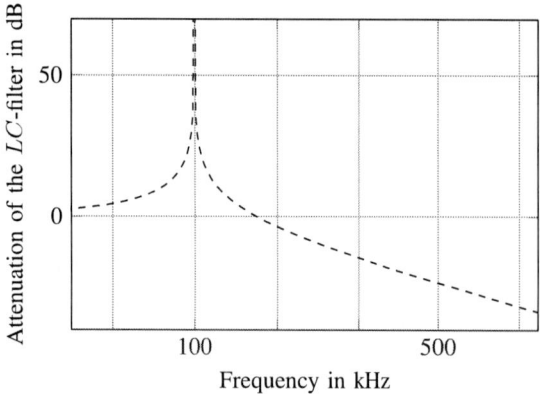

Fig. 6: Bode Diagram of the realized LC-filter

TABLE II: Dimensions of both transformers

	Width	Length	Depth	Volume
50 kHz-Transformer	72 mm	38.5 mm	124 mm	344 cm³
500 kHz-Transformer	52 mm	17 mm	65 mm	58 cm³

which results in $L_\sigma \approx 5\,\mu\mathrm{H}$. To achieve the desired stray inductance an additional inductor has to be connected in series to the transformer which will be discussed in the next subsection.

At the operating frequency the main inductance L_{m} is much higher than the stray inductance which makes the main inductance negligible for calculations. Additionally, the reactive current to magnetize L_{m} is small.

The measured coupling capacitance C_{coupling} between the primary and secondary side is in the range of $60\,\mathrm{pF}$. This capacitance results in common mode current and should be minimized to reduce stress on components [15] and for safety issues.

In the last measurement plot the influence of the skin- and proximity effect can be seen since the series resistance R_{s} increases with an increase of the switching frequency. However, for the realized transformer R_{s} is still below $0.5\,\Omega$ at a frequency of $10\,\mathrm{MHz}$. Thus, there are low losses caused by low order switching harmonics.

C. Series Inductor

To obtain an efficient DAB converter in a wide operating range a stray inductance of $5\,\mu\mathrm{H}$ is necessary. Unfortunately, the described compact transformer only has a stray inductance of about $200\,\mathrm{nH}$. Otherwise, the resulting losses could not be cooled at that small volume. Thus, an additional series inductor is needed to privide the desired inductance of $5\,\mu\mathrm{H}$. For demonstrating issues and due to linearity air coils have been used to realize the additional inductance by using equation 16.

$$L \approx \frac{\mu_0 N^2 A}{l} \tag{16}$$

The number of turns is given by N, l is the length of the air coil and A is the cross section area. Since this does not result in a compact air coil, a proper design for planar based inductors could be done. This would also improve emitted EMI due to better shielding properties and power density. Nevertheless, all demonstrations and design consideration proposed by this paper are still valid and are verified in the section measurement results. The measurement result of the resistance of the air coil is given in 9 whereas a figure of the air coils is depicted in Fig. 11a. The resistance keeps below $0.2\,\Omega$ for frequencies below $2\,\mathrm{MHz}$ which is suitable for an operation of $500\,\mathrm{kHz}$.

IV. Thermal Design

In simulations based on MATLAB/Simulink models, the maximum losses for each switch have been simulated to $15\,\mathrm{W}$. The used GaN switches have a cooling area of $20\,\mathrm{mm^2}$ which results in a thermal flow density of about $0.75\,\mathrm{W/mm^2}$. Thus,

The 2018 International Power Electronics Conference

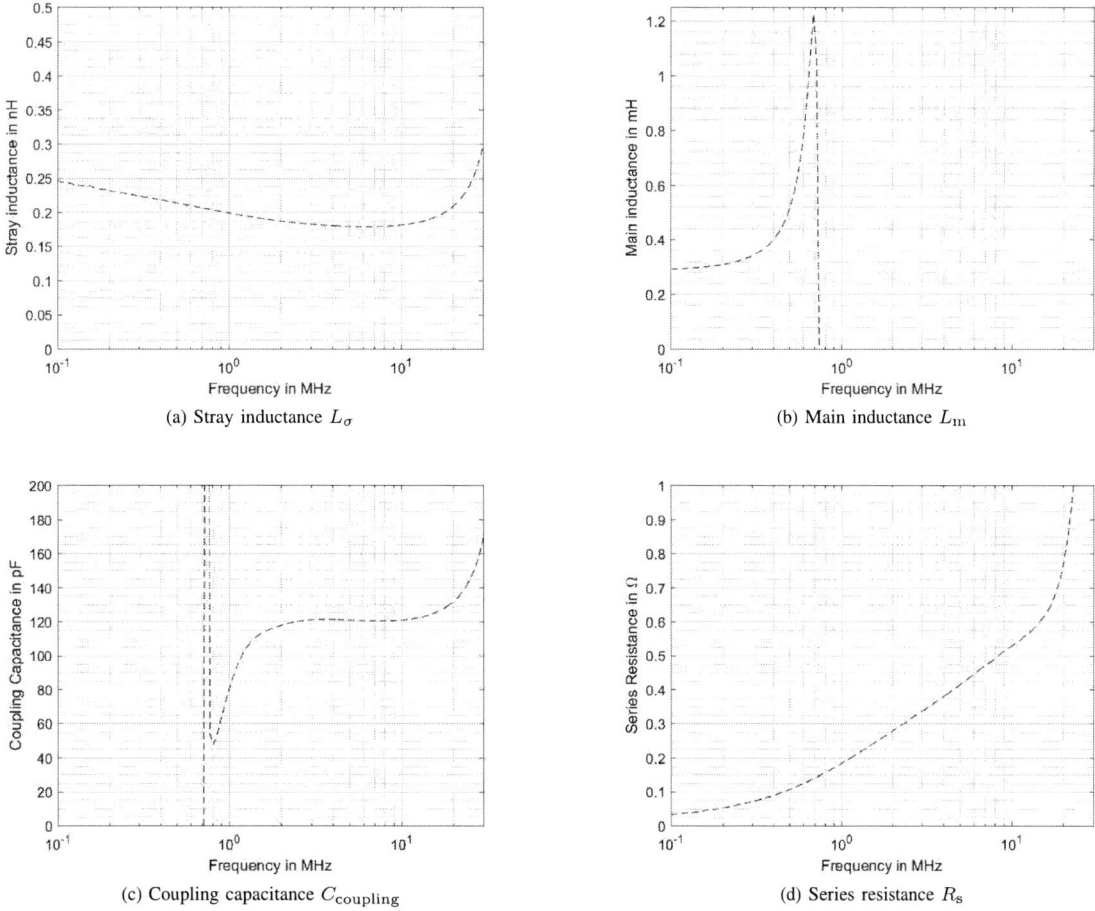

(a) Stray inductance L_σ

(b) Main inductance L_m

(c) Coupling capacitance $C_{coupling}$

(d) Series resistance R_s

Fig. 7: Measurement results of the designed high frequency transformer

Fig. 8: Comparison of a $50\,kHz$ transformer with a $500\,kHz$ transformer in the same power range of $4\,kW$

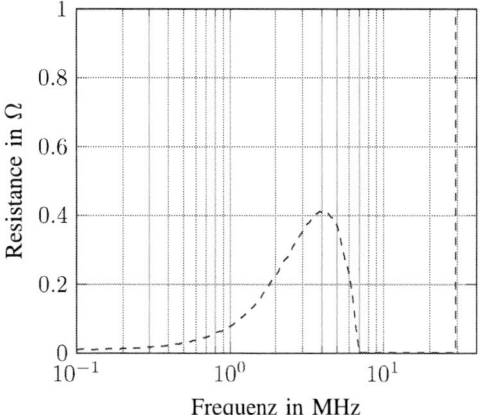

Fig. 9: Resistance of the realized air coil

an effective and efficient cooling is needed to transfer the heat from the switches to the heat sink. Different cooling methods have been investigated and the results are shown in Fig. 10. The simulations are based on the simulation principle given in [16]. For every simulation the four switches of one full bridge are simulated. On top of the switches different cooling structures are mounted. Each cooling structure ensures electrical isolation by gap fillers or ceramics. In the first

393

The 2018 International Power Electronics Conference

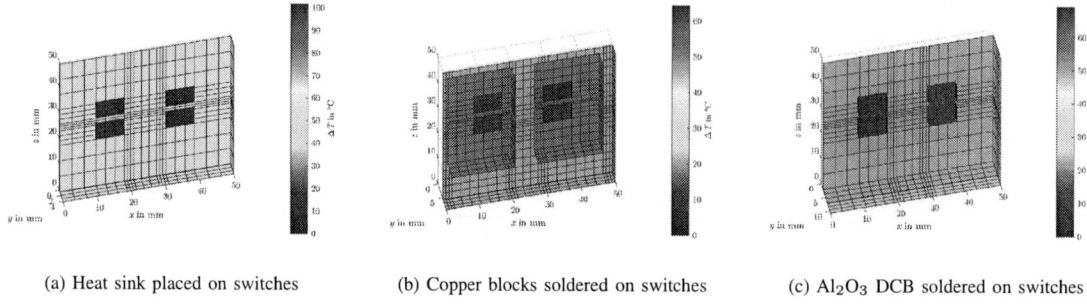

(a) Heat sink placed on switches   (b) Copper blocks soldered on switches   (c) Al$_2$O$_3$ DCB soldered on switches

Fig. 10: Different methods for cooling with 15 W losses in each switch.

simulation result, the heat sink with a gap filler on the bottom side for electrical isolation is placed directly on the four switches of one full bridge. This results in a temperature rise of about 100 °C. As a second example, copper blocks are soldered on the switches which results in a more efficient heat spreading and a higher thermal capacitance. The heat sink with a gap filler is mounted on the copper blocks. Unfortunately, the potential of the copper blocks is identical to the source potential which increases EMI due to jumping potentials. In comparison to the first example the temperature rise can be reduced to about 62 °C. In the third example an Al$_2$O$_3$ DCB has been soldered on the switches which already includes electrical isolation to the heat sink. With a DCB, the temperature of the whole system is very homogeneous and the temperature rise is nearly the same than with the copper blocks but the EMI will be less. For the realized prototype the version with the Al$_2$O$_3$ DCB has been chosen.

The losses of the transformer are rated with 28 W in nominal operating mode. Thus, a standard heat sink will be mounted on the transformer to ensure sufficient cooling.

V. MEASUREMENT RESULTS

A single phase dual-active bridge converter based on the proposed concept has been realized and is depicted in Fig. 11a.

Additionally, the whole test bench is shown in Fig. 11b with the control board on the right side and a passive load on the left side. This converter includes a cooling system, voltage and current sensors as well as gate driver supply and has a total size of only 0.61 l resulting in a power density of 6.1 $^{kW}/_l$ with air cooling.

One measurement result of the realized converter operating in the nominal operating point of 400 V at a transferred power of 3.7 kW is given in Fig. 12. The efficiency in the nominal operating point has been measured to 95.7 %. It can be seen that both full bridges operate at a duty cycle of 50 % phase shifted to each other. The voltage slopes as well as the transformer current i_p are smooth without significant overshoot. For this specific operation point the voltage slope is about 50 $^{kV}/_{\mu s}$. Even at the high switching frequency of 500 kHz and fast switching no oscillations can be observed on the voltage and current waveforms. This demonstrates a low inductive and well damped design. As a second example the transformer current i_p and the clamp voltages v_p and v_s are given in Fig. 13 at a voltage of 350 V on both sides operating at an output power of $P = 3.7$ kW. During the phase shift δ the current changes from approximately -15 A to 15 A linearly without any oscillation. At the clamp voltages small and high ferquency oscillations can be observed which are cause by

 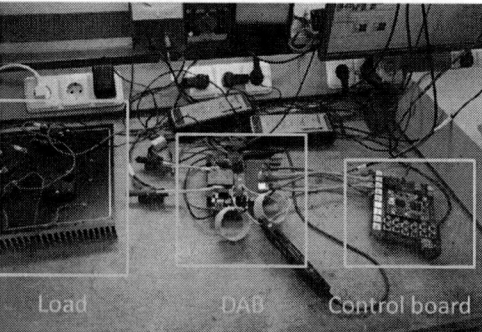

(a) DAB with cooling, sensors and air coils   (b) Hardware setup

Fig. 11: Hardware realization

394

The 2018 International Power Electronics Conference

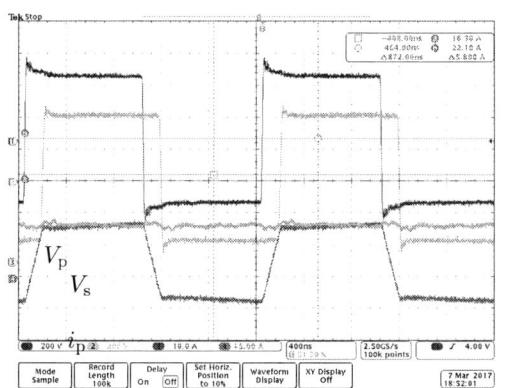

Fig. 12: Measurement results at $V_{dc} = 400\,\text{V}$ and nominal power of $3.7\,\text{kW}$

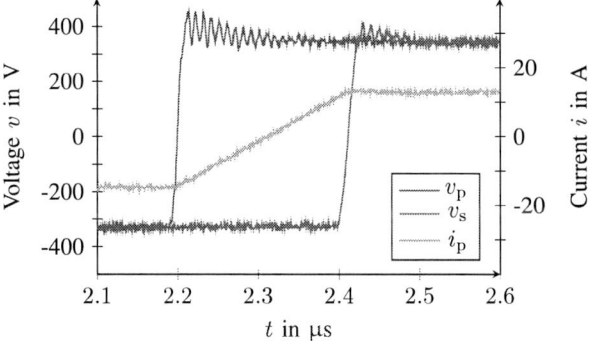

Fig. 13: Switching behavior during transition at $V_{dc} = 350\,\text{V}$ and $P = 3.7\,\text{kW}$

parasitic elements like it has been demonstrated in this paper.

VI. CONCLUSION

This paper shows a design procedure for a fast switching single phase Dual Active Bridge converter which can be designed in the frequency domain to ensure safe operation at high switching frequencies and suitable low pass filter on the input and output sides. This demonstrated design considerations include input and output filter, transformer design as well as thermal aspects. The design concept has been verified with measurements at a rated power of $3.7\,\text{kW}$, a switching frequency of $500\,\text{kHz}$ and a voltage of $400\,\text{V}$ at both full bridges of the converter with an efficiency of about $96\,\%$. At those specifications it is possible to realize very compact converters with high power densities. The realized converter has a volume of only $0.61\,\text{l}$ resulting in a power density of $6.1\,\text{kW/l}$ with air cooling. Current and voltage sensors as well as the cooling structure are included to this small volume.

VII. ACKNOWLEDGEMENT

We would like to thank the company Standex Meder for supporting us with the transformer and the additional inductance.

REFERENCES

[1] A. Stippich et al. "Key components of modular propulsion systems for next generation electric vehicles". In: *CPSS Transactions on Power Electronics and Applications* 2.4 (2017), pp. 249–258. ISSN: 2475-742X. DOI: 10.24295/CPSSTPEA.2017.00023.

[2] Johannes Schmalstieg et al. "From accelerated aging tests to a lifetime prediction model: Analyzing lithium-ion batteries". In: *2013 World Electric Vehicle Symposium and Exhibition*, pp. 1–12. DOI: 10.1109/EVS.2013.6914753.

[3] Philipp Schülting, Martin Rosekeit, Sarriegui Garikoitz, Matthias Biskoping and Rik De Doncker. "Potential of Using GaN Devices Within Air Cooled Bidirectional Battery Chargers for Electric Vehicles". In: *2015 IEEE 6th International Symposium on Power Electronics for Distributed Generation Systems (PEDG)*, pp. 1–6. DOI: 10.1109/PEDG.2015.7223071.

[4] Howard Tweddle. "GS66516T DS Rev 161007: GaN Systems". In: (2016).

[5] Jan Riedel et al. "ZVS Soft Switching Boundaries for Dual Active Bridge DC–DC Converters Using Frequency Domain Analysis". In: *IEEE Transactions on Power Electronics*. Vol. 32, pp. 3166–3179. DOI: 10.1109/TPEL.2016.2573856.

[6] Jan Riedel et al. "Active Suppression of Selected DC Bus Harmonics for Dual Active Bridge DC-DC Converters". In: *IEEE Transactions on Power Electronics*, p. 1. DOI: 10.1109/TPEL.2016.2647078.

[7] G. Bertotti. "General properties of power losses in soft ferromagnetic materials". In: *IEEE Transaction on Magnetics*. Vol. 24, pp. 621–630. DOI: 10.1109/20.43994.

[8] M. H. Kheraluwala and R. W. De Doncker. "Single phase unity power factor control for dual active bridge converter". In: *Conference Record of the 1993 IEEE Industry Applications Conference Twenty-Eighth IAS Annual Meeting*. Vol. 2, pp. 909–916. DOI: 10.1109/IAS.1993.299007.

[9] Robert W. Erickson and Dragan Maksimovic. *Fundamentals of Power Electronics: Second Edition*. New York: Springer Science+Business Media, 2001. ISBN: 978-1-4757-0559-1.

[10] D. Grahame Holmes and Thomas A. Lipo. *Pulse Width Modulation for Power Converters: Principles and Practice*. Piscataway, NJ 08854: IEEE Press, 2003. ISBN: 0-471-20814-0.

[11] P. Hillenbrand, S. Tenbohlen, C. Keller, K. Spanos. "Understanding conducted emissions from an automotive inverter using a common-mode model". In: *2015 IEEE International Symposium on Electromagnetic Compatibility (EMC)*.

[12] P. Schuelting, D. Kubon, and R. W. De Doncker. "Electrical design considerations for a 4 kW buck converter with normally-off GaN devices at a Dc-link voltage of 400 V". In: *2016 IEEE 2nd Annual Southern Power Electronics Conference (SPEC)*. 2016, pp. 1–6. DOI: 10.1109/SPEC.2016.7846124.

[13] Sascha Stegen and Junwei Lu. "Structure Comparison of High-Frequency Planar Power Integrated Magnetic Circuits". In: *IEEE Transaction on Magnetics*. Vol. 47, pp. 4425–4428. DOI: 10.1109/TMAG.2011.2158071.

[14] Ziwei Ouyang, Ole C. Thomsen, Michael A. E. Andersen. "Optimal Design and Tradeoffs Analysis for PlanarTransformer in High Power DC-DC Converters". In: *IEEE Transaction on Industrial Electronics*. Vol. 59, pp. 2800–2810.

[15] Chen Liu et al. "Wideband Mechanism Model and Parameter Extracting for High-Power High-Voltage High-Frequency Transformers". In: *IEEE Transactions on Power Electronics*. Vol. 31, pp. 3444–3455. DOI: 10.1109/TPEL.2015.2464722.

[16] C. H. van der Broeck et al. "Spatial Electro-Thermal Modeling and Simulation of Power Electronic Modules". In: *IEEE Transactions on Industry Applications* 54.1 (2018), pp. 404–415. ISSN: 0093-9994. DOI: 10.1109/TIA.2017.2757898.

The 2018 International Power Electronics Conference

Exploration of the Design and Performance Space of a High Frequency 166 kW / 10 kV SiC Solid-State Air-Core Transformer

Piotr Czyz, Thomas Guillod, Florian Krismer and Johann W. Kolar
Power Electronic Systems Laboratory (PES), ETH Zurich
8092 Zurich, Switzerland
czyz@lem.ee.ethz.ch

Abstract— With the availability of 10 kV SiC MOSFETs with low Zero Voltage Switching (ZVS) losses, Medium-Voltage (MV) converters, e.g., Solid-State Transformers (SST), capable of operation at very high switching frequencies become feasible. However, the optimization of MV and Medium-Frequency (MF) transformer of the dc–dc converter stage of a high power SST reveals that only limited improvements in efficiency and weight result for switching frequencies exceeding 50 kHz. Therefore, air-core transformers are expected to enable a realization with lower weight and, at the same time, simplified insulation coordination. This paper presents a comprehensive exploration of the design and performance spaces (efficiency, mass, volume) of a conventional, i.e., a magnetic-core based, and an air-core transformer employed in a resonant dc–dc converter with input and output voltages of 7 kV and a rated power of 166 kW. As a result, comparable efficiencies are achievable for both transformers (99.3% and 99.0%), but the SST with air-core transformer at a switching frequency of 103.6 kHz features 41% of the mass (10.3 kg) of a conventional transformer (24.9 kg at a switching frequency of 48.5 kHz). Accordingly, air-core transformers are of special interest for future weight-critical SST applications, e.g., in More Electric Aircraft and More Electric Ships.

Keywords—Air-Core Transformer, Power Electronics, Medium-Frequency, Medium-Voltage, Solid-State Transformers, Inductive Power Transfer.

I. INTRODUCTION

Medium-Voltage and Medium-Frequency (MV/MF) transformers represent core elements of Solid-State Transformers (SSTs) and enable galvanic isolation and high step-down or step-up voltage ratios at high conversion efficiencies, high power densities, and low mass. Thus, SSTs are of special interest in aerospace applications [1] (particularly in the field of commercial transport aircraft), in maritime applications [2] and, due to advanced control capabilities, also in general smart grid applications [3]. The trend for More Electric Aircraft (MEA) already exists in the industry for years, resulting in aircraft with hydraulic and pneumatic systems, partially exchanged by electric systems (e.g., in Airbus A380 or in Boeing 787). More recently, thanks to emerging advanced technologies in composites, electric batteries, and motors, further new design concepts regarding the propulsion architecture are developed, i.e., Electrical Propulsion (EP) architectures. EP cover all-electric, hybrid and Turboelectric Distributed Propulsion systems (TeDP). An all-electric aircraft still, due to the limitation of batteries, is feasible only for small, short-haul aircraft. Therefore, in the area of long-haul commercial aircraft only hybrid and TeDP architectures are considered. In hybrid propulsion systems one or more of the gas turbines of a conventional aircraft is replaced by an electric motor with fan, as e.g., in the hybrid-electric flight demonstrator *E-Fan X* [4]. However, the vast majority of the thrust is still generated by gas turbines. TeDP architectures, apart from the gas turbogenerators, utilize an array of motor driven fans, as shown in **Fig. 1**. In this architecture the primary function of the turbogenerators is to generate the electric power which is

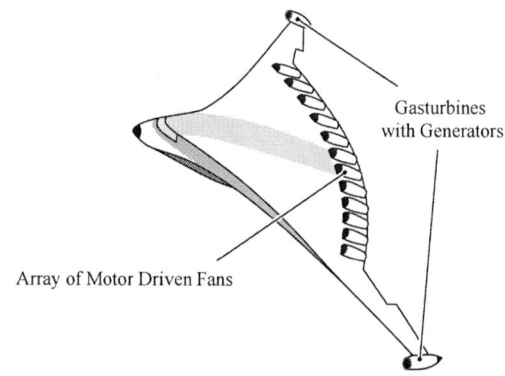

Fig. 1. Concept of TeDP aircraft, based on [5]. Two turbogenerators, i.e., turbine engines driving electrical generators, are located on the wingtips. The electric power from the turbogenerators is distributed to an array of motor driven fans located on the trailing edge of the aircraft body.

then transferred to distributed electric fans, where the thrust is generated. The power rating of the motors driving the fans might reach up to 4 MW [5], therefore, initial studies imply using superconducting electric components. However, due to challenges with lightweight, efficient, and reliable cryocoolers, recent studies also focus on conventional ambient temperature systems [5]. To reduce the losses and masses of the components used in conventional electric systems, a HVDC electric bus is considered. Depending on which parameter is minimized (electrical system's mass or losses) the recommended optimum HVDC bus operating voltage is ranging from ± 3 kV to ± 4.5 kV [6]. A basic concept of a TeDP aircraft electric power system architecture with the ± 3.5 kV bus voltage is presented in **Fig. 2**. In the context of safety and redundancy considerations, such a system is expected to incorporate a circuit breaker between the two dc buses to facilitate continued operation with the working system in case of a severe failure. Advanced state-of-the-art solutions are hybrid dc circuit breakers [7] that incorporate a parallel connection of a mechanical circuit breaker and a string of IGBTs. Pure solid-state dc circuit breakers (SSCB) featuring low losses which could be realized with MV SiC MOSFETs, are, however, still subject to research [8]. Due to the limited flexibility of circuit breakers, also a light-weight air-core SST could be used instead, which provides increased reliability by reason of galvanic isolation, is tolerant with regard to different dc bus voltages, and features overcurrent limitation. In addition to this, further applications of SSTs are highly likely to appear in future TeDP aircraft, e.g., to realize power conditioning units for battery energy storages.

State-of-the-art high efficiency power converters typically feature gravimetric power densities up to 5 kW/kg (cf. **Fig. 3**) [9]. TeDP aircraft applications, however, rather require values greater than $\gamma = 10$ kW/kg, e.g., $\gamma = 14$ kW/kg at an efficiency of $\eta = 99\%$ for the combination of a dc

The 2018 International Power Electronics Conference

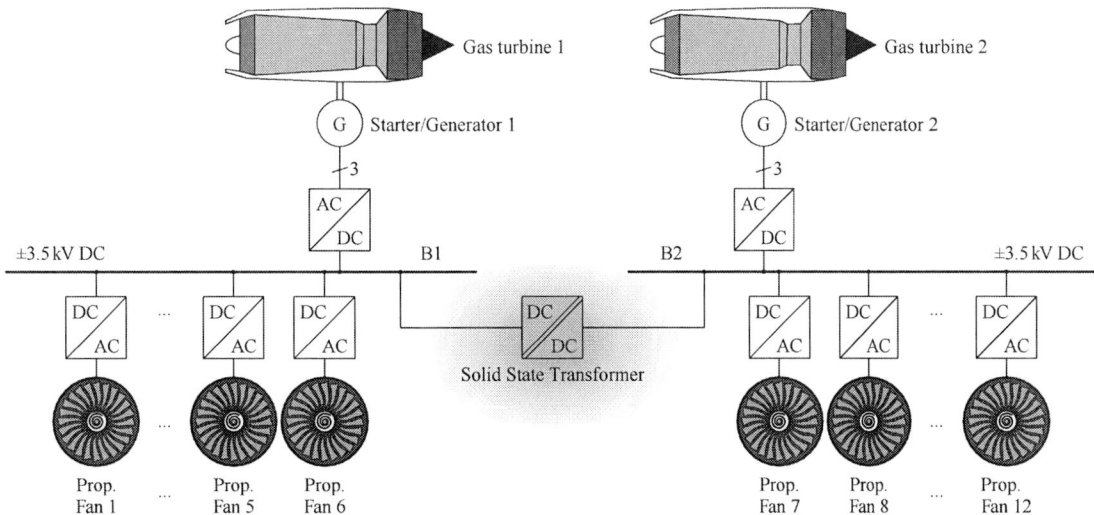

Fig. 2. Draft concept of the electric power system architecture of a TeDP aircraft. In such application the SST could be used as a power electronic link between the two HVDC buses and also fulfill the function of a dc SSCB. A SST, which is formed by parallel connection of SST cells with 166 kW of power, provides increased reliability by reason of galvanic isolation, is tolerant with regard to different dc bus voltages, and features overcurrent limitation.

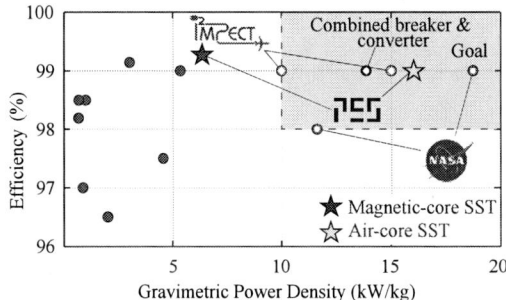

Fig. 3. The gravimetric power density and efficiency of power electronic converters for MEA applications. State-of-the-art high efficiency power converters (blue markers) are presented together with goals set by Integrated, Intelligent Modular Power Electronic Converter (I2MPECT) project [9] and The National Aeronautics and Space Administration (NASA). Proposed SST utilizes the functionality of combined circuit breaker and power converter (purple marker) [10]. The air-core and magnetic core SST designs presented in this paper are shown with stars.

circuit breaker and a power converter [10]. The air-core SST design presented in this paper, cf. **Fig. 3**, could meet the high requirements of minimum gravimetric power density and system efficiency, e.g., in turboelectric aircraft.

In order to take advantage of the favorable properties of MV/MF transformers and to implement SST-specific beneficial control features, SSTs require MV power converters for generation of MF voltages suitable for the MF/MV transformer, and output voltage regulation [11]. An isolated dc–dc converter which allows for high efficiency and high power density, provides the isolation tasks of the SST. In this regard, recent literature documents that high efficiencies are achievable with Dual Active Bridge (DAB) converters and Series Resonant Converters (SRC), due to their Zero Voltage Switching (ZVS) and/or Zero Current Switching (ZCS) properties [12]–[14]. The SRC, in addition, features reduced currents at the switching instants, effectively reduces high frequency harmonic components in the currents of the MV transformer and is therefore particularly interesting for the selected application.

With the availability of MV SiC MOSFETs, very low switching losses can be achieved for ZVS operation [15], [16], which, in principle, enables MV converters to be operated with very high switching frequencies exceeding 100 kHz. In conventional

MV/MF transformers, however, a substantial part of the core window is required for insulation (between the coils and between coils and core) [17]. Due to the associated low filling factor of the core window and core material limits, limited benefits are expected at high switching frequencies [18]. In contrast, the challenging insulation coordination is reduced in air-core transformers (cf. **Fig. 4**), however, lower coupling factors result. Though, recent literature reveals that high efficiency operation can still be achieved for systems with loosely coupled coils [19], [20].

This paper, therefore, explores the design and performance spaces concerning the efficiency, power density, and power to mass ratio achievable with dc–dc SRCs (cf. **Fig. 5**) employing MV and MF to High Frequency (HF) transformers with air-cores and magnetic-cores. **Table I** summarizes the specifications of the considered SRC, which is part of a multi-cell SST. **Section II** describes the model and the optimization procedure applied to the SRC with MV/MF air-core transformer and **Section III** summarizes the approach considered for the SRC with conventional MV/MF transformer. Subsequently, **Section IV** evaluates the obtained optimization results. The converter with air-core transformer is found to achieve a maximum gravimetric power density of $\gamma = 16.1\,\text{kW/kg}$ at a switching frequency of $f_s = 103.6\,\text{kHz}$, whereas the optimal

TABLE I. Specifications of the prototype system

Electric specifications		
P	166 kW	output power
$V_{1,\text{dc}}$	7 kV(± 3.5 kV)	input side dc-link voltage
$V_{2,\text{dc}}$	7 kV(± 3.5 kV)	output side dc-link voltage
V_{iso}	10 kV	isolation voltage between coils of prim. and sec. sides
Litz wire		
d_{litz}	71 μm	single strand diameter
k_{litz}	28%	total fill factor of the windings (incl. the litz wire fill factor)
Thermal specifications		
$p_{\text{v,max}}$	0.25 W/cm^2	surface related power loss density
Fixed dimensions		
w_{iso}	4 mm	isolation distance magnetic-core
w_{iso}	16 mm	isolation distance air-core

397

The 2018 International Power Electronics Conference

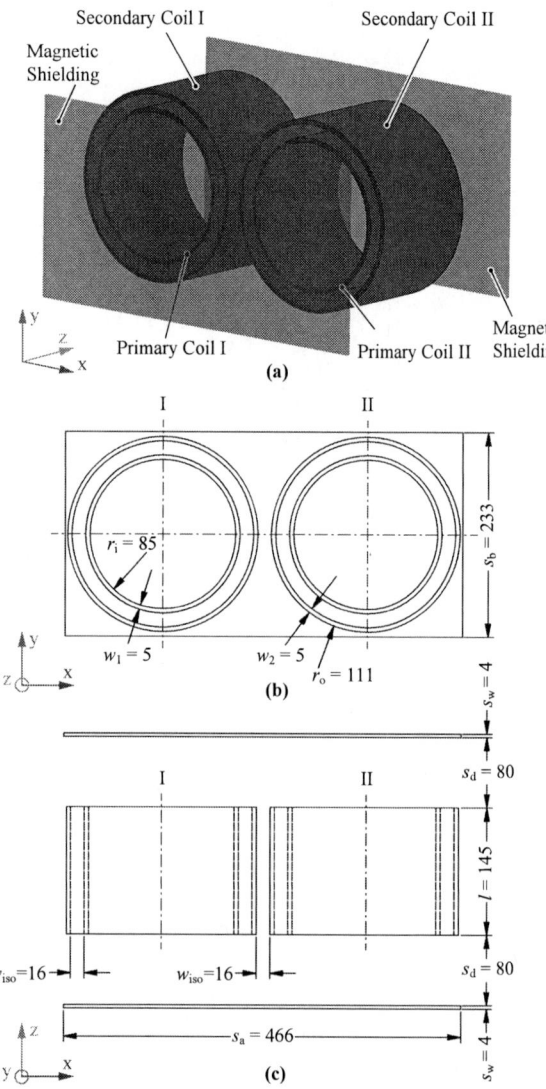

Secondary Coil I — Secondary Coil II
Magnetic Shielding
Primary Coil I — Primary Coil II — Magnetic Shielding

(a)

$r_1 = 85$
$w_1 = 5$ — $w_2 = 5$ — $r_o = 111$
$s_b = 233$
$s_w = 4$

(b)

$s_d = 80$
$l = 145$
$w_{iso} = 16$ — $w_{iso} = 16$ — $s_d = 80$
$s_a = 466$
$s_w = 4$

(c)

Fig. 4. (a) Schematic representation of the MV/HF air-core transformer with primary and secondary coils divided into two equal parts to provide guidance of the magnetic flux along the whole geometric path, i.e. similar to a toroidal arrangement in order to minimize the external magnetic field. Magnetic shielding plates provide the functions of back irons for stray fields at the ends of the transformer. (b) Projection from front. (c) Projection from top. Projections are in scale and all dimensions are given in mm.

solution with magnetic-core transformer is approximately one half of this gravimetric power density at a switching frequency of f_s = 48.5 kHz. The presented η-γ-Pareto-optimal [21] results show that a further increase of the gravimetric density of a transformer with magnetic-core is limited (γ = 8.0 kW/kg at η = 99.1%), whereas the design with air-core transformer can achieve comparable efficiency (η = 99.0%) with gravimetric densities higher than γ = 16.1 kW/kg.

II. AIR-CORE MV/MF TRANSFORMER

A. Design of the SRC

The converter design involves the calculation of the transformer's self inductances, L_1 and L_2, its mutual inductance, M, and the resonant capacitances. The inductances are obtained from the transformer geometry and the resonance capacitances are calculated in order to obtain high efficiency

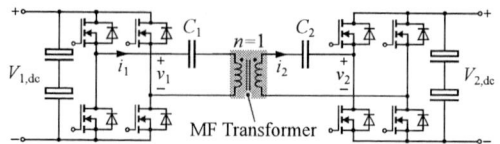

Fig. 5. Schematic drawing of the considered SRC.

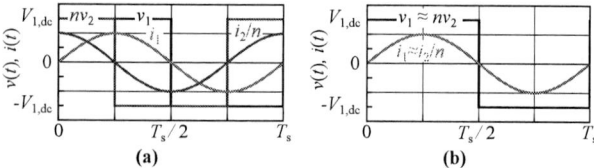

(a) **(b)**

Fig. 6. AC current and voltage waveforms characterizing the operation of the dc-dc converter's MF transformer of the SRC topology: (a) waveforms resulting with the compensation technique applied to transformers with low coupling factor, $k < 0.7$; (b) waveforms resulting for the technique employed for transformers with high coupling factor, $k > 0.7$.

converter operation. Depending on the coupling factor of the investigated air-core transformer,

$$k = \frac{M}{\sqrt{L_1 L_2}}, \quad (1)$$

different compensation techniques are suitable. For low coupling factors, as e.g., in inductive power transfer [22], the SRC achieves highest efficiency if C_1 and C_2 compensate the impedances of the self inductances L_1 and L_2, respectively [22],

$$C_1 = \frac{1}{\omega_s^2 L_1}, \qquad C_2 = \frac{1}{\omega_s^2 L_2}, \quad (2)$$

and for operation with a phase shift of v_1 and v_2 of 90° (cf. **Fig. 6(a)**). In this study it is found that all feasible designs achieve relatively high coupling factors, $k > 0.7$, and as no tolerance to misalignment of coils has to be considered [22], it is therefore, more reasonable to apply the commonly used series-resonant compensation technique, i.e., to set the resonance frequency of C_1, C_2, and the leakage inductance equal to the switching frequency,

$$C_1 = \frac{1}{\omega_s^2 (L_1 - M)}, \qquad C_2 = \frac{1}{\omega_s^2 (L_2 - M)}, \quad (3)$$

with v_1 and v_2 approximately in phase (**Fig. 6(b)**).

B. Transformer Set-up

The air-core transformer is implemented with two coaxially arranged primary and secondary windings (cf. **Fig. 4**). Both windings are arranged as cylindrical solenoids that are realized with HF litz wire conductors. Single-layer windings are assumed for both primary and secondary coils. The results presented in **Section IV-B** are based on 2–D finite-element method (FEM) simulations and selected results are verified by means of 3–D FEM simulations in **Section IV-D**. The 3–D FEM simulations are conducted without and with magnetic shielding plates that are used to close the flux path and/or limit the magnetic stray field.

C. 2–D FEM-Based Design and Optimization

Different approximations of the transformer configuration depicted in **Fig. 4** can be considered. One approximation is to run separate 2–D FEM simulations for the two sets of coils, I and II, and determine the final values for inductances and couplings from the electric series connection of both sets of coils. Another 2–D approximation considers the placement of coils I and II coaxially in series such that a single set of coils with a length of $2l$ results. The first approach fully decouples coils I and II and underestimates the coupling,

398

The 2018 International Power Electronics Conference

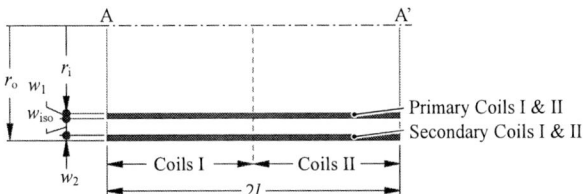

Fig. 7. Schematic representation of the 2–D axisymmetric model (in r–, z– spatial coordinates) of one set of the coils as used for the FEM simulation.

whereas the second approach overestimates the coupling. In this paper, the second approach has been selected with the corresponding axisymmetric 2–D FEM model shown in **Fig. 7.** A detailed investigation of the accuracies achieved with both approaches is subject to further research. The employed 2–D model considers no shielding plates. Furthermore, the primary coil (inner solenoid) is separated from the secondary coil (outer solenoid) by an isolation distance of width w_{iso}.

Fig. 8 depicts the design procedure and optimization of the transformer, which utilizes analytical calculations and FEM simulations. The system specifications listed in **Table I** are providing the input parameters. Second, for the investigated winding arrangements the space of geometry dimensions is defined in **Table II**. For each geometry a FEM simulation is performed with values of self-inductances and mutual inductance normalized to the number of turns, i.e. a single turn inductor. In addition, the values of external magnetic fields and current densities are extracted for the calculation of litz wire losses. Finally, all obtained results are stored in a look-up table, which then is used for analytical calculations. In the next loop, a sweep through switching frequency and number of turns is realized. In this loop the normalized values of parameters obtained from FEM are scaled and the converter compensation method is chosen based on the magnetic coupling value. Using the presented formulas (1)–(3) and an electric model of the SRC (cf. **Fig. 9(a)**), the resonant capacitances, the primary and secondary currents, and losses in the litz wire windings are calculated. The additional copper losses due to skin and proximity effects in each litz wire conductor are calculated based on the peak current and magnetic field values obtained

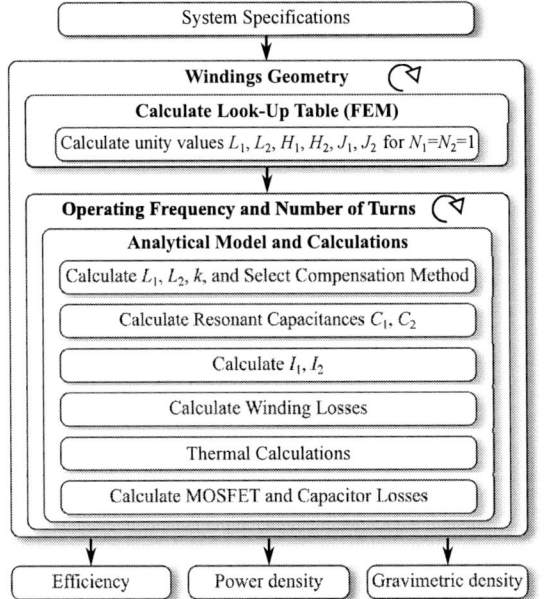

Fig. 8. Flowchart of the implemented optimization procedure for the MV/HF air core transformer.

TABLE II. Design space for the optimization

Var.	Min.	Max.	# Points	Description
2–D FEM Sweep				
f_s	20 kHz	200 kHz	15	switching frequency
N	30	200	170	number of turns
r_i	35 mm	105 mm	15	internal diameter
w_1	5 mm	60 mm	15	width of primary winding
w_2	5 mm	60 mm	15	width of secondary winding
l	60 mm	350 mm	15	length of transformer
3–D FEM Sweep				
s_d	50 mm	80 mm	4	shielding plate distance from end of transf. coils
s_w	4 mm	9 mm	15	thickness of shielding plate
d	1.05	1.20	4	shielding plate to transformer cross section ratio[1]

[1] $s_a = d \cdot 4r_o$, $s_b = d \cdot 2r_o$, cf. **Fig. 4**

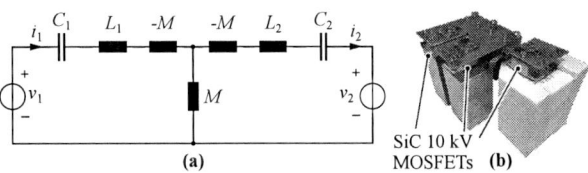

Fig. 9. (a) Lossless electric model of the SRC. (b) 10 kV SiC MOSFET based half-bridge setup for calorimetric measurements of zero voltage switching (ZVS), i.e. soft-switching, losses [16].

from FEM simulation (superposition of scaled vectors of magnetic fields and integration of the loss density according to the models presented in [22]). Furthermore, to ensure thermal feasibility, a thermal model is considered. The transformer designs which exceed a surface related power loss density of $p_{v,max} = 0.25$ W/cm^2 are removed from the calculated designs. The limit value for thermal calculation is based on [22] where the transformer is assumed to have active cooling, i.e., fans, and both, inner and outer, surfaces are considered as participating in the heat transfer.

D. 3–D FEM-Based Design and Optimiz. w/ Shielding Plates

The 3–D FEM simulations of the air-core transformer are conducted for the actual physical configuration with splitted primary- and secondary-side coils, i.e., two sets of coils, such that the external magnetic fields cancel. The simulations are conducted with and without shielding plates, for closing the flux path and/or for limiting the magnetic stray field (back iron for limiting stray fields), for three selected, optimal 2–D designs. The results are presented and discussed in **Section IV-D**.

Ferrite shielding plates (material N97) with rectangular cross sections are assumed to be two equal cores, placed at both ends of the splitted coils according to **Fig. 4**. For the 3–D models parameter sweeps for different geometries of plates are conducted, similarly to the procedure in **Fig. 8**. However, only one optimization loop is realized since for one design the operating frequency and the number of turns are constant. Therefore, instead of the windings geometry sweep, the distances between coils and the plates, thicknesses, widths, and lengths of the plates are varied (cf. **Table II**). For each configuration, the model is solved and all performance indexes are calculated and stored. The modeled transformer has three planes of symmetry, thus, to reduce the computational effort, mirroring is used in all planes to simplify the model.

III. CONVENTIONAL MV/MF TRANSFORMER

In case of a conventional MV/MF transformer (magnetic core based transformer, MCT), the design of the SRC requires

399

The 2018 International Power Electronics Conference

the values of leakage and magnetizing inductances, which are obtained from the transformer geometry. Furthermore, it involves the calculation of the resonant capacitances, which are calculated in order to obtain a resonance frequency equal to the switching frequency. In the SRC with leakage inductance compensation, the magnetizing inductance provides a basic inductive load to the full bridges on the primary and secondary sides. With this, ZVS, i.e. complete charging and discharging of all MOSFETs' output capacitances, is achieved for the complete power range.

In this study, the MCT is used as a reference and for comparison with the air-core transformer (ACT) for the same application and specifications as presented in **Table I**. The design and optimization of the MCT transformer is well documented in the literature, therefore, in this paper, the MCT is designed and optimized based on the routine and models presented in [23]. The considered MCT consists of E-cores (ferrite N95) and shell-type windings. The winding loss model includes skin and proximity losses in the litz wire conductors [23]. For the calculation of the ferrite core losses the improved Generalized Steinmetz Equation (iGSE) is used [24]. Furthermore, similarly to the ACT, the thermal model is based on the surface related power loss density of the transformer. Finally, the value of magnetizing current required for ZVS is set by introducing an air gap of suitable length.

IV. EVALUATION

To complete the analysis, the remaining elements of the converter, such as transistors, resonant and dc-link capacitors, and the cooling system, are evaluated. In the final stage of the optimization, losses, volumes, and masses of those components

are calculated. Finally, only the Pareto-optimal designs with respect to efficiency η and gravimetric power density $\gamma\,(\mathrm{kW/kg})$ and volumetric power density $\rho\,(\mathrm{kW/dm^3})$ are stored and used for evaluation and comparison.

A. Considered Components

The conduction losses of the SiC MOSFETs are calculated assuming an on-state resistance of $400\,\mathrm{m\Omega}$ as given for a junction temperature of $100°$ [16]. Furthermore, experimental data from the characterization of the soft-switching losses of the SiC MOSFETs, obtained with the setup presented in **Fig. 9(b)**, are used to calculate the switching losses for $7\,\mathrm{kV}$ dc input and $7\,\mathrm{kV}$ dc output voltage. In addition, the number of parallel dies per switch is limited to 3 and the maximum losses are set to $100\,\mathrm{W}$ per die. For resonant and dc-link capacitors the losses are estimated using the dissipation factors specified in the data sheets, i.e., 0.05% and 0.1% respectively. The capacitor models of volume and mass use the scaled energy density factors of existing designs, i.e., for the resonant capacitors $2\,\mathrm{J/dm^3}$ and $1\,\mathrm{J/kg}$ [22] and for the dc-link capacitors $40\,\mathrm{J/dm^3}$ and $30\,\mathrm{J/kg}$ [16]. Furthermore, the peak voltage of resonant capacitors is limited to $10\,\mathrm{kV}$. Volume, mass, and losses of the cooling system are based on experimental data from [16].

B. η-γ-ρ-Pareto Fronts for ACT and MCT Converters (2–D)

The η-γ-ρ-performance spaces (planes) and the η-γ-ρ-Pareto fronts that result from the 2–D FEM simulations for ACT and MCT transformers with the respective dc–dc converters are shown in **Figs. 10(a)–(d)**. From the η-γ-performance planes shown in **Figs. 10(a)** and **(c)** it becomes apparent that the converter with MCT due to thermal limitation cannot

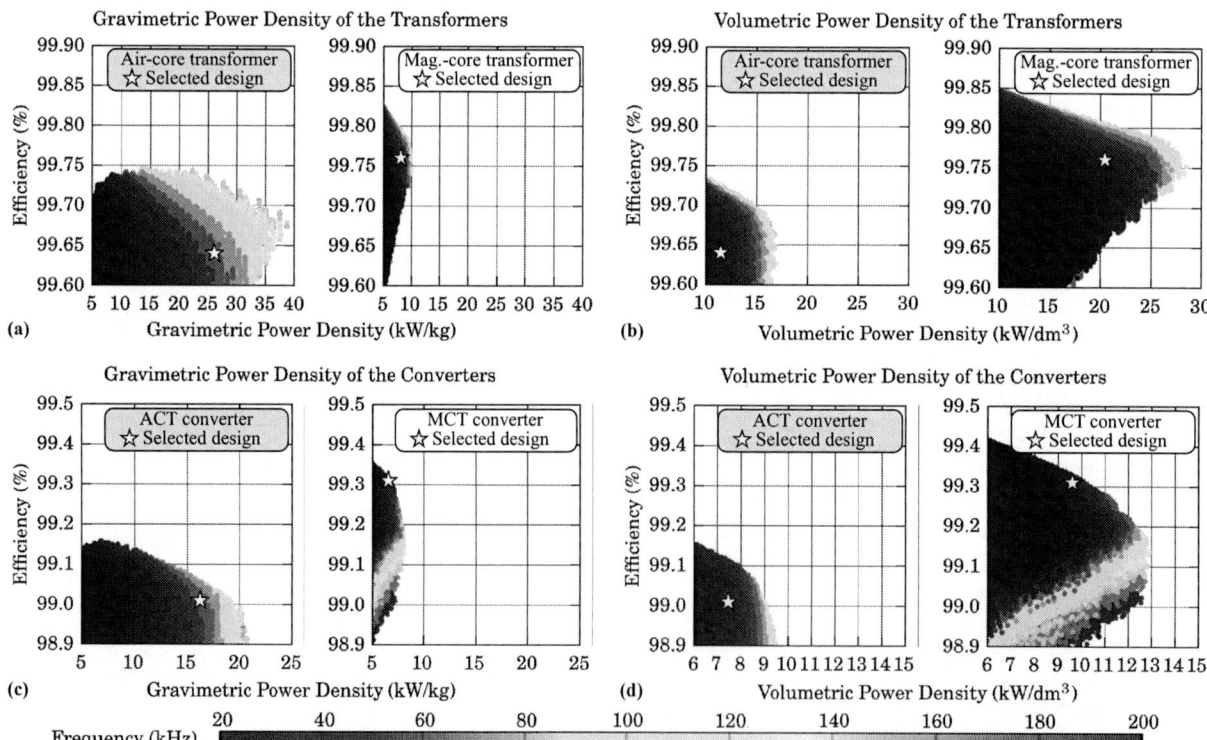

Fig. 10. Results of the 166 kW / 7 kV transformer multi-objective optimization (2–D FEM for the ACT, numerical approx. without FEM for the MCT). η-γ-ρ-performance planes: (a) gravimetric power density γ of the air-core transformer (ACT) vs. the magnetic-core transformer (MCT); (b) volumetric power density ρ of the ACT vs. the MCT; (c) gravimetric power density γ of the overall ACT converter vs. the overall MCT converter (d) volumetric power density ρ of the overall ACT converter vs. the overall MCT converter.

400

The 2018 International Power Electronics Conference

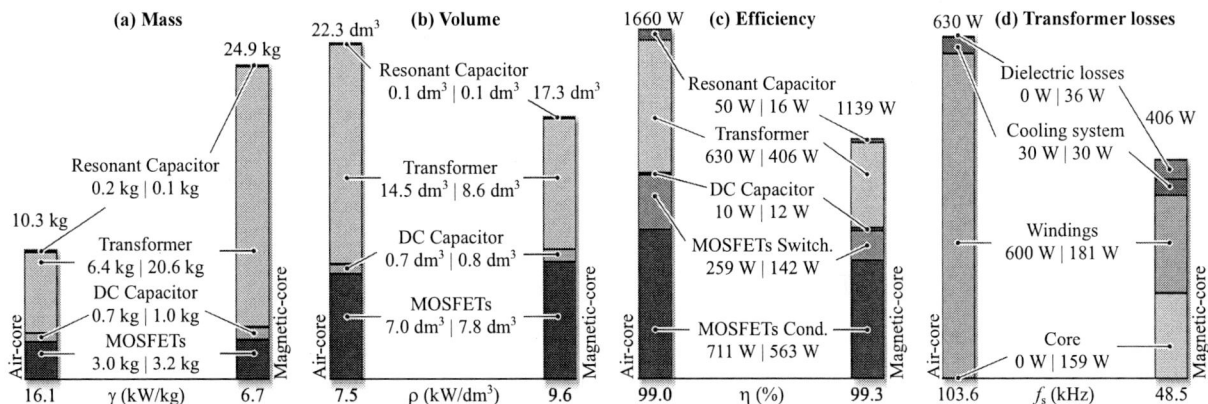

Fig. 11. Break-down of the calculated (a) masses, (b) volumes, (c) losses of the components employed in the two chosen designs. (d) Detailed transformer losses including losses for the cooling system operation.

achieve a power density of more than $8\,\mathrm{kW/kg}$, whereas the converter with ACT can achieve densities higher than $21\,\mathrm{kW/kg}$, while still maintaining a comparably high efficiency of 99.0%. It is worth to point out that, even though there are high-efficient (99.7%) air-core transformer designs (cf. **Fig. 10(a)**) with relatively high switching frequency up to 200 kHz, due to high switching losses of the transistors those designs have poor overall system performance (cf. **Fig. 10(c)**). From the η-ρ-performance plane shown in **Figs. 10(b)** and **(d)** it can be seen that the maximum achievable volumetric power density for an ACT ($17\,\mathrm{kW/dm^3}$) is approximately half of the maximum achievable density of a MCT ($29\,\mathrm{kW/dm^3}$). The larger volume of the ACT can be explained by the fact that either a higher number of turns (long transformer) or a larger diameter of the solenoid windings is needed to obtain effectively the same inductances in comparison to the MCT. According to **Fig. 10(d)**, however, the disproportion of the maximum volumetric power density of the converter is not that significant ($13\,\mathrm{kW/dm^3}$ to $9.5\,\mathrm{kW/dm^3}$) and is in favor of the MCT. For a more detailed comparison one design from each system type is selected for analysis, as indicated by yellow stars in **Fig. 10**. The designs were selected by introducing the following converter performance criteria: efficiency >99.0% and the highest achievable gravimetric power density. For the ACT the selected design is not located on the front of the Pareto plane because such design is not optimal after adding shielding plates, which is explained in more detail in **Section IV-D**. **Figs. 11(a)–(c)** show detailed information about the converter elements' share in overall mass, volume, and losses for the two selected designs, i.e., {99.0%, 103.6 kHz} for the ACT and {99.3%, 48.5 kHz} for the MCT. In addition, in **Fig. 11(d)** the breakdown of the losses of the corresponding transformers including cooling systems is presented. It is interesting to notice, that the design with ACT has almost three times higher gravimetric density with a trade-off of only $\approx 0.3\%$ concerning efficiency than a design with MCT, which constitutes a considerable advantage for a weight-critical application.

C. Investigations of Isolation Distances and Insulation in ACT

Further exploration of the design space of the ACT focuses on the inter-winding isolation distance and insulation material. In the analyzed system for the operating voltage of $7\,\mathrm{kV}$ the required withstand voltage for the transformer is chosen to be $10\,\mathrm{kV}$. It can be achieved either with dry-type insulation for small distances or air insulation provided that the distance is at least equal to the air clearance distance for the required withstand voltage and given that no creepage path exists.

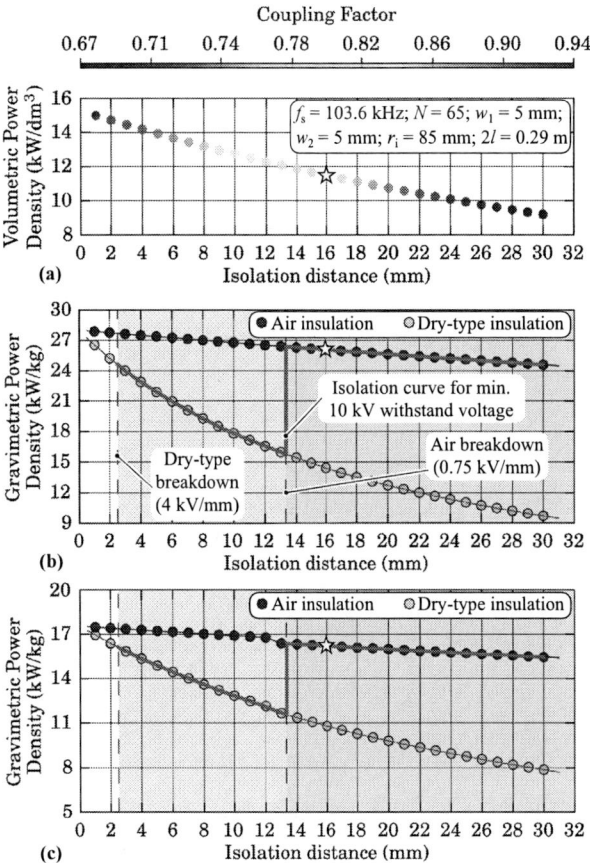

Fig. 12. Volumetric and gravimetric power density for the selected converter with ACT for different isolation distance w_{iso} between the windings and different insulation materials, i.e., air and silicone (from 2–D FEM simulations without shielding plates). (a) Volumetric power density of transformer. (b) and (c) gravimetric power density of transformer and complete converter system, respectively. For the chosen 10 kV withstand voltage, the breakdown boundaries and the resulting functional isolation curve for the system are plotted.

For the selected ACT design ({99.0%, 103.6 kHz}) additional 2–D FEM simulations were carried out for different isolation distances between the windings $w_{\mathrm{iso}} \in \{1, 2, ..., 30\}$ mm and both insulation types: dry-type (silicone) and air. In case of dry-type insulation, the additional mass of the insulation material was accounted for in the gravimetric power density. In **Fig. 12(a)** the volumetric power density for ACTs without shielding plates is presented along with the information about the coupling factor which varies from 0.67 to theoretically 0.94 for small isolation distances. **Fig. 12(b)** and **(c)** clearly shows that feasible designs start from the dry-type insulation breakdown boundary (2.25 mm), which is the only possibility up to the air breakdown boundary (13.33 mm). As expected it is more beneficial to use air insulation as significantly higher gravimetric densities can be achieved, since air insulation designs have higher gravimetric power density than dry-type insulation designs with same isolation distance. Furthermore, the selected design ($w_{\mathrm{iso}} = 16$ mm) with air insulation allows to avoid using additional insulation materials and associated challenges such as dielectric losses, and partial discharges [17].

D. η-γ-Pareto Fronts for ACT with Shielding Plates (3–D)

Fig. 13 shows the η-γ-performance space for a converter employing an ACT, achieved from 2–D FEM simulations (without shielding plates, cf. **Fig. 10(c)**). The designs are colored according to the lengths of the transformers. Additionally, three selected designs are shown for reference, for three different transformer lengths $2l \in \{17.5, 22.5, 29.0\}$ cm and efficiencies of 99.0%. Finally, for those three designs the corresponding Pareto optimal results with optimized shielding plates from 3–D FEM simulations are presented considering the shielding plates. The 3–D FEM optimal results are selected with respect to gravimetric power density and from the subset of designs, in which the shielding plates are not in saturation, i.e., the maximum magnetic flux density in plates is below 300 mT.

From **Fig. 13** it can be seen that choosing the optimal design without shielding plates ($2l = 17.5$ cm) does not lead to the

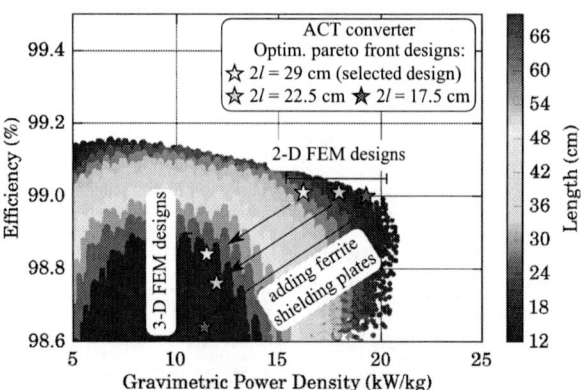

Fig. 13. Results of the multi-objective optimization of the ACT for the ACT converter (166 kW / 7 kV) without shielding plates (2–D FEM simulation) and with shielding plates (3–D FEM simulation) for three designs with different transformer lengths. The mapping of 99% efficient 2–D designs calculated without considering shielding plates (**Fig. 10(c)**) into corresponding optimal 3–D designs with plates is indicated with arrows.

TABLE III. Performance indexes of the selected optimal design ($2l = 29$ mm) of the ACT converter: 2–D, 3–D without shielding plates and 3–D with shielding plates

Solution	Gravimetric density (kW/kg)	Volumetric density (kW/dm³)	System efficiency (%)	Magnetic flux density - stray field (mT)
2–D	16.2	7.47	99.01	-
3–D without shielding plates	16.2	7.47	98.94	3.0
3–D with shielding plates	11.5	3.97	98.84	0.5

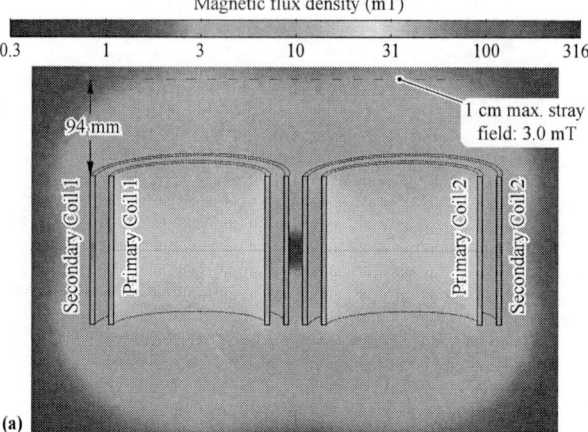

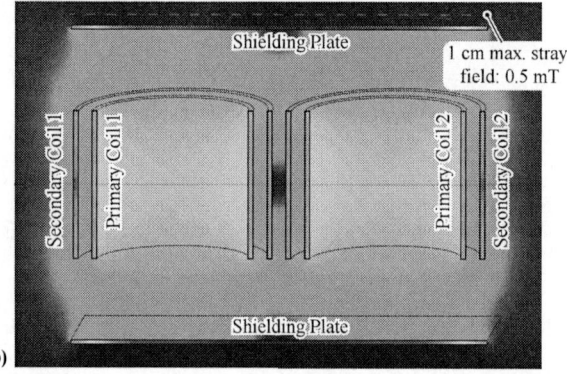

Fig. 14. Results of the 3–D FEM simulation for the selected design: magnetic flux density in the xz–plane. Air-core transformer (a) without shielding plates and (b) with shielding plates. The maximum values of the magnetic flux density in 1 cm proximity from shielding plates are provided.

best design after adding the shielding plates. In fact, the best performance with shielding plates is achieved for the longest transformer that features the lowest gravimetric density without shielding plates. This relation can be explained by the fact that long transformers are also characterized by smaller outer diameters, which determine the sizes of the shielding plates. More detailed performance indexes of the selected optimal design ($2l = 29$ cm) for all models, i.e., 2–D, 3–D without shielding plates, and 3–D with shielding plates, are shown in **Table III**. Adding the ferrite shielding plates generates additional losses, which causes a reduction of the efficiency of approx. 0.1% and a drop of the gravimetric density from 16.2 kW/kg to 11.5 kW/kg. On the other hand, the shielding plates realize the function of back iron for stray fields, which otherwise could cause eddy current losses in surrounding conductive elements or generate EMI perturbations. In the presented case the shielding plates are reducing the value of the magnetic flux density in axial distance of 94 mm from the end of the coils (1 cm from shielding plates) from 3.0 mT to 0.5 mT (cf. **Fig. 14**), which is below typical values specific for MV magnetic core transformers [17].

V. CONCLUSION

This paper evaluates MV transformers with air-core (ACT) and magnetic-core (MCT) for a 7 kV / 166 kW series-resonant dc–dc converter employed in a multi-cell SST, based on a comprehensive design space exploration. With the use of MV SiC MOSFETs featuring ZVS and a blocking voltage of 10 kV, the converter with ACT can achieve gravimetric power densities higher than 21 kW/kg, while still maintaining comparably high efficiency of 99.0%, whereas a converter system employing a MCT cannot achieve a gravimetric power density of more than $8\,kW/kg$, due to thermal limitation of the MCT. The selected Pareto optimal design of the ACT itself is found to enable a lower mass ($10.3\,kg$ vs. $24.9\,kg$) and a comparable efficiency of 99.0% at a switching frequency of $103.6\,kHz$.

One of the challenges in designing MV/MF transformers is the insulation. It is shown that air-core transformers allow to avoid using additional insulation materials and/or the associated challenges (dielectric losses, partial discharges). From the comparison of the air and silicone insulation for the required withstand voltage of $10\,kV$ for the transformer it is clear that it is more beneficial to use air insulation as higher gravimetric power densities can be achieved.

The ACT employs splitted coils to provide guidance of the magnetic flux along the whole geometric path, and magnetic shielding plates as back iron for limiting the external magnetic stray field. Results from 3–D FEM simulations show that shielding plates are reducing the value of magnetic flux density in an axial distance of $94\,mm$ measured from the end of the coils from $3.0\,mT$ to $0.5\,mT$, which is below typical values specific for MV magnetic core transformers. This solution, however, comes with a significant reduction of gravimetric and volumetric power density, and it is therefore beneficial to use ACTs without shielding plates, if possible.

In summary, the ACT represents an attractive alternative to an MCT for weight-critical applications, e.g., future Turboelectric Distributed Propulsion aircraft. Furthermore, the proposed construction of the ACT is advantageous with regard to insulation coordination and is expected to allow for more effective active cooling compared to MCTs. A scaled hardware prototype with a rated power of $25\,kW$ is currently under construction in order to verify the presented calculated results, examine the residual external magnetic field, and study the effectiveness of the proposed shielding plates. Besides that, future research will focus on the identification of accuracies of 2–D FEM simulations of advanced configurations when compared to the results of 3–D FEM simulations and/or experimental results, e.g., to enable computationally efficient Pareto-optimizations of configurations with different types of shielding (e.g. plates, bars) and alternative configurations of ACTs. A respective example is the optimization of an arrangement with balanced primary- and secondary-side transformer currents, where primary coil I is placed inside secondary coil I and secondary coil II is placed inside primary coil II.

ACKNOWLEDGMENT

The authors are very much indebted to the Swiss Centre for Competence in Energy Research on the Future Swiss Electrical Infrastructure (SCCER-FURIES) and the National Research Programme on the Energy Turnaround (NRP 70) of the Swiss National Science Foundation (SNSF) for the support of the research in the area of Solid-State Transformer Technology at the ETH Zurich.

REFERENCES

[1] B. Sarlioglu and C. T. Morris, "More Electric Aircraft: Review, Challenges, and Opportunities for Commercial Transport Aircraft," *IEEE Transactions on Transportation Electrification*, vol. 1, no. 1, pp. 54–64, 2015.

[2] S. Castellan, R. Menis, A. Tessarolo, and G. Sulligoi, "Power Electronics for All-Electric Ships with MVDC Power Distribution System: An Overview," in *Proc. of International Conference on Ecological Vehicles and Renewable Energies (EVER)*, 2014, pp. 1–7.

[3] X. She, A. Q. Huang, and R. Burgos, "Review of Solid-State Transformer Technologies and Their Application in Power Distribution Systems," *IEEE Journal of Emerging and Selected Topics in Power Electronics*, vol. 1, no. 3, pp. 186–198, 2013.

[4] Airbus, "Airbus, Rolls-Royce, and Siemens Team up for Electric Future Partnership Launches E-Fan X Hybrid-Electric Flight Demonstrator." [Online]. Available: http://www.airbus.com/newsroom/press-releases/en/2017/11/airbus-rolls-royce-and-siemens-team-up-for-electric-future-par.html

[5] N. Madavan, "NASA Investments in Electric Propulsion Technologies for Large Commercial Aircraft," in *Proc. of Electric and Hybrid Aerospace Technology Symposium*, 2016.

[6] M. Armstrong, "Superconducting Turboelectric Distributed Aircraft Propulsion," in *Proc. of Cryogenic Engineering Conference / International Cryogenic Materials Conference*, 2015.

[7] M. Callavik, A. Blomberg, J. Häfner, and B. Jacobson, "The Hybrid HVDC Breaker. An Innovation Breakthrough Enabling Reliable HVDC Grids," *ABB Grid Systems, Technical Paper*, 2012.

[8] C. Gu, P. Wheeler, A. Castellazzi, A. J. Watson, and F. Effah, "Semiconductor Devices in Solid-State/Hybrid Circuit Breakers: Current Status and Future Trends," *Energies*, vol. 10, no. 12, p. 495, 2017.

[9] M. Guacci, "Analysis and Design of a 1200 V All-SiC Planar Interconnection Power Module for Next Generation More Electrical Aircraft Power Electronic Building Blocks," *CPSS Transactions on Power Electronics and Applications*, vol. 2, no. 4, pp. 320–330, Dec 2017.

[10] A. T. Isikveren, A. Seitz, P. C. Vratny, C. Pornet, K. O. Plötner, and M. Hornung, "Conceptual Studies of Universally-Electric Systems Architectures Suitable for Transport Aircraft," in *Proc. of 61th Deutscher Luft- und Raumfahrtkongress*, 2012.

[11] J. E. Huber, D. Rothmund, and J. W. Kolar, "Comparative Evaluation of Isolated Front End and Isolated Back End Multi-Cell SSTs," in *Proc. of IEEE International Power Electronics and Motion Control Conference (IPEMC-ECCE Asia)*, 2016, pp. 3536–3545.

[12] N. Soltau, H. Stagge, R. W. D. Doncker, and O. Apeldoorn, "Development and Demonstration of a Medium-Voltage High-Power DC-DC Converter for DC Distribution Systems," in *Proc. of IEEE International Symposium on Power Electronics for Distributed Generation Systems (PEDG)*, 2014, pp. 1–8.

[13] C. Zhao, D. Dujic, A. Mester, J. K. Steinke, M. Weiss, S. Lewdeni-Schmid, T. Chaudhuri, and P. Stefanutti, "Power Electronic Traction Transformer – Medium Voltage Prototype," *IEEE Transactions on Industrial Electronics*, vol. 61, no. 7, pp. 3257–3268, 2014.

[14] M. S. Agamy, M. E. Dame, J. Dai, X. Li, P. M. Cioffi, R. L. Sellick, and R. K. Gupta, "Resonant Converter Building Blocks for High Power, High Voltage Applications," in *Proc. of IEEE Applied Power Electronics Conference and Exposition (APEC)*, 2015, pp. 2116–2121.

[15] D. Han, J. Noppakunkajorn, and B. Sarlioglu, "Efficiency Comparison of SiC and Si-Based Bidirectional DC-DC Converters," in *Proc. of IEEE Transportation Electrification Conference and Expo (ITEC)*, 2013, pp. 1–7.

[16] D. Rothmund, D. Bortis, J. Huber, D. Biadene, and J. W. Kolar, "10kV SiC-based Bidirectional Soft-Switching Single-Phase AC/DC Converter Concept for Medium-Voltage Solid-State Transformers," in *Proc. of IEEE International Symposium on Power Electronics for Distributed Generation Systems (PEDG)*, 2017, pp. 1–8.

[17] T. Guillod, F. Krismer, and J. W. Kolar, "Electrical Shielding of MV/MF Transformers Subjected to High dv/dt PWM Voltages," in *Proc. of IEEE Applied Power Electronics Conference and Exposition (APEC)*, 2017, pp. 2502–2510.

[18] F. Kieferndorf, U. Drofenik, F. Agostini, and F. Canales, "Modular PET, Two-Phase Air-Cooled Converter Cell Design and Performance Evaluation with 1.7kV IGBTs for MV Applications," in *Proc. of IEEE Applied Power Electronics Conference and Exposition (APEC)*, 2016, pp. 472–479.

[19] R. Bosshard, U. Iruretagoyena, and J. W. Kolar, "Comprehensive Evaluation of Rectangular and Double-D Coil Geometry for 50 kW/85 kHz IPT System," *IEEE Journal of Emerging and Selected Topics in Power Electronics*, vol. 4, no. 4, pp. 1406–1415, 2016.

[20] S. Mao, C. Li, T. Song, J. Popovic, and J. A. Ferreira, "High Frequency High Voltage Generation with Air-Core Transformer," in *Proc. of IEEE International Workshop on Integrated Power Packaging (IWIPP)*, 2017, pp. 1–5.

[21] J. W. Kolar, F. Krismer, Y. Lobsiger, J. Mühlethaler, T. Nussbaumer, and J. Miniböck, "Extreme Efficiency Power Electronics," in *Proc. of International Conference on Integrated Power Electronics Systems (CIPS)*, 2012, pp. 1–22.

[22] R. Bosshard, J. W. Kolar, J. Mühlethaler, I. Stevanovi, B. Wunsch, and F. Canales, "Modeling and η-γ-Pareto Optimization of Inductive Power Transfer Coils for Electric Vehicles," *IEEE Journal of Emerging and Selected Topics in Power Electronics*, vol. 3, no. 1, pp. 50–64, 2015.

[23] M. Leibl, G. Ortiz, and J. W. Kolar, "Design and Experimental Analysis of a Medium-Frequency Transformer for Solid-State Transformer Applications," *IEEE Journal of Emerging and Selected Topics in Power Electronics*, vol. 5, no. 1, pp. 110–123, 2017.

[24] K. Venkatachalam, C. R. Sullivan, T. Abdallah, and H. Tacca, "Accurate Prediction of Ferrite Core Loss with Nonsinusoidal Waveforms Using Only Steinmetz Parameters," in *Proc. of IEEE Workshop on Computers in Power Electronics*, 2002, pp. 36–41.

The 2018 International Power Electronics Conference

Novel Calculation Method of Iron Loss of Gapped Inductors Using Loss Map

Yoshihiro Miwa[1*] and Toshihisa Shimizu[1]

1 Electrical and Electric Engineering, Tokyo Metropolitan University, Tokyo, Japan

*E-mail: miwa-yoshihiro@ed.tmu.ac.jp

Abstract—In recent years, the volume of magnetic components have become dominant in power electronics circuits, and their losses cannot always be neglected. Therefore, design of magnetic components based on an accurate estimation of their losses is important to achieve high power density of power electronics circuits. However, conventional calculation method for the iron loss of a gapped inductor requires much measurement or complex electromagnetic simulation. This paper proposes a novel calculation method for iron loss of gapped inductors using the loss map method. An advantage of the proposed method is that the iron loss of gapped inductor can be calculated without electromagnetic simulation and with less measurement effort than conventional method. Experimental verification has been conducted with gapped inductors where MnZn ferrite core and iron dust core are used as the core material.

Keywords— Air gap, Buck converter, Inductor, Iron loss

I. INTRODUCTION

Recently, the volume of magnetic components have become dominant in power electronics circuits, and their losses cannot always be neglected [1]. Therefore, decreasing the volume and losses of magnetic components are some of the most important challenges to overcome in order to increase the power density of power electronics circuits. Since heat generated from losses in magnetic components limits their volume, accurate estimation of their losses is necessary when designing magnetic components.

However, the problem of estimation of their losses is not entirely solved. Magnetic components are classified into transformers and inductors, depending on the function. Iron losses of transformers can be calculated with the Steinmetz equation or iGSE (Improved Generalized Steinmetz Equation) [2]. Nevertheless, these equations have a drawback: the equations neglect the influence of dc premagnetization on iron losses. Iron losses vary with the dc component of magnetic field (referred to as dc premagnetization), and the influence cannot be neglected [3]-[5]. Therefore, it is difficult to calculate iron losses of inductors with these equations if the magnetic field has dc premagnetization.

Some approaches to estimate iron losses of inductors have been proposed. One of these approaches is the loss map method [3] [6] [7]. As discussed in detail later in this paper, the loss map method is an iron loss calculation method using a loss map. The loss map storages measured iron loss data for the excited condition including dc premagnetization. Because one parameter of the loss map is dc premagnetization, the iron loss of an inductor can be calculated while considering the influence of dc premagnetization. Another approach is the improved iGSE [5]. In this approach, the iron loss characteristic for dc premagnetization is described by representing the exponential part in iGSE as a dc premagnetization variable. The variables are determined with measured iron loss data, i.e. a loss map.

Nevertheless, both approaches have the same drawback due to using a loss map. As discussed in section 4, the loss map varies with air gap length. In other words, it is difficult to calculate the iron loss of a gapped inductor with a loss map measured from a gapless inductor. In order to know the iron loss characteristic of a gapped inductor, fabrication of the gapped inductor and measurement of the iron loss are necessary. Fabricating the gapped inductor and measuring the iron loss are time consuming and make optimizing the air gap length difficult. Using the FEM by electromagnetic analysis, the iron loss of the gapped inductor can be calculated [8]. However, such a simulation is complex and requires significant simulation time.

This paper proposes a method to calculate the loss map of a desired air gap length from a loss map measured with a gapless inductor. With the proposed method, acquiring the loss map for various air gap lengths is easier than using the conventional method, and the iron loss of the gapped inductor can be calculated with the loss map method. Therefore, the proposed method contributes to iron loss estimation for a gapped inductor and the optimization of airgap length. This paper is organized as follows. The loss map and loss measurement method are explained in Sections 2 and 3, respectively. Section 4 introduces the novel calculation method of iron loss of gapped inductors. The proposed method is experimentally verified using both ferrite and iron dust cores in Section 5. Section 6 concludes the paper.

II. LOSS MAP METHOD

A. Iron loss of inductors

First, the iron loss characteristic of inductors is discussed using an inductor for a general buck converter application. Fig. 1 (a) and (b) show a general buck converter circuit and the inductor waveforms, respectively. As shown in Fig. 1 (b), the inductor current includes the dc component $I_{L,dc}$. Therefore, the magnetic field includes dc premagnetization in core $H_{c,dc}$ and the flux density includes the dc component $B_{c,dc}$. As a result, the BH curve is positioned in the first quadrant in the B-H plane, as shown in Fig. 2. This curve is called a minor loop. The area of the minor loop corresponds to iron loss per unit volume Q [J/m³]. The shape and the area vary not only with ripple flux density in core ΔB_c and switching frequency f_{sw}, but also with $H_{c,dc}$. As an example, the iron loss characteristic for the excited condition (referred to as loss map) of MnZn ferrite (PC40 TDK) and iron dust (SK Toho Zinc) are shown in Fig. 3(a) and (b), respectively. The values of iron loss on the loss maps are measured data, and the procedure to derive the loss map is introduced in Section 3. As shown in Fig. 3(a) and (b), the iron loss of ferrite and iron dust increase when increasing $H_{c,dc}$. In addition, the iron loss of ferrite increases more sharply with increasing $H_{c,dc}$ than that of iron dust. The iron loss characteristic curve depends heavily on the magnetic material. Therefore, it is necessary to consider the influence of $H_{c,dc}$ on iron losses. when calculating the iron loss of inductors.

B. Loss Map Method

The loss map method is an iron loss calculation method for inductors used for power electronics circuits. This paper shows the procedure to calculate the iron loss of an inductor used for a general buck converter application. First, the excited condition, such as $H_{c,dc}$ and ΔB_c, are calculated. According to Ampere's law, $H_{c,dc}$ is calculated as

$$H_{c,DC} = \frac{N I_{L,dc}}{l_c} \qquad (1)$$

Where N is the number of windings and l_c is the flux path length of the core. According to Faraday's law, ΔB_c is calculated as

$$\Delta B_c = \frac{1}{N A_c} \int_0^{dT_s} v_L dt = \frac{d^2 V_{in}}{N f_{sw} A_c} \qquad (2)$$

Where A_c is the effective cross-sectional area of the core, d is the duty ratio, T_s is the switching cycle, v_L is the inductor voltage, V_{in} is the input voltage, and f_{sw} is the switching frequency. Second, the value of Q corresponding to the values of $H_{c,dc}$ and ΔB_c is obtaind with the loss map. Finally, the iron loss P_i [W] is calculated as

$$P_i = Q V_c f_{sw} \qquad (3)$$

Where V_c is the core volume. With the loss map method, not only can the iron loss of an inductor used for a buck chopper be calculated, but the iron loss of an inductor used for a PWM inverter can also be calculated [6][7]. In conclusion, once the loss map has been measured once, the iron losses under various excited conditions can be calculated.

III. IRON LOSS MEASUREMENT SYSTEM

Fig. 4 shows a measurement system for iron losses and loss maps. The system mainly consists of a B-H Analyzer (SY-8219 IWATSU) and a DC-bias source (SY-961 IWA

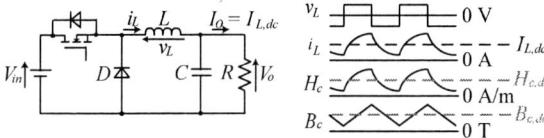

(a) Buck converter circuit (b) The inductor waveforms
Fig. 1. Inductor used for buck converter.

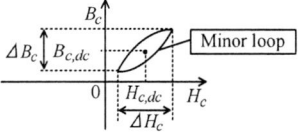

Fig. 2. Minor loop in B-H plane.

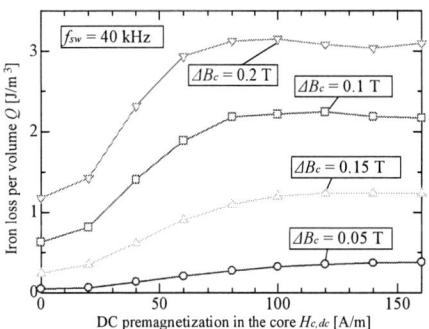

(a) Ferrite core.

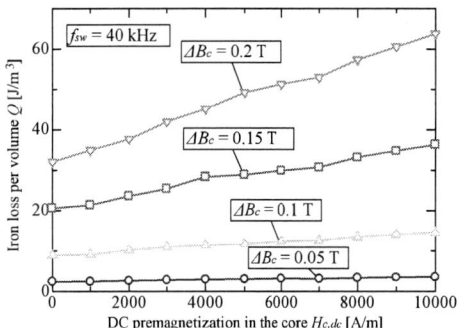

(b) Iron dust core.
Fig. 3. Loss map of MnZn ferrite (PC40 TDK).

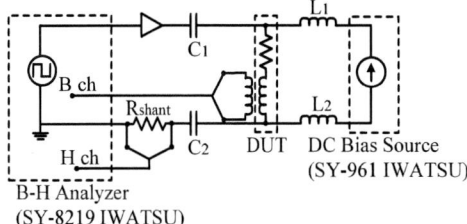

Fig. 4. Iron loss measurement system.

TABLE 1.
SPECIFICATION OF MEASUREMENT SYSTEM.

Ripple inductor current	6 A
DC bias current	30 A
Inductor voltage	±200 V
Switching frequency	10 kHz~1 MHz

TSU) and a DC-bias tester (SY-961 IWATSU). The B-H Analyzer excites a inductor under measurement with a rectangular voltage with the desired switching frequency and ripple component. Additionally, the DC bias source provides the desired dc current to the inductor under measurement. As a result, the inductor under measurement is excited with the desired rectangular voltage including a dc current similar to Fig. 1 (b). Then, the B-H Analyzer measures the inductor current i_L and the inductor voltage v_L and calculates the iron loss. v_L is measured with a secondary winding wound to the inductor being measured, and the measured iron loss does not include copper loss. As a result,

The iron loss under a desired value of $H_{c,dc}$, ΔB_c, and f_{sw} can be measured. One measurement period is approximately 10 seconds. Therefore, it is assumed that the temperature of the core has been kept at room temperature ($T_C = 27$ ℃). The specification of this measurement system is shown in Table 1.

The phase error between measurement of i_L and that of v_L has much influence on iron loss measurement accuracy [3] [5]. In this system, the phase error correction system installed in the B-H analyzer automatically corrects for phase error. The principle is introduced in detail in [3]. Residual phase error after correction is ± 0.15 deg. (sin f_{sw}=10 kHz~10 MHz).

IV. IRON LOSS CALCULATIN METHOD FOR GAPPED INDUCTORS

As discussed in detail later, loss map varies with air gap length l_g as shown in Fig. 5. This paper proposes a method to calculate the loss map for a desired air gap length from a loss map measured with a gapless inductor. Fig. 6 shows the outline of the proposed method. The loss map of the gapped inductor is obtained from some input factors and a conversion equation (10).

A. Conversuib equationt

To obtain the conversion equation, the inductor models shown in Fig. 7 (a) and (b) are used. In the gapped inductor model shown in Fig. 7 (a), the magnetic field in core H_c is lower than the magnetic field in air gap H_g. On the other hand, the equivalent inductor model shown in Fig. 7 (b) has no air gap, and the virtual magnetic field H_v is the same along the flux path. The equivalent inductor model describes the increase in reluctance due to the air gap in the gapped inductor model as a decrease in the virtual permeability. Therefore, the virtual permeability of the equivalent inductor model is lower than the original permeability of core. Applying Ampere's law to both models, $H_{c,dc}$ and the dc component of the virtual magnetic field $H_{v,dc}$ can be determined as

$$H_{c,dc} = \frac{NI_{L,dc}}{l_c} - H_{g,dc}\frac{l_g}{l_c} \tag{4}$$

$$H_{v,dc} = \frac{NI_{L,dc}}{l_c} \tag{5}$$

By eliminating N and $I_{L,dc}$ from (4) and (5), the relationship between $H_{v,dc}$ and $H_{c,dc}$ is described as

$$H_{v,dc} = H_{c,dc} + H_{g,dc}\frac{l_g}{l_c} \tag{6}$$

In (6), the equation shows that the loss map of a gapped inductor described as $H_{v,dc}$ varies with air gap as shown in Fig. 7 under the same ΔB_c and f_{sw} conditions. In other words, the loss map of gapped inductor described as $H_{v,dc}$ is obtained by rewriting the loss map measured from the gapless inductor described as $H_{c,dc}$ with $H_{v,dc}$. In order to eliminate the unknown parameter $H_{g,dc}$ from (6), (7) ~ (9) have been used.

$$H_{c,dc} = \frac{B_{c,dc}}{\mu_o\mu_{r,dc}} \tag{7}$$

$$H_{g,dc} = \frac{B_{g,dc}}{\mu_o} \tag{8}$$

$$B_{g,d}A_g = B_{c,dc}A_c \tag{9}$$

Where $B_{c,dc}$ is the dc component of flux density in the core, $B_{g,dc}$ is the dc component of flux density in the air gap, μ_o is space permeability, and $\mu_{r,dc}$ is relative permeability for $H_{c,dc}$ and $B_{c,dc}$. A_g is the effective cross-sectional area of the air gap. (7) and (8) are the relational expressions between magnetic field and flux density in the core and air gap, respectively. (9) is an equation that shows that total flux content is same in the core and in the air gap. By eliminating $H_{g,dc}$, $B_{c,dc}$, and $B_{g,dc}$ from (6)-(9), the conversion equation (10) is obtained.

$$H_{v,dc} = \left(1 + \mu_{r,dc}\frac{l_g A_c}{l_c A_g}\right)H_{c,dc} \tag{10}$$

With (10), the loss map of a gapless inductor described as $H_{c,dc}$ is converted to a loss map measured with a gapless inductor. In (10), A_g and $\mu_{r,dc}$ are unknown parameters. The method for determining the values of these unknown parameters is shown next.

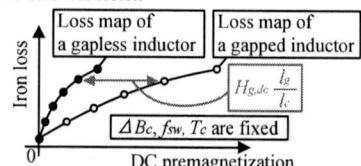

Fig. 5. Air gap length dependency of loss map.

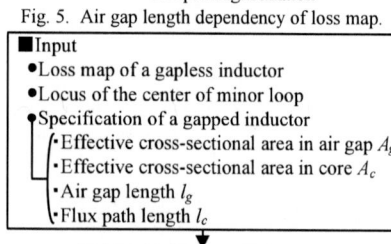

Fig. 6. Outline of the proposed method.

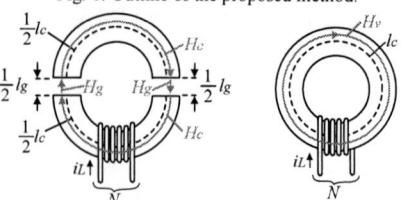

(a) Gapped inductor   (b) Equivalent inductor

Fig. 7. Inductor model.

406

B. Determination of A_g

The value of A_g is generally larger than the value of A_c resulting from the fringing flux. A popular equation to calculate A_g is the following equation [9].

$$A_g = \left(l_w + l_g\right)\left(t + l_g\right) \qquad (11)$$

Where l_w is the width of the core, and t is the thickness of the core. (11) is satisfied in a limited condition $(10 l_g < l_w,\ t)$. In this paper, experimental verification of the proposed method has been conducted with gapped inductors which meet the condition. Therefore, A_g has been calculated with (11).

C. Determination of $\mu_{r,dc}$

As shown in Fig. 8, $\mu_{r,dc}$ is the relative permeability for $H_{c,dc}$ and $B_{c,dc}$. Therefore, the position of the center of the minor loop in the B-H plane needs to be determined in order to determine the value of $\mu_{r,dc}$. In this paper, it is assumed that the top of the minor loop is tangent to the initial magnetization curve shown in Fig. 8 according to [10] and [11]. The center of the minor loop, i.e. the value of $H_{c,dc}$ and $B_{c,dc}$, has been measured with a B-H analyzer. Fig. 9(a) and (b) show measurement results of the locus of the minor loop of ferrite core and iron dust core ,respectively, when increasing $H_{c,dc}$ under the same ΔB_c and f_{sw}. In addition, Fig. 10(a) and (b) show the value of $\mu_{r,dc}$ for the $H_{c,dc}$ of ferrite core and iron dust core, respectively, which is obtained from Fig. 9. The value of $\mu_{r,dc}$ for each $H_{c,dc}$ is Fig. 10(a) and (b) is used in (10).

As this paper focuses on iron loss characteristics of the powder or the ferrite cores for dc premagnetization and air gap length, an increase in iron loss due to fringing loss is neglected. Iron loss calculations for the sheet cores that need to take fringing loss into consideration are reserved for future work. The experimental verification neglecting fringing loss is discussed in Section 5.

V. EXPERIMENTAL VERIFICATION

The experiment has been conducted with inductors where MnZn ferrite, as a high permeability material, and iron dust, as a low permeability material, are used for the core.

A. Experimental condition

All experiments have been conducted with the iron loss measurement system shown in Fig. 4 under room temperature ($T_C = 27\ ^\circ$C). Table 2 shows the experimental conditions. The switching frequency has been chosen based on the general switching frequency range of Si devices. The value of ΔB_c and the range of $H_{c,dc}$ were determined on the basis of the general operational range due to the saturation flux density $B_{c,sat}$ (Ferrite: $B_{c,sat} = 0.5$ T and iron dust: $B_{c,sat} = 1.6$ T)

B. Experimental inductor

Fig. 11 and Fig. 12 show the exterior view of the experimental inductors using the ferrite core and the iron dust core, respectively. All inductors are wound with the exciting winding and the secondary winding which is need for iron loss measurement. The experiment was conducted with two gapped inductors with different air gap lengths

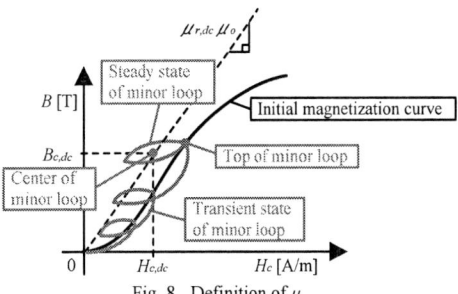

Fig. 8. Definition of $\mu_{r,dc}$.

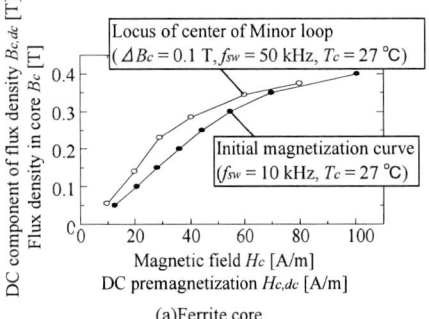

(a)Ferrite core

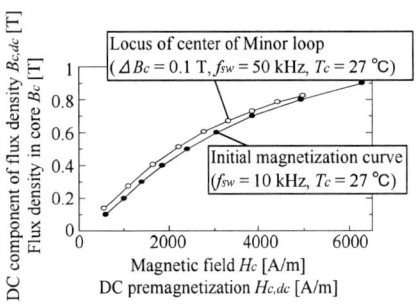

(b)Iron dust core

Fig. 9. Locus of the center of minor loop.

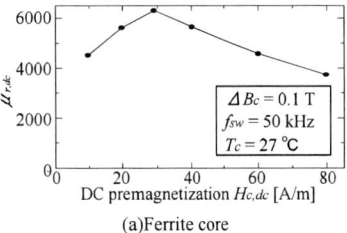

(a)Ferrite core

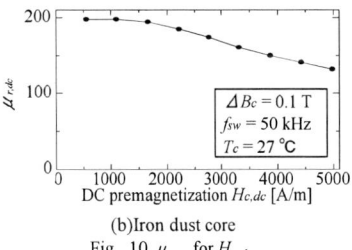

(b)Iron dust core

Fig. 10. $\mu_{r,dc}$ for $H_{c,dc}$.

(the ferrite core: $l_g = 0.2$, 0.4 mm and the iron dust core: $l_g = 0.62$, 1.17 mm) shown in Fig. 11 (b), (c), Fig. 12 (b) and (c). The gapless inductors shown in Fig. 11(a) and Fig. 12(a) ware used to measure the loss map of the gapless inductor. The cores used for the gapless inductor and the gapped inductors are the same model (the ferrite core: PC40 T51×13×31 TDK l_c =124 mm, A_c=127 mm^2 the iron dust: SK 28M Toho zinc l_c =108 mm, A_c=121.5 mm^2). The cores used for the gapped inductors have been cut in a similar fashion as Fig. 7 (a), and the value of l_g shown in Fig. 11(b), (c), Fig. 12(b) and (c) are the sum of the two air gap lengths.

Ferrite and powder core have high electrical resistance, and fringing losses resulting from eddy currents in the core surface are minimal. Moreover, the windings of the gapped inductors have been wound avoiding the air gaps in order to prevent interlinkage of fringing flux and windings, as shown in Fig. 11 (b), (c), Fig. 12(b) and (c). Therefore, it is assumed that an increase in iron loss due to fringing loss is negligible.

C. Experimental result and calculation result

Fig. 13 (a) and (b) show the calculation and experimental results of ferrite and iron dust, respectively. In order to clearly show that the loss maps of the gapped inductor are converted from that of the gapless inductor, all loss maps have been drawn in the same picture and the horizontal axis is composed of $H_{c,dc}$ and $H_{v,dc}$. Additionally, all loss maps are described as iron loss P_i [W] not as iron loss per unit volume Q [J/m^3] from the iron loss calculation with (3) because the purpose of this paper is to calculate iron loss P_i [W].

As shown in Fig. 13(a), the experimental results and calculation results of ferrite core coincide well and the maximum error in the results is approximately 6%. The error is thought to be due to the individual differences of the cores. As shown in Fig. 13(b), the maximum error between experimental results and calculation results of iron dust core is 10%. This error is expected to be caused by the calculation error of the effective cross-sectional area of the air gap, A_g. In the inductor specification of the inductor used in this experiment, about 30 % calculation error of A_g results in 10 % error of iron loss calculation. The value of l_g of the inductor used in the experiment meets the condition discussed in Section 4 Nevertheless, the permeability of the iron dust is low. Therefore, the practical fringing flux is expected to large, i.e., the value of the effective cross-sectional area of the air gap is expected to be higher than the value calculated with (11). From this result, A_g must be calculated accurately when using low permeability magnetic materials, such as iron dust, as a core, and this is left for future work.

TABLE 2.
EXPERIMENTAL CONDITIONS.

Ripple flux density in core ΔB_c	0.1 T
Switching frequency f_{sw}	50 kHz
Duty ratio d	0.5
DC premagnetization in core $H_{c,dc}$	0~80 A/m (Ferrite core) 0~5000 A/m (Iron dust core)

VI. CONCLUSION

This paper proposes a method to calculate the loss map of a gapped inductor, which has a desired air gap length from a loss map measured with a gapless inductor. The proposed method has been verified experimentally with a MnZn ferrite core and iron dust core. Nevertheless, when using low permeability materials, such as iron dust, as the core in an inductor, calculation error is expected to be caused. To overcome this issue, the method to calculate accurately the effective cross-sectional area in the air gap is left for future work.

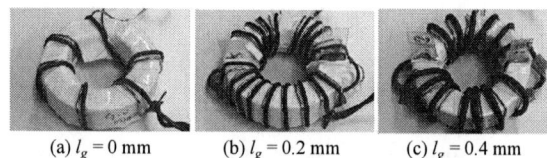

(a) $l_g = 0$ mm   (b) $l_g = 0.2$ mm   (c) $l_g = 0.4$ mm

Fig. 11. Exterior view of experimental inductors (Ferrite core).

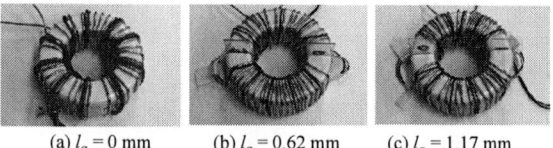

(a) $l_g = 0$ mm   (b) $l_g = 0.62$ mm   (c) $l_g = 1.17$ mm

Fig. 12. Exterior view of experimental inductors (Iron dust core).

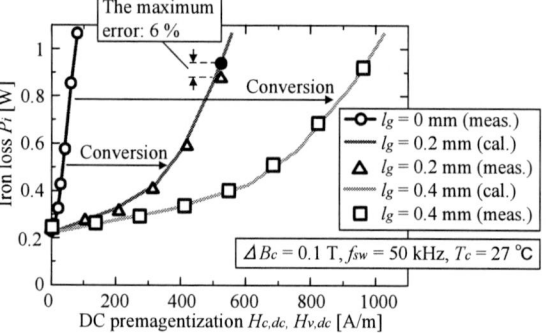

(a)Ferrite core.

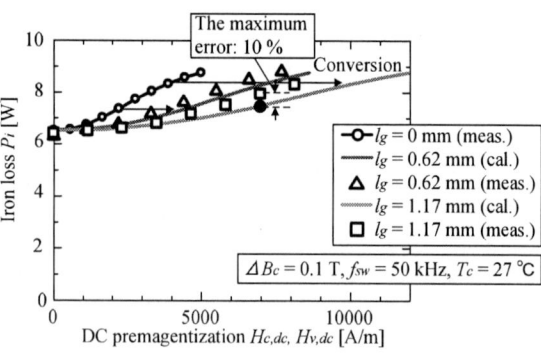

(b)Iron dust core.

Fig. 13. Calculation results and experimental results.

REFERENCES

[1] D. Reusch, J. Strydom, and J. Glaser, "Improving High Frequency DC-DC Converter Performance with Monolithic Half Bridge GaN ICs," ECCE, pp. 381-387, Montreal, Canada, 2015.

[2] K. Venkatachalam, C. Sullivan, T. Abdallah, and H. Tacca, "Accurate prediction of ferrite core loss with nonsinusoidal waveforms using only Steinmetz parameters," IEEE Workshop on Computers in Power Electronics, pp. 36-41, 2002.

[3] S. Iyasu, T. Shimizu and K. Ishii, "A Novel Inductor Loss Calculation Method on Power Converters Based on Dynamic Minor Loop," *IEEJ Transactions on Industry Applications*, Vol. 126, No. 7, pp. 1028-1034, 2006.

[4] A. Kring and J. Soulard, "Overview and Comparison of Iron Loss Models for Electrical Machines," MC2D & MITI, EVER MONACO, 2011.

[5] J. Muhlethaler, J. Biela, J. Kolar, A. Ecklebe, "Core Losses Under the DC Bias Condition Based on Steinmetz Parameters," *IEEE Transaction on Power Electronics*, Vol. 27. pp. 953-963, 2012.

[6] T. Shimizu, K. Kakazu, T. Takano, and H. Ishii, "Verification of Iron Loss Calculation Method using a High-Precision Iron Loss Analyzer," *IEEJ Trans. on Industry Applications*, Vol. 133, No. 1, p84-p93, 2013.

[7] H. Matsumori, T. Shimizu, K. Takano, and H. Ishii, "Iron Loss Calculation of AC Filter Inductor for Three-phase PWM Inverters," *IEEJ Trans. on Industry Applications*, Vol. 133, No. 10, pp. 1009-1021, 2013.

[8] H. Sato, T. Shimizu, "Study on an accurate iron loss calculation method considering the non-uniformity of the magnetic flux density," ECCE, pp. 3032-3039, Montreal, Canada, 2015.

[9] N. Mohan, T. Underland and W. Robbins, "Power electronics : theory, design and applications, (Third Edition)," Wiley, pp. 758, 2003

[10] P. Tenant, J. J. Rousseau, "Dynamic Model of Magnetic Materials Applied on Soft Ferrites," *IEEE Trans. on Power Electronics*, Vol. 13, No. 2, pp. 372-379, 1998.

[11] W. G. Hurley and W. H. Wolfle, "Transformers and inductors for power electronics: theory, design and applications," Wiley, pp. 16, 2013

The 2018 International Power Electronics Conference

Verification of the Reduction of the Copper Loss by the Thin Coil Structure for Induction Cookers

Morimasa Hataya*, Koki Kamaeguchi, Eiji Hiraki, Kazuhiro Umetani, Takayuki Hirokawa, Makoto Imai, Hideki Sadakata

1. Graduate School of Natural Science and Technology, Okayama University, Okayama, Japan
2. Electronic Control Technology Group, Panasonic Corporation, Kusatsu, Japan
*E-mail: pyjv3v5x@s.okayama-u.ac.jp

Abstract- **Litz wire is commonly employed as the heating coil of induction cookers. In order to realize further low cost and profile, the solid wire with simple construction and high space factor is required. However, the solid wire is may suffer from the large copper loss increased by the skin and proximity effect. Then, the previous study proposed the novel coil structure, which can suppress these effects, only by the FEM simulation. Therefore, the purpose of this paper is to verify this structure experimentally in comparison with the Litz wire coil. The result revealed that the proposed structure can have similar AC resistance and the similar height with the same surface area and the same number of turns. Moreover, the experimental result showed a possibility to further height reduction by optimization of the magnetic and winding isolation design. Consequently, the experiment supported practical effectiveness of the proposed structure for induction heating.**

Keywords— copper loss, foil wire, induction cooker, proximity effect.

I. Introduction

Recently, the induction heating is employed in many applications due to its advantages such as fast heating, high efficiency, cleanness, and safety. Particularly, the induction heating is widely utilized for the induction cookers. These cookers are commonly operated above 20kHz in order to suppress the acoustic noise. The operating frequency can reach further higher to approximately 100kHz for heating the pan of non-iron metals.

However, the high frequency operation may cause intense skin and proximity effects at the heating coil. These effects leads to concentration of the current distribution in the wire cross-section [1]–[7], generating large copper loss. Therefore, the heating coil is generally made of the Litz wire to suppress these effects [8]–[11].

The Litz wire is a special wire made of thin isolated wire strands. These strands have the far smaller dimension than the skin depth; and these strands are generally twisted or woven so that all of the strands pass all points of the cross-section of the Litz wire. Because of the symmetry of the electromagnetic condition among the strands, these strands carry the same current. Therefore, the Litz wire can achieve the uniform current distribution. As a result, the Litz wire is free from the skin and proximity effects and can suppress the copper loss.

However, the Litz wire tends to be expensive because of its complicated construction. In addition, the Litz wire coils may have large winding height due to low space factor of the Litz wire. Therefore, the solid copper wire is intensely required to replace the Litz wire for reduction of the cost and height of the heating coil because the solid copper wire tends to have simple construction and higher space factor than Litz wire.

Certainly, the solid copper wire tends to suffer from large copper loss in high frequency operation due to the skin and proximity effects. Therefore, a special magnetic structure suppressing the skin and proximity effects are essential for applications to the induction cookers. As a probable candidate of this magnetic structure, a novel heating coil structure of the copper foil, which is a thin solid copper wire, has been proposed in the previous study [12].

In this heating coil structure, a simple magnetic structure was employed to suppress the skin and proximity effects, as reviewed in the next section. In combination with this magnetic structure, the copper foil with the thickness less than the skin depth is wound to form the heating coil. As a result, the AC current is distributed uniformly inside the copper foil, thus reducing the copper loss.

This previous study discussed only the theoretical principle of this heating coil structure. Actually, this study has verified suppression of the skin and proximity effect, as well as resultant reduction of the copper loss only by the FEM analysis and in comparison with the thick solid rectangular copper wire. Therefore, experimental evaluation is needed to verify the effectiveness of this heating coil structure in comparison with the Litz wire heating coil.

The purpose of this paper is to verify the effectiveness of this proposed heating coil structure of the copper foil both by simulation and experiment. Section II briefly reviews the structure of the proposed structure and explains the theoretical principle how the skin and proximity effect can be suppressed in the structure. Then, section III presents the simulation and the experiment carried out to verify the effectiveness of the proposed structure. In addition, section □ presents an experiment which compares the copper loss between the proposed structure and the Litz wire heating coil.

II. PROPOSED STRUCTURE

Figure 1 shows the proposed heating coil structure with copper foil. The structure employs two strategies to suppress the inhomogeneity of the current distribution in vertical and horizontal directions respectively. One is that the wire thickness is designed to be less than the skin depth; the other is that the ferrite core is placed next to the wire edges. Below, the two strategies are explained.

A. Wire Thickness Smaller Than Skin Depth

First, the conventional problem of the solid copper wire is discussed with respect to the inhomogeneous current distribution in the vertical direction. For this purpose, we analyze the AC current distribution inside the thick solid copper wire. Figure 2 shows the cross section of the thick wire supplied with the high frequency AC current. The AC current is confined at the wire surface within the skin depth due to the skin effect. The skin depth δ is defined as following equation,

$$\delta = \sqrt{2\rho / \omega\mu} \tag{1}$$

where ρ is the resistivity of the material of the wire, μ is the permeability of the material of the wire and ω is the angular frequency.

We apply Ampere's law along the dotted line as shown Fig.2. Because the AC magnetic field does not penetrate the conductor through the skin depth, we can neglect the integration of the magnetic field along the top side of the dotted line. In addition, the integration of the magnetic field along the vertical sides can also be neglected because the magnetic field is perpendicular to the vertical sides. As a result, the integration of the magnetic field is contributed only by the bottom side, obtaining

$$\Delta I = H_s \tag{2}$$

where ΔI is the surface current per unit length and H_s is the surface magnetic field of the wire.

Equation (2) indicates that the surface current per unit length equals to the surface AC magnetic field. In other words, the AC current flowing the surface of wire is proportional to the surface magnetic field. Therefore, the surface magnetic field should be distributed uniformly for the uniform current distribution on the wire surface.

Now, we consider the heating coil of the solid rectangular wire with the thickness far greater than the skin depth, as shown in Fig. 3 (Left figure). This heating coil is constructed on the ferrite plate as is common in many practical heating coil to avoid the electromagnetic interference with the inverter, which is commonly disposed at the bottom of the heating coil.

The bottom wire surface of the lowest layer carries no AC current because the surface magnetic field is small due to large permeability of the ferrite plate. Therefore, all the AC current of the wire of the lowest layer flows in the top wire surface.

We apply again Ampere's law to the closed path passing through the center of the wire of the lowest layer

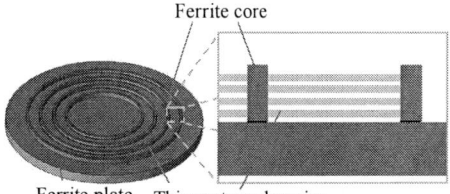

Fig.1 Proposed coil structure

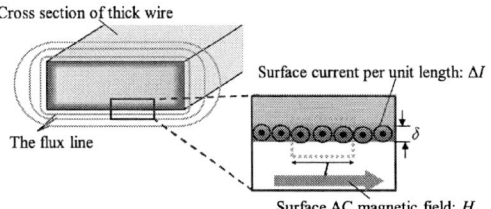

Fig.2 Cross section of thick wire

and the center of the wire of the next layer, as shown in Fig. 3. Because the magnetic field vanishes inside the wire at the depth larger than the skin depth, the horizontal side of the closed path does not contribute to the integration of the magnetic field. In addition, the vertical side of the closed path does not also contribute the integration because the vertical side is perpendicular to the magnetic field. Therefore, no AC current must flow through the closed path, indicating that the bottom wire surface of the next layer carries the AC current flowing oppositely to but having the same amplitude as that of the top wire surface of the lowest layer. As a consequence, the top surface of the next layer must carry twice as large AC current as the top surface of the lowest layer.

According to the similar discussion, the winding layer of the higher level must carry larger AC current at the bottom and top surfaces flowing in the opposite directions each other. This opposite AC current between the top and bottom surfaces tends to have much larger amplitude than the total AC current flowing in the wire, resulting in large copper loss.

In order to overcome this problem, the proposed structure employs the copper foil thinner than twice of the skin depth. As mentioned above, the AC current flows at the surface within the skin depth. Therefore, if the wire thickness is greater than twice of the skin depth, no AC current flows in the inner region of the wire; and therefore, the opposite current flows at the top and bottom surfaces without canceling each other. However, if the wire is thinner than twice the skin depth, the inner region of the wire also can carry the AC current and there the opposing current surface can cancel each other. This corresponds to suppressing the vertical inhomogeneity of the AC current distribution, which results in reduction of the copper loss.

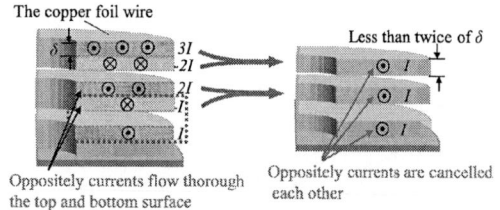

Fig.3 Strategy A for suppressing the vertical inhomogeneity of the AC current distribution

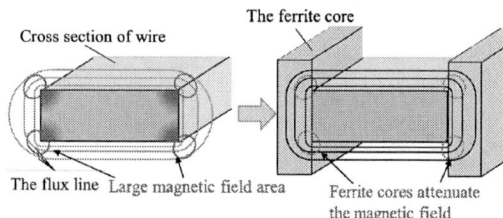

Fig. 4 Strategy B for suppressing the horizontal inhomogeneity of the AC current distribution

B. Ferrite Core Wall at Wire Edge

Next, the other strategy, which is the ferrite core is placed at the wire edge, is discussed to explain how it can dissolve the horizontal inhomogeneity of the AC current distribution [13].

When the AC current flows through in the wire, the flux path is formed to surround the wire. As for the solid rectangular wire including the copper foil, the flux path curves at the wire edge. The curving flux path generally generates the inhomogeneity of the magnetic flux density because the inner side of the flux path is shorter than the outer side. As a result, the inner side has larger flux density than the outer side [14][15]. Consequently, intense magnetic field tends to occur near the wire edge. As discussed above, the AC current is distributed to be proportional to the surface magnetic field. Therefore, the AC current is concentrated at the wire edge, generating large copper loss.

In order to suppress this inhomogeneous AC current distribution at the wire edge, this strategy places vertical walls of the ferrite core in adjacent to the wire edge, as shown in Fig. 4. Ferrite can greatly reduce the magnetic field owing to its high permeability. Therefore, by covering the wire edge by the ferrite walls, the magnetic field at the wire edge can be reduced to avoid the AC current concentration at the wire edge.

III. VERIFICATION OF PROPOSED COIL STRUCTURE

Simulation and experiment were carried out to verify that the proposed structure can reduce the copper loss. In order to verify the proposed structure, the AC resistance was compared among the three coils shown in Fig. 5. Photograph of the experimental prototype for coil C is presented in Fig. 6.

Coil A is the conventional coil structure with the solid rectangular wire, which has large cross sectional area to

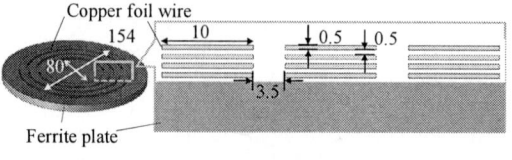

(a) coil A(conventional structure)

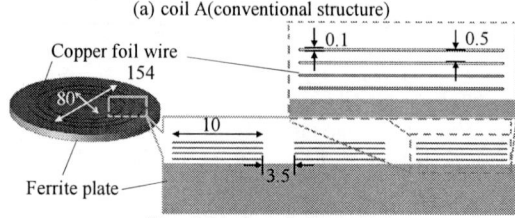

(b) coil B(reflected strategy 1)

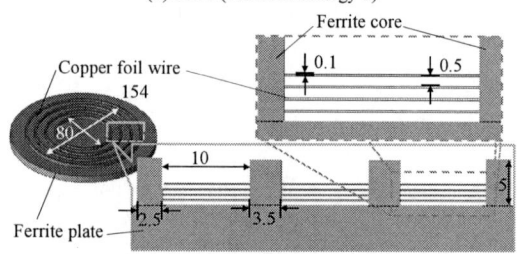

(c) coil C(reflected strategy 1 and 2)
Fig.5 The magnified view of three coils

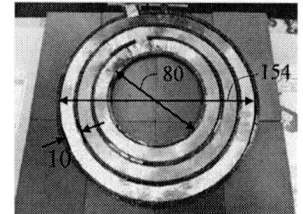

Fig.6 The coil C for the experimental verification of the proposed structure

TABLE I
SPECIFICATIONS OF THREE COILS

	Coil A	Coil B	Coil C
AC current	80kHz, 1Arms		
Number of turns [T]	12	12	12
Wire thickness [mm]	0.5	0.1	0.1
Insulation sheet thickness [mm]	0.5	0.5	0.5
Coil thickness [mm]	4	2.4	5

TABLE II
MATERIALS OF THREE COILS

Component	Coil A	Coil B	Coil C
The copper foil	Hikari Corp. HC0526	Hikari Corp. HC133T	
The insulation sheet	Artec Corp. 20512		
Ferrite plate	FDK Corp. 6H60		
Ferrite core			Laird Corp. 33P2098-0M0

decrease the DC resistance. The wire thickness is designed to be far larger than the skin depth. On the other hand, Coil B is the coil structure with the copper foil, which is thinner wire than the skin depth. Hence, coil B employs only strategy A for suppressing the vertical inhomogeneity of the AC current distribution. Coil C is

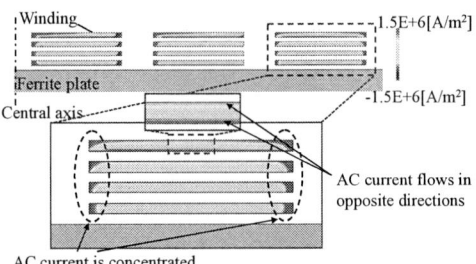

Winding
Ferrite plate
Central axis
1.5E+6[A/m²]
-1.5E+6[A/m²]
AC current flows in opposite directions
AC current is concentrated

(a) coil A(conventional coil)

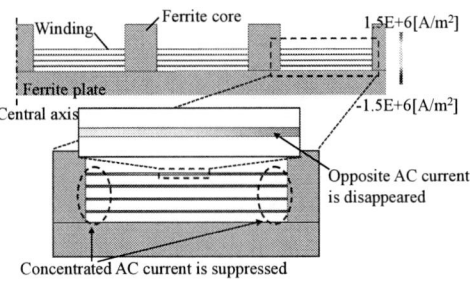

Winding
Ferrite plate
Central axis
1.5E+6[A/m²]
-1.5E+6[A/m²]
Opposite AC current is disappeared
AC current is concentrated

(b) coil B(reflected strategy A)

Winding   Ferrite core
Ferrite plate
Central axis
1.5E+6[A/m²]
-1.5E+6[A/m²]
Opposite AC current is disappeared
Concentrated AC current is suppressed

(c) coil C(reflected strategy A and B)
Fig. 7 Simulation result of current density distribution

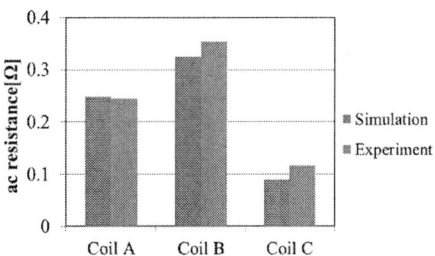

Fig. 8 The simulation and experiment result of ac resistance at 80kHz

between the top and bottom surface of the wire in coil A. Particularly, this opposite current flows intensely in the wire near the top of the winding, indicating the intense vertical inhomogeneity. Furthermore, the AC current was found to be concentrated at the wire edge, indicating the intense horizontal inhomogeneity.

This vertical inhomogeneity is suppressed in coil B, indicating the effectiveness of strategy A. However, the concentration of the AC current at the wire edge still remains in coil B. Because the thin wire reduced the cross-sectional area for the current flowing at the wire edge in coil B, the AC resistance of coil B was found to be larger than that of coil A, as can be seen in Fig. 8.

Next, coil B and coil C are compared to verify the effectiveness of strategy B. As seen Fig. 7, the AC current concentration at the wire edge was found to be successfully suppressed in coil C. As a result, the AC current is distributed more uniformly in coil C, resulting in reduction of the AC resistance. By suppressing the vertical and horizontal inhomogeneity, coil C exhibits the least AC resistance among the three coils. Particularly, coil C reduced the AC resistance by 52.2% compared with coil A, according to the experimental result.

Figure 7 indicates that the proposed structure, i.e. coil C, can achieve uniform AC current distribution without using the Litz wire. Therefore, this result implies that the proposed structure can have similar AC resistance as the Litz wire coil but with higher space factor. In the next section, the proposed structure is compared with the Litz wire coil.

IV. COMPARISON BETWEEN PROPOSED STRUCTURE COIL AND LITZ WIRE COIL

A. Prototype Design

The purpose of this section is to verify that the proposed structure can actually be utilized in replace of the Litz wire heating coil. Therefore, we designed the prototype of the proposed structure so that this prototype has the same number of turns and the same coil diameter as the prototype of the Litz wire heating coil.

The prototypes of the proposed structure and the Litz wire coil were shown in Fig. 9. Tabel III and Table IV show the specifications and materials of two heating coils. The outer and inner diameter of the coil was set at 175mm and 88mm respectively, excluding the ferrite core at wire edges. In addition, the number of turns of the coils

the proposed structure with copper foil, which further has the ferrite core walls in adjacent to the wire edge in addition to coil B.

Table I and Table II show the specifications and materials of the prototypes. The coils were wound to have 12 turns with 4 winding layers on the ferrite plate. The outer and inner diameter of the coil is 154mm and 80mm . The width of copper foil is 10mm. The gap between the horizontally adjacent wires is 3.5mm. For the insulation between the vertically adjacent wires, the polypropylene sheets (Artec Corp. 20512) with thickness 0.05mm are inserted between the winding turns. Coil C has ferrite walls with the thickness of 5mm.

The AC current of 80kHz was applied to these coils. Therefore, the skin depth was 0.23mm. The wire thickness of the coil A was designed to be 0.5mm (approximately twice as large as the skin depth), whereas the wire thickness of coil B and coil C was designed to be 0.1mm (approximately a half of the skin depth).

Figure 7 shows the simulation result of current density distribution. Figure 8 shows the simulation and experiment results of the AC resistance. First, coil A and coil B were compared to verify the effectiveness of the strategy A. As shown in Fig. 7, opposite current flows

413

The 2018 International Power Electronics Conference

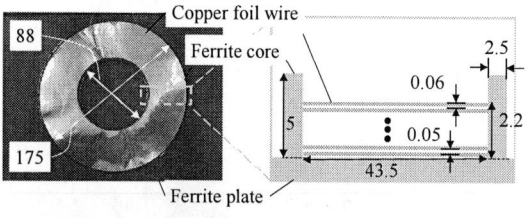

(a) Proposed structure coil

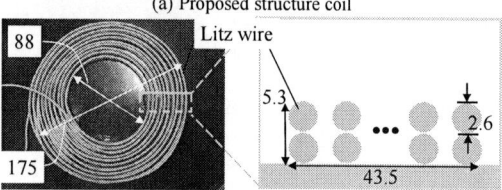

(b)Litz wire coil

Fig. 9 Prototype coil

TABLE III
SPECIFICATIONS OF EXPERIMENT

	Proposed structure	Litz wire
AC current	80kHz, 1Arms	
Number of turns [T]	20	20
Wire thickness [mm]	0.06	2.6
Wire width [mm]	43.5	2.6
Insulation sheet thickness [mm]	0.05	
Coil thickness [mm]	5	5.3

TABLE IV
THE COMPONENTS

Component	Proposed structure	Litz wire
The copper wire	Shim&gauge Corp. 10M×100×0.06	2UEWSTC
The insulation sheet	TGK Corp. 638-17-97-01	
Ferrite plate	FDK Corp. 6H60	
Ferrite core	Laird Corp. 33P2098-0M0	

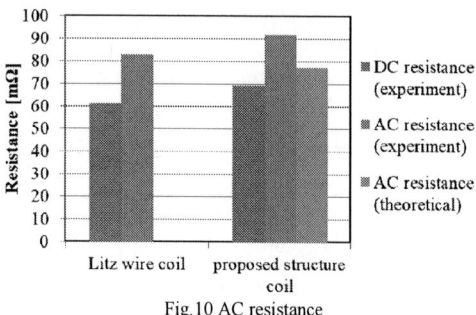

Fig.10 AC resistance

was set at 20. These coils were placed on the ferrite plate (FDK Corp., 6H60). The relative permeability of ferrite plate was 3000. The thickness of ferrite plate was 10mm.

The Litz wire employed for the prototype was 2UEWSTC. The diameter of the Litz wire was 2.6mm. The Litz wire incorporates 308 strands with the diameter of 0.1mm. This coil was designed to have two winding layers. As a result, the winding height of the Litz wire heating coil was 5.3mm.

On the other hand, the winding of the proposed

structure was designed to have large width in order to make sufficiently large cross-section area of the copper foil. Therefore, the wire width the proposed structure has designed cover the whole coil area. As a result, the proposed structure has 20 winding layers.

The thickness of the copper foil was designed to minimize the AC resistance. In the proposed structure, the winding is located above the ferrite plate, and its edges adjacent to the ferrite core. Therefore, the magnetic field only exists at the winding layer and the space between the adjacent wires because the magnetic field into the high permeability material is very small. Furthermore, as mentioned at previous section, the flux line curves around the wire in the ferrite plate or ferrite core, and passes through straightly into the wire and space. Therefore, the magnetic field distribution is one-dimensional. Under this situation, the copper loss of proposed structure is calculated by one-dimensional analysis. The AC resistance of the proposed structure was calculated based on the theoretical analysis presented in [16]. According to this analysis, the AC resistance can be calculated using the one-dimensional electromagnetic field analysis according to the following equation:

$$R_{ac\_N} = \frac{R_{dc\_N}}{2} \frac{t}{\delta} \left[\frac{\sinh\left(\frac{t}{\delta}\right) + \sin\left(\frac{t}{\delta}\right)}{\cosh\left(\frac{t}{\delta}\right) - \cos\left(\frac{t}{\delta}\right)} + \frac{4N^2-1}{3} \frac{\sinh\left(\frac{t}{\delta}\right) - \sin\left(\frac{t}{\delta}\right)}{\cosh\left(\frac{t}{\delta}\right) + \cos\left(\frac{t}{\delta}\right)} \right] \quad (3)$$

where R_{ac_N} is the ac resistance of the coil, R_{dc_N} is the dc resistance of the coil, t is the wire thickness, δ is the skin depth, and N is the number of turns.

If the copper foil has excessively large thickness, the intense skin and proximity effects increases the AC resistance. If the copper foil has excessively small thickness, the wire does not have sufficient cross-sectional area for the AC current flow. Therefore, this case also increases the AC resistance. Consequently, there is the optimal thickness that minimize the AC resistance.

This optimal thickness can be determined by differentiating (3) with respect to the wire thickness t. As for the prototype of the proposed structure, the optimal thickness was determined as 68μm, given δ=0.23mm and N=20. Therefore, we employed the copper foil with the thickness of 60μm for constructing the prototype of the proposed structure. For the insulation between the vertically adjacent wires, the insulation sheets (TGK Corp. 638-17-97-01) with thickness 0.05mm which is made of PTFE are inserted between the winding turns.

The proposed structure has the ferrite walls that covers the outer and inner edge of the copper foil wire to suppress the horizontal inhomogeneity of the AC current distribution. The height of ferrite cores was set at 5.0mm, which determined the total height of the coil. Accordingly, the proposed structure reduced the winding height by 6% compared with the Litz wire coil.

Certainly, in this experiment, the proposed structure had similar height as the Litz wire coil. However, the height of the proposed structure was determined by the

414

ferrite wall, which was designed to be sufficiently high to cover the whole winding, which is only 2.2mm high. Therefore, the net winding height is much reduced in the proposed structure. If the ferrite wall is optimized to reduce the height, the proposed structure probably exhibits effective reduction in height, which will be elucidated in the future study.

B. AC Resistance Measurement

The AC resistance of the prototypes were measured using the LCZ meter (NF Corp., NF2340). Figure 10 shows the experiment results of the AC resistance at 80kHz. The AC resistance of the proposed structure coil was slightly larger than that of Litz wire coil, although both of the prototypes exhibited the similar values for the AC resistance.

Consequently, the prototype of the proposed structure was found to have the similar height as well as the similar AC resistance as the prototype of the Litz wire coil. This indicates that the proposed structure can replace the Litz wire heating coil, suggesting the effectiveness of the proposed structure.

In this paper, the ferrite walls and the ferrite plate were implemented separately. However, in the manufacturing process, these ferrite elements can be integrated into one piece because the ferrite is commonly produced by the mold. This may probably reduce the cost of implementing the additional ferrite walls.

V. CONCLUSION

The Litz wire has been commonly utilized for the heating coils for the induction heating. However, the comparatively complicated construction as well as the low space factor of the Litz wire leads to high cost and large winding height of the heating coil. In order to mitigate these drawbacks, this paper investigated the feasibility of the recently proposed heating coil structure with the copper foil winding.

This paper verified the effectiveness of this structure by simulation and experiment. As a result, this structure was found to suppress the skin and proximity effect without utilizing the Litz wire. Owing to this attractive feature, the proposed structure reduced the AC resistance compared with the heating coil of the solid rectangular wire. In addition, comparison of the prototype heating coils between the proposed structure and the Litz wire coil revealed that the proposed structure can have similar AC resistance and similar height. The results supported that the proposed structure can replace the Litz wire heating coil. The results also suggested that optimization of the ferrite wall may further reduce the height of the proposed structure, which will leads to the effective reduction of the heating coil using the proposed structure.

REFERENCES

[1] S. E. Schwarz, "Electrodynamics," in *Electromagnetics for engineers*, Orland, FL, USA: Sounders College Publishing, 1990, pp. 214–246.

[2] H. Hämäläinen, J. Pyrhönen, J. Nerg, J. Talvitie, "AC resistance factor of Litz-wire windings used in low-voltage high-power generators," *IEEE Trans. Ind. Electron.*, vol. 61, no. 2, pp. 693–700, Feb. 2014.

[3] H. Shinagawa, T. Suzuki, M. Noda, Y. Shimura, S. Enoki, and T. Mizuno, "Theoretical analysis of ac resistance in coil using magnetoplated wire," *IEEE Trans. Mag.*, vol. 45, no. 9, pp. 3251–3259, Sep. 2009.

[4] S. L. M. Berleze and R. Robert, "Skin and proximity effects in nonmagnetic conductors," *IEEE Trans. Educ.*, vol. 46, no. 3, pp. 368–372, Aug. 2003.

[5] N. H. Kutkut and D. M. Divan, "Optimal air-gap design in high-frequency foil windings," *IEEE Trans. Ind. Electron.*, vol. 13, no. 5, pp. 942–949, Sep. 1998.

[6] M. K. Kazimierczuk and R. P. Wojda, "Foil winding resistance and power loss in individual layers of inductors," *Intl. J. Electron. Telecommunications*, vol. 56, no. 3, pp. 237–246, Sept. 2010.

[7] A. Roßkopf, E. Bar, and C. Joffe, "Influence of inner skin- and proximity effects on conduction in litz wires," *IEEE Trans. Power Electronics*, vol. 29, no. 10, pp. 5454–5461, Oct. 2014.

[8] J. Acero, P. J. Hernández, J. M. Burdío, R. Alonso, and L. A. Barragán, "Simple resistance calculation in Litz-wire planar windings for induction cooking appliances," *IEEE Trans. Magn.*, vol. 41, no. 4, pp. 1280–1288, April. 2005.

[9] J. Acero, R. Alonso, J. M. Burdío, L. A. Barragán, and D. Puyal, "Frequency-dependent resistance in Litz–wire planar windings for domestic induction heating appliances," *IEEE Trans. Power Electron.*, vol. 21, no. 4, pp. 856–866, July. 2006.

[10] I. Lope, J. Acero, and C. Carretero, "Analysis and optimization of the efficiency of induction heating applications with Litz-wire planar and solenoidal coil," *IEEE Trans. Power Electron.*, vol. 31, no. 7, pp. 5089–5101, July 2016.

[11] J. Acero, R. Alonso, J. M. Burdío, L. A. Barragán, and C. Carretero, "A model of losses in twisted-multistranded wires for planar windings used in domestic induction heating appliance," *IEEE Applied Power Electronics Conf. Rec.*, pp.1247–1253, 2007,

[12] M. Hataya, Y. Oka, K. Umetani, E. Hiraki, T. Hirokawa and M. Imai "Novel thin heating coil with reduced copper loss for high frequency induction cookers." *2016 International Conference on Electrical Machines and System*, 2016.

[13] I. Sasada, "Alternating current loss reduction for rectangular busbars by covering their edges with low permeable magnetic caps," *J. Appl. Phys.*, vol. 115, 17A343, 2014.

[14] K. Umetani, "Improvement of saturation property of iron powder core by flux homogenizing structure," *IEEJ Trans. Elect. Electron. Eng.*, vol. 8, no. 6, pp. 640–648, Sep. 2013.

[15] K. Umetani, Y. Itoh, and M. Yamamoto, "A detection method of DC magnetization utilizing local inhomogeneity of flux distribution in power transformer core," in *Proc. IEEE Energy Conversion Congr. Expo.*, Pittsburgh, PA, pp. 3739–3746, 2014.

[16] M. P. Perry, "Multiple layer series connected winding design for minimum losses." *IEEE Trans. Power App. Syst.*, vol. PAS-98, pp.116–123, Jan./Feb. 1979.

The 2018 International Power Electronics Conference

Condition Monitoring of Electrolytic Capacitor based on ESR Estimation and Thermal Impedance model using Improved Power Loss Computation

Sundararajan Prasanth[1*], Mohamed Halick Mohamed Sathik[1], Firman Sasongko[1], Tan Chuan Seng[1], Peng Yaxin[1] and Rejeki Simanjorang[2]

1 Rolls-Royce @ NTU Corporate Lab, Nanyang Technological University, Singapore
2 Applied Technology Group, Rolls-Royce Singapore Pte. Ltd., Singapore
*E-mail: prasanth006@e.ntu.edu.sg

Abstract— **Aluminium Electrolytic Capacitors (AEC) are indispensable in applications where high capacitance per unit volume and low cost are preferred. However, it is also one of the most failure-prone components in a power converter system. As AEC is an electrochemical device, which follows Arrhenius law, the most important life-deciding factor of an AEC is its operating temperature [1]. In this paper, a condition monitoring system for AECs is proposed based on the accurate estimation of equivalent series resistance (ESR) and operating temperature. Operating temperature is estimated using the thermal impedance model based on improved power loss calculation methodology. The proposed method computes power loss using the capacitor voltage and current in contrast to the existing methods [4] - [7] which uses capacitor current and ESR to estimate the power loss. Thereby, it avoids the issue of ESR variation with ageing and temperature which may give erroneous results with ageing in case of stored look-up tables for ESR. ESR estimation is based on the method proposed in [14]. As ESR is the most widely used degradation indicator, the ESR and operating temperature can be used to estimate the present health status of an AEC.**

Keywords—Aluminium electrolytic capacitors, Condition monitoring, ESR estimation, temperature estimation.

I. INTRODUCTION

Lifetime of an AEC is largely dependent on its core temperature. Core is the hottest spot at about the center of the capacitor. Generally, lifetime doubles for every 10 °C decrease in core temperature [1]. Therefore, core temperature of the capacitor is one of the critical parameters to predict capacitor's life. AEC thermal models are known to be used widely to estimate the core/hotspot temperature of a capacitor based on its electrical and thermal stresses [2]-[7]. When the mission profile of a power converter is known, electrical and thermal stresses can be obtained to calculate the hotspot temperature of the capacitor in a power converter, which in turn reflects the long-term thermal stress on the capacitor. Using these signatures, thermal stress based life models shall be developed to predict capacitor's life [8].

Most of the existing condition monitoring systems for capacitors involves estimation of ESR and capacitance [9]. If the estimated ESR of a capacitor is greater than twice its initial value, it is considered as capacitor's end-of-life. However, the ESR and capacitance values depend on temperature and operating frequency of a capacitor, which makes condition monitoring difficult without the core temperature information [10]-[11]. Several methods can be employed to model thermal characteristics of a capacitor. Physics-based thermal models using finite element analysis (FEA) can be developed, if the geometry and properties of all the constituent materials in a capacitor are known [2]. Temperature sensitive electrical parameter (TSEP) based approach is proposed in [3], and lumped-parameter RC thermal models based on Cauer and Foster-based networks are discussed in [4] - [7]. In this paper, an improved thermal model has been developed for condition monitoring applications based on Foster network representation of thermal impedance and improved power loss estimation methodology. ESR estimation is based on method proposed in [14]. The core temperature estimated using the proposed method along with the ESR can be used to determine the present health status of a capacitor.

II. PROPOSED CORE TEMPERATURE ESTIMATION METHOD

In any thermal model, there is a heat source, which heats up the object. In case of a capacitor, the power loss due to the capacitor's ESR generates heat. The amount of heat generated depends on the magnitude of ripple current and ESR [12]. In existing thermal models, heat source is assumed to be at the center of the capacitor, which is the hottest part of the capacitor during its operation. The generated heat spreads from the center towards the case and to the ambient air via conduction and convection. Thermal characteristics of a capacitor can be modeled as a lumped thermal network consisting of a thermal resistance from the core to the case and a thermal capacitance. Fig. 1 shows the Foster network representation of capacitor's thermal impedance.

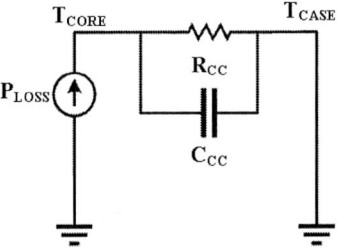

R_{CC} – THERMAL RESISTANCE FROM CORE TO CASE
C_{CC} – THERMAL CAPACITANCE
T_{CORE} – CORE TEMPERATURE
T_{CASE} – CASE TEMPERATURE

Fig. 1. Foster network representation of capacitor's thermal impedance

In Foster network representation, power loss is analogous to a current source in an electrical network, whereas thermal resistance and thermal capacitance are analogous to an electrical resistance and capacitance. Thermal resistance determines the steady state thermal characteristics. The combination of thermal resistance and thermal capacitance determines the transient characteristics. The mathematical equation to calculate the core temperature from case temperature derived from the Foster-network representation is shown in (1).

$$\frac{T_{CORE}(t) - T_{CASE}(t)}{P_{LOSS}(t)} = R_{CC}\left(1 - e^{-\frac{t}{R_{CC}C_{CC}}}\right) \qquad (1)$$

where T_{CORE} is the core temperature (°C), T_{CASE} is the case temperature(°C), P_{LOSS} is the power loss (W), R_{CC} is the thermal resistance from core to case (°C/W), and C_{CC} is the associated thermal capacitance (J/°C). Based on the system requirements, Foster network representation can be extended to estimate the core temperature from ambient temperature, by adding case to ambient thermal resistance and capacitance of the surrounding air. However, the case to ambient thermal resistance depends on the environmental conditions, which may introduce error in core temperature estimation. Therefore, the thermal model considered in this work, estimates core temperature by using core to case thermal characteristics.

If the power loss and case temperature are known, the core temperature can be obtained using (2).

$$T_{CORE}(t) = T_{CASE}(t) + $$
$$P_{LOSS}(t) \times R_{CC}\left(1 - e^{-\frac{t}{R_{CC}C_{CC}}}\right) \qquad (2)$$

The calculation of power loss is one of the challenging tasks in estimating the core temperature of a capacitor. Many of the existing RC thermal models lack in estimating power loss accurately as the variation of ESR with ageing is not considered. The ESR represents all the losses in a capacitor. The equation for calculating the power dissipated by a resistor can be expressed as shown in (3).

$$P_{LOSS} = I_{RMS}^2 \times R \qquad (3)$$

Even though (3) looks simple, the calculation of power loss for a capacitor operating in the power converter is a complex task because the ESR of the capacitor varies with the frequency and temperature. If the fundamental and significant harmonic components of the capacitor current in a power converter are known, then the value of ESR at corresponding frequencies can be used to calculate the total power loss as shown in (4).

$$P_{LOSS} = (I_f^2 \times ESR_f) + \sum_{h=1}^{n} I_{fh}^2 \times ESR_{fh} \qquad (4)$$

where I_f is the RMS value of fundamental component of the capacitor current, I_{fh} represents the RMS value of dominant harmonic components with h=1, 2, 3...n, ESR_f and ESR_{fh} with h=1, 2, 3...n, are the corresponding ESR values.

In [7], the power loss is calculated using stored look-up tables for various operating conditions. This method is better than the other methods since it considers the dominant harmonic components and corresponding ESR values. But, there are two main drawbacks in this method:

1. ESR is a temperature dependent parameter and the use of (4) to predict the core temperature in ripple current reconstruction method, requires accurate ESR values at all operating conditions. But this problem is eliminated in direct look-up table based loss calculation method as it directly gives the power loss for present operating conditions.

2. As the capacitor ages, the ESR value changes and it must be updated periodically. Otherwise, it will lead to large error in loss calculations. The methods proposed in [7] do not consider the effect of ageing.

Therefore, for condition monitoring applications, where capacitors are monitored until its end-of-life, a power loss calculation method which takes the effect of ageing into account is imperative. In this paper, a power loss calculation method which gives accurate estimation of power loss independent of ageing effects is proposed. The total active power drawn by the capacitor can be calculated using (5) as shown in [13].

$$P_{LOSS} = \frac{1}{T}\int_0^T v_c(t)i_c(t)\,dt \qquad (5)$$

where P_{LOSS} is the total power loss, T is the integration period, v_c is the instantaneous value of capacitor voltage and i_c is the instantaneous value of capacitor current. The capacitor voltage and current are both periodic waveforms, and can be expressed in Fourier series as the sum of fundamental and harmonic components as in (6) and (7).

$$v_c(t) = V_{DC} + V_{f1m}\sin(\omega_1 t)$$
$$+ V_{f2m}\sin(\omega_2 t) + \cdots \qquad (6)$$
$$i_c(t) = I_{DC} + I_{f1m}\sin(\omega_1 t + \emptyset_1)$$
$$+ I_{f2m}\sin(\omega_2 t + \emptyset_2) + \cdots \qquad (7)$$

where V_{DC} and I_{DC} are the DC components of capacitor voltage and current, V_{fnm} ($n = 1,2,3\ldots$) and I_{fnm} ($n = 1,2,3\ldots$) represents the peak value of the harmonic components of capacitor voltage and current respectively.

At operating voltages less than the rated voltage, the power loss due to DC leakage current is negligible and it can be neglected. Therefore, the total power loss is the sum of power loss due to individual AC components expressed as in (8).

$$P_{LOSS} = V_{f1}I_{f1}\cos\emptyset_1 + V_{f2}I_{f2}\cos\emptyset_2 + \cdots \qquad (8)$$

The power loss expressed in (8) shows that the phase information of capacitor voltage and current is important for calculating the power loss. Therefore, the phase shift introduced by the voltage and current sensors must be considered to calculate power loss accurately. The integration operation (5) is easy to implement in a microcontroller as it involves simple multiplication and addition. The power loss and thermal network parameters facilitate the estimation of core temperature. This paper focuses on improving the core temperature estimation methodology for developing a reliable condition monitoring solution for AECs as core temperature is one of the key factors in deciding capacitor's life. Also, core temperature is necessary to validate the most commonly used degradation indicator, ESR, which is temperature sensitive. The block diagram of the proposed method is shown in Fig. 2.

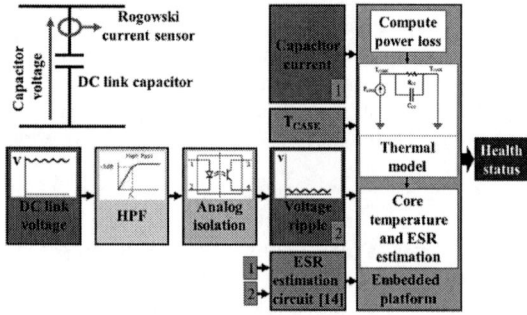

Fig. 2. Block diagram of the proposed core temperature estimation method

III. THERMAL NETWORK PARAMETERS EXTRACTION

The thermal resistance R_{CC} and thermal capacitance C_{CC} of an AEC can be extracted experimentally. Fig. 3 shows the experimental setup for extracting the thermal resistance and thermal capacitance of an AEC. The capacitor part number 500R112M500BC2B from Cornell Dubilier is used in the experiment.

The AEC is connected to a programmable voltage source to supply DC (70 V) + AC (10V, 120 Hz). The capacitor has a thermocouple embedded in its core to measure the core temperature and a thermocouple placed on its case to measure the case temperature. The capacitor current and temperature data are logged using dSPACE 1104. The waveforms of the measured core temperature and case temperature are shown in Fig. 4. The value of ESR at different core temperatures are measured using an LCR meter. Fig. 5 shows the thermal impedance graph for the measured data obtained by applying equation (1). The value of thermal resistance and thermal capacitance are extracted by applying curve-fitting technique on the thermal impedance graph in MATLAB.

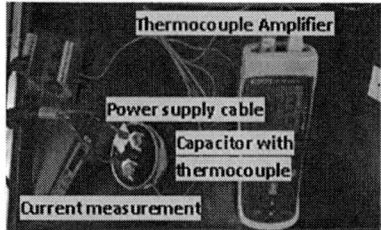

Fig. 3. Experimental Setup to extract the thermal resistance and thermal capacitance of an AEC

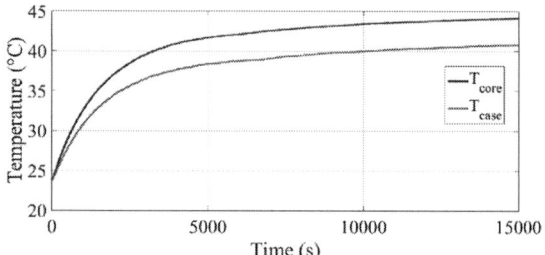

Fig. 4. Core temperature and case temperature of the capacitor under test

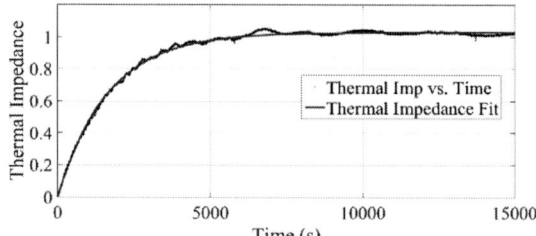

Fig. 5. Thermal impedance parameters extraction

The value of thermal resistance R_{CC}=0.84 °C/W and the thermal capacitance C_{CC}=1623 J/°C. If the power loss and case temperature are known, the core temperature can be obtained using (9).

$$T_{CORE}(t) = T_{CASE}(t) + \\ P_{LOSS}(t) \times 0.84 \left(1 - e^{-\frac{t}{0.84 \times 1623}}\right) \quad (9)$$

IV. SIMULATION RESULTS

The existing power loss calculation method used for core temperature estimation is compared with the proposed method to analyze the advantages and disadvantages of the proposed method. For this purpose, a front-end rectifier fed three-phase inverter system shown in Fig. 6 is simulated in PSpice with input voltage of 100V. The Spice model of DCMC102T500BC2B capacitor from Cornell Dubilier is used as DC link capacitor. The switching frequency of the inverter is 2.5 kHz. The capacitor voltage and current data collected from the simulation is used to calculate the power loss. The capacitor voltage and current are shown in Fig. 7 and Fig. 8 respectively. The calculated power loss using the proposed method is compared with power loss obtained from the other methods which use either ESR or look-up table to estimate power loss.

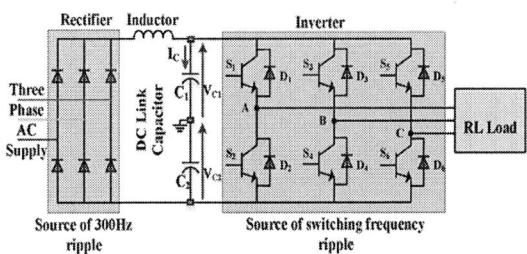

Fig. 6. Front-end rectifier fed three-phase inverter

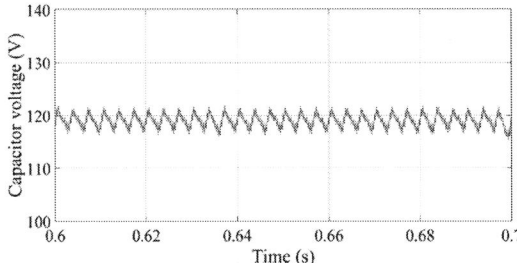

Fig. 7. Capacitor voltage

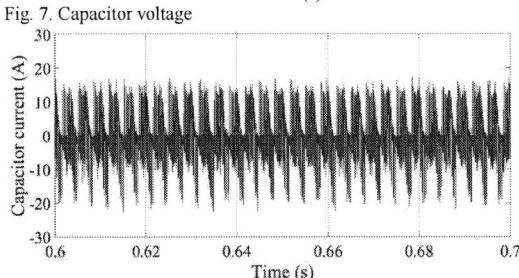

Fig. 8. Capacitor current

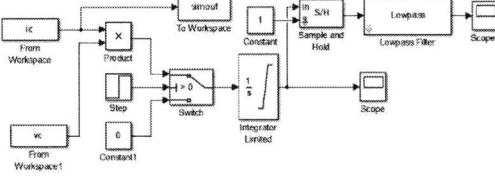

Fig. 9. MATLAB Simulink model to calculate power loss

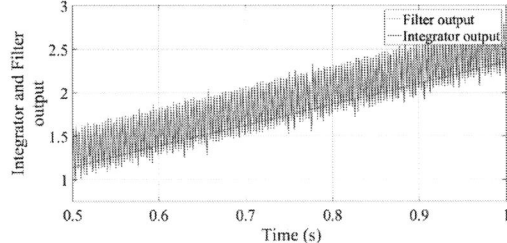

Fig. 10. Simulation results

To calculate power loss using (4), MATLAB code is written to compute FFT of capacitor current. The ESR values for different frequencies for the SPICE model are taken from the manufacturer's impedance modelling tool and the calculations were performed for 20%, 40% and 60% increase in ESR and compared with power loss calculated for actual ESR. To calculate power loss using the proposed method, a Simulink model is developed as

shown in Fig. 9. The voltage and current data from PSpice simulation are multiplied and integrated. The integrator's output is filtered, and the slope is calculated to get the power loss. The outputs of the integrator and filter are shown in Fig. 10. The summary of results from both the methods are shown in Table I.

TABLE I
RESULTS SUMMARY

ESR change	Error in power loss computation	
	Existing method	Proposed Method
20%	18%	1.54%
40%	29%	1.83%
60%	39%	1.68%

The error obtained from simulation is negligible for the proposed method as phase shift introduced by the sensors is not considered. In practical applications, the phase shift introduced by the sensors should be corrected. Even if the ESR can be estimated, it cannot be estimated without considerable amount of error and it will be reflected in power loss calculations. Direct look-up table based method requires ESR values for all possible operating conditions and the present health status to calculate power loss accurately. From the results, it is clear that the proposed method can perform well for condition monitoring applications.

V. EXPERIMENTAL RESULTS

The proposed power loss estimation method is validated experimentally using a front-end rectifier fed three-phase inverter test rig. The AEC from Cornell Dubilier 500R112M500BC2B is used as a DC-link capacitor. The capacitor has a thermocouple in its core to measure core temperature. The capacitor current is measured using Rogowski coil based current sensor and the capacitor voltage is measured using a custom made circuit based on HCNR201. The switching frequency of the inverter is 10 kHz.

The selection of voltage and current sensors play an important role in deciding the accuracy of proposed power loss estimation method. As discussed in section II, phase response of voltage and current sensors is considered carefully for the experiment. For achieving better accuracy, the voltage and current components up to four times the switching frequency are considered.

Fig. 11 shows the phase response of a commercially available voltage sensor with a bandwidth of 20 kHz. It is clear that the phase information of the sensed voltage starts to deviate significantly from 0° as the frequency increases above few kilohertz. Therefore, if this sensor is used for frequencies above few kHz, it will introduce error in phase, thereby reducing the accuracy of power loss estimation. Also, the commercially available sensors measure both DC and AC voltage. But, the power loss depends largely on the AC content. The presence of DC content reduces the room for AC content thereby affecting its resolution. Therefore, a voltage sensor circuit which filters DC and measures only AC with a bandwidth of 650 kHz is designed for power loss estimation.

The phase response of the custom-made voltage sensor

is shown in Fig. 12. The phase response shows that the phase change is negligible up to few tenths of kilohertz. Therefore, it can be used for switching frequencies up to 15 kHz. The capacitor current is measured using Rogowski-based current sensor with a bandwidth of 20 MHz. The phase response of current sensor is also verified to ensure the accuracy of estimation.

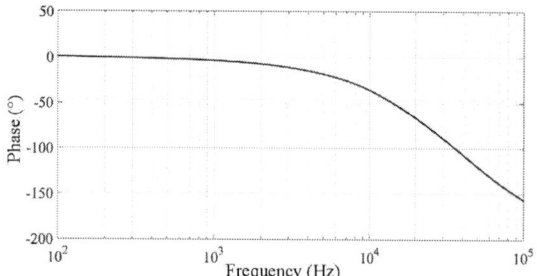

Fig. 11. Phase response of a commercial voltage sensor (Bandwidth = 20 kHz)

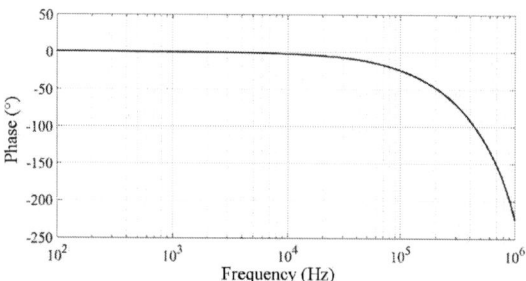

Fig. 12. Phase response of custom-made voltage sensor for power loss estimation (Bandwidth = 650 kHz)

The experimental setup of front-end rectifier fed three-phase inverter test rig is shown in Fig.13. The power loss estimation algorithm is implemented in PE-Expert4 digital control system. The outputs from the voltage and current sensors are sampled at a frequency of 1 MHz. The power loss is estimated by computing the product of each sample followed by discrete integration and division.

The estimated power loss is then used to calculate the core temperature of the capacitor based on (9). The case temperature of the capacitor is measured using LMT70 based temperature sensor.

The results of power loss estimation and subsequent temperature estimation using the thermal model are shown in Fig. 14 and Fig. 15. The ESR estimation results based on [14] are shown in Fig. 16. The proposed method estimates core temperature with error less than 6%. It is validated by comparing the estimated core temperature with the core temperature measured using the thermocouple.

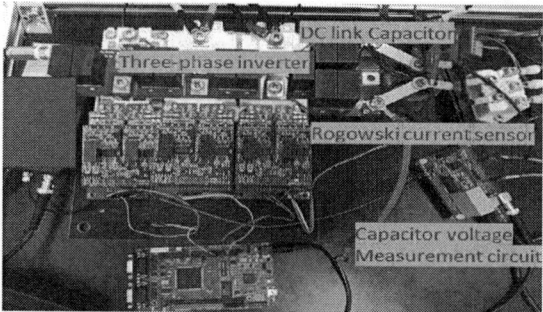

Fig. 13. Experimental Setup

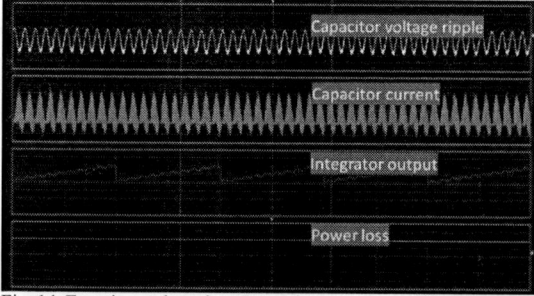

Fig. 14. Experimental results – Power loss

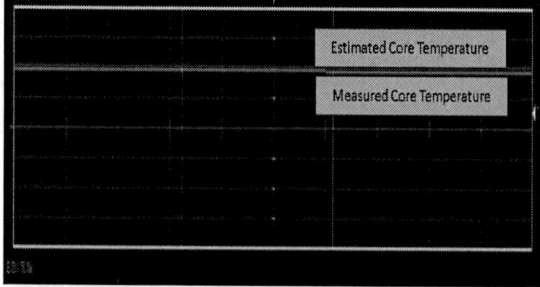

Fig. 15. Experimental results – Temperature estimation

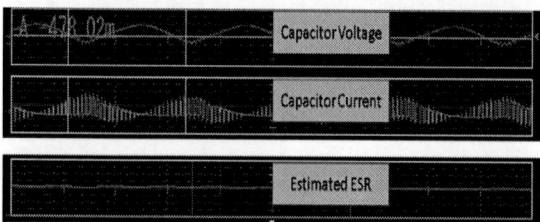

Fig. 16. Experimental results – ESR

VI. CONCLUSION

The core/hotspot temperature of the capacitor is one of the principal factors in deciding capacitor's life. The proposed method enables accurate estimation of capacitor's core temperature. The core temperature along with the capacitor's ESR can be used to monitor the health of an AEC. Therefore, the capacitor can be replaced when it nears the end-of-life preventing unexpected shutdown/ failure of the entire system.

ACKNOWLEDGEMENT

This study was conducted within the Rolls-Royce@NTU Corporate Lab with support from the National Research Foundation (NRF) Singapore under the Corp Lab@University Scheme.

REFERENCES

[1] Application Guide, Aluminum Electrolytic Capacitors, Cornell Dubilier, Liberty, SC, USA. [Online]. Available: *http://www.cde.com/catalogs/AEappGUIDE.pdf*

[2] S. G. Parler, "Thermal modeling of aluminum electrolytic capacitors," Conference Record of the 1999 IEEE Industry Applications Conference. Thirty-Forth IAS Annual Meeting (Cat. No.99CH36370), Phoenix, AZ, pp. 2418-2429 vol.4, 1999.

[3] Z. Sarkany and M. Rencz, "A way for measuring the temperature transients of capacitors," *2016 IEEE 18th Electronics Packaging Technology Conference (EPTC), Singapore*, pp. 818-822, 2016.

[4] T. Furukawa, D. Senzai and T. Yoshida, "Electrolytic capacitor thermal model and life study for forklift motor drive application," *2013 World Electric Vehicle Symposium and Exhibition (EVS27), Barcelona*, pp. 1-6, 2013.

[5] J. N. Davidson, D. A. Stone and M. P. Foster, "Required Cauer network order for modelling of thermal transfer impedance," in *Electronics Letters*, vol. 50, no. 4, pp. 260-262, February 13, 2014.

[6] P. Freiburger, "Transient thermal modeling of aluminum electrolytic capacitors under varying mounting boundary conditions," *2015 21st International Workshop on Thermal Investigations of ICs and Systems (THERMINIC), Paris*, pp. 1-5, 2015.

[7] Y. Yang, K. Ma, H. Wang and F. Blaabjerg, "Instantaneous thermal modeling of the DC-link capacitor in PhotoVoltaic systems," *2015 IEEE Applied Power Electronics Conference and Exposition (APEC), Charlotte, NC*, pp. 2733-2739, 2015.

[8] D. Zhou, H. Wang, and F. Blaabjerg, "Lifetime estimation of electrolytic capacitors in a fuel cell power converter at various confidence levels," in *2016 IEEE 2nd Annual Southern Power Electronics Conference (SPEC)*, pp. 1–6, Dec 2016.

[9] H. Soliman, H. Wang and F. Blaabjerg, "A Review of the Condition Monitoring of Capacitors in Power Electronic Converters," in *IEEE Transactions on Industry Applications*, vol. 52, no. 6, pp. 4976-4989, Nov.-Dec. 2016.

[10] Prymak, John, et al. "Why that 47 uF capacitor drops to 37 uF, 30 uF, or lower." *Proc. of the CARTS USA conference*, Mar. 2008.

[11] L. Young, "Models for ionic conduction in anodic oxide films," *Journal of the Electrochemical Society*, v 126, n 5, p 765-8, May 1979.

[12] R. S. Alwitt, "Contribution of spacer paper to the frequency and temperature characteristics of electrolytic capacitors," *Journal of the Electrochemical Society*, v 116, n 7, p 1024-7, July 1969.

[13] M. A. Vogelsberger, T. Wiesinger, and H. Ertl, "Life-cycle monitoring and voltage-managing unit for dc-Link electrolytic capacitors in PWM converters," *IEEE Trans. Power Electron.*, vol. 26, no. 2, pp. 493–503, Feb. 2011.

[14] S. Prasanth, M. H. M. Sathik, F. Sasongko, T. C. Seng, M. Tariq and R. Simanjorang, "Online equivalent series resistance estimation method for condition monitoring of DC-link capacitors," *2017 IEEE Energy Conversion Congress and Exposition (ECCE), Cincinnati, OH*, pp. 1773-1780, 2017.

The 2018 International Power Electronics Conference

Test Setup for Characterisation of Biased Magnetic Hysteresis Loops in Power Electronic Applications

Min Luo and Drazen Dujic
Power Electronics Laboratory
École Polytechnique Fédérale de Lausanne (EPFL)
Lausanne CH-1015, Switzerland
min.luo@epfl.ch, drazen.dujic@epfl.ch

Jost Allmeling
Plexim GmbH
Zürich CH-8005, Switzerland
allmeling@plexim.com

Abstract—Hysteresis effect of core materials contributes significantly to the power loss and nonlinearity of the transformers and filter inductors in power electronic applications. For design or modeling of the magnetic components, information of the magnetic material's characteristic under the desired operation condition is usually required. In many common types of power converters the magnetic components undertake biased excitation, which leads to hysteresis loop with DC-offset on both magnetic field strength and flux density. This work proposes a test setup combining both linear amplifier and switching cells, which is able to conveniently generate magnetic hysteresis loops at arbitrary biased levels.

I. INTRODUCTION

The core material of isolation transformers or filter inductors used in power electronic converters exhibits hysteresis effect, which significantly contributes to the power loss and nonlinear inductivity of the magnetic components. In order to properly design the magnetic component in terms of core dimension, winding configuration or thermal condition, the core material characteristic on the B-H plane, or in other words, the hysteresis loops are usually required. For the measurement of hysteresis loops, the V-I approach introduced in the work of [1] is commonly adopted nowadays. This approach is improved in terms of measurement accuracy by several publications like [2] and [3] for power loss of symmetrical hysteresis loops.

In many types of power converters, unsymmetrical hysteresis loops with bias are also present, as in the case of a flyback DC-DC converter shown in Fig. 1(a) or a voltage source DC-AC inverter shown in Fig. 1(b). In order to characterise the biased hysteresis loops of the core material using a core sample, one option is to introduce an extra DC excitation. Authors of [4] installed a third winding supplied by DC voltage source and authors of [5] made use of two identical core samples. Aiming to imitate the operation conditions in power electronic converters under PWM excitation, [6] used a H-bridge converter and the DC-biased excitation was generated via applying different duty cycles to the gate signals. In these publications, only the DC-offset on the field strength H direction was considered and explicitly controlled during the measurement. However the bias of the hysteresis loops is also present on the flux density direction (vertical direction of the B-H plane in Fig. 1(a) and Fig. 1(b)). Authors of [7]

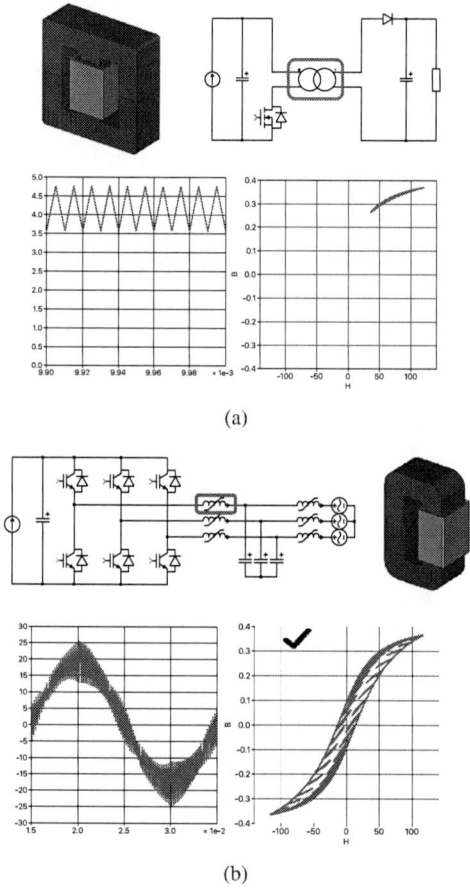

Fig. 1: Biased magnetic hysteresis loop as well as the corresponded magnetisation current waveform in power electronic applications (a) Simplified flyback converter; (b) Voltage source inverter.

and [8] have observed the "noncongruency" effect present in the real magnetic materials, that is: Even the minor hysteresis loops having the same range of field strength H exhibit different shapes if they are located at different flux density

The 2018 International Power Electronics Conference

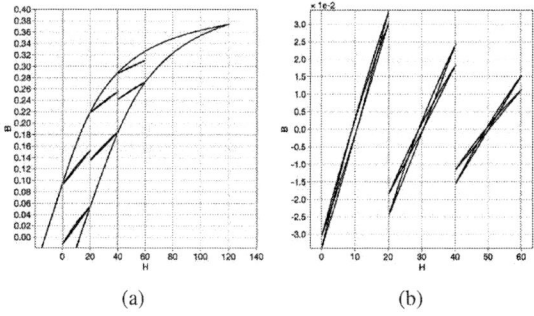

(a)   (b)

Fig. 2: Demonstration of noncongruency in reality of hysteresis loops in the same H range and different B positions (a) Biased hysteresis loop on their original B positions; (b) Shape comparison after the bias of B is artificially removed.

position on the B-H plane, depending on the magnetisation history, as has been demonstrated in Fig. 2(a) of the ferrite material N87. The difference is not only related to the enclosed loop area (determines the power loss) but also the equivalent permeability, which directly affect the dynamic behaviour of the magnetic component and need to be taken into account during design phase.

In order to characterise the hysteresis loops at different B positions, authors [9] and [10] proposed to overlap a fundamental frequency- with a harmonic frequency sinusoidal voltage excitation. The fundamental frequency excitation accounts for generating a symmetrical major hysteresis loop and the harmonic frequency one results in biased minor loops attached on the major loop, so that the B position of each biased loops can be identified. However using sinusoidal harmonic excitation it is not easy to quantitatively control the B position of the biased loops, also in power electronic applications the biased loops are initiated by PWM- instead of sinusoidal excitation. This work aims to facilitate the characterisation of biased hysteresis loops by means of accurately controlling the B position as well as equivalent switching frequency for power electronic applications. For that purpose, a highly flexible low power characterisation setup is developed, as described next, allowing for great degree of freedom in driving a magnetic core sample into a desired operation point.

II. PROPOSED APPROACH

The basic circuit configuration of the proposed test setup is depicted in Fig. 3(a), which is constructed based on the V-I method introduced by [1]. The core sample is equipped with two windings: The primary winding is connected to the excitation and the current I_p is measured, which is converted into the field strength H. The secondary winding is left open and the voltage V_s is obtained, which is integrated in time domain and converted into the flux density B.

Different from the existing publications, the primary winding in the proposed test setup is supplied by a sinusoidal-

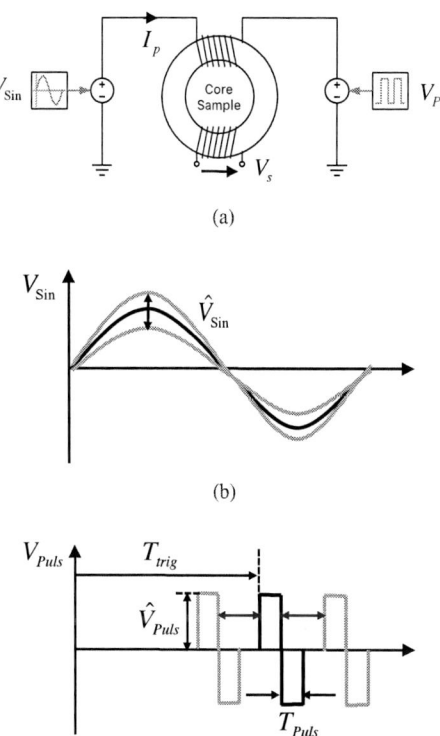

(a)

(b)

(c)

Fig. 3: Basic configuration of the proposed test setup (a) Schematic; (b) Voltage waveform of the sinusoidal excitation; (c) Voltage waveform of the pulsed excitation.

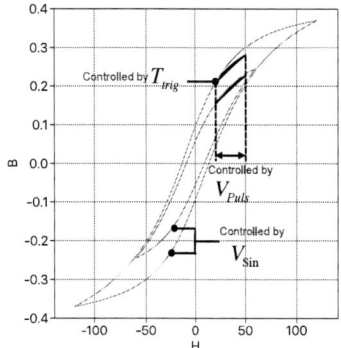

Fig. 4: Biased hysteresis loop located at desired field strength range H and flux density B position via adjusting the amplitude of the sinusoidal- and pulsed excitation, as well as the phase delay of the pulsed excitation.

and a square-wave pulsed voltage source, which allows to setup the biased hysteresis loop at arbitrary operation point: When the primary winding is supplied solely by the sinusoidal

423

voltage source (Fig. 3(b)), a symmetrical hysteresis loop with large amplitude is generated. If the pulsed voltage source is applied at the same time (Fig. 3(c)), a biased hysteresis loop is present attaching on the symmetrical loop. As demonstrated in Fig. 4, on a symmetrical loop generated by the sinusoidal excitation, the H position where the biased loop is initiated can be explicitly controlled by the phase delay T_{trig} (Fig. 3(c)) of the pulsed excitation, with respect to the sinusoidal excitation. The width of the pulsed voltage T_{puls} can be configured to setup the desired equivalent switching frequency for the biased loop. With given T_{Puls}, the field strength H range of the biased loop can be adjusted by tunning the amplitude T_{Puls} of the pulsed voltage. If another amplitude $\hat{V}_{Sin}$ of the sinusoidal excitation is applied which changes the symmetrical hysteresis loop's size, the biased loop in the same H range can be located at any desired B positions. In this way, the proposed test setup allows to characterise the materials's B-H property at arbitrary operating conditions.

Please note that it is important to setup the period of the pulsed voltage excitation to be equal to that of the sinusoidal one and keep both excitations always synchronised, so that the biased loop stay at the same location on the B-H plane.

III. HARDWARE REALISATION

The hardware realisation of the proposed test setup is illustrated in Fig. 5(a) and Fig. 5(b), which is composed of the following parts:

- A power amplifier of type LM3886 to generate the sinusoidal excitation.

- A single phase T-shape MOSFET bridge to provide three-level square-wave pulsed excitation, the DC side of which is supplied by another two power amplifiers.

- The primary current is measured by a shunt resistor in combination with fully differential instrumentation amplifier.

- The secondary voltage is measured using resistive voltage divider.

- A control unit (PLECS RT-Box) is equipped to provide reference signal for the power amplifiers, generate gate signal for the MOSFET bridge as well as process the measurements.

Since the test setup is supposed to characterise the material characteristic using small core samples, only low voltage output in the range $\pm 24V$ and low current up to $6A$ are required. The MOSFETs of the T-shape bridge are driven by carrier based PWM, in order to generate the pulsed voltage waveform shown in Fig. 3(c), the carrier as well as modulation index are configured as demonstrated in Fig. 6. The four switches are divided into two groups - $Q1\&Q3$ and $Q2\&Q4$, both of which have the same saw-tooth carrier. The two switches inside each group have the same modulation index while opposite polarity. The modulation index m_1 of the group $Q1\&Q3$ and m_2 of $Q2\&Q4$ fullfills $m_1 = 1 - m_2$.

IV. EXPERIMENTAL RESULTS

For demonstration purpose, test result has been produced for ferrite material N87 from TDK, where the toroidal core of size code "R 41.8x26.2x12.5" is taken as sample. The turns number of the primary- and secondary windings are both eight. Two test schemes has been carried out, where the amplitude of the sinusoidal excitation has been configured to make the amplitude of the symmetrical hysteresis loop to be $\hat{H} = 45A/m$ and $\hat{H} = 60A/m$, respectively. The frequency of the sinusoidal excitation has been configured to be $200Hz$. The phase delay T_{trig} and amplitude of the pulse excitation are adjusted so that the biased minor loop in both test schemes are within the same field strength range $20\text{~}40A/m$. The pulse width T_{Puls} is configured to be $0.25ms$, yields $2kHz$ equivalent switching frequency. The measured hysteresis loop (including the symmetrical- and biased loops) as well as the time domain waveform of the primary winding current and

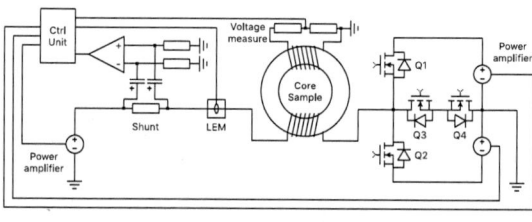

(a)

(b)

Fig. 5: Implementation of the proposed test setup (a) Hardware setup; (b) Circuit schematic.

The 2018 International Power Electronics Conference

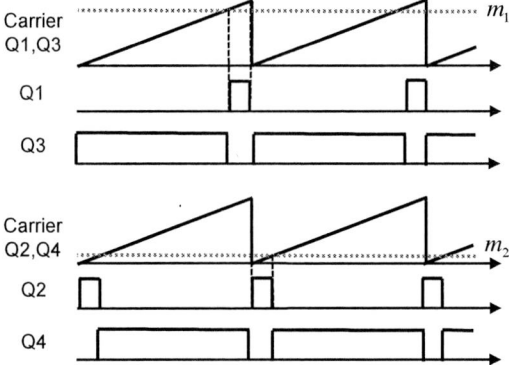

Fig. 6: Saw-tooth carrier based PWM generation of the T-shape MOSFET bridge.

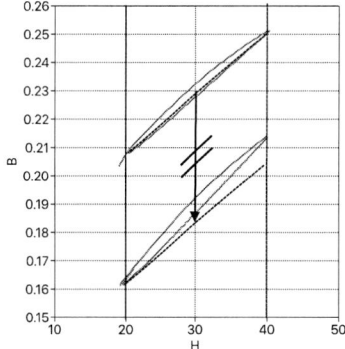

Fig. 9: Comparison of the two biased loop at different flux density B positions, where obvious difference on equivalent permeability can be observed.

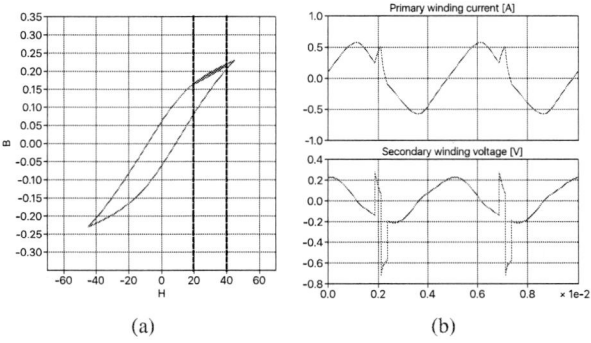

(a) (b)

Fig. 7: Biased minor loop in the range $20\sim40A/m$ with symmetrical loop's amplitude $45A/m$ and equivalent switching frequency $2kHz$ (a) B-H characteristic; (b) Time domain waveform of the primary current and secondary voltage.

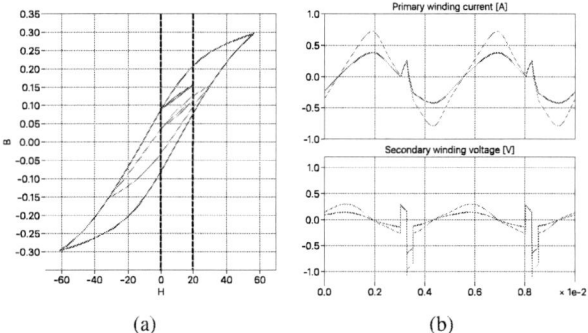

(a) (b)

Fig. 10: Biased minor loops within $0\sim20A/m$ with symmetrical loop's amplitude $60A/m$ and $30A/m$, equivalent switching frequency $2kHz$ (a) B-H characteristic; (b) Time domain waveform of the primary current and secondary voltage.

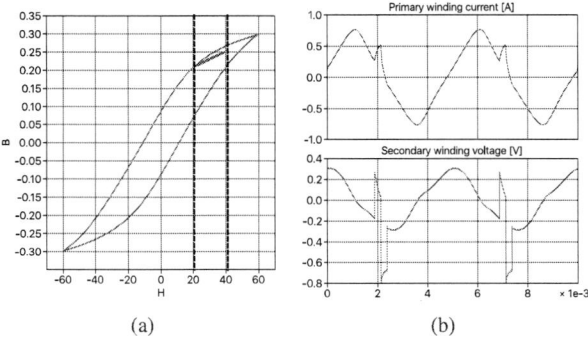

(a) (b)

Fig. 8: Biased minor loop in the range $20\sim40A/m$ with symmetrical loop's amplitude $60A/m$ and equivalent switching frequency $2kHz$ (a) B-H characteristic; (b) Time domain waveform of the primary current and secondary voltage.

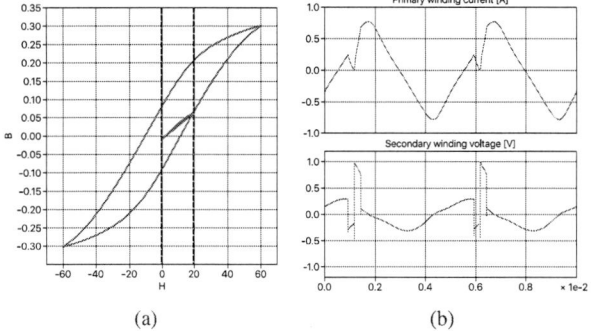

(a) (b)

Fig. 11: Biased minor loop within $0\sim20A/m$ initiated on the ascending branch of a symmetrical loop with amplitude equal to $60A/m$ (a) B-H characteristic; (b) Time domain waveform of the primary current and secondary voltage.

425

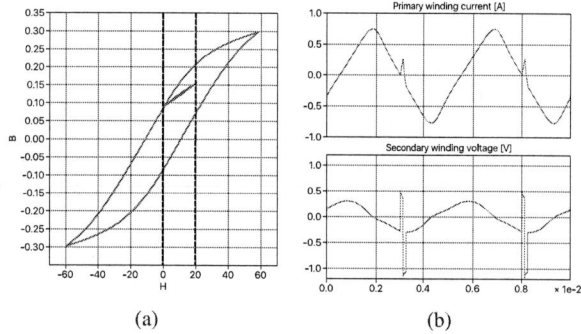

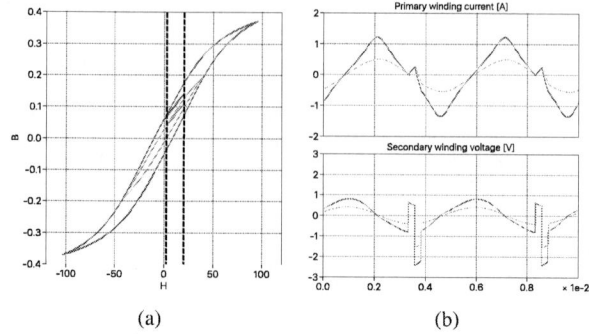

(a)　　　　　　　(b)　　　　　　　　　　　　　(a)　　　　　　　(b)

Fig. 12: Biased minor loop in the range 0~20A/m with equivalent switching frequency 4kHz (a) B-H characteristic; (b) Time domain waveform of the primary current and secondary voltage.

Fig. 13: Biased minor loops in the range 0~20A/m of ferrite material 3C81, equivalent switching frequency 2kHz (a) B-H characteristic; (b) Time domain waveform of the primary current and secondary voltage.

secondary voltage in both test schemes are shown in Fig. 7(a) and Fig. 8(a), respectively. The minor loop on the B-H plane corresponds to the triangular peak on the primary winding current waveform as an result of the pulsed-excitation provided by the T-bridge, which can be observed on the secondary winding voltage waveform. In Fig. 9, the two biased hysteresis loops obtained at different flux density B positions are compared with each other, obvious difference can be observed on the equivalent permeability, or in other words, the slope of the virtual straight line connecting the two peaks of the loops.

By adjusting the phase delay of the gate signals T_{trig} with respect to the sinusoidal excitation voltage, the H range of the minor loops can be changed. In Fig. 10(a) and Fig. 10(b), the measured minor loops within the field strength range 20~40A/m with different flux density bias are presented together with the corresponding time domain waveforms. Please note that the previous results are obtained when the minor loops are initiated on the descending branch of the symmetrical loop, if lower flux density bias is desired, the minor loop should be initiated on the ascending branch. For this purpose, the gate signals for the two switch groups $Q1\&Q3$ and $Q2\&Q4$ are exchanged. The test result of a minor loop initiated on the ascending branch of the symmetrical loop is presented in Fig. 11(a) and Fig. 11(b).

Higher equivalent switching frequency can be achieved via applying shorter pulse-width T_{Puls} and higher DC voltage V_{Puls}. The measured minor loop within the field strength range 0~20A/m with a equivalent switching frequency of 4kHz as well as the corresponding primary winding current and secondary winding current are demonstrated in Fig. 12(a) and Fig. 12(b), respectively. In comparison to the secondary voltage waveform in Fig. 10(b), the duration of the pulse component is one half while the voltage is approximately doubled. If a duty cycle other than 50% is desired, the summation of the modulation indices m_1 and m_2 becomes higher than 1, while the DC supply voltages need to be trimmed as well so that the

voltage-second products of the positive- and negative impulse remain the same.

At the end, another core sample $TX51/32/19$ made of ferrite material 3C81 from Ferroxcube is characterised using the proposed approach. The minor loops in the field strength range 0~20A/m with different flux density bias are shown in Fig. 13(a). In comparison to the material N87, the large symmetrical hysteresis loop of 3C81 is significantly narrower. Considering the fact that the flux density range of the minor loops does not exceed the boundary of a large symmetrical major loop, the minor loops of 3C81 have limited range in the B direction, so that the impact of the flux density bias on the shape of the minor loops is also lower.

V. CONCLUSION

This work proposes a new approach to characterise the biased hysteresis loop of core materials in power electronic applications. This approach combines continuous sinusoidal- and switched PWM excitation together to imitate arbitrary operating conditions that take place in real power converter systems. The test can be easily carried out under low-voltage and low-current. Using this approach, the offset of the magnetic hysteresis loop on not only the field strength H but also that on the flux density B are accurately controlled, so that the influence of the historical magnetisation on the core loss and equivalent permeability can be clearly investigated. Also the equivalent switching frequency of the biased loop can be configured in a convenient way. Based on the measured material characteristic, the design and modelling of the magnetic component can be carried out in a more accurate way, in terms of meeting the THD requirement as well as optimising the system efficiency.

ACKNOWLEDGMENT

This project has been supported in the frame of the ECPE Joint Research Programme.

REFERENCES

[1] D. Tan, J. L. Vollin, and S. M. Cuk, "A practical approach for magnetic core-loss characterization," in *IEEE Transactions on Power Electronics*, vol. 10, no. 2, 1995, pp. 124–130.

[2] M. Mu, Q. Li, D. J. Gilham, F. C. Lee, and K. D. T. Ngo, "New core loss measurement method for high-frequency magnetic materials," in *IEEE Transactions on Power Electronics*, vol. 29, no. 8, 2014, pp. 4374–4381.

[3] D. Hou, M. Mu, F. C. Lee, and Q. Li, "New high-frequency core loss measurement method with practial cancellation concept," in *IEEE Transactions on Power Electronics*, vol. 32, no. 4, 2017, pp. 2987–2994.

[4] J. Reinert, A. Brockmeyer, and R. W. A. A. D. Doncker, "Calculation of losses in ferro- and ferritemagnetic materials based on the modified steinmetz equation," in *IEEE Transactions on Industry Applications*, vol. 37, no. 4, 2001, pp. 1055–1061.

[5] Y. Han and Y. Liu, "A practical transformer core loss measurement scheme for high-frequency power converter," in *IEEE Transactions on Industrial Electronics*, vol. 55, no. 2, 2008, pp. 941–948.

[6] J. Muehlethaler, J. Biela, J. W. Kolar, and A. Ecklebe, "Core losses under the dc bias condition based on steinmetz parameters," in *IEEE Transactions on Power Electronics*, vol. 27, no. 2, 2012, pp. 953–963.

[7] G. Kadar and E. D. Torre, "Hysteresis modeling: I. noncongruency," in *IEEE Transactions on Magnetics*, vol. 23, no. 5, 1987, pp. 2820–2822.

[8] M. Marracci and B. Tellini, "Hysteresis losses of minor loops versus temperature in mnzn ferrite," in *IEEE Transactions on Magnetics*, vol. 49, no. 6, 2013, pp. 2865–2869.

[9] M. Ibrahim, "M. ibrahim and p. pillay," in *Core loss prediction in electrical machine laminations considering skin effect and minor hysteresis loop*, vol. 49, no. 5, 2013, pp. 2061–2068.

[10] T. Taitoda, Y. Takahashi, and K. Fujiwara, "Iron loss estimation method for a general hysteresis loop with minor loops," in *IEEE Transactions on Magnetics*, vol. 51, no. 11, 2015.

A Fast Open-Circuit Fault Diagnosis Scheme for Modular Multilevel Converters with Model Predictive Control

Dehong Zhou[1], Shunfeng Yang[2], Yi Tang[2*]

1 Maritime Institute @ NTU, Nanyang Technological University, Singapore
2 School of Electrical and Electronic Engineering, Nanyang Technological University, Singapore
*E-mail: yitang@ntu.edu.sg

Abstract—**Fault diagnosis is a crucial way to increase the reliability of modular multilevel converters (MMCs) which consist of a large number of submodules (SMs). This paper proposes a fast fault detection and isolation (FDI) method to identify single open-circuit faults for MMCs with model predictive control (MPC). The proposed FDI method is simply implemented by comparing the applied arm voltage obtained in the former control cycle and the estimated arm voltage calculated by the load and circulating current models. No additional transducer or measurement is required. Experimental results show that an open-circuit fault in the MMC can be accurately detected and isolated in a very short time.**

I. Introduction

Modular multilevel converters (MMCs) have emerged as one of the most attractive topology for high voltage applications [1], such as high-voltage direct-current (HVDC) transmission [2], medium voltage motor drives [4] and energy storage systems [5] thanks to the advanced benefits of modularity, flexible expandability, transformer-less configuration, ease of assembling, scalability, and so on.

The primary concern of the MMC is the reliability issue, especially in the HVDC application. The MMC operation may be interrupted by the potential failure in the power semiconductor switches. Power semiconductor switches are considered as one of the most fragile components. It is estimated that about 38% of the faults in the power conversion system are caused by the semiconductor failures [6], [7]. This is particularly true in MMC since there are always hundreds (or even thousands) of insulated-gate bipolar transistors (IGBTs) in one MMC. Broadly, the power semiconductor switch failures can be categorized into two groups: short-circuit faults and open-circuit faults [8]. Short-circuit fault protection is often provided by hardware solutions [9]. The open-circuit fault, however, does not damage the SM immediately. But its danger cannot be neglected because it degrades the system performance and extensive operation with open-circuit fault may lead to a potential secondary fault or catastrophic failure of the whole MMC system. Therefore, it is crucial to detect and isolate these faults in MMCs immediately after their occurrence.

Fault diagnosis schemes for IGBT open-circuit faults of MMCs have been a hot-spot to improve the reliability of

MMCs [10]–[16]. A fast fault diagnosis method based on adding extra hardware to each SM was proposed in [10]. However, this solution was very expensive and volume-consuming. A fault detection method based on unconformity information between SM output voltage and switching signals was proposed in [14]. But with the output voltage of each SM measured instead of capacitor voltage of each SM capacitor, potentially instability may be introduced into the control system. In [13], a sliding-mode observer was proposed to detect the open-circuit fault in MMCs. The same observer was improved in [11] with increased robustness and reduced detection time. The detection method was based on the errors of the observed circulating current and the measured one. This method is robust and without additional sensors. However, the locating time was relatively slow with a minimum locating time of 50 ms. [12] proposed a state observer based fault detection method. The minimum required locating time was also slow with a minimum localization time of 50 ms. The main drawbacks of these observer-based methods were the assumption-verification process which was very time-consuming.

In an MMC system, efforts have also to be devoted to SM capacitor voltage balancing or circulating currents suppression simultaneously in addition to controlling the output powers or current [17]. Therefore, it is very favorable to use model predictive control (MPC) since all the control variables control can be managed in a single cost function. The MPC are widely used in the MMC control because of its advantages including fast dynamic response, easy inclusion of nonlinearities and constraint. Compared with linear control scheme with pulse-width modulation (PWM), the MPC has the known and unchanged switching state in a sampling period which can be utilized for fast open-circuit fault location.

Although considerable research work has been devoted to the implementation of MPC for MMCs recently, to our best knowledge, a little effort has been made to develop fault diagnosis methods for the MMC with MPC. To enhance the reliability of MMC with MPC, a fault detection and isolation (FDI) method to identify single open-circuit faults is proposed in this paper. The FDI method is simply implemented by comparing the applied arm voltage obtained in the former control cycle and the estimated arm voltage obtained by the

The 2018 International Power Electronics Conference

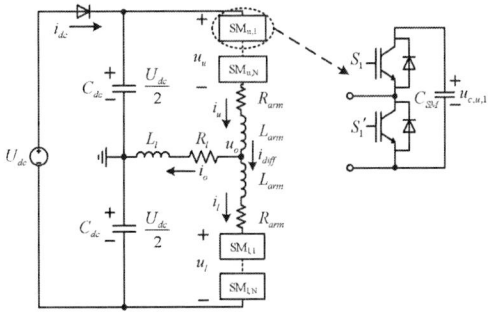

Fig. 1. Structure of a single-phase MMC inverter.

load and circulating current models. No extra measurement is required and it is possible to detect and isolate an open-circuit fault in several sampling periods (less than 1 ms). With only one parameter to be tuned, the proposed FDI method is very easy to be implemented by avoiding tedious factor tuning process. Experimentally obtained data are presented to verify the effectiveness of the proposed FDI scheme of MMCs.

II. System Description and MPC for MMCs

The mathematical model and MPC of an MMC are reviewed. The notations which is introduced in this section is used throughout this paper. The circuit configuration of a single phase MMC is shown in Fig. 1.

A. Mathematical Model

Each SM consists of a capacitor C_{SM} and two complementary IGBT (i.e., S_x and S'_x). For each SM, the capacitor voltage appears in the terminal when S_x is on, and the capacitor voltage is bypassed when S'_x is on. Therefore, the terminal voltage in the upper arm and the lower arm can be, respectively, expressed as

$$u_{\text{sm},u,i} = S_{u,i} \times u_{c,u,i} \qquad (1)$$

$$u_{\text{sm},l,i} = S_{l,i} \times u_{c,l,i} \qquad (2)$$

where $S_{u,i}$ and $S_{l,i}$ are the binary switching functions of the ith ($i \in [1, 2, ..., N]$) SM in the upper and lower arms. $u_{c,l,i}$ and $u_{c,u,i}$ denote, respectively, the capacitor voltages of the ith SM in the upper and lower arms.

Each arm is equipped with an arm inductor L_a to limit the arm currents in the case of DC-link short-circuit. R_a is the equivalent resistance of the arm inductor. u_u and u_l are the voltages of the upper and lower arm, respectively. i_u and o_l are the currents of the upper and lower arm, respectively. u_o is the output voltage and i_o is the output current. L_l, R_l are the load inductor and resistance, respectively. U_{dc} is the dc source voltage. $u_{\text{sm},u,i}$ and $u_{\text{sm},l,i}$ are the ith SM capacitor voltages of the upper arm and the lower arm, respectively. N is the number of SMs.

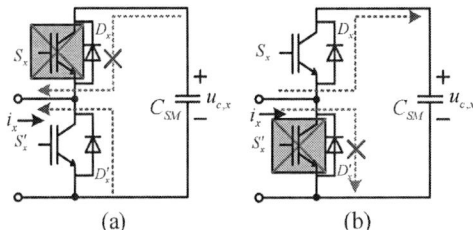

Fig. 2. Current path of the open-circuit failure in one SM (a) upper switch fault (b) lower switch fault

TABLE I.   SM x Output Voltage Deviation,
$\Delta u_{c,x} = u_{expected} - u_{actual}$

Fault Location	Current Direction i_x	Switching signal 1	0
S_i	> 0	0	0
	< 0	$u_{c,x}$	0
S'_i	> 0	0	$-u_{c,x}$
	< 0	0	0

The dynamic equations of the MMCs can be written as

$$\frac{di_c}{dt} = -\frac{R_a}{L_a} i_c + \frac{U_{dc} - u_u - u_l}{2L_a} \qquad (3)$$

$$\frac{di_o}{dt} = -\frac{R_a + 2R_l}{L_a + 2L_l} i_o + \frac{-u_u + u_l}{L_a + 2L_l}. \qquad (4)$$

B. MPC for MMC

The core idea of MPC can be summarized as the following: (1) Measure the current state of the MMC formed by capacitor voltage, circulating current, and load current; (2) Predict the state at the next sampling instant for all the valid switching patterns; (3) The difference of all the state vectors and their references are calculated and normalized with weighting factors to formulate the cost function. The switching combination with the lowest cost function is applied in the next control period.

The cost function for MMCs control can be written as

$$J = \lambda_1 (i_c^{ref} - i_c^{k+1})^2 + \lambda_2 (i_o^{ref} - i_o^{k+1})^2 \qquad (5)$$

$$+ \sum_{i=1}^{N} \left((u^{ref} - u_{c,u,i}^{k+1})^2 + (u^{ref} - u_{c,l,i}^{k+1})^2 \right),$$

where λ_1 and λ_2 are weighting factors that determine the importance of different objectives when selecting the optimal switching combination, i_c^{ref} is the reference value for the circulating current, i_o^{ref} is the reference value for the load current, u^{ref} is the reference value for capacitor voltage of each SMs. The detailed control is presented in [18]. In the end of each control period, the switching combination is obtained. With the known applied switching combination, it is very easy to locate the open-circuit switch fault in MPC of MMCs.

III. Behavior Analysis of The MMC under Open-Circuit Faults

Two possible open-circuit faults (S_x fault and S'_x fault), as shown in Fig. 2, may occur in each SM. If the upper switch

The 2018 International Power Electronics Conference

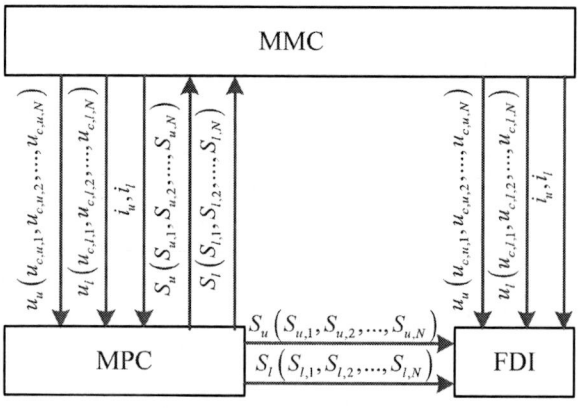

Fig. 3. The proposed open-circuit fault FDI scheme for MMCs with MPC.

TABLE II. Voltage Characteristics Under Faulty Conditions

State	Voltage error		Faulty switch location
Normal	$\varepsilon_{sum} \approx 0,$	$\varepsilon_{diff} \approx 0.$	Healthy
Faulty	$\varepsilon_{sum} > U_{th},$	$\varepsilon_{dif} < -U_{th}.$	S_1 fault in $SM_{u,x}$
	$\varepsilon_{sum} < -U_{th},$	$\varepsilon_{dif} > U_{th}.$	S_1' fault in $SM_{u,x}$
	$\varepsilon_{sum} > U_{th},$	$\varepsilon_{dif} > U_{th}.$	S_1 fault in $SM_{l,x}$
	$\varepsilon_{sum} < -U_{th},$	$\varepsilon_{dif} < -U_{th}.$	S_1' fault in $SM_{l,x}$

$$u_{dif,e} = (L_a + 2L_l)\frac{i_o(k) - i_o(k-1)}{T_s} + i_o(k)(R_a + 2R_l). \quad (7)$$

The expected or measured sum and difference of the upper and the lower arm voltage can be obtained from the former control cycle

$$u_{sum,m} = \sum_{i=1}^{N}(S_{u,i} \cdot u_{c,u,i}) + \sum_{i=1}^{N}(S_{l,i} \cdot u_{c,l,i}) \quad (8)$$

$$u_{dif,m} = -\sum_{i=1}^{N}(S_{u,i} \cdot u_{c,u,i}) + \sum_{i=1}^{N}(S_{l,i} \cdot u_{c,l,i}). \quad (9)$$

The voltage errors between the two voltages can be utilized to locate the possible fault primarily.

$$\Delta u_{sum} = u_{sum,m} - u_{sum,e} \quad (10)$$

$$\Delta u_{dif} = u_{dif,m} - u_{dif,e}. \quad (11)$$

To simplify the analysis, each SM's capacitor voltage is assumed to be perfectly balanced. In order to make the FDI independent to SM capacitor voltage, these errors can be normalized as

$$\varepsilon_{sum} = \frac{N\Delta u_{sum}}{U_{dc}} \quad (12)$$

$$\varepsilon_{dif} = \frac{N\Delta u_{dif}}{U_{dc}}. \quad (13)$$

Both the two errors can be utilized for fault detection. If the difference between the expected and actual voltage satisfies $|\varepsilon_{sum}| > U_{th}$ or $|\varepsilon_{dif}| > U_{th}$, the fault can be detected. However, a transient spike caused by estimation errors, measurement noise or electromagnetic interference may lead to erroneous diagnosis. Although these errors are very small and within the predefined threshold in the normal case, they may, however, occasionally exceed the threshold and cause a false alarm. In order to eliminate this situation, the fault detection is modified. If the difference between the expected and actual voltage satisfies $|\varepsilon_{sum}| > U_{th}$ or $|\varepsilon_{dif}| > U_{th}$, and this condition persists for five control cycles. Then, an open-circuit fault occurs and the FDI scheme enters the fault isolation mode; otherwise, the FDI scheme stays in the fault detection mode.

fault in an SM, the output voltage of the SM should be $u_{c,x}$ when the faulty switch S_x should conduct current. However, the current does not flow through S_x due to the open-circuit fault and this current path is blocked by the anti-parallel diode, and the actual current flows through D_x'. Therefore, the actual voltage of the SM is 0. The SM's output voltage is affected by the faulty SM's state and the current direction under open-circuit fault conditions, because the anti-parallel diode is assumed to be unaffected by the open-circuit fault. The relationship the output voltage of SM x with the current direction and switching state is presented in Table I. Table I show that the S_i fault can be observed when $i_x < 0$ and switching state of S_i is on, the S_i' fault can be observed when $i_x > 0$ and switching state of S_i' is on. The proposed open-circuit fault detection scheme presented in this paper is based on these observations.

IV. FAULT DETECTION AND ISOLATION METHOD

In this section, the proposed open-circuit switch fault detection and isolation with model predictive control is presented. The proposed technique takes advantage of the measurements and calculations performed for the MPC and requires no extra hardware or measurement. The whole diagram of the FDI scheme for MMCs with MPC is presented in Fig.3. The proposed scheme is carried out in three steps: 1) fault detection 2) possible open-circuit fault location 3) fault isolation. In the fault detection stage, health state of MMCs is monitored. Unless the fault is detected, the fault location process will not be implemented. In the second stage, the possible faulty switch location can be narrowed into one of the four groups by checking the voltage error polarity. In the last stage, the faulty switch is isolated by directly checking the switching states.

A. Fault Detection

The actual sum and difference of the upper and the lower arm voltage are estimated by

$$u_{sum,e} = U_{dc} - 2L_a\frac{i_c(k) - i_c(k-1)}{T_s} - 2i_c(k)R_a \quad (6)$$

430

B. Possible Open-Circuit Fault Location

These two normalized errors between the expected and actual voltage can be utilized for fault isolation. Taking the open-circuit fault at S_1 for example, whenever S_1 should carry a current, the voltage errors between the sum and difference of the upper and the lower arm voltage are as follows according to (12) (13)

$$\varepsilon_{sum} \approx 1 \tag{14}$$

$$\varepsilon_{dif} \approx -1. \tag{15}$$

Based on this feature, the possible open-circuit fault switch can be narrowed down by introducing a threshold U_{th}. When ε_{sum} surpasses the positive threshold and ε_{dif} surpasses the negative threshold, the possible faulty switch can be narrowed down to S_1 fault in $SM_{u,x}$. Other types of the open-circuit fault can be narrowed in the same way and the whole narrowed table for possible open-circuit switch fault is given in Table. II.

C. Faulty Switch Isolation

Due to the discrete characteristic of MPC, the switching state of each SM is known, which makes it easy for fault isolation. Unlike checking the capacitor voltages in prior approaches, the proposed approach directly checks the switching states of SMs for fault isolation. After the possible faulty switch is narrowed into one of the four groups, for example, if S_x fault is in the upper arm, the potential faulty switch location set can be represented as

$$S_F = \{S_{u,1}, S_{u,2}, ..., S_{u,N}\}. \tag{16}$$

Once an open-circuit is detected, S_F must be made to have a single element.

When there exist errors between the measured and estimated voltages, it indicates that one of the inserted SMs in S_F fails and the ones not inserted in S_F are in healthy conditions. By taking this feature into consideration, the fault can be fast isolated by introducing a counter to each potential faulty switch. For example, if the voltage errors between the estimated voltages and the measured ones are not zero, the counters of the possible faulty switches in S_F increase by one, and the counters of the healthy switches in S_F decrease by one. This idea can be denoted as follows

$$if(S_i == 1), T_{cnt,i} += 1, \tag{17}$$

$$if(S_i == 0), T_{cnt,i} -= 1. \tag{18}$$

As long as one of the counters has the largest value, the open-circuit fault can be isolated.

TABLE III. PARAMETERS

Parameters	Values
DC bus voltage: U_{DC}	240 V
DC bus capacitance: C_{DC}	3.9 mF
Arm inductance: L_a	5 mH
Arm resistance: R_a	0.2 Ω
SM capacitance: C_{SM}	940 uF
Load inductance: L_l	2 mH
Rated output frequency:	50 Hz
No. of SM in each arm: N	3
Sampling frequency $1/T_s$	10 kHz

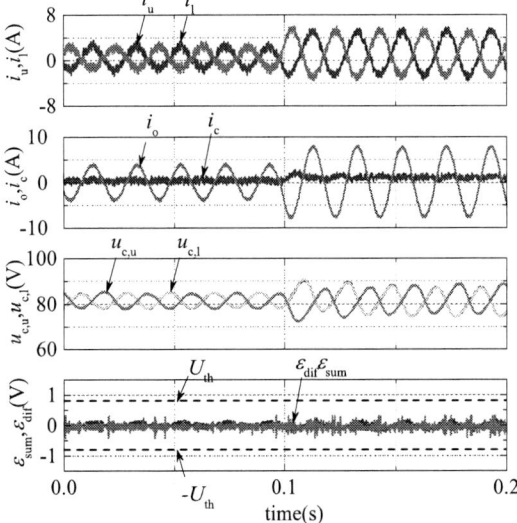

Fig. 4. Experimental results of immunity to transient process

V. EXPERIMENTAL RESULTS

The detailed parameters of the single-phase MMC inverter experimental setup are listed in TABLE III, was built to validate the effectiveness of the FDI method. A dSPACE MicroLabBox DS1202 was adopted as a controller and a slave Xilinx was utilized to generate the gate signals. The voltage and current quantities required by the proposed control strategy were sampled every T_s.. A resistive load of 5 Ω was adopted in the experiments. In the following experiments, the IGBT open-circuit faults were generated by permanently inhibiting the corresponding gate signals to emulate the open-circuit fault .

Step change of output current change is tested to prove the robustness of the proposed diagnostic method, as shown in Fig. 4. The waveforms of arm currents, output current, circulating current, six SMs capacitor voltages and voltage errors are presented. As shown, the output current and circulating current track their references and the SM capacitor voltage is balanced. During the transient, the two normalized voltage errors stay within the predefined threshold $U_{th} = 0.8$. No false alarm occurs during the test, indicating the robustness of the FDI

The 2018 International Power Electronics Conference

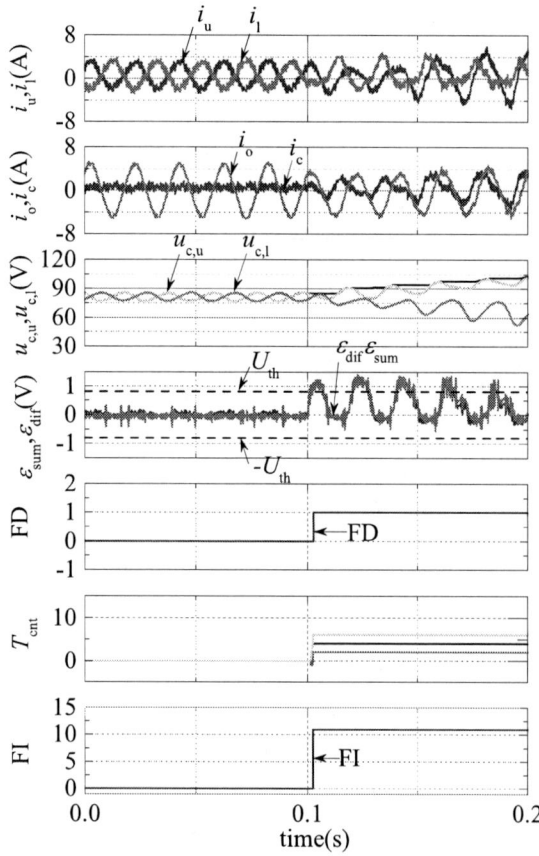

Fig. 5. Experimental Results of open-circuit fault at S_1 in the $SM_{l,3}$

the proposed scheme. As shown in the experimental results, the normalized errors are lower than 0.3 in normal condition. Therefore, the threshold with a value higher than 0.3 can be adopted to avoid false alarms. A higher value of U_{th} will decrease the probability of false alarm caused by transient disturbances and parameter variations. On the other hand, when the capacitor voltage is perfectly balanced, then the normalized error will become 1 at the open-circuit fault case. However, this voltage error is not a perfect dc component but dc has an ac component with fundamental frequency. In addition, in the open-circuit fault case, the capacitor voltage will deviate from the balanced state. Therefore, a conservative threshold value of 0.8 is adopted to avoid false alarm in this paper.

VI. CONCLUSIONS

In this paper, a fast single open-circuit fault FDI has been proposed to enhance the reliability of MMCs with MPC. The fault in one the of SMs is detected by comparing the theoretically expected voltage obtained from the former control cycle with the estimated voltage obtained by the load and circulating current models. The proposed diagnosis method is robust to transient disturbance and is free from tedious threshold tuning. Unlike other fault detection methods in the literature, no additional hardware is required and it is possible to detect and isolate an open-circuit fault in several sampling periods (less than 1 ms). Experimental results demonstrate the feasibility and effectiveness of the proposed FDI method.

ACKNOWLEDGEMENT

This research is funded by the Singapore Maritime Institute under the Maritime Research between Norway and Singapore (MNS) R&D Programme - Project SMI-2015-MA-15.

method to output current step change.

Experiments with the proposed FDI method were conducted to verify the validity of the analyzed behavior of the MMC under open-circuit fault conditions and the developed fault diagnosis method. Fig. 5 shows the experimental waveforms with S_1 fault in $SM_{l,3}$. Initially, it can be seen that the observed voltage errors stay around zero, indicating the measured voltages well coincided with the estimated ones. But when the open-circuit fault occurs, the normalized voltage errors deviate from zero. ε_{sum} surpasses the positive threshold and the ε_{dif} surpasses the negative threshold which is in agreement with the theoretical relationship in Table II. When one of the voltage error surpasses the threshold and this situation persists for 0.5 ms, the fault detection signal changes from 0 to 1, indicating the fault is detected. After the fault is detected, the fault isolation method is implemented. The open-circuit fault is isolated within 0.5 ms when the counter value of $T_{cnt,3}$ becomes the largest among the three counters where $T_{cnt,1} = 2$, $T_{cnt,2} = 4$, and $T_{cnt,3} = 6$. Other open-circuit fault can be isolated in the same way.

One of the important issue for the proposed fault diagnosis method is the threshold tuning. There is only one threshold in

REFERENCES

[1] H. Abu-Rub, J. Holtz, J. Rodriguez, and G. Baoming, "Medium-voltage multilevel converters—state of the art, challenges, and requirements in industrial applications," *IEEE Trans. Ind. Electron.*, vol. 57, no. 8, pp. 2581–2596, 2010.

[2] A. Nami, J. Liang, F. Dijkhuizen, and G. D. Demetriades, "Modular multilevel converters for hvdc applications: Review on converter cells and functionalities," *IEEE Trans. Power Electron.*, vol. 30, no. 1, pp. 18–36, 2015.

[3] M. T. Bina *et al.*, "A transformerless medium-voltage statcom topology based on extended modular multilevel converters," *IEEE Trans. Power Electron.*, vol. 26, no. 5, pp. 1534–1545, 2011.

[4] H. Akagi and R. Kondo, "A transformerless hybrid active filter using a three-level pulsewidth modulation (pwm) converter for a medium-voltage motor drive," *IEEE Trans. Power Electron.*, vol. 25, no. 6, pp. 1365–1374, 2010.

[5] M. Vasiladiotis and A. Rufer, "Analysis and control of modular multilevel converters with integrated battery energy storage," *IEEE Trans. Power Electron.*, vol. 30, no. 1, pp. 163–175, 2015.

[6] D. Zhou, Y. Li, J. Zhao, F. Wu, and H. Luo, "An embedded closed-loop fault-tolerant control scheme for nonredundant vsi-fed induction motor drives," *IEEE Trans. Power Electron.*, vol. 32, no. 5, pp. 3731–3740, May 2017.

[7] D. Zhou, J. Zhao, and Y. Liu, "Independent control scheme for nonredundant two-leg fault-tolerant back-to-back converter-fed induction motor drives," *IEEE Trans. Ind. Electron.*, vol. 63, no. 11, pp. 6790–6800, Nov. 2016.

[8] M. Ciappa, "Selected failure mechanisms of modern power modules," *Microelectron. Reliab.*, vol. 42, no. 4, pp. 653–667, 2002.

[9] U.-M. Choi, F. Blaabjerg, and K.-B. Lee, "Study and handling methods of power igbt module failures in power electronic converter systems," *IEEE Trans. Power Electron.*, vol. 30, no. 5, pp. 2517–2533, 2015.

[10] K. Bi, Q. An, J. Duan, L. Sun, and K. Gai, "Fast diagnostic method of open circuit fault for modular multilevel dc/dc converter applied in energy storage system," *IEEE Trans. Power Electron.*, vol. 32, no. 5, pp. 3292–3296, 2017.

[11] S. Shao, A. J. Watson, J. C. Clare, and P. W. Wheeler, "Robustness analysis and experimental validation of a fault detection and isolation method for the modular multilevel converter," *IEEE Trans. Power Electron.*, vol. 31, no. 5, pp. 3794–3805, 2016.

[12] B. Li, S. Shi, B. Wang, G. Wang, W. Wang, and D. Xu, "Fault diagnosis and tolerant control of single igbt open-circuit failure in modular multilevel converters," *IEEE Trans. Power Electron.*, vol. 31, no. 4, pp. 3165–3176, 2016.

[13] S. Shao, P. W. Wheeler, J. C. Clare, and A. J. Watson, "Fault detection for modular multilevel converters based on sliding mode observer," *IEEE Trans. Power Electron.*, vol. 28, no. 11, pp. 4867–4872, 2013.

[14] S. Haghnazari, M. Khodabandeh, and M. R. Zolghadri, "Fast fault detection method for modular multilevel converter semiconductor power switches," *IET Power Electronics*, vol. 9, no. 2, pp. 165–174, 2016.

[15] F. Deng, Z. Chen, M. R. Khan, and R. Zhu, "Fault detection and localization method for modular multilevel converters," *IEEE Trans. Power Electron.*, vol. 30, no. 5, pp. 2721–2732, 2015.

[16] D. Zhou, S. Yang, and Y. Tang, "A voltage-based open-circuit fault detection and isolation approach for modular multilevel converters with model predictive control," *IEEE Trans. Power Electron.*, vol. PP, no. 99, p. 1, 2018.

[17] L. Harnefors, A. Antonopoulos, S. Norrga, L. Angquist, and H.-P. Nee, "Dynamic analysis of modular multilevel converters," *IEEE Trans. Ind. Electron.*, vol. 60, no. 7, pp. 2526–2537, 2013.

[18] L. Ben-Brahim, A. Gastli, M. Trabelsi, K. A. Ghazi, M. Houchati, and H. Abu-Rub, "Modular multilevel converter circulating current reduction using model predictive control," *IEEE Trans. Ind. Electron.*, vol. 63, no. 6, pp. 3857–3866, 2016.

The 2018 International Power Electronics Conference

An Online Open-Circuit Fault Diagnosis and Fault Tolerant Scheme for Three-Phase AC-DC Converters with Model Predictive control

Dehong Zhou[1], Yi Tang[2*]

1 Maritime Institute @ NTU, Nanyang Technological University, Singapore
2 School of Electrical and Electronic Engineering, Nanyang Technological University, Singapore
*E-mail: yitang@ntu.edu.sg

Abstract—This paper proposes an open-circuit fault diagnosis and fault-tolerant control to enhance the reliability of the AC-DC converter using model predictive control (MPC). A fault diagnosis method, which takes advantage of the known and unchanged switching state is proposed to identify single open-circuit fault. By introducing the healthy leg variables (HLVs), a generalized model including both normal and faulty operations is proposed and the fault-tolerant scheme can be integrated into the normal MPC scheme. With the proposed generalized model and fault-tolerant scheme, the AC-DC converter can autonomously recover from open-circuit faults and the continuous operation of three-phase AC-DC converters is maintained. Both the fault diagnosis and fault-tolerant schemes are based on the stationary frame, the phase-locked loop (PLL) is eliminated. Experimental results are presented to validate the effectiveness of the proposed integrated fault diagnosis and tolerant scheme.

I. INTRODUCTION

Two-level three-phase AC-DC pulse-width modulation (PWM) converters are extensively employed in many mission-critical applications such as data centers, more electric aircrafts and marine electric propulsion systems, due to their advanced merits including controllable power factor, control of dc output voltage and sinusoidal input current [1]–[4]. However, the failure in power devices such as the insulated-gate bipolar transistors (IGBTs) of this sort of converter can be hazardous. It is estimated that 38% of the failures in power conversion systems are due to the power devices fault [5], [6]. This is not desirable in many applications such as aerospace and military applications, where halt of the system after a fault will result in immeasurable economic losses.

To prevent the unscheduled shutdown, real-time fault-tolerant operation must be implemented. Fault diagnosis is the prerequisite of fault-tolerant operation. Extensive researches have been conducted to identify open-circuit fault with conventional modulation strategies [7]–[11]. However, the fault diagnosis methods proposed in these work are based on the conventional control scheme using linear controller with modulation techniques. To our best knowledge, little work is published on the fault diagnosis for AC-DC converter with model predictive control (MPC). The model predictive control is a nonlinear control scheme without modulation techniques. The switching state is directly calculated and the switching state is known in each control period. This feature

can be utilized for fast isolation of the open-circuit fault. To enhance the reliability of the AC-DC converter with MPC, a fault diagnosis method to identify single open-circuit fault is desired.

After the fault is isolated, the corresponding fault-tolerant schemes need to be conducted for post-fault operation. Extensive works on fault-tolerant scheme of the three-phase AC-DC converters have been published in the past literature. A parallel redundant leg was introduced for post-fault operation in [12]. However, this solution is volume-consuming and will increase the system cost. A fault-tolerant control scheme by simply revising the pulse-width modulation (PWM) switching pattern was proposed in [7]. However, the three-phase current was distorted and unbalanced in fault-tolerant mode. A three-phase four-switch topology which connects the corresponding faulty phase to the mid-point of dc-link was presented in [13]–[15]. However, the effort of these works was to improve the control performance AC-DC converters under fault-tolerant operation conditions. The control schemes for the normal operation and fault-tolerant operation were designed separately. The smooth transition from the faulty operation to the fault-tolerant operation are not considered in these works. It is impossible for the converter system with these control schemes to recover from the open-circuit fault autonomously [16] and the continuous operation of the converter cannot be maintained online after an open-circuit fault occurs. Therefore, the applications of these fault-tolerant schemes are limited.

To overcome these issues, a real-time fault diagnosis and tolerant scheme for three-phase AC-DC converters is proposed in this paper. An open-circuit fault diagnosis scheme, which takes advantages of the known and unchanged switching state in each control period in MPC, is proposed in this paper. The proposed fault diagnosis method is directly implemented by checking the mismatch between the converter voltages and the estimated ones. The fault diagnosis scheme is straightforward and no additional measurement is required. As long as the faulty switch conducts current, the open-circuit fault can be isolated very fast within several sampling periods. Generalized models of AC-DC converter models including voltage vectors, switching states and cost function for both healthy and faulty operations are proposed by introducing the healthy leg variables (HLVs). Experimental obtained data demonstrate

The 2018 International Power Electronics Conference

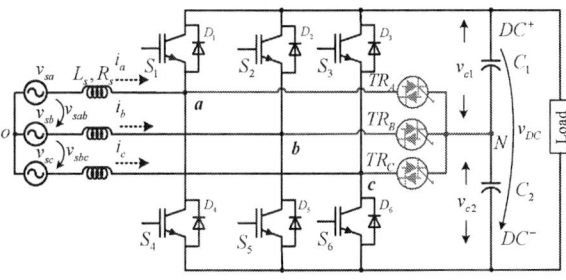

Fig. 1. Topology of a reconfigurable fault-tolerant three-phase AC-DC converter

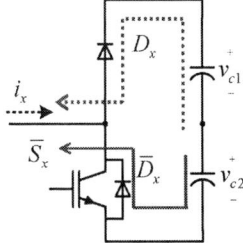

Fig. 2. Illustration of the equivalent circuit for the open-circuit failure in the switch S_x

the effectiveness of the proposed scheme.

II. FAULT DIAGNOSIS METHOD

Fig. 1 shows the fault-tolerant AC-DC PWM converter without redundancy. It consists of three source voltages (v_{sa}, v_{sb}, v_{sc}), three inductances L_s with resistances R_s, six IGBTs ($S_1, S_2, ..., S_6$) with anti-parallel diodes ($D_1, D_2, ..., D_6$), the dc-link capacitors C_1 and C_2. The two capacitor values are assumed to be identical $C_1 = C_2 = C$. The two-phase currents i_a, i_b and full dc-link voltage v_{DC} are measured directly. Each leg is connected to the mid-point of dc-link by bidirectional switches (e.g. relays). When an open-circuit fault occurs, the fault is initially identified by the fault diagnosis scheme. Then the faulty part is isolated from the normal part by removing the gate signal of the faulty leg switches. In fault-tolerant operation, the corresponding bidirectional switch (relay) will be fired to connect the phase with faulty switch to the midpoint of dc-link.

After an IGBT open-circuit fault occurs in phase $x(x \in \{a, b, c\})$, the equivalent circuit for the corresponding switching leg is illustrated in Fig. 2. As shown, the faulty IGBT is replaced with an anti-parallel diode. When the switching state is "1", the expected or measured pole voltage is v_{c1} and the current path is supposed to be in the dotted line direction. However, this current path is in the solid line because of the anti-parallel diode. The estimated switching state is "0" and the estimated pole voltage becomes $-v_{c2}$. Therefore, the supposed pole voltage in the faulty phase will differ from its actual value whenever the fault switch is supposed to conduct

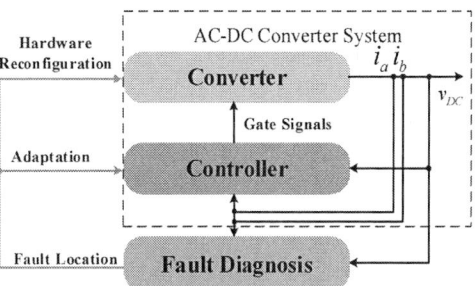

Fig. 3. Illustration of the fault-tolerant AC-DC converter system without redundancy

current. This feather can be utilized to locate the open-circuit fault in AC-DC converters with MPC.

The instant voltage error between the measured and estimated pole voltages in phase x is given by

$$\Delta u_{xN} = u_{mxN} - u_{exN} \tag{1}$$

In order to avoid additional hardware, the measured voltages are obtained directly by the switching combination and measured dc-link voltage. In the MPC scheme, the switching state is known in each sampling period. Therefore, the measure pole voltage can be written as

$$u_{mx} = \begin{cases} v_{c1}, & S_x = 1 \\ -v_{c2}, & S_x = 0 \end{cases} \tag{2}$$

The estimated pole voltage are obtained by the measured current and source voltage. However, the accuracy of phase-to-neutral voltage is affected by the common mode voltage. Therefore, the error of line-to-line voltage $xy(xy \in \{ab, bc, ca\})$ is utilized to perform the analysis of the pole voltage error. The line-to-line voltage errors are estimated as

$$\Delta u_{xy} = (u_{mx} - u_{my}) - \left(v_{xy} - (i_x - i_y) \cdot R_s - L_s \cdot \frac{d(i_x - i_y)}{dt} \right) \tag{3}$$

where the first term is the measured value of the line-to-line voltage. The second term is the estimated line-to-line voltage derived from rectifier model. This error is normalized to make the algorithm independent from the dc-link voltage

$$\varepsilon_{xy} = \frac{\Delta u_{xy}}{v_{DC}} \tag{4}$$

These three normalized instant voltage errors can be utilized for fault location. The line-to-line voltage of the unaffected phase is always zero while the variables of the two affected phase are equal to $\frac{\Delta u_{xy}}{v_{DC}}$ or $-\frac{\Delta u_{xy}}{v_{DC}}$. Based on this feature, the open-circuit faulty phase can be located by introducing a threshold T_d. ε_{xy} approximates 0 in normal conditions while approximate 1 or -1 in open-circuit conditions. Therefore, T_d can be set to a value near 1 to locate the fault (0.9 is adopted in this paper).

435

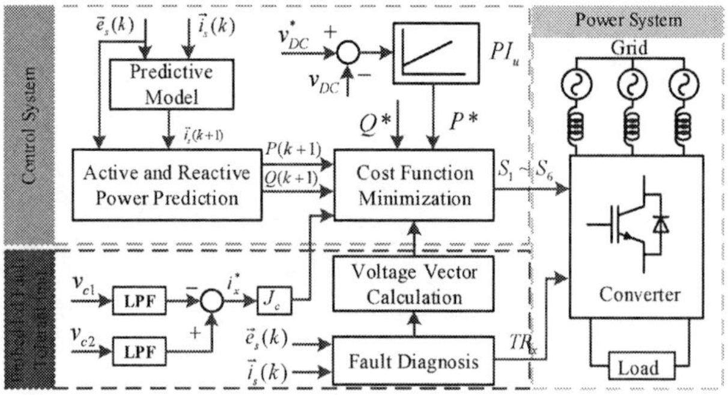

The 2018 International Power Electronics Conference

Fig. 4. Control diagram of the proposed fault diagnosis and tolerant scheme

III. INTEGRATED FAULT-TOLERANT SCHEME

The fault-tolerant AC-DC converter system is illustrated in Fig. 3. This system consists of three parts, i.e., fault diagnosis, hardware reconfiguration, and control scheme adaptation. The fault is first located by the fault diagnosis method. Then, the converter is reconfigured by connecting the faulty phase to the mid-point of dc-link for fault-tolerant operation according to the fault location information.

The integrated fault-tolerant control scheme for AC-DC PWM rectifiers is mainly composed of five parts, i.e., generalized voltage vectors, generalized switching combinations, active and reactive power prediction, capacitor voltage balancing and generalized cost function design.

1) Voltage Vector Calculation: In order to carry out the generalized model for both normal and fault-tolerant operation, the HLVs can be introduced as

$$h_x = \begin{cases} 1, & x \text{ phase is healthy} \\ 0, & otherwise \end{cases}, (x \in \{a, b, c\}). \quad (5)$$

In normal operation, there are eight switching combinations and voltage vectors while only four are available in fault-tolerant operation. By introducing HLVs, the generalized pole voltage of each phase can be given as

$$v_{xo} = h_x \cdot (S_x \cdot v_{c1} - \bar{S}_x \cdot v_{c2}) \quad (6)$$

where v_{c1} is the upper capacitor voltage and v_{c2} is the lower capacitor voltage. The voltage vectors for both normal and fault-tolerant operations can be obtained by the Clarke transform.

2) Switching Combinations: When an open circuit occurs, the faulty leg is isolated by removing the gate signal of the faulty leg switches. Therefore, the generalized switching command of each phase can be given as

$$\begin{cases} \gamma_x = h_x \cdot S_x \\ \bar{\gamma}_x = h_x \cdot \bar{S}_x \end{cases} \quad (7)$$

where γ_x and $\bar{\gamma}_x$ ($x \in \{a, b, c\}$) are the generalized switching commands for the upper and lower switch of each phase, respectively.

3) Active and reactive power prediction: By using the three-phase to two-phase transformation, the predictive model of the AC-DC PWM rectifier can be calculated as

$$\vec{i}_s(k + 1) = (1 - \frac{R_s T_s}{L_s})\vec{i}_s(k) + \frac{T_s}{L_s}\vec{v}_s(k) - \frac{T_s}{L_s}\vec{u}_s(k) \quad (8)$$

where $\vec{v}_s, \vec{i}_s, \vec{u}_s$ are the grid voltage vector, current vector, rectifier voltage vector, respectively. T_s is sampling time.

Considering the input voltage and current vectors in orthogonal coordinates, the predicted instantaneous active and reactive power can be calculated as

$$P(k + 1) = 1.5\text{Re}\{\vec{v}_s(k + 1)\bar{\vec{i}}_s(k + 1)\} \quad (9)$$

$$Q(k + 1) = 1.5\text{Im}\{\vec{v}_s(k + 1)\bar{\vec{i}}_s(k + 1)\} \quad (10)$$

4) Capacitor voltage balancing: In the fault-tolerant operation, the faulty phase current flows through the midpoint of the dc-link,which will lead the unbalanced capacitor voltage. The unbalanced capacitor voltage will reduce the dc-link voltage utilization and degrade the performance of the converter.

The cost function component for capacitor voltage balancing can be written as

$$J_c = \frac{\left| i_{cap}^{ref} - i(k + 1) \right|}{I_{nom}} \quad (11)$$

where $i_{cap}^{ref} = k_p(\bar{v}_{c2} - \bar{v}_{c1})$, $\bar{v}_{c1}$ and $\bar{v}_{c2}$ are the filtered values of upper and lower capacitor voltage, respectively. k_p is an adjustable coefficient and I_{nom} is the rated phase current.

The generalized model for capacitor voltage balancing can be given as

$$J_c = \frac{\left| i_{cap}^* - \Sigma((1 - h_x) \cdot i_x(k + 1)) \right|}{I_{nom}}, x \in \{a, b, c\} \quad (12)$$

By adopting this generalized model, J_c will be zero when the converter is in health state.

5) Cost Function Design: In the predictive power control scheme, no internal control loops and no modulator are required. The currents are directly regulated by controlling the input active and reactive power. All the voltage vectors are evaluated to control active and reactive power to track the reference value of P^{ref} and Q^{ref}. Therefore, the cost function component for the active and reactive power control can be given as

$$J_r = \frac{\left| P^{ref} - P(k + 1) \right| + \left| Q^{ref} - Q(k + 1) \right|}{P_{nom}} \quad (13)$$

where P_{nom} is the rated output power.

Taking both the normal operation and fault-tolerant operation into consideration, the cost function for both modes can be written as

$$J = J_r + \lambda_c J_c \quad (14)$$

where λ_c is the weighting factor.

436

The 2018 International Power Electronics Conference

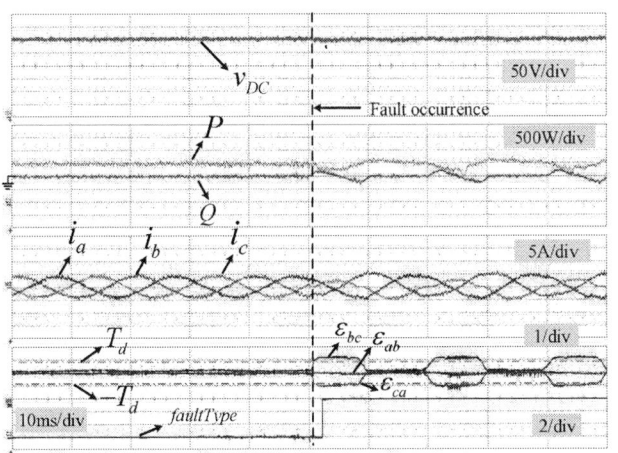

Fig. 5. Fault diagnosis results in open-circuit fault case. From top to bottom, dc-link voltage, active and reactive power, three-phase current, instant voltage error, fault type

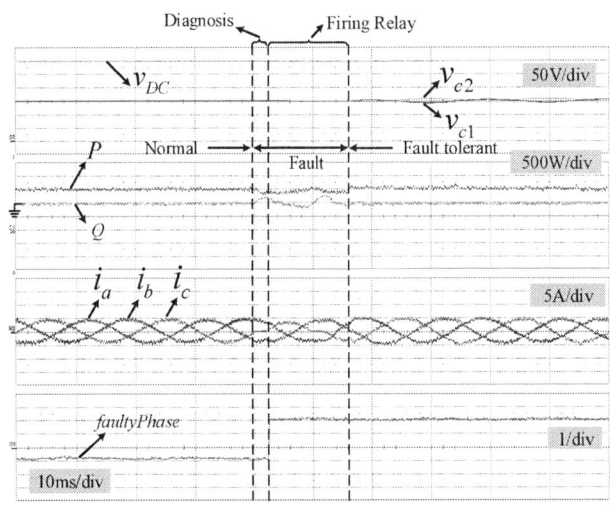

Fig. 6. Fault-tolerant results. From top to bottom, dc-link voltage, C_1 and C_2 voltage, active and reactive power, three-phase current, faulty phase

IV. EXPERIMENTAL RESULTS

In this section, experimentally obtained data are presented to validate the the proposed open-switch fault diagnosis and tolerant scheme of the AC-DC converter. The proposed fault diagnosis and fault tolerant schemes are implemented in DSP TMS320F28335. The triacs utilized for topology reconfiguration are relays (finder AC250V20A). The filter inductance for the experimental test is 10 mH with equivalent resistance 0.5 Ω. The source voltage is 50 V with frequency of 50 Hz. The sampling frequency for both schemes is 20 kHz. The capacitance of C_1 and C_2 is 1410 μF.

1) Fault Diagnosis Results: Experimental results of an open-circuit fault in S_6 are shown in Fig.5. The instant voltage errors stay around zero until the fault occurs at $t = 0.05$ s. The voltage errors ε_{bc} and ε_{ca} deviate to the opposite direction while ε_{ab} remains unchanged. ε_{bc} surpasses the predefined upper threshold and ε_{ca} surpasses the predefined lower threshold which indicate the fault of S_6. In this way, the fault type can be identified. As shown in Fig. 5, the currents are distorted and significant ripples are shown in active and reactive power after the open-circuit fault occurs.

2) Fault Tolerant Results: The control performance of the integrated fault-tolerant control scheme is presented in Fig.6. In the figure, it can be clearly seen that the topology is reconfigured to the fault tolerant topology (connecting the phase with faulty switch to the mid-point of the DC-link) very quickly by firing the corresponding relay. After the fault identification, the control scheme is adapted in the next sampling period, which means that the adaptation of the control scheme can be implemented seamlessly. However, it takes several sampling periods to adapt to the fault tolerant mode due to the relay adopted for hardware reconfiguration is a relatively slow actuator. The transient from faulty state to postfault state takes about 20 ms, 2 ms for fault diagnosis and 18 ms for firing the relay. The fault tolerant control scheme can achieve almost the same control performance of dc-link voltage, active and reactive power tracking as that in normal operation. As shown in Fig.6, the control scheme for normal converter can be adjusted to fault-tolerant one seamlessly.

V. CONCLUSION

To improve the reliability of AC-DC rectifiers, an integrated open-circuit fault diagnosis and tolerant scheme based on MPC has been proposed in this paper. A robust open-circuit fault diagnosis scheme based on instant voltage error is integrated into MPC for online fault diagnosis. By introducing the HLVs, a generalized model including both healthy and faulty operations is proposed and the fault-tolerant scheme can be integrated into the normal control scheme. With the proposed fault diagnosis and tolerant scheme, the three-phase AC-DC converter can autonomously recover from open-circuit faults and continuous operation of the converter is maintained. Experimental results showed the fast and seamless transition from the fault occurrence to fault-tolerant operation. The proposed scheme provides an easy, compact and real-time solution to the reliability issue of AC-DC converters.

ACKNOWLEDGEMENT

This research is funded by the Singapore Maritime Institute under the Maritime Research between Norway and Singapore (MNS) R&D Programme - Project SMI-2015-MA-15.

REFERENCES

[1] J. Hu and Z. Q. Zhu, "Improved voltage-vector sequences on dead-beat predictive direct power control of reversible three-phase grid-connected voltage-source converters," *IEEE Trans. Power Electron.*, vol. 28, no. 1, pp. 254–267, Jan. 2013.

[2] D. Zhou, P. Tu, and Y. Tang, "Multi-vector model predictive power control of three-phase rectifiers with reduced power ripples under nonideal grid conditions," *IEEE Trans. Ind. Electron.*, vol. PP, no. 99, p. 1, 2018.

[3] S. C. Shin, H. J. Lee, Y. H. Kim, J. H. Lee, and C. Y. Won, "Transient response improvement at startup of a three-phase AC/DC converter for a DC distribution system in commercial facilities," *IEEE Trans. Power Electron.*, vol. 29, no. 12, pp. 6742–6753, Dec. 2014.

[4] Y. Suh and T. A. Lipo, "Control scheme in hybrid synchronous stationary frame for PWM AC/DC converter under generalized unbalanced operating conditions," *IEEE Trans. Ind. Appl.*, vol. 42, no. 3, pp. 825–835, May 2006.

[5] D. Zhou, Y. Li, J. Zhao, F. Wu, and H. Luo, "An embedded closed-loop fault-tolerant control scheme for nonredundant vsi-fed induction motor drives," *IEEE Trans. Power Electron.*, vol. 32, no. 5, pp. 3731–3740, 2017.

[6] D. Zhou, J. Zhao, and Y. Liu, "Independent control scheme for nonredundant two-leg fault-tolerant back-to-back converter-fed induction motor drives," *IEEE Trans. Ind. Electron.*, vol. 63, no. 11, pp. 6790–6800, Nov. 2016.

[7] W. S. Im, J. M. Kim, D. C. Lee, and K. B. Lee, "Diagnosis and fault-tolerant control of three-phase AC –DC PWM converter systems," *IEEE Trans. Ind. Appl.*, vol. 49, no. 4, pp. 1539–1547, Jul. 2013.

[8] L. Tian, F. Wu, and J. Zhao, "Current kernel density estimation based transistor open-circuit fault diagnosis in two-level three phase rectifier," *Electron. Lett.*, vol. 52, no. 21, pp. 1795–1797, 2016.

[9] I. Jlassi, J. O. Estima, S. K. E. Khil, N. M. Bellaaj, and A. J. M. Cardoso, "Multiple open-circuit faults diagnosis in back-to-back converters of pmsg drives for wind turbine systems," *IEEE Trans. Power Electron.*, vol. 30, no. 5, pp. 2689–2702, May 2015.

[10] F. Wu and J. Zhao, "Current similarity analysis-based open-circuit fault diagnosis for two-level three-phase pwm rectifier," *IEEE Trans. Power Electron.*, vol. 32, no. 5, pp. 3935–3945, 2017.

[11] L. M. A. Caseiro and A. M. S. Mendes, "Real-time IGBT open-circuit fault diagnosis in three-level neutral-point-clamped voltage-source rectifiers based on instant voltage error," *IEEE Trans. Ind. Electron.*, vol. 62, no. 3, pp. 1669–1678, Mar. 2015.

[12] S. Karimi, A. Gaillard, P. Poure, and S. Saadate, "FPGA-based real-time power converter failure diagnosis for wind energy conversion systems," *IEEE Trans. Ind. Electron.*, vol. 55, no. 12, pp. 4299–4308, Dec. 2008.

[13] Z. Zeng, W. Zheng, R. Zhao, C. Zhu, and Q. Yuan, "Modeling, modulation, and control of the three-phase four-switch PWM rectifier under balanced voltage," *IEEE Trans. Power Electron.*, vol. 31, no. 7, pp. 4892–4905, Jul. 2016.

[14] T. S. Lee and J. H. Liu, "Modeling and control of a three-phase four-switch PWM voltage-source rectifier in d-q synchronous frame," *IEEE Trans. Power Electron.*, vol. 26, no. 9, pp. 2476–2489, Sep. 2011.

[15] D. Zhou, X. Li, and Y. Tang, "Multiple-vector model predictive power control of three-phase four-switch rectifiers with capacitor voltage balancing," *IEEE Trans. Power Electron.*, vol. PP, no. 99, p. 1, 2017.

[16] W. Zhang, D. Xu, P. N. Enjeti, H. Li, J. T. Hawke, and H. S. Krishnamoorthy, "Survey on fault-tolerant techniques for power electronic converters," *IEEE Trans. Power Electron.*, vol. 29, no. 12, pp. 6319–6331, Dec. 2014.

The Lifetime Assessment of a Micro-Inverter for PV Applications

Tohihiro Shimao[1*], Koji Kato[1], Youichi Ito[1], Akio Iwabuchi[1], Yongheng Yang[2] and Frede Blaabjerg[2]

1 Sanken Electric Co. Ltd., Saitama, Japan
2 Department of Energy Technology, Aalborg University, Aalborg, Denmark
*E-mail: tshimao@sanken-ele.co.jp

Abstract— **Recently, introducing grid-tied PV systems is recommended in many countries. The inverter lifetime in PV applications is affected by their operating conditions. That is affected by the thermal loading of the power devices and capacitors. Especially, the lifetime of electrolytic capacitors at the DC-link to compensate the power ripple decreases when the ambient temperature increases. To improve the lifetime, a micro-inverter with an active buffer at the DC-link is introduced in this paper. In this micro-inverter, the DC-link electrolytic capacitor is replaced with a ceramic capacitor that has small capacitance. However, the reliability performance of the entire micro-inverter has not been evaluated. This paper thus assesses the lifetime of the micro-inverter considering the operational conditions (i.e., ambient temperature and solar irradiance, also referred to as mission profiles). A comparison between the proposed circuit with an active buffer and the conventional circuit based on a boost chopper is performed in this paper. It is revealed that the lifetime of the micro-inverter is very long. This is due to the over-designed power devices.**

Keywords— Lifetime estimation, Micro inverter, PhotoVoltaic

I. INTRODUCTION

The photovoltaic (PV) technology has a potential to become a major energy source in the future, and it has experienced a high growth rate during the last decades. As more PV systems have been installed and connected to the grid, their reliability and lifetime are gaining more and more attention [1]–[5]. With the recent technology, the lifetime of PV panels is normally warranted at 20-25 years, while the PV inverter lifetime is usually limited to 10-20 years. On the other hand, such short lifetime can be improved by redesigning the inverters with advanced power semiconductors and capacitors.

With this background, a micro-inverter is designed to achieve a continuous service period of more than 25 years in this paper. The micro-inverter adopts an active buffer at the DC-link [6], which compensates power ripples, to replace the conventional electrolytic capacitors. The active buffer enables using small ceramic capacitors, and thus, it possibly contributes to extended lifetime .

For an efficient design, the lifetime of the micro-inverter should be estimated considering the degradation of components. Thus, a lifetime estimation of power devices in the micro-inverter is also presented in this

paper. Failure mechanisms including solder fatigue and bond-wire liftoff are considered. Those are closely related to the thermal loading on the power devices and also the operating conditions (referred to as mission profiles). However, due to the lack of experimental lifetime testing data and reliability models, this paper evaluates the lifetime of the micro-inverter through its thermal performance. In the assessment, mission profiles are also considered. Finally, the total lifetime estimation including other components and total lifetime are compared between the conventional boost chopper-based circuit and the proposed active buffer-based circuit.

II. SYSTEM DESCRIPTION

Fig. 1 shows one of the conventional PV inverters. The conventional PV inverter is based on the boost chopper and the voltage source inverter. The inverter requires a large electrolytic capacitor for compensation of the power fluctuation with twice of the grid frequency. Electrolytic capacitors have shorter lifetime than film capacitors or ceramic capacitors due to fatigue of their sealing cover.

Fig. 2 shows the micro-inverter with an active buffer. This inverter consists of a resonant isolated DC/DC converter, an active buffer, and a current source inverter (CSI). Due to the active buffer, the power oscillation between DC and AC can be actively compensated. As a result, a small capacitor (denoted as C_{ab}) can be adopted. Additionally, the CSI can be switched with a high frequency (above 100 kHz), leading to an increased

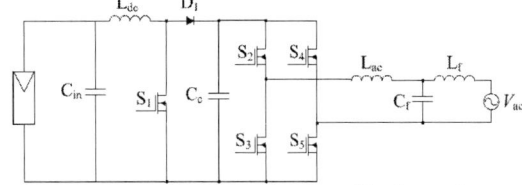

Fig. 1. A conventional micro-inverter with a boost chopper.

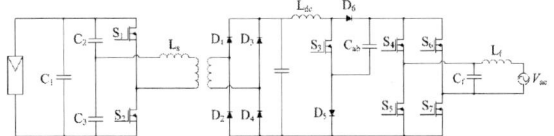

Fig. 2. A micro-inverter with an active buffer for PV applications.

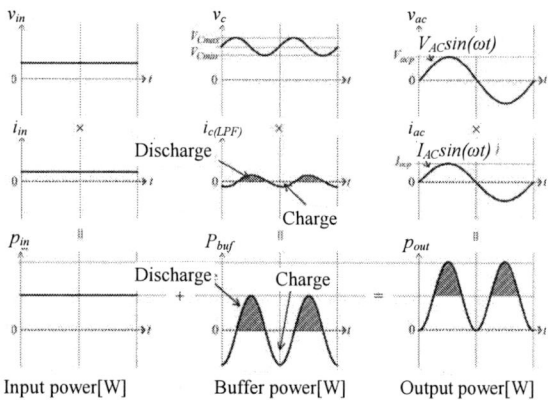

Fig. 3. Principle of the power ripples compensation.

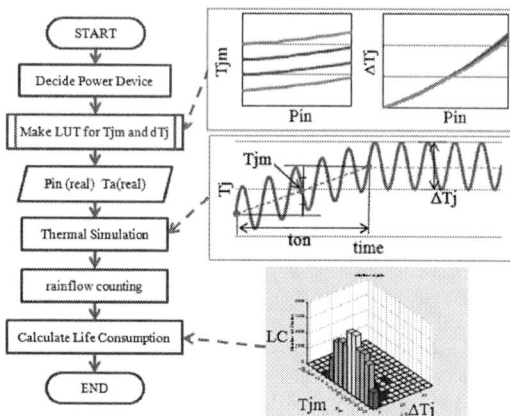

Fig. 4. Flowchart of the lifetime estimation for power devices.

power density.

More details of this micro-inverter are given as follows. Firstly, the resonant isolated DC/DC converter achieves a zero-current switching operation with arm capacitors C_1, C_2 and the inductor L_s. Noted that, the switching frequency is set at the resonant frequency, and the primary side devices S_1 and S_2 operates with a duty cycle of 50%. A high resonant frequency (and also the switching frequency) is designed in order to reduce the transformer volume, and thus the entire system volume. The active buffer compensates the power ripple with the small capacitor C_{ab} and rectifies the high frequency resonant current. The power devices of the active buffer and the CSI, i.e., S_3-S_7, are controlled coordinately to achieve both power decoupling and the Maximum Power Point Tracking (MPPT). All capacitors in the inverter are ceramic, which is benefited from the active buffer.

Fig. 3 shows the principle of the power ripple compensation. When both the grid voltage v_{ac} and the output current i_{ac} are sinusoidal at the unity power factor operation, the instantaneous output power p_{out} is also a sinusoidal wave with twice the fundamental grid frequency, as shown in Fig. 3. Thus, the instantaneous power of the active buffer p_{buf} should be the same as the output power p_{out}. However, a negative offset whose value is the same as the input power p_{in} is added to p_{buf}. By doing so, the power oscillation is mitigated.

III. LIFETIME ESTIMATION PROCEDURE

A procedure to estimate the lifetime of PV inverters is presented in this section considering mission profiles. The lifetime of the critical components in PV inverters such as power devices and capacitors is related to temperature variations [3], [4]. Therefore, the mission profile should be considered when evaluating the thermal performance of PV inverters [5]. Then, the required data by the lifetime model such as the mean junction temperature T_{jm} and the cycle amplitude ΔT_j can be extracted from the thermal loading profile by the rainflow counting algorithm. Finally, the specific lifetime model can be applied, and the lifetime can be estimated.

Fig. 4 shows the evaluation flowchart. Notably, the lifetime models for power devices may be different.

A. Mission Profile Translation to Thermal Loading

First, the inverter input power P_{in} and the ambient temperature T_a can be translated to the thermal profiles (T_{jm} and ΔT_j) by simulations with a Look-Up Table (LUT). This LUT is generated from the power device characteristic and the thermal impedance given in the datasheet in order to assist a long-term simulation [7].

B. Rainflow Cycle Counting

The junction temperature variation obtained from the previous step is an irregular profile according to the mission profile. In order to apply a junction temperature profile to the lifetime model, a counting algorithm such as a rainflow counting method is needed [7]-[9]. The rainflow counting algorithm is used in the lifetime and stress analysis related to the thermal cycling. This algorithm can partition the irregular profile into several regular cycles. Each regular cycle can be categorized according to the cycle amplitude, its average value, and the cycle period, as it is shown in Fig. 4. Then, the number of cycles n_i is obtained according to the categorized data (e.g., the cycle amplitude ΔT_j, mean junction temperature T_{jm}, and cycle period t_{on}).

C. Lifetime Model of Power Devices

The lifetime model of the power device is given as

$$
N_f = A\Delta T_j^{-\alpha} \cdot \exp\left(\frac{E_a}{k_B T_{jm}}\right) \cdot ar^{\beta_1 \Delta T_j + \beta_0} \cdot \left(\frac{C + (t_{on})^\gamma}{C}\right) \cdot f_d
$$

(1)

which is based on the Semikron model [9]. Here, N_f is the number of cycles to failure, T_{jm} is the mean junction temperature, ΔT_j is the cycle amplitude, and t_{on} is the cycle period. The other parameters of the lifetime model in (1) are given in Table I.

TABLE I
PARAMETERS OF THE LIFETIME MODEL

Symbol	Value
A	3.4368×10^{14}
α	-4.923
β_1	-9.012×10^{-3}
β_1	1.942
C	1.434
γ	-1.208
f_d	0.6204
E_a	0.06606eV
k_B	8.6173×10^{-5}eV/K
ar	0.29

The Lifetime Consumption (LC) is then calculated according to the Miner's rule [11] as

$$LC = \sum_i \frac{n_i}{N_{fi}} \qquad (2)$$

where n_i is the number of cycles obtained from the rainflow analysis and N_{fi} is the number of cycles to failure calculated from (1) at that specific stress condition. The LC can be used to indicate how much the life of the power device is consumed during operation. For instance, if the number of cycles n_i is counted from an annual mission profile, the LC calculated in (2) will represent a yearly LC of the power device. The lifetime of the power device is then determined when the LC accumulates to the unity, which is when the device reaches its end of life.

D. Lifetime Model of an Electrolytic Capacitor

Fig. 5 shows the diagram of this evaluation process for the electrolytic capacitor. The temperature increment ΔT is calculated from the power loss is given as

$$\Delta T = \frac{I_R{}^2 R}{\beta A} = \frac{I_R{}^2 \tan \delta}{\beta A \omega C} \qquad (3)$$

where I_R is the ripple current through the capacitor, R is the Equivalent Series Resistance (ESR), $\tan\delta$ is the dissipation factor at 120 Hz, β is the radiation coefficient obtained from datasheet, A is surface area of the capacitor, ω is the frequency (120 Hz), and C is the capacitance.

The lifetime model of an electrolytic capacitor is based on the Arrhenius equation as

$$L = L_0 \times 2^{\frac{T_0 - T_x}{10}} \times 2^{\frac{\Delta T_0 - \Delta T}{5}} \qquad (4)$$

in which L_0 is the lifetime at the upper limit of the category temperature range and at the rated ripple current superimposed to a DC voltage, T_0 is the upper limit of the category temperature range, T_x is ambient temperature, ΔT_0 is the rise of the hotspot temperature due to the rated ripple current and ΔT is the internal temperature rise due to the actual ripple current.

Eqs. (3) and (4) are from the technical note of Nippon Chemi-con [12].

E. Monte Carlo Simulation and Weibull Distribution

The ideal lifetime is calculated by lifetime estimation

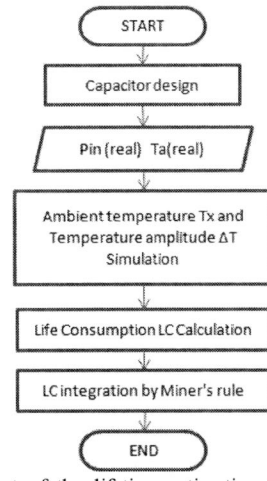

Fig. 5. Flowchart of the lifetime estimation for electrolytic capacitors. [10]

procedures as aforementioned. However, there will be deviations in the parameters caused by the manufacturing process or aging. For a more realistic lifetime evaluation, the deviations are introduced in the procedure through the Monte Carlo simulations [13]-[16] T, which is an effective mean to the component parameter-drifting. In this paper, 1000 samples are calculated with 5% of parameter variations in Eqs. (1) and (4). After that, the Probability Density Function (PDF) is obtained with a Weibull distribution [24] from the 1000 samples. The PDF is given as

$$f(x) = \frac{\beta}{\eta^\beta} x^{\beta-1} \exp\left(-\left(\frac{x}{\eta}\right)^\beta\right) \qquad (5)$$

where β is the shape parameter, η is the scale parameter, and x is the operation time. The unreliability can be evaluated by considering the Cumulative Density Function (CDF) of the Weibull distribution. This Weibull CDF $F(x)$ is normally referred to as the unreliability function, which can be obtained as

$$F(x) = \int_0^x f(x)dx \qquad (6)$$

Finally, the total unreliability is estimated with the integration of the unreliability of all components as

$$F_{total}(x) = 1 - \prod_{i=1}^n (1 - F_i(x)) \qquad (7)$$

Thus, the B_x lifetime can be obtained. Here, B_x gives the time at which $x\%$ of the units in the population will have failed. In practice, the B_{10} lifetime, which indicates lifetime when the unreliability reaches 10%, is used as an assessment standard.

IV. LIFETIME EVALUATION OF EACH STAGE

The lifetime evaluation of the proposed micro-inverter is discussed in this section. The lifetime estimation results of the inverter with various T_{jm} and ΔT_j are compared.

441

The 2018 International Power Electronics Conference

TABLE II
PARAMETERS OF THE MICRO-INVERTERS

Output power	P_{out}	300 W
Output voltage	V_{ac}	200 V
Switching frequency of DC/DC converter in proposed circuit	f_{sw1}	200 kHz
Switching frequency of active buffer and inverter in proposed circuit	f_{sw2}	40 kHz
Trans turn ratio	$N1:N2$	1:1
Switching frequency of active buffer and inverter in conventional circuit	f_{sw}	40 kHz

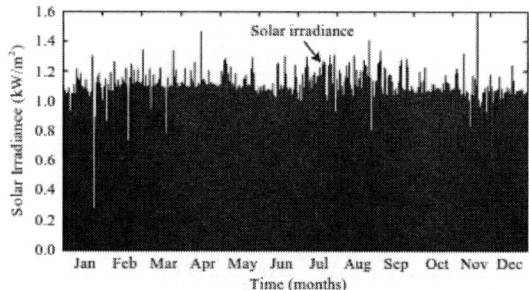

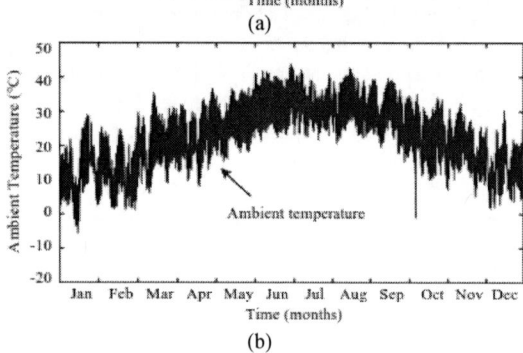

Fig.6. Annual mission profile in Arizona: (a) solar irradiance and (b) ambient temperature.

A. Estimation Conditions

Table II shows each parameters of the micro inverter. A yearly mission profile of Arizona is adopted, which is shown in Fig. 6. The annual data (ambient temperature and solar irradiance) was recorded in Arizona. The LUT of T_{jm} and ΔT_j is shown in Fig. 7. Three characteristics are generated in the LUT by changing the thermal impedance considering uncertainties. The existing characteristic is the calculated temperature based on the conduction loss and the switching loss of components with Plexim simulations. The other characteristics are calculated temperatures based on the existing characteristic for comparison.

Fig. 8 shows the LUT data of each component. It can be implied in Fig. 8 that ΔT_j depends on the conduction loss and the switching loss of each component. However, ΔT_j of the power device in the DC/DC is 0.05°C because the conduction loss is reduced with a high switching frequency and the switching loss is almost zero due to the ZCS. Furthermore, the ΔT_j characteristic of the active buffer is similar to that of the chopper. In terms of

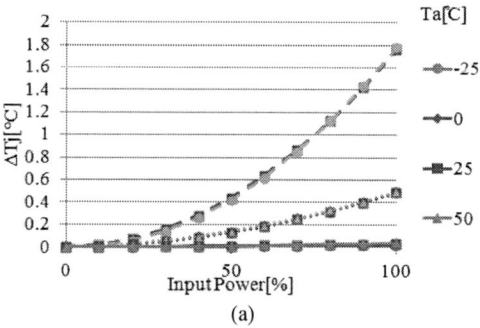

(a)

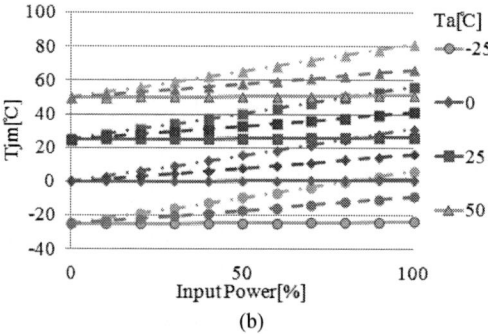

(b)

Fig. 7. LUT data of the inverter switches of the proposed circuit for lifetime evaluation: (a) ΔT_j (solid line: the existing characteristic, dash line: 3 times of the existing, dot-dash line: 5 times of the existing) and (b) T_{jm} (solid line: the existing characteristic, dash line: added 15 °C to the existing, dot-dash line: added 30 °C to the existing).

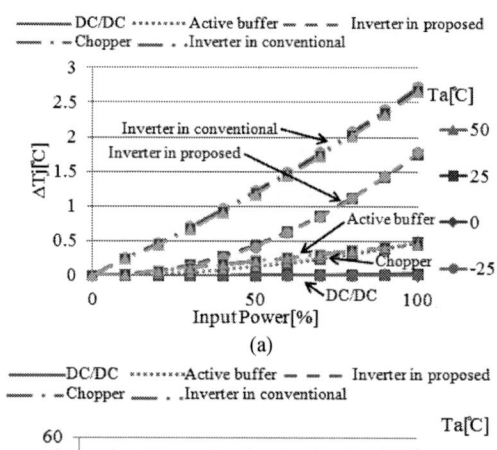

(a)

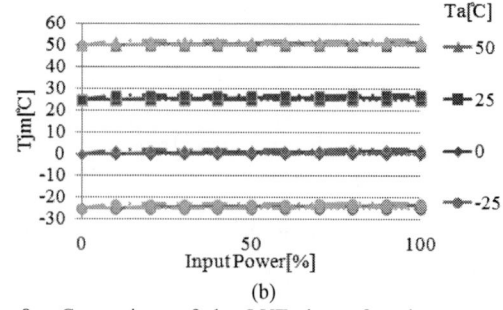

(b)

Fig. 8. Comparison of the LUT data of each component between the proposed circuit and the conventional circuit at the existing condition: (a) ΔT_j and (b) T_{jm}.

442

The 2018 International Power Electronics Conference

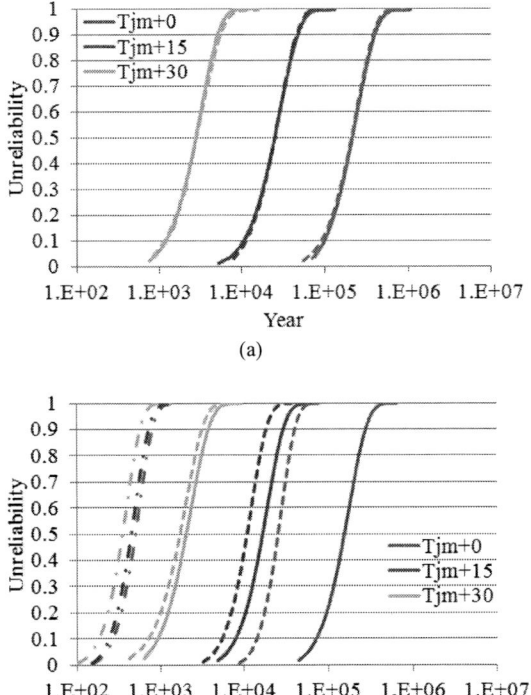

(a)

(b)

Fig. 9. Unreliability of the power devices of the conventional circuit by the Monte-Carlo simulations: (a) unreliability of the DC/DC switch (solid line: existing characteristic, dash line: 3 times of the existing, dot-dash line: 5 times of the existing) and (b) unreliability of the inverter switch (solid line: existing characteristic, dash line: 3 times of the existing, dot-dash line: 5 times of the existing).

inverter switches, ΔT_j of the proposed circuit is 1-°C less than that of the conventional circuit. This is because the reduced voltage fluctuation with the active buffer leads to a loss reduction in the proposed circuit. Meanwhile, the characteristics of T_{jm} are similar, as shown in Fig. 8(b).

B. Simulation Results of Switching Devices

Lifetime estimation results are shown in Figs. 9 and 10. It can be observed that the original B_{10} lifetime is significantly high, because the devices have a large design margin (100 times) in terms of current and voltage. That is, the power devices in the analysis are for high power applications.

Nevertheless, the lifetime of the power devices decreases along with the increase of T_{jm} in the DC/DC stage, active buffer stage and chopper stage. Their characteristics have few deviations. However, in the inverter stage, the lifetime becomes dependent on ΔT_j. The lifetime of inverter switches in the conventional circuit is significantly different when its ΔT_j is 3 times of the existing. In addition, the lifetime of the inverter switches in the proposed circuit is also different when its ΔT_j is 5 times of the existing. Thus, the effective value of ΔT_j is conjectured at 7.5°C in their lifetime model.

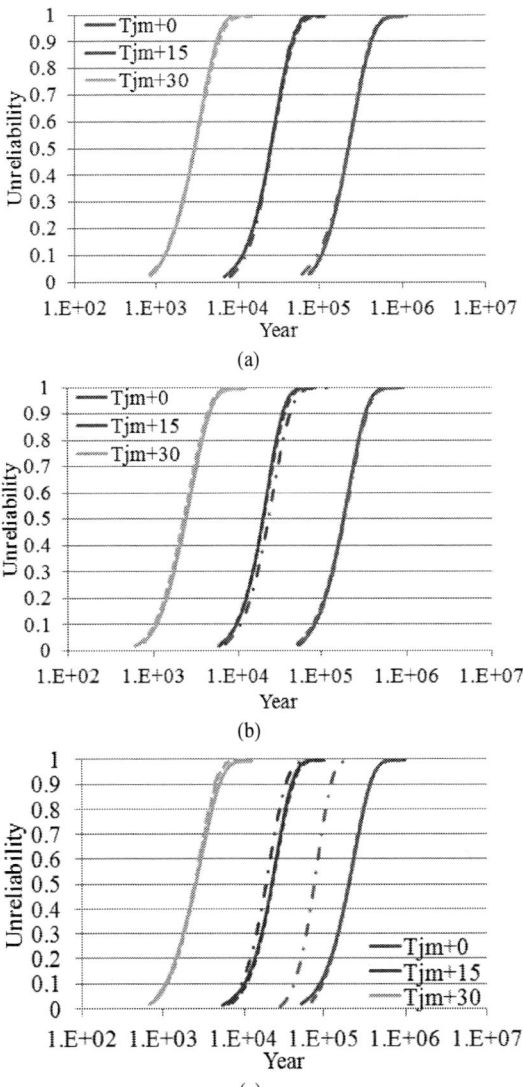

(a)

(b)

(c)

Fig. 10. Unreliability of the power devices of the proposed circuit by the Monte-Carlo simulations: (a) unreliability of the DC/DC switch (solid line: existing characteristic, dash line: 3 times of the existing, dot-dash line: 5 times of the existing), (b) unreliability of the active buffer switch (Solid line: existing characteristic, dash line: 3 times of the existing, dot-dash line: 5 times of the existing), and (c) unreliability of the inverter switch (Solid line: existing characteristic, dash line: 3 times of the existing, dot-dash line: 5 times of the existing).

Fig. 11 shows the total lifetime estimation results obtained from Eq. (6). The B_{10} lifetime with the existing condition is very long in both circuits due to overdesign.

However, the total lifetime of the conventional circuit depends on T_{jm} rather than ΔT_j due to the shorter lifetime of the inverter. The total lifetime of the conventional circuit is tenth of that of the proposed circuit in the higher ΔT_j. Thus larger heat sink is needed to equalize the lifetime of the conventional circuit to the lifetime of the proposed circuit. This indicates that the proposed circuit has long lifetime but reduced requires of heatsink. It is

443

The 2018 International Power Electronics Conference

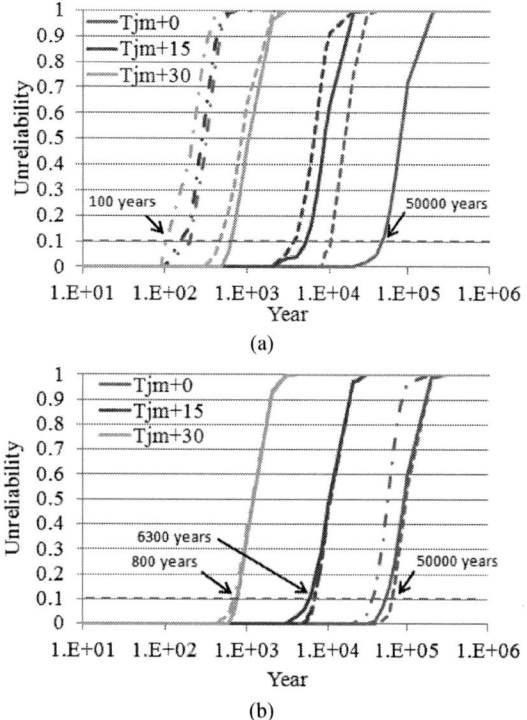

(a)

(b)

Fig. 11. Lifetime estimation results with all switching devices: (a) unreliability of the conventional circuit (solid line: existing characteristic, dash line: 3 times of the existing, dot-dash line: 5 times of the existing) and (b) unreliability of the proposed circuit (Solid line: existing characteristic, dash line: 3 times of the existing, dot-dash line: 5 times of the existing).

TABLE III
SPECIFICATIONS FOR THE DC-LINK CAPACITOR DESIGN.

Rated voltage V_{dc}	< 300V
Capacitance C	< 120uF
Rated ripple current I_R	< 2.11A_{rms}
Voltage ripple rate	2.5%

TABLE IV
CHARACTERISTICS OF THE ELECTROLYTIC CAPACITORS.

No.	V_{dc}[V]	C[uF]	I_R[A_{rms}]	β	tanδ	A[cm²]
1		330	2.11			36.9
2		390	2.37			45
3	450	470	2.66	1.58	0.2	44.8
4		560	2.93			49.4
5		680	3.21			59.1
6		820	3.60			64.6

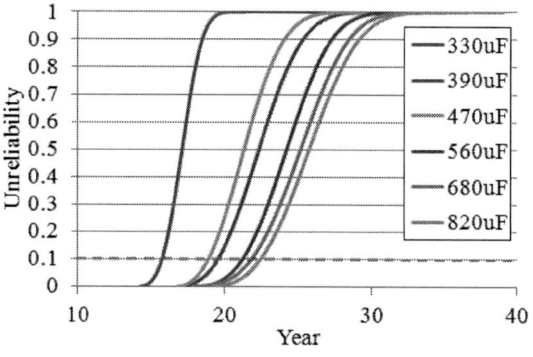

(a)

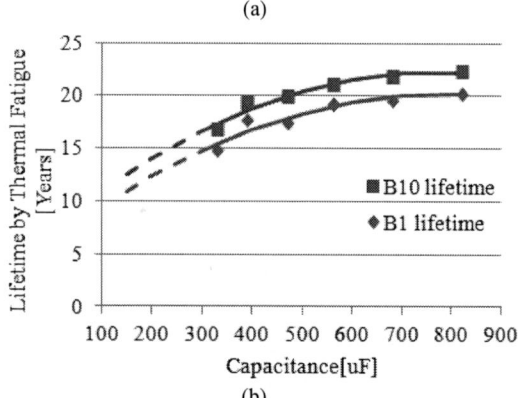

(b)

Fig. 12. Lifetime estimation result of the DC-link capacitor: (a) unreliability of the DC-link capacitor and (b) lifetime characteristics with different capacitance.

worth mentioning that although the total lifetime of the switching devices is very long in both circuits, which is far more than 100 years. The total lifetime of the inverter depends on the presence or absence of the DC-link capacitor.

C. Capacitor design and lifetime evaluation

Table III shows the specifications of the DC-link capacitor for the conventional circuit. The capacitor is designed for ripple currents of L_{dc} same as that of proposed circuit. Some capacitors are chosen from the electrolytic capacitor SMR series made by Nippon Chemi-Con for comparison. Table IV shows their characteristics. 2000 years of lifetime at 85°C and at the rated ripple current superimposed to a DC voltage is guaranteed in this series.

Fig. 12 shows the lifetime estimation result of the DC-link capacitor. The lifetime range is from 10 to 40 years. The larger capacitance extends their lifetime. However, the lifetime of 390 uF is longer than that of 470 uF due to its larger surface area.

Nonetheless, the DC-link capacitor is the dominant lifetime-limiting component of the conventional micro-inverter. In addition, a lifetime of sealing cover of the electrolytic capacitor is defined as 15 years even if the

lifetime estimation result is long enough. Longer lifetime is obtained without the capacitors.

V. CONCLUSION

This paper presented the lifetime estimation considering mission profiles. As a result of the estimation, the existing micro-inverter has very long lifetime due to the over-designed power devices. In addition, the lifetime

444

of the DC-link capacitor in the conventional circuit is dominant in the total lifetime evaluation.

As a future plan, more accurate lifetime of the micro-inverter will be estimated. The accuracy of lifetime estimation will be increased by considering a temperature increment with the radiation or including the lifetime of other components such as solder joints. In addition, more experimental tests will be performed in order to obtain an accurate lifetime model and thermal impedance characteristics of the power devices. With this, the selection of power devices and thermal design can be done with reasonable design margins.

References

[1] H. Hu, S. Harb, N. Kutkut, I. Batarseh, and Z. Shen, "A review of power decoupling techniques for micro-inverters with three different decoupling capacitor locations in PV systems," IEEE Trans. Power Electron., vol. 28, no. 6, pp. 2711–2726, Jun. 2013.

[2] C. Rodriguez and G. Amaratunga, "Long-lifetime power inverter for photovoltaic ac modules," IEEE Trans. Ind. Electron., vol. 55, no. 7, pp. 2593–2601, Jul. 2008

[3] G. Petrone, G. Spagnuolo, R. Teodorescu, M. Veerachary, and M. Vitelli,"Reliability issues in photovoltaic power processing systems," IEEE Trans. Ind. Electron., vol. 55, no. 7, pp. 2569–2580, Jul. 2008.

[4] Y. Song and B. Wang, "Survey on reliability of power electronic systems," IEEE Trans. Power Electron., vol. 28, no. 1, pp. 591–604, Jan. 2013.

[5] H. S.-H. Chung, H. Wang, F. Blaabjerg, and M. Pecht, Reliability of Power Electronic Converter Systems. IET, 2015.

[6] Y. Ohnuma, K. Orikawa, J. Itoh, "A Single-Phase Current-Source PV Inverter With Power Decoupling Capability Using an Active Buffer" IEEE Trans. Ind. Applications, vol. 51, pp. 531 – 538, no. 1, Jan. 2015

[7] Y. Yang, H. Wang, F. Blaabjerg, and K. Ma, "Mission profile based multi-disciplinary analysis of power modules in single-phase transformerless photovoltaic inverters," in Proc. of EPE, pp. 1–10, Sep. 2013.

[8] H. Huang and P. A. Mawby, "A lifetime estimation technique for voltage source inverters," IEEE Trans. Power Electron., vol. 28, no. 8, pp. 4113–4119, Aug. 2013.

[9] A. Sangwongwanich, Y. Yang, D. Sera, F. Blaabjerg, "Lifetime Evaluation of Grid-Connected PV Inverters Considering Panel Degradation Rates and Installation Sites", IEEE Trans Power Electron, early access, 2017

[10] Yanfeng Shen, Huai Wang, Frede Blaabjerg, "Reliability Oriented Design of a Grid-Connected Photovoltaic Microinverter" IEEE 3rd International Future Energy Electronics Conference and ECCE Asia (IFEEC 2017 - ECCE Asia), pp.81-86, 2017

[11] H. Huang, P. A. Mawby, "A Lifetime Estimation Technique for Voltage Source Inverters", IEEE Trans. Power Electron., vol. 28, no. 8, August 2013

[12] Nippon Chemi-con, "Judicious Use of Aluminum Electrolytic Capacitors"

[13] P. D. Reigosa, H. Wang, Y. Yang, and F. Blaabjerg, "Prediction of bond wire fatigue of IGBTs in a PV inverter under a long-term operation," IEEE Trans. Power Electron., vol. 31, no. 10, pp. 7171–7182, Oct. 2016.

[14] Y. Shen, H. Wang, Y. Yang, P. D. Reigosa, and F. Blaabjerg, "Mission profile based sizing of IGBT chip area for PV inverter applications," in Proc. of PEDG, pp. 1–8, Jun. 2016.

[15] D. Zhou, H. Wang, F. Blaabjerg, S. K. Kaer, and D. Blom-Hansen, "System-level reliability assessment of power stage in fuel cell application," in Proc. of ECCE, pp. 1–8, Sep. 2016.

[16] K. Ma, H. Wang, and F. Blaabjerg, "New approaches to reliability assessment: Using physics-of-failure for prediction and design in power electronics systems," IEEE Power Electron. Mag., vol. 3, no. 4, pp. 28–41, Dec. 2016.

The 2018 International Power Electronics Conference

Online Health Monitoring of Multiple MOSFETs in a Grid-Tied PV Inverter using Spread Spectrum Time Domain Reflectometry (SSTDR)

Sourov Roy and Faisal Khan
Department of Computer Science and Electrical Engineering,
University of Missouri, Kansas City, MO, USA
Email: srdh9@mail.umkc.edu

Abstract- **In this paper, Spread Spectrum Time Domain Reflectometry (SSTDR) based condition monitoring technique for large PV inverter is proposed, which is capable of characterizing the degradation levels in all four MOSFETs in a single phase inverter using a single measurement. To achieve this, SSTDR test signal was applied across a node pair inside the PV inverter, and the corresponding peak responses were captured and processed to determine the level of aging and the location of the aged MOSFET within the inverter. The ON-state channel resistance increases with MOSFET aging, and the SSTDR reflections contain the information regarding this impedance change even though the inverter is live. That being said, the presence of PWM signal only makes the degradation detection more challenging. A new algorithm has been introduced in this paper to solve the problem mentioned above, and finally, experimental results have been included to demonstrate the validity of the proposed algorithm to locate individual degraded MOSFET with their corresponding level of aging. Although the initial test results are obtained from a single phase inverter, the proposed technique is equally applicable to three-phase inverters as well.**

Keywords— Condition monitoring, degradation, live health monitoring, PV inverter, reflectometry, reliability.

I. INTRODUCTION

With increased penetration of renewable energy sources, the failure-free operation of power converter circuits in PV systems, wind turbines and motor drives are gaining importance. Among various power converters used in PV based power systems, inverters contribute to more challenging reliability issues because of their complicated switching schemes and the use of different component types with dissimilar aging characteristics [1]. PV modules usually have much higher life (more than 20 years) than PV inverters [1], and 36% of lost energy is occurred due to inverter failures as opposed to 5% loss occurred by the PV module failures [2]. Therefore, making the PV inverter more reliable is the key to prevent the unwanted loss of potential energy production in PV power systems. Among the failure-prone components used in the inverter system, power semiconductor devices such as IGBTs and MOSFETs are considered to be the most vulnerable components followed by the capacitors and gate drivers. According to an industry based survey, IGBT incurs around 80% of the failure cost of the total power converter system [3]. Therefore, estimating the health of

power switches are essential when considering PV inverter reliability.

A considerable amount of literature has been reported on developing suitable reliability accessing techniques for PV inverters. These methods can be roughly divided into two categories- reliability modeling and condition monitoring [4]. Reliability modeling gives the tool for analyzing the reliability of PV inverters in the form of failure rates of the switching devices. For instance, real-field mission profile (RFMP) and Markov reliability model have been developed to access the mean-time-to-failure (MTTF) of PV inverters/converters in [5] and [6], respectively. In [7], taking into account the variations in IGBT parameters, a statistical approach based on Monte-Carlo simulations to predict the lifetime consumption of IGBT bond wires has been proposed. In a similar way, authors of [8] have estimated the failure rate of the capacitors and the power MOSFETs of a PV inverter considering different quality factors (π_Q) based on MIL-217F N2 method. Although reliability modeling gives a tool to attain a detailed inverter lifetime assessment that helps in the design stage, it cannot prevent the failure from happening.

Condition monitoring technique guarantees reliability in real-time operation of power converters. Condition monitoring is defined as the real-time estimation of the physical states of a component or the entire system so that any deviation from the healthy states can be used to perform maintenance before breakdown occurs [3]. A variety of methods have been used to monitor the real-time health of the power converters by measuring changes in the failure precursors of switching devices such as MOSFETs and IGBTs with the help of additional sensors or circuits [9]-[13]. Among these precursors, ON-state channel resistance, $R_{DS(ON)}$, $V_{GS(TH)}$ are the most significant factors associated with MOSFET degradation. Furthermore, the ON-state collector-emitter voltage, $V_{CE(ON)}$ is considered to be prominent failure precursor for IGBTs. However, these methods demand additional sensors and monitoring circuits, which may not be cost-effective. Moreover, variations in $R_{DS(ON)}$ and $V_{CE(SAT)}$ are negligible compared to the OFF-stage resistance and voltage, demanding highly sensitive sensors for high-resolution measurement. In contrast, model and system-identification based condition monitoring methods measure the degradation level of the switching devices as

446

well as the converters by comparing the measured response to the predicted variable of the healthy model. A few of them are proposed in [14]-[17]. However, these methods suffer from huge computational burden with a large number of training data, and they have a poor estimation of specific variables in different operating conditions. Moreover, all of the above techniques were applied to monitor the health of the different types of dc-dc converters rather than the PV inverters, where the later require a relatively higher number of switches along with complicated control and switching schemes.

In order to overcome the above-mentioned limitations, spread spectrum time domain reflectometry (SSTDR) based condition-monitoring methods of power devices have been proposed in [18]-[21]. The ON-state channel resistance, $R_{DS(ON)}$ increases with MOSFET aging, and the SSTDR reflection characterizes the aging level based on this resistance variation. Therefore, SSTDR's performance does not depend on the circuit's operating conditions, and this unique feature makes SSTDR an excellent candidate for condition monitoring process. In [18], SSTDR was applied to individual components while the power converter was running in a static condition (constant input at the gate) instead of real-life operating condition (PWM switching). Also, SSTDR based degradation monitoring methods proposed in [19], [20] was carried out in DC-DC converter where usually the PWM switching scheme is not as complicated as it is in the inverter. Since SSTDR test results depend on the impedance variations in its propagation path, different switching schemes will have different impacts on the SSTDR test results.

In [21], SSTDR has been used to monitor the degradation associated with MOSFETs in a live grid-tied PV inverter. However, this technique requires separate accessible nodes to connect the SSTDR test probes for detecting degradation level in each MOSFET, which may not always be available in real life applications. Moreover, an additional resistance was placed in series with the MOSFET under test to emulate the natural degradation process (increased $R_{DS(ON)}$), which does not represent the real-life aging precursor as well. In contrast, in this paper, we propose a new SSTDR based algorithm that characterizes the degradation levels in all four MOSFETs in the H-bridge PV inverter from a single measurement. Therefore, the proposed method can save additional computation/processing time as well as reducing the number of accessible nodes for performing the health monitoring for all four individual MOSFETs. Moreover, active power cycling was carried out to degrade the MOSFETs that was quantified by the aging precursor, $R_{DS(ON)}$.

II. SPREAD SPECTRUM TIME DOMAIN REFLECTOMETRY (SSTDR)

Reflectometry methods have been successfully used for locating and detecting faults in transmission lines [22], [23], and photovoltaic (PV) arrays [24]. A high-frequency electrical signal (V_0^+, I_0^+) is sent down the wire and a portion of the signal (V_0^-, I_0^-) is reflected back from wherever it faces impedance discontinuity. The reflection

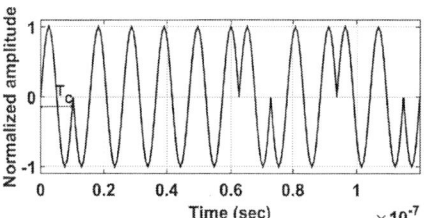

Fig.1. Sine modulated PN sequence with carrier frequency of 96 MHz. Here, T_C= length of a chip.

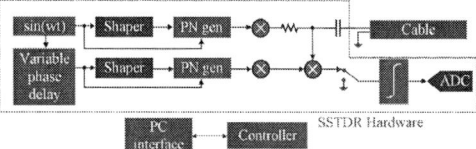

Fig.2. Schematic diagram of SSTDR implementation process [23].

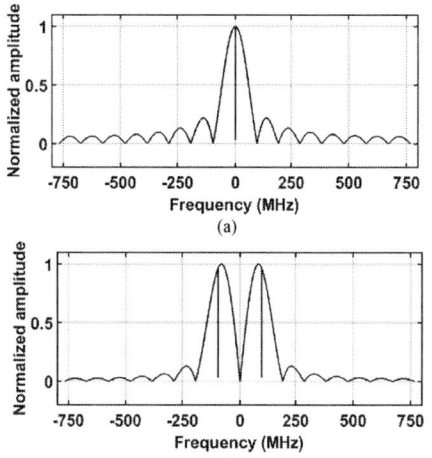

Fig.3. Frequency domain (FFT) representation of (a) PN sequence (chip rate, f_C= 96 MHz) and (b) sine modulated PN sequence with carrier frequency of 96 MHz.

coefficient that represents ratio between the reflected signal and the incident signal, is defined as follows [22], [23]:

$$\rho = \frac{V_0^-}{V_0^+} = \frac{z_L - z_0}{z_L + z_0} \quad (1)$$

Here, Z_0 is the characteristics impedance of the cable interconnecting the source and the load/ device under test (DUT), and Z_L is the impedance of the load/ the DUT.

Based on the incident signal used, there are several types of reflectometry based fault detection methods. Among them, spread spectrum time domain reflectometry (SSTDR) uses a direct sequence spread spectrum (DSSS) binary phase shift keyed (BPSK) signal as incident signal which is obtained by modulating pseudo-noise (PN) sequence with a high-frequency sine wave (please see Fig.1).

The 2018 International Power Electronics Conference

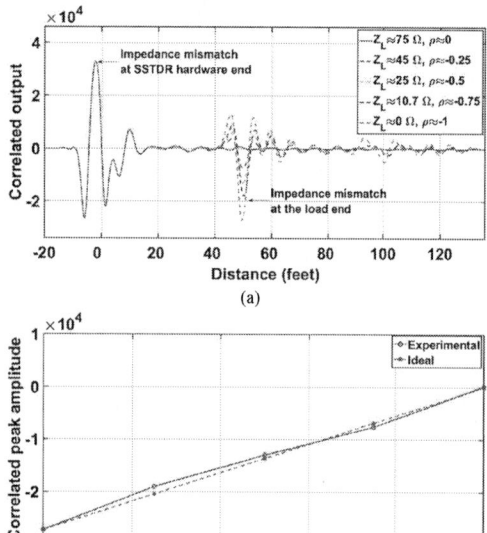

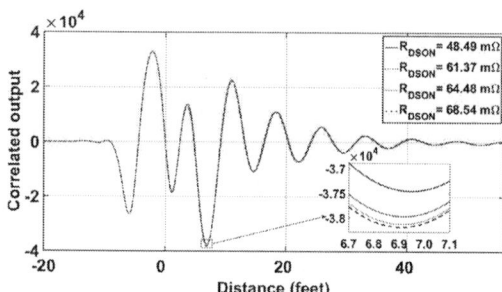

Fig. 5. Correlated amplitude variation for different aging levels (corresponding to different values of $R_{(DSON)}$) of power MOSFETs in a DC-DC buck converter.

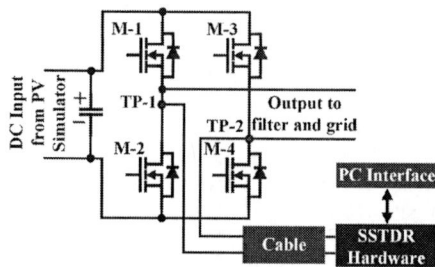

Fig. 6. SSTDR test diagram for detecting component degradation of single phase grid-tied PV inverter [21].

Fig. 4 (a). Variation in correlated amplitude for the different values of reflection coefficients ($\rho \leq 0$) and load impedances ($Z_L \leq Z_0$) (for 50 feet long, 75 Ω co-axial cable), (b) Variation in the correlated peak amplitudes for different reflection coefficients ($\rho \leq 0$) (obtained from Fig. 4 (a) [19]).

A schematic diagram of SSTDR implementation process is shown in Fig. 2. In general, PN sequence consists of binary random numbers (1s and 0s), and each 1 or 0 is denoted as a chip. This sine modulated PN sequence is sent down the wire, and the signal is reflected back if it finds any impedance discontinuity against the characteristic impedance of the propagation path. This reflected signal is cross-correlated with the delayed copies of the incident signal with the help of variable phase delay, and a lobe is generated at a time delay in the auto-correlation plot that corresponds to the distance from the source terminal to the impedance mismatch at the load terminal. It is worth to mention that the main lobe of the Fast Fourier Transform (FFT) of PN sequence is twice of its chip rate (T_C) and it is centered on 0 (zero) Hz (please see Fig. 3 (a)). Modulating the PN sequence with high-frequency carrier sine wave shifts the main lobe away from 0 (zero) Hz, and the amount of shift is proportional to the carrier frequency (please see Fig. 3(b)). Since the noise power is centered on 0 (zero) Hz, this sine modulated PN sequence produces less power in the cross-correlation with the noise signal, which will eventually increase the signal-to-noise ratio (SNR) of the system. Having best SNR along with its low voltage characteristics leads to its excellent performance in the live circuit among other reflectometry methods.

From (1), it is evident that the reflection coefficient is negative ($-1 \leq \rho \leq 0$) for $Z_L \leq Z_0$ and positive ($0 \leq \rho \leq 1$) for $Z_L \geq Z_0$. Fig. 4 (a) shows the different correlation curves that have been generated applying 48 MHz SSTDR signal passing through a 50 feet co-axial cable having a characteristic impedance of 75 Ω where reflection coefficient is negative. Fig. 4(b) shows the linear relationship between the correlated peak amplitudes and the corresponding reflection coefficient obtained from Fig. 4(a). From these two plots, it is evident that, when $\rho \leq 0$, an increased value in Z_L leads to less negative auto-correlated amplitude (or lower magnitude). Since the ON state channel resistance ($R_{DS(ON)}$) of power switches is significantly less than that of the SSTDR test cable and the lumped network of the PV inverter, the auto-correlated peak amplitude will be negative ($-1 \leq \rho \leq 0$) and the increased $R_{DS(ON)}$ of an aged MOSFET will lower the magnitudes of the auto-correlated amplitudes (Please see Fig.5). Fig. 5 was obtained applying a 48 MHz SSTDR signal across the drain-source of power MOSFETs with different aging levels (corresponding to different values of R_{DSON}) in a DC-DC buck converter. This test was done during static condition meaning no PWM was applied.

III. EQUIVALENT CIRCUIT OF THE INVERTER AND VARIOUS IMPEDANCE PATHS

In Fig. 6, a test diagram has been shown where two accessible points, TP-1 and TP-2 (across the AC output) of the single-phase H-bridge PV inverter has been used to connect the SSTDR hardware. There are two operating states of the inverter along with two sub-states for each of them depending on the ON/OFF state of the MOSFETs (Table I), and the corresponding equivalent impedance paths for the four sub-states have been shown in Fig. 7. In

448

TABLE I
OPERATING STATES OF THE INVERTER (M1, M2, M3 & M4 REPRESENT THOSE FOUR MOSFETs)

State	Sub-State	M1	M2	M3	M4
1	a	ON	OFF	OFF	PWM (OFF)
	b	ON	OFF	OFF	PWM (ON)
2	a	OFF	PWM (OFF)	ON	OFF
	b	OFF	PWM (ON)	ON	OFF

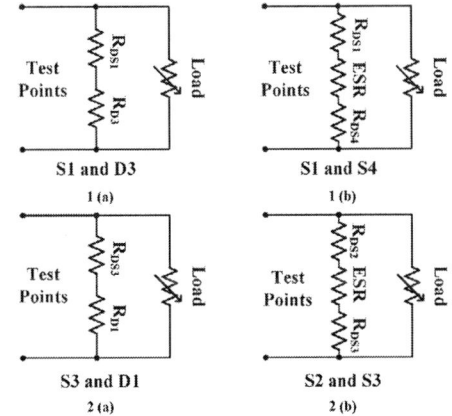

S1 and D3
1 (a)

S1 and S4
1 (b)

S3 and D1
2 (a)

S2 and S3
2 (b)

Fig. 7. Various equivalent impedance paths inside the inverter.

addition to these four sub-states, two additional impedance paths are possible: combination of the impedance paths shown in state 1(a), 1(b) and combination of the impedances paths shown in state 2(a) and 2(b). Here, the equivalent series resistance (*ESR*) of the capacitor is included as SSTDR signals will go through the capacitor in sub-states 1(b) and 2(b).

It is well understood that $R_{DS(ON)}$ of the MOSFET increases with the higher level of aging, and this increase in $R_{DS(ON)}$ of any MOSFET will increase equivalent impedance paths of corresponding sub-states. Therefore, each sub-state will have an impact on the SSTDR propagation. For instance, degradation in MOSFET 1 will affect sub-state 1(a), 1(b) and 2 (a) (provided R_{D1} increases). In the same way, change in $R_{DS(ON)}$ of MOSFET 2 will affect sub-state 2(b), and aging in MOSFET 3 will affect sub-states 1(a) (provided R_{D3} increases), 2(a) and 2(b). Sub-state 1(b) will be the only sub-state that will be influenced by the change in the value of $R_{DS(ON)}$ of MOSFET 4.

IV. EXPERIMENTAL SETUP FOR THE AGING PROCESS OF THE MOSFETs

Multiple MOSFETs were aged with power cycle to apply electro-thermal stress to an N-channel power MOSFET. This power cycling was done with a temperature gradient of 20°C where maximum and minimum temperature thresholds were maintained at 230°C and 210°C, respectively. The aging process continued for 2844 cycles /19 hours with the duty cycle of

TABLE II
EFFECT OF AGING ON MOSFET ON-RESISTANCE, $R_{DS(ON)}$ AND BODY DIODE RESISTANCE, R_D

	Before Aging	After Aging
$R_{DS(ON)}$	70.4 mΩ	108.9 mΩ
R_D	341.6 mΩ	386.84 mΩ

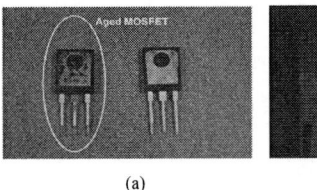
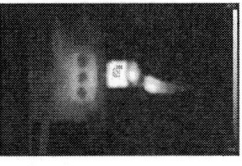

(a)   (b)

Fig.8. (a) MOSFET (DUT) before and after aging process (b) thermal image during aging process.

50%, and a constant drain current of 12.8 A. After 2844 cycles, we found that the DUT experienced 38.5 mΩ rise in ON-state channel resistance ($R_{DS(ON)}$) and 45.24 mΩ rise in body diode resistance (R_D) (Table II). This higher value of $R_{DS(ON)}$ along with R_D is a clear indication of the aging of the DUT. Fig. 8 shows the photograph of aged MOSFET as well as the thermal image during the aging process.

V. EXPERIMENTAL RESULTS

The experimental test set-up is shown in Fig. 9, and the SSTDR hardware was connected across the AC output of a 700 W commercial PV inverter. Five group of tests were conducted to verify the proposed SSTDR algorithm, and we considered the first group as a baseline, when, all MOSFETs used in the circuit were new (no aging). Another four groups were created where each new MOSFET was replaced by one aged MOSFET, one at a time. Thus in group 1, 2, 3 and 4, an aged MOSFET is used to replace the MOSFET 1, MOSFET 2, MOSFET 3 and MOSFET 4, respectively. Ten readings were taken in each group using 24 MHz SSTDR signal. During each measurement, SSTDR hardware continuously sent test signals to the power circuit, and autocorrelation peaks were generated approximately at the rate of 1200 times per second. After a series of data filtering procedures, only the peak values which represent the full conduction of each MOSFET were collected. Interestingly, the

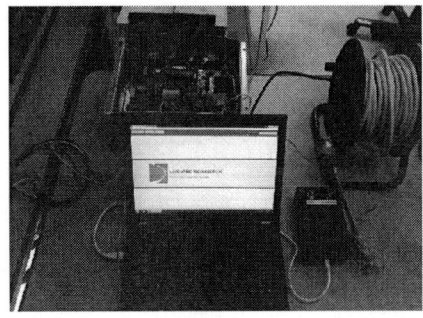

Fig. 9. Experiment test set-up for condition monitoring of PV inverter [21].

449

autocorrelation peaks generated by the SSTDR hardware in each reading for the same setup were not identical, and it was due to the continuous changes in the impedance paths due to the changes in the switching states. Moreover, temperature variation, system power variation, system noise, hardware measurement error due to small reflection, etc. will slightly vary the auto-correlated peak amplitudes in each reading [22]. Therefore, the averaged peaks cannot be taken as an indicator of the degradation level of the MOSFETs.

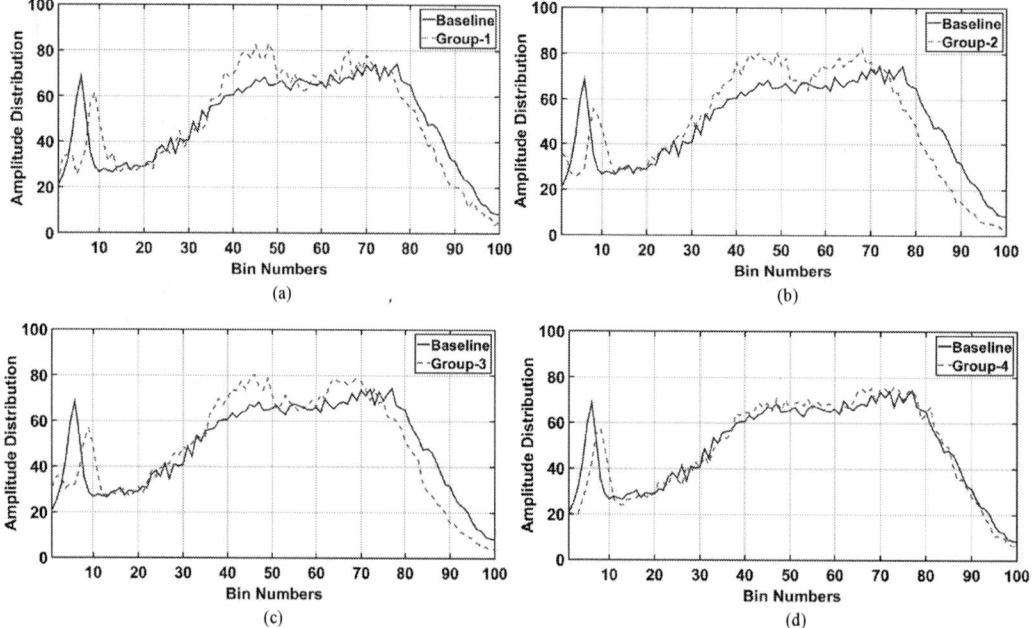

Fig. 10. Amplitude distribution-comparison between baseline and (a) aged MOSFET 1 (Group 1) (b) aged MOSFET 2 (Group 2); (c) aged MOSFET 3 (Group 3); (d) aged MOSFET 4 (Group 4).

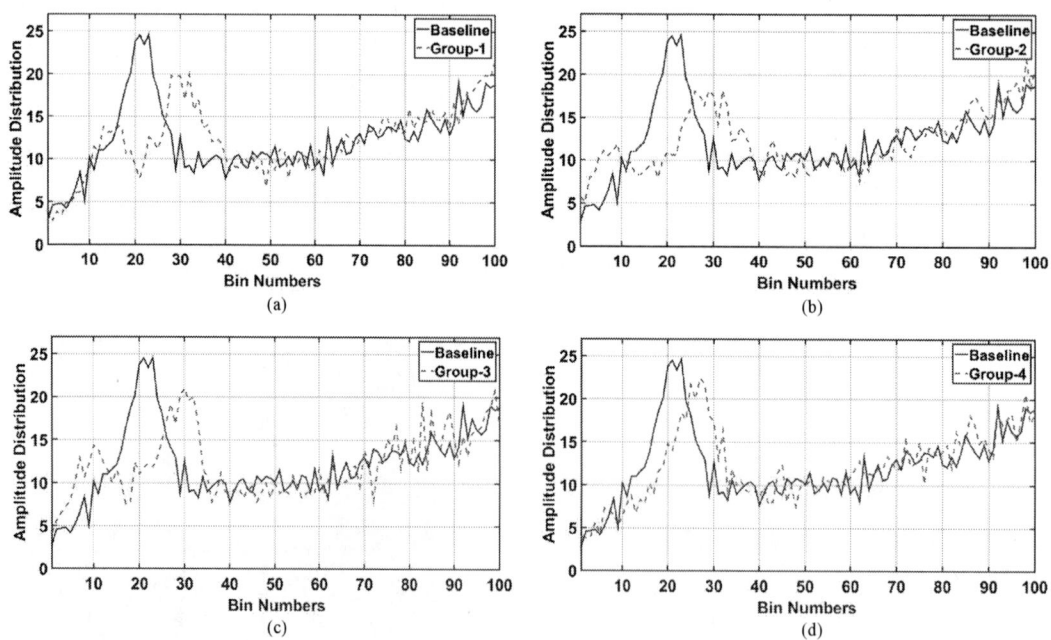

Fig. 11. Amplitude distribution-comparison (by taking only top 20 % data) between baseline and (a) aged MOSFET 1 (Group 1) (b) aged MOSFET 2 (Group 2); (c) aged MOSFET 3 (Group 3); (d) aged MOSFET 4 (Group 4).

To overcome this drawback, the entire dataset during any test/ reading was divided into one hundred (100) bins based on their cross-correlation peak amplitudes of descending order (smallest to largest negative peak values). The frequency of each bin (frequency distribution of the peak amplitudes) for the ten readings were averaged and plotted as an indicator of impedance allocation (Please see Fig. 10). To avoid confusion, we named this frequency distribution of the peak values as **amplitude distribution**. We applied amplitude distribution to count the number of points located within a specific range of autocorrelation peak values. There was a rightward shift from the baseline for each group at the left side of the bin plots. The rightward shift at the left side of the bin plot indicates that the correlated peak amplitudes with lower magnitudes (after a specific range) have higher counts for the aged MOSFET compared to the baseline. This implies that the $R_{DS(ON)}$ has increased in an aged MOSFET, thus making the magnitudes of the correlated peak amplitudes smaller ($-1 \leq \rho \leq 0$). Thus, the analysis can be focused on the top 20% of the data (the left side of the bin plots) that are plotted in Fig. 11. Because the physical distances from each of these four MOSFETs to the test points are different, a small variation exists in their distributed amplitude locations. This observation can be primarily used to determine the location of the aged MOSFET in the H-bridge PV inverter, and detailed analysis will be provided in future publications.

VI. CONCLUSION AND FUTURE WORK

A novel SSTDR based condition-monitoring technique for single-phase live grid-tied PV inverter has been presented in this paper. SSTDR signal was applied across a node pair inside the PV inverter, and the corresponding peak responses have been recorded and analyzed. By analyzing the amplitude distribution of SSTDR peak values and their locations, it is possible to measure the individual degradation in all four MOSFETs from a single measurement. The authors are working on the detailed analysis of the proposed method that bridges the SSTDR data with the different equivalent impedance paths of corresponding switching states along with the location information. This proposed method will be a breakthrough to determine the level of degradation and the physical location of the aged MOSFETs or IGBTs without affecting the normal operation of the inverter.

REFERENCES

[1] J. Flicker and S. Gonzalez, "Performance and reliability of PV inverter component and systems due to advanced inverter functionality," 2015 IEEE 42nd Photovoltaic Specialist Conference (PVSC), New Orleans, LA, 2015, pp. 1-5. doi: 10.1109/PVSC.2015.7355978.

[2] A. Golnas, "PV system reliability: An operator's perspective," 2012 IEEE 38th Photovoltaic Specialists Conference (PVSC) PART 2, Austin, TX, USA, 2012, pp. 1-6. doi: 10.1109/PVSC-Vol2.2012.6656744.

[3] S. Yang, A. Bryant, P. Mawby, D. Xiang, L. Ran and P. Tavner, "An Industry-Based Survey of Reliability in Power Electronic Converters," in IEEE Transactions on Industry Applications, vol. 47, no. 3, pp. 1441-1451, May-June 2011. doi: 10.1109/TIA.2011.2124436S.

[4] Yang, D. Xiang, A. Bryant, P. Mawby, L. Ran and P. Tavner, "Condition Monitoring for Device Reliability in Power Electronic Converters: A Review," in IEEE Transactions on Power Electronics, vol. 25, no. 11, pp. 2734-2752, Nov. 2010. doi: 10.1109/TPEL.2010.2049377.

[5] N. C. Sintamarean, F. Blaabjerg, H. Wang, F. Iannuzzo and P. de Place Rimmen, "Reliability Oriented Design Tool For the New Generation of Grid Connected PV-Inverters," in IEEE Transactions on Power Electronics, vol. 30, no. 5, pp. 2635-2644, May 2015.doi: 10.1109/TPEL.2014.2361918.

[6] E. Hofreiter and A. M. Bazzi, "Single-stage boost inverter reliability in solar photovoltaic applications," 2012 IEEE Power and Energy Conference at Illinois, Champaign, IL, 2012, pp. 1-4. doi: 10.1109/PECI.2012.6184601.

[7] P. D. Reigosa, H. Wang, Y. Yang and F. Blaabjerg, "Prediction of Bond Wire Fatigue of IGBTs in a PV Inverter Under a Long-Term Operation," in IEEE Transactions on Power Electronics, vol. 31, no. 10, pp. 7171-7182, Oct. 2016. doi: 10.1109/TPEL.2015.2509643.

[8] F. Obeidat and R. Shuttleworth, "Reliability prediction of PV inverters based on MIL-HDBK-217F N2," 2015 IEEE 42nd Photovoltaic Specialist Conference (PVSC), New Orleans, LA, 2015, pp. 1-6. doi: 10.1109/PVSC.2015.7356277S.

[9] Dusmez, M. Bhardwaj, L. Sun and B. Akin, "In Situ Condition Monitoring of High-Voltage Discrete Power MOSFET in Boost Converter Through Software Frequency Response Analysis," in IEEE Transactions on Industrial Electronics, vol. 63, no. 12, pp. 7693-7702, Dec. 2016. doi: 10.1109/TIE.2016.2595482.

[10] X. Ye, C. Chen, Y. Wang, G. Zhai and G. J. Vachtsevanos, "Online Condition Monitoring of Power MOSFET Gate Oxide Degradation Based on Miller Platform Voltage," in IEEE Transactions on Power Electronics, vol. 32, no. 6, pp. 4776-4784, June 2017. doi: 10.1109/TPEL.2016.2602323.

[11] F. Stella, G. Pellegrino, E. Armando and D. Dapra, "On-line Junction Temperature Estimation of SiC Power MOSFETs through On-state Voltage Mapping," in IEEE Transactions on Industry Applications, vol. PP, no. 99, pp. 1-1. doi: 10.1109/TIA.2018.2812710.

[12] P. Sun, C. Gong, X. Du, Q. Luo, H. Wang and L. Zhou, "Online Condition Monitoring for Both IGBT Module and DC-Link Capacitor of Power Converter Based on Short-Circuit Current Simultaneously," in IEEE Transactions on Industrial Electronics, vol. 64, no. 5, pp. 3662-3671, May 2017. doi: 10.1109/TIE.2017.2652372.

[13] U. M. Choi, F. Blaabjerg, S. Jørgensen, S. Munk-Nielsen and B. Rannestad, "Reliability Improvement of Power Converters by Means of Condition Monitoring of IGBT Modules," in IEEE Transactions on Power Electronics, vol. 32, no. 10, pp. 7990-7997, Oct. 2017. doi: 10.1109/TPEL.2016.2633578.

[14] M. Heydarzadeh, S. Dusmez, M. Nourani and B. Akin, "Bayesian remaining useful lifetime prediction of thermally aged power MOSFETs," 2017 IEEE Applied Power Electronics Conference and Exposition (APEC), Tampa, FL, 2017, pp. 2718-2722. doi: 10.1109/APEC.2017.7931083.

[15] S. Mohagheghi, R. G. Harley, T. G. Habetler and D. Divan, "Condition Monitoring of Power Electronic Circuits Using Artificial Neural Networks," in IEEE Transactions on Power Electronics, vol. 24, no. 10, pp. 2363-2367, Oct. 2009. doi: 10.1109/TPEL.2009.2017806.

[16] J. M. Anderson and R. W. Cox, "On-line condition monitoring for MOSFET and IGBT switches in digitally controlled drives," 2011 IEEE Energy Conversion Congress and Exposition, Phoenix, AZ, 2011, pp. 3920-3927. doi: 10.1109/ECCE.2011.6064302.

[17] M. A. Eleffendi and C. M. Johnson, "In-Service Diagnostics for Wire-Bond Lift-off and Solder Fatigue of Power Semiconductor Packages," in IEEE Transactions on Power Electronics, vol. 32, no. 9, pp. 7187-7198, Sept. 2017. doi: 10.1109/TPEL.2016.2628705.

[18] M. S. Nasrin, F. H. Khan and M. K. Alam, "Quantifying Device Degradation in Live Power Converters Using SSTDR Assisted Impedance Matrix," in IEEE Transactions on Power Electronics, vol. 29, no. 6, pp. 3116-3131, June 2014. doi: 10.1109/TPEL.2013.2273556.

[19] S. Roy and F. Khan, "Live condition monitoring of switching devices using SSTDR embedded PWM sequence: A platform for

intelligent gate-driver architecture," 2017 IEEE Energy Conversion Congress and Exposition (ECCE), Cincinnati, OH, 2017, pp. 3502-3507. doi: 10.1109/ECCE.2017.8096625.

[20] S. Roy and F. Khan, " State of Health (SOH) Estimation of Multiple Switching Devices Using a Single Intelligent Gate Driver Module," accepted and presented at 2018 IEEE Applied Power Electronics Conference and Exposition (APEC), San Antonio, TX, USA, March 2018.

[21] Q. Li and F. H. Khan, "Identifying natural degradation/aging in power MOSFETs in a live grid-tied PV inverter using spread spectrum time domain reflectometry," 2014 International Power Electronics Conference (IPEC-Hiroshima 2014 - ECCE ASIA), Hiroshima, 2014, pp. 2161-2166. doi: 10.1109/IPEC.2014.6869888.

[22] C. Furse, Y. C. Chung, C. Lo, and P. Pendayala, "A critical comparison of reflectometry methods for location of wiring faults," J. Smart Structures Syst., vol. 2, no. 1, pp. 25–46, 2006.

[23] P. Smith, C. Furse and J. Gunther, "Analysis of spread spectrum time domain reflectometry for wire fault location," in IEEE Sensors Journal, vol. 5, no. 6, pp. 1469-1478, Dec. 2005. doi: 10.1109/JSEN.2005.858964.

[24] S. Roy, M. K. Alam, F. Khan, J. Johnson and J. Flicker, "An Irradiance Independent, Robust Ground Fault Detection Scheme for PV Arrays Based on Spread Spectrum Time Domain Reflectometry (SSTDR)," in IEEE Transactions on Power Electronics, vol. PP, no. 99, pp. 1-1. doi: 10.1109/TPEL.2017.2755592.

An Improved Equivalent Model for a Long PV String under Partial Shading Conditions

Xiaoyang Wang*, Huiqing Wen*, Xingshuo Li*

*Dept. of Electrical & Electronic Engineering, Xian Jiaotong-Liverpool University, Suzhou, China
Email: Xiaoyang.Wang14@student.xjtlu.edu.cn Email: Huiqing.Wen@xjtlu.edu.cn

Abstract—**An accurate model for a long Photovoltaic (PV) string is essential to predict the position of maximum power point under partial shading conditions. Normally it divides a PV string into three parts, namely a diode forward voltage drop, an equivalent PV source and a linear voltage source, under different operating stages. However, significant error may occur when this model is applied on a long PV array or under low solar irradiance condition. Therefore, a new algorithm is proposed to improve the performance of this equivalent model. The proposed method collects and takes advantage of irradiance information in calculation, together with curve fitting method to obtain PV model of higher frequency. To validate this algorithm, error analysis is presented on results of both original and proposed algorithm. It illustrates that there is little deviation on each calculated MPP by applying the proposed algorithm even in a long PV string under low irradiance condition.**

I. INTRODUCTION

Mathematical model on photovoltaic (PV) device is crucial for the PV system since it can be used to express the behavior of a PV string current and power with respect to voltage under different weather conditions. Under uniform irradiance condition, the P-V curve of the PV string usually presents a single peak known as MPP [1–13].

However, when the PV string is partially shaded, multiple peaks will be exhibited in the P-V curve. This will result that some conventional maximum power point tracking (MPPT) methods cannot distinguish the local maximum power point (LMPP) and global maximum power point (GMPP). Furthermore, it is also difficult to analytically express the I-V and P-V curves and accurately predict the GMPPP under this partial shading condition (PSC).

So far, many PV modeling methods have been proposed to express the I-V and P-V curve under the

PSC, such as $0.8V_{oc}$ model [14, 15] and PV equivalent circuit [16–18].

$0.8V_{oc}$ model assumes that the peaks of a P-V curve under the PSC occur nearly at multiples of $0.8V_{oc}$ [14], where V_{oc} is open circuit voltage of PV module. This model can help some conventional MPPT methods, such as perturb and observe (P&O) method, to track the GMPP. Furthermore, it can also increase the tracking speed for these methods by limiting searching region to $0.8V_{oc}$. Although this method has been validated efficiently, Jubaer Ahmed [15] have found that this assumption may fail to predict MPP in a long PV string such as a 10-module PV string. In [15], it was found that large deviation occurs between accurate MPP and calculated locations. In addition, errors from $0.8V_{oc}$ further increase as the number of illumination levels getting larger.

PV equivalent circuit is another way to describe a PV string under the PSC [16–18]. It divides a PV string into three parts, namely a diode forward voltage drop, an equivalent PV source and a linear voltage source, under different operating stages. The equivalent PV voltage is calculated according to this model, then the whole PV string will be converted into several equivalent PV modules. The effectiveness of this modeling method has been validated in [16, 17] and it is also used to develop global MPPT (GMPPT) method, such as the modified Beta method [18]. However, this method in [18] may also fail in a long PV string under very low solar irradiance.

In this paper, an algorithm is proposed to improve performance of the PV equivalent model under aforementioned conditions. Firstly, source of error is analyzed for this model used by modified Beta method. Then, an explicit expression of error calculation is derived and then used to evaluate model performance. The proposed method senses irradiance information from string current to determine linear voltage sources in equivalent circuit.

The 2018 International Power Electronics Conference

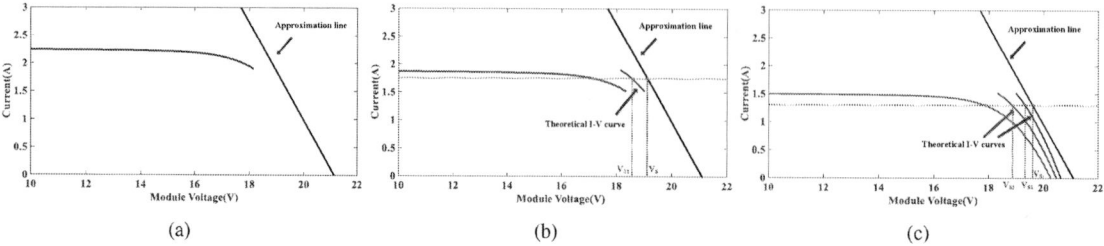

(a)   (b)   (c)

Fig. 1. Error analysis on curves for each operation stage in a 3-module PV string under PSC. (a) Error analysis on stage I. (b) Error analysis on stage II. (c) Error analysis on stage III.

More accurate voltage sources are selected according to irradiance levels, while the model in modified Beta method only takes one approximation for various irradiance conditions. By applying the proposed algorithm, accuracy of equivalent voltage calculation on key module in modified Beta algorithm is improved significantly. The improvement is more obvious in long PV string under low irradiance. Simulation results on both 3-module and 5-module PV string under different shading conditions are presented to verify the advantage of proposed algorithm.

II. EFFECT OF LOW IRRADIANCE LEVELS ON EQUIVALENT MODEL

A. Error analysis on 3-Module PV String

The modified beta algorithm [18] traces GMPP in a PV string under PSC, which is based on the PV equivalent model [17]. The model describes a PV string as series connected linear voltage sources, diodes and an equivalent PV source. Diode has a forward voltage drop of 0.8V and each linear voltage source can be calculated as [18].

$$V_S \approx \frac{V_{MPP,stc} - V_{oc,stc}}{I_{MPP,stc}} \times I_{String} + V_{oc,stc} \quad (1)$$

where I_{String} is the string current and all other parameters are obtained under standard condition. This approximation of linear voltage source is applied to calculate equivalent voltage V_{eq} on key PV source [18] as shown in equation 2.

$$V_{eq} = V_{String} - (n-1) \times V_S + (m-n) \times V_d \quad (2)$$

where V_{String} is PV string voltage, m is total number of modules, n refers to operation stage number and V_d is diode voltage drop.

The result of this algorithm is verified to be effective under high irradiance levels. However, error on V_{eq}

Algorithm 1 Error calculation of modified Beta algorithm in a N-module array

Input: $I[1...N]$: N vectors for measured current values for N stages; $V_{Sth}[1...N]$: Theoretical V_S for N stages; $I_{Sth}[1...N]$: Current values corresponding to V_{Sth}; N: Total module number
Output: $Error[1...N]$: N vectors representing errors in N stages
1: $V_S[1...N] \leftarrow 0$
2: $Error[1...N] \leftarrow 0$
3: $pos \leftarrow 0$
4: **for** $i = 1 \rightarrow N$ **do**
5:     **for** $j = 1 \rightarrow length(I[i])$ **do**
6:         // V_S from modified Beta algorithm
7:         Calculate $V_S[i,j]$ with $I[i,j]$
8:         **for** $k = 1 \rightarrow length(I[i])$ **do**
9:             // Obtain theoretical V_S under I[i,j]
10:             **if** $I_{Sth}[i,k] == I[i,j]$ **then**
11:                 $pos \leftarrow k$
12:                 $break$
13:             **end if**
14:             // Calculate absolute error
15:             $Error[i,j] \leftarrow |V_{Sth}[i,pos] - V_S[i,j]|$
16:         **end for**
17:     **end for**
18: **end for**
19: **return** $Error[1...N]$

calculation can be obvious under low irradiance levels. Fig. 1 illustrates error analysis of a 3-module PV string. A single approximation line in equation 1 is used to calculate V_S for all three stages, while the theoretical V_S should be V_{S1} and V_{S2} in Fig. 1(b) and (c). Those values are obtained form real I-V curves not approximation. Thus theoretical equivalent voltage V_{th} on stage n in a should be calculated as in equation 3

$$V_{th} = V_{String} - (V_{S1} + V_{S2} + ... + V_{S(n-1)}) + (m-n) \times V_d \quad (3)$$

By comparison, errors between V_{eq} in equation 2 and V_{th} is from approximation of V_S. Thus equation for error on stage n can be obtained as equation 4. Note that on error appears on the first stage because no V_S is involved in calculation as shown in both equation 2 and 3.

$$Error = |(n-1) \times V_S - (V_{S1} + V_{S2} + ... + V_{S(n-1)})| \quad (4)$$

The 2018 International Power Electronics Conference

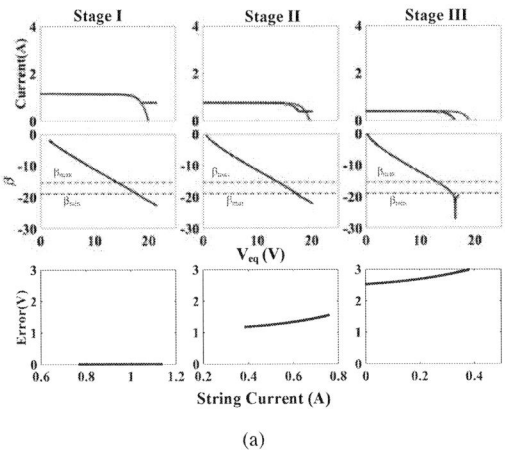
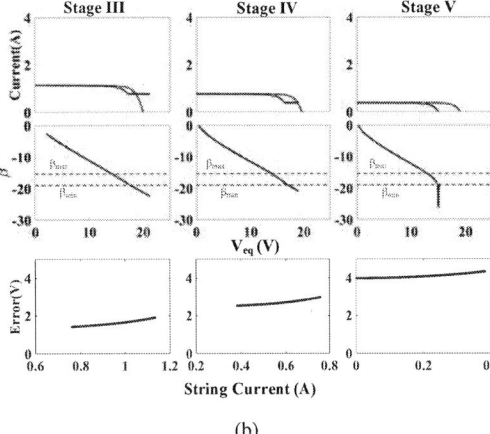

(a)　　　　　　　　　　　　(b)

Fig. 2. Direct comparison and errors between equivalent voltage and its theoretical value for each operation stage. (a) 3-module PV string under 300 W/m^2, 200 W/m^2, and 100 W/m^2 irradiance levels. (b) The last three operation stages of 5-module PV string under 500 W/m^2, 400 W/m^2, 300 W/m^2, 200 W/m^2, and 100 W/m^2 irradiance levels.

The overall procedure of error calculation is illustrated in algorithm 1. This is applied in following parts to evaluate cases of low irradiance levels of 3-module PV string and 5-module PV string.

B. Effect of low irradiance on equivalent model

In Fig. 2, three operation stages of a 3-module PV string is illustrated. Calculated equivalent voltage in each stage is shown by blue lines in first row of each subplot, while green lines demonstrate theoretical I-V curves of each key module. Direct comparison between voltages can be made from those two lines. Last row of subplots shows errors varing with string current. Fig. 2 shows that beta algorithm cause large errors when PV String is working under low irrdiance. In Fig. 2(a), errors in the last stage in a 3-module array can be larger than 2.5 volts. In a 5-module array, errors will further increase around 4 volts as shown in Fig. 2(b). In such case, beta algorithm may fail to track GMPP.

III. PROPOSED METHOD TO IMPROVE ACCURACY OF CALCULATED EQUIVALENT VOLTAGE

A. More accurate approximation lines

Previous examples shows that beta algorithm fails to obtain accurate equivalent voltage. The problem is that only one approximation line cannot fully describe all I-V curve sections under different irradiance levels. Thus, next step is to generalize a new algorithm that can apply different approximation lines for corresponding

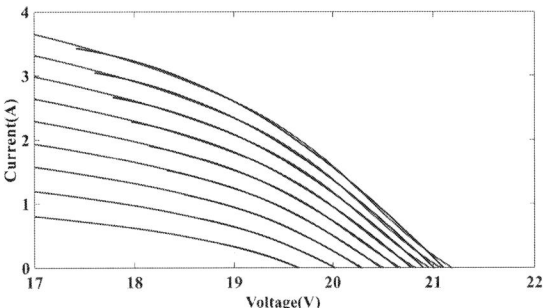

Fig. 3. I-V curve of 1000 W/m^2 to 100 W/m^2 with step of 100 W/m^2 from simulation

irradiance conditions. To obtain expressions of those lines, first step is to get simulation results of single module under 10 illumination conditions.

Next step is to apply Matlab Curve Fitting Tool with simulation data to get equation descriptions. Note that only parts of data will be fitted because only the approximately linear parts in a I-V curve will be involved in calculation. Considering fitting performance and calculation efficiency, polynomial with highest degree of 3 is chosen to fit those curves. Fitting results are illustrated in Fig. 3, where blue lines are simulation results and green lines are fitting results. All expressions are functions of current and detailed parameters are shown in table I.

B. Proposed Algorithm

Main function of proposed algorithm is illustrated by pseudocode in algorithm 2. When main function is

455

The 2018 International Power Electronics Conference

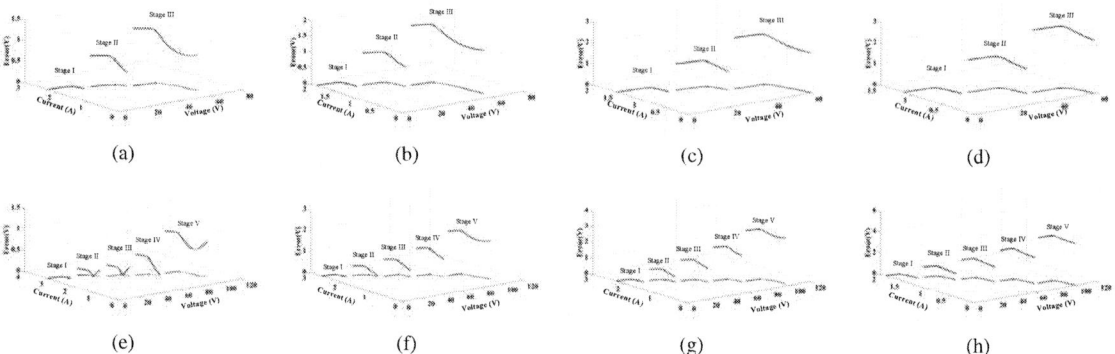

Fig. 4. Error comparison between beta method and proposed method. (a)-(h) are correspond to case 1-8 in table

TABLE I. PARAMETERS OF FITTED CURVES UNDER DIFFERENT IRRDIANCE LEVELS

Fitted Equation	$V = p_1 I^3 + p_2 I^2 + p_3 I + p_4$									
Irradiance (W/m^2)	1000	900	800	700	600	500	400	300	200	100
p_1	-0.09709	-0.1205	-0.1532	-0.2004	-0.2718	-0.3874	-0.5908	-0.998	-2.01	-83.28
p_2	0.313	0.3339	0.3553	0.3759	0.39	0.3887	0.3435	0.1723	-0.4278	25.5000
p_3	-0.998	-1.009	-1.022	-1.041	-1.066	-1.106	-1.175	-1.312	-1.644	-5.863
p_4	21.18	21.07	20.96	20.82	20.67	20.50	20.29	20.02	19.65	19.10

executed, it calls algorithm 3 that obtain illumination information and 4 that chooses approximation expressions accordingly. First step is to initialize total module numbers M in the PV string and N the stage number. Then, this algorithm will sense initial current in each region to determine stage irradiance level. For example, an initially sensed current value is around 3.8A, which is close to short circuit current under 1000 W/m^2. Then $Flag[1]$ changes from 0 to 1 indicating stage irradiance level is 1000 W/m^2. This irradiance information will be used to determine approximation line to calculate V_S. When this process is iterated, V_{EQ} for each stage will be obtained and finally V_{EQ} for all stages will be calculated.

IV. VALIDATION OF NEW ALGORITHM

Validation mainly focuses on PV string of 3 and 5 modules under low irradiance conditions. To evaluate results of proposed method, errors are calculated in the way mentioned in Section II. Note equivalent voltage calculated from beta algorithm as V_β and that from proposed method as V_P. Theoretical voltage values V_{th} are simulation result of key modules in each stage. Then errors of beta method are obtained by point to point comparison of V_β and V_{th} as $|V_\beta - V_{th}|$ shown in blue lines in Fig. 4 and that of proposed method are $|V_P - V_{th}|$ denoted by green lines. To see the effect on MPPT, actual

MPP locations are marked on error plots illustrating deviation on actual MPPs for both beta and proposed methods. First 4 cases in Fig. 4 and Table II illustrates the improvement of proposed method on 3-module PV string. In the first case where PV string is under 600, 500, and 400 W/m^2 irradiance levels, MPP deviation on stage III of beta algorithm is 1.247V while that of proposed algorithm is only 0.003V. In following case 2, 3 and 4 with irradiance decreases, errors from beta algorithm keep increasing and maximum error reach 2.953V. However, errors from proposed algorithm remain less than 0.001V. The last 4 cases illustrates comparisons on 5-module string and similar pattern can be observed.

V. CONCLUSION

In this paper, an explicit error analysis process is carried out to evaluate results of PV equivalent model in modified Beta method under PSC. By comparing simulation result of key module and calculated result, large errors are found under low illumination condition, which indicate that beta method may fail to predict MPP locations accurately. Then, to improve its accuracy in equivalent voltage calculation section, a new algorithm is proposed to apply fitted curves in calculation. Choice of curves involved in calculation is also determined by algorithm according to irradiance. By applying proposed

The 2018 International Power Electronics Conference

Algorithm 2 Equivalent voltage calculation in a 5-module PV array

---

Input: $I_{IN}[1..5]$: Measured initial current of each stage; $I_{ST}[1...5]$: 5 vectors for measured stage current; $V_{ST}[1...5]$: 5 vectors for measured stage voltage; V_D: Diode voltage drop
Output: $V_{EQ}[N]$: Calculated equivalent voltage for N stages
1: **function** VEQCALCULATION($I_{IN}[1..5], I_{ST}[1...5], V_{ST}[1...5], V_D$)
2:　　$M \leftarrow 5$ // Total module number
3:　　$N \leftarrow 1$ // Operation stage number
4:　　$Flag[1...10] \leftarrow 0$ // Irradiance information
5:　　$V_S[1...N] \leftarrow 0$ // N vectors representing approximation voltage sources
6:　　$V_{EQ} \leftarrow 0$ // N vectors representing N segments of equivalent voltage
7:　　// Calculation of the first stage does not require Vs
8:　　$V_{EQ}[1] \leftarrow V_{ST}[1] + (M - N)V_D$
9:　　$V_S[1] \leftarrow$ **VsCalculation**($Flag[1...10], I_{ST}[1]$)
10:　　**for** $N = 2 \to M$ **do**
11:　　　$Flag[1...10] \leftarrow$ **FlagSet**($I_{IN}[N], I_{SC}[1...10]$)
12:　　　$V_S[N] \leftarrow$ **VsCalculation**($Flag[1...10], I_{ST}[N]$)
13:　　　$V_{EQ}[N] \leftarrow V_{ST}[N] - (V_S[1] + V_S[2] + ...V_S[N - 1]) + (M - D)V_D$ // Vectorized calculation of equivalent voltage in stage N
14:　　　$Flag[1...10] \leftarrow 0$ // Reset flag to store irradiance information of next stage
15:　　**end for**
16:　　**return** $V_{EQ}[1...10]$
17: **end function**

TABLE II.　　COMPARISON ON 5-MODULE PV STRING

Case	Irradiance level on each module(W/m^2)	MPP Deviation of proposed algorithm(V)					MPP Deviation of beta algorithm(V)				
		I	II	III	IV	V	I	II	III	IV	V
1	600 500 400	-0.001	0.009	0.003	-		-0.001	0.602	1.247	-	-
2	500 400 300	-0.001	0.0001	0.007	-	-	-0.001	0.898	1.816	-	-
3	400 300 200	-0.001	0.002	0.005	-	-	-0.001	1.193	2.385	-	-
4	300 200 100	-0.001	0.002	0.003	-	-	-0.001	1.487	2.953	-	-
5	800 700 600 500 400	-0.003	0.011	0.019	0.007	0.043	-0.003	0.006	0.100	0.384	0.953
6	700 600 500 400 300	-0.003	0.013	0.017	0.013	0.030	-0.003	0.303	0.673	1.231	2.072
7	600 500 400 300 200	-0.003	0.007	0.003	0.020	0.012	-0.003	0.600	1.245	2.075	3.192
8	500 400 300 200 100	-0.003	0.0001	0.008	0.015	0.011	-0.003	0.900	1.816	2.919	4.310

algorithm, equivalent voltage for each operation stage can be more accurate compared with that in previous model, permitting better MPPT performance.

VI. ACKNOWLEDGEMENT

This work was supported by the Research development fund of XJTLU (RDF-16-01-10), the Jiangsu Science and Technology Programme (BK20161252), and the Suzhou Prospective Application programme (SYG201723).

REFERENCES

[1] W. D. Soto, S. Klein, and W. Beckman, "Improvement and validation of a model for photovoltaic array performance," *Solar Energy*, vol. 80, no. 1, pp. 78 – 88, 2006.

[2] H. Tian, F. Mancilla-David, K. Ellis, E. Muljadi, and P. Jenkins, "A cell-to-module-to-array detailed model for photovoltaic panels," *Solar Energy*, vol. 86, no. 9, pp. 2695 – 2706, 2012.

[3] S. Jain and V. Agarwal, "A new algorithm for rapid tracking of approximate maximum power point in photovoltaic

systems," *IEEE Power Electron. Lett.*, vol. 2, no. 1, pp. 16–19, Mar. 2004.

[4] X. Li, H. Wen, and C. Zhao, "Improved beta parameter based mppt method in photovoltaic system," in *Proc. IEEE 9th Int. Power Electron. ECCE Asia Conf.*, Jun. 2015, pp. 1405–1412.

[5] X. Li, H. Wen, L. Jiang, Y. Hu, and C. Zhao, "An improved beta method with auto-scaling factor for photovoltaic system," *IEEE Trans. Ind. Appl.*, vol. 52, no. 5, pp. 4281–4291, Sep. 2016.

[6] X. Li, H. Wen, L. Jiang, W. Xiao, Y. Du, and C. Zhao, "An improved mppt method for pv system with fast-converging speed and zero oscillation," *IEEE Trans. Ind. Appl.*, vol. 52, no. 6, pp. 5051–5064, Nov. 2016.

[7] L. X. W. H. . X. W. Du, Y., "Perturbation optimization of maximum power point tracking of photovoltaic power systems based on practical solar irradiance data," *2015 IEEE 16th Workshop on Control and Modeling for Power Electronics (COMPEL)*, pp. 1–5, 2015.

[8] W. H. . H. Y. Li, X., "Evaluation of different maximum power point tracking (mppt) techniques based on practical meteorological data," *2016 IEEE International Conference on Renewable Energy Research and Applications*

(ICRERA), pp. 9–17, 2016.

[9] W. H. J. L. L. E. G. D. Y. . Z. C. Li, X., "Photovoltaic modified -parameter-based mppt method with fast tracking," *Journal of Power Electronics*, vol. 16, no. 1, pp. 1–5, 2016.

[10] N. K. H. Y. S. J. W. H. . Y. D. Xu, R., "Analysis of the optimum tilt angle for a soiled pv panel," *Energy Conversion and Management*, vol. 148, pp. 100–149, 2017.

[11] W. H. L. X. J. L. . H. Y. Luo, H., "Synchronous buck converter based low-cost and high-efficiency sub-module dmppt pv system under partial shading conditions," *Energy Conversion and Management*, vol. 126, pp. 473–487, 2017.

[12] D. Y. . W. H. Chen, X., "Forecasting based power ramp-rate control for pv systems without energy storage," *2017 IEEE 3rd International Future Energy Electronics Conference and ECCE Asia (IFEEC 2017-ECCE Asia)*, pp. 733–738, 2017.

[13] W. H. . H. Y. Chu, G., "Control method for flyback based submodule integrated converter with differential power processing structure," *2016 IEEE International Conference on Renewable Energy Research and Applications (ICRERA)*, pp. 684–689, 2016.

[14] H. Patel and V. Agarwal, "Maximum power point tracking scheme for pv systems operating under partially shaded conditions," *IEEE Trans. Ind. Electron*, vol. 55, no. 4, pp. 1689–1698, Apr. 2008.

[15] J. Ahmed and Z. Salam, "An improved method to predict the position of maximum power point during partial shading for pv arrays," *IEEE Trans. Ind. Informat.*, vol. 11, no. 6, pp. 1378–1387, Dec. 2015.

[16] E. I. Batzelis, I. A. Routsolias, and S. A. Papathanassiou, "An explicit pv string model based on the lambert w function and simplified mpp expressions for operation under partial shading," *IEEE Trans. Sustain. Energy*, vol. 5, no. 1, pp. 301–312, Jan. 2014.

[17] E. I. Batzelis, G. E. Kampitsis, S. A. Papathanassiou, and S. N. Manias, "Direct mpp calculation in terms of the single-diode pv model parameters," *IEEE Trans. Energy Convers.*, vol. 30, no. 1, pp. 226–236, Mar. 2015.

[18] X. Li, H. Wen, Y. Hu, L. Jiang, and W. Xiao, "Modified beta algorithm for gmppt and partial shading detection in photovoltaic systems," *IEEE Transactions on Power Electronics*, vol. PP, no. 99, pp. 1–1, 2017.

APPENDIX

Algorithm 3 Update irradiance level flags

Input: I_{IN}: Region initial current; $I_{SC}[1...10]$: Pre-load short circuit current values

Output: $Flag[1...10]$: Irradiance level information

1: **function** FLAGSET($I_{IN}, I_{SC}[1...10]$)
2:     **if** $I_{IN} - I_{SC}[1] < 0.1$ **then** $Flag[1] \leftarrow 1$
3:     **else if** $I_{IN} - I_{SC}[2] < 0.1$ **then** $Flag[2] \leftarrow 1$
4:     **else if** $I_{IN} - I_{SC}[3] < 0.1$ **then** $Flag[3] \leftarrow 1$
5:     **else if** $I_{IN} - I_{SC}[4] < 0.1$ **then** $Flag[4] \leftarrow 1$
6:     **else if** $I_{IN} - I_{SC}[5] < 0.1$ **then** $Flag[5] \leftarrow 1$
7:     **else if** $I_{IN} - I_{SC}[6] < 0.1$ **then** $Flag[6] \leftarrow 1$
8:     **else if** $I_{IN} - I_{SC}[7] < 0.1$ **then** $Flag[7] \leftarrow 1$
9:     **else if** $I_{IN} - I_{SC}[8] < 0.1$ **then** $Flag[8] \leftarrow 1$
10:    **else if** $I_{IN} - I_{SC}[9] < 0.1$ **then** $Flag[9] \leftarrow 1$
11:    **else if** $I_{IN} - I_{SC}[10] < 0.1$ **then** $Flag[10] \leftarrow 1$
12:    **end if**
13:    **return** $Flag[1...10]$
14: **end function**

Algorithm 4 Determine the correct approximation of voltage source in each operation stage

---

Input: $Flag[1...10]$: information of irradiance levels; I_{ST}: Measured current values of certain operation stage

Output: V_S: the approximation of current stage irradiance level

1: **function** VSCALCUALTION($Flag[1...10], I_{ST}$)
2:     // All equations below refer to Table I
3:     // Eg. eq.1000 refers to parameters under $1000\ W/m^2$
4:     **if** $Flag[1] == 1$ **then** $V_S \leftarrow eq.1000$ with I_{ST}
5:     **else if** $Flag[2] == 1$ **then** $V_S \leftarrow eq.900$ with I_{ST}
6:     **else if** $Flag[3] == 1$ **then** $V_S \leftarrow eq.800$ with I_{ST}
7:     **else if** $Flag[4] == 1$ **then** $V_S \leftarrow eq.700$ with I_{ST}
8:     **else if** $Flag[5] == 1$ **then** $V_S \leftarrow eq.600$ with I_{ST}
9:     **else if** $Flag[6] == 1$ **then** $V_S \leftarrow eq.500$ with I_{ST}
10:    **else if** $Flag[7] == 1$ **then** $V_S \leftarrow eq.400$ with I_{ST}
11:    **else if** $Flag[8] == 1$ **then** $V_S \leftarrow eq.300$ with I_{ST}
12:    **else if** $Flag[9] == 1$ **then** $V_S \leftarrow eq.200$ with I_{ST}
13:    **else if** $Flag[10] == 1$ **then** $V_S \leftarrow eq.100$ with I_{ST}
14:    **end if**
15:    **return** V_S
16: **end function**

Optimized Flux-Weakening Control of Induction Motor for Torque Enhancement in Voltage Extension Region

Zhen Dong, Yong Yu*, Bo Wang, Qinghua Dong and Dianguo Xu

Power Electronics & Electrical Drives (PEED), Harbin Institute of Technology, Harbin, China

*E-mail: yuyong@hit.edu.cn

Abstract— It is known that higher torque capability and less torque ripple cannot be obtained at the same time for flux-weakening control of induction motor in voltage extension region (outside the inscribed circle but within the hexagon). Therefore, a tradeoff strategy for maximum toque and suppressed torque ripple is necessary. Meanwhile, in many previous research studies, VCFS (voltage closed-loop flux-weakening scheme) based voltage extension methods rely on the reconstructed voltage, which means the control performance is affected by the accuracy of reconstruction algorithm and the stability of DC-Link voltage. In this paper, an optimized flux-weakening controller is proposed with the following features: a proper balance between torque capability and torque ripple, an enhanced DC-Link overvoltage rejection and a better torque performance for load change conditions. Simulation and experiment results verify the validity of the proposed scheme.

Keywords—Flux-weakening control, induction motor, torque enhancement, voltage extension.

I. INTRODUCTION

The high speed induction motor (IM) drives are extensively used in electrical vehicles (EVs) and tool spindle drives applications. Under constrains of maximum current limit and maximum voltage limit, the crucial task is to achieve high torque performance in high speed, including maximum output torque and ideal dynamic responses. Different from current limit which cannot change under certain conditions, voltage limit is determined by the output capability of inverter with coordination of SVPWM. When the voltage-limit trajectory migrates out of the inscribed circle, the instantaneous torque is improved because of the higher utilization of DC-Link voltage. It is significant for flux-weakening control to operate in over-modulation region in terms of fast acceleration capability and larger torque capability [1]. However, compared with hexagon operation, six-step operation in over-modulation region II brings in deteriorated sub-problems such as torque ripple, acoustic noise and inaccurate flux-orientation [2], it is not considered in this paper where high-performance control is mainly discussed.

Till now, high speed control schemes can be divided into two parts, direct torque control (DTC) [3] and field-oriented vector control (FOC) [4], and more information on the classification can be referred in [5]. Considering the intrinsic demerit of hysteresis loops in DTC, Deadbeat-Direct Torque and Flux control (DB-DTFC) [6] and Finite-Settling-Steps Direct Torque and Flux control (FSS-DTFC) [7] are developed. As potential techniques, they promise the hexagon operation for high speed control. However, these techniques are still full of challenges. In these methods, the voltage is calculated and manipulated based on motor model directly. The sensitivity of parameters requires the coordination of advanced current and flux observer and parameter estimation techniques to ensure the control performance [8]. While DTFC-based high speed control still needs to be further optimized, indirect FOC (IFOC) has been widely applied for high speed control. The classical method is "$1/\omega_r$" [9], which cannot achieve maximum torque capability. The voltage closed-loop flux-weakening scheme (VCFS) proposed in [10] is a successful method to ensure maximum torque and high-performance of high speed control [11]. Since the voltage reference in flux-weakening controller is set as the inscribed circle, the utilization of DC-Link voltage is limited. Schemes in [12-14] can be regarded as VCFS-based schemes, which are further developed to explore the limit capability of inverter, but they narrowly take torque ripple caused by over-modulation into consideration. In [15], considering the tradeoff between maximum torque and torque ripple suppression, a flux-weakening controller with selective quasi-hexagon voltage reference is designed. However, since the reconstructed voltage is used to build voltage reference, negative effects may be caused by the inaccuracy of reconstructed algorithm and fluctuation of DC-Link voltage.

In the following part, first, the VCFS and VCFS-based schemes to explore the maximum utilization of DC-Link voltage will be introduced briefly. Then, the quantitative analysis on the relationship among extended voltage, torque and torque ripple is given. Further, torque ripple brought by DC-Link overvoltage and the selection of

---

This paper and its related research are supported by grants from the Power Electronics Science and Education Development Program of Delta Group (DREK2017002)

operating point for torque ripple alleviation in load change condition are discussed. Based on the analysis, an optimized flux-weakening control strategy for torque enhancement in voltage extension region is proposed. Finally, the simulation and experimental results show the feasibility and the superiority of the proposed method with regard to the speed acceleration time, torque ripple suppression and DC-Link overvoltage rejection.

II. MAXIMUM TORQUE CONTROL OF IM IN HIGH SPEED REGION

A. Ideal Maximum Torque Control for High Speed Operation of IM

Combining the simplified IM voltage equation and torque equation under the maximum current and voltage limit, the limitations in voltage form and current form are expressed as

$$\begin{cases} u_{sd}^2 + u_{sq}^2 \le u_{s\max}^2 \\ |u_{sd}| \le u_{s\max}/\sqrt{2} \\ \left(\dfrac{u_{sd}}{\omega_e \sigma L_s}\right)^2 + \left(\dfrac{u_{sq}}{\omega_e L_s}\right)^2 \le i_{s\max}^2 \end{cases} \quad (1)$$

$$\begin{cases} i_{sd}^2 + i_{sq}^2 \le i_{s\max}^2 \\ u_{sd}^2 + u_{sq}^2 \le u_{s\max}^2 \\ i_{sd} \ge \sigma i_{sq} \end{cases} \quad (2)$$

The ideal voltage (segment OCD in left half plane) and current (segment ABO in right half plane) vector trajectory for maximum torque is shown in Fig. 1, where both u_{smax} and i_{smax} are treated as the constant values. Thus the whole speed operation is divided into three regions: the base speed region (the constant-torque region), segment OC and point A; the flux-weakening I region (the constant-power region), segment CD and segment AB; the flux-weakening II region (the constant-voltage region), point D and segment BO.

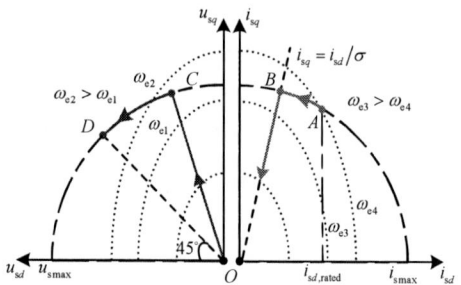

Fig. 1. Ideal voltage (left half plane) and current (right half plane) vector trajectory for maximum torque.

B. Analysis of D- and Q-axis Current Distribution in Existing VCFS-based Methods

1) D-axis current distribution

Fig.2 shows the structures of flux-weakening controller in VCFS and VCFS-based schemes in previous research studies. Fig. 2(a) is VCFS presented in [10]. The voltage reference is set to $U_{dc}/\sqrt{3}$, the radius of inscribed circle in SVPWM. Fig. 2(b) extends the voltage reference by adjusting the coefficient k_{utl}, and satisfies $k_{utl} \le 1$ [16].

Fig.2(c) uses the error between switching time and active switching time to build the flux-weakening controller. This method guarantees the hexagon operation in high speed. In Fig. 2(d), the reconstructed voltage after over-modulation is used as voltage reference. Fig. 2(e) utilizes the adjustable voltage reference by comparing the value of adjusted extended circle and the reconstructed voltage. The features of the schemes above are summarized in Table I. An optimized scheme should have adjustable DC-Link voltage utilization so that the torque and torque ripple are adjustable. Meanwhile, it is better to generate voltage reference without using reconstructed voltage and with the rejection of DC-Link overvoltage.

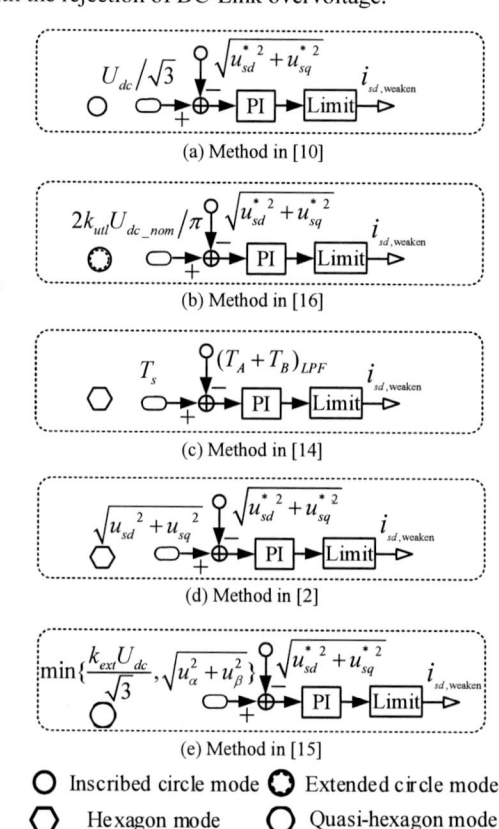

Fig. 2. Block diagram of VCFS and VCFS-based schemes.

TABLE I
PERFORMANCES COMPARISON

Schemes	Utilization of DC-Link Voltage	DC-Link overvoltage Rejection	Torque ripple	Voltage reconstruction
(a)	Low	No	Low	No
(b)	Adjustable	Yes	High	No
(c)	High	No	High	No
(d)	High	No	High	Need
(e)	Adjustable	No	Adjustable	Need

2) Q-axis current distribution

For VCFS-based methods, the maximum voltage limit is used as close loop, and therefore, q-axis current is limited as $\sqrt{i_{s\max}^2 - i_{sd}^2}$ in the constant-torque region and the constant-power region. (If the maximum voltage acts as a limit, the maximum current limit is for close-loop control, seen in [17]). While in the constant voltage region, q-axis current needs to be reduced due to the maximum slip constraint under the maximum voltage limit. There are three equivalent conditions for q-axis current limit if stator resistance is neglected: (1) current/flux inequality: $i_{sd} \geq \sigma i_{sq}$ or $\lambda_{dr} \geq \sigma L_m i_{sq}$ [4],[18]; (2) voltage inequality: $|u_{sd}| \leq u_{s\max}/\sqrt{2}$ [10]; (3) the angle between the input voltage vector and the rotor flux vector is no more than $3\pi/4$ [19].

C. Torque Analysis in Hexagon Region

When IM operates in hexagon region to further extend the utilization of DC-Link voltage, the circumstance is much more complicated than that shown in Fig. 1. The detailed quantitative analysis can be found in our previous work shown in [15] and here the final result is shown directly. Fig. 3 describes the possible voltage vector trajectory at a certain operating time. The operating point could be randomly at any points on segment AB, causing the torque ripple. The quantitative relationship between torque and voltage at a certain speed ω_e^* in hexagon region can be derived as

$$T_{e\max, \omega_e^*} = \min\{T_{e\max\_\beta_1, \omega_e^*}, T_{e\max\_\beta_2, \omega_e^*}\} \qquad (3)$$

where:

$$T_{e\max\_\beta_x, \omega_e^*} = G(\omega_e^{*2})[-k_1 k_2 u_{sd}^2 + (k_1 b_2 - k_2 b_1)u_{sd} + b_1 b_2]$$

$$\omega_e^* = \sqrt{\frac{U_{dc}^2(\cos^2\alpha^* + \sigma^2\sin^2\alpha^*)}{3i_{s\max}^2\sigma^2 L_s^2}}$$

$$G(\omega_e^{*2}) = \frac{3n_p L_m^2}{2L_r(R_s + \omega_e^{*2}\sigma L_s^2)^2}$$

$$k_1 = R_s - \frac{\cos\beta_x}{\sin\beta_x}\omega_e^*\sigma L_s, b_1 = \omega_e^*\sigma L_s \frac{U_{dc}}{\sqrt{3}\sin\beta_x}$$

$$k_2 = \frac{\cos\beta_x}{\sin\beta_x}R_s + \omega_e^*L_s, b_2 = \frac{U_{dc}R_s}{\sqrt{3}\sin\beta_x}$$

$$x = 1 \ for \ l_{\beta_1}; \quad x = 2 \ for \ l_{\beta_2}$$

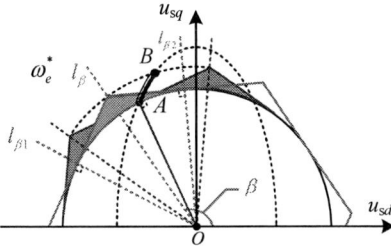

Fig. 3. Possible operating points at a certain operating time.

Fig. 4 shows the torque trend of hexagon operation mode and inscribed circle operation mode at different speeds. It can be seen that the extended voltage brings in the increase of maximum torque and severer torque ripple. However, different extended voltage values make different contributions to maximum torque improvement and torque ripple. In the other word, as limit voltage extends towards the hexagon, it brings more torque ripple but less torque improvement. Though the torque ripple cannot be removed completely, as a tradeoff, it can be

suppressed if the torque trend is manipulated by setting a proper k_{ext} value to remove the top part.

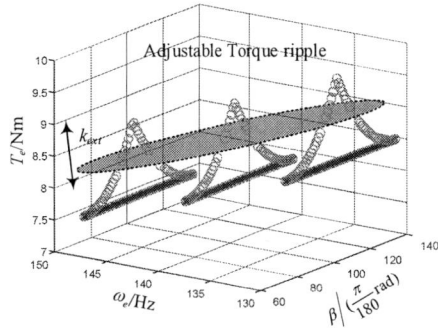

Fig. 4. Torque trend of hexagon mode (red curve) and inscribed circle mode (blue curve) at different speeds.

1) Torque ripple analysis under the DC-Link overvoltage

It is known that the DC-Link overvoltage usually happens due to the energy feedback during brake and the case is worse especially when the step deceleration from a high speed is required. For example, it is assumed that the overvoltage of DC-Link is 130%, which is the protection threshold value that is often set in commercial inverter. According to (3), Tab. II shows the changes of maximum torque and torque ripple of inscribed circle mode and hexagon operation mode in varying frequencies under 100% and 130% DC-Link voltage respectively.

TABLE II
TORQUE COMPARISON

$\frac{(\Delta)T_e}{U_{dc}}\diagdown\omega_e$	180Hz		200Hz		220Hz	
	○	⬡	○	⬡	○	⬡
1.0p.u.	6.03—17.19%→7.07		5.35—17.69%→6.29		4.77—18.08%→5.64	
	33.21%	30.89%	34.67%	32.19%	36.20%	34.08%
	55.57%	58.33%				
1.3p.u.	8.04—15.14%→9.26		7.20—15.52%→8.32		6.50—16.24%→7.56	

$N \cdot m$

From Tab. II, the following conclusions can be yielded:

(1) The change of torque (maximum torque or torque ripple) caused by the change of DC-Link voltage is severer as ω_e increases, no matter in the inscribed circle mode or in the hexagon mode.

(2) At a certain ω_e, the higher DC-Link voltage always shows a better tolerance to torque ripple in the hexagon mode.

(3) In high speed region, 130% DC-Link overvoltage results in more than 50% torque ripple in the hexagon mode, while in the inscribed circle mode the torque ripple is about 30%, and this difference is larger as ω_e increases.

It indicates that hexagon operation for flux-weakening control shows weaker robustness to DC-Link voltage fluctuation, which is a serious factor that results in deteriorated system performance and even instability. Therefore, even though torque ripple caused by the feedback energy is inevitable and cannot be controlled by flux-weakening methods, it still will be an improvement for system stability if the fluctuated DC-Link voltage is removed from the control loop, especially for hexagon operation mode.

2) Torque ripple alleviation during load change

In voltage extension region, since the hexagon or quasi-hexagon voltage limit is used for closed-loop flux-weakening control, the operating point always moves along the fluctuated voltage boundary, resulting in undesirable torque ripple regardless of the load condition. As shown in Fig.5, with the reduction of torque from T_1 to T_2, the operating point O will possibly move along the red curve in transient state and keep varying between Point A and Point B in steady state. Considering that the maximum torque control is only needed in some cases such as step acceleration/deceleration and full-load operation, torque ripple can be averted by adjusting the operating point in non-maximum torque control. For example, if point D is selected, the influence of torque ripple can be well eliminated as d- and q-axis voltage shrinks rapidly, however, the great reduction may bring in undesired instability. Point C is proved as a proper point for dynamic torque response with additional control strategy to ensure a constant flux [10], while Point B can be easily achieved by adjusting the voltage reference in flux-weakening controller. Besides, the operating point selection can be further discussed if other requirements e.g. minimum current amplitude, are taken into consideration.

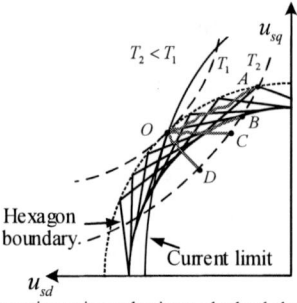

Fig. 5. Operating points selection under load change conditions

III. PROPOSED FLUX-WEAKENING CONTROL METHOD

Based on the summary discussed in Section II, this part proposes a novel flux-weakening controller by designing the voltage reference in a direct way. To extend the maximum torque while suppress the torque ripple, the voltage reference in flux-weakening controller is determined by the minimum value of $k_{ext}U_{dc\_nom}/\sqrt{3}$ and an ideal hexagon boundary to form a quasi-hexagon with rounded edges, as shown in Fig.6. As for the construction of the hexagon, in two phases stationary reference frame, synchronous reference frame and the voltage vector are rotating and maintaining the relative position shown in Fig. 1 as ω_e increases. It is important to ensure the phase consistency of reference and feedback voltage vector so that the feedback voltage vector can always follow the reference to slide alongside the quasi-hexagon boundary, as shown in Fig. 7. In the proposed

controller, the nominal value of U_{dc} is applied. Meanwhile, the zero-phase-error over-modulation ensure the accuracy of the constructed voltage reference. The ideal hexagon is expressed as (4), and since the entire controller is only related to θ, it is robust and independent with the accuracy of reconstructed voltage applied in [2], [15].

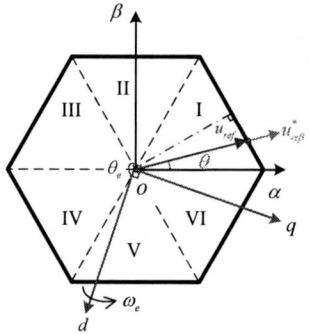

Fig. 6. Block diagram of reference adjustable flux-weakening controller with direct constructed boundary.

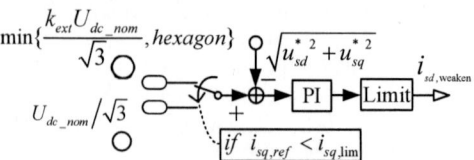

Fig. 7. Formation of the direct constructed voltage reference.

$$
hexagon = \begin{cases}
\text{Sec.I:}\ \dfrac{U_{dc\_nom}}{\sqrt{3}}/(\dfrac{\sqrt{3}}{2}\cos\theta + \dfrac{1}{2}\sin\theta) \\[2mm]
\text{Sec.II:}\ \dfrac{U_{dc\_nom}}{\sqrt{3}}/\sin\theta \\[2mm]
\text{Sec.III:}\ \dfrac{U_{dc\_nom}}{\sqrt{3}}/(-\dfrac{\sqrt{3}}{2}\cos\theta + \dfrac{1}{2}\sin\theta) \\[2mm]
\text{Sec.IV:}\ \dfrac{U_{dc\_nom}}{\sqrt{3}}/(-\dfrac{\sqrt{3}}{2}\cos\theta - \dfrac{1}{2}\sin\theta)\ (4) \\[2mm]
\text{Sec.V:}\ -\dfrac{U_{dc\_nom}}{\sqrt{3}}/\sin\theta \\[2mm]
\text{Sec.VI:}\ \dfrac{U_{dc\_nom}}{\sqrt{3}}/(\dfrac{\sqrt{3}}{2}\cos\theta - \dfrac{1}{2}\sin\theta)
\end{cases}
$$

where: $\sin\theta = \dfrac{u_\beta^*}{\sqrt{u_\alpha^{*2} + u_\beta^{*2}}}$; $\cos\theta = \dfrac{u_\alpha^*}{\sqrt{u_\alpha^{*2} + u_\beta^{*2}}}$

Further, for a better torque performance under load change conditions, the operating point should be manipulated to shrink out of the voltage extension region. Here, point B in Fig. 5 is taken as an example and the control strategy is modified as Fig. 8 shows. The error of q-axis current limit and reference is used as the judgment.

$$\min\{\frac{k_{ext}U_{dc\_nom}}{\sqrt{3}}, hexagon\}$$

$$U_{dc\_nom}/\sqrt{3}$$

$$\sqrt{u_{sd}^{*2} + u_{sq}^{*2}} \rightarrow \text{PI} \rightarrow \text{Limit} \rightarrow i_{sd,weaken}$$

$$if\ i_{sq,ref} < i_{sq,\lim}$$

Fig. 8. Block diagram of flux-weakening controller for operating point manipulation.

IV. SIMULATION AND EXPERIMENTAL RESULTS

Following simulation and experimental results are given to verify the features of adjustable torque and torque ripple and the rejection of DC-Link overvoltage respectively. Parameters of the induction motor is shown in Table. III.

TABLE III
PARAMETERS OF THE INDUCTION MACHINE

Parameter	Value	Parameter	Value
Rated Power	3.7kW	Rated Torque	23.6N·m
Rated Speed	1500r/min	Stator Resistance	1.142Ω
Rated Voltage	380V	Rotor Resistance	0.825Ω
Rated Current	8.9A	Mutual Inductance	118.9mH
Rated frequency	50Hz	Stator Inductance	124.4mH
Inertia	0.0123kg·m²	Rotor Inductance	124.4mH

A. Adjustable Maximum Torque and Torque Ripple

In this part, k_{ext} is set as 1.15, 1.05 and 1.00, corresponding to hexagon (the same as Scheme (c) and Scheme (d) in Fig. 2), quasi-hexagon and inscribed circle (the same as Scheme (a) in Fig. 2) respectively. Fig. 9 shows that the maximum torque increases as k_{ext} is adjusted to larger values, therefore shorting the step acceleration time. In Fig. 10 and Fig. 11, more detailed information on torque ripple caused by different k_{ext} is shown. It can be seen that when k_{ext} is set from 1.05 to 1.15, the increase of maximum torque is so limited that the acceleration time is hardly shorten, but the torque ripple is more severe. Though the increasing maximum torque and the more severe torque ripple are syngeneic, we can still find a proper value $k_{ext} \in [1.05, 1.10]$ as a tradeoff. The experimental results are shown in Fig. 12.

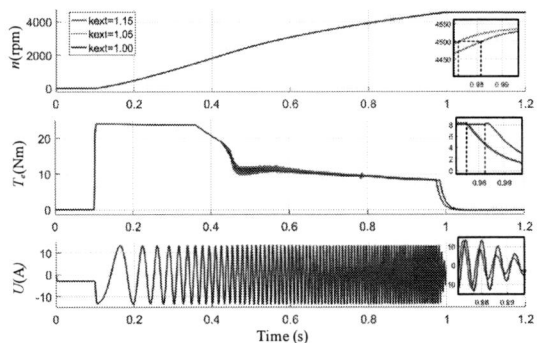

Fig. 9. Simulation results: Speed, torque and phase current responses of speed step command under various k_{ext}.

B. Torque Ripple Alleviation under Load Change

To verify the effectiveness of the operating point manipulation method shown in Fig. 8, comparison experiment shown in Fig. 13 is undertaken. A step speed command from 0rpm to 4500rpm is given, and after a short acceleration time when the system is in the steady state, a step change of 10% rated load each time is given until the system tends to be instable. It can be seen that under the same load condition, without operating point manipulation, torque response suffers with severer torque ripple even in steady state, while with operating point manipulation, torque ripple is greatly alleviated in non-maximum output torque conditions.

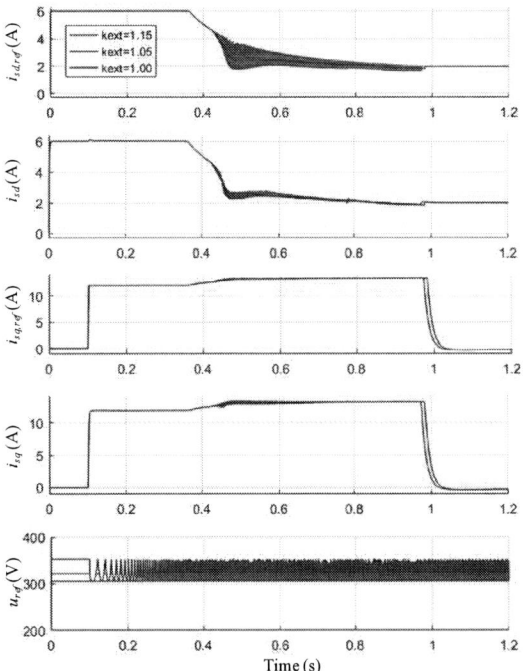

Fig. 10. Simulation results: D-q axis current references and responses and voltage reference of speed step command under various k_{ext}.

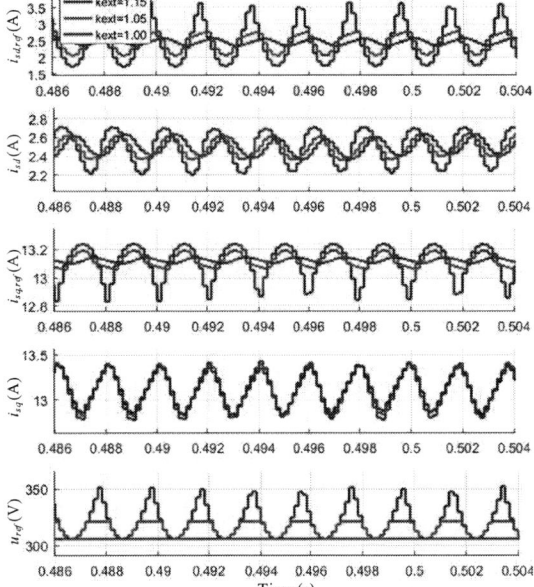

Fig. 11. Simulation results (enlarged): D-q axis current references and responses and voltage reference of speed step command under various k_{ext}.

C. DC-Link Overvoltage Rejection

In this part, conventional method is Scheme (e) shown in Fig. 2. In Fig. 14, DC-Link voltage is manipulated from 537V to 577V. It is seen from d-axis current that the proposed method has strong robustness for DC-Link overvoltage. The experimental condition in Fig. 15 is step deceleration from 4500rpm to 3000rpm. The wave of terminal voltage indicates the high speed operation in quasi-hexagon mode. The instantaneous increase of DC-

The 2018 International Power Electronics Conference

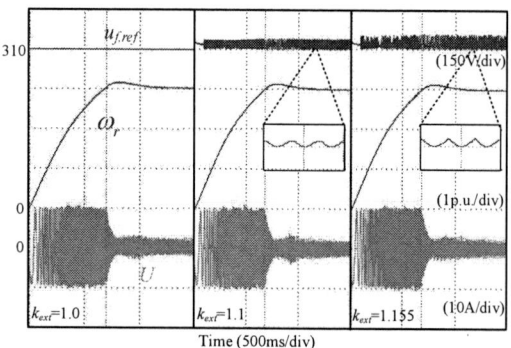

Fig. 12. Experimental results: voltage reference in flux-weakening controller, speed response and phase current of speed step command from 0 to 4500rpm under various k_{ext}.

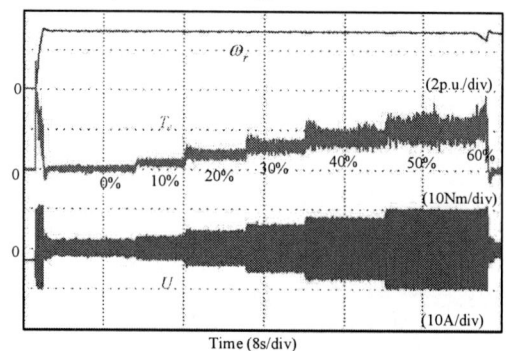

(a) Without operating point manipulation

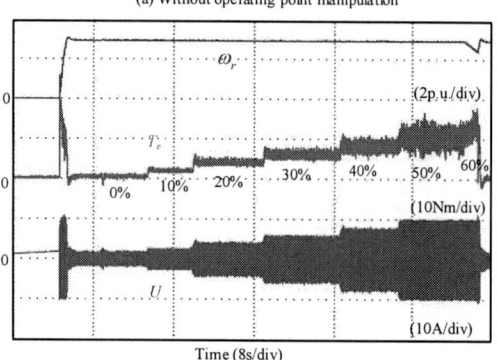

(b) With operating point manipulation

Fig. 13. Experimental results: Comparison of torque response under load change conditions at 4500rpm.

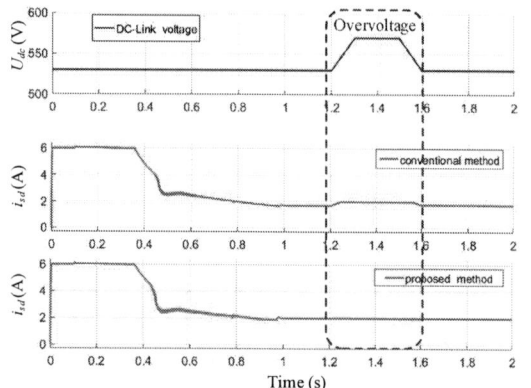

Fig. 14. Simulation results: Comparison of d-axis current responses under overvoltage of DC-Link voltage; k_{ext}=1.05.

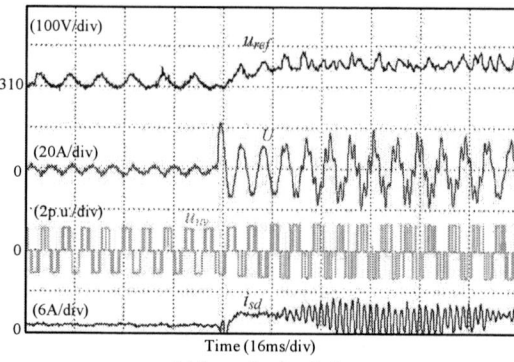

(a) Conventional method

(b) Proposed method

Fig. 15. Experimental results: Comparison between conventional method and proposed method under overvoltage of DC-Link voltage; k_{ext}=1.10.

Link voltage is caused by the energy feedback. The smoother phase current and more stable d-axis current response shown in Fig. 15(b) prove the superiority of proposed method.

V. CONCLUSION

This paper focuses on two issues on flux-weakening controller design of high speed IM control system in voltage extension region: the tradeoff between maximum torque increase and torque ripple suppression and DC-Link voltage overvoltage rejection. A novel method for flux-weakening controller design is proposed in which the voltage reference is constructed directly instead of using the reconstructed voltage. Meanwhile, torque ripple analysis under DC-Link overvoltage and operating point selection for torque ripple alleviation under non-maximum output torque condition are presented. Verification on proposed scheme has been taken on by simulation and experiment on a commercial 3.7kW platform. The results show that higher torque with relatively lower torque ripple can be obtained by adjusting coefficient k_{ext}, and a strong overvoltage rejection of DC-Link voltage is ensured at the same time. Besides, with an operating point manipulation, torque performance is further enhanced.

REFERENCES

[1] L. Ping-Yi and L. Yen-Shin, "Novel voltage trajectory control for field-weakening operation of induction motor drives," *IEEE Trans. Ind. Appl.*, vol. 47, no. 1, pp. 122–127, Jan./Feb. 2011.

[2] Y. Kwon, S. Kim and S. K. Sul, "Six-Step Operation of PMSM With Instantaneous Current Control" *IEEE Trans. Ind. Appl.*, vol. 50, no. 4, pp. 2614–2625, Jul./Aug. 2014.

[3] A. Tripathi, A. M. Khambadkone, and S. K. Panda, "Dynamic control of torque in overmodulation and in the field weakening region," *IEEE Trans. Power Electron.*, vol. 21, no. 4, pp. 1091–1098, Jul. 2006.

[4] S. H. Kim and S. K. Sul, "Maximum torque control of an induction machine in the field weakening region," *IEEE Trans. Ind. Appl.*, vol. 31, no. 4, pp. 787–794, Jul./Aug. 1995.

[5] L. Zarri, M. Mengoni, A. Tani, G. Serra, D. Casadei, and J. O. Ojo, "Control schemes for field weakening of induction machines: A review," *In Proc. IEEE Workshop Electr. Mach. Design, Control, Diagnosis*, 2015, pp. 146–155.

[6] C.-H. Choi, J.-K. Seok, and R. D. Lorenz, "Wide-speed direct torque and flux control for interior PM synchronous motors operating at voltage and current limits," *IEEE Trans. Ind. Appl.*, vol. 49, no. 1, pp. 109–117, Jan./Feb. 2013.

[7] S. H. Kim and J. K. Seok, "Finite-settling steps direct torque and flux control (FSS-DTFC) for torque-controlled interior PM motors at voltage limits," *IEEE Trans. Ind. Appl.*, vol. 50, no. 5, pp. 3374–3381, Sep./Oct. 2014.

[8] S. Kim and J. K. Seok, "Hexagon voltage manipulating control (HVMC) for AC motor drives operating at voltage limit," *IEEE Trans. Ind. Appl.*, vol. 51, no. 5, pp. 3829–3837, Sep./Oct. 2015.

[9] X. Xu and D. W. Novotny, "Selection of the flux reference for induction machine drives in the field weakening region," *IEEE Trans. Ind. Appl.*, vol. 28, no. 6, pp. 1353–1358, Nov./Dec. 1992.

[10] S. H. Kim and S. K. Sul, "Voltage control strategy for maximum torque operation of an induction machine in the field-weakening region," *IEEE Trans. Ind. Electron.*, vol. 44, no. 4, pp. 512–518, Aug. 1997.

[11] L. Harnefors, K. Pietilainen, and L. Gertmar, "Torque-maximizing field-weakening control: design, analysis, and parameter selection," *IEEE Trans. Ind. Electron.*, vol. 48, no. 1, pp. 161–168, Feb. 2001.

[12] T. S. Kwon and S. K. Sul, "Novel antiwindup of a current regulator of a surface-mounted permanent-magnet motor for flux-weakening control," *IEEE Trans. Ind. Appl.*, vol. 42, no. 5, pp. 1293–1300, Sep./Oct. 2006.

[13] T.-S. Kwon, G.-Y. Choi, M.-S. Kwak, and S.-K. Sul, "Novel flux-weakening control of an IPMSM for quasi-six-step operation," *IEEE Trans. Ind. Appl.*, vol. 44, no. 6, pp. 1722–1731, Nov./Dec. 2008.

[14] L. Ping-Yi and L. Yen-Shin, "Voltage control technique for the extension of DC-link voltage utilization of finite-speed SPMSM drives," *IEEE Trans. Ind. Electron.*, vol. 59, no. 9, pp. 3392–3402, Sep. 2012.

[15] Z. Dong, Y. Yu, W. Li, B. Wang and D. Xu, "Flux-weakening Control for Induction Motor in Voltage Extension Region: Torque Analysis and Dynamic Performance Improvement," *IEEE Trans. Ind. Electron.*, vol. 65, no. 5, pp. 3740–3751, May. 2018.

[16] B. J. Siebel, T. M. Rowan, and R. J. Kerkman, "Field-oriented control of an induction machine in the field-weakening region with dc-link and load disturbance rejection," *IEEE Trans. Ind. Appl.*, vol. 33, no. 6, pp. 1578–1584, Nov./Dec. 1997.

[17] S. D. Sudhoff and H. J. Hegner, "A flux-weakening strategy for current regulated surface-mounted permanent-magnet machine drives," *IEEE Trans. Energy Convers.*, vol. 10, no. 3, pp. 431–437, Sep. 1995.

[18] B. Wang, Y. Zhao, Y. Yu, G. Wang, D. Xu, and Z. Dong, "Speed-sensorless induction machine control in the field-weakening region using discrete speed-adaptive full-order observer," *IEEE Trans. Power Electron.*, vol. 31, no. 8, pp. 5759–5773, Aug. 2016.

[19] M. Mengoni, L. Zarri, A. Tani, G. Serra, and D. Casadei, "Stator flux vector control of induction motor drives in the field-weakening region," *IEEE Trans. Power Electron.*, vol. 23, no. 2, pp. 941–949, Mar. 2008.

The 2018 International Power Electronics Conference

Improved Performance of CFTC-based Direct Torque Control of Induction Machines by Increasing Torque Loop Bandwidth

Ibrahim Mohd Alsofyani[1], June-Hee Lee[1], Byung-Moon Han[2], and Kyo-Beum Lee[1*]

1 Department of Electrical and Computer Engineering, Ajou University, Korea
2 Department of Electrical Engineering, Myong-ji University, Korea
*E-mail: kyl@ajou.ac.kr

Abstract— Constant frequency torque controller-based direct torque control (CFTC-DTC) of AC motor drives is known to have limitation with torque loop bandwidth, since it uses sampling frequency available from a digital signal processor for generating triangular carrier-based waveform. The switching frequency of the control system is determined by the carrier frequency of CFTC. The worst case of CFTC-DTC drive exists at very low speed under light load where the zero voltage vectors are dominant, and the radial component of active voltage vectors are very short leading to flux droop at sector transitions. This results in the distortion of motor phase current and aggravation of speed waveform. This paper proposes a modification on the frequency carriers of CFTC to increase the torque loop bandwidth by replacing the triangular carrier-based with ramp carrier-based waveform. The proposed method is proven by the simulation results showing excellent performance of motor flux, phase current, and speed at low speed.

Keywords— *Direct torque control, Flux regulation, Induction motor.*

I. INTRODUCTION

Since the introduction of direct torque control (DTC) of AC motor drives in 1985 [1] and being marketed by ABB in 1996, DTC has established an increasing popularity in industrial applications due to its simple structure and excellent torque performance. The common problems associated with the original DTC are variable switching frequency, large torque ripples, flux droop at low speed. Over the last three decades, significant efforts have been made to eliminate these inherent problems associated with the DTC . Some efforts have focused on the modification of the switching table [2-4], while others have exploited high-computational methods such as space vector modulation [5], and model predictive torque control [6]. However, these proposed algorithms deteriorate the simplicity and robustness of original DTC.

However, there are some variations to the original construction of DTC were proposed to solve these drawbacks in the hysteresis based DTC. Among these methods, authors [7-8] have been carried out by injecting high-frequency triangular signals into the torque errors;

this method is called the dithering technique. Even though it is a simple method, an unpredictable switching frequency is generated since the torque slopes that govern the frequency of the torque controller varies based on different operating conditions.

Recently, constant frequency torque controller-based direct torque control (CFTC-DTC) has received increased attention in both Induction Motor (IM) [9-10] and Permanent-Magnet Synchronous Motors (PMSM) [11] drives due to its simplicity and easy implementation. The configuration of CFTC-DTC is the simplest of other variations of direct torque control (DTC), since it retains the same components of the original DTC except for the replacement of torque hysteresis comparator. In addition, CFTC drive can be utilized in 2-level inverter and multi-level inverter topologies [11]. When compared to the original DTC, the CFTC-DTC can provide constant switching frequency operation, improved flux regulation, and reduced torque ripple. However, although CFTC-DTC has a good capability to regulate the flux at low and zero speed region, it still requires some improvement during the sector transition of flux linkage vector. An analysis of the flux droop for the original DTC is presented in [9]. It has been investigated that the flux droop problem is highly influenced by the long duration of null voltage vector at low speed, whereby the motor stator resistance effect is inevitable. Due to the constraint of torque loop bandwidth in CFTC-DTC, the null vector is dominant and the active voltage vector (i.e. radial voltage component) is too short at very low motor speed under light load causing a droop in stator flux magnitude at each sector transition. One way to solve this problem is to ensure higher torque loop bandwidth.

This paper proposes a ramp carrier-based waveform for the CFTC-DTC of induction machine, which has twice switching frequency of the triangular wave-based carrier. In this way, the effect of stator resistance on the reduction of the flux magnitude will be smaller, and at the same time, the number of radial components of voltage vectors will increase. Consequently, the flux droop at low speed will be eliminated without extending another FPGA controller as in [12]. In addition, further reduction in the torque ripple will be achieved, the current distortions will be reduced, and the speed waveform will be enhanced. The main advantage of this proposed method is its simplicity while retaining the same structure of the conventional CFTC controller. Some experimental results are presented to show the effectiveness of the new carrier on CFTC-DTC

II. CONVENTIONAL CFTC-DTC AND ITS LIMITATION

Fig. 1(a) shows the and complete structure of CFTC-DTC of induction machine [13] consisting mainly of a switching table, two-level flux hysteresis comparator, and conventional CFTC controller (see Fig. 1(b)), which consists of two triangular wave-based carriers in the upper and lower sides of the CFTC controller, a PI controller and two comparators to select torque error status (T_{stat}) which are indicated by: 1 for forward active voltage, -1 for reverse active voltage, and 0 for zero voltage vector. The general rules for the selection of the switching states ($S_{a,b,c}$) for active and zero voltage vectors are available in Table I.

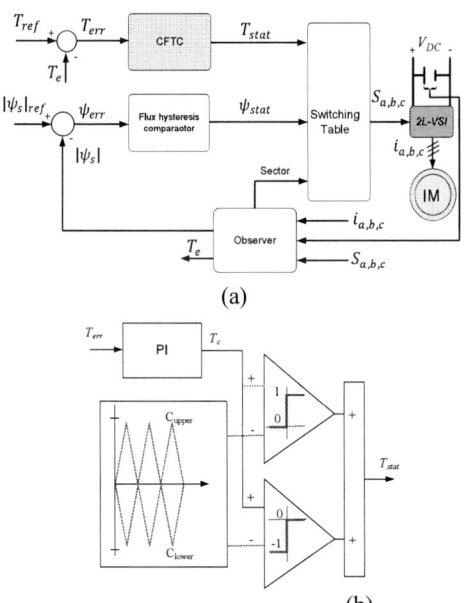

(a)

(b)

Fig.1 (a) Constant frequency torque controller (CFTC). (b) Complete structure of Conventional CFTC-DTC.

The complete description of the control system is available in [9] where the estimation of d-q components of flux and calculated electrical torque can be found. The output rules for the CFTC are expressed as:

$$T_{stat} = \begin{cases} 1 & for & T_c \geq C_{upper} \\ 0 & for & C_{lower} < T_c < C_{upper} \\ -1 & for & T_c < C_{lower} \end{cases} \quad (1)$$

where T_{stat} is the torque error status and the indexes; 1, 0, -1 indicate forward, null, and reverse voltage vectors,

respectively. T_c is the output of PI controller. Cupper and C_{lower} are upper and lower triangular-based carriers, respectively. In the discrete implementation of the DTC, the stator flux change in each sampling time (T_s) is given by

$$\Delta\psi_s = (v_s - i_s R_s)T_s \quad (2)$$

where vs is the stator voltage vector, is i_s the stator current vector, Rs is the stator resistor, and ψ_s is the stator flux vector. The stator resistance effect ($-i_s R_s$) in (2) is normally neglected when estimating the flux in the medium and high speed. However, the effect of resistance becomes more pronounced at very low speed under light load as the duration of the null voltage vector becomes the longest. The basic DTC is highly influenced by the resistance [14, 15] effect leading to a failure of flux regulation. Unlike original DTC, the stator flux in CFTC-DTC does not fall completely at very low speed owing to the nature of constant switching operation as shown in Fig. 2. Due to the limitation of switching frequency of carriers, it is noticed there is a droop at the beginning of each sector causing distortion of the motor phase current. This is because of radial component (to the circular flux locus) of voltage vector responsible for increasing flux.

TABLE I
SELECTION OF VOLTAGE VECTORS

Flux Demand (ψ_{stat})	Torque Demand (T_{stat})	Voltage Selection
1	1	Forward Active voltage vectors
	-1	Reverse active voltage vectors
	0	Zero voltage vectors
0	1	Forward Active voltage vectors
	-1	Reverse active voltage vectors
	0	Zero voltage vectors

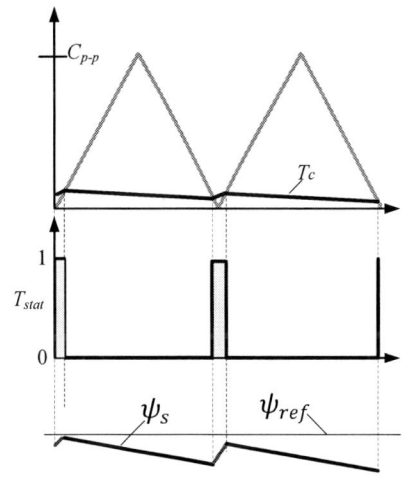

Fig. 2: Flux regulation using upper carrier conventional CFTC of at low speed.

The 2018 International Power Electronics Conference

Fig. 3 shows the space vector of stator flux linkage moving counterclockwise in sector (k+1). In this sector, there are two optimized voltage vectors to increase and reduce the flux linkage represented by v_{k+2} and v_{k+3}, respectively. Focusing only in voltage vector of increasing flux (due to the space limitation), it is shown there are three positions of flux linkage; ($\psi_{s,0}$) in the beginning, ($\psi_{s,1}$) in the middle, and ($\psi_{s,2}$) at the end of sector. It can be noticed the radial component ($d^{v_{k+2}}$) is the shortest at the beginning of sector, and keeps incrementing until it becomes the largest at the end of sector. Therefore, the long zero vector duration and weak radial component are the main reasons for flux droop during sector transitions.

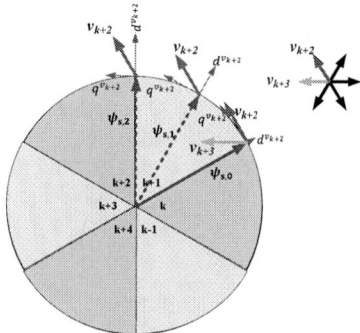

Fig. 3: Variations of radial component of voltage vector through sector k+1

III. IMPROVEMENT OF CFTC

To enhance the performance of the CFTC-DTC, the torque loop bandwidth should be increased. In the proposed method, the rules and parts of the conventional CFTC are retained the same except for the replacement of the triangular-based with a ramp-based carrier waveform (see Fig. 4). It is shown in Fig. 5 that the new carrier frequency (solid lines) is twice that of the conventional one (dashed lines), thus resulting in the increase of switching frequency. As such, the problem of flux regulation at sector transition can be simply solved by reducing the duration of null vectors and increasing the number of active vectors as indicated by T_{stat} shown in Fig. 5.

IV. SIMULAITON RESULTS

To validate the effectiveness of the proposed flux regulation method, simulations are carried out for conventional and proposed CFTC-DTC methods using a PSIM simulation tool. The system parameters and DTC values are as shown in Table II. Figs. 6 and 7 show the experimental results for the conventional and proposed CFTC-DTC drives, respectively. From top to bottom, the waveforms are the speed, stator flux, electrical torque, and phase current. It is obvious from [16] that the original DTC has large ripple in torque and current waveforms. It is also shown when the motor speed steps to lower speed,

the flux completely falls to zero and as a result all the variables drop to zero. On the other hand, Fig. 6 shows the performance of the CFTC-DTC with basic carrier and at a motor speed of 15 rad/s. The problem of poor flux regulation at sector transition is clearly shown with conventional carrier. However, when the ramp carrier is applied to CFTC-DTC, the flux becomes well-regulated and the performance of speed and current is enhanced as seen in Fig. 7.

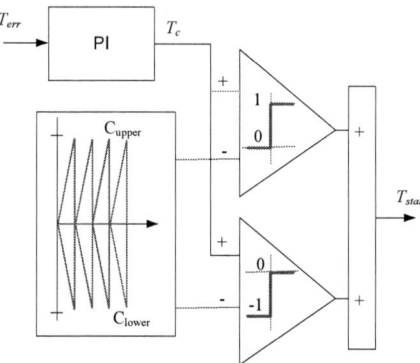

Fig. 4: Proposed CFTC controller

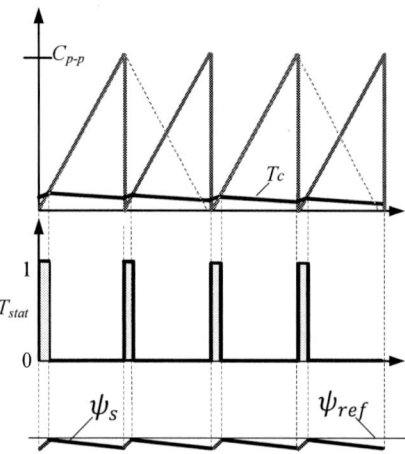

Fig. 5: Effect of proposed upper carrier frequency on Flux regulation

TABLE II
INDUCTION MOTOR PARAMETERS

Rated power	3.7 kW
Rated current	8.28 A
Rated speed	1750 rpm
Rated Torque	20.36 Nm
Stator resistance	0.934 ohm
Rotor resistance	1.225 ohm
Stator inductance	146.213 mH
Rotor inductance	146.213 mH
Mutual inductance	139.516 mH
Pole pairs	2

The 2018 International Power Electronics Conference

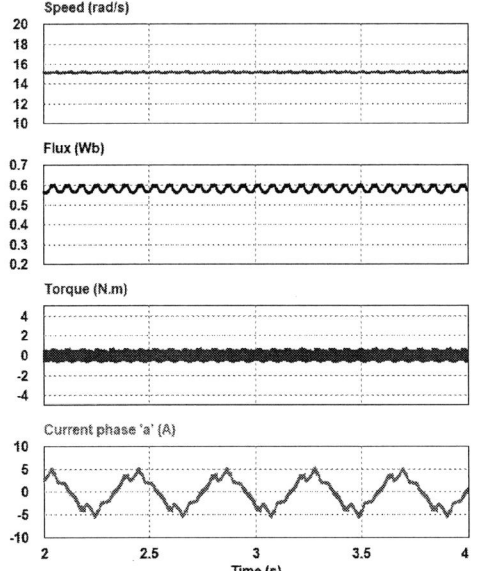

Fig.6. Simulation results for CFTC-DTC with the conventional carrier

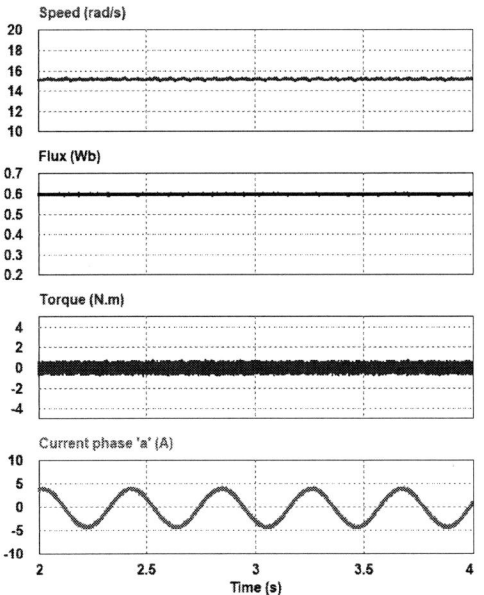

Fig.6. Simulation results for CFTC-DTC with the proposed carrier

V. CONCLUSION

In this paper, a new frequency carrier is proposed for the CFTC-DTC drive of Induction machine to increase the torque loop bandwidth. The proposed carrier solved the problem of flux droop at sector transitions at low speed region, hence, improving the performance of speed and phase current, in addition to the reduction of torque ripple. The effectiveness of the proposed methods was demonstrated in simulations and compared with the original DTC and conventional CFTC-DTC at low speed region. In the future work, mathematical analysis and descriptions of system will be provided, as well as more investigations will be carried out to test the effectiveness of the proposed algorithm. Furthermore, the systematic design for the PI controller used in the proposed carrier will be discussed.

ACKNOWLEDGMENT

This work was supported by the Basic Science Research Program through the National Research Foundation of Korea (NRF) funded by the Ministry of Science, ICT & Future Planning [2016R1A2B4010636].

REFERENCES

[1] I. Takahashi and T. Noguchi, "A New Quick-Response and High-Efficiency Control Strategy of an Induction Motor," IEEE Trans. Ind. App., vol. IA-22, no. 5, pp. 820-827, 1986.

[2] W.S.H. Wong and D. Holliday, "Minimisation of flux droop in direct torque-controlled induction motor drives," Electric Power Applications, vol. 151, pp. 694 - 703, 2004.

[3] P. Vas, Sensorless Vector and Direct Torque Control. New York: Oxford Univ. Press, 1998.

[4] K. B. Lee, J.H. Song, I. Choy, and J.Y. You, "Improvement of low-speed operation performance of DTC for three-level inverter-fed induction motors," IEEE Transactions on Industrial Electronics, vol. 48, no. 5, pp. 1006-1014, 2001.

[5] L. Harnefors and M. Hinkkanen, "Stabilization Methods for Sensorless Induction Motor Drives—A Survey," IEEE JOURNAL OF EMERGING AND SELECTED TOPICS IN POWER ELECTRONICS, vol. 2, pp. 132-142, 2014.

[6] M. Habibullah, D. D. C. Lu, D. Xiao, and M. F. Rahman, "Finite-State Predictive Torque Control of Induction Motor Supplied From a Three-Level NPC Voltage Source Inverter," IEEE Transactions on Power Electronics, vol. 32, no. 1, pp. 479-489, 2017.

[7] M. P. Kazmierkowski and A. Kasprowicz, "Improved direct torque and flux vector control of PWM inverter-fed induction motor drives," *IEEE Trans. Ind. Electronics,* vol. 42, pp. 344–350, Aug. 1995.

[8] T. Noguchi, M. Yamamoto, S. Kondo, and I. Takahashi, "High frequency switching operation of PWM inverter for direct torque control of induction motor," in *Proc. IEEE Industry Applications Conf.,* 1997, pp. 775-780 vol.1.

[9] I. M. Alsofyani and N. R. N. Idris, "Simple Flux Regulation for Improving State Estimation at Very Low and Zero Speed of a Speed Sensorless Direct Torque Control of an Induction Motor," IEEE Transactions on Power Electronics, vol. 31, no. 4, pp. 3027-3035, 2016.

[10] I. M. Alsofyani and N. R. N. Idris, "Look-up Table-Based DTC of Induction Machines With Improved Flux Regulation and Extended Kalman Filter State Estimator at Low Speed Operation," IEEE Transactions on Industrial Informatics, vol. PP, no. 99, pp. 1-1, 2016.

[11] D. Mohan, X. Zhang, and G. Foo, "Three Level Inverter fed Direct Torque Control of IPMSM with Constant Switching Frequency and Torque Ripple Reduction," IEEE Transactions on Industrial Electronics, vol. PP, no. 99, pp. 1-1, 2016.

[12] C. L. Toh, N. R. N. Idris, and A. H. M. Yatim, "Constant and high switching frequency torque controller for DTC drives," IEEE Power Electronics Letters, vol. 3, no. 2, pp. 76-80, 2005.

[13] N. R. N. Idris and A. H. M. Yatim, "Direct torque control of induction machines with constant switching frequency and reduced torque ripple," IEEE Trans. Ind. Electronics, vol. 51, no. 4, pp. 758-767, 2004.

[14] K.-B. Lee and F. Blaabjerg, "Simple power control for sensorless induction motor drives fed by a matrix converter," *IEEE Trans. Energy Convers.,* vol. 23, no. 3, pp. 781–788, Sep. 2008.

[15] Y. Cho, Y. Bak and K. B. Lee, "Torque-Ripple Reduction and Fast Torque Response Strategy for Predictive Torque Control of Induction Motors," in *IEEE Transactions on Power Electronics,* vol. 33, no. 3, pp. 2458-2470, March 2018.

[16] I. M. Alsofyani, N. R. N. Idris and K. B. Lee, "Dynamic Hysteresis Torque Band for Improving the Performance of Lookup-Table-Based DTC of Induction Machines," in *IEEE Transactions on Power Electronics*, vol. PP, no. 99, pp. 1-1.

μ-Analysis Evaluation of A Novel Combined Current-and-Speed Control for Induction Motors via ILQ Design Method

Shuto Omori[1], Hiroshi Takami[1], Masashi Nakamura[2]
1 Shibaura Institute of Technology, Tokyo, Japan
2 Toshiba Mitsubishi-Electric Industrial Systems Corporation, Tokyo, Japan
takami@sic.shibaura-it.ac.jp

Abstract- **We proposed combined electric and dynamic system based on the ILQ optimal control, which has high robust performance for induction motor drive. This method realizes an excellent speed control, even if the response of current-control loop is more than three times of the response of speed-control loop. When the motor parameters are varied, the ILQ optimal controller can hold more robust condition than previous method. This paper presents quantitative evaluations for robust stability and robust performance via μ-analysis. Generalized state equations for μ-analysis are derived and compare the ILQ control method with proposed PI method.**

I. INTRODUCTION

A variable-speed control of the induction motor based on the vector control is widely used such as steel rolling machine and servo drive for factory automation, and so on. Generally, variable-speed control consists of the current minor loop, controlling motor's flux and torque, and the speed major loop. However, the cutoff frequency of the current minor loop should be not less than three times as much as that of the major speed loop for stable control[1]. An optimal state feedback control strategy based speed control for synchronous motor, decoupled d-axis and q-axis currents by nonlinear state feedback and collaborated the current and speed control loops, had been proposed[1].

The optimal state feedback control is solved as LQ (Linear Quadratic) problem. The LQ problem issued to minimize the cost function for evaluation, represented by secondary form, trading off between control errors and manipulating inputs. The optimal control has these three advantages: (1) gain margin is infinity, (2) decrement of gain margin is less than 50%, (3) phase margin is less than $\pm 60°$ [2]. However, the LQ problem has disadvantage such as: we have to trial and error many times by simulation to determine the optimal gain, thus it is practically awkward.

Proposed ILQ (Inverse Linear Quadratic) design method is inverse approach solution without trial and error process in previous LQ problem. Moreover, it's solution is given as analytical form and transfer function of each reference input to output is set independently when gain adjusting parameter of the ILQ servo system converges to

infinity[3]. Thus, it is easy to design the optimal servo system in practical use.

Some of authors had proposed *ILQ combined current-and-speed (ILQ-CCS) controller* based on the ILQ design method for speed control of induction motor[4] . In numerical simulations and experiments, proposed controller has lead to superior control more robust than conventional PI control. However, it is difficult for numerical simulations and experiments to verify the robustness under simultaneous parameter variations of the motor.

One of the authors clarified superiority of only robust current control of the synchronous motor based on the ILQ design method via μ-analysis, evaluating quantitatively control-robustness for all of the parameter variations[5].

This paper extends and generalizes the state equations of induction motor control system by proposed *the ILQ-CCS controller* and conventional PI controller for μ-analysis. Furthermore, robustness of the proposed ILQ control is evaluated, and compared with conventional PI control.

II. COMBINED CURRENT-AND-SPEED CONTROLLER

Equations (1) and (2) show the state and output equations, respectively, which combined the voltage equation on the dq-axis reference frame, rotating on the angular velocity ω_g and the kinetic equation of the induction motor.

$$
p\begin{bmatrix} i_{sd} \\ i_{sq} \\ i_{0d} \\ \omega_r \end{bmatrix} = \begin{bmatrix} -\dfrac{r_s+r_r}{l_s} & \omega_g & \dfrac{r_r}{l_s} & 0 \\ -\omega_g & -\dfrac{r_s}{l_s} & -\dfrac{L_0}{l_s} & 0 \\ \dfrac{r_r}{L_0} & 0 & -\dfrac{r_r}{L_0} & 0 \\ 0 & \dfrac{P_0^2}{4J}L_0 i_{0d} & 0 & -\dfrac{D}{J} \end{bmatrix} \begin{bmatrix} i_{sd} \\ i_{sq} \\ i_{0d} \\ \omega_r \end{bmatrix}
$$
$$
+ \begin{bmatrix} \dfrac{1}{l_s} & 0 \\ 0 & \dfrac{1}{l_s} \\ 0 & 0 \\ 0 & 0 \end{bmatrix} \begin{bmatrix} v_{sd} \\ v_{sq} \end{bmatrix} + \begin{bmatrix} 0 \\ \omega_g \\ 0 \\ 0 \end{bmatrix} L_0 i_{0d} \quad (1)
$$

$$y = \begin{bmatrix} 1 & 0 & 0 & 0 \\ 0 & 0 & 0 & 1 \end{bmatrix} \begin{bmatrix} i_{sd} & i_{sq} & i_{0d} & \omega_r \end{bmatrix}^{\mathrm{T}} \tag{2}$$

where ω_g is the angular velocity of the secondary number of flux interlinkage, v_{sd}, i_{sd} and v_{sq}, i_{sq} represent the d-axis of the primary voltage, current and the q-axis of the primary voltage, current, respectively. i_{0d} is the excitation current converted to secondary side, ω_r is the electrical angular velocity of the rotor. The primary and secondary resistance are represented by r_s and r_r, respectively. l_s is the leakage inductance and L_0 is the excitation inductance. P_0 represents the number of poles. J and D are the inertia and the dumping coefficient. Superscript "T" means transpose operator.

Figure 1 shows the optimal combined current-and-speed control law including decoupling control for dq-axis, derived from the ILQ design method in (1).

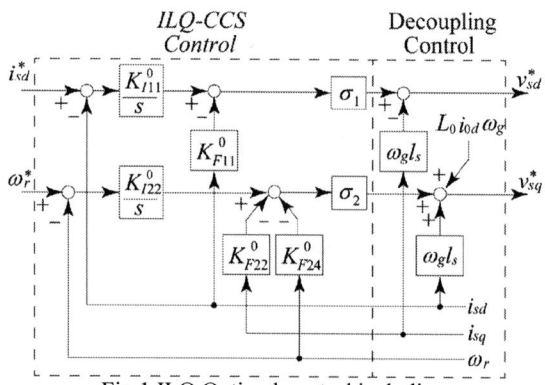

Fig.1 ILQ Optimal control including decoupling control.

where K_{F11}^0, K_{F22}^0 and K_{F24}^0 represent basic state feedback gains, and K_{I11}^0 and K_{I22}^0 are basic integral gains of servo controller, and σ_1 and σ_2 refer to parameters for adjusting gains.

Equations (3) and (4) give basic optimal gain matrixes K_F^0, K_I^0, and (5) gives optimal conditions for *the ILQ-CCS control* by ILQ design method[6] in Fig.1.

$$K_F^0 = \begin{bmatrix} K_{F11}^0 & 0 & 0 & 0 \\ 0 & K_{F22}^0 & 0 & K_{F24}^0 \end{bmatrix}$$
$$= \begin{bmatrix} l_s & 0 & 0 & 0 \\ 0 & l_s & 0 & -\dfrac{4l_s(2Js_2^* + D)}{P_0^2 L_0 i_{0d}} \end{bmatrix} \tag{3}$$

$$K_I^0 = \begin{bmatrix} K_{I11}^0 & 0 \\ 0 & K_{I22}^0 \end{bmatrix} = \begin{bmatrix} -l_s s_1^* & 0 \\ 0 & \dfrac{4l_s Js_2^{*2}}{P_0^2 L_0 i_{0d}} \end{bmatrix} \tag{4}$$

$$\sigma_1 > \underline{\sigma_1} = -\frac{2(r_s + s_1^* l_s)}{l_s}$$

$$\sigma_2 > \underline{\sigma_2} = -\frac{2(Jr_s + 2Js_2^* l_s + l_s D)}{l_s} \tag{5}$$

where s_1^* and s_2^* represent assigned poles specifying

response of current and speed, respectively. $\underline{\sigma_1}$ and $\underline{\sigma_2}$ is lower limits of σ_1 and σ_2 guaranteeing optimal control, respectively.

Equations (6) and (7) show the transfer functions of *the ILQ-CCS controller* and PI controller, respectively.

$$G_R^{ILQ}(s) = \frac{K_I^0}{s}(r - y_B) - K_F^0 x \tag{6}$$

$$G_R^{PI}(s) = (K_P + \frac{K_I}{s})(r - y_B) \tag{7}$$

where K_P is proportional gain matrix, and K_I is integral gain matrix, $r = \begin{bmatrix} i_{sd}^* & \omega_r^* \end{bmatrix}^{\mathrm{T}}$ is reference vector and $y_B = \begin{bmatrix} i_{sd} & \omega_r \end{bmatrix}^{\mathrm{T}}$ is output vector for servo control, and superscript * represents command value.

III. μ–ANALYSIS FOR EVALUATION OF ROBUSTNESS

The μ-analysis is a powerful tool for evaluating robust stability and robust performance quantitatively, when we control a plant with structured uncertainties of parameters, which parameter variations vary simultaneously[7].

A. Definition of Parameter Variations and Derivation of Generalized Plant

Parameter variations in (1) are given in the primary resistance r_s, the secondary resistance r_r, the leakage inductance l_s and inertia J. Defining additive parameter variations e_1, e_2, e_3 and e_4 from a nominal operating point of r_s^0, r_r^0, l_s^0, J^0, and separating variations from a nominal operating point, motor parameters are given as:

$$\begin{cases} r_s = r_s^0(1+e_1) = r_s^0 + r_s^0 \bar{e}_1 \delta_1 \\ r_r = r_r^0(1+e_2) = r_r^0 + r_r^0 \bar{e}_2 \delta_2 \\ l_s = l_s^0(1+e_3) = l_s^0 + l_s^0 \bar{e}_3 \delta_3 \\ J = J^0(1+e_4) = J^0 + J^0 \bar{e}_4 \delta_4 \end{cases} \tag{8}$$

where $\bar{e}_1, \bar{e}_2, \bar{e}_3, \bar{e}_4$ are the supremum of variations, and $|\delta_1|, |\delta_2|, |\delta_3|, |\delta_4| < 1$ are normalized variations. Moreover structured block variation is also defined as:

$$\Delta_p = \text{block-diag}(\delta_1 I_2, \delta_2 I_2, \delta_3 I_2, \delta_4) \tag{9}$$

where I_2 is a 2×2 identity matrix and 0_2 is a 2×2 zero matrix.

Finally, (10) to (13) are given the state and output equations extracting Δ_p out from (1) and (2) as (14).

$$\dot{x} = Ax + B_1 w_p + B_2 u \tag{10}$$

$$z_p = C_1 x + D_{11} w_p + D_{12} u \tag{11}$$

$$y_{B1} = C_2 x + D_{21} w_p + D_{22} u \tag{12}$$

$$y_{B2} = C_3 x + D_{31} w_p + D_{32} u \tag{13}$$

and

$$w_p = \Delta_p z_p \tag{14}$$

where $x = \begin{bmatrix} i_{sd} & i_{sq} & i_{0d} & \omega_r \end{bmatrix}^{\mathrm{T}}$ is the state variable vector, $u = \begin{bmatrix} v_{sd} & v_{sq} \end{bmatrix}^{\mathrm{T}}$ represents manipulated input vector, w_p and z_p are input and output vectors, respectively. $y_{B1} = [\omega_r]$ is output for servo control, $y_{B2} = [i_{sd}]$ in *the ILQ-CCS control*, or $y_{B2} = \begin{bmatrix} i_{sd} & i_{sq} \end{bmatrix}^{\mathrm{T}}$ in the PI control. Constant coefficients of (10) to (13) are represented in the following doyle form as:

$$P(s) = \left[\begin{array}{c|c|c} A & B_1 & B_2 \\ \hline C_1 & D_{11} & D_{12} \\ \hline C_2 & D_{21} & D_{22} \\ \hline C_3 & D_{31} & D_{32} \end{array}\right]$$

$$= \left[\begin{array}{cccc|c} a_1^0 + a_3^0 & \omega_g & -a_3^0 & 0 & \bar{e}_1 \\ -\omega_g & a_1^0 & -\dfrac{L_0}{l_1^0} & 0 & 0 \\ -a_2^0 & 0 & a_2^0 & 0 & 0 \\ 0 & \dfrac{P_0^2}{4J^0}L_0 i_{0d} & 0 & -\dfrac{D}{J^0} & 0 \\ \hline a_1^0 & 0 & 0 & 0 & 0 \\ 0 & a_1^0 & 0 & 0 & 0 \\ a_3^0 & 0 & -a_3^0 & 0 & 0 \\ -a_2^0 & 0 & a_2^0 & 0 & 0 \\ -a_1^0 - a_3^0 & -\omega_g & a_3^0 & 0 & -\bar{e}_1 \\ \omega_g & -a_1^0 & \dfrac{L_0}{l_1^0} & 0 & 0 \\ 0 & -\dfrac{P_0^2}{4J^0}L_0 i_{0d} & 0 & \dfrac{D}{J^0} & 0 \\ \hline 0 & 0 & 0 & 1 & 0 \\ c_{31} & c_{32} & 0 & 0 & 0 \\ c_{33} & c_{34} & 0 & 0 & 0 \end{array}\right] *$$

$$* \left[\begin{array}{cccccc|cc} 0 & \bar{e}_2 & \bar{e}_2 & \bar{e}_3 & 0 & 0 & b_1^0 & 0 \\ \bar{e}_1 & 0 & 0 & 0 & \bar{e}_3 & 0 & 0 & b_1^0 \\ 0 & \bar{e}_2 & \bar{e}_2 & 0 & 0 & 0 & 0 & 0 \\ 0 & 0 & 0 & 0 & 0 & \bar{e}_4 & 0 & 0 \\ \hline 0 & 0 & 0 & 0 & 0 & 0 & 0 & 0 \\ 0 & 0 & 0 & 0 & 0 & 0 & 0 & 0 \\ 0 & 0 & 0 & 0 & 0 & 0 & 0 & 0 \\ 0 & 0 & 0 & 0 & 0 & 0 & 0 & 0 \\ 0 & -\bar{e}_2 & -\bar{e}_2 & -\bar{e}_3 & 0 & 0 & -b_1^0 & 0 \\ -\bar{e}_1 & 0 & 0 & 0 & -\bar{e}_3 & 0 & 0 & -b_1^0 \\ 0 & 0 & 0 & 0 & 0 & -\bar{e}_4 & 0 & 0 \\ \hline 0 & 0 & 0 & 0 & 0 & 0 & 0 & 0 \\ 0 & 0 & 0 & 0 & 0 & 0 & 0 & 0 \\ 0 & 0 & 0 & 0 & 0 & 0 & 0 & 0 \end{array}\right]$$

(15)

where $a_1^0 = -r_s^0/l_s^0$, $a_2^0 = -r_r^0/L_0$, $a_3^0 = -r_r^0/l_s^0$ and $b_1^0 = 1/l_s^0$. c_{31} to c_{34} represent the components of output matrix improved by the generalized servo controller such as *the ILQ-CCS control* and PI control, namely, the former by $\{c_{31}, c_{32}, c_{33}, c_{34}\} = \{1,0,0,0\}$, the latter by $\{c_{31}, c_{32}, c_{33}, c_{34}\} = \{1,0,0,1\}$.

B. Derivation of Generalized Servo System

We introduce a generalized servo system in Fig.2, which applied to both *the ILQ-CCS control* and PI control, and set up in each controller as:

1) *The ILQ-CCS controller*

References $r_1 = \begin{bmatrix} i_{sd}^* & \omega_r^* \end{bmatrix}^{\mathrm{T}}$, $r_2 = 0_2$, gains $K_{P1} = 0_2$, $K_{P2} = I_2$, $K_{I1}^0 = K_I^0$, $\bar{K}_{I1}^0 = K_I^0$, $K_{I2}^0 = 0_2$, $K_{F2}^0 = 0_4$ and gain adjusting parameter $\Sigma = \mathrm{diag}(\sigma_1, \sigma_2)$.

2) *PI controller*

References $r_1 = \omega_r^*$, $r_2 = \begin{bmatrix} i_{sd}^* & i_{sq}^* \end{bmatrix}^{\mathrm{T}}$, gains $K_{P1} = K_{F1}^0$ $= K_{PS}$, $K_{P2} = K_{F2}^0 = K_{PC}$, $K_{I1}^0 = K_{IS}$, $K_{I2}^0 = K_{IC}$ and gain adjust parameter $\Sigma = I_2$.

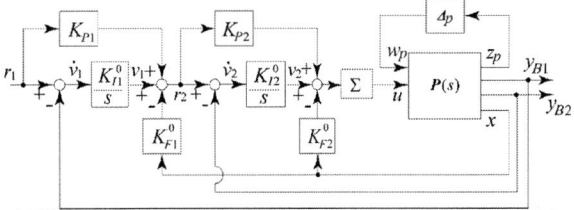

Fig. 2 Generalized servo system for μ-analysis.

Thus new temporary variables v_1 and v_2 are defined as $\dot{v}_1 = r_1 - y_{B1} = w_r - y_{B1}$, $\dot{v}_2 = r_2 - y_{B2}$. The control inputs represents as:

$$r_2 = \begin{bmatrix} -K_{F1}^0 & K_{I1}^0 \end{bmatrix}\begin{bmatrix} x \\ v_1 \end{bmatrix} + \begin{bmatrix} 0_{27} & K_{P1} \end{bmatrix}\begin{bmatrix} w_p \\ w_r \end{bmatrix} \quad (16)$$

$$u = \begin{bmatrix} -K_{F2}^0 & K_{I2}^0 \end{bmatrix}\begin{bmatrix} x \\ v_2 \end{bmatrix} + \begin{bmatrix} 0_{27} & K_{P2} \end{bmatrix}\begin{bmatrix} w_p \\ r_2 \end{bmatrix} \quad (17)$$

where 0_{27} means 2×7 zero matrix.

The generalized servo system is derived from substituting (16) and (17) to (10), (11), (12) and (13).

$$\begin{bmatrix} \dot{x} \\ \dot{v}_1 \\ \dot{v}_2 \end{bmatrix} = \begin{bmatrix} A_{bk} - B_2 K_{P2} K_{F1}^0 & B_2 K_{P2} K_{I1}^0 \\ -C_{3k} & 0_{21} \\ -C_{3k} - I_{dk} K_{F1}^0 & I_{dk} K_{I1}^0 \end{bmatrix} *$$

$$* \begin{bmatrix} B_2 K_{I1}^0 \\ 0_{21} \\ -D_{22} K_{I2}^0 \end{bmatrix}\begin{bmatrix} x \\ v_1 \\ v_2 \end{bmatrix} + \begin{bmatrix} B_1 & B_2 K_{P2} K_{P1} \\ 0_{27} & I_{dk} \\ -D_{31} & I_{dk} K_{P1} \end{bmatrix}\begin{bmatrix} w_p \\ w_r \end{bmatrix} \quad (18)$$

$$\begin{bmatrix} z_p \\ y_{B1} \\ y_{B2} \end{bmatrix} = \begin{bmatrix} C_{1k} - D_{12} K_{P2} K_{F1}^0 & D_{12} K_{P2} K_{I1}^0 \\ C_{2k} - D_{22} K_{P2} K_{F1}^0 & D_{22} K_{P2} K_{I1}^0 \\ C_{3k} - D_{32} K_{P2} K_{F1}^0 & D_{32} K_{P2} K_{I1}^0 \end{bmatrix} *$$

$$* \begin{bmatrix} D_{12} K_{I2}^0 \\ D_{22} K_{I2}^0 \\ D_{32} K_{I2}^0 \end{bmatrix}\begin{bmatrix} x \\ v_1 \\ v_2 \end{bmatrix} + \begin{bmatrix} D_{11} & D_{12} K_{P2} K_{P1} \\ D_{21} & D_{22} K_{P2} K_{P1} \\ D_{31} & D_{32} K_{P2} K_{P1} \end{bmatrix}\begin{bmatrix} w_p \\ w_r \end{bmatrix} \quad (19)$$

where $A_{bk} = A - B_2 K_{F2}^0$, $C_{1k} = C_1 - D_{12} K_{F2}^0$, $C_{2k} = C_2$ $-D_{22} K_{F2}^0$, $C_{3k} = C_3 - D_{32} K_{F2}^0$, $I_{dk} = I_2 - D_{22} K_{P2}$.

C. Derivation of Integrated State Equation

In Fig.3 we derive extended system $M(s)$, consisted of plant $P(s)$, servo controller $G_R^{ILQ}(s)$ or $G_R^{PI}(s)$, nominal plant $T(s)$ and evaluation function $W_T(s)$, for evaluating robustness and robust performance below. The robustness is evaluated via μ-analysis of $M(s)$ closed by Δ_p . On the other hand, the robust performance is evaluated via μ-analysis of $M(s)$ closed by virtual variation Δ_r . Δ_r is given by the differences between the generalized plant with parameter variations and the nominal plant. The cost function, which evaluates the robust performance quantitatively, is given in (20), introducing evaluation function $W_T(s)$. This is chosen so as to be directly proportion to the cost function.

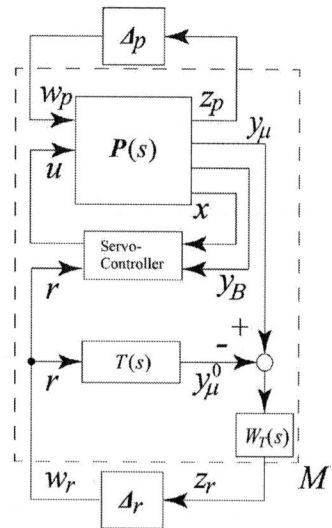

Fig.3 Extended system M for evaluation of the robustness and the robust performance

$$z_r(s) = W_T(s)\Delta y = W_T(s)\left(y - y^0\right)$$
$$= W_T(s)\{T(s, \Delta_p) - T(s)\}w_r(s)$$
$$(20)$$

where y^0 represents output from the nominal plant, $\Delta y = y - y^0$, $T(s, \Delta_p)$ is the transfer function from input ω_r to output y_μ including Δ_p, and $T(s)$ is the transfer function of nominal plant with no Δ_p for evaluating the robust performance.

A minimal realization of $W_T(s)$ is given as:
$$\dot{x}_r = A_r x_r + B_r \Delta y$$
$$\dot{z}_r = C_r x_r + D_r \Delta y \qquad (21)$$

where x_r is the state variable of the cost function, z_r is the output of the cost function, and A_r, B_r, C_r and D_r are constant coefficient matrixes.

The state equation of the nominal plant is derived from similar procedure by generalized plant. Appending the state variables x^0, v_1^0 and v_2^0 of nominal plant to those of (18), the integrated state equation $M(s)$ is obtained as following (23) and (24).

$$\begin{bmatrix} \dot{x} \\ \dot{v}_1 \\ \dot{v}_2 \\ \dot{x}^0 \\ \dot{v}_1^0 \\ \dot{v}_2^0 \\ \dot{x}_r \end{bmatrix} = \begin{bmatrix} A_{bk} - B_2 K_{P2} K_{F1}^0 & B_2 K_{P2} K_{I1}^0 & B_2 K_{I1}^0 \\ -C_3 & 0_{22} & 0_{22} \\ -C_{3k} - I_{dk} K_{F1}^0 & I_{dk} K_{I1}^0 & -D_{22} K_{I2}^0 \\ 0_{44} & 0_{42} & 0_{42} \\ 0_{24} & 0_{22} & 0_{22} \\ 0_{24} & 0_{22} & 0_{22} \\ B_r C_{4k} & B_r D_{22} K_{P2} K_{I1}^0 & B_r D_{22} K_{I2}^0 \end{bmatrix} *$$

$$\begin{matrix} 0_{44} & 0_{42} & 0_{42} \\ 0_{24} & 0_{22} & 0_{22} \\ 0_{24} & 0_{22} & 0_{22} \\ * A_{bk} - B_2 K_{P2} K_{F1}^0 & B_2 K_{P2} K_{I1}^0 & B_2 K_{I1}^0 & * \\ -C_3 & 0_{22} & 0_{22} \\ -C_{3k} - I_{dk} K_{F1}^0 & I_{dk} K_{I1}^0 & -D_{22} K_{I2}^0 \\ -B_r C_{4k} & -B_r D_{22} K_{P2} K_{I1}^0 & -B_r D_{22} K_{I2}^0 \end{matrix}$$

$$\begin{bmatrix} 0_{41} \\ 0_{21} \\ 0_{21} \\ *0_{41} \\ 0_{21} \\ 0_{21} \\ A_r \end{bmatrix} \begin{bmatrix} x \\ v_1 \\ v_2 \\ x^0 \\ v_1^0 \\ v_2^0 \\ x_r \end{bmatrix} + \begin{bmatrix} B_1 & B_2 K_{P2} K_{P1} \\ 0_{27} & I_{dk} \\ -D_{31} & I_{dk} K_{P1} \\ B_1 & B_2 K_{P2} K_{P1} \\ 0_{27} & I_{dk} \\ -D_{31} & I_{dk} K_{P1} \\ B_r D_{21} & 0_{12} \end{bmatrix} \begin{bmatrix} w_p \\ w_r \end{bmatrix}$$
$$(22)$$

$$\begin{bmatrix} z_p \\ z_r \end{bmatrix} = \begin{bmatrix} C_{1k} - D_{12} K_{P2} K_{F1}^0 & D_{12} K_{P2} K_{I1}^0 & * \\ D_r C_{4k} & D_r D_{22} K_{P2} K_{I1}^0 \end{bmatrix}$$

$$\begin{matrix} * & D_{12} K_{I2}^0 & 0_{74} & 0_{72} & * \\ & D_r D_{22} K_{I2}^0 & -D_r C_{4k} & -D_r D_{22} K_{P2} K_{I1}^0 \end{matrix}$$

$$\begin{matrix} * & 0_{72} & 0_{71} \\ & -D_r D_{22} K_{I2}^0 & C_r \end{matrix} \begin{bmatrix} x \\ v_1 \\ v_2 \\ x^0 \\ v_1^0 \\ v_2^0 \\ x_r \end{bmatrix}$$
$$(23)$$

where $C_{4k} = C_{2k} - D_{22} K_{P2} K_{F1}^0$ and in this case, each coefficients are $A_r = B_r = C_r = 0$ and $D_r = 1$.

D. Introducing of Evaluating Robustness and Robust Performance Indexes

In the definition of $\mu_{\Delta_e}(M)$, there is an uncertainty structure Δ_e of a prescribed set of block diagonal matrixes, on which whole variations in the sequel depends, and $\mu_{\Delta_e}(M)$ for M is defined as[8]:

$$\mu_{\Delta_e}(M) = \frac{1}{\min\left\{\bar{\sigma}(\Delta_e) : \Delta_e \in \Delta, \det(I - M\Delta_e) = 0\right\}} \quad (24)$$

Thus on a basis of small gain theorem, the stability margin $\wp_{stb}$ for evaluating the robustness of the system M is defined as inverse of the supremum value of $\mu_{\Delta_e}(M)$:

$$\wp_{stb} = \frac{1}{\sup_\omega \mu_{\Delta_e}(M(s))} \quad (25)$$

where the maximum perturbation matrix Δ is represented by $\bar{\sigma}(\Delta_e)$, and function "sup" means supremum operator.

Moreover the evaluation function $W_T(s)$ is chosen so as to be proportion to the virtual variation Δ_r, thus $W_T(s)$ is defined as:

$$W_T(s) = \{W_T > 0 \mid W_T \text{ is a scolar and a constant}\} \quad (26)$$

Consequently, we can introduce the robust performance index $\wp_{rob}$ for evaluating the robust performance as[9]:

$$\wp_{rob} = \max\left[W_T \mid \sup \mu_\Delta \{M(j\omega)\} \le 1\right] \quad (27)$$

When the supremum of $\mu_{\Delta_e}(M)$ is equal to 1.0, $\wp_{rob}$ is given as maximum value of the evaluation weight W_T.

IV. ROBUSTNESS COMPARISON ILQ OPTIMAL CONTROL WITH PI CONTROL

In the servo controller in Fig.3, *the ILQ-CCS control* and the conventional PI control are evaluated to their robust stability and robust performance when structured variations occur. Table 1 shows the rating and parameters of the tested three-phase induction motor for evaluation.

The assigned poles s_1^* and s_2^*, which determine the responsiveness of the exciting current and speed in *the ILQ-CCS control*, are both set to $s_1^* = s_2^* = -60$. The PI controller is configured whose zero cancels the pole of the plant, appending a first order time lag of the time constant T_d. Condition $T_d = -1/s_1^* = -1/s_2^*$ is adopted to almost correspond speed response of the conventional PI control to that of *the ILQ-CCS control* at the nominal parameters.

Table 1 Rating and parameters of induction motor.

Rated Output	1.5	kW
Rated Speed	1500	min⁻¹
Rated Torque	8.0	N-m
Resistance-Stator Winding: r_s	0.889	Ω
Resistance-Stator Reducing: r_r	0.669	Ω
Stator Leaked Inductance: l_s	0.011	H
Excitation Inductance: L_0	0.084	H
Inertia Moment: J	0.011	kg-m²
Viscosity Coefficient: D	0.004	N-m/rad/s

According to (8), the parameter variations are varied continuously from –50% to 50%. We carry out to evaluate both the robust stability and the robust performance at each motor speed: low speed of 0.1pu=150min⁻¹, middle speed of 1.0pu=1500 min⁻¹ and high speed of 2.0pu=3000 min⁻¹. Fig.3 to Fig.5 illustrate μ-plots for evaluating the robust stability, comparing *the ILQ-CCS control* with the PI control. The horizontal axis is frequency of *the frequency transfer-functions* in the closed loop. Fig.6 shows the stability margin of (25), comparing proposed ILQ and conventional PI controls. Fig.7 shows the robust performance of (27). The conventional PI control is slightly good in the stability margin, but *the ILQ-CCS control* is extremely higher in the robust performance at the especially low speed region. The response fluctuation in PI control is larger than that in *the ILQ-CCS control*. This is due to setting zero-pole cancelation in PI control. In contrast, *the ILQ-CCS control* leads to robust response in parameter fluctuation due to cooperative state feedback.

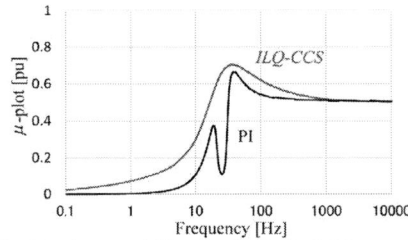

Fig.3 Robust stability in *the ILQ-CCS control* and PI control (0.1pu).

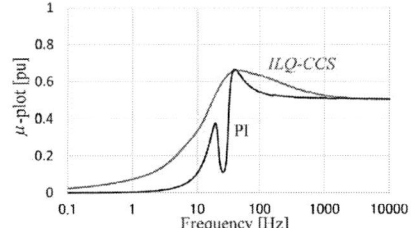

Fig.4 Robust stability in *the ILQ-CCS control* and PI control (1.0pu).

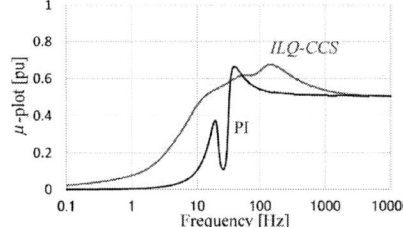

Fig.5 Robust stability in *the ILQ-CCS control* and PI control (2.0pu).

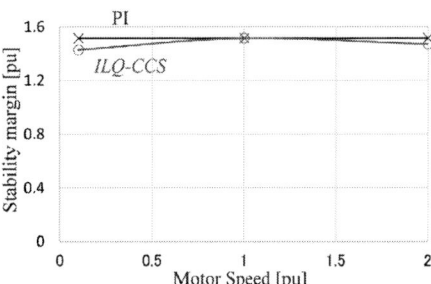

Fig.6 Stability margin in *the ILQ-CCS control* and PI control.

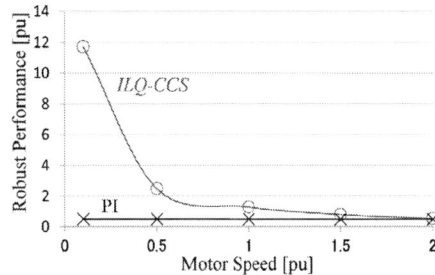

Fig.7 Robust performance in *the ILQ-CCS controller* and PI controller.

V. VERIFICATION OF NUMERICAL SIMULATION AND EXPERIMENT

The results of μ-analysis in this paper are verified by numerical simulation and experiment indicating the superiority of *the ILQ-CCS control*. The robustness is verified by given speed step command with 100 min⁻¹ at low speed, where the parameter variation is set. Fig.8 shows a block diagram for verifying μ-analysis evaluations in chapter IV by numerical simulation and experiment. The controller is composed of *the ILQ-CCS controller* or PI controller. They are including decoupled control law. A symbol PSS is speed sensor.

The 2018 International Power Electronics Conference

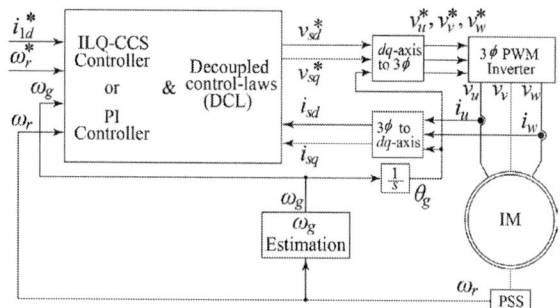

Fig.8 Block diagram for speed control of IM.

A. Verification by The Numerical Simulation

Figure 9 and Fig.10 show torque current response and motor speed response when the inertia fluctuation, which affects speed response, is given and the variation is ± 50% in *the ILQ-CCS control* and the PI control, respectively. The results show that the speed responses of *the ILQ-CCS control* are not much affected by the inertia fluctuation, and it is found that have high robustness, which is confirmed by μ-analysis result.

B. Verification by Experiment

For verification of μ-analysis and the numerical simulation, the experiments are carried out by comparing deviations of the rise time of the motor speed response in *the ILQ-CCS control* with those in the PI control. The rise time is defined as the time taken by a height of indicial response to change from 10% to 90%.

Figure 11 and Fig. 12 illustrate both torque current and speed responses in *the ILQ-CCS control* and the PI control, respectively. The inertia fluctuation is given similar to the numerical simulation. Table 2 show comparison of maximum deviations in *the ILQ-CCS control* and the PI control. Each rise time are rewritten from Fig.11 and Fig.12.

Table 2 Comparison of maximum deviations in
the ILQ-CCS control and the PI control.

		Fluctuation of inertia			Maximum deviation
		-50%	0%	+50%	
Rise time	*ILQ-CCS control*	32.5 ms	33.0 ms	33.5 ms	0.5ms
	PI control	27.5 ms	33.0 ms	48.0 ms	15.0ms

Both maximum deviations of *the ILQ-CCS control* and PI control can be calculated by subtracting the rise time in the nominal condition from that in the fluctuation of +50% (see half tone of table 2). Thus maximum deviation of *the ILQ-CCS control* is given as 33.5ms−33.0ms =0.5ms, and that of PI control is given as 48.0ms−33.0ms =15.0ms. Therefore, the PI control is much deviated from the nominal speed response in the inertia fluctuation. On the other hand, *the ILQ-CCS control* leads to robust response of 30 times against inertia fluctuation. It proves the superiority of *the ILQ-CCS control*, and verifies the

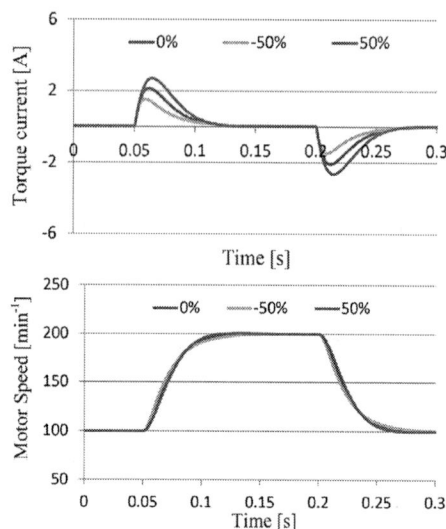

Fig.9 Speed and torque current responses with parameter error in *the ILQ-CCS control*.

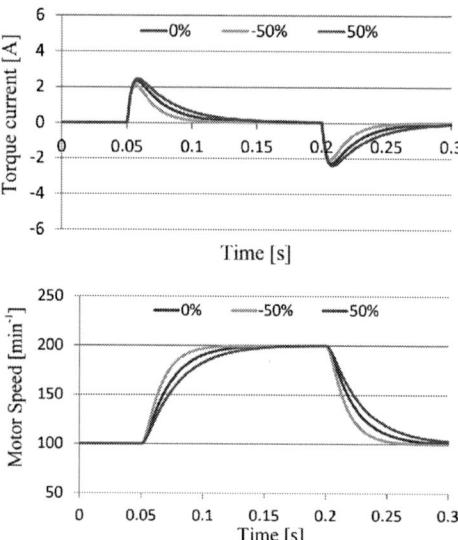

Fig.10 Speed and torque current responses with parameter error in PI control.

accuracy of μ-analysis and numerical simulation.

VI. CONCLUSION

In this paper, a new method for evaluating robustness by μ-analysis is presented. Also, *the ILQ-CCS control* is compared with PI control at minor loop. As a result, it was quantitatively clarified that *the ILQ-CCS control* has a high robust performance. The introduced μ-analysis for servo system, which can deal with multiple parameter fluctuations, could be a powerful tool for evaluating robustness.

In the future, it is necessary to evaluate the robustness when the speed fluctuates dynamically. In addition, it is necessary to derive the generalized state equation including the disturbance factor to verify robustness against disturbance.

The 2018 International Power Electronics Conference

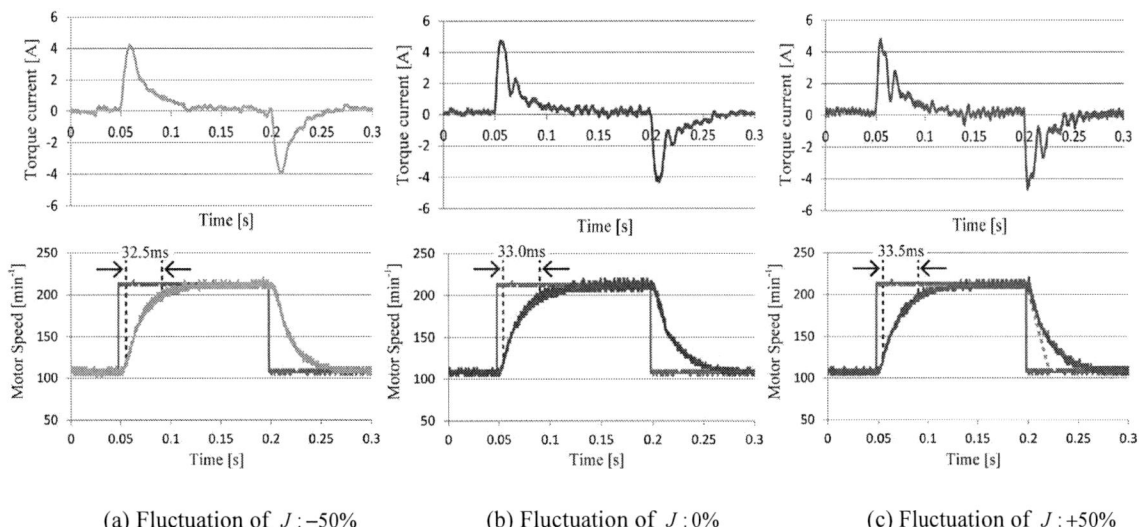

(a) Fluctuation of $J : -50\%$ (b) Fluctuation of $J : 0\%$ (c) Fluctuation of $J : +50\%$

Fig.11 Experimental results of speed and torque current responses with parameter error in *the ILQ-CCS control.*

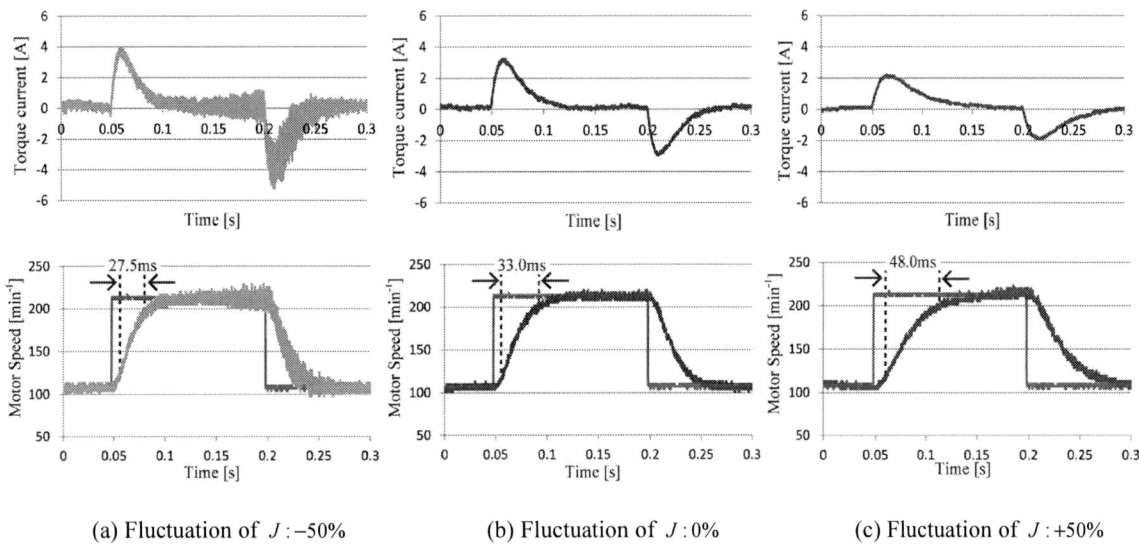

(a) Fluctuation of $J : -50\%$ (b) Fluctuation of $J : 0\%$ (c) Fluctuation of $J : +50\%$

Fig.12 Experimental results of speed and torque current responses with parameter error in PI control.

REFERENCES

[1] Y. Kuroe, T. Maruhashi, and K.Okamura, "Linearizing Control of Synchronous Motors through Decoupling of d-q Axes and its Application to Design of Optimal Speed Servo Systems," *IEEJ Trans.* IA, Vol. 109, no. 11, pp. 817-823, 1989

[2] M. Ikeda, Y. Fujisaki "Control of Multivariable Systems," CORONA PUBLISHING CO. , 2010

[3] T. Fujii and N. Mizushima, "A New Approach to LQ Design Application to the Design of Optimal Servo Systems," *Trans. on the Society of Instrument and Control Engineers* Vol. 23 , no. 2, pp. 129-135, 1987

[4] Y. Nakamura, H.Takami, M.Nakamura, T. Okamoto, "A Method of ILQ Optimal Speed Sensorless Control for Three-Phase induction Motor," *Trans. Industry Applications Society Conference IEEJ*, no. 1, pp. I429-I432, 2011

[5] H. Takami, T. Tsujino, "Robust Stability and Performance Evaluation of an ILQ Optimal Current-Control System for Permanent Magnet Synchronous Motors via μ-Analysis –Robust

Analysis Including State Feeedback Control for Decoupling d- and q-axis Subsystems–," Trans. on the Society of Instrument and Control Engineers, Vol. 42, no. 5, pp. 510-519, 2006

[6] T. Fujii, T. Tsujino "Practical design method for optimal control –designation of controller via ILQ design method and ILQ applicational example– ," Morikita Established, 2012

[7] J. C. Doyle, J. E. Wall and G.Stein, "Performance and Robustness Analysis for Structured Uncertainty," *Proc. of the 21st IEEE Conference on Decision and Control*, pp. 629-636, 1982

[8] Kemin Zbou, John C. Doyle, "ESSENTIAL OF ROBUST CONTROL," PRENTICE HALL, 1998

[9] H.Takami, T.Tsujino, T.Fujii, "Robust Stability and Performance Evaluation of an ILQ Optimal Current-Control System for Permanent Magnet Synchronous Motors via μ-Analysis" *Trans. on the Society of Instrument and Control Engineers* Vol. 39, no. 9, pp. 808-816, 2003

Loss Minimization Control of Sensorless Scalar-Controlled Induction Motor Drives Considering Iron Loss

Nguyen Anh Tan and Dong-Choon Lee
Department of Electrical Engineering, Yeungnam University
280 Daehak-Ro, Gyeongsan, Gyeongbuk, 38541 Korea
nguyenanhtan91bk@gmail.com, dclee@yu.ac.kr

Abstract— **In this paper, a novel method of loss minimization control for sensorless scalar-controlled induction motor (IM) drives is proposed. In this work, a simple estimation scheme of the motor speed is developed based on the difference between the reference back-EMF and the estimated back-EMF. The loss minimization control is achieved through an optimized slip speed which is determined from the total power loss equation. Since only one current sensor is required, the cost of this sensorless scalar control system can be reduced. The validity of the proposed method is verified by the simulation results.**

Keywords— *iron loss, loss minimization, optimal slip speed, sensorless scalar control*

I. INTRODUCTION

The open-loop scalar-controlled IM drives are widely applied in the industrial applications including fans, pumps, compressors, and etc. where the accurate speed and torque control is not required. The installation of speed sensors can cause the whole system to be more expensive and more cumbersome. Thereby, the sensorless scalar control in the IM drive has been studied in [1] and [2].

In recent years, with the increases of the energy price and the requirements for the clean environment, the loss minimization in electric machines is receiving the great attentions, especially under light load conditions. In the IM drive systems, the loss minimization control can be classified into two types, they are the Search Control (SC) and the Loss-Model-Based Control (LMC). The SC is applied without requiring the knowledge of the motor loss model [3]-[5]. However, the SC has the drawbacks of the long convergence time, ripple torque at steady state and poor performance at transient state. Compared with the SC methods, the LMC method offers the faster convergence time as well as the better performance at steady state and transient state [6]-[10]. However, the LMC methods require the knowledge of the motor loss model which is not always available in practice. In this study, the method of loss minimization by applying the optimal slip speed in sensorless scalar-controlled IM drived is proposed. Since only one single current sensor is utilitied, the cost of presented sensorless scalar control system can be reduced.

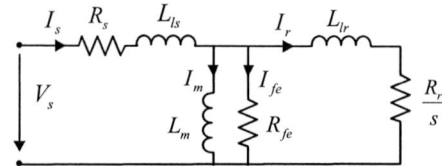

Fig. 1. Per phase equivalent circuit of induction motor.

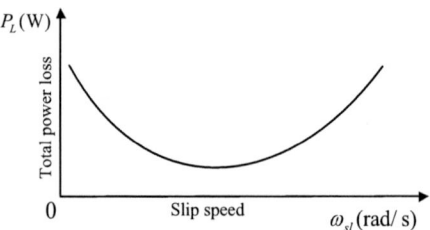

Fig. 2. Relation of the total power loss and slip speed under certain speed and torque condition.

II. LOSS MINIMIZING CONTROL

Fig. 1 shows the equivalent circuit of the IM considering the iron loss resistance connected in parallel to the magnetizing branch. In the IM drive systems, by retaining the motor speed and load condition, it can be noted that the total power losses are the convex function of the slip speed as shown in Fig. 2. Hence, there exists an optimal slip speed value which gives the minimum power losses. The power loss components of the IM are stator copper loss P_{sc}, core loss P_c, rotor copper loss P_{rc}, friction and windage loss P_{fw}, and stray loss P_{str} [6], [12], [13]. The power losses P_{sc}, P_c, and P_{rc} account for about 80% of total power loss. Thus, the friction and windage loss and stray loss are neglected in this analysis. The total power losses in the three-phase IM is expressed as follows:

$$P_L = 3\left(R_s I_s^2 + R_r I_r^2 + \frac{E^2}{R_{fe}} \right) \qquad (1)$$

where P_L is the total power losses, I_s, I_r are the stator and rotor currents, respectively. E is the back-EMF. R_s, R_r, R_{fe} are the stator, rotor and iron loss resistances, respectively. $R_s I_s^2$ and $R_r I_r^2$ represents the stator and rotor copper losses, respectively. E^2/R_{fe} means the core loss.

The rotor current can be given by [14]

$$I_r^2 = \frac{E^2}{\left(\dfrac{R_r}{s}\right)^2 + \left(\omega_e L_{lr}\right)^2} \tag{2}$$

And the electromagnetic torque can be expressed as

$$T_e = \frac{3pI_r^2 R_r}{\omega_r}\frac{(1-s)}{s} \tag{3}$$

where ω_{sl} is the slip speed, ω_e is the synchronous speed, and slip $s=\omega_{sl}/\omega_e$, T_e is the electromagnetic torque and p is the number of pole pairs.

Since $\left(\omega_e L_{lr}\right)^2 \ll \left(R_r/s\right)^2$, from (2) and (3),

$$E^2 \approx \frac{T_e \omega_e^2 R_r}{3p\omega_{sl}} \tag{4}$$

The magnetizing current I_m is given by

$$I_m^2 = \frac{E^2}{\left(L_m \omega_e\right)^2} \tag{5}$$

Assume that the iron loss current I_{fe} is low enough to be neglected. Thus, the stator current I_s is expressed as,

$$I_s^2 \approx I_m^2 + I_r^2 \approx \frac{T_e}{3p}\left(\frac{R_r}{\omega_{sl}L_m^2} + \frac{\omega_{sl}}{R_r}\right) \tag{6}$$

Hence, the total power loss can be expressed as

$$P_L = \frac{R_s T_e}{p}\left(\frac{R_r}{\omega_{sl}L_m^2} + \frac{\omega_{sl}}{R_r}\right) + \frac{T_e \omega_{sl}}{p} + \frac{T_e \omega_e^2 R_r}{p\omega_{sl}R_{fe}} \tag{7}$$

By taking derivative of the total power losses P_L with respect to slip speed ω_{sl}, then setting it zero, the optimized slip speed can be obtained as

$$\omega_{sl\_opt} = \sqrt{\frac{R_r R_{fe}}{R_r^2 + R_s R_{fe} + R_r R_{fe}}\left(\frac{R_r R_s}{L_m^2} + \frac{R_r \omega_r^2}{R_{fe}}\right)} \tag{8}$$

From (8), it can be noted that the optimized slip speed ω_{sl_opt} is the function of shaft speed ω_r and the iron loss resistance R_{fe} but not the torque. If iron loss resistance is not considered, the optimized slip speed can be expressed as [6],

$$\omega_{sl\_opt\,[6]} = \sqrt{\frac{R_s R_r^2}{R_s\left(L_{lr} + L_m\right)^2 + R_r L_m^2}} \tag{9}$$

In (9), the optimal slip speed is fixed at variable speed conditions.

III. SENSORLESS SCALAR CONTROL ALGORITHM

In the IM, the stator flux can be expressed as,

$$\lambda_{s\alpha} = \int \left(V_s \cos(\omega_e t) - R_s i_{s\alpha}\right)dt \tag{10}$$

$$\lambda_{s\beta} = \int \left(V_s \sin(\omega_e t) - R_s i_{s\beta}\right)dt \tag{11}$$

$$\lambda_s = \sqrt{\left(\lambda_{s\alpha}\right)^2 + \left(\lambda_{s\beta}\right)^2} \tag{12}$$

where V_s is the stator voltage, $i_{s\alpha}$ and $i_{s\beta}$ are the α- and β-axis stator currents, $\lambda_{s\alpha}$ and $\lambda_{s\beta}$ are the α- and β-axis stator fluxes, and λ_s is the stator flux.

The air gap power P_g can be expressed as,

$$P_g = P_{in} - P_{sc} - P_i = \frac{3R_r I_r^2}{s} \tag{13}$$

where P_{in} is the input power, P_{sc} is the stator copper loss, P_i is the stator core loss and I_r can be obtained from (2).

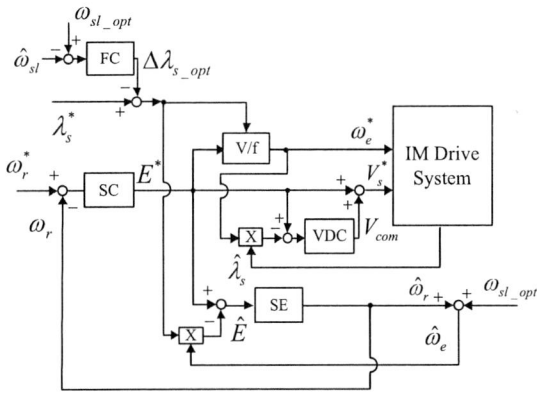

Fig. 3. Block diagram of proposed sensorless scalar control system with loss minimization (SC: Speed controller; SE: Speed estimator; VDC: Voltage drop compensator; FC: Flux compensator).

If the rotor leakage inductance is neglected, from (2) and (13),

$$\omega_{sl} \approx \frac{P_g \omega_e R_r}{3E^2} \tag{14}$$

The block diagram of the proposed sensorless scalar control system with loss minimization is shown in Fig. 3. The optimized slip speed ω_{sl_opt} from (8) is applied to express the estimated synchronous speed $\hat{\omega}_e$. The estimated shaft speed $\hat{\omega}_r$ should be varied so that the estimated back-EMF $\hat{E}$ can follow the reference back-EMF E^*. Due to the stator resistance voltage drop, a voltage compensator should be added. The estimated stator flux $\hat{\lambda}_s$ from (12) is used for this compensator. For loss minimizing control, the optimized slip speed ω_{sl_opt} needs to be applied to change the stator flux. All controllers in Fig. 3 are PI controllers. To mitigate the ripples caused by the stator flux variations, the working period of stator flux compensator should be longer than that of other controllers, such as speed controller, speed estimator and voltage drop compensator.

IV. STATOR CURENT ESTIMATION

From (10) and (11),

$$\hat{\lambda}_{s\alpha} = \left(V_s \sin(\omega_e t) - R_s i_{s\beta}\right)/\omega_e \tag{15}$$

$$\hat{\lambda}_{s\beta} = -\left(V_s \cos(\omega_e t) - R_s i_{s\alpha}\right)/\omega_e \tag{16}$$

where $\hat{\lambda}_{s\alpha}$ and $\hat{\lambda}_{s\beta}$ stand for the estimated state of α- and β-axis stator fluxes, respectively. In this work, the stator currents can be estimated by using only phase-A current sensor. Instead of (10) and (11), this method uses (15) and (16) to mitigate the estimation error caused by the integration and the derivation.

Fig. 4 shows the stator flux vector in a stationary reference frame. It can be noted that stator flux vector $\vec{\lambda}_s$ has the speed of ω_e and β-axis stator flux $\lambda_{s\beta}$ begins to change from the positive value to the negative one at stator flux angle $\theta_f = \pi$ or $\theta_f = -\pi$. Thereby, the estimated stator flux angle $\hat{\theta}_f$ can be obtained by integrating the

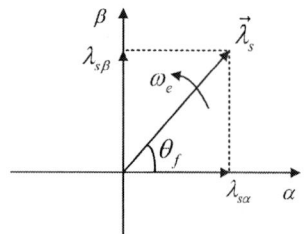

Fig. 4. Stator flux vector with α- and β- axis components.

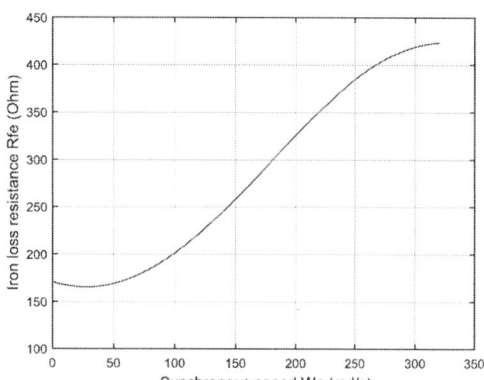

Fig. 5. Estimation of iron loss resistance R_{fe}

TABLE I
MOTOR PARAMETERS

Symbol	Meaning	Value
P_{out}	Output power	3 kW
ω_r	Rated shaft speed	1430 rpm
T_e	Rated motor torque	19 Nm
I_s	Rated phase current	10.9 A
p	Number of poles	4
λ_s	Rated stator flux	0.43 V/(rad/s)
L_m	Magnetizing inductance	90 mH
L_{ls}	Stator leakage inductance	3 mH
L_{lr}	Rotor leakage inductance	3 mH
R_s	Stator resistance	0.833 Ω
R_r	Rotor resistance	0.596 Ω
R_{fe}	Iron loss resistance	422 Ω

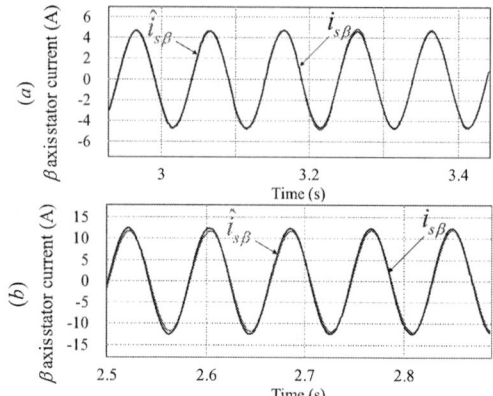

Fig. 6. β-axis stator current at 300 rpm. (a) no load (b) 70% load.

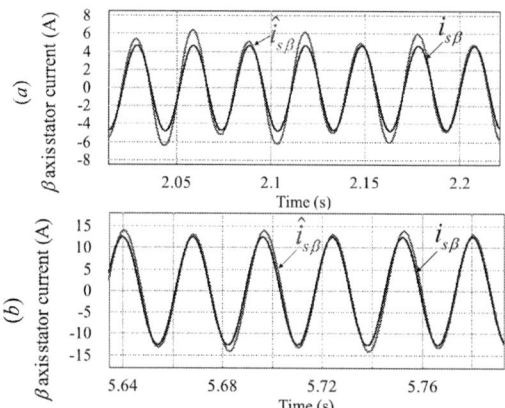

Fig. 7. β-axis stator current at 1000 rpm. (a) no load (b) 70% load.

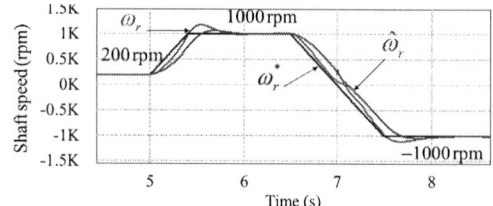

Fig. 8. Shaft speed in transient condition.

TABLE II
ESTIMATION ERROR (*) AT VARIOUS SPEED AND LOAD CONDITIONS

ω_r (rpm) / T_{load}	No load	30% load	50% load
200	0.3	2.3	3.5
400	0.3	2.5	3.9
800	1	2.6	4.2
1200	1	3	4.6

(*) in rpm

synchronous speed ω_e and be set as $-\pi$ whenever the stator flux $\hat{\lambda}_{s\beta}$ in (16) changes from positive to negative.

Assuming that the stator resistance voltage drop is well compensated, we have

$$\hat{\lambda}_{s\alpha} \approx \lambda_s^* \cos(\hat{\theta}_f) \qquad (17)$$

From (15) and (17), the estimated β-axis stator current $\hat{i}_{s\beta}$ can be obtained as,

$$\hat{i}_{s\beta} = \frac{V_s \sin(\omega_e t) - \hat{\lambda}_{s\alpha} \omega_e}{R_s} \qquad (18)$$

V. SIMULATION RESULTS

The performance of the proposed method has been verified and evaluated in PSIM simulation. In the simulation, the motor model is built with the equivalent circuit parameters listed in Table I. The iron loss resistance R_{fe} which is almost linearly dependent on the synchronous speed is shown in Fig. 5 [11]. Fig. 6 shows that at the low speed of 300 rpm, the current estimation method offers the good performances both at no load and

high load conditions. However, at high speed of 1000 rpm, since the stator resistance voltage drop is much less than the phase voltage in (15), the estimated stator current is susceptible to the small error of the current estimation method. Hence, the estimated β-axis stator current is not accurate at the high speed as shown in Fig. 7. Fig. 8 shows the performance of sensorless speed control under transient conditions and Table. II summarizes the estimation error of the proposed sensorless control at various speed and load conditions. In the Table. II, under higher load conditions, the rotor

480

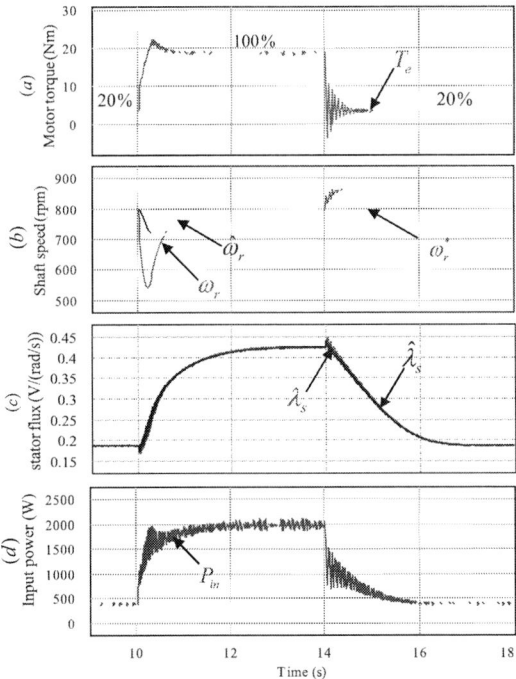

Fig. 9. Proposed loss minimization method under transient load condition. (a) motor torque (b) shaft speed (c) stator flux (d) input power.

current calculation error in (2) is higher, and this causes the estimation error to be increased. Figs. 9 and 10 show the performances of the loss minimization control under transient conditions. In Fig. 9, the changes of load (from 20% to 100% and back to 20%) are applied at speed of 800 rpm. After the load changes, the convergence time is about 3s. The change of speed (from 800 rpm to 1200 rpm and back to 800 rpm) under 30% load condition is shown in Fig. 10. These results prove that the proposed algorithm can give the good performance under transient conditions of the speed and the load. Fig. 11 shows the comparison of the proposed method and SC method on the convergence time. After applying the loss minimization control at t=6s under 10% load condition, the convergence time of the SC method is about 21s, whereas that of proposed method is about 5s. After 30% load is applied at t=30s, the SC method takes about 13s to reach the steady state, whereas the proposed method takes only 2s. Table III lists the efficiency improvements of the proposed method and existing method in [6] under different conditions.

VI. CONCLUSION

In this work, the method of loss minimization in sensorless scalar-controlled IM drives has been proposed. Based on the difference between the estimated back-EMF and the reference one, the speed can be easily estimated. For loss minimization, the optimized slip speed has been obtained in a closed form from the total power losses equation. With the optimized slip speed, at 1200 rpm and 10% load conditions, the convergence time is about 5s and efficiency improvement can reach 8.5%. In addition,

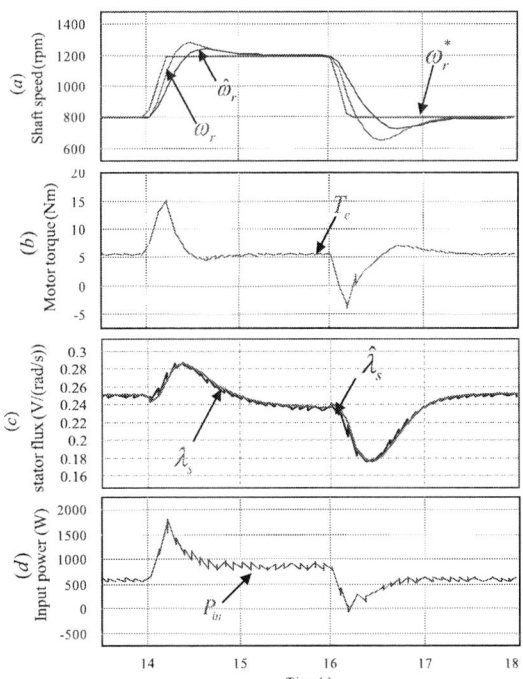

Fig. 10. Proposed loss minimization method under transient speed condition: (a) shaft speed (b) motor torque (c) stator flux (d) input power.

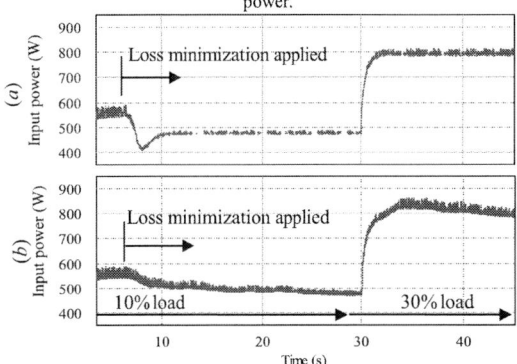

Fig. 11. Input power. (a) proposed method and (b) search control method.

TABLE III
$\Delta \eta$ (%) OF PROPOSED METHOD AND EXISTING METHOD IN [6]

Speed (rpm)	10% load		20% load		25% load	
	Prop.	[6]	Prop.	[6]	Prop.	[6]
300	4.6	3.7	1.5	0.8	1	0.7
800	6.75	5.2	2.1	1.3	1.1	0.7
1200	8.5	6.8	2.14	1.5	1.3	0.8

the stator current estimation method has been suggested based on the desired stator flux which is available in scalar control.

ACKNOWLEDGMENT

This work was supported by the Korea Institute of Energy Technology Evaluation and Planning(KETEP) and the Ministry of Trade, Industry & Energy(MOTIE) of the Republic of Korea (No. 20173030024770).

REFERENCES

[1] C. C. Wang and C. H. Fang, "Sensorless Scalar-Controlled Induction Motor Drives With Modified Flux Observer", *IEEE Trans. On Energy Conversion*, vol. 18, no. 2, pp. 181-186, 2003.

[2] C. O. Adiuku, A. R. Beig and S. Kanukollu, "Sensorless Closed Loop *V/f* Control of Medium-Voltage High-Power Induction Motor with Synchronized Space Vector PWM", in *Proc. 8th IEEE GCC Conference and Exhibition*, 2015, pp. 1-6.

[3] P. Famouri and J. J. Cathey, "Loss Minimization Control of an Induction Motor Drive", *IEEE Trans. On Industry Applications*, vol. 27, no. 1, pp. 32-37, 1991.

[4] I. Kioskeridis and N. Margaris, "Loss Minimization in Scalar-Controlled Induction Motor Drives with Search Controllers", *IEEE Trans. On Power Electronics*, vol. 11, no. 2, pp. 213-220, 1996.

[5] J. M. Eguilaz, M. Cipolla, J. Peracaula and P. J. da Costa Branco, "Induction Motor Optimum Flux Search Algorithms with Transient State Loss Minimization using a Fuzzy Logic based Supervisor" in *28th Annual IEEE Power Electronics Specialists Conference*, vol. 2, 1997, pp. 1302-1308.

[6] A. Kusko and D. Galler, "Control Means for Minimization of Losses in AC and DC Motor Drives", *IEEE Trans. On Industry Applications*, vol. IA-19, no. 4, pp. 561-570, 1983.

[7] S. Chen and S. N. Yeh, "Optimal Efficiency Analysis of Induction Motors Fed by Variable-Voltage and Variable-Frequency Source", *IEEE Trans. On Energy Conversion*, vol. 7, no. 3, pp. 537-543, 1992.

[8] I. Kioskeridis and N. Margaris, "Loss Minimization in Induction Motor Adjustable-Speed Drives", *IEEE Trans. On Industrial Electronics*, vol. 43, no. 1, pp. 226-231, 1996.

[9] R. H. A. Hamid, A. M. A. Amin, R. S. Ahahmed and Adel. A. A. El-gammal, "New Technique for Maximum Efficiency of Induction Motors Based on Particle Swarm Optimization (PSO)", in *IEEE International Symposium on Industrial Electronics*, 2006, pp. 2176-2181.

[10] A. Consoli, G. Scarcella, G. Scelba and M. Cacciato, "Energy Efficient Sensorless Scalar Control for Full Speed Operating Range IM Drives", in *Proc. 14th European Conference on Power Electronics and Applications*, 2011, pp. 1-10.

[11] E. Levi, M. Sokola, A. Boglietti and M. Pastorelli, "Iron Loss in Rotor-Flux-Oriented Induction Machines: Identification, Assessment of Detuning, and Compensation", *IEEE Trans. on Power Electronics*, vol. 11, no. 5, pp. 698-709, 1996.

[12] K. H. Nam, *AC Motor Control and Electric Vehicle Applications*. Boca Raton, FL: CRC Press, 2010.

[13] R. H. Engelmann and W. H. Middendorf, *Handbook of Electric Motors*. New York: Marcel Dekker, 1995.

[14] S. K. Sul, *Control of Electric Machine Drive Systems*. Hoboken, NJ: Wiley, 2011.

Tuning of Induction Motor Drive with Torque Sensor

Hajime Kubo[1]*, Yugo Tadano[1]

MEIDENSHA CORPORATION

Numazu, Japan

*Email: kubo-ha@mb.meidensha.co.jp

Abstract—To achieve precise torque control of an induction motor drive, the control parameters are tuned with torque sensor before shipping the drive system. In this paper, a method to estimate the rotor resistance and the magnetizing inductance with measured test data which consist of torque, current amplitude and slip frequency is proposed. Then a torque control with look up tables (LUTs) which are built from test data is also proposed. The LUTs decide the current amplitude and the voltage amplitude based on the operating point which is defined as the condition of the motor speed and the torque command.

Keywords—Induction motor drive, Parameter tuning, Torque control

I. Introduction

The torque generated by an induction motor (IM) driven by a voltage source inverter (VSI) can be controlled by using vector control. The vector control is realized based on the equivalent circuit model of the motor and the motor parameters of the model. Estimated motor parameters are set to the controller of the VSI to use the vector control mode. The performance of the torque control largely depends on the precision of the parameter estimation because the torque control is an open-loop control based on the equivalent circuit model without torque sensor. The motor parameters are usually identified by no-load test and locked rotor test conducted before the shipping of the motor. Even after the installation of the motor, the motor parameters can be identified with several self-commissioning methods which are implemented in VSIs as auto-tuning function [1]–[5]. These parameter identification methods use only electrical information such as stator current, stator voltage and output frequency.

Some applications require fine torque control whose precision cannot be achieved with parameters tuned by using the electrical information. For such application, a torque sensor is used to tune the control parameters during the test before shipping or installation of the drive system. Furthermore, look up tables (LUTs) built from the test data are used to bridge the gap between the simplified equivalent circuit model and the actual motor.

In this paper, a method to estimate the motor parameters used for the torque control based on the measured electrical data and torque data is proposed. The torque control with LUTs is also proposed. The proposed methods are evaluated in experiments.

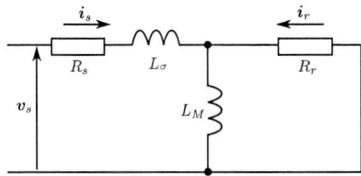

Fig. 1. Equivalent circuit model of IM

II. Methodology

A. Overview of the basic IM model and vector control

First, the basics of the IM model and the vector control are overviewed to use them in the later section. The equivalent circuit model of IM is shown in Fig 1. In this paper, the equivalent model called T-I or inverse-Γ model [6] is used. The voltage equation of the model is expressed as follows.

$$\boldsymbol{v}_s = R_s \boldsymbol{i}_s + \omega_e (L_M + L_\sigma) \boldsymbol{J} \boldsymbol{i}_s + p\boldsymbol{i}_s + \omega_e L_M \boldsymbol{J} \boldsymbol{i}_r + p\boldsymbol{i}_r \quad (1)$$

$$0 = \omega_{sl} L_M \boldsymbol{J} \boldsymbol{i}_s + p\boldsymbol{i}_s + R_r \boldsymbol{i}_r + \omega_{sl} L_M \boldsymbol{J} \boldsymbol{i}_r + p\boldsymbol{i}_r \quad (2)$$

$$\begin{bmatrix} \phi_s \\ \phi_r \end{bmatrix} = \begin{bmatrix} L_M + L_\sigma & L_M \\ L_M & L_M \end{bmatrix} \begin{bmatrix} \boldsymbol{i}_s \\ \boldsymbol{i}_r \end{bmatrix} \quad (3)$$

Where $\boldsymbol{v}_s$ is the stator voltage, $\boldsymbol{i}_s$ is the stator current, $\boldsymbol{i}_r$ is the rotor current, ϕ_s is the stator flux, ϕ_r is the rotor flux, R_s is the stator resistance, R_r is the rotor resistance, L_σ is the leakage inductance, L_M is the magnetizing inductance, ω_e is the output frequency, ω_{sl} is the slip frequency and p is time differential operator. The voltage equation is also expressed another form which uses stator current and rotor flux.

$$\boldsymbol{v}_s = R_s \boldsymbol{i}_s + \omega_e L_\sigma \boldsymbol{J} \boldsymbol{i}_s + L_\sigma p\boldsymbol{i}_s + \omega_e \boldsymbol{J} \phi_r + p\phi_r \quad (4)$$

$$0 = -R_r \boldsymbol{i}_s + R_r / L_M \phi_r + \omega_{sl} \boldsymbol{J} \phi_r + p\phi_r \quad (5)$$

The torque T_e is expressed by the product of stator current $\boldsymbol{i}_s$ and rotor flux ϕ_r as follows.

$$T_e = \text{PP} \phi_r \times \boldsymbol{i}_s \quad (6)$$

$$= \text{PP} (\boldsymbol{J} \phi_r)^T \boldsymbol{i}_s \quad (7)$$

Where PP is the number of the pole pair.

The block diagram of the typical IM drive system is shown in Fig. 2. The current $\boldsymbol{i}_{abc}$ and the speed ω_m are measured and controlled by closed loop PI control. The VSI has two different operating modes: speed control mode and torque control mode. One of the two operating modes is selected by the mode selector. The torque command T_e^* comes from the speed control block in the speed control mode or from the

The 2018 International Power Electronics Conference

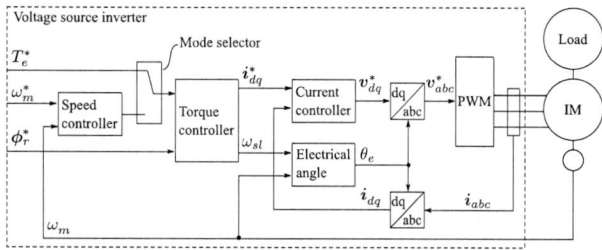

Fig. 2. Overview of IM drive system

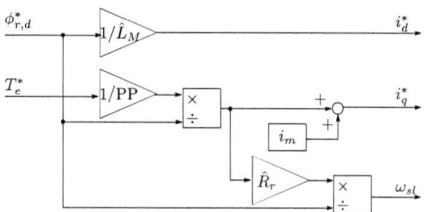

Fig. 3. Block diagram of the torque control

external command in the torque control mode. As shown in the figure, no torque feedback is given to the torque control block.

The block diagram of the torque control is shown in Fig. 3. The torque command T_e^*, d-axis rotor flux command $\phi_{r,d}^*$ and measured motor speed ω_m are converted into the current command i_d^* i_q^* and the slip frequency ω_{sl}. The constant value i_m is added to the q-axis current command to compensate the torque error due to iron loss. As shown in the figure, the torque control uses the estimated rotor resistance $\hat{R}_r$ and the estimated magnetizing inductance $\hat{L}_M$. Therefore, the precision of the open-loop torque control depends on the estimation of these two parameters.

B. Parameter tuning with torque sensor

The proposed method estimates the two parameters, $\hat{R}_r$ and $\hat{L}_M$, from the measured data which consist of torque, stator current amplitude, output frequency and rotor speed. The estimation method is derived from the motor mathematical model. The relation between stator current i_s and rotor flux ϕ_r in steady state is expressed by using R_r and L_M as follws by substituting zero for the time differential terms of (5) .

$$0 = R_r i_s - R_r/L_M \phi_r - \omega_{sl} J \phi_r \qquad (8)$$

By multiplying J to (8), the following equation is obtained.

$$0 = R_r i_s - R_r/L_M J \phi_r + \omega_{sl} \phi_r \qquad (9)$$

The outer products of (8) and (9) with $PP i_s$ result in following two equations respectively by substituting T_e for $PP \phi_r \times i_s$.

$$0 = -R_r/L_M T_e - PP \omega_{sl} J \phi_r \times i_s \qquad (10)$$

$$0 = -PP R_r/L_M |i_s|^2 - PP J \phi_r \times i_s + \omega_{sl} T_e \qquad (11)$$

By eliminating the term of $J \phi_r \times i_s$ with (10) and (11), the following equation is obtained.

$$0 = -(R_r/L_M)^2 T_e + PP \omega_{sl} R_r i_A^2 - \omega_{sl}^2 T_e \qquad (12)$$

Where i_A is the stator current amplitude which is equal to $|i_s|$. The equation (12) is transformed into the following form.

$$\left[T_e, \quad -PP \omega_{sl} i_A^2 \right] \begin{bmatrix} (R_r/L_M)^2 \\ R_r \end{bmatrix} = -\omega_{sl}^2 T_e \qquad (13)$$

Since the two parameters R_r and L_M are expressed only by the measurable variables T_e, ω_{sl} and i_A, they can be estimated from the data of $[T_e, \omega_{sl}, i_A]$ sampled at $N(N \geq 2)$ points. The data matrix X and the data vector y are constructed from N samples of data as follows.

$$X = \begin{bmatrix} T_{e,1}, & -PP \omega_{sl,1} i_{A,1}^2 \\ \cdots, & \cdots \\ T_{e,N}, & -PP \omega_{sl,N} i_{A,N}^2 \end{bmatrix} \qquad (14)$$

$$y = \begin{bmatrix} -\omega_{sl,1}^2 T_{e,1} \\ \cdots \\ -\omega_{sl,N}^2 T_{e,N} \end{bmatrix} \qquad (15)$$

Then the relation between the parameter vector $w(w = [(R_r/L_M)^2, R_r]^T)$ and the N samples of the data is expressed by the following equation.

$$X w = y \qquad (16)$$

Then the least square estimation of w is calculated as follows.

$$\hat{w} = (X^T X)^{-1} X y \qquad (17)$$

Thus the two parameters are tuned to be fit to the measured torque data. Note that the three variables T_e, ω_{sl} and i_A are purely measurable without any assumption or approximation while the variables on the synchronous reference frame such as i_d, i_q v_d and v_q require the estimated orientation of the rotor flux.

C. Torque control with LUTs

Even if the parameters of the equivalent circuit model are turned to fit the torque data, torque error due to the gap between the simplified equivalent circuit model and the actual motor remains. This torque error is reduced or eliminated by using LUTs based on the measured data.

The torque T_e in steady state is uniquely determined by the three variables which are the voltage amplitude v_A, the current amplitude i_A and the motor speed ω_m. This is explained as follows. First, the following equation is obtained from the inner product of (1) and stator current i_s and (7).

$$v_s^T i_s = R_s i_A^2 + \frac{\omega_e}{PP} T_e \qquad (18)$$

This equation can be transformed into the expression which consists of only the four variables: T_e, v_A, i_A and ω_m. The first term of the right side of the equation only contains one of the four variables. The second term of the right side has two variables: ω_e and T_e. Since the slip frequency ω_{sl} is the function of i_A and T_e as shown in (12), $\omega_e(\omega_e = PP\omega_m + \omega_{sl})$

484

is also the function of three variables: ω_m, i_A and T_e. The left side of (18) is the form of the power so it is expressed with v_A, i_A and phase angle θ_{PF}.

$$\boldsymbol{v}_s^T \boldsymbol{i}_s = v_A i_A \cos\theta_{PF} \tag{19}$$

In steady state, the voltage equations, (1) and (2), become following forms.

$$\boldsymbol{v}_s = (R_s \boldsymbol{I} + (L_M + L_\sigma)\omega_e \boldsymbol{J}) \boldsymbol{i}_s + L_M\omega_e \boldsymbol{J} \boldsymbol{i}_r \tag{20}$$
$$\boldsymbol{0} = L_M\omega_{sl}\boldsymbol{J}\boldsymbol{i}_s + (R_r\boldsymbol{I} + L_M\omega_{sl}\boldsymbol{J})\boldsymbol{i}_r \tag{21}$$

By eliminating rotor current $\boldsymbol{i}_r$, the stator voltage is expressed as linear transformation of the stator current.

$$\boldsymbol{v}_s = \boldsymbol{A}\boldsymbol{i}_s \tag{22}$$

$$\begin{aligned}\boldsymbol{A} = {}& R_s\boldsymbol{I} + (L_M + L_\sigma)\omega_e\boldsymbol{J} \\ & -L_M\omega_e\boldsymbol{J}\left(R_r\boldsymbol{I} + L_M\omega_{sl}\boldsymbol{J}\right)^{-1} L_M\omega_{sl}\boldsymbol{J} \end{aligned} \tag{23}$$

Since $\boldsymbol{A}$ contains two variables: ω_e which is a function of ω_m, i_A and T_e and ω_{sl} which is a function of i_A and T_e, the phase angle θ_{PF} is also a function of ω_m, i_A and T_e. From the above, all the three terms in (18) are expressed as function of the four variables : T_e, v_A, i_A and ω_m. Therefore the torque T_e can be controlled by v_A, i_A and ω_m. The voltage amplitude and current amplitude are controlled by the torque controller while the motor speed is given from the outside. Therefore the voltage amplitude and current amplitude are decided by LUTs to control the torque precisely.

The proposed torque control with LUTs is shown in Fig. 4. The torque offset LUT decides the current amplitude by giving the offset to the torque command T_e^*. Voltage amplitude LUT outputs the voltage amplitude command v_A^*. The input to both the two LUTs is the operating point op which is defined as $[\omega_m, T_e]$. The role of the torque offset LUT is to adjust the torque command at each operating point. The voltage amplitude LUT compensates fluctuation of R_r due to temperature change. In the voltage amplitude control block, the estimated rotor resistance $\hat{R}_r$ is updated to reduce the error between the voltage amplitude command v_A^* and the voltage amplitude of that control cycle as follows.

$$\hat{R}_{r,k+1} = \hat{R}_{r,k} - K_e(v_A^* - v_A) \tag{24}$$

Whrere K_e is the estimation gain. The relation between the stator current and the rotor flux in steady state (8) is transformed into the following form and depicted in Fig.5.

$$L_M \boldsymbol{i}_s = \boldsymbol{\phi}_r + \frac{L_M}{R_r}\omega_{sl}\boldsymbol{J}\boldsymbol{\phi}_r \tag{25}$$

While the amplitude of the stator current $L_M|\boldsymbol{i}_s|$ is kept constant by the current PI controller, the magnitude of rotor flux $|\boldsymbol{\phi}_r|$ increases (or decreases) if that of the second term of right side of (25) decreases (or increases). For example, if the rotor resistance R_r increases along with the temperature rise, the magnitude of rotor flux also increases. This results in the larger voltage amplitude. In a similar way, the voltage amplitude decreases if the slip frequency ω_{sl} increases with the estimated rotor resistance $\hat{R}_r$. The voltage amplitude controller keeps

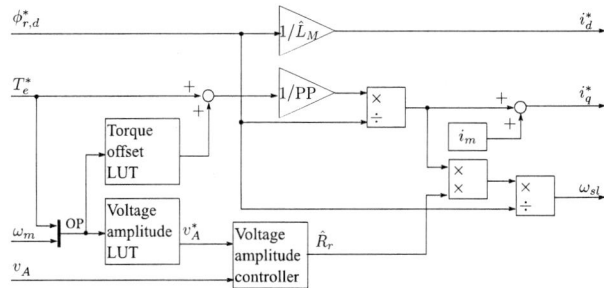

Fig. 4. Torque control with LUTs

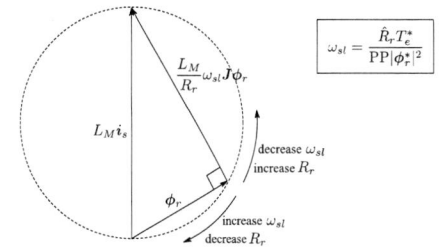

Fig. 5. Rotor flux monotonically increases as slip frequency decreases

the voltage amplitude to the value of the voltage amplitude command v_A^* by adjusting the estimated rotor resistance $\hat{R}_r$, while the actual rotor resistance R_r changes with temperature. From (6), the torque is proportional to the area of the triangle in the figure. The area of the triangle is kept constant by current controller and the voltage amplitude controller because $\boldsymbol{i}_s$ and $\boldsymbol{\phi}_r$ are kept constant by these two controllers.

The test to build the two LUTs is conducted as follows. The IM under the test is driven in torque control mode and the coupled motor is driven in speed control mode. At an operating point $[\omega_m, T_e^*]$, the torque is controlled with the torque controller shown in Fig. 3 and an integral controller of torque offset. The integral controller updates the torque offset to reduce the torque error as follows.

$$T_{eo,k+1} = T_{eo,k} + K_I(T_e^* - T_e) \tag{26}$$

Where $T_{eo,k}$ is the torque offset at control cycle k, K_I is the integral gain and T_e is the torque measured with a torque sensor. The torque command with offset $T_e^* + T_{eo}$ is entered into the torque controller. When the torque error $T_e^* - T_e$ becomes zero, the torque offset and the voltage amplitude at the moment are recorded as the table data for the operating point.

III. EXPERIMENT

A. Experimental setup

The diagram of the experimental setup is shown in Fig. 6. The rating values of the IM are shown in Table I. The IM are coupled with the load machine. The IM is driven in the torque control mode while the load machine is driven in the speed control mode. The torque sensor is in between the two motors.

The 2018 International Power Electronics Conference

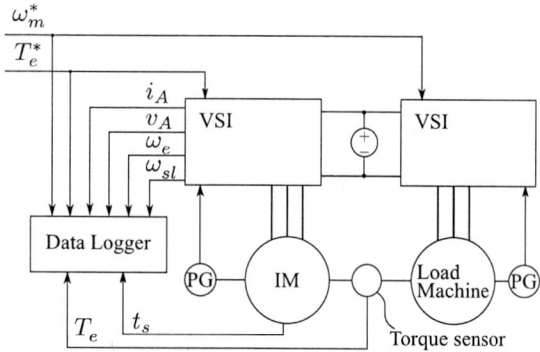

Fig. 6. Diagram of the experimental setup

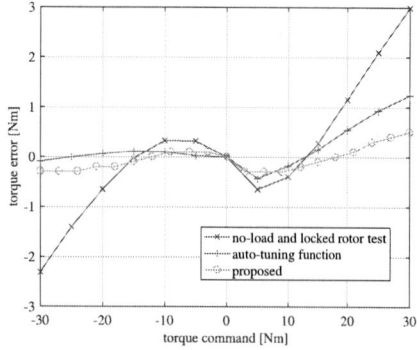

Fig. 7. Torque errors of the three methods.

The temperature of the stator coil-end t_s is measured with a thermocouple.

B. Evaluation of the parameter tuning

The rotor resistance and the magnetizing inductance of the motor are estimated by no-load and locked rotor test, the auto-tuning function of the VSI and the proposed method. The estimated values are shown in Table II.

The torque errors of the torque control with these three estimations are compared in Fig. 7. The proposed method has the smallest torque error of the three.

TABLE I. SPECIFICATION OF THE IM

Power	POLE	Voltage	Frequency	Current	Speed
5.5 kW	4	160 V	51.5 Hz	28 A	1500 rpm

TABLE II. ESTIMATED MOTOR PARAMETERS

	R_r	L_M
no-load and locked rotor	95.5 mΩ	24.5 mH
auto-tuning function	84.6 mΩ	24.1 mH
proposed method	93.8 mΩ	27.3 mH

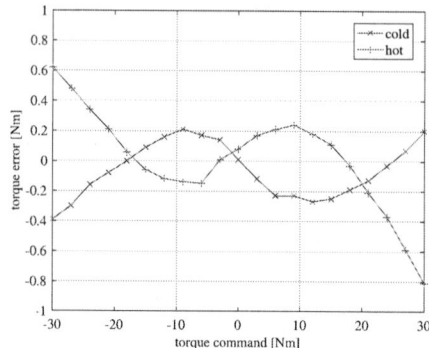

Fig. 8. Torque error without voltage amplitude LUT

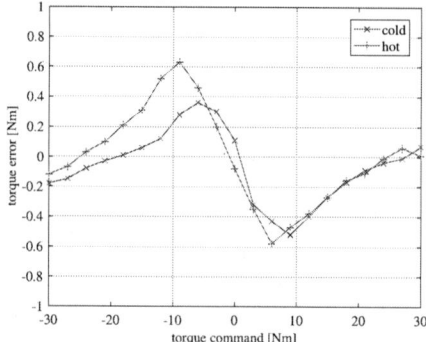

Fig. 9. Torque error with voltage amplitude LUT

C. Evaluation of the torque control with LUTs

To evaluate the temperature fluctuation compensation with the voltage amplitude LUT and the voltage amplitude control, the torque control with LUT was tested in two different temperature conditions. First, the torque control was tested before the IM was warmed up. The stator coil-end temperature t_s is from 9.7 K to 11.1 K lower than that in which the LUTs were made. Then the torque control was tested after the IM was warmed by running in hours. The stator coil-end temperature t_s is from 7.2 K to 9.8 K higher than that in which the LUTs were made. The results are shown in Fig. 8 and Fig. 9. Fig. 8 is the result of the torque control without the voltage amplitude LUT and the voltage amplitude control. In this case, torque error occurs due to the temperature changes. Fig. 8 is the result of torque control with the voltage amplitude LUT and the voltage amplitude control. The torque error in high torque region is eliminated by controlling the voltage.

IV. CONCLUSION

To achieve precise torque control of IM, a method to tune the two control parameters, the rotor resistance and the magnetizing inductance, with torque sensor is proposed. The method uses data which consist of slip frequency, current amplitude and torque. The two parameters are tuned to fit the IM model to the torque data. The result of the experiment shows that the proposed method achieves more precise torque control than the conventional no-load and locked rotor test or auto-tuning function of VSI which only uses electrical information. The torque control using LUTs is also proposed. The torque offset LUT and the voltage amplitude LUT are used in the proposed torque control. The LUTs are built based on the test data at each operating point which is defined by motor speed and torque. The result of the experiment shows that the proposed method can control the torque precisely even if the temperature of the motor changes. The improvement of the torque control in low torque region is the future work.

REFERENCES

[1] D. Telford, M. W. Dunnigan, and B. W. Williams, "Online identification of induction machine electrical parameters for vector control loop tuning," *IEEE Transactions on Industrial Electronics*, vol. 50, no. 2, pp. 253–261, Apr 2003.

[2] H. A. Toliyat, E. Levi, and M. Raina, "A review of rfo induction motor parameter estimation techniques," *IEEE Transactions on Energy Conversion*, vol. 18, no. 2, pp. 271–283, June 2003.

[3] K. Wang, J. Chiasson, M. Bodson, and L. M. Tolbert, "A nonlinear least-squares approach for identification of the induction motor parameters," *IEEE Transactions on Automatic Control*, vol. 50, no. 10, pp. 1622–1628, Oct 2005.

[4] Y. He, Y. Wang, Y. Feng, and Z. Wang, "Parameter identification of an induction machine at standstill using the vector constructing method," *IEEE Transactions on Power Electronics*, vol. 27, no. 2, pp. 905–915, Feb 2012.

[5] W. M. Lin, T. J. Su, and R. C. Wu, "Parameter identification of induction machine with a starting no-load low-voltage test," *IEEE Transactions on Industrial Electronics*, vol. 59, no. 1, pp. 352–360, Jan 2012.

[6] G. R. Slemon, "Modelling of induction machines for electric drives," *IEEE Transactions on Industry Applications*, vol. 25, no. 6, pp. 1126–1131, Nov 1989.

Quasi-Two-Level Converter for overvoltage mitigation in medium voltage drives

F. Bertoldi[1], M. Pathmanathan[1], R. S. Kanchan[1], K. Spiliotis[2] and J. Driesen[2]
1 Corporate Research, ABB, Västerås, Sweden
2 ESAT-ELECTA, KU Leuven, Leuven, Belgium
Email: federico.bertoldi@se.abb.com, konstantinos.spiliotis@kuleuven.be

Abstract - This work focuses on a Quasi-Two-Level (Q2L) converter topology for medium voltage drives (MVDs). It targets applications where the drive and the motor are connected via a long cable. In such systems a *dv/dt* filter is usually placed between the drive and motor to increase the rise time of inverter output voltage pulses, and thereby reduce the motor terminal overvoltage. The Q2L converter allows for an intrinsic reduction in the motor terminal overvoltage by dividing the voltage transitions in the inverter output voltage waveform into multiple smaller steps. Simulation results obtained with a distributed parameter cable model are presented, which show that the Q2L converter can offer lower converter capacitor energy and reduced *dv/dt* filter component values for a given motor terminal overvoltage requirement compared to the 5LANPC, a conventional MVD topology, at the cost of worse WTHD and efficiency.

Keywords – Long cable phenomena, overvoltage mitigation, quasi-Two-Level, dv/dt, reflection wave.

I. INTRODUCTION

Medium voltage electrical machines are widely used in high-power industrial applications. Medium voltage drives are often used together with these machines due to their ability to offer energy savings compared to a direct-on-line (DOL) configuration.

In some applications, due to space constraints, a long cable is present between the medium voltage drive and electrical machine [1, 2]. The schematics of such an application is shown in Fig. 1.

When connecting a converter to an electrical machine via a long cable, overvoltage due to wave reflection can appear at the motor terminals due to the fast switching process of the semiconductor devices, causing risk of premature insulation failure [3]. The voltage at the motor terminals can be up to double the voltage step applied by the converter [4].

Medium voltage drives are commonly constructed using either a two-level (2L) configuration formed from series-connected power semiconductors, or multilevel converters. 2L solutions can be adopted by connecting switches in series and operating them simultaneously with high precision gate drivers. Higher voltage rated devices can also achieve a 2L operation. However, the *dv/dt* applied by the Voltage Source Inverter (VSI) is steep, which is bad for the wave reflection overvoltage. Typically, *dv/dt* filters are used to mitigate the overvoltage. For a 2L inverter, they have to be dimensioned over the DC-link voltage to limit the percentage overshoot at the motor terminals below the insulation limit.

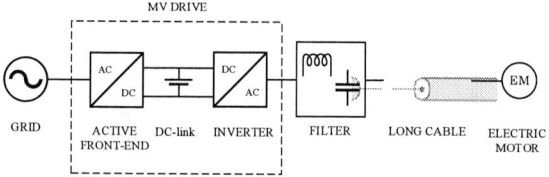

Fig. 1. Schematics of the MV drive application of this work.

When multilevel converters are used, the overvoltage problem at the motor terminals is less of a concern because of the smaller steps applied by the converter [5, 6].

Beside their low output THD, multilevel converters also offer low voltage stress and low *dv/dt* [7]. For this reason, when more level are present, smaller *dv/dt* filters can be used to avoid risky overvoltage at the motor terminals.

Increasing of the number of levels comes along with increase complexity, in non-modular topologies as Neutral Point Clamped (NPC), or increase in the capacitor energy as it happens for Modular Multilevel Converter (MMC) or Flying Capacitor (FC) topologies. This results in harder voltage scalability of the former, and in increase of the total converter capacitance of the latter, driving up the costs and the footprint of the drive system.

In this work, the use of a Quasi-Two-Level (Q2L) converter as a low capacitance and low *dv/dt* solution for MVDs is analyzed based on simulations, aiming at low converter and filter footprint and costs associated with the components energy.

A square modulated Q2L converter topology was firstly proposed in [8] for HVDC application to reduce the *dv/dt* at the terminals of the insulation transformer. Recently, a PWM based Q2L operation of an MMC was introduced in [9, 10] to enable the use of lower voltage components and reduce the converter energy while keeping low the *dv/dt*.

The Q2L has the modular structure of an MMC (as depicted in Fig. 2., which results in easy voltage scalability and good reliability, but mainly works on two levels. It is composed by n cells per arm (here half-bridge have been used), and two arms make up a leg of the three-phase converter. A typical line-to-line voltage waveform is shown in Fig. 3.: on the fundamental period zoom the is no difference from that of a traditional 2L converter, but when zoomed in, each pulse is made up by a staircase of n steps, fired with a dwell time t_D delay one after another. By varying t_D and n the effective *dv/dt* applied by the converter can be changed.

In this work, dwell time, dv/dt and n are related by the following formula:

$$t_D = \frac{1}{dv/dt} \cdot \frac{1}{n+1} \qquad (1)$$

While the voltage of each Q2L step is given by:

$$V_{step} = \frac{V_{DC}}{n} \qquad (2)$$

II. REFLECTION WAVE PHENOMENON

When a long cable is placed between the converter and the motor, wave reflection phenomena can occur. This phenomenon comes as a result of the impedance mismatch of the cable, motor and drive, and the fast rising time of the switching devices. The overvoltage problem can then be tackled either by matching the impedances of the cable and the motor (with impedance matching RC circuits at the cable end) or by increasing the effective rise time of the applied voltage step, thereby decreasing the dv/dt [11].

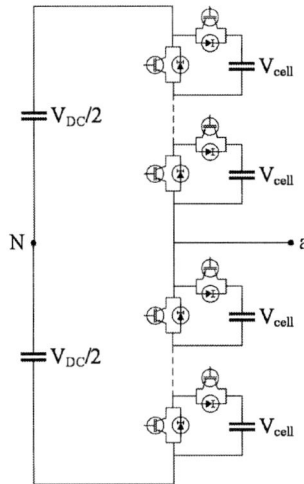

Fig. 2. Single-phase schematics of a Q2L converter.

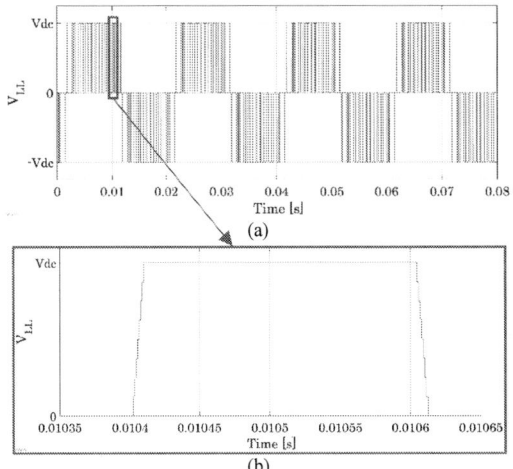

Fig. 3. Q2L line-to-line output voltage waveform: a) fundamental period zoom, b) zoomed in pulse view.

RLC dv/dt filters at the output terminals follow this second principle, and are the most common choice in industry.

When a cable connects the motor to the drive, a reflection coefficient appears at the cable ends, given by:

$$\Gamma_M = \frac{Z_M - Z_C}{Z_M + Z_C} \qquad (3)$$

where Z_M and Z_C represent the motor and cable characteristic impedances, respectively. Typical values for Γ_M are between 0.6 and 1, since the motor impedance is generally much larger than the cable's. Analogously, a reflection coefficient at the converter terminals Γ_D can be derived. Because the converter impedance is usually very low, Γ_D has usually values close to -1.

Then, a full reflection phenomenon with reflection coefficient Γ_M occurs when the rise time of the pulse applied by the converter is lower than three times the wave propagation time from one cable end to the other, defined here as t_t. Thus, for a reflection as such the overvoltage at the motor terminals V_M is given by:

$$V_M = V_{step}(1 + \Gamma_M) \qquad (4)$$

being V_{step} the voltage step applied by the VSI.

It is important to notice that, as the device turn on and off processes become faster, the overvoltage appears for shorter cable lengths [12].

Fig. 4. shows a simplified example of how cable and switching parameters influence the overvoltage oscillation at the motor terminals. The voltage at the motor terminals oscillates with a peak value that depends on the rise time of the applied voltage step, and a frequency dependent on the cable parameters, namely the per-unit-length inductance and capacitance, and the cable length. Therefore, a slower pulse rise time and a smaller step magnitude can be applied at the converter side to decrease the overvoltage peak.

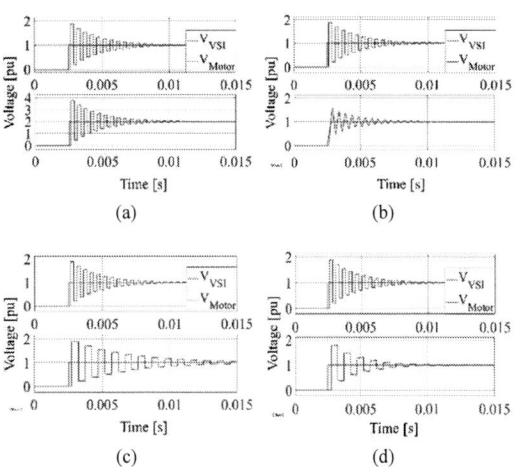

Fig. 4. Overvoltage at the motor terminals: (a) doubling the voltage step; (b) increasing the switching rise time; (c) halving the cable propagation speed; (d) doubling the cable length.

The first is done by appropriately selecting the *dv/dt* filter cut-off frequency, the second is intrinsically given by the converter. In addition, a longer cable or a slower propagation speed result in a lower oscillation frequency that will be cut by the filter, thereby increasing the *LC* component values.

In contrast to *sine* filters, *dv/dt* filters only cut-off the frequency range related to the wave reflection, and are therefore smaller and cheaper. Then, for a given overshoot, filters are designed based on the cable parameters and the reflection coefficient due to impedances mismatch [13, 4]. When THD requirements are not strict, they can be used for cable lengths up to a few hundred meters.

III. Converter and Filter Design

In this section an overview of the design of the Q2L VSI and *dv/dt* filter used in this work is given.

A. Q2L IGBT choice and cell capacitor design

The IGBT module to be used in a Q2L converter can be based on the desired V_{step}, which is derived from the *dv/dt*, t_D and V_{DC} requirements and by using (1) and (2). While the nominal current is the same, the voltage rating decreases according to (2).

The second converter parameter to be chosen is the cell capacitance C_{cell}. In this work, the following approach has been followed.

During a transition from the upper level to the lower, the current commutates from the upper arm to the lower, charging or discharging the capacitors. The delta energy of on arm can be written as the difference between its final and initial value:

$$\Delta E_{arm} = E_{fin} - E_{init} = \frac{1}{2}\frac{C_{cell}}{n}V_{DC}^2(2\delta + \delta^2) \quad (5)$$

Where δ is the target cell voltage deviation over the nominal value.

During an upper to lower voltage level transition, the upper arm passes from providing 0 voltage and full phase current I_{ph} (all cells bypassed), to provide full V_{DC} voltage and 0 current (all cells inserted), in a time nt_D. If, as in this case, both quantities have a linear slope, the energy involved in the transition is:

$$\Delta E_{arm} = \frac{V_{DC} I_{ph} n t_D}{4} \quad (6)$$

Then, by equalizing (5) with (6), and considering the worst case when the phase current is at its peak, it is found:

$$C_{cell} = \frac{I_{pk} n^2 t_D}{2 V_{DC}(2\delta + \delta^2)} \quad (7)$$

Each arm is parallelized to the DC link every two transitions, when cells are inserted again, charging or discharging the cells with a current circulating in the phase

leg. A sorting algorithm then takes care that cells are kept balanced.

From (7) it can be notice that, for a given DC voltage, the capacitance increases inversely to the ideal *dv/dt* of the Q2L pulse flank given in (1); i.e. a higher *dv/dt* requires less capacitor energy.

B. RLC filter design

The filter is made by an RLC star connected circuit, where the inductance L_F is connected in series at the converter output and the $R_F C_F$ components are in shunt configuration. The filter resistance R_F is set equal to the cable characteristic impedance to absorb the reflected wave travelling backwards [11]. The filter inductance and capacitance (L_F and C_F, respectively) are derived with the method in [4], by solving by comparison with a second order transfer function with damping equal to 1 the filter-cable transfer function:

$$H(s) \approx \frac{\dfrac{Z_C}{L_F}s + \dfrac{1}{L_F C_F}}{\dfrac{2}{1+\Gamma_M}s^2 + \dfrac{Z_C}{L_F}s + \dfrac{1}{L_F C_F}} \quad (8)$$

where Z_C is the cable characteristic impedance and Γ_M the reflection coefficient at the motor terminals defined in (3).

However, the filter does not have to be dimensioned for the whole transition from the bottom to the top level of the Q2L flank (i.e. a *0* to V_{DC} transition). Instead, it can be dimensioned in such a way that the overshoot related to the last step is kept within the desired limit. This allows for smaller L_F and C_F components value, and also results in shorter settling time. By choosing a dwell time equal to the settling time of the filter-cable system it can be ensured that the overvoltage at the motor terminals is kept below the insulation level.

IV. Methodology

A. Simulation setup

For an accurate representation of the drive system a PLECS-PSCAD (EMTDC) co-simulation was set up. PLECS is a software used for converter modeling, since it provides well defined switching events needed for the Q2L staircase modulation and good losses estimation. At the same time, the cable model for high frequencies studies cannot be modeled as a single-frequency Bergeron model [14]. The *r, l, c* (and *g,* usually neglected) per-unit-length cable parameters should be frequency dependent. One way of obtaining such parameters is by measuring the cable over a frequency range and find an equivalent circuit [15]. The PSCAD EMTDC module provides a geometry-based cable model that include all the conducting and insulation layers, deriving the impedance and admittance matrices to be computed by PSCAD. This geometry based cable model was preferable, since measured data of the cable used in this study was not available.

The 2018 International Power Electronics Conference

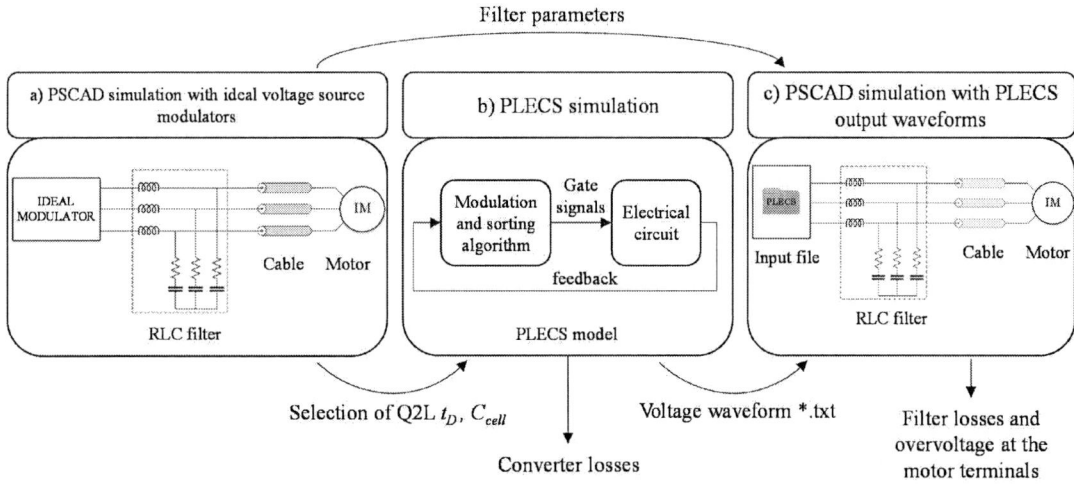

Fig. 5. Simulation process and outputs of each step.

The simulation process of the drive system is divided in three main steps:

a) The drive system is firstly modeled in PSCAD, with the VSI modeled as an ideal voltage source based modulator, to tune the Q2L flank dv/dt and the filter accordingly to the desired overshoot.

b) Then, the Q2L cell capacitor value are defined for a given ripple using (7), and a PLECS simulation is conducted to calculate losses and obtain the converter output waveforms which are saved.

c) The converter waveforms are loaded into PSCAD, where the motor terminal overvoltage and filter losses can be evaluated.

The PLECS model of the converter in Fig. 5. b) makes use of a thermal description of the IGBT modules to evaluate the switching and conduction losses. The converter input reference is a simple sine PWM. The cell capacitors voltage is measured and fed back to a control algorithm that keeps the capacitors balanced.

For the PSCAD simulation in Fig. 5. a) and Fig. 5. c), a simulation step of one order of magnitude smaller than the cable traveling time was used.

The motor is modeled as a RL load connected at the cable end for simulations Fig. 5.a) and Fig. 5. c) and at the inverter output for the simulation in Fig. 5. b). The current reaching the motor is found to be the same in all the three steps, with only a small difference due to the paths that enable the current to flow through the cable capacitance.

In [16] equivalent circuits for low voltage induction motors are derived to derive the motor behavior at high frequencies. Such models can be obtained by experimental measurements of the machine. In this work however, only the low frequency RL load is used, since it results in the largest reflection coefficient, which in turn causes the highest motor terminal overvoltage.

B. Simulated scenarios and converter designs

The simulation model was set up for the application case shown in TABLE I

The system is made of a VSI with a dv/dt filter connected at the output, a 50 and 100 m cable and the RL load at the cable's end.

The Q2L was benchmarked versus a 5LANPC multilevel converter presented in [17], which represents a state of the art converter technology for medium voltage drives.

The 5LANPC was modulated with a hybrid Phase Shifted Carriers (PSC) scheme. Two Q2L configurations with four and eight cells per arm are chosen, the former having the same voltage step as the 5LANPC.

All converters are modulated to have the same device switching frequency at 540 Hz. The converter is tested for a 60 Hz fundamental frequency and three different amplitude modulation indexes, 0.1, 0.5 and 1, in order to check the impact on the converter and filter losses.

TABLE I
APPLICATION CASE DESCRIPTION

Apparatus	Model	Description
Drive	5LANPC Q2L ($n=4,8$)	S = 1.25 MVA V_{DC} = 6 kV Fundamental frequency: 60 Hz Device switching frequency: 540 Hz Amplitude modulation index: 0.1, 0.5, 1
Filter	RLC dv/dt filter	
Cable	CU105G30-010 Southwire Canada	Coaxial Shielded Cable Lengths: 50, 100 m Insulation: 7.2 kV
Motor	ABB HXR500LK8	Induction Motor S = 1.25 MVA V_{LL} = 4.16 kV

491

The 2018 International Power Electronics Conference

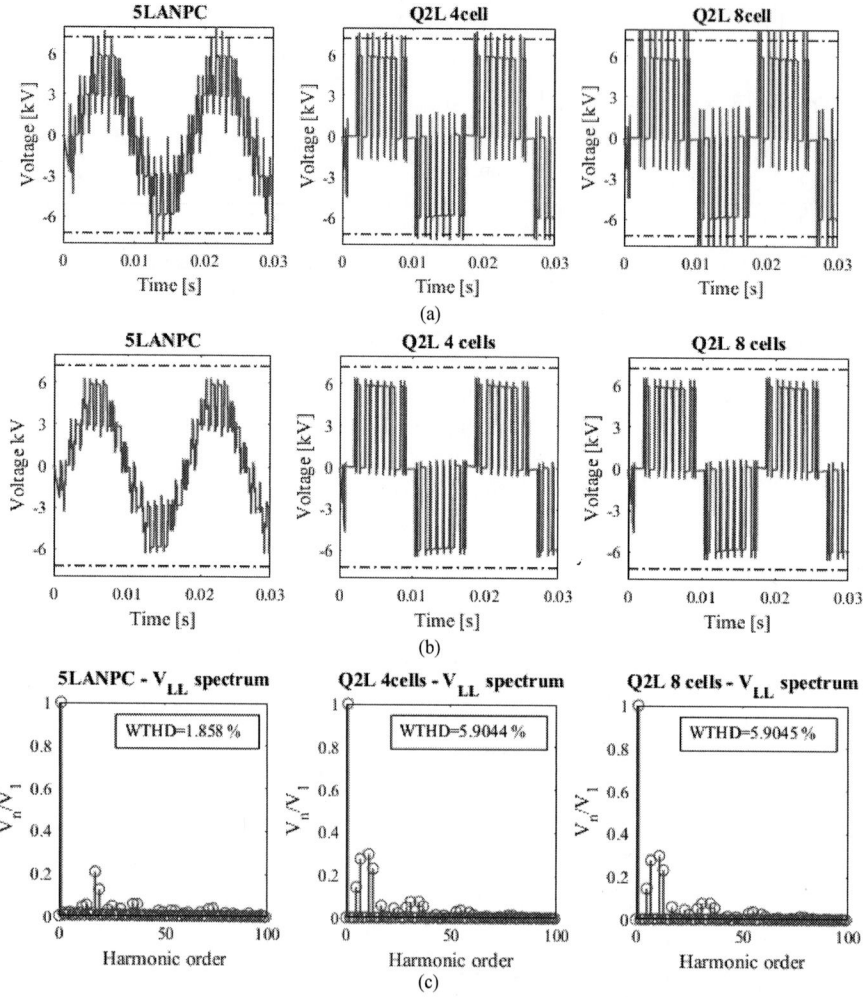

Fig. 6. Voltage waveforms at the motor terminals for a 100 m cable with insulation 7.2 kV ($-\cdot-\cdot$). (a) when no filter is applied, (b) when filters are applied; (c) the converters line-to-line voltage spectrum.

V. SIMULATION RESULTS

Fig. 6. shows the voltage waveforms at the motor terminals without and with the dv/dt filter. The multilevel waveform of the 5LANPC clearly shows a benefit in terms of weighted total harmonic distortion (WTHD), as it can be observed by looking at the waveform spectra in Fig. 6. (c). If low WTHD is a requirement, the multilevel property is beneficial, accompanied by a *sine* filter to smooth out output voltage and currents. However, this would come with greater converter capacitance and filter components.

When no filter is applied, uncontrolled overvoltage spikes threaten the cable's insulation. The filters are tuned so that the overvoltage at the motor terminal is kept below 5% of the DC-link voltage.

Since the step of the *eight* cells Q2L is half of that of the *four* cells Q2L and 5LANPC, in order to have the same overvoltage at the motor terminals, a smaller dv/dt filter can be dimensioned. This can be appreciated by looking at the chart in Fig. 8. (b), where the L_F and C_F component energy is halved for the *eight* cells Q2L.

Having smaller LC filter components value allows the filter to have a higher natural frequency and a faster settling time (R_F however, is always equal to the cable characteristic impedance). This results in the *four* and *eight* cells Q2L configurations having the same dv/dt, as shown in Fig. 7. which results in roughly the same converter capacitor energy for a given cable length, as illustrated in Fig. 8.(a).

In the case of the 5LANPC the converter energy does not change with the cable length, only the filter components value increases as the cable becomes longer. In contrary, the Q2L capacitance must be tuned for a given cable length. As the cable length becomes longer, the settling time of the system comprising the filter and cable becomes greater. The dwell time of the Q2L converter needs to be greater than the settling time in order to avoid a summation of overvoltages at the motor terminals. Thus the Q2L cell capacitance must increase as a function of cable length, as shown in Fig. 7.

On the other hand, as the cable length decreases, the Q2L capacitance decreases accordingly. Therefore, for the application case presented here, the Q2L requires 36% and

492

72% the capacitor energy of the 5LANPC, for the 50 and 100 m setup, respectively.

In Fig. 9. the loss split in the converter and filter for 50 and 100 m cable length for the three different modulation indexes is illustrated. The 5LANPC converter losses are lower than the Q2L in both configurations.

The filter losses depend on the applied step, the filter size and the number of transitions the filter is attenuating.

Because the filter losses are proportional to the square of the applied voltage, in the *eight* cells Q2L configuration they are around one quarter those in the *four* cells Q2L and half those with the 5LANPC.

In addition, filter losses are not affected by the modulation index and represent a constant loss contribution for a given converter switching frequency, as the number of level transitions does not vary significantly with different modulation indexes. For this reason, the *eight* cells Q2L drive configuration can achieve better efficiency at lower modulation indexes, where the filter loss makes up the larger part of the total VSI-filter system.

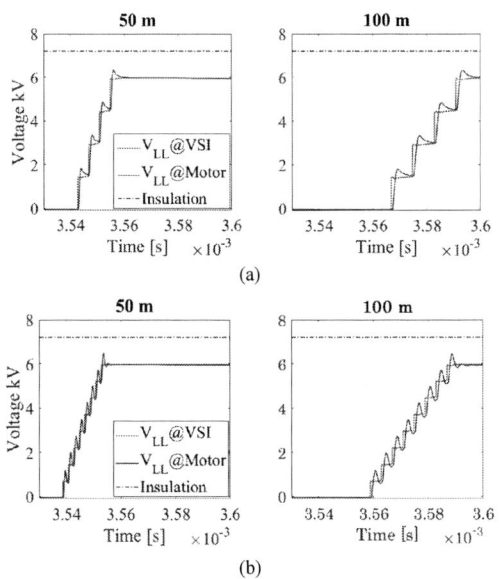

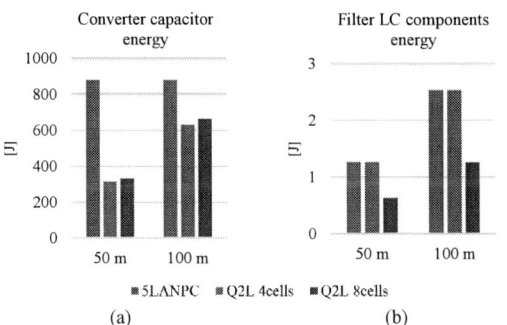

Fig. 7. Q2L flank, line-to-line voltage at the VSI and motor terminals for 50 and 100 m cable length configurations: (a) *four* cells Q2L, (b) *eight* cells Q2L.

Fig. 8. VSI and filter total energy: (a) total converter capacitor energy, (b) total filter L_F and C_F components energy, based on nominal voltage and current.

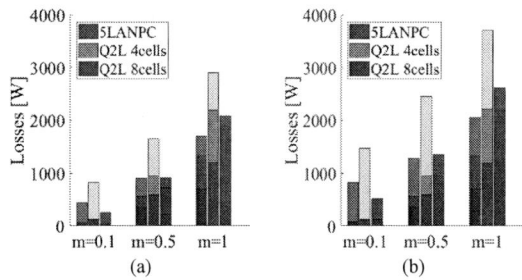

Fig. 9. Converter and filter losses split: switching (darkest, bottom), conduction (medium dark, middle), and filter (brighter, top). (a) For a 50 m cable, (b) for a 100 m cable.

VI. CONCLUSIONS

In this paper a Q2L converter is proposed as a solution for medium voltage drives, and it benchmarked against a 5LANPC. The converters have been modeled in PLECS, using commercially available IGBT modules. The interaction with the cable and filter was implemented in PSCAD, where an appropriate cable model was realized.

The ability if the Q2L converter to generate a line voltage waveform with a controllable *dv/dt* flank made it an attractive drive choice in cases where the cable length was long enough for motor terminal overvoltages to be a problem. Compared to the 5LANPC, the eight cell Q2L converter was found to have lower converter capacitor and filter capacitor requirements for a given motor terminal overvoltage requirement. These benefits came at the cost of an increased line voltage WTHD and losses.

The Q2L converter can become a more attractive solution as faster switching devices such SiC become adopted, since faster switch rise times will cause the problem of motor terminal overvoltage to be present at shorter cable lengths. Moreover, the modular cell-based design of the Q2L converter inherently results in small commutation loops, which manifests itself in lower converter switching losses compared to non-modular designs.

REFERENCES

[1] L. Lobianco and W. Wardani, "Electrical submersible pumps for geothermal applications," *Geothermal Energy for Power Production*, 2000.

[2] G. Skibinski and S. Breit, "Line and load friendly drive solutions for long length cable applications in electrical submersible pump applications," *IEEE*, no. PCIC-2004-30, pp. 269–278, 2004.

[3] J. Rodriguez, J. Pontt, C. Silva, R. Musalem, P. Newman, R. Vargas, and S. Fuentes, "Resonances and overvoltages in a medium voltage fan motor drive with long cables in an underground mine," *IEEE Transactions on Industry Applications*, vol. 42, no. 3, pp. 856–863, 2006.

[4] S. Lee and K. Nam, "Overvoltage suppression filter design methods based on voltage reflection theory," *IEEE Transactions on Power Electronics*, vol. 19, no. 2, pp. 266–271, 2004.

[5] F. Endrejat and P. Pillay, "Resonance Overvoltages in Medium Voltage Multilevel Dive Systems," *IEEE Transactions on Industry Applications*, vol. 45, no. 4, pp. 1199–1209, 2009.

[6] J. Oliver and G. Stone, "Implications for the Application of Adjustable Speed Drive Electronics to Motor Stator Winding Insulation ," *IEEE Electrical Insulation Magazine*, pp. 32–36, 1995.

[7] J. Rodríguez, L. G. Franquello, S. Kouro, J. I. León, R. C. Portillo, M. A. Martín Prats, and M. A. Pérez, "Multilevel converters: An enabling technology for high-power applications," *IEEE*, vol. 97, no. 11, pp. 1786–1917, 2009.

[8] I. A. Gowaid, G. P. Adam, A. M. Massoud, S. Ahmed, D. Holliday, and B. W. Williams, "Quasi Two Level operation of a modular multilevel converter for use in a High-Power DC transformer with DC fault isolation capability," *IEEE Transaction on Power Electronics*, vol. 30, no. 1, pp. 108–123, 2015.

[9] A. Mertens and J. Kucka, "Quasi Two Level operation of an MMC phase leg with reduced module capacitance," *IEEE Transactions on Power Electronics*, vol. 31, no. 10, pp. 6765–6769, 2016.

[10] J. Kucka and A. Mertens, "Control for quasi-two-level PWM operation of modular multilevel converter," *IEEE*, pp. 448–453, 2016.

[11] K. Kuen-Faat Yuen and H. Chung Shu-Hung, "A Low-Loss "RL-plus-C" filter for overvoltage suppression in inverter-fed drive system with long motor cable," *IEEE Transaction on Power Electronics*, vol. 30, no. 4, pp. 2167–2181, 2015.

[12] T. Mukundan, "Calculation of voltage surges on motor fed from PWM drives - a simplified approach," *IEEE Transactions on Energy Cconversion*, vol. 19, no. 1, pp. 223–225, 2004.

[13] A. Von Jouanne and P. Enjeti, "Design considerations for an inverter output filter to mitigate the effects of long motor leads in ASD applications," *IEEE Transaction on Industry Applications*, vol. 33, no. 5, pp. 1138–1145, 1997.

[14] G. Skibinski, R. Kerkman, D. Leggate, J. Pankau, and D. Schlegen, "Reflected wave modeling technique for PWM AC motor drives," *APEC '98 Thirteenth Annual Applied Power Electronics Conference and Exposition*, vol. 2, pp. 1121–1129, 1998.

[15] A. Moreira, T. Lipo, G. Venkataramanan, and S. Bernet, "High-Frequency modeling for cable and induction motor overvoltage studies in long cable drives," *IEEE Transactions on Industry Applications*, vol. 38, no. 5, pp. 1297–1306, 2002.

[16] B. Mirafzal, G. Skibinski, R. Tallam, D. Schlegel, and R. Lukaszewski, "Universal induction motor model with low-to-high frequency response characteristics," *IEEE Transactions on Industry Applications*, vol. 43, no. 5, pp. 1233–1246, 2007.

[17] P. Barbosa, P. Steimer, J. Steinke, M. Winkelnkemper, and N. Celanovic, "Active-neutral-point-clamped (ANPC) multilevel converter technology," *European Conference on Power Electronics Applications*, pp. 1–10, 2005.

The 2018 International Power Electronics Conference

A Medium-Voltage Three-Phase AC-DC Converter Consisting of Cascaded Three-level Boost-type Rectifiers and an Open-End Winding Transformer

Ryoji Tsuruta[1]*, Hiromitsu Suzuki[2] and Ritaka Nakamura[2]

1 Advanced Technology R&D Center, Mitsubishi Electric Corp., Hyogo, Japan

2 Power Electronics System Division, Toshiba Mitsubishi-Electric Industrial Systems Corp. (TMEIC), Tokyo, Japan

*E-mail: Tsuruta.Ryoji@ea.MitsubishiElectric.co.jp

Abstract- This paper proposes a three-phase AC-DC converter topology used for a front-end converter of medium-voltage (MV) motor drives suitable for inverter with a common DC-link bus, such as modular multilevel converters (MMCs) or others. A multilevel cascaded H-bridge inverter is traditionally used for the MV drives which has distributed DC-link buses. For the purpose of the reduction of harmonics and the production of the distributed DC-voltage sources, the multilevel cascaded H-bridge inverter is generally used with a phase-shifted multi-winding transformer and diode rectifiers. However, the phase-shifted multi-winding transformer has many secondary windings, resulting in complex structure, bulky and costly. The MMC is one of the most expected topologies as a next-generation inverter for the MV drives. However, the front-end converter topology still consists of the phase-shifted multi-winding transformer and the diode rectifiers. Besides, the diode rectifiers have no capabilities of boosting DC-link voltage and suppressing the influence of the voltage sags. By contrast, cascaded single-phase three-level boost-type rectifiers and an open-end winding transformer can provide less harmonics, high functionality and capabilities of boosting DC voltages and suppressing the influence of the voltage sags. This paper presents the circuit configuration, the operation principle, and the simulation results with the input voltage of 6.6 kVAC and the output voltage of 12 kVDC, where a novel control method by a third-order harmonic current injection to reduce the harmonics is proposed. The simulation results verify the validity of the proposed configuration and the proposed control method.

I. INTRODUCTION

For traditional medium-voltage (MV) drives, a multilevel cascaded H-bridge inverter shown in Fig. 1 is widely used [1]-[3]. The input-side converter consists of a phase-shifted multi-winding transformer and six-pulse diode rectifiers, and the output-side inverter consists of series-connected H-bridge inverters. The configuration is characterized by low harmonics of the source current, and the multilevel output voltage by the series-connected H-Bridge inverters. On the other hand, the phase-shifted multi-winding transformer is bulky and costly because of a number of secondary windings with complicated structure.

Fig. 2 shows a conventional MV drive system with a modular multilevel converter (MMC). The input-side

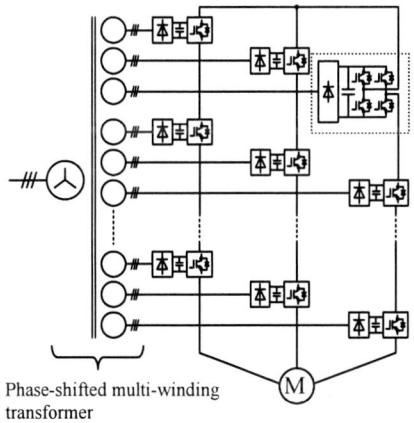

Fig. 1. The traditional MV drive system with multilevel cascaded H-bridge inverter [3].

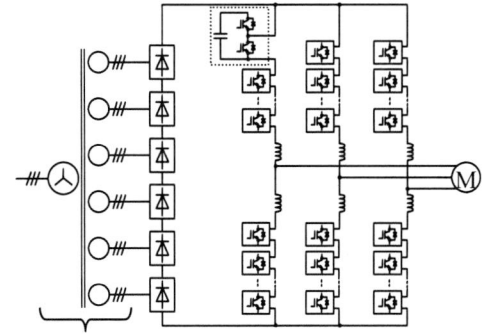

Fig. 2. The conventional MV drive system with MMC [4].

converter consists of a phase-shifted multi-winding transformer and diode rectifiers, and the output-side inverter consists of the MMC. Recently, the MMC is expected as a high-voltage and large-capacity DC-AC or AC-DC converter for an application of high-voltage direct current (HVDC) transmission especially [4]-[6]. The MMC is also applicable for the MV drives, and one of the features is a cell-based modularity design that

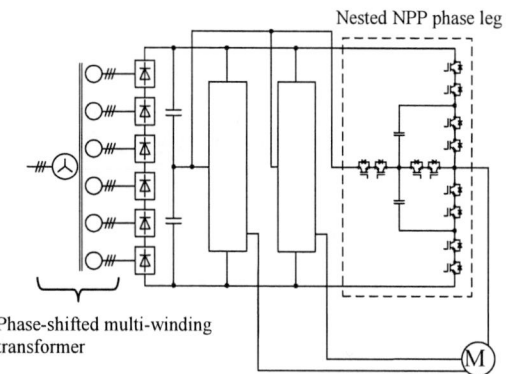

Fig. 3. The conventional MV drive system with a nested NPP inverter [10].

enables to obtain ensuring redundancy and ease of maintenance [7]-[11]. The MMC has one common DC-link bus in contrast to the multilevel cascaded H-bridge inverter. However, for the purpose of the reduction of harmonics in the source current, the input-side converter consists of multi-pulse diode rectifiers with the phase-shifted multi-winding transformer. The use of two MMCs with back-to-back connection enables transformer-less configuration [12], but there are problems in terms of the cost in case of non-regenerative load and the isolation of the motor.

Fig. 3 shows another conventional MV drive system with a nested neutral point piloted (Nested NPP) inverter [13]. The input-side converter consists of the phase-shifted multi-winding transformer and diode rectifiers, and the output inverter consists of the Nested NPP whose each phase leg has two floating capacitors, and the output voltage of the each phase leg is five-level waveform. The Nested NPP inverter also has one common DC-link bus as well as the MMC, so the use of two Nested NPP inverters with the back-to-back connection can also realize transformer-less configuration [14].

As mentioned above, the MV inverters have a trend of requiring the common DC-link bus. However, the realistic solution is the use of the bulky and costly phase-shifted multi-winding transformer. Regarding other solution, Fig. 4 shows the MV drive system consisting of the MMC with a star-delta transformer and a hybrid active filter [15] [16]. Applying the hybrid active filter enables to reduce the harmonics of the source current and to simplify the structure of the transformer with the 12-pulse diode rectifier, whereas this system requires a large LC filter. In addition, the diode rectifiers also disable to boost the DC-link voltage and to suppress the influence of the voltage sags.

With the focuses of the low-cost converter topology to obtain the simple transformer, the reduction of harmonics in the source current, and the fault-ride-through capability, this paper proposes a three-phase AC-DC converter consisting of cascaded single-phase three-level boost-type rectifiers and a non-phase-shifted open-end winding transformer to connect to MV inverters requiring

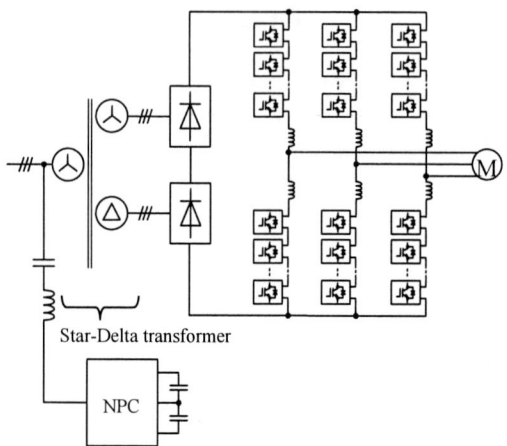

Fig. 4. The MV drive system applied for hybrid filter with MMC [16].

the common DC-link bus [17] [18]. The operation principle and a novel control method by a third-order harmonic current injection to reduce the harmonics are represented.

II. CIRCUIT CONFIGURATION

Fig. 5 shows the proposed MV drive system. The input-side converter consists of an open-end winding transformer with primary three-phase windings, secondary non-phase-shifted windings, and series-connected converter cells. A single-phase three-level boost-type rectifier is used as the converter cell. In this paper, k means the number of stages of the three-phase open-end secondary windings. In Fig.5, k is 2 and the open-end winding transformer provides six single-phase AC-voltage sources. Each secondary winding is connected to the AC-input terminal of the converter cell. The DC-output terminals of the converter cells are connected in series one another. Fig. 6 shows the configuration of the single-phase three-level boost-type rectifier as the converter cell where the converter cell requires four semiconductor switching devices. Compared to the diode-clamped three-level neutral-point-clamped (NPC) topology, the number of semiconductor switching devices becomes half [19]. The series connection of the DC-output terminals provide a common DC-link bus. Table I shows the comparison of the number of components and, functions in varies AC-DC

TABLE I
A COMPARISON OF STRUCTURES OF TRANSFORMER AND AC TO DC CONVERTER

	Proposed Fig.5	Conventional Fig. 2	Traditional Fig.1
Secondary windings	6	18	45
IGBTs	24	0	0
Diodes	48	36	75
DC-link bus	common	common	distributed
DC-voltage boost capability	Yes	No	No
Harmonic mitigation	Voltage Source Converter	Multi-pulse transformer	Multi-pulse transformer

496

The 2018 International Power Electronics Conference

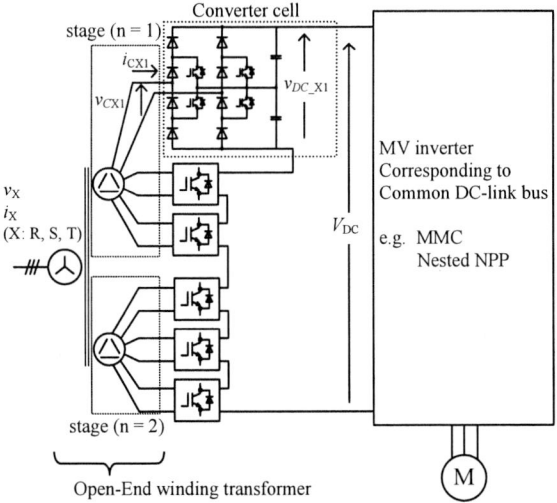

Fig. 5. The proposed MV drive system with an inverter that has common DC-link bus (k =2).

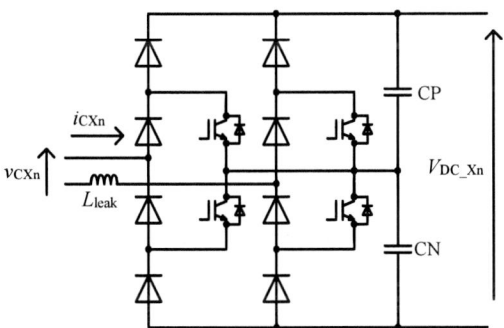

Fig. 6. The single-phase three-level boost-type rectifier as the converter cell.

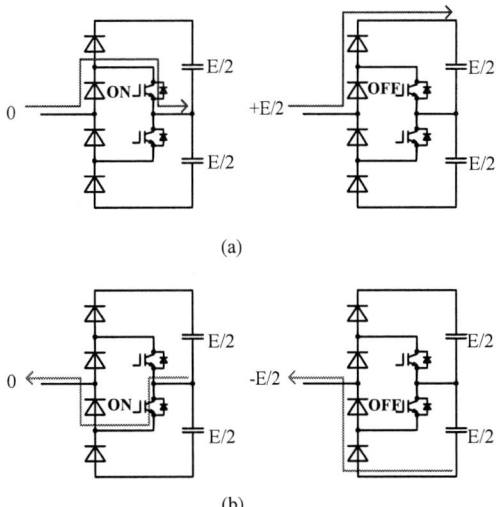

Fig. 7. The switching modes of the converter cell.
(a) $i_{CXn} > 0$. (b) $i_{CXn} < 0$.

+E/2 or 0. On the other hand, when $i_{cXn} < 0$, the converter cell can output only –E/2 or 0. These mean that the converter cell can operate only under the unity power factor. By the reason, if the power factor is not 1, and the converter cell is impossible to output the desired voltage. Moreover, the proposed topology requires the DC-capacitor voltage balancing control of the cascaded converter cells, which may affects the power factor, resulting in distortion of i_{CXn} [20].

III. BASIC CONTROL SCHEME

Fig. 8 shows the control block diagram for the proposed topology. This control block is applied to the three converter cells connected to same number of the stage indicated by n ($1 \leq n \leq k$) as shown in Fig. 5. For the purpose to control the DC-link voltage, the DC-capacitor voltage of each converter cell, and the source current to follow the referenced value, the control block is divided into the following control schemes;

(A) averaging control,

(B) phase balancing control, and

(C) AC current control.

A. Averaging Control

In Fig. 8, block (A) represents the averaging control of the DC-link voltage V_{DC} by adjusting the active-power flow from the source into the coverter cell. Basically, the source voltage is a constant amplitude and a constant frequency. Therefore, the converter cells control I_{qn} and I_{dn} that are obtained by the abc/dq transformation of each phase converter-cell current i_{CRn}, i_{CSn}, and i_{CTn}. The q-axis current I_{qn} represents the active-power component, and the d-axis current I_{dn} represents the reactive-power component. In the averaging control, the error between V^*_{CDC} and V_{DCn_ave} is provided to the Proportional and

conversion topologies under the assumption of 6.6 kV line to line voltage in Fig. 1, Fig.2 and Fig. 5. In Table I, the rated voltage of IGBTs is assumed as 1.7 kV. The proposed configuration uses six converter cells because the number of the three-phase open-end secondary windings k is set to 2. While, the conventional configuration uses 36-pulse diode rectifier and the traditional configurations use 30-pulse diode rectifier. Using semiconductor switching devices, the proposed configuration enables to reduce the number of the secondary windings, and to boost the DC-link voltage.

In this paper, the polarity of the converter-cell current i_{CXn} (X = R, S, T), ($1 \leq n \leq k$) is defined that the direction flowing from the AC-input terminal into the converter cell is positive. v_{CXn} is defined as the AC-terminal voltage of the converter cell. V_{DC_Xn} represents the DC-capacitor voltage of each converter cell, and the summation of each V_{DC_Xn} equals the DC-link voltage V_{DC}. v_X represents the phase voltage of the source, and i_X represents the source current. Two series-connected capacitors, CP and CN are used in the DC-output terminals of the converter cell. L_{leak} indicates the leakage inductance of the transformer.

Fig. 7 shows the four switching modes of the converter cell. When $i_{cXn} > 0$, the converter cell can output only

497

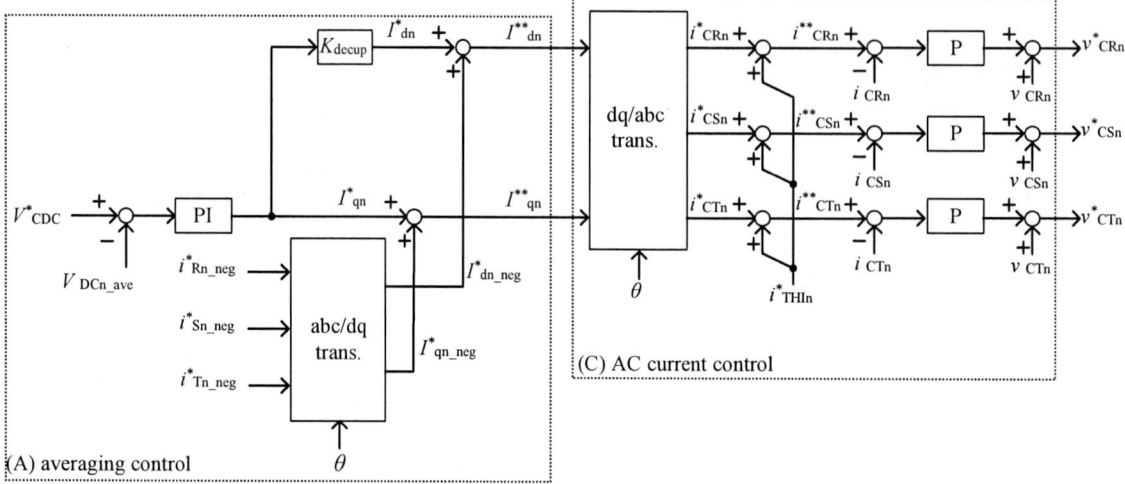

Fig. 8. The block diagram of the overall of control for the proposed topology.

Integral (PI) controller. I^*_{qn} represents the reference of the q-axis positive-sequence current. Under ideal operation with unity power factor, and the reactive-power component I_{dn} is zero. However, in order to decuple the interference between I_{qn} and I_{dn}, I^*_{dn} is given by $K_{decup} \times I^*_{qn}$. I^*_{dn} means the reference of the d-axis positive-sequence current. I^*_{dn} and I^*_{qn} are added to the reference of the negative-sequence current $I^*_{dn\_neg}$ and $I^*_{qn\_neg}$. As the result, the references of the converter current I^{**}_{dn} and I^{**}_{qn} are derived. In Fig. 8, V^*_{CDC} means the reference of one converter-cell DC-capacitor voltage. The DC-link voltage V_{DC} can be calculated by following equation (1) ideally.

$$V_{DC} = 3^{phase} \times k \times V^*_{CDC} . \tag{1}$$

While, V_{CDCn_ave} can be calculated by following equation (2).

$$V_{DCn\_ave} = (V_{DCnR} + V_{DCnS} + V_{DCnT}) \div 3 . \tag{2}$$

θ is the phase angle of the source voltage. In this paper, the line frequency is set to 50 Hz.

B. Phase Balancing Control

Fig. 9 shows the control block diagram of phase balancing control. To balance the voltage of the converter cell between three phases (X: R, S, T) in same stage n, the control method of the proposed topology uses the negative-sequence current. In Fig. 9, the references of the three-phase negative-sequence currents are calculated. At first, the moving average value of the DC-capacitor voltages of the converter cells connected to each phase V_{DCXn_MA} are calculated. For example, with focus on R-phase, the error between V_{DCRn_MA} and the average value of "V_{DCSn_MA} and V_{DCTn_MA}" is provided to the PI controller, and the output of the PI controller is the amplitude reference of the negative-sequence current based on R-phase $I^*_{Rn\_neg}$. The three-phase negative-sequence currents are given by multiplying $I^*_{Rn\_neg}$ by the

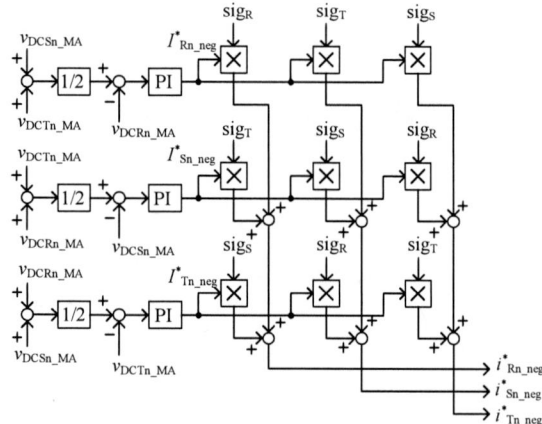

Fig. 9. The block diagram of phase balancing control (B).

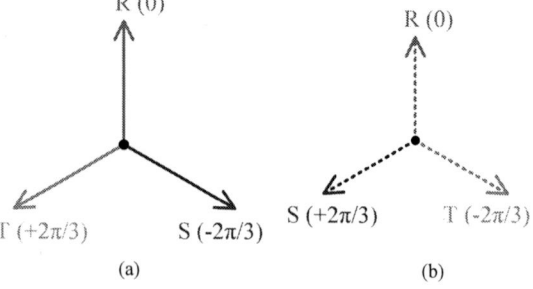

Fig. 10. Phasor diagrams. (a) AC voltage. (b) Negative sequence current based on R phase.

phase information sig_X to produce the negative-sequence current based on each phase as shown in Fig.9. The phase information sig_X is represented by following equation (3). In this equation, ω represents the angular velocity synchronized with the line frequency. With the same way, the references of the negative-sequence current based on S-phase and T-phase are calculated as shown in Fig. 9.

$$\begin{cases} sig_R = \sin \omega t \\ sig_S = \sin\left(\omega t - \dfrac{2\pi}{3}\right) \\ sig_T = \sin\left(\omega t + \dfrac{2\pi}{3}\right) \end{cases} \quad (3)$$

Fig.10 shows the example of the phasor diagram of the positive-sequence AC voltage and the negative-sequence current based on R-phase. In this case, the active power of R-phase P_{Rn_neg} is given by following equation (4), therefore the active power flows into the converter cell to increase the DC-capacitor voltage of the converter cell. V_C represents the amplitude of the AC-terminal voltage of the converter cell in steady state and T represents the period of one cycle in line frequency.

$$\begin{aligned} P_{Rn\_neg} &= \frac{1}{T}\int_0^T V_C \sin(\omega t) \times I_{Rn\_neg}\sin(\omega t)dt \\ &= \frac{1}{T}\int_0^T \frac{V_C I_{Rn\_neg}\left(1 - \cos(2\omega t)\right)}{2}dt \\ &= \frac{V_C I_{Rn\_neg}}{2} \end{aligned} \quad (4)$$

On the other hand, the active power of S-phase P_{Sn_neg} and T-phase P_{Tn_neg} are given by following equations (5) and (6), therefore the active power flow out from converter cells and the DC-capacitor voltage of the converter cell decreases.

$$\begin{aligned} P_{Sn\_neg} &= \frac{1}{T}\int_0^T V_C \sin(\omega t - \frac{2\pi}{3}) \times I_{Rn\_neg}\sin(\omega t + \frac{2\pi}{3})dt \\ &= \frac{1}{T}\int_0^T \frac{V_C I_{Rn\_neg}\left(\cos(-\frac{4\pi}{3}) - \cos(2\omega t)\right)}{2}dt \\ &= -\frac{V_C I_{Rn\_neg}}{4} \end{aligned} \quad (5)$$

$$\begin{aligned} P_{Tn\_neg} &= \frac{1}{T}\int_0^T V_C \sin(\omega t + \frac{2\pi}{3}) \times I_{Rn\_neg}\sin(\omega t - \frac{2\pi}{3})dt \\ &= \frac{1}{T}\int_0^T \frac{V_C I_{Rn\_neg}\left(\cos(\frac{4\pi}{3}) - \cos(2\omega t)\right)}{2}dt \\ &= -\frac{V_C I_{Rn\_neg}}{4} \end{aligned} \quad (6)$$

The negative-sequence current calculated in Fig. 9 is provided to the abc/dq transformer as shown in Fig.8, and transformed to $I^*_{dn\_neg}$ and $I^*_{qn\_neg}$. The references of the converter-cell current I^{**}_{dn} and I^{**}_{qn} are given by adding $I^*_{dn\_neg}$ and $I^*_{qn\_neg}$ to I^*_{dn} and I^*_{qn}.

C. AC Current Control

The references of the converter-cell current I^{**}_{dn} and I^{**}_{qn} calculated by the averaging control and the phase balancing control are provided to the dq/abc transformer, and the references of the three-phase converter-cell current i^*_{CXn} are given. Adding the third-order harmonic current reference i^*_{THI} to i^*_{CXn}, the true reference of the converter-cell current i^{**}_{CXn} are given. The third-order harmonic current injection is proposed control method to reduce the distortion of the source current. The detail of the third-order harmonic current injection will be

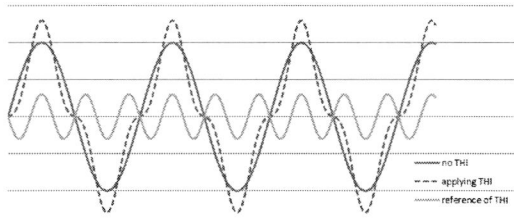

Fig. 11. The difference between added i^*_{THIn} and not added i^*_{THIn} of the converter-cell current.

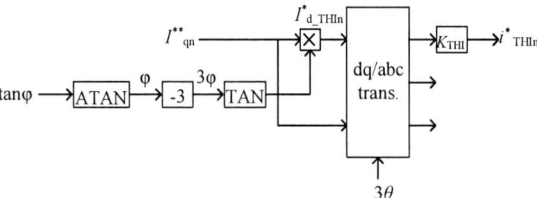

Fig. 12. The block diagram of i^*_{THIn} calculation.

mentioned in the next section.

The error between i^{**}_{CXn} and the current i_{CXn} is provided to P controller. Adding AC-terminal voltage of the converter cell v_{CXn} to the output value of the P controller, the references of the converter-cell voltage v^*_{CXn} are given. After considering unitization and including PN balance component for balancing the voltages of two series connected capacitors of each converter cell, the pulse-width modulation is carried out by comparing the references and a triangular wave carrier. Each stage of the three-phase converter cells has different carrier waves whose phase is shifted by π/k one another. Applying the phase-sifted carriers, the equivalent switching frequency of primary windings becomes k times of the individual switching frequency of the secondary windings.

IV. THIRD-ORDER HARMONIC CURRENT INJECTION FOR REDUCTION OF CURRENT DISTORTION

As shown in Fig.7, the AC-output voltage of the single-phase three-level boost-type rectifier depends on the polarity of the converter-cell current. Because of the dependence, the distortion of the source current increases near the zero-crossing point. To suppress the distortion, this paper proposes the third-order harmonic current injection (THI) control method. When the polarities of the converter-cell current and the reference of converter-cell voltage are different, the converter cell is impossible to output the referenced voltage. As the result, the error between the real value and referenced value of the converter-cell current increases and a harmonic current increases. The THI control decrease the error between the reference of the converter-cell current and the current of each converter cell near the zero-crossing point, and the distortion of source current is reduced. As shown in Fig.8, the third-order harmonic current reference i^*_{THIn} is added to the converter-cell current reference i^*_{CXn}. Fig.11 shows the difference between the reference of the converter-cell current added i^*_{THIn} and the reference of the converter-cell current not added i^*_{THIn}.

Fig. 12 shows the control block diagram of the calculation of i^*_{THIn}. In Fig.12, $\tan\varphi$ is given by the following equation (7).

$$\tan\phi = \frac{I^{**}_d}{I^{**}_q} \tag{7}$$

φ means the error of initial phase angle from zero radian of the fundamental wave of i^*_{CXn} including the negative-sequence current. The frequency of i^*_{THIn} is three times higher than the line frequency, and the error of initial phase angle from zero radian has to be three times larger than φ. To make the gap of the initial phase between the line frequency component i^*_{CXn} and the third-order harmonic component i^*_{THIn} zero, the reference of the third-order harmonic d-axis current $I^*_{d\_THIn}$ is given by multiplying $\tan(3\varphi)$ by I^{**}_{qn}. $I^*_{d\_THIn}$ and I^{**}_{qn} are provided to the dq/abc transformer whose rotation speed is three times faster than the fundamental wave. The a-phase output of the dq/abc transformer is multiplied by K_{THIn} to correct the amplitude, and i^*_{THIn} is given. By this method, the initial phase angles of i^*_{CXn} and i^*_{THIn} become same each other even if the amplitude imbalance of i^*_{CXn} exists in three phase because of the negative-sequence current.

The third-order harmonic current is canceled because of the star connection on the primary windings. By the reason, the third-order harmonic current of each converter cell does not interfere with the source current.

V. Simulation Result

To verify the proposed control method for regulating the DC-link voltage and reducing the distortion of the source current by the third-order harmonic current injection control, the simulation of the cascaded three-level boost-type rectifiers and an open-end winding transformer is carried on. Table II shows the circuit parameter of the simulation. Fig.13 shows the

TABLE II
PARAMETER OF SIMULATION

	Parameters of the simulation
Source voltage (line to line)	6.6 kV
Line frequency	50 Hz
DC-link voltage	12 kV
Power rating	3 MVA
AC voltage of converter cell	1.27 kV
DC voltage of converter cell	2 kV
Number of stage of secondary windings (k)	2
Leakage inductance of primary windings	1.39 mH (3% : based on 6.6 kV, 3MVA, 50Hz)
Leakage inductance of secondary windings	1.23 mH (12% : based on 1.27 kV, 500kVA, 50Hz)
Capacitance of converter cell	2.4 mF (H = 9.6 ms)
Carrier frequency	900 Hz

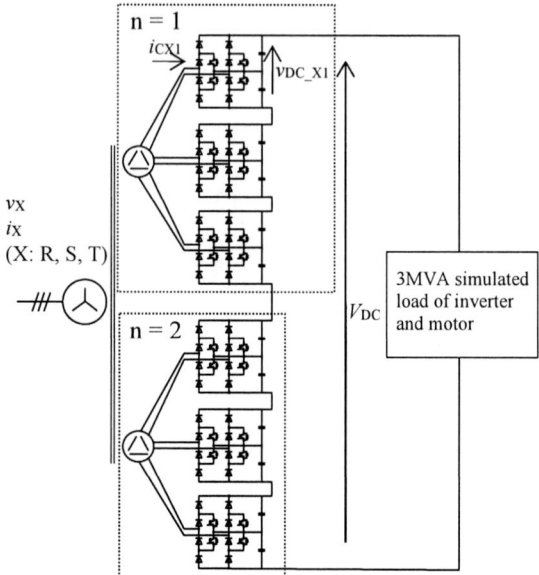

Fig. 13.   The configuration of simulated circuit.

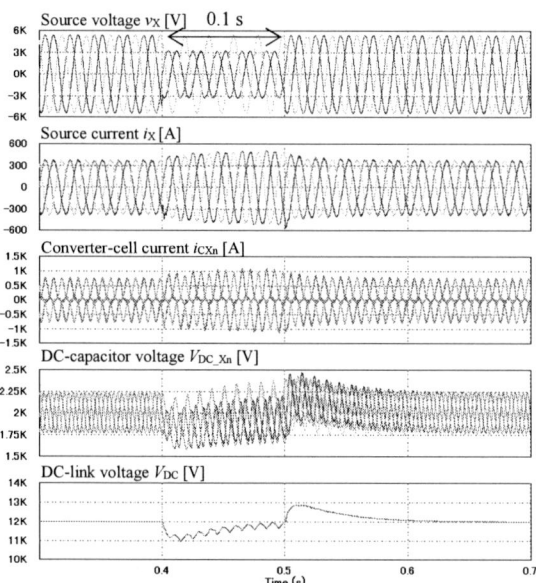

Fig. 14.   The simulation result under the voltage sag.

configuration of the simulating circuit. In this simulation, k is set to 2 and there are two stages of the three-phase converter cells. The number of cascaded converter cells is six.

Fig. 14 shows the simulation result under the voltage sag. In Fig. 14, the voltage sag with the depth of 60 % occurs for 0.1 s in R-phase and S-phase. Although the input voltage is imbalanced and the negative-sequence current increase, the control method suppresses the DC-voltage fluctuation and produces the sinusoidal source current.

Fig. 15 shows the comparison of the waveforms without THI and with THI under the imbalanced source

The 2018 International Power Electronics Conference

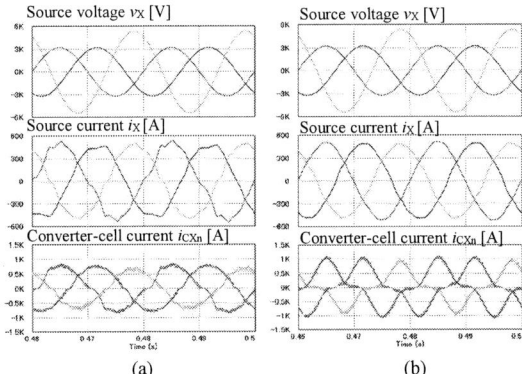

(a)   (b)

Fig. 15.   The simulation result under the imbalanced source voltage condition. (a) no THI. (b) applying THI.

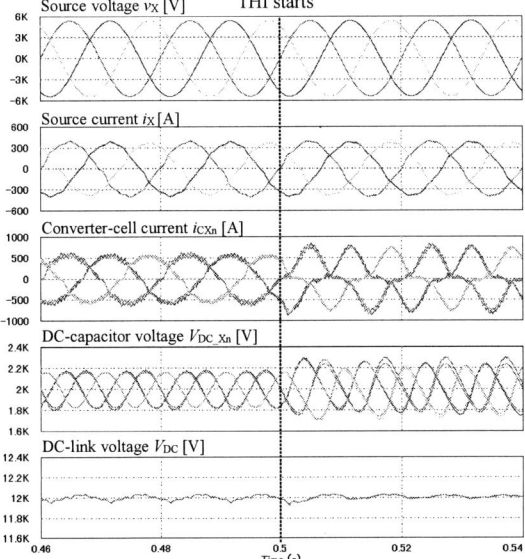

Fig. 16.   The simulation result of the imbalance loss condition.

voltage. With this situation in Fig. 15 (a), the increased negative-sequence component makes the zero-crossing point of the source current shifted. This causes the different polarities of the converter-cell current and the converter-cell voltage. As the result, the distortion of the source current increases. On the other hand, applying THI in Fig. 15 (b), the distortion of the source current is suppressed even if the imbalanced source voltage causes the increased negative-sequence current.

Fig.16 shows the simulation result under the imbalanced converter-cell loss condition. In Fig.16, the transition from no THI to applying THI is verified when the extra loads are connected to the DC-output terminals of the R1-converter cell (10% of rated power of the converter cell) and the S1-converter cell (5%) as the imbalanced loss. Applying THI starts without any fluctuation and reduces the harmonics of the source current. This means that the proposed method is able to compensate the harmonics immediately when the negative-sequence current increase or the power factor become not 1.

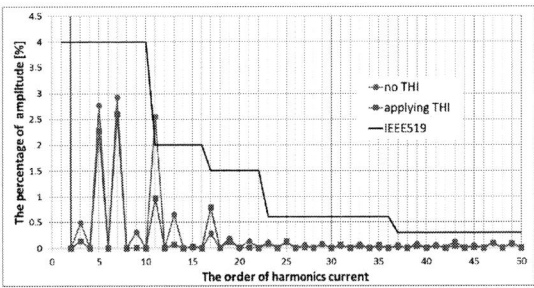

Fig. 17.   The comparison of the harmonics of the source current between no THI and applying THI.

Fig. 17 shows the comparison of the harmonics spectrum of the source current between no THI and applying THI of the situation in Fig.16. The graph shows that THI reduces 11th harmonics especially.

VI. CONCLUSION

This paper proposes a topology of a MV three-phase AC-DC converter intended for a MV drives which consists of cascaded single-phase three-level boost-type rectifiers and an open-end winding transformer. The topology can reduce the number of the secondary windings, boost the DC-link voltage, and continue the operation with low harmonics during the voltage sags. The third-order harmonic current injection control method proposed in this paper reduces the distortion of the source current. The simulation results show the stability operation and the reduction of the harmonics by the proposed method.

REFERENCES

[1] R. Teodorescu, F. Blaabjerg, J. K. Pedersen, E. Cengelci, and P. N. Enjeti "Multilevel Inverter by Cascading Industrial VSI," *IEEE Trans. on Industrial Electronics*, vol. 49, no. 4, pp. 832-837, 2002.

[2] X. Liang, and J. He, "Load Model for Medium Voltage Cascaded H-Bridge Multi-Level Inverter Drive Systems," *IEEE Power and Energy Technology Systems Journal*, vol. 3, no. 1, pp. 13-23, 2016.

[3] K. Thantirige, S. Kumar Panda, A. Kumar Rathore, S. Mukherjee, M. Adam Zagrodnik, A. Kumar Gupta, "Fault-Tolerant Cascaded Multi-level Inverter with Improved Output Quality," *in Conf. Rec. ICSET 2016*, pp. 332-337.

[4] M. Mehrasa, E. Pouresmaeil, S. Zabihi, and J. P. S. Catalrao, "Dynamic Model, Control and Stability Analysis of MMC in HVDC Transmission Systems," *IEEE Trans. on Power Delivery*, vol. 32, no. 3, pp. 1471-1482, 2017.

[5] U. N. Gnanarathna, A. M. Gole, and R. P. Jayasinghe, "Efficient modeling of modular multilevel HVDC converters (MMC) on electromagnetic transient simulation programs," *IEEE Trans. on Power Delivery.*, vol. 26, no. 1, pp. 316–324, Jan. 2011.

[6] K. Friedrich, "Modern HVDC PLUS application of VSC in Modular Multilevel Converter Topology," *in Conf. Rec. ISIE 2010.*

[7] M. Glinka, and R. Marquardt, "A New AC/AC-Multilevel Converter Fatmily Applied to a Single-phase Converter," *in Conf. Rec. PEDS2003*, vol.1, pp. 16-23

[8] P. Himmelmann, M. Hiller, D. Krug, and M. Beuermann, "A new Modular Multilevel Converter for Medium Voltage High Power Oil & Gas Motor Drive Applications," *in Conf. Rec. EPE 2016.*

[9] G. Mondal, and S. Nielebock, "Control of M2C Direct Converter for AC to AC Conversion with Wide Frequency Range," *in Conf. Rec. EPE 2016.*

[10] M. Hiller, D. Krug, R. Sommer, S. Rohner, "A New Highly Modular Medium Voltage Converter Topology for Industrial Drive Applications," *in Conf. Rec. EPE 2009.*

[11] S. Sau, S. Karmakar, and B. G. Fernandes, "Reduction of Capacitor Ripple Voltage and Current in Modular Multilevel Converter based Variable Speed Drives," *in Conf. Rec. IFEEC2017*, pp. 1451-1456.

[12] H. Akagi, "Classification, Terminology, and Application of the Modular Multilevel Cascade Converter (MMCC)," *IEEE Trans. on Power Electronics*, vol. 26, no. 11, pp. 3119-3130, 2011.

[13] J. Li, J. Jiang, and S. Qiao, "A Space Vector Pulse Width Modulation for Five-Level Nested Neutral Point Piloted Converter," *IEEE Trans. on Power Electronics*, vol. 32, no. 8, pp. 5991-6004, 2017.

[14] M. Mechlinski, S. Schröder, J. Shen, and R. W. D. Doncker, "Grounding Concept and Common-Mode Filter Design Methodology for Transformerless MV Drives to Prevent Bearing Current Issues," *IEEE Trans. on Industry Applications*, IEEE Early Access, 2017.

[15] H. Akagi, "Multilevel Converters Fundamental Circuits and Systems," *Proceedings of the IEEE*, 2017.

[16] Y. Okazaki, W. Kawamura, M. Hagiwara, H. Akagi, T. Ishida, M. Tsukakoshi, R. Nakamura, "Which is More Suitable for MMCC-Based Medium-Voltage Motor Drives, a DSCC Inverter or a TSBC Converter?," *in Conf. Rec. ICPE-ECCE Asia2015*, pp. 1053-1060.

[17] Y. Zhao, Y. Li, and T. A. Lipo, "Force Commutated Three Level Boost Type Rectifier," *IEEE Trans. on Industry Applications*, vol. 31, no. 1, pp. 155-161, 1995.

[18] B. Lin, and H. Lu, "Single-phase three-level PWM rectifier," *in Conf. Rec. PEDS 1999*. pp. 63-68.

[19] A. Nabae, I. Takahashi, and H. Akagi, "A new neutral-point-clamped PWM inverter," *on Industry Applications*, vol. 17, no. 5, pp. 518–523, 1981.

[20] J. C. Salmon, "Circuit topologies for pwm boost rectifiers operated from 1-phase and 3-phase ac supplies and using either single or split dc rail voltage outputs," *in Conf. Rec. APEC1995*, vol. 1, pp. 473-479.

The 2018 International Power Electronics Conference

A Fault Tolerant Control Strategy for the Delta-Connected Cascaded Converter

Ping-heng Wu[1*], *Student Member, IEEE*, and Po-tai Cheng[1], *Fellow, IEEE*

[1]Center for Advanced Power Technologies, Department of Electrical Engineering
National Tsing Hua University, Hsinchu, Taiwan
*E-mail: super497415008@gmail.com

Abstract—As the increased reliability requirement, the fault tolerant operation as the bridge-cell failures is an important issue for the modular multilevel cascaded converter (MMCC). This paper proposes an effective fault tolerant strategy for the delta-connected cascaded converter. During the occurrence of the bridge-cell fails, the proposed open-delta operation enables the converter continuous operation and maintains the dc capacitor voltages balanced. This fault tolerant technique does not require extra hardware circuit installation. Laboratory tested results are given to validate the proposed control approach.

Index Terms—Modular multilevel cascaded converter based on single-delta bridge cells (MMCC-SDBC), fault tolerant control, dc capacitor voltage balancing control, open-delta operation

I. INTRODUCTION

Over the past years, the modular multilevel cascaded converter (MMCC) has drawn considerable attention for its proper applications in the medium-voltage (MV) utility grid. The MMCC with the configuration of single-delta bridge cells (MMCC-SDBC) [1] is a prominent circuit topology that has been applied to various grid-connected applications such as static compensator (STATCOM) and interface converter in photovoltaic (PV) system [2]–[6].

The cascaded converter has the control potential of maintaining operation when one or more of its H-bridge modules fail, which is called fault tolerant control. This feature increases the reliability and availability of the STATCOM, and several papers have presented in this issue [7]–[12]. The papers [7], [8] describe the PWM redundant space vector selection methods which are applied to the motor drive system. In [9], a strategy of H-bridge building block redundancy in a STATCOM system is proposed. The method effectively advances the reliability, but also raises the construction cost. The control methods in [10]–[12] are achieved by injecting the zero sequence voltage to redistribute each phase modulation. These fault tolerant approaches allow the converter maintaining operation without requiring extra redundant modules. However, all of the discussions are based on the star-connected cascaded converter, none of the papers address the fault tolerant control strategy of MMCC-SDBC.

This paper presents a fault tolerant control strategy of MMCC-SDBC without installing extra redundancy bridge cells. The proposed open-delta operation is able to maintain dc voltages balancing even during the bridge cells are faulty. The controller of the open-delta operation is based on average

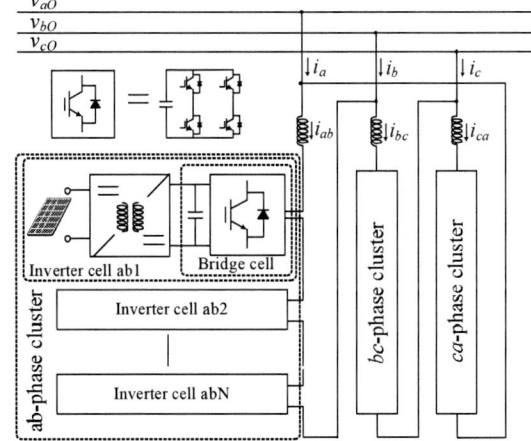

Fig. 1. The MMCC-SDBC in the utility-scale PV system.

power analysis. The proposed control strategy is verified by using a 7-level MMCC-SDBC in the PV system application.

II. SYSTEM CONFIGURATION AND THE EFFECT OF THE FAULTY CELL

Fig. 1 shows the MMCC-SDBC applied a PV inverter in the power system, which N H-bridge modules are cascaded in one cluster. The grid voltages and output line currents are presented by the forms of positive (V_q^p, V_d^p, I_q^p, I_d^p) and negative sequence components (V_q^n, V_d^n, I_q^n, I_d^n) in this paper. These definitions are shown in (1). Besides, the phase voltages and currents are paid attention which the relations with line components are derived as

$$v_{ab} = v_{aO} - v_{bO}, v_{bc} = v_{bO} - v_{cO}, v_{ca} = v_{cO} - v_{aO} ;$$
$$i_{ab} = \frac{(i_a - i_b)}{3} + i_z, i_{bc} = \frac{(i_b - i_c)}{3} + i_z, i_{ca} = \frac{(i_c - i_a)}{3} + i_z \quad (2)$$

where i_z represents the zero sequence current, and it is defined as $I_o \cos(\omega t + \gamma)$ in this paper. Since the dc voltages contain twice the grid frequency ripple, a moving-average filter is used to obtain the average dc voltage value (V_{dcmn}, $m \in \{a, b, c\}$, $n \in \{1...N\}$).

503

The 2018 International Power Electronics Conference

$$
\begin{bmatrix} v_{aO} \\ v_{bO} \\ v_{cO} \end{bmatrix} = \begin{bmatrix} 1 & 0 \\ -\frac{1}{2} & -\frac{\sqrt{3}}{2} \\ -\frac{1}{2} & \frac{\sqrt{3}}{2} \end{bmatrix} \begin{bmatrix} v_\alpha \\ v_\beta \end{bmatrix} = \begin{bmatrix} 1 & 0 \\ -\frac{1}{2} & -\frac{\sqrt{3}}{2} \\ -\frac{1}{2} & \frac{\sqrt{3}}{2} \end{bmatrix} \left(\begin{bmatrix} \cos \omega t & \sin \omega t \\ -\sin \omega t & \cos \omega t \end{bmatrix} \begin{bmatrix} V_q^p \\ V_d^p \end{bmatrix} + \begin{bmatrix} \cos \omega t & -\sin \omega t \\ \sin \omega t & \cos \omega t \end{bmatrix} \begin{bmatrix} V_q^n \\ V_d^n \end{bmatrix} \right)
$$

$$
\begin{bmatrix} i_a \\ i_b \\ i_c \end{bmatrix} = \begin{bmatrix} 1 & 0 \\ -\frac{1}{2} & -\frac{\sqrt{3}}{2} \\ -\frac{1}{2} & \frac{\sqrt{3}}{2} \end{bmatrix} \begin{bmatrix} i_\alpha \\ i_\beta \end{bmatrix} = \begin{bmatrix} 1 & 0 \\ -\frac{1}{2} & -\frac{\sqrt{3}}{2} \\ -\frac{1}{2} & \frac{\sqrt{3}}{2} \end{bmatrix} \left(\begin{bmatrix} \cos \omega t & \sin \omega t \\ -\sin \omega t & \cos \omega t \end{bmatrix} \begin{bmatrix} I_q^p \\ I_d^p \end{bmatrix} + \begin{bmatrix} \cos \omega t & -\sin \omega t \\ \sin \omega t & \cos \omega t \end{bmatrix} \begin{bmatrix} I_q^n \\ I_d^n \end{bmatrix} \right)
$$

(1)

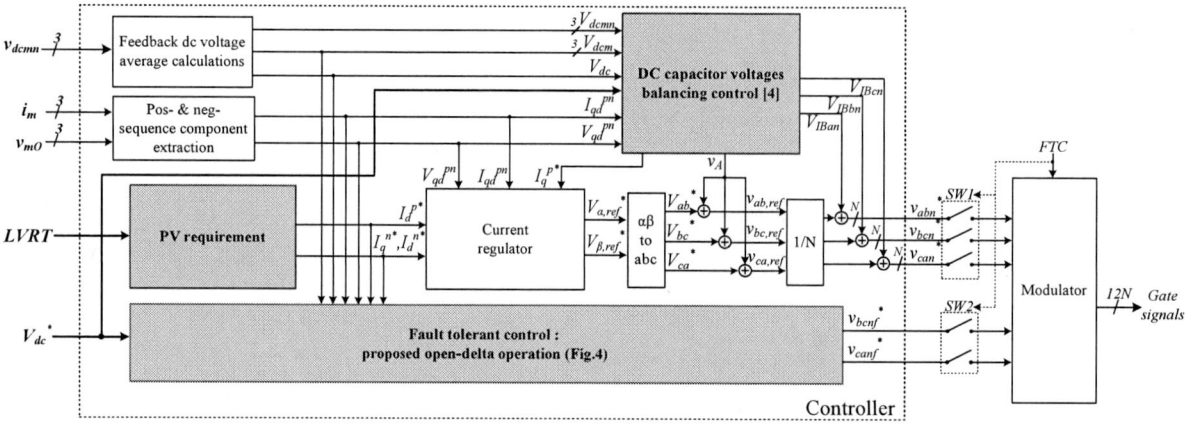

Fig. 2. The overall control block diagram of MMCC-SDBC.

Fig. 2 shows the control block diagram of MMCC-SDBC. The whole control is divided into the following subcontrols:

- dc capacitor voltage balancing control
- PV requirement control
- fault tolerant control

The balancing control of dc capacitor voltages of H-bridge cells depends on the average active power of the entire system. In [4], a precise zero sequence current injection based on the detailed power analysis to manage the dc bus voltages of all modules has been presented. The dc capacitor voltage balancing control in Fig. 2 is based on [4], which includes three-layer controls: overall, phase cluster, and individual voltage control. The overall voltage control is to regulate the average value of all modules dc bus voltages (V_{dc}) at dc bus command (V_{dc}^*). The phase cluster voltage control is to regulate the average dc bus voltages of each cluster (V_{dcm}, $m = ab, bc, ca$) at V_{dc}. The individual voltage control is to maintain the dc bus voltage of each bridge cell at V_{dcm}. The experimental results in [4], [6] proved that all the modules of dc voltages are supported balanced, even during the unbalanced load compensation and imbalanced dc powers.

The cascaded H-bridge modules of the MMCC-SDBC are capable of redundancy operation when some modules fail. However, if the number of faulty modules increases, then the performance of the redundant operation will degrade because the remaining H-bridge modules cannot produce sufficient output voltages. This paper proposes a fault tolerant operation strategy to extend the fault operation capability for a large number of faulty modules.

Assuming the system is lossless and the filter inductor is neglected, the modulation indexes of the $3N$ converter cells get equal for normal operation. Fig. 3(a) shows its phasor diagram. When the module is detected to be faulty, the bridge cell is bypassed immediately by short-circuiting its ac terminals [10]. Fig. 3(b) shows the diagram which the bridge cells $ab1$ to aby are faulty and bypassed, and their required output voltage is produced by the remaining bridge cells. The increase in ac voltages in the ab-phase puts an additional burden on the $N - y$ healthy bridge cell in the ab-phase, thus may result in the bridge cell in over-modulation operation.

To avoid this problem, one of the solutions is to increase the dc bus voltages. However, it may increase the volume and the design complexity of the converter capacitor. The other method is to redistribute the phase cluster modulation by the appropriate control algorithm [10]–[12]. Unlike the star-connected cascaded converter, the MMCC-SDBC cannot shift the potential of neutral point to regulate each phase modulation. Only the zero sequence current can be managed which is not helping with the modulation redistribution.

The dc bus voltage balancing control of MMCC-SDBC have already been implemented in [4], [6]. This paper mainly focuses on the fault tolerant operation of MMCC-SDBC. The proposed technique is operated when the fault is detected. After the fault occurred, the control strategy is no longer the conventional method. The core of the fault tolerant control strategy is the open-delta operation, which is discussed in

The 2018 International Power Electronics Conference

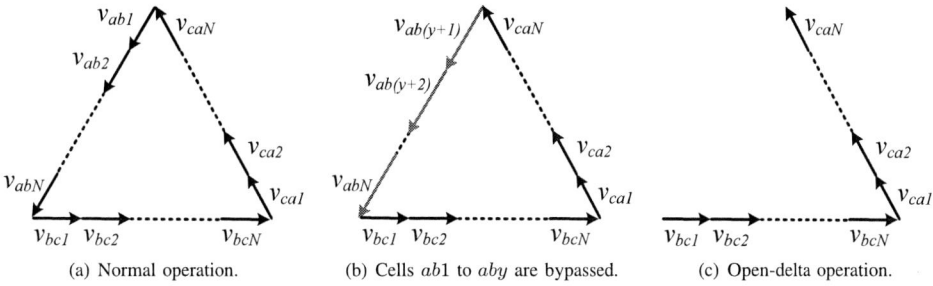

(a) Normal operation. (b) Cells $ab1$ to aby are bypassed. (c) Open-delta operation.

Fig. 3. The diagram of the voltage distribution for dc bus voltages in MMCC-SDBC.

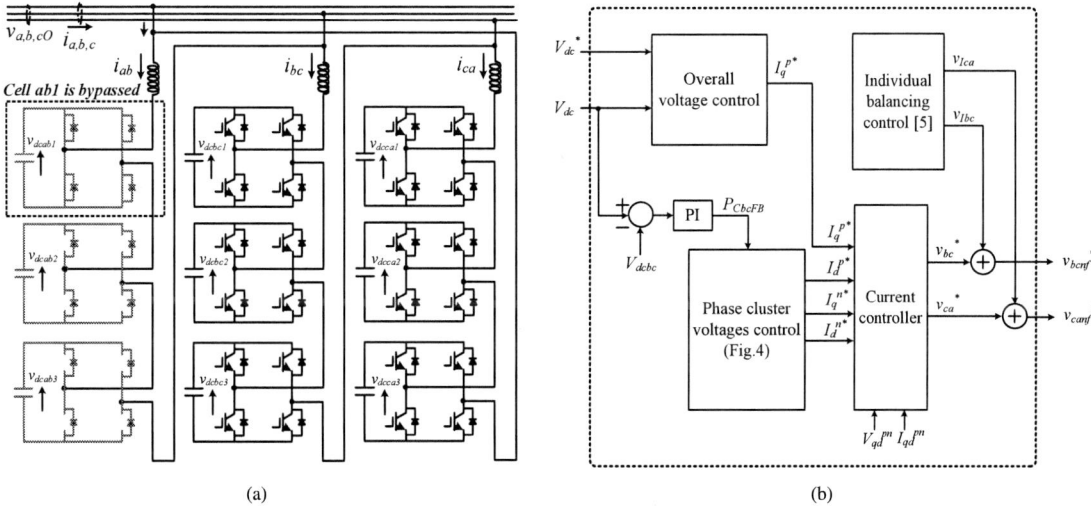

(a) (b)

Fig. 4. The equivalent circuit and control method of MMCC-SDBC in open-delta operation. (Assumption of cell $ab1$ is bypassed) (a)The configuration of MMCC-SDBC. (b)The dc voltages balancing control of open-delta operation.

detail in the following section.

III. Fault Tolerant Control Strategy

In order to maintain continuous operation during the bridge cells are faulty, the open-delta operation, which is to open circuit other healthy converter cells in the same phase, is proposed in this paper. In this way, as shown in Fig. 3(c), the system can successfully avoid operating in over-modulation.

Fig. 4(a) shows the system configuration in this paper, where the grid voltage is 220 V, and the cascaded number is three ($N = 3$). Assuming the faulty bridge cell is cell $ab1$, which is bypassed by turning on the upper switches of each bridge leg. In order to achieve proposed open-delta operation, other healthy bridge cells in ab-phase are open-circuit by turning off all of the switches. Since the grid is based on a three-phase three-wire system, it is possible to output three-phase currents by only two cluster converters.

A. Power flow analysis of open-delta operation

The dc bus voltage balancing control is an important issue for the cascaded converter. The proposed method is based on average power analysis. From the equivalent circuit of the open-delta operation in Fig. 5, the phase cluster currents i_{bc} and i_{ca} can be expressed as

$$i_{bc} = i_b \; ; \; i_{ca} = i_c + i_{bc}. \tag{3}$$

Note that only bc- and ca-phase exist in the system during the open-delta operation. Therefore, the calculations of phase average power are calculated as follows [13]:

$$P_{bc} = \frac{\omega}{2\pi} \int_0^{\frac{2\pi}{\omega}} v_{bc} \cdot i_{bc} dt \; ; \; P_{ca} = \frac{\omega}{2\pi} \int_0^{\frac{2\pi}{\omega}} v_{ca} \cdot i_{ca} dt. \tag{4}$$

The summation of P_{bc}, P_{ca} then leads to the overall power P_T of the system, and it can be calculated with (1):

$$P_T = P_{bc} + P_{ca} = \frac{3}{2} \left(V_q^p I_q^p + V_d^p I_d^p + V_q^n I_q^n + V_d^n I_d^n \right). \tag{5}$$

Since the unbalanced sharing of power among phase clusters cause the dc capacitor voltages unbalanced [4], the power

The 2018 International Power Electronics Conference

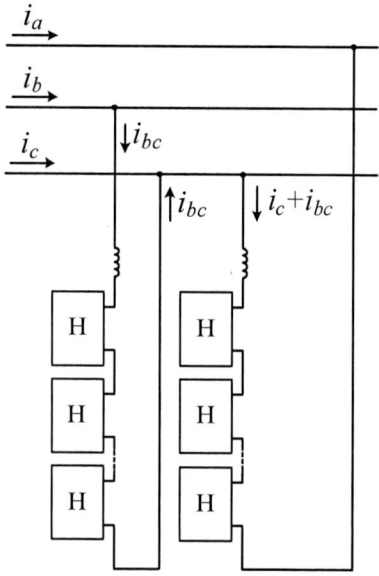

$$P_{Ccaq}^{p} = \left(-\frac{\sqrt{3}}{4}V_d^p + \frac{3}{4}V_q^n - \frac{\sqrt{3}}{4}V_d^n \right) I_q^p$$

$$P_{Ccad}^{p} = \left(\frac{\sqrt{3}}{4}V_q^p - \frac{\sqrt{3}}{4}V_q^n - \frac{3}{4}V_d^n \right) I_d^p$$

$$P_{Ccaq}^{n} = \left(\frac{3}{4}V_q^p - \frac{\sqrt{3}}{4}V_d^p + \frac{\sqrt{3}}{4}V_d^n \right) I_q^n \tag{8}$$

$$P_{Ccad}^{n} = \left(-\frac{\sqrt{3}}{4}V_q^p - \frac{3}{4}V_d^p + \frac{\sqrt{3}}{4}V_q^n \right) I_d^n$$

The phase cluster powers P_{Cbc} and P_{Cca} are unbalanced because of the injection of compensating currents I_{dL}^p, I_{qL}^n and I_{dL}^n during the unbalanced load compensation or the imbalanced cluster generation powers in the PV system. As the results, the balancing of each phase dc bus voltages is affected by these unbalanced phase cluster powers.

Fig. 4(b) shows the dc voltages balancing control block diagram of the open-delta operation of SDBC. The overall voltage control is to manage the V_{dc} by a PI regulator. The output of the PI regulator is used as the active current command I_q^{p*} because controlling the average dc bus voltages of all modules (V_{dc}), requires overall active power as in (5). Therefore, the leftover control freedom is the regulation of the reactive current command I_d^{p*} and negative sequence current commands I_q^{n*}, I_d^{n*} to eliminate the unbalanced cluster powers. Note that the cluster voltage balancing controller is based on the information in bc-phase (V_{dcbc}), the PI regulator is applied to generate the feedback command P_{CbcFB} in order to balance the cluster voltages precisely.

B. The control strategy in PV system (balanced grid)

Fig. 6(a) shows the proposed cluster voltages balancing controller when MMCC-SDBC operates as an interface converter in the PV system. In the PV system, the balanced output grid currents are required, which restrict the injection of the negative sequence currents. For designing the cluster balancing control, the cluster powers in the PV system are calculated as:

$$
\begin{aligned}
P_{Cbc} &= P_{CbcPV} + P_{Cbcloss} \\
&= P_{Cbcq}^{p} + P_{Cbcd}^{p} + P_{Cbcq}^{n} + P_{Cbcd}^{n} \\
P_{Cca} &= P_{CcaPV} + P_{Ccaloss} \\
&= P_{Ccaq}^{p} + P_{Ccad}^{p} + P_{Ccaq}^{n} + P_{Ccad}^{n}
\end{aligned} \tag{9}
$$

The proposed control strategy in this application is to re-arrange the total output powers of bc- (P_{CbcPV}) and ca-phase (P_{CcaPV}) converter. Therefore, the cluster powers are balanced when two cluster converters output the same powers ($P_{CbcPV} = P_{CcaPV}$):

$$P_{Cbc} = P_{Cbcloss} \text{ when } P_{bcPV} = P_{caPV}. \tag{10}$$

Note that there are still some power losses P_{Cmloss} existed among each bridge cell, which influences the balancing of

---

Fig. 5. The equivalent circuit of the cascaded converter in open-delta operation. (Assumption of cells $ab1$ to aby are bypassed and ab-phase is open-circuit.)

balancing of each phase cluster is necessary. The phase cluster power is defined as:

$$
\begin{aligned}
P_{Cbc} &= P_{bc} - \frac{1}{2}P_T = P_{Cbcq}^{p} + P_{Cbcd}^{p} + P_{Cbcq}^{n} + P_{Cbcd}^{n} \\
P_{Cca} &= P_{ca} - \frac{1}{2}P_T = P_{Ccaq}^{p} + P_{Ccad}^{p} + P_{Ccaq}^{n} + P_{Cbcd}^{n}
\end{aligned} \tag{6}
$$

where P_{Cmq}^{p}, P_{Cmd}^{p} ($m \in \{bc, ca\}$) represent the cluster powers induced by voltages and positive sequence currents (I_q^p, I_d^p), and P_{Cmq}^{np}, P_{Cmd}^{n} ($m \in \{bc, ca\}$) represent the cluster powers induced by voltages and negative sequence currents (I_q^n, I_d^n). Note that the cluster powers P_{Cbc} and P_{Cca} are dependent ($P_{Cbc} = -P_{Cca}$), which implies that the phase cluster powers contribute no overall power in the system ($P_{Cbc} + P_{Cca} = 0$) and the balancing control only have to focus on one of the phase clusters. The detailed calculations are listed as:

$$P_{Cbcq}^{p} = \left(\frac{\sqrt{3}}{4}V_d^p - \frac{3}{4}V_q^n + \frac{\sqrt{3}}{4}V_d^n \right) I_q^p$$

$$P_{Cbcd}^{p} = \left(-\frac{\sqrt{3}}{4}V_q^p + \frac{\sqrt{3}}{4}V_q^n + \frac{3}{4}V_d^n \right) I_d^p$$

$$P_{Cbcq}^{n} = \left(-\frac{3}{4}V_q^p + \frac{\sqrt{3}}{4}V_d^p - \frac{\sqrt{3}}{4}V_d^n \right) I_q^n \tag{7}$$

$$P_{Cbcd}^{n} = \left(\frac{\sqrt{3}}{4}V_q^p + \frac{3}{4}V_d^p - \frac{\sqrt{3}}{4}V_q^n \right) I_d^n$$

506

The 2018 International Power Electronics Conference

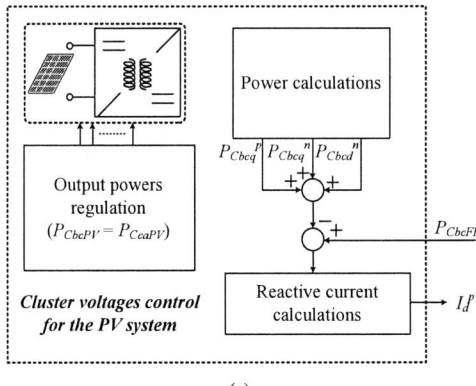

(a)

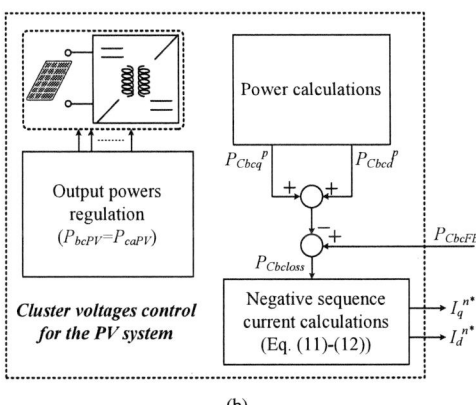

(b)

Fig. 6. The phase cluster voltages balancing control of open-delta operation in the PV system applications. (a)Normal grid voltages. (b)Grid fault.

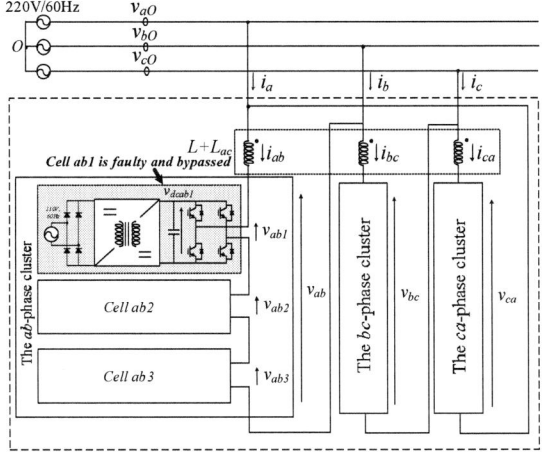

Fig. 7. The test-bench of 7-level MMCC-SDBC (for PV system).

the dc voltages. Based on the cluster powers in (6), the unbalance can be solved by injecting fewer reactive current to regulate the power P_{Cbcd}^p. Although the unbalanced power losses among phases induce the reactive current injection, this amount current is difficult to exceed grid code voltage magnitude variation requirement (10%) [14].

C. Discussion of the reactive current polarity

The controller in Fig. 6(a) provides a strategy for MMCC-SDBC achieving fault tolerant control without injecting amounts of extra currents by sacrificing the maximum power point tracking (MPPT). If the local grid code permits injecting the reactive current, the dc side can maintain MPPT and accomplish cluster balancing during the open-delta operation. Based on the (6)-(8), the relation between reactive current and cluster power can be arranged as:

$$P_{Cbc} = P_{bc} - \frac{1}{2}P_T > 0, \ I_d^p < 0$$
$$P_{Cbc} = P_{bc} - \frac{1}{2}P_T < 0, \ I_d^p > 0 \qquad (11)$$

As the (11) shown, the capacitive reactive current is injected to balance the cluster voltages when bc-phase cluster power is positive. On the other hand, the inductive reactive current is injected when bc-phase cluster power is negative.

D. The control strategy in PV system (grid fault)

In the utility-grid PV system, the converter injects the required reactive current (I_d^p) to meet the LVRT requirements [15], [16]. This section discusses the dc voltage balancing control strategy of the open-delta operation during the grid fault period.

In the section III-B, the cluster voltage balancing is achieved by adjusting the PV output powers of each cluster converter to be equal. During the LVRT, the negative sequence grid voltages contribute the unbalanced cluster powers P_{Cmq}^p, P_{Cmd}^p based on the (6). The related cluster powers are displayed as follows:

$$(P_{Cbcq}^p + P_{Cbcd}^p) - P_{Cbcloss} = P_{Cbcq}^n + P_{Cbcd}^n. \qquad (12)$$

Note that the P_{CbcPV} is zero since the output powers in bc- and ca-phase are regulated as equal ($P_{bcPV} = P_{caPV}$). Since the reactive current I_d^p is injected to meet the grid code requirement, only the control freedom of the negative sequence currents (I_q^n, I_d^n) could be used to regulate the cluster powers:

$$-P_{Cbcloss} = P_{Cbcq}^n + P_{Cbcd}^n. \qquad (13)$$

From (4)-(13), the negative sequence current commands I_q^{n*}, I_d^{n*} can be calculated. Note that the current phase angle is designed as an inductive negative sequence current to support the grid [17].

507

TABLE I
SYSTEM HARDWARE PARAMETERS (BASED ON 220 V AND 1 KVAR)

Variables	Symbol	Value
Line-to-line rms voltage	v_g	220(V)
Rated active power	P_R	3.0(kW)
Cascaded cell number	N	3
AC filter inductor	L_{ac}	10(mH)
		7.8%
Couple inductor	L	2.6(mH)
		2%
Nominal dc voltage	V_{dc}^*	120.0(V)
dc bus capacitor	C_{mn}	840(μF)
Unit capacitor constant [18]	H	54.4ms
Switching frequency	f_{sw}	2(kHz)
Sampling frequency	f_{sp}	12.0(kHz)
Switching dead time	T_{dt}	1.0(μsec)

TABLE II
CONTROLLER PARAMETERS

Variables	Symbol	Value
Moving averaging filter	MAF	33points at $4kHz$
Overall dc bus voltage control	K_{pTB}	0.2(A/V)
	K_{iTB}	2.0(A/V·sec)
Cluster balancing control	K_{pCB}	60.0(A/V)
	K_{iCB}	1200(A/V·sec)
Individual balancing control	K_{IB}	0.07(V/V)
Circulating current control	K_{pZ}	50(V/V)

IV. TEST RESULTS

The control verification for the PV system is built in the "PSCAD/EMTDC" simulation environment, which is shown in Fig. 7. Assuming the cell $ab1$ is faulty and bypassed to test the fault tolerant performance of the SDBC converter. TABLE I shows the parameters of the MMCC-SDBC, and TABLE II shows its controller parameters. The unit capacitor constant is $H = 54.4$ ms at 120 V [18].

The proposed fault tolerant strategy for PV system is verified in the simulation environment in Fig. 7. A MMCC-SDBC with three bridge cells per cluster is combined with nine isolated dc-power converters. Each of the nine dc-power converters consists of a unidirectional galvanically isolated dc-to-dc converter and a front-end rectifier that is fed from a single-phase ac mains. The dc-power converter operates in constant-power mode emulating an MPPT operation in the actual system. This provides the control flexibility to adjust the dc output power by controlling the power commands of the dc-power converters [5].

A. Verification for open-delta operation in normal grid

The simulation results in Fig. 8 are designed as three steps. The first step is in normal operation without broken cells. Each cluster operates in the unbalanced power outputs ($P_{abPV} = P_{caPV} = -1$ kW, $P_{bcPV} = -450$ W). Note that the minus power means the direction of the generated power is from dc to ac side. In the second step, the cell $ab1$ is faulty and bypassed. The cluster powers are maintained unbalanced. The

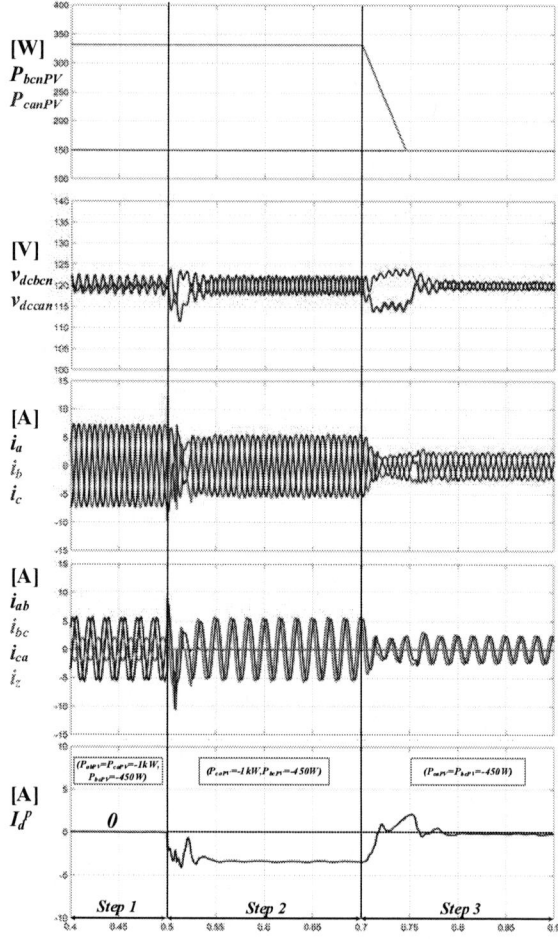

Fig. 8. Simulation results for verifying proposed fault tolerant control strategy in PV system.

proposed open-delta operation balances the dc bus voltages by injecting the reactive current I_d^p. Note that the cluster power in bc-phase P_{Cbc} is positive, therefore, the injected reactive current is capacitive. The proposed control strategy regulates the equal powers in bc- and ca-phase in step three. Therefore, the dc voltages maintain balanced and only injecting the minor reactive current to compensate the power loss of the converter.

B. Verification for open-delta operation in grid fault

To verify the proposed open-delta operation during the grid fault conditions. The simulation results under single-phase and two-phase voltage sags are displayed in Fig. 9 and Fig. 10. The simulations are designed as four steps. The first step is in normal operation without broken cells. Each cluster operates in the unbalanced power outputs ($P_{abPV} = P_{caPV} = -1$ kW, $P_{bcPV} = -450$ W). In the second step, the cell $ab1$ is faulty and bypassed. The cluster powers are maintained unbalanced. The proposed open-delta operation balances the dc bus voltages by injecting the reactive current I_d^p. In step three, the system occurred in the grid fault. In this period, the reactive current are injected to meet the LVRT requirement ($I_d^p = -3.711A$). Therefore, the cluster voltages balancing control is achieved by negative sequence current injection. Note that the PV output powers in bc- and ca-phase converters are not equal in this period. The proposed control strategy regulates the equal powers in bc- and ca-phase in step three. The output powers of bc- and ca-phase converters are regulated as the same value. In this way, the unbalanced cluster powers are contributed by the negative sequence voltage. The dc voltages maintain balanced and by injecting the corresponding negative sequence currents (I_q^n, I_d^n).

Fig. 9 shows the tested results under the a-phase 100% voltage sag. In step one, the converter achieves dc voltages balancing by injecting the zero sequence current. The output currents are balanced in step two since the cluster balancing is achieved by injecting the reactive current without the grid fault. From the figure, the dc voltages are maintained balanced even the during the single-phase voltage sag.

On the other hand, the Fig. 10 shows the tested results under the a-phase and b-phase 50% voltage sag. The output currents are balanced in step two since the cluster balancing is achieved by injecting the reactive current without the grid fault. In Fig. 10, the dc voltages are maintained balanced even the during the two-phase voltage sag. As the results, the proposed fault tolerant control method, which is called open-delta operation is able to achieve the stable operation and dc voltages balancing under dc and ac side unbalanced conditions.

V. CONCLUSION

This paper has presented a fault tolerant control scheme for MMCC-SDBC. During the occurrence of the bridge cell fault, the fault phase bridge of the converter is open-circuit to prevent the over-modulation problem, which is called open-delta operation. The proposed dc voltages balancing control in the open-delta operation of MMCC-SDBC is based on the average power analysis. This control approach has been verified by using the MMCC-SDBC in the application of the PV system. The unbalanced cluster powers are regulated by adjusting the PV cluster powers. The dc voltages balancing feedback control is achieved by injecting reactive current in the normal grid voltage. On the other hand, the reactive current injection is required by the grid code during the grid fault. In this condition, the balancing control is achieved by injecting the proper negative sequence currents. The simulation results show that the dc voltages are maintained balanced and the converter is stable operation even one of the bridge modules fails.

VI. ACKNOWLEDGMENT

This research is funded by the Ministry of Science and Technology of Taiwan under grant MOST-104-2221-E-007-045-MY3.

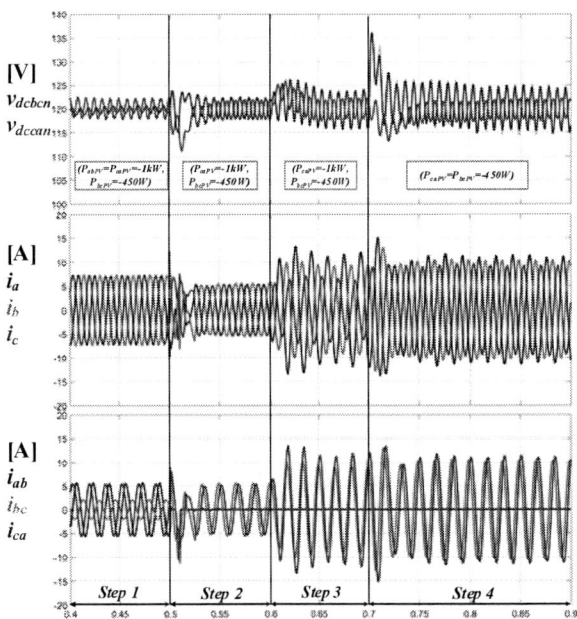

Fig. 9. Simulation results for verifying proposed fault tolerant control strategy in PV system under single-phase 100% sag.

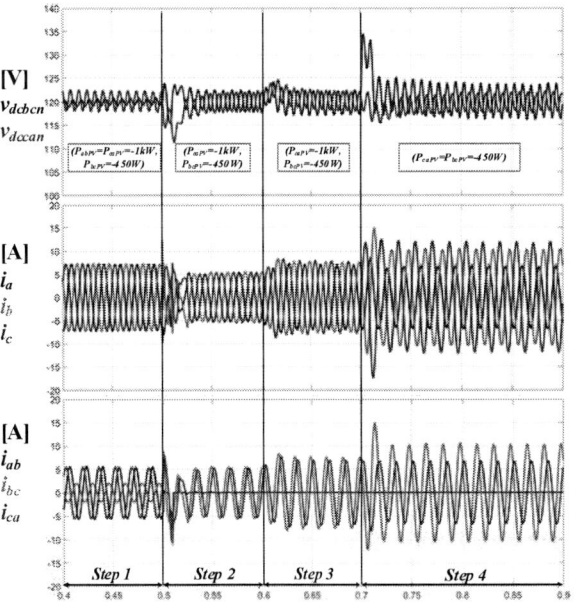

Fig. 10. Simulation results for verifying proposed fault tolerant control strategy in PV system under two-phase 50% sag.

The 2018 International Power Electronics Conference

REFERENCES

[1] H. Akagi, "Classification, terminology, and application of the modular multilevel cascade converter (mmcc)," *Power Electronics, IEEE Transactions on*, vol. 26, no. 11, pp. 3119–3130, Nov 2011.

[2] M. Hagiwara, R. Maeda, and H. Akagi, "Negative-sequence reactive-power control by a pwm statcom based on a modular multilevel cascade converter (mmcc-sdbc)," *Industry Applications, IEEE Transactions on*, vol. 48, no. 2, pp. 720–729, March 2012.

[3] S. Du, J. Liu, J. Lin, and Y. He, "A novel dc voltage control method for statcom based on hybrid multilevel h-bridge converter," *IEEE Transactions on Power Electronics*, vol. 28, no. 1, pp. 101–111, Jan 2013.

[4] P. Wu, H. Chen, Y. Chang, and P. Cheng, "Delta-connected cascaded h-bridge converter application in unbalanced load compensation," *IEEE Transactions on Industry Applications*, vol. PP, no. 99, pp. 1–1, 2016.

[5] P. Sochor and H. Akagi, "Theoretical comparison in energy-balancing capability between star- and delta-configured modular multilevel cascade inverters for utility-scale photovoltaic systems," *Power Electronics, IEEE Transactions on*, vol. 31, no. 3, pp. 1980–1992, March 2016.

[6] P. Wu, Y. Chen, and P. Cheng, "The delta-connected cascaded h-bridge converter application in distributed energy resources and fault ride through capability analysis," *IEEE Transactions on Industry Applications*, vol. PP, no. 99, pp. 1–1, 2017.

[7] P. Correa, M. Pacas, and J. Rodriguez, "Modulation strategies for fault-tolerant operation of h-bridge multilevel inverters," in *2006 IEEE International Symposium on Industrial Electronics*, vol. 2, July 2006, pp. 1589–1594.

[8] M. Ma, L. Hu, A. Chen, and X. He, "Reconfiguration of carrier-based modulation strategy for fault tolerant multilevel inverters," *IEEE Transactions on Power Electronics*, vol. 22, no. 5, pp. 2050–2060, Sept 2007.

[9] W. Song and A. Q. Huang, "Fault-tolerant design and control strategy for cascaded h-bridge multilevel converter-based statcom," *IEEE Transactions on Industrial Electronics*, vol. 57, no. 8, pp. 2700–2708, Aug 2010.

[10] L. Maharjan, T. Yamagishi, H. Akagi, and J. Asakura, "Fault-tolerant operation of a battery-energy-storage system based on a multilevel cascade pwm converter with star configuration," *IEEE Transactions on Power Electronics*, vol. 25, no. 9, pp. 2386–2396, Sept 2010.

[11] Y. Yu, G. Konstantinou, B. Hredzak, and V. G. Agelidis, "Operation of cascaded h-bridge multilevel converters for large-scale photovoltaic power plants under bridge failures," *IEEE Transactions on Industrial Electronics*, vol. 62, no. 11, pp. 7228–7236, Nov 2015.

[12] C. Lee, H. Chen, P. Wu, C. Wang, C. Yang, and P. Cheng, "A fault tolerant operation technique for statcoms based on star-connected cascaded h-bridges multilevel converter," in *2015 IEEE Applied Power Electronics Conference and Exposition (APEC)*, March 2015, pp. 995–1001.

[13] H. Akagi, Y. Kanazawa, and A. Nabae, "Instantaneous reactive power compensators comprising switching devices without energy storage components," *IEEE Transactions on Industry Applications*, vol. IA-20, no. 3, pp. 625–630, May 1984.

[14] "En 50160, voltage characteristics of electricity supplied by public distribution systems," 1999.

[15] "Grid code high and extra high voltage," E.ON Netz GmbH, Bayreuth, April 2006. [Online]. Available: http://www.eon-netz.com

[16] "The grid code, issue 4 revision 2," National Grid Electricity Transmission plc, Great Britain, March 2010. [Online]. Available: http://www.nationalgrid.com/uk

[17] S. Chou, C. Lee, H. Ko, and P. Cheng, "A low-voltage ride-through method with transformer flux compensation capability of renewable power grid-side converters," *Power Electronics, IEEE Transactions on*, vol. 29, no. 4, pp. 1710–1719, April 2014.

[18] H. Fujita, S. Tominaga, and H. Akagi, "Analysis and design of a dc voltage-controlled static var compensator using quad-series voltage-source inverters," *IEEE Transactions on Industry Applications*, vol. 32, no. 4, pp. 970–978, Jul 1996.

510

Cooling Performance Improvement of Heat Sink by Oscillating Heat pipe Addition and Design for Environment of Oscillating Heat Pipe Refrigerant

Kuan-Chung Tey[1] and Kenichiro Suzuki[1*]

1 Basic & Core Technology Research Laboratories, Meidensha Corporation, Numazu, Japan

*E-mail: suzuki-ken@mb.meidensha.co.jp

Abstract— Oscillating heat pipe addition to the heat sink was considered for the cooling performance improvement of heat sink. Hydrofluoroolefins (HFO) was chosen as the refrigerant to be encapsulated into the heat pipe, as it is non-flammable and has low global warming potential (GWP) index. HFO is proven to have the equal performance as the existing refrigerant of flammable butane. The optimum liquid volume ratio of the refrigerant encapsulated into the heat pipe is about 40%. By insulating the heat between the heat source and the chassis, the oscillating heat pipe could improve the cooling performance by 26.4%.

Keywords— CFC alternative, HFO, oscillating heat pipe, refrigerant

I. INTRODUCTION

The Paris Agreement was adopted during COP21 on December 2015 as the international effort to avoid the risk of global warming. In Japan, the goal of the global warming countermeasures is to reduce the greenhouse gases emission by 26% before year 2030 for midterm target and 50% before 2050 for long term target. For our operations we developed the guiding principle and environmental philosophy of top-down management system based on Meiden Group President's environmental management policy statement as below.

A. Philosophy

With our basic environmental philosophy: "Contribute to people, society and the global environment," Meiden Group aims to help build a sustainable society and to realize the growth of the Group and actively implement environmental management to tackle important issues: mitigating climate change, efficient use of resources (building a recycling society) and conserving biodiversity.

B. Policies

① By promoting the development of new products and innovative technologies and providing such products to wider global markets, we endeavor to positively contribute to the society.

② We strive to design and develop environmentally conscious products by conducting environmental impact evaluation for the product's life cycle, from initial material procurement to final disposal.

③ We strive to promote environmentally conscious business processes with green initiatives: promoting energy saving, promoting the 3Rs (reduce, reuse and recycle) and reducing the release of hazardous material to reduce the environmental impact from our business activities.

④ After establishing our internal guidelines, we endeavor to comply with the related environmental laws, regulations, rules and other required matters and strive to prevent the pollutions from our operations.

⑤ After establishing an environmental management system, we strive to maintain and improve it through the QC (quality control) tool of the PDCA (Plan-Do-Check-Act) Cycle and achieve our environmental goals.

⑥ We strive to implement initiatives including environmental education and PR (public relation) activities in order to increase all of our employees' understanding of environmental management and environmental protection and in so doing, we shall activate our environmental programs.

⑦ We endeavor to publicize our environmental initiatives both within the Group and to society and promote broader communication with our stakeholders.

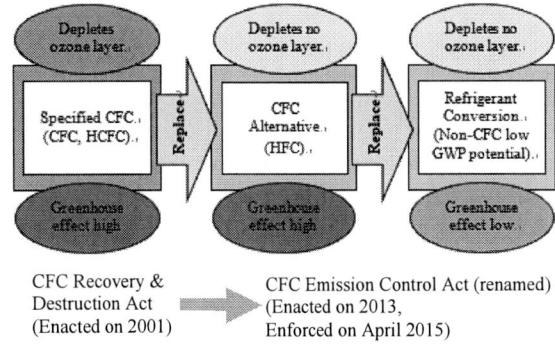

Fig. 1. Legal reform of CFC emission control in Japan [1]

II. DESIGN FOR ENVIRONMENT OF OSCILLATING HEAT PIPE

Figure 1 shows about the legal reform of CFC (chlorofluorocarbon) emission control in Japan. In 1987, The Montreal Protocol was signed to protect the ozone layer, while production and export of 5 specified CFCs (11, 12, 11, 114 and 115) are being regulated since 1988 in accordance with the ozone protection law. The usage of ozone depleting substances except HCFCs (Hydrochlorofluorocarbons) has been abolished since 2005, and HCFC itself by 2020. As the alternative for the specified CFCs, there is HFC (Hydrofluorocarbon) that does not deplete ozone layers but it has greenhouse gas effect that is 100 to 10,000 times more than carbon dioxide. It is necessary to have the alternative of non-CFC, low global warming potential (GWP) refrigerant like HFO (Hydrofluoroolefins) or butane gases that has even lower greenhouse gas effect. However, with safety in mind HFO is preferred as the alternative than flammable butane. Table I shows the specifications of major refrigerants available.

Table I. Refrigerant specification comparisons

	Refrigerant	Ozone layer depletion	GWP	Flammable
Specified CFC	CFC HCFC	Yes	100 to 10,000	No
CFC alternative	HFC	No	100 to 10,000	No
Low GWP ①	Butane	No	15	Yes
Low GWP ②	HFO	No	Up to 5	No

III. COOLING PERFORMANCE OF HEAT SINK BY OSCILLATING HEAT PIPE ADDITION

Power conversion equipment like inverters has about 97% efficiency with the remaining 3% being emitted as heat. The heat is well dissipated by the heat sink cooler, and it can be categorized into passive cooling and forced air cooling with blower type depending on the equipment rating capacity. High capacity equipment that forced air cooling could not handle are usually cooled by fitting water cooled heat sink. However, water cooling requires coolant circulation system that would increase the cost and size, and it is necessary to increase the performance of forced air cooling. Figure 2 explains the wind speed – thermal resistance curve of the commonly used heat sink. It is mentioned earlier that the thermal resistance has reached a saturation point about 0.02K/W, but in detail near to 0.02K/W is the region where the thermal resistance would not decrease even when the wind speed increases at high wind speed region shown in Figure 2. Therefore we made the cooling device prototypes that aim to increase the cooling performance by adding the heat pipe that assists the heat dissipation from the force air cooling heat sink under saturated condition. There are 3 types of heat pipe: gravity, capillary and oscillating type. Gravity type

heat pipes lack the freedom on mounting orientation, while capillary type heat pipes have small amount of heat transport due to the single direction flow of refrigerant.

In this work, for the heat pipe to be added to the heat sink we adopted the oscillating type heat pipe that has circulating refrigerant flow and high degree of freedom of mounting orientation. In order to verify the effect of the heat pipe addition and its orientation, we evaluated the heat pipe in cross orientation (Figure 3) and longitudinal orientation (Figure 4) of the heat sink. To optimize the right amount of the refrigerant liquid volume encapsulated into the heat pipe, we performed the evaluation on the encapsulation mount change. The definition of the refrigerant encapsulation ratio is defined in Figure 5. Encapsulation amount is defined as the liquid volume ratio of the all mixture content of vapor and liquid.

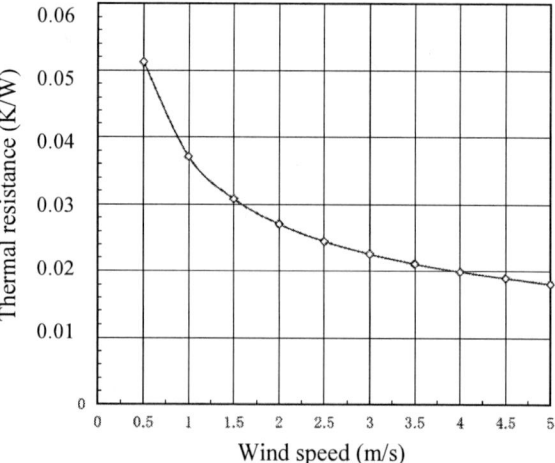

Fig. 2. Thermal resistance vs. Wind speed curve

Fig. 3. Heat pipe added heat sink (cross orientation)

Fig. 4. Heat pipe added heat sink (longitudinal orientation)

The 2018 International Power Electronics Conference

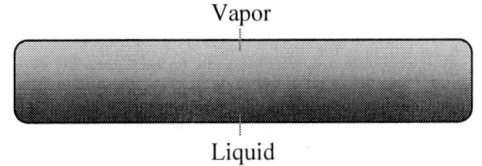

Fig. 5. Refrigerant encapsulated content inside heat pipe

IV. COOLING PERFORMANCE EVALUATIONS & THE RESULTS

A. The Effect of Heat Pipe Orientation

Figure 7 and 8 are the evaluation results for the heat pipe as oriented in Figure 4 and 5 respectively with 30% and 50% encapsulated liquid volume of butane refrigerant. X-axis refers to the frontal wind speed (m/s) at the heat sink while Y-axis refers to the temperature difference (rise) between the heat source and ambient surrounding. The heat sink was heated by a 300 watts heat source shown in Figure 6.

Fig. 6. Heating position

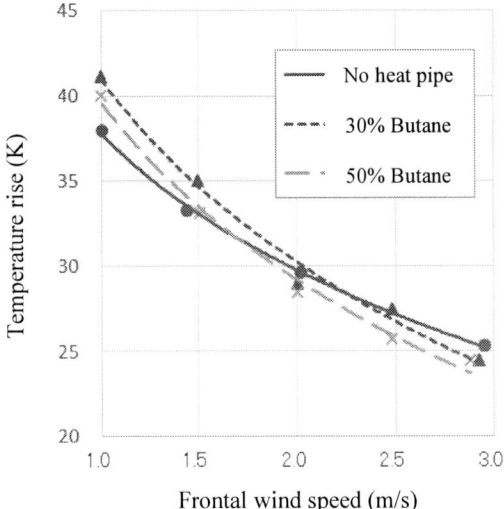

Fig. 7. Heat sink with cross oriented heat pipe

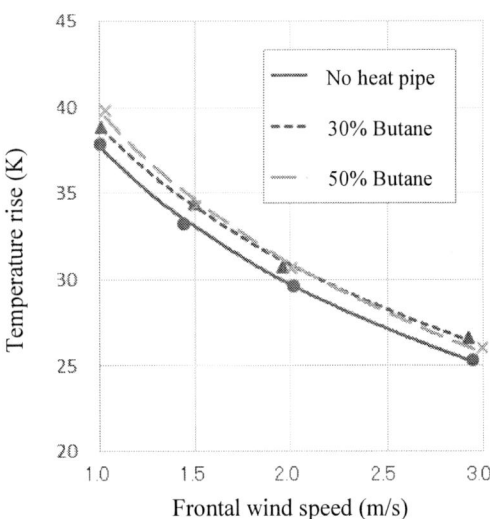

Fig. 8. Heat sink with longitudinal oriented heat pipe

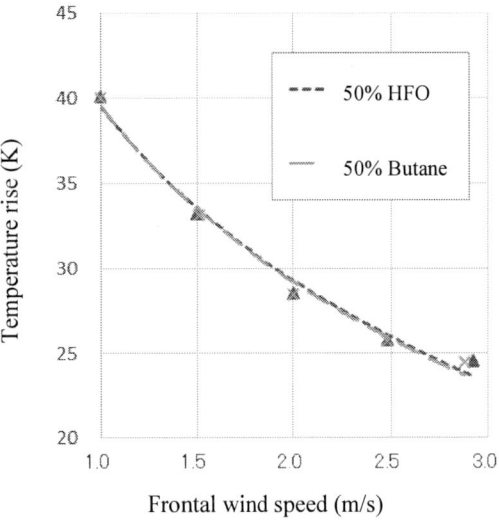

Fig. 9. Heat sink with longitudinal oriented heat pipe (Butane & HFO)

We understand that the cooling performance shows 7% improvement for the case of the heat pipe with cross orientation (Figure 7), while the cooling performance deteriorates for the case of the heat pipe with longitudinal orientation (Figure 8). Also, the result shows that in the case of longitudinal orientated heat pipe the performance was not affected by the encapsulated liquid volume of the refrigerant. Next, we evaluated the refrigerant performance difference between flammable butane and inflammable HFO, and confirmed that there are no significant difference between two as shown in Figure 9. In the next section, we further investigated about how the encapsulated refrigerant liquid volume would affect the cooling performance regardless the heat pipe orientation.

513

B. The Effect of Refrigerant Liquid Volume Encapsulated Inside Heat Pipe

Inside the heat pipe conduit the refrigerant exists in the form of mixture between vapor and liquid form, shown in Figure 5 above. From the results in Section A, since we know that the optimum encapsulated liquid volume ratio of the refrigerant is about 50%, we re-evaluated the cooling performance under the liquid volume ratio condition of 40% and 50%. Also, both butane and HFO are used in order to confirm the reproducibility of the earlier results in Section A. Heat sink overview, heat pipe overview and evaluation apparatus setup are shown in Figure 10, 11 and 12 respectively. Heat amount of 125 watts was applied by placing the heater on the center of the heat sink. The chassis that heat sink was mounted was cooled passively. In order to boost the operating efficiency of the oscillating heat pipe, the rectangular spiral shape and orientation was adopted to generate the pressure (temperature) difference.

Fig.10. Overview of the cooling device consisting of heat pipe mounted under the heat sink base

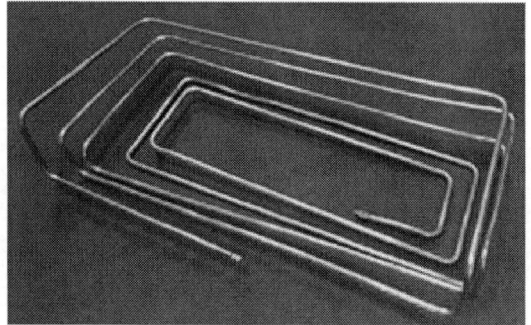

Fig.11. Overview of the heat pipe

Fig.12. Evaluation apparatus setup

Table. 2. Effect of encapsulated liquid volume comparison between butane and HFO

Refrigerant	Liquid volume ratio (%)	Temperature rise (K) ※
Butane	40	42.2
	50	43.4
HFO	40	43.1
	50	43.3

※Temperature rise = heat source temperature - ambient temperature

Table 2 shows the comparison on the effect of the encapsulated liquid volume between butane and HFO. As the temperature rise is calculated as the difference between the heat source temperature and the ambient temperature, we understand that the encapsulated liquid volume ratio of 40% is the optimum amount due to the slightly lower temperature rise. However, in the case of the metal chassis shown in Figure 10, we presume that the operating efficiency of the heat pipe decreased due to the thermal diffusion effect to the chassis. Therefore, we investigated the effect of the heat source insulation on the heat pipe operating efficiency.

C. The Effect of Heat Insulation on Heat Source

From the section B evaluation results, in order to investigate the effect of heat insulation we compare and evaluate the heat pipes that were winded on two different chassis materials: Bakelite plastic made chassis for the thermal insulation between heat source and chassis (Case 1), and aluminum metal made chassis for the thermal conduction between heat source and chassis (Case 2). Also, on these two cases the heat pipe winding in cross and longitudinal orientation were considered in order to investigate the effect of the way heat pipe being winded on the chassis. The heat pipes being winded on the Bakelite plastic made chassis in cross and longitudinal orientation are shown in Figure 12 and 13 respectively, while the heat pipes being winded on the aluminum metal made chassis in cross and longitudinal orientation are shown in Figure 14 and 15 respectively. Heat source are placed on the back base of the heat sink as seen on Figure 12 to 15 for applying 600 watts of heat on it. As for the cooling, the heated heat sink was forced air cooled with the wind speed of 1.5m/s, while the chassis was cooled passively. These were compared with the setup without heat pipe as well.

Fig.12. Thermal insulated chassis (Bakelite) with heat pipe winded in cross orientation

Fig.13. Thermal insulated chassis (Bakelite) with heat pipe winded in longitudinal orientation

Fig.14. Thermal conducted chassis (Aluminum) with heat pipe winded in cross orientation

Fig.15. Thermal conducted chassis (Aluminum) with heat pipe winded in longitudinal orientation

Table. 3. Evaluation results for thermal insulated chassis (Bakelite)

Heat pipe	Heat pipe orientation	Fig.	Temperature rise (K) (※)	Cooling performance improvement (%)
None	Cross		89.7	
	Longitudinal		89.7	
With	Cross	12	70.9	21.0
	Longitudinal	13	66.0	26.4

※Temperature rise = heat source temperature - ambient temperature

Table. 4. Evaluation results for thermal conducted chassis (Aluminum)

Heat pipe	Heat pipe orientation	Fig.	Temperature rise (K) (※)	Cooling performance improvement (%)
None	Cross		72.1	
	Longitudinal		71.9	
With	Cross	14	66.8	7.4
	Longitudinal	15	60.4	16.0

※Temperature rise = heat source temperature - ambient temperature

The 2018 International Power Electronics Conference

The evaluation results for the Case 1 of thermal insulated Bakelite made chassis is tabulated in Table 3. Compared to the setup without heat pipe, the setup that heat pipe winded in longitudinal orientation achieved 26.4% improvement in cooling performance. Next, the evaluation results for the Case 2 of thermal conducted aluminum made chassis is tabulated in Table 4. Compared to the setup without heat pipe, the setup that heat pipe winded in longitudinal orientation achieved 16.0% improvement in cooling performance. From these results, we understand that the heat sink with heat pipe winded on thermal insulated chassis in longitudinal orientation has the best cooling performance. We presume that longitudinal oriented heat pipe manages to perform better in cooling, as the refrigerant in the cross oriented heat pipe accumulates at the bottom of heat pipe due to the gravitational effect, resulting to the operating efficiency decrease.

V. CONCLUSIONS

From all the results, we confirmed the applicability of environmentally friendly inflammable HFO as the encapsulated refrigerant substitute for the heat pipe. By adding the heat pipe to the forced air cooling heat sink, longitudinal orientation shows no cooling performance improvement but 7% improvement is achieved by positioning the heat pipe in cross orientation. Contrarily, we presumed that the only small percentage (7%) of cooling performance improvement is due to the high dependency of the heat sink cooling performance in the case of forced air cooling. Also, we investigated that encapsulated refrigerant liquid volume ratio of 50% is the optimum point as seen on Figure 7, and thus further investigated about the effect of the refrigerant encapsulated liquid volume on the cooling performance. Lastly, we understand that by insulating the heat between the heat source and the chassis, the pressure (temperature) difference between the evaporator and condenser sections of heat pipe increases and this would affect the cooling performance improvement. For the future work, we consider investigating the effect of the heat pipe length on the cooling performance.

ACKNOWLEDGMENTS

We would like to express our upmost gratitude to the vendor partner who wish not to be named for supplying the oscillating heat pipe, as well as evaluating the assembled prototypes of heat pipe and heat sink that contributed to this work.

REFERENCES

[1] Ministry of Environment, Ministry of Economy, Trade & Industry, Ministry of Land, Infrastructure, Transport & Tourism

Compact large capacity gas turbine static starter

Hironori Kawaguchi, Shigeyuki Nakabayashi, Akinobu Ando, Hiroshi Ogino,
Yasuaki Matsumoto, Ikuto Udagawa, Takahiro Ohta

Toshiba Mitsubishi-Electric Industrial Systems Corporation, TMEIC
1-1-2, Wadasaki-cho, Hyogo-ku, Kobe, 652-8555, JAPAN

Abstract- **Today, the demand for gas turbine combined cycle (GTCC) power plants is increasing owing to the need to increase efficiency and reduce impact on the environment. The device used for starting a gas turbine is called a static frequency converter (SFC). Owing to the increasing size of gas turbines in GTCC plants, gas turbines require more output power from the SFC. Recently, larger capacity SFC (7 MW) products are available; however, they require a larger footprint for installation and additional work on-site for wiring between the SFC cubicles. This paper describes the development of a compact and large capacity SFC to address these new challenges.**

Keywords— Gas turbine starter, Load commuated inverter (LCI), Static starting device, Thyristor converter

I. INTRODUCTION

Static frequency converters (SFCs) have been employed to start gas turbines in high-efficiency gas turbine combined cycle (GTCC) power plants, and blast furnace blowers in steel mills. In recent years, GTCC power plants are widely used because they offer the advantages of energy conservation and environmental conservation.

In GTCC plants, combustion air is compressed by a rotating gas turbine, mixed with fuel, and the energy obtained by igniting this mixture is used to generate electric power. Because the gas turbine cannot generate energy at start-up, an external device is used to accelerate the turbine. Until recently, the gas turbine was started using a starting motor with a single-shaft arrangement. A large gas turbine requires a large starting motor. In recent years, static frequency converters (SFCs), which are load commutated inverters (LCIs), have been employed to start large gas turbines in GTCC power plants, instead of using starting motors. SFCs are variable-voltage, variable-frequency inverters that drive a generator such as a synchronous generator so that it accelerates up to the

gas turbine's self-sustaining speed. A GTCC plant with an SFC is compact and reliable owing to the absence of a starting motor [1]. SFCs can be applied not only to GTCC plants, but also to blast furnace blowers in steel mills, and other applications. The configuration of an SFC system, including the generator and load, is shown in Fig. 1. An SFC consists of four components. The rectifier is a thyristorized 3-phase, 6-pulse bridge connection circuit in which a 3-phase AC voltage is converted into a variable DC voltage, and the DC current is controlled using phase control thyristors. The DC reactor smoothens the DC current ripples from the rectifier. The inverter is also a thyristorized 3-phase, 6-pulse bridge connection circuit using phase control thyristors. The inverter outputs AC current to the stator of the gas turbine generator along its rotor position, which is detected by the advanced position sensor (APS) [1], so that the inverter can convert DC to AC while synchronizing with the generator motor rotation. The inverter controls its output to synchronize with the rotor position. The control panel controls and monitors the devices.

II. SUBSTANCE

TABLE I
RATINGS OF NEW SFC AND CONVENTIONAL SFC

	New SFC	Conventional SFC
Input/Output Power	7 MW	7 MW
Input AC Voltage	2.9 kV	3.9 kV
Input/Output Current	2041 A (rms)	1429 A (rms)
Output AC Voltage	2.7 kV (for 60 Hz sites) 2.6 kV (for 50 Hz sites)	3.6 kV (for 60 Hz sites) 3.5 kV (for 50 Hz sites)
DC Link Voltage	2.8 kV	4.0 kV
DC Link Current	2500 A	1750 A
Thyristor Configuration	1S1P6A ×2 (Rectifier + Inverter)	2S1P6A ×2 (Rectifier + Inverter)
Footprint	4.3m²	8.5m²

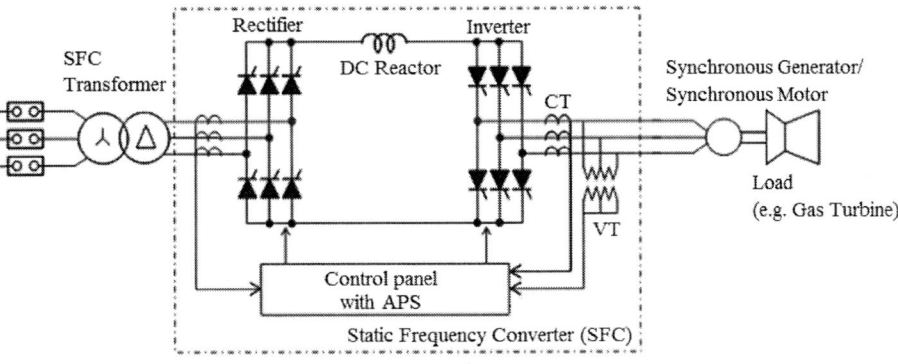

Fig. 1. SFC overview.

517

A. Electrical Design

1. Change in electrical design

Table I presents the ratings of the 7 MW conventional SFC and the new SFC. The input AC voltage and output AC voltage of the new SFC were decreased by 25% to minimize the insulation distance and scale down the cubicle. However, the current rating was increased by 43% because the required output capacity was not changed.

2. Change in voltage rating

The input AC voltage rating was changed from 3.9 kV to 2.9 kV, while the DC link voltage rating was changed from 4.0 kV to 2.8 kV in order to decrease the required insulation clearance, the creepage distance in accordance with the IEC-61800 standards, and the number of series thyristors. The number of series thyristors was decreased from 2 to 1, and the converter dimensions were reduced.

3. Change in current rating

The DC current rating was changed from 1750 A to 2500 A. The losses dissipated from a single thyristor element increased. Therefore, the cooling performance of the thyristors should be improved. The cooling fans were changed to a larger capacity type, and the wind velocity around the fins approximately doubled. In addition, a new cooling fin with a low heat resistance was constructed and tested. However, the number of cooling fans could be decreased to half because the number of thyristors was reduced, and the surge absorber circuits were changed.

4. Snubber circuit

The snubber circuits were optimized. The number of snubber circuits was decreased from 24 to 12 because the number of series thyristors was decreased. The capacitance of the snubber circuits in the new SFC was optimized. The losses dissipated from the snubber circuit can be calculated as given in (1) [2]. The parameters of the snubber circuits were decided such that the surge voltage will be restricted within the thyristor's withstand voltage rating and the losses from the snubber is minimized. The total losses from all the snubber circuits were reduced.

$$Psn = 4 \times f \times C \times \left(Eu \times \sin(\alpha + u)\right)^2 \tag{1}$$

where *Psn* is the loss dissipated from the snubber circuit, *f* is the circuit frequency, *C* is the capacitance of the snubber capacitor, *Eu* is the AC voltage, α is the trigger delay angle, and *u* is the angle of overlap (commutation angle).

5. Surge absorber

To withstand against a surge of voltage from the feeder, surge absorbers were installed in the input circuit of the SFC. In a conventional SFC, RC-type surge absorbers are used. RC-type surge absorbers generate high losses and require a large space for installation. In the new SFC, ZNR arrestors were used owing to the benefits of reduced losses and installation space. This approach contributed to the scaling down of the space and higher efficiency.

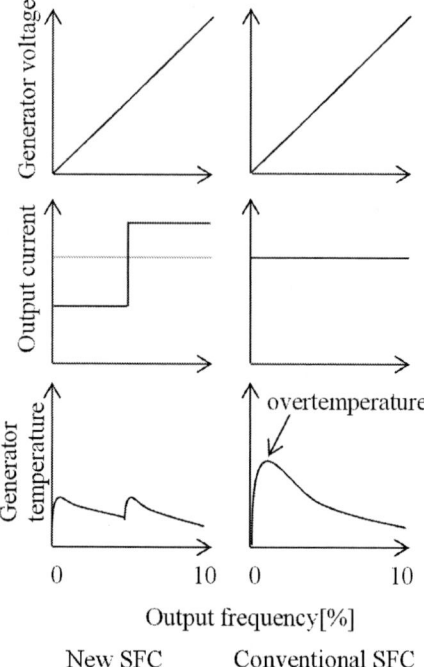

Fig. 2. Change of output current ($f < 0.1$ pu.).

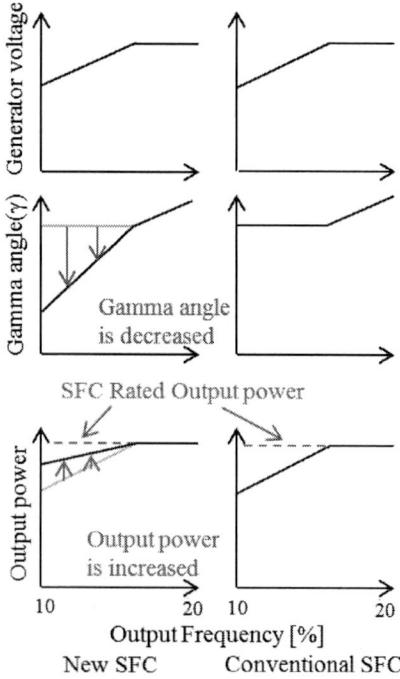

Fig. 3. Variable gamma control and resulting output power (approximately 10%< f <20%).

B. New Functions

1. Change in current control

If the voltage rating is decreased, the current rating should be increased. However, at a lower rotating speed, a large current cannot be applied to the generator stator windings because the generator lacks suitable cooling performance. On the other hand, at lower rotating speeds, the resistance torque of the gas turbine is smaller, so that a large power output is not required from the SFC. A new function known as the "2-step pulse mode current control" was applied. Fig. 2 shows the current and voltage curves versus the output frequency (f < 10%). At lower speeds, the new SFC outputs less current and can increase the output current when the rotating speed is sufficiently high for the generator to create appropriate cooling. This can prevent the generator from overheating due to high current, even if the generator cooling performance is low at the beginning of startup.

2. Change in gamma angle control

The gamma angle (γ) control was changed for output power at medium rotating speeds. Fig. 3 shows the gamma angle and output power curves versus output frequency for medium range (approximately 10% < f < 20%). In the conventional SFC, the gamma angle is constant at medium rotating speeds, while the gamma angle can be varied at medium rotating speeds in the new SFC. As a result, the new SFC can improve its output power factor, and output more power at medium speeds.

C. Mechanical Design

1. Effect of decrease in number of thyristors

The number of thyristors was decreased. Fig. 4 shows the circuit configuration of the conventional SFC, while Fig. 5 shows that of the new SFC. The conventional SFC has two series thyristors in one arm, with a total of 24 thyristor elements. The new SFC has one thyristor in one arm, with a total of 12 thyristor elements. This decrease in the number of the thyristors contributed to reducing the peripheral circuits, such as snubbers and gate circuits, and to simplifying structures, such as stacks, cooling fins, bus bars, and frames. The conventional SFC consists of separate cubicles: a rectifier cubicle, a DC reactor cubicle, an inverter cubicle, and a control cubicle. However, the new SFC can accommodate all circuits, including the rectifier, DC reactor, inverter, and controls within a single cubicle. This new structure does not require on-site wiring between SFC cubicle interfaces. The removal of the interface terminals contributes to scaling down.

2. Mechanical design strength

CAE (computer-aided engineering) is a computer technology that offers useful assistance for many engineering tasks, including 3D CAD, thermo-fluid analysis, mechanical strength analysis, and vibration analysis. Static analyses (such as stress and strain analyses) and dynamic analyses (such as vibration and impact analyses) can be carried out using CAE [3]. In this study, CAE was used for the mechanical design of the new SFC. We carried out virtual assembly using CAE,

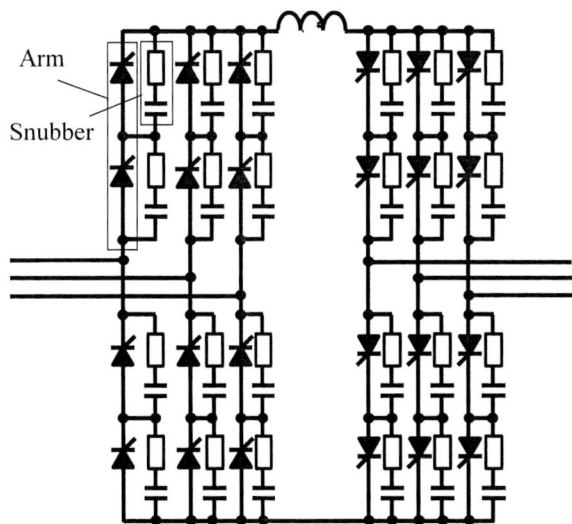

Fig. 4. Thyristor configuration of a conventional SFC.

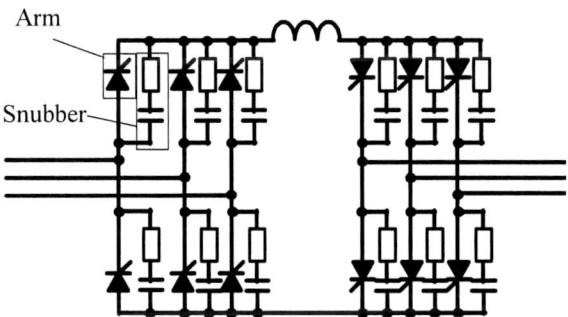

Fig. 5. Thyristor configuration of a new SFC.

Fig. 6. A model of the CAE analysis.

which demonstrates that designers can virtually assemble and disassemble products on a computer to examine the compact mounting. From the vibration analysis, we completed a natural frequency response analysis using only the CAE simulation capabilities. In this study, we confirmed that the new SFC cubicle has sufficient mechanical strength.

3. Thermo-fluid analysis

Furthermore, we carried out thermo-fluid analysis using CAE. Fig. 7 shows the results of the temperature rise of the thyristor. In addition, Table II presents a comparison of the temperature resistance results obtained from the CAE calculations and values measured from the actual device. The CAE calculation results were consistent with measured values from the actual device.

III. RESULTS

Owing to these changes, we obtained the following excellent results.

A. Smaller Footprint

By decreasing the number of thyristors, applying a single-cubicle structure, optimizing the snubber circuit, and decreasing the insulation distance, the amount of installation space per unit of the output power is 0.61 m²/MW compared to that of the conventional SFC (1.21 m²/MW). Therefore, the new SFC achieved a reduction of 50%.

B. Easy Installation

The single-cubicle structure ensured a relatively easy installation compared to that of the conventional setup. Fig. 8 shows the required site works. No on-site rewiring work or re-assembly work is required.

C. Higher Efficiency

The reduction in the number of thyristors, the optimized snubber circuit, the surge absorber, and the development of a low-loss DC reactor improved the efficiency of the entire SFC system. Compared to the conventional SFC, the new SFC achieved a 40% reduction in losses. This reduction also lowered the power consumption of the air-conditioning system, contributing to environmental protection.

REFERENCES

[1] H. Ogino, S. Tamai, Y. Hosokawa, and A. Ando, "Static starting device for the start-up of the gas turbine using position sensorless control method," *Proc. of ICPE 2011-ECCE Asia*, pp. 1649-1654.

[2] J. Waldmeyer and B. Backlund, "Application note Design of RC Snubbers for Phase Control Applications," pp. 12, 2008.

[3] S. Nakabayashi, T. Koga, Y. Hosokawa, and A. Ando, "CAE application to the power converters development," *Annual Conference of IEEE of Japan, Industry Applications Society*, pp. I-309-310, 2008.

TABLE II
COMPARISON OF COOLING FIN TEMPERATURE RESISTANCE

Cooling fin temperature resistance	CAE calculation	Measured value, test on actual device
	0.0161 K/W	0.0161 K/W

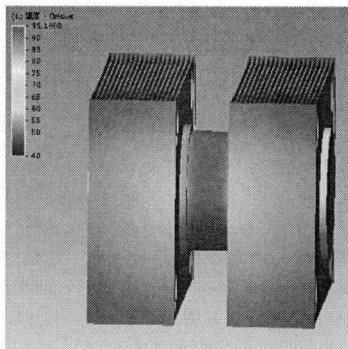

Fig. 7. Thermo-fluid analysis of the thyristor.

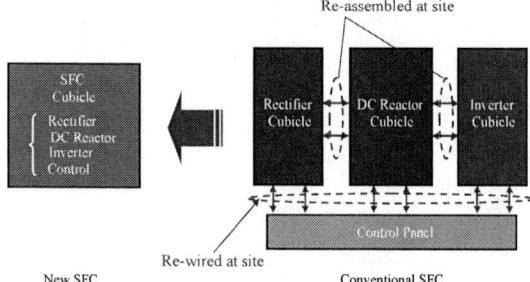

Fig. 8. Comparison of the new SFC and conventional SFC structures.

Fig. 9. Outline of new SFC.

Voltage Reference Modification Scheme for Resonance Suppression in LCL-filtered Inverters with Discontinuous PWM method

Hyeon-Sik Kim, and Seung-Ki Sul

Department of Electrical and Computer Engineering, Seoul National University, Seoul, Korea

hyeonsik@eepel.snu.ac.kr, sulsk@plaza.snu.ac.kr

Abstract—This paper analyzes *LCL* resonance problem provoked by discontinuous PWM (DPWM) through frequency and time domain analysis. Moreover, a voltage reference modification algorithm is proposed to alleviate *LCL* resonance problem while maintaining advantages of DPWM scheme. In the proposed method, pole voltage references at the edge of offset voltage are slightly modified to reduce a current oscillation. It can be easily implemented by adopting perturb and observe (P&O) technique. The effectiveness of the proposed method is verified by computer simulation and experimental test. Through the proposed algorithm, the oscillation could be minimized while keeping the merits of DPWM, which decreases effective switching frequency under the grid harmonics regulation.

Keywords—Discontinuous PWM (DPWM), LCL filter, active damping, resonance suppression.

I. INTRODUCTION

Voltage source inverters (VSIs) have been widely used for grid connection of renewable energy source (RES), distributed generation (DG), and energy storage system (ESS). On the output terminals of inverters, passive filter should be placed to attenuate high frequency harmonics generated by a pulse-width modulation (PWM) switching.

Compared with *L* filter, *LCL* filter has better harmonic attenuation performance, which not only minimizes size, weight, and cost of the filter inductor, but also improves the dynamics of the system [1]-[2]. However, the resonance of *LCL* filter could provoke unexpected high frequency current ripples and even instability.

There have been lots of researches to suppress *LCL* resonance, which could be classified into two groups: passive damping [3]-[4] and active damping methods [5]-[7]. Passive damping methods are quite simple and robust, but result in losses from damping resistor and reduced filter attenuation ratio at high frequency range. As an alternative, active damping methods have been proposed to simulate virtual resistor or avoid resonant frequency by modifying a current controller, which could eliminate losses on the filter. However, the majority of literature did not cover a resonant problem provoked by discontinuous modulation schemes itself.

In respect of an efficient operation, the selection of PWM method is crucial to minimize power losses under the constraints of grid standard, such as IEEE 519 or BDEW [8]-[9]. In [10], the impacts of PWM methods on

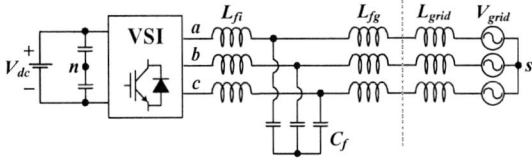

Fig. 1. Configuration of three-phase VSI with LCL filter.

LCL-filtered two-level three-phase inverter were compared in terms of power loss and power density. It shows 60° discontinuous PWM (DPWM) has higher performance over continuous PWM (CPWM) in most aspects, i.e., lower power losses while keeping the harmonic standard limits. However, it also did not mention a current oscillation induced by abrupt variation of offset voltages.

In [11], the resonant problem come from DPWM was identified by a frequency domain analysis. Also, a hybrid PWM method mixed with DPWM and CPWM was proposed as an alternative to alleviate the cause of the resonant problem. In [12], an adaptive DPWM method was proposed where the ratio of CPWM is increased when the modulation index is low, which could be classified as one of hybrid PWM methods. However, the hybrid PWM proposed in [11] and [12] not only increases switching loss compared to DPWM, but also cannot remove current harmonics near the resonant frequency completely.

In this paper, the resonance issue due to DPWM is analyzed by *LCL* filter modeling in the frequency domain and harmonic spectra analysis according to modulation schemes. In addition, a voltage reference modification scheme is proposed to alleviate the resonance problem while preserving DPWM scheme. Voltage references at the edge of offset voltage are moved to near the optimal point by using perturb and observe (P&O) algorithm. Lastly, simulation and experimental results are provided to verify the validity of the proposed method.

II. ANALYSIS OF RESONANCE PROBLEM

A. LCL filter Modeling

Fig. 1 shows the circuit topology of a three-phase VSI with *LCL* filter, where inverter-side inductance, grid-side inductance and filter capacitance are defined as L_{fi}, $L_g = L_{fg} + L_{grid}$, and C_f, respectively. The filter admittances

The 2018 International Power Electronics Conference

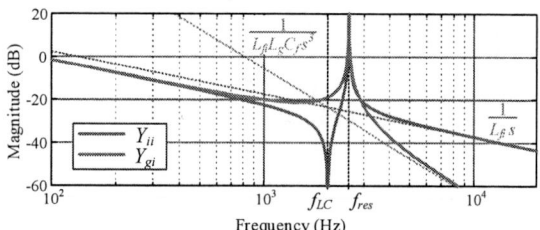

Fig. 2. Bode plot of LCL filter admittance when $L_{grid} = 0$.

TABLE I
SYSTEM PARAMETERS

Parameter	Value
Grid voltage, V_{grid}	220 $V_{l-l,rms}$, 60Hz
Filter inductance on inverter side, L_{fi}	1.2 mH
Filter capacitance, C_f	9 μF
Filter inductance on grid side, L_{fg}	0.7 mH
Resonant frequency, $f_{res}(L_{grid}=0)$	2.52 kHz
Switching frequency, f_{sw}	7.20 kHz
Sampling frequency, f_{samp}	7.20 kHz

from inverter voltage, v_{inv}, to inverter current, i_{inv}, and grid current, i_{grid}, can be defined as (1) and (2), where f_{res} is the resonant frequency and f_{LC} is LC resonant frequency of grid-side.

$$Y_{ii}(s) = \frac{i_{inv}(s)}{v_{inv}(s)} = \frac{L_g C_f s^2 + 1}{L_{fi} L_g C_f s^3 + (L_{fi} + L_g)s} = \frac{1}{L_{fi}} \frac{s^2 + \omega_{LC}^2}{s(s^2 + \omega_{res}^2)}. \quad (1)$$

$$Y_{gi}(s) = \frac{i_{grid}(s)}{v_{inv}(s)} = \frac{1}{L_{fi} L_g C_f s^3 + (L_{fi} + L_g)s} = \frac{1}{L_{fi} + L_g} \frac{\omega_{res}^2}{s^2 + \omega_{res}^2} \quad (2)$$

, where $\omega_{res} = 2\pi f_{res} = \sqrt{\dfrac{L_{fi} + L_g}{L_{fi} L_g C_f}}$, $\omega_{LC} = 2\pi f_{LC} = \sqrt{\dfrac{1}{L_g C_f}}$.

Fig. 2 shows the Bode plot of filter admittances where *LCL* filter parameters are given in Table I. Y_{gi} has 20 dB/dec attenuation in lower frequency range, i.e., $f < f_{res}$, whereas 60 dB/dec attenuation in higher frequency range, i.e., $f_{res} < f$. The resonant peak occurs at f_{res}, which means that a small amount of v_{inv} around f_{res} range could provoke large i_{inv} and i_{grid} distortion and would violate harmonic regulation. Therefore, f_{res} should be far away from switching frequency, f_{sw}, and its sideband.

B. Harmonic Spectra Analysis

Harmonic characteristics vary depending on the PWM methods, which gives different voltage harmonics, v_{har}, at each frequency. Thus, the harmonic spectra analysis of different PWM schemes is required to improve a system efficiency for given operating conditions. The harmonic spectra of PWM can be derived by an analytical method as a technique using Double Fourier integration. It is widely applied to calculate the magnitude of each v_{har} as (3), where *m* and *n* are carrier and baseband integer indices, respectively [13].

$$C_{mn} = \frac{V_{dc}}{2\pi^2} \int_{-\pi}^{\pi} \int_{x_r}^{x_f} e^{j(mx+ny)} dx dy . \quad (3)$$

Fig. 3 shows the PWM waveforms and harmonic spectra of *a*-phase phase voltage, v_{as}, when modulation index (MI) and f_{sw} are set to 0.9 and 7.2 kHz, respectively. Here, MI is defined as the ratio between a fundamental component of the phase voltage, v_{s1}, and a half of dc-link voltage like (4).

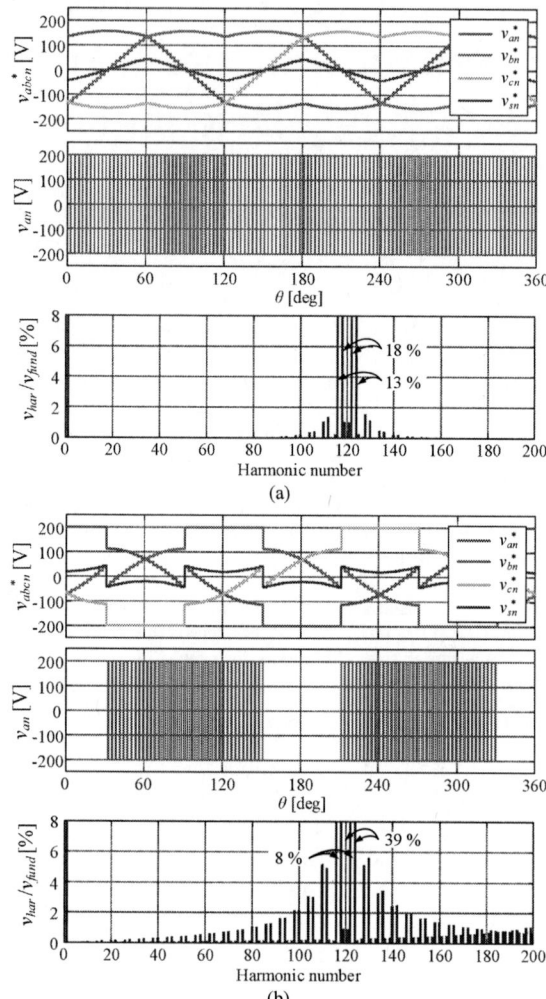

Fig. 3. PWM waveforms and harmonic spectra of v_{as} when MI = 0.9, f_{sw} = 7.2 kHz. (a) CPWM. (b) DPWM.

$$MI = \frac{v_{s1}}{0.5 V_{dc}} . \quad (4)$$

Offset voltage of CPWM, v_{sn}^*(CPWM), is defined as (5), and it is known to be equivalent to SVPWM [14]. Compared to CPWM, DPWM has larger and widespread sideband harmonics at near f_{sw}. Effective f_{sw} of DPWM is two-thirds of that of CPWM, and switching loss of DPWM is almost a half of that of CPWM thanks to no switching at near maximum current. Therefore, f_{sw} of DPWM could be increased up to two times of CPWM depending on the phase angle [15]. In spite of increasing f_{sw}, sideband harmonics around f_{sw} are enlarged when DPWM is applied. In this case, the sideband harmonics of f_{sw} and resonant frequency band, near f_{res}, might be overlapped, which may provoke severe harmonic oscillation [11]. Thus, v_{har} components located near f_{res} should be avoided by changing PWM schemes or revising other control part.

$$v_{sn}^*(\text{CPWM}) = -0.5(v_{max}^* + v_{min}^*) . \quad (5)$$

The offset voltage of DPWM, v_{sn}^*(DPWM), can be defined as (6), where v_{max}^*, v_{min}^* are defined as the

522

The 2018 International Power Electronics Conference

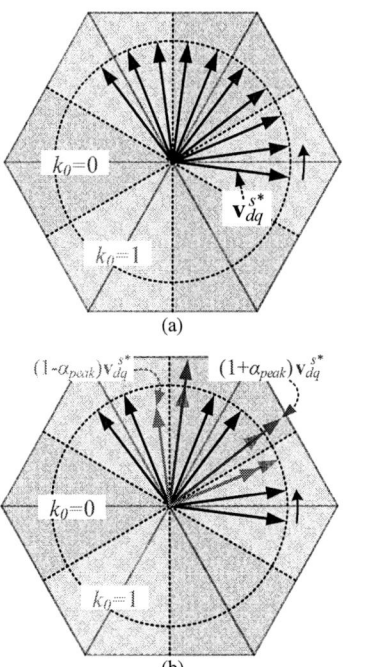

(a)

(b)

Fig. 4. Traces of $\mathbf{v}_{dq}{}^{s*}$ at steady-state in space vector diagram. (a) Conventional method. (b) Proposed method.

maximum and minimum pole voltage reference, respectively. It shows abrupt variation of $v_{sn}{}^*$ from $v_{sn,pos}{}^*$ to $v_{sn,neg}{}^*$ or vice versa, 6 times per an electrical period. Fig. 3(b) shows that a pulse width of v_{an} abruptly varies at rising and falling edge of $v_{an}{}^*$ contrary to the case of SVPWM in Fig. 3(a), which induces more sideband harmonics around f_{res}.

$$v_{sn}^*(\text{DPWM}) = \begin{cases} +0.5V_{dc} - v_{\max}^* \equiv v_{sn,pos}^*, & (v_{\max}^* + v_{\min}^* \geq 0) \\ -0.5V_{dc} - v_{\min}^* \equiv v_{sn,neg}^*, & (v_{\max}^* + v_{\min}^* < 0) \end{cases}. \quad (6)$$

In the aspect of time domain, this phenomenon could be comprehended that harmonic oscillation occurs at each $v_{sn}{}^*$ edge, which induces oscillation not only in inverter currents but also in grid currents. Consequently, the grid currents starts to oscillate at an interval of 60° in case of DPWM and that cannot be easily suppressed with conventional methods.

III. PROPOSED SCHEME

A. Concept of Proposed Scheme

To suppress harmonic oscillation incurred by DPWM, v_{har} near f_{res} should be avoided. v_{har} spectra is determined by pole voltage references, $\mathbf{v}_{abcn}{}^*$, which means $\mathbf{v}_{abcn}{}^*$ should be revised to change v_{har} near f_{res}. In the view point of controller, $\mathbf{v}_{abcn}{}^*$ is set up by d-q voltage references, $\mathbf{v}_{dq}{}^*$, and offset voltage, $v_{sn}{}^*$, where $\mathbf{v}_{dq}{}^*$ and $v_{sn}{}^*$ are set by the current controller and PWM method, respectively. However, $v_{sn}{}^*$ cannot be used to suppress sideband harmonics near f_{res} because a transition from DPWM to other PWMs would degrade the system efficiency. Alternately, $\mathbf{v}_{dq}{}^*$ could be modified to suppress harmonic oscillation while current control performances are not much impaired.

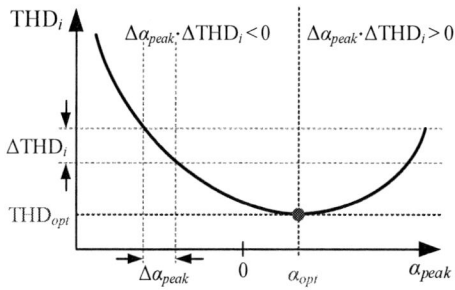

Fig. 5. Relationship between α_{peak} and THD$_i$.

Fig. 4(a) shows the traces of d-q voltage references in stationary reference frame, $\mathbf{v}_{dq}{}^{s*}$, at steady-state under the conventional current controller, which shows that the magnitude of $\mathbf{v}_{dq}{}^{s*}$ remains constant at $v_{sn}{}^*$ edge moments. The bandwidth of the current controller is too low to catch up current harmonics near f_{res}, which is about one-third of f_{sw}. To improve damping performance, a feedforward voltage defined as $\alpha_{peak}\cdot\mathbf{v}_{dq}{}^{s*}$ could be added or subtracted at the output of the current controller as shown in Fig. 4(b). Because the dominant voltage harmonics, i.e., v_{har} near f_{res}, are invoked at $v_{sn}{}^*$ edge moment, $\alpha_{peak}\cdot\mathbf{v}_{dq}{}^{s*}$ should be imposed at the edge of $v_{sn}{}^*$ in the direction of damping LCL resonance. To keep an output voltage of the current controller on the average, the magnitude of the feedforward voltage before the edge of $v_{sn}{}^*$ should be compensated after the edge of $v_{sn}{}^*$. Proposed scheme is called as Edge Voltage Modification (EVM) algorithm hereafter.

Fig. 5 shows the conceptual relations between the ratio of an edge voltage, α_{peak}, and the magnitude of harmonic currents represented by THD$_i$ when other conditions are identical, where THD$_i$ means total harmonic distortion of i_{grid}. α_{opt}, optimal α_{peak} to minimize harmonic distortion at $v_{sn}{}^*$ edge, depends on the filter admittance and harmonic spectra of PWM. It is difficult to find an analytical solution of α_{opt} in real time due to heavy computational burden. Instead, a premade 1-D look-up table (LUT) can be employed to obtain α_{opt} according to MI when f_{grid} and f_{res} are fixed, i.e., $\alpha_{opt} = f(\text{MI})$, which can be extracted by an experimental test in advance.

However, the harmonic spectra of PWM is not only the function of MI and f_{sw} but also varied with operating conditions such as grid voltage and frequency. Likewise, the filter admittance is also affected by manufacturing tolerance and grid impedance. Therefore, a searching algorithm could be a practical solution to find α_{opt} in real time. As a result, instead of the premade LUT, EVM scheme can be easily implemented by combining the real time searching algorithm.

Meanwhile, active damping algorithm, e.g., filter capacitor current- or voltage- feedback based algorithm, is not sufficient to suppress the resonant problem incurred by DPWM because damping gain is limited by sampling noise and digital time delay. It only minimizes current harmonics after $v_{sn}{}^*$ edge moments have passed, which cannot eliminate the root cause of LCL resonance, i.e., v_{har} near f_{res}. However, it can be utilized to suppress current harmonics provoked by the other causes such as grid harmonics. If necessary, active damping algorithms

523

The 2018 International Power Electronics Conference

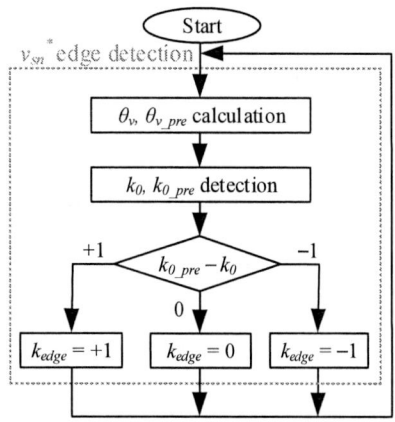

Fig. 6. Flowchart of $v_{sn}{}^{*}$ edge detection algorithm.

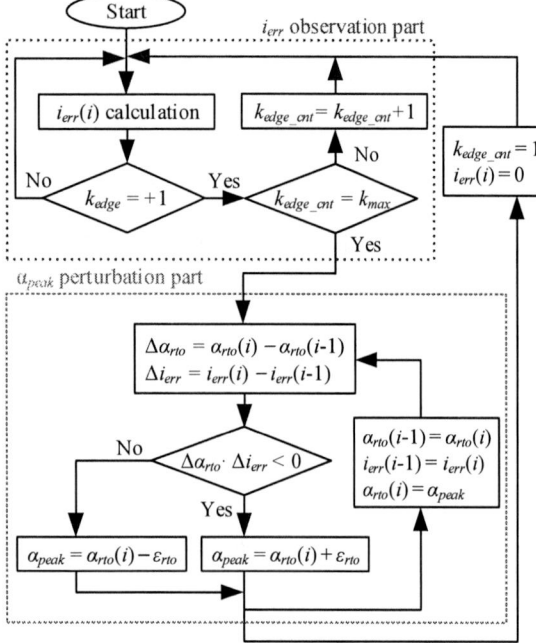

Fig. 7. Flowchart of proposed α_{peak} P&O algorithm.

can be used in conjunction with the proposed EVM in parallel. e.g., under a weak grid condition.

B. Implementation of Proposed Scheme

To implement the proposed scheme, $v_{sn}{}^{*}$ edge detection and α_{opt} searching algorithms should be devised.

Fig. 6 shows the flowchart of $v_{sn}{}^{*}$ edge detection algorithm, where $k_0 = 0$ means that $v_{sn}{}^{*}$ is set to $v_{sn,pos}{}^{*}$, otherwise, $k_0 = 1$ means that $v_{sn}{}^{*}$ is set to $v_{sn,neg}{}^{*}$. Firstly, voltage reference angles at present and next operating point, θ_v and θ_{v_pre}, are calculated by an arctangent function as (7), where θ_{v_pre} is predicted under the assumption of a steady-state condition. Based on the voltage reference angles, the each status of $v_{sn}{}^{*}[n]$ and $v_{sn}{}^{*}[n+1]$, equivalent to k_0 and k_{0_pre}, is decided. Positive and negative edge of $v_{sn}{}^{*}$ are distinguished by the difference of k_0 and k_{0_pre}, which is stored to k_{edge} variable. Consequently, k_{edge} is used to trigger the operation of EVM scheme and searching algorithm, where $k_{edge} = +1$

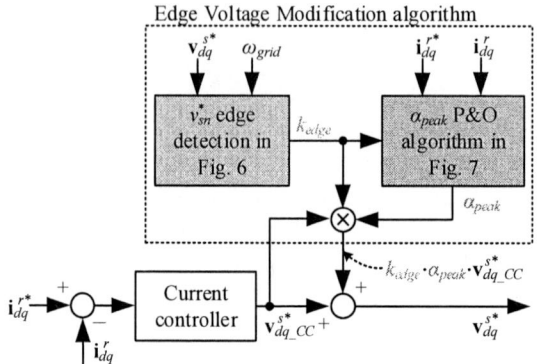

Fig. 8. Overall control block diagram of proposed algorithm.

means positive $v_{sn}{}^{*}$ edge, whereas $k_{edge} = -1$ means negative $v_{sn}{}^{*}$ edge.

$$
\begin{aligned}
\theta_v &= \mathrm{atan}2(v_{qs}^{s*}, v_{ds}^{s*}), \\
\theta_{v\_pre} &= \theta_v + \omega_{grid} T_{samp}.
\end{aligned}
\tag{7}
$$

α_{opt} searching algorithm is based on P&O method, also referred to as hill-climbing method [16]-[17], which is commonly used for MPPT in photovoltaic applications due to easy implementation. The basic principle is to perturb α_{peak} and observe a current error, i_{err}, where i_{err} is measured twice, and compared.

Fig. 7 shows the flowchart of α_{peak} P&O algorithm. Firstly, i_{err} has been extracted until the number of $v_{sn}{}^{*}$ edge is sufficient to neglect other disturbances, which is defined as i_{err} observation part. After that, the direction of α_{peak} is determined by comparing measured i_{err} in α_{peak} perturbation part. The principle of α_{peak} perturbation part is as follows: If measured i_{err}, $i_{err}(i)$, is less than i_{err} measured at one sample before, $i_{err}(i-1)$, α_{peak} keeps the present moving direction, $\Delta\alpha_{rto} \equiv \alpha_{rto}(i) - \alpha_{rto}(i-1)$. Otherwise, α_{peak} is adjusted in the opposite direction to $\Delta\alpha_{rto}$. It can be implemented shortly by comparing $\Delta\alpha_{rto}$ with Δi_{err} as (8), where ε_{rto} means a perturbation step size of α_{peak}. As a result, α_{peak} is updated and directly applied at next $v_{sn}{}^{*}$ edge. In parallel, α_{rto} and i_{err} at present operating point are stored, which is utilized during the next calculation step.

$$
\alpha_{peak} =
\begin{cases}
\alpha_{rto}(i) - \varepsilon_{rto}, & (\Delta\alpha_{rto} \cdot \Delta i_{err} < 0) \\
\alpha_{rto}(i) + \varepsilon_{rto}. & (\Delta\alpha_{rto} \cdot \Delta i_{err} \geq 0)
\end{cases}
\tag{8}
$$

Fig. 8 shows the overall control block diagram of the proposed algorithm. Firstly, k_{edge} is extracted by the $v_{sn}{}^{*}$ edge detection algorithm. Then, α_{peak} is calculated based on the P&O algorithm, where α_{peak} is updated when the number of k_{edge} is matched to k_{max}. k_{edge} and α_{peak} are utilized to modify the output voltages of the current controller, $v_{dq\_CC}^{s*}$. Finally, d-q voltage references, v_{dq}^{s*}, are revised as (9) after the EVM algorithm is applied.

$$
v_{dq}^{s*} = (1 + k_{edge}\alpha_{peak})v_{dq\_CC}^{s*}.
\tag{9}
$$

The tuning factors of the proposed algorithm are k_{max} and ε_{rto}. The disturbances such as inverter nonlinearities and grid harmonics influence on i_{err}, i.e., i_{err} observation part, which could provoke undesirable fluctuations of α_{peak}. In that case, k_{max} should be increased, which makes the P&O algorithm more robust to the disturbances.

524

The 2018 International Power Electronics Conference

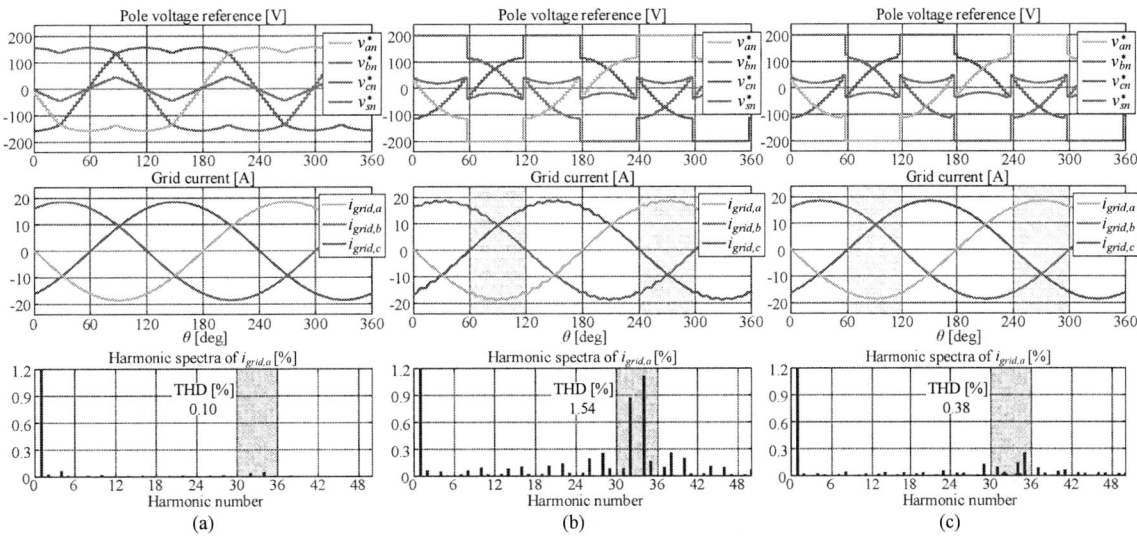

Fig. 9. Simulation 1: Waveforms and harmonic spectra of $i_{grid,a}$ when V_{dc} = 400 V. (a) CPWM. (b) DPWM without EVM. (c) DPWM with EVM.

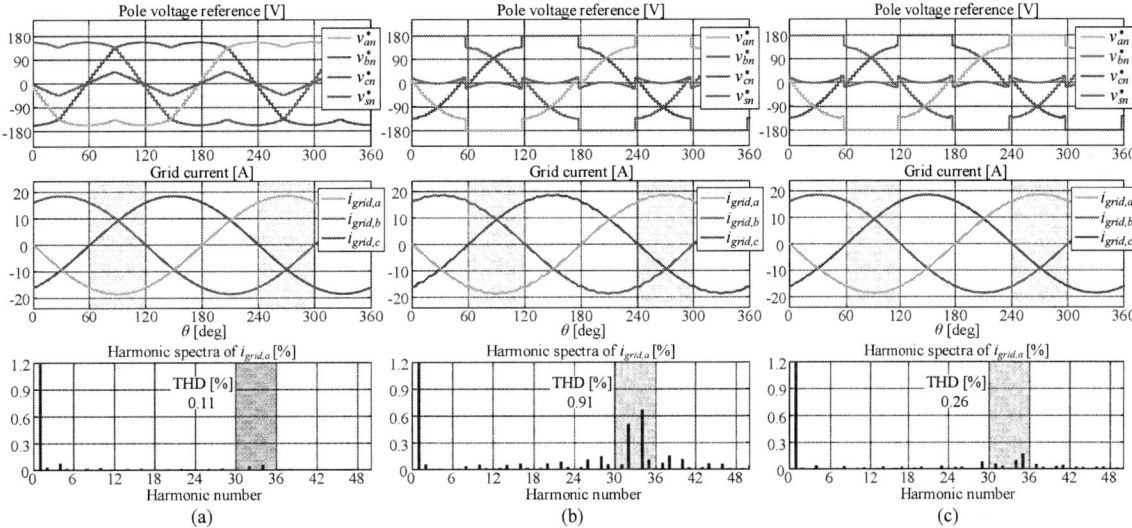

Fig. 10. Simulation 2: Waveforms and harmonic spectra of $i_{grid,a}$ when V_{dc} = 360 V. (a) CPWM. (b) DPWM without EVM. (c) DPWM with EVM.

However, the dynamic response is getting sluggish as k_{max} is getting larger. Likewise, small ε_{rto} minimizes oscillation of α_{peak}, and enhances the steady-state performance, while slows down the transient response. Thus, k_{max} and ε_{rto} should be adjusted in consideration of the tradeoff between dynamic responses in the transient and oscillatory responses in the steady-state. A current observer and dynamic step size may be used to improve the performances of the proposed algorithm further.

IV. SIMULATION AND EXPERIMENTAL RESULTS

A. Simulation Results

Simulation was carried out to evaluate the proposed scheme before experiments. The simulation results are valuable in that performances of the proposed scheme can be evaluated under ideal conditions without the effect of disturbances such as inverter nonlinearities and grid harmonics. 5 kW two-level inverter with *LCL* filter and three-phase grid with 5 % grid impedance as shown in

Fig. 1 are simulated in MATLAB/Simulink with PLECS where the system parameters are set as Table. I. When L_{grid} has 5 % per unit impedance, i.e., 0.8 mH, f_{res} is located at 2.03 kHz, near 34th harmonics. EVM controller gains, k_{max} and ε_{rto}, are set to 3 and 0.0005, respectively.

Fig. 9 shows the waveforms and harmonic spectra of a-phase grid current, $i_{grid,a}$, at rated current and unity power factor conditions, i.e., $P^* = 5$ kW and $Q^* = 0$ kW, where the dc-link voltage, V_{dc}, is fixed to 400 V. In Fig. 9(a), where CPWM is applied, there is no harmonic oscillation provoked by v_{sn}^* jumps. However, harmonic current near f_{res} arises in $i_{grid,a}$ when DPWM is applied as shown in Fig. 9(b) because $\mathbf{v}_{abcn}^*$ abruptly vary at v_{sn}^* edge moments. It can excess harmonic current limits near f_{res} band, e.g., IEEE 519. After applying the proposed EVM scheme, harmonics near f_{res} are obviously reduced as shown in Fig. 9(c), which satisfy harmonic regulation still keeping DPWM. It shows that effective f_{sw} can be reduced by adopting DPWM with the proposed EVM

525

The 2018 International Power Electronics Conference

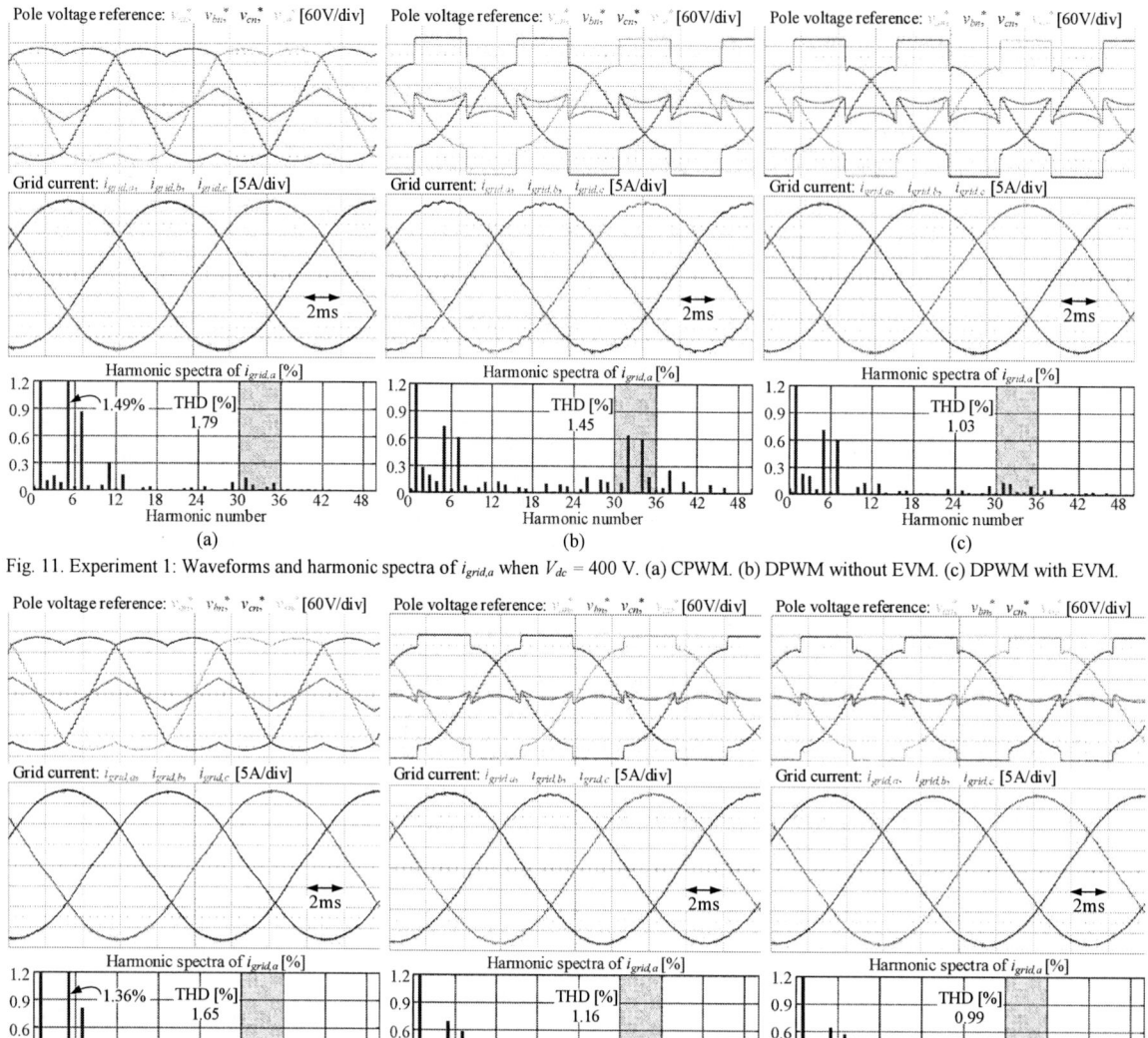

Fig. 11. Experiment 1: Waveforms and harmonic spectra of $i_{grid,a}$ when $V_{dc} = 400$ V. (a) CPWM. (b) DPWM without EVM. (c) DPWM with EVM.

Fig. 12. Experiment 2: Waveforms and harmonic spectra of $i_{grid,a}$ when $V_{dc} = 360$ V. (a) CPWM. (b) DPWM without EVM. (c) DPWM with EVM.

scheme when compared to CPWM under the same conditions. Above all, it should be noted that the switching loss of PWM converter with CPWM shown in Fig. 9(a) would be two times of that with DPWM in Fig. 9(c).

Fig. 10 shows the simulation results when P^*, Q^* and V_{dc} are set to 5 kW, 0 kW, and 360 V, respectively. Dc-link voltage has been decreased by 10 % when compared to the case of Fig. 9, which reduces the effects of v_{sn}^* jumps when DPWM is applied. Fig 10(a) shows that LCL resonance does not incurred when CPWM is applied as Fig. 9(a). However, harmonic oscillation occurs at v_{sn}^* edge moments when DPWM is applied, which is clearly shown in the harmonic spectra of $i_{grid,a}$ at Fig. 10(b). It reveals that harmonic regulation cannot be achieved under some v_{sn}^* jump without proper damping schemes. Applying the EVM scheme, current harmonics near f_{res} are remarkably reduced as shown in Fig. 10(c). LCL

resonance could be minimized by using the proposed α_{peak} P&O algorithm under various V_{dc} conditions.

B. Experimental Results

The proposed and conventional methods were also tested in a practical system. All control algorithms are implemented digitally in a digital signal processor (DSP), TMS320F28335. The system parameters of experimental setup are identical with those in the simulation. EVM controller gains, k_{max} and ε_{rto}, are the same as those of the simulation. The dead time is set to 2 μs and dc-link voltage is changed by a dc supply.

Fig. 11 shows the experimental results during steady-state operation under the same condition with those in Fig. 9. Contrast to simulation results, low-order harmonics such as 5th and 7th are provoked by the nonlinearities of inverters. In Fig. 11(a), where CPWM is applied, there is a few harmonics near f_{res} due to the smooth transition of

526

v_{sn}^*. In contrast, harmonic distortion occurs at the edge of v_{sn}^* when DPWM is applied as shown in Fig. 11(b), which is easily seen in the harmonic spectra of $i_{grid,a}$. Thanks to the proposed EVM scheme, harmonic oscillation is sufficiently suppressed by shifting $\mathbf{v}_{abcn}^*$ at the edge of v_{sn}^* as shown in Fig. 11(c). It has similar harmonic spectra near f_{res} compared to that of CPWM.

Fig. 12 shows the experimental results under the same condition with those in Fig. 10. Similar to the case of Fig. 11(a), harmonic oscillation is not provoked when CPWM is applied as shown in Fig. 12(a). On the contrary, harmonic distortion occurs at v_{sn}^* edge moments induced by DPWM as shown in Fig. 12(b). It could be inhibited by the current controller or an internal resistance of the filter, which is not enough to satisfy harmonic regulation without further damping methods. After applying the proposed EVM scheme, harmonic oscillation provoked by DPWM could be efficiently minimized as shown in Fig. 12(c). It satisfies harmonic regulation through simple modification of PWM algorithm in the software without sacrificing the system efficiency, e.g., without adding damping resistor or increasing effective switching frequency.

V. CONCLUSIONS

In this paper, *LCL* resonance problem incurred by DPWM has been analyzed by the frequency and time domain analysis. Also, the voltage reference modification scheme has been proposed to alleviate *LCL* resonance problem while keeping the advantages of DPWM scheme. By applying the proposed EVM method, the pole voltage references at the edge of offset voltage could be slightly revised to minimize the current oscillation. Finally, the simulation and experimental results are provided to verify the effectiveness of the proposed method. Through the proposed EVM algorithm, *LCL* resonance by DPWM has been conspicuously minimized, which decreases effective switching frequency of PWM converter as much as possible while satisfying grid harmonics regulation.

REFERENCES

[1] M. Liserre, F. Blaabjerg, and S. Hansen, "Design and Control of an *LCL*-Filter-Based Three-Phase Active Rectifier," *IEEE Trans. Ind. Appl.*, vol. 41, no. 5, pp. 1281–1291, Sep. 2005.

[2] K. Jalili and S. Bernet, "Design of *LCL* filters of active-front-end two-level voltage-source converters," *IEEE Trans. Ind. Electron.*, vol. 56, no. 5, pp. 1674–1689, 2009.

[3] R. Peña-Alzola, M. Liserre, F. Blaabjerg, R. Sebastián, J. Dannehl, and F. W. Fuchs, "Analysis of the Passive Damping Losses in *LCL*-Filter-Based Grid Converters," *IEEE Trans. Power Electron.*, vol. 28, no. 6, pp. 2642–2646, Jun. 2013.

[4] R. N. Beres, X. Wang, F. Blaabjerg, M. Liserre, and C. L. Bak, "Optimal Design of High-Order Passive-Damped Filters for Grid-Connected Applications," *IEEE Trans. Power Electron.*, vol. 31, no. 3, pp. 2083–2098, Mar. 2016.

[5] J. Dannehl, M. Liserre, and F. W. Fuchs, "Filter-Based Active Damping of Voltage Source Converters With *LCL* Filter," *IEEE Trans. Ind. Electron.*, vol. 58, no. 8, pp. 3623–3633, Aug. 2011..

[6] S. G. Parker, B. P. McGrath, and D. G. Holmes, "Regions of Active Damping Control for *LCL* Filters," *IEEE Trans. Ind. Appl.*, vol. 50, no. 1, pp. 424–432, Jan. 2014.

[7] V. Miskovic, V. Blasko, T. M. Jahns, A. H. C. Smith, and C. Romenesko, "Observer-Based Active Damping of *LCL* Resonance in Grid-Connected Voltage Source Converters," *IEEE Trans. Ind. Appl.*, vol. 50, no. 6, pp. 3977–3985, Nov. 2014.

[8] *IEEE Recommended Practices and Requirements for Harmonic in Electrical Power Systems*, IEEE Standard 519, 1992.

[9] *Technical Guideline Generating Plants Connected to the Medium-Voltage Network*, BDEW Standard, 2008.

[10] K.-B. Park, F. D. Kieferndorf, U. Drofenik, S. Pettersson, and F. Canales, "Weight Minimization of *LCL* Filters for High-Power Converters: Impact of PWM Method on Power Loss and Power Density," *IEEE Trans. Ind. Appl.*, vol. 53, no. 3, pp. 2282–2296, May 2017.

[11] J.-H. Park and K.-B. Lee, "Performance Improvement for Reduction of Resonance in a Grid-Connected Inverter System Using an Improved DPWM Method," *Energies*, vol. 11, no. 1, p. 113, Jan. 2018.

[12] F. Liu, K. Xin, and Y. Liu, "An adaptive Discontinuous Pulse Width Modulation (DPWM) method for three phase inverter," in *2017 IEEE Applied Power Electronics Conference and Exposition (APEC)*, 2017, pp. 1467–1472.

[13] D. G. Holmes and T. A. Lipo, *Pulse Width Modulation for Power Converters: Principles and Practice*. New York, NY, USA: Wiley, 2003.

[14] D.-W. Chung, J.-S. Kim, and S.-K. Sul, "Unified voltage modulation technique for real-time three-phase power conversion," *IEEE Trans. Ind. Appl.*, vol. 34, no. 2, pp. 374–380, 1998.

[15] D.-W. Chung and S.-K. Sul, "Minimum-loss strategy for three-phase PWM rectifier," *IEEE Trans. Ind. Electron.*, vol. 46, no. 3, pp. 517–526, Jun. 1999.

[16] T. Esram and P. L. Chapman, "Comparison of Photovoltaic Array Maximum Power Point Tracking Techniques," *IEEE Trans. Energy Convers.*, vol. 22, no. 2, pp. 439–449, Jun. 2007.

[17] M. A. G. De Brito, L. Galotto, L. P. Sampaio, G. De Azevedo Melo, and C. A. Canesin, "Evaluation of the main MPPT techniques for photovoltaic applications," *IEEE Trans. Ind. Electron.*, vol. 60, no. 3, pp. 1156–1167, 2013.

Parametric Robustness Analysis for Parallel Feedforward Compensation Based Active Damping of LCL Grid Connected Inverter

Muhammad Talib Faiz[1], Muhammad Mansoor Khan[2], Xu Jianming[3], Muhammad Ali[4], Houjun Tang[5]

[1,2,4,5] SEIEE, Shanghai Jiaotong University, Shanghai, China.

[3] Changzhou Power Supply Company, Changzhou, Jiangsu, China

[1] talib_faiz@sjtu.edu.cn

Abstract- **In this paper a Parallel Feedforward Compensation method is considered to actively damp the resonance of single phase grid interfaced inverter with LCL filter. In this method, a compensator is added with filter plant in parallel manners and sum of capacitor voltage and compensator output is feedback at reference input voltage. Further, the difference voltage signal is feed to proportional controller to complete the damping loop. A bandpass section (BPS) is used as a compensator in damping loop. This method is attractive owing to a simple and relatively easy design procedure for a compensator. The stability and robustness of PFC method is analyzed under compensator and LCL filter parameter variations through bode plot tool. A Proportional (PR) Controller is used to regulate injected grid. Finally, the robustness of PFC method is validated by MATLAB simulation results.**

Keywords- Active Damping, Grid Connected Inverter, LCL-Filter, Parallel Feedforward Compensation

I. INTRODUCTION

Renewable energy resources are integrated with grid through voltage source converter that results high frequencies harmonics. It requires a low pass filter to attenuate such harmonics to follow the grid interaction codes. A high order low pass LCL filter is preferred over simple L filter due to high attenuation characteristics and economic benefits [1]. However, inherent resonance phenomena associated with LCL filter can destabilize the system at certain harmonics frequencies. Active damping (AD) solutions are preferred to mitigate the resonance issue due to lower power losses as compared to passive damping methods that are simple to implement but high-power losses [2].

Several methods have been used in literature for active damping depends upon measurement of filter variable such as capacitor voltage, capacitor current or grid current [3], [4], [5]. Due to low cost of voltage sensor, the filter capacitor voltage damping technique is more economical for MW power application range. It needs a differential compensator in additional inner feedback damping loop [6]. However, the design procedure for differential compensator used in a feedback compensation loop is relatively complicated and more analytical and computational efforts are needed to achieve damping.

In this paper, a compensator is added across LCL filter following parallel feedforward compensation (PFC)

structure where a bandpass section is used as a compensator. This approach was initially introduced in simple adaptive control (SAC) [7], [8] to enhance system robustness against variation in system parameters. This robustness property of PFC method is exploited in active damping of LCL filter with sever filter parameter variations. In [9], this method is used to achieve damping by grid current feedback with hysteresis control for grid current regulation.

Firstly, the basics of PFC method is described in section II. Secondly, LCL filter modelling, design of compensator and current control loop for grid current regulation is discussed in Section III. Thirdly, the stability and robustness of PFC method with designed compensator is analyzed under compensator and filter parameter variation in section IV. Simulation results are demonstrated in section V to verify the robustness. Finally, the conclusion is given in section VI.

II. REVIEW OF PARALLEL FEEDFORWARD COMPENSATION METHOD

Parallel Feedforward compensation configuration is used to transform a non-minimum phase or poorly minimum-phase system into a minimum phase system to enhance system stability [7]. This transformation is achieved by adding a compensator parallel with considered plant and feedback at reference signal. A high proportional controller is used to achieve required stability characteristics as shown in fig. 1 (a).

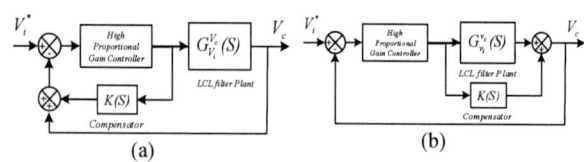

Fig. 1. Proposed control method (a) Parallel Feedforward Compensation (b) Equivalent control diagram

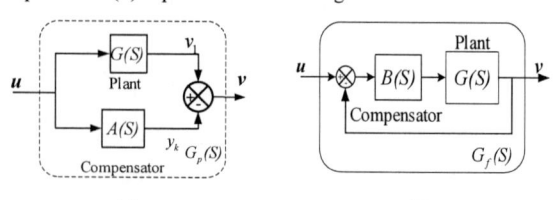

Fig. 2. Duality Property (a) Parallel feedforward compensation (b) Negative Feedback compensation

A LCL filter transfer function (TF) between capacitor voltage and inverter output voltage is considered as plant G(S) here. Fig. 1(b) is equivalent to fig. 1(a) from stability point of view where the output of closed loop is sum of capacitor voltage and compensator output.

This method adds zero when a compensator is placed in parallel with plant and results the poles shifting of compensated system towards left side when feedback loop is completed. This is equivalent to negative feedback compensated to achieve the damping of LCL filter plant and enhance system stability. It can be realized by duality property existing between parallel feedforward compensation method and negative feedback compensation method shown in fig. 2 [10]. The mathematical realization of duality property is given as follows.

Assume a strictly proper transfer function (TF) $G(s) = \frac{n_G}{d_G}$ for plant and $A(S) = \frac{n_A}{d_A}, B(S) = \frac{n_B}{d_B}$ for compensator used in fig. 2(a)-(b). Parallel Feedforward compensated system can be formulated as following:

$$G_p(S) = G(S) + A(S) = \frac{n_G}{d_G} + \frac{n_A}{d_A} = \frac{n_G d_A + n_A d_G}{d_A d_G} \quad (1)$$

Negative feedback compensated system can be calculated as below:

$$G_f(S) = \frac{V(S)}{U(S)} = \frac{B(S)G(S)}{1 + B(S)G(S)}$$

$$= \frac{\frac{n_B}{d_B} \cdot \frac{n_G}{d_G}}{1 + \frac{n_B}{d_B} \cdot \frac{n_G}{d_G}} = \frac{n_B n_G}{n_B n_G + d_B d_G} \quad (2)$$

If $A(S) = B^{-1}(S)$, then numerator of $G_p(S)$ in equation (1) becomes equal to denominator of $G_f(S)$ in equation (2). This is the fundamental of duality property. This property helps to convert an undamped system into a damped system. However, compensator design need special attention to have relative degree one in a way that $G_p(S) = G(S) + A(S)$ should be minimum-phase system to achieve asymptotically stable system.

III. SINGLE PHASE GRID CONNECTED INVERTER MODELLING AND CONTROL STRUCTURE DESIGN

A single-phase, H bridge voltage source inverter connected with grid is given in fig. 3 that contains input DC capacitor, conversion unit and LCL filter section. The equivalent series resistances of filter inductors are ignored for worst case. The resonance frequency f_{res} of LCL filter can be computed from following relation.

$$f_{res} = \frac{1}{2\pi} \sqrt{\frac{L_1 + L_2}{L_1 L_2 C}} \quad (3)$$

The linearized averaged model of inverter control structure with PFC method is depicted in fig. 4. It

includes inner damping loop and outer current control loop. The damping loop is formed by sum of filter capacitor voltage and compensator output and feedback at inverter reference voltage. The difference signal of current controller o/p and a feedback damping signal is feed to high proportional gain. Since, the compensator bandpass section K_{BPS} is added across the conversion unit and filter plant, that's why, this structure is called parallel feedforward compensation (PFC) method. A proportional resonant controller G_{PR} is used to regulate grid current error in current control loop. The reference grid current I_2^* is calculated from voltage control loop that is not discussed here. By simplifying the control structure, the grid injected current I_2 can be computed from equation (4) as below:

$$i_2(s) = \frac{P G_{PR}(S).M(S).N(S)}{1 + P[K_{BPS}(S) + M(S) + G_{PR}(S).M(S).N(S)]} i_2^*(s)$$

$$- \frac{N(S).[1 + P(K_{BPS}(S) + M(S))]}{1 + P[K_{BPS}(S) + M(S) + G_{PR}(S).M(S).N(S)]} v_g(s) \quad (4)$$

Where,

$$M(S) = \frac{G_d(S).G_{inv}}{L_1 C S^2 + 1}, \quad N(S) = \frac{L_1 C S^2 + 1}{C L_1 L_2 S^3 + (L_1 + L_2)S}$$

The capacitor voltage measured for damping loop can be calculated as follows:

$$v_c(s) = \frac{P M(S)}{1 + P(K_{BPS}(S) + M(S))} \Delta v \quad (5)$$

It is revealed from equation 5 that damping can be achieved using filter capacitor voltage by appropriate selection of compensator $K_{BPS}(S)$.

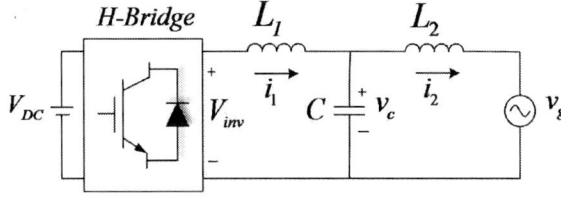

Fig. 3. Schematic Diagram of LCL type Single Phase Grid Connected Voltage Source Inverter

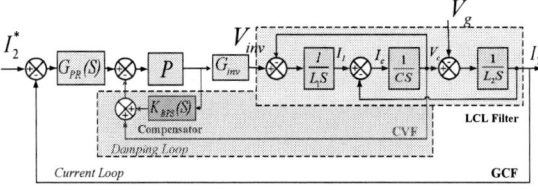

Fig. 4. Control Structure of Inverter with proposed damping Method

A. Compensator Design

A BPS is employed to provide an alternate path for specific frequency range around resonance frequency according to damping characteristics requirements. It is a best choice due to low order compensator that offers better control performance and easy design procedure. The transfer function (TF) of BPS is given by equation (6) as following:

$$K_{BPS}(S) = \frac{\dfrac{\omega_c}{Q}S}{S^2 + \dfrac{\omega_c}{Q}S + \omega_c^2} \qquad (6)$$

Where 'Q' is quality factor and 'ω_c' is the center frequency of bandpass section. The frequency response of BPS is given in fig. 5(a) where it has maximum magnitude in magnitude plot at the center frequency of BPS and 90^0 phase shift in phase plot of bode plots. When BPS is added in parallel with filter system it provides alternate path to damp the resonance frequency range. This damping ability is influenced by quality factor 'Q' parameter of BPS. For higher value of 'Q', the bandwidth of BPS decreased, and sharpness of curve increased as shown in fig. 5(b).

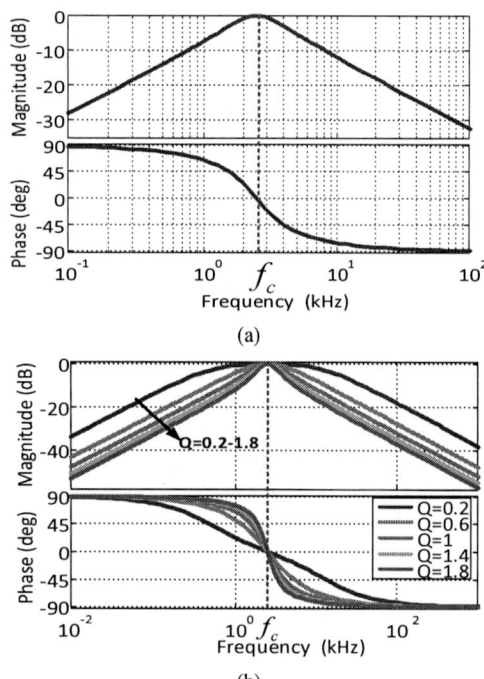

(a)

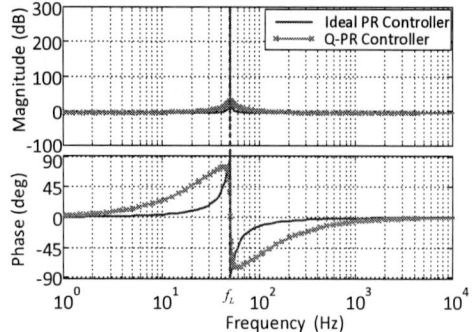

(b)

Fig. 5. Frequency response for (a) BPS (b) BPS with varying quality factor 'Q'

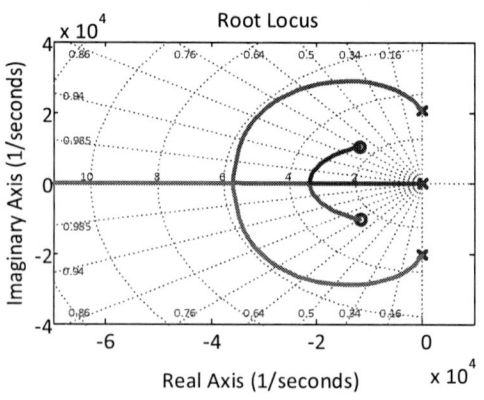

Fig. 6. Frequency Response for Proportional Resonant controller

B. Current Regulation

PR controller is preferred for sinusoidal grid current regulation over a Proportional Integral controller in current control loop for high control performance. It offers highest

gain at line frequency that eliminate the steady state error and rise and settling time are also reduced. Moreover, it is easy to implement and need less signal processing work. Owing to such advantages, a PR control is used for grid current regulation in current control loop. However, realization of ideal PR controller is difficult due to lossless system and Quasi PR controller is proposed in practical systems. [11].The mathematical form of Q-PR controller is given by equation (7):

$$G_{PR}(S) = K_p + \frac{2K_r\omega_{PR}S}{S^2 + 2\omega_{PR}S + \omega_L^2} \qquad (7)$$

Where 'K_p' is proportional gain, 'K_r' is resonant gain and 'ω_{PR}' is bandwidth of controller. The bode plot for simple PR controller and Quasi PR controller is given in fig. 6. It revealed that the magnitude plot for Q-PR controller shows a small output gain for resonant frequency region in comparison with PR controller that lessen the sensitivity level against grid current frequency variation.

IV. STABILITY ANALYSIS

Fig. 7(a)-(b) shows the root-locus of open damping loop without compensator and with BPS compensator. A well damped system is achieved by adding one imaginary zero to LHS and a pole at origin for desired control performance as shown in figure 7(b).

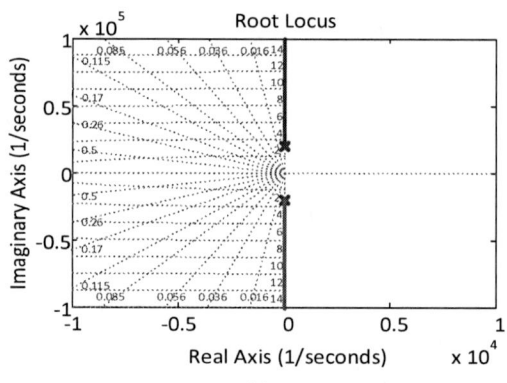

(a)

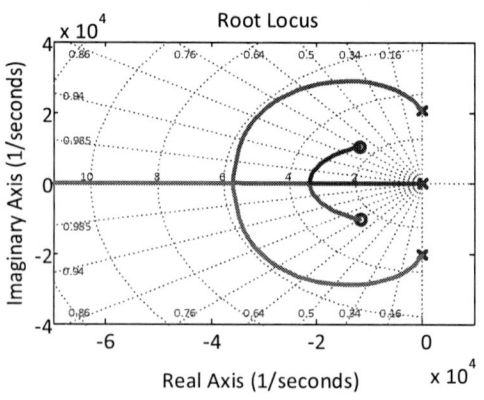

(b)

Fig. 7. Root Locus of open damping loop of LCL filter (a)Without Compensation (b) with parallel feedforward Compensation

A. Compensator Parameter Variations

A current loop is design by using PR controller for current regulation followed by damping loop. It is necessary to analyze the impact of BPS compensator parameter variation on current control loop to get a specific range for inverter secure operation. BPS has two parameters i.e. quality factor 'Q' and a center frequency 'f_c'. The range of these BPS parameters is defined by bode plot to ensure a well damped and stable operation of system. Fig. 8(a)-(b) show the bode plot of current control loop of compensated system for different values of quality factor and center frequency while stability margins (i.e. phase margin and gain margin) are tabled in table I.

The minimum value considered for phase margin is 30° and for gain margin is 6dB for better control performance and system robustness. By following this criterion, the value of quality factor 'Q' should not be less than '0.55' and the center frequency should not exceed than '2.9kHz'. These values can vary across a range according to specific control performance requirement. As the value of Q increases, the stability margins and bandwidth (BW) increases while reducing the phase disturbance around line frequency 'f_L' and vice versa. On the other hand, it reduces attenuation capability of high frequency ripples. The center frequency of BPS should always less than the resonance frequency 'f_{res}'. The stability margins of current loop decreased as the center frequency increases and approach towards resonance frequency with slightly increment in BW. Frequency response of current loop with varying 'f_c' is shown in figure 14(b) when system becomes unstable at '$f_c = f_{res}$'.

Fig. 8. Bode plot of open current loop with compensator parameter variation(a) With Quality Factor Q (b) With Center frequency fc of BPS

TABLE I
STABILITY MARGINS WITH BPS PARAMETER VARIATIONS

Sr.No.	Q	GM	PM	BW	fc(kHz)	GM	PM	BW
1	0.15	Unstable System			2.0	16.7	35.7	432
2	0.45	3.27	3.74	396	2.3	13.4	37.9	450
3	0.75	8.53	40.2	479	2.6	10	39.9	476
4	1.05	12.1	45.8	544	2.9	5.9	41.7	496
5	1.5	17	52.1	604	3.5	Unstable System		

B. LCL Filter Parameter Variations

The Stability and robustness of PFC method is tested under variation in LCL filter parameters as given in fig. 8(a)-(d). Frequency response of current control loop show the effective damping and good stability margins at rated filter parameters as depicted by bode plot in fig. 9(a). PFC method shows good control performance and stability margins with ±50% variation in filters inductance either inverter-side or grid side which is shown in fig. 9 (b) and (c). Even the system is stable for any filter inductor value < ±100%. However, control performance and system stability are reduced in grid side inductor variation. The high frequency harmonics attenuation capability is also reduced considerably in both cases for such high variation. Stability margins and bandwidth with varying filter inductance values is given in table II. The robustness can be increased by reducing the center frequency or bandwidth of BPS on cost of low dynamic performance. Current control loop is also stable for variation in capacitor value upto ±25% as can be observed in fig. 9(d) and stability margins are given in table III.

These observations show an effective resonance damping, better control performance with the excellent stability and robustness of a capacitor voltage damping with PFC configurations and designed compensator.

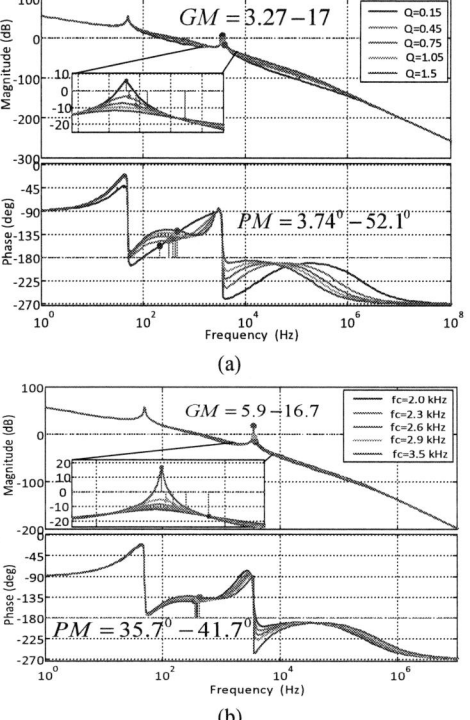

(a)

(b)

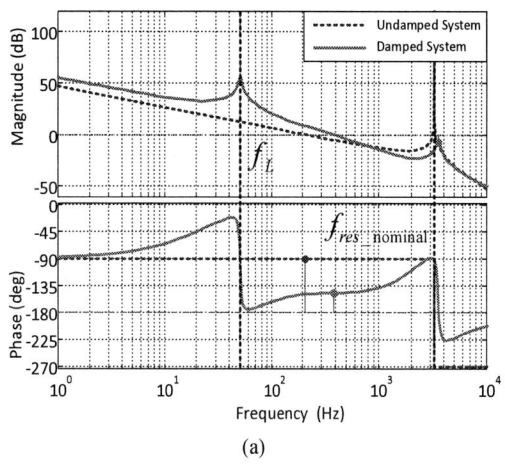

(a)

The 2018 International Power Electronics Conference

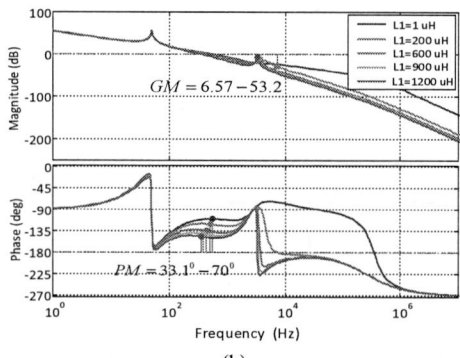

(b)

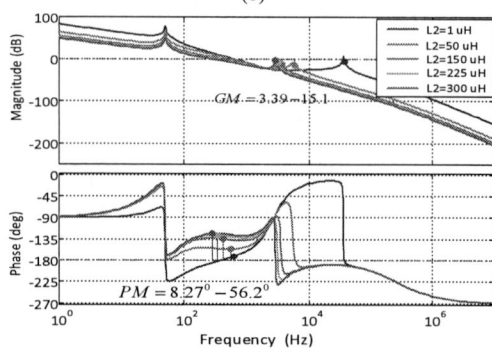

(c)

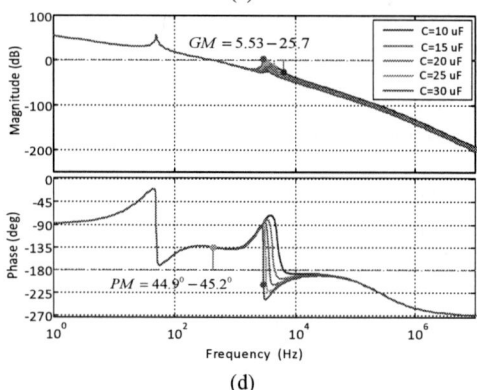

(d)

Fig. 9. Frequency response with BPS (a)Undamped and damped system (b) Variation in L1<L1 ±100% at L2=150uH, C=20uF (c) Variation in L2 <L2±100% at L1=600uH, C=20uF (d) Variation in C ±50% at L1=600uH, L2=150uH.

TABLE II
STABILITY MARGINS WITH VARIATION IN INVERTER-SIDE AND GRID-SIDE INDUCTANCE

Sr.No.	L1(uH)	GM	PM	BW	L2(uH)	GM	PM	BW
1	1	53.2	70	735	1	6.02	8.27	730
2	200	28.2	59.2	651	50	15.1	23.9	674
3	600	11.6	45.1	537	150	11.6	45.1	537
4	900	8.46	38.2	475	225	8.23	53	445
5	1200	6.57	33.1	437	300	3.39	56.2	368

TABLE III
STABILITY MARGINS WITH VARIATION IN FILTER CAPACITOR

Sr.No.	L1(uF)	GM	PM	BW
1	10	25.7	44.9	735
2	15	17.7	45.0	651
3	20	11.6	45.1	537

4	25	5.53	45.2	475
5	30		Unstable System	

V. SIMULATION VERIFICATION

A MATLAB Simulink model for single phase LCL filtered grid connected inverter with PFC damping loop is developed to verify the stability and robustness. Equivalent series resistance of inductor and control loop delay are ignored while ideal IGBT switches are assumed. The SPWM technique is used for gating signal generation. The detail simulation parameters [3] are given in Table 1.

TABLE I
SYSTEM SIMULATION PARAMETERS

Parameter	Symbol	Value
DC-Link voltage	V_{DC}	500V
Grid Voltage	V_g	220V
Output Power	P_o	6kW
Filter parameters	L_1, L_2, C	0.6 mH, 0.15mH, 20uF
Line Frequency	f_L	50 Hz
Switching Freq.	f_{sw}	10 kHz
Sampling Freq.	f_s	10 kHz
Resonance Freq.	f_{res}	3.25 kHz

As fig. 10(a)-(b) indicate the steady-state response of a system with and without damping loop where voltage at point of common coupling and grid current are in phase. It can be observed that the resonance frequency harmonics is eliminated effectively with proposed damping method. The transient response of system is analyzed under 50% change in reference current given in fig. 11 where grid injected current follows the reference current accordingly. The current overshoot is less than 10% that can be reduced by increasing center frequency of BPS compensator. However, it will reduce the robustness of proposed PFC method against increment in filter parameters as it reduces the resonance frequency.

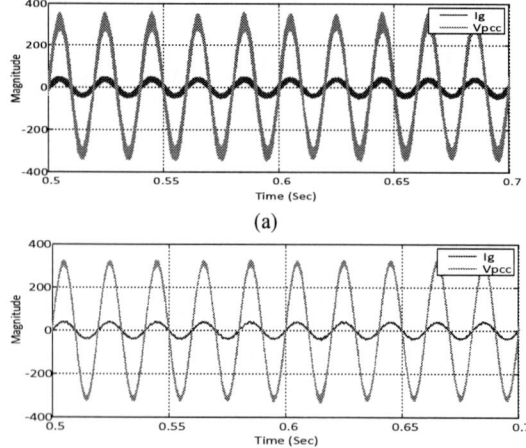

(a)

(b)

Fig. 10. Steady-state response (a) Without damping loop(b) With PFC damping method

532

The 2018 International Power Electronics Conference

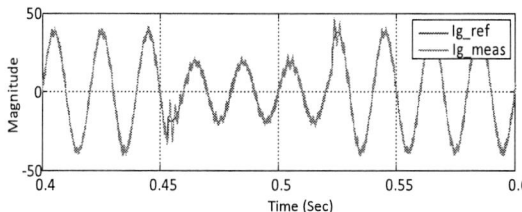

Fig. 11. Transient response of Inverter Under change in reference current PFC method

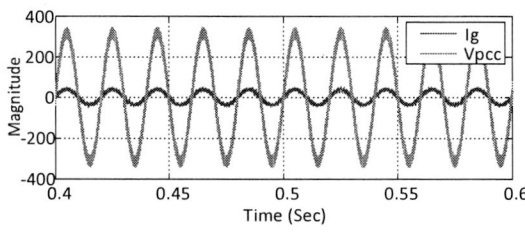

(a)

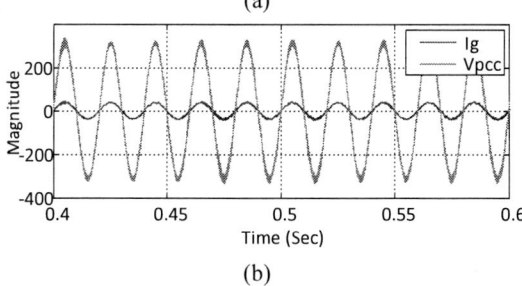

(b)

Fig. 12. Grid injected voltage and current under LCL filter variation (a) 50%L1, 50%L2, 80%C (b) 150%L1, 150%L2, C

The robustness of proposed method against filter parameter variations is shown in Fig. 12 (a)-(b), where the system is stable with reduced high harmonics attenuation capability.

VI. CONCLUSION

A Parallel Feedforward Compensation (PFC) method is developed for active damping of resonance issue in single phase LCL grid connected inverter. The bandpass section (BPS) is designed as a compensator in analytical way and connected across LCL filter plant in parallel feed forward compensation manners. The stability and control performance are analyzed of current control loop with proposed design method. The stability plots show effective damping with good dynamic response. The robustness of considered method is examined under compensator and filter parameter variation that shows a good robustness of proposed method. Lastly, the damping effectiveness and robustness of PFC method is validated by simulation results.

REFERENCES

[1] F. Blaabjerg, R. Teodorescu, S. Member, M. Liserre, A. V Timbus, and S. Member, "Overview of Control and Grid Synchronization for Distributed Power Generation Systems," vol. 53, no. 5, pp. 1398–1409, 2006.

[2] R. Peña-Alzola, M. Liserre, F. Blaabjerg, R. Sebastían, J. Dannehl, and F. W. Fuchs, "Analysis of the passive damping losses in lcl-filter-based grid converters," *IEEE Trans. Power Electron.*, vol. 28, no. 6, pp. 2642–2646, 2013.

[3] D. Pan, X. Ruan, C. Bao, W. Li, and X. Wang, "Capacitor-current-

feedback active damping with reduced computation delay for improving robustness of LCL-type grid-connected inverter," *IEEE Trans. Power Electron.*, vol. 29, no. 7, pp. 3414–3427, 2014.

[4] X. Wang, F. Blaabjerg, and P. C. Loh, "Virtual RC Damping of LCL -Filtered Voltage Source Harmonic Compensation," *IEEE Trans. Power Deliv.*, vol. 30, no. 9, pp. 4726–4737, 2015.

[5] C. C. Gomes, A. F. Cupertino, and H. A. Pereira, "Damping techniques for grid-connected voltage source converters based on LCL filter: An overview," *Renew. Sustain. Energy Rev.*, vol. 81, no. January 2017, pp. 116–135, 2018.

[6] Z. Xin, P. C. Loh, X. Wang, F. Blaabjerg, and Y. Tang, "Highly Accurate Derivatives for LCL-Filtered Grid Converter With Capacitor Voltage Active Damping," *IEEE Trans. Power Electron.*, vol. 31, no. 5, pp. 3612–3625, 2016.

[7] I. Bar-Kana, "On Parallel Feedforward and Simplified Adaptive Control," in *IFAC Adaptive Systems in Control and Signal Processing*, 1986, pp. 99–104.

[8] Z. Iwai, I. Mizumoto, and Mingcong Deng, "A parallel feedforward compensator virtually realizing almost strictly positive real plant," *Proc. 1994 33rd IEEE Conf. Decis. Control*, vol. 3, no. December, pp. 2827–2832, 1994.

[9] G. R. C.Santiago, "Hysteretic control of grid-side current for a single-phase LCL grid-connected voltage source converter," *Math. Comput. Simul.*, vol. 130, pp. 194–211, 2016.

[10] I. Rusnak and I. Barkana, "The duality of parallel feedforward and negative feedback," in *2012 IEEE 27th Convention of Electrical and Electronics Engineers in Israel, IEEEI 2012*, 2012, pp. 1–4.

[11] N. Bianchi and M. Dai Pre, "Proportional-resonant controllers and filters for grid-connected voltage-source converters," *IEE Proceedings-Electric Power Appl.*, vol. 150, no. 2, pp. 750–762, 2006.

The 2018 International Power Electronics Conference

Open-loop-based Island-mode Voltage Control Method for Single-phase Grid-tied Inverter with Minimized LC Filter

Satoshi Nagai and Jun-ichi Itoh
Department of Energy and Environment Science Engineering,
Nagaoka University of technology, Niigata, Japan
*E-mail: satoshi_nagai@stn.nagaokaut.ac.jp, itoh@vos.nagaokaut.ac.jp

Abstract— This paper proposes an island-mode voltage control method by using an open-loop control applying a high-gain disturbance observer (DOB) for a single-phase inverter with a low capacitance output filter. The output voltage is distorted with the low capacitance by the conventional method which consists of a PID regulator for an automatic voltage regulator (AVR) and a PI regulator for an automatic current regulator (ACR). In order to compensate the output voltage distortion, the high-gain DOB for the output voltage is applied in the open-loop-based voltage control. DOB is implemented into a field-programmable gate array (FPGA) with high-speed sampling. In the LC filter of a 1-kW prototype, the impedance of the inductor is minimized to 1.0% of the normalized inverter impedance, whereas the admittance of the capacitor is reduced to 0.25% of the normalized inverter admittance. By the proposed method, the inverter output voltage total harmonic distortion (THD) is reduced by 82.4% even with the diode rectifier load compared to the conventional voltage control, whereas the constant output voltage is achieved regardless of any conditions of load.

Keywords— Grid-tied inverter, Island-mode, Minimized LC filter, Open-loop-based voltage control

I. INTRODUCTION

In recent years, grid-tied inverters, e.g. photovoltaic systems, fuel cell systems, and wind turbine systems, have been actively studied for energy saving [1]-[3]. The inverter is required to have a small volume in order to increase the power density of the system [4]-[5]. Generally, interconnected inductors occupy a majority of the inverter size. Therefore, the interconnected inductor is highly required to reduce the size. By reducing the inductance, it is possible to reduce the volume of the inductor. A high switching frequency with SiC or GaN devices is applied in term of a same current ripple in order to reduce the inductance of the interconnected inductor. Moreover, the capacitance of the filter capacitor is possible to reduce due to a filter design of high cutoff frequency. Therefore, it is possible to apply the high resonance frequency output filter for the inverter by increasing the switching frequency. On the other hand, the inverter output current overshoot rate becomes high during the voltage sag when the inductance of the interconnected inductor is reduced. Thus, the authors propose the high-speed gate-block method and the design method of LC or LCL filter in order to meet the fault-ride-through (FRT) requirements [6]-[7]. In particular, in order to meet the FRT requirements, a filter capacitor with low capacitance is desirably designed to reduce the resonance.

Meanwhile, when a black out in a long period that is exceeding the FRT requirements occurs, the grid-tied inverter is required to perform the island mode operation in order to supply the power to the load [8]-[10]. In order to output the voltage, it is necessary to implement the voltage controller for the inverter. However, the disturbance suppression performance of the voltage controller worsens due to the low capacitance of the filter capacitor in the minimized LC filter. The inverter output voltage is distorted when the island mode operation with the low disturbance suppression performance is connected to nonlinear-load such as a rectifier load. As one of the solution for this problem, the control response of the voltage controller and the current controller are increased in order to enhance the disturbance suppression performance. However, the increase in the control response is limited to the detection delay time and the sampling frequency. Moreover, in order to improve the output voltage distortion, the voltage control methods applied in such as Uninterruptible Power Supply (UPS) systems are proposed in [11]-[12]. The operation of UPS is not necessary to consider the FRT operation. Thus, the filter capacitor of the output filter is possible to increase the capacitance. The voltage control methods of [11]-[12] are not considered with the output filter of the low capacitance. Therefore, in the grid-tied inverter, the voltage control with the low capacitance has to be considered. Moreover, it is necessary to implement a complex control or sampling method for the control methods of [11]-[12].

This paper proposes new open-loop-based island-mode voltage control method equipped with a high-gain disturbance suppression in order to achieve a low total

534

harmonic distortion (THD) of the output voltage. The originality in this paper is that the island-mode voltage control with the minimized LC filter is possible to compensate an output error voltage for the load current without a current feedback loop. Moreover, it is not necessary to consider the interference between the voltage controller and the current controller. This control is defined as the VOLTAGE DISTURBANCE COMPENSATION OPEN-LOOP CONTROL (VDCOLC) in this paper. The proposed VDCOLC compensates the load current disturbance by using the output voltage detection value to calculate the disturbance through a high-gain disturbance observer (DOB) [13]-[15]. In order to compensate the disturbance with the high-speed sampling, the high-gain DOB is implemented in field-programmable gate array (FPGA). By using this proposed VDCOLC, a reduction of THD for the inverter output voltage is confirmed with a 1-kW prototype.

II. PROBLEM OF ISLAND-MODE VOLTAGE CONTROL WITH MINIMIZED LC FILTER

Figure 1 shows a circuit configuration of a single-phase grid-tied inverter with an LC filter. In this paper, an H-bridge single-phase two-level inverter is employed due to its simplicity. In order to reduce the volume, the LC filter is minimized. In the island mode operation, the output voltage is controlled by configuring the voltage controller with the filter capacitor of the output LC filter as the control object. The disturbance gain of the output current i_{out} $G_D(s)$ when the voltage controller is constructed by PI controller is expressed as

$$G_D(s) = \frac{s}{s^2 C_f + sK_p + \frac{K_p}{T_i}} \quad (1),$$

where C_f is the capacitance of the filter capacitor, K_p is the proportional gain, T_i is the integral time, s is the Laplace operator, and the current control gain is one. The disturbance suppression performance worsens due to the increase of the disturbance gain, when the capacitance of the filter capacitor C_f is decreased in (1). Thus, it is necessary to improve the disturbance suppression performance, in order to achieve the island mode operation with the minimized LC filter.

III. OPEN-LOOP-BASED VOLTAGE CONTROL FOR ISLAND MODE OPERATION

A. Current Detection Problem of Filter Capacitor

Figure 2 shows a conventional closed-loop-based island-mode voltage control with current controller composed by PI regulator. In general, an island-mode voltage control is carried out by a voltage controller that has inner loop of the current controller. Note that, the disturbance suppression performance is improved by using the filter capacitor current control [16]. The disturbance for the voltage controller such as the load current is included in the filter capacitor current. Thus, the disturbance compensation for the output voltage controller is not

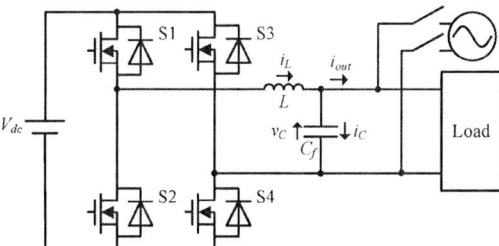

Fig. 1. Circuit diagram for island mode. In this paper, the output LC filter is minimized.

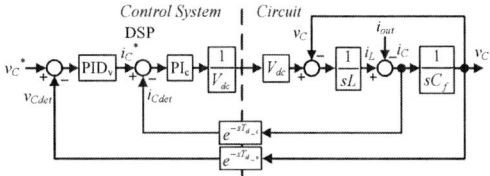

Fig. 2. Control block diagram of conventional closed-loop-based voltage control for island mode operation. In order to improve the disturbance suppression performance, the control response has to be increased.

necessary by the filter capacitor current feedback control. Moreover, it is necessary to increase the control response, which is limited due to the controller hardware, in order to achieve the island mode with the minimized LC filter.

Figure 3 shows the filter capacitor current detection value under the same output voltage with different detection delay time of the filter capacitor current, where the impedance of the interconnected inductor is 1.0% of the normalized inverter impedance, the admittance of the filter capacitor is 0.25% of the normalized inverter admittance in the 1-kW system. The detection of the filter capacitor average current for one switching period is required as shown in Fig. 2. However, the filter capacitor current has a large current ripple when the low capacitance of the filter capacitor is applied. Because the switching frequency becomes higher than the fundamental frequency component of the output current. Thus, it is impossible to detect the filter capacitor average current per the switching period due to the current detection delay time. In Fig. 3 (a) and (b), it is possible to detect the average filter capacitor current as a sinusoidal wave, when the detection delay time is 0 s or half period of carrier (6.25 μs). However, the detection point of the output current is different from the average value, i.e. the filter capacitor current detection value becomes non-sinusoidal wave, when the detection delay time of the filter capacitor current becomes longer as shown in Fig. 3 (c) as 10 μs. Thus, it is impossible to detect the filter capacitor average current due to the above problem. As a result, the inverter output voltage THD becomes worse. It is necessary to apply the multi-rate sampling for the average capacitor current detection [11]-[12], in order to solve above problem.

B. Proposed Open-loop-based Island-mode Voltage Control

The resonance frequency of the filter is shifted to the high frequency range when the LC filter is minimized.

The 2018 International Power Electronics Conference

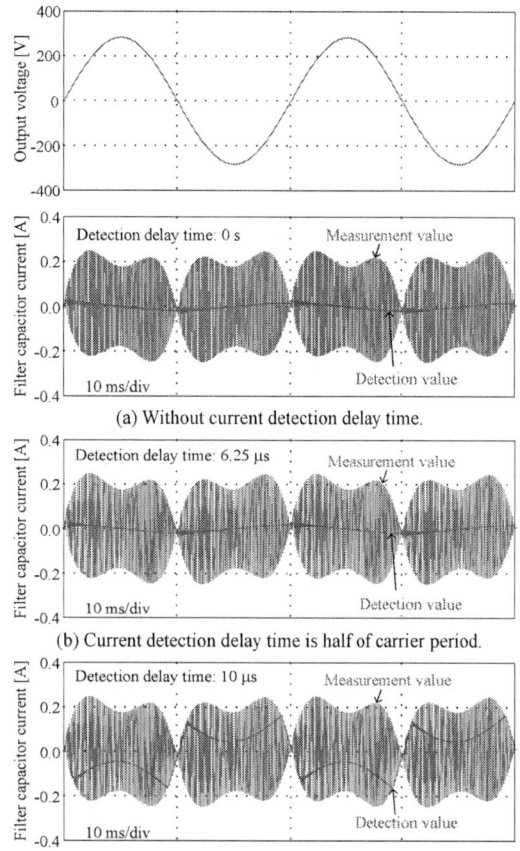

(a) Without current detection delay time.

(b) Current detection delay time is half of carrier period.

(c) Current detection delay time is not half of carrier period.

Fig. 3. Comparison of filter capacitor current detection with different current detection delay time.

The transfer function for the output voltage command with the open-loop operation $G_{com\_OL}(s)$ is expressed as

$$G_{com\_OL}(s) = \frac{1}{s^2 LC_f + 1} \qquad (2).$$

The transfer function becomes wide bandwidth for the output voltage command v_c^* with the minimized LC filter by (2). Therefore, an open-loop-based voltage control can be employed without the occurrence of the output voltage resonance. Nevertheless, the output voltage with the open-loop-based voltage control varies highly dependently on the load, because the output voltage is not feedback and is not regulated by the voltage controller. Furthermore, odd number harmonic components occur in the output voltage due to the nonlinear output current, when a diode rectifier load is connected to the inverter. Therefore, it is necessary to compensate the disturbance of the output current in order to obtain the stable and constant output voltage regardless of the load.

Figure 4 shows the conventional open-loop-based island-mode voltage control with the typical dead-time compensation. The output voltage error due to the dead-time error voltage is compensated by using this method. However, the harmonic components due to the nonlinear-load such as the diode rectifier cannot be compensated.

Figure 5 shows the proposed VDCOLC. In order to achieve the high-gain compensation, the compensator is

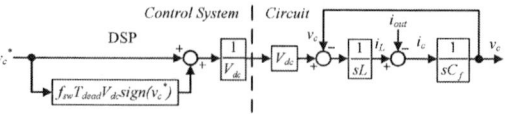

Fig. 4. Control block diagram of conventional open-loop-based island-mode voltage control for island mode operation.

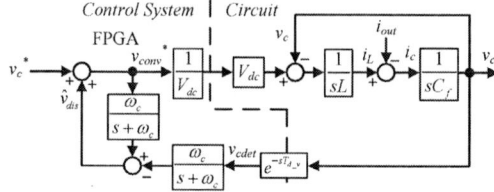

Fig. 5. Control block diagram of proposed VDCOLC for island mode operation. In order to compensate output voltage in the high-gain, the DOB is implemented in FPGA.

implemented by FPGA. Moreover, the compensator is operated as a DOB that compensates the disturbance due to the output current i_{out}. The disturbance estimation value $\hat{v}_{dis}$ is expressed by

$$\hat{v}_{dis} = \frac{\omega_c}{s + \omega_c} v_{conv}^* - \frac{\omega_c}{s + \omega_c} v_{cdet} \qquad (3),$$

where ω_c is the cutoff angular frequency of the DOB, v_{conv}^* is the voltage command after the disturbance compensation, and v_{cdet} is the detection value of the filter capacitor voltage.

C. Appling Area of Open-loop-based Voltage Control for Island Mode

In the disturbance compensator of (3), the Low-pass-filter (LPF) is implemented into DOB in order to suppress the resonance components of the LC filter for the voltage command in the voltage controller. The resonance frequency of the LC filter f_{res_LC} is expressed as

$$f_{res\_LC} = \frac{1}{2\pi \sqrt{LC_f}} \qquad (4).$$

In order to reduce the resonance component of the LC filter by using LPF, the cutoff frequency of LPF is designed to less than 1/10 to 1/2 for the resonance frequency of the LC filter f_{res_LC}. Thus, it is possible to reduce the effect of the LC filter resonance. However, in order to increase the performance of the disturbance compensator, it is necessary to increase the cutoff frequency of LPF. Due to the output current disturbance caused by the diode rectifier load, the odd-order harmonic components are included in the inverter output voltage. For instance, 11th order harmonic wave is 550 Hz; where fundamental wave is 50 Hz, when considering the reduction of the harmonics to 11th order for the fundamental wave of the output voltage. Thus, it is necessary to increase the LPF cutoff frequency of DOB higher than 550 Hz. Therefore, in order to consider the suppression of the harmonics to k order, the proposed method is applied as follows condition

$$kf_{out} < f_c < \frac{f_{res\_LC}}{n}, \quad \left(n : 2 \sim 10, f_c = \frac{\omega_c}{2\pi} \right) \qquad (5),$$

536

where f_{out} is the frequency of the output voltage, n is the ratio for the resonance frequency of the LC filter f_{res_LC}, which implies the DOB cutoff frequency from 1/10 to 1/2 of the resonance frequency of the LC filter. In this paper, the resonance frequency of the LC filter is approximately 10 kHz, where L is 1.29 mH (%Z = 1.0%), and C_f is 0.2 µF (%Y = 0.25%); therefore, if k is eleven, it is possible to meet (5). Furthermore, in order to compensate 11th order harmonic without interfering with the resonance frequency of the LC filter, the DOB cutoff frequency is set to 2 kHz.

Figure 6 shows the comparison of the disturbance gain characteristics of the conventional closed-loop-based voltage control and the proposed VDCOLC. The controller of the conventional method is similarly constructed as Fig. 2, where the angular frequency of the voltage controller is 3000 rad/s, and the angular frequency of the current controller is 30000 rad/s. Moreover, the disturbance suppression gain is calculated from the transfer function, where the output current i_{out} is the input, and the output voltage v_c is the output. Furthermore, the disturbance suppression performance is analyzed by sweeping the frequency of the disturbance suppression gain from 1 Hz to 100 kHz as Fig. 6. In Fig. 6, the disturbance gain is reduced with the proposed control from 30 Hz to 1 kHz of the disturbance frequency, which includes the harmonic components from first to 20th for the fundamental frequency of the output voltage. Thus, it is possible to reduce the output voltage distortion by using the proposed method.

IV. SIMULATION RESULT FOR PROPOSED VDCOLC

Table I shows the simulation conditions for the proposed method. The LC filter parameter and the DOB cutoff frequency are changed in the simulation results. Figure 7 shows the simulation results of the proposed method with the diode rectifier load. In Fig. 7 (a), it is possible to suppress the inverter output voltage THD less than 5.0%, when the resonance frequency of the LC filter is 10 kHz (L = 1.29 mH; %Z = 1.0%, C_f = 0.2 µF; %Y = 0.25%), and the DOB cutoff frequency is 2 kHz. The LC filter parameter is designed in order to suppress the output current overshoot with the low inductance during the voltage sag [7]. In order to reduce the resonance of the output current at the voltage fluctuation, the capacitance of the filter capacitor is designed to small. The inverter output voltage THD is 1.56%. Moreover, the design of the DOB cutoff frequency lower than the resonance frequency of the LC filter is possible to reduce the resonance component in the inverter output voltage. Thus, the resonance component of the inverter output voltage in Fig. 7 (a) does not occur, i.e. the condition of Fig. 7 (a) meets (5). In Fig. 7 (b), the resonance frequency of the LC filter and the DOB cutoff frequency are 2 kHz. The inverter output voltage is higher than 5%; in addition, the resonance occurs in the inverter output voltage. This is because the condition of Fig. 7 (b) does not meet (5), i.e. the LC filter resonance component is

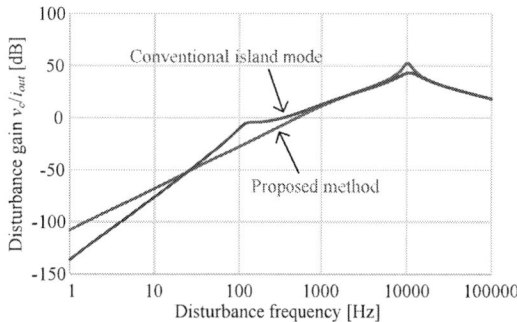

Fig. 6. Characteristics of disturbance gain from output current i_{out} to output voltage v_c. The resonance due to the controller is effected to the output voltage distortion with the conventional method.

TABLE I. SIMULATION CONDITIONS.

Output power	P_{out}	1 kW	Carrier fre.	f_{cry}	80 kHz
DC link vol.	V_{dc}	380 V	Samp. fre. of DOB	f_{so}	80 kHz
Output volt. command	v_c^*	200 V_{rms}	Crest factor of load		3.0

included in the voltage command. In Fig. 7 (c), the resonance frequency of the LC filter is 10 kHz, and the DOB cutoff frequency is 200 Hz. The inverter output voltage THD is higher than 5.0% in this condition. This is because that the disturbance suppression performance of the DOB decreases due to the low DOB cutoff frequency. Thus, the inverter output voltage THD is worse due to the disturbance for the diode rectifier load. In addition, the dead-time error voltage compensation by DOB becomes not effective enough due to the low DOB cutoff frequency. In Fig. 7 (d), the resonance frequency of the LC filter is 2 kHz, and the DOB cutoff frequency is 200 Hz. The inverter output voltage THD is also higher than 5.0% in this condition. The inverter output voltage includes the distortion due to the low resonance frequency of the LC filter. Furthermore, the disturbance compensation value becomes not effective enough for the diode rectifier load due to the low DOB cutoff frequency. Thus, the inverter output voltage THD in Fig. 7 (d) is higher than the condition of Fig. 7 (c). In conclusion, the proposed method can only be applied in the area of (5).

V. EXPERIMENTAL RESULTS

In the experimental verification of the island mode operation, three cases of load test are considered. The load conditions are as follows: (A) No-load, (B) Linear load such as resistance load, (C) Nonlinear-load such as diode rectifier load.

Table II shows the experimental conditions. By considering the frequency of the voltage command as 50 Hz, the angular frequency of the voltage controller for the conventional closed-loop-based island-mode voltage control is designed to ten times of the voltage command frequency. Thus, the angular frequency is set to 3000 rad/s. Moreover, in order to prevent the interference, the angular frequency of the current controller is designed to ten times for the angular frequency of the voltage controller. Thus, the angular frequency of the current

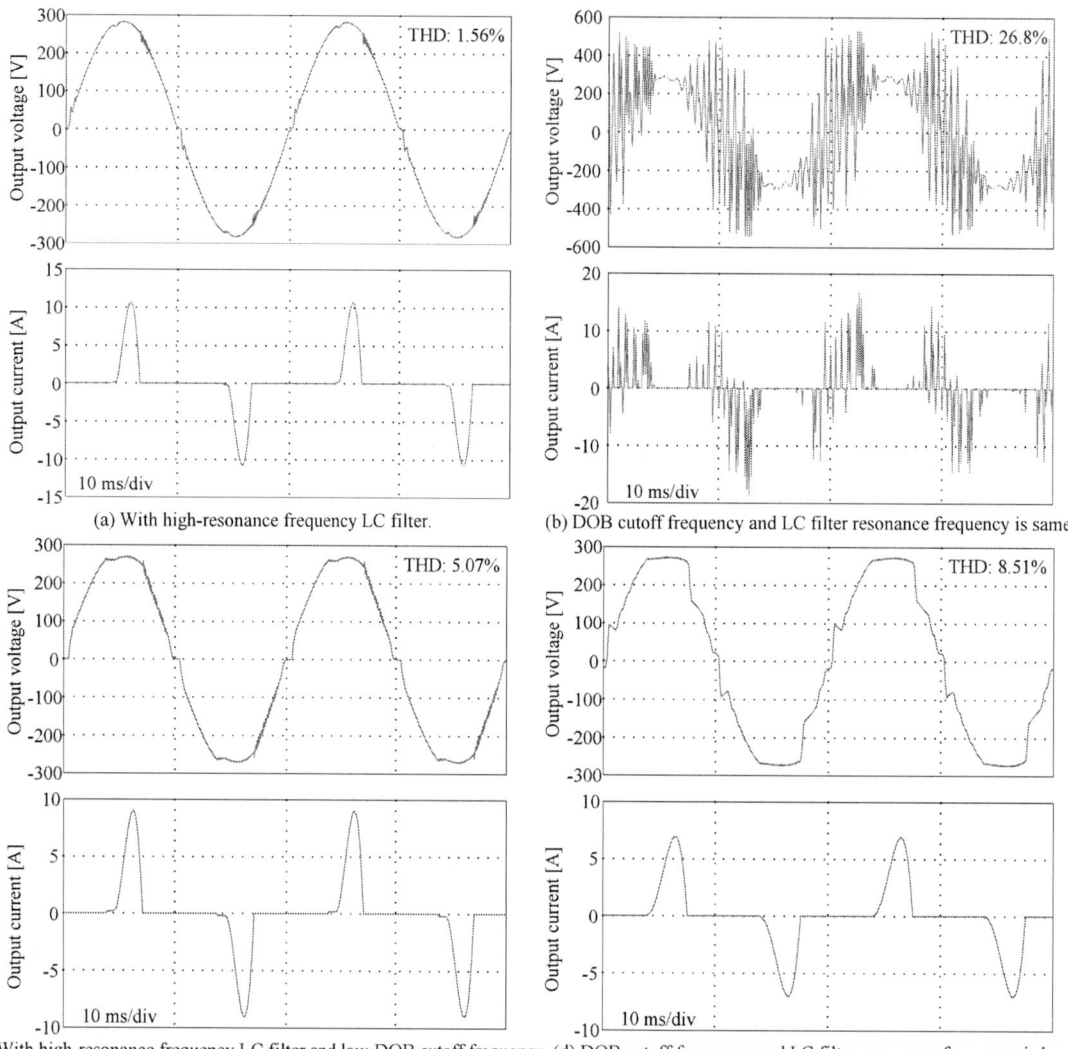

(a) With high-resonance frequency LC filter. (b) DOB cutoff frequency and LC filter resonance frequency is same.

(c) With high-resonance frequency LC filter and low-DOB cutoff frequency. (d) DOB cutoff frequency and LC filter resonance frequency is low.

Fig. 7. Simulation results for any condition of resonance frequency of LC filter and DOB cutoff frequency with the proposed VDCOLC. The inverter output voltage THD is suppressed to less than 5% when (5) is met.

controller is set to 30000 rad/s. Note that the minimized LC filter parameter is designed in order that the FRT requirements are met during the voltage sag [7]. The island mode operations are compared with the conventional closed-loop-based voltage control, the conventional open-loop-based island mode with the dead-time error compensation, and the proposed VDCOLC.

A. No-load Operation

Figure 8 shows the experimental result for the no-load operation. In Fig. 8 (a), (b) and (c), the output voltage THD of three methods are suppressed to less than 5.0%. However, the inverter output voltage THD in Fig. 7 (a) is high in the conventional closed-loop-based voltage control. This is because the filter capacitor current detection has the delay time that is different from the half period of the carrier as shown in Fig. 3 (c). Thus, it is impossible to detect the average value for the filter capacitor current. Therefore, the inverter output voltage is distorted in Fig. 8 (a). In particular, the distortion of the

inverter output voltage with the proposed method is reduced by 70.2% compared with the conventional closed-loop-based voltage control. In addition, the output voltage in Fig. 8 (a) is higher than the voltage command due to the high gain of the current controller. In order to regulate the output voltage in Fig. 8 (a), it is necessary to reduce the gain of the current controller. Moreover, the output voltage in Fig. 8 (b) is also higher than the voltage command due to the typical dead-time compensation. The dead-time error voltage does not occur in the no-load operation. Thus, the output voltage increases by the dead-time compensation value. On the other hand, the output voltage in Fig. 8 (c) is almost same compared with the voltage command.

B. Linear-load Operation

Figure 9 shows the experimental result for the linear load operation. In Fig. 9 (a), (b) and (c), the output voltage THD at the rated power with three methods are also suppressed to less than 5.0%. The inverter output

The 2018 International Power Electronics Conference

TABLE II. EXPERIMENTAL CONDITIONS.

Output power	P_{out}	1 kW	Carrier fre.	f_{cry}	80 kHz
DC link vol.	V_{dc}	380 V	Samp. fre. of DSP	f_s	20 kHz
Output volt. command	$v_c{}^*$	200 V_{rms}	Angl. fre. of AVR	ω_{avr}	3000 rad/s
Inter. Induc. (%Z)	L	1.29 mH (1.0%)	Angl. fre. of ACR	ω_{acr}	30000 rad/s
Filter cap. (%Y)	C_f	0.2 μF (0.25%)	Samp. fre. of DOB	f_{so}	80 kHz
			Cutoff fre. of DOB	f_c	2 kHz

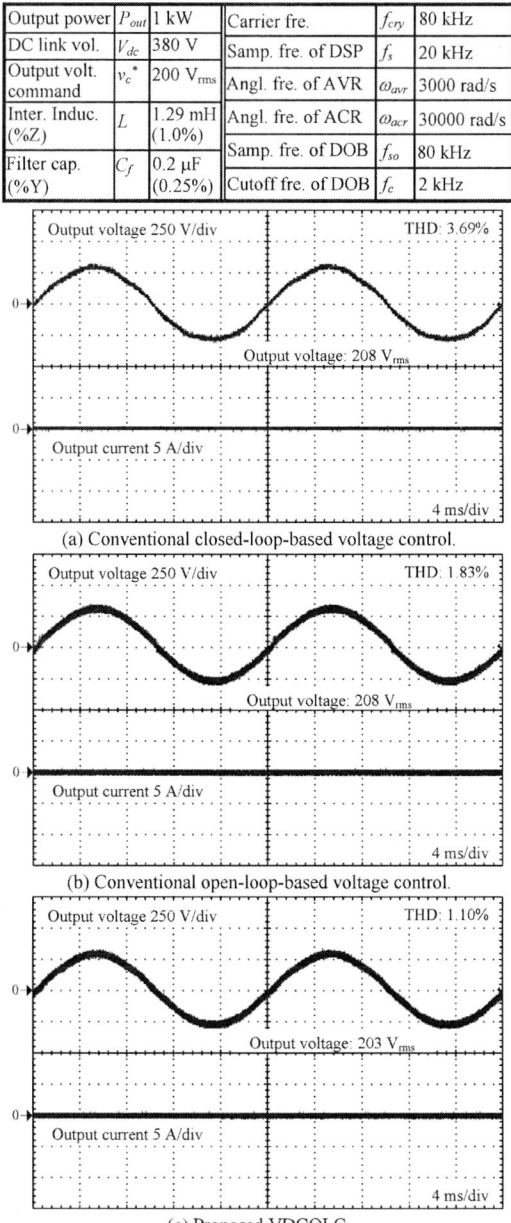

(a) Conventional closed-loop-based voltage control.

(b) Conventional open-loop-based voltage control.

(c) Proposed VDCOLC.

Fig. 8. Experimental result for no-load operation. The output voltage THD is suppressed less than 5.0% in both the conventional method and the proposed method.

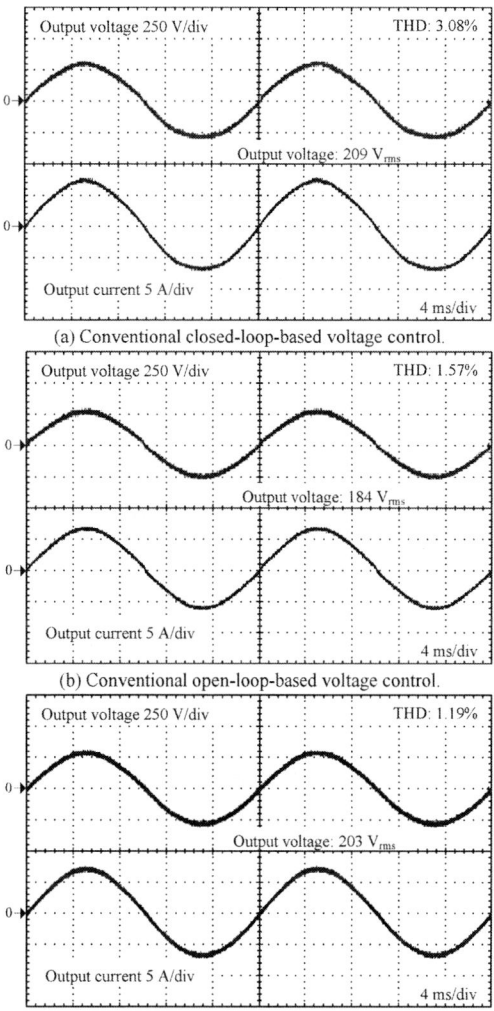

(a) Conventional closed-loop-based voltage control.

(b) Conventional open-loop-based voltage control.

(b) Proposed VDCOLC.

Fig. 9. Experimental result for linear load operation at rated load. The output voltage THD is suppressed to less than 5.0% in both the conventional method and the proposed method.

voltage THD in Fig. 9 (a) is high compared with both method of Fig. 9 (b) and (c), due to the same reason in Fig. 8 (a). In particular, the inverter output voltage THD with the proposed method is improved by 61.4% compared with the conventional closed-loop-based voltage control at the rated load. In addition, the output voltage in Fig. 9 (a) is higher than the voltage command due to the same reason in Fig. 8 (a). Moreover, the output voltage in Fig. 9 (b) decreases due to the load current compared with Fig. 8 (b). On the other hand, the output voltage in Fig. 9 (c) is almost same compared with the voltage command.

Figure 10 shows the output voltage THD characteristic and the output voltage variation against the linear load. In Fig. 10 (a), the output voltage THD with the proposed method is reduced more effectively than with the conventional closed-loop-based voltage control. The maximum improvement of the output voltage THD is 80.8% at load of 0.9p.u.. Moreover, the inverter output voltage THD characteristics with the conventional open-loop-based island mode or the proposed method are approximately constant compared with the conventional closed-loop-based voltage control in the load range from 0p.u. to 1.0p.u.. The cause of difference for the inverter output voltage THD with the conventional closed-loop-based voltage control is that the variation of the load current causes the error of the filter capacitor average current detection value. Due to the detection error of the filter capacitor average current, the inverter output voltage becomes non-sinusoidal wave. Thus, the operation of the conventional closed-loop-based voltage control becomes unstable. On the other hand, in Fig. 10 (b), the inverter

539

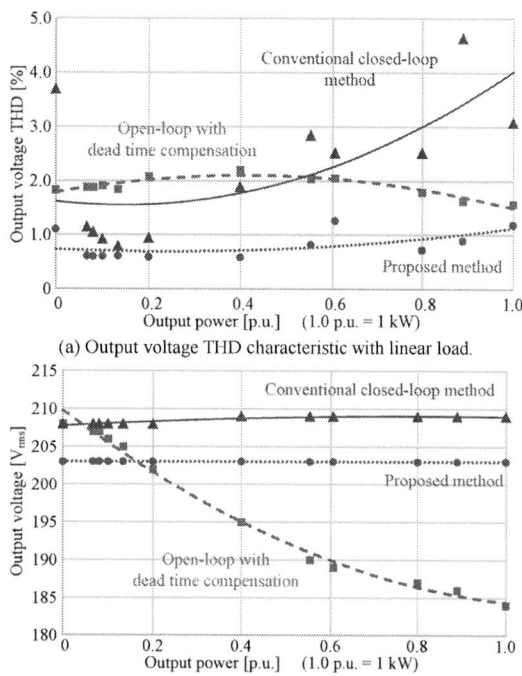

(a) Output voltage THD characteristic with linear load.

(b) Output voltage variation with linear load.

Fig. 10. Relationship between output voltage and resistive load in 1-kW system. By the proposed method, the output voltage THD is better than other two method. In addition, the output voltage is constant.

output voltage with the conventional closed-loop-based voltage control is almost constant in the load range from 0p.u. to 1.0p.u.. This is because the voltage controller regulates the inverter output voltage. However, the output voltage with the conventional open-loop-based voltage control varies significantly depending on the load. This is because that the inverter output voltage decreases due to the decrease of the filter capacitor current by the load current. On the other hand, the output voltage with the proposed method is constant. In the proposed control, DOB uses the detection value of the output voltage in order to estimate the output current and then compensate for the disturbance of the output current. On other perspective, the output voltage is indirectly regulated by a feedback loop in DOB. Therefore, it is possible to achieve that the inverter outputs the constant voltage during the island mode with the linear load in the proposed method.

C. Nonlinear-load Operation

Figure 11 shows the experimental result for the nonlinear-load operation. The diode rectifier is applied as nonlinear load. The crest factor of the diode rectifier load is approximately 3.0. The output voltage THD is higher than 5.0%, when using the conventional closed-loop-based voltage control and the conventional open-loop-based island mode as in Fig. 11 (a) and (b). The inverter output voltage with the conventional closed-loop-based voltage control distorts due to the reduction of the disturbance suppression performance by the low capacitance of the filter capacitor and the error of the filter

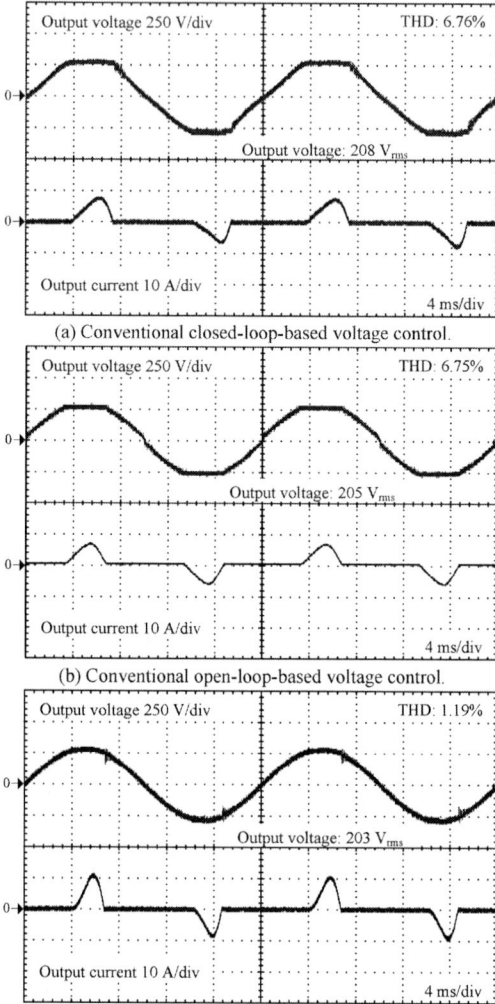

(a) Conventional closed-loop-based voltage control.

(b) Conventional open-loop-based voltage control.

(c) Proposed VDCOLC.

Fig. 11. Experimental result for diode rectifier-load operation. By using the proposed method, the output voltage THD is reduced to 1.19% with the nonlinear-load.

capacitor average current detection. In particular, the disturbance compensation is not effective enough for the output current control with the nonlinear load. Moreover, the inverter output voltage with the conventional open-loop-based island mode also distorts due to the non-sinusoidal output current which flows during the output voltage peak. On the other hand, by using the proposed method as in Fig. 11 (c), the output voltage THD is less than 5.0%. It is possible to compensate the drop voltage due to the open-loop-based operation with the high-gain DOB in the proposed method. Thus, the inverter output voltage becomes sinusoidal wave even with the diode rectifier load. Therefore, it is possible to suppress the output voltage THD by using the proposed method.

Figure 12 shows the THD characteristic and the output voltage variation against the nonlinear-load. In Fig. 12 (a), the output voltage THD with the proposed method is reduced more effectively than with the conventional closed-loop-based voltage control or the conventional open-loop-based voltage control. The maximum improvement of the output voltage THD with the proposed method is 82.4% compared with the conventional closed-loop-based voltage control at the heavy nonlinear load. Moreover, in Fig. 12 (b), the output voltage with the conventional open-loop-based island mode decreases greatly at the heavy nonlinear-load. On the other hand, the output voltage with the conventional closed-loop-based voltage control and the proposed method is constant value. Thus, it is possible to achieve that the inverter outputs the constant voltage during the island mode with the nonlinear-load in the proposed method. Therefore, it is confirmed that by using the proposed method, the output voltage is constant regardless of any conditions of load, i.e. linear/nonlinear load, or light/heavy load.

VI. CONCLUTION

This paper proposed the open-loop-based island-mode voltage control with the high-gain disturbance compensation when the LC filter is minimized. Moreover, the applying area of the proposed method is clarified by the DOB cutoff frequency and the resonance frequency of the LC filter. By using the proposed method, the output voltage is constant in the island mode during regardless of any conditions of load. Furthermore, the output voltage THD with the proposed method is suppressed to less than 5.0%. In particular, the output voltage THD was improved by 82.4% compared with the conventional closed-loop-based voltage control when the nonlinear-load was connected. Therefore, the proposed method was confirmed to the utility for the island mode with the minimized LC filter.

REFERENCES

[1] Hung-I Hsieh, and Jiaxin Hou, "Realization of Interleaved PV Microinverter by Quadrature-Phase-Shift SPWM Control", *IEEJ J. Ind. Appl.*, Vol.4, No.5, pp.643-649, 2015.

[2] Luis Valverde, Carlos Bordons, and Felipe Rosa, "Integration of Fuel Cell Technologies in Renewable-Energy-Based Microgrids Optimizing Operational Costs and Durability", *IEEE Trans. Ind. Electron.*, Vol. 63, No. 1, pp. 167-177, 2016.

[3] Zhenbin Zhang, Hui Fang, Feng Gao, José Rodríguez, and Ralph Kennel, "Multiple-Vector Model Predictive Power Control for Grid-Tied Wind Turbine System With Enhanced Steady-State Control Performance", *IEEE Trans. Ind. Electron.*, Vol. 64, No. 8, pp. 6287-6298, 2017.

[4] Min Huang, Xiongfei Wang, Poh Chiang Loh, Frede Blaabjerg, and Weimin Wu, "Stability Analysis and Active Damping for LLCL-Filter-Based Grid-Connected Inverters", *IEEJ J. Ind. Appl.*, Vol.4, No.3, pp.187-195, 2015.

[5] R. Peña-Alzola and M. Liserre,"LCL-Filter Design for Robust Active Damping in Grid-Connected Converters", *IEEE Trans. Ind. Info.*, Vol. 10, No. 4, pp. 2192-2203, 2014.

[6] S. Nagai, K. Kusaka, J. Itoh, "FRT Capability of Single-phase Grid-connected Inverter with Minimized Interconnected Inductor", *in Proc. IEEE Appl. Power Electron. Conf. and Expo. (APEC)* 2017, No. 1800, pp. 2802-2809, 2017.

[7] S. Nagai, K. Kusaka, J. Itoh, "ZVRT Capability of Minimized-LCL-filter-based Single-phase Grid-tied Inverter with High-speed Gate-block", in Proc. IEEE Energy Conversion Congress Expo. (ECCE) 2017, No. 847, pp. 1757-1764, 2017.

[8] Rong-Jong Wai, Wen-Hung Wang, and Chung-You Lin, "High-Performance Stand-Alone Photovoltaic Generation System", *IEEE Trans. Ind. Electron.*, Vol. 55, No. 1, pp. 240-250, 2008.

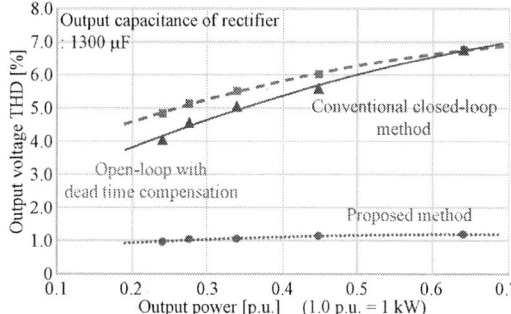

(a) Output voltage THD characteristic with nonlinear-load.

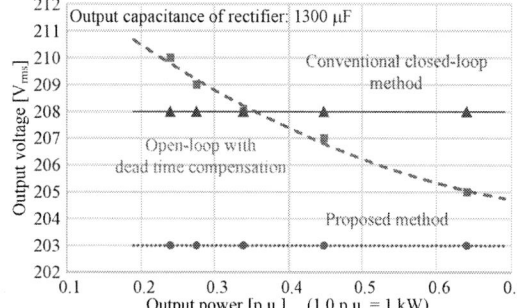

(b) Output voltage variation with nonlinear-load.

Fig. 12. Relationship between output voltage and rectifier load in 1-kW system. By using the proposed method, the output voltage THD is suppressed to less than 50%. Moreover, the output voltage is constant.

[9] Kenta Sayama, Shohei Anze, Kiyoshi Ohishi, Hitoshi Haga, and Takayuki Shimizu, "Robust and fine sinusoidal voltage control of self-sustained operation mode for photovoltaic generation system", *IECON 2014 - 40th Annual Conference of the IEEE Industrial Electronics Society*, pp. 1760-1765, 2014.

[10] César Trujillo Rodríguez, David Velasco de la Fuente, Gabriel Garcerá, Emilio Figueres, and Javier A. Guacaneme Moreno, "Reconfigurable Control Scheme for a PV Microinverter Working in Both Grid-Connected and Island Modes", *IEEE Trans. Ind. Electron.*, Vol. 60, No. 4, pp. 1582-1595, 2013.

[11] Mitsutoshi Ito, Reo Fujiwara, Ryunosuke Araumi, Takuma Yoshino, and Tomoki Yokoyama, "Robust digital control of single phase PWM inverter using 3MHz multi sampling method with FPGA based hardware controller", *in Proc. IEEE International Power Electron. and Motion Control Conf. - ECCE Asia*, No. P103 (2016).

[12] Atsuo Kawamura, Hiroshi Fujimoto, and Tomoki Yokoyama, "Survey on the real time digital feedback control of PWM inverter and the extension to multi-rate sampling and FPGA based inverter control", IECON 2007 - 33rd Annual Conference of the IEEE Industrial Electronics Society, Vol. TPC3-24, pp. 2044-2051.

[13] S. Nagai, H. N. Le, T. Nagano, K. Orikawa, J. Itoh, "Minimization of Interconnected Inductor for Single-Phase Inverter with High-Performance Disturbance Observer", *in Proc. IEEE International Power Electron. and Motion Control Conf. - ECCE Asia*, No. Wb8-06 (2016).

[14] T. Yamaguchi, Y. Tadano, and N. Hoshi, "Using a Periodic Disturbance Observer for a Motor Drive to Compensate Current Measurement Errors", *IEEJ J. Ind. Appl.*, Vol.4, No.4, pp.323-330, 2015.

[15] N. Hoffmann, M. Hempel, M. C. Harke and F. W. Fuchs, "Observer-based Grid Voltage Disturbance Rejection for Grid Connected Voltage Source PWM Converters with Line Side LCL filters", *IEEE Energy Conversion Congress and Expo.*, 2012, pp. 69-76.

[16] Walid R. Issa, Mohammad A. Abusara, and Suleiman M. Sharkh, "Control of Transient Power During Unintentional Islanding of Microgrids", *IEEE Trans. Power Electron.*, Vol. 30, No. 8, 2015, pp. 4573-4584.

The 2018 International Power Electronics Conference

Experimental Validation of Adaptive Current Injecting Method for Grid-synchronization Improvement of Grid-tied REGS during Short-circuit Fault

Shaokang Ma[1*], Hua Geng[1], Geng Yang[1], and Bo Liu[2]

1 Department of automation, Tsinghua University, Beijing, China
2 Zongheng Electro-mechanical technology development co., Beijing, China
*E-mail: mashaokang1111@163.com

Abstract-During short-circuit fault, the power converter based grid-tied renewable energy generation system (REGS) is supposed to inject active current in addition to the reactive current stipulated by the grid code to prevent loss of synchronization between the REGS and the power grid. In this paper, an adaptive current injecting method (ACIM) is applied to improve the grid-synchronization stability of REGS. The method adjusts the active and reactive currents according to the frequency estimated by the phase-locked-loop (PLL), ensuring that a stable operating point of the REGS exists during the short-circuit fault. An experimental system is built to validate the effectiveness of the ACIM.

Keywords—Synchronization, renewable energy gerenation system, phase-locked-loop, current control, fault ride-through.

I. INTRODUCTION

Renewable energy is a promising way to address the environmental issues and energy crisis. Many kinds of the REGSs are based on power converters, such as the variable speed wind turbine system and the photovoltaic (PV) system, which have been widely used all over the world. Unlike the synchronous generator (SG), power converter usually connects to the power grid based on the PLL, which estimates the phase and frequency of the grid voltage for the vector control[1]. Although the virtual SG technique without PLL has been developing, the PLL based REGS still dominates.

Grid-synchronization, which is the basis of the upper level controls, means that the output frequency of PLL of the REGS is identical with the grid voltage frequency[2]. When the REGS connects to a "strong" power grid, the grid-synchronization would not be a problem. However, when the power grid becomes "weak", i.e. the short-circuit ratio (SCR) at the point of common connection (PCC) of the REGS becomes low, the REGS may encounter the grid-synchronization issues[3, 4]. Low SCR means a high line impedance or a low amplitude of grid voltage[5]. The current injected by the REGS will impact the terminal voltage of the REGS itself under weak connection. The terminal voltage of REGS is the input of the PLL and the PLL estimates the frequency and phase of the grid voltage for power converter vector control, which will in turn affect the REGS current[6]. The interaction between the injected current and the line

impedance aggravates the grid-synchronization stability of the REGS[7, 8].

Under short-circuit grid fault, terminal voltage of the REGS becomes very low. Hence the equivalent SCR becomes very low and the connection between the REGS and the power grid is very weak. As a result, the terminal voltage of REGS is vulnerable to the current injected by the REGS. The grid code stipulates that the REGSs are supposed to inject reactive current to support the power grid during the fault[9]. Thus, the interaction between the REGS current and the line impedance may lead to loss of grid-synchronization during the short-circuit fault. The output frequency of PLL will deviate from the frequency of the grid voltage and the REGS will trip off and the desired reactive current can be never injected into the grid correctly.

The grid-synchronization problem is caused by the interaction between the REGS and the grid impedance. Hence improving the tracking capability of PLL only, like [10], will not help to solve the grid-synchronization issues. In [11], a reduced-order model is proposed to describe and analyze the grid-synchronization problem. It also proposes a necessary condition for the power converter to maintain grid-synchronization. The same issues in permanent magnet synchronous wind generator (PMSG) system is studied under severe grid fault in [2], and a frequency feed-back control scheme is proposed to improve the grid-synchronization stability of the wind generation system. The method regulates output current of PMSG to maintain the frequency of PLL at its rated value. However, in real world, the grid frequency does not strictly equal to its rated value and is hard to be known for the REGS controller. As a result, the method is ineffective since it has to set a target frequency which should be equal to the actual grid frequency.

During grid fault, the REGS is supposed to inject not only reactive current stipulated by the grid code, but also active current to ensure grid-synchronization[2]. In this paper, an adaptive current injecting method (ACIM), which regulates the active and reactive current of the REGS according to the output frequency of PLL, is applied to improve the grid-synchronization stability of the power converter based REGS during grid fault. An experimental system, which is a two-level three-phase grid-tied converter system, is established to validate the

542

proposed method. Three cases are examined, indicating that the ACIM is effective in improving the grid-synchronization stability of the REGS.

The rest of the paper is organized as follow: Section II describes the simplified model of the REGS. Section III discusses the ACIM. Section IV shows the experimental results and Section V gives the conclusion of the paper.

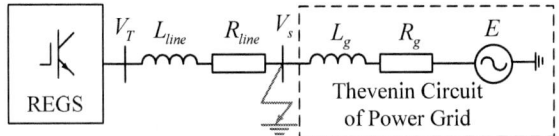

Fig. 1. Grid-tied REGS.

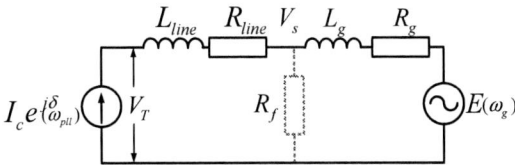

Fig. 2. Equivalent circuit of grid-tied REGS.

II. MODELLING OF REGS

Fig.1 shows the basic structure of a grid-tied REGS. The power grid is represented by a thevenin equivalent circuit E serialized with inductance L_g and resistance R_g. L_{line} and R_{line} represent the transmission line and V_T is the terminal voltage of the REGS. The REGS is controlled by power converter. Under grid fault, the REGS is controlled to inject the desired current according to the grid code. Since the dynamic of current loop is much faster than the PLL, the REGS is modeled as a controlled current source during grid fault. The electromagnetic transient of the impedance is also omitted, as shown in Fig.2.

In Fig.2, $I_c = I_d + jI_q$, I_d is the d-axis component and I_q is the q-axis component of the injected current of the REGS. I_d and I_q are on the PLL frame, so the velocity of I_c is ω_{pll}. E is on the grid voltage frame, so the velocity of E is ω_g. δ is the angle between the PLL frame and the grid voltage frame, i.e. $\delta = \theta_{pll} - \theta_g$. The typical structure of PLL refers to [11].

In Fig.2, a small resistance R_f is used to represent the short-circuit fault. Regarding E as the reference vector, the terminal voltage of REGS under fault situation can be expressed as

$$V_T = Z_i\left(\omega_{pll}\right)\left|I_c\right|e^{j(\phi_c+\delta)} + K_v\left(\omega_g\right)E \quad (1)$$

where

$$
\begin{cases}
\phi_c = \arctan\dfrac{I_q}{I_d} \\[2mm]
Z_i\left(\omega_{pll}\right) = R_{line} + j\omega_{pll}L_{line} + \dfrac{\left(R_g + j\omega_{pll}L_g\right)R_f}{R_f + R_g + j\omega_{pll}L_g} \\[3mm]
K_v\left(\omega_g\right) = \dfrac{R_f}{R_f + R_g + j\omega_g L_g}
\end{cases} \quad (2)
$$

On the PLL frame, the q-axis component of (1), which is shown in (3), is the input of the PLL.

$$
\begin{aligned}
v_q = &\left|Z_i\left(\omega_{pll}\right)I_c\right|\sin\left(\phi_c + \angle Z_i\left(\omega_{pll}\right)\right) \\
&+\left|K_V\left(\omega_g\right)E\right|\sin\left(\angle K_V\left(\omega_g\right)-\delta\right)
\end{aligned} \quad (3)
$$

Hence, according to the PLL structure[12], we can obtain

$$
\begin{cases}
\dfrac{d\lambda}{dt} = K_i v_q \\[2mm]
\dfrac{d\delta}{dt} = K_p v_q + \lambda
\end{cases} \quad (4)
$$

where λ is the value of the integrator of the PLL. K_p and K_i are the parameters of the PLL.

The relationship between the state variable δ and the input v_q reflect the interaction between the REGS and the impedance of the power grid.

III. IMPROVEMENT OF GRID-SYNCHRONIZATION

A. Analysis on grid-synchronization

At the equilibrium operating point of the differential equation (4), both of the λ and the v_q equal to 0. According to (3), the v_q can equal to zero only if the following condition is met.

$$\left|Z_i\left(\omega_g\right)I_c\sin\left(\phi_c + \angle Z_i\left(\omega_{pll}\right)\right)\right| < \left|K_v\left(\omega_g\right)E\right| \quad (5)$$

(5) is the necessary condition in [11]. From (5), it can be obtained that severe grid fault make the value of the right side term lower. If the injected current is set to zero, the left side term will be zero and hence there will be no grid-synchronization problem. However, according to the grid code, the REGS is required to inject reactive current. If condition (5) cannot be met, the v_q can never be zero, which means that the differential equation (4) has no equilibrium operating point. As a result, the REGS will lose grid-synchronization. Note that if the left side term of (5) is close to the right side term, even though equilibrium operating point exists, the dynamic behaviors of PLL may still lead to loss of grid-synchronization[12].

B. The ACIM

A feasible way to regulate the value of the left side term of (5) is to regulate the value of ϕ_c. Under grid fault condition, since R_f is small, we assume that

$$
\begin{cases}
Z_i\left(\omega_{pll}\right) \approx R_{line} + j\omega_{pll}L_{line} \\[2mm]
\angle K_v\left(\omega_g\right) \approx -\arctan\dfrac{\omega_g L_g}{R_g}
\end{cases} \quad (6)
$$

If the d- and q-axis current of REGS are set as

$$\frac{I_q}{I_d} = -\tan\angle Z_i\left(\omega_{pll}\right) = -\frac{\omega_{pll}L_{line}}{R_{line}} \quad (7)$$

then the left side term of (5) equals to zero. Hence (5) can be always met. With (7), (4) can be rewritten as

$$
\begin{cases}
\dfrac{d\lambda}{dt} = -K_i\left|K_V\left(\omega_g\right)E\right|\sin\left(\delta - \angle K_V\left(\omega_g\right)\right) \\[2mm]
\dfrac{d\delta}{dt} = -K_p\left|K_V\left(\omega_g\right)E\right|\sin\left(\delta - \angle K_V\left(\omega_g\right)\right) + \lambda
\end{cases} \quad (8)
$$

A Lyapunov function is expressed in equation (9).

543

$$V(\lambda,\delta)=0.5\lambda^2$$
$$+K_i\left|K_v\left(\omega_g\right)E\right|\left[1-\cos\left(\delta-\angle K_v\left(\omega_g\right)\right)\right] \quad (9)$$

The value of the Lyapunov function $V(\lambda,\delta)$ is no less than zero while the value of the differential of $V(\lambda,\delta)$ is no greater than zero. Consequently, system (8) is globally stable, which means the ACIM can ensure that the REGS has a stable operating point under short-circuit fault situation. Hence, the PLL of the REGS is capable to track the grid voltage.

During the grid fault, the grid code requires the REGS to inject reactive current. According to the ACIM in (7), the active current should also be injected. From (9), we can obtain that even if the grid voltage frequency, i.e. the ω_g, does not equal to its rated value, the ACIM can still improves the grid-synchronization stability of the REGS.

Note that I_d and I_q having different signs does not mean that the REGS has to absorb active power while injecting inductive reactive power. Instead, because the apparent power is the multiply of the voltage and the conjuncture of the current, I_d and I_q having different signs means that the REGS injects both active power and inductive reactive power into the grid.

IV. EXPERIMENTAL VALIDATION

Fig.3 illustrates the experimental system. The three phase power converter connects to the power grid via a voltage sag generator. The voltage sag generator is used to emulate the short-circuit fault. The control board of power converter is a DSP-FPGA platform and the proposed ACIM is integrated in the controller of the power converter. When the voltage sag is detected, the proposed ACIM is activated and both active and reactive current are injected into the grid according to (7). The amplitude of the injected current is 1p.u.. The parameters of the experimental system is shown in Table.I. Fig.4 shows the actual experimental system.

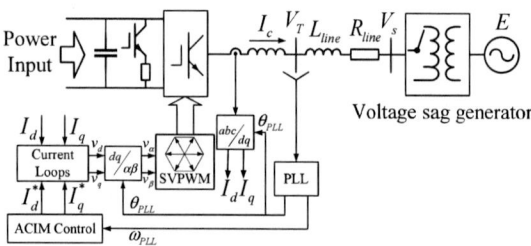

Fig. 3. Experimental system of grid-tied REGS.

TABLE I
PARAMETERS OF THE EXPERIMENTAL SYSTEM

Symbol	Meaning	Value
V_{DC}	DC voltage (V)	425
E_p	Peak value of grid voltage E (V)	300
I_n	Rated current (A)	10
f_n	Frequency of grid voltage (Hz)	50
f_s	Switching frequency (Hz)	8000
R_{line}	Line resistance (ohm)	3.5
L_{line}	Line inductance (H)	0.02
K_p	K_p of PLL	90
K_i	K_i of PLL	220

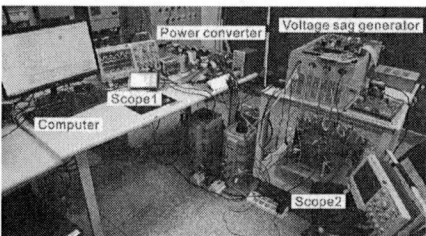

Fig. 4. Illustration of the experimental system.

The voltage sag generator reduces the voltage to 0.09 times of its rated value to emulate the short-circuit fault. Before the grid fault, the output current of REGS is set to 1p.u. active current, i.e. $I_c=1+j0$(p.u.). Three cases are examined and illustrated in the following.

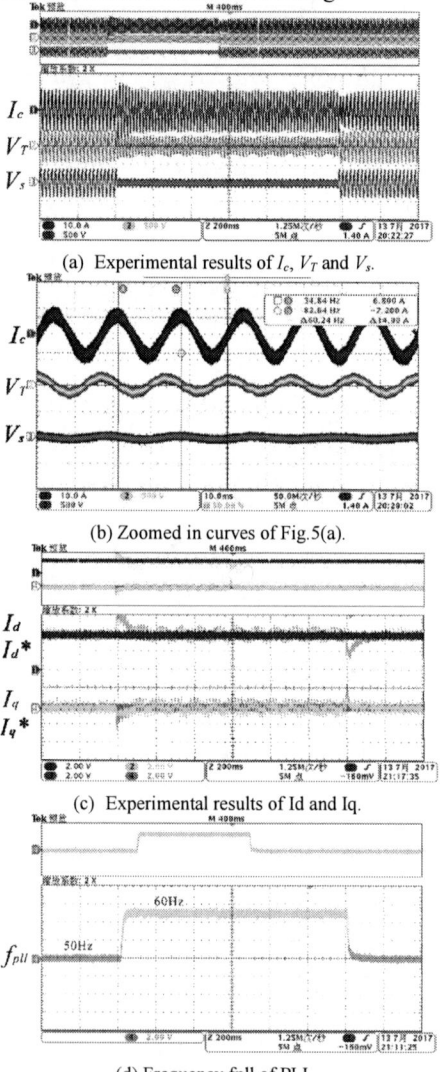

(a) Experimental results of I_c, V_T and V_s.

(b) Zoomed in curves of Fig.5(a).

(c) Experimental results of Id and Iq.

(d) Frequency fpll of PLL.
Fig. 5. Experimental results of case one.

In the first case, the current of power converter during the voltage sag is unchanged, i.e. $I_c=1+j0$(p.u.). Fig.5(a) shows the waveforms of I_c, V_s and V_T. Fig.5(b) is the zoomed in figure of Fig.5(a) during the voltage sag. Fig.5(c) shows current commands and real currents on the d- and q-axis and Fig.5(d) is the PLL frequency. In this case, the left side term of (5) is 50V and the right side term of (5) is 30V. Hence there is no equibilium operating point for the REGS during the voltage sag and the system loses grid-synchronization. Besides, the v_q is always greater than zero. According to (4), the PLL frequency accelerates untill it reaches its upper limitation, as shown in Fig.5(d). Fig.5(b) shows the frequency of the current at certain time during the voltage sag, which is 60Hz and does not equal to the grid freqeucny.

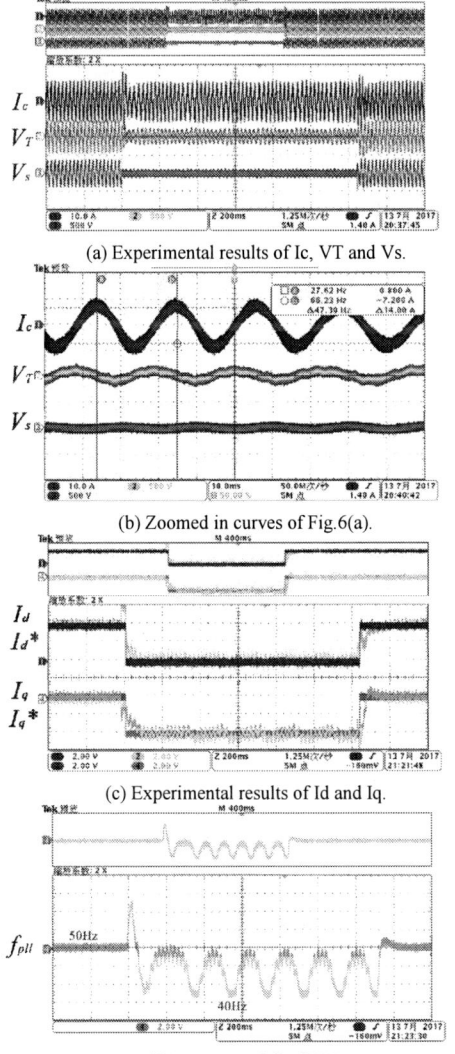

(a) Experimental results of Ic, VT and Vs.

(b) Zoomed in curves of Fig.6(a).

(c) Experimental results of Id and Iq.

(d) Frequency fpll of PLL.

Fig. 6. Experimental results of case two.

In the second case, the REGS injects 1p.u. inductive reactive current, i.e. $I_c = 0 - j1(\text{p.u.})$, during the voltage sag. The left side term of (5) is -28V, whose amplitude is smaller than but close to that of the right side term of (5), i.e. 30V. As shown in Fig.6(d), the PLL still cannot track the grid voltage because of the dynamic process of the PLL[12]. Fig.6(c) shows that the converter can track the current commands during the voltage sag. Fig.6(b) shows that the frequency of the injected current at certain time during the voltage sag is 47.39Hz.

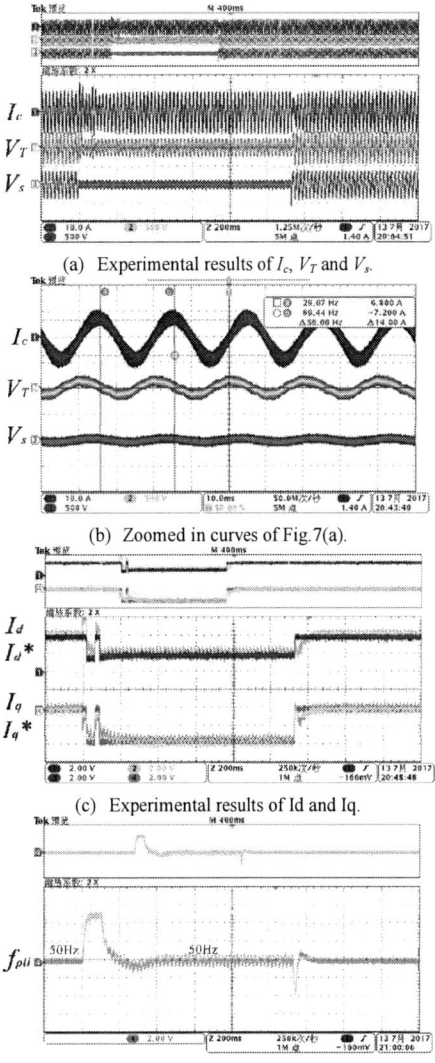

(a) Experimental results of I_c, V_T and V_s.

(b) Zoomed in curves of Fig.7(a).

(c) Experimental results of Id and Iq.

(d) Frequency fpll of PLL.

Fig. 7. Experimental results of case three.

In the third case, the ACIM is applied. From Fig.7(d), the frequency of PLL can strictly track the grid voltage, indicating that with the proposed ACIM, the power converter has stable equilibrium operating point and the stability of the system is guaranteed. Consequently, the power converter can keep synchronous with the power grid during the voltage sag.

V. CONCLUSION

This paper applies the ACIM for the power converter based grid-tied REGS. Under short-circuit grid fault situation, in addition to the reactive current stipulated by the grid code, the ACIM injects active current according to the PLL frequency to improve the grid-synchronization stability of the REGS. The experimental results show that under short-circuit fault, the REGS has risk to lose the grid-synchronization and the ACIM can keep the REGS synchronous with the grid voltage.

The 2018 International Power Electronics Conference

ACKNOWLEDGMENT

This work is supported by National Key Research and Development Program under Grant 2016YFB0900300, National Natural Science Foundation of China (NSFC) under Grant 61273045, 61722307, U1510208, 51711530235, 51361135705, Technology project of State Grid Corporation of China under Grant NY71-16-052, and UK China grant EP/L014343/1: Stability and Control of Power Networks with Energy Storage (STABLE-NET) (P48270).se place an eventual Acknowledgment here, before the References. Put sponsor acknowledgments in an unnumbered footnote on the first page.

REFERENCES

[1] B. Wen, D. Dong, D. Boroyevich, R. Burgos, P. Mattavelli, and Z. Y. Shen, "Impedance-Based Analysis of Grid-Synchronization Stability for Three-Phase Paralleled Converters," *IEEE Transactions on Power Electronics,* vol. 31, pp. 26-38, Jan 2016.

[2] O. Goksu, R. Teodorescu, C. L. Bak, F. Iov, and P. C. Kjaer, "Instability of Wind Turbine Converters During Current Injection to Low Voltage Grid Faults and PLL Frequency Based Stability Solution," *IEEE Transactions on Power Systems,* vol. 29, pp. 1683-1691, Jul 2014.

[3] L. D. Zhang, L. Harnefors, and H. P. Nee, "Power-Synchronization Control of Grid-Connected Voltage-Source Converters," *IEEE Transactions on Power Systems,* vol. 25, pp. 809-820, May 2010.

[4] X. Z. Xi, H. Geng, and G. Yang, "Enhanced model of the doubly fed induction generator-based wind farm for small-signal stability studies of weak power system," *IET Renewable Power Generation,* vol. 8, pp. 765-774, Sep 2014.

[5] J. Z. Zhou, H. Ding, S. T. Fan, Y. Zhang, and A. M. Gole, "Impact of Short-Circuit Ratio and Phase-Locked-Loop Parameters on the Small-Signal Behavior of a VSC-HVDC Converter," *IEEE Transactions on Power Delivery,* vol. 29, pp. 2287-2296, Oct 2014.

[6] D. Jovcic, "Phase locked loop system for FACTS," *IEEE Transactions on Power Systems,* vol. 18, pp. 1116-1124, Aug 2003.

[7] D. Dong, B. Wen, P. Mattavelli, D. Boroyevich, and Y. Xue, "Grid-synchronization modeling and its stability analysis for multi-paralleled three-phase inverter systems," in *Applied Power Electronics Conference and Exposition (APEC)*, pp. 439-446.

[8] B. Wen, D. Boroyevich, P. Mattavelli, Z. Shen, and R. Burgos, "Influence of phase-locked loop on input admittance of three-phase voltage-source converters," in *Applied Power Electronics Conference and Exposition (APEC)*, 2013, pp. 897-904.

[9] S. Xiao, G. Yang, H. L. Zhou, and H. Geng, "An LVRT Control Strategy Based on Flux Linkage Tracking for DFIG-Based WECS," *IEEE Transactions on Industrial Electronics,* vol. 60, pp. 2820-2832, Jul 2013.

[10] P. Rodriguez, J. Pou, J. Bergas, J. I. Candela, R. P. Burgos, and D. Boroyevich, "Decoupled double synchronous reference frame PLL for power converters control," *IEEE Transactions on Power Electronics,* vol. 22, pp. 584-592, Mar 2007.

[11] D. Dong, B. Wen, D. Boroyevich, P. Mattavelli, and Y. S. Xue, "Analysis of Phase-Locked Loop Low-Frequency Stability in Three-Phase Grid-Connected Power Converters Considering Impedance Interactions," *IEEE Transactions on Industrial Electronics,* vol. 62, pp. 310-321, Jan 2015.

[12] S. M. a. H. G. a. L. L. a. G. Y. a. B. C. Pal, "Grid-Synchronization Stability Improvement of Large Scale Wind Farm During Severe Grid Fault," *IEEE Transactions on Power Systems,* vol. PP, pp. 1-1, 2017.

Adaptive Control of Grid-Voltage Feedforward for Grid-Connected Inverters based on Real-Time Identification of Grid Impedance

Roni Luhtala[1]*, Tuomas Messo[2], Tomi Roinila[1]

1 Laboratory of Automation and Hydraulics, Tampere University of Technology, Tampere, Finland
2 Laboratory of Electrical Energy Engineering, Tampere University of Technology, Tampere, Finland
*E-mail: roni.luhtala@tut.fi

Abstract—**Time-varying grid impedance has a major effect on the stability and operation of grid-connected systems. The stability of such systems can be assessed by impedance-based stability analysis. Recent studies have shown methods for online stability assessment and adaptive control of grid-connected inverters to ensure the system stability in varying grid conditions. This paper extends the previous studies, and presents a method in which a grid-voltage feedforward is adaptively controlled by applying real-time grid-impedance measurements. The measurements are performed by using pseudo-random binary sequences and Fourier techniques. In parallel with the adaptive control, the system stability is assessed in real-time. Experimental results based on a three-phase grid-connected inverter are presented and used to demonstrate the effectiveness of the proposed methods.**

Keywords—*Adaptive Control, Grid-Connected inverters, Grid-Voltage Feedforward, Real-Time Stability Assessment*

I. INTRODUCTION

The electric power systems will become increasingly complex in the near future. Such systems will include great amount of environmentally friendly renewable energy, energy-storage systems and controllable loads, such as electric cars' charging devices [1]. Rapid replacement of traditional synchronous generators by power-electronic-connected renewable energy have already started to affect the grid dynamics [2], and have caused excessive oscillations [3]. The route for power flow from single source to consumption varies more rapidly over the time as the utilization of energy-storage systems (batteries) and controllable loads increase. As a consequence, the dynamics of the power grid have become highly time variant [4]. This trend has had a major effect on the stability and operation of grid-connected devices.

The stability of grid-connected systems can be assessed by impedance-based stability analysis [5] which states that system will remain stable if the ratio between the grid impedance and inverter output impedance satisfies the Nyquist stability criterion [6]. The impedance-based stability is ensured if both impedances remain passive or the inverter output impedance is always greater than the grid impedance. [7] However, the passivity of a three-phase inverter output impedance is impossible to attain at all frequencies. In addition, it is difficult to ensure that the inverter output impedance is greater

than the grid impedance, particularly under weak grid conditions. [8] As the grid impedance highly varies over time, the conventional off-line impedance measurements have become insufficient to guarantee stability as the grid impedance highly varies over time [4].

Recent studies have provided various online-stability assessment methods based on broadband injections and Fourier techniques for grid-connected inverters. Such methods include, for example, impulse response [9], multi-tone injections [10], sine sweeps [11], and maximum-length binary sequence (MLBS) [12]. As the grid conditions are strongly time varying, adaptive control of a grid-connected inverter, based on real-time measurements of the grid impedance, will be the most efficient method to guarantee the stability of the grid-connected system. By using adaptive control system, the inverter output impedance can be re-shaped in order to avoid impedance-based stability issues. [13]

The feedforward proportional of grid voltage is a widely used concept to mitigate the effect of grid voltage harmonics. The grid-voltage feedforward increases the magnitude of the inverter output impedance, but due to time delays, also causes significant phase-lag which may lead the inverter to lose its passivity. [14] This paper introduces a practical implementation of an adaptively controlled grid-voltage feedforward for the grid-connected converters. By adaptively adjusting the feedforward gains based on grid-impedance information acquired in real time, the stability issues can be effectively avoided [15]. Experimental results using a grid-connected inverter under rapidly varying grid conditions are provided.

The remainder of the paper is organized as follows: Simplified impedance-based stability analysis for grid-connected devices and grid-voltage feedforward control scheme are reviewed in Section II. Section III introduces the adaptive control of grid-voltage feedforward gains, grid-inductance identification methods based on maximum-length binary sequence (MLBS), and real-time stability-assessment methods. Section IV presents a power hardware-in-the-loop test bench, practical implementation of the proposed methods, and experimental results. Section V draws conclusions.

II. THEORY

A. Impedance-Based Stability Analysis

A three-phase system can be transformed to a synchronous reference frame (dq domain) where a three-phase AC system can be presented as a model that includes two DC-valued components, direct (d) and quadrature (q). The analysis in the dq domain has attractive features as the models can be linearized around the steady-state operating point which simplifies the complex inverter models and allows observing the small-signal stability. The dq transformation introduces the cross-coupling terms (dq and qd) between the individual components. [16] Thus, the analysis of dq-domain impedances must include individual impedances for both components (dd and qq) and also cross-coupling impedances (dq and qd) between them. The dq-domain impedance can be represented as a matrix form as

$$\mathbf{Z}_{DQ} = \begin{bmatrix} Z_{dd} & Z_{dq} \\ Z_{qd} & Z_{qq} \end{bmatrix} \tag{1}$$

Fig. 1 shows a small-signal equivalent circuit of the grid-connected system at the point of common coupling (PCC) which is used to visualize the impedance-based stability in the dq domain. In the presentation, the inverter model is described by impedance matrix $\mathbf{Z}_{o\text{-}DQ}$ representing the output impedance, and grid as $\mathbf{Z}_{g\text{-}DQ}$ describing non-ideal load effects. The stability of such system can be analyzed by examining the return ratio of these two impedance matrices $\mathbf{Z}_{g\text{-}DQ}(s)/\mathbf{Z}_{o\text{-}DQ}(s)$. The ratio must satisfy the generalized Nyquist stability criterion. [6]

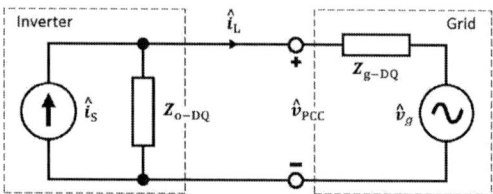

Fig. 1: Equivalent small-signal circuit of grid-connected system.

Studies have shown that the inverter output impedance matrix is clearly diagonally dominant, meaning that the cross-coupling impedances of $\mathbf{Z}_{o\text{-}DQ}$ are small and can be thus neglected from the impedance-based stability analysis when unity power factor is desired [17]. It is emphasized, however, that this assumption does not hold when the power factor is decreased. In such case, the cross-coupling must be included in the analysis to avoid inaccuracy.

The d and q component of the conventional grid impedance are most commonly modeled at frequencies below 1 kHz by using resistance and inductance [18]. It can be also assumed that the resistance and inductance are the same for both components as the grid impedance is not shaped by any active control loops. Thus, the grid can be represented by a single impedance as the assumption

$$Z_g(s) = Z_{g\text{-}dd}(s) = Z_{g\text{-}qq}(s) = R_g + j\omega L_g \tag{2}$$

holds, where R_g is the resistance and L_g is the inductance of the grid.

The stability analysis can be conducted by analyzing separately transfer functions

$$G_{dd}(s) = \frac{1}{1 + Z_g(s)/Z_{o\text{-}dd}(s)} \tag{3}$$

$$G_{qq}(s) = \frac{1}{1 + Z_g(s)/Z_{o\text{-}qq}(s)} \tag{4}$$

from which the grid impedance $Z_g(s)$ is measured and the inverter output impedances $Z_{o\text{-}dd}$ and $Z_{o\text{-}qq}$ are analytically modeled. The analytical model of the inverter output impedance changes as the operating conditions or control parameters vary, and thus, the model has to be updated when assessing the impedance-based stability.

B. Grid-Voltage Feedforward

The grid-voltage feedforward is used to mitigate the effect of low-order harmonics to the output currents. The feedforward path is implemented by adding the measured grid voltages to the duty ratios of d and q components through appropriate feedforward gains K_{ffd} and K_{ffq}. The feedforward significantly affects the inverter output impedance by increasing the magnitude, but also causes significant phase-lag due to control delays. [19]

The ideal gains for grid-voltage feedforward can be derived from an average model (introduced in [16]) by adding desired effect to the duty ratios that would compensate the grid voltage variations. In the case of a two-level three-phase inverter, the ideal feedforward gain for both components are $1/V_{DC}$ where V_{DC} is the DC-side voltage which is estimated to remain constant. [14] However, the operating conditions in DC and AC side vary over time and the ideal feedforward gains are not the same all the time. High gains provide sufficient mitigation of harmonics but too high gains may cause stability issues especially with inductive grid due to produced phase-lag to the inverter output impedance. Reference [15] provides further analysis of the grid-voltage feedforward and derivation of the accurate impedance-model used in this paper.

III. METHODS

A. Adaptive Control of Grid-Voltage Feedforward

The ideal feedforward gains ($1/V_{DC}$) are based on the steady-state value of the DC-side voltage and the same gains are not suitable when the operating conditions change. Additionally, the grid impedance has a major effect on the inverter operation, especially when grid-voltage feedforward is utilized. Fig. 2 shows the d component of the inverter output impedance with high (red) and low (blue) grid-voltage feedforward gains, when applying an impedance model derived in [15]. Due to significant phase lag between 60 and 600 Hz, the high gains may cause stability issues when the inverter is connected to a high-impedance grid. The stability issues can be avoided by boosting the phase of the inverter impedance. The phase boost is achieved by reducing the gains. However, the gains should be increased close to their ideal values, in order to ensure sufficient mitigation of harmonics and good performance during transients.

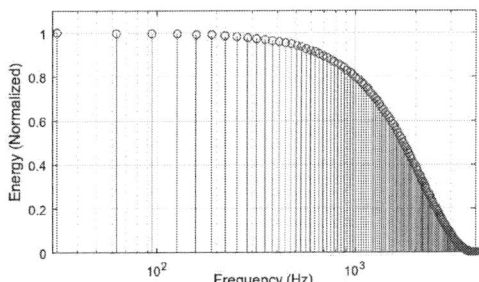

Fig. 3: normalized power spectrum of MLBS.

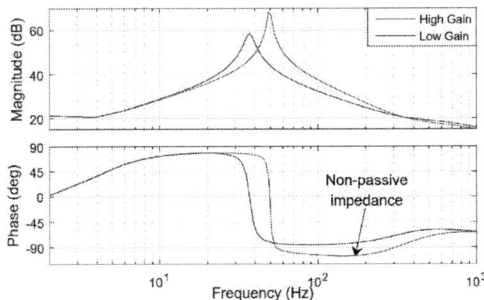

Fig. 2: Inverter output impedances with two different grid-voltage feedforward gains.

B. Maximum-Length Binary Sequences

Maximum-length binary sequence (MLBS) is the most common form of the pseudo-random sequences, and has proven to be effective in online measurements of the grid-connected systems for example in [20] and [21]. The MLBS is a periodic binary signal which can be straightforwardly generated by using a shift register with an XOR-feedback loop [22]. Due to deterministic binary form, the MLBS has the lowest possible peak factor. Therefore, the measurements can be averaged over many periods, allowing the use of relatively low signal amplitude of perturbations which do not disturb the normal system operation and power quality too much during the identification [13].

Fig. 3 shows a normalized power spectrum of the MLBS with a sequence length of 127 bits generated at 4 kHz (f_{gen}). The power spectrum of the MLBS follows $sinc^2$ function [22] and the power drops to zero at the generation frequency. The power spectrum is usually considered flat to $0.45f_{gen}$ which is typically considered as the measurable bandwidth.

C. Grid Inductance Identification

According to (2), the inductance value L_g can be identified from the imaginary part of complex-plane grid impedance as

$$L_g = \frac{imag[Z_g(j\omega)]}{\omega} = \frac{imag[R_g + j\omega L_g]}{\omega} = \frac{\omega L_g}{\omega} \quad (5)$$

from which $Z_g(j\omega)$ is measured in real-time as a ratio of MLBS-perturbed grid current and voltages.

D. Real-time Stability Assessment

As the adaptive control of the feedforward requires real-time data of the grid inductance, the MLBS is used to perturb the grid-side waveforms. The perturbed grid currents and voltages can be applied also for identifying the grid impedance for stability assessment, in which the grid impedance is computed from Fourier transformed measurements of the perturbed grid currents and voltages [23]. For the impedance-based stability analysis, the analytical model of the inverter output impedance is applied. The stability assessment can be visualized by plotting the Nyquist contours of both components continuously based on the ratio of the analytical inverter output impedance and the measured grid impedance. The Nyquist stability criterion states that system is unstable if Nyquist curve encircles the critical point (-1,0) clockwise in the complex plane [24].

IV. EXPERIMENTS

A. System Under Study

Fig. 4 shows the inverter control system which applies grid-voltage feedforward gains K_{ffd} and K_{ffq}. The gains are automatically adjusted, based on real-time measurements of the grid impedance. The grid-voltage feedforward may lead the inverter output impedance to lose its passivity (decrease phase below -90 degrees) in certain frequency band which may cause stability issues if the inverter is connected to a very inductive grid. In that case, the grid-voltage feedforward gains must be re-adjusted to avoid stability issues. L_g represents the measured grid inductance, and is used to adjust suitable feedforward gains in this implementation.

The inverter itself it configured to inject the MLBS continuously to the output current d component. The perturbed grid currents and response from voltages are sampled using the inverter's own sensors. No external signal generators are required to perturb the grid currents nor sensors to collect the grid-side waveforms. Thus, the real-time data of L_g can be provided easily for adaptive control system without using any external measurement or signal generation devices. The impedance-based stability assessment for the grid-connected inverter, performed in real time, requires relatively heavy computational effort and may need the use of external computation devices.

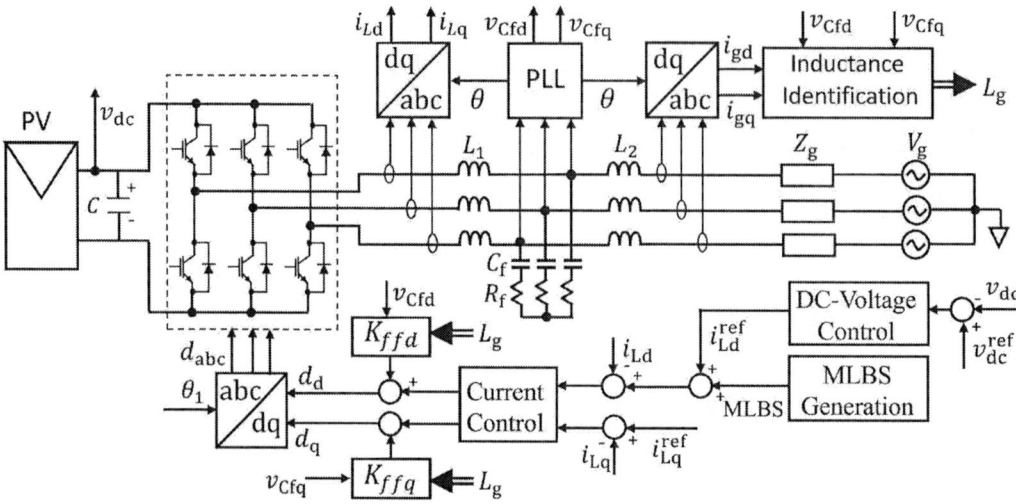

Fig. 4: Three-phase inverter with adaptive grid-voltage feedforward gains.

In this paper, the real-time stability assessment is performed by external computational device which utilizes the MLBS used for adaptive control.

B. Experimental setup

The presented methods are experimentally verified by a photovoltaic grid-connected inverter. Fig. 5 shows a setup of a three-phase IGBT inverter in which the internal control system is bypassed and replaced by real-time simulator dSPACE model 1103. The PV simulator is operated at 3 kW. The grid emulator acts as a three-phase voltage sink, repeating stiff reference waveforms generated by the dSPACE. The real-time impedance-based stability assessment is performed in parallel with the adaptive control using a measurement card NI USB-6363 and PC (MATLAB/Data Acquisition Toolbox). The variation in the grid impedance is implemented by using an inductor which can be bypassed using relay.

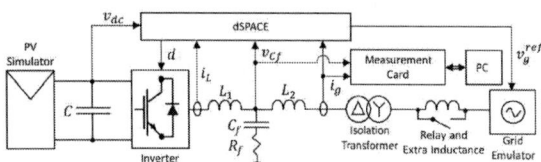

Fig. 5: Experimental setup.

Three sets of experiments are performed to demonstrate the efficiency of the presented methods. The first set analyzes how the MLBS injection affects the grid currents, and shows that the implemented inductance measurements using such injections do not disturb the produced power quality too much. In the second experiment, the adaptive control of the grid-voltage feedforward is implemented. The results will show the effectiveness of adaptive control in order to avoid stability issues

in high-impedance grids. The last set of experiments shows the real-time stability assessment, performed in parallel to adaptive control. In the stability assessment, the inverter output impedance is updated by taking the adaptive control into account.

C. Experiment 1

In the first experiment, a 127-bit-length MLBS was injected into the d-component output-current reference with a generation frequency of 4 kHz, and the grid currents were measured. The measurement setup allowed adjusting the injection amplitude during the measurements. Higher amplitude provides more accurate results as the signal-to-noise (SNR) ratio increases. In the online- and especially real-time implementations, the perturbation effect to the produced power quality has to be considered because the injections are added on the top of the grid currents. The SNR can be improved by averaging the results over a multiple injection periods, but this increases the measurement time. That limits the number of averaging periods in real-time implementations. Hence, the selection of the amplitude is a trade-off between measurement accuracy and minimal disturbance to grid currents.

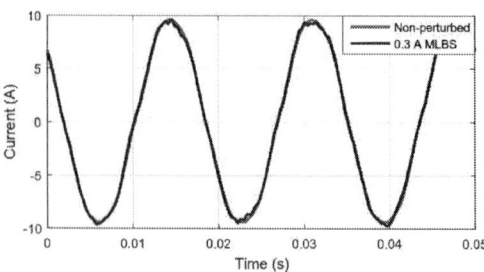

Fig. 6: Effect of the perturbation signal to the output current.

The MLBS was applied with an amplitude of 0.3 A, which provided a sufficiently low measurement time and reliable inductance values. Fig. 6 shows the effect of the applied MLBS to phase current. Fig. 7 shows the normalized power spectra of the non-perturbed (red) and perturbed (blue) phase currents. It should be noted that the power values are presented on a logarithmic scale. The conventional total harmonic distortion (THD) analysis (only harmonics of 60 Hz) hides the effect of the MLBS, because the MLBS has energy at many frequencies, not only at grid-voltage harmonics. To analyze the harmonics, interharmonics, and the effect of the MLBS injection based on the shown power spectra, the share of total distortion (STD) is derived in this paper as

$$\text{STD} = \frac{\int_0^{4000} E(f)df - E(60)}{E(60)} \qquad (6)$$

where the energy contents $E(f)$ of all frequencies (excluding the fundamental) is divided by the energy at the fundamental frequency (60 Hz). Here the analysis is limited to the frequencies below 4 kHz which is the generation frequency of the MLBS. For the current flowing to grid without perturbation, the STD value is 0.0011, and with the MLBS only 0.0005 higher (0.0016). For comparison, 5 % harmonic in phase current increases the STD value by 0.0025 (=0.05^2), and thus has five times higher impact than the applied MLBS.

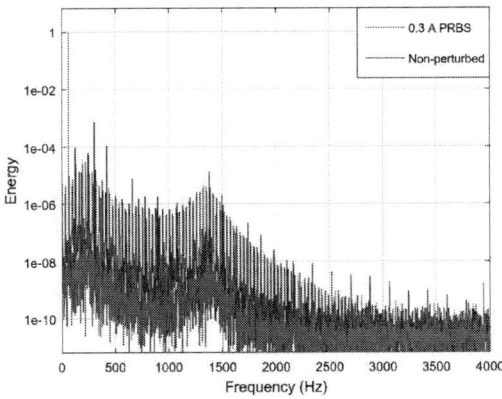

Fig. 7: Spectral analysis of perturbed and non-perturbed current.

D. Experiment 2

In the second experiment, the adaptive control of the grid-voltage feedforward gains is implemented. The MLBS generated in Experiment 1 is applied for the real-time measurements of the grid inductance. The analytical models of the inverter output impedances $Z_{o\text{-}dd}$ and $Z_{o\text{-}qq}$ presented in [15] are used. The adaptive control of grid-voltage feedforward is implemented by setting the gains to zero when the grid inductance value is increased above 5 mH and back to ideal value ($1/V_{\text{DC}}$) when the inductance is decreased below the same limit. Therefore, the limit of 5 mH is representing change from low- to high-impedance grid. It was observed that the phase margins became

insufficient when the inverter with ideal feedforward gains was connected to the grid with inductance of 5 mH or more. Thus, the inverter impedance is re-shaped in order to provide sufficient phase margin. The desired phase boost is provided by reducing the feedforward gains. In this implementation the gains are reduced directly to zero. By using this implementation, the stability issues with high-impedance grid are effectively avoided.

One measurement cycle takes 0.03175 s. The measurements are averaged over 10 injection periods in order to increase the signal-to-noise ratio. The feedforward gains are immediately changed when the limit of 5 mH is crossed. The measured grid voltage are added to the duty ratios of d and q components through the feedforward gains. Using the ideal gain ($1/V_{\text{DC}} = 0.00214$), the value added to the duty ratios in balanced (60 Hz, 120 V) grid are: 0.4096 for d component and zero for q component. The steady-state value for duty ratio d component is 0.4122. Hence, if the feedforward gains are adaptively set to zero as inductance has increased above the limit, the duty ratio d component drops immediately from its steady-state value to almost zero (0.4122 - 0.4096 = 0.0026).

Fig. 8 shows the inverter output current when too drastic change in duty ratios occurs at 2.4 s. The inverter control system can not compensate the change before the DC-side voltage starts to increase. This activated the protection system of the PV-emulator. The DC-side protection system disconnects the PV emulator from the inverter approximately at 2.6 s. This issue is solved by adding the ramp-rate limiters to feedforward paths in order to avoid too rapid changes in duty ratios.

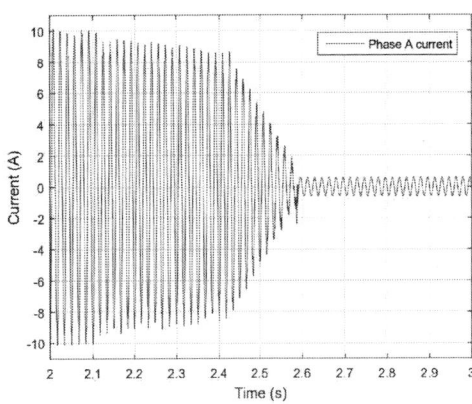

Fig. 8: Too fast changes in feedforward cause issues at the DC side.

Fig. 9 shows the internal trigger signal for the adaptive control of the feedforward gains, the actual value of the d component added to duty ratio through feedforward path, and the measured grid inductance during the increase in grid impedance approximately at 2 s. The grid impedance is suddenly increased by switching off the bypass relay of the extra inductance (shown in Fig. 5). The measured grid inductance do not rise immediately to its steady-state value due to averaging. Once the measured inductance

551

The 2018 International Power Electronics Conference

value is crossed the 5 mH limit, the feedforward gains are set to zero. Due to ramp-rate limiter for falling edges in feedforward path, the decrease of d-component value from the ideal (0.4096) to zero takes approximately 0.3 s.

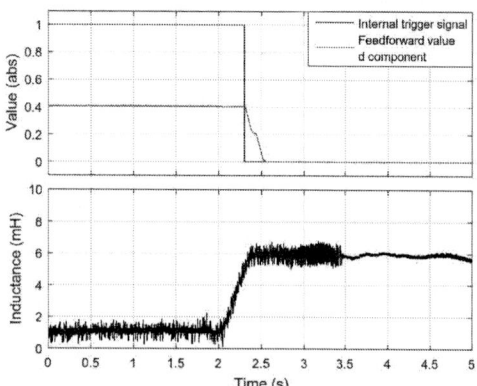

Fig. 9: Adaptive control of feedforward when grid inductance is increasing.

Fig. 10 shows the same parameters as Fig. 9 but the feedforward gains are set to the ideal value from zero. The measured grid inductance decreases below the limit of 5 mH at 1.2 s, and the feedforward gains are set to their ideal values. The d-component value, added to duty ratios through the feedforward path, starts to increase towards its ideal value of 0.4096. The slower ramp-rate limiter for rising edges is used for avoiding instability at output currents due to duty-ratio references rise over their maximum values.

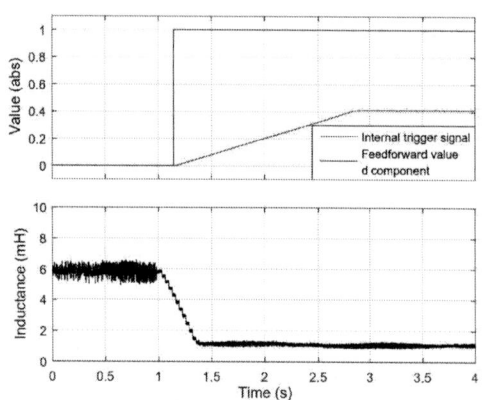

Fig. 10: Adaptive control of feedforward when grid inductance is decreasing.

E. Experiment 3

This experiment shows an impedance-based stability assessment of the studied system. The assessment is performed in real time, under the same operating conditions

and MLBS injection as applied in Experiment 2. The measurement card continuously collects the perturbed d-component voltages and currents from the grid side. From the sampled waveforms, the d component of the grid impedance can be computed in real-time and used for the impedance-based stability assessment with the analytical model of the inverter output impedance. To produce accurate stability assessment in real time, the varying output impedance of the inverter, caused by adaptive controllers, must be updated in the analytical models. According to approximation in (2), the impedance-based stability assessment can be performed for both d and q components by measuring only the d component of the grid impedance. In this experiment, a highly accurate grid-impedance estimation is desired in order to study the stabilization effect of the proposed adaptive control methods. Thus, the grid-impedance results are averaged over 100 periods, assessing the impedance-based stability once in every 3.175 s.

Fig. 11 shows Bode plot of the measured Z_g in a low-impedance grid, and the inverter output impedances Z_{o-dd} and Z_{o-qq} when ideal feedforward gains are used. The grid impedance do not overlap neither of the inverter impedances in magnitude. This provides sufficient stability margins which is confirmed by the Nyquist contour of the impedance ratio.

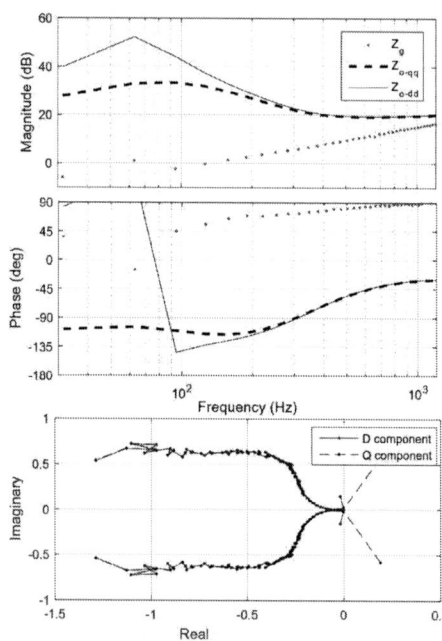

Fig. 11: Bode plot and Nyquist diagram of the measured impedances when applying optimal feedforward gains in low-impedance grid.

Fig. 12 shows a visualization of the real-time stability assessment in a case when the feedforward gains are adaptively set to zero due to a high grid impedance. The grid impedance overlaps both d and q components of the inverter output impedance approximately at 220 Hz. Due to reduced feedforward gains, and thus, additional phase

552

The 2018 International Power Electronics Conference

boost, the phase margin is sufficient for both components: 64 degrees for the d component and 54 degrees for the q component.

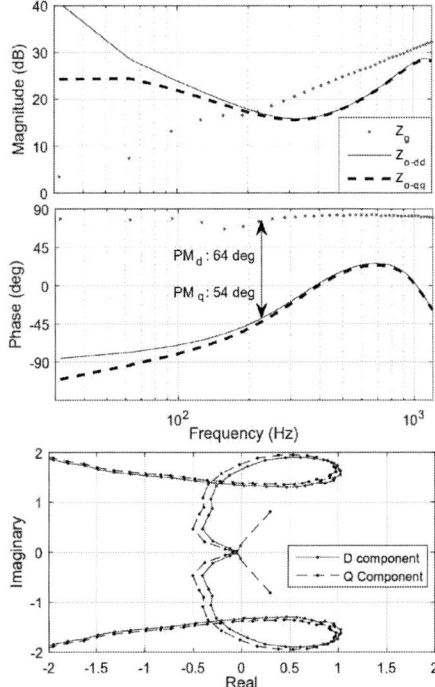

Fig. 12: Bode plot and Nyquist diagram of the measured impedances applying zero feedforward gains.

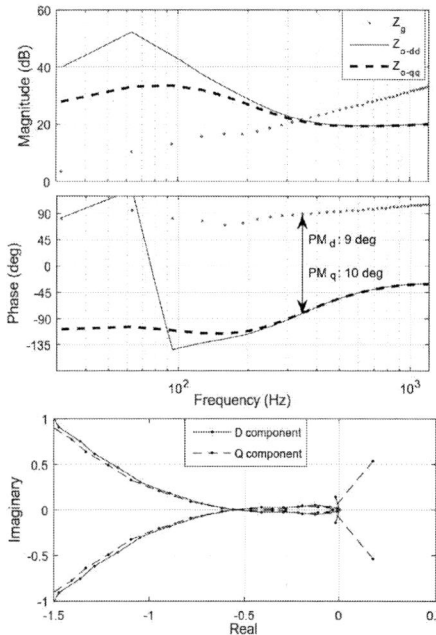

Fig. 13: Bode plot and Nyquist diagram of the measured impedances applying overly high feedforward gains in high-impedance grid.

Fig. 13 shows a stability analysis when the inverter is connected to a high-impedance grid. The feedforward gains are not adaptively changed and they remain at constant value $(1/V_{dc})$. Due to overly high feedforward gains, the magnitudes of the inverter output impedances are increased and the grid impedance overlaps both of them approximately at 350 Hz. The feedforward gains produce significant phase-lag to inverter output impedances, resulting very low phase margins for both components: 9 degrees for the d component and 10 degrees for the q component. The Nyquist contour confirms the low stability margins. An inverter with this low stability margins is very sensitive for external disturbances and a small increase in grid inductance may lead the inverter to lose its stability. [15] Such issues can be effectively avoided by the presented adaptive control of the grid-voltage feedforward gains.

V. CONCLUSION

Grid-voltage feedforward is used to mitigate the effects of the grid voltage harmonics but it may cause impedance-based stability issues when the inverter is connected to a high-impedance grid. Real-time identification of time-varying grid impedance and continuous stability assessment provide effective tools for adaptive control of the grid-connected inverters in order to avoid the stability issues caused by the grid-voltage feedforward.

This paper introduced a practical implementation of an adaptive control of grid-voltage feedforward based on real-time estimation of the grid inductance. By adaptively adjusting the feedforward parameters, the stability issues can be effectively avoided under varying grid conditions.

ACKNOWLEDGEMENT

This work is supported by the Academy of Finland.

REFERENCES

[1] B. K. Bose, "Global energy scenario and impact of power electronics in 21st century," *IEEE Transactions on Industrial Electronics*, vol. 60, no. 7, pp. 2638–2651, July 2013.

[2] J. L. Agorreta, M. Borrega, J. Lpez, and L. Marroyo, "Modeling and control of n-paralleled grid-connected inverters with lcl filter coupled due to grid impedance in pv plants," *IEEE Transactions on Power Electronics*, vol. 26, no. 3, pp. 770–785, March 2011.

[3] C. Li, "Unstable operation of photovoltaic inverter from field experiences," *IEEE Transactions on Power Delivery*, no. 99, pp. 1–1, 2017.

[4] L. Jessen and F. W. Fuchs, "Modeling of inverter output impedance for stability analysis in combination with measured grid impedances," in *2015 IEEE 6th International Symposium on Power Electronics for Distributed Generation Systems (PEDG)*, June 2015, pp. 1–7.

[5] T. Messo, R. Luhtala, R. Roinila, D. Yang, X. Wang, and F. Blaabjerg, "Real-time impedance-based stability assessment of grid converter interactions," in *the Eighteenth IEEE Workshop on Control and Modeling for Power Electronics, IEEE COMPEL 2017*, 2017.

[6] J. Sun, "Impedance-based stability criterion for grid-connected inverters," *IEEE Transactions on Power Electronics*, vol. 26, no. 11, pp. 3075–3078, Nov 2011.

[7] L. Harnefors, X. Wang, A. G. Yepes, and F. Blaabjerg, "Passivity-based stability assessment of grid-connected vscs - an overview," *IEEE Journal of Emerging and Selected Topics in Power Electronics*, vol. 4, no. 1, pp. 116–125, March 2016.

[8] T. Messo, J. Jokipii, A. Mkinen, and T. Suntio, "Modeling the grid synchronization induced negative-resistor-like behavior in the output impedance of a three-phase photovoltaic inverter," in *2013 4th IEEE International Symposium on Power Electronics for Distributed Generation Systems (PEDG)*, July 2013, pp. 1–7.

[9] M. Cespedes and J. Sun, "Online grid impedance identification for adaptive control of grid-connected inverters," in *2012 IEEE Energy Conversion Congress and Exposition (ECCE)*, Sept 2012, pp. 914–921.

[10] A. Rygg and M. Molinas, "Real-time stability analysis of power electronic systems," in *2016 IEEE 17th Workshop on Control and Modeling for Power Electronics (COMPEL)*, June 2016, pp. 1–7.

[11] J. Jokipii, T. Messo, and T. Suntio, "Simple method for measuring output impedance of a three-phase inverter in dq-domain," in *2014 International Power Electronics Conference (IPEC-Hiroshima 2014 - ECCE ASIA)*, May 2014, pp. 1466–1470.

[12] T. Roinila, M. Vilkko, and J. Sun, "Broadband methods for online grid impedance measurement," in *2013 IEEE Energy Conversion Congress and Exposition*, Sept 2013, pp. 3003–3010.

[13] R. Luhtala, T. Messo, T. Roinila, T. Reinikka, J. Sihvo, and M. Vilkko, "Adaptive control of grid-connected inverters based on real-time measurements of grid impedance: Dq-domain approach," in *2017 IEEE Energy Conversion Congress and Exposition (ECCE)*, Oct 2017, pp. 69–75.

[14] T. Messo, A. Aapro, T. Suntio, and T. Roinila, "Design of grid-voltage feedforward to increase impedance of grid-connected three-phase inverters with lcl-filter," in *2016 IEEE 8th International Power Electronics and Motion Control Conference (IPEMC-ECCE Asia)*, May 2016, pp. 2675–2682.

[15] T. Messo, R. Luhtala, A. Aapro, and R. Roinila, "Accurate impedance model of grid-connected inverter for small-signal stability asssessment in high-impedance grids," in *IPEC-Niigata 2018 -ECCE ASIA*, 2018.

[16] T. Suntio, T. Messo, and J. Puukko, *Power Electronic Converters: Dynamics and Control in Conventional and Renewable Energy Applications.* Wilwy-VCH, 2017.

[17] B. Wen, R. Burgos, D. Boroyevich, P. Mattavelli, and Z. Shen, "Ac stability analysis and dq frame impedance specifications in power-electronics-based distributed power systems," *IEEE Journal of Emerging and Selected Topics in Power Electronics*, vol. 5, no. 4, pp. 1455–1465, Dec 2017.

[18] M. Liserre, R. Teodorescu, and F. Blaabjerg, "Stability of photovoltaic and wind turbine grid-connected inverters for a large set of grid impedance values," *IEEE Transactions on Power Electronics*, vol. 21, no. 1, pp. 263–272, Jan 2006.

[19] T. Messo, J. Jokipii, and T. Suntio, "Effect of conventional grid-voltage feedforward on the output impedance of a three-phase photovoltaic inverter," in *2014 International Power Electronics Conference (IPEC-Hiroshima 2014 - ECCE ASIA)*, May 2014, pp. 514–521.

[20] T. Roinila, R. Luhtala, T. Reinikka, T. Messo, A. Aapro, and J. Sihvo, "dspace implementation for real-time stability analysis of three-phase grid-connected systems applying mlbs injection," in *9th EUROSIM Congress on Modelling and Simulation, EUROSIM2016*, Sept 2016.

[21] T. Roinila, T. Messo, and E. Santi, "Mimo-identification techniques for rapid impedance-based stability assessment of three-phase systems in dq domain," *IEEE Transactions on Power Electronics*, vol. 33, no. 5, pp. 4015–4022, May 2018.

[22] K. R. Godfrey, *Perturbation Signals for System Identification.* Prentice Hall, UK, 1993.

[23] Z. Sharif and A. Z. Sha'ameri, "The application of cross correlation technique for estimating impulse response and frequency response of wireless communication channel," in *2007 5th Student Conference on Research and Development*, Dec 2007, pp. 1–5.

[24] R. Dorf and R. Bishop, *Modern Control Systems.* Prentice-Hall, 2000.

Model Based Tuning of Proportional Resonant Controllers for Voltage Source Inverters

Stefan Almér, Thomas Besselmann and Mario Schweizer
ABB Corporate Research
Segelhofstrasse lK, 5405 Baden-Dattwil, Aargau, SWITZERLAND
Email: stefan.almer@ch.abb.com

Abstract—**The paper considers the optimal choice of gains of proportional-resonant controllers applied to control voltage source inverters. An inverter, controlled in closed loop, but without proportional resonant controllers, is modeled as a second order transfer function from reference to output. The system is augmented with proportional-resonant controllers by feeding the error between reference and output back through a set of N proportional resonant controllers to subsequently alter the reference. The root locus of the closed loop system is considered as a function of the proportional-resonant gains. To find the optimal choice of gains, we maximize the damping of the mode with smallest damping. This corresponds to solving a nonlinear min-max problem. After linearization, the problem is stated as a linear program.**

I. INTRODUCTION

Uninterruptible power source (UPS) systems are used in industrial processes in order to decouple loads partially from the grid. Short power outages are compensated and the load is supplied with a clean voltage waveform. Furthermore, UPS systems mitigate the injection of current harmonics to the utility grid that originate from high power non-linear loads. Consequently, converters for UPS applications are required to have a very high output voltage quality even in presence of highly non-linear loads such as diode rectifiers.

The system efficiency is a key aspect of such systems. Usually, very low semiconductor switching frequencies in the range of 2-4 kHz are employed in order to limit the switching losses and keep the efficiency high. Passive filter components such as inductors and capacitors are minimized such that further power losses are avoided. The filtering performance of these passive filters is usually poor for low order harmonics of non-linear loads. Therefore, the output voltage quality has to be ensured by means of proper control.

Due to the low switching frequency, the closed loop voltage control bandwidth is limited and usually not sufficient to cope with non-linear loads [1]. Additional means of compensating the voltage harmonics are required such as harmonic compensators tuned at the specific harmonic frequencies. Harmonic compensators can be implemented e.g. with proportional resonant (PR) controllers suggested in [2]–[8].

Although the performance of proportional resonant controllers in compensating harmonics was investigated extensively, only few publications deal with the proper selection

and tuning of the gains of the PR controllers. In [9] and [10] analytical parameter tuning rules are provided, but only for a system with a single PR controller tuned at the fundamental frequency. For systems containing several PR controllers tuned at the harmonic frequencies, approximate and empiric parameter tuning rules are given in [2] and [5]. In [7], it is suggested to investigate the bode-plot of the open loop transfer function and design for phase margin. However, this approach should only be applied to closed-loop systems that can be represented as a second order system. Due to the introduction of the harmonic compensators, additional poles are introduced and the system is turned into a high order system. Designing for phase margin can lead to unexpected closed loop system behavior in that case.

To Summarize, no systematic parameter tuning approach considering the interactions of the individual PR controllers and the impact on the damping of the resonant modes is given.

In this paper, a method is presented to optimally choose the gains of the PR controllers: First the inverter is considered without PR controllers and the damping of the inverter is computed. We then decide on the amount of decrease of the damping, which is caused by the introduction of PRs, we can accept. We then maximize the damping of the least damped PR controller, while respecting that the damping of the inverter does not fall below the specified limit. This corresponds to solving a nonlinear min-max problem. After linearization, the problem is stated as a linear program which can be solved efficiently.

The paper is outlined as follows: Section II introduces the model of the inverter and PR controllers. Section III formulates the PR gain design problem as an optimization problem, which is then approximated and solved in Section IV. The method is applied to a numerical example and evaluated in simulation in Section V. Finally, conclusion and outlook to further work are given in Section VI.

II. CONVERTER MODEL

Our starting point is to consider a voltage source inverter (VSI) which is assumed to operate in closed loop, but without PR controllers. The controlled inverter is modeled as a second order transfer function which maps the (sinusoidal) reference

The 2018 International Power Electronics Conference

to the output,

$$y = G(s)y_{\text{ref}}, \quad G(s) = \frac{\omega^2}{s^2 + \xi\omega s + \omega^2}, \quad (1)$$

where ω is the natural frequency and ξ is the damping of the controlled inverter. One example of a system which can be modeled on the form (1) is a VSI with LC filter, controlled in abc frame by a cascaded voltage-current control system comprising proportional control. We note that, with properly designed control, a VSI is expected to behave as second order systems in closed loop. Thus, assuming a system model of the form (1) is not restrictive.

A. Proportional-Resonant Control

To achieve offset free tracking of the sinusoidal reference y_{ref} and to reduce harmonics in the output, we consider adding PR controllers [11] to the system (1). The PR controllers are added in an outer loop (see Fig. 1) and adjust the reference according to

$$\tilde{y}_{\text{ref}} = y_{\text{ref}} + \sum_{n \in \{1,3,5,...,N\}} H_n(s)(y_{\text{ref}} - y),$$

where

$$H_n(s) = \frac{\lambda_n s}{s^2 + (n\omega_0)^2},$$

where the fundamental frequency ω_0 is the frequency of the reference (typically 50 or 60 Hz), and where λ_n are the feedback gains of the PR controllers. The gains λ_n are tuning parameters which affect the transient response of the closed loop system.

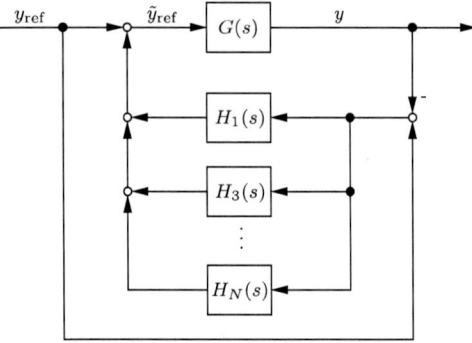

Fig. 1.   Control structure of converter model with PR controllers.

B. Closed Loop System

The order of the closed loop system dynamics is $2 + 2N_{\text{PR}}$ where N_{PR} is the number of PR controllers added in the outer loop. The resulting closed-loop system can be stated as

$$y = \frac{G(s)\left(I + \sum_n H_n(s)\right)}{I + G(s)\sum_n H_n(s)} y_{\text{ref}}, \quad (2)$$

and by changing the gains λ_n, we influence the location of the poles and zeros of the closed-loop system.

C. Closed Loop Poles

The PR gains λ_n affect the poles of the closed loop system. In the design approach outlined below, we seek to maximize the damping of the (complex) pole pair which has the lowest damping.

To clarify the approach we consider an example: Consider the case where two PR controllers (with harmonics number 1 and 3) are included in the control loop. For this case the system has 6 poles, and their position in the complex plane is determined by two gains λ_1, λ_3. We enumerate different combinations of gains λ_1, λ_3 and plot the resulting poles on the complex plane; the result is shown in Fig. 2. In this figure we also plot the poles obtained when both gains are close to zero (blue circles), and the poles obtained with one particular choice of higher gains (red circles).

From Fig. 2 it can be seen that one pole pair moves to the right, closer to the imaginary axis (and unstable domain), while the other two pole pairs move left. For sufficiently high gains, one of the pole pairs turn into two purely real poles, one of which moves left and the other moves right, towards the unstable domain.

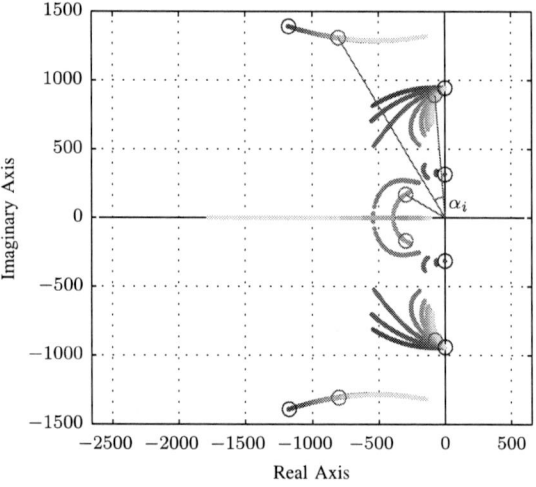

Fig. 2.   Poles of the closed loop system with $N_{PR} = 2$ PR controllers: The green points show poles for various combinations of gains λ_1, λ_3. Blue circles show the poles for low gains. Red circles show poles for high gains.

Since changes in one gain affects all poles, it is not obvious how to choose the gains optimally. Increase in one particular gain may make one pole pair "more stable", but may have negative effects on another pole pair.

III. PROBLEM FORMULATION

To address the problem of how to choose the PR gains, we propose to formulate a max-min optimization problem: We first consider the damping of the transfer function G, representing the inverter without PR controllers. We decide on a bound on how much we can accept the damping to decrease, and we then maximize the damping of the least damped mode of the PR controllers.

Let α_0 be the angle between the pole and the imaginary axis (assuming the pole is in the open left half plane) of the second order transfer function G in (1), i.e., α_0 is the damping of the system without PRs in the loop;

$$\alpha_0 = \tan^{-1}(-\text{real}(p_0)/\text{imag}(p_0))$$

where

$$p_0 = -\frac{\xi\omega}{2} + \sqrt{\left(\frac{\xi\omega}{2}\right)^2 - w^2}.$$

Adding PRs to the control loop will inevitably decrease the damping of G. We decide on the amount of decrease of damping we are willing to accept and define

$$\alpha_{\text{tol}} = \kappa \cdot \alpha_0$$

where $\kappa \in (0, 1)$.

We now add the PRs to the control loop as illustrated in Fig. 1. The number α_{tol} is used as a bound on the damping of the mode corresponding to the second order transfer function in Fig. 1. Denoting this damping α_1, the problem we ideally want to solve is

$$\begin{aligned}
\max_{\lambda_1, \lambda_3, \ldots, \lambda_N} \min_{i \in \{2, \ldots, i_{\max}\}} & \quad \alpha_i(\lambda_1, \lambda_3, \ldots, \lambda_N) \\
s.t. & \quad \alpha_1(\lambda_1, \lambda_3, \ldots, \lambda_N) \geq \alpha_{\text{tol}}
\end{aligned} \tag{3}$$

with $i_{\max} = (N + 3)/2$. We note that the angles α_i are dependent on the PR gains λ_n, and that as the gains vary, different angles take on the role of being "the least damped". We also note that the gains also have to be chosen to keep the closed loop system stable.

Figure 3 shows $\min_{i \in \{1,2,3\}} \alpha_i$; the smallest damping of the three pole pairs as a function of the gains. High gains for both PR controllers push one pole into the right half plane, resulting in an unstable system.

IV. PROBLEM APPROXIMATION

To obtain a tractable optimization problem, we proceed to approximate $\alpha_i(\lambda_1, \lambda_3, \ldots \lambda_N)$ with affine functions of the gains λ_n: That is, the angles are approximated by

$$\tilde{\alpha}_i(\lambda_1, \lambda_3, \ldots, \lambda_N) = a_i^T \lambda + b_i \tag{4}$$

where $\lambda = \begin{bmatrix} \lambda_1 & \lambda_3 & \ldots & \lambda_N \end{bmatrix}^T$ is a vector containing the gains and where $a_i \in \mathbb{R}^{N_{\text{PR}}}$, $b_i \in \mathbb{R}$ are constant vectors obtained by a sampling and least squares fitting procedure:

Values of the angles α_i are sampled for a number of gain values λ_j; we thus obtain a set of sampling points $\{\alpha_i(\lambda_j)\}_{j=1}^M$, $i = 1, \ldots, 2 + 2N_{\text{PR}}$. The vectors a_i, b_i are chosen to solve the least squares fitting problem

$$\min_{a_i, b_i} \sum_{j=1}^M \left(\alpha_i(\lambda_j) - (a_i^T \lambda_j + b_i)\right)^2.$$

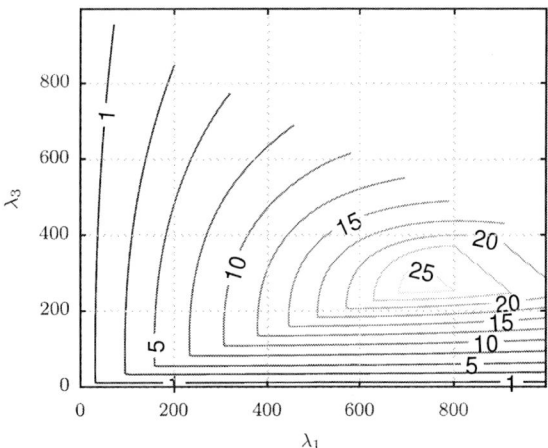

Fig. 3. Minimum of the damping α_i of the pole pairs of the closed loop system with $N_{\text{PR}} = 2$ PR controllers as a function of the PR controller gains λ_1 and λ_3. The labels on the contour lines indicate damping in degrees.

The problem above is an unconstrained quadratic problem which can be solved by soling a linear system of equations.

By describing the angles in (3) with the approximation (4), we obtain a max-min problem with affine cost function:

$$\begin{aligned}
\max_{\lambda_1, \lambda_3, \ldots, \lambda_N} \min_{i \in \{2, \ldots, i_{\max}\}} & \quad a_i^T \lambda + b_i \\
s.t. & \quad a_1^T \lambda + b_1 \geq \alpha_{\text{tol}}
\end{aligned} \tag{5}$$

This problem can be equivalently formulated a linear program (LP) according to

$$\min c^T x, \quad s.t. \quad Ax \leq b \tag{6}$$

with matrices

$$A = \begin{bmatrix} 0 & -a_1^T \\ 1 & -a_2^T \\ \vdots \\ 1 & -a_{i_{\max}}^T \end{bmatrix}, \quad b = \begin{bmatrix} b_1 - \alpha_{\text{tol}} \\ b_2 \\ \vdots \\ b_{i_{\max}} \end{bmatrix}, \quad c = \begin{bmatrix} -1 \\ 0 \\ \vdots \\ 0 \end{bmatrix}.$$

Linear programs can be solved efficiently with readily available software.

V. NUMERICAL EXAMPLE

The PR gain design approach outlined above was applied to a VSI with LC filter used in UPS applications. The VSI considered is a four-wire topology where the dynamics of the three phases are decoupled, due to the connection between filter and DC side neutral point. Because of the decoupling, we consider each of the three phases individually and design stabilizing controllers with proportional feedback. The closed loop system thus becomes a second order transfer function from voltage reference to output voltage on the form (1).

The inner control loop is augmented with $N_{\text{PR}} = 4$ PR controllers as described in Fig. 1. We thus have five

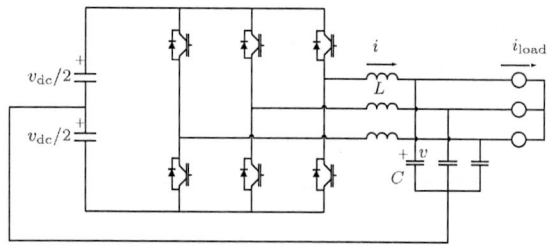

Fig. 4. Voltage source inverter with LC filter.

angles α_i which are functions of four gains λ_i. The nonlinear functions α_i are sampled over a grid of gain values and a linear approximation of the nonlinear functions is made by least squares fitting. We choose the parameter $\kappa = 0.9$ and thus allow for a 10% decrease of damping;

$$\alpha_{\mathrm{tol}} = 0.9 \cdot \alpha_0.$$

The resulting LP (5) is solved. The optimal solution (in per unit) is

$$\lambda_{1,\mathrm{opt}} = 0.34, \quad \lambda_{2,\mathrm{opt}} = 1.04, \lambda_{3,\mathrm{opt}} = 1.16, \quad \lambda_{4,\mathrm{opt}} = 1.33.$$

We note that the gain increase for higher order PRs.

The poles of the system, using the optimal gains, is shown by the red circles in Fig. 5. The red lines illustrate the limit α_{tol} and the optimal solution value α_{opt} of the LP (5).

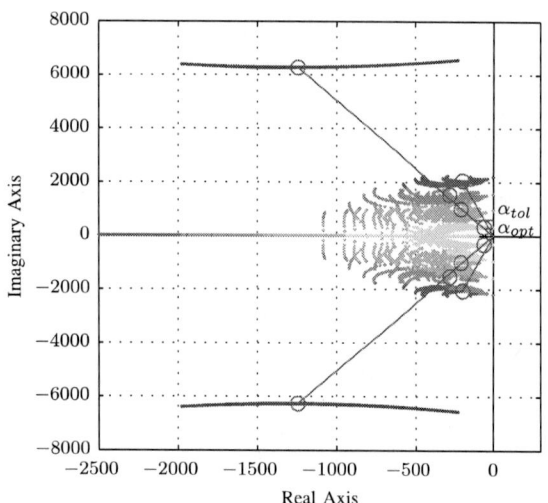

Fig. 5. Poles of the closed loop system with $N_{PR} = 4$ PR controllers: The green points show poles for various combinations of gains λ_1, λ_3, λ_5 and λ_7. Red circles show poles for the optimal gains.

The system is simulated with a nonlinear load; a diode rectifier bridge. The dynamic response and the steady state behavior with and without PR controllers is evaluated in simulation. All values are in per unit.

When the system is at steady state the load is switched out. The transient response of the output voltage is shown in Fig. 6.

It can be seen that the introduction of the PR controllers (blue line) only causes minor changes in the transient peak value. However, the PRs introduce a slow oscillation which takes two grid periods to damp out.

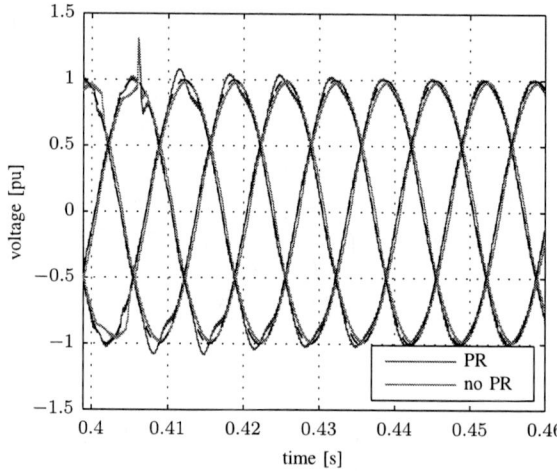

Fig. 6. Transient response of the output voltage when the load (a diode rectifier bridge) is switched out.

The steady state output voltage and load current are shown in Fig. 7 and 8. The harmonics of the steady state voltage are shown in Fig. 9. Both figures verify that the PR controllers result in closer tracking of the sinusoidal reference, as well as reduction of harmonic content. From Fig. 9, it can be seen that the PR controllers (blue circles) reduce the third, fifth and seventh harmonics to less than -60 dB. Without the PR controllers, the THD is 6.2%. Adding the PR controllers reduces the THD to 2.7%.

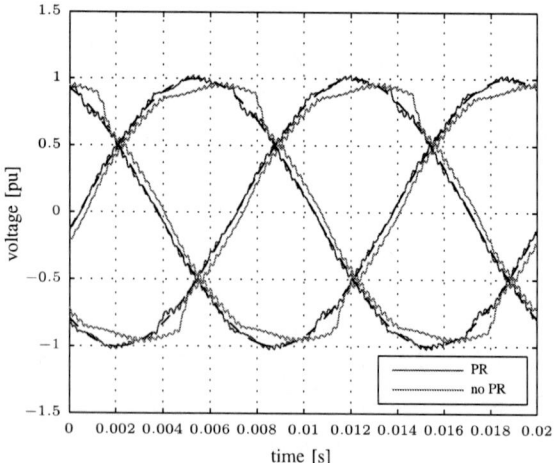

Fig. 7. Steady state output voltage with diode rectifier bridge as load.

The 2018 International Power Electronics Conference

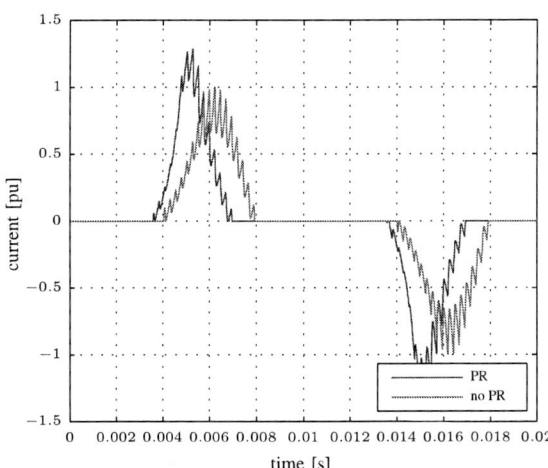

Fig. 8. Steady state load current of phase a with diode rectifier bridge as load.

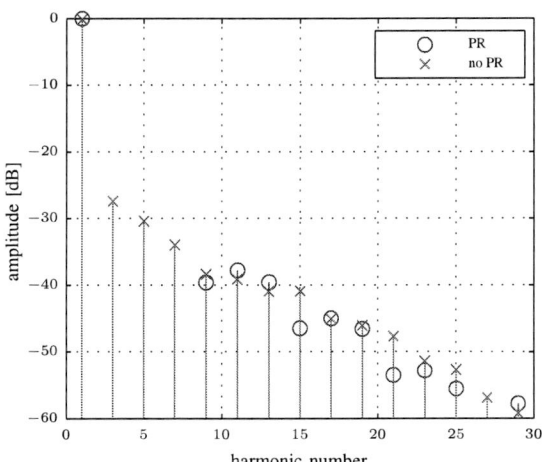

Fig. 9. Harmonics of the steady state output voltage with diode rectifier bridge as load.

VI. CONCLUSIONS

The design of PR gains was formulated as a min/max optimization problem: The inverter, without PR controllers, was modeled as a second order transfer function. It is noted that the damping of this transfer function is inevitably decreased by the introduction of PR controllers. We decide on a limit on how much we are willing to decrease the damping, and maximize the damping of the PR controllers, while respecting the bound on the decrease of damping of hte original transfer function. This problem is approximated as a linear program which can be solved efficiently. The method is verified in simulation.

REFERENCES

[1] S. Almér and U. Jönsson. Harmonic analysis of pulse-width modulated systems. *Automatica*, 45(4):851–862, 2009.

[2] P. Mattavelli. Synchronous-Frame Harmonic Control for High-Performance AC Power Supplies. *IEEE Transactions on Industry Applications*, 37(3):864–872, May 2001.

[3] E. Demirkutlu, S. Cetinkaya, and A. M. Hava. Output Voltage Control of A Four-Leg Inverter Based Three-Phase UPS by Means of Stationary Frame Resonant Filter Banks. In *2007 IEEE International Electric Machines Drives Conference*, volume 1, pages 880–885, May 2007.

[4] M. Monfared, S. Golestan, and J. M. Guerrero. Analysis, Design, and Experimental Verification of a Synchronous Reference Frame Voltage Control for Single-Phase Inverters. *IEEE Transactions on Industrial Electronics*, 61(1):258–269, Jan 2014.

[5] A. Vidal, F. D. Freijedo, A. G. Yepes, P. Fernandez-Comesana, J. Malvar, O. Lopez, and J. Doval-Gandoy. Assessment and Optimization of the Transient Response of Proportional-Resonant Current Controllers for Distributed Power Generation Systems. *IEEE Transactions on Industrial Electronics*, 60(4):1367–1383, April 2013.

[6] X. Yuan, W. Merk, H. Stemmler, and J. Allmeling. Stationary-Frame Generalized Integrators for Current Control of Active Power Filters With Zero Steady-State Error for Current Harmonics of Concern Under Unbalanced and Distorted Operating Conditions. *IEEE Transactions on Industry Applications*, 38(2):523–532, Mar 2002.

[7] A. G. Yepes, F. D. Freijedo, J. Doval-Gandoy, O. Lopez, J. Malvar, and P. Fernandez-Comesana. Effects of Discretization Methods on the Performance of Resonant Controllers. *IEEE Transactions on Power Electronics*, 25(7):1692–1712, July 2010.

[8] R. A. Gannett, J. C. Sozio, and D. Boroyevich. Application of Synchronous and Stationary Frame Controllers for Unbalanced and Nonlinear Load Compensation in 4-Leg Inverters. In *IEEE Applied Power Electronics Conference and Exposition (APEC)*, volume 2, pages 1038–1043 vol.2, 2002.

[9] D. G. Holmes, T. A. Lipo, B. P. McGrath, and W. Y. Kong. Optimized Design of Stationary Frame Three Phase AC Current Regulators. *IEEE Transactions on Power Electronics*, 24(11):2417–2426, Nov 2009.

[10] F. O. Martinz, K. C. M. de Carvalho, N. R. N. Ama, W. Komatsu, and L. Matakas. Optimized Tuning Method of Stationary Frame Proportional Resonant Current Controllers. In *International Power Electronics Conference (IPEC - ECCE ASIA)*, pages 2988–2995, May 2014.

[11] S. Fukuda and T. Yoda. A Novel Current-Tracking Method for Active Filters Based on a Sinusoidal Internal Model. *IEEE Transactions on Industry Applications*, 37(3):888–895, 2001.

The 2018 International Power Electronics Conference

An SoC-based platform for integrated multi-axis motion control and motor drive

Yongping Sun[1*], Ming Yang[1], Yangyang Chen[1], Wangpin He[1] and Dianguo Xu[1]

1 State Key Laboratory of Robotics and System, Harbin Institute of Technology, Harbin, China
*E-mail: sunyongping000@163.com

Abstract— **This paper presents a Field Programmable Gate Array (FPGA) -based platform for integrated multi-axis motion control and motor drive. As the development of FPGA, the integrated high-performance ARM processor on the FPGA has become the direction of FPGA development. This paper chooses the Xilinx Zynq-7020 programmable system-on-chip(SoC) which integrated dual-core ARM CPU and FPGA as a High-performance hardware platform. This platform is suits for the requirements of integrated multi-axis motion control and motor drive well, one ARM CUP completes the multi-axis position loop algorithm, speed loop algorithm and multi-axis trajectory generation, the other completes the human-computer interaction function. 4-axis current loop pipeline control and double sampling double updating current loop algorithm are accomplished by FPGA which can expand the current loop bandwidth and save FPGA logic resources. This platform's architecture will improve the overall bandwidth of the system, achieve multi-axis synchronization accuracy at nanosecond level which makes position trajectory more accurate.**

Keywords— *servo system; current loop ; System-on-chip; multi-axis control system*

I. INTRODUCTION

With the development of industrial demand, multi-axis motion control products are required for miniaturization, low cost, high reliability and flexibility[1]. For motor driver, it is required to complete advanced algorithms[2-4], such as resonance suppression algorithm, terminal jitter suppression parameter identification algorithm, it also can improve the performance and reliability of servo system. FPGA has been widely used in motion control and motor driver[5], using FPGA high speed parallel computing capability that can accomplished some advanced algorithms and robust algorithms. Xilinx Zynq-7000, as full programmable system-on-chip(SoC), combined FPGA and ARM is very suitable for the application of integrated motion control and motor drive [6,7].

Compared with distributed structure multi-axis motion control which transfers data by industrial Ethernet [8,9], the FPGA based platform's data exchange through shared memory and high-speed internal bus, the speed of data transmission can reach Gb/s level. The reduction of the external bus reduces the volume and cost of the system, and improves the reliability of the system.

In this paper, The Zynq-7020-based solution for 4-axis control systems is proposed, one ARM CUP completes the multi-axis position loop algorithm, speed loop algorithm and multi-axis trajectory generation, the other ARM CUP complete the function of human-computer interaction, 4-axis current loop pipeline control. 4-axis current loop pipeline control and double sampling double updating current loop algorithm are accomplished by FPGA. Thanks to the parallel computation, the algorithm of 4-axis current regulator can be completed in several hundreds nanoseconds, which can expand the current loop bandwidth, which also can achieve high synchronization of 4-axis speed.

II. DESIGN AND ANALYSIS OF THE ZYNQ-7020 SoC BASED PLAT FORM ARCHITECTURE

A. Architecture of the Zynq-7020 based multi-axis control system

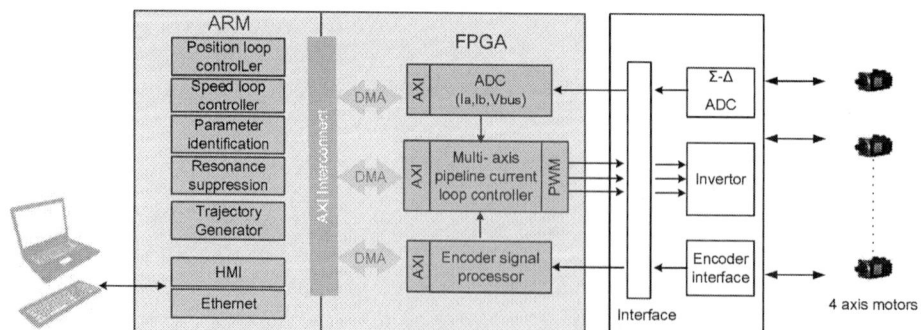

Fig.1. Architecture of the Zynq-based multi-axis control system

Architecture of the Zynq-based multi-axis control system is in Fig.1, the core of the main control chip uses Xilinx Zynq-7000 series SoC, which integrates two high-performance ARM CPU and FPGA, one ARM CUP completes the multi-axis position loop algorithm, speed loop algorithm and multi-axis trajectory generation, and also can completes the servo advanced algorithm, such as resonance suppression algorithm, parameter identification algorithm, the other ARM CPU handles the data transfer between the FPGA board and host PC

In terms of motor drive, the FPGA-based current loop algorithm is used to reduce the computational delay and improve the current loop bandwidth of the system without increasing the overall complexity of the algorithm. At the same time, considering the limited resources of FPGA, in order to accomplish the multi-axis cooperative control, the time-divisional multiplexing algorithm is used to build the multi-axis hardware current loop of the pipeline architecture to reduce the hardware resource occupancy by a small amount of computing time. Encoders and AD signals are also handled by the FPGA to facilitate the timing of the planning and signal segmentation. Advanced extensible interface (AXI) is bus on-chip bus which fulfill the high-speed information exchange between PL and PS.

B. Design of block diagram of the 4-axis control systems

The 4-axis control system can be shown in Fig.2. The processing logic (PL) section complete current signal sampling and angle signal sampling through IP core, the

trajectory generator gets the feedback angle value, then calculates the 4-axis position commands by trajectory generator. The value of the position deviation is given to the position loop controller and speed loop controller output current command value. The current command value is transmitted to the PL section via the AXI bus. Multi-axis pipeline current loop controller completes the SVPWM operation and output PWM signals after obtaining current command value.

Single-axis controller system is shown in Fig.3, The position loop adopts the proportional controller $W_{APR}(s) = K_P$, the speed loop adopts the proportional integral controller $W_{ASR}(s) = K_P + \dfrac{K_I}{s}$, and the current loop adopts the proportional integral controller $W_{ASR}(s) = K_P + \dfrac{K_I}{s}$.

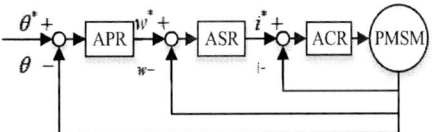

Fig.3. Block diagram of the single-axis control system

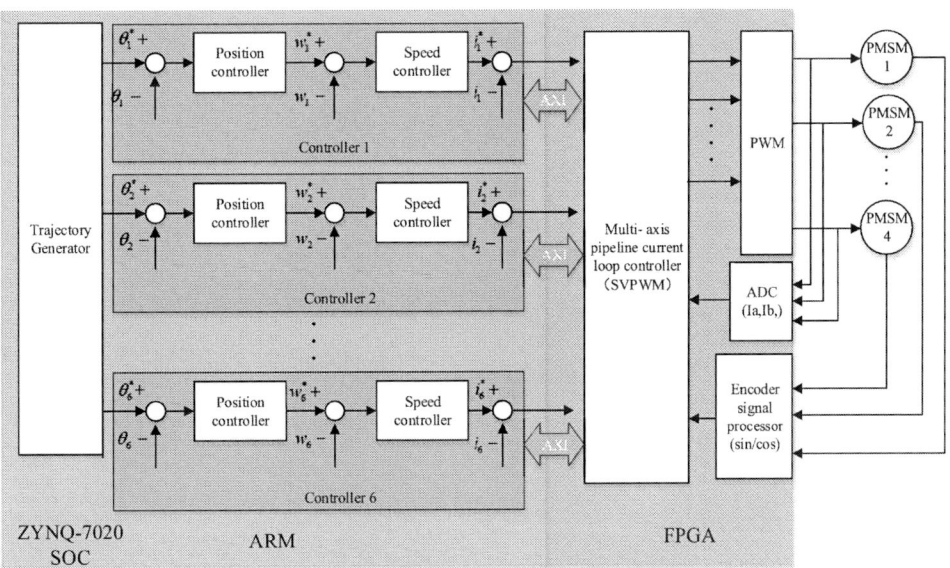

Fig.2. Block diagram of the 4-axis control system

current sampling is obtained AD7401 that it can provide high-resolution current signal, The IP core of Multi-axis pipeline current loop is also completed in the PL. The processing system (PS) section gets the angle value and current value from the PL by the AXI bus. The

III. IMPLEMENTION OF MULTI-AXIS CURRENT LOOP

A. Multi-axis current loop scheduling algorithm based on pipeline

The 2018 International Power Electronics Conference

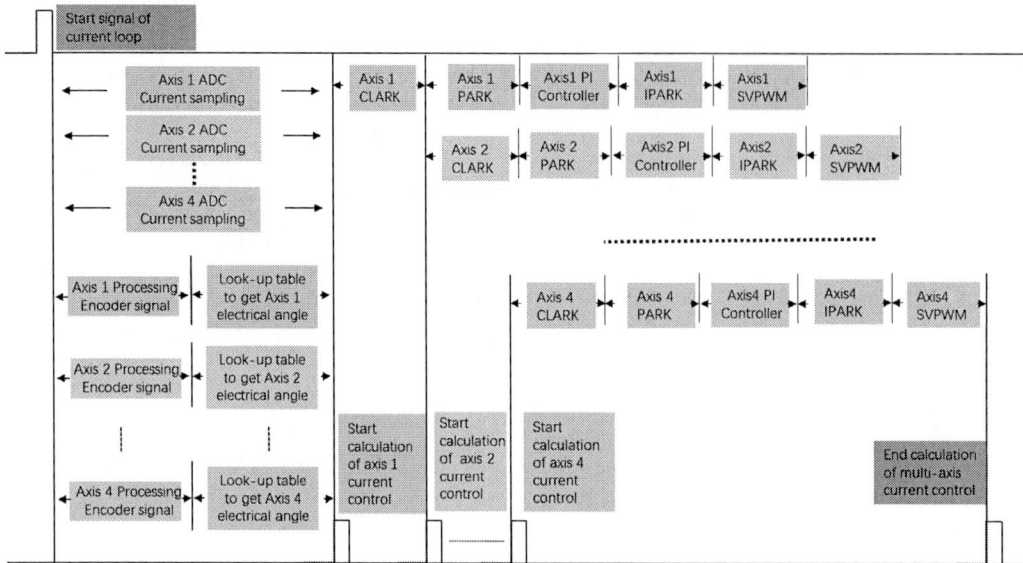

Fig.4. Diagram of multi - axis current loop pipeline structure

At present, multi-axis synchronous control servo control is obtained by industrial Ethernet technology. In order to achieve multi-axis synchronous control that needs to deal with distributed clock operations in the network protocol. The SoC platform can achieve better multi-axis synchronization than the Ethernet technology. In order to save FPGA resources, using pipeline structure and time-sharing reuse ideas, multi-axis current controller can use single-axis' hardware logic resources, Let the idle arithmetic module put into the next axis' arithmetic module in time to save the FPGA resources.

Multi - axis current loop pipeline structure is shown in Fig.4 PL gets 4-axis' current values through Σ-Δ ADC chips and 4-axis electrical angle values through Look-up sin/cos table. According to the idea of time division multiplexing, one function module can only control one motor at the same time, When the Clark transform module of the 1 axis is finished, the 1 axis' PARK transform module starts, and so on until the Space Vector pulse width modulation (SVPWM) module starts to operate on the 1 axis. When the 1 axis starts the PARK transform, and then the 2 axis starts the CLARK transform, When the 2 axis finished inverse park transformation (IPARK) transform, and the 1 axis finished SVPWM module then 2 axis start SVPWM module, and so on, until the 4th-axis crurent loop is finished.

B. FPGA-based PWM Strateges

FPGA-based motor control algorithm has been widely studied, the dual-sampling and dual-update PWM strategies is shown in Fig. 5 is proposed in [4] to extend the current bandwidth . Dn(k) is the nth PWM duty cycle update in the kth cycle, in (k) is the nth current sampling in the kth cycle (N = 1, 2), where T$_s$ (50μs) is the period of sampling current, and represents the operating cycle of the current loop, where T$_c$ (100μs)is the period of carrier .. As explained in [10], the execution time is very short so

the PWM updating can occur immediately. As the current sampling uses AD 7401, which the decimation rate is 128, the resulting 16-bit word rate is 125kHz at a 16 MHz external clock frequency. The delay of current sampling is 8μs, therefore T$_{sc}$ is set to 9μs between current sampling and PWM updating.

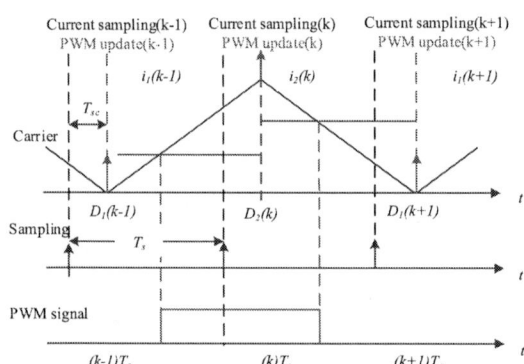

Fig. 5. Time sequence of current sampling and PWM duty cycle update

C. Design of Current Regulator

The control object is Surface Mounted Permanent Magnet Synchronous Machine (SMPMSM). The voltage equation [11] of the electrical model of the machine in the rotor reference frame is shown in (1)

$$u_{dqs}^r = R_s i_{dqs}^r + L_s \frac{d i_{dqs}^r}{dt} + j\omega_r \psi_{dqs}^r \quad (1)$$

The current regulator uses PI control, which is known to be robust to the parameter. The reference voltage in the synchronous rotation coordinate system is expressed in [2].

$$u_{dqs}^{r*} = K_P (i_{dqs}^{r*} - i_{dqs}^r) + K_I \int (i_{dqs}^{r*} - i_{dqs}^r) dt + j\omega_r \psi_{dqs}^r \quad (2)$$

where, $K_p = L_s \omega_{cb}^*$, $K_I = R_s \omega_{cb}^*$.

562

ω_{cb}^* is the desired bandwidth of the current regulator. After compensating back KMK and the cross-coupling, the transfer function of the current closed loop can be shown in (3)

$$G_{cl}(s) = \frac{i_{dqs}^r}{i_{dqs}^{r*}} = \frac{K_p s + K_I}{L_s s^2 + (K_p + R_s)s + K_I} = \frac{\omega_{cb}^*}{s + \omega_{cb}^*} \quad (3)$$

D. Design of Speed Regulator

The equation of the mechanical model can be described as follow

$$T_{em} = J\frac{d\omega_r}{dt} + B\omega_r + T_{load} \quad (4)$$

Where J represents the inertia and B represents the viscous friction of the mechanical system. The Proportional and Intergal (PI) controller is adopted in the speed regulator. The transfer function of speed regulator and the close loop system are shown in (5) and (6) respectively

$$T_{em}^* = \left(K_{p\_sr} + \frac{K_{I\_sr}}{s}\right)(\omega_r^* - \omega_r) \quad (5)$$

$$\frac{\omega_r}{\omega_r^*} = \frac{K_{p\_sr}s + K_{I\_sr}}{Js^2 + (B+K)s + K_{I\_sr}} \quad (6)$$

where $K_{p\_sr} = J\omega_{c\_sr}^*$, $K_{I\_sr} = K_{v\_sc}K_{p\_sc}\omega_{c\_sc}$.

K_{v_sc} is set to 0.2 [10].

IV. KEY PARAMETERS OF MUTIL-AXIS SYSTEM

A. Sampling Period

The SoC platform of multi-axis system has A/D converter block and encoder block, the current sampling chip adopt AD7401, which takes 9μs to finish current sample. The encoder uses Tamagawa absolute encoder, that completing angle sampling needs about 24μs. All of these are executed every 50μs and PWM switching frequency is 10kHz. The frequency of speed loop and the position loop are 4kHz.

B. Dvice utilization summary

The architecture of the Zynq-7020-based multi-axis control system is completed in Zynq-7020, which has abundant logic resources is suitable for multi-axis current loop algorithm. In this paper, the multi-axis platform adopts pipeline structure and time-sharing reuse ideas to save FPGA resources and completed 4-axis' current loop, it has ability to complete 8-axis current loop operation. Table I presents the hardware resource utilization in Figure 2.

TABLE I
HARDWARE RESOURCE UTILIZATION

Logic untilization	Used	Available	Util%
Slice Luts	14279	53200	26.8
Slice Register	32812	106400	38.82
DSP	14	220	6.36
Block RAM	1.5	140	1.07

V. EXPERIMENT

The experimental platform is shown in Fig.6, it is consisted of four 750w Tamagawa motors and the SoC multi-axis platform integrated control motion control and motor drive.

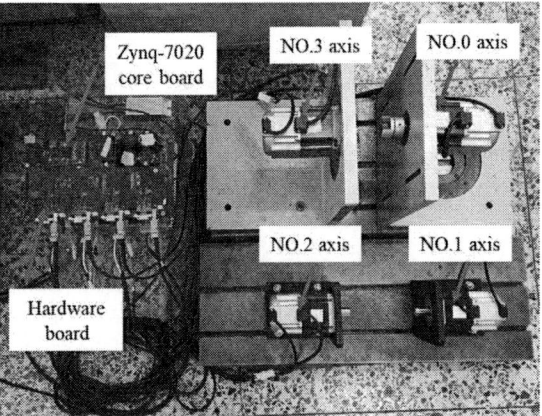

Fig. 6. Experimental setup with 4 750W PMSMs and SoC platform

A. The performance of current loop

In the current loop bandwidth test experiment, The id-axis reference current is given by chirp which the frequency increases, the amplitude of the signal is 0.3A and the offset of the signal is 0.5A, the SoC platform sends the id-axis reference current data and feedback current data to the PC through the serial port. The Bode plots of current loop is drawn through MATLAB and is shown in Fig 7. When the amplitude decays to -3dB, the frequency is 1550Hz. While the phase delays is 45°, the frequency is 990.3Hz, so the current loop bandwidth is 990.3Hz

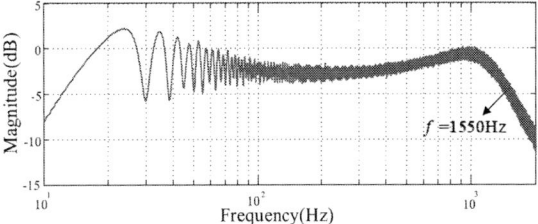

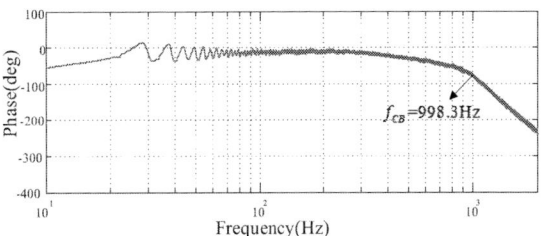

Fig.7. Bode diagrams of id chirp Chirp sweep frequency

B. The performance of speed loop synchronization

Thanks to high-speed communication between PS and

PL, PS can complete the calculation of the 4-axis speed loop and sends the speed loop output directly to the PL. Because there is no communication delay, the multi-axis system has a high degree of speed synchronization. Fig.8 shows experimental result of the speed synchronization, the speed reference of 0 axis and 1 axis is a ramp which rise to 1000 r/min at 2.5s and then change to constant speed at 1000r/min, after 2s the speed drops to 0r/min in a ramp way within 2s. The speed curve of 2 axis and 4 axis is the opposite to 0 axis and 1 axis. The data of 4-axis' speed feedback is transferred to the PC through the serial port. The result shows there is no significant difference between 2-axis.

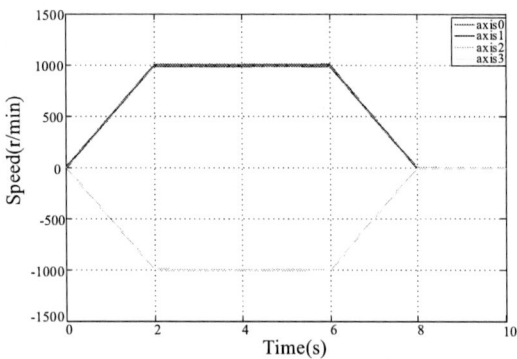

Fig.8. Experiment of 4-axis speed synchronization

VI. CONCLUSIONS

This paper designs a Zynq-7020 based for multi-axis motion control and motor drive which has advantages such as low cost, high reliability and flexibility, fully use of FPGA resources to achieve multi-axis synchronization control. The multi-axis pipeline current loop is used, which can expand current loop bandwidth, and save the FPGA logic resources. The information of position, speed, current in PL can be transferred to PS simultaneously, it can improve the speed synchronization of multi-axis. ARM resource realizes advanced servo algorithm, and accomplishes multi-axis trajectory planning. Zynq-7020 SoC is more suitable for efficient system integration than conventional ones. Future works will focus on the use of the proposed solution on SCARA(Selective Compliance Assembly Robot Arm).

ACKNOWLEDGMENT

This work is supported by The National Key Research and Development Program of China (NO.2017YFB1300801), Performance Optimization of Industrial Robot Servo Motor and Driving Product.

REFERENCES

[1] Monmasson E, Idkhajine L, Cirstea M N, et al. FPGAs in industrial control applications[J]. IEEE Transactions on Industrial informatics, 2011, 7(2): 224-243.

[2] Jeppesen B P, Crosland A, Chau T. An FPGA-based platform for integrated power and motion control[C]//Industrial Electronics Society, IECON 2016-42nd Annual Conference of the IEEE. IEEE, 2016: 2684-2689.

[3] Bartsch A, Senicar F, Soter S. Design of a scalable FPGA based inverter for complex drive systems[C]//Industrial Technology (ICIT), 2012 IEEE International Conference on. IEEE, 2012: 1086-1091

[4] Wang, Hongjia, et al. "Current-loop bandwidth expansion strategy for permanent magnet synchronous motor drives." Industrial Electronics and Applications IEEE, 2010:1340-1345.

[5] Itoh J I, Araki T. "Volume evaluation of a PWM inverter with wide band-gap devices for motor drive system". Ecce Asia Downunder. IEEE, 2013:372-378.

[6] Gu Q, Li Y, Niu P. SoC-based solution for multi-axis control systems using high-level synthesis[J]. Proceedings of the Institution of Mechanical Engineers, Part I: Journal of Systems and Control Engineering, 2015, 229(1): 63-73.

[7] Zhong G, Shao Z, Deng H, et al. Precise Position Synchronous Control for Multi-Axis Servo Systems[J]. IEEE Transactions on Industrial Electronics, 2017, 64(5): 3707-3717.

[8] Liu J, Chen S, Zhang G, et al. The development of a novel servo motor controller based on EtherCAT and FPGA[C]//Control and Decision Conference (CCDC), 2016 Chinese. IEEE, 2016: 3174-3179.

[9] Cena G, Bertolotti I C, Scanzio S, et al. Evaluation of EtherCAT distributed clock performance[J]. IEEE Transactions on Industrial Informatics, 2012, 8(1): 20-29.

[10] Jung E, Lee H J, Sul S K. FPGA-based motion controller with a high bandwidth current regulator[C]//Power Electronics Specialists Conference, 2008. PESC 2008. IEEE. IEEE, 2008: 3043-3047.

[11] Briz F, Degner M W, Lorenz R D. Analysis and design of current regulators using complex vectors[J]. IEEE Transactions on Industry Applications, 2000, 36(3): 817-825.

Variable Switching Frequency Strategy for Enhanced Settling Performance of Position Control within Inverter Loss Limit

Choongin Lee[1] and Jung-Ik Ha[1*]

1 Electrical & Computer Engineering, Seoul National University, Seoul, Republic of Korea

E-mail: soscucu@snu.ac.kr, jungikha@snu.ac.kr

Abstract— **This paper proposes variable switching frequency strategy for improving settling performance of position control. The proposed algorithm increases the switching frequency by using the amount of conduction loss reduction in the settling region. By increasing sampling frequency due to enhanced switching frequency, control performance limited by distortion from digital implementation is improved. The influence of increased sampling frequency is analyzed through the closed loop system pole analysis in z domain. The settling performance enhancement is verified through two experiments with 1kW SPMSM. In the experiment using the same PID gain set, overshoot was suppressed, and in the small step reference test (0.2rad) with same overshoot limit, 17% settling time reduction was achieved.**

Keywords—Position Control; Settling Performance; Switching Loss; VSFPWM;

I. INTRODUCTION

Performance improvement of position control is essential issue for productivity enhancement. Through the calculated speed, current, and voltage feedforward references by the position profile, the target position can be reached with high performance. The feedforward reference values allow each controller to control only parameter variation and to compensate the unexpected disturbance. However, after the given profile is ended, the settling period starts where the performance of the position controller directly acts. In this period, because the previously calculated feedforward references cannot affect the performance of position control, it is required to improve the performance of the position controller.

The PID controller has existed as a position controller which occupies the most of the industrial field. The performance of the position controller is determined by how the PID gains are tuned, but the tuning optimal gain set is somewhat intuitive and tricky. The controller with PID gain set for fast response induces overshoot, and the controller with the gain sets of high damping for preventing overshoot has a slow response. Therefore, in the context of this trade-off, in order to improve the control performance, it is necessary to solve the fundamental limitations of implemented controller, not simply to find the optimum PID gain set.

There are two factors which limit the performance of position control. The first element is the inner controller bandwidth, and the second one is a distortion of controller due to digital implementation with discretization, PWM delay, and zero order hold effect. If the bandwidth is not high enough for the inner controllers to follow the output command of position controller, the phase delay from it induces a position control oscillation, and it impairs the position control performance. And, controller distortion also has potential to causes oscillation which prevents the PID gain from setting the direction with high performance.

In order to minimize the interference of inner controller bandwidth, conventional studies for improving the current control bandwidth is effective[1-5]. The inner controllers of the position control are normally cascade structure of speed controller and current controller or simple PI current controller. Conventional studies are focused on improving current controller bandwidth. [2, 3] are based on predictive control strategy, and the direct torque control (DTC) method is used in [1, 4, 5].

The limits related to controller distortion were discussed in [6, 7]. [6] mentioned the problem that the maximum available pulse width ratio is limited from time delay including AD conversion. To avoid this problem, synthesizing PWM pulses one sample after AD conversion instant is used. However, in this case, system becomes unstable at high fundamental frequency. Stability problem in a current control considering the delay is analyzed in [7]. In [7], discrete time complex vector synchronous frame PI current regulator using Tustin transform is proposed. Time delay effect is compensated by time delay compensator. Previous studies have mainly focused on the current controller distortion, but this effects also exist on the position controller as well.

Although those problems can be solved by applying all of the exiting algorithms mentioned above, there is a solution which commonly weakens both problems. It is increasing the sampling frequency. Increasing sampling frequency reduces negative effects from the sampling delay and digital implementation. In particular, the maximum bandwidth of the PI structured current controller is determined from the sampling frequency, so that increasing sampling frequency can improve the bandwidth of the inner controller. Also, the performance improvement of the position controller with increased sampling frequency has been mentioned in [8].

However, because the sampling frequency is synchronized with the switching frequency, high sampling frequency inevitably leads to high inverter switching losses. Therefore, it is limited to a certain frequency considering the inverter rating. To improve sampling frequency in this environment, existing variable switching frequency PWM (VSFPWM) algorithm which is used to reducing inverter loss, are effective [9]. In [9], the switching frequency is determined considering the load torque and current ripple, so that the loss is reduced compared to the conventional constant switching frequency PWM (CSPWM) method.

In this paper, the VSFPWM is applied to the position control. The settling period is an effective period to adapt VSPWM. That's because the conduction loss is low in settling period, so that switching frequency can be increased, and also there is no pre-calculated feedforward reference, so that the performance of the controller directly affects. The inverter loss is measured in advance according to the load current and switching frequency, and then stored in the form of a table. In the section where the position profile is given, it operates at 20kHz which is the nominal switching frequency of the IGBT, and after profile ends, the switching frequency starts increasing as a function of current so that it does not exceed the rating loss and earns performance enhancement.

In section II, two positive effects obtained by increasing the sampling frequency is analyzed. The improvement of the maximum inner controller bandwidth by reducing the distortion of the current controller will be shown first. And the decreasing position controller's oscillatory response from the controller distortion effect will be seen with the system pole analysis. Next, the relationship between the load current and the switching frequency in inverter loss will be analyzed, and measurement result with power meters will be shown in section III. The measurement result supports the effectiveness of the variable switching operation which use load current level as variable. In section IV, the variable switching frequency position control method will be introduced with each control block, and it will be verified through the experimental results of section V.

II. EFFECT OF HIGH FREQUENCY SAMPLING ON SERVO DRIVE

In the case of implementing a control algorithm with a DSP, the sampling frequency acts as a factor that causes the system to deviate from the intended system in continuous time domain. The deviation leads to an oscillatory response at the controller. Therefore, it limits maximum bandwidth of current control, and it prevents the PID gain set of position controller setting in the direction of high performance.

A. Increased Maximum Current Controller Bandwidth

Assuming continuous time domain, PI type current controller which is the most widely used in industrial fields has an infinite bandwidth. However, the distortion caused by PWM delay for calculation time and zero order hold confines the maximum bandwidth of current controller. Zero-order hold effect considered SPMSM plant model shown in (1) is from [7].

$$G_{plant,electtric}(z) = \frac{\left(1 - e^{-(R/L)T_s}\right)}{R \cdot \left(z - e^{-(R/L)T_s}\right)} \tag{1}$$

With PWM delay and backward transformed PI controller, open loop transfer function expressed in z domain is as (2).

$$G_{open}(z) = \frac{\omega_{cc}\{z(L + RT_s) - L\}}{z - 1} \frac{1}{z} \frac{\left(1 - e^{-(R/L)T_s}\right)}{R \cdot \left(z - e^{-(R/L)T_s}\right)} \tag{2}$$

The parameters of PI controller, K_p and K_i, are set in the usual way ($K_p = \omega_{cc} \cdot L$, $K_i = \omega_{cc} \cdot R$) which makes the closed loop transfer function of current control system 1st order low pass filter in the continuous time domain. The variable ω_{cc} means the cut-off frequency of the 1st order low pass filter and the bandwidth of the current control. The closed-loop transfer function for analyzing the system stability can be expressed as (3).

$$G_{closed}(z) = \frac{G_{open}(z)}{1 + G_{open}(z)} = \frac{K\omega_{cc}}{z^2 - z + K\omega_{cc}} \tag{3}$$
$$\left(K = (L + RT_s)\left(1 - e^{-(R/L)T_s}\right) / R\right)$$

Pole-zero cancellation between the pole ($z = L/(L+RT_s)$) and zero ($z = e^{-(R/L)T_s}$) in (2) changes the characteristic equation of the closed loop system to 2nd order system so that it allows an intuitive observation of the divergent point of the system.

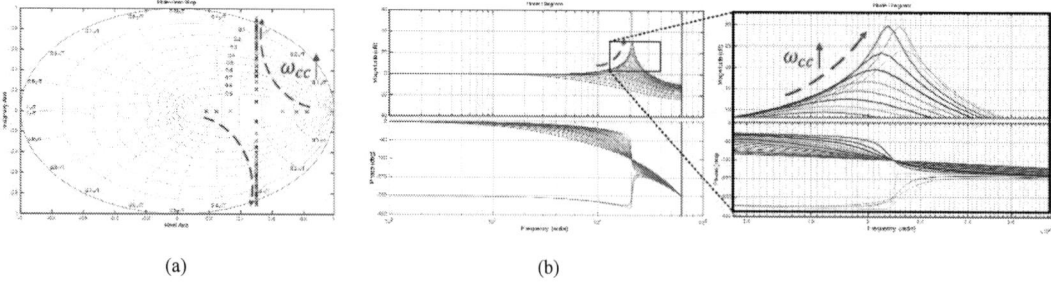

(a)   (b)

Fig. 1.(a) pole-zero map (b) bode plot of closed loop current control system by bandwidth increased from 0.4kHz to 2.3kHz at the 20kHz fixed sampling.

Fig. 1 shows the variation of the system in the pole-zero map and the bode plot as increasing the intended system bandwidth (ω_{cc}). As the pole has an imaginary value, the system starts to show an oscillatory response, which eventually causes the system to diverge. Because the q-axis current oscillation generates a torque ripple, it degrades the position control accuracy. In addition, since the magnitude of the current determines the sampling frequency in the variable switching operation, instant variation from the current oscillation degrades system stability. For the position current accuracy and the stable operation, current controller bandwidth is set under the frequency which occurs system oscillation shown in (4). $\omega_{cc,oscillation\_start}$ is the value at which the roots of the denominator in (3) begins to have an imaginary component.

$$\omega_{cc} < \omega_{cc,oscillation\_start} = \frac{R}{4\left(L + RT_s\right)\left(1 - e^{-(R/L)T_s}\right)} \quad (4)$$

In (4), the maximum current control bandwidth which can be represented by $\omega_{cc,oscillation\_start}$ increases as sampling frequency increases. Thus, increasing the sampling frequency has the effect of increasing the maximum bandwidth in the PI type current controller.

B. Decresed Oscillatory Response on Position Control

The PID type controller has an oscillatory response, and it can cause overshoot. Position control is an overshoot sensitive system, so that it is important to reduce the oscillatory response.

In this paper, the target system is a system where the structure of the entire system is designed without speed controller in order to minimize the influence of the bandwidth of the low level controller, and the position controller and current controller are both implemented with the synchronized sampling in a single DSP.

For analyzing the effect of insufficient sampling on the position control, zero order hold and PWM delay is considered in the plant model of (6). Even with an ideal current controller which produces the desired torque in one sample, a delay model should be added to analyze the system.

$$G_{plant,mech}(s) = \frac{1}{s(Js + B)} \quad (5)$$

$$G_{plant,mech}(z) = \left(1 - z^{-1}\right) Z \left\{ L^{-1} \left[\frac{G_{plant,mech}(s)}{s} \right] \right\} \quad (6)$$

The block diagram of the designed position control system is shown in Fig. 2. The PID position controller in the block is implemented in z domain which is transformed with backward transformation ($s = (z-1)/(T_s \cdot z)$).

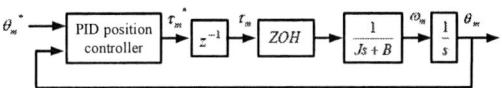

Fig. 2. Block diagram of the closed loop position control system with the infinite torque control bandwidth.

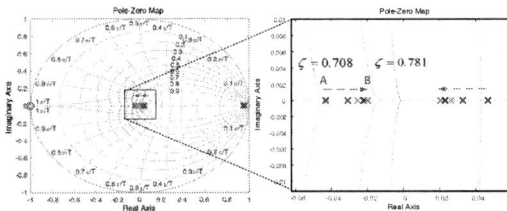

Fig. 3. Pole-zero map of the closed loop system by sampling frequency increased from 10kHz to 50kHz.

It can be shown that the oscillation affecting the overshoot of the system can be shown from analyzing the pole location of the enlarged figure in Fig. 3. The pole in the oscillatory region at Fig. 3 migrates from point A to point B during the sampling frequency changing from 10kHz to 50kHz in the same PID gain. It shows that the damping (ζ) increases as the sampling frequency increases. As a results, the frequent sampling reduces the oscillatory response of the system.

III. INVERTER LOSS ANALYSIS

As mentioned, the higher the sampling frequency, the better the control performance. However, in the environment where control algorithm is implemented in DSP, sampling instant is synchronized to switching carrier, so increasing sampling frequency causes additional loss at the inverter. A typical SPMSM drive circuit is shown in Fig. 4 which is consists of 3 leg IGBT inverter. The inverter loss is the loss occurring in the switching devices and diodes. In [10], the inverter loss is expressed as (7) by dividing the conduction loss (P_{cond}), and the switching loss (P_{sw}).

$$P_{loss} = P_{cond} + P_{sw} \quad (7)$$

$$P_{cond} = P_{cond\_IGBT}(I) + P_{cond\_FWD}(I) \quad (8)$$

$$P_{sw} = P_{sw\_IGBT}\left(f_{sw}, E_{on}(I), E_{off}(I)\right) + P_{sw\_FWD}\left(f_{sw}, E_{rec}(I)\right) \quad (9)$$

The detail expression of the inverter loss is presented in [10]. However, this paper focuses only on the position settling area at almost zero speed, so that the loss can be regarded as a function of phase current magnitude and switching frequency as shown in (8) and (9).

TABLE. I.
PARAMETERS OF SWITCHING DEVICE

PARAMETER	VALUE
SWITCHING DEVICE	IGBT(IRGIB6B60KD)
$V_{(BR)CES}$	600V
I_{C_MAX} @ T=25°C	11A

TABLE. II.
PARAMETERS OF SPMSM

PARAMETER	VALUE
Rated Power	1kW
Rated Current	5.37 A
DC link Voltage	100 V
Stator Resistance	0.5 Ω
Stator Self-inductance	3 mH
Back EMF Constant	85 mWb

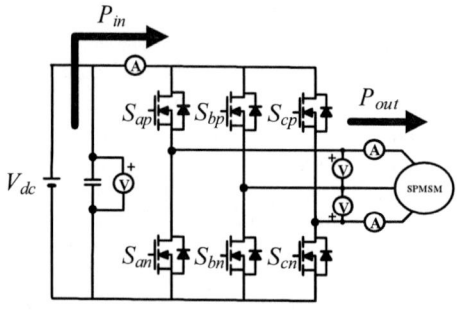

Fig. 4. Three phase inverter circuit with power meters.

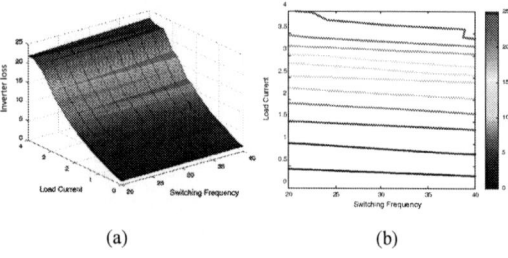

(a)   (b)

Fig. 5. Inverter loss by load current and switching frequency.

In this paper, the inverter loss is measured at each current and switching frequency using three power meters as shown in Fig. 4. One power meter is for measuring the input power (P_{in}), and with the remaining two power meters, the power consumed by the motor (P_{out}) is measured. Because the settling performance is focused, the loss value is measured with d-axis current control at zero speed in the target motor. The conditions

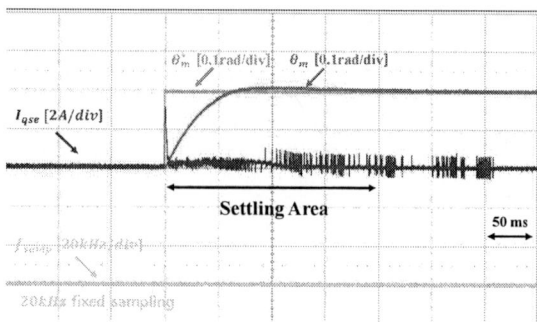

Fig. 6. Small step reference position control test with fixed switching frequency operation.

for the IGBT devices and the target SPMSM motor are given in TABLE I and TABLE II.

The measured results are shown in Fig. 5. The results show that the inverter loss increases as the switching frequency and phase current rise. Fig. 5 (b) is the contour plot which represents the same loss with the same color

IV. VARIABLE SWITCHING FREQUENCY POSITION CONTROL

Conventional variable switching frequency PWM strategies have been studied to change the switching frequency in purpose of reducing loss [9]. However, in this paper, because it is identified that the high sampling frequency has numerous advantages in reducing position control oscillatory response and improving the current control bandwidth, the variable frequency switching technique is used to improve the performance of the position control in the settling area.

The current waveform in a typical position control in the settling area is shown at the blue line of Fig. 6. Fig. 6 is the waveform of the position control result of the small step reference (0.2 rad) under the conditions of TABLE I and II. The small step reference is applying for only simulating the settling area without acceleration and deceleration.

As it is shown, in the most settling areas where the performance needs improvement, the load current is almost zero. It means that there is a remaining space for improving the switching frequency in the settling area. The increased switching frequency eventually leads to an increase in the sampling frequency. And this also leads to improved position control performance due to the two mechanisms mentioned in section II. The block diagram

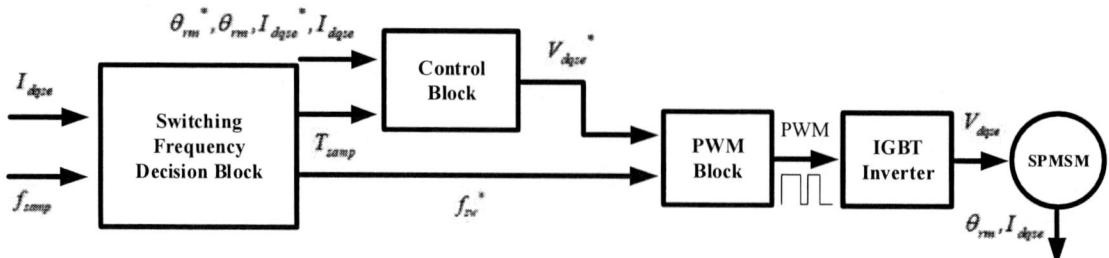

Fig. 7. Control block diagram of variable swiching frequency operation.

568

Control Block

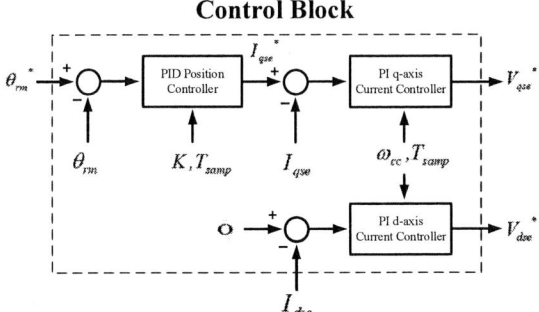

Fig. 8. Structure of control block.

of the overall control algorithm is shown in Fig. 7.

A. Switching Frequency Decisigon Block

Variable switching block is the most key block in the proposed algorithm. In general, the switching frequency is set to a certain fixed value so that the inverter loss does not exceed the rated power even at the rated current. However, as shown in Fig. 5 (b), the equal-loss line has a straight line shape. Therefore, if the switching frequency is increased along this straight line, the switching frequency can be increased without the extra burden on switching devices.

The rated loss of the switching devices is arbitrarily determined in advance according to the switching device and the heat dissipation condition. Therefore, before starting the operation, the specific straight line should be selected. The line is approximated by a first-order linear equation, which determines the switching frequency according to the load current. In this paper, since the 13W is set to the rated loss, the equation of the switching frequency (f_{sw}) by load current (I_{load}) is expressed as (10)

$$f_{sw} = -1.0 \cdot 10^5 I_{load} + 3.0 \cdot 10^5 \quad (10)$$

$$\left(f_{fixed} \leq f_{sw} \leq f_{samp,max} \right)$$

In the settling period where the inverter current is close to zero, the switching frequency is possible to be set to extremely large value, however the switching instant is synced with sampling instant, so the maximum value is limited by the processor's calculation capability.

B. Control Block

The sampling frequency, the output of the previous block, determines the gains of the two controllers in the control block. First, the PI gain values of the current controller are set through (6). As mentioned, the problems caused by the digital implementation of the current controller appear at higher bandwidth. One of the problems is the oscillation, which generates a torque ripple and interferes with position settling, so the bandwidth is set below the oscillation start point ($\omega_{cc,oscillation\_start}$) in (6). Once ω_{cc} is determined, the gain values are set to $K_p = \omega_{cc,oscillation\_start} \cdot L$, $K_i = \omega_{cc,oscillation\_start} \cdot R$ so that the zero of the PI controller is compensated with the pole of the plant.

Next step is to determine the PID position controller gains. In this paper, the ratio of each PID is set according to the method of Ziegler-Nichol[11], and the position control performance is adjusted by defining a constant (K) which is commonly multiplied by each gain. As mentioned above, the entire control structure without speed controller is a form of cascade connection between the PID position controller and the PI current controller as shown in Fig. 8. The common constant (K) is set to the largest value under the constraints that do not cause overshoot for fast position control settling. Systems with improved sampling frequency can reach the target position faster with larger K value.

V. Experimental Results

The experimental sets are composed of three power meters, servo driver, and PMSM. One power meter is for measuring the input power to the inverter, and two power meters measures the output power. Fig. 9 (a) shows the entire experimental sets and Fig. 9 (b) shows the servo driver. The experimental conditions are listed in Table I&II at section II.

All experiments are small step position movement test (0.2rad) which models settling period with short accelerating period. The experiment of Fig. 10(a) was done with fixed switching operation at 20kHz which is a generally recommended rating switching frequency of IGBT, and the experiment of Fig. 10(b) was conducted with the proposed variable switching operation under arbitrary set rating loss limit(13W). The switching frequency were increased along the equal loss line defined by the limit value (13W). Due to the limitations of AD conversion time and DSP calculation capability, the maximum switching frequency was limited to 40kHz in the experiments.

The controller parameters of those two experiments in Fig. 10 were set with the same value. Due to the high damping obtained by the increased sampling frequency in the settling area, the rotor position of Fig. 10(b) settled down to the reference position without overshoot. In the same experimental condition, 4% overshoot occurred with the fixed sampling operation (20kHz) as shown in Fig 10(a). From the experimental results in Fig. 10, it is verified that the high sampling frequency achieved from the proposed operation suppresses overshoot.

The experiments in Fig. 11 were conducted within the overshoot limit. In the experiments, the common gain factor K at the position controller was set to largest value until the system response met the overshoot limit (3 pulse, $2.0 \cdot 10^{-3}$ rad). As a result of the damping effect obtained by the high sampling frequency, higher K could be set in the proposed operation. Therefore, as shown in Fig. 11, the variable switching operation had better performance, and it meets in-position 15ms earlier than the fixed switching operation.

VI. Conclusion

This paper adapts variable switching frequency PWM (VSPWM) on position control. VSPWM is used to maximize the position control performance in settling

The 2018 International Power Electronics Conference

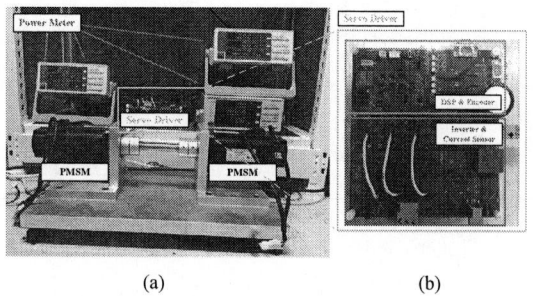

(a)   (b)

Fig. 9. Experimental sets. (a) motor drive system with power meters, (b) motor drive circuit.

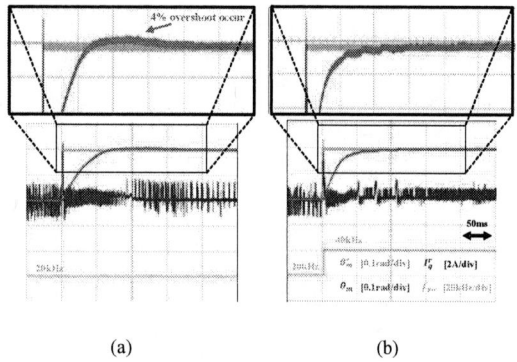

(a)   (b)

Fig. 10. Experimental results with same PID gain. (a) fixed sampling operation at 20kHz, (b) variable sampling operation.

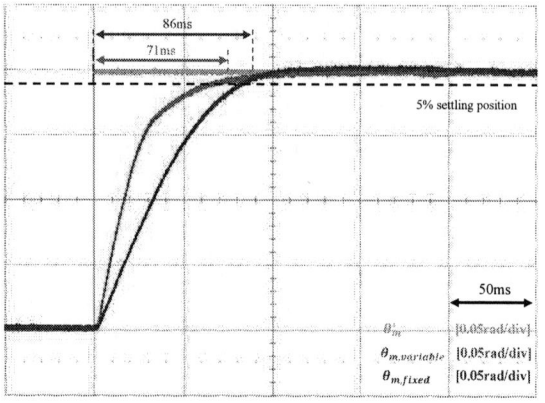

Fig. 11. Experimental results using PID gain set determined from the same overshoot limit (3 encoder pulse).

period rather than reducing inverter loss. In order to apply VSPWM, the inverter loss according to the load current and switching frequency was measured by three power meters. The improvement of the sampling frequency is obtained by improving the switching frequency in the settling area where the performance of the controller is most dominant, and conduction loss is low. In this paper, by increasing sampling frequency, the reducing the distortion phenomenon of the controller implemented in the DSP is analyzed with closed loop pole analysis in z domain. This improves the maximum bandwidth in the current controller and the damping in

the position controller. The settling performance enhancement was verified through two small step reference position control experiments with 1kW SPMSM. With the first experiment which compared fixed switching operation and variable switching operation in same PID gain set, the proposed algorithm suppressed the overshoot 4%. In the second experiment with limited overshoot level, the variable switching operation reached the target position 15ms earlier than fixed switching operation.

ACKNOWLEDGMENT

This work was supported by the Industrial Strategic technology development program funded by the Ministry of Trade, Industry & Energy (MI, Korea) [10060121, Development of Reconfigurable Articulated Robot Manipulators and All-in-One Motion Control System for Manufacturing of Mobile IT Products]

REFERENCES

[1]   J. S. Lee and R. D. Lorenz, "Robustness Analysis of Deadbeat-Direct Torque and Flux Control for IPMSM Drives," *IEEE Transactions on Industrial Electronics*, vol. 63, no. 5, pp. 2775-2784, 2016.

[2]   L. Niu, M. Yang, and D. g. Xu, "Deadbeat predictive current control for PMSM," in *2012 15th International Power Electronics and Motion Control Conference (EPE/PEMC)*, 2012, pp. LS6b.1-1-LS6b.1-6.

[3]   T. Türker, U. Buyukkeles, and A. F. Bakan, "A Robust Predictive Current Controller for PMSM Drives," *IEEE Transactions on Industrial Electronics*, vol. 63, no. 6, pp. 3906-3914, 2016.

[4]   L. Weijie, L. Dongliang, W. Qiuxuan, C. Lili, and Z. Xiaodan, "A novel deadbeat-direct torque and flux control of IPMSM with parameter identification," in *2016 18th European Conference on Power Electronics and Applications (EPE'16 ECCE Europe)*, 2016, pp. 1-9.

[5]   L. Zhong, M. F. Rahman, W. Y. Hu, and K. W. Lim, "Analysis of direct torque control in permanent magnet synchronous motor drives," *IEEE Transactions on Power Electronics*, vol. 12, no. 3, pp. 528-536, 1997.

[6]   D. Heng, R. Oruganti, and D. Srinivasan, "PWM methods to handle time delay in digital control of a UPS inverter," *IEEE Power Electronics Letters*, vol. 3, no. 1, pp. 1-6, 2005.

[7]   H. Kim, M. W. Degner, J. M. Guerrero, F. Briz, and R. D. Lorenz, "Discrete-Time Current Regulator Design for AC Machine Drives," *IEEE Transactions on Industry Applications*, vol. 46, no. 4, pp. 1425-1435, 2010.

[8]   C. Du, Y. Zhang, A. Kong, and Z. Yuan, "High-Precision and Fast Response Control for Complex Mechanical Systems—Servo Performance of Dedicated Servo Recording Systems," *IEEE Transactions on Magnetics*, vol. 53, no. 3, pp. 1-5, 2017.

[9]   W. Cao, F. Wang, and D. Jiang, "Variable switching frequency PWM strategy for inverter switching loss and system noise reduction in electric/hybrid vehicle motor drives," in *2013 Twenty-Eighth Annual IEEE Applied Power Electronics Conference and Exposition (APEC)*, 2013, pp. 773-780.

[10] T. Jalakas, D. Vinnikov, and J. Laugis, "Light load operation of 6.5 kV 200 A IGBTs in half-bridge configuration," in *2008 International Symposium on Power Electronics, Electrical Drives, Automation and Motion*, 2008, pp. 1373-1378.

[11] B. Jaganathan, R. Sharanya, S. K. Devi, and S. K. Sah, "Ziegler-Nichol's method of online tuning of PMSM for improved transient response," in *2010 International Conference on Power, Control and Embedded Systems*, 2010, pp. 1-4.

The 2018 International Power Electronics Conference

Two-Wheel Cane for Walking Assistance

Phi Van Lam*, Yasutaka Fujimoto**
Electrical and Computer Engineering, Yokohama National University, Yokohama, Japan
*E-mail: www.pvl.vn@gmail.com
**E-mail: fujimoto@ynu.ac.jp

Abstract—**A hardware design and controller of two-wheel cane has been proposed in this paper. Our two-wheel cane has been designed to assist users in maintaining balance as they move. We validated the performance of the controller designed to stabilize the inverted pendulum's balance based on the test results with the designed robot modeling. Experimental results show that two-wheel cane can be most useful to support elderly persons maintain balance.**

Keywords—*Two-wheel cane, robotic cane, inverted pendulum, non-linear disturbance observer, assist devices.*

I. INTRODUCTION

There are a lot of researches on robots that serve the elderly or disabled people to move, especially the compact type and flexible for the users.

For example, three-wheels robots that serve disabled people to move in the rehabilitation centers studied by the author [1]. However, its design is rather cumbersome and less flexible, the ability to apply it in practice is limited, it is hard to use when the users walk on rugged terrain.

In addition, there are some researches on more compact assistant devices like robotic canes to help users stand or move but their results are just limited on details about robot system design [2] or simulations [3], [4] without experimental results to validate the controller's performance.

In papers [5], [6], experimental results of actual robotic cane modeling were presented. However, its mechanical structure is quite complex by using inverted pendulum model, consequently users may feel uncomfortable when using it.

To overcome the drawbacks from the above studies, we propose a two-wheel cane based on an inverted pendulum model. Parallel with the design of the robot model and the controller are the actual test results with the users.

This paper is composed of four sections: In section II, the two-wheel cane hardware is thoroughly presented. In section III, two-wheel cane mathematical equations including Lie algebra method and nonlinear disturbance observer are solved to evaluate the controller parameters. In section IV, the controller performance discussed on based on experimental results. Finally, the conclusion and future works are described.

II. HARDWARE OF THE TWO-WHEEL CANE

In this section, two-wheel cane hardware is presented. Its design is as shown in Fig. 1.

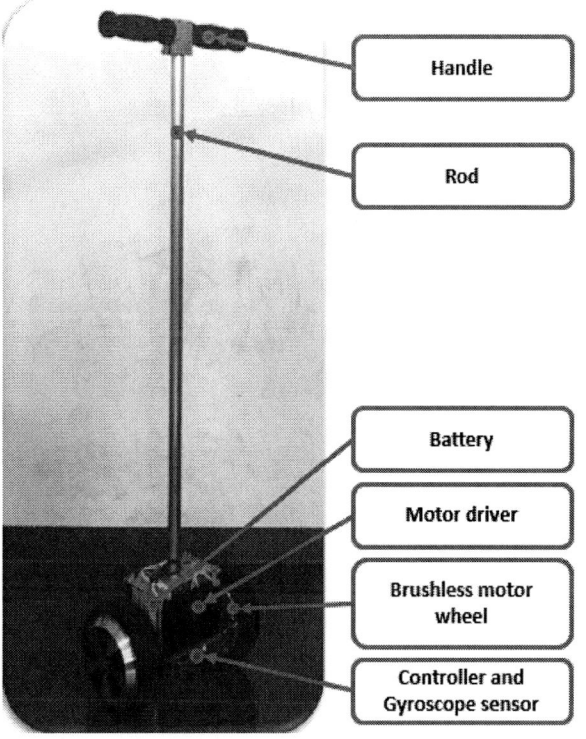

Fig. 1. Two-wheel cane for walking assistance.

It has a handle on the top of the cane, connected to the frame of the two-wheel cane by a rod. This frame is printed by a computer numerical control (CNC) machine. This frame is connected with two natural rubber tire electric wheels (Fig. 2) on the left and right sides. There are two motor drivers on the top of the frame can control the output current of 10 A. At the center of the frame these are Li-ion batteries with capacity of 36V/4400 mAh. These batteries supply power for the two-wheel cane to work in a long time without charging. At the bottom these is a raspberry pi zero controller and a bridge circuit connected to an accelerometer and gyroscope sensor to calculate and control motion of the cane based on our algorithm.

Natural rubber tire electric wheel includes a brushless motor and hall sensors as shown in Fig. 2. Thanks to this structure, the two-wheel cane works well in any environments, and details about them are shown in Table I. Due to the special structure of the wheel without any gearbox types, the two-

The 2018 International Power Electronics Conference

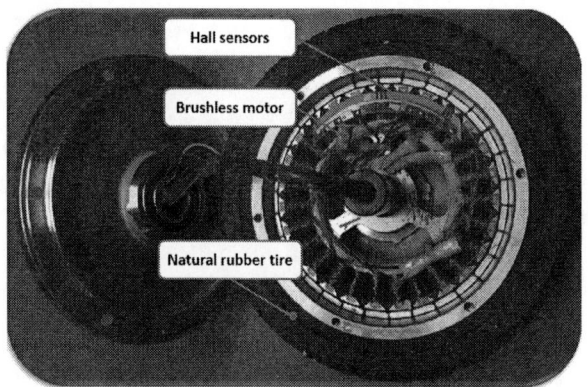

Fig. 2. Natural rubber tire electric wheel included a brushless motor and hall sensors.

TABLE I.   PARAMETER OF NATURAL RUBBER TIRE ELECTRIC WHEEL

Symbol	Meaning	Value	Unit
V	Voltage	36	V
P	Output power	250	W
H	Hall sensor	90	p/r
S	Size (thickness x diameter nominal)	46 x 168	mm

wheel cane work without any noises.

Moreover, MPU-6050 accelerometer and gyroscope sensor is used at relatively cheap price. The MPU-6050 devices combine with a 3-axis gyroscope - selectable range up to ±2000 degree/s and a 3-axis accelerometer - selectable range up to ±8 g and 400 kHz, fast mode I^2C (Inter-Integrated Circuit) to communicate with all registers to easily connect with raspberry pi zero w controller.

III. MATHEMATICAL MODEL AND DESIGN OF THE CONTROLLER SYSTEM OF TWO-WHEEL CANE

A. Mathematical model of two-wheel cane using an inverted pendulum model

The two-wheel cane in Fig. 3 is basically based on an inverted pendulum model. The rod is l in length, the wheel is r in radius, the mass of wheel is M, and mass of rod is m. By analyzing this system, we use the Lagrange equation to determine the motion of an inverted pendulum as follows:

$$\frac{d}{dt}\left(\frac{\partial L}{\partial \dot{\phi}}\right) - \frac{\partial L}{\partial \phi} + \frac{\partial F_{fr}}{\partial \dot{\phi}} = 0 - d_1 \qquad (1)$$

$$\frac{d}{dt}\left(\frac{\partial L}{\partial \dot{\theta}}\right) - \frac{\partial L}{\partial \theta} + \frac{\partial F_{fr}}{\partial \dot{\theta}} = \tau - d_2 \qquad (2)$$

The analysis of the inverted pendulum model was presented by authors in [5], [6]. Similarly, the system motion equation is as follows.

$$\begin{bmatrix} H_{11} & H_{12} \\ H_{21} & H_{22} \end{bmatrix} \begin{bmatrix} \ddot{\phi} \\ \ddot{\theta} \end{bmatrix} + \begin{bmatrix} b_1 \\ b_2 \end{bmatrix} = \begin{bmatrix} -d_1 \\ \tau - d_2 \end{bmatrix} \qquad (3)$$

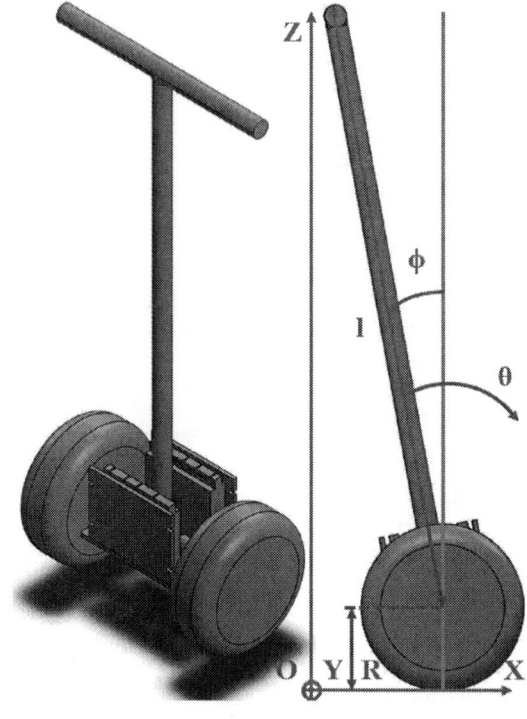

Fig. 3. Two-Wheel Cane System coordinates using Inverted Pendulum.

where,

$$H_{11} = J_\theta + (M + m)\, r^2 + 2mrl\cos\phi + J_\phi + ml^2 \qquad (4)$$

$$H_{12} = H_{21} = -J_\theta - (M + m)\, r^2 - mrl\cos\phi \qquad (5)$$

$$H_{22} = J_\theta + (M + m)\, r^2 \qquad (6)$$

$$b_1 = -\dot{\phi}^2 mrl\sin\phi - mgl\sin\phi + D_\phi\dot{\phi} \qquad (7)$$

$$b_2 = \dot{\phi}^2 mrl\sin\phi + D_\theta\dot{\theta} \qquad (8)$$

The two-wheel cane needs a force to be applied to the axis motor to move and help users maintain balance. From (3), the torque on the motor axis is given the equation as below:

$$\tau = \left(H_{22} - \frac{H_{12}H_{21}}{H_{11}}\right)u - \frac{H_{21}b_1}{H_{11}} + b_2 - d_2 \qquad (9)$$

B. Linearization of nonlinear system by Lie algebra method

From the motion equation of two-wheel cane, we expand it in (10). We easily recognize that this system is a nonlinear system.

$$\dot{x} = \begin{bmatrix} \dot{\phi} \\ \ddot{\phi} \\ \dot{\theta} \\ \ddot{\theta} \end{bmatrix} = \begin{bmatrix} \dot{\phi} \\ -\frac{b_1}{H_{11}} \\ \dot{\theta} \\ 0 \end{bmatrix} + \begin{bmatrix} 0 \\ -\frac{H_{12}}{H_{11}} \\ 0 \\ 1 \end{bmatrix} u \qquad (10)$$

The Lie algebra method to linearize the nonlinear system is one of the best selections to control the inverted pendulum model. This method is defined as a function given in (11) and (12):

$$\dot{x} = f(x) + g(x)u \qquad (11)$$

$$y = h(x) \qquad (12)$$

After expanding them to find the y function is a linear equation, its derivatives from the first order to the third order can be calculated by (13)-(16).

$$y = \int_0^\phi \frac{H_{11}}{H_{12}} d\phi + \theta \qquad (13)$$

$$\dot{y} = \frac{H_{11}}{H_{12}} \dot{\phi} + \dot{\theta} \qquad (14)$$

$$\ddot{y} = \frac{\partial}{\partial \phi} \frac{H_{11}}{H_{12}} \dot{\phi}^2 - \frac{b_1}{H_{12}} \qquad (15)$$

$$y^{(3)} \simeq \frac{\partial^2}{\partial \phi^2} \frac{H_{11}}{H_{12}} \dot{\phi}^3 - \frac{\partial}{\partial \phi} \frac{b_1}{H_{12}} \dot{\phi} - 2(\frac{\partial}{\partial \phi} \frac{H_{11}}{H_{12}}) \frac{b_1}{H_{11}} \dot{\phi} \qquad (16)$$

Obviously, these equations are the linear equations of the nonlinear system to show the motion of the two-wheel cane.

The higher derivatives are, the smoother the system is. But depending on the controller, whether it has enough speed to calculate the motion equations of the system or not. In this case, we use the derivative with the fourth order, which meets the speed to process a basic feedback loop controller to controls this system, and the input value u is given by (17).

$$u = \frac{v - L_f^4 h(x)}{L_g L_f^3 h(x)} \qquad (17)$$

where,

$$L_g L_f^3 h(x) = -3 \frac{\partial^2}{\partial \phi^2} \frac{H_{11}}{H_{12}} \frac{H_{12}}{H_{11}} \dot{\phi}^2 + \frac{\partial}{\partial \phi} \frac{b_1}{H_{12}} \frac{H_{12}}{H_{11}}$$
$$+ 2 \frac{\partial}{\partial \phi} \frac{H_{11}}{H_{12}} \frac{H_{12}}{H_{11}^2} b_1 \qquad (18)$$

$$L_f^4 h(x) = -\frac{\partial^2}{\partial \phi^2} \frac{b_1}{H_{12}} \dot{\phi}^2 + \frac{\partial^3}{\partial \phi^3} \frac{H_{11}}{H_{12}} \dot{\phi}^4$$
$$- 5(\frac{\partial^2}{\partial \phi^2} \frac{H_{11}}{H_{12}}) \frac{b_1}{H_{11}} \dot{\phi}^2 - 2 \frac{\partial}{\partial \phi} \frac{b_1}{H_{12}} \dot{\phi}^2$$
$$+ \frac{\partial}{\partial \phi} \frac{b_1}{H_{12}} \frac{b_1}{H_{11}} + 2 \frac{\partial}{\partial \phi} \frac{H_{11}}{H_{12}} \left(\frac{b_1}{H_{11}}\right)^2 \qquad (19)$$

IV. EXPERIMENTAL RESULTS

A. Stabilization of two-wheel cane by itself

The two-wheel cane is designed with the ability of balance with or without being held by the users. This is a basic working mode of an inverted pendulum. Experiments of the controller on two-wheel cane are recorded by videos, and a frame is cut from the video as shown in Fig. 4.

The angle of two-wheel cane has a relationship with the angle of the cane, which is determined by gyroscope sensor in Fig. 5, the result shows that the two-wheel cane achieves

Fig. 4. Two-wheel cane stability by itself.

a stability point after only 1.5 s with a big tilted angle of the rod around 0.18 rad and remain unchanged until the end of the period.

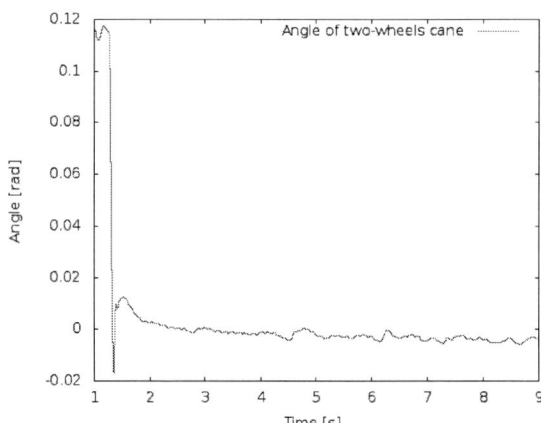

Fig. 5. Angle of two-wheel cane when stability by itself.

This result shows that by using Lie algebra method we can linearize the nonlinear system not only around the zero points as the linear-quadratic regulator (LQR) method but also big tilted angles of the rod. Therefore, this method is really good to control the two-wheel cane to help users maintain balancing when the tilted angle is equal or greater than zero degree.

B. Support users maintain balancing

In the second test, the efficiency of the two-wheel cane is shown by the angle and the feedback position of the two-wheel cane at the same time in Fig. 6.

The 2018 International Power Electronics Conference

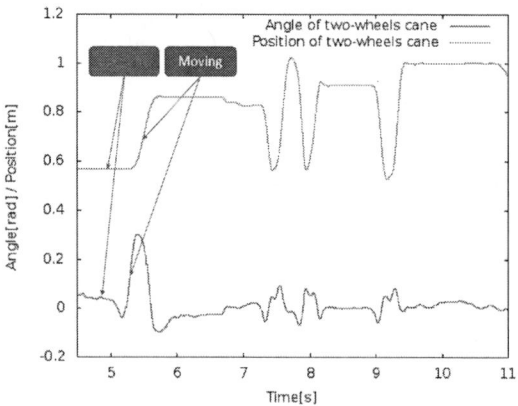

Fig. 6. Angle and position of two-wheel cane when it help users maintain balancing.

We present two cases: the first case is when users need to remain stable at the stand upright position (Fig. 7), at that time the two-wheel cane helps users achieve the balancing point through the handle; the second case is when the cane supports users to move (Fig. 8). In this case, the two-wheel cane is based on the angle of the rod compared to the stability point to get to a suitable position to support users.

Fig. 7. Two-wheel cane helps users to stand.

Fig. 8 shows that when the user needs to go ahead, the cane is tilted at an angle of ϕ compared to the vertical axis. Then, an output torque is applied to the motor axis, this torque calculated by (9) helps the two-wheel cane change the position with the corresponding angle of θ.

The results are shown in Fig. 6, with the angle of ϕ changes from 0 rad to around 0.3 rad, the two-wheel cane automatically changes its position from around 0.58 m to 0.88 m before achieving a balancing point to help users stand from the sixth second.

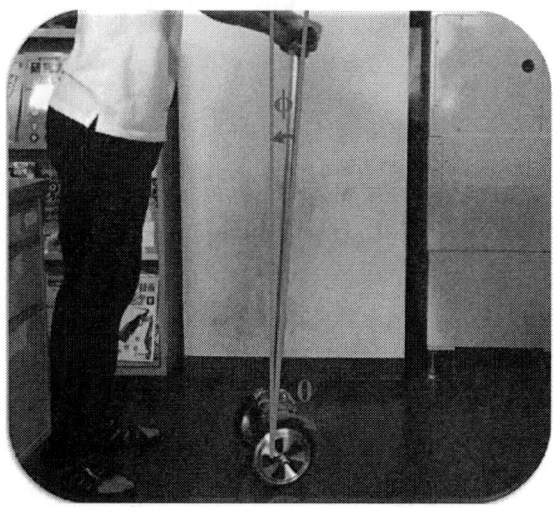

Fig. 8. Two-wheel cane help users to move.

V. Conclusion and Future Work

We designed the hardware of the two-wheel cane based on a natural rubber tire electric wheel with a high-speed processing controller for walking assistance. From the experimental results on the hardware, the two-wheel cane is more effective to help the users to maintain balance. In future work, we intend to reduce the size of the system while increasing capacity of the battery to increase the running time of the two-wheel cane without charging. Moreover, we will test the two-wheel cane at the rehabilitation center to get more experimental results and improve it.

Acknowledgment

We would like to thank Otsuka Toshimi Scholarship Foundation for doctoral research funding.

References

[1] S. Nakagawa, Y. Hasegawa, T. Fukuda, I. Kondo, M. Tanimoto, P. Di, J. Huang, and Q. Huang, "Tandem stance avoidance using adaptive and asymmetric admittance control for fall prevention," *IEEE Transactions on Neural Systems and Rehabilitation Engineering*, vol. 24, DOI 10.1109/TNSRE.2015.2429315, no. 5, pp. 542–550, May. 2016.

[2] Y. Ota, M. Ryumae, and S. Keiichi Sato, "Robotic cane devices," *United States Patent Application Publication*, no. US 2013/0041507 A1, 2013.

[3] K. Shimizu and Y. Fujimoto, "A robotic cane for walking support using two control modes," *IEEJ International Workshop on Sensing, Actuation, Motion Control, and Optimization*, pp. TT9–5, Mar. 2016.

[4] K. Shimizu, S. A. Issam, and Y. Fujimoto, "A robotic cane for walking assistance'," *IEEJ International Power Electronics Conference (IPEC)*, DOI 10.1109/IPEC.2014.6869857, pp. 1968–1973, May. 2014.

[5] P. V. Lam and Y. Fujimoto, "Building and test a controller of the robotic cane for walking assistance," *The IEEJ International Workshop on Sensing, Actuation, Motion Control, and Optimization*, vol. 3, pp. SS2–6, Mar. 2017.

[6] P. V. Lam and Y. Fujimoto, "Completed hardware design and controller of the robotic cane using the inverted pendulum for walking assistance," *2017 IEEE 26th International Symposium on Industrial Electronics (ISIE)*, vol. 19-21 June 2017, pp. 2163–5145, Aug. 2017.

574

Fall Prevention and Vibration Suppression of Wheelchair Using Rider Motion State

Isseki Takahashi[1*], Toshiyuki Murakami[1**]

1 Department of Science and Technology, Keio University, Yokohama City, Japan
*E-mail: takahashi@sum.sd.keio.ac.jp
**E-mail: mura@sd.keio.ac.jp

Abstract—**In recent years, with the declining birthrate and the aging of the population, the demand for nursing care support equipment is increasing. Among them, wheelchairs are already common as nursing care support equipment. Therefore, demand for wheelchair research is high, and electric wheelchair is required to have higher functionality. In this paper, prevention of falling of the wheelchair as a part of enhancing the functionality of the electric wheelchair is described. Several studies on prevention of rollover of electric wheelchairs have already been reported, but since rider motion state is not taken into account in the calculation of rollover conditions, there is a problem that the rollover conditions does not conform to the actual situation. Therefore, the purpose of this study is to estimate the state of the rider by IMU sensor and to calculate accurate rollover condition. Several simulation results are shown to confirm the proposed approach.**

Keywords—Wheelchair, Kalman filter, Fall prevention

I. INTRODUCTION

Due to the extension of the average life expectancy and the decline of the birthrate, population aging has progressed on a global scale. In advanced countries, the number of elderly people already exceeds the number of children in 1998, and it will be the first time in 2045 for the world. The world population of over aged 60 years is 3.5 times from 205 million to 737 million in 1950 to 2009. Thereafter, it will continue to increase at a rate of 1.2% per year, and it is estimated that it will be three times by 2050. In developed countries the population of over aged 60 years will be around 30% in 2050 from 20% in 2009 from 60 years old or more. On the other hand, in developing countries, it is expected that it will increase from 8% in 2009 to 20% same as present developed countries in 2050[1].

One of the problems in the aging society is the increase in the number of people who need nursing care and the decrease in caregivers. The shortage of caregiver has already become a big problem in many countries, and it is urgent to deal with these problems.

To address the above issues, one solution is the development of nursing care equipment. As the declining birthrate and the aging population progresses, it is necessary to support people who need nursing care to have independent living. It is important not only from the viewpoint of lack of caregivers, but also from the viewpoint of the quality of life of the care recipient, that the elderly can live independent. Support that does nursing

more than necessary and support that does not reflect the intention of the care recipient is counterproductive[2][3].

From the above it can be seen that there is a need for a wheelchair that can support independent living of care recipients. Among them, electric wheelchairs have been used for a long time as a means by which a care recipient can easily move.

However, wheelchair accidents also occur frequently[4]. Older people account for a large proportion of care recipients, and those who are declining in cognitive ability often operate. As a result, accidents are caused by erroneous operations or delayed reactions. Not only that, there are many places where wheelchairs are not supposed to pass even after the necessity of barrier-free has been proposed. In such places it is easy to get accidents due to wheelchairs. Therefore, it is desired to make electric wheelchairs more sophisticated and safer, and research facilities and companies are pursuing the issue. Gina[5] *et al.* simulated the shock when an electric wheelchair was hit by an obstacle and made a consideration on safety. David[6] *et al.* estimated the load applied to the wheelchair and used it for wheelchair control. Katsura[7] *et al.* proposed a method for generating the trajectory of a wheelchair to avoid obstacles.

In this way, many functions and safety enhancement of wheelchairs are being pursued, therefore in this paper rolling of a wheelchair is studied. Many accidents caused by wheelchairs are caused by rollover of a wheelchair. Although this rollover seems not to occur easily at first glance, in reality it occupies a lot of wheelchair accidents. Much of the rollover occurs at the slope. Speed tends to high in the slope, and the wheelchair itself is inclined, so the risk of rollover is high. Furthermore, when a wheelchair travels on an inclined road, the wheelchair often curves even if the rider wishes to go straight ahead. For these reasons, there are many studies pursuing safety on a slope[8][9].

However, there are not many papers mentioned about rolling wheelchairs. Furthermore, there are very few paper which considers the existence of riders. Among the total weight of the wheelchair, the mass of the rider occupies a lot, and the rider is inclined at the curve and the slope. Therefore, the moment changes and the rollover condition also fluctuates, and it is indispensable to know the state of the rider to calculate the accurate rollover condition.

The 2018 International Power Electronics Conference

In this paper, calculation of rollover condition considering rider's state is proposed. The IMU sensor is used to estimate the state of the rider. Furthermore, in order to predict the rider's state fluctuation while going on a curve or a slope, the Kalman filter are used. Moreover, a person is modeled by a spring-mass-damper system, and it is regarded that the coefficient of the spring-mass-damper system is related to the reactivity to the curve and slope. By identifying each coefficient appropriately, the rider's states in a slope and a curve are predicted, and the rollover condition is calculated.

Determining the parameters of the body and changing the command value for the wheelchair, it is possible to reduce the acceleration of the body and improve the ride comfort. This means that the proposed approach is also effective for ride comfort improvement using body parameters.

The paper is composed of six sections. In section II, modeling of wheelchair and its rider is described. In section III, system for estimating rider's state. In section IV, control system is described. In section V, simulation conditions and results are mentioned. This paper is concluded in section VI.

II. MODELING

In this section, the two-wheeled wheelchair to be controlled and its rider are modeled. First, these models are shown and then, kinematics and dynamics are described.

In this paper, modeling is carried out assuming a two-wheeled electric wheelchair which is inferior drive compare with four wheel wheelchair. In the established model, when the pitch angle of the wheelchair is fixed at 0, the model is the same as the general wheelchair of the four wheels.

A. Modeling of two-wheeled wheelchair and its rider

In this subsection, models of two-wheeled wheelchair and rider are explained. Fig. 1 (a) and (b) shows top view and front view of a two-wheeled wheelchair respectively. In this study, it is regarded that the vibration characteristics of the rider can be modeled by the spring-mass-damper system, and the center of gravity of the upper body of the vehicle body and rider is connected by a spring and a damper. P is the control reference which is the center of gravity of the wheelchair. Furthermore, G is the center of gravity of the lower body and exists just above the control reference point. Furthermore, the center of gravity H of the upper body of the rider is present at a position inclined by δ directly above the center of gravity of the lower body. Symbols are summarized in Table I. The state vector in the workspace is defined as (1), and the state vector in the joint space is defined as follows.

$$\boldsymbol{X} = \begin{bmatrix} \theta_p & \delta & \kappa + \varphi & d & \varphi \end{bmatrix}^{\mathrm{T}} \quad (1)$$

$$\boldsymbol{\theta} = \begin{bmatrix} \theta_p & \delta & \kappa & \theta_r & \theta_l \end{bmatrix}^{\mathrm{T}} \quad (2)$$

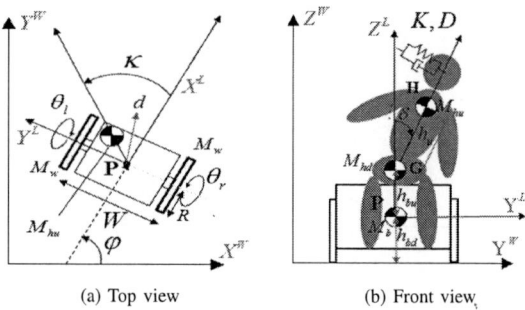

| (a) Top view | (b) Front view |

Fig. 1. Modeling of two-wheeled wheelchair and its rider.

TABLE I. SYMBOLS USED IN THIS PAPER

Variables	Explanation
P	Control reference point and wheelchair center of gravity
G	Lower center of mass
H	Upper body center of gravity
$\bigcirc^W$	World coordinate system
$\bigcirc^L$	Local coordinate system
W	Distance between right wheel and left wheel (tread)
R	Wheel radius
d	Progressive distance
θ_p	Pitch angle of car body
θ_r	Right wheel rotation angle
θ_l	Left wheel rotation angle
φ	Azimuth angle
δ	Body tilt angle
κ	Body tilt azimuth angle
X, Y	Position of wheelchair in world coordinate system
M_w	Weight of both wheels
M_b	Mass of car body
M_{hd}	Mass of lower body
M_{hu}	Weight of upper body
h_{bd}	Distance between ground and P
h_{bu}	Distance between P and G
h_u	Distance between G and H
J_w	Moment of inertia around the axle
J_y	Moment of inertia in the direction of rotation of motorcycle chair
K	Spring constant of rider
D	Viscosity constant of rider
r	Turning radius (positive when turning clockwise)
g	Acceleration of gravity

B. Kinematics

In deriving kinematics, the following two constraints are set.

- The wheelchair does not move in the direction of the axle.

- The left and right wheels do not slip.

(3) is derived from first constraint condition.

$$\dot{X}\sin\varphi - \dot{Y}\cos\varphi = 0 \quad (3)$$

From the second condition, (4), (5) are derived.

$$\dot{X}\cos\varphi + \dot{Y}\sin\varphi + \frac{W}{2}\dot{\varphi} = R\dot{\theta}_r \quad (4)$$

$$\dot{X}\cos\varphi + \dot{Y}\sin\varphi - \frac{W}{2}\dot{\varphi} = R\dot{\theta}_l \quad (5)$$

Also, the turning radius r is formulated as (6).

$$r = W\frac{\dot{\theta}_l + \dot{\theta}_r}{2\left(\dot{\theta}_l - \dot{\theta}_r\right)} \quad (6)$$

C. Dynamics

Considering Lagrangian formulation, the motion equation of wheelchair system with the rider is given by (7).

$$M(\theta)\ddot{\theta} + H(\theta, \dot{\theta}) + G(\theta) + F(\theta, \dot{\theta}) = T \quad (7)$$

Here, $M(\theta)$ is the inertia matrix, $H(\theta, \dot{\theta})$ is the centrifugal force / Coriolis force, $G(\theta)$ is the gravitational term, $F(\theta, \dot{\theta})$ represents the damping force by the elastic force of the spring. The components of the vector are indicated by (8)-(12), respectively.

$$M = \begin{bmatrix} M_{11} & M_{12} & M_{13} & M_{14} & M_{15} \\ M_{21} & M_{22} & M_{23} & M_{24} & M_{25} \\ M_{31} & M_{32} & M_{33} & M_{34} & M_{35} \\ M_{41} & M_{42} & M_{43} & M_{44} & M_{45} \\ M_{51} & M_{52} & M_{53} & M_{54} & M_{55} \end{bmatrix} \quad (8)$$

$$H = \begin{bmatrix} H_1 & H_2 & H_3 & H_4 & H_5 \end{bmatrix}^T \quad (9)$$

$$G = \begin{bmatrix} G_1 & G_2 & G_3 & G_4 & G_5 \end{bmatrix}^T \quad (10)$$

$$F = \begin{bmatrix} 0 & F_2 & 0 & 0 & 0 \end{bmatrix}^T \quad (11)$$

$$T = \begin{bmatrix} \tau_p & \tau_\delta & \tau_\kappa & \tau_r & \tau_l \end{bmatrix}^T \quad (12)$$

III. ESTIMATION OF SYSTEM PARAMETERS

In this section, estimation of the state of the rider using the extended Kalman filter(EKF) and calculation of rollover condition using the estimated rider state are performed. First, state estimation of the rider is described, then calculation of rollover condition is shown.

A. Estimation of body parameters by EKF

First, the state matrix and observation matrix in EKF are defined, and the body parameters are estimated while removing noise by EKF.

1) Definition of state matrix and observation matrix: In this paragraph, the state matrix and the observation matrix are defined. The symbols are shown in Table II.

TABLE II. SYMBOLS USED IN ESTIMATION SYSTEM BY EKF

Variables	Explanation
x	State matrix
$\dot{x}$	Differentiation of state matrix
y	Observation matrix
M_δ	Transformation matrix
Φ_δ	Transformation matrix
ω	Natural frequency
β	Attenuation constant
δ	Body tilt angle
r	Turning radius
$\dot{d}$	Wheelchair speed
$\hat{x}$	Estimated value of state matrix
Δ	Sampling time

The equations of motion of the upper body are formulated as (13) using the natural frequency ω, the attenuation ratio β, the body inclination angle δ, the turning radius r and the speed $\dot{d}$.

$$\ddot{\delta} + 2\beta\omega\dot{\delta} + \omega^2\delta = -\frac{\dot{d}^2}{r}\sin\kappa - g\sin\delta \quad (13)$$

Here, x, $\dot{x}$, y, M_δ, Φ_δ are defined as (14)-(18).

$$x = \begin{bmatrix} x_1 \\ x_2 \\ x_3 \\ x_4 \end{bmatrix} = \begin{bmatrix} \delta \\ \dot{\delta} \\ \beta \\ \omega \end{bmatrix} \quad (14)$$

$$\dot{x} = \begin{bmatrix} x_2 \\ -2x_3x_4x_2 - x_4^2x_1 - \frac{\dot{d}^2}{r}\sin\kappa - g\sin x_1 \\ 0 \\ 0 \end{bmatrix} \quad (15)$$

$$M_\delta = \begin{bmatrix} 0 & 1 & 0 & 0 \end{bmatrix} \quad (16)$$

$$y = M_\delta x \quad (17)$$

$$\Phi = \begin{bmatrix} 1 & \Delta & 0 & 0 \\ \Phi_{21} & \Phi_{22} & \Phi_{23} & \Phi_{24} \\ 0 & 0 & 1 & 0 \\ 0 & 0 & 0 & 1 \end{bmatrix} \quad (18)$$

$$\Phi_{21} = -\Delta\hat{x_4}^2$$

$$\Phi_{22} = -2\Delta\hat{x_2}\hat{x_4}$$

$$\Phi_{23} = 1 - 2\Delta\hat{x_3}\hat{x_4}$$

$$\Phi_{24} = -2\Delta(\hat{x_2}\hat{x_3} + \hat{x_1}\hat{x_4})$$

2) Extended Kalman filter algorithm: In this paragraph, the extended Kalman filter algorithm is described. The symbols are shown in Table III.

TABLE III. SYMBOLS USED IN EKF ALGORITHM

Variables	Explanation
$K(k)$	Kalman gain
$P^-(k)$	Predicted (a priori) estimate covariance
$P(k)$	A Posteriori error covariance matrix
$\hat{x}$	Estimated value of state matrix
$\hat{x}^-$	Predicted (a priori) state estimate
σ_w^2	Process noise
σ_v^2	Observation noise

Extended Kalman filter algorithm is formulated as (19)-(23)

$$x^-(k) = x(k-1) + \dot{x}(k-1)\Delta \quad (19)$$

$$P^-(k) = \Phi P(k-1)\Phi^T \quad (20)$$

$$K(k) = \frac{P^-(k)M_\delta}{M_\delta^T P^-(k)M_\delta + \sigma_w^2} \quad (21)$$

$$\hat{x} = x^-(k) + K(k)\left(y - M_\delta x^-(k)\right) \quad (22)$$

$$P(k) = \left(1 - K(k)M_\delta\right)^2 P^-(k) + \sigma_v^2 \quad (23)$$

By calculating the state matrix $\hat{x}$ using this algorithm, it is possible to calculate the rider's state such as angle δ, angular velocity $\dot{\delta}$, natural frequency ω, and attenuation constant β.

B. Calculation of rollover condition considering rider's state

In this section, calculation of the rollover condition considering the state of the rider are be explained.

1) Calculation of moment: Because the rolling condition of the wheelchair is same on the right and left wheels, in this section the rollover in the direction of the right wheel is taken into consideration. Here, set the inclination of the ground to ζ (positive when the left wheel is high) and set the moment on the left wheel while turning to the left is Mo (clockwise is positive). Mo is given by (24).

$$
\begin{aligned}
\frac{Mo}{2} = &\ (M_b + M_w)(W\cos\zeta - 2h_{bd}\sin\zeta) \\
&+ M_{hd}\{W\cos\zeta - 2(h_{bd} + h_{bu})\sin\zeta\} \\
&+ M_{hu}\{W\cos\zeta - 2(h_{bd} + h_{bu} + h_u)\sin\zeta \\
&- 2h_u\sin(\zeta + \delta)\} \\
&- 2K\delta(h_{bd} + h_{bu} + h_u\cos\delta)\sin\kappa \\
&- \frac{2\dot{d}^2}{r}\{(M_b + M_{hd})h_{hd} + M_{bu}h_{bu}\} \quad (24)
\end{aligned}
$$

Here, the rollover condition is described as (25).

$$
Mo = 0 \quad (25)
$$

2) Calculation of rollover condition: In this paragraph, the maximum body inclination angle of a person is predicted from the current speed and rotation radius, natural frequency by EKF, and the speed limit value is calculated by the moment method. Since the damping constant in the body model is greater than 1, when the traveling trajectory of the wheelchair is a curve, the maximum estimated value of the body tilt angle $\hat{\delta}_{MAX}$ is given by (26).

$$
\hat{\delta_{MAX}} = \frac{\dot{d}^2}{(\omega^2 - g)r} \quad (26)
$$

However, in practice, the natural frequency calculated by the Kalman filter is inaccurate, and there is a possibility that the estimated maximum value of the body tilt angle becomes smaller than the actual body tilt angle. Therefore, the maximum value of the body tilt angle is determined as (27).

$$
\delta_{MAX} = \begin{cases} \hat{\delta}_{MAX} & if\ \ \hat{\delta}_{MAX} > \delta \\ \delta & else \end{cases} \quad (27)
$$

The velocity limit value v_{Me} that a wheelchair does not overturn is expressed by (28).

$$
v_{Me}^2 = \frac{rg}{2\dot{d}^2}\frac{S_1}{S_2} \quad (28)
$$

Here, S_1 and S_2 are expressed as follows.

$$
\begin{aligned}
S_1 = &\ (M_b + M_w)(W\cos\zeta - 2h_{bd}\sin\zeta) \\
&+ M_{hd}\{W\cos\zeta - 2(h_{bd} + h_{bu})\sin\zeta\} \\
&+ M_{hu}\{W\cos\zeta - 2(h_{bd} + h_{bu} + h_u) \\
&- 2h_u\sin\delta_{MAX}\sin(\zeta + \delta_{MAX})\sin\kappa\} \\
= &\ (M_b + M_w)(W\cos\zeta - 2h_{bd}\sin\zeta) \\
&+ M_{hd}\{W\cos\zeta - 2(h_{bd} + h_{bu})\sin\zeta\} \\
&+ M_{hu}\{W\cos\zeta - 2(h_{bd} + h_{bu} + h_u)\sin\zeta \\
&- 2h_u\sin\kappa\sin\left(\zeta + \frac{\dot{d}^2}{(\omega^2 - g)r}\right)\} \quad (29) \\[4pt]
S_2 = &\ (M_b + M_{hd})h_{hd} + M_{bu}h_{bu} \\
&+ M_{hu}(h_{bd} + h_{bu} + h_u\cos\delta_{MAX}) \\
= &\ (M_b + M_{hd})h_{hd} + M_{bu}h_{bu} \\
&+ M_{hu}\left(h_{bd} + h_{bu} + h_u\cos\frac{\dot{d}^2}{\omega^2 r}\right) \quad (30)
\end{aligned}
$$

C. Control to suppress vibration

By getting the body parameters and changing the command value for the wheelchair, the acceleration of the body is suppressed. The rotation radius corresponding to the speed command is r_{cmd}, and the rotation radius for vibration suppression while corresponding to the speed input is r_{prv} which are given by (31), (32).

$$
r_{prv} = \frac{r_{cmd}}{A_{prv}} \quad (31)
$$

$$
r_{prv} = v^2\frac{M_h}{KB_{prv}\delta} \quad (32)
$$

Here, A_{prv}, B_{prv} are represented by (33) and (34).

$$
A_{prv} = \begin{cases} K_{prv}(t - t_1)^2 & (t_1 < t < t_1 + 0.1) \\ 1 & (else) \end{cases} \quad (33)
$$

$$
B_{prv} = \begin{cases} 1 - K_{prv}(t - t_2)^2 & (t_2 < t < t_2 + 0.1) \\ 0 & (else) \end{cases} \quad (34)
$$

Here, t_1 and t_2 indicate the rotation start time and the rotation end time, respectively. K_{prv} indicates suppress coefficient determined by empirically.

By changing the rotation radius command value, it is possible to suppress the acceleration of the human model and to improve the riding comfort.

IV. Control system[10]

This section describes the control system. As for the control system, the paper which already exists[10] is referred to. Since the movement of people is not taken into account when constructing the control system, riders are not considered in modeling.

A. Model of Two-Wheel Mobile Robot

Kinematics is derived as (35) and (36) which assumes that wheels never slips.

$$\dot{\boldsymbol{x}}_w = \begin{bmatrix} \dot{\theta}_p \\ \dot{d} \\ \dot{\varphi} \end{bmatrix} = \begin{bmatrix} 1 & 0 & 0 \\ 0 & \frac{R}{2} & \frac{R}{2} \\ 0 & \frac{R}{W} & -\frac{R}{W} \end{bmatrix} \begin{bmatrix} \dot{\theta}_p \\ \dot{\theta}_r \\ \dot{\theta}_l \end{bmatrix} = \boldsymbol{J}_{aco}\dot{\boldsymbol{\theta}}_w \tag{35}$$

$$\dot{\boldsymbol{\theta}}_w = \begin{bmatrix} \dot{\theta}_p \\ \dot{\theta}_r \\ \dot{\theta}_l \end{bmatrix} = \begin{bmatrix} 1 & 0 & 0 \\ 0 & \frac{1}{R} & \frac{W}{2R} \\ 0 & \frac{1}{R} & -\frac{W}{2R} \end{bmatrix} \begin{bmatrix} \dot{\theta}_p \\ \dot{d} \\ \dot{\varphi} \end{bmatrix} = \boldsymbol{J}_{aco}^{-1}\dot{\boldsymbol{x}}_w \tag{36}$$

The relationship between torque vector of joint space $\boldsymbol{\tau}$ and that of work space $\boldsymbol{f}$ is expressed as (37).

$$\boldsymbol{\tau} = \boldsymbol{J}_{aco}^T \boldsymbol{f} \tag{37}$$

Dynamics in the work space is derived as (38).

$$\boldsymbol{m}\left(\boldsymbol{x}_h\right)\ddot{\boldsymbol{x}}_h + \boldsymbol{h}\left(\boldsymbol{x}_h, \dot{\boldsymbol{x}}_h\right) + \boldsymbol{g}\left(\boldsymbol{x}_h\right) = \boldsymbol{f} \tag{38}$$

$$\boldsymbol{m} = \begin{bmatrix} m_{11} & m_{12} & 0 \\ m_{21} & m_{22} & 0 \\ 0 & 0 & m_{33} \end{bmatrix}, \boldsymbol{h} = \begin{bmatrix} h_1 \\ h_2 \\ h_3 \end{bmatrix},$$

$$\boldsymbol{g} = \begin{bmatrix} g_1 \\ 0 \\ 0 \end{bmatrix}, \boldsymbol{f} = \begin{bmatrix} 0 \\ f_d^{ref} \\ \tau_\phi^{ref} \end{bmatrix}$$

B. Velocity-Based Controller

Velocity-based controller is designed extending [11] and adapting rotational motion additionally. This controller is designed based on sliding mode control. The sliding surface $\boldsymbol{\sigma}$ is determined as (39).

$$\boldsymbol{\sigma} = \begin{bmatrix} \sigma_1 \\ \sigma_2 \end{bmatrix}$$
$$= \begin{bmatrix} c_1\left(\theta_p^{res} - \theta_p^{cmd}\right) + c_2\left(\dot{\theta}_p^{res} - \dot{\theta}_p^{cmd}\right) + \left(\dot{d}^{res} - \dot{d}^{cmd}\right) \\ \left(\dot{\phi}^{cmd} - \dot{\phi}^{res}\right) \end{bmatrix}. \tag{39}$$

Here, c_1 and c_2 are positive values which are defined empirically, and θ_p^{cmd} and $\dot{\theta}_p^{cmd}$ are set to zero because pitch angle should converge to zero. The time derivative of $\boldsymbol{\sigma}$ is expressed as (40).

$$\dot{\boldsymbol{\sigma}} = \begin{bmatrix} c_1\dot{\theta}_p^{res} + c_2\ddot{\theta}_p^{res} + \left(\ddot{d}^{res} - \ddot{d}^{cmd}\right) \\ \left(\ddot{\phi}^{cmd} - \ddot{\phi}^{res}\right) \end{bmatrix} \tag{40}$$

Here, the disturbance added to the model is also considered and $\bigcirc^{dis}$ means disturbance. (40) is rewritten as (41) using dynamics.

$$\dot{\boldsymbol{\sigma}} = \begin{bmatrix} c_1\dot{\theta}_p^{res} - A_n f_d^{ref} + A_n\tilde{f}_d^{dis} - B_n\tilde{\tau}_p^{dis} - \ddot{d}^{cmd} \\ \ddot{\phi}^{cmd} - \frac{1}{m_{n33}}\left(\tau_\phi^{ref} - \tilde{\tau}_\phi^{dis}\right) \end{bmatrix} \tag{41}$$
$$A_n = \frac{c_2 m_{n12} - m_{n11}}{m_{n11}m_{n22}}, B_n = \frac{c_2}{m_{n11}}$$

The total disturbance, $\tilde{\tau}_p^{dis}$, $\tilde{f}_d^{dis}$, and $\tilde{\tau}_\phi^{dis}$ is estimated by DOB [12] and PADO [13] as (42)-(44).

$$\hat{f}_d^{dis} = \frac{g_d}{s + g_d}\left(f_d^{ref} + g_d m_{n22}\dot{d}^{res}\right) - g_d m_{n22}\dot{d}^{res} \tag{42}$$

$$\hat{\tau}_\phi^{dis} = \frac{g_\phi}{s + g_\phi}\left(f_d^{ref} + g_\phi m_{n33}\dot{\phi}^{res}\right) - g_\phi m_{n33}\dot{\phi}^{res} \tag{43}$$

$$\hat{\tau}_p^{dis} = \frac{g_p^2}{s + g_p}\left(m_{n11}\dot{\theta}_p^{res} + m_{n12}\dot{d}^{res}\right) \\ - g_p\left(M_{n11}\dot{\theta}_p^{res} + M_{n12}\dot{\theta}_w^{res}\right) \tag{44}$$

The equivalent control input is derived as (45) by setting $\dot{\boldsymbol{\sigma}}$ to zero and combining estimated disturbance in (42)-(44).

$$\boldsymbol{f}_{eq} = \begin{bmatrix} \frac{1}{A_n}\left(c_1\dot{\theta}_p^{res} + A_n\hat{f}_d^{dis} - B_n\hat{\tau}_p^{dis} - \ddot{d}^{cmd}\right) \\ \hat{\tau}_\phi^{dis} + m_{n33}\ddot{\phi}^{cmd} \end{bmatrix} \tag{45}$$

The control input in the work space is described as (46): sum of the equivalent control input $\boldsymbol{f}_{eq}$ and the nonlinear control input $\boldsymbol{f}_{nl}$.

$$\boldsymbol{f}^{ref} = \boldsymbol{f}_{eq} + \boldsymbol{f}_{nl} \tag{46}$$

In order to determine $\boldsymbol{f}_{nl}$ such that the convergence into sliding surface is guaranteed, Lyapunov function is set as (47).

$$V = \frac{1}{2}\boldsymbol{\sigma}^T\boldsymbol{\sigma} \tag{47}$$

The time derivative of V is calculated as (48) by using (40), (45), and (46).

$$\dot{V} = \begin{bmatrix} -\sigma_1 A_n & -\frac{\sigma_2}{m_{n33}} \end{bmatrix} \boldsymbol{f}_{nl} \tag{48}$$

$\boldsymbol{f}_{nl}$ is designed as (49), which leads $\dot{V}$ to be negative as (50), and guarantees $\boldsymbol{\sigma} \to \boldsymbol{0}$.

$$\boldsymbol{f}_{nl} = \begin{bmatrix} \frac{1}{A_n}\left(k_{11}\text{sgn}\left(\sigma_1\right) + k_{12}\sigma_1\right) \\ m_{n33}\left(k_{21}\text{sgn}\left(\sigma_2\right) + k_{22}\sigma_2\right) \end{bmatrix} \tag{49}$$

$$\dot{V} = -k_{11}\frac{\sigma_1^2}{|\sigma_1|} - k_{12}\sigma_1^2 - k_{21}\frac{\sigma_2^2}{|\sigma_2|} - k_{22}\sigma_2^2 < 0 \tag{50}$$

Here, k_{11}, k_{12}, k_{21}, and k_{22} are control gains and they concern the adjustment of wheelchair behavior in control. $\boldsymbol{f}_{nl}$ is modified as (51) to avoid chattering phenomenon because of the discontinuous terms.

$$\boldsymbol{f}_{nl} = \begin{bmatrix} \frac{1}{A_n}\left(k_{11}\frac{\sigma_1}{|\sigma_1|+\delta_1} + k_{12}\sigma_1\right) \\ m_{n33}\left(k_{21}\frac{\sigma_2}{|\sigma_2|+\delta_2} + k_{22}\sigma_2\right) \end{bmatrix} \tag{51}$$

Here, δ_1 and δ_2 are smooth function parameters. The chattering is suppressed by adjusting these parameters. The block diagram of the total system is shown in Fig. 2.

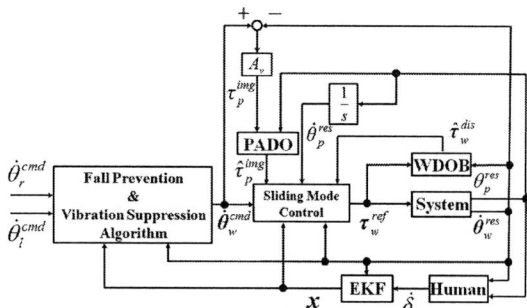

Fig. 2.   Block diagram of total system

V.   SIMULATION

Three simulations are performed in this paper. In the first simulation, the validity of estimating the angle and state of the rider using the extended Kalman filter is verified. Assuming the case where the IMU sensor is actually used, noise is included in the observation value. In the second simulation, it is verified how the rollover condition of the wheelchair changes as the rider's state changes. In the third simulation, the effectiveness of the rider's damping control is verified.

A. Verification of EKF estimation

In this section, human model estimation by extended Kalman filter is described. Verification is conducted in the presence of noise. The wheelchair runs on a bumpy road at a constant speed in this simulation. The rotation radius is given by a random number. The actual vibration and vibration estimation of the human model in this case are compared. Observed value δ_o is the sum of true value δ_m and noise which is given by (52).

$$\delta_o = \delta_m + n_w \qquad (52)$$

Note that n_w is a value that gives the maximum of its absolute value $|n_w|$ as 10 deg/s in radians.

TABLE IV.   SIMULATION CONDITION

Variables	Values	Description
v	5.0 km/h	Wheelchair speed
ω	7.97	Person model natural frequency
β	1.1	Person model attenuation constant
ω_{init}	10.7	Person model natural frequency initial value
β_{init}	1.7	Person model attenuation constant initial value
h_u	0.4 m	Upper body center of gravity height

Simulation results are shown in Fig. 3-6. As shown in Fig. 3, although the initial value of the damping constant and natural frequency greatly differs from the true value greatly, estimated value almost agrees with the true value. It is found that the estimated values are in agreement. As shown in Fig. 4, it is understood that the observation value gradually agrees with the true value even when noise is heavily loaded. As you can see from Fig. 3, the value obtained by simply integrating the observation value is far away from the true value, On the other hand, the estimated value gradually agrees with the true value. Furthermore,

as can be seen from Fig. 5, 6, the convergence of the attenuation constant and natural frequency to the true value is made.

From the above, the effectiveness of human model estimation by EKF is confirmed.

B. Calculation of rollover condition

In this section, the rollover condition is calculated which is taking into account the model of the rider. The extent to which the speed limit value varies depending on the natural frequency of the human model is verified by simulation. This verification is carried out by changing the rotation radius, the mass of the wheelchair main body, and the speed command value. Simulation conditions are shown in Table V. Depending on the simulation, parameters are changed, but unless otherwise noted, it is assumed to follow Table IV.

TABLE V.   SIMULATION CONDITION

Variables	Values	Description
v	5.0 km/h	Wheelchair speed
r	5.0 m	Turning radius
g	9.8 m/s^2	Gravitational acceleration
W	0.4 m	Wheelchair width
ζ	5.0 deg	Ground inclination
h_u	0.4 m	Upper body center of gravity height
h_{bu}	0.30 m	Lower center of gravity height
h_{bd}	0.30 m	Wheelchair center of gravity height
M_b	20.0 kg	Wheelchair weight
M_w	10.0 kg	Wheel weight
M_{hd}	30.0 kg	Lower body weight
M_{hu}	30.0 kg	Upper body weight

Simulation results are shown in Fig. 7, 8. Fig. 7 shows the rollover condition when the turning radius is $r = 3.0, 5.0, 8.0$ m. Fig. 8 shows the rollover condition when the ground inclination is $\zeta = 0.0, 5.0, 10.0$ deg. Other parameters follow Table V. As shown in Fig. 7, 8, the difference between the speed limit value in the conventional method and the speed limit value in the proposed method is large. Especially when there is a tilt in the horizontal direction on the ground, the difference of the speed limit value is particularly large. As can be seen from these, the rollover condition in the conventional method is incomplete, and the wheelchair tends to collapse and the speed limit value becomes lower when the rider's state is considered. From the above, the effectiveness of the proposed method is shown.

C. Verification of ride comfort improvement control

In this subsection, the usefulness of the vibration suppression algorithm is simulated. First, the wheelchair travels straight at 5.0 km/h for 0.5 sec, and then it is given a command to rotate 90 deg at the radius of 5 m to the right. The magnitude of the vibration of the human model is compared from the viewpoint of angular acceleration. Simulation conditions are shown in Table VI.

Simulation results are shown in Fig. 9-11. Fig. 9 shows the body tilt angle and angular velocity. In the case of the presence of the algorithm, it is understood

The 2018 International Power Electronics Conference

TABLE VI. PHYSICAL PARAMETERS USED IN SIMULATION

Variables	Values	Description
v	5.0 km/h	Wheelchair speed
ω	7.97	Person model natural frequency
β	1.1	Person model attenuation constant
r	5.0 m	Turning radius
g	9.8 m/s^2	Gravitational acceleration
W	0.4 m	Wheelchair width
ζ	5.0 deg	Ground inclination
h_u	0.4 m	Upper body center of gravity height
h_{bu}	0.30 m	Lower center of gravity height
h_{bd}	0.30 m	Wheelchair center of gravity height
M_b	20.0 kg	Wheelchair weight
M_w	10.0 kg	Wheel weight
M_{hd}	30.0 kg	Lower body weight
M_{hu}	30.0 kg	Upper body weight

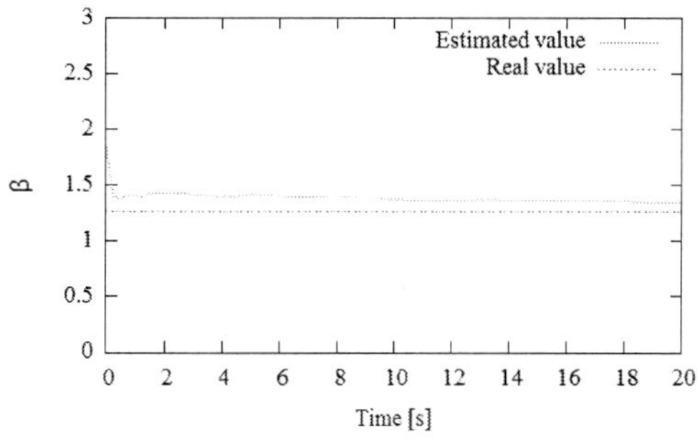

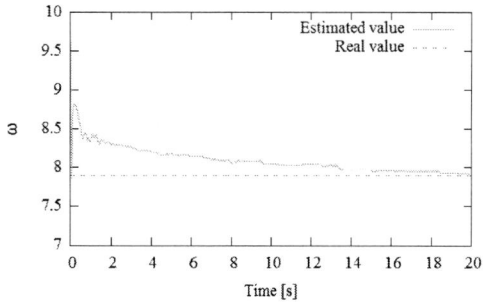

Fig. 5. Estimation value of attenuation constant

that the body tilt angle is slightly delayed as compared with the case where there is no algorithm. Fig. 10 shows the body tilt angular acceleration. Comparing the body tilt angular acceleration with the presence or absence of the algorithm, it can be seen that the maximum value of angular acceleration is reduced to about 60 % compared to the absence of the algorithm. From this, it can be seen that damping has been achieved. However, as shown in Fig. 11, there is a problem that the trajectory deviates slightly from the command value.

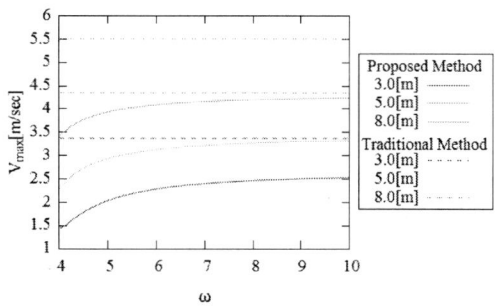

Fig. 6. Estimation value of natural frequency

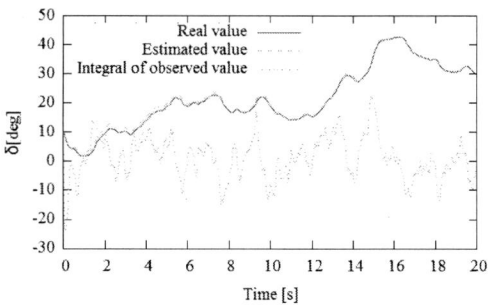

Fig. 3. Estimation value of body tilt angle

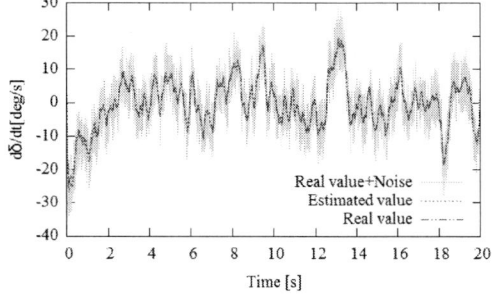

Fig. 4. Estimation value of body tilt angular velocity

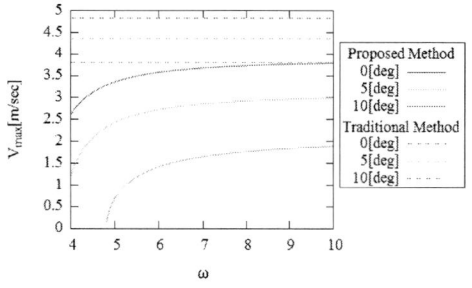

Fig. 7. Rollover condition ($r = 3.0,\ 5.0,\ 8.0$)

Fig. 8. Rollover condition ($\zeta = 0.0,\ 5.0,\ 10.0$)

581

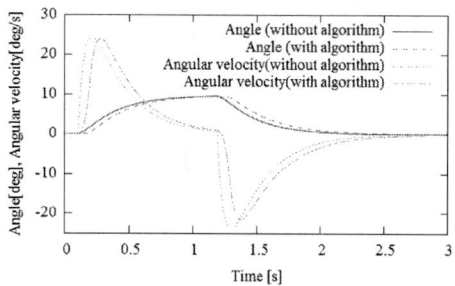

Fig. 9.   Body tilt angle and angular velocity

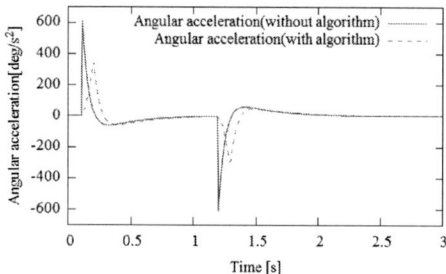

Fig. 10.   Body tilt angular acceleration

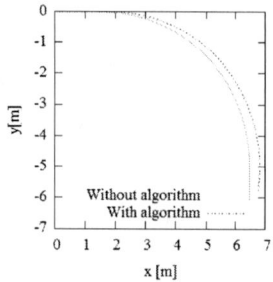

Fig. 11.   Travel locus

VI. CONCLUSION

In this paper, the purpose is to calculate accurate rollover condition by considering human state. The body tilt angle is measured by the IMU sensor, and the body parameters are estimated while removing the noise by the Kalman filter. Based on this estimated parameter, rollover condition is calculated. By the simulation, the following two points are found out. 1) Physical parameters can be estimated even when there is noise in the observed value. 2) The rollover condition changes greatly by considering the human body parameters. Furthermore, it is found that further improvement is required for the enhancement of ridding comfortableness.

One of the issues to be addressed in the future is that it is necessary to verify with the actual machine and human body.

ACKNOWLEDGEMENT

This work is supported in part by a Grant-in-Aid for Scientific Research(15H02235).

REFERENCES

[1] United nations(UN), "World Population Ageing 2009" 19 December 2009, New York.:UN

[2] WALTER M. BORTZ, "The Disuse Syndrome," THE WESTERN JOURNAL OF MEDICINE, vol. 5, no. 141, pp. 691-694, 1984.

[3] H. Wakabayashi, H. Sashika, "Association of Nutrition Status and Rehabilitation Outcome in the Disuse Syndrome: a Retrospective Cohort Study," Arch Phys Med Rehabil Vol 92, June 2011.

[4] W. Chen, Y. Jang, J. Wang, W. Huang, C. Chang, H. Mao, Y. Wang, "Wheelchair-Related Accidents: Relationship With Wheelchair-Using Behavior in Active Community Wheelchair riders." A Treatise on Electricity and Magnetism, 3rd ed., vol. 2. Oxford: Clarendon, 1892, pp. 68-73.

[5] Gina E. Bertocci, Douglas A. Hobson, Kennerly H. Digges, "Development of a Wheelchair Occupant Injury Risk Assessment Method and Its Application in the Investigation of Wheelchair Securement Point Influence on Frontal Crash Safety," IEEE TRANSACTIONS ON REHABILITATION ENGINEERING, VOL. 8, NO. 1, MARCH 2000.

[6] David P. Vansickle, Rory A. Cooper, Rick N. Robertson, Michael L. Boninger "Determination of Wheelchair Dynamic Load Data for Use with Finite Element Analysis," IEEE TRANSACTIONS ON REHABILITATION ENGINEERING, VOL. 4, NO. 3, SEPTEMBER 1996.

[7] S. Katsura, K. Ohnishi, "Semiautonomous Wheelchair Based on Quarry of Environmental Information," IEEE TRANSACTIONS ON REHABILITATION ENGINEERING, VOL. 8, NO. 1, MARCH 2000.

[8] S. Oh, and Y. Hori, "Disturbance Attenuation Control for Power-Assist Wheelchair Operation on Slopes," IEEE TRANSACTIONS ON CONTROL SYSTEMS TECHNOLOGY, VOL. 22, NO. 3, MAY 2014.

[9] H. Seki, K. Ishihara, A. Tadakuma, "Novel Regenerative Braking Control of Electric Power-Assisted Wheelchair for Safety Downhill Road Driving," IEEE TRANSACTIONS ON REHABILITATION ENGINEERING, IEEE TRANSACTIONS ON INDUSTRIAL ELECTRONICS, VOL. 56, NO. 5, MAY 2009.

[10] S. Amagai, M.Kamatani, T. Murakami: "A Comparison Study of Velocity and Torque Based Control of Two-Wheel Mobile Robot for Human Operation" 2017 24th International Conference on Mechatronics and Machine Vision in Practice (M2VIP),21-23 Nov. 2017, New Zealand

[11] S. Amagai, T. Murakami: "A stabilization Control of Two Wheels Driven Wheelchair" IECON 2016 - 42nd Annual Conference of the IEEE Industrial Electronics Society, 23-26 Oct. 2016, Forence, Itary

[12] K. Ohnishi: "Robust Motion Control by Disturbance Observer", Journal of the Robotic Society of Japan, Vol. 11, No. 4, pp. 486-493, 1993 (in Japanese)

[13] A. Nakamura, T. Murakami: "A stabilization Control of Two Wheels Driven Wheelchair" The 2009 IEEE/RSJ International Conference on Intelligent Robots and Systems, October 11-15, 2009 St.Louis, USA

The 2018 International Power Electronics Conference

Stabilization Method for Residential DC System Based on Passivity Criterion

Hiroaki Kakigano[1*]

1 College of Science and Engineering, Ritsumeikan University, Kusatsu, Japan
*E-mail: kakigano@fc.ritsumei.ac.jp

Abstract— In this paper, we assume a dc distribution board that can be attached and detached to dc sources and loads flexibly. In previous researches, stable criteria based on the impedance ratio conditions were proposed. These criteria can deal with one-way power transfer from a power source to a load. However, if a system has a secondary battery, it is difficult to apply the impedance-based conditions due to its bi-directional power flow. We study a stabilization method using a passivity-based stability criterion in a residential dc system. A virtual admittance was added with an input admittance of a converter in parallel using a feedforward control. Experimental results showed input bus voltage of a buck converter was stable when it satisfied the passivity condition.

Keywords— *dc power sypply, stability, passivity, feedforward control, input admittance.*

I. INTRODUCTION

To reduce greenhouse gas emissions and fossil fuel usages, many photovoltaic (PV) systems for residential houses have been installed widely because of feed-in tariff programs. However, the outputs of the PV systems are not stable due to the variations of both the solar radiation and the temperature. This unstable output limits the connectable capacity of PV systems to each utility grid. To solve this problem, it can be common that a secondary battery consists of the PV system not to affect a utility grid and to utilize the PV output power for the local consumption. Since PV panels and secondary batteries are dc output, dc power supply system is gaining attention to be a next power supply system instead of a conventional ac system. In this case, grid-tied inverters can be merged into one unit. In addition, the dc system can connect with fuel cells and/or electric vehicles which are expected to increase the installations.

If the dc systems are applied to residential houses, it could be demanded to change the composition of the system along with family structure flexibly. In this case, a dc distribution board is favorable to be attached and detached with the dc output equipment as a common platform. The image of the dc distribution board is shown in Fig. 1. The dc distribution board can connect with any dc-dc converters which are satisfied with the specification of the dc distribution board. As one of the issues, we should handle with the instability caused by the negative

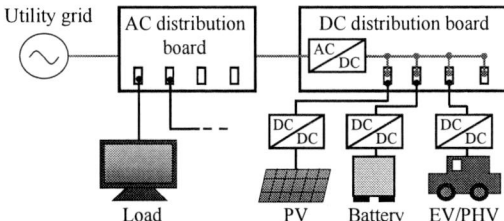

Fig. 1. DC system with a dc distribution board.

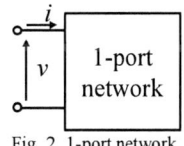

Fig. 2. 1-port network.

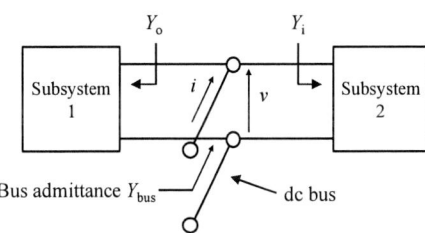

Fig. 3. Two subsystems connected with a dc bus.

impedance characteristic of the dc-dc converters when the converters are regarded as constant power loads [1].

In Ref. [2], a stable criterion was proposed using input-output impedance ratio, and there are some derived stable criteria based on the impedance ratio have been proposed [3] – [5]. They assume one-way power transfer from a power source side to a load side. However, if system has equipment that can change the power flow bidirectionally like secondary batteries, it is difficult to use the stable conditions based on the impedance ratio.

As the other stable criterion for dc system, passivity-based stability criterion was proposed. The criterion can be acceptable to the bidirectional power flow. To satisfy the passivity, positive feedforward controls were added to converter controls [6] – [8]. However, the controls in the previous researches were too simple to apply a practical converter.

583

The 2018 International Power Electronics Conference

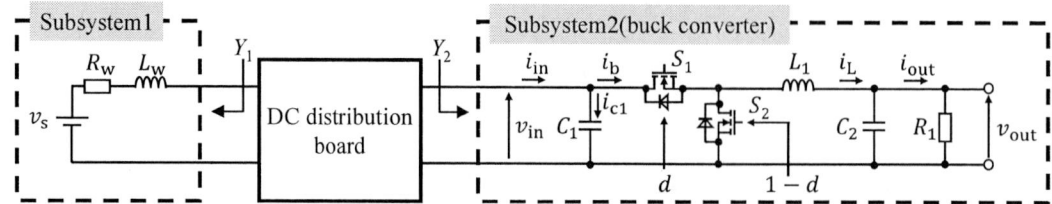

Fig. 4. Test system for stability analysis.

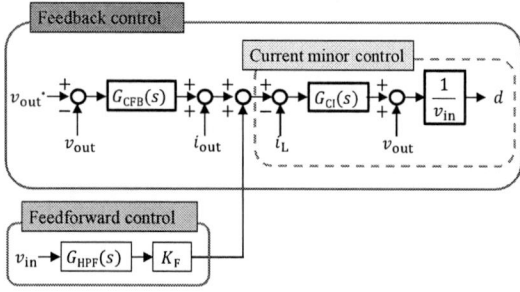

Fig. 5. Control block diagram of the buck converter.

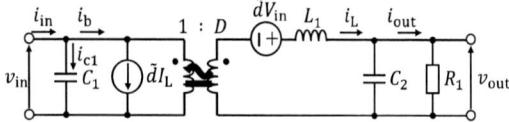

Fig. 6. Buck converter equivalent average linear circuit.

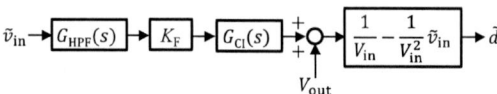

Fig. 7. Control block diagram of a buck converter
(perturbations of v_{in} and d are considered).

In this paper, to solve the above problem, we apply the passivity-based stability criterion for the target dc system, and propose a modified practical feedforward control to satisfy the passivity.

II. PASSIVITY-BASED STABILITY CRITERION

A circuit of 1-port network connected to a dc system is shown in Fig. 2. If and only if it satisfies (1) for any T, the 1-port circuit satisfies passivity and it is stable.

$$\int_{-\infty}^{T} v(t)i(t)dt \geq 0 \qquad (1)$$

Substituting $+\infty$ for T, the equation becomes

$$\int_{-\infty}^{+\infty} v(t)i(t)dt$$
$$= \frac{1}{2\pi}\int_{-\infty}^{+\infty} V(j\omega)I^*(j\omega)d\omega$$
$$= \frac{1}{2\pi}\int_{-\infty}^{+\infty} Z(j\omega)|I(j\omega)|^2 d\omega$$
$$= \frac{1}{\pi}\int_{0}^{+\infty} \mathrm{Re}[Z(j\omega)]|I(j\omega)|^2 d\omega \qquad (2)$$

If an impedance $Z(s)$ has a positive real part at any frequency, (2) satisfies passivity. In addition, if an impedance has a positive real part, the admittance also has a positive real part. Therefore, when an admittance has a positive real part at any frequency, it also satisfies passivity.

Fig. 3 shows a system example where two sub-systems are connected to a dc bus. If the input admittances of both Subsystem1 Y_o $(j\omega)$ and Subsystem2 Y_i $(j\omega)$ have positive real parts, which can be expressed as follows:

$$\mathrm{Re}[Y_o(j\omega)] \geq 0 \qquad (3)$$
$$\mathrm{Re}[Y_i(j\omega)] \geq 0 \qquad (4),$$

each system satisfies passivity. Then, the bus admittance Y_{bus} $(j\omega)$ satisfies the following equation:

$$\mathrm{Re}[Y_{bus}(j\omega)] = \mathrm{Re}[Y_o(j\omega)] + \mathrm{Re}[Y_i(j\omega)] \geq 0 \qquad (5).$$

Therefore, the system is stable because the $Y_{bus}(j\omega)$ satisfies passivity.

As shown the above, dc system stability can be confirmed when the real part of the input admittance of each converter connected to the system is always positive. It means all converters satisfy passivity, the system is stable. However, most converters do not satisfy passivity especially when they are regarded as constant power loads. One of the solution is to add an additional feedforward control to satisfy passivity.

III. INPUT ADMITTANCE OF BUCK CONVERTER

A. Proposed Feedforward Control

A test dc system for stability analysis is shown in Fig. 4. Fig. 5 shows the control block diagram of the buck converter in Fig. 4. The test system consists of the buck converter, a dc source and a line resistance and inductance. The input admittance Y_1 of Subsystem1 satisfies passivity as shown in the following equation:

$$Y_1(s) = \frac{1}{R_w + sL_w} \qquad (6).$$

If the input admittance Y_2 of Subsystem2 (buck converter) satisfies passivity, the total bus admittance satisfies passivity, which means the test system is stable. $G_{CFB}(s)$ and $G_{CI}(s)$ in Fig. 5 are the transfer functions of PI controller of voltage control and current minor loop control, respectively. K_F and $G_{HPF}(s)$ are gain and the transfer function of high pass filter (HPF) for the additional feedforward control shown in Fig. 5.

In previous researches [6] – [8], similar feedforward controls were proposed for a converter to achieve passivity. However, the study did not include the output voltage v_{out} and the input voltage v_{in} in the current minor loop control. In this case, it is difficult to set appropriate PI gains of the current minor loop control, and the current control cannot work in a practical condition. Therefore, we propose a more practical control considering the output voltage v_{out} and the input voltage v_{in} in the current minor loop control.

584

TABLE I
System parameters.

DC voltage source, v_s	V	48
Wire resistance, R_w	Ω	0.2
Wire inductor, L_w	mH	5
Input capacitor, C_1	μF	470
Inductor, L_1	mH	5.6
Output capacitor, C_2	μF	470
Load resistance, R_1	Ω	3.8
Cut of frequency(HPF), f_c	Hz	10

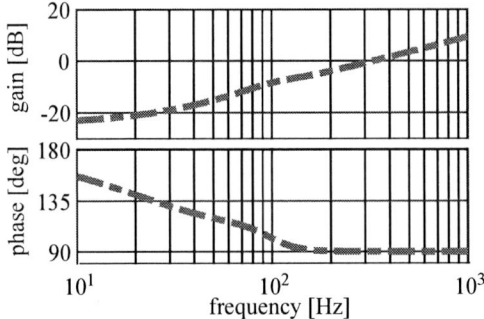

Fig. 8. Input admittance of buck converter (without PFF control).

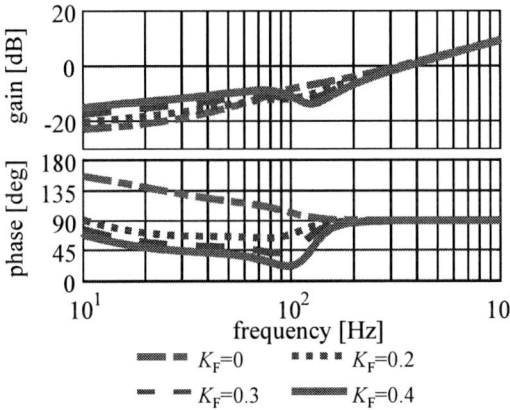

- - - K_F=0   ▪▪▪▪ K_F=0.2
- - K_F=0.3   ▬▬ K_F=0.4

Fig. 9. Input admittance of buck converter (without and with PFF control: f_c=10 Hz).

The buck converter equivalent average linear circuit is shown in Fig. 6. The perturbations of i_b can be expressed with the duty d as

$$\tilde{i}_b(s) = \tilde{d}I_L \qquad (7).$$

The tilde and the capital letter denote the perturbation and the steady value of the variables, respectively. V_{out} is expressed as follows:

$$V_{out}(s) = \frac{DV_{in}R_1}{s^2 C_2 L_1 R_1 + sL_1 + R_1} \qquad (8).$$

I_L is expressed by

$$I_L(s) = \frac{V_{out}(s)}{R_1}(1 + sC_2R_1) = \frac{DV_{in}(1+sC_2R_1)}{s^2 C_2 L_1 R_1 + sL_1 + R_1} \qquad (9).$$

By substituting (9) into (7), (7) is rewritten as shown in

$$\tilde{i}_b(s) = \frac{DV_{in}(1+sC_2R_1)}{s^2 C_2 L_1 R_1 + sL_1 + R_1}\tilde{d} = G_{id\_OL}(s)\tilde{d} \qquad (10).$$

To consider the perturbation and the steady value, $1/v_{in}$ can be rewritten to $1/(V_{in} + \tilde{v}_{in})$. This term is nonlinear because the denominator includes the perturbation

component. To linearize this term, $1/(V_{in} + \tilde{v}_{in})$ is expanded by Maclaurin series as shown in

$$\frac{1}{V_{in} + \tilde{v}_{in}} = \frac{1}{V_{in}} - \frac{1}{V_{in}^2}\tilde{v}_{in} + \frac{1}{V_{in}^3}\tilde{v}_{in}^2 + \cdots$$
$$\approx \frac{1}{v_{in}} - \frac{1}{v_{in}^2} \qquad (11).$$

To consider the perturbations of both v_{in} and d in the control block shown in Fig. 7, the following equation is derived

$$\tilde{d}(s) = \left(\frac{-V_{out}}{V_{in}^2} + \frac{K_F G_{HPF}(s) G_{CI}(s)}{V_{in}}\right)\tilde{v}_{in}(s) \qquad (12).$$

By substituting (12) into (10), (13) is derived as follows:

$$\tilde{i}_b(s) = G_{id\_OL}(s)\left(\frac{-V_{out}}{V_{in}^2} + \frac{K_F G_{HPF}(s) G_{CI}(s)}{V_{in}}\right)\tilde{v}_{in}(s) \qquad (13).$$

The input current i_{in} of the buck converter is expressed as the following equation

$$\tilde{i}_{in}(s) = sC_1\tilde{v}_{in}(s) + \tilde{i}_b(s)$$
$$= \left\{sC_1 + G_{id\_OL}(s)\left(\frac{-V_{out}}{V_{in}^2} + \frac{K_F G_{HPF}(s) G_{CI}(s)}{V_{in}}\right)\right\}\tilde{v}_{in}(s) \qquad (14).$$

The input admittance Y_2 of the buck converter is expressed as follows:

$$Y_2(s) = \frac{\tilde{i}_{in}(s)}{\tilde{v}_{in}(s)}$$
$$= sC_1 + G_{id\_OL}(s)\left(\frac{-V_{out}}{V_{in}^2} + \frac{K_F G_{HPF}(s) G_{CI}(s)}{V_{in}}\right) \qquad (15).$$

From (15), The input admittance Y_2 can be divided into the admittance $Y_{in}(s)$ without PFF control and a virtual admittance $Y_{damp}(s)$ that is added by the feedforward control. The equations are as follows:

$$Y_{in}(s) = sC_1 + G_{id\_OL}(s)\frac{-V_{out}}{V_{in}^2} \qquad (16),$$

$$Y_{damp}(s) = G_{id\_OL}(s)\frac{K_F G_{HPF}(s) G_{CI}(s)}{V_{in}} \qquad (17).$$

Therefore, the input admittance of the buck converter can satisfy passivity by adjusting $Y_{damp}(s)$.

B. Frequency response of input admittance

To use the parameters of Table I, frequency response of the input admittance of the buck converter is calculated using MATLAB/Simulink. The transfer function of the high pass filter $G_{HPF}(s)$ is expressed as

$$G_{HPF}(s) = \frac{s}{s + 2\pi f_c} \qquad (18).$$

Since the resonance frequency of L_w and C_1 is about 103 Hz, and the cut off frequency f_c is set to 10 Hz.

The frequency response of the input admittance without feedforward control and with feedforward control are shown in Fig. 8 and Fig. 9, respectively. The phase of the input admittance without feedforward control is not within between $+90°$ and $-90°$ at any frequency as shown in Fig. 8, and Y_2 does not satisfy passivity. On the other hand, the phase of the input admittance with the feedforward control is within between $+90°$ and $-90°$ as shown in Fig. 9, and Y_2 satisfies passivity. As K_F is larger, the phase is within between $+90°$ and $-90°$ in the lower frequency range.

The 2018 International Power Electronics Conference

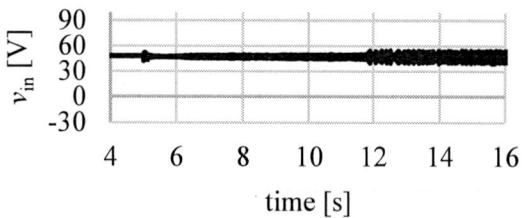

Fig. 10. Experiment of input voltage without PFF control (long range).

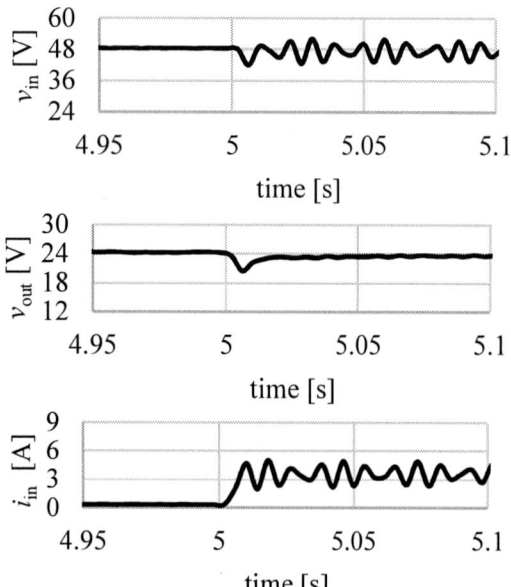

Fig. 11. Experimentation results without PFF control.

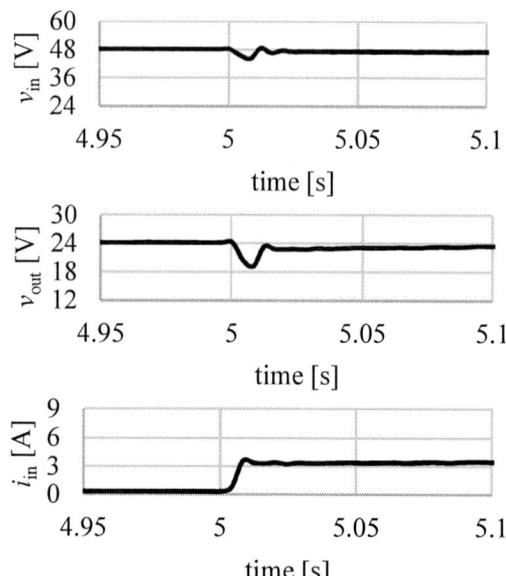

Fig. 12. Experiment results with PFF control (f_c=10 Hz, K_F=0.2).

V to 58 V after the load change. The waveforms shown in Fig. 11 are focused on the range from 4.95 s to 5.1 s. The input current also had the continuous oscillation after the load change. The output voltage was controlled constantly after the load change, while the output voltage had a very small oscillation by the oscillation of the input voltage.

The input voltage, current and the output voltage with the feedforward control are shown in Fig. 12. The waveforms were stable after the load change. However, the output voltage was dropped at the load change, and it was deeper than the result of no feedforward control. However, the influence of the feedforward control was relatively small.

IV. EXPERIMENTAL VERIFICATION

A. Experimental condition

The effect of the feedforward control was verified by experiment. The circuit, the control block and the parameters are shown in Fig. 4, Fig. 5 and Table I, respectively. The dead time and switching frequency are set to be 0.5 μs and 20 kHz, respectively. Regarding the measuring instruments, the oscilloscope is MDO3024 (Tektronix). The voltage source of Subsytem1 is PWR1600H (KIKUSUI), and MOSFET is Si-MOSFET: TK20N60W (TOSHIBA). TMS320F28335 (TI) is used as the controller. The cut off frequency f_c of the HPF is set to be 10 Hz, and the feedforward control gain K_F is set to be 0.2. The load was step-changed from 38 Ω (power consumption 15 W) to 3.8 Ω (power consumption 150 W) at 5 s.

B. Experimental results

The waveforms without the feedforward control are shown in Fig. 10 and Fig. 11. The waveforms with the feedforward control show in Fig. 12.

The input voltage without the feedforward control was oscillated continuously. The amplitude was varied from 40

V. CONCLUSION

In this paper, we studied a dc system stability with a dc distribution board. For the system with bidirectional power flow, the passivity-based stability criterion is applicable. To satisfy passivity, a feedforward control is used, and the admittance with a practical control is analyzed. To add the virtual admittance generated by the feedforward control to the input admittance of the buck converter in parallel, it is shown the converter satisfies passivity from the bode diagram. The phase of the input admittance of the buck converter with the feedforward control is within between +90° and −90°. The experimental results show the input bus voltage of the buck converter becomes stable when it satisfies passivity with the feed forward control.

REFERENCES

[1] N. O. Sokal, "System oscillations from negative input resistance at power input port of switching-mode regulator, amplifier, DC/DC converter, or DC/DC inverter", IEEE Power Electronics Specialists Conference, p.138-140 (1973)

[2] R. D. Middlebook, "Input Filter Considerations in Design and Application of Switching Regulators", IEEE IAS Annual Meeting, pp.366-382 (1976)

[3] C. M. Wildrick, F. C. Lee, B. H. Cho, B. Choi, "A method of defining the load impedance specification for a stable distributed power system", IEEE Trans. Power Electron., Vol. 10, No. 3 pp. 280–285 (1995)

[4] X. Feng, Z. Ye, K. Xing, F. C. Lee, D. Borojevic, "Impedance specification and impedance improvement for dc distributed power system", IEEE PESC, Vol. 2, pp. 889–894 (1999)

[5] S. D. Sudhoff, S. F. Glover, P. T. Lamm, D. H. Schmucker, D.E. Delisle, "Admittance space stability analysis of power electronic systems" IEEE Trans. Aerosp. Electron. Syst., Vol. 36, pp. 965–973 (2000)

[6] A. Riccobono and E. Santi, "Comprehensive review of stability criteria for DC power distribution systems" IEEE Trans. Ind. Applicat., Vol. 50, No. 5 pp.3525 - 3535 (2014)

[7] Y. Gu, W. Li, X. He, "Passivity-Based Control of DC Microgrid for Self-Disciplined Stabilization", IEEE Trans. Power Electron, Vol. 30, pp.2623-2632 (2015)

[8] A. Riccobono, "Stabilizing controller design for a dc power distribution system using a passivity-based stability criterion," Ph.D. dissertation, University of South Carolina (2013)

[9] R. W. Erickson, D. Maksimovic, "Fundamentals of Power Electronics, 2nd edition", Springer (2001)

A Novel Control Approach to Multi-Terminal Power Flow Controller for Next-Generation DC Power Network

Kenji Natori*, Yuta Nakao, and Yukihiko Sato
Department of Electrical and Electronic Engineering, Chiba University, Chiba, Japan
*E-mail: knatori@chiba-u.jp

Abstract— This paper studies a novel control approach to multi-terminal power flow controller (MTPFC) for next-generation DC power network. We have proposed and validated bidirectional power flow controller (BPFC) and MTPFC as promising power flow controllers for next-generation DC power networks. In BPFC, total power imbalance possibly causes voltage fluctuation of the inter-stage link capacitor. The fluctuation affects performance of the power flow control. Then, in MTPFC, a compensation node that keeps the inter-stage link capacitor voltage at a predetermined constant value, plays an important role for precise operation of MTPFC. However, the introduction of the compensation node needs additional costs. In this paper, we propose a novel control approach that properly controls power flows even in case that there are fluctuations of the voltage of the inter-stage link capacitor. The control approach is realized by using measured values of the voltages for the control system. The proposed approach is validated by experimental results.

Keywords— *Next-generation DC power network, BPFC, MTPFC, control system design.*

I. INTRODUCTION

For realization of sustainable society with electrical energy (power), renewable energy sources, such as photovoltaic and wind power have been expected to be massively installed as future energy sources. In addition, intelligent usage of available energy based on highly-developed ICT (information and communication technology) has been receiving much attention as a promising technology for future energy systems. One of such technologies or concepts is so-called smart grid that enables energy-efficient and smart operation of electrical energy systems or grids.

Since most of renewable energy sources and energy storage devices (ESDs) have DC input/output terminals, many researches have studied DC smart grids or DC microgrids [1-3] (an example is shown in Fig. 1). A bidirectional resonant converter for DC microgrids has been studied in [4]. A multiport-type DC/DC converter for DC networks has been proposed in [5]. In [6], stability of dual-active-bridge (DAB) converters for DC microgrids has been studied.

In DC power systems or grids, generally, only the controllable variables are voltages of the nodes (energy

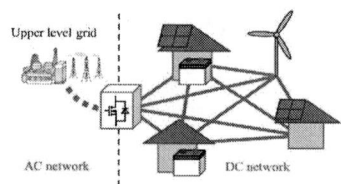

Fig. 1. Example of expected future power network.

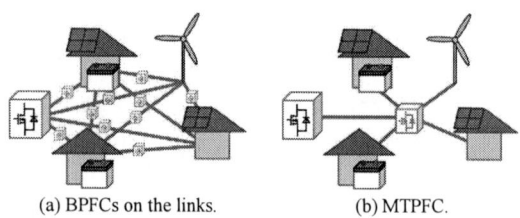

(a) BPFCs on the links. (b) MTPFC.

Fig. 2. Power network with power flow controllers.

sources, ESDs, and loads), and the power flows among the nodes are determined according to the relationship of the voltages. Therefore, it is generally impossible to arbitrarily control all the power flows among all the nodes simultaneously.

To overcome the problem, we have proposed a bidirectional power flow controller (BPFC) that is implemented on the link between the nodes [7]. Since it intentionally makes voltage differences on the links, it enables to arbitrarily control the power flow on the link. In case BPFCs are implemented on the all links as shown in Fig. 2 (a), it is possible to control all power flows on the all links independently. On the other hand, a multi-terminal power flow controller (MTPFC) achieves the same control goal by using only one control unit [8]. In other words, it enables to drastically reduce the number of the power conversion stages by using MTPFC as shown in Fig. 2 (b). For those controllers, researches on improvement of the efficiencies have been also studied [9] [10].

In proposal of MTPFC, we have also proposed a concept of a compensation node. In the power flow control among multiple nodes, total power imbalance due to losses and various kinds of disturbances occurs, because control of input/output power of each node is conducted

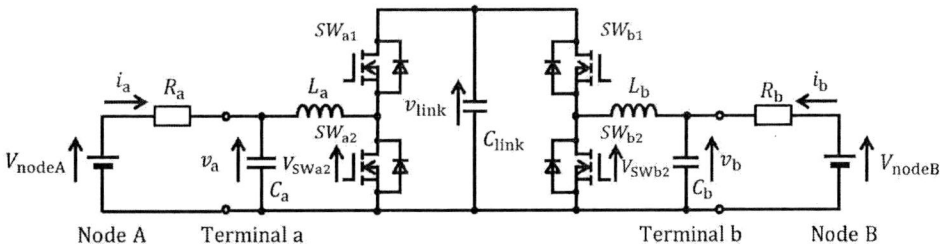

Fig. 3. Circuit of BPFC.

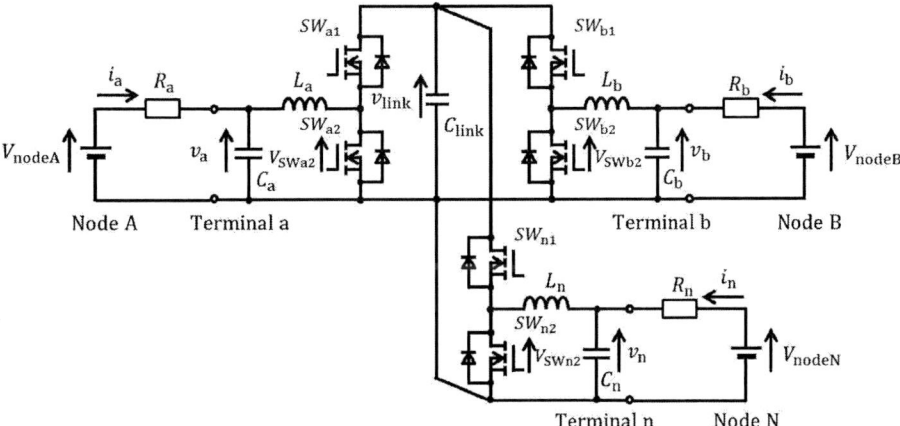

Fig. 4. Circuit of MTPFC.

independently. Then, the power imbalance causes voltage fluctuation of the inter-stage link capacitor. The fluctuation possibly deteriorates the power flow control. The compensation node has been therefore introduced to compensate the power imbalance and to keep the voltages of the inter-stage link capacitor at a predetermined constant value. However, the introduction of the compensation node needs additional costs. In addition, the compensation node needs to have ability to both send and receive certain amount of power in order to compensate total power imbalance of the power network.

In this paper, our purpose is to realize precise power flow control without the compensation node. For achieving the purpose, we need to implement a control system that properly control power flows even in case that there are voltage fluctuations of the inter-stage link capacitor. Then, we propose a control system that uses measured values of the voltages. The effectiveness and feasibility of the proposed control approach are verified by experimental results.

This paper is organized as follows: MTPFC is introduced in section II. Then, control systems implemented for MTPFC are described in section III. In section IV, we introduce the proposed control approach. Section V validates the proposed controller by experimental results. Finally, we conclude this paper in section VI.

II. MTPFC

This section introduces concept and topology of MTPFC. First, BPFC and MTPFC are described. Then, the function of the compensation node in MTPFC is explained.

A. BPFC and MTPFC

Fig. 3 shows a circuit of BPFC [7][8]. The circuit topology is required to have ability to realize bidirectional power flows between two nodes of different voltages. Therefore, we have adopted a bidirectional buck-boost converter that is composed of two-quadrant choppers connected with an inter-stage link capacitor [11] [12]. The relationship between the terminal voltage V_a and the inter-stage link capacitor voltage V_{link} is given as follows:

$$V_{link} = \frac{1}{D_a} V_a \qquad (1)$$

On the other hand, the relationship between the terminal voltage V_b and V_{link} is described as

$$V_b = D_b V_{link} \qquad (2)$$

Therefore, the voltage ratio V_b/V_a is controlled by the ratio of duty ratios as follows:

$$\frac{V_b}{V_a} = \frac{D_b}{D_a} \qquad (3)$$

Since the power flow is controlled by the ratio V_b/V_a [7], the power flow is controlled by the ratio of the duty ratios.

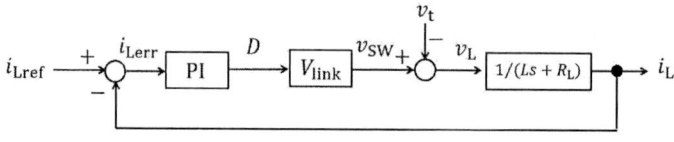

(a) Control system for power-sending/receiving node.

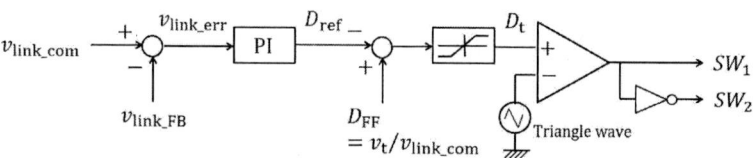

(b) Current control system.

Fig. 5. Conventional control approach for power-sending/receiving node.

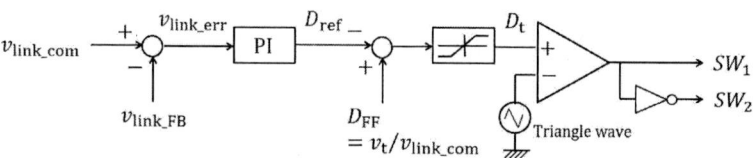

Fig. 6. Control system for compensation node.

When the number of the nodes is more than 3, we call the controller MTPFC as shown in Fig. 4 [8]. In MTPFC, the power flow is realized in a similar manner of BPFC.

B. Function of Compensation Node

Even in case that the input power of one node is equal to the output power of the other node in Fig. 3, total power imbalance occurs since there exist losses and various disturbances. As described above, the total power imbalance causes fluctuations of the voltage of the inter-stage link capacitor v_{link}. As (1) and (2) indicate, the power flow control is affected by the fluctuation. It is therefore necessary to keep the voltage v_{link} at a predetermined constant value. In order to keep the voltage at the value, we have introduced the compensation node (Node N) as shown in Fig. 4. The compensation node works to keep the voltage v_{link} at the constant value. As described above, the compensation node needs to have ability of supplying and absorbing certain amount of power to compensate the total power imbalance.

III. CONVENTIONAL CONTROL SYSTEMS

This section presents controllers for MTPFC. The controller for power-sending/receiving node is firstly introduced. Then, the controller for the compensation node is described.

A. Control System for Power-Sending Node

Fig. 5 (a) shows a block diagram of the control system for the power-sending node. P_{com} is a command value of the sending power, i_{ref} is a current reference value, i_{FB} is a feedback of the current values i_a or i_b in Figs. 3 and 4, i_{err} is an error of the current, D_{ref} is the reference value of the duty ratio D_t (D_t is D_a or D_b). It should be noticed that the duty ratio is determined as follows:

$$D_t = 1 - D_{ref} \qquad (4)$$

For design of the PI current controller, we consider a transfer function that governs the inductor current i_L. The block diagram of the current controller considering the transfer function is depicted as Fig. 5 (b). i_{Lref} is a reference value of the inductor current, i_{Lerr} is an error of the current, v_{SW} is a voltage of the lower switch, v_t is v_a or v_b in Figs. 3 and 4, and v_L is the voltage of the inductance L_t (t is a or b). The proportional and integral control gains are designed to make the closed-loop control system a first-order system with the time constant T_d. The designed controller gains are described as follows:

$$K_p = \frac{L}{T_d V_{link}}, \quad K_i = \frac{R_L}{T_d V_{link}} \qquad (5)$$

B. Control System for Power-Receiving Node

In case the sign of the command value of the sending power is minus, the control system of the power-sending node (Fig. 5 (a)) works as the controller for the power-receiving node. The PI controller gains are designed in the same manner as the power-sending node.

C. Control System for Compensation Node

Fig. 6 shows a block diagram of the control system for the compensation node. v_{link_com} is a predetermined

The 2018 International Power Electronics Conference

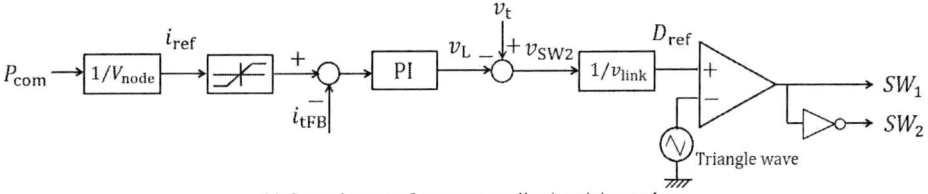

(a) Control system for power-sending/receiving node.

(b) Current control system.

Fig. 7. Proposed control approach for power-sending/receiving node.

constant value of the inter-stage link capacitor voltage v_{link}. v_{link_FB} is the feedback of v_{link}, v_{link_err} is the error, and D_{ref} is the reference value of the duty ratio D_t (D_t is D_a or D_b). The purpose of the control system is to keep the voltage of the inter-stage link capacitor v_{link} at the predetermined constant value v_{link_com} by feedback of v_{link}. The duty ratio is determined as below

$$D_t = D_{FF} - D_{ref} \tag{6}$$

where D_{FF} is a feedforward term to achieve faster response. It is derived considering the similar relationship as (1) or (2) (D_t is D_n that is a duty ratio of the compensation node). By introducing the compensation node, the voltage is able to be controlled independently from other nodes.

IV. PROPOSED CONTROL SYSTEM

In MTPFC, the compensation node has been introduced to keep the voltage of the inter-stage link capacitor. However, for implementing the compensation node, additional cost is required. Moreover, as the compensation node, it should have ability to both send and receive certain amount of power in order to compensate the total power imbalance of the power network.

This paper proposes a control approach that achieves required power flow even when there are voltage fluctuations of the inter-stage link capacitor. The control approach is based on the usage of the measured voltages. Fig. 7 (a) shows the proposed control system. P_{com} is the command value of the sending/receiving power, i_{ref} is the current reference value, i_{tFB} is the feedback of the current values i_a or i_b in Figs. 3 and 4, v_L is the voltage of the inductance L_t (t is a or b), v_t is v_a or v_b in Figs. 3 and 4, D_{ref} is the reference value of the duty ratio D_t (D_t is D_a or D_b). v_{SW2} is given as follows:

$$v_{SW2} = D_t v_{link} \tag{7}$$

In the proposed control approach, by using the measured voltage values v_t and v_{link}, the power flow control works properly even when there are fluctuations of v_{link}.

In the case of power-receiving node, the same control system with minus power command value works as the control system for the power-receiving node.

For designing PI current controller gains, a transfer

TABLE I
PARAMETERS FOR EXPERIMENT

Node voltages	V_{node_A}, V_{node_B}	30V
Link voltage	V_{link}	40V
Line resistances	R_a, R_b	0.6 Ω
Load resistance	R_{load}	30 Ω
Capacitances	C_a, C_b	100uF
Link capacitance	C_{link}	1880uF
Inductances	L_a, L_b	1mH
Dead time	DT	1000ns

function that governs the inductor current i_L is considered as section III-A. The block diagram of the current controller that considers the transfer function is shown as Fig. 7 (b). i_{Lref} is the reference value of the inductor current, i_{Lerr} is the error of the current, v_L is the voltage of the inductance L_t (t is a or b). The controller gains are designed in the same manner as section III-A, and obtained as follows:

$$K_p = \frac{L}{T_d}, \quad K_i = \frac{R_L}{T_d} \tag{8}$$

where T_d is the time constant of the designed closed-loop control system.

V. EXPERIMENTAL VALIDATION

In this section, we present experimental results that validate the proposed control approach. The circuits shown in Figs. 3 and 4 are used. Parameters used for the experiment are shown in TABLE I. The power command of the sending power of the node A is changed from 10W to 30W as a step function. The transient responses are studied in 2 cases. One case is that v_{link} is constant at 40V, and the other case is that v_{link} changes as

$$v_{link} = 40 + \sin 10\pi t \text{ [V]}$$

In other words, the voltage fluctuates with the amplitude of 1V at 5 Hz. In this experiment, just for verification of the proposed control approach, the voltage v_{link} is controlled by using additional equipment.

Experimental results in the case of $v_{link} = 40$ V (constant) are shown in Fig. 8. In both control approaches, the required power flows are stably achieved.

The 2018 International Power Electronics Conference

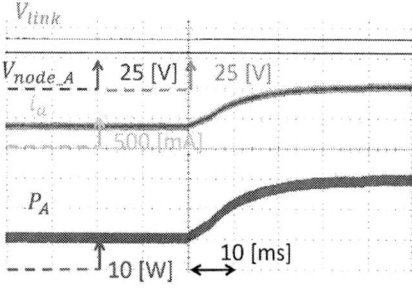

(a) Conventional control approach.

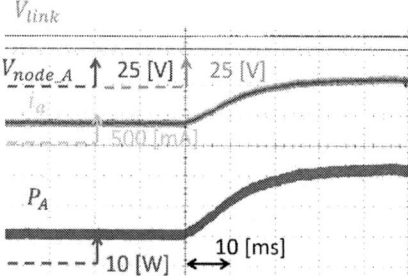

(b) Proposed control approach.

Fig. 8. Experimental results (v_{link} is constant).

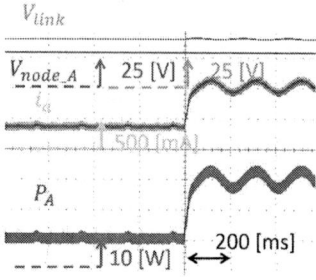

(a) Conventional control approach.

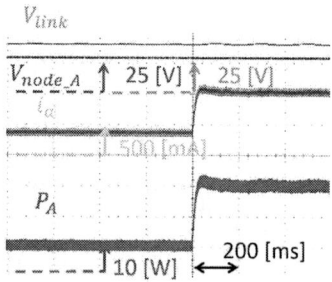

(b) Proposed control approach.

Fig. 9. Experimental results (v_{link} is fluctuating).

Next, experimental results in the case of $v_{\text{link}} = 40 + \sin10\pi$ [V] (fluctuating) are depicted in Fig. 9. In the result of the conventional control, there exists fluctuations of the current and the power flow due to the fluctuation of v_{link}. On the other hand, in the case of the proposed control approach, the required power flow is properly achieved and there are no fluctuations of the current and the power flow.

The experimental results demonstrate that the proposed control approach works properly even when there are fluctuations of v_{link}.

VI. CONCLUSIONS

In this paper, we proposed a novel control approach for MTPFC. The proposed control approach properly controls the power flows even when there are fluctuations of the inter-stage link capacitor voltage. The validity and feasibility of the proposed control approach were verified by experimental results.

REFERENCES

[1] L. Meng, Q. Shafiee, G. Ferrari, H. Karimi, D. Fulwani, X. Lu, and J. M. Guerrero, "Review on Control of DC Microgrids and Multiple Microgrid Clusters," *IEEE Journal of Emerging and Selected Topics in Power Electronics*, vol. 5, no. 3, pp. 928-948, 2017.

[2] T. Dragičević, X. Lu, J. C. Vasquez, and J. M. Guerrero, "DC Microgrids-Part I: A Review of Control Strategies and Stabilization Techniques," *IEEE Transactions on Power Electronics*, vol. 31, no. 7, pp. 4876-4891, 2016.

[3] T. Dragičević, X. Lu, J. C. Vasquez, and J. M. Guerrero, "DC Microgrids-Part II: A Review of Power Architectures,

Applications, and Standardization Issues," *IEEE Transactions on Power Electronics*, vol. 31, no. 5, pp. 3528-3549, May 2016.

[4] H. Wu, S. Ding, K. Sun, L. Zhang, Y. Li, and Y. Xing, "Bidirectional Soft-Switching Series-Resonant Converter with Simple PWM Control and Load-Independent Voltage-Gain Characteristics for Energy Storage System in DC Microgrids," *IEEE Journal of Emerging and Selected Topics in Power Electronics*, vol. 5, no. 3, pp. 995-1007, 2017.

[5] M. Corti, E. Tironi, and G. Ubezio, "DC Networks Including Multiport DC/DC Converters: Fault Analysis," *IEEE Transactions on Industry Applications*, vol. 52, no. 5, pp. 3655-3662, 2016.

[6] Q. Ye, R. Mo, and H. Li, "Low-Frequency Resonance Suppression of a Dual-Active-Bridge DC/DC converter Enabled DC Microgrid," *IEEE Journal of Emerging and Selected Topics in Power Electronics*, vol. 5, no. 3, pp. 982-994, 2017.

[7] K. Natori, H. Obara, K. Yoshikawa, B. C. Hiu, and Y. Sato, " Flexible Power Flow Control for Next-Generation Multi-Terminal DC Power Network", *Proceedings of IEEE Energy Conversion Congress and Exposition (ECCE)*, pp. 778-784, 2014.

[8] Y. Takahashi, K. Natori, and Y. Sato, "A Multi-Terminal Power Flow Control Method for Next-Generation DC Power Network", *Proceedings of IEEE Energy Conversion Congress and Exposition (ECCE)*, pp. 6223-6230, 2015.

[9] T. Tanaka, Y. Takahashi, K. Natori, and Y. Sato, "High-Efficiency Floating Bidirectional Power Flow Controller for Next-Generation DC Power Network," *IEEJ Journal of Industry Applications*, vol. 7, no. 1, pp. 29-34, 2018.

[10] K. Natori, T. Tanaka, Y. Takahashi, and Y. Sato, " A Study on High-Efficiency Floating Multi-Terminal Power Flow Controller for Next-Generation DC Power Networks ", *Proceedings of IEEE Energy Conversion Congress and Exposition (ECCE)*, pp. 2631-2637, 2017.

[11] British Patent GB2376357B - Power converter and method for power conversion.

[12] A. Maclaurin, R. Okou, P. Barendse, M.A. Khan, and P. Pillay, "Control of a flywheel energy storage system for rural applications using a Split-Pi DC-DC converter," *Proceedings of IEEE International Electric Machines and Drives Conference (IEMDC)*, pp. 265-270, 2011.

DC Microgrid for Telecommunications Service and Related Application

Keiichi Hirose

Data center business Headquarters, NTT FACILITIES, INC., Tokyo, Japan

E-mail: hirose36@ntt-f.co.jp

Abstract— Facilities for telecommunications can also be said to be DC microgrids with storage batteries and backup power supplies. Telecommunication facilities around the world have been using DC 48 V for more than a century. The reason is that it is highly reliable and it is easy to integrate with the storage battery and DC-input communication equipment and maintain it. Expansion of the use of ICT has become a trigger to reconsider the primary energy source of the facility, and in particular promotes the introduction of renewable energy. Photovoltaic power generation and fuel cells have direct current output, and have high affinity for integration with a power supply system for telecommunications. By connecting the DC output directly to the power system in telecom site/data center, it is easy to construct a system with higher efficiency and higher reliability. This paper reports overview of telecommunication facility, recent topics, examples of adopting renewable energy, etc., and reports on future trends. In order to expand the use of the ICT system, the load density has increased, and various difficulties have arisen with the conventional DC 48 V as it is. One solution to this problem is the use of higher voltages such as DC 380V. Several practical examples are also included.

Keywords— DC power, Communications Energy System, Microgrid, 380 Vdc.

I. INTRODUCTION

In recent years, the role and importance of ICT services in industry, finance, and the daily lives of individuals has been increasing. Providing good ICT service requires a stable and high-quality power supply. From the viewpoint of the reliability and quality supplied by power companies, deregulation in the power industry and environmental problems have created various problems. The three factors described below have major impact on telecom power [1] [2].

A. Power market deregulation

Deregulation of utilities, including the power industry, began in the 1980s in Europe and North America. Japan, too, has seen deregulation of the wholesale power market since 2000. There is concern that the constraints on investment in the electric power system due to the price competition that comes with deregulation may lower the reliability of the commercial power supply. In fact, large-scale electric power failures such as the major power blackout in North America and the California power crisis in the United States have occurred. Furthermore, the

reluctance to disclose information on power quality in the name of competition is cause for concern by users.

B. Introduction of renewable energy

The introduction of distributed power production using renewable energy sources such as solar cells and wind power has been increasing, and large-capacity facilities are experiencing power quality problems such as voltage instability and harmonics. As a result, instability in the electric power system is expected to increase even beyond current levels in the future.

C. Increase in disaster risk

Climate changes due to global warming are expected to increase harm from large-scale, long-period power outages due to heavy rain, hurricanes, typhoons, and other such weather phenomena. Also, the great difficulty of predicting earthquakes requires that the occurrence of earthquakes be assumed in the operation of telecom power facilities in regions where earthquakes occur frequently, such as Japan and California. For example, the northeastern part of Japan, including Sendai, was hit by a devastating earthquake on March 11, 2011. The tsunami triggered by the earthquake severely damaged coastal areas, but not the city center. This natural disaster caused the power supply trouble for approximately 7.9 million kW (about 60% of the earthquake demand is outage) and maximum number of blackout was about 4.66 million customers.

As noted above, present power reliability may gradually decrease due to a variety of factors whose future increase is difficult to take into account (power market deregulation, increased use of natural energy sources, global warming induced climate change, terrorism, etc.).

To cope with such decrease in system power reliability, the user requires their own self-defense capability in the form of tools that allow 'visualization' of power reliability based on a database and reliability evaluation. Furthermore, the ability to simply simulate the power system configuration to match the reliability required by the load would be greatly useful in optimizing investment.

II. TELECOM POWER SUPPLY [3]

Conventional communication equipment generally requires -48 VDC input power for telecom services. The power supply system consists of multiple parallel redundant rectifiers that convert AC power to -48 VDC power supply, charge the storage battery and supply power to the communications equipment. The converters or inverters are used to supply other necessary voltages from the -48 VDC power supply system. Long battery support time is required to support the equipment in case of AC main or rectifier failures. Most of telecom operators make use of the engine -generator set to supplement AC power when the power outage is continuing. Battery discharge times are from at least 1 hour to over 24 hours, typical times are 3, 8, 10, and 24 hours.

Figure 1 (top right) shows a typical communication power system using a rectifier that supports important load equipment and a -48 VDC battery system.

Conventional IT equipment has a single phase AC of 100, 120, 200 V, or 240 V in a country of 60 Hz, a single phase AC of 220 V to 240 V in a country of 50 Hz and AC input power matching the configuration of commonly available AC power supply is required. These power supply configurations use AC UPS systems with batteries that are sized to provide the time required for shutdown of IT equipment and startup of the standby engine generator. In effect, all critical IT facilities include a permanently deployed engine generator system (and associated automatic transfer switch), which can prevent AC power failure. Figure 1 (top left) contains a typical IT power configuration that uses an AC UPS system to support critical loads.

Traditional communication and IT facilities generally include other important support equipment such as lighting and air conditioner. The equipment generally tolerates brief blackouts without adversely affecting communication or IT equipment operation.

Communication systems without a permanently deployed engine generator generally supply power to a support device with an inverter connected to a -48 VDC power system. IT systems operate such equipment with AC power. IT systems without a permanently placed engine generator provide an orderly shutdown time and systems with engine generators provide long runtime. The convergence of communication and IT equipment may require both -48 VDC power and one or more AC supply voltages for the critical electronic load of either facility. In many cases, the site's system can not operate unless both DC and AC power loads are running. This co-dependency requires that the reliability and availability of DC and AC power systems be equally high.

Fig. 1 also illustrates the comparison of three configurations of an AC power distribution system, a 48VDC power distribution system, and a 380VDC power distribution system. In an AC power distribution system, AC/DC conversions and DC/AC conversions are required inside the AC UPS in order to charge the embedded battery.

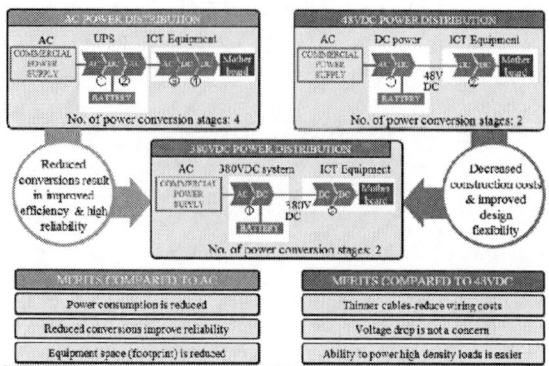

Fig. 1. AC and DC power supply systems for ICT facilities.

Also, AC/DC conversion at the input of ICT equipment is necessary because the components in the ICT equipment (e.g., CPU) require DC power. Consequently, four stages of power conversion would be required from the input of the AC UPS to the input of the CPU inside the ICT equipment, which results in decrease of total efficiency due to the loss in each conversion stages.

A 380VDC power distribution system is a much simplified system with having only two stages of power conversion. It was originally proposed based on the architecture of the 48VDC power distribution system traditionally used in telecommunication buildings. One of the merits is its high efficiency and reliability because there are fewer stages of power conversion. Moreover, DC power also has in its favor that it is highly compatible with many renewable energy and power storage systems because of direct connection without AC/DC conversions. Since data centers are increasingly required to take measures on energy conservation and renewable energy, we therefore assume that the need for efficiency improving technologies like 380VDC power supply systems will increase.

Moreover, renewable energy sources (like solar power and fuel cells) provide DC output. Interconnect of the energy sources and the DC power distribution system is easy with only the voltage controlled. So their compatibility with DC power distribution systems is high, and they can be utilized efficiently. Thus, the DC power distribution system, applications, and future prospects are expected to be utilized in a smart grid. In Fig. 2, we show an example of a smart grid using the DC power distribution system. The power sources (such as solar cells, fuel cells, and batteries) and the load (such as electric vehicles and consumer electronics) are connected to the bus line of a 380VDC, and is considered to be achieved by monitoring and controlling the state.

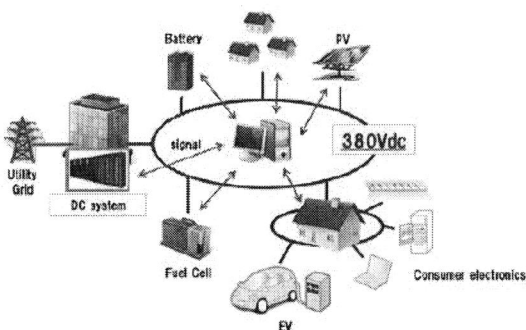

Fig. 2. Example of a smart grid using the DC power distribution system.

III. DC MICROGRID

Fig. 3 shows an example of a typical ICT power supply infrastructure in telecom site/data center. The power supply system consists of rectifier, dc-dc converter, dc-ac inverter, AC-UPS, storage battery, buck-up generator-set, monitoring/control system, and so on. In recently, many sites deploy the distributed energy resources (DERs) to use renewable energy more.

A dc power plant in ICT site is characterized by its technical simplicity. It consists of a number of paralleled rectifiers that connect to two or more battery strings, which are also connected in parallel. In the event of a power outage or rectifier failure, the load continues to operate from the batteries without switching or interruption.

It is relatively easy to connect rectifiers and batteries in parallel for reliable load sharing, since there is no need to consider phasing. Those attributes allow for power supply systems with modular components and interconnections-an imperative for simple and inexpensive maintenance by people with limited training on power systems. While there is a trend toward modularity in UPS systems employing "hot-swappable" battery and inverter modules, it is not universal and typically involves proprietary components.

During operation, parallel rectifiers in a dc system provide the current consumed by the load, the float current for the batteries and the additional current for recharging the batteries after a utility failure. Redundant rectifiers fill two needs: battery recharging after an outage and continued operation if one rectifier fails.

The typical ICT power supply system has some essential technologies originaly below;

· Grid-interconnection/disconnection (Islanding/microgrid) methods,

· Optimized control for backup gen-set, distributed generator, and power converters,

  · Management of Storage batteries,

  · Providing higher power quality and reliability,

  · Demand response,

  · High density mounting/packaging,

· Operation & maintenance technics,

· Wiring, and protection,

· Cost-benefit analysis, etc.

These technologies are essential for microgrid as well as independent and many types of power supply systems. In other words, ICT power supply infrastructure has a good potential to work as microgrid without any modification. Telecommunications operators have used a conventional 48VDC power plant for ICT survices more than a century. As mentioned, the 380 VDC power supply systems have been deployed to replace the 48 VDC power plants and AC UPSs.

The 380VDC power system fits for battery systems, renewable energy resouces, and many appliances, and it is very easy to intagrate the optimized power sub-system or DC microgrid.

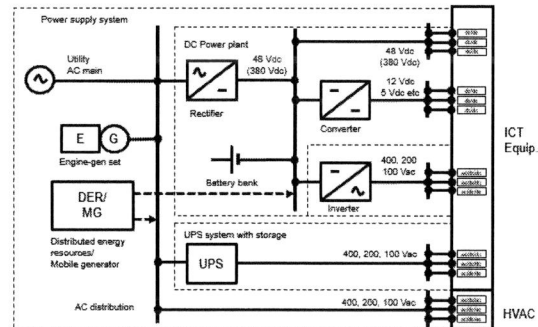

Fig. 3. Example of a typical ICT power supply infrastructure, like DC microgrid.

IV. DC MICROGRID DEMONSTRATIONS

Recently, microgrid demonstrations and deployments are expanding in US power systems and around the world. Although goals are specific to each site, these microgrids have demonstrated the ability to provide higher reliability and higher power quality than utility power systems and improved energy utilization, etc.

Table I shows DC power demonstrations which carried out by NTT FACILITIES. Many lessons for DC microgrid operations and design learned from each demonstration have reported to the IEEE INTELECs, many conferences, workshops, and forums, and so on.

Three demonstrations are described briefly as follows:

A. Sendai Microgrid

The report of Sendai microgrid discribes on the development of a multiple power quality level supply system that provides five kinds of high quality electric power at the same time. The system utilizes both renewable energy and utility and supplies consumers with various kinds of electric power of high quality. It gives consumers the advantages of low cost and reduces the amount of space needed for existing such power converters as UPSs. It also contributes to promoting the

TABLE I
DC MICROGRID DEMONSTRATIONS BY NTT FACILITIES

Site	Futures	Dome years
Sendai Microgrid	AC and DC multiple power supply system	2005 - 2008
AIT microgrid	AC and DC hybrid power supply system	2009 - present
Tsukuba	380 VDC for data center	2010-2012
Aomori	Container data center with wind turbine	2011
Obihiro city office	380 VDC microgrid for office	2012 – 2014
Yamagata	DC power exchange	2012 – 2014
University of Texas, Austin, TACC	380 VDC for super computer with 200 kW PV panels	2015 – 2016

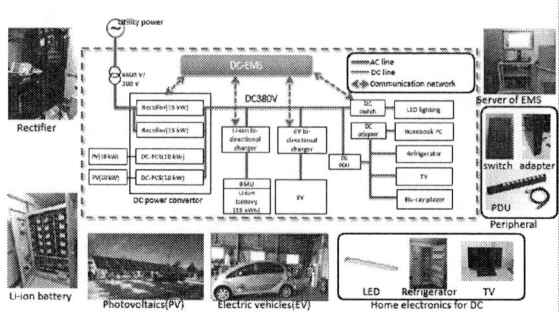

Fig. 4. 380 VDC microgrid in Obihiro city office, Hokkaido, Japan.

introduction of renewable energy. For DC power, the system has 48 VDC and 300 VDC distributions.

Loads for 48 VDC are not telcom but IT servers, fluorescent lamp ballasts, and air flow fans. 300 VDC power system contributes the early stage of development for 380 VDC today. During demonstration, many studies of intagration of dc-dc converters, PV panels, lead-acid batteris, semi-conducter circut breaker, high-resister mid point grounding, operation of 400 VDC input survers had carried out.

B. Obihiro DC microgrid

This report shows the experimental study of a DC microgrid for an office building that constructed in Obihiro City, Hokkaido, Japan. The objective of this study is to develop a self-sustained distributed energy system by combining distribution energies, batteries, and appliances with DC power. This new DC energy system not only reduces the environment load and improves energy efficiency but also forms a community energy system that can become independent from utility grids and resistant to natural disasters. We have found that compared with AC power supply, the DC system, which uses power generated by solar panels as is, increases the efficiency by 3.2 percent and decreases commercial power to be purchased by 4.2 percent. Fig. 4 shows the power configuration of DC microgrid in Obihiro city office.

C. 380 VDC power supply for super computer

The international team between Japan and US demonstrated a 380VDC power distribution system interconnected with a solar power generation system in Texas, USA. The purpose of this demonstration was to show that a 380VDC power supply system saves more energy than an AC power supply system, and to show how much carbon dioxide emissions can be reduced by integrating a solar power generation system. This demonstration resulted in an approximate 17% energy reduction compared with an AC power supply system having the same level of reliability. Also, an evaluation using Data center Performance Per Energy (DPPE) as a performance index of the efficiency of data centers was

carried out. The results showed that Power Usage Effectiveness (PUE), one of the sub-metrics of DPPE, improved with the 380VDC power supply system compared with the AC power supply system. Loads of 380 VDC power system are not only a super computer, but also cooling units and LED lighting systems.

V. INTERNATIONAL STANDARDS

International Standards for telecom/data centers create an integrated, open platform for power, infrastructure, peripheral device and control applications to facilitate the hybrid use of AC and DC power within data centers and telecom central offices as well as promoting microgrid applications.

The battery voltage of most telecom switching equipment is -48 V. That voltage is a universal standard for telecommunications equipment and is well-defined by some international standards bodies. For AC power, on the other hand, there are 14 different voltage and distribution systems defined across the world.

In the existing infrastructure of the telecommunications industry, millions of highly dependable DC power installations are already in use. In addition to these experiences, the 380 VDC makes the power system more effective and reliable.

Since 2017, the IEC established the new system committee on low voltage DC up to 1500 Vdc, short for the SyC LVDC [4]. The IEC SyC LVDC works actively to develop many international standards for DC power applications including DC microgrids.

VI. NEXT CHALLENGE

Direct current 380 V power distribution system already has sufficient deployments as communication buildings, data centers and microgrids both in Japan and overseas, but in order to promptly incorporate advanced technology and its results, to further improve convenience and economy, there are still some issues to overcome [5].

It is necessary to reduce losses and improve reliability by decreasing the number of components by improving the density and size of facilities and equipment and devising circuit schemes. One of the challenges is to avoid costs by avoiding special parts dedicated to direct current and to share costs as much as possible with general-purpose AC specifications.

In addition, in order to reduce costs, it is also necessary to consider the common use of devices and parts that conform to standards and standards such as the IEC, and common specifications and utilization beyond the boundaries of a wide range of DC applications.

Furthermore, because digital control of power conversion devices used for DC 380 V power distribution and distribution system can be expected to further improve efficiency of power distribution and distribution, high reliability, and support for flexible and advanced operation mode, Further expansion of application is required.

From the viewpoint of safety protection, it is technically possible to deal with overcurrent, overvoltage, earth fault and earth leakage, and equivalent products are also being expanded to the market. Recently, with the introduction of photovoltaic power generation system, in order to prevent heat generation / fire accidents caused by loosening of connection portion of DC circuit which has been considered as a problem, construction failure, etc., for reliable DC arc detection protection technology, Market establishment is required.

It is necessary to rectify false recognition due to lack of understanding and urge correct understanding through appropriate education and education as a common problem in the whole of direct current system in which use expansion expands in various fields.

In order to further activate the market, the role of international standardization plays a minor role. Great introduction of useful introduction examples is applicable to international standards.

VII. Summary

Telecom and IT equipment that requires AC and -48VDC power provides new challenges for system designers. Conventional configurations may not be the most appropriate choice for powering this type of equipment, particularly for Internet applications with high-load power densities, codependency of the AC- and DC-powered load equipment, and a mixture of loads that require AC input power.

Alternative, by using 380 VDC power system, cost-effective configurations can create systems that have the necessary reliability, availability, and maintainability to meet today's nonstop processing requirements.

As a power supply infrastructure for ICT services for over 10 years systematically organized the process of examining the DC 380 V power distribution system developed simultaneously in many areas such as Japan, Europe and the United States, and a number of technical outcomes and development cases.

As mentioned in the next challenge, further technical development, issuance of timely and strategic international standards and appropriate education / awareness activities will further expand the introduction of DC 380 V power distribution system, eventually to protect the global environment and living social system.

It is expected that this will contribute to improving the convenience of users. In addition to ICT applications, DC power distribution systems are expanding also for applications such as micro grids, smart grids, commercial buildings and factories, and it is highly expected that they will be applied in developing countries other than developed countries in future.

In the 1880's, T. Edison started the electricity business by the direct current in New York in the United States, but at that time it was not able to step up and down the voltage, and gradually replaced with an alternating current system that could freely convert voltage using a transformer.

However, the semiconductor power conversion technology superbly overcomes the weak point, it is possible to effectively use direct current such as integration / interconnection with renewable energy and accumulator, easy to manage total including direct current load I lay the foundation for the arrival of the times.

In the future, the achievements and examples introduced in this paper should be useful because not only a simple confrontation structure of AC versus direct current but also the right place for DC technology / application matching the needs and era background is increasingly required.

REFERENCES

[1] N. Takeuchi, K. Sakamoto, K. Tsutsui and M. Ogasawara, "The Great Hanshin Earthquake Brings Many Lessons," INTELEC 1995.

[2] S. Iwai, H. Iwamoto, T. Aidsu and Y. Kawagoe, "Business Continuity Management of NTT FACILITIES, INC.," INTELEC 2008.

[3] Thomas M. Gruzs, "Powering Telecom and Info Technology Systems," POWER QUALITY MAGAZINE ARCHIVE, Apr 01, 2001.

[4] IEC System Committee LVDC website: http://www.iec.ch/dyn/www/f?p=103:7:0::::FSP_ORG_ID:20447

[5] E. Pritchard, D. C. Gregory, and S. Srdic, "The dc Revolution", IEEE Electrification Magazine / June 2016.

MVDC Distribution Grids for Electric Vehicle Fast-Charging Infrastructure

Marco Stieneker, Benedict J. Mortimer, Arne Hinz, Adolf Müller-Hellmann and Rik W. De Doncker

Institute for Power Generation and Storage Systems

E.ON Energy Research Center

RWTH Aachen Univeristy, Aachen, Germany

Email: post_pgs@eonerc.rwth-aachen.de

Abstract—The electrification of the mobility sector is steadily increasing. Besides established electrical vehicles like railways, also the number of full-electric busses with on-board batteries grows. Moreover, the acceptance and hence the share of electric cars in traffic is continously rising. To further accelerate the electrification of public and individual traffic, large-scale usual and fast-charging infrastructures should be developed. This paper presents different approaches to ensure the proper integration into existing electrical grids. Also, the power supply with medium-voltage direct-current (MVDC) distribution grids that lowers the costs for loss energy and components is presented.

I. INTRODUCTION

The share of electric vehicles (EV) in the mobility sector is steadily increasing. Thereby, the number of electrical cars and the amount of electric busses in service is steadily growing. Taking the capital of the Netherlands as an example, the target of Amsterdam is the electrification of the entire public transport by 2026 [1]. This goal is necessary to achieve climate targets, the reduction of noise, as well as to establis a sustainable transport systems. With this change in the transportation sector, the need for large-scale implementation of high-power and fast-charging infrastructure arises.

In reaction to the increasing demand for charging infrastructure, the Federal Republic of Germany announced a 300 million euro program to support 5,000 fast-charging stations (charging power $P_{chrg} > 22\,kW$) and 10,000 standard charging stations ($P_{chrg} \leq 22\,kW$) [2]. Also the utilities in North America (USA and Canada) plan to install more than 5,500 EV standard-charging and fast-charging stations resulting in a more than one billion dollar investment over the next five years [3]. These numbers emphazise the future (peak) power demand caused by the mobility sector.

In Europe, where four-wire $400\,V$ are common, the most preferred location for charging private EVs is at home, where convenient over-night charging with e.g. $P_{chrg} = 3.7\,kW$ can be easily realized. According to [4], $70\,\%$ of the users of cars have access to a garage, carport or car park on private properties. Therefore, the possiblity for installing charging spots in low-voltage distribution grids is given. The acceptance for EVs can further be increased by implementing charging facilities at work.

Fast charging ($P_{chrg} > 22\,kW$) of EVs is required at least for long-distance travelling, in particular distances that

cannot be reached with the installed battery capacity. But also if low-power over-night charging is not sufficient to increase the state-of-charge (SOC) of the battery for the next day's driving distance, fast-charging stations become necessary. Suitable locations to implement direct-current fast-charging (DCFC) stations can be found at highways, transport hubs (e.g. trainstations, airports, car parks, gas stations), supermarkets, theaters, cinemas and at work. Depending on the location of the charging infrastructure, however, the maximum acceptable time for charging the EV might be different than for occasional charging.

TABLE I: Acceptance of fast-charging time [5]

Charging Time	5 min	10 min	15 min	30 min
Acceptance	14 %	23 %	37 %	25 %
Cumulative acceptance	99 %	85 %	62 %	25 %

According to the data given in [5], the maximum fast-charging times accepted by people considering electric cars are given in Table I. For $62\,\%$ of the potential users, a fast-charging time t_{fchrg} of 15 min is acceptable. Reducing t_{fchrg} down to 10 min by increasing P_{chrg}, the expectations for fast-charging stations of $85\,\%$ of the interviewed persons would be fulfilled.

Different charging power levels for on-board and off-board battery charger topologies are presented in [6]. In the following section the minimum P_{chrg} that is required to fulfill these user expectations is discussed.

II. REQUIRED CHARGING POWER LEVELS

Achieving charging times with a high acceptance rate accelerates the electrification of individual traffic, since more people consider purchasing an EV. However, the required charging power P_{chrg} has to be determined in accordance to the battery capacity E_{bat} of (future) electric cars. The expected development of E_{bat} in electric cars within the next decade is shown in Fig. 1 and is derived from data presented in [7]. It can be seen that the difference in E_{bat} between premium and mass market vehicles tend to be $40\,kWh$ in future. However, this does not necessarily result in different driving distances of the two classses of EVs due to varying specific energy consumption of each particular electric car.

The 2018 International Power Electronics Conference

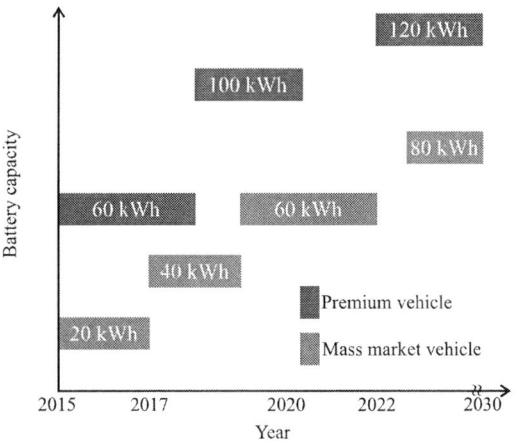

Fig. 1: Expected development of E_{bat} in EV according to [7]

TABLE II: Required charging power P_{chrg} in kW depending on the battery capacity E_{bat} in kWh

$t_{\mathrm{fchrg}} \backslash E_{\mathrm{bat}}$	20	40	60	80	100	120
5 min	187	373	560	747	933	1120
10 min	93	187	280	373	467	560
15 min	62	124	187	249	311	373

Although increasing E_{bat} results in higher driving ranges, the length of t_{fchrg}, hence the length of an unintentionally but necessary stop has still to fulfill the expectations according Table I. The required (future) P_{chrg} in dependency of E_{bat} for the different targeted charging times can be calculated with (1).

$$P_{\mathrm{chrg}} = \frac{E_{\mathrm{bat}} \Delta \mathrm{SOC}}{t_{\mathrm{fchrg}} \eta_{\mathrm{chrg}}} \qquad (1)$$

The calculation takes an conservative efficiency of the charging process η_{chrg} of 90 % into account. The batteries' state of charge (SOC) at the beginning of the charging process is considered to be 10 %. Finally, depending on the battery technology, the charging process stops at SOC = 80 % that results in a difference ΔSOC of 70 %. However, due to the required change of the charging strategy at SOC = 80 % (from constant-current to constant-voltage charging), P_{chrg} is reduced. Hence, the time to further increase the SOC up to 100 % takes too long compared with the gain in driving distance. The results of the calculation for different E_{bat} and different t_{fchrg} are summarized in Table II.

Besides the discussed power ratings for P_{chrg} of 150 kW and 350 kW, future E_{bat} require even higher power ratings if the expectations of t_{fchrg} for fast-charging stations of the users will not change. But also the electric system on-board the EV as well as the charging equipment has to be capable for such high power ratings.

Today's charging plug with a standardized voltager rating of 1 kV and a current rating of 350 A allows a maximum power of 350 kW. It can be derived from the results presented in Table II that the expectation of all scenarios will not be fulfilled, if no further development also of auxiliary equipment is pushed.

Clearly, with respect to the data shown in Table II, existing low-voltage distribution grids are not capable to supply future fast-charging infrastructure, especially when several charging spots have to be supplied simultaneously. Large-scale installations of normal charging spots will exceed the current installed power capabilities resulting from established low-voltage distribution grid structures.

III. STRUCTURE OF LOW-VOLTAGE DISTRIBUTION GRIDS

Existing low-voltage distribution grids are connected to the feeding medium-voltage distribution grid via a step-down transformer with a typical power rating of 250 kVA or 630 kVA in Germany. As an example, Fig. 2 shows the schematic of an urban distribution grid [8]. In total, 65 households are fed by three branches. Branch A connects 21 households with a total maximum power demand of 98 kVA, branch B feds 34 households with a total maximum power of 129 kVA whereas maximum 68 kVA are supplied via branch C to 10 households. In total, all branches can be fed by the 250 kVA transformer since the maximum load of the three branches do not occur at the same time.

Considering a scenario where all of these households are using at least one electric car, over-night charging with 3.7 kW results in an additional power demand of 240 kW, hence 96 % of the installed power. The impact of EV charging on existing distribution grids is discussed in detail in [9]. It is shown that even under low EV penetration significant voltage deviations and additional transformer load can be expected. Hence, to avoid excessive expansion of the low-voltage grid, coordinated charging strategies like priorization of EVs should be developed and implemented [9, 10].

Moreover, the power supply of lumped loads, like fast-charging stations or large-scale installations of charging points in e.g. public parking garages, pose a challenge for existing distribution grids. In the following, different concepts are presented to integrate future (fast-) charging infrastructures into existing electricity grids.

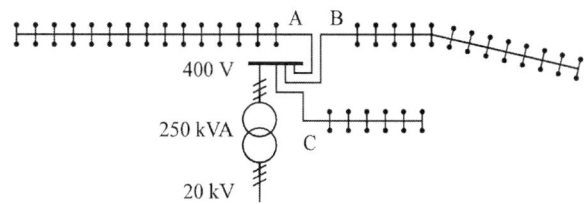

Fig. 2: Schematic of an urban distribution grid

The 2018 International Power Electronics Conference

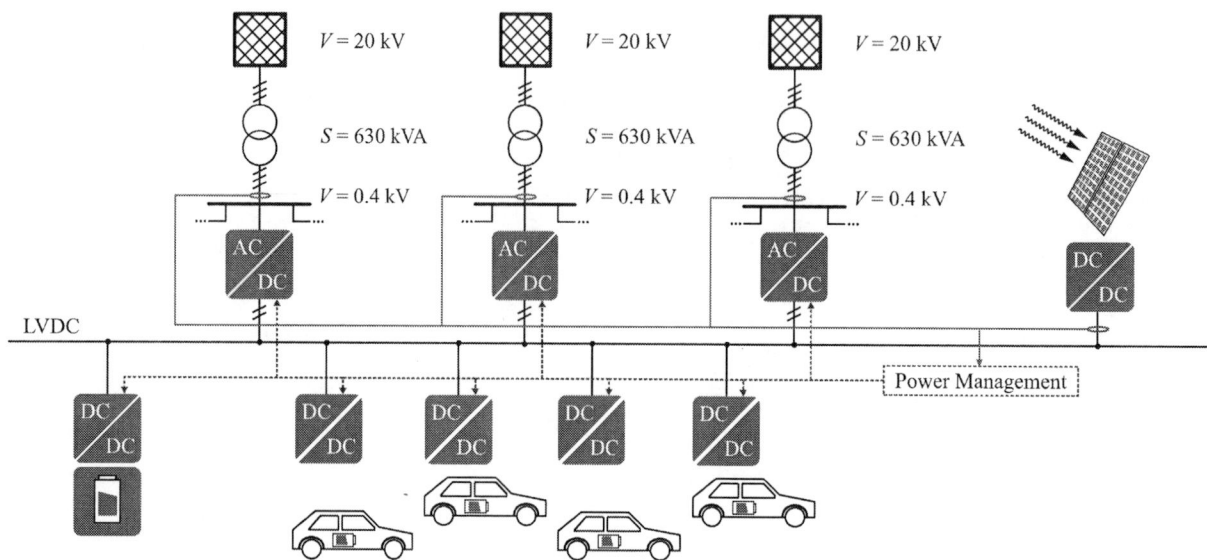

Fig. 3: Schematic of possible charging infrastructures supplied by several existing low-voltage transformers

IV. GRID INTEGRATION OF CHARGING INFRASTRUCTURES

The installation of fast-charging infrastructures, especially if several charging spots are supposed to be installed at a single location, lead to a high peak power demand in the supplying grid, which might exceed installed capacities. Considering charching spots with a maximum power of 150 kW each, only four cars can be charged at maximum power simultaneously, if a 630 kVA transformer is installed, while only 30 kVA remains for other loads.

As investigated in [11], the installed apparent power of the distribution transformers is the limiting factor in most cases and usually not the supplying medium-voltage ac grid. Furthermore, the low-voltage grid is not capable to carry the additional current to enable high charging powers (≥ 100 kW). Therefore, an approach to bundle existing grid capacities to directly feed a fast-charging infrastructure is proposed and presented in this paper.

Fast-charging infrastructure can be integrated into the supply structures of railway systems as well. In this approach, existing railway substations or the catenary can be used to deliver electrical energy.

In case the power rating of existing infrastructure is not sufficient, additional components have to be installed to ensure the energy supply. As investigated in [11], medium-voltage cables in the supplying distribution grid can be the bottleneck. Instead of installing a new ac cable, a co-infrastructure based on medium-voltage dc (MVDC) can be implemented leading to further advantages from a system perspective.

A. Bundling of Existing Low-Voltage AC Capacities

On the one hand, distribution-grid transformers are typically not fully utilized as it is presented in [12] for the city of Gothenburg, Sweden. It can be derived from this case study, that in average 25 % of the installed capacity in Gothenburg is not used and therefore available to supply EVs. Further, it can be assumed that other cities offer similar idle power capabilities.

On the other hand, the remaining power capabilities might not be sufficient to supply high lumped loads like fast-charging infrastructures. A distribution transformer (10 kV/0.4 kV) in [12] has a power rating of 200 kVA to 800 kVA. Hence, an average spare capacity of 50 kVA to 200 kVA at each existing connection point to the medium-voltage distribution grid can be used for charging. Regarding Table II, this power is not sufficient to allow fast-charging within 5 min for $E_{bat} > 20$ kWh, which is the preferred t_{fchrg} according to Table I. Therefore, to increase P_{chrg}, one solution is to bundle residual power capacities of several distribution transformers.

The most obvious measure is the parallel connection of transformers in different low-voltage grids on the secondary side to supply the charging spots. But this can lead to circulating currents within the ac grid and to increased short-circuit currents. Futhermore, the power load is distributed among both of the transformers according to the impedances and not to the spare power capacity. Hence, over-load of a particular transformer can occur while the other transformers have still remaining capacities.

An alternative to supply a load with parallel transformers is a dc grid infrastructures that inherently enables power flow control. In Fig. 3, an exemplarily implementation of such a dc grid is shown. Active bi-directional rectifiers (active-front ends,

600

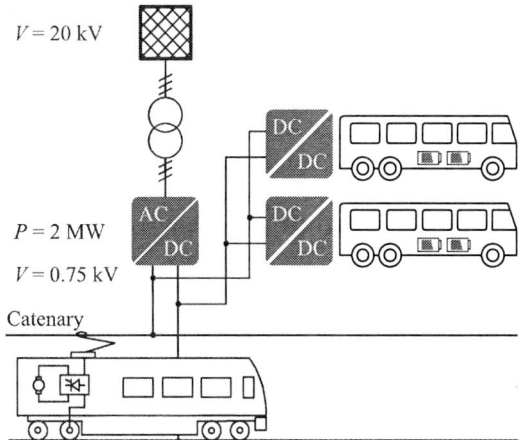

Fig. 4: Dual-use of light-rail infrastructures [15]

AFE) are connected to the secondary side of a transformer to rectify the 400 V ac voltage. The output of the AFEs are linked by a local dc grid. This dc grid can be used to supply different dc-dc converters (charging spots) to fast charge EVs.

Furthermore, renewable energies like photovoltaics and wind energy as well as battery energy storage systems can be integrated which is also addressed in [13, 14]. From a system perspective, this approach can reduce peak loads for the supplying ac grid. Since the peak power demand also determines the energy costs, this approach can reduce the costs for the power supply.

With dc infrastructures, inverters and lossy grid filters can be omitted resulting in lower investment costs and conversion losses compared with conventional ac-based approaches. Since less components are required, a higher availability of the installations can be expected.

The implementation of a dc-based energy supply also enables further services for the feeding ac distribution grid. The AFE can provide reactive power to support the ac-grid voltage stability. Possible unsymmetries in the grid voltge can be balanced and the implementation of active filtering can mitigate harmonics, hence the AFEs contribute to the power quality.

B. Dual-Use of Railway and Light Rail Infrastructure

Due to the low utilization of railway supply infrastructure, fast-charging stations can also be supplied with existing infrastructure that is installed to supply dc railways. Hence, new installations especially in cities and urban areas can be avoided. Not only investment costs, but also the footprint of overall installations can be reduced.

Light rails are usually operated with a dc voltage of 750 V via a catenary, which is supplied by substations that are built with a transformer and a three-phase diode rectifier. Since the power rating can reach several megawatts (typically 1 MW to

2 MW), the substations are typically connected to a medium-voltage ac distribution grid. Another important aspect, for use as a power source for fast-charging stations, is that substations are designed to deliver continously twice the nominal power [16]. Hence, it is feasible to operate fast-charging stations, although a certain peak power has to be delivered for the railway system. Power management systems can be installed to ensure a reliable operation of the railway system while maximum power can be delivered for fast charging.

To access the available dc power in railway substations, a dc-dc converter can be connected to the catenary of the railway system to provide a suitable dc voltage for fast charging. The advantage of the use of the catenary voltage is that no investment costs for the distribution grid connection occur. However, it has to be taken into account that additional ohmic losses occure in the catenary and the current capability of the catenary may limit the charging power.

Due to ohmic losses, fast-charging stations should be placed at the railway substation directly as it is shown in Fig. 4. Since substations are placed 4 km to 6 km apart from each other in 750 V based light rail systems [17], there is a high potential to supply numerous fast-charging spots in urban areas. Recently, in the EU-project ELIPTIC the functionality of catenary-bound fast-charging stations is demonstrated by an implementation at the railway station Sterkrade in Oberhausen, Germany. Here, three fast-charging spots with a power rating of 50 kW each for individual vehicles and two fast-charging spots with a power rating of 220 kW for public electric busses are installed. The power supply is ensured by the catenary of the municipal light rail system [18].

Municipal light rail systems usually cover a significant area of a city and connect traffic hubs. Consequently, the integration of fast-charging stations into urban railway systems not only makes use of otherwise idle power capacities. This concept also offers the opportunity to place fast-charging infrastructure in areas with high demand for charging power without major modifications of the public grid.

In some countries like the Netherlands, France, Belgium, Italy, and Spain, intercity railways are also operated with dc voltages, but with 1500 V or 3000 V dc [19]. This approach enables the power supply of fast-charging stations in rural areas, e.g. at highways. However, this concept is not discussed in this paper due to the similarity to the fast-charging infrastructure integration into light-rail systems.

C. Medium-Voltage DC Grids

With increasing installed (fast-) charging infrastructure, the power demand for EVs might exceed already installed power capabilities even if existing transformers are bundled. Beside the installation of a new ac distribution grid that might be required in different application scenarios according to e.g. [11], a (co-) infrastructure based on MVDC grids with a voltage of e.g. ± 5 kV, ± 10 kV or ± 20 kV is a promising solution.

The 2018 International Power Electronics Conference

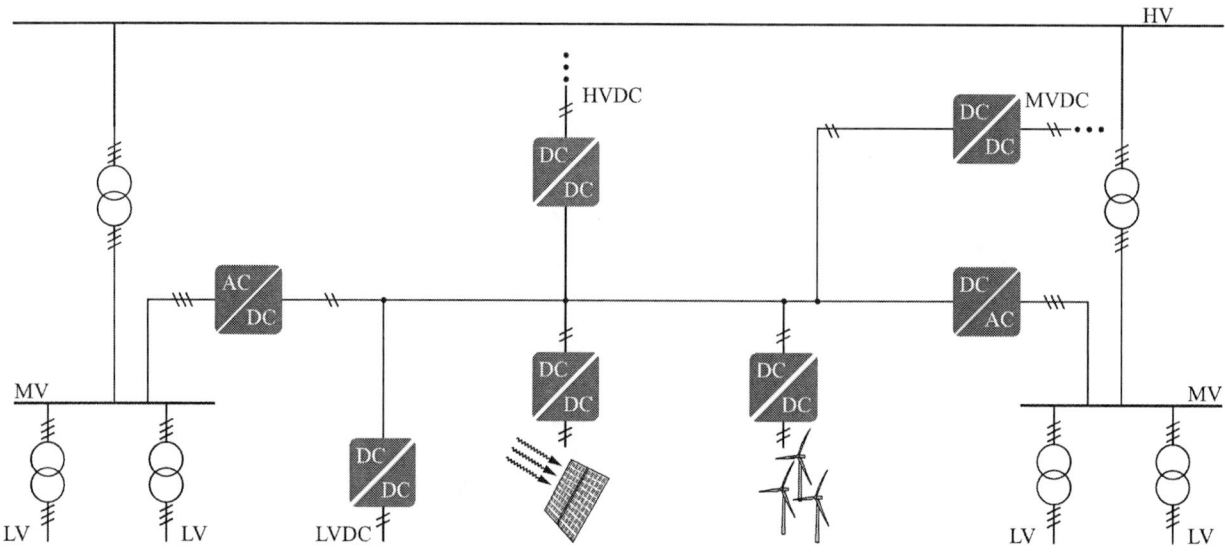

Fig. 5: MVDC co-infrastructure to supply fast-charging stations

Figure 5 shows the concept of MVDC grids as co-infrastructure embedded into an existing ac grid. The MVDC connection allows active power exchange between the two MVAC grids. Besides active power flow control, the power electronic converters can provide services like reactive power flow control and active filtering.

Since wind turbines and photovoltaics operate with a dc voltage internally, the implementation of a dc grid makes pulse-width modulated inverters, lossy grid filters and bulky 50 Hz grid transformers obsolete. Hence, conversion losses can be reduced and the system efficiency of electrical grids can be improved. Due to charging with dc, the efficiency of fast-charging infrastructure also benefits from the integration into MVDC grids.

At present, standards for MVDC distribution grids are under development. In preliminary studies, due to commercially available power electronic converters, the $\pm 5\,\mathrm{kV}$, $10\,\mathrm{kV}$ and $\pm 10\,\mathrm{kV}$ are promising voltage levels. Figure 6 shows

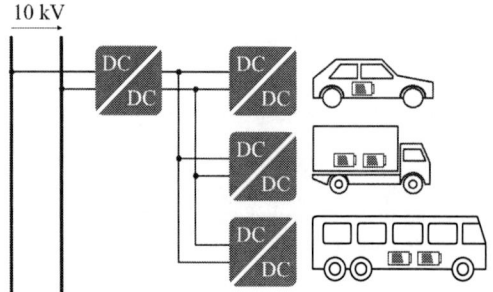

Fig. 6: Two-stage dc charging infrastructure

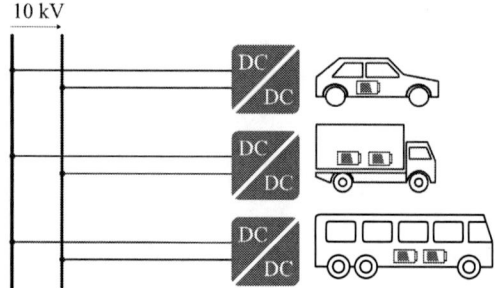

Fig. 7: One-stage dc charging infrastructure

an approach to integrate fast-charging stations into the grid. The grid voltage is stepped down with the dc-dc converter to e.g. $1\,\mathrm{kV}$. Afterwards, galvanically-isolated dc-dc converters are implemented to charge the EV batteries with constant current. The adavantage of this approach is that only one power electronic converter has to be isolated against the grid voltage. Furthermore, due to the high power rating that is required to supply all of the connected charging spots, the utilization of the dc-dc converter can be improved. However, the integration of fast-charging spots without a second voltage level (c.f. Fig. 7) is also a feasible solution.

Bipolar ($\pm 5\,\mathrm{kV}$) MVDC distribution grids lead to lower voltage stress on insulation material during normal operation while the power capability is kept constant compared to equivalent monopolar voltages ($10\,\mathrm{kV}$). In case a (grounded) neutral conductor is implemented, components can be connected either to $5\,\mathrm{kV}$ or $10\,\mathrm{kV}$ according to the power rating of the consumer (e.g. fast-charging station) or producer (e.g. photovoltaics). Hence, based on a trade-off between insulation demand and

602

conduction losses, the most suitable voltage level can be chosen.

This is illustrated in Fig. 8 by a high power charging station for public busses connected to both dc grid poles, hence operating with an input voltage of $10\,\text{kV}$. In contrast to this, the fast-charging stations for cars and light trucks are connected to one dc grid pole operating at $5\,\text{kV}$ and $-5\,\text{kV}$ respectively.

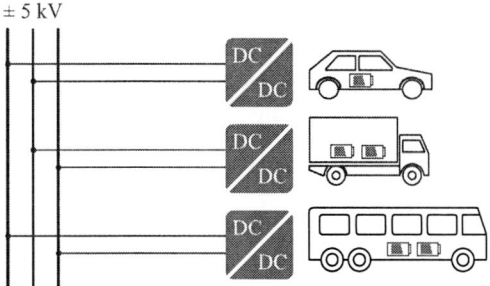

Fig. 8: Bipolar power supply

Also established substations for light-rail systems and dc intercity railways can be integrated into MVDC grids. Instead of the transformer to adapt the voltage level and the rectifier to convert ac into dc, a galvanically-isolated dc-dc converter can be used to supply the catenary with electrical energy.

The effort to implement railway substations for a MVDC grid connection is higher than for conventional MVAC distribution grids. Also the efficiency of the dc-dc converter at full power is slightly lower than that of transformer and diode rectifier [15]. In contrast, the advantages of the substation shown in Fig. 12 are the bidirectional power flow capability and the enhanced controllability. Due to the bidirectional power flow capability, recuperation energy of trains can be fed into the supplying grid or used to charge electric vehicles, instead of being burned in braking resistors. Hence, the system-level energy consumption can be reduced significantly.

Additionally, the galvanically isolated dc-dc converter can be set into a standby mode if no power is demanded by trains or EVs. In this operation mode, only the power demand of the control units has to be covered. Thus the standby losses are lower than those of a convensional ac substations [15].

V. FUTURE ON-BOARD CHARGERS

The integration of battery chargers into the onboard system of EVs promises a simplification of the fast-charging infrastructure, since only a defined constant voltage is required. Further, there is no need for communication between EV and charging spot. This also leads to a more flexible fast-charging infrastructure as one power electronic converter can supply several vehicles as shown in Fig. 10. Consequently, the dc-dc converter providing the constant charging voltage has not to be specialized regarding specific vehicles, but can supply power regardless whether the connected vehicle is a bus, a car or a truck [20]. Moreover, less converter are necessary to establish

a universal fast-charging station as the comparison of Fig. 10 with Fig. 8 shows.

The onboard chargers can be formed by re-arranging the drivetrain converter as the company Continental presented at the Internationale Automobilaustellung (IAA) 2017 in Frankfurt [21]. However, to obtain reasonable charging times, the drivetrain converter has to be designed regarding the required charging power. Hence, this approach leads to overdimensioned drivetrain converters with respect to the power demand during driving, especially in case of small vehicles with small traction power.

VI. CONVERTER TOPOLOGIES FOR MVDC DISTRIBUTION GRIDS

The galvanically-isolated dual-active bridge (DAB) dc-dc converter [22] is a promising topology for utility applications. The DAB is built with two power electronic converters linked with a medium-frequency transformer, as shown in Fig. 11. Considering the power rating and voltage level, the DAB can be implemented with one phase or with three phases.

Besides the capability of stepping up or down the output voltage, the winding ratio of the transformer dictates the static voltage ratio between input and output. In addition, bidirectional power flow capability and the ability to isolate failures in the grid is given inherently.

The DAB can be designed to link MVDC with LVDC grid [23]. As it is discussed in [24] and [8], $5\,\text{MW}$ DAB converters can be designed to build a modular converter system. This approach allows the easy adaption of the power and voltage rating according to the application by series and parallel-connection of converters.

But also the parallel connection of several DABs with the same voltage specifications to increase the power capability can lead to advantages compared with a single converter for rated power. For substations supplying light-rail systems

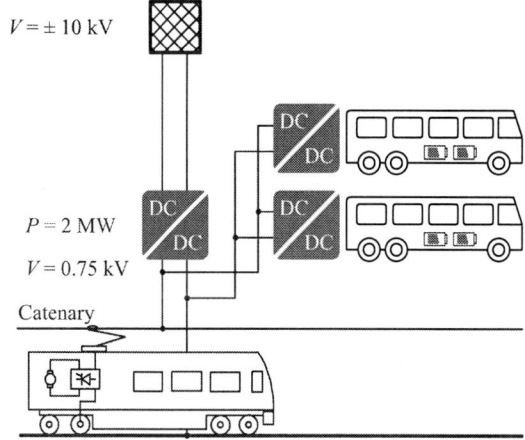

Fig. 9: Dual-use of a dc-supplied light-rail infrastructure [15]

as shown in Fig. 12, four DABs are connected in parallel. The modular approach allows to imnprove the partial-load efficiency.

Figure 13 gives the efficiency of the parallel connection over power transfer for two different operation modes. In case of equal power sharing, each of the four parallel-connected converters transfers the same power. In contrast, the operation with load-dependent power sharing adjusts the number of converters according to the power demand. As the simulation results shown in Fig. 13 point out, the efficiency for low-load operation of the substation can be improved significantly.

VII. BENEFITS OF MVDC GRID SUPPLY

Due to the steadily decreasing costs for power electronics [25], MVDC distribution grids become economical competitive with established ac distribution grids. The galvanically isolated dual-active bridge (DAB) dc-dc converter [22] is a promising topology for grid applications. Besides a high efficiency [26] and a good controllability [27], costs of $45\,{}^{\text{EUR}}/\text{kW}$ [8] contribute to the economic feasibility of dc systems.

In addition, the size and weight of a transformer in medium-voltage high-power DABs operated with a switching frequency of $1\,\text{kHz}$ is reduced by the factor of 10 compared with conventional $50\,\text{Hz}$ grid transformers [28]. Hence, MVDC distribution grids lead to advantages regarding material costs and power density compared with a conventional ac distribution grid supply. This allows to save valuable space, especially in high-populated and commercial areas.

Furthermore, local renewable power sources within distribution level distance, railway substations [15] and rural distribution grids to supply households [8] can also be linked with each other via MVDC grids. The amount of conversion steps (ac to dc and vice versa) can be reduced, hence the system efficiency can be improved.

VIII. CONCLUSION

This paper discusses the required power rating for future fast-charging stations according to the expectations of electric vehicle users. It is pointed out that the utilization of existing

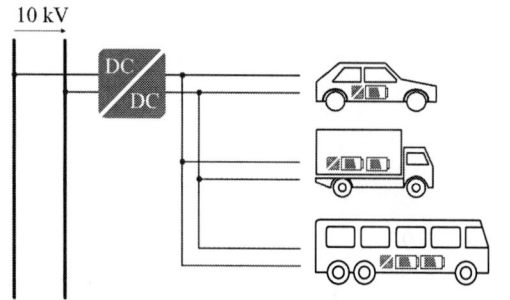

Fig. 10: Fast-Charging station for EVs with on-board chargers

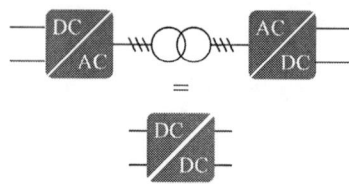

Fig. 11: Schematic of a three-phase DAB dc-dc converter

infrastructure can be increased to cover parts of the power demand for charging, but new concepts and coordinated charging have to be implemented. Further, the dc power distribution allows the efficient integration of renewable energy sources and battery energy storage systems. In addition, it is shown that medium-voltage direct-current (MVDC) grids are a promising solution to integrate fast-charging infrastructures into medium-voltage ac distribution grids with several advantages from system perspective.

REFERENCES

[1] Gemeente Amsterdam. "Duurzaam Amsterdam". In: (2015).

[2] Federal Ministry of Transport and Digital Infrastructure, Germany. "Bekanntmachung: Foerderrichtlinie Ladeinfrastruktur fuer Elektrofahrzeuge in Deutschland". In: (2017).

[3] D. Bowermaster, M. Alexander, and M. Duvall. "The Need for Charging: Evaluating utility infrastructures for electric vehicles while providing customer support". In: *IEEE Electrification Magazine* 5.1 (Mar. 2017), pp. 59–67.

[4] Fraunhofer ISI. "Markthochlaufszenarien fuer Elektrofahrzeuge". In: (Sept. 2013).

[5] ING Economics Department. "Breakthrough of Electric Vehicle Threatens European Car Industry". In: (July 2017).

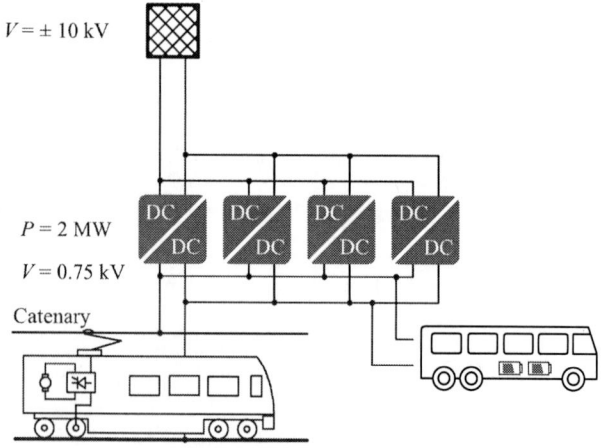

Fig. 12: Modular dc-supplied light-rail infrastructure

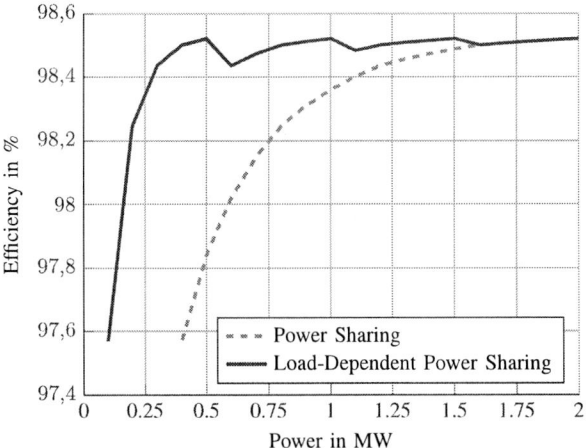

Fig. 13: Efficiency of modular parallel-connected DABs

[6] M. Yilmaz and P. T. Krein. "Review of Battery Charger Topologies, Charging Power Levels, and Infrastructure for Plug-In Electric and Hybrid Vehicles". In: *IEEE Transactions on Power Electronics* 28.5 (May 2013), pp. 2151–2169.

[7] J. B. Moreau. "The EV Charging Market: Present and Future Outlook". In: *Presentation at Automotive Power Electronics Conference (APE), Paris, France*. Apr. 2017.

[8] M. Stieneker and R. W. De Doncker. "Medium-voltage DC distribution grids in urban areas". In: *2016 IEEE 7th International Symposium on Power Electronics for Distributed Generation Systems (PEDG)*. June 2016, pp. 1–7.

[9] P. S. Moses, M. A. S. Masoum, and S. Hajforoosh. "Overloading of distribution transformers in smart grid due to uncoordinated charging of plug-In electric vehicles". In: *2012 IEEE PES Innovative Smart Grid Technologies (ISGT)*. Jan. 2012, pp. 1–6.

[10] A. Zerres. "Elektromobilitaet - Stand der Diskussion, Regulierungsbedarf und andere Weiterungen". In: *Presentation at netconomica, Bonn, Germany*. May 2011.

[11] M. Kurth, M. Gödde, A. Schnettler, A. Probst, and D. Pieper. "Determination of the integration and influencing potential of rapid-charging systems for electric vehicles in distribution grids". In: *1st E-Mobility Power System Integration Symposium*. 2017.

[12] D. Steen and L. A. Tuan. "Fast charging of electric buses in distribution systems". In: *2017 IEEE Manchester PowerTech*. June 2017, pp. 1–6.

[13] S. Rivera and B. Wu. "Electric Vehicle Charging Station With an Energy Storage Stage for Split-DC Bus Voltage Balancing". In: *IEEE Transactions on Power Electronics* 32.3 (Mar. 2017), pp. 2376–2386.

[14] Clemente Capasso, Sebastian Riviera, Samir Kouro, and Ottorino Veneri. "Charging Architectures Integrated with Distributed Energy Resources for Sustainable Mobility". In: *Energy Procedia* 105.Supplement C (2017). 8th International Conference on Applied Energy, ICAE2016, 8-11 October 2016, Beijing, China, pp. 2317–2322.

[15] A. Hinz, M. Stieneker, and R. W. De Doncker. "Impact and opportunities of medium-voltage DC grids in urban railway systems". In: *2016 18th European Conference on Power Electronics and Applications (EPE'16 ECCE Europe)*. Sept. 2016, pp. 1–10.

[16] "IEEE Standard for Uncontrolled Traction Power Rectifiers for Substation Applications Up to 1500 V DC Nominal Output". In: *IEEE Std 1653.2-2009* (Jan. 2009), pp. 1–48.

[17] R. D. White. "DC electrification supply system design". In: *Railway Electrification Infrastructure and Systems (REIS 2013), 6th IET Professional Development Course on*. 2013, pp. 57–85.

[18] Stadtwerke Oberhausen GmbH. "Schnellladetankstelle fuer Elektroautos eingeweiht". In: (2017).

[19] A. Steimel. "Electric railway traction in Europe. A survey of the state-of-the-art". In: *Proceedings of IEEE International Symposium on Industrial Electronics*. Vol. 1. June 1996, 40–48 vol.1.

[20] A. Mueller-Hellmann and M. Schmitz. "Ueberlegungen zu zukuenftigen Batteriebus-Systemen". In: *Der Nahverkehr* 9 (Sept. 2016).

[21] Continental AG. "AllCharge Technology from Continental Makes EVs Fit for Any Type of Charging Station". In: (2017).

[22] R.W.A.A. De Doncker, D.M. Divan, and M.H. Kheraluwala. "A three-phase soft-switched high-power-density DC/DC converter for high-power applications". In: *Industry Applications, IEEE Transactions on* 27.1 (Jan. 1991), pp. 63–73.

[23] M. Neubert, A. Gorodnichev, J. Gottschlich, and R. W. De Doncker. "Performance analysis of a triple-active bridge converter for interconnection of future dc-grids". In: *2016 IEEE Energy Conversion Congress and Exposition (ECCE)*. Sept. 2016, pp. 1–8.

[24] M. Stieneker and R. W. De Doncker. "Dual-active bridge dc-dc converter systems for medium-voltage DC distribution grids". In: *2015 IEEE 13th Brazilian Power Electronics Conference and 1st Southern Power Electronics Conference (COBEP/SPEC)*. Nov. 2015, pp. 1–6.

[25] R. W. De Doncker. "Power electronic technologies for flexible DC distribution grids". In: *Power Electronics Conference (IPEC-Hiroshima 2014 - ECCE-ASIA), 2014 International*. May 2014, pp. 736–743.

[26] N. Soltau, J. Lange, M. Stieneker, H. Stagge, and R. W. De Doncker. "Ensuring soft-switching operation of a three-phase dual-active bridge DC-DC converter applying an auxiliary resonant-commutated pole". In: *2014 16th European Conference on Power Electronics and Applications*. Aug. 2014, pp. 1–10.

[27] S. P. Engel, N. Soltau, H. Stagge, and R. W. De Doncker. "Dynamic and Balanced Control of Three-Phase High-Power Dual-Active Bridge DC-DC Converters in DC-Grid Applications". In: *Power Electronics, IEEE Transactions on* 28.4 (2013), pp. 1880–1889.

[28] N. Soltau, H. Stagge, R. W. De Doncker, and O. Apeldoorn. "Development and demonstration of a medium-voltage high-power DC-DC converter for DC distribution systems". In: *2014 IEEE 5th International Symposium on Power Electronics for Distributed Generation Systems (PEDG)*. June 2014, pp. 1–8.

Review of Resonant Gate Driver in Power Conversion

Bainan Sun, Zhe Zhang and Michael A.E. Andersen
Department of Electrical Engineering,
Technical University of Denmark,
2800 Kgs. Lyngby, Denmark
baisun@elektro.dtu.dk, zz@elektro.dtu.dk, ma@elektro.dtu.dk

Abstract-**Resonant gate driver is a vital trend of research topic along with the development of high electron mobility transistor (HEMT). Compared with conventional gate driver, resonant gate driver achieves much lower power dissipation during switching transient and widely viewed as one essential technique for high frequency power conversion. This paper provides a state-of-art review and thorough comparison of different resonant gate driver topologies. Case study of two representative topologies is carried out. Application of resonant gate driver in Gallium Nitride (GaN) HEMT is discussed.**

Keywords— Resonant gate driver, GaN HEMT, power conversion.

I. INTRODUCTION

GaN HEMT is widely considered as a promising candidate for the next generation of power transistor [1]. Compared with Silicon MOSFET, GaN HEMT has a higher semiconductor band gap, which leads to lower on-resistance, lower leakage current and higher operating temperature [2-5]. All these salient features enable the GaN HEMT's application in high frequency power conversion. Benefit from fast switching capability, GaN HEMT based high frequency converter can effectively achieve volume reduction and high power density integration [6].

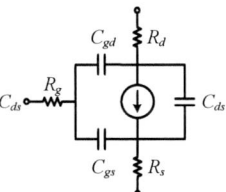

Fig. 1. Equivalent circuit of transistor.

Efficiency is the chief consideration in power converter design. For high frequency switching application, transistor parasitic capacitor and loop impedance will result in considerate power loss. Equivalent circuit of the transistor is shown as Fig. 1 [7]. Two major sources of power loss during the switching process are: transistor switching loss and gating loss [8]. Switching loss is related to the transistor output capacitor ($C_{oss}=C_{DS}+C_{GD}$) [9]. Soft-switching techniques has been widely investigated for decades to minimize this part of energy dissipation [10]. Gating loss is related to transistor's input capacitor ($C_{iss}=C_{GS}+C_{GD}$). Charging and discharging process of transistor C_{iss} during switching

transient lead to energy dissipated in the gate loop impedance, which is also known as CV^2 loss. Equation for CV^2 loss calculation is given as (1):

$$P_{Gate} = V_G \cdot Q_G \cdot f_s = V_G^2 \cdot C_{iss} \cdot f_s \qquad (1)$$

, where Q_G is the total gate charge, f_s is the switching frequency and V_G is the gate voltage [11]. Furthermore, loss dissipation on the gate driver transistor also contributes to the overall gate driver loss.

For conventional gate driver, CV^2 loss is totally dissipated in gate loop impedance during the switching transient. According to (1), gating loss is proportional to the transistor switching frequency, which highly deteriorates the converter efficiency in high frequency application [12]. Resonant gate driver is introduced to recycle this part of energy. With resonant tank added in the gating path, part of the gate energy can be recycled and gating loss is significantly reduced [13].

In this paper, review of resonant gate driver topologies is given in section II. In section III, case study of two representative resonant gate driver is carried out. Detailed analysis of gating performance and loss evaluation is given according to the simulation results. Section IV concludes the paper.

II. REVIEW OF RESONANT GATE DRIVER

Conventional gate driver is the push-pull (totem pole) topology, shown in Fig. 2(a). PMOS and NMOS are in series configured as half bridge structure to generate square wave gate signal [14]. Full-bridge configuration is aiming to provide negative gate voltage for the transistor effectively gated-off, which is shown in Fig. 2(b). All kinds of resonant gate driver design are based on these two fundamental topologies. Difference lays in the resonant tank design and control strategy.

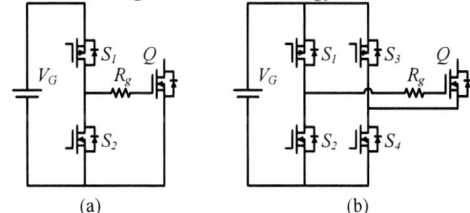

Fig. 2. Conventional gate driver. (a) Half-bridge configuration (b) Full-bridge configuration

One criterion to classify resonant gate drivers is the initial inductor current [15]. For zero initial inductor current topology, gate charging process is simultaneous with the charging process of resonant inductor. For non-zero initial inductor current topologies, gate charging is initialed with a start-up current, which helps to shorten the switching transient. Different resonant gate driver topologies are classified and reviewed according to this criterion in this paper.

A. Zero initial inductor current gate driver

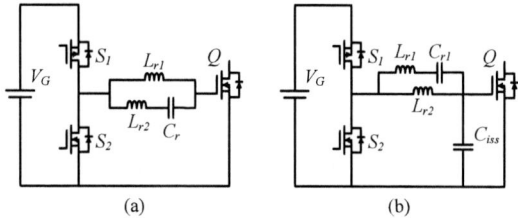

(a)   (b)

Fig. 3. Multi-resonant tank topology. (a) External resonant tank (b) Parasitic included resonant tank

A multi-resonant gate driver designed for Class-E power amplifier is proposed in [16], which is shown as Fig.3 (a). The resonant tank, L_{r1}, L_{r1} and C_r, are designed to pass through 1st and 3rd harmonic component. A quasi-square gate signal is generated for efficiency consideration. Gate transistors S_1 and S_2 both operate at soft switching, which further helps to minimize the gate driver loss. For Si MOSFET (FDMC86248) in a 32W Class-E amplifier operating at 20MHz, gating loss is reported to be lowered from 1.8W to 720mW, compared with the hard gating method by gate driver IC (LM5114).

Further research into this idea for GaN HEMT application is carried out in [17], which is shown as Fig.3 (b). Input capacitor of GaN HEMT is considered as part of the resonant tank. The proposed gate driver is tested to drive GaN-HEMT (TPH3066PS) based 13.56MHz Class-E inverter. Gating loss is reported to be lowered from 880mW to 800mW compared with the conventional totem pole gate driver.

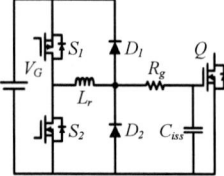

Fig. 4. Efficient power recovery topology.

Resonant gate driver topology with efficient power recovery is first proposed in [11], which is shown as Fig. 4. Resonant inductor L_r is connected to the totem pole output, where freewheeling diodes D1 and D2 are installed to provide returned current path for excessive energy recovery. The proposed topology is tested based on MOSFET switching at 500kHz. Gating loss is reported to be 209mW, which reduces 55% of the conventional loss. This topology is further researched in [18] to be applied for GaN HEMT. Gating loss is reported to be 198mW in a 10MHz IC design simulation. Application of this topology in driving e-mode GaN transistor (EPC2001) is researched in [19]. During 10MHz operation, gating loss is reported be 200mW,

compared with 450mW consumed by conventional gate driver.

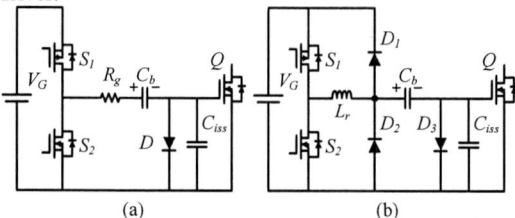

(a)   (b)

Fig. 5. Negative gate voltage topology. (a) Resonant tank included (b) Freewheeling diodes included

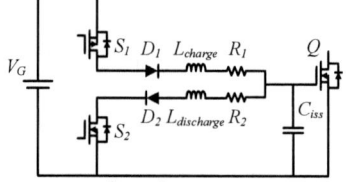

Fig. 6. Separate gate path topology.

Resonant gate driver with negative voltage driving capability is proposed in [20], which is shown as Fig.5 (a). Capacitor C_b and diode D_3 are installed in the gate path to provide negative driving voltage. A normally-on GaN transistor is fabricated in this paper to be applied in a 5.67W Class-E power amplifier operating at 13.56MHz. Based on the similar idea, energy recovery resonant gate driver with negative gate voltage is given in [21] and [22], which is shown as Fig.5 (b).

The resonant gate driver topology with controllable slew rate is proposed in [23] and further explored in [24] to fit for GaN HEMT application, which is shown as Fig. 6. The totem pole output is separated into charging and discharging path, where different resonant inductor are installed to provide asymmetric output. The topology is tested to drive the 600V commercial GaN HEMT. With 200V applied on the transistor, the gating loss is reported to steadily rise from 250mW to 1.04W, when switching frequency rises from 200kHz to 1MHz.

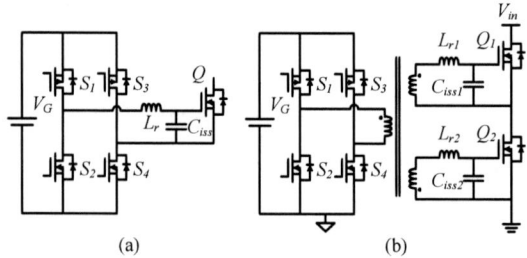

(a)   (b)

Fig. 7. Full bridge resonant gate driver. (a) For single transistor (b) For half bridge module

A full bridge configured resonant gate driver is proposed in [25], which is shown as Fig.7 (a). The proposed gate driver is tested in a 360kHz and 1kW MOSFET inverter, which helps to realize the overall efficiency of 99.1%. Compared with the conventional gate driver, gating loss is reported to be lowered from 9.3W to 0.9W. A further research of this topology is carried out in [26], which is shown as Fig.7 (b). Transformer with two secondary windings is integrated in

608

the gate driver to provide half-bridge driving capability. Converter overall efficiency is reported to be improved by 0.5% for 500kHz, 1kW full bridge inverter.

B. Non-zero initial inductor current gate driver

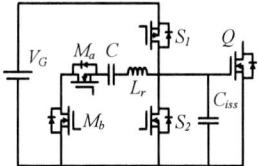

Fig. 8. Pulse resonant gate driver.

A pulse resonant gate driver, with online controllable slew rate, is proposed in [27], shown as Fig. 8. The basic idea is to generate initialized current to charge or discharge the gate capacitor C_{iss}. M_a and M_b are ancillary switches, which is in reverse series connection to provide separate path for charging and discharging LC resonant tank. Both driving switches pair and ancillary switches pair operate in complementary conduction mode, where phase shift control is utilized to control the slew rate. PSPICE simulation result is given for 12V 1MHz switching condition. Switching loss is reported to be 330mW when rise/fall time is 25ns and 420mW when rise/fall time is 20ns.

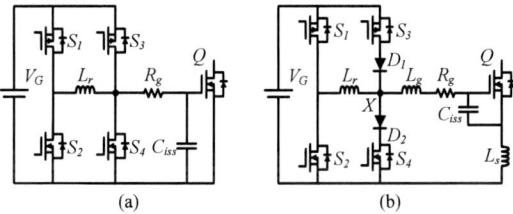

(a)   (b)

Fig. 9. Gate power recovery topology. (a) Two totem pole topology (b) Gate impedance insensitive topology

The resonant gate driver with power recovery capability is proposed in [28], shown as Fig.9 (a). A small resonant inductor is connected between two totem pole transistor pairs. Excessive inductor stored energy is returned to the line voltage source V_G and power MOSFET Q is charged with a non-zero initial current. 6.5% overall efficiency improvement is reported to be achieved in a 1MHz 12V Boost converter. The same idea is studied in [29] to provide dual-channel output, which is reported to reduce the gating loss by 67% compared with conventional non-resonant gate driver. A further research into this topology is given in [15], shown as Fig.9 (b). Schottkey diodes D1 and D2 are installed to recovery excessive gate power as well as clamp the gate voltage during this process, which helps the resonant tank design less sensitive to the parasitic component on the gate path.

Self-oscillating resonant gate driver is one special category in this classification. Different from totem pole based gate driver, additional gate transistor is not needed in the design. Drain source voltage of the transistor is fed back to the resonant tank and thus a close loop is formed to maintain self-oscillating. As a result, volume of the gate driver can be minimized to achieve high power density. Furthermore, switching frequency of the gate signal is no longer limited by the gate transistors, which

can be largely increased to meet for very-high-frequency application.

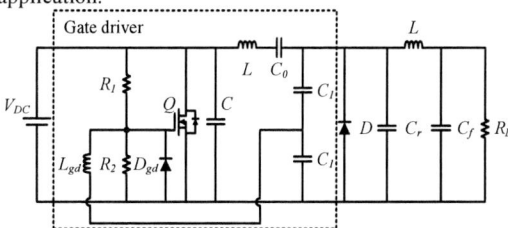

Fig. 10. Self-oscillating resonant gate driver.

One application of self-oscillating gate driver is researched in [30]. A VHF Class E DC-DC converter is built to operate in the whole VHF band width from 30MHz to 300MHz. Self-oscillating gate driver is utilized in the Class-E inverter part, which is shown as Fig. 10. Gating loss is reported to be 290.4mW and switching loss is reported to be 933.7mW at 100MHz according to the simulation result based on transistor MRF6V2010NR1. Other application of self-oscillating resonant gate driver can be found in [31] for Class-Φ2 inverter application and [32] for boost converter application.

The salient drawback of the self-oscillating gate driver is that impedance matching must be considered in the converter output. The inverter output must be followed by a rectifier stage designed accordingly to generate the desired self-oscillating gate signal [30]-[32]. As a result, application of self-oscillating gate driver is limited to DC-DC converter design, which is widely adopted in the audio amplifier.

III. CASE STUDY

A. Summary of literature review

From literature review in the last chapter, several conclusions can be made:

1) General idea for resonant gate driver is the construction of resonant tank.

2) General idea for resonant gate driver with negative voltage driving capability is the resonant capacitor in series installed to gate path.

3) Controllable slew rate is one popular research topic, which is generally achieved by separate gate path or phase shift control scheme.

4) Efficient power recovery is one popular research topic, which is generally achieved by additional diodes or ancillary switches to provide current returned path.

5) Additional gate transistors are essential for non-zero initial inductor current gate driver.

6) Self-oscillating gate driver requires impedance match which is optimal for VHF DC-DC converter design.

Several problems need further research include:

1) Different topologies are researched in different application circumstances. A general evaluation method for resonant gate driver is absent.

2) In many researches, only overall efficiency improvement of the converter is given to validate the feasibility of the proposed topology. Detailed loss analysis is essential.

3) Many topologies are researched based on the application in conventional silicon MOSFET. Performance in the GaN HEMT application need to be further evaluated.

B. Case Analysis

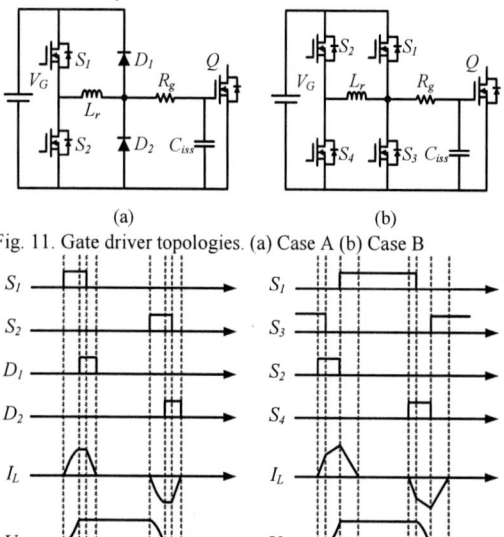

(a)   (b)

Fig. 11. Gate driver topologies. (a) Case A (b) Case B

(a)   (b)

Fig. 12. Ideal waveform. (a) Case A (b) Case B

According to analysis above, case study of two resonant gate drivers is carried out. Case A is the zero initial inductor topology proposed in [11] and Case B is the non-zero initial inductor current topology proposed in [28]. Two topologies are similar in structure and both topologies implement the idea of energy recovery, which is optimal for comparison. Topologies of Case A and Case B are shown in Fig. 11.

The ideal operating waveform of two topologies is shown in Fig. 12. For intuitive comparison, all the transistors in both topology are selected as n-mos. Topology of Case A is composed of a conventional totem pole, a resonant inductor and two schottky diodes. Input capacitor of Q is charged during $t_0 \sim t_1$, along with charging of the resonant inductor. Energy recovery takes place during $t_2 \sim t_3$ and the conventionally wasted energy is returned to the gate voltage source via the path of S_2 body diode, L_r and D_1. In Case B, two schottky diodes are replaced with another totem pole. Accordingly, the resonant inductor is pre-charged via the path of S_2, L_r and S_3. After that, the inductor current charges the input capacitor of Q during $t_1 \sim t_2$. Energy recovery takes place during $t_2 \sim t_3$ via the path of S_4 body diode, L_r and S_1.

According to the analysis above, gate energy can be recycled in both Case A and Case B, via a similar conduction path. The extra totem pole in Case B provides a pre-charging stage of the resonant inductor and thus results in fast gate charging/discharging capability. Shortened switching transient of the driven transistor

shall reduce the switching loss and further improve the power conversion efficiency.

C. Test Benchmark Specification

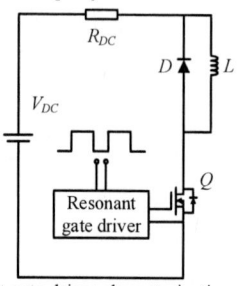

Fig. 13. Resonant gate driver characterization circuit.

The general method for transistor loss characterization is double pulse tester, which is widely adopted in [33]-[35]. However, in case of resonant gate driver, the first pulse, which is conventionally used for current forming, will interfere with the charging state of the resonant inductor and thus cannot be directly applied. As a result, based on the idea of conventional double pulse tester, topology for resonant gate driver characterization is proposed in this paper, which is shown in Fig. 13. Repetitive gate signal with fixed duty ratio is generated by the resonant gate driver. Resistor R_{DC} is installed in the power loop to adjust the switching current of the driven transistor. Schottkey diode is installed in parallel to the inductor L to provide a circulating current path during the transistor Q gated off

TABLE I

PARAMETERS OF DRIVEN TRANSISTOR

	IPW60R045CP	GS66516B
Type	MOSFET	GaN HEMT
V_{DS}	650V	650V
I_D	60A	60A
Q_g	150nC	12.1nC
V_{GS}	±20V	-10V to +7V

Conventional totem pole gate driver and resonant gate driver in both cases are tested. Transistor Q is selected as MOSFET and GaN HEMT respectively to evaluate the loss of two resonant gate drivers. Parameters of each transistor are shown in Table I. Two transistors have the same power rate. Benefit from the wide band gap characteristic of GaN HEMT, GS66516 enjoys a much smaller gate charge (Q_g) and lower gate voltage (V_{GS}). Transistor in gate driver is selected as IRF7309N for both cases. Resonant inductor in Case A and Case B are designed according to the design rules specified in each paper. Calculation equation from Case A and Case B are shown in (2) and (3) respectively:

$$L_R \le \frac{1}{C_G} \times \left(\frac{2 \times t_{rise}}{\pi} \right)^2 \qquad (2)$$

$$L_R = \frac{V_{cc} t_{rise}}{Q_G} \times \left(\frac{t_{rise}}{4} + t_d \right) \qquad (3)$$

, where t_{rise} represents the gate voltage rising time ($t_0 \sim t_1$ in Case A and $t_1 \sim t_2$ in Case B) and t_d represents the inductor pre-charging time ($t_0 \sim t_1$ in Case B).

D. Simulation results

1) Gate waveform

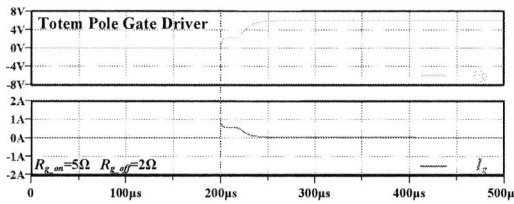

Fig. 14. Switching on transient of totem pole gate driver applied in GaN HEMT.

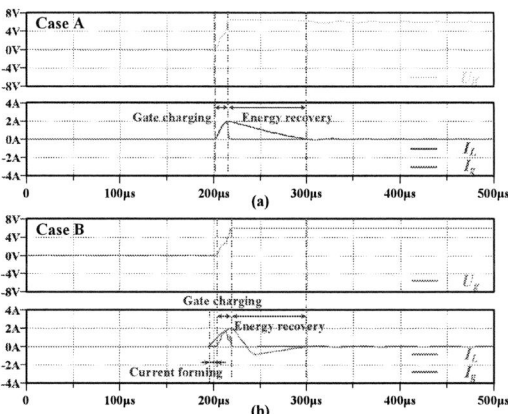

Fig. 15. Switching on transient of resonant gate driver applied in GaN HEMT. (a) Case A (b) Case B

Both of the two resonant gate drivers are originally proposed for silicon MOSFET application. As a result, gate waveform will be given for their GaN HEMT application. Simulation is carried out in LTSpice and transistor models are obtained from each manufacturer. Switching frequency is 100 kHz. As a comparison, simulation results of the conventional totem pole gate driver is given in Fig. 14.

Gate waveform of two resonant gate drivers applied for GaN HEMT is given in Fig. 15. U_g is the gate voltage and both of the two cases show good gate driving capability in GaN HMET application. I_L and I_g are the resonant inductor current and gate charging current respectively. The gate charging current rises along with the resonant inductor current in Case A. While in Case B, there is a current forming stage in the resonant inductor before the gate capacitor is charged. Also, there's a noticeable current ringing in the resonant inductor during energy recovery for Gate B application, which is resulted from the oscillation of resonant inductor and the parasitic capacitor of gate transistors. From simulation results, it can be concluded that resonant gate drivers from both Case A and Case B show well capability in GaN HEMT application. Waveform validates the energy recovery process researched in two cases.

2) Gate charging capability

In conventional totem pole gate driver, gate charging current is controlled by the gate resistor. Gate-on resistor is 5 Ω and gate-off resistor is 2 Ω, which is the typical value specified in GS66516B application note from the manufacturer. For resonant gate driver, gate charging current is related to the resonant inductor. In Case A, the gate charging speed is solely determined by the time constant of the resonant tank (L_r and C_{iss}). In Case B, the gate charging speed is determined by the resonant inductor current as well as the gate charging time. As a result, benefit from the two extra gate transistors, resonant gate driver in Case B can achieve online changeable slew rate.

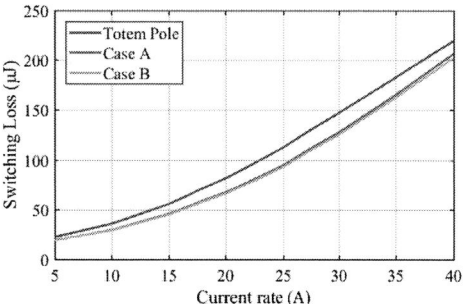

Fig. 16. Switching loss characterization.

From the simulation result, it should be noted that the peak gate charging current of conventional totem pole gate driver is less than 1A and decays quickly. For resonant gate driver, the peak gate current is around 1.9A in two cases. A faster gate charging speed will result in a faster transistor switching transient, which will thus lead to a lower switching loss in the driven transistor. To validate the analysis, loss characterization of GaN HEMT is carried out in simulation, which is shown in Fig. 16. Conventional totem pole gate driver and resonant gate drivers in two cases are tested respectively to obtain the switching loss. GS66516B is tested to switch at fixed drain-source voltage (V_{ds}=400V) and a large range of source current (I_{ds}=5A~40A). According to the simulation results, switching loss characterization is pretty identical in Case A and Case B, which is resulted from the similar gate charging current. Compared with conventional totem pole gate driver, there's a noticeable reduction of switching loss when resonant gate driver is applied. At 40A source current, switching loss can be reduced by 7.7% compared with conventional totem pole gate driver.

3) Gating loss evaluation

TABLE II
GATE LOSS ANALYSIS IN SILICON MOSFET APPLICATION

	Totem pole	Case A	Case B
Gating loss	2.86 μJ	1.62 μJ	1.37 μJ
Gate driver E_{sw}	0.29 μJ	0.37 μJ	0.51 μJ
Overall gate driver loss	3.15 μJ	1.99μJ	1.88 μJ

One salient advantage of resonant gate driver is the reduction of gating loss. The overall gate driver loss can be categorized into two parts: the driven transistor gating loss and the gate transistor switching loss (E_{sw}). Gate loss analysis in Silicon MOSFET application is shown in TABLE II. Compared with the conventional totem pole gate driver, gating loss is significantly reduced by 44.4% and 52.1% in Case A and Case B respectively. On the other hand, extra components in resonant gate driver introduces additional gate driver E_{sw} in both cases. The overall gate driver loss is reduced by 36.8% and 40.3% in Case A and Case B respectively.

TABLE III
GATE LOSS ANALYSIS IN GAN HEMT APPLICATION

	Totem pole	Case A	Case B
Gating loss	55.4nJ	25.4nJ	23.4nJ
Gate driver E_{sw}	98.7nJ	102.94nJ	113.1nJ
Overall gate driver loss	154.1nJ	128.34nJ	136.5nJ

Gate loss analysis in GaN HEMT application is shown in TABLE III. Benefit from small input capacitor, GaN HEMT has a much lower gating loss compared with its Silicon MOSFET counterpart. Resonant gate driver in Case A and Case B reduces 54.2% and 57.8% of the conventional gating loss respectively. It should be noted that gate driver switching loss is the majority of overall gate driver loss in the case of GaN HEMT application. As a result, the overall gate driver efficiency is merely improved by 16.7% in Case A and 11.4% in Case B.

IV. CONCLUSIONS

In this paper, a thorough review of resonant gate driver for switching application is given. Two general categories, zero initial inductor current topology and non-zero initial inductor current topology, are analyzed and compared in details. According the case study, several conclusions can be made:

(1) Both of the two categories of resonant gate driver can effectively reduce the gating loss in Silicon MOSFET and GaN HMET application.

(2) Resonant gate driver is capable of forming large gate charging current, which helps to reduce the switching loss of driven transistor.

(3) For the zero initial inductor current category, topology and control is rather simple, which is highly reliable and capable of integrating the commercial half bridge gate driver.

(4) For the non-zero initial inductor current category, gating loss can be further reduced. Gate charging current is online changeable with the extra gate transistors. However, complicated controller design limits its application in the multi-transistor topologies.

(5) Reduction of overall loss in resonant gate driver is not evident for GaN HEMT application. Switching loss of the transistors in gate driver contributes to the majority of gate driver loss.

REFERENCES

[1] Lidow, Alex, Johan Strydom, Michael De Rooij, and David Reusch. *GaN transistors for efficient power conversion*. John Wiley & Sons, 2014.

[2] H. Wang, J. Wei, R. Xie, C. Liu, G. Tang and K. J. Chen, "Maximizing the Performance of 650-V p-GaN Gate HEMTs: Dynamic RON Characterization and Circuit Design Considerations," in *IEEE Transactions on Power Electronics*, vol. 32, no. 7, pp. 5539-5549, July 2017.

[3] T. Ishibashi *et al.*, "Experimental Validation of Normally-On GaN HEMT and Its Gate Drive Circuit," in *IEEE Transactions on Industry Applications*, vol. 51, no. 3, pp. 2415-2422, May-June 2015.

[4] T. Ishibashi, M. Okamoto, E. Hiraki, T. Tanaka, T. Hashizume and T. Kachi, "Resonant gate driver for normally-on GaN high-electron-mobility transistor," *2013 IEEE ECCE Asia Downunder*, Melbourne, VIC, 2013, pp. 365-371.

[5] J. Lautner and B. Piepenbreier, "Analysis of GaN HEMT switching behavior," *2015 9th International Conference on Power Electronics and ECCE Asia (ICPE-ECCE Asia)*, Seoul, 2015, pp. 567-574.

[6] K. Kruse, M. Elbo and Z. Zhang, "GaN-based high efficiency bidirectional DC-DC converter with 10 MHz switching frequency," *2017 IEEE Applied Power Electronics Conference and Exposition (APEC)*, Tampa, FL, 2017, pp. 273-278.

[7] M. Orabi and A. Shawky, "Proposed Switching Losses Model for Integrated Point-of-Load Synchronous Buck Converters," in *IEEE Transactions on Power Electronics*, vol. 30, no. 9, pp. 5136-5150, Sept. 2015.

[8] J. R. Warren, K. A. Rosowski and D. J. Perreault, "Transistor Selection and Design of a VHF DC-DC Power Converter," in *IEEE Transactions on Power Electronics*, vol. 23, no. 1, pp. 27-37, Jan. 2008.

[9] W. Eberle, Y. F. Liu and P. C. Sen, "A Resonant Gate Drive Circuit with Reduced MOSFET Switching and Gate Losses," *IECON 2006 - 32nd Annual Conference on IEEE Industrial Electronics*, Paris, 2006, pp. 1745-1750.

[10] M. R. Ahmed, R. Todd and A. J. Forsyth, "Predicting SiC MOSFET Behavior Under Hard-Switching, Soft-Switching, and False Turn-On Conditions," in *IEEE Transactions on Industrial Electronics*, vol. 64, no. 11, pp. 9001-9011, Nov. 2017.

[11] Yuhui Chen, F. C. Lee, L. Amoroso and Ho-Pu Wu, "A resonant MOSFET gate driver with efficient energy recovery," in *IEEE Transactions on Power Electronics*, vol. 19, no. 2, pp. 470-477, March 2004.

[12] T. Lopez, G. Sauerlaender, T. Duerbaum and T. Tolle, "A detailed analysis of a resonant gate driver for PWM applications," *Applied Power Electronics Conference and Exposition, 2003. APEC '03. Eighteenth Annual IEEE*, Miami Beach, FL, USA, 2003, pp. 873-878 vol.2.

[13] J. V. P. S. Chennu and R. Maheshwari, "Study on Resonant Gate Driver circuits for high frequency applications," *2016 IEEE 6th International Conference on Power Systems (ICPS)*, New Delhi, 2016, pp. 1-6.

[14] Kaiwei Yao and F. C. Lee, "A novel resonant gate driver for high frequency synchronous buck converters," in *IEEE Transactions on Power Electronics*, vol. 17, no. 2, pp. 180-186, Mar 2002.

[15] X. Zhou, Z. Liang and A. Huang, "A new resonant gate driver for switching loss reduction of high side switch in buck converter," *2010 Twenty-Fifth Annual IEEE Applied Power Electronics Conference and Exposition (APEC)*, Palm Springs, CA, 2010, pp. 1477-1481.

[16] L. Gu, W. Liang and J. Rivas-Davila, "A multi-resonant gate driver for Very-High-Frequency (VHF) resonant converters," *2017 IEEE 18th Workshop on Control and Modeling for Power Electronics (COMPEL)*, Stanford, CA, 2017, pp. 1-7.

[17] F. Hattori, H. Umegami and M. Yamamoto, "Multi-resonant gate drive circuit of isolating-gate GaN HEMTs for tens of MHz," in *IET Circuits, Devices & Systems*, vol. 11, no. 3, pp. 261-266, 5 2017.

[18] B. Wang, N. Tipirneni, M. Riva, A. Monti, G. Simin and a. E. Santi, "An Efficient High-Frequency Drive Circuit for GaN Power HFETs," in *IEEE Transactions on Industry Applications*, vol. 45, no. 2, pp. 843-853, March-april 2009.

[19] Y. Long, W. Zhang, D. Costinett, B. B. Blalock and L. L. Jenkins, "A high-frequency resonant gate driver for enhancement-mode GaN power devices," *2015 IEEE Applied Power Electronics Conference and Exposition (APEC)*, Charlotte, NC, 2015, pp. 1961-1965.

[20] M. Okamoto, T. Tanaka, K. Matuzaki, T. Hashizume and H. Yamada, "13.56-MHz Class-E RF power amplifier using normally-on GaN HEMT," *IECON 2014 - 40th Annual Conference of the IEEE Industrial Electronics Society*, Dallas, TX, 2014, pp. 982-987.

[21] M. Okamoto, T. Ishibashi, H. Yamada and T. Tanaka, "Resonant Gate Driver for a Normally ON GaN HEMT," in *IEEE Journal of Emerging and Selected Topics in Power Electronics*, vol. 4, no. 3, pp. 926-934, Sept. 2016.

[22] T. Ishibashi, M. Okamoto, E. Hiraki, T. Tanaka, T. Hashizume and T. Kachi, "Resonant gate driver for normally-on GaN high-electron-mobility transistor," *2013 IEEE ECCE Asia Downunder*, Melbourne, VIC, 2013, pp. 365-371.

[23] I. D. de Vries, "A resonant power MOSFET/IGBT gate driver," *APEC. Seventeenth Annual IEEE Applied Power Electronics Conference and Exposition (Cat. No.02CH37335)*, Dallas, TX, 2002, pp. 179-185 vol.1.

[24] Y. Yan, A. Martinez-Perez and A. Castellazzi, "High-frequency resonant gate driver for GaN HEMTs," *2015 IEEE 16th Workshop on Control and Modeling for Power Electronics (COMPEL)*, Vancouver, BC, 2015, pp. 1-6.

[25] H. Fujita, "A Resonant Gate-Drive Circuit Capable of High-Frequency and High-Efficiency Operation," in *IEEE Transactions on Power Electronics*, vol. 25, no. 4, pp. 962-969, April 2010.

[26] Fei-Fei Li, Z. Zhang and Y. F. Liu, "A novel dual-channel isolated resonant gate driver to achieve gate drive loss reduction for ZVS full-bridge converters," *Proceedings of The 7th International Power Electronics and Motion Control Conference*, Harbin, China, 2012, pp. 936-940.

[27] S. Pan and P. K. Jain, "A new pulse resonant MOSFET gate driver with efficient energy recovery," *2006 37th IEEE Power Electronics Specialists Conference*, Jeju, 2006, pp. 1-5.

[28] W. Eberle, Y. F. Liu and P. C. Sen, "A Resonant Gate Drive Circuit with Reduced MOSFET Switching and Gate Losses," *IECON 2006 - 32nd Annual Conference on IEEE Industrial Electronics*, Paris, 2006, pp. 1745-1750.

[29] Z. Yang, S. Ye and Y. F. Liu, "A New Dual-Channel Resonant Gate Drive Circuit for Low Gate Drive Loss and Low Switching Loss," in *IEEE Transactions on Power Electronics*, vol. 23, no. 3, pp. 1574-1583, May 2008.

[30] T. M. Andersen, S. K. Christensen, A. Knott and M. A. E. Andersen, "A VHF class E DC-DC converter with self-oscillating gate driver," *2011 Twenty-Sixth Annual IEEE Applied Power Electronics Conference and Exposition (APEC)*, Fort Worth, TX, 2011, pp. 885-891.

[31] J. M. Rivas, O. Leitermann, Y. Han and D. J. Perreault, "A very high frequency dc-dc converter based on a class Φ2 resonant inverter," *2008 IEEE Power Electronics Specialists Conference*, Rhodes, 2008, pp. 1657-1666.

[32] R. C. N. Pilawa-Podgurski, A. D. Sagneri, J. M. Rivas, D. I. Anderson and D. J. Perreault, "Very-High-Frequency Resonant Boost Converters," in *IEEE Transactions on Power Electronics*, vol. 24, no. 6, pp. 1654-1665, June 2009.

[33] Yuancheng Ren, Ming Xu, Jinghai Zhou and F. C. Lee, "Analytical loss model of power MOSFET," in IEEE Transactions on Power Electronics, vol. 21, no. 2, pp. 310-319, March 2006.

[34] L. Xue, D. Boroyevich and P. Mattavelli, "Switching condition and loss modeling of GaN-based dual active bridge converter for PHEV charger," 2016 IEEE Applied Power Electronics Conference and Exposition (APEC), Long Beach, CA, 2016, pp. 1315-1322.

[35] Z. Chen, D. Boroyevich, R. Burgos and F. Wang, "Characterization and modeling of 1.2 kv, 20 A SiC MOSFETs," 2009 IEEE Energy Conversion Congress and Exposition, San Jose, CA, 2009, pp. 1480-1487.

A Low Profile High Frequency LED Driving System Based on Aircore Planar Inductor

Yueshi Guan*, Xihong Hu, Shu Zhang, Yijie Wang, Dianguo Xu and Wei Wang
Department of Electrical Engineering, Harbin Institute of Technology, Harbin, China
*E-mail: hitguanyueshi@163.com

Abstract-This paper proposes a high frequency high power density LED driving system. Based on the optimal design method, the soft-switching characteristics of switch and diode can be achieved, meanwhile, the diode stress is reduced. The values of the inductors and capacitors are greatly reduced in high frequency condition. For small values, the aircore planar inductors with low profile can be adopted. To reduce the resistance and corresponding loss, an optimal structure and design method is proposed. To verify the feasibility of the proposed LED driving system, a 20MHz prototype is built.

Keywords— High Frequency, LED Driving System, Planar Inductor.

I. INTRODUCTION

With the advantages of long lifetime and high efficacy, LEDs have become the mainstream of lighting sources. With the fast development of LED production and packaging process, LEDs have been widely used in wide applications, such as street lighting, indoor lighting, plant factory, and aerospace lighting situations [1-7].

Some deep researches have been done for suitable topologies of LED driving systems. For small power conditions, the Flyback topology is widely used, which owns simple structure and small components number [3, 5]. It is also suitable for multi-output conditions. For large power condition, the LLC topology is widely adopted. The primary side switch and secondary side diode can operate in zero voltage switching (ZVS) condition and zero current switching (ZCS) condition respectively. However, for abovementioned converters, they usually operate around tens or hundreds of kilohertz, and the system volume is quite large.

To reduce the volume and improve the power density of LED driving system, the most effective approach is to promote the operating frequency. In recent years, the very high frequency (VHF) converters have been gradually developed [8-10]. In tens of megahertz conditions, the values of the inductors and capacitors can be greatly reduced. In such high switching conditions, the most change aspect is the high switching loss. Thus, the soft-switching characteristics of switch and diode must be guaranteed to reduce the corresponding loss.

The increment in operating frequency can greatly reduce value of magnetic components. Thus, in high frequency conditions, the magnetic core can be avoided and aircore inductors can be used [11-14]. Meanwhile, in some harsh environments, such as aerospace applications, the extreme temperature greatly affect the operating characteristics and life time of LED driving system. The characteristics of magnetic material vary a lot in different ambient temperature conditions. Thus, the LED driving system with aircore inductors can keep a well operating situation in harsh environments.

For aircore inductors, the planar ones can use the copper tracks on PCB as windings. It can help to achieve a low profile system, and make the LED driving system more compact and flexible. Also, the mass of the aircore inductors can be greatly reduced with the absence of magnetic core. Thus, the payload can be improved for the aerospace equipment. However, the winding resistance of inductors also increases a lot in high frequency condition. Thus, the optimal winding structure is necessary to reduce the copper loss.

In this paper, a 20MHz LED driving system is proposed. The detailed design method of the LED driving system is proposed, which makes the switch and diode operating in soft switching modes. Meanwhile, the diode voltage stress is also reduced. The aircore planar inductor is adopted. With an optimal winding structure, the winding loss can be greatly reduced. For the proposed LED driving system, the profile is greatly reduced and power density is greatly improved.

The paper is organized as follows. In Section II, the operating mode and design method of the high frequency circuit are analyzed. The optimal winding structures for aircore planar inductors are presented in Section III. The design method of corresponding parameters is given in detail. A prototype is built and the experimental waveforms are given in Section IV. The conclusion of this paper is given in Section V.

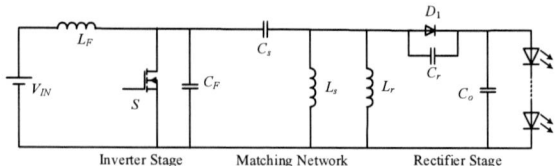

Fig. 1. The topology of proposed high frequency LED driving system.

II. TOPOLOGY OF THE PROPOSED CONVERTER

Fig. 1 shows the topology of proposed high frequency LED driving system. It can be seen that it consists of three parts, namely inverter stage, matching network stage and rectifier stage. The inverter stage transfers the DC component into high frequency AC component and the rectifier stage regulates the AC component into DC output voltage. The matching network stage is adopted here to adjust the equivalent impedance of the rectifier stage to guarantee the desired output power.

During the design procedure, we start from the rectifier stage. Because once the rectifier stage is determined, it can be approximately represented by a certain impedance in fundamental component condition. Then, the inverter stage and matching network stage can be designed based on equivalent rectifier impedance to guarantee the switch ZVS characteristics.

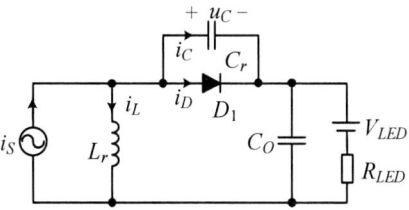

Fig. 2. The circuit of rectifier stage with LED as load.

The diagram of the rectifier stage is shown in Fig. 2. To simply the design procedure of rectifier stage, the forward conduction voltage of the diode is ignored. The parasitic capacitance of diode is absorbed by the resonant capacitor. Meanwhile, the voltage ripple of output voltage is ignored. The input voltage of rectifier stage is the output voltage the inverter stage, which can be approximately represented by a sinusoidal current source $i_S(t) = I_{IN} \sin(\omega t + \varphi)$. The operating modes of the rectifier can be divided into two parts. During $0 < t < (1-D)\,T$, the diode is off and during $(1-D)\,T < t < T$, the diode is on. Here, T is operating period and D is the diode duty ratio when it is on.

During the diode turn-off procedure, based on KVL, the voltage across the diode can be represented by:

$$\frac{d^2 u_C(t)}{dt^2} + \frac{u_C(t)}{L_r C_r} = \frac{di_S(t)}{C_r dt} - \frac{V_{LED} + I_O R_{LED}}{L_r C_r} \quad (1)$$

According to the soft-switching characteristics, the boundary conditions of equation (1) can be obtained.

$$\begin{cases} u_C(t)|_{t=0} = u_C(t)|_{t=(1-D)T} = 0 \\ i_C(t)|_{t=0} = C_r \dfrac{du_C(t)}{dt}\bigg|_{t=0} = 0 \end{cases} \quad (2)$$

Substituting (2) into (1), the expression of diode voltage can be obtained as (3) shown, where $\omega_r = 1/\sqrt{L_r C_r}$. Also, the equation (4) and (5) can be obtained.

$$\left(V_{LED} + I_O R_{LED}\right)\left[\cos\left(\omega_r(1\text{-}D)T\right) - 1\right] + \frac{I_{IN}\omega L_r}{1 - \omega^2/\omega_r^2} \cdot$$
$$\cos\varphi\cos\left(\omega DT\right) - \frac{I_{IN}\omega L_r}{1 - \omega^2/\omega_r^2}\cos\varphi\cos\left(\omega_r(1\text{-}D)T\right) = 0 \quad (4)$$

$$\sin\varphi = 0 \quad (5)$$

Then, during the diode turn-off and turn-on procedures, according to the diode voltage expression, the current of inductor can be represented by (6) and (7).

$$i_{L,off}(t) = I_{IN}\sin(\omega t + \varphi) + \frac{I_{IN}\omega L_r}{1 - \omega^2/\omega_r^2}C_r\omega\sin(\omega t + \varphi) +$$
$$\left(V_{LED} + I_O R_{LED} - \frac{I_{IN}\omega L_r}{1 - \omega^2/\omega_r^2}\cos\varphi\right)C_r\omega_r\sin(\omega_r t) \quad (6)$$

$$i_{L,on}(t) = i_{L,off}(t)\big|_{t=(1-D)T} + \frac{V_{LED} + I_O R_{LED}}{L_r}\left[t - (1-D)T\right] \quad (7)$$

During one period, for the voltage seconds balance rule of inductor and the current seconds balance rule of capacitor, the equations (8) and (9) can be obtained.

$$\left(V_{LED} + I_O R_{LED} - \frac{I_{IN}\omega L_r}{1 - \omega^2/\omega_r^2}\cos\varphi\right)\frac{1}{\omega_r}\sin\left(\omega_r(1-D)T\right) +$$
$$\left(V_{LED} + I_O R_{LED}\right)DT - \frac{I_{IN}L_r}{1 - \omega^2/\omega_r^2}\sin\left(\omega DT\right)\cos\varphi = 0 \quad (8)$$

$$\frac{I_{IN}}{\omega}\cos\varphi\left[1 - \cos\left(\omega DT\right)\right] - \frac{\left(V_{LED} + I_O R_{LED}\right)D^2 T^2}{2L_r} + I_O T = 0 \quad (9)$$

From above analysis, it can be seen that there are nine parameters in these equations, namely I_O, V_{LED}, R_{LED}, I_{IN}, ω, L_r, C_r and D, φ. Once the load parameters, operating frequency and diode duty cycle are determined, the other four parameters can be calculated. The expressions of L_r and I_{IN} are shown as follows.

$$L_r = \frac{\left[K\left(1 - \omega^2/\omega_r^2\right)\left(1 - \cos(2\pi D)\right) + 2\pi^2 D^2\right]\left(V_{LED} + I_O R_{LED}\right)}{2\pi\omega I_O} \quad (10)$$

$$I_{IN} = \frac{2\pi K\left(1 - \omega^2/\omega_r^2\right)I_O}{\left[K\left(1 - \omega^2/\omega_r^2\right)\left(1 - \cos(2\pi D)\right) + 2\pi^2 D^2\right]\cos\varphi} \quad (11)$$

Here, K is defined as (12) shown.

$$K = \frac{\sin\left(2\pi\dfrac{\omega_r}{\omega}(1-D)\right) + 2\pi D\dfrac{\omega_r}{\omega}}{\sin\left(2\pi\dfrac{\omega_r}{\omega}(1-D)\right) + \dfrac{\omega_r}{\omega}\sin(2\pi D)} \quad (12)$$

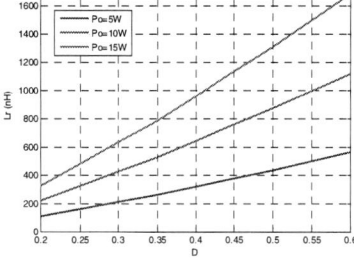

Fig. 3. The curve of resonant inductor in different load and duty cycle conditions.

$$u_C(t) = \begin{cases} \left(V_{LED} + I_O R_{LED} - \dfrac{I_{IN}\omega L_r}{1 - \omega^2/\omega_r^2}\cos\varphi\right)\cos(\omega_r t) + \dfrac{I_{IN}\omega L_r}{1 - \omega^2/\omega_r^2}\cos(\omega t + \varphi) - \left(V_{LED} + I_O R_{LED}\right), & 0 \le t < (1-D)T \\ 0, & (1-D)T \le t \le T \end{cases} \quad (3)$$

According to (4), (5), and (8), the value of ω_r / ω is determined only by the specific value of duty cycle D. Thus, the parameter K is only related with parameter D.

Then, based on above equations, the resonant inductor and capacitor value can be calculated in different duty cycle and output power conditions. The curve of L_r is shown in Fig. 3. From Fig. 3, it can be seen that for a certain output power, the inductance forms a proportional relationship with the duty cycle. On the other hand, at the same duty cycle condition, the inductance increases in larger output power condition. It can be concluded that a larger duty cycle is not expected in the LED driver, because that large duty cycle means large inductance, which increases the loss.

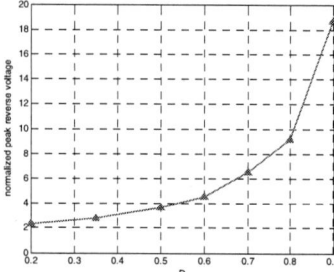

Fig. 4. The curve of normalized diode voltage stress in different duty cycle conditions.

Another very important property of the rectifier stage is the diode voltage stress. According to (3), the diode normalized voltage stress can be represented by (13).

$$U_{C,n} = \frac{U_{C\text{-max}}}{V_O} = \max\left(\left|(1-K)\cos(\omega_r t) + K\cos(\omega t) - 1\right|\right) \quad (13)$$

Fig. 4 shows the curves of the normalized diode voltage stress in different duty cycle conditions. It can be seen that small duty cycle can also help to reduce the diode voltage stress. With a certain duty cycle, inductance and capacitance, the equivalent impedance Z_R of rectifier stage can be obtained based on the fundamental component analysis.

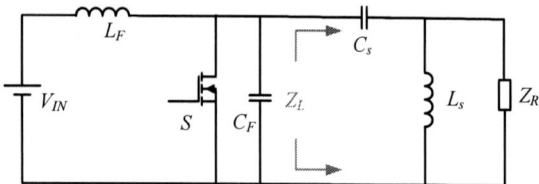

Fig. 5. The equivalent circuit of the inverter and matching network stages.

Fig. 5 shows the equivalent circuit of the inverter and matching network stages. The matching network is composed of C_S and L_S. With the help matching network, the rectifier impedance Z_R can be adjusted to be Z_L, which meets the output power requirement. For the inverter stage, the impedance across the switch decides the shape of voltage waveform, where the switch is expected to operate in ZVS condition. Thus, the impedance is expected to be inductive and with a phase angle θ around 30 to 60 degree. The phase angle forms a relationship with other parameters as shown in (14). With the help of above equations, the system parameters can be finally decided.

$$\tan(\theta) = \frac{Z_L}{\omega L_F} - C_F \omega Z_L \quad (14)$$

III. OPTIMAL DESIGN OF PLANAR INDUCTOR

In high frequency condition, the aircore planar inductor can be adopted, which greatly reduces the profile of the system and improves system power density. To reduce the resistance of the winding, an optimal winding structure is proposed in this paper, and the diagram is shown in Fig. 6. It can be seen that the track width gradually increases from inner to outer. Here, we start from the DC condition. Fig. 6(a) shows the simplified winding structure, where it keeps a common ratio $a=(r_{iw}+w_1)/r_{iw}$ between adjacent tracks. Thus, (15) can be obtained.

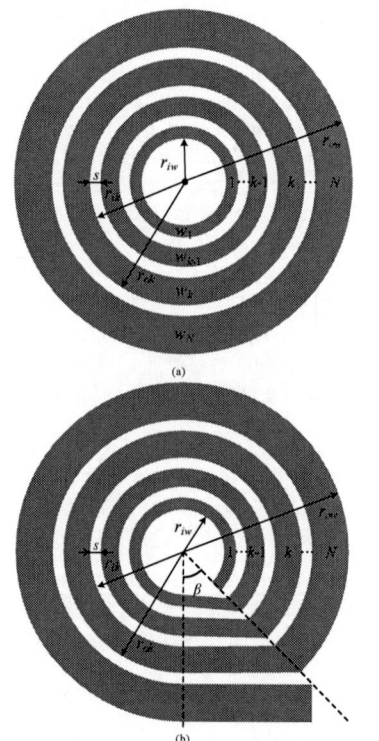

Fig. 6. The winding structure of the proposed planar inductor.

$$\begin{cases} a^N + a_s \sum_{i=1}^{N-1} a^i = \dfrac{r_{ow}}{r_{iw}} \\ a = \dfrac{r_{ok}}{r_{ik}}, \quad a_s = \dfrac{s}{r_{iw}} \end{cases} \quad (15)$$

However, taking into consideration the actual condition, a transition straight conductor is needed to connect the two turns, as shown in Fig. 6(b). The angle β should be calculated to achieve the minimum resistance value. The DC resistance of the arc in k-th turn can be calculated as:

$$R_{k\_(2\pi-\beta)} = \frac{(2\pi-\beta)\rho}{t \ln a} \quad (16)$$

Also, the DC resistance of the chord in k-th turn is:

$$R_{k\_\beta} = \frac{\rho\sqrt{1+(a+a_s)^2 - 2(a+a_s)\cos\beta}}{t\ln\left(\dfrac{a+1}{2}\right)} \quad (17)$$

Thus, the DC resistance of the whole winding can be obtained as:

$$R_{dc} = \frac{N\rho}{t}\cdot\left(\frac{2\pi-\beta}{\ln a} + \frac{\sqrt{1+(a+a_s)^2 - 2(a+a_s)\cos\beta}}{\ln\left(\dfrac{a+1}{2}\right)}\right) \quad (18)$$

By solving the derivative of equation (18), the extreme points of the winding resistance can be calculated as:

$$\begin{cases} \beta_1 = \arccos\dfrac{\ln^2\left(\frac{a+1}{2}\right) + \sqrt{\ln^4\left(\frac{a+1}{2}\right) - \ln^2 a\left[\left(\ln^2\left(\frac{a+1}{2}\right) - \ln^2 a\right)(a+a_s)^2 + \ln^2\left(\frac{a+1}{2}\right)\right]}}{(a+a_s)\ln^2 a} \\[4mm] \beta_2 = \arccos\dfrac{\ln^2\left(\frac{a+1}{2}\right) - \sqrt{\ln^4\left(\frac{a+1}{2}\right) - \ln^2 a\left[\left(\ln^2\left(\frac{a+1}{2}\right) - \ln^2 a\right)(a+a_s)^2 + \ln^2\left(\frac{a+1}{2}\right)\right]}}{(a+a_s)\ln^2 a} \end{cases}$$
$$(19)$$

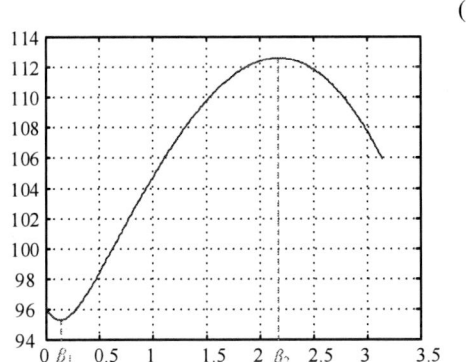

Fig. 7. The resistance curve in different β conditions.

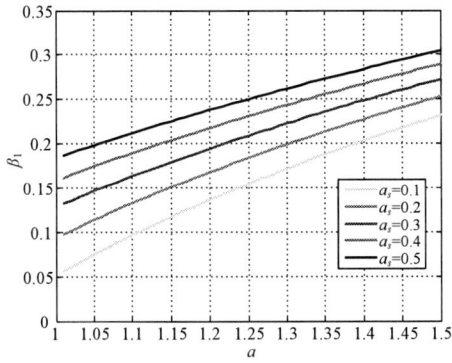

Fig. 8. The resistance curve in different a and β conditions.

Based on above equations, Fig. 7 shows the resistance curve with different β when r_{iw}=2mm, r_{ow}=6mm, N=5 and s=0.2mm (a=1.1886, a_s=0.1) to verify the correctness of the above analysis. Also, it is necessary to slightly adjust β_1 to promise the fabrication of PCB, although sacrificing part of quality factor. Fig. 8 shows the value of β_1 with different a and a_s, it can be seen that β_1 forms an approximately proportional relationship with a, and β_1 also increases with the increase of a_s.

TABLE I
THE SYSTEM PARAMETERS

Symbol	Description	Value
C_F	Resonant capacitor	390pF
L_F	Resonant inductor	56nH
C_s	Resonant capacitor	330pF
S	Switch	SI7454
D	Diode	STPS2H100A
C_r	Resonant capacitor	150pF
L_{rec}	Parallel of L_r and L_s	110nH

IV. EXPERIMENTAL RESULTS

Based on the proposed topology, a 9 W LED driving system is built in this paper. The input voltage is 12V and output voltage is 27 V. The operating frequency is set to be 20 MHz. Based on the design method of the proposed converter, the system parameters can be calculated, which are shown in TABLE I.

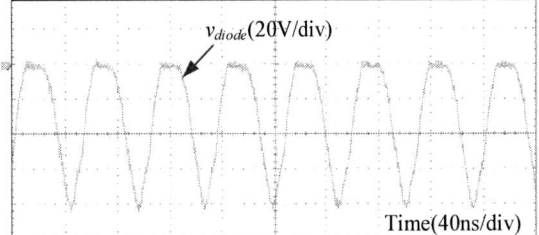

Fig. 9. The diode voltage waveforms of rectifier stage.

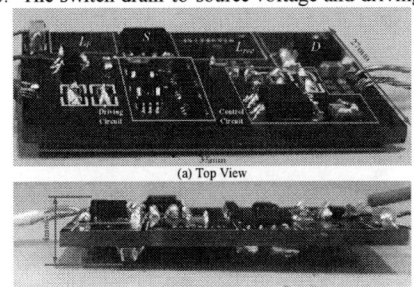

Fig. 10. The switch drain-to-source voltage and driving voltage.

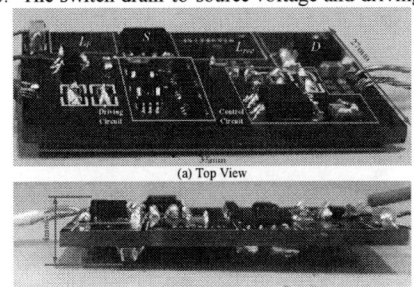

(a) Top View

(b) Side View

Fig. 11. The top and side view of the prototype.

Fig. 9 shows the diode voltage waveforms of the rectifier stage. The duty cycle of the diode is about 0.35. From the Fig. 9, it can be seen that the diode operates in ZCS turn-off condition, which reduces the switching loss. Fig. 10 shows the switch drain-to-source voltage and the driving voltage. It can be seen that the switch can operate in ZVS condition. Fig. 11 shows the top and side view of the prototype, the height of the LED driver is 4 mm, the power density is as high as 45 W/inch³, and the efficiency is 80.2%.

617

V. CONCLUSION

This paper proposes a low profile high frequency LED driving system which is based on aircore planar inductors. The parameters design method is derived in detail. An optimal winding structure is proposed, which can effectively reduce the corresponding resistance. A 20 MHz prototype is built which verifies the feasibility of the proposed LED driving system. The power density is 45 W/inch^3 and the efficiency is 80.2%.

ACKNOWLEDGEMENT

This work is supported by the National Key Research and Development Program of China under Grant 2017YFB0402800.

REFERENCES

[1] J. M. Alonso, M. S. Perdigão, M. A. Dalla Costa, G. Martínez and R. Osorio, "Analysis and Experiments on a Single-Inductor Half-Bridge LED Driver With Magnetic Control," *IEEE Trans. Power Electron.*, vol. 32, no. 12, pp. 9179-9190, Dec. 2017.

[2] G. M. Soares, J. M. Alonso and H. A. C. Braga, "Investigation of the Active Ripple Compensation Technique to Reduce Bulk Capacitance in Off-line Flyback-Based LED Drivers," *IEEE Trans. Power Electron.*, vol. PP, no. 99, pp. 1-1.

[3] X. Xie, J. Wang, C. Zhao, Q. Lu, and S. Liu, "A novel output current estimation and regulation circuit for primary side controlled high power factor single-stage flyback LED driver," *IEEE Trans. Power Electron.*, vol. 27, no. 11, pp. 4602–4612, Nov. 2012.

[4] J. M. Alonso, M. S. Perdigão, G. Z. Abdelmessih, M. A. Dalla Costa and Y. Wang, "SPICE Modeling of Variable Inductors and Its Application to Single Inductor LED Driver Design," *IEEE Trans. Industr. Electron.*, vol. 64, no. 7, pp. 5894-5903, July 2017.

[5] Y. C. Chuang, Y. L. Ke, H. S. Chuang, and C. C. Hu, "Single-stage power factor-correction circuit with flyback converter to drive LEDs for lighting applications," in *Proc. Rec. IEEE IAS Annu. Meeting*, 2010, pp.1–9.

[6] J. M. Alonso, M. Perdig£o, M. A. Dalla Costa, G. Martínez and R. Osorio, "Analysis and design of a novel variable-inductor-based LED driver for DC lighting grids," in *Proc. 2016 IEEE Industry Applications Society Annual Meeting*, Portland, OR, 2016, pp. 1-8.

[7] S. Moon, G. B. Koo, and G. W. Moon, "A new control method of interleaved single-stage flyback ac–dc converter for outdoor LED lighting systems," *IEEE Trans. Power Electron.*, vol. 28, no. 8, pp. 4051–4062, Aug. 2013.

[8] J. Warren, K. Rosowski, and D. Perreault, "Transistor selection and design of a VHF DC-DC power converter," *IEEE Trans. on Power Electron.*, vol. 23, pp. 27–37, 2008.

[9] J. M. Burkhart, R. Korsunsky, and D. J. Perreault, "Design methodology for a very high frequency resonant boost converter," *IEEE Trans. on Power Electron.*, vol. 28, no. 4, pp. 1929–1937, 2013.

[10] J. Hu, "Design of a low-voltage low-power dc-dc hf converter", M.S. thesis, Dept. Elect. Eng. Comput. Sci.,Massachusetts Institute of Technology (MIT), Cambridge, 2008.

[11] S. Michelis, F. Faccio, P. Jarron and M. Kayal, "Air core inductors study for DC/DC power supply in harsh radiation environment," in *Proc. 2008 Joint 6th International IEEE Northeast Workshop on Circuits and Systems and TAISA Conference*, Montreal, QC, 2008, pp. 105-108.

[12] R. C. N. Pilawa-Podgurski, A. D. Sagneri, J. M. Rivas, D. I. Anderson and D. J. Perreault, "Very-High-Frequency Resonant Boost Converters," *IEEE Trans. on Power Electron.*, vol. 24, no. 6, pp. 1654-1665, June 2009.

[13] T. Akagi, S. Abe, M. Hatanaka and S. Matsumoto, "An isolated DC-DC converter using air-core inductor for power supply on chip applications," in *Proc. 2015 IEEE International Telecommunications Energy Conference (INTELEC)*, Osaka, 2015, pp. 1-6.

[14] W. Liang, L. Raymond and J. Rivas, "3-D-Printed Air-Core Inductors for High-Frequency Power Converters," *IEEE Trans. on Power Electron.*, vol. 31, no. 1, pp. 52-64, Jan. 2016.

The 2018 International Power Electronics Conference

Analysis and Compensation of Dead-Time Effect in SiC-Device-based High-Switching-Frequency Inverters

Qingzeng Yan[1,2], Xibo Yuan[1*], Xiaojie Wu[2] and Yiwen Geng[2]

1 Department of Electrical and Electronic Engineering, University of Bristol, Bristol, United Kingdom
2 School of Electrical and Power Engineering, China University of Mining and Technology, Xuzhou, China
*E-mail: xibo.yuan@bristol.ac.uk

Abstract—The ultra-fast-switching silicon carbide (SiC) devices enable power inverters to achieve extremely high switching frequencies (e.g. 100kHz). However, as a limiting factor at high switching frequencies, the dead-time of the conventional pulse width modulation (PWM) can cause serious low-frequency current/voltage harmonics and reduce the dc-link voltage utilization. It is timely to investigate the aggravated dead-time effect on the performance of high-switching-frequency inverters with the increasing application of wide-bandgap devices. In this paper, the dead-time effect and its compensation at high switching frequencies are investigated. The voltage loss and the maximum linear modulation index are derived considering both the dead-time and the switching frequency. The increased amplitude of modulation wave by the dead-time compensation is illustrated, indicating a significantly decreased linear modulation region in high-switching-frequency conditions. A dead-time elimination PWM scheme is therefore adopted, which is equivalent to the conventional PWM with dead-time compensation while has a larger linear modulation region but a slight lower efficiency. The efficiency difference can be very small or even negligible at high switching frequencies and high operating powers. The analyses are finally validated by the experimental results at 100kHz on a three-phase inverter with SiC MOSFETs and SiC Schottky diodes.

Keywords—Dead-time, high-switching-frequency, pulse width modulation (PWM), silicon carbide (SiC).

I. INTRODUCTION

With the increasing commercial availability, silicon carbide (SiC) devices have been further proved to be promising alternatives of Si devices, though there is still a long way before their widespread application [1, 2]. The fast switching speed of SiC devices can enhance the switching frequency of the inverter up to hundreds of kHz, reduce the size and weight of passive filters and improve the power density. Besides, a high switching frequency can also provide lower control delay, higher control bandwidth, and better accuracy [3]. However, the dead-time in the conventional pulse width modulation (PWM) can limit the high-switching-frequency operation of SiC-device-based inverters, though it can be very short, e.g. 200ns [4]. The dead-time effect can cause voltage losses, generate low-frequency (mainly 5th and 7th) current/voltage harmonics, and decrease the linear

modulation region with a lower dc-link voltage utilization [5]. The impact of dead-time effect on inverters will be further aggravated if the switching frequency becomes higher [3, 6].

Numerous methods have been proposed for compensating the dead-time effect [3-12]. Since the voltage errors caused by dead-time are closely related to the current polarity, most of the well-known dead-time compensation methods are implemented by adjusting the amplitude of the modulation wave with current polarities [7, 8]. The dead-time compensation performance therefore heavily relies on the precision of the current-polarity detection method, which is challenging for an extremely small current with noise and ripple at zero crossing. Several current-polarity detection methods have been proposed, e.g. detecting the current polarity from filtered currents [7], indirectly obtaining the current polarity by detecting the turn-on/-off state of switching devices with auxiliary circuits [9], estimating the current polarity from the load electrical angle [8], etc. Apart from the dead-time compensation methods depending on current polarities, there are also many other methods, e.g. the adaptive feed-forward dead-time compensation [10], the dead-time compensation using the integrator output of the synchronous d-axis current PI controller [5], the on-line dead-time compensation method using the disturbance observer [11], etc. To compensate the low-frequency-harmonic currents caused by the dead-time effect, the proportional-resonant (PR) controller and the repetitive controller which have high gains at harmonic frequencies can be adopted [12]. However, all the aforementioned dead-time compensation methods will cause the amplitude increment of modulation wave in inverter mode, which has not been addressed in previous studies. As analyzed in this paper, the amplitude increment of the modulation wave can be negligible at low-switching-frequencies, while it is remarkable at high switching frequencies, thus seriously decreasing the linear modulation region. The dead-time elimination PWM in [9] can be a possible solution for avoiding the dead-time issue at high switching frequencies, which distributes the gate drive pulses according to the current polarity without dead-time.

This work was supported in part by the Newton Research Collaboration Programme under Project NRCP/1415/138 and the Natural Science Foundation of Shandong Province under Grant ZR2017BEE019.

The 2018 International Power Electronics Conference

Due to the requirement of current polarity, the implementation of dead-time elimination PWM is more complex, especially at the current zero-crossing, where the distortion issue is evident [13-17]. The current hysteresis control [13] and the control distributing the gate drive pulses according to current references [14] have been presented for the practical implementation of the dead-time elimination PWM. Regarding the minimization of the current zero-crossing distortion, serval solutions have been proposed. The precision of the current zero-crossing measurement was enhanced in [9] and [15] by detecting the conduction of the switching devices, thus suppressing the current zero-crossing distortion. A mixed modulation was proposed in [16] to avoid the current zero-crossing distortion issue, which alternates the conventional PWM and the dead-time elimination PWM respectively in and out of the current zero-crossing area. In [17], the cascade dual buck inverter with phase-shift control was presented, which can eliminate the current zero-crossing distortion in the dead-time elimination PWM. In this paper, the dead-time elimination PWM is implemented by using the modulation wave and the estimated load power factor angle, which is similar to the method in [14] distributing the gate drive pulses according to current references. The current zero-crossing distortion can be minimized by adjusting the estimated load power factor angle. And the high-switching-frequency performance of the dead-time elimination PWM will be focused.

The main contributions of this paper can be given as follows in accordance with the structure of the paper. The influence of dead-time on the maximum linear modulation index is derived in this paper. The increased amplitude of modulation wave by the dead-time compensation is revealed, which will decrease the linear modulation region and the dc-link voltage utilization especially at high switching frequencies. The conventional PWM with dead-time compensation and the dead-time elimination PWM, as two possible solutions for the dead-time effect at high switching frequencies, are compared regarding output current harmonics and inverter efficiency.

II. DEAD-TIME EFFECT ANALYSIS AT HIGH SWITCHING FREQUENCIES

A. Dead-Time Effect Analysis at High Switching Frequencies

The topology of the two-level three-phase voltage-source inverter is shown in Fig. 1, where Q_x^+ and Q_x^- are respectively the upper and the lower SiC MOSFETs in Phase x ($x=a, b, c$). Considering the body diode of the SiC MOSFET usually has relatively larger forward voltage drops and higher reverse-recovery losses compared with the purpose-designed single diode, in practice anti-paralleling a better performance SiC Schottky diode is preferred [18].

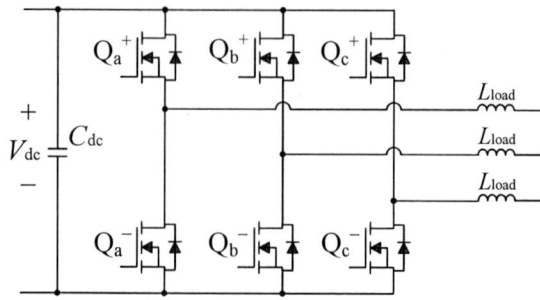

Fig. 1. Topology of the two-level three-phase voltage-source inverter.

The conventional PWM for the inverter in one phase is presented in Fig. 2, where the carrier frequency is ten times that of the modulation wave to achieve a clear figure. When the modulation wave is larger than the triangle carrier, the upper switching device will be turned on, otherwise the lower switching device will be turned on. The dead-time is indispensable in the conventional PWM in order to avoid the shoot-through failure. During the dead-time period, the output voltage is generated by the current freewheeling through diodes, thus causing voltage-error pulses related to the output current polarity, and generating low-frequency harmonics in the output current/voltage contain [5].

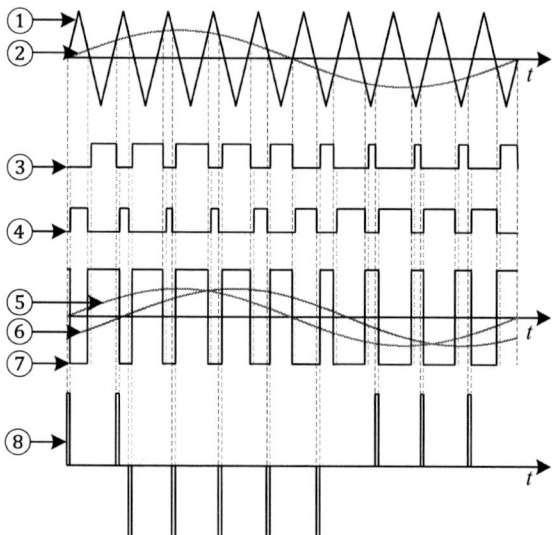

Fig. 2. Modulation mechanism of the conventional PWM [①Carrier, ②Sinusoidal modulation wave, ③Upper drive pulse, ④Lower drive pulse, ⑤Fundamental output voltage, ⑥Output current, ⑦Output voltage, ⑧voltage-error pulse].

Note that, the output voltage in Fig. 2⑦ is the voltage between the midpoint of the dc-link and the output of the phase leg. Therefore, the amplitude of the output voltage equals to half of the dc-link voltage $V_{dc}/2$. To simplify the analysis, the turn-on/-off time and the device voltage drop are neglected. With the dead-time of T_d, one voltage-error pulse with the amplitude of V_{dc} and the width of T_d can be generated in each switching period as seen in Fig.2⑧. When the current is positive (i.e. flowing out of the inverter), negative voltage-error pulses will be generated, making the amplitude of the output voltage lower. Assuming the power factor angle is zero, in the positive-half period of the fundamental voltage/current, there are overall $f_s/(2f_1)$ voltage-error pulses, where f_s and f_1 are the switching frequency and fundamental frequency, respectively. The corresponding voltage-loss area (S_{V_loss}) caused by the voltage-error pulses can be expressed as

$$S_{V\_loss} = V_{dc}T_d \frac{f_s}{2f_1}. \tag{1}$$

Without the dead-time, ideally, the fundamental component of the output voltage should be a sinusoidal wave with an amplitude of $MV_{dc}/2$ (M is the modulation index). The envelope area of the positive-half period of the sinusoidal wave can be given by

$$S_{half} = \int_0^{\frac{T_1}{2}} \frac{MV_{dc}}{2}\sin(\omega_1 t)dt = \frac{MV_{dc}}{2\pi f_1} \tag{2}$$

where T_1 and ω_1 are the fundamental period and angle frequency, respectively. Define the ratio of S_{V_loss} to S_{half} as the dead-time voltage-loss ratio η, which can be derived from (1) and (2) as

$$\eta = \frac{S_{V\_loss}}{S_{half}} = \frac{\pi T_d f_s}{M} \times 100\%. \tag{3}$$

As seen in (3), the voltage loss caused by dead-time is directly proportional to the dead-time T_d and switching frequency f_s, while reversely proportional to the modulation index M. Fig. 3 is plotted according to (3) with varying T_d and f_s (M=0.9). As seen, the voltage loss increases with the increasing dead-time and switching frequency. At the switching frequency of 10kHz, the voltage losses caused by the dead-time of 1μs is 3.5% of the ideal output voltage. However, the voltage loss will reach up to 34.9% at 100kHz switching frequency. In addition, if the modulation index becomes smaller, the voltage loss caused by dead-time can be further aggravated at high switching frequencies.

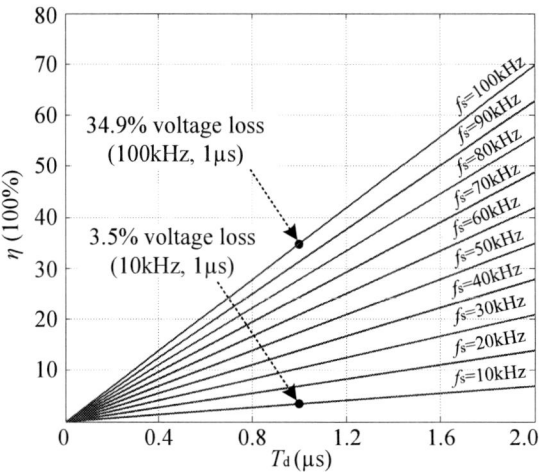

Fig. 3. η with varying T_d and f_s (M=0.9).

B. Influence of Dead-Time and Switching Frequency on the Linear Modulation Region

The modulation of one switching cycle at the top of modulation wave is shown in Fig. 4, where S_a^+, S_a^-, and u_a^* are the drive pulses and the modulation wave in Phase A; T_s refers to the switching period; T_i is the time between the intersections of the modulation wave and the triangle carrier; u_{rm} and u_{cm} are the amplitudes of the modulation wave and the triangle carrier, respectively. The modulation index M can be given as

$$M = \frac{u_{rm}}{u_{cm}}. \tag{4}$$

According to properties of similar triangles, the relationship in (5) can be obtained as

$$\frac{T_i}{T_s} = \frac{u_{cm} - u_{rm}}{2u_{cm}}. \tag{5}$$

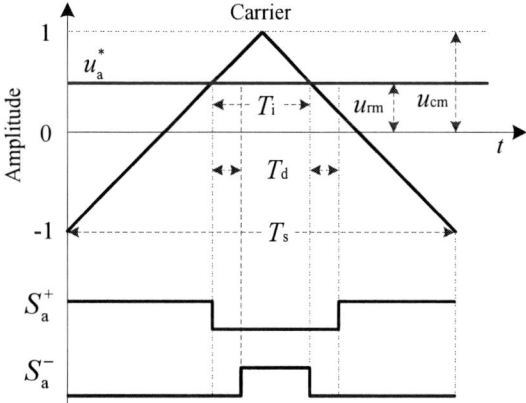

Fig. 4. The modulation of one switching cycle at the top of the modulation wave.

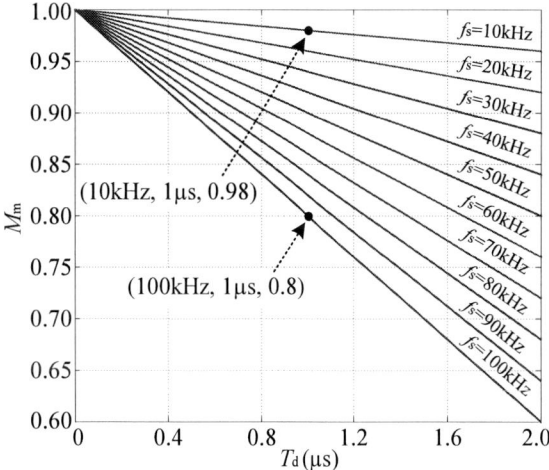

Fig. 5. M_m with varying T_d and f_s.

As illustrated in Fig. 4, if T_i is shorter than T_d, the lower drive pulse S_a^- will keep in low-state, presenting a nonlinear modulation. The maximum linear modulation

621

index M_m can be obtained when $T_i=T_d$, and derived from (4) and (5) as

$$M_m = 1 - 2f_sT_d. \tag{6}$$

Based on (6), the maximum linear modulation index M_m with varying dead-time T_d and switching frequency f_s can be plotted in Fig. 5. As seen, M_m decreases when T_d and f_s increase. At high switching frequencies, a larger impact of T_d on M_m is shown compared with that at low switching frequencies, e.g. the M_m with T_d of 1μs is 0.98 at 10kHz switching frequency, but is remarkably reduced to 0.8 at 100kHz switching frequency. It is verified that, the linear modulation region can be significantly reduced by the dead-time at high switching frequencies.

III. INCREASED AMPLITUDE OF THE MODULATION WAVE BY THE DEAD-TIME COMPENSATION

The dead-time compensation can effectively eliminate the voltage error pulses caused by dead-time and solve the dead-time effect. However, the modulation wave amplitude will be increased by the dead-time compensation, which will be explained in this section.

Taking Phase A for example and assuming a positive current (flowing out of the inverter), the dead-time compensation mechanism is illustrated in Fig. 6. As seen, to compensate the negative voltage-error pulse with the width of T_d (caused by dead-time), the amplitude of Δu needs to be increased in the modulation wave, to make the falling edge of the upper drive pulse (S_a^+) delay by $T_d/2$, while the rising edge move ahead by $T_d/2$. And the edges of the lower drive pulse (S_a^-) change correspondingly as well. Consequently, the positive width of the output voltage is prolonged by T_d, and the negative voltage-error pulse is thus compensated.

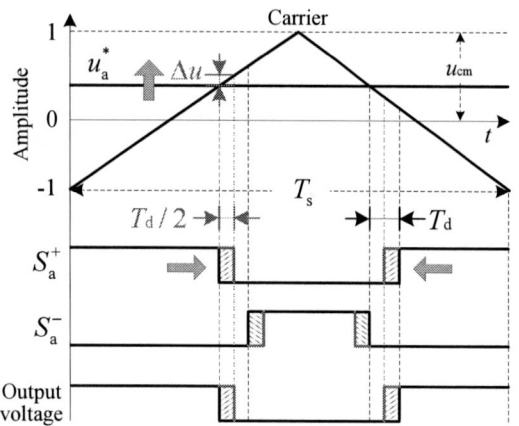

Fig. 6. Dead-time compensation in Phase A.

According to the properties of similar triangles, the increased amplitude Δu of the modulation wave can be derived as

$$\Delta u = \frac{2u_{cm}}{T_s}T_d = 2u_{cm}T_df_s. \tag{1}$$

According to (1), assuming $u_{cm}=1$ and $T_d=1$μs, Δu can be calculated as 0.02 at 10kHz, while at 100kHz Δu

calculated as 0.2. The calculation results indicate that, the increased amplitude of modulation wave caused by the dead-time compensation is small at low switching frequencies, while it can be quite significant at high switching frequencies, which is highly possible to lead to overmodulation and significantly affect the linear modulation region.

Assuming the three-phase inverter operates at 100kHz with an unity power factor, the modulation waves before and after dead-time compensation are shown in Fig. 7. Note that, the adopted dead-time compensation is a commonly-used method presented in [19] which adjusts the modulation wave amplitude according to the current direction. As seen in Fig. 7, due to the amplitude of the carrier is 1, the modulation index M equals to the modulation wave amplitude. The *nonlinear modulation region* is composed of the *overmodulation region* and the *nonlinear modulation region caused by dead-time*. The well-known *overmodulation region* is in the interval of $M>1$, while the *nonlinear modulation region caused by dead-time* is in the interval $0.8<M<1$ ($T_d=1$μs and $f_s=100$kHz) as analyzed in Section II-B. With $M=0.9$, the original modulation waves have already been in the *nonlinear region caused by the dead-time* ($0.8<M<1$). After the dead-time compensation is employed, the amplitude of the modulation wave is increased by 0.2 to compensate the dead-time of 1μs. However, the modulation nonlinearity is further intensified to the *overmodulation region* ($M>1$). To avoid the nonlinearity of the modulation, the modulation index must be reduced which however leads to a lower dc-link voltage utilization.

In addition, it should be noted that, in the rectifier, due to the polarity of the voltage-error pulse caused by dead-time is reversed with the current polarity, the dead-time compensation can make the modulation wave amplitude decrease, providing a larger linear modulation region. The dead-time compensation therefore will not cause overmodulation in the rectifier.

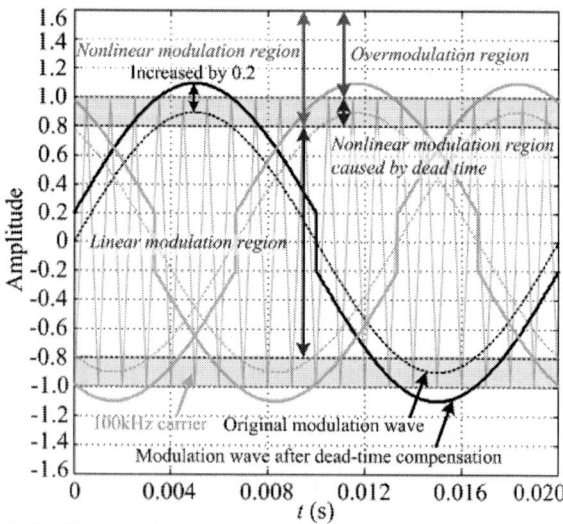

Fig. 7. Modulation waves before and after the dead-time compensation ($T_d=1$μs, $M=0.9$, $f_s=100$kHz).

IV. A DEAD-TIME ELIMINATION PWM SCHEME

The dead-time elimination PWM scheme in [9] can be a solution for avoiding the dead-time effect issue in high switching frequency conditions. The modulation mechanism of the dead-time elimination PWM is illustrated in Fig. 8. As seen, with the dead-time elimination PWM, the drive pulses in half of the fundamental period are enabled or disabled according to the current polarity. Taking the positive-current period as an example, the upper drive pulses are enabled, while the lower drive pulses are disabled. In this period, the negative output voltage are generated by the current freewheeling in body diodes, to guarantee the principle of equivalent accumulated impulse. Due to the alteration of the upper and the lower drive pulses, the dead-time can therefore be eliminated in this modulation scheme. And the output current/voltage distortion can be essentially avoided even in high-switching-frequency conditions. Besides, the abandon of dead-time also reduces the shoot-through possibility of the inverter. A drawback of the dead-time elimination PWM is the requirement of current polarity, which can affect the output voltage/current zero-crossings. The existing current-polarity-detection methods [7-9] used in the dead-time compensation can be directly employed in the dead-time elimination PWM.

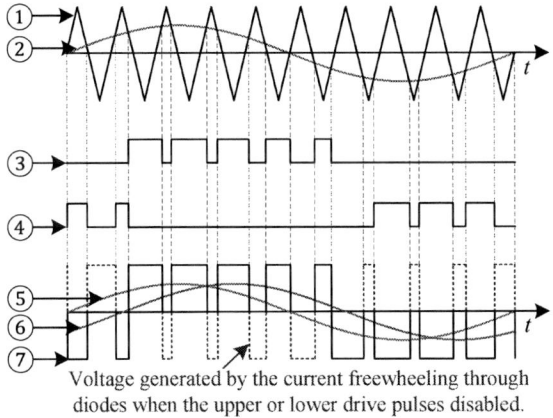

Voltage generated by the current freewheeling through diodes when the upper or lower drive pulses disabled.

Fig. 8. Modulation mechanism of dead-time elimination PWM [① Carrier, ②Sinusoidal modulation wave, ③Upper drive pulse, ④Lower drive pulse, ⑤Fundamental output voltage, ⑥Output current, ⑦Output voltage].

One issue which should be noted is that, in the standard two-level voltage-source inverter with the conventional PWM and MOSFETs, there can be synchronous rectification with reduced conduction losses and improved inverter efficiency [20]. However, no synchronous rectification exists in the dead-time elimination PWM due to the upper and the lower drive pulses are enabled alternatively. Thus the inverter efficiency with the dead-time elimination PWM may be lower than that with the conventional PWM. The output current harmonics and the efficiencies with different PWM schemes will be presented in next section.

V. EXPERIMENTAL RESULTS

A standard two-level three-phase voltage-source inverter has been used to demonstrate the dead-time effect, the PWM with dead-time compensation, and the dead-time elimination PWM. SiC MOSFETs (C2M0080120D, 20A, 1200V) are adopted as the switching devices; SiC Schottky diodes (C4D20120A, 20A, 1200V) are used as anti-parallel diode to bypass the body diodes of the SiC MOSFETs, which have relatively large on-state resistance and high reverse-recovery current. The power from the dc power supply (600V) is transferred via the inverter to the three-phase R-L load. The experiments are carried out at the switching frequency of 100kHz with the load resistance of 30Ω and the inductance of 6.2mH. Apart from the modulation schemes, the experimental conditions are the same. The three-phase currents and their THD using different PWM schemes are presented in Fig. 9. Note that, the parasitic capacitance is indispensable in practical loads, which can be charged/discharged by the very high dv/dt caused by the high-switching speed of SiC devices [21], thus generating the high-frequency harmonics in the currents of Fig. 9. The THD is calculated up to 300kHz which can include both high-frequency harmonics caused by the power-device switching and low-frequency harmonics for reflecting the impact of various modulation schemes.

As seen in Fig. 9(a) with the conventional PWM, the three phase currents are seriously distorted by the voltage-error pulses generated by dead-time. Therefore, the current THD in Fig. 9(a) is the largest among the results in Fig. 9.

With the dead-time compensation, the current THD shown in Fig. 9(b) has been effectively reduced in comparison with that in Fig. 9(a). However, due to the dead-time compensation can cause amplitude increment of modulation wave (analyzed in Section III), the overmodulation happens generating the three-phase currents with flattened top in Fig. 9(b).

Due to the abandon of dead-time, no voltage-error pulses will be generated by the dead-time elimination PWM, and the linear modulation region is extended as well. Therefore with the dead-time elimination PWM, the three-phase currents in Fig. 9(c) have the minimum distortion with the lowest THD.

Besides, with the same experimental conditions and parameters, the output powers of the three PWM schemes are measured as 1.72kW, 2.93kW, and 3.25kW, respectively, verifying a larger dc-link voltage utilization in dead-time elimination PWM.

In order to compare the efficiencies when employing the conventional PWM without dead-time compensation and the dead-time elimination PWM, experiments are carried out at various power levels by changing the load resistance with the same dc-link voltage of 600V. The measured efficiencies are shown in Fig. 10. Due to the compared modulation schemes have different dc-link voltage utilizations, different power samples are shown in Fig. 10 to make the efficiency comparison in the same power range. As seen, at low operating power, the efficiencies with the conventional PWM are much higher

The 2018 International Power Electronics Conference

than those with the dead-time elimination PWM. As the operating power increases, the efficiencies with the two PWM schemes tend to be very close. However, the efficiencies with the conventional PWM are still higher than those with the dead-time elimination PWM.

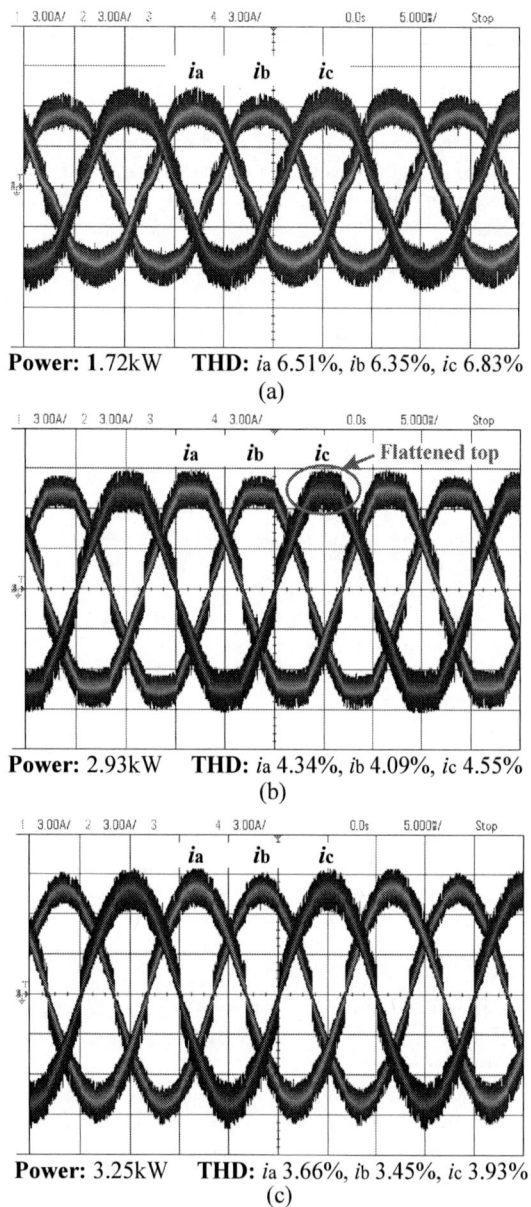

Power: 1.72kW   THD: *i*a 6.51%, *i*b 6.35%, *i*c 6.83%

(a)

Power: 2.93kW   THD: *i*a 4.34%, *i*b 4.09%, *i*c 4.55%

(b)

Power: 3.25kW   THD: *i*a 3.66%, *i*b 3.45%, *i*c 3.93%

(c)

Fig. 9. Three-phase currents and their THD: (a) using conventional PWM (T_d=1μs), (b) using conventional PWM with dead-time compensation, and (c) using dead-time elimination PWM.

The phenomenon can be explained as follows. As aforementioned in Section IV, the synchronous rectification mode exists in the conventional PWM but there is no synchronous rectification with the dead-time elimination PWM. During the synchronous rectification mode, the SiC MOSFET channel bypasses the SiC Schottky diode thus sharing the freewheeling current and

reducing the conduction loss. At low operating powers, due to the synchronous rectification in the conventional PWM, the efficiencies are higher than those with the dead-time elimination PWM. As the operating power becomes larger, the nonlinear conduction characteristic of the diode makes a larger proportion of freewheeling current flows through the SiC Schottky diode, and the proportion of the decreased loss by the synchronous rectification will reduce in the total losses of the inverter. However, the synchronous rectification in the conventional PWM is still beneficial to the efficiencies, thus the efficiencies are close to but always higher than those with the dead-time elimination PWM.

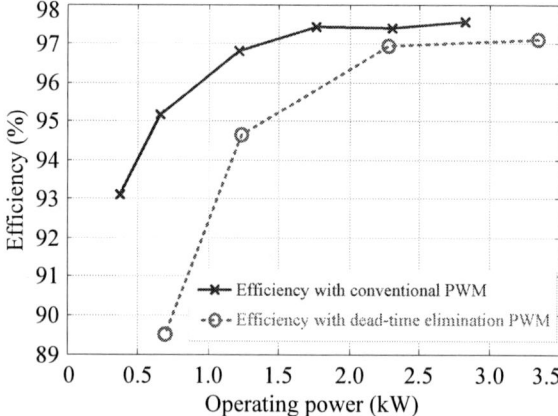

Fig. 10. Efficiencies at different operating powers with the conventional PWM and the dead-time elimination PWM (f_s=100kHz and M=0.9).

VI. CONCLUSIONS

From this paper, it can be seen the dead-time issue for high-switching-frequency operation should be paid careful attention in the inverter design with SiC devices. The investigation at high switching frequencies indicates that, apart from the well-known low-frequency harmonics, the dead-time can also remarkably decrease the linear modulation region as well as the dc-link voltage utilization. Although the dead-time compensation can effectively mitigate low-frequency harmonics, due to the amplitude increment of modulation wave, the linear modulation region can be further reduced at high switching frequencies.

In high switching frequency conditions, the dead-time elimination PWM can be a preferred solution for avoiding the dead-time effect. The experimental results on the SiC-device-based inverter with 100kHz switching frequency show that, the dead-time elimination PWM has less output harmonics, higher linear modulation region, and larger dc-link voltage utilization compared with both the conventional PWM and the conventional PWM with dead-time compensation. However as a main drawback, the synchronous rectification mode is not achievable in the dead-time elimination PWM, which can decrease the efficiency of the inverter with MOSFETs.

624

REFERENCES

[1] J. Biela, M. Schweizer, S. Waffler, and J. W. Kolar, "SiC versus Si—Evaluation of potentials for performance improvement of inverter and DC–DC converter systems by SiC power semiconductors," *IEEE Trans. Ind. Electron.*, vol. 58, no. 7, pp. 2872–2882, Jul. 2011.

[2] X. Yuan, N. Oswald, and P. Stark, "Superjunction MOSFETs in voltage-source three-level converters: experimental investigation of dynamic behavior and switching losses," *IEEE Trans. Power Electron.*, vol. 30, no. 12, pp. 6495–6501, Dec. 2015.

[3] T. Mannen and H. Fujita, "Dead-time compensation method based on current ripple estimation," *IEEE Trans. Power Electron.*, vol. 30, no. 7, pp. 4016–4024, Jul. 2015.

[4] Z. Zhang, F. Wang, D. J. Costinett, L. M. Tolbert, B. J. Blalock, and H. Lu, "Dead-time optimization of SiC devices for voltage source converter," in *Proc. IEEE Appl. Power Electron. Conf. Expo.*, Charlotte, NC, Mar. 2015, pp. 1145–1152.

[5] S.-H. Hwang and J.-M. Kim, "Dead-time compensation method for voltage-fed PWM inverter," *IEEE Trans. Energy Convers.*, vol. 25, no. 1, pp. 1–10, Mar. 2010.

[6] A. C. Oliveira, C. B. Jacobina, and A. M. N. Lima, "Improved dead-time compensation for sinusoidal PWM inverters operating at high switching frequencies," *IEEE Trans. Ind. Electron.*, vol. 54, no. 4, pp. 2295–2304, Aug. 2007.

[7] A. R. Munoz and T. A. Lipo, "On-line dead-time compensation technique for open-loop PWM-VSI drives," *IEEE Trans. Power Electron.*, vol. 14, no. 4, pp. 683–689, Jul. 1999.

[8] H. Mese and A. Ersak, "Compensation of nonlinear effects in three-level neutral-point-clamped inverters based on field oriented control," in *Proc. IEEE 25th Power Electron. Motion Control Conf.*, Novi Sad, Sep. 2012, pp. DS2c.2-1–DS2c.2-8.

[9] L. Chen and F. Peng, "Dead-time elimination for voltage source inverters," *IEEE Trans. Power Electron.*, vol. 23, no. 2, pp. 574–580, Mar. 2008.

[10] M. A. Herran, J. R. Fischer, S. A. Gonzalez, M. G. Judewicz, and D. O. Carrica, "Adaptive dead-time compensation for grid-connected PWM inverters of single-stage PV systems," *IEEE Trans. Power Electron.*, vol. 28, no. 6, pp. 2816–2825, Jun. 2013.

[11] H.-S. Kim, H.-T. Moon, and M.-J. Youn, "On-line dead-time compensation method using disturbance observer," *IEEE Trans. Power Electron.*, vol. 18, no. 6, pp. 1336–1345, Nov. 2003.

[12] Y. Yang, K. Zhou, H. Wang, and F. Blaabjerg, "Harmonics mitigation of dead-time effects in PWM converters using a repetitive controller," in *Proc. IEEE Appl. Power Electron. Conf. Expo.*, Charlotte, NC, Mar. 2015, pp. 1479–1486.

[13] Z. Yao, L. Xiao, and Y. Yan, "Dual-buck full-bridge inverter with hysteresis current control," *IEEE Trans. Ind. Electron.*, vol. 56, no. 8, pp. 3153–3160, Aug. 2009.

[14] P. Sun, C. Liu, J.-S. Lai, C.-L. Chen, and N. Kees, "Three-phase dual-buck inverter with unified pulsewidth modulation," *IEEE Trans. Power Electron.*, vol. 27, no. 3, pp. 1159–1167, Mar. 2012.

[15] Y.-K. Lin and Y.-S. Lai, "Dead-time elimination of PWM-controlled inverter/converter without separate power sources for current polarity detection circuit," *IEEE Trans. Ind. Electron.*, vol. 56, no. 6, pp. 2121–2127, Jun. 2009.

[16] Y. Wang, Q. Gao, and X. Cai, "Mixed PWM for dead-time elimination and compensation in a grid-tied inverter," *IEEE Trans. Ind. Electron.*, vol. 58, no. 10, pp. 4797–4803, Oct. 2011.

[17] P. Sun, C. Liu, J.-S. Lai, and C.-L. Chen, "Cascade dual buck inverter with phase-shift control," *IEEE Trans. Power Electron.*, vol. 27, no. 4, pp. 2067–2077, Apr. 2012.

[18] Q. Yan, X. Yuan, Y. Geng, A. Charalambous, and X. Wu, 'Performance evaluation of split output converters with SiC MOSFETs and SiC Schottky Diodes', *IEEE Trans. Power Electron.*, vol. 32, no. 1, pp. 406–422, Jan. 2017.

[19] L. Ben-Brahim, "The analysis and compensation of dead-time effects in three phase PWM inverters," in *Proc. IEEE IECON'98*, Aachen, 1998, pp. 792–797.

[20] W. Feng, F. C. Lee, P. Mattavelli, and D. Huang, "A universal adaptive driving scheme for synchronous rectification in LLC resonant converters," *IEEE Trans. Power Electron.*, vol. 27, no. 8, pp. 3775–3781, Nov. 2012.

[21] Z. Zhang, F. Wang, L. M. Tolbert, B. J. Blalock, and D. J. Costinett, "Evaluation of switching performance of SiC devices in PWM inverter-fed induction motor drives," *IEEE Trans. Power Electron.*, vol. 30, no. 10, pp. 5701–5711, Oct. 2015.

Control and Performance of New Asymmetrical Operation for Switched-Capacitor-based Resonant Converters

Hadi Setiadi* and Hideaki Fujita

Department of Electrical and Electronic Engineering, Tokyo Institute of Technology, Tokyo, Japan
*E-mail: setiadi.h.aa@m.titech.ac.jp

Abstract—This paper proposes a new asymmetrical control method for a switched-capacitor-based resonant converter consisting of two half-bridge circuits using four switching devices. While the existing control method only adjusts the switching frequency and phase-shift angle, the proposed method additionally controls the duty ratio of each half-bridge circuit independently. The proposed method makes it possible to adjust the average voltage across the switched capacitor although the resonant current has an asymmetric waveform. The proposed control method also enables to operate the SCRC at a frequency closer to the resonant frequency between the resonant inductor and the switched capacitor than the existing control method. Moreover, the switching devices in three of the four arms are turned off at their minimum current requirement for ZVS operation. Thus, it is possible to reduce the switching power loss significantly in a wide power range. A 350-kHz, 3-kW prototype is constructed and tested in this paper. The experimental result shows an improved power conversion efficiency as high as 98% even under various output voltage conditions.

Keywords—*asymmetrical operation, duty ratio, and switched-capacitor-based resonant converter.*

I. INTRODUCTION

The concept of a switched capacitor was proposed several decades ago [1] which led to a switched-capacitor converter (SCC) topology [2]. Various types of SCCs have been proposed and examined in many papers [3]–[6]. SCCs are mostly found in low-power applications because it has an inherent problem of a spike current at a turn-on transition [7]. The spike current quickly charges or discharges the switched capacitor. This causes a non-negligible power loss.

One of the solutions for the spike current problem is to add a small inductor in series with the switched capacitor [8]. This circuit is now called as a switched-capacitor-based resonant converter (SCRC). Many control methods have been proposed to improve performance of the SCRCs. Looking back at the early developments such as blanking time [9] and duty-cycle control [10], these control methods relied on the zero-current-switching (ZCS) without considering the difference among the supply, load, and switched capacitor voltages. The voltage difference may cause a high-frequency ringing which would bring an additional power loss. On the other hand, a frequency control method has proposed to solve the

problem of the ringing [11]. It can achieve zero-voltage switching (ZVS) to reduce the switching power loss. However, it may suffer a large resonant current and a large on-state power loss because its power factor is considerably low.

The authors have proposed a phase-shift control method to improve the power factor [12]. Furthermore, this control method has the capability of controlling the load voltage to be higher than half the input voltage. However, the control method suffers a high turn-off current.

To solve this problem, a new control method has also been proposed which adjusts the phase-shift angle and the operating frequency at the same time [13]. It can reduce the turn-off current, and improve the efficiency when the load voltage is close to half of the supply voltage. However, the required switching frequency becomes too high when the load voltage deviates from a half of the input voltage, especially at a light load condition. Therefore, this method still causes a large switching power loss in the switching devices as well as a non-negligible core loss in the resonant inductor.

This paper proposes a new asymmetrical control method for an SCRC consisting of two half-bridge circuits using four active switches totally. The proposed control method adjusts not only the switching frequency and phase-shift angle but also the duty ratio of each half-bridge circuit independently. The proposed method makes it possible to adjust the average voltage across the switched capacitor although the resonant current has an asymmetric waveform. Then, the proposed control method makes it possible to achieve ZVS turn-on and turn-off in the all four switches. Especially, three of four arms can be turned off at their minimum current requirement for ZVS operation. The proposed method requires a lower switching frequency than the existing method. As a result, it is possible to reduce the switching power loss in a wider power and voltage range. Moreover, the proposed method can also reduce the core loss in the resonant inductor because it has a low voltage stress across the inductor.

A 3-kW 350-kHz prototype is constructed and tested to evaluate the operating performance of the proposed method in experiments. The existing method [13] and the proposed method are implemented and examined using

The 2018 International Power Electronics Conference

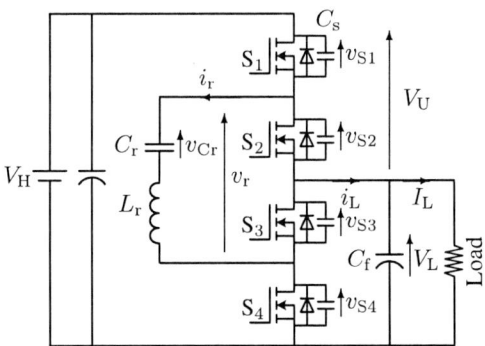

Fig. 1. Switched-capacitor-based resonant converter.

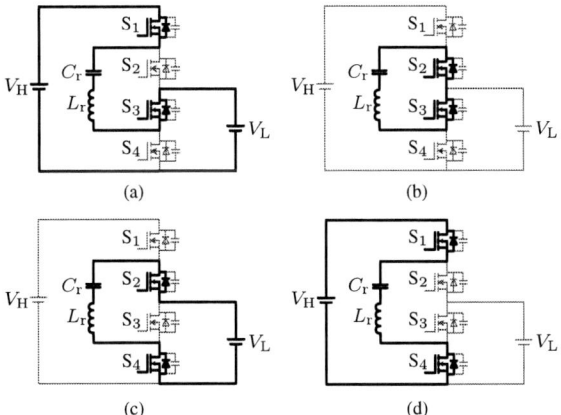

Fig. 2. Switching modes in the SCRC, (a) mode 1, (b) mode 2, (c) mode 3, and (d) mode 4.

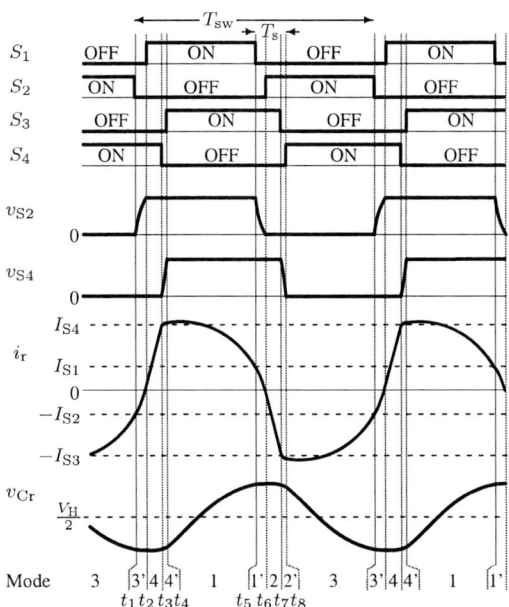

Fig. 3. Waveforms of the SCRC with the frequency and phase-shift control method at $V_L > V_H/2$.

the same main power circuit under various power/voltage conditions. As a result, the proposed asymmetrical control method exhibits an improved conversion efficiency of more than 98% in a wider operating range.

II. SWITCHED-CAPACITOR-BASED RESONANT CONVERTER

A. Circuit Configuration

Fig. 1 illustrates the circuit configuration of an SCRC. The SCRC consists of two half-bridge circuits connected in series. A dc power supply V_H is connected with the series-connected two half-bridge circuits, while a dc load is connected with the lower half-bridge circuit. The dc voltage applied across the upper half-bridge circuit is $V_U = V_H - V_L$, where V_L is the load voltage. A series resonant circuit is connected between the ac terminals of the half-bridge circuits, which is composed of a switched capacitor C_r and a resonant inductor L_r. An external snubber capacitor C_{ext} can be connected between the drain and source terminals of each MOSFET. Considering the drain-to-source output capacitance C_{oss}, the total capacitance value C_s is used in the following discussion, which is represented by $C_s = C_{ext} + C_{oss}$.

B. Control Method

Fig. 2 shows four possible operating modes of the SCRC. The SCRC provides four different voltage levels to the series resonant circuit, namely V_U, 0, V_L, and V_H, according to the switching modes 1, 2, 3, and 4, respectively. Fig. 3 illustrates the voltage and current waveforms of the SCRC with the existing control method, assuming the load voltage is $V_L > V_H/2$. There are additional four modes, which are 1', 2', 3', and 4', representing the operation during the dead time of every switching device. Fig. 4 shows the circuit operation these additional modes. The duration of these modes are adjusted so that each of the switching devices achieve ZVS operation. In general, the existing control method adjusts the phase-shift time between the upper and lower half-bridge circuits, T_s, as well as the switching frequency $f_{sw} = 1/T_{sw}$. Since the duty ratio is fixed at 0.5, the durations of modes 1, 1', 2, and 2' are the same with those in modes 3, 3', 4, and 4', respectively. Therefore, the resonant current i_r has a symmetrical waveform.

In this control method, all four switching devices are turned off with their minimum current requirement to allow the ZVS when $V_L = V_H/2$. This results in a great reduction of the switching power losses [13]. However, only two of them can be turned off with their minimum current requirement for the ZVS transition, and the other two have to be turned off with a higher current, when the load voltage is deviated from $V_H/2$. For example, the turn-off current in the lower half-bridge circuit is much larger than that in the upper one when $V_L > V_H/2$ as shown in Fig. 3. Then, the lower half-bridge circuit should suffer a higher switching power loss. Moreover, a higher switching frequency is required to reduce the excessive turn-off current. This also increases the switching power

627

The 2018 International Power Electronics Conference

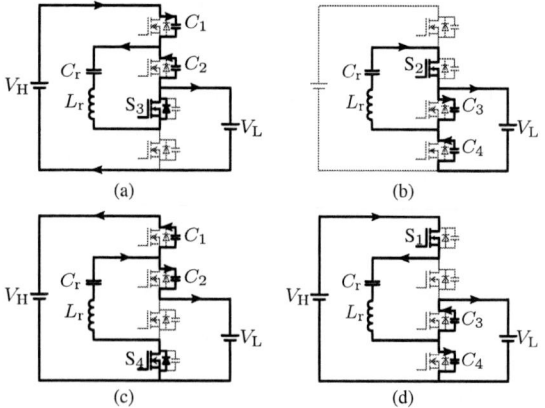

(a)   (b)

(c)   (d)

Fig. 4. Operation of the SCRC during (a) mode 1', (b) mode 2', (c) mode 3' and (d) mode 4'.

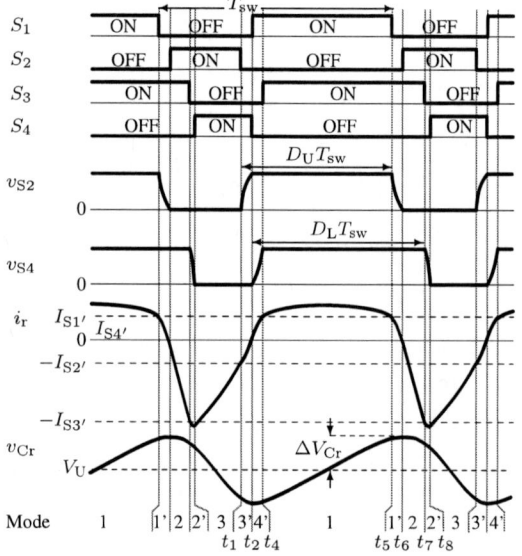

Fig. 5. Waveforms of the SCRC with the asymmetrical control method at $V_L > V_H/2$.

loss. Furthermore, in this case also the voltage stress across the inductor causes a non-negligible core loss.

III. ASYMMETRICAL OPERATION

A. Operating Principle

Fig. 5 shows the voltage and current waveforms of the proposed asymmetrical control method, which assumes that the load voltage is $V_L > V_H/2$. This control method additionally adjusts the duty ratios of the upper and lower half-bridge circuit, D_U and D_L, as well as the operating frequency and the phase-shift time.

The duty ratio control enables to adjust the average switched-capacitor voltage V_{Cr}. The proposed control method controls V_{Cr} at the lower voltage of the two, V_U and V_L. Thus, V_{Cr} is controlled at V_U in the case of Fig. 5. Then, it enables S_1, S_2, and S_4 to be turned off with their minimum current requirement for ZVS operation as

shown in Fig. 5. Therefore, only S_3 has a larger turn-off current than the other switches, which is in comparable level to the peak value of the current in the existing control method.

In mode 1, the applied voltage across the resonant circuit is V_U as shown in Fig. 5 and is close to the V_{Cr}. Since the inductor current i_r has a small change in this mode, it is possible to expand the duration of mode 1 and to decrease the switching frequency. As a result, this control method makes it possible to reduce the switching power loss. Moreover, it is also possible to weaken the voltage stress across the resonant inductor because the applied voltage is quite low in mode 1. Thus, the proposed method enables to reduce the core loss in the resonant inductor.

B. Switching Sequence

As shown in Fig. 5, switches S_2 and S_4 are conducting in mode 3. The mode 3' starts from t_1 when switch S_2 enters the off state with the turn-off current equal to its minimum current requirement for ZVS operation. During mode 3', the resonant current discharges and charges the snubber capacitors of C_1 and C_2 connected in parallel with S_1 and S_2. The SCRC continues the mode 3' until the capacitor voltages across C_1 and C_2 reach zero and V_U, respectively. Since the mode 3' starts with the minimum turn-off current for achieving ZVS operation, the resonant current becomes zero when the snubber capacitor voltage across C_1 reaches zero. At this time, switches S_1 and S_4 are set to enter the on and off state, which makes the SCRC operate with mode 4' from t_2. During mode 4', the snubber capacitors of C_3 and C_4 are discharged and charged. The SCRC continues the mode 4' until the capacitor voltages across C_3 and C_4 reach zero and V_L, respectively. At this time, switch S_3's body diode conducts, indicating that the SCRC starts operating with mode 1 at t_4. During this mode, switch S_3 is turned on with ZVS operation. The mode 1' starts from t_5 when switch S_1 enters the off state with the turn-off current equal to its minimum current requirement for ZVS operation. During the mode 1', the resonant current charges and discharges the snubber capacitors C_1 and C_2. The SCRC continues the mode 1' until the capacitor voltage across C_1 and C_2 reaches V_U and zero, respectively. Since the mode 1' starts with the minimum turn-off current for achieving ZVS operation, the resonant current becomes zero when the snubber capacitor voltage across C_2 reaches zero. At this time, switch S_2 is set to enter the on state, which makes the SCRC operate with mode 2 from t_6. The mode 2 ends when switch S_3 is entering the off state. At this moment, the SCRC starts operating with mode 2'. During mode 2', the snubber capacitors of C_3 and C_4 are charged and discharged. The SCRC continues the mode 2' until the capacitor voltages across C_3 and C_4 reach V_L and zero, respectively. Once the voltage across C_4 reaches zero, the S_4 body diode starts conducting, which indicates the SCRC operates with mode 3. During mode 3, switch S_4 is turned on with ZVS operation.

628

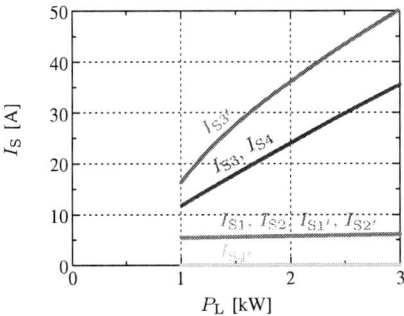

Fig. 6. Comparison of the turn-off current on every MOSFET between the existing and asymmetrical control method at $V_L = 200$ V.

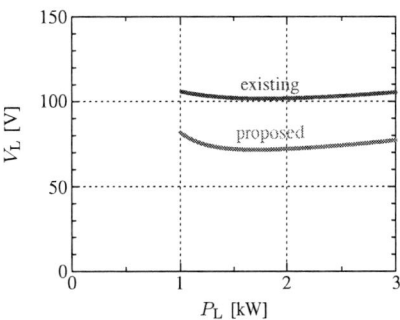

Fig. 8. Comparison of the rms voltage across the resonant inductor between the existing and asymmetrical control method at $V_L = 200$ V.

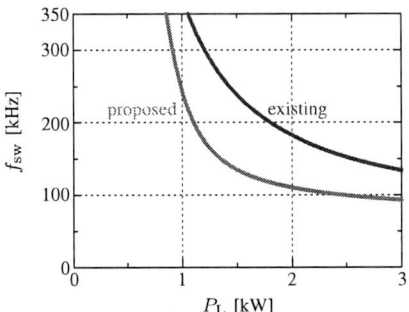

Fig. 7. Comparison of the switching frequency between the existing and asymmetrical control method at $V_L = 200$ V.

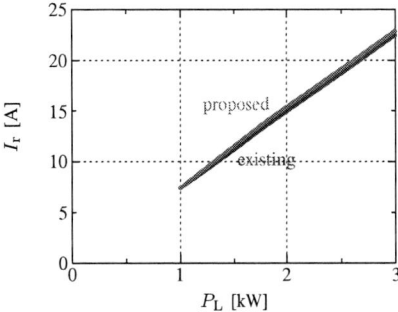

Fig. 9. Comparison of the rms resonant current between the existing and asymmetrical control method at $V_L = 200$ V.

The most important difference between the existing and asymmetrical control method is the existence of mode 4. In the existing method, S_1 and S_4 are conducting at the same time in mode 4. The asymmetrical control method does not apply the mode 4. Without the mode 4, the SCRC starts operating with mode 4' after it operates with mode 3'. The purpose of the elimination of mode 4 is to operate with the average switched-capacitor voltage V_{Cr} at a lower value between V_L and V_U.

C. Estimated Power Loss

Table I shows the circuit parameters for the experiment. No external capacitor is installed in parallel with the MOSFET because its output capacitance is effectively utilized as a snubber capacitor. The loss analysis is performed based on these circuit parameters at the highest load voltage of 200 V to show the advantages of the asymmetrical control method over the existing one.

Fig. 7 shows the comparison of the turned-off current for all of the four MOSFETs on the SCRC. $I_{S1'}$, $I_{S2'}$,

TABLE I.   SCRC's PARAMETERS

Symbol	Parameter	Value
V_H	High-side voltage	300 V
V_L	Load voltage	100–200 V
L_r	Resonant inductor	5.2 μH
C_r	Resonant capacitor	1.2 μF
C_s	Snubber capacitor (parasitic)	2.45 nF
f_{sw}	Switching frequency	87.5–350 kHz

$I_{S3'}$, and $I_{S4'}$ are the turn-off currents in the corresponding MOSFET during the proposed asymmetrical operation. On the other hand, the turn-off currents when the SCRC operates under the existing frequency and phase-shift control method [13] are presented as I_{S1}, I_{S2}, I_{S3}, and I_{S4}. Only one MOSFET S_3 suffers from a high turn-off current under the proposed control method compared with two MOSFETs S_3 and S_4 in the existing control method. The turn-off current for S_4 can be kept zero, and all MOSFETs can achieve ZVS operation under the proposed control method. Even though the turn-off current of S_3 is increasing, the asymmetrical control method has a smaller total turn-off current compared with the total turn-off current under the existing control method. Furthermore, the proposed control method reduces the required switching frequency at the whole operating power range, as shown in Fig. 7. Therefore, the proposed asymmetrical control method has a lower switching loss compared with the existing frequency and phase-shift control method.

Fig. 8 shows the comparison of the rms voltage applied across the resonant inductor. The proposed control method reduces the voltage stress across the resonant inductor at the whole operating range. Since the inductor core loss is proportional to the square of the rms voltage across its terminal, the asymmetrical operation effectively suppresses the resonant inductor core loss compared with the existing control method.

With the reduction of the switching loss and inductor core loss, the remaining unanalyzed loss component on the circuit is the power losses induced by the ESR in all

The 2018 International Power Electronics Conference

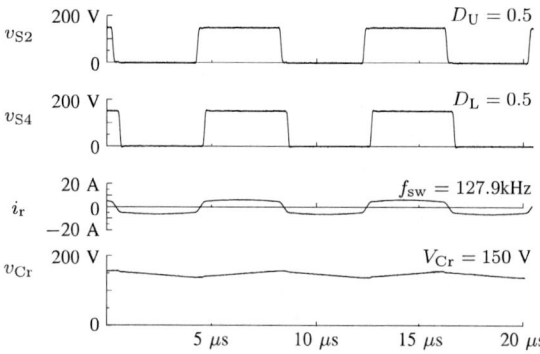

Fig. 10. Experimental waveforms at $V_L = 150V$, $V_H = 300V$ and $I_L = 5A$.

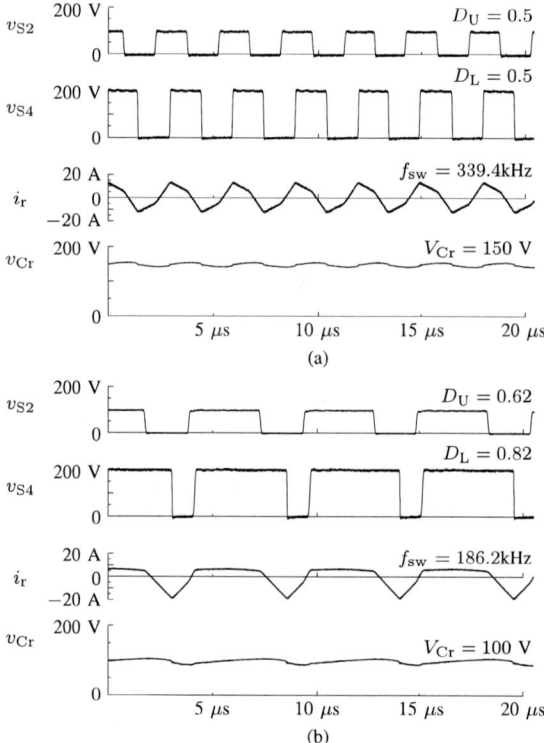

Fig. 11. Experimental waveforms at $V_L = 200V$, $V_H = 300$ V and $I_L = 5$ A, (a) the existing method and (b) the proposed asymmetrical method.

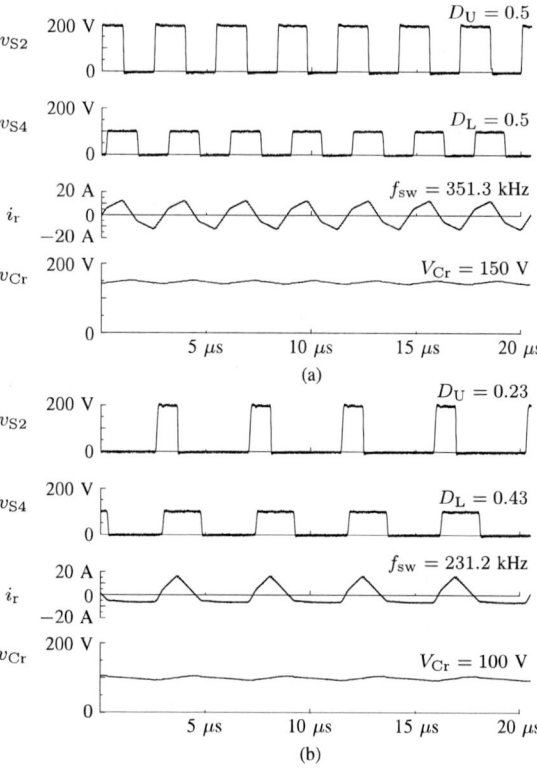

Fig. 12. Experimental waveforms at $V_L = 100$ V, $V_H = 300$ V and $I_L = 5$ A, (a) the existing method and (b) the proposed asymmetrical method.

IV. EXPERIMENTAL RESULTS

Fig. 10 shows experimental waveforms of the proposed control method measured at a voltage ratio of $V_L/V_H = 0.5$. The proposed method adjusted the duty ratio at $D_U = D_L = 0.5$ according to the voltage ratio under this operating condition. Thus, the voltage and current waveforms are almost the same with those in the existing method.

Fig. 11 shows experimental waveforms of the existing and proposed methods measured at a voltage ratio of $V_L/V_H = 200/300$ and a load current of $I_L = 5$ A. The existing method increased the switching frequency to 340 kHz to reduce the peak value of i_r. However, it reached 14 A as shown in Fig. 11(a). On the other hand, the proposed asymmetrical control method increased the duty ratios to $D_L = 0.62$ and $D_U = 0.82$. Then, the average switched-capacitor voltage was well controlled at $V_{Cr} = 100$ V which was almost equal to V_U. Although the peak value of i_r reached 19 A, which was moderately higher with that in Fig. 11(a), the three switching devices were turned off at their minimum current requirement for ZVS operation. Moreover, the switching frequency was also reduced to 190 kHz.

Fig. 12 was measured under the condition of $V_L = 100$ V and $I_L = 5$ A. The proposed method also controlled V_{Cr} at 100 V similarly to Fig. 11(b) by adjusting the duty ratios to $D_U = 0.23$ and $D_L = 0.43$. Accordingly, a positive peak appeared in i_r in Fig. 12(b).

components and tracks. These losses strongly depend on the circulating current inside the SCRC. Fig. 9 shows the comparison of the rms value of the circulating current (or resonant current) when the SCRC operates under the proposed asymmetrical and existing control methods. There is no significant differences between these two rms currents. In other words, the ESR losses between the two control methods are identical. In summing up, the proposed asymmetrical control method is capable to reduce the total power loss compared with the existing control method.

630

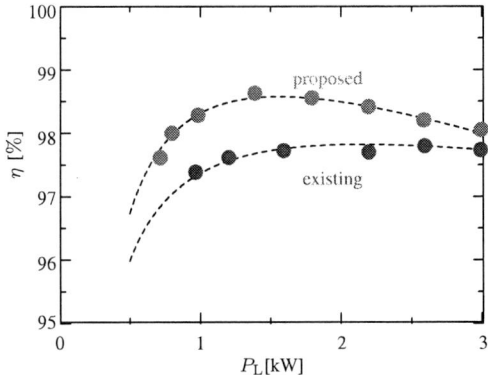

Fig. 13. Power conversion efficiency measured at $V_H = 300$ V and $V_L = 200$ V.

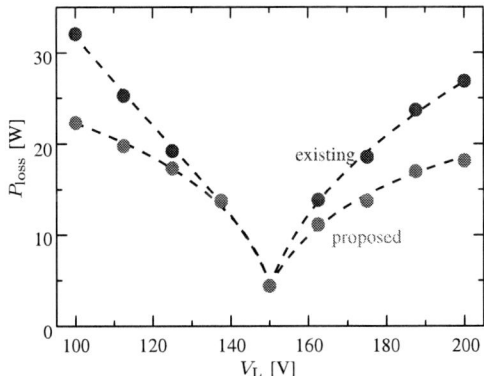

Fig. 14. Measured power loss at $V_H = 300$ V and $I_L = 5$ A.

The waveform was quite similar to i_r in Fig. 11(b) but the polarity was opposite. The switching frequency was also maintained at a low frequency of 230 kHz in the proposed method.

Fig. 13 shows the power conversion efficiency measured at $V_H = 300$ V and $V_L = 200$ V. The existing method had a maximum power efficiency of 97.8% at a load power of $P_L = 2.5$ kW while the proposed method achieved 98.2% in the same load power condition. On the other hand, the highest efficiency reached 98.6% at $P_L = 1.4$ kW in the proposed method compared with 97.6% in the existing one. This means that 1% efficiency improvement is achieved around the middle power range. Fig. 14 is the measured power loss P_{loss} against the load voltage change under a condition of $V_H = 300$ V and $I_L = 5$ A. The proposed asymmetrical control method exhibits a lower power loss than the existing method over the measured range. In other words, the proposed asymmetrical control method makes it possible to operate the SCRC with a high efficiency in a wider load voltage range.

V. CONCLUSION

This paper discussed about the new asymmetrical control method for the SCRC. This control method not only adjusts the switching frequency and phase-shift time, similar to the existing control method [13], but also adjusts the duty ratio of the upper and lower half-bridge circuits

independently. The analysis showed that the asymmetrical control method is capable to reduce the total power loss compared with the existing one by reducing the switching loss in MOSFETs and the core loss in the resonant inductor. The experiment confirmed the operation of the proposed asymmetrical control method with a 3-kW, 350-kHz SCRC prototype. The prototype operated at a wide input-to-output voltage ratio with input voltage of 300 V and output voltage in the range of 100–200 V. The power loss measurement during the experiment proved the loss reduction at the whole operating power and voltage ratio range. As a result, the conversion efficiency can be maintained as high as 98% under the various operating conditions.

REFERENCES

[1] Z. Singer, A. Emanuel, and M.S. Erlicki "Power regulation by means of a switched capacitor," *in Proc. the Institution of Electrical Engineers 1972*, Feb., vol. 119, no. 2, pp. 149–152.

[2] D. Midgley and Z. Sigger "Switched capacitors in power control," *in Proc. the Institution of Electrical Engineers 1974*, Jul., vol. 121, no. 7, pp. 703–704.

[3] H. P. Le, S. R. Sanders and E. Alon, "Design techniques for fully integrated swithed-capacitor DC-DC converters," *IEEE Jurnal of Solid-State Circuits*, vol. 46, pp. 2120–2131, Sep. 2011.

[4] T. M. Andersen, F. Krismer, J. W. Kolar, T. Toifl, C. Menolfi, L. Kull, T. Morf, M. Kossel, M. Brandli, P. Buchmann, P. A. Francese, "4.7 A sub-ns response on-chip switched capacitor DC-DC voltage regulator delivering 3.7W/mm2 at 90% efficiency using deep-trench capacitors in 32nm SOI CMOS," *in Proc. ISSCC 2014*, CA, USA Feb. 2014, pp. 90–91.

[5] D. Kilani, M. Alhawari, B. Mohammad, H. Saleh and M. Ismail, "An efficiency switched-capacitor DC-DC buck converter for self-powered wearable electronics," *IEEE Trans. Circuit and System*, vol. 63, no. 10, pp. 1557–1566, Oct. 2016.

[6] S. Xiong, S. C. Wong, S. C. Tan, and C. K. Tse, "A family of exponential step-down switched-capacitor converter and their applications in two-stage converters," *IEEE Trans. Power Electron.*, vol. 29, no. 4, pp. 1870–1879, Apr. 2014.

[7] Y. Lei and R. C. N. Pilawa-Podgurski, "A general method for analyzing resonant and soft-charging operation of switched-capacitor converters" *IEEE Trans. Power Electron.*, vol. 30, no. 10, pp. 5650–5664, Oct. 2015.

[8] K. W. E. Cheng, "New generation of switched capacitor converters," *in Proc. PESC '98.*, Fukuoka, Japan, May. 1998, vol.2, pp. 1529–1535.

[9] Y. C. Lin and D. C. Liaw, "Parametric study of a resonant switched capacitor DC-DC converter," *in Proc. IEEE TENCON 2001*, Singapore, Aug. 2001, vol. 2, pp. 710–716.

[10] D. Qiu and B. Zhang, "Duty ratio control of resonant switched capacitor DC-DC converter," *in Proc. ICEMS 2005*, Nanjing, China, Sep. 2005, vol. 2, pp. 1138–1141.

[11] M. Shoyama and T. Ninomiya, "Output voltage control of resonant boost switched capacitor converter," *in Proc. PCC-Nagoya 2007*, Nagoya, Japan, Apr. 2007, LS3-4-4, pp. 899–903.

[12] K. Sano and H. Fujita, "Performance of a high-efficiency switched-capacitor-based resonant converter with phase shift control," *IEEE Trans. Power Electron.*, vol. 26, no. 2, pp. 344–354, Feb. 2011.

[13] H. Setiadi and H. Fujita, "Combined frequency and phase-shift control for switched-capacitor-based resonant converter," *in Proc. IPEMC-ECCE Asia 2016*, Hefei, China, May 2016.

The 2018 International Power Electronics Conference

High-Frequency Resonant Converter with Synchronous Rectification for High Conversion Ratio and Variable Load Operation

Lei Gu, Kawin Surakitbovorn, and Juan Rivas-Davila

Department of Electrical Engineering, Stanford University, Stanford, USA

E-mail: {leigu, north, jmrivas}@stanford.edu

Abstract—Applying RF circuit design techniques to dc-dc power conversion has pushed the switching frequency of power converters using discrete components well beyond 10 MHz. This paper proposes a simple solution to implement synchronous rectification for resonant dc-dc power converters operating at above 10 MHz, by using a symmetric bandpass matching network between the inverter and rectifier. The bandpass matching network enables the synchronous switch to be driven by the same clock signal as the inverter switch. Furthermore, this family of high frequency converter is able to maintain zero-voltage-switching operation over a wide load range. A 200 W 13.56 MHz 210 V-to-30 V bidirectional converter with a peak efficiency of 90% is demonstrated.

Keywords—*Class E, Resonant Converter, Synchronous rectification, VHF*

I. INTRODUCTION

A fundamental solution to improving power density, transient response, and realizing higher level of integration of state-of-art power converters is to increase the switching frequency [1]. Resonant power converters with soft-switching characteristics can greatly reduce switching losses and in some configurations have the ability to absorb and utilize some of the components parasitics, so that they can operate in the High Frequency (HF) / VHF (3~300 MHz) range. Fig. 1 shows a basic structure of a resonant dc-dc converter, which consists of a resonant inverter, a transformation stage, and a resonant rectifier. At VHF frequencies, RF power amplifiers like the Class E, Class F, and Class EF or EF^{-1} [2]–[7], with single ground-referenced switch are prefered in dc-dc converters due to their zero-voltage-switching (ZVS) and/or zero-current-switching (ZCS) characteristics and simplicity of driving a ground referenced gate. Previous work has demonstrated successful dc-dc power conversion at switching frequencies greater than 10 MHz, and even beyond 100 MHz [8]–[13]. These works demonstrated some of the expected benefits of VHF power conversion, such as drastically reduced passive component size, much higher power density, increased loop-gain bandwidth, and improved load transient. [14] summarized the fundamental considerations for typical VHF power conversion circuit including component stress factor, control/regulation techniques, and gate driver design. Detail design considerations of HF/VHF dc-dc converter are discussed in [15]. Along the path, efforts to further improve the performances and

include galvanic isolation such as reducing component counts, transformer synthesis, and integrated transistor optimization have been demonstrated in [16]–[19].

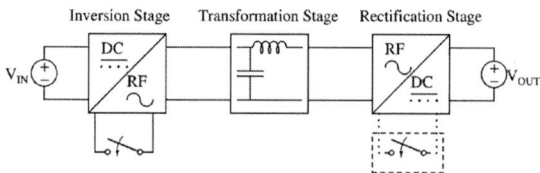

Fig. 1: Structure of a resonant dc−dc converter

In addition to the research efforts mentioned above, synchronous rectification in HF/VHF dc-dc converters is also an interesting topic worth investigating for the design of radio frequency dc-dc converters. In these HF/VHF resonant converters, Schottky diodes are normally used in the rectifier stage, either hard- or soft-switched. Synchronous rectification can reduce conduction loss under certain conditions [20], [21]. Self-driven based gating approaches of synchronous rectifier have been demonstrated in low power and low voltage HF/VHF dc-dc power converters in [22], [23]. Another method of introducing a fixed time delay of the synchronous rectifier gating signal has been demonstrated in a higher power and higher voltage bidirectional converter [24]. However, these approaches need extra resonant circuitry and pose difficulties to precisely drive the synchronous rectification switch, especially at above 10 MHz.

One feature of this type of HF/VHF resonant converter is that the dc-dc gain varies significantly with load [25]. Firstly, this is because the gain of the transformation stage (i.e. a narrow-band matching network), is generally dependent on the load. On the other hand, the design of conventional Class E or higher order tuned Class E power amplifier variations relies on the load resistance to achieve ZVS [2], [5], [7]. As a result, the ZVS characteristic cannot be maintained when the load resistance varies more than 2 times.

In this paper, we focus on the challenge of designing a resonant converter with synchronous rectification operating at above 10 MHz given that the commercially available gate driver ICs have 10-20 ns propagation delay. We presents a

HF/VHF resonant converter topology with synchronous rectification that is capable of maintaining a fixed dc-dc gain across a wide load range. The circuit utilizes a symmetric bandpass matching network for voltage conversion, which maintains zero phase shift between the inverter and rectifier. As a result, the rectifier switch can be driven by the same gate signal as the inverter switch across a wide range of operation. It greatly reduces the difficulty of accurately adjusting the delay of the gating signal for the synchronous switch at above 10 MHz under different load conditions. Compared to the voltage/current sensing based synchronous gating approaches in [22], [23], no extra sensing circuit is required to track the zero crossing of the rectifier voltage or current waveform. To allow design flexibility, both the inverter switch and rectifier switch are designed to maintain ZVS turn-on across a wide load range, but not necessarily zero dv/dt turn-on as usually demonstrated in a conventional Class E power amplifier design.

The rest of this paper is organized as follows. Section II explains the basic operation principles of the proposed VHF power conversion architecture with a single-phase circuit example. Section III shows a two-phase interleaving configuration which can improve the transient performance of a single-phase circuit. The experimental results of a 13.56 MHz prototype are presented in Section IV. Section V concludes the paper.

II. CONVERTER OPERATION

As shown in the Fig. 1, a Radio Frequency (RF) resonant converter consists of three parts: an inverter, an impedance transformation stage, and a rectifier. The inverter generates an RF voltage/current waveform at a switching frequency f_s, while the resonant rectifier converts the ac voltage/current back to dc. It is usually assumed that most of the power is delivered at the fundamental frequency [26], while the harmonic components deliver a negligible amount of power to the load. The transformation stage sets the gain, matches the impedance to deliver the desired amount of power, and can also be used to adjust the phase between the ac voltage and current waveform to achieve ZVS/ZCS [15]. A transformer is normally used to provide a large conversion gain in a resonant converter operating at frequency below 1 MHz [27], [28]. At higher switching frequency (i.e. above 10 MHz), a narrowband matching network is preferred to set the conversion gain due to the limited availability of suitable core materials [29], [30].

Fig. 2a shows the structure of the VHF resonant converter that is the focus of this paper. It consists of a ZVS inverter, a bandpass matching network consisting of two cascaded L matching stages, and a ZVS rectifier. Both the inverter and rectifier have a single ground-referenced switch. The choice of the inverter circuit can be either a Class E, F, or higher order tuned amplifier provided that it is also used in the rectifier stage to maintain symmetry. Fig. 2b shows the schematic of the converter studied here. In this paper, a Φ_2 amplifier were selected. It can generate a high frequency RF signal with ZVS but lower voltage stress on the switch compared with Class E [6].

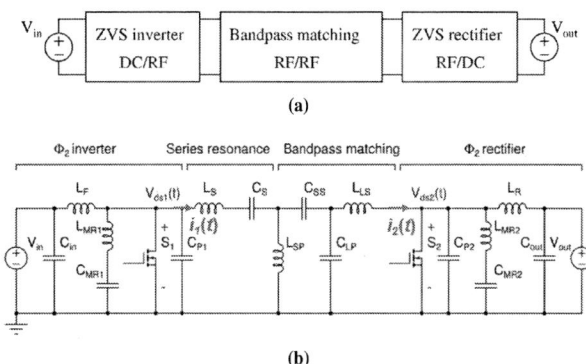

Fig. 2: Proposed converter structure and single phase converter circuit. (a) VHF resonant converter with bandpass matching network; (b) Single phase Φ_2^2 converter with bandpass matching network

One of the reasons of selecting the Φ_2 inverter/rectifier and the bandpass matching network here is due to its lower sensitivity to load variation. In a DC/RF power amplifier with a single ground-referenced switching device, the impedance looking across the semiconductor determines the time domain waveform. For example, Fig. 3 shows the general structure of a power amplifier using a single MOSFET generating an RF signal from a DC supply. The impedance $Z_{DS}(\omega)$ across the drain and source determines the MOSFET's steady-state time domain waveform $V_{DS}(t)$ and $I_{DS}(t)$ [4].

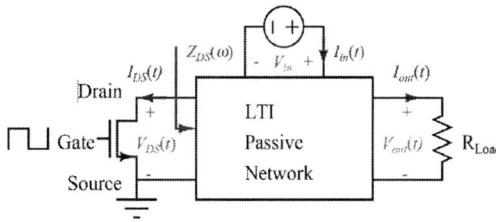

Fig. 3: Power amplifier with single ground-referenced MOSFET

As the MOSFET is being switched on and off periodically at f_s, the impedance Z_{DS} across the drain and source at the fundamental frequency and its harmonic components determines the time domain waveform [5]. Fig. 5 shows an example of a 50 MHz Φ_2 inverter and its ability to main ZVS independent of its output power. Modified from the design guidelines outlined in [31], the general rules to design a Φ_2 inverter in Fig. 4 that can maintain ZVS across a wide resistive load range are:

1) The Φ_2 inverter can be separated into two parts, the multi-resonant network Z_{MR} and the loading network Z_L. The loading network Z_L should always look resistive at f_s.
2) L_S and C_S should resonate at f_s if it is necessary to block DC voltage, otherwise can be eliminated for efficiency improvement. The nominal R_L value is determined by the maximum output power.

3) The impedance magnitude of Z_L should be larger than the multi-resonant ZVS network's impedance Z_{MR} at f_s and its harmonic components at different load conditions.
4) The impedance of Z_{MR} at the fundamental frequency f_s is inductive in phase, while at the third harmonic $3f_s$ is capacitive. At the second harmonic $2f_s$, Z_{MR} is small due to the series resonance between L_{MR} and C_{MR}.

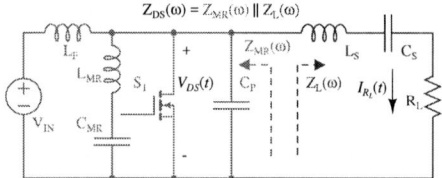

Fig. 4: Φ_2 DC-RF circuit

As shown in Fig. 5a, the ZVS operation can be maintained across a variation in the resistive load of about 10 times the nominal value. [32] explains similar desgin principles in more details and with more practical examples. Fig. 2b also uses a Φ_2 circuit for rectifier stage. If a Φ_2 DC/RF inverter is designed using the principles above, when it is running as a rectifier to convert RF signal back to DC, its effective input impedance will look roughly resistive, i.e., the fundamental component of rectifier input voltage $V_{ds2}(t)$ is roughly in phase with the input current $i_2(t)$. Therefore, the rectifier effectively acts as a resistance at the fundamental frequency. When the dc output power P_{out} changes, the rectifier maintains resistive at the input but effective resistance changes inversely proportional to the power.

A step-down bandpass matching network is connected between the inverter and rectifier. Configured this way, the converter is capable of bidirectional power delivery. At the resonate frequency, a bandpass matching network is working as a narrow-band transformer. When used for resistive impedance matching, it maintains a fixed voltage/resistance conversion ratio independent of the load resistance value. If modeling the rectifier in Fig. 2b as a variable resistive load at the fundamental frequency, we can analyze the RF voltage transfer function between the drain of the two switches using a simplified circuit shown in Fig. 6a. Fig. 6b shows that regardless of the variation of the load, the voltage gain of bandpass matching is always constant at resonant frequency, and the phase shift between the $V_{ds1}(\omega_s)$ and $V_{ds2}(\omega_s)$ maintains zero. In the converter shown in Fig. 2b, because the Φ_2 circuit is used for both the DC/RF and RF/DC conversion, the two drain-to-source voltage $V_{DS}(t)$ would look like a quasi-square waveform as shown in Fig. 5a, and these two voltages are forced to be in phase by the bandpass matching network. As a result, such in-phase relationship causes the zero value of the drain-to-source voltage $V_{DS,S1}(t)$ and $V_{DS,S2}(t)$ to occur during the same time. So S_1 and S_2 can be driven by the same gate clock signal. While being driven by the same gate clock signal, S_1 and S_2 can still maintain ZVS operation across a wide load range. This provides a simple solution to synchronously drive these

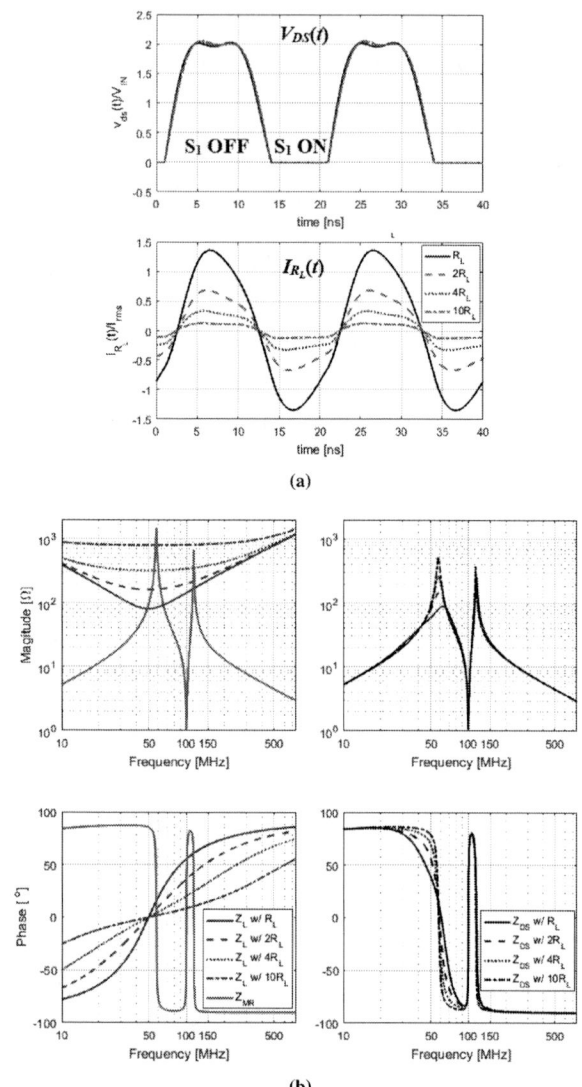

Fig. 5: A Φ_2 inverter operating under variable resistive load. L_F = 80 nH, L_{MR} = 169 nH, C_{MR} = 15 pF, C_P = 73 pF. (a) $V_{ds1}(t)$ waveform, ZVS is achieved when R_L is changed from nonminal load to 10 times large. (b) The impedance of multi-resonant network Z_{MR} and the loading network Z_L on the left, the total impedance Z_{DS} on the right.

two switches at above 10 MHz while still maintaining ZVS. While the matching network is used to match the impedance and convert the voltage, the parasitic resistance in the inductor and the capacitor would generate power losses. The losses of a single-stage L-matching-network is roughly proportional to the voltage conversion ratio [33]. In a case where a high voltage conversion ratio is required, using multiple stages of matching networks reduces the conversion ratio of each stage, which would affect the total losses in the inductors and capacitors. For certain voltage conversion ratio, this two-stage configuration provides the highest conversion efficiency [34].

Therefore, by using a two-stage bandpass matching network with a symmetric Φ_2 inverter/rectifier, this bidirectional VHF dc-dc converter in Fig. 2b can maintain a near-constant high conversion ratio and high efficiency for a wide load range of operation.

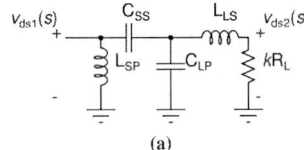

(a)

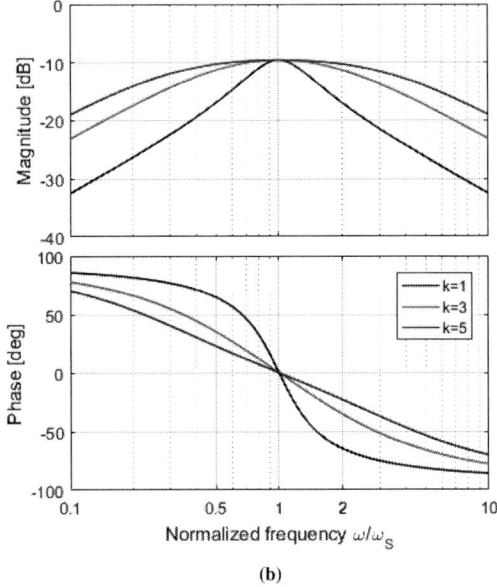

(b)

Fig. 6: Bandpass matching example with variable resistive load, nominal $\frac{V_{ds2}}{V_{ds1}} = \frac{1}{3}$ (a) simplified circuit (b) voltage gain $\frac{v_{ds2}(j\omega)}{v_{ds1}(j\omega)}$ vs frequency ω

III. Implementation

The work in [35] demonstrated a single-phase 64 MHz 36 V-to-12 V prototype converter of the circuit in Fig. 2b. Two eGaN FETs are used in the prototype and driven by the same gate clock. The demonstration experimentally verifies that S_1 and S_2 can maintain ZVS and be driven by the same clock across a wide load range. However, one of the disadvantage in the single-phase configuration is that the current ripple in the input inductor L_F and output inductor L_R is large. As a result, large capacitance is necessary at the input and output, which slows down the load transient. A method to overcome this issue is connecting two of the same converter in Fig. 2b in parallel at the input and output, but two phases are operating 180 degree out-of-phase. One example of this push-pull implementation is shown in Fig. 7.

The components labeled with subscript a belongs to Phase a, and vice versa for Phase b. The two switches in the same

phase are still driven by the same clock signal. However, Phase a and b are operating 180° out-of-phase. The key waveforms of a simulation are shown in Fig. 8. It can be seen that the ripples of input current $i_{in}(t)$ and output current $i_{out}(t)$ are greatly reduced compared with a single phase converter, which reduces the required input/output bypass capacitance and improve the transient performance. Moreover, as the interleaving operation creates differential ground, some of the components can be combined [36]. In Fig. 7, C_{SP} are combined components of the original C_{SPa} and C_{SPb}, which is same case for L_{LP}. The low-Q series resonant circuit L_S-C_S in Fig. 2b now can be eliminated, which was used to provide DC blocking in the single-phase converter. A 2-stage bandpass matching network consists of a low-pass L matching and high-pass L matching. Depending on the order of the low-pass and high-pass, there are two different circuits representation, as shown in Fig 9. These two bandpass matching networks have the same electric behavior.

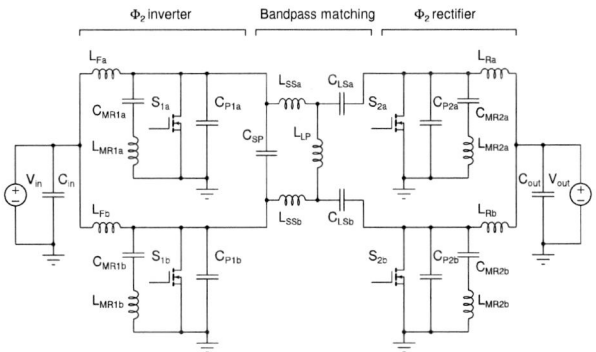

Fig. 7: Two-phase interleaved Φ_2^2 converter schematic

IV. Experimental Results

To experimentally demonstrate the operation principles of the push-pull Φ_2^2 converter in Fig. 7, a 13.56 MHz 200 W 200 V-to-30 V prototype was designed and built. The components of the prototype are listed in Table I. Fig. 10 shows the photograph of this prototype. All the resonant passive components and the gate-drive circuits are placed on one side of the PCB, while the GaN FETs are placed on the other side for better heat extraction.

When the converter is delivering power from the high-voltage to the low-voltage side, the measured drain-to-source voltage waveforms $V_{DS}(t)$ of all the switches are shown in Fig 11. We can see that for the two switches in Phase a, $V_{DS,S1a}(t)$ is in phase with $V_{DS,S2a}(t)$, shown on the top of Fig. 11. The same in-phase relationship is found in the Phase b $V_{DS}(t)$ waveform, shown on the bottom of Fig. 11. Phase a and b are operating 180° out of phase. This phase relationship between the $V_{DS}(t)$ of all the switches holds true under different output power conditions.

In this case, looking at the drain-to-source voltage $V_{DS}(t)$ and gate-to-source voltage $V_{GS}(t)$ of the inverter switches S_{1a}

The 2018 International Power Electronics Conference

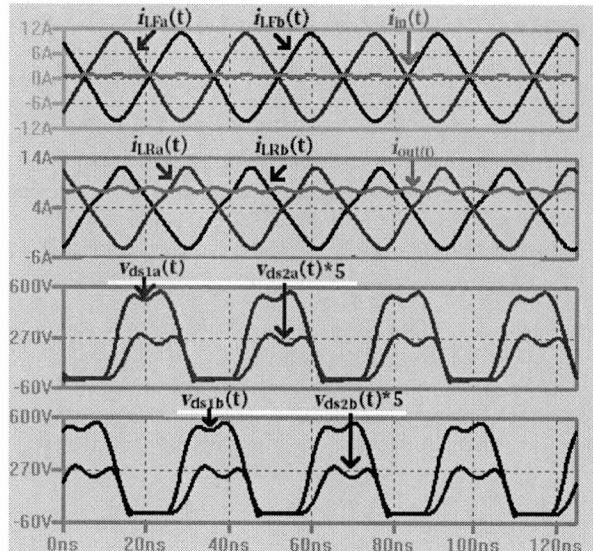

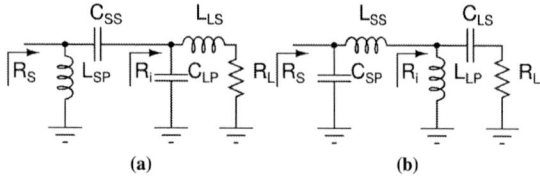

Fig. 8: Key waveforms of a push-pull Φ_2^2 converter, V_{in}=270 V, V_{out}=28 V, P_{out}=200 W

(a)

(b)

Fig. 9: Two bandpass matching network circuits

and S_{1b} in Fig. 12, we are clear that both switches are operating with ZVS.

When the converter is delivering power reversely from the low-voltage to the high-voltage side, the measured drain-to-source voltage waveforms of all the switches $V_{DS}(t)$ are shown

TABLE I: Push-pull Φ_2^2 Prototype BOM

Parameter	Value
Gate drive	UCC27516
$S_{1a,b}$	GS66502B
$S_{2a,b}$	GS61004B
f_s	13.56 MHz
$L_{Fa,b}$	660 nH
$L_{MR1a,b}$	500 nH
$C_{MR1a,b}$	69 pF
$C_{P1a,b}$	56 pF
C_{SP}	78 pF
$L_{SSa,b}$	1550 nH
L_{LP}	476 nH
$C_{LSa,b}$	680 pF
$C_{MR2a,b}$	940 pF
$L_{MR2a,b}$	123 nH
$C_{P2a,b}$	280 pF
$L_{Ra,b}$	50 nH

in Fig 13. It can also be seen that for the two switches in Phase a, $V_{DS,S1a}(t)$ is in phase with $V_{DS,S2a}(t)$, shown on the top of Fig. 13. The same in-phase relationship is found in the Phase b $V_{DS}(t)$ waveform, shown on the bottom of Fig. 13. Phase a and b are operating 180° out of phase. This phase relationship between the $V_{DS}(t)$ of all the switches holds true under different output power conditions.

In the case of delivering power in the reverse direction, looking at the drain-to-source waveform $V_{DS}(t)$ and gate-to-source waveform $V_{GS}(t)$ of the inverter switches S_{2a} and S_{2b} in Fig. 14, we are clear that both switches are operating with ZVS.

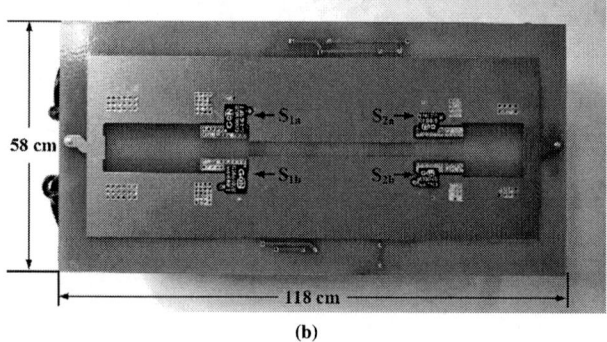

(a)

(b)

Fig. 10: Photograph of 13.56 MHz push-pull Φ_2^2 prototype converter. (a) top of the converter, including all the resonant passives and gate-drive circuits. (b) bottom of the converter, 4 GaN FETs.

V. CONCLUSION

This paper proposes to utilize a bandpass matching network as a impedance transformation stage in a VHF dc-dc converter. The bandpass matching network maintains a close–to–zero phase shift between the inverter and rectifier voltage, enabling the synchronous switch to be driven by the same clock signal as the inverter switch. It provides a simple solution to implement synchronous rectification for HF/VHF dc-dc power conversion at above 10 MHz. This approach also extends the ZVS operation range of all the switches. A two-phase interleaving operation is proposed to improve the efficiency and transient performance of a previously demonstrated single-phase circuit. A 13.56 MHz 210 V-to-30 V prototype converter

636

The 2018 International Power Electronics Conference

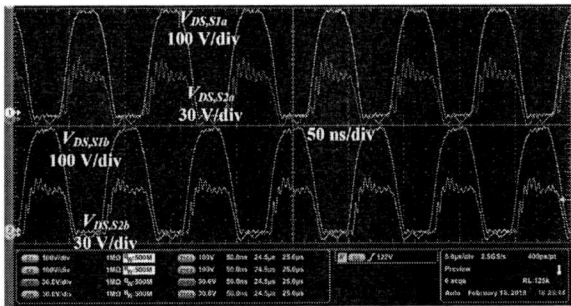

Fig. 11: Forward direction, drain-to-source waveform, V_{in} = 210 V, V_{out} = 30 V, R_{out}=10 Ω, CH1 $V_{DS,S1a}(t)$, CH2 $V_{DS,S1b}(t)$, CH3 $V_{DS,S2a}(t)$, CH4 $V_{DS,S2b}(t)$.

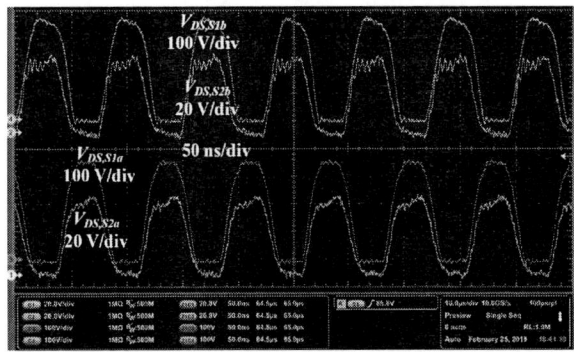

Fig. 13: Reverse direction, drain waveform, V_{in} = 30 V, V_{out} = 193 V, R_{out}=333 Ω, CH1 $V_{DS,S1a}(t)$, CH2 $V_{DS,S1b}(t)$, CH3 $V_{DS,S2a}(t)$, CH4 $V_{DS,S2b}(t)$.

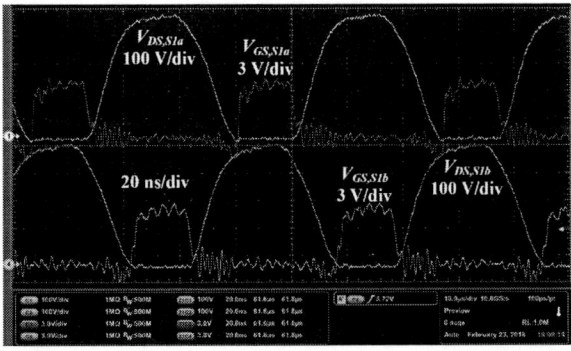

Fig. 12: Forward direction, drain and gate waveform of the high voltage switches, V_{in} = 210 V, V_{out} = 30 V, R_{out}=10 Ω, CH1 $V_{DS,S1a}(t)$, CH2 $V_{DS,S1b}(t)$, CH3 $V_{GS,S1a}(t)$, CH4 $V_{GS,S1b}(t)$.

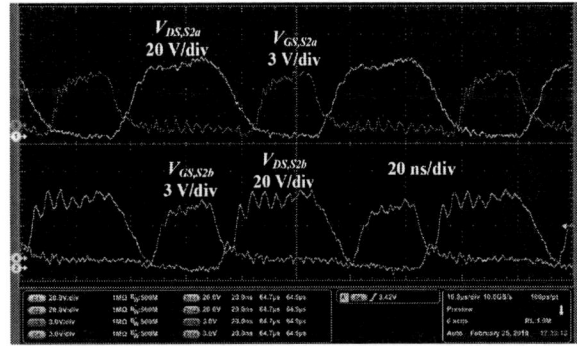

Fig. 14: Reverse direction, drain and gate waveform of low voltage switches, V_{in} = 30 V, V_{out} = 193 V, R_{out}=333 Ω, CH1 $V_{DS,S2a}(t)$, CH2 $V_{DS,S2b}(t)$, CH3 $V_{GS,S2a}(t)$, CH4 $V_{GS,S2b}(t)$.

with peak efficiency of 90% at 200 W is demonstrated.

ACKNOWLEDGMENT

The authors would like to thank Texas Instruments Kilby Labs for their support in this project through the funding provided to the Stanford SystemX Alliance FMA (Faculty, Mentor, Advisor) Research program.

REFERENCES

[1] D. J. Perreault, J. Hu, J. M. Rivas, Y. Han, O. Leitermann, R. C. N. Pilawa-Podgurski, A. Sagneri, and C. R. Sullivan, "Opportunities and challenges in very high frequency power conversion," in *2009 Twenty-Fourth Annual IEEE Applied Power Electronics Conference and Exposition*, Feb 2009, pp. 1–14.

[2] N. Sokal and A. Sokal, "Class E-a new class of high-efficiency tuned single-ended switching power amplifiers," *IEEE Journal of Solid-State Circuits*, vol. 10, pp. 168–176, 1975.

[3] F. Raab, "Class-F power amplifiers with maximally flat waveforms," *IEEE Transactions on Microwave Theory and Techniques*, vol. 45, no. 11, pp. 2007 – 2012, 1997.

[4] S. D. Kee, "The class E/F family of harmonic-tuned switching power amplifiers," Ph.D. dissertation, California Institute of Technology, 2002.

[5] S. D. Kee, I. Aoki, A. Hajimiri, and D. Rutledge, "The class-e/f family of zvs switching amplifiers," *IEEE Transactions on Microwave Theory and Techniques*, vol. 51, no. 6, pp. 1677–1690, June 2003.

[6] J. Rivas, Y. Han, O. Leitermann, A. Sagneri, and D. Perreault, "A high-frequency resonant inverter topology with low-voltage stress," *Power Electronics, IEEE Transactions on*, vol. 23, no. 4, pp. 1759–1771, July 2008.

[7] S. Aldhaher, D. C. Yates, and P. D. Mitcheson, "Modeling and Analysis of Class EF and Class E/F Inverters With Series-Tuned Resonant Networks," *IEEE Transactions on Power Electronics*, vol. 31, no. 5, pp. 3415–3430, May 2016.

[8] R. J. Gutmann, "Application of RF circuit design principles to distributed power converters," *IEEE Transactions on Industrial Electronics and Control Instrumentation*, vol. IECI-27, no. 3, pp. 156–164, 1980.

[9] R. Redl and N. Sokal, "A 14-MHz 100-Watt Class E resonant converter: Principles, design considerations and measured performance," in *17th Annual IEEE Power Electronics Specialists Conference Proceedings*, 1986.

[10] A. F. Goldberg and J. G. Kassakian, "The application of power mosfets at 10mhz," in *1985 IEEE Power Electronics Specialists Conference*, June 1985, pp. 91–100.

[11] J. Jozwik and M. Kazimierczuk, "Analysis and design of class E^2 dc/dc converter," *IEEE Transactions on Industrial Electronics*, vol. 37, pp. 173–183, 1990.

637

The 2018 International Power Electronics Conference

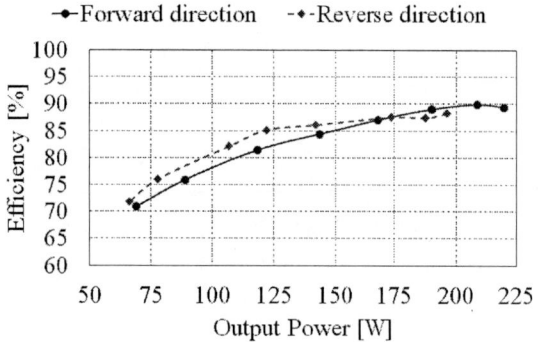

Fig. 15: Measured efficiency vs output power

[12] J. M. Rivas, O. Leitermann, Y. Han, and D. J. Perreault, "New architectures for radio-frequency dc/dc power conversion," *IEEE Transactions on Power Electronics*, vol. 21, no. 2, pp. 380 – 393, 2006.

[13] R. Pilawa-Podgurski, A. Sagneri, J. Rivas, D. Anderson, and D. Perreault, "Very-high-frequency resonant boost converters," *Power Electronics, IEEE Transactions on*, vol. 24, no. 6, pp. 1654–1665, June 2009.

[14] R. Redl, "Fundamental considerations for very high frequency power conversion," in *Int. Workshop Power Supply on Chip Power Power-SoC'08*, Sept 2008.

[15] J. M. Rivas, "Radio frequency dc-dc power conversion," Ph.D. dissertation, Massachusetts Institute of Technology, 2006.

[16] Y. Han, O. Leitermann, D. A. Jackson, J. M. Rivas, and D. J. Perreault, "Resistance compression networks for radio-frequency power conversion," *IEEE Transactions on Power Electronics*, vol. 22, no. 1, pp. 41–53, Jan 2007.

[17] J. M. Burkhart, K. R., and D. J. Perreault, "Design methodology for a very high frequency resonant boost converter," *IEEE Transactions on Power Electronics*, vol. 28, no. 4, pp. 1929 – 1937, April 2013.

[18] A. D. Sagneri, D. I. Anderson, and D. J. Perreault, "Transformer synthesis for VHF converters," in *Power Electronics Conference (IPEC), 2010 International*, June 2010.

[19] A. Sagneri, D. Anderson, and D. Perreault, "Optimization of integrated transistors for very high frequency dcdc converters," *Power Electronics, IEEE Transactions on*, vol. 28, no. 7, pp. 3614–3626, July 2013.

[20] M. Jovanovic, M. Zhang, and F. Lee, "Evaluation of synchronous-rectification efficiency improvement limits in forward converters," *Industrial Electronics, IEEE Transactions on*, vol. 42, no. 4, pp. 387–395, Aug 1995.

[21] R. Blanchard and P. E. Thibodeau, "The design of a high efficiency, low voltage power supply using mosfet synchronous rectification and current mode control," in *Power Electronics Specialists Conference, 1985 IEEE*, June 1985, pp. 355–361.

[22] J. Pedersen, M. Madsen, A. Knott, and M. Andersen, "Self-oscillating galvanic isolated bidirectional very high frequency dc-dc converter," in *Proc. 30th Annual IEEE Applied Power Electronics Conf. and Exposition (APEC)*, 2015, pp. 1974–1978.

[23] X. Ren, Y. Zhou, D. Wang, X. Zou, and Z. Zhang, "A 10-MHz isolated synchronous class−ϕ_2 resonant converter," *Power Electronics, IEEE Transactions on*, vol. 31, no. 12, pp. 8317–8328, Jan 2016.

[24] L. Gu, W. Liang, L. C. Raymond, and J. Rivas-Davila, "27.12 MHz GaN bi−directional resonant power converter," in *Control and Modeling for Power Electronics (COMPEL), 2015 IEEE 16th Workshop on*, June 2015.

[25] J. M. Rivas, D. Jackson, O. Leitermann, A. D. Sagneri, Y. Han, and D. J. Perreault, "Design considerations for very high frequency dc-dc converters," in *2006 37th IEEE Power Electronics Specialists Conference*, June 2006, pp. 1–11.

[26] R. W. Erickson and D. Maksimovic, *Fundamentals of Power Electronics*, 2nd ed. Springer Science & Business Media, 2007.

[27] M. N. Kheraluwala, R. W. Gascoigne, D. M. Divan, and E. D. Baumann, "Performance characterization of a high-power dual active bridge dc-to-dc converter," *IEEE Transactions on Industry Applications*, vol. 28, no. 6, pp. 1294–1301, Nov 1992.

[28] B. Yang, F. Lee, A. Zhang, and G. Huang, "Llc resonant converter for front end dc/dc conversion," in *Applied Power Electronics Conference and Exposition, 2002. APEC 2002. Seventeenth Annual IEEE*, March 2002.

[29] Y. Han, G. Cheung, A. Li, C. R. Sullivan, and D. J. Perreault, "Evaluation of magnetic materials for very high frequency power applications," *IEEE Transactions on Power Electronics*, vol. 27, no. 1, pp. 425–435, Jan 2012.

[30] A. J. Hanson, J. A. Belk, S. Lim, C. R. Sullivan, and D. J. Perreault, "Measurements and Performance Factor Comparisons of Magnetic Materials at High Frequency," *IEEE Transactions on Power Electronics*, vol. 31, no. 11, pp. 7909–7925, Nov 2016.

[31] J. M. Rivas, O. Leitermann, Y. Han, and D. J. Perreault, "A very high frequency dc–dc converter based on a class Φ_2 resonant inverter," *IEEE Transactions on Power Electronics*, vol. 26, no. 10, pp. 2980–2992, 2011.

[32] L. Roslaniec, A. S. Jurkov, A. A. Bastami, and D. J. Perreault, "Design of Single-Switch Inverters for Variable Resistance/Load Modulation Operation," *IEEE Transactions on Power Electronics*, vol. 30, no. 6, pp. 3200–3214, June 2015.

[33] Y. Han and D. J. Perreault, "Analysis and design of high efficiency matching networks," *IEEE Transactions on Power Electronics*, vol. 21, no. 5, pp. 1484–1491, Sept 2006.

[34] P. A. Kyaw, A. L. F. Stein, and C. R. Sullivan, "Analysis of high efficiency multistage matching networks with volume constraint," in *2017 IEEE 18th Workshop on Control and Modeling for Power Electronics (COMPEL)*, July 2017, pp. 1–8.

[35] L. Gu, W. Liang, and J. R. Davila, "Design of very-high-frequency synchronous resonant dc-dc converter for variable load operation," in *2017 IEEE Energy Conversion Congress and Exposition (ECCE)*, Oct 2017, pp. 3447–3454.

[36] J. S. Glaser and J. M. Rivas, "A 500 W push-pull dc-dc power converter with a 30 MHz switching frequency," in *Proc. Twenty-Fifth Annual IEEE Applied Power Electronics Conf. and Exposition (APEC)*, 2010, pp. 654–661.

The 2018 International Power Electronics Conference

Smart PV Inverters for Smart Grid Applications

Cheng-Jhen Yang[1*], Terng-Wei Tsai[1], Yi-Chan Li[1], Cheng-Yu Tang[2], Yaow-Ming Chen[1] and Yung-Ruei Chang[3]

1 Dept. of Electrical Engineering, National Taiwan University, Taipei, Taiwan
2 Dept. of Electrical Engineering, Feng Chia University, Taichung, Taiwan
3 Institute of Nuclear Energy Research, Atomic Energy Council, Taoyuan, Taiwan
E-mail: r05921024@ntu.edu.tw

Abstract—For a micro-grid system, the increasing distributed generations and unbalanced loads may easily cause unbalanced grid voltages and reduce the power system stability. Therefore, a smart photovoltaic (PV) inverter capable of transferring demanded unbalanced power of each phase will be needed to help to increase the stability and power quality of the micro-grid system. To address these issues, this paper proposed a per-phase current control strategy for the PV power system. With the proposed control strategy, the output current and power of each phase can be directly controlled and determined by the user according to the load requirement. It is worth mentioning that complex dq-axis transformations and extra circuits or components are not needed. The thorough mathematical derivations and operation principles will be presented in this paper. Also, a 5kVA prototype circuit is implemented to help to demonstrate the feasibility of the proposed control strategy.

Keywords— PV inverter, Micro-grid, Per-phase control, Unbalanced.

I. INTRODUCTION

Recently, in order to deal with fossil fuel usage and carbon dioxide emission problems, renewable energy such as photovoltaic (PV) energy, wind energy and tidal energy have received lots of attention [1]. Among all the renewable energy, photovoltaic (PV) energy might be the most promising one in the future due to its cost and easy implementation. Therefore, PV power system has been rapidly developed and integrated with the micro-grid systems for supplying the electric power [2]. To transfer PV energy to the power system, the grid-connected PV inverter plays an important role [1]-[5].

For a grid-connected PV inverter, boost converter is usually adopted as the front-end maximum power point (MPPT) tracker in virtue of the linearity of its input current [6]-[8]. Moreover, the step-up mechanism of boost converter can also help the PV array, which usually has relatively low output voltage, be utilized to match with high voltage transmission of the grid side. Various kinds of MPPT algorithm can be found in the literature to help overcome difficulties tracking the real maximum power point under different irradiations or partially shaded conditions [9]-[13]. Besides, to further decrease the input current ripple, the interleaved boost converter topology is usually adopted [13].

Due to the increasing presence of single-phase distributed generations and unbalanced load in the electric power system, which may lead to unbalanced three phase voltage problem especially for micro-grid systems, the three-phase four-wire split-capacitor topology is adopted in this paper as the rear-end inverter. Researches have pointed out the advantages of the three-phase four-wire split-capacitor inverter topology [14]. Owing to the fact that micro-grid systems are easily tend to be unstable under fault incidents or unbalanced load conditions, it is not suitable for the PV inverter to keep transferring balanced three-phase output power to the grid under such conditions, which may further intensify the unbalanced problem.

On the contrary, the PV inverter should be able to transfer unbalanced output power according to the load condition of each phase to mitigate the unbalanced situation and help to increase the stability of the micro-grid system. Fig. 1 shows the conceptual block diagram of the system that the three-phase four-wire grid-tied PV inverter is used to compensate for the unbalance created by the single-phase distributed generations and unbalanced loads.

Usually, to achieve the goal, detection of the voltage sags, active and reactive power oscillation, current reference calculation and dc-link voltage balancing are the key issues to be concerned for the PV power system. Among them, the current reference calculation plays the most important role of the proper operation. Therefore, various current control strategies dealing with unbalanced conditions and load requirements can be found in literature [15]-[18]. However, complex transformations in dq-axis reference frame and the theory of symmetrical components are usually needed in these control methods.

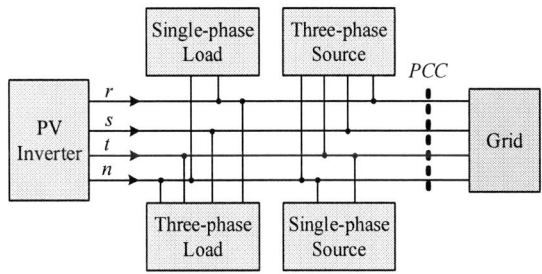

Fig. 1. Conceptual block diagram of the unbalanced three-phase power system.

The 2018 International Power Electronics Conference

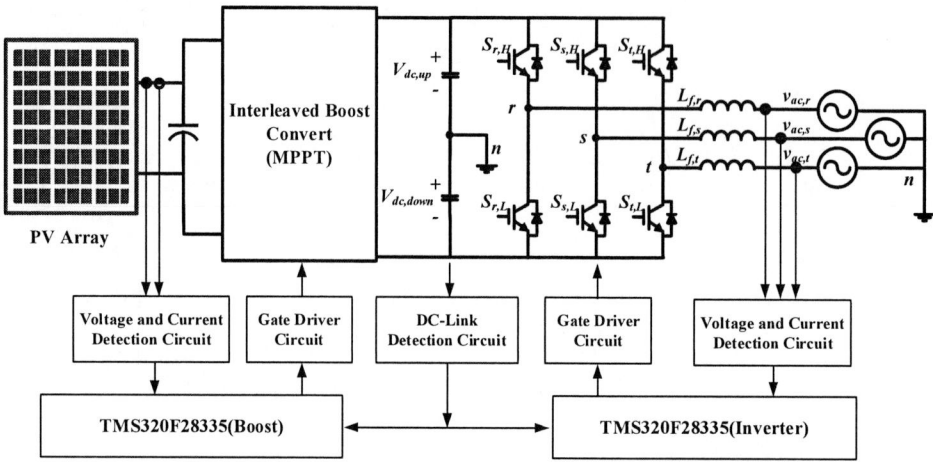

Fig. 2. The circuit diagram and control blocks of the PV system.

As a result, the associated control scheme will become complex and the calculation may take longer time to be done by using Digital Signal Processor (DSP) in practices, which will even result in the restriction of the switching frequency.

To simplify the control strategies, the PV inverter with the per-phase current control strategy is proposed in this paper. With the proposed current control strategy, the output current of each phase can be individually controlled without any transformation to meet different load requirements. The simulation results as well as the hardware experimental results will be presented to verify the performance of the proposed PV inverter with per-phase current control strategy.

II. OPERATION PRINCIPLE AND MATHEMATICAL DERIVATIONS

The circuit diagram and control blocks of the PV power system are shown in Fig. 2. A two stage topology of the front-end interleaved boost converter integrates with the rear-end three-phase four-wire split-capacitor inverter is utilized. The TMS320F28335 microcontroller from TI are used as the controllers of the front-end converter and the rear-end inverter. In general, the dc-dc interleaved boost converter will act as the MPPT to extract the maximum power of the PV array. Meanwhile, the rear-end three-phase four-wire split-capacitor inverter will transfer the entire solar energy extracted by the MPPT to the power system. As a result, the DC-link voltage between the front-end converter and the rear-end inverter would be properly balanced.

Because of the non-linear characteristic of the PV array, effective MPPT algorithms are of great concern to help maximize the power extraction of the PV array. By taking the stability and simplicity into account, the algorithm of the perturbation and observation (P&O) method is adopted in this paper. The control flow chart of the P&O method is shown in Fig. 3. The voltage V_{PV} and the current I_{PV} of the PV array will be sensed and then be multiplied with each other to get the PV power P_{PV} every cycle. By comparing the result of V_{PV} and P_{PV} under the present cycle with the

result of the previous cycle, the PV current reference I_{ref} can be determined to be increased or decreased so as to achieve the function of MPPT.

To transfer the power extracted by the MPPT, as Fig. 2 shows, the three-phase four-wire split-capacitor inverter topology is adopted to be the rear-end inverter. The Sinusoidal Pulse Width Modulation (SPWM) technique is used to generate the gate signal for the switches of the inverter. As a result, a proper sinusoidal reference signal in phasor form $\overrightarrow{V_{ref,x}}$ must be derived to achieve the operation of the per-phase current control strategy.

For the purpose of the analysis of the proposed current control strategy, the inverter can be regarded as three independent single-phase half-bridge inverter sharing a common dc bus. Fig. 4 and Fig. 5 show the xth phase single-phase half-bridge inverter and the SPWM scheme, where the parameter x can be r, s or t. According to Fig. 4, it can be seen that each phase of the inverter can be analyzed individually. The mathematical expression of the relation between inverter switching node voltage $\overrightarrow{v_{xn}}$ and grid voltage $\overrightarrow{v_{ac,x}}$ can be derived as (1).

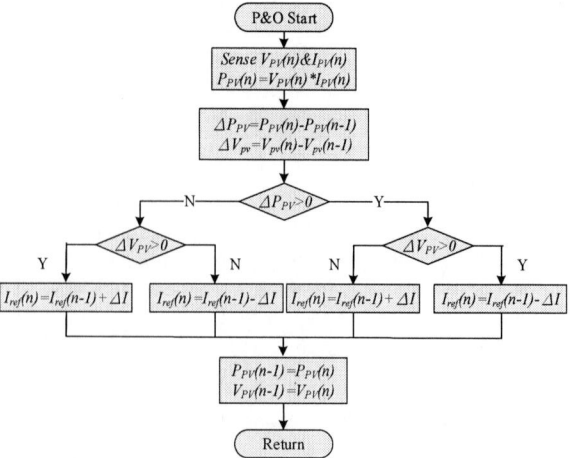

Fig. 3. Control flow chart of the P&O method.

640

$$\overrightarrow{v_{xn}} = \overrightarrow{i_{ac,x}} \times j\omega L + \overrightarrow{v_{ac,x}} \tag{1}$$

where $\overrightarrow{i_{ac,x}}$ is the line current and ω is the line frequency. By assuming that the voltage of the split capacitors are balanced, the voltage $\overrightarrow{v_{xn}}$ can be expressed as (2), where V_{dc} is the total dc-bus voltage, due to the complementary duty ratio of $S_{x,H}$ and $S_{x,L}$, which will be D and ($1-D$), respectively.

$$\overrightarrow{v_{xn}} = (2D - 1) \times \frac{V_{dc}}{2} . \tag{2}$$

Also, according to Fig. 5, which is the conceptual diagram for SPWM scheme, the duty ratio D can be further derived as (3):

$$D = \frac{1}{2} + \frac{1}{2} \times \frac{\overrightarrow{V_{ref,x}}}{V_{tri,peak}} \tag{3}$$

where $\overrightarrow{V_{ref,x}}$ is the sinusoidal reference signal and $V_{tri,peak}$ is the peak value of the triangular reference signal. The time-domain sinusoidal reference signal for SPWM can be expressed as:

$$v_{ref,x}(t) = \left|\overrightarrow{V_{ref,x}}\right| \times \cos(\omega t + \theta_{V_{ref,x}}) \tag{4}$$

where $\theta_{V_{ref,x}}$ is the phase angle of the reference signal for xth phase. Finally, by substituting equations (2) and (3) into (1), the reference signal in phasor form can be obtained:

$$\overrightarrow{V_{ref,x}} = \left(\overrightarrow{v_{ac,x}} + \overrightarrow{i_{ac,x}} \times j\omega L\right) \times \frac{2V_{tri,peak}}{V_{dc}} . \tag{5}$$

From the equation above, the relation between the sinusoidal reference signal and the grid voltage is derived. Moreover, according to equation (5), the amplitude and phase of $\overrightarrow{V_{ref,x}}$ can be further expressed and obtained respectively as follows:

$$\left|\overrightarrow{V_{ref,x}}\right| = \frac{2V_{tri,peak}}{V_{dc}} [|\overrightarrow{v_{ac,x}}|^2 + |\overrightarrow{i_{ac,x}}|^2 \times \omega^2 L^2 -$$
$$2\omega L \times |\overrightarrow{v_{ac,x}}||\overrightarrow{i_{ac,x}}| \sin(\theta_{i_{ac,x}} - \theta_{v_{ac,x}})]^{\frac{1}{2}} \tag{6}$$

$$\theta_{V_{ref,x}} = \tan^{-1}\left(\frac{|\overrightarrow{v_{ac,x}}| \sin\theta_{v_{ac,x}} + |\omega L \times \overrightarrow{i_{ac,x}}| \cos\theta_{i_{ac,x}}}{|\overrightarrow{v_{ac,x}}| \cos\theta_{v_{ac,x}} - |\omega L \times \overrightarrow{i_{ac,x}}| \sin\theta_{i_{ac,x}}}\right) \tag{7}$$

According to equations (6) and (7), it can be seen that once the amplitude and phase of $\overrightarrow{i_{ac,x}}$ are determined, the sinusoidal reference signal $\overrightarrow{V_{ref,x}}$ can be derived since the information of $\overrightarrow{v_{ac,x}}$ can be directly obtained by sensing the grid voltage. As shown in Fig. 4, $\overrightarrow{i_{ac,x}}$ is the xth phase sinusoidal output current, which can be manipulated by the user. By assuming that the output current of each phase are well-regulated, $\overrightarrow{i_{ac,x}}$ will be approximately equal to the xth phase sinusoidal output current reference $\overrightarrow{i_{ref,x}}$. To generate the current reference $\overrightarrow{i_{ref,x}}$, the amplitude and phase of $\overrightarrow{i_{ref,x}}$ can be determined as follows:

$$\left|\overrightarrow{i_{ref,x}}\right| = \sqrt{\frac{2(P_{cmd,x}^2 + Q_{cmd,x}^2)}{v_{ac,x,rms}^2}} \tag{8}$$

$$\theta_{i_{ref,x}} = \theta_{v_{ac,x}} - \tan^{-1}\frac{Q_{cmd,x}}{P_{cmd,x}} \tag{9}$$

where $P_{cmd,x}$ and $Q_{cmd,x}$ are the power commands of xth phase, which can be directly determined by the user.

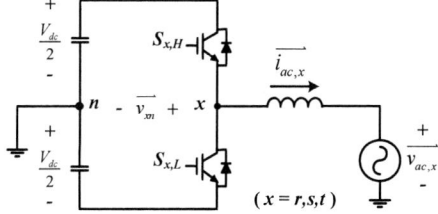

Fig. 4. Single-phase half-bridge inverter.

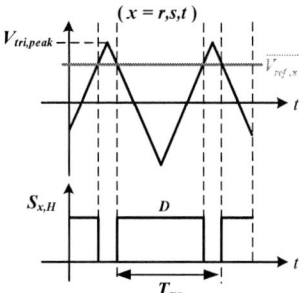

Fig. 5. The conceptual diagram for SPWM scheme.

Besides, the zero-crossing detection technique will be used to synchronize the phase angle between $\overrightarrow{i_{ref,x}}$ and $\overrightarrow{v_{ac,x}}$.

According to equation (8) and (9), by adjusting the power commands of each phase, the xth phase output current reference $\overrightarrow{i_{ref,x}}$ will be individually derived. Besides, due to the fact that $v_{ac,x,rms}$ is simultaneously updated with the grid, even if the grid voltage is changing, the xth phase output current reference $\overrightarrow{i_{ref,x}}$ can still be correctly derived to transfer the demanded output power. To conclude, the xth phase output current can be flexibly controlled to meet different load requirements under either balanced or unbalanced voltage condition. Finally, by substituting equation (8) and (9) into equation (6) and (7), the proper sinusoidal reference signal $\overrightarrow{V_{ref,x}}$ can be derived to achieve the operation of per-phase current control. Figure. 6 shows the conceptual control block diagram of xth phase for the proposed current control strategy, where the Current reference generator is composed of equation (8) and (9), the SPWM reference generator is composed of equation (6) and (7), and the term $i_{fb,x}$ represents the feedback signal of the xth phase ac current.

According to the analysis, it is worth mentioning that the proposed per-phase current control strategy doesn't need any complex transformation or calculation of symmetrical components, which may require large computational power in practical. Moreover, no additional circuits or devices will be required to help achieving the proposed control strategy.

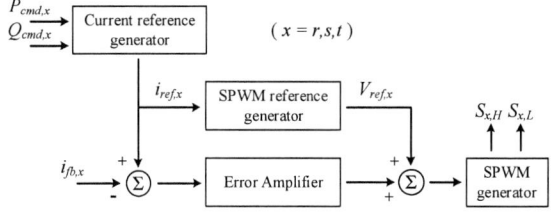

Fig. 6. The conceptual control block diagram of xth phase inverter current.

III. SIMULATION AND EXPERIMENTAL RESULTS

To verify the feasibility and performance of the proposed per-phase current control strategy, computer simulation software Matlab/Simulink is used to simulate the PV power system in this paper. Moreover, a 5kVA prototype circuit is also implemented on the basis of the circuit specifications shown in Table. I.

A. Simulation Results

The simulation results of the PV current, I_{pv}, PV power, P_{pv}, dc-link voltage, V_{dc} and three phase output current, I_r, I_s, I_t are shown in Fig. 7. A built-in PV array model, which has 5kW maximum power output under $1000\ W/m^3$ irradiation and 25°C temperature condition, is used to be the source of the PV power system. From the figure, the interleaved boost converter can be properly operated to achieve the function of MPPT. Besides, the dc bus voltage can be seen to be well regulated at around 760V under steady-state. Moreover, it can be seen that the amplitude of r-phase current can be controlled to be twice the s-phase current and the t-phase current, verifying that the proposed per-phase current control strategy can indeed control the output current of each phase individually to have different output power for unbalanced load requirement.

B. Experimental Results

Fig. 8(a) shows the 5kVA prototype circuits of the front-end converter, which consisting of a three-channel interleaved boost converter and its DSP based control circuit. In addition, the 5kVA prototype circuits of the rear-end inverter are also shown in Fig. 8(b). The control circuit and dc bus is on the left and the filter inductor is on the right, while the driver circuit is on top of the heat sink and the power board. A programmable PV Simulator from Chroma is used to imitate the PV source. The TMS320F28335 microcontroller from TI is used here as the controller of both the front-end converter and the rear-end inverter.

TABLE I
CIRCUIT SPECIFICATIONS

Symbol	Value
C_{dc} (DC-link capacitance)	1.4 mF
L_B (Boost inductor)	1.6 mH
L_f (Inductance of the LC filter)	2 mH
V_{MPP} (PV voltage at MPP)	300 V
P_{MPP} (PV power at MPP)	5 kW
V_{dc} (Total dc-link voltage)	760 V
V_{ac} (Line-to-neutral ac voltage)	220 V_{rms}

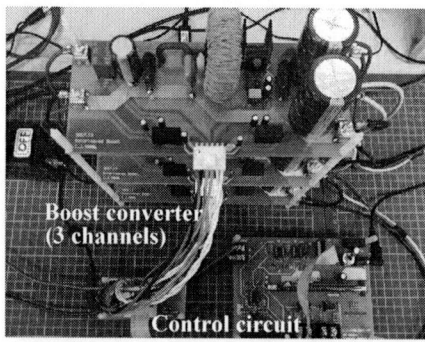

(a). The interleaved boost converter.

(b). The rear-end inverter.

Fig. 8. Photographs of the prototype circuits.

To verify the function of MPPT, Fig. 9(a) and Fig. 9(b) show the current and voltage waveforms of the front-end boost converter. As shown in Fig. 9(a), it is the interleaved operation of the front-end boost converter, in which a 120 degree out of phase shift exists between the inductor current of each channel. The total input current can be seen to have a very small ripple under interleaved operation. As for Fig. 9(b), it can be seen that the interleaved boost converter can help tracking the maximum power point (MPP) to harvest the maximum power from the PV simulator, which is set to be 4kW in this case. It should be mentioned that the dc bus voltage is pre-charged by the grid and therefore it will remain at about 600V in the beginning before the function of MPPT is started up.

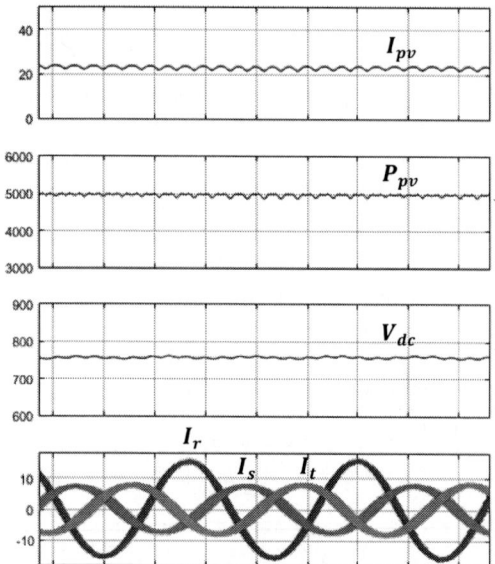

Fig. 7. Simulation waveforms of steady-state operation.

The 2018 International Power Electronics Conference

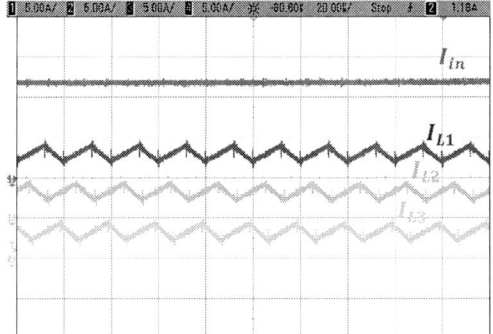

Fig. 9(a). Inductor current under interleaved operation.

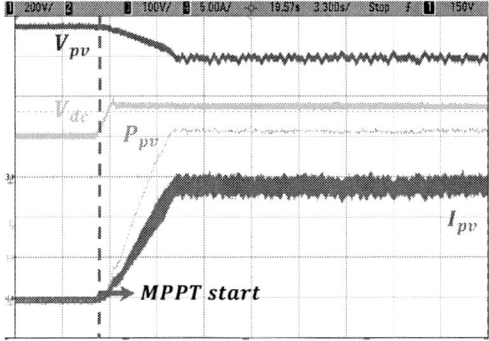

Fig. 9(b). Waveforms of the function of MPPT.

As for Fig. 10(a), this is the overview results of the whole operation of the rear-end inverter, which can be divided into five time intervals. It can be noted that the front-end converter will not start extracting the power from the PV simulator to the dc bus until time interval II.

At the beginning of time interval II, the front-end converter will start the function of MPPT so that the dc bus voltage can be seen to keep rising up in this time interval. Once the dc bus voltage reaches the voltage level, which is set to be 760V, the rear-end inverter will start transferring the power to the grid so as to regulate the dc bus voltage, as shown in time interval III. As for time interval III to time interval V, the output current of each phase will be controlled and changed with different ratio according to the power command determined by the user. The ratio of the power command of r phase, s phase and t phase in these time intervals are determined to be 1:1:1, 3:1:2 and 1:0:1, respectively.

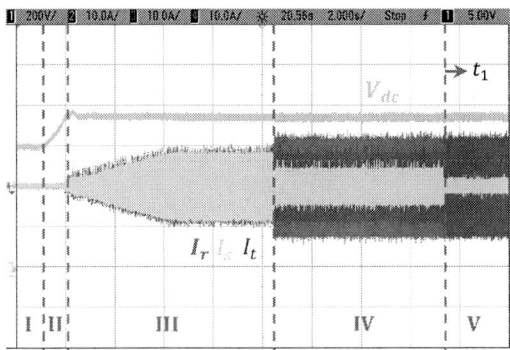

Fig. 10(a). Overview results of the whole operation.

To have a more clear view of the output current in these time intervals, Fig. 10(b) and Fig. 10(c) are the partially enlarged views of time interval III and time interval IV. From the figures, the output current of each phase can be seen to be well controlled to be balanced or unbalanced. Furthermore, Fig. 10(d) shows the waveforms of transient response at t_1. It can be seen that it only takes less than one 60-Hz ac mains cycle for the transient-state to be completed. Besides, from the following experimental results, it can also be seen that the dc bus voltage can be well regulated at 760V under either steady-state or transient-state, which can help verify the feasibility of the proposed control strategy.

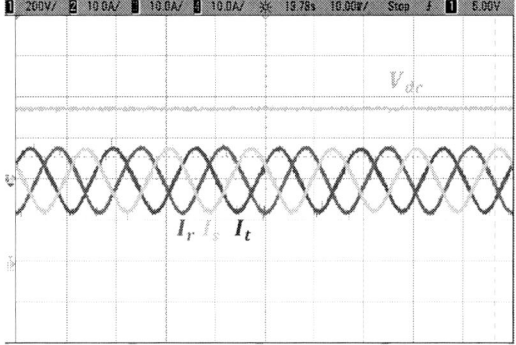

Fig. 10(b). A partially enlarged view of time interval III.

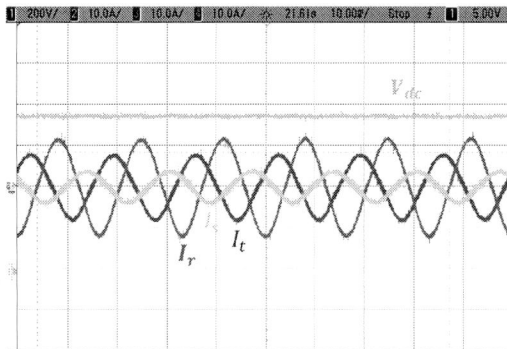

Fig. 10(c). A partially enlarged view of time interval IV.

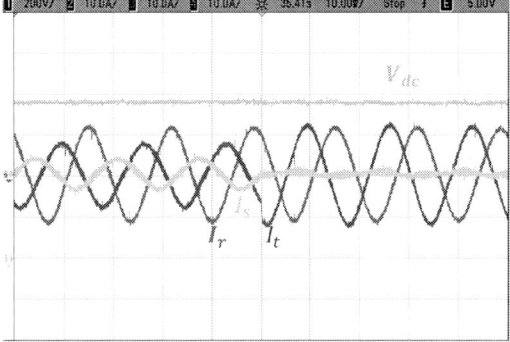

Fig. 10(d). Waveforms of transient-state response at t_1.

643

The 2018 International Power Electronics Conference

IV. CONCLUSIONS

Due to the fact that micro-grid systems are easily tend to be unstable under unbalanced load conditions, keep transferring balanced output power to each phase under these conditions may further intensify the unbalanced problem such as unbalanced grid voltage. To mitigate the unbalanced condition, a per-phase current control PV power system is proposed.

With the proposed control strategy, the output current of each phase can be individually controlled without using complex transformations. Also, the command of each phase can be directly determined and changed by the user according to the load requirements to improve the stability of the micro-grid. Both the operation principle and thorough mathematical derivations are presented in this paper. A 5kVA prototype circuit has been implemented. The simulation results as well as the experimental results presented to verify the performance of PV inverter with the proposed control strategy.

ACKNOWLEDGMENT

This work is partially supported by a research grant from the Institute of Nuclear Energy Research, Taiwan.

REFERENCES

[1] L. B. G. Campanhol, S. A. O. da Silva, A. A. de Oliveira and V. D. Bacon, "Dynamic Performance Improvement of a Grid-Tied PV System Using a Feed-Forward Control Loop Acting on the NPC Inverter Currents," in *IEEE Transactions on Industrial Electronics*, vol. 64, no. 3, pp. 2092-2101, March 2017.

[2] N. C. Sintamarean, F. Blaabjerg, H. Wang, F. Iannuzzo and P. de Place Rimmen, "Reliability Oriented Design Tool For the New Generation of Grid Connected PV-Inverters," in *IEEE Transactions on Power Electronics*, vol. 30, no. 5, pp. 2635-2644, May 2015.

[3] M. O. Badawy and Y. Sozer, "Power Flow Management of a Grid Tied PV-Battery System for Electric Vehicles Charging," in *IEEE Transactions on Industry Applications*, vol. 53, no. 2, pp. 1347-1357, March-April 2017.

[4] Y. Shi, R. Li, Y. Xue and H. Li, "High-Frequency-Link-Based Grid-Tied PV System With Small DC-Link Capacitor and Low-Frequency Ripple-Free Maximum Power Point Tracking," in *IEEE Transactions on Power Electronics*, vol. 31, no. 1, pp. 328-339, Jan. 2016.

[5] B. Yang, W. Li, Y. Zhao and X. He, "Design and Analysis of a Grid-Connected Photovoltaic Power System," in *IEEE Transactions on Power Electronics*, vol. 25, no. 4, pp. 992-1000, April 2010.

[6] J. Kim and C. Kim, "A DC–DC Boost Converter With Variation-Tolerant MPPT Technique and Efficient ZCS Circuit for Thermoelectric Energy Harvesting Applications," in *IEEE Transactions on Power Electronics*, vol. 28, no. 8, pp. 3827-3833, Aug. 2013.

[7] H. C. Chen and W. J. Lin, "MPPT and Voltage Balancing Control With Sensing Only Inductor Current for Photovoltaic-Fed, Three-Level, Boost-Type Converters," in *IEEE Transactions on Power Electronics*, vol. 29, no. 1, pp. 29-35, Jan. 2014.

[8] M. Tampubolon *et al.*, "A study and implementation of three-level boost converter with MPPT for PV application," *2017 IEEE 3rd International Future Energy Electronics Conference and ECCE Asia (IFEEC 2017 - ECCE Asia)*, Kaohsiung, 2017, pp. 1143-1148.

[9] G. Escobar, S. Pettersson, C. N. M. Ho and R. Rico-Camacho, "Multisampling Maximum Power Point Tracker (MS-MPPT) to Compensate Irradiance and Temperature Changes," in *IEEE Transactions on Sustainable Energy*, vol. 8, no. 3, pp. 1096-1105, July 2017.

[10] T. K. Soon and S. Mekhilef, "A Fast-Converging MPPT Technique for Photovoltaic System Under Fast-Varying Solar Irradiation and Load Resistance," in *IEEE Transactions on Industrial Informatics*, vol. 11, no. 1, pp. 176-186, Feb. 2015.

[11] T. Esram and P. L. Chapman, "Comparison of Photovoltaic Array Maximum Power Point Tracking Techniques," in *IEEE Transactions on Energy Conversion*, vol. 22, no. 2, pp. 439-449, June 2007.

[12] C. W. Chen and Y. M. Chen, "Analysis of the series-connected distributed maximum power point tracking PV system," *2015 IEEE Applied Power Electronics Conference and Exposition (APEC)*, Charlotte, NC, 2015, pp. 3083-3088.

[13] C. Y. Tang, Y. T. Chen and Y. M. Chen, "PV Power System With Multi-Mode Operation and Low-Voltage Ride-Through Capability," in *IEEE Transactions on Industrial Electronics*, vol. 62, no. 12, pp. 7524-7533, Dec. 2015.

[14] W. T. Franke, C. Kürtz and F. W. Fuchs, "Analysis of control strategies for a 3 phase 4 wire topology for transformerless solar inverters," *2010 IEEE International Symposium on Industrial Electronics*, Bari, 2010, pp. 658-663.

[15] A. Junyent-Ferre, O. Gomis-Bellmunt, T. C. Green and D. E. Soto-Sanchez, "Current Control Reference Calculation Issues for the Operation of Renewable Source Grid Interface VSCs Under Unbalanced Voltage Sags," in *IEEE Transactions on Power Electronics*, vol. 26, no. 12, pp. 3744-3753, Dec. 2011.

[16] N. A. Ninad and L. A. C. Lopes, "Unbalanced operation of per-phase vector controlled four-leg grid forming inverter for stand-alone hybrid systems," *IECON 2012 - 38th Annual Conference on IEEE Industrial Electronics Society*, Montreal, QC, 2012, pp. 3500-3505.

[17] H. Cao, H. Zhang, W. Jiang and S. Wei, "Research on PQ Control Strategy for PV Inverter in the Unbalanced Grid," *2012 Asia-Pacific Power and Energy Engineering Conference*, Shanghai, 2012, pp. 1-3.

[18] N. A. Ninad and L. A. C. Lopes, "Per-phase vector (dq) controlled three-phase grid-forming inverter for stand-alone systems," *2011 IEEE International Symposium on Industrial Electronics*, Gdansk, 2011, pp. 1626-1631.

High-voltage Bi-directional Half-bridge Three-level Series Resonant Converter with Frequency Modulation Control

Sih-Yi, Lee[1], Jynu-Jhe, Jhang[1], Jing-Yuan, Lin[1], Yao-Ching, Hsieh[2] and Haung-Jen, Chiu[1]

[1]Department of Electronic and Computer Engineering, National Taiwan University of Science and Technology,

Taipei, Taiwan

[2]Department of Electrical Engineering National Sun Yat-Sen University, Kaohsiung, Taiwan

E-mail: d10302012@mail.ntust.edu.tw

Abstract- **This paper aims to develop a bi-directional half-bridge three-level series resonant converter for high-voltage DC micro-grid. The topology of three-level circuit can reduce the device voltage stresses and electromagnetic interference. The clamped diode and capacitor unit can provide an appropriate current path to achieve zero-voltage switching on power switches. By conventional frequency modulation control, the resonant tank can be operated in SRC region and achieve bi-directional power conversion. Since the proposed control method is developed in this paper, the circuit has a characteristic of transformer decoupling called LLC-SRC region. It achieves quasi-zero-current switching to reduce the switching loss. Synchronous rectifier is also implemented to reduce conduction loss on the secondary side. A laboratory prototype of bi-directional half-bridge three-level series resonant converter is designed and verified in high-voltage applications. The circuit specifications are 3-kW rated power, 1-kV input voltage, and 3 A output current. A digital signal processor (DSP) chip is used to realize the digital controller of this converter. The measured efficiency can be achieved 96% under different load conditions.**

I. INTRODUCTION

Since the rapid development of industry and commerce, the energy demand is increasing. The energy becomes shortage since the power generation is insufficient, therefore, how to effectively reduce the power transmission losses and convert the energy effectively are the main challenge which we face to. In order to improve the above problems, the direct solution to the power system is to raise the voltage to extremely high voltage that can reduce the long-distance transmission loss on the transmission line. Besides, many countries have promoted renewable energy, such as: solar energy, tidal energy, wind power, in this application, the power capacity becomes gradually high power capability and the power delivery is connected to DC gird. Fig.1 shows the simple structure of micro-grid system. Compared to traditional AC transmission system, It is more reliable and higher efficiency since long-distance transmission on the power generation side to the load side has no skin effect on the line and don't need to add line compensation unit.

Based on the above-mentioned argument, the transmission voltage is designed to cope with long-distance, high-power application and improves efficiency on energy transmission, However, the voltage stress of the component is also increased at the same time, so the multi-level converter architecture has been proposed [1]-[3], the multi-level converter distribute the voltage stress evenly by using unique modulation control method, besides, the multi-level converter also can effectively reduce the electromagnetic interference (EMI) and other issues. Therefore, this paper studies the architecture of high voltage, high power and high efficiency half-bridge third-order series resonant converter in topological architecture. In addition to improving the above problems [4]-[5], and achieving zero-voltage switching on power switches, quasi-zero current cut-off and bi-directional demands [6], and finally verify the feasibility of the proposed architecture.

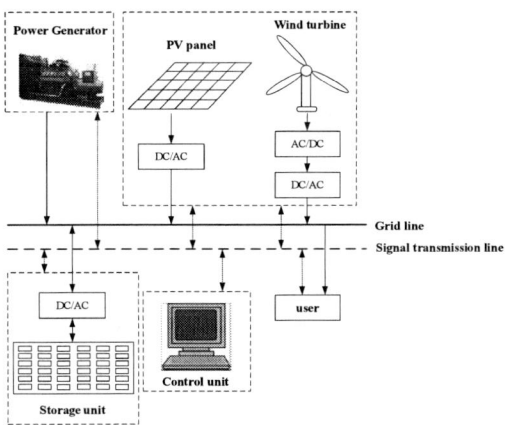

Figure 1. simple structure of micro-grid system

II. SYSTEM ARCHITECTURE AND ACTION PRINCIPLE

A. System architecture

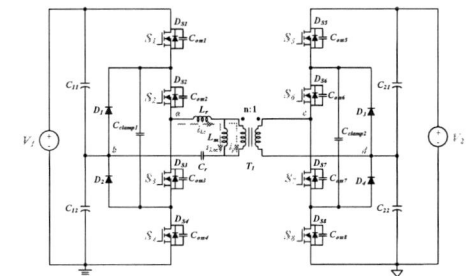

Figure 2. System architecture diagram

Figure 2 shows the system of the bi-directional half bridge three-level series resonant converter, as shown in Figure 2, As the primary side and the secondary side are two switches in series, so the power stress of each power switch is half of the input and output voltage; since the input and output capacitors are in series, each input and output capacitor is only half of the input and output voltage. Based on this half bridge architecture, the transformer is only half the input and output voltage, therefore, it is suitable for high voltage and low current application. In the conventional control, through the power switch S_2 and clamp diode D_1 conduction, the V_{ab} of the cross-voltage clamped on the zero voltage level; when S_1 and S_2 conduction, the V_{ab} can be pressed across the compression in half of the input Voltage, and when S_3 and S_4 are turned on, the cross-voltage of V_{ab} can be clamped to half the reverse input voltage, so the third-order voltage level can be achieved [7]-[8]..

Resonant tank includes the resonant capacitor C_r, resonant inductance L_r, magnetizing inductance L_m, two resonant frequencies f_{r1} and f_{r2} of the series resonant converter can be divided into three intervals, and the value of f_{r1} and f_{r2} are shown as follows (1) and (2).

$$f_{r1} = \frac{1}{2\pi\sqrt{L_r C_r}} \quad (1)$$

$$f_{r2} = \frac{1}{2\pi\sqrt{(L_r + L_m)C_r}} \quad (2)$$

The three intervals can be divided into the first interval (Region-1), the second interval (Region-2) and the third interval (Region-3), the gain curve of the relationship between three intervals is depicted in Figure 3.

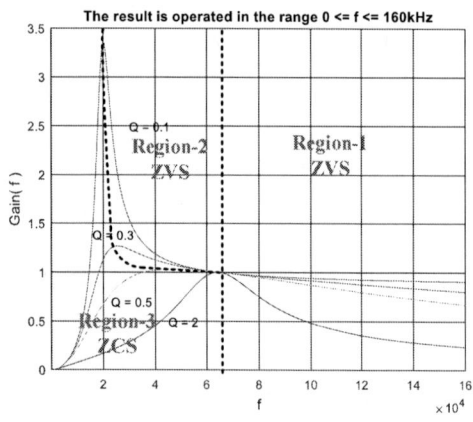

Figure 3. Gain curve of series resonant converter

The first interval (Region-1) is the SRC resonant mode whose operating frequency is greater than the first resonant frequency f_{r1} and is the zero-voltage switching (ZVS) region where the maximum gain ratio is 1, since the gain curve in both power flow is almost the same, it is very suitable to adopt in bi-directional application, therefore, in this paper, bi-directional half bridge three-

level series resonant converter is operated in the first interval (Region-1).

B. Operation principle of half-bridge third-level series resonant converter

The power stage of this paper is a bi-directional half-bridge third-order series resonant converter. Since the operation principle in forward and reverse power flow is almost the same, the operation principle of forward action is discussed. The intervals of the operation are shown in Figure 4.

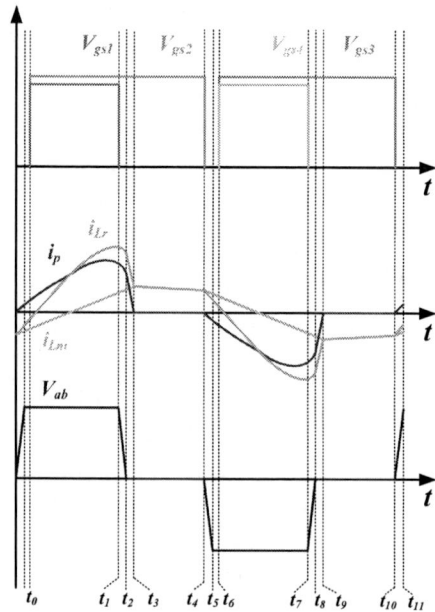

Figure 4. Third-level series resonant converter operation timing diagram

The structure of this paper is operated in the first interval (Region-1), the intervals of bi-directional half-bridge third-order series resonant converter can be divided into 12 stages, but the positive half operation and negative half operation are the same, this paper only illustrates the operation intervals in positive half ($t_0 \sim t_6$), assuming $C_{out} = C_{11} = C_{12} = C_{21} = C_{22} \gg C_r \gg C_{oss1} = C_{oss2} = C_{oss4} = C_{oss4} = C_{oss5} = C_{oss6} = C_{oss7} = C_{oss8} = C_{oss}$, $C_{clamp1} = C_{clamp2} \gg C_{oss}$, $L_m \gg L_r$, transformer turns ratio n.

Stage 1 ($t_0 \sim t_1$):

As shown in Fig. 5, the resonant current i_{Lr} is driven by the bulk diodes D_{s1}, D_{s2}, D_{s5} and D_{s6} flows through the switches S_1, S_2, S_5 and S_6 for the freewheeling, the switches S_1, S_2, S_5 and S_6 are turned on and zero voltage switching (ZVS) is achieved at t_0. Other switches and clamp diodes are cut-off, so the V_{ab} is $+V_{in}/2$. The primary current i_p delivers energy through the transformer T_1 to the output. Since the voltage across magnetizing inductance L_m is half of the output voltage, the magnetizing inductance current i_{Lm} linear rises, the resonant current i_{Lr} also gradually increase, when the switch S_1 cut off, this stage is end.

646

The 2018 International Power Electronics Conference

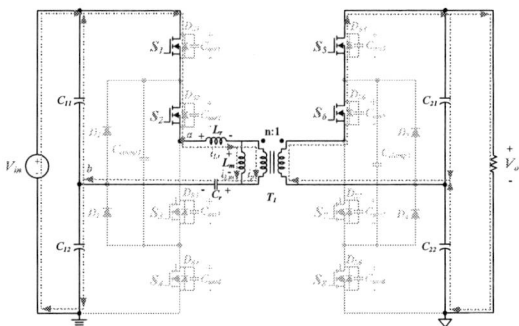

Figure 5. Stage 1

Stage 2 ($t_1 \sim t_2$):

The conduction path is shown in Fig. 6, it can be seen that only the switch S_2 is turned on, the other switches and the clamp diodes are turned off. the resonant current i_{Lr} charges the C_{oss1} and discharge the clamping capacitor C_{clamp1} and C_{oss4}, so V_{ab} decrease, the primary current i_p flow through the transformer T_1 and S_5, S_6 of the body diodes D_{s5}, D_{s6} and delivers energy to the output. Since the the voltage across magnetizing inductance L_m is half of the output voltage, the magnetizing inductance current i_{Lm} is still linear rise, the resonant current i_{Lr} and transformer primary current i_p gradually decrease. When the C_{oss1} Charge to half the input voltage $V_{in}/2$, C_{oss4} discharge to zero voltage.

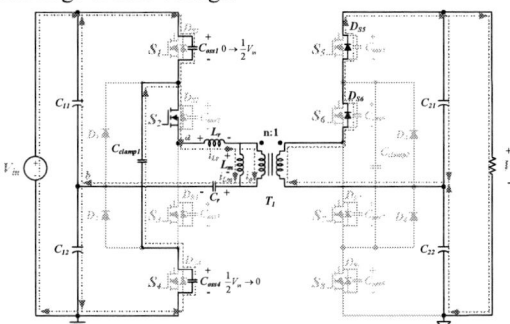

Figure 6. Stage 2

Stage 3 ($t_2 \sim t_3$):

As shown in Fig. 7, only the switch S_2 is turned on and the other switches are off-state. the resonant current i_{Lr} flows through the power switch S_2, resonant inductance L_r, transformer T_1, resonant capacitor C_r and clamp diode D_1, so that V_{ab} is zero voltage, the primary current i_p flows through the transformer T_1 and S_5, S_6 of the body diodes D_{s5}, D_{s6} to the output. Since the voltage across magnetizing inductance L_m is half of the output voltage, the magnetizing inductance current i_{Lm} still linear rises, but the resonant capacitor C_r voltage can't be changed, the resonant inductor produce a larger negative voltage which make i_{Lr} drop sharply, when i_{Lr} drops to i_{Lm}, this stage is end.

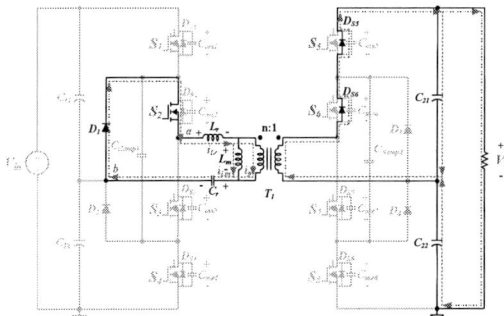

Figure 7. Stage 3

Stage 4 ($t_3 \sim t_4$):

As shown in Fig. 8, the switch S_2 is turned on and the transformer T_1 is decoupled. The resonant current i_{Lr} is same as magnetizing inductance current i_{Lm} and flows through the resonant capacitor C_r, the clamp diode D_1 and the power switch S_2, V_{ab} is zero voltage. As the value of magnetizing inductance L_m is very large, resonant current i_{Lr} becomes very small in this interval, S_2 can achieve quasi-zero current switching. However, the voltage of C_{oss5}, C_{oss6}, C_{oss7} and C_{oss8} is not clamped, so it resonates with resonant inductor L_r. When the switch S_2 is off, this stage is end.

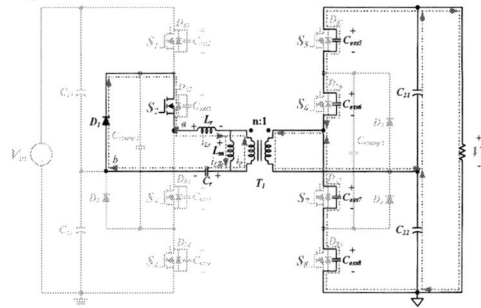

Figure 8. Stage 4

Stage 5 ($t_4 \sim t_5$):

As shown in Fig. 9, all the power switches are turned off. The resonant current i_{Lr} is maintained in this period, the C_{oss2} is charged, the C_{oss3} is discharged, the current of the clamped diode D_1 gradually decreases and the current which flows through the body diode D_{s4} of the switch S_4 is gradually increases. V_{ab} decreases from zero voltage to $-V_{in}/2$. C_{oss5} and C_{oss6} are charged to $+nV_o/2$, C_{oss3}, C_{oss7} and C_{oss8} are discharged to zero voltage when C_{oss2} is charged to $+V_{in}/2$.

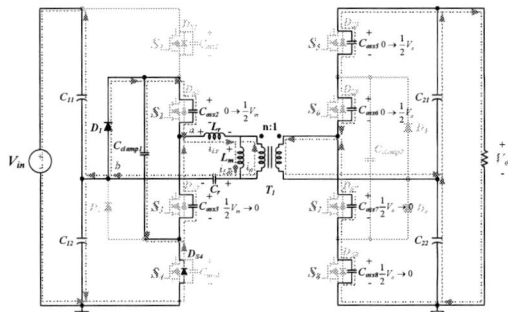

Figure 9. Stage 5

647

Stage 6 ($t_5 \sim t_6$):

The conduction path shown in Fig. 10, C_{oss2}, C_{oss5} and C_{oss6} has been charged to $+V_{in}/2$, and C_{oss3}, C_{oss7} and C_{oss8} has been discharged to zero voltage. the resonant current i_{Lr} is turned on for the body diodes D_{S3}, D_{S4}, D_{S7} and D_{S8} of the switches S_3, S_4, S_7 and S_8. At this time the magnetizing inductance L_m cross-pressure $-nV_o/2$, so the magnetizing inductance current i_{Lm} also linearly decline, when the switch S_3, S_4, S_7 and S_8 conduction, the end of this stage.

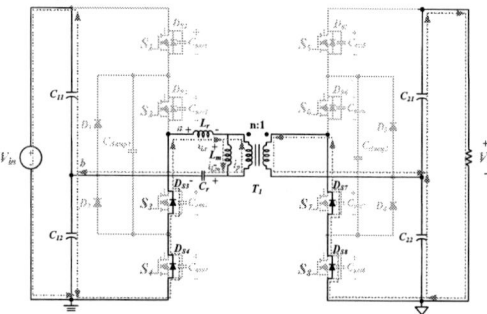

Figure 10. Stage 6

C. Resonant tank design

Since the structure of this paper is operated in the SRC interval, the resonant period and the effective duty cycle are shown in Fig. 11. The time of its half cycle must be greater than the time of the effective duty cycle. The relationship is shown as follow (3).

$$\pi \cdot \sqrt{L_r \times C_r} \geq DT_s \quad (3)$$

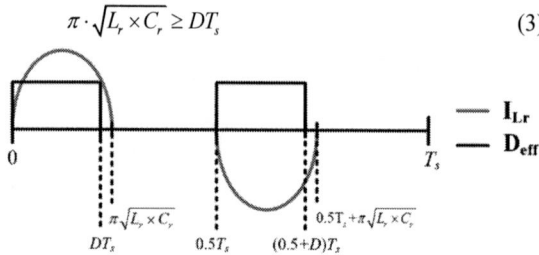

Figure 11. diagram of resonant period and effective duty cycle

III. VERIFICATION OF SIMULATION AND IMPLEMENTATION

This paper actually developed a 3 kW bi-directional half-bridge third-order series resonant converter, its specifications are shown as follows:

- Input voltage V_{in} : 1 kV
- Output voltage V_o : 1 kV
- Output current I_o : 3 A
- Output power P_o : 5 kW

According to the previous circuit parameter design, the simulation results are verified by PSIM circuit simulation software. The relevant parameters are set as follows: input voltage V_{in} = 1 kV, switching frequency f_s = 50 kHz, resonant capacitor C_r = 500 nF, resonant inductance L_r = 7.20 µH, transformer turns ratio n is 1, magnetizing inductance L_m = 4.80 mH, each Switch capacitor C_{oss} = 80 pF, clamp capacitor C_{clamp} = 880 nF, the output capacitor C_{out} = 170 µF.

Fig. 12 shows the gate signal in all the switches, the switches S_1, S_5 and S_6 signals are the same which are shorter than switches S_2, S_4, S_7 and S_8 signals are the same which are shorter than switch S_3 and the switches S_2 and S_3 signal is complementary.

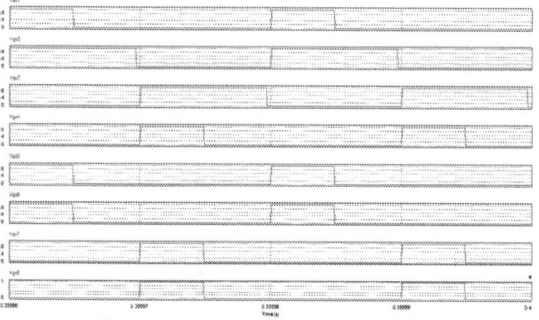

Figure 12. Gate-singals timing waveform

Fig. 13 shows the waveforms of V_{ab} and i_{Lr} at full load, V_{ab} is the three-level square wave, and the resonant current i_{Lr} operates in first resonant interval. Fig. 14 shows the waveforms of V_{gs} and V_{ds} of switches S_1 and S_4, it is clear that the switches S_1 and S_4 are zero voltage switching.

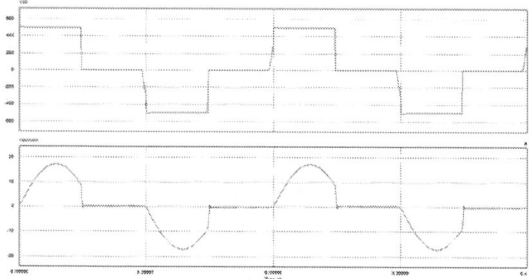

Figure 13. V_{ab} and i_{Lr} waveforms at full load

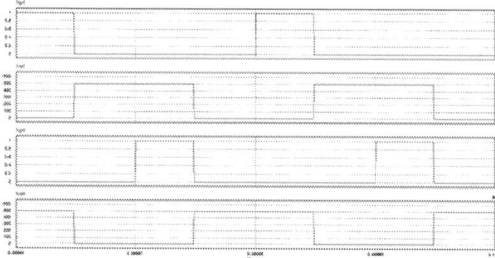

Figure 14. V_{gs} and V_{ds} of S_1 and S_4 waveforms

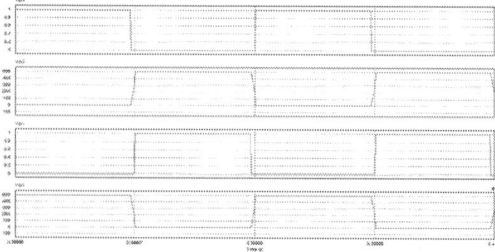

Figure 15. V_{gs} and V_{ds} of S_2 and S_3 waveforms

Fig. 15 shows the waveforms of V_{gs} and V_{ds} of switches S_2 and S_3. It can be seen that the energy stored in the magnetizing inductance is not enough to achieve zero-voltage switching on switches S_2 and S_3, because the magnetizing inductance L_m is designed to achieve quasi-zero current switching on switches S_2 and S_3, the magnetizing inductance current i_{Lm} is much smaller than the resonant current i_{Lr} of the transfer energy interval.

Using the previously designed component parameters and simulation results, the laboratory prototype is implemented and verified. The following results show the experimental waveforms on forward mode and reverse mode. Fig. 17 shows the experimental waveforms on forward mode at full load. Fig. 18 shows the experimental waveforms on reverse mode at full load.

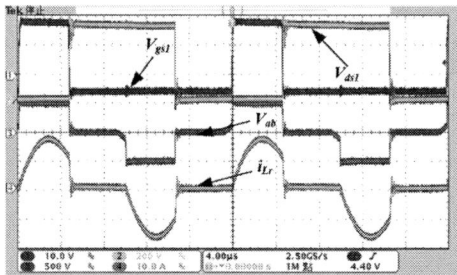

Figure 17. Forward mode waveform at full load

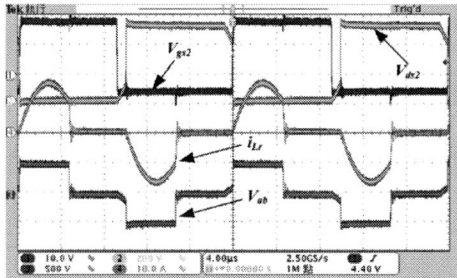

Figure 18. Forward mode waveform at full load

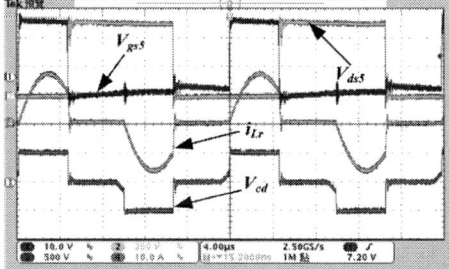

Figure 19. Reverse mode waveform at full load

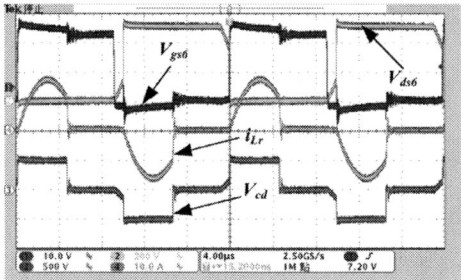

Figure 20. Reverse mode waveform at full load

As shown in Fig. 17 and Fig. 19, S_1 and S_5 both achieve the zero voltage switch in both forward and reverse

directions, and both V_{ab} and V_{cd} are the third-level square wave and achieves the voltage balance. As shown in Fig. 18 and Fig. 20, although the S_2 and S_6 don't reach the zero voltage switching, they can achieve the zero-current switching, so that the loss on the switching can be effectively reduced. The efficiency curves are shown in Fig. 21.

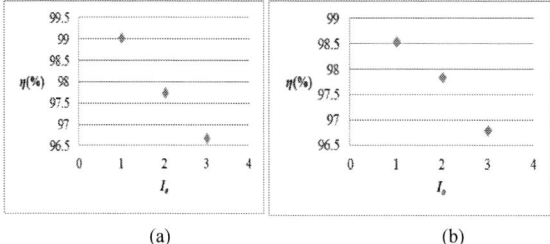

(a)   (b)

Figure 21. Efficiency curve (a) Forward mode (b) Reverse mode

IV. CONCLUSION

In order to improve the transmission efficiency in response to long-distance, high-power and high-efficiency bi-directional power transmission, the transmission voltage is increased to improve the transmission efficiency, but the voltage stress of the component is also increased. Therefore, the multi-level converter architecture has been proposed.

This paper focuses on the development of a bi-directional half-bridge third-order series resonant converter, the third-order architecture can reduce the voltage stress on power switch effectively, by using the diode and capacitor clamp circuit, the switch can achieve zero voltage switching, therefore, the switching loss can be reduced effectively, the switching frequency can be increased to pursue higher power density. In the secondary side, adding synchronous rectification to further reduce the conduction loss of the switch. At the end of this paper, a high-voltage bidirectional isolated third-order resonant converter with input voltage of 1 kV, output current of 3 A and output power of 3 kW is completed by using digital signal processor chip TMS320F28035, conversion efficiency are more than 96% in the light, medium and full load.

REFERENCES

[1] J. S. Lai, and F.Z. Peng, "Multilevel Converters – A New Breed of Power Converter," IEEE Transactions. Industry Application, vol. 32, no. 3, pp. 509-517, May/Jun. 1996.

[2] K. Jin and X. B. Ruan, "Hybrid Full-Bridge Three-Level LLC Resonant Converter—A Novel DC–DC Converter Suitable for Fuel-Cell Power System," in Proc. Power Electronics Specialists Conference, 2005, pp.361-367.

[3] X. Wang, G. Z. Wang, Y. B. Wang, X. W. Sun and Z. J. Ou, "Three-Level Half-Bridge LLC Converter with Phase Shift and Frequency Modulation Control," in Proc. Annual Conference of the IEEE, 2016, pp.1429-1434.

[4] Y. Y. Cai, C. Wang, F. Zhao and R. Dong, "Design of A High-frequency Isolated DTHB CLLC Bidirectional Resonant DC-DC Converter," in Proc. IEEE Conference and Expo Transportation Electrification Asia-Pacific, 2014, pp.1-6.

[5] F. Canales, P. M. Barbosa, J. M. Burdio, and F. C. Lee, " A Zero Voltage Switching Three-Level DC/DC Converter," in Proc.

International Telecommunications Energy Conference, 2000, pp. 512-517

[6] L. Corradini, D. Seltzer, D. Bloomquist, R. Zane, D. Maksimović and B. Jacobson, "Zero Voltage Switching Technique for Bi-Directional DC/DC Converters," in Proc. IEEE Energy Conversion Congress and Exposition, 2011, pp.2215-2222.

[7] C. Wang, F. Zhao, Q. M. Gao, Y. Y. Cai, and H. Cheng, "Performance Analysis of High-Frequency Isolated Dual Half-Bridge Three-Level Bi-Directional DC/DC Converter," in Proc. IEEE Conference and Expo Transportation Electrification Asia-Pacific, 2014, pp.1-6.

[8] F. Canales, P. M. Barbosa and F. C. Lee, "A Zero-Voltage and Zero-Current Switching Three-Level DC/DC Converter," IEEE Transactions on Power Electronics, vol. 17, no. 6, pp. 898-904, Nov. 2002.

The 2018 International Power Electronics Conference

A Control Strategy for Flying-Start of Shaft Sensorless Permanent Magnet Synchronous Machine Drive

Zih-Cing You and Sheng-Ming Yang*

Electrical Engineering, National Taipei University of Technology, Taipei, Taiwan

*E-mail: smyang@ntut.edu.tw

Abstract—**Starting a spinning shaft sensorless controlled permanent magnet synchronous machine drive is difficult due to the lack of rotor position and speed information. This information is needed for the current and speed controllers when they are activated initially. Particularly, current controller requires back-EMF to calculate the decoupling voltage to prevent drive overcurrent. In this paper, a restarting strategy for back-EMF based shaft sensorless drive is proposed. The existing position and speed estimator for shaft sensorless control is used and no additional algorithm is required. The current controller is set to a particular condition so the back-EMF can be estimated and decoupled. With the proposed strategy the regeneration current is suppressed, and the motor can start from various spinning conditions with either positive or negative speed.**

Key words: **shaft sensorless control, flying start, PMSM**

I. INTRODUCTION

Shaft position sensor is generally used to detect the rotor position for implementation of the vector control for permanent magnet synchronous machine (PMSM). However, these sensors not only increase the cost but also degrade reliability of the motor drive. Elimination of the shaft position sensor can be achieved by using the machine itself as a sensor. Many shaft position sensorless control strategies have been developed in recent years. These strategies generally fall into two groups: (1) saliency-based, for example high frequency voltage injection (HFVI) [1-3], and (2) back-EMF based [4-5]. Because these two approaches have complimentary speed range limitations, two different sensorless control algorithms are generally mixed to achieve full speed range operations.

A stable start-up of shaft sensorless controlled PMSM can be accomplished with saliency-based control algorithms. However, starting a shaft sensorless controlled PMSM drive when it is spinning, which is known as the flying start, is difficult due to the lack of position and speed information. The main challenge is in the presence of the back-EMF voltage. This voltage cannot be decoupled from the current controller before the rotor position and speed are identified. Consequently, when the inverter starts to synthesize terminal voltages in a spinning condition, the back-EMF induces regeneration current due to the absence of the decoupling voltage. The regeneration current may cause unexpected motor rotation and rapid rise of the dc-bus voltage. For safety concern, a restarting strategy that

can suppress regeneration current is essential for PMSM drives.

Several restarting strategies have been developed for shaft sensorless PMSM drives in recent years. Most of these methods apply zero-voltage vector pulses to identify the initial rotor position and speed and avoid regeneration current [6-8]. In [9], additional zero-voltage vector pulses are applied to reduce speed estimation error resulting from the limited time interval between the two zero-voltage vector pulses. To mitigate the sensitivity to the parameter and speed variations, an adjustment procedure for the time duration and interval of zero-voltage vector pulse is developed in [10] based on the methods described in [7-9]. Although these methods are feasible for implementation in shaft sensorless PMSM drives, they are generally complex and sensitive to speed variations and is redundant to the existing sensorless control algorithms.

This paper presents a restarting strategy for the back-EMF based shaft sensorless PMSM drives. The existing position and speed estimator for shaft sensorless control is used and no additional algorithm is required. With the proposed strategy the regeneration current produced by the back-EMF is suppressed.

II. SHAFT SENSORLESS CONTROL SYSTEM

This paper implements a mixed sensorless control strategy for IPMSM with the saliency- and the back-EMF based sensorless techniques. Figure 1 shows the block diagram of the control system. A saliency-based algorithm with square wave HFVI estimates shaft position at low speed, while a back-EMF based algorithm estimates shaft position at high speed. A transition algorithm combines the estimator results for the shaft position when the motor is operating in the transition region.

The stator voltage for interior permanent magnet PMSM in the rotor reference frame can be expressed as

$$\begin{bmatrix} v_{qs}^r \\ v_{ds}^r \end{bmatrix} = \begin{bmatrix} r_s + L_{qs}s & \omega_r L_{ds} \\ -\omega_r L_{qs} & r_s + L_{ds}s \end{bmatrix} \begin{bmatrix} i_{qs}^r \\ i_{ds}^r \end{bmatrix} + \begin{bmatrix} \omega_r \lambda_m \\ 0 \end{bmatrix} \quad (1)$$

where v_{qs}^r, v_{ds}^r, i_{qs}^r, and i_{ds}^r are the q- and d-axis voltages and currents, respectively, L_{qs} and L_{ds} are the q- and d-axis inductance, respectively, r_s, ω_r and λ_m are the phase resistance, rotor speed and magnet flux, respectively, and s is the differential operator.

The 2018 International Power Electronics Conference

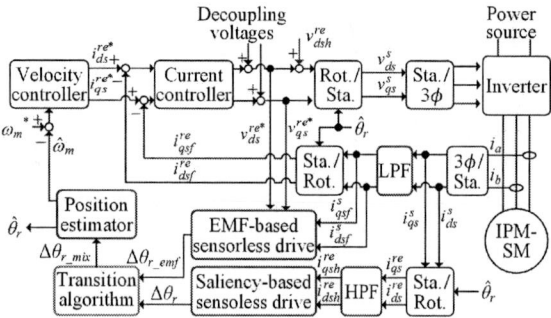

Fig. 1. Control system for the shaft sensorless speed drive.

A. Saliency-based shaft sensorless control

The saliency-based sensorless drive implemented in this paper is similar to that reported in [3]. A square-wave voltage is injected in the estimated d-axis, and the saliency spatial signal is extracted from the induced q-axis difference current. Let v_{inj} denotes the magnitude of the injection voltage, the induced difference currents are

$$\begin{bmatrix} \Delta i_{qsh}^{re} \\ \Delta i_{dsh}^{re} \end{bmatrix} = \text{sign}(\pm v_{inj}) \cdot \frac{\pm v_{inj} \cdot \Delta T}{(L_\Sigma^2 - L_\Delta^2)} \begin{bmatrix} L_\Delta \sin(2\Delta\theta_r) \\ L_\Sigma + L_\Delta \cos(2\Delta\theta_r) \end{bmatrix} \quad (2)$$

and

$$L_\Sigma = (L_{qs} + L_{ds})/2, \; L_\Delta = (L_{qs} - L_{ds})/2 \quad (3)$$

$$\Delta\theta_r = \theta_r - \hat{\theta}_r \quad (4)$$

where subscript 'h' denotes high-frequency (HF) quantities, θ_r is the rotor position, and $\hat{\theta}_r$ is the rotor position seen from the estimated frame. Note that '$\wedge$' denotes estimated quantities. As shown in Eq. (2), as the estimated rotor frame does not aligned with the actual rotor frame a $2\Delta\theta_r$ position-dependent current signal is induced in both d- and q- axis currents. The $\pm$ sign compensation is required due to the square-wave voltage injection. Also, in order to calculate differential currents a high-pass filter (HPF) is implemented to remove the low frequency contents in the current.

When the position error is sufficiently small, the position-error-dependent current signal in the q-axis is linearized to

$$\Delta i_{qsh}^{re} \approx \frac{v_{inj} \cdot \Delta T \cdot L_\Delta}{(L_\Sigma^2 - L_\Delta^2)} \cdot 2\Delta\theta_r = k_{err} \cdot \Delta\theta_r \quad (5)$$

Express the position error as a function of the current error as

$$\Delta\theta_r \approx \Delta i_{qsh}^{re} / k_{err} \quad (6)$$

Then, rotor position can be estimated by manipulating $\Delta\theta_r$ to zero with a closed loop estimator.

B. EMF-based shaft sensorless control

Motor position can also be estimated based on the extended back-EMF voltage [4-5]. Rewrite Eq. (1) as

$$\begin{bmatrix} v_{qs}^s \\ v_{ds}^s \end{bmatrix} = \begin{bmatrix} r_s + L_{ds}s & P\omega_m L_\Delta \\ -P\omega_m L_\Delta & r_s + L_{ds}s \end{bmatrix} \begin{bmatrix} i_{qs}^s \\ i_{ds}^s \end{bmatrix} + \begin{bmatrix} e_{qs} \\ e_{ds} \end{bmatrix} \quad (7)$$

where $e_{qs}=E_b \cos(\theta_r)$ and $e_{ds}=-E_b \sin(\theta_r)$ are known as the extended back-EMF, $E_b = L_\Delta(pi_{qs}^r - \omega_r i_{ds}^r) + \omega_r \lambda_m$, and P is the number of rotor poles. With the knowledge of parameters the extended back-EMF is calculated as

$$\begin{bmatrix} \hat{e}_{qs} \\ \hat{e}_{ds} \end{bmatrix} = \hat{E}_b \begin{bmatrix} \cos\theta_{r\_emf} \\ -\sin\theta_{r\_emf} \end{bmatrix} = \begin{bmatrix} v_{qsf}^{s*} \\ v_{dsf}^{s*} \end{bmatrix} - \begin{bmatrix} \hat{r}_s + \hat{L}_{ds}s & P\hat{\omega}_m \hat{L}_\Delta \\ -P\hat{\omega}_m \hat{L}_\Delta & \hat{r}_s + \hat{L}_{ds}s \end{bmatrix} \begin{bmatrix} i_{qsf}^s \\ i_{dsf}^s \end{bmatrix} \quad (8)$$

where the subscript 'f' denotes fundament-frequency quantities, '$*$' denotes the command value, ω_m is the mechanical speed and, and θ_{r_emf} is the rotor position estimated from the back-EMF. A low-pass filter (LPF) is used to remove the high frequency components in motor currents to avoid high frequency noise. Then, a position error dependent signal ($\Delta\theta_{emf}$) is extracted from the following vector product,

$$\Delta\theta_{emf} = (-\hat{e}_{qs}\sin\hat{\theta}_r - \hat{e}_{ds}\cos\hat{\theta}_r)\text{sign}(\hat{\omega}_m)/|\hat{E}_b|$$
$$= \text{sign}(\hat{\omega}_m)\frac{E_b}{|\hat{E}_b|}\sin(\theta_{r\_emf} - \hat{\theta}_r) \quad (9)$$

Note that the estimated back-EMF is normalized with its magnitude, and the sign of the estimated speed is required to correct the phase error for negative speeds. Figure 2 shows the calculations of $\hat{e}_{qs}$ and $\hat{e}_{ds}$. When the position error is sufficiently small $\Delta\theta_{emf}$ can be approximated as

$$\Delta\theta_{emf} \approx \theta_{r\_emf} - \hat{\theta}_r \quad (10)$$

Thus, rotor position can be estimated by manipulating $\Delta\theta_{emf}$ to zero with a closed loop estimator.

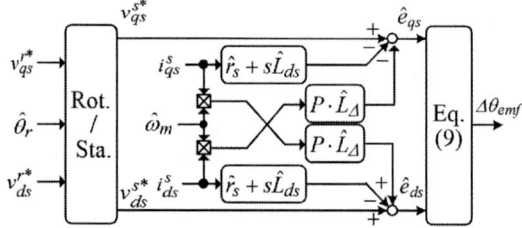

Fig. 2. Extended back-EMF estimator.

C. Transition period control

In the transition period, a speed dependent weighting function mixes the position error generated by the saliency-based ($\Delta\theta_r$ in (6)) and the back-EMF based ($\Delta\theta_{emf}$ in (10)) algorithms as follows,

$$\Delta\theta_{r\_mix} \approx (1-G_\omega) \cdot \Delta\theta_r + G_\omega \cdot \Delta\theta_{r\_emf} \quad (11)$$

where G_ω is the weighting function. Then, the position estimator illustrated in Fig. 3 estimates the rotor position and speed through the convergence of $\Delta\theta_{r\_mix}$. The mixed position error signal $\Delta\theta_{r\_mix}$ is the input to the estimator. The estimator is designed according to the following mechanical model,

$$T_e = Js \cdot \omega_m + B \cdot \omega_m + T_L \quad (12)$$

where J is the inertia combined rotor and load, B is the frictional coefficient, and T_L is the load torque.

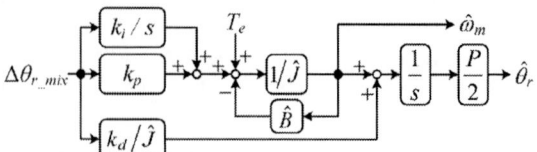

Fig. 3. Rotor position and speed estimator with the mixed position error signal.

III. FLYING-START WITHOUT POSITION AND SPEED INFORMATION

As shown in Fig. 1, the conventional PMSM rotor frame current controller requires rotor position for vector control and motor speed to calculate the decoupling voltages. However, when restarting without position and speed information, the controller is equivalent to a stationary frame controller with no feedforward voltage, as shown in Fig. 4. Because of no rotor position information, the frame transformation is disabled and the voltage commands become stationary frame quantities. The back-EMF becomes a disturbance to the current controller due to the absence of the decoupling voltage.

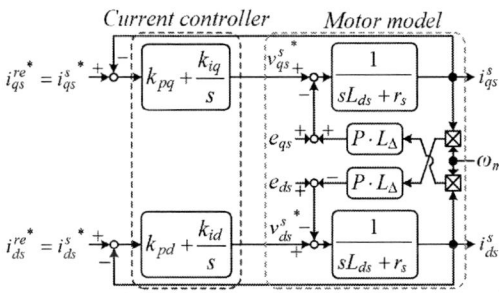

Fig. 4. Equivalent stationary frame current controller.

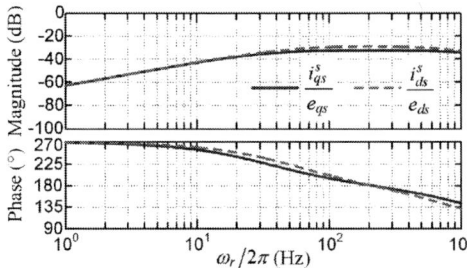

Fig. 5. Frequency response between stator current and back-EMF.

Because the cross-coupling voltages, $P\omega_m L_\Delta i_{qs}^s$ and $P\omega_m L_\Delta i_{ds}^s$, are smaller in comparisons to the back-EMF, they are ignored in the following analysis. Moreover, the current command is set to zero to deliver zero torque during the restart condition. Therefore, from Fig. 4, the approximated transfer functions between current and back-EMF are

$$\frac{i_{qs}^s}{e_{qs}} \approx -\frac{s}{L_{ds}s^2 + \left(r_s + k_{pq}\right)s + k_{iq}} \tag{13}$$

$$\frac{i_{ds}^s}{e_{ds}} \approx -\frac{s}{L_{ds}s^2 + \left(r_s + k_{pd}\right)s + k_{id}} \tag{14}$$

where k_{pq}, k_{iq} and k_{pd}, k_{id} are the proportional and integral gains for the q- and d-axis current controller, respectively. For convenience, the controller gains are designed based on the pole-zero cancelation method [11], and the bandwidth of both controllers are set at 1 kHz. Figure 5 shows the frequency response of these two functions. The parameters of the PMSM used in the calculations are given in Appendix A. As shown in this figure, for both functions the phases between regeneration current and back-EMF are all within 135° and 270°. Consequently, the currents

generate average and pulsating power back to the inverter. Moreover, the magnitude of the average and pulsating power are functions of rotor electrical speed.

By solving Eqs. (13) and (14) through inverse Laplace transform the steady-state regeneration currents can be found as

$$i_{qs}^s(t) = 2\left(E_b/Z_{eq}\right)\cdot\cos\left(\omega_r t + \phi_q\right) \tag{15}$$

$$i_{ds}^s(t) = -2\left(E_b/Z_{ed}\right)\cdot\sin\left(\omega_r t + \phi_d\right) \tag{16}$$

where ϕ_q and ϕ_d are the phase between the EMF voltage, Z_{eq} and Z_{ed} are the equivalent impedance, and they are

$$\begin{bmatrix} Z_{eq} \\ Z_{ed} \end{bmatrix} = \begin{bmatrix} \omega_r\Big/\sqrt{4\omega_r^2\left(r_s + k_{pq}\right)^2 + 4\left(k_{iq} - L_{ds}\omega_r^2\right)^2} \\ \omega_r\Big/\sqrt{4\left(L_{ds}\omega_r^2 - k_{id}\right)^2 + 4\omega_r^2\left(r_s + k_{pd}\right)^2} \end{bmatrix} \tag{17}$$

$$\begin{bmatrix} \phi_q \\ \phi_d \end{bmatrix} = \begin{bmatrix} \sin^{-1}\left(-2\left(k_{iq} - L_{ds}\omega_r\right)/Z_{eq}\right) \\ -90° - \sin^{-1}\left(-2\omega_r\left(r_s + k_{pd}\right)/Z_{ed}\right) \end{bmatrix} \tag{18}$$

The braking torque (T_{eb}) produced by the regeneration current can be calculated by substituting currents into the following expression:

$$T_{eb} = \frac{3}{4}P\lambda_m\cdot\left(\cos\theta_r\cdot i_{qs}^s - \sin\theta_r\cdot i_{ds}^s\right) \\ -\frac{3}{2}PL_\Delta\cdot\left(\sin\theta_r\cdot i_{qs}^s + \cos\theta_r\cdot i_{ds}^s\right)\left(\cos\theta_r\cdot i_{qs}^s - \sin\theta_r\cdot i_{ds}^s\right) \tag{19}$$

Because an analytic solution for T_{eb} is complicated, numerical analysis of T_{eb} is performed instead. Figure 6 shows T_{eb} for various speeds. It can be seen that significant average and pulsating torque exist in the machine at high speeds. The average torque represents the average power generated by the machine. Large average power may cause rapid rise of the dc bus and a potential damage to the power supply. On the other hand, the torque ripple represents the pulsating power produced by the machine, and they may cause unexpected motor rotations.

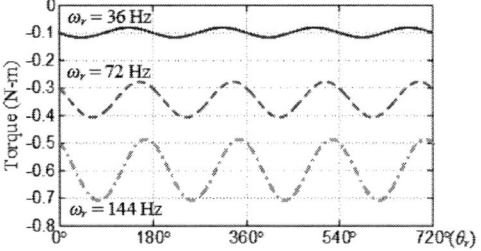

Fig. 6. Torque produced by the regeneration current.

IV. PROPOSED FLYING-START STRATEGY

The proposed restarting strategy uses the existing position and speed estimator. The relationships between the estimated and the actual back-EMF can be derived from substituting Eq. (7) into (8), and the results are

$$\hat{e}_{qs} = G_1\cdot e_{qs} + v_{qs}^{s*} - G_1\cdot v_{qs}^s + P\left(G_1\omega_m L_\Delta - \hat{\omega}_m\hat{L}_\Delta\right)i_{ds}^s \tag{20}$$

$$\hat{e}_{ds} = G_1\cdot e_{ds} + v_{ds}^{s*} - G_1\cdot v_{ds}^s + P\left(\hat{\omega}_m\hat{L}_\Delta - G_1\cdot\omega_m L_\Delta\right)i_{qs}^s \tag{21}$$

where $G_1 = \left(s\hat{L}_{ds} + \hat{r}_s\right)/\left(sL_{ds} + r_s\right)$. Note that the subscript 'f' in the variables are neglected for convenience. It can be seen that variation in parameters such as in L_{ds} and r_s causes amplitude and phase error between the estimated and the actual back-EMF voltages. Moreover, inverter nonlinearity such as dead-time effects also contributes errors to the estimated back-EMF. To mitigate these errors, motor parameters are measured with reasonable accuracy and the inverter dead-time effects are compensated. Other uncompensated errors are treated as disturbances to the back-EMF estimator.

Because rotor speed is not identified yet, the estimated speed $\hat{\omega}_m$ is set to zero. Similarly, current commands are also set to zero. Consequently, $G_1 = 1$, and Eqs. (20) and (21) become

$$\hat{e}_{qs} \approx e_{qs} + P\omega_m L_\Delta i_{ds}^s \tag{22}$$

$$\hat{e}_{ds} \approx e_{ds} - P\omega_m L_\Delta i_{qs}^s \tag{23}$$

Substituting Eqs. (13) and (14) into (22) and (23) and eliminate stator currents, the relationships between the estimated and the actual extended back-EMF can be expressed as

$$\frac{\hat{e}_{qs}}{e_{qs}} = \frac{\left(2 - L_{qs}/L_{ds}\right)s^2 + \left(r_s + k_{pq}\right)s/L_{ds} + k_{iq}/L_{ds}}{s^2 + \left(r_s + k_{pq}\right)s/L_{ds} + k_{iq}/L_{ds}} \tag{24}$$

$$\frac{\hat{e}_{ds}}{e_{ds}} = \frac{\left(2 - L_{qs}/L_{ds}\right)s^2 + \left(r_s + k_{pd}\right)s/L_{ds} + k_{id}/L_{ds}}{s^2 + \left(r_s + k_{pd}\right)s/L_{ds} + k_{id}/L_{ds}} \tag{25}$$

It can be seen that both functions have two poles and two zeros. At low frequencies their magnitudes approach 1, and at high frequencies their magnitudes are depend on the q- and d-axis inductance ratio. Figure 7 depicts their frequency responses for various saliency ratio. The current loop bandwidth is set at 1 kHz. It can be seen that the amplitude and phase error generally increases as rotor speed increases. However, for the machines with saliency ratio less than 2, the amplitude and phase error are not as severe when rotor speed is lower than the bandwidth of the current controller. This implies that $\hat{e}_{qs} \approx e_{qs}$ and $\hat{e}_{ds} \approx e_{ds}$.

Subsequently, back-EMF can be estimated with good accuracy under the above-mentioned conditions without position and speed information.

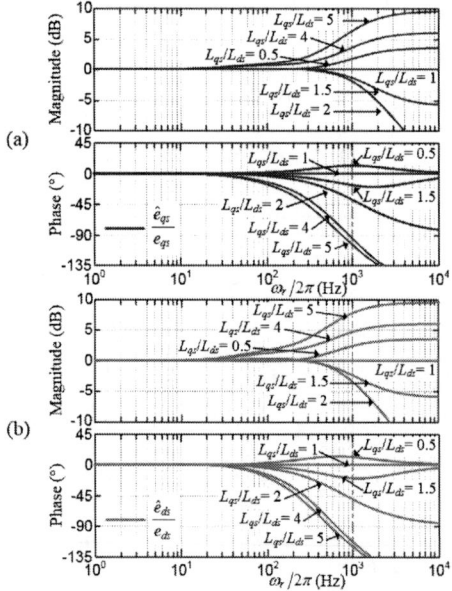

Fig. 7. Frequency response of (a) $\hat{e}_{qs}/e_{qs}$, (b) $\hat{e}_{ds}/e_{ds}$.

Figure 8 shows the current controller and the back-EMF estimator when the motor is restarting. Motor back-EMF is estimated right after the inverter and the current controller are activated. The estimated back-EMF is applied to decouple the actual back-EMF immediately after they are calculated. After the decoupling, the stator current produced by the back-EMF is suppressed and its influences are mitigated. Normal shaft sensorless control is then followed. Note that during restarting the back-EMF estimator is effectively in the stationary frame and contains no integrator, the settling time is inherently zero. Therefore, the transient current is very limited. Figure 9 shows the flowchart for the proposed flying start strategy. Note that the proposed strategy is valid regardless of motor rotating direction.

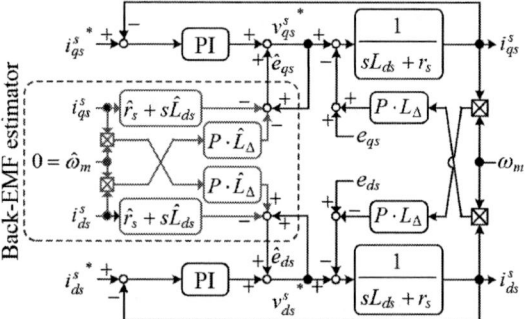

Fig. 8. Current controller with the decoupling voltage from the etimated back-EMF when restarting.

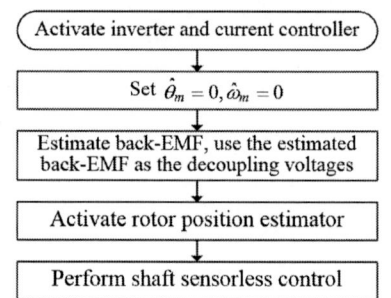

Fig. 9. Flowchart for the proposed flying start strategy.

V. EXPERIMENTAL RESULTS

An IPMSM is used for experimental verifications of the proposed restarting strategy. Appendix A gives its parameters. The shaft sensorless speed drive is implemented with a TMS320F28335 digital signal processor. The sampling rate for the current control and velocity controller are set to 18 kHz and 2.2 kHz, respectively. The bandwidth of the velocity control loop is set to 25 Hz. The poles of the position estimator are all located at 50 Hz. The dc power supply is 300 V. For the saliency-based sensorless control, the frequency and magnitude of the injection square-wave voltage is 9 kHz and 60 V, respectively. The transition speed zone between the saliency-based and the back-EMF based sensorless control algorithm is between 600 to 900 rpm. A load motor provides external load to the test motor. The actual rotor position and speed is monitored by an encoder with 2500 pulse/rev. Figure 10 shows the experimental system.

The 2018 International Power Electronics Conference

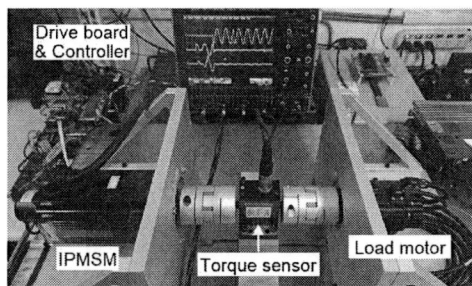

Fig. 10. Experimental system.

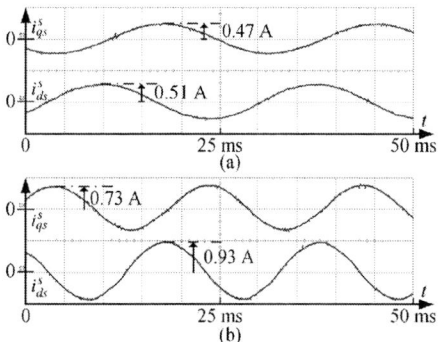

Fig. 11. Motor current waveforms when command is set to zero for both axes and the back-EMF is not decoupled, (a)1080 rpm, (b)1500 rpm.

Figure 11 shows the motor current waveforms when the current controllers are activated and the back-EMF is not decoupled, both current commands are set to zero. Motor speed is regulated by the load motor at 1080 rpm and 1500 rpm, respectively, when the experiments are conducted. It can be seen that current cannot be regulated effectively without the decoupling voltage. Figures 12(a) and (b) compare the calculated and the measured braking torque at 1080 rpm and 1500 rpm, respectively. The braking torque is measured with a torque meter. To match the measurement results, the calculated value is multiplied by a minus sign. It can be seen that the frequency of the torque ripple is twice of the rotor speed. In addition, the measured torque is highly consistent with the calculated values. Note that the peak torque is approximately 68% of the rated torque. These results verified that regeneration current produces undesirable braking torque.

Figure 13 compares the current responses with and without the decoupling voltage, speed is regulated at 1080 rpm. The back-EMF is estimated with the proposed technique. It can be seen that large stator currents appeared before $t = 0.05$s because the controller has no decoupling voltage. At $t = 0.05$s the decoupling voltage is applied, and the current ripple is suppressed effectively. This result verified that back-EMF can be estimated correctly without motor speed information. In addition, the estimated back-EMF is not affected by the sudden addition of the decoupling voltage.

Figure 14 shows the results of starting the motor as the load motor regulates speed at -3000 rpm. Current commands are set to zero. The inverter and current controller is activated at $t0$. It can be seen that significant regeneration currents appear as expected. The estimator activates and enables the decoupling voltage at $t1$. Then, the decoupling voltage suppresses the regeneration current almost instantaneously. Because the motor is running at a negative speed, a position error of 180° appears in the estimated position even the speed is correctly estimated. The controller runs normally after the phase correction.

Figure 15 shows the experimental results when the motor is restarted at -1500 rpm. The load motor turned off at $t0$, and the PMSM is coasting down during time interval $t0$–$t1$. The restarting control is activated at $t1$ and completed at $t2$. After the motor is restarted, the shaft sensorless speed control algorithm is enabled and drives the motor speed upward with a constant torque. After motor speed reached the pre-set command (3000 rpm), the controller switches to regular speed regulation.

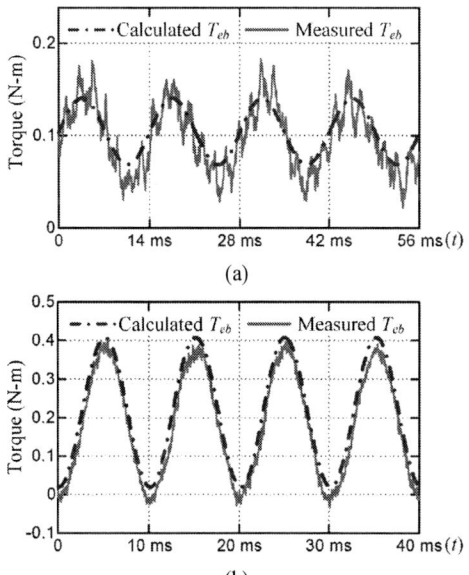

Fig. 12. Comparison of measured and calculated torque at (a) 1080 rpm, (b) 1500 rpm.

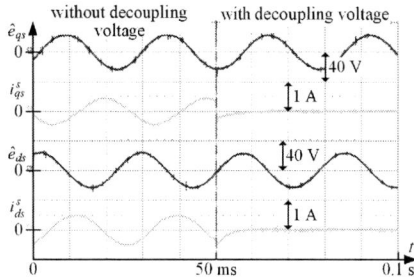

Fig. 13. Comparison of the current responses with and without voltage decoupling, motor speed 1080 rpm.

Figure 16 shows the motor restarts from an unexpected error occurrence with the rated load torque. In the beginning, the PMSM is running with sensorless speed control and subjected to the rated load. Then, the inverter is turned off at $t0$ to emulate an error has occurred. The motor coasting down rapidly during $t0$–$t1$ due to the large load torque. Then, the restarting procedure activates at $t1$. After the motor is restarted, the sensorless control is implemented with constant current command to drive

655

The 2018 International Power Electronics Conference

motor speed upward. Finally, the system switches to the regular sensorless speed control after the speed reached the pre-set command (3000 rpm).

VI. CONCLUSION

This paper presents a restarting strategy for back-EMF based shaft sensorless drive when the motor is in spinning condition. In the proposed scheme, the existing back-EMF based position and speed estimator is used and no additional algorithm is required. The analytic results show that motor back-EMF can be estimated with good accuracy without rotor position and speed information when restarting. The experimental results verified that the regeneration current can be suppressed effectively, and the motor can restart successfully from various spinning conditions.

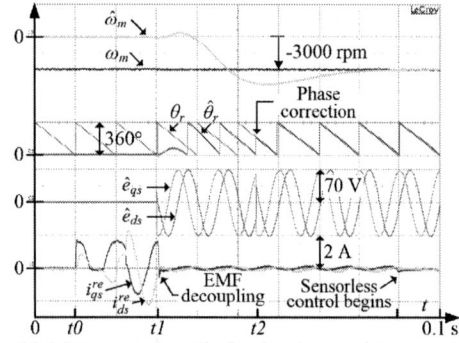

Fig. 14. Motor restarts as the load motor regulates speed at -3000 rpm.

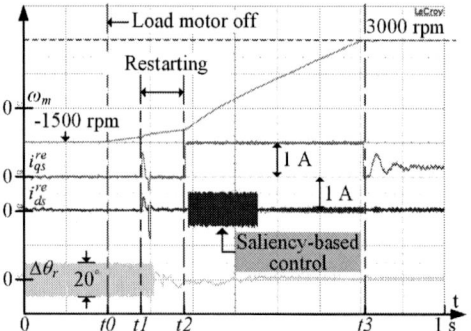

Fig. 15. Motor restarts from free running condition.

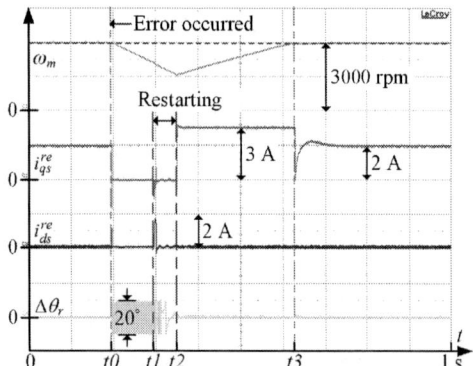

Fig. 16. Motor restarts with rated load.

APPENDIX A   MOTOR PARAMETERS

Parameter	Value	Unit
Rated power	370	W
Rated speed/pole	6000/4	rpm
Magnet flux (λ_m)	0.11	Wb-turns
q-axis inductance (L_{qs})	7.1	mH
d-axis inductance (L_{ds})	4.8	mH

REFERENCES

[1] J. M. Liu, and Z. Q. Zhu, "Sensorless Control Strategy by Square-Waveform High-Frequency Pulsating Signal Injection into Stationary Reference Frame," *IEEE Journal of Emerging and Selected Topics in Power Electronics*, vol. 2, pp. 171-180, Jun. 2014.

[2] Y. D. Yoon, S. K. Sul, S. Morimoto, and K. Ide, "High-Bandwidth Sensorless Algorithm for AC Machines Based on Square-Wave-Type Voltage Injection," *IEEE Transactions on Industry Applications*, vol. 47, pp. 1361-1370, May/Jun. 2011.

[3] S. C. Yang, S. M. Yang, and J. H. Hu, "Robust Initial Position Estimation of Permanent Magnet Machine with Low Saliency Ratio," *IEEE Access*, vol. 5, pp. 2685-2695, Feb. 2017.

[4] J. M. Liu and Z. Q. Zhu, "Novel Sensorless Control Strategy with Injection of High-Frequency Pulsating Carrier Signal into Stationary Reference Frame," *IEEE Transactions Industry Applications*, vol. 50, pp. 2574-2583, Jul./Aug. 2014.

[5] Z. Chen, M. Tomita, S. Doki, and S. Okuma, "An Extended Electromotive Force Model for Sensorless Control of Interior Permanent-Magnet Synchronous Motors," *IEEE Transactions on Industrial Electronics*, vol. 50, pp. 288-295, Apr. 2003.

[6] H. Yoo, J. H. Kim, and S. K. Sul, "Sensorless Operation of a PWM Rectifier for a Distributed Generation," *IEEE Transactions on Power Electronics*, vol. 22, pp. 1014-1018, May 2007.

[7] Y. C. Son, S. J. Jang, and R. D. Nasrabadi, "Permanent Magnet AC Motor Systems and Control Algorithm Restart Methods," U.S. Patent US8054030B2, 2011.

[8] Y. C. Son, J. Jang, B. A. Welchko, N. R. Patel, and S. E. Schulz, "Method and System for Initiating Operation of an Electric Motor," U.S. Patent US8319460B2, 2012.

[9] S. Taniguchi, S. Mochiduki, T. Yamakawa, S. Wakao, K. Kondo, and T. Yoneyama, "Starting Procedure of Rotational Sensorless PMSM in the Rotating Condition," *IEEE Transactions on Industry Applications*, vol. 45, pp. 194-202, Jan./Feb. 2009.

[10] K. Lee, S. Ahmed, and S. M. Lukic, "Universal Restart Strategy for High-Inertia Scalar-Controlled PMSM Drives," *IEEE Transactions on Industry Applications*, vol. 52, pp. 4001-4009, Sept./Oct. 2016.

[11] Y. Inoue, K. Yamada, S. Morimoto, and M. Sanada, "Effectiveness of Voltage Error Compensation and Parameter Identification for Model-Based Sensorless Control of IPMSM," *IEEE Transactions on Industry Applications*, vol. 45, pp. 213-221, Jan. 2009.

Contactless EV Power Track System with Segment-Excited Inductively Coupled Structure

Jia-You Lee, Yu-Chi Wang and Chih-Yi Liao*

Department of Electrical Engineering, National Cheng Kung University, Tainan, Taiwan
*E-mail: N26054312@mail.ncku.edu.tw

Abstract— **This paper is aimed to utilize the technology of contactless power transmission to design and implement contactless electric vehicle power track system with segment-excited inductively coupled structure array. To increase the tolerances of lateral and longitudinal displacement that between vehicle and the track, the simulation software have analyzed the track coil of magnetic field and selected the spiral structure with a uniform magnetic field distribution. The size of pickup coil has also been determined to let the system receive power efficiently. The overall track consists of pad arrays. For reducing power loss caused by turning on the track power simultaneously, the segment-excited control has been set up in the system. Based on the experimental results, the maximum output power of system is 910.6 W and the highest efficiency is 75% with 160 V input dc supply voltage and load resistance 80.7 Ω under 12 cm air-gap.**

Keywords—Contactless power track system, inductively coupled structure array, segment-excited control system.

I. INTRODUCTION

In recent years, large consumption of fossil fuels makes air pollution considerably worse and makes nature resources around the world gradually reduce. To i mprove these problem, people have begun finding the way to replace fossil fuels. Electric vehicle (EV) which utilizes electrical energy to provide motive power with zero carbon emissions is the most common research in the aspect of transport. If the electric-powered public transport had operated a long distance, it should be recharged frequently. Therefor, the method of transfering power to EVs by the power track have been considered to reduce battery bank capacities and range anxiety. The power supply track for EVs could divide into traditional contact power and contactless inductive power transmission as shown in Fig. 1 [1]. A large air-gap between the power track and EV in contactless power transfer track system would result in reducing power transmission efficiency, but it is more safe and convenient than traditional contact power transmission because of wireless charging. We could design the inductively coupled structure with a uniform magnetic field distribution [2] and choose the LCCL resonant circuit to make the power transfer ability of the system become better [3]. In addition, the track consists of many arrays and segment-excited control is utilized for improving the power transmission efficiency. The power of the track array only turned on and transferred power when EVs passed above it.

(a)　　　　　　　　(b)

Fig. 1. The way of power transmission for EV [1]. (a) Contact power transmission. (b) Contactless inductive power transmission.

II. ANALYSIS OF CONTACTLESS INDUCTIVE POWER TRANSFER SYSTEMN

When the contactless inductively coupled system transfer power, the induced voltage is generated on both the primary and secondary system as shown in Fig. 2. Besides, the equivalent secondary impedance would reflect to the primary and cause the situation about the interference of the primary resonant circuit. To understand these problems, First, the reflected impedance from the secondary to the primary is

$$Z_r = \frac{-j\omega M i_s}{i_p} = \frac{\omega^2 M^2}{Z_s} \tag{1}$$

where the the subscripts "s" and "p" stand for primary and secondary, Z_s represents the impedance of the secondary circuit, M is the mutual inductance between primary and secondary.

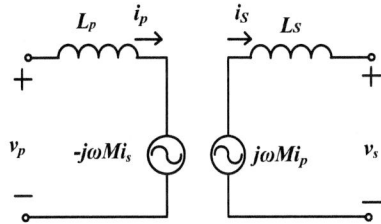

Fig. 2. The mutual inductance coupling model with dependent source.

If the secondary circuit is an open circuit, the open-circuit voltage induced in pickup coil L_s is

$$V_{oc} = j\omega M i_p \tag{2}$$

and the short-circuit current is given by

$$I_{sc} = \frac{V_{oc}}{j\omega L_s} = \frac{M i_p}{L_s} \tag{3}$$

The maximum uncompensated power of the pickup S_u is the product of the short circuit current I_{sc} and the open circuit voltage V_{oc}. To improve the power transfer ability to the pickup, the inductor with capacitors would provide a resonant frequency which is the same as the frequency of track system. Therefore, the maximum output power P_{max} [4] is the product of S_u and the quality factor of pickup circuit Q

$$P_{\max} = S_u \cdot Q = \omega \cdot \frac{M^2}{L_s} \cdot i_p^2 \cdot Q \tag{4}$$

The contactless power supply track system is different from the traditional transformer. It has a large distance between the track and EV, the efficiency would reduce because there is a large number of leakage flux loss in the inductively coulped structure. To improve the overall power transmission efficiency, the resonant circuit is added on both track and pickup structure. When resonant frequency is equal to operating frequency of the system, output voltage and current of inverter are in phase, inductive and capacitive reactance would offset each other then equivalent impedance is regard as a resistance.

The track coils on the primary with high voltage and current could enhance power transmission efficiency of system. Analyzing resonant circuit and choosing appropriate topology to let the track coil provide a large magnetic field. The high voltage and low current is the characteristics of series LC circuit on primary for the form of power supply. The disadvantage is that output-voltage regulation at no load Z_r is poor. The parallel LC resonant circuit with low voltage, high current and load regulation is suitable for the input current supply. To improve this shortcoming, we chose the LCCL circuit on track structure. If the series LC circuit is used on the primary as shown in Fig. 3, the voltage gain, equivalent impedance and quality factor Q_p are

$$\frac{v_r}{v_I} = \frac{1}{(\frac{r_p + Z_r}{Z_r}) + j \cdot Q_p \cdot (\frac{\omega}{\omega_o} - \frac{\omega_o}{\omega})} = \frac{Z_r}{r_p + Z_r} \bigg|_{\omega = \omega_o = \frac{1}{\sqrt{L_p \cdot C_1}}} \tag{5}$$

$$Z_I = r_p + Z_r + j \cdot Q_p \cdot (\frac{\omega}{\omega_o} - \frac{\omega_o}{\omega}) = r_p + Z_r \bigg|_{\omega = \omega_o = \frac{1}{\sqrt{L_p \cdot C_1}}} \tag{6}$$

$$Q_p = \frac{\omega_o \cdot L_p}{Z_r} \tag{7}$$

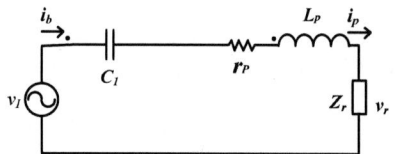

Fig. 3. Circuit diagram of the series LC circuit on primary (track side).

When Z_r is a real impedance and r_P is not 0, Fig. 4 (a) shows the primary output voltage V_r at $\omega = \omega_o$ with a fixed input voltage V_I would change with Z_r. At the same time,

the sensitivity of operating frequency ω increase because of higher Q_p, the small offset of ω may cause the output voltage reduce dramatically. For the primary series resonant circuit, the variation of load would obviously affect the system stability. In addition, if $Z_r=0$, Q_p would become greater resulting in the high inverter output current, it would cause a problem about the damage of inverter and power supply as shown in Fig. 4 (b). Therefore, we have considered another resonate circuit to avoid these disadvantage.

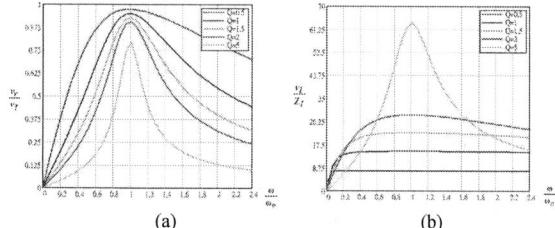

(a)   (b)

Fig. 4. The characteristics of series LC circuit on the primary. (a) Voltage gain versus normalized operating frequency. (b) Inverter current versus normalized operating frequency.

The LCL resonant circuit is used on the primary as shown in Fig. 5, the current gain, equivalent impedance and quality factor Q_p are

$$\frac{i_p}{v_I} = \frac{\frac{1}{\omega_o \cdot L_b}}{(1 - \frac{\omega^2}{\omega_o^2})(\frac{r_p}{\omega_o \cdot L_b} + \frac{1}{Q_p}) + j[\frac{\omega}{\omega_o} + \frac{L_p}{L_b}(\frac{\omega}{\omega_o} - \frac{\omega^3}{\omega_o^3})]} \tag{8}$$

$$= \frac{1}{j\omega \cdot L_b} \bigg|_{\omega = \omega_o = \frac{1}{\sqrt{L_b \cdot C_1}}}$$

$$Z_I = \frac{j\omega L_b + j\omega L_p (1 - \omega^2 L_b C_1) + (r_p + Z_r)(1 - \omega^2 L_b C_1)}{(1 - \omega^2 L_b C_1) + j\omega C_1(r_p + Z_r)}$$

$$= \frac{(\omega L_b)^2}{(r_p + Z_r)} \bigg|_{\omega = \omega_o = \frac{1}{\sqrt{L_p \cdot C_1}} = \frac{1}{\sqrt{L_b \cdot C_1}}} \tag{9}$$

$$Q_p = \frac{\omega_o \cdot L_p}{Z_r} \tag{10}$$

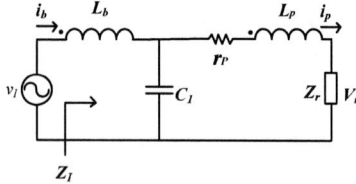

Fig. 5. Circuit diagram of the LCL circuit on on primary (track side).

If track coil L_P is same as L_b of the LCL circuit, Fig. 6(a) shows output current i_p at $\omega = \omega_o$ would only change with input voltage V_I, it is independent of load. When $Z_r=0$, the inverter output current is 0 as shown in Fig. 6(b). By designing the appropriate ω_o and L_b, Z_I presents characteristic of high equivalent impedance could reduce the situation about damage of inverter and power supply.

658

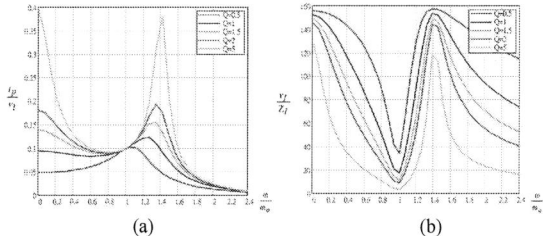

Fig. 6. The characteristics of LCL circuit on the primary. (a) Current gain versus normalized operating frequency. (b) Inverter current versus normalized operating frequency.

In the LCL circuit, the inductance of power track coil L_P is designed too small, the value would affect the power transfer ability. Besides, if the L_b is too large, the magnetic field may cause the influence on the system. Therefore, adding a capacitor to adjust the LCL circuit and become the LCCL resonant circuit as shown in Fig. 7, the reactance of each branch is

$$X_1 = \omega L_1 = \omega L_b = \frac{1}{\omega C_1} = \omega L_p - \frac{1}{\omega C_{1s}} \quad (11)$$

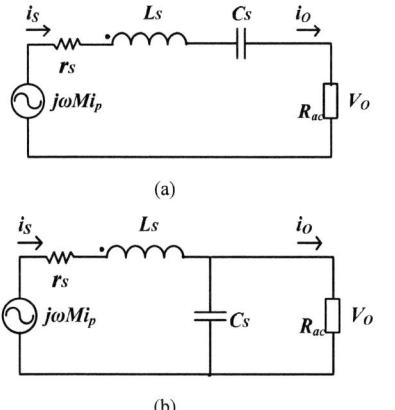

Fig. 7. Circuit diagram of the LCCL circuit on primary (track side).

To reduce the loss on the secondary circuit components, the pickup resonant circuit would select the simple LC circuit as shown in Fig. 8. By considering the system applications, we choose the parallel LC circuit as the pickup resonant circuit.

(a)

(b)

Fig. 8. Diagram of the resonant circuit on the secondary (pickup side).(a) Series LC circuit. (b) Parallel LC circuit.

III.   SIMULATION AND ANALYSIS OF INDUCTIVELY COUPLED STRUCTURE

The contactless power track system transfer power by electromagnetic induction. When the current flow through the track coil and generate alternating magnetic field, the pickup coil receives the induced power by a changing magnetic field. A large distance between the track and EV cause the power transmission efficiency get worse. To improve this problem, we need to pay more attention for designing the inductive coupled structure. First, the simulation software of magnetic field Maxwell is used to analyze the track coil and pickup coil respectively. The appropriate track coil have been selected to have a uniform magnetic flux distribution and enhance the lateral and longitudinal displacement tolerances between the inductive coupled structure. The pickup coil could be chose to receive power effectively and stably under a large distance by the simulation result.

The coil size of a single power supply track is about 30 cm × 30 cm for increasing the implementation of the results. In the aspect of designing the shape of track coil, the uniform magnetic flux distribution and strong magnetic fields are mainly considered. To minimize the situation about the magnetic flux offset, do not select the complex winding structure. In this paper, the vertical distance of the coupling structure is set at 12 cm. The simulation result of track coil can be obtained by the analysis plane and line in Maxwell is shown in Fig. 9. Fig. 10 shows the three different coil configurations with the same length of wire. It is obvious that the type of 0.5 cm between wires has larger magnetic field in the middle, but the magnetic flux distribution is not better than the type of single and double ring coil.

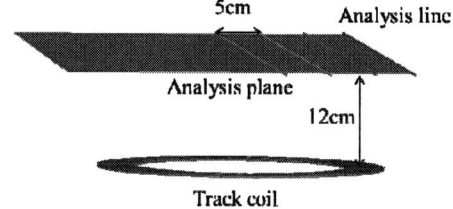

Fig. 9. Diagram of magnetic field simulation of track coil.

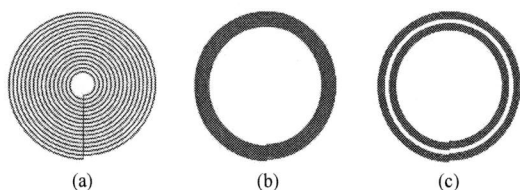

Fig. 10. Different spiral coil configuration with the same length of wire. (a) 0.5 mm between wires. (b) 0 mm between wires (single ring). (c) 0 mm between wires (double ring).

The magnetic field of single ring coil is slightly larger than the double ring type as shown in Fig.11 and Fig. 12. Therefore, to enhance the lateral and longitudinal displacement tolerances between the EV and track, we have chosen the single ring spiral configuration by this analysis.

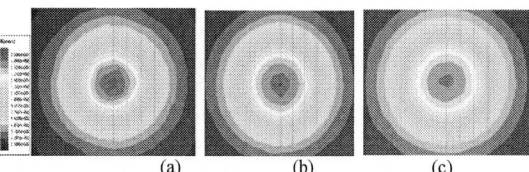

Fig. 11. Simulation of the magnetic flux distribution of Fig. 10 (top view). (a) 0.5 mm between wires. (b) 0 mm between wires(single ring). (c) 0 mm between wires(double ring).

The 2018 International Power Electronics Conference

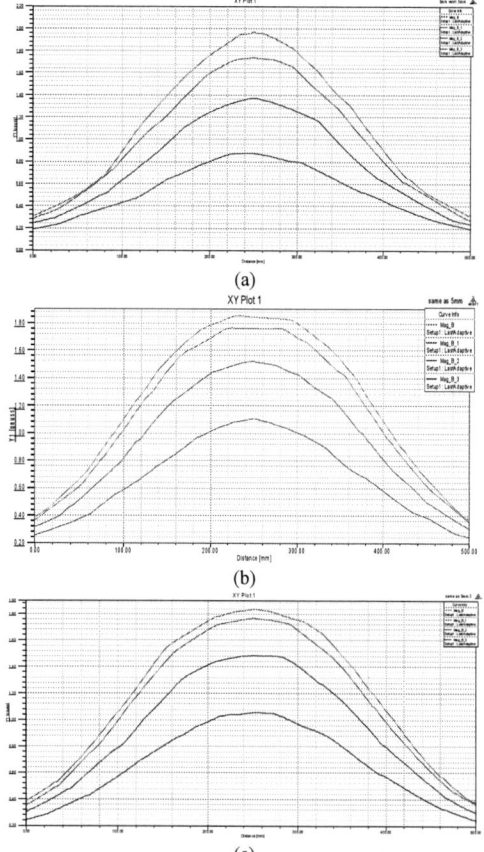

(a)

(b)

(c)

Fig. 12. Simulation curve of the magnetic flux distribution of Fig. 10. (a) 0.5 mm between wires. (b) 0 mm between wires (single ring). (c) 0 mm between wires(double ring).

The size of contactless EV power track array block is about 40 cm × 40 cm. Fig. 13(a) shows the track coil used in actual. The outer diameter and inner diameter of spiral coil is 30 cm and 14 cm respectivel. According to Fig. 13(b), it can be seen that the simulation of this configuration has a uniform magnetic flux distribution and large magnetic filed.

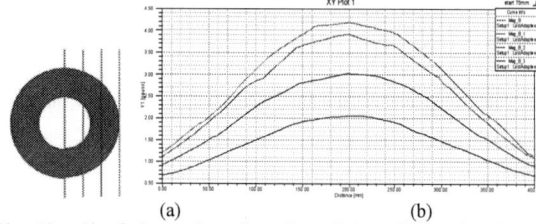

(a) (b)

Fig. 13. Simulation and configuration of the spiral track coil. (a) Configuration. (b) Simulation curve of magnetic flux distribution.

The secondary coil needs to be considered the pickup system receive power stably and effectively form track when EV is moving. The relationship between different diameter of the pickup coil and coupling factor under 12 cm air-gap analyze by simulations is shown in Fig. 14(a). If the size of the pickup coil is smaller, the surface which the magnetic flux pass through would become smaller. When the radius of the pickup is greater than the track

coil, there will be offset the magnetic flux at zero horizontal misalignment, resulting in the decrement of total magnetic flux through the surface of pickup coil as shown in Fig. 14(a). According to Fig. 14(b), the pickup coil size is selected the radius of 15 cm with highest coupling factor at 0 horizontal misalignment.

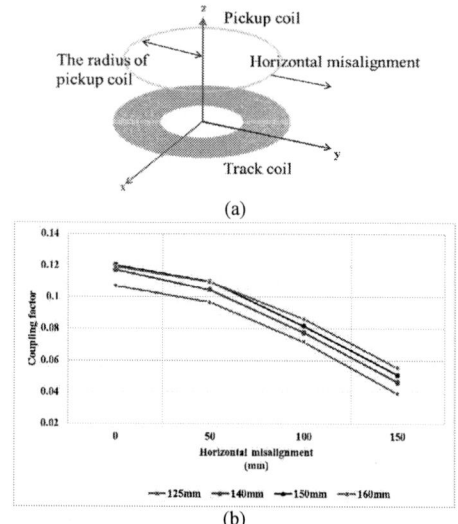

(a)

(b)

Fig. 14. Simulation of relationship between pickup radius and coupling factor. (a) Configuration. (b) Simulation curve.

To ensure the effect of improving the magnetic flux linkage, as shown in Fig. 15, we investigated the effects for the coil structure without ferrite and with ferrite, respectively. Fig.16 shows the stray magnetic field is concentrate by the ferrite. For the track coil, the ferrite can concentrate the magnetic field to make the magnetic flux path reach a position which is higher than the coil structure without ferrite, and it is beneficial for the contactless power transmission at a large distance. The pickup coil with ferrite can prevent the magnetic field emitting to a position higher than the inductive coupled structure, and passed through the pickup coil intensively.

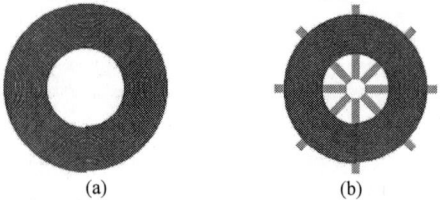

(a) (b)

Fig. 15. Configuration of the spiral track coil. (a) Without ferrite. (b) With ferrite.

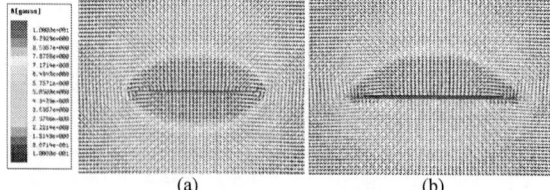

(a) (b)

Fig. 16. Simulation of the magnetic flux distribution of the spiral track coil (side view). (a) Without ferrite. (b) With ferrite.

660

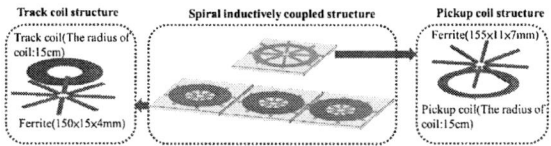

Fig. 17. Configuration of the inductively coupled structure.

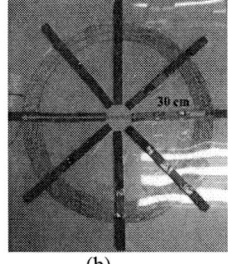

(a)   (b)

Fig. 18. Actual inductively coupled structure. (a) Primary power array track. (b) Secondary pickup.

IV. CONTACTLESS POWER ARRAY TRACK SYSTEM

The overall system of contactless EV power supply track is shown in Fig. 19. It could be divided into the structure of power supply track, the structure of power pickup circuit, inductively coupled structure array, and segment-excited control system.

The system supplied by the DC power supply, the full bridge inverter provide high frequency square wave for the primary track coil to generate a high frequency alternating magnetic field and transfer power to the pickup coil. The operating frequency of the system could be the same as resonant frequency by adjusting the switch of full-bridge inverter. Therefore, the AC power received from the track becomes DC to the load by rectifier circuit. Segment-excited control system utilizes the way of infrared (IR) transmission and reception. First, installed the IR emitter module on the pickup and send a signal continuously with Arduino. If the IR receiver module on the track receives the signal from the emitter, Arduino would determine whether the received information is correct and whether the switch is open to achieve the purpose of the segment-excited control.

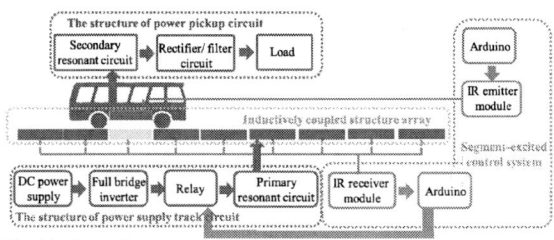

Fig. 19. Contactless power array track system.

When the EV is moving, the power transmission only operate in the area near the track coil that pickup coil covered. It is unnecessary that the overall track turn on the power. Therefore, we have added the segment-excited control to avoid too much power loss in the contactless transfer system. Because the magnetic field generated by the contactless power track system is much stronger than the segment-excited system, it is difficult to detect EV with the magnetic field. Therefore, we choose the infrared sensor to implement segment-excited control which could detect EV in a certain distance and the sensing distance of the device is no more than 120 cm.

Infrared (IR) transmission of the principle is the IR LED convert electrical power into Infrared and flash with a specific frequency to send the signal. The IR with wavelengths from 700 nm to 1 mm belongs to an invisible light. The IR receiver contains the photodiode and IC. Photodiode would receive the optical signal, then the signal been processed by the IC and out of the IR receiver device.

The segment-excited control system implement by the way of IR transmission and reception. The IR emitter module with Arduino is used to send a set of signals. When the receiver get this signal and determine the code successfully, turned on the switch in the track system. At this time, the current flow through track coil to transfer power to the power picker as shown in Fig. 20. The response time of infrared transmission is about 500 ms. The sensing distance of module could obtain by a simple measurement, the horizontal displacement can receive the signal is more than 15 cm at vertical height of 12 cm.

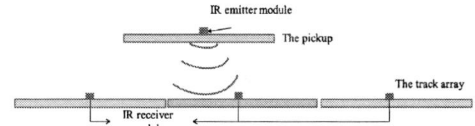

Fig. 20. Configuration of Segment-excited control system.

V. EXPERIMENTAL RESULTS

The measurement combines with segment-excited control and power track system. The contactless electric vehicle power track system with segment-excited inductively coupled structure array is shown in Fig. 21.

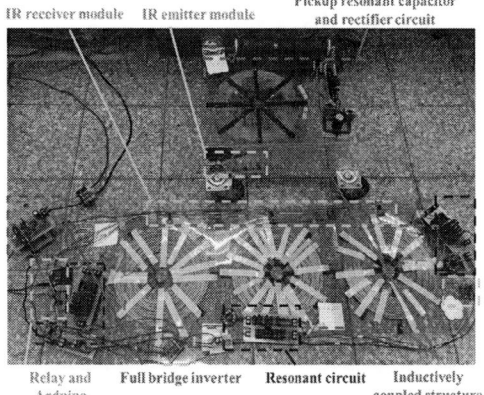

Fig. 21. Experiment setup.

The circuit diagram of the power track system include full-bridge inverter, LCCL resonant circuit on the track and simple parallel circuit, rectifier, load on EV as shown in Fig. 22, measurement specification of the power track system is shown in Table I and circuit parameters of the proposed system is shown in Table II.

661

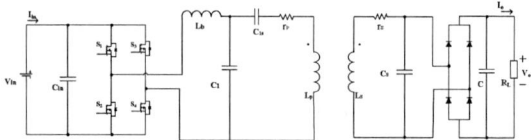

Fig. 22. Circuit diagram of the power track system.

TABLE I
MEASUREMENT SPECIFICATIONS OF THE POWER TRACK SYSTEMS

Input voltage	60-160 V	Frequency	40 kHz
Output voltage	100-300 V	Gap	12 cm
Load	80.7 Ω	EV speed	0.12 m/s

TABLE II
CIRCUIT PARAMETERS OF THE PROPOSED SYSTEM

Input DV voltage	160 V	Secondary resonant inductance L_S	111.4 µH
Inductance L_b	40 µH	Secondary resonant capacitance C_S	142.1 nF
Primary resonant capacitance C_1	396.4 nF	Resistance of pickup coil r_S	0.54 Ω
Primary resonant inductance L_P	483.9 µH	Mutual inductance M	32.7 µH
Capacitance C_{1s}	37 nF	Resistance of track coil r_P	0.75 Ω

Measuring the efficiency and output power of single track array system with 160 V input voltage, when EV moved under 12 cm air-gap as shown in Fig. 23. When the input voltage is 160 V at zero horizontal misalignment, the maximum output power is 910.6 W and the highest power transmission efficiency is 75%, when load impedance is 80.7 Ω as shown in Fig. 24. The proposed structure of this paper provides uniform magnetic field distribution been confirmed.

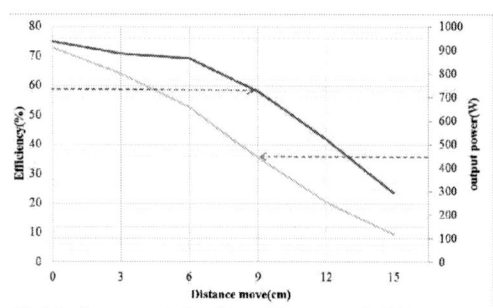

Fig. 23. Misalignment relationships with power and efficiency.

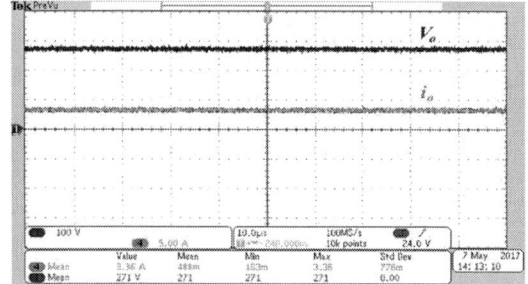

Fig. 24. Experimental waveform of output voltage and current.

The measurement method of overall system which included power transfer system with three track array and segment-excited control as shown in Fig. 25. Fig. 26 shows the waveform of the track array current and voltage on the load. According to the results, it can be seen that the voltage is unstable because of EV movement but the pickup could receive a certain voltage when the power of the track coil turned on.

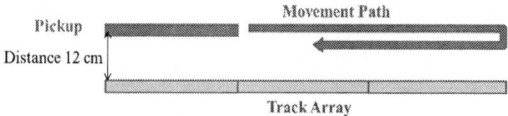

Fig. 25. Movement path of pickup.

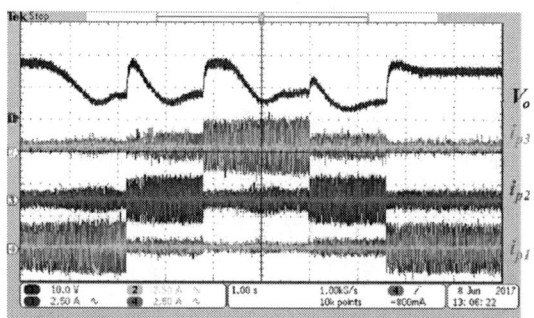

Fig. 26. Experimental waveform of voltage and current for track array system.

VI. CONCLUSIONS

This paper proposes the contactless power transmission for implementing contactless power array track with segment-excited overlapping-circle inductive coupled structure, which using overlapping type and adding appropriate ferrite to secondary to improve the efficiency of vertical and horizontal offset. In addition to this, the segment-controlled can be utilized for improving the power transmission ability and efficiency. According to experimental results with distance of 15 cm, the maximum efficiency comes up to 89.9% with 360 W output power, and the maximum output power of overall system is 759 W with efficiency of 82.95%.

REFERENCES

[1] O. C. Onar, J. M. Miller, S. L. Campbell, C. Coomer, C. P. White, and L. E. Seiber, "A novel wireless power transfer for in-motion EV/PHEV charging," in *Proc. IEEE APEC*, pp. 3073–3080, 2013.

[2] A. Zaheer, H. Hao, G. A. Covic, and D. Kacprzak, "Investigation of multiple decoupled coil primary pad topologies in lumped IPT systems for interoperable electric vehicle charging," *IEEE Trans. Power Electron.*, vol. 30, no. 4, pp. 1937–1955, Jan. 2014.

[3] J. M. Miller, L. M. Li, O. C. Onar, and P. T. Jones, "ORNL experience and challenges facing dynamic wireless power charging of EV's," *IEEE Circuits Syst. Mag.*, vol. 15, no. 2, pp. 40–53, May 2015.

[4] J. T. Boys, G. A. Covic, and A. W. Green, "Stability and control of inductively coupled power transfer systems," *IEEE Proc. Electric Power Appl.*, vol. 147, no. 1, pp. 37–43, Jan. 2000.

[5] H. H. Wu, A. Gilchrist, D. Bronson, and K. D. Sealy, "A high efficiency 5 kW inductive charger for EVs using dual side control," *IEEE Trans. Ind. Informat.*, vol. 8, no. 3, pp. 585–595, Apr. 2012.

Driving Test Evaluation of Sensorless Vehicle Detection Method for In-motion Wireless Power Transfer

Katsuhiro Hata[1*], Kensuke Hanajiri[1], Takehiro Imura[1], Hiroshi Fujimoto[1], Yoichi Hori[1],
Motoki Sato[2], Daisuke Gunji[3]

1 Department of Electrical Engineering, The University of Tokyo, Kashiwa, Japan
2 Engineering Research Division, Toyo Denki Seizo K.K., Yokohama, Japan
3 Powertrain Technology Development Department, NSK Ltd., Fujisawa, Japan
*E-mail: hata.katsuhiro13@ae.k.u-tokyo.ac.jp

Abstract—In-motion wireless power transfer (WPT) has the capability to drastically increase a cruising distance of electric vehicles (EVs). A vehicle detection technique is important for a road facility to reduce standby power consumption and to prevent an unnecessary magnetic field leakage. A sensorless vehicle detection method using voltage pulses has been proposed and fundamental experiments have been demonstrated with small-scale equipment. In this paper, a full-scale in-motion WPT system is implemented and a test vehicle is developed with the second generation wireless in-wheel motor (W-IWM2). The sensorless vehicle detection method is applied to the implemented in-motion WPT system and the feasibility of the proposed system is verified by the driving experiment with the test vehicle.

Keywords—Electric vehicle, Wireless power transfer, In-motion charging, Sensorless vehicle detection

I. INTRODUCTION

Electric vehicles (EVs) have gathered much attention not only for their environmental performance but also for their motion controllability based on an electric motor drive. Additionally, EVs with in-wheel motors (IWMs) have advantages such as a driving range extension based on independent control of each wheel and weight saving due to reduction of drive part components [1]. However, the IWM has a wire connection to a chassis of EVs and its reliability is one of limiting factors for commercialization. In order to solve this problem, the first generation wireless in-wheel motor (W-IWM1) has been developed [2]. A driving test has been demonstrated by the test vehicle with the W-IWM1 units on the rear wheels [3].

Meanwhile, the mileage per charge of EVs is limited due to the capacity of the on-board battery. Although previous research has proposed efficient motor driving [4] and range extension autonomous driving [5], in-motion wireless power transfer (WPT) is expected to radically improve the driving range of EVs [6], [7]. In this paper, a novel in-motion WPT system is proposed for EVs with the IWMs considering efficient use of energy.

The concept of the proposed system is not to charge the on-board battery first but to deliver the transferred power to the IWMs. The test vehicle is implemented with the second generation W-IWM (W-IWM2) units

Fig. 1. Test vehicle with W-IWM2.

on the front wheels and the road facility is designed based on a sensorless vehicle detection system [8]. In this paper, an implementation methodology of the in-motion WPT system is described and a driving test result of the sensorless vehicle detection method is discussed.

II. TEST VEHICLE WITH W-IWM2

A. Concept of in-motion WPT for EVs with IWMs

Most of previous research has assumed EVs with an on-board motor as standard and proposed in-motion WPT systems to charge the on-board battery through the receiving coil placed at the bottom surface of the vehicle.

In this study, a novel in-motion WPT system is proposed for EVs with IWMs. Since the road facility provides energy to the IWMs directly, the proposed in-motion WPT system has the following advantages compared to the conventional in-motion WPT systems.

1) High energy efficiency can be achieved by powering the IWMs without the need to charge/discharge the on-board battery.

2) The clearance margin can be minimized because the gap between the road coil and the receiving coil placed at the IWM is nearly unchanged regardless of the suspension movement.

3) The output power of the road coil can be reduced because each IWM can receive power from each road coil.

TABLE I. SPECIFICATION OF THE TEST VEHICLE

Max. motor power	12 kW
Max. motor torque	76.4 Nm
Reduction ratio	4.407
Number of motors	2 (4)
Total power	24 kW (48 kW)
Total wheel torque	672 Nm (1344 Nm)

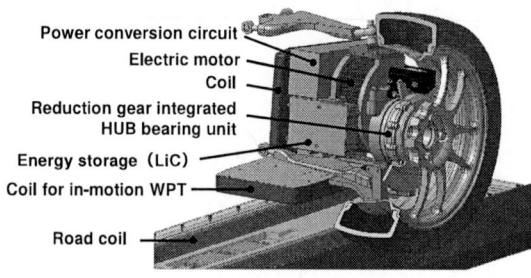

Fig. 2. Configuration of the W-IWM2.

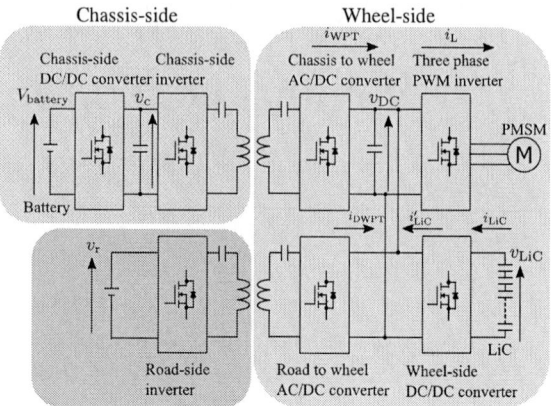

Fig. 3. Circuit configuration of the W-IWM2.

B. Configuration of W-IWM2

The test vehicle with the W-IWM2 units on the front wheels is shown in Fig. 1 and its specification is described in Table I. Although the current test vehicle employs two W-IWM2 units on the front wheels, the total power and wheel torque with the four-wheel W-IWM2 units reach the same level as the driving performance of the base commercial EV.

The configuration of the W-IWM2 is shown in Fig. 2. The W-IWM2 is designed as the electro-mechanical integrated IWM and composed of the electric motor, the power conversion circuits, and the two receiving coils for WPT. Each coil is placed to avoid any contact with the suspension arm even if the coil displacement is caused by the steering action and the suspension movement. The coil gap between the chassis side and the wheel side is set to 100 mm and the distance from the road surface to the receiving coil placed at the IWM is adjusted to 100 mm. The motor output is delivered to the wheel through the reduction gear integrated hub bearing unit.

C. Functional features of W-IWM2

The W-IWM2 has the following functional features.

1) Bidirectional WPT between the chassis side and the wheel side
2) In-motion WPT from the road facility to the IWM
3) Power management with the energy storage device integrated into the IWM

The circuit configuration of the W-IWM2 is illustrated in Fig. 3. The W-IWM2 employs a series-series (SS) compensated circuit topology of WPT via magnetic resonance coupling [9]. Then, the transmitter and receiver coils are connected to the resonant capacitors in series. Their resonance frequency and the operating frequency of the inverters are set to 85 kHz.

The bidirectional WPT system between the chassis side and the wheel side is the same configuration as the

W-IWM1 [2]. In-motion WPT from the road facility to the IWM is appended in the W-IWM2. The receiving coil for in-motion WPT is connected to the wheel-side DC link through the AC/DC converter. Additionally, the energy storage devices, the lithium-ion capacitors (LiCs), are integrated into the W-IWM2 and connected to the wheel-side DC link through the DC/DC converter.

In order to ensure safe and reliable performance, the W-IWM2 has to handle power flow in the IWM and to stabilize the wheel-side DC link voltage. The power management of the W-IWM2 has been proposed based on the wheel-side DC link voltage control with the wheel-side DC-DC converter and LiCs [10].

III. ROAD FACILITY FOR IN-MOTION WPT

A. Configuration of road facility

Previous research has proposed a variety of the road facility configurations for in-motion WPT [7]. These are broadly divided into the following systems.

1) Long road coils driven by high-capacity inverters
2) A number of short road coils driven by lower-capacity inverters

Although the former system does not require a large number of the inverters, their capacity has to be increased because of the possible existence of multiple EVs on the road coil. Furthermore, to reduce the losses of the road coil and to prevent magnetic flux leakage become more difficult because the road coil excited by the high-frequency AC current covers a much larger area compared to the latter system. If the road coil length is shorter than the vehicle overall length, the problems described above could be eased. Consequently, the latter system is employed in this study.

B. Implementation of in-motion WPT system

The implemented full-scale in-motion WPT system is shown in Fig. 4. The in-motion WPT lane is 20 m length and the total installed capacity is designed to be

664

Fig. 4. Implemented full-scale in-motion WPT system.

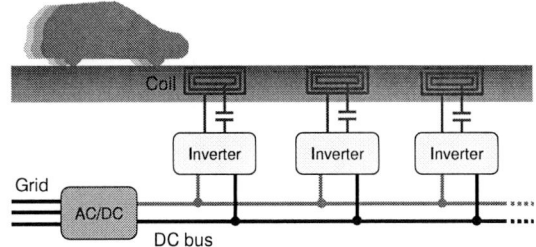

Fig. 5. Configuration of road facility.

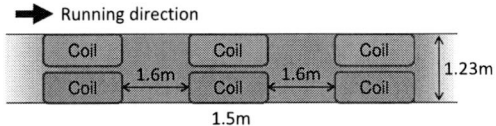

Fig. 6. Arrangement of road coils.

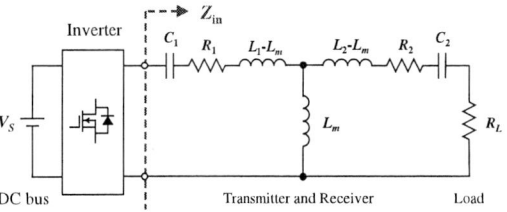

Fig. 7. Input impedance of the in-motion WPT system.

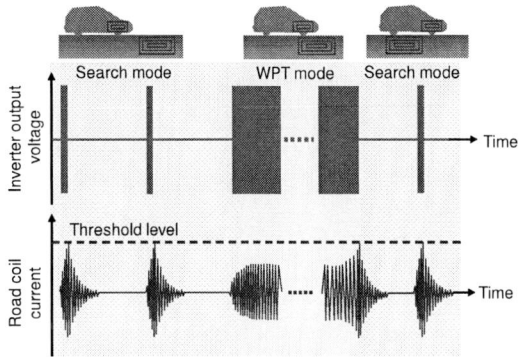

Fig. 8. Concept of vehicle detection operation.

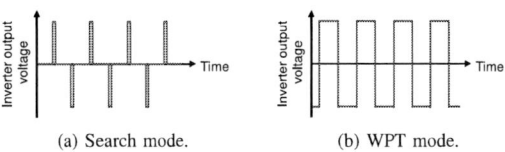

(a) Search mode. (b) WPT mode.

Fig. 9. Waveforms of inverter output voltage.

up to 50 kVA. In the current setup, the rated power of in-motion WPT from the road to the IWM is up to 9 kW, nonetheless, the test vehicle equipped with the four-wheel W-IWM2 units can receive 36 kW from the road facility.

The configuration of the road facility is illustrated in Fig. 5. The DC bus is generated from the grid with the AC/DC converter and each inverter drives each road coil. In this study, 6 road coils are placed in the in-motion WPT lane and the arrangement of the road coils is indicated in Fig. 6. The road coils are placed in 2 lines for the right and left wheels and their intervals are set to 1.6 m.

C. Sensorless vehicle detection method

In terms of energy saving and safety issues, the in-motion WPT system should be executed when the vehicle exists above the road coil and a vehicle detection system should be installed into the road facility. In this study, a sensorless vehicle detection method using voltage pulses [8] is employed to simplify the road-side components. This method estimates strength of magnetic coupling between the road coil and the vehicle-side receiving coil based on the input impedance of the system.

Figure 7 shows the T-type equivalent circuit of the in-motion WPT system and indicates the input impedance of the system Z_{in}. The equivalent load resistance R_L includes the power conversion circuits and drive motor on the vehicle side. When the operating angular frequency ω_0 is equated to the resonance frequency of the transmitter and receiver, Z_{in} is expressed as follows:

$$Z_{\mathrm{in}} = R_1 + \frac{(\omega_0 L_m)^2}{R_2 + R_L}. \qquad (1)$$

Since Z_{in} changes according to the mutual inductance L_m between the transmitter and receiver coils, the approaching vehicle can be detected based on (1). However, the value of R_L is different for each of the vehicles. Then, if the receiver is shorted with the AC-DC converter on the vehicle side, R_L is assumed to be 0 and it permits the same design strategy of the detection method to be applied to every vehicle.

D. Transmission control based on impedance detection

Figure 8 shows the concept of the vehicle detection operation for in-motion WPT. Firstly, the inverter of the road facility outputs the 3-level voltage wave, which is indicated in Fig. 9(a), to reduce the power consumption during the search mode. Its frequency is set to be the same as the resonance frequency and Z_{in} is estimated based on the current envelop amplitude of the road coil I_{1env}.

If the vehicle does not appear above the road coil, I_{1env} is steeply increased because Z_{in} is quite small. As

The 2018 International Power Electronics Conference

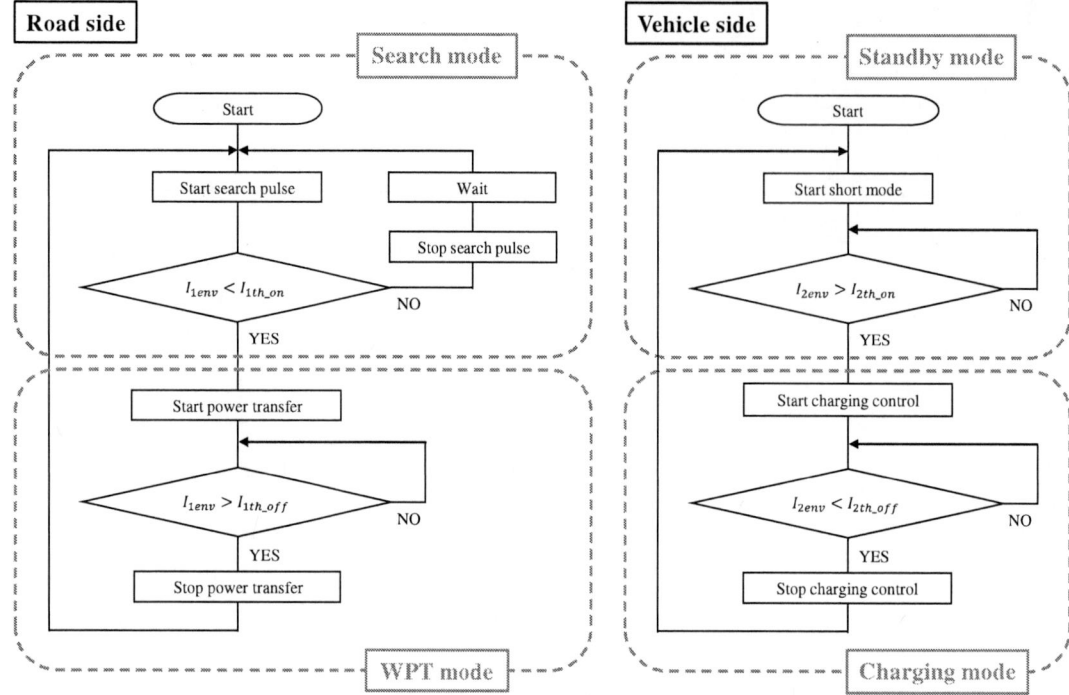

Fig. 10. Flowchart of the in-motion WPT system.

the vehicle approaches the road coil, Z_{in} is increased and the climb rate of I_{1env} is gradually decreased. As a result, if I_{1env} does not exceed the designed on-threshold during the detection time, the inverter operation is changed from the search mode to the WPT mode and then the square voltage wave is generated as shown in Fig. 9(b).

When the vehicle passes over the road coil, Z_{in} is decreased again. Therefore, the inverter can stop then by comparing I_{1env} to the designed off-threshold. After that, the inverter repeats the foregoing searching sequence.

E. Flowchart of the in-motion WPT system

The flowchart of the in-motion WPT system is indicated in Fig. 10. When the vehicle is not above the road coil, the road side and the vehicle side are operated as the search mode and the standby mode, respectively. Then, the receiver is shorted by the wheel-side AC/DC converter to apply the detection method to every vehicle.

If I_{1env} does not exceed the on-threshold I_{1th_on} during the detection time, the road-side inverter operation is changed from the search mode to the WPT mode. Then, the vehicle side detects receiving power based on the current envelop amplitude of the receiving coil I_{2env} and the state of the wheel-side AC/DC converter is switched by comparing I_{2env} to the on-threshold I_{2th_on}.

When the vehicle passes over the road coil, the road side changes the inverter operation based on a comparison between I_{1env} and the off-threshold I_{1th_off}. And then, the vehicle side goes back to the standby mode using I_{2env} and the off-threshold I_{2th_off}.

(a) Inverter.

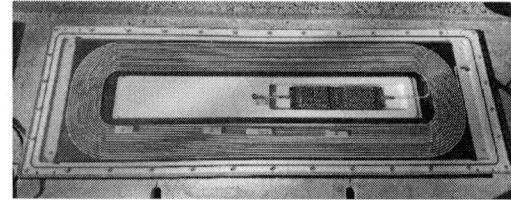

(b) Road coil.

Fig. 11. Experimental equipment of road facility.

IV. EXPERIMENT

A. Experimental condition

The experimental equipment is shown in Fig. 11. Each inverter box includes two inverters and the full-bridge circuit topology is employed. The overall size of the road coil is 1.5 m × 0.5 m. The resonance capacitor is placed at the center of the coil and connected to the coil in series.

The experimental condition is indicated in Table II. The search pulse voltage V_{search} is manipulated by the

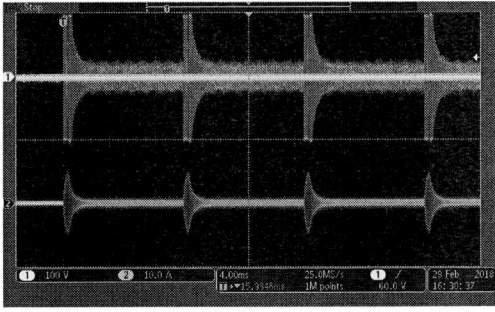
(a) Overall view

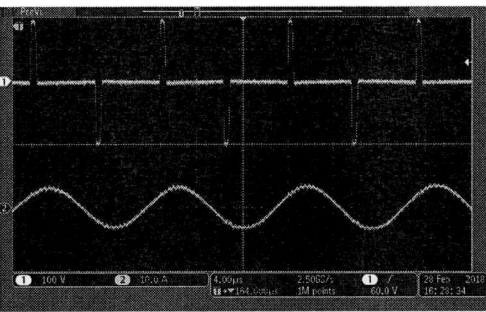
(b) Enlarged view

Fig. 12. Road-side voltage and current waveforms in search mode.

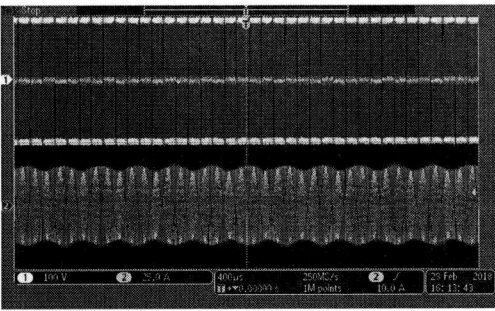
(a) Overall view

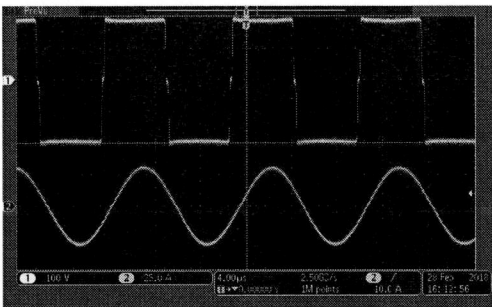
(b) Enlarged view

Fig. 13. Road-side voltage and current waveforms in WPT mode.

TABLE II. PARAMETERS OF THE EXPERIMENTAL SETUP

Operating frequency f_0	89 kHz
Road-side DC bus voltage V_S	200 V
Wheel-side DC link voltage V_{DC}	500 V
Road coil inductance L_1	429.0 μH
Road coil resistance R_1	342.5 mΩ
Road coil quality factor Q_1	700.4
Receiving coil inductance L_2	377.7 μH
Receiving coil resistance R_2	383.3 mΩ
Receiving coil quality factor Q_2	551.0
Searching period T_{search}	10 ms
Search pulse voltage V_{search}	50 V
Detection time T_{det}	400 μs

duty ratio of the road-side inverters and each of the thresholds was designed by trial and error. In this experiment, the detection method can assess the travel distance of the vehicle less than 3.33 cm, because the driving speed is slower than 12 km/h.

B. Road-side voltage and current waveforms

Figure 12 shows the road-side voltage and current waveforms in the search mode. The yellow and blue lines indicate the road-side voltage V_1 and current I_1, respectively. During the search mode, the 3-level voltage wave was generated properly as shown in Fig. 12(b). When the vehicle did not exist above the road coil, I_1 was steeply increased. After the detection time T_{det}, the road-side inverter stopped to inject the searching voltage pulses because I_{1env} exceeded I_{1th_on}. The searching operation was conducted with the searching period T_{search}.

Figure 13 illustrates the waveforms in the WPT mode. The square voltage wave was generated by the road-side inverter and the continuous charging was achieved. However, I_{1env} was fluctuated with several kHz and its reason will be investigated and analyzed as a future work.

C. Driving test result of in-motion WPT

The driving test was demonstrated by the test vehicle with the W-IWM2 and the constructed in-motion WPT system. The torque commands of the right and left IWMs were equalized and generated by the driver input with the accelerator pedal. Although it is possible to demonstrate in-motion WPT for the right and left wheels simultaneously, in this paper, the test result of in-motion WPT only for the right wheel is indicated.

The diving test result is shown in Fig. 14. Note that time series of the vehicle-side data (Fig. 14(a)–(c)) and the road-side data (Fig. 14(d), (e)) do not accord because these results were independently measured by the sensors in the test vehicle and the road facility. Additionally, Fig. 14(d), (e) indicate the results of the second road coil.

The test vehicle was gradually accelerated as shown in Fig. 14(a) and the in-motion WPT was successfully received as shown in Fig. 14(b). Then, the state of the wheel-side AC/DC converter was switched based on I_{2env} and the thresholds. From Fig. 14(c), though it was assumed that the vehicle was displaced from directly above the first road coil, the mode change was properly conducted based on I_{2env} and the thresholds.

The 2018 International Power Electronics Conference

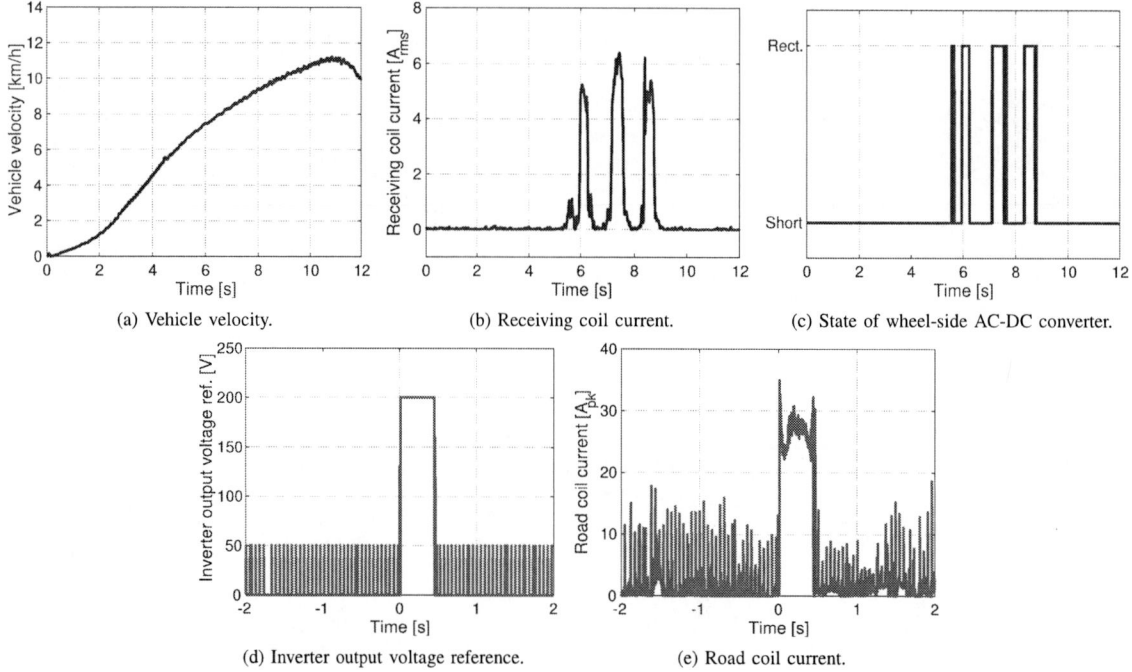

(a) Vehicle velocity.

(b) Receiving coil current.

(c) State of wheel-side AC-DC converter.

(d) Inverter output voltage reference.

(e) Road coil current.

Fig. 14. Driving test result of in-motion WPT.

Figure 14(d) indicates the inverter output voltage reference V_{1ref}^*. In this experiment, $V_{1ref}^* = 50$ V and $V_{1ref}^* = 200$ V mean the search mode and the WPT mode, respectively. As shown in Fig. 14(e), I_{1env} was gradually decreased during the search mode according to the approximation of the vehicle to the road coil. And then, the inverter operation was successfully changed based on I_{1env} and the thresholds. Furthermore, when the vehicle passed over the road coil, the road facility detected it and went back the search mode properly.

V. CONCLUSION AND FUTURE WORK

This paper presented an implementation methodology of an in-motion WPT system for EVs with IWMs. A test vehicle was developed with the W-IWM2 units on the front wheels and the road facility was implemented based on a sensorless vehicle detection method. The driving test result demonstrated the feasibility of the detection method with the full-scale in-motion WPT system.

Future works are to optimize the designed parameters of the detection method and to evaluate energy efficiency of the implemented in-motion WPT system. Additionally, the cause of the current fluctuation during the charging operation will be investigated in detail.

ACKNOWLEDGMENT

This research was partly supported by JSPS KAK-ENHI Grant Number 26249061 and JST CREST Grant Number JPMJCR15K3. The authors would like to express their deepest appreciation to the Murata Manufacturing Co., Ltd. for providing the laminated ceramic capacitors used in these experiments.

REFERENCES

[1] S. Murata, "Innovation by in-wheel-motor drive unit", *Vehicle Syst. Dynamics, Int. J. Vehicle Mech. Mobility*, vol. 50, no. 6, pp. 807–830, Jun. 2012.

[2] M. Sato, G. Yamamoto, D. Gunji, T. Imura, and H. Fujimoto, "Development of wireless in-wheel motor using magnetic resonance coupling", *IEEE Trans. Power Electron.*, vol. 31, no. 7, pp. 5270–5278, Jul. 2016.

[3] H. Fujimoto, M. Sato, D. Gunji, and T. Imura, "Development and driving test evaluation of electric vehicle with wireless in-wheel motor", in *Proc. EVTeC & APE Japan*, 2016.

[4] A. Kawamura, G. Guidi, Y. Watanabe, Y. Tsuruta, N. Motoi, and T. W. Kim, "Driving performance experimental analysis of series chopper based EV power train", *J. Power Electron.*, vol. 12, no. 6, pp. 992–1002, Nov. 2012.

[5] Y. Ikezawa, H. Fujimoto, Y. Hori, D. Kawano, Y. Goto, M. Tsuchimoto and K. Sato, "Range extension autonomous driving for electric vehicles based on optimal vehicle velocity trajectory generation and frontrear drivingbraking force distribution with time constraint", *IEEJ J. Ind. Appli.*, vol. 5, no. 3, pp. 228–235, May 2016.

[6] G. A. Covic and J. T. Boys, "Modern trends in inductive power transfer for transportation application,", *IEEE J. Emerg. Sel. Topics Power Electron.*, vol. 1, no.1, pp. 28–41, Mar. 2013.

[7] C. C. Mi, G. Buja, S. Y. Choi, and C. T. Rim, "Modern advances in wireless power transfer systems for roadway powered electric vehicles", *IEEE Trans. Ind. Electron.*, vol. 63, no. 10, pp. 6533–6545, Oct. 2016.

[8] D. Kobayashi, T. Imura, H. Fujimoto and Y. Hori, "Sensorless vehicle detection Using Voltage Pulses in Dynamic Wireless Power Transfer system", in *Proc. EVS29*, 2016.

[9] A. Kurs, A. Karalis, R. Moffatt, J. D. Jonnopoulos, P. Fisher, and M. Soljacic, "Wireless power transfer via strongly coupled magnetic resonance", *Science*, vol. 317, no. 5834, pp. 83–86, Jun. 2007.

[10] T. Takeuchi, T. Imura, H. Fujimoto and Y. Hori, "Power management of wireless in-wheel motor with dynamic-wireless power transfer", in *Proc. EVS30*, 2017.

A System Design Method of High-Frequency Class-D Inverter for Wideband Current Control

Hiroki Kurumatani[1], Seiichiro Katsura[1*]

1 Department of System Design Engineering, Keio University, Yokohama, Japan

*E-mail: kurumatani@katsura.sd.keio.ac.jp

Abstract—**This paper reports a system design method of high-frequency class-D inverter for wideband current control for robotics in human society. There is a problem that conventional inverters are restricted by a switching frequency and hence the motor can not drive in a high-frequency range. Since a control system requires a severe condition as for a phase lag, such restriction should be removed. To overcome this problem, the paper designs a system with providing three pillars: a controller to attain a robustness, hardware improvement and a modulation technique to increase a bandwidth, and design policy for a processor to treat a high-frequency signal. Then, a GaN-HEMT-based class-D inverter and an FPGA-based controller are prepared to support these pillars. The paper proves an importance of the integrated design.**

I. INTRODUCTION

Robotics has been improving even in supporting the human society. Power electronics certainly guarantees a stable operation of motors [1]. Current control for permanent magnet synchronous motors offers precise output-torque control and precise servoing. As a performance of the torque control becomes better, a control theory of the motors has been developed [2]. Along systematization of a robust control theory, control designs on an upper layer, such as a command generation using an artificial intelligence, get to be a hot topic [3], [4]. Such cascade-layer system allocating proper roles on each layer is one of the solutions for complicated systems. Here, it should be noted that an upper layer can not elicit a performance exceeding a lower layer. Namely, the total performance of the system is restricted by an inner control-system. Now, the performance of the robot is restricted by the drivability of the machines, which is important capability in contact motions or crisis prevention. It is the time of re-examining and enhancing the inner-loop system.

One of the biggest challenges in the robotics is to work in human society. A safety is quite important and hence the machines should ensures a high back-drivability over wideband [5]. However, the robots is difficult to contact with a hard object and deal with a collision because it requires a severe condition as for a phase-lag [6]. For example, the law of action-reaction requires instantaneous force-balance at any point. This means that the robot must follow to an contact object as fast as possible. A hard object reacts quickly and lag in a control loop of the robot causes destabilization of the system. This problem can be seen remarkably in teleoperation system. Reasons are a calculation speed, an insufficient bandwidth of a controller limited by sensor noise, a lack

of drivability due to electrical systems, and so on. Since the phase lag is accumulated through a round loop, all factors which leads the phase lag should be removed.

To remove these factors, many efforts have beed devoted. Owing to improvement of processors, heterogeneous computing has begun to spread. CPU, GPU and FPGA, which show fine performance at operation flexibility, calculation parallelism and processing speed, are integrated into one system [7]. In the cascaded control, each layer has each required specification and hence they requires appropriate processors. Then, the calculation speed is not a big problem compared with the other factors. As for the sensor noise, high-fidelity sensors and observer techniques have been emerging and a phase lag due to filtering a sensor signal is reduced. Also, fusion of external sensors remarkably increase a sensing bandwidth [8]. An application specific circuit for the rotary encoder has been reported and monitoring a high-speed motion is achieved [9]. Considering the aforementioned techniques, a lack of drivability is dominant problem in recent robotics in the factors. This is because most classical technologies have been focusing on an energy saving. In such case, switching inverter is used while being conservative to a performance improvement. To break through above situation, the research [11] has designed a class-G linear amplifier. It showed a good performance while requiring a proper heat radiation and a large power because the linear amplifier is forced to have a large DC-link voltage for high-speed drive. For this problem, a switching inverter using emerging power devices will be a solution [12]. The emerging power devices, such as GaN or SiC transistors, have low energy-consumption comparing Si devices and enable the inverter to operate in the high-frequency domain. Even though a class-D inverter has switching noise, enhancing of the bandwidth would be possible by controlling it properly.

The paper designs a system with providing three pillars: a controller to attain a robustness, hardware improvement and a modulation technique to increase a bandwidth, and design policy for a processor to treat a high-frequency signal. The robust control system can not be established by only one of them, but be achieved by combination of them. This is because these components are closely related to the others. Then, the paper prepares a GaN-HEMT-based class-D inverter and an FPGA-based controller to realize above requirements. Calculation circuits on the FPGA is designed to elicit the system performance. The paper shows these pillars and evaluates a validity of the designed system by simulation.

The 2018 International Power Electronics Conference

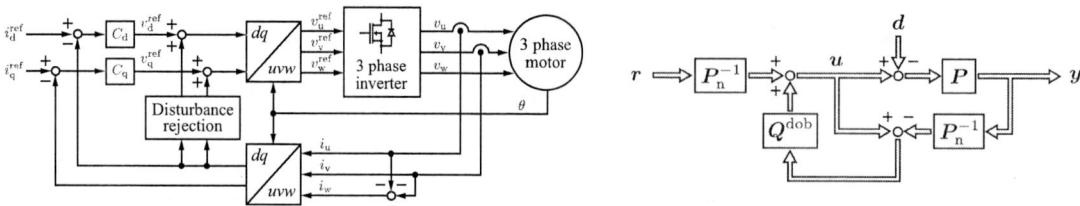

Fig. 1. Block diagram of the current controller.

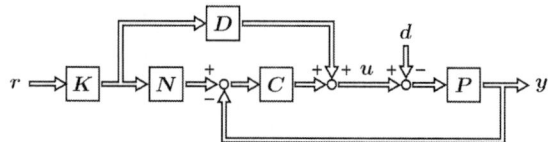

Fig. 2. Block diagram of the DOB.

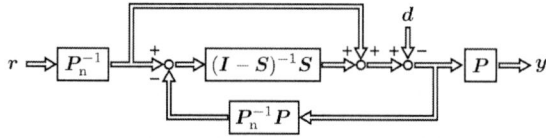

Fig. 3. Block diagram of generalized 2-DOF controller.

Fig. 4. Equivalent block diagram of generalized 2-DOF controller.

II. CONTROLLER DESIGN

The section shows a control architecture of an armature current of a three-phase motor. A block diagram of a classical current controller is shown in Fig. 1. Here, v, i and θ stand for a voltage, a current and an electrical angle. A superscript $\bigcirc^{\text{ref}}$ and subscripts $\bigcirc_{\text{d}}$, $\bigcirc_{\text{q}}$ and $\bigcirc_{\text{u}}$, $\bigcirc_{\text{v}}$, $\bigcirc_{\text{w}}$ mean a reference signal, parameters on a d-q coordinate and a uvw coordinate. In the paper, sets of these parameters are expressed by a vector format as $\bigcirc_{\text{dq}}$ and $\bigcirc_{\text{uvw}}$. Controllers to be designed are (i) a tracking controller and (ii) a sensor controller. The controller is designed on a d-q coordinate to control a generated torque. Many classical controllers have a PI controller and a disturbance rejection controller to compensate a voltage drop due to a coil resistance, effect of back electromotive force and interface terms between d-q axes. In this time, the paper introduces a 2-DOF control structure using a disturbance observer (DOB) [13]. Since the DOB has a role of a type-1 servo, it works like I controller while not paying attentions to a pole-zero assignment problem [14]. Furthermore, the DOB also suppress the other disturbances by setting a cutoff frequency of the DOB high, though it can not satisfy the inner model principle. As for a distortion of an output voltage, it is difficult to compensate as it is in very a high-frequency domain. Then, its suppression is addressed by an inverter design.

A. Disturbance Observer

The DOB is good tool to attain the robustness. To enrich an understanding for the DOB, this section shows an equivalent 2-DOF controller. Figs. 2 and 3 show the DOB and a generalized 2-DOF controller, where P, C, Q^{dob}, N, D and K are a plant, a feedback controller, a Q filter of the DOB, an irreducible numerator and denominator of P given by

$$P = ND^{-1} = \tilde{D}^{-1}\tilde{N}, \tag{1}$$

and a tracking controller which makes N and D proper, respectively. Subscript $\bigcirc_{\text{n}}$ denotes a nominal model. The controllers which make the system stable can be expressed by using free parameters X, Y and Q as

$$C = (Y - Q\tilde{N})^{-1}(X + Q\tilde{D}), \tag{2}$$

which fulfill a following Bézout's lemma

$$YD + XN = I. \tag{3}$$

r, d, u and y denote a command, a disturbance, an input to the plant and an outputs of this system, respectively. In this architecture, a following regulation can be found.

$$
\begin{aligned}
u &= DKr + (Y - Q\tilde{N})^{-1}(X + Q\tilde{D})(NKr - y) \\
&= DKr + Y^{-1}X(NKr - y) + Y^{-1}Q(\tilde{N}u - \tilde{D}y) \\
&= Y^{-1}(Kr - Xy + Q(\tilde{N}u - \tilde{D}y)).
\end{aligned} \tag{4}
$$

When setting parameters as

$$\begin{bmatrix} X & Y & K \end{bmatrix} = \begin{bmatrix} O & D^{-1} & N^{-1} \end{bmatrix}, \tag{5}$$

The control architecture becomes equal to that of the DOB. Here, the output of this controller is expressed as

$$
\begin{aligned}
y &= r + P(DQ(\tilde{N}u - \tilde{D}y) + d) \\
&= r - P(I - DQ\tilde{N})d.
\end{aligned} \tag{6}
$$

By setting Q as

$$Q = D^{-1}S\tilde{N}^{-1}, \tag{7}$$

a desired disturbance-rejection performance can be attained, where S is a free parameter, correspond to a Q filter of the DOB. Here, the controller is expressed as

$$C = (I - S)^{-1}SP^{-1}. \tag{8}$$

Then, we obtain an equivalent block diagram as shown in Fig. 4. This result shows that the DOB is a type of the series 2-DOF controllers. By giving disturbance-generating-polynomials to S, the robust control will be attained. Comparing PI controllers, the DOB simplifies the control design as there is no need to consider the pole-zero assignment.

670

B. Current Controller

This paper adopted a combination of a P controller and the DOB. Each controller is used for specification of a control bandwidth and disturbance suppression, respectively. To implement the DOB, the paper introduces a system model. A governing equation of the three-phase motor on the d-q coordinate is expressed as

$$v_{\mathrm{dq}} = \begin{bmatrix} R_{\mathrm{d}} + sL_{\mathrm{d}} & -\omega L_{\mathrm{q}} \\ \omega L_{\mathrm{d}} & R_{\mathrm{q}} + sL_{\mathrm{q}} \end{bmatrix} i_{\mathrm{dq}} + \begin{bmatrix} 0 \\ \omega K_{\tau} \end{bmatrix} \quad (9)$$

$$=: A i_{\mathrm{dq}} + B\omega \quad (10)$$

$$v_{\mathrm{dq}} \triangleq D v_{\mathrm{uvw}}, \ i_{\mathrm{dq}} \triangleq D i_{\mathrm{uvw}} \quad (11)$$

where D is a coordinate-transformation matrix from uvw coordinate to d-q coordinate defined as

$$D \triangleq \sqrt{\frac{2}{3}} \begin{bmatrix} \cos\theta & \cos(\theta - \frac{2}{3}\pi) & \cos(\theta - \frac{4}{3}\pi) \\ -\sin\theta & -\sin(\theta - \frac{2}{3}\pi) & -\sin(\theta - \frac{4}{3}\pi) \end{bmatrix} \quad (12)$$

and R, L and K_τ are a resistance, an inductance and a torque constant. To suppress the resistance components, the non-diagonal terms in the first term and the second term of (9), the DOB is implemented on the inner loop. An estimate process is expressed as

$$\hat{v}_{\mathrm{dq}}^{\mathrm{dis}} = Q_{\mathrm{dq}}(v_{\mathrm{dq}}^{\mathrm{ref}} - sL_{\mathrm{dqn}} i_{\mathrm{dq}}), \quad (13)$$

where Q denotes a Q filter of the DOB. A superscript $\hat{\bigcirc}$ and a subscript $\bigcirc_{\mathrm{n}}$ means an estimated value and a nominal value. The DOB has a role of inner model control and the plant behaves as it is nominal model designed by the DOB. Hence, the series current controller is able to be designed with only considering a bandwidth of the current controller, without paying attentions to pole-zero assignment problem which determines disturbance-suppression performance. This is one of the advantage to use the DOB-based 2-DOF controller. By using a desired bandwidth of the current control ω_c [rad/s], the current controllers are designed as

$$C_{\mathrm{dq}} = \mathrm{diag}(\omega_c L_{\mathrm{dn}}, \ \omega_c L_{\mathrm{qn}}). \quad (14)$$

A control architecture using these controllers is shown in Fig. 5. The reference signals on the uvw coordinate are

$$v_{\mathrm{uvw}}^{\mathrm{ref}} = D^{-1}(v_{\mathrm{dq}}^{\mathrm{ref}} + \hat{v}_{\mathrm{dq}}^{\mathrm{dis}}) \quad (15)$$

$$v_{\mathrm{dq}}^{\mathrm{ref}} = C_{\mathrm{dq}}(i_{\mathrm{dq}}^{\mathrm{cmd}} - i_{\mathrm{dq}}). \quad (16)$$

Here, there are many residual process in this calculation. This causes a waste of logic elements and lengthen a processing time and a phase asynchrony between a real plant and a controller space. The paper shows a minimal realization with several pretreatments. The first and second term of (15) is able to be transformed as

$$D^{-1} v_{\mathrm{dq}}^{\mathrm{ref}} = D_{\mathrm{dq}}^{-1} C_{\mathrm{dq}}(i_{\mathrm{dq}}^{\mathrm{cmd}} - D i_{\mathrm{uvw}}) \quad (17)$$

$$D^{-1} \hat{v}_{\mathrm{dq}}^{\mathrm{dis}} = D^{-1} Q_{\mathrm{dq}} \left(D v_{\mathrm{uvw}}^{\mathrm{ref}} - sL_{\mathrm{dqn}} D i_{\mathrm{uvw}} \right). \quad (18)$$

By imposing following conditions

$$C_{\mathrm{dq}} = \mathrm{diag}(C_{\mathrm{q}}) \quad (19)$$

$$Q_{\mathrm{dq}} = \mathrm{diag}(Q_{\mathrm{q}}) \quad (20)$$

$$L_{\mathrm{dqn}} = \mathrm{diag}(L_{\mathrm{qn}}) \quad (21)$$

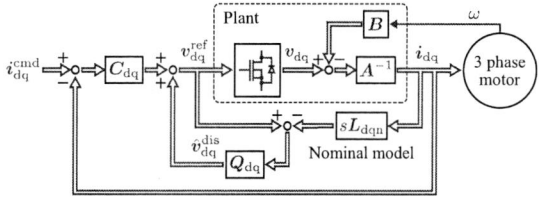

Fig. 5. Control architecture using the P controller and the DOB.

on the design, the process (17) and (18) are simplified as

$$D^{-1} v_{\mathrm{dq}}^{\mathrm{ref}} = C_{\mathrm{q}}(D_{\mathrm{dq}}^{-1} i_{\mathrm{dq}}^{\mathrm{cmd}} - i_{\mathrm{uvw}}) \quad (22)$$

$$D^{-1} \hat{v}_{\mathrm{dq}}^{\mathrm{dis}} = Q_{\mathrm{q}} \left(v_{\mathrm{uvw}}^{\mathrm{ref}} - sL_{\mathrm{qn}} i_{\mathrm{uvw}} \right). \quad (23)$$

These equation shows that the feedback controllers can be designed only on the uvw coordinate without d-q transformation. A point to notice is that a nominal model of the DOB equivalently designed on the uvw coordinate is the parameter of the d-q coordinate. All designed parameters are designed with focusing on the q axis, which has a dominant effect on a generated torque. Since the nominal parameter of the DOB affects on robust stability and performance, it should be careful when there exists a large salient ratio. Compared with the classical model-based feedback control, the d-q transformation process are removed and it achieves a much reduction of a computational complexity.

C. Sensor Controller

Most of the controllers requires to use differential signals of sensor outputs to ensure a stability margin. A point to notice is that a performance of the differentiation deeply depends on the sampling rate of the differentiator. Since we could not obtain future values of the sensor outputs, a pseudo-differentiator should be embedded on the system. This pseudo-differentiation is realized using an IIR filter in many case. Here, the IIR filter uses past values of a system input and a filter output and then an operation speed of the differentiator is important index to specify a performance of the differentiation.

An IIR filter which ensures a differential characteristic within bandwidth ω_{pd} is described as

$$D(s) = \frac{\omega_{\mathrm{pd}}}{s + \omega_{\mathrm{pd}}} s. \quad (24)$$

By discretizing using the bilinear transformation, we obtain

$$\hat{x}[k] = N_1(1 - z^{-1})x[k] + N_2 z^{-1}\hat{x}[k], \quad (25)$$

where N_1 and N_2 are regression coefficients such that

$$N_1 = (2\omega_{\mathrm{pd}})/(2.0 + \omega_{\mathrm{pd}} T_{\mathrm{s}}) \quad (26)$$

$$N_2 = -(\omega_{\mathrm{pd}} T_{\mathrm{s}} - 2.0)/(2.0 + \omega_{\mathrm{pd}} T_{\mathrm{s}}), \quad (27)$$

and x denotes an input and $\hat{\bigcirc}$ means estimated value. The bandwidth of differentiation should be designed with considering noise sensitivity. By saving these coefficients in BRAMs, this calculation process becomes executable. Since this process has a few calculation, the differentiator has a capability of high-speed operation.

Fig. 6. Three phase 2 level GaN-HEMT-based class-D inverter.

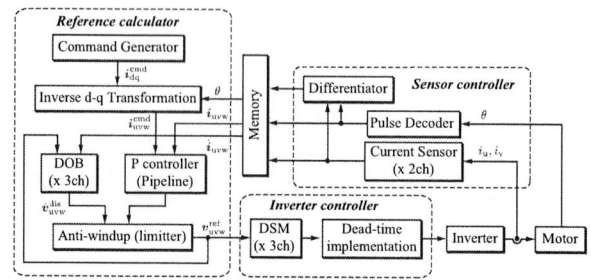

Fig. 7. Functional block-diagram of the FPGA

III. INVERTER DESIGN

The voltage inverter located on the feedforward pass plays an important role to ensure a tracking performance and a disturbance rejection control. Main requirements are (i) low-distortion and (ii) wide-band voltage feeding. To satisfy the both, the paper introduced a PDM inverter using a delta-sigma modulation (DSM). It is true that the DSM is effective for noise suppression of low-frequency domain and usually used for precise measurement. From a different viewpoint, it simultaneously denotes that the high-frequency switching inverter is able to elicit a performance of the DSM. This is realizable because an operation frequency of the inverter is quite higher than a drive frequency which is restricted by a mechanical disturbance and available energy.

One of the advantage to use the PDM inverter is to ensure a number of switchings even when a reference voltage is small. This is difficult to achieve by the PWM inverter, though multi-leveling of the inverter improves an output characteristics on the low-voltage area. In relation to a quantization noise, 2 level PDM inverter and PWM inverter have almost same noise level while the PDM inverter has a larger number of switching [15]. This result denotes a superiority of the PDM inverter in a frequency domain. An operation of the PDM inverter is expressed in a discrete domain as

$$Y[k] = X[k] + (1 - z^{-1})^L Q[k], \qquad (28)$$

where Y, X, L and Q stand for an output, an input, an order of the delta-sigma modulator, and the quantization noise and operator z is a shift operator. By this modulation characteristics, a signal-to-noise ratio (SNR) is calculated as [16]

$$\mathrm{SNR} \approx 20 \log(M - 1) - 1.25 + A_L + A_K \qquad (29)$$

$$A_L = -15.96L + 10 \log(2L + 1) \qquad (30)$$

$$A_K = (2L + 1)10 \log K, \qquad (31)$$

where M and K denote a number of output stages of the inverter and an ratio of an control bandwidth and a switching frequency, called as oversampling rate. Each term A_L and A_K express contribution from the DSM and oversampling. It should be noted that the DSM enhances the contribution of the oversampling. As for the PWM inverter, a carrier wave works like a dither signal and make noise characteristics uniform. An operation and a SNR of the PWM inverter is equal to a case when

substituting a condition $L = 0$ to (28) and (29)–(31). Hence, the in-band performance is difficult to improve even when using the high-frequency switching inverter since the contribution from the oversampling is not large. It can also be construed as the PDM inverter enhances the in-band performance by increasing the number of switching.

To elicit the performance of the DSM, the inverter having a capability to operate in a high frequency domain is required. The GaN-HEMT is a good tool as it can perform high speed switching and has a small conduction loss [17], [18]. These characteristics provides remarkable reduction of a dead-time and a distortion following to the switching and heat generation. Hence, this power device enables the inverter to be operated in the high frequency domain where is difficult to achieve by silicon devices. The paper uses a three phase 2 level GaN-HEMT-based class-D inverter shown in Fig. 6.

IV. CIRCUIT DESIGN

The paper designed circuits for each control system on the FPGA to elicit a system performance. Signal processing treating a high-frequency signal is important to realize a wideband control. Since the control system is composed of several parts which have different sampling rates, it is better to allocate appropriate time-schedulers for each operation. For example, the sensory system operates continuously without waiting the other process. Differentiation, which determines a control performance in a transient state, is always a big problem in the controller design as it has possibility to excite an unexpected oscillation. To obtain a low-noise differential value, the oversampling is a good technique. Since a differentiator can be designed as an independent module for most control systems, the oversampling rate is easy to ensure; A calculation process of the reference signals does not include the differentiation process but just orders the differentiator to feeding a differential value when the calculation starts. Similarly, although a control period of the inverter is determined by an available operation frequency of the power devices, there is no need for the other controllers to wait for a next switching period. Namely, a control module is established by fetching a decimated outputs of the other modules while operating them continuously. This design policy also applies to a

672

The 2018 International Power Electronics Conference

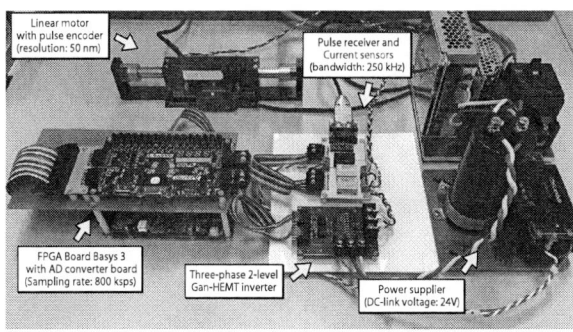

Fig. 8. Experimental setup.

TABLE I. LATENCIES OF THE MODULES.

Module	Function	Latency
Reference calculator		
Inv. dq Trans.	Transform signal from uvw to dq coordinate	100 cycles.
P controller	Calculate control signal	6 cycles
DOB	Estimate disturbances on the dq-axes	20 cycles
Anti-windup	Prevent the DSM output from oscillating	3 cycles
Inverter controller		
DSM	Output modulated binary for gate signal	10 cycles
Dead-time Imp.	Fix all gate signals to '0' in predetermined period	–
Sensor controller		
Differentiator	Output differentiated signal	10 cycles
Current sensor	Measure-then-output armature current	125 cycles
Pulse decoder	Measure-then-output position and electrical angle	10 cycles

controller for external sensors such as a pulse decoder or an A/D converter. To design such multirate blocks, an FPGA which makes any private circuits on a chip is good tool. The paper shows a functional block-diagram of the FPGA in Fig. 7. Even though a slow operation of the controller causes an aperture effect, the drive bandwidth does not so much affected because it is mainly restricted by the bandwidth of the switching inverter. The data type in the calculation process is double-precision floating-point according to IEEE Standard 754 in order to ensure a dynamic range. The differentiator has a high-sensitivity to a rounding error and it is better to ensure a precision when treating a wideband signal. This structure is able to elicit a performance of the peripheral devices.

V. SIMULATION

The paper conducted simulation of the current controller to check an ideal performance of the designed system as a pre-test. An object is to confirm the disturbance suppression performance using the wideband control system.

A. Setup

1) Hardware: For the simulation, the paper used a parameters in an actual platform. A target is a linear motor (S120T; GMC Hillstone). Variations of a rotor angle and an electrical angle are measured by a pulse encoder (RGH24H15A30A; Renishaw), which has a resolution of 50 nm. The power is supplied through a switching AC-DC converter (PLA100F-24; Cosel) which outputs 24 V. To control the inverter, peripherals are prepared as shown in Fig. 8: (i) an FPGA board (Basys3; Digilent) which has a Xilinx XC7A35T and clock source of 100 MHz, (ii) a pulse receiver and current sensors (CDS4006; Sensitec) whose sensing bandwidth is 250 kHz, (iii) an AD converter board and its controller (ADD-16B8 and AX-Card7/100C2; Prime Systems) which have a Xilinx XC7A100T and 8ch pairs of an AD converter AD7980, an ADC buffer ADA4841 and an isolator ADUM3441. This module is able to perform conversion at 824 Hz with 16-bit resolution.

2) Controller: Minimum latencies of the modules in Fig. 7 are listed in TABLE. I. In the reference calculator, the inverse dq transformation and the P controller should

be executed sequentially but the DOB is able to run in parallel. Then, the reference calculation finishes in almost 110 cycles. The inverter controller is operated at switching frequency. In the sensor controller, the differentiator runs along each target module since the differentiator is an IIR filter requiring recursive values. Then, the differentiators for the position and the armature current is operated at 10 MHz and 800 kHz.

3) Parameters: Since the prepared power source have low voltage, the system is easily suffered from saturation. In addition, the DSM works like an IIR filter, a large input lead to sticking of gate signals to one side and make its output oscillatory. Then, a rely model is introduced as

$$v_{\mathrm{uvw}}^{\mathrm{ref}} = \begin{cases} 12 & \text{when } v_{\mathrm{uvw}}^{\mathrm{ref}} > 12 \\ -12 & \text{when } v_{\mathrm{uvw}}^{\mathrm{ref}} < -12 \\ v_{\mathrm{uvw}}^{\mathrm{ref}} & \text{when others.} \end{cases} \quad (32)$$

Now, the other anti-windup controller is not implemented. The controller gains are designed as

$$C_{\mathrm{dq}} = \mathrm{diag}(31400,\ 31400) \quad (33)$$

$$Q_{\mathrm{dq}} = \mathrm{diag}\left(\frac{62800}{s+62800},\ \frac{62800}{s+62800}\right). \quad (34)$$

Here, the current controller and the DOB have bandwidths of 5 kHz and 10 kHz. It should be noted that a high gain P controller easily causes the input saturation and hence its gain is restricted. The model values of the DOB are set at nominal values. The switching frequency of the inverter is set at 500 kHz. The command values are given as

$$i_{\mathrm{dq}}^{\mathrm{cmd}} = \begin{bmatrix} 0 & 0.5(1 - e^{-1000(t-0.002)}) \end{bmatrix}^{\mathrm{T}}. \quad (35)$$

B. Result

Outputs of the current controller is shown in Fig.9. This results shows that a disturbances on the dq coordinate, such as voltage drop due to the coil resistance, back-electromotive-force and magnetic field interference, are able to be suppressed by only the DOB, which has wideband estimation-capability. It also be confirmed that the wideband inverter certainly feeds the estimated disturbance to the plant. Owing to the fast processing and the wideband transmission, the robustness is attained. Simultaneously, the fast processor is able to elicit the performance of the inverter and then widen the bandwidth. It can be expected that experiment shows good results.

673

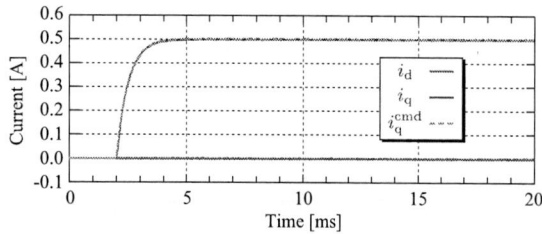

Fig. 9.   Simulation result: currents on the d-q coordinate.

VI.   CONCLUSION

This paper aims to design the wideband inverter for high back-drivable system. The three pillars, the controller design, the hardware design and the design policy, is introduced and the design method which conforms to these pillars is considered. Then, the simple controller, the inverter controller and the hardware structure on the FPGA to realize the wideband system are derived. An experimental prototype is under development and we will conduct the experiment.

ACKNOWLEDGEMENT

This work was partially supported by JSPS KAKENHI.

REFERENCES

[1] H. Kaimori and K. Akatsu, Behavior Modeling of Permanent Magnet Synchronous Motors Using Flux Linkages for Coupling with Circuit Simulation, *IEEJ Journal of Industry Applications,* vol. 7, no. 1, pp. 56-63, Jan. 2018.

[2] T. Oomen, Advanced Motion Control for Precision Mechatronics: Control, Identification, and Learning of Complex Systems, *IEEJ Journal of Industry Applications,* vol. 7, no. 2, pp. 127-140, Feb. 2018.

[3] Y. Fujimoto, T. Murakami, and R. Oboe, "Advanced motion control for next-generation industrial applications", *IEEE Transactions on Industrial Electronics,* vol. 63, no. 3, pp. 1886-1888, 2016.

[4] T. Tsuji, K. Kutsuzawa and S. Sakaino, Optimized Trajectory Generation based on Model Predictive Control for Turning Over Pancakes, *IEEJ Journal of Industry Applications,* vol. 7, no. 1, pp. 22-28, Jan. 2018.

[5] A. Z. Shukor and Y. Fujimoto, "Direct-Drive Position Control of a Spiral Motor as a Monoarticular Actuator", *IEEE Transactions on Industrial Electronics,* vol. 61, no. 2, pp. 1063-1071, 2014

[6] S. D. Eppinger and W. P. Seering, "Three Dynamic Problems in Robot Force Control," *IEEE Transactions on Robotics and Automation,* vol. 8, no. 6, 751-758, Dec. 1992.

[7] S. McIntosh-Smith T. Wilson, A, Í. Ibarra and J. Crisp, R. B. Sessions, Benchmarking Energy Efficiency, Power Costs and Carbon Emissions on Heterogeneous Systems, *The Computer Journal,* vol. 55, no. 2, pp. 192–205, Feb. 2102.

[8] S. Katsura, Y. Matsumoto, K. Ohnishi, "Analysis and Experimental Validation of Force Bandwidth for Force Control," *IEEE Transactions on Industrial Electronics,* vol. 53, no. 3, pp. 922-928, Jun. 2006.

[9] S. Katsura, K. Irie, K. Ohishi, Wideband Force Control by Position-Acceleration Integrated Disturbance Observer, *IEEE Transactions on Industrial Electronics,* vol. 55, no. 4, pp. 1699–1706, Apr. 2008.

[10] K. Ohyama, K. Fujii, H. Fujii and K. Uehara, Application of FPGA to Rotor Position Detection of SR Motor Using Rotary Encoder, *IEEJ Transactions on Industry Applications,* vol. 125, no. 5, pp. 473-481, Aug. 2005.

[11] Y. Yokokura and K. Ohishi, "FPGA-Based Broadband Current Control of a Linear Motor with Class-G Power Amplifiers", *IEEE International Conference on Mechatronics,* pp. 528-533, 2013.

[12] Y. F. Wu, D. Kebort, J. Guerrerol, S. Yea, J. Honeal, K. Shirabe and J. Kang, "High-Frequency, GaN Diode-Free Motor Drive Inverter with Pure Sine Wave Output", *Power Transmission Engineering,* pp. 40-43, 2012.

[13] K. Ohishi, K. Ohnishi and K. Miyachi, "Torque-Speed Regulation of DC Motor Based on Load Torque Estimation", *International power electronics conference,* pp. 1209-1218,1983.

[14] T. Mita, M. Hirata and K. Murata, "Theory of H ∞ Control and Disturbance Observer", *IEEJ Transactions on Electronics, Information and Systems,* vol. 115-C, no. 8, pp. 1002-1011, 1995.

[15] H. Kurumatani and S. Katsura, "GaN-HEMT-based three level T-type NPC inverter using reverse-conducting mode in rectifying", *IEEE International Symposium on Industrial Electronics,* pp. 1941-1946, 2017.

[16] H. Kurumatani and S. Katsura, "FPGA-Based Voltage Control with PDM Inverter for Wideband Drive System Using T-type Neutral-Point-Clamped Topology", *IEEE International Conference on Mechatronics,* pp. 272-277, 2017.

[17] J. Millan, P. Godignon, X. Perpina, A. Perez-Tomas and J. Rebollo, "Asurvey of Wide Bandgap Power Semiconductor Devices", *IEEE Transactions on Power Electronics,* vol. 29, no. 5, pp. 2155-2163, 2014.

[18] W. Zhang, Z. Xu, Z. Zhang, F. Wang, L. M. Tolbert and B. J. Blalock, "Evaluation of 600 V Cascode GaN HEMT in Device Characterization and All-GaN-Based LLC Resonant Converter", *IEEE Energy Conversion Congress and Exposition,* pp. 3571-3578, 2013.

Analysis of Interior Permanent Magnet Two Degrees of Freedom Motor Based on Cross-Coupled Structure

Yoshiyuki Hatta[1,2]*, Tomoyuki Shimono[2,3]**

1 Graduate School of Engineering, Yokohama National University, 79-5 Tokiwadai, Hodogaya, Yokohama, Japan
2 Kanagawa Institute of Industrial Science and Technology, 3-2-1 Sakado Takatsu, Kawasaki, Japan
3 Faculty of Engineering, Yokohama National University, 79-5 Tokiwadai, Hodogaya, Yokohama, Japan
*E-mail: hatta-yoshiyuki-ws@ynu.jp
**E-mail: shimono-tomoyuki-hc@ynu.ac.jp

Abstract—This paper proposes an interior permanent magnet (IPM) two degrees of freedom (2-DOF) motor based on cross-coupled structure. This paper describes a structure of the IPM 2-DOF motor and an analysis of the IPM 2-DOF motor with Finite Element Analysis (FEA). The relationship between electrical angles, amplitude of current, thrust force and torque is considered by the analysis. Moreover, an analysis for the comparison with the conventional 2-DOF motor and the proposed 2-DOF motor is conducted.

Keywords—Cross-coupled motor, 2-DOF motor, Interior permanent magnet, Multi degrees of freedom

I. INTRODUCTION

Recently, robots have been more and more important in the industrial field, because the robots have been seen as an alternative to human workers. The robots will play more important role in aging society. They will make up for the labor in a lot of fields. In order to realize that, robots need to realize human's motion. Human's motion is realized by the multi-joint structure. Therefore, the robots generally need to be composed of multi degrees of freedom (multi-DOF) system in order to realize complicated motion such as human-like motion. Many kinds of the multi-DOF system have been researched and developed.

The number of DOF, which a robot has, is increased as motion of the robot becomes more complicated. One DOF generally consists of one motor. Because of that, the number of motors is the same as the number of DOF. The number of motors is increased as motion of the robot becomes more complicated. The robot would be big and heavy with an increase in the number of motors.

In order to prevent the number of motors from being increased, spherical motors have been proposed [1]–[3]. Each spherical motor has two or three DOF so that each of them rotates about multi axes. By applying the spherical motors to the multi-DOF systems, it is possible to decrease the number of motors in the multi-DOF systems.

Additionally, a 2-DOF actuator which generates thrust force in one axial direction and torque about the axis has been proposed [4]. The actuator realizes not only rotational motion but also linear motion. Furthermore, the

actuator realizes two kinds of the motion simultaneously. However, the actuator has nine kinds of windings. It is necessary that current of each winding is controlled independently. Because of that, the controller could become big and expensive.

Then 2-DOF actuators having two kinds of 3-phase windings have been proposed [5], [6]. One kind of 3-phase windings generates thrust force and the other set generates torque. However, one permanent magnet (PM) is opposed to either of two kinds, but not both. Because of that, even if magnetic flux of one PM is improved, only thrust force or torque, which is generated by the kind of 3-phase windings opposed to the PM, is improved. It is difficult that thrust force volume ratio and torque volume ratio are improved.

In order to improve thrust force volume ratio and torque volume ratio, a cross-coupled 2-DOF direct drive motor has been proposed [7]. The 2-DOF motor is the surface permanent magnet (SPM) motor, which has a mover composed of a shaft and PMs. In the 2-DOF motor, PMs are arranged on the surface of the shaft. A stator of the 2-DOF motor has two kinds of 3-phase windings. The 2-DOF motor generates each of thrust force and torque by combination of Lorentz force generated by two kinds of 3-phase windings. Additionally, one PM is opposed to both of two kinds of 3-phase windings. Because of that, if magnetic flux of one PM is improved, Lorentz force generated by both of two kinds of 3-phase windings is improved. Therefore, it is possible that thrust force volume ratio and torque volume ratio are improved.

Recently, in order to improve torque of the rotational motors, interior permanent magnet (IPM) motors have been researched and developed [8]. The IPM motor has a shaft composed of the ferromagnetic material. PMs of the IPM motor are not arranged on a surface of the shaft but implanted in the shaft. Therefore, it is possible that the PMs are molded into the arbitrarily shape. The PMs are molded into the optimal shape so that the magnet torque is improved. Additionally, IPM motors are able to generate not only magnet torque but also reluctance torque. Therefore, it is possible that torque of motors is improved by applying IPM structure to the motors. In

The 2018 International Power Electronics Conference

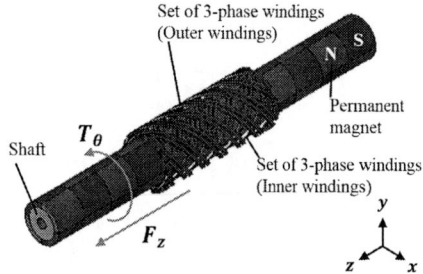

Fig. 1.   Conventional 2-DOF motor.

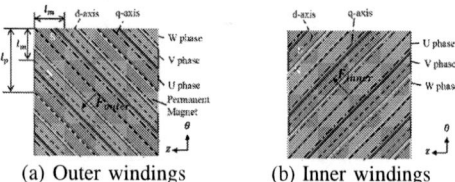

(a) Outer windings (b) Inner windings

Fig. 2.   A development view of the 2-DOF motor

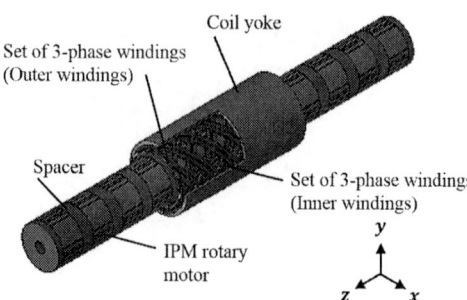

Fig. 3.   Proposed 2-DOF motor.

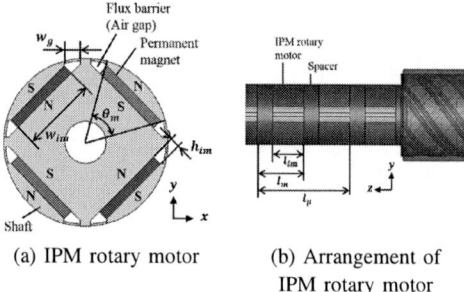

(a) IPM rotary motor (b) Arrangement of
IPM rotary motor

Fig. 4.   Structure of the IPM rotary motor in the proposed 2-DOF motor

order to improve thrust force and torque of the cross-coupled 2-DOF direct drive motor, this paper proposes an interior permanent magnet (IPM) two degrees of freedom (2-DOF) motor based on cross-coupled structure.

This paper is organized as follows. The conventional cross-coupled 2-DOF direct drive motor is described in section II. In section III, the structure of the IPM 2-DOF motor is described. Analysis results of the IPM 2-DOF motor are shown in section IV. Finally, this paper is summarized in section V.

II.   Cross-coupled 2-DOF direct drive motor

The structure of the conventional 2-DOF motor based on cross-coupled structure is shown in Fig. 1. The conventional 2-DOF motor has a mover, which is composed of a shaft and PMs, and a stator, which is two kinds of 3-phase windings. In the 2-DOF motor, one kind of 3-phase windings is composed of inner windings. The other kind of 3-phase windings is composed of outer windings. Inner windings and outer windings are orthogonally arranged. Because of that, these windings are called cross-coupled windings.

In the mover, PMs are arranged on a surface of the shaft. Each PM is formed into arc shape. An arrangement of PMs is shown in Fig. 2 (a) and (b). Fig. 2 (a) and (b) indicate a development view of the mover, outer windings and inner windings. The PMs are arranged not only in z direction but also in θ direction, which is a rotational direction about z direction. A surface of the PM is S pole or N pole. These poles are alternately placed to z direction and θ direction, like squares on a chessboard.

Inner and outer windings are wired along diagonals of the PMs. When current is flowed to inner and outer windings, Lorentz force F_{outer} and F_{inner} shown in Fig. 2 (a) and (b) is generated. A direction of F_{outer}

is different from a direction of F_{inner}. The sum of z directional components of F_{outer} and F_{inner} produces thrust force F_z shown in Fig. 1. The difference between θ directional components of F_{outer} and F_{inner} produces torque T_θ.

In the conventional 2-DOF motor, d axes are arranged on diagonals of the PMs and q axes are arranged between consecutive d axes. In order to maximize thrust force and torque of the conventional 2-DOF motor, current is controlled so that d-axis current I_d is 0 A.

III.   Structure of the proposed 2-DOF direct drive motor

A. Basic structure of the proposed motor

The structure of the proposed 2-DOF motor is shown in Fig. 3. A stator of the proposed 2-DOF motor is composed of two kinds of 3-phase windings and a coil yoke, which covers two kinds of 3-phase windings. The mover has multi IPM rotary motors. Additionally, the mover has spacers arranged between the IPM rotary motors. The spacers are composed of non-magnetic material such as aluminum.

The IPM rotary motor is shown in Fig. 4 (a). The IPM rotary motor has a shaft composed of the ferromagnetic material, such as iron, and the PMs which are implanted in the shaft. In order to prevent magnetic flux from shorting, flux barrier is formed in the IPM rotary motor. The flux barrier is composed of materials whose magnetic resistance is high, such as air gap. A mechanical angle of each IPM rotary motor is different from a mechanical

676

The 2018 International Power Electronics Conference

angle of consecutive IPM rotary motors by 90° as shown in Fig. 4 (b).

B. Reluctance torque

In the conventional 2-DOF motor, thrust force and torque generated by one kind of 3-phase winding are calculated as shown in (1)–(4)

$$F_z = -P_N \frac{1}{2} \boldsymbol{I}^T \frac{\partial \boldsymbol{L}}{\partial z} \boldsymbol{I} = -P_N \frac{1}{2} \boldsymbol{I}_{dq}^T \boldsymbol{C} \frac{\partial \boldsymbol{L}}{\partial z} \boldsymbol{C}^T \boldsymbol{I}_{dq} \quad (1)$$

$$T_\theta = -P_N \frac{1}{2} \boldsymbol{I}^T \frac{\partial \boldsymbol{L}}{\partial \theta} \boldsymbol{I} = -P_N \frac{1}{2} \boldsymbol{I}_{dq}^T \boldsymbol{C} \frac{\partial \boldsymbol{L}}{\partial \theta} \boldsymbol{C}^T \boldsymbol{I}_{dq} \quad (2)$$

$$\boldsymbol{I} = \begin{bmatrix} I_U & I_V & I_W & I_M \end{bmatrix}^T \quad (3)$$

$$\boldsymbol{I}_{dq} = \begin{bmatrix} I_d & I_q & 0 & I_M \end{bmatrix}^T \quad (4)$$

where P_N indicates the number of 3-phase windings in one kind. In the conventional 2-DOF motor shown in Fig. 1, P_N is 2. I_U, I_V and I_W indicate U-phase, V-phase and W-phase current and I_M indicates magnetizing current. I_d and I_q indicate d-axis current and q-axis current. $\boldsymbol{L}$ and $\boldsymbol{C}$ indicate a inductance matrix and a transformation matrix. These parameters are defined as shown in (5)–(13)

$$I_U = I_{as} \sin\left(\alpha + \alpha_0\right) \quad (5)$$

$$I_V = I_{as} \sin\left(\alpha + \alpha_0 - \frac{2}{3}\pi\right) \quad (6)$$

$$I_W = I_{as} \sin\left(\alpha + \alpha_0 + \frac{2}{3}\pi\right) \quad (7)$$

$$I_d = I_a \sin\left(\alpha_0\right) \quad (8)$$

$$I_q = I_a \cos\left(\alpha_0\right) \quad (9)$$

$$I_a = \sqrt{\frac{3}{2}} I_{as} \quad (10)$$

$$\alpha = 2\pi \left(\frac{z}{2l_m}z + \frac{P_\theta \theta}{2\pi}\right) \quad (11)$$

$$\boldsymbol{L} = \begin{bmatrix} L_U & M_{UV} & M_{WU} & M_{MU} \\ M_{UV} & L_V & M_{VW} & M_{MV} \\ M_{WU} & M_{VW} & L_W & M_{MW} \\ M_{MU} & M_{MV} & M_{MW} & L_M \end{bmatrix} \quad (12)$$

$$\boldsymbol{C} = \sqrt{\frac{2}{3}} \begin{bmatrix} \cos(\beta) & \cos(\beta - \frac{2}{3}\pi) & \cos(\beta - \frac{2}{3}\pi) & 0 \\ -\sin(\beta) & -\sin(\beta - \frac{2}{3}\pi) & -\sin(\beta - \frac{2}{3}\pi) & 0 \\ \frac{1}{\sqrt{2}} & \frac{1}{\sqrt{2}} & \frac{1}{\sqrt{2}} & 0 \\ 0 & 0 & 0 & \sqrt{\frac{3}{2}} \end{bmatrix} \quad (13)$$

where I_{as} indicates peak current. α_0 and α indicate the current phase angle and the electrical angle calculated from z and θ. z and θ indicate linear position and rotational position of the mover. l_m indicates the length per one pole in z direction. P_θ indicates the number of magnetic pole pairs in θ direction and equivalent to P_N. L_U, L_V, L_W and L_M indicate self-inductance and M_{UV}, M_{VW}, M_{WU}, M_{MU}, M_{MV} and M_{MW} indicate mutual inductance.

In common IPM motors, which are 1-DOF rotary motors, self-inductance and mutual inductance are defined

as shown in (14)–(19)

$$L_U = l_a + L_a - L_{as} \cos\left(2\alpha\right) \quad (14)$$

$$L_V = l_a + L_a - L_{as} \cos\left(2\alpha + \frac{2}{3}\pi\right) \quad (15)$$

$$L_W = l_a + L_a - L_{as} \cos\left(2\alpha - \frac{2}{3}\pi\right) \quad (16)$$

$$M_{UV} = \frac{1}{2}L_a - L_{as} \cos\left(2\alpha - \frac{2}{3}\pi\right) \quad (17)$$

$$M_{VW} = \frac{1}{2}L_a - L_{as} \cos\left(2\alpha\right) \quad (18)$$

$$M_{WU} = \frac{1}{2}L_a - L_{as} \cos\left(2\alpha + \frac{2}{3}\pi\right) \quad (19)$$

where l_a indicates the leakage inductance per one phase winding. L_a and L_{as} indicate the average and the vibration amplitude of the effective inductance per one phase winding.

If reluctance torque of the common IPM motors is calculated, torque is calculated based on $L_M = M_{UV} = M_{VW} = M_{WU} = M_{MU} = M_{MV} = M_{MW} = 0$. Therefore it is assumed that reluctance force F_r and reluctance torque T_r of the proposed 2-DOF motor are T_θ and F_z calculated based on $L_M = M_{UV} = M_{VW} = M_{WU} = M_{MU} = M_{MV} = M_{MW} = 0$. F_r and T_r are calculated as shown in (20) and (23)

$$F_r = P_N \frac{\pi}{l_m} \left(L_q - L_d\right) I_d I_q \quad (20)$$

$$T_r = P_N P_\theta \left(L_q - L_d\right) I_d I_q \quad (21)$$

$$L_d = l_a + \frac{3}{2}\left(L_a - L_{as}\right) \quad (22)$$

$$L_q = l_a + \frac{3}{2}\left(L_a + L_{as}\right) \quad (23)$$

where L_d and L_q indicate d-axis inductance and q-axis inductance.

IV. ANALYSIS OF THE PROPOSED MOTOR

A. Thrust force and torque of the proposed IPM motor

In order to verify that the proposed 2-DOF motor generates thrust force and torque, an analysis of the proposed 2-DOF motor is conducted. In order to simplify the analysis and to improve computation accuracy of magnetic flux, an analytical model shown in Fig. 5 (a) is used for 3-D FEA. The PMs and the ferromagnetic material are composed of neodymium magnets and magnetic steel sheet. The analysis parameters are shown in Table I.

In the analysis, F_z and T_θ in liner motion is analyzed. The mover moves 63.0 mm in z direction, which is equivalent to the length of two PMs. Additionally, F_z and T_θ in rotational motion is analyzed. The mover rotates 180°, which is equivalent to 360° electrical angle.

The analysis results are shown in Table II. In the analysis, I_{as} is 1.414 A. Table II indicates the average of thrust force and torque. In linear motion, thrust force is generated and torque is suppressed. In rotational motion, torque is generated and thrust force is suppressed. It turns out that linear motion and rotational motion is realized.

677

The 2018 International Power Electronics Conference

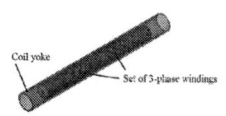
(a) First analytical model

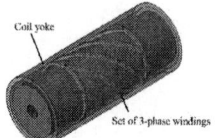
(b) Second analytical model

Fig. 5. An analytical model

TABLE I. ANALYSIS PARAMETERS

Outer diameter of the mover	D_{mo}	40.0 mm
Inner diameter of the mover	D_{mi}	10.0 mm
Length of a pole	l_m	31.5 mm
Length of a PM	l_{im}	21.0 mm
Hight of a PM	h_{im}	2.50 mm
The number of sets of 3-phase windings	P_N	2
The number of magnetic pole pairs in θ direction	P_θ	2
Pole opening angle	θ_{im}	60.0°
Width of flux barrier	w_g	4.00 mm
Outer diameter of the coil yoke	D_{co}	52.0 mm
Inner diameter of the mover	D_{ci}	46.0 mm
Width of air gap between the windings and the mover	w_{ag}	1.00 mm
Material of the shaft		35JN250
Material of the PMs		NEOREC45SH
Winding number		65

TABLE II. ANALYSIS RESULTS

	Linear motion	Rotational motion
Thrust force [N]	14.92	3.38×10^{-3}
Torque [Nm]	5.70×10^{-5}	0.31

B. Relationship between torque, thrust force and current phase

Generally, common IPM motors, which are 1-DOF rotary motors, have the relationship among current phase angle α_0, magnet torque T_m and reluctance torque T_r as shown in (24) and (25).

$$T_m = T_{m0} I_a \cos(\alpha_0) \qquad (24)$$
$$T_r = T_{r0} I_a^2 \sin(2\alpha_0) \qquad (25)$$

where T_{m0} and T_{r0} indicate magnet torque and reluctance torque in the case where I_a is 1 A. In order to confirm that the proposed 2-DOF motor has the same relationship as the common IPM motors, the analysis of the proposed 2-DOF motor is conducted.

In order to simplify the analysis and to analyze F_z and T_θ generated by one kind of 3-phase windings, an analytical model shown in Fig. 5 (b) is used. The d axis and q axis of the proposed 2-DOF motor are defined in the same as the d axis and q axis shown in Fig 2 (a) and (b). In the analysis, the average of F_z and T_θ in the case where the mover rotates $180°$ is analyzed. I_{as} is 1.414 A.

The analysis results are shown in Fig. 6–8. Fig. 6 indicates F_z and T_θ which change in response to α_0. Fig. 7 indicates magnet force F_m and magnet torque T_m, which are contained by F_z and T_θ shown in Fig. 6. Wave shapes of F_m and T_m are close to cosine curves. Therefore, the proposed 2-DOF motor has the same

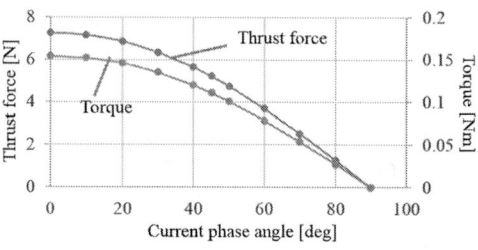

Fig. 6. Relationship among the current phase angle, force and torque

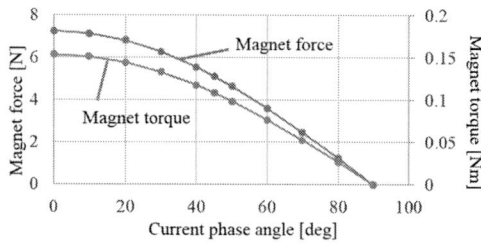

Fig. 7. Relationship among the current phase angle, magnet force and magnet torque

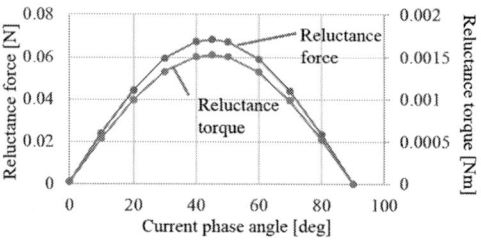

Fig. 8. Relationship among the current phase angle, reluctance force and reluctance torque

relationship between F_m, T_m and α_0 as common IPM motors. Fig. 8 indicates reluctance force F_r and torque T_r, which are contained by the F_z and T_θ shown in Fig. 6. Wave shapes of F_r and T_r are close to sine curves of two order component. Therefore, the proposed 2-DOF motor has the same relationship between F_r, T_r and α_0 as common IPM motors.

However, F_r and T_r are smaller by far than F_m and T_m. Because of that, F_z and T_θ are close to F_m and T_m.

C. Relationship between torque, thrust force and amplitude of current

The analysis to confirm the relationship between thrust force F_z, torque T_θ and current I_{as} of the proposed 2-DOF motor is conducted. The analysis results are shown in Fig. 9 and Fig. 10. Fig. 9 indicates F_m of the proposed 2-DOF motor in the case where α_0 is 0°. And this figure indicates F_r of the proposed 2-DOF motor in the case where α_0 is 45°. In the analysis, F_m and F_r in rotational motion are analyzed.

678

The 2018 International Power Electronics Conference

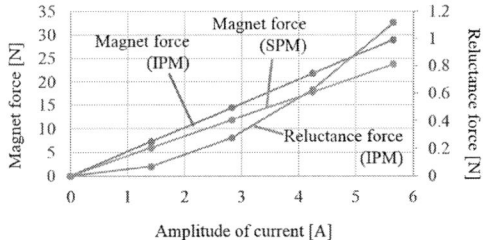

Fig. 9.   Relationship between current and force

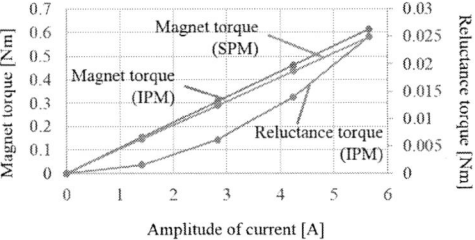

Fig. 10.   Relationship between current and torque

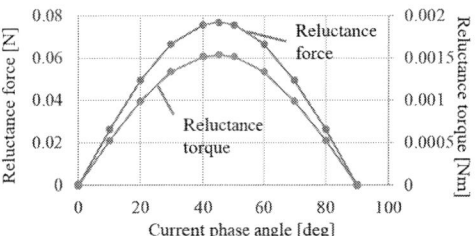

Fig. 11.   Relationship among the current phase angle, reluctance force and reluctance torque based on the calculation model

and L_q are calculated as shown in (26)–(29)

$$L_d = \frac{\Phi_d}{I_d} \tag{26}$$

$$L_q = \frac{\Phi_q}{I_q} \tag{27}$$

$$\Phi_d = \sqrt{\tfrac{3}{2}}\Phi_{as} \quad \text{when } I_d = I_a, I_q = 0 \tag{28}$$

$$\Phi_q = \sqrt{\tfrac{3}{2}}\Phi_{as} \quad \text{when } I_d = 0, I_q = I_a \tag{29}$$

where Φ_d and Φ_q indicate d-axis flux linkage and q-axis flux linkage. Φ_{as} indicates vibration amplitude of U-phase flux linkage. Therefore, L_d and L_q are calculated with the analysis result of $\alpha_0 = 0°$ and $\alpha_0 = 90°$. From the analysis result, it turns out that L_d is 6.05 mH and L_q is 7.02 mH.

Fig. 11 shows F_r and T_r calculated based on L_d, L_q, (20) and (21) Fig. 11 is close to Fig. 8. However, F_r in Fig. 11 is mildly larger than F_r in Fig. 8. In section IV. B, rotational motion is analyzed. The calculation model is applied to T_r better than F_r because L_d and L_q were calculated with U-phase flux linkage of the analysis in section IV. B. Because of that, T_r in Fig. 11 is closer to T_r in Fig. 8 than F_r. It is necessary that L_d and L_q are calculated with U-phase flux linkage in linear motion in order to improve the calculation for F_r.

From the results, it is possible to calculate F_r and T_r of the proposed 2-DOF motor based on L_d and L_q. And it is confirmed that F_r and T_r are generated by the difference between L_d and L_q. Therefore, it turns out that F_r and T_r are smaller by far than F_m and T_m because the difference between L_d and L_q is too small.

E. Analysis of the magnetic flux

In common IPM motors, L_d is small and L_q is high. That is because a path of the d-axis magnetic flux which passes PMs and flux barriers is long. And a path of the q-axis magnetic flux which passes PMs and flux barriers is short. The analysis of d-axis magnetic flux is conducted.

Fig. 12–14 show the magnetic density generated by I_d in the proposed 2-DOF motor. In other words, the figures indicate the magnetic density generated by 3-Phase windings, but not by PMs. Analysis condition is

From this figure, it turns out that F_r is proportional to squared I_{as} while F_m is proportional to I_{as}. Fig. 10 indicates T_m and T_r of the proposed 2-DOF motor. As well as F_m and F_r shown in Fig. 9, T_r is proportional to squared I_{as} while T_m is proportional to I_{as}. Therefore, F_z and T_θ of the proposed 2-DOF motor has the same feature as common IPM motors.

Moreover, Fig. 9 and Fig. 10 indicate F_z and T_θ of the conventional 2-DOF motor in the case where the current phase angle is 0°. F_z and T_θ of the conventional 2-DOF motor are composed of only magnet force and magnet torque. Quantity of PMs which the conventional 2-DOF motor has is equivalent to quantity of PMs which the proposed 2-DOF motor has. Although the conventional 2-DOF motor has the different mover from the mover of the proposed 2-DOF motor, the conventional 2-DOF motor has the same stator as the proposed 2-DOF motor.

From Fig. 9 and Fig. 10, it turns out that F_m and T_m of the proposed 2-DOF motor is larger than F_m and T_m of the conventional 2-DOF motor. One of the reasons is that magnetic flux is improved thanks to changing the shape of PMs. In other words, the PMs are molded into the optimal shape so that the magnet torque be improved.

D. Verification of the calculation model for inductance force and torque

The calculation model for inductance force and torque is derived as described in section III. B Verification of the calculation model is conducted. At first, L_d and L_q is calculated from the analysis result in section IV. B. L_d

The 2018 International Power Electronics Conference

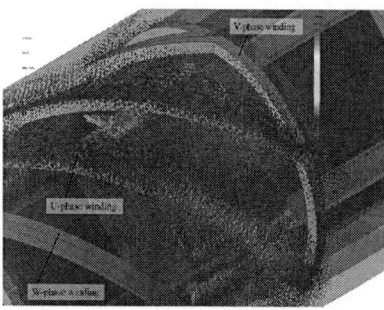

Fig. 12. Perspective view magnetic flux line

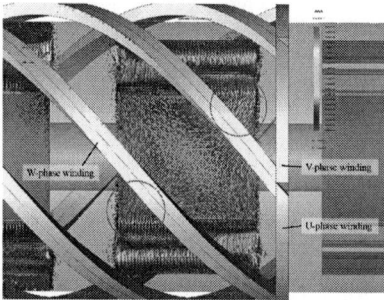

Fig. 13. Side view magnetic flux line

shown in (30)–(32)

$$\alpha_0 = 90° \tag{30}$$

$$\theta = 0 \tag{31}$$

$$z = 0. \tag{32}$$

Fig. 12 and 13 indicate the magnetic flux line. The magnetic flux flowing out of U-phase winding flows into the shaft. And the magnetic flux flows from the shaft to V-phase and W-phase windings. However, the magnetic flux almost does not pass PMs and spacers.

Fig. 14 indicates the contour plot viewed from the same viewpoint as the viewpoint in Fig. 13. The contour plot in Fig. 14 indicates radial component of the magnetic density. In the figure, magnetic density flowing to the center of the shaft is described as a positive value. Magnetic density flowing outward is described as a negative value. It turns out that the magnetic flux flowing into the shaft flows from the points as shown in red circles to V-phase and W-phase windings. Because of that, the magnetic flux almost does not pass PMs and spacers and d-axis inductance is not small. Therefore, it is necessary that the shape of the shaft is changed so that d-axis inductance is decreased.

V. CONCLUSION

In this paper, the interior permanent magnet two degrees of freedom motor based on cross-coupled structure was proposed. The analysis results showed that the relationship between current, thrust force and torque of the proposed 2-DOF motor was close to the relationship of common IPM motors. Additionally, by the analysis

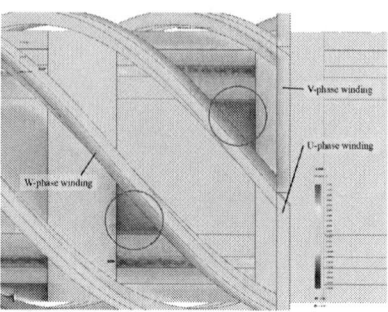

Fig. 14. Contour plot of magnetic density

results, it was verified that thrust force and torque were improved by applying IPM structure to the 2-DOF cross-coupled motor.

However, the difference of torque and thrust force between the proposed 2-DOF motor and the conventional 2-DOF motor is small. Reluctance force and torque generated by the proposed 2-DOF motor is smaller by far than magnet force and torque. Therefore, it is necessary that the shape of the shaft is changed so that reluctance force and torque is increased.

ACKNOWLEDGEMENT

This research was supported in part by Japan Society for the Promotion of Science under Grant-in-Aid for Scientific Research (B), 17H03211.

REFERENCES

[1] W. Wang, J. Wang, G. W. Jewell and D. Howe, "Design and control of a novel spherical permanent magnet actuator with three degrees of freedom", *IEEE/ASME Transactions on Mechatronics*, vol. 8, no. 4, pp. 457–468, Dec. 2003.

[2] B. Li, G. D. Li and H. F. Li, "Magnetic Field Analysis of 3-DOF Permanent Magnetic Spherical Motor Using Magnetic Equivalent Circuit Method", *IEEE Transactions on Magnetics*, vol. 47, no. 8, pp. 2127–2133, Aug. 2011.

[3] J. F. P. Fernandes and P. J. C. Branco, "The Shell-Like Spherical Induction Motor for Low-Speed Traction: Electromagnetic Design, Analysis, and Experimental Tests", *IEEE Transactions on Industrial Electronics*, vol. 63, no. 7, pp. 4325–4335, July 2016.

[4] G. Krebs, A. Tounzi, B. Pauwels, D. Willemot and F. Piriou, "Modeling of A Linear and Rotary Permanent Magnet Actuator", *IEEE Transactions on Magnetics*, vol. 44, no. 11, pp. 4357–4360, Nov. 2008.

[5] T. T. Overboom, J. W. Jansen, E. A. Lomonova and F. J. F. Tacken, "Design and Optimization of a Rotary Actuator for a Two-Degree-of-Freedom $zphi$-Module", *IEEE Transactions on Industry Applications*, vol. 46, no. 6, pp. 2401-2409, Nov.-Dec. 2010.

[6] J. Si, H. Feng, L. Ai, Y. Hu and W. Cao, "Design and Analysis of a 2-DOF Split-Stator Induction Motor", *IEEE Transactions on Energy Conversion*, vol. 30, no. 3, pp. 1200–1208, Sept. 2015.

[7] S. Tanaka, T. Shimono, and Y. Fujimoto, "Development of a Cross-Coupled 2DOF Direct Drive Motor", *Proceedings of the IEEE International Conference on Industrial Electronics*, pp. 508–513, 2014.

[8] Y. Honda, T. Higaki, S. Morimoto and Y. Takeda, "Rotor design optimisation of a multi-layer interior permanent-magnet synchronous motor", *IEE Proceedings - Electric Power Applications*, vol. 145, no. 2, pp. 119–124, Mar 1998.

Study comparison between firefly algorithm and particle swarm optimization for SLAM problems

Mounia Janah[1*], Yasutaka Fujimoto[1**]

1 Yasutaka Fujimoto Laboratory, Electrical and computer engineering departement, Yokohama National University, Yokohama, Japan
*E-mail: mounjanah@gmail.com
**E-mail: fujimoto@ynu.ac.jp

Abstract—**Simultaneous localization and mapping abbreviated in SLAM is a procedure used to elaborate autonomous mobile robots that can construct a map of an unknown surroundings and simultaneously use that map to calculate its own position. There is many implementations of methods for solving the SLAM problem. In this context, Firefly Algorithm is inspired by fireflies behavior in nature and is one of the latest models same as PSO that is the abbreviation for Particle Swarm Optimization which is inspired by bird flocking or fish schooling and described as population depend on optimization process. We conducted a sequence of experiments using each algorithm and the results of those experiments were evaluated and compared to find the best solutions.**
Keywords—**FA, LRF, PSO, SLAM**

I. INTRODUCTION

The Simultaneous Localisation and Mapping problem request for a mobile robot if it is possible to construct a consistent map of an unknown surroundings and determine it's own location within the same map simultaneously. Over the past decade, The solution of the SLAM problem has been one of the remarkable successes of the robotics community.

SLAM has been developed and determined as a abstract problem in a number of different models. However, consequential problems continue in basically accomplishing more general SLAM solutions and particularly as part of a SLAM algorithm in building and using maps [1]. This paper aims to compare the firefly algorithm with Particle Swarm Optimisation, there are test problems that can be used to compare between them, experiments will be presented applying these algorithms for solving SLAM problem and comparing the results.

II. PARTICLE SWARM OPTIMIZATION (PSO) AND FIREFLY OPTIMIZATION ALGORITHM (FA)

Particle Swarm Optimisation (PSO) constructed a set of possible problem solutions as particles that are moved through a problem area, a particle updates its velocity and finally its position in the search area Fig. 1.

(1) is the velocity and (2) is the position of the ith particle at the kth step :

$$v_i^k = wv_i^{k-1} + c_1r_1(p_i - x_i^{k-1}) + c_2r_2(p_n x_i^{k-1}) \quad (1)$$

$$x_i^k = x_i^{k-1} + v_i^k \quad (2)$$

where v_i^k and x_i^k describe velocity and position for the ith particle in the kth step of the PSO algorithm, p_i represents the local best location of the ith particle and p_n represents the global best location of all particles. the acceleration coefficients parameters are c_1 and c_2, and the arbitrary numbers taken from an homogeneous distribution are r_1, r_2 [2]. The parameter w is called inertia weight and was initiated in [3] to manage how much devotes the current velocity of the particle to its velocity in the next iteration. It plays the role of adjusting the balance between the global and local searches [2].

We set the following parameters:
- Number of particle: $m = 100$
- Weight coefficients: $w = 0.729$, $c_1 = 1.4955$, $c_2 = 1.4955$
- Number of iteration: $K_{max} = 500$

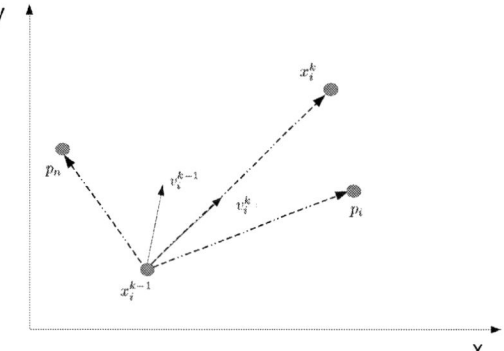

Fig. 1: PSO process

A. Procedure of the first test

To find the right parameters for particle swarm optimization algorithm we test our algorithm for different values of weight coefficients :

$w = 0.6$; 0.7; 0.729
$c_1 = 0.8$; 2.8 ; 1.4955
$c_2 = 0.8$; 1.3 ; 1.4955

We test the parameters by using sphere function (5) with :
$-5.12 \leq x_i \leq 5.12 \Rightarrow$
$-5.12 \leq x \leq 5.12$; $-5.12 \leq y \leq 5.12$; $1 < i < n$

The results of Fig. 2 and Fig. 3 shows when we set w = 0.729, c_1 =1.4955, c_2 = 1.4955, the particle swarm optimization algorithm converge better

The 2018 International Power Electronics Conference

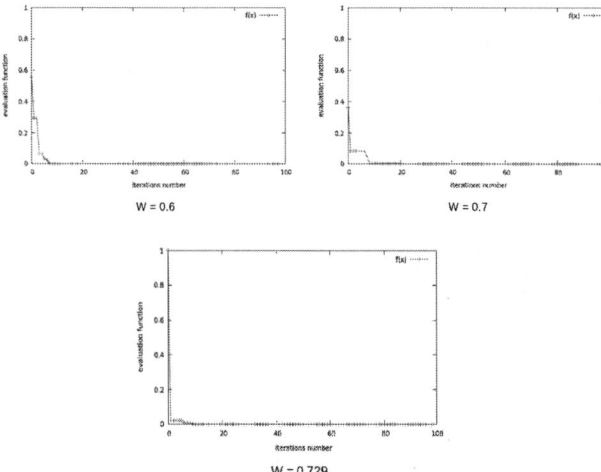

W = 0.6   W = 0.7

W = 0.729

Fig. 2: Results from first test when we change the value of w

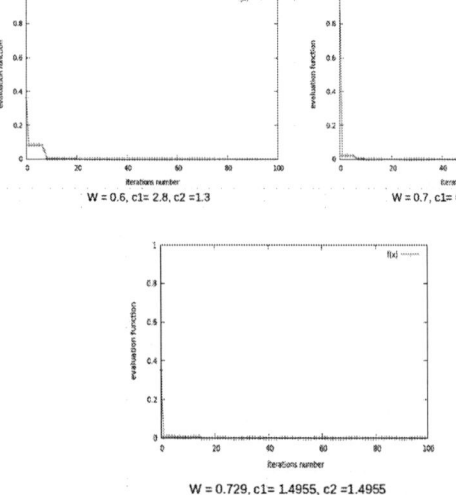

W = 0.6, c1= 2.8, c2 =1.3   W = 0.7, c1= 0.8, c2 =0.8

W = 0.729, c1= 1.4955, c2 =1.4955

Fig. 3: Results from first test when we change the values of w, c_1 and c_2

In other hand, Firefly Algorithm (FA) depends on the glowing paradigm of fireflies and the attractiveness of a firefly is compatible to its lightness, and this decreases with the observed distance Fig. 4. The fluctuation of attractiveness with distance r is shown as below :

$$\beta = \beta_0 e^{-\gamma r^2} \qquad (3)$$

where γ is the luminous abstraction coefficient of the medium and β_0 is the attractiveness at distance $r = 0$. The movement of ith firefly to a brighter j^{th} firefly which is attracted to is

$$x_i^{t+1} = x_i^t + \beta_0 e^{-\gamma r_{ij}^2}(x_j^t - x_i^t) + \alpha_t \epsilon_i^t \qquad (4)$$

where respectively x_i^t and x_j^t are the positions of fireflies i and j at time t and α_t is the randomisation parameter that can be related to rate the random element and ϵ_i^t is a vector

of random numbers [2]. This random vector is mostly peaked from either Gaussian or uniform distribution, but can also be applied to other probability distributions. The second term is used for the attraction to the lighter neighbour and the third term is dedicated to randomisation.

We set the following parameters:
- Number of firefly: $m = 100$
- Weight coefficients: $\beta_0 = 0.2$, $\gamma = 6.5$, $\alpha = \text{rand}(0,1)$
- Number of iteration: $K_{max} = 500$

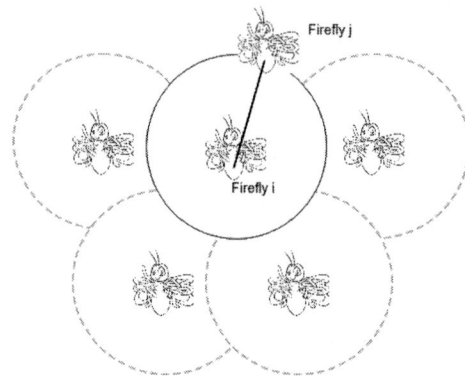

Fig. 4: FA process

B. Procedure of the second test

To find the right parameters for Firefly optimization algorithm we test our algorithm for different values of weight coefficients :

$\beta_0 = 0.2, 1$

$\gamma = 0.8, 1, 2.5, 4.5, 6.5, 7$

$\alpha = 0.2, \text{rand}(0,1)$

We test the parameters by using sphere function (5)
With: $-5.12 \le x_i \le 5.12 \Rightarrow$
$-5.12 \le x \le 5.12; -5.12 \le y \le 5.12; 1 < i < n$

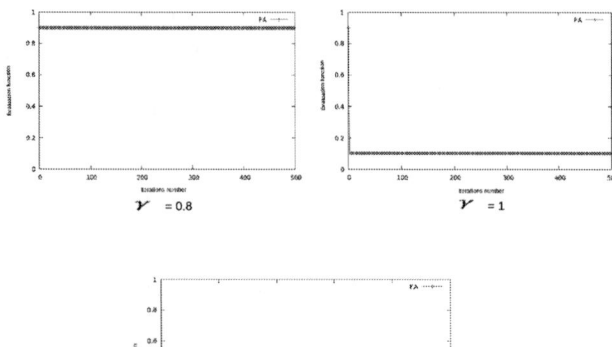

$\gamma = 0.8$ $\gamma = 1$

$\gamma = 2.5$

Fig. 5: Results from second test when we change the value γ = 0.8; 1; 2.5

682

The 2018 International Power Electronics Conference

TABLE I
TABLE OF EXECUTION TIME OF BOTH PROCESS

Functionsl	PSO	FA
Sphere fct	0.1s	0.45s
Maytas fct	0.4s	2.88s
Schaffer fct	0.5s	7.49s
Michalewicz fct	0.16s	0.76s

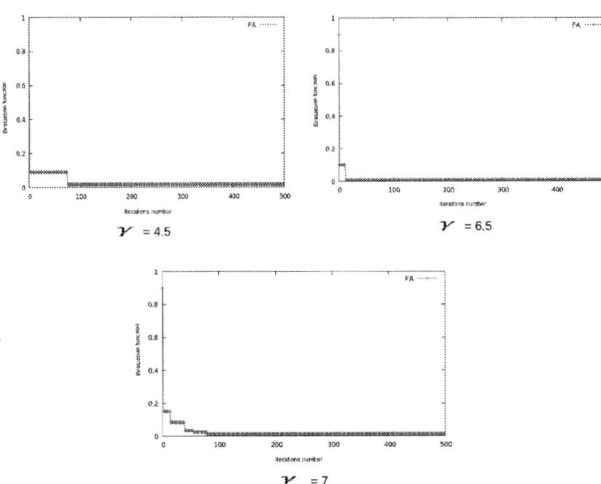

$\gamma = 4.5$ $\gamma = 6.5$

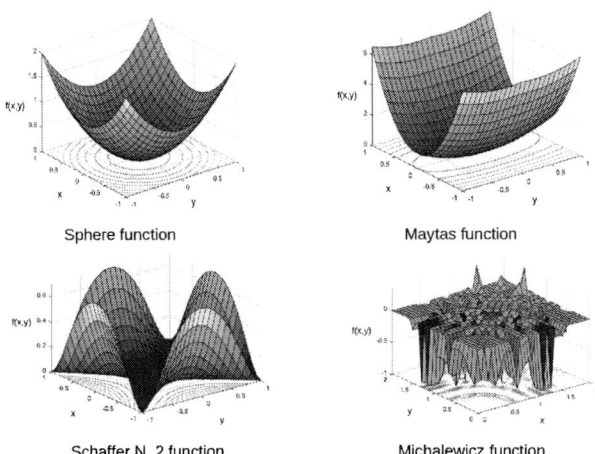

$\gamma = 7$

Fig. 6: Results from second test when we change the value γ = 4.5; 6.5; 7

The results of Fig. 5 and Fig. 6 shows, when we set $\gamma = 6.5$; $\beta_0 = 0.2$; $\alpha = \text{rand}(0, 1)$ the Firefly optimization algorithm converge better

Testing on Benchmark functions Fig. 7:

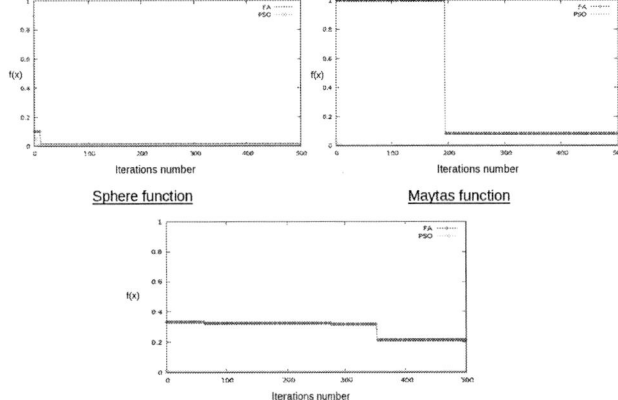

Sphere function   Maytas function

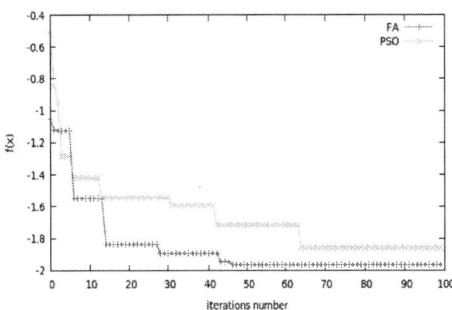

Schaffer function N. 2

Fig. 8: Results of testing on Benchmark functions

- Michalewicz function :

$$f(x) = -\sum_{i=1}^{d} \sin(x_i)[\sin(\frac{ix_i^2}{\pi})]^{2n} \qquad (8)$$

with n = 10 and d = 1, 2, ...; in the region $10 \leq x \leq 0, 0 \leq y \leq 10$.

Fig. 9: Results of testing on Michalewicz function

From TABLE 1 and Fig. 8, PSO seems to be better in terms of speed of convergence and to find the global optimum.

III. SOLVING SLAM PROBLEM

In SLAM without the priority of knowing the location, the trajectory and the location are estimated SLAM is a procedure for an autonomous robot to construct simultaneously a map of an surroundings and use that map to estimate its position [4].

Sphere function   Maytas function

Schaffer N. 2 function   Michalewicz function

Fig. 7: 3D plotting of Benchmark functions

- Sphere function :

$$f(x) = \sum_{i=1}^{n} x_i^2 \qquad (5)$$

- Maytas function:

$$f(x) = 0.6(x^2 + y^2) - 0.48xy \qquad (6)$$

- Schaffer function N. 2:

$$f(x) = 0.5 + \frac{\sin^2(x^2 - y^2) - 0.5}{[1 + 0.001(x^2 + y^2)]^2} \qquad (7)$$

The 2018 International Power Electronics Conference

A grid map is rebuild from the information obtained from an LRF that belongs to the environment platform Fig. 17.

We modify the basic data extract from the Laser Range Finder sensor onto an occupied grid-map [5], the cells size of the grid-map is appointed at $20 \times 20\text{mm}^2$, and the cell value is described as $\text{MAP}(x, y) \in \{0, 1\}$ at the coordinates (x, y) and is initiated by 0. When MAP(x, y) = 1 that means there is an obstacle.

Respectively, the position and head angle of the robot are determined as x_0, y_0 and θ_0. The robots movement is described as $\delta x, \delta y, \delta\theta$ in a sampling period δt. Sampled data extracted from an LRF include movement information. hence, we can evaluate the robots movement by coordinating the latest data to a map built from past data. After the robot try to minimize Eq. (9) by using the PSO algorithm .

Depending on the number of coordinated points, the rate of points that have been matched $f(z)$ decided how well the local-map matches each of two the current-map or the global-map [5]:

$$f(z) = \frac{N_{valid} - N_{fit}(z)}{N_{valid}} \quad (9)$$

where $z = (\delta x, \delta y, \delta\theta)$ define the robots movement and N_{valid} describe the number of samples extracted from the LRF in each scan and detection distance is from 30 mm to 30,000 mm and N_{fit} is described as below [5]:

$$N_{fit} = \sum_{i=1}^{n_{max}} MAP(x_i, y_i) \quad (10)$$

$$x_i = d_i \cos(\theta_0 + \delta\theta) + x_0 + \delta x \quad (11)$$

$$y_i = d_i \sin(\theta_0 + \delta\theta) + y_0 + \delta y \quad (12)$$

where d_i is directly the distance from the sensor to the obstacle extracted from the LRF, (x_i, y_i) is the point coordinates on the map equivalent to the distance d_i and n_{max} is the number of valid points (719).

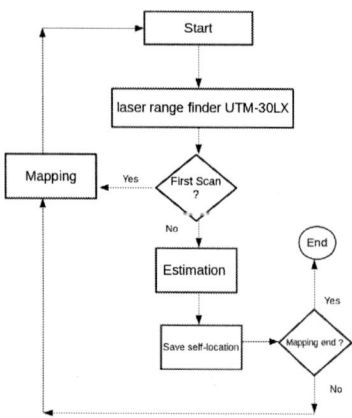

Fig. 10: System flow chart [6]

A. Procedure of the third test

We will test in different environment the simulator tool that we will use is Gazebo we will perform in ideal map Fig. 11

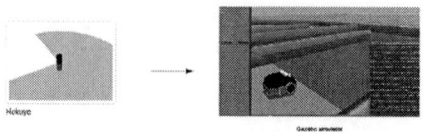

Fig. 11: Our environment in the simulator Gazebo

We can obviously add noise to the data extracted from Gazebo's sensors that by default examine the world completely to introduce a more practical environment because in the real world sensors display noise.

We will test on the case when we do a simple translation using the laser sensor with adding the gaussian noise : A mean of 0.0m and stddev of 0.02m, and the angular resolution by default equal to 1 deg.

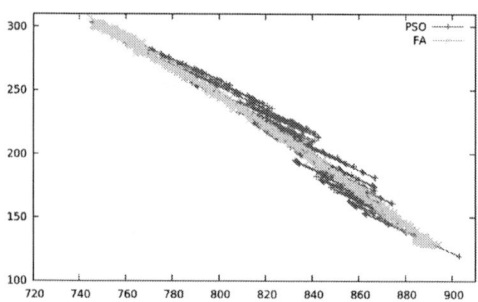

Fig. 12: trajectory of both (PSO/FA)

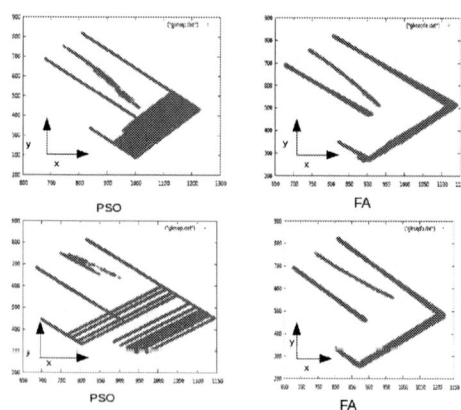

Fig. 13: results of the third test

From the simulation results Fig. 12, Fig. 13 we can see that FA outperfoms PSO.

When we conduct tests in terms of angular resolution and in terms of standard deviation of Gaussian noise we got the following results Fig. 14

684

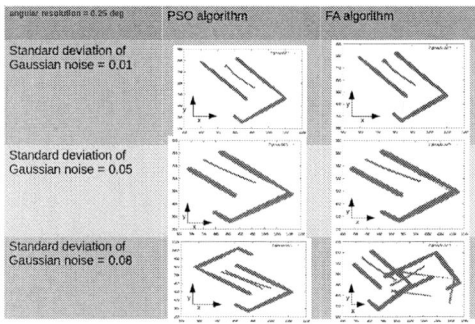

Fig. 14: results of the third test

B. Procedure of the fourth test

Testing in real world in two diffrent cases: The first and second examples Fig. 18 and Fig. 20 will test on the case when we do a simple translation, the third example Fig. 23, Fig. 24 will test on the case when we do (translation, rotation)

About the platform i am using 2D laser range finder UTM-30LX, iRobot Roomba 600, Raspberry Pi 3 and Power Bank Fig. 17

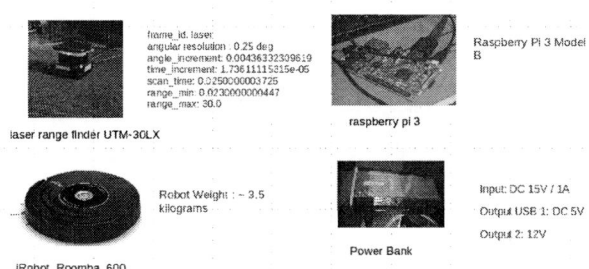

Fig. 15: Environment platform :2D laser range finder UTM-30LX + iRobot Roomba 600 + Raspberry Pi 3 + Power Bank

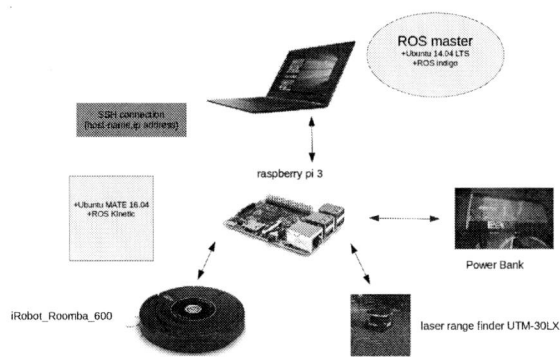

Fig. 16: Architecture of the environment platform

Extending the comparison between two algorithms for SLAM problem

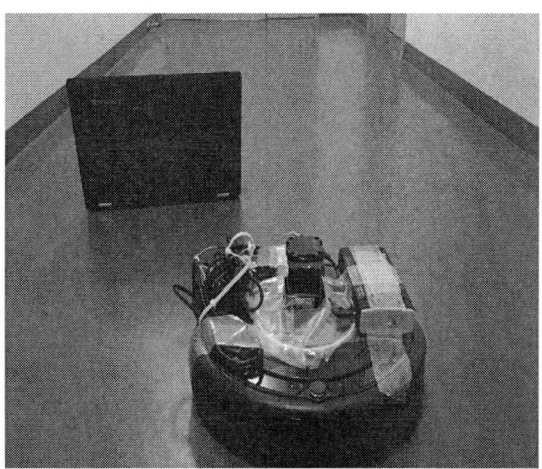

Fig. 17: Environment platform :2D laser range finder UTM-30LX + iRobot Roomba 600

TABLE II
SPECIFICATIONS OF UTM-30LX

provision	specification
Size (WxDxH)	60 mm x 60 mm x 87 mm
Current	under 0.7 A
detection distance	3030,000 mm
angle of the scan	270 deg
angular resolution	0.25 deg
scanning time	25 ms/scan

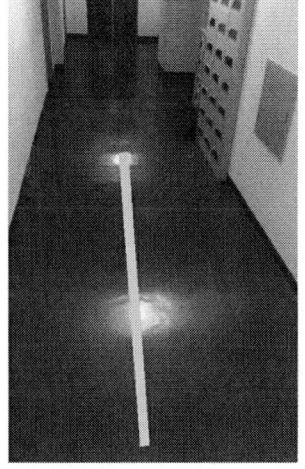

Fig. 18: Example I environment

Number of iterations : we fixed the number of fire-flies/particles to 40 and we varied the number of iterations from 5 to 500 Fig. 21

Number of particles/ fireflies : we fixed the number of iter-ations to 500 and we varried the number of fireflies/particles from 5 to 100 Fig. 22

In the benchmark functions, we have seen the PSO achieves better results than the FA Fig. 8, and does so in a small

The 2018 International Power Electronics Conference

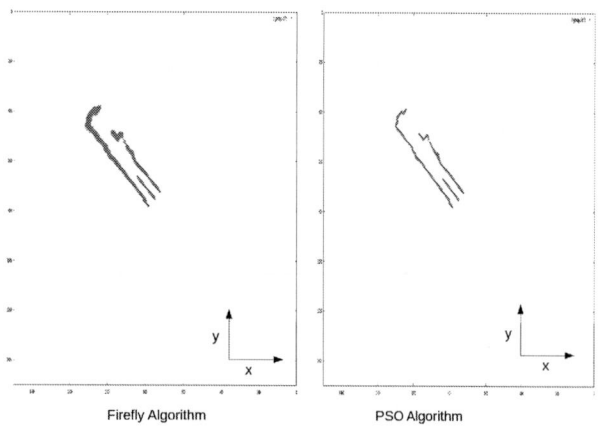

Firefly Algorithm   PSO Algorithm

Fig. 19: Result of the first case (translation)

Fig. 20: Example II environment

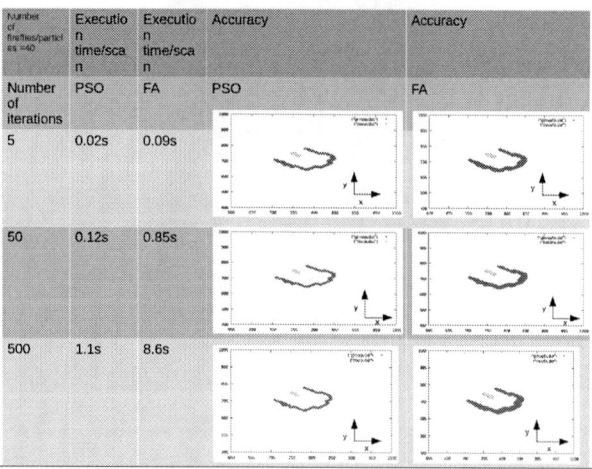

Fig. 21: Results of the second case (translation) with varying the number of iterations

fraction of the time TABLE. 1, but in Fig. 9 firefly algorithm

outperforms PSO but PSO still faster than FA in all cases . The results demonstrated in Fig. 19, Fig. 21, Fig. 22 and Fig. 25 The PSO and FA results are compared in Fig. 19, Fig. 21, Fig. 22 and Fig. 25. It appears that the PSO is the superior algorithm when it comes to solve SLAM problem. And when time is of the essence, the PSO returns results extremely fast.

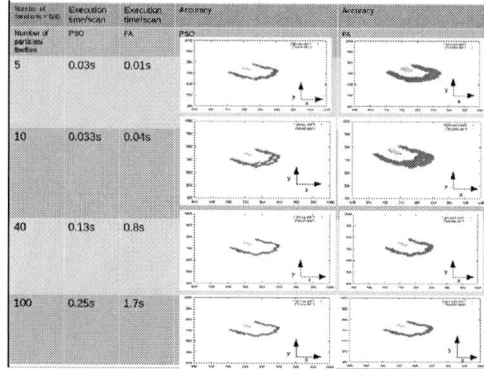

Fig. 22: Results of the second case (translation) with varying the number of fireflies/particles

Fig. 23: Example III environment

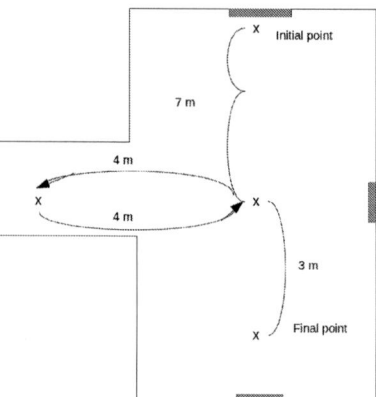

Fig. 24: Trajectory plan in Example III

686

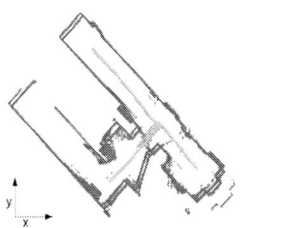 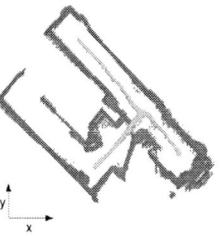

Fig. 25: Results of third case (translation,rotation)

IV. CONCLUSIONS

In this work,we compare between two procedures particle swarm optimization (PSO) algorithm and FireFly algorithm (FA), investigated their resemblances and inequalities. simulation results for determining the global optima of benchmark functions propose that particle swarm usually exceed Firefly algorithm in general cases and for solving Simultaneous Localisation and Mapping problems.

ACKNOWLEDGEMENT

This research was supported by Yasutaka Fujimoto Laboratory, we would like to thank the reviewers for their insights and we are also immensely grateful for their comments on an earlier version of the manuscript.

REFERENCES

[1] Hugh Durrant-Whyte, Fellow, IEEE, and Tim Bailey "1Simultaneous Localisation and Mapping (SLAM): Part I The Essential Algorithms," 2006.

[2] Michael Lohrer, "A comparison between the firefly algorithm and particle swarm optimization," Honors College Theses, Oakland University, pp. 11-16, Diss. 2013.

[3] Yuhui Shi, and Russell Eberhart, "A modified particle swarm optimizer," Proc.IEEE Int. Conf. on Evolutionary Computation, DOI:10.1109/ICEC.1998.699146, 1998.

[4] Jongdae Jung, Seung-Mok Lee, and Hyun Myung, "Indoor Mobile Robot Localization Using Ambient Magnetic Fields and Range Measurements," Robot Intelligence Technology and Applications 2. Springer, Cham, 137-143, 2014.

[5] Yudai Hasegawa, and Yasutaka Fujimoto, "Experimental Verification of Path Planning with SLAM," IEEJ Journal of Industry Applications, vol.5, no.3, pp. 253–260, 2016.

[6] Kazuaki Okada, and Yasutaka Fujimoto, "Grid-based localization and mapping method without odometry information," IEEE Industrial Electronics Society, proc. IEEE Industrial Electronics Society Annual Conference (IECON), PP.159-164, 2011.

Bandwidth Limitations in Force Control of a Series Elastic Actuator with Backlash and Quantization

Hanul Jung, Chan Lee and Sehoon Oh
Department of Robotics Engineering, DGIST, Daegu, Korea
E-mail: (jungsky14, chanlee, sehoon)@dgist.ac.kr

Abstract—Series Elastic Actuator has been widely utilized in various robotic applications, where precise and compliant force control is required. In SEA, the amount of spring deformation, which is mainly controlled for the compliant force control of SEA, is not very large, and thus the feedback control of this spring deformation is very subject to various nonlinearity such as backlash and quantization. In this paper, the effect of this nonlinearity on the force control of SEA is investigated. In particular, how the feedback controller gains are affected by the nonlinearity is examined through experiments, and a guideline for the gain determination is suggested.

Keywords—*Series Elastic Actuator, Force Control, Backlash, Quantization error, Encoder resolution.*

I. INTRODUCTION

Series Elastic Actuator (SEA) has been emerging as a potential actuator that can provide high torque and compliance at the same time. Unlike rigid actuators, SEA has an elastic element which leads to benefits such as impact tolerance, precise and stable force control and elastic energy storage [1]. Taking advantages of these features, SEA is widely applied to various service robots including rehabilitation and legged robotics [2], [3].

The force control is mainly applied to SEA applications to provide desired forces to the load and environments. The deformation of the spring is measured and considered proportional to the force output based on Hooke's law. The measured (or estimated) force is fed back to the controller so that the motor torque can be controlled for the spring force output to track the desired force values.

Many researchers have conducted much research to analyze and improve SEA's force control performance [4]–[9]. Eppinger et al. analyzed that there is a gain limit depending on the location of the sensor, which is related to the spring deformation measurements [6]. Pratt proposed a PD controller that feedback the load acceleration [1]. Most of these approaches are dealing with the design of force control to improve performance of the force control. However, the most significant restriction of the SEA force control is considered due to the stability condition imposed by its flexibility and phase delay.

In addition to this stability condition, there is nonlinearity that substantially affects the control performance [8]; nonlinear elements such as backlash, friction and quantization increase the error of the control loop and limit the accuracy and control bandwidth of the SEA force control. To tackle this issue caused by nonlinearity, this paper investigates the relationship between the nonlinearity of SEA and its effect on control performance.

In summary, the main contributions of this paper are as follows:

1) Investigation of the effect of backlash and quantization on SEA force control.
2) Gain selection method considering the influence of backlash and quantization.

This paper is organized as follows. The dynamic model of the SEA is derived assuming the SEA is in contact with stiff environment, and the force controller is designed and analyzed based on the model. The types of nonlinearity in an SEA system are introduced and its effect on the control performance is analyzed in Sec. II. The influence of backlash and quantization on force control analyzed in Sec. II is verified in Sec. III through experiments. In particular, an experimental analysis of how the PD controller should be set up when there is a nonlinearity effect. In Sec. IV, conclusions are given.

II. EFFECT OF NONLINEARITY ON SEA FORCE CONTROL

A. Dynamics of Series Elastic Actuator

SEA is modeled as a two mass system which consists of two second order system connected via a elastic component, which results in a fourth order system [9]. In that case, the elastic element of SEA, which is a spring, is considered connected to a load, which can move freely (Dynamic case).

Meanwhile, SEA is also frequently modeled as a second order system, where the spring is considered to be connected to a very stiff environment, which does not move at all (Static case). In this paper, this model is employed and analyzed, as it is simple but still very practical [10].

Figure 1 is the model of SEA utilized for the forthcoming analysis, where there is no position/angle deviation from the environment, as it is fixed. Two nonlinear

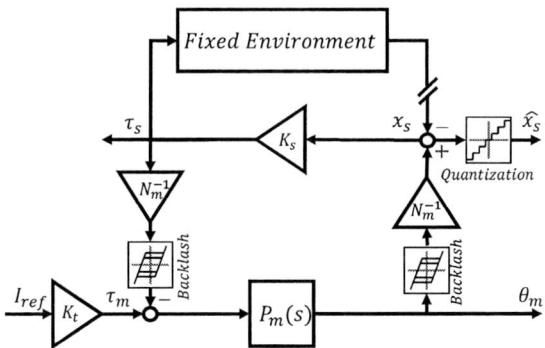

Fig. 1: Fixed load environment model.

TABLE I: Parameters of SEA utilized in this paper

Symbol	Meaning	Value
J_m	Motor Inertia	8.5×10^{-6} kg m^2
B_m	Motor Damping	8.0×10^{-5} Nm s / rad
K_s	Spring constant	180000 Nm / rad
N_m	Gear Ratio	7854
K_t	Current Coefficient	0.0276 Nm / A

elements - backlash and quantization are included in this model, and will be analyzed. In the block diagram of Fig. 1, $P_m(s)$ represents the motor dynamics ($P_m(s) = \frac{1}{J_m s^2 + B_m s}$), K_s is the spring constant, N_m is the reduction ratio of gear train, τ_m is motor torque, x_s is the spring deformation, θ_m is the motor angle, τ_s is output torque, and K_t is the torque constant. The transfer function from the current reference for the motor to the output torque of SEA can be derived as follows.

$$P(s) = \frac{\tau_s(s)}{I_{ref}(s)} = \frac{K_t K_s N_m^{-1}}{J_m s^2 + B_m s + K_s N_m^{-2}} \quad (1)$$

The parameters for this model are identified from an experimental setup in Fig. 4, and listed in Table I.

B. Force Feedback Controller Design

The goal of SEA force control is to control the forces that are transmitted to the outside (which is fixed in this paper). The controller for this is designed as shown in Fig. 2: feedback control and feedforward control (or feedforward filter).

A Proportional-Derivative (PD) controller is adopted as the feedback controller, and in addition to it a feedforward filter is added to compensate for the steady-state error.

As the system to be controlled is a second order system, the PD controller gains can arbitrarily place the closed loop poles. Based on this idea, the PD gains are determined to place the closed loop poles in desired places. The feedback controller is given as follow

$$G_c(s) = K_d s + K_p, \quad (2)$$

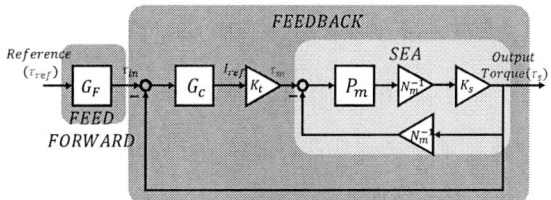

Fig. 2: Block diagram of controller.

where K_p is the P gain, and K_d is the D gain. The transfer function of the closed loop under the $G_c(s)$ feedback control is given as

$$G_{cl}(s) = \frac{\tau_s(s)}{\tau_{ref}(s)} = \frac{G_F(s) G_c(s) P(s)}{1 + G_c(s) P(s)}, \quad (3)$$

where $G_F(s)$ is the feedforward controller in Fig. 2.

The controller $G_c(s)$ and $G_F(s)$ are designed to modulate the closed loop system $G_{cl}(s)$ as (4), which is a standard second order system. The characteristic of $G_{cl}(s)$ is specified by the natural frequency ω and the damping factor ζ.

$$G_{cl}(s) \approx \frac{\omega^2}{s^2 + 2\zeta\omega s + \omega^2} \quad (4)$$

To render the closed loop system characteristic as (4), the controller gains K_p and K_d should be determined as follows.

$$K_p(\omega) = \frac{J_m \omega^2 - K_s N_m^{-2}}{K_t K_s N_m^{-1}}, \quad (5)$$

$$K_d(\omega, \zeta) = \frac{2 J_m \zeta \omega - B_m}{K_t K_s N_m^{-1}} \quad (6)$$

It is widely-accepted that the damping factor ζ as unity demonstrates a good transient performance, which is also know as the critically-damped system [11]. With this unity ζ, K_p and K_d are determined only based on one variable ω.

With this gain decision, the PD controller can determine the control bandwidth ω, but it does not guarantee the low frequency range to be unity to reduce the steady state error. Therefore, in this paper, a feedforward controller is designed to improve the steady state tracking performance. The feedforward controller $G_F(s)$ is designed as follows to fit the numerator of (3) to (4).

$$G_F(s) = \frac{K_s N_m^{-2} + K_p K_t K_s N_m^{-1}}{(K_d s + K_p) K_t K_s N_m^{-1}} \quad (7)$$

C. Effect of Backlash and Quantization on SEA Force Control

The effect of the backlash and quantization of the system on the force control performance is analyzed here. Backlash and quantization are added in a MATLAB Simulink® simulation as in Fig. 1. Figure 3 shows the simulation results where a step-wise force reference is

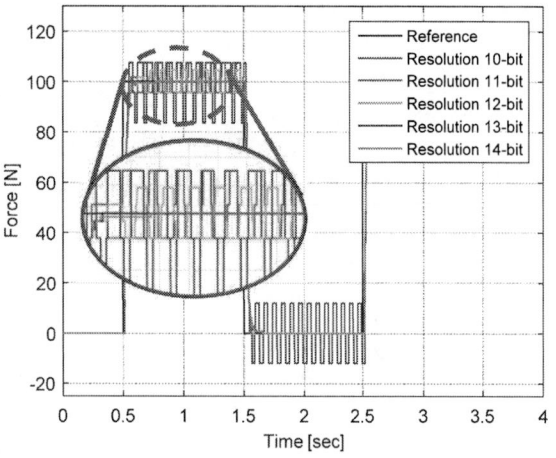

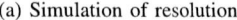

(a) Simulation of resolution

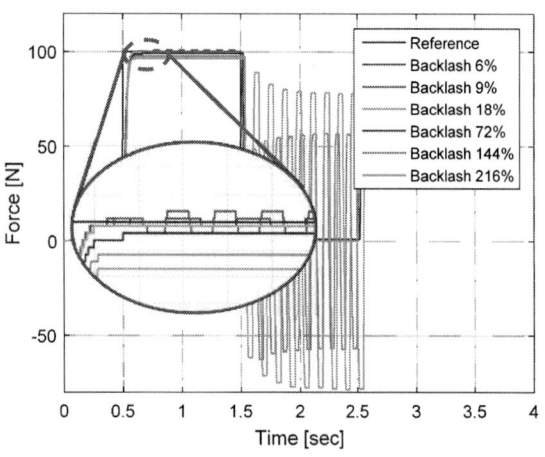

(b) Simulation of backlash

Fig. 3: Simulation of nonlinearity.

Fig. 4: Experimental setup of an SEA.

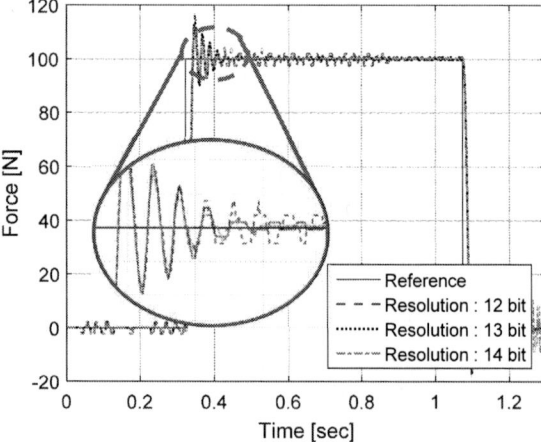

Fig. 5: Experiment of comparison of sensor resolution.

given with different level of nonlinearity; varying levels of quantization are simulated in Fig. 3a, and varying levels of backlash are simulated in Fig. 3b. It is noteworthy that there is a certain level of nonlinearity from which the performance is deteriorated significantly.

Derivative controller with the gain K_d determines the transient characteristic ζ as seen in (6). However, Derivative controller is considered to be affected by the nonlinearity as the numerical difference method is largely affected by quantization and backlash.

Therefore, this paper investigates how these nonlinearities affect the control performance, and how the gains are to be adjusted to overcome the nonlinearity. For this purpose, the gain set with ζ set to unity is considered as the nominal gain set , that is $K_{p0} = 0.12182$ and $K_{d0} = 0.00248$. The actual gains K_p, K_d are designed R_p/R_d times the nominal gains. Equation (8) shows the design of actual gain K_p, K_d, where R_p and R_d are the varying ratio.

$$K_p = R_p K_{p0}, \quad K_d = R_d K_{d0} \tag{8}$$

The control performance is examined with respect to R_p and R_d.

The maximum control bandwidth is also examined with different nonlinearity level and different ζ values. To this end, the relationship between the control gain sets (K_p and K_d) and the corresponding control bandwidth ω and damping ζ are given as follows.

$$\omega = \frac{K_s N_m^{-2} + K_p K_t K_s N_m^{-1}}{J_m}, \tag{9}$$

$$\zeta = \frac{B_m + K_d K_t K_s N_m^{-1}}{2 J_m \omega} \tag{10}$$

III. EXPERIMENTAL VERIFICATION

A. Experiment of SEA Force Control with Nonlinearity

In order to verify the effect of nonlinearity on SEA force control, the experimental set-up shown in Fig. 4 is

The 2018 International Power Electronics Conference

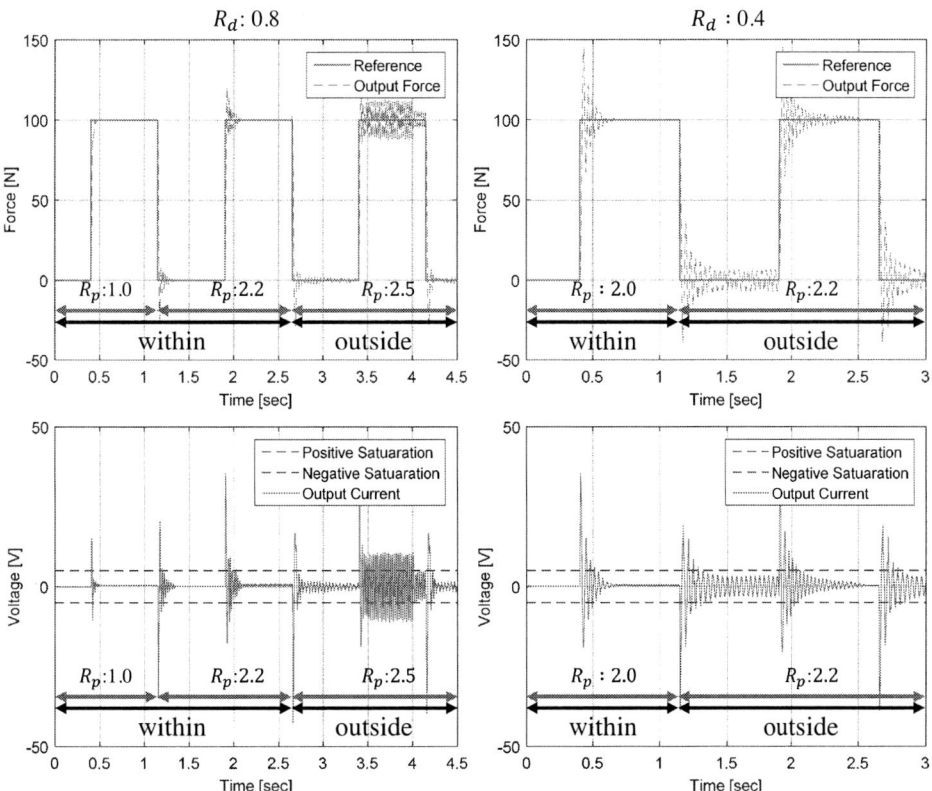

Fig. 6: Control performance evaluation.

utilized. Maxon EC-4 pole 200W BLDC motor with a 14-bit pulse-per-revolution encoder is employed as the torque source, and the timing belt and pulley are connected to the motor to transmit the torque to the ball screw. In the spring part, a rotary encoder with a 13-bit pulse-per-revolution and a timing belt are used to measure the spring deformation.

Figure 5 shows the experimental results of force control with step-wise reference. To verify the effect of quantization, different levels of encoder-resolution are applied for the spring deformation feedback, and its effect is shown in Fig. 5. This result seems similar to the simulation results in Fig. 3 (a), which shows that quantization has an influence on force control performance.

B. Evaluation Force Control Performance

Experiments were conducted while changing control gains to analyze control performance. The (maximum permissible) gain values are obtained evaluating the force control performance under various levels of quantization.

Figure 6 is the experimental result, where K_d is fixed (to 0.8 in the left figures and 0.4 in the right figures), while K_p increases (from 1.0 to 2.5 in the left figures and 2.0 to 2.2 in the right figures).

With K_p increasing, the rising time becomes faster, however, 1) the overshoot increases, 2) the output be-

comes oscillating but converges, and 3) finally the output keeps oscillating.

The state 3) where the oscillations do not stop is not acceptable as the control output. The state 2) can be considered as a marginally acceptable state. In this paper, the marginal gain sets are selected as the gain sets that control the system to reach the marginal control performance 2).

As described in Sec. II-C, these marginal gain sets depend on backlash and quantization level. In this paper, the experiments shown in Fig. 6 are conducted with various levels of quantization level, and the marginal gain sets are found for each quantization level.

C. Experiment Result and Discussion

The marginal gain sets found in the experiments are depicted in Fig. 7, which plots K_p and K_d of each marginal gain set in the xy plane. As the control bandwidth ω is uniquely determined by K_p as in (9), ω converted from K_p based on (9) is utilized in Fig. 7.

In this figure, the solid, dash, and dash-dotted lines represent resolution 14, 13, and 12 bits, respectively. Since the low pass filter utilized for the time derivative of the encoder measurement, the cut-off frequency of the filter is also a tunable factor. Different cut-off frequencies are utilized in the experiments, and the red, blue and black

The 2018 International Power Electronics Conference

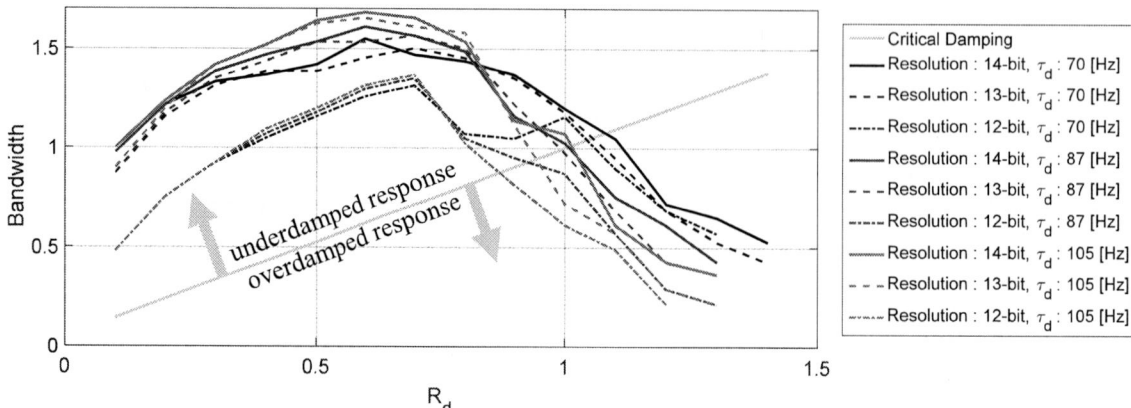

Fig. 7: Marginal gain sets obtained through experiments.

lines indicate that the cutoff frequency if 105, 87, 70 Hz, respectively.

Experimental results show that the bandwidth decreases as the resolution increases from 14 bits to 12 bits. The bandwidth increases to a certain level as K_d increases. However, when K_d is set larger than a certain level, the control bandwidth decreases. The maximum bandwidth was achieved when K_d was around 0.6 at almost all the resolution conditions. The decrease in the control bandwidth becomes more drastic, when the cutoff frequency is set too high. Smaller cut-off frequency achieves the smallest bandwidth when K_d is set low, but it achieves the highest bandwidth when K_d is set high.

IV. CONCLUSIONS

In this paper, the effect of nonlinearity on the SEA force control was investigated by simulations and experiments. Moreover feedback controller gains to achieve stable control performance under nonlinearity is analyzed. Theoretical analysis and different controller design approach to overcome the nonlinearity are the future works.

ACKNOWLEDGEMENT

This work was supported by the DGIST Start-up Fund of the Ministry of Science, ICT and Future Planning (2018010078) and by the Korea government of the Ministry of Trade, Industry & Energy (10080547)

REFERENCES

[1] G. Pratt and M. Williamson, "Series elastic actuators," in *Intelligent Robots and Systems 95. 'Human Robot Interaction and Cooperative Robots', Proceedings. 1995 IEEE/RSJ International Conference on*, vol. 1, Aug 1995, pp. 399–406 vol.1.

[2] Q. Zhu, Y. Mao, R. Xiong, and J. Wu, "Adaptive torque and position control for a legged robot based on a series elastic actuator," *International Journal of Advanced Robotic Systems*, vol. 13, no. 1, p. 26, 2016.

[3] H. Yu, S. Huang, G. Chen, Y. Pan, and Z. Guo, "Human–robot interaction control of rehabilitation robots with series elastic actuators," *IEEE Transactions on Robotics*, vol. 31, no. 5, pp. 1089–1100, 2015.

[4] W. S. Newman, "Stability and performance limits of interaction controllers," *TRANSACTIONS-AMERICAN SOCIETY OF MECHANICAL ENGINEERS JOURNAL OF DYNAMIC SYSTEMS MEASUREMENT AND CONTROL*, vol. 114, pp. 563–563, 1992.

[5] E. Colgate and N. Hogan, "An analysis of contact instability in terms of passive physical equivalents," in *Robotics and Automation, 1989. Proceedings., 1989 IEEE International Conference on*. IEEE, 1989, pp. 404–409.

[6] S. D. Eppinger and W. P. Seering, "Three dynamic problems in robot force control," in *Robotics and Automation, 1989. Proceedings., 1989 IEEE International Conference on*. IEEE, 1989, pp. 392–397.

[7] S. P. Buerger and N. Hogan, "Complementary stability and loop shaping for improved human–robot interaction," *IEEE Transactions on Robotics*, vol. 23, no. 2, pp. 232–244, 2007.

[8] G. Wyeth, "Control issues for velocity sourced series elastic actuators," in *Proceedings of the Australasian Conference on Robotics and Automation 2006*. Australian Robotics and Automation Association Inc, 2006.

[9] S. Oh and K. Kong, "High-precision robust force control of a series elastic actuator," *IEEE/ASME Transactions on Mechatronics*, vol. 22, no. 1, pp. 71–80, 2017.

[10] N. Paine, S. Oh, and L. Sentis, "Design and control considerations for high-performance series elastic actuators," *IEEE/ASME Transactions on Mechatronics*, vol. 19, no. 3, pp. 1080–1091, 2014.

[11] Y. Zhao, N. Paine, S. J. Jorgensen, and L. Sentis, "Impedance control and performance measure of series elastic actuators," *IEEE Transactions on Industrial Electronics*, 2017.

The 2018 International Power Electronics Conference

Rotor Shape Optimization of Interior Permanent Magnet Synchronous Motors with Concentrated Windings by Considering End-Leakage Flux

Katsumi Yamazaki[*] and Hiroki Narushima

Dept. of Electrical, Electronic and Computer Engineering, Chiba Institute of Technology, Narashino, Japan

*E-mail: yamazaki.katsumi@it-chiba.ac.jp

Abstract— **In this paper, we propose an optimization procedure to determine the shapes and orientations of permanent magnets in interior permanent magnet synchronous motors with concentrated armature windings. In the proposed procedure, 2D finite element analysis coupled with armature voltage equation is used to take into account the leakage flux at the end windings. The end leakage inductance in the equation is determined by 3D finite element analysis in advance. By using this procedure, the rotor shape of the motor can be optimized within practical computational time. First, the validity of the analysis is confirmed by comparing measured and calculated characteristics of the motor. Then, the optimal shape and orientation of the permanent magnets in the motor are discussed from the results of the proposed optimization procedure. It is revealed that the consideration of the end-leakage flux is indispensable to estimate the characteristics of the motors with concentrated windings whose core length is relatively short as compared to the diameter, particularly under flux weakening control. An optimized rotor that reduces the iron loss at high speeds without considerable deterioration of the other important characteristics is also obtained by the proposed optimization procedure.**

Keywords— *Optimizations, Permanent magnet motors, Losses, Variable Speed Drives.*

I. INTRODUCTION

Interior permanent magnet motors (IPMSMs) with concentrated armature windings are widely used for industry applications because of many advantages [1]-[5]. In these motors, each armature coil is wound onto one stator tooth. As a consequence, the end windings can be shorter than those of the distributed (overlapping) winding motors. Therefore, this motor is particularly useful to reduce the core length as compared with the distributed winding motor according to the decrease in the size of the end windings.

On the other hand, one of the disadvantages of these motors is the generation of large rotor losses caused by the wide stator-slot pitch [3]-[5]. In reference [5], the shapes of stator and rotor of an IPMSM with concentrated windings is optimized in order to reduce the magnet eddy current loss by using automatic iterative

calculations of 2 dimensional (2D) time-stepping finite element analysis (FEA) with an optimization method. In this optimization, the amplitude and phase angle of armature current are fixed in each motor designs. However, more practical optimization by considering the actual driving condition is desired for advanced motor designs, in particular, under flux weakening control at high speeds. In this case, the armature current angle is determined according to the maximum inverter voltage. Therefore, for the accurate characteristics estimation, FEA should be iteratively carried out to determine the armature current angle at each motor design. Furthermore, 3 dimensional (3D) FEA is strongly required to estimate the accurate armature voltage because considerable end-leakage flux is generated in the IPMSMs with concentrated windings, whose core length is relatively short as compared with the diameter. However, the application of the 3D time-stepping FEA to this optimization is practically impossible because it requires vast computational time according to the very large number of FEA iterations to determine both the motor design and the appropriate current angle.

From these viewpoints, we propose an optimization procedure using combination of 2D and 3D FEAs. In the proposed method, the end-leakage flux is taken into account in the optimization by coupling 2D FEA and the armature voltage equation including the end-leakage inductance. This inductance is calculated by static 3D FEA in advance under the assumption that the end-leakage inductance is constant with the driving condition and detailed motor designs. In this case, the 3D time-stepping FEA is needless. As a consequence, the iterative calculations to determine the motor design and the appropriate current angle can be carried out within practical computational time.

First, the validity of the FEA and the necessity of the consideration for end-leakage flux are discussed by comparing the measured and calculated characteristics of an IPMSM with concentrated windings. Then, the shape and the orientation of the permanent magnets (PMs) in the IPMSM are optimized by the proposed procedure to decrease the loss and to increase the torque.

The 2018 International Power Electronics Conference

TABLE I
MOTOR SPECIFICATION.

Phases and poles	3 phases, 16 poles
Maximum rotational speed	7000 r/min
Maximum phase current	320 A
Diameter of stator, Core length	255 mm, 55 mm
Number of stator slot	24 (Concentrated)
Magnet type, Magnetization	Nd-Fe-B, 1.24 T
Thickness of electrical steel sheet	0.35 mm

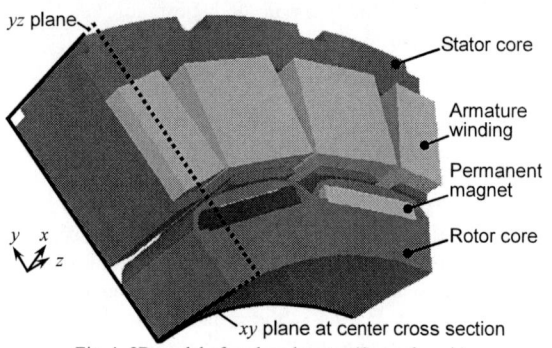

Fig. 1. 3D model of analyzed motor (One pole pair).

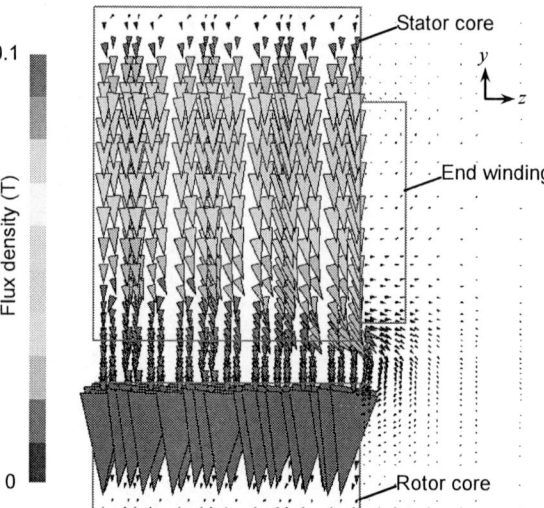

Fig. 2. Flux density distribution on *yz* plane by 3D FEA.

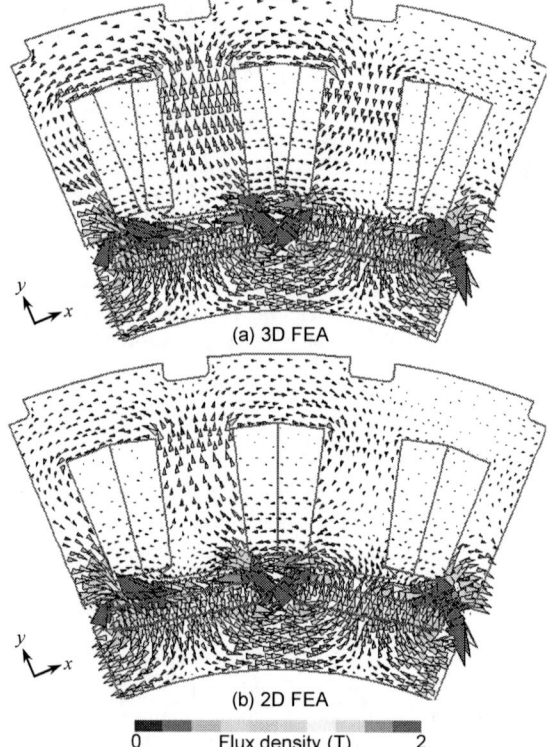

Fig. 3. Flux density distribution on *xy* plane.

II. ANALYZED MOTOR AND CALCULATION METHOD

A. Analyzed Motor

Table I lists the specification of the initial IPMSM. Fig. 1 shows the 3D model per pole pair. The motor has one interior magnet per pole. It is a sintered magnet whose remnant flux density is 1.24 T. One magnet is subdivided into 7 pieces along the axial length to prevent the eddy currents. The stator has 24 slots with concentrated windings. Both the stator and rotor cores are laminated. The inverter is an insulated gate bipolar transistor (IGBT) type PWM inverter whose carrier frequency is 10 kHz.

B. Proposed Method to Consider End-Leakage Flux

Fig. 2 shows the flux density distribution calculated by the 3D FEA on the *yz* plane in Fig. 1 under flux weakening control with maximum armature current. The figure indicates considerable flux at the end winding. The 3D analysis is indispensable to estimate this flux. On the other hand, Fig. 3 shows the flux density distribution on the *xy* plane at the center cross section. This distribution can be obtained by both the 2D and 3D FEAs. It is observed that the results of the 2D and 3D FEAs are nearly identical to each other.

From these results, we propose following method, which approximately estimates the end-leakage flux in the 2D FEA.

1) First, the total armature flux linkage is calculated by both the 3D and 2D FEAs, as follows:

$$\Phi_{3D} = \iiint_{Va} N\boldsymbol{A} \cdot \boldsymbol{n} \, dv / S_a \tag{1}$$

$$\Phi_{2D} = L \iint_{S_a} NA_z n_z \, dS / S_a \tag{2}$$

where Φ_{3D} and Φ_{2D} are the total armature flux linkage calculated by the 3D and 2D FEAs, respectively, V_a and S_a are the volume and the cross sectional area of the armature winding, respectively, $\boldsymbol{n}$ and n_z are the unit vector and unit z component that are parallel to the armature current, respectively, $\boldsymbol{A}$ and A_z are the magnetic vector potentials used in the 3D and 2D FEAs, respectively, N is the number of armature-winding turns, L is the core length.

2) Then, the end-leakage flux Φ_{end} is defined as the difference between Φ_{3D} and Φ_{2D}, as follows:

$$\Phi_{end} = \Phi_{3D} - \Phi_{2D}. \tag{3}$$

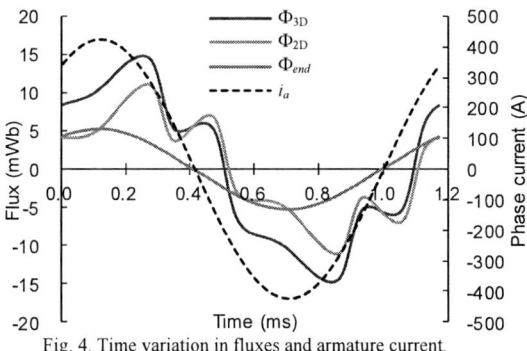

Fig. 4. Time variation in fluxes and armature current.

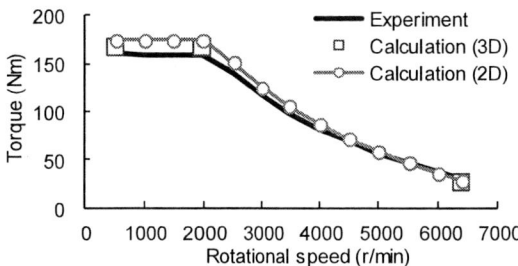

Fig. 5. Experimental and calculated torques (300 A).

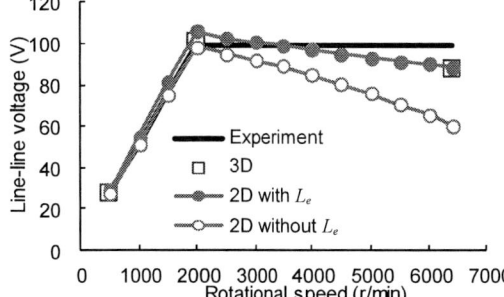

Fig. 6. Experimental and calculated armature voltages (300 A).

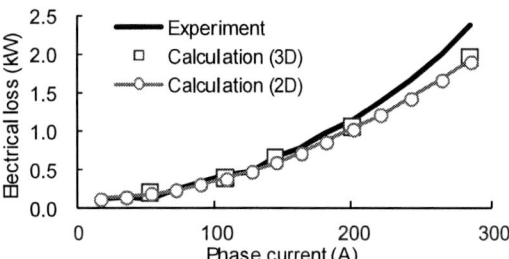

Fig. 7. Experimental and calculated electrical losses (2000 r/min).

3) It is considered that Φ_{end} is mainly produced by the armature current i_a. Therefore, the end-leakage inductance L_e is defined, as follows:

$$L_e = \Phi_{end} / i_a . \qquad (4)$$

4) The armature voltage of the motor v_a is estimated by following armature voltage equation including Φ_{2D} and L_e:

$$v_a = \frac{d\Phi_{2D}}{dt} + R_a i_a + L_e \frac{di_a}{dt} \qquad (5)$$

where R_a is the armature resistance.

Fig. 4 shows the calculated time variation in Φ_{3D}, Φ_{2D}, Φ_{end}, and i_a. It is confirmed that Φ_{3D} is larger than Φ_{2D}. In addition, Φ_{end} and i_a show sinusoidal variation with an identical phase, whereas Φ_{3D} and Φ_{2D} are not sinusoidal. This is because Φ_{end} is mainly produced only by i_a, whereas Φ_{3D} and Φ_{2D} are produced by both i_a and residual flux density of the PMs.

As Φ_{end} is nearly proportional to i_a, the end-leakage inductance L_e can be determined by the static solution when i_a is maximum. Therefore, time-stepping 3D FEA is needless. In addition, it is considered that L_e is not considerably affected by the core shape and driving conditions because it is mainly determined by the magnetic resistance at the air region. Therefore, L_e can be fixed in the optimization procedure.

III. Experimental Verification

Next, the validity of the proposed method for end-leakage flux is confirmed by the measurement of initial motor. In the calculation, the amplitude and phase of the armature current are set to be identical to those in the measurement. The end leakage inductance L_e is determined to be 12.23 µH by using the calculated fluxes under the maximum speed condition.

Fig. 5 shows the experimental and calculated torque-speed curves with maximum armature current. Both the results of 3D and 2D FEAs are found to be in good agreement with the experimental result. The slight overestimation by the calculation must have been caused by the effect of temperature rise of the PMs.

Fig. 6 shows the experimental and calculated armature voltages. The result calculated by the 2D FEA without L_e considerably underestimates the armature voltage at the constant voltage region under flux

weakening control. It is considered that the ratio of the end-leakage flux to the total flux linkage becomes large under the flux weakening control because only the flux in the core is reduced by the flux weakening. On the other hand, the result of the 2D FEA with L_e is found to be in good agreement with the result of the 3D FEA and the experiment.

Fig. 7 shows the experimental and calculated electrical losses. The stator/rotor core losses and PM eddy current losses are calculated by the method described in [4]. In the 2D FEA, the conductivity of the PMs is modified by Russell and Norsworthy coefficient [5]. The slight underestimation must have been caused by the neglect of the harmonic losses caused by inverter carrier.

From these results, it can be stated that the 2D FEA with the proposed end-leakage inductance gives acceptable results of the motor characteristics.

IV. Rotor Magnet Optimization

A. Conditions of Optimization

The shape and orientation of PMs in the IPMSM are automatically optimized by using the proposed procedure.

Fig. 8 shows the design variables. The PM region is subdivided by 6 regions. By expecting the development

695

The 2018 International Power Electronics Conference

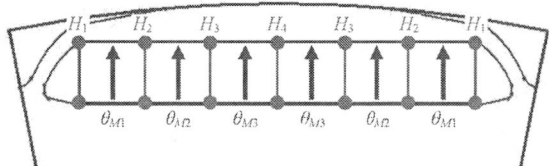

Fig. 8. Design variables for optimization.

TABLE II
ESTIMATED DRIVING CONDITIONS.

Driving conditions	A : Max torque	B : High speed
Rotational speed (r/min)	500	6000
Armature phase current i_a (A)	330	140
Current angle β (deg)	Peak torque point	Max. voltage point
Line-line voltage V_{line} (V)	Variable	99.0

of PM technologies in the near future, it is assumed that the magnet orientation can be changed at each region and the shape of the PM is not restricted to the rectangle. In this case, the design variables are selected, as follows:

1) The orientation (direction of magnetization vector) at each region θ_{M1}, θ_{M2}, and θ_{M3}
2) The position of the PM surface H_1, H_2, H_3, and H_4

The total magnet volume is set to be constant by moving the bottom position of the PM according to H_1 to H_4.

Table II lists the estimated driving conditions. The condition A is under the maximum torque control, whereas the condition B is under the flux weakening control.

Two optimizations are carried out, as follows:

(a) Maximum torque increase (T_{max} up)
The objective is maximizing the torque T_{max} under the condition A. The constraint condition is imposed on the iron loss (Sum of the core loss and the PM eddy current loss) W_{iron} under the condition B, which should be less than 105% of the initial motor.

(b) Iron loss reduction at high speeds (W_{iron} down)
The objective is minimizing W_{iron} under condition B. The constraint condition is imposed on the torque T_{max} under the condition A, which should be larger than 97% of the initial motor.

Fig. 9 shows the procedure of the optimization. First, the end-leakage inductance L_e is determined due to (1)-(4) by using the 2D and 3D static FEAs in advance. As explained in the previous section, this inductance is assumed to be constant with the driving conditions and rotor-design change in the optimization procedure. Rosenbrock's method [5] is selected for the optimization method because this method often gives acceptable results within relatively short computational time. At each optimization step, the current angle β under the conditions A and B are determined by the iterative calculations of the 2D time stepping FEA. The L_e is not required to search β under the condition A because the torque is not related to L_e when the armature current is given. On the other hand, the L_e is indispensable to search β under the condition B because the armature voltage is considerably affected by L_e, as shown in Fig. 6.

Table III shows the specification of the optimization calculation. An adaptive finite element method with permissive error coefficient [6] is applied to the 2D FEAs

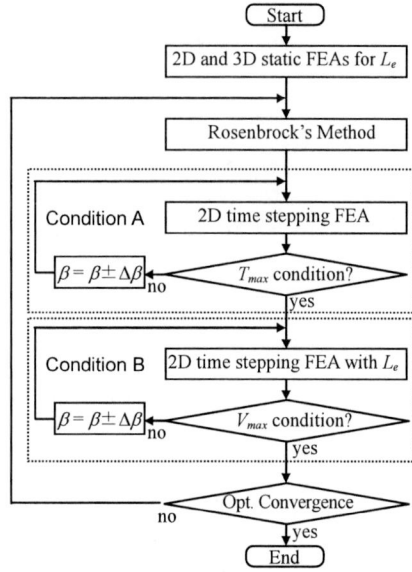

Fig. 9. Procedure of optimization.

TABLE III
SPECIFICATION OF OPTIMIZATION CALCULATION.

	T_{max} up	W_{iron} down
Number of optimized iteration	193	102
Average number of elements	7413	6955
Time steps / period	128 step / period	
Total calculation time	122 hour	62 hour
Permissive error coefficient of FEA	10 %	

Core i7 3.5 GHz

TABLE IV
CALCULATED MOTOR CHARACTERISTICS.

		Initial	(a) T_{max} up	(b) W_{iron} down
Angle β (deg)	Con.A	21.3	21.6	14.3
	Con.B	72.7	78.9	69.0
Torque (Nm)	Con.A	188.7	198.6 (+5.2%)	185.4 (-1.8%)
	Con.B	29.9	19.7 (-34.3%)	30.3 (+1.2%)
Iron loss (W)	Con.B	330.7	336.0 (+1.6%)	275.3 (-16.8%)

in order to make the finite element mesh automatically according to the rotor shape determined by Rosenbrock's method.

B. Resuls of Optimization

Table IV lists the characteristics of the initial and optimized motors. Fig. 10 shows the variation in shape and orientation according to the optimization iteration.

In the case of the T_{max}-up optimization, the center part of the PM is shifted to inside of the rotor, whereas the ends of the PM are shifted to air-gap side. As a consequence, a "partial V-shape" PM is obtained. The direction of the magnetization vector also inclines to d-axis, particularly at the V-shape part. In this case, the maximum torque increases by 5.2%. This increase is attributed to an increase in the PM flux by the optimization. However, this flux increase results in a considerable decrease in the torque under the condition B (-34.3%) according to an increase in the current angle β. It can be stated that this motor is not suitable for variable

696

The 2018 International Power Electronics Conference

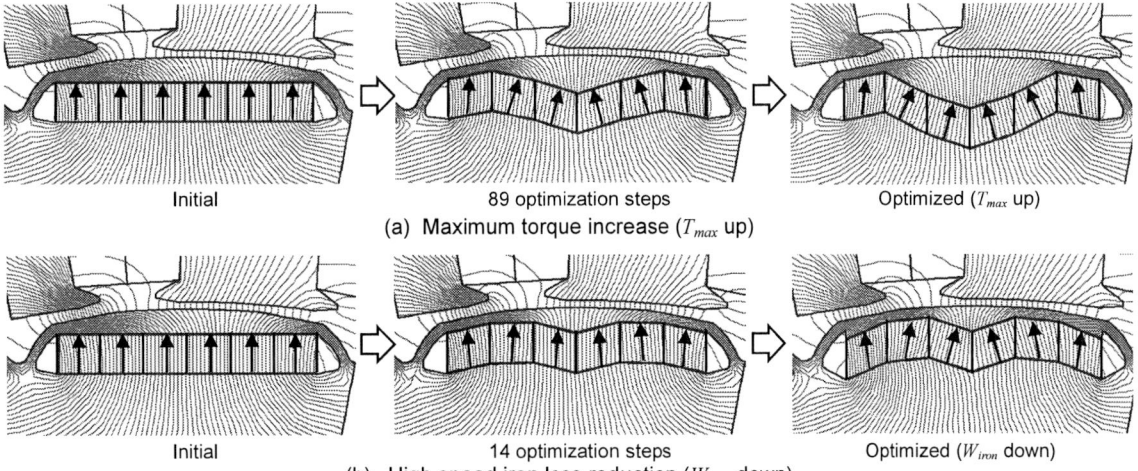

(a) Maximum torque increase (T_{max} up)

(b) High speed iron loss reduction (W_{iron} down)

Fig. 10. Variation in PM shape and magnetizing direction according to the optimization iterations.

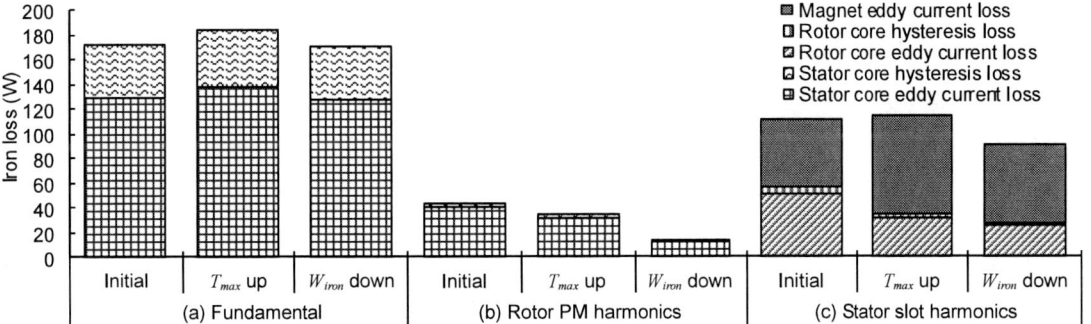

Fig. 11. Iron losses classified according to origins (Driving condition B, High speed, flux weakening).

speed applications.

In the case of the W_{iron}-down optimization, the center part of the PM is also shifted to inside of the rotor. However, this displacement is smaller than that in the case of the T_{max}-up optimization. The ends of the PM are also shifted to inside. As a consequence, a "gull-wing" shape PM is obtained. In this case, the iron loss under the condition B is reduced by -16.8%. In addition, variations in the torque under both the conditions A and B are very small (-1.8% and +1.2%). It can be stated that the iron loss is significantly reduced, whereas the other characteristics are not considerably deteriorated by this optimization.

C. Mechanizm of Iron Loss Reduction by Optimization

Then, let us discuss why the iron loss is reduced by the W_{iron}-down optimization.

The iron loss of IPMSMs with 3 slots per pole pair can be classified into following major components [4]:
(a) Stator core loss caused by the fundamental rotational field
(b) Stator core losses caused by odd harmonic magnetomotive forces of PMs
(c) Rotor core and PM losses caused by stator slot harmonics whose order is $3i$ (i is integer)
(d) Carrier harmonic losses caused by inverters.

Fig. 11 shows the calculated iron loss components of the initial and optimized motors under the driving condition B. The loss component (d) is neglected because

it is not dominant in this condition [4]. The figure indicates that the harmonic loss components (b) and (c) are significantly reduced by the W_{iron}-down optimization.

Fig. 12 shows the detailed harmonic loss components. The harmonic stator core losses are significantly reduced by the optimizations. In particular, all the harmonic losses are reduced by the W_{iron}-down optimization. It is considered that the gull-wing shape PM is very effective to reduce the harmonic magnetomotive forces of PMs. On the other hand, the tendency of the harmonic rotor core losses and PM eddy current losses are different from each other. The harmonic rotor core losses are also reduced by the optimizations, whereas the harmonic PM eddy current losses increases, particularly, in the case of the T_{max}-up motor.

To understand these rotor-loss variations, the loss density and permeability distributions under the driving condition B are investigated. Figs. 13 and 14 show the results, respectively. Fig. 13 indicates that the loss density at the ends of the PM in the T_{max}-up motor considerably increases. This increase must have been caused by an increase in the slot harmonic effects because the PM ends were shifted to the air-gap side. On the other hand, the PM ends of the W_{iron}-down motor were shifted to inside. As a consequence, the increase in the PM eddy current loss is suppressed as compared with the T_{max}-up motor. It is also observed that the saturated area at the rotor-core surface increases in the W_{iron}-down motor as compared with the initial and T_{max}-up motor in

697

The 2018 International Power Electronics Conference

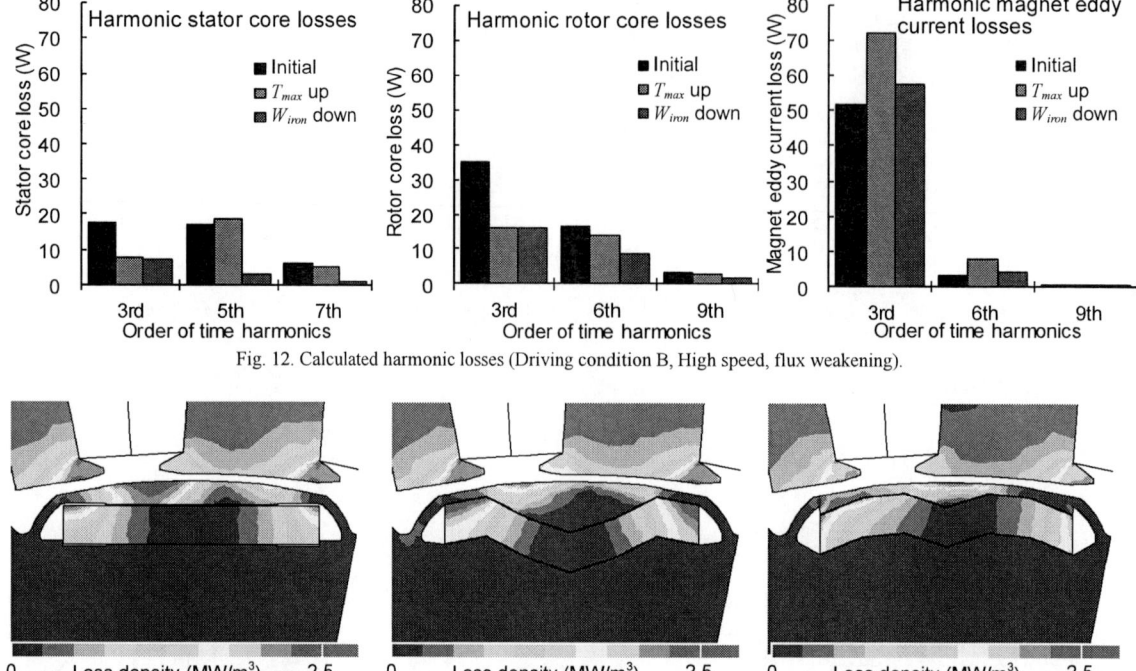

Fig. 12. Calculated harmonic losses (Driving condition B, High speed, flux weakening).

Fig. 13. Loss density distributions (Driving condition B, High speed, flux weakening).

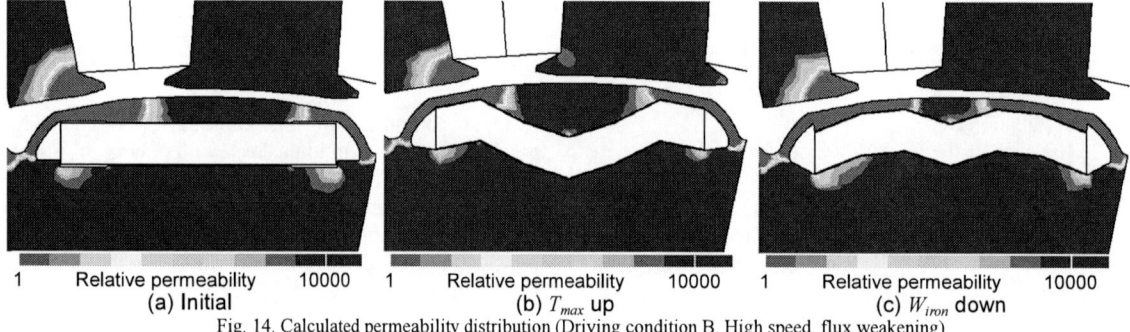

Fig. 14. Calculated permeability distribution (Driving condition B, High speed, flux weakening).

Fig. 14. As a consequence, the amplitude of the slot harmonic magnetic fields at the rotor surface is reduced in this motor. This is the reason why the harmonic rotor core loss of the W_{iron}-down motor becomes the smallest.

V. CONCLUSIONS

A shape optimization procedure that considers the end-leakage flux in IPMSMs with concentrated windings is proposed. The validity and necessity of the proposed procedure is confirmed by experimental results. The proposed procedure is applied to the optimization of shape and orientation of rotor PMs by expecting the development of PM technologies in the near future. Then, a novel rotor design with gull-wing shape PMs, which can reduce the iron loss at high speeds without considerable deterioration of the other characteristics, is obtained. Further work will be required to manufacture the prototype of the optimized motor.

REFERENCES

[1] J. Cros and P. Viarouge, "Synthesis of high performance PM motors with concentrated windings," *IEEE Trans. on Energy Conversion*, vol. 17, no. 2, pp. 248-253, 2002.

[2] A. M. El-Refaie, T. M. Jahns, P. J. McCleer, and J. W. McKeever, "Experimental verification of optimal flux weakening in surface PM machines using concentrated windings," *IEEE Trans. on Ind. Appl.*, vol. 42, no. 2, pp. 443-453, 2006.

[3] S. H. Han, T. M. Jahns, and Z. Q. Zhu, "Analysis of rotor core eddy-current losses in interior permanent magnet synchronous machines," *IEEE Trans. on Ind. Appl.*, vol. 46, no. 1, pp. 196-205, 2010..

[4] K. Yamazaki, Y. Fukushima, and M. Sato, "Loss analysis of permanent-magnet motors with concentrated windings –variation of magnet eddy current loss due to stator and rotor shapes," *IEEE Trans. on Ind. Appl.*, vol. 45, no. 4, pp. 1334-1342, 2009.

[5] K. Yamazaki, Y. Kanou, Y. Fukushima, S. Ohki, A. Nezu, T. Ikemi, and R. Mizokami, "Reduction of magnet eddy current loss in interior permanent magnet motors with concentrated windings," *IEEE Trans. Ind. Appl.*, vol. 46, no. 6, pp. 2434-2441, 2010.

[6] K. Yamazaki, S. Watari, and A. Egawa, "Adaptive finite element meshing for eddy current analysis of moving conductor," *IEEE Trans. Magn.*, vol. 40, no. 2, pp. 993-996, 2004.

The 2018 International Power Electronics Conference

Loss Analysis of Permanent-Magnet Synchronous Machines Considering In-plane Eddy Current in Electrical Steel Sheets

Hideki Ohguchi[1], Satoshi Imamori[1*], Katsumi Yamazaki[2], Haiyan Yui[2], and Masao Shuto[1]

1 Fuji Electric Co., Ltd., Hino-City, Tokyo, Japan
2 Chiba Institute of Technology, Narashino-City, Chiba, Japan
*E-mail: imamori-satoshi@fujielectric.com

Abstract— The quantification of the stray load losses such as the in-plane eddy-current loss in electrical steel sheets, and circulating-current loss in the armature windings is important in order to develop rotating machines with high efficiency. We investigate the losses caused by the in-plane eddy current in the stator cores of permanent-magnet synchronous machines by using three-dimensional finite element method. The influences of the coil-end structure, skew structure, and carrier harmonics of a pulse-width-modulated inverter on the in-plane eddy-current loss in the stator core are described. The circulating-current loss in the parallel connected armature windings is analyzed by using two-dimensional finite element method. The accuracy of the analyses is confirmed by the comparison with experimental results.

Keywords— *circulating-current loss, finite element method, in-plane eddy current loss, permanent-magnet synchronous machine*

I. INTRODUCTION

Permanent-magnet synchronous machines are widely used because of their higher efficiency compared to induction machines. In order to further improve the efficiency of permanent-magnet synchronous machines, the losses need to be estimated with high accuracy. In order to calculate the iron loss accurately, some papers have reported an improved analysis method that uses two-dimensional (2D) finite element method (FEM) to consider, for example, the carrier harmonics and mechanical stress [1]. However, 2D FEM is inapplicable when the magnetic flux flow is in the stacking direction. In other words, three-dimensional (3D) FEM is necessary.

In order to achieve higher efficiency, the iron loss analysis needs to be improved, and stray load losses such as the in-plane eddy-current loss in the stator core and circulating-current loss in the armature windings need to be quantified. Some papers have examined the in-plane eddy-current loss [2, 3]. However, few papers have reported on the influences of the coil-end structure, skew structure, and carrier harmonics of a pulse-width-modulated (PWM) inverter on the in-plane eddy-current loss in the stator core to the best of the authors' knowledge. Some papers have focused on using homogenization technique to evaluate the loss at each armature wire [4–7]. However, few papers have reported on the influence of the position of each wire in the stator

slot on the estimated circulating-current loss in parallel connected armature windings.

The purpose of this study is to clarify the behavior of the losses caused by the in-plane eddy current in the stator cores of permanent-magnet synchronous machines by using 3D FEM and the loss caused by the circulating current in the parallel connected armature windings by using 2D FEM. First, the in-plane eddy-current loss for two coil-end structures is analyzed. The conventional model assumes that the coil ends are perpendicular to the symmetry boundary to simplify the calculation. With this assumption, however, leakage flux of the coil ends is not sufficiently considered. For example, the leakage flux of the wires passing over the stator core is ignored. However, the leakage flux is accounted for if the coil ends are connected from slot to slot. Second, two skew structures are compared: a stator continuously skewed by one slot pitch and a step-skewed rotor. Third, the influence of the carrier harmonics induced by the PWM inverter is examined. Fourth, the circulating-current loss in the parallel connected armature windings when each wire in the stator slot is considered is estimated by using 2D FEM. Finally, the total electrical losses obtained by the analysis and experimental results are compared to confirm the validity of the proposed model.

II. INFLUENCE OF THE COIL-END STRUCTURE

In order to accurately evaluate the in-plane eddy-current loss, the lamination structure and skin effect need to be considered in the analysis model. However, it takes a long time to create a model and generate elements. Then, the electrical conductivity in the in-plane direction is set to the catalog data, and that in the stacking direction is set to 1/5000 of the catalog data to improve the convergence stability.

Table I summarizes the specifications of the analyzed machine. Fig. 1(a) and (b) show the coil-end structures of the conventional coil-end model straight to the symmetric boundary and that considering the connection between slots, respectively. Fig. 2 shows one part of the coil ends in Fig. 1(b), and region B is the wire passing over the stator core. Table II summarizes the analysis specifications.

The 2018 International Power Electronics Conference

TABLE I
MACHINE SPECIFICATIONS

Item	Value
Phases and poles	3 phases, 8 poles
Rated output power	105 kW
Rated rotational speed	4,000 r/min
Diameter of stator, stack length	300 mm, 287 mm
Stator skew	1 slot pitch
Number of stator slots	48 (distributed)
Magnet type, magnetization	Nd–Fe–B, 1.2 T
Thicknesses of electrical steel sheets	Stator: 0.35 mm Rotor: 0.5 mm
Inverter	V_{dc} = 690 V, f_c =5 kHz

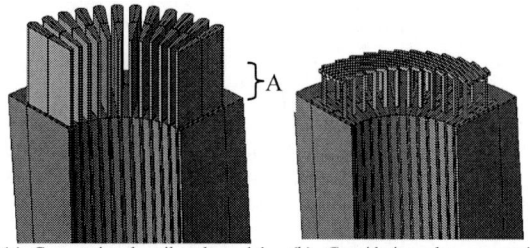

(a) Conventional coil-end model straight to the symmetric boundary

(b) Considering the connection between slots

Fig. 1. Coil-end structures.

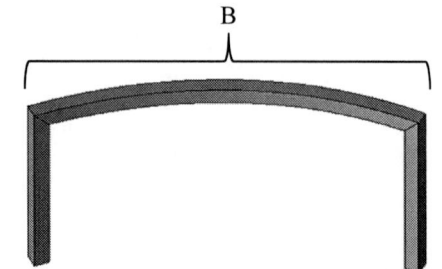

Fig. 2. One part of the coil ends.

TABLE II
DISCRETIZATION DATA

Number of time steps/period	Number of elements		Computational time	
	(a)	(b)	(a)	(b)
48	1,677,136	1,744,838	47 h 16 min	10 h 51 min

Xeon 2.3 GHz, 8 parallel cores

Fig. 3 compares the in-plane eddy-current losses of the two models when the rated sinusoidal current flows in the armature windings. The total in-plane eddy-current loss in Fig. 3(a) is set to 1 p.u. Fig. 4(a) and (b) show the contour plots of the Joule loss density. The amount of the in-plane eddy-current loss is almost the same for the coil-end structures of Fig. 1(a) and (b), and the Joule loss densities are similar. The in-plane eddy-current loss at the edge part of the stator core appears to be caused by the leakage flux of the coil ends near the stator core (region A in Fig. 1). In other words, the influence of the wires passing over the stator core on the in-plane eddy-current loss is ignored.

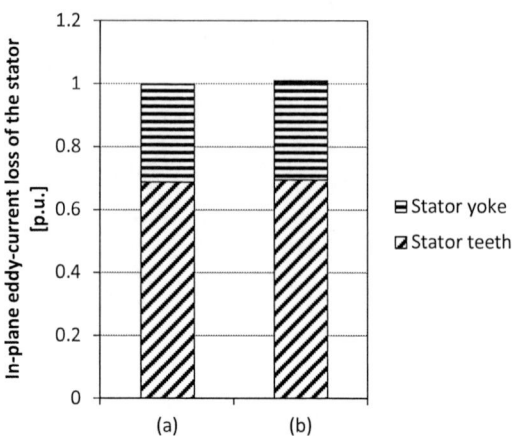

Fig. 3. Comparison of the in-plane eddy-current losses.

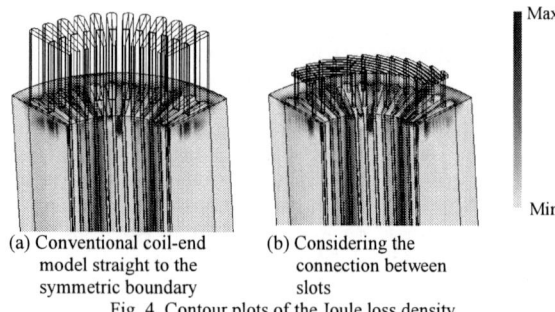

(a) Conventional coil-end model straight to the symmetric boundary

(b) Considering the connection between slots

Fig. 4. Contour plots of the Joule loss density.

III. INFLUENCE OF THE SKEW STRUCTURE

Fig. 5(a) and (b) show the skew structures of the continuously skewed stator core and step-skewed rotor core, respectively. The step-skew angle is 3.75°. The conventional coil-end structure is applied to the step-skewed rotor machine. Table III summarizes the analysis specifications.

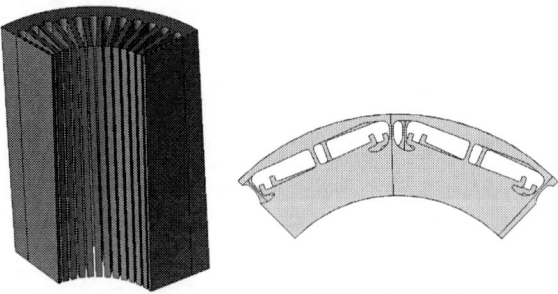

(a) Skewed stator

(b) Step-skewed rotor

Fig. 5. Skew structures.

TABLE III
DISCRETIZATION DATA

Number of time steps/period	Number of elements	Computational time
48	1,706,400	27 h 46 min

Xeon 2.3 GHz, 8 parallel cores

700

Fig. 6 compares the in-plane eddy-current losses of the two models. Fig. 7(a) and (b) show the contour plots of the Joule loss density. The in-plane eddy-current loss is almost the same for the skewed stator (Fig. 1(a)) and step-skewed rotor. The in-plane eddy-current loss appears to be dominant at the edge part of the stator core compared with inside.

IV. INFLUENCE OF THE CARRIER HARMONICS OF THE INVERTER

Table IV summarizes the analysis specifications. The current waveforms are simulated with a circuit simulator. Fig. 8 shows the simulated current waveforms of the armature windings. The amplitude of the fundamental component is set to 1 p.u. The model considering the connection between the slots, as shown in Fig. 1(b), is adopted.

Fig. 9 compares the in-plane eddy-current losses for the sinusoidal and PWM currents. The in-plane eddy-current loss for the PWM current is 1.8 times that for the sinusoidal current. Fig. 10 shows the contour plot of the Joule loss density at the time step when the maximum Joule loss is generated. The loss of the edge part of the stator core is high compared with that of the inside. This seems to be due to the leakage flux of the carrier harmonic component at the coil ends.

TABLE IV
DISCRETIZATION DATA

Number of time steps/period	Computational time
300	27 h 22 min
	Xeon 2.3 GHz, 8 parallel cores

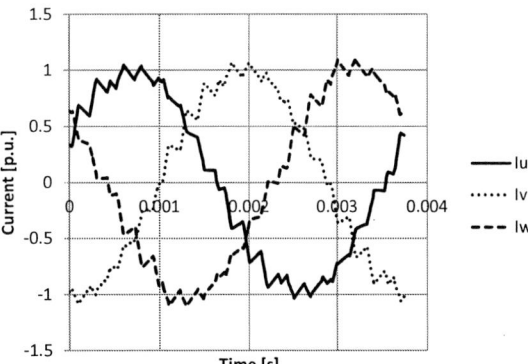

Fig. 8. Simulated current waveforms of the armature windings.

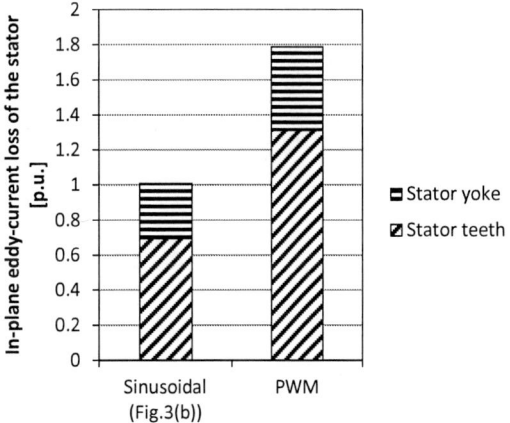

Fig. 9. Comparison of the in-plane eddy-current losses.

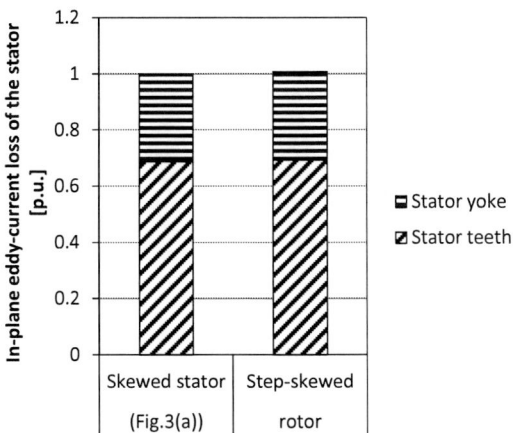

Fig. 6. Comparison of the in-plane eddy-current losses.

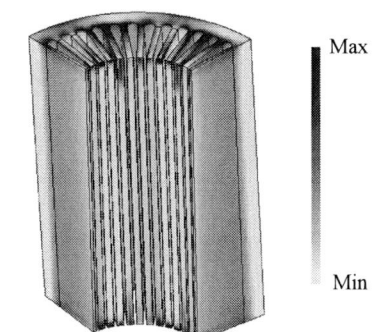

Fig. 10. Contour plot of the Joule loss density for the PWM current.

V. CIRCULATING-CURRENT LOSS IN THE PARALLEL CONNECTED ARMATURE WINDINGS

The circulating-current loss in the parallel connected armature windings is analyzed by using 2D FEM.

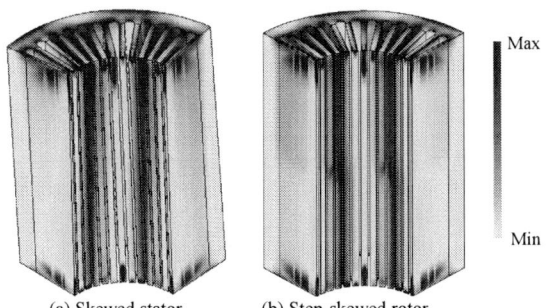

(a) Skewed stator   (b) Step-skewed rotor
Fig. 7. Contour plots of the Joule loss density.

701

Fig. 11 shows the finite element model for the FEM analysis, and Fig. 12 shows the wire position. Fig. 13 indicates the wire connection in the armature windings. There are 10 turns per slot, and each turn consists of 26 wires. In Fig. 12, wires with the same number are connected in series, whereas wires with the different number are connected in parallel.

The circulating-current loss in the parallel connected armature windings is analyzed considering the driving by the PWM inverter.

Fig. 14 shows the classical armature copper loss, which is calculated by using the armature current, armature resistance, and circulating-current loss in the parallel connected armature windings. The classical armature copper loss is set to 1 p.u. The circulating-current loss in the parallel connected armature windings is about twice the usual copper loss. Furthermore, the fundamental and harmonic circulating-current losses are the same. In other words, the circulating-current generated by the carrier harmonics of the PWM inverter cannot be ignored.

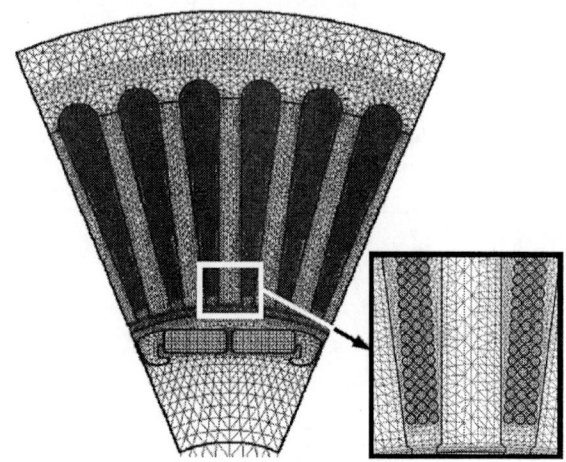

Fig. 11. Finite element model.

Fig. 12. Wire position.

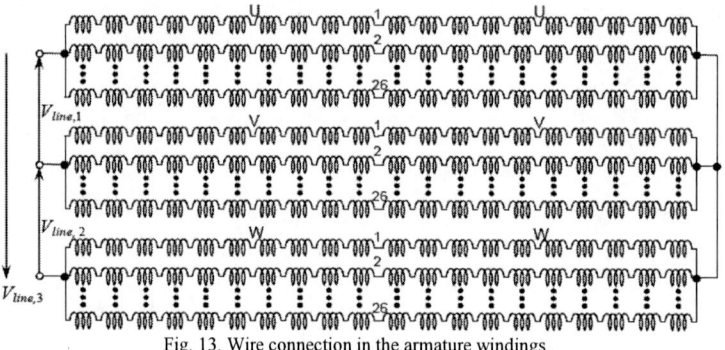

Fig. 13. Wire connection in the armature windings.

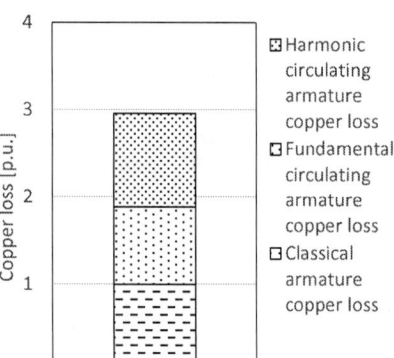

Fig. 14. Classical armature copper loss and circulating-current loss.

VI. TOTAL ELECTRICAL LOSSES

Fig. 15 compares the total electrical loss obtained from the experiment and the analysis [8]. The iron losses of the stator and rotor cores are calculated from the time variation in the flux density by using the method described in [9]. The experimental and analysis values are almost the same. The in-plane eddy-current loss in the stator core and the circulating-current loss in the parallel connected armature windings are about 15% and about 21%, respectively.

The 2018 International Power Electronics Conference

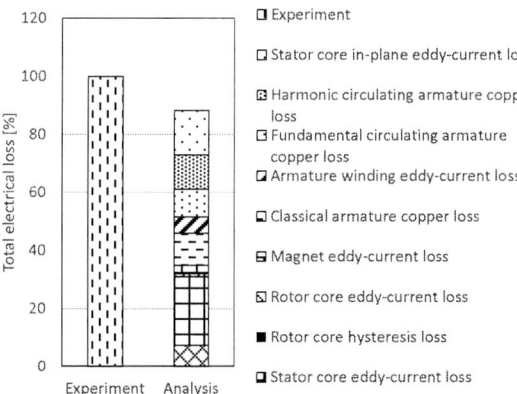

Fig. 15. Comparison of the total electrical loss obtained by the experiment and analysis.

VII. CONCLUSIONS

This study examined the influences of the coil-end structure, skew structure, and carrier harmonics of a PWM inverter on the in-plane eddy-current loss in a stator core, and the circulating-current loss in the parallel connected armature windings. The findings are as follows:

1) The in-plane eddy-current loss is almost the same for the conventional coil ends and coil ends considering the connection between slots.
2) The in-plane eddy-current loss is almost the same for the skewed stator and step-skewed rotor.
3) The in-plane eddy-current loss for the PWM current is 1.8 times that for the sinusoidal current.
4) The circulating-current loss in the parallel connected armature windings is about twice that of the usual copper loss, which is calculated from the armature current and the armature resistance.
5) The analysis is validated by a comparison to experimental results.

Loss reduction is planned for future work.

REFERENCES

[1] K. Yamazaki and Y. Kato, "Iron Loss Analysis of Interior Permanent Magnet Synchronous Motors by Considering Mechanical Stress and Deformation of Stators and Rotors," *IEEE Trans. Magn.*, vol. 50, no. 2, pp. 909-912, 2014.
[2] K. Yamazaki, Y. Yamato, H. Mogi, C. Kaido, A. Nakahara, K. Takahashi, K. Ide, and K. Hattori, "In-Plane Eddy Current Analysis for End and Interior Stator Core Packets of Turbine Generators," *2008 18th International Conference on Electrical Machines*, pp. 1-6, 2008.
[3] K. Yamazaki, Shin Tada, H. Mogi, Y. Mishima, C. Kaido, S. Kanao, K. Takahashi, K. Ide, K. Hattori, and A. Nakahara, "Eddy Current Analysis Considering Lamination for Stator Core Ends of Turbine Generators," *IEEE Trans. Magn.*, vol. 44, no. 6, pp. 1502–1505, 2008.
[4] J. Gyselinck and P. Dular, "Frequency-domain homogenization of bundles of wires in 2-D magneto dynamic FE calculations," *IEEE Trans. Magn.*, vol. 41, no. 5, pp. 1416-1419, 2005.
[5] X. Nan and C. R. Sullivan, "Eddy-Current-Effect Homogenization of sindings in harmonic-balance finite element models," *IEEE Trans. Ind. Appl.*, vol. 45, no. 2, pp. 854-860, 2009.
[6] H. Igarashi, "Semi-analytical approach for finite element analysis of multi-turn coil considering skin effects," *IEEE Trans. Magn.*, vol. 53, no. 1, 7400107, 2017.

[7] R. V. Sabariego1 and J. Gyselinck, "Eddy-current-effect homogenization of windings in harmonic-balance finite-element models," *IEEE Trans. Magn.*, vol. 53, no. 6, 7206304, 2017.
[8] K. Yamazaki, H. Yui, and H. Ohguchi, "Loss Analysis of Permanent Magnet Synchronous Machines by Considering Circulating Currents of Armature Windings," *The Papers of Joint Technical Meeting on Motor Drive, Rotating Machinery and Vehicle Technology, IEE Japan*, MD-17-074/RM-17-057/VT-17-011, pp. 25-30, 2017.
[9] K. Yamazaki and Y. Seto, "Iron loss analysis of interior permanent magnet synchronous motors -Variation of main loss factors due to driving condition," *IEEE Trans. on Ind. Appl.*, vol. 42, no. 4, pp. 1045-1052, 2006.

Study on Influence of Difference in Structure of Concentrated Winding IPMSMs Obtained by Automatic Design

A. Ura, M. Sanada, S. Morimoto and Y. Inoue
Graduate School of Engineering
Osaka Prefecture University
Sakai, Japan
E-mail: sxb01034@edu.osakafu-u.ac.jp

Abstract— Interior permanent magnet synchronous motors (IPMSMs) are widely used because of their high efficiency and high power. However, since their design has a high degree of freedom, application-specific design of IPMSMs is difficult. In this study, we investigated the characteristics of the structure of models obtained by an automatic design system using a genetic algorithm (GA) and examined the structural factors useful in the design of IPMSMs that reduce eddy current loss in the stator and have high efficiency at high speeds. Then, we applied the obtained design guidelines to a typical motor and confirmed the reduction of the eddy current loss in the stator at high speed.

Keywords— automatic design, genetic algorithm (GA), interior permanent magnet synchronous motor (IPMSM), eddy current loss.

I. INTRODUCTION

Interior permanent magnet synchronous motors (IPMSMs) are widely used in various applications due to their high efficiency and high power. However, their design has a high degree of freedom, and structural differences in IPMSMs have a considerable influence on the motor characteristics [1]. Thus, application-specific design of an IPMSM requires experimental knowledge or many iterative calculations. Therefore, the simpler design methods and a shorter design process are being pursued [2].

To generalize the design of IPMSMs, design methods using various optimization algorithms have been proposed [3], [4]. In particular, genetic algorithms (GAs) have many applications in the design of electromagnetic devices [5] and are widely used in the design of IPMSMs [6]-[8].

In this study, a finite element method (FEM) was used for electromagnetic analysis of IPMSMs. The FEM is often used for an electromagnetic field analysis of motors because its consideration of magnetic saturation allows detailed analysis. However, FEM has the problem of generally requiring long computation times. To decrease design periods, coarse-mesh FEM, which uses a relatively small number of mesh elements, was proposed. In

IPMSM analysis, the coarse-mesh FEM shows almost similar results compared to those obtained with fine-mesh FEM [9].

In this paper, first we describe the design of IPMSM structures that have large high efficiency driving area and fulfill torque condition. The design was performed with an automatic design system using a combination of GA and coarse-mesh FEM. Then, we examine the structures of the motors obtained by automatic design to investigate which part of the structure affects the motor characteristics, especially for loss and efficiency.

II. DESIGN MODEL AND CONDITION

The structure of the design model is shown in Fig. 1. The model has a concentrated winding stator with four poles and six slots. A total of nine design variables were evaluated, eight of which are shown in Fig. 1, and the additional variable was x_5 (mm^3), the volume of each permanent magnet per pole. Since the stator slot size varies with changes in the stator structure, such as the tooth width x_7 and the yoke width x_8, the number of winding turns was changed to be approximately equal to the reference winding space factor shown in Table I. The thickness of the flux barriers was the same as the width of the permanent magnets.

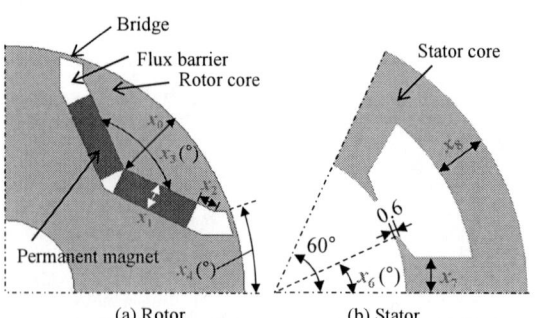

(a) Rotor   (b) Stator
Fig. 1. Structure and design variables of IPMSM model
(units: mm).

TABLE I
COMMON SPECIFICATIONS OF THE IPMSM MODEL

Item (unit)	Value
Stator diameter (mm)	112.2
Rotor diameter (mm)	60
Rotor inner diameter (mm)	16
Stack length (mm)	60
Air gap length (mm)	0.5
Bridge width (mm)	0.5
Permanent magnet coercive force (kA/m)	915
Winding resistance (Ω/km)	23.33
Reference winding space factor (%)	50
Current limit (A)	4.4
Voltage limit (V)	100
Steel grade	35H300

TABLE II
DESIGN VARIABLES

Item (unit)	Value	Number of Patterns	Step size
x_0 (mm)	5 – 20	16	1
x_1 (mm)	2.0 – 3.5	16	0.1
x_2 (mm)	1 – 8	8	1
x_3 (°)	30 – 180	16	10
x_4 (°)	10 – 25	16	1
x_5 (mm³)	1560 – 2460	16	60
x_6 (°)	16 – 23	8	1
x_7 (mm)	3 – 10	8	1
x_8 (mm)	4 – 11	8	1

Table I shows the common specifications of the design model. The design variables were changed over the range of values shown in Table II. Some combinations of variable values produced a motor structure that was impossible to make. In such cases, the fitness of this individual was set to zero and it was not carried over to the next generation.

To shorten the computation time, coarse-mesh FEM was used. The number of finite element meshes was set so that the rotor had approximately 3,000 elements and the stator had approximately 3,200 elements, both relatively small numbers.

In this study, the purpose of automatic design was to expand high efficiency driving area as much as possible while satisfying maximum torque condition. The GA fitness was set to the number of grid points where the efficiency was 93% or more in the efficiency map, shown in Fig. 2. The torque range was 0 to 3.5 Nm, and the torque step size was 0.025 Nm, which corresponds to

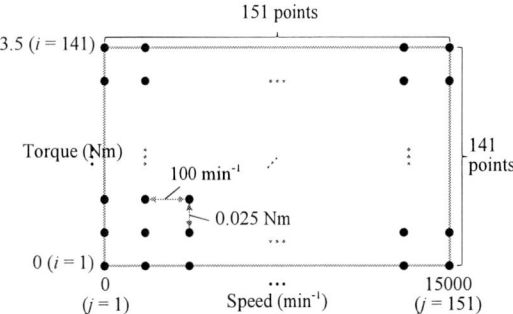

Fig. 2. Grid points in the efficiency map.

TABLE III
GA PARAMETERS

Item	Value
Population size	32
Crossover method	Uniform crossover
Crossover rate	60%
Mutation rate	15%
Selection method	Tournament selection + elite selection
Elite population size	2
Termination condition	50 generations

$i = 1$ to 141. Similarly, the speed range was 0 to 15,000 min⁻¹, and the speed step size was 100 min⁻¹, which corresponds to $j = 1$ to 151. Therefore, the total number of grid points was 21,291. The maximum torque T_{max} of the design model was required to satisfy the constraint condition of exceeding 2.0 Nm. Thus, the fitness f of the GA and the constraint condition were as follows:

$$\text{maximize} \quad f = \sum_{i=1}^{141} \sum_{j=1}^{151} P_{ij} \tag{1}$$
$$\text{subject to} \quad T_{max} \geq 2.0$$

where P_{ij} is defined as follows:

$$P_{ij} = \begin{cases} 1, & \eta_{ij} \geq 93 \\ 0, & \eta_{ij} < 93 \end{cases} \tag{2}$$

and η_{ij} is the efficiency at point (i, j) in the efficiency map (%).

In this analysis, the GA was used as an optimization method in the automatic design of the motor structure. The combinations of the nine design variables of motor structure were considered gene sequences. Five variables had gene lengths of 4 bits and four variables had gene lengths of 3 bits. Therefore, the total length of each gene was 32 bits and the total number of variable combinations was 2^{32}. Genetic operations were performed on those gene sequences to create new motor structures.

The GA parameters for this design process are shown in Table III. To give an adequate diversity to the first generation, the population size per generation was set to 32 individuals. The selection methods were elite selection and tournament selection, to avoid the disappearance of high-fitness individuals in each generation. To obtain a solution within a practical amount of time, the GA analysis was terminated when it reached 50 generations.

III. DESIGN RESULTS AND CONSIDERATIONS

The results of the automatic design are presented in this section. At the last generation, the maximum fitness f is 1,606. The model that showed the highest fitness is called the GA-A model, and its rotor and stator structures are shown in Fig. 3.

For comparison, we consider one other structure obtained by automatic design, although it exhibited relatively low fitness. This model is called GA-B, and its structures are shown in Fig. 4. The fitness of the GA-B model is 1,258. Both of the models have a V-shaped rotor, but the shape of their flux barrier is different.

Table IV compares the model parameters for GA-A and B. Torque, current phase, armature flux linkage, and inductance values are for the maximum torque. Figure 5 shows efficiency maps for both models. Especially in the high-speed range, the operable and high-efficiency areas of the GA-A model are wider than those of GA-B.

To analyze the efficiency during operation, we set two points P_1 and P_2 on the efficiency map. Point P_1 is at high torque and low speed ($T = 2.0$ Nm and $N = 2,000$ min^{-1}), and point P_2 is low torque and high speed ($T = 0.5$ Nm and $N = 6,000$ min^{-1}). The loss details for both models at both points are shown in Fig. 6. At point P_1, there is no significant difference in loss between the two models. In contrast, at point P_2, copper and iron losses of the GA-A model were small compared to those of the GA-B model. Due to a smaller number of winding turns, the winding resistance of the GA-B model was lower, but it had a phase current of 3.25 A for driving at point P_2, which was larger than the 2.79 A seen in the GA-A model. Therefore, the copper loss of the GA-B model was larger.

Details of iron loss in both models are shown in Fig. 7. At point P_1, no significant difference is seen between both models, but at point P_2, the stator eddy current loss accounts for the largest iron loss in both models. The GA-B stator eddy current loss had the largest increase in comparison with the GA-A model. To investigate the position where the stator eddy current loss occurs, the distributions of eddy current loss are shown in Fig. 8. The stator eddy current loss is particularly large at the pole horn at the end of the stator tooth indicated by circles in the figure.

To consider whether high-efficiency operation at high speed in GA-A is mainly caused by the rotor or stator structure, we created models in which the rotor and stator of GA-A and GA-B were swapped. The model that combined the rotor of GA-A and the stator of GA-B is called AR-BS, and the model that combined the rotor of GA-B and the stator of GA-A is called BR-AS.

Table V compares the parameters of these additional models with those of GA-A and GA-B. As in the previous case, torque, current phase, armature flux linkage, and inductance values are at the maximum torque. The efficiency maps of all four models are shown in Fig. 9. In the AR-BS, the maximum torque was greatly decreased, but the high-efficiency operation area expanded in the high-speed, low-torque range. In contrast, in the BR-AS model, the maximum torque was larger,

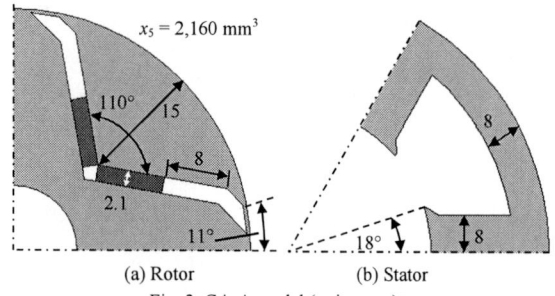
(a) Rotor   (b) Stator
Fig. 3. GA-A model (units: mm).

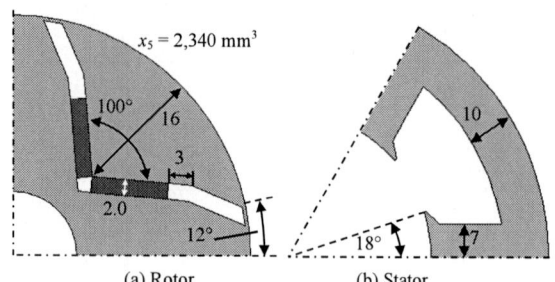
(a) Rotor   (b) Stator
Fig. 4. GA-B model (units: mm).

TABLE IV
COMPARISON OF CHARACTERISTICS OF GA-A AND GA-B

Item (Unit)	Value	
	GA-A	GA-B
Maximum torque T_{max} (Nm)	2.05	2.01
Magnet torque T_m (Nm)	1.71	1.73
Reluctance torque T_r (Nm)	0.34	0.29
Current phase β (°)	22.1	20.4
Armature flux linkage Ψ_o (Wb)	0.162	0.151
d-axis inductance L_d (mH)	10.58	8.45
q-axis inductance L_q (mH)	20.40	16.05
$L_q - L_d$ (mH)	9.82	7.60
Minimum d-axis flux linkage Ψ_{dmin} (Wb)	0.04	0.05
Winding turns (per tooth)	126	114
Winding resistance (Ω)	0.624	0.575
Fitness f	1606	1258

The 2018 International Power Electronics Conference

and the high-efficiency area expanded in the low-speed, high-torque range, but efficiency decreased at high speed. From this, we concluded that the rotor of the GA-A contributes significantly to the high-efficiency area expansion at high speed. In addition, the GA-A model shows that the high-efficiency area expanded in both the high-speed and low-speed ranges, making it a well-balanced model.

Figure 10 shows the details of losses at point P_2. In the AR-BS model, in which the rotor of GA-B was replaced by that of GA-A, both copper and iron losses decreased. In contrast, in the BR-AS model, in which the stator of GA-B was replaced by that of GA-A, iron loss was almost unchanged and copper loss increased. The change in copper loss is due to the reduction in the amount of current required to drive the same point as in the GA-A and GA-B models. To investigate the cause of the decrease in iron loss by the rotor structure, the details of iron loss were examined. Figure 11 provides details of the iron loss at point P_2, and it shows that, among the iron losses, the rotor change reduced the stator's eddy current loss the most.

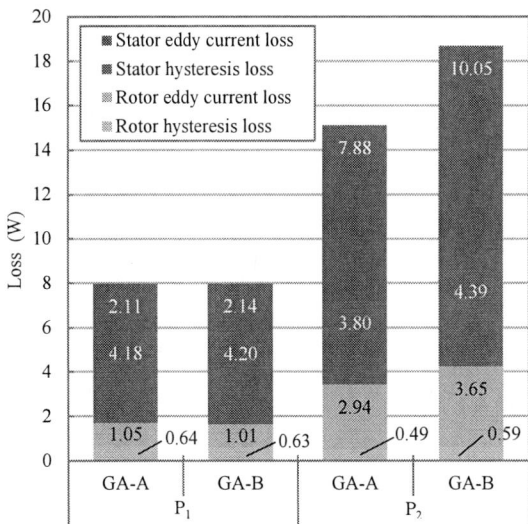

Fig. 7. Details of iron loss at points P_1 and P_2 in GA-A and GA-B.

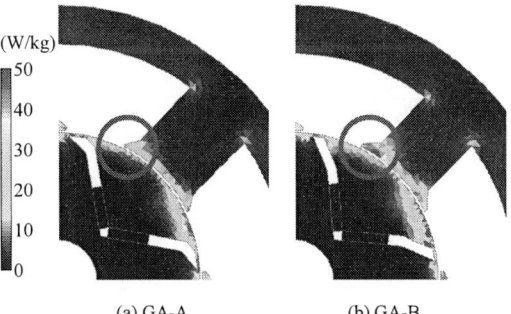

(a) GA-A   (b) GA-B

Fig. 8. Eddy current loss distributions at point P_2 in GA-A and GA-B.

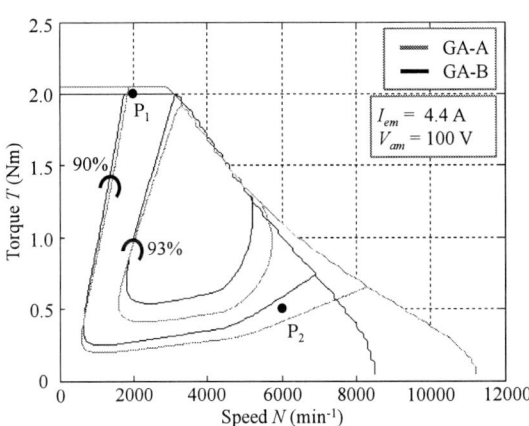

Fig. 5. Efficiency maps for GA-A and GA-B.

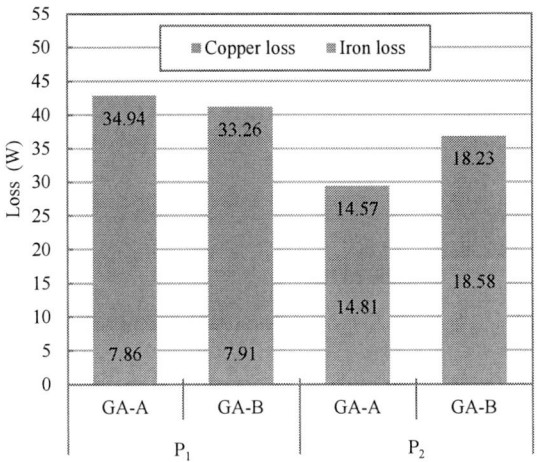

Fig. 6. Details of losses at points P_1 and P_2 in GA-A and GA-B.

TABLE V

COMPARISON OF CHARACTERISTICS OF ALL FOUR MODELS

Item (Unit)	Value			
	GA-A	GA-B	AR-BS	BR-AS
Maximum torque T_{max} (Nm)	2.05	2.01	1.82	2.27
Magnet torque T_m (Nm)	1.71	1.73	1.55	1.89
Reluctance torque T_r (Nm)	0.34	0.29	0.26	0.38
Current phase β (°)	22.1	20.4	20.6	22.0
Armature flux linkage Ψ_o (Wb)	0.162	0.151	0.140	0.176
d-axis inductance L_d (mH)	10.58	8.45	8.36	10.53
q-axis inductance L_q (mH)	20.40	16.05	15.32	19.98
$L_q - L_d$ (mH)	9.82	7.60	6.95	9.44
Minimum d-axis flux linkage Ψ_{dmin} (Wb)	0.040	0.053	0.042	0.050
Winding turns (per tooth)	126	114	114	126
Winding resistance (Ω)	0.624	0.575	0.575	0.624
Fitness f	1606	1258	1481	1340

707

The 2018 International Power Electronics Conference

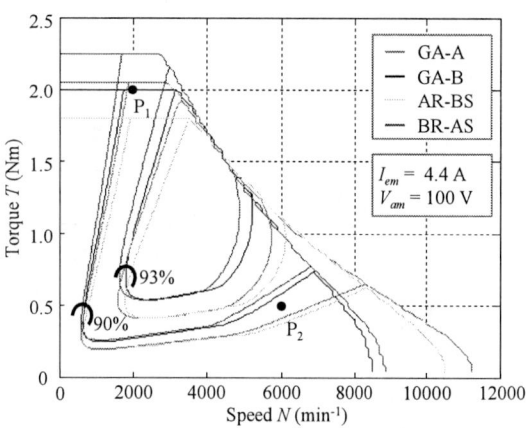

Fig. 9. Efficiency maps for all four models.

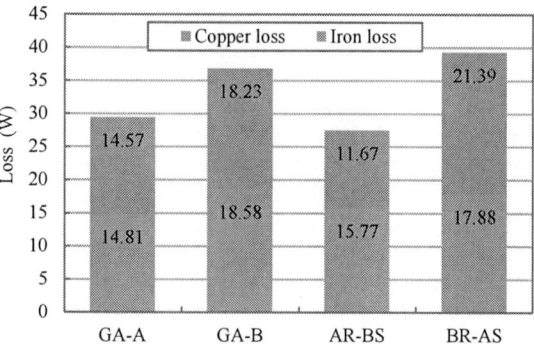

Fig. 10. Details of losses at point P₂ in all four models.

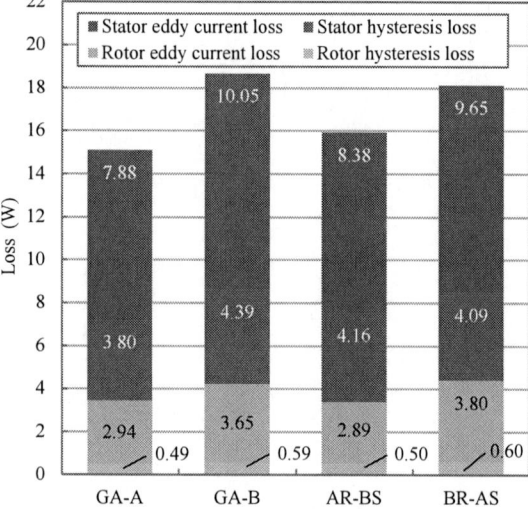

Fig. 11. Details of iron loss at point P₂ in all four models.

Next, we examined how the difference in rotor structure affects stator eddy current loss. The stator eddy current losses calculated for each harmonic component for two models with different rotors, GA-B and AR-BS, are shown in Fig. 12. The figure shows that in the stator eddy current loss, the fundamental wave component is the largest and the fifth, seventh, and ninth harmonic components are greatly reduced due to the change of rotor structure. The eddy current loss is expressed by the following equation [10], [11]:

$$W_e = \sum_{i=1}^{e}\left[m_i \sum_{k=1}^{n}\left\{ k_e \left(kf\right)^2 B_{rk-i}^{\ 2} + k_e \left(kf\right)^2 B_{\theta k-i}^{\ 2} \right\} \right], \quad (3)$$

where W_e is the eddy current loss (W), e is the number of elements in the region, m_i is the mass of the i-th element (kg), k_e is the eddy current loss coefficient, k is the harmonic order, n is the maximum harmonic order, f is the basis frequency (Hz), B_{rk-i} is the amplitude of the k-th harmonic order of the radial magnetic flux density of the i-th element (T), and $B_{\theta k-i}$ is the amplitude of the k-th harmonic order of the tangential magnetic flux density of the i-th element (T).

Equation (3) shows that a sharp change in the magnetic flux density caused the large harmonic component of the eddy current loss.

To observe how the magnetic flux density changes due to the difference in rotor structure, we focused on the flow of magnetic flux during rotation. Figure 13 shows the magnetic flux passing through the pole horn when the flux barrier of the rotor overlaps the position of the pole horn of the stator in the same two models. The change of the magnetic flux in AR-BS is gentle compared to that of GA-B. Magnetic flux diagrams for a rotor position of 45° are shown in Fig. 14. In GA-B, the magnetic flux from the rotor toward the stator rapidly flows into the stator pole horns intensively. In contrast, in AR-BS, the flux barriers bend outwardly with respect to the magnets, so that the magnetic flux toward the stator gradually flows to the pole horns. This reduces the eddy current loss in the stator pole horn.

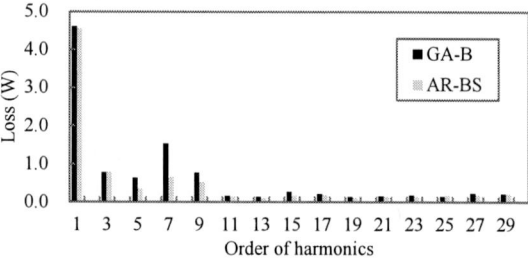

Fig. 12. Eddy current loss harmonics at point P₂ in GA-B and AR-BS.

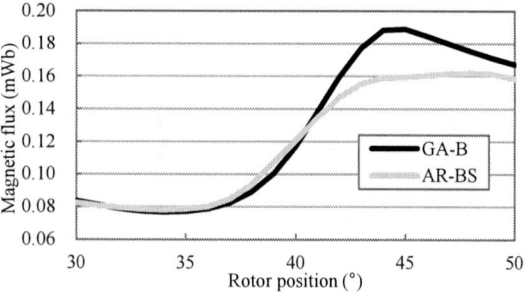

Fig. 13. Magnetic flux passing through pole horn at point P₂ in GA-B and AR-BS.

708

The 2018 International Power Electronics Conference

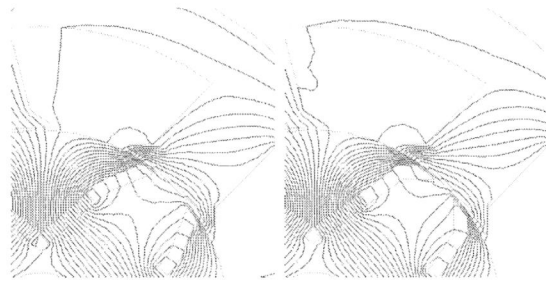

(a) GA-B   (b) AR-BS

Fig. 14. Magnetic flux diagrams at point P₂ in GA-B and AR-BS.

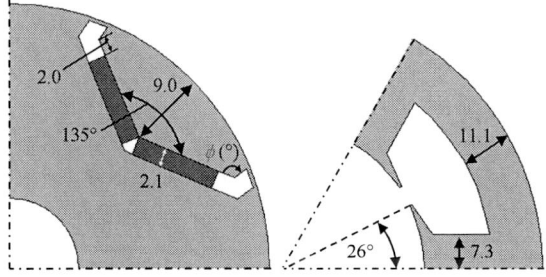

(a) Rotor   (b) Stator

Fig. 15. Standard model (units: mm).

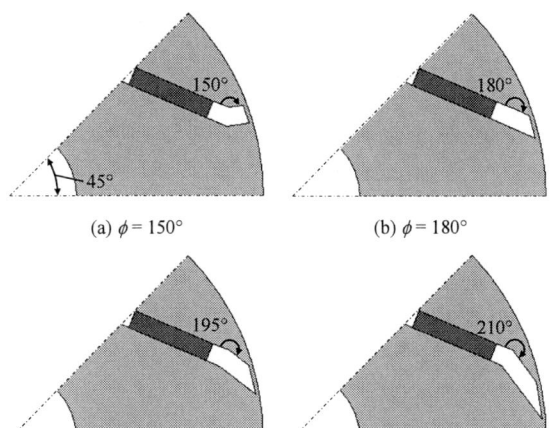

(a) $\phi = 150°$ (b) $\phi = 180°$

(c) $\phi = 195°$ (d) $\phi = 210°$

Fig. 16. Additional models with flux barrier variations.

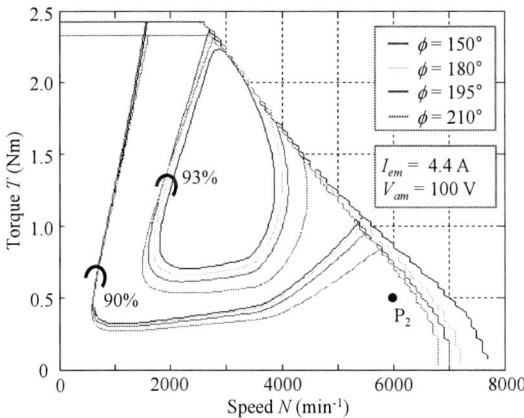

Fig. 17. Efficiency maps for flux barrier variations.

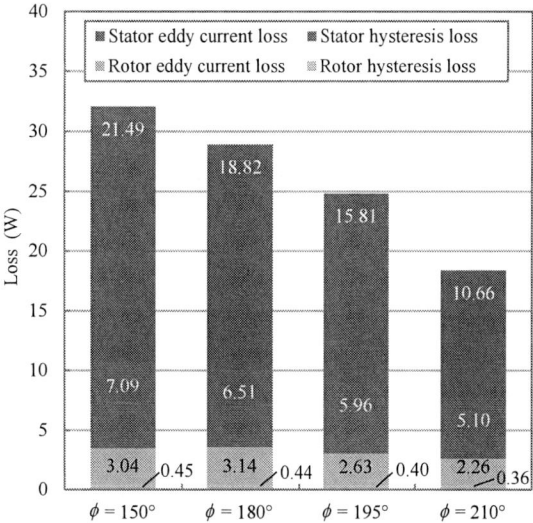

Fig. 18. Details of iron loss at point P₂ in flux barrier variation models.

IV. APPLYING DESIGN METHOD TO STANDARD IPMSM

In the previous section, we showed that the stator eddy current loss at high speed can be reduced by bending the flux barrier of the rotor outward with respect to the magnet. We applied this rotor structure to a standard IPMSM shape and confirmed that the stator eddy current loss is reduced.

The standard model to which this design was applied is shown in Fig. 15. It has a concentrated winding stator with four poles and six slots. The rotor is V-shaped, and the volume of the permanent magnet per pole is 2,520 mm³. The parameters for motor size, magnet, steel grade, current, and voltage are the same as those in the automatic design model (Table I). The winding has 117 turns per tooth. We developed four models by changing the flux barrier bending angle ϕ as shown in Fig. 16.

The efficiency maps for these models are shown in Fig. 17. As the angle ϕ increased, the high-efficiency area in the high-speed range expanded. The details of iron loss at point P₂ are shown in Fig. 18. The eddy current loss of the stator decreased as the angle ϕ increased, and the eddy current loss at $\phi = 210°$ was approximately half of the eddy current loss at $\phi = 150°$.

These results confirmed that stator eddy current loss is reduced by increasing the angle of the flux barrier of the rotor.

V. CONCLUSION

In this study, we investigated the rotor and stator structure of IPMSMs to expand the high-efficiency operation area by automatic design using GA. To evaluate the obtained models, we focused on improvement of efficiency, especially at high speed, and detailed investigation was carried out on the factors that reduce stator eddy current loss. The results showed that certain rotor flux barrier shapes suppress rapid changes in magnetic flux density at the stator pole horns, which reduces eddy current loss. In addition, it was shown that this design improved the efficiency at high speed.

REFERENCES

[1] K. C. Kim, J. Lee, H. J. Kim, and D. H. Koo, "Multiobjective Optimal Design for Interior Permanent Magnet Synchronous Motor," *IEEE Trans. on Magnetics*, vol. 45, no. 3, pp. 1780-1783, 2009.

[2] F. Parasiliti, M. Villani, S. Lucidi, and F. Rinaldi, "Finite-Element-Based Multiobjective Design Optimization Procedure of Interior Permanent Magnet Synchronous Motors for Wide Constant-Power Region Operation," *IEEE Trans. on Industrial Electronics*, vol. 59, no. 6, pp. 2503-2514, 2012.

[3] C. Lu, S, Ferrari, and G. Pellegrino, "Two Design Procedures for PM Synchronous Machines for Electric Powertrains," *IEEE Trans. on Transportation Electrification*, vol. 3, no. 1, pp. 98-107, 2017.

[4] P. Zhang, G. Y. Sizov, M. Li, D. M. Ionel, N. A. Demerdash, S. J. Dtretz, A. W. Yeadon, "Multi-Objective Tradeoffs in the Design Optimization of a Brushless Permanent-Magnet Machine With Fractional-Slot Concentrated Windings," *IEEE Trans. on Industry Applications.*, vol. 50, no. 5, pp. 3285-3294, 2014.

[5] J. A. Vasconcelos, R. R. Saldanha, L. Krahenbuhl, and A. Nicolas, "Genetic algorithm coupled with a deterministic method for optimization in electromagnetics," *IEEE Trans. on Magnetics*, vol. 33, no. 2, pp. 1860-1863, 1997.

[6] W. Zhao, F. Zhao, T. A. Lipo, and B. I. Kwon, "Optimal Design of a Novel V-Type Interior Permanent Magnet Motor with Assisted Barriers for the Improvement of Torque Characteristics," *IEEE Trans. on Magnetics*, vol. 50, no. 11, 8104504, 2014.

[7] T. Nakata, M. Sanada, S. Morimoto, and Y. Inoue, "Automatic Design of IPMSMs Using a Genetic Algorithm Combined with the Coarse-mesh FEM for Enlarging the High-Efficiency Operation Area," *IEEE Trans. on Industrial Electronics*, vol.64, no. 12, pp. 9721-9728, 2017.

[8] D. Lee, S. Lee, J. W. Kim, C. G. Lee and S. Y. Jung, "Intelligent Memetic Algorithm Using GA and Guided MADS for the Optimal Design of Interior PM Synchronous Machine," *IEEE Trans. on Magnetics*, vol. 47, no. 5, 2011.

[9] T. Nakata, M. Sanada, S. Morimoto, and Y. Inoue, "Automatic Design of IPMSMs Using a GA Coupled with the Coarse-mesh Finite Element Method," *in Proc. ICEMS* 2016, DS3G-1-11, 2016.

[10] S. Hashimoto, M. Sanada, S. Morimoto, and Y. Inoue, "Basic Study on the Suitable Structure of a Permanent Magnet Synchronous Motor with a Powder Magnetic Core," *in Proc. IPEC* 2014, 21P5-3, 2014.

[11] K. Yamazaki, Y. Seto, "Iron Loss Analysis of Interior Permanent-Magnet Synchronous Motors—Variation of Main Loss Factors Due to Driving Condition," *IEEE Trans. on Industry Applications.*, vol. 42, no. 4, pp. 1045-1052, 2006.

Carrier Harmonic Loss Reduction Technique on Dual Three-Phase Permanent-Magnet Synchronous Motors with Phase-Shift PWM

Yoshihiro Miyama[1,2], Haruyuki Kometani[1] and Kan Akatsu[2]

1 Advanced Technology R&D Center, Mitsubishi Electric Corporation, Amagasaki, Hyogo, Japan

2 Department of Electrical Engineering, Shibaura Institute of Technology, Tokyo, Japan

Abstract—This work investigates a method to reduce the carrier harmonic current and the carrier harmonic losses of a permanent-magnet synchronous motor (PMSM) with dual three-phase windings. The motor input impedance of the carrier harmonics is increased to reduce the carrier harmonic current by reinforcing the carrier harmonic gap flux density of the dual three-phase windings. Our study is carried out based on a theoretical approach by calculating the gap flux density, including the space, time, and carrier harmonics. The result of our theoretical approach is confirmed by finite element analysis (FEA) based on a 12-slot, 10-pole phase-shift winding dual three-phase PMSM. The measured result of our manufactured motor also reveals that the target harmonic current was reduced by applying our proposed technique.

Keywords—*Carrier harmonics, Carrier phase-shift PWM, Dual three-phase windings, Input impedance, Iron loss, Permanent magnet synchronous motor.*

I. INTRODUCTION

Motors require higher efficiency and fewer emissions for sustainable industrial activity. To improve a motor's efficiency, loss calculations (including the iron loss calculation) have been scrutinized [1]–[6]. The following manufacture processing effects have also been investigated because they increase the iron loss: the stress in the laminated steel core by press fit [2]–[4], the stress by punching [5], [6], and the stress and the conduction between laminated steel cores by a swage. Recently, a pulse width modulation (PWM) voltage-source inverter fed motor is widely being used for variable speed operation. However, since the iron loss of carrier harmonics is not negligible from the viewpoint of loss reduction, loss calculation techniques have been studied for motor applications [7]–[11].

We previously proposed a carrier harmonics iron loss reduction technique [12] on a dual three-phase PMSM with core deactivation that partially used the stator core by stopping the energization of one of the dual three-phase windings in a low torque region.

Recently, we developed another carrier harmonics control method and applied a carrier harmonics vibration reduction technique [13] to a dual three-phase PMSM. In this paper, we proposed a carrier harmonics control method to reduce the losses generated by the PWM carrier harmonics based on the gap magnetic flux density distribution control of a dual three-phase PMSM by controlling the PWM phases of dual winding.

In this manuscript, we discuss the concept and the theory of a carrier harmonic iron loss reduction technique in Section II. We focus on the relationship of the carrier phase difference, the space phase difference, and the time phase difference of the gap magnetic flux density of dual three-phase windings. We formulated the reduction conditions for four types of dual three-phase PMSM configurations. Our theoretical approach is confirmed by FEA in Section III. In Section IV, we apply a theoretical condition to a manufactured 12-slot, 10-pole phase shift concentrated dual three-phase winding PMSM and the measured results reveal a halved carrier harmonic current in the primary component.

II. REDUCTION CONCEPT AND THEORY OF CARRIER HARMONICS IRON LOSS

On a PWM voltage-source inverter-fed motor, the carrier harmonic iron loss is generated by the magnetic flux density fluctuation of the iron core in the carrier harmonic frequency. Magnetic flux density fluctuation is generated by the carrier harmonic current that is excited by the carrier harmonic voltage. To reduce the carrier harmonic iron loss, one effective approach is to reduce the magnetic flux density fluctuation or the PWM carrier harmonic currents. From Ohm's law, for a given voltage, the motor's input impedance must be increased in the range of the carrier harmonic frequency. Simultaneously, the motor's input impedance in the range of the fundamental frequency must be maintained to avoid a decrease in the high-speed output power. Inductance is a dominant factor in the input impedance, and the winding's inductance is greatly affected by the air gap flux density. Increasing the air gap flux density effectively improves the input impedance. To increase the carrier harmonic inductance without changing the fundamental inductance, the gap flux density of the carrier harmonic component was designed by a dual three-phase winding PMSM property.

For a dual three-phase winding PMSM, the radial magnetic flux density in air gap B_{gr} is summed by the air gap flux density generated from first winding B_{gr1}, and from second windings B_{gr2}, including space, time, and carrier harmonics. B_{gr} can be expressed by the sum of these harmonics [13]:

$$B_{gr}(k,n,m) = B_{gr1}(k,n,m) + B_{gr2}(k,n,m)$$
$$= \sum_{a=1,2}\sum_{k}\sum_{n}\sum_{m} A_{gr}(k,n,m)\cos\{k(\theta-\alpha_a)-n(\omega t-\beta_a)-m(\omega_c t-\gamma_a)+\varphi(k,n,m)\}$$
$$, (1)$$

where k, n, and m are respectively the harmonic order of space, time, and carrier. A and φ are the amplitude and the phase of each harmonics. θ is the rotational space position in the mechanical angle, ω is the electric angular frequency of a fundamental component, ω_c is the angular frequency of the carrier component, t is the time, and α_1, β_1, γ_1, α_2, β_2, γ_2 are the space, time, and carrier phases of the first and second windings, respectively. Fig. 1 shows the phase difference of the PWM triangle carrier waves between the first and second windings of the dual three-phase PMSM. Fig. 2 shows the construction of a dual three-phase VSI fed PMSM.

We respectively determine the phase differences of the space, time, and carrier of the first and second windings as α, β, and γ. By controlling phase differences α, β, and γ for harmonic orders k, n, and m, the larger impedance disturbs the carrier harmonics flow to both the stator and rotor sides. For larger impedance, the air gap flux density from the first and second windings must be reinforced. This means the carrier harmonic gap flux densities of the first and second windings are in-phase, and the following condition is required:

$$k\alpha - n\beta - m\gamma = 0. \quad (2)$$

For a PWM voltage-source inverter-fed motor, the phase current waveform includes carrier harmonics and carrier sideband harmonics of $n\omega \pm m\omega_c$ as a result of sine-triangle modulation [14]-[16]. The following is the relationship between n and m when m is an odd number,

$$n = \pm(6j + 3 \pm 1), \quad (3)$$

and when m is an even number,

$$n = \pm(6j \pm 1), \quad (4)$$

where $j = 0, 1, 2, \ldots$, and double-sign corresponds in (3) and (4).

The time harmonics of the air gap magnetic flux density, which is a positive sequence for n, is plus, and a negative sequence for n is minus. On the three-phase PMSM, the fundamental component is positive ($n = 1$), the second harmonic is a negative sequence ($n = -2$), and the third harmonic does not exist. Since the carrier harmonics does not include space distribution, the space harmonic coefficient must be 1 ($k = 1$). Then carrier harmonics around the carrier frequency ($m = -1$ or 1)

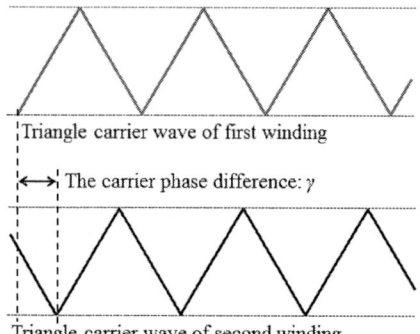

Fig. 1. Phase difference of triangle carrier wave between first and second windings of dual three-phase PMSM.

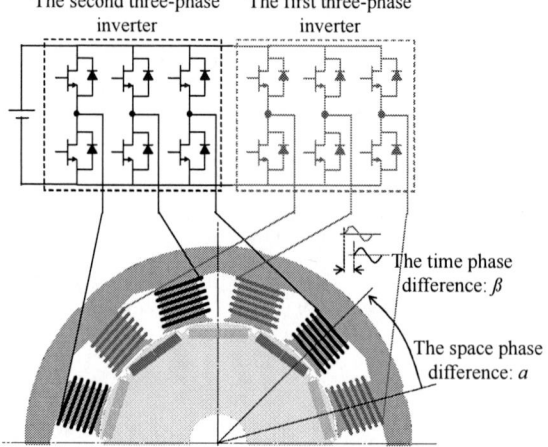

Fig. 2. Construction of dual three-phase PMSM fed by VSI.

exists at $-2\omega \pm \omega_c$, $4\omega \pm \omega_c$. . . . Carrier harmonics around the second carrier frequency ($m = -2$ or 2) exists at $\omega \pm 2\omega_c$, $-5\omega \pm 2\omega_c$. . . . The following relationship is established at the nearest sideband of the carrier harmonics and the second carrier harmonics.

A. *At around the carrier frequency ($k = 1$, $n = -2$, $m = \pm 1$)*

1) $m = -1$

$$\gamma = -(\alpha + 2\beta). \quad (5)$$

2) $m = 1$

$$\gamma = \alpha + 2\beta. \quad (6)$$

B. *At around second carrier frequency ($k = 1$, $n = 1$, $m = \pm 2$)*

1) $m = -2$

$$\gamma = -(\alpha - \beta)/2. \quad (7)$$

2) $m = 2$

$$\gamma = (\alpha - \beta)/2. \quad (8)$$

For a given α and β from (5)-(8), γ is specified to reduce the carrier harmonic iron loss for each harmonic order. This is discussed in Section III related to motor configuration.

For a specified α, to reduce both the carrier sidebands of the carrier harmonic, β and γ follow from (5) and (6) around the carrier harmonic frequency:

$$\beta = -\alpha/2,\ \gamma = 0 \text{ or } \beta = (\pi - \alpha)/2,\ \gamma = \pi. \quad (9)$$

From (7) and (8) at around the second carrier harmonic frequency,

$$\beta = \alpha,\ \gamma = 0 \text{ or } \beta = (\alpha - \pi),\ \gamma = \pi/2. \quad (10)$$

III. COMPARISON CARRIER HARMONICS FOR WINDING STRUCTURES

Dual three-phase winding motors have two different types of winding structures: phase-shifted and in-phase (Fig. 3). Figs. 3(a) and (c) illustrate a 48-slot 8-pole, 2 per pole per slot distributed winding motor: (a) is the first winding and the second phase is in an alternate slot; (c) is a winding in the first and second phases in alternate pole pairs. In (a), the first and second phase windings have a $\pi/6$ rad space phase difference in their electric angles; in (c) the first and second phase windings do not have a space phase difference. Similarly, Figs. 3(b) and (d) show a 24-slot 20-pole concentrated winding motor; (b) is the winding in the first and second phases in alternate teeth, and (d) is the winding in the first and second phases in alternate pole pairs. In (b), the first and second phase windings have a $\pi/6$ rad space phase difference in electric angles; in (d), the first and second phase windings do not have any space phase differences.

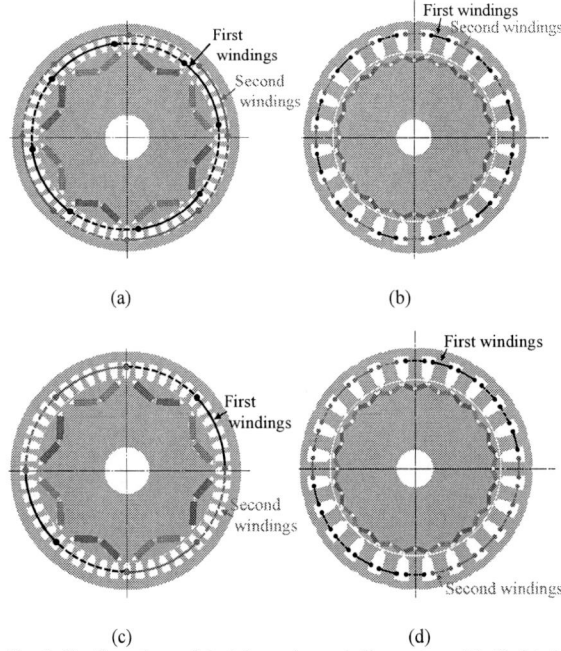

Fig. 3. Configurations of dual three-phase winding motors: (a) 48-slot, 8-pole, 2 per pole per slot, distributed winding motor with phase-shifted windings; (b) 24-slot, 20-pole concentrated winding motor with phase shifted windings; (c) 48-slot, 8-pole, 2 per pole per slot, distributed winding motor with in-phase windings; (d) 24-slot, 20-pole concentrated winding motor with in-phase windings.

In each winding structure, space phase difference α is specified when the winding structure is fixed. Current time phase difference β and carrier phase difference γ can be arbitrarily changed to reduce the carrier harmonic currents. A small phase difference between α and β is desired for higher current versus torque characteristics.

TABLE I

CARRIER PHASE DIFFERENCE γ RELATED TO SPACE PHASE DIFFERENCE α AND TIME PHASE DIFFERENCE β TO REINFORCE OR CANCEL CARRIER HARMONIC GAP FLUX DENSITY OF FIRST AND SECOND WINDINGS AROUND FIRST AND SECOND CARRIER HARMONICS

		Phase-shift windings		In-phase windings	
		Reinforce conditions	Cancel conditions	Reinforce conditions	Cancel conditions
Space phase difference α		$\dfrac{\pi}{6}$		0	
Time phase difference β		$\dfrac{\pi}{6}$		0	
Carrier phase difference γ	$-2f+f_c$	$\dfrac{\pi}{2}$	$\dfrac{3\pi}{2}$	0	π
	$-2f-f_c$	$\dfrac{3\pi}{2}$	$\dfrac{\pi}{2}$	0	π
	$f-2f_c$	$0, \pi$	$\dfrac{\pi}{2}, \dfrac{3\pi}{2}$	$0, \pi$	$\dfrac{\pi}{2}, \dfrac{3\pi}{2}$
	$f+2f_c$	$0, \pi$	$\dfrac{\pi}{2}, \dfrac{3\pi}{2}$	$0, \pi$	$\dfrac{\pi}{2}, \dfrac{3\pi}{2}$

TABLE II

TIME PHASE DIFFERENCE β AND CARRIER PHASE DIFFERENCE γ TO SIMULTANEOUSLY REDUCE BOTH SIDEBAND OF FIRST OR SECOND CARRIER HARMONIC IRON LOSS

		Phase-shift windings				In-phase windings		
Space phase difference α		$\dfrac{\pi}{6}$				0		
Time phase difference β		$\dfrac{23\pi}{12}$	$\dfrac{5\pi}{12}$	$\dfrac{\pi}{6}$	$\dfrac{7\pi}{6}$	0	$\dfrac{\pi}{2}$	π
Carrier phase difference γ	$-2f\pm f_c$	0	π	-		0	π	-
	$f\pm 2f_c$	-		0	$\dfrac{\pi}{2}$	0	-	$\dfrac{\pi}{2}$

TABLE III
MOTOR SPECIFICATIONS

Maximum output	70 kW
Maximum rotational speed	10,000 min⁻¹
Maximum torque	150 Nm
Poles	10
Number of magnet per pole	2
Number of slots	12
Stator outer diameter	200 mm
Stator inner diameter	122 mm
Rotor outer diameter	120 mm
Air-gap length	1.0 mm
Stack length	100 mm
Magnet width	15 mm
Magnet thickness	5 mm
Stator and rotor core materials	30JNE
Material of permanent magnet	N37UZ
Remanent of magnet	1.2 T
Number of turns per coil	20

the iron loss for the first and second carrier harmonics is the same in $\gamma = 0$. This is a big advantage for in-phase windings.

Table II shows time phase difference β and carrier phase difference γ for simultaneously reducing both the sideband of the first or second carrier harmonic iron losses from (9), (10). No condition reduces both the first and second carrier harmonic iron losses for the phase-shift windings, and it is possible to switch the control to reduce either of the carrier harmonics because β and γ can be controlled by an inverter.

Finite element analysis is carried out for the 12-slot, 10-pole, dual three-phase winding PMSM with concentrated phase-shift windings to verify the proposed theory in Table III at a 400 V DC voltage, a carrier frequency of 5 kHz, a motor rotation speed of 3000 min⁻¹. Fig. 4 shows the analyzed result for carrier harmonic currents of $-2f\pm f_c$ and $f\pm 2f_c$ versus γ at $\alpha = \pi/6$, $\beta = \pi/6$. For each carrier harmonic component, reasonable agreement with the proposed theory (Table I) is shown. Fig. 5 shows the analyzed result for carrier harmonic currents of $-2f\pm f_c$ and $f\pm 2f_c$ versus β at $\alpha = \pi/6$, $\gamma = 0$. For each carrier harmonic component, reasonable agreement with the proposed theory (Table II) was shown. The proposed theoretical approach's validity is confirmed from these results.

IV. VERIFICATION OF CARRIER HARMONIC CURRENT REDUCTION EFFECT

A 12-slot, 10-pole dual three-phase winding PMSM with concentrated phase-shift windings was made (Fig. 6). The stator and rotor structure is identical as the FEA model in Figs. 6(a) and (b). All phase windings are connected to a parallel Y. The first and second windings are separated (Fig. 2), and six leading lines are wired in a radial fashion to keep the same line impedance (Fig. 6(c)).

A measurement system for the tested motor is shown in Fig. 7, and the model numbers of the applied instruments are shown in Table IV. The first and second inverters are operated by one controller to synchronize the carrier phase.

Table I shows carrier phase difference γ that is related to space phase difference α and time phase difference β for the carrier harmonic gap flux density of the first and the second windings that reinforce each other around the first and second carrier harmonic from (5)-(8). $\alpha = \pi/6$ represents the phase-shift windings, and $\alpha = 0$ represents the in-phase windings. In Table I, carrier phase differences γ that cancels the gap flux density are also shown. In the phase-shift windings, the carrier phase difference that reduces the iron loss of the first carrier harmonics is different and is switched to cancel the conditions. In these conditions, if the iron loss of one sideband is reduced, the iron loss of the other sideband is increased. The carrier phase difference to reduce the iron loss for the second carrier harmonics is the same, and both sideband iron losses are reduced simultaneously. For in-phase windings, the carrier phase difference to reduce

The 2018 International Power Electronics Conference

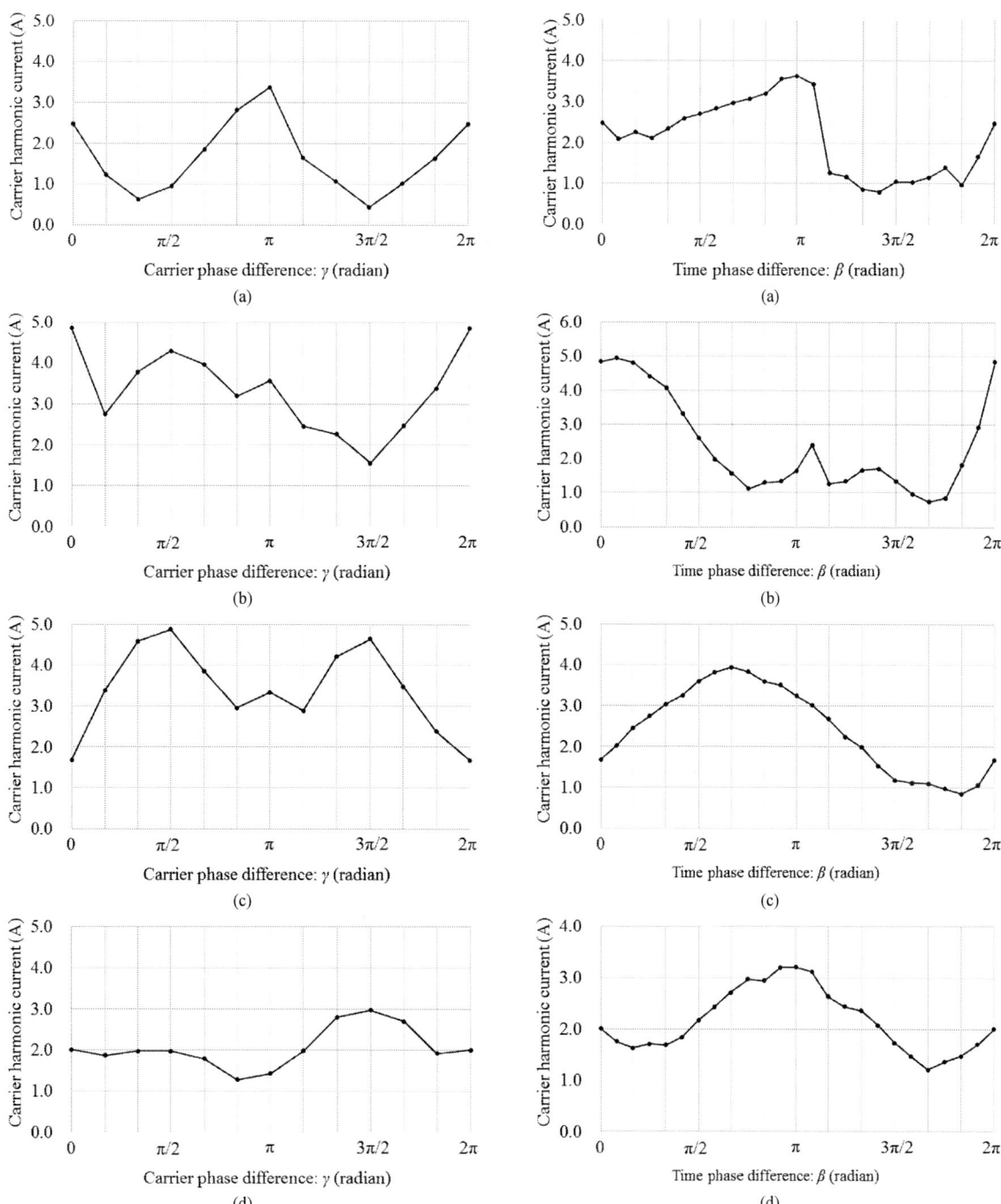

Fig. 4. Analyzed result for carrier harmonic currents versus γ at $\alpha = \pi/6$, $\beta = \pi/6$. (a) - $2f + f_c$. (a) - $2f - f_c$. (c) $f - 2f_c$. (d) $f + 2f_c$.

Fig. 5. Analyzed result for carrier harmonic currents versus β at $\alpha = \pi/6$, $\gamma = 0$. (a) - $2f + f_c$. (b) - $2f - f_c$. (c) $f - 2f_c$. (d) $f + 2f_c$.

Figure 8 shows the measured results of the first carrier harmonic sideband current at 1,000 min^{-1}, 15 Nm. The DC voltage is 400 V, the carrier frequency is 5 kHz, space phase difference α is $\pi/6$, time phase difference β is $\pi/6$, and carrier phase difference γ is 0, $3\pi/2$. In this motor, as shown in Table I, when $\alpha = \pi/6$, $\beta = \pi/6$, and $\gamma = 3\pi/2$, the gap flux density of the first carrier harmonic in the lower sideband component, -$2f - f_c$, generated by

dual three-phase windings, is reinforced, and the input impedance becomes higher, reducing the current. In Fig. 8, when $\gamma = 0$, the first carrier harmonic sideband (-$2f + f_c$, -$2f - f_c$) current has almost the same amplitude. When $\gamma = 3\pi/2$, the -$2f + f_c$ current is slightly increased, and the -$2f - f_c$ current is halved compared to the current when $\gamma = 0$. From this result, increasing the input impedance by reinforcing the dual, three-phase carrier harmonics

715

The 2018 International Power Electronics Conference

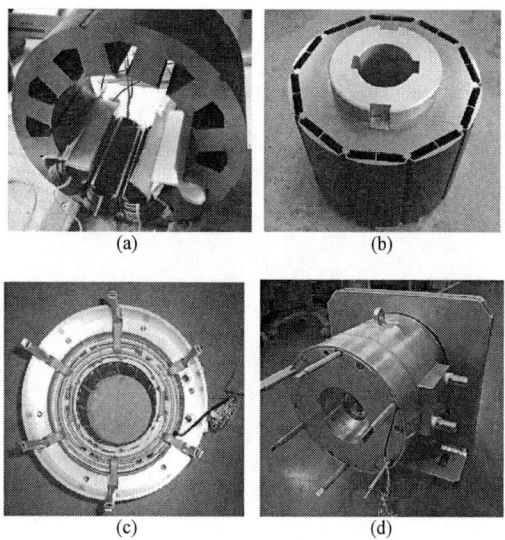

(a)　　　　　(b)

(c)　　　　　(d)

Fig. 6. Manufactured 12-slot 10-pole dual three-phase winding PMSM with concentrated phase-shift windings: (a) stator structure, (b) rotor structure, (c) winding connection structure, (d) and motor assy.

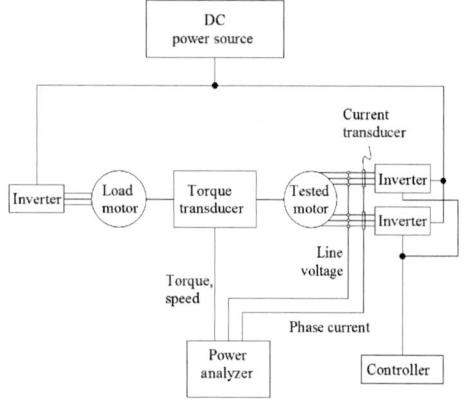

Fig. 7. Measurement system of tested PMSM

magnetic flux density effectively reduces the carrier harmonics current, verifying the proposed method's availability.

In this case, only the carrier harmonics in the lower sideband component current are reduced, and the second carrier harmonics currents are increased [13]. No carrier harmonics iron loss was reduced. To reduce the carrier harmonic currents at both the sidebands of the carrier frequency, the phase differences (α, β, γ) must satisfy (4) = (5):

$$\beta = -\alpha/2. \tag{11}$$

For the manufactured motor, when $\alpha = \pi/6$, $\beta = -\pi/12$, and $\gamma = 0$, both carrier harmonic sideband currents are reduced.

TABLE IV
INSTRUMENTS OF TESTED MOTOR MEASUREMENT SYSTEM

Instrument	Manufacturer	Model number
DC power source	Matsusada Precision	KRS650
Inverters	Myway plus	MWINV-5022B
Controller	Myway plus	PE-Expert 3
Torque transducer	Magtrol	TMHS312/11
Power analyzer	Yokogawa	WT3000
Current transducer	LEM	LEM ITN600-S Ultrastab

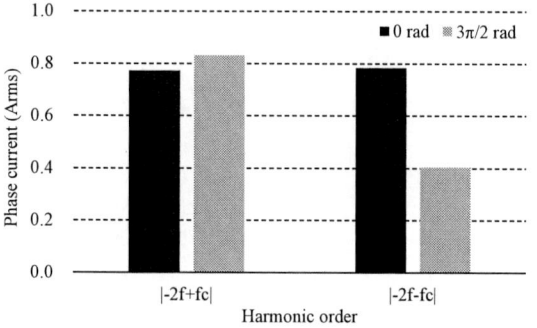

Fig. 8. Measured results of first carrier harmonic current of sideband frequency at motor is driven at 1,000 min^{-1}, 15 Nm. DC voltage is 400 V, carrier frequency is 5 kHz, and carrier phase difference is 0, $3\pi/2$.

V. CONCLUSIONS

We proposed a carrier harmonic current reduction technique for dual three-phase windings PMSM fed by VSI with a carrier phase-shift. Our study was based on the formulation of gap flux density, including space, time, and carrier harmonics. FEA results and our measured result confirmed that the target harmonic current was reduced by applying the proposed technique. Future work will continue to seek an effective loss reduction approach.

REFERENCES

[1] K. Yamazaki and N. Fukushima, "Experimental validation of iron loss model for rotating machines based on direct eddy current analysis of electrical steel sheets," in *Proc. IEEE Int. Electr. Mach. Drives Conf.*, 2009, pp. 851-857.

[2] A. Daikoku, M. Nakano, S. Yamaguchi, Y. Tani, Y. Toide, H. Arita, T. Yoshioka, and C. Fujino, "An accurate magnetic field analysis for estimating motor characteristics taking account of stress distribution in the magnetic core," *IEEE Trans. Ind. Appl.*, vol. 42, no. 3, pp. 668-674, May/Jun. 2006.

[3] D. Miyagi, N. Maeda, Y. Ozeki, K. Miki, and N. Takahashi, "Estimation of iron loss in motor core with shrink fitting using FEM analysis," *IEEE Trans. Magn.*, vol. 45, no. 3, pp. 1704-1707, Mar. 2009.

[4] N. Takahashi and D. Miyagi, "Effect of stress on iron loss of motor core," in *Proc. IEEE Int. Electr. Mach. Drives Conf.* 2011, pp. 469-474.

[5] L. Vandenbossche, S. Jacobs, F. Henrotte, and K. Hameyer, "Impact of cut edges on magnetization curves and iron losses in e-machines for automotive traction," in *Proc. 25th EVS*, 2010.

[6] W. M. Arshad, T. Ryckebusch, F. Magnussen, H. Lendenmann, B. Eriksson, J. Soulard, and B. Malmros, "Incorporating Lamination Processing and Component Manufacturing in Electrical Machine Design Tools," in *Proc. IEEE Ind. Appl. Annu. Meet.*, 2007, pp. 94-102.

The 2018 International Power Electronics Conference

[7] M. Amar and R. Kaczmarek, "A general formula for prediction of iron losses under nonsin," *IEEE Trans. Magn*, vol. 31, no. 5, pp. 2504-2509, Sep. 1995.

[8] E. Barbisio, F. Fiorillo, and C. Ragusa, "Predicting loss in magnetic steels under arbitrary induction waveform and with minor hysteresis loops," *IEEE Trans. Magn.*, vol. 40, no. 4, pp. 1810-1819, Jul. 2004.

[9] A. Boglietti, A. Cavagnino, M. Lazzari, and M. Pastorelli, "Predicting iron losses in soft magnetic materials with arbitrary voltage supply: An engineering approach," *IEEE Trans. Magn.*, vol. 39, no. 2, pp. 981-989, Mar. 2003.

[10] K. Yamazaki, Y. Seto, and M. Tanida, "Iron Loss Analysis of IPM Motor Considering Carrier Harmonics," *IEEJ Trans. Ind. Appl.*, vol. 125, no. 7, pp. 758-766, 2005.

[11] K. Akatsu, K. Narita, Y. Sakashita, and T. Yamada, "Impact of flux weakening current to the iron loss in an IPMSM including PWM carrier effect," in *Proc. IEEE Energy Convers. Congr. Expo.*, 2009, pp. 1927-1932.

[12] Y. Miyama, M. Hazeyama, S. Hanioka, N. Watanabe, A. Daikoku, and M. Inoue," PWM carrier harmonic iron loss reduction technique of permanent-magnet motors for electric vehicles," *IEEE Trans. Ind. Appl.* vol. 52, No. 4, Jul./Aug. 2016.

[13] Y. Miyama, M. Ishizuka, H. Kometani, and Kan Akatsu, "Vibration reduction by applying carrier phase-shift PWM on dual three-phase windings permanent-magnet synchronous motor," in *Proc. IEEE Int. Elect. Mach. Drives Conf. (IEMDC'17)*, 2017.

[14] S. R. Bowes and B. M. Bird, "Novel approach to the analysis and synthesis of modulation processes in power convertors," *IEE Proceedings*, Vol. 122, No. 5, pp. 507-513, May, 1975.

[15] D. G. Holmes, "A general analytical method for determining the theoretical harmonic components of carrier based PWM strategies," in *Conf. Rec. IEEE-IAS Annu. Meeting*, 1998, pp. 1207-1214.

[16] D. G. Holmes, "Opportunities for harmonic cancellation with carrier-based PWM for two-level and multilevel cascaded inverters," *IEEE Trans. Ind. Appl.* vol. 37, No. 2, Mar./Apr. 2001.

Flux Intensifying PM-Motor with Variable Leakage Magnetic Flux Technique

Masahiro Aoyama[1*] and Toshihiko Noguchi[2]

1 Electric Component Development Department, SUZUKI Motor Corporation, Hamamatsu, JAPAN

2 Graduate School of Science and Technology, Shizuoka University, Hamamatsu, JAPAN

*E-mail: aoyamam@hhq.suzuki.co.jp

Abstract— **This paper describes a flux intensifying PM-motor with variable magnetic flux. The unique feature of this proposed motor is the ability to passively adjust the magnetic flux linkage into the armature windings in proportion to the armature magnetomotive force and/or armature current phase. The magnetic circuit topology of the flux intensifying PM-motor and the passive variable leakage magnetic flux are determined through FE-analysis. Then, the driving performance is experimentally elucidated through comparison with that of a reverse pole type (flux weakening) PM-motor without variable leakage magnetic flux.**

Keywords— *Permanent magnet type synchronous motor, variable leakage magnetic flux, flux intensifying.*

I. INTRODUCTION

Recently, strengthening of global environmental regulations, are accelerating the EVs (Electric Vehicles) development for the zero-emission society in the transportation equipment sector. In general, as traction motor for EVs, an IM (Induction Machine) and a PMSM (Permanent Magnet Synchronous Motor) is employed mainly [1-4]. These motors are selected as the suitable motor by a target and the common use drive domain of the vehicle. In the case of PMSM, high torque density can be realized, and high efficiency drive can be realized in the range of low-load to maximum load while low-speed and medium-speed range. On the other hand, in the extremely low-speed range, the efficiency is lower than IM due to the iron loss caused by the magnetic flux of the PMs (Permanent Magnets). At higher loads in the high speed range, iron loss increases due to large distortion of the gap magnetic flux which is caused by flux weakening control, resulting in lower efficiency than IM.

In view point of the above problems, studies on variable magnetic flux motors have been actively studied in recent years for the purpose of expanding the PMSM's high efficiency area and output. As representative studies of today, 1) the memory motor type that makes the PM flux of the PMSM variable [5-8], 2) the type to adjust the rotor skew angle [9], 3) the type to adjust the iron-pole magnetization amount of the consequent pole structure [10-13], and 4) the type that utilizes the leakage magnetic flux [14-18]. Particularly in the case of 4), the variable magnetic flux function is realized only by devising the

magnetic circuit design of the rotor while applying the conventional vector control as it is without requiring the variable-magnetized magnets and extra additional devices. It has a simple structure that passively controls the short magnetic flux path in the rotor by the armature magnetomotive force and the armature magnetic flux vector which is greatly different from the conventional variable magnetic flux technique. This technique is superior in terms of cost, control, and robustness. However, there is a problem in Ref. [14] and [15], that is difficult to utilize the reluctance torque because the variable magnetic flux range is narrow and the leakage magnetic flux amount tends to increase as the armature current vector advances. On the other hand, in Ref. [16] and [17], since the PMs are arranged in the salient pole portions, the salient pole ratio is low, which is a factor of lowering the torque density.

In view point of the above problems of the prior studies, the authors have been studied the possibility of a passive variable magnetic flux range and improve the reluctance torque by combining field adjustment by variable leakage magnetic flux and flux intensifying effect by the $+d$-axis armature magnetic flux [19],[20]. The following three points differ from the prior studies,

A) Study of possibility of utilization of reluctance torque by designing the forward saliency circuit which attempted to improve the forward saliency ratio by separating the iron core magnetic path and the PM-magnetic path.

B) A magnetic circuit design in which the leakage magnetic flux path of PM-flux is formed in an iron core magnetic path provided between PM-magnetic paths by applying the concept of a consequent pole structure. The leakage magnetic flux is induced by the magnetic shielding effect on the magnetic flux on the magnet by the q-axis armature magnetic flux.

C) Place the salient pole on the d-axis and alternately arrange the PM-magnetic path and the iron core magnetic path between the d-axis and the q-axis. Under the flux intensifying control, i.e., $+d$-axis armature magnetic flux is so arranged as to face the same direction, the d-axis armature magnetic flux of the iron core magnetic path becomes the forward direction with respect to the PM-magnetic flux and performs flux intensifying. The

The 2018 International Power Electronics Conference

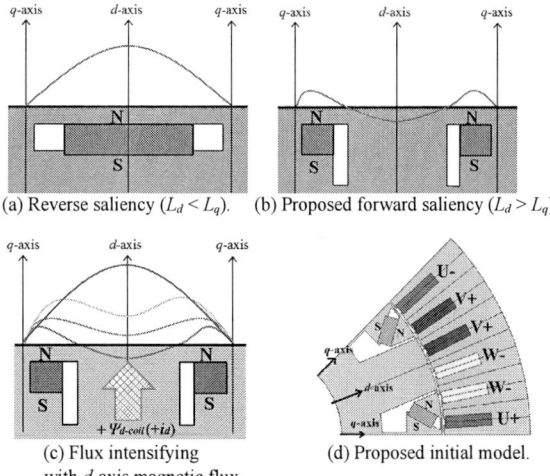

(a) Reverse saliency ($L_d < L_q$).   (b) Proposed forward saliency ($L_d > L_q$).

(c) Flux intensifying with d-axis magnetic flux.   (d) Proposed initial model.

Fig. 1. Concept of proposed flux intensifying PM-motor.

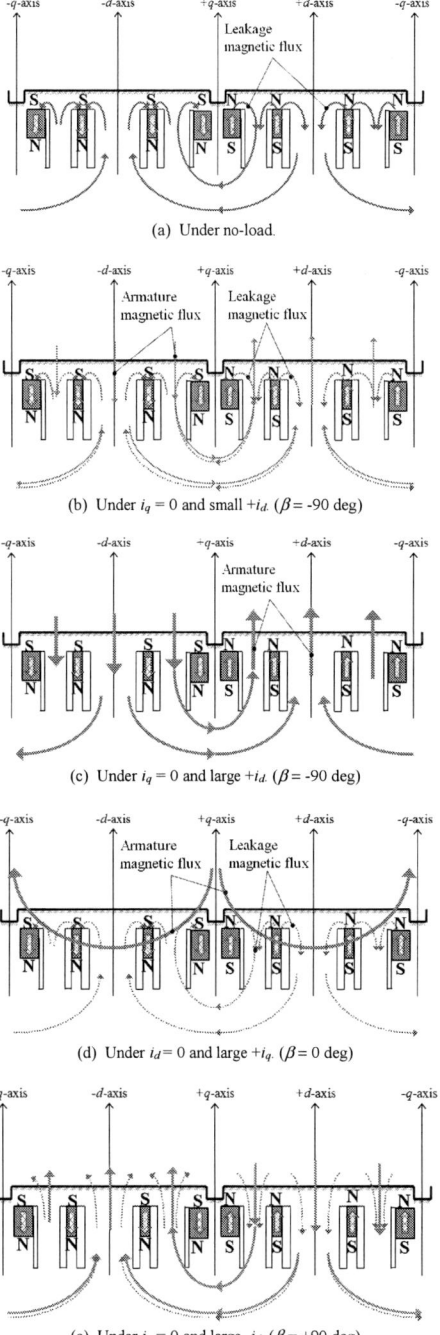

(a) Under no-load.

(b) Under $i_q = 0$ and small $+i_d$. ($\beta = -90$ deg)

(c) Under $i_q = 0$ and large $+i_d$. ($\beta = -90$ deg)

(d) Under $i_d = 0$ and large $+i_q$. ($\beta = 0$ deg)

(e) Under $i_q = 0$ and large $-i_d$. ($\beta = +90$ deg)

Fig. 2. Leakage PM-flux and armature magnetic flux with respect to armature magnetomotive force and armature current vector.

amount of flux intensifying becomes variable according to the d-axis armature magnetic flux amount.

In this paper, the concept of magnetic circuit design for achieving the variable magnetic flux principle having the above characteristics is described. Then, the prototype for principle verification and comparing the drive performance with that of the reverse salient pole type PM-motor of the same core size without variable leakage magnetic flux function is experimentally demonstrated. In addition, verification of the possibility of the proposed variable magnetic flux PM-motor, drive characteristics, and problems will be elucidated.

II. PRINCIPLE OF PROPOSED VARIABLE MAGNETIC FLUX

Generally, as shown in Fig. 1 (a), PM magnetomotive force is designed so that magnetic flux is distributed sinusoidally, but the proposed motor is intentionally designed such that a iron core is provided between the magnets and the waveforms with superimposing the positive third harmonic as shown in Fig. 1 (b). On the other hand, under the load condition, the third harmonic wave superimposed on the PM-magnetic flux is canceled by the $+d$-axis armature magnetic flux, and the combined magnetic flux of the PM-flux and the armature magnetic flux is distributed in the sinusoidal wave as shown in Fig. 1 (c). Applying this concept, it is possible to reduce the no-load core loss under no-load, reduce the iron loss due to the PM-flux at the time of extremely low load, and adjust the amount of flux intensifying by adjusting the $+d$-axis armature magnetic flux amount as the load increases. In the prior research of Refs. [14-17], flux bypass magnetic path are provided on both sides of magnets arranged on the d-axis to passively change armature flux linkage amount by armature magnetomotive force and current vector. Although the leakage magnetic flux of the proposed motor also the same with conventional technique, however, applying the concept of the consequent pole structure as shown in Fig. 2, the iron core magnetic flux path between PM-flux path, the leakage magnetic flux formed by the iron core

magnetic flux path is variable with the $+d$-axis armature magnetic flux amount and the armature current vector. As a result, the change of the armature interlinkage flux number is passively increased by the armature magnetomotive force and the current vector. On the other hand, the iron core magnetic flux path becomes a magnetic path of the armature magnetic flux and contributes to the generation of the reluctance torque.

719

Table I. Specifications of prototype.

	Benchmark	Proposed
Number of rotor poles	8	
Number of stator slots	48	
Stator core outer diameter	200 mm	
Air-gap length	0.7 mm	
Axial length of core	108 mm	
Maximum armature magnetomotiveforce	1060 A_{rms}T (28.1 A_{rms}/mm^2), 60 s	
Number of armature coil-turn	12	
Armature winding connection	2 series-4 parallel	
Armature winding resistance	12.76 mΩ/phase	
Core material	35H-EA (Nippon steel & Sumitomo metal)	
Magnet material	N39UH (Shinetsu chemical)	N52AS-GF (Shinetsu chemical)

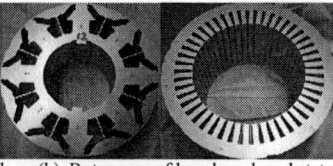

(a) Rotor core of proposed   (b) Rotor core of benchmark and stator core.

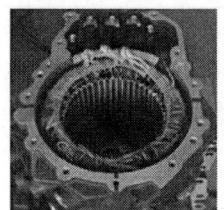
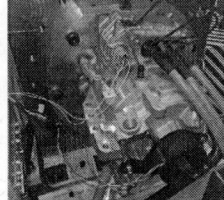

(c) Stator assembly (Left-side) and experimental setup (Right-side).

Fig. 6. Actual prototype.

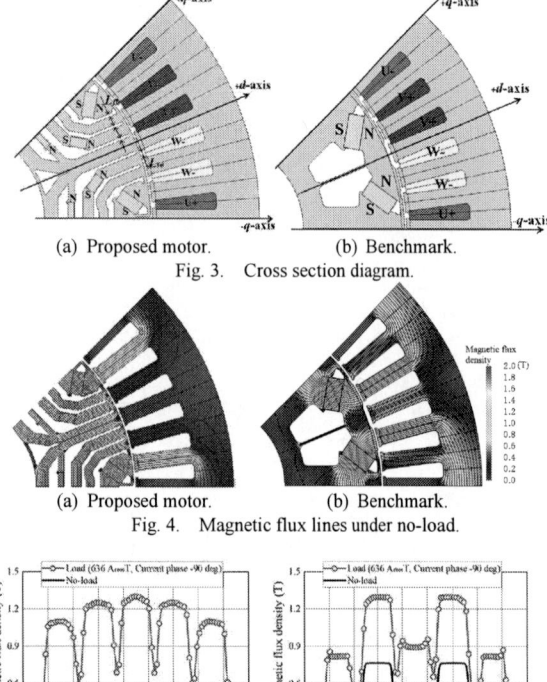

(a) Proposed motor.   (b) Benchmark.

Fig. 3. Cross section diagram.

(a) Proposed motor.   (b) Benchmark.

Fig. 4. Magnetic flux lines under no-load.

(a) Proposed motor.   (b) Benchmark.

Fig. 5. Air-gap magnetic flux waveforms.

III. PROTOTYPE MACHINE

Magnetic circuit and structural design of the proposed principle verification machine were performed with the specifications in Table I. The advantages of proposed motor will be revealed with comparing that of benchmark model. The benchmark and the proposed motor are designed with a common stator, the same air-gap length, the same magnet volume, the same core material and the same core stack length. For the magnet, a magnet suitable for the necessary coercive force is selected from the

difference of the magnetic circuits of both motors, respectively. Figure 3 shows the cross section diagram of magnetic circuit design of proposed motor and benchmark. Figure 4 shows the magnetic flux density and flux lines under no-load and Fig. 5 shows the magnetic flux density waveform in air-gap (intermediate position between stator and rotor) under no-load (black solid line) and load with +d-axis armature current (red dotted line). As can be seen in these figures, the proposed motor leaks PM-flux into the iron core magnetic flux path on the d-axis (rotor position 22.5 deg is +d-axis) and a short-circuited magnetic path is formed inside the rotor. On the other hand, in the benchmark, the magnetic resistance of the both end bridges on the outside of the magnetic pole and the d-axis center bridge is higher than the magnetic resistance of the air-gap, so the larger number of PM-fluxes are linked to the stator than the proposed motor. Next, when comparing the air-gap magnetic flux density waveform at the flux intensifying state (current phase -90 deg) due to the +d-axis armature magnetic flux in Fig. 5, it can be confirmed that the slot harmonic is superimposed on the proposed motor, but it is possible to obtain a sinusoidal gap magnetic flux waveform. In the current phase, the +q-axis is the phase reference and the CCW direction is positive. Here, in order to facilitate the formation of a short-circuit magnetic flux in the rotor according to the condition of the armature magnetomotive force and the current phase, the width of the flux barrier (L_{fb} in the Fig. 3 (a)) separating the iron core magnetic flux path and the PM-flux path from the slot open width (L_{so} in the Fig. 3 (a)) of the stator is designed narrow. That is, the magnetic circuit is designed so that the magnetic resistance of the leakage flux magnetic path becomes small when the armature magnetic flux is perpendicular to the PM-flux.

Figure 6 shows the prototype of the benchmark and the proposed motor. Both motors has the distributed winding stator structures with 8 poles-48 slots (number of phase slots per pole q = 2), and the armature winding adopts AW round wire with H-type heat resistant class. Both rotors are not skewing. Peripheral parts such as shaft and end-plate of both rotors are prototyped with common parts. The stator is attached to a water-cooled motor case by shrink fitting. Performance evaluation of both motors is performed by replacing the rotor of Fig. 6 (a) and (b) and assembling.

IV. EXPERIMENTAL TEST

A. Current Phase-vs.-Torque Characteristics

A universal inverter manufactured by Myway Plus Inc. was used to motor drive and set the carrier frequency to 8 kHz. The torque measurement was carried out with speed control on the motor bench side and torque control with the inverter for driving the test motor. It is controlled by sinusoidal wave PWM driving, the DC-bus voltage is set at 245-V_{dc}, and the upper limit of the voltage utilization rate with respect to DC-bus voltage is set within the range of 95-96 %. Figure 7 shows the no-load induced voltage waveform at 1000 r/min and its harmonic analysis results. According to the figure, magnets with the same volume per pole and high residual magnetic flux density B_r are adopted from Table I, but as shown in Fig. 2, the proposed motor is short-circuited with the PM flux in the rotor at no-load, the armature interlinkage magnetic flux number decreases, and as a result, the induced voltage becomes extremely lower than the benchmark.

Figure 8 shows the current phase-vs.-torque characteristics of the benchmark and the proposed motor. The +q-axis is defined as the current phase reference and the -d-axis direction is defined as the advanced angle direction. In measurement of the reluctance torque in the figure, measurement is made using unmagnetized magnets. For the convenience of the measurement bench, measurement was made with a load of 70 % of design value as the upper limit. Comparing the reluctance torques of both motors, under the armature magnetomotive force 750 $A_{rms}T$, the proposed motor stays as only 66.3 % of the benchmark. In the total torque, the proposed motor is only about 59.2 % of the benchmark. It can be cited whether or not the armature reaction torque can be used as a main cause that the forward salient pole PM-motor is much lower in torque density than the reverse salient pole PM-motor. As shown in Fig. 9, in the case of the forward salient pole type PM-motor, it is necessary to drive the flux intensifying with saliency of $L_d > L_q$. It is necessary to design the magnetic circuit so that the dq-axis magnetic path does not interfere around the MTPA point. Meanwhile, since the reverse salient pole type PM-motor is driven by flux weakening, a permanent magnet having sufficient coercive force is arranged on the d-axis to have reverse saliency of $L_d < L_q$, the armature reaction torque can be utilized. As a result, the torque generation surface of proposed motor is enlarged more than the forward salient pole type PM-motor, realizing high torque density. In the case of flux intensifying PM-motor, it is necessary to design the magnetic circuit so that the dq-axis magnetic path does not interfere. Therefore, if the same volume of magnet is embedded in the rotor of the same core size, the core magnetic path inevitably narrows in width and becomes easy to be magnetically saturated, so that the

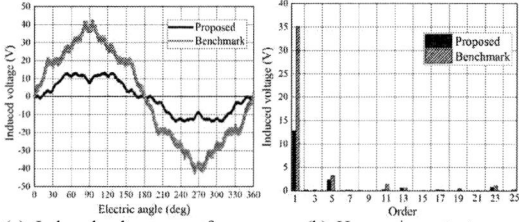

(a) Induced voltage waveforms.   (b) Harmonic contents.
Fig. 7. Induced voltage waveforms and its harmonic contents.

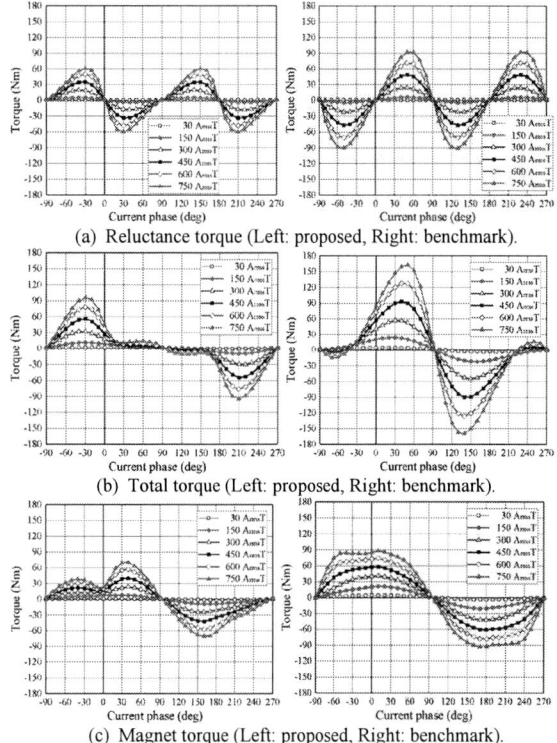

(a) Reluctance torque (Left: proposed, Right: benchmark).

(b) Total torque (Left: proposed, Right: benchmark).

(c) Magnet torque (Left: proposed, Right: benchmark).
Fig. 8. Current phase-vs.-torque characteristics.

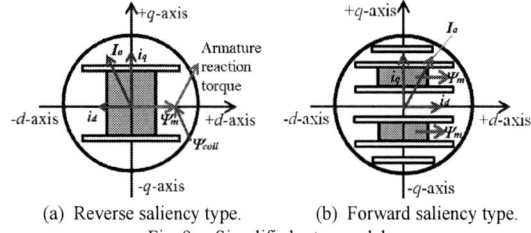

(a) Reverse saliency type.   (b) Forward saliency type.
Fig. 9. Simplified rotor model.

reluctance torque of the forward salient pole PM-motor is inferior to that of the reverse salient pole PM-motor.

Next, in order to confirm the passive variable magnetic flux effect of the proposed motor, the magnet torque was approximated by the difference between the total torque and the reluctance torque shown in Fig. 8 (a) and (b). Figure 8 (c) shows the current phase-vs.-magnet torque characteristics. From this figure, the magnet torque of the benchmark is the theoretical $\cos\beta$-function, but it can be confirmed that the magnet torque of the proposed motor drops greatly when the current phase is

The 2018 International Power Electronics Conference

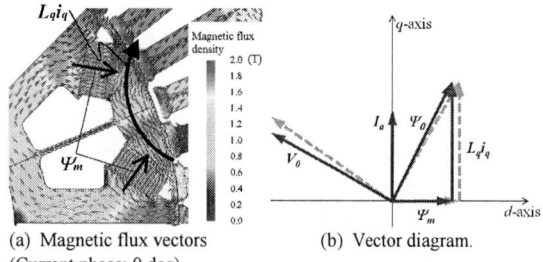

(a) Magnetic flux vectors
(Current phase: 0 deg).

(b) Vector diagram.

Fig. 10. Magnetic flux vectors and vector diagram under high-load and current phase 0 deg. (Benchmark).

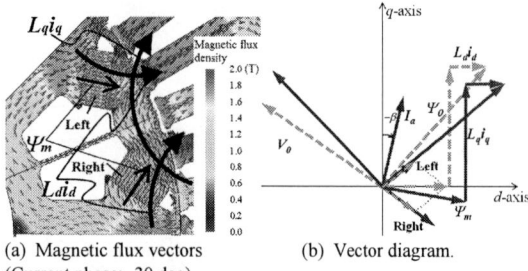

(a) Magnetic flux vectors
(Current phase: -30 deg).

(b) Vector diagram.

Fig. 11. Magnetic flux vectors and vector diagram under high-load and current phase -30 deg. (Benchmark).

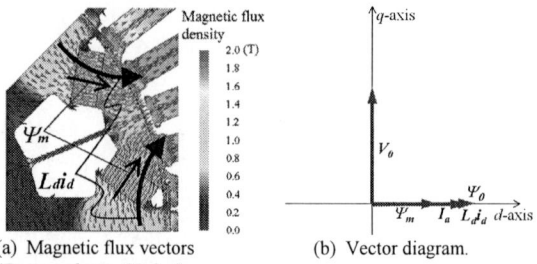

(a) Magnetic flux vectors
(Current phase: -90 deg).

(b) Vector diagram.

Fig. 12. Magnetic flux vectors and vector diagram under high-load and current phase -90 deg. (Benchmark).

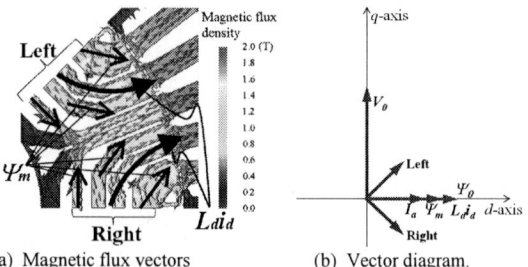

(a) Magnetic flux vectors
(Current phase: -90 deg).

(b) Vector diagram.

Fig. 13. Magnetic flux vectors and vector diagram under high-load and current phase -90 deg. (Proposed).

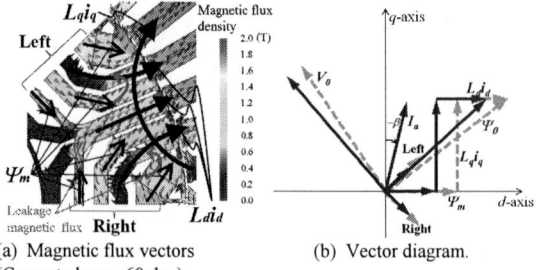

(a) Magnetic flux vectors
(Current phase: -60 deg).

(b) Vector diagram.

Fig. 14. Magnetic flux vectors and vector diagram under high-load and current phase -60 deg. (Proposed).

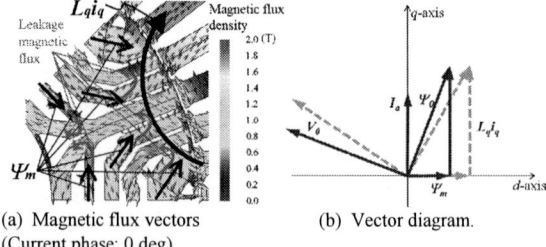

(a) Magnetic flux vectors
(Current phase: 0 deg).

(b) Vector diagram.

Fig. 15. Magnetic flux vectors and vector diagram under high-load and current phase 0 deg. (Proposed).

near 0 deg. As described in the previous chapter, when β = 0 deg, the flux intensifying effect does not effects due to the $+d$-axis armature magnetic flux $\Psi_{+d\text{-}coil}$ becomes zero. As a result, the leakage magnetic flux linking to the iron core magnetic flux path increased and the magnetic flux linking the stator decreased as shown in Fig. 2 (d). Furthermore, the $+q$-axis armature magnetic flux $\Psi_{+q\text{-}coil}$ is orthogonal to the magnetic flux vector. The larger the $+\Psi_{+q\text{-}coil}$, the more the magnetic shielding effect on the PM-flux vector works and the magnetic flux becomes a flow of magnetic flux that forms leakage magnetic flux in the iron core magnetic flux path.

Next, it is considered using vector distribution simulated FE-analysis and vector diagram to push forward an analysis about variable magnetic flux function of the proposed motor. First, consider the magnet torque characteristics of the benchmark in Fig. 8 (c). When the armature magnetomotive force is high, with respect to the theoretical characteristic of the magnet torque of the ideal $\cos\beta$ function, the magnet torque decreases near the current phase β = 0 deg. Especially when the current phase is 0 deg, the magnetic shielding effect occurs due to the influence of the q-axis armature reaction $L_q i_q$ orthogonal to the PM-flux Ψ_m as shown in Fig. 10, the d-axis magnetic resistance increases and the PM-flux Ψ_m becomes the demagnetization action. As a result, it changes from the dotted line vector diagram (without considering reduction of Ψ_m) to the solid line vector diagram (with consideration for decreasing Ψ_m) in Fig. 10 (b). Here, although the $L_q i_q$ magnetic path exists also on the inner diameter side of the rotor, it contributes less to the reduction of Ψ_m, it is not indicated by a black line arrow. On the other hand, when the current phase is other than 0 deg, for example, in the case of -30 deg shown in Fig. 11, the flux intensifying effect differs between the

right side and the left side of the magnet arranged in the V-shape in the magnetic pole. As a result, a phase shift of the PM-flux vector occurs with respect to the d-axis in the case of the current phase of -90 deg in Fig. 12. Here, the phase of Ψ_m is an image diagram. The q-axis armature reaction $L_q i_q$ is orthogonal to Ψ_m and the magnetic resistance of the d-axis increases to act as a demagnetization action. In addition, the d-axis armature reaction $L_d i_d$ in the same direction (flux intensifying direction) as the PM flux vector also orthogonally acts so

722

The 2018 International Power Electronics Conference

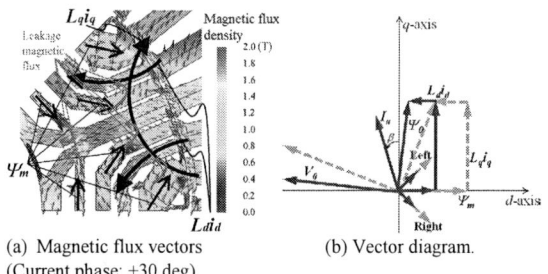

(a) Magnetic flux vectors
(Current phase: +30 deg).

(b) Vector diagram.

Fig. 16. Magnetic flux vectors and vector diagram under high-load and current phase +30 deg. (Proposed).

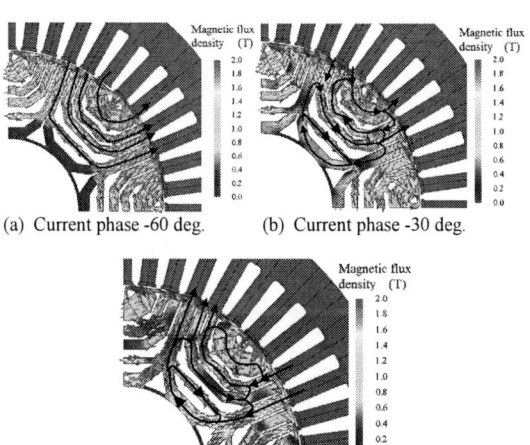

(a) Current phase -60 deg.

(b) Current phase -30 deg.

(c) Current phase +60 deg.

Fig. 17. PM-flux vectors with respect to current phase.

that the flux intensifying of the right magnet becomes strong in the case of CCW rotation. Thus, this is due to interference of the magnetic paths between the *dq*-axes, and since this influence is not taken into account in the mathematical model of the theoretical *dq*-axis, there is a difference between the magnet torque of the theoretical cos*β*–function of the mathematical model and the actual measurement value.

Next, consider the magnet torque of the proposed motor. Figure 13 to 16 show the magnetic flux vector distribution and the *dq*-axis vector diagram when the current phase is changed. The vector diagram of each figure is drawn in the image diagram. Comparing the figures, as the current phase advances from -90 deg (flux intensifying) to +*q*-axis direction, the flux intensifying effect due to the +*d*-axis armature magnetic flux decreases and the leakage magnetic flux increases due to the magnetic shielding effect by the *q*-axis armature magnetic flux, therefore the amount of magnetic flux linking the armature winding decreases. As a result, in the vector diagram of each figure (b), a dotted line vector (without considering the leakage magnetic flux) is changed to a solid line vector (with leakage magnetic flux). On the other hand, Fig. 16 shows a diagram showing only the DC magnetic flux in the rotor by using the harmonic analysis function of the electromagnetic field analysis software (JMAG-Designer ver.15). As shown in Fig. 17 (a) and (b), as the current phase *β* advances from +*d*-axis to +*q*-axis direction, the leakage

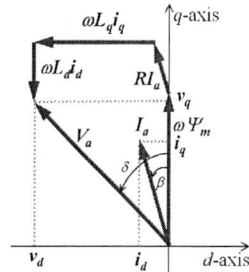

Fig. 18. Vector diagram of PM-motor.

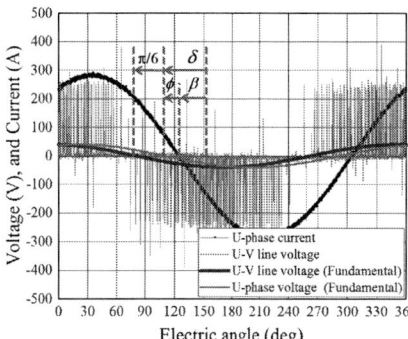

Fig. 19. Measured voltage and current.

magnetic flux increases as indicated by an arrow through an iron core magnetic flux path provided between the PM-flux paths. On the other hand, as shown in Fig. 17 (c), according to the current phase *β* advancing from the +*q*-axis to the -*d*-axis direction (flux weakening region), as explained in Fig. 2 (e), -*d*-axis armature magnetic flux forms a closed magnetic flux path between the iron core magnetic flux path and the stator via an air-gap, the leaking PM-flux will not be short-circuited within the rotor and will be linked to the armature winding, resulting in an increase in armature flux linkage. Considering the actual driving, basically it is driven at the MTPA point, so it is driven in the flux intensifying region of the forward salient pole type proposed motor, and it can be inferred that the influence of this driving characteristic on the performance is small.

B. dq-axis Voltage Ellipse

Next, the drive performance is considered by visualizing the trajectory of the *dq*-axis voltage ellipse. As shown in the vector diagram of Fig. 18, if the value of the voltage phase *δ* is known, v_d and v_q can be obtained from Eq. (1) [21].

$$v_d = -V_1 \sin \delta, \quad v_q = V_1 \cos \delta \qquad (1)$$

Voltage phase *δ* is measured by operating the test motor at a constant speed with arbitrary armature magnetomotive force and current phase, measuring UV-line voltage and U-phase current by an oscilloscope, and deriving the fundamental wave with an FFT analyzer. As an example, Fig. 19 shows the measured UV-line voltage and U-phase current, the UV-line voltage fundamental wave and the U-phase voltage fundamental wave calculated by postprocessing at a rotation speed of 500 r/min, the armature magnetomotive force of 600 A$_{rms}$T,

723

The 2018 International Power Electronics Conference

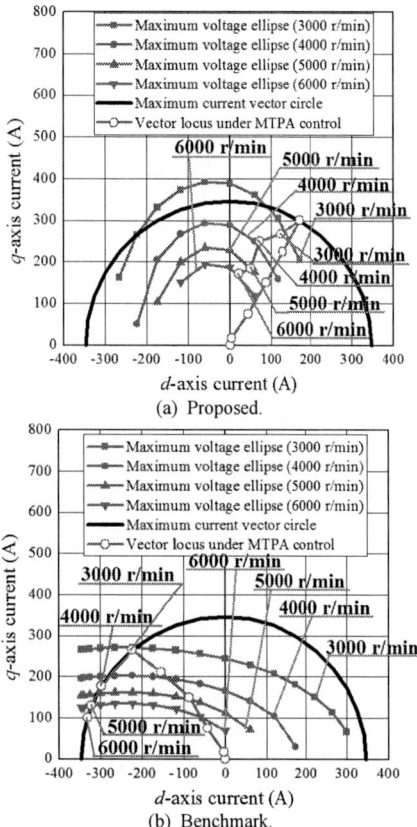

(a) Proposed.

(b) Benchmark.

Fig. 20. Vector locus under MTPA control.

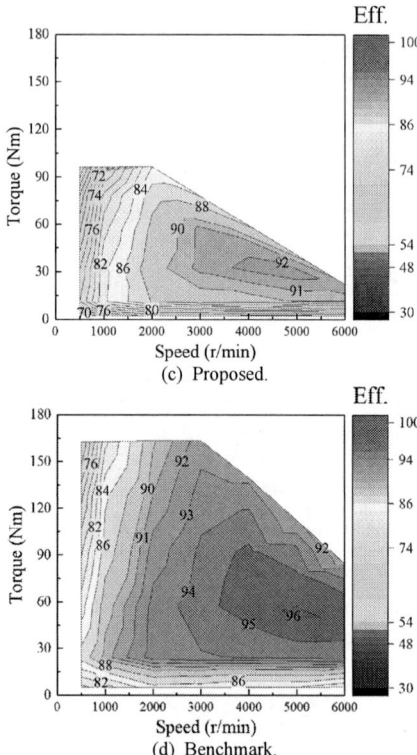

(c) Proposed.

(d) Benchmark.

Fig. 21. Efficiency maps in motoring under 750 A_{rms}T limitation. (Limited under 70 %-load.)

and the current phase of +30 deg. From this figure, the voltage phase δ is obtained.

Figure 20 shows the current-limited circle, the dq-axis voltage ellipse, and the current vector trajectory in MTPA control, where v_d and v_q are obtained from actually measured voltage phase and current phase information [21]. As shown in this figure, the proposed motor has forward saliency, so that $v_q > v_d$, and the i_q-axis becomes the long axis radius of the voltage ellipse. As a result, when it comes to voltage limitation, it can not be driven within the voltage threshold unless it is a current vector trajectory that reduces the armature current norm rather than the current phase advance. On the other hand, since the benchmark is a reverse salient pole, the i_d-axis becomes the long axis radius of the voltage ellipse. As a result, when it comes to voltage limitation, the current norm is not changed and the current phase is advanced to drive within the voltage threshold. From these results, it is clear from the current vector trajectory that the leakage magnetic flux effect due to the current phase advance of the proposed motor can not be utilized to improve the drive characteristics.

C. Efficiency Map

Figure 21 shows the motor efficiency map in motoring. From the figure, the motor efficiency is significantly lower than the benchmark. In particular, the peak efficiency was about 4 % lower. From the measurement result of the torque characteristic, it was possible to

demonstrate the variable flux function, but the difference in performance from the magnet type reverse salient pole motor due to the essential problem of the magnet type forward salient pole motor structure rather than the performance improvement effect by the variable magnetic flux was revealed. In the case of the magnet type forward salient pole structure, as mentioned in the previous section, the low torque density due to the fundamental magnetic circuit by the non-interference design of the dq-axis magnetic path and the current norm adjustment type variable speed drive due to the major axis radius of the voltage ellipse on the i_q-axis appears remarkably in efficiency drops.

V. CONCLUSION

In this paper, applying the feature of a consequent pole structure, the study was presented on a motor featuring enhanced salient pole ratio, variable leakage magnetic flux function, and flux intensifying by d-axis armature magnetic flux by arranging the PM-flux path and the iron core magnetic flux path alternately. As a result of the actual machine evaluation, the variable magnetic flux effect described above could be demonstrated. By visualizing the vector locus, it was also clarified that the variable magnetic flux effect due to the current phase advance was not able to contribute to performance improvement. It is because that the feature of the variable speed drive characteristic by the current-norm adjustment type drive of the forward salient pole motor. As a result of this study, it was clarified that in the case of realizing passive variable magnetic flux due to variable leakage magnetic flux characteristics in the magnet type, it is not possible to fully utilize the variable magnetic flux effect for improving drive characteristics with forward salient pole type.

According to the results of this study, it is preferable to design the variable leakage magnetic flux motor with a reverse salient pole structure with a low salient pole ratio and designed, so that the magnet torque can be utilized to the utmost. And it can be considered that the magnetic circuit design in which the leakage magnetic flux increases with the d-axis armature magnetic flux above the nominal speed is more suitable from the viewpoint of extending the adjustable speed drive characteristic. The future work is to investigate the magnetic circuit of the variable leakage magnetic flux motor with the magnetic circuit design which can be utilized to the utmost magnet torque and increase leakage magnetic flux in proportion to negative d-axis armature magnetic flux increase.

REFERENCES

[1] https://www.tesla.com/jp/blog/induction-versus-dc-brushless-motors

[2] Y. Sato, S. Ishikawa, T. Okubo, M. Abe, and K. Tamai: "Development of High Response Motor and Inverter System for the Nissan LEAF Electric Vehicle", *SAE Technical Paper*, No. 2011-01-0350 (2011).

[3] K. Handa, H. Yoshida: "Development of Next-Generation Electric Vehicle i-MiEV", *Mitsubishi Motors Technical Review*, No. 9, pp. 65-69 (2007) (in Japanese).

[4] F. Momen, K. Rahman, Y. Son, and P. Savagian: "Electrical Propulsion System Design of Chevrolet Bolt Battery Electric Vehicle", *IEEE Energy Conversion Congress and Expo* (ECCE) (2016).

[5] Ostovic, V.: "Memory Motors", *IEEE Industry Applications Magazine*, vol. 9, pp.52-61 (2003).

[6] Ostovic, V. : "Memory Motors – a New Class of Controllable Flux PM Machines for a True Wide Speed Operation", *Proc. of IEEE Industry Applications Society Conference*, 2001, vol. 4, pp.2577-2584 (2001).

[7] K. Sakai, K. Yuki, Y. Hashiba, N. Takahashi, K. Yasui, and L. Kovudhikulrungsri: "Principle and Basic Characteristics of Variable Magnetic-Force Memory Motors", *IEEJ Trans. on IA.*, vol. 131, No. 1 pp.53-60 (2011) (in Japanese).

[8] T. Kato, N. Limsuwan, C. Y. Yu, K. Akatsu, and R. D. Lorenz: "Rare Earth Reduction Using a Novel Variable Magnetomotive Force, Flux Intensified IPM Machine", *IEEE Trans. on IA.*, vol. 50, No. 3, pp.1748-1756 (May/June, 2016).

[9] T. Nonaka, S. Oga, and M. Ohto: "Consideration about the Drive of Variable Magnetic Flux Motor", *IEEJ Trans. on IA.*, vol. 135, No. 5, pp. 451-456 (2015) (in Japanese).

[10] T. Mizuno, K. Nagayama, T. Ashikaga, and T. Kobayashi: "Basic Principles and Characteristics of Hybrid Excitation Type Synchronous Machine", *IEEJ Trans. on IA.*, vol. 115, No. 11, pp.1402-1411 (1995) (in Japanese).

[11] J. A. Tapia, F. Leonardi, and T. A. Lipo: "Consequent-Pole Permanent-Magnet Machine with Extended Field-Weakening Capability", *IEEE Trans. on IA.*, vol. 39, No. 6, pp.1704-1709 (2003).

[12] M. Namba, K. Hiramoto, and H. Nakai: "Novel Variable-Field Motor with a Three-Dimentional Magnetic Circuit", *IEEJ Trans. on IA.*, vol. 135, No. 11, pp.1085-1090 (2015) (in Japanese).

[13] T. Ogawa, T. Takahashi, M. Takemoto, H. Arita, A. Daikoku, and S. Ogasawara: "The Consequent-Pole Type Ferrite Magnet Axial Gap Motor with Field Winding for Traction Motor Used in EV", *SAEJ Proc. of EVTeC & APE Japan 2016*, No. 20169094 (2016).

[14] T. Kato, M. Minowa, H. Hijikata, and K. Akatsu: "High Efficiency IPMSM Effectively Utilizing Variable Leakage Flux Characteristics", *IEEJ JIASC 2014*, No. 3-13, pp. 139-142 (2014) (in Japanese).

[15] T. Kato, and K. Akatsu: "Magnet Operating Point Characteristics of Variable Leakage Flux Interior Permanent Magnet Motor", *IEEJ JIASC 2015*, No. 3-1, pp. 65-70 (2015) (in Japanese).

[16] A. Athavale, T. Fukushige, T. Kato, C.Y. Yu, and R. D. Lorenz: "Variable Leakage Flux (VLF) IPMSMs for Reduced Losses over a Driving Cycle while Maintaining the Feasibility of High Frequency Injection-Based Rotor Position Self-Sensing", *IEEE Energy Conversion Congress and Exposition* (ECCE), (2014).

[17] M. Minowa, H. Hijikata, K. Akatsu, and T. Kato: "Variable Leakage Flux Interior Permanent Magnet Synchronous Machine for Improving Efficiency on Duty Cycle", *International Power Electronics Conference* (IPEC-Hiroshima 2014 –ECCE ASIA).

[18] I. Urquhart, D. Tanaka, R. Owen, Z. Q. Zhu, J. B. Wang, and D. A. Stone: "Mechanically Actuated Variable Flux IPMSM for EV and HEV Applications", *Proc. of EVS27 International Battery, Hybrid and Fuel Cell Vehicle Symposium 2013*, pp. 0684-0695 (2013).

[19] M. Aoyama, K. Nakajima, and T. Noguchi: "Preliminary Study of Flux Intensifying PM Motor with Variable Leakage Magnetic Flux Technique", *IEEJ Annual Meeting*, No. 5-001, pp. 1-2 (2017) (in Japanese).

[20] M. Aoyama, K. Nakajima, and T. Noguchi: "Driving Performance of Flux Intensifying PM Motor with Variable Leakage Magnetic Flux Technique", *IEEJ JIASC*, No. 3-42, (2017) (in Japanese).

[21] M. Morimoto, Y. Takeda, and T. Hirasa: "Parameter Measurement of PM Motor in dq Equivalent Circuit", *IEEJ Trans. on IA.*, vol. 113, No. 11, pp.1330-1331 (1993) (in Japanese).

The 2018 International Power Electronics Conference

Continuous Operation Control of PMSM in the case of DC Power Supply Loss

Jongwon Heo[1*], Keiichiro Kondo[1]
1 Dept. of Electrical and Electronic Eng., Chiba University, Chiba, Japan
*E-mail: jheo@chiba-u.jp

Abstract— **In the PMSM sensorless drive system, when the power supply is lost, the position of rotor cannot be estimated by the usual control method. And the system stops due to the Filter Condenser (FC) voltage or the rotation speed decrease. To solve this problem, we proposed effective control methods at the interruption of power supply. There are three methods depending on inertial loads of the application and duration of the power interruption. In this paper, we explain the structure and characteristics of the three control methods and compare them. And we propose an appropriate continuous operation control depending on application and usage environment.**

Keywords— *Continuous Operation, PMSM, Power Supply Loss, Sensorless*

I. INTRODUCTION

The Permanent Magnet Synchronous Motor (PMSM) is widely used in electric vehicles, industrial applications, and home appliances because it is compact, lightweight, and high efficient compared to the Induction Motor (IM) [1]. In order to control the PMSM easily, the vector control in the rotating coordinate system based on the magnetic pole position of the rotor is generally used, but in order to detect the position of the rotor, a position sensor such as an encoder or resolver is necessary. However, those sensors have some disadvantages such as cost and reliability due to breakdown or disconnection. Therefore, the position-sensorless control is also widely used [2].

One of the problems of the position-sensorless control is the continuous operation during the power interruption. In the inverter system for the PMSM driving shown in Fig. 1, when the input power source is lost due to the power supply side, the Filter Condenser (FC) voltage abruptly drops due to the load under normal control method. And the inverter is forcibly gate-off due to the low voltage protection. As a result, the position estimation becomes impossible, and the system must be stopped.

In applications with large inertial load, even the rotor is rotated by inertia after the inverter is stopped, it is difficult to restart immediately because the position information is unknown. And, the FC voltage may exceed the limit value due to an induced voltage of PMSM.

And, depending on the application, it may be difficult to restart immediately once it stops. For example, in a compressor for air conditioning, if it is attempted to restart immediately after stopping by power interruption, a large starting torque is generated while the discharge pressure of the compressor remains high. Therefore, when the compressor stops, it takes about a few minutes and decreases the discharge pressure before restarting the compressor.

When the power supply is lost, it is impossible to supply energy from the outside. So, it must be operated using the internal energy (electric energy of FC, kinetic energy of inertia) of the system. As a method of continuous operation, the following three control methods have been proposed depending on the magnitude of the kinetic energy and duration of the power interruption.

First, in applications where the kinetic energy is sufficient, it is possible to maintain the FC voltage and continue operation by regenerating the kinetic energy slightly just for several seconds or less because we assume an instantaneous power interruption in this study.

On the other hand, in applications where the inertial loads are very small and the kinetic energy is small, regeneration causes the rotational speed to drop abruptly to the lower limit value even if the FC voltage is maintained, and the system may stop. Also, since the total energy is small, it is necessary to extend the operation time by effectively utilizing the charged energy of FC. Furthermore, when the duration of power interruption is longer than the allowable operating time by control method using charged energy of FC, it is necessary to minimize the loss of the system and extend the operation time as long as possible.

As concrete realizations of these controls, the FC voltage control method has been proposed when the kinetic energy is large [3], and the control method using FC energy has been proposed when the kinetic energy is small [4]. Also, a method to maximize the operation time even in the case of the longer power interruption has been proposed [5]. In this paper, we introduce specific structure and how to realize these three control methods including experimental results using 1-kW class PMSM. And we propose an appropriate continuous operation control according to the application and usage environment by comparing the three control methods.

II. Configuration and Characteristics of Continuous Operation Controls

A. FC voltage control [3]

First, the FC voltage control method which is effective for applications with high inertial loads is described. In the PMSM drive system like Fig. 1, when the power supply of the inverter is lost, if the positive torque current reference value is given, the FC voltage decreases rapidly. Then, the system must be stopped even if the rotation speed of the motor is sufficiently high. In order to prevent stopping the system, it is necessary to supply energy to the FC from the kinetic energy of the inertial loads. In an application in which the inertial loads are high and the kinetic energy is sufficiently higher than the

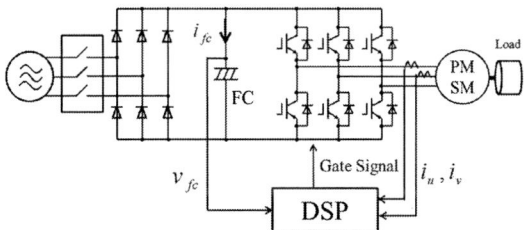

Fig.1. Circuit configuration of PMSM drive system

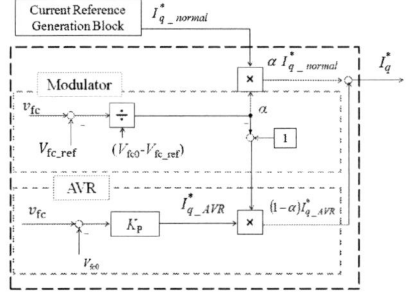

Fig.2. Block diagram of FC voltage control

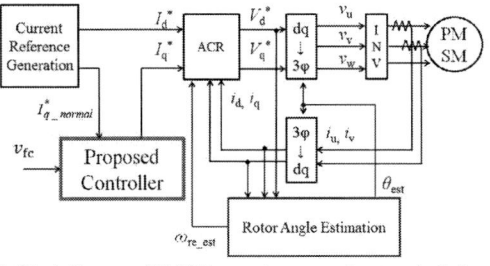

Fig.3. Block diagram of PMSM sensorless control system including FC voltage control block

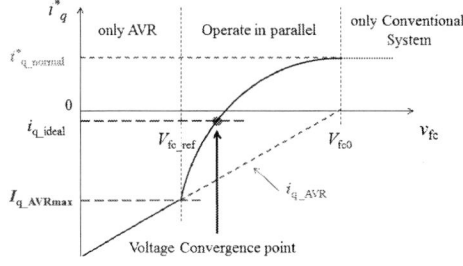

Fig.4. Action of q axis current reference by the proposed method

energy of FC, by regenerating the kinetic energy slightly, FC voltage can be kept over low-voltage limit and the gate-off of the inverter can be prevented.

Fig. 2 shows the block diagram of the FC voltage controller. V_{fc0} is the FC voltage in the normal state. V_{fc_ref} is the voltage at which the output of the q-axis current reference value is maximum by the automatic voltage regulator (AVR) of the proposed controller. Fig. 3 shows a block diagram of the PMSM sensorless control system including the FC voltage controller.

The operation principle of the FC voltage control method is described. The lower part of Fig. 2 is AVR. In this part, the q-axis current reference value $i^*_{q\_AVR}$ is generated by

$$i^*_{q\_AVR} = -(V_{fc0} - V_{fc})K_p \tag{1}$$

where, V_{fc0} is the power supply voltage as the target value and K_p is the proportional gain.

The middle part in Fig. 2 is a modulator. The modulator multiplies the q-axis current reference value $i^*_{q\_normal}$ from the current reference generation block by modulation variable α, and the q-axis current reference value $i^*_{q\_AVR}$ from AVR by (1-α) respectively. The modulation variable α is calculated by

$$\alpha = \frac{V_{fc} - V_{fc\_ref}}{V_{fc0} - V_{fc\_ref}} \tag{2}$$

according to the value of V_{fc} where $V_{fc\_ref} \leq V_{fc} \leq V_{fc0}$.

Finally, the q-axis current reference value i^*_q is expressed by

$$i^*_q = \begin{cases} i^*_{q\_normal} & (V_{fc0} \leq V_{fc}) \\ \alpha i^*_{q normal} + (1-\alpha)i^*_{q\_AVR} & (V_{fc\_ref} \leq V_{fc} \leq V_{fc0}) \\ i^*_{q\_AVR} & (V_{fc0} \leq V_{fc\_ref}) \end{cases} \tag{3}$$

The modulator performs three kinds of operations as shown in (3). First, when $V_{fc} \geq V_{fc0}$ during the power supply exists, i^*_q is equal to $i^*_{q\_normal}$ obtained from the current reference generation block. And the output of the AVR becomes 0, so there is no voltage adjustment function. Next, when V_{fc} decreases because of power supply loss, the modulation variable α is gradually decreased from 1 by the modulator. As a result, $\alpha i^*_{q\_normal}$ decreases, and $(1-\alpha)i^*_{q\_AVR}$ output by AVR is increased conversely. Finally, when V_{fc} becomes V_{fc_ref}, α becomes 0, and i^*_q is equal to $i^*_{q\_AVR}$ regardless of $i^*_{q\_normal}$. It means that it is regenerated with the maximum negative current reference value.

Fig. 4 shows the i^*_q calculated by (3) versus V_{fc}. The steady-state value of i^*_q converges to the value at which the losses and the regenerative power are balanced during the power interruption. If the inertial loads are large enough, i^*_q converges with a negative value close to 0 because the losses are extremely small compared to the kinetic energy. And V_{fc} can be kept at a value higher than the lower limit voltage.

Table I shows the specifications of the PMSM and control parameters. Table II shows the experimental

TABLE I
PMSM Specifications and Control Parameters

Symbol	Meaning	Value
P_{rated}	Rated Power	1 kW
p	Pole pairs	4
R_m	Stator coil resistance	1.1 Ω
L_d	d-axis inductance	12.0 mH
L_q	q-axis inductance	14.0 mH
ϕ_f	PM flux	0.21 Wb
ω_{rated}	Rated Speed	2000 rpm
I_{max}	Rated current	3.7 A
C_{fc}	Capacitance of FC	1320 μF
F_s	Carrier frequency	5000 Hz

TABLE II
Experimental Conditions of FC Voltage Control

Symbol	Meaning	Value
J_m	Inertial load	0.744 kgm²
V_{fc0}	Power supply voltage	150 V
I_d^*	d axis current reference	0 A
I_{q_normal}	q axis current reference	4 A
V_{shut}	Low voltage protection	95 V
V_{fc_ref}	Terminal voltage of change control	100 V
K_P	Proportional gain of AVR	0.1

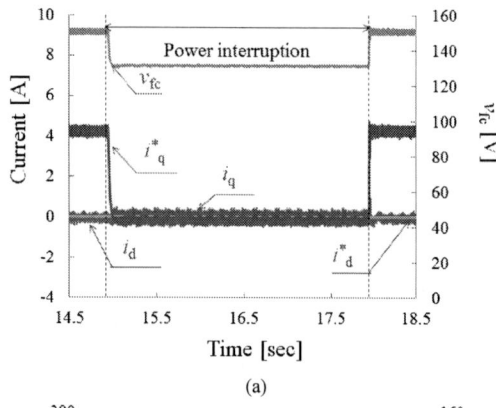

(a)

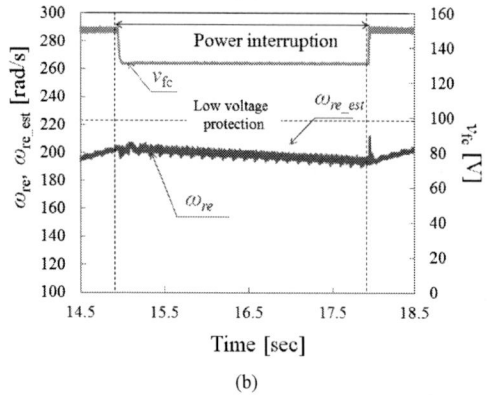

(b)

Fig.5. Experiment result of FC voltage control. (a) d-q axis current response. (b) Real and estimated rotational speed.

conditions of FC voltage control. And Fig. 5 shows the experimental results. From Fig. 5, it can be seen that the FC voltage V_{fc} decreases during the power interruption. But the q-axis current reference value calculated by (3) becomes a negative value close to 0, and V_{fc} is stable at a voltage higher than lower limit voltage by regenerating

slightly. Therefore, FC voltage control method can avoid the gate-off during the power interruption. And the q-axis current immediately follows the q-axis current reference value by ACR, and V_{fc} also responds without any vibration. Also, it can be confirmed that the estimated speed ω_{re_est} follows without error.

B. Control Method Using FC Energy [4]

Next, the control method using FC energy, which is effective for applications with small inertial loads like a compressor for air conditioning, is described. In the case where the inertial loads are small and the kinetic energy of inertia is not enough large, if the regeneration is performed as in the FC voltage control method, the rotational speed suddenly decreases and the system will stop. So it is necessary to continue operation by supplying the FC energy to the inertia effectively.

Fig. 6 shows a block diagram of the continuous operation control method using FC energy and Fig. 7 shows a block diagram of the PMSM sensorless control system including the continuous operation controller.

First, in order to estimate the amount of energy, the electrical energy of FC and the kinetic energy of the inertia are calculated as

$$Energy_C = \frac{1}{2}C_{fc}\left(V_{fc}^2 - V_{fc\_min}^2\right) \tag{4}$$

$$Energy_M = \frac{1}{2}J_m(\omega^2 - \omega_{min}^2) \tag{5}$$

where, C_{fc} is the capacity of the FC, V_{fc_min} is the lower limit voltage of FC, J_m is the moment of inertia of the rotating body including the load, ω is the rotational speed of the motor, ω_{min} is the minimum operating speed that can be operated.

The energy difference $\Delta Energy$ is expressed as

$$\Delta Energy = Energy_C - Energy_M. \tag{6}$$

And current reference value i_{aE}^* is calculated by

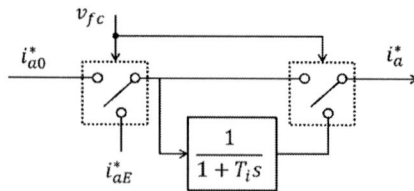

Fig.6 Block diagram of control using FC energy

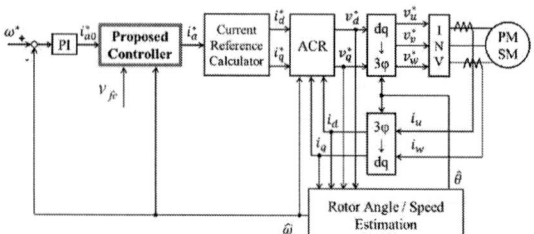

Fig.7 Block diagram of PMSM sensorless control system including control using FC energy

$$i_{aE}^* = \begin{cases} 0 & (\Delta Energy \leq 0) \\ i_{a0}^* \times \dfrac{\Delta Energy}{Energy_C} & (\Delta Energy > 0) \end{cases} \qquad (7)$$

where, i_{a0}^* is the current reference value calculated from the speed PI controller just before the power supply loss.

At the time of power supply loss, if the motor rotation speed is sufficiently high and the kinetic energy is equal to or more than the electrical energy of FC ($\Delta Energy \leq 0$), the current reference value i_{aE}^* is set to zero. Then, the rotation speed is decelerated quickly because of load torque. When the kinetic energy of the inertia becomes smaller than the electrical energy of FC ($\Delta Energy > 0$), i_{aE}^* is proportional to the value obtained by dividing the energy difference $\Delta Energy$ by the FC energy $Energy_C$. And the maximum value thereof is i_{a0}^*. It is possible to drive the motor to the minimum speed in accordance with the energy of the FC without generating torque more than

TABLE III
EXPERIMENTAL CONDITIONS OF CONTROL USING FC ENERGY

Symbol	Meaning	Value
J_m	Inertial load	0.028 kgm²
V_{fc0}	Power supply voltage	300 V
i_{a0}^*	Current reference value before power interruption	3.2 A
V_{fc_min}	Minimum FC voltage	75 V
ω_{min}	Minimum speed	200 rpm
T_i	Time constant of LPF of current reference	100 ms

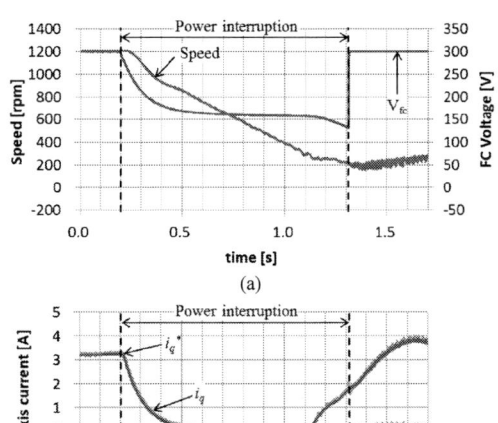

(a)

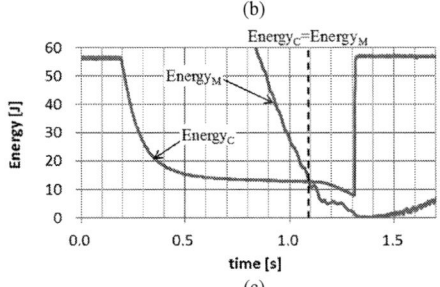

(b)

(c)

Fig.8 Experimental Results of control using FC energy. (a) Speed & FC Voltage. (b) d-q Axis Current. (c) Electric Energy of Filter Capacitor & Kinetic Energy of Motor

the load torque by calculating the current reference value from (7). As a result, both the electrical energy of the FC and the kinetic energy of the inertia gradually decrease, so that the rotation speed can be gradually reduced to the minimum speed until both energy is exhausted.

In a compressor load, a sudden fluctuation in the current reference value may damage the compression section. Therefore, by providing a first-order lag filter with a time constant T_i in the current reference value generation section, a sudden change in the current reference value is prevented.

When it returns to the speed control by power recovery, internal integral variable of the speed PI control needs to be reversely calculated from magnitude of current value and reset in order to prevent abrupt change of current.

The results of experiments under the experimental conditions as shown in Table III are shown in Fig. 8 based on PMSM specifications and control conditions as shown in Table I. An IM load rated at 2.2 kW was used as a load. When the power supply is lost, the current reference value is gradually decreased by the first-order lag filter and the rotation speed is gradually lowered because of $Energy_M > Energy_C$. When the current reference value becomes completely zero, no energy is supplied from the FC to the motor. So, the voltage of FC drops by the internal losses. When the rotation speed of the motor decreases and the kinetic energy becomes smaller than the electric energy of FC (around 1.1 seconds), the current reference value is calculated from the second line of (7). As the energy difference $\Delta Energy$ increases, the current reference value gradually increases, and the FC voltage decreases by applying energy to the motor. In this control method, since the current reference value is calculated on-line from the rotational speed ω and the FC voltage V_{fc}, it is possible to slowly decelerate to the minimum speed depending on situation of ω and V_{fc}. Also, even if switching to the speed control by power recovery, it is found that abrupt fluctuation of the current value does not occur by resetting the internal integral variable of the speed PI control (around 1.3 seconds).

C. Maximum Continuous Operation Time Control [5]

If the duration of power interruption is longer than the allowable operating time by control method using FC energy of the previous section, it is necessary to minimize losses of the whole system and extend the operation time as long as possible. In this section, it is explained the control method that maximizes the operation time using the calculus of variations.

Since energy cannot be supplied from the outside when the power supply is lost, the energy that the system has E_{all} coincides with the time integration of losses of whole system P_{loss} and power transmitted to the load side from FC Energy and/or kinetic energy P_{load} until the system stops as follows:

$$E_{all} = \int (P_{loss} + P_{load}) dt. \qquad (8)$$

And, P_{loss} and P_{load} can be defined as follows:

$$P_{loss} = R_m i_q{}^2 + K_{IL}(p\omega)^3\left(\phi_f{}^2 + L_q{}^2 i_q{}^2\right) + K_{me}\omega^2$$
$$+V_{fc}{}^2/R_d + K_{on}|i_q| + K_{sw}|i_q|V_{fc} \qquad (9)$$

$$P_{load} = \tau_L \times \omega \qquad (10)$$

where, i_q is q-axis current, ω is mechanical rotational speed, R_d is equivalent parallel resistance of FC including discharge resistor, and $K_{IL}, K_{me}, K_{on}, K_{sw}$ are coefficients of each losses, respectively. In (9), from the first term on the right side, they mean copper loss, iron loss, and mechanical loss of motor, loss of equivalent resistance of FC, conduction loss, and switching loss of inverter, in the order respectively.

From the time derivative of the electrical energy of the capacitor and the kinetic energy of the rotating body, and the power transmitted to the load side are in equilibrium, and ignoring the losses those are small with respect to the energies, (11) is obtained.

$$C_{fc}V_{fc}\frac{dV_{fc}}{dt} + J_m\omega\frac{d\omega}{dt} + \omega\tau_L = 0 \qquad (11)$$

If the loss power from the time $t = 0$ to $t = t_a$ is set to the evaluation function as (12), and (11) is set to the boundary condition, the Lagrangian function L is obtained as (13) and (14).

$$E = \int_0^{t_a}(P_{loss} + P_{load})\,dt = \int_0^{t_a} P\,dt \qquad (12)$$

$$L(V_{fc}, V_{fc}, t) = P(V_{fc}, V_{fc}, t)$$
$$+\lambda\left(C_{fc}V_{fc}\frac{dV_{fc}}{dt} + J_m\omega\frac{d\omega}{dt} + \omega\tau_L\right) \qquad (13)$$

$$\frac{\partial L(V_{fc}, V_{fc}, t)}{\partial V_{fc}} - \frac{d}{dt}\left(\frac{\partial L(V_{fc}, V_{fc}, t)}{\partial V_{fc}}\right) = 0 \qquad (14)$$

where, λ is Lagrangian multiplier.

From the Lagrange function L does not have any solution, the energy loss E in (12) is monotonous increase or decrease according to the $V_{fc}(t)$. And, E is the minimum when $V_{fc}(t)$ is the maximum in the assumed operation condition of the compressor load of air conditioner.

That $V_{fc}(t)$ is the maximum means that $V_{fc}(t)$ is maintained the initial value until ω reaches the lower limit ω_{min}. After then, $V_{fc}(t)$ goes down to the lower limit V_{fcmin} to keep $\omega = \omega_{min}$. During $\omega = \omega_{min}$, $V_{fc}(t)$ can be calculated as (15) from (11).

$$C_{fc}V_{fc}\frac{dV_{fc}}{dt} = -\omega_{min}\tau_L \qquad (15)$$

From (15), the operation time becomes longer as higher V_{fc}. And V_{fc} can be as high as possible by maintaining the initial value before power interruption.

Therefore, the maximum continuous time control is that current reference value is set to zero to keep V_{fc} constant until ω reaches ω_{min} and after then, current reference value is set to the value before power interrupt to keep $\omega = \omega_{min}$.

The results of simulation and experiment under the experimental conditions as shown in Table IV are shown in Fig. 9(a) and Fig. 9 (b) respectively. And, the results of simulation under the compressor conditions as shown in Table V are shown in Fig. 9(c). In the simulation, the above loss model and power balance model are used. From Fig. 9(a) and Fig. 9(b), the validity of the models can be confirmed. And, from Fig.9 (c), it can be confirmed that the continuous operation time becomes about 40% longer under the assumed conditions of the compressor.

III. COMPARISON OF THE THREE CONTROL METHODS

In applications with large inertial loads, kinetic energy of inertia is relatively large, and the total energy of the system is large. Therefore, by applying the FC voltage control method during the power interruption, a negative current reference value close to 0 is given to the inverter. Then, the FC voltage can be maintained over the low limit and the rotation speed decreases just a little by regenerating slightly. On the other hand, in applications with small inertial loads, the total energy of the system is relatively small. So, it is necessary to make effective use of both energies and gradually lower the rotation speed so that it can quickly converge to the original rotation speed at power recovery by applying the control method using FC energy utilization method.

In addition, when the inertial loads are large, the FC voltage control method can cope well with the power interruption within a few seconds, and it is not necessary to study for extending the continuous operation time. On the other hand, when the inertial loads are small, since the total energy is small, the continuous operation time is very short when the power supply is lost. In that case, it is necessary to study the maximum continuous operation time control method that can continue the operation as long as possible considering the internal losses during the power interruption. But, if it is expected that the duration of power interruption is shorter than the allowable operating time by control method using charged energy of FC, it is suitable for control method using charged energy in FC. Because it gradually reduces the rotational speed, the method can be quickly converged to the original rotation speed after power recovery.

For example, from the simulation result in Fig. 9(c), in case of power interruption of less than 0.2 second, the operation method using charge energy in FC is advantageous because it maintains a higher rotation speed in the simulation conditions. On the other hand, in case of power interruption of 0.2 seconds or more, the maximum continuous operation time control is advantageous because it can continue operation until about 0.28 seconds.

Therefore, it is necessary to select an appropriate continuous operation control method depending on the conditions such as the magnitude of the inertia load and the time of power interruption that occurs frequently.

TABLE IV
SPECIFICATIONS AND CONTROL PARAMETERS OF THE EXPERIMENT

Symbol	Meaning	Value
ω_0	Speed before Power interruption	126 rad/s (1200rpm)
ω_{min}	Minimum Speed	21 rad/s (200rpm)
τ_L	Load Torque	1.5 Nm
C_{fc}	Capacitance of FC	1320 μF
J_m	Inertial Load	0.028 kgm²
V_{fc0}	DC Power supply voltage	300 V
V_{fcmin}	Minimum FC voltage	75 V

TABLE V
SPECIFICATIONS AND CONTROL PARAMETERS OF THE COMPRESSOR

Symbol	Meaning	Value
ω_0	Speed before Power interruption	377 rad/s (3600rpm)
ω_{min}	Minimum Speed	126 rad/s (1200rpm)
τ_L	Load Torque	2.0 Nm
C_{fc}	Capacitance of FC	1020 μF
J_m	Inertial Load	0.00075 kgm²
V_{fc0}	DC Power supply voltage	380 V
V_{fcmin}	Minimum FC voltage	100 V

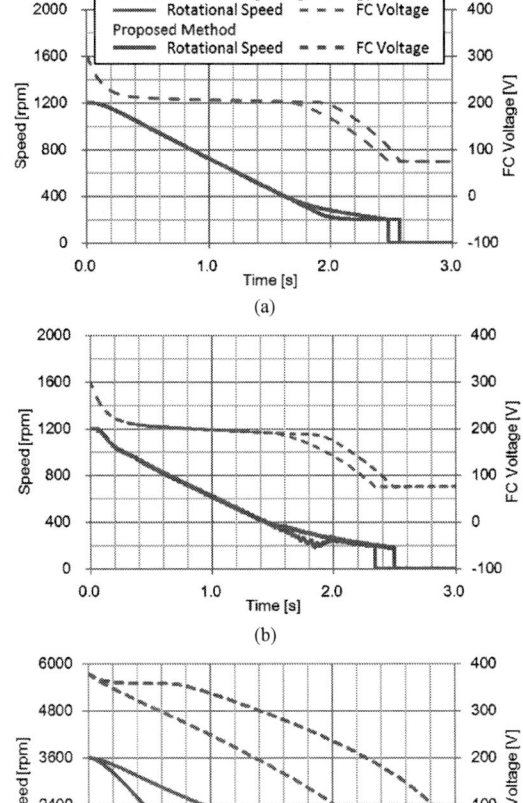

Fig. 9. Simulation and Experiment Results: (a) simulation results at the experiment conditions (Table IV), (b) experimental results, (c) simulation results at the compressor conditions (Table V)

IV. CONCLUSIONS

In this paper, we explained the FC voltage control method, control method using FC energy, and maximum continuous operation time control in detail as the continuous operation control in the case of power supply loss for the PMSM driving system. And it is clarified that the FC voltage control method is proper when the inertial loads are large and the control method using FC energy is proper when the inertial loads are small respectively. And, if the duration of power interruption is longer than the allowable operating time by control method using FC energy, it is suitable for maximum continuous operation time control. So, it is necessary to select a suitable control method for each applications and usage environments.

By applying these control methods to the operating environment where the power supply is not stable, it is expected that it is possible to prevent the system from stopping during the power interruption of the input power supply and save energy and time for restart.

REFERENCES

[1] K. Kondo, H. Kubota, "Innovative Application Tech-nologies of AC Motor Drive Systems," IEEJ J. Industry Applications, Vol.1, No.3, pp.132-140, 2012

[2] Seung-Ki Sul, Sungmin Kim, "Sensorless Control of IPMSM : Past, Present, and Future," IEEJ J. Industry Applications, Vol.1, No.1, pp.15-23, 2012

[3] S. Kurita, T. Horie, and K. Kondo, "Study on Continued Operation for the Direct-Current Power-Supply Loss for a Rotation-Angle Sensorless PMSM," IEEJ Transactions on IA, Vol. 135, No. 6, pp.671-678, 2015 (in Japanese)

[4] J. Heo, K. Matsuo, K. Natori, and K. Kondo, "Continuous Operation Control of PMSM with Compressor Loads of Air Conditioners in Case of DC Power Supply Loss," IEEJ Transactions on IA, Vol. 137, No. 9, pp.687-695, 2017 (in Japanese)

[5] J. Heo, K. Kondo, "Maximum Continuous Operation Time Control of PMSM with Compressor Loads of Air Conditioners in case of DC Power Supply Loss," IEEJ J. Industry Applications, Vol.7, No.3, 2018

The 2018 International Power Electronics Conference

Model Predictive Control for Multiphase Motor Drives – a Technology Status Review

A. Tenconi, S. Rubino and R. Bojoi

Dipartimento Energia "G. Ferraris", Politecnico di Torino, Torino, 10129, Italy

Abstract-The application of model predictive algorithms for three-phase electric drives has gained an impressive attention in the last decade. Nowadays, the model predictive controls (MPC) represent one of the most competitive solutions in alternative to the conventional feedback control schemes. At the same time, multiphase motor drives have become a feasible solution for high power/high current applications and most likely they will be employed in the applications like transportation electrification and energy production. Although the technical literature contains many references regarding the application of MPC for three-phase drives, there are only few attempts about the application of predictive algorithms for multiphase machines. The paper provides a comprehensive analysis and survey on the main MPC schemes applied to multiphase drives. In addition, experimental results on the application of a MPC scheme for a six-phase asymmetrical induction machine are provided.

Keywords— multphase machines, model predictive control, multiphase drives.

I. INTRODUCTION

In the recent years, the application of model predictive controls (MPC) for electric drives has been strongly developed [1]-[2]. In fact, by using the computational power of the modern microcontrollers, it is possible to develop more complex control algorithms to improve the drive performance. In this context, the predictive torque control of the electrical machines represents an interesting evolution for next future.

The use of MPC schemes is justified by multiple reasons which depend on the considered application. In high power/high current systems it represents a smart solution for the reduction of the power converter's switching frequency [3]. For this kind of applications, the dynamic requirements are less significant with respect to the reduction of the power switches commutations. Therefore, the use of MPC schemes allows at increasing the efficiency of the power converter while making possible the use of the consolidated and reliable Gate Turn Off Thyristor (GTO) technologies.

Nevertheless, the MPC schemes can be also used to improve the dynamic torque response, generally better than traditional feedback control schemes and without any change in the control hardware [4]. In this case, the MPC represents a competitive solution for the implementation of high performance electrical drives where the dynamic requirements have the priority on every other aspect.

The above-mentioned examples are the proof how the MPC schemes can be used to improve the drives performance for multiple aspects. Furthermore, the MPC generally requires a less demanding control tuning procedure, so it becomes particularly interesting for the application engineers to reduce the time-to-market.

The technical literature contains a lot of examples about the application of the MPC for three-phase electrical drives and some of these solutions have been already implemented in industry [3]. Nevertheless, with the continuous development of the transportation electrification and energy production, together with the advent of the power electronics, also the multiphase electrical systems are gaining an impressive attention [5]-[6]. In fact, when the power range of the application increases to MW levels, the use of multiphase systems allows at keeping the phase currents at acceptable limits that can be handled with today's fast power electronic components. Furthermore, the multiphase machines possess inherent fault-tolerant capability due to their redundant structure [7]. This feature makes them a promising solution for safety-critical applications and with a strategic role in the process of transportation electrification [5].

In this context, in the recent years it has been started a promising research on the application of the MPC on the multiphase drives to combine their advantages [10-43]. Currently, only some solutions have been proposed due the complexity of the problem. Nevertheless, the proposed solutions are fairly diversified to justify the need of a survey related to the state of the art in this field of the research. Therefore, the goal of the work is to provide an exhaustive analysis and survey of the multiphase MPC schemes. The paper is organized as follows. Section II contains an overview of the multiphase topology where the MPC has been successfully implemented. Section III deals with the employed multiphase machine typologies. Section IV provides a comprehensive analysis of the multiphase MPC control solutions. Experimental results for a six-phase MPC drive are provided in Section V.

II. MPC MULTIPHASE DRIVE TOPOLOGIES

Three-phase MPC drives are generally referred to a machine's structure that consists of a conventional three-phase winding with isolated neutral point. With respect to the three-phase counterpart, the multiphase topologies present more degrees of freedom.

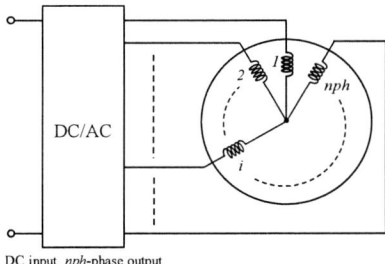

DC input, *nph*-phase output

Fig. 1. Multiphase topology with single neutral point.

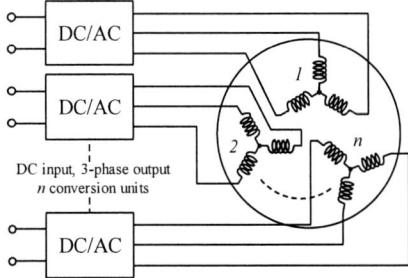

DC input, 3-phase output
n conversion units

Fig. 2. Multiphase topology with multiple three-phase units.

TABLE I.   MPC MULTIPHASE DRIVE TOPOLOGIES

Topology	References
5-Phase	[15-21], [23-25], [28-32], [34-37], [39-40]
6-Phase	[10-14], [26-27], [33], [38], [43]
9-Phase	[22]

In some configurations they consist of a simple extension of the three-phase case (Fig. 1). Nevertheless, when the phase number becomes a multiple of three, the stator can be configured as independent three-phase windings with isolated neutral points, as shown in Fig. 2 [5]. In terms of implementation of MPC schemes, both structures present huge complications related the exponential increment of the power converter's discrete states [8]-[9].

The main multiphase structures used for the implementation of MPC schemes are reported in Table I. It is possible to note how the most employed configurations are the 5-phase machine and the 6-phase one in double three-phase configuration. Table I is quite representative of the current technological development. For historical reasons, the 5-phase machine is the most analyzed configuration. For a given power and rated voltage, it allows a strong reduction of the phase currents (near to 40%) with respect to the three-phase counterparts. Furthermore, it can be easily controlled with conventional microcontrollers as the increment of power converter's discrete states is still limited (32 against the conventional 8 ones of the three-phase case). When the number of phase increases, the complexity of a MPC scheme becomes very high, as it will be described in Section IV. A relevant attempt has been done in [22] with a 9-phase PM machine using a single neutral point configuration.

The 6-phase machine in double three-phase configuration represents the simplest case of multiple three-phase machine.

TABLE II.   MPC MULTIPHASE MACHINE TYPES

Typology	References
IM	[10-21], [23-24], [26], [28], [33-38], [43]
PM	[25], [27], [40]
BLDC	[29-32]
PM-FS	[22]
SRM	[41-42]

In this configuration, multiple independent three-phase windings are employed, as shown in Fig. 2. The phase current amplitude is reduced by a factor n (where n is the number of three-phase sets) with respect to the three-phase counterparts (for a given power and rated voltage). Furthermore, these machines use the well consolidated power electronics three-phase technology [5]. This feature most likely will make the multiple three-phase configurations preferred by the industry [5-6].

Similarly to the 5-phase machine, the 6-phase configuration has been deeply analyzed and relevant attempts have been done into defining MPC schemes for it, as shown in Table I. Nevertheless, for multiple three-phase machines dedicated mathematical approaches can be defined to make straightforward the electric drive [5]. For example, the Multi-Stator (MS) approach allows at defining MPC schemes able to manage a high number of three-phase units in parallel despite the exponential increment of the power converter's discrete states [7]-[38].

III.   MPC MULTIPHASE MACHINES

With respect to the three-phase MPC drives, where almost all machine types have been analyzed, in the multiphase MPC ones only some machine types have been considered. In particular, the isotropic structures like the Induction Machines (IM) and Permanent Magnet (PM) machines have been addressed. The latter category includes also the Brushless DC machines (BLDC) and the PM-Flux Switching (PM-FS) ones. Furthermore, there are some interesting attempts of the MPC on Switched Reluctance machines (SRM). Table II reports the multiphase machine typologies where the MPC has been implemented.

It is interesting to note the absence of Synchronous Reluctance (SyR) and Interior Permanent Magnets (IPM) machines. Conversely, the IM results as the most analyzed machine. This trend is mainly related to the historical development of the three-phase drives. In fact, only in recent times the synchronous machines have started to replace the conventional IM drives. Consequently, the development of multiphase synchronous machines is in progress and only feedback control strategies have been mainly defined. A relevant exception are the BLDC drives [29-32] which is a consolidated and robust technology. Some attempts have been done in SPM motor drives, as shown in Table II. However, it is quite evident how the multiphase MPC schemes have been mainly developed for IM drives.

Further limitations on the application of predictive algorithms on the synchronous machines are their state equations.

733

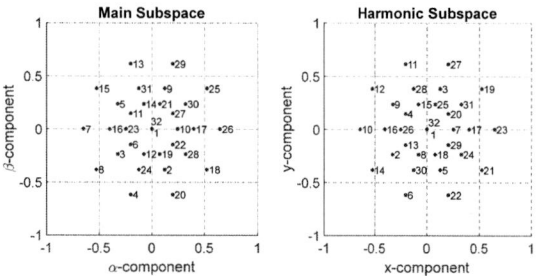

Fig. 3. Normalized 5-phase VSI power converter's discrete states.

TABLE III. MPC MULTIPHASE DRIVE TYPOLOGIES

Typology		References
FCS-MPC	SCC	[10-14], [16], [18-21], [23-24] [26], [28], [32-36], [39], [43]
	DCPC	[27], [29]
	FOC	[17], [22], [25], [37], [40]
	DTC	[15]
CCS-MPC	FOC	[30-31], [40]
	DFVC	[38]

In fact, in the synchronous machines the evolution of the variables depends on the rotor position. Conversely, the dynamic behavior of IM depends exclusively from the rotor speed. Since the predictive algorithms are model-based, it is necessary to predict also the mechanical variables for both cases. However, the changes of the mechanical speed in a single sample time can be usually neglected (inertial dynamic). Conversely, the rotor position can have huge variations that must be evaluated for a proper operation of the MPC scheme. This factor make necessary the use of model predictive motional estimators which are difficult to implement.

IV. MPC CONTROL SOLUTIONS

The implementation of MPC control schemes can be performed in different ways. The choice of one strategy with respect to another one depends by the requirements of the system. For this reason, in three-phase drives it is not possible to provide a clear comparison between the control strategies. In this paper, it will be provided the main classifications of the MPC multiphase drives with a full description of the advantages and disadvantages of each control strategy.

A. Reference Voltages Selection

The main classification of the MPC multiphase schemes is related to the selection method of the reference voltages. The most part of the MPC schemes select the reference voltages among the instantaneous power converter's discrete states. This category includes the Finite Control Set MPC (FCS-MPC) schemes. Nevertheless, there are MPC schemes where the reference voltages are selected between all possible average voltage vectors which the power converter can apply. These control strategies belong to the category of Continuous Control Set MPC schemes (CCS-MPC). In Table III is provided a survey of both solutions together a classification of the machine control type.

1) Finite Control Set MPC (FCS-MPC)

In the FCS-MPC, the reference voltages are selected among the instantaneous power converter's discrete states according with the minimization of a cost function. This operation is quite easy to implement in three-phase FCS-MPC drives where the number of all possible discrete states is only eight (2-level inverters). Conversely, the extension of the FCS-MPC algorithms to multiphase drives is quite complicated as the number of power converter's discrete states increase exponentially.

In addition, these control strategies usually use the Vector Space Decomposition (VSD) approach to reduce the complexity of the state equations of the machine [9]. The direct consequence of the VSD approach is the creation of harmonic subspaces. In normal operation, they should not be excited to avoid not useful currents that reduce the efficiency of the system. Therefore, the design of the cost function must take in account not only the terms related to the electromechanical conversion but also the harmonic subspaces ones which assume the meaning of penalty factors. To understand the complexity of the FCS-MPC in multiphase drives, the power converter's discrete states of a 5-phase machine are shown in Fig. 3. For this simple case, there is only the main subspace and one harmonic subspace. However, it is possible to note the relevant number of states where the cost function must be evaluated. When the number of phases increases, the evaluation of the cost function in each single state is not viable [8]-[9]. Consequently, the literature contains some examples of optimization algorithms to reduce the computational efforts. For example, [13] proposes an interesting optimization method called as "restrained search predictive control" able to reduce the number of discrete states according to the predefined constraints.

In analogy with the conventional three-phase MPC drives, the cost function can be defined to optimize currents, flux, torque, reactive power etc. Therefore, the FCS-MPC schemes allows high degrees of freedom in terms of variables that is possible to optimize. For these reasons, with the FCS algorithms it is quite simple to implement the conventional drive schemes as Field Oriented Control (FOC) with current control in synchronous reference frame, Direct Torque Control (DTC), Direct Flux Vector Control (DFVC). An interesting approach is the FOC based on the Stationary Current Control (SCC). The SCC is convenient in multiphase machines since it is valid for all machine subspaces. There are also some attempts for the Direct Control of the Phase Currents (DCPC). In Table III, in the section related to FCS-MPC schemes, a full classification upon the base of the machine control type is provided.

To understand the versatility of FCS algorithms, [16] proposes a predictive control strategy where the cost function is designed to reduce the common mode voltage which compromises the windings' insulation and the motor bearings. In addition, also the electromagnetic noise is reduced. However, it is quite important to point out the increment of currents distortion when secondary variables are optimized together with the ones related to the electromechanical conversion. This factor cannot be neglected because already in normal conditions the FCS algorithms present problems of current distortion.

In fact, the performance of FCS-MPC schemes depends on the machine's parameters with a strong increment of the phase currents distortion for low impedance machines. Furthermore, as the machine control is based on the VSD modelling approach, there are relevant difficulties to limit the current's components of the harmonic subspaces. In the technical literature some solutions have been proposed to solve these issues [11-12], [25-26]. Large part of them introduce sub-modulations between the power converter's discrete states to reduce the current's derivatives. For example, the solution proposed in [25-26] use the concept of "virtual voltage vector modulation" to delete/reduce the low order harmonics of the stator currents introduced by the harmonic subspaces.

Another important aspect of to the FCS-MPC multiphase drives is their performance under machine's parameters detuning conditions. Some contributions have been carried out from [35, 43]. In conclusion, the main advantage of the FCS-MPC is the reduction of the switching frequency of the power converter. As declared in the introduction, this feature makes them a smart solution for high power/high voltage drive where the dynamic requirements are not so important compared to the efficiency and reduction of power switches commutations.

2) Continuos Control Set MPC (CCS-MPC)

In the CCS-MPC schemes the voltage range corresponds with all possible average voltage vectors which the power converter can apply. In this case, the reference voltages are computed by using the inverse machine model according with the reference commands and the predicted variables (torque, fluxes and currents). These control schemes use a PWM modulator and they are executed at constant switching frequency and thus the sampling frequency is also constant and synchronized with the switching frequency. This is an important difference with respect to the FCS-MPC schemes where the switching frequency is variable.

The CCS-MPC schemes do not use any cost function to design and the number of power converter's discrete states is not so important (conventional PWM strategies are employed). Furthermore, the CCS-MPC allows at using alternative mathematical approaches like the Multi-Stator (MS) one, which is very useful for multiple three-phase machines (Fig. 2). In this way, it is possible to define modular control schemes with an independent control of each three-phase winding set [7]-[38].

The use of CCS-MPC schemes avoid the distortion of the phase currents with respect to the FCS-MPC schemes. In addition, they allow at obtaining high dynamic drive performance. In fact, the CCS-MPC schemes use their predictive nature to improve the dynamic of the torque response [38] which is an important feature especially in some applications like aircraft etc. Therefore, the goal of CCS-MPC schemes is not the optimization of the switching frequency like the FCS-MPC ones.

In three-phase MPC drives, the CCS-MPC category contains the well-known Dead-Beat Direct Flux and Torque Control [4]-[44-45]. In Table III, in the section related to CCS-MPC schemes is provided a full classification upon the base of the machine control type.

The CCS-MPC can be used to implement the conventional drive schemes like FOC, DFVC and DTC. It must be emphasized here that the CCS-MPC requires more complex schemes with respect to the FCS-MPC ones. Furthermore, they offer less degrees of freedom with respect to the FCS-MPC schemes where, in theory, it is possible to optimize each variable through an appropriate design of the cost function [16]-[27].

An important advantage related to the use of CCS-MPC schemes is the possibility to manage in easy way the operation of drive under voltage and current limitations. Indeed, the operation in flux-weakening is one of the less addressed topic in MPC drives. This issue can be found also in conventional three-phase MPC drives and this factor strongly hinders the implementation of MPC schemes on the industrial drives. Currently, the most important contributions in three-phase MPC drives have been obtained with the Dead-Beat Direct Torque and Flux Control (DB-FTC) [44-45]. The results achieved with the DB-FTC have shown how the direct stator flux regulation is a key-factor for the robustness and feasibility of a MPC scheme able to face with the full speed range of a motor drive. This aspect has been pointed out also in the conventional feedback control schemes [46].

According with the best authors' knowledge, in the field of multiphase MPC drives the [38] is the only work who has shown the full operation of the machine under voltage and current limitations. The control scheme defined in [38] regards the application of a CCS-MPC algorithm on the basic structure of a DFVC scheme used for multiple three-phase induction motor drives. It represents the experimental proof on how the direct stator flux regulation easily guarantees the maximum torque production under current and voltage limitations able to provide Maximum Torque per Volt (MTPV) operation. More details are reported in Section V related to the experimental results.

The control scheme defined in [38] has been developed by using the MS-approach to deal with a straightforward control of each three-phase winding set. The use of MS approach is not mandatory for the full control of the single three-phase sets. Similar results can be obtained with the well-known VSD combined with the torque/current sharing equations, as described in [47].

Independently of the mathematical approach, it is quite evident that the performance of the CCS-MPC schemes are not affected by the number of power converter's discrete states. All PWM modulation techniques defined for multiphase machines can be still used being the cost function removed. This factor makes the CCS-MPC the only practical solution to design MPC multiphase drives able to manage high phase order machines. Therefore, the CCS-MPC algorithms could represents a key-factor to define general MPC schemes for multiphase machines.

B. Post-Fault MPC Drive Operation

Another important aspect related to the multiphase machines is the post-fault drive operation. This feature in not so relevant for three-phase drives while is quite important for the multiphase ones to satisfy the fault-tolerance requirements. In this survey, only open-winding fault events are considered.

TABLE IV.   MPC MULTIPHASE DRIVES WITH POST-FAULT OPERATION

Typology	References
Single-phase open-winding fault	[18-20], [27-30], [32], [36]
Three-phase open-winding fault	[38]

For multiphase machines with single neutral point shown in Fig. 1, the fault event corresponds with the single-phase open-winding fault. In multiple three-phase machines (Fig. 2) the fault events can also include the three-phase open-winding fault [7]-[38].

To become competitive control solutions for multiphase drives, the MPC schemes must guarantee the post-fault operation, i.e. they must be fault ride-through compliant. These operations are not easy to perform, especially for FCS-MPC schemes. Table IV reports the MPC schemes where efficient open-winding post-fault operations of the drive are provided. The fault-events are classified according to their typology.

It can be noted how only some works have provided the post-fault operation of the MPC scheme. Large part of them are focused on the 5-phase machine where the number of power converter's discrete state is limited. In fact, the reconfiguration of a FCS-MPC scheme after an open-winding fault event is not straightforward. Furthermore, the current distortion usually increases as the cost function must be optimized with a lower number of available states (for each single phase open-winding fault 2^{x-1} states are lost, where 'x' is the active phases number before the fault). An interesting contribution has been carried out in [27] for a 6-phase axial flux PM machine. However, it is quite clear how the FCS-MPC schemes have low management capability of post-fault open winding events, especially with high phase order machines.

Another important advantage of CCS-MPC drives is the possibility to use the fault tolerant strategies defined for the conventional multiphase feedback control schemes. In practice, in case of fault it is necessary to modify the reference variables (currents, fluxes, torque etc.) without changing the reference voltage selection system. This operation is quite easy in multiple three-phase machines where the only post-fault action to perform is the redistribution of the total torque production on the remained healthy three-phase sets [5]-[7]-[38].

In conclusion, the development of MPC schemes able to satisfy the fault-tolerance requirements is still premature and relevant improvement are necessary for the FCS-MPC algorithms. At present only, the CCS-MPC schemes seems to provide efficient solutions as they can recycle the solutions defined in the technical literature for conventional multiphase feedback controls.

V.   EXPERIMENTAL RESULTS

As example, experimental results are provided for a CCS-MPC applied to a six-phase asymmetrical induction drive [38]. The proposed CCS-MPC is implemented on the basic structure of the DFVC scheme and uses the MS-approach to deal with an independent control of each three-phase winding set. The goal of the proposed control is the improvement of the dynamic torque response with respect to the traditional feedback control schemes. This solution has been proposed as Started/Alternator for open-rotor aircraft application [48].

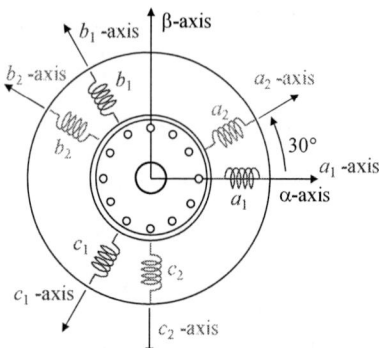

Fig. 4.   Asymmetrical 6-phase induction machine configuration (2x3ph).

Fig. 5.   Induction machine prototype.

TABLE V.   CHARACTERISTICS OF THE MACHINE

Starter/Generator	
Number of phases	6 (2x3-phase, star connected)
Rated phase voltage	230 Vrms
Continuous output power	10 kVA
Base speed	6000 rpm
Overload capability	150%, 5 min.

The stator has 6 phases with two slot/pole/phase, forming a double-three-phase winding with relative shift of 30 electrical degrees among the two three-phase sets ($a_k b_k c_k$), k=1,2 as shown in Fig. 4. The main characteristics of the machine are shown in Table V. The machine has been mounted on a test rig for development purposes, as shown in Fig. 5. The shaft of the machine is coupled with a driving machine acting as prime mover with speed control. The power converter consists of two independent three-phase inverter IGBT power modules fed by a single DC power source of 550 V. The digital controller is the dSpace DS1103 development board. The sampling frequency and the inverters' switching frequency have been set at 6 kHz.

The experimental results are related to the drive operation in healthy and open-winding fault conditions, as well as open loop torque and closed loop speed operation. The experimental work focused on the drive in both healthy and faulty conditions. The faulty condition means open phases after sudden shut-off of one inverter power modules due to a failure in power electronics. More details about the cited control solution are reported in [38].

A.   Torque Control

The machine has been tested with torque control at a negative speed of -6000 rpm (about 200 Hz of electrical frequency), imposed by the prime mover. All tests have been performed with the machine's rated flux (0.23 Vs).

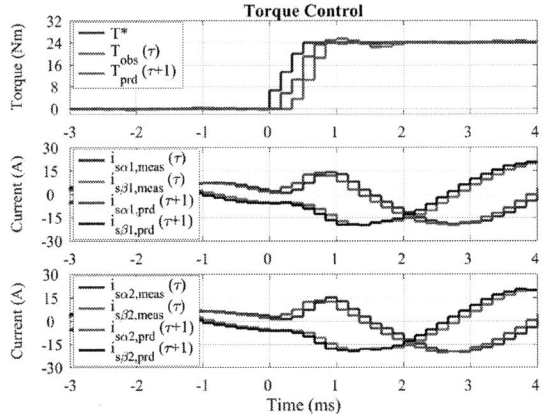

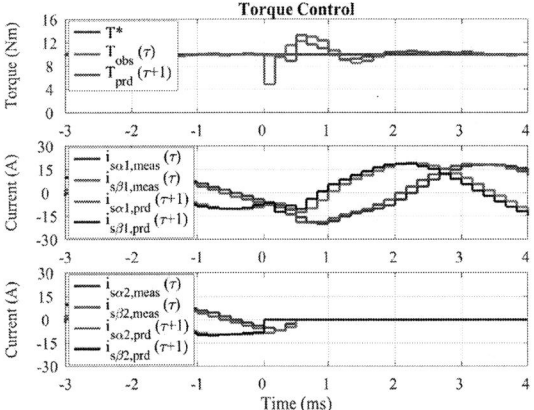

Fig. 6. Fast torque transient from no-load up to 150% rated torque (24Nm) at -6000 rpm. From top to bottom: reference, observed and predicted torque (Nm); Set 1 - (α,β) measured and predicted currents (A); Set 2 - (α,β) measured and predicted currents (A).

Fig. 8. Inverter 2 shut off during generation mode at -6000 rpm and 10Nm. From top to bottom: reference, observed and predicted torque (Nm); Set 1 - (α,β) measured and predicted currents (A); Set 2 - (α,β) measured and predicted currents (A).

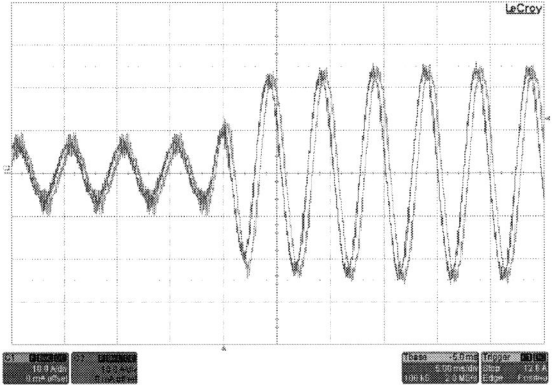

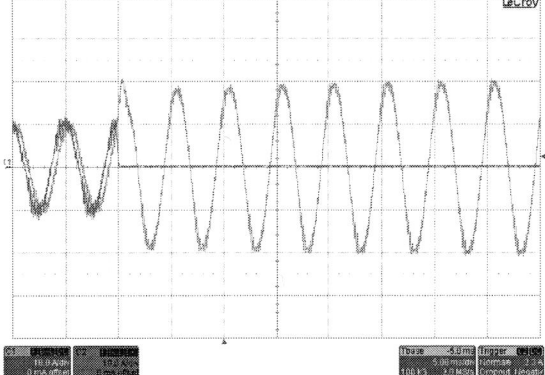

Fig. 7. Fast torque transient from no-load up to 150% rated torque (24Nm) at -6000 rpm. Ch1: i_{sa1} (10 A/div), Ch2: i_{sa2} (10 A/div), Time scale: 5 ms/div.

Fig. 9. Inverter 2 shut off during generation mode at -6000 rpm and 10Nm. Ch1: i_{sa1} (10 A/div), Ch2: i_{sa2} (10 A/div), Time scale: 5 ms/div.

1) Torque control in healthy conditions

The drive has been tested in generating operation being the machine mainly designed as induction generator. A fast reference torque transient (40 Nm/ms) from no-load up to +24 Nm (150% of the rated value) has been imposed for a speed of -6000 rpm and the results are shown in Figs.6-7. Each stator set produce half of the total machine torque.

It can be clearly noted the one-step ahead prediction of currents, as well as the dead-beat torque response, despite the high slew-rate of the torque reference that corresponds to a mechanical power transient from 0 to 15 kW (150% rated value) performed in less than 1 ms. The two three-phase current sets are kept balanced at both no-load and load-conditions, as can be seen in Fig. 7.

The results demonstrate how the proposed CCS-MPC scheme is stable and it works properly despite the lowly ratio between sampling/switching frequency (6 kHz) and fundamental frequency (200 Hz).

2) Torque control in faulty condtions

The control "fault ride-through" capability when one inverter unit is suddenly disabled is shown in Figs. 8-9 (inverter 2 off) for generation operation at -6000 rpm and 10 Nm (60% of rated value).

The healthy unit exhibits sinusoidal current that increases within the allowed limits with the attempt at keeping the same torque and machine flux. The torque and currents of the healthy set exhibit slight overshoots due to turn-off dynamics of the faulty set that acts as disturbance on the predictive algorithm, as shown in Fig. 8.

The open-winding fault event does not add any distortion on the currents waveforms, as shown in Fig. 9. In fact, being a CCS-MPC scheme MS-based, each three-phase set is independently controlled with the own PWM modulator. Therefore, the loss of 56 power converter's discrete states (64 before the fault event) is not relevant and the re-configuration of the control scheme is not required.

The 2018 International Power Electronics Conference

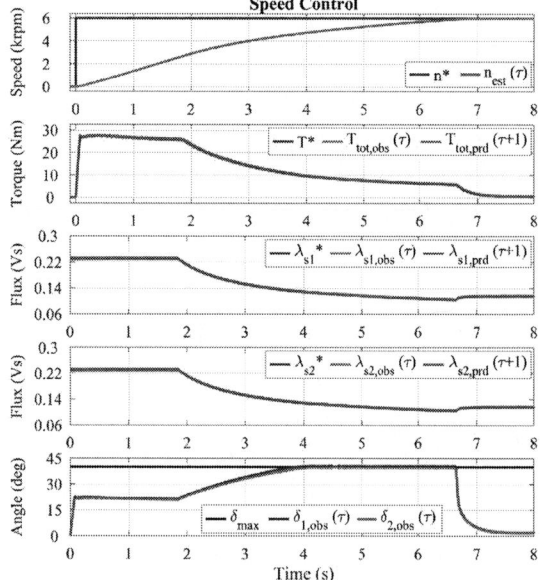

Fig. 10. Speed control with inertial load from 0 to 6000 rpm. From top to bottom: reference and estimated speed (krpm); reference, observed and predicted torque (Nm); Set 1 - reference, observed and predicted stator flux (Vs); Set 2 - reference, observed and predicted stator flux (Vs); maximum and single sets observed load angle (deg).

Conversely, in a FCS-MPC scheme it would have been necessary to change the reference voltages selection method and in any case with lower performance due the increment of the currents distortion (similar performance to a three-phase DTC executed at low sampling frequencies are obtained).

B. Speed Control

Due to the mechanical limitations of the test rig, the maximum speed has been limited at ±6000 rpm. Therefore, the DC source voltage has been reduced from 550V to 275V to test the flux-weakening and MTPV operation below the speed limit of test rig.

The speed control has been implemented with a PI controller whose output is the reference torque provided to the DFVC scheme. The obtained results for a step reference speed from zero up to 6000 rpm are shown in Fig. 10.

At low speed, the torque is limited only by the power converter current limit (24A). The flux-weakening becomes active for a speed that is close to 3000 rpm. The stator flux and stator currents of the single sets are perfectly controlled. This is the experimental proof on how a CCS-MPC scheme with a direct stator flux regulation guarantees high performance at high speed/frequency under limited voltage.

The MTPV limitation becomes active when the maximum load angle is reached, at a speed of about 4500 rpm. For safety, the maximum load angle has been set at 40 electrical degrees to avoid pull-out. As can be seen in Fig. 10, at MTPV operation the load angle of each single three-phase set is properly limited so obtaining the full machine controllability.

The experimental results clearly demonstrate how a CCS-MPC scheme allows a straightforward control of multiple three-phase machines in all operating conditions (post fault operation included).

VI. CONCLUSION

The paper provides a survey of the main MPC algorithms applied to multiphase drives to provide a comprehensive state of the art in this field of the research. Exhaustive classifications upon base of multiphase topologies, machine types and control solutions are provided. Furthermore, also the post-fault operation of the MPC drives has been included. Experimental results are provided for a CCS-MPC scheme applied to a six- phase asymmetrical IM in double three-phase configuration.

REFERENCES

[1] P. Correa, M. Pacas, J. Rodriguez, "Predictive Torque Control for Inverter-Fed Induction Machines", *IEEE Trans. Ind. Electron.*, vol. 54, no. 2, 2007, pp. 1073-1079.

[2] P. Cortes, M. P. Kazmierkowski, R. M. Kennel, D. E. Quevedo and J. Rodriguez, "Predictive Control in Power Electronics and Drives," in *IEEE Trans. Ind. Electron.*, vol. 55, no. 12, pp. 4312-4324, 2008.

[3] T. Geyer, "Algebric Weighting Factor Selection for Predictive Torque and Flux Control", ECCE 2017, pp.1 – 8.

[4] Y. Wang, S. Tobayashi, and R.D. Lorenz, "A Low-Switching-Frequency Flux Observer and Torque Model of Deadbeat–Direct Torque and Flux Control on Induction Machine Drives", *IEEE Trans. Ind. Applicat.*, vol. 51, no. 3, 2015, pp. 2255-2267.

[5] R. Bojoi, S. Rubino, A. Tenconi, S. Vaschetto, "Multiphase electrical machines and drives: A viable solution for energy generation and transportation electrification ", EPE 2016, pp. 632 – 639.

[6] W. Cao, B. Mecrow, G. Atkinson, J. Bennett, and D. Atkinson, "Overview of Electric Motor Technologies Used for More Electric Aircraft (MEA)", *IEEE Trans. Ind. Electron.*, 2011, vol. 59, no. 9, pp. 3523-3531.

[7] S. Rubino, R. Bojoi, A Cavagnino, and S. Vaschetto, "Asymmetrical Twelve-Phase Induction Starter/Generator for More Electric Engine in Aircraft", Conf. Rec. IEEE-ECCE, 2016, pp. 1-8.

[8] J. W. Kelly, E. G. Strangas, and J. M. Miller, "Multi-phase inverter analysis", Conf. Rec. IEEE-IEMDC, 2001, pp. 147–155.

[9] Y. Zhao and T.A. Lipo, "Space vector PWM control of dual three-phase induction machine using vector space decomposition", *IEEE Trans. Ind. Appl.*, 1995, vol. 31, no. 5, pp. 1100-1108.

[10] F. Barrero, M. R. Arahal, R. Gregor, S. Toral and M. J. Duran, "A Proof of Concept Study of Predictive Current Control for VSI-Driven Asymmetrical Dual Three-Phase AC Machines", *IEEE Trans. Ind. Electron.*, vol. 56, no. 6, pp. 1937-1954, June 2009.

[11] F. Barrero, M. R. Arahal, R. Gregor, S. Toral and M. J. Duran, "One-Step Modulation Predictive Current Control Method for the Asymmetrical Dual Three-Phase Induction Machine", *IEEE Trans. Ind. Electron.*, vol. 56, no. 6, pp. 1974-1983, June 2009.

[12] R. Gregor, F. Barrero, S.L. Toral, M.J. Duran, M.R. Arahal, J. Prieto, J.L. Mora, "Predictive-Space Vector PWM Current Control Method for Asymmetrical Dual Three-Phase Induction Motor Drives", *IET Electric Power Applications*, vol. 4, no. 1, pp. 26-34, January 2010.

[13] M. J. Duran, J. Prieto, F. Barrero and S. Toral, "Predictive Current Control of Dual Three-Phase Drives Using Restrained Search Techniques", *IEEE Trans. Ind. Electron.*, vol. 58, no. 8, pp. 3253-3263, Aug. 2011.

[14] F. Barrero, J. Prieto, E. Levi, R. Gregor, S. Toral, M.J. Duran, M. Jones, "An Enhanced Predictive Current Control Method for Asymmetrical Six-Phase Motor Drives", *IEEE Trans. Ind. Electron.*, vol. 58, no. 8, pp. 3242-3252, Aug. 2011.

The 2018 International Power Electronics Conference

[15] J. A. Riveros, F. Barrero, E. Levi, M. J. Durán, S. Toral and M. Jones, "Variable-Speed Five-Phase Induction Motor Drive Based on Predictive Torque Control", *IEEE Trans. Ind. Electron.*, vol. 60, no. 8, pp. 2957-2968, Aug. 2013.

[16] M. J. Duran, J. A. Riveros, F. Barrero, H. Guzman and J. Prieto, "Reduction of Common-Mode Voltage in Five-Phase Induction Motor Drives Using Predictive Control Techniques", *IEEE Trans. on Ind. Applic.*, vol. 48, no. 6, pp. 2059-2067, Nov.-Dec. 2012.

[17] C. S. Lim, E. Levi, M. Jones, N. A. Rahim and W. P. Hew, "FCS-MPC-Based Current Control of a Five-Phase Induction Motor and its Comparison with PI-PWM Control", *IEEE Trans. Ind. Electron.*, vol. 61, no. 1, pp. 149-163, Jan. 2014.

[18] H. Guzman, M. J. Duran, F. Barrero, B. Bogado and S. Toral, "Speed Control of Five-Phase Induction Motors With Integrated Open-Phase Fault Operation Using Model-Based Predictive Current Control Techniques", *IEEE Trans. Ind. Electron.*, vol. 61, no. 9, pp. 4474-4484, Sept. 2014.

[19] H. Guzman, F. Barrero and M. J. Duran, "IGBT-Gating Failure Effect on a Fault-Tolerant Predictive Current-Controlled Five-Phase Induction Motor Drive", *IEEE Trans. Ind. Electron.*, vol. 62, no. 1, pp. 15-20, Jan. 2015.

[20] H. Guzman, M.J. Duran, F. Barrero, L. Zarri, B. Bogado, I.G. Prieto, M.R. Arahal, "Comparative Study of Predictive and Resonant Controllers in Fault-Tolerant Five-Phase Induction Motor Drives", *IEEE Trans. Ind. Electron.*, vol. 63, no. 1, pp. 606-617, Jan. 2016.

[21] C. Martín, M. R. Arahal, F. Barrero and M. J. Durán, "Five-Phase Induction Motor Rotor Current Observer for Finite Control Set Model Predictive Control of Stator Current", *IEEE Trans. Ind. Electron.*, vol. 63, no. 7, pp. 4527-4538, July 2016.

[22] M. Cheng, F. Yu, K. T. Chau and W. Hua, "Dynamic Performance Evaluation of a Nine-Phase Flux-Switching Permanent-Magnet Motor Drive With Model Predictive Control", *IEEE Trans. Ind. Electron.*, vol. 63, no. 7, pp. 4539-4549, July 2016.

[23] J. Rodas, F. Barrero, M. R. Arahal, C. Martín and R. Gregor, "Online Estimation of Rotor Variables in Predictive Current Controllers: A Case Study Using Five-Phase Induction Machines", *IEEE Trans. Ind. Electron.*, vol. 63, no. 9, pp. 5348-5356, Sept. 2016.

[24] J. Rodas, C. Martín, M. R. Arahal, F. Barrero and R. Gregor, "Influence of Covariance-Based ALS Methods in the Performance of Predictive Controllers With Rotor Current Estimation", *IEEE Trans. Ind. Electron.*, vol. 64, no. 4, pp. 2602-2607, April 2017.

[25] C. Xue, W. Song and X. Feng, "Finite control-set model predictive current control of five-phase permanent-magnet synchronous machine based on virtual voltage vectors", *IET Electric Power Applications*, vol. 11, no. 5, pp. 836-846, 5 2017.

[26] I. Gonzalez-Prieto, M. J. Duran, J. J. Aciego, C. Martin and F. Barrero, "Model Predictive Control of Six-phase Induction Motor Drives Using Virtual Voltage Vectors", *IEEE Trans. Ind. Electron.*, vol. PP, no. 99, pp. 1-1.

[27] H. Lu, J. Li, R. Qu, D. Ye and L. Xiao, "Reduction of Unbalanced Axial Magnetic Force in Post-Fault Operation of a Novel Six-Phase Double-Stator Axial Flux PM Machine Using Model Predictive Control", *IEEE Trans. Ind. Electron.*, vol. PP, no. 99, pp. 1-1.

[28] I. Gonzalez-Prieto, M. J. Duran, N. Rios-Garcia, F. Barrero and C. Martin, "Open-switch Fault Detection in Five-phase Induction Motor Drives using Model Predictive Control", *IEEE Trans. Ind. Electron.*, vol. PP, no. 99, pp. 1-1.

[29] M. Salehifar, M. Moreno-Equilaz, "Fault diagnosis and fault-tolerant finite control set-model predictive control of a multiphase voltage-source inverter supplying BLDC motor", *ISA Transactions*, Vol. 60, 2016, pp. 143-155.

[30] R. S. Arashloo, M. Salehifar, L. Romeral, V. Sala, "A robust predictive current controller for healthy and open-circuit faulty conditions of five-phase BLDC drives applicable for wind generators and electric vehicles", *Energy Conversion and Management*, vol. 92, 2015, pp. 437-447.

[31] J.L.R. Martinez, R.S. Arashloo, M. Salehifar, J.M. Moreno, "Predictive current control of outer-rotor five-phase BLDC generators applicable for off-shore wind power plants", *Electric Power Systems Research*, vol. 121, 2015, pp. 260-269.

[32] M. Salehifar, M. M. Eguilaz, G. Putrus, P. Barras, "Simplified fault tolerant finite control set model predictive control of a five-phase inverter supplying BLDC motor in electric vehicle drive", *Electric Power Systems Research*, Vol. 132, 2016, pp. 56-66.

[33] M.R. Arahal, F. Barrero, S. Toral, M. Duran, R. Gregor, "Multi-phase current control using finite-state model-predictive control", *Control Engineering Practice*, Vol. 17, 2009, pp. 579-587.

[34] C. Martín, M.R. Arahal, F. Barrero, M.J. Durán, "Multiphase rotor current observers for current predictive control: A five-phase case study", *Control Engineering Practice*, Vol. 49, 2016, pp. 101-111.

[35] C. Martín, M. Bermúdez, F. Barrero, M.R. Arahal, X. Kestelyn, M.J. Durán, "Sensitivity of predictive controllers to parameter variation in five-phase induction motor drives", *Control Engineering Practice*, Vol. 68, 2017, pp. 23-31.

[36] H. Guzmán, M. J. Durán and F. Barrero, "Speed control of five-phase induction motor drives with an open phase fault condition and predictive current control methods", *IECON*, Montreal, QC, 2012, pp. 3647-3652.

[37] C. S. Lim, E. Levi, M. Jones, N. Abdul Rahim and W. P. Hew, "Experimental evaluation of model predictive current control of a five-phase induction motor using all switching states", *(EPE/PEMC)*, Novi Sad, 2012, pp. LS1c.4-1-LS1c.4-7.

[38] S. Rubino, R. Bojoi, S.A. Odhano, P. Zanchetta, "Model Predictive Direct Flux Vector Control of Multi Three-Phase Induction Motor Drives", ECCE, 2017, pp. 1-8.

[39] A. Iqbal, R. Alammari, M. Mosa and H. Abu-Rub, "Finite set model predictive current control with reduced and constant common mode voltage for a five-phase voltage source inverter", *ISIE*, Istanbul, 2014, pp. 479-484.

[40] Cheng Xue, Wensheng Song and Xiaoyun Feng, "Model predictive current control schemes for five-phase permanent-magnet synchronous machine based on SVPWM", *IPEMC-ECCE*, Hefei, 2016, pp. 648-653.

[41] D. Winterborne and V. Pickert, "Improving direct instantaneous torque control of switched reluctance machines with predictive flux constraints", *PEMD 2016*, Glasgow, 2016, pp. 1-6.

[42] H. J. Brauer, M. D. Hennen and R. W. De Doncker, "Multiphase Torque-Sharing Concepts of Predictive PWM-DITC for SRM", *7th International Conference on Power Electronics and Drive Systems*, Bangkok, 2007, pp. 511-516.

[43] B. Bogado, F. Barrero, M. R. Arahal, S. Toral and E. Levi, "Sensitivity to electrical parameter variations of Predictive Current Control in multiphase drives", IECON 2013, Vienna, 2013, pp. 5215-5220.

[44] C. H. Choi, J. K. Seok and R. D. Lorenz, "Wide-Speed Direct Torque and Flux Control for Interior PM Synchronous Motors Operating at Voltage and Current Limits," in *IEEE Trans. Ind. Appl.*, vol. 49, no. 1, pp. 109-117, Jan.-Feb. 2013.

[45] J. S. Lee and R. D. Lorenz, "Deadbeat Direct Torque and Flux Control of IPMSM Drives Using a Minimum Time Ramp Trajectory Method at Voltage and Current Limits," in *IEEE Trans. Ind. Appl.*, vol. 50, no. 6, pp. 3795-3804, Nov.-Dec. 2014.

[46] G. Pellegrino, R. Bojoi, and P. Guglielmi, "Unified Direct-Flux Vector Control for AC Motor Drives", *IEEE Trans. Ind. Appl.*, 2011, vol. 47, no. 5, pp. 2093-2102.

[47] I. Zoric, M. Jones and E. Levi, "Arbitrary Power Sharing Among Three-Phase Winding Sets of Multiphase Machines", in *IEEE Trans. Ind. Electron.*, vol. 65, no. 2, pp. 1128-1139, Feb. 2018.

[48] R. Bojoi, A. Cavagnino, A. Tenconi and S. Vaschetto, "Control of Shaft-Line-Embedded Multiphase Starter/Generator for Aero-Engine," in *IEEE Tran. on Ind. Electron.*, vol. 63, no. 1, pp. 641-652, Jan. 2016.

Influence of fast switching semiconductors on the winding insulation system of electrical machines

Kay Hameyer, Andreas Ruf and Florian Pauli
Institute of Electrical Machines (IEM)
RWTH Aachen University, Germany
Email: Kay.hameyer@iem.rwth-aachen.de

Abstract-Variable speed and low voltage electrical drives are commonly operated by frequency converters. According to recent developments [1-2], there is a trend in the area of semi-conductors, that switching frequency and voltage slew rate will increase significantly. The aim of these semiconductors is to reduce the switching losses and to increase the switching frequency, which enables to reduce the size of passive components in the power-electric circuit. This results in less material effort and lower cost, for the power electronic component. However, electric motors operated by large slew rate inverters show particular problems in the winding insulation, which have to be analyzed. Such problems are well known for high voltage machines. Due to the increasing slew rate, this problematic occurs in low voltage machines now as well. Here, the influence of fast switching semiconductors on the winding insulation system is studied, using accelerated ageing tests with fast switching high-voltage generators.

Keywords— Life time, winding insulation system, partial discharge, SiC semiconductors

I. INTRODUCTION

According to recent developments in power semiconductor technology, particularly in the area of wide bandgap semiconductors, the switching frequency and the slew rate of the voltage will increase significantly. State of the art silicon carbide SiC MOSFETs generate slew rates of $du/dt \approx 20$ kV/µs, which stimulate harmonics in the range of approx. 10 MHz [3-4]. From the state-of-the-art silicon-based IGBT inverter systems, the following three parasitic effects, which might endanger the drive system are already well known [5-6]:

1. Travelling wave phenomena, which result in high overvoltage at the terminals of the machine (cp. fig. 1),
2. Parasitic high frequency currents caused by the high du/dt of the "common-mode" inverter output voltage and
3. Line-conducted and radiated HF electromagnetic signals in the MHz-range.

The motors supply voltage represented in the frequency domain consists of three parameters: basic frequency, switching frequency and the frequency components due to the voltage slew rate. The insulation system of electrical machines as the most important part for the reliability experiences a deterioration by a combination of ageing mechanisms. The dominant ageing mechanisms are thermal, mechanical, ambient- and electrical loads [7].

In low voltage machines, which are operated by sinusoidal

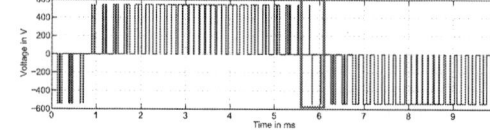

a) Idealized PWM voltage (4 kHz): line-to-line.

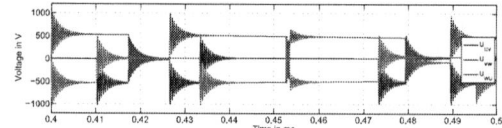

b) Simulated PWM voltage (16 kHz): line-to-line.

Fig. 1. Simulated PWM voltages for a two-level three-phase inverter with 540 V DC link voltage at a star-connected three phase winding system.

voltages or low switched dc voltages, deterioration by temperature is the dominant ageing factor for the winding system. Therefore, ageing by high electric fields (e.g. partial discharges) is usually not considered. The insulation systems for low voltage machines are typically defined as being not partial discharge (PD) resistant. For this reason, the machine designer and manufacturer must avoid partial discharges during the entire service life of the drive system, which is discussed in the standards [8] and [9] for different winding types. Standard [8] describes a methodology to calculate the maximum voltage amplitude for an example of a two-level inverter operating on a three-phase grid (cp. TABLE I). According to the standard the stress by overvoltage can be separated into stress categories depending on the overshoot OF = u_{pk}/u_{dc} and the impulse rise time t_r. When compared to the state of the art semiconductors the stress category must be assumed to be extreme (D) for a rise time $t_r<0.1$ µs of modern SiC semiconductor.

TABLE I.   STRESS CATEGORIES FOR INSULATION SYSTEMS [8].

Stress category	Overshoot factor ($p.u. u_{pk}/u_{dc}$)	Impulse rise time t_r in µs
A-Benign	OF ≤ 1.1	$1 < t_r$
B-Moderate	$1.1 < OF ≤ 1.5$	$0.3 ≤ t_r < 1$
C-Severe	$1.5 < OF ≤ 2.0$	$0.1 ≤ t_r < 0.3$
D-Extreme	$2.0 < OF ≤ 2.5$	$0.03 ≤ t_r < 0.1$

The combination of high overvoltage and high safety factors according to the standards lead to high test-voltages, which increase the requirements on the insulation system being free of partial discharges. The technological improvements of switching losses by fast switching semiconductors, the increase of dc link voltages and the development of corona resistant wire enamels with inorganic nano-particles and composites [10] moves the topic of electrical ageing by partial discharges into focus for low voltage machines. The technological question which must be answered is, whether the existing insulation systems must be improved to be PD free or PD resistant. To answer this question, the influence of the high loads due to fast switching has to be studied and analyzed in detail.

II. METHODOLOGY

Figure 2 presents the measurement configuration to investigate the electrical ageing of winding insulation probes. For this, a special high slew rate pulse voltage generator (cp. fig. 7) is developed at IEM, the Institute of Electrical Machines of RWTH Aachen University, which allows to vary the parameters of slew rate, switching frequency and the voltage magnitude, to adjust the desired load. To detect the PD-characteristics during the electric load, a high frequency oscilloscope is used with the antenna of a commercial winding test system for end-of-line tests (Schleich MTC 3). This device captures the electromagnetic waves which are caused by PD. Using the measurement configuration, the voltage threshold where PDs start to occur is determined for different probes of enameled wire.

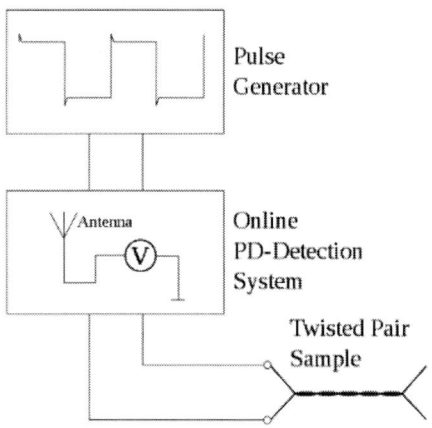

Fig. 2. Measurement configuration consisting of a pulse generator, an online PD-Detection System and a twisted pair probe of enameled wires.

III. DESIGN OF SPECIMENS

To imitate the setup of the interturn insulation of machine windings different models of enameled copper wires can be used. The most common model is a twisted pair (cp. fig. 3) of enameled copper wires for a better comparability

to standard tests [11]. Depending on the nominal diameter of an enameled wire, standards define different mechanical and electrical requirements.

To consider that thicker wires may be wound with a larger force, the specimens are twisted with different pulling forces during their set-up (TABLE II. This results into higher pressure forces between the wires with increasing diameter.

TABLE II.   LOAD AND NUMBER OF TWISTS FOR TWISTED PAIRS [11].

Nominal diameter in mm	Load in N	Number of twists
0.50 − 0.71	7	12
0.71 − 1.06	13.5	8
1.06 − 1.40	27	6
1.40 − 2.00	54	4

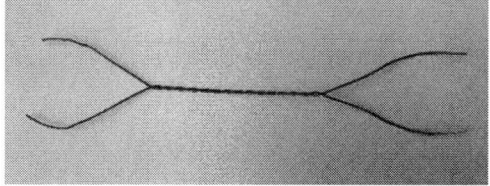

Fig. 3. Geometry of investigated twisted wire probes.

$$n \geq \frac{\left(z_{1-\alpha/2} \cdot \sigma\right)^2}{e^2} \qquad (1)$$

Requiring a maximum error e of 5 %, a confidence level of 95 % ($z_{1-\alpha/2} = 1.96$) and assuming a standard deviation σ of 10 %, the minimum number of samples n must be 15. In this study, $n=20$ samples of each enameled wire are chosen to increase the confidence level of the results. The specimens are tested with different voltage shapes under defined conditions (20 °C (68 °F), 50 % r.H.). For measurements with sinusoidal voltage with a frequency of 50 Hz (slew rate ≈ 1 V/µs) and unipolar surge voltages (slew rate ≈ 20 kV/µs) a commercial stator analyzer according to standard [12] is utilized. The different voltage shapes are displayed in fig. 4

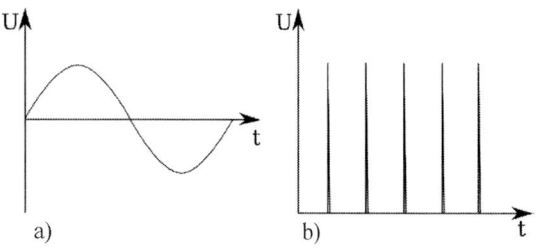

Fig. 4. Shape of test voltage of standard winding tester a) sinusoidal test voltage and b) unipolar surge voltage.

According to IEC 60034-18-41 [8] both voltage shapes are suitable to evaluate the performance of insulation systems for inverter driven electrical machines. Effects such as voltage overshoot and reflection of electromagnetic waves are considered by a series of experience based factors that are multiplied by the dc-link voltage.

The source of the standard winding test system provides up to 5 pulses per second in the unipolar surge voltage mode.

In this study, two types of wires are studied. The wire's properties are listed in TABLE III. Both, wire enamels consist of a polyester imide basecoat and a polyamide imide topcoat. For the grade 3 wire nano-particles made of Cr_2O_3 are added to the topcoat to improve the PD-resistance. The samples are chosen to study the influence of an increased coating layer and to take electrical aging of PD-resistant materials into consideration.

IV. INITIAL MEASUREMENTS

For sinusoidal voltages, the partial discharge inception voltage (PDIV) is measured by exposing a specimen to the output voltage of the measuring system. The amplitude of the voltage is increased until PD is detected. The lowest voltage that leads to PD is the PDIV. This procedure is performed for the 20 specimens of each wire. The results for this measurement are collected in fig. 5. Here, the mean partial discharge inception voltage (PDIV) is 1458 V for the wire with a grade 2 enamel and 2039 V for the grade 3 enamel, which is about 40% higher. Also the lowest measured PDIV for the grade 3 wire (1680 V) is significantly higher than the minimum PDIV for the grade 2 wire (968 V).

Regarding the repetitive partial discharge inception voltage (RPDIV) for the pulsed unipolar surge voltage measurement (The lowest voltage where at least half of the pulses of the same amplitude cause PD) the increase of the mean RPDIV is less significant: from 2029V for grade 2 to 2131 V for grade 3 which is only 5 %. Also the minimum RPDIV only changes from 1407 V to 1461 V. While for the sinusoidal test-voltage an increase of the grade of the insulation yields significantly higher PDIVs. For the pulsed unipolar surge voltage there is no reliable

correlation between increase of the PDIV and insulation grade to be identified (cp. fig. 6). These results lead to the conclusion that considering effects that are due to a pulse shaped voltage with a high du/dt by using standardized measurement systems with sinusoidal voltages with increasing magnitude or pulsed unipolar surge voltage, is not a valid option for a reliable analysis of the winding insulation system.

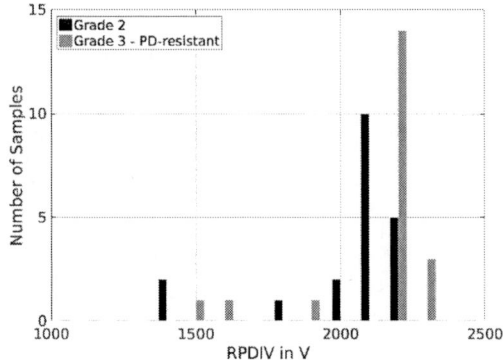

Fig. 6. Measured RPIDV for pulsed unipolar surge voltages of a grade 2 enameled wire and a corona resistant grad 3 enameled wire.

TABLE III.   PROPERTIES OF ENAMELLEL WIRES

Grade	Nominal diameter in mm	Corona resistant
Grade 2	0.56	no
Grade 3	1.0	yes

Applying this approach would mislead to the conclusion that also for pulse voltages a grade 3 enameled wire yields significantly higher PDIVs than a grade 2 wire. Inverters employing SiC-semiconductors can provide slew rates of more than 50 kV/µs and significantly higher switching frequencies as the standard measurement test system. As SiC-inverters are considered to be an important component in future drive systems, the behavior of enameled wires under such an inverter load has to be studied. For this purpose, a pulse voltage generator is developed and a test procedure is set-up.

V. HIGH SLEW RATE PULSE VOLTAGE GENERATOR

The voltage generator that is developed at IEM is displayed in fig 7. It can provide unipolar and bipolar pulses. Amplitude, switching frequency and slew rate are displayed in TABLE IV. The parameters can be varied independently. For initial tests the generator is operated with a twisted pair specimen. For an output voltage of 1200 V and a switching frequency of 20 kHz the voltage and current output is measured during the test and displayed in figures 8 and 9.

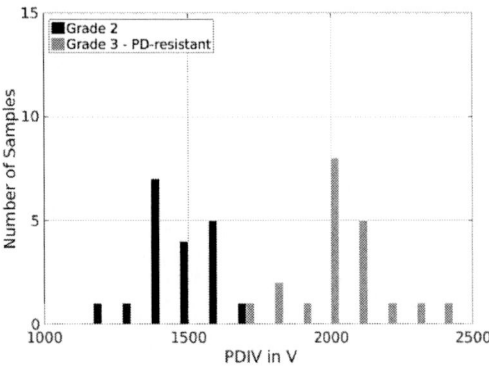

Fig. 5. Measured PIDV for sinusoidal test voltage of a grade 2 enameled wire and a corona resistant grade 3 enameled wire.

The 2018 International Power Electronics Conference

TABLE IV.   PARAMETERS OF THE PULSED VOLTAGE OF THE IEM-PULSE-VOLTGAE-GENERATOR

Parameter	Minimum value	Maximum value
Voltage Amplitude	-1200 V	1200 V
Switching Frequency	1 Hz	100 kHz
Slew rate	10 kV/ μs	50kv/ μs

Fig. 7.   Developed SiC high slew rate pulse voltage generator.

It can be seen that each voltage pulse causes a current pulse which decays within 20 ns. From fig. 9 can be taken, that the pulse starts as soon as the voltage starts to increase. This is caused by the capacitances of the specimen. It can be detected that the specimen is operated under the condition of PD. Under the influence of partial discharge, the insulation ages rapidly, causing at the end an electric breakdown. Such a breakdown is finally characterized by a short circuit between the two wires which occurs after PD breaks the insulation material. At this point the remaining life time is zero.

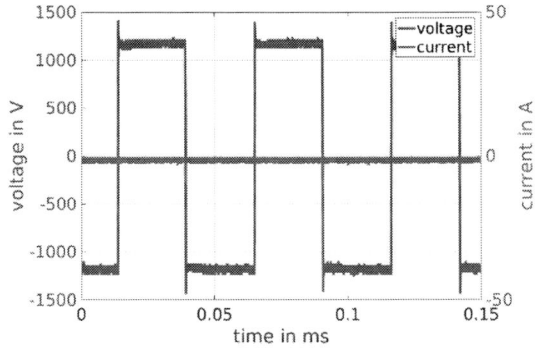

Fig. 8.   Measured voltage an current of the high slew rate pulse voltage generator, operated with a twisted pair specimen.

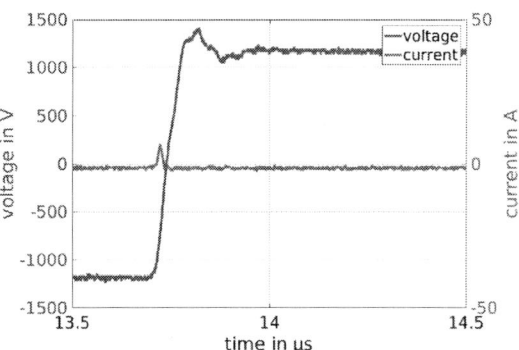

Fig. 9.   Measured voltage step and current of the high slew rate pulse voltage generator.

In this case the current rises very strong (fig. 10). The breakdown occurs at 200 μs. The current rises to more than 50 A, which is the maximum current that can be detected by the measuring system used.

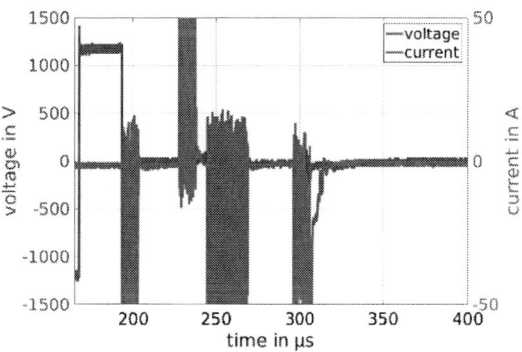

Fig. 10. Measured voltage step and current of the high slew rate pulse voltage generator, operated with a twisted pair specimen.

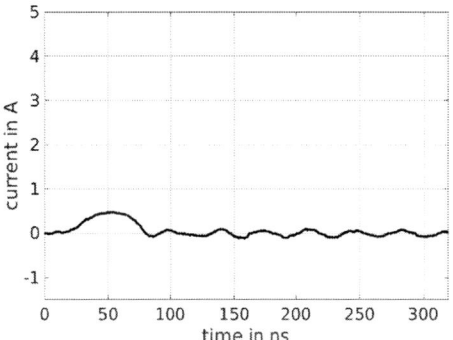

Fig. 11. Current for a PD-free specimen.

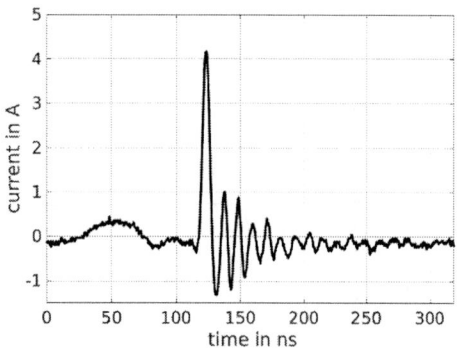

Fig. 12. Current for a specimen where PD occurs at 120 ns.

The developed measuring system is equipped with a current sensor, which provides a bandwidth of 120 MHz to detect PD. In fig 11 and 12 the measured currents for a specimen that is PD-free and a specimen where PD occurs are displayed. At the time 0 s, the voltage slope is initialized. For the specimen that is PD-free (fig. 11) a capacitive current with a rise time of 35 ns is measured. For the specimen which is exposed to PD (fig. 12), this current is superposed with high current peaks with a significantly lower rise time of 8 ns. The magnitude of the current peaks that indicates PD is limited to 4… 5 A and therefore significantly lower than the aforementioned currents of more than 50 A that occur at a breakdown (cp. fig 10).

VI. ELECTRICAL AGEING

Besides evaluating PDIVs the setup can be used to perform accelerated electrical ageing tests. It is well known that the service life of conventional enameled wires under the influence of PD is reduced to a few hours. Therefore, winding systems of machines equipped with standard enameled wires are always designed to be PD-free.

TABLE V.   LIFE EXPECTANCIES OF ENAMELLED WIRES UNDER THE INFLUENCE OF PD [13]

Enameled wire	Life expectancy in h
Grade 2 Standard	0.7
Grade 3 standard	4
Grade 2 PD-resistant	221
Grade 3 PD-resistant	>500

New insulation materials are mixed with inorganic nano-particles made of Cr_2O_3. The life-time reduction of these wires due to PD is not as severe as for standard wires. In TABLE V the life expectancy of standard wires and PD-resistant wires is compared. The data are taken from a datasheet of a manufacturer of enameled wires. However, the data are only provided for one point of operation. Here, at 16 kHz and 2.4 V peak to peak and an unknown slew rate. To estimate if machines with PD- resistant wires can be operated at voltages that exceed the PDIV, the ageing

TABLE VI.   PARAMETERS OF THE TESTING VOLTAGE FOR THE MEASURING SEREIS WITH A VARIED FREQUENCY [13]

Parameter	Value
Voltage amplitude	Bipolar voltage of 1200V (ΔV=2400V)
Slew rate	45 kV/ μs
Frequencies	10, 20, 40 kHz

behavior of these new generation enameled wires under the influence of PD must be studied.

For a first measuring series the dependency of the frequency on the time to failure is examined. For this purpose, a standard non-PD-resistant enameled wire is studied. In TABLE VI the parameters of the pulsed testing voltage are displayed. Voltage amplitude and slew rate are kept constant, while the frequency is varied. In our experiment 5 twisted pair specimen are exposed to the test voltage for each measured frequency. The resulting time to failure is displayed in fig. 13. Due to the overshoot most of the PD occur during the voltage slope. Therefore, it is expected that the time to failure decreases reciprocally in the proportion the frequency increases. The measured data however suggest that the lifetime decreases super-linear. This may be accounted to the fact that the PD creates a loss power which leads to an increase of the specimen's temperature.

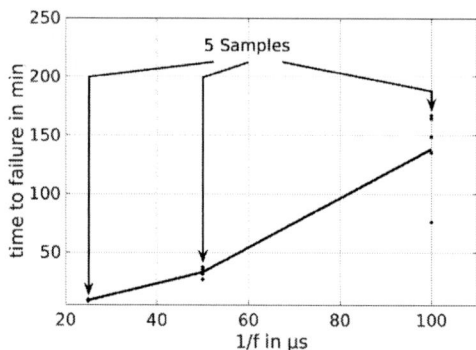

Fig. 13. Time to failure of the specimen for a variation of switching frequency.

The high local temperatures lead to a further decrease of the time to failure. The temperatures of the specimen are measured using an infrared camcorder. The results are displayed in TABLE VII.

TABLE VII.   TEMPERATURE OF THE SPECIME IN DEPENDANCE ON THE FREQUENCY

Frequency in kHz	Temperature in °C
10	60
20	115
40	250

For a Frequency of 40 kHz the temperature is significantly higher than the index temperature of the enamel of 200 °C. Therefore, it can be stated that the insulation material decays not only due to PD but also due to temperature effects. The influence of a varying voltage amplitude and slew rate can be studied by employing the developed high slew rate pulse voltage generator. As for the detection of the PDIV, the parameters of the output voltage are varied separately. Qualitatively the results from TABLE VIII are expected.

TABLE VIII.   EXPECTED INFLUENCE OF THE PULSE-VOLTAGE-PARAMETERS ON THE LIFE EXPECTANCY OF ENAMELED WIRES

Parameter variation	Expected influence on the life time
Increase of the switching frequency	Decrease of the service life
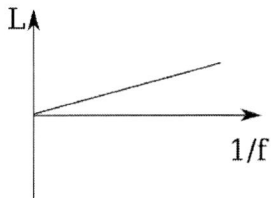	
Increase of the voltage amplitude	Decrease of the service life/ below the PDIV the lifespan is significantly larger than above it.
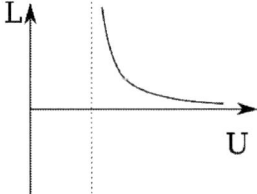	
Increase of the slew rate	Decrease of the service life
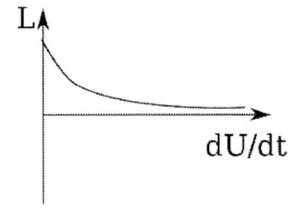	

I.   RESULTS

In this paper it is shown that for sinusoidal voltages the PD-inception voltage increases significantly when a grade 2 wire is substituted with a grade 3 wire. However, for the standard pulsed unipolar surge voltage there is no reliable correlation between insulation grade and RPIDV. The standard sinusoidal voltage test as well as the pulsed unipolar surge voltage are not a suitable means to identify the PDIV for inverter operated electrical machines.

The developed high slew rate pulse voltage generator is capable to load a winding system for a test varying the relevant parameters such as frequency, slew rate and peak voltage.

To guaranty long-life time for the winding system a PD-resistant system can be designed. To characterize the ageing behavior of the winding specimens the high slew rate pulsed voltage generator can be employed. PWM signals can be analyzed which promise better PD resistivity. The methodology discussed in this paper can be applied to answer the question if PD-resistant materials are a suitable approach to design machines for the operation with fast switching SiC-inverters.

REFERENCES

[1] B. Wrzecionko, J. Biela, J. W. Kolar, "SiC Power Semiconductors in HEVs: Influence of Junction Temperature on Power Density, Chip Utilization and Efficiency", Industrial Electronics IECON, 2009.

[2] D. Bortis, B. Wrzecionko, J. W. Kolar, "A 120 °C ambient temperature forced air-cooled normally-off SiC JFET automotive inverter system", Applied Power Electronics Conference and Exposition (APEC), 2011.

[3] T. Köneke, A. Mertens, D. Domes, P. Kanschat, "Highly Efficient 12kVA Inverter with Natural Convection Cooling Using SiC Switches", PCIM Europe, Nuremberg, 2011.

[4] J. Biela, M. Schweizer, S. Waffler, J. W. Kolar, "SiC versus Si-Evaluation of Potentials for Performance Improvement of Inverter and DC-DC Converter Systems by SiC Power Semiconductors", IEEE Transactions on Industrial Electronics, vol. 58, no. 7, 2011.

[5] X. Gong, J.A. Ferreira, "Comparison and Reduction of Conducted EMI in SiC JFET and Si IGBT-Based Motor Drives", IEEE Transactions on Power Electronics, vol. 29, no. 4, April 2014.

[6] Andreas Ruf, Jörg Paustenbach, David Franck and Kay Hameyer, A methodology to identify electrical ageing of winding insulation systems, International Electric Machines and Drives Conference, IEMDC 2017, pages CD-ROM, 2017.

[7] IEC 60034-18-1:2010 -Rotating electrical machines - Part 18-1: Functional evaluation of insulation systems,2010.

[8] IEC 60034-18-41 ed. I, Rotating electrical machines - Part 18-41: Partial discharge free electrical insulation systems (Type I) used in rotating electrical machines fed from voltage converters – Qualification and quality control tests, 2014.

[9] IEC 60034-18-42, Rotating electrical machines - Part 18-42: Qualification and acceptance tests for partial discharge resistant electrical insulation systems (Type II) used in rotating electrical machines fed from voltage converters, 2008.

[10] H. Kikuchi and H. Hanawa, "Inverter surge resistant enameled wire with nanocomposite insulating material", IEEE Transactions on Dielectrics and Electrical Insulation, vol. 19, no. 1, pp. 99-106, February 2012.

[11] IEC 60317-0-1, Specifications for particular types of winding wires – Part 0-1: General requirements - Enameled round copper wire, 2013.

[12] IEC 61934, Electrical insulating materials and systems – Electrical measurement of partial discharges (PD) under short rise time and repetitive voltage impulses, 2011.

[13] LWW-Group: Dahrentrad Daprest 200 Round enamelled conductor of copper, corona resistant, class 200 datasheet.

The 2018 International Power Electronics Conference

Centralized Control of Modular Multi Rectifier for Motor Drive Applications under Unbalanced Grid

Yipeng Song*, Pooya Davari and Frede Blaabjerg
Department of Energy Technology, Aalborg University, Aalborg, Denmark
*E-mail: yis@et.aau.dk

Abstract — **A centralized control scheme is proposed for modular multi rectifier system as a popular front-end rectifier solution for the industrial application where multiple motors are supplied by the Silicon Controlled Rectifier (SCR). Normally under conventional balanced grid voltage, the input current distortions can be improved by implementing the appropriate SCR firing angles in each unit. However, the three-phase grid voltage may be either amplitude or phase angle unbalanced, which deteriorate the performance of the conventional harmonic mitigation method. In order to remedy this issue, the three-phase total input currents 3rd harmonic components are extracted and included in the closed-loop control using PI controller, and a reset unit with the input of inductor current is adopted to reset the integrator. The proposed control strategy is able to obtain the appropriate SCR firing angle compensation, and then consequently achieve harmonics improvement in the input current under unbalanced grid voltage. Simulation results based on PLECS and experimental results have validated the effectiveness of the proposed improved control strategy under unbalanced grid voltage.**

Keywords — **Multi-drive system; unbalanced grid voltage; centralized control; harmonic mitigation; firing angle compensation.**

I. INTRODUCTION

The Adjustable Speed Drive (ASD) has been widely implemented in the field of motor control due to its performance advantages such as variable magnitude and frequency of the output control voltage. From the point view of less cost and higher capacity, the Silicon Controlled Rectifier (SCR), rather than the full-controllable power electronics device, is adopted for the industrial ASD applications.

Fig. 1 shows the typical configuration of a multi-drive application with multiple SCR units [1]-[3], where three-phase full-bridge SCR converters and the Electronic Inductor (EI) [1] are adopted. The EI consists of one inductor, one diode and one IGBT, and performs as a boost converter. The output capacitor in each unit is used to smooth the output voltage fluctuation, and a resistive load is presented to act as load which is often an inverter plus a motor in practical. Since the SCR-based ASD will unfortunately inject significant amount of low order harmonic current into the grid, it is essential to mitigate these harmonic components in order to improve the performance of the entire ASD system and also ensure good grid voltage quality.

As a cost-effective solution reported in [1]-[3], the appropriate SCR firing angles can be applied to each single unit, and the harmonics in the total input current can be improved. Note that this control strategy is proposed under three-phase balanced grid voltage condition, however the grid voltage in practice may behave as either amplitude unbalance, or phase angle unbalance, or even both amplitude and angle unbalance. Fig. 2 shows the phasor diagram of the grid voltage unbalance cases which will be discussed in this paper, i.e., Fig. 2(a) shows the amplitude unbalance in phase b, and Fig. 2(b) and (c) shows the capacitive and inductive angle unbalance in phase b. It is assumed that the maximum unbalance is 10% according to the definition of Voltage Unbalance Factor (VUF). Then, the amplitude unbalance in Fig. 2(a) is set as 230 V, 165.5 V and 230 V in phase a, b and c respectively; while the angle unbalance in Fig. 2(b) and Fig. 2(c) is set as 0°, -120°±17.1° and +120° in phase a, b and c respectively.

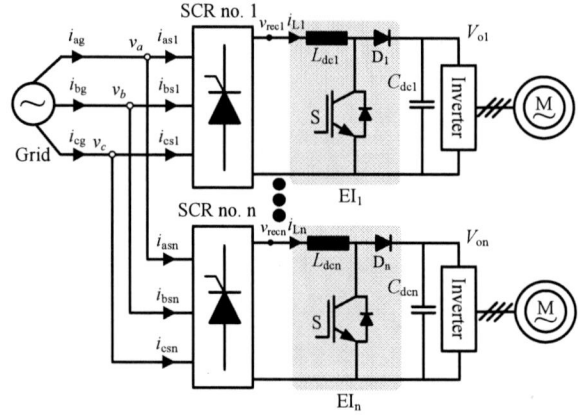

Fig. 1. Configuration of a multi-drive application with multiple SCR units

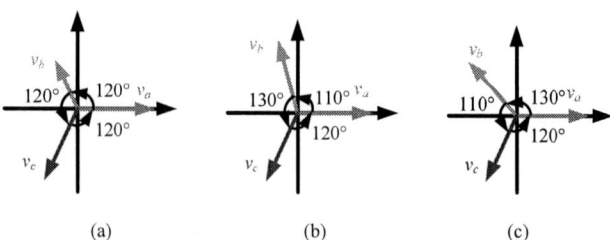

Fig. 2. Grid voltage unbalance in phase b, (a) amplitude unbalance; (b) angle unbalanced (capacitive); (c) angle unbalanced (inductive).

Due to the grid voltage unbalance, the SCR may fail to commute at the instance of firing pulse, and no equal conduction periods among three phases can be achieved, and finally results in the deterioration of input current harmonics distortion. In order to solve this problem, it is necessary to introduce the SCR firing compensation angle so as to ensure 120° conduction angle for each phase and then consequently improve the input current harmonics. For the purpose of achieving this improvement, the three-phase total input currents 3rd harmonic components are extracted and included in the closed-loop control using PI controller, and a reset unit with the input of inductor current is also adopted to reset the integrator, which will be explained in the following discussion.

This paper proposed a centralized control of modular multi rectifier for motor drive applications under unbalanced grid. The performance deterioration of multi-drive system under unbalanced grid voltage is demonstrated in Section II. The proposed improved control strategy, which consists of closed-loop control of output voltage and inductor current, as well as the firing angle compensation method based on the input current 3rd harmonics, are illustrated in Section III. The simulation validation and experimental results are provided in Section IV and V respectively to verify the effectiveness of the improved control strategy.

II. System Performance under Balanced and Unbalanced Grid

The conventional harmonic mitigation strategy is able to achieve certain low order harmonic mitigation under balanced grid voltage [1] when the equal-space firing pulse method, which generates the firing pulse every 60°,

is adopted for the triggering of the SCR. Nevertheless, its performance can be deteriorated when the grid unbalance occurs.

Fig. 3 shows the FFT analysis of the total input current of two SCR units with firing angle of $\alpha_{f1} = 30(0+30)°$ and $\alpha_{f2} = 66(36+30)°$ under (a) balanced grid voltage; (b) amplitude unbalanced grid voltage in phase b; (c) angle unbalanced (capacitive) grid voltage in phase b; (d) angle unbalanced (inductive) grid voltage in phase b. Note that the firing angles for SCR unit 1 and unit 2 $\alpha_{f1} = 30°$ and $\alpha_{f2} = 66°$ are designed as for the purpose of eliminating the 5th harmonics of input current. It can be seen from Fig. 3(a) that the three phase total input currents are balanced under balanced grid voltage. Due to the firing angle difference of 36° between unit 1 and unit 2, the 5th harmonics current can be eliminated, while the 7th, 11th, 13th, 17th and 19th harmonics remain.

Nevertheless, once the grid amplitude unbalance occurs with lower amplitude in phase b in Fig. 3(b), the 5th harmonic current can not be well suppressed, and most importantly, the 3rd harmonics are produced in the total input current as a consequence. Similar unbalanced conduction period can be seen for the angle unbalanced (capacitive) grid in Fig. 3(c), the 3rd harmonics current also appears in this case.

Moreover, for the case of the angle unbalance (inductive) grid in phase b as shown in Fig. 3(d), the appropriate commutation and balanced three phase input current can still be ensured in the three phase SCR bridge due to the fact that the firing pulse is phase lagging compared with the three-phase grid voltage intersection moment, thus the 5th harmonics can still be suppressed, and no 3rd harmonic current can be seen, which is quite similar to the performance under balanced grid voltage.

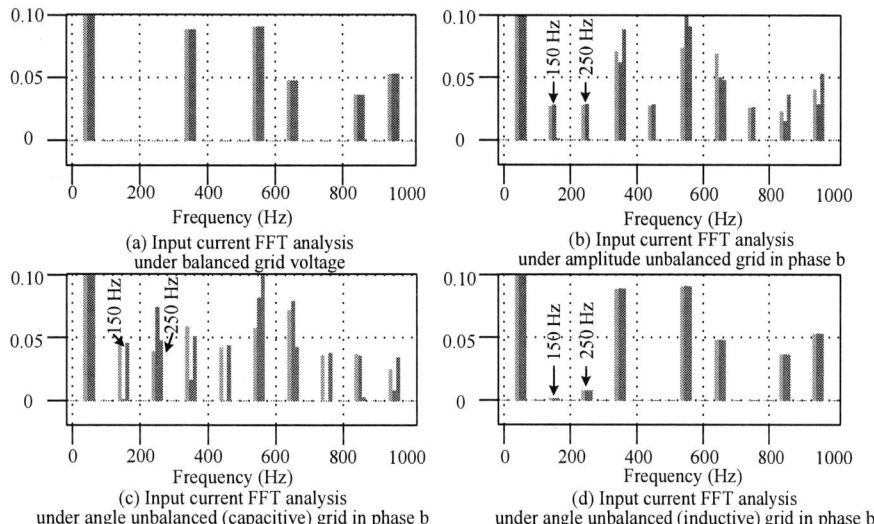

Fig. 3. FFT analysis of the total input current of two SCR units with firing angle of $\alpha_{f1} = 30(0+30)°$ and $\alpha_{f2} = 66(36+30)°$ under (a) balanced grid voltage; (b) amplitude unbalanced grid voltage in phase b; (c) angle unbalanced (capacitive) grid voltage in phase b; (d) angle unbalanced (inductive) grid voltage in phase b.

Based on above descriptions on Fig. 3, it can be found out that the 3rd harmonics of total input current may occur consequently when the grid voltage unbalance happens. Thus, the total input current 3rd harmonics can be chosen as the feedback of the closed-loop control, which is able to achieve the appropriate firing angle compensation for the SCR units. As long as the firing angle compensation is applied, the multi-drive system under unbalanced grid voltage is able to achieve the same current harmonics performance as the one under balanced grid voltage.

III. PROPOSED CONTROL STRATEGY UNDER UNBALANCED GRID

In this section, the proposed improved control strategy, which is able to achieve the firing angle compensation under unbalanced grid voltage, will be demonstrated. Besides, output voltage outer closed-loop control and the inductor current inner closed-loop control will also be illustrated.

As shown in blue box in Fig. 4, the output voltage closed-loop control is implemented to regulate the output voltage V_o according to the reference V_o^* using the PI controller. The PI controller output is considered as the inductor current closed-loop control reference i_L^*. Thereafter, the closed-loop control of the load current i_L is implemented with the reference i_L^* using the hysteresis controller, and its output switching signal is delivered to the IGBT.

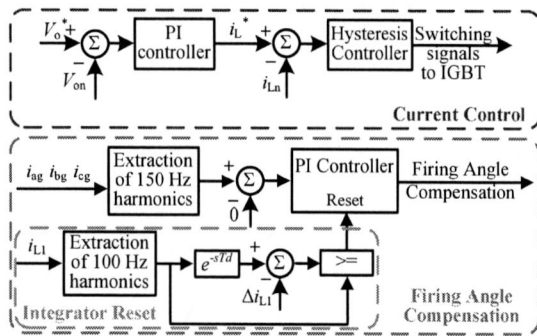

Fig. 4. Overall system control block diagram

On the other hand, the firing angle compensation method is presented in red box in Fig. 4. Based on the total input current waveforms and THD analysis shown in Fig. 3, the 3rd harmonics of total input current i_{ag}, i_{bg} and i_{cg} will appear if the grid voltage unbalance occurs. Thereafter, the first step of the firing angle compensation method is to extract the 3rd harmonics and obtain its amplitude information.

Since it is always preferred to have zero 3rd current harmonics, the reference value of the 3rd current harmonics closed-loop control is set as zero. The PI controller is adopted here to output the firing angle compensation value, which can be added to the conventional firing angle of each SCR unit.

Most importantly, it should be pointed out that the amplitude information of the 3rd current harmonics is always a positive value, thus the PI controller output can not decrease due to the integrator and the non-negative input error value. As a consequence, once the grid voltage changes from unbalanced to balanced, the firing angle compensation

can not go back to zero. This means the unnecessary firing angle compensation may still be applied under balanced grid voltage, and cause the decreasing of the system power factor.

In order to solve this problem, the integrator reset unit is implemented in the green box in Fig. 4. Since the inductor is connected between the SCR output rectified voltage V_{rec} and the output voltage V_o, while the V_{rec} is related to the three phase grid voltage V_{abc}, thus the inductor current i_L can be used to detect the grid voltage unbalance by extracting its 100 Hz harmonics component.

Note that the amplitude value of the i_L 100 Hz harmonics is proportional to the extent of the grid voltage unbalance. In order to give out a reset signal when the grid voltage unbalance extent becomes smaller, the i_L 100 Hz amplitude information is compared with a flexible threshold value. This flexible threshold value is obtained by delaying the i_L 100 Hz harmonics a small period of time e^{-sTd} and subtract it with a small positive value $\varDelta i_L$. The logic comparator, with the inputs of i_L 100 Hz harmonics amplitude and the flexible threshold, can output the reset signal for the integrator.

IV. SIMULATION VALIDATION

In order to validate the proposed centralized improved control strategy for multi-drive system under unbalanced grid voltage, the simulation model based on PLECS is built, and the two SCR units are demonstrated as an example, while simulations with more units can be similarly conducted and will not be described in detail here. The simulation system block diagram and the control strategy have been described in detail in Fig. 4. The simulation parameters are shown in Table I.

TABLE I. PARAMETERS OF THE SIMULATED SYSTEM

Output voltage V_o	700 V	Output cap C_{dc}	470 μF
Firing angle α_{f1}	30°	Firing angle α_{f2}	66°
Rated power (each)	5 kW	Input voltage	230 V
K_p for V_o	0.01	K_i for V_o	1.5
K_p for i_{abcg}	8	K_i for i_{abcg}	20
Hysteresis band	±1.5 A	Inductor	2 mH
Delay T_d	50 ms	$\varDelta i_L$	5e-4 A

A. Simulation of outer closed-loop control of output voltage and inner closed-loop control of inductor current

Fig. 5 shows the simulation of the outer closed-loop control of output voltage and the inner closed-loop control of inductor current. It can be seen that the output voltage reference (red in the top) is set as 700 V, and the output voltage feedback (green in the top) is able to track precisely the reference due to the PI controller. Thereafter, the PI controller output is regarded as the reference of the inductor current control (red in the bottom), and the inductor current feedback (green in the bottom) follows the reference with the hysteresis band of ±1.5 A.

The 2018 International Power Electronics Conference

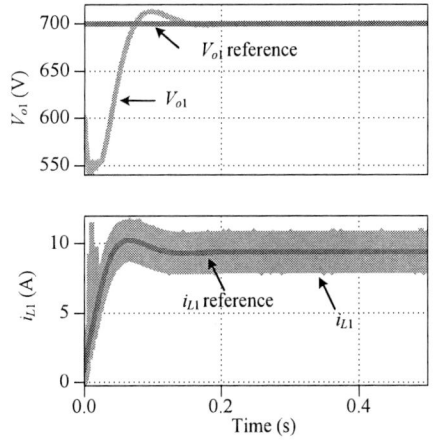

Fig. 5. Simulation of outer closed-loop control of output voltage and inner closed-loop control of inductor current

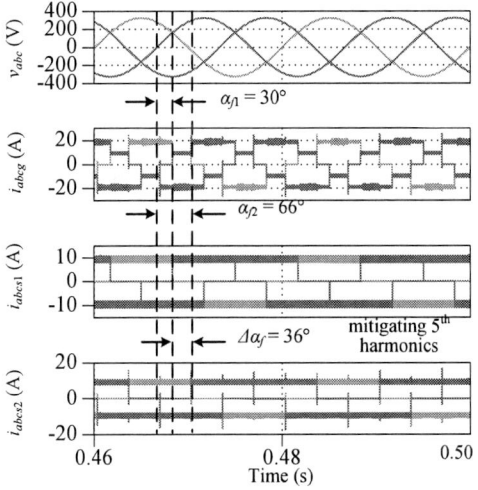

Fig. 6. Simulation of the overall system performance with two SCR units under balanced grid

B. Simulation of the firing angle compensation method

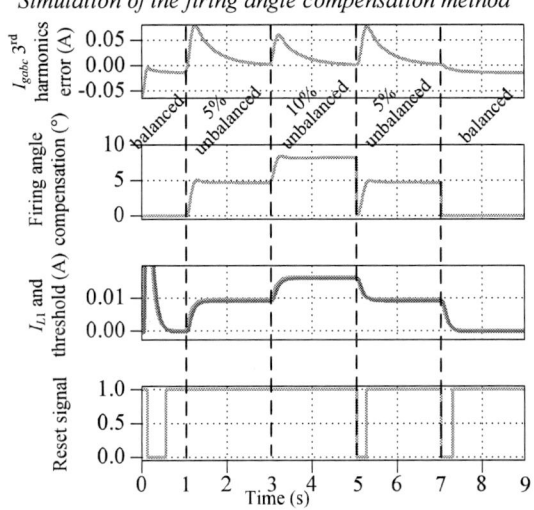

Fig. 7. Simulation of the firing angle compensation method under 5% and 10% unbalanced grid

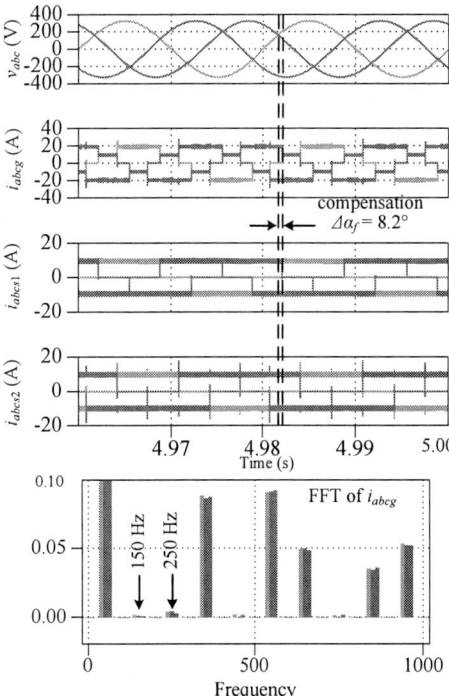

Fig. 8. Simulation of overall system performance with two SCR units under 10% unbalanced grid and the firing angle compensation is applied

Fig. 6 shows the simulation of the overall system performance with two SCR units under balanced grid. It can be seen that the first unit has the firing angle $\alpha_{f1} = 30°$, thus its current commutation happens at the intersection point of v_a and v_b. While the second unit has the firing angle $\alpha_{f2} = 66°$. The firing angle difference of 36° results in the mitigation of 5th harmonics in the total input current i_{abcg}.

Thus, based on the simulation shown in Fig. 5 and Fig. 6, the effective control performance of both output voltage and the inductor current can be validated, and the balanced three phase total input current with the mitigation of 5th harmonic can be verified.

Fig. 7 shows the simulation of the firing angle compensation method under 5% and 10% unbalanced grid. It can be seen that during 0 – 1s the firing angle compensation is zero under balanced grid; then during 1 – 3s when the 5% grid voltage unbalance happens, the i_{abcg} 3rd harmonics control error becomes larger, and the firing angle compensation comes to the steady state of 4.8°; similarly, the firing angle compensation becomes 9.2° when the grid voltage unbalance increases up to 10% during 3 – 5s.

On the other hand, when the grid voltage unbalance decreases from 10% to 5% during 5 – 7 s, the integrator reset signal is generated at 5 s, and the firing angle compensation can be reset to 0° first, and then comes to the steady state output of 4.8°. Similarly, when the grid voltage unbalance decreases from 5% to balanced during 7 – 9 s, the firing angle compensation can be reset to zero at 7 s.

Fig. 8 shows the simulation of overall system performance with two SCR units under 10% unbalanced grid. It can be seen that the firing angle compensation $\Delta\alpha_f = 8.2°$ are applied in both SCR units. As a result, the 150 Hz

749

and 250 Hz harmonics in the total input current can be well suppressed as shown in Fig. 8, and the FFT analysis is almost the same as the one under balanced grid voltage as shown in Fig. 3(a).

Therefore, based on the simulation results shown in Fig. 7 and Fig. 8, the effectiveness of the firing angle compensation method can be validated. The limitation of the firing angle compensation method is that the integrator reset unit always resets the firing angle compensation to zero, thus the slow dynamics can be a consequence.

V. EXPERIMENTAL VALIDATION

Besides the simulation validation, an experimental setup is also built up with two drive units in parallel connection. The grid voltage amplitude unbalance of 10% in phase a or b is implemented during the experiment.

The parameters of the experiment setups are listed as shown in Table II. The input voltage is set as 110 V RMS per phase, and the 10% amplitude unbalance results in the phase voltage decreasing to 79.1 V. As a results, the two cases of grid voltage amplitude unbalance are set as, 1) $v_a = 79.1$ V, $v_b = 110$ V and $v_c = 110$ V; 2) $v_a = 110$ V, $v_b = 79.1$ V and $v_c = 110$ V. The output dc voltage is set at 300 V, with the rated power of each unit is 1.3 kW. The other parameters are available in Table II.

TABLE II. PARAMETERS OF THE EXPERIMENT SETUPS

Input voltage (Phase RMS)	110 V	Output cap C_{dc}	470 μF
Firing angle α_{f1}	30°/40°	Firing angle α_{f2}	60°/70°
Rated power (each)	1.3 kW	Output voltage	300 V
K_p for V_o	0.01	K_i for V_o	1.5
K_p for i_{abcg}	8	K_i for i_{abcg}	20
Hysteresis band	±1.5 A	Inductor	2 mH

Fig. 9 shows the experimental results of multi-drive system with two SCR units under balanced grid ($v_a = 110$ V, $v_b = 110$ V and $v_c = 110$ V), the firing angle $\alpha_{f1} = 30 (0+30)°$ and $\alpha_{f2} = 60 (30+30)°$. As it is shown, under balanced grid, the three-phase total input currents have symmetric waveforms, and the FFT analysis results show that no 3rd harmonics are generated in this case, a relatively low THDi can be achieved consequently, i.e., i_a: 16.17%, i_b: 15.94%, i_c: 16.40%.

Fig. 10 shows the experimental results of multi-drive system with two SCR units under 10% amplitude unbalanced grid in phase a ($v_a = 79.1$ V, $v_b = 110$ V and $v_c = 110$ V), the firing angle $\alpha_{f1} = 30 (0+30)°$ and $\alpha_{f2} = 60 (30+30)°$. Due to the grid unbalance, the 3rd harmonic sequence occurs in the total input current, and the THDi also becomes worse compared with the case under balanced grid, i.e., i_a: 18.44%, i_b: 15.71%, i_c: 18.28%.

Fig. 11 shows the experimental results of multi-drive system with two SCR units under 10% amplitude unbalanced grid in phase a ($v_a = 79.1$ V, $v_b = 110$ V and $v_c = 110$ V), the firing angle $\alpha_{f1} = 30+10°$ and $\alpha_{f2} = 60 +10°$, the firing angle compensation of 10°. Compared with Fig. 10, the firing angle compensation value of 10° is implemented,

as a result the THDi of total input current can be improved, i.e., i_a: 15.83%, i_b: 15.56%, i_c: 17.50%.

Besides the case of grid unbalance in phase a, the experimental results with grid unbalance in phase b are also similarly shown in Fig. 12 and Fig. 13. Similar performance can be achieved with the improved THDi by adopting the firing angle compensation. Due to the sake of simplicity, no detailed explanations will be provided here.

In conclusion, the THDi of total input current under both balanced grid and unbalanced grid in phase a/b can be summarized as shown in Table III. By comparing the THDi data under different cases, it can be verified that the firing angle compensation can help to improve the total input current distortion performance under unbalanced grid voltage condition.

TABLE III. THDI OF TOTAL INPUT CURRENT

THDi	i_a	i_b	i_c
Balanced grid	16.17%	15.94%	16.40%
Unbalance in phase a Without compensation	18.44%	15.71%	18.28%
Unbalance in phase a With compensation	15.83%	15.56%	17.50%
Unbalance in phase b Without compensation	18.07%	18.94%	15.78%
Unbalance in phase b With compensation	18.40%	16.15%	15.63%

Moreover, the discussion regarding the total input 3rd harmonics needs to be conducted. In the theoretical section, the 3rd harmonics based firing angle compensation method is illustrated. The authors would like to validate the effectiveness of this method, nevertheless the thyristor triggering pulse board RT380S from Semikron fails to implement the appropriate equal-pulse-space (EPS) firing method under unbalanced grid condition, which indicates that the 60 degree firing pulse is no longer available.

Since the EPS firing method serves as fundamental role in the system, as a consequence, the proposed 3rd harmonics based firing angle compensation method can not be validated in this setup. As it can be seen from the experimental results, the 3rd harmonics always exists even when the firing angle compensation is applied. Instead, the TDHi of total input currents are compared and summarize in order to verify the effectiveness of the firing angle compensation in respect to the improvement of the input current quality.

Furthermore, although the electronic inductor emulates an infinite inductor and provides a constant DC load current, the grid voltage unbalance will inevitably cause 100 Hz oscillation in the DC link. Thereafter, the interaction between the DC-link second-order harmonic current and the front-end rectifier switching functions will produce the third-order harmonic in the line currents. Thus in practice, the multi-drive system always injects the triple harmonics into the grid under grid voltage unbalanced conditions.

The 2018 International Power Electronics Conference

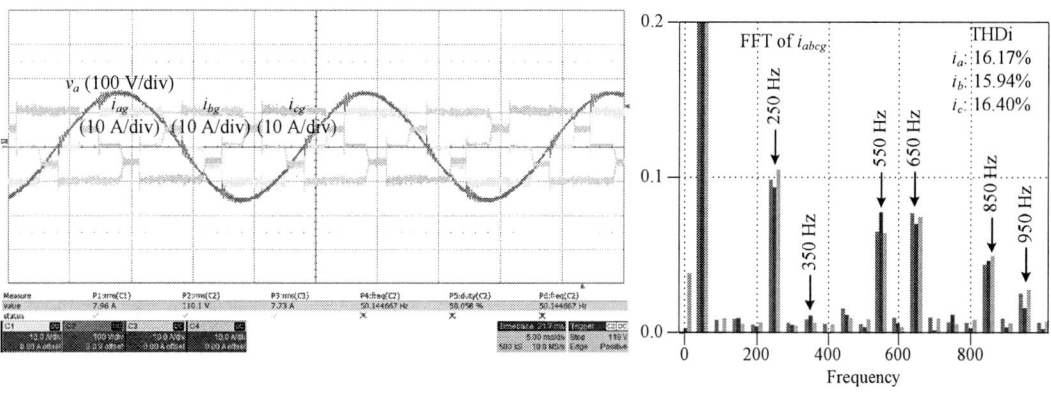

Fig. 9. Experimental results of multi-drive system with two SCR units under balanced grid (v_a = 110 V, v_b = 110 V and v_c = 110 V), the firing angle α_{f1} = 30 (0+30)° and α_{f2} = 60 (30+30)°.

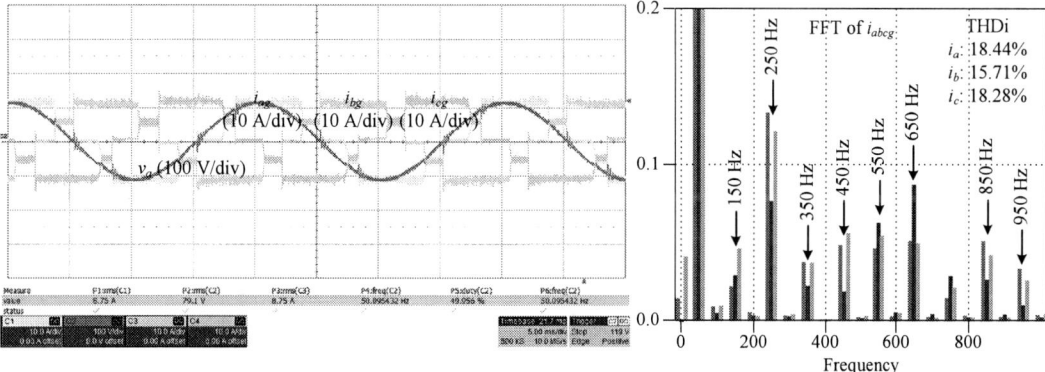

Fig. 10. Experimental results of multi-drive system with two SCR units under 10% amplitude unbalanced grid in phase a (v_a = 79.1 V, v_b = 110 V and v_c = 110 V), the firing angle α_{f1} = 30 (0+30)° and α_{f2} = 60 (30+30)°.

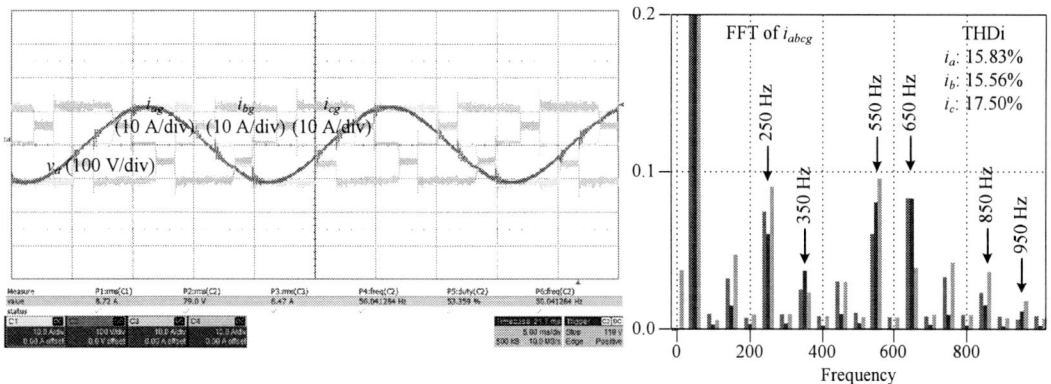

Fig. 11. Experimental results of multi-drive system with two SCR units under 10% amplitude unbalanced grid in phase a (v_a = 79.1 V, v_b = 110 V and v_c = 110 V), the firing angle α_{f1} = 30+10° and α_{f2} = 60 +10°, the firing angle compensation of 10°.

The 2018 International Power Electronics Conference

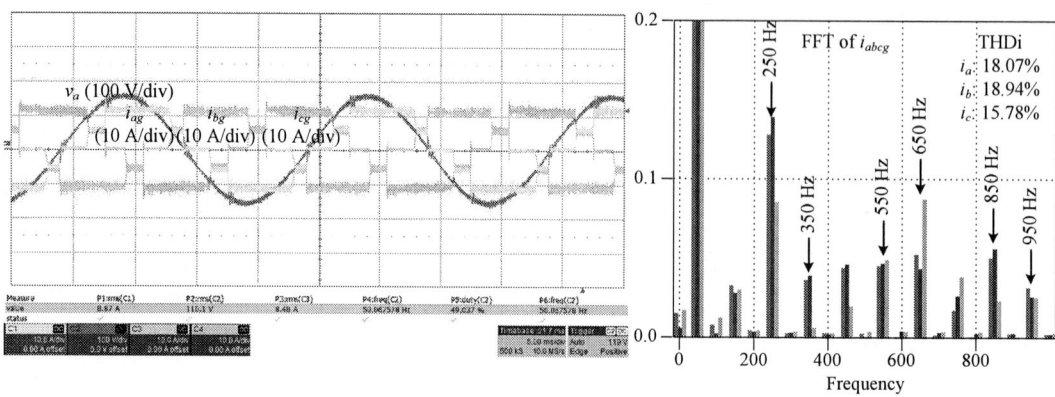

Fig. 12. Experimental results of multi-drive system with two SCR units under 10% amplitude unbalanced grid in phase b (v_a = 110 V, v_b = 79.1 V and v_c = 110 V), the firing angle α_{f1} = 30 (0+30)° and α_{f2} = 60 (30+30)°.

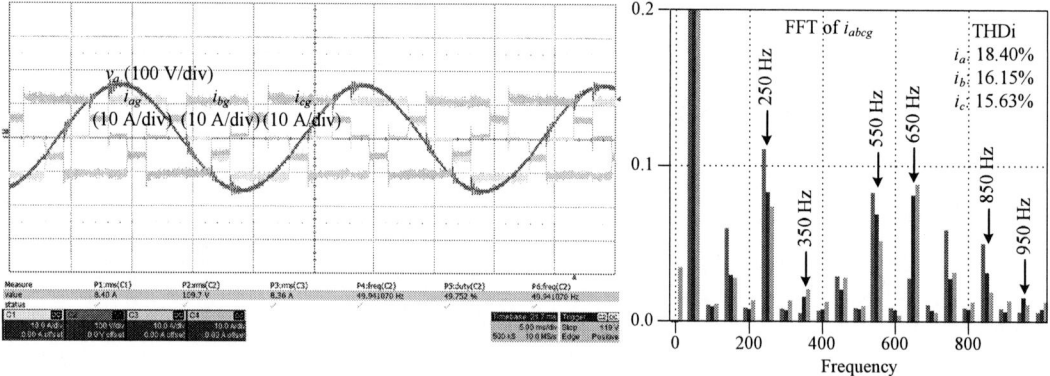

Fig. 13. Experimental results of multi-drive system with two SCR units under 10% amplitude unbalanced grid in phase b (v_a = 110 V, v_b = 79.1 V and v_c = 110 V), the firing angle α_{f1} = 30+10° and α_{f2} = 60 +10°, the firing angle compensation of 10°.

VI. CONCLUSIONS

This paper investigated a centralized control scheme for modular multi rectifier system under unbalanced grid voltage. The firing angle compensation method is introduced under unbalanced grid voltage to guarantee identical overall input current harmonics as the case under balanced grid voltage. The total input current 3rd harmonic component is included as the closed-loop control feedback to generate the firing angle compensation value. The proposed control strategy has been validated by simulation results based on PLECS and experimental results in respect to the THDi of the total input current.

REFERENCE

[1] P. Davari, Y. Yang, F. Zare, and F. Blaabjerg, "A Multipulse Pattern Modulation Scheme for Harmonic Mitigation in Three-Phase Multimotor Drives," *IEEE Journal of Emerging and Selected Topics in Power Electronics*, vol. 4, no. 1, pp. 174 – 185, March 2016.

[2] P. Davari, F. Zare, and F. Blaabjerg, "Pulse Pattern-Modulated Strategy for Harmonic Current Components Reduction in Three-Phase AC–DC Converters," *IEEE Trans. Ind. Appl.*, vol. 52, no. 4, pp. 3182-3192, July/Aug. 2016.

[3] P. Davari, Y. Yang, F. Zare, and F. Blaabjerg, "Predictive Pulse-Pattern Current Modulation Scheme for Harmonic Reduction in Three-Phase Multidrive Systems," *IEEE Trans. Ind. Electron.*, vol. 63, no. 9, pp. 5932-5942, Sept. 2016.

[4] H. Soltani, P. Davari, F. Blaabjerg, and F. Zare, "Harmonic Distortion Performance of Multi Three-Phase SCR-Fed Drive

Systems with Controlled DC-Link Current under Unbalanced Grid," in *Proc. IECON 2017*, pp. 1210-1214, 2017.

[5] F. Zare, P. Davari, and F. Blaabjerg, "A Modular Active Front-End Rectifier With Electronic Phase Shifting for Harmonic Mitigation in Motor Drive Applications," *IEEE Trans. Ind. Appl.*, vol. 53, no. 6, pp. 5440-5450, Nov/Dec. 2017.

[6] J. W. Gray, and F. J. Haydock, "Industrial power quality considerations when installing adjustable speed drive systems," *IEEE Trans. Ind. Appl.*, vol. 32, no. 3, pp. 646–652, May/June, 1996.

[7] D. Basic, "Input current interharmonics of variable-speed drives due to motor current imbalance," IEEE Trans. Power Del., vol. 25, no. 4, pp. 2797–2806, Oct. 2010.

[8] Y. Yang, P. Davari, F. Zare, and F. Blaabjerg, "A DC-link modulation scheme with phase-shifted current control for harmonic cancellations in multidrive applications," IEEE Trans. Power Electron., vol. 31, no. 3, pp. 1837–1840, Mar. 2016.

Vector Control of Magnetically Modulated Motor for Power Splitting of HEV Application

Toshihiko Noguchi[1*], Sawanth Krishna Machavolu[1], Masahiro Aoyama[1] and Yuto Motohashi[1]
1 Graduate School of Integrated Science and Technology, Shizuoka University, Hamamatsu, Japan
*E-mail: noguchi.toshihiko@shizuoka.ac.jp

Abstract— **This paper describes a vector control method of a magnetically modulated motor. The motor has a quite different structure from a standard permanent magnet synchronous motor (PMSM) because of the two rotors with two different mechanical output shafts, i.e., an inner PM rotor and an outer magnetic-flux modulator composed of only iron and air. Investigation on the vector control method using a real machine is still unexplored, thus, the paper discusses the vector control algorithm of the motor. In addition, several experimental tests have been conducted to examine the control performance of the mechanical output power delivered by the two rotors, using a prototype machine. It has been verified that the discussed algorithm makes it possible to control the two rotor output power as well as the stator input power like a conventional planetary-gear-based HEV system, including the motoring and regenerating operations.**

Keywords— *Hybrid vehicle, magnetically modulated motor, power split, vector control.*

I. INTRODUCTION

In order to enhance integration of a hybrid electric vehicle (HEV) power train with a power split mechanism, a magnetically modulated motor (MMM) is supposed to be one of the solutions and has been studied intensively in recent years. There are two rotors in the MMM, i.e., an inner rotor with permanent magnets (PMs) and an outer rotor which works as a magnetic flux modulator [1], [2]. The mechanical power is basically transferred to the drive shaft by the outer rotor, and the mechanical engine power is provided through the inner rotor in a standard application. The electric power is supplied to the stator by the external power converter such as an inverter, and is converted to the mechanical power like a conventional motor. The MMM is possible to synthesize the mechanical engine power and the electrical motor power given by an inverter through the stator, and is capable to deliver the synthesized mechanical power to the drive shaft through the modulator. In addition, it is possible to regenerate the surplus mechanical power from the engine to the inverter if the drive shaft does not require higher power than the engine output. The outer rotor modulates the stator rotating magnetic flux, and the inner PM rotor receives the modulated magnetic flux to generate the electromagnetic torque. Various high-frequency components are generated through the magnetic flux modulation process, which do not contribute to generate

the effective mechanical power. These extra harmonic components of the magnetic flux cause additional power losses, e.g., an eddy current loss in the PM and extra iron losses in the rotor [3]. Therefore, optimization of the motor geometry is significantly important to reduce the power losses in the MMM, where an FEM based electromagnetic analysis plays an important role for the optimization [4]. However, technical discussion on the control scheme of the MMM looks relatively inactivated by researchers in the motor drive field, probably due to its electrical and mechanical complexity [5].

The voltage equation of the MMM have been investigated because the mathematical motor model is the basis of the vector control algorithm. The voltage equation has been derived from a simplified magnetic circuit of the MMM [6]. In this paper, this fundamental investigation is described, and the operation principle is verified with a prototype of the MMM.

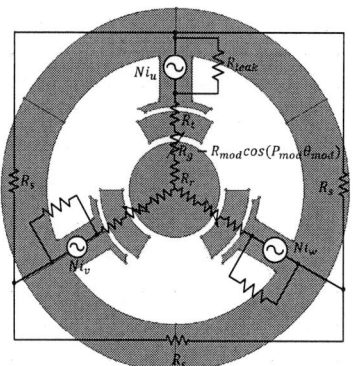

Fig. 1. Simplified MMM model and magnetic circuit.

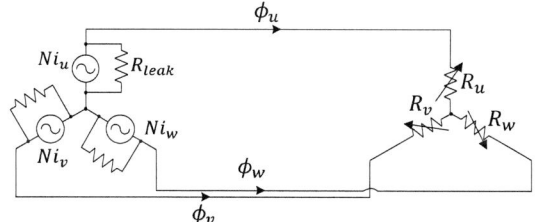

Fig. 2. Three-phase magnetic circuit of MMM.

II. DERIVATION OF VOLTAGE EQUATION

It is assumed that the MMM discussed in the paper has $P_s : P_{pm} : P_{mod} = n : 2n : 3n$ (n is a positive integer) relationship, where P_s is the pole pair number of the stator, P_{pm} is the pole pair number of the inner PM rotor, and P_{mod} is the number of the modulator iron cores. Fig. 1 shows the most fundamental geometry and its magnetic circuit of the MMM, where the least combination of $P_s : P_{pm} : P_{mod} = 1 : 2 : 3$ ($n = 1$) is assumed. Although the MMM is a kind of synchronous machines, it has different pole pair numbers between the stator and the rotor. In the magnetic circuit, the following reluctances are considered; the stator back yoke reluctance $\mathcal{R}_s$, the stator tooth reluctance $\mathcal{R}_t$, the air gap reluctance $\mathcal{R}_g$, the modulator reluctance $\mathcal{R}_{mod}$, the inner PM rotor core reluctance $\mathcal{R}_r$, and the reluctance corresponding to the total leakage flux $\mathcal{R}_{leak}$. N is the number of the winding turns for each phase.

The total reluctance of the three phases $\mathcal{R}_u$, $\mathcal{R}_v$, and $\mathcal{R}_w$ are expressed by the following equation because the three-phase reluctance variations caused by the modulator are in phase with each other:

$$
\begin{aligned}
&\mathcal{R}_u = \mathcal{R}_v = \mathcal{R}_w = \mathcal{R}_{dc} - \mathcal{R}_{mod}\cos(P_{mod}\theta_{mod}) \\
&\because \mathcal{R}_{dc} = \mathcal{R}_g + \mathcal{R}_t + \mathcal{R}_r + \mathcal{R}_s/3
\end{aligned}, \tag{1}
$$

where θ_{mod} is a position of the modulator. The total number of flux linkage at the phase U coil is calculated as

$$
\begin{aligned}
\psi_u = &\left\{\frac{N^2}{\mathcal{R}_{leak}} + \frac{N^2}{\mathcal{R}_{dc}} + \frac{\mathcal{R}_{mod}N^2}{\mathcal{R}_{dc}^2}\cos(P_{mod}\theta_{mod})\right\}i_u \\
&-\frac{1}{2}\left\{\frac{N^2}{\mathcal{R}_{dc}} + \frac{\mathcal{R}_{mod}N^2}{\mathcal{R}_{dc}^2}\cos(P_{mod}\theta_{mod})\right\}i_v \\
&-\frac{1}{2}\left\{\frac{N^2}{\mathcal{R}_{dc}} + \frac{\mathcal{R}_{mod}N^2}{\mathcal{R}_{dc}^2}\cos(P_{mod}\theta_{mod})\right\}i_w
\end{aligned}. \tag{2}
$$

The self-inductance L of each phase and the mutual-inductance M can be derived from (2) as follows:

$$
\begin{aligned}
&L = \ell + L_{dc} + L_{ac}\cos(P_{mod}\theta_{mod}) \\
&M = -\frac{1}{2}L_{dc} - \frac{1}{2}L_{ac}\cos(P_{mod}\theta_{mod}) \\
&\because \ell = \frac{N^2}{\mathcal{R}_{leak}},\ L_{dc} = \frac{N^2}{\mathcal{R}_{dc}},\ L_{ac} = \frac{\mathcal{R}_{mod}N^2}{\mathcal{R}_{dc}^2}
\end{aligned}. \tag{3}
$$

One of the most significant different features of the MMM from common PM motors is that the number of the stator pole pair is different from that of the rotor. Therefore, the voltage equation of the MMM should be derived by taking the pole pair number difference into account. The magnetomotive force of the rotor PM for each phase $\mathcal{F}_u$, $\mathcal{F}_v$, $\mathcal{F}_w$ can be expressed as

$$
\begin{aligned}
&\mathcal{F}_u = \mathcal{F}\cos(P_{pm}\theta_{pm}) \\
&\mathcal{F}_v = \mathcal{F}\cos\left(P_{pm}\theta_{pm} - \frac{2\pi P_{pm}}{3P_s}\right), \\
&\mathcal{F}_w = \mathcal{F}\cos\left(P_{pm}\theta_{pm} - \frac{4\pi P_{pm}}{3P_s}\right)
\end{aligned} \tag{4}
$$

where θ_{pm} is a position of the inner PM rotor and $\mathcal{F}$ is the maximum amplitude of the equivalent PM magnetomotive force.

In general, a three-phase voltage equation of the PM motor can be expressed as follows on the stationary reference frame:

$$
\begin{bmatrix} v_u \\ v_v \\ v_w \end{bmatrix} = \begin{bmatrix} R & 0 & 0 \\ 0 & R & 0 \\ 0 & 0 & R \end{bmatrix}\begin{bmatrix} i_u \\ i_v \\ i_w \end{bmatrix} + \frac{\mathrm{d}}{\mathrm{dt}}\begin{bmatrix} L & M & M \\ M & L & M \\ M & M & L \end{bmatrix}\begin{bmatrix} i_u \\ i_v \\ i_w \end{bmatrix} + \frac{\mathrm{d}}{\mathrm{dt}}\begin{bmatrix} \psi_u \\ \psi_v \\ \psi_w \end{bmatrix}, \tag{5}
$$

where R is the stator winding resistance, and ψ_u, ψ_v, and ψ_w are the number of flux linkages for each phase. In the MMM case, it is necessary to consider the number of flux linkage after modulation, which has a relationship such as $\psi_u = L\mathcal{F}_u/N$.

The voltage equation on the γ-δ rotating reference frame can be obtained by applying coordinate transformations to (5) as

$$
\begin{bmatrix} v_\gamma \\ v_\delta \end{bmatrix} = \begin{bmatrix} R + pL & -\omega L \\ \omega L & R + pL \end{bmatrix}\begin{bmatrix} i_\gamma \\ i_\delta \end{bmatrix} + \begin{bmatrix} -E_\gamma \\ \omega\sqrt{\dfrac{3}{8}}\dfrac{L_{ac}\mathcal{F}}{N} - E_\delta \end{bmatrix}. \tag{6}
$$

$$
\because \omega = P_{mod}\omega_{mod} - P_{pm}\omega_{pm}
$$

It should be noted that the following rotational angle between the stationary reference frame and the rotating reference frame was used through the rotational coordinate transformation:

$$
\theta = \omega t = P_{mod}\omega_{mod}t - P_{pm}\omega_{pm}t = \theta_{mod} - \theta_{pm}, \tag{7}
$$

where ω_{pm} and ω_{mod} are mechanical angular frequencies of the inner PM rotor and the modulator, respectively. As is shown in (6), the voltage equation of the MMM is similar to that of the common PM motor whereas some extra electromotive forces are generated by the asynchronous frequency components as follows:

$$
\begin{aligned}
E_\gamma = &\ P_{pm}\omega_{pm}\sqrt{\frac{3}{2}}\frac{(\ell + L_{dc})\mathcal{F}}{N}\sin(P_{mod}\theta_{mod}) \\
&+ \left(P_{mod}\omega_{mod} + P_{pm}\omega_{pm}\right)\sqrt{\frac{3}{8}}\frac{L_{ac}\mathcal{F}}{N}\sin(2P_{mod}\theta_{mod}) \\
E_\delta = &\ P_{pm}\omega_{pm}\sqrt{\frac{3}{2}}\frac{(\ell + L_{dc})\mathcal{F}}{N}\cos(P_{mod}\theta_{mod}) \\
&+ \left(P_{mod}\omega_{mod} + P_{pm}\omega_{pm}\right)\sqrt{\frac{3}{8}}\frac{L_{ac}\mathcal{F}}{N}\cos(2P_{mod}\theta_{mod})
\end{aligned}. \tag{8}
$$

The voltage equation has been discussed so far, and a relationship of the electromagnetic torque between the modulator and the inner PM rotor is investigated as follows:

$$
\begin{aligned}
P_i &= v_\gamma i_\gamma + v_\delta i_\delta = R(i_\gamma^2 + i_\delta^2) + \omega\sqrt{\frac{3}{8}}\frac{L_{ac}\mathcal{F}}{N}i_\delta \\
&= R(i_\gamma^2 + i_\delta^2) + \omega_{mod}\tau_{mod} + \omega_{pm}\tau_{pm}.
\end{aligned} \tag{9}
$$

$$
\because \tau_{pm} = -P_{pm}\sqrt{\frac{3}{8}}\frac{L_{ac}\mathcal{F}}{N}i_\delta,\ \tau_{mod} = P_{mod}\sqrt{\frac{3}{8}}\frac{L_{ac}\mathcal{F}}{N}i_\delta
$$

The above equation is an input electric power, where the first term represents a copper loss of the windings and the second term corresponds to the mechanical output power. Therefore, the following torque relationship can be derived from the result of (9):

$$
\tau_s = \frac{P_s}{P_{pm}}\tau_{pm} = -\frac{P_s}{P_{mod}}\tau_{mod}. \tag{10}
$$

TABLE I
SPECIFICATIONS OF PROTOTYPE MMM.

Number of stator pole pairs P_s	4
Number of rotor pole pairs P_{pm}	8
Number of modulator cores P_{mod}	12
Stator outer diameter	120 mm
Rotor diameter	61.2 mm
Axial length of core	49.5 mm
Air gap length	0.7 mm
Winding connection	4 series-2 parallel
Maximum current	150 A_{rms}
Armature winding resistance R	33.3 mΩ
Inductance on rotating reference frame L	0.27 mH
Stator flux linkage ψ_a	3.8 mWb

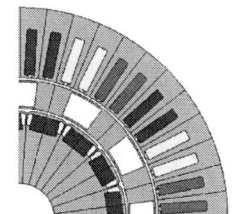

(a) Radial cross section of MMM.

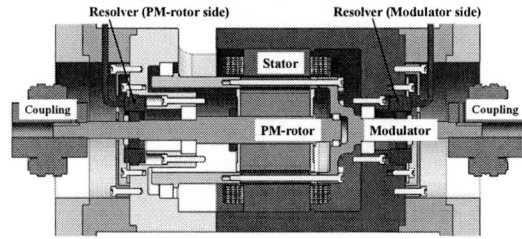

(b) Axial cross section of MMM.
Fig. 3. Configuration of MMM.

The above equation shows that the MMM works like a planetary gear used as a mechanical power splitter of hybrid vehicles and that the mechanical power can be distributed among the electric power fed by the inverter, mechanical power of the inner PM rotor, and another mechanical power of the modulator.

III. OUTLINE OF PROTOTYPE AND CONFIGURATION OF EXPERIMENTAL SETUP

A small scale prototype of the MMM has been developed to evaluate validity of the proposed vector control technique. TABLE I shows major specifications of a prototype MMM. N39UH (Nd-Fe-B, B_r = 1.22 T, H_{cb} = 965.7 kA/m@293 K) neodymium PM was employed in the inner rotor. A radial cross section and an axial cross section are shown in Fig. 3, where how the two rotating parts are installed in the prototype is illustrated. In general, the inner PM rotor is mechanically connected to the combustion engine, while the magnetic modulator is connected to the drive shaft in a typical HEV application. The electromagnetic torque is delivered by the MMM according to (10) in either a motoring operation or a regenerating operation.

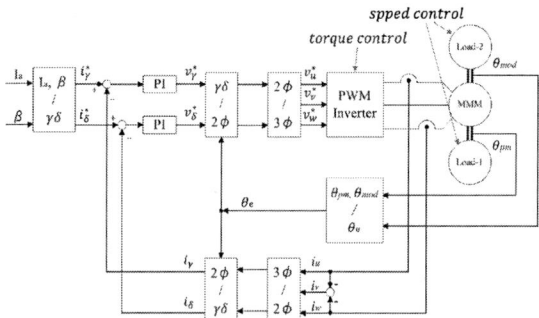

Fig. 4. Block diagram of vector control system for MMM.

Fig. 5. Experimental setup.

TABLE II
EXPERIMENTAL TEST CONDITIONS.

Inverter DC-bus voltage	80 V
Inverter switching frequency	10 kHz
Dead time	4 μs
Sampling interval of torque	0.5 s

Fig. 4 is a block diagram of the vector control system for the MMM. As can be seen in the figure, two of the resolvers are used to detect the rotational positions of the inner PM rotor and the modulator. The electric angle for rotational coordinate transformation is calculated by (7), and is used for the stator current control which is carried out on the $\gamma\delta$ stator reference frame. For the convenience of the tests, the current commands are given to the controller in a polar coordinate manner, i.e., a current amplitude I_a and a current phase angle β measured from δ-axis.

Fig. 5 shows the experimental setup, where the two load machines are directly connected to the inner PM rotor and the modulator, respectively. The two load machines control their rotation speeds independently. On the other hand, the prototype of the MMM does not have a speed control loop; thus the torque control is only achieved by the controller indicated in Fig. 4. Two of the mechanical output powers from both shafts are measured with two torque meters. The input electric power is also measured with a power meter to evaluate the total power flow. A positive direction for both of the torque and the speed is defined as a CCW rotation of the inner PM rotor.

IV. EXPERIMENTAL TEST RESULTS

A. Experimental Test Conditions

The experimental test conditions are indicated in TABLE II. An inverter was used to operate the prototype MMM, where the DC-bus voltage was 80 V, the switching frequency was 10 kHz, and the dead time was

The 2018 International Power Electronics Conference

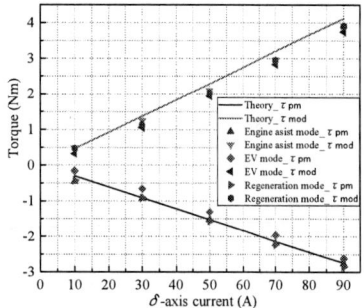

Fig. 6. Output torque and δ-axis current characteristic.

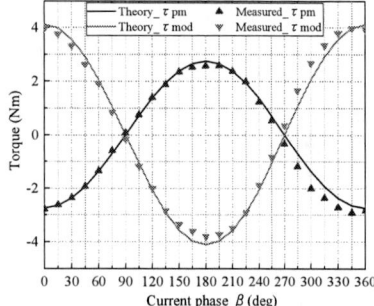

Fig. 7. Output torque and current phase angle β characteristic.

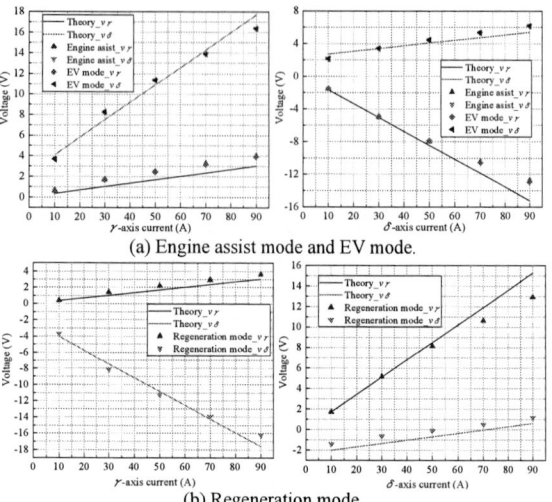

(a) Engine assist mode and EV mode.

(b) Regeneration mode.

Fig. 8. Confirmation of voltage equation.

set at 4 μs, respectively. The sampling period of torque measurement was 0.5 s, and the average torque for 15 s was calculated. Many of the compensation control techniques are employed to make the vector control as precise as possible such as a dead time compensation, a discrete time error compensation, an on-voltage compensation of the IGBTs, and so on [7].

B. Output Torque and Equivalent Gear Ratio

A relationship between the torque and the δ-axis current i_δ with the constant $i_\gamma = 0$ is shown in Fig. 6, where i_δ is changed every 20 A over the range from 10 to 90 A. The output torque of the inner PM rotor τ_{pm} and that of the modulator τ_{mod} are delivered in proportional to i_δ regardless of the driving modes. The theoretical relationship between τ_{mod} and τ_{pm} given by (10) is indicated by the solid lines, and the test result agrees with the theoretical relationship of $\tau_{pm} = -2\,\tau_{mod}/3$.

C. Output Torque and Current Phase Angle

Fig. 7 shows a torque characteristic where the current phase angle β is varied every 15 deg from 0 to 360 deg (electric angle) with constant current vector amplitude of 90 A under the engine assist mode. As can be seen in the figure, the torque characteristic similar to an SPMSM is confirmed because of the surface PM structure of the prototype rotor. However, τ_{mod} and τ_{pm} are in opposite phase with each other and their ratio is kept at $\tau_{pm} = -2\,\tau_{mod}/3$.

D. Confirmation of Voltage Equation

Consistency was examined between the two-axis voltages v_γ and v_δ calculated by (6) and their actual command values in the controller to confirm the validity of the voltage equation. Either i_γ or i_δ was varied in the

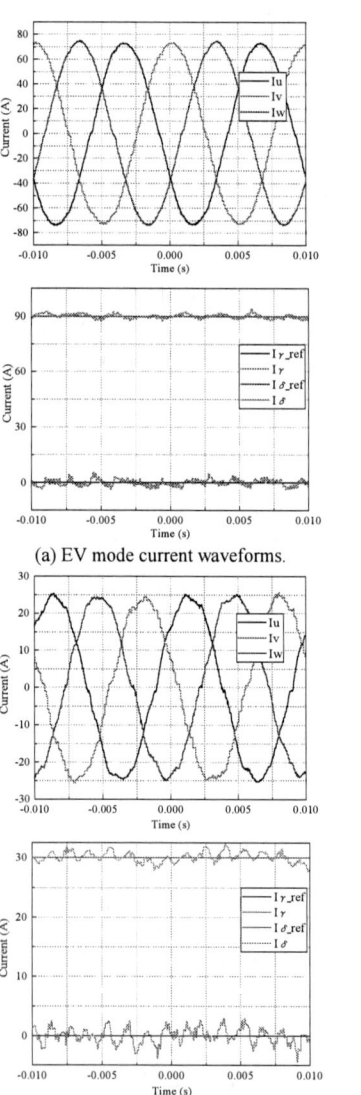

(a) EV mode current waveforms.

(b) Regeneration mode current waveforms.

Fig. 9. Current waveforms in EV and regeneration modes.

756

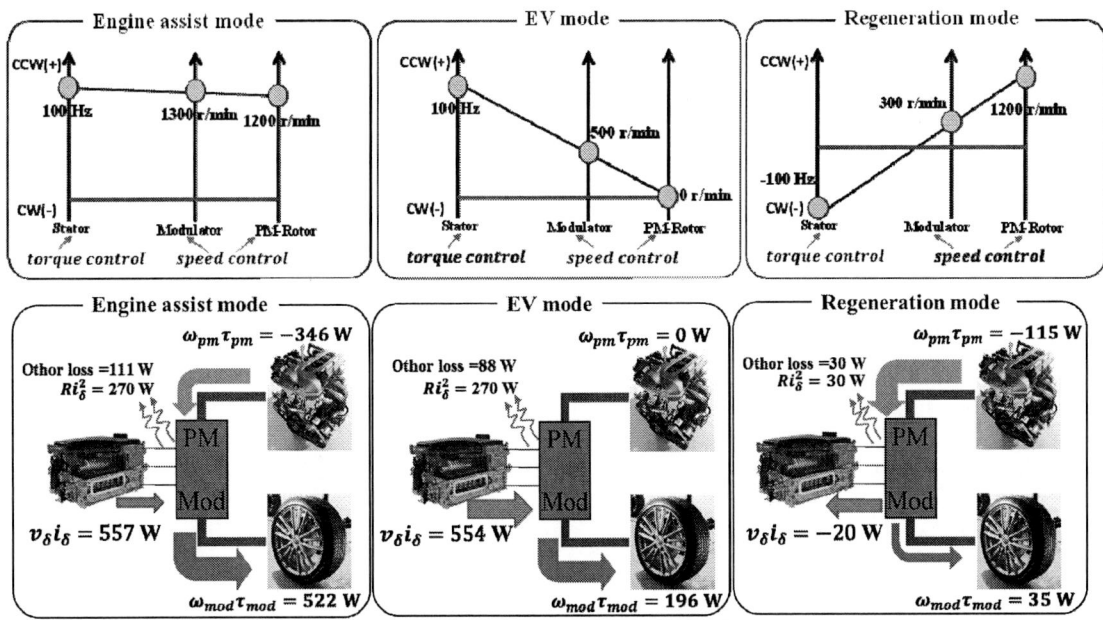

Fig. 10. Three driving modes, collinear charts, and power flow of MMM.

range from 10 to 90 A, and the rest one was kept constant at zero in the tests. The measured and the theoretical results of v_γ and v_δ are shown in Fig. 8, where all the three driving modes are checked. Since the phase sequence and the inverter frequency are identical in the engine assist and the EV modes, both of the characteristics of v_γ and v_δ satisfy the same voltage equation.

$v_\gamma = Ri_\gamma$ and $v_\delta = \omega(Li_\gamma + \psi_a)$ from (6) when $i_\delta = 0$; hence, both of the voltages are proportional to i_γ. In a similar way, if $i_\gamma = 0$ is substituted into (6), $v_\gamma = -\omega Li_\delta$ and $v_\delta = Ri_\delta + \omega\psi_a$ are obtained, which means the voltages are proportional to i_δ. As shown in Fig. 8 (a), actual voltages v_γ and v_δ vary almost linearly, which agrees with their theoretical values indicated by the solid lines. The driving modes, i.e., the engine assist and the EV modes, hardly have influence on the relationships between the voltages and the currents. However, Fig. 8 (a) indicates a slight non linearity due to the magnetic saturation because v_γ or v_δ includes inductance L. Fig. 8 (b) shows v_γ and v_δ in the regenerative operation mode, where either $i_\delta = 0$ or $i_\gamma = 0$ condition is similarly given. The actual values of the voltages agree with their theoretical values for any conditions of the two-axis currents. Therefore, consistency between the mathematical voltage equation of the MMM and the real machine characteristics is appropriately confirmed.

The three-phase currents and the two-axis currents in the EV and the regeneration modes are shown in Fig. 9. The EV mode test condition was $i_\gamma = 0$ and $i_\delta = 90$ A, and the regeneration mode test condition was $i_\gamma = 0$ and $i_\delta = 30$ A. The balanced and sinusoidal three-phase current waveforms are observed in Fig. 9 (a), and the two-axis currents i_γ and i_δ are properly controlled to follow their command values with small deviations. On the other hand, the current waveforms seem to have some ripples

in the regeneration mode shown in Fig. 9 (b), but the vertical axis scale of the graphs is 1/3 of Fig. 9(a). Therefore, the current control in the regeneration mode is properly achieved. As described previously, the regenerating operation makes a phase sequence of the three-phase currents negative.

E. Collinear Chart and Power Flow

Fig. 10 shows collinear charts of the three driving modes that simulate HEV system, i.e., the engine assist mode, the EV mode, and the regeneration mode. The engine assist mode and the EV mode were operated at $i_\gamma = 0$ and $i_\delta = 90$ A, while the regeneration mode was examined at $i_\gamma = 0$ and $i_\delta = 30$ A. The modulator (the drive shaft) speed can be adjustable by changing the inverter frequency and/or the inner PM rotor (the engine) speed in the engine assist mode. On the other hand, the EV mode is achieved by fixing the inner PM rotor (the engine) speed at zero, and the modulator (the drive shaft) speed can be changed by controlling only the inverter frequency. The regeneration mode is achieved by a negative frequency of the inverter.

The test results shown in Fig. 6 indicate that $\tau_{pm} < 0$ and $\tau_{mod} > 0$ when $i_\gamma = 0$ and $i_\delta = 90$ A in the engine assist or the EV mode. The collinear chart depicted in Fig. 10 also indicates that the MMM is operated under the $\omega_{pm} > 0$ and $\omega_{mod} > 0$ condition. Therefore, the mechanical power of the inner PM rotor is negative $\tau_{pm}\omega_{pm} < 0$, i.e., the rotor receives the mechanical power from the load machine (the engine), while the mechanical power of the outer modulator is positive $\tau_{mod}\omega_{mod} > 0$, i.e., the modulator delivers its mechanical power to the load machine (the drive shaft). The supplemental electric power is provided by the inverter, which is converted to the assist mechanical power for the engine in the MMM.

It is found from Fig. 6 that $\omega_{pm} = 0$ and $\omega_{mod} > 0$, and that $\tau_{mod} > 0$ when $i_\gamma = 0$ and $i_\delta = 90$ A in the EV mode. Therefore, the mechanical power of the inner PM rotor is zero, whereas that of the modulator is positive. This operating condition means that the MMM gives the mechanical power to the load machine through the modulator with standstill inner PM rotor (engine). The inverter provides the MMM with positive electric power in the case; hence, the MMM simply converts the received electric power to the mechanical power.

The surplus mechanical power of the inner PM rotor can be regenerated to the inverter through the stator windings with transferring the requested mechanical power to the modulator (the drive shaft) at the same time. Negative electric power of the inverter $v_\gamma i_\gamma + v_\delta i_\delta < 0$ is a requirement of the regeneration. As described previously in (9), $v_\delta i_\delta < 0$ is the condition of the regeneration mode because of $i_\gamma = 0$. This condition is satisfied in the range of i_δ from 10 to 50 A. In the actual experimental test, the regeneration mode has been achieved when $i_\gamma = 0$ and $i_\delta = 30$ A because the electric power of the inverter is measured negative.

The voltage equation (6) does not include any iron losses and mechanical loss; thus, these losses are residual quantities calculated by the difference between the measured value and the theoretical value. The total of the iron losses and the mechanical loss represented as other losses seems to be more than 15 % of the total input power.

V. CONCLUSIONS

The voltage equation and the torque equation of the magnetically modulated motor have been investigated in the paper. On the basis of the mathematical model, the vector control algorithm has been derived from the voltage equation. In order to confirm the feasibility of the derived vector control algorithm, a prototype was developed and an experimental system was set up. Assuming the three HEV driving modes, i.e., the engine assist, EV, and regeneration modes, various experimental tests were conducted with the prototype motor.

As a result, feasibility of the derived voltage equation was confirmed by comparing the theoretical values calculated from the equation with the measured voltages. The torque equation of the motor is also feasible because it satisfies a relationship of the equivalent gear ratio between the inner PM rotor and the outer magnetic modulator. This characteristic can be represented by collinear charts, which corresponds to a planetary gear actually used as a power splitter of the HEV system.

The power flow of the motor was also investigated through the three driving mode tests. Proper operation was confirmed in every driving mode. However, because the iron loss was dominant in the miniature prototype, high efficiency was not achieved. This iron loss was likely caused by the harmonics of the magnetic flux modulated by the outer modulator, which is a future work to improve the motor efficiency.

REFERENCES

[1] W. Kaewjinda and M. Konghirun, "A DSP-Based Vector Control of PMSM Servo Drive Using Resolver Sensor," *2006 IEEE Region 10 Conference TENCON.*

[2] C. C. Mi, G. R. Slemon, and R. Bonert, "Minimization of Iron Losses of Permanent Magnet Synchronous Machines," *IEEE Trans. on Energy Conversion.*, vol. 20, no. 1, pp. 121-127, 2005.

[3] T. Tonari, H. Kato, and H. Matsui, "Study on Iron Loss of Flux Modulated Type Dual-Axis Motor," *IEEJ RM Tech. Meet.*, RM-13-142, pp. 101-105, 2013.

[4] M. Fukuoka, K. Nakamura, H. Kato, et al., "A Consideration of the Optimum Configuration of Flux-Modulated Type Dual-Axis Motor," *IEEJ RM Tech. Meet.*, RM-13-141, pp. 95-100, 2013.

[5] Y. Takeuchi, H. Kato, M. Tago, S. Ogasawara, et al., "Operating Principle and Control Method of the Magnetic Modulated Motor," *IEEJ Conf. Rec.*, no.5-041, pp. 73-74, 2013.

[6] M. Aoyama, Y. Kubota, T. Noguchi, et al., "Prototype Design of Permanent-Magnet-Free Magnetic Geared Motor for HEV Application," *IEEJ Ind. Appl. Soc. Conf.*, 3-8, pp. 97-100, 2015.

[7] J. Kudo, T. Noguchi, M. Kawakami, and K. Sano, "Mathematical Model Errors and Their Compensations of IPM Motor Control System," *IEEJ SPC Tech. Meet.*, SPC-08, pp. 25-30, 2008.

AUTHOR INDEX

Aapro, Aapo 3156
Abdollahi, Hessamaldin.........................1719
Abe, Kazuyuki....................................1567
Abe, Kensho.......................................767
Abe, Kodai..................................1741, 3890
Abe, Seiya.................................2360, 2370
Abe, Takashi.....................................2176
Abrishamifar, Adib..............................2854
Abuogo, James....................................1125
Acharya, Anirudh Budnar 2630
Acharya, Sayan...................................3564
Adachi, Masakazu................................2237
Afsharian, Jahangir 1537, 3797
Agarwal, Vivek...................................3471
Agelidis, Vassilios G............................3215
Agostinelli, Matteo 3140
Ahmad, Hamzeh J.................................3273
Aiso, Kohei......................................3186
Akagi, Hirofumi..................................2352
Akahane, Masashi................................2774
Akama, Yousuke..................................1741
Akao, Naoki......................................1217
Akatsu, Kan...............................711, 3186
Alatise, O 1149
Alenius, Henrik 1704, 4205
Ali, Muhammad....................................528
Ali, Murad.......................................2317
Allmeling, Jost............................422, 2199
Almér, Stefan....................................555
Alsofyani, Ibrahim Mohd 466
Alvarez, S.......................................4009
Amano, Koki.......................................94
Amei, Kenji......................................3182
Amin, Mohammad...................................759
Amrhein, Wolfgang................................3640
An, Ronghui............... 957, 1524, 3251, 3692, 3924
An, Zheng..4001
Andenna, M.......................................3596
Andersen, A. E. Michael..........................1351
Andersen, Michael A. E....................607, 4066
Ando, Akinobu.....................................517
Ando, H..3665
Ando, Masato.....................................1919
Ando, Takashi....................................3658
Ang, Simon S......................................153
Antivachis, Michael...............................181
Antonini, Giulio.................................3588
Antonopoulos, Antonios 2335
Anurag, Anup.....................................3564
Anyapo, Chan.....................................3332
Aoyagi, Kazuki...................................2237
Aoyama, Masahiro............................718, 753
Arai, Takuro.....................................1997
Araumi, Ryunosuke.........................1877, 3658
Arimatsu, Kenji..................................1370

Arita, Hideaki.............................2796, 2820
Arrua, Silvia....................................1719
Asada, Kazunori..................................3658
Asama, Junichi...................................4016
Ashizaki, Yusuke.................................3450
Ashourloo, Mojtaba...............................2380
Aso, Shinji......................................3086
Aware, Mohan.....................................1730
Ayano, Hideki....................................1080
Azad, A N M Wasekul..............................2416
Azegami, Kazuya..................................3723
Azuma, S...3665
Baba, Teppei.....................................2283
Babasaki, Tadatoshi...............................207
Bach, Hoang Linh.................................2410
Baek, Jae-Il...........108, 2365, 3100, 3533, 3538
Baek, Miran......................................1141
Bahat-Treidel, Eldad.............................3607
Bai, Baodong.....................................2638
Baik, Jeong Min..................................3063
Bak, Yeongsu...............................1736, 4104
Bakran, Mark-M...................................2476
Bandyopadhyay, Soumya 1426
Barrena, Jon Andoni...............................759
Barrera-Cardenas, Rene...........................3431
Bauer, Pavol...............................1426, 2630
Bauer, Walter....................................3640
Bayer, Christoph Friedrich 2410
Bellini, M.......................................4009
Berg, Matias...............................963, 4205
Bergveld, H.J.....................................267
Bertoldi, F.......................................488
Besselmann, Thomas................................555
Bezha, Minella...................................3170
Bhattacharya, Subhashish 3564, 3993
Bhowate, Apekshit................................1730
Bhumkittipich, Krischonme 2430
Biela, J...1896
Biela, Jürgen...............1103, 1509, 2301, 3734
Bilal, Ahmad.....................................2193
Bilsalam, A......................................1622
Bin, Zhao..2692
Bixel, Paul.......................................238
Blaabjerg, Frede...................... 439, 746, 1183,
     1246, 1711, 1788, 2512, 2604, 2743, 3123,
     3164, 3357
Blanes, José M...................................1435
Böcker, Jan......................................3607
Bojoi, R...732
Bonyadi, R.......................................1149
Boroyevich, Dushan 790, 3705, 3749, 3985
Bortis, D..4080
Bortis, Dominik...................................181
Boynov, K.O......................................161
Braun, Michael.............................2848, 3074

AUTHOR INDEX

Büdel, Johannes 3034
Bui, M.X. 4174
Bunlaksananusorn, Chanin 2490
Burgos, Rolando 790, 3705, 3749, 3985
Cai, Kejun 3965
Cai, Panpan 3495
Cai, Xu 1004, 1491, 2245, 4162, 4220
Canales, F. 4009
Cao, Hu 1816, 3484
Cao, Pengpeng 2973
Cao, Qi 100
Cao, Wu 3002, 3010, 3015
Cardenas, Rene Alexander Barrera 1111
Carvalho, Kelly C. M. 3785
Castellazzi, Alberto 130, 2932
Ceballos, Salvador 3117
Celik, Mustafa 1680
Cha, Honnyong.............927, 1046, 2619, 3134
Chae, Beomseok 1977
Chailloux, Thibaut 2153
Chang, Chen-Wei 1617
Chang, Chien-Hsuan 2860
Chang, Liuchen 815, 1472, 1793, 2505
Chang, Yung-Ruei 639, 883
Chanmontree, P. 1622
Chao, Yi-Hao 1145
Charalambous, Apollo 1634
Charoensuksirikul, Supanut 2113
Chattopadhyay, Ritwik........................3564
Chazal, Hervé 2158
Chen, Ang-Tung 2102
Chen, Bo 1397
Chen, C. 142
Chen, Ching-Chen 1617
Chen, Ching-Jan 2086
Chen, Chuantong 1598
Chen, Dezhi 2638
Chen, Guan-Jung 1341
Chen, Guo 370
Chen, Hao 3112
Chen, I-Lin 2107
Chen, Jiangnan 1157, 1167
Chen, Jiann-Fuh 2653
Chen, Jie 1015, 1177
Chen, Kai-Hui 3081
Chen, Ke 1391
Chen, Kun-Feng 1341
Chen, Min 878
Chen, Minwu 2547
Chen, Nan 2335
Chen, Pingping 1118
Chen, Shen-Li 1145
Chen, Song 2153
Chen, Tang-Jung 1617
Chen, Tao..1872

Chen, Wan-Jung 3544
Chen, Wenjia 4213
Chen, Wenjie 1062, 2854, 3329
Chen, Wu 1504, 2496
Chen, Xiliang 3329
Chen, Xin 1015, 1177
Chen, Xingxing 1051, 3129, 3439
Chen, Yang 2785
Chen, Yangyang 560
Chen, Yaow-Ming 639, 883
Chen, Yenan 1118
Chen, Yen-Wen 2576
Chen, Yufeng 3383
Chen, Yu-Jen 275
Chen, Zhe 1758, 2708
Chen, Zhi 2997
Chen, Zhigang 3040
Cheng, Ching-Hsiang 2086
Cheng, Chun-An 2860
Cheng, Hung-Liang 2860
Cheng, Nie 2625
Cheng, Po-Tai.............503, 1038, 2462, 3549
Cheng, Ran 3877
Cheng, Xiangpeng 2435, 3934
Chengbi, Zeng 2718
Chi, Yongning 1491
Chiba, Akira 3627
Chien, Lin-Hao 2102
Chiu, Huang-Jen 2092, 3151
Chiu, Hui-Lung 123
Chiu, Yi-Hao 1145
Cho, Geum-Bae 2145
Cho, In-Ho 3323
Cho, Shin-Young 1530
Cho, Young Joon 137
Cho, Younghoon 1403
Choe, Chanyang 1598
Choi, Byungcho 1465
Choi, Hyun-Jun 383
Choi, Jae Hyuk 1336
Choi, Jaeho 803
Choi, Joon-Ho 982, 1799
Choi, Seung-Hyun 4049
Choi, Sewan 256
Choi, Sung-Jin 1409
Choi, Youn-Ok.....................................2145
Chou, Shih-Feng 1711
Chou, T.-C. 1912
Choudhury, Abhijit..............................3401
Chuai, Guoming 3025
Chung, Daewoong.................................1141
Chung, Henry S. H. 917
Chunkag, V. 1622
Collins, Caspar 1931
Cortes, Camilo 2193

AUTHOR INDEX

Corvasce, C. 3596
Cucala, Asuncion P. 2534
Cui, Shenghui 2250, 2484
Cui, Xiang 1125
Cvetkovic, Igor 790, 3985
Czyz, Piotr 396
D'arco, Salvatore 782, 2003
Da Silva, C. 267
Dahidah, Mohamed S A 3215
Dai, T. 1149
Dai, Wenjing 1015
Daikoku, Akihiro 2796
Danqing, Liu 1376
Dao, Ngoc Dat 1212
Dauphin, Benjamin 3644
Davari, Pooya 746
Davletzhanova, Z 1149
De Doncker, Rik W. 375, 388, 598,
 1073, 2250, 2484, 2768, 3729, 3979
Decker, Simon 2848, 3074
Delaforge, Timothé 2158, 3820
Deng, Fujin 1758, 2708
Deng, Jinxin 2992
Deshpande, Prathamesh Pravin 4186
Dieckerhoff, Sibylle 3607
Dimarino, Christina 3985
Din, Zakiud 2262
Ding, Yong 815
Dinh, Nguyen Duy 363
Diouf, Fatou 2078
Dirksen, Daniel 2410
Divan, Deepak 4001
Doki, Shinji 1032, 1223, 1228, 1295, 1747, 2224
Dong, Hanjing 987
Dong, Mi 1771
Dong, Qinghua 459
Dong, Xiaofeng 4168
Dong, Zhen 459
Dong, Zheng 3768
Driesen, J. 488
Du, Chao 2204
Du, Xiaotong 1167, 2780
Du, Xizhou 1491
Du, Yan 1472, 2877
Du, Zhijiang 84
Duarte, J. L. 946, 1067, 2697
Duarte, Jorge L. 1447, 3840
Dugal, F. 3596
Dujic, Drazen 422, 1484, 1498, 2170
Duong, Truong-Duy 982
Duque, C. A. 1067
Eberle, Wilson 927
Ekman, Jonas 3588
Elbaset, Adel A. 3945
Endegnanew, Atsede G. 2003

Endo, Hiroaki 4151
Endres, Tobias Maximilian 2410
Engelmann, Georges 3979
Enomoto, Bruno Yukio 3785
Eto, Haruhi 2097
Faiz, Muhammad Talib 528
Fajri, Poria 3223
Fan, Dongchen 3002, 3010, 3015
Fan, Shengwen 977, 3040
Fan, Weiyan 1386, 1421
Fang, Jingyang 337, 3910
Fang, Ran 4213
Fangfang, Luo 1282
Farkas, Gabor 137
Fayyaz, Asad 130
Felderer, Niklaus 2199
Feng, Chao 2058
Feng, Wei 3678
Ferdowsi, Mehdi 3223
Fernandez, Gabriel 3209
Fernandez-Cardador, Antonio 2534
Fischer, F. 3596
Foo, Gilbert 1724
Formentini, A. 4034
Freijedo, Fracisco D. 1498
Friedrichs, Peter 3584
Fuchs, Simon 2301
Fujii, H. 1253
Fujii, Kansuke 3711
Fujii, Keisuke 1189
Fujii, Toshiyuki 2540, 3578
Fujimoto, Hiroshi 77, 663
Fujimoto, Kazuki 2047
Fujimoto, Yasutaka 571, 681
Fujimura, Akira 1080
Fujita, Atsushi 296
Fujita, Goro 363
Fujita, Hideaki 626, 1854, 3813, 3940
Fujiwara, Hajime 1381
Fujiwara, Kazuya 3773
Fukuda, Hiroto 2938
Fukuda, Kenji 2558
Fukui, Tomoya 860
Fukuoka, T. 1240
Fukushima, Kentarou 2176
Fukushima, Takafumi 3478
Funabiki, Shigeyuki 2449
Funaki, Tsuyoshi 309, 2181, 3092
Funato, Hirohito 94, 2036
Funato, Hiroki 2073
Furukawa, Keita 3349
Furukawa, Kimihisa 3572
Furukawa, Yudai 4193
Furusho, Yasuaki 3711
Gan, Yiliang 1391

AUTHOR INDEX

Ganisetti, V. K. 2907
Gao, Feng...........................2016, 3383, 3965
Gao, Xiaonan 1661
Gao, Zhuo 3455
Garrigós, A. 1435
Gasim, Abdulaziz 2836
Gehlot, Deepak 3471
Geng, Hua 542
Geng, Yiwen 619
Gerada, C. 4034
Gheonjian, Anna 2078
Gietler, Harald 3140
Gohara, Hiromichi 2764
Gondo, Ryota 3490
Gong, Bing 3797
Gong, Chunying 1015, 1177
Gong, Z. 267
Gorodnichev, Anton 375
Goto, Akihisa 2449
Goto, Hiroki 3192
Goto, Kazuya 1315
Goto, Yasuyuki 809
Gou, Yating 1157, 1167
Grimm, Ferdinand 2895
Grossner, Ulrike 3588
Gruber, Wolfgang................3632, 3640, 4028
Gu, Lei 632
Gu, Qing 2963
Guajardo, Cristian Andres Garces.......1854
Guan, Bo 1032
Guan, Yajuan 2668, 3678
Guan, Yueshi 614, 3780
Guangzhu, Wang 1376
Guerrero, Josep M. 1498, 2668, 3112
Guerrero, M. Josep 3678
Gui, Yonghao.......................................2668
Guidi, Giuseppe.............................782, 2003
Guillod, Thomas 396
Gunji, Daisuke.....................................663
Guo, Leilei 904
Guo, Yanjie 3338
Guozhao, Duan 2625
Gupta, K. 267
Gurpinar, Emre 130
Gutiérrez, R. 1435
Ha, Jung-Ik..................................565, 2500
Ha, Sang-Hyun 3466
Haga, Hitoshi................................1370, 3890
Hagiwara, Makoto 3273
Hahashi, Yuji 4059
Haider, M. 4080
Halamicek, Michael 831
Halick, Mohamed 416
Hamabe, Yasumasa 1276
Hamada, Shizunori 227

Hamaguchi, Takumi 3507
Hamasaki, S. 1240
Hamasaki, Shin-Ichi.................1217, 1276, 2938, 3237
Hameyer, Kay 740
Han, Byung-Moon 466
Han, Jung-Kyu 3107, 3533, 4049, 4054
Han, Pengcheng 1027, 2714
Han, Yang 3112
Hanajiri, Kensuke 663
Hanamoto, Tsuyoshi 1315, 1698
Hancioglu, Oguz Kaan 1680
Handa, Hiroyuki 3762
Handa, Yuuichi 4059
Hane, Yoshiki..2426
Hang, Lijun 1391, 2866
Hanju, Cha 1985
Hao, Liu 3484
Hao, Xiang 1478
Harnefors, Lennart 3684
Hartmann, S. 3596
Haruna, Junnosuke 94, 2036
Hasegawa, Kazunori 1938
Hasegawa, M. 3665
Hasegawa, Ryuta 2011
Hashempour, Mohammad M. 4198
Hashimoto, Kazuki 3757
Hasler, Jean-Philippe 3684
Hata, Katsuhiro 663
Hata, Ryotaro 2149
Hatakeyama, Tomoyuki 1991
Hataya, Morimasa 410
Hatipoglu, E. 3805
Hatsumi, Takuya 94
Hatta, Yoshiyuki 675
Hattori, Fumiya 2738
Hattori, Keisuke 3286
Haung-Jen, Chiu 645
Hayashi, Nobuo 866
Hayashi, Yuji 356
He, Wangpin 560
He, Xiaokun 1504, 2496
He, Xiaoqiong 1027, 2714
He, Yigang...2317
He, Yingjie 3439
Hendrix, M. A. M. 946, 2697
Heo, Jongwon 726
Hidaka, Yuki 2820
Higuchi, Keiichi 2764
Higuchi, Masato......................................3952
Higuchi, Shinichi 2216
Hikaru, Naruse 3418
Hikihara, T. 3665
Hikihara, Takashi 3654, 3757
Hiller, Marc 3074
Hillers, A. 1896

AUTHOR INDEX

Hillers, André 2301
Hilt, Oliver 3607
Hinz, Arne 598
Hirahara, Hideaki 1960
Hiraki, Eiji 410, 1602, 1610
Hirao, Takashi............................2082, 2137
Hirase, Yuko.......................................767
Hirayama, Katsutoshi 4193
Hirayama, Tadashi 3406
Hirokawa, Masahiko...................1543, 4133
Hirokawa, Takayuki 296, 410
Hiromoto, Masayuki 3644
Hirose, Keiichi.............................593, 822
Hirose, Naoki 3791
Hiroshi, Tadano 3431
Hiroshige, Shinichi 3369
Hirota, Takashi 3952
Hoang, Tuan V. 1752
Hoda, Isao..2073
Hofmann, Viktor 2476
Hofmann, Wilfried 3243
Hojo, Masahide 3369
Holenstein, Thomas...........................3619
Holmes, D. G. 3670
Hong, Miao 2718
Hongpeng, Liu.....................1442, 2969
Honjo, Satoshi 2066
Hori, Motohito 3396
Hori, Yoichi 77, 663
Horie, Shunsuke 809
Horikoshi, Takahiro 1997
Hoshi, Nobukazu 971, 2660, 3855
Hou, Chung-Chuan 1617
Hou, Lijun 2901
Houran, Mohamad Abou 1062, 2854
Hsieh, Guan-Chyun 123
Hsieh, Hung-I 123
Hsieh, Yao-Ching 3151, 3544
Hsu, Chi-Hsuan 2653
Hu, Jiewen 3985
Hu, Jingxin 1073, 2250, 2484
Hu, Sheng 3052
Hu, Song 370
Hu, Xihong................................614, 3780
Hu, Xing 2262
Huang, Bing-Siang 2092
Huang, Bo-Jia 3528
Huang, Chien-Chun 3151
Huang, Huazhen................................1125
Huang, Jingjin 2980, 4157
Huang, Jingjing.............1004, 2688, 2692
Huang, Jun-Xian 1626, 3081
Huang, Lang......................................1478
Huang, Pin Yu 2165
Huang, Ta-Wei 1626

Huang, Wen-Mei 2576
Huang, Xianjin 1131, 2051
Huang, Xiaoliang..................................84
Huang, Xuehao 3455
Huang, You-Chun 275
Huemer, Mario 3140
Hui, S. Y. Ron 889, 2552
Hung, Chun-Yao 2576
Hung, Shun-Kang 1575
Huo, Chongcan 987
Huo, Junya 1206, 1234
Hussein, Abdallah............... 130, 2932
Huynh, Dang Minh 3086
Hwang, Duck-Hwan 1403
Hwang, Seon-Ik 3323
Hwu, K.I. 851
Hyakutake, Y. 1253
Hyodo, Takashi 2589
Hyunsung, An 1985
Iannuzzi, Diego 2527
Ibuchi, Takaaki 309
Ichinose, H. 1240
Ide, Yuji.. 3896
Iijima, Ryuji 313, 1111
Iioka, Daisuke 2278
Ikari, Yuki 148
Ikeda, Hidehiro 1315
Ikeda, Yoshinari 3396
Ilves, Kalle 2335
Imai, Kazu 3363
Imai, Makoto 296, 410
Imamori, Satoshi 699
Imaoka, Jun 1087, 1095, 1554, 3773
Imoto, R. 2808
Imtiaz, Abu Saleh 2416
Imura, Takehiro 77, 663
Inaba, Tsuyoshi 4114
Inomata, Kentaro 3952
Inoue, Daisuke 2764
Inoue, Kaoru 1264, 2186, 4151
Inoue, Kent 348
Inoue, Masamichi 1228
Inoue, Takatoshi 1276
Inoue, Y. 704, 2808
Inoue, Yukinori 1189, 1289, 1329, 2802, 2814, 3197
Irino, Yusuke 244
Ise, Toshifumi 775, 2393, 3762, 3902
Ishibashi, Mikiya 1370
Ishibashi, Naoyuki 1543
Ishibashi, Taku 2292
Ishigaki, Shingo 227
Ishiguro, Takahiro 1997, 2011, 3304
Ishihara, Masataka 1610
Ishii, Y. 1834
Ishii, Yuki....................................... 1196

AUTHOR INDEX

Ishikawa, Hiroki 2176, 3412
Ishikawa, Kohsuke 2725
Ishikura, Yuki 1087, 1095, 3717
Isobe, Eisuke 2042
Isobe, Takanori 313, 1111, 3375, 3431
Isozaki, Keisaku 1364
Itaya, Yohei 3450
Ito, Kazuhiko 2540
Ito, Yasuaki 1586, 2324
Ito, Yoichi 3086
Ito, Youichi 439
Itoh, Gimpei 1289
Itoh, Jun-Ichi 69, 348, 534, 896, 1567,
          2229, 2237, 2519, 2596, 3349, 3797
Iwabuchi, Akio 439
Iwai, Akinobu 2066
Iwaji, Yoshitaka 1301
Iwasaki, Makoto 1666
Iwasaki, Tetsuya 3490
Iwata, Hiroki 3896
Iyasu, Seiji 4059
Iyoda, Isao 2914
Jacobs, Keijo 3292
Jaffar, Hanis Afiqah Binti 2956
Jain, Prashant 3471
Janah, Mounia 681
Jang, Duekjin 2619
Jang, Yu-Jin 1655, 3466
Jang, Yun 1736
Jangs, Yujin 1562
Jarutus, Neerakorn 2121
Jehle, Andreas 1509
Jennings, M 1149
Jeong, Seog Y 2564
Jeong, Si-Hoon 289
Jeong, Yeonho 838, 2365, 2376
Jhang, Ying-Yi 3884
Jhou, Yu-Lin 1145
Ji, Guyuan 2921
Jia, Haiyang 998
Jia, Pengyu 977, 3040
Jia, Xu 3025
Jiacheng, Wang 2986
Jiajie, Zang 2986
Jiajie, Zhou 1442
Jian, Jun-Min 2653
Jiang, Jinhai 84
Jiang, Shuai 987
Jiang, Siyue 4168
Jiang, Yanfeng 2058
Jiang, Yongbin 3863
Jianhua, Wang 1282
Jianming, Xu 528
Jianqiao, Zhou 2986
Jianwen, Zhang 2986

Jiaxing, Liu 1376
Jikumaru, Takehiro 177
Jimichi, T. 1834
Jimichi, Takushi 3729
Jin, Nan 904
Jin, Zheming 2668
Jing, Lei 878
Jing, Lyu 2692
Jing, Yang 3383
Jingyu, Song 1282
Jing-Yuan, Lin 645
Jinjun, Liu 4181
Jinshui, Zhang 4181
Jisaki, Jun 3182
Joebges, Philipp 375
Jongudomkarn, Jonggrist 3902
Jonishi, Akihiro 2774
Joryo, Satoshi 1202
Joseph, Anto 1358
Jumayev, S. 161
Jung, Hanul 688
Jung, Hyun-Sam 911
Jung, Jae-Jung 3557
Jung, Jee-Hoon 289, 383
Jung, Jun-Hyung 3323
Jung, Si-Hoon 383
Jungmayr, Gerald 3640
Junior, Lourenço Matakas 3785
Jynu-Jhe, Jhang 645
Kada, Haruya 3890
Kadota, Mitsuhiro 3572
Kai, Masahiko 1803
Kaicheng, Ding 4181
Kaipia, T. 2948
Kaishakuji, Hikaru 2360
Kakigano, Hiroaki 583, 2956
Kamaeguchi, Koki 410
Kamakura, Kousuke 2756
Kamejima, Takayoshi 3286
Kamiya, Naoki 1673
Kamiyama, Naosumi 1955
Kamoshida, Naoki 1111
Kampeerawar, Warayut 3257
Kanai, Naoyuki 3396
Kanaya, Kazuhisa 2011
Kanazawa, Yasuki 2789
Kanchan, R. S. 488
Kandula, Prasad 4001
Kaneko, Satoshi 3396
Kanetani, Kaisei 207
Kang, Dong-Hun 3030
Kang, Feel-Soon 2376
Kang, Kyoung-Suk 922
Kang, Tahyun 1977
Kang, Yong 2997

AUTHOR INDEX

Kanno, Junya 3299
Kano, Fumihisa 2036
Kanoda, Akihiko 3572
Kanzian, Marc 3140
Kapisch, E. B. 1067
Karami, Bagher..2854
Karppanen, J. 2948
Kasai, Yuji 2036
Kashihara, Tatsuki 1741
Katayama, Tatsuji 1346
Kato, Hideaki 1580, 1586, 2324
Kato, Hirokazu 3478
Kato, Koji 439, 1370, 3086
Kato, Toshiji................................1264, 2186, 4151
Katoh, Kaoru...233
Katoh, Shinji 2176
Katsuki, Akihiko 1543
Katsura, Seiichiro 669
Katsura, Shogo 767
Katsushi, Terazono 3431
Kawabata, Naoki 2887
Kawabata, Shuma 3406
Kawagoe, Natsuki 3490
Kawaguchi, Hironori 517
Kawaguchi, Jun'ichiro 1828
Kawaguchi, Yuki 3572
Kawakami, Masaki 2756
Kawakami, Noriko 1346
Kawamura, Atsuo 318, 1649, 1687, 3916
Kawamura, Itsuo 3396
Kawamura, Kazuki 1567
Kawanishi, Kota 169
Kawashita, Jun 2042
Kayashima, Kazuya 1315
Kaymak, Murat 3729
Kazmi, Syed Muhammad Raza 4168
Ke, Junji 1125
Kennel, Ralph 1661, 2895, 3965
Kezuka, Nobutaka 227
Khan, Ashraf Ali 927
Khan, Faisal...446, 2416
Khan, Muhammad Mansoor............................528
Khan, Usman Ali 927
Khomfoi, Surin 1460
Khubchandani, Vasudha 845
Kiatsookkanatorn, Paiboon...........................2581
Kida, Masahiro..................................1586, 2324
Kido, Tatsuya 329
Kikuchi, Ryosuke 1877
Kikuchi, Takaaki 2292
Kikuchi, Takeshi 3578
Kikuma, Toshiaki 3299
Kim, Byeongwoo...256
Kim, Chong-Eun 108, 3538
Kim, Dong-Kwan 1655, 3466, 3538

Kim, Gun-Woo 838
Kim, Hansang 1465
Kim, Heung-Geun 927, 1046, 2619, 3134
Kim, Hideaki 207
Kim, Hyeon-Sik 521
Kim, In-Dong 3229
Kim, Jae-Kuk 3100
Kim, Jang-Mok 3323
Kim, Jin-Hak 1530
Kim, Jin-Young 3229
Kim, Jong-Woo 3107, 4049
Kim, Kangsan 256
Kim, Katherine A. 2092, 3063
Kim, Keon Young 4104
Kim, Keon-Woo.................108, 1562, 1655, 2365, 2376
Kim, Ki-Mok 2365
Kim, Myong Hwan 2500
Kim, Sanghun 2619
Kim, Sunju 3833
Kim, Yeonjung 1465
Kimura, Hideki 2036
Kimura, Mamoru 1991, 1997
Kimura, Noriyuki 1202, 1259, 2558, 2887, 2914
Kinoshita, Masahiro 3929
Kishimoto, Toshihiko 261
Kishita, Ken 1301
Kitagawa, Wataru 1847, 3507
Kitamura, Akio 2764
Kitamura, Toshinori 2660
Kiyoshi, Ohishi 1673
Kiyota, Kyohei 3182
Klammer, Bianca 3632
Ko, Chien-Tzu 2107
Kobayashi, Hiroyasu 2527, 3490
Kobayashi, Koji 1741
Kobayashi, Marika 2802
Kodaka, Wataru 2589
Kogai, Naoki 1364
Koizumi, Hirotaka 4114
Kolar, J. W. 3805, 4080
Kolar, Johann W. 181, 396, 3619
Kolb, Johannes 2848
Komaru, Yuma 1329
Komatsu, Hiroyoshi 1346
Komatsu, Taiga 2820
Komatsu, Wilson 3785
Komeda, Shohei 3813
Kometani, Haruyuki 711
Kondo, Keiichiro 726, 2047, 2527, 3490
Kondo, Shota 1295
Kondo, Takeshi 4114
Kong, Wei 3460
Kongjeen, Yuttana 2430
Konishi, Akihiro 1602
Konno, Junya 1692

AUTHOR INDEX

Konstantinou, Georgios 3117
Kopta, A. 3596
Kosaka, Takashi 3418
Koseki, K. 1162
Koseki, Takafumi 2042, 2309, 3257
Koshikizawa, Hiroyuki 1567
Kostov, Konstantin 2732
Kouketsu, Masaju 227
Kouno, Yusuke 2176
Kovacevic-Badstübner, Ivana 3588
Kowatari, Hiroki 2660
Koyama, Yushi 2011
Krismer, F. 3805
Krismer, Florian 396
Kubo, Hajime 483
Kubota, Hisao 1196
Kucka, Jakub 1904
Kumada, Keishirou 3396
Kumagai, Shuta 1264
Kumar, Ashish 3993
Kumar, Rajesh 2456
Kumar, S. Gautam 3471
Kumsuwan, Yuttana 2113, 2121
Kunomura, Ken 1803
Kuo, Chun-Ting 1145
Kuraishi, Daigo 3896
Kuraku, Nagendra Vara Prasad 2317
Kuring, Carsten 3607
Kurisaka, Masakatsu 4151
Kurita, Naoyuki 1991
Kurita, Nobuyuki 3640
Kurokawa, Fujio 826, 2097, 2283, 4193
Kurosawa, Nobuhito 1810
Kurumatani, Hiroki 669
Kusaka, Keisuke 69, 348, 2237, 3349
Kusumah, Ferdi Perdana 3870
Kuwata, Gen 177
Kwon, Min-Jun 114
Kyyrä, Jorma 2193, 3870
Lai, Jih-Sheng 3107, 4049
Lai, Jui-Hung 3081
Lan, Yuanliang 1167
Lana, A. 2948
Le, Hanh-Phuc 213
Le, Hoai Nam 2519
Lee, Byoung-Hee 838
Lee, Byung-Kwon 3030
Lee, Chan 688
Lee, Choongin 565
Lee, Dong-Choon 478, 1212
Lee, Hong-Hee 1752
Lee, Hyong Gun 1336
Lee, Il-Oun 1530
Lee, Jae-Bum 3100
Lee, Jia-You 657, 2107

Lee, Joon-Hee 3557
Lee, Junbae 1141
Lee, June-Hee 466
Lee, Jung-Yong 1403
Lee, Jun-Young 3030
Lee, Jusuk 1336
Lee, Kyo-Beum 466, 1736, 4104, 4109
Lee, Kyoung-Won 2145
Lee, Kyung-Hwan 2500
Lee, Min-Su 108
Lee, Minsub 1141
Lee, Nayoung 1562
Lee, Song-Kai 2102
Lee, T. L. 4198
Lee, Tzung-Lin 2576
Lee, Woo-Cheol 114
Lee, Woo-Seok 1530
Lee, Young-Dal 3466, 3538
Lehn, Peter W. 3203
Lei, Qin 2400, 3742
Leng, Darith 1764
Leubner, Martin 3243
Li, Bodong 878
Li, Chi 790, 3705
Li, Dongsheng 1301
Li, Fei 2611
Li, Fujian 2944
Li, Guanglei 1455
Li, Haijin 2270
Li, Haisi 3040
Li, Haoyu 2901
Li, Hong 2058
Li, Hongchang 337, 3910
Li, Jhih-Sian 3081
Li, Jia 2073
Li, Jianfeng 130
Li, Kaiyuan 1517, 1592
Li, Lei 1172
Li, Li 1771
Li, Ming 2973
Li, Mingshen 2668, 3678
Li, Pengcheng 3698
Li, Shufan 3338
Li, Sinan 889, 2552
Li, T.-Y. 1912
Li, Xiaodong 370
Li, Xiaolu Lucia 3768
Li, Xiaoqiang 3910
Li, Xingshuo 453
Li, Xinying 2646
Li, Yan 2245
Li, Yang 795, 1478
Li, Yangman 2901
Li, Yi-Chan 639, 883
Li, Yongdong 1010, 2386

AUTHOR INDEX

Li, Yong-Jyun 275
Li, Yunwei 3958
Li, Yunwei Ryan 1537
Li, Yuze 2997
Li, Zhenjie 84
Li, Zhenwei 998
Li, Zhiqing 100
Liang, Daniel 1943
Liang, Junrui 4122
Liang, Ning 1157
Liang, Wencai 1131
Liao, Chenglin 3338
Liao, Chih-Yi 657
Liao, Hsuan 2653
Liao, Jian-Tang 4233
Liao, Mengyan 1386, 1421
Liaw, C. M. 2907
Lim, Cheon-Yong 1655, 2376, 3533
Lim, Dae-Sik 1212
Lim, Kyungbae 803
Lim, Young-Cheol 982, 1799
Lin, Chang-Hua 1341, 1777
Lin, Cheng-Hung 2092
Lin, Fei 1131, 1816, 2051, 2058, 3484, 3495
Lin, Jin 3460
Lin, Jing-Yuan 3151
Lin, K.-E. 1912
Lin, Min 4133
Lin, Xiang 3460
Lin, Xiaolan 1027
Lin, Xuerui 1537
Lin, Yu-Hsiu 1575
Lin, Yu-Lin 1145
Lisha, Chen 3958
Liske, Andreas 2848
Liu, Baojin 1051, 2944, 3924
Liu, Bi 1872
Liu, Bo 542, 878
Liu, Chao 2245
Liu, Chunhui 3742
Liu, Cuicui 1157, 1167
Liu, Dong 1758, 2708
Liu, Fang 2611, 2992
Liu, Furong 3052
Liu, He 3215
Liu, Hwa-Dong 1341, 1777
Liu, Jia 775, 3902
Liu, Jiaxin 2016
Liu, Jinjun 957, 1051, 1524, 2435, 2646, 2681, 3129, 3176, 3251, 3439, 3692, 3924, 3934
Liu, Junwen 3863
Liu, Kangli 3010, 3015
Liu, Nianzhou 1010
Liu, Ning 2877

Liu, Pang-Jung 2102
Liu, Ruofei 2547
Liu, Shu 3052
Liu, Siqi 1491
Liu, Tao 1478
Liu, Teng 2681, 3176, 3934
Liu, Wei 3164
Liu, Wenzhao 3678
Liu, Xiaosheng 934
Liu, Xicai 1661, 3965
Liu, Xinbo 3455
Liu, Yifu 2400, 3742
Liu, Yu-Chen 2092
Liu, Yuping 1816
Liu, Zeng 957, 1524, 2435, 2681, 3176, 3251, 3692, 3749, 3924
Liu, Zhiyuan 3495
Liu, Zipeng 2681, 3176
Lo, Jen-Hao 1145
Lomonova, E.A. 161
Lopez-Lopez, Alvaro J. 2534
Lotfi, Nima 3223
Lovison, Giorgio 77
Lu, David H. 2404
Lu, David Hongfei 3390
Lu, Kaiyuan 1183, 1246, 2842
Lu, M. Z. 2907
Lu, Shengli 3145
Lu, Shuai 3698
Lu, Y. 267
Luhtala, Roni 547, 2470, 3156
Lunglmayr, Michael 3140
Luo, Min 422, 2199
Luo, Rui 3129, 3439
Luo, Y. 267
Luong, Hoan-Tien 2145
Lyu, Jing 1004, 4162, 4220
Ma, Baohui 2882
Ma, Jie 1118
Ma, Ke 3877
Ma, Shaokang 542
Ma, Tianshu 2703
Ma, Yue 3717
Ma, Zhixun 917, 2688, 2692, 4157, 4162
Mabuchi, Yuichi 3572
Machavolu, Sawanth Krishna 753
Machida, Yuuki 2449
Maharjan, Laxman 1840
Makishima, Shingo 2047
Mannen, Tomoyuki 1414, 1866
Mantooth, H. Alan 153
Mao, Meiqin 815, 1472, 1793, 2505
Mariéthoz, Sébastien 2158, 3820
Marinescu, Radu-Florin 1822
Marroquí, D. 1435

AUTHOR INDEX

Martinez, Wilmar 2193
Maruta, Hidenori 826
Maruyama, Kouji 3396
März, Martin 2410
Masuda, Eisuke 309
Masuda, Mitsuru 88
Masuko, Toshitake 3723
Matsubayashi, Tatsushi 207
Matsuda, Akihiro 2329
Matsuda, Tomohiro 1972
Matsudate, Koki 2022
Matsui, Nobumasa 826, 2283
Matsui, Nobuyuki 3418
Matsui, Teruhisa 1803
Matsui, Yoshihiro 1080
Matsui, Yuto 1847, 3791
Matsuki, Yosuke 2224
Matsumori, Hiroaki 3357
Matsumoto, Satoshi 2360, 2370
Matsumoto, Takashi 2404
Matsumoto, Toshiaki 2011, 3304
Matsumoto, Yasuaki 517
Matsumoto, Yohei 233
Matsumura, Toshiro 809
Matsuo, Keisuke 169
Matsuse, Kouki 169
Mattsson, A. 2948
Mawby, P 1149
Mcgrath, B. P. 3670
Meng, Xin 957, 1549, 3251
Menzi, David 181
Mertens, Axel 1904
Messo, Tuomas 547, 963, 1704, 2470, 3156, 4205
Michihira, Masakazu 992, 3058
Michikoshi, Hisato 2558
Milovanovic, Stefan 1484
Min, Geon-Hong 2500
Minami, Masataka 992, 3058
Mino, Kazuaki 3717
Mira, Maria C. 1351
Mishima, Tomokazu 329, 872
Misra, Mitradatta 3884
Mitsantisuk, Chowarit 3332
Miura, Yushi 775, 2393, 3762
Miwa, Yoshihiro 404
Miyajima, Hiroki 1803
Miyama, Yoshihiro 711
Miyawaki, Satoshi 2738
Miyazaki, Toshimasa 1673
Mizumoto, Yuki 1810
Mizuno, Takayuki 169
Mizuno, Yuji 2283
Mizushima, Takuya 1543
Mocevic, Slavko 3985
Mochidate, Sae 1972

Mogorovic, Marko 2170
Moiannou, Tom 831
Mok, Hyung Soo 1336
Molinas, Marta 759
Moo, Chin-Sien 275, 3544
Moon, Gun-Woo 108, 838, 1562, 1655, 2365, 2376, 3100, 3466, 3533, 3538, 4049, 4054
Mori, Kazuhisa 233
Morimoto, Hiroaki 2540, 3265
Morimoto, S. 704, 2808
Morimoto, Shigeo ...1189, 1289, 1329, 2802, 2814, 3197
Morimoto, Shinya 2210
Morishima, Naoki 2540, 3450
Moriyama, Hiroyuki 1580, 1586, 2324
Morizane, Toshimitsu 1202, 1259, 2558, 2887, 2914
Mortimer, Benedict J. 598
Motegi, Shin-Ichi 992, 3058
Motohashi, Yuto 753
Motoyama, Hiromasa 356
Mouawad, Bassem 130
Mukaiyama, Naoki 2558
Müller-Hellmann, Adolf 598
Muni, Bishnu Prasad 3471
Murakami, Toshiyuki 575
Nabetani, Yoichi 2404
Nada, Kaho 3578
Nagai, Sakahisa 1687
Nagai, Satoshi 534
Nagao, S. 142
Nagaoka, Naoto 3170
Nagaoka, Shingo 118, 4139
Nagasaka, Kuniaki 1692
Nagashima, Takumi 3490
Nagira, Yoshiki 4016
Naina, Sagar 3046
Nakabayashi, Shigeaki 1692
Nakabayashi, Shigeyuki 517
Nakagawa, Hidehiko 767
Nakahara, Kengo 3237
Nakahara, Mizuki 3572
Nakai, Masanobu 3182
Nakajima, Mizuki 2750
Nakajima, Tatsuhito 1997, 3299
Nakamura, Fuminori 2329
Nakamura, Hideyo 1137
Nakamura, Kenji 2426
Nakamura, Kimikazu 4059
Nakamura, M. 201
Nakamura, Masashi 471
Nakamura, Ritaka 495
Nakano, Hayato 2764
Nakano, Shigeki 2370
Nakao, Hiroshi 196
Nakao, Kazushige 148, 2914

AUTHOR INDEX

Nakao, Yuta 588
Nakashima, Yoshiyasu 196
Nakatsu, Kinya 2082
Nakazawa, Haruo 2404
Nakazawa, Y. 1253
Nakazawa, Yuji 244
Namba, Akihiro 2082
Nanamori, Kimihiro 2789
Naradhipa, Adhistira M. 3833
Narita, Takayoshi 1580, 1586, 2324
Narushima, Hiroki 693
Nashida, Norihiro 1137
Nasr, Miad 2380
Natori, Kenji 588, 1860
Nawaz, Muhammad 2335
Nazib, A. A. 3670
Nee, Hans-Peter 2732, 3292, 3684
Neubert, Markus 3979
Ngamroo, Issarachai 2287
Ngo, Tung 1724
Nguyen, Bang Le-Huy 1046, 3134
Nguyen, Hong-Quan 3426
Nguyen, Minh-Khai 982, 1799, 2145
Nguyen, Tien-The 1046, 3134
Nho, Eui-Cheol 922
Nicolae, Ileana-Diana 1822
Nicolae, Petre-Marian 1822
Nie, Jintong 2963
Niki, Toru 856
Nimura, Takumi 1295
Ninomiya, Tatsuya 2836
Nishikata, Shoji 4227
Nishimura, Yoshitaka 1137
Nishino, Taisei 1364
Nishiyama, Shigeki 2149
Nishizawa, Koroku 2229
Nishizawa, Shin-Ichi 1938
Niu, Haonan 3025
Niyomsatian, K. 4096
Noah, Mostafa 1087, 1095
Noda, Taku 2176
Noda, Yujiro 324
Noguchi, Toshihiko 718, 753
Noh, Seungjun 1598
Nomura, Naofumi 2216
Nomura, Shinichi 2022
Nonogaki, Midori 2292
Noro, Osamu 767
Norrga, Staffan 3292
Norum, Lars 2630
Noto, Yasuyuki 3711
Notohara, Yasuo 1301
Nuchnoi, S. 4096
Nugroho, Dannisworo S. 3855
Nussbaumer, Thomas 3619

Nuutinen, P. 2948
Obara, Hidemine 1649
Oda, Yoshiho 1586, 2324
Ogasawara, Satoshi 2589, 2725, 2796, 3315
Ogawa, Eri 2768
Ogawa, Kazuki 1580
Ogawa, Takuro 866
Ogawa, Tomoyuki 1828, 3265
Ogawa, Toru 2796
Ogino, Hiroshi 517
Oh, Sehoon 688
Ohashi, Hidetomo 2774
Ohdera, Fumiya 1322
Ohguchi, Hideki 699
Ohishi, Kiyoshi 1741, 3332, 3890, 3896
Ohji, Takahisa 3182
Ohnishi, Haruna 3273
Ohno, Takanobu 971
Ohno, Tatsuki 1649
Ohnuma, Naoto 233
Ohnuma, Takumi 1223
Ohnuma, Yoshiya 2738
Ohta, Kazuki 1223
Ohta, Takahiro 517
Ohtake, Asuka 3286
Ohyama, K. 1253
Ohyama, Kazuhiro 2921
Ohyama, Kazunobu 244
Oi, Kazunobu 1890
Oishi, Kazuki 3644
Oiwa, Takaaki 157, 4042
Oka, Toshiomi 2370
Okamoto, Kenkichiro 1095
Okazaki, Yuhei 2335
Okazawa, Toshio 2066
Oki, Yusuke 1828
Okitsu, Takashi 169
Okuda, Takafumi 3654, 3757
Okuno, Kengo 1586, 2324
Okuyama, Ryota 3450
Omori, Hideki 1202, 1259, 2558, 2887
Omori, Shuto 471
Omura, Ichiro 1938
Onishi, Hiroyuki 4139
Onishi, Masami 2082
Ono, Y. 4080
Onozawa, Yuichi 2768
Ooshima, Masahide 3613
Orikawa, Koji 2589, 2725, 3315
Ortiz-Gonzalez, J. 1149
Osawa, Akihiro 2764
Oshima, Takuya 4088
Osman, Ilham 3971
Ota, Ryosuke 3855
Ouaida, Rémy 2153

AUTHOR INDEX

Ouchi, Takayuki.................................250
Ouyang, Shaodi.....................1051, 3129
Ouyang, Ziwei................................4066
Owaki, Daiki....................................809
Paiboon, Supakorn...........................1642
Pairindra, Worapong.........................1460
Pan, Pengpeng.................................1504
Pan, Xuewei....................................1172
Panda, Sanjib Kumar.........................4186
Pang, Hui.......................................2343
Papadopoulos, C..............................3596
Papini, L.......................................4034
Paramalingam, Jan............................2329
Parashar, Sanket..............................3993
Park, Hwa-Pyeong.............................289
Park, Jin-Hyuk................................4104
Park, Jun H....................................2564
Park, Kwon-Sik................................922
Park, Moo-Hyun.............1562, 3100, 3533
Park, Mu-Hyun.................................838
Park, Sang Uk.................................1336
Park, Sanghyeon..............................282
Partanen, J.....................................2948
Pasterczyk, Robert............................2158
Patel, Prashant................................3046
Patel, Utsav....................................3046
Pathmanathan, M...............................488
Patwa, Premal..................................3046
Pauli, Florian..................................740
Pecharroman, Ramon R.......................2534
Pei, Xuejun....................................2997
Peltoniemi, P...................................2948
Peng, Jinjie....................................939
Peng, Xu.......................1027, 2714, 3020
Pengxiang, Zeng...............................4181
Pham, N. Ha...................................1414
Pidaparthy, Syam Kumar.....................1465
Pinomaa, P.....................................2948
Polmai, Sompob.......................1764, 2490
Pou, Josep.....................................3117
Prabowo, Yos...................................3564
Prasanth, Sundararajan.......................416
Prodic, Aleksandar.............................831
Promyoo, Adisak...............................2871
Pueschel, Tilo..................................190
Pyrhonen, J.....................................161
Qi, Wenlong....................................889
Qian, Cheng...................................1472
Qian, Qinsong.................................3145
Qiao, Liang....................................3329
Qin, Zian.......................................1925
Qiu, Maohang..................................878
Qiu, Zhifeng....................................939
Rabkowski, Jacek..............................2129
Radman, Karlo.................................3632

Radwan, Hamdy...............................3945
Rahimo, M.............................3596, 4009
Rahman, Ahmad Arif Bin Abd................2956
Rahman, Faz...................................3971
Rahmati, Abdolreza............................2854
Ramirez-Elizondo, Laura.....................1426
Ramos, Niño Christopher.....................3092
Ran, L...1149
Ran, Li..1931
Rao, Eswar....................................3471
Rathore, Akshay Kumar.................342, 2456
Reinikka, Tommi...............................1704
Remus, Nico...................................3243
Ren, Haijun....................................2714
Ren, Yu..3329
Rencz, Marta...................................137
Rengarajan, Satish............................3564
Riar, Baljit..............................4074, 4145
Rietmann, Stefan..............................2301
Rim, Chun T...................................2564
Risseh, Arash Edvin...........................2732
Rivas-Davila, Juan..............282, 632, 3848
Robert, Mickaël................................2158
Rodriguez-Diaz, Enrique......................1498
Roes, M. G. L..........................946, 2697
Roinila, Tomi.........547, 1704, 1719, 2470, 3156, 4205
Romano, Daniele...............................3588
Roy, Sourov.............................446, 2416
Ruan, Liheng...........................3010, 3015
Rubino, S......................................732
Ruf, Andreas..................................740
Rygg, Atle.....................................759
Sadakata, Hideki..............................410
Sagawa, Kouhei...............................2036
Saha, Tarak.............................4074, 4145
Saito, Tatsuhito.......................1828, 3265
Saito, Yota....................................1782
Saitoh, Hiroumi................................2278
Sakabe, Tomoki................................3058
Sakai, Kazuto..................................2826
Sakai, Ryosuke................................2832
Sakai, Yoshikazu..............................4114
Sakawaki, Atsushi.............................244
Sakimoto, Kenichi..............................767
Sakiyama, Taiki................................2186
Sakoda, Kenichi................................860
Sakr, Nadim...................................2078
Sakuma, Kensuke..............................3522
Sakuraba, Tomokazu..........................2153
Sakurai, Seiya.................................3412
Samanta, Suvendu.............................342
Samermurn, S..................................4096
Samizadeh, Mehdi.....................1062, 2854
Sanada, M...............................704, 2808
Sanada, Masayuki ..1189, 1289, 1329, 2802, 2814, 3197

AUTHOR INDEX

Sangwongwanich, Ariya 2512
Sangwongwanich, S...................................4096
Sangwongwanich, Somboon 1642, 2581
Sannomiya, Kenta 1259
Sano, Kenichiro 3299
Sano, Toshiki 3896
Santi, Enrico 1719
Sasaki, Masahiro 2774
Sasaki, Masato 3344
Sasongko, Firman 416
Sathik, Mohamed 416
Sato, Fumihiro 250
Sato, Keisuke 3265
Sato, Kenji 3478
Sato, Mitsuru 118
Sato, Motoki 663
Sato, Takashi 3644
Sato, Yasuhiro 2042
Sato, Yukihiko 588, 1860, 1972, 3514, 3522
Satoh, Nobuo 2750
Sayed, Mahmoud A. 3945
Schanen, Jean-Luc 2158
Schletz, Andreas 2410
Schülting, Philipp 388
Schweiker, Daniel 2848
Schweizer, Mario 555
Schwendemann, Rüdiger 3074
See, Kye Yak 2296
Sekiba, Yoichi 2176
Sekimoto, Morimitsu 866
Sekisue, Takayuki 2176
Sekiya, Hiroo 3650, 4127
Semwal, R. R. 1358
Senanayake, Thilak 313
Seng, Tan Chuan 416
Seo, Byuong-Jun 922
Seo, Gab-Su 213
Sera, Dezso 2512
Setiadi, Hadi 626
Settels, Sjef J. 3840
Severson, Eric L. 4020
Sewergin, Alexander 3979
Sha, Yilin 3329
Shabib, G. 3945
Shamseh, Mohammad Bani 3916
Shan, Zhenyu 977
Shang, Gao 1282
Shao, Chi 2866
Shao, Riming 1793
Sharma, Avinash 2456
Sharma, Sohit 1730
Shen, Yanfeng 1788, 1925
Shen, Yatao 815
Shen, Yecheng 2842
Shen, Zhan 1788, 1925

Sheng, Caiwang 1167
Shi, Gang 4220
Shi, Haixu 4168
Shi, Xiangyue 939
Shi, Yong 2877
Shibata, Naoya 3929
Shigeeda, Hidenori 2540
Shigematsu, Koichi 2176
Shigeuchi, Koji 3514, 3522
Shijo, Takuya 324
Shimada, Takae 250
Shimakage, Toyonari 2292
Shimamoto, Keita 2210
Shimao, Tohihiro 439
Shimaoka, Masahiro 1747
Shimizu, Toshihisa 302, 404, 2137, 2165, 3309, 3357
Shimizu, Toshimasa 1803
Shimomura, Shoji 2836
Shimono, Tomoyuki 675
Shimosato, Noboru 261, 3514, 3522
Shimoyama, A. 142
Shin, Sungyong 3418
Shinohara, Atsushi 1308, 1322
Shinohara, Hiroshi 1840
Shinshi, Tadahiko 4016
Shintani, Michihiro 3644
Shirai, Ryo 3309
Shirata, Kento 1137
Shiyuan, Yin 2625
Shoyama, Masahito 1095, 1554, 3773
Shujiang, Duan 2718
Shunsuke, Ohasi 3363
Shuto, Masao 699
Si, Yunpeng 2400, 3742
Sihvo, Jussi 2470
Sih-Yi, Lee 645
Silber, Siegfried 4028
Silventoinen, P. 2948
Simanjorang, Rejeki 416, 2296
Singh, Amit Kumar 4186
Singh, Vijay Kumar 1698
Son, Yung-Deug 3323
Song, Hongyu 3825
Song, Injong 803
Song, Kai 84
Song, Seung-Min 3229
Song, Shuguang 1051, 3129, 3924
Song, Wensheng 1872
Song, Yang 3698
Song, Yipeng 746
Song, Yubo 3877
Soong, Boon-Hee 1517, 1592
Soong, Theodore 3203
Soontorntaweesub, Kittichot 1764
Spiliotis, K. 488

AUTHOR INDEX

Stieneker, Marco 598, 2484
Stock, Alexander 3034
Stojadinovic, Miloš 1103
Su, Huiling 795
Su, Jianhui 2877
Su, Yu-Chen 1038, 3549
Sudo, K. 1240
Suetake, A. 142
Suetsugu, Tadashi 4193
Sueuchi, Yuki 1955
Sugahara, Satoshi 2756
Sugahara, T. 142
Suganuma, K. 142
Suganuma, Katsuaki 1598
Sugihara, Yusuke 2789
Sugimoto, Hiroya 3627
Sugimoto, Kazushige 767
Sugiyama, Takashi 3578
Suh, Yongsug 1977
Sul, Seung-Ki 521, 911, 3557
Sumida, Hitoshi 2774
Sun, Bainan 607
Sun, Chuan 370
Sun, Haotian 2780
Sun, Jianning 2963
Sun, Kai 3460, 4168
Sun, Lejia 2882
Sun, Peng 1125
Sun, Shumin 1455
Sun, Weifeng 3145
Sun, Xiangdong 2204
Sun, Yongping 560
Sun, Yuchong 3650, 4127
Sung, Kyungmin 1364
Suntio, Teuvo 963
Supanyapong, S. 1622
Surakitbovorn, Kawin 632, 3848
Surinkaew, Tossaporn 2287
Suul, Jon Are 782, 2003
Suwa, Hiroshi 1997
Suwankawin, S. 4096
Suwankawin, Surapong 2871
Suzuki, Akio 1840
Suzuki, Dai 157
Suzuki, Hiromitsu 495
Suzuki, Kazuma 1847, 3501, 3507
Suzuki, Kenichiro 511
Suzuki, Toshiki 1586, 2324
Suzuki, Yuhei 3390
Suzumori, Hirofumi 2066
Tabata, Yoichiro 329
Tada, Makoto 1580
Tadano, Hiroshi 313, 1111, 3375
Tadano, Yugo 483, 1890
Taguchi, Masashi 826

Taguchi, Yoshiaki 3280
Taiyuan, Yin 2625
Tajima, Katsubumi 2832
Tajyuta, Toshihisa 1840
Takahashi, Akihiko 3896
Takahashi, Akiko 2449
Takahashi, Arata 1270
Takahashi, Isseki 575
Takahashi, Masaki 3186
Takahashi, R. 3665
Takahashi, Shotaro 3315
Takahashi, Tomohira 2796
Takahashi, Toshimichi 227
Takahashi, Yuki 3375
Takakura, Shotaro 1270
Takami, Hiroshi 471
Takamura, Kenya 1381
Takano, Sho 3390
Takasho, Kenta 1890
Takatori, Koji 4139
Takayanagi, Ryohei 3396
Takeda, Kodai 2309
Takemoto, Masatsugu 2589, 2725, 2796, 3315
Takenaka, Hiroshi 3304
Takeno, K. 201
Takenoiri, Shunji 2764
Takeshita, Takaharu 356, 1847, 3501,
 3507, 3791, 3945, 4088
Takeuchi, Norikazu 2292
Takeuchi, Yoko 1828, 3265
Takiguchi, Masashi 3723
Takimoto, Kazuyasu 3304
Takishima, Kenta 2826
Takubo, Hiromu 3390
Takuma, Shunsuke 2596
Takuno, Tsuguhiro 3578
Tamate, Michio 3315
Tan, Nguyen Anh 478
Tan, Siew-Chong 889
Tanaka, Akira 1960
Tanaka, Takaaki 2604
Tanaka, Takahide 2774
Tanaka, Toshihiko 324, 1381
Tanaka, Tsuguhiro 3929
Tanaka, Y. 1162
Tanemo, Masamichi 2022
Tang, Cheng-Yu 639
Tang, Houjun 528
Tang, Ye 3705
Tang, Yi 337, 428, 434, 3910
Taniguchi, Katsumi 3396
Taniguchi, Katsunori 1202
Taniguchi, Tomoisa 866
Tatsumi, Kazuto 1202
Tatsuta, Fujio 4227

AUTHOR INDEX

Tatte, Yogesh 1730
Tausif, Ali 3833
Tcai, Anatolii 4109
Techama, Pantarote 2490
Teerakawanich, Nithiphat 3332
Teigelkötter, Johannes..........................3034
Tenconi, A......................................732
Teraoka, Kenji 3086
Tey, Kuan-Chung.................................511
Thai, Van X. 2564
Thummala, Prasanth 4066
Tian, Mofan 998, 2785
Tian, Wei 1661
Tian, Xiaoyu 1771
Tian, Yanjun 1397
Tibola, Gabriel 1447
Tikka, V. 2948
Toba, Akio 1840
Toi, Takato 2229
Tokumaru, Syohei................................2938
Tokusaki, Hiroyuki...............................2589
Tominaga, Isamu 1692
Tomita, Mutuwo.................................1295
Tong, Anping..............................1391, 2866
Tran, Hai N......................................3833
Tran, Tan-Tai 1799, 2145
Trescases, O.....................................267
Trescases, Olivier 2380
Tripathi, Ravi Nath...............................1698
Troppenz, Maria 3607
Trung, Tran Vu 1666
Tsai, Chang-Lin 3151
Tsai, Meng-Jiang.................................2462
Tsai, Men-Shen..................................1575
Tsai, Terng-Wei 639, 883
Tsai, Tsung-Lin 3151
Tsai, Yue-Ting...................................4198
Tse, Chi K......................................3768
Tseng, King Jet.............................1517, 1592
Tseng, Wei-Jing..................................1626
Tsuchiya, Taichiro 2329
Tsuji, Hitoshi 3717
Tsuji, M...1240
Tsuji, Mineo...........................1217, 1276, 2938
Tsukakoshi, Masahiko............................238
Tsumura, Akihiko 3490
Tsuno, Masahito.................................2558
Tsuruta, Ryoji 495
Tsuruta, Yukinori 318
Tsutsumi, Hirohiko...............................3723
Tu, Yiming................2435, 2681, 3176, 3439, 3934
Tuji, Mineo 3237
Tumerdem, Ugur 1680
Tumurbaatar, Anudari 1972
Uchida, Junichi..................................1955

Uchida, Yuuki 2750
Uchino, Yuki 324
Uda, Ryosuke 3578
Udagawa, Ikuto 517, 1692
Ueda, Tetsuzo...................................3762
Uehara, H. 1253
Uematsu, Takeshi 118, 4139
Uemura, Takamasa 860
Ueno, Tsutomu 4151
Uesugi, Yuma 3412
Ueta, Hiroaki 1883
Umeda, Takashi 2814
Umetani, Kazuhiro.................410, 1602, 1610
Unamuno, Eneko 759
Uno, Masatoshi 1782, 2030
Unterrieder, Christoph 3140
Ura, A. 704
Urabe, Shinichi 1782
Urata, Kazuki 302
Ute, Ryo..3773
Valente, G. 4034
Van De Ven, B.A.C. 267
Van Duivenbode, Jeroen..........................3840
Van Lam, Phi 571
Vasquez, C. Juan 3678
Vasquez, Juan C..................................1498
Vass-Varnai, Andras 137
Veerachary, M. 845
Vemulapati, U. 3596, 4009
Vobecky, J. 3596
Vukadinovic, Nenad 831
Vyacheslav, Shkodyrev 1966
Wachi, Tsuneshisa 1997
Wada, Haruhisa..................................3286
Wada, Keiji...........1414, 1866, 1919, 2137, 4059
Wakimoto, Hiroki................................2404
Wang, Beibei 795
Wang, Bo 459
Wang, Can 1172
Wang, Chao 2386, 2901
Wang, Congling 3112
Wang, Dong 1183, 1246
Wang, Feng 1157, 1167, 2882
Wang, Fusheng 2611, 2992
Wang, Gaolin 1206, 1234
Wang, Guoxin 1206
Wang, Hanyu 2997
Wang, Hao 2270
Wang, Haoyu 100, 3825
Wang, Hechao 1183, 1246
Wang, Hongjie 4074, 4145
Wang, Huai 1021, 1788, 1925, 2604, 2743, 3123
Wang, Huiying 1234
Wang, Jianing 2611
Wang, Jizhe...............................826, 2097

AUTHOR INDEX

Wang, Jun.................................3749, 3985
Wang, Kui.............................1010, 2386
Wang, Laili...........................2785, 3863
Wang, Liang................................3958
Wang, Lifang................................3338
Wang, Liwei..................................927
Wang, Meng..................................2992
Wang, Naizeng.........................998, 2785
Wang, Panrui................................3383
Wang, Po-Wei................................1617
Wang, Qiusheng..............................2421
Wang, Shike...........................1524, 3692
Wang, Shinn-Shyong.........................2086
Wang, Shitao................................2866
Wang, Shunyu................................3002
Wang, Wei...............................614, 3780
Wang, Wenjie..........................1391, 2866
Wang, Xiaolei..............................1455
Wang, Xiaoqing..............................878
Wang, Xiaoyang..............................453
Wang, Xiongfei..............1711, 2673, 3164, 3357, 3684
Wang, Yanbo...........................1758, 2708
Wang, Yangyang..............................2505
Wang, Yi.........................1027, 1397, 3495
Wang, Yijie...................614, 934, 3780, 3825
Wang, Youyun................................2204
Wang, Yu-Chi................................657
Wang, Yue.............................1455, 3863
Wang, Yuncheng..............................1177
Wang, Zhongxu.........................2743, 3123
Watanabe, Hiroki............................896
Watanabe, Shoichiro........................2042
Wei, Baoze.................................3678
Wei, Feng.............................1517, 1592
Wei, Jianzhao...............................2630
Wei, Juan..................................1131
Wei, Shilei................................1397
Wei, Wang.............................1442, 2969
Wei, Xiaoguang..............................2343
Wei, Xiuqin...........................3650, 4127
Wei, Zhang.................................2969
Wellawatta, Thusitha Randima...............1409
Wen, Huiqing................................453
Wen, Po-Hsiang.............................3544
Wenbing, Li................................1282
Wickramasinghe, Harith R...................3117
Wijaya, Febry Pandu........................3490
Wikström, T................................3596
Winter, Christian...........................388
Wolf, Mihaela..............................3607
Wolski, Kornel.............................2129
Wu, Bin....................................3797
Wu, Heng...................................2673
Wu, Hongfei................................4168
Wu, Min....................................3863

Wu, Pei-Lin................................1145
Wu, Ping-Heng.........................503, 3549
Wu, T.-F...................................1912
Wu, Tsai-Fu................................3884
Wu, Tsung-Hsi..............................3544
Wu, Xiaojie.................................619
Wu, Xiaojun...........................3010, 3015
Wu, Ya'nan.................................2496
Wu, Zhiqian................................1549
Würfl, Joachim.............................3607
Wyss, Jonas................................3734
Xia, Meng..................................3484
Xia, Yongming..............................2842
Xiao, Chanjuan.............................1131
Xiao, Dan..................................3971
Xiao, Guochun.........................1549, 2944
Xiao, Jianfang.............................4157
Xiao, Xi...................................1966
Xiaoxi, Liu................................2969
Xie, Jingwen...............................3069
Xie, Shaofeng..............................2547
Xie, Xiaogao................................987
Xie, Zhen.............................2611, 2992
Xiong, Wei.................................939
Xu, Binci..................................2270
Xu, Cai....................................2986
Xu, Dehong............................1118, 2270, 2569
Xu, Dewei David.......................1537, 3797
Xu, Dianguo.......................459, 560, 614, 934, 1206, 1234, 3780, 3825, 4213
Xu, Guangzhao...............................998
Xu, Huadian................................2877
Xu, Jin...........................261, 3514, 3522
Xu, Peng...................................1478
Xu, Sheng..................................3002
Xu, Shuang.................................1793
Xu, Yin-Chi................................3884
Xu, Yue....................................3985
Xuan, Yang.................................1478
Xuanjie, Gao...............................2718
Xue, Danhong...............................2435
Yabuuchi, Tatsushi..........................233
Yada, Tomoharu.............................1381
Yamada, Hiroaki.......................324, 1381
Yamada, Koji................................169
Yamaguchi, Daiki...........................3940
Yamaguchi, Koji............................1972
Yamaji, Masaharu...........................2774
Yamamoto, Aoto.............................2558
Yamamoto, Hidekazu.........................2750
Yamamoto, Kichiro.....................1308, 1322
Yamamoto, Masaya......................1782, 2030
Yamamoto, Masayoshi.....1087, 1095, 2738, 2789, 3344
Yamamoto, Ryo..............................4016
Yamamoto, Shu.........................1949, 1960

AUTHOR INDEX

Yamamoto, Yuuto 3197
Yamanaka, Daisuke 2329
Yamanaka, Kenji 3369
Yamashita, Hiroki 1196
Yamashita, Yoshinori 3490
Yamazaki, Katsumi...........................693, 699
Yamazaki, Masahiro..........................207
Yan, Qingzeng 619
Yan, Y.T. 851
Yan, Zhang 4181
Yanagisawa, Yuta 3762
Yang, Chang-Jun 3884
Yang, Cheng-Jhen 639, 883
Yang, Daoshu 1549
Yang, Dongsheng 3357
Yang, Geng 542
Yang, Hong-Tzer 4233
Yang, Hui-Chen 2296
Yang, Mei..3958
Yang, Ming 560
Yang, Peng 1966
Yang, Ping 3112
Yang, Renxin 4220
Yang, Sheng-Ming 651, 3426
Yang, Shunfeng 428
Yang, Shuying 2611
Yang, Xu 998, 1062, 1478, 2785, 2854, 3329
Yang, Ying 2973
Yang, Yongheng.............. 439, 1021, 1788, 2512, 2743
Yang, Yugang 2703
Yang, Zebin 1157, 1167
Yang, Zhichang 2058
Yang, Zhihua.......................................3797
Yang, Zhiqing......................................1073
Yang, Zhongping............. 1131, 1816, 2058, 3484, 3495
Yano, Junya 3723
Yao-Ching, Hsieh 645
Yaoqin, Jia 2441
Yasuda, Takumi 992
Yasuda, Yusuke 2082
Yaxin, Peng...416
Ye, Han 1504, 2496
Yeh, Shun-Hao 4233
Yelaverthi, Dorai Babu.........................4066
Yen, Chih-Ying 1145
Yenchamchalit, Kulsomsup...................2430
Yi, Hao 2780, 2882
Yijie, Hou 2441
Yin, Shiyuan 1455
Yin, Taiyuan 1455
Yin, Zhijian 1021
Yin, Zhonggang....................................2204
Yingchun, Xu.......................................2441
Yokokura, Yuki...................... 1673, 1741, 3890, 3896
Yokoyama, T. 3665

Yokoyama, Tomoki 1270, 1877, 1883, 2914, 3363, 3658
Yonezawa, Y. 3603
Yonezawa, Yu 196
Yoon, Bo-Kyung 3063
Yoshida, Souichi 2764
Yoshida, Yukihiro 2832
Yoshihara, Hidemasa 219
Yoshihara, Tohru 1997
Yoshikawa, Gaku 3280
Yoshimi, Daisuke 3952
Yoshimura, Eiji 767
Yoshino, Takuma 3363
Yoshino, Teruo 1692, 3916
Yoshioka, Yusuke 4151
Yoshizawa, Daisuke............................ 238
You, Jiang 1386, 1421
You, Zih-Cing 651
Yu, Yong 459
Yuan, Huawei 889
Yuan, Liqiang 2963
Yuan, Xibo 619, 1634
Yuan, Yiqin 977, 3040
Yue, Wang.. 2625
Yui, Haiyan 699
Yukita, Kazuto.................................... 809
Zaijun, Wu 1282
Zaitsu, Toshiyuki 118, 4139
Zaman, Mohammad Shawkat 2380
Zanchetta, P. 4034
Zane, Regan 4066, 4074, 4145
Zdanowski, Mariusz 2129
Zeng, Pengxiang 2646
Zhang, Chen 4220
Zhang, Feili 1315
Zhang, Guoqiang................................. 1206, 1234
Zhang, H. 142
Zhang, Hailong 3863
Zhang, Hao 1131, 1598
Zhang, Hongyang 3684
Zhang, Jianwen 1004
Zhang, Jianzhong 2262
Zhang, Le 3145
Zhang, Lei 3383
Zhang, Lifei 2703
Zhang, Meng 1966
Zhang, Qianfan 3025
Zhang, Runze 1816
Zhang, Shichong 2638
Zhang, Shu 614, 934, 3780
Zhang, Shuai 2944
Zhang, Tengfei 2980
Zhang, Wang 2625
Zhang, Xiaofang 2547

AUTHOR INDEX

Zhang, Xin 917, 953, 1004, 2688,
    2692, 2980, 4157, 4162
Zhang, Xinan 1724
Zhang, Xing 2973, 2992
Zhang, Xueguang 4213
Zhang, Y. 946, 2697
Zhang, Yan 2646
Zhang, Yang 1177
Zhang, Yanping 2204
Zhang, Yaqian 2262
Zhang, Yi 2743, 3123
Zhang, Zhe.......................................607, 1351, 3460
Zhang, Zhenbin 1661, 2895, 3965
Zhang, Zhigang 1157, 1167
Zhao, Chongyan 904
Zhao, Fangzhou.............................1549, 2944
Zhao, Fei 1172
Zhao, Jianfeng.......................3002, 3010, 3015
Zhao, Juan 2051
Zhao, Shengnan 795
Zhao, Tianshu 3020
Zhao, Tianyang 1172
Zhao, Yuanliang...3698
Zhao, Zhengming 2963
Zhao, Zhibin 1125
Zhao, Zhiqing 2714
Zheng, Deyou...2611
Zheng, Xuemei 2901
Zheng, Zedong.............................1010, 2386
Zhong, Wenxing 1118, 2569
Zhou, Dao 1758
Zhou, Dehong 428, 434
Zhou, Fulin 3257
Zhou, Jiuyang...2462
Zhou, Lei 2505
Zhou, Sheng-Zhi...370
Zhou, Victor 1943
Zhou, Yan 934
Zhou, Yimin 2547
Zhu, Cailing 3052
Zhu, Chunbo 84
Zhu, Helin 1336
Zhu, Junjie...3145
Zhu, Lianghong 1206, 1234
Zhu, Qingwei 3338
Zhu, Yanlin 2780
Zhu, Ye 2270
Zhujian, Ou 1376
Zhuo, Fang..............................1157, 1167, 2780, 2882
Zhuyong, Li 2986
Zischler, Sigrid 2410
Zou, Yaohan 3455

IEEE
445 Hoes Lane
Piscataway, NJ 08854-4141

ISBN 978-1-5386-4190-3